S0-BBU-882

ENCYCLOPEDIA OF

FOOD SCIENCE AND TECHNOLOGY

VOLUME 3

WILEY ENCYCLOPEDIA OF FOOD SCIENCE AND TECHNOLOGY, Second Edition

Editor-in-Chief
Frederick J. Francis
University of Massachusetts, Amherst

Associate Editors
Christine M. Bruhn
Center for Consumer Research
Pavinee Chinachoti
University of Massachusetts, Amherst
Fergus M. Clydesdale
University of Massachusetts, Amherst
Michael P. Doyle
University of Georgia
Kristen McNutt
Consumer Choices, Inc.
Carl K. Winter
University of California, Davis

Editorial Staff
Publisher: **Jacqueline I. Kroschwitz**
Associate Editor: **Glenn Collins**
Managing Editor: **John Sollami**
Editorial Assistants: **Susan O'Driscoll, Hugh Kelly**

ENCYCLOPEDIA OF

FOOD SCIENCE AND TECHNOLOGY
Second Edition

VOLUME 3

Frederick J. Francis
University of Massachusetts
Amherst, Massachusetts

A Wiley-Interscience Publication
John Wiley & Sons, Inc.
New York / Chichester / Weinheim / Brisbane / Singapore / Toronto

This book is printed on acid-free paper. ♾

Published simultaneously in Canada.

For ordering and customer service, call 1-800-CALL-WILEY.

Library of Congress Cataloging-in-Publication Data:

Wiley encyclopedia of food science and technology.—2nd ed. / [edited by] Frederick J. Francis.

p. cm.

Rev. ed. of: Encyclopedia of food science and technology / Y.H. Hui, editor-in-chief. c1992.

Includes bibliographical references.

ISBN 0-471-19285-6 (set : cloth : alk. paper).—ISBN 0-471-19255-4 (v. 1 : cloth : alk. paper).—ISBN 0-471-19256-2 (v. 2 : cloth : alk. paper).—ISBN 0-471-19257-0 (v. 3 : cloth : alk. paper).—ISBN 0-471-19258-9 (v. 4 : cloth : alk. paper)

1. Food industry and trade Encyclopedias. I. Francis, F. J. (Frederick John), 1921- . II. Encyclopedia of food science and technology.

TP368.2.E62 2000

664′.003—dc21 99-29003

CIP

Printed in the United States of America.

10 9 8 7 6 5 4 3 2

INTEGRATED PEST MANAGEMENT.

See POSTHARVEST INTEGRATED PEST MANAGEMENT in the Supplement section.

INTERNATIONAL DEVELOPMENT

INTERNATIONAL DEVELOPMENT: AUSTRALIA

Australia is an island continent similar in area to mainland United States, but with a much smaller population (16.3 million). Because it lies between latitudes 10° and 43.5° south, spanning temperate and tropical zones and corresponding to a climatic range in the Americas from northern California to Costa Rica, Australia is able to produce a wide range of food crops.

However, food production in Australia is greatly constrained by the limits of arable land and rainfall. Intensive agriculture is possible only on the eastern seaboard, in the southwestern corner, and in some inland irrigation districts. Progressively inland from the coast there is a band of country suitable for dryland farming of cereal crops and then large areas fit only for open grazing of cattle and sheep; half of the continent is desert.

Nevertheless, Australia has exportable surpluses of many foods, notably meat, wheat, rice, cane sugar, dairy products, fruits, and seafoods. Australia is second in world ranking, after the European Community, in exports of beef and veal, second after New Zealand in exports of mutton and lamb, third after the United States and Canada in exports of wheat and third after Cuba and Europe in exports of sugar. The major export markets for Australian food commodities are Japan (25%), other Asian countries (23%) and the United States (12%) (1).

Although the Australian aborigines lived off the land, the only indigenous plant product that has achieved commercial production is the macadamia nut. Of the native animals, only the kangaroo is used for meat, mainly for pet food, but the introduced and now-feral rabbit and buffalo are also used for human food.

THE FOOD INDUSTRY

Among the manufacturing industries in Australia the food industry stands first and accounts for about 20% of employment, turnover, and value added. The larger enterprises are mostly owned completely or partially by multinational corporations. However only 7.5% of the turnover is exported and, as for the primary commodities, the major markets for Australian processed foods are Japan (30%), other Asian countries (25%), and the United States (12%) (2).

Meat

Meat is the most important food commodity produced in Australia in terms of value, with a total production mostly beef, around 1.5 million tons per year of which about one-half is exported. Australia has a large population of sheep (150 million), but production of sheep meat (0.5 million t) is secondary to wool production.

Because cattle are raised mainly on the open range, the beef produced tends to be low in fat and high in protein and is most suitable for manufactured meat products such as hamburgers and sausages. Thus Australia has developed valuable export markets, notably in the United States and Japan, for frozen boneless beef and mutton and chilled vacuum-packed boneless beef. Consumption of meat in Australia is around 236 lb per head per year, which may be the highest in the world, and is made up of 94 lb beef, 50 lb lamb and mutton, 40 lb pork, and 52 lb poultry.

Innovative techniques developed in the Australian meat industry for converting live cattle into beef save manpower and eliminate many of the physically demanding and unpleasant tasks on the slaughter floor. Electric stimulation of beef carcases is widely applied to increase tenderness especially in conjunction with hot boning. Spraying carcases with hot water around 80°C for 10s removes 90–99% of contaminating bacteria including food poisoning organisms. Packaging of meat cuts under vacuum or in modified atmospheres in plastic films of the right barrier properties is now widely practiced to extend the storage life in chilled storage and distribution (3).

The poultry industry has moved from frozen whole chickens to marketing 85% of its production (approaching 0.5 million tons) in the chilled fresh form. The range of cut-up chicken products in raw, semiprocessed, and cooked forms has greatly increased. Modern plants with computerized control typically process 8,000 birds per hour (4).

Cereals

Wheat is the major cereal crop in Australia with annual production ranging up to 13 million t of which 80% is exported. Domestic usage of wheat is about 1 million t mainly for bread. A few large companies dominate the cereals industry with integrated operations embracing flour milling, computerized feed formulation for stock and poultry, and bread manufacture (5).

Australia is pioneering a modern application of an ancient technique for control of insect infestation in stored grains by creating an atmosphere of carbon dioxide, which effectively asphyxiates the insects but leaves no chemical residue. This technique is being applied on a large scale to the storage and transport of wheat, rice, and cocoa beans in silos, storage sheds, and intermodal containers.

Most Australian bread is made by an activated dough process developed by the Bread Research Institute. Consumer interest in dietary fiber has encouraged production of whole meal and multigrain breads and breakfast foods of the muesli type and those containing wheat, oat, and rice brans. Small speciality bakeries produce croissants, French pastries, rye breads, and Arabic breads.

An Australian company is third in the world in yeast manufacture and pioneered distribution of bulk cream yeast in refrigerated tankers. Some 20% of the flour milled in Australia is used for the manufacture of starch and gluten. Australia is one of the few countries in the world with a starch industry based on wheat rather than on maize or potatoes.

Rice growing under irrigation is a highly developed industry, which claims to achieve the highest yields of rice anywhere in the world. Of the annual production of 0.5 million t, 85% is exported. One feature of technology in this industry is the drying of rice in silos by the circulation of large volumes of ambient air.

A new grain legume crop, developed and commercially pioneered in Australia, is the sweet lupin, *Lupinus angustifolius*, which is extremely well adapted to the infertile sandy soils of Western Australia. Annual production is now around 1 million t of lupin seed, of which half is exported. Seeds of the sweet lupin are low in alkaloids and contain about 30% protein but very little starch. They are particularly useful for stock feeding and are beginning to be exploited for human food as sprouts, in fermented oriental foods, and in bread (6,7).

Dairy Products

Milk production in Australia amounts to 6 million t per year of which 29% is consumed directly, 20% go to butter manufacture, 22% is used in cheese manufacture, and the remainder goes to production of dried milks and casein. Most of the dried milks produced are exported.

Pasteurized milk is distributed in 1- and 2-L cartons and plastic bottles, but pressure from the green movement is encouraging a return to returnable and recyclable glass bottles. A range of modified milks is available: low fat, calcium fortified, and low lactose produced by hydrolysis of lactose by immobilized enzyme technology. Long-life milks are produced by aseptic processing and the same lines are used for shelf-stable cream and custards. Cultured dairy products, especially yogurt, have become popular items (8).

Cheese manufacture is largely mechanized and the inauguration of a continuous curd-making plant for Cheddar cheese in 1987 was a world first for Australia. Many small companies are making specialty cheeses in French. Swiss, and Italian styles. With an ice cream consumption of 18 L per head per year, Australia is second only to the United States.

Australia has been a leader in the application of membrane processes in the dairy industry. Ultrafiltration is being used to produce protein concentrates from whey and cheese base concentrates for cheese manufacture. Milk concentrated by reverse osmosis to twice the solids content is being exported in frozen form to Southeast Asian countries (9).

Fruits and Vegetables

Because of its wide climatic range, Australia is able to produce a great variety of fruits and vegetables. Grapes, apples, citrus fruits, peaches, pears, apricots, pineapples, and bananas are the principal fruit crops. Minor crops include other stone fruits, berry fruits, and tropical fruits such as mango, avocado, and passion fruit as well as coffee and tea. These fruits provide the raw materials for the fruit canning, fruit juice, sun drying, and wine industries all of which export substantial proportions of their production. A recent development is the canning of fruits with no added sugar, using a fruit juice, particularly pineapple or pear juice, as the packing medium.

The annual crop of grapes in Australia approaches 0.8 million t, of which 62% are used for wine making; 33%, for drying into raisins, sultanas, and currants; and 5%, for table fruit. The Granny Smith apple, which originated in Australia as a chance seedling, is reputed to be the best of all apple varieties in keeping quality, keeping for a year or longer in controlled-atmosphere storage with reduced oxygen and increased carbon dioxide contents.

Among vegetables, potatoes, tomatoes, and green peas are grown in greatest amounts and, with many other vegetables, are distributed in fresh, frozen, and canned forms. Australia led the world in the introduction of free-flowing frozen peas packed in polyethylene film bags (10).

Fats and Oils

For the manufacture of margarines and edible oils, Australia uses 100,000 t/yr of locally produced oils (mainly sunflower, safflower, cottonseed, groundnut and maize oils) together with a similar quantity of imported oils (mainly palm oil with some canola, coconut, and soybean oils) and 70,000 t of edible tallow. A recent achievement by plant breeders is a linseed variety yielding an edible oil that has been called linola. Polyunsaturated margarines are widely consumed with polyunsaturated to saturated ratios 2:1 or 3:1 (ratios of *cis*-methylene-interrupted polyunsaturated fatty acids to saturated fatty acids).

Seafoods

In relation to the length of the coastline of the continent, the Australian fishing industry is small, but there are significant exports of several high-value seafoods, notably rock lobsters, prawns, and abalone. Australia is the largest producer of abalone in the world. Commercial aquaculture of Atlantic salmon and rainbow trout is being successfully practiced in Tasmania (11).

Sugar

The cane sugar industry in Australia produces 3 million t/yr of which 2 million tons are exported. Australian consumption of sugar directly and in manufactured foods and beverages is consistently ca 112 lb per head per year. Confectionery consumption averages 20 lb per head per year, which is considerably less than Switzerland and the UK but similar to consumption in the United States.

Beverages

Beer is the most popular alcoholic beverage in Australia with an annual consumption of 115 L per head; wine consumption is 20 L. Two companies dominate the brewing industry but there are many boutique brewers producing specialty brews. The major products are of the lager type with ales as the minor segment. Low-alcohol beers claim a significant share of the market, typically with 2.2% alcohol by direct brewing or 0.9% alcohol with alcohol removed by vacuum distillation; excise is related to alcohol content (12).

There are more than 500 wineries in Australia, but 80% of the wine is produced by 12 companies. Recent advances in wine technology include partial or complete clarification of grape juice for white wine before fermentation, temperature control and oxidation control with inert gas blanketing in tank storage, more rational use of sulfur dioxide and other antioxidants, and centrifugation and membrane filtration technology. There has been a large increase in wine sales in 5-L collapsible bag-in-box packages (13).

A wide variety of fruit juices is available in several processed forms: chilled (49%) and aseptically processed (28%) in gable-top paperboard containers or high-density polyethylene bottles, pasteurized (14%) in 2-L plastic bottles, and canned (9%). The cost of metal cans is currently about six times that of equivalent packs in polyethylene.

An Australian company has pioneered the development of cross-flow microfiltration technology with self-cleaning by gas back-flushing. This technology finds application in the wine and fruit juice industries for sterilization of grape juice prior to fermentation, for sterile filtration of wines prior to bottling, and for production of sterile, bright fruit juices; extension to the beer industry to replace diatomaceous earth filtration is envisaged. A further application is for replacement of conventional water-treatment technology in the soft drinks industry (14). Other recent developments are a countercurrent extractor for fruit juices with intermittent reversal of the screw, which greatly improves efficiency, and a spinning cone distillation column for recovery of fruit flavors and removal of alcohol to produce nonalcoholic wines (15).

EDUCATION

Degree courses in food science and technology requiring three or four years of postsecondary study to attain the bachelor's degree and one or two years in addition to the master's degree, are offered in a number of tertiary educational institutions in Australia: University of New South Wales, Sydney; University of Western Sydney, Hawkesbury, New South Wales; University of Queensland, Gatton; Deakin University, Geelong, Victoria; Royal Melbourne Institute of Technology, Victoria; Western Institute, St. Albans, Victoria; Ballarat College of Advanced Education, Victoria; and Bendigo College of Advanced Education, Victoria. The University of New South Wales offers a master's course in food engineering, and the Curtin University of Technology, Perth, Western Australia, a course in nutrition and food science.

Degree courses are also offered in a number of areas peripheral to food science and technology: biotechnology at the universities of New South Wales and Western Sydney; food microbiology at the South Australian Institute of Technology, Adelaide; wine science at Charles Sturt University, Wagga Wagga, New South Wales, and Roseworthy Agricultural College, South Australia; home economics at the University of Western Sydney, Brisbane College of Advanced Education, Queensland and Western Australian College of Advanced Education, Perth; hospitality studies at the University of Queensland, Gatton, Footscray Institute of Technology, Victoria and Wodonga Institute of Tertiary Education, Victoria; and applied toxicology at Royal Melbourne Institute of Technology. Graduate diploma courses are offered in food and drug analysis at the University of New South Wales, in food packaging at the University of Queensland, Gatton, and in human nutrition at Deakin University. Three universities (University of New South Wales, University of Western Sydney, and University of Queensland) accept candidates for the doctor of philosophy degree in food science and technology, which requires at least three years of study and research beyond the bachelor's degree. The annual output of graduates from all the Australian courses is around 100, and the job opportunities in the food industry are such that these people are readily absorbed.

In all the Australian states, in the capital cities, and in many regional centers, there are colleges of technical and further education which offer diploma courses in most of the areas already listed and also certificate or trade courses in baking, bread making, butchery, small goods manufacture, meat inspection, dairy products manufacture, food processing, food laboratory techniques, packaging, hospitality, and catering and commercial cookery.

RESEARCH AND DEVELOPMENT

Expenditure on food research and development in the food industry amounts to only 0.1–0.2% of turnover (16). Most of this work is development work related directly to the production and marketing plans of the particular company. The companies that have overseas affiliations share the benefits of research and development in their parent companies but it is seldom possible for the results of overseas research and development to be applied directly in Australia. Additional development steps are generally necessary because Australian raw materials and food regulations are different and Australian consumers make different demands.

The Commonwealth (Federal) government makes a major contribution to food research and development through the Commonwealth Scientific and Industrial Research Organization (CSIRO). Among the 40 divisions of CSIRO, those that engage in some research related to food science and technology are the divisions of Food Processing, Plant Industry, Human Nutrition, Horticulture, and Entomology. The Division of Food Processing has a staff of 120 professional scientists and 150 support staff and an annual budget of $15 million (Australian). Its headquarters are at the CSIRO Food Research Laboratory, North Ryde, New South Wales, which conducts research on food microbiology, food engineering, food packaging and transport, food acceptance, and the functional properties of food constituents. The division also embraces the CSIRO Meat Research Laboratory, Cannon Hill, Queensland which conducts research on meat quality, storage, and processing and the CSIRO Dairy Research Laboratory, Highett, Victoria, which works on the separation and utilization of milk components and the microbiology and technology of dairy products.

The Division of Plant Industry maintains the CSIRO Wheat Research Unit at North Ryde, New South Wales, which has projects on the proteins of wheat and the utilization of gluten. The research program of the Division of Human Nutrition includes investigations on the role of dietary fiber. The Division of Horticulture works on the postharvest handling and storage of fruits and vegetables, and the Division of Entomology has a project on the control of insect pests in stored grains. The Commonwealth Department of Community Services and Health is responsible for the collection of data on the composition of Australian foods and has recently published revised composition tables and a related computer data base (17). The Department of Defence Materials Research Laboratory at Scottsdale, Tasmania, undertakes food research and development to meet specific needs of the armed forces.

Some state governments support food-related research, notably the Victorian Department of Agriculture and Rural Affairs, which maintains the Food Research Institute at Werribee, Victoria, and the Queensland Department of Primary Industries, which has the International Food Institute of Queensland at Hamilton, Queensland. Most of the universities with food science and technology programs undertake contract research and development for the food industry and the University of New South Wales has the Food Industry Development Centre.

Commodity research trust funds in which the federal government matches amounts raised by compulsory levies have been set up for the meat, pig meat, poultry, egg, dairy, wheat, grain legume, dried fruit, fisheries, honey, and oilseed industries. These funds are used in part to support competitive grants for food research and development projects.

Three branches of the food industry have created research associations for carrying out cooperative research and development the Bread Research Institute of Australia, North Ryde, New South Wales; the Australian Wine Research Institute, Adelaide, South Australia; and the Australian Sugar Research Institute, Mackay, Queensland.

ORGANIZATIONS

The professional organization representing food science and technology in Australia is the Australian Institute of Food Science and Technology (AIFST), which began in 1950 as a regional section of the Institute of Food Technologists, the first regional section to be commissioned outside of the United States. In 1967, AIFST became an independent organization and now has 1,850 members in several grades: Fellows, Associates, graduate members, student members, licentiates, and affiliates, according to defined requirements of professional qualification and food industry experience. AIFST is the Australian representative body in the International Union of Food Science and Technology. AIFST has six state branches which come together in an annual convention; there are also technical interest groups in the areas of food microbiology, food processing and engineering, food service, seafood technology, and nutrition. The technical journal *Food Australia* is the joint official organ of AIFST and the Council of Australian Food Technology Associations (CAFTA).

CAFTA is an organization of companies in the food industry and its membership of 233 companies embraces the majority of food and beverage manufacturers, food ingredient suppliers, and some retailers. Its principal function is to represent the food industry on government committees concerned with food standards and regulations. CAFTA also conducts, in conjunction with AIFST, the Food Information Service, which seeks to inform the media, educators, and consumers about food safety, quality, and nutrition.

Other technical organizations serving the general area of food science and technology are the Cereal Chemistry Group of the Royal Australian Chemical Institute, the Australian Society for Viticulture and Oenology, the Dairy Industry Association of Australia, the Australian Society of Sugar Cane Technologists, and the Australian and New Zealand Branch of the Institute of Brewing (UK).

REGULATION

Regulatory control of the food industry in Australia is in the hands of the six states and two territories, each of which has a pure food act or an equivalent and accompanying regulations. Substantial uniformity in these regulations has been achieved through the Australian Food Standards Committee administered by the Bureau of Consumer Affairs in the federal attorney-general's department (18). The Food Standards Committee recognizes a commitment to harmonize Australian standards as far as possible with the international standards of the Codex Alimentarius Commission.

BIBLIOGRAPHY

1. S. W. Gunner, "Food Technology for the Export Market," *Food Australia* **41**, 946–951 (1989).
2. W. N. Goff, "Food Export Strategy," *Food Australia* **41**, 827–828 (1989).
3. D. J. Walker, "Trends in the Australian Meat Industry," *Food Australia* **40**, 418–421 (1988).
4. J. G. Fairbrother, "The Poultry Industry," *Food Australia* **40**, 456–462 (1988).
5. A. J. Meers, "Trends in the Cereal Industry," *Food Australia* **40**, 356–358 (1988).
6. J. S. Gladstones, "Development of Lupins as a New Crop Legume," *Food Australia* **42**, 270–272 (1990).
7. D. S. Petterson and G. B. Crosbie, "Potential of Lupins as Food for Humans," *Food Australia* **42**, 266–268 (1990).

8. L. L. Muller, "Advances in Dairy Processing," *Food Australia* **40**, 414–416 (1988).
9. J. G. Zadow, "Extending the Shelf Life of Dairy Products," *Food Australia* **41**, 935–937 (1989).
10. P. C. Thompson, "Refrigerated Food Industry," *Food Australia* **40**, 361–367 (1988).
11. P. G. Taylor, "Australian Seafoods," *Food Australia* **40**, 505–507 (1988).
12. D. C. O'Donnell, "Brewing in Australia 1970–1988," *Food Australia* **41** (Suppl.), 1–5 (1989).
13. B. Rankine, "The Australian Wine Industry," *Food Australia* **40**, 449–451 (1988).
14. S. Paterson and R. Wale, "New Developments in Membranes for Beverage Processing," *Food Australia* **41**, 852 (1989).
15. D. J. Casimir, "Food Process Engineering Over the Last Twenty-One Years," *Food Australia* **40**, 521–523 (1988).
16. J. R. Vickery, *Food Science and Technology in Australia*, Commonwealth Scientific and Industrial Research Organisation, Canberra, 1990.
17. K. Cashel, J. Lewis, and R. English, *Composition of Foods, Australia*, Australian Government Publishing Service, Canberra, 1989; *NUTTAB 89, Nutrient Data Base*, Australian Government Publishing Service, Canberra, 1989.
18. R. A. Edwards, "Food Legislation in Australia," *Food Australia* **40**, 369–375 (1988).

J. F. KEFFORD
North Ryde, New South Wales
Australia

INTERNATIONAL DEVELOPMENT: CANADA

SIGNIFICANCE

The food and beverage processing sector is a major contributor to the Canadian economy as a supplier of food, a market for agricultural production, and a source of economic growth. Since the 1960s the sector experienced substantial growth due to rising population and income levels, and lifestyle changes that emphasized convenience. However, increasing global economic integration since the mid-1980s, particularly in North America, compelled the sector to shift its focus from the domestic market and develop a more international, export orientation. This sector has significantly contributed to the provision of reasonably priced food to consumers, a remarkable achievement given the geographic distances, the diversity of markets and the climatic hurdles in Canada. In fact, relative to other developed nations, Canadian spending on food as a share of personal disposable income ranked second lowest after the United States.

Canada's food and beverage processing sector, in terms of its relative size, trade, and processing intensity, is similar to those in other large industrial countries. Domestically, food and beverage processing is the third largest manufacturing sector, accounting for 11.8% of manufacturing gross domestic product (GDP) and about 2% of total GDP in 1996 (Figure 1). The sector produced shipments valued at $52 billion in 1996 and provided 234,000 jobs (Table 1). The statistics used to describe the size and significance of the Canadian food and beverage processing sector are derived from Statistics Canada, the only source of information that could provide consistent and comparable data across Canada. The data presented may vary from provincial sources of information primarily because of differences in how food and beverage processing firms are defined.

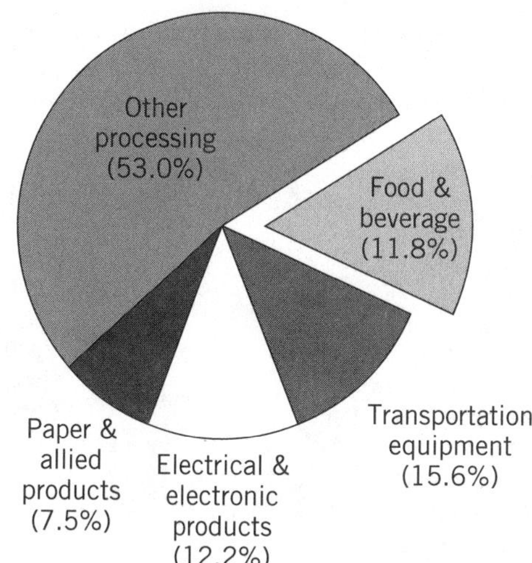

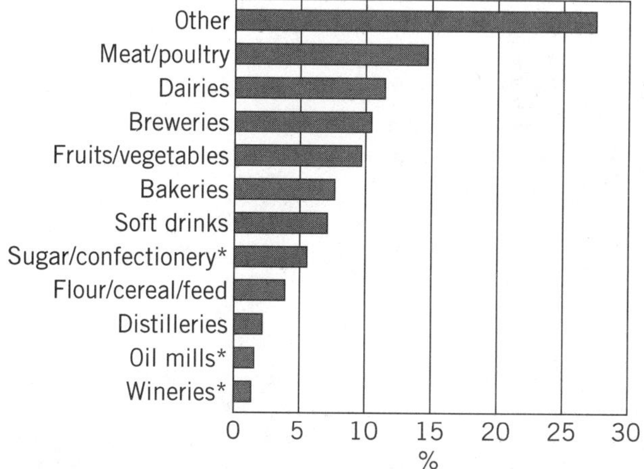

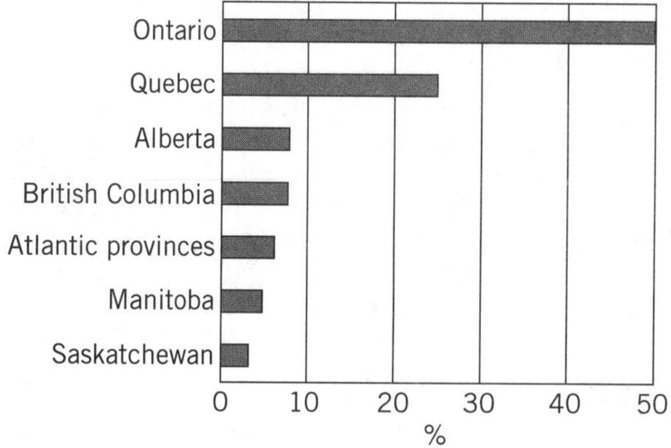

Figure 1. Share of GDP for manufacturing sectors and for the food and beverage processing sector by subsector and by region, 1996. *Source:* Ref. 1.

Table 1. Principal and Trade Statistics for the Food and Beverage Processing Sector (excluding fish), 1988–97

	1988	1991	1995	1996[d]	1997[d]
		Principal statistics			
GDP ($millions 1988)	11,108	10,924	11,952	12,225	—
Share of manufacturing GDP (%)	11.8	12.7	11.7	11.8	—
Establishments	3,147	2,976	2,790	—	—
Shipments ($millions)	40,196	41,418	48,791	51,793	—
Employment[a] (000)	235	223	232	234	—
Investment ($millions)					—
Buildings and construction	260	285	196	371	—
Machinery and equipment	1,083	1,182	1,207	1,450	—
Value added[b] ($millions)	14,309	16,301	17,698	—	—
Net profit margin[c] (%)	3.93	2.45	2.26	2.76	—
Return on assets[c] (%)	6.60	3.64	3.23	4.13	—
Return on equity[c] (%)	14.42	8.45	8.20	11.07	—
		Trade statistics			
Exports ($million)	3,523	3,838	7,129	8,454	9,557
Imports ($million)	4,205	5,274	7,861	8,423	9,526
Balance of trade ($million)	(682)	(1,435)	(731)	31	31

[a]Labour Force Survey.
[b]Value added in the value of the product sold by a firm or industry less the value of the materials purchased and used by the firm or industry to manufacture the products.
[c]Includes fish.
[d]Preliminary for 1996 principal statistics and 1997 trade statistics.
Source: Ref. 1.

About 40% of agricultural production is exported in raw form, and 12% is either sold directly to consumers or sold for nonfood uses. The remaining 45% is marketed as processed food through the food and beverage processing sector, which supplies more than 80% of the food consumed in Canada (Fig. 2). The sector accounted for 44% of agricultural and agrifood product exports in 1996, up from only 32% in 1988. Increasingly, international and domestic markets are characterized by trade in higher-valued products.

As a result of increased market globalization and freer trade, Canadian food and beverage processing firms, which were characterized as being big in a small market in the 1980s, are now rather small in a global market.

STRUCTURE

The food and beverage processing sector remains oriented to the domestic market. However, the share of production that Canada exports has increased from about 9% to about 16% between 1988 and 1996. Imports have also increased from about 10% of the domestic market in 1988 to about 16% in 1996. In 1997, based on preliminary data, Canadian processed food exports reached $9.6 billion, while imports increased to $9.5 billion. Because exports have increased at a faster rate than imports, Canada has enjoyed a positive trade balance of $31 million in both 1996 and 1997. The trade relationship with the United States continues to strengthen, accounting for about 73% of exports and 62% of imports in 1997, up from 60% and 47%, respectively, in 1988.

Although the bulk of food and beverage processing sector activity occurs in Ontario and Quebec, the economic significance in terms of the share of regional manufacturing activity is greatest in the Atlantic and Prairie provinces (Table 2). The influence of world trade in processed food and beverage products varies significantly by region. Export orientation is greatest for the Prairie provinces and lowest in Quebec. Plant locations also vary; red meat and fruit and vegetable processing plants tend to be located near sources of farm production, while fluid milk and bakery product plants tend to be clustered near large centers of population. Activity in the major subsectors occurs in every region. Meat processing is generally the most important activity in each region, particularly in the Prairie provinces. Dairy and meat products are the major products in Quebec, with the share of total shipments of dairy products being significantly higher than in most other regions.

The food and beverage processing sector is fragmented and diverse and is dominated by 67 large and very large firms. The two largest subsectors, meat and dairy products, account for about one-third of the total value of shipments in 1996. Import penetration and export orientation vary widely among the subsectors. For example, export orientation ranges from less than 10% in the supply-managed subsectors, canned and preserved fruits and vegetables, sugar, snack food, soft drink, and brewery subsectors, to more than 60% in the distillery products subsector.

The food and beverage processing sector is composed of relatively large Canadian-owned multinational enterprises (MNEs), foreign-owned MNEs, large and small cooperatives, and small and medium-sized enterprises

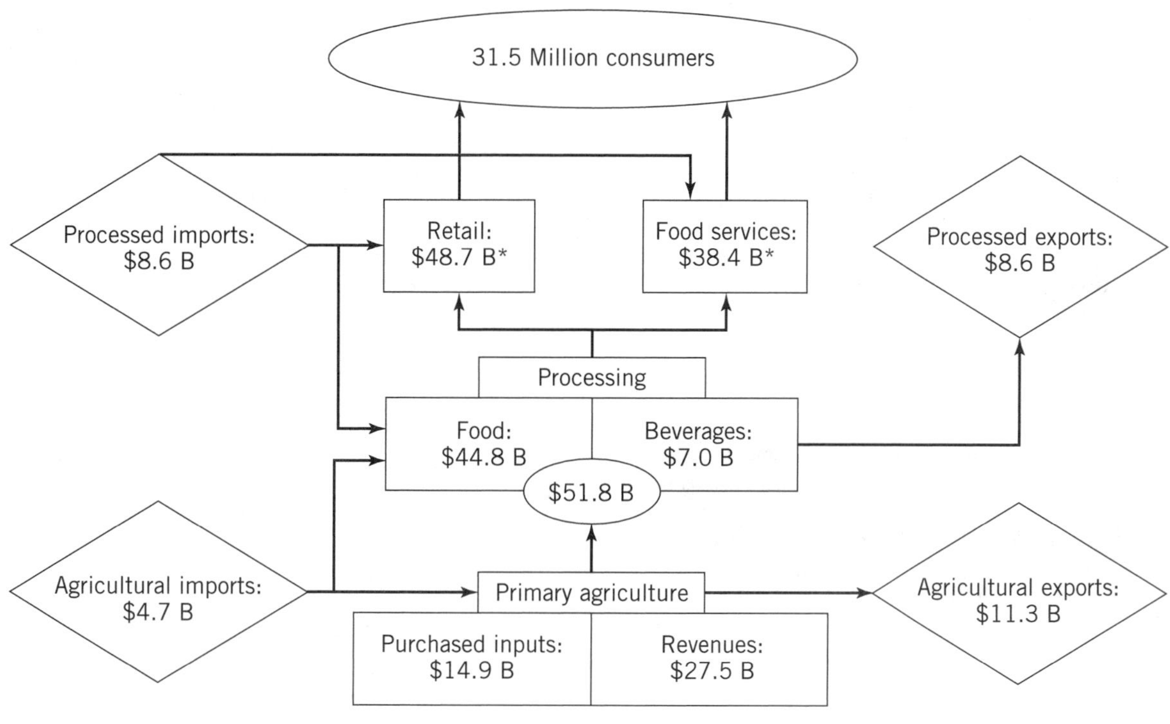

Figure 2. Value of transactions between various components of the Agri-food system, 1996. *Source:* Ref. 1.

Table 2. Food and Beverage Processing Sector by Region, 1996

	Atlantic Provinces	Quebec	Ontario	Prairie Provinces	BC
Shipments ($million)	2,838	12,332	22,951	9,933	3,740
Share of Canadian shipments	5.5%	23.8%	44.3%	19.2%	7.2%
Share of provincial manufacturing	17.5%	12.8%	10.9%	22.1%	11.4%
Imports ($millions)	248	1,321	4,875	797	1,161
Exports ($millions)	415	1,488	3,720	2,316	515
Trade balance ($millions)	167	167	−1,155	1,519	−666
Export Orientation[a]	14.6%	12.1%	16.2%	23.3%	13.8%

[a]Exports divided by shipments.
Source: Ref. 1.

(SMEs), both Canadian and foreign owned. Although some Canadian food sector firms are relatively large, in general, Canadian firms are smaller and less diversified than many of their major U.S. competitors.

In 1996, the food and beverage processing sector was structured as follows:

- A core of 12 very large firms (annual sales of over $1 billion) represent about 35% of shipments; 8 of them are Canadian owned.
- Fifty-five large firms (annual sales between $100 million and $1 billion) produce about 25% of shipments; about 44% of them are foreign owned.
- Approximately 2000 SMEs (annual sales under $100 million) account for the remaining 40% of shipments—about three-quarters of these firms had sales of less than $10 million annually. Based on other sources of information, this number is underestimated and could reach an estimated 4000 SMEs.

A number of the large and medium-sized firms are publicly listed companies, while the small enterprises tend to be privately owned. Most foreign-owned firms are privately held by their parent corporations. Cooperatives, which account for about 20% of the value of shipments, play an important role in the Canadian food and beverage processing sector. Aside from their core role in marketing and providing inputs and services to farmers, cooperatives operate mostly in the dairy and poultry subsectors.

The market share of foreign-controlled firms has increased since 1989, to about 40%. These firms tend to be significantly larger and produce higher value-added products than domestic firms. MNE-dominated subsectors include beef, fruits and vegetables, prepared flour mixes and breakfast cereals, biscuits, confectionery, and breweries. Subsectors dominated largely by domestically controlled firms include pork, poultry, and bakery.

To strengthen and maintain competitiveness in the international market, industrialized nations have placed technological capacity high on their political and industrial agendas. Scientific research increasingly underpins the economic vitality of nations by providing the foundation for industrial renewal and growth.

FOOD RESEARCH AND DEVELOPMENT

Research and development is the ultimate source of technological improvements. Food research in Canada is conducted in several types of organizations and institutions distributed across the country; involves several hundred scientific/technical personnel; is highly variable in nature; is planned and coordinated to varying degrees; is funded federally, provincially, and privately; and is influenced in several ways by the characteristics of the Canadian economy, society, and infrastructure. Three basic kinds of organizations are responsible for food research in Canada: government establishments, universities, and private sector laboratories.

Private Sector

The private sector laboratories engage mainly in applied research because they are obliged to show results to shareholders and because they must be able to maintain a position in the marketplace where demands shift rapidly. The proprietary nature of most industrial research limits the information available in this area. In many cases industry implements the results of research and development conducted by the publicly funded organizations.

Canadian industry food research organizations range from a relatively few, fairly large and comprehensive programs consisting of several professionals, to a fairly large number of small programs with less than a full person-year committed to the activity. Many of the latter are integrated with the companies' quality assurance programs.

Canada's effort in private sector food research appears much smaller in comparison to other industrial nations. This may be partly due to Canada's branch-plant economy with its reliance on technology developed elsewhere. Research and development by multinational firms is centralized in the parent company and then the technology is adapted as seems necessary for branch-plant operations.

Government

Federal government food research is conducted by several departments. Agriculture and AgriFood Canada (AAFC) concentrates its research on agricultural products. Fisheries and Oceans Canada focuses on fish quality issues, and Health Canada is involved in the area of food safety and nutrition.

AAFC plays a major role of in Canadian agrifood research and development, maintaining a critical mass of fundamental, long-term, high-risk research activity, focusing on areas that the industry is unable to conduct on its own and to develop and maintain a communications/coordination/information infrastructure involving university, private sector, and provincial food research agencies. The Research Branch mission is "to improve the on-going competitiveness of the Canadian agrifood sector through the development and transfer of innovative technologies." The food research program contributes toward this by focusing on increasing the market value (quality and value-added processing) and the utilization (new uses and value-added processing) of agricultural products. In the Research Branch of Agriculture Canada, food research is conducted at six establishments across the country (Table 3).

The Health Protection Branch of Health and Welfare Canada also maintains a research program primarily at its central Ottawa laboratory, but with some regional activity. Its efforts relate to the Branch's mandate for ensuring food safety and preventing consumer fraud.

Several provinces have or are developing technology transfer facilities to work with food processing firms in their provinces (Table 4). Provincial food research endeavors are frequently oriented to enhancing industrial productivity in trade. They generally do not undertake fundamental research but work with industry in application and development work. The provincial facilities support the food processing sector through:

Development of food products and processing technologies incorporated the following:

- Value-added opportunities
- Unexploited or underused raw materials

Table 3. Agriculture and Agrifood Canada Food Research Establishments

Kentville Research Station, Kentville, Nova Scotia
Food Research and Development Centre, Saint Hyacinthe, Quebec
Southern Crop Protection and Food Research Program, London, Ontario
Saskatoon Research Centre, Saskatoon, Saskatchewan
Lacombe Research Centre, Lacombe, Alberta
Pacific Agri-Food Research Centre, Summerland, British Columbia

Table 4. Provincial Food R&D Establishments

Alberta Special Crops & Horticulture Research, Brooks, Alberta
Food Process Development Centre, Leduc, Alberta
POS Pilot Plant Corporation, Saskatoon, Saskatchewan
Canadian Food Products Development Centre, Portage La Prairie, Manitoba
Geulph Food Technology Centre, Guelph, Ontario
Service de Recherche sur les Aliments, MAPAQ St. Hyacinthe, Québec
Centre de Recherche sur les Aliments, Moncton, New Brunswick
New Brunswick Research and Productivity Council, Fredericton, New Brunswick
P.E.I. Food Technology Centre, Charlottetown, Prince Edward Island
The Marine Institute, St. John's, Newfoundland

- By-products of agriculture
- Provision of analytic and technical services for food products and food-processing systems
- Identification of domestic and foreign technologies appropriate to local regions
- Provision of technical information directed to improvement and development of food products and processes

One unique organization, representing a coalition of industry (49 companies) and government (federal and provincial) interests, is the Protein, Oilseeds and Starch (POS) Pilot Plant located in Saskatoon. This organization was established in 1976 to facilitate grain and oilseed technology transfer from the laboratory bench to commercial production.

Universities

Several universities across the country have food science programs (Table 5), some as freestanding departments, others in association with nutrition and home economics programs. The trend toward integration is increasing as universities seek to reduce administrative costs and minimize overhead.

The Canadian Council of University Food Science Administrators (CCUFSA) ensures that there is academic consistency in the food science programs across the country. The programs at the various universities follow guidelines established by the Education Committee of the Canadian Institute of Food Science and Technology.

Although the primary role of universities is education, a considerable amount of research is being conducted. As resources at universities become more scarce, future research will require close collaboration among disciplines and among sectors. Research collaboration between universities and the private sector, is an important facet of the Canadian food research and development scene. Scientific collaboration between the universities and the industry is encouraged through the AAFC/Natural Sciences and Engineering Research Council (NSERC) Partnerships Program.

Table 5. Universities with Food-Related Programs

University of British Columbia, Vancouver, British Columbia
University of Alberta, Edmonton, Alberta
University of Saskatchewan, Saskatoon, Saskatchewan
University of Manitoba, Winnipeg, Manitoba
University of Guelph, Guelph, Ontario
University of Toronto, Toronto, Ontario
Macdonald College of McGill University, Montreal, Quebec
Université Laval, Laval, Quebec
Acadia University (undergraduate program only), Wolfeville, Nova Scotia
Daltech Dalhousie University (graduate program only), Halifax, Nova Scotia

COORDINATION OF FOOD RESEARCH

Government, university, and private sector establishments have fundamentally different mandates and responsibilities. However, for at least the past two decades there has been a general realization of the value of coordination and cooperation in many areas of research. Linkages have been fostered in a number of ways.

In 1932 a major national provincial network of agricultural services personnel was established. This network included research scientists and other technical personnel from federal and provincial governments and universities. The organization, now known as the Canadian Agri-Food Research Council, has evolved into a complex, diversified organization with its membership drawn from industry, government, and academia as well as representation from the consumers' association. Within the network are provincial and federal food science committees, which provide a meeting ground for members to discuss food research needs and to develop and recommend strategies to meet these needs.

A second avenue that assists in achieving a coordinated and efficient approach to food research is the annual publication of the *Inventory of Canadian Agri-Food Research.* This provides a project-by-project compilation of food research being undertaken in the reporting establishments. This information is reasonably current and comprehensive, and therefore valuable for planning by research directors, coordinators, and individual scientists.

Perhaps the most effective mechanism for coordination is individual contact. An annual opportunity for this is provided by the national conference of the Canadian Institute of Food Science and Technology (CIFST), the membership of which includes people from all areas of food research and development. CIFST also publishes a bimonthly scientific journal as a means of communicating among scientists.

SCIENTIFIC AND TECHNICAL ACHIEVEMENTS

Many Canadian scientists in industry, universities, and government have made major contributions in the area of food science. The results of many of the early commercial successes are still having positive effects on the Canadian economy. William Heeney made a tremendous contribution with regard to the commercialization of frozen foods. His early work in frozen food technology is still as current today as he foresaw it before World War II. Joe Yarem, with the assistance of Percy Gittelman, commercialized the processing of mechanically deboned poultry meat in the late 1960s and early 1970s. This technical development currently generates in excess of $60 million in sales worldwide, in addition to several international joint ventures.

The development work done at Labatt's by Dr. Norman Singer with regard to synthetic low-calorie artificial fat substitutes during the 1970s was a major breakthrough in reduced-calorie foods. Dr. René Riel made a great contribution in the area of aseptic packaging, and Dr. Doug Emmons is credited for his work in acid coagulation for continuous production of cottage cheese. Dr. Wayne Modler developed a process and the equipment for continuous production of ricotta-type cheeses that is now a commercial reality. Subsequent development of this equipment has seen its application to continuous tofu manufacture.

The early engineering development work of Dr. Gordon Timbers and Dr. Moustafa Aref led to the freezing of liquid and semiliquid foods known as Cryogran in the late 1960s and early 1970s. This process as developed received several international awards and is now an international commercial success. In addition, Dr. Timbers, along with Dr. Robert Stark and Dr. Daniel Cumming, made a significant contribution during the 1970s and 1980s to the world's most energy-efficient ABCO vegetable-blanching system. These energy-efficient blanching systems are now installed and are used in commercial production worldwide.

Jim Squires and David Kennedy developed the first shelf-stable nonfrozen french fry, one of the first successful modified gas-packaged vegetable products. The evolution of this process took more than 10 years to bring to commercial reality in the early 1980s.

The commercial development and process engineering in Canada of the Sous Vide cooking method was undertaken by Ron White and Ray Mitchell. Although initially developed in France, it was their engineering and development work during the mid 1980s that made it commercially viable for large-scale products.

MDS Nordion has gained a worldwide reputation for its work in the development of food irradiation technology and manufactures and ships equipment meeting the highest standards for the application of this very important technology.

Percy Gitelman and his team comprised of Nora Martin, Dr. Paul Fedec, and Dr. John D. Jones made a commercial breakthrough in deheated ground mustard in the mid-1970s. This continues to be a commercial success with sales around the world. Donald Grier and Jeff Dyson during the mid- to late 1970s developed an energy-efficient impulse combustion-drying system that received worldwide patent protection and commercial installation. This system has one of the widest ranges of application in the drying of food, from liquids to solids.

Engineering developments applied to the food industry continue to make their mark. Recent installations of robotic systems for the separation of ribs from the meat of pork sides illustrate the high technology trends within the industry. This unit, and the associated computer algorithms, were developed by the Conseil de recherche Industrielle de Québec and have the capability to handle 1400 sides per hour. Also in the meat industry, a carcass pasteurizer developed by Dr. Colin Gill and associates provides an increased margin of safety for consumers. A process for the fractionation of oats developed by Drs. Burrows, Paton, Collins and Woods has renewed interest in this traditional crop with diverse applications in health (β-glucans), cosmetics, anti-irritants and other food and nonfood products. A primary focus is now toward the fractionation of various crops for the production of nutraceuticals and functional foods with their potential for impact on the health status of the population.

Biotechnology will play an increasing role in future of the Canadian agrifood industry as developments from the R&D community are adopted. Examples include enzymes for accelerated ripening of cheese (Dr. Byong Lee) and *bioingredients* (Dr. François Cormier). Biotechnology is taking its place beside conventional plant breeding as various plants, including the Canadian *Cinderella crop*, Canola, are being further tailored to meet food industry requirements. Of the highest priority in any national food supply is its safety. Canada is well regulated as to sanitary and safe food processing, handling, and distribution; new hazards arise from the evolution of knowledge, organisms, practices, and regulations. Examples include investigations concerning the effects of ionizing radiation on foods and microorganisms, as well as on food packaging materials and the development of handling and processing protocols for newly identified pathogens implicated in food-poisoning cases.

The demand for convenience features in foods continues to grow as discretionary family income grows, as an increasing proportion of families have more than a single breadwinner, and as society becomes more leisure-oriented. Much research is therefore focused on quality retention, sensory properties, product uniformity, and convenience features. This, in turn, has led to the need for a Sophisticated understanding of the molecular level changes in food undergoing various processes and the development of highly controlled processes.

BIBLIOGRAPHY

1. *Statistics Canada*, Agriculture and Agri-Food Canada, Food Bureau, Ottawa, Canada, 1997.

GORDON TIMBERS
CLAUDE JANELLE
Agriculture and AgriFood Canada
Ottawa, Ontario
Canada

INTERNATIONAL DEVELOPMENT: FINLAND

Finland became a member of the European Union (EU) in 1995, which has contributed to the national economy and food industry over the past few years. By the EU membership, the Finnish companies now operate in the Common Market. As a consequence, the share of food in the final consumption expenditure per household decreased from 16 to 13% during one and a half years. A significant reduction in prices was observed for pork, eggs, and oil plants. To hasten the adaptation to the new situation, large structural changes in the food industry have been undertaken. Food industry has grown slowly. Food companies from other EU countries have invaded the Finnish market by increasing their ownership of Finnish companies and extending their product selection. Finland became a member of the European Monetary Union and among the first countries to accept European currency in 1999.

The predicted gross national product per capita has increased from $17,000 to $21,000 between 1991 and 1997 with the population growth from 5.0 to 5.1 million. The gross value of all Finnish industry in 1997 was $80 billion; the share of food industry was 10% (Fig. 1). Since 1992 the Finnish food industry has employed more than 40,000 people. The most significant branches in the Finnish food industry include slaughter and meat production, dairy production, and beverages. A proportion of 3.5% ($280 million) of gross value in food industry was allocated to research and development.

The export and import values of food industry were $1,020 million and $1,900 million, respectively, in 1997. Most export was directed to the Eastern European countries (Fig. 2). As a result of the difficult economic situation in Russia, the export from Finland to Russia has markedly decreased in 1997–1998, while export to other Eastern European countries has increased. The share of the Common Market was 30%. Most imports came from the Common Market (Fig. 3). The most important Finnish export and import products are presented in Table 1.

CHARACTERISTICS OF FINNISH FOOD CONSUMPTION

The food expenditure per household was $3,000 in 1997. Meat and meat products (26%), milk, cheese and eggs (19%), bread and cereals (17%), and fruits and vegetables (15%) formed the major food expenditures per capita. Con-

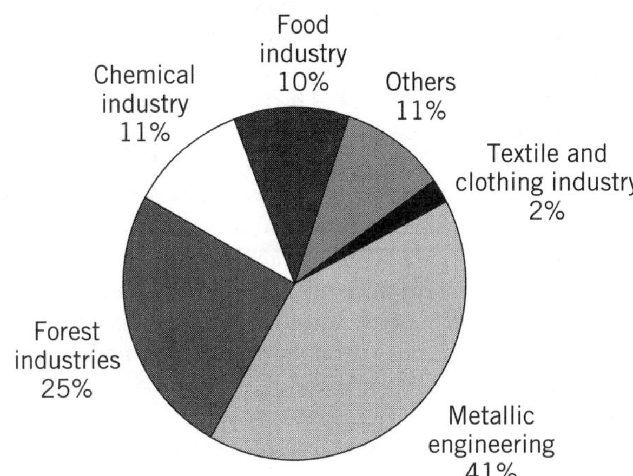

Figure 1. Gross value of production by branch of industry in 1997. Total gross value of production was $80 billion.

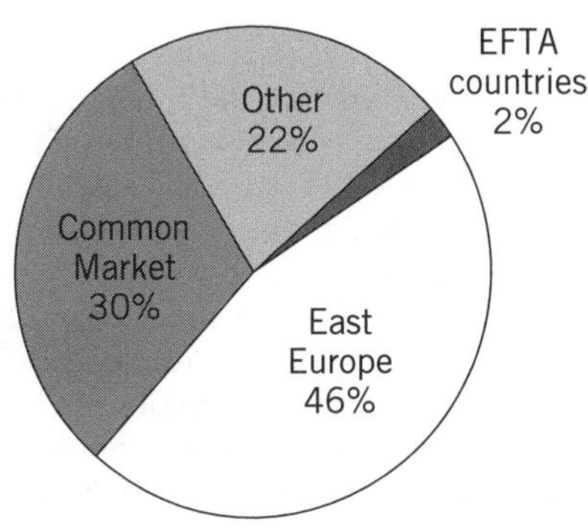

Figure 2. Food export by sales area in 1997.

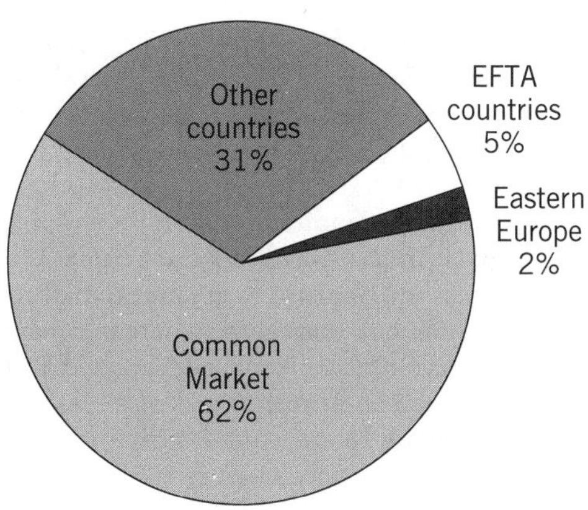

Figure 3. Food import by sales area in 1997.

Table 1. Finnish Export and Import in 1997

	Export value		Import value	
Product	$ million	%	$ million	%
Industrial products				
Cheese	99	9.7	67	3.5
Chocolate	78	7.6	38	2.0
Butter and other milk fats	57	5.6	—	—
Alcoholic beverages	52	5.1	125	6.6
Margarine	50	4.9	—	—
Sucrochemical products	49	4.8	—	—
Pork	38	3.7	29	1.5
Coffee	36	3.5	—	—
Malt	31	3.0	—	—
Milk powder	30	2.9	—	—
Confectionery	29	2.8	—	—
Beer	21	2.1	—	—
Starch derivatives	—	—	70	3.7
Canned fish	—	—	46	2.4
Juices	—	—	40	2.1
Canned fruits	—	—	36	1.9
Other	304	29.8	547	28.8
Total	*872*	*85.5*	*1036*	*54.5*
Primary products				
Oats	47	4.6	—	—
Barley	27	2.6	—	—
Edible fats and oils	13	1.3	42	2.2
Eggs	11	1.1	—	—
Raw coffee	—	—	220	11.6
Fruits	—	—	194	10.2
Vegetables	—	—	87	4.6
Oil seeds	—	—	72	3.8
Raw sugar	—	—	23	1.2
Other	27	2.6	108	5.7
Total	*124*	*12.2*	*747*	*39.3*
Animal feed				
Total	*23*	*2.3*	*118*	*6.2*

sumption of milk and fermented milk (199 L/year/per capita) was highest, followed by fruits and vegetables (85 kg per capita), beer (84 L per capita), and vegetables (70 kg per capita). A total of 66 kg of meat (one-half of pork, one-third of beef, and one-sixth of poultry), and 60 kg of potatoes were consumed per capita. Consumption of coffee was among the highest in the world (9 kg per capita).

During the past few years the consumption of ready-to-eat and frozen foods has increased, whereas consumption of canned foods has slightly decreased. Organically farmed products have gained their own segment of consumption. Nearly 3% of cultivated area is under organic farming in Finland, while the European mean is 1%.

PUBLIC HEALTH AND FUNCTIONAL FOODS

Health aspect is considered one of the main selection criteria of foods. Cardiovascular diseases form a significant public health hazard in Finland. High sodium chloride level in the diet is considered one of the most important risk factors. Therefore, a marked reduction in salt intake has become a desirable trend. The level of dietary salt has been decreased by lowering the sodium chloride concentration in food products and by replacing sodium with magnesium and potassium. One such product is the Finnish innovation Pansalt, which is showing beneficial health effects (1,2). This product has been patented in 22 countries. A similar trend has been observed in consumption of fat. Food industry has reacted to the recommendations of health authorities by launching various low-fat and nonfat products into the market. Consumption of nonfat milk was one-fourth of total milk consumption in 1998.

In addition to lowering the salt and fat content of different foods, much research has been done on foods that promote human health or prevent diseases. Technological use of applications of medicine and biosciences has become a trend in food production. New products and processes are developed using enzymes and the natural bioactivity of plants. Probiotic lactic acid bacteria are studied as a means of influencing intestinal microecology, infection resistance, and gut barrier mechanisms.

FOOD SAFETY

High hygienic quality in primary production and effective food control guarantee the food safety in Finland. Significant infectious diseases in animals do not occur, and the rate of zoonoses is very low.

A lot of emphasis has been given to prevention of *Salmonella*. One reason for the low prevalence of *Salmonella* in poultry is competitive exclusion, a widely used method invented by Nurmi and coworkers in the 1970s (3). On the basis of the very low rate of salmonelloses in Finland, the EU commission accepted the National Salmonella Control Programme in Finland in 1994. The purpose of the programme is to ensure the safety of foods of animal origin by limiting the number of salmonelloses in Finnish food animals and in foods of animal origin to internationally low levels. Cattle, pigs, and poultry, as well as meat and eggs, are being controlled according to the program. The prevalence of *Salmonella* in food animals and foods of animal origin is expected to be below 5% in each slaughterhouse and cutting plant, whereas the total prevalence in Finland may not exceed 1%. Furthermore, imported beef, pork, and poultry meat, as well as living poultry and eggs, are expected to be completely free from *Salmonella*. A total of 0.1% of lymph node samples and surface swap samples in beef and pork were positive for *Salmonella*, while in living poultry the proportion of positive samples was 0.3% according to the results of the National Veterinary and Food Research Institute. The share of *Salmonella* positive cattle and pig herds was below 0.3% and 0.04%, respectively. Altogether 29,255 samples were investigated.

Finnish foodstuffs are generally considered pure. The use of drugs and pesticides is widely controlled in Finland. The National Veterinary and Food Research Institute is responsible for testing residues and contaminants in meat, milk, egg, and fish samples. According to the national residue-monitoring program, no prohibited growth promoters in meat production were detected. Residues of antimicrobial drugs over the maximum residue level were found from pork (0.05%) and milk (0.81%). The content of pesticides, heavy metal, and other contaminants was low. The number of investigated samples was 26,110. Cold winters and strict hygienic control obviously cause the relatively low rate of environmental contamination in foods compared with several other industrial countries. Therefore, Finnish food meets well the consumers' expectations of foods with low environmental contamination levels.

There is an extensive surveillance system for food poisoning outbreaks in Finland. The average rate of reported food poisoning outbreaks has been 0.7 outbreaks/100,000 inhabitants in the 1990s. The average number of cases in these outbreaks has been 32 cases/100,000 inhabitants. The surveillance system yields valuable information about the infectious agents and other factors leading to the epidemics. This information is used in the controlling of epidemics by the food control authorities that in cooperation with health authorities, veterinarians, and medical doctors are responsible for the investigations of food poisoning epidemics.

A lot of research on the entire production chain has been done in Finland to improve the safety of foods. Molecular biological methods have been applied in tracing the origin of epidemics (4) and outlining the controlling measures (5).

FOOD CONTROL

The organization of food control in Finland is presented in Figure 4. The division of labor at each administrative level is distinct, although the structure of central administration may seem complex. At the municipal level the food control authorities act under the direction of the pertinent ministry, which makes the municipal administration in Finland effective. In each of 452 municipalities in Finland an effective food control system is usually run by municipal veterinary officers. In the 1990s, the in-house control system based on Hazard Analysis of Critical Control Points (HACCP) has become a common practice in the food industry.

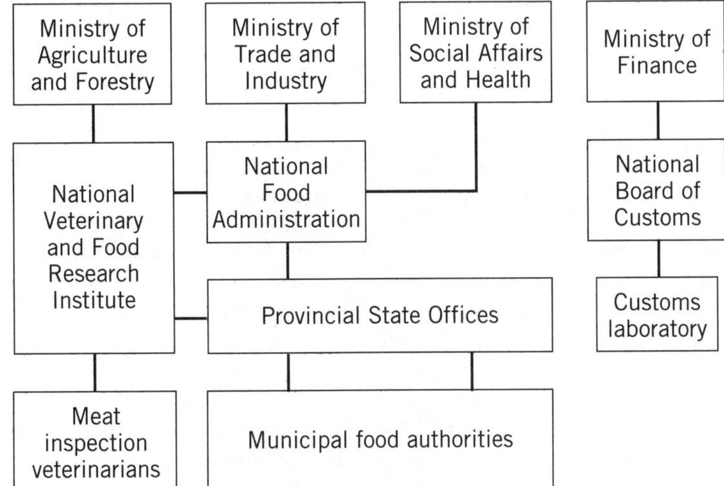

Figure 4. Food control in Finland.

FOOD INDUSTRY IN FINLAND

Dairy Industry

The share of dairy industries in total value-added production was 16% in 1997. The dominating dairy processors are Valio Ltd. and Ingman Foods Oy Ab. Emmental cheese is the most important export product of the Finnish dairy industry. Due to the strict control and high hygienic quality of primary production, the quality of milk is high. Of raw milk, almost 91% was ranked as class E in 1998, which means that the geometric means of bacterial and cell counts were below 250,000/mL and 50,000/mL, respectively.

Milk and milk products are the main source of calcium in the diet of most Finnish people. In addition to normal milk, Valio Ltd. produces milk with 50% higher calcium content than in normal milk. HYLA products contain less than 1 g/100 g lactose.

The probiotic effect of lactic acid bacteria on the human gastrointestinal tract has been intensively studied in Finland. *Lactobacillus rhamnosus* GG is one of the most investigated probiotic strains in the world. The latest studies on *L. rhamnosus* GG have shown its stabilizing effect on gut permeability and suppression of allergic reactions in food hypersensitivity (6). Products containing *L. rhamnosus* GG include milk, sour milk, juices, yogurts, whey drinks, and capsules. Valio Ltd. has the worldwide commercial rights for the strain and is sublicensing the exclusive rights for its use to other countries. Products containing *L. rhamnosus* GG are on the market in 23 countries, and one of these products was officially accepted as the first probiotic food (Food for Specific Health Use) in Japan.

Meat Industry

The meat industry is the largest branch of the food industry in Finland. The gross value of the meat industry was $2,140 million in 1997, with the total value-added production being $440 million. A total of 31 kg of meat products were consumed per capita. The majority of these products were cooked sausages. The largest Finnish meat companies are Atria Oyj, HK Ruokatalo Oyj, and Saarioinen Oy. Finnish meat research includes food safety, development of process technology, and studies on product shelf lives. As a Finnish speciality, elk and reindeer meat are produced on a small scale. Elks are wild animals found in all parts of Finland, whereas semidomesticated reindeer are kept in North Finland.

Bakery

The baking industry is one of the largest branches of the food industry in Finland. A typical feature in the baking industry is a high demand of labor compared with other branches of the food industry; the bakery branch is the biggest employer in the food industry. The largest baking companies in Finland are Fazer Bakeries Ltd. and Vaasan & Vaasan Oy. In addition to these major bread producers, there are more than 1,000 small and medium-sized bakeries in Finland. Traditionally, freshness of the product has been considered the most important quality factor in bakery products. To fulfill consumer demand for fresh, tasteful, and easy-to-use bread products, production and use of frozen and prebaked products has increased heavily during the last years.

In addition, the nutritional value of whole-meal breads has obtained a growing interest. For example, commonly consumed rye bread is an excellent source of dietary fiber, vitamins, and minerals. Recent studies indicate the positive influences of rye on colon fermentation and suggest that the lignans located in the bran may play a role in preventing certain types of cancer.

Beverages

The inland sale of beers and soft drinks in 1997 was 725 million liters. More than half of the volume consisted of beer, and soft drinks and mineral water formed the rest. The volume of annual export was 53 million liters and import 11 million liters. Hartwall Plc and Oy Sinebrychoff Ab are the dominating beer and soft drink producers in Finland. Of the strong liquors, the most famous export article is Finlandia Vodka, whose production was initiated by Primalco Ltd. in 1970. By the late 1970s it had propagated worldwide.

The latest research on brewing and malting technology includes the entire production chain from barley to beer. Research in close cooperation between VTT Biotechnology and Research and industry has aimed at improvement of fermentation process of yeast, using immobilized yeast for fermentation, starter technology in malting, endogenous enzymes in malt, rapid microbiological control methods, and hygiene of dispensing systems. Novel investigations on transgenic barley producing β-glucanase have been initiated.

Margarine Industry

High concentration of LDL cholesterol in the diet has been considered as the main risk factor for atherosclerotic vascular disease. The incidence has been relatively high in Finnish people. Therefore, there is a tendency to introduce low-fat, low-salt, and low-cholesterol diets to consumers in order to minimize the risk of atherosclerotic vascular disease. According to recent research (7), margarine enriched with fat-soluble sitostanol, when substituting part of the daily fat intake in individuals with mild hypercholesteremia, was effective in lowering serum total cholesterol and LDL cholesterol. One of the most noteworthy innovations in the Finnish food industry has been the application of the aforementioned plant stanol ester into food products known as Benecol® (8). The first Benecol® product, a full-fat margarine, was launched into the market in 1995 by Raisio Group Plc, followed by a low-fat margarine. Both Benecol® products have been reported to effectively reduce the serum total cholesterol, especially the LDL cholesterol, as a part of a low-fat, low-cholesterol diet.

Candy Industry

The Finnish innovation xylitol, added to chewing gum, has been reported to prevent caries by inhibiting the growth of *Streptococcus mutans* (9,10). Moreover, the repairing effect of xylitol on primary damage in the dental enamel has been proved. The Finnish company Huhtamäki Leaf Oy launched a xylitol-containing chewing gum; thereafter, many other products with xylitol as a sweetener have come to the market. Since 1988 the Finnish Dental Association has thus recommended the use of xylitol to improve the dental quality of Finnish people. According to the recent research, daily use of xylitol has also been shown to inhibit adhesion of otopathogenic bacteria, therefore decreasing the incidence of acute otitis media in small children by 40% (11).

Oy Karl Fazer Ab is the market leader of confectionery in Scandinavia. The share of import of the whole turnover of Oy Karl Fazer Ab is 60%. Two-thirds of the sales of Fazer Confectionery Ltd. is generated abroad. The most important brand is Karl Fazer Milk Chocolate.

FOOD PROCESSING AND PACKAGING

New technologies, effectiveness, and preparedness to quickly react to consumers' demands are essential in the harsh competition of modern food processing. Purity and nutritive and sensory quality of foods, as well as lack of preservatives, have become important selection criteria for consumers. VTT Biotechnology and Food Research has intensively studied minimal processing techniques for improving the sensory quality and nutritional value of foods. Minimal processing, vacuum packaging, and extended chilled storage of such food products, however, possess certain microbiological risks that are currently being studied by VTT Biotechnology and Food Research and in the Department of Food and Environmental Hygiene, Faculty of Veterinary Medicine, University of Helsinki.

New processing techniques and the yield of food products with extended shelf lives are reflected through the increasing need of vacuum packaging and modified atmosphere packaging (MAP). Research on smart and active packaging such as quality and leak indicators, edible films, oxygen removers, and biodegradable materials is currently being done by VTT Biotechnology and Food Research (12,13). Such packaging technologies seem promising with respect to sensory quality and product safety.

Of the traditional packaging material producers, Stora-Enso Oyj is one of the world's leading companies in the production of carton packages for liquid foods. Other significant cardboard and carton producers are Metsä-Serla Corporation and UPM Kymmene Corporation. The largest packaging film producers are Wihuri Wipak Oy and UPM Kymmene Corporation.

EDUCATION AND RESEARCH

Food sciences education is shared among four universities. The Faculty of Agriculture and Forestry at the University of Helsinki is concentrated on food technology, food chemistry, microbiology, and nutrition; the Department of Food and Environmental Hygiene in the Faculty of Veterinary Medicine, the University of Helsinki, focuses on food hygiene. Other branches are food chemistry in the University of Turku, nutrition in the University of Kuopio, and food technology and chemistry in the Helsinki University of Technology. Research on food science and technology is carried out in the preceding universities as well as in four governmental research institutes and two private institutes. The governmental institutes include the Agricultural Research Centre of Finland (MTT), National Public Health Institute (KTL), National Veterinary and Food Research Institute (EELA), and VTT Biotechnology and Food Research. The private institutes are the Meat Research Institute and Potato Research Institute. Universities and governmental institutes have organized postgraduate education, and in 1994 the Finnish Graduate School on Applied Bioscience: Bioengineering, Food & Nutrition, Environment (ABS) was founded. A total of 154 graduate students registered in the ABS Graduate School in 1998.

BIBLIOGRAPHY

1. E. Mervaala et al., "Beneficial Effects of a Potassium- and Magnesium-Enriched Salt Alternative," *Hypertension* **19**, 535–540 (1992).
2. U.S. Pat. 4,931,305 (June 5, 1990), H. Karppanen and P. Karppanen (to Pharmaconsult Ltd.).

3. E. Nurmi and M. Rantala, "New Aspects of *Salmonella* Infection in Broiler Production." *Nature* **241**, 210–211 (1973).
4. M. K. Miettinen et al., "Molecular Epidemiology of an Outbreak of Febrile Gastroenteritis Caused by High Numbers of *Listeria monocytogenes* in Cold-Smoked Rainbow Trout," *J. Clin. Microbiol.* **37**, 2358–2360 (1999).
5. T. Autio et al., "Sources of *Listeria monocytogenes* Contamination in a Cold-Smoked Rainbow Trout Processing Plant Detected by Pulsed-Field Gel Electrophoresis Typing," *Appl. Environ. Microbiol.* **65**, 150–155 (1999).
6. M. Saxelin, "*Lactobacillus* GG—A Human Probiotic Strain With Thorough Clinical Documentation," *Food Rev. Int.* **13**, 293–313 (1997).
7. T. A. Miettinen et al., "Reduction of Serum Cholesterol With Sitostanol-Ester Margarine in a Mildly Hypercholesterolemic Population," *New Engl. J. Med.* **333**, 1308–1312 (1995).
8. U.S. Pat. 5,502,045 (March 26, 1996), T. Miettinen, H. Vanhanen, and I. Wester (to Raisio Ltd., Finland).
9. A. Scheinin, K. K. Mäkinen, and K. Ylitalo, "Turku Sugar Studies. V. Final Report on the Effect of Sucrose, Fructose and Xylitol Diets on the Caries Incidence in Man," *Acta Odontol. Scand.* (Suppl. 70) **33**, 337–343 (1975).
10. A. Scheinin et al., "Turku Sugar Studies. XVIII. Incidence of Dental Caries in Relation to 1-year Consumption of Xylitol Chewing Gum," *Acta Odontol. Scand.* (Suppl. 70) **33**, 307–316 (1975).
11. M. Uhari et al., "Xylitol Chewing Gum in Prevention of Acute Otitis Media: Double Blind Randomized Trial," *Br. Med. J.* **313**, 1180–1184 (1996).
12. M. Smolander, E. Hurme, and R. Ahvenainen, "Leak Indicators for Modified-Atmosphere Packages," *Trends Food Sci. Technol.* **8**, 101–106 (1997).
13. T. Mattila-Sandholm "Oxygen-Sensitive Colour Indicator for Detecting Leaks in Gas-Protected Food Packages," European Patent 0 666 977 B1.

Miia Lindström
Hannu Korkeala
University of Helsinki
Helsinki, Finland
Raija Ahvenainen
VTT Biotechnology and Food Research
VTT, Finland

INTERNATIONAL DEVELOPMENT: GERMANY

The Federal Republic of Germany (FRG), since the reunification on October 3, 1990, consists of 16 federal states (Bundesländer) covering an area of 357,030 square kilometers. The population density varies of almost 4000 persons/km^2 in the city state of Berlin and around 80 persons/km^2 in the northeastern states of Brandenburg and Mecklenburg-Vorpommern.

Germany is situated in the center of Europe at the point of intersection of many cultural regions. People in Germany—their culture, social behavior, and, last but not least, their eating habits (food production, processing, and preparation)—therefore have been formed by a great many influences. The broad spectrum of food and beverages was influenced in south Germany to some extent by southern Europe, but traces of influences of the adjoining countries also may be found in the west, north, and east of Germany. This can be exemplified by viticulture, which the Romans introduced into the regions north of the Alps 2000 years ago; by the pasta dishes, typical for parts of southern Germany, that came from Italy; and by many desserts from eastern Europe. Varieties of bread found in Scandinavia are also found in northern Germany. This development has been continuing; the food supply of today contains many products foreign workers brought into Germany.

In the many local regions of Germany different nutritional behaviors developed over the centuries. The great variety of sausages and meat dishes, the high number of bread varieties (mostly based on rye), and many kinds of beer have been known all over the world. Important cereal products have been unripe spelt grain and oat flakes. In the mountains, milk production and dairying have been the basis of income and nutrition. Fruit and vegetable processing also go back to the Middle Ages. Important products are sauerkraut, dried fruit, and cider.

In the Middle Ages, the qualities of food products such as bread, meat, and beer, produced in small craftsmen's establishments, were protected by guild laws and rules of the sovereign. One of these was the so-called pureness rule of 1516, governing the production of beer. Water, malt, and hops were the only ingredients allowed for beer. Similar, usually local, rules existed for other food items like sausages and flour. These rules and regulations may be regarded as the precursors of today's food laws. Due to the long tradition of crafts in Germany, food still is produced today mostly in small-sized plants.

The essential impetus for the industrialization of food production in central Europe, and thus in Germany, originated from the continental blockade Napoleon I imposed in 1806. Blocked off from the world market, Europe had to substitute for several essential foods or start food production by itself. Thus, industrial production of sugar was developed. Because coffee was no longer available, substitutes had to be developed, and industrial production soon started in Germany. At the end of the nineteenth century, the first industrially produced convenience food followed Liebig's invention of meat concentrate. Such production included dry soups and vegetable protein hydrolysates, which were processed into so-called seasonings for soups.

There was a need for preserved food in the French-German War of 1870/1871, which led to the development of condensed soups. Typical food manufacturing companies were founded in the late nineteenth century, and when surviving the intensified competition, today they partly exist as affiliates of large international food producing companies.

In the following sections, the demographic structure of the German population will be addressed, and determining data of the food industry in Germany will be presented. Subsequent sections will be devoted to the education of persons working in the food industry and to research institutions and developments in the field of food industry.

STRUCTURE OF THE POPULATION AND DEMOGRAPHIC DEVELOPMENT OF RELEVANCE FOR FOOD PRODUCTION AND PROCESSING

Important for determining the quantities of food production is the number of people to be supplied, whereas the age structure, size of private households, and kinds of jobs are essential for determining the kind of product spectrum. Another important factor influencing the product spectrum is available income. The population statistics of all European countries show a stagnation or a slight decrease of population numbers. This is true for the German population within the new Bundesländer and also for the former Federal Republic, but foreign people applying for the German citizenship exceed the decreasing number (Table 1). In 1995, 7.2 million foreigners in Germany accounted for 8.8% of the total population, the percentage in the new Bundesländer was 1.7% at the same time.

The low birth rate and the further extended life expectancy (1995: boys, 72.8; girls, 79.3) led to an increased proportion of older people, and to a lowered percentage of younger people (Table 2).

FOOD CONSUMPTION

The consumption behavior of people may, besides the distribution according to age and country of origin, also be influenced by the fact that the number of small households and households without children is continuously increasing (status of 1995 shown in Table 3). Despite the increasing number of unemployed (especially in the new Bundesländer, Table 4) the average consumption per household still increases (Table 5). The development after 1994 slowed down but kept the trend.

Total quantities of food sold in Germany will remain on an almost constant level. But demand will develop to more convenient products to serve older people or fulfill the needs of small households. Special food items are created to attract health-conscious consumers, and the exotic and ethnic products supply follows the needs of foreigners and experiences made on vacations all over the world. A new dimension arises, when extrinsic properties like environmentally friendly production and sustainable farming as well as social responsibility of companies become quality criteria. A consumer-oriented food development is growing in Germany, relatively new in the branches of fresh produce. Sensory science and consumer research assess the known and unknown desires of the consumer and replace the sole view on a product's properties (3–6).

Table 1. Development of the German Population from 1950 to 1995 in Millions of People

Year	Germany	Former Federal Republic	New Bundesländer and East Berlin
1950	68,724	50,336	18.388
1960	72,937	55,785	17.188
1970	78,069	61,001	17.068
1980	78,397	61.651	16,740
1990	79,753	63,726	16.028
1995	81,817	66,342	15.476

Source: Ref. 1.

Table 2. Development of Age Structure in Germany from 1955 to 1995

		Age		
Year	Population (millions)	Below 20 (%)	20–59 (%)	Above 60 (%)
1955	70.945	30.2	54.2	15.6
1965	75.591	28.6	52.7	18.6
1975	78.882	29.1	50.4	20.5
1985	77.709	24.2	55.9	19.9
1995	31.539	21.5	57.8	20.7

Source: Ref. 1.

AGRICULTURAL AND COMMERCIAL PRODUCTION

Agricultural and commercial food production are important economic factors in Germany. The net production value of agriculture and food industry was 34.3 and 48.8 billion DM respectively, compared with 2,759.6 billion of the German economy. General structural and economic data of Germany, the German food industry, and its comparison to other industry branches are presented in Tables 6 and 7.

Agricultural Production

In 1997, 522,000 enterprises cultivated 117,000 km^2 in the former Federal Republic, and only 33,145 enterprises produced on 56,000 km^2 within the new Bundesländer; together they covered 54.1% of the total area of Germany.

The number of employees and the total area is decreasing year by year, but production and turnover have been fairly stable within the last years. The most important produce, in terms of quantity, are cereals, sugar beets, potatoes, and meat (Table 8). Self-sufficiency depends on the crop (Table 8), only 18% in case of fruits and 41% in case of vegetables. Of the main agricultural crops, the self-supply is above 100%.

Commercial Food Production

Commercial food producing enterprises in Germany largely comprise small- to medium-sized companies. Of the companies in the German food industry, 72% employ less than 100 people, and only 3% of all enterprises employ more than 500 people (Tables 9 and 10). In addition to the producing food industry with a total number of 20 or more employees per enterprise, in 1995 there were 44,464 small craftsmen's shops with 1 to 19 employees per enterprise; 19,610 bakery shops; and 20,210 butcher shops producing and selling their products. In 1994 they employed 314,472 people (baker; 145,632; butcher: 138,019) and had a turnover of 31,204 million DM (baker: 11,241; butcher: 16,612).

In the food industry most of the employees work in the baking sector, slaughterhouse and meat processing industry, brewing industry, and dairy, and confectionery industries. These are the branches with the highest turnover and a low number of employees per enterprise, both indicating the traditional importance of the produced food items.

The total turnover, number of employees, and total payroll in the food industry is typical for the producing indus-

Table 3. Number of Households With and Without Children in 1995

Household, no. of members	Households (millions)	(%)	Without children (%)	With children (%)
		Former Federal Republic		
1	10.800	36.0	100	—
2	9.612	31.9	88.4	11.6
3	4.571	15.2	4.5	95.5
4	3.618	12.0	0.6	99.4
5 or more	1.518	5.0	0.3	99.7
Total	*30.119*	*100*	*64.8*	*35.2*
		New Bundesländer and East Berlin		
1	2.065	30.4	100	—
2	2.246	33.1	85.4	14.7
3	1.276	18.8	3.4	96.6
4	0.979	14.4	0.3	99.7
5 or more	0.227	3.3	0.0	100
Total	*6.793*	*100*	*59.3*	*40.7*

Source: Ref. 1.

Table 4. Population, Potential Working and Unemployment

Year	Potentially working (millions)	Percentage of total population	Working (millions)	Unemployed (millions)
		Former Federal Republic		
1991	30.662	47.9	28.973	1.689
1994	30.872	46.9	28.316	2.556
1997	30.512	45.8	27.491	3.021
		New Bundesländer and East Berlin		
1991	8.503	53.4	7.590	0.913
1994	7.798	50.1	6.656	1.142
1997	7.748	50.4	6.385	1.363

Source: Ref. 2.

Table 5. Structure of Consumption of Private Households in 1991, 1994, and 1997

	Former Federal Republic		New Bundesländer and East Berlin		Germany
Private consumption	1991	1994	1991	1994	1997
Million DM	1,446,940	1,647,070	183,390	255,790	2,086,200
Food, beverages, tobacco (%)	21.3	20.1	30.4	27.6	18.0
Clothing, shoes	7.7	7.1	7.3	5.5	6.5
Housing	15.9	18.3	4.2	12.7	19.9
Energy (no fuel)	4.1	3.6	5.0	4.4	3.6
Furniture, appliances	8.8	8.7	10.8	10.3	8.2
Health and body care	5.9	6.2	4.3	5.8	5.6
Mobility and communication	17.3	16.4	20.8	16.0	16.7
Education, entertainment, leisure	11.2	11.0	11.9	11.4	9.5
Personal goods	7.8	8.6	5.2	6.3	8.1

Source: Refs. 1 and 2.

try in Germany: slowly increasing turnover, slowly decreasing number of employees, and almost constant total payroll. The export share (foreign turnover divided by total turnover, Table 10) has steadily increased from 8.9% in 1993 to 11.1% in 1997. This is a small figure compared to 33.2% export share of the total producing industry, but there is considerable variation among the different sectors of food industry (Table 10). The value of imports was 71,534 million DM in 1996; the value of exports was in the same year 41,939 million DM. Over the last five years imports and exports slowly grew at the same rate.

The percentage of payroll in the total turnover of food industry (1996) was low (10.9%) compared with the producing industry of other branches (18.5%), whereas the

Table 6. General Structural and Economic Data of Germany, 1995 and 1997

	Germany 1995	Germany 1997
Population	81,700,000	82,100,000
Persons employed	34,800,000	33,900,000
agriculture and forestry	1,001,000	903,000
food industry	524,500	502,700
Private households	36,900,000	37,457,000
Gross national product (in present prices)	3,443 billion DM	3,624 billion DM
Private consumption	1,979 billion DM	2,096 billion DM
Expenditures for food, beverages, and tobacco	372 billion DM	376 billion DM
Net income per employed person and month	2,610 DM	2,710 DM
Total cost-of-living index (1993 = 100)	104.5	108.0
Food price index (1993 = 100)	102.4	104.7
Gross salary (wage) index (1993 = 100)	106.1	109.7
Net salary (wage) index (1993 = 100)	101.4	105.2
Foreign trade		
Total imports	664 billion DM	756 billion DM
Food imports	68 billion DM	72 billion DM
Total exports	750 billion DM	887 billion DM
Food exports	38 billion DM	42 billion DM

Source: Refs. 1 and 2.

Table 7. Comparative Structural Data, 1995

Industry	Number of enterprises	Number of employed	Turnover (billion DM)	Export (%)
Total producing industry	46,398	6,593,000	2,033	28.8
Food industry	5,085	525,000	221	9.8
Petrochemical industry	86	27,000	111	2.8
Chemical industry	1,717	536,000	219	41.6
Metallurgy	1,128	295,000	102	32.8
Metal products	6,498	1,044,000	251	42.7
Equipment for power plants	2,341	495,000	123	32.0
Vehicles and parts	1,047	689,000	262	47.6

Source: Ref. 1.

Table 8. Agricultural Production, 1997

Produce	Quantity (million tons)	Self-sufficiency (%)
Cereals (wheat, rye) feed and industry excluded	23.2	120
Sugar (as white sugar)	4.0	150
Potatoes	12.5	102
Fruit	0.948	18
Vegetables	2.561	41
Wine (million hL)	8.311	46
Meat (slaughter weight)	6.2	85
Beef	1.5	127
Pork	3.5	80
Cow's milk	28.7	
Eggs	13,890 million	72

Source: Ref. 2.

material value was high (50.8% compared with 39.2%). This indicates the high degree of automation that is supported by the above average investments made in the food industry. In 1994, 10,305 million DM were invested by the food industry, 86,321 million DM of the total producing industry of Germany. The energy consumption increased from 1996 to 1997 by 2.18% to 26,558 million kWh in the case of gas and by 1.24% to 12,378 million kWh in the case of electric current. Consumption of coals and fuel oils reduced in the same period by 9.8% to 17,424.2 million kWh.

EDUCATION AND RESEARCH

Education

Cultural and educational policies are within the competencies of the federal states of Germany. However, the federal government exerts coordinating functions providing for framework and financing laws. The German educational system therefore varies to some extent from state to state. The mandatory education period is 9 (in several states, 10) years of school attendance.

In 1999, 935,000 pupils finished school education, with the following breakdown (7):

With no qualification	8%
After successful elementary school	25%
Realschule (intermediate level)	40%
With admission to university or technical college (Ing.)	27%

Table 9. Branches of Food Industry, Number of Companies and Persons Employed, 1995

	Number of companies employing						Total number of companies	Total employed persons
	1–19 persons	20–49 persons	50–99 persons	100–499 persons	500–999 peresons	1000+ persons		
Commercial food production	505	1,988	1,148	1,281	120	35	5,077	534,000
Slaughterhouse and meat processing industry	85	520	248	275	18	5	1,151	105,900
Fish processing industry	28	32	16	28	4	1	109	12,100
Fruit and vegetable industry	56	95	65	70	5	2	293	28,600
Potato products	8	16	11	14	1	—	84	6,400
Fat and oil industries	2	5	10	19	2	1	39	8,500
Dairies, cheese producing plants (fresh and keeping)	34	85	83	109	12	4	324	49,600
Spices and sauces	9	26	10	24	4	3	76	15,400
Mills and starch	12	42	31	19	3	3	107	9,300
Baking products	71	650	334	325	24	3	1418	129,200
Sugar industry	1	2	5	37	1	—	46	8,800
Confectionery	8	28	45	59	14	7	161	37,900
Coffee and tea producing industry	3	9	15	20	3	—	50	7,900
Mineral water, soft drinks	31	73	50	86	4	—	244	24,900
Breweries and malting plants	46	206	117	107	11	4	491	49,300
Wineries	—	13	12	7	1	—	33	3,100
Spirits	29	51	21	16	2	—	119	7,600

Source: Ref. 2.

Table 10. Branches of Food Industry, Index of Net Production, Turnover, Total Payroll, and Export Share, 1997

	Index of net production (1995 = 100)				Turnover	Total payroll	Export share
Branch of industry	1993	1994	1996	1997	(million DM)[a]	(% of turnover)	(%)
Commercial food production	94.1	95.8	101.9	102.7	225,981	10.9	11.1
Slaughterhouse and meat processing industry	97.5	99.0	102.7	104.9	39,839	10.3	5.0
Fish processing industry	96.0	97.6	102.3	97.2	3,604	13.3	15.5
Fruit and vegetable industry	82.2	86.5	100.4	102.4	12,285	10.5	12.6
Potato products					1,691	15.6	9.1
Fat and oil industries	110.4	104.5	95.2	99.4	8,541	4.7	29.1
Dairies, cheese producing plants (fresh and keeping)	90.1	90.0	99.7	97.1	39,396	6.4	16.5
Spices and sauces					6,612	16.3	7.9
Mills and starch	82.8	80.4	106.4	112.4	6,345	9.6	22.4
Baking products	76.1	77.9	103.8	105.8	18,533	27.7	2.8
Sugar industry					6,383	7.5	15.9
Confectionary	105.5	108.3	103.1	99.3	14,893	11.0	18.2
Coffee and tea producing industry					9,472	4.8	12.8
Mineral water, soft drinks	91.4	97.7	100.5	103.1	11,770	12.1	1.8
Breweries and malting plants	101.8[b]	103.5	97.7	96.8	19,972	15.0[b]	6.0[b]
Wineries					2,675	5.8	
Spirits	110.6	107.7	97.2	89.4	6,668	5.0	3.3

[a]Without tax.
[b]Only breweries.
Source: Ref. 2.

Professional qualification can be performed in multiple ways; the main levels of professional qualification are listed in Table 11.

After the successful attendance of 9 (or 10) years of the elementary school, a professional education of 2 to 3 years may follow. In addition to the training within the company, there is parttime school education. In some cases the complete education is conducted in specialized schools. It can be a choice between an industry and a smaller craftsman shop's environment education. Further qualification after several years of job experience can lead to a master or technician certificate. The first enables education of pupils within the industry's own branch.

Universities are institutions of the dual system of teaching and research. In 1998/99, 1,824,000 students were registered at German universities, 271,575 persons, 48.4% women, in the first semester. In 1996 the student numbers were lower, for the first time, than in the year before. In

Table 11. Levels of Professional Qualifications in Germany

School	Professional education	Course of qualification	Number of educated persons in 1996	Skills
12–13 years secondary school or gymnasium	4–5 years university academic		Agriculture 7,260[a] Food Sciences 4,870[b]	Work scientifically Self-responsible Management duties
12–13 years secondary school or gymnasium	4 years technical college Ing. graduate		Agriculture 2,692[a] Food Sciences 3,144[b]	Self-employed, transfer of scientific knowledge into practical problem solving, technical management
10 years Realschule	2 years school technical assistant			Implementation of technical work independently
9–10 years elementary school	3 years dual education within an enterprise and/or school education	Master additional school or courses experience		Performing specific work without instruction, instructing other workers, educating
9–10 years elementary school	3 years dual education within an enterprise and/or school education	2 years school technician experience		Planning of investment and operation of machinery, managing supply and labor requirements
9–10 years elementary school	3 years dual education within an enterprise and/or school education		Several professions thereof: Baker 17,642 Butcher 9,086	Technical skills, capable of performing difficult specific work
9–10 years elementary school	On-job training			Specific work

[a]Horticulture collage and university: 2871 students.
Source: Refs. 1, 2, and 8.

the food sector, students specialize in food technology (Universities: Berlin, Bonn, Hohenheim, Munich), food chemistry (Berlin, Bonn, Braunschweig, Dresden, Erlangen-Nümberg, Frankfurt/M, Hamburg, Hohenheim, Kaiserslautern, Karlsruhe, Munich, Münster, Stuttgart, Wuppertal, Würzburg), and nutrition science and nutrition economy (Bonn, Gießen, Hohenheim, Jena, Kiel, Munich, Potsdam). Horticultural universities are located in Berlin, Hannover, and Munich; agricultural branches are in Berlin, Bonn, Gieβen, Göttingen, Halle-Wittenberg, Hohenheim, Kassel, Kiel, Munich, and Rostock. Additional disciplines taught at German universities are wine production, beverage technology, dairy economy, packaging technology, production technology, and automation and process technology. The German food industry also engages graduates from other disciplines, such as chemistry, physics, and pharmacy, thus providing transfer of knowledge and encouraging interdisciplinary concepts. It should be noted here that there are no restrictions or specific educational requirements to a position in the food industry. One exception is the official inspection of foods, where food chemists or veterinarians are required.

Research

Total expenditures for research and development in Germany amounted in 1996 to 88,600 million DM. Industry and trade spent 67% of research funds, 18% were spent by universities, and 15% by noncommercial governmental or private institutions.

Research at Universities. The dual system of research and teaching is traditionally practiced at German universities and technical universities. They are exclusively affiliated with the federal states. Scientists (professors) appointed to a chair for an unlimited period may conduct self-responsible research within their working field. The budget is financed by the federal government concerned. Additional money sources are industry, private foundations, the Deutsche Forschungsgemeinschaft (mainly fundamental research), or "Innovationskompetenz mittelständischer Unternehmen" by the Federal Ministry for Education and Research (applied research, development). Other sources of funding are the research projects supported by the EU in the FAIR, FLAIR-FLOW, and COST projects. More details can be found at the Web sites *www.exp.ie / flair.html* or *www.kowi.de*. Information on the fields of the German universities and addresses can be found at the Web sites *www.hochschulkompass.hrk.de*, *www.bmbf.de*, or *www.studienwahl.de*.

Federal Ministry for Nutrition and Agriculture. The Federal Ministry for Nutrition and Agriculture spends 500 million DM annually on research (9). The most important institutions are the 10 federal research centers, the Central Institution for Agricultural Documentation and Information (*www.dainet.de* (/zadi)) and 7 institutions (blue list), which are cofinanced with federal states. The topics of investigation are sustainable use of ecosystems and resources, protection of the natural environment and quality production, enhancement of food quality, further development of renewable resources, social economy, and information and documentation. The knowledge produced serves governmental decisions in the field of nutrition, agriculture and forestry, and consumer policy. It also is intended to support small- and medium-sized enterprises in answering technical questions and as an information source for the public.

Federal Ministry of Education and Research and the Commercial Environment. The Federal Ministry of Education and Research also funds a variety of projects within the food sector. As in the case of the Agricultural Ministry, the cooperation between university and other nonprofit research institutions and commercial R&D initiatives should be facilitated. Actual information can be obtained from *www.bmbf.de*. Searchers about actual developments within the commercial sector can be started from *www.behrs.de* or *www.fnii.ifis.org*.

MAIN FIELDS OF RESEARCH AND DEVELOPMENT

The developments in Germany are not independent of the developments in other countries. The aim of the substantial support by the European research programs is the transfer of ideas, knowledge, and techniques. Besides the traditional fields of technological and conceptional evolution, the following topics have emerged as fields of interest:

Preservation and modification of food by new technologies like high hydrostatic pressure, high-pressure freezing, high-intensity electric field pulses, and supercritical carbon dioxide.

Use of rapid and online measurements (ultrasound, NIR, NMR, electronic noses) to facilitate and document good manufacturing practice. Certification and accreditation have become increasingly necessary tools in quality concepts (quality assurance).

Assessment of allergic potential of food additives, novel food, or genetically engineered products.

Development of improved packaging, reduction of material. Recycling and elimination costs are extremely high in Germany and Austria compared with other countries.

Identification implementation, and promotion of food items with human health promoting functions (antioxidants, bioactive substances, secondary plant components, probiotic microorganisms, fiber). The claim of specific food effects on human health is illegal in Germany.

Use of sensory methodologies and other results of consumer research to implement consumer orientated product development, combining convenience, sustainable production, minimal processing, and new experiences.

BIBLIOGRAPHY

1. Statistisches Bundesamt, ed., *Datenreport*, Bundeszentrale für politische Bildung, Bonn, Germany, 1997.
2. Statistisches Bundesamt, ed., *Statistisches Jahrbuch über Ernährung, Landwirtschaft und Forsten 1998*, Landwirtschaftsverlag, Münster-Hiltrup, Germany, 1999.
3. Z. Andani and H. J. H. MacFie, "Consumer Preference," in R. L. Shewfelt and B. Brückner, eds., *Fruit and Vegetable Quality: An Integrated View*, Technomic Publishing, Lancaster, Pa., 1999.
4. B. Brückner and H. Auerswald, "Instrumental Data—Consumer Acceptance," in R. L. Shewfelt and B. Brückner, eds., in *Fruit and Vegetable Quality: An Integrated View*, Technomic Publishing, Lancaster, Pa., 1999.
5. K. G. Grunert, "Food and Science. Wissenschaft im Dienst der Ernährung," *Proceedings of the Symposium 25th Anniversary "Euro Research and Development" CPC Europe, 35th Anniversary "Institute for Research and Development" CPC Deutschland*, 47–63, 1997.
6. H. R. Moskowitz, B. Krieger, and D. Cohen, "Meeting the Food Market Challenge: Creating Winning Food Concepts for Consumers in a 'Real Time' Mode," *Health and Pleasure at the Table. A Symposium held at the University of Montreal*, Montreal, Canada, May, 1994.
7. *Bundesanstalt für Arbeit. Beruf aktuell: Ausgabe 1998/99*, MEDIALOG Verlag, Mannheim, Germany, 1998.
8. *BLK für Bildungsplanung und Forschungsförderung, Studien- & Berufswahl*, BW Bildung und Wissen Verlag und Software, Nürnberg, Germany, 1998.
9. *Forschungsrahmenplan. Bundesministerium für Ernährung, Landwirtschaft und Forsten*, Referat 624, Bonn, Germany, 1998.

B. Brückner
Institute of Vegetable and Ornamental Crops
Grossbeeren/Erfurt, Germany

INTERNATIONAL DEVELOPMENT: IRELAND

The island of Ireland lies west of Britain off the northwest coast of Europe. It is compact, measuring about 220 km east to west, 480 km north to south. It lies in the north temperate zone but is warmed by the Gulf Stream. Mean 24-h temperature is 8 to 11°C, and rainfall is about 1080 mm/year.

Low (ca 1000-m maximum) mountain ranges lie behind the coastal plains, except on the central east coast where rivers draining the central limestone plain enter the Irish Sea. The Shannon River (384 km) drains the western region into the Atlantic. The climax vegetation is deciduous woodland, which covered about 12% of the land in A.D. 1600. Now about 8% is in forestry. Deep peat bogs have developed under the wet climate in some areas. Grass grows well everywhere, and the long grazing season, together with grass conservation for winter feed, is the foundation for agriculture.

The Republic of Ireland, with 3.6 million people, comprises 70,000 km^2 and achieved political independence in 1921. Northern Ireland, in the northeast of the country and part of the UK (qv), with 1.5 million people, comprises 14,000 km^2. Most of this article will deal with the Republic.

HISTORIC PERSPECTIVE

Given the large fertile land area relative to population, an abundance of temperate foodstuffs have been exported since the Viking age. Hides, wool, grain, butter, and fish were exported through the trading towns, such as Dublin, Cork, and Galway, to Britain and Europe (1).

In 1700 Ireland was similar to other western European countries; her population was about 4 million and most of them worked on the land. From the middle of the eighteenth century, the demand for agricultural products grew sharply. In a century, beef exports had quadrupled, pork increased eightfold, and the carefully supervised butter trade doubled. This supply of low-cost foods was critical in releasing British agricultural labor for factory work, thus contributing to the British industrial revolution. Ireland's population, supported by increased potato cultivation, reached 8 million by 1800, when the Act of Union joined Ireland's Parliament to that of the UK (2).

Much of this population lived in poverty as farm tenants and agricultural laborers, with most of the fertile land owned by the Anglo-Irish gentry in great estates. The Great Famine, 1845 to 1849, was a calamity in which about 1 million died and a further million emigrated (3). It arose from the devastation of the staple food, the potato, through blight. Land tenure was transformed in subsequent decades to family-owned and -operated farms, which now predominate (4).

MODERN ECONOMY

Ireland's joining of the European Union (EU) in 1973 accelerated economic development in both agriculture and industry. Economic prosperity has followed, with Ireland's gross national product (GNP) per head rising from 62% of the EU average in 1980 to 85% in 1998, with some projections suggesting equality by 2003 (5). Factors involved are external, such as trade access to the single EU market and EU structural fund transfers, and internal, such as consistent macroeconomic policy, major investment in education, and industrial strategy. The latter has involved concentration on selected industrial sectors, such as information technology, financial services, chemicals, pharmaceuticals, and food.

With economic development, primary agriculture has been a declining sector, with its share of gross domestic product (GDP) falling to 5.2% in 1998. The overall agrifood sector is still of major importance in the economy, however, accounting for nearly 13% of GDP in 1998 (6). Furthermore, the low import content of food exports means that the share of the agrifood sector in net foreign earnings is nearly 33% (ie, foreign earnings adjusted for imports of materials and repatriation of profits by foreign-owned companies). Agricultural employment has correspondingly fallen to about 9% of the total. Although the food industry is still dominated by meat and dairy products, advanced food ingredients and consumer products sectors have also emerged. The drinks sector is also very strong with global brands such as Guinness and Bailey's. The food industry is strongly export oriented with about half of all food manufacture being exported. Food industry structure has also changed rapidly, being dominated by about 10 leading companies that have expanded rapidly through merger and foreign direct investment, primarily in the UK, the EU, and the United States.

THE COOPERATIVE MOVEMENT

In 1889 Sir Horace Plunkett started the producer cooperative movement. The Co-op Societies grew rapidly in numbers and prospered, especially in the dairy and pigmeat sectors (7). Today the cooperative movement continues to flourish, with some of the larger cooperatives, particularly in dairying, being very successful international food business organizations. Since the mid-1980s a few of these larger cooperatives have adopted a rather unique cooperative–public company joint ownership structure. The linkage by the cooperatives to a public company structure has provided finance for international growth. Cooperative types involve both primary cooperatives, in which membership consists of farmer suppliers, and secondary cooperatives, in which particular functions are performed on behalf of the primary cooperatives that constitute the membership. The Irish Dairy Board, which markets dairy products internationally on behalf of primary dairy cooperatives under the brand name Kerrygold, is a good example of a successful secondary cooperative. The Irish Cooperative Organisation Society (I.C.O.S.) is the national representative and promotional agency for cooperatives in Ireland, while the Centre for Co-operative Studies in University College Cork conducts research and provides education on all aspects of cooperatives.

GOVERNMENT AND AGRIFOOD

The Department of Agriculture and Food is responsible for all aspects of supervision and development of the agrifood industry. For food, its goals include the further development of high-quality international markets and the fostering of added-value food exports rather than basic commodity trading. Related activities include the operation and implementation of EU schemes and regulations, the provision of grants for farm improvements, the operation of measures to improve livestock and horticultural production, the control and elimination of animal disease, ensuring quality control in food processing and marketing, and the formulation and operation of land policy (8).

The Department of the Marine and Natural Resources is a comparatively new government department, being first established as a Department of Fisheries, separate from Agriculture, in 1977, with subsequent functional and title changes. Ireland as an island has great scope for sea fisheries development and also is famous for freshwater fishing, both game and coarse. The Department has responsibility for all aspects of regulation, protection, and development of the marine resources. The Department is also responsible for all aspects of forest policy (8).

IMPORTANT STATE AGENCIES

Teagasc—the Agriculture and Food Development Authority—is the national body providing advisory, research, education, and training services to the agricultural and food industry. It was established in September 1988 and replaced ACOT, the agricultural advisory and training body, and An Foras Talúntais, the agriculture and food research organization. Of particular importance to Teagasc is the education and training for young farmers, research in the food industry, together with farm management, economics, marketing of agricultural products, and rural development (8). Teagasc has two major food research centers, the Dairy Products Research Center in Cork and the National Food Center in Dublin.

Bord Bia, the Irish Food Board, is the market development and promotional body established by the government to assist the Irish food and drink industry. Its role is to work in partnership with Irish companies, bringing them together with prospective buyers to build sales of Irish food and drink in export markets. It provides a range of services, including market information support, product promotion, inward buyer programs, quality assurance initiatives, and marketing enhancement programs for individual companies. It has a network of international offices, in Dusseldorf, London, Madrid, Milan, Moscow, New York, and Paris.

An Bord Glas, the Horticulture Development Board, was established in March 1990 as a statutory semistate body. The primary function of the board is to assist and encourage all aspects of horticulture so as to increase output, improve the domestic market, and thereby create employment and contribute more significantly to GNP. BIM/ Irish Sea Fisheries Board is the state agency with primary responsibility for the sustainable development of the Irish seafish and aquaculture industry both at sea and ashore and the diversification of the coastal economy so as to enhance the employment, income, and welfare of people in coastal regions and their contribution to the national economy. It provides a wide range of financial, technical, educational, marketing, resource development, and ice supply services for the production sector through to the processing and marketing sectors.

UNIVERSITIES AND COLLEGES

Ireland has many universities, institutes of technology, and regional technical colleges that offer undergraduate and postgraduate food science and technology programs of relevance to the agrifood sector.

University College Cork has a faculty dedicated to food science and technology and is internationally recognized for the excellence of its education, research, and training programs. It is unique in Ireland for its commitment to food science and technology, food business, and food biotechnology. The Faculty of Food Science and Technology offers five undergraduate degree programs: B.S. in Food Business (offered jointly with the Commerce faculty), B.S. in Food Science, B.S. in Food Technology, B.S. in Nutritional Sciences, and B.E. in Food Process Engineering (offered jointly with the Engineering faculty). The faculty also offers a range of postgraduate opportunities, which include M.S. and Ph.D. by research, M.S. by a combination of research and coursework, and also higher diplomas. The Faculty occupies a dedicated Food Science and Technology Complex, including an extensive Food Processing Hall and specialized laboratory facilities where the National Food Biotechnology Center is incorporated. University College Dublin offers a B.S. in Food Science within the Faculty of Agriculture and provides both full-time and parttime taught master's programs in Food Science. In addition, graduates may be admitted for research programs leading either to master's or Ph.D. qualifications in Food Science.

The University of Limerick offers a B.S. in Food Technology associated with the Biological Sciences Department and also provides postgraduate programs. Trinity College Dublin offers undergraduate and postgraduate courses in nutrition and dietetics. The Dublin Institute of Technology offers a B.S. in Human Nutrition and Dietetics. In addition, a range of certificate and diploma courses is offered by the Institutes of Technology in Carlow, Cork, Dundalk, Letterkenny, Sligo, Tralee, and Waterford.

In Northern Ireland, the University of Ulster at Coleraine provides a B.S. in Human Nutrition, as well as postgraduate opportunities in the area. Loughry College also offers a B.S. in Food Technology, as well as postgraduate courses.

Extensive research programs in the food science and technology area are under way in most of the universities just listed. Funding comes from the European Commission, government agencies, and national and international industry.

BIBLIOGRAPHY

1. R. F. Foster, *Modern Ireland 1600–1972*, Allen Lane Penguin Press, London, United Kingdom, 1988.
2. L. M. Cullen, *The Formation of the Irish Economy*, Mercier Press, Cork, Ireland, 1969.
3. C Woodham-Smith, *The Great Hunger*, Hamish Hamilton, London, United Kingdom, 1962.
4. F. S. L. Lyons, *Ireland Since the Famine*, Collins/Fontana, London, United Kingdom, 1975.
5. *Agenda 2000, Implications for Ireland*, Institute for European Affairs, Dublin, Ireland, 1999.
6. Department of Agriculture and Food, *Annual Review and Outlook for Agriculture and the Food Industry*, Dublin, Ireland, 1998.
7. R. A. Anderson, *With Plunkett in Ireland: The Co-op Organiser's Story*, Irish Academy Press, Dublin, Ireland, 1983.
8. *I. P. A. Administration Year Book & Diary*, Institute of Public Administration, Dublin, Ireland, 1999.

Michael Keane
University College Cork
Cork, Ireland

INTERNATIONAL DEVELOPMENT: NEW ZEALAND

The characteristics of New Zealand and its economy, with special reference to food production and processing, will be discussed. The new science structures and funding are mentioned, and then some detail of food science and research providers is given.

New Zealand, with its nearest neighbor the continent of Australia, is located in the southwest Pacific Ocean. New Zealand consists of two main islands and is similar in size to Japan and to the UK. It is a diverse land, changing from subtropical in the far north to cool temperate in the south. The population of New Zealand is approximately 3.7 million (1996 census).

Seasons are opposite to the Northern Hemisphere, with January and February being the warmest months and July the coldest. The climate is temperate with averages from 8°C (46.4°F) in July to 17°C (62.6°F) in January, but summer temperatures occasionally reach the low 30s (mid 80s°F to low 90s°F) in many inland and eastern regions.

The mean average rainfall varies widely, from less than 400 mm (15.7 in.) in the lower central part of the South Island to over 12,000 mm (470 in.) in the Southern Alps—which form the backbone of the South Island. For most of the North Island and the northern South Island, the driest season is summer. However, for the west coast of the South Island and much of inland Canterbury, Otago, and Southland, winter is the driest season.

New Zealand is a food producing and exporting nation with food and beverage products comprising 48% of New Zealand's exports in 1998. Over the past 10 years there has been a decline in the earnings from unprocessed primary produce (particularly wool) and an increase in the earnings from processed products. The New Zealand economy is heavily dependent on overseas trade. Until the late 1980s New Zealand had an almost complete dependence on dairy, meat, and wool exports, but since that time increases in value of exports from forestry, fishing, horticulture, and the manufacturing industries have been very significant. Tourism has become a very important earner for the country.

The major trading partners that together account for about 70% of the country's exports are Australia, Japan, EC, the United States, Korea, and China (including Hong Kong). The EU has been the major market for food and beverage products, accounting for 28% of the value, but the patterns for different food items are very varied. North America is the primary market for beef, but Europe is the primary market for sheepmeat products. Milk powder is primarily targeted at Southeast Asia; butter and cheese at Europe, although North Asia is also a high importer of cheese; and more than half of the casein production is exported to North American markets. Fish and fish preparations markets tend to be in North Asia, although there are also significant markets in the North Americas.

Kiwifruit are the major earner in the horticultural sector, followed reasonably closely by apples and pears with vegetables third. The major markets for exported fruit are in Europe (58% of the value), whereas North Asia is more important for vegetables, accounting for some 35% of the export value. Australia is the largest market for fruit and vegetable preparations.

The value of totally processed primary produce is almost three times that of the unprocessed primary produce. Since the major contributors to the unprocessed primary produce are wood pulp and wool, it is obvious that most food products are in the processed primary products category.

NEW ZEALAND SCIENCE STRUCTURES

The government science agencies were completely reorganized in 1992. The activities of the Department of Scientific and Industrial Research (DSIR), the Ministry of Agriculture and Fisheries (MAF), the Forest Research Institute (FRI), and other government research organizations were reorganized into new business enterprises called Crown Research Institutes (CRIs). The CRI structure was established to provide an open and flexible framework for the management of science with the expectation that it would create better collaboration between the public and private sectors in areas of research and development (R&D) and technology transfer. Each institute is based around a productive sector of the economy or a grouping of natural resources and in general provides a vertical integration from production through to market.

The CRIs are as follows:

- New Zealand Pastoral Agriculture Research Institute Ltd. (AgResearch)
- The New Zealand Institute for Crop and Food Research Ltd. (Crop & Food)
- Horticulture and Food Research Institute of NZ (HortResearch).
- Industrial Research Ltd. (IRL)
- Institute of Environmental Science & Research Limited (ESR)
- Forest Research (FRI)
- Institute of Geological & Nuclear Sciences Ltd. (IGNS)
- Manaaki Whenua–Landcare Research (Landcare)
- National Institute of Water & Atmospheric Research (NIWA)

The first three on the list have a significant focus on food; the fourth, IRL, contributes significantly to processing of biological materials and in equipment development for the food industry and the others can have some involvement providing specific skills for food projects.

RESEARCH FUNDING

The New Zealand government, via the Foundation for Research Science and Technology, makes available just under $NZ300 million for *public good science* each year. Research organizations bid for these funds, which are distributed across 17 output areas that embrace everything from animal industries (production and processing) to research on Antarctica. There is no specific output for food and food science and technology; instead, it is an integral part of many of the outputs, especially those focused on animal industries, dairy industries, horticulture, arable and other food and beverage industries, manufacturing, and industrial technologies. Funding for these output classes collectively account for about half of the public good science funding. Since the funding is focused on the production aspects of the raw material as well as the food and other products related to the sector, it is difficult to get a measure of the government funding for food science and technology programs. Government funding provides the most obvious research funding source, but individual sector businesses and industries also fund universities and research institutes for R&D projects to adapt generic science or to undertake specific development projects. The dairy and meat industries support, to some degree, dedicated research institutes, but there is no overall food research institute.

FOOD SCIENCE

New Zealand has a number of significant providers of food science research: New Zealand Dairy Research Institute, MIRINZ Food Technology and Research Ltd., three of the CRIs (AgResearch, HortResearch, and Crop & Food), and two of the universities (Massey University and Otago University). Other CRIs and two other universities play some role in food science research.

New Zealand Dairy Research Institute

The New Zealand Dairy Research Institute (NZDRI) is the key provider of strategic product and process research and development to New Zealand's cooperative dairy industry. NZDRI has close links with AgResearch and the Dairy Research Corporation (a joint venture between the NZ Dairy Board and AgResearch) for production research and with Massey University for production, processing, and product research. Since R&D is integrated with marketing and manufacturing strategies, NZDRI works closely with the New Zealand Dairy Board's development laboratories around the world and with individual companies.

Research carried out by NZDRI primarily focuses on new milk products and processes for the export markets, with emphasis on recombined products and longer-life consumer products rather than fresh liquid milk products. Fundamental studies on the properties of milk components and how these can be influenced by production and manipulated during processing feeds research to provide products with specific attributes for defined-ingredient market niches. The research draws on specialized knowledge on separation and evaporation as well as basic protein and fat chemistry. The research on milk proteins has led to a wide range of milk powder products designed for specific bakery applications as well as some for meat products. The research on milkfats and how they can be manipulated has led to the development of butters that are spreadable at refrigeration temperatures. Considerable effort is also directed at cheese production so that the New Zealand industry can meet the demand for continuous supplies of high-quality cheese products with consistent flavor, the development of which is a slow, complex biological process. The two concerns are loss of starter strains (and of product) due to bacteriophage infection, and achievement of the correct balance of microflora for the development of required flavors. NZDRI aims to provide the strategic science to develop starter strains with improved resistance to bacteriophage infection. Knowledge of the physiological and enzymological systems of lactic acid bacteria that contribute to critical flavor compound development from the breakdown of protein and fat will provide the critical science necessary to predict and control flavor development in cheese.

Flavor is another critical component of milk-based products, and the research program aims to enable the industry to provide products with consistent and desired flavor attributes. The investigations focus on the relationships between the sensory characteristics of dairy products, and the concentrations of critical flavor compounds that differentiate the flavor of New Zealand products from that of competitors. Sensory characteristics are being defined using both consumer acceptability and descriptive analysis techniques and related to the concentrations of critical compounds accurately measured using stable isotope dilution assays. Concentrations of critical compounds in products will be related to processing procedures in the factory and to seasonal, dietary, and breed factors affecting milk characteristics on the farm. This information will be used to develop integrated flavor management strategies to meet market requirements for specific dairy flavors.

Much of NZDRI's R&D is confidential, providing products for specific end uses meeting specific applications requirements, but information can be obtained on research undertaken with public good science funding.

MIRINZ Food Technology and Research

Formerly known around the world as the Meat Industry Research Institute of New Zealand (Inc), (MIRINZ), institute has been regarded as the independent central R&D agency servicing the New Zealand meat processing industry. Established in 1955 to study all aspects of meat processing from farm gate to the consumer, the organization has been the primary driver for many of the technological innovations in the world meat industry. The organization was recently restructured with AgResearch taking a 30% shareholding in the new MIRINZ Food Technology and Research Limited while the other 70% remains with MIRINZ (Inc). At the same time the scope of activities was broadened so that MIRINZ Food Technology and Research can apply its skills to other foods.

MIRINZ has been at the forefront of research into meat science, methods, and systems to ensure eating quality; humane electrical stunning systems for sheep, cattle, and deer; packaging developments to maintain product safety over longer periods; and predictive microbiology techniques to alert processors to potential risks. Research continues in these areas to reduce costs, reduce variability, and integrate processes and procedures into pasture-to-plate management systems.

Modeling of biological processes responsible for the conversion of muscle to meat continues to play an important role in the improvement of methods to maintain quality of meat. As a consequence of the predictive models, new chilling and freezing processes have been developed to reduce energy wastage while improving potential product qualities. Mechanization of the dressing and boning processes have been significant areas of endeavor and continue to be important with machines and ergonomic tools being developed that simplify and improve processing of high value products.

Product development, especially of products with nutritional and health benefits, is a crucial area of research to increase the profitability of the industry. The minimization and treatment of wastes from animal and meat processing remains an important area of R&D, since the protection of the environment is crucial to maintenance of a sustainable industry.

MIRINZ Food Technology and Research provides a wide range of research and development activities—with its skills in meat science, physiology, biochemistry, microbiology, measurement technologies, electronics, environmental science, and engineering—to address the needs of the meat industry and, increasingly, the broader food industry (MIRINZ was absorbed into AgResearch in August 1999). The drive is to move the meat industry from being a commodity supplier to a supplier of specific tailored products.

New Zealand Pastoral Agriculture Research Institute Ltd. (AgResearch)

AgResearch rightly sees itself as providing "the bridge between the food, fiber and biotechnology products of New Zealand's pastoral industries and the consumer trends of the 21st century." Although largely focused on production there are three streams of food-related research at AgResearch. First, there is the research aimed at new foods or improving product quality; second, that aimed at allaying customer fears on food safety; and, finally, that aimed at ensuring food is grown in a healthy environment. Research in the first category includes attempts to tailor meat-based products to market specification; alter milkfat composition and casein production; enhance milk flavor; and provide velvet antler products with proven nutritional and health properties. Much of the product research is conducted by association with other groups such as NZDRI, MIRINZ, and Massey University so that the production-modified products can be processed to utilize the improved characteristics. Research to allay consumer fears of food safety has focused on animal disease control, residue elimination or reduction, and detection of natural toxins. The assurance that food is grown in a healthy environment is supported by the programs on animal welfare and development of sustainable production systems from soil management to the final product.

AgResearch has recently established the Agresearch Centre for Pastoral Food Research in Palmerston North as a joint venture with Massey University to develop innovative and differentiated food products with added value derived from forage and forage-fed animals. AgResearch has research capabilities in physiology, genetics, reproduction, nutrition, and management of grazed livestock, biotechnology, and environmental impacts on agriculture.

The New Zealand Institute for Crop and Food Research Ltd. (Crop & Food)

Crop & Food undertakes vertically integrated research and technology development programs to support the development of commercially successful products from the cereal, vegetable, flower, forestry, and seafood industries. For the vegetable and cereal products, there is a strong focus on safe and environmentally sustainable production, but there is also a link with the companies processing the products. Increasingly, Crop & Food supplies information on ways to grow, process, and use foods and their nutritional value for a better lifestyle. One area of nutritional research where Crop & Food has been involved is the relationship between dietary lipids and cardiovascular disease. Animal models are being developed that enable appropriate end points reflecting thrombosis and atherosclerosis to be developed.

Crop & Food is the primary provider of seafood product research. The group has considerable expertise in preservation of the quality of fish from catching to the marketplace, both as "live" fish and as processed product. Crop & Food seafood focus is development of technologies that enhance the consistency of product and product safety, thereby increasing its market value.

Crop & Food has recently established the Food Industry Science Centre (FISC) to link its activities with those of researchers at Massey University, NZDRI, and food companies operating in the North Island with the aim of developing a center of excellence for food, nutrition, and health. Research projects include increasing the storage life of vegetables, expanding the use of maize and oats in food products and the use of vegetables and fruits to promote healthy eating, as well as exploring consumer perceptions of food. The skills of the Institute's staff include expertise in plant breeding and biotechnology, food technology, nutrition, biochemistry, postharvest technology, agronomy, plant physiology, and pest and disease control.

Horticulture and Food Research Institute of NZ (HortResearch)

HortResearch is effectively the primary research arm of the horticulture industry working in partnership with horticulture and food operations. HortResearch is involved in research to assist growers of fruit, some vegetable crops, and flowers in developing new varieties, new environmentally friendly methods of crop production and protection, as well as novel harvesting and processing technologies underpinned by fundamental research. HortResearch has relaunched its food science capabilities under a Food + Innovation banner, to create food research solutions for business built on three key strengths: sensory and consumer science, food technology, and industry foresight with a strong commercial focus.

HortResearch provides research, development, and technology to help bring new and improved products and processes to the marketplace. New varieties of kiwifruit and apples tailored to meet the consumer requirements have resulted from HortResearch's efforts. Noninvasive measurement methods are being developed by HortResearch to enable consistent quality product to be delivered to the marketplace. Changing consumer trends toward convenience and "healthier" foods means that demand for ready-to-eat or use foods is increasing. Such items as peeled apples, sliced rock melon, and kiwifruit and prepared raw vegetables that stay fresh in the refrigerator for up to six months are the possibilities being contemplated by the latest research. Enzyme-peeled fruit technology does away with labor-intensive hand-peeling methods. HortResearch has already developed safe, high-quality clarified juice concentrates, with extended shelf life, that have the flavor and appearance of fresh juices. HortResearch pioneered the technology behind kiwifruit wine, clarified kiwifruit juices and concentrates, fruit nectars, and tamarillo juice products. It has also been involved with enzyme peeling of fruit, heat pump technology, and improved processing of kiwifruit. Processed kiwifruit is naturally tangy but no longer irritates the throat, thanks to new processing methods.

New preservation methods are being developed that reduce the use of artificial preservatives. The use of heat pump drying means manufacturers can reduce the use of preservatives. An example is a new vanilla drying process that improved product quality, increased vanilla content by up to four times, and reduced processing time from three months to a matter of weeks.

HortResearch's interests encompass forestry, environment, forage, animal & biomedical products, pipfruit, summerfruit, viticulture and ornamentals, kiwifruit, berryfruit, citrus, and exotic products as well as the food manufactured from them. Its skill base spans molecular biology, plant and tree breeding, crop production, food processing, fruit storage and transport, and the evaluation of consumer preferences. HortResearch also has a strong technology development unit supplying engineering solutions to assist the biological science developments.

Industrial Research Limited

Industrial Research Limited (IRL), although primarily involved with the manufacturing industries, services natural products processing industries and therefore has a significant role in food research in New Zealand. IRL is heavily involved in process equipment development and biotechnological processing. Key science disciplines are biochemistry, chemical and biological engineering, physics, organic and inorganic chemistry, mathematics, electrical and electronic engineering, mechanical engineering, and information science and technology.

New Zealand Universities

Two of New Zealand's universities have significant programs with a food focus and two others have small food science interests.

Massey University. At Massey University, food science, technology, and engineering can be part of undergraduate degrees in Applied Science, Science, Technology, and Veterinary Science and part of postgraduate degrees and diplomas. Although research into food is primarily conducted within the Institute for Food, Nutrition and Human Health (IFNHH), which is one of eight institutes within the College of Science, there is significant food product and food process research undertaken by other institutes within the College of Science. The IFNHH's scope includes food engineering, food science, food technology, agribusiness, and human and animal nutrition, as well as physiology and biochemistry, as they all relate to lifestyle science. Food science is an integral part of the degree in Food Technology, focusing on the application of sciences and engineering to food processing and food presentation. The emphasis is technological, with a major input in areas of food engineering, biochemistry, biophysics, food microbiology, product and process development, and quality assurance.

Research areas are as varied as the types of food and processes in the industry. Dairy technology research includes studies on casein, cheese manufacture, ultrafiltration and membrane processing, dairy fat characteristics, and uses of whey proteins. Separation technologies are key to many new food product developments. Food rheology is important research not only in terms of dairy products but also in terms of other products and the linkage of rheological characteristics with sensory attributes. Product and process development research involves the total process from product idea generation to product launching using a systematic method for product development, including linear programming, experimental designs, multidimensional scaling, and analysis and profile testing.

Research into food processing and packaging includes wine and fruit juice processing as well as investigation of the changes that occur during processing and storage. Research programs in food microbiology, nonenzymatic browning, and nutritional composition are also undertaken. There is also an increasing emphasis on nutrition and its role in sports and general health.

Within the IFNHH are two research centers with special roles in food research, the Milk and Health Research Centre and the Food Technology Research Centre. The former is a joint venture with the New Zealand Dairy Research Institute and the New Zealand Dairy Board undertaking research into the health-promoting properties of dairy products. The Food Technology Research Centre is a commercial center funded by grants and private research. Its research areas are new cereal and horticultural products, texture, shelf-life extension, consumer research, and product management.

Although there is food-related research undertaken by groups other than the IFNHH, generally it is undertaken by multidisciplinary teams. Examples are research on refrigeration processes from both an energetic efficiency view as well as from a product quality viewpoint and research on packaging technology, which encompasses the technology of the package and its component materials and performance and work on performance of packaging in relation to the product. The research on impacts of processing on quality will use food science and technology expertise often from the IFNHH. The multidisciplinary teams provide a strong resource to tackle almost any food processing problem: functional foods, supply chain management, food safety, food preservation, food formulation, consumer science, and sensory evaluation.

Otago University. At Otago University, a degree in food science is undertaken in the Faculty of Applied and Consumer Sciences. Food science is studied in relation to the chemistry of food, food processing sciences, food rheology, market acceptance, and sensory evaluation. Research is carried out on horticultural crops, especially concerning the composition, performance, and postharvest changes. Flavor research focuses on the impact of processing on flavor precursors. Some research has been conducted on venison. Some of the flavor studies are undertaken in a Plant Extracts Unit, which is a joint venture with Crop & Food. This group examines extracts from a wide range of natural products for uses as flavorings and as health products.

Otago University has two commercial research centers. The Food Product Development Research Centre undertakes development work of both a problem-solving and strategic nature. The center offers a complete service for the development and refinement of new and existing food products. It undertakes comprehensive applied research programs to improve the commercial profitability of food products and assists with in-house food product development programs.

The Sensory Science Research Centre undertakes fundamental and applied research on sensory perceptions and preferences. It investigates the senses that are important in the perception of consumer products such as foods and personal care products and provides training at both undergraduate and postgraduate levels; it also runs short courses on sensory science for industry.

Otago University has a strong nutritional science program that relates the chemistry department with health sciences in examining not only the food but also its impacts on human health and well-being.

Some food research is undertaken at Lincoln University through the Animal and Food Sciences Division and at Auckland University through the Chemistry Department.

FUTURE DIRECTIONS

The future role of food science and technology in New Zealand remains to deliver products that compete in terms of quality, service, and innovation against all others in the world market. Sustainable safe production supported by innovative processing to give long-life natural products that are free from residues and toxins but with characteristics demanded by the high-value customers and consumers is the key focus of the New Zealand food industry and its researchers. Because of the distance from markets, new and improved preservation methods will continually be sought so that New Zealand can satisfy the growing emphasis for convenience and health products. Innovative food products will be required to satisfy the potential demand for home meal replacements as well as the functional food interests.

Biotechnology will have an increasing influence on the direction of the New Zealand food industry leading to new products and new processes. Not only will researchers need to understand the molecular biological means of manipulating foods, they will also need to prove to a skeptical consumer that products are safe.

FURTHER INFORMATION

Further information on the food science and technology research providers and access to current information is available on the Internet from the following Web addresses or the E-mail address for NZDRI:

New Zealand Pastoral Agriculture Research Institute Ltd.: *http://www.agresearch.cri.nz*

The New Zealand Institute for Crop and Food Research Ltd.: *http://www.crop.cri.nz*

Horticulture and Food Research Institute of NZ Ltd.: *http://www.hort.cri.nz*

Industrial Research Ltd.: *http://www.irl.cri.nz*

New Zealand Dairy Research Institute: *Anne.Podd@nzdri.org.nz*

Massey University: *http://www.massey.ac.nz*

Otago University: *http://www.otago.ac.nz*

Lincoln University: *http://www.lincoln.ac.nz/ansc*

B. B. CHRYSTALL
Massey University–Albany
Auckland, New Zealand

INTERNATIONAL DEVELOPMENT: SOUTH AFRICA

The economy of South Africa has the distinct features of both a First World and a Third World sector. Since 1994 the annual growth rate has fallen short of the annual population increase; the population increased from 29 million in 1988 to 41 million in 1998. The result has been increasing unemployment, which has been aggravated by deregulation and globalization of the economy since 1994. From a food supply point of view South Africa nevertheless remains self-sufficient and a net exporter of agricultural produce.

The agricultural achievements may be regarded as remarkable, because the country is not endowed with high rainfall or good arable land. Annual precipitation is 465 mm, compared with the world average of 857 mm. Only 12% of the 86 million ha utilized by agriculture is arable, the vast majority being rangeland used for livestock production. Climatic regions vary from semidesert, with only 125 mm rainfall annually, on the Atlantic Coast to more than 1000 mm annually on the eastern escarpment. About 86% of the area receives almost exclusively summer rains, the exception being the southwestern Cape area, which has a Mediterranean climate and winter rains. The vegetation can be divided into five biomes: desert and semidesert, Mediterranean plants and shrubs, forests, savanna, and open grasslands.

The principal cropping and mixed farming regions are the 1500-m plateau of the Mpumalanga, Gauteng, and Free State area, as well as the northern and midlands of KwaZulu-Natal and the southwestern Cape. The summer rainfall regions constitute mixed farming, with production of corn, sorghum, groundnuts, and potatoes together with dairying and sheep, cow, pig, and poultry production. Sheep farming includes both wool and mutton production, although the major wool and also mohair production is found to the south in the vast expanse of the arid and semiarid Karoo and adjacent regions. The southwestern Cape, with its winter rainfall, is the primary region for the production of winter cereals, deciduous fruit, and wine. Since 1995, wine and grape production for the export market has increased dramatically, the bulk of the increase being produced in the Orange River irrigation scheme. The subtropical low-lying savanna area of Mpumalanga bordering the escarpment is an important region for citrus, subtropical fruit, and vegetables. These commodities also extend toward the frost-free coastal belt of KwaZulu-Natal and the eastern Cape, with additional crops being sugar and pineapples. The far northern and northwestern savanna, the open grasslands, and the area bordering the Kalahari Desert north of the Orange River are major beef producing areas.

AGRICULTURAL PRODUCTION AND EXPORT

The relative importance of agriculture is illustrated by the sectoral analysis of the gross domestic product (Table 1). Agriculture has declined since 1970 but has become more stable since 1990. Real growth rates in agricultural produce have increased by 1.7% per annum since 1960, with the respective sectors being livestock, 1.9%; field crops, 1.0%; and horticulture, 2.9%. In terms of gross value, animal products constitute 42%, field crops 33%, and horticultural products 25% of agricultural produce. As an export commodity, agricultural produce increased from 3.76 billion rand in 1988 to 11.983 billion rand in 1996, which represents a share of the total export of 7.6% and 9.5%, respectively. Animal products constitute a declining share of agricultural produce export, from 34% in 1988 to 11.4% in 1996. In contrast, field crops and horticulture, the major contribution being the latter, increased from 66% in 1988 to 89% in 1996.

Field and Horticultural Crops

Field crops are mostly produced without irrigation. Quantitatively, corn and wheat are the major crops, but significant quantities of sunflower seed, grain sorghum, groundnuts, and barley are also produced. South African corn is recognized worldwide for its quality and stringent control tests. Cultivars are established for yield under dryer conditions, disease resistance, persistence under poor climatic conditions, and lysine and methionine content. In the case of wheat, breeding lines are screened for baking qualities. Specific efforts to eliminate the Russian aphid have been successful. In total, 10 to 15 million t of field crops are produced annually.

The horticultural crop consists of some 30 types of deciduous and subtropical fruit, together with citrus and a number of vegetable species. Grapes are the most important fruit crop, with more than 80% finding their way into wine production. Wine is largely exported. Significant quantities of oranges, apples, pears, peaches, bananas, and other subtropical fruits are also exported. Fruit accounts for about 60% of the gross value of all horticultural products, vegetables and potatoes some 30%, and flowers, tea, and other products the remainder.

Animal Products

The gross value of white meat in 1997 was 5.651 billion rand compared with 4.907 billion rand for red meat and 4.699 billion rand for corn, the most important field crop. Due to the intensive nature and relative efficiency of poultry (white) meat production, the industry has made rapid strides in capturing a substantial share of the meat market. This share is further boosted by meat demand, because South Africa is a net importer of meat. The percentage composition of the different animal products produced is shown in Table 2.

A salient feature in Table 2 is the proportional decline in the market of red meat and fiber (mainly wool) production. Although South Africa still produces of the best-quality fine wool and mohair in the world, the decline in market share, as well as export, is noteworthy.

By far the majority of dairy produce is fresh milk, followed by ice cream, condensed milk, cheese, butter, and fermented products such as yogurt and fruit-based drinks. Dairy produce has lost share of the market (Table 2) since 1980, whereas egg production and ostrich products have shown a steady increase.

Food Consumption Patterns

The per capita consumption of food commodities comprises about 120 kg grain, 65 kg vegetables, 55 kg dairy produce, 35 kg sugar, and 30 kg red meat. Because of price, 30% of the food budget is spent on meat, 20% on bread and grain products, 18% on vegetables and fruit, 9% on dairy products, and the remainder on miscellaneous commodities including sugar, preserved products, and jam. The demographic character of South Africa has a major influence on food consumption patterns, as does urbanization. These

Table 1. Gross Domestic Product

Year	Agriculture (%)	Mining (%)	Manufacturing (%)	Commerce (%)	Other (%)	GDT[a] ($\times$R10^6)
1970	8.1	10.0	23.3	15.1	44	12,037
1980	7.0	22.0	21.6	11.6	38	58,056
1990	5.0	9.2	24.0	21.5	40	247,315
1997	5.2	8.0	23.2	21.8	42	268,142

[a]Values in nominal terms (average inflation 10%); R, rand, GDT, gross domestic total.

factors present exciting, innovative challenges to the food manufacturers.

MANUFACTURED FOOD PRODUCTS

The food and beverage industry is one of the largest industries in the country—in fact, the major manufacturing industry—and a major employment multiplier. The employment multiplier indicates the number of jobs generated in the economy for each new job generated in a specific sector. Table 3 shows the multiplier effect in the animal product and processing sectors. Generally, the multiplier for the primary livestock sectors is small, whereas the processing sectors have the highest multipliers. This indicates that the driving force behind employment creation in the livestock industry, as in the other agricultural sectors, lies within the processing sectors and not within the primary production sectors.

Foods and beverages are produced in fruit and vegetable factories, grain mills, bakeries, breakfast cereal plants, dairy factories, meat and fish plants, sweet and cold drink producers, food oil companies, and many others. The total production of manufactured food products in 1995 exceeded 40 million t, with the proportional contribution being about 81% grain and bakery products, 5.7% sugar and candy, 3.3% fat and oils, 0.6% fish products, 2.0% fruit and vegetable products, and 0.9% meat, dairy, coffee, and tea products (Table 4). The industry is modern, well organized, and growing to comply with the needs and demands of diverse local consumers as well as with those of countries that import manufactured foods and beverages.

Grain Products

The milling industry in South Africa processed 2.5 million t of wheat in 1995. Brown and whole wheat bread production amounted to 575,000 t and white bread 622,000 t. A unique feature of the grain products industry is the fact that corn products produced from white corn make up such a high proportion. This is because corn porridge is a popular food among the majority of the population. Yellow corn is primarily used in the animal feed industry, as is grain sorghum, although a substantial portion is also utilized to brew traditional sorghum beer.

Table 2. Gross Value of Animal Products Produced, %

Product	1980	1990	1997
Meat			
Red	46.0	39.9	29.0
White	15.5	22.3	33.4
Dairy products	18.7	16.6	16.6
Eggs	6.5	8.1	12.2
Fiber	11.9	9.8	4.1
Ostriches	0.5	0.9	1.8
Other	0.9	2.4	2.9

Table 3. Employment Multipliers of the Animal Industry for the Total Workforce, 1985

Sector	Multiplier
Meat products	11.7
Animal feeds	10.0
Dairy products	8.24
Meat animals	1.92
Milk	1.51
Poultry	1.48
Animal fiber	−1.17

Table 4. Manufactured Foods and Beverages, 1995

Product	Tons (thousands)	Percentage of total
Canned and prepared meats	121	0.3
Dairy products	206	0.5
Canned fruit and vegetables	804	2.0
Fish products and similar foods	234	0.6
Vegetable and animal oils and fats[a]	1,343	3.3
Grain mill products[a]	6,117	15
Bakery products	26,807	66
Sugar	2,183	5.4
Chocolates, sugar confectionery, and cocoa	136	0.3
Roasted peanuts and other nuts	351	0.9
Coffee roasting, tea blending, and packing	50	0.1
Other food products	113	0.3
Balanced animal feeds	4,261	
Alcoholic and nonalcoholic beverages	2,256 kiloliters	5.5

[a]Include by-products used in animal feeds.

Sugar Products

In 1995, 2.2 million t of sugar was produced, of which 60% was refined, 3.5% was golden brown, and 13.5% was exported as raw sugar. In the same year chocolates, sugar confectionery, and cocoa amounted to 136,000 t (Table 4).

Oils, Fats, and Fish Products

The production of oils and fats in 1995 amounted to 1.34 million t (Table 4), of which 10%was manufactured into margarine and 20% into edible oils and fats. Fish products for human consumption equaled 234,000 t, of which 25% were canned fish, 28% frozen fish, and 2.5% frozen crayfish. In contrast to many countries, the per capita consumption of fish compared with that of red and white meat is low (<15%).

Fruit and Vegetable Products

The total production of fruit and vegetable products in 1995 was 804,000 t (Table 4). Of this, jams amounted to 30,700 t, canned fruit and fruit salad 183,785 t, fruit juices and drinks 418,400 kL, and canned and frozen vegetables 150,185 t. The industry is expanding rapidly, with a growth rate of 5% per year.

Dairy and Meat Products

The combined production of manufactured products in 1995 amounted to 327,000 t (Table 4). Dairy products consisted of 9.5% butter, 24% cheeses, 25% condensed and powdered milk, and 36% ice cream and related products. Centralization in the dairy industry resulted in the virtual control of the market by four companies. Meat products consisted of 8% bacon, 4.5% ham, 8% patties, 29% polony, and 28% sausage types. As with the dairy industry, the largest proportion of the manufactured meat industry is controlled by three companies. From a consumption point of view, South Africans eat much more fresh meat than meat products.

QUALITY CONTROL AND GRADING

The Department of Agriculture is responsible for compiling specifications for the sorting and classification of most of the processed and unprocessed agricultural products. In the metropolitan areas, vegetables and fruit must comply with certain packaging, grading, and marketing conditions. Local and export regulations apply for canned vegetables and fruit, dehydrated vegetables, dried fruit, and frozen vegetables and fruit. Red meat carcasses are graded according to fatness, lean yield, and meat tenderness. Grading is compulsory for all carcasses auctioned in urban areas. The grading and classification of eggs in specific weight and quality groups apply to eggs for local consumption as well as for export. Poultry carcasses are graded in four classes according to their appearance, fleshing, market readiness, and moisture content in the case of frozen carcasses.

A parastate institution, the South African Bureau of Standards (SABS), operates an extensive testing service for a range of food products with regard to compulsory specifications for canned meat, frozen rock lobster, canned fish, other fish, mollusks, and frozen prawns and crab. It also voluntarily tests for certain canned meat products, carbonated soft drinks, grain sorghum beer, frozen fish, and edible gelatin. The SABS mark of approval guarantees the quality of the product, which is to the advantage of both the manufacturer and the consumer.

AGRICULTURE AND FOOD SCIENCE AND TECHNOLOGY RESEARCH

Research and development needs in a country with both First and Third World sectors are diversified and an enormous challenge. Research activities have to comply with the needs of both the resource poor rural communities, as well as the standards set by countries importing primary and secondary food products from South Africa.

Several universities, government institutions, semistate organizations, and private sector companies are involved in research and technology development in the agriculture and food and beverage industries. In agricultural research the Agricultural Research Council (ARC) of South Africa, a parastate organization, contributes about 65% of the agricultural research capacity, the remainder being from university departments and government stations in the provinces. The ARC, with 13 institutes appropriately situated across the country, caters to the needs of all agricultural commodity research, including in some instances food product and processing technology development. Other institutions with food and beverage research capacities include the Division of Food Science and Technology (Foodtek) of the Council for Scientific and Industrial Research (CSIR) and a number of food science and technology departments at universities and technicons. In addition, many multinational companies in the food and beverage industry operating in South Africa, as well as local meat, dairy, fruit, vegetable, bakery, and beverage manufacturers, have sufficient infrastructure to support product-oriented research, development, and quality control programs. These companies and the government-supported institutions employ food scientists and technologists with both generic and specific training.

To effectively operate as scientists and specialists, they found common ground in the South African Association for Food Science and Technology (SAAFoST). Different scientific societies also promote their sciences and professions and publish research results in disciplinary journals such as the *South African Food Review* and *Food Industries of South Africa*, to name only two of the major ones.

It would be impossible in a contribution of this scope to give an overview of research programs or achievements. Suffice it to mention that new cultivars of fruit with consumer-specific attributes and longer shelf life are introduced to the export market virtually every year, that the potentially problematic product cheese whey is reutilized to the extent of 70% in various dairy products, that South Africa is a world leader in irradiated food technology and nutritive enrichment, that an internationally recognized breakthrough was made with the isolation of fungi and mycotoxin on cereal grains and oilseeds, that a guava puree treated with pectolytic enzymes to remove stone cells has captured the world market, that South African wines compete successfully on the international market with leading wine producing countries, that a controlled-atmosphere packaging technique was perfected to facilitate marketing of deciduous fruit throughout the year, that tender lean beef with less than 15% fat is high in demand with tourists, and that various molecules and substances isolated from indigenous and other plants find their way into international pharmaceutical products. These examples show that the food and beverage industry and its research support are on par with those of the developed world.

GENERAL REFERENCES

D. J. Bosman, "The South African Livestock Industry," *Fundepec International Meeting*, Brazil, August 20–21, 1998.

Manufacturing: Products Manufactured (Food, Beverages and Tobacco Products), Statistical Release p 3051.1, Central Statistical Service, Pretoria, South Africa, 1996.

M. R. Mokoena, *Impact Assessment of Livestock Production Research in South Africa*, ARC-Animal Improvement Institute, Irene, South Africa, 1998.

R. T. Naudé, "International Development: South Africa," in Y. H. Hui, ed., *Encyclopedia of Food Science and Technology*, 1st ed., 1988, John Wiley & Sons, New York, 1992.

H. H. Meissner
ARC—Animal Nutrition and Animal Products Institute
Irene, South Africa

INTERNATIONAL DEVELOPMENT: SPAIN

AGRICULTURE AND FISHERIES

Agriculture

Spain, with an area of 505,990 km^2 and a population of 39.8 million in 1999, is one of five large countries of the European Union (EU) Geographic and climatic differences between regions of continental Spain, with mean annual rainfall ranging from 200 to 1600 mm and annual mean temperatures ranging from 9 to 18°C, are responsible for a great variety in the agricultural production. The particular characteristics of the Balearic and, especially, the Canary Islands contribute to the diversity of Spanish agricultural systems.

Vegetables, with a production of 10.62 million t, citrus fruits, with 4.89 million t, and noncitrus fruits, with 3.08 million t, are among the most characteristic agricultural Spanish productions. Annual grape production raised to 3.35 million t, out of which 12% was table grapes. Olive production in 1996 reached 4.47 million t, table olives representing approximately 4% of total production. The production of grain cereals and pulses was 11.57 and 0.19 million t, respectively, and that of sugar beet and potatoes 7.63 and 4.18 million t, respectively (1).

Meat produced in Spain raised to 4.13 million t in 1996, with pork (56.1%), poultry (21.3%), and beef (13.7%) standing out as the major productions. Cows' milk production, mainly in northern Spain, was 5.56 million t, and significant amounts of ewes' milk and goats' milk (0.23 and 0.30 million t, respectively) were also produced, mainly in central and southern Spain. Egg production reached 0.75 million t per year (1).

The importance of agriculture and food industry for the Spanish economy is shown by the fact that agricultural exports contributed to 15.3% of total exported goods in 1997, for a value of 14,034 million euros (ME), out of which 50.5% were products from the food industry. The value of final agricultural production was 26,772 ME, and that of agricultural imports 11,704 ME. Other countries forming part of the EU constitute the major market (67.5% of the total value) for Spanish agricultural exports, and Spain, on its turn, receives 66.7% of total agricultural and food imports from its EU partners (1,2).

Population active in agriculture was 1.254 million in 1997, which added to 0.437 million active in the food industry sector accounted for 10.5% of total active population.

Fisheries

Fisheries and aquaculture are economic activities of a major importance for Spain, the second country in the world in per capita consumption of fish (more than 35 kg per year). Catches and production of fish and seafood raised to a total amount of 1.21 million t in 1995 (Secretaría de Pesca, unpublished data, 1999). Catches of fish species of marine origin accounted for 75.9% of the total tonnage and were marketed as fresh, frozen, or salted fish (77, 22, and 1%, respectively), with commercial values of 1203 ME, 230 ME, and 20 ME, respectively. Catches of crustaceans represented 2.1% of the total tonnage, with a commercial value of 191 ME, whereas catches of mollusks accounted for 10.9% of the total tonnage, with a commercial value of 218 ME per year.

Fish farming has increased exponentially during the present decade, both in marine and continental waters, to a production with an estimated value of 210 ME per year. Marine aquaculture yielded 0.11 million t in 1995, equivalent to 9.2% of the total tonnage of fish and seafood (Secretaría de Pesca, unpublished data, 1999). Mussels are the product with the highest tonnage. However, fish species such as sea bass, sole, bream, and turbot and seafood species such as oyster, clam, cockle, and prawn are gaining importance. Continental aquaculture produced 0.02 million t in 1995, equivalent to 1.8% of the total tonnage, with trout, tench, and eel as the main freshwater species.

THE FOOD INDUSTRY

The food industry is the most important manufacturing sector in Spain in terms of value of the final production (20% of total industrial production), contributing to 15% of added value and employing 17% of the workforce (3). The Spanish food sector ranks fifth within the EU, accounting for 9.9% of the total value. Total sales of the food industry in 1996 raised to 49,555 ME and total consumption of raw materials to 29,415 ME (2).

The high number of companies (37,857) and of manufacturing plants (42,264) is a characteristic of the Spanish food industry. The reduced volume of production of many small to medium companies results in 40% of the market being shared by a few large companies. Companies of a large size are especially found in the beer and sugar sectors. Bakery (39.9%), meat products (11.3%), and winery (9.6%) are the food industry sectors with the highest number of companies. Because of the considerable differences between company size of the various sectors, the highest total sales corresponded to meat products (19.8%), dairy products (10.8%), oils and fats (9.9%), and bakery (8.4%). Sales of the main Spanish food industry sectors in 1996 are shown in Table 1.

Andalusia (17.6%), Catalonia (13.9%), and Valencia (10.3%) are the Spanish autonomous regions (Comunidades Autónomas) with the highest number of food industry companies (2). Autonomous regions with the highest sales were Catalonia, Andalusia, Castilla-León, Madrid, and Valencia, with 22.0, 18.1, 9.1, 8.0, and 6.7%, respectively, of total sales of the Spanish food industry.

Sixty-four associations of different sectors of the food industry are grouped under the Spanish Federation of Food and Beverages Industries (FIAB). The FIAB is the representative organization of the food industry in committees for regulatory affairs and also offers different services to associates such as commercial and technical information, technology transfer, links between private companies with public research organisms, and so on.

FOOD CONSUMPTION

Food habits have changed considerably during the last three decades in Spain, with a higher proportion of meals

Table 1. Total Sales of the Main Sectors of the Spanish Food Industry, 1996

Food sector	Sales (ME)	%
Meat and meat products	9802	19.78
Milk and dairy products	5347	10.79
Oils and fats	4881	9.85
Bakery	4153	8.38
Canned vegetables	3469	7.00
Wines	3365	6.79
Nonalcoholic drinks	2879	5.81
Milling	1962	3.96
Chocolate and cocoa	1794	3.62
Beer and malt	1710	3.45
Fish and fish products	1630	3.29
Sugar	976	1.97
Spirits	738	1.49

Table 2. Expenditure per Capita for Selected Groups of Foods in Spain, 1998

Group of foods	Euros per capita	%
Meat and meat products	238.2	25.3
Fish and fish products	118.3	12.6
Bread	71.2	7.6
Fresh fruits	66.6	7.1
Dairy products	64.2	6.8
Fresh vegetables	55.1	5.9
Milk	50.1	5.3
Bakery	33.2	3.5
Oils	26.4	2.8
Processed fruits and vegetables	20.4	2.2
Nonalcoholic drinks	19.0	2.0
Eggs	15.0	1.6
Chocolate and cocoa	12.2	1.3
Total foods	939.2	100

out of homes and a decrease in the time employed for cooking at home. A significant increase in the consumption of products manufactured by the food industry with respect to unprocessed foods has occurred. The pattern of food consumption has also suffered from changes in lifestyle, although in general terms it still remains close to the Mediterranean diet. Consumption per capita of some selected groups of foods and beverages were 143.4 kg for fresh fruits and vegetables, 20.6 kg for processed fruits and vegetables, 116.8 L for milk, 28.5 kg for dairy products, 65.5 kg for meat and meat products, 28.4 kg for fish and fish products, 68.6 kg for bread and bakery products, 11.6 kg for rice and pasta, 34.8 L for wine, and 53.4 liters for beer (2).

A long-term tendency to the increase in the consumption of fresh and processed fruits, milk and dairy products, meat, and fish has been recorded during the last three decades, whereas the consumption of bread, pasta, potatoes, fresh and processed vegetables, eggs, and oil decreased during the same period. In the short term, the consumption of fish, olive oil, fruits, vegetables, and wine has increased; that of bread and eggs tends to keep constant; and that of sugar, potatoes, beer, and spirits decreases, indicating the good acceptance of the Mediterranean diet by the Spanish consumer.

Expenditure in foods by Spanish consumers in 1997 reached the total amount of 50,162 ME, out of which 73% was consumed at homes, 25% at public premises, and only 2% at collective facilities. Food consumption at homes was influenced by variables such as town size, age and working status of housewife, social class, and, very strongly, by the number of persons at home. Some data on per capita expenditure for selected groups of foods in 1998 (4) are presented in Table 2.

Food policy in Spain is oriented both to the safety and the quality of products. Food safety is under the competency of the Ministry of Health and the governments of the autonomous regions. Regulations concerning food safety are gradually being implemented at European level by EU directives, although some national regulations are still valid for particular products. Harmonization of national and international standards for different food products is under way.

Food quality programs, under the competency of the Ministry of Agriculture, Fisheries and Food and the governments of autonomous regions, have especially focused on the promotion and protection of high-quality traditional products under European regulations for Appellation d' Origine or Geographic Indication products. Agricultural and food industry products nowadays under these regulations are 111, with a total sales value of 2,799 ME in 1997 (2). The most important products are wines, with 53 Appellations d' Origine and 69% of total value, followed by spirits, with 10 Appellations d' Origine and 19% of total value, and cheeses, with 12 Appellations d' Origine and 2% of total value.

Products from ecological agriculture have considerably increased during the last decade, gaining market both in Spain and for export. Oppositely, the introduction of genetically modified foods in the marketplace is suffering from the rejection of a growing sector of Spanish consumers.

EDUCATION

University studies on food science and technology are relatively new in Spain. Those studies were comprised in other careers, mainly veterinary, pharmacy, and agricultural engineering, and have been redesigned as a second-cycle degree during the present decade. In the new plan of studies, the first cycle consists of the three first academic years of a disciplinary career such as chemistry, biology, biochemistry, and so on or of a more technical career such as veterinary, pharmacy, or agricultural engineering, after which the university graduate degree is attained. Afterward, food science and technology studies, the second cycle, consists of two academic years, although in some cases, depending on previous studies, some complementary subjects may be needed to obtain a five-university-years degree of *licenciado*. The third cycle to attain the philosophy doctor degree usually takes four years, one year of Ph.D. courses, and three years of research work. Twenty faculties all over Spain offer now the second cycle of food science and technology studies, with 50 to 75 new admissions per year in each faculty.

Technical and further education oriented to the different sectors of the food industry are also found in all autonomous regions of Spain, generally focused on the areas of the greatest importance for each particular region.

RESEARCH AND DEVELOPMENT

The Organization of Research

Spain has a complex public research and development system, under the competency of various ministries. Organisms funding research projects on food science and technology are the Inter-ministerial Commission for Science and Technology (CICYT); the Ministry of Agriculture, Fisheries and Food; the Ministry of Industry; and the governments of the autonomous regions. A Science and Technology Bureau (OCYT), reporting directly to the presidency of the government, was created in 1998 to plan, coordinate, and evaluate the research carried out in Spain under public funding.

The CICYT, working closely with the Ministry of Education, is the organism responsible for the National Plan of Science and Technology. Within the National Plan there is a National Program of Food Technology (which has recently changed its name to Agro-Food Resources and Technologies). During the period 1995 to 1998, the National Program of Food Technology distributed 61% of its budget for funding public research projects, 24% for projects in collaboration with private companies, 11% for scientific equipment, and 5% for special actions (CICYT, unpublished data, 1999).

Projects funded by the National Program of Food Technology, oriented both to basic knowledge on food science and to the development of products and technologies, were mostly carried out at university departments (53% of the National Program budget for projects) and at public research organisms such as the CSIC (National Council for Scientific Research), with several institutes actively working on food science and technology (38% of the National Program budget for projects).

The main objectives of the current National Program of Food Technology are (*1*) study of the modification of food components and functional properties of foods by physiological and industrial processes; (*2*) biotechnological processes occurring in or applied to fermented foods; (*3*) development of industrial equipment, processes, and products; (*4*) food safety, with particular reference to toxicological aspects and detection of alergenic compounds in foods; (*5*) nutrition, in particular methodology for the evaluation of the real nutritive value of foods; (*6*) evaluation of food quality, by means of the development of new analytical techniques for sensory grading and equipment for online determinations; and (*7*) improvement of raw materials for the food industry.

Three strategic actions within the area of food science and technology will be launched in the near future, on the following subjects: new species in aquaculture, control of food quality and safety, and improvement of wine quality and competitiveness. Strategic actions are oriented to increase the research effort carried out by private companies in dynamic sectors of the economy. Participation in a project of one or more private companies, generally together with a public research organism, is wished, with funding of the project from both public and private sources.

The Ministry of Agriculture, Fisheries and Food runs its own Sectorial Research Program, managed by the INIA (National Institute for Agricultural and Food Research and Technology), which funds projects on food science and technology, as well as on agriculture and forestry. These projects are mostly carried out at research centers dependent on the regional governments, where a more product-oriented research is done, although universities and other organisms may also compete for this type of funding. Current priorities of this Sectorial Research Program related to food science and technology are (*1*) development of raw materials for the agroindustry, (*2*) postharvest physiology and storage of raw materials, (*3*) applications of biotechnology to the improvement of agroindustry processes, (*4*) improvement of quality and safety of food, (*5*) improvement of artisanal food products, and (*6*) consumer acceptance and market tendency studies.

Particular types of research projects are integrated projects, such as the one already under way on olive oil. In these projects, with a vertical design, all steps from the production of the raw material, including genetics, irrigation, plant protection, and so on to technological processes, final product characteristics, consumer acceptance, marketing, and so on are covered. Multidisciplinary teams of a large size are thus intended to achieve a more global approach to the problems of a sector and to the most promising trends in research and innovation.

The Ministry of Industry, through its Center for Technological and Industrial Development (CDTI), funds concerted projects, jointly carried out by private companies and public research organisms, and development projects, carried out by private companies on themselves. Funding of projects by CDTI consists in no-interest or low-interest loans to private companies involved in the projects. There are no scientific priorities with regard to the subject of the projects funded by CDTI, technological innovation being the main criterion for the selection of projects. During the last three years, the meat, dairy, fruit, and wine industries have shown to be the most innovative sectors, taking into consideration the number of funded projects.

Participation of Spanish researchers in food science and technology projects funded by the EU has gradually increased throughout the successive European Commission Framework Programs, under ECLAIR, FLAIR, AAIR, and FAIR Programs. Research topics in which Spanish groups are involved in the area of Nutrition and Consumer Well-Being of FAIR projects are quality policy and consumer behavior, mealiness in fruits, impact of dietary fat reduction, foodborne glucosinolates, wine and cardiovascular disease, conjugated linoleic acid, food allergens of plant origin, and bioactive constituents in new food plant varieties. In the area of New and Optimized Food Materials, Spanish groups participate in projects on topics such as *N*-3 polyunsaturated fatty acids, molecular markers for olives, production of oral vaccines in plants, genetic manipulation of melon ripening, molecular improvement of strawberry quality, and genetic manipulation of vegetable shelf life. In the area of Advanced Technologies and Processes, current projects deal with the evaluation of fruits' and vegetables' internal quality, physical and biochemical markers of meat quality, high-pressure treatment of liquid foods, osmotic treatment of foods, sourdough starters, frozen food quality, onion wastes, dry sausages ripening, olives' texture, robotics technology, active and intelligent packaging concepts, and improvement of natural resistance of fruit. Finally, in the area of Generic Food Science the research of Spanish groups is focused on natural antimicrobial systems, quality of thermally processed foods, detection of pesticide residues in foods, whey proteins, risk assessment of pathogenic spore formers, and prevention of mold spoilage in bakery products (5).

The Development of New Products

As in any other European country, a considerable variety of new products appears in the Spanish food marketplace every year. Because of the increasing circulation of products within the EU, and also due to the fact that most large Spanish food production plants belong to multinational companies, part of these new products enter simultaneously in the global European market. Low fat content, high fiber content, vitamin enriched, calcium fortified, omega-3 fatty acids enriched, and so on form part of the labeling culture of the new products, which in most cases are compositional modifications of existing products oriented to a growing healthy-food market. Probiotics added to fluid milk, milk powder, fermented milks, and cheese have been extensively introduced in Spain in the last five years, being accepted as a nutritional benefit by consumers of these dairy products. Yogurt cannot be labeled as such if it has been heat treated, but dairy products based on heat-treated fermented milks are more and more common.

Prepared dishes, generally under frozen form, some of recent development but many others following traditional Spanish cuisine recipes, appear continuously in Spanish supermarkets. Traditional Spanish desserts under new commercial presentations are also gaining place in the supermarkets. More original is refrigerated pasteurized gazpacho, a new food industry product that eliminates time-consuming preparation at home of the cold vegetable soup abundantly consumed in hot summer months. The fish products industry has shown to be particularly innovative, with different fish block products, caviarlike products, and especially baby eel-like products (baby eel is a very appreciated, scarce, and expensive fish) developed by means of a sophisticated surimi technology.

The wine industry has considerably enlarged the scenario of Spanish wines, based on the conjunction of new wine-making technologies with the introduction of foreign grape varieties (Chardonnay, Cabernet Sauvignon, Merlot, Syrah) and the revalorization of local grape varieties. A strong promotional effort of moderate wine consumption on the basis of its beneficial health effects is being carried out, regaining market with respect to beer. Olive oil, another characteristic Spanish product, is also regaining market share against other oils and fats in spite of its higher price, on the basis of its nutritional advantages and favored by new attractive commercial presentations, especially of extra-virgin (cold-pressed) olive oil.

BIBLIOGRAPHY

1. *Agricultural Statistics 1997 Handbook*, Spanish Ministry of Agriculture, Fisheries, and Food, Madrid, Spain, 1998.
2. *Facts and Figures of the Spanish Agro-Food Sector in 1998*, Spanish Ministry of Agriculture, Fisheries, and Food, Madrid, Spain, 1999.
3. *The Spanish Food Industry in 1998*, Spanish Federation of Foods and Beverages Industries, Madrid, Spain, 1999.
4. *Food Consumption Panel in 1998*, Spanish Ministry of Agriculture, Fisheries, and Food, Madrid, Spain, 1999.
5. *FAIR Agriculture and Fisheries, Food Projects Synopses*, Directorate General XII, European Commission, Office of Official Publications of the European Communities, Luxembourg, 1998.

M. Nuñez
M. Medina
National Institute for Agricultural and Food Research and Technology (INIA)
Madrid, Spain

INTERNATIONAL DEVELOPMENT: SWITZERLAND

COMMERCIAL FOOD PROCESSING

Environment and Structure

Switzerland is a country with an area of 41,293 km^2; its population was 7.08 million in 1996. It is entirely surrounded by land, has no natural resources, and, except for agricultural and forestry products, does not produce any raw material.

Currently, domestic agricultural production is providing 50 to 55% of the total energy that is consumed as food. Thus a considerable share of raw material for food processing and of processed foods is imported. The self-subsistence of Swiss agriculture varies from food group to food group. Consumption of Swiss products is reaching 100% for dairy products and is around 90% for potatoes; 80% for bread wheat, meat, and pome and stone fruits; 40% for eggs and crystal sugar; and 20% for vegetable oils.

In spite of the high dependence on food imports, the Swiss food and beverage industry has been able to develop into one of the most important industrial sectors of the national economy (see Table 1). Part of this success has been due to the strong presence of Switzerland on the export market (see Table 2) not only with traditional products such as chocolate and cheese but also with food additives (eg, flavors and vitamins) and other auxiliary

Table 1. Total Expenditure for Food and Beverages in Switzerland

Year	Swiss francs (billions)
1990	33.15
1995	36.33
1996	36.25

Table 2. Exports by Swiss Commercial Food Processing Companies (1997)

Product	Metric tons	Percentage of total production
Hard cheese	61,400	60
Chocolate and confectionery	66,782	49
Soups, sauces, and seasonings	39,036	65
Sweets	12,610	59
Process cheese	10,274	50
Canned foods	12,900	15
Pasta products	8,349	14
Baby foods	14,588	85

chemicals, food machinery, and licenses for individual food products and for whole food process operations. Several renowned food machinery companies are exporting more than 90% of their production. Today some Swiss food processing companies belong to the largest ones operating in Europe or the world.

Therefore, the ties of the Swiss food processing industry to other industrial sectors, particularly to the mechanical and chemical industries, have become as important as its traditional relation to agriculture. Nevertheless, agricultural cooperatives are still involved in some commercial food processing (eg, for dairy products, meat products, fruit juices, and wine). A substantial part of the food and beverage production lies in the hands of entrepreneurs with small tradelike operations. These entrepreneurs are professionally trained and federally licensed food specialists. A large proportion of cheese, bakery goods, and meat products are manufactured in this way.

In Switzerland, commercial food processing and retail markets are integrated to a considerable degree. This integration is particularly due to the existence of two large consumers' cooperative organizations that operate their own food processing plants or use plants as copackers. Their combined share on the retail food market is reaching almost 40%. Other parts of the tertiary food sector (eg, catering and restaurants) are mostly independent of the food processing companies.

On the retail market, a high living standard and a high purchasing power maintain the demand for quality and diversity of domestically produced and imported foods. Quality standards are met by an equally high standard of professional training on all levels of education. Diversity is enhanced by the cultural variety based on the four national languages of Switzerland (German, French, Italian, and Rhaeto-Romanic). In turn, the high standards for domestic food quality ensure a lead in quality on the export market, where Switzerland as a small nation would be unable to compete with low-priced bulk products.

Statistical Information

The number of persons involved in commercial food processing (estimated in 1995) is 10% of all persons involved in industrial production.

The number and size of commercial food processing operations was estimated in 1997. There are approximately 9000 small-scale tradelike operations, half of which have fewer than 10 employees. There are 1400 cheese plants, 4000 bakeries, and 2900 butcheries. Switzerland has 134 industrial food processing companies with approximately 190 production sites and approximately 22,000 employees. Of the 134 companies, 13 employ more than 500 people, 39 employ between 100 and 499, and 82 employ fewer than 100. There are 300 plants for beverage production (mineral water, soft drinks, beer, wine, and distilled beverages) with approximately 6800 employees. (Small wine farms with their own vinification and sales are not included in this estimate.)

EDUCATION IN FOOD SCIENCE AND TECHNOLOGY

University Level

University education is preceded by high school education and the Swiss Federal Maturity Examination (baccalaureate).

Swiss Federal Institute of Technology (Eidgenoessische Technische Hochschule, ETH), Zurich. Courses in food science and technology include food engineering, introductory nutrition, and biotechnology. The M.Sc. degree (diploma) is a nine-semester curriculum with 24 months of practical industrial training. It takes 3 to 4 years of graduate research work to earn the D.Sc. or Ph.D. degree. There is a 1-year postgraduate curriculum in nutrition as well.

Universities of Basel, Bern, Lausanne, and Zurich. Selected courses at the undergraduate and graduate levels in food chemistry (particularly food analysis), food microbiology, and food technology are part of the curricula in chemistry and veterinary medicine.

Swiss Federal Office of Public Health. The Swiss Federal Office of Public Health is responsible for licensing official state chemists after postgraduate study courses taught at the Swiss Federal Institute of Technology, Zurich, or the University of Basel, Bern, Lausanne, or Zurich.

College Level (*Fachhochschulen*)

Education on the college level is usually preceded by formal vocational training in the food industry (see "Vocational Training Level"). All colleges award a 4-year curriculum (1 year of general and 3 years of professional education) and a degree equivalent to a B.Eng. or a B.Sc. (Ingenieur FH).

Hochschule Waedenswil. There are two separate curricula in food technology and biotechnology.

Hochschule Sion. The curriculum in food technology includes courses in general biotechnology.

Education in Special Areas: Commodities. Additional college-level curricula are available in beverage technology (Changins), dairy technology (Zollikofen), and biotechnology (Winterthur).

Vocational Training Level

Education under the Swiss Federal Legislation on Professional Training. Programs include 3- and 4-year vocational training courses that are a combination of on-the-job learning at the production site and theoretical education in the classroom (1 to 2 days per week). The courses lead to federally recognized licenses for food technologists (in the areas of bakery technology, chocolate and confectionery technology, and soup and sauce manufacturing), bakers, baker-confiseurs, brewers, butchers, chemical and microbiological laboratory technicians, cheese processors, dairy technologists, fruit and wine processors, and milling technologists.

Further education of graduates from vocation training is possible at the Swiss School of Food Technology, Posieux (1-year course), and for milling technologists at the Swiss School of Milling, St. Gallen (2-year course). There are also advanced extension courses leading to the professional masters diploma in some of these fields. They are organized by the different Swiss associations of professional masters. For food technologists, there is also the possibility of advanced training in food technology and management.

RESEARCH AND DEVELOPMENT

Expenditure for Food Research and Development (1996)

Swiss Food Processing Industry. The Swiss food processing industry spends 0.5% of the total turnover, or approximately 150 million Swiss francs, for research and development. This amount is primarily used for in-company research (93%); the rest is split for contract research at universities (2.5%), joint ventures between industrial and public research (2.5%), and private research institutions (2%).

In addition, corporate research of several worldwide food companies is carried out in Switzerland. Expenditures by chemical industry for research on flavors and vitamins and by machinery industry on process and packaging equipment are also substantial.

Federal Government. The federal government spends 170 million Swiss francs for agricultural research and extension service, approximately 20% of which is allocated for food processing research. Therefore, public funding for food research amounts to approximately 20 million Swiss francs. Some aspects of the research projects are operated on a joint venture basis with industry.

Public International Research Projects. Switzerland is actively collaborating in public international research projects. Examples of these projects are the European Cooperation in Scientific and Technical Research (COST) projects and the Framework Programs of the European Union.

Public Establishments for Food Research

Public establishments for food research include the following:

- Swiss Federal Institute of Technology (ETH) Zurich: Institute of Food Science (basic and applied research in food chemistry, technology, engineering, microbiology, biotechnology, and nutrition)
- Hochschulen Waedenswil and Sion (applied research in food technology and biotechnology)
- Swiss Agricultural Research Stations, Bern (dairy technology), Changins (plant products including wine), Posieux (meat products), Reckenhoz (plant products), and Waedenswil (fruits, vegetables, and wine)
- Swiss Federal Office of Public Health, Bern
- University of Berne (section for food chemistry), University of Lausanne (group for food chemistry), and University of Zurich (Institute of Veterinary Hygiene)

ORGANIZATIONS

Industrial Level

Industrial-level organizations include the following:

- Federation of Swiss Food Industries (Federation des Industries Alimentaires Suisse, FIAL), Berne (comprising subgroups for various food products)
- Federation of Swiss Machinery Industry (Verein Schweizerischer Maschinenindustrien, VSM), Zurich
- Swiss Society of Chemical Industries (Schweizerische Gesellschaft der Chemischen Industrie, SGCI), Zurich
- Several commodity-oriented organizations

Professional and Scientific Level

Professional- and scientific-level organizations include the following:

- Swiss Society of Food and Environmental Chemistry
- Swiss Society of Food Hygiene
- Swiss Society of Food Science and Technology
- Swiss Society of Nutrition Research
- Swiss Association of Agricultural and Food Engineers and Scientists

The first four professional- and scientific-level organizations listed are represented in the Swiss National Committee to the International Union of Food Science and Technology (IUFoST) and to the European Federation of Food Science and Technology (EFFoST). Switzerland also has delegations in the International Union of Food Science and Technology (IUNS), the Division of Food Chemistry of the Federation of European Chemical Societies, in the Food Working Party of the European Federation of Chemical Engineering, in the International Association of Engineering and Food, and in several commodity-oriented food research organizations.

GENERAL REFERENCES

Lebensmittel und Getraenkeindustrie der Schweiz [Swiss Food and Beverage Industry], 17th ed., Orell Fuessli, Zurich, Switzerland, 1998.

J. Solms, "Directory of Swiss Educational and Research Establishments," [Compiled for the European Federation of Food Science and Technology, Department of Food Science, Swiss Federal Institute of Technology Zurich], *Lebensmittel-Technologie* **24**, 53–56 (1991).

Statistisches Jahrbuch der Schweiz [Statistical Yearbook of Switzerland], Neue Zuercher Zeitung, Zurich, Switzerland, 1998.

Swiss Association of Food Industries, *Jahresstatistik* [Annual Statistical Report], Bern, Switzerland, 1998.

Swiss Farmers Association, *Monatsstatistik* [Monthly Statistical Reports], Brugg, Switzerland, 1998.

FELIX E. ESCHER
Swiss Federal Institute of Technology (ETH)
Zurich, Switzerland

INTERNATIONAL DEVELOPMENT: TAIWAN

Taiwan is a tobacco-leaf-shaped island located due east of Fu-Jian Province of China across the Taiwan Strait. It has a surface area of 36,000 km^2, or about the size of the Netherlands, but almost three-fourths of the Island is covered by a steep mountain range running from the north to the south, in which more than 100 peaks are higher than 3000 m. The eastern end of the monsoon belt reaches the island, and the Tropic of Cancer dissects the middle. Thus the climate is subtropic in the plain area but cool temperate up in the hills, and the entire Island has abundance of rainfall. Rather broad and very narrow alluvial plain belts suitable for agricultural exploitation are found on the western and eastern coastal zones, respectively. Rivers originate at the central mountain range and run westward or eastward reaching the Taiwan Strait or the Pacific Ocean, respectively. They are short in length and rapid in flow, so the water is little retained before being discharged into the sea. Nevertheless, intensive and extensive tropical to subtropical agriculture is practiced on coastal plains. Some temporal agricultural activities are also found at the scattered mountain farms where mainly temporal fruits and vegetables are grown. Although forests are dense in mountains, lumbering is almost prohibited for land conservation reasons. Because of the prevalence of the wide range of tropical, subtropical, and temporal climatic conditions, a very rich fauna is found on the island. Similarly, in addition to the main cash crops such as rice, sugar cane, sweet potato, tea, pineapple, banana, and so on, diverse agricultural crops are produced in Taiwan.

Although Taiwan is rich in rainfall and favorable in temperature for agricultural operation, it often suffers from the visits of typhoons and the torrential rainfalls that accompany them. Damages caused by severe wind and excessive rainfall are not limited to agriculture but also extend to many social functions. The development and implementation of technologies for food processing, storage, transportation, and so on are needed for not only enhancing the usefulness and value of food materials but also meeting emergencies caused by natural disasters.

Among 21 million inhabitants, currently Taiwan's agricultural population is stabilized at about 18% of the total in about 800,000 households. The number fluctuates more or less within 5% according to changes in economic situations, because the agricultural economy serves as the buffering domain for accommodating fluctuation in the nonagricultural employment. The size of an average farm is 1.1 ha. The low economic efficiency of ordinary farming is attributed to the small farm size and has provided an incentive to establish containment facilities for the production of higher-value-added products such as medicinal plants, ornamental flowers, off-season fruits and vegetables, and so on, in recent years. The income of farming families is the lowest among all categories of households as seen from the fact that agricultural productions earn only 2.78% of total domestic productions in terms of the dollars.

Up to early sixties, the export of rice and bananas and some processed agricultural products such as refined cane sugar, canned pineapple, mushrooms and asparagus, and frozen fisheries products had played a significant role in earning foreign exchanges needed for industrial development. In 1968 rice and cane sugar alone still represented 11% of total national exports. Although the share of raw and processed food and feed products dropped from 31.6% of total export value in 1968 to 5.5% in 1988 and then 2.8% in 1996, the value increased from $250 million to $3.32 billion and then to $4.08 billion in the same periods. Up to 1996, the total export value of food and feed-related products exceeded that of imports, meaning that the agricultural productivity and capacity of the food industry of Taiwan had been more than self-sustaining in feeding the population in terms of commodity exchange values until recently (Table 1). This situation is now reversed; Taiwan is a net importer of agricultural products by continuously expanding imports and diminishing exports simultaneously. The affluence of the society has caused significant changes in food consumption habits, such as shifting from the rice-based diet to the wheat-based one, and the consumption of more expensive formerly exported and now imported foods and beverages, such as fisheries products, dairy products, beef, and wines and liquors.

DOMESTIC FOOD CONSUMPTION AND ITS EFFECTS ON TRENDS OF THE FOOD INDUSTRY AND INTERNATIONAL TRADE

The per capita food intakes in recent years are shown in Table 2. Besides keeping a constant intake level of carbohydrate, intakes of the other two main dietary ingredients, protein and fat, are increasing steadily, contributing to increases in calorie intake. In particular, the consumption of animal protein has almost quadrupled in 40 years. Daily intakes of fat and carbohydrate were 40 g and 432 g, re-

Table 1. International Trades of Foods ($ million)

Year	Export	% of Total Export	Import	% of Total Import
1993	3484	4.0	2502	3.28
1994	3808	3.7	2906	3.37
1995	4284	3.4	3196	3.18
1996	4083	3.1	3372	3.29
1997	2551	1.8	2931	2.92

Table 2. Per Capita Food Intake

Year	1991	1992	1993	1994	1995	1996	1997
Calories	2857	2875	2930	2979	2971	2972	3045
Protein	91.45	93.64	96.89	96.06	97.72	97.85	101.14
Fat (g)	114.04	114.93	120.38	124.72	125.94	126.80	131.07
Carbohydrate (g)	373.80	373.24	372.51	375.23	369.47	367.45	372.97

spectively, in 1957. We may see a significant degree of substitution of carbohydrate-derived energy by that derived from fat during the past 40 years.

The overintake of food calories is now making obesity a national epidemic. The data in Table 3 illustrate the population's current trend of food habits. These data again illustrate the increase of protein and fat intake in the forms of meat, dairy products, eggs, and cooking oils and also the need for very active international trade of food commodities to satisfy the national appetite.

Among food ingredients, about one-half of cereals and almost all of cooking oils are derived from imported wheat and soybean, respectively. The little availability of arable land, the diversion of land for industrial use, and the growth of the population from about 6 million in 1945 to 21 million in 1997 have made it impossible to produce even a self-sustainable amount of staple food. In 1997 the imports of cereals and their products, mainly wheat and maize, and those of oil seeds and their flours, mainly soybean, were 7.202 million t (metric tons) and 2.818 million t, respectively. Wheat supplements rice as the staple food. Besides being used as the source of cooking oil, soybean is used for the manufacturing of various forms of soy protein foods, such as tofu, various forms of vegetarian meat, and seasoning agents such as miso paste, soy sauce, and so on. Maize, wheat bran, and oil seed flours are used as feeds in swine and poultry productions.

Summarizing the preceding section, we may say that agricultural production in Taiwan is aimed at a self-sustainable supply of foods, yet the former trend of exporting some specialized items for earning foreign exchanges still remains. Formerly such items needed rather low technology for production. Now the trend shifts to the development and manufacturing of high-value-added processed foods. They are for domestic use as well as for export to exchange raw materials that need large land areas for production, and also high-quality and even exotic foods for satisfying increasingly sophisticated social demands.

Food Industry

Most food industrial operations in Taiwan are located in rural villages for the convenience of collection and transportation of raw materials from farms. Production of canned, frozen, and salted foods is closely related to the rural economy and farmers' income. The total number of registered food processing plants in Taiwan has diminished from 6805 in 1987 to 6023 in 1997. The total number of employees is more than 110,000. Several large incorporated companies that have overseas manufacturing branches in mainland China and southeast Asia produce a full range of products such as cooking oils, seasonings, frozen foods, pickles, instant foods, candies, crackers, biscuits and other types of snacks, vegetable and fruit juices, soft drinks, and health foods. However, most factories are small or family businesses specializing in the production of one type of food, and 95% of them have a combined capital of less than $200,000 per operation. The frozen food industry is the largest, with a capital of $0.3 to 2 million and a workforce of 100 to 200 people for each operation. The canned food industry comes to the second, with a capital around $0.3 to $1 million and 50 to 150 employees per plant; the salted food industry is the third, with an approximate capital of $60,000 to $300,000 and 20 to 50 employees per plant. The other industries are worth $30,000 to $60,000 each, and each one employs only 5 to 20 people. The total production value of the food industry was $1.57 billion in 1988, which grew to $14.15 billion in 1997, showing a formidable ninefold increase in nine years. During the same period, per capita national income doubled from $5,829 to $12,019 at current prices, and the percentage of the consumption of food, beverages, and tobacco to the

Table 3. Per Capita Intake of Food Items (g)

Year	1991	1992	1993	1994	1995	1996	1997
Cereals	273.42	275.05	272.91	279.50	275.78	266.18	263.56
Tubers	58.03	60.08	59.42	58.00	49.49	53.93	64.11
Sugar and honey	72.03	71.17	69.42	68.83	67.18	68.14	66.98
Kernels and oil seeds	80.41	70.41	80.09	82.87	6.89	86.99	86.71
Vegetables	259.53	266.28	268.73	255.70	278.08	297.41	296.63
Fruits	379.97	354.70	396.11	374.08	378.98	379.32	411.17
Meats	176.73	183.51	192.68	198.33	199.34	205.03	211.80
Eggs	36.58	37.49	38.93	40.79	44.47	48.01	53.23
Fisheries products	108.82	120.08	130.02	105.12	105.08	104.91	116.03
Dairy products	46.91	50.04	55.29	58.25	63.10	62.92	64.20
Cooking oil	64.97	64.35	68.25	71.02	71.32	71.34	73.27

household expenditure decreased from about 35 to 25%. These trends show clearly that the demands for processed and prepared foods, although at prices much higher than the raw food materials available at markets, have increased in recent years. The main reasons for the shift in food habits may be listed as follows: More women are moving into the job market; more modern home appliances are available; the younger generation has become less interested to learn cooking; more outdoor activities are accepted as the lifestyle, and so on. Instant noodles (made from wheat or rice) are one of the fastest-growing products in this category. A wide range of baked goods, frozen foods, packaged dairy products and fruit and vegetable juice drinks, soft drinks, snack foods, and so on have a large share of food consumed in Taiwan. Besides using much of imported cereals, oil seeds, starches, fruits, and so on, the food industry depends largely on the domestic agricultural industry for raw materials. Thus, the food industry is considered to be a great booster to the rural economy.

The government policy of market liberalization and import tariff reduction, together with appreciation of the New Taiwan dollar in recent years, has also boosted dramatically the increase of importation of processed foods. The value increased from $679 million in 1987 to $2.40 billion in 1997. The flourishing economy and rising national income have led to a big change in consumer habits for foods and drinks. The trend of consumer taste is toward high-quality, semiprepared and flavored, or prepared and flavored foods because such products are much more convenient. The partial lifting of the monopoly policy over production and trade of alcoholic beverage has witnessed a sharp increase in the import of beers, grape wines, cereal and fruit liquors, and so on.

Cereal and Soybean Products

The import of wheat from the United States to Taiwan started in the fifties. Wheat products such as bread and noodles have replaced a large part of the traditional staple food rice, and bread and noodles in various forms have a big share of the domestic staple food market now. The instant noodle alone had a sales value of $26 million in 1977, which quadrupled to $110 million in 1987, and further increased to $346 million in 1997, all along with a proportionate increase in volume. As wheat products have increasingly supplanted the traditional staple food rice, although the brown rice production decreased from the peak value of 2.713 million t in 1976 to 1.663 million t in 1997, and the population increased from 16.508 millions to 21.683 millions in the same period, the domestic rice supply has been suffering from a surplus. The development and manufacturing of rice products, such as rice crackers and instant rice foods, are encouraged by the government to alleviate this situation. Rice crackers are popular; some are imported from Japan, and some are produced locally in collaboration with Japanese firms. However, they use mainly glutinous rice, which is not common for daily use, as the raw material. Imports of breakfast cereals and biscuits from Western countries are increasing. These facts also contribute to the deepening of the grave situation of rice surplus.

Besides producing some vegetable soybean that is consumed domestically and also exported to Japan, the supply of soybean in Taiwan depends almost exclusively on the import from the United States. As mentioned before, imported soybean is used mainly for oil extraction and the residual flour for feed formulation, and for manufacturing of various forms of vegetarian meats, including tofu and tofu products; it is also important in the soy sauce industry. In 1997 the production of soy sauce amounted to 1.034 million t and was valued at $95 million. Most soy sauce is consumed domestically, because it is not generally used in Western cuisine.

Meat and Poultry

Pork is the major meat consumed in Taiwan. Annual per capita pork consumption increased from 20.6 kg in 1967 to 33.86 kg in 1987 and then stabilized at around 39 to 40 kg since 1993. However, the ratio of pork consumption to total meat consumption has declined in recent years due to the rise in poultry and beef consumption.

Poultry ranks second as the preferred meat. The poultry industry, similar to the swine industry, has shown a tremendous growth, largely because of technological improvements, a high degree of specialization of the industry, and an ample supply of feeds through the very active international trade and development of the food industry, which yielded many by-products suitable for feed use. The per capita poultry consumption increased steadily from 5.12 kg in 1967 to 20.12 kg in 1988 and 33.49 kg in 1997, with its share in total meat consumption increased from 19.5 to 35.4 and 43.3%. Poultry consumption is expected to grow more.

Meat imports, including beef and mutton, rapidly grew with the quick development of the fast-food industry in recent years. Domestic production of cattle is limited, and the number raised is decreasing. The population of sheep is almost nonexisting. Frozen meats are imported from Australia, New Zealand, and the United States. There was 38,349 t of beef imported in 1988, which increased to 56,785 t in 1997, valued at $129 million and $169 million, respectively. The imported beef satisfied nearly 80% of the consumption. The import volume of mutton is about one-third that of beef. The per capita consumption of mutton was 0.86 kg in 1993 and 1.30 kg in 1997, or equivalent to 17,200 t and 27,300 t of annual import, respectively. We may expect that more imported beef and mutton will supplement the consumption of domestic pork.

Pork is the main item of meat Taiwan exports. The market is Japan. Annually nearly $500 million worth of pork had been exported, but, due to the outbreak of the foot and mouth disease in 1997, it has been halted since then.

Dairy Products

Taiwan launched the Dairy Farmer Support Program in 1958. Dairy farmers got reasonable profits from their milk production, and the program was further extended with the result that dairy farms increased from 804 in 1978 to 1,322 in 1988. Milk production increased from 44,615 to 144,390 t during the same period. Most milk produced is consumed fresh (eg, 105,703 t of the 1988 production was consumed fresh), but due to the variations in milk supply and market demand, surplus milk in winter is processed into dry milk and sweetened condensed milk. Fresh milk consumption and milk powder production reached the peak values of 215,799 t in 1995 and 9,603 t in 1992, respectively, but both declined to 179,567 t and 6,907 t in 1997. Although the self-supply ratio of dairy products increased and imports decreased, dairy products are still leading in imports. The main forms of imported dairy products are milk solids and butter with cheese and curd sharing about 10% of the total. The import value increased from $144 million in 1983 to $363 million in 1997. The average yearly milk consumption per capita comprises both locally produced milk and imported products. When converting their food values into full-cream milk equivalent, the total milk consumption per capita was 1.50 kg in 1957, which increased to about 50 kg in 1987. Because local production of fresh milk now meets only about 20% of the market demand, there is much room for the domestic dairy industry to develop.

Beverages

Besides the local brands of various types of soft drinks, either carbonated or noncarbonated, various kinds of soft drinks bearing international brands, such as Coca-Cola and Pepsi-Cola, are bottled in Taiwan.

Fruit juices and vegetable juices, either single or mixed, produced from local raw materials are abundant. But fruit juices, such as orange, grapefruit, peach, and so on, reconstituted from imported concentrates are also having a large market share. A very interesting feature of Taiwan's fruit and vegetable juice industry is its highly diversified products. You may find juices and juice drinks bearing the names of asparagus, guava, carambora, coconut, winter melon, plum, sugar cane, mango, pomelo, tangerine, litchi, water melon, passion fruit, honey dew, lotus root, and kiwifruit in addition to the names ordinary found in Western markets. Sport drinks aimed at replenishing blood salt balance and providing a quick energy supply grew fast at the initial phase of their introduction from 1985 to 1990 but are now stabilized. Because most people consume soft drinks and juices for quenching thirst, the seasonal variation in consumption is large.

Owing to the monopoly policy over alcoholic beverages, local products dominated the market until recently. Now, imports are taking a share of about 25% of the total consumption of $1.92 billion.

Locally produced beverages consumed in Taiwan in 1997 are (in kiloliters and $ million, respectively) alcoholic, excluding beer, 241,365 and 791; beer, 365,406 and 637; fruit and vegetable juices, 260,194 and 195; carbonated soft drinks, 456,620 and 311; sports drinks, 171 and 773. Other types of nonalcoholic beverages, including coffee and tea drinks, mineral water, traditional beverages, and so on, had a total sales value of $593 million.

Fruits and Vegetables

The consumption of fruits and vegetables is still increasing, although at a slow pace, as shown in Table 3. Most are consumed fresh. Canned, frozen, dehydrated, or preserved fruits and vegetables are not as important in the domestic market as in the foreign market. But manufacturers are paying more attention to them in the domestic market, because these items are now less competitive in the international market, and changes in the social situation have enhanced their domestic consumption.

The frozen food items for domestic consumption, except ice cream, include frozen vegetables (mainly mixed vegetables), pork dumplings, hamburger patties, pork balls, and fish balls. Frozen poultry and seafood have not yet received a welcome from consumers in Taiwan, but the government had launched a sales promotion campaign. To help consumers identify excellent frozen food, and to help producers enhance quality and sanitation in their products, the Council of Agriculture implemented an "Excellent Frozen Food Mark" system in 1988. Establishment of wholesale-type chain stores and supermarket facilities islandwide has overcome the difficulties experienced in the early stage of the promotion campaign, which were due mainly to the lack of cold chain facilities.

Canned pineapple, mushroom, and asparagus played an important role as the foreign exchange earner in the development of the Taiwan economy. The canned pineapple industry was initiated during the colonial era. It had a systematic operation from cultivation, harvesting, and processing to exports handled by a big company. The other two items, the cultivation of which was started in the fifties on small farms, were processed by small-scale factories by using raw materials collected locally. Exports of pineapple and asparagus have declined to almost nil, but supplies such as fresh fruit and vegetables are still available in the domestic market. Various kinds of canned mushrooms still have foreign markets, the export of which amounted to $283 million in 1996. Bamboo shoots have found a niche in the international market in the meantime, and the export of canned products amount to $531 million in 1996. Tomatoes have been grown in Taiwan as far back as the nineteenth century. Originally, they were treated as fruit, to be eaten raw. It was not until 1967 that local farmers began growing tomatoes specially for canning. At the peak time, Taiwan's canned tomato products, ranging from 2 million to 3 million standard cases a year, accounted for 2% of the world's canned tomato production. It is now rapidly declining, with the annual export values of $71.4 million in 1995, $12.9 million in 1996, and $3.5 million in 1997.

Frozen Food

The frozen food industry has been rapidly developing during the past two decades. Frozen food became a leading export item among the processed products in the early 1980s. Export value reached more than $2.25 billion in 1996. In spite of the appreciation of the New Taiwan dollar and keen competitions in the international market, frozen food processors in Taiwan have enjoyed a steady growth of exports in recent years. And yet they have been able to maintain outstanding quality, achieve significant reduction in production costs, and enhance production efficiency. Among the factors contributing to such an excellent performance are the advancement in quality control technology, the installation of modern production facilities, and the assistance provided by the related government organizations. The major frozen food items and their production values in 1996 (in metric tons and $ million) are vegetables, 32,326 and 59.05; fruits, 1461 and 2.52; seafood, 359,303 and 838.93; meat and offal, 163,227 and 4016.4; prepared foods, 16,541 and 199.2; and eggs and egg products, 13 and 0.70. The majority of frozen food factories are located close to their raw material supplying areas in central and southern Taiwan. However, frozen seafood producers choose their plant sites in areas close to the two main seaports, Keelung and Kaohsiung, located on the northern and southern tips of Taiwan, respectively. About 98% of Taiwan's frozen foods are exported to Japan, the Federal Republic of Germany, the United States, Hong Kong, and Canada. The three biggest items are pork, grass shrimp, and eel. The most important frozen vegetables are soybeans, pea pods, and string beans. Prepared foods in frozen form, including roasted eel and Chinese-style convenience foods, now enjoy a booming market in Japan.

Dehydrated and Preserved Foods

The dehydrated and preserved food industries are important sources of income for farmers who grow fruits and vegetables on slope land. The main items are ginger, prunes, eggplant, mustard, cucumber, mushrooms, and bulbous onions. The total export values for dehydrated and salted foods in 1996 were $60.6 million and $74.4 million, respectively; Japan imports 80% of Taiwan's salted foods. Because most products are in semifinished form, the profit is limited.

Others

Among the great changes in international trade that Taiwan has witnessed during the past half-century, we should mention the disappearance of cane sugar from the list of top export items, and the emergence of seasoning agents, mainly monosodium glutamate, a more-advanced product of fermentation technology, which enjoy a significant position continuously in the list of total productions and exports. In 1996 sugar and seasoning agents had total export values of $14.8 million and 144.9 million, respectively. These figures again signify that the era of agricultural development to support industrial takeoff has already become a part of Taiwan's history.

RESEARCH AND DEVELOPMENT AND FUTURE TRENDS

Taiwan has identified food science and technology as one of the major fields of national research and development (R&D) programs in the midseventies. The policy was formulated primarily not to enhance the international trade values of agricultural products, but to improve the nutrition and safety of foods as well as to enhance the economic values of agricultural resources so as to improve the rural economy. Since then the R&D capabilities at academic and commercial institutions have been strengthened along with the establishment of the Institute of Food Science and Technology (IFoST), an academic society supported by all related institutions and private scholars and technologists, to promote and coordinate all food-related activities. The number of students enrolled at both undergraduate and graduate levels in teaching institutions, the number of teaching and research staffs, the investment in R&D facilities and research program funding, and the number of R&D personnel holding advanced academic degrees have greatly expanded. We can see the outcomes of these efforts, first, in the continually growing food industry and, second, from the fact that the annual meeting of the IFoST is attended by more than 1000 participants. At the annual meeting, several hundred original research papers are presented and a rather large-scale food show is staged.

For practical on-the-job training, the Food Industry Research and Development Institute (FIRDI) located at Hsinchu (northern Taiwan), which was originally established under the sponsorship of the Canned Food Exporters' Association in the early sixties but is now a government-funded foundation, plays a major role. In addition, FIRDI provides contractual services on research and training for satisfying both government and industry needs. The Food and Drug Inspection Bureau of the Ministry of Health located in Taipei mainly researches food safety inspection technology. One of the major material bases for biotechnological R&D, the Type Culture Collection (microbial, animal, and plant cell lines), is also operated by FIRDI. Gene banks for supporting agricultural research are located at the Provincial Agricultural Research Institute in Taichung (central Taiwan) and the Asian Vegetable Research Institute in Shanhua, Tainan (southern Taiwan). These establishments started as rather modest operations but now have been greatly expanded to serve the international community as well.

The era of the development of specific type of foods at academic institutions is almost gone. Their research is now more concentrated on the basic aspects of food science and the development of modern biotechnology. So besides the Council of Agriculture–supported, mission-oriented practical aspects of technology development at government as well as private institutions, funding from the National Science Council is approved on a competitive basis and based on the scientific merits for future advancement rather than the prospect of immediate benefits.

We may see changes in the trends by citing the policy promulgated by the relevant government agencies as follows. The policy on the development of food technology includes:

1. Securing the sources of food supply and strengthening the production and marketing systems
2. Leveling-up the technology and improving the infrastructure of the industry
3. Leveling-up the food quality and safety
4. Enhancing competitiveness to promote domestic and international sales
5. Improving the environment for investment
6. Promoting an international cooperation mechanism in which the root remains in Taiwan
7. Streamlining of policy formulation and implementation

The strategy for development includes:

1. Streamlining a collaboration mechanism among government agencies related to food research, production, marketing, safety regulation, environmental concerns, and so on
2. Totally upgrading products
3. Strengthening research activities aimed at new products development
4. Rationalizing and enhancing productivity
5. Strengthening training
6. Modernizing production management and marketing systems
7. Promoting the interchange and cooperation between food industries on both sides of the Taiwan Strait

GENERAL REFERENCES

FIRDI Statistics Database, Food Industry Research and Development Institute, Taiwan, ROC, 1999.

Food Industry Statistics of the Republic of China, Food Industry Research and Development Institute, Taiwan, ROC, 1998.

Monthly Reports on Industry Statistics, Taiwan, ROC, Council for Economic Planning and Development, Taiwan, ROC, 1999.

Taiwan Statistical Data Book, Council for Economic Planning and Development, Taiwan, ROC, 1998.

JONG-CHING SU
National Taiwan University
Taipei, Taiwan

INTERNATIONAL DEVELOPMENT: UNITED KINGDOM

Services and industry are the major employers in the UK, and agriculture employs a relatively small workforce of half a million. This is equivalent to ~2% of the total civilian workforce, which is low by comparison with all other European countries (1). The food industry, on the other hand, employs more than 2.5 million. Agriculture and the food industry make a considerable contribution to UK exports. There is a high level of trade in food commodities: in 1997, food, beverage, and tobacco accounted for 6.5% of gross domestic product (GDP) as exports and 8.8% of GDP as imports (1).

THE REGULATORY ENVIRONMENT

The main piece of primary food legislation in Britain is the Food Safety Act (1990). Northern Ireland has equivalent legislation: the Food Safety (Northern Ireland) Order (1991). A feature of the Food Safety Act is that in addition to enabling legal rule-making, it also provides for defenses in law, including due diligence defense. Thus, manufacturers, importers, retailers, or individuals can use due diligence as a defense, if it can be shown that all reasonable precautions and due diligence were used to avoid the commission of an offense.

Central responsibility for the food supply in the UK is shared between the Ministry of Agriculture, Fisheries and Food (MAFF) and the Department of Health (DoH). Some responsibility in Wales, Northern Ireland, and Scotland is delegated to, for example, the Welsh Office, the Scottish Office Agriculture and Fisheries Department, the Department of Health and Social Services for Northern Ireland, and the Department of Agriculture for Northern Ireland. MAFF also maintains an executive agency, the Central Science Laboratory (CSL), at sites in Norwich and York. The CSL is responsible for the protection of consumers, food supplies, and the environment by means of food surveillance and research programs. Other executive agencies of MAFF include the Pesticides Safety Directorate (PSD; responsible for licensing and monitoring pesticides with respect to safety, effectiveness, and use) and the Veterinary Medicines Directorate (VMD; responsible for licensing and control of veterinary medicines and surveillance of residues in foods). The executive agency responsible for advising MAFF on land management, food production, waste disposal, and some aspects of food processing and retail, the Agricultural Development Advisory Service (ADAS), was privatized in 1997.

A number of independent advisory committees are established to advise government departments on aspects of food policy and safety. The main task of the Food Advisory Committee (FAC) is to advise on the safety of chemicals in foods, food additives, and food surveillance. However, policy on food labeling, including nutritional and health claims, is also determined to a large extent by the work of FAC. The assessment of novel foods, including genetically modified foods, is conducted by the Advisory Committee on Novel Foods and Processes (ACNFP). More specific toxicological expertise is available on the Committee on Toxicity of Chemicals in Foods, Consumer Products and the Environment (COT), to which specific questions may be referred by the FAC or the ACNFP. Similarly, the government has access to expert nutrition advice on the Committee on Medical Aspects of Food and Nutrition Policy (COMA). In the case of genetically modified organisms, two additional committees are involved in the research and development stages: the Advisory Committee on Genetic Modification (ACGM) and the Advisory Committee on Releases to the Environment (ACRE). The Health and Safety Executive, as advised by ACRE, is responsible for enforcing legislation concerned with occupational safety and the safety of members of the public exposed to agents in the environment. The Advisory Committee on the Microbiological Safety of Foods (ACMSF) advises the government on the risks to humans of microorganisms used or occurring as contaminants in foods. Other advisory committees that deal with the food supply include the Advisory Committee on Pesticides (ACP), the Consumer Panel, the Spongiform Encephalopathy Advisory Committee (SEAC), and the Veterinary Products Committee (VPC).

Enforcement of food legislation in the UK is largely decentralized. At a local level, food control is the responsibility of environmental health officers (EHOs) and trading standards officers (TSOs) employed by local authorities (of which there are 589 in the UK). The former deal with issues of safety and hygiene, the latter with food quality, composition, labeling, and claims. There is some overlap between the activities of EHOs and TSOs and, in some cases (eg, contaminants), either or both may be involved in an action. Local authority food inspectors are responsible for enforcing the Food Safety Act. Public analysts support the activities of Environmental Health and Trading Standards departments. In addition, samples may be referred to other laboratories, such as the Public Health Laboratory Service (PHLS) or the Central Science Laboratory. Coordination of the activities of individual EHOs and TSOs is provided by the Local Authorities Coordinating Body on Food and Trading Standards (LACOTS), the PHLS, DoH, and MAFF.

Medicinal claims are only permitted for products licensed under the Medicines Act (1968). The FAC recently published draft guidelines on health claims on foodstuffs, which is the basis for a code of practice proposed by a working group drawn from industry, local authority representatives, and consumer representatives: the Joint Health Claims Initiative. The voluntary code of practice proposes two main types of health claims associated with all foods marketed (including supplements): generic health claims based on well-established knowledge, and innovative health claims where substantiation is required in accordance with proposed criteria. The market for organic foods in the UK is small compared with Germany, Austria, or Denmark but is growing rapidly. Most certification of organic foods in the UK is conducted by the Soil Association, which was established in 1946 and contributed to the formulation of the first EC organic food regulations in 1974.

The government published a white paper in 1998 containing proposals for the establishment of an independent Food Standards Agency for the UK (2). This has been the subject of much discussion among the interested parties, including food industry, professional societies, and the relevant government departments. The starting point for such an agency would no doubt be the current Joint Food Safety and Standards Group set up by MAFF and DoH. Various templates have been proposed for such an agency, including the current Health and Safety Commission and Health and Safety Executive (HSE). The overall mission of the proposed agency would be to provide objective advice to government on the food supply and to protect public health. More specifically, the following activities have been proposed (2): policy formulation; drafting secondary legislation; representing the UK internationally (eg, at EU, WTO, and Codex Alimentarius Commission); providing information and education, food research, and surveillance, and food law enforcement; and issuing licenses and approvals. It is proposed that some of the funding for the agency will come from a levy imposed on all food premises in the UK ($\sim$600,000 manufacturing plants, shops, and catering facilities). This controversial proposal has provoked much debate.

THE INDUSTRIAL ENVIRONMENT

Mergers and acquisitions have had an impact on UK food companies as elsewhere. The most notable of such mergers was between Grand Metropolitan (parent of Pillsbury and Burger King) and Guinness at the end of 1997, creating Diageo, the world's biggest alcoholic drinks company and the UK's ninth biggest company. Following Diageo, the other major UK food companies are Albert Fisher Group Plc, Associated British Foods, Booker Plc, CPC (UK) Ltd, Cadbury Schweppes Plc, Nestle Plc, Northern Foods Plc, RHM Ltd, Unigate Plc, Unilever Plc, and United Biscuits (Holdings) Ltd.

RECENT ISSUES AFFECTING THE UK FOOD CHAIN

In recent years, the UK food system has been characterized by a series of high-profile events concerned with food safety and the acceptability of new products by consumers: the bovine spongiform encephalopathy (BSE) crisis, outbreaks of food poisoning, and the introduction of genetically modified plant foods.

BSE was first recognized in cattle in the UK in 1986, although there are suggestions that it existed earlier. The disease resembles other transmissible spongiform encephalopathies (TSEs) such as Creutzfeldt-Jakob disease (CJD) and scrapie. The disease was attributed to the use of infected meat and bonemeal in cattle feed, although maternal transmission was suspected in some cases (3). The ban on such feeding practices led to a dramatic reduction in the incidence of the disease. However, the discovery of a variant form of CJD in British patients in 1996 (4) led to the suspicion that eating infected beef was responsible. The ensuing crisis resulted in a ban on exports of British beef and the slaughter and destruction of thousands of cattle. The immunological detection of the prion protein in the tonsils and appendices of patients with variant CJD may offer the potential to better assess the number of people likely to develop the disease (5,6). To date, all cases of the variant form of CJD have been in the UK, with only one exception, and all those affected have been homozygous for methionine at codon 129 of the PrP gene, which opens the possibility that there may be a genetic predisposition to the disease (4,6).

The incidence of the variant form of CJD is now being monitored by the National Creutzfeld-Jacob Disease Surveillance Unit in Edinburgh. Some evidence, based on postmortem data collected since 1995, indicates that the incidence may be increasing (7). However, an unusually high number of deaths from the disease in the last quarter of 1998 is currently complicating statistical analysis. Continued data collection over several years will be necessary before any firm conclusions are drawn about the etiology of the disease.

Several outbreaks of microbiological food poisoning have led to illness and death in the UK in the past 10 years. These events and the publicity surrounding them, in addition to the BSE crisis, have been major driving forces in the UK government's plans to establish a Food Standards Agency with executive responsibility for the food system. An infectious disease surveillance system is operated by the Communicable Disease Surveillance Centre (CDSC) of the Public Health Laboratory Service (PHLS). The most notorious outbreak occurred in Lanarkshire, Scotland, in November and December 1996 due to cooked meat contaminated with *Escherichia coli* 0157. Almost 500 people were affected and 18 died, the second highest number of deaths associated with an outbreak of *E. coli* 0157 in the world: The age range of those who died was 69 to 93 years. The outbreak was traced to a single butcher's shop, the owner of which was subsequently successfully prosecuted in a criminal trial. The butcher's shop was closed at the peak of the outbreak, but efforts to prevent infection were complicated by the diverse distribution chain from the shop at the source of the outbreak: a total of 85 outlets throughout central Scotland were eventually identified. An expert group was set up to investigate the outbreak under the chairmanship of Professor Hugh Pennington, University of Aberdeen. Some of the recommendations of the Pennington Group can be summarized as follows (8):

1. Better education and training of farm workers, slaughterhouse staff, and food handlers.
2. Hazard Analysis and Critical Control Point (HACCP) principles should be enshrined in legislation governing slaughter houses and meat transportation; accelerated implementation and enforcement of HACCP contained in existing legislation.
3. Licensing should make provision for HACCP implementation and record keeping, including temperature control and monitoring.
4. Inclusion of food hygiene on school curricula.

Although the consequences of the Lanarkshire outbreak were serious, verocytotoxin-producing *E. coli* is a relatively minor cause of food poisoning outbreaks in the UK. In 1997 and 1998, the main contributor to foodborne disease outbreaks was *Salmonella enteritidis* (9), which is closely associated with eggs and poultry. *Camplyobacter* infections are the major cause of laboratory-confirmed enteric infections (outbreaks plus sporadic) (9). The parasitic protozoan *Cryptosporidium* has been associated with waterborne enteric infections in the south of England. The most notable outbreak had 345 confirmed cases in the North Thames area in 1997 (10). The multi-drug-resistant strain, *S. typhimurium* DT 104 is increasing in importance as a cause of food poisoning (more than 4000 cases in 1996). Most outbreaks of enteric infections in the UK in recent years have been associated with hospitals and residential institutions.

Some preliminary evidence indicates that some cases of Crohn's disease may be linked to infection with *Mycobacterium paratuberculosis*, which is present in some dairy cows (11). The organism has been the subject of several studies in the UK because of the possibility of some cells surviving conventional pasteurization when levels of contamination are high. MAFF has commissioned a survey of raw and pasteurized cow's milk to determine the incidence of contamination with viable *M. paratuberculosis* cells.

One of the most significant scientific events in the UK in recent years was the world's first cloning of a sheep at the Roslin Institute near Edinburgh. The lamb, named Dolly, was cloned from an udder cell derived from an adult sheep (12,13). The same group has reported success in the use of transgenic sheep for the manufacture of human proteins such as the clotting Factor IX (14), which opens up the possibility of so-called pharming for the manufacture of pharmaceutical products.

RESEARCH INFRASTRUCTURE

Much UK research is supported through a series of autonomous research councils, funded by government through the Office of Science and Technology/Department of Trade and Industry. Those principally involved in food and nutrition research include the Biotechnology and Biological Sciences Research Council (BBSRC) and the Medical Research Council (MRC); the Engineering and Physical Sciences Research Council (EPSRC) and the Natural Environment Research Council (NERC) have a smaller impact on the food and agricultural sciences. The past decade has seen a major restructuring of the research infrastructure in the UK, changing the way in which food research is conducted and funded. The former Agricultural and Food Research Council (AFRC) was merged with the biological sciences program of the former Science and Engineering Research Council (SERC) in 1994 to establish the BBSRC. The BBSRC supports several institutes that contribute to food research, the most important of which is the Institute of Food Research (IFR) based at Norwich. Other BBSRC institutes that contribute to food research include the Institute of Arable Crops Research (Rothamsted, Long Ashton, and Broom's Barn), Institute of Animal Health (Newbury and Pirbright), John Innes Centre (Norwich), Roslin Institute (Edinburgh), and the Silsoe Research Institute (Bedford). On average, about half of the income of such institutes is derived from the BBSRC, usually a significant amount from MAFF, with the balance coming from research grants and industrial contracts.

Research associations are a major source of support for industry in the UK and also contribute to training and research output. In recent years, all of the UK associations concerned with food research have had considerable success in developing international markets. The three main associations are Leatherhead Food Research Association (Leatherhead, Surrey), Campden and Chorleywood Food Research Association (Chipping Campden, Gloucestershire), and the British Industrial Biological Research Association (Carshalton, Surrey; now more commonly known as BIBRA International).

GOVERNMENT-FUNDED RESEARCH PROGRAMS

MAFF funds food research to support its mission that is largely concerned with protecting the public and stimulating improved economic performance. The 1997/1998 budget was ~£130 million. Collaborative research is encouraged. The LINK program is a collaborative research program, encouraging applied research initiatives in a variety of areas in which 50% of the funding is provided by government and 50% by industry. The MAFF-led programs in the LINK scheme are concerned with food processing sciences, agrofood quality, and advanced and hygienic food manufacturing. As mentioned, the UK research councils (in particular the BBSRC) are major sources of funding for food research in the UK, through a variety of schemes.

The Foresight Programme was launched in 1994 by the Office of Science and Technology of the Department of Trade and Industry. The activities of the Programme are broad and are concerned with the identification of future challenges and opportunities for UK companies. For example, thematic panels address, *inter alia*, food chain and crops for industry; health care; education, and energy. Panels draw together relevant contributions to the knowledge pool from industry, government, academe, and professional societies. The output of the Foresight Programme influences government policy, spending decisions, and research programs and initiatives.

EDUCATION

The number of universities in the UK has grown considerably in recent years as a consequence of the award of charters to many former polytechnics, which were intermediate between universities and colleges of further education. Historically, the major centers of food science education in the UK were the University of Reading (also home to the National College of Food Technology, formerly based at Weybridge, Surrey), University of Nottingham, University of Leeds, and The Queen's University, Belfast. Currently, these universities offer both BSc and thought postgraduate courses (certificate, diploma, or MSc level) in food-related subjects. Primary degrees are also offered by Bournemouth University; Glasgow Caledonian University; Loughborough University of Technology; Manchester Metropolitan University; Oxford Brookes University; Queen Margaret University, Edinburgh; Robert Gordon University, Aberdeen; Scottish Agricultural College; Sheffield Hallam University; South Bank University; University of Huddersfield; University of Lincolnshire and Humberside; University of Newcastle; University of North London; University of Plymouth; University of Strathclyde; University of Surrey; University of Teeside; University of Ulster; and University of Wales Institute Cardiff. Postgraduate courses in food-related subjects are offered by the Glasgow College of Food Technology; Manchester Metropolitan University; Oxford Brooks University; Queen's University Belfast; Robert Gordon University; South Bank University; and the Universities of Birmingham, Bristol, Huddersfield, Lincolnshire and Humberside, Leeds, London (Wye College), Reading, Strathclyde, and Wales Institute Cardiff. Many of the preceding institutions also award Ph.D. degrees in food research. Nutrition degrees are offered by the Universities of London, Newcastle, Sheffield, Southampton, Surrey, and Ulster. A number of UK universities, such as Surrey, now offer modular degrees that enable students to chose from a selection of courses shared with other degree programs, such as toxicology, nutrition, physiology, and molecular biology. A number of UK colleges of further education offer courses in food science to Higher National Certificate (HNC) and Higher National Diploma (HND) level. Some of the UK professional societies are also involved in educational initiatives. For example, the Institute of Food Science and Technology (IFST) is an examining body for the Higher Certificate in Food Premises Inspection. The European Commission is considering plans that may lead to the accreditation of courses in food science throughout Europe.

PROFESSIONAL SOCIETIES

Most UK food scientists are members of the Institute of Food Science and Technology. The London-based society was established in 1959 and currently has more than 3000 members of which almost 600 are fellows of the institute. The IFST also acts as a professional qualifying body for food science in the UK, the grade of membership reflecting the professional competence of the individual. The Institute publishes a newsletter for members (*Keynote*) and two journals for members and subscribers (*Food Science & Technology Today* and the *International Journal of Food Science & Technology*). The activities of the institute broadly reflect the activities of members and the needs of UK consumers, regulators, and industry. A number of subject interest groups address areas of specific interest including food control, food law enforcement, and the Internet. The IFST publishes regular position statements on topical issues and in 1994 formed the Professional Food Microbiology Group reflecting interest in this area.

The Nutrition Society was established in 1941 and maintains registers of accredited nutritionists and public health nutritionists. In addition to a newsletter for members, the society publishes four journals: *Proceedings of the Nutrition Society, Nutrition Research Reviews, Public Health Nutrition*, and the *British Journal of Nutrition*. Special interest groups meet to discuss specific aspects of nutrition such as clinical nutrition and metabolism, micronutrients, animal nutrition and metabolism, and nutrition and behavior. The society hosts several conferences each year; the proceedings are published in paper or in the form of abstracts in the *Proceedings of the Nutrition Society*.

The Royal Society of Chemistry (RSC) is the professional body for chemists in the UK with 46,000 members, including food chemists and technologists. The RSC serves the needs of food scientists mainly through the Food Chemistry Group, which has more than 800 members. The Society is a major publisher and publishes many books in the field of food chemistry and nutrition. Most of the society's journals are devoted to chemistry, although the monthly magazine, *Chemistry in Britain*, covers general developments including news of developments in food science. The Society of Chemical Industry serves members in the chemical, pharmaceutical, and food industries. The Society has 20 subject groups, including Agriculture and Environment, Biotechnology, Crop Protection, Food Commodities and Ingredients, Food Engineering, Oils and Fats, and Consumer and Sensory Research. The Society publishes the *Journal of the Science of Food and Agriculture*, the fortnightly magazine *Chemistry and Industry*, and a range of books on food science.

Some UK food scientists are also members of several other professional societies such as the Biochemical Society, Institute of Biology, British Dietetic Association, Institute of Physics, Society of Dairy Technology, Society for General Microbiology, Association of Public Analysts, Institution of Chemical Engineers, Royal Society of Health, and the British Toxicology Society.

The Royal Society (London) is perhaps the most prestigious scientific academy in the UK, promoting both the basic and applied sciences. Founded in 1660, the Society comprises more than 1000 elected fellows. Previous fellows include Sir Isaac Newton, Sir Humphry Davy, Lord Lister, and Lord Kelvin. The activities of the Society are broad, including a program to support the public understanding of science, awarding of small research grants, hosting of conferences and meetings (including subjects relevant to food research), publishing reports, and acting as a forum for independent expert discussions on issues affecting British society.

BIBLIOGRAPHY

1. *OECD Economic Surveys 1997–1998: United Kingdom*, Organisation for Economic Cooperation and Development, Paris, France, 1998.
2. *The Food Standards Agency—A Force for Change*, The Stationery Office, London, United Kingdom, 1998.
3. R. M. Anderson et al., "Transmission Dynamics and Epidemiology of BSE in British Cattle," *Nature* **382**, 779–788 (1996).
4. J. Collinge et al., "Molecular Analysis of Prion Strain Variation and the Aetiology of 'New Variant' CJD," *Nature* **383**, 685–690 (1996).
5. D. A. Hilton et al., "Prion Immunoreactivity in Appendix Before Clinical Onset of Variant Creutzfeldt-Jakob Disease," *Lancet* **352**, 703–704 (1998).
6. R. Smith, "Early Identification of Variant Creutzfeldt-Jakob Disease," *Br. Med. J.* **316**, 563–564 (1998).
7. R. G. Will et al., "Deaths From Variant Creutzfeldt-Jakob Disease," *Lancet* **353**, 979 (1999).
8. The Pennington Group, *Report on the Circumstances Leading to the 1996 Outbreak of Infection With* E. coli *O157 in Central Scotland, the Implications for Food Safety and the Lessons to be Learned*, The Stationery Office, Edinburgh, Scotland, 1997.
9. H. S. Evans et al., "General Outbreaks of Infectious Gastrointestinal Diseases in Scotland and Wales: 1995 and 1996," *Comm. Dis. Pub. Health* **1**, 165–171 (1998).
10. L. Willcocks et al., "A Large Outbreak of Cryptosporidosis Associated With a Public Water Supply From a Deep Chalk Borehole," *Comm. Dis. Pub. Health* **1**, 239–243 (1998).
11. R. J. Chiodini, "Crohn's Disease and the Mycobacterioses: A Review and Comparison of Two Disease Entities," *Clin. Microbiol. Rev.* **2**, 90–117 (1989).
12. K. H. Campbell et al., "Sheep Cloned by Nuclear Transfer From a Cultured Cell Line," *Nature* **380**, 64–66 (1996).
13. I. Wilmut et al., "Viable Offspring Derived From Fetal and Adult Mammalian Cells," *Nature* **385**, 810–813 (1997).
14. A. E. Schneike et al., "Human Factor IX Transgenic Sheep Produced by Transfer of Nuclei From Transfected Fetal Fibroblasts," *Science* **278**, 2130–2133 (1997).

JOHN O'BRIEN
University of Surrey
Guildford, United Kingdom

INTERNATIONAL FOOD INFORMATION SERVICE (IFIS)

STRUCTURE AND AIMS

The International Food Information Service (IFIS) is a not-for-profit organization that provides information products and services and commissions research and provides education in information science for the international food science, food technology, and human nutrition community. It comprises two companies, IFIS Publishing in the UK and IFIS GmbH in Germany. Specifically, IFIS Publishing (Charity No. 1068176, Limited Company No. 3507902) is concerned with providing information products and services; IFIS GmbH is concerned with commissioning basic and applied research and providing education and training programs in information science. IFIS Publishing houses the headquarters of the organization at Shinfield near Reading (some 40 mi west of London). IFIS Publishing has 36 staff members (three are based at the British Library at Boston Spa in West Yorkshire) and its primarily concerned with producing and marketing the database *Food Science & Technology Abstracts (FSTA)*. IFIS GmbH is based in Frankfurt and has one part-time staff member who holds the title of managing director of IFIS GmbH. IFIS also has a Japanese representative based in Tokyo. Funding for IFIS is derived solely from product sales; the organization is fully self-supporting.

IFIS is governed by a board of governors, comprising two members each from CAB International in the UK; the Bundesministerium für Landwirtschaft, Ernährung, und Forsten (BML) (represented by Deutsche Landwirtschaft-Gesellschaft e.V. [DLG]) in Germany; the Institute for Food Technologists (IFT) in the United States; and PUDOC (Centrum voor Landbouwpublikaties en Landbouwdocumentatie) in the Netherlands. The governing organizations constitute a unique consortium of one intergovernmental organization, two government ministries, and one professional association; they represent the four organizations (or their direct descendants) that were responsible for establishing IFIS in 1968. The board of governors meets twice yearly. Day-to-day responsibility is delegated to the general manager, who is based at Shinfield; the general manager also has the title managing director to describe his role within IFIS Publishing.

DATABASES AND COVERAGE

Information in *FSTA* is provided in the form of abstracts, irrespective of the mechanism of access. Abstracts are faithful descriptions of the original work. All abstracts are written in English but are prepared from source material written in more than 40 languages. Detailed author and subject indexes are provided to facilitate searching the database. *FSTA* is available in printed form, on-line, on magnetic tape, on compact disk (CD-ROM) and via the Internet (*Food Science Alerts, fsa*); it has been available in both electronic and printed formats since its launch in 1969. *FSTA* has always been available in the state-of-the-art formats, and the appearance of the database on the Internet further demonstrates the commitment of IFIS in this direction. The database can be accessed via seven hosts (DataStar, Dialog, DIMDI, EINS, Orbit, Questel, and STN). The database contains approximately 530,000 records covering the period 1969–1999 inclusive and is updated monthly (except for the CD-ROM, which has quarterly updates). *FSTA* is growing at a rate of approximately 20,000 records per year. Approximately 1800 different primary journals are scanned regularly for articles of relevance to *FSTA*. Several hundred other journals and relevant literature are also scanned on an intermittent basis. *FSTA* contains records on basic sciences relevant to food (eg, chemistry, biochemistry, physics, microbiology, biotechnology, hygiene, and toxicology) as well as food processing, food products, packaging, economics, and legislation. Of the 18 different sections into which the database is subdivided, 12 contain records on specific foods or food groups as follows: speciality and multicomponent foods; alcoholic and nonalcoholic beverages; fruits, vegetables, and nuts; cocoa, chocolate, and sugar confectionery products; sugars, syrups, and starches; cereals and bakery products; fats, oils, and margarine; milk and dairy products (including butter); eggs and egg products; fish and marine products; meat, poultry, and game; and additives, spices, and condiments. There is an ongoing policy to review coverage to reflect developments in food science, food technology, and human nutrition as they occur. This policy is evident following increased coverage of biotechnology and related disciplines since the beginning of the 1989 volume of *FSTA*.

ADDITIONAL PRODUCTS AND SERVICES

A compact disk (POLTOX) has been produced in association with Cambridge Scientific Abstracts (CSA) and the U.S. National Library of Medicine. POLTOX contains relevant abstracts taken from *FSTA* relating to food safety. A second compact disk (Human Nutrition on CD-ROM) has also been produced in association with *CSA* and the U.S. National Library of Medicine; this disk gives one source access to 14 international databases including *FSTA*. The disk gives comprehensive coverage of research on human nutrition and its direct impact on human health.

An additional service to *FSTA* users is provided by food science profiles (FSP); these consist of sets of records specially selected from *FSTA*'s current input and cover any food-related subject from aseptic packaging to viscosity. FSPs give monthly updates in print or electronic format; 30 standard titles covering "hot topics" are routinely produced, but personalized searches are available on request. Current awareness services and retrospective database searches are also available to complement the FSP service.

In June 1998 a new service, the Food and Nutrition Internet Index (FNII), was launched by IFIS. FNII is a fully searchable Web site describing and indexing food and nutrition resources available on the Internet. The main focus is on food science, food technology, and human nutrition, although there is also coverage of food business and company information.

The 2000 edition of the *FSTA* companion thesaurus will soon be available; it is an invaluable user aid and is available in conjunction with the CD-ROM, with *fsa*, and from some on-line hosts. It is also available separately as a product in its own right in both print and electronic formats.

S. Hill
International Food Information Service
Reading, United Kingdom

INTERNATIONAL UNION OF FOOD SCIENCE AND TECHNOLOGY

The International Union of Food Science and Technology (IUFoST) is a voluntary, nonprofit organization that promotes worldwide exchange of ideas and experience in those scientific disciplines and technologies relating to the expansion, improvement, distribution, and conservation of the world's food supply. Its members, "adhering bodies" in the terminology of international scientific organizations, consist of one representative food science and technology organization or committee from each country. IUFoST also has associate members, who are individuals, and who receive the Union newsletter, but have no vote. IUFoST is a scientific member of the International Council of Scientific Unions (ICSU).

OBJECTIVES

The chief aims of IUFoST are the encouragement and fostering of

- international cooperation and exchange of knowledge and ideas among food scientists and technologists;
- further development of and support for food research;
- progress in the fields of theoretical and applied food science for improvements in the processing, manufacturing, preservation, and distribution of food;
- the education and training of food scientists and technologists; and
- development of both individual professionalism and professional organization among food scientists and technologists.

Increasingly, IUFoST sees its role as that of providing services that implement the preceding objectives to its various constituencies, including its adhering bodies, international organizations, national governments, and individual food scientists. Currently, IUFoST pursues its objectives primarily through stimulation and sponsorship of international and regional congresses, conferences, short courses, and symposia.

In these activities IUFoST subscribes to the principles of ICSU, which promote the participation by all bona fide scientists without regard to race, religion, political philosophy, ethnic origin, citizenship, language, or sex.

The Union establishes such committees, commissions, and working parties as are required to cope with international issues concerning the food supply. Its activities are coordinated to supplement the efforts of other similarly interested international scientific and technological groups. A number of these activities are carried out jointly with the International Union of Nutritional Sciences (IUNS), a closely related organization. It develops projects with the World Health Organization (WHO), and has official Non-Governmental Organization (NGO) status with the Food and Agriculture Organization (FAO).

ADHERING BODIES TO IUFOST

There are currently 54 adhering bodies in the Union; several countries have membership on a nonfee-paying basis. The complete current list is as follows: Albania, Argentina, Australia, Austria, Belgium, Brazil, Canada, Chile, China (Beijing), China (Taipei), Czech Republic, Denmark, Egypt, Finland, France, Germany, Greece, Hungary, Indonesia, Ireland, Italy, Japan, Kazakhstan, Korea, Kuwait, Lesotho, Lithuania, Macedonia, Malawi, Malaysia, Mexico, Mozambique, Netherlands, New Zealand, Nigeria, Norway, Philippines, Poland, Portugal, Qatar, Russia, Saudi Arabia, Singapore, Slovenia, South Africa, Spain, Sweden, Switzerland, Thailand, Uganda, United Kingdom, USA, Zambia, Zimbabwe.

HISTORY

Discussion of the need to establish an international organization of food science and technology to advance food availability, food safety, and nutrition began in Britain and North America in 1959–1960. These early conversations resulted in the First International Congress of Food Science and Technology held in London, England, in 1962. At that time an International Committee of Food Science and Technology was formed to plan further congresses and a more formal international union. Four years later, in 1966, the Second Congress was held in Warsaw.

The IUFoST was formally inaugurated during the Third Congress held in Washington, D.C., USA, in 1970. Subsequent congresses, their dates, and venues were Madrid, Spain, 1974; Kyoto, Japan, 1978; Dublin, Ireland, 1983; Singapore, 1987; Toronto, Canada, 1991; and Budapest, Hungary, 1995. The Tenth International Congress will be held in Sidney, Australia, in 1999. IUFoST decided that, beginning with the Eleventh Congress in Seoul, Korea, in 2001, it will hold International Congresses every two years. The Twelfth Congress will be held in Chicago, Illinois, USA, in 2003.

ORGANIZATION

General Assembly

The Union is governed by a General Assembly, which meets at the international congresses of the Union, and by an Executive Committee, which exercises the authority of the General Assembly between congresses.

The General Assembly develops and controls IUFoST policy and action. Its responsibilities include: the consideration of proposals involving international cooperative scientific programs and activities, the approval of methods to finance Union activities and reports of expenditures, the development of rules and procedures governing Union activities and the election of officers, and the Executive and Finance Committees.

Each member nation is allocated a number of voting delegates to the General Assembly according to its annual fee assessment; the maximum number of delegates that a country can have is five. The Executive Committee has the power to admit temporarily an adhering body to nonvoting membership without fee.

At the Ninth Congress in Budapest, the General Assembly of the Union unanimously adopted the Budapest Declaration in response to the 1992 International Conference on Nutrition in Rome. The Declaration gave special attention to those areas in which food science and technology can make a major contribution to the quality and availability of food, including reduction in postharvest losses, improvements in food quality and safety, and adaptation and improvement of traditional foods and food processing.

Fees

There is an initial entrance fee of US$100, payable on acceptance into membership. An annual assessment is made on each adhering body according to the number of votes it has in the General Assembly, with one vote per delegate, as shown in Table 1.

Union Officers

The president, secretary general, and treasurer, and three vice-presidents are elected to four-year terms by the General Assembly. Not more than two officers may be citizens of the same country. The mailing address of the secretary general is the official headquarters of the Union. The president is the principal executive officer, and the secretary general/treasurer the principal administrative officer of the Union. The vice-presidents act for the president in his absence, lead major activities of the Union as determined by the Executive Committee, and represent the Union as determined by the president.

The Union plans to establish a permanent secretariat in order to expand its activities and further develop the services it can offer. The Swiss journal *Lebensmittel Wissenschaft und Technologie* (*LWT*, or *Food Science and Technology*) is the official journal of the Union. The Union publishes its own newsletter, *Newsline*, circulated three times yearly to adhering bodies, associate members, and subscribing libraries.

Table 1. Schedule of Assessment

	U.S. dollars (US$)	
No. of Delegates	1998	1999
1	310	310
2	1,240	1,280
3	2,480	2,560
4	3,720	3,840
5	6,200	6,400

Regional Groupings

IUFoST has several regional groupings whose constitutions and purposes are consistent with those of the Union, and whose programs are specifically oriented to the interests of their regions. These include: The European Federation of Food Science and Technology (EFFoST); The Federation of Food Science and Technology in ASEAN (FIFSTA—southeast Asia); the Eastern, Central, and Southern Africa Association of Food Science and Technology (ECSAFoST); and the Latin American and Caribbean Association of Food Science and Technology (ALACCTA).

Union Committees

The Executive Committee is comprised of the IUFoST officers, including the immediate past president, as ex officio members, and seven members elected by delegates to the General Assembly from among their number. The Executive Committee is empowered to execute the policies of the General Assembly, and to act for it in conducting the affairs of the Union between Assembly meetings. It approves budgets and major activities, selects the time and place of international congresses and meetings, and acts on applications from new adhering bodies. Insofar as possible, the Executive Committee meets in developing countries timed to coincide with locally organized scientific events. This allows for a broader international participation and provides speakers without incurring additional costs to the conference organizers.

The Finance Committee provides advice on financial matters and appropriate control procedures and recommends the selection of outside auditors.

The Congress Advisory Committee serves as consultant to the organizing committee of the country sponsoring an international congress.

The Scientific Activities Committee plans, organizes, and sponsors symposia, conferences, and other scientific programs in keeping with IUFoST objectives, often in association with other interested organizations. IUFoST symposia, typically three or four per year, have been held on a wide variety of topics including food safety, sensory evaluation, food biotechnology, food composition, nutritional quality, and preservation technology.

The International Liaison Committee works for cooperation between IUFoST and other international organizations. It has been active in helping to meet the needs of developing countries, especially in Africa where it has successfully promoted the establishment of ECSAFoST, mentioned earlier.

In 1995 the General Assembly established the International Academy of Food Science and Technology. Election as a Fellow of the Academy honors those throughout the world who have made outstanding contributions to the field.

D. E. Hood
Dublin, Ireland
R. L. Hall
Baltimore, Maryland

IRRADIATION OF FOODS

THE EVOLVING NATURE OF FOOD PROCESSING TECHNOLOGY

Food is one of the basic needs of humanity. Its production and processing constitute one of the major economic pillars of all societies. And the ways in which that production and processing are done reflect the general level of technological advancement within society at any given point in time. Current food processing techniques constitute a snapshot of a dynamic system in continuous, expansive change. Driven by ever more demanding requirements of consumers, food processors continually seek ways to improve their way of doing things. Even casual reflection allows identification of several relatively recent additions to the arsenal of food processing technologies. Familiar examples of such additions include microwave cooking, ultrafilteration, ohmic heating, modified atmosphere packaging, freeze drying, high pressure processing, blast freezing, supercritical fluid extraction, reduced-fat and -calorie formulations, myriad flavor extracts and enhancers, enzymic processing, and a host of functionality modifiers. Thus, technological change and innovation is no stranger to the food industry. It is in this context of continuous change, driven by changing needs, that radiation processing of food is emerging to join the food processors' arsenal.

THE CONCEPT OF FOOD IRRADIATION HAS A LONG HISTORY

The concept of using ionizing energy to process food has a long history. Soon after the discovery of roentgen rays in 1895, a number of visionaries suggested that this new form of energy could be of benefit to food preservation. Indeed, one of the first patents based on this idea dates back to the early years of this century (1). However, as history subsequently showed, practical limitations precluded early industrial and commercial development and application of these concepts. It is only much more recently that progress has advanced to the stage where significant industrial utilization of food irradiation is becoming widespread. This follows several decades of research and development, in government, university, military, and industrial laboratories around the world, which put in place the technical and scientific foundation for the safe and effective utilization of radiation for the processing of food.

CURRENT APPLICATIONS OF RADIATION TECHNOLOGY IN AGRICULTURE

There are a number of ways, other than food processing, in which radiation and related technology are utilized in the service of food and agriculture. Examples include the following: sterile male technique for insect pest control, mutation induction for plant breeding purposes, radiotracers for studies on agrichemical pathways in the environment, studies on utilization of fertilizers and other plant nutrients by plants cultivated as human food crops, radioimmunoassays for animal physiology studies, animal husbandry research, and vaccine production to protect animals against parasites. Collectively, these techniques constitute an important component of the toolbox available to agricultural scientists in their never-ending quest for improvements to the world's food production system(s).

GROWING INTEREST IN FOOD IRRADIATION

Acceptance of any new technology by society depends on there being a genuine need for the benefits offered by that technology. In recent years there has been a veritable explosion of consumer interest in, and concern with, foodborne illness. There is a growing realization that such illness constitutes a very real problem around the world, with significant costs inflicted on the personal as well as national levels. Food irradiation is increasingly viewed as a technology that has much to offer in the ongoing battle against foodborne illness. Responsible public health agencies are urging adoption of irradiation as an additional means of protecting the safety of our food supply. The net effect of all these developments is that there now is an unprecedented level of interest in food irradiation.

TECHNICAL ASPECTS

Definition of Food Irradiation

Food irradiation is a process involving the exposure of food to a field of ionizing energy (radiation) for the purpose of effecting some desired benefit. Hence, in respect to purpose, radiation processing of food is similar to the processing of food by more conventional methods.

Energy Considerations. Food processing of any type involves energy transfer into, or out of, the food substance being treated. The form of energy involved can differ and can include thermal, chemical, mechanical, microwave, pressure, or ionizing energy (synonymous with radiation). Thus, at this basic level, food irradiation differs from more conventional forms of food processing in the form of energy involved. However, it should be noted that at this level, the conventional forms of food processing differ from one another as well.

Types of Ionizing Energy. Ionizing energy is any form of energy whose individual quanta (discrete packets of energy) are energetic enough to create ions by ejecting electrons from the atoms within a material absorbing that energy. Such energy can be photonic (pure electromagnetic energy, with no physical particles involved) or particulate (real particles involved). Examples of the former type include γ-rays, spontaneously given off by certain radioactive elements (like cobalt-60 or cesium-137) and X rays (produced by X-ray generating machines). Beams of high-energy electrons, generated by special machines called electron accelerators, constitute the most common form of particulate radiation. Any of these types of ionizing energy can be used for processing materials. However, to avoid potential induction of radioactivity in the treated material, electron energies must be kept below 10 MeV, while photons must not exceed 5 MeV.

Radiation as Processing Energy. Ionizing energy, in its various forms, possesses several characteristics that make it extremely useful as a form of processing energy (2). These include its versatility in effecting a variety of technical end points in a wide variety of materials, its ability to penetrate through the bulk of the product being treated, its ability to effect treatment without significantly increasing the temperature of the material being treated, its controllability, its flexibility, its convenience, its low cost, its environmental friendliness and its lack of residues in the treated product. Figure 1 illustrates the general concept of radiation processing. As shown in Figure 1, radiation processing of any material gives rise to change(s) in the physical, chemical, and biological properties of the treated substance. One or more of these changes in material properties constitute the purpose of the treatment. It is worth noting that radiation processing is no stranger to our daily lives. Although most consumers are generally not aware of it, there already is widespread use of consumer products whose manufacture is associated with radiation processing in one way or another, reflecting the great utility and versatility of this type of industrial processing. Some familiar examples of such products include electrical insulation, automobile tires, personal hygiene products, bandages, cosmetics, plastic films and coatings, baby soothers, packaging materials for juices and other liquids, advanced composites used in aircraft manufacture, cellulosic products (eg, fabrics) derived from plant biomass, medical devices, and even jewelry.

Process of Food Irradiation

Conceptually, the process of food irradiation is very simple. The essential elements include a shielded chamber containing a field of ionizing energy, and appropriate equipment for conveying foodstuffs through the facility. In operation, the food material is transported into the irradiation chamber, kept there for an appropriate length of time during which the desired amount of ionizing energy is absorbed, and then removed from the chamber. Upon removal from the chamber, the food material is ready for immediate utilization, which could involve consumption, further processing, or storage, as appropriate. In actual practice, the irradiation facility operator must pay attention to a lot of detail to ensure that the process is carried out properly, but that detail is invisible to both the facility client and the end user of the treated product.

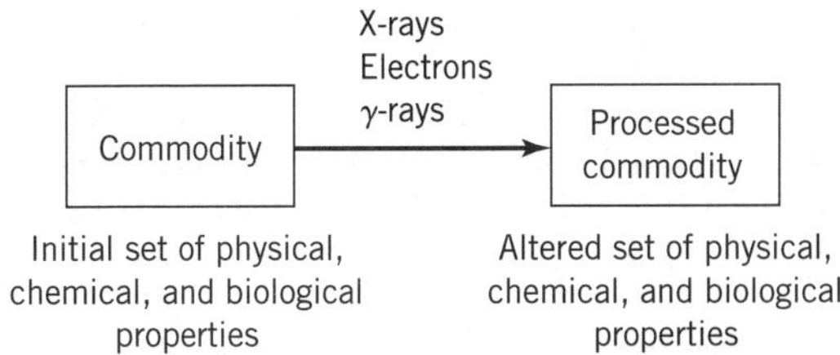

Figure 1. Illustration of principle of radiation processing. Prior to irradiation, the commodity has an initial set of properties. Following treatment, the processed commodity has an altered set of properties.

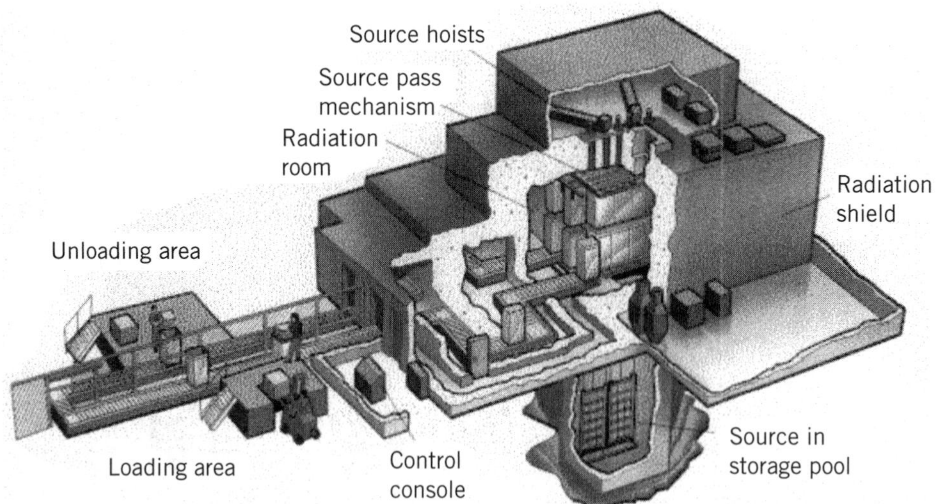

Figure 2. Cutaway view of typical γ-radiation facility, such as could be used for processing food.

Figure 2 illustrates a typical industrial irradiation plant, which can be used for treating food. The illustration shows the essential elements of such a facility, including the warehouse area, the biological shield, the maze (required to allow transfer of product through the shield while not permitting escape of radiation to the outside), the product carriers, the cobalt-60 source assembly, and the storage pool (needed for safe storage of the cobalt-60 when product is not being treated). It should be noted that, unlike the case with microwave cooking, food irradiation is an industrial process and cannot be miniaturized for use in home kitchens.

Mechanism of Action

Exposure of the food product to the field of ionizing energy within the radiation chamber results in a specified, desired amount of energy being absorbed by the food material. The amount of energy absorbed per unit mass of the absorbing material is termed the absorbed *dose* of radiation and is measured in units called *grays*. The practical unit for measuring radiation dose used in food processing is called the *kilogray*. It is the *absorbed* energy that gives rise to the effects ascribed to the treatment. Energy that passes through the product without being absorbed does not affect the exposed material. A central question is how the absorbed energy leads to the observed effects, which constitute the benefits and detriments of the process.

Thermal Effects Are Negligible. Table 1 presents data on the theoretical maximum temperature rise associated with irradiation of food products treated to effect the specified technical end points. This illustrates that, for all practical purposes, the maximum temperature rise is generally negligible. From this it follows that the benefits of the treatment are effected by nonthermal mechanisms.

Action Cascade as Mechanism. Food is a biological system, and the effects of irradiation on food constitute a particular subset of the more general effects of irradiation on any such system. Biological effects of irradiation result from a complex sequence of reactions (3) that are well described by the term *action cascade*. Figure 3 illustrates the essential features of this sequence. The action cascade consists of a series of stages, beginning with energy absorption (a physical, discrete process that occurs at random throughout the irradiated material). Through ionization and excitation pathways the absorbed energy generates a variety of primary reactive species (free radicals, ions, excited molecules) that serve to propagate the effects through the subsequent chemical, biochemical, physiological, and biological stages. The final result consists of physiological and biological effects that constitute the observable benefits and detriments deriving from the treatment. Note that each successive stage operates on a longer time scale than the preceding one.

Basis for Beneficial Effects. Given that the initiating event (energy absorption) of the action cascade is unavoidably random in its occurrence within the treated material, it is instructive to examine the basis for beneficial effects of food irradiation. Intuitively, it is not obvious how random acts of molecular damage can give rise to a net benefit. The secret lies in the fact that different functional entities within biological systems exhibit large intrinsic differences with respect to their sensitivity to inactivation by ionizing radiation. Empirical characterization of the individual dose response curves for inactivation of specific biological functions of interest allows exploitation of these differential dose responses for our benefit, in favorable cases. Thus, *differential dose response underlies the benefits of the process*. This is illustrated schematically in Figure 4. With reference to Figure 4, it can be seen that, because the individual dose response curves are separated along the dose axis, it is generally possible to select a treatment dose suitable for effecting a desired technical end point (such as insect killing) without at the same time inducing significant detriment to the nutritional value, taste, or texture of the treated food.

Benefits

Technical End Points of Irradiation. Empirical observation has demonstrated a variety of possible technical

Table 1. Temperature Rise Associated with Food Irradiation

Dose (kGy)	Typical application	Energy Absorbed (J/kg)	Temperature rise (°C.; water)
0.1	Sprout inhibition	1×10^2	0.024
0.3	Insect disinfection	3×10^2	0.072
1	Ripening delay	1×10^3	0.239
3	Meat and poultry pasteurization	3×10^3	0.716
10	JECFI limit[a]	1×10^4	2.390
50	Sterilization	5×10^4	11.94

[a]The Joint Expert Committee on Food Irradiation (JECFI) in 1981 set an upper limit of 10 kGy on the permitted dose for food irradiation, pending further evaluation of the data on safety and wholesomeness at higher doses. That limit has since been removed.
Source: Ref. 8.

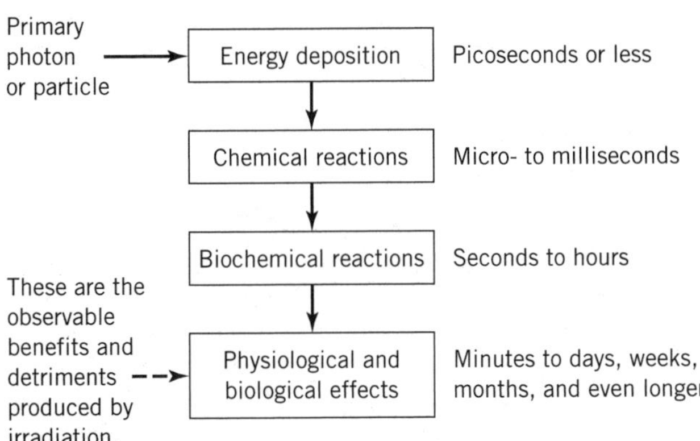

Figure 3. Schematic representation of the action cascade that underlies the biological effects of irradiation.

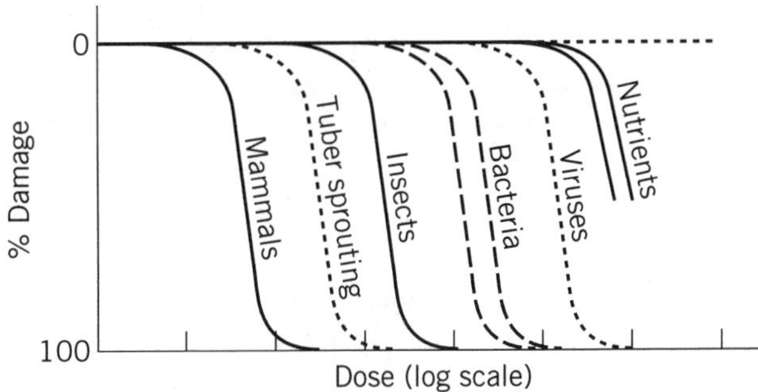

Figure 4. Schematic representation of dose response curves for inactivation of viability or other functionality in various biological systems. Differential dose response underlies the benefits of food irradiation.

benefits (4). These include microbial inactivation, insect disinfestation, sprout inhibition, maturation delay, and functional modification (eg, viscosity reduction, accelerated rehydration of dried preparations, reduced cooking time). Particular beneficial effects are associated with particular foods, and their achievement requires that an appropriate treatment protocol be followed. Only certain foods can be treated beneficially for any particular desired technical end point. It is generally not possible to elicit arbitrarily specified benefits in arbitrarily chosen foods, although some technical end points (eg, microbial inactivation) are less restrictive in this regard than others (eg, maturation delay).

Tangible Benefits for the Consumer. To the consumer, irradiation offers the benefits of food that is safer, fresher, more convenient, and therefore of higher value than the unirradiated counterpart. Of course these tangible consumer benefits derive from the specific technical end points just listed.

Limitations

As is the case with any food processing method, there can be detriments as well as benefits arising from irradiation. Negative effects of the radiation processing of food can include undesirable alteration of sensory properties (color, odor, taste, texture), loss of desired functional properties, and loss of some nutrient value. These negative effects tend to increase with increasing dose. To have a successful treatment, it is essential that unacceptable levels of detriment be avoided.

Avoidance of Detrimental Effects. To ensure a favorable result, irradiation must be carried out according to a proper treatment protocol, which has been validated by empirical testing. As with any processing method, improper application of the process will generally lead to unacceptable results. A treatment protocol is simply a 'recipe' wherein the process variables affecting the outcome are controlled within an appropriate range. Every product requires its own treatment protocol. In this sense irradiation is no different than conventional baking or cooking, which is done according to a proven recipe. There are many proven treatment protocols available, but there is ample opportunity for further development in this area.

SAFETY AND WHOLESOMENESS

Confidence in the safety and wholesomeness of irradiated food is critical for consumer acceptance of such products. It is instructive to examine the scientific approaches that have led to the current understanding in this vital area.

Approaches Used to Evaluate Safety and Wholesomeness

Two principal approaches are used to determine the safety and wholesomeness of food treated in any manner, including by irradiation. The first is based on chemical analysis and involves the identification and toxicological characterization of any new chemicals generated by the treatment in question. The second approach is one based on biological function and involves feeding the treated food to test animals, which are subsequently carefully observed for any manifestations of harm resulting from ingestion of that food. These two approaches are complementary, and together they constitute an extremely powerful paradigm for evaluating the overall safety and wholesomeness of irradiated food.

Summary of Findings from Chemical Approach

Chemical analysis of irradiated foods has led to the detection and identification of something in excess of 100 compounds referred to as radiolytic products (5). Evaluation of the toxicological significance of these identified compounds can be done by reference to available toxicological data for the compounds in question, or by comparisons with the same and similar compounds present in various foods either naturally or as the result of conventional processing. For the vast majority of identified radiolytic products the data reveal that the same compounds are present in various foods, either naturally or after conventional processing, and generally at much higher concentrations than are generated by irradiation. The results of these evaluations overwhelmingly support the conclusion that there is no indication of harm from consumption of irradiated foods (6).

Summary of Findings from Feeding Tests

There is a very large body of published literature describing toxicological studies on various animals fed a variety of irradiated foods (7). These tests have ranged from short term to long term, multigenerational in duration. In the various studies, the state of health of the test animals has been evaluated by a multiplicity of tests, examining many critical aspects of their function, including life span, reproductive ability, cancer incidence, birth defects, blood chemistry, physiology, chromosome aberrations, and even feed conversion efficiency. The vast body of literature describing these studies has been examined by several groups of experts (8,9,10). Results of these expert evaluations indicate that there are no reproducible findings of harm to humans and other mammalian species resulting from consumption of irradiated foods, in respect to dose, biological function monitored, test species, or variety of test foods.

Public Health Institutions Endorse the Safety and Wholesomeness of Irradiated Food

The available information on safety and wholesomeness of irradiated foods has been examined by many of the most prestigious and respected public health institutions around the world. Without exception these evaluations have led to unconditional reaffirmation of the absence of hazard attributable to irradiation. Such endorsements have been extended by the World Health Organization, American Medical Association, U.S. Food and Drug Administration, American Dietetic Association, American Gastroenterological Association, Mayo Clinic Health Letter, the UK Advisory Commission on Irradiated and Novel Foods, plus many others.

OVERVIEW OF IMPLEMENTATION PROGRESS

The Needs Driving Implementation

Several unmet needs are driving implementation of food irradiation. The two major ones include food safety concerns arising from foodborne pathogens and parasites, and the need for effective quarantine security treatment of agricultural commodities in interregional trade. Of lesser impact are the desire for extended shelf life of foods, reduction of food losses due to spoilage, and particular functional modifications.

Food Safety Concerns

Foodborne Illness Is Increasing. Most experts agree that foodborne illness is on a long-term upward trend in countries around the globe. This is in contrast to the dramatic decline of infectious disease overall (not foodborne), resulting from improvements in health care and in particular the advent of modern antibiotics and vaccines. The rise in foodborne illness has occurred in parallel to widespread, extensive changes in the structure of our food system, suggesting a cause-and-effect relationship. Over the last several decades, this system has evolved into its present form, markedly different from that which existed previously. In particular it has acquired new attributes that are essential for provision of a plentiful supply of affordable food to a largely urban population. These essential features include mass production of farm animals reared for food, central processing, widespread distribution of products, a diversity of sources of both raw and finished products, increasing trade between previously isolated regions, and recycling of nonedible portions of animals. These features make the system very efficient, but unfortunately they also make it very vulnerable to dissemination of infectious agents. Given that the social changes (principally urbanization) driving this evolution are effectively irreversible, it follows that the essential features of our current food system will not change. Thus, the negative effects will also persist, unless some intervention can be introduced to reduce or eliminate them without affecting the desired features.

Social Costs of Foodborne Illness. Estimates of the impact of foodborne illness on society are indirect and are based on incomplete data and imperfect methodology. Nevertheless, available data indicate that annually the financial cost in the United States alone is in the billions of dollars, with millions of cases of illness, and some thousands of deaths (11). There may be some disagreement on the exact values for these three categories of cost, but there is agreement that the magnitudes are reasonably accurate. From this it is evident that foodborne illness constitutes a serious burden on society, even in the developed countries. It can be expected that the problem is much worse in developing countries that lack the developed food system infrastructure central to the provision of safe and affordable food for large concentrations of people.

Irradiation and HACCP. Recent times have seen the introduction of new and more effective measures to protect the safety of our food supply. The Hazard Analysis and Critical Control Point (HACCP) approach represents the most advanced and comprehensive attempt in this regard. HACCP constitutes a rational plan for ensuring food safety by first identifying all possible hazards (biological, chemical, physical) associated with the production of any given food, and then introducing appropriate barriers or critical control points (CCPs) to prevent those hazards from reaching the consumer. For foods derived from animals (red meat, poultry, seafood) the microbial hazards (salmonella, *Escherichia coli* 0157:H7, campylobacter, listeria, vibrio, toxoplasma, other) are of greatest concern. HACCP plans for such foods require one or more CCPs of proven effectiveness against microbial pathogens to reduce these hazards. Irradiation is a true kill step and can reliably achieve several logs of kill (12,13) of all the common microbial pathogens responsible for the majority of foodborne illness in the United States and other countries. Inclusion of irradiation as a CCP against microbial hazards greatly enhances the effectiveness of any HACCP plan. Conversely, irradiation works best as part of an overall HACCP program. In this sense, irradiation and HACCP are natural partners in protecting the safety of our food.

Quarantine Security The need for effective quarantine security treatment constitutes the second major driver behind implemention of food irradiation.

Role in International Trade. An extremely important consideration in the trading of agricultural commodities between different regions involves protecting the importing region against the introduction of pests from the exporting region. Unwitting introduction of pests into a region where they are not yet present poses substantial risk of damage to the ecology and economy of the receiving region. For this reason such trade is contingent on the availability and use of effective quarantine treatment against pests of concern. In general, every importing jurisdiction has a system in place to enforce this requirement and prevent entry of products that have not been adequately treated. Historically, such treatment has involved the use of poison gas, such as methyl bromide, to fumigate the shipments of product. The need for treatment is pervasive, and essentially every traded product of agricultural origin is fumigated at least once, and in some cases several times, before reaching its final destination. Without availability of an approved and effective quarantine security treatment, international trade in agricultural products would be dramatically curtailed.

Loss of Methyl Bromide. For decades methyl bromide has served as the universal fumigant for quarantine security treatment of agricultural products. Unfortunately, methyl bromide is an ozone-depleting chemical and as such has been designated by international agreement (14) to be removed from service. The current phaseout schedule calls for this fumigant to be taken out of use in the United States early in the first decade of the twenty-first century. Loss of methyl bromide is expected to create a major disruption in trade of the affected commodities. Irradiation is a good alternative to methyl bromide for disinfestation of agricultural commodities for quarantine security purposes and may well be the only treatment available for many products when fumigation is no longer an option.

Costs of Food Irradiation

Cost considerations are critical in the determination of whether a particular process will be adopted by industry. Radiation processing is a capital-intensive technology and as such requires outlay of considerable funds before there can be any incoming flow of revenue.

Capital Costs. Capital costs for irradiation facilities depend on a number of variables, which will have different values for different enterprises. Main components common to all facilities include the biological shield, material conveyance system, source of ionizing energy, ventilation system, computer and operating software for the facility, and ancillary equipment (water purification and cooling). Depending on the required throughput capacity, type of plant, and degree of plant sophistication desired, capital cost for the irradiator could range between a low of perhaps $2 million (US) to possibly as high as $10 million with a median value of about $5 million. In addition, there will be a cost for land and warehouse requirements, which are site specific.

Operating Costs. The cost model for operating a food irradiation facility involves a large number of financial variables, including capital amortization and depreciation, labor, utilities, administration, taxes, insurance, radiation safety officer, license and permits, supplies, radioisotope replenishment (for a gamma plant), repair and maintenance, as well as process-related ones like dose, product density, radiation utilization efficiency, actual annual throughput, and plant utilization (% of maximum design capacity). Values for each of these variables will differ between individual operations, but in general the unit processing cost for a typical plant will be a few cents a pound, ranging from perhaps 2 cents to possibly 10 cents a pound, depending on the particular situation. The lower end of the cost range would apply to large capacity plants operated near maximum capacity, processing in excess of about 100 million lb per year.

Service Options

Radiation processing of food represents a fledgling industry, where the commercial infrastructure is not yet well developed. Thus, the delivery of service to potential users of this technology faces obstacles such as availability and accessibility of processing capacity, and additional costs arising from transport to and from existing facilities, which may be at some distance from the food manufacturing plant. It can be expected that as significant demand for service develops, the marketplace will implement solutions to these shortcomings. The ultimate structure of the industry will undoubtedly include a variety of irradiator business configurations, tailored to respond to different business needs.

Type of Facility. Irradiators can differ in regard to the type of source providing the required processing energy. This can either be radioisotope based, using cobalt-60 or, to a lesser extent, Cs-137 (both of which emit γ-rays), or it could be machine source based, utilizing electrons or X-rays. Each source option has particular advantages and disadvantages, and selection of the best choice for a particular application requires careful technical and economic evaluation in relation to the particular needs of that application.

Facilities can also differ in respect to their processing flexibility and relationship to the primary manufacturing location for the product requiring irradiation. Dedicated plants can be optimized for the particular product to be treated, providing both economic and product quality advantages. This type of plant would be integrated into the overall manufacturing process, both physically and operationally. It would be suitable for large processors who produce enough product to utilize the entire capacity of the irradiator. This option would not be suitable for the many smaller manufacturers whose requirements do not justify a dedicated facility. To service these smaller users, contract service providers using multipurpose facilities are the logical answer. Contract service facilities require greater processing flexibility than is present in the dedicated type and would be stand-alone enterprises, separate from the manufacturing site. Business considerations will ensure that such contract service facilities are optimally located and configured to provide access to efficient and economical radiation processing on a regional basis.

Since the existing irradiation facilities were mostly designed to treat nonfood items, their suitability for treating food, which has some very stringent processing requirements, is less than optimal. Several companies have initiated new designs intended to rectify the shortcomings of the existing facilities.

Business Arrangements. Needs relating to facility ownership, financing, location, and operation will be met by particular arrangements that reflect the unique circumstances of each business enterprise. The marketplace will ultimately decide on the most advantageous arrangements.

Regulations

Food irradiation can only be carried out in accordance with appropriate regulations in the country of use.

Philosophy of Regulatory Approach. In the majority of countries around the world, food irradiation is regulated by means of a blanket prohibition, with provision for exemptions, or clearances, which are granted for specific food items, for a specified purpose. Thus, only those food items that are explicitly named within the law as being exempt from the prohibition can legally be processed by irradiation. Origins of this regulatory approach date back to the early days of development of food irradiation technology when information about the safety and wholesomeness of irradiated foods was lacking. This prudent approach reflected the novelty of the process at that time, and the desire to ensure to the maximum extent possible the safety of a nation's food supply while leaving an avenue for industry to introduce useful new technology.

Clearances. The first clearance for irradiation of a food item was that for grain, granted in the former USSR in 1959 for insect disinfestation purposes. Currently there are some 227 specific clearances, in 39 countries (15), and the list continues to grow. South Africa leads the way, with 82 specific clearances. The effective list of clearances is much larger than the 227 items just noted, because many of the clearances are for categories, such as cereal grains, fish, seafood, fruits and vegetables, dehydrated mixtures and condiments, which are composed of a large number of individual food items within the category. At present there is a clear need to harmonize clearances internationally, especially between countries that are trading partners, to avoid development of nontariff trading barriers. This latter requirement is a special concern of the International Consultative Group on Food Irradiation (ICGFI), which is a United Nations–based organization mandated to coordinate food irradiation initiatives around the world.

Labeling Requirements. Current regulations governing food irradiation specify that irradiated product must be clearly labeled as having been treated in this manner. The exact wording of the label can vary somewhat from country to country, but in most cases it must include both the radura logo (shown in Figure 5) and text explicitly stating that the product has been irradiated. Most jurisdictions permit the inclusion of an additional message of an educational nature, conveying to the consumer the benefit effected by the process. Acceptable text on labels could be "Treated by irradiation to control harmful bacteria" or "Treated with radiation to control harmful bacteria." The phrase describing the benefit can be adjusted to accurately describe the purpose and benefit, as long as it is not misleading. The intent of these labeling requirements is to ensure that the individual consumer has the opportunity to make an informed choice in regard to selecting irradiated product. In the United States the prominence of the label text must be similar to that of the ingredient list.

Logistical Considerations

Implementation of food irradiation poses some difficult logistical challenges, stemming from the requirement to match scarce processing resources with the emerging needs in a regionally disseminated food industry. These

Figure 5. Radura symbol. This is the internationally designated symbol to identify food that has been treated by irradiation.

difficulties are most pronounced during the early stages of the implementation process and will diminish as new facilities are brought into service.

Limited Availability of Processing Capacity. At present few irradiation facilities are dedicated to food irradiation, anywhere in the world. Where food processing is being done, it generally utilizes surplus capacity in plants designed primarily to serve the sterilization needs of the medical device industry. Thus, aside from a few exceptions, little unused capacity is available for treating food. In the United States at present, one commercial irradiator, the Food Technology Services Incorporated plant in Florida, is dedicated exclusively to processing food, and at least two facilities have the capability for treating food along with other products. There is some spare capacity in existing contract irradiation plants, but far short of what would be required to handle even a small fraction of the country's potential food processing requirements. The same situation holds for most other regions of the world.

Geographic Distribution and Access to Facilities. The general lack of availability of radiation processing capacity for food is compounded by the fact that the location of the available capacity does not match with the location of the meat and poultry processors, which constitute the main base of potential clients for the technology in the United States. This reflects the fact that the existing irradiation facilities were located to serve needs unrelated to the needs of the food industry, and this lack of correspondence can only be resolved by building new facilities in more suitable locations. Such new facilities will be located by reference to both food manufacturing plants and existing geographic flow patterns of product. In general, irradiation plants are most efficient if located at node points (such as warehouses, distribution centers, manufacturing plants) in the product flow patterns, since such points are natural sites for product concentration. High throughput of product is essential for achieving favorable processing economics.

Time to Build. Experience has shown that from concept to commissioning of a new plant requires approximately 15 to 24 months. This time frame could be significantly shortened if a standardized design were adopted. In addition, several projects could be developed concurrently, so that after an initial lag period, new capacity could be coming on stream very rapidly.

Seasonality of Some Products. An additional complication arises from the fact that many agricultural products requiring treatment are seasonal. This is especially true for produce (fruits and vegetables). Thus, for part of the year there would be unused capacity, unless some secondary product(s) could fill in the low usage periods. This seasonal variation in availability of product for treatment impacts negatively on the processing economics.

CURRENT STATUS

State of Public Readiness to Accept Food Irradiation

Indications are that in North America there is an unprecedented willingness to accept irradiated food. This positive attitude cuts across a broad cross section of society, including regulators, politicians, public health professionals, academics, food processors, food industry associations, and the majority of consumers. Also, mass media has become more supportive. This favorable environment has developed relatively recently, and in a rather short time. The driving force behind these latest developments has been the explosive growth in public awareness of foodborne illness. This was unleashed by an outbreak of foodborne illness in the Pacific Northwest region of the United States in 1993, when several children died as a result of eating contaminated hamburgers in a fast-food chain outlet. Since that time there have been a continuing series of additional, highly publicized recalls of contaminated products, in some cases accompanied by plant closures and business failures. The net result is that consumers have become highly sensitized to the issue of foodborne pathogens like *E. coli* 0157:H7, listeria, salmonella, campylobacter, vibrio, and the like. The heightened level of consumer awareness and demand for safer food has led to implementation of new, more stringent food safety regulations in the United States, with specific pathogen reduction standards, increased testing and mandatory HACCP plans.

Along with the consumer awareness factor, there is a growing realization among food manufacturers of the limitations of end-product testing to combat foodborne disease, and of the need for an effective kill step to eliminate pathogens from their product. Radiation pasteurization of those foods that are at high risk of being contaminated is increasingly recognized as a safe and effective intervention technology to help solve this problem (16,17). The 1998 consumer survey conducted by the Food Marketing Institute in the United States clearly reveals that the majority of consumers are willing to purchase irradiated foods (18). This survey finding confirms the positive results obtained with real, albeit limited (but growing), market experience in the United States in the last few years. In addition, it is becoming more evident that as consumers learn more about the process, their willingness to accept it increases.

FUTURE OUTLOOK

From numerous indications, it is clear that there is a rising ground swell of support for implementation of food irradiation in the United States. It appears increasingly likely that a major breakthrough in terms of significant commercial and industrial use may happen relatively soon. When that comes to pass consumers will at long last have the opportunity to choose and pass judgment on the promise of this technology.

In summary, the current status is perhaps best described by a quotation from the late Adlai Stevenson:

> That which seems the height of absurdity in one generation . . . often becomes the height of wisdom in another.

Perhaps the age of wisdom in respect to food irradiation has finally arrived. The next few months and years will be critical.

BIBLIOGRAPHY

1. J. Appleby and A. J. Banks, British Patent No. 1609 (Jan. 26, 1905), cited in J. F. Diehl, *Safety of Irradiated Foods*, 2nd ed., Marcel Dekker, New York, 1995.
2. *Proceedings of the 10th International Meeting on Radiation Processing, Radiation Physics and Chemistry* **52**, Numbers 1–6, June 1998.
3. C. von Sonntag, *The Chemical Basis of Radiation Biology*, Taylor and Francis, London, United Kingdom, 1987.
4. "Practical Applications of Food Irradiation," in *Food Irradiation. A Technique for Preserving and Improving the Safety of Food*, World Health Organization, Geneva, Switzerland, 1988, pp. 33–43.
5. C. Merritt, Jr., "Qualitative and Quantitative Aspects of Trace Volatile Components in Irradiated Foods and Food Substances," *Rad. Res. Rev.* **3**, 353–368 (1972).
6. *Evaluation of the Health Aspects of Certain Compounds Found in Irradiated Beef*, Report AD-A045716, Life Sciences Research Office, Federation of American Societies for Experimental Biology, Bethesda, Md., 1977 (Supplement in 1979).
7. J. Barna, "Compilation of Bioassay Data on the Wholesomeness of Irradiated Food Items," *Acta Alimentaria* **8**, 205–315 (1979).
8. Joint Expert Committee on Food Irradiation, *Report of a Joint FAO / IAEA / WHO Expert Committee*, Technical Report No. 659. World Health Organization, Geneva, Switzerland, 1981.
9. *Ionizing Energy in Food Processing and Pest Control: I. Wholesomeness of Food Treated With Ionizing Energy*, Report No. 109, Council for Agricultural Science and Technology, Ames, Iowa, July 1986.
10. U.S. Food and Drug Administration, "Irradiation in the Production, Processing and Handling of Food. Final Rule," *Federal Register*, Friday April 18, 1986.
11. J. C. Buzby et al., *Bacterial Foodborne Disease: Medical Costs and Productivity Losses*, Agric. Econ. Report No. 741, U.S. Dept. of Agriculture, Washington, D.C., 1996.
12. T. Radomyski et al., "Elimination of Pathogens of Significance in Food by Low-Dose Irradiation: A Review," *J. Food Protect.* **57**, 73–86 (1994).
13. D. W. Thayer, "Use of Irradiation to Kill Enteric Pathogens on Meat and Poultry," *Journal of Food Safety* **16**, 181–192 (1995).
14. United Nations Environment Program, *Proc. 4th Meeting of the Parties to the Montreal Protocol on Substances that Deplete the Ozone Layer*, Copenhagen, Denmark, November 23–25, 1992.
15. Joint FAO/IAEA Division of Nuclear Techniques in Food and Agriculture, *Clearance of Item by Country*, Food Irradiation Newsletter Supplement 1, International Atomic Energy Agency, Vienna, Austria, December 1996.
16. C. Bruhn, "Consumer Attitudes and Market Response to Irradiated Food," *J. Food Protect.* **58**, 175–181 (1995).
17. L. M. Crawford, "Food Irradiation's Advantages Will Not Escape Public Attention," *Food Technol.* **52**, 55 (1998).
18. *Consumers' Views on Food Irradiation*, Food Marketing Institute and Grocery Manufacturers of America, Washington, D.C., 1998.

GENERAL REFERENCES

J. F. Diehl, *Safety of Irradiated Foods*, 2nd ed., Marcel Dekker, New York, 1995.

J. Farkas, *Irradiation of Dry Food Ingredients*, CRC Press, Boca Raton, Fla., 1988.

Ionizing Energy in Food Processing and Pest Control: II. Applications, Task Force Report No. 115, Council for Agricultural Science and Technology, Ames, Iowa, 1989.

E. S. Josephson and M. S. Peterson, eds., *Preservation of Food by Ionizing Radiation*, Vol. 1–3, CRC Press, Boca Raton, Fla., 1982.

E. A. Murano, ed., *Food Irradiation: A Sourcebook*, Iowa State University Press, Ames, Iowa, 1995.

M. Satin, *Food Irradiation: A Guidebook*, 2nd ed., Technomic, Lancaster, Pa., 1996.

W. M. Urbain, *Food Irradiation*, Academic Press, Orlando, Fla., 1986.

World Health Organization in collaboration with the Food and Agriculture Organization of the United Nations, *Food Irradiation: A Technique for Preserving and Improving the Safety of Food*, World Health Organization, Geneva, Switzerland, 1988.

JOSEPH BORSA
MDS Nordion, Inc.
Kanata, Ontario
Canada

K

KINETICS

The engineering of food materials, that is, the industrial production and processing of food materials, requires the knowledge of an engineer and a good understanding of the food system in terms of its chemical, physical, and biological properties. In other words, it requires the knowledge of an engineering and scientific nature. Among the said knowledge, one subject stands out that most closely links a scientist and an engineer; that is the topic of kinetics. Generally speaking, kinetics is the study of motion. Because motions are time dependent, kinetics is the study of rates of changes. Traditionally, scientists have been interested in knowing the nature or the mechanism of reaction, and engineers have been interested in the efficiency of the reactions and that of reactors. These two subject matters can be rephrased as molecular (chemical) kinetics and reactor (chemical engineering) kinetics, respectively. Both subjects are important in enabling a food engineer to arrive at an optimal design for a food process.

MATHEMATIZATION OF KINETIC BEHAVIOR

The ultimate goal of science is the formulation of laws, principles, or models, through which the behavior of a system can be deduced without the necessity of experimenting with and constructing the system. The extent to which this goal is reached is an objective measure of the scientific success achieved by the scientist or engineer. Economically, such achievement can help cut a great deal of the cost due to system construction from trial-and-error experimentation. Practically, such achievement is reflected in the predictability of the system derived from the use of the model. Several prerequisites are important in striving for predictability. The first is mathematization. This involves more than just fitting numbers into an equation. With the advent of computers, mathematical modeling has become a more powerful tool than in the past. A successful mathematization of a system, and predictability exerted by the model, depends on the understanding of the mechanistic aspect of the system involved and the identification of the pertinent variables of the system. Experimental design and statistical analysis are two powerful tools for gathering such information in the analytical approach to the system.

The mathematician in the scientist or engineer can say whether a given model is adequate or not, and if such and such conditions are met, a certain behavior will follow. However, it is the scientist or engineer in the mathematician who must judge the adequacy or appraise the predictability of the behavior. Mathematization is a necessity in forming models and making use of their predictability; sound experiments and good judgment are equally necessary in the formulation of foul proof models. Not to take any credit from the usefulness of a pure stochastic, statistical, or synthetic approach to rather unanalyzable problems, the subject matter in this article calls for the analytical, deterministic approach to problem solving.

Chemical kinetics is the study of rate, changes occurred per unit time, and mechanism by which one chemical species is converted to another. Mechanism is the elemental process of motion, that is, collisions among the molecules, atoms, radicals, or ions that take place simultaneously or consecutively in producing the observed overall rate. Overall rate depends on the nature of the participants and the environment surrounding them. The overall rate, in some cases, is controlled by the slowest step of the simultaneous or consecutive reactions. If the overall reaction involves a heterogeneous environment, where, for example, concentration is not uniform, the rate of the transport of the materials involved in the reaction can also control the overall rate of the reaction. In such a case, it is usually considered to be an overall rate controlled by transport (mass transfer), a physical change, not a chemical reaction. Similar arguments can be said for reaction systems with temperature variations. Overall rates can be controlled by heat-transfer rate in some situations.

The changes food materials undergo during processing, preparation, or storage are the results of complex physical changes, chemical reactions, and biological activities. Microscopically or macroscopically biological materials represent systems of heterogeneous natures in terms of the chemicals and biochemicals involved and in terms of the physical separations of these molecules into compartments in the system. Again, the overall rate of a change can be controlled either by the slowest step of a chemical reaction or by a physical change such as mass transfer of an essential substrate. Such a step is called the rate-limiting step or the rate-controlling step. The identification of the said step is necessary for the formulation of the correct mathematical model, which is useful for the design of efficient processes.

BASIC CHEMICAL KINETICS

General Considerations

The kinetics of chemical reactions are usually studied within a deterministic framework; that is, the considerations of reacting chemical species are assumed to be single-valued, continuous functions of time and position, uniquely determined by their initial values and a complete knowledge of the temperature, pressure, and so on throughout the system. The rate of homogeneous simple reaction (one corresponding to the actual molecular event) is assumed to follow the law of mass action (ie, to be proportional to the product of the concentrations of the reacting species).

For example, the rate of the unimolecular reactions A to P and bimolecular reaction A + B to P are expressed as

$$\frac{dp}{dt} = k_1 a - k_{-1} p \tag{1}$$

$$\frac{dp}{dt} = k_2 ab - k_{-2} p \tag{2}$$

respectively, where k_1 and k_2 are the rate constants for forward reaction and k_{-1} and k_{-2} are those for backward reactions; a, b, and p are the molar concentrations of chemical species A, B, and P, respectively. Most chemical reactions are reversible, to a certain extent. It is customary to write a chemical reaction left to right with the forward reaction rate constants (k_1 and k_2) larger than those of backward reactions (k_{-1} and k_{-2}). In measuring the initial rate of reaction $(dp/dt)_{\text{initial}}$, that is, in the early stage of a reaction, because product concentration p is rather small, equations 1 and 2 are reduced to equations 3 and 4:

$$\left(\frac{dp}{dt}\right)_{\text{initial}} = k_1 a \tag{3}$$

$$\left(\frac{dp}{dt}\right)_{\text{initial}} = k_2 ab \tag{4}$$

Order of Reaction

Equation 3 states that the initial rate of product formation, $(dp/dt)_{\text{initial}}$ is proportional to the concentration of the substrate A, namely a, to the first power. Hence, this reaction is called a first-order reaction. By the same token, equation 4 represents a second-order reaction. Experimentally, the order of a reaction can be determined by fitting experimental data to the following general equation:

$$\frac{dp}{dt} = ka^n \tag{5}$$

The value of n found could be fractions or whole numbers. Order is strictly an empirical concept. When the stoichiometric equation truly represents the mechanism of the reaction, the order and the molecularity both have the same value. In other words, on a molecular scale the reaction occurs exactly as written. These reactions are called elementary reactions. Many problems in food processing involve the studies of nonelementary reactions. Overall rates are usually measured without knowing the mechanism of the changes, including the steps of chemical reactions and physical changes such as mass and heat transfer.

FIRST-ORDER RATE PROCESSES

Macroscopic changes in the concentration (or other extensive quantity) of a substance tend to be exponential kinetically; in other words, first order in rate expression. This is exemplified by the phenomena of bacterial growth, radioactive decay, enzyme decay, pharmacokinetics (ie, kinetics of drug, absorption, etc), and many other chemical reactions and food processes. This is simply because the behavior of any one bacterium (atom, pair of reactive molecules) is governed by the law of chance. For any one bacterium and such there is a well-defined probability that it will change during a given period of observation. Or more simply, not one changing atom knows whether any one of the other unstable atoms in the sample population has decayed or not. The following example will illustrate how this condition leads to exponential decay. Consider a classroom containing 100 students; each one of them tosses a coin once a minute. Suppose that anyone who obtains a head is asked to leave the room. After the first toss, there will be approximately 50 people leaving; after two tosses, about 25 people leaving; and the number of people remaining in the room will decrease exponentially with time. Mathematically, the number of people, N, in the room, varies with time, according to the following equation (the negative sign recognizes the number is decreasing):

$$-\frac{dN}{dt} = kN \tag{6}$$

Separating the variables and integrating:

$$-\int_{N_0}^{N} \frac{dN}{N} = \int_0^t kdt;\ \ln N_0 = kt + \ln N \tag{7}$$

The rate of decrease of people in the room (decay of radioactivity, enzyme activity, or food quality, and microbial death rate in sterilization process), that is, $-dN/dt$, is dependent on the number of people at that time. The rate constant k, though, is independent of both time and the number of people (or other quantity such as enzyme concentration, radioactivity, etc). The half-life ($t_{1/2}$) of an exponential decay function is the time required for the population to become half of its original number; namely,

$$\ln \frac{N_0}{N} = \ln 2 = kt_{1/2};\ t_{1/2} = 0.693/k$$

Instead of decay, for growth such as cell culture after the lag phase, the rate is increasing (positive instead of negative in equation 6), and so the population doubling time is

$$\ln \frac{N}{N_0} = \ln 2 = kt_{1/2};\ t_{1/2} = 0.693/k$$

the same expression as that for half-life. It is worth noting that k, the first-order rate constant, or the exponential growth (or decay) constant, has a dimension of reciprocal of time.

Microbial cell growth rate at exponential phase and cell death rate in a thermal sterilization process are both first order. The specific growth rate and death rate are customarily denoted by μ and D (the decimal reduction time), respectively. They are shown in the following equations. For cell growth rate, x being cell concentration:

$$\frac{dx}{dt} = \mu x \text{ or } \frac{d \ln x}{dt} = \mu \tag{8}$$

For cell death rate:

$$\frac{dx}{dt} = -kx, \frac{d \ln x}{dt} = -k \text{ or } \frac{d \log x}{dt} = -1/D \quad (9)$$

ENZYME- AND CELL-CATALYZED REACTIONS

Mechanistic mathematization of a biological phenomenon is the first step toward quantifying, solving, and controlling (quantitatively) it. The difficulty faced is not that it cannot be solved, but rather that the result is of such algebraic complexity as to render little conceptual understanding of the process involved. Not to condemn the total empirical approach (which is usually limited in its applicability) to the analysis of data (such as curve fitting), a rigorous mechanistic approach yielding monumental amounts of mathematical expressions fail, usually, just as badly in terms of its usefulness in actual process operation. In such cases, a midway approach seems to be beneficial. The development of simplified kinetics of enzyme-catalyzed reactions provides one of the best examples of the judicious use of approximations to render a complicated system-equation intuitively understandable and practically useful. As discussed in the following, the steady-state and equilibrium assumptions in enzyme kinetics achieve exactly that.

Most biological reactions, those catalyzed by enzymes, follow Michaelis-Menten kinetics (equation 10), which represents mixed-order reactions. Enzyme-catalyzed reactions follow a first-order kinetic at low value of substrate to enzyme concentration ratio, and they become zero order when values of the said ratios are high (in that case, enzyme molecules are saturated with substrates). The reaction rate in this region (zero order) is then independent of substrate concentration. Enzymes play important roles in the growth and maintenance of cells. The design of such enzyme-catalyzed reactions as the functioning unit of the metabolic machinery in biological systems is of primary importance to the economy of the cell. The growth and maintenance of the cell require a highly integrated interplay of anabolism and catabolism. The subtle and precise metabolic controls are achieved through the regulations of the rate of enzyme formation (induction) and of enzyme activities (activation or inhibition).

Cell growth involves concurrent and serial reactions using enzyme as catalysts. The overall specific growth rate of cells, in either microbial, animal, or plant cell culture, has been modeled by using Monod equation (equation 11), where S denotes the limiting substrate concentration.

$$\frac{dp}{dt} = v = V_{max}S/(K_M + S) \text{ where}$$
$$V_{max} = k_3 e_0$$
$$K_M = (k_2 + k_3)/k_1 \quad (10)$$

$$\left(\frac{1}{x}\right)\frac{dx}{dt} = \mu = \frac{\mu_{max}S}{K_S + S} \quad (11)$$

K_M and K_S are saturation constants. When S/E is small, or $K_M \gg S$, equation 10 can be reduced to

$$v = \frac{V_{max}S}{K_M} = kS; \text{ where } \frac{V_{max}}{K_M} = k$$

(a first-order reaction)

At the other extreme, $S \ll K_M$, equation 10 becomes

$$v = \frac{V_{max}S}{S} = V_{max}$$

where the reaction is going at maximum velocity and is independent of substrate concentration (a zero-order reaction). In measuring the total activity of an enzyme preparation, the substrate concentration used should be large enough to saturate all the active sites of the enzyme molecules during the course of measurements (to follow the zero-order kinetics); whereas if a standard enzyme preparation is used to titrate the concentration of a substrate, substrate concentration should be diluted to make sure that first-order kinetic is followed ($v = kS$, velocity is a measure of substrate concentration).

Many food materials are composed of live cells (fresh food), many others are made by using the activities of live cells (such as fermentation processes) or by using enzymes (such as glucose from starch), and all food materials are affected by foreign live cells (food spoilage). From a food processing application point of view, two types of enzyme are currently important. These are hydrolytic enzymes (no cofactor required) and oxidoreductases (cofactors required).

The kinetic behavior of an enzyme system is not unique to the system. Other chemical kinetic analyses have found similar phenomena. For example, the concept of stationarity or a steady state is common in enzyme kinetics, in kinetic studies involving free-radical intermediates, and in those involving a stationary activated complex concentration.

The first satisfactory mathematical analysis of the diphasic activity curve was carried out by Michaelis and Menten in 1913. They assumed that the intermediate complex ES was reversibly formed according to the mass action law:

$$\text{E} + \text{S} \underset{k_2}{\overset{k_1}{\rightleftarrows}} \text{ES} \xrightarrow{k_3} \text{E} + \text{P} \quad (12)$$

and that the rate of breakdown of ES to form product P was small in relation to the rate of establishment of the equilibrium described by k_1 and k_2. The constant that they derived is, therefore, the dissociation constant of ES complex. Twelve years later, the concept of steady-state approximation to enzyme kinetics was introduced (1). It was believed that the catalyzed reaction may deplete ES complex at a substantial rate and that the Michaelis constant K_M measured experimentally from kinetic curves is in fact $(k_2 + k_3)/k_1$, which can be derived from the steady-state solution (equation 10) to the rate equations for mechanisms describable by equation 12. In addition to the steady-state assumption, it should be noted that equation 10 has been derived for negligible P, the product concentration. In other words, in using equation 10, initial rate of reaction should be used. A practical definition of the Michaelis constant is that it is the substrate concentration at half maximum velocity. Under carefully defined conditions of temperature, pH, ionic strength of buffer, and so on, for an enzyme-substrate pair, K_M is a constant. It approximates the affinity of an enzyme for its substrate. In general, the affinities of respiratory enzymes (oxidoreductases) for their substrates are higher than those of hydrolytic enzymes for their substrates. As shown in Table 1, the K_M values for respiratory enzymes are lower than that of hydrolases. There are a few reports on the K_M values of immobilized enzymes (3,4). Because of the possible complications of external and internal diffusional resistance, the true values of K_M for immobilized enzymes are difficult to obtain. Experimentally, a high substrate-to-enzyme concentration ratio (zero-order region of Michaelis-Menten kinetics) in a batch reactor with high velocity of flow (high agitation rate in stirred reactor) would minimize the external mass-transfer effect; the use of ultrafine particles (or ultrathin membrane) as enzyme carriers would minimize the internal mass-transfer effect.

The activities of enzymes are affected by temperature and pH. Because of the protein nature of an enzyme, thermal denaturation of the enzyme protein is evident in the high end of the temperature range. Generally speaking, up to perhaps 45°C, the predominant effect will be an increase in reaction rate and have a temperature dependence of the Arrhenius form. Above 45°C, the opposing factor, namely thermal denaturation, becomes increasingly important. Around 55°C, rapid denaturation usually destroys the catalytic function of the enzyme protein.

The effect of pH, as that of temperature, on enzymatic activity is typified by a bell-shaped curve with a relatively narrow plateau. The plateau is usually called the optimal pH, or optimal temperature point. These optimal points are to be maintained if maximum enzymatic activities are desired. After an enzyme is immobilized on a carrier, the pH and temperature optimal points may be different from that of the free-enzyme counterpart. In addition to pH and temperature stabilities, operational stability and storage stability of immobilized enzymes are also important from practical points of view. Storage stability of 50 immobilized enzyme systems were reviewed (5). Of these, 30 exhibited greater storage stability, 8 exhibited less than the free-enzyme counterparts. Operational stability of immobilized enzymes depends on the enzyme itself, method of immobilization, operational conditions, and so on. Table 2 illustrates the half-life of various immobilized-enzyme systems, which range from a few days to more than a year.

The kinetic behavior of enzymes may be modified by the following factors when they are immobilized: (*1*) change in enzyme conformation, (*2*) steric effects, (*3*) microenvironmental effects, and (*4*) external and internal diffusional effects. Theoretically, the change in enzyme conformation would cause a change in innate activity of the enzyme, that is, the K_M value. Factors 2 to 4 are ineffective in modifying the true catalytic reaction per se, but nevertheless, obscure the kinetics of the reaction. If enzymes are bound to nonporous systems, the reaction rate can be controlled by one of the following steps: (*1*) diffusion of substrate from the bulk solution to the surface of the immobilized enzyme, (*2*) enzymatic reaction at the surface, or (*3*) diffusion of the reaction products back into the bulk solution. Assuming a linear gradient of substrate concentration across a Nernst diffusion layer, a near-stagnant layer of fluid around the immobilized enzyme particle surface, it has been suggested that the approximate behavior of such systems can be represented by:

$$N_S = -\frac{dS}{dt} = \frac{V_M S_B}{S_B + K_M + \left(\frac{V_M}{D_S}\right)\Delta x} \tag{13}$$

where N_S is the molar flux of substrate at the boundary layer, V_M is the maximum activity of the enzyme, S_B is the bulk concentration of substrate, Δx is the thickness of the Nernst layer, and D_S is the substrate diffusivity (12).

The values of K_M and V_M from a Lineweaver-Burk plot ($1/v$ vs $1/s$) of data obtained using immobilized enzymes are usually dependent on flow conditions and particle size (or membrane thickness) of the immobilized enzyme used. If existing mass-transfer resistances are neglected in immobilized enzymes experiments, the apparent K_M obtained is usually higher in value than that of the free enzyme counterpart, unless the enzyme carriers have high affinity for the substrate. Because the thickness of the diffusion layer is inversely proportional to agitation rate, the apparent K_M value can be expected to decrease with increased substrate flow over the solid catalysts, provided bulk diffusional resistance is the principal factor affecting the kinetics of the process. As indicated before, an apparent K_M value close to the true K_M of the immobilized enzyme can be obtained by experiments where possible bulk diffusional effect is limited by using a high substrate-to-enzyme concentration ratio and by maintaining high substrate flow over the catalyst surface, and internal diffusional effect is minimized by using fine particles or thin membranes. True K_M values of immobilized enzymes are important parameters for the purpose of reactor scale-up and control for commercial purposes. On the other hand, because apparent K_M is dependent on mass-transfer effects and thus on reactor size, throughput and configuration, it is of little use in this connection. Thus the determination of true K_M is an important task. Basically there are three groups of experimental methods for evaluating the intrinsic kinetic parameters, such as K_M and V_M, in heterogeneous catalysis. These are as follows (13,14):

1. A certain type of immobilized-enzyme reactor configuration of a single size and a single loading of enzymes are used; the variable is substrate-surface concentration.

Table 1. Michaelis-Menten Constants K_M for Various Enzymes

Enzyme	Substrate	K_M, M
Maltase	Maltose	2.1×10^{-1}
Invertase	Sucrose	2.8×10^{-2}
Lactase (yeast)	Lactose	4.0×10^{-2}
Lactic dehydrogenase	Pyruvate	3.5×10^{-2}

Source: Ref. 2.

Table 2. Operational Stability of Some Immobilized Enzymes

Enzyme-carrier	Operation temperature (°C)	Operation time (days)	Estimated half-life (days)[a]	Reference
Glucoamylase–porous glass (zirconium-coated)	45	21	645	6
Glucoamylase–porous glass	60	15	4.2	6
Invertase–porous glass	23	28	42.5	7
Invertase–cellulose	23	16	3.7	7
Invertase–collagen	30	15	200	8
Glucose isomerase–porous glass	60	46	14.4	9
Glucose isomerase in cells–collagen	70	55	50	10
Acylase–DEAE-sephadex	50	35	65	11

[a]Half-lives are estimated as the time required for the loss of 50% of the initial activity during operation, assuming an unimolecular decay kinetics.

2. The variable is the loading of the enzyme or different-size enzyme particles.
3. An enzymatic membrane is inserted between two reservoirs with different substrate concentrations.

To understand the effects and the control of the live cells, kinetics studies of the enzyme-catalyzed reaction, namely, the functioning unit of metabolic machinery, are necessary as starters. Such studies are usually performed *in vitro*. The extrapolations of the results of *in vitro* studies to the *in vivo* situations are sometimes useful, yet are limited in most cases. Structural factors bearing on the relation of the organization of the cell to enzymatic action should be considered if such extrapolations are to be on the safe side. Most enzymes are quite soluble in water. The *in vitro* studies on such enzymes can be considered to be in a homogeneous catalytic environment. However, within cells, enzymes are not totally free to move, especially the so-called membrane-bound enzymes. Due to such considerations, sometimes the actual effect of an enzyme-catalyzed reaction in a cell cannot be explained by the chemical reaction studied *in vitro* alone. The coupling of mass-transfer effect to chemical reaction is necessary to explain the global effect of a reaction on a system with heterogeneous nature. While the advantages of immobilized enzymes in food manufacturing (eg, high fructose syrup production) are evident from an economical point of view, the research on immobilized enzymes will also help the food manufacturers in general, through better understanding of the regulatory machinery in the cells.

A typical microbial growth curve involves a lag phase, an exponential phase, a stationary phase, and a decline phase. In most industrial microbial processes, cells or products accumulated by cells are harvested in the late exponential phase or early stationary phase. During the exponential phase of growth for unicellular organisms undergoing binary fission, the growth rate dx/dt can be expressed in terms of a growth rate constant (or specific growth rate) (and the cell concentration x [equation 11]).

Most microorganisms, if grown properly, follow this first-order, exponential growth. There are exceptions. A linear growth model, namely, dx/dt = constant, was found in some hydrocarbon cultures where limitation is caused by the rate of diffusion of substrate from the surface of oil droplets that have essentially a constant surface area during cultivation. The growth of yeast and filamentous organisms do not follow the first-order kinetics of growth in some cases. In these organisms, growth occurs from the tip, but nutrients diffuse throughout the cellular tissue; sometimes fractional reaction orders are obtained in kinetic analyses.

The specific growth rate μ is essentially a first-order reaction rate constant with a dimension of reciprocal of time, that is, h^{-1}. The constancy of μ is limited. It is a constant under specified conditions of temperature, pH, type of substrate, substrate concentration, and so on. When different μ values are obtained by changing substrate concentration in the experiments, an expression for μ, the so-called Monod equation, is obtained (equation 11).

The form of this equation is exactly the same as that of the Michaelis-Menten equation for enzyme-catalyzed reactions. However, this equation is entirely empirical, whereas that of Michaelis-Menten is mechanistic. This equation usually fits experimental data when cell growth is limited only by a single limiting substrate in a pure culture situation. When there are inhibitors or mixed substrates in the medium, a modified Monod equation can be used to fit the data.

In addition to cell tissue, there are three types of product that can be produced by industrial fermentation processes. These are primary metabolites (eg, monosodium glutamate and food acidulants such as citric acid), secondary metabolites (eg, antibiotics), and enzymes. For primary metabolites and some enzymes, the rate of product formation is proportional to the rate of cell growth. The following equation has been proposed for organic acid fermentations (15):

$$\frac{dp}{dt} = \alpha\left(\frac{dx}{dt}\right) + \beta x$$

or

$$\left(\frac{1}{x}\right)\frac{dp}{dt} \equiv \nu_p = \alpha\mu + \beta \qquad (14)$$

The production model expressed by equation 14 is called a growth-associated model. The kinetics for the production of secondary metabolites and enzymes (when gratuitous inducer is used) are such that they are accumulated or induced after the cells have advanced into late exponential phase or early stationary phase of growth where cell growth rate has started to decline. A nongrowth-associated model is appropriate in these cases:

$$\frac{dp}{dt} = \alpha x$$
$$\left(\frac{1}{x}\right)\frac{dp}{dt} \equiv \nu_p = \alpha \quad (15)$$

The following growth-associated model has been proposed for the kinetics of enzyme induction (16):

$$\left(\frac{1}{x}\right)\frac{dE}{dt} = \alpha\mu - kE \quad (16)$$

where the term kE on the right-hand side of equation 16 represents the monomolecular (first-order) decay of the enzyme synthesized *in vivo*. A more complete model on the kinetics of enzyme induction includes the effects of inducer and repressor concentrations on the kinetics, and the mechanism of protein synthesis can be used as basis for the construction of the model.

STARCH CONVERSION

When the starch granule is heated in the presence of an adequate or excess amount of water, gelatinization of starch takes place. Several steps are associated with gelatinization, such as hydration and swelling of granules, heat absorption, and the loss of crystallinity as judged by the disappearance of both optical birefringence and X-ray diffraction pattern. In the extrusion cooking of starch containing materials, starch is heated under high pressure, shear force, and a limited amount of water, to say less than 40%. At a high enough temperature and low moisture content (17,18), starch granules undergo both gelatinization and melting processes. Thus, starch conversion during food processing is not an elementary reaction as discussed before. The conversion itself is a complex phenomenon that is further complicated by heat, mass, and momentum transfer in a reactor like an extruder. The kinetics studies on starch conversion, and other similar processes, take the advantage of an overall rate approach and use the parameter percent conversion instead of concentration for the formulation of kinetics equations. Starch conversion is an endothermic reaction. The degree of conversion of starch during processing due to heat or shear can be evaluated by the decrease of the differential scanning calorimeter (DSC) endotherm area. This area is directly proportional to the mass of unconverted starch in the sample. The degree of conversion X can be defined as follows:

$$X = 1 - \frac{\Delta H_\mathrm{i}}{\Delta H_0} \quad (17)$$

where ΔH_0 is the enthalpy change of the DSC endotherm without conversion, and ΔH_i that with conversion. Thus X is equivalent to the degree of conversion and is defined as

$$X = \frac{M_0 - M_t}{M_0 - M_f} \quad (18)$$

where M_0 is the initial amount of unconverted starch; M_f, the final amount of unconverted starch; and M_t, the amount of unconverted starch at time t. Thus the rate of the disappearance of native (unconverted) starch during a process can be written in the following form:

$$\frac{d(1 - X)}{dt} = -k(1 - X)^n \quad (19)$$

There are two general approaches to analyzing kinetics data, namely, the differential and integral methods. In the differential method of test, the differential at each specific time is calculated, and after taking the logarithm, the data are plotted. The reaction order is the slope of a plot of log ($d[1 - X]/dt$) versus log ($1 - X$), and the reaction-rate constant can be obtained from the intercept. In the integral method, a particular form of rate expression is assumed. After appropriate integration and mathematical manipulation, a certain concentration function versus time is plotted. If a reasonably good fit is obtained, then the rate equation is adopted to describe the reaction under study. There are advantages and disadvantages to each method. The advantages of the differential method are the ease of obtaining the appropriate reaction order and reaction rate, and the disadvantage is the magnification of experimental errors. The advantage of the integral method is that experimental data can be used directly, while the disadvantage is that a trial-and-error procedure must be followed. In general, it is suggested that the differential analysis be attempted first, then the integral analysis is performed to check if the obtained rate expression is appropriate or not. Most investigators apply the integral approach to the studies of starch gelatinization with an excess amount of water. The kinetics of starch gelatinization in the cooking of rice grains with 94% moisture content has been studied (19). The kinetics of the water diffusion and starch gelatinization at 91.4% moisture content has been reported (20). The kinetics of cooking of rice starch also in an excess amount of water has been studied (21). The gelation kinetics of barley starch in a diluted system (1% solid) has been investigated (22). A first-order kinetics equation has been employed by these investigators to describe the gelatinization of starch in an excess amount of water. The kinetics of starch conversion in a limited amount of water (40% or less) has not been as well studied, and that reaction order has been reported to be zero order (18,23).

In the presence of an excess amount of water (ie, >60% water for Amioca), different sources of starch gelatinize at different temperatures. The range of temperature is between 50 and 80°C. If a limited amount of water is used, the temperature at which gelatinization and melting (as detected in DSC thermogram) happen is elevated. Figure 1 shows the dependence of peak temperature T_p of the DSC thermogram on the water content of the Amioca samples. At water content higher than 60%, T_p stays constant, whereas at water content lower than 60%, there are two peaks in the thermogram. The second peak temperature, T_p, increases with decreasing water content according to

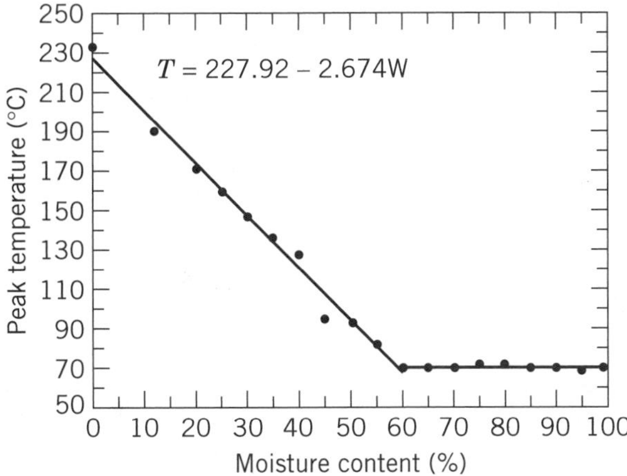

Figure 1. Peak temperatures from DSC endotherms for amioca with various moisture contents.

the linear equation $T_p = 227.92 - 2.674$ W, where W is the water content in wet wt %. To consider the combined effects of heat, moisture content, and the source of starch on the conversion of starch, a dimensionless temperature parameter T/T_p was developed (18). This parameter correlated well with the rates of conversion of starch. Figure 2 shows the plot of k, the zero-order rate of starch conversion versus T/T_p, T is the operating temperature and T_p, the second endotherm peak temperature, is a constant for a specific starch at a specified water content. In Figure 2, data from different studies are presented (18,19,24). Excluding data from Reference 19, which show obvious mass-transfer resistance (diffusion controlling), the rest of the data can be represented by the following equation:

$$k = \frac{x}{74.97x^2 - 167.21x + 93.38} \tag{20}$$

where $x = T/T_p$ and the limit for x is $0.63 < x < 1.06$, for the equation to be meaningful.

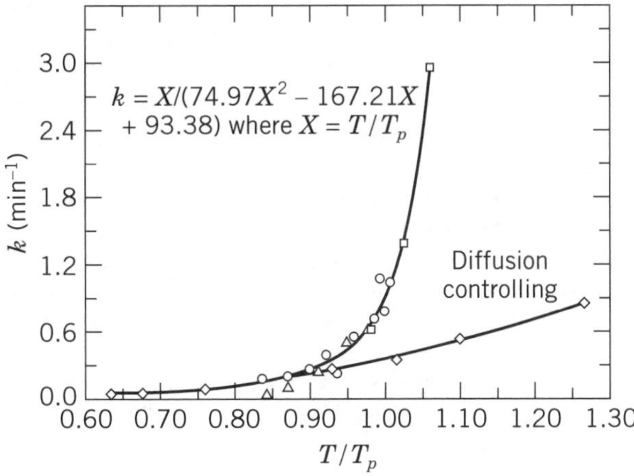

Figure 2. A general correlation between starch conversion rate constant k, and T/T_p, a dimensionless temperature parameter. T_p is peak temperature of DSC thermogram. *Source:* ○, Ref. 18; □, Ref. 24; △, Ref. 21; ◊, Ref. 19.

This equation is useful for calculating reaction rates of various starch conversion at various water contents and various temperatures. The source of starch and water content determines the value of T_p, which is measured by using DSC endotherms.

EFFECTS OF ENERGY INPUTS

It is generally correct to say that some form of energy input is required to cause a reaction to happen. A classic way of supplying the said energy is by heating up the system. As a rule of thumb, for every increment of 10°C, the reaction rate will double. That is the Q_{10} principle. The temperature effect can be visualized by looking at the increased kinetic energy of the molecules that makes the molecules move faster and so increases the probability of colliding with other molecules in the system. Such reasoning is good for explaining temperature effects in gaseous and solution chemistry. For food systems that are solid or semisolid, the resistances of heat and mass transfer and the fact that the mobility of macromolecules are relatively small cause the temperature effect to deviate from the simple Q_{10} principle. Another idiosyncrasy of biological reactions is that many of them are catalyzed by enzymes that are unstable at high temperatures. This causes nonlinearity of Arrhenius plot for biological and biochemical reactions. Arrhenius derived the following equation to fit experimental data obtained from studying the effect of temperature on the kinetics of sucrose inversion:

$$k = \alpha e^{-\Delta E/RT} \tag{21}$$

The logarithmic form of the preceding equation is shown as follows:

$$\ln k = \ln \alpha - \frac{\Delta E}{RT} \tag{22}$$

Equation 22 states that the natural log of reaction rate constant ln k varies linearly with the reciprocal of absolute temperature $1/T$ with a constant negative slope, $-\Delta E/R$. The evaluated value of the said slope provides a way to calculate ΔE, the activation energy of the reaction for the temperature range that the said slope was evaluated on. If 1.987 is used for the value of R, the gas constant in calculating ΔE, the dimension for the calculated value of ΔE is cal/g mol. For some reactions involved in food processing, different temperature ranges may result in different calculated activation energy. As shown in Table 3, the thermal activation energy for the conversion of several starches at different temperature ranges varies from about 10 to 230 kcal/mol. These differences can be rationalized by looking at the different granular sizes of the native starches from rice, corn, and potato; the different moisture contents in the samples; and possibly different mechanisms of conversion at different temperature ranges. In dealing with biological or biochemical reactions where activities of the catalysts (enzyme or cells) involved are also affected by temperature, a composite Arrhenius plot can be obtained. The study of the effect of temperature on the

Table 3. Activation Energy for Starch Conversion Due to Thermal Effect

Materials	Temperature (°C)	E_a (kcal/mol)	Reference
Amoica, 75%	145–160	44.25	18
Amoica, 80%	150–170	33.78	18
White rice	75–100	19.8	19
(65% water)	100–150	8.8	
Brown rice	50–85	18.5	28
(>45% water)	85–120	10.5	
Rough rice	50–85	24.7	28
(>45% water)	85–120	9.6	
Rice starch	70–85	24.7	29
(90% water)			
Potato starch	<65	230	29
(90% water)			
Potato starch	<67.5	196	24
(82% water)	<67.5	58.2	

catalytic breakdown of hydrogen peroxide by beef liver catalase uncovered the nonlinearity in an Arrhenium plot (ln k vs $1/T$) as shown in Figure 3 (25). The effect of temperature on the rate of production of penicillin with *Penicillium chrysogenum* was reported to show a similar effect (Fig. 4). (26). The data points on these two figures fall into two groups, one of which is related to the activation of the substrate(s) and the enzyme molecules (catalase, or other enzymes in the cell of *P. chrysogenum*), while the other is indicative of the inactivation of the enzyme or the cell at higher temperature range. The maximum of the roughly bell-shaped curve of Figure 3 and the intercept of the two lines in Figure 4 are corresponding to the critical temperatures causing the denaturation of the enzyme (53°C for catalase) and the inactivation of the cell activity. In studying activation energies for biological or biochemical reactions, care must be taken to avoid such high temperatures. It has

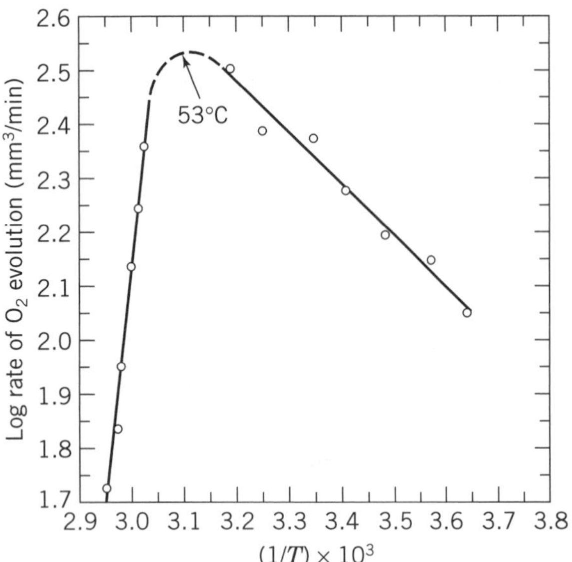

Figure 3. Nonlinearity of Arrhenius plot for breakdown rate of hydrogen peroxide in the presence of catalase. *Source:* Ref. 25.

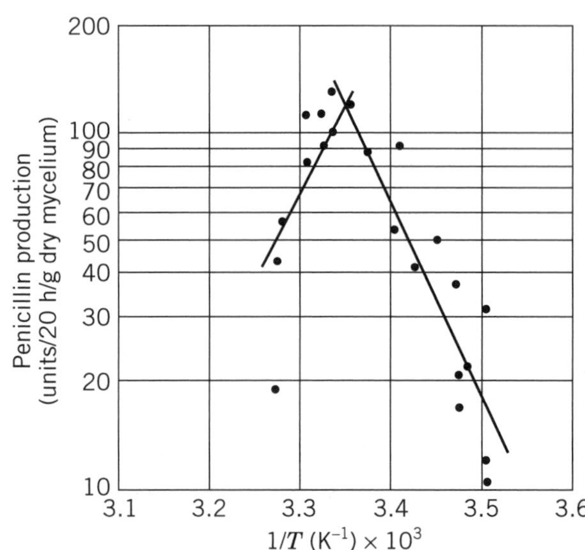

Figure 4. Nonlinearity of Arrhenius plot for penicillin production rate with *Penicillium chrysogenum*. *Source:* Ref. 26.

been shown that the activation energy for cell growth is sufficiently different from that for penicillin production. This difference suggests that the enzymes involved in each sequence are different. (Penicillin, being a secondary metabolite, is produced by mature culture of the *Penicillium*.) Studies such as these are helpful in elucidating the kinetic mechanisms of reactions. To determine whether a biological or a biochemical reaction is kinetically controlled or mass-transfer controlled, the Arrhenius plot and the calculation of activation energy can be used. The activation energy of a catalyzed biological or biochemical conversion is in the range of 8 to 10 kcal/mol or higher, whereas that of diffusion is smaller than that (27).

The activation energy calculation according to Arrhenius only deals with one type of energy source: heat or thermal. It is known that there are reactions that can be activated by other types of energy source, such as photoenergy, electrical energy, shear energy, and pressure-related energy. The process of photosynthesis is initiated by photoenergy. The mechanism of the transduction of light energy into chemical energy has been a challenge to the scientific community. Electrochemical reactions are initiated by electrical energy. Manufacturing of chlorine and sodium hydroxide are examples. Shear energy has been shown to initiate free-radical formation. What is the efficiency of these different energy source in causing the reactions to happen? Activation energy can be looked on as the energy barrier to the initiation of a reaction. What are the activation energy requirements for photochemical, electrical, or shear energy induced reactions? It is possible to look at the activation energy as an index for the efficiency of a particular energy source. The smaller the value, the higher the efficiency. Using starch conversion as an example, starch can be converted (based on DSC thermogram) by supplying thermal energy or shear energy in an extruder. It is known that starch can also be converted (based on DSC thermogram) by shear energy alone (30–32) that is without external heating or dissipated heat. The efficiency of shear energy in causing this conversion is two to three orders of magnitude higher than that by thermal energy. The thermal activation energies shown in Table 3 are in the range of kilocalories per mole, whereas those of shear, as shown in Table 4, are in the range of calories per mole. This is an important reason why extruders, due to their capability of providing high shear, are efficient reactors in which to carry out starch conversion for food processors.

BIOREACTOR

This section deals with the selection of the best types of reactor and operation for a particular biochemical or biological process. When faced with a design problem such as this, there are two previously fixed factors that must be considered to make a proper selection. These are, first, the scale of operation (production rate) and, second, the kinetics of the reaction involved. Invariably the criterion that is used to make the selection is a monetary one. Design and economics are two sides of the same coin. To be responsible to society as a whole, industries must also pay attention to safety and environmental and ethical problems in making decisions on the overall design of a process. This is much easier said than done. Engineering design and economics are quantitative, whereas social benefits and values are difficult to access. At the present time, the subject matter dealt here is limited to the scientific aspect of the problem that can be expressed clearly in quantities.

BIOLOGICAL REACTOR

In general, reactors have been classified in two ways, one according to the type of operation and the other according to design features. Based on the types of operation, there are batch, continuous, and semicontinuous reactors. Due to the relatively high productivity of a continuous process as compared with a batch one, when a process is reaction cost intensive, it is advisable to use the former rather than the latter. On the other hand, if the process is recovery cost intensive, high concentration of product instead of high productivity is desirable, and a batch process is usually preferred. Whether the process is a batch or a continuous one, the reactor is the heart of the process. The efficiency of the reactor frequently, if not always, determines the economics and, hence, the commercial feasibility of the process. To achieve high efficiency of the reactor, it must be designed according to the specific reaction system concerned, namely, the kinetics of the reaction.

Table 4. Activation Energy for Shear Effect on 65% Amioca at Different Temperatures

Temperature (°C)	Activation Energy (cal/mol)	Preexponent (k_{so})
21	108.01	6.99×10^7
25	91.77	8.60×10^3
35	71.60	4.29×10^4
40	37.32	4.79×10^3
50	24.42	7.45×10^2
70	20.52	9.77
80	12.97	8.10
90	10.44	9.48

Source: Ref. 30.

Chemical reactors have also been classified according to the design features. There are tank reactors, tubular reactors, tower reactors, and fluidized reactors. Most of these reactors have found their uses in biological and biochemical processes. The most popular reactor used in biological and biochemical processes has been the tank reactor. The majority of the industrial fermentation process, waste products–treatment processes, and food-cooking processes use tank reactors. For factories making a repertoire of hundreds of products each at a small scale, the tanks are versatile and flexible, and batch operation is the one of choice. However, a batch process is inherently harder to control than a continuous one, even with the advent of computer-coupled control systems. The right decision between batch and continuous operation depends on the relative magnitude of labor and capital costs. What is best for a developed country is not necessarily good for a developing one.

In the following, the performances of distinctive reactors in continuous operation are discussed. For the sake of simplicity in mathematical analyses of the reactor performances, these reactors can be categorized into two ideal reactors: well-mixed (back-mixed or stirred tank) and plug-flow reactors, according to the flow patterns in the reactors. In a well-mixed reactor, it is assumed that there is no concentration distribution in the reactor; if the process is continuous, there is no distribution of residence time among the elements of fluid (or substrate) leaving the reactor tank. Under such a condition (complete mixing), the rate is constant throughout the reactor, and there is no need for a differential equation describing the change in concentration or conversion between the inlet and outlet streams. A simple material balance is sufficient. In a plug-flow reactor (wherein no longitudinal mixing but complete radial mixing is assumed), the concentration of the substrate varies longitudinally and so does the rate of the reaction (unless the reaction is of zero order with respect to the substrate concentration). In such a case, a differential equation is needed to express the performance of the reactor, and the performance of the reactor can be obtained by integrating the equation. Due to such differences in reactor features for well-mixed and plug-flow reactors, other things being equal, when a reaction is substrate inhibitory, it is desirable to use a well-mixed reactor. On the other hand, if the reaction is product inhibitory, it is desirable to use a plug-flow reactor. In most occasions, the selection of either a well-mixed or a plug-flow reactor depends mainly on the kinetics of the reaction concerned. In general, if a first-order enzymatic reaction is free of substrate and product inhibition, the amount of enzyme required for 99% conversion in a well-mixed reactor is 21 times that required for a plug-flow reactor. The ratio of the amount of the enzyme required (ECSTR/EPFR) becomes four to one when 90% conversion is desired. Consequently, other things being equal, for a first-order enzymatic reaction, it is advisable to use a plug-flow reactor rather than a well-mixed reactor.

The growth of cells is a self-catalytic process. The plot of cell growth rate versus cell concentration follows a bell-shaped curve with an optimum. In such processes, if the reaction rate in the reactor never reaches the maximum rate, theoretically the use of one single well-mixed reactor is superior to the use of a plug-flow reactor. This is based on the comparison of the efficiency of the reactors in carrying out a specific conversion under identical environmental conditions. If the cultivation of cell tissue is the purpose of a fermentation process, either in a batch or continuous process, the rate of cell growth in the fermenter usually does not exceed the maximum rate during the process. This is one of the reasons why a mechanically stirred-tank reactor with air dispersed into liquid medium by sparging is the most typical industrial fermenter, although the production of secondary metabolites and enzymes from fermentation processes usually includes the growth of microorganisms beyond the exponential phase. This type of process usually warrants a two-stage process, and theoretically the use of a stirred-tank reactor followed by a plug-flow reactor is most desirable (33). The most obvious difficulty of using a plug-flow reactor is the maintenance of the flow pattern, especially under aeration requirements.

Another aspect of the discussion of reactor kinetics is the use of immobilized enzymes in various reactor types for biocatalytic conversions. Free enzymes can be used in these reactors, except that a rather large amount of energy will have to be spent on the retention of enzymes usually through the use of ultrafilters. Basically, enzymes are expensive materials; because they are catalysts, their reusability is logical, provided it is economical to do so. Immobilization of enzymes seems to be the answer to this question. Although some of the immobilization processes are quite elaborate and potentially expensive, many are quite simple and effective (8,10).

The efficiency of a reactor used for immobilized enzymes depends on the following factors: mass-transfer resistances (external and internal), enzyme-loading factor, enzyme stability, substrate contact efficiency, and so on. In selecting reactors to house immobilized enzymes, these factors must be considered. As an example, the selection of a reactor for glucose isomerase is discussed. There are three types of immobilized glucose isomerase: immobilized cell-free glucose isomerase (8,9,34,35), immobilized cells with intracellular glucose isomerase (10), and free cells with intracellular glucose isomerase (36,37). As has been pointed out for immobilized glucose isomerase reactors (35) and as a general rule, for higher-order reactions or Michaelis-Menten kinetics with a high value of K_M, when high conversion is desired, much more enzyme is required in the continuous stirred-tank reactor than in the plug-flow reactor. The choice of a plug-flow reactor is also partly due to the fact that the intrinsic reaction kinetics there are equilibrium controlled. The Keg value of the isomerization reaction is close to the value of one at most operating temperatures. The concentration of product cannot be neglected in the rate expression. Applying reversible Michaelis-Menten kinetics to glucose-fructose conversion, the resulting rate expression is

$$v = \frac{V_M(S - P/K)}{K_M + S + (K_M P/K_P)} \quad (23)$$

where K is equilibrium constant (P/S at equilibrium), S is the glucose concentration, P is the fructose concentration, K_M and K_P are constants, v is the reaction rate, and V_M is equal to $(E)_{\text{total}}$ times k, the maximum reaction rate. This rate expression can be looked upon as one derived from the product-inhibition model, and, as discussed before, the use of a plug-flow reactor is more efficient for this type of reaction. In published data on immobilized glucose isomerase, mass-transfer resistances were not found to be limiting factors. Both internal and external resistances were found to be negligible. In accordance with this, it has been found that the activation energy of immobilized and soluble glucose isomerase are close, 15.7 and 14.5 kcal/g ~ mol, respectively (35). Due to the lack of mass-transfer resistances in this case, it is advisable to use an immobilized whole-cell system instead of an immobilized enzyme system, provided that the cells are fully induced and are high in glucose isomerase concentration (8).

In batch reactors, concentrations of reactant(s) and product(s) are a function of time. So batch processes are transient processes. Continuous processes, on the other hand, can be in either transient or steady state. There are two types of steady state in continuous operations: volumetric steady state and concentration steady state. Other things being equal, a volumetric steady state is a prerequisite for reaching a concentration steady state. Mathematically, steady state means that the variable (volume or concentration) is not changing with respect to time, dV/dt or $dC/dt = 0$. In operation, volume invariance can be achieved mechanically; for biological systems, concentration steady state, such as constant cell concentration, is difficult to achieve. In reality, $dC/dt = +$ or -10% of the mean is a good criterion for reaching the steady state of cell cultures (38).

Deviations from ideal reactors are common in practice. Many real reactors behave in between a well-mixed and a plug-flow reactor. A column reactor is closer to a plug-flow reactor, a tank reactor to a well-mixed reactor, and a fluidized-bed reactor is a hybrid of the two ideal reactors. A series of well-mixed reactors can be operated to perform like a plug-flow reactor. An extruder is closer to a plug-flow than a well-mixed reactor. These arguments can be evaluated by looking at the residence time distributions of various reactors.

An extruder is like a pump. It is a versatile reactor, capable of performing other operations in concert with its pumping function. For instance, extruders can be used as a mixer and as reactors to make pasta, pet foods, instants, confectioneries, and snacks and can be used to determine spices. Extrusion is a typical continuous process that has the potential of steady-state operation with high productivities. The engineering analysis of an extrusion process is a difficult task. Chemical and physical changes of materials that occur in extruders are the results of solid (powdery) or semisolid interactions under limited-moisture contents. The initiation, or activation, of such changes can be done by thermal energy input or mechanical energy input. In addition to changes at the molecular level, extruders are also most effective in rendering size reductions to affect changes at the particle level. All these changes are important in determining the product quality, especially in texture-related features. The kinetics of chemical and physical changes of starch and other food ingredients in extrusion process are difficult to study because of the highly nonideal environment, as opposed to gas or dilute solution, and the complication of transport processes (heat, mass, and momentum transfers) coupling with reaction kinetics. The idea of predicting the performance of an extruder by using transport coupled kinetic models is a desirable one. The task of correlating product qualities, such as specific texture, flavor, or color, to the performance of an extruder is a difficult one. Certainly innovative ideas are needed to achieve such a goal.

FOOD STORAGE AND PREDICTION OF SHELF LIFE

Quality deterioration in the form of decrease of nutritional value, color change, development of off-flavor, and textural change may occur in foods during storage. The purpose of food preservation is to minimize such quality deterioration. To provide adequate but not excess protection, the studies of environmental factors affecting quality deterioration are of prime importance. To predict quality changes, or to predict shelf life of foods, kinetics studies of essential parameters, such as product characteristics, storage environment, and container protection are necessary. Computer-aided simulation of models derived from such kinetics studies can provide information for adequate design of a preservation scheme for foods and also for prediction of the shelf life of foods (39–43). The success of such design and prediction depends mainly on the understanding of the nature of the chemical and physical changes of the food system involved and the subsequent derivation of a correct kinetics model from such understanding. A direct mathematization of an overall change without mechanistic understanding of such changes may result in an inadequate kinetics model that has limited application. Earlier studies on the quality deterioration of foods (44,45) mostly depend on such technique. The effects of storage temperature and time on the retention of thiamin and ascorbic acid in canned foods has been studied (44). Nomographs for ascorbic acid and thiamin retention based on first-order decay kinetics of these vitamins were constructed. By extrapolation, these homographs permit prediction of retention values beyond the conditions of the actual experiments. Similar techniques, were used in another study that went further to calculate activation energy and frequency factors (ΔE and α of equation 21) by studying the first-order decay reaction at different temperatures. By using these constants (ΔE and α), it would be also possible to calculate the effect of fluctuating temperature on the storage stability of the foods studied. Although both these studies found first-order decay of ascorbic acid in frozen spinach and various canned foods, respectively, another study (43) reported that the rate of oxidation of ascorbic acid is not only dependent on ascorbic acid concentration, but also on dissolved oxygen (DO) concentration if it is below a critical value, such as 8.71 ppm in that study. Unless there is a continuous supply of oxygen to the packaged food system, the DO value will decrease from an initial saturation value (solubility of oxygen in water is about 7–8 ppm at 25°C) to a limiting value during storage. The consideration of DO concentration in the rate expression of ascorbic acid degradation seems to be logical under such circumstances. As proposed, the following bimolecular kinetics equation is appropriate in describing the rates of ascorbic acid oxidation in infant foods (43).

$$-\frac{d(A)}{dt} = k(A)(B) \tag{24}$$

where (A) and (B) are concentrations of ascorbic acid and DO, respectively. It was also found that the rate constant k increases linearly with light intensity up to 1756 (transmitted) lux.

Shelf life, or storage stability of foods, depends on a large number of factors such as those of environmental origin—temperature, equilibrium relative humidity, oxygen partial pressure, light, intensity, and so on—and those of container characteristics—package permeabilities, package configurations, and so on. Studies of the effect of the critical factor or factors among those listed on the stability of food make it possible to derive kinetics equations suitable for modeling the changes and predicting the shelf life of foods. However, for the purpose of package design and optimization (overprotection is expensive), it is necessary to consider the deteriorative reactions that depend on the characteristics of the package. In other words, the kinetics model or the formulas derived from it will be useful for package design and optimization if the packing parameters are explicitly included in the formulas (42).

As an example of optimization of package design, the results of one study (40) are summarized here. Two critical factors affecting the shelf life of potato chips were identified as (*1*) transfer of moisture into the package, causing a textural change, and (*2*) the initial presence and continued transfer of oxygen, which causes oxidative rancidity. If shelf life is defined as the time to reach maximum allowable moisture content or the maximum allowable extent of oxidation, the optimized design of packaged potato chips is that in which these two limits are reached simultaneously. This requirement determines the selection of suitable film for packaging. A lamination of two films with one high in oxygen permeability but low in water permeability and the other high in water permeability but low in oxygen permeability would be the choice for such a case.

Kinetics models are useful in describing the rates of changes related to food quality, including spoilage. A more important aspect of kinetics model development is its ability to be used for predicting the spoilage before the foods are consumed. The traditional microbiological approach to determining the stability and safety of foods is retrospective. Information is obtained several days after the samples are collected and analyzed in the laboratories. The use of predictive microbiology, within which the growth rates of the microbes of concern in foods would be modeled with respect to critical environmental factors, such as temperature, pH, and water activity, has been proposed (46). Models relevant to broad categories of foods can be used to predict the spoilage rates of foods. More studies are needed to learn how much information is required to produce such a workable model. Nevertheless, modeling by using relevant kinetics equations offers the most cost-effective approach to the understanding and control of microbial growth in foods (46).

BIBLIOGRAPHY

1. G. E. Briggs and J. B. S. Haldane, "A Note on the Kinetics of Enzyme Action," *Biochem. J.* **19**, 338–339 (1925).
2. J. B. Neilands and P. K. Stumpf, *Outlines of Enzyme Chemistry*, John Wiley & Sons, New York, 1958.
3. G. B. Borglum and M. Z. Sternberg, "Properties of a Fungal Lactase," *J. Food Sci.* **37**, 619–623 (1972).
4. H. H. Weetall, "The Preparation of Immobilized Lactase and Its Use in the Enzymatic Hydrolysis of Acid Whey," *Biotechnol. Bioeng.* **16**, 295–313 (1974).
5. G. J. H. Melrose, "Insolubilized Enzymes: Biochemical Applications of Synthetic Polymers," *Rev. Pure Appl. Chem.* **21**, 83–119 (1971).
6. H. H. Weetal and N. B. Havewala, "Continuous Production of Dextrose From Corn Starch: Reactor Parameters Necessary for Commercial Application," *Biotechnol. Bioeng. Symposium* **3**, 241–256 (1972).
7. R. D. Mason and H. H. Weetall, "Invertase Covalently Coupled to Porous Glass: Preparation and Characterization," *Biotechnol. Bioeng.* **14**, 637–645 (1972).
8. S. S. Wang and W. R. Vieth, "Collagen-Enzyme Complex Membranes and Their Performance in Biocatalytic Modules," *Biotechnol. Bioeng.* **14**, 93–101 (1972).
9. G. W. Strandberg and K. L. Smiley, "Glucose Isomerase Covalently Bound to Porous Glass Beads," *Biotechnol. Bioeng.* **14**, 509–513 (1972).
10. S. S. Wang, W. R. Vieth, and A. Constantinedes, "Complexation of Enzymes or Whole Cells With Collagen," in E. K. Pye and L. B. J. Wingard, eds., *Enzyme Engineering*, Vol. 2, Plenum Press, New York, 1974, pp. 123–129.
11. I. Chibata et al., *Fermentation Technology Today*, Society of Fermentation Technology, Osaka, Japan, 1972.
12. M. D. Lilly and A. K. Sharp, *Chem. Eng.* (London) CE12 (Jan./Feb. 1968).
13. J. Engasser and C. Howath, "Effect of Internal Diffusion in Heterogeneous Enzyme Systems: Evaluation of True Kinetic Parameters and Substrate Diffusivity," *J. Theor. Biol.* **42**, 137–155 (1973).
14. Y. Kobayashi and K. J. Laidler, "Kinetic Analysis for Solid Supported Enzymes," *Biochim. Biophys. Acta* **302**, 1–12 (1973).
15. R. Leuedeking and E. L. Piret, *J. Biochem. Microbiol. Technol. Eng.* **1**, 393–405 (1959).
16. A. Shinmyo, M. Okagaki, and G. Terui, "Kinetic Studies on Enzyme Production by Microbes. III: Process Kinetics of Glucoamylase Production by *Aspergillus niger*," in D. Perlman, ed., *Fermentation Advances*, Academic Press, Orlando, Fla., 1969, pp. 337–367.
17. J. W. Donovan, "Phase-Transition of the Starch-Water System," *Biopolymers* **18**, 263–275 (1979).
18. S. S. Wang, W. C. Chiang, A. I. Yeh, B. L. Zhao, and I. H. Kim, "Kinetics of Phase-Transition of Waxy Corn Starch under Extrusion Temperature and Moisture Content," *J. Food Sci.* **64**, 1298–1301 (1989).
19. K. Suruki et al., "Kinetic Studies on Cooking Rice," *J. Food Sci.* **41**, 1180–1183 (1976).
20. K. Kubota et al., "Studies of the Gelatinization Rate of Rice and Potato Starches," *J. Food Sci.* **44**, 1394–1397 (1979).
21. D. B. Lund and M. Wirakartakusumah, "A Model for Starch Gelatinization Phenomena," in B. M. McKenna, ed., *Engineering and Food, Vol. 1: Engineering Sciences in the Food Industry*, 1984, pp. 425–432.
22. C. Mok et al., "Kinetic Study on the Gelatinization of Barley Starch," *Korean Journal of Food Science Technology* **17**, 409–414 (1985).
23. M. Bhattacharya and M. A. Hannai, "Kinetics of Starch Gelatinization During Extrusion Cooking," *J. Food Sci.* **52**, 764–766 (1987).
24. C. I. Pravisani, A. N. Califano, and A. Calvilo, "Kinetics of Starch Gelatinization in Potato," *J. Food Sci.* **50**, 657–660 (1985).
25. I. W. Sizer, "Temperature Activation and Inactivation of the Crystalline Catalase-Hydrogen Peroxide System," *J. Biol. Chem.* **154**, 461–473 (1944).
26. C. T. Calam, N. Driver, and R. H. Bowers, "Studies in the Production of Penicillin, Respiration and Growth of *Penicillium chrysogenum* in Submerged Culture, in Relation to Agitation and Oxygen Transfer," *J. Appl. Chem.* **1**, 209–216 (1951).
27. P. Schneider and J. M. Smith, "Chromatographic Study of Surface Diffusion," *AIChE J.* **14**, 886 (1968).
28. A. S. Bakchi and R. P. Singh, "Kinetics of Water Diffusion and Starch Gelatinization During Rice Parboiling," *J. Food Sci.* **45**, 1387–1392 (1980).
29. Kubota et al., "Studies of the Gelatinization Rate of Rice and Potato Starches," *J. Food Sci.* **44**, 1394–1397 (1979).
30. S. S. Wang et al., "Development of an Energy-Equivalent Concept for the Formulation of Extrusion-Kinetic Equations," *Fiftieth IFT Annual Meeting*, Chicago, Ill., June, 1989.
31. X. Zheng and S. S. Wang, "Shear Induced Starch Conversion During Extrusion" *J. Food Sci.* **59**, 1137–1142 (1994).
32. S. S. Wang, and X. Zheng, "Tribological Shear Conversion of Starch," *J. Food Sci.* **60**, 520–522 (1995).
33. K. B. Bischoff, "Optimal Continuous Fermentation Reactor Design," *Can. J. Chem. Eng.* **44**, 281–284 (1966).
34. Ger. Pat. 2,303,872 (1973), D. Dinelli, F. Morisi, S. Giovenco, and P. Pansolli.
35. N. B. Havewala and W. H. Pitcher, "Immobilized Glucose Isomerase for the Production of High Fructose Syrup," in E. K. Pye and L. B. J. Wingard, eds., *Enzyme Engineering*, Vol. 2, Plenum Press, New York, 1974, pp. 315–328.
36. U.S. Pat. 3,623,953 (November 30, 1971), W. P. Cotter, N. E. Lloyd, and C. W. Hinman (to Standard Brands, Inc.).
37. U.S. Pat. 3,694,314 (September 26, 1972), N. E. Lloyd, L. T. Lewis, R. M. Logan, and D. N. Patel (to Standard Brands, Inc.).
38. C. K. Jin and S. S. Wang. "Steady State Analysis of the Enhancement in Productivity of a Continuous Fermentation Process Employing Protein Phospholipid Complex as a Protecting Agent," *Enzyme Microb. Technol.* **3**, 249–257 (1981).
39. T. P. Labuza, S. Mizrahi, and M. Karel, "Mathematical Models for Optimization of Flexible Film Packaging of Foods for Storage," *Trans. Am. Soc. Agricultural Eng.* **1G**, 150–155 (1972).

40. D. G. Quast and M. Karel, "Computer Simulation of Storage Life of Foods Undergoing Spoilage by Two Interacting Mechanisms," *J. Food Sci.* **37**, 679–683 (1972).
41. Y. S. Henig and S. G. Gilbert, "Computer Analysis of the Variables Affecting Respiration and Quality of Produce Packaged in Polymeric Films," *J. Food Sci.* **40**, 1033–1035 (1975).
42. K. I. Hayakaw, Y. S. Henig, and S. G. Gilbert, "Formulas for Predicting Gas Exchange of Fresh Produce in Polymeric Film Package," *J. Food Sci.* **40**, 186–191 (1975).
43. R. P. Singh, D. R. Heldman, and J. R. Kirk, "Kinetics of Quality Degradation: Ascorbic Acid Oxidation in Infant Formula During Storage," *J. Food Sci.* **41**, 304–308 (1976).
44. M. Freed, S. Brenner, and V. O. Wodicka, "Prediction of Thiamine and Ascorbic Acid Stability in Stored Canned Foods," *Food Technol.* **3**, 148–151 (1949).
45. A. L. Brody, K. Bedrosian, and C. O. Ball, "Low Temperature Handling of Sterilized Foods, V: Biochemical Changes in Storage," *Food Technol.* **14**, 552–556 (1960).
46. T. A. Roberts, "Combinations of Antimicrobials and Processing Methods," *Food Technol.* **43**, 156–163 (1989).

SHAW WANG
Rutgers University
Piscatawav, New Jersey

KOSHER FOODS AND FOOD PROCESSING

The kosher dietary laws determine which foods are "fit or proper" for consumption by Jewish consumers who observe these laws. The laws are biblical in origin, coming mainly from the original five books of the Holy Scriptures. Over the years, the details have been interpreted and extended by rabbis around the world to protect the Jewish people from violating any of the fundamental laws and to address new issues and technologies. For example, the rabbis are currently dealing with issues related to biotechnology (see later section).

Why do Jews follow the kosher dietary laws? Many explanations have been given. The following by Rabbi Grunfeld is possibly the best written explanation and probably summarizes the most widely held ideas about the subject (1):

> "And ye shall be men of a holy calling unto Me, and ye shall not eat any meat that is torn in the field" (Exodus XXII:30). Holiness or self-sanctification is a moral term; it is identical with . . . moral freedom or moral autonomy. Its aim is the complete self-mastery of man.
>
> To the superficial observer it seems that men who do not obey the law are freer than law-abiding men, because they can follow their own inclinations. In reality, however, such men are subject to the most cruel bondage; they are slaves of their own instincts, impulses and desires. The first step towards emancipation from the tyranny of animal inclinations in man is, therefore, a voluntary submission to the moral law. The constraint of law is the beginning of human freedom. . . . Thus the fundamental idea of Jewish ethics, holiness, is inseparably connected with the idea of Law; and the dietary laws occupy a central position in that system of moral discipline which is the basis of all Jewish laws.
>
> The three strongest natural instincts in man are the impulses of food, sex, and acquisition. Judaism does not aim at the destruction of these impulses, but at their control and indeed their sanctification. It is the law which spiritualizes these instincts and transfigures them into legitimate joys of life.

These laws are not health laws. For a more complete discussion, see Ref. 2.

THE KOSHER MARKET

The kosher market covers more than 30,000 products in the United States. In dollar value, about $35 billion worth of products have a kosher marking on them. The actual consumers of kosher food, that is, those who specifically look for the kosher mark, are estimated to be about 6 to 8 million Americans, and they are purchasing almost $2 billion worth of kosher product each year. Only about one-third of the kosher consumers are Jewish; other consumers include Muslims, Seventh Day Adventists, vegetarians, people with various types of allergy—particularly dairy, grain, and legume—and other consumers who value the quality of kosher products. "We report to a higher authority" was the ad claim for Hebrew National hot dogs. *AdWeek Magazine* has called kosher "the Good Housekeeping Seal for the 90s." By undertaking kosher certification, food companies can incrementally expand their market by opening up new markets.

Although limited market data are available, the most dramatic data about the impact of kosher have been provided by Coors when they went kosher. According to their market analysis, their share of the Philadelphia market went up 18% on going kosher. Somewhat less dramatic increases were observed in other cities in the Northeast.

THE KOSHER DIETARY LAWS

The kosher dietary laws predominantly deal with three issues, all in the animal kingdom:

1. Allowed Animals
2. Prohibition of Blood
3. Prohibition of Mixing of Milk and Meat

However, for the week of Passover (in late March or early April), restrictions on "chometz," the prohibited grains and the rabbinical extensions of this prohibition, lead to a whole new set of regulations, focused in this case on the plant kingdom.

In addition, a separate set of laws deals with grape juice, wine, and alcohol derived from grape products, which must be handled by sabbath-observing Jews. However, if the juice is pasteurized (heated or "mevushal" in Hebrew), then this juice can be handled like any other kosher ingredient.

ALLOWED ANIMALS AND PROHIBITION OF BLOOD

Ruminants with a cloven hoof, most domestic birds, and fish with fins and removable scales are generally permitted. Pigs, wild birds, sharks, dogfish, catfish, monkfish and similar species along with all crustacean and molluscan shellfish are prohibited. Insects are also prohibited so that carmine and cochineal (natural red pigments) are not used in kosher products. Civet, ambergesis, and castoreum are the major flavors that are automatically not kosher.

Furthermore, ruminants and fowl must be slaughtered according to Jewish law by a specially trained religious slaughterer. These animals are also subsequently inspected by the rabbis for various defects. In the United States, the desire for more stringent meat inspection requirements has led to the development of a kosher meat conforming to a stricter inspection requirement, referred to as "glatt kosher." The term "glatt," or smooth, refers to the condition of the lung. The highest standard requires that no adhesions be present; that is, the lung is totally smooth. Most glatt kosher meat is limited to a few such adhesions. For regular kosher meat, the religious meat inspectors remove all of the adhesions, and if the lung can still be inflated, the meat is kosher.

The meat and poultry must be further prepared by properly removing certain veins, arteries, prohibited fats, blood, and the sciatic nerve. In practical terms this means that only the front quarter cuts of meat are generally used because of the difficulty in removing the sciatic nerve. To remove the blood, the meat is soaked and salted within a specified time period. Furthermore, any materials that might be derived from animal sources are generally prohibited because they are difficult to obtain from strictly kosher animals. Thus, many products that might be used in the food industry, such as emulsifiers, stabilizers, and surfactants, particularly those that are fat-derived, need careful rabbinical supervision to ensure that no animal-derived ingredients are used. Almost all such materials are also available in a kosher form derived from plant oils.

PROHIBITION OF MIXING OF MILK AND MEAT

"Thou shalt not seeth the kid in its mother's milk." This passage appears three times in the Torah (the first five books of the Holy Scriptures) and is thus taken religiously as a very serious admonition. The meat side of the equation has been extended to include poultry. The dairy side includes all milk derivatives.

To keep meat and milk separate requires that the processing and handling of all products that are kosher will fall into one of three categories:

1. Meat products
2. Dairy products
3. Pareve (Parve) or neutral products

The last includes all products that are not classified as meat or dairy. All plant products as well as eggs, fish, honey, and lac resin (shellac)—which is used as a coating for some confections and fruits—are pareve. These pareve foods can be used with either meat products or dairy products, except that fish cannot be mixed directly with meat. Once a pareve product is mixed with either meat or dairy products, it takes on the status of meat or dairy, respectively.

Some kosher-observant Jews are concerned about the possible adulteration of kosher milk with the milk of other animals (eg, mare's milk) and as such require that the milk be watched from the time of milking. This "Cholev Yisroel" milk and products derived from milk are required by some of the stricter kosher supervisory agencies for all dairy ingredients, so that dairy products would have to meet these requirements.

To ensure the complete separation of milk and meat, all equipment, utensils, and so on must be of the proper category. Thus, if plant materials (eg, a fruit juice) is run through a dairy plant, it would become a dairy product religiously. Some kosher supervisory agencies do permit such a product to be listed as "dairy equipment (D.E.)" rather than "dairy." The D.E. tells the consumer that it does not contain dairy but was made on dairy equipment (see allergy discussion). With the D.E. listing, the consumer can use the product immediately after a meat meal (on dairy dishes); ordinarily a significant wait would be required to consume a dairy product. In either case, the dishes would be switched from meat dishes to dairy dishes.

Kosher-observant Jews must wait a fixed time between meat and dairy consumption. Customs vary, but the wait after meat before consuming dairy is generally much longer (three to six hours) than the wait from dairy to meat (zero to one hour). However, when a hard cheese (defined as a cheese that has been aged for more than six months) is eaten, the wait is the same as that for meat. Thus, most companies producing cheese for the kosher market age their cheese for less than six months.

If one wants to make the product truly pareve, the dairy plant can usually be made pareve by the process of equipment kosherization (see later section).

PASSOVER

During this holiday, which occurs in the spring, all products made from the five prohibited grains (Hebrew: chometz)—wheat, rye, oats, barley, and spelt—cannot be used except for the specially supervised production of unleavened bread (Hebrew: matzos) that is prepared especially for the holiday. Special care is taken to ensure that the matzos do not have any time to "rise." In addition, products derived from corn, rice, legumes, sesame seeds, mustard seed, buckwheat, and some other plants (Hebrew: kitnyos) are prohibited for Jews whose origins are from central Europe. Thus, items like corn meal, corn syrup, corn starch, would be prohibited. Many rabbis permit kitnyos oils. Some rabbis permit liquid derivatives of kitnyos such as corn syrup. The major source of sweeteners and starches generally used for Passover production of "sweet" items is either real sugar or potato-derived products. Some potato syrup is also used. Passover is a time of large family gatherings. Overall, 40% of kosher sales for the traditional "kosher" companies occurs during the week of Passover.

EQUIPMENT KOSHERING

There are three ways to make equipment kosher and/or to change its status, depending on the equipment's prior production history. Note: After a plant (or a line) has been used to produce kosher pareve products, it can be switched to either kosher dairy or kosher meat without a special equipment kosherization step.

The simplest equipment kosherization occurs with equipment made of materials that can be koshered and that have only been handled cold. These require a good caustic/soap cleaning. However, materials such as ceramics, rubber, earthenware, and porcelain cannot be koshered. If these materials are found in a processing plant, new materials may be required for production; switching between different status conditions will be difficult.

Most food processing involves a heat treatment, generally above 115 to 120°F, which is defined rabbinically as "cooking." However, the exact temperature for "cooking" depends on the rabbi. To kosher these items, the equipment must be thoroughly cleaned with caustic/soap. The equipment must be left idle for 24 h and then the equipment must be flooded with boiling water (defined between 190 and 212°F) in the presence of a kosher supervisor.

In the case of ovens or equipment that use "fire," kosherization involves heating the metal until it glows. Again, the rabbi will generally be present while this process is taking place.

The procedures that must be followed for equipment kosherization can be quite extensive, so the fewer status conversions, the better. Careful formulating of products and good production planning can minimize the inconvenience and cost.

INDUSTRIAL JEWISH COOKING

Depending on what is being cooked, it may be necessary for the rabbi to "do" the cooking. In practical terms this is often accomplished by having a rabbi light the pilot light, which is then left on continuously.

In the case of cheese making, a similar concept usually requires the rabbi to add the coagulating agent into the vat. If the ingredients used during cheese making are all kosher but a rabbi has not added the coagulant, then the whey derived from such cheese would be considered kosher as long as the curds and whey have not been heated above about 120°F before the whey is drained off. As a result, there is much more kosher whey available than kosher cheese.

DEALING WITH KOSHER SUPERVISION AGENCIES

Kosher supervision is taken on by a company to expand its market opportunities. It is a business investment that, like any other investment, should be examined critically. It is appropriate for companies to look carefully at how they select a kosher supervisory agency. The agency's name recognition is only one important consideration. Other questions to be raised include: (*1*) How responsive is the agency? (*2*) How willing are the rabbis to work with the company on problem solving? (*3*) How willing are they to explain their kosher standards and their fee structure? (*4*) Is the "personal" chemistry right? and finally (*5*) What are the agency's religious standards and do they meet the company's needs in the marketplace?

One of the hardest issues for the food industry is the existence of so many different kosher supervisory agencies and standards. The trend in the mainstream kosher community today is toward a more stringent standard.

One can generally divide the kosher supervisory agencies into three broad categories. First there are the large organizations that dominate the supervision of larger food companies: the OU (Union of Orthodox Jewish Congregations), the OK (Organized Kashruth Laboratories), the Star-K, and the Kof-K—all of which are nationwide and "mainstream." Two of these, the OU and the Star-K are communal organizations: that is, they are part of a larger community religious organization. This provides them with a wide base of support but also means that the organizations are potentially subject to the other priorities and needs of the organization. The Kof-K and the OK are private companies whose only function is to provide kosher supervision. In addition to these national companies, there are smaller private organizations and many local community organizations that provide equivalent religious standards of supervision on a smaller scale. As such, products accepted by any of these mainstream organizations will generally be accepted by all other mainstream organizations. Local organizations may have a bigger stake in the local community; they may be more accessible and easier to work with. Although they often have less technical expertise, the smaller agencies are often backed up by one of the national organizations. For a company that nationally markets, a limitation of using a local kosher certifying agency may be name recognition of their kosher symbol elsewhere in the United States. With the advent of *KASHRUS* magazine, and its yearly review of symbols, this has become somewhat less of a problem. *KASHRUS* magazine (Box 204, Brooklyn, NY 11204) does not try to "evaluate" the standards of the various kosher supervisory agencies, but simply "reports" their existence. It is the responsibility of the local congregational rabbi to define his or her standards for the congregation. A remote and lesser-known certification organization may be difficult to recommend.

The second category of kosher supervision includes individual rabbis, generally associated with the "Hassidic" communities. These are often affiliated with the ultra-orthodox communities of Williamsburg and Borough Park in Brooklyn, Monsey, New York, and Lakewood, New Jersey. There are special food brands that cater to these communities, oftentimes providing continuous rabbinical supervision rather than the occasional supervision used by the mainstream organizations. The symbols of the kosher supervisory agencies representing these consumers are not as widely recognized beyond these communities as those of the major mainstream agencies in the kosher market. The rabbis will often do special supervisions of products using a facility that is normally under mainstream supervision—often without any changes, but sometimes with special needs for custom production.

The third level of supervision is done by individual rabbis who are more "lenient" than the mainstream standard. Many of these rabbis are Orthodox; some may be Conservative. Their standards are based on their interpretation of the kosher laws. More lenient standards may cut out some of the "mainstream" and stricter markets. Each company must weigh such issues when deciding about kosher supervision.

Ingredient companies are very much encouraged to use a "mainstream" kosher supervisory agency so that they can sell to as many customers as possible. Unless an ingredient is acceptable to the mainstream, it is almost impossible to gain the benefit of kosher certification.

With respect to interchangeability between kosher supervisory agencies: a system of certification letters is used to provide information among certifying rabbis about products they have approved. A supervising rabbi certifies that a particular plant produces kosher products, or that only products with certain labels or certain codes are kosher under his supervision. Such letters should be renewed each year and should be dated with both a starting and ending date. These letters are the mainstay of how companies establish the kosher status of ingredients as these ingredients move in commerce. Obviously a kosher supervisory agency will only "accept" letters from agencies they consider acceptable. Consumers may also ask to see such letters.

In addition, the kosher symbol of the certifying agency or rabbi usually appears on the packaging. (In some industrial situations, where kosher and nonkosher products are similar, some sort of color coding of products may also be used.) Most of these symbols are "trademarks" that are duly registered. In a few cases, the trademark is not registered, and more than one rabbi has been known to use the same kosher symbol.

With respect to kosher markings on products, two issues need to be highlighted:

1. It is the responsibility of the food company to show its labels to its kosher certifying agency prior to printing to ensure that the labels are marked correctly. This responsibility includes both the agency symbol and the documentation establishing its kosher status (eg, meat, dairy, or pareve). Many agencies currently do not require that "pareve" be marked on products; others do not use the "dairy" marking.
2. The labels for private label products marked with specific agency symbols cannot be moved easily between plants. This is why some companies—both private label and others—use the generic "K," which can continue to be used even if the kosher supervisory agency changes. Increasingly, sophisticated kosher consumers are questioning the use of this generic symbol.

GELATIN

Gelatin is probably the most controversial of all modern kosher ingredients. Gelatin can be derived from pork skin, beef bones, or beef skin. In recent years, some fish gelatins have also appeared. As a food ingredient, fish gelatin has many similarities to beef and pork gelatin. Depending on the species from which the fish, skins are obtained, its melting point can vary over a much wider range of melting points than beef or pork gelatin. This may offer some unique opportunities.

Currently available gelatins—even if called "kosher"—are not acceptable to the mainstream kosher supervisory organizations. A recent production of gelatin from the hides of kosher-slaughtered cattle has been available in limited supply at great expense and is accepted by the mainstream kosher supervisory agencies and some of the stricter ones.

Attitudes about gelatin certification vary greatly among the lenient kosher supervisory agencies. The most liberal view holds that gelatin, being made from bones and skin, is not being made from a food (flesh). Further, the process used to make the product goes through a stage where the product is so "unfit" for consumption that it is not edible by man or dog, thus becoming a new entity. (Rabbis holding this view may accept pork gelatin.) Most gelatin desserts with a generic "K" follow this ruling.

Other rabbis only permit gelatin from beef bones and hides but not pork. Other rabbis will only accept "India dry bones" as a source of beef gelatin. These bones, found naturally in India (because of the Hindu custom of not using cattle) are aged for over a year and are "dry as wood"; additional religious laws exist for permitting these materials. Again, none of these products is accepted by the "mainstream" kosher supervisions, rendering them unacceptable to a significant portion of the kosher community.

BIOTECHNOLOGY

Rabbis currently accept products like chymosin (rennin) that are made by simple genetic engineering. The production conditions in the fermentors must still be kosher; that is, the ingredients and the fermentors and any subsequent processing must use kosher equipment and ingredients of the appropriate status. A product produced in a dairy medium would be dairy; any product produced by cattle by excretion in the milk is defined as dairy. The rabbis continue to consider the status of more complex genetic manipulations. Note: A natural rennin extracted from the dried stomach of a kosher-killed and inspected calf is kosher and is pareve.

FEDERAL AND STATE REGULATIONS

Making a claim of kosher on a product is a "legal" claim. Federal regulation 21CFR101.29 has a paragraph indicating that such a claim must be appropriate. Approximately 20 states and a number of counties and cities have laws specifically regulating the claim of "kosher." Many of these laws refer to "Orthodox Hebrew Practice" or some variant of this term, and their legality in the 1990s is subject to further court interpretation.

New York State probably has the most extensive set of kosher laws, including a requirement to register kosher products with the Kosher Enforcement Bureau of the Department of Agriculture and Markets (55 Hanson Pl., Brooklyn, NY 11217). However, the laws in New Jersey—having been written after the state's original laws were declared unconstitutional by the state supreme court—probably have the clearest focus and, to date, no Constitutional issue. The regulations focus specifically on "consumer right to know" and "truth in labeling" issues. They avoid having the state of New Jersey define kosher. Rather, the rabbis providing supervision declare the information—adhered to by the food manufacturers—that consumers need to make an informed decision.

ALLERGIES

Many consumers use the kosher markings as a guideline for determining whether products might meet their special health needs. There are also limitations that the particularly sensitive consumer must consider.

With respect to all kosher products, two important limitations need to be recognized:

1. A process of equipment kosherization is used to convert equipment from one status to another. This is a well-defined religious procedure but may not lead to 100% removal of previous materials run on the equipment.
2. Religious law does permit certain ex post facto (after the fact) errors to be negated. Thus, trace amounts (less than 1/60 by volume under very specific conditions) can be nullified. In deference to a company's desire to minimize negative publicity, a kosher supervisory agencies may *not* announce when it has used this procedure to make a product acceptable.

Products that one might surmise to be made in a dairy plant—for example, pareve substitutes for dairy products and some other liquids like teas and fruit juices—may be produced in plants that have been kosherized, but they may not meet a very critical allergy standard. Another product that can be problematic is chocolate: many plants make both milk chocolate and pareve chocolate. Getting every last trace of dairy out of the pareve chocolate can be difficult.

Dairy and meat equipment: A pareve product may be produced on a dairy or meat line without any equipment kosherization. Again, religiously insignificant traces of dairy may remain that cannot be tolerated by a very allergic consumer. (The product can be used in a kosher home without the normal waiting period.)

Fish: In a few instances where pareve or dairy products contain small amounts of fish (eg, anchovies in Worcestershire sauce), this ingredient may be marked as part of the kosher supervision symbol. Many certifications, however, do not specifically mark this.

For Passover, there is some dispute about "derivatives" of both chometz and kitnyos materials. A few rabbis permit items like corn syrup, soybean oil, peanut oil, and similarly derived materials from these extensions. More generally, "proteinaeous" parts of these materials are not used. Thus, people with allergies to these items should purchase these special Passover products from supervisory agencies that do not permit kitnyos derivatives. With respect to "equipment kosherization", supervising rabbis tend to be very strict about the clean-up of the prohibited grains (wheat, rye, oats, barley, and spelt) but may not be as critical with respect to the extended prohibition of kitnyos.

Consumers should not assume that kosher markings ensure the absence of trace amounts of the ingredient to which they are allergic.

BIBLIOGRAPHY

1. I. Grunfeld, *The Jewish Dietary Laws*, The Soncino Press, London, United Kingdom, 1972.
2. J. M. Regenstein, "Health Aspects of Kosher Foods," *Activities Report and Minutes of Work Groups & Sub-Work Groups of the R & D Associates* **46**, 77–83 (1994).

GENERAL REFERENCES

M. M. Chaudry and J. M. Regenstein, "Implications of Biotechnology and Genetic Engineering for Kosher and Halal Foods," *Trends Food Sci. Technol.* **5**, 165–168 (1994).

J. M. Regenstein and C. E. Regenstein, "An Introduction to the Kosher (Dietary) Laws for Food Scientists and Food Processors," *Food Technol.* **33**, 89–99 (1979).

J. M. Regenstein and C. E. Regenstein, "The Kosher Dietary Laws and Their Implementation in the Food Industry," *Food Technol.* **42**, 86, 88–94 (1988).

JOE M. REGENSTEIN
Cornell University
Ithaca, New York
CARRIE E. REGENSTEIN
University of Rochester
Rochester, New York

KRILL PROTEIN PROCESSING

Krill belongs to the order Euphausiacea. There are 2 families, 11 genera, and 84 species. It is distributed in both shallow and deep seawaters. The appearance of krill is much like that of small shrimp. Among them the Antarctic krill (*Euphausia superba*) is most important and most abundant. It is distributed south of 60°S around the South Pole. The stock of krill is estimated at 360–1,400 million tons (1). Krill receives much attention because of its potential use as a protein resource for human foods. The use of Antarctic krill for human food is outlined here. More detailed reports concerning krill use are given in Refs. 2 and 3. Figure 1 shows Antarctic krill.

CHEMICAL COMPOSITION

The chemical composition of Antarctic krill varies slightly because of differences in the size, age, and sex of the sample and the fishing season. Average values of general composition of whole body of krill are as follows: moisture, 77.9–83.1%; crude protein, 11.9–15.4%; chitin and glucide, 2%; crude fat, 0.5–3.6%, and crude ash, as 3% (4,5). The composition of the muscle after removing the shell is more protein and less ash compared to the whole body of krill.

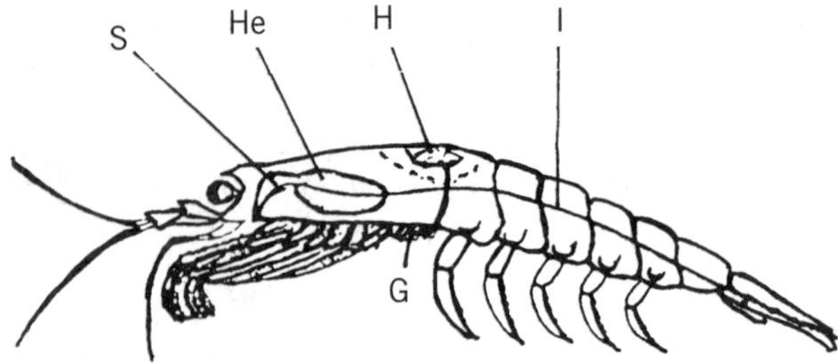

Figure 1. Antarctic krill. G, gill; S, stomach; He, hepatopancreas; H, heart; I, intestine.

Table 1 shows the chemical composition of whole krill body taken in the period December through February. It can be seen that the chemical composition of krill is well balanced in foodstuff. The lipids of krill have a high iodine value of 110–190. About 70% of the whole lipids of krill is unsaturated fatty acid. Concentration of oleic acid, eicosapentaenoic acid, and docosahexaenoic acid are high. Krill lipids are, however, not easily oxidized, which may be due to their high concentration of vitamin E in the oil as antioxidant. Cholesterol levels of shellfish are higher than that of fish; krill is not an exception. Krill tissue contains 62.4–71.6 mg/100 g cholesterol. Krill contains large amounts of vitamin A, D, and the B group complex. Most of the vitamin A is astaxanthin in the whole body, which is highly concentrated in the eyes.

About 80% of the nitrogen compounds in krill meat is protein-nitrogen. Myofibrillar protein makes up 60–70% of the total protein (average 9.4% in muscle) (6). The value is similar to fish meats. Krill myofibrillar protein is mainly composed of myosin, actin, and paramyosin, making of similar to that of other invertebrate meats. The protein composition of krill meat is given in Table 2. Amino acid composition of krill protein is similar to that of common crustaceans, such as prawn or shrimp. Sulfur-containing amino acids are low compared to those in whole-egg protein; on the other hand, lysine and threonine are relatively abundant in krill (7). The amino acid score for krill protein ranges from 85 and 100. The discrepancy of the values obtained by different scientists are probably because of different analytical methods. Nutritive values of krill protein are lower than those of whole-egg protein but are similar to or higher than those of casein (8).

Table 1. Chemical Composition of Krill (% in Whole Body)

Date of fishing	Moisture	Crude protein	Crude fat	Crude ash	Chitine and glucide
Dec. 12	83.7	11.8	0.48	3.24	1.15
Dec. 26	81.8	13.8	0.60	3.19	1.19
Jan. 5	80.4	14.2	1.48	3.00	1.05
Jan. 15	80.8	13.5	1.90	3.07	1.08
Jan. 25	79.7	14.2	2.50	3.07	1.03
Feb. 5	78.8	14.2	3.61	3.08	1.09
Feb. 15	80.0	13.4	3.37	2.60	1.09

Table 2. Protein Contents in Krill Meat

Sample no.	Nonprotein-N (mg N/g)	Protein-N % in total protein-N			
		Sp	Mf	A-S	Stroma
1	4.72	14.2	70.0	14.1	1.6
2	4.58		84.0	14.7	1.5

Note: Sp. sarcoplasmic protein; Mf, myofibrillar protein; A-S, alkali-soluble protein.

PROCESSING

Raw Material Properties

Because of the characteristics of krill, processing into the final products must be accomplished as soon as possible after catch on board. The body of krill is so small (average body weight is 0.68 g) that its muscle is located close to its digestive organs containing protease, and muscle protein is easily affected by protease (9). Also the level of autoproteolysis varies considerably among different season catches from the same area (11). The speed of denaturation of myofibrillar protein of krill is much faster than it is for other fish and shellfish. Deterioration starts taking place immediately after death of krill. As krill fishing is not done continuously, harvested krill is stored in a pool until processing starts. The main proteases responsible for the autolysis are trypsinlike proteases at neutral pH and cathepsins (B, H, and L types) at acidic pH (10). Rolling and pitching of the ship causes damage to the bodies.

Fresh-Frozen Products

Fresh-frozen products of krill are made by quick freezing immediately after catch by means of a contact freezer at −40°C or individual quick freezing (IQF). The key point of this procedure is to freeze and store the material below −20°C while it is still fresh. Temperature control of krill on board is important; if the temperature is insufficient, blacking of krill bodies takes place. The cause of such deterioration has been found to be the enzyme (tyrosinase) contained in the hepatopancreas. Myofibrillar protein in krill muscle aggregates and its molecular size enlarges during freezing-storage. The freeze-denaturation of krill protein occurs more quickly than for other fish and shellfish (6). Frozen krill should be stored at −30°C; otherwise, unpalatable texture due to protein denaturation occurs. The specific odor of frozen krill is found to be caused by dimethylsulfide (DMS) (12). DMS evokes a flavor specific to crustacea when its concentration is low. If the concentration becomes over 1 ng/g tissue, the aroma becomes unpleasant. Storage at −30°C is desirable to prevent formation of DMS.

Peeled Meat

As krill has many unpalatable appendages, peeled meat is more palatable. There are two kinds of peeled krill products, boiled peeled meat and raw peeled meat.

To make boiled peeled meat, whole krill body is cooked in water at 90°C for 5 min, then excess water is removed by centrifugation, then the krill is quickly frozen. The frozen material is used for removing unpalatable parts, such as appendages and shells A shot-blast, vibration, screw-grinding, or air-blast method is employed for the removing process (2). The peeled meat is stored at −20°C.

To make raw peeled meat, the muscle of the tail of the krill is collected mechanically by a peeler (Fig. 2) (13,14). A pair of rollers, parallel to each other, rolls inward and then turns back. During this process, the shell of the krill is torn and the meat is pushed on the rollers and picked up by a jet stream of brine. This seawater is absorbed and makes the water content of the collected meat as high as 80–90%. The raw peeled meat is stored at −30°C. The yield of the raw peeled meat is very low, about 10–15% by weight of whole krill body. Improvements of the peeler design are needed to increase productivity per hour and yield rate of meat from raw material. The raw peeled krill is produced in Japan to be used as an ingredient in many kinds of processed foods.

Dried Products

Drying is an easy way of processing krill on board. Half-dried and full-dried krill have been produced. The half-dried krill is better tasting than the full-dried krill because the fully dried krill are more fragile and have an unpalatable texture. The water content of full-dried krill is 5–25%; half-dried is 40–50%. Half-dried krill must be refrigerated.

Protein Paste

Krill paste has been developed and commercial produced mostly in the former USSR. Fresh krill is pressed and a juice containing protein is collected. The coagulated protein is obtained by heating and pressing. The obtained coagulated protein is kneeded to make paste and then frozen. A krill paste-manufacturing machine for use on board ship has been developed in the former USSR (15). The krill paste-manufacturing machine consists of parts for peeling, heating, protein-separating, and kneeding. The krill paste is pink in color and has the appearance of soft cheese. It is used as an ingredient in cheese and butter and various other foods, such as sausage, clroquette, and priozhki (15).

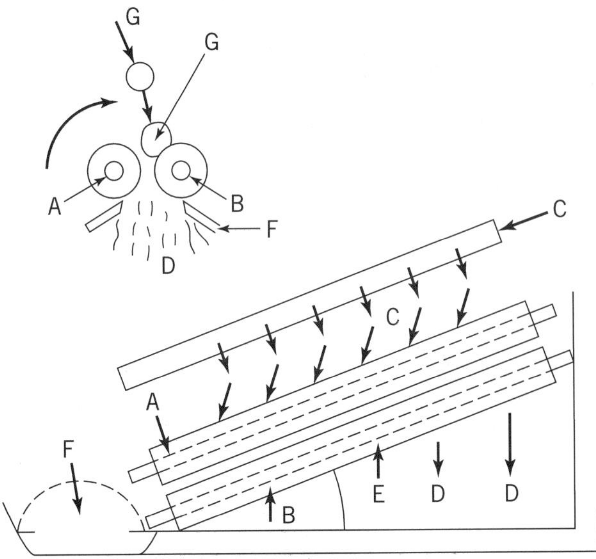

Figure 2. Roller-type peeling machine. A, B, roller; C, seawater; D, shell; E, krill meat in the getter of the peeler; F, krill meat; G, whole krill; H, knife.

Frozen Surimi

Frozen surimi from krill meat have been developed in processing on board. There are two types of krill surimi red surimi and white surimi. The method for preparing red surimi is to remove the internal organs by a centrifuge and collect the meat by a drum-type bone separater. The red surimi is contaminated red pigments from krill eyes. The starting material of white surimi is raw, peeled krill meat. Water-washing, which is always used in making surimi from other fish, is not used in either type of krill surimi processing because of difficulties in removing excess water after water-washing, owing to the strong water-holding ability of krill protein and also due to the fact that myofibrillar protein is quite soluble, and some is lost during washing (16). Five to 10% sorbitol and 0.3% polyphosphate are mixed to prevent protein denaturation of red and white surimi. Also, denaturated soybean protein, beef blood plasma, or egg white is added to diminish the effect of the residual protease; these nonmuscle proteins act as coenhancers of gelation. The gel-forming ability of red surimi is weaker than that of white surimi. Red surimi is rich in vitamin A and red pigment. White surimi is rich in taste compound and strong gel-forming ability similar to surimi made from fish meat. The Japan Fisheries Agency was engaged in a project beginning in 1977 to develop the use of Antarctic krill as food for humans. Krill surimi developed by this project is not yet commercially successful (17).

Various Processed Krill

Development of highly processed food with added value has received attention. For example, in Japan prepared

Table 3. Yield of Krill Products

Product	Yield (%[a])	References
On the market		
Frozen raw whole	100	2
Frozen raw peeled	10–28	3
Frozen boiled whole	105	4,18
Frozen boiled peeled	80–90	4,18
Frozen tail meat	20–30	4,18
Dried whole (15% M[b])	25	3
Dried whole (40% M[b])	35	3
Paste	10–40	2,15
Being developed		
Red surimi	20–50	3
White surimi	10–20	3
Meat-textured fish protein concentrate	15	3,19
Imitation crabmeat	20	20
Isolated protein	10–15	2

[a]Yield, % = product weight/whole krill body.
[b]M = moisture.

foods sold at delicatessens are made by mixing krill with vegetables and hamburger meat. Peeled, raw krill that has been boiled and canned has been officially adopted in the school lunch program in Japan. Because krill is rich with free amino acids and with shrimplike flavor, a method is being developed to produce a natural seasoning by liquefying krill with enzymes in a bioreactor, filtering the liquid to remove fragments of shell, and concentrating the result by vacuum or spray-drying. Table 3 lists the many kinds of krill products developed as human foods.

BIBLIOGRAPHY

1. I. Yamanaka, "Interaction Among Krill, Whales and Other Animals in tl:e Antarctic Ecosystem," *Proceedings of the BIOMASS Colloquium in 1982, Memoir of the National Institute of Polar Research* Special issue No. 27, 220–232 (1983).
2. T. Suzuki and N. Shibata, "The Utilization of Antarctic Krill for Human Food" *Food Revue International* **6**, 119–147 (1990).
3. T. Suzuki, *Fish and Krill Protein: Processing Technology*, Applied Science Publishers, London, 1981.
4. G. J. Grantham, "The Utilization of Krill," Food and Agriculture Organization, Southern Ocean Fisheries Survey Programme, Rome, 1977.
5. M. Mori and S. Yasuda, *Bulletin Central Research Laboratory*, Nippon Suisan Co. LTD. **11**, 1–5 (1976).
6. N. Shibata, "Food Biochemical Study on Utilization of Muscular Proteins in Antarctic Krill for Human Consumption," *Bulletin of the Tokai Regional Fishery Research Laboratory* No. 11, 63–141 (1983).
7. T. Kinumaki, in *Comprehensive Studies on the Effective Utilization of Krill Resources in Antarctic Ocean*, The Science and Technology Agency, Japanese Government, 1980, pp. 135–140 (Text in Japanese).
8. Y. Obatake, Ref. 7, pp. 46–51.
9. S. Konagaya, "Protease Activity and Autolysis of Antarctic Krill," *Nippon Suisan Gakkai-shi* **46**, 175–183 (1980).
10. Y. Kawamura et al., "Effects of Protease Inhibitors on the Autolysis and Protease Activities of Antarctic Krill," *Agric. Biol. Chem.* **48**, 923–930 (1984).
11. K. Osnes and V. Mohr, "Peptide Hydrolases of Antarctic Krill *Euphasia superba*," *Comp. Biochem. Physiol.* **82B**, 559–606 (1985).
12. T. Tokunaga and co-workers, "Formation of Dimethyl Sulfide in Antarctic Krill, Euphausia superba," *Nippon Suisan Gakkai-shi* **43**, 1209–1217 (1977).
13. A. Kawakami and M. Tone, Jap. Pat. (Open No 55-31911).
14. M. Yanagimoto and T. Kobayashi, in Ref. 7, pp. 154, 174, 182–183.
15. T. Masuda and co-workers, "Utilization and Processing of Fish in USSR," Fisheries Agency, Japanese Government, 1980. (Text in Japanese).
16. P. Montero and M. C. Gomez-Guillen, "Recovery and Functionality of Wash Water Protein from Krill Processing," *J. Agric. Food Chem.* **46**, 3300–3304 (1998).
17. "Development of Technology of Antarctic Krill for Human Foods" Fisheries Agency, Japanese Government, 1980, 1981 (Text in Japanese).
18. H. Ozaki, "Peeling Technology of Krill and a Screw Grinder Method" *Shokuhin Kogyou* **24**, 34–44 (1981) (Text in Japanese).
19. T. Suzuki and co-workers, "Manufacturing of Meat-textured Krill Protein Concentrate," **48**, 105–111 (1982).
20. H. Takizawa and co-workers, "Manufacturing of Crab Meat-Textured Protein Fiber from Antarctic Krill," *Nippon Suisan Gakkai-shi* **52**, 1243–1248 (1986).

TANEKO SUZUKI
Nihon University
Kanagawa, Japan

L

LABORATORY ROBOTICS AND AUTOMATION

Robotics is not new to the general public but is a recent addition to the analytical laboratory. The history of robotics spans more than a half century, with the first use of the term in a 1921 play *Rossums Universal Robots*. The term robot is derived from the Czechoslovakian word "robota," which can be defined as one who performs compulsory labor much like an automaton (1). Science fiction literature and films are full of robots in various forms, ranging from humanoid such as Maria in the movie *Metropolis* to R2D2 and C3PO in *Star Wars* and a cerebral HAL in *2001, A Space Odyssey*. Probably the first exposure of many persons in the written literature came in the form of Robbie in *I, Robot* by Isaac Asimov (2). This volume also introduced the Three Laws of Robotics and the first robot psychometrician: (*1*) A robot may not injure a human being or through inaction allow a human being to come to harm; (*2*) A robot must obey orders given by human beings except where such would conflict with the First Law; (*3*) A robot must protect its own existence as long as the protection does not conflict with the First and Second Law.

Commercial laboratory robotics was introduced to the scientific community in 1982 in the form of Zymark Corporation's Zymate 1. Although there was a substantial number of organizations doing robotic research for the laboratory, this introduction allowed those who were interested in the technology to purchase it off the shelf. Laboratory automation before robotics was divided into two general categories: high performance liquid chromatography (HPLC)/gas chromatographic (GC) autosamplers and data capture and analysis units. The category of autosamplers is self-explanatory. The second category included units that ranged from instruments as simple as an integrator to as complex as in integrated laboratory information management system (LIMS). At that time, inefficient sample preparation was a substantial bottleneck to laboratory operations, so it was sensible and prudent that the early robots concentrated on sample preparation. Robotics was billed as filling the missing link in the laboratory. This article cannot hope to cover laboratory robotics in its entirety, but it provides an introduction to its uses in the laboratory and some additional references for further study.

ROBOT TYPES

Three general types of robots are used in the laboratory. They are the Cartesian or cylindrical, articulating arm or revolute, and X, Y, Z. Each has advantages and disadvantages. The most widely used and first to be commercially available is the Zymate series (Fig. 1), which is a cylindrical type. It has an established user base of about 1,500. *Forbes* dubbed this unit the "One Armed Chemist" (3). Other cylindrical robots are being marketed for the laboratory market, but they tend to resemble a reverse-engineered Zymate.

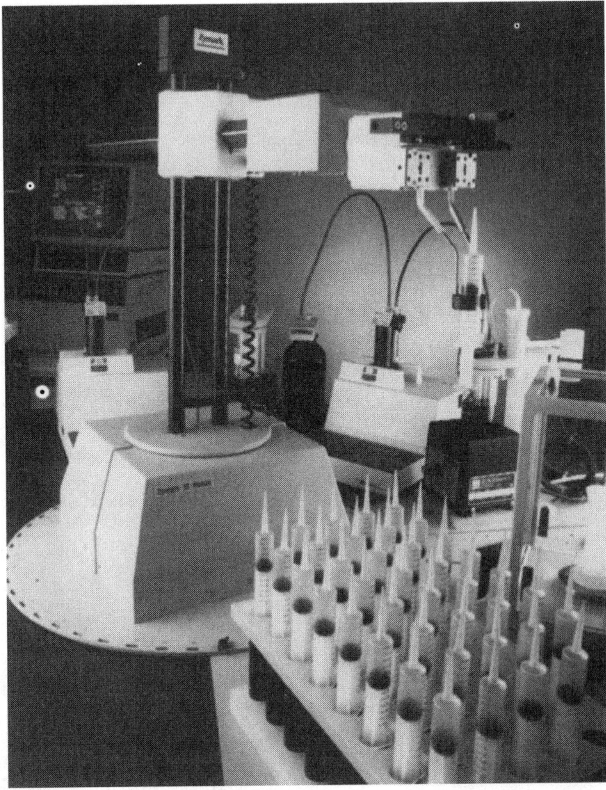

Figure 1. The Zymate series is a cylindrical type of robot and the first to be commercially available.

The articulating arm or revolute robot is exemplified by a number of units, including the second commercial additions to the technology from Perkin Elmer and Fisher Scientific. These units were based on the Mitsubishi RM501 Movemaster and were available for only a limited time. A number of additional ventures, based on the CRS-Plus/CRS-One series of robots, use this type of robot.

The final of the three types is the X, Y, Z robot, named so because it moves along three axes. Examples of this type include the MilliLab, the Biomek, and an offering by Source for Automation. Figures 2 through 4 illustrate several of these units. It is not within the scope of this article to discuss robotics in general or examine all the various robot types in detail, but these examples illustrate the diversity in the area of laboratory robotics. Finally, with the new and additional offerings from a variety of vendors, there will be a blurring of the distinctions among the robot types and probably a loss of distinction among robots and robotic and automated workstations.

ROBOT PROGRAMMING

Because robots are a computer-based system, they have a unique language that is used to program them and allows

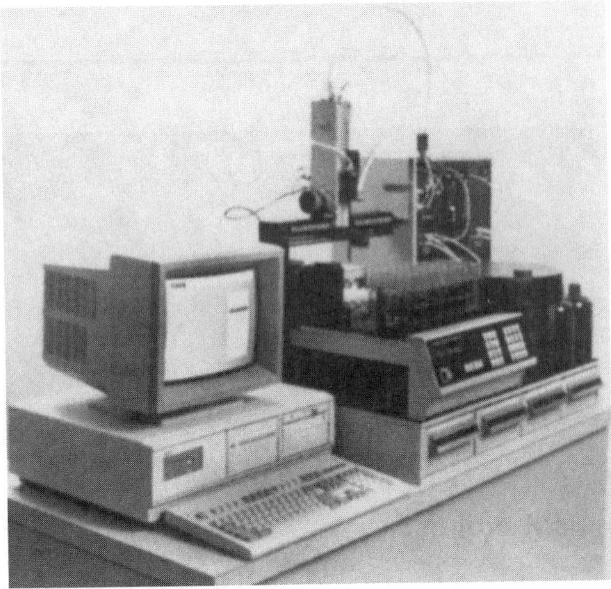

Figure 2. Window-based systems such as the MilliLab make X,Y,Z robots more user-friendly.

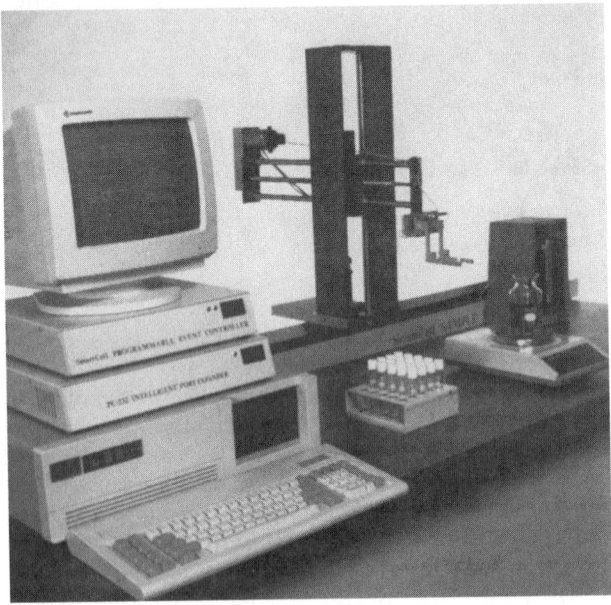

Figure 3. The Biomek, and X,Y,Z robot, communicates with peripheral devices.

them to communicate with other instruments or peripheral devices. The early generation of robots was difficult to program, not owing to a difficult robot language but largely to the fact that in some instances each robot movement had to be taught and then stored in the robot memory. The current generation of robots has eliminated this bottleneck with many systems being able to run samples within hours after their installation. This is due to a number of factors, including the introduction of PyTechnology by Zymark, where certain common laboratory operations had the necessary movements preprogrammed. Additional and also very important factors allowed for the use of standard MS-DOS computers with various window-based systems such as the MilliLab or icon-based system to make the units more user-friendly.

Figure 4. An X,Y,Z robot by Source for Automation.

APPLICATIONS

Early robots were purchased for a variety of reasons, including what could be called an evaluation of the technology, which was a very acceptable justification. This could be called the era of gee-whiz robotics. Because laboratory robotics is well established, the gee-whiz era has passed and users are more interested in an application-based system. These types of systems can be divided into two broad categories: (*1*) task based and (*2*) assay based. The task-based robot is configured to accomplish a particular laboratory operation or discrete series of laboratory operations. Laboratory unit operations are (4): weighing, grinding, manipulation, liquid handling, conditioning, measurement, control, data reduction, and documentation.

These various operations are common to laboratories and not particular to any specific segment. Examples of this might be liquid-liquid extraction, centrifugation, weighing, or solid-phase extraction. A laboratory might do enough of these operations to use a robot or robotic workstation. Additionally, a unit might be procured to accomplish a number of these operations, such as solid phase extraction (SPE) and HPLC injection. As one of the major uses of robotics is HPLC sample preparation, it is reasonable to assume that commercial units have been developed to concentrate in this area. The ASPEC and ASTED by Gilson and the Benchmate by Zymark are examples of these units that will accomplish several of the laboratory unit operations. Figures 5 and 6 illustrate two of these units. One could of course use a robotic system to accomplish operations, but it would not be used to the fullest capacity.

A second group of task-based units could be classified as robotic autosamplers for HPLC. A substantial number of vendors have this type of unit. In addition to acting as an HPLC autosampler, these types of units also have

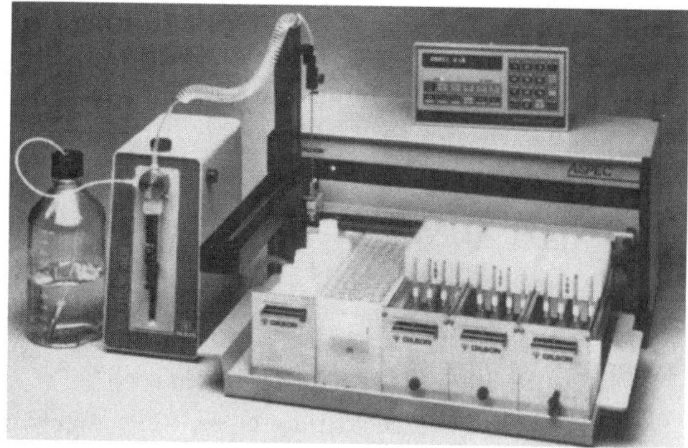

Figure 5. One of the major uses of robotics is HPLC sample preparation.

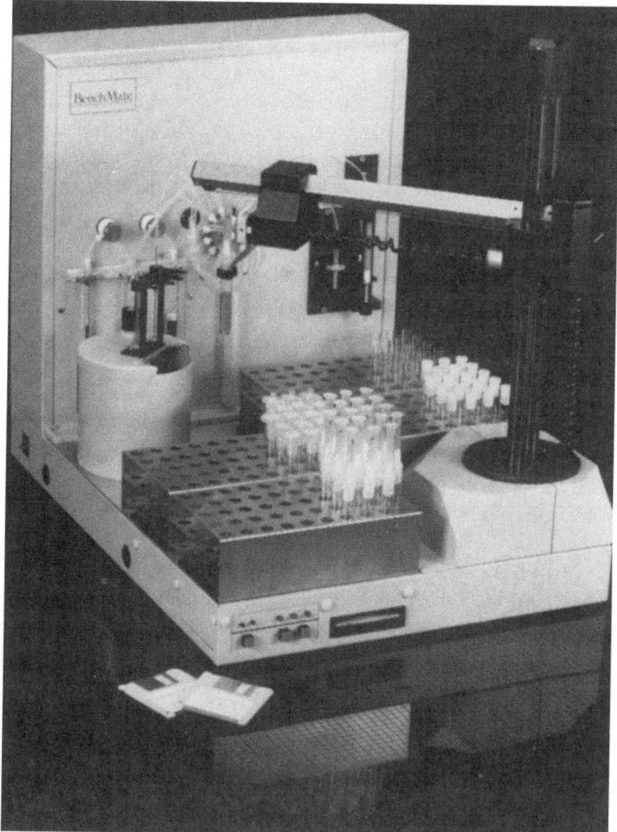

Figure 6. The Benchmate by Zymark accomplishes several of the laboratory unit operations.

the capability of making dilutions of samples and do automated derivative formation. One of the largest uses of this capability is the ability to form amino acid derivatives in a serial fashion. For example, the opa derivative for amino acids is subject to decay immediately after it is formed, so these cannot be made in a batch fashion. Using one of the robotic autosamplers, it is now possible to do the derivatization in a serial fashion so each derivative is injected onto the HPLC immediately after it is formed. This eliminates the data variability due to derivative formation and derivative decay. A final type of task-based instrument is typified by the AASP by Varian, which allows on-line solid-phase extraction and subsequent HPLC injection using a prepacked Solid Phase Extraction cassette. This unit has seen limited success in the food industry but has loyal following in a number of life sciences laboratories.

The other general category of robots that are in use could be classified as assay based. A prime example that is in use and likely one of the largest uses of robotics is in pharmaceutical organizations accomplishing drug dissolution testing. This particular assay is labor intensive and repetitive, making it very suitable for robotic implementation. A similar assay that is in use in the food industry is the Mojonnier assay. This assay is used for the determination of percentage fat content and is also labor intensive and repetitive. An organization can presently purchase a turnkey system through a partnership between a major robot vendor, Zymark, and a custom programming and device manufacturer, Forcoven Products. A number of these systems are in place throughout the world producing data that are equivalent or superior to that obtained by manual techniques (4).

Another type of assay-based system that is more generic in nature is a system capable of sample preparation for HPLC, injection onto the HPLC, and the transmission of the data to an external computer for subsequent analysis. Other examples of the use of robotics in the food industry include: Karl Fischer titration, particle size analysis, vitamin analysis, pesticide residue analysis, flavor compounding, pH analysis, and total dietary fiber.

This information provides some selected uses of the various forms of robotics in food industry laboratories. There is, of course, substantial overlap among many application areas, so it is not easy to segment them as an application used in the food industry, such as HPLC sample preparation, could easily apply to the petrochemical industry or a life sciences laboratory.

Sometimes a laboratory might have a custom application for robotics that does not fit easily into either the task-based or assay-based unit, so a custom robot is the obvious choice. When a custom application is needed, a user then has two ways to implement this procedure. This can be accomplished either through the use of internal resources or through the use of a system house. Each has its series of attributes, which will not be outlined in this article.

IMPLEMENTATION

When a robotic or other such automated system is chosen for a certain assay or procedure, the next logical phase involves the implementation. This phase starts with the installation and is completed when the system is integrated into the daily laboratory operations. In many of these components such an installation is straightforward. Likely an often encountered roadblock does not revolve around the hardware or the software; it involves a series of management issues such as personnel, management, and organizational support. Key questions to answer are

Do I have proven chemistry?

Do I have the correct person?

Do I have the correct application?

Do I have organizational commitment?

Do I have the right vendor?

Laboratory robotics presents an exciting opportunity to persons interested in laboratory automation. The technology is continuing to develop through evolution in laboratory and equipment design. Additionally, several organizations are actively involved in intelligent robots with feedback, knowledge engineering, and data-based systems. This article has provided an introduction to laboratory robotics and has overviewed the technology. Developments are occurring at a rapid pace; the adage "science is a moving target" certainly applies.

BIBLIOGRAPHY

1. I. Asimov and K. A. Frenkel, *Robots-Machines In Man's Image*, Harmony Books, New York, 1985.
2. I. Asimov, *I, Robot*, Ballantine Books, New York, 1950.
3. "The One-Armed Chemist," *Forbes* **135**, 116, 1985.
4. Zymark Corporation, *Laboratory Unit Operations*, Hopkinton, Mass., 1984.
5. R. Bradley, *Laboratory Robotics and Automation* **1**, 17, 1989.

GENERAL REFERENCES

G. L. Hawk and J. Strimaitus, eds., *Advances in Laboratory Automation-Robotics 1984–1988*. Zymark Corp, Hopkinton, Mass.

These volumes are the Proceedings of The International Symposium on Laboratory Robotics and are an excellent compilation of applications and advances in the technology.

W. J. Hurst, and J. W. Mortimer, *Laboratory Robotics, A Guide to the Planning, Programming and Application*. VCH Publishers, New York, 1986.

This book provides an introduction in the technology.

W. JEFFREY HURST
ROBERT A. MARTIN
Hershey Foods Technical Center
Hershey, Pennsylvania

LACTOSE

Lactose (4-*O*-*β*-D-galactopyranosyl-D-glucose), formerly called milk sugar, is a disaccharide in which a *β*-D-galactopyranosyl unit is attached to D-glucose by a (1 → 4) glycosidic linkage (1–6) (see the article CARBOHYDRATES: CLASSIFICATION, CHEMISTRY, LABELING).

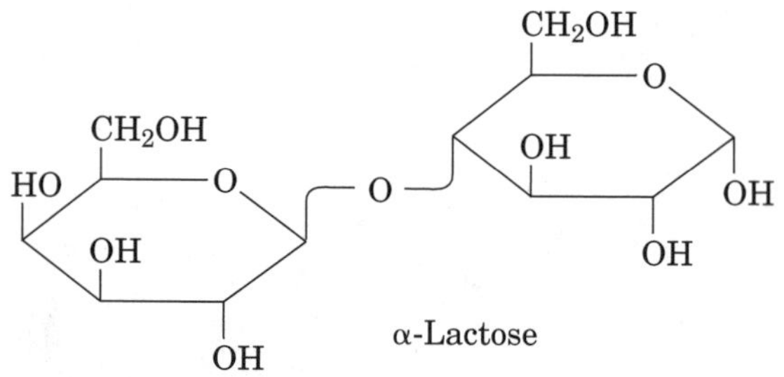

α-Lactose

SOURCE

Lactose is the major carbohydrate present in the milk of most mammals. The amount varies from species to species. Cow and goat milks contain approximately 4.5% lactose; human milk contains approximately 7%.

Lactose is produced commercially from cow's milk that has had the casein coagulated (the first step in the manufacture of cheese) by adjustment of the pH of the milk to the isoelectric pH of casein (4.5–4.7) and heating or by use of rennin. The resulting sweet whey, the solids of which are approximately 70% lactose, is subjected to ultrafiltration to remove remaining proteins (7). Then minerals are removed by ion exchange, and the solution is concentrated to 50 to 65% solids so that the lactose can be crystallized or precipitated. The recovered lactose is redissolved, decolorized with carbon, and recrystallized as *α*-lactose monohydrate (see the article CARBOHYDRATES: CLASSIFICATION, CHEMISTRY, LABELING), the most commonly prepared form of lactose and the one that crystallizes from supersaturated solutions at temperatures below 93.5°C (1–6). For every kilogram (2.2 lb) of cheese produced, approximately 9 kg (19 lb) of whey is recovered (3). Because whey contains about 4.7% lactose, about 450 g (1 lb) is potentially available as a by-product of the manufacture of each kilogram of cheese. However, little commercial use is made of lactose.

Milk also contains 0.3 to 0.6% lactose-containing oligosaccharides, many of which are important as energy sources for growth of a specific variant of *Lactobacillus bifidus*, which as a result is the predominant microorganism of the intestinal flora of breast-fed infants.

PROPERTIES

α-Lactose monohydrate dissolves slowly, which means that it cannot be used in instant food products. Three forms of anhydrous *α*-lactose are known (1–6). A stable anhydrous form is prepared by heating *α*-lactose monohydrate in air at 130°C. *β*-Lactose is more soluble than *α*-lactose. It is prepared by heating *α*-lactose · H_2O in an alcohol in the presence of a base. *α*-Lactose is also sweeter than is *α*-lactose, but each anomer will form the same equilibrium anomeric mixture in solution. The equilibrated solution is 15 to 30% as sweet as a sucrose solution of the same concentration.

LACTOSE DIGESTION AND LACTOSE INTOLERANCE

The enzyme lactase, located in the small intestine, catalyzes the hydrolysis of lactose into its monosaccharide constituents, D-glucose and D-galactose. Both are rapidly absorbed and enter the bloodstream.

$$\text{lactose} \xrightarrow{\text{lactase}} \text{D-glucose} + \text{D-galactose}$$

Only monosaccharides can be absorbed from the small intestine. Therefore, if for any reason, ingested lactose is only partially hydrolyzed (digested) or is not hydrolyzed at all, a clinical syndrome results (8–10). Its symptoms are abdominal distention, cramps, flatulence (gas), and diarrhea. When there is an insufficient amount of lactase, some lactose is not absorbed from the small intestine and remains in the lumen. The presence of lactose draws fluid into the intestinal lumen by osmosis. It is this fluid that leads to the abdominal distention, cramps, and diarrhea. From the small intestine, the lactose passes into the large intestine (colon) where bacteria effect its anaerobic fermentation to short-chain fatty acids. Production of more molecules results in still greater retention of fluid. The acidic products of fermentation lower the pH and irritate the lining of the colon, leading to an increased movement of the contents. Diarrhea is caused by the retention of fluid and the increased movement of the intestinal contents. The gaseous products of fermentation (carbon dioxide, hydrogen, methane, and oxygen) cause bloating.

This syndrome resulting from an absence or deficiency of lactase is called lactose intolerance. Lactose intolerance is not usually seen in children until after about 6 years of age. At this point, the incidence of lactose-intolerant individuals begins to rise. There are varying degree of lactose intolerance. The difference in incidence among populations of different genetic backgrounds can be large. By 12 years of age, 45% of blacks develop the symptoms of lactose intolerance; among teenage blacks, the incidence climbs to 70% and by adulthood, 90% of the black population in the United States shows symptoms of lactose intolerance. Lactose intolerance is also high among Orientals. Among the whites of western European ancestry, the peak incidence in adulthood is about 15%.

Humans consume lactose in milk and other nonfermented dairy products. Cheese, cottage cheese, and most yogurts contain less lactose because some lactose is converted into lactic acid during fermentation and/or removed with the whey. Milk from which much of the lactose has been removed is available.

USES

Because of its low solubility and the common occurrence of lactose intolerance in adults, food uses of crystalline lactose are limited; but it must be kept in mind that lactose constitutes about 70% of the solids of sweet whey, so where whole liquid or dried whey is used as an ingredient, some functionality of lactose is realized. Crystalline lactose is used to a small extent in toppings, icings, pie fillings, confections, and ice creams to provide body and texture while contributing less sweetness than the same amount of sucrose would (3). It is used as a carrier for flavors, because enhancement of flavor and colors is one of its attributes (11–15). Lactose provides bulk and rapid disintegration and dissolution of pharmaceutical tablets. It is present in about 20% of tableted prescription drugs and in about 6% of over-the-counter drugs.

PRODUCTS FROM LACTOSE

Lactose is a reducing disaccharide with the aldose D-glucose at its reducing end. Reduction (hydrogenation) yields the disaccharide alditol (polyol), lactitol (1,4) (see the article CARBOHYDRATES: CLASSIFICATION, CHEMISTRY, LABELING). Lactitol can be crystallized as either a monohydrate or a dihydrate, both of which are nonhygroscopic. Therefore, it can be used in the manufacture of products such as chocolate that require that there be no moisture pickup during processing and bakery products that should remain crisp. Lactitol provides a clean, sweet taste (30–40% of that of sucrose) and provides foods with a bulk and texture similar to that provided by sucrose. Its solubility is slightly less than that of sucrose. Its heat of solution is slightly higher than that of sucrose and much below that of sorbitol. Lactitol is not acted on by human digestive enzymes, so it does not effect an increase in blood glucose or insulin levels and, thus, is safe for diabetics. Neither is it cariogenic. It is, however, fermented by colonic microorganisms, being converted into carbon dioxide and volatile fatty acids. There is, therefore, a limit to tolerance and daily consumption as with lactose in lactose-intolerant individuals.

Lactulose, 4-*O*-β-D-galactopyranosyl-D-fructose, is made by alkali isomerization of lactose, which converts the reducing end to the corresponding ketose, D-fructose (1,4) (see the article CARBOHYDRATES: CLASSIFICATION, CHEMISTRY, LABELING). Lactose is quite soluble; a 77% solution can be made at 30°C (86°F). Its sweetness is 48 to 62% that of sucrose, depending on concentration. Lactulose is not acted on by human digestive enzymes, but is readily fermented by colonic microorganisms, which accounts for its major use for and treatment of chronic constipation. Lactulose is a bifidus factor; in other words, it is effective in promoting the proliferation of bifidobacteria in the intestine, increasing the level of the organism in the feces of bottle-fed infants to that in breast-fed infants.

PROPERTIES

Many physical properties of lactose have been determined (6). These include solubility in water over a wide temperature range, solubility in salt solutions and organic solvents, heat of combustion, heat of solution, heat of dilution, heat capacity, and specific heat. Also determined have been the density, viscosity, refractive index, osmotic pressure, vapor pressure, freezing-point depression, and interfacial tension with organic liquids of lactose solutions and the rate of crystallization of lactose from solution.

Crystalline lactose has adsorptive properties and can be used as a carrier for aromas and flavors (11–15). The chemistry of lactose has been rather extensively examined (1,3), in part because of an interest in converting the rather abundantly available lactose into a more useful product.

ANALYSIS

Methods for the determination of lactose have been reviewed (4,16).

BIBLIOGRAPHY

1. J. R. Clamp et al., "Lactose," *Advances in Carbohydrate Chemistry* **16**, 159–206 (1961).
2. T. A. Nickerson, "Lactose Sources and Recovery," in G. G. Birch and R. S. Shallenberger, eds., *Developments in Food Carbohydrate-1*, Elsevier Applied Science, Barking, United Kingdom, 1977, pp. 77–90.
3. L. A. W. Thelwall, "Lactose," in C. K. Lee, ed., *Developments in Food Carbohydrate-2*, Elsevier Applied Science, Barking, United Kingdom, 1980, pp. 275–326.
4. L. W. Doner and K. B. Hicks, "Lactose and the Sugars of Honey and Maple: Reactions, Properties, and Analysis," in D. R. Lineback and G. E. Inglett, eds., *Food Carbohydrates*, AVI, Westport, Conn., 1982, pp. 74–112.
5. J. G. Zadow, "Lactose: Properties and Uses," *J. Dairy Sci.* **67**, 2654–2679 (1984).
6. V. H. Holsinger, "Lactose," in N. P. Wong, ed., *Fundamentals of Dairy Chemistry*, 3rd ed., Van Nostrand Reinhold, New York, 1988, pp. 279–342.
7. J. L. Short, "Prospects for the Utilization of Deproteinated Whey in New Zealand: A Review," *New Zealand J. Dairy Sci. Technol.* **13**, 181–194 (1978).
8. D. M. Paige and T. M. Bayless, *Lactose Digestion: Clinical and Nutritional Implications*, Johns Hopkins University Press, Baltimore, Md., 1981.
9. A. Lutkic and A. Votava, "Enzymic Deficiency and Malabsorption of Food Disaccharides," in C. K. Lee and M. G. Lindley, eds., *Developments in Food Carbohydrates-3*, Elsevier Applied Science, Barking, United Kingdom, 1982, pp. 183–211.
10. J. Delmont, *Milk Intolerances and Rejections*, S. Karger, Basel, Switzerland, 1983.
11. S. L. McMullin et al., "Heats of Adsorption of Small Molecules on Lactose," *J. Agric. Food Chem.* **23**, 452–458 (1975).
12. T. A. Nickerson, "Lactose Chemistry," *J. Agric. Food Chem.* **27**, 672–677 (1979).
13. K. F. Ehler et al., "Heats of Adsorption of Small Molecules on Various Forms of Lactose, Sucrose, and Glucose," *J. Agric. Food Chem.* **27**, 921–927 (1979).
14. F. W. Parish et al., "Demineralization of Cheddar Whey Ultrafiltrate with Thermally Regenerable Ion-Exchange Resins: Improved Yield of α-Lactose Monohydrate," *J. Food Sci.* **44**, 555–557 (1979).
15. J. W. Marvin et al., "Interactions of Low Molecular Weight Adsorbates on Lactose," *J. Dairy Sci.* **62**, 1546–1557 (1979).
16. W. J. Harper, "Analytical Procedures for Whey and Whey Products," *New Zealand J. Dairy Sci. Technol.* **14**, 156–171 (1979).

J. N. BEMILLER
Purdue University
West Lafayette, Indiana

See also CARBOHYDRATES: CLASSIFICATION, CHEMISTRY, LABELING; CARBOHYDRATES: FUNCTIONALITY AND PHYSIOLOGICAL SIGNIFICANCE.

LECITHINS

Lecithin is a tan to amber viscous liquid or solid primarily derived from the soybean (Fig. 1). It is a complex mixture of phospholipids combined with triglycerides, fatty acids, phytoglycolipids, sterols, and minor neutral lipid components (1). Fourcray presented evidence of its existence in 1793, and Gobley first identified it from egg yolk lipids in 1846 (2). The applications and uses have expanded for this natural surfactant since its discovery, with more than 300 million pounds now available for commercial application.

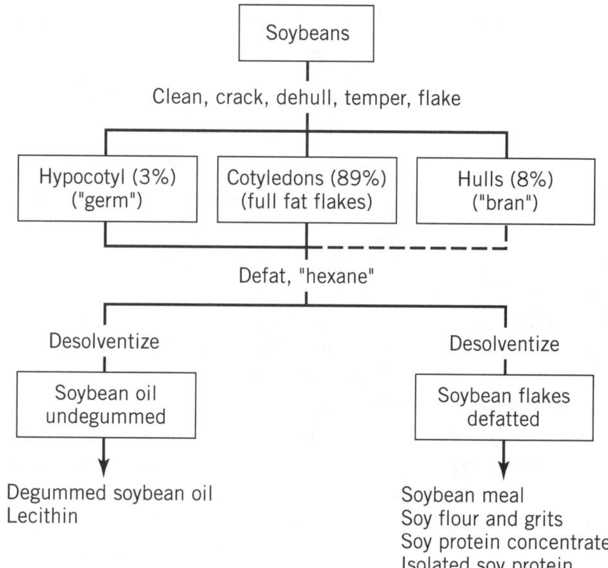

Figure 1. Processing of soybeans by solvent extraction. *Source:* Courtesy of Central Soya Company, Inc.

Lecithin is present from animal and vegetable sources. The classic definition of lecithin referred to phosphatidylcholine and not the complex mixture commercially available today. Chemical phosphatidylcholine is known as 1,2-diacyl-*sn*-glycero-3-phosphorylcholine according to IUPAC nomenclature (3). The archaic term Lecithin is no longer used for phosphatidylcholine.

COMMERCIAL LECITHIN

Commercial lecithin is a complex mixture of phospholipids, primarily phosphatidylcholine, phosphatidylethanolamine, phosphatidylinositol, and sometimes phosphatidylserine (Fig. 2). Various amounts of triglycerides and fatty acids are present, depending on consistency and end use.

Composition

Phospholipids are the major components in commercial lecithin. The phospholipids can range from 50 to 95%, depending on the neutral lipid content (Table 1). Triglycerides are the second major component and have fatty acids similar to the source oil. Fatty acids are natural components found in the lecithin manufacturing process, although additional fatty acids may be added as fluidizing agents. Again, the fatty acids resemble the source oil. Phospholipid fatty acids are peculiar to their source. Each vegetable or animal source can change the resulting fatty acid composition.

Phytoglycolipids are present in vegetable lecithins whereas animal lecithins may contain sphingomyelin, car-

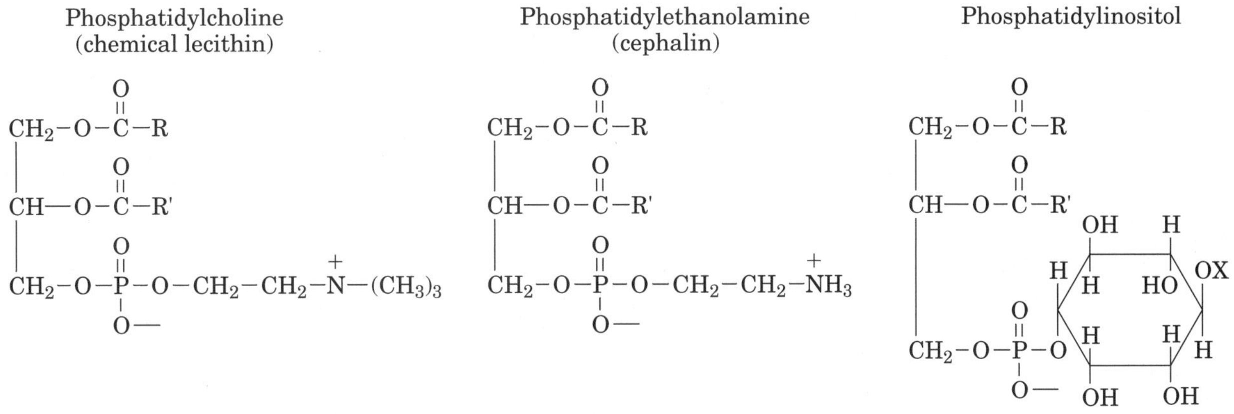

Phosphatidylcholine (chemical lecithin)

Phosphatidylethanolamine (cephalin)

Phosphatidylinositol

Figure 2. Three principal components of soybean lecithin. *Source:* Ref. 2.

diolipids, and other minor phospholipids. Minor amounts of sterols and sugars are carried through the manufacturing process. Sitosterol and stigmasterol are found in vegetable lecithin; cholesterol is found in animal sources.

Sources

Vegetable and Animal. The primary source of lecithin is from the plant kingdom, with the highest concentrations in seeds from soy, corn, rapeseed, and sunflower (4). Egg and brain are the major sources from the animal kingdom (5).

Microbial and Synthetic. Phospholipids are available from the extraction of microbes (6), and chemically derived and semisynthetic products (7) are available for medical research and cosmetics.

Manufacture

Production. Vegetable lecithins are prepared simply from the hydration of the crude oil source with centrifuging and drying (Fig. 3). Additives for viscosity and function may be added in the process (9). Animal products are usually prepared by solvent extraction of the tissue. Synthetic products usually are recovered from the reaction mixture by solvent extraction.

Analysis. Product composition and quality usually are determined by the methods established in the American Oil Chemists' Society Official Handbook of Methods, Section J (10). Tests include acetone insolubles (AI), acid value (AV), hexane insoluble matter (HIM), peroxide value (PV), phosphorus (P), composition by HPLC, viscosity, and moisture (Table 2).

Functional Uses

Lecithins are multifunctional products that have a wide range of applications at low use levels of 0.5–3.0% (2).

Emulsification. As surface-active agents they have wide usage in foods as emulsifiers. Their ability to bring two dissimilar liquids together make them important in water/oil (W/O) and oil/water (O/W) surfactants. These amphoteric surfactant properties make them ideal in food systems. The estimated HLB values for lecithin products range from 2–12 when compared to nonionic surfactants.

Release. Lecithins are good release or parting agents. Their polar groups tend to bind to metallic surfaces, and the nonpolar fatty acids act as slipping agents.

Instantizing. Most problems with dispersion deal with the rapid hydration of the dispersing item. Lecithin will reduce the solution rate and allow for easy dispersion. It also helps to disperse fat in high fat systems.

Anti-Spatter. In fat systems that contain water as an ingredient, lecithin keeps the moisture dispersed for slow evaporation under heat processing. Moisture will foam off rather than spatter.

Lubrication. The fatty acids present in the nonpolar part of lecithin give the added lubricity to prevent sticking in food-processing equipment. Because of its hydratability, the product is easier to wash off surfaces and it is completely edible.

Multifunctional. Lecithin users usually find that it is the combination of functional properties that make it a beneficial ingredient. Rarely do the functions counteract in a finished food system. Emulsification and release go well in baking systems. Anti-spatter and release are perfect frying systems. Instantizing and emulsification work together in beverage mixes.

Applications

The following are from Ref. 11.

Margarine. One of the first uses for lecithin was in regular margarine around 1940. The emulsifying properties were combined with anti-spatter properties to make margarines more stable and better for sauteing. Diet margarines with the high water content use special lecithins to retain the tight emulsion and hold the moisture.

Table 1. Lecithin Nutritional Profile/Typical Analysis

	Granular lecithin	Typical liquid lecithin
	Nutritional summary[a]	
Phosphatides, acetone insoluble %	95% min	60% min
Natural soybean oil	2–3%	39%
Moisture	1%	0.7%
Fat, per 100 g product	90 g	93 g
Monounsaturated (oleic)	9.2%	17.9%
Polyunsaturated (linoleic, linolenic)	65.9%	60.7%
Saturated (palmitic, stearic)	24.9%	20.3%
Calories, per 100 g product	700	790
Carbohydrates, per 100 g product	8 g	5 g
Cholesterol, per 100 g product	0 g	0 g
	Approximate composition	
Fatty acid content, g/100 g product	60 g	66 g
Fatty acid content, relative composition		
Linoleic	58.9%	54.0%
Linolenic	7.0%	6.7%
Oleic	9.2%	17.9%
Palmitic	20.3%	15.6%
Stearic	4.6%	4.7%
Other fatty acids	0.0%	1.1%
Total	100.0%	100.0%
	Primary acetone insolubles	
Phosphatidylcholine, g/100 g product	23 g	15 g
Phosphatidylethanolamine g/100 g product	20 g	12 g
Phosphatidylinositol, g/100 g product	14 g	9 g
	Elemental analysis, mg / 100 g product	
Calcium		
Without flow agent	65 mg	40 mg
With flow agent (tricalcium phosphate)	745 mg	
Iron	2 mg	1 mg
Magnesium	90 mg	60 mg
Phosphorus		
Without flow agent	3,000 mg	2000 mg
With flow agent (tricalcium phosphate)	3,400 mg	
Potassium	800 mg	440 mg
Sodium	30 mg	10 mg

Source: Courtesy of Central Soya Company, Inc.
[a]The nutritional profile of lecithin depends, in part, on the amount and type of oil used. The breakdown shows a deoiled, granular lecithin and liquid lecithin containing a typical level of natural soybean oil.

Confections. Chocolate is also one of the oldest uses for lecithin. It was found that lecithin could reduce the amount of cocoa butter needed for chocolate handling and also retarded the chocolate bloom from cocoa butter crystallization. Lecithin helps to reduce the viscosity of chocolate manufacturing and gives better flow-coating properties.

In chewing gum it prevents the drying out and cracking of the sticks on shelf storage. It also provides softness to the chicle gum base for better bite.

Caramel-coated products will not stick together when lecithin is added to the sugar. This gives the consumer a better-handling product with less staling.

Baking. Commercial lecithins use many of their functional properties in baking areas.

Breads. Lecithin distributes other emulsifiers in dough, gives less sticking in proofing, and retards bread staling on the shelf.

Cakes. The distribution of the cake and added moisture give quick hydrating and less lump formation in the batter for creamy smooth consistent cakes.

Cookies. Similar to cakes, lecithin provides better mixing of the batter and soft, smooth consistency. Lecithin can give-good chewiness for freshlike texture. It can also reduce dry-out.

Icings. Smooth creamy texture is needed for icing performance. Lecithin, when added to the fat system, will allow for better mouthfeel and good spreading qualities.

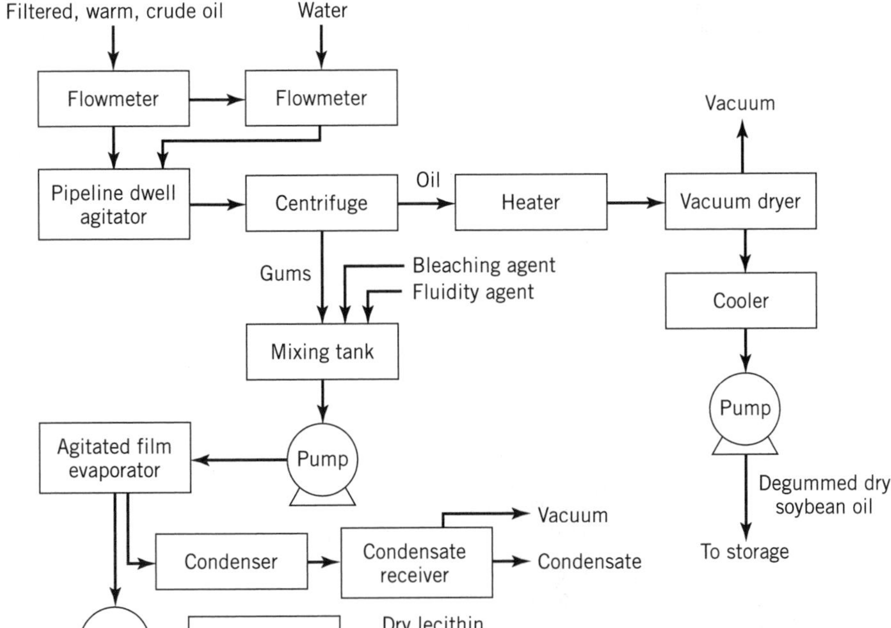

Figure 3. Flow sheet for degumming soybean oil and crude lecithin production. *Source:* Ref. 8.

Table 2. Food Grade Soybean Lecithin Specifications

Acetone-insoluble matter	Not less than 50.0%
Acid value	Not more than 36
Hexane-insoluble matter	Not more than 0.3%
Water	Not more than 1.5%
Peroxide value	Not more than 100
Limits of impurities	
Arsenic (as As)	Not more than 3 parts per million (0.0003%)
Heavy metals (as Pb)	Not more than 40 parts per million (0.004%)
Lead	Not more than 10 parts per million (0.001%)

Source: Ref. 10.

Beverages. Instant beverages are prepared by the careful application of lecithin to the powdered surface. Lecithin prevents the rapid hydration of the sugar or hydrophylic components and gives better dispersion.

Food Systems. Food systems are combinations of several ingredients that are used for covering food entrees such as meat or pastas or vegetable products in salads. Lecithin is a fine emulsifier for sauces and gravies that brings fat sources in contact with water. Lecithin keeps the oil-in-water systems intact even under heated conditions. In dressings, lecithin is a good emulsifier for high-stress systems where acidity and salt can be a problem. It provides a method of stabilizing the emulsion with a natural food amphoteric surfactant.

Health and Nutrition

Commercial lecithin and phosphatidylcholine have been extensively evaluated in health and disease areas (12). Lecithin may be effective in treating certain neurologic disorders such as tardive dyskinesia and some types of Alzheimer's disease. Most likely, lecithin plays an important role in health and nutrition as a precursor for choline (13) and may effect membrane fluidity as it effects membrane emulsion systems in food applications.

BIBLIOGRAPHY

1. B. F. Szuhaj and E. F. Sipos, "Food Emulsifiers from the Soybean," p. 115, in G. Charalambous and G. Doxastakis, G, eds., *Food Emulsifiers*, Elsevier, New York, 1989.
2. W. E. Prosise. "Commercial Lecithin Products: Food Use of Soybean Lecithin," in B. F. Szuhaj and G. R. List, eds., Lecithins, *American Oil Chemists' Society Monograph* **12**, 163–182, 1985.
3. C. R. Scholfield, "Occurrence, Structure, Composition, and Nomenclature," in Ref. 2, pp. 1–20.
4. J. P. Cherry, "Plant Sources of Lecithin," in B. F. Szuhaj, ed., *Lecithins: Sources, Manufacture and Uses*, American Oil Chemists' Society, 1989, pp. 16–31.
5. Kuksis, A., "Animal Sources of Lecithin," in Ref. 4, pp. 32–71.
6. C. Ratledge, "Microbiological Sources of Phospholipids," in Ref. 4, pp. 72–96.
7. M. Ghyczy, "Synthesis and Modification of Phospholipids," in Ref. 4, pp. 131–144.
8. R. Brian, *J. AOCS* **53**, 27 (1976).
9. G. R. List, "Commercial Manufacture of Lecithin," in Ref. 4, pp. 145–161.
10. R. A. Lantz, "Industrial Methods of Analysis," in Ref. 4, pp. 162–173.
11. G. L. Dashiell, "Lecithin in Food Processing Applications," in Ref. 4, pp. 213–234.
12. S. H. Zeisel, "Lecithin in Health and Human Nutrition," in Ref. 4, pp. 225–236.

13. S. H. Zeisel, "Phosphatidylcholine: Endogenous Precursor of Choline," in *Lecithin: Technological, Biological, and Therapeutic Aspects*, vol. 33 in I. Hanin and G. B. Ansell, eds., *Advances in Behavioral Biology*, Plenum Press, New York, 1987, pp. 107–120.

BERNARD F. SZUHAL
Central Soya Company, Inc.
Fort Wayne, Indiana

See also EMULSIFIER TECHNOLOGY IN FOODS.

LICORICE CONFECTIONERY

Licorice (also liquorice) is the name given to both the flavor extracted from the root of the plant *Glycyrrhiza glabra* and to a popular confection whose flavor is derived from the extract. The word licorice comes from the Greek words "glykys" (meaning sweet) and "rhiz" (meaning root). In ancient times the root was valued not only for its sweet taste, but also for its perceived medicinal value, both as a treatment for illness and as a substance for preserving youth and beauty.

The licorice plant is native to an extensive region, which includes the Mediterranean, Asia Minor, and southern Asia (see Fig. 1). It has also been grown in Pontefract, England, and Bamberg, Germany. The prime sources of supply of licorice root currently are Iran, Turkey, China, Afghanistan, Syria, and Pakistan. Extracts from Spain and Italy are also available in limited quantities but at higher prices because they are less bitter in flavor.

The licorice plant takes about 4 years to grow to a height of 2–4 feet. At this age the root can be as thick as 1 inch but averages about a half inch. The lateral roots can be as long as 25 feet. The roots are harvested in autumn and spring and transported to baling stations where they are inspected, dried, and baled. The bales are shipped to extraction plants where the root is shredded and immersed in hot water. The extraction process can take place under pressure or atmospherically. The resulting liquor is then transferred to evaporators where it is concentrated into three basic forms:

1. *Syrup* containing up to 30% moisture
2. *Paste* or block licorice with moisture content between 14 and 22%
3. *Powder* with moisture content around 4.5%.

The different processes result in slightly different flavor profiles.

Licorice extract is used in the following industries:

Tobacco as flavor, to increase the mildness and as a moisture retainer.

Pharmaceutical, to mask the bitter flavor of some drugs and as a flavor enhancer.

Food and beverage as flavor, sweetness enhancer and foaming agent.

BY-PRODUCTS

As in most industries, the desire to maximize productivity and find uses for by-products has led to useful materials:

1. A foaming agent to put out fires, created via a second extraction process.
2. An insulation made from the fibers of the root results in a product with high structural strength.

CANDY

Commercial licorice candy as known today started in England in the second half of the 18th century and in the United States in the 19th century. Y&S Candies (the current market leader in sales of licorice candies) was founded

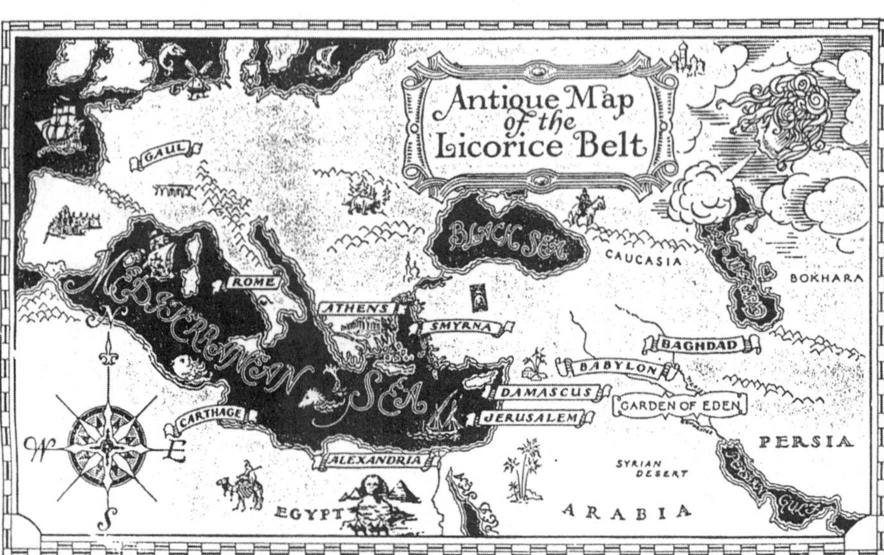

Figure 1. This curious old map shows the ancient lands in which licorice flourished thousands of years ago. This area, the Cradle of Civilization, is the principal source of licorice today.

in Brooklyn in 1845 by J. S. Young and C. A. Smylie. In 1902 they changed the name of the company to National Licorice and in 1968 changed the name back to Y&S Candies. Other well-known companies, such as Switzer and American Licorice, were founded early in the 20th century.

TYPES OF LICORICE

Licorice extract is added to gums, lozenges, and hard candies as a flavor and a demulcent for cough relief, but the most common form of candy is where the extract is added to a paste or dough and extruded into

1. Sheets, when combined with sheets of cream paste in sandwich form provide the main component pieces for licorice allsorts.
2. Strips of different shapes to form whips, braids, twists, shoelaces, reels, ribbons, etc.
3. Tubes, both hollow and filled.
4. Novelties, such as pipes and cigars.
5. Centers, for coating with a hard sugar shell and known as torpedoes, comfits, Good and Plenty®, and Goodies®.

INGREDIENTS

Licorice paste is made of wheat flour, crude sugars, molasses, gelatin, flavoring (which includes licorice extract), and color (see Table 1).

Flour

Wheat flour is normally used at levels between 25 and 40%. A strong flour, rich in elastic gluten, used to be recommended for licorice. Today, flour of low gluten content, which is less water absorbent, is preferred. More emphasis is placed on the starch content of the flour (approximately 69%), which contains 25% amylose and 75% amylopectin. A change in this percentage can regulate the gelatinization temperature and the degree of gelatinization (viscosity) before, during, and after cooking. Such changes in proportion can be obtained by the replacement of certain quantities of flour by native or modified maize, wheat, potato, rice, or other starches. The starch industry has made such advances in the last few years that it is possible to obtain a tailor-made product for each individual requirement. A hard wheat flour has a high protein content and is strong in gluten. Soft wheat flour is low in protein with little gluten. The gluten in soft flours is of a different character than that in hard flours and is more elastic. These variations in content and type of gluten give rise to differences in the texture of the licorice paste.

Table 1. Typical Formulas

Ingredient	Short texture	Long texture	Red color
Wheat flour	26.0	26.0	19.0
White sugar			18.0
Brown sugar	23.0	13.0	
Molasses	10.0		
Caramel color	4.0	13.0	
Dextrose	3.0		3.0
Corn syrup		7.0	12.0
Block licorice	3.0	4.0	
Salt	0.5		0.1
Gelatin (low bloom)	0.3	4.0	
Red color			0.3
Black color	0.1		
Glyceryl monostearate	0.1		
Fruit flavor			q.s.[a]
Anise oil	q.s.	q.s.	
Citric acid			0.6
Water	30.0	33.0	47.0

[a]q.s., quantity sufficient.

Sugar

Sugars used in licorice paste can be low grade provided that they meet food safety standards. Reducing sugars (present in low-grade sugars) are beneficial in retaining gloss and moisture of the paste. Low-grade or crude sugars also contribute to flavor. Molasses are used for the same reasons. Corn syrups are used in red licorice to replace the molasses; they have the same function, ie, that of inhibiting sugar crystallization, which in turn results in a loss of gloss. High dextrose or high fructose corn syrups are generally preferred.

Gelatin

Gelatin is sometimes necessary as a binder and also retards moisture loss. It contributes to gloss and inhibits cracking. Low-bloom quality is used at levels between 0 and 5%.

Flavor and Color

Flavors include (for black licorice) licorice extract, salt, molasses, and oil of anise, and (for red product) fruit flavors and acids. Color is contributed by licorice extract, molasses, caramel, and black (or red) dyes.

BATCH PROCESS MANUFACTURE

Premix

With the exception of flavors and acid, most of the ingredients mentioned in the above formulas are mixed in a slurry premixer. The practice of preparing a thick paste of flour with a little water in the cooker and then converting the paste into a creamlike slurry by gradually adding more water is not very satisfactory. It is slow, and small lumps of dry or damp flour often remain in the slurry. These not only spoil the finished paste, but also provide a lot of trouble in the extrusion process by blocking nozzles. An emulsifier or slurry mixer will prevent these shortcomings and produce a smooth, homogenous slurry. The correct quantity of water in relationship to flour and sugars is of paramount importance and is often neglected. Short textured licorice requires less water (30–32%) than long texture, which can go as high as 47%.

Sugars inhibit the swelling of starch in a water system and thus retard the gelatinization of starch. This factor can be used to advantage by adding all or part of the sugars at the premix stage and by adding the balance at some later stage of the cooking process.

Cook

Conventionally, licorice paste is manufactured in 1/2 to 1 ton batches in open steam-jacketed kettles. The mixing paddles must efficiently sweep the heated paste from the kettle walls as the paste has poor heat-transfer properties. Cooking times can vary from 60 to 220 minutes. Steam pressures around 60 psi are adequate.

Licorice paste has many unusual rheologic properties resulting in large variations in observed elasticity. These largely occur through variations in starch gelatinization. Factors that influence the degree of gelatinization are quantity of flour, quality of flour, batch boiling time, concentration of sugars, and available water.

The progress of gelatinization in the batch can be followed microscopically by alteration in appearance of starch granules. As boiling continues, the granules swell and eventually rupture. (This change can also be clearly seen under polarized light when the characteristic cross on the surface of the granules disappears when the granule rupture occurs.) Count lines, shoelaces, reels, etc (long textured product), will show starch that is fully swollen or ruptured whereas sandwich paste and centers for Good & Plenty® (short texture) should show only partial gelatinization of starch.

The gloss of licorice is dependent primarily on moisture content which typically in the final product is 16–20%. Too high moisture can lead to mold growth, so sometimes preservatives are added. At the end of the cooking process, moisture is reduced to 23–32%. After cooking, the licorice paste is partially cooled then extruded or stamped into its final shape and placed on trays in racks, after which it goes into a warm, dry room known as a stove where more moisture is removed down to its final moisture content. The time of stoving will vary depending on the thickness and shape of the final piece and the amount of moisture to be removed. If this process is rushed, a hard crust will form on the surface of the licorice, which inhibits the moisture loss from the center of the product. When the product is dried in this manner, the moisture from the center will ultimately migrate to the surface and cause it to be sticky. Good air circulation within the stove is essential. Stove temperatures should be 100–125°F, with relative humidities of 50–60%.

Figure 2 outlines the various steps in both the batch and continuous processes, along with an indication of the relative process times. A semicontinuous final moisture process is also included for comparative purposes.

CONTINUOUS PROCESS MANUFACTURE

Slurry to be cooked in a continuous mode, whether heated by steam injection or steam jacketing, must contain less flour and consequently more sugars than normally used for the batch process. This formula change offsets the higher cooking temperatures and tends to repress gelatinization. Because little or no evaporation takes place, the moisture content before and after cooking is substantially the same; therefore, the water used in the slurry must be reduced.

As is seen from Figure 2, the advantages of the semicontinuous and continuous processes are shorter process time, less floor space, reduced labor, and reduced amount of material in process. One negative is that it is difficult to get the same texture from a continuous process compared to the batch.

DOMESTIC AND FOREIGN STANDARDS AND REGULATIONS

- Standards and regulations given here are as of April 1989 as reported by the British Food Manufacturing Industries Research Association.
- *United States*. The standard for licorice as an ingredient for foodstuffs is covered by Title 21, part 184.1408, of the *Code of Federal Regulations*. Specific limitations (based on glycyrrhizin content) for the use of licorice in certain foodstuffs are laid down.
- *Argentina*. The Food Code lays down standards for licorice confectionery, namely licorice pastilles. These are pastilles (containing sugar, glucose, flavored distilled water, permitted natural or synthetic essences and colors, edible gums, and gelatin with or without maximum 5% starch or dextrin) to which at least 4% licorice extract or juice has been added. Modified or unmodified starches may be used in necessary proportions when gums or gelatin have not been used.
- *Austria*. Regulations on sugar products state that licorice is manufactured from a mixture of at least 5% sweet licorice extract or a corresponding quantity of dry extract with sugar, flour, and also the use of authorized gelling agents (as given in the 1988 Emulsifiers Regulations). The use of colors in licorice confectionery is prohibited.
- *Belgium, Denmark, Finland, Greece, Norway, and Sweden*. These countries have no standards of composition for licorice confectionery, although the use of additives in confectionery is controlled by the national regulations of each of the countries. Licorice confectionery should be of a safe and suitable composition such that it is not harmful to the consumer, and the label should not mislead the consumer as to the nature of the product.
- *Bolivia, and Dominican Republic*. Licorice is permitted as a color for foodstuffs unless otherwise restricted or prohibited by regulations for a specific type of food.
- *France*. The French standards are a little confusing, but appear to be broken down into three categories: pure licorice, licorice, and starched licorice. (1) The name licorice, with or without the name pure, is reserved for the product obtained by extraction of all or part of the soluble substances contained in the licorice root and containing no more than 15% water.

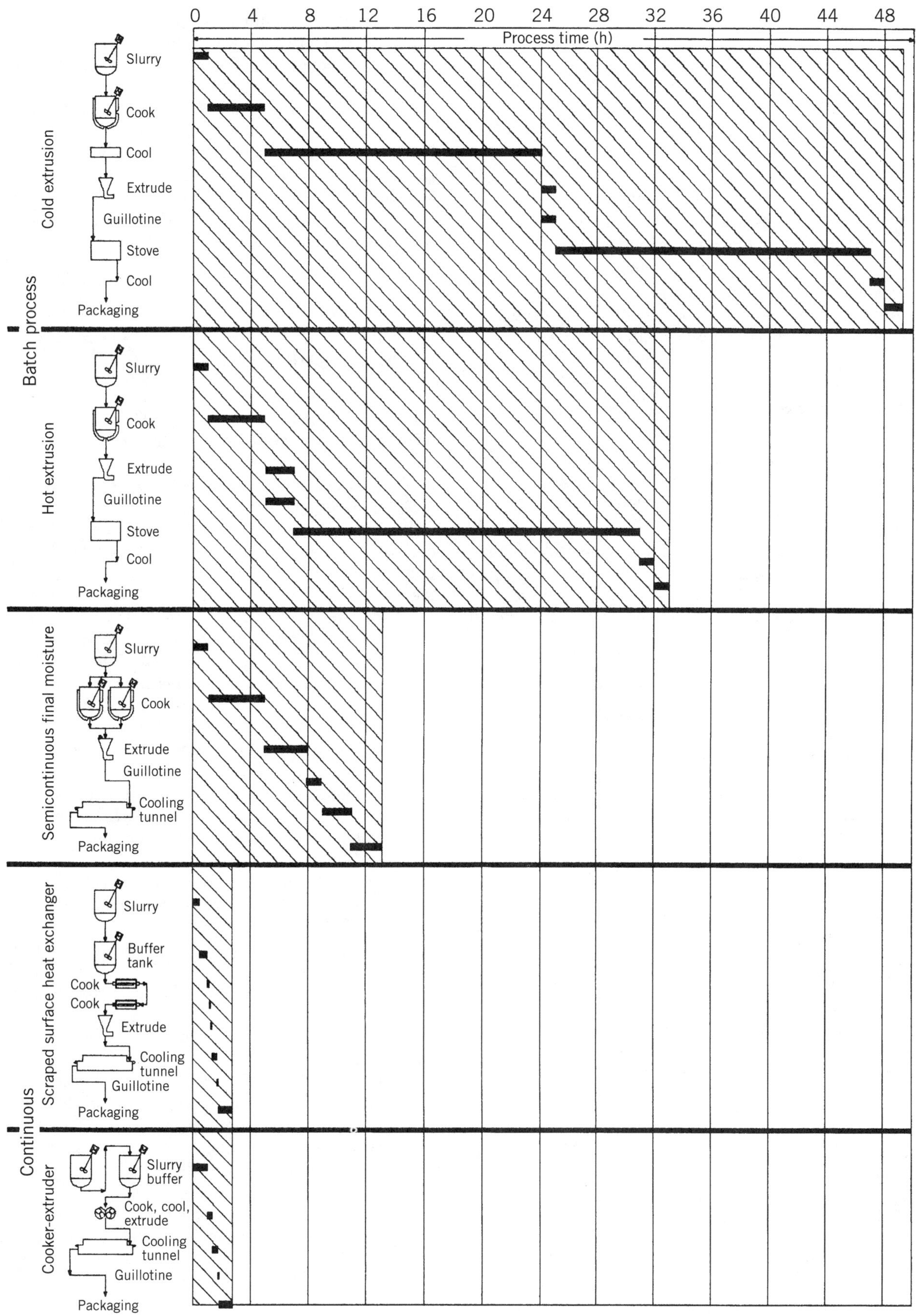

Figure 2. The various steps in both batch and continuous processes.

This product can be qualified as pure. (2) It is permissible for the addition of edible sweetened matter or gum, as long as the product still contains 6% glycyrrhizin; however, this product cannot be labeled pure. (3) Starched licorice (the most common classification) permits the addition of edible sweetened matter, gum, starch matter, and dextrin. This category must contain at least 1.5% glycyrrhizin.

- *Italy*. There is no standard of composition laid down for licorice. However, the use of additives in confectionery products (which one assumes includes licorice) is covered by the additives list.
- *Netherlands, Ireland, and United Kingdom*. There are no standards of composition on licorice confectionery. Products that are made to a standard recipe and fulfill consumer expectation of licorice should be acceptable.
- *Peru*. Substances based on licorice root are permitted as foaming agents for food use.
- *Spain*. Spain has no specific regulation or compositional standards on licorice as such; however, the list of additives permitted for confectionery products covers licorice.
- *Switzerland*. There is no standard of composition laid down for licorice, but there are regulations on confectionery and bakery products. The use of additives in licorice and confectionery are covered by the additives list.
- *Germany*. The guidelines are quite lengthy but can be simply stated as any product labeled as licorice must contain a minimum of 5% licorice juice in the dried form as customarily traded. Interpretation of the dried form probably means the powder form; however, if one takes the as customarily traded statement into consideration it could include block licorice. Although the guidelines do not have the force of law, they are an indication of trade and consumer expectation and should be complied with.

Although this article has described the black-colored licorice-flavored candy product, a similar textured product containing fruit flavors and pastel colors is made and sold throughout the world. In the United States, this particular type of product (red-colored and strawberry or cherry flavored) outsells the black variety by about 10 to 1.

J. Frank Smullen
Hershey Foods Corporation
Hershey, Pennsylvania

LIMONIN

One of the most characteristic flavor attributes associated with grapefruit (*Citrus paradisi* Macf) juice is bitterness. This is a quality factor that, present in a moderate degree, is responsible for much of the bite or zestiness of the juice. This bite or zestiness is expected and appreciated by many grapefruit juice consumers. However, if present at too high a level, excess bitterness will result in low approval ratings from even people who like grapefruit juice. Bitterness is also one of the chief reasons why most consumers dislike the overall flavor of grapefruit juice (1–6), with only about 29% of all U.S. households purchasing grapefruit juice in 1983 (7). The two major constituents in grapefruit juice responsible for bitterness are limonin and naringin. Of the two, limonin is considered to be the more significant contributor. This may be because bitterness due to limonin is perceived sooner and lingers longer than bitterness from naringin (8).

HISTORY OF LIMONIN IN CITRUS JUICES AND ITS DETERMINATION

Limonin was first reported (9) in 1938 as the cause of bitterness in Washington navel orange juice. It was not until 1961 that the complex chemical structure of limonin was finally worked out (10) and 1965 when Maier and Dryer (11) first reported the presence of limonin in commercial grapefruit juice. Limonin is a highly oxygenated triterpene derivative whose structural features include a furan ring, two lactone rings, a five-member ether ring, and an epoxide (Figure 1). Freshly squeezed citrus juices do not contain limonin. After extraction, limonin forms slowly over a period of several hours from a nonbitter precursor present in the fruit due to enzyme activity (12). Limonin forms rapidly if the juice is heated. The limonin precursor is limonoic acid A-ring lactone; being formed by closure of the D-ring as shown in Figure 1. An in-depth review of limonin and related structures has been presented by Maier et al. (13).

When Maier and Dryer (11) first discovered the presence of limonin in commercial grapefruit juice, thin-layer chromatography (TLC) was used. In 1970 Maier and Grant (14) developed a specific TLC assay for limonin. Scott (15), using an adaptation of their assay, reported limonin in many Florida citrus cultivars, including grapefruit. Tatum and Berry (16) in 1973 developed a quicker TLC method, which was adopted commercially.

Limonin today is generally assayed by high-performance liquid chromatography (HPLC), and a number of methods have been published. The real challenge with limonin determination by HPLC is one of detection. Although limonin is a major contributor to bitterness in citrus juices, it is only present at the low part per million level. Limonin does not contain a good UV absorbing chromophore, does not fluoresce, and is not easily reduced or oxidized using electrochemical detection. This limits LC methods to refractive index or UV detection at the lower wavelengths unless limonin is derivatized to enhance UV absorbance or fluorescence.

One of the first HPLC methods was developed by Fisher (17) and utilized refractive index detection. Subsequent HPLC methods (18–21) have utilized the more sensitive detection method of UV absorbance at lower wavelengths of 205 to 215 nm. Because limonin occurs in citrus juices in the low part per million range and most substances absorb in the UV below 220 nm, preparation of the sample

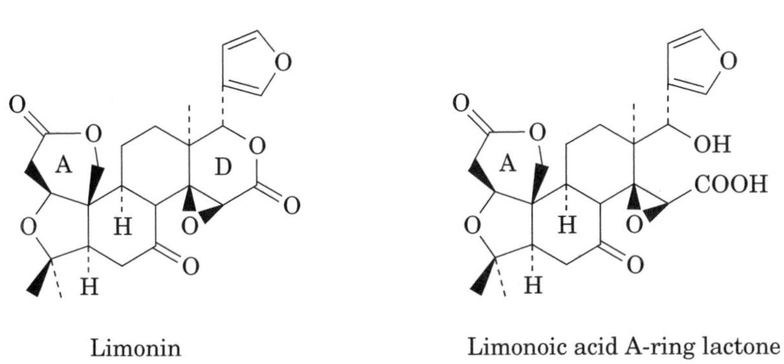

Limonin

Limonoic acid A-ring lactone

Figure 1. Structures of limonin and limonoic acid A-ring lactone, the precursor to limonin.

to obtain an extract free of interferences upon analysis is demanding. Rouseff and Fisher (18) developed an accurate method using chloroform extraction to isolate limonin. Analysis with a cyano column by normal phase chromatography with hexane:isopropanol as the mobile phase and UV detection at 214 nm produced excellent results. Analysis of the same chloroform extracts under reversed phase conditions with acetonitrile:water on a cyano column resulted in a good separation but short analytical column life.

A number of reversed phase chromotography methods that utilize the more robust octyl (C8) and octydecyl (C18) analytical columns have also been published. Sample cleanup to remove interfering components was accomplished by solid phase extraction (19–21). Solid phase extraction (SPE) offers several advantages over solvent extraction. It is less time and labor intensive. Less solvent is required for preparation of each sample. Reversed phase chromatography is also generally preferred over normal phase chromatography as water comprises most of the mobile phase solvent and can be produced relatively inexpensively. For limonin analysis, one problem with methods that utilize C8 and C18 reversed phase columns is that the mobile phase conditions must be changed depending on the type of citrus juice being analyzed. Some grapefruit juice samples require different analysis conditions than those that provide a good separation of orange juice extracts. The mobile phases required also generally consist of mixtures of methanol:acetonitrile:water or methanol:tetrahydrofuran:water. Mobile phases containing methanol:acetonitrile:water give low broad peaks reducing detectability. Tetrahydrofuran tends to form epoxides that absorb strongly in the low UV requiring diligent use of fresh solvents.

In 1991 Widmer (22) developed a general citrus juice analysis method with sample preparation by SPE, analysis on a cyano column under reversed phase conditions using acetonitrile:water, and detection by UV at 214 nm. Unlike chloroform extracts of citrus juices, the SPE extract preparation was optimized so integrity of the cyano analytical column under reverse phase conditions was maintained, resulting in less cost per sample analysis. Limonin recovery using SPE from spiked grapefruit and model juices ranged from 95 to 108%. Results from 25 samples analyzed by the new method and the method of Rouseff and Fisher (18) also showed good agreement between the two methods with differences in the range of 5 to 13%. The method has since been modified for analysis of citrus juices and citrus peel extracts by direct injection with an in-line automated sample cleanup (23). Both limonin and naringin are determined simultaneously with an analysis time of 20 min. Sample preparation for citrus juice requires only heating and (1:1) dilution with 40% aqueous acetonitrile. Preparation of peel extracts is accomplished by sonication of the peel in aqueous acidified acetonitrile.

Other novel methods for limonin analysis that have been developed include the rapid radioimmunoassay (RIA) developed by Mansell and Weiler (24) and the enzyme linked immunoassay (EIA) developed by Jourdan et al. (25). The EIA method was commercialized and available as a rapid test kit (26) for a number of years. Carter and coworkers (27) in 1985 compared the RIA and EIA methods to HPLC methods available at the time. Samples analyzed by EIA tended to have limonin values slightly higher than those analyzed by HPLC. Widmer and Rouseff (28) reported on a collaborative study done to assess the commercial EIA method. Problems with reproducibility in citrus quality assurance labs (private communications), sample analysis costs that were higher than anticipated, and subsequent development of rapid and reliable HPLC methods likely resulted in the loss of interest by the citrus industry for the commercial EIA method.

TASTE THRESHOLD LEVELS

Guadagni et al. (29) reported a threshold of 1 μg limonin/mL distilled water, while for naringin the threshold was 20 μg/mL. In orange juice they found the limonin threshold was dependent on juice pH. At a juice pH of 3.2 to 3.5 the limonin threshold was approximately 3.5 μg/g. The threshold increased to 6.5 μg/g at pH 3.8 and then decreased to about 4 to 4.5 μg/g at a juice pH greater than 4. This suggests natural orange juice components are effective in neutralizing or masking limonin bitterness to a certain degree and are pH dependent. Among the 27 tasters used in the study, the most sensitive was able to detect limonin at 0.5 μg/g in orange juice and the least sensitive required 32 μg/g. This range illustrates the wide variation in individual palates to limonin bitterness. Similar results were found by Widmer (unpublished data) in panel tests for limonin bitterness using model juice solutions containing 8% sucrose and 0.7% citric acid. Sensitivity to limonin ranged from 1 μg/g to greater than 16 μg/g limonin between individuals.

Limonin sensitivity in grapefruit juice is difficult to measure due to the presence of naringin and other bitter components. Work by Guadagni et al. (30), however, has shown that the relative bitterness contributions of limonin and naringin are additive.

REGULATORY ASPECTS FOR LIMONIN IN GRAPEFRUIT JUICE

Until recently there were no federal regulations for limonin content of grapefruit juice. Regulations issued by the state of Florida Department of Citrus (31) were in effect for many years to ensure the quality of juice produced in Florida. Recent changes in federal regulations, however, prohibit regulations at the state level more strict than federal regulations. Both grapefruit juice not from concentrate and grapefruit juice from concentrate products are covered under a federal standard of identity (32) with no mention of a maximum amount of limonin that may be present. Therefore, these may not be regulated at the state level for limonin content. Grapefruit juice concentrate does not have a federal standard of identity and therefore may still be regulated at the state level for limonin and naringin content. The Florida regulations (31) regarding limonin content in frozen concentrated grapefruit juice specify:

> For the period August 1 to December 1 of each season: (a) Grade "A" finished product shall meet at least one of the following requirements: (1) contains less than 5.0 ppm [5.0 μg/mL] limonin, measured by high pressure liquid chromatography, or (2) contains less than 600 ppm [600 μg/mL] naringin, measured by the Davis Test. (b) Grade "B" finished products shall meet at least one of the following requirements: (1) contains less than 7.0 ppm [7.0 μg/mL] limonin, measured by high pressure liquid chromatography, or (2) contains less than 750 ppm [750 μg/mL] naringin, measured by the Davis Test. (c) All products failing to meet the permissible limits of subsection (b) for Grade "B" product, shall be labeled Substandard Frozen Concentrated Grapefruit Juice, and may be held in appropriately marked bulk containers for subsequent blending into Frozen Concentrated Grapefruit Juice for Manufacturing only. (d) Maximum naringin and limonin requirements set forth in subsections (a) and (b) shall not apply: (1) to finished product produced solely from concentrate processed between December 2 and July 31, or (2) to finished product produced from concentrate or bulk single strength juice processed between December 2 and July 31, provided any raw juice blended therewith meets the permissible limits of naringin or limonin for Grade "A" product respectively.

LIMONIN CONTENT OF GRAPEFRUIT JUICE

Results of studies in which limonin levels in fresh grapefruit juice are reported follow. In an early study, Scott (15) found from 1 to 5 μg/mL limonin in juice from Marsh and Duncan grapefruit harvested in Florida at maturity during the 1967, 1969, and 1970 seasons; a TLC method was used. Mansell and McIntosh (33), using the RIA method, reported limonin contents of fresh grapefruit juice samples derived from Marsh, Duncan, and Ruby Red cultivars harvested from mid-December 1979 to early March 1980 and juiced at three state test houses in Florida. A mean limonin content of 6.76 μg/mL was found for 1058 samples analyzed, with a range of 1.35 to 13.8.

Ting and McAllister (4), using HPLC, determined the limonin contents for 156 samples of Florida-packed, canned grapefruit juice obtained from U.S. supermarkets. The overall mean was 5.6 μg/mL with a range of 1.5 to 10.9. Dougherty and Fisher (34) reported limonin contents (HPLC) for canned single-strength grapefruit juices obtained from surveys of Florida production during two seasons. For the 1975–1976 season the overall mean for 117 samples was 3.0 μg/mL, range 0.3 to 7.4. For the 1976–1977 season the overall mean for 99 samples was 4.7 μg/mL, range 0.6 to 12.8. Results by EIA for the 1985–1986 (87 samples) and 1986–1987 (61 samples) seasons showed means and ranges of 9.4 μg/mL, 2.8 to 18.9 and 9.4 μg/mL, 2.8 to 16.6, respectively (Barros, unpublished data). Grapefruit juice values in more recent years have shown this same type of variation for limonin content.

EFFECT OF SEVERAL FACTORS ON LIMONIN CONTENT OF GRAPEFRUIT JUICE

Mansell and McIntosh (33) reported limonin contents (EIA) for a large number of samples of the major Florida grapefruit cultivars. Samples from three Florida state test houses collected from mid-December 1979 through early March 1980 revealed the following limonin concentrations in the cultivars (mean and range in μg/mL): Duncan—394 samples, 6.03, 1.35 to 12.10; Marsh—537 samples, 7.15, 1.78 to 13.80; pink seedless—127 samples, 7.47, 2.41 to 12.00. The mean limonin content for Duncan was significantly lower ($P < 0.05$) than that for the Marsh or pink seedless.

In the same study, limonin contents were also reported for the three different test houses utilized (overall mean and range in μg/mL): test house 1—344 samples, 7.52, 2.38 to 12.30; test house 2—488 samples, 6.09, 1.35 to 12.30; test house 3—272 samples, 7.05, 1.78 to 13.80. The means for the three test houses differed significantly at the 95% confidence level.

Maturity

Generally, as grapefruit mature through the season the limonin content decreases. Tatum et al. (35) used a TLC method to analyze samples at about weekly intervals from two Florida plants from late November 1971, when the limonin level was 10 μg/mL, through May 1972, when the level had decreased to 2 μg/mL. The decline in limonin was gradual over the study period. Dougherty et al. (36), working with Marsh grapefruit from Florida packinghouse eliminations (fruit that was not suitable for marketing fresh, primarily because of skin blemishes, but was of processing quality), reported limonin contents by HPLC of juice from five separate batches of fruit harvested between September 12 and December 17, 1974, at irregular intervals. Using the mean limonin levels for the juice obtained with soft and hard extractor settings (to produce soft and hard squeezes) for each batch of fruit, the limonin content was found to decline gradually from 7.5 μg/mL in the first juice extracted to 2.5 μg/mL in the last. A gradual decrease in limonin content was again noted over the time period.

Until late 1989 it was generally believed that decreasing limonin as fruit matured was the result of transport of limonin from the fruit to developing seeds. Grapefruit seeds contain very high concentrations of limonin (37). However, Hasagawa et al. (38) discovered an enzyme that bound the limonin as a nonbitter glucoside as the fruit matured. Early season grapefruit contain few limonoid glucosides while mature grapefruit contain approximately 240 μg/g limonoid glucosides on average.

Extractor Pressure and Extractor Types

Dougherty et al. (36) also showed that there was an average 65% increase (range, 17–167%) in limonin content of juice obtained using a soft versus hard squeeze using an FMC Corp. (Lakeland, Fla.) extractor. This was based on values obtained from five batches of Marsh grapefruit harvested between September 12–17, 1974, and processed at both soft and hard squeeze settings. Similar results were obtained, but not presented, for the Duncan cultivar. In the 1975–1976 season, the same authors used FMC Corp. and Automatic Machinery Corp. (Winter Haven, Fla.) extractors to produce juice from three batches with each machine of Florida grapefruit at about monthly intervals beginning in October for both Marsh and Duncan cultivars. The data again showed more limonin extracted with the hard squeeze than with the soft squeeze and for both extractor types. In addition, one extractor produced an overall average of 34% (ranging from 6 to 63%) more limonin than did the other extractor for the same batches of fruit.

These results are not surprising as limonin is most prevalent in the seeds at levels in excess of 0.5% wet weight (37). Limonoic A-ring lactone, the nonbitter precursor to limonin occurs at highest concentrations in the fruit core, followed by the segment membranes, and peel (39). Any process that tends to crack the seeds or cause more peel and pulp constituents to be incorporated into the juice product will result in higher limonin values. Grapefruit core and segment membranes contain 500 to 1500 μg/g limonin; and the peel, 12 to 234 μg/g (39).

Season

Dougherty and Fisher (34) presented data showing season-to-season variation in limonin contents (HPLC) for Florida-canned, single-strength grapefruit juice. The mean of 117 samples from the 1975–1976 season with 3.9 μg/mL limonin while that for 99 samples from the 1976–1977 season was 4.7 μg/mL. During the latter season there was a freeze during mid-January, and the authors speculated that the freeze caused the peel of the fruit to break down, making limonin in the peel more available for extraction.

EFFECT OF LIMONIN IN GRAPEFRUIT JUICE ON SENSORY FLAVOR QUALITY

Studies with Small Experienced Taste Panels

Between September 12 and December 17, 1974, five batches of Florida Marsh grapefruit were harvested on an irregular basis (36). From each batch of fruit, packinghouse eliminations were obtained for quality evaluation. Juices were extracted using FMC equipment set for soft, medium, hard, and state test house pressure squeezes. Values for the four extraction pressures were averaged for limonin (HPLC) and flavor (nine-category hedonic scale as judged by 10–12 experienced panelists). During the period of the study, flavor increased from 5.0 (neither like nor dislike) to 5.9 (like slightly) while limonin decreased from 7.0 to 2.3 μg/mL. Naringin (by the Davis test) remained about the same, ranging from 600 to 700 μg/mL. The correlation between flavor and limonin was significant at the 99.9% confidence level ($r = -0.91$).

Flavor and several physical and chemical quality characteristics, including limonin content, have been determined in commercial Florida-produced canned and chilled single-strength grapefruit juices in a survey by the Citrus Research and Education Center, Lake Alfred, Florida, during each citrus season starting with 1973–1974. Samples were collected twice monthly from processing plants (14 initially; 7 during the 1986–1987 season). The number of samples collected annually varied between 61 and 168. Flavor was determined using 10 to 12 panelists judging samples on a nine-point hedonic scale. Details of the study and results covering two seasons' data were presented by Dougherty and Fisher (34). Over the first 10 seasons (1973–1974 to 1982–1983) in which the study was under way, limonin concentration (HPLC) in the juice was highly correlated with flavor at the 95% or better confidence level in 8 of the seasons (Barros and Dougherty, unpublished data). The best correlation coefficient was achieved for the 1978–1979 season for 113 samples from 11 plants ($r = -0.53$), explaining almost 28% of the variation in flavor due to limonin. Four individual plants that season showed significant correlations ($P < 0.05$) between flavor and limonin. The two seasons showing no significant correlations were the 1974–1975 and 1976–1977 seasons.

EIA and RIA procedures were used at the Citrus Research and Education Center to measure limonin concentration in juices for the 1983–1984 and 1984–1985 seasons (Barros, unpublished data). There were low but significant correlation coefficients ($P < 0.05$) for limonin content (both EIA and RIA) and flavor score for 1983–1984. For the 1984–1985 season, $r = -0.01$ for EIA limonin versus flavor was not significant ($P > 0.05$), while $r = -0.27$ for RIA limonin versus flavor was at the 99% level. It would have been interesting to have information to calculate r for HPLC limonin versus flavor. One plant showed significant correlation ($P < 0.05$) between limonin concentration and flavor for no fewer than 7 of the 12 study seasons (1973–1974 through 1984–1985); 5 seasons had limonin analysis by HPLC, 1 season with EIA and RIA, and 1 season with RIA methodology. On the other hand, another plant showed no significant correlation ($P > 0.05$) between limonin (HPLC) and flavor for the 10 years the plant was included in the study. Every other plant showed significant correlation ($P < 0.05$) between limonin (HPLC from 1973–1974 to 1982–1983; EIA and RIA the last 2 seasons) and flavor for 1 or more years. However, no significant correlations ($P > 0.05$) were noted between limonin (HPLC) and flavor for any of the 12 plants surveyed during the 1974–1975 season.

The two most recent years of the study, using only EIA methodology, showed highly significant correlations ($P < 0.01$) between limonin and flavor: for 135 samples during 1985–1986, $r = -0.28$; for 61 samples during 1986–1987, $r = -0.40$.

Fellers and Ting (unpublished data) surveyed the U.S. retail market to determine quality of three types of Florida grapefruit juice. There were 10 frozen concentrated grapefruit juices, 11 grapefruit juices from concentrates (the USDA term for grapefruit juices reconstituted from concentrates) and 14 canned (direct) grapefruit juices. Flavor was determined by 12-member experienced panels grading samples on a nine-point hedonic scale. Among the several chemical and physical analyses run on the juices was limonin by HPLC. No significant correlation ($P > 0.05$) was found between flavor score and limonin for any of the product types or for all samples combined ($r = -0.18$). The correlation coefficients were negative for the canned and reconstituted products, but positive for the concentrate. As has already been discussed, it is well established that one of the reasons grapefruit juice users like grapefruit juice is because of its zestiness due at least in part to bitterness. The limonin contents of the 10 concentrates were quite low with a mean of 3.0 μg/mL (range 1.3–7.5), apparently a concentration viewed more or less favorably by the taste panelists. The mean limonin contents for the canned and reconstituted products, respectively, were 4.2 μg/mL (range 1.3–7.5) and 5.2 μg/mL (range 2.5– 10.0). The mean for the reconstituted products was significantly higher ($P < 0.05$) than the mean for the concentrates but not for the canned juices. It should also be pointed out that concentrate also has a higher Brix:acid ratio requirement than does the canned or reconstituted products. The degree Brix refers to the weight percent of soluble solids (which is mostly sugar) that are present in the juice. The Brix:acid ratio is a measure of the ratio of soluble solids to total acidity present and is an indicator of fruit maturity. Generally juice with a higher Brix:acid ratio will be perceived as sweeter and is produced from fruit that is more mature.

Study Utilizing a Small Consumer Panel

Ting and McAllister (4) reported on flavor and several other quality factors, including limonin content of 156 Florida-processed canned single-strength grapefruit juice samples obtained from the U.S. retail market. Juices were judged by 12 to 14 consumers (mostly housewives) from the Winter Haven, Florida area. Samples were graded on a nine-point hedonic scale. Limonin content (HPLC) correlated negatively and significantly with flavor ($r = -0.166$; $P < 0.05$). The mean limonin content for all samples was 5.6 μg/mL. Panelist's comments of bitterness were significantly correlated with limonin content ($r = 0.233$; $P < 0.01$) and with flavor ($r = -0.434$; $P < 0.01$). The average number of bitterness comments by the panel per sample was 2.19, the largest number for any unfavorable type of comment. Age of the sample from the date of packing also correlated significantly with limonin content ($r = -0.233$; $P < 0.01$).

Large Consumer Panel Studies

In a study by Fellers et al. (40), 30 Florida-processed grapefruit juices were obtained from the U.S. retail market during 1983 for flavor and chemical and physical analyses of quality factors, including limonin by the EIA method. Six juices of each of five types were used: frozen concentrated, grapefruit juice from concentrate packed in glass, cartons, and cans and grapefruit juice (canned directly from freshly harvested fruit). Ten products were evaluated by grapefruit juice users at each of three metropolitan test locations in the United States. Two of the five juice types were tested at each location. Each sample was tested by 108 consumers with each testing no more than two samples. Consumers evaluated samples on a nine-point hedonic scale (1 = dislike extremely, 9 = like extremely) for overall flavor and on a five-point scale (-2 = not at all enough; 0 = just right; $+2$ = much too much) for the attributes of sweetness, tartness, bitterness, aroma, and color.

Correlation analysis revealed no significant association ($P > 0.05$) between overall flavor and limonin content (EIA). However, the bitterness attributed, overall and for each product type, was somewhat too much ($P < 0.05$) and correlated significantly ($P < 0.01$) with limonin concentration.

A serious problem apparently arose from the inability of consumers to differentiate bitterness from tartness. There was a significant positive correlation between tartness and bitterness ($r = -0.66$; $P < 0.01$) as judged by the consumers and negative correlation between sweetness and bitterness ($r = 0.55$; $P < 0.01$); in addition, tartness correlated directly ($P < 0.05$) with both limonin and naringin content. However, bitterness, as measured by the consumer, also correlated directly at the 99% confidence level with limonin and naringin content.

In another study (6), the effect on flavor of limonin addition to grapefruit juice was determined. Using a single commercial Florida frozen concentrate, limonin levels (HPLC) of 2.3, 7.1, 8.8, 9.2, and 11.0 μg/mL were obtained in the respective reconstituted juices at 10.0° Brix with a Brix:acid ratio of 8.5. The target limonin levels were 2, 5, 7, 10, and 15 μg/mL. A stock solution of limonin in distilled water (obtained by slow addition of limonin to boiling water) was used to spike the juice. Both grapefruit juice users and nonusers at shopping malls in three large U.S. cities participated in the large consumer study. An overall flavor score was obtained on each product using a nine-point hedonic scale (1 = dislike extremely; 9 = like extremely); a five-point scale (-2 = not at all enough; 0 = just right; $+2$ = much too much) was used to rate bitterness, sweetness, tartness, aroma, and color; finally, after each panelist had finished rating their two samples, they were asked which of the two samples was preferred. $P < 0.05$ was chosen as the level for significance.

Overall flavor means for all panelists (users and nonusers) varied within an acceptable range from the low end of the like moderately flavor category to the top end of the like slightly category (Table 1). The product having the highest limonin level (11.0 μg/mL) was assessed as having a flavor significantly inferior to the two products having

Table 1. Mean Flavor Score Values and Preference Ratings by Consumer Panels for Grapefruit Juice with Varying Limonin Contents

	Limonin Concentration (μg/mL by HPLC)				
Consumers	2.3	7.1	8.8	9.2	11.0
		Overall flavor score[a]			
Nonuser[b]	6.49[c]	6.33[c,d]	6.26[c,d]	6.21[c,d]	6.08[d]
Users[b]	7.07[c,d]	7.20[a]	6.97[c,d]	7.04[c,d]	6.91[d]
All Panelists[b]	6.77[c]	6.76[c]	6.62[c,d]	6.62[c,d]	6.49[d]
		Preference ration (%)			
Nonusers[b]	58[c]	50[c]	52[c]	44[d]	32[e]
Users[b]	51[c]	46[c]	51[c]	44[c]	36[d]
All Panelists[b]	55[a]	48[d]	51[c]	44[d]	34[e]

Source: Ref. 6.

[a]Based on a nine-point hedonic scale: 1 = dislike extremely; 9 = like extremely.

[b]Means in the same row sharing the same letter c, d, e, are not significantly different at the 95% confidence level.

the smallest amounts of limonin, but not inferior to the products with 9.2 or 8.8 μg/mL. There was no significant difference between the four products having the lowest limonin levels.

Significant flavor differences existed between users and nonusers for every product, the users in each case rating the product higher. Within the nonuser group there was a steady downward trend in flavor scores with increasing limonin concentration. Significant differences existed between the products having the lowest and highest limonin levels, with no significant difference noted between the products having from 7.1 to 9.2 μg/mL limonin. Among users, the highest rated product (7.2 flavor score) was the sample with the second lowest limonin levels (7.1 μg/mL). The only significant difference in flavor scores among users was between this highest rated product having 7.1 μg/mL limonin and the product having the most limonin (11.0 μg/mL) and the lowest score (6.91).

The data in Table 1 also show the panelists' preference measurements. For all panelists the sample with the least limonin (2.3 μg/mL) was significantly preferred over samples having 7.1, 9.2, and 11.0 μg/mL, but not the sample with 8.8 μg/mL limonin. The sample with the most limonin (11.0 μg/mL) was significantly least preferred of all.

When preference measurements by users were compared with those by nonusers, there was no difference in their preference for each product. Within the nonuser group the most highly preferred product had the least limonin and was significantly preferred to the products having the two highest limonin levels. The least preferred product had the highest limonin level and had only a 32% preference, 45% less than the most preferred product (Table 1). Among users, the product with the highest amount of limonin was again rated significantly lower than all other products. No significant difference in preference was noted among users between the samples having the lowest limonin contents. Apparently, a clear-cut consensus existed among both users and nonusers of grapefruit juice that grapefruit juice having ~11.0 μg/mL limonin (HPLC) was less preferred to grapefruit juice containing less limonin.

For those specific attributes of sweetness, tartness, and bitterness that were also rated (Table 2), there was a general tendency for the level of sweetness perceived by consumers to decrease with increasing limonin concentration in the juice. Simultaneously, the perception of tartness and bitterness intensity increased with increasing limonin content. Because the only differences between the five samples was limonin content, bitterness was perceived as tartness or reduced sweetness to a measurable degree, a phenomenon also noted in a study described earlier (40). All five samples were judged by consumers as being somewhat not sweet enough and somewhat too bitter and tart.

Correlation analysis revealed a significant negative correlation ($r = -0.973$; $P < 0.01$) between overall flavor score and limonin level. Differences in limonin levels explained over 94% of the variation in flavor. In addition, overall flavor score correlated significantly with two of the quality attributes judged by the consumers: directly with sweetness ($r = 0.852$; $P < 0.01$) and negatively with bitterness ($r = -0.962$; $P < 0.01$). Tartness did not quite make the 95% confidence level.

INTERACTION BETWEEN BRIX:ACID RATIO AND LIMONIN CONTENT IN AFFECTING SENSORY FLAVOR OF GRAPEFRUIT JUICE

There have been two studies showing that the Brix:acid ratio affects the perception of bitterness caused chiefly by limonin and naringin in grapefruit juice. Tatum et al. (35) studied seasonal effects on the flavor perception of limonin and naringin in grapefruit juice. Results from paired comparison taste panels led the authors to observe "when both naringin and limonin were higher and Brix:acid was higher, it was always judged least bitter." They concluded that the results suggest that "in many cases the Brix:acid ratio may be more important than naringin and limonin content in determining quality of grapefruit juice."

In the second study (Fellers and Carter, unpublished data), an experiment was done to determine the effect of

Table 2. Means and Comparison of Means for Selected Quality Attributes, as Evaluated by Consumers, of Grapefruit Juice Having Varying Limonin Contents

	Limonin Concentration (μg/mL by HPLC)				
Attributes	2.3	7.1	8.8	9.2	11.0
Sweetness	-0.28^{a}	$-0.41^{a,b}$	$-0.37^{a,b}$	$-0.36^{a,b}$	-0.47^{b}
Tartness	0.19^{a}	$0.30^{a,b}$	$0.29^{a,b}$	0.19^{a}	0.42^{b}
Bitterness	0.32^{a}	$0.38^{a,b}$	$0.39^{a,b}$	$0.41^{a,b}$	0.51^{b}

Source: Ref. 6.
Note: Each mean value based on about 300 judgments by consumers using a five-point scale: −2 = not at all enough of the attribute; 0 = just right; +2 = much too much of the attribute.
^{a,b}Mean values in the same row sharing the same letter are not significantly different at the 95% confidence level.

Brix:acid ratio on the flavor of grapefruit juice varying in the amount of added grapefruit pulpwash containing high levels of limonin and naringin. The three Brix:acid ratios were 8.0, 9.5, and 11.0. Approximate limonin (HPLC) and naringin (Davis test) contents were, respectively, for each of the pulpwash levels: 0% pulpwash, 2.5, 660 μg/mL; 7% pulpwash, 10.0, 1010 μg/mL; 14% pulpwash, 14.4, 1280 μg/mL. For each pulpwash level, 72 panelists drawn from USDA personnel in Washington, D.C., judged samples having each of the three ratios. The mean hedonic scores for the three Brix:acid ratios, when averaged across the three pulpwash levers, were, respectively, 6.04, 6.20, and 6.69. Analysis of variance indicated that the Brix:acid ratio had a significant effect on taste preferences. The effect of Brix:acid ratio at 11.0 was found to be significantly different from the effect at both 9.5 and 8.0 at the 99.9% confidence level; the effects at 8.0 and 9.5 did not significantly differ from one another ($P < 0.05$).

That the Brix:acid ratio had an effect on limonin (bitterness perception in grapefruit juice) was not totally unexpected. Guadagni et al. (29), using model systems containing sucrose, citric acid, and combinations of these constituents, increased the thresholds of limonin (and naringin) severalfold. In an orange juice, these workers were able to increase the limonin threshold from about 6.2 to 8.5 μg/g by increasing the Brix:acid ratio from 10 to 16, while keeping the pH at a constant 3.65. No tests were done using grapefruit juices.

METHODS FOR REDUCING THE LIMONIN CONTENT OF GRAPEFRUIT JUICE

To meet government regulations and customer quality specifications, juices are often blended. Orange juices are blended to produce the desired color, flavor, Brix, and Brix:acid ratio in the final product. Processed Florida grapefruit juice products are also blended to produce the desired aspects listed above and to control the level of bitterness (limonin and naringin) in the juice.

Although a great deal of research has been done in developing methods for bitterness reduction in citrus juices for more than 40 years, control of juice extraction conditions and juice blending were the only ways permissible for controlling bitterness until 1990. In response to a petition filed to allow commercial debittering of citrus juices, the Food and Drug Administration Department of Health and Human Services (FDA) sent a letter to the State of Florida Department of Citrus. In this letter the FDA withdrew previous objections to using treatments for reducing the levels of limonin and naringin in citrus juices using food-grade polymeric adsorption resins. This was provided the resin used conformed to food additive regulations, nutritional content of the juice was not altered, other organoleptic properties were not altered, and the limonin and naringin content were not lowered below those ordinarily found in a good-quality juice. There are now processors in the United States and throughout the world who utilize adsorption resins commercially to control the bitterness content in citrus juices that meet these requirements.

Early research was focused on reducing the limonin content of navel orange juice because delayed bitterness caused by excessive limonin was a real problem. Bitterness reduction really consists of two processes. Particulates must first be removed so adsorption materials are not fouled. Second is the process of bitter component adsorption or their conversion to nonbitter components. The debittering systems now commercially in use (41) for both orange and grapefruit consist of an ultrafiltration step to remove particulates, and a debittering chamber containing nonfunctionalized cross-linked hydrophobic styrene divinylbenzene polymer resin (SDVB) to adsorb limonin and naringin from the clarified serum. The oil content of the juice must also be low to prevent premature fouling of the system. After debittering, the clear serum and removed pulp and cloud particulates (or retentate) are recombined to obtain a juice with reduced bitterness. Some bitter components are still present in the final juice because the retentate portion, which comprises 10 to 20% of the whole juice, has not been treated. Natural flavor components may also be added to the treated juice, or it may then be blended with nontreated juice to obtain product with the desired levels of bitterness and other flavor characteristics.

The first commercial system in the United States was installed at Sunkist Growers, Inc., in California for the treatment of navel orange juice in 1990. Commercial systems for debittering citrus juices in Florida were available soon after (41). These systems were available through a partnership between Mitco Water Laboratories, Inc. (Winter Haven, FL), and Romicon (now owned by Koch Membrane Systems, Inc., Wilmington, MA). The two companies have since become independent. Most recently, Sepragen, Inc. (Haywood, CA), has introduced a commercial system for citrus juice debittering that utilizes a novel radial flow chamber for holding the debittering adsorption resin allowing for improved juice flow during the debittering step (42). How advantageous this novel design will be remains to be seen as the method for citrus juice processing is still being evaluated.

The ability of SDVB adsorbent resins for debittering citrus juices was first demonstrated by Puri who received a patent in 1984 for the process (43). Limonin and naringin levels could be reduced by 90 and 80%, respectively, in grapefruit juice. In navel orange juice, limonin reduction was 85%. Vitamin C losses were negligible. Norman (44) received a patent utilizing SDVB resins that were uniquely modified with hydrophillic amine groups. These resins acted as ion-exchange resins and were considered to not meet the restrictions specified by the FDA for adsorption resins and debittering. The most recent patent for debittering citrus juices awarded in 1998 to Sepragen Corp, Inc., utilizes SDVB resins with a smaller particle size for increased efficiency (42).

The earliest attempts at debittering were conducted in the late 1950s by Pritchett (45) and Swisher (46), who used lengthy selective extraction techniques. Years later in 1968, Chandler et al. (47) successfully used polyamides to adsorb significant quantities of limonin from Washington navel orange juice. Johnson and Chandler also tried cellulose acetate gels for limonin removal in orange juice (48), and cellulose acetate, various nylon-based materials, porous polymers, and ion-exchange resins to reduce bitterness and acidity in grapefruit juice (49). Variable success was achieved with each material and with certain combinations of materials. Cellulose acetate and two of the porous polymers were found to have powerful affinity for limonin. They were not as efficient as the styrene-divinyl benzene resins now being used, however. A comprehensive article by Johnson and Chandler (50) discusses the results of the Australian program from 1968 to 1988 in the field of adsorptive removal of bitter principals (limonin and naringin) from citrus juices.

Konno et al. (51,52) used a soluble β-cyclodextrin monomer to reduce the bitter taste of limonin and naringin in the juices of grapefruit, Iyo orange, and *Citrus natsudaidai*. Using this information, Shaw et al. (53) were able to utilize insoluble β-cyclodextrin polymer to remove limonin and naringin from aqueous solution and from filtered orange and grapefruit juices, employing both batch and continuous-flow column treatments. The commercial possibilities were enhanced with the successful use of β-cyclodextrin polymer in pilot-scale experiments (54).

Barmore et al. (55) showed that Florisil (activated magnesium silicate) added to grapefruit juice acts to significantly reduce limonin, naringin, narirutin, and total acid without reducing vitamin C, sugars, or flavor. In organoleptic testing, the flavor of Florisil-treated (30–60 g/L) juice was improved significantly ($P < 0.01$).

Utilizing a different approach, Hasegawa et al. (56) utilized an enzyme, limonoate dehydrogenase (no EC number assigned) of *Arthrobacter globiformis*, to prevent the development of, or removal of, limonin bitterness in citrus juice. They went on to develop a process using immobilized *A. globiformis* cells (57) and *Corynebacterium fascians* (58) where the enzyme system in the immobilized cells actively converted 80 to 85% of the limonin to nonbitter products. Maier et al. (59) used brief treatments of fruits for 3 h with 20 μg/mL ethylene to accelerate limonoid metabolism in navel oranges, lemons, and grapefruit to reduce bitterness in the juice expressed from the fruit.

Kimball (60) successfully removed an average of 25% of the limonin from California Washington navel orange juice using carbon dioxide at pressures between 21 and 41 MPa and temperatures between 30 and 60°C for 1 h. A 60% reduction in limonin was effected using 4-h runs. Unfortunately no mention was made of the effect of the process on flavor, and apparently no other citrus juices, such as grapefruit juices, were studied. Tamaki et al. (61) demonstrated that pressure treatments alone at 195 MPa for 10 m could reduce limonin content by 35 to 55% in orange and grapefruit juices. Acidity, brix, and vitamin C content were not affected. High-pressure treatments are expensive, however.

A recent process by Van Eikeren et al. (62) utilizes liquid/liquid/liquid extraction with a thin liquid membrane film immobilized and supported by a hydrophobic polymer membrane. Limonin and naringin in citrus juice are extracted by the thin immobilized liquid, diffuse across the liquid film and into a basic stripping liquid (pH 12–13) where the limonin and naringin are ionized and cannot diffuse back across the membrane into the citrus juice. Limonin reduction was demonstrated to be 85% in orange juice with no efficiency listed for naringin. Development of this new process on a commercial scale has not yet been accomplished.

SUMMARY, CONCLUSIONS, AND RECOMMENDATIONS

Limonin has proved to be a consistent quality factor in grapefruit juice acceptance. Apparently, for grapefruit juice users its presence in moderate amounts contributes to the overall zesty flavor that characterizes this juice and is liked by this consumer group. However, even for users, excessive amounts of limonin results in reduced acceptance. For nonusers of grapefruit juice, limonin, with its perceived bitterness-tartness, is a key factor in grapefruit juice flavor leading to low acceptance scores. Research efforts should be continued to improve ways to modulate the negative effect of bitter limonin on consumer acceptance of grapefruit juice.

Limonin levels currently are only regulated in processed Florida grapefruit concentrate by the State of Florida Department of Citrus (21) for early season fruit processed harvested prior to December 1. Limonin levels can still be rather high in December and January, thus an extension of limonin caps for one or preferably two months would help ensure against products with excessive limonin (bitterness) being packed. There would also be another advantage in that juice having other flavor problems associated with maturity in the early season and high limonin values, such as green, immature flavor that are only measured subjectively, would also be reduced.

It is likely that the public would be very receptive to an alternative to current regular grapefruit juices; namely, a natural grapefruit juice low in limonin (<3 μg/mL by HPLC) and naringin (300–350 by Davis Test), and having a natural Brix:acid ratio > 9.5. This would likely increase the market to individuals who find moderate levels of bitterness in grapefruit unacceptable. With the availability of commercial debittering capability, there is no reason why such a product could not be offered along with regular grapefruit juice. The product could carry special identification and likely a premium price. But it should serve to satisfy the taste preferences of a significant number of people who currently are nonusers of grapefruit juice. It is also likely to satisfy some established grapefruit juice users who prefer lower bitterness levels in the grapefruit juice that is available later in the season when fruit is more mature, less tart, and less bitter.

ACKNOWLEDGMENTS

This article has been adapted and updated from P. J. Fellers, "A Review of Limonin in Grapefruit (*Citrus paradisi*) Juice, Its Relationship to Flavour, and Efforts to Reduce It," *Journal of the Science of Food and Agriculture* **49**, 389–404 (1989). © 1989 Society of Chemical Industry.

BIBLIOGRAPHY

1. United States Department of Agriculture, *Consumers' Use of and Opinions About Citrus Products*, Bureau of Agricultural Economics, Division of Special Surveys, USDA, Washington, D.C., 1950.
2. United States Department of Agriculture, Market Research Report No. **243**, Agricultural Marketing Service, Market Research Division, USDA, Washington, D.C., 1958.
3. R. E. Branson, C. Price, and H. V. Courtenay, *Consumer Panel Tests of Texas Fortified Red Grapefruit Juice*, Agric and Mech Coll Texas, Texas Agricultural Experimental Station, College Station, Tex., 1961.
4. S. V. Ting and J. W. McAllister, "Quality of Florida Canned Grapefruit Juice in Supermarket Stores of the United States," *Proceedings of the Florida State Horticultural Society* **90**, 170–172 (1977).
5. S. M. Barros et al., "Interrelationships of °Brix, Brix:Acid Ratio, Naringin and Limonin and Their Effect on Flavor of Commercial, Canned, Single-Strength Grapefruit Juice," *Proceedings of the Florida State Horticultural Society* **96**, 316–318 (1983).
6. P. J. Fellers, R. D. Carter, and G. de Jager, "Influence of Limonin on Consumer Preference of Processed Grapefruit Juice," *J. Food Sci.* **52**, 714–743, 746 (1987).
7. State of Florida Department of Citrus, *Orange and Grapefruit Consumer Profile Report Dec 1, 1982–Dec 30, 1983*, Market Research Report, State of Florida, Department of Citrus, Lakeland, Fla., 1984.
8. R. L. Rouseff, W. E. Lee III, and C. A. Huefner, "Time-Intensity Studies of Citrus Bitter Compounds," in G. Charalambous, ed., *Flavors and Off-Flavors '89: Proceedings of the 6th International Flavor Conference, Rethymnon, Crete, Greece 5–7 July 1989*, Elsevier, New York, 1990, pp. 213–223.
9. R. Higby, "Bitter Constituents of Navel and Valencia Oranges," *J. Am. Chem. Soc.* **60**, 3013–3018 (1938).
10. D. H. R. Barton et al., "Triterpenoids. Part XXV. The Constitutions of Limonin and Related Bitter Principles," *J. Chem. Soc.* 255–275 (1961).
11. V. P. Maier and D. L. Dryer, "Citrus Bitter Principles IV. Occurrence of Limonin in Grapefruit Juice," *J. Food Sci.* **30**, 874–875 (1965).
12. V. P. Maier and D. A. Margileth, "Limonoic Acid A-Ring Lactone, A New Limonin Derivative in *Citrus*," *Phytochemistry* **8**, 243–248 (1969).
13. V. P. Maier et al., "Limonin and Limonoids: Chemistry, Biochemistry, and Juice Bitterness," in S. Nagy and J. Attaway, eds., *Citrus Nutrition and Quality*, American Chemical Society, Washington, D.C., 1980, pp. 63–82.
14. V. P. Maier and E. R. Grant, "Specific Thin Layer Chromatography Assay of Limonin, A Citrus Bitter Principle," *J. Agric. Food Chem.* **18**, 250–252 (1970).
15. W. C. Scott, "Limonin in Florida Citrus Fruits," *Proceedings of the Florida State Horticultural Society* **83**, 270–277 (1970).
16. J. H. Tatum and R. E. Berry, "Method for Estimating Limonin Content of Citrus Juices," *J. Food Sci.* **38**, 1244–1246 (1973).
17. J. F. Fisher, "Quantitative Determination of Limonin in Grapefruit Juice by High-Pressure Liquid Chromatography," *J. Agric. Food Chem.* **23**, 1199–1200 (1975).
18. R. L. Rouseff and J. F. Fisher, "Determination of Limonin and Related Limonoids by High Performance Liquid Chromatography," *Anal. Chem.* **52**, 1228–1233 (1980).
19. P. E. Shaw and C. W. Wilson, "Use of a Simple Solvent Optimization Program to Improve Separation of Limonin in Citrus Juices," *J. Chromatogr. Sci.* **24**, 364–366 (1986).
20. P. E. Shaw and C. W. Wilson, "Quantitative Determination of Limonin in Citrus Juices by HPLC Using Computerized Solvent Optimization," *J. Chromatogr. Sci.* **26**, 478–481 (1988).
21. Van Beek and A. Blaakmeer, "Determination of Limonin in Grapefruit and Other Citrus Juices By High Performance Liquid Chromatography," *J. Chromatogr.* **464**, 375–386 (1989).
22. W. W. Widmer, "Improvements in the Quantitation of Limonin in Citrus Juice by Reversed Phase High Performance Liquid Chromatography," *J. Agric. Food Chem.* **39**, 1472–1476 (1991).
23. W. W. Widmer and S. F. Martin, "Analysis of Limonin in Citrus Juices by Direct Injection and On-Line Sample Clean-Up," *Proceedings of the Forty-Fifth Annual Citrus Processors' Meeting* **45**, 21–24 (1994).
24. R. L. Mansell and E. W. Weiler, "Radioimmunoassay for the Determination of Limonin in Citrus," *Phytochemistry* **19**, 1403–1407 (1980).
25. P. S. Jourdan et al., "Competitive Solid Phase Enzyme-Linked Immunoassay for the Quantification of Limonin in Citrus," *Anal. Biochem.* **138**, 19–24 (1984).
26. *Limonin Standards*, Bitterdetek kit, Idetek, Inc., San Bruno, Calif., 1986.
27. B. A. Carter, D. G. Oliver, and L. Jang, "A Comparison of Methods for Determining the Limonin Content of Processed California Navel Orange Juice," *Food Technol.* **39**, 82–86, 97 (1985).
28. W. W. Widmer and R. L. Rouseff, "Quantitative Analysis of Limonin in Grapefruit Juice Using an Enzyme-Linked Immunoassay: Interlaboratory Study," *J. Assoc. Off. Anal. Chem.* **74**, 513–515 (1991).
29. D. G. Guadagni, V. P. Maier, and J. G. Turnbaugh, "Effect of Some Citrus Juice Constituents on Taste Thresholds for Limonin and Naringin Bitterness," *J. Sci. Food Agric.* **24**, 1277–1288 (1973).

30. D. G. Guadagni, V. P. Maier, and J. G. Turnbaugh, "Effect of Subthreshold Concentrations of Limonin, Naringin and Sweeteners on Bitterness Perception", *J. Sci. Food Agric.* **25**, 1349–1354 (1974).
31. State of Florida Department of Citrus, *Official Rules Affecting the Florida Citrus Industry*, State of Florida Department of Citrus, Lakeland, Fla., 1982.
32. U.S. Food and Drug Administration, "Grapefruit Juice," *21 CFR 146.132*, Office of the Federal Register National Archives and Records Administration, Washington, D. C., 1992.
33. R. L. Mansell and C. A. McIntosh, "An Analysis of Limonin, Brix and Acid Content in Grapefruit Samples Collected from Three State Test Houses," *Proceedings of the Florida State Horticultural Society* **93**, 289–293 (1980).
34. M. H. Dougherty and J. F. Fisher, "Quality of Commercial, Canned, Single-Strength Grapefruit Juice Produced in Florida During the 1975–76 and 1976–77 Citrus Seasons," *Proceedings of the Florida State Horticultural Society* **90**, 168–170 (1977).
35. J. H. Tatum, J. C. Lastinger, Jr., and R. E. Berry, "Naringin Isomers and Limonin in Canned Florida Grapefruit Juice," *Proceedings of the Florida State Horticulture Society* **85**, 210–213 (1972).
36. M. H. Dougherty et al., "Grapefruit Juice Quality Improvement Studies, Introduction and Scope of the Study; and the Effect of Processing Variables, Temperature and Duration of Storage on the Quality of Grapefruit Juice," *Proceedings of the Florida State Horticultural Society* **90**, 165–167 (1977).
37. S. Hasagawa, R. D. Bennett, and C. P. Verdon, "Limonoids in Citrus Seeds: Origin and Relative Concentration," *J. Agric. Food Chem.* **28**, 922–925 (1980).
38. S. Hasagawa et al., "Limonoid Glucosides in Citrus," *Phytochemistry* **28**, 1717–1720 (1989).
39. R. L. Mansell and C. A. McIntosh, "Three Dimensional Distribution of Limonin, Limonoate A-Ring Lactone, and Naringin in Fruit From Three Grapefruit Cultivars," *Proceedings of the Thirty-Fifth Annual Citrus Processors' Meeting* **35**, 12–14 (1984).
40. P. J. Fellers et al., "Quality of Florida-Packed Retail Grapefruit Juices as Determined by Consumer Sensory Panels and Chemical and Physical Analyses," *J. Food Sci.* **51**, 417–420 (1986).
41. M. Wethern, "Citrus Debittering With Ultrafiltration/Adsorption Combined Technology," in *Transactions of the 1991 Citrus Engineering Conference*, American Society of Mechanical Engineers, 1991, Lakeland, Fla., pp. 48–65.
42. U.S. Pat. 5,817,354 (October 6, 1998), Z. Mozaffar, Q. R. Miranda, and V. Saxena (to Sepragen Corp., Inc.).
43. U.S. Pat. 4,439,458 (March 27, 1984), A. Puri (to Coca Cola Co.).
44. U.S. Pat. 4,965,083 (October 23, 1990), S. I. Norman, R. T. Stringfield, and C. C. Gopsill (to Dow Chemical Co.).
45. U.S. Pat. 2,816,033 (Dec. 10, 1957), D. E. Pritchett.
46. U.S. Pat. 2,834,687 (May 13, 1958), H. E. Swisher.
47. B. V. Chandler, J. F. Kefford, and G. Ziemelis, "Removal of Limonin From Bitter Orange Juice," *J. Sci. Food Agric.* **19**, 83–86 (1968).
48. B. V. Chandler and R. L. Johnson, "Cellulose Acetate as a Selective Sorbent for Limonin in Orange Juice," *J. Sci. Food Agric.* **28**, 875–884 (1977).
49. R. L. Johnson and B. V. Chandler, "Reduction of Bitterness and Acidity in Grapefruit Juice by Adsorptive Processes," *J. Sci. Food Agric.* **33**, 198–293 (1982).
50. R. L. Johnson and B. V. Chandler, "Adsorptive Removal of Bitter Principles and Titratable Acid From Citrus Juices," *Food Technol.* **42**, 130–137 (1988).
51. A. Konno et al., "Bitterness Reduction of Citrus Fruits by β-Cyclodextrin." *Agric. Biol. Chem.* **45**, 2341–2342 (1981).
52. A. Konno et al., "Bitterness Reduction of Naringin and Limonin by β-Cyclodextrin," *Agric. Biol. Chem.* **46**, 2203–2208 (1982).
53. P. E. Shaw, J. H. Tatum, and C. W. Wilson III, "Improved Flavor of Navel Orange and Grapefruit Juices by Removal of Bitter Components with β-Cyclodextrin Polymer," *J. Agric. Food Chem.* **32**, 832–836 (1984).
54. C. J. Wagner, C. W. Wilson III, and P. E. Shaw, "Reduction of Grapefruit Bitter Components in a Fluidized β-Cyclodextrin Polymer Bed," *J. Food Sci.* **53**, 516–518 (1988).
55. C. R. Barmore et al., "Reduction of Bitterness and Tartness in Grapefruit Juice With Florisil," *J. Food Sci.* **51**, 415–416, 439 (1986).
56. S. Hasegawa, L. C. Brewster, and V. P. Maier, "Use of Limonoate Dehydrogenase of *Arthrobacter globiformis* for the Prevention or Removal of Limonin Bitterness in Citrus Products," *J. Food Sci.* **38**, 1153–1155 (1973).
57. S. Hasegawa, V. A. Pelton, and R. D. Bennett, "Metabolism of Limonoids by *Arthrobacter globiformis II*: Basis for a Practical Means of Reducing the Limonin Content of Orange Juice by Immobilized Cells," *J. Agric. Food Chem.* **31**, 1002–1004 (1983).
58. U.S. Pat. 4,447,456 (May 8, 1984), S. Hasegawa.
59. V. P. Maier, L. C. Brewster, and A. C. Hsu, "Ethylene-Accelerated Limonoid Metabolism in Citrus Fruits: A Process for Reducing Juice Bitterness," *J. Agric. Food Chem.* **21**, 490–495 (1973).
60. D. A. Kimball, "Debittering of Citrus Juices Using Supercritical Carbon Dioxide," *J. Food Sci.* **52**, 481–582 (1987).
61. U.S. Pat. 5,049,402 (September 17, 1991), Y. Tamaki, O. Mutsushika, and H. Mieda (to Pokka Corp.).
62. U.S. Pat. 5,263,409 (November 23, 1993), P. van Eikeren and D. J. Brose (to Bend Research, Inc.).

W. W. Widmer
P. J. Fellers
State of Florida Department of Citrus
Lake Alfred, Florida

LIPIDS: NUTRITION

Fat, carbohydrate, and protein are the three macronutrients found in the diet that provide energy to sustain life. Each macronutrient class is comprised of many diverse types of constituents that have important biological functions. With respect to fat, there are many compounds that vary widely in structure and function. This chemical heterogeneity is important for numerous biological processes and contributes to sensory and functional characteristics of foods. Fats, commonly referred to as lipids, are found in animals or plants and are generally insoluble in water and soluble in organic solvents. There are three major classes of lipids: the neutral lipids, the glycerophosphatides, and the sphingolipids. The neutral lipids, referred to as triglycerides, are comprised of glycerol (a 3-carbon molecule) and three fatty acids. There are short-chain (≤6-carbon atoms), medium-chain (8- to 12-carbon atoms), and long-chain (>12- to 22-carbon atoms) fatty acids. The long-chain fatty acids predominate in the diet. Two or three different fatty acids are typically present in a triglyceride molecule. Because of the diversity of fatty acids found in nature, fats are a mixture of different triglycerides.

Chemically, fatty acids are categorized as either saturated or unsaturated. Saturated fatty acids (SFA) contain no double bonds, whereas the unsaturated fatty acids contain one or more double bonds varying in isomeric configuration. Fatty acids with one double bond are referred to as monounsaturated fatty acids (MUFA) and those with two or more double bonds are called polyunsaturated fatty acids (PUFA). With regard to MUFA and PUFA, further distinctions are made on the basis of the position of the first double bond from the CH_3 terminus (yielding the designations omega-9, omega-6, or omega-3) and the confirmation of the double bond (yielding the designations *cis* or *trans*). In *cis* fatty acids the hydrogen atoms are located on the same side of the double bond, whereas in *trans* fatty acids the hydrogen atoms are located on opposite sides of the double bond. The chain length of a fatty acid as well as the number and type (ie, *cis* or *trans*) of double bonds in the molecule determine its physical, chemical, and biochemical properties.

Fat is an essential nutrient in the diet as are protein and carbohydrate. A unique characteristic of fat is its energy density. Fat supplies more than twice as much energy as protein or carbohydrate on a mass basis (9 kcal/g versus 4 kcal/g) and provides approximately one-third of dietary energy. In addition, fat provides essential fatty acids that are required for normal growth and development. Fat also promotes absorption of important fat-soluble vitamins such as vitamins A, D, E, and K. Moreover, fat contributes to satiety after consumption of food and delays the onset of hunger after a meal due to slowing of gastric emptying. Fat also contributes importantly to sensory attributes of foods. For example, it contributes to the unique textures of certain foods (ie, flakiness of pastries, creamy mouthfeel of desserts and salad dressings, crunchy texture of snack foods and cookies) and imparts distinctive flavors to foods.

Fat has numerous biological, physical, and chemical properties. With respect to its biological effects, much work has been done to elucidate how they affect various physiological processes that affect risk of the major chronic diseases observed in developed countries. The ensuing discussion reviews our present understanding of the beneficial and adverse effects that total fat, fat classes, and individual fatty acids have on risk and incidence of obesity, cardiovascular disease (CVD), and cancer.

FAT EFFECTS ON HEALTH

Impressive strides have been made in increasing the understanding of how total fat, fat classes, and individual fatty acids influence risk of chronic diseases. The diversity of effects of total fat, fat class, and individual fatty acids on risk of chronic disease is remarkable. Because of this, when discussing the role of fat in chronic disease, both fat quantity and quality must be considered. Importantly, the type and amount of fat can affect risk of chronic diseases in different ways.

Total Fat

Several issues are being debated about the ideal quantity of total fat in the diet. The first relates to the amount of total fat that most favorably affects risk of CVD. Second, there is still uncertainty about the relationship between the quantity of fat in the diet and weight control. In addition, there are questions about the relationship between fat intake and risk of certain cancers.

Inherent to the discussion about how much total fat should be in the diet is that saturated fat and cholesterol are low (ie, <10% of calories and <300 mg/day, respectively). The key question that follows is whether MUFA or carbohydrate should replace SFA calories. Scientists who favor a high MUFA diet (1) contend that replacing SFA with carbohydrate leads to a decrease in high-density lipoprotein (HDL) cholesterol and an increase in plasma triglyceride levels, both of which increase risk of CVD. In contrast, when MUFA replaces SFA, plasma triglyceride levels are not increased and HDL cholesterol is not decreased compared with a high-carbohydrate diet. Thus, advocates of a high-MUFA, low-SFA diet believe that it results in a more favorable overall CVD risk profile than a low-SFA, high-carbohydrate diet. Advocates of a high-carbohydrate, low SFA diet, however, argue that this diet will result in weight loss because fewer calories are consumed and, hence, elicit a beneficial effect on HDL cholesterol and plasma triglyceride levels (2). Intertwined in the debate is the question of whether a high-MUFA, low-SFA diet (a higher-fat diet) leads to an increase in energy consumption, thereby leading to weight gain resulting in overweight/obesity. Scientists who favor the position that increasing dietary fat leads to weight gain because of the higher caloric value of fat (3) maintain that calorie control is not regulated sufficiently, leading to overconsumption of energy. On the other hand, well-respected scientists oppose this conclusion and cite epidemiologic evidence that shows little association between total fat intake and the incidence of overweight/obesity (4). Furthermore, fat may play an important role in satiety and, thus, help control calorie intake and body weight. Irrespective of the ongoing discussion, the fact remains that control of energy intake, regardless of macronutrient source, will result in achieving and maintaining a healthy body weight (5).

Some epidemiologic evidence supports a relationship between dietary fat and the incidence of breast, prostate, colon, and lung cancers (6). However, because of the difficulties in evaluating this association due to limitations in assessment methodologies and confounding variables such as total energy and micronutrient intake as well as physiological and environmental factors (such as age at menarche and level of physical activity) the results are discrepant. Thus, with respect to total fat it is not possible to reach any meaningful conclusion about the relationship between total fat intake and risk of certain cancers. It is important to note, however, that there is a larger database for individual fatty acids and risk of certain cancers (discussed in the following).

As discussed, total fat is a mixture of individual fatty acids. Likewise, different food sources of fat differ markedly in their fatty acid profile. For example, canola oil, a fat that is very low in saturated fat, is comprised of more than 90% unsaturated fat, whereas coconut oil contains more than 95% saturated fat! The other fats in our food supply contain, in general, a prominent fatty acid class but also contain varying amounts of the other fatty acid classes. For instance, soybean oil contains about 60% PUFA and about 18% SFA and 22% MUFA. In contrast, lard contains approximately 42% SFA, 44% MUFA, and the remainder is other fatty acids including PUFA. Because of the considerable variation in fatty acid composition of fats found in the diet, scientists have begun to investigate how individual fatty acids affect risk factors for chronic diseases.

Saturated Fatty Acids

The Seven Countries Study (7) was a landmark epidemiologic investigation that played a seminal role in establishing a relationship between diet and the incidence of coronary heart disease (CHD). Moreover, this study also provided evidence that diet affected serum cholesterol levels and that an elevation in cholesterol increased risk of coronary disease. This marked the beginning of the diet-heart hypothesis era, a time during which numerous studies were conducted to examine the effects of different dietary factors on risk factors for coronary disease. Many of these studies evaluated the relationship between the type of fat in the diet and serum cholesterol levels.

An important finding of the Seven Countries Study was that saturated fat intake (as a percentage of calories) was significantly correlated with serum cholesterol levels; 80% of the variability was due to differences in dietary SFA among the populations. Moreover, SFA intake was also correlated with five-year incidence of CHD. This finding has been confirmed in numerous subsequent studies. This epidemiologic evidence provided the impetus for an era in which carefully controlled clinical studies were conducted initially to evaluate the effects of fat classes and then to assess the impact of individual fatty acids on plasma lipids and lipoproteins. The results from many of the well-controlled clinical studies were used to develop blood cholesterol predictive equations for estimating the changes in total cholesterol (TC) in response to changes in type of fat and amount of dietary cholesterol. The original equations developed by Keys et al. (8) and Hegsted et al. (9) demonstrated that saturated fat was markedly hypercholesterolemic, whereas PUFA lowered blood cholesterol levels. Saturated fat was found to be twice as potent in raising blood cholesterol levels as PUFA were in lowering them. Both groups of investigators reported that MUFA had a neutral effect and that dietary cholesterol raised the blood cholesterol level but less so than saturated fat. More recently, predictive equations have been developed for low-density lipoprotein (LDL) and HDL cholesterol. The LDL cholesterol response mimics that for total cholesterol. All fatty acid classes and dietary cholesterol increase HDL cholesterol.

These studies have been followed by investigations that evaluated the effects of individual fatty acids on plasma lipids and lipoproteins. A recent summary of the literature evaluating the effects of individual fatty acids is shown in Figure 1 (10). It is evident that the effects observed are quite divergent when comparisons are made among the different fatty acids and even within a fatty acid class. For example, when comparisons are made among the long-chain SFAs, it is apparent that there are pronounced differences in potency. Specifically, myristic acid (C14:0) is twice as potent as lauric acid (C12:0) in raising total and LDL cholesterol. However, stearic acid (C18:0) is uniquely different; it has either a neutral or slight cholesterol-lowering effect. The most potent cholesterol-lowering fatty acid is linoleic acid (C18:2). Oleic acid has effects that are intermediate to those of linoleic acid and the cholesterol-raising SFA. The *trans* isomer of oleic acid has effects that are quite different from the *cis* isomer; *trans* 18:1 raises serum total and LDL cholesterol and may decrease HDL cholesterol.

In contrast to the extensive literature on fat classes and individual fatty acids on blood lipid/lipoprotein responses, much less is known about the effects of dietary fat on thrombosis. Some epidemiologic evidence from the Atherosclerosis Risk in Communities (ARIC) Study indicates that a high intake of total fat, SFA, and cholesterol is associated with higher levels of Factor VII and fibrinogen, two hemostatic factors that play a role in blood clot formation. Likewise, in the Dietary Effects on Lipoproteins and Thrombogenic Activity (DELTA) Study, a well-controlled multicenter feeding study, a reduction in saturated fat decreased Factor VII. However, a reduction in total fat increased fibrinogen levels. It is important to note that the magnitude of response for both was modest (ie, 2–3%).

There has been great debate about the association between fat intake and breast cancer. However, a pooled analysis of seven major cohort studies from four countries did not find any association between total fat intake and incidence of breast cancer (11). Population studies have shown a relationship between SFA intake, especially from animal products, and risk of colorectal adenomas. In addition, there is some evidence that SFA increases risk of ovarian cancer (12) and prostate cancer (13).

Unsaturated Fatty Acids

Monounsaturated Fatty Acids. MUFAs are a unique class of fatty acids that provides great flexibility in diet planning. They can be used to replace SFAs or carbohydrate calories or both. Depending on the substitution made, there can be a variable change in the total fat content of the diet, varying from essentially no or little change to an approximate twofold increase. The impact of these scenarios has already been discussed. In brief, however, diets high in MUFA (that are low in SFA and cholesterol) will lower total and LDL cholesterol and plasma triglycerides and minimize any potential decrease in HDL cholesterol levels (14). In individuals with diabetes, MUFAs improve the glycemic profile (plasma glucose and insulin levels) (15). There is some evidence that MUFAs may decrease susceptibility of LDL particles to oxidative modification (which is an important initiating event in the development of atherosclerosis), thereby reducing their atherogenic potential (16).

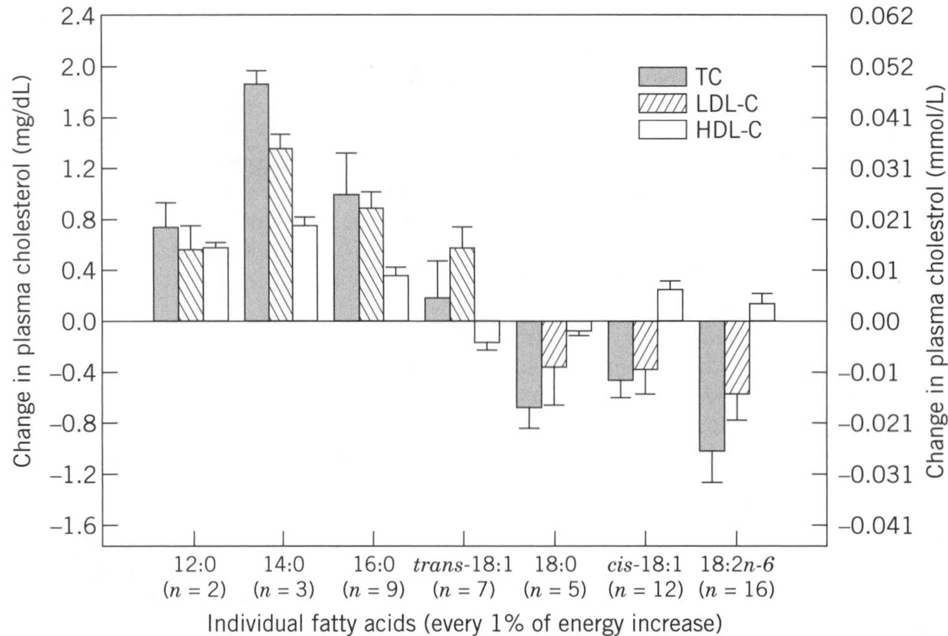

Figure 1. The effects of individual fatty acids compared with oleic acid. TC, plasma total cholesterol; LDL-C, low-density lipoprotein cholesterol; and HDL-C, high-density lipoprotein cholesterol. The results reported for oleic acid are based on comparisons with carbohydrate. The results reported for the other fatty acids are compared with oleic acid and carbohydrate.

***Trans* Fatty Acids.** Several epidemiologic studies have examined the relationship between *trans* fatty acid intake and both incidence of coronary disease and risk factors for CHD, notably lipids and lipoproteins (reviewed in Ref. 17). Evidence from some of these studies, but not all, has suggested a positive relationship. There are a number of limitations, however, of epidemiologic studies that prevent making cause-and-effect conclusions. These include the challenges associated with accurately assessing *trans* fatty acid intake due to limitations in diet assessment methodologies and an incomplete nutrient database for *trans* fatty acids. Moreover, there is a question of how accurately measures of recent food intake approximate long-term food intake, the latter of which is important relative to the chronic nature of the CHD. Despite the limitations of the epidemiologic studies conducted to date, enough have reported a positive association between *trans* fatty acid intake and CVD risk to warrant further evaluation in controlled clinical studies.

From a historical perspective, numerous early controlled feeding studies demonstrated that hydrogenated fats elicited a blood cholesterol response that was intermediate to that observed for unhydrogenated oils and saturated fats (reviewed in Ref. 18). More recently, studies have been conducted to evaluate the plasma lipid and lipoprotein effects of *trans* fatty acids. These studies have consistently shown that *trans* fatty acids increase plasma total and LDL cholesterol relative to unsaturated fatty acids (reviewed in Ref. 18). Compared with SFAs *trans* fatty acids elicit a similar or slightly lower cholesterol-raising effect (reviewed in Ref. 18). In addition, *trans* fatty acids lower HDL cholesterol, resulting in a worsening of the LDL:HDL ratio, which in turn increases CHD risk.

Polyunsaturated Fatty Acids ***Omega-6 Fatty Acids.*** The evidence that certain SFAs raise blood cholesterol levels whereas PUFAs lower them was a justification for several major diet trials designed to lower CHD risk by feeding a high-fat diet (about 40% of calories) that was low in SFA and very high in PUFA (16 to 20% of calories; Oslo Diet Heart Study, 19; VA Diet Heart Study, 20; Finnish Mental Hospital Study, 21). These studies all showed that this diet lowered CHD risk by 24 to 53% concomitant with a reduction in serum cholesterol of 13 to 15%. Because of the perception that these diets might be associated with a higher incidence of cancer and also because this diet is not consumed by population groups worldwide to meaningfully assess its safety, it has not widely been recommended for the prevention and treatment of CHD.

Recent studies have reported that while PUFA has a greater total and LDL cholesterol-lowering effect versus MUFA, the differences are not as great as once believed. Thus, for practical purposes, MUFA or PUFA will elicit effects that are quite similar when incorporated in a diet that meets current recommendations for total fat (≤30% of calories) and PUFA (<10% of calories). Some experts advocate, however, that PUFA not exceed approximately 7% of total energy (22). Collectively, the results of many studies indicate that while high levels of dietary PUFA seem to beneficially affect CHD, within the range of recommended intake, PUFA and MUFA have effects on plasma lipids and lipoproteins that are quite similar. Thus, either can be incorporated into the diet resulting in comparable effects.

Although some of the early studies suggested that high intakes of PUFA were associated with an increased risk of certain cancers, a recent meta-analysis of the literature concluded that a high intake of PUFA did not increase the risk of various cancers (23). This analysis did not rule out the possibility of a small increase in risk, however.

Omega-3 Fatty Acids. The early 1970s marked the beginning of a new era of investigation into the role of omega-3 fatty acids in the development of CVD. The seminal studies of Dyerberg et al. (24) showed that serum cholesterol and β-lipoprotein levels were lower in Greenland Eskimos compared with Danish subjects who consumed a high SFA diet. Interestingly, coronary atherosclerotic disease was rare in the Eskimos and prevalent in the Danes. These scientists attributed this difference in the incidence of CHD to the high intake of marine oils and in particular, two predominant constituent highly unsaturated fatty acids, eicosapentaenoic acid (EPA, C20:5) and docosahexaenoic acid (DHA, C22:6). During the past 30 years numerous studies have demonstrated that these fatty acids have diverse biological effects. Collectively, they confer impressive cardioprotective effects via multiple mechanisms that involve antiarrhythmic actions, sudden death, thrombosis, growth of atherosclerotic plaques, lipids, and lipoproteins.

Restenosis is the disproportionate activation of one or several normal wound-healing responses that occur as the result of coronary artery bypass surgery or other invasive procedures. The clinical significance of this is that while angioplasty may initially increase lumen size by 20 to 40% and restore blood flow, a significant number of individuals (approximately 30 to 40%) undergoing the procedure will experience restenosis leading to luminal narrowing within 3 to 6 months, which requires another intervention procedure. Meta-analyses established that overall, restenosis was reduced by 14 to 29% by supplemental fish oils (25). Interestingly, when angiography was used to define coronary restenosis, there was a direct relationship between dose of omega-3 fatty acids and the reduction in restenosis. A recent large clinical trial, however, failed to find a beneficial effect of fish oil on restenosis after coronary angioplasty (26).

Impressive evidence indicates that omega-3 fatty acids reduce sudden death (27,28). In the initial report of the Lyon Diet Heart Trial (27), patients who had suffered a myocardial infarction (MI) followed an American Heart Association (AHA) Step 1 "Mediterranean-type" diet rich in linolenic acid versus a prudent Western-type diet. There was a striking reduction in coronary events and, in particular, no sudden deaths in the treatment group (versus 8 in the control group) despite no improvement in lipids, lipoproteins, and body mass index (BMI). These results have been corroborated in a population-based case-control study (28). In this study, as little as one fatty fish meal per week (ie, 1.4 g of omega-3 fatty acids per week) decreased the risk of primary cardiac arrest by 50%. More recently, the final report from the Lyon Diet Heart Trial (29) found that a Mediterranean dietary pattern (high in linolenic acid) reduced all cardiac death and nonfatal MI by approximately 70% and all coronary events measured by about 50%.

Fatty fish consumption has been shown to prevent cardiac arrest from ventricular fibrillation (28). In addition, omega-3 fatty acids have been shown to favorably affect susceptibility to premature ventricular contractions (30). These data are significant because they suggest that fish intake affects risk of CHD via events that are distinct from the effects of fish consumption on plasma lipids.

There is convincing evidence that a major beneficial effect of omega-3 fatty acids is on the prevention of thrombosis (31). Omega-3 fatty acids reduce platelet aggregation (including reactivity and adhesion) and vasoconstriction. In addition, omega-3 fatty acids favorably affect hemostasis. These effects are the result of enhanced fibrinolysis and reduced blood clot formation.

Fish oil has a marked hypotriglyceridemic effect in both normotriglyceridemic and hypertriglyceridemic ($\geq$2 mmol/L) individuals. The addition of approximately 9 to 13 g/day of fish oil (eg, 1.1 to 7 g/day of omega-3 fatty acids) resulted in a plasma triglyceride decrease of about 20 to 25% in normotriglyceridemic individuals and a decrease of about 26 to 33% in hypertriglyceridemic individuals. These levels of fish oil have a modest LDL cholesterol-raising effect (eg, 4 to 5%) in normotriglyceridemic individuals. In comparison, LDL cholesterol is elevated approximately 5 to 11% in hypertriglyceridemic individuals, and even more so (30%) in some individuals with familial hyperlipidemia (Type IV/V) (32). Thus, fish oil supplements can be an effective treatment for some patients with hypertriglyceridemia, although close monitoring by a physician is essential to ensure that there is not a concurrent significant increase in LDL cholesterol.

Conjugated Linoleic Acid. Conjugated linoleic acid (CLA) is a collective term given to a group of linoleic acid isomers in which the double bonds are conjugated instead of being in the typical methylene interrupted configuration. CLA levels are higher in animal products than in plant products and, in general, CLA levels are higher in ruminant than nonruminant tissues. CLA has attracted great interest recently because of the evidence that it favorably affects several major chronic diseases, most notably cancer, obesity/overweight, and CVD. The evidence about the biological effects of CLA to date has originated from cell culture and experimental animal studies. Providing 0.1 to 1% CLA in the diet has been reported to suppress tumor incidence in models using carcinogens that require or do not require metabolic activation, suggesting that more than one metabolic mechanism may account for the anticarcinogenic properties of CLAs. At least one study has shown that CLA supplementation to normal human mammary cells or to MCF-7 human mammary tumor cells did not lead to increased intracellular lipid peroxidation, whereas linoleic acid did (33). Thus, it is possible that the effect of CLA relates to differences in the ability to enhance free radical formation.

With respect to CVD there is limited but interesting evidence that CLA may beneficially affect plasma lipids and lipoproteins as well as atherosclerosis in rabbits (34) and hamsters (35). In both studies, the CLA treatment groups had markedly lower ($\approx$20%) total and LDL cholesterol levels and triglyceride levels. HDL cholesterol levels were unaffected by CLA. Based on the study of Nicolosi et al. (35), CLA had more potent effects than linoleic acid. Collectively, these studies are highly suggestive that CLA may have a more potent cardioprotective effect beyond that of the predominant fatty acids in the diet. In addition, while CLA has antioxidant effects, little is known about whether CLA protects against oxidative modification of LDL.

FAT AND FATTY ACID INTAKE

Data from the 1995 USDA Continuing Survey of Food Intake of Individuals (CSFII) indicates that total fat consumption is 32 to 33% of calories for males and females 19 years of age and greater. Saturated fat intake is approximately 11%, MUFA is 12 to 13% and PUFA is 6 to 7% of calories. As shown in Table 1, palmitic acid is the most predominant SFA in the diet, accounting for 5 to 6% of calories. Likewise, oleic acid is the major MUFA, and linoleic acid is the most abundant PUFA in the diet. *Trans* fatty acids comprise approximately 2.6% of total energy intake. The intake of omega-3 fatty acids in the United States is approximately 1.6 g/day (about 0.7% of energy intake). α-Linoleic acid is the predominant omega-3 fatty acid in the diet, accounting for 1.4 g/day. Intake of EPA and DHA is quite low (only 0.1 to 0.2 g/day). Otherwise, intake of CLA is estimated to be approximately 0.2 g/day in participants of the Nationwide Food Consumption Survey.

Total fat intake has declined from approximately 36 to 37% of calories in 1987–1988 to 32 to 33% of calories in 1995, which represents an approximate 10% reduction. Likewise, SFA intake has decreased from 13% of calories to 11% of calories during this same period of time (36). Present intake of total and saturated fat is approaching current dietary recommendations (30% of calories from total fat and <10% of calories from SFA).

Table 1. Fat and Fatty Acid Intake in Adult U.S. Population (≥19 Years of Age)

Fat/fatty acid[a]	Men	Women
Total fat	*33.3*	*32.4*
Total SFA	*11.2*	*10.7*
C12:0	0.3	0.3
C14:0	0.9	0.9
C16:0	5.2	5.9
C18:0	2.9	2.8
Total MUFA	*12.9*	*12.2*
C18:1 *cis*	11.9	11.4
Total trans	*2.6*	
Total PUFA	*6.6*	*6.8*
C18:2	5.9	6.0
C18:3	0.6	0.6
EPA + DHA	*	*

[a] % of energy.

* Value of less than 0.05 but greater than 0.

SUMMARY

Impressive progress has been made in understanding the role of the amount and type of fat in the diet on health. There is general agreement that certain SFAs in the diet be reduced to lower risk of CVD. However, debate continues about what nutrient source should replace SFA calories in the diet. Some scientists advocate that SFA be replaced with carbohydrate, resulting in a high-carbohydrate, low-SFA diet. Others believe that MUFA is the preferable replacement for SFA calories, resulting in a diet higher in total fat. There is an increasing consensus that *trans* fatty acids be reduced in the diet to decrease CVD risk. In contrast, compelling emerging evidence suggests that omega-3 fatty acids elicit cardioprotective effects and that CLA may confer a protective effect against certain cancers. There is a growing sense that intake of omega-3 fatty acids needs to be increased. The small database for CLAs is exciting; however, further studies are needed to define their health benefits.

The evidence we have about the effects of the amount and type of fats on chronic disease risk has enabled us to develop contemporary dietary recommendations that reduce risk of these diseases. As additional evidence is accumulated it is not unreasonable to speculate that further changes in the amount and type of fat in the diet can be made that will further enhance health.

BIBLIOGRAPHY

1. M. B. Katan, S. M. Grundy, and W. C. Willet, "Should a Low-Fat, High Carbohydrate Diet Be Recommended for Everyone? Beyond Low-Fat Diets," *N. Engl. J. Med.* **337**, 563–566 (1997).
2. W. E. Connor and S. Connor, "Should a Low-Fat, High Carbohydrate Diet Be Recommended or Everyone? The Case for a Low-Fat, High-Carbohydrate Diet," *N. Engl. J. Med.* **337**, 562–563 (1997).
3. G. A. Bray and B. M. Popkin, "Dietary Fat Intake Does Affect Obesity!" *Am. J. Clin. Nutr.* **68**, 1157–1173 (1998).
4. W. C. Willett, "Dietary Fat and Obesity: An Unconvincing Relation," *Am. J. Clin. Nutr.* **68**, 1149–1150 (1998).
5. R. L. Leibel et al., "Energy Intake Required To Maintain Body Weight Is Not Affected by Wide Variation in Diet Composition," *Am. J. Clin. Nutr.* **55**, 350–355 (1992).
6. L. H. Kuller, "Dietary Fat and Chronic Diseases: Epidemiologic Overview," *Journal of the American Dietetic Association* **97** (Suppl), S9–S15 (1997).
7. A. Keys, "Coronary Heart Disease in Seven Countries," *American Heart Association Monograph*, No. 29, I-1–I-211 (1970).
8. A. Keys, J. T. Anderson, and F. Grande, "Serum Cholesterol Response to Changes in the Diet. IV. Particular Saturated Fatty Acids in the Diet," *Metabolism* **14**, 776–787 (1965).
9. D. M. Hegsted et al., "Quantitative Effects of Dietary Fat on Serum Cholesterol in Man," *Am. J. Clin. Nutr.* **17**, 281–295 (1965).
10. P. M. Kris-Etherton and S. Yu, "Individual Fatty Acid Effects on Plasma Lipids and Lipoproteins: Human Studies," *Am. J. Clin. Nutr.* **65** (Suppl.), 1628S–1644S (1997).

11. D. J. Hunter et al., "Cohort Studies of Fat Intake and the Risk of Breast Cancer—A Pooled Analysis," *N. Engl. J. Med.* **334**, 356–361 (1996).
12. H. A. Risch et al., "Dietary Fat Intake and Risk of Epithelial Ovarian Cancer," *J. Nat. Cancer Inst.* **86**, 1409–1415 (1994).
13. J-R. Zhou and G. L. Blackburn, "Bridging Animal and Human Studies: What Are the Missing Segments in Dietary Fat and Prostate Cancer," *Am. J. Clin. Nutr.* **66** (Suppl.), 1572S–1580S (1997).
14. R. P. Mensink and M. B. Katan, "Effect of a Diet Enriched With Monounsaturated or Polyunsaturated Fatty Acids on Levels of Low-Density and High-Density Lipoprotein Cholesterol in Healthy Women and Men," *N. Engl. J. Med.* **321**, 436–441 (1989).
15. A. Garg et al., "Comparison of a High-Carbohydrate Diet With a High-Monounsaturated-Fat Diet in Patients With Non-Insulin-Dependent Diabetes Mellitus," *N. Engl. J. Med.* **319**, 829–834 (1988).
16. P. D. Reaven et al., "Effects of Oleate-Enriched and Linoleate-Enriched Diets on the Susceptibility of Low Density Lipoprotein to Oxidative Modification in Hypercholesterolemic Subjects," *J. Clin. Invest.* **91**, 668–676 (1993).
17. A. H. Lichtenstein, "Trans Fatty Acids, Plasma Lipid Levels, and Risk of Developing Cardiovascular Disease: A Statement for Healthcare Professionals From the American Heart Association," *Circulation* **95**, 2588–2590 (1997).
18. M. A. Denke, "Serum Lipid Concentrations in Humans," *Am. J. Clin. Nutr.* **62**, 693S–700S (1995).
19. P. Lernen, "The Effect of Plasma Cholesterol Lowering Diet in Male Survivors of Myocardial Infarction. A Controlled Clinical Trial," *Acta Medica Scandinavica* (Suppl.) **466**, 1–92 (1966).
20. S. Dayton et al., "A Controlled Clinical Trial of a Diet High in Unsaturated Fat in Preventing Complications of Atherosclerosis," *Circulation* (Suppl.) **XL**, 1–63 (1969).
21. M. Miettinen et al., "Effect of Cholesterol-Lowering Diet on Mortality From Coronary Heart-disease and Other Causes. A Twelve Year Clinical Trial in Men and Women," *Lancet* **2**(7782), 835–838 (1972).
22. S. M. Grundy, "What Is the Desirable Ratio of Saturated, Polyunsaturated and Monounsaturated Fatty Acids in the Diet?" *Am. J. Clin. Nutr.* (Suppl.) **66**, 988S–990S (1997).
23. P. L. Zock and M. B. Katan, "Linoleic Acid Intake and Cancer Risk: A Review and Meta-Analysis," *Am. J. Clin. Nutr.* **68**, 142–153 (1998).
24. J. Dyerberg, H. O. Bang, and N. Hjørne, "Fatty Acid Composition of the Plasma Lipids in Greenland Eskimos," *Am. J. Clin. Nutr.* **28**, 958–966 (1975).
25. J. P. Gapinski et al., "Preventing Restenosis With Fish Oils Following a Coronary Angioplasty. A Meta-Analysis," *Archives of Internal Medicine* **153**, 1595–1601 (1993).
26. A. Leaf et al., "Do Fish Oils Prevent Restenosis After Coronary Angioplasty?" *Circulation* **90**, 2248–2257 (1994).
27. M. de Lorgeril et al., "Mediterranean Alpha-Linoleic Acid-Rich Diet in Secondary Prevention of Coronary Heart Disease," *Lancet* **343**, 1454–1459 (1994).
28. D. S. Siscovick et al., "Dietary Intake and Cell Membrane Levels of Long-Chain n-3 Polyunsaturated Fatty Acids and the Risk of Primary Cardiac Arrest," *JAMA—J. Am. Med. Assoc.* **274**, 1363–1367 (1995).
29. M. de Lorgeril et al., "Mediterranean Diet, Traditional Risk Factors, and the Rate of Cardiovascular Complications After Myocardial Infarction: Final Report of the Lyon Diet Heart Study," *Circulation* **99**, 779–785 (1999).
30. A. Sellmayer et al., "Effects of Dietary Fish Oil on Ventricular Premature Complexes," *American Journal of Cardiology* **76**, 974–977 (1995).
31. W. E. Connor, "The Beneficial Effects of Omega-3 Fatty Acids: Cardiovascular Disease and Neurodevelopment," *Current Opinion in Lipidology* **8**, 1–3 (1997).
32. W. W. Harris, "Fish Oils and Plasma Lipid and Lipoprotein Metabolism in Humans: A Critical Review," *J. Lipid Res.* **30**, 785–807 (1989).
33. D. C. Cunningham, L. Y. Harrison, and T. D. Shultz, "Proliferative Responses of Normal Human Mammary and MCF-7 Breast Cancer Cells to Linoleic Acid, Conjugated Linoleic Acid and Eicosanoid Synthesis Inhibitors in Culture," *Anticancer Res.* **17**, 197–203 (1997).
34. K. N. Lee, D. Kritchevsky, and M. W. Pariza, "Conjugated Linoleic Acid and Atherosclerosis in Rabbits," *Atherosclerosis* **108**, 19–25 (1994).
35. R. J. Nicolosi et al., "Dietary Conjugated Linoleic Acid Reduces Plasma Lipoproteins and Early Aortic Atherosclerosis in Hypercholesterolemic Hamsters," *Artery* **22**, 266–277 (1997).
36. A. H. Lichtenstein et al., "Dietary Fat Consumption and Health," *Nutrition Reviews* **56**, S3–19 (1998).

PENNY M. KRIS-ETHERTON
TERRY D. ETHERTON
Penn State University
University Park, Pennsylvania

LIVESTOCK FEEDS

Animals have been part of the human food chain since Neanderthals hunted them about 50,000 B.C. (1). Today, in those countries with a high standard of living, foods from animals (eg, meat, milk, and eggs) provide about 25 to 30% of the total caloric intake, whereas in developing countries it is a much lower proportion. High-quality protein is the primary nutrient supplied by animal products. In countries with adequate diets, half or more of the total protein intake comes from animal products, whereas in countries with inadequate diets, animal protein intake is 6 to 35% of the total protein intake. Protein in animal foods is high quality; the essential amino acids are present in high and well-balanced quantities. Thus, the consumption of animal foods supplements and complements the lower-quality protein found in many plant foods (eg, grains, vegetables, and fruits). A deficiency of animal protein in small children results in the disease kwashiorkor, a well-known problem in developing countries. Table 1 shows the edible protein consumed from different animal products on a worldwide basis (2).

Animal products are rich sources of required minerals (especially calcium, phosphorus, magnesium, potassium, iron, zinc) and vitamins (especially vitamin A, thiamin, riboflavin, niacin, vitamins B_6 and B_{12}) as shown in Table 2 (3). Because animal products contain protein, minerals, and vitamins in higher amounts relative to their caloric content in comparison to human dietary requirements, their consumption facilitates a balanced diet (4).

The efficiency with which animals convert feed resources into human food varies depending on the type of

Table 1. Edible Protein from Animal Sources Consumed Annually by the World Population

Source	Protein, ×1000 metric tons
Milk and milk products	18,550
Eggs	2,849
Meat	
Cattle	7,591
Buffalo	198
Sheep	979
Goat	282
Horse	70
Pork	5,859
Poultry	4,464
Other[a]	2,916
Edible offal	1,381
Fish	7,589
Total	*52,728*

[a]Includes camelidae, yak, deer, elk, reindeer, antelope, kangaroo, rabbit, and other small animals.

Table 2. Contribution of Animal-Derived Foods to the Nutrient Availability in the United States (1994)

	% of total nutrient intake			
Nutrient	Meat[a]	Dairy products[b]	Eggs	Total
Calories	14	9	1	25
Protein	39	19	4	62
Calcium	3	73	2	78
Phosphorus	25	33	4	62
Magnesium	13	16	1	30
Potassium	17	18	1	36
Iron	16	2	2	20
Zinc	42	19	3	64
Vitamin A	21	17	4	42
Thiamin	19	6	1	26
Riboflavin	18	31	6	55
Niacin	38	1		39
Folate	7	7	5	19
Vitamin B_6	36	10	2	48
Vitamin B_{12}	73	21	4	98

[a]Beef, pork, poultry, and fish.
[b]Excluding butter.

feed consumed by a particular class of livestock. Table 3 shows the percentage of energy and protein in the feed of animals that is recovered as edible animal products in the United States (5). These gross conversion efficiencies do not consider that livestock consume feeds of no value to humans (eg, range grass, pastures, hays, silage, straw, stover, stubble) or are not readily consumed by humans, such as feed grains (sorghum, barley, oats, millet, triticale) and by-products from the food industry (bran, gluten feeds, hulls, oilseed meals, tankage, meat and bone meal, feather meal, blood meal, whey, brewers and distillers grains, citrus and sugar-beet pulp, inedible fats). Table 4 shows the consumption of concentrates and forages, and the percentage of each used in the diets of U.S. livestock (5). China

Table 3. Gross Efficiency of Converting Animal Feed into Edible Products for Humans

	Percentage recovered	
Animal products	Calories	Protein
Monogastric animals		
Broilers (meat)	11	23
Turkeys (meat)	9	22
Hens (eggs)	18	26
Swine (meat)	14	14
Ruminant animals		
Cattle (milk)	17	25
Cattle (meat)	3	4
Sheep (meat)	3	4

Note: See text for discussion of net efficiency.

has the world's largest population of swine (33% of the total number of pigs), which is fed largely on forages, by-products, and wastes. Of the total land in the world, 11% is arable (suitable for crop production); 22% is range, pasture, and meadow; and 66% is not suitable for agricultural use. If animal efficiency is based on the calories that can be utilized directly by humans, their net efficiencies are nearly equal (6). Considering only the feed protein consumed by animals that could be utilized directly by humans, edible protein from animal products is produced with an efficiency of 60 to over 100% (7). Until all available land that can be used for photosynthesis yields food completely consumed directly by humans without any loss of plant material, including the supporting portions of the plant and by-products from the processing of plant materials for human consumption, animals will make a positive contribution to the supply of human food (8). Because the relationships between ruminants (cattle, sheep, goats) and human needs are complex and interdependent (Fig. 1), quantitative efficiency measurements are difficult (9).

The use of grains for food animals varies greatly throughout the world. In advanced countries, approximately 70% is used for animal feed and 30% for human food, whereas in developing countries, only about 8% is used as animal feed. Except for wheat, the principal end use of grains in the United States is for livestock feed. The current interest in oats as a source of dietary fiber in the human diet and corn for the production of fructose sweeteners and fuel alcohol has changed somewhat the usage for these two grains as well as the quantity of by-products available from these food industries.

Animals constitute a tremendous reservoir of human food, thereby reducing humankind's vulnerability to periods of crop shortages due to droughts, disease, and political events (2). Animals are found throughout the world, whereas grain reserves are found only in certain areas.

Animals also provide a host of other human needs. Wool is used for clothing and hides are used for shoes and other leather products. Many pharmaceutical products are isolated from animal glands. Fats and other lipids are used to manufacture soap and other products. Animals are used in many parts of the world for transportation and draft

Table 4. Consumption of Feed from Various Sources by Different Classes of Livestock in the United States

	Grains and high-protein concentrates		Forage	
Class of livestock	×1000 Metric tons	% of Diet	×1,000 Metric tons	% of Diet
Dairy cattle	25,602	37	43,739	63
Beef cattle	56,963	27	152,168	73
Sheep and goats	974	11	7,610	89
Hogs	51,396	86	8,178	14
Hens and pullets	20,499	97	734	3
Broilers	12,104	100		0
Turkeys	5,091	95	272	5
All livestock	185,105	46	218,879	54

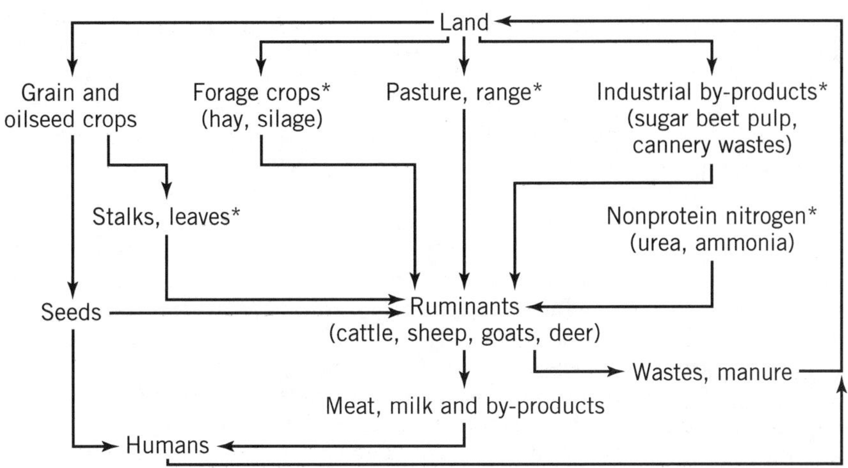

Figure 1. The Role of ruminants in human nutrition. Items marked with an asterisk are converted by ruminants but are not consumed directly as food by humans.

power, especially in agricultural production. Animal wastes are valuable fertilizer, and animal manure is used as a source of heating fuel and building material.

FARMING UNCONVENTIONAL ANIMALS

Although small in magnitude, the commercial production of unconventional animals is increasing. Aquaculture for the production of fish and crustaceans is definitely on the increase. There is a small commercial rabbit industry. Farmers and ranchers are increasingly interested in the propagation of deer, other wild ruminants, and game birds for private hunting and for exotic meats. Proper nutrition and feeding of these new farm animals is of interest (10).

COMPANION ANIMALS

In addition to food-producing animals, feeds are used to support a large number and array of companion animals (horses, dogs, cats, and other pets). Horse owners in the United States use about 4.3 million metric tons (t) of mixed feed each year (11) to feed about 5.25 million horses (12). In addition, approximately this same amount is also fed as hay and pasture.

The dog and cat population in the United States is estimated to be 52 and 56 million, respectively (13); 60% of the households in the United States have pets. These pets consume about 5,133,500 t of food each year (about 73% by dogs and 26% by cats) resulting in $9.9 billion in pet food sales (14). Six companies account for 87% of the dog food, and five companies account for 87% of the cat food marketed through major food stores. Additionally, pet foods are sold as private label products through pet specialty stores, veterinarians, feed stores, and direct distribution. The trend has been toward more dry food products (88% of the total), less canned foods, and more specialty products (eg, high energy, high protein to reduce defecation and life-cycle products for growing and aging pets).

Other specialty feeds for rabbits, fur-bearing animals, birds, laboratory, and other animals total about 20% of that produced for dogs and cats. Feeding wild animals in captivity (zoos and game parks) is of increasing interest. Diet monotony is a problem in feeding zoo animals; adapting wild animals to diets and supplements formulated by humans is a challenge.

NUTRIENT REQUIREMENTS

Research continues on the nutrient requirements and interactions in all classes of livestock, companion animals, fish, and laboratory animals. These are summarized in a publication series *Nutrient Requirements of Domestic Animals* by the National Academy of Science–National Research Council (NAS–NRC) that are revised periodically (15). This series incorporates computer models to predict nutrient requirements based on genetic variance, environmental conditions, and the use of various production technologies. These are used by feed companies, animal practitioners, consultants, producers, and teachers as guides to formulating diets that will meet the nutritional requirements of animals, generally with some safety margin, from feeds typically consumed and therefore palatable to a particular animal. Many animal feeding guides are also available from state extension offices, animal associations, feed companies, and veterinarians.

COMPUTERS

During the past 35 years, the feed business has been revolutionized by computers. Diet formulation, least-cost diet formulas, process and inventory control, and accounting have been impacted by the introduction of computers. Nutritionists can not only formulate diets from palatable feeds that meet the nutritional requirements of animals, they can also formulate diets from ingredients that minimize the total cost of the formula. Data describing the physical characteristics of each ingredient, restrictions on the amount of a given ingredient, and current prices can now all be handled in great detail using linear programming and least-cost formulation (16).

Computers are also being used to simulate livestock production systems to study inputs and their impact on production output and net returns. These programs are more sophisticated than older measures of profit potential (17).

THE FEED MANUFACTURING INDUSTRY

Feed manufacturing is one of the 25 leading industries in the United States (18); it employs about 97,000 people (11). Feed purchases historically have been the largest U.S. farm expenditure, surpassing rent, interest, fertilizer, and energy; in 1984, feed purchases totaled $18.3 billion (11). Feed cost in the production of meat, milk, and eggs accounts for 50 to 75% of the total cost of production. The early development of this industry and the later application of science, engineering, and merchandising to feed manufacturing provides an interesting success story in the development of U.S. agriculture (19). The challenge facing the industry is to formulate, manufacture, and distribute feeds that will enable livestock, poultry, and fish farmers to produce quality products at the lowest possible cost (16).

Many feeds used in feed manufacturing were once considered waste materials from the food industry that were disposed of in ways that would be considered environmentally unacceptable by today's standards. In 1947, it was noted: "Sixty years or more ago, the flour mills in Minneapolis dumped wheat bran into the river because nobody wanted to buy it" (20). Feed mixing moved from the scoop shovel to a mechanized operation, and many companies entered the feed manufacturing business between 1895 and 1930. During this time, feeds were sold more for their mystical qualities than for demonstrated production and health responses in animals.

Many feed companies founded during this early period are still in business today. The Ralston Purina Company (now Purina Mills, Inc.) was founded in 1894; the Checkerboard trademark was copyrighted in 1900. A recent survey provides a listing of more than 800 existing feed companies, and poultry and livestock integrators (21). Of these companies, 95% produce complete feeds, 66% produce feed supplements, and 39% produce premixes. Most manufacture feed for hogs (76%) and beef cattle (73%), followed by feed for dairy cattle (68%) and laying hens (51%). Approximately 4% of the companies produce more than 227,000 t of feed annually, 30% produce between 18,000 and 227,000 t a year, and 66% produce less than 18,000 t. Farmer cooperatives account for about 22% of the formula feed tonnage. About 6700 feed mills were in operation in 1984 (11). In 1988, the U.S. feed manufacturing industry produced an estimated 94 million t of primary feed (feed mixed from individual ingredients). Secondary feed (feed mixed with one or more ingredients and a supplement) totaled 9 million t. Table 5 lists the livestock uses of the primary feed total (12).

Today's feed-manufacturing industry is a highly complex, scientifically based, quality-controlled, and heavily regulated industry (Fig. 2). It combines a variety of fields ranging from chemistry, nutrition, computer programming, engineering, materials handling, and transportation to personnel, inventory control, economic analysis, advertising, and law. Patents related to manufacturing processes, ingredients, nutrient forms, and drugs are many. It is a progressive and competitive industry that contributes advanced technology to the U.S. animal industry and enhances the supply, quality, and nutritional value of food for humans.

Several trade and related associations service the industry, including the American Feed Industry Association (AFIA; Arlington, VA), Association of American Feed Control Officials (AAFCO; West Lafayette, IN), Animal Health Institute (Alexandria, VA), American Association of Feed Microscopists (Sacramento, CA), National Grain and Feed Association (Washington, DC), several state grain and feed associations, and allied industry associations (22).

FEEDING SYSTEMS USED IN LIVESTOCK PRODUCTION

Poultry

For the most part, poultry (chickens, turkeys, ducks, geese) are fed in houses on complete mixed diets. Chickens are

Table 5. Livestock Use of Manufactured Feed in the United States (1997)

Livestock	Percentage of total
Starter, grower, layer, and breeder hens	12.7
Broilers	31.1
Turkeys	7.7
Dairy cattle	13.2
Beef cattle and sheep	15.4
Hogs	13.1
Other	6.8

Note: Primary feed mixed from individual ingredients as complete feed, supplement, or concentrate.

Figure 2. Feed Manufacturing plant with ingredient storage, feed elevators, warehouse, delivery trucks, and office. *Source:* Courtesy of Ezell-Key Grain Co., Inc., Snyder, Texas.

fed either for meat (broilers) or eggs (laying hens). Breeds and strains of chickens differ for these two purposes. Turkeys are also produced primarily in houses, although some range or pasture is still used. The poultry industry is very different from that of 40 years ago (23). Many broiler production units are vertically integrated (24) from hatching, production, and feed preparation through shipment of processed broilers to grocery stores (Fig. 3). Nearly all (98%) of the broiler feed is fed to the manufacturer's own chickens (11). The eight largest companies produce 55% of the total U.S. broiler production. Modern poultry production firms are well capitalized, take advantage of the economies of large-scale operations, and rapidly adopt new biological and technological advances. As a result, the relative price of poultry products has steadily decreased, which has greatly increased the demand and therefore the production of broilers and turkeys.

Swine

Swine are also largely produced in houses on complete mixed diets. The swine industry is becoming more vertically integrated (11). Many independent swine producers, large (>5000 head) and small (<100 head), purchase feeds and other inputs and sell finished pigs to meat-processing companies for final distribution of pork products. Many swine producers are farrow-to-finish, meaning they breed, produce, and market finished pigs. A few producers sell feeder pigs to others for finishing. Swine producers effectively use new technology in nutrition, feeding, and other management practices (Fig. 4).

Dairy Cattle

Milk production and the production of dairy cattle is still largely by individual producers, although many of these

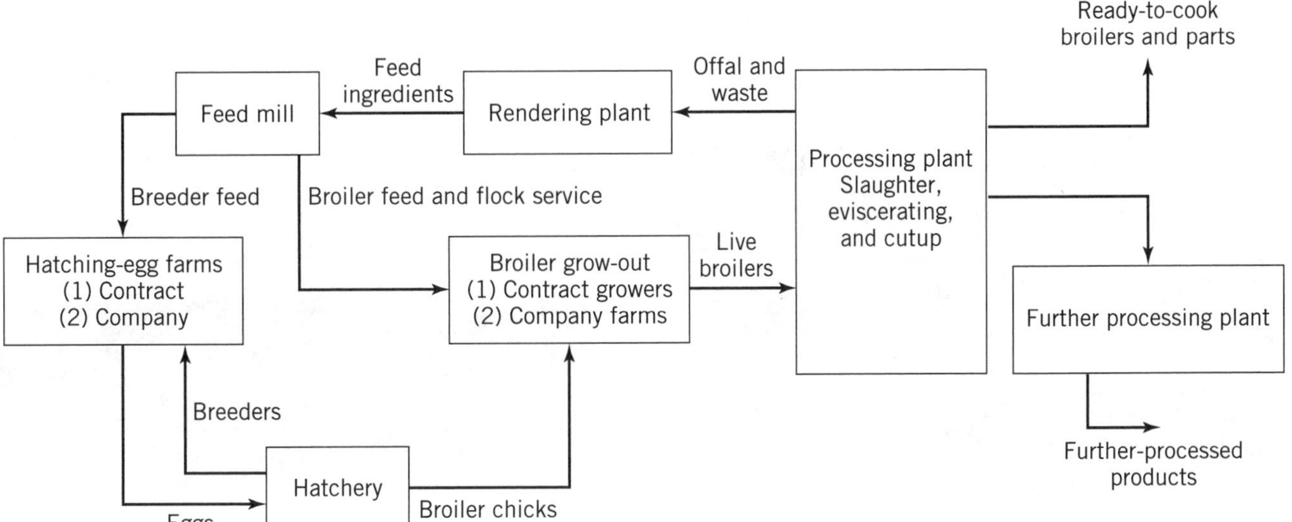

Figure 3. Many modern broiler firms control production from chick to consumer product.

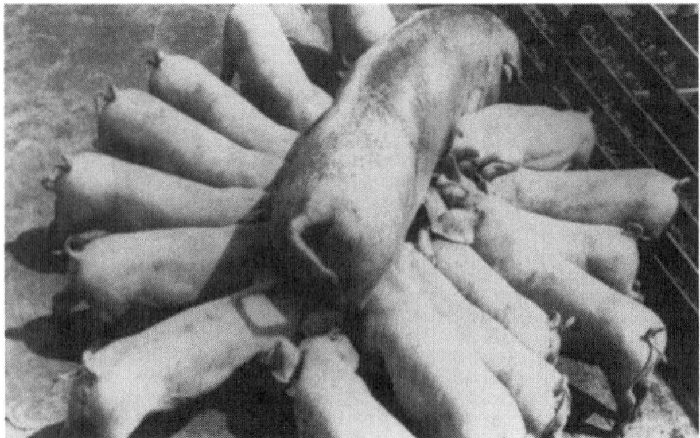

Figure 4. Swine production; sow and her baby pigs. *Source:* Courtesy of *National Food Review*.

producers are quite large (eg, 300 to 3500 cows). Some have integrated into producer-processor operations. Most producers also farm and produce a major portion of their own feed, especially roughage (eg, hay, silage, and pasture) and purchase a commercial concentrate feed. Large dairy units may purchase all of the required feed. With continuing increases in milk production per cow, due largely to genetic, feeding, and reproductive technology, the number of dairy producers continues to decrease but milk production per cow steadily increases (Fig. 5).

Beef Cattle

The beef cattle industry is made up of two distinct segments. The cow-calf industry is land based, largely in the western and southeastern areas of the United States, relying on range and pasture as the major feed input. Because it is land based, requiring a major capital investment, most cow-calf producers are small in size (eg, 50 cows), although a few are very large (eg, 20,000 cows). The feedlot industry, on the other hand, is made up of large feedlots (5000 to 200,000 head onetime capacity) located primarily in seven states (Texas, Nebraska, Kansas, Colorado, Iowa, California, and Arizona). These feedlots are often integrated with grain or feed companies (24). Some are integrated feedlot and meat-processing companies, and a few are integrated feed, feedlot, and meat-processing companies. Many feedlots custom-feed cattle on contract for absentee owners and investors. A third, less important, segment backgrounds cattle between weaning and a yearling age on pasture, crop stubble, and aftermath (eg, corn and sorghum stalks) and on various by-product feeds. Because of the diversified nature of the beef cattle industry and because many cow-calf producers derive a significant portion of their income from other farming enterprises or off-farm income, genetic and other new technologies are slow in becoming established. On the other hand, nutrition, health, and other technologies are rapidly applied by the feedlot industry, similar to the poultry industry (Fig. 6).

Sheep

Much of what characterizes the beef cattle industry is also true for the sheep industry, but on a smaller scale. Many of the producers are relatively small. Feedlot production of lambs is limited. In addition to meat, wool is an important source of income to a sheep producer (25). Sheep production in the United States has steadily declined over a long period of time.

Goats

The production of goats is similar to that described for beef cattle and sheep (25) but is a relatively small industry. Mohair from Angora goats and milk are the major products, although there is some consumption of goat meat. The Boer goat has been introduced into the United States primarily for meat production. Roughage (range, pasture, hay) is the primary feed input for goat production.

Fish

Fish and other aquatic animals are produced in commercial ponds and tanks or separated areas of lakes and seacoasts. For the most part, they are fed commercially prepared diets, although they also obtain food material from plankton in these waters. Catfish, trout, and salmon are the major fish, and shrimp and crawfish are the major crustaceans commercially produced in the United States

Figure 5. Dairy farm: dairy cows, milking parlor, barn, silo, and granary. *Source:* Courtesy of *National Food Review*.

Figure 6. Cattle feedlot: cattle on feed and feed mill. *Source:* Courtesy of Texas Cattle Feeders Assoc., Amarillo, Texas.

(26). Catfish production increased 330% between 1980 and 1987. The catfish industry is largely vertically integrated. Because of its rapid growth, aquaculture will use increasing amounts of feed and improved technology to produce aquatic animals.

FEED INGREDIENTS

Basic Feed Ingredients

Roughages (range, pasture, hay, and silage) are characterized as high-fiber feeds (usually more than 18% crude, 30% acid detergent, or 45% neutral detergent fiber). They are important feeds for ruminant animals. In the rumen, anaerobic microorganisms degrade a large portion of the fiber and other organic constituents, thereby making the fiber available as an energy source. A minimum amount of roughage in the diet of ruminants is desirable for maintaining normal rumen function.

Range and Pasture. As mentioned earlier, 22% of the world's land surface is range, pasture, and meadow, two-thirds of which is not suitable for crop production. To utilize this land for the production of human food requires ruminant animals (cattle, sheep, goats, and wild ruminants). Rangeland is permanent pasture, in that it is not seeded periodically, although interseeding is sometimes practiced as an improvement method. Desirable native species of grasses and forbes are maintained and sustained largely through range management practices that include stocking rate, rotation of animals on range areas, complementarity between grazing animals, location of animal water supplies, fertilization, and the use of supplemental feeds. The major determinant of range production is the amount of precipitation.

Pasture is largely improved pasture because land is periodically reseeded to improve pasture varieties, often with the use of fertilizer and irrigation. Grazing of small grain crops, especially wheat, during the vegetative stage is an important source of pasture. Improved pastures are often part of a crop rotation system to improve soil fertility and humus and to enhance soil conservation practices. Genetic programs provide varieties and strains of plants for improved pastures. These pastures are used in the production of all ruminants but are especially important in dairy production.

Meadows are water drainage areas that are permanent grasslands between arable land areas and in stream valleys. These are grazed periodically, primarily by beef cattle and sheep, and are sometimes used for hay.

Harvested Roughages. There are many kinds of harvested roughages. Hays are forages that are cut and dried in the field and gathered relatively dry (85–90% dry matter) as baled or chopped hay. Both grass (eg, timothy, bromegrass, orchard grass) and legumes (alfalfa, clover) are used for hay. Approximately 136 million t of hay are fed to U.S. livestock each year, 57% of which is alfalfa (27). Alfalfa meal, often dehydrated, is an important specialty feed (1.76 million t; 11) in several diet formulas.

Crops can also be harvested as relatively high-moisture material (25–45% dry matter) and stored in oxygen-limiting structures (upright and bunker silos) where the bacteria present on the harvested material or added intentionally at the time of ensiling convert part of the carbohydrate in the crop to various organic acids (especially lactic acid), utilizing the oxygen present, thereby preserving the acidified product as silage. Corn, small grains (barley, oats), alfalfa, and other crops can be harvested as the entire plant for silage.

Grains. All grains can be used to feed all classes of livestock. The major feed grains in the United States are corn, sorghum grain (milo), barley, and oats. Small quantities of millet, rye, and triticale are also fed, and wheat is used when the price makes it competitive. Because grain is produced in the United States in excess of that required for human food, feeding grain to livestock is an important factor in maintaining the price of grain. About 114 million t of corn, 14 million t of sorghum grain, 7.6 million t of wheat and rye, 5.2 million t of barley, and 3.1 million t of oats are fed to U.S. livestock (12).

Protein Feeds. These feeds are largely by-products of the food-processing industry; most are the result of vegetable oil production. Soybean meal is the major protein feed (18.1 million t) followed by cottonseed meal (1.5 million t), canola meal (0.25 million t), sunflower meal (0.22 million t), peanut meal (0.15 million t), and linseed meal (0.10 million t). Corn gluten feed and meal (1.32 million t) are by-products of the corn starch and sweetener industries. Protein feeds derived from the processing of animals for food include tankage and meat meal (3.1 million t), fish meal and solubles (0.74 million t), milk products such as whey (0.36 million t), hydrolyzed feather meal (0.24 t), and smaller quantities of blood meal (11,12).

By-product Feeds. By-products used as livestock feeds are too numerous to list here; a table of by-product and unusual feed composition (28) lists 355 feeds. Those available in greatest quantities include wheat mill feeds (5.79 million t), molasses (1.66 million t), distillers grains (0.97 million t), fats and oils (0.86 million t), beet pulp (0.60 million t), rice mill feeds (0.58 million t), hominy feed (0.48 million t), brewers grains (0.34 million t), and miscellaneous by-product feeds (0.80 million t) (11,12). Other by-product feeds include almond hulls, apple pomace, bakery waste, citrus pulp, cannery waste, cottonseed hulls, cotton-gin trash, grape pomace, oat hulls, cull peas, potato wastes, soybean hulls, and many others.

Supplements

Minerals. Mineral additions to livestock diets are common practice, either to complete diets and supplements or as free-choice ingredients. Salt (NaCl) has been used for a long time; early pioneers sought out salt licks where horses and cattle obtained salt and other minerals. Commonly supplemented are calcium and phosphorus as limestone, calcium carbonate, bone meal, various calcium phosphates, sodium and ammonium phosphates, and phosphoric acid. Many of these sources are mined products from natural deposits, where fluorine is often present and must be reduced (defluorinated phosphate) to acceptable limits to prevent toxicity to livestock.

Other major minerals required by livestock include potassium, magnesium, and sulfur, which are supplemented as potassium chloride and potassium ammonium sulfate, magnesium oxide, flowers of sulfur, sulfate salts, and organic sulfur (methionine).

Livestock require eight trace minerals (cobalt, copper, iron, iodine, manganese, molybdenum, selenium, zinc) and perhaps several others (chromium, fluorine, silicon, nickel, vanadium). The required trace minerals are commonly added to complete diets, trace mineral supplements, and trace mineralized salt. Most are added as inorganic salts. There is increasing interest in the bioavailability of various organic forms of these trace minerals (eg, ethylenediamine dihydroiodide; zinc methionine, selenium methionine, and other mineral-amino acid complexes; iron and copper gluconate; and various chelated trace minerals). Some of these trace minerals have wide toxicity to requirement ratios (zinc) whereas others have very narrow ratios (selenium). As a result, trace minerals are seldom added as single ingredients to complete diets or supplements; they are formulated into multimineral premixes and then added as dry or liquid premixes.

Vitamins. All the required vitamins can be supplemented in livestock diets. These vitamins are chemically synthesized and are relatively inexpensive. Vitamin A is commonly added to all diets. The other fat-soluble vitamins—D, E, and K—are used in special situations (vitamin D for animals housed indoors, vitamin E to enhance the immunological competence, and vitamin K where bacterial synthesis in the intestine may be limited).

Water-soluble vitamins are commonly added to the diets of monogastric animals (swine, poultry, dogs, cats). Although these vitamins are synthesized by bacteria in the gastrointestinal tract, only in the case of ruminants where synthesis takes place at the beginning of the tract (rumen) is there opportunity for absorption eliminating the need for supplementation except in special situations. Thiamin, riboflavin, niacin, pyridoxine, pantothenic acid, choline, and B_{12} are commonly added to diets for monogastric animals; biotin and folic acid (folate) are added in some cases. Vitamin C is required by fish but not required by livestock, although it may be useful in heat stress, especially in laying hens for maintaining egg shell strength.

Amino Acids. Methionine and methionine hydroxy analogue (MHA) have been used for S-amino acid supplementation for several years. Fermentive synthesis of lysine has made this often limiting amino acid an economical addition to many monogastric diets. Arginine, threonine, and tryptophan are also available.

Urea. Ruminant animals can make use of nonprotein nitrogen (NPN) through conversion to microbial protein in the rumen and subsequent digestion in the lower gastrointestinal tract. This conversion results in high-quality animal protein (milk and meat) for human consumption from an inexpensive, nonprotein source. Urea is the major source of NPN used in ruminant diets (355,500 t in 1984; 11). Biuret and various ammoniated feed products are also sources of NPN. Use of urea in ruminant diets is limited to about 1% of the total diet (one-third of the total required protein) for effective conversion to microbial protein. An excess of NPN can result in acute ammonia toxicity due to more rapid microbial conversion to ammonia in the rumen, compared with its incorporation into microbial protein.

Additives. Various other compounds are added to livestock diets for purposes other than nutritional supplements. Sales of animal health products totaled $3.6 billion in 1997 (29). Several antibiotics are added for growth promotion, improved feed utilization, and disease prevention and treatment. Chlortetracycline (aureomycin), oxytetracycline (terramycin), penicillin, virginiamycin, monensin, and lasalocid are the primary antibiotics used in livestock diets. Sulfa drugs are also used as bacteriostats. Coccidiostats, anthelmentics, and mold inhibitors are also used. These additives result in healthier animals. Use of these compounds is highly regulated by the U.S. Food and Drug Administration (FDA) both in terms of efficacy for a stated claim as well as safety for the target animal and for humans consuming foods from animals. Research costs can total $8 to $15 million to achieve final approval for one of these additives. The U.S. Food Safety and Inspection Service (FSIS) constantly monitors animal food products for the absence of residues. Use of additives constantly changes; the most current reference source on their approved use is the Feed Additive Compendium (30), which is updated annually plus monthly supplements.

Grading and Definitions

Many feeds, especially grains, are graded for quality factors (moisture, broken kernels, damage, foreign material) that are used in the trade as purchasing specifications (31). Many of these visual standards are defect-type factors that may or may not relate to the nutritional value of the feed. End-user factors such as moisture, protein, oil, starch, fiber, hardness, and amino acid ratios are important economically to the needs of specific grain users (32). Hay grading standards have also been proposed, but their use is limited.

Official processing and composition definitions for various feed ingredients have been established and are constantly updated by the Association of American Feed Control Officials (AAFCO) (33). These definitions are used by state and national feed control officials in regulatory aspects of the feed industry. Also listed are the names and addresses of state feed control officials, state feed registration requirements, the uniform state feed bill, related associations, and other useful information on livestock feeds.

Analyses

The National Research Council (NRC) has published an atlas of feed composition (34) and a more recent table of feed composition (35) tabulating the composition of feeds. There are also many other tables of feed composition (28,36,37). Feed composition tables, along with actual feed analyses, are the basis for formulating nutritionally complete diets for livestock.

Feeds can be analyzed for many constituents. It is convenient to divide these into chemical and biological constituents (37). Chemical analyses for nutrients, antinutritional factors, toxicants, contaminants, and residues can be performed, but these results do not necessarily indicate the bioavailability of these constituents when consumed by animals. Thus, biological constituents are measured using a biological response such as digestibility, absorption, metabolic efficiency, or some biochemical response. Gross energy (GE or heat of combustion) indicates the total energy value of a feed but only a portion is digestible (digestible energy, DE), less is available for metabolism (ME), and the net energy (NE) value of a feed is the energy available for maintenance of body tissues and growth (meat), wool, milk, and egg production. Similar bioavailability considerations apply to protein, amino acids, minerals, vitamins, and other constituents. Feeds are commonly analyzed for moisture, crude protein, ash, fiber (crude, acid detergent, neutral detergent fiber), and crude fat. Additional analyses for individual amino acids, minerals, and vitamins can be performed.

Regulations

Many agencies and regulations impact the feeding of livestock and the manufacture of feeds, including state feed regulations, the FDA, the Environmental Protection Agency, and the Occupational Safety and Health Agency (19,38,39). In 1909, AAFCO was organized; its objective is "... to promote uniformity in legislation, definitions, and rulings, and the enforcement of laws relating to the manufacture, sale, and distribution of feeds and livestock remedies in the continent of North America." Good Manufacturing Practice Regulations for Medicated Feeds (30) were published by the FDA in 1976 and revised in 1986.

FEED PROCESSING

Most feeds must be processed before consumption by livestock. Only the free-grazing animal consumes feed that has not been processed.

Roughages

Hay is cut, field-dried, and generally baled into rectangular or round bales that weigh 25 to 300 kg each. The moisture content of baled material must be below 14% for proper storage. Hay can also be field chopped and hauled to storage structures or chopped and pressed into cubes (approximately 3 cm^3) either in the field or at hay storage sites. Dehydrated alfalfa is cut fresh, sometimes partially dried in the field, chopped, hauled to dehydrating facilities, and pelleted (0.6–1.3 cm). Forages can be cut and chopped in the field, hauled, and fed fresh to livestock (green chop). Forages and other plant materials can be chopped and ensiled in oxygen-limiting structures and fed as silage.

Roughages can also be treated with various chemicals to enhance their digestibility. Treatment with alkali (NaOH) has been practiced for a long time, especially in northern Europe. Use of anhydrous ammonia for enhancing the digestibility of straws and silage is a more recent practice. The use of alkaline peroxide appears especially promising for high-fiber roughages (40). Pressure treatment of acidified roughage material also enhances digestibility but is generally not cost effective. Treatment with ozone also increases the digestibility of roughages (41).

Roughages that are finely chopped (screen size less than 1.3 cm) and/or pelleted are generally consumed in greater amounts by ruminants and have a faster passage rate through the rumen, resulting in a lower digestibility of the fiber fraction. Because of greater intake, however, greater animal productivity generally occurs. Lactating dairy cows fed large amounts of concentrate feed should also be fed a minimum quantity of unchopped roughage (3 kg) to prevent abnormally low milkfat due to insufficient acetic acid production by rumen microorganisms.

Grains

Grains and other seeds can be dehulled, ground, cracked, rolled, steamed, steam-flaked, micronized, popped, extruded, roasted, and pelleted before feeding. In some cases, grains (corn) and other seeds (cotton and soybeans) are fed unprocessed (whole) to livestock. The need for processing depends on the grain, animal to be fed, feed mixture, potential improvement in utilization, and processing cost.

Swine, sheep, and goats chew their feed more thoroughly than do cattle; therefore, improvements in feed utilization through processing are comparatively small. Sorghum grain (milo), because of the small seed, must be processed for all livestock. Corn grain, on the other hand, can be fed whole to most classes of livestock except poultry, although improvements in utilization may be obtained with steam-flaking, owing to partial gelatinization of the starch granule. Concentrate ingredients may require processing to achieve a somewhat uniform particle size in complete mixed diets to prevent sorting and separation by animals. This is especially true for poultry and swine diets, and dairy and other concentrate mixes.

Grains can also be stored and fed as high-moisture or reconstituted grain. High moisture refers to grain that is harvested while the moisture content is still 20 to 30% and stored in oxygen-limiting structures. Reconstitution refers to grain that is relatively dry to which water is added to bring the moisture content to about 27% and then stored in oxygen-limiting structures. The stored grains are generally rolled when removed from the storage structures before feeding.

Complete Diets and Supplements

Common forms of formula feeds are meal or mash, pellets, crumbles and cubes, blocks, and liquids (11). Complete diets are sometimes pelleted before feeding to keep the diet from separating, to facilitate mechanical handling, and to decrease feed wastage by livestock. The pelleting process involves grinding the feed to a relatively small particle size; adding moisture, often as steam; and forcing the feed through metal dies with openings that vary from 0.3 to 1.6 cm in diameter, which generates considerable heat. The compressed feed is cut or broken into various lengths. As the compressed feed cools, it tends to set as pellets, with reasonable durability for mechanical handling. The starch content of grains enhances the pelleting process through partial gelatinization, and various pellet binders can be added (eg, lignosulfonates) to enhance the pelleting process and improve pellet durability.

Feeding pelleted diets to swine and poultry is relatively common. Larger-size pellets (cubes) are used to provide supplemental feed to grazing ruminants. Large blocks (ca 25–30 cm^3) of supplemental feeds are used for pasture and range livestock where the animals chew off portions of these blocks over time, enabling infrequent supplemental feeding of these animals. Tubs with solidified molasses plus supplemental nutrients are also convenient, allowing animals to consume the supplement over time by licking.

DIET FORMULATION AND EXAMPLE DIETS

Nutritionally complete diets for livestock can be formulated using calculators or elaborate computer programs. The NRC nutrient requirement series (15) give example methods for calculating complete diets. Because ruminant animals have fewer dietary requirements than monogastric animals, it is easier to formulate complete diets for these animals. A complex poultry diet may require balancing for 25 or more nutritional attributes.

The science of animal nutrition allows the formulation of complete diets from a multitude of feed ingredients. Therefore, there is not a best or ideal diet for any animal. Table 6 shows intakes per animal and example diets simply to illustrate what could be fed to these livestock.

Table 6. Intakes per Animal and Example Diets

Ingredient	g/kg Diet
Growing chickens, 20–160 g / d	
Corn, ground	580
Soybean meal	350
Methionine, DL	2
Animal/vegetable fat	30
NaCl	5
$CaCO_3$	10
$CaHPO_4 \cdot 2H_2O$	20
Vitamin premix	1
Trace mineral premix	1
Antimicrobial premix	1
Growing swine, 1–3 kg / d	
Corn, ground	810
Soybean meal	140
Meat and bone meal	40
NaCl	3
$CaCO_3$	3
$CaHPO_4 \cdot 2H_2O$	1
Vitamin premix	1
Trace mineral premix	1
Antimicrobial premix	1
Lactating dairy cattle, 15–28 kg / d	
Alfalfa hay	180
Corn silage	500
Brewers grains, dried	100
Corn and cob meal	209
Urea	5
NaCl	3
$CaHPO_4 \cdot 2H_2O$	2
Trace mineral premix	1
Feedlot cattle, 5–14 kg / d	
Corn, whole or steam-flaked	637
Wheat middlings	80
Corn silage	120
Cottonseed hulls	70
Cottonseed meal	10
Corn gluten meal	10
Urea	5
Molasses, cane	40
Animal/vegetable fat	10
KCl	2
NaCl	3
$CaCO_3$	10
Trace mineral premix	1
Vitamin premix	1
Antimicrobial premix	1

FUTURE DEVELOPMENTS

The potential is great for improving our knowledge to optimize the nutrition of livestock, understanding the interactions between nutrients, quantitating the productive responses to varying nutrient inputs, and improving the economic return from the feeding of livestock. With genetic changes in livestock and the use of additives, the potential exists to produce livestock with improved efficiency that are better suited to consumer desires. Progress has already been made in increasing the leanness of animals with less waste fat (42). Through better understanding of metabolic processes, it may be possible to produce foods of animal origin with enhanced nutritional value. The potential impact of advanced technologies on animal productivity has been estimated (43).

Better and faster analysis of feed ingredients will enable more accurate formulation of livestock diets. Near-infrared reflectance spectroscopy is a promising technique in this regard. Enzyme assay may give rapid estimates of nutrient bioavailability in feeds. New and improved methods of processing feeds will improve their utilization, resulting in less animal waste. Through plant breeding, introduction of new strains, and biotechnology (44), feed materials will be improved to better meet the nutritional needs of livestock or lower the content of natural toxicants. High-lysine corn and canola meal are examples of what can be accomplished in this regard.

New strains of livestock will have different nutritional requirements that must be determined to maximize genetic improvements. The use of production- and efficiency-enhancing compounds will also result in altered nutritional requirements. An increased lysine concentration in the diet of swine whose growth rate and leanness has been improved through the use of recombinant porcine growth hormone or other repartitioning agents is but one example.

Genetically engineered microorganisms that will predigest feed materials, better preserve ensiled crops, or enhance feed utilization in the rumen or large intestine of animals could have a major impact in livestock feeding. Genetically engineered organisms may also be effective in decreasing gastrointestinal problems and diseases.

The science of feeding livestock has an exciting future, but acceptance of scientific findings by producers and consumers will depend on a better understanding and appreciation of science in general. The current public backlash to the introduction of scientific improvements in general poses a major threat to the competitive position of the United States and to the competitive position of livestock in the human food chain.

BIBLIOGRAPHY

1. S. M. Garn and W. R. Leonard, "What Did Our Ancestors Eat?" *Nutrition Reviews* **47**, 337–345 (1989).
2. T. J. Cunha, "The Animal as a Feed Resource for Man," *Feedstuffs* **64**, 18–32 (1982).
3. S. Gerrior and L. Bente, *Nutrient Content of the U.S. Food Supply, 1909–94*, Home Economics Res. Rpt. No. 53, USDA Center for Nutrition Policy and Promotion, Washington, D.C., 1987.
4. R. G. Hansen, "An Index of Food Quality," *Nutrition Reviews* **31**, 1–7 (1973).
5. W. F. Wedin, H. J. Hodgeon, and N. L. Jacobson, "Utilizing Plant and Animal Resources in Producing Human Food," *Journal of Animal Science* **41**, 667–686 (1975).
6. K. L. Blaxter, "Efficiency of Food Conversion by Different Classes of Livestock in Relation to Food Production," *Federation Proceedings* **20**, 268–274 (1961).
7. R. L. Preston, "Let's Feed People and Animals," *Advance* **12**, 6–7, 11 (1975).
8. R. L. Preston, "Land Use—A Time for Reconsideration," *Journal of Animal Science* **46**, 1467–1468 (1977).
9. H. A. Fitzhugh et al., *The Role of Ruminants in Support of Man*, Winrock International, Petit Jean Mountain, Ark., 1978.
10. K. L. Blaxter, "Conventional and Unconventional Farmed Animals," *Proceedings of the Nutrition Society* **34**, 5156 (1975).
11. M. Ash, W. Lin, and M. D. Johnson, "The U.S. Feed Manufacturing Industry, 1984," *USDA-ERS Statistical Bulletin* No. 768 (1988).
12. "Feed Marketing and Distribution," *Feedstuffs* **70**, 6–20 (1998).
13. J. Corbin, "Pet Foods and Feeding," *Feedstuffs* **70**, 82–87 (1998).
14. R. A. Schoeff and P. Lobo, "Most Feed Potentials Up, Except—Surprise—for Broilers," *Feed Management* **49**, 7–16 (1998).
15. NAS-NRC, *Nutrient Requirements of Domestic Animals*, National Academy Press, Washington, D.C., ("Horses" [1989]; "Dairy Cattle" [1989]; "Beef Cattle" [1996]; "Goats" [1981]; "Poultry" [1994]; "Sheep" [1985]; "Swine" [1998]; "Cats" [1986]; "Dogs" [1985]; "Laboratory Animals" [1995]; "Mink and Foxes" [1982]; "Rabbits" [1977]; "Fishes" [1993]; "Nonhuman Primates" [1978]).
16. R. W. Schoeff, "The Formula Feed Industry," in H. B. Pfost and D. Pickering, eds., *Feed Manufacturing Technology*, American Feed Manufacturing Association, Arlington, Va., 1976, pp. 1–17.
17. W. L. Alsmeyer, "A Multiple Regression Model for Net Income Estimates from 100-Sow Production Units," *Journal of Animal Science* **40**, 6–12 (1975).
18. W. Anderson, "The U.S. Formula Feed Industry Today," in R. R. McEllhiney, ed., *Feed Manufacturing Technology III*, American Feed Industry Association, Arlington, Va., 1985, pp. 9–12.
19. R. W. Schoeff, "History of the Formula Feed Industry," in R. R. McEllhiney, ed., *Feed Manufacturing Technology III*, American Feed Industry Association, Arlington, Va., 1985, p. 28.
20. L. Wherry, *The Golden Anniversary of Scientific Feeding*, Business Press, Milwaukee, Wisc., 1947.
21. "Survey Gives Updated View of Feed Industry," *Feedstuffs* **61**, 28–51 (1989).
22. "Feedstuffs 1998 Reference Issue," *Feedstuffs* **70**, 171–174 (1998).
23. "The Poultry Industry Today," *National Food Review* **12**, 9–12 (1989).
24. "The Integrators: The 'In-House' Feed Manufacturers," *Feedstuffs* **61**, 27 (1989).
25. C. S. Menzies et al., "The U.S. Sheep and Goat Industry: Products, Opportunities and Limitations," Council for Agriculture Science and Technology Report No. 94, CAST, Ames, Iowa.
26. D. Harvey, "Aquaculture: Meeting Fish and Seafood Demand," *National Food Review* **11**, 10–13 (1988).
27. "Hay: Production, Harvested Acreage, Yield, Prices Received by Farmers," *Feed Situation and Outlook Report* **305**, 53 (1988).
28. D. Bath et al., "Byproducts and Unusual Feedstuffs," *Feedstuffs* **70**, 32–38 (1998).
29. "Survey Shows Animal Health Product Sales Rose to $3.6 Billion in 1997," *Feedstuffs* **70**, 7 (1998).
30. S. Muirhead, ed., *1998 Feed Additive Compendium*, The Miller Publishing Co., Minnetonka, Minn., 1997.
31. *The Official United States Standards for Grain*, USDA Federal Grain Inspection Service, Washington, D.C., 1978.
32. C. R. Hurburgh, Jr., "Grain Quality, Defined by End-use Values, Moving into Market," *Feedstuffs* **61**, 1–56 (1989).
33. *Official Publication*, Association of American Feed Control Officials, Atlanta, Ga., 1989.
34. E. W. Crampton and L. E. Harris, *Atlas of Nutritional Data on United States and Canadian Feeds*, National Academy of Science, Washington, D.C., 1972.
35. NAS-NRC, *United States-Canadian Tables of Feed Composition*, 3rd rev., National Academy Press, Washington, D.C., 1982.
36. N. Dale, "Ingredient Analysis Table: 1998 Edition," *Feedstuffs* **70**, 24–31 (1998).
37. R. L. Preston, "Typical Composition of Feeds for Cattle, 1998," *Beef* **34**, FC1–FC15 (1998).
38. L. H. Boyd, "Feed Regulation," *Feedstuffs* **70**, 126–128 (1998).
39. L. H. Boyd, "Medicated Feed Regulation," *Feedstuffs* **70**, 129–132 (1998).
40. "Alkaline Hydrogen Peroxide Unlocks Energy in High-fiber Lignified By-products," *Nutrition Reviews* **44**, 251–252 (1986).
41. L. D. Bunting, C. R. Richardson, and R. W. Tock, "Digestibility of Ozone-Treated Sorghum Stover by Ruminants," *Journal of Agricultural Science (Camb)* **102**, 747–750 (1984).

42. V. W. Hays and R. L. Preston, "Nutrition and Feeding Management to Alter Carcass Composition of Pigs and Cattle," in H. D. Hafs and R. G. Zimbelman, eds., *Low-Fat Meats Design Strategies and Human Implications*, Academic Press, San Diego, Calif., 1994, pp. 13–33.
43. R. L. Preston, "Livestock Technologies," in B. C. English, R. L. White, and L. H. Chuang, eds., *Crop and Livestock Technologies, RCA III Symposium*, Iowa State University Press, Ames, Iowa, 1997, pp. 115–126.
44. B. K. Symonds and J. Orendorff, "Agriculture to be Changed by Emerging Biotechnology Age," *Feedstuffs* **70**, 12–18 (1998).

R. L. Preston
Texas Tech University
Lubbock, Texas

LOBSTER: BIOLOGY AND TECHNOLOGY

The animals referred to as lobsters are representative of a number of families of the decopod crustacea. The colors of lobsters are due to carotenoid pigments bound to proteins and deposited in the exoskeleton. These pigments are released from the proteins when the proteins are heat denatured during cooking, thus giving cooked lobster the characteristic reddish color (1).

There are more than 30 commercially exploited species of spiny lobsters alone. Description of distinguishing characteristics of species is beyond the scope of this article; refer to a review for guidance in lobster identification and world description (2). The families Galatheidae (squat lobster) and Scyllaridae (slipper or flat lobster) are minor contributors to world lobster catch (11 and 1%, respectively). On the other hand, the family Nephropinae (clawed lobsters) accounts for 71% of world commercial catch of lobster.

HOMARUS AMERICANUS

The American lobster has a carapace length of up to 210 mm (3) and is dark green with dark spotting above and yellowish color on the underside. These lobsters inhabit the shallow water of the Atlantic Coast of North America from shore to a depth of 480 m. Most commonly, they are found at depths of 4 to 5 m (4,5). These lobsters typically inhabit rocky areas but also make burrows under solid objects or in clay banks (4). They range from the Strait of Belle Isle, Newfoundland, to Cape Hatteras, North Carolina (4,6). References 3 and 7 give a thorough treatment of the biology of the species. This species accounts for 36% of the world lobster catch. *H. gammarus* (the European lobster) is biologically similar to *H. americanus* in all respects except distribution. It is a minor contributor in the world fishery, accounting for only 2% of the catch.

NEPHROPS NORVEGICUS

The Norway lobster (also known as scampi or Dublin Bay prawn) has an average carapace length of 30 mm for males and 20 mm for females at maturity; individuals grow as large as 70 mm (8). They live in burrows in muddy bottoms at depths of 15 to 820 m (9). Their range is from Iceland to the coast of Norway and from the Adriatic Sea south to the coast of Morocco. They are also found in the western Mediterranean (4). There are related species on the Brazilian coast and in the Indo-Pacific areas (10). In total, these species account for 34% of the world catch.

PALINURIDAE

The family Palinuridae (langouste or spiny lobster) is characterized by strong spines on the carapace and/or on the abdominal segments. Pincers, when present on the front legs, are small. European species belong to the genera *Panulirus* and *Jasus* (9,10). They are distributed throughout tropical and subtropical oceans (see Reference 1 for a map). For a discussion of geographic and vertical distribution of Palinurid lobsters, see Reference 11. Phyllosoma larvae hatch from eggs carried by females. These larvae are extremely flat, with long divided legs and delicate mouthparts. After about 12 molts while drifting in the ocean, the juveniles become puerile larvae; they swim actively and approach the coast. These larvae subsequently molt into the juvenile form of the adult at about 12 months of age (12). For biological details on one species (*P. argus*) see Reference 5. Palinuridae account for 17% of the world catch. The annual catch is about 80,000 t, mostly in Australia (12,000 t), Cuba (11,000 t), and Brazil (9000 t) (13).

CATCHING METHODS

American lobsters (*H. americanus*), European lobsters (*H. gammarus*), and spiny lobsters (*P. argus*) are primarily caught by traps (also called pots or creels) (14–17), but trawling of lobsters is legal in U.S. waters, except for in Maine (18). The recently observed practice of displaced New England cod fishermen turning to trawling for lobster is causing a great concern for the future state of the lobster resource (18,19). Some spiny lobsters are also caught by trawl or by diving (13). Norway lobsters (*N. norvegicus*) are primarily caught by trawl (20).

Traditionally, in both Europe and North America, half-round (semicylindrical) traps, made with wooden laths, were the common type of lobster traps used (15,17,21,22). Regardless of its shape, each trap contains one or more funneled entrances made of wood or netting (21,22), with the spacing of the laths set to allow undersized lobsters to escape (23). Because European lobstermen have learned that the traditional North American parlor pot is much more productive than their traditional single-chamber pots, they have introduced parlor pots into the European lobster fishery (13,24); some of these parlor pots are extremely large and made of galvanized metal (25). Similarly, in addition to the traditional wooden traps, metal traps are now also used by North American lobstermen (15,26).

A lobster trap made of wood with synthetic twine, metal and wire, or plastic does not biodegrade after the trap becomes lost on the fishing grounds. Therefore, these traps ghost fish much longer than the conventional wooden lobster traps with cotton netting at the ends (15,27). Use of nonbiodegradable traps has led to state regulations regarding both biodegradable panels and escape vents (14,23).

Before being lowered to the bottom, each lobster trap is weighted with a flat stone or concrete and baited with fish such as herring, mackerel (28), or flatfish (29). At least with spiny lobsters, traps with live sublegal-sized lobsters used as decoys caught three times as many lobsters as did traps without lobster decoys (16). All lowered traps are attached to brightly painted wooden or plastic floats on the surface. The length of time the traps are left on the bottom depends on the location, the number of traps a fisherman has, the rate of deterioration of the bait, and weather (21). Maximum productivity is obtained from daily lifting of traps (14), and mechanical hauling is often a necessity (30).

In addition to being caught by standard baited wooden or plastic traps lowered to the bottom (31), by trawls, or by divers using spears (13), spiny lobsters are also caught commercially by Caribbean fishermen using concrete slabs kept a few inches off the bottom (32). Seeking shelter, the lobsters go underneath the concrete slabs. The fisherman dives to the bottom, taps the concrete slab with a gaff, and grabs the lobsters as fast as they crawl out from underneath the concrete (32).

With Norway lobsters, which live mainly in or on muddy bottoms (20), the spread of the trawl is a more important catching factor than the trawl's headline height (33). Consequently, the towing of two trawls simultaneously from a single stern trawler has proved to be a very efficient method of catching Norway lobsters (33).

As soon as lobsters are caught, the length of the carapace is measured (15,23,34) and the sublegal-sized lobsters are returned to the ocean. With lobsters that have large claws, the claws are immobilized using twine, string, or rubber bands, not wooden pegs (35,36).

On small boats on which live holding is not practical, North American (or European) lobsters are kept as cool and damp as possible so that the lobsters are all alive when the fishermen return to port (36) a few hours later. Larger North American lobster boats often have tanks of seawater that will keep the lobsters alive until they return to port (37). Similarly, a large proportion of spiny lobsters (*P. argus*) are also kept onboard in wells of seawater and landed alive when the vessels return to port (38). Large Norway lobsters (≦30 lobster/kg) are often landed alive and sold whole, whereas those that are too small or damaged to be sold whole are tailed onboard the vessel and then washed and stored in ice (39).

LIVE HOLDING

North American lobsters, European lobsters, spiny lobsters, and large Norway lobsters may be held live onboard commercial fishing vessels (37–39). Spiny lobsters and some Norway lobsters are usually frozen before distribution (13,19,40,41). However, the vast majority of North American and European lobsters are sold live (at the retail level) (13,41). Several systems may be used to hold live lobsters (19,41,42). For example, at the first tier of distribution, the systems range from inexpensive tidal ponds with little control over the conditions to Clearwater Fine Food's $13.5 million facility at Arachat, Nova Scotia, which holds 1.5 million lb of lobsters, with each lobster in its own environmentally controlled compartment (26). Individual compartments prevent disease and cannibalism, while controlled environmental systems promote hibernation and reduce the need for food (26). Apart from storing the animals, live holding also allows for the depuration of taint from lobster meat (43).

Regardless of the system used to hold North American and European lobsters live, one should always (*1*) check animals daily and remove weak or damaged lobsters; (*2*) turn over stock on a rotating system; (*3*) avoid toxic material in construction and operation; (*4*) band or tie claws securely; (*5*) check salinity and temperature regularly; (*6*) keep water temperature at 4.5 to 10°C; (*7*) change water frequently or use an efficient filter system; (*8*) keep tanks shaded and away from bright lights; (*9*) when preparing artificial seawater, use correct weight and chemical composition; and (*10*) where possible, keep lobsters in a single layer (41). In contrast, regardless of the system used to hold North American and European lobsters live, one should never (*1*) attempt to store soft, weak, or damaged lobsters; (*2*) handle lobsters roughly; (*3*) feed lobsters in tanks, which creates more problems than it solves; (*4*) cut or peg claws to prevent fighting; (*5*) cause sudden changes in temperature or salinity; (*6*) expose lobsters to extremes of temperatures; (*7*) overload storage systems; or (*8*) overfill tanks with water, because shallow water is adequate (41).

SHIPPING

Spiny lobsters and some Norway lobsters are usually shipped frozen (13,19,44) and should be kept at or below −18°C during primary (wholesale) distribution and at or below −15°C during final distribution to smaller retail stores (45). Norway lobsters shipped fresh should be adequately chilled before being shipped and surrounded with ice inside open plastic boxes or inside covered insulated boxes (45,46). The distribution of high-quality fresh or frozen lobsters requires an integrated system of chilled or cold storage distribution centers, refrigerated vehicles, and, at the retail level, refrigerated cabinets (47).

During the 1980s, the shipment of live lobsters (American, European, and some Norway) great distances increased dramatically (47–49). This shipment is accomplished either by consumers, at their point of departure, purchasing a live lobster that is packed with seaweed and dry ice in a styrofoam container, protected by a double corrugated box, and transported live by the consumer to its final destination (48), or commercially when live lobsters are bought and transported live via commercial passenger or freight airlines to their final destination, where they are sold live commercially (47,49). For live lobsters to be of good quality when sold at their final destination, it is necessary to get lobsters from a fisherman; keep the lobsters in a holding tank for at least 24 h; separate the weak and injured lobsters; shortly before shipment, store the strong and healthy lobsters in an insulated, leakproof, lightweight, but strong container along with gel ice (a frozen chemical solution in plastic bags) and Kem-Pak (a crepe cellulose fiber blanket) or wooden shavings; store the unsealed container in a cold storage room for at least 1 h; insulate, seal, and secure the container with double strapping; ship the container; and, on reaching the final destination, remove dead or weak lobsters and place the lobsters that are in good condition in tanks of recirculating seawater (47,49). Critical factors in the distribution of live lobsters are the condition of the lobsters before shipment, maintenance of a low storage temperature of 1 to 7°C, maintenance of at least 70% relative humidity, and prevention of injury to lobsters (49). The assurance of high quality standards within the commercial lobster industry involves the utilization of production plans based on hazard analysis critical control point (HACCP) (50,51,52).

PROCESSING

American and European lobsters are primarily sold live, with some sold frozen, a few sold fresh, and a very small number canned (19), but the production of lobster product frozen with or without brine has been increasing dramatically (18,53). The quality of frozen lobster has been reported to be good for at least 9 months of frozen storage (54). Spiny lobsters are often marketed in the United States as frozen raw tails and in Asia as live or fresh whole cooked spiny lobster (13,55). In the United Kingdom, Norway lobsters are almost always sold as frozen peeled tails, whereas in the rest of Europe, they are sold whole (live, freshly cooked, or cooked and frozen) (44).

Because live lobster is normally cooked well immediately before it is eaten, it does not cause any health problems. Canned lobster products that are thermally processed inside hermetically sealed containers are also safe. However, any lobster products that are not normally cooked before being consumed present a potential health problem because of postprocessing contamination with bacteria, particularly *Listeria monocytogenes* (56,57). Consequently, the Canadian Inspection Services Branch has published *Good Manufacturing Practices* (58) for frozen ready-to-eat lobster products, which will help ensure the acceptability of such products (59). Anyone who produces any ready-to-eat lobster products should follow these or similar good manufacturing practices. References 50 through 52 provide more detailed information concerning the safety of seafood and utilization of HACCP principles.

International codes of practice regarding the handling of lobsters at sea and onshore have been published (60), as have international standards for quick-frozen lobster (61) and general processing procedures for Norway lobster (62).

More than 12 alternatives to live lobster (such as fillets, medallions, tails, cocktail claws, whole frozen, and whole cooked) are offered by producers in Maine and Canada (19). In addition, lobster paté or bites have been developed from extracted lobster meat (63), lobster Mornay is being commercially produced in Australia (64), and a lobster bisque has been developed in Denmark (65).

AQUACULTURE

Juvenile and adult Palinuridae exhibit little aggression or cannibalism even under high densities. However, at the early stage (phyllosoma), they are difficult to rear. Scientists in New Zealand have recently succeeded in rearing one species of rock lobster (*J. edwardsii*) to the puerile stage in the laboratory (66), thus increasing hopes of aquaculture of the species. Western rock lobsters have been reared from puerile to commercial size, and optimum conditions for growth have been established (67). The pond culture of juvenile rock lobsters caught in the wild has been reported (68).

Homarid lobster larvae are easily cultured, have a short development time, and can be hatched at any time of the year. They are, however, cannibalistic and must be segregated, thereby raising the costs. Nutrition is also a major problem. Lobsters normally eat a varied diet of seaweed, small mollusks, and crustacea. Such a diet is too expensive and impractical to provide on a commercial basis (69). For techniques of homarid lobster culture, see References 12, 70, 71, and 72.

CRAYFISH

Members of the crayfish group are all freshwater species distributed worldwide but are most abundant in temperate climates (73,74). More than 500 species and subspecies are recognized to belong to three families (Ascidae, Parastacidae, and Cambaridae). They range in size from 2 to 50 cm total length (73). More than 300 species are found in North America. In Europe, native crayfish were decimated by a fungal disease introduced around 1880. Subsequently, disease-resistant North American species were successfully introduced into affected areas (74).

The signal crayfish (*Pacifastacus leniusculus*), which originated in California, is believed to have the most potential for farming in temperate climates (74), but other crayfish, such as the red claw crayfish (*Cherax quadricarinutus*), are reported to have good potential for aquaculture (75–77). However, presently, the single most commercially exploited species is the red swamp crayfish, *Procambarus clarkii*. This species accounts for about 85% of world production (78). Most commercial crayfish are produced in Louisiana in culture ponds and natural habitats (79). *Procrambarus* spp. burrow in the mud of pond or swamp banks during the summer as water levels recede. Females lay their eggs and attach them to their pleopods, where they remain until hatching in 2 to 3 weeks. In the fall, when water levels rise, adults and the young of the year emerge from the burrows and into the water. The young of the year grow rapidly and can attain marketable size (65 mm total length) in 2 to 3 months (80). More than 60,000 ha of culture ponds are estimated to be in production in the United States, mostly in Louisiana. The annual natural and cultivated production in the United States is about 66,000 t, and an additional 10 t are produced in Canada. This combined total is about 80% of total world production (78).

Both farmed crayfish (November to June) and wild crayfish (March to June) are caught (usually 4 to 6 days per week) using stand-up pillow, pyramid, or barrel traps baited with natural (fish) bait or formulated artificial bait (81–83). Once harvested, 40 to 50 lb of crayfish are tightly packed inside onion sacks, protected against both excess heat and dehydration, and then transported to the processing plant, where they are processed immediately or stored live in coolers (82). The main crayfish products are live crayfish (unwashed, washed, or purged by being starved for 12 to 48 h to rid them of waste), fresh whole boiled crayfish, fresh crayfish tail meat (fat-on or washed tail meat), frozen whole boiled crayfish, frozen fat-on or washed tail meat, and frozen soft-shell crayfish (84). Information on harvesting and processing crayfish, including soft-shell crayfish, is found in References 82 through 85.

BIBLIOGRAPHY

1. B. F. Phillips, J. S. Cobb, and R. W. George. "General Biology" in J. S. Cobb and B. F. Phillips, eds., *The Biology and Management of Lobsters*, Vol. 1, Academic Press, New York, 1980, pp. 1–82.
2. A. B. Williams, "Lobsters—Identification, World Distribution and the U.S. Trade," *Marine Fisheries Revue* **48**, 1–36.
3. F. H. Herick, "The American Lobster: A Study of Its Habits and Development," *Bulletin of the U.S. Fisheries Commission*, 15 (1895).
4. L. B. Holthius, "Lobsters of the Superfamily Nephrppidae of the Atlantic Ocean (Crustacacae: Decopoda)," *Bulletin of Marine Science* **24**, 732–884 (1974).
5. A. B. Williams, *Shrimps, Lobsters and Crab of the Atlantic Coast of the Eastern United States, Maine to Florida*, Smithsonian Institution Press, Washington, D.C., 1984.
6. A. B. Williams and D. M. Williams, "Carolinian Records for American Lobster *Homarus americanus* and Tropical Swimming Carb, *Callinectus bocourti*. Postulated Means of Dispersal," *Fishery Bulletin* **79**, 192–198 (1981).
7. F. H. Herrick, "Natural History of the American Lobster," *Bulletin of the U.S. Bureau of Fisheries* **29**, 149–408 (1911).
8. A. S. Farmer, "Synopsis of Biological Data on the Norway Lobster *Nephrops norvegicus*," *FAO Fisheries Synopsis* **112**, 1–97 (1975).
9. J. S. Cobb and D. Wang, "Fisheries Biology of Lobster and Crayfishes," in A. J. Provenzano, ed., *The Biology of Crustacea*, Vol. 10 Academic Press, New York, 1985, pp. 168–230.
10. M. Burton, *The New Larousse Encyclopedia of Animal Life*, Bonanza Books, New York, 1981.
11. R. W. George and A. R. Main, "The Evolution of Spiny Lobster (*Palinuridae*): A Study of Evolution in the Marine Environment." *Evolution* **21**, 803–821 (1967).
12. P. R. Walne, "The Potential for the Culture of Crustracea in Salt Water in the United Kingdom," *Laboratory Leaflet* **40**, Ministry of Agriculture, Fisheries and Food, Directorate of Fisheries Research, Lowestoft, United Kingdom, 1977.
13. "Spiny Lobster. 1996 Buyer's Guide," *Seafood Leader* **16**, 197–204 (1996).
14. D. B. Bennett and S. R. J. Lovewell, "The Effects of Pot Immersion Time on Catches of Lobsters (*Hommarus gammarus*) in the Welsh Coast Fishery," *Fisheries Research Technical Report No. 36*, Ministry of Agriculture, Fisheries and Food, Directorate of Fisheries Research, Lowestoft, United Kingdom, 1977.
15. N. Griffen, "Closeup on Lobster," *National Fisherman* **69**, 18–21 (1988).
16. D. W. Heatwole, J. H. Hunt, and F. S. Kennedy, Jr., "Catch Efficiencies of Live Lobster Decoys and Other Attractants in the Florida Spiny Lobster Fishery," *Florida Marine Research Publication No. 44*, Florida Department of Natural Resources, Bureau of Marine Research, St. Petersburg, Fla., 1988.
17. A. Spence, "Creels, Pots, Their Treatment and Materials Used in Construction," in A. Spence, ed., *Crab and Lobster Fishing*, Fishing News Books, London, United Kingdom, 1989, pp. 73–90.
18. "American Lobster. 1995 Buyer's Guide," *Seafood Leader* **15**, 155–166 (1995).
19. "American Lobster. 1997 Buyer's Guide," *Seafood Leader* **17**, 98–107 (1997).
20. A. Bjordal, "The Behaviour of Norway Lobster Towards Baited Creels and Size Selectivity of Creels and Trawl," *Fish. Dir. Skr. Ser Hav. Unders.* **18**, 131–137 (1986).
21. J. T. Everett, "Inshore Lobster Fishing," U.S. Department of Commerce, National Oceanic and Atmospheric Administration, National Marine Fisheries Service, *Fishery Facts* **4**, 1972.
22. R. Stewart, *A Living From Lobster*, Fishing News Books, London, United Kingdom, 1971.
23. *Atlantic Fisheries Regulations*, Canada Department of Fisheries and Oceans, Ottawa, Ontario, 1985.
24. S. R. Loveweil, A. E. Howard, and D. B. Bennett, "The Effectiveness of Parlor Pots for Catching Lobsters (*Hommarus gammarus*) and Crabs (*Cancer pagurus*)," *J. Cons. Int. Explor. Mer.* **44**, 247–252 (1988).
25. H. Allen, "Giant Pots From West Coast Man . . . ," *Fishing News* 32–33 (Aug. 25, 1989).
26. J. Barnett, "Make it a Lobster Summer," *Seafood Business* 90–106 (July/August 1988).
27. P. Lazarus, "Resistance to Lobster Escape Panels Slows Down Legislation," *National Fisherman* **69**, 69–71 (1988).
28. *The Atlantic Lobster Story*, Clearwater Lobster Ltd., Halifax, Nova Scotia, Canada.
29. A. Spence, "Bait, Different According to Season," in A. Spence, ed., *Crab and Lobster Fishing*, Fishing News Books, London, United Kingdom, 1989, pp. 116–121.
30. A. Spence, "Hauling Equipment and Other Auxiliary Controls," in A. Spence, ed., *Crab and Lobster Fishing*, Fishing News Books, London, United Kingdom, 1989, pp. 56–63.
31. C. Piatt, "Florida Fisherman Are Slow to Adopt to Plastic Traps," *National Fisherman* **69**, 66–68 (1988).
32. W. E. Garrett, "LaRuta Maya," *National Geographic* **176**, 424–479 (1989).
33. T. Wray, "Stern Trawlers Prove Multi-rigs," *Fishing News International*, 26 (June 1989).
34. *Guide to the Federal Management Regulations for Lobster*, The New England Fishery Management Council, Saugus, Mass., 1989.
35. H. D. DeBoer and J. D. Castell, "A Study of Various Devices Which Can Be Used to Immobilize Lobster Claws," *Canadian Industry Report of Fisheries and Aquatic Sciences* **129** (1981).
36. A. Spence, "Marketing, Care of the Catch," in A. Spence, ed., *Crab and Lobster Fishing*, Fishing News Books, London, United Kingdom, 1989, pp. 126–136.
37. M. Crowley, "Shafmaster's Log Low Lobster Boats Are Steady as a Rock," *National Fisherman* **67**, 34–36 (1986).
38. B. H. Lamadrid and A. W. Blanco, "Handling of Live Lobsters in Cuba," *Infofish Marketing Digest* **6**, 44–46 (1986).

39. "Nephrops Rising to Fame," *Seafood International* **3**, 50–55 (1988).
40. "Live Tanks Nourish Profits," *Seafood Leader* 171–181 (Fall 1987).
41. P. A. Ayres and P. C. Wood, "Live Storage of Lobsters," *Laboratory Leaflet No. 37*, Ministry of Agriculture, Fisheries and Food, Directorate of Fisheries Research, Lowestoft, United Kingdom, 1977.
42. "Selling Seafood Live," *Seafood International* **3**, 37–43 (1988).
43. U. P. Williams et al., "Tainting and Depuration of Taint by Lobsters (*Homarus americanus*) Exposed to Water Contaminated With a No. 2 Fuel Oil: Relationships with Aromatic Hydrocarbon Content in Tissue," *J. Food Sci.* **54**, 240–243, 257 (1989).
44. "Norway Lobster (*Nephrops norvegicus*), Buyer's Guide," *Seafood Leader* **7**, 214–220 (1987).
45. "Seafood on the Move," *Seafood International* **2**, 33–35 (1987).
46. "More Seafood Takes to the Road . . . as Specialist Distribution Steps Up," *Seafood Processing and Packaging* **2**, 19–25 (1988).
47. "Keep It Cool, Clean and Moving," *Infofish International* **6**, 24–28 (1988).
48. "Lobsters Take-off at U.S. Airports," *Seafood International* **2**, 17–18 (1987).
49. H. Lisac, "A Note on the Live Shipment of Live Lobsters," *Infofish Marketing Digest* **4**, 38–40 (1986).
50. S. Subasinghe, "Handling and Marketing Aquaculture Products," *Infofish International* **3**, 44–51 (1996).
51. E. S. Garrett, M. J. Jahncke, and J. M. Tennyson, "Microbiological Hazards and Emerging Food-Safety Issues Associated With Seafood," *J. Food Prot.* **60**, 1409–1415 (1997).
52. National Advisory Committee on Microbiological Criteria for Foods, "Hazard Analysis and Critical Control Point Principles and Application Guidelines," *J. Food Prot.* **61**, 762–775 (1998).
53. "Lobster Update," *Seafood Leader* **16**, 211 (1996).
54. "Freezing in a Flash," *Seafood Leader* **13**, 98–100 (1993).
55. "Spiny Lobster. 1993 Buyer's Guide," *Seafood Leader* **13**, 157–162 (1993).
56. R. Fitzgerald, "Listeria hysteria," *Seafood Leader* **6**, 232–233 (1988).
57. S. D. Weagant et al., "The Incidence of *Listeria* Species in Frozen Seafood Products," *J. Food Prot.* **51**, 655–657 (1988).
58. Department of Fisheries and Oceans, Inspection Services Branch, *Good Manufacturing Practices (GMP)—Lobster Processing*, Document 890410-58-G-I-L, Ottawa, Ontario, Canada, 1988.
59. H. H. Huss, "Microbiological Quality Assurance in the Fish Industry," *Infofish International* **5**, 36–37 (1989).
60. Codex Alimentarius Commission, Joint FAO/WHO Food Standards Programme, *Thirteenth Session of the Codex Committee on Fish and Fishery Products, Bergen, Norway, May 7–11, 1979. Recommended International Code of Practice for Lobsters*, 1979.
61. Codex Alimentarius Commission, Joint FAO/WHO Food Standards Programme, "Codex Standard for Quick Frozen Lobsters," Codex Standard 95, in *Codex Alimentarius*, Vol. V, FAO/WHO, Rome, Italy, 1981.
62. J. C. Early, "Process Norway Lobsters," *Torry Advisory Note No. 29*, UK Ministry of Technology, Torry Research Station, 1965.
63. D. D. Duxbury, "Lobster Paté or Bites for Seafood Hors d'oeuvres," *Food Processing* **47**, 68–70 (1986).
64. "Constantia Food's Shellfish Microwave Appetisers," *Seafood International* **3**, 23 (1988).
65. P. Moustgaard, "Denmark's Changing Markets," *Seafood International* **4**, 24–29 (1989).
66. J. Illingworth et al., "Upwelling Tank for Culturing Rock Lobster (*Jasus edwardsii*) Phyllosomas," *Mar. Freshwater Res.* **48**, 911–914 (1997).
67. R. G. Chittleborough, "Review of Prospects for Rearing Rock Lobsters," *Australian Fish*, 4–8 (April 1974).
68. D. O. C. Lee and J. F. Wickens, *Crustacean Farming*, Blackwell Scientific Publications, Oxford, United Kingdom 1992.
69. E. Edwards, "The Future for Farmed Lobster," *Seafood International* **4**, 33–34 (1989).
70. T. W. Beard, P. R. Richards, and J. F. Wickins, "The Techniques and Practicability of Year-round Production of Lobsters, *Homarus gammarus* (**L**), in Laboratory Recirculation Systems," *Fisheries Research Technical Report 79*, Directorate of Fisheries Research, Lowestoft, United Kingdom, 1985.
71. A. J. Provenzano, Jr., "Commercial Culture of Decapod Crustacians," in A. J. Provenzano, ed., *The Biology of Crustacea*, Vol. 10, Academic Press, New York, 1985, pp. 269–323.
72. J. C. VanOlst, J. M. Carlberg, and J. T. Hughes, "Aquaculture," in J. S. Cobb and B. F. Phillips, eds., *The Biology and Management of Lobsters*, Vol. 2, Academic Press, New York, 1980, pp. 333–384.
73. H. H. Hobbs, Jr., "Crayfish Distribution, Adaptive Radiation and Evolution," in D. M. Holdich and R. S. Lowery, eds., *Freshwater Crawfish: Biology, Management, and Exploitation*, Croom Helm, London, United Kingdom, 1988, pp. 52–82.
74. R. E. Groves *The Crayfish: Its Nature and Nurture*, Fishing News Books Ltd, Surrey, United Kingdom, 1985.
75. M. C. Rubino, "Economics of Red Claw (*Cherax quadricarinatus*, von Martens, 1868) Aquaculture," *Journal of Shellfish Research* **11**, 157–162 (1992).
76. J. V. Huner, "An Overview of the Status of Freshwater Crawfish Culture," *Journal of Shellfish Research* **14**, 539–543 (1995).
77. D. B. Rouse, "Australian Crayfish Culture in the Americas," *Journal of Shellfish Research* **14**, 569–572 (1995).
78. J. V. Huner, "Overview of International and Domestic Freshwater Crawfish Production," *Journal of Shellfish Research* **8**, 259–265 (1989).
79. L. W. de la Bretonne, Jr., and R. P. Romaire, "Commercial Crawfish Cultivation Practices: A Review," *Journal of Shellfish Research* **8**, 267–275 (1989).
80. J. W. Avault, Jr. and J. V. Huner, "Crawfish Culture in the United States," in J. V. Huner and E. E. Brown, eds., *Crustacian and Mollusk Aquaculture in the United States*, AVI, Westport, Conn., 1985, pp. 1–54.
81. "Crawfish, *Cambaridae* and *Astaciadae*," *Seafood Leader* **9**, 262–270 (1989).
82. M. W. Moody, "Processing of Freshwater Crawfish: A Review," *Journal of Shellfish Research* **8**, 293–301 (1989).
83. R. P. Romaire, "Overview of Technology Used in Commercial Crawfish Aquaculture," *Journal of Shellfish Research* **8**, 281–286 (1989).
84. "Crayfish. Seafood Buyer's Guide," *Seafood Leader* **14**, 213–221 (1994).

85. D. D. Cully and L. Duobins-Gray, "Soft-shell Crawfish Production Technology," *Journal of Shellfish Research* 8, 287–291 (1989).

J. R. Botta
Moncton Food Quality Investigation, Inc.
Moncton, New Brunswick
Canada

J. W. Kiceniuk
Fernleigh House
Halifax, Nova Scotia
Canada

LOW-ACID AND ACIDIFIED FOODS

A low-acid food is a food that has a pH value greater than 4.6; an acid food is a food with a natural pH of 4.6 or below; and an acidified food is a low-acid food that has been treated to reduce the pH to 4.6 or below. The pH of a food is one of the most important parameters in determining the types of microorganisms that are capable of growing in the product. This in turn determines the types of processes that are appropriate to preserve that food. Although the terms *low-acid food* and *acidified food* do not in the strictest sense refer to canned foods, the terms are most frequently applied by the food industry to canned food products, and this will be the focus here. It should be noted that pH is not the only factor involved in preservation of canned foods. Reduced water activity, preservatives, and other factors may also be used in combination with heat to preserve foods; for some foods a heat treatment is not required because of the combinations of preservation. However, this article focuses on traditional low-acid and acidified canned foods.

PH OF VARIOUS FOODS

The pH of selected fresh, fermented, and commercially canned foods is shown in Table 1. In general, the pH of most fruits falls in the acid range (notable exceptions include bananas and cantaloupes), whereas the pH of meat, poultry, fish, and vegetables falls in the low-acid range. Fermentation may be used to reduce the pH of a low-acid food, as is done with sauerkraut, pickles, and yogurt. For other foods the pH is lowered by direct addition of acid (marinated vegetables such as mushrooms or three-bean salad). A variety of foods are formulated with low-acid and acid components. In such cases the acid component may be used to acidify the low-acid component. An example of this might be a spaghetti sauce in which the tomato component (tomatoes, tomato paste, and/or tomato sauce) acidifies the low-acid components (meat, onions, peppers).

MICROBIOLOGY OF LOW-ACID AND ACIDIFIED FOODS

All raw foods contain microorganisms. These organisms will grow and spoil foods unless the organisms are properly controlled. Even more important, among the microorganisms that can be found on raw foods are pathogens. Thus,

Table 1. pH Range of Selected Fresh, Fermented, and Commercially Canned Foods

Food	Approximate pH range
Apples	2.9–3.5
Apple juice	3.3–3.5
Apple sauce	3.4–3.5
Apricots	3.5–4.0
Asparagus	5.0–6.1
Bananas	4.5–5.2
Beans	
Baked	4.8–5.5
Green	4.9–5.5
Lima	5.4–6.5
Soy	6.0–6.6
Beans, with pork	5.1–5.8
Beef	5.3–6.2
Beef, corned, hash	5.5–6.0
Beers	4.0–5.0
Beets, whole	4.9–5.8
Blackberries	3.0–4.2
Blueberries	3.2–3.6
Boysenberries	3.0–3.3
Bread	
White	5.0–6.0
Date and nut	5.1–5.6
Broccoli	5.2–6.0
Brussel sprouts	6.3–6.6
Butter	6.1–6.4
Cabbage	5.2–6.3
Cantaloupe	6.2–6.5
Carrots	4.9–6.3
Catfish	6.6–7.0
Cauliflower	6.0–6.7
Celery	5.7–6.0
Cheese (most)	5.0–6.1
Camembert	6.1–7.0
Cottage	4.1–5.4
Parmesan	5.2–5.3
Roquefort	4.7–4.8
Cherries	3.2–4.7
Chicken	5.5–6.4
Chicken with noodles	6.2–6.7
Chop Suey	5.4–5.6
Cider	2.9–3.3
Clams	5.9–7.1
Codfish (Canned)	6.0–6.1
Corn	
On the cob	5.9–6.8
Cream style	5.9–6.5
Whole grain	
Brine packed	5.8–6.5
Vacuum packed	6.0–6.4
Crab	6.8–8.0
Crab apples—spiced	3.3–3.7
Cranberry	
Juice	2.5–2.7
Sauce	2.3
Currant juice	3.0
Dates	6.2–6.4
Duck	6.0–6.1
Dry sausages	4.4–5.6
Egg white	7.6–9.5
Egg yolk	6.0–6.3
Fermented vegetables	3.9–5.1
Figs	4.9–5.0
Frankfurters	6.2

Table 1. pH Range of Selected Fresh, Fermented, and Commercially Canned Foods *(continued)*

Food	Approximate pH range
Fruit cocktail	3.6–4.0
Ginger ale	2.0–4.0
Gooseberries	2.8–3.1
Grapefruit	
Juice	2.9–3.4
Pulp	3.4
Sections	3.0–3.5
Grapes	3.3–4.5
Haddock	6.2–6.7
Halibut	5.5–5.8
Ham, spiced	6.0–6.3
Hominey, lye	6.9–7.9
Honey	6.0–6.8
Huckleberries	2.8–2.9
Jam, fruit	3.5–4.0
Jellies, fruit	3.0–3.5
Lemons	2.2–2.4
Juice	2.2–2.6
Lettuce	6.0–6.4
Limes	1.8–2.0
Juice	2.2–2.4
Loganberries	2.7–3.5
Mackerel	5.9–6.2
Maple syrup	6.5–7.0
Mayonnaise	3.8–4.0
Milk	
Cow, whole	6.0–6.8
Evaporated	5.9–6.3
Molasses	5.0–5.4
Mushrooms	6.0–6.5
Olives, ripe	5.9–7.3
Onions	5.3–5.8
Oranges	2.8–4.0
Juice	3.0–4.0
Oysters	5.9–6.7
Peaches	3.1–4.2
Pears	3.4–4.7
Peas	5.6–6.8
Pickles	
Dill	2.6–3.8
Sour	3.0–3.5
Sweet	2.5–3.0
Pimentos	4.3–5.2
Pineapple	
Crushed	3.2–4.0
Sliced	3.5–4.1
Juice	3.4–3.7
Plums	2.8–4.6
Pork	5.3–6.4
Potatoes	
White, whole	5.4–6.3
Sweet	5.3–5.6
Potato salad	3.9–4.6
Prune juice	3.7–4.3
Pumpkin	5.2–5.5
Raspberries	2.9–3.7
Rhubarb	2.9–3.3
Salmon	6.1–6.5
Sardines	5.7–6.6
Sauerkraut	3.1–3.7
Juice	3.3–3.4
Scallops	6.8–7.1
Shrimp	6.8–8.2

Table 1. pH Range of Selected Fresh, Fermented, and Commercially Canned Foods *(continued)*

Food	Approximate pH range
Soda crackers	6.5–8.5
Soups	
Bean	5.7–5.8
Beef broth	6.0–6.2
Chicken noodle	5.5–6.5
Clam chowder	5.6–5.9
Duck	5.0–5.7
Mushroom	6.3–6.7
Noodle	5.6–5.8
Oyster	6.5–6.9
Pea	5.7–6.2
Tomato	4.2–5.2
Turtle	5.2–5.3
Vegetable	4.7–5.6
Spinach	4.8–6.8
Squash	5.0–5.3
Strawberries	3.0–4.2
Tomatoes	3.7–4.7
Juice	3.9–4.7
Tuna	5.9–6.1
Turkey	5.6–6.0
Turnips	5.2–5.6
Turnip greens	5.4–5.6
Vegetable	
Juice	3.9–4.3
Mixed	5.4–5.6
Vinegar	2.4–3.4
Walnuts	5.4–5.5
Water	
Distilled, CO_2	6.8–7.0
Mineral	6.2–9.4
Sea	8.0–8.4
Whiting	6.2–7.1
Wines	2.3–3.8
Yogurt	3.8–4.2

Source: Reprinted from Ref. 1 with permission from *Dairy, Food and Environmental Sanitation*. Copyright held by the International Association of Milk, Food and Environmental Sanitarians, Inc. Gravani is with Cornell University.

to ensure foods are safe, these pathogens must be destroyed or reduced to levels at which they pose no hazard to public health. In some instances, controlling the growth of pathogens that may be present will render the food safe to eat. The thermal processes designed for low-acid foods, and the pH along with a heat treatment for acid and acidified foods, destroy pathogens and spoilage organisms capable of growing in the product under normal distribution conditions. The can or jar prevents recontamination of the food after it has been heat treated. Thus these products are rendered shelf stable, and they are considered *commercially sterile*. They cannot be called *sterile* because some of these products may contain viable microorganisms that cannot grow under normal conditions. For example, low-acid canned vegetables may contain spores of thermophilic bacteria that cannot grow unless temperatures exceed 100°F for several days. Canned fruits may contain the spores of *Bacillus* or *Clostridium* species that cannot grow because the pH of the product is too low. Knowledge of the microorganisms that may be present on a product and their characteristics is necessary to establish an appropriate process to prevent health hazards and spoilage of canned foods.

It is not possible to cover the field of food microbiology here. Readers should consult additional texts for detailed information on this subject (2–4). This article briefly covers, in general terms, those organisms of concern in low-acid and acidified foods.

Low-Acid Canned Food Microbiology

While low-acid foods may be contaminated with viruses, yeast, molds, parasites, and bacteria, it is the spore-forming bacteria that are of greatest concern from the standpoint of sterilization. The bacterium most important for low-acid canned foods is *Clostridium botulinum*, which produces spores that survive boiling. When the vegetative cells of *C. botulinum* grow, they produce potent neurotoxins (Types A, B, and E are of greatest significance) that can cause death if consumed. *C. botulinum* Type A and proteolytic strains of Type B have decimal reduction (D) values at 250°F (121°C) of 0.10 to 0.20 minutes (5), the D value being the time in minutes at the specified temperature to destroy 1 log, or 90%, of the population. Thus, thermal processes for low-acid canned foods must be designed to destroy the spores of this organism. By convention, a process equivalent to 12D is considered adequate to protect the public health. However, there are other clostridia important in food spoilage that produce more heat-resistant spores. To produce shelf-stable products, processes are therefore designed to destroy the spores of organisms such as PA 3679, an organism similar to *Clostridium sporogenes*, that has a $D_{250°F}$ value of 0.50 to 1.50 minutes (5). The heat treatments required to render low-acid foods commercially sterile will destroy viruses, yeast, molds, parasites, and bacterial pathogens such as *Salmonella* and *Escherichia coli* O157:H7.

Acid and Acidified Food Microbiology

Since *C. botulinum* spores cannot germinate and grow in foods at pH values below approximately 4.8 (regulations setting pH 4.6 as the cutoff between low-acid and acid/acidified include a safety factor), heat treatments for acid and acidified foods need not destroy the spores of this organism. Heat treatments for products with pH values <4.0 are designed to inactivate yeast, molds, and lactic acid bacteria that can cause spoilage. The most resistant of these organisms have $D_{150°F}$ values on the order of 1.0 min. A notable exception are the ascospores of the heat-resistant molds such as *Byssochlamys, Neosartorya*, and *Talaromyces* (6–8). These organisms may have $D_{194°F}$ values of 1 to 12 min, which allow them to survive commercial processes; fortunately, such spoilage is rare.

For products with pH values between 4.0 and 4.6, there are spore-forming bacteria that can be important in spoilage. Of greatest importance are *Bacillus coagulans* and the butyric anaerobes such as *Clostridium pasteurianum* and *C. butyricum*. Products in this pH range are usually given more severe heat treatments to prevent spoilage from these organisms. Recently, spoilage in fruit juices has been caused by spore formers of the newly recognized genus *Alicyclobacillus* (9). These organisms can grow and spoil products at pH 3.5, and they can survive the commercial processes for juice products. Studies to characterize and control these organisms are under way; as with heat-resistant molds, spoilage of this type is not common.

Although the ability of pathogens such as *Salmonella* and *E. coli* O157:H7 to grow in acid and acidified foods depends on the food, the pH, the acidulent, and other factors, it is not always necessary for these pathogens to grow to make someone sick. Thus it is important that these organisms be destroyed by a heat treatment. The pasteurization treatments given acid and acidified foods are more than adequate to kill these pathogens, as well as organisms such as the enteric protozoan parasite *Cryptosporidium*. In recent years, all three of these pathogens have caused outbreaks of foodborne illness as the result of consumption of contaminated, unpasteurized fruit juices (10).

THERMOBACTERIOLOGY

To properly design a thermal process for a product, one must know the heat resistance of the organisms of concern; these organisms were just discussed. Procedures for determining the heat resistance of a microorganism are similar, regardless of whether the organism is a spore former or a vegetative cell.

Heating Medium

The heat resistance of a microorganism is frequently determined in a buffer such as phosphate buffer because it is a homogeneous, easily reproducible liquid. Moreover, when the heat resistances of microorganisms are determined in the same buffer, they can be compared. However, it is well known that foods may affect microbial heat resistance; heat resistance may be either higher or lower in a food compared with phosphate buffer. For example, fat may be protective of organisms, resulting in increases in resistance. Heat resistance can increase with reduced water activity, and it will be affected by the pH of the heating medium. Thus, determining the heat resistance of a microorganism in a food more accurately predicts the thermal process requirements. However, food products are more difficult to work with than buffer.

Heating Methods

A variety of methods have been used for determining the heat resistance of a microorganism:

1. For liquids being heated at 100°C or below, a three-neck flask can be used. A thermometer is included in one neck of the flask, a stirrer in another, and the third is used to add microorganisms and to remove samples. This is a convenient, rapid method for determining the heat resistance of vegetative cells. Heating lag times are minimized because small volumes of inoculum are added to a large volume of preheated liquid, but splashing or flocculation may give erroneous results.

2. One of the most common methods uses thermal death time (TDT) tubes to which a given volume (usually 1–2 mL) of product containing the microorganism is added, and the tube is sealed and heated in an oil or water bath (depending on the heating temperatures). The procedure is useful for liquids and liquefied food homogenates. There are heating and cooling lag times (since heat must penetrate into the tube and heat the product) that must be considered, but a thermocouple in a replicate sample can be used to measure the temperature during heating and cooling to account for these lags in the TDT calculations. It is easy to heat many replicates at one time. Usually 10 tubes per time interval and five or six time intervals are heated at the same time at a single temperature, and the procedure is repeated at additional temperatures. In general, the TDT tubes will need to be opened and the contents subcultured in an appropriate growth medium or the food product. The procedure is suitable for vegetative cells and spores.
3. A similar method uses capillary tubes, which are much smaller versions of TDT tubes. Because of the small diameter and thickness of capillary tubes, heating and cooling lags are negligible. This procedure is limited to liquids, and the contents must be subcultured. The external surface of the tube must be sterilized prior to subculturing the contents, because the tube is generally crushed to release its contents. The procedure is suitable for vegetative cells and spores.
4. A specialized piece of equipment known as a thermoresistometer has been used to determine the heat resistance of microorganisms at very high temperatures (121–138°C in saturated steam and 121–260°C in superheated steam) for very short time intervals (seconds). The procedure is suitable for liquids or diluted food homogenates. Heating and cooling lag times are negligible, and the system allows for very precise timing, which is critical with high temperature–short time studies. There is a high initial cost for the equipment, and product must be subcultured for survivors. This procedure is suitable for spores.
5. Another specialized piece of equipment for TDT studies is the thermal death time retort. These retorts are used to heat either TDT tubes (containing liquid or food homogenates) or TDT cans (containing ground or homogenized foods) at temperatures of 100 to 121°C. Use of TDT cans closely simulates canning practices, and the cans can be incubated directly. Spoilage by gas producers is detected by swelling of the can. Heating and cooling lag times are substantial, and it is essential to account for the lethality that occurs by measuring the temperatures with a thermocouple inserted in replicate cans. The incubation time can be lengthy—several months to ensure all injured organisms are recovered. This procedure is suitable for spores.

Recovery of Survivors

The apparent heat resistance of an organism depends on the recovery medium, as well as the substrate in which it is heated. When testing the heat resistance of a microorganism in a particular food, recovery in the product is preferred, as this will reflect the environment the injured organisms would see after processing. This procedure measures whether or not there are organisms that can grow in the product after the heat treatment, not how many there are. Alternatively, the heated product can be subcultured to determine survivors. This may be either a qualitative procedure, where the heated product is transferred to a broth medium to determine if there are organisms that can grow, or a quantitative procedure, where the survivors in the heated product are enumerated. Use of a culture medium has several advantages. The media used generally allow better recovery of survivors, as the medium can be formulated to favor growth of the specific organism being tested, and components that enhance the recovery of injured organisms can be added. Also, organisms generally recover more quickly in the more favorable environment.

Thermal Death Time Calculations

An in-depth discussion of TDT calculations is not possible here. Readers are referred to works such as Ref. 5. The following formula is used to calculate the D value:

$$D_{\mathrm{T}} = \frac{t}{\log \mathrm{A} - \log \mathrm{B}}$$

where t is the heating time (corrected for the come up time) at temperature T, A is the initial number of organisms, and B is the number of survivors. Alternatively, the D value can be determined by plotting the number of survivors at several time intervals against time on semilogarithm paper. (The D value is calculated from the slope of the best-fit line.) D values determined at several temperatures are plotted on semilogarithm paper against temperature (called a phantom TDT curve), and the z value (the change in the thermal death rate with temperature) is calculated from the slope of the line. Knowing the D and z values for an organism allows the calculation of the D value at any other temperature. The F value, or the number of minutes to destroy a given number of organisms at a specific temperature, can also be calculated. The minimum health sterilizing value for a low-acid food is generally a 12D value for *C. botulinum* (12 times the D value determined for *C. botulinum* spores in the specific food); the commercial sterility value is generally 5D for *C. sporogenes* (PA3679) spores.

REGULATIONS GOVERNING LOW-ACID AND ACIDIFIED FOODS

The U.S. Department of Agriculture's (USDA) Food Safety and Inspection Service (FSIS) has responsibility for meat, poultry, and egg products. The Food and Drug Administration (FDA) has responsibility for all other foods, including seafood and shell eggs. For details on the regulations, see the article titled CANNING: REGULATORY AND SAFETY CONSIDERATIONS.

FDA Regulations

The FDA's general regulatory requirements for foods ("Current Good Manufacturing Practice in Manufacturing, Packing, or Holding Human Food") are found in Title 21 of the *Code of Federal Regulations (CFR)* Part 110. Requirements for "Thermally Processed Low-Acid Foods Packaged in Hermetically Sealed Containers" are found in 21 *CFR* 113, and requirements for "Acidified Foods" in 21 *CFR* 114. Foods that have a water activity of 0.85 or less and those that are refrigerated or frozen are exempt from the low-acid and acidified food regulations, as are carbonated and alcoholic beverages, jams, jellies, preserves, dressings, and condiments. The FDA has recently mandated Hazard Analysis and Critical Control Point (HACCP) regulations for seafood and seafood products (21 *CFR* 123), although canned foods complying with 21 *CFR* 113 or 114 do not have to include critical control points for microbiological hazards. In addition, 21 *CFR* 108 describes the emergency permit system that can be imposed on processors who fail to comply with the regulations.

The regulations in 21 *CFR* 113 define low-acid foods and commercial sterility. They specify the proper design, controls, and instrumentation for all the common retorting systems and the practices necessary in the operation of these systems to ensure safety. The regulations require that records be kept of all coding, processing, and container closure inspections and that these be reviewed prior to shipment of product. The regulations also describe the actions to take when a process deviation has occurred. They require that the heat process be designed by qualified persons having expert knowledge of thermal processing requirements (processing authority). The term *processing authority* is not explicitly defined in the regulations, there are no specific criteria for qualifications, nor does FDA (or USDA) maintain a list of processing authorities. However, certain organizations, such as the National Food Processors Association (NFPA), are widely recognized by government agencies and the food industry as having the experience and expertise to serve in this capacity (11). Some food processing firms have individuals on staff who can serve as the processing authority, and a number of consultants are recognized as processing authorities. Another requirement is that all thermal processing operations be conducted under the supervision of an individual who has satisfactorily completed an FDA-approved course of instruction; these Better Process Control Schools, sponsored jointly by the FDA, the Food Processors Institute (FPI) (12), and key universities, are held around the country every year. Occasionally such courses are given to specific companies or overseas.

The regulations in 21 *CFR* 114 define acidified foods and describe procedures for acidification (described in the next section). The regulations require that the process be established by a processing authority, and they define the procedures to be followed in the event of a deviation from the scheduled process or the pH exceeds 4.6. They describe the methods to determine pH or acidity for acidified foods and the records that must be kept. Supervisors with responsibility for pH control and other critical factors must receive instruction through an FDA-approved course.

USDA FSIS Canning Regulations

In general, foods containing more than 2% cooked (3% raw) meat or poultry fall under the jurisdiction of the USDA. The regulations for canned meat products are found in 9 *CFR* 318.300 and for canned poultry products in 9 *CFR* 381.300; they are essentially identical. Requirements are very similar to those for FDA products. As with the FDA, the USDA has also mandated HACCP regulations for meat and poultry products (9 *CFR* 417), although canned foods complying with 9 *CFR* 318.300 or 381.300 do not have to include critical control points for microbiological hazards.

ESTABLISHING A PROCESS

Acidified Foods

Processes for acidified foods are based on a relationship between the pH of the product and temperature. Such processes are to be established by a qualified person having expert knowledge in acidification and processing of acidified foods.

Acidification Procedures. Five methods are used in the industry to acidify a low-acid product (21 *CFR* 114.80):

1. *Blanching of the food ingredients in acidified aqueous solutions.* An effective way to acidify large particles, this method will require control of the time and temperature of the blanch, as well as the type and concentration of the acid.
2. *Immersion of blanched food in acid solution.* This two-step acidification process requires blanching the low-acid product in a normal steam or water blancher, followed by dipping the blanched product into an acid solution. The product is removed from the acid solution prior to filling into the containers. The blanch procedure, concentration of the acid solution, and length of time the product is in the acid solution need to be controlled to ensure proper acidification.
3. *Direct batch acidification.* This method is used for liquid products or other products mixed in batch kettles. Acid is added to the batch in predetermined levels that will achieve the desired lowering of pH. The pH level is confirmed prior to releasing the batch for filling.
4. *Direct addition of predetermined amount of acid in individual containers.* Liquid acid solutions or acid pellets are added to each container of product. Diligent control is necessary to ensure that each individual container receives the proper addition of acid. Control is also necessary to ensure that the proper solid-to-liquid ratio is obtained to achieve adequate acidification of the low-acid components.
5. *Addition of acid foods to low-acid foods.* Proper attention to formulation is necessary for this final

method of acidification that relies on the acid from the acid food to acidify the low-acid component. The proportion of acid and low-acid components will be critical to ensure a proper final equilibrium pH.

These five methods of acidification are all acceptable; each requires a certain degree of control for proper acidification. Some products may require more than one method to guarantee consistent acidification.

Pasteurization Process. Pasteurization processes for acidified foods are designed to destroy any vegetative pathogens and spoilage organisms that could grow in the reduced pH environment. The processes are established to either deliver a final product temperature at the end of heating or a specific F value for the microorganism of concern. The heating rate of the product needs to be determined for processes established to deliver a specific F value. Product heating rates for acidified foods are determined with heat penetration studies. These studies will be discussed in detail in the following section on low-acid foods. Processes that rely on a final product temperature may be monitored with maximum reading thermometers or by actual temperature measurements from containers periodically sampled at the end of the final heating section.

Low-Acid Foods

Establishing a thermal process for low-acid foods requires knowledge of both the heat resistance of microorganisms, which has been discussed previously, and the heating rate of the products. The heating rate of a product is determined by conducting heat penetration studies, which measure the changes in product temperature during processing. In addition to establishing a thermal process, the thermal process system must be operated in an appropriate manner to ensure the production of a commercially sterile product. Temperature distribution studies are conducted to establish the operating procedures for the thermal processing systems. The regulations previously discussed require that thermal processes and thermal processing system operating procedures be established by a processing authority.

Heat Penetration. Heat penetration studies involve measuring the product temperature inside test containers by placing one temperature sensor in each container. A series of tests are conducted to locate the slowest heating zone within the container and then to confirm the heating rate in the slowest heating zone. A processing authority will analyze the data from all of the heat penetration tests to determine the heating rate of the product and calculate thermal processes. The heat penetration tests should be designed to collect data under the "worst case" conditions that could be reasonably encountered in commercial production. Test design should consider product preparation, temperature sensor location, container fill procedures, container geometry and orientation, product initial temperature, retort temperature, and other possible critical factors.

Products typically heat by conduction, convection, or a combination of both. Conduction heating involves particle-to-particle heat transfer with no gross particle movement. Products that heat by conduction are typically viscous, with little free liquid—pumpkin, stews, corned beef hash, and condensed soups are examples. Convection heating involves particle-to-particle heat transfer with particle movement. Products that heat by convection are typically thin or contain free liquid—brine-packed products such as mushrooms and green beans are examples.

Copper-constantan (Type T) thermocouples are usually the temperature sensor of choice for heat penetration testing. The time- and temperature-recording equipment used for heat penetration tests normally consists of a multichannel recording potentiometer or computer modified for data collection. The thermocouples should be connected to the recording equipment with copper-constantan thermocouple wires. The test containers may be processed in a commercial retort or in a pilot retort simulator. The thermocouple wires are normally installed through the retort wall using a stuffing box or packing gland to seal the opening into the retort.

The data must be analyzed by a processing authority. The time-temperature data from all heat penetration containers should be analyzed unless there is a problem noted with a specific temperature sensor. The slowest heating thermocouple located at the cold zone is used to define the product heating rate for process establishment. The processing authority will use this information, along with the thermal resistance data, to calculate a thermal process. Several mathematical methods have been developed to calculate a thermal process, but the scope of this chapter does not allow further discussion.

Temperature Distribution. Thermal processing systems (retorts) must be operated in such a way that the process timing does not start until the temperature-indicating device (mercury-in-glass thermometer [MIG] or equivalent) reaches the process temperature and uniform temperatures are achieved throughout the retort. Processing authorities conduct temperature distribution tests to develop the operating procedures that are necessary to establish uniform temperatures throughout the retort. For steam retorts, the operating procedure is commonly referred to as the venting procedure or vent schedule. Venting procedures are designed to remove the air present in the retort and replace it with saturated steam. Vent schedules have at least two critical components: time and temperature.

The timing of the retort process cannot begin until the scheduled process temperature has been achieved and the prescribed operating procedures have been completed. Properly designed operating procedures will ensure that the retort temperature, as indicated by the temperature-indicating device, is uniform in the retort environment.

The processing authority needs to be aware of the many variables that may affect the development of an adequate operating procedure. A few examples of these variables include container loading configuration, percent open area of the baskets and divider plates, vent pipe sizes, or pump speed.

The distribution temperature sensors (typically a continuous Type T thermocouple wire with the copper and constantan wire tips connected to form the temperature sensor) are distributed among the containers to monitor the temperatures throughout the retort load. In addition, at least one sensor is typically located next to the temperature-indicating device to serve as a reference against which all the other sensors can be compared. The temperature-recording equipment used for distribution tests is similar to that used for heat penetration tests.

The temperature distribution data are analyzed by the processing authority to establish the operating procedure, which will provide a reproducible relationship between the temperature-indicating device and the thermocouples located throughout the load of containers. Once established, the operating procedure needs to be followed each time the retort is operated to ensure adequate delivery of the scheduled process.

TYPES OF CONTAINERS

The types of containers used for low-acid and acidified foods have expanded quite a bit in the past 20 years. The metal can has proven to be a versatile and well-established container, but the industry is also producing low-acid and acidified products in glass, semirigid, and flexible containers.

The metal can commonly used is either a three-piece steel can or two-piece steel or aluminum can. The can designation is determined by the number of metal pieces needed to form the can. Three-piece cans are formed from three pieces of metal: top, bottom, and body. The cylindrical body is formed from one piece of metal rolled and joined by welding a side seam. Two-piece cans are formed from two pieces of metal: a top-end and can body that includes the bottom. The body and bottom are formed from one flat, circular piece of metal by either a drawing and ironing process or a drawing and redrawing process. The can tops (and bottoms for three-piece cans) are attached to the can bodies in a complex, double-seaming operation. In addition to the familiar cylindrical can, the industry also uses oblong, rectangular, and half steam table tray metal cans.

Glass packages (jars or bottles) are another common container and are probably more readily used for acid and acidified food than the metal can. The glass container is sealed with a metal closure to ensure the integrity of the package. The closure may be one of several different designs. The lug or twist cap and the PT (Press-on Twist-off) cap are commonly used for low-acid food products and will withstand the retort pressure process. The plastisol-lined, continuous thread cap is often used for acidified foods.

Semirigid and flexible containers are gaining wider acceptance in the marketplace. Some semirigid and flexible containers, such as the retort pouch, plastic containers with double seamed metal ends, and some semirigid containers with heat-sealed lids, will withstand the pressures of a retort process and are suitable for low-acid foods. Other containers (paperboard, flexible pouches, and semirigid containers with heat-sealed lids) will not withstand retort processing, but low-acid foods can be packaged acceptably into them with aseptic processing and packaging techniques. All of these semirigid and flexible containers have been adapted for use with acidified and acid products that do not require a severe heat process.

DEVIATION MANAGEMENT

A thermal process deviation is defined as a thermal process that is less than scheduled and/or in which one or more of the critical factors has not been satisfied. The corrective action taken and the specific procedures required to handle the deviation will depend on the regulatory agency involved. The FDA regulations concerning the handling of low-acid deviations are located in 21 *CFR* part 113.89. The corresponding regulations for the USDA are found in 9 *CFR* 318.308 and 381.308.

Deviations that are discovered prior to or during the thermal process may be corrected, thus eliminating the need to handle the thermal process as a deviation. All deviations must be completely documented on the thermal processing records. If it is not possible to correct a problem on the spot, or if the deviation is discovered after the fact, all product involved with the deviation must be placed on hold with one of the following options undertaken:

1. The deviant product may be reprocessed if a suitable reprocess is available from a processing authority.
2. The deviant process could be evaluated by a processing authority.
3. The deviant product could be destroyed. Records must be kept that delineate the disposition of the destroyed product.

All thermal processing, product disposition, and deviation evaluation records pertaining to a specific deviation need to be kept in a separate file. Alternatively a log may be kept that identifies all deviations.

The best way to manage deviations is to prevent them from occurring. This may be accomplished by training the retort operators and employees assigned to monitor critical factors to recognize the potential for the occurrence of a deviation and be familiar with corrective action.

PROCESSING SYSTEMS FOR LOW-ACID CANNED FOODS

There are many systems for the thermal processing of low-acid canned foods. Traditional thermal processing methods utilize retorts (pressure vessels) for processing products. The food is filled into containers, the containers are sealed, and then containers are placed into the retort for processing. Many styles of retorts have been developed for use in the low-acid canned foods industry. The retorts can be categorized by container handling methods, type of processing medium, the use of overpressure, and the use of agitation.

Container Handling Methods

Retorts may be designed to use discontinuous (batch) container handling or continuous container handling. Generally, batch container handling systems require the processor to place the containers into metal crates or baskets to facilitate loading the retort. In a few systems, referred to as *crateless*, the cans are loaded directly into the retort. Batch retorts require an operator to close and tighten a door or lid to seal the pressure vessel prior to introducing the heating medium and/or bringing the retort up to the required temperature and pressure.

Continuous container handling systems have two self-sealing openings, which allow the retort to be constantly at the required temperature and pressure and still allow the introduction and discharge of containers. The self-sealing openings are either mechanical valves or columns of water depending on the system. The containers travel through the retort during the thermal process by either a rotating reel or a carrier chain.

Type of Processing Medium

The first retorts used steam at high temperatures and pressures as the processing medium for processing low-acid canned foods. Circulating water, steam/air mixtures and steam/water/air mixtures have also been used successfully as the processing medium. All of the processing media mentioned provide an efficient means of heat transfer to the containers. Regardless of the type of processing medium utilized, a system must be in place to circulate the medium to allow for uniform temperatures within the retort.

Use of Overpressure

Overpressure is a term used to describe the additional pressure supplied to the retort that exceeds that supplied by steam at a given temperature. Depending on the retort system, overpressure may be supplied in the form of air or steam. Overpressure is used to maintain the integrity of containers that have limited resistance to internal pressures (ie, glass jars, semirigid plastic containers, and flexible pouches).

Use of Agitation

Agitating the contents of the containers during processing may allow for a faster heating rate, thus resulting in a shorter process time for some products. The agitation of the containers acts to "stir" the contents during heating, allowing for faster and more uniform heat transfer. Several systems provide agitation—either intermittent or continuous—of the containers during processing.

PROCESSING ACIDIFIED FOODS

Because high temperatures and pressures are not required for processing acidified foods, the methods for thermal processing may be less complex. Processing acidified foods is often referred to as pasteurization. Pasteurization temperatures are generally at or lower than boiling water temperatures at atmospheric pressures. Pasteurizers, in the form of water baths, atmospheric cookers, or retorts, may be used for processing acidified foods. When a pasteurizer is used, hot or cold product is filled into the containers that are then sealed. The sealed containers are heated in the pasteurizer for a given period of time or until a specific temperature is achieved. Pasteurizers may be batch systems (such as a water bath) or continuous systems. Continuous pasteurizers may use steam, hot water, or cascading hot water as the processing medium. A moving belt or other mechanism is used to move the containers through the processing medium for heating and through cold water or sprays for cooling. The speed of the belt will determine how long the containers are in the pasteurizer. It is essential to operate pasteurizers properly to ensure that all containers achieve the same degree of heating. Improper speed or temperature settings or overloaded belts may have a negative impact on product heating.

Hot-fill-hold procedures are another method of processing commonly used for acidified foods. Hot-fill-hold procedures involve heating the product to a specified temperature, filling hot product into the containers, and holding the filled and sealed container at a minimum temperature for a period of time prior to cooling. Generally the containers will be inverted or tipped after sealing and prior to cooling to allow the hot product to contact and sterilize the inside surface of the lid.

ASEPTIC PROCESSING AND PACKAGING

Low-acid and acidified foods can also be aseptically processed and packaged. For details on how products are processed aseptically, see the article titled ASEPTIC PROCESSING: OHMIC HEATING.

SPOILAGE OF CANNED FOODS—DETERMINATION OF CAUSE

There are two major causes of spoilage of canned foods: postprocess contamination (leakage) and insufficient thermal processing. Postprocessing contamination usually results in a mixed microbial flora of vegetative cells and spore formers, generally with little heat resistance. There is very little risk that such contamination will present a public health hazard. Underprocessing of low-acid foods usually is characterized by pure cultures of spore-forming bacteria; there is the risk that *C. botulinum* and its toxins may be present. In addition, foods may spoil if stored at temperatures above 43°C (109°F) because of growth of thermophilic spore formers, which present no health hazard. In acid and acidified foods, underprocessing results in the growth of organisms that are not a health hazard; however, improper acidification can result in the potential for spores of *C. botulinum* to germinate and grow.

Detailed procedures for evaluating the cause of spoilage can be found in Ref. 13, along with methods for the microbiological examination of a variety of other foods, including meat and poultry products, milk and milk products, eggs, seafood, and fruits and vegetables. There are also procedures for isolating and identifying indicator organisms and foodborne pathogens. Cultural methods are being supplemented or replaced by techniques to identify specific proteins or DNA sequences, such as ELISA (enzyme-linked immunosorbent assay) methods, PCR (polymerase chain reaction) technology, and fingerprinting methods such as ribotyping and pulsed field gel electrophoresis. Such techniques are being used to determine not only what organisms caused the problem but also where the organism came from.

CONCLUSION

The pH of foods is one of the key factors in determining what microorganisms can grow in the food, which in turn defines how the product must be processed to ensure safety and prevent spoilage. An understanding of food microbiology, coupled with knowledge of the parameters critical to delivery of a thermal process, is essential for establishing an adequate thermal process. This expertise forms the basis for a processing authority, who will also be knowledgeable in the applicable regulations that apply to these products. Assistance in assessing the microbial inactivation requirements and how to deliver a process can be found through trade associations and universities, as well as contract laboratories.

BIBLIOGRAPHY

1. Robert B. Gravani, "Food Science Facts for the Sanitarian: The pH of Foods (continued)," *Dairy Food Sanit.* **6**, 112–114 (1986)
2. M. P. Doyle, L. R. Beuchat, and T. J. Montville, eds., *Food Microbiology, Fundamentals and Frontiers*, ASM Press, Washington, D.C., 1997.
3. International Commission on Microbiological Specifications for Foods, *Microorganisms in Foods 5: Microbiological Specifications of Food Pathogens*, Blackie Academic & Professional, London, United Kingdom, 1996.
4. International Commission on Microbiological Specifications for Foods, *Microorganisms in Foods 6: Microbial Ecology of Food Commodities*, Blackie Academic & Professional, London, United Kingdom, 1998.
5. C. R. Stumbo, *Thermobacteriology in Food Processing*, 2nd ed., Academic Press, New York, 1973.
6. H. G. Bayne and H. D. Michener, "Heat Resistance of *Byssochlamys* Ascospores" *Appl. Env. Microbiol.* **37**, 449–453 (1979).
7. L. R. Beuchat, "Spoilage of Acid Products by Heat-Resistant Molds," *Dairy Food Env. Sanit.* **18**, 588–593 (1998).
8. V. N. Scott and D. T. Bernard, "Heat Resistance of *Talaromyces flavus* and *Neosartorya fischeri* Isolated From Commercial Fruit Juices," *J. Food Protect.* **50**, 18–20 (1987).
9. I. Walls and R. Chuyate, "*Alicyclobacillus*—Historical Perspective and Preliminary Characterization Study," *Dairy Food Env. Sanit.* **18**, 499–503 (1998).
10. M. E. Parish, "Public Health and Nonpasteurized Fruit Juices," *Crit. Rev. Microbiol.* **23**, 109–119 (1997).
11. *Thermal Processes for Low-Acid Foods in Metal Containers*, Bulletin 26-L, 13th ed., National Food Processors Association, Washington, D.C., 1996.
12. A. Gavin and L. M. Weddig, eds., *Canned Foods: Principles of Thermal Process Control, Acidification and Container Closure Evaluation*, 6th ed., Food Processors Institute, Washington, D.C., 1995.
13. C. Vanderzant and D.F. Splittstoesser, eds., *Compendium of Methods for the Microbiological Examination of Foods*, 3rd ed., American Public Health Association, Washington, D.C., 1992.

V. N. SCOTT
L. M. WEDDIG
National Food Processors Association
Washington, D.C.

See also FOOD PROCESSING.

MACHINE VISION SYSTEMS: PROCESS AND QUALITY CONTROL

Many processes and inspection systems in manufacturing operations were done manually until the early 1980s, when the normalized grayscale correlation (NGC) was adapted to vision systems. With the introduction of new technologies, increased emphasis on improved production quotas, labor costs and shortages, and the introduction of new products such as computer chips, the need to develop the machine vision system (MVS) for processing and quality/inspection of components or whole products was necessary. Machine vision has an additional advantage in that it can provide 100% inspection.

This powerful system can be compared to the human vision system (Fig. 1) in that it uses a lens (eye) with proper illumination that sends a signal (optic nerve) to a computer (brain) that analyzes the image and sends information to a controller (arm) that then proceeds with a programmed task.

MVS offers many advantages over manual and mechanical systems:

1. Accuracy and precision
2. Automation
3. Better quality
4. Improved production
5. Higher throughput

Machine vision may be defined as "the mechanism of gathering information optically, quickly and nondestructively to control machines or processes" (1). Others define it as "the automatic acquisition of images by noncontact means and their automatic analysis to extract needed data for controlling a process or activity" (2).

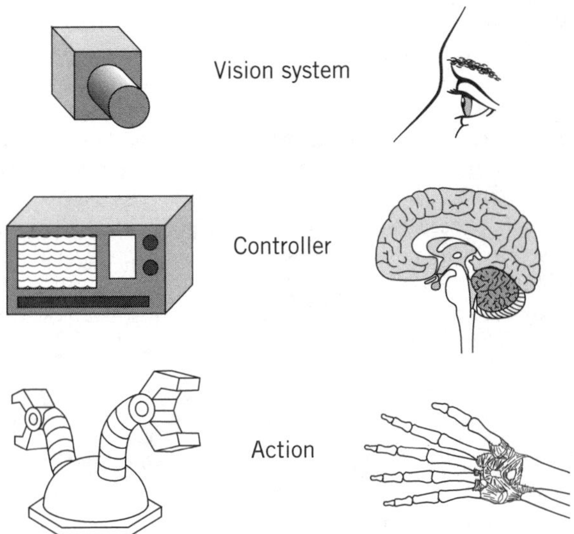

Figure 1. Analogy between human and machine vision.

To understand MVS, we have to look at its components and how they work; applications; and functions and types of MVS, including descriptions of hardware, software, and future developments.

THE SYSTEM AND ITS COMPONENTS

A machine vision system is designed to obtain, automatically, an electronic or digital image of an object through a camera lens. The object through a camera lens. The object is illuminated with a set light (light source) at a set angle such that the image can be recognized and processed by the system (image formation) (3). The camera extracts from the object the feature to analyze, and a board captures the image, digitizes it, and divides it into small sections called pixels or picture elements. A number is given to each pixel based on a binary scale, a gray scale, or a color scale. In binary images, the number represents either dark or light. Color processing uses information from the red, blue, and green color spectra to detect and differentiate shades of color (4). The gray scale number is proportional to the level of light intensity of the corresponding area represented by the pixel. The digitized image is placed into the vision-processor memory where a decision using vision algorithms is used to decide the output. The output is used to make a decision about the object in terms of processing or rejection (Fig. 2).

The MVS (Fig. 3) consists of an image-sensing device, an illuminator (not shown), an image-processing device, a control device, an operating interface, and an output device.

Image-Sensing Device

This part of the system is able to locate the object and positions itself so as to capture or acquire the measured feature in the orientation it was originally programmed to find (5). The image-sensing device captures a light intensity and changes this to an electrical signal, which is sent to the processor. Image-sensing devices are usually video

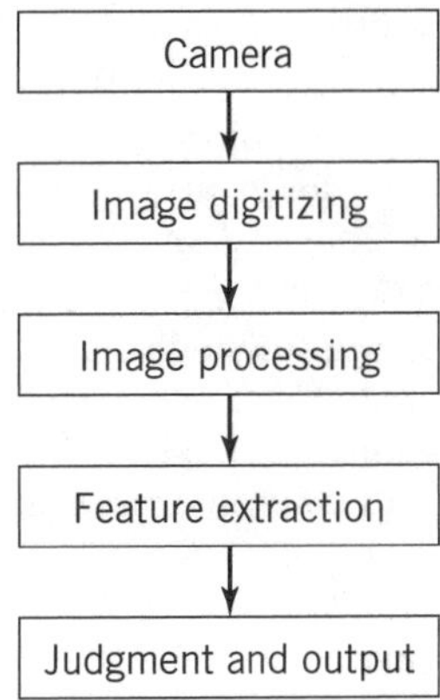

Figure 2. Flowchart for a machine vision system.

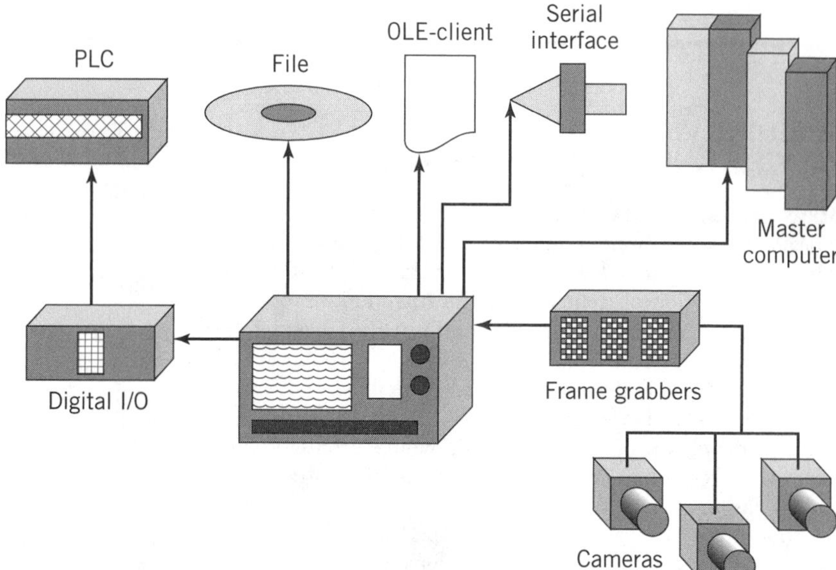

Figure 3. Components of a machine vision system.

cameras or solid-state cameras (charge-couple devices [CCD] or charge-injection devices [CID]); infrared [IR], ultraviolet (UV) or X rays; laser scanners, and ultrasonic sensing devices. These devices can be very small and sealed in housings, have low or high resolution, and be monochrome or color. The small, high-resolution and/or color cameras are suitable where detail is needed but are more costly and take longer to process the image.

Illuminator

Lighting is critical to an MVS in that it allows for accurate image capture. Broadband sources, fiberoptic illuminators, or lasers can be used as light sources. The wavelength can be from the IR to the UV to the X-ray regions. Light can be provided in the shape of a ring, a line, a point, or multiple points, or it can cover an area (1) and can be continuous or pulsed (flashes). The latter is used for rapidly moving objects. There are three basic lighting configurations, depending on the features to be highlighted (4,6):

1. Front lighting (front light), in which objects are lit from the top or front of the object but possibly at an angle. This is used for extracting surface features.
2. Back lighting (backlight), in which the light is placed below or behind the object, providing an outline of it. This is suitable for nontransparent objects where high contrast is needed and is usually coupled with a binary-image sensor device.
3. Structural or multidimensional lighting, in which light is directed at various angles so as to extract a two- or three-dimensional image. These may be used to measure a dimension or find shape defects in an object.

Image-Processing Device

This device processes the image depending on the application. An analogic-to-digital transform module changes the captured analogic image and transforms it to a digital image for processing in the computer or for display. The image-storage module stores the digitized image for further processing and analysis. An image pretreatment module filters the captured image for enhancement of it or its features. The image analysis module calculates the gray-level distribution of the image and counts the area for gray level.

Operating Device

This device simply translates the captured image to usable bits. Speed is necessary here so that a response is given quickly. The speed will depend on the capacity of the computer, the complexity of the image, and the type and quantity of information desired from it. The system and software should be optimized for the specific application so that the necessary frame grabber programs, algorithms, accelerators that interface with the CPU, and application software are adequate. Among the software available are NeuroCheck, a general-purpose image-processing system for industrial quality control; Eurosys, suitable for industrial machine vision; HACCON, which is flexible and gives high performance for research and education; and WiT, for image-processing development (2). WiT, along with VPE, ProtoPIPE, VPM, and Concept Vi, allow the user to pick the image-processing icon, drag it to the workspace canvas, connect I/Os using links, and define operator-specific parameters (7).

Choosing specifications for an MVS requires answers to the following questions (4):

1. How many parts must be processed or inspected per unit time?
2. Which and how many distinguishable features are there?
3. What environmental factors are present?
4. What level of technical knowledge is available on the floor?

5. Who will install, program, maintain, and integrate the system?
6. Will any other machines or components be interfaced with the vision system?
7. What are the upgrade and flexibility requirements?
8. What is the budget?

On the other hand, lack of need for precision of part appearance, poor visual access, or poor visibility due to the environment are key factors in not choosing an MVS (8).

Control Device

The control device, a board or, more commonly, a computer, is the heart of the system. It controls the imaging device and stores the digitized image, does image analysis and pattern recognition, and transmits the results or command to a controller. The size of this device will determine the speed of the process and the precision of the measurement or task(s).

Output Device

This device obtains the translated information and acts on it accordingly. In a manual system, the information (display) is passed to an operator who will make a decision and proceed accordingly. In an automated system, the processor makes a decision based on the data evaluated and sends a signal to some part of the process to act accordingly. This signal could trigger an alarm, stop a line, reject or accept the product, or send a signal to a robot to perform a task. In a closed-loop system, the feedback from the statistics process is provided to the controller to reduce process deviations.

Image recognition and inspection is done via image comparison (sample image vs. standard image) or feature extraction (sample image features vs. standard image features). These are named "something-somewhere" inspections, which are usually quantitative. A more flexible, qualitative image-inspection system based on the "anything-anywhere" principle works on the principle of example parts or illustrations. An example of the "anything-anywhere" principle is the system named Mentor®, which uses a filter generated from a neural network that creates and refines the filter during the set-up process and creates different algorithms to analyze different area parts (9).

APPLICATIONS FOR MVS

There are eight general categories for MVS applications (2):

1. Machine monitoring
2. Material handling
3. Process control
4. Quality control/assurance and inspection
5. Robot guidance
6. Safety
7. Sorting and grading
8. Test and calibration

Quality control/assurance and inspection has the function of removing defective products from the manufacturing process. This can be coupled with statistical process control (SPC) to increase information and efficiency. It is the single largest application because it requires repetitive tasks and constant attention. It may also be an important tool in the monitoring of critical control points (CCP) in hazard analysis and critical control points (HACCP) systems. One such application is the detection of metal particles. Bottling lines, poultry and fish processing lines, and confectionery lines are candidates for this system (Fig. 4).

Sorting or grading is a process whereby objects can be classified by size, dimension, color, or other physical attributes. Roasted nuts could be classified according to degree of browning and size for specific applications. Fresh shrimp are graded by size (count) to be able to be processed correctly and into the right category.

Food Industry Applications

Many repetitive, measured processes or quality/inspection-control functions could be performed by MVS. These applications range from sorting and inspection of raw product to product reduction or grading to inspection of final product, package, or both. Some of these applications are:

1. Process control
   a. Monitoring and controlling cooking operations
   b. Monitoring color of foods
   c. Detecting abnormal conditions
   d. Monitoring product fill
   e. Monitoring product size and shape for size reduction

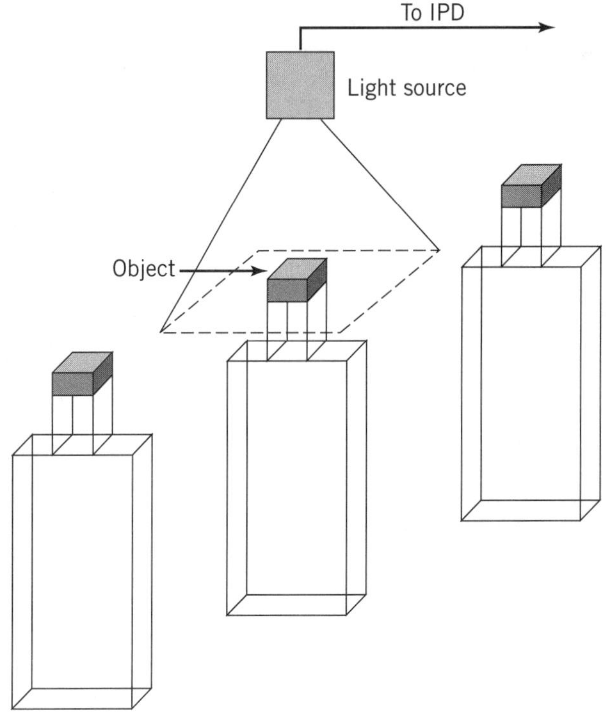

Figure 4. The frontlight (top) illumination method.

   f. Monitoring fermentation process through color development
2. Quality control/assurance
   a. Recognition of product size and variety
   b. Sorting of vegetables
   c. Egg inspection
   d. Inspection of food containers
   e. Inspection of packaging label and codes
   f. Detection of missing or different products placed in trays
   g. Detection of faulty or missing seal
   h. Detection and classification of defective product
   i. Detection of abnormal conditions

Process Control Applications

Color machine vision (CMV) can be used to monitor color of incoming and final products, extent of cooking process, and pigment concentration. Strong correlations were found between a colorimeter and CMV systems (CMVS) and colors as perceived by humans (10). Among the applications of a CMVS are the monitoring and measurement of product color, extent of retort, and other thermal processes through chemical markers. One such application is the monitoring of baked and roasted materials as an indication of quality or degree of cooking (11). Baked muffins (doneness), roasted peanuts, and pizzas (mapping) are some of the specific products (Fig. 5). Minimum description analysis is used when the quantity of irrelevant colors (e.g., blueberries in muffins, raisins with peanuts, etc.) vary between samples (12).

A system to predict crawfish molting based on color ratio was used to produce soft-shelled crawfish (13). The system used an IBM PC/AT to control a Versa Module Eurocard (VME) image-processing system. A Pulmix CCD color camera with a polarizing filter was used to acquire the image. A system consisting of an optical scanner, a computer and image-processing system, and a robot water jet cutter was designed to produce fat-free steaks (14). The system uses a 3D laser scanner to obtain the image profile with up to 0.1 pixel. A digital signal processor (DSP) program enables the image-processing system to receive the product's volume and thus its mass. Another similar system using rotating blades instead of water jets combines image processing with a scale to define the shape, orientation, and mass of product to be cut into portions (Fig. 6). This process allows for maximum efficiency and minimum waste. These systems not only save on labor and add efficiency but also are more effective in controlling hazards from handlers and cutting utensils.

Quality Control/Assurance Applications

The Kernel Extension Rapid New Evolution Level (KERNEL) was developed to grade morphological features and color defects in turkey carcasses (15). The turkey carcasses suspended on a chain are appraised for pin feathers, visceras, and filets/cuts yield. Different appearance (color) characteristics were extracted using color segmentation methods. A CMVS was developed to identify the reddish comb of each chicken and thus be able to count the chickens before placing them in the crate automatically. Another system was developed to measure count, uniformity ratio, color, and melanosis and to detect foreign objects in processing white shrimp (16). Shrimp area was used to estimate weight and calculate count and uniformity ratio. Color and melanosis were quantified by correlating with a trained inspector.

A CMVS was developed to inspect frozen dinners for proper food amount and ratio in each compartment (17). The image captured is analyzed by color-connectivity analysis where it is compared against predetermined tolerance limits and proximity of one food to another to check for

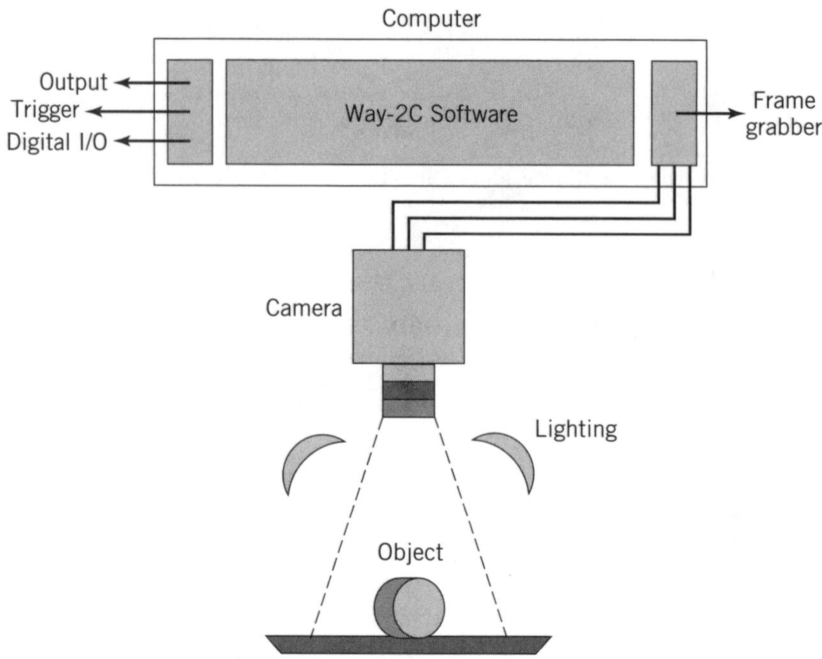

Figure 5. Color measurement using an MVS.

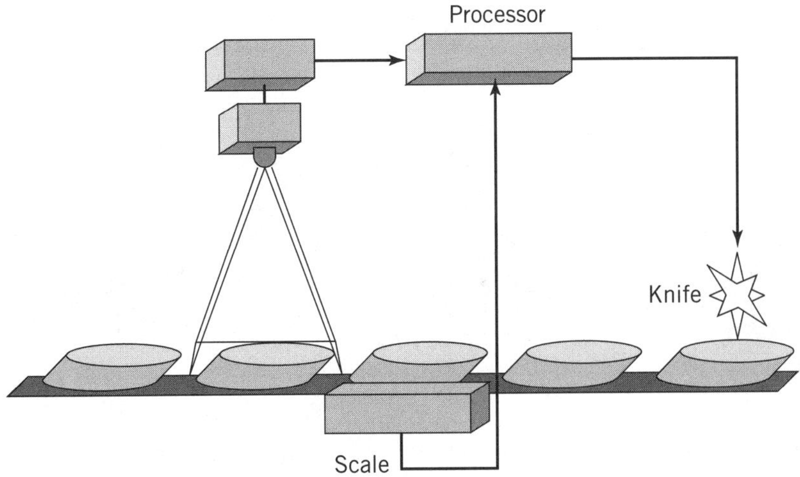

Figure 6. Product cutting/portioning using an MVS, a scale, and a cutting tool.

spills. A method to characterize the color of fruits and vegetables at different ripening states, describe color changes during storage of mushrooms, distinguish varieties by shape, and recognize shape defects was developed calculating the average red, green, and blue (RGB) color components of the pixels belonging to each product (18). Utilizing hue, saturation, and intensity, an (HSI) MVS was trained to distinguish between good and greened potatoes and yellow and green apples (19). The CMVS used multivariate discriminate techniques to hue histograms of representing features and increased accuracy by reducing the number of hue bins by selecting significant features or summing groups of hue bins. X-ray technology has been developed to detect motals, glass, minerals, stone, polyvinyl chloride (PVC), dense rubber, and other materials within a package or in a conveyor belt (20). The system consists of an X-ray generator, a detector, and a processor. Differences in density are detected, and the sensor converts the X-ray signal into light, then into an electrical signal before it is converted into a video signal. It can also be used for morphology processing and for detection of underfilled or broken packages.

FUNCTIONS OF MACHINE VISION SYSTEMS

MVS is capable of performing many functions, depending on the type of information extracted during image analysis. Functions are gauging, verification, flow detection, identification, recognition, finding position, and tracking (Fig. 7).

Gauging refers to the process of making measurements. These measurements can be made on-line and off-line. Online inspects or sorts parts, provides adjustment to machines, calibrates, or gathers SPC information. An off-line system could be used to adjust or coordinate various machines or machine parts in a process.

Verification refers to the process whereby an activity is assured to happen correctly. This could be the correct positioning of a label or of an object prior to packaging. Flaw detection is the most common function of an MVS. It refers to the detection of any defect in an object, such as a cracked bottle, a defective can, an underfilled container, or a defective seal. Identification is usually accomplished by optical character recognition and bar coding. This is used where multiple products are packaged in the same container but the code reveals their identity, such as in warehouses to move the oldest manufactured product first (first-in, first-out [FIFO]), and for inventory (21).

Recognition, as compared to identification, utilizes characteristic features of the image to determine the object's identity. This is usually used in combination with robotics to identify an object or its position.

Finding position refers to the determination of the location and orientation of an object or objects. This allows for rapid and precise working of the object while reducing waste. Tracking usually precedes a position-finding function in an MVS, since the part position needs to be known before the tracking function begins (22).

CONFIGURATION AND TYPES OF MVS

Three configurations of MVS are available, depending on environment and requirements. Stand-alone systems are well suited for factory automation. They use application specific integrated circuits (AISCs) that deliver accelerated processing power. They are compatible with other factory automation devices [programmatic language controllers (PLCs), photoelectric sensors, and radio frequency identification (RF/ID) systems] and withstand adverse environments (temperature, vibration, electrical interference). The second type are PC-based systems, which consist of a computer and a motherboard or CPU with dedicated-vision ASICs. These are lower in price than stand-alone systems and take advantage of microprocessor technology, but they are not as resistant to environmental and other pressures as stand-alone systems. The third type are VME systems, which are robust yet flexible. The VME is an Institute of Electrical and Electronic Engineers (IEEE) bus standard, where system components are connected via pin-and-socket connections to the bus. These systems are more complex and costly than the others (23).

Machine vision systems can be classified into eight categories. Each has advantages and disadvantages (Table 1) and can be suitable to a particular operation. Turnkey

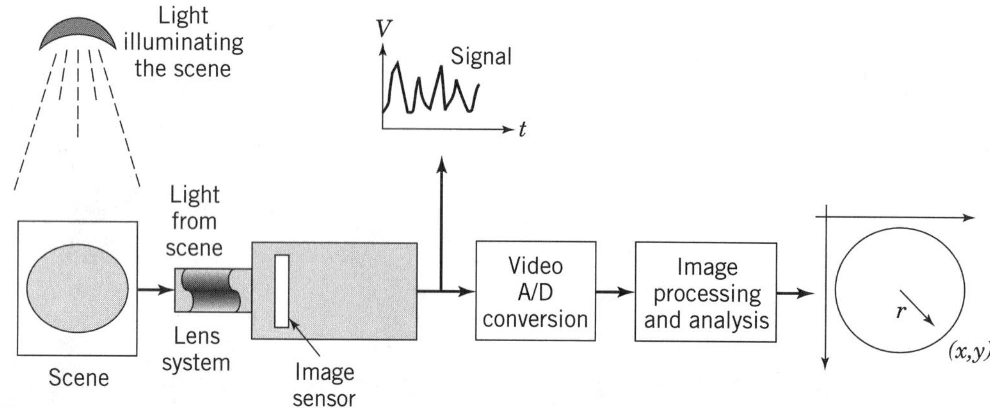

Figure 7. Typical machine vision system and its processes.

Table 1. Advantages$^{(+)}$ and Disadvantages$^{(-)}$ of Machine Vision System Types

	MVS type							
Characteristic	Turnkey standard	Embedded	Turnkey custom	General purpose	Vision engines	Board level	Modular	Custom
Low engineering cost	+	+	+/−	−	−	−	+	+
Long-term support	+	+	+/−	+	+	+/−	+	+/−
"Customer-friendly"	+	+	+/−	−	−	−	+/−	−
Low project risk	+	+	+/−	+/−	+/−	+/−	+	−
Lead time (short)	+/−	+/−	−	−	−	−	+	−
Low cost	−	+/−	+/−	+/−	+/−	+	+	−
Maximum flexibility	−	−	+	+	+	+	−	+

Source: Ref. 2.

standard MVS is designed as a complete package for a specific application but with a broad market basis. Embedded MVS is similar to the turnkey standard MVS except that the system is part of a larger machine. Turnkey custom integrated MVS is customized to the client's requirements. The general-purpose MVS is flexible in that it is designed for the end user as well as the system integrator. It is noted for its easy-to-use controls. Vision engines are designed for the original equipment manufacturer (OEM) or sophisticated system integrator and are noted for their high performance. Board-level MVS is designed as peripheral boards plugged onto a computer. Modular MVS integrates the camera and the processor into one piece. Custom MVS is characterized by uniquely engineered components, offering the highest flexibility but also the highest cost and risk.

BIBLIOGRAPHY

1. D. Zankowsky, *Laser Focus World* **4**, 127–137 (1996).
2. R. Grosklaus, *Sensors* **14**, 22–27 (1997).
3. E. Kubel, *Manuf. Eng.* **121**, 42–44 (1998).
4. M. D. Wheeler, *Photonics Spectra* **32**, 112–114 (1998).
5. J. R. Ggorki, *Mach. Des.* **2**, 42–48 (1994).
6. C.-T. Yang, "Quality Control: Machine Vision System," in Y. H. Hui, ed., *Encyclopedia of Food Science and Technology*, 1st ed., John Wiley and Sons, New York, 1992, pp. 2220–2225.
7. D. LaCroix and B. Smyth, *Photonics Spectra* **28**, 129–130 (1994).
8. A. Davis, *Advanced Imaging* **10**, 52–56 (1995).
9. K. Harding, *Optics and Photonic News* **7**, 29–33 (1996).
10. P. P. Ling et al., *ACS Symposium Series* **631**, 253–256 (1997).
11. R. K. McConnell, Jr., and H. H. Blau, Jr., *Proc. FPAC IV Conf., ASAE*, Chicago, Ill., November 3–5, 1995.
12. R. K. McConnell, Jr., and H. H. Blau, Jr., *Proc. Electron. Imaging Int. Mtg.*, Boston, Mass., September 30–October 3, 1991.
13. T. Timmermans, F. E. Sistler, and T. B. Lawson, *J. World Aquac. Soc.* **26**, 234–239 (1995).
14. J. Jacobs, *Advanced Imaging* **11**, 21–23 (1996).
15. P. Marty-Mahé and P. Marchal, *C. R. Acad. Agric. Fr.* **83**, 121–136 (1997).
16. D. A. Luzuriaga, M. O. Balaban, and S. Yeralan, *J. Food Sci.* **62**, 113–118 (1997).
17. S. Montellese, *Photonics Spectra* **29**, 107–108 (1995).
18. J. Felfoldi, F. Firtha, and E. Gyori, *Acta Alimentaria* **26**, 289–292 (1997).
19. Y. Tao et al., *Trans. ASAE* **38**, 1555–1561 (1995).
20. L. Mancini, *Food Eng.* **70**, 89–96 (1998).
21. D. Braggins, *Sens. Rev.* **18**, 152–153 (1998).
22. J. Testa, *Industrial Robot.* **22**, 13–15 (1995).
23. T. Seitzler, *Advanced Imaging* **11**, 56–58 (1996).

Juan L. Silva
Roberto S. Chamul
Mississippi State University
Mississippi State, Mississippi

MALTS AND MALTING

Malts are cereal grains that have been germinated to a limited extent to alter their physical and biochemical states. Usually they are dried and partially cooked (cured), and the rootlets are removed before use (1–10). There are many types of malt. Barley malts vary in color from a pale straw-yellow to nearly black. The darker malts have characteristic colors and aromas and they give particular flavors to products made from them, but in contrast to pale malts, they contain few or no enzymes. Most malts are prepared from barley, but smaller amounts are made from wheat, rye, possibly triticale, oats, sorghum, and millet. Barley is often preferred for malting because the adherent husk protects the grain during handling and malting whereas naked grains, like wheat and sorghum, are readily damaged and crushed. Barley can be induced to germinate evenly and generate useful levels of hydrolytic enzymes (including cytases, proteases, and amylases), and varieties are available for making many kinds of malts. Also, the relatively low gelatinization temperature of large barley starch granules (about 60–65°C [140–149°F]) renders them easily converted when the malt is mashed. Malts are used in making beers, African-style opaque beers, whiskeys, malt vinegars, and syrupy extracts and in some bakery products, confectionery, breakfast cereals, baby foods, malt coffee, and malted-milk beverages (1–10).

Small amounts of malt flours are used in baking. However, for most purposes malts must be mashed, sometimes after they are mixed with other starch-containing materials, to produce a sugary solution or extract (1,3,4). In mashing, ground malt is mixed in a chosen ratio with warm water and the mixture is heated, in a controlled way, to progressively higher temperatures. After a time the liquid or sweet wort, which contains the extract materials dissolved from the malt, is separated from the residual undissolved solids, known as spent grains or draff. The latter is usually used to feed animals, but other uses are being sought. The sweet wort is processed further to produce beer, whiskey, vinegar, extract, diastase, and so on (1–10). However, mashing to prepare opaque beers is somewhat different (1,3). If ground malt is extracted with cold, slightly alkaline water to prevent enzyme activity, then about 18 to 21% of the malt solids (the cold-water extract) dissolve. If the malt is mashed with hot water (e.g., 65°C; 149 °F) then the hot-water extract is about 77 to 83% of the dry matter of the malt, depending on the quality of the malt and the mashing procedure used. Mashing processes vary in the degree of grinding of the malt, the liquid-to-solids ratio, the temperatures used, and the durations of the different stages. The hot-water extract is much larger than the cold-water extract because during the hot mash, enzymes act to hydrolyze insoluble materials to give soluble products (1,2,4,9,10). Some degradation of proteins and cell-wall polysaccharides occurs, but the main change is the conversion of insoluble starch into a complex mixture of soluble sugars and dextrins. The increase in the density of wort above that of water is a measure of the amount of extract in solution (1,4). For most purposes the yield of hot-water extract and its quality (color, level of fermentable sugars, contents of amino acids and proteins, etc.) are the major measures of malt quality. In making beers, malt vinegars, and whiskeys it is the simple sugars that yeasts convert to ethanol and carbon dioxide (1,2,3,5,9). Malts are usually analyzed by one of three internationally recognized systems (1,4).

CHANGES OCCURRING IN MALTING GRAINS

Carefully selected grain, usually barley, is first hydrated, or steeped (1–4,6–11). Usually this is achieved by immersion in water, but sometimes water is sprayed onto the grain. At intervals the grain-water mixture is aerated by blowing compressed air in at the base. At the end of each immersion the water is drained to waste, and air (which may, to advantage, be cooled and saturated with water vapor) is sucked down through the grain. During each air-rest period the downward ventilation, or carbon dioxide extraction, cools the grain, provides oxygen, removes carbon dioxide, and accelerates germination. Typically the grain may be successively immersed in three batches of clean water. Later, water is sometimes sprayed on to adjust the moisture content.

After steeping, the grain, which has started to germinate, continues to grow surrounded by humid air (1–4,6–10). Rootlets (culms, sprouts) grow from one end of each grain and the embryonic coleoptile, also known as the shoot or acrospire, grows beneath the husk. The grain respires, using oxygen and producing carbon dioxide and heat. Hydrolytic enzymes are formed in the scutellum and later, partly in response to gibberellin hormones from the embryo, in the aleurone layer. These enzymes are released into the starchy endosperm, where they degrade the cell walls, some of the reserve proteins, and some of the starch. At the same time some sugars and simple nitrogenous substances accumulate. Collectively these changes are termed modification. When modification is sufficiently advanced, usually when the acrospire is about 0.8–0.9 the length of the grain (and certainly before it has grown out from beneath the husk) and endosperm cell-wall degradation is nearly complete, the malt is ready to be dried. This green malt (green means undried; it is not green in color) is now kilned or sometimes it is cooked and dried in a drum. In a very few cases the green malt is used directly (it cannot be stored), such as in grain distilleries where large amounts of starchy materials must be broken down in mashes and therefore maximum enzyme survival is needed (1,2). In kilning, the malt is dried and cooked, or cured, to a limited extent in a stream of heated air. This stops germination, produces a stable product, removes undesirable flavors, and gives the malt character (desirable color, flavor, and aroma). When pale malts are being made, the initial airflow is rapid and the air temperature is low to maximize enzyme survival and minimize color formation. As the malt dries, the air temperature is raised and the airflow is reduced without major effects on the enzymes, which are more stable in drier grain. When colored malts are being prepared, the temperatures used are higher and the airflows are less. At first the grain is held warm and moist, conditions that encourage first enzyme formation and then, as the temperature rises, enzyme inactivation, but that also favor the accumulation of sugars and amino acids (1,3,4,8). Later, as the temperature is raised, the sugars and amino acids interact to form colored melanoidins and flavorful, aromatic substances. Because darker malts contain lower levels of enzymes, they must be mashed with pale malts, which provide the enzymes to hydrolyze the starch.

After kilning the cooled malts are dressed, that is, sieved to remove the rootlets, which are sold for animal feed, sieved to remove malt and grain dust and thin grains (1–4,6–10). Barley malts lose only rootlets, but malts from naked (huskless) grains, such as wheat, rye, triticale, and sorghum, also lose shoots, because they are not protected by adherent husks (1). After kilning, pale malts are often stored for about 6 weeks when, for reasons that are not understood, their performance in mashing improves. In contrast, darker malts should be used soon after preparation, so that they do not lose their special flavors and aromas.

Roasted and caramel malts are prepared in roasting drums (1,3,4,8), which fall into two classes: (*1*) those in which kilned malts are roasted (brown, chocolate, and black malts, which are friable) and (*2*) those in which green, undried malts are processed (caramel malts in which the endosperms are replaced by hard, barley sugar–like masses).

Thus malting converts tough, raw-tasting, enzyme-poor grain into (usually) readily crushed, flavorful, and, at least in the case of pale malts, enzyme-rich materials that can easily be stored and transported and that when mashed give worts that are suitable for a range of uses, such as the manufacture of beer. However, matter is lost during this process. During the preparation of pale malts, made without additives, the total loss (dry weight) is about 10%, 1% in steeping, 4% as rootlets, and 5% as respiratory losses) (1–4,6–10). The losses incurred in making colored malts can be much greater. Losses may appear to be greater until it is realized that the initial moisture content of the barley going into steep may be 8 to 16%, and the moisture of the finished malt 1.5 to 6%.

MALTING TECHNOLOGIES

Barley or other grain may be delivered to a malting factory, or maltings, by road, rail, or barge (1–4,6–10). Before a load is accepted, each batch of grain will be checked, usually for viability, varietal purity, freedom from insect pests, moldy grains (which may lead to poorly flavored malt and carry mycotoxins and agents that cause beers to gush, or overfoam) (1,5), freedom from physical damage and premature sprouting, the appropriate protein ($N \times 6.25$) content, moisture content, and screenings (thin grains). If the load is substandard it may be rejected or accepted at a reduced price. The grain may be moved pneumatically, by drag-flight or helical screw conveyors, or by bucket elevators. The grain will quickly be precleaned, by sieving and aspiration, to remove most of the coarse and fine impurities and will be passed over magnets to remove fragments of iron and steel. If necessary the grain will later be dried to less than 12% moisture for prolonged storage. It will also be thoroughly cleaned to remove grains of other cereals, dust, weed seeds, small stones, and broken grains. The grain may also be graded, by sieving, to separate it into width classes, which may be malted separately. Usually the thinnest grains, or screenings, will be rejected and used for animal feed, but occasionally they are made into low-extract, enzyme-rich malt. The grain will be stored, at least until any dormancy has declined, under conditions that prevent its deterioration and the development of pest infestations.

The malting process itself may be carried out in many different sorts of plants. Old-fashioned floor maltings are still in limited use for making small quantities of specialist malts. Most malt is made in pneumatic maltings, which may operate in batch, continuous, or semicontinuous fashions. Most malting is carried out in batches, and steeping, germination, and kilning operations are carried out in different vessels. However, vessels are in use that combine steeping and germination (SGV), germination, kilning, and steeping (SGKV), and germination and kilning (GKV). Modern steeps are either cylindrical with conical bottoms or circular with flat bases. The latter have perforated false floors on which the grain rests, and they have mechanical leveling and unloading devices. Each is equipped to add and remove water, to blow compressed air into the base of the steep when the grain is underwater, and to suck air down through the vessel for CO_2-extraction between immersions. Sometimes provision is made to wash the grain as it is conveyed to the steep. When the grain is sufficiently moist (e.g. 44–46% fresh weight for many barley brewing malts and up to 50% for some distilling malts) it is drained and transferred to a germination compartment. Where their use is allowed, additives such as gibberellic acid and (though rarely) a bromate salt may be applied to the grain during the transfer by spraying on a solution (1,4). Gibberellic acid supplements the grains' own gibberellins and accelerates the formation of many enzymes, thus speeding the rate of modification. Bromates reduce rootlet growth, respiration, and therefore total malting losses (e.g., to 6% dry basis) and the levels of soluble protein released when the treated malt is mashed. Treated malts contain traces of bromide ions.

Traditionally, barley is germinated cool (10–14°C; 50–57°F) on smooth tiled or concrete floors. It used to be raked or turned by workers with shovels to control the temperature and prevent the rootlets from matting together. These processes are now partly mechanized (1,4). If the batch of barley (piece) becomes too warm, it is spread more thinly, for example, to 3 in., to allow it to cool. Conversely, if it is too cool, it is piled thicker, to 6 in. or more, to retain the heat. In modern plants grain temperatures and ventilation are achieved pneumatically, that is, by passing a variable flow of temperature-controlled and water-saturated air through a comparatively thick bed of grain (2 ft in the older plants, up to 8 ft in modern plants; about 4 ft is the present average). Maximum airflow rates may be 0.14 to 0.2 m^3/ton of barley steeped/s (11). Mechanical devices are used, about three times per day, to lighten the bed, to turn it, and to prevent the rootlets from matting. Various temperature regimes are used depending on the grain and the type of malt being produced, usually in the range of 13 to 18°C (54.4–64.0°F). In drums, horizontal metal cylinders, turning is achieved by intermittent rotation about the long axis. For most purposes drum batch sizes (average about 50 tons) are uneconomically small. Most malt is germinated in compartment vessels, usually of up to 300 tons capacity (of original barley). However, one Scottish combination steeping, germination, and kilning unit has a batch size of 520 tons of barley. Most compartments, like the early Saladin boxes, are rectangular, but newer plants are usually circular. In compartments grain is supported on a perforated deck through which the fan-driven stream of cooled and humidified air passes. Turning is achieved by a row of vertical, rotating helical screws, or augers, that move slowly through the grain mass, lifting it and separating the roots. In rectangular boxes the turners are supported from a carriage that moves slowly over the bed of grain. With circular compartments the set of turners, mounted below their support, may rotate around the center of the static bed of grain, or the bed of grain may be rotated and thus moved through a static row of turners.

When it is ready the green malt is usually transferred to a kiln, where it will be dried and slightly cooked in a stream of warm air. The temperatures used depend on the type of malt being made (1–4,6–11). For pale malts the initial air-on temperature may be 50 to 60°C (122–140°F), rising to 80 to 105°C (176–221°F). The temperatures of the grain and the air leaving the grain bed are less as the air is cooled by evaporating moisture from the malt. The grain is supported on a perforated deck, and the air moves through it from below. Kilns may be rectangular or circular in cross-section and may be of single-, double-, or, rarely, triple-deck construction. In modern kilns the grain beds may be 2 to 6 ft deep, and outside North America they are not turned. In double kilns the wetter batch, on the top deck, receives some air from the earlier, drier batch of malt below. This arrangement saves heat, and a similar effect is achieved by linking kilns with air ducts (1,4). Usually air-to-air heat exchangers or other devices are used to abstract heat from the warm and wet outgoing kiln air and convey it to the dry-air inlet, saving heat and reducing fuel usage. Sometimes, toward the end of the curing period when the malt is nearly dry, a large proportion of the hot air is recirculated, also to save heat. Kilns may be directly fired, that is, furnace gasses pass directly through the bed of malt. Alternatively, the furnace heats a heat exchanger that in turn heats the air. With such indirect heating the furnace gasses do not come into contact with the malt. Sometimes sulphur dioxide is added to the kiln airstream to reduce the color of the malt, to make the malt more acid so that the pH of the mashed malt is reduced, to increase the levels of soluble nitrogenous substances in the wort, and to prevent the formation of carcinogenic nitrosamines on the surfaces of the malt grains. The quantities of oxides of nitrogen that reach the malt during kilning are also limited, to minimize the formation of nitrosamines. Malt for making Scotch whiskey is flavored with peat smoke, applied during kilning, and *rauchmalz*, used in the making of some special German beers, is flavored with hardwood smoke (1–3,8,9). At the end of kilning the malt is cooled and dressed—that is, the sprouts or rootlets, which are now brittle, are broken by agitation, and they and dust are removed by sieving and aspiration. These, together with thin grains and grain dust, are used for animal feed, often after being pelletized. Some special malts are finished in roasting drums, commonly of 0.5–5 tons capacity (1,8). Brown, chocolate, amber, and black malts are made from plump, lightly kilned pale malts. In the United Kingdom roasted barley is regarded as a special malt. It imparts a sharper, drier flavor than true roasted malts. The malt is loaded into a horizontal metal cylinder that rotates around its long axis and, for this purpose, is heated from the outside. The temperature is carefully raised until the correct coloration is achieved. For making black malts, when the highest temperatures are used, heating begins at about 75°C (167°F) and the temperature is raised to about 225°C (437°F). The products are quickly cooled, to fix their composition and reduce the risk of fire. They should not be charred. Any substantial amount of charcoal in these products makes them worthless. They deteriorate on storage and therefore are used as soon as possible. During roasting objectionable fumes are formed, and these must be destroyed, for example, by being consumed in an afterburner.

Crystal and caramel malts are prepared differently. Wet malt, either green malt or a pale malt that has been rewetted, is warmed in a closed cylinder, with no drying allowed, to 65 to 70°C (149–158°F). At this temperature the endosperms of the grains are mashed and the contents liquefy to a sugary mass. When all the grains have liquid contents, hot air is passed through the drums to dry the grains and to develop the desired color. At lower temperatures pale, luscious-flavored products are made, whereas at higher temperatures progressively darker and more strongly flavored materials are produced. When the malts are cooled the endosperm contents set solid and become hard, with a brittle, barley sugar–like texture. These are the only dry malts that should not be friable.

When malts are stored, care is taken to ensure that they are not mixed with unmalted grain; do not become damp; and do not become infested with insects, rodents, or other pests. Normally malt is transported in bulk, but for some purposes it is stored and moved in laminated, moisture-proof sacks of up to a 1-ton capacity.

MALT TYPES

The number of malt types in use is very large (1,4,7,8,10). Analyses of seven examples are given in Table 1. Malts vary with the cereal used, the quality and variety of the grain, the detail of the malting process, and the kilning and/or the drum heating process. It follows that the potential number of malt types is almost infinite. Undoubtedly the most common malts are pale with moderate enzyme levels (as in Pilsner and pale-ale brewing malts), with higher enzyme levels (as in North American brewing malts used with high levels of starch-rich adjuncts in mashes), or with exceptionally high enzyme levels (as in some distilling malts that must convert very large amounts of adjuncts during mashing). Usually the more enzyme-rich malts are made from protein-rich barleys that yield less extract than samples of the same variety that contain less protein. German kilned malts form a series of increasingly dark colors: (approximate; °EBC) Pilsner lager, 3.0; light lager, 3.5; Vienna-style malt, 7; and Munich-style malt, 17

Table 1. Analyses of Some Examples of Different Malts

Type of malt	Pale ale malt (UK)	Brewer's malt (N. America)	Grain distillers malt (Europe)	Lactic malt (Europe)[a]	Caramel malts (Europe)	Chocolate and black malts (UK)	Pale wheat malts (Europe)
Barley type (2- or 6-row)	2	2 or 6[b]	2 or 6[b]	2	2	2	n/a[c]
Hot-water Extract (% d.b)[d]	79–83	77–81	78–80	77	76–78	66–72	82–85
Protein (N × 6.25; % d.b)	8.8–10.3	10.5–13.5	12–13	11	11.0–12.5	9.4–10.6	11.6–13.6
Soluble/Total protein ratio (%)	38–42	38–48	50–55	50	31–34	—	36–50
Color							
(°EBC)	4–6	—	3–4	4–5	2–300	500–1,600	4–8
(°Lovibond)	—	1.5–2.0	—	—	—	—	—
Diastatic power							
(°Institute of Brewing)	30–45	80–130	180	65	0	0	100
(°Lintner)	35–50	90–145	200–250	70	0	0	110
(W.-K)	100–140	230–360	480–600	190	0	0	280
α-Amylase (DU)	30–55	33–49	105	25	0	0	47–55
Moisture (% fr. wt)	2–4	3.6–4.3	6.5	5–6	5–8	2	6–8

Note: Many types of malts are prepared to meet different specifications. Because malts are analyzed by different methods and often there are no true conversion factors, the example values given are approximate. Data mostly from reference 1. d.b-dry basis; fr. wt.-wet weight basis. I thank S. M. Sole of the Crisp Malting Group for much help.

[a]Lactic acid content 2–4%.

[b]In general malts from 6-row barleys tend to have higher enzyme and protein contents and lower extracts than malts from 2-row barleys. Wheats, being husk-free, tend to give malts with higher extracts than barley, because the husk yields no extract.

[c]n/a = not applicable.

[d]Hot-water extracts determined on finely ground samples. Different laboratory mashing methods are used.

(1,8). There is also a series of caramel malts: carapils, 2–5; carahell, 20–33; and caramuench, 50–300. Common British malts usually have colors of about:- lager, 3; pale ale, 4–6; mild ale, 7–8, amber, 40–85; brown, 100–200; chocolate, 500–1,100; and black, 1,200–1,600 (1,4). A full range of caramel malts is also available. These color ranges are not legally defined, and malt may be made to any reasonable specification. In addition to color, these malts impart special characteristics to beers and some other products. They are of no value in making distilled products. Colored and special malts, lacking enzymes, are used in small proportions in mashes mixed with paler malts that do contain enzymes. Less usual barley malts include undergrown or chit, malts that have the advantages of raw grain adjuncts in mashes, are less costly to make than normal malts, and can be used legally where the use of unmalted grains is illegal. Proteolytic, enzymic, and lactic malts contain a proportion (e.g., 2–4%) of lactic acid. These malts are used to adjust mash pH values where other methods are not permitted (1,4,8). Mashes made with a proportion of these malts yield a good extract, and the wort is enriched with N-containing yeast nutrients. Wheat beers are made from mashes prepared mainly with wheat malts (1,8). These malts are sometimes used in small proportions in mashes making barley beers, to improve the beer foam. Smoked, caramel, and roasted wheat malts are also used to a small extent. Sorghum malts are increasingly being used in Africa to make lager-type beers as well as traditional opaque beers (1). Distillery malts are of two main classes. Those in which the malt is the main or only component of the mash must have high extract yields and need only moderate enzyme contents, but those used to convert large amounts of starchy adjuncts in the mash must be rich in enzymes, although their extract yield is not so critical. These may be green (undried) malts (1,2).

BIBLIOGRAPHY

1. D. E. Briggs, *Malts and Malting*, Blackie Academic and Professional, London, United Kingdom, 1998.
2. G. N. Bathgate and R. Cook, "Malting of Barleys for Scotch Whiskies," in J. R. Piggott, R. Sharp, and R. E. B. Duncan, eds., *The Science and Technology of Whiskies*, Longman Scientific and Technical/John Wiley and Sons, New York, 1989, pp. 19–63.
3. D. E. Briggs, *Barley*, Chapman and Hall, London, United Kingdom, 1978.
4. D. E. Briggs et al., *Malting and Brewing Science*, 2nd. ed., Vol. 1, Chapman and Hall, London, United Kingdom, 1981.
5. J. S. Hough et al., *Malting and Brewing Science* 2nd ed., Vol. 2, Chapman and Hall, London, United Kingdom, 1982.
6. W. C. Burger and D. E. La Berge, "Malting and Brewing Quality," in D. C. Rasmusson, ed., *Barley*, American Society of Agronomy, Madison, Wis., 1985, pp. 367–401.
7. A. H. Cook, ed., *Barley and Malt*, Academic Press, London, United Kingdom, 1962.
8. L. Narziss, *Die Bierbrauerei (6 Auflage), Band I, Die Technologie der Malzbereitung*, Ferdinand Enke Verlag, Stuttgart, Germany, 1976.
9. G. H. Palmer and G. N. Bathgate, "Malting and Brewing," in Y. Pomeranz, ed., *Advances in Cereal Science and Technology*, Vol. 1, American Association of Cereal Chemists, St. Paul, Minn., 1976, pp. 237–324.
10. R. E. Pyler and D. A. Thomas, "Malted Cereals: Production and Use," in K. J. Lorenz and K. Kulp, eds., *Handbook of Cereal: Science and Technology*, Marcel Dekker, New York, 1991, pp. 815–832.
11. G. Gibson, "Malting Plant Technology," in G. H. Palmer, ed., Aberdeen University Press, Aberdeen, Scotland, pp. 279–325.

D. E. Briggs
University of Birmingham
Birmingham, England

MARGARINE

Margarine is a fatty food emulsion manufactured in semblance of butter. The appearance, color, texture, flavor, and nutritional content are equivalent to those of butter. The primary difference between margarines and butter is that the fat in margarine is not primarily milkfat. In turn, many lower-fat spread products have been invented and marketed in semblance of margarine. Margarines and lower-fat spreads share common attributes that are designed to resemble those of butter: product color is typically yellow; flavor is reminiscent of, although not identical to, that of butter; salt and acidity are comparable; functional utility is similar except for some cooking operations; and vitamin content is usually equivalent. The manufacturing processes are different, however, and shelf life length favors margarine over butter.

HISTORICAL MILESTONES

Margarine was invented in 1869 in France by Hippolyte Mege-Mouries in response to a need for a satisfactory butter substitute for military personnel. The original invention was prepared from beef fat, artificial gastric juice, salt, and milk. The resultant product was described as having a pearl-like luster, so Mege-Mouries named the material margarine, after the Greek *margarite*, or pearl-like. The inventor further described the product as a variety of true butter and later as artificial butter, butterine, or margarine.

The manufacture and sale of margarine was permitted by 12 April 1872 by action of the Council of Hygiene of Paris, which authorized its sale if not described as butter. Margarine manufacture expanded throughout Europe from 1872 to 1874. The first U.S. patent for margarine was granted in December 1873.

In the United States, the first Oleomargarine Act was passed in 1886. Amendments followed in 1902 and 1930. These actions placed a tax of $0.02/lb on margarine and required licensure of manufacturers and distributors to prevent adulteration of butter. The first definition of margarine was prescribed in 1941, and the 1950 Margarine Act removed the special taxation. The last state tax on margarine, in Minnesota, was abolished in 1975.

The history of margarine formulation in the United States is a series of incremental changes in processes and composition, all directed at attaining more favorable economics.

1890	Cultured milk used as a flavorant.
1917	Coconut oil used as a substitute for animal fat, most likely as a reaction to shortages of animal fat caused by World War I.
1923	Vitamin A fortification popularized. Butter contains about 15,000 IU/lb of this vitamin. Most margarines and spreads in the United States today are fortified to 16,000 IU/lb to permit labeling claims of 10% daily reference value intake per serving.
1934	Hydrogenated vegetable oils replaced coconut oil.
1941	U.S. federal Standard of Identity for margarine adopted.
1947	Beta-carotene (provitamin A) replaced coal tar colorants.
1950	Margarine Act passed by U.S. Congress, abolishing federal taxes on the product.
1952	First soft margarine in tub form appeared. One product was marketed as "Emde," at a price of over $3.00/lb and available only by prescription of a physician.
1956	Blended vegetable oil and butterfat products appeared.
1957	U.S. per capita consumption of margarine exceeded that of butter for the first time.
1957	Whipped sticks (6 per pound) first appeared.
1964	Soft tub margarine distributed nationally as mainstream grocery item.
1966	Diet margarine (40% fat) introduced.
1967	Fluid, pourable margarine developed and introduced.
1975	Lower-fat spreads (typically 60% fat) introduced.
1981	Butter/margarine blends introduced to national market.
1981–1997	Lower-fat spreads proliferated.
1991	U.S. Nutritional Labeling and Education Act passed by U.S. Congress.
1991	No-fat margarines appeared.

TYPES OF PRODUCTS

The products within the margarine and yellow fat spread category may be classified in any of several groups. Most relevant is grouping by physical form, which follows function.

Solid

Solid products are bulk cubes of 50 lb or 25 kg (55 lb), packed for institutional food processing or food service consumption. These products are typically formulated specifically for the intended use and generally are of firmer structure and higher solid fat content than retail margarine. Typical use is in general baking, with additional usage in layered pastries. Other types of solid margarine are brick and stick.

Brick margarine is typically of 1 lb net weight, wrapped in paper or foil laminate, and packed 24 to 35 lb per case. Customary uses are in general cooking, shallow pan frying, basting, and the like. The products are similar to retail formulations in texture, flavor, and consistency.

Stick margarines are made for household kitchen and table use and are typically packaged four sticks to the pound. These products are packaged first in paper or laminated foil, then in a carton of paperboard. Cases contain 12 to 36 lb. The lower-weight cases are recent developments to address inventory-management programs among grocers, who count case turnover in lieu of actual poundage. In the United States, two styles of stick margarine are found. The eastern stick is sold east of the Rocky Mountains and is packed either in flat units or in two-over-two cartons. The western-style product is shorter in length but larger in width and breadth. The two styles arose from availability of packaging equipment in the respective market areas. Solid margarine is also packaged in individual servings of 5 to 14 g, distributed via food service operators.

Soft

Soft products are generally intended for direct use, as for spreading on breads. Retail products are packed in twin cups of 8 oz net weight, sleeved two per pound, or in 16-oz tubs. Margarine that conforms to the U.S. Standard of Identity may not be sold in packages greater than 1 lb at retail. Many lower-fat spreads are packaged for retail sale in units of 2 or 3 lb or more. Soft products for food service use include portion-control cups of 5, 7, 10, or 14 g. Soft product is occasionally seen in single-serving pouches.

Whipped products are soft or stick products to which nitrogen is injected at manufacture, with the primary intent being expansion in volume for economy of serving. Typical volume expansion is about 50%, yielding 6 sticks per pound, or 24 fluid ounces at net weight of 16 oz.

Liquid margarines are typically formulated to be resistant to solidification at refrigerated temperatures while maintaining structure of the emulsion. Typical uses are as a topping for hot vegetables, waffles, pancakes, and popcorn.

THE REGULATORY IDENTITY OF MARGARINE

In the United States, margarine and butter are two foods that are defined by action of the U.S. Congress. Other foods are defined by regulations promulgated by the U.S. Food and Drug Administration, the U.S. Department of Agriculture, or other regulatory agencies.

Title 21, Section 166 of the Code of Federal Regulations (21 CFR 166) defines margarine labeling and composition. The labeling provision requires that the word "margarine" or "oleomargarine" appear on the principal display panel in type or size at least as large as any other type or lettering. Other regulations require a listing of ingredients in the product, with the sole exception that proprietary flavorant substances need only be identified as "artificial flavor."

The composition provision requires that the product contain at least 80% fat and be made from edible fats, oils, or mixtures thereof whose origins are vegetable, rendered animal carcass fats, or any form of oil from a marine species that has been affirmed as generally recognized as safe (GRAS) or listed as a food additive.

The product must also contain one or more of the following:

- Water and/or milk and/or milk products
- Suitable edible protein, including whey, lactose-reduced whey, non-lactose-containing whey components, albumin, casein, caseinate, vegetable proteins, or soy protein isolate, in amounts not greater than that reasonably required to accomplish the desired effect.
- Any mixture of 1 and 2. This must be pasteurized. Any mixture of this may be subjected to the action of harmless bacterial starters.
- Vitamin A of at least 15,000 IU/lb concentration.

Margarine may contain any of these optional ingredients:

- Vitamin D. If added, product must contain no less than 1,500 IU/lb.
- Salt, which typically ranges from about 1.25% to 2.0% by weight.
- Nutritive carbohydrate sweeteners
- Emulsifiers
- Preservatives, including but not limited to the following:
  - Sorbic acid, benzoic acid, and their sodium, potassium, and calcium salts, individually 0.1% or in combination, 0.2%
  - Calcium disodium EDTA, 0.0075%
  - Gallates, BHT, BHA, ascorbyl palmitate, and ascorbyl stearate, individually or in combination, 0.02%
  - Stearyl citrate, 0.15%; isopropyl citrate mixture, 0.02%
- Color additives (Beta-carotene or annatto or annatto/turmeric extracts)
- Flavoring substances
- Acidulants and/or alkalizers

MARGARINE FATS

Margarine is a water-in-oil emulsion, which is a semistable suspension of dispersed aqueous materials within a continuous matrix of fat. The fat typically is refined, mixed triglycerides of vegetable or animal carcass origin. The predominant vegetable oil in the United States is soybean oil, which is a by-product of soybeans. The soybean is grown to produce proteins for animal and human foodstuffs; soybean oil is about 18% of the soybean. Additional vegetable oils used in margarines are corn, safflower, sunflower, and palm. Lauric oils are seldom seen. Marine-origin oils that are GRAS are also permitted. Lower-priced margarines, especially in private-label segments, may contain lard, beef tallow, or combinations of these with soybean oil.

The structure of the fat or blend of fats is the most significant factor in margarine formulation. The triglycerides present in the oils are reaction products of 1 molecule of glycerol with 3 molecules of fatty acids, with mixed triglycerides predominant in nature. The triglycerides provide structure, lubricity, caloric density, and a satiety to the product.

Physical properties of margarines and spreads, especially texture, spreadability, color, appearance, and melting profile, are functions of the structure of the fat and the processing conditions used in manufacture. Margarines and spreads are composed of liquid oil, fat crystals, and the aqueous phase. The crystals give margarine the required consistency and stabilize the water droplets. The fat crystals are in effect frozen oil. Hydrogenation of the liquid fat is used to raise the melting point of the fat to provide the structure needed. Hydrogenation is the addition of hydrogen across the unsaturated carbon-carbon double bond.

Soybean oil margarine and/or spread is typically a blend of two or more partially hydrogenated components mixed in appropriate proportions for functional effect. The functional effect can be predicted from the measurement of solid fat index (SFI), which is the ratio of solid fat to liquid fat at specified temperatures. Measurement of SFI is done by dilatometry. A similar measurement, solid fat content, may be performed instrumentally via nuclear magnetic resonance. Table 1 demonstrates the differences in SFI among different types of margarines and spreads, with reference to butter.

The significance of the temperatures used in the test protocol are that 50°F is a typical temperature of processing, 70°F may describe the stability of the product outside the cold-storage area or refrigerator, 80°F is a typical temperature found in a commercial bakery or kitchen, 92°F is reflective of mouth temperatures, and 104°F is a typical bakery proofing chamber temperature. The relevance of the SFI temperatures is usage specific.

The process temperature of 50°F is typically the temperature at which product is packaged. For solids, bricks, and sticks, the product itself provides structure to the package. SFI content of about 25% is required for efficient packing of such product. For soft margarine, the solids content of only about 11% assures the product is still fluid and pourable when the containers are filled. The soft product will become dramatically firmer after resting in the package for as few as 5 min. Liquid margarine SFI of about 3% at 50°F indicates the product will remain fluid at refrigeration/cold-store temperatures.

At 70°F, the SFI indicates resistance to ambient temperature slump or partial melt. In the United States, stick margarines are occasionally offered for sale in supermarket aisles rather than from the refrigerated dairy case. This practice is caused in part by the economic phenomenon of slotting allowances that retailers require manufacturers to pay. Space in the dairy case is premium compared to dry shelf display, which is premium compared to aisle display. The practice of offering margarines for sale in nonrefrigerated areas is not condoned by the manufacturers, however.

The 92°F temperature in SFI is significant in that it is a quantitative measure of waxiness in the mouth. If the value is greater than about 3%, the product will melt slowly in the mouth and impart a slight coating or waxlike aftertaste. Significance of 80°F and 104°F were described previously.

Margarine fats are generally blends of oils. Hydrogenation is energy intensive and also produces *trans* fat isomers. Its use in margarine is generally minimized for economic as well as functional effect. For example, stick margarine fat may be a blend of 50% of partially hydrogenated soybean oil of iodine value 68 and 50% of refined, bleached, and deodorized liquid soy oil of iodine value 135. Tub margarine may be a blend of the same components, with the sole difference being the ratio of liquid oil to hard fat at 2:1 rather than 1:1 as for stick product. The liquid product may be a blend of about 98% of liquid soy oil and 2% of fully hydrogenated soybean oil. Bakery products typically combine partially hydrogenated oil and liquid oil with fully hydrogenated fat to maintain structure in warm bakeries.

Table 1. Typical Solid Fat Indices

Percent Solid fat at	Margarine type				Butter
	Solid	Soft	Liquid	Bakery	
50°F = 10°C	25–29	9–13	2–4	28–32	29–32
70°F = 21.1°C	13–17	6–8	1–3	16–20	11–14
80°F = 26.7°C	8–11	4–6	1–3	11–15	6–8
92°F = 33.3°C	1–3	1–2.5	1–3	6–10	1–2.5
104°F = 40°C	0	0	1	2–6	0

OTHER FATTY TYPE INGREDIENTS

Fat-soluble margarine ingredients include soy lecithin, monoglycerides, derivatives of monoglycerides, vitamin A, beta-carotene, flavorants, and some preservatives.

Lecithin is phosphatidylcholine, a by-product of refining soybean oil. The addition of lecithin to margarine is to promote the frying performance of the margarine. Addition of about 0.2% of lecithin will dramatically reduce the tendency of margarines to "spatter" in frying. Lecithin prevents the coalescence of large droplets of water that would become steam at frying temperatures.

Monoglycerides and distilled monoglycerides are emulsifiers, or surface active agents that promote stability of the oil and water mixture by linking the naturally immiscible fats and water-laden ingredients at the fat-water interface. Use of 0.10 to 0.25% of monoglycerides also promotes creaminess in mouth-feel.

Derivatives of monoglycerides include polyglycerol esters, propylene glycol monoesters, and other such materials. Primary use of these materials is as whipping/creaming agents for expanding volume of the margarine or spread by incorporation of nitrogen, creating a three-phase foam of gas in water-in-oil.

Vitamin A palmitate is the palmitic acid ester of retinoic acid. Regulations require addition to total fortification of 15,000 IU/lb, or about 4,500 μg per pound, which equates to about 10 ppm. Vitamin A palmitate is the fat-soluble market form of vitamin A, as opposed to vitamin A citrate, a water-soluble form typically found in multivitamin tablets. Although no regulation requires spreads (nonstandard products) to be fortified, market forces have caused all nonmargarine spreads in the United States to be fortified with vitamin A at the same concentrations as those of margarine.

Vitamin D (calciferol) is an optional ingredient that may be added to margarine and spreads at 1,500 IU/lb.

Beta-carotene is a double molecule of retinoic acid and is used as the colorant of choice in the majority of margarines marketed in the United States. It is also called provitamin A because it is metabolized within the liver to become vitamin A. In practice, margarine is fortified with about 5 ppm of beta-carotene for optimal coloring. This permits the manufacturer to reduce the amount of vitamin A palmitate added, as both vitamin A palmitate and beta-carotene contribute to the total vitamin A activity in the product.

Annatto or annatto-turmeric extracts are also used as margarine colorants, typically at concentrations of about 10 to 20 ppm. These are vegetable extracts.

Preservatives may be added to protect the oil from oxidative rancidity or to retard the growth of microorganisms.

Flavorants typically added to margarine and/or spreads are generally mixtures of synthesized or extracted chemicals. Principal flavorant substances are 2,3 butadione (diacetyl); lactic acid; short-chain fatty acids such as butyric, caproic, or capryllic; lactones; keto acids; some aldehydes; indole; methyl indole; or similar materials. Typical usage is about 5.0 to about 300 ppm, depending on desired strength of flavor.

AQUEOUS COMPONENTS

The dispersed phase typically contains droplets of 1 to 20 μm in size. The primary aqueous material is water, to which may be added skim milk, salt, and water-soluble preservatives and/or flavorants. Lightly salted margarine is preferred in the United States. Typical salt concentration is about 1.5%.

Skim milk, whey, or other edible proteins are added to enhance the release of the flavor profiles and to permit the margarine or spread to perform in semblance of butter in shallow pan frying. The milk solids and protein will brown in the pan, leaving a golden color for frying. Water-based margarines that contain no milk products simply overheat and blacken in the pan. The flavor profiles are often described as deficient or not well rounded.

Lactic acid, if added, is both a flavorant and an acidulant. Addition of about 0.05% to 0.1% of lactic acid will reduce the typical serum phase pH to less then 5.5, which increases the effect of the preservatives.

The typical margarine and spread preservative system is a mix of potassium sorbate and sodium benzoate, each added at about 0.10% by weight. These preservatives are added to sustain the resistance to microbial spoilage after the package has been opened in the home.

MANUFACTURING PROCESS

The margarine manufacturing process is a series of unit operations that transform a blend of ingredients into the product. The five unit operations are emulsification, cooling, working, resting, and packaging.

Margarine is mixed either in batches or by continuous methods. In batching, discrete amounts of the ingredients are measured, either by weight or by volume; blended; and transferred to processing steps. Batching is the procedure of choice when the products are of a variety of recipes. Continuous mix simultaneously measures all ingredients and mixes them in a continuous output stream. One method of continuous mix uses a multiple-headed adjustable calibration pump operated from a common drive shaft. This method is useful if the type of product is constant.

Emulsification is achieved by measuring appropriate amounts of oils, emulsifiers, vitamin A, coloring, and oil-soluble flavorings. Aqueous-phase ingredients are blended separately, then combined with the oil phase at a temperature slightly above the melting point of the oil. Agitation of the mix is required to promote the formation of the emulsion.

Mixed emulsion at a temperature slightly higher than the melting point of the mixed fat (typically about 100–105°F) is pumped through a filter to a scraped surface heat exchanger cooling system such that the output temperature of the chilled margarine or spread is about 50 to 55°F, which is well below the melting point of the oils. This attribute of supercooling by rapid chilling is employed to reduce the size of fat crystals to a fine dispersion. If the emulsion were cooled via simple refrigeration, the fat crystals would be excessively large. Large crystals promote grainy texture and instability of emulsion, which are both undesirable. Nitrogen gas is injected either before or after scraped surface heat exchange but prior to pinworker (see next paragraph). Chilling and crystallizing of molten product yields an increase in production line pressure to as much as 250 to 300 psig, depending on the specific composition of product and equipment configuration. The scraped surface heat exchange equipment is designed to tolerate pressure slightly in excess of this. Contact the manufacturer for specifics.

The chilled product is agitated by pinworker units that reduce the size of the fat crystals by mechanical working. This further softens the texture of the product and fully disperses the injected nitrogen gas. Working also lengthens the time for the product to fully set.

Supercooled fluid product is allowed to rest in crystallizing chambers before packaging. Stick and brick product can be packed by either of two processes: (*1*) form the stick, then apply wrap, collect, and carton; or (*2*) insert cut wrapping material into a mold, then inject the fluid product into the mold cavity, wrap, collect, and carton. Soft product is dispensed into preformed and printed cups or tubs, capped, then sleeved. Many soft products are packaged with a plastic shrink-wrap band for package surety.

Packed units are collected into cases, stacked on pallets, and then moved to cold storage temporarily until quality testing is completed.

QUALITY TESTING

Ingredients are tested for conformance to requirements. The increasing tendency is for suppliers to test and validate requirements by providing certified testing results with delivery of the ingredient. Primary testing of the oils is conducted at the plant. Significant attributes in routine testing include flavor, aroma, color, peroxide value, free fatty acids, moisture, refractive index or iodine value, and SFI, when applicable. Other ingredients are tested for microbiological quality as appropriate.

Manufacture processes are monitored by recording instruments, visual observation, and analysis of production efficiencies for conformance to requirements. Typical process temperatures are product specific, but general approximations are:

As mixed	ex SSHE	ex pinworker	As packed	10 min postpack
105°F	53°F	56°F	58°F	64°F

Product quality includes nominal chemical characteristics of moisture %, salt %, curd/milk solids %, with fat % determined by difference. Other characteristics of product that are determined at time of production include pH of the aqueous phase, color, flavor, appearance, and net weight. Several package characteristics, including completeness of wrap, presence of lids, and legibility of code dating are checked also. Microbiological assays for standard plate count (SPC), yeasts, molds, and coliforms are determined routinely.

Typical limits are SPC 3,000/g, yeasts, molds, and coliforms negative by analysis. Periodic sampling for additional quality characteristics is also required, including vitamin A assay and selected pathogens. Quality begins with clearly stated product requirements and is built into the product via instructions for formulation and manufacture.

GENERAL REFERENCES

A. J. C. Andersen and P. N. Williams, *Margarine*, 2nd ed., Pergamon Press, New York, 1965.

M. M. Chrysam, "Table Spreads and Shortenings," in T. H. Applewhite, ed., *Bailey's Industrial Oil and Fat Products*, 4th ed., Vol. 3, John Wiley & Sons, New York, 1965, pp. 41–126.

Code of Federal Regulations, Title 21 CFR Part 166, Secs. 401, 701, U.S. Government Printing Office, Washington, D.C., 1999.

D. Melnick, "Development of Organoleptically and Nutritionally Improved Margarine Products," *J. Home Economics* **6**, 793–798 (1968).

A. J. Haighton, "Blending, Chilling and Tempering of Margarines and Shortenings," *J. Am. Oil Chem. Soc.* **53**, 357–399 (1978).

Handbook of Soy Oil Processing and Utilitization, American Soybean Association, St. Louis, Mo., and American Oil Chemists Society, Champaign, Ill., 1980.

Y. H. Hui, ed., *Bailey's Industrial Oil and Fat Products*, 5th ed., John Wiley & Sons, New York, 1996.

Margarine Statistics Report, National Association of Margarine Manufacturers, Washington, D.C., 1988.

S. F. Riepma, *The Story of Margarine*, Public Affairs Press, Washington, D.C., 1970.

L. H. Weiderman, "Margarine and Margarine Oil: Formulation and control," *J. Am. Oil Chem. Soc.* **55**, 823–829 (1978).

JOHN BUMBALOUGH
Land O'Lakes, Inc.
Arden Hills, Minnesota

See also FATS AND OILS: PROPERTIES, PROCESSING TECHNOLOGY, AND COMMERCIAL SHORTENINGS.

MARINE ENZYMES

Marine animals possess the same functional classes of enzymes as other living organisms, which enable them carry out virtually the same metabolic activities. These enzymes are present in the digestive glands and the muscle tissues of the animals and may be recovered in active and stable forms for commercial use. There is considerable demand for enzymes throughout the world for food, biomedical, and other commercial applications. Worldwide sales of commercial enzymes were estimated at $1.5 billion for 1997 (1). Traditional enzymes such as pepsins, rennets, trypsins, and lipases are derived from animal tissues, whereas bromelain, ficin, papain, lipoxygenase, and amylases are derived from plants. Plant and animal enzymes represent a small fraction of commercial enzymes, with the greatest diversity of commercial enzymes coming from microorganisms that have been stringently evaluated and certified as safe. Examples of microbial enzymes are glucose oxidase, pectinesterases, cellulase, and glucose isomerase. Microbial enzymes are used as replacements or substitutes for homologous enzymes from animals or plants because they are relatively easier and cheaper to produce. Other contributory factors to the decline in the use of plant and animal enzymes are related to political as well as agricultural and economic policies of various governments that regulate food and agricultural practices. The marine environment presents an excellent opportunity for supplying commercial enzymes to help meet the demands for these compounds. In several of the major fish-producing countries, the by-products of seafood harvesting comprise about 50% of the entire harvest. These materials are largely underutilized and discarded as waste. However, this abundant material also includes the enzyme-rich digestive organs, and the enzymes may be recovered in various forms to suit a range of commercial applications. It is estimated that less than half of the fish offal generated in fish plants is converted into value-added products such as pet food, fish meal, and compost, while the remainder poses problems related to disposal and environmental pollution. This discussion focuses on enzymes from fish and shellfish tissues that either have the potential for commercial application or constitute a source of concern for food-processing operations, especially extracellular enzymes produced by digestive organs and intracellular enzymes from fish muscle tissues. Enzymes from digestive organs have the greatest potential for commercial application because of their high tissue concentration. However, the recovery of enzymes present at low concentration, such as intracellular enzymes, is also of interest, because recent advances in biotechnology will facilitate the cloning of the genes for those enzymes and their subsequent production by fermentation. The use of immobilized forms of these enzymes will also allow use of expensive enzymes in clinical or biochemical applications.

ENZYME TYPES, DEFINITIONS, SOURCES, AND SOME PROPERTIES

Marine animals carry out essentially the same types of metabolism as land animals. Thus, it is to be expected that marine animals will contain all six functional classes of enzymes (i.e., oxidoreductases, transferases, hydrolases, lyases, isomerases, and ligases). The scope of this article will be limited to those enzymes from fish and shellfish that either have the potential for commercial application or constitute a source of concern for food processing operations. They include representatives from the oxidoreductases, transferases, and hydrolases.

Oxidoreductases

Oxidoreductases are the group of enzymes that catalyze oxidation-reduction reactions. Examples include polyphenol oxidases, peroxidases, and the dehydrogenases. Polyphenol oxidases (PPO) catalyze the oxidation of phenolic compounds in the presence of molecular oxygen to form dark-colored melanins. These enzymes are found in the particulate and in the soluble fraction of skin homogenates of fish and shellfish. Examples of marine animals from which PPOs have been isolated are lobster (2,3), shrimp (4,5), crab (6), the common mussel (*Mytilus edulis*) (7), goldfish (8) and otopus (9). PPOs are metalloenzymes and have copper ions in their catalytic sites. They lose activity when the copper ions are removed or replaced by other metal ions. PPOs are active over a broad pH range from 5.0 to 8.5, and they are most stable within a weakly acidic to alkaline pH range of 6 to 10 (4,7,8). The enzymes also have different molecular weights such as 30 to 40 kDa for shrimp PPOs, 64 kDa for the lobster enzyme, 120 kDa for the enzyme from *Mytilus edulis*, and 125 kDa to 205 kDa for cephalopods (2–4,7,9). The enzymes are inhibited by p-amino benzoic acid, oxygen competitors like cyanide and N2, and metal chelators like diethyldithiocarbamate and EDTA.

Another group of oxidoreductases found in fish tissue is the lipoxygenases. These enzymes are of considerable economic importance to the fishing industry because of the adverse effects it causes with regard to fish texture, odor, and color during storage. These enzymes have been isolated from microsomal fractions of various fish species including herring (10), trout (11,12), flounder (13), Pacific rockfish (14), mackerel (14), gurnard, grouper, and tilefish (14). Lipoxygenases from marine animals catalyze the oxidation of polyunsaturated fatty acids, similar to plant lipoxygenases. These enzymes are most active between pH 6 and 7 and remain active at subzero temperatures (10). Microsomal lipid oxidation enzymes have been implicated in fish quality deterioration during iced or frozen storage. For example, lipoxygenases associated with fish skin promote postharvest bleaching of carotenoid pigments in fish skin and/or flesh (15).

Peroxidases are also oxidoreductases. Peroxidases with molecular weights of 35.5 kDa, 76.5 kDa, and 147 kDa have been isolated from the hepatopancreas of crayfish (16). These enzymes catalyze the incorporation of halogens into organic molecules, for example, halogenation of tyrosine to thyroxine, the thyroid hormone responsible for the prevention of goiter in humans.

The glycolytic pathway includes a number of dehydrogenases that participate in the removal of hydrogen atoms from various substrate molecules. The glycolytic enzymes remain active and catalyze their respective reactions in the postmortem animal. The degree to which glycolysis proceeds in the postmortem animal has important ramifications concerning the postharvest quality of fish and shellfish. Several of the glycolytic enzymes have been characterized from muscles of fish and shellfish such as lobster and cod (17), tuna and halibut (18), shrimp (19), and trout (20). Examples of the enzymes are lactate dehydrogenase (LDH), an oxidoreductase that catalyzes the interconversion between pyruvate and lactate under limited supply of oxygen. Lactate dehydrogenases from marine species exist as tetramers with subunits of 36 kDa and are inhibited in a noncompetitive fashion by 3-acetyl-pyridine NAD. The muscle of some invertebrates like squid does not contain lactate dehydrogenase, and pyruvate is metabolized by a different mechanism during anoxia.

Glutamate dehydrogenases have been isolated from various shellfish, for example, *Mytilus edulis* (21), mollusks (22), and *Modiolus demissus* (23). The enzymes catalyze the oxidative deamination of glutamate to α-ketoglutarate, thereby serving to increase concentration of TCA cycle intermediates (so-called anaplerotic reactions) to allow an increase in rate of oxidation of two-carbon units:

$$\text{glutamate} + \text{NAD}^+ + \text{H2O} \leftrightarrow \alpha\text{-ketoglutarate} + \text{NH}_3 + \text{NADH} + \text{H}^+$$

The same reaction may run in reverse, draining off TCA cycle intermediates for biosynthetic purposes. Glutamate dehydrogenases from marine animal tissues were shown to be most active from pH 7.4 to 9.5.

Another dehydrogenase of the glycolytic pathway that has been isolated from marine animal tissues is muscle-type glyceraldehyde-3-phosphate dehydrogenase. The enzyme catalyzes the substrate level oxidative phosphorylation of glyceraldehyde-3-phosphate (G3P) to the high-energy compound glyceraldehyde-1,3-diphosphate (1,3-DPG), whose high-energy group transfer potential is subsequently used to synthesize ATP. The same enzyme drives the reverse reaction, that is, the formation of glyceraldehyde-3-phosphate from 1,3-diphosphoglycerate during gluconeogenesis. Glyceraldehyde-3-phosphate dehydrogenase has been isolated from lobster and cod muscle and shown to exhibit maximum activity within pH range 8.5 to 9.0.

Uricase, also known as urate oxidase, is an oxidoreductase and catalyzes the oxidation of uric acid to allantoin prior to further degradation and excretion. The enzyme is usually located in peroxisomes and is used as a marker enzyme for this organelle and has a potential commercial value as a diagnostic enzyme. Uricase has been isolated from the livers of a number of fish, such as trout, mackerel, catfish, shark, and tilapia (24). Uricases from various marine animals are active over a broad pH range, from pH 7.0 to 9.5 with a maximum at 8.8. Uricase is an oligomeric enzyme containing subunits of 32.5 kDa. It is also a metalloenzyme, requires Cu^{2+} for activity, and is inhibited by chelating agents.

Transferases

The transferases are a group of enzymes that remove groups (other than H) from various substrates and then transfer them to acceptor molecules. Examples of the transferases are transglutaminases and the kinases. Transglutaminases (TGases) catalyze the Ca^{2+}-dependent acyl-transfer reactions between the γ-carboxyamide groups of protein-bound glutamine residues and the primary amino groups in a variety of compounds (including ϵ-NH_2 of lysine). TGases have the unique ability to modify the functional properties of protein molecules via covalent cross-linking. TGase activity has been described in the tissues of various marine animals such as carp (25), rainbow trout (26), mackerel (26), lobster (27), scallop, shrimp, and squid (26). The TGases have molecular weights ranging from 80 to 200 kDa and temperature optima ranging from 40 to 50°C. They are active over a broad pH range of 6.0 to 9.5. The kinases catalyze the phosphorylation of substrates in reactions that invariably require ATP as phosphate donor. Pyruvate kinase (PK) is an important regulatory enzyme in the glycolytic pathway. It catalyzes the reaction that prepares pyruvate for subsequent utilization in the TCA cycle, that is, the irreversible conversion of pyruvate to acetyl CoA. PK has been purified from various shellfish including *Mytilus edulis* (28), *Crassostrea gigas* (29), and *Cardium tuberculatum* (30). It is a tetrameric protein molecule with 70 kDa subunits (30) that requires monovalent and divalent cations for activity. It is activated by fructose-1,6-bisphosphate and inhibited by L-alanine, ATP, D-lactate, and citrate.

Phosphofructokinase (PFK) is another transferase, and it catalyzes the conversion of fructose-6-phosphate (F6P) to fructose-1,6-diphosphate (F1,6DP) in the glycolytic pathway. PFK is allosterically activated by F6P and AMP but is inhibited by ATP and citrate. The enzyme has been isolated from trout muscle (31).

Glycogen phosphorylase is a transferase and catalyzes the first step in the utilization of glycogen. The enzyme has been isolated from lobster muscle and shown to exist as a dimer with a molecular weight of 170 kDa. Lobster glycogen phosphorylase is similar to its mammalian counterpart in having two forms: the inactive dephosphorylated form known as phosphorylase b, and the phosphorylated active form known as phosphorylase a.

Glutathione-S-transferases are a group of multifunctional proteins that are involved in the detoxification of xenobiotics in animals. The enzyme has been isolated from the liver, kidney, and gills of rainbow trout and shown to exist as a dimer whose subunits have molecular weights ranging from 23.9 to 25 kDa (32). The enzyme is inhibited by 1-chloro-2,4-dinitrobenzene.

Aspartate transcarbamylase, also known as aspartate carbamoyltransferase, catalyzes the first committed reaction in the biosynthesis of pyrimidine bases. This involves the transfer of a carbamoyl group directly to L-aspartate without the formation of any intermediate products. The enzyme has been isolated from the mantle of *Mytilus edulis* (33) and is most active within pH range of 8.5 to 8.9. It is strongly inhibited by *p*-hydroxymercuribenzoate and Cu^{2+} but is strongly stimulated by organic solvents like dimethylsulfoxide and dimethylformamide.

Hydrolases

Hydrolases are the group of enzymes that catalyze the cleavage of covalent bonds in substrates with the participation of water molecules as reactants. The hydrolytic enzymes of importance in this regard include the proteases that catalyze hydrolytic cleavage of peptide bonds in proteins and polypeptides to produce low-molecular-weight peptides and/or amino acids, and lipolytic enzymes that catalyze the hydrolysis of ester bonds of triglycerides and phospholipids to produce fatty acids and other products. Proteolytic enzymes include those found in fish muscle as lysosomal proteases and those from digestive glands or fluids, namely, digestive proteases. Lysosomal proteases, known as cathepsins, have been detected in various species of fish such as carp (34), albacore (35), cod (36), salmon (37), Pacific sole (38), squid (39,40), and tilapia (41). Cathepsins from marine animals occur as single polypeptide chains with different molecular weights, for example, 13.6 kDa for squid cathepsin B (40) and 25 kDa for cathepsin C from squid hepatopancreas (39). Some cathepsins exhibit maximum activity within pH 3.5 to 7.0, whereas others are most active at alkaline pH (8.0). They are activated by Cl^- ions and usually require thiol reagents as well as chelating agents for activity and are believed to play an important role in postmortem deteriorative changes in fish texture prior to processing or storage (42). According to Siebert and Schmitt (43), fish muscle contains about 10 times more catheptic enzyme activity than mammalian muscle.

The most studied digestive proteases from marine animals are the pepsin(ogen)s, trypsin(ogen)s, and chymotrypsin(ogen)s. Pepsin(ogen)s are acidic proteases secreted by the gastric mucosal glands, whereas trypsin(ogen)s and chymotrypsin(ogen)s are secreted by pancreatic tissue and pyloric ceca. A number of exopeptidases are also secreted in the gut, where they act to degrade peptides into free amino acids. Pepsin(ogen)s have been isolated from several fish species, including tuna (44,45), salmon (46), dogfish (47), bonito (48), trout (49,50), sardine (51), hake (52), capelin (53), cod (54–56), and smelt (57). The carnivorous fish have been shown to have the highest levels of pepsins in their stomachs (58,59). Pepsin activity has not been detected in digestive tracts of stomachless fish (60,61). The pH optima of fish pepsins range from 2.5 to 3.0, which is slightly more alkaline than pH optima for mammalian pepsins (51). The molecular weights of pepsins isolated from various fish species range between 36 and 41 kDa. Gastricsin(ogen)s are acidic proteases present in the gastric mucosa of fish. Gastricsin(ogen)s have been purified from *Merluccius gayi* with molecular weights of 27 to 28 kDa (36). They exhibit maximum proteolytic activity at pH 3.0 and are generally stable up to pH 10.0, the point beyond which they rapidly lose activity. Marine gastricsins catalyze the clotting of milk similar to fish and mammalian pepsins but differ from pepsins by showing activation with sodium chloride (56). Trypsin(ogen)s and chymotrypsin(ogen)s have also been recovered from the digestive tracts of several fish species such as sardine (62), goldfish (61), capelin (63), catfish (64), rainbow trout (65), salmon (66), green chromide (67), cod (68–71), cunner (72), crayfish (73), shrimp (74), and starfish (75–78). They are generally most active within pH range 7 to 8.5 and are also most stable at alkaline pH, unlike trypsin(ogen)s from mammals (69). They are single polypeptide chains with relatively lower molecular weights, ranging from 21 kDa to 26 kDa.

Chymotrypsins have been isolated and characterized from marine species such as anchovy (79), Atlantic cod (79–80), capelin (81), herring (81), rainbow trout (82), and spiny dogfish (83,84). In general, these enzymes are single polypeptide molecules with molecular weights ranging between 25 and 28 kDa. They are most active within pH range 7.5 to 8.5 and are most stable at around pH 9.0 (82). Chymotrypsins from marine animals have a higher catalytic activity and hydrolyzed more peptide bonds in various protein substrates (casein, collagen, and bovine serum albumin) at subdenaturation temperatures than mammalian chymotrypsins. In general, the marine chymotrypsins were more heat-labile than mammalian chymotrypsins (80). Dogfish chymotrypsin was more active toward soy protein isolate from 5°C to 35°C than bovine chymotrypsin (84) and could also clot milk. Chymotrypsins from herring and capelin exhibited greater hydrolytic activities toward the synthetic substrate BTEE than the bovine enzyme (81).

Collagenases are capable of degrading the polypeptide backbone of native collagen under conditions that do not denature the protein. Collagenolytic enzymes have been isolated and characterized from the digestive glands of a variety of teleost fish (85) and crab (86–89). Marine collagenases resemble their mammalian counterparts in their mode of action, that is, they act on native collagen in solution and cleave the helical structure at loci 75%, 70%, and 67% from the NH_2-terminal of the molecule. However, they differ from mammalian collagenases in exhibiting trypsin- and chymotrypsin-like specificities toward synthetic substrates for these enzymes. They are also subject to inhibition by serine protease inhibitors such as PMSF and DIPF. In this regard they are described as serine proteases, unlike mammalian collagenases, which are mostly metalloproteinases. Marine collagenases show optimum activity within a pH range of 6.5 to 8.0 and are inactivated at pH values below 6.0. They are also inhibited by well-known serine protease inhibitors like diisopropylfluorophosphate, phenyl methyl sulfonyl fluoride, soybean trypsin inhibitor, and chicken ovomucoid. A collagenase purified from crab was shown to be a single polypeptide molecule with a molecular weight of ~25 kDa. Collagenases from hepatopancreas tissue have been implicated in postharvest texture deterioration in prawns during ice storage (90). Although true collagenases have not been isolated from the muscle of fish, there is evidence that they contribute to flesh softening in postharvest Pacific rockfish (91).

Alkaline proteases is the general term used to describe proteolytic enzymes present in fish muscle that show maximum activity as well as stability within the alkaline pH range. These enzymes have been isolated from muscles of various fish species such as filefish, croaker, anchovy, rockfish, yellowtail, shrimp, and garfish (92). They show optimum activity within pH range 7.9 to 8.1, are stable within a broad pH range (from 4.5 to 10.0), and generally are high-molecular proteins with molecular weights ranging from 780 kDa to 920 kDa (92). These enzymes require relatively high temperatures for activity; however, they show differences in their thermostabilities. For example, incubation of white croaker alkaline protease for 10 min at 60°C destroyed about 95% of original enzyme activity, whereas carp alkaline protease lost less than 5% of its original activity when subjected to the same time and temperature treatment (92). Alkaline proteases have been implicated in postmortem textural deteriorations in fish and fish products.

The enzyme urease catalyzes the breakdown of urea to ammonia and carbon dioxide during excretion in most fish species. The flesh of elasmobranchs contains much more nonprotein nitrogen of ureic origin than teleost fish, and this imparts a peculiar odor and sour-bitter taste to these fish species. During ice storage of meats from elasmobranchs like shark, a large amount of ammonia is generated due to the high content of urea. Some of the predominant microflora associated with skins of freshly caught sharks (i.e., *Pseudomonas, Micrococcaceae*, and *Moraxella*) produce urease that promotes ammonia formation from urea during storage, which is detrimental to the quality and acceptability of shark meats. This undesirable development in the product can be minimized by improved sanitary conditions during handling; soaking meat in water, lactic acid, salt, or urease solution prior to storage; blanching; or heat sterilization.

Lipolytic enzymes that have been characterized from fish include lipases from digestive organs or juices from both vertebrate and invertebrate fish, and phospholipases from microsomal fraction from muscle tissues of various fish such as skate (93), cod (94,95), trout (96,97), crayfish (98), pollock (99), and lobster (100). These lipolytic enzymes are single polypeptide chains with a broad pH-activity range (from 6 to 10) and are generally very heat labile. They are inhibited by thiol reagents, surface active agents, heavy metals, and oxygen. In general, the activities of lipolytic enzymes from marine species are lower than those from mammals. However, this is probably due to lower levels of enzymes present rather than the lower specific activities of the enzymic proteins.

Acid phosphatases are a group of heterogenous sialic acid–containing glycoenzymes that catalyze removal of phosphate groups from their substrates, and this reaction is important in the formation of triglycerides from phosphatidates. An acid phosphatase with molecular weight of 123 kDa has been isolated from liver of carp (101). The enzyme is stable over a broad pH range, between 3.5 and 5.5, and is most active between pH 4.5 and 5.0. It is strongly inhibited by L-(+) tartrate, molybdate, fluoride, urea, and mercuric ions.

In addition to the enzymes described previously, all the enzymes of the TCA cycle, with the exception of succinyl CoA synthetase, have been detected in the liver of rainbow trout (102). Citrate synthetase, α-ketoglutarate dehydrogenase, and succinate dehydrogenase were detected in the mitochondrial fraction, whereas the rest of the enzymes were detected in the cytosolic fraction. The intracellular distribution of TCA enzymes in trout was similar to that of mammalian species.

Comparative Biochemistry

Marine enzymes are similar to their mammalian counterparts in several of their characteristics. Generally, their pH-activity profiles, substrate specificities, molecular weights, and response to inhibitors or activators resemble those of their mammalian counterparts (44,56,70,84). However, they may differ from mammalian enzymes in a number of ways. For example, digestive proteases from marine organisms have been shown to be generally less stable under acid conditions, have higher catalytic activity, digest undenatured protein substrates more efficiently, and are heat labile but more stable at cold temperatures (44,69,70,84,103). Some of these distinctive properties of marine enzymes could be exploited successfully in various commercial operations.

APPLICATIONS, CURRENT USE, AND IMPACT OF BIOTECHNOLOGY

Biotechnology as it pertains to food production has been practiced for several years and has been or area of prolific growth in science and engineering within the past two decades. In the food and allied industries, major importance is now being attached to the use of enzymes for enhancing product quality, by-product utilization, increasing yields of extractive processes, product stabilization, and flavor. For example, fresh seafoods exhibit delicate aromas and flavors that are quite distinct from those usually evident in "commercially fresh" seafood. A group of enzymatically derived aldehydes, ketones, and alcohols has been found to be responsible for the characteristic aroma of very fresh seafood (104), and these compounds are similar to the C-6, C-8, and C-9 compounds produced by lipoxygenases. Chen and Li (105) isolated proteolytic enzymes from the cephalothorax of grass shrimp and demonstrated that enzymes like carboxypeptidase A and B, trypsin, chymotrypsin, cathepsin, and collagenase are important in seafood flavor production. They further demonstrated that these proteolytic enzymes from grass shrimp heads could tenderize meats just as well as papain or bromelain.

The use of enzymes in food processing and their potential for future applications have been extensively covered in publications by Godfrey and Reichelt (106), Schwimmer (107), and Reed (108). Studies carried out by various researchers have also demonstrated potential applications of certain marine enzymes as processing aids, based on some of their unique properties described earlier in this article. For example, pepsins from cold temperature–adapted fish have been used as a rennet substitute in preparation of cheddar cheese (55), while fish and shellfish trypsins and cathepsins have been applied to facilitate fermentation of capelin, squid, and herring (109–111). Cathepsins have also been shown to play an important role in patis fermentation (112). Enzyme preparations from digestive organs of herring, mackerel, sardine, and sprats have been used to refine the taste of mackerel preserves and to accelerate ripening of herring fillets (57). Similarly, trypsins from cod and cunner fish have been successfully applied to recover carotenoprotein from crustacean waste for incorporation into fish feeds and as flavorants, colorants, or both in surimi-based seafood analogs (72,113). Digestive proteolytic enzymes from stomachless marine species like cunner, crayfish, and mullet appear to be better suited to digestion and/or inactivation of native proteins than their counterparts derived from species with a functional stomach like mammals (73,114–116). These enzymes are currently being investigated with regard to their ability to inactivate PPOs, pectinesterases, or both in fruit juices. For example, apple juice develops a brown color very rapidly due to the presence of PPOs. To overcome the browning process, the substrates responsible for this phenomenon may be chemically altered to render them resistant to oxidation by PPOs. Another approach is to use a hot clarification process to inactivate the enzymes responsible for the discolorations, which adds to the energy cost as well as destroys some desirable heat-labile volatile components in fruit juices. An alternative approach that could offer an even greater appeal to consumers eager for "natural" food products may be to use digestive proteases from stomachless marine species described previously to inactivate both PPOs and pectinesterases in fruit juices. The successful application of such enzymes to inactive PPO could have a major impact on the shrimp-processing industry, which is facing the problem of finding effective alternative(s) to sulfiting agents used in some countries like the United States to control dark discolorations in these economically important marine species. Also, a potential interesting use of marine enzymes is the immobilization of lipolytic enzymes for catalyzing interesterification of fish oils to produce glycerides enriched with ω-3 polyunsaturated fatty acids (PUFAs). The consumption of ω-3 PUFAs from fish have been associated with a number of beneficial health effects (117). The use of marine enzymes for the interesterification reactions is suggested because the intended substrates are more natural to these enzymes than microbial or plant lipases. There are few instances where marine enzymes are applied on a commercial scale. For example, the Icelandic Fisheries Laboratory (IFL) has developed a process for utilizing trypsin-like enzymes from cod viscera for removal of skins, membranes, and scales from fish. The fish skins, membranes, and scales thus produced are used to prepare fish gelatin, which differs from mammalian gelatin in that it has a relatively low gelling temperature due to its lower proline and hydroxyproline content (118). At the present time, fish frames are hydrolyzed in Norway with cold-adapted pepsins from cod viscera to produce marine peptones for subsequent use as media for commercial cultivation of microorganisms for production of microbial proteins or fertilizer (119). Marine enzymes are also used in fish feed and for the production of immune stimulants for fish reared in captivity (118). For more information on the current and potential uses of marine enzymes in industry, there are excellent reviews on the subject by Haard (120) and Raa (118).

The major obstacles to using marine by-products as sources of industrial enzymes include their limited availability due to seasonal harvest; variability in levels and/or activity of enzymes due to nutritional status, seasonal variation, and spawning: and the highly perishable nature of the raw material. However, attitudes toward the use of marine enzymes in industry are slowly changing, and the potential profits to be derived from their commercial production may be enhanced by the use of recombinant DNA techniques for cloning enzyme protein genes from the marine animals into suitable bacteria or yeasts to facilitate production of the recombinant enzymes by fermentation.

Several intracellular enzymes from fish and shellfish have been studied, as shown earlier in the text. In general, large-scale production of intracellular enzymes from marine animals is limited to only a few enzymes, and sales from these enzymes represent only a small fraction of the total volume and value of commercial enzyme production and sales. However, with the rapid advances in enzyme technology in recent years, production of intracellular enzymes is very feasible for a number of reasons. For example, intracellular enzymes are widely used in clinical research, and their production is crucial to the furtherance of much biochemical research—for example, uricase is used as a diagnostic enzyme; however, the pH sensitivity of uricase presents a problem in clinical use for the quantification of serum uric acid, which is conducted around pH 7.1. The observation that trout uricase retains stability and activity at pH 7.0 may indicate its usefulness for this clinical application. Different kinds of fish enzymes have been cloned and expressed by microorganisms. Examples of such fish enzymes include tissue-type TGase from sea bream, which was successfully cloned and expressed in *E. coli* (121). In fish, TGases have been studied in relation to the elucidation of the molecular mechanisms for gel formation in fish mince products. It has been suggested that endogenous tissue-type TGases catalyze the setting reaction that involves the cross-linking of myosin heavy chains in fish paste during processing. This setting reaction is thought to be important with respect to the viscoelastic properties of fish paste (122). Another interesting fish enzyme that has been cloned is the cytochrome P450c17, which is a key enzyme in determining the synthesis of sex steroids (androgens and estrogens) and progestins. The enzyme is responsible for the sequential 17 α-hydroxylase and $C_{17,20}$-lyase reactions in the synthesis of the aforementioned compounds. The expression of an active form of shark P450 was achieved by Trant (123) using a *Pichia pastoris* GS115 yeast expression system and the expression vector pHiL-D2. Male et al. (124) described the nucleotide sequences and the structural models of five trypsins variants from Atlantic salmon, whereas Klein et al. (125) studied the trypsin gene expression in shrimp (*Penaeus vannamei*) and detected the cDNAs encoding five isoforms of trypsin. Gudmunsdottir et al. (126) also isolated cDNAs encoding two different forms of trypsinogen from Atlantic cod. LeBoulay et al. (127) reported the cloning and expression of cathepsin L from shrimp hepatopancreas, whereas Van Wormhoudt and Sellos (128) cloned and sequenced three cDNAs for amylase in *P. vannamei*, and Su et al. (131) constructed a cDNA library of the phospholipase from *Artemia* and expressed it in *E. coli* BL21(DE3). As more basic information on the distinctive properties of marine enzymes becomes available through research, it is expected that those marine enzymes with potential for industrial use would be produced in commercial quantities by gene cloning technology.

BIBLIOGRAPHY

1. Novo Nordisk, 1998.
2. O. J. Ferrer et al., "Phenoloxidase Levels in Florida Spiny Lobster (*Panulirus argus*): Relationship to Season and Molting Stage," *Comp. Biochem. Physiol.* **93B**, 595–599 (1989).
3. K. A. Savagaon and A. Sreenivasan, "Purification and Properties of Latent and Isoenzymes of Phenoloxidase in Lobster (*Panulirus homarus* Linn.)," *Ind. J. Biochem. Biophys.* **12**, 94–99 (1978).
4. B. K. Simpson, M. R. Marshall, and W. S. Otwell, "Phenoloxidases From Pink and White Shrimp: Kinetic and Other Properties," *J. Food Biochem.* **12**, 205–217 (1988).
5. M. E. Bailey, E. A. Fieger, and A. F. Novak, "Physicochemical Properties of the Enzymes Involved in Shrimp Melanogenesis," *Food Res.* **25**, 557–563 (1959).
6. N. M. Summers, "Cuticle Sclerotization and Blood Phenol Oxidase in the Fiddler Crab, *Uca pugnax*," *Comp. Biochem. Physiol.* **23**, 129–138 (1967).
7. J. H. Waite, "Catechol Oxidase in the Byssus of the Common Mussel, *Mytilus edulis* L," *J. Mar. Biol. Ass. U.K.* **65**, 359–371 (1985).
8. K-H. Kim and T. T. Tchen, "Tyrosinase of the Goldfish *Carassius auratus* L. I. Radio-assay and Properties of the Enzyme," *Biochim. Biophys. Acta.* **59**, 569–576 (1962).
9. G. Prota et al., "Occurrence and Properties of Tyrosinase in the Ejected Ink of Cephalopods," *Comp. Biochem. Physiol.* **68B**, 415–419 (1981).
10. B. M. Slabyj and H. O. Hultin, "Lipid Peroxidation by Microsomal Fractions Isolated From Light and Dark Muscles of Herring (*Clupea harengus*)," *J. Food Sci.* **47**, 1395–1398 (1982).
11. R. J. Hsieh and J. E. Kinsella, "Lipoxygenase-catalyzed Oxidation of N-6 and N-3 Polyunsaturated Fatty Acids. Relevance to Activity in Fish Tissue," *J. Food Sci.* **51**, 940–945, 996 (1986).
12. T.-J. Han and J. Liston, "Lipid Peroxidation and Phospholipid Hydrolysis in Fish Muscle Microsomes and Frozen Fish," *J. Food Sci.* **52**, 294–296, 299 (1987).
13. R. E. McDonald and H. O. Hultin, "Some Characteristics of the Enzymic Lipid Peroxidation System in the Microsomal Fraction of Flounder Skeletal Muscle," *J. Food Sci.* **52**, 15–21, 27 (1987).
14. N. Tsukuda, "Discoloration of Red Fishes," *Bull. Tokai Reg. Rish. Research Lab.* **70**, 103–174 (1972).
15. N. Tsukuda, "Studies on the Discoloration of Red Fishes. VI. Partial Purification and Specificity of Lipoxygenase-like Enzymes Responsible for Carotenoid Discoloration in Fish Skin During Refrigerated Storage," *Bull. Jap. Soc. Sci. Fish.* **36**, 725–733 (1970).
16. R. H. Ilgner and A. E. Woods, "Purification, Physical Properties and Kinetics of Peroxidases From Freshwater Crayfish (Genus *orconectes*)," *Comp. Biochem. Physiol.* **82B**, 433–440 (1985).
17. C. B. Cowey, "Comparative Studies on the Activity of D-Glyceraldehyde-3-Phosphate Dehydrogenase From Cold and Warm-blooded Animals With Reference to Temperature," *Comp. Biochem. Physiol.* **23**, 969–976 (1967).
18. P. S. Low, J. L. Bada, and G. N. Somero, "Temperature Adaptation of Enzymes: Roles of the Free Energy, the Enthalpy, and the Entropy of Activation," *Proc. Nat. Acad. Sci.* **70**, 430–432 (1973).
19. M. T. Thebault, A. Bernicard, and J. F. Lennon, "Lactate Dehydrogenase From the Caudal Muscle of the Shrimp *Palaemon serratus*: Purification and Characterization," *Comp. Biochem. Physiol.* **68B**, 65–70 (1981).
20. T. Henry, and A. Ferguson, "Kinetic Studies on the Lactate Dehydrogenase (LDH-5) Isozymes of Brown Trout, *Salmo trutta L*," *Comp. Biochem. Physiol.* **82B**, 95–98 (1985).

21. A. R. Ruano et al., "Some Enzymatic Properties of NAD+-dependent Glutamate Dehydrogenase of Mussel Hepatopancreas (*Mytilus edulis* L.)—Requirement of ADP," *Comp. Biochem. Physiol.* **82B**, 197–202 (1985).

22. M. A. Wicker and R. P. Morgan II, "Effects of Salinity on Three Enzymes Involved in Amino Acid Metabolism From the American Oyster *Crassostrea virginia*," *Comp. Biochem. Physiol.* **53B**, 339–343 (1976).

23. P. M. Reiss, S. K. Pierce, and S. H. Bishop, "Glutamate Dehydrogenases From Tissues of the Ribbed Mussel, *Modiolus demissus*: ADP Activation and Possible Physiological Significance," *J. Exp. Zool.* **202**, 253–258 (1977).

24. J. E. Kinsella, B. German, and J. Shetty, "Uricase From Fish Liver: Isolation and Some Properties," *Comp. Biochem. Physiol.* **82B**, 621–624 (1985).

25. H. Kishi, H. Nozawa, and N. Seki, "Reactivity of Muscle Transglutaminase on Carp Myofibrils and Myosin B," *Nippon Suisan Gakkaishi* **57**, 1203–1210 (1991).

26. H. Nozawa, S. Mamagoshi, and N. Seki, "Partial Purification and Characterization of Six Transglutaminases From Ordinary Muscles of Various Fishes and Marine Invertebrates," *Comp. Biochem. Physiol.* **118B**, 313–317 (1997).

27. R. Myhrman and J. Bruner-Lorand, "Lobster Muscle Transpeptidase," *Meth. Enzymol.* **19**, 765–770 (1970).

28. D. A. Holwerda et al., "Regulation of Mussel Pyruvate Kinase During Anaerobiosis and in Temperature Acclimation by Covalent Modification," *Molec. Physiol.* **3**, 225–234 (1983).

29. T. Mustafa and P. W. Hochachka, "Catalytic and Regulatory Properties of Pyruvate Kinase in Tissues of a Marine Bivalve," *J. Biol. Chem.* **246**, 3196–3203 (1983).

30. J. Chrispeels and G. Gade, "Purification and Characterization of Pyruvate Kinase From the Foot Muscle of the Cockle, *Cardium tuberculatum*," *Comp. Biochem. Physiol.* **82B**, 163–172 (1985).

31. S. P. J. Brooks and K. B. Storey, "Revaluation of the Glycolytic Complex: A Multitechnique Approach Using Trout Muscle," *Arch. Biochem. Biophys.* **267**, 13–22 (1988).

32. I. A. Nimmo and C. M. Spalding, "The Glutathione S-transferase Activity in the Kidney of Rainbow Trout (*Salmo gairdneri*)," *Comp. Biochem. Physiol.* **82B**, 91–94 (1985).

33. M. Mathieu, "Partial Characterization of Aspartate Transcarbamylase From the Mantle of the Mussel *Mytilus edulis*," *Comp. Biochem. Physiol.* **82B**, 667–674 (1985).

34. K. Hara, A. Suzumatsu, and T. Ishihara, "Purification and Characterization of Cathepsin B From Carp Ordinary Muscle," *Nihon Suisan Gakkai-shi* **54**, 1243–1252 (1988).

35. H. S. Groninger, Jr., "Partial Purification of a Proteinase From Albacore Muscle," *Arch. Biochem.* **108**, 175–182 (1964).

36. A. Schmitt and G. Siebert, "Distinguishing Aliphatic Dipeptidases From Cod Muscle," *Z. Physiol. Chem.* **348**, 1009–1016 (1967).

37. C.-Y. Ting, M. Montgomery, and A. F. Anglemier, "Partial Purification of Salmon Muscle Cathepsins," *J. Food Sci.* **33**, 617–621 (1968).

38. G. M. Geist and D. L. Crawford, "Muscle Cathepsins in Three Species of Pacific Sole," *J. Food Sci.* **39**, 548–551 (1974).

39. K. S. Hameed and N. F. Haard, "Isolation and Characterization of Cathepsin C From Atlantic Short-finned Squid, *Illex illecebrosus*," *Comp. Biochem. Physiol.* **82B**, 241–246 (1985).

40. T. Inaba, N. Shindo, and M. Fuji, "Purification of Cathepsin B From Squid Liver," *Agric. Biol. Chem.* **40**, 1159–1165 (1976).

41. S. V. Sherekar, M. S. Gore, and V. Ninjoor, "Purification and Characterization of Cathepsin B From the Skeletal Muscle of Fresh Water Fish, *Tilapia mossambica*," *J. Food Sci.* **53**, 1018–1023 (1988).

42. S. B. K. Warrier, V. Ninjoor, and G. B. Nadkarni, "Involvement of Hydrolytic Enzymes in the Spoilage of Bombay Duck (*Harpodon nehereus*). Harvest and Postharvest Technology of Fish," *Soc. Fish. Technol. (India)*, pp. 470–472 (1985).

43. G. Siebert and A. Schmitt, "Fish Tissue Enzymes and Their Role in Determination Changes in Fish," in R. Kreuzer, ed., *The Technology of Fish Utilization*, Fishery News Books, Ltd., London, United Kingdom, pp. 47–52, 1965.

44. M. Tanji, T. Kageyama, and K. Takahashi "Tuna Pepsinogens and Pepsins. Purification, Characterization and Amino-terminal Sequences," *Eur. J. Biochem.* **177**, 251–259 (1988).

45. E. R. Norris and J. C. Mathies, "Preparation, Properties and Crystallization of Tuna Pepsin," *J. Biol. Chem.* **204**, 673–680 (1953).

46. J. S. Fruton and M. Bergmann, "The Specificity of Salmon Pepsin," *J. Biol. Chem.* **136**, 559–560 (1940).

47. T. G. Merrett, E. Bar-Eli, and H. Van Vunakis, "Pepsinogens A, C, and D From the Smooth Dogfish," *Biochemistry* **8**, 3696–3702 (1969).

48. M. Kubota, A. Ohnuma, and I. Karube, "Kinetics of Protease Inactivation by Gamma Irradiation," *J. Tokyo Univ. Fish.* **55**, 9–20 (1970).

49. T. G. Owen and A. J. Wiggs, "Thermal Compensation in the Stomach of the Brook Trout (*Salvelinus fantinalis* Mitchill)," *Comp. Biochem. Physiol.* **40B**, 465–473 (1971).

50. S. S. Twining et al., *Comp. Biochem. Physiol.* **75B**, 109–112 (1983).

51. M. Nora and K. Murakami, "Purification and Characterization of Two Acid Proteinases From the Stomach. Studies on Proteinases From the Digestive Organs of Sardine," *Biochim. Biophys. Acta.* **658**, 27–34 (1981).

52. L. Sanchez-Chiang, and O. Ponce, "Gastricsinogens and Gastricsins From *Merluccius gayi*—Purification and Properties. *Comp. Biochem. Physiol.* **68B**, 251–257 (1981).

53. A. Gildberg and J. Raa, "Purification and Characterization of Pepsins From the Arctic Fish Capelin (*Mallotus villosus*)," *Comp. Biochem. Physiol.* **75A**, 337–342 (1983).

54. K. Arunchalam and N. F. Haard, "Isolation and Characterization of Pepsin Isoenzymes From Polar Cod (*Boreogadus saida*)," *Comp. Biochem. Physiol.* **80B**, 467–473 (1985).

55. P. Brewer, N. Helbig, and N. F. Haard, "Atlantic Cod Pepsin—Characterization and Use as a Rennet Substitute," *Can. Inst. Food Sci. Technol. J.* **17**, 38–43 (1984).

56. E. J. Squires, N. F. Haard, and L. A. W. Feltham, "Gastric Proteases of the Greenland Cod *Gadus ogac*. II. Structural Properties," *Can. J. Biochem. Cell Biol.* **64**, 205–222 (1986).

57. N. F. Haard et al., "Modification of Proteins With Proteolytic Enzymes From the Marine Environment," in R. L. Feeney, J. R. Whitaker, eds., *Modification of Proteins*. Advances in Chemistry Series 198, American Chemical Society, Washington, D.C., pp. 223–224, 1982.

58. B. G. Kapoor, H. Smit, and I. A. Verighina, "The Alimentary Canal and Digestion in Teleost," *Adv. Mar. Biol.* **13**, 109–239 (1975).

59. Y-I. Hsu and J-L. Wu, "The Relationship Between Feeding Habits and Digestive Proteases of Some Fresh Water Fishes," *Bull. Inst. Zool. Acad. Sinica.* **18**, 45–53 (1979).

60. A. Nilsson and R. Fange, "Digestive Proteases in the Cyclostome *Myxina glutinosa* (L)," *Comp. Biochem. Physiol.* **32**, 237–250 (1970).
61. K. D. Jany, "Studies on the Digestive Enzymes of the Stomachless Bonefish *Carassius auratus gibelio* (bloch): Endopeptidases," *Comp. Biochem. Physiol.* **53B**, 31–38 (1976).
62. K. Murakami and M. Noda, "Studies on Proteinases From the Digestive Organs of Sardine—Purification and Characterization of Three Alkaline Proteinases From the Pyloric Ceca," *Biochem. Biophys. Acta.* **65B**, 17–26 (1981).
63. K. Hjelmeland and J. Raa, "Characteristics of Two Trypsin Type Isozymes Isolated From the Arctic Fish Capelin (*Mallotus villosus*)," *Comp. Biochem. Physiol.* **71B**, 557–562 (1982).
64. R. Yoshinaka et al., "Enzymatic Characterization of Anionic Trypsin of the Catfish *Parasilurus asotus*," *Comp. Biochem. Physiol.* **77B**, 1–6 (1984).
65. M. Kitamikado and S. Tachino, "Studies on the Digestive Enzymes of Rainbow Trout—II. Proteases," *Bull. Jap. Soc. Scient. Fish.* **26**, 685–694 (1960).
66. C. B. Croston, "Endopeptidases of Salmon Ceca: Chromatographic Separation and Some Properties," *Arch. Biochem. Biophys.* **112**, 218–223 (1965).
67. S. Sundaram and P. S. Sarma, "Purification and Properties of a Protease From the Gut of *Etroplus suratensis*," *Biochem. J.* **77**, 465–471 (1960).
68. B. K. Simpson, M. V. Simpson, and N. F. Haard, "On the Mechanism of Enzyme Action: Digestive Proteases From Selected Marine Organisms," *Biotechnol. Appl. Biochem.* **11**, 226–234 (1989).
69. B. K. Simpson and N. F. Haard, "Trypsin From Greenland Cod (*Gadus ogac*). Isolation and Comparative Properties," *Comp. Biochem. Physiol.* **79B**, 613–627 (1984).
70. B. K. Simpson and N. F. Haard, "Purification and Characterization of Trypsin From the Greenland Cod (*Gadus ogac*). I. Kinetic and Thermodynamic Characteristics. *Can. J. Biochem. Cell Biol.* **62**, 894–900 (1984).
71. J. Overnell "Digestive Enzymes of the Pyloric Ceca and the Associated Mesentery in the Cod (*Gadus morhua*)," *Comp. Biochem. Physiol.* **46B**, 519–531 (1973).
72. B. K. Simpson and N. F. Haard, "The Use of Proteolytic Enzymes to Extract Carotenoproteins From Shrimp Wastes," *J. Appl. Biochem.* **7**, 212–222 (1985).
73. G. Pfleiderer, R. Zwilling, and H. Sonneborn, "Eine Protease von Molekulargewicht 11,000 und eine Trypsinahniche Fraktion aus *Astacus fluviatilis*," *Z. Physiol. Chem.* **348**, 1319–1331 (1967).
74. B. T. Gates and J. Travis, "Isolation and Comparative Properties of Shrimp Trypsin," *Biochemistry* (N.Y.) **8**, 4483–4489 (1968).
75. C-S. Chen, T-R. Yan, and H-Y. Chen, "Purification and Properties of Trypsin-like Enzymes and a Carboxypeptidase A From *Euphasia superba*," *J. Food Biochem.* **21**, 349–366 (1978).
76. Z. Camacho, J. R. Brown, and G. B. Kitto, "Purification and Properties of Trypsin-like Proteases From the Starfish *Dermasterias imbricata*," *J. Biol. Chem.* **245**, 3964–3972 (1970).
77. H. F. Bundy and J. Gustafson, "Purification and Comparative Biochemistry of a Protease From the Starfish *Pisaster giganteus*," *Comp. Biochem. Physiol.* **44B**, 241–251 (1973).
78. E. P. Kozlovskaya and L. A. Elyakova, "Purification and Properties of Trypsin-like Enzymes from the Starfish *Lysastrosoma anthocticta*," *Biochim. Biophys. Acta.* **371**, 63–70 (1974).
79. M. S. Heu, H. R. Kim, and J. H. Pyeum, "Comparison of Trypsin and Chymotrypsin From the Viscera of Anchovy, *Eugraulis japonica*," *Comp. Biochem. Physiol.* **112B**, 557–568 (1995).
80. B. Asgiersson and J. B. Bjarnasson, "Structural and Kinetic Properties of Chymotrypsin From Atlantic Cod (*Gadus morhua*). Comparison With Bovine Chymotrypsin," *Comp. Biochem. Physiol.* **99B**, 327–335 (1991).
81. J. Kalac, "Studies on Herring (*Clupea harengus* L.) and Capelin (*Mallotus villosus* L.) Pyloric Ceca Proteases. III. Characterization of Anionic Fractions of Chymotrypsins," *Biologia (Bratislava)* **33**, 939–945 (1978).
82. M. M. Kristjansson and H. H. Nielson, "Purification and Characterization of Two Chymotrypsin-like Proteases From the Pyloric Ceca of Rainbow Trout (*Oncorhynchus mykiss*)," *Comp. Biochem. Physiol.* **101B**, 247–253 (1992).
83. W. F. Racicot, *A Kinetic and Thermodynamic Comparison of Bovine and Dogfish Chymotrypsins*, Ph.D. thesis, Univ. of Massachusetts, Amherst, 1984.
84. M. Ramakrishna, H. O. Hultin, and M. T. Atallah, "A Comparison of Dogfish and Bovine Chymotrypsins in Relation to Protein Hydrolysis," *J. Food Sci.* **52**, 1198–1202 (1987).
85. R. Yoshinaka, M. Sato, and A. Ikeda "Distribution of Collagenase in the Digestive Organs of Some Teleost," *Bull. Jap. Soc. Scient. Fish.* **44**, 263–268 (1978).
86. G. A. Grant, J. C. Sacchettini, and H. G. Welgus, "A Collagenolytic Serine Protease With Trypsin-like Specificity From the Fiddler Crab *Uca pugilator*," *Biochemistry* **22**, 354–358 (1983).
87. A. Z. Eisen and J. J. Jeffrey, "An Extractable Collagenase From Crustacean Hepatopancreas," *Biochim. Biophys. Acta* **191**, 517–520 (1969).
88. A. Z. Eisen et al., "A Collagenolytic Protease From the Hepatopancreas of the Fiddler Crab, *Uca pugilator*. Purification and Properties," *Biochemistry* **12**, 1814–1822 (1973).
89. H. Welgus et al., "Substrate Specificity of the Collagenolytic Serine Protease From *Uca pugilator*," *Biochemistry* **21**, 5183–5189 (1982).
90. W. K. Nip, C. Y. Len, and J. H. Moy, "Partial Characterization of a Collagenolytic Enzymes Fraction From the Hepatopancreas of the Freshwater Prawn, *Macrobranchium rosenbergii*, " *J. Food Sci.* **50**, 1187–1188 (1985).
91. R. Cepeda et al., *An Immunological Method for Measuring Collagen Degradation in the Muscle of Fish. Seafood Biotechnology Workshop. August 31–September 1, 1989*, St. John's, Newfoundland, Canada, 1990.
92. K. Iwata, K. Kabashi, and J. Hase, "Studies on Muscle Alkaline Protease—III. Distribution of Alkaline Protease in Muscle of Freshwater Fish, Marine Fish and in Internal Organs of Carp," *Bull. Jap. Soc. Sci. Fish.* **40**, 201–209 (1974).
93. H. Brockerhoff and J. R. Hoyle, "Hydrolysis of Triglycerides by the Pancreatic Lipase of a Skate," *Biochim. Biophys. Acta* **98**, 435–436 (1965).
94. P. Chawla and R. F. Ablett, "Detection of Microsomal Phospholipase Activity in Myotomal Tissue of Atlantic Cod (*Gadus morhua*)," *J. Food Sci.* **52**, 1194–1197 (1987).
95. H. Brockerhoff, "Digestion of Fat by Cod," *J. Fish. Res. Bd. Can.* **23**, 1835–1839 (1966).
96. J. E. Geromel, *Release of Lipase From Liposomes of Rainbow Trout* (Salmo gairdneri) *Muscle Subjected to Low Pressures*, Informativo Anual, Faculdade de Engenharia de Alimentos e Agricola, Universidad Estadual de Campinas, No. 8, 57–60, 1980.

97. C. Leger, "Essai de Purification de la Lipase du Tissu Intercaecal de la Truite (*Salmo gairdneri* Rich)," *Ann. Biol. Anim. Biochim. Biophys.* **12**, 341–345 (1970).

98. D. L. Berner and E. G. Hammond, "Phylogeny of Lipase Specificity," *Lipids* **5**, 558–562 (1970).

99. M. A. Audlrey, K. J. Shetty, and J. E. Kinsella, "Isolation and Properties of Phospholipase A From Pollock Muscle," *J. Food Sci.* **43**, 1771–1775 (1978).

100. H. Brockerhoff, "On the Function of Bile Salts and Proteins as Cofactors of Lipase," *J. Biol. Chem.* **246**, 5828–5831 (1971).

101. A. Kubicz, E. Dratewka-Kos, and R. Zygmuntowicz, "Multiple Molecular Forms of the Acid Phosphatase From *Cyprinus carpio* liver: Isolation and Comparison With Those of *Rana esculenta*," *Comp. Biochem. Physiol.* **68B**, 437–443 (1981).

102. M. J. Walton, "Intracellular Distribution of Tricarboxylic Acid Cycle Enzymes in Liver of Rainbow Trout *Salmo gairdneri*," *Comp. Biochem. Physiol.* **82B**, 87–90 (1985).

103. A. Ooshiro, "Studies on a Proteinase in the Pyloric Ceca of Fish. II. Some Properties of Proteinase Purified From the Pyloric Ceca of Mackerel," *Bull. Jap. Soc. Scient. Fish.* **34**, 847–854 (1968).

104. D. B. Josephson, R. C. Lindsay, and D. A. Sturber, "Identification of Compounds Characterizing the Aroma of Fresh Whitefish (*Coregonus clupeaformis*)," *J. Agric. Food Chem.* **31**, 326–330 (1983).

105. H. Y. Chen and C. F. Li, "Isolation, Partial Purification and Application of Proteinases From Grass Shrimp Heads," *Food Sci. (China)* **15**, 230–243 (1988).

106. T. Godfrey and J. Reichelt, *Industrial Enzymology*, Nature Press, New York, 1983.

107. S. Schwimmer, *Source Book of Enzymology*, AVI, Westport, Conn., 1981.

108. G. Reed, *Enzymes in Food Processing*, Academic Press, New York, 1975.

109. N. Raksakulthai, Y. Z. Lee, and N. F. Haard, "Effect of Enzyme Supplements on the Production of Fish Sauce From Male Capelin (*Mallotus villosus*)," *Can. Inst. Food Sci. Technol.* **19**, 28–33 (1986).

110. B. K. Simpson and N. F. Haard, "Trypsin From Greenland Cod as a Food-processing Aid," *J. App. Biochem.* **6**, 135–143 (1984).

111. Y. Z. Lee, B. K. Simpson, and N. F. Haard, "Supplementation of Squid Fermentation With Proteolytic Enzymes," *J. Food Biochem.* **6**, 127–134 (1982).

112. R. R. del Rosario and S. M. Maldo, "Biochemistry of Patis Formation. I. Activity of Cathepsins in Patis Hydrolysis," *Philippines Agriculturists* **67**, 167–175 (1984).

113. A. Cano-Lopez, B. K. Simpson, and N. F. Haard, "Extraction of Carotenoprotein From Shrimp Process Wastes With the Aid of Trypsin from Atlantic Cod," *J. Food Sci.* **52**, 503–504, 506 (1987).

114. M. A. Kyei, *Trypsin and Trypsin Like Enzyme From the Stomachless Cunner—Preparation and Applications*, M.Sc. thesis, McGill University, Montreal, Canada, 1996.

115. B. K. Simpson and N. F. Haard, "Trypsin and a Trypsin-like Enzyme From the Stomachless Cunner. Catalytic and Other Physical Characteristics," *J. Agric. Food Chem.* **35**, 652–656 (1987).

116. N. Guizani et al., "Isolation and Characterization of a Trypsin From the Pyloric Ceca of Mullet (*Mugil cephalus*)," *Comp. Biochem. Physiol.* **98**, 517–521 (1991).

117. J. E. Kinsella, "Dietary Fats and Cardiovascular Disease," in *Seafoods and Fish Oils in Human Health and Disease*, Marcel Dekker, New York, pp. 1–24, 1987.

118. J. Raa, "Biotechnology in the Fish Processing Industry—A Success Story in Norway," in *Advances in Fisheries Technology for Increased Profitability, Proceedings of the Atlantic Fisheries Technological Conference, August 27–30, 1989*, St. John's, Newfoundland, Canada, 1990.

119. K. Almas, "Utilization of Marine Biomass for the Production of Microbial Growth Media and Biochemicals," in *Advances in Fisheries Technology for Increased Profitability, Proceedings of the Atlantic Fisheries Technological Conference, August 27–30, 1989*. St. John's, Newfoundland, Canada, 1990.

120. N. F. Haard, "Specialty Enzymes From Marine Organisms," *Food Technol.* **52**, 64–67 (1998).

121. H. Yasueda et al., "Tissue-type Transglutaminase From Red Sea Bream, Sequence Analysis of the cDNA and Functional Expression in *Escherichia coli*," *Eur. J. Biochem.* **232**, 411–419 (1995).

122. G. G. Kamath et al., "Nondisulfide Covalent Cross-linking of Myosin Heavy Chain in Setting of Alaska Pollock and Atlantic Croaker Surimi," *J. Food Biochem.* **16**, 151–172 (1982).

123. J. M. Trant, "Functional Expression of Recombinant Spiny Dogfish Shark Cytochrome P450c17 in Yeast *Pichia pastoris*," *Arch. Biochem. Biophys.* **326**, 8–14 (1996).

124. R. Male et al., "Molecular Cloning and Characterization of Anionic and Cationic Variants of Trypsin From Atlantic Salmon." *Eur. J. Biochem.* **232**, 677–685 (1995).

125. B. Klein et al., "Molecular Cloning and Sequencing of Trypsin cDNAs From *P. vannamei*: Use in Assessing Gene Expression During the Moult Cycle," *Int. J. Biochem. Biol.* **28**, 551–563 (1996).

126. A. Gudmundsdottir et al., "Isolation and Characterization of cDNAs From Atlantic Cod Encoding Two Different Forms of Trypsinogen," *Eur. J. Biochem.* **217**, 1091–1097 (1993).

127. C. LeBoulay, A. Van Wormhoudt, and D. Sellos, "Cloning and Expression of Cathepsin L-like Proteinases in the Hepatopancreas of Shrimp *Penaeus vannamei* During Intermolt Cycle," *J. Comp. Physiol.* **166B**, 310–318 (1996).

128. A. Van Wormhoudt and D. Sellos, "Cloning and Sequencing Analysis of Three Amylase cDNAs in the Shrimp *Penaeus vannamei*: Evolutionary Aspects," *J. Mol. Evol.* **42**, 43–551 (1996).

129. X. Su, F. Chen, and L. E. Hokin, "Cloning and Expression of a Novel, Highly Truncated Phosphoinositide-specific Phospholipase C cDNA From Embryos of the Brine Shrimp, *Artemia*, " *J. Biol. Chem.* **269**, 12925–12931 (1994).

GENERAL REFERENCES

S. A. Assaf and D. J. Graves, "Structural and Catalytic Properties of Lobster Muscle Glycogen Phosphorylase," *J. Biol. Chem.* **244**, 5544–5555 (1969).

N. F. Haard, "A Review of Proteolytic Enzymes From Marine Organisms and Their Application in the Food Industry," *J. Aq. Food Product. Tech.* **1**, 17–35 (1992).

C. de Haen, K. A. Walsh, and H. Neurath, "Isolation and Amino Acid Terminal Sequence Analysis of a New Pancreatic Trypsinogen of the African Lungfish *Protopterus aethiopicus*," *Biochemistry* (N.Y.) **16**, 4421–4425 (1977).

K. A. Hansen, "Processing of Cod in Scandinavia 1990," in *Advances in Fisheries Technology for Increased Profitability, Proceedings of the Atlantic Fisheries Technological Conference, August 27–30, 1989*, St. John's, Newfoundland, Canada, 1989.

E. R. Norris and D. W. Elam, "Preparation and Properties of Crystalline Salmon Pepsin," *J. Biol. Chem.* **134**, 443–454 (1940).

R. L. Shewfelt, R. E. McDonald, and H. O. Hultin, "Effect of Phospholipid Hydrolysis on Lipid Oxidation in Flounder Muscle Microsomes," *J. Food Sci.* **46**, 1297–1301 (1981).

BENJAMIN K. SIMPSON
AMARAL SEQUEIRA-MUNOZ
McGill University
Ste. Anne de Bellevue, Quebec
Canada

NORMAN F. HAARD
University of California
Davis, California

MARINE TOXINS

A very wide variety of biotoxins occur with varying frequencies in fish, shellfish, mammals, reptiles, algae, and other animals and plants existing in the marine environment (1). Toxic organisms have been collected from most regions of the world, and many of them have caused illness in humans following ingestion. Ciguatera and scombroid poisoning together account for more than 80% of all seafood outbreaks and more than 50% of all outbreaks caused by the consumption of muscle protein with confirmed etiologies reported to the US Centers for Disease Control (2). This article emphasizes the similarities among the most frequently encountered or potentially encountered biotoxins in United States commercial fisheries, including imports. Significant differences between toxins or illnesses are also presented. Information is presented to assist in identifying major disorders and their circumstances. There are seven common and well characterized disorders—paralytic, neurotoxic, amnesic, and diarrhetic shellfish poisonings, ciguatera, scombroid poisoning, and pufferfish poisoning (poisonings or disorders refer to illnesses whereas toxins refer to specific chemical complexes) and these are summarized in Table 1. Three additional disorders that may increase in occurrence as international commerce of fishery products increases are briefly described at the end of the article: clupeotoxicity, hallucinogenic mullet poisoning, and shark liver poisoning.

BIOTOXINS

Intoxications are caused by a variety of biotoxins or families of biotoxins. Those common toxins that are chemically characterized are often grouped according to their chemical solubility. Brevetoxins (neurotoxic shellfish poisoning), okadaic acids (diarrhetic shellfish poisoning), and ciguatoxin (ciguatera) are lipid soluble, whereas, saxitoxins (paralytic shellfish poisoning), domoic acid (amnesic shellfish poisoning), maitotoxin (ciguatera), scombrotoxin (scombroid poisoning), and tetrodotoxin (pufferfish poisoning) are water soluble. All are low molecular weight (111–3425 D) containing no known sulfur or phosphorus groups. All but scombrotoxin and tetrodotoxin are produced by marine algae. Both scombrotoxin and tetrodotoxin are pro-

Table 1. Marine Biotoxins

Disorder	Toxin(s)	Agent(s)	Chemical structure	Sources	Assays	Controls	References
Paralytic shellfish poison	Saxitoxins (12)	*Protogonyaulax calenella, P. tamarensis*	$C_{10}H_{17}N_7O_3$, water soluble, 299 MW	Oysters, clams, scallops	Mouse, bioassay HPLC	Harvest restrictions	1,3,4,13,16, 17,18,19
Neurotoxic shellfish poison	Brevstoxins (5 or more)	*Ptychodiscus brevis*	$C_{50}H_{70}O_{14}$, ether soluble, 894 MW	Oysters, clams, coquina	Mouse bioassay, HPLC	Harvest restrictions	1,3,5,6,16,17, 18,19,20
Amnesic shellfish poison	Domoic acid	*Nitzchia pungens*	$C_{15}H_{21}O_6N$, water soluble, 311 MW	Mussels	Mouse, bioassay, HPLC	Harvest restrictions	1,3,7,16,17, 18,19
Diarrhetic shellfish poison	Okadaic acid and derivatives	*Dinophysis fortii, D. acuminata*	$C_{44}H_{70}O_{13}$, ether soluble, 804 MW	Mussels, scallops	Mouse and rat bioassays, HPLC, ELISA	Harvest restrictions, remove digestive gland	1,3,8,9,16,17, 18,19
Cigustera	Cigustoxin	*Gambierdiscus toxicus*	$C_{83}H_{77}O_{24}N$ or $C_{54}H_{78}O_{24}$, ether soluble, 1111 MW	Barracuda, grouper, snapper, amberjeck	Mouse, cat, and guinea pig bioassays, EIA, RIA	Avoid endemic areas, remove organs	1,3,10,16,17, 18,19
	Maltoxin	*G. toxicus*	?, water soluble, 3425 MW	Surgeonfish	Mouse bioassay	Avoid endemic areas, remove organs	1,3,10,16,17, 18,19
Scombroid poison	Histamine	Bacteria species	$C_8H_8N_3$, water soluble, 111 MW	Tuna, skipjack, mackerel, mahi-mahi	Ouines pig bioassay fluorometry, HPLC	Refrigeration lot rejection	1,3,11,14,16, 17,18,19
Pufferfish poison	Tetrodotoxin	*Vibrio alginolyticus*	$C_{11}H_{17}O_8N_3$, water soluble, 319 MW	Pufferfish	Mouse bioassay, HPLC	Fugu dealers and chefs	1,3,12,15,16, 17,18,19

duced by bacteria. These groups of toxins cause very similar symptoms (neurologic and/or gastrointestinal) characterized by rapid onset. Four of these groups of toxins are found in molluscan shellfish consumed in the United States: saxitoxins, brevetoxins, domoic acid, and okadaic acids; and four in finfish: ciguatoxin, maitotoxin, scombrotoxin, and tetrodotoxin (3).

OCCURRENCE

Shellfish Poisoning

Shellfish toxins fall into three specific groups; paralytic, amnesic, and diarrhetic (20). None of these toxins are produced by shellfish themselves but are obtained from microalgae such as phytoplankton naturally present in the oceans.

Paralytic. Of the molluscan shellfish-associated disorders, three are found in temperate areas and one in semitropical. Paralytic shellfish poisoning (saxitoxins) is found in shellfish such as mussels, clams, oysters, and scallops that have fed on the toxic dinoflagellates (21), including several species of the genus *Alexandrium*, previously *Protogonylauax catenella* and *P. tamarensis* (4). The toxins produced by these dinoflagellates tend to persist in shellfish for varying periods of time, depending on shellfish species and tissues involved. Paralytic shellfish poisoning is reported from all temperate oceans of the world; *P. catenella* on the west coast of North America and *P. tamarensis* on the east coast. In most cases the water temperature must be 5–8°C. The three families of toxins involved (saxitoxins, neosaxitoxins, and gonyautoxins) are all water soluble and heat stable. The first European case was documented in Norway in 1962 (22). Between 1977 and 1981, the United States had 126 cases reported from nine outbreaks.

Amnesic. Amnesic shellfish poisoning (domoic acid) has been reported in blue mussels from Canada and the United States (Maine). The source of the toxin is the marine diatom *Pseudonitzschia* (formerly *Nitzschia*) (7,23). In 1987, more than 100 illnesses and three deaths were reported from the Prince Edwards Island area of Canada.

Diarrhetic. Diarrhetic shellfish poisoning (okadaic acid and derivatives) is found in shellfish, especially mussels, that have fed on toxic marine dinoflagellates and then concentrated the lipid soluble toxins in their hepatopancreas. It appears that the edible muscle tissues remain free of toxin. The causative dinoflagellates (*Dinophysis fortii* and *Dinophysis acuminata*) are widespread in occurrence (8). Europe (Netherlands, Spain, France, Sweden), Southeast Asia (Japan, Thailand), and South America (Chile) have all reported outbreaks of diarrhetic shellfish poisoning. As many as 1,300 cases were reported in Japan from 1976 to 1982; and during 1981, 5,000 cases were reported in Spain (9,24).

Neurotoxic. Neurotoxic shellfish poisoning (brevetoxins) is found throughout the Gulf of Mexico, the Atlantic coast of the southern United States, and the Carribbean Sea. The causative organism is the dinoflagellate *Ptychodiscus brevis* (5). It is most frequently found on the western coast of Florida but has been confirmed from Texas to North Carolina. Eleven people in three episodes were intoxicated in Florida in 1973–1974 (6). Cases continue to be reported. Oysters, clams, and coquina are primary transvectors for the toxins to humans.

Ciguatera Poisoning

About 50,000 cases of ciguatera poisoning are reported annually worldwide; more than 2,000 cases in the United States (mostly Hawaii, Puerto Rico, and Florida). The toxins occur throughout the tropical regions of the world (between 35°N and 34°S) in shallow reef areas, mostly around islands (3). The apparent causative organism is the dinoflagellate *Gambierdiscus toxicus* (10). Fish that cause human illnesses include mostly barracuda, grouper, snapper, and amberjack. Other affected fishes include mackerel and surgeonfish.

Scombroid Poisoning

Many species of time–temperature abused fresh, canned, salted, or dried fish have been associated with scombroid poisoning (11). Scombroid fish include tuna, skipjack, mackerel, bonito, albacore, bluefish, saury, butterfly kingfish, and seerfish. Nonscombroid fish include mahi-mahi, sardines, pilchards, anchovies, herring, black marlin, and kahawai. These fish normally contain high levels of free histidine that is converted to histamine by a variety of bacteria species. Optimum decarboxylation occurs at 20–25°C, pH 2.5–6.5. Chemical potentiators that are also present in flesh, ie, putrescine and cadaverine, enhance the oral toxicity. There were 78 outbreaks of scombroid poisoning reported to the US Centers for Disease Control between 1983 and 1987 (2).

Puffer Poisoning

Tetrodotoxin is found in the flesh of pufferfish (Tetraodontidae). Not all species of puffer are equally toxic. The most notoriously toxic species, *Fugu rubripes* (tiger puffer), is also said to be the tastiest. Toxin concentration, principally in the ovaries and to a lesser extent in the liver and intestines, is highest during the winter months before spawning. Representatives of the tetraodontids are found throughout a broad circumglobal belt extending from latitudes of 47°N to 47°S. Recently, the primary source of the toxin has been determined to be several marine bacteria including *Vibrio alginolyticus* (12). In Japan, the only major country where pufferfish (fugu) are a delicacy, tetrodotoxin caused an average of 84 deaths a year between 1886 and 1963; having a fatality rate of 59% of all reported intoxications (1).

HEALTH EFFECTS

All but one of these disorders (diarrhetic shellfish poisoning) are characterized by similar neurologic symptoms (particularly tingling of the extremities). Some also elicit gastrointestinal (nausea, diarrhea, abdominable pain) and/or cardiovascular (pulse rate and blood pressure changes) symptomology. All biotoxin disorders are characterized by rapid onset—within a few hours. The disorders associated with mulluscan shellfish—paralytic shellfish poisoning, neurotoxic shellfish poisoning, amnesic shellfish poisoning, and diarrhetic shellfish poisoning—all elicit nausea, diarrhea and abdominal pain. All but diarrhetic shellfish poisoning cause tingling in the extremities of consumers.

For paralytic shellfish poisoning, the toxins contain a tetrahydropurine skeleton with around 20 forms of basic, water-soluble saxitoxins identified (25). The saxitoxins act by binding to the sodium channels in nerve cell membranes and blocking nerve transmission. Symptoms occur within 1 hour, initially with a tingling sensation of the lips, gums, tongue, and face. Death occurs as a result of respiratory paralysis, usually within a period of 12 hours.

Symptoms of neurotoxic shellfish poisoning include reverse sensation of hot and cold temperatures, muscular incoordination and pain, and lowered pulse rate. Neurotoxic shellfish poisoning is caused by a family of about six chemically related toxins known as brevetoxins. Symptoms begin within about 3 hours and stop in 2–3 days. No deaths have been reported.

Amnesic shellfish poisoning is due to the potent neurotoxin domoic acid, which causes gastroentiritis and mental confusion (1,26). Diarrhetic shellfish poisoning symptoms occur from 30 minutes to 12 hours following ingestion of shellfish. The symptoms usually disappear within 1–3 days and produce no long-term effects. Diarrhetic shellfish poisoning can be distinguished from paralytic shellfish poisoning by the absence of neurologic symptoms.

The disorders associated with finfish: ciguatera, scombroid poisoning and pufferfish poisoning; all elicit nausea, diarrhea, abdominal pain, and tingling of the extremities (1). Other ciguatera symptoms include reversed sensation of hot and cold temperatures, headache, itching skin, muscle and joint pain, metallic taste, malais, anxiety, chills, convulsions, paralysis, hallucinations, lost equilibrium, changes in pulse rate, dilated eyes, and reduced blood pressure. Onset occurs in 2–12 hours. Severe itching, temperature reversal, and tingling of the extremities are the most distinctive symptoms, which may last many months. Morbidity in the Pacific region averages five per thousand (3).

Scombroid poisoning symptoms include rash, flushing, burning of the mouth and throat, and heart palpitation, in addition to nausea, diarrhea, abdominal pain, and tingling of the extremities. Onset occurs in a few minutes to a few hours and the duration is 4–12 hours (11,14,27).

Pufferfish poison is a potent vasodepressor that selectively blocks the sodium channel in excitable membranes (15). The symptoms of intoxication start with almost immediate tingling sensations (usually within minutes but up to 3 hours). In severe cases, respiratory failure and death occur. Because pufferfish poison is found in the highest concentration in the gonads, liver, and intestine, the fish are most dangerous to eat immediately before and during the reproductive season.

DETECTION

All of the more common biotoxins can be measured by bioassay, immunologic, or chemical techniques (16,17). It is generally more cost-effective to survey or monitor with relatively nonspecific bioassay methods and confirm with more specific methods. However, specific, rapid, low-cost, field methods are under development and will greatly improve controls of illness in the future. All the common toxins (except scombroid poison) can be detected by mouse bioassay, differing in sample preparation and extraction. Mouse bioassay is the most common regulatory method in use. Ciguatera is also measured by cat, mongoose, various aquatic invertebrates, and guinea pig ileum bioassay techniques. A method of measuring diarrhetic shellfish poison in common use in the Netherlands is the rat fecal consistency assay. The toxins of paralytic, neurotoxic, amnesic, and diarrhetic shellfish poisonings, scombroid poisoning and pufferfish poisoning are measured by relatively sophisticated analytic high-performance liquid chromatography (HPLC) methods. The toxins of ciguatera are not yet measured using HPLC. It is reported that ciguatoxin can be detected using a radioimmunoassay or an enzyme immunoassay.

Scombroid poison is detected by measurement of histamine using guinea pig ileum bioassay, fluorometry, colorimetry, enzyme isotopic assays, or thin-layer chromatography. Toxic samples contain 100–4,000 mg histamine per 100 g fish. The toxic threshold is affected by levels and types of potentiators and microflora present, spoilage conditions, and fish species.

Immunoassay methods have considerable potential for detecting marine toxins. The sensitivity and specificity of this method would quickly eliminate the negative samples so that only the positive sample will require further testing. A number of immunoassay-based kits are available for okadaic acid (28), saxitoxins (29), cigautoxins, brevetoxins, and diarrhetic shellfish toxins (30).

CONTROLS

Foods contaminated with marine toxins cannot be identified without exhaustive analyses. Cooking will not inactivate these toxins. Controls are largely dependent on restrictions on harvests and prompt recognition of illnesses (18,19). Harvests of all contaminated shellfish are usually restricted based on toxin levels in edible tissues. Systematic processes to screen fish and shellfish are not in place except for monitoring of paralytic and neurotoxic shellfish poisons in shellfish-growing waters under the National Shellfish Sanitation Program. The National Shellfish Sanitation Program is described in greater detail in another article. Harvests are prohibited when concentrations of saxitoxins exceed 80 μg/100 g tissue for paralytic shellfish poison and 20 mouse units of brevetoxins per 100 g tissue for neurotoxic shellfish poison. Because neurotoxic shellfish poison correlates with levels of causative dinoflagellates in the water column, many states prohibit harvesting shellfish when concentrations exceed 5,000 cells *P. brevis* per liter of seawater (31). These areas are kept closed until edible tissue levels of toxin fall below FDA tolerances specified above.

Harvests of amnesic shellfish poison-contaminated shellfish can be controlled by monitoring of the causative organism in the growing area, but this is not done routinely. Cell monitoring is not effective in controlling paralytic and diarrhetic shellfish poisonings. Ciguatera is controlled, in part, by avoiding harvests from reefs implicated in frequent intoxications. It is controlled by avoiding large specimens of frequently toxic species from suspect areas because the toxin is bioaccumulated as fish prey on small herbivores.

Scombroid poisoning can be completely controlled by rapid and constant refrigeration of susceptible species immediately following harvest and throughout all processing and handling steps. Lots suspected of time—temperature abuse can be analyzed for the presence of histamine and rejected when levels exceed 50 mg/100 g.

Some of the toxins can be controlled during processing, handling, and meal preparation. Removal of certain internal and reproductive organs of some fish and shellfish (scallops) will remove a majority of paralytic and diarrhetic shellfish poisons, ciguatera, and pufferfish poison. Pufferfish poisoning is largely controlled, however, by proper meal preparation by trained fugu dealers and cooks.

OTHER DISORDERS

Some other seafood toxins such as clupeotoxin, hallucinogenic mullet, or fish poison and shark liver poison occur infrequently but may have severe consequences (1,11). Clupeotoxin poisoning, which is caused by contaminated herrings (Clupeidae) and anchovies (Engraulidae), is found sporadically in the tropical Atlantic Ocean, Pacific Ocean, and Caribbean Sea. The ingestion of certain types of reef fish, five families including Acanthuridae (tang) and Mugilidae (mullet), that occur in the tropical Pacific and Indian oceans, sometimes cause hallucinogenic mullet poisoning. The causative agents for clupeotoxicity and hallucinogenic mullet poisoning are unknown but thought to be dinoflagellates or bluegreen algae.

The symptoms of clupeotoxicity are both gastrointestinal and neurologic, occur quickly, and are usually violent in nature. Mortality rates as high as 45% have been reported. Hallucinogenic mullet poisoning symptoms include those of the nervous system such as dizziness, loss of equilibrium, and hallucinations. The symptoms appear quickly (within 2 hours) and can persist for up to 24 hours. However, hallucinogenic mullet poisoning is relatively mild with no long-term effects.

Several species of fish have been reported to have toxic livers, especially some of the sharks. Shark liver poisoning is thought to be vitamin A toxicosis. Symptoms begin within 30 minutes of ingestion and include nausea, diarrhea, abdominable pain, headache, rapid pulse, cold sweats, and tingling and burning of the tongue, throat, and esophagus. Recovery occurs in 1 day to 2 weeks. Mortality is high.

Natural marine toxins are low molecular weight compounds usually produced by dinoflagellates that ascend the food chain. Fish and shellfish that contain the toxins cannot be identified by casual inspection. Generally, handling, processing, and cooking do not affect levels of toxins in tissues. The disorders are characterized by rapid onset and usually cause gastrointestinal and/or neurologic symptoms. Control of illnesses usually requires harvest restrictions and prompt recognition of symptoms. Bioassays are most often used to detect the presence of toxins in fish.

BIBLIOGRAPHY

1. B. W. Halstead, *Poisonous and Venomous Marine Animals of the World*, 2nd rev. ed., Darwin Press, Princeton, N.J., 1988.
2. United States Centers for Disease Control, *Foodborne Disease Surveillance Annual Summaries*, US Department of Health and Human Services, Atlanta, Geo.
3. E. P. Ragelis, ed., *Seafood Toxins*, American Chemical Society, Washington, D.C., 1984.
4. E. J. Schautz, Ref. 3, pp. 99–111.
5. D. G. Baden, "Brevetoxins: Unique Polyether Dinoflagellate Toxins," *FASEB Journal* **3**, 1807–1817 (1989).
6. W. H. Hemmert, "The Public Health Implications of *Gymnodinium breve* Red Tides, A Review of the Literature and Recent Events," in *Proceedings of the First International Conference on Toxic Dinoflagellate Blooms*, Boston, Mass., Nov. 4–6, 1974.
7. D. V. Subba Rao, M. A. Quillian, and R. Pocklington. "Domoic Acid—A Neurotoxic Amino Acid Produced by the Marine Diatom *Nitzschia pungens* in Culture," *Canadian Journal of Fish. Aquatic Science* **45**, 2076–2079 (1988).
8. T. Yasumoto, T. Oshima, W. Sugaware, Y. Fukuyo, H. Oguri, T. Igarashi, and N. Fujita. "Identification of *Dinophysis fortii* as the Causative Organism of Diarrhetic Shellfish Poisoning," *Bulletin of the Japanese Society of Sci. Fish.* **46**, 1405–1411 (1980).
9. E. Stammon, D. A. Segar, and P. G. Davis, "A Preliminary Assessment of the Potential for Diarrhetic Shellfish Poisoning in the Northeast United States," *NOAA Technical Memorandum* NOS OMA 34, Rockville, Md., 1987.
10. T. Yasumoto, I. Nakajima, R. Bagnis, and R. Adachi. "Finding of a Dinoflagellate as a Likely Culprit of Ciguatera," *Bulletin of the Japanese Society Sci. Fish.* **43**, 1021–1026 (1977).
11. H. Dymsza et al., "Poisonous Marine Animals," in H. D. Graham, ed., *The Safety of Foods*, 2nd ed., Avi Publishing Co., Westport, Conn., 1980, pp. 625–651.
12. U. Simidu, T. Noguchi, D. F. Hwang, Y. Shida, and K. Hashimoto. "Marine Bacteria which Produce Tetrodotoxin," *Applied and Environmental Microbiology* **53**, 1714–1715 (1987).
13. J. J. Sullivan and W. T. Iwaoka, "High Pressure Liquid Chromatographic Determination of Toxins Associated with Paralytic Shellfish Poisoning," *Journal of the Association of Official Analytical Chemists* **66**, 297–303 (1983).
14. S. L. Taylor, J. Y. Hui, and D. E. Lyons, Ref. 3, pp. 417–430.
15. S. L. Hu and C. Y. Kao, "The pH Dependence of the Tetrodotoxin-Blockade of the Sodium Channel and Implications for Toxin Binding," *Toxicon* **24**, 25–31 (1986).
16. E. F. McFarren, "Assay and Control of Marine Biotoxins," *Food Technology* **25**, 234–244 (1971).
17. S. Hall and Y. Shimizu, "Toxin Analysis and Assay Methods," in D. M. Anderson, A. W. White, and D. G. Baden, eds., *Toxic Dinoflaggelates*, Elsevier, New York, pp. 545–548.

18. E. L. Bryan, "Epidemiology of Foodborne Diseases Transmitted by Fish, Shellfish and Marine Crustaceans in the United States, 1970–1978," *Journal of Food Protection* **43**, 859–876 (1980).
19. J. M. Hughes and M. H. Merson, "Current Concepts, Fish and Shellfish Poisoning," *Medical Intelligence* **93**, 1117–1120 (1976).
20. J. L. C. Wright, "Dealing with Seafood Toxins: Present Approaches and Future Options," *Food Research International* **28**, 347–358 (1995).
21. K. A. Steidinger, "Some Taxonomic and Biologic Aspects of Toxic Dinoflagellates," in I. R. Falconer, ed., *Algal Toxins in Seafood and Drinking Water*, Academic Press, San Diego, Cal., 1993, pp. 1–28.
22. H. P. van Egmond et al., "Paralytic and Diarrhoeic Shellfish Poisons: Occurrence in Europe, Toxicity, Analysis and Regulation," *Journal of Natural Toxins* **2**, 41–83 (1993).
23. S. S. Bates et al., "Pennate Diatom *Nitzchia pungens* as the Primary Source of Domoic Acid, A Toxin in Shellfish from Eastern Prince Edward Island, Canada," *Canadian Journal of Fisheries and Aquatic Sciences* **46**, 1203–1215 (1989).
24. T. Aune and M. Yndestad, "Diarrhetic Shellfish Poisoning," in I. R. Falconer, ed., *Algal Toxins in Seafood and Drinking Water*, Academic Press, San Diego, Cal., 1993, pp. 87–104.
25. Y. Oshima, S. Blackburn, and G. M. Hallegraff, "Comparative Study on Paralytic Shellfish Toxin Profile of the Dinoflagellate *Gymnodinium catenateum* from Three Different Countries," *Marine Biology* **95**, 217–220 (1993).
26. E. C. D. Todd, "Domoic Acid and Amnesic Shellfish Poisoning—A Review," *Journal of Food Protection* **56**, 69–83 (1993).
27. B. Bartholomew et al., "Scombrotoxic Fish Poisoning in Britain: Features of Over 250 Suspected Incidents Between 1976 and 1986," *Epidemiological Infections* **99**, 775–782 (1986).
28. J. D. Chin et al., Screening for Okadaic Acid, A New Probe for the Study of Cellular Regulation," *TIBS* **15**, 98–100 (1995).
29. E. Usleber, E. Schneider, and G. F. Terplan, "Direct Enzyme Immunoassay in Microtitration Plate and Test Strip Format for the Detection of Saxitoxin in Shellfish," *Letters in Applied Microbiology* **13**, 275–277 (1991).
30. U.S. Patent 4,816,392 (March 28, 1989), Y. Hokama.
31. Food and Drug Administration, *National Shellfish Sanitation Program Manual of Operations*, Part I, U.S. Department of Health and Human Services, Washington, D.C., 1988.

See also TOXICANTS, NATURAL.

MASS TRANSFER AND DIFFUSION IN FOODS

Mass transfer occurs in a number of food-processing operations such as dehydration, evaporation, concentration, distillation, solvent extraction, packaging, peeling, and leaching. Mass transfer is a common and important phenomenon in food processing and preservation. During osmotic drying of fruits, for example, sugar from a concentrated solution moves into the fruit while water leaves the fruit and migrates into the surrounding solution. By definition, mass transfer is the migration of matter from one location to another due to concentration or partial pressure gradient. According to this definition, the motion of air due to wind or the transport of water through a pipe by a pump is not mass transfer, because these motions occur due to either the presence of a total pressure difference (wind) or the application of mechanical work (pump).

Mass transfer subdivides into molecular mass transfer and convective mass transfer. For the former, the term diffusion is often used. Diffusion deals with the random molecular migration of matter through a medium, whereas convective mass transfer involves the migration of matter from a surface into a moving fluid or a stream of gas. Diffusion further subdivides into molecular diffusion, transitional diffusion, Knudsen diffusion, and Eddy diffusion. Among these, molecular diffusion is the most important phenomenon. It exists in most processes of mass transfer and often has a major influence when other kinds of diffusion are also present.

From the theory of molecular kinetics, it is known that when a randomly moving molecule diffuses through a capillary tube, it can move freely for some distance before a collision with another molecule or with the wall of the capillary tube occurs. If a molecule would collide with other molecules more often than it would with the wall, molecular diffusion occurs. If molecule-wall collisions are predominant, it is Knudsen diffusion. If both molecule-molecule and molecule-wall collisions are important, it is transition-type diffusion. The specific state can be determined by evaluating Knudsen's number (1). When bulk fluid motion is involved, the term Eddy diffusion applies.

Diffusion of a substance may occur in gases, liquids, or solids. Within the range of validity of the ideal gas law, the rate of diffusion in gases is affected primarily by temperature and pressure. When diffusion in liquids is considered, however, the effect of pressure is usually negligible because liquids are incompressible. Diffusion in solids is far more complex than diffusion in gases or liquids, because (1) the substance diffusing through a solid may actually be diffusing through a liquid or gas contained within the pores of the solid; and (2) many solids, such as crystals, polymeric films, and solids with capillaries, are anisotropic, in which case the molecules have a preferential direction of movement.

The general equation for all types of mass transfer is

$$\text{mass transfer rate} = \frac{\text{driving force}}{\text{resistance}}$$

where the driving force is the partial pressure difference or the concentration gradient (depending on the transferred matter); and the resistance is a function of the properties of the medium through which the matter is transferred.

Mass transfer can occur with steady-state or unsteady-state conditions. In a steady state the concentration or partial pressure, the resistance, and the transfer rate are constant over time. In an unsteady state, all these properties vary with time. Obviously, the mathematical treatment of steady-state problems is simpler than that involving unsteady-state conditions. For this reason, the following sections shall first deal with the simple steady-state cases. It is important to remember, however, that in reality every process begins with unsteady-state conditions, which over time change to steady state.

MOLECULAR DIFFUSION

Molecular diffusion of matter is analogous to the conductive transfer of heat. In heat conduction, energy moves from a region of high temperature to a region of low temperature due to the random motion of gas or liquid molecules or the vibration of solid molecules. Similarly, in molecular diffusion, matter moves from a region of high concentration to a region of low concentration due to the random motion of the molecules of that substance. Therefore, molecular diffusion can be defined as the net transport of matter on a molecular scale due to a concentration or partial pressure gradient through a medium, which is either stagnant or has a laminar flow with a direction perpendicular to that of the concentration gradient. For example, water will evaporate from an open surface into still surrounding air, and a piece of sugar will dissolve in a cup of coffee and spread without stirring.

Fick's First Law

Fick first recognized the analogy between heat conduction and mass diffusion in 1855. Following Fourier's equation for heat conduction, Fick quantitatively expressed the rate of diffusion of a substance, through an isotropic medium with a unit surface area, as being proportional to the concentration gradient measured as a vector normal to the surface:

$$j = -D\frac{dC}{dy} \qquad (1)$$

where j is the diffusion rate, C is the concentration of the substance being transferred, y is the distance, and D is a measure of the resistance against the diffusion of matter, usually called the diffusivity or diffusion coefficient. The rate j is usually expressed in mol/m$^2\cdot$ s; C, in mol/m^3; y, in m; and D, in m^2/s. The negative sign in Fick's law indicates that the flow of matter is in the direction of decreasing concentration. It should be noted that equation 1 applies equally well to diffusion in gases, liquids, and solids (assuming isotropic properties).

Diffusion in Gases

From the standpoint of molecular kinetics, gas molecules contain energy and are in a state of continuous random motion. During such movement, the molecules bounce against each other or against other surfaces, continuously changing their direction. Consequently, in a region of constant concentration, the probability of an individual molecule moving in any direction is the same. This phenomenon is called the random walk. When considering a homogeneous mixture of two gases, for example, and imaging a certain number of molecules traveling in one direction, there must be an equal number of molecules traveling in the opposite direction and no net mass transfer is taking place. If a concentration gradient present were, however, then the number of molecules traveling from the high to the low concentration region would be larger than that traveling in the opposite direction. Therefore, net mass transfer would take place from the high to the low concentration region.

Assuming that a gas A follows the ideal gas law, the concentration term in Fick's law can be expressed as:

$$C^A = \frac{n_A}{V} = \frac{P_A}{RT} \qquad (2)$$

where n_A is the amount of gas A in moles, V is the total volume, P_A is the partial pressure of A, R is the gas constant, and T is the absolute temperature. Thus, Fick's law becomes

$$j_A = -\frac{D}{RT}\frac{dP_A}{dy} \qquad (3)$$

In a more general case, the medium is moving in a laminar flow with a bulk speed v_b while the component is diffusing through the medium. N_A, the total flux of component A, is the sum of diffusion and bulk flow of the mixture

$$N_A = C_A v_d + C_A v_b \qquad (4)$$

where v_d is the velocity of A due to true diffusion and $C_A v_d$ is the number of molecules being diffused. The term $C_A v_d$ can be expressed by Fick's law and is equivalent to j_A. Hence the total flux of A is

$$N_A = j_A + C_A v_b \qquad (5)$$

For a binary mixture of gases A and B, the total mass flux is the sum of the fluxes of the two gases

$$N = N_A + N_B = C v_b \qquad (6)$$

and solving for v_b gives

$$v_b = \frac{N_A + N_B}{C} \qquad (7)$$

where C is the total concentration of the mixture ($A + B$). By substituting equation 7 into equation 5, and using $C_A = P_A/RT$, and $C = P/RT$, gives

$$N_A = j_A + \frac{C_A}{C}(N_A + N_B) \qquad (8)$$

For ideal gases where $C_A/C = P_A/P$, Fick's law gives the following general equation for diffusion in gases

$$N_A = -\frac{D_{AB}}{RT}\frac{dP_A}{dy} + \frac{P_A}{P}(N_A + N_B) \qquad (9)$$

Applications of equation 9 in two specific cases will be discussed in the following section.

Equimolecular Gas Counterdiffusion. Consider two gases A and B in two jars connected by a tube as shown in Figure 1. When the total pressure in each jar is the same, a molecular diffusion through a stationary medium will occur due to the presence of partial pressure difference between each individual component. Thus

$$N_A = j_A \text{ and } N_B = j_B$$

As long as the partial pressure $P_{A1} > P_{A2}$ and $P_{B2} > P_{B1}$, diffusion will continue to occur. The number of molecules of A diffusing from jar 1 to jar 2 is equal to the number of molecules of B diffusing in the opposite direction (to maintain the equal total pressure in both jars)

$$N_A = -N_B \text{ or } j_A = -j_B$$

Setting $N_A = -N_B$ in equation 9, we have

$$N_A = j_A = -\frac{D_{AB}}{RT}\frac{dP_A}{dy} = -j_B = -\left(-\frac{D_{BA}}{RT}\frac{dP_B}{dy}\right) \quad (10)$$

For steady-state conditions, integration of equation 10 gives

$$N_A = \frac{D_{AB}}{RTY}(P_{A1} - P_{A2} \quad (11)$$

and

$$N_B = \frac{D_{BA}}{RTY}(P_{B1} - P_{B2}) \quad (12)$$

where Y is the total distance of diffusion. Because $P = P_A + P_B =$ constant, it follows that $dP = dP_A + dP_B = 0$ and therefore $dP_A = -dP_B$. Substituting this expression into equation 10 gives

$$D_{AB} = D_{BA} \quad (13)$$

In other words, in the case of diffusion of ideal gases, the diffusivities of each individual component are identical.

Diffusion of Gas *A* Through a Stagnant Layer of Gas *B*. Consider water evaporating from an open flask into the air, as shown in Figure 2. In this case, water vapor A is diffusing through a stagnant layer of air B at the top of the water surface. We know that air is only slightly absorbed by water, whereas water vapor does escape from the water's surface through the still air layer and into the bulk stream outside the flask. Therefore

$$N_B = 0$$

and the general equation 9 becomes (by rearranging)

$$N_A dy = -\frac{D_{AB}}{RT}\frac{dP_A}{1 - \frac{P_A}{P}} \quad (14)$$

When dealing with a dilute system at constant temperature T and pressure P, D_{AB} can be considered constant, and thus integration of equation 14 gives

$$N_A = \frac{D_{AB}P}{RTY}\ln\left(\frac{P - P_{A2}}{P - P_{A1}}\right) \quad (15)$$

By introducing the concept of logarithmic mean value

$$P_{m\ln} = \frac{(P - P_{A2}) - (P - P_{A1})}{\ln\left(\frac{P - P_{A2}}{P - P_{A1}}\right)} = \frac{P_{A1} - P_{A2}}{\ln\left(\frac{P - P_{A2}}{P - P_{A1}}\right)} \quad (16)$$

we can rewrite equation 15 as

$$N_A = \frac{D_{AB}}{RTY}\frac{P}{P_{m\ln}}(P_{A1} - P_{A2}) \quad (17)$$

When the difference between P_{A1} and P_{A2} is small, say $P - P_{A1}/P - P_{A2}) < 1.5$, using an arithmetic mean P_m instead of a logarithmic mean introduces an error of less than 1.5%, and equation 16 becomes

$$P_{m\ln} = P_m = \frac{(P - P_{A1}) + (P - P_{A2})}{2} \quad (18)$$

Furthermore, when the partial pressure of the diffusing matter is low compared to the total pressure, then $P - P_{A1} \approx P - P_{A2} \approx P$ and $P_m \approx P$, and equation 17 becomes

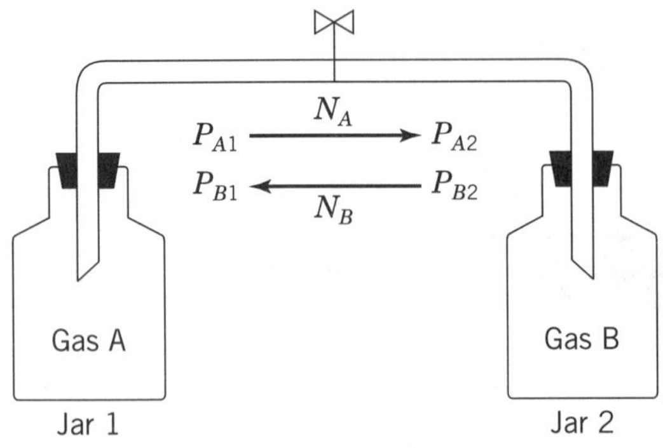

Figure 1. Equimolecular counterdiffusion.

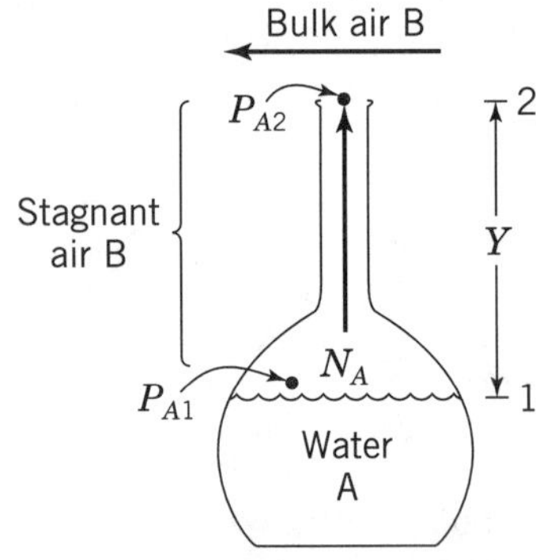

Figure 2. Diffusion of water vapor through stagnant air.

$$N_A = \frac{D_{AB}}{RTY}(P_{A1} - P_{A2}) \tag{19}$$

Diffusion in Liquids

Diffusion in liquids can be observed when a drop of food color, for example, is put into a cup of water. The color diffuses into the surrounding water due to the concentration difference until a homogeneous solution is formed. The mechanism of diffusion in liquids is similar to that of diffusion in gases. However, because liquids are denser than gases, the resistance to diffusion in a liquid is much greater and the diffusion rates are lower. Because liquids are incompressible, the total pressure has very little influence on the diffusion. In addition, because concentration in liquids is considerably higher and some chemical changes of the solute may occur, diffusion is often concentration dependent.

The general equation 8, derived for diffusion in gases, can be directly adopted here. Substituting j_A from equation 1 gives

$$N_A = -D_{AB}\frac{dC_A}{dy} + \frac{C_A}{C}(N_A + N_B) \tag{20}$$

This is the general equation for diffusion in liquids.

Equimolecular Liquid Counterdiffusion. Similar to the situation in gases, equimolecular diffusion in liquids has the feature of $N_A = -N_B$. Thus equation 20 can be simplified as

$$N_A = -D_{AB}\frac{dC_A}{dy} = D_{AB}\frac{(C_{A1} - C_{A2})}{Y} \tag{21}$$

When the difference in the concentration between two points is small, an average concentration ($\bar{C}_A$) can be defined as

$$\bar{C}_A = \frac{C_{A1} + C_{A2}}{2} \tag{22}$$

If we set $\bar{C}_{A1} \approx \bar{C}_A X_{A1}$ and $\bar{C}_{A2} \approx \bar{C}_A X_{A2}$, equation 22 becomes

$$N_A = \frac{D_{AB}\bar{C}_A}{Y}(X_{A1} - X_{A2}) \tag{23}$$

where X_{A1} and X_{A2} are the molecular fractions of A at point 1 and point 2, respectively. Similarly, for liquid B, we have

$$N_B = \frac{D_{BA}\bar{C}_B}{Y}(X_{B1} - X_{B2}) \tag{24}$$

Because the diffusivity D_{AB} is usually concentration dependent, an average concentration $\bar{C}_A$ should be used when calculating the value of D_{AB}.

Diffusion of Liquid A through a Stagnant Layer of Liquid B. In this case, $N_B = 0$, and equation 17 can be rewritten by substituting $C_{A1} = \bar{C}_A X_{A1} = P_{A1}/RT$, $C_{A2} = \bar{C}_A X_{A2} = P_{A2}/RT$, and $X_{m\ln} = P_{m\ln}/P$ so that

$$N_A = \frac{D_{AB}}{YX_{m\ln}}(C_{A1} - C_{A2}) = \frac{D_{AB}\bar{C}_A}{YX_{m\ln}}(X_{A1} - X_{A2}) \tag{25}$$

where $X_{m\ln}$ is the logarithmic mean of the molecular fractions. For a small difference between X_{A1} and X_{A2}, the arithmetic mean (X_m) can be used. For a dilute solution, $1 - X_{A1} \approx 1 - X_{A2}$ and $X_m = 1.0$, giving

$$N_A = \frac{D_{AB}}{Y}(C_{A1} - C_{A2}) \tag{26}$$

Diffusion in Solids

Diffusion of gases and liquids in solids are phenomena occurring in many processes such as drying, packaging, catalytic reactions, leaching, and membrane separations. The two classifications of diffusion in solids are structure independent and structure dependent. In the former, the diffusing substance is dissolved in the solid to form a homogeneous solution, and thus Fick's law can be directly applied. In the latter, pores, capillaries, and other interconnected voids in the solid allow some flow of the diffusing substance. Because this porous solid is anisotropic, the diffusion path of molecules will be different than that described by Fick's law and an overall, apparent, or effective diffusivity coefficient may be used.

Structure-Independent Diffusion in Solids. A homogeneous solution in the solid is assumed here; therefore, diffusion is independent of the actual structure of the solid. The solute concentration gradient is the actual driving force, and the general equation 20 for binary diffusion can be applied directly. Usually the concentration C_A is small compared to the total concentration C, and C_A/C can be neglected. By further assuming that D_{AB} is constant, we can integrate to get

$$N_A = \frac{D_{AB}}{Y}(C_{A1} - C_{A2}) \tag{27}$$

When dealing with diffusion of gases in a solid, the solubility S of a gas solute A is usually expressed as (vol of A at 0°C and 1 atm)/(vol of solid · pressure of A). To convert this into concentration C_A (kmol of A/vol of solid), the following equation can be used.

$$C_A = \frac{SP_A}{22.414} \tag{28}$$

where 22.414 is the volume (in m^3) of 1 kmol of an ideal gas at 0°C and 1 atm.

In the case of diffusion through a solid cylinder in the radial direction, the total diffusing area is not constant. For a cylinder with an inner radius r_1, outer radius r_2, and length l, we have

$$N_A = \frac{\dot{N}_A}{A} = -D_{AB}\frac{dC_A}{dr} \quad (29)$$

Because $A = 2\pi rl$ for the cylinder, substituting this expression into equation 29, rearranging, and integrating gives

$$\dot{N}_A = D_{AB} \cdot 2\pi l \frac{C_{A1} - C_{A2}}{\ln\left(\frac{r_2}{r_1}\right)} \quad (30)$$

where $\dot{N}_A$ is the total flux in mol/s. It should be noted that in diffusion in solids, $D_{AB} \neq D_{BA}$ and both D_{AB} and D_{BA} are independent of the partial pressure of the gas or liquid outside the solid. However, the solubility in the solid is directly proportional to the partial pressure.

Structure-Dependent Diffusion in Solids. The presence of interconnected pores and capillaries in porous solids does not allow direct application of Fick's law, because the diffusing substance follows a tortuous path, which is not equal to the distance between two points. To correct for that tortuous path, it is necessary to define a factor τ, called the tortuosity, and rewrite the diffusion equation 27 as

$$N_A = \frac{\epsilon D_{AB}}{\tau Y}(C_{A1} - C_{A2}) \quad (31)$$

where ϵ is the porous fraction of the solid. Usually the porous fraction and tortuosity factor are combined into an effective diffusivity as

$$D_{AB_{\text{eff}}} = \frac{\epsilon}{\tau} D_{AB} \quad (32)$$

When equation 31 is used to describe diffusion of gases in solids, and because $C_A = P_A/RT$,

$$N_A = \frac{\epsilon D_{AB}}{\tau YRT}(P_{A1} - P_{A2}) \quad (33)$$

If the pores of the solid are very small compared to the mean free path of the gas, other types of diffusion may occur (2,3).

Diffusion in Nonideal Systems

Nonideal systems play an important role in food, biological, and life processes of humans, animals, plants, and microorganisms. In the presence of macromolecules in the system, interactions may occur between them and smaller molecules. Such phenomena cause the system to deviate from its ideal behavior, even though a true solution (e.g., colloids) might be formed. As a result, the mobility of the diffusing molecule changes unpredictably, even in dilute solutions. Diffusion in nonideal systems is complex and difficult to describe by theoretical models. Generally, it is assumed that the differential form of Fick's law applies

$$\frac{\partial C}{\partial t} = \frac{\partial}{\partial y}\left(D\frac{\partial C}{\partial y}\right) \quad (34)$$

where D is concentration dependent. Solution of this equation requires knowledge of the actual function of D and the corresponding boundary conditions. Later, D will be expressed as a function of concentration, temperature, viscosity of solution, and the properties of the medium.

DIFFUSION WITH BULK FLOW AND CHEMICAL REACTIONS

In some practical diffusion problems involving the presence of catalysts, enzymes, or other reagents, the diffusing substance A may undergo a chemical reaction. In such a case, the general equation 8 is still valid. To solve this equation, the relationship between N_A and N_B, which depends on the type of chemical reaction, must be known. In the following sections it is demonstrated how equation 8 can be solved for some chemical reactions commonly observed in food systems.

Reactions $A + Z \rightarrow B$

In this case, if 1 mol of A is diffusing toward the reacting surface, there must be 1 mol of B diffusing away in the opposite direction (as long as Z is adequate on that surface). Substituting $N_A = -N_B$ into equation 8 and observing that

$$j_A = -D_{AB}\frac{dC_A}{dy} = -CD_{AB}\frac{dX_A}{dy}$$

the following equation is obtained.

$$N_A = -CD_{AB}\frac{dX_A}{dy} \quad (35)$$

Equation 35 is identical to equation 21 for equimolecular counterdiffusion. Solving equation 35 gives

$$N_A = \frac{CD_{AB}}{Y}(X_{A1} - X_{A2}) \quad (36)$$

If the reaction is instantaneous, no A exists on the reacting surface and $X_{A2} = 0$. Equation 36 then becomes

$$N_A = \frac{CX_{A1}}{(Y/D_{AB})} \quad (37)$$

If the reaction is first order, N_A in the reacting surface can be expressed as

$$N_A|_{y=Y} = \mu C_{A2} = \mu CX_{A2} \quad (38)$$

where μ is the first-order reaction rate constant in mol/m^2 s. Solving equation 38 for X_{A2} gives

$$X_{A2} = \frac{N_{A\,|y=Y}}{\mu C} \tag{39}$$

By substituting equation 39 into equation 36 and assuming steady-state conditions ($N_A = N_{A\,|y=Y}$), the following is finally obtained.

$$N_A = \frac{CX_{A1}}{\left(\dfrac{Y}{D_{AB}}\right) + \dfrac{1}{\mu}} \tag{40}$$

In equations 37 and 40, the terms Y/D_{AB} and $Y/D_{AB} + 1/\mu$ provide a measurement of the resistance to diffusion in systems with instantaneous or first-order chemical reaction of the form $A + Z \rightarrow B$, respectively.

Reaction $A + Z \rightarrow 2B$

In this case, 1 mol of A produces 2 mol of B if Z is adequate on the reacting surface. Substituting $N_B = -2N_A$ into equation 8, noting that $C_A/C = X_A$ and rearranging, gives

$$N_A dy = -CD_{AB} \frac{dX_A}{1 + X_A} \tag{41}$$

For constant C, D_{AB}, and N_A, it is possible to integrate both sides of equation 41 so that

$$N_A \int_0^Y dy = -CD_{AB} \int_{X_{A1}}^{X_{A2}} \frac{dX_A}{1 + X_A} \tag{42}$$

which, after solving for N_A, gives

$$N_A = \frac{CD_{AB}}{Y} \ln\left(\frac{1 + X_{A1}}{1 + X_{A2}}\right) \tag{43}$$

For an instantaneous reaction with $X_{A2} = 0$ equation 43 can be simplified to

$$N_A = \frac{CD_{AB}}{Y} \ln(1 + X_{A1}) \tag{44}$$

For a first-order reaction, substituting equation 39 into equation 43 gives

$$N_A = \frac{CD_{AB}}{Y} \ln\left(\frac{1 + X_{A1}}{1 + \dfrac{N_A}{\mu C}}\right) \tag{45}$$

Equation 45 can be solved by trial and error, but it is appropriate to point out that the reaction rate in equation 45 is lower than that in equation 44, because $(1 + N_A/\mu C) > 1$.

UNSTEADY-STATE MASS TRANSFER

In the previous sections, only steady-state mass transfer phenomena were dealt with. In practice, mass transfer always begins with an unsteady-state condition. Steady-state conditions may be established only after some time has elapsed. In unsteady-state mass transfer, the concentrations or partial pressures and consequently the mass transfer rate are functions of time and position. Fick's second law can be used to describe the concentration changes with time and position. For one dimension, Fick's second law can be written as

$$\frac{\partial C}{\partial t} = D \frac{\partial^2 C}{\partial y^2} \tag{46}$$

To solve equation 46, a set of initial and boundary conditions must be given. A simple example is the diffusion through a membrane with known initial and boundary conditions (4,5), stated as

$$\begin{aligned} &\text{at } t = 0,\ C = C_0, \text{ for } 0 < y < Y \\ &\text{at } y = 0,\ C = C_1, \text{ for } t > 0 \\ &\text{at } y = Y,\ C = 0, \text{ for } t > 0 \end{aligned} \tag{47}$$

The solution of equation 46 is then

$$C = C_1\left(1 - \frac{y}{Y}\right) + \frac{2}{\pi} \sum_{n=1}^{\infty} \frac{C_1}{n} \sin \frac{n\pi y}{Y} \exp\left(-n^2\pi^2 \frac{Dt}{Y^2}\right) \tag{48}$$

The total amount of substance diffused through the membrane in time t, Q_t, is

$$Q_t = \frac{DC_1 t}{Y} - \frac{YC_1}{6} - \frac{2YC_1}{\pi^2} \sum_{n=1}^{\infty} \frac{(-1)^n}{n^2} \exp\left(-n^2\pi^2 \frac{Dt}{Y^2}\right) \tag{49}$$

When t approaches infinity, the exponential terms vanish and equation 49 simplifies to a linear form

$$Q_t = \frac{DC_1 t}{Y}\left(t - \frac{Y^2}{6D}\right) \tag{50}$$

A plot of Q_t versus t gives an intercept L on the t-axis (where $Q_t = 0$) as

$$L = \frac{Y^2}{6D} \tag{51}$$

where L is the so-called time lag. Equation 50 can be used to determine D from experimental data. With more complex initial and boundary conditions, analytical solutions of equation 46 become more difficult and sometimes even impossible. In practice, numerical methods are often used (1,6).

MASS TRANSFER IN TWO-PHASE SYSTEMS

In the previous discussion, the mass transfer of one phase was considered. The concentration gradient of the other phase was assumed to be zero (the system was assumed either well stirred or very dilute). In most practical problems, however, concentration gradients may exist on both sides of the interface of the system. For example, consider that solute A is being transferred from a gas to a liquid phase. In such a case, A must travel from the bulk volume into the boundary layer of the gas phase, through the interface, into the boundary layer, and finally into the bulk volume of the liquid phase. The concentration profile of A is shown in Figure 3. Defining P_A as the partial pressure of A in the gas phase, C_A as the concentration in the liquid phase, P_{Ai} as the partial pressure of A in the gas boundary layer, and C_{Ai} as the concentration in the liquid boundary layer, the rate of transfer of A from the gas phase to the interface is

$$N_A = k_g(P_A - P_{A1}) \tag{52}$$

The rate of transfer of A away from the interface and into the liquid phase is

$$N_A = k_l(C_{Ai} - C_A) \tag{53}$$

where k_g and k_l are transfer coefficients for the gas and liquid boundary layer, respectively. At steady state, the rate of transfer to the interface must be equal to the rate of transfer away, so that

$$\frac{k_l}{k_g} = \frac{(P_A - P_{Ai})}{(C_{Ai} - C_A)} \tag{54}$$

If $k_g \gg k_l$, then $(C_{Ai} - C_A) \gg (P_A - P_{Ai})$ and the mass transfer is controlled by the liquid boundary layer. Increasing k_l increases the overall transfer rate. If $k_l \gg k_g$, the process is controlled by the gas boundary layer, and the overall rate is greatly affected by k_g.

For equimolecular counterdiffusion, comparing equation 11 with equation 52, and equation 21 with equation 53 obtains the following.

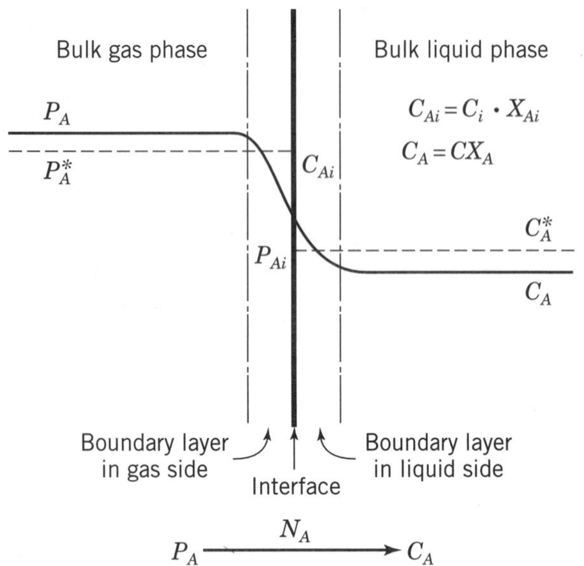

Figure 3. Concentration profile of a solute in a two-phase system.

$$k_g = \frac{D_{AB}}{RTY_g} \tag{55}$$

$$k_l = \frac{D_{AB}}{Y_l} \tag{56}$$

For a substance A diffusing through a stagnant layer of B, comparing equation 17 with equation 52, and equation 25 with equation 53 gives

$$k_g = \frac{D_{AB}}{RTY_g}\frac{P}{P_{m\ln}} \tag{57}$$

and

$$k_l = \frac{D_{AB}}{Y_l}\frac{1}{X_{m\ln}} \tag{58}$$

where Y_g and Y_l are the thicknesses of the gas and liquid boundary layers, respectively, and $P_{m\ln}$ and $X_{m\ln}$ are given by

$$P_{m\ln} = \frac{P_A - P_{Ai}}{\ln\left(\frac{P - P_{Ai}}{P - P_A}\right)} \tag{59}$$

and

$$X_{m\ln} = \frac{X_{Ai} - X_A}{\ln\left(\frac{1 - X_A}{1 - X_{Ai}}\right)} \tag{60}$$

The measurement of the boundary layer thicknesses, Y_g and Y_l, is difficult. For that reason, it is convenient at times to use the boundary layer coefficients for calculations.

DETERMINATION OF DIFFUSIVITY AND OVERALL MASS TRANSFER COEFFICIENT

So far, it has been assumed that the transfer coefficients are known constants. In reality transfer coefficients in food systems are unknown and not constant. Constant transfer coefficients can be used only in special situations and under restricted conditions. A number of experimental methods (7) and several empirical equations (1) have been developed to determine and predict the mass transfer coefficients. In the following subsections the theory that underlines some of the simpler experimental methods used is presented.

Diffusivity in Gases

One of the simplest methods for determining the diffusivity of a vapor or gas into another gas is to evaporate pure liquid A using a graduated cylinder with small diameter, as shown in Figure 4. The vapor A diffuses through an inert gas B, which is gently passed over the opening of the cylinder. The gas B moves slowly so that the layer at the top of liquid A (in the cylinder) can be assumed stagnant. The whole apparatus should be maintained at constant

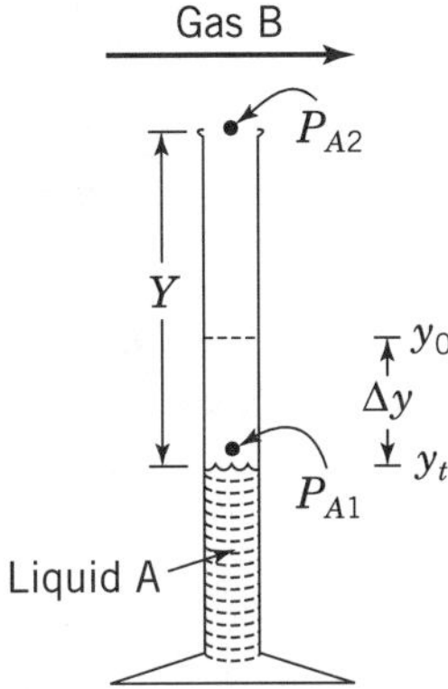

Figure 4. Evaporation of a liquid.

temperature and pressure. The drop of the liquid's level is measured over time and used to determine the diffusivity. The amount of liquid A evaporated in a small time interval is

$$N_A = \frac{\rho_A dy}{M_A dt} \tag{61}$$

Substitution of equation 61 into equation 17 gives

$$\frac{\rho_A dy}{M_A dt} = \frac{D_{AB}}{RTY}\frac{P}{P_{m\ln}}(P_{A1} - P_{A2}) \tag{62}$$

Rearranging and observing that only y and t are variables under steady-state conditions gives

$$D_{AB}\int_0^t dt = \frac{RTP_{m\ln}\rho_A}{PM_A(P_{A1} - P_{A2})}\int_{y_0}^{y_t} y dy \tag{63}$$

and solving for D_{AB}

$$D_{AB} = \frac{RTP_{m\ln}\rho_A(y_t^2 - y_0^2)}{2PM_A t(P_{A1} - P_{A2})} \tag{64}$$

where P_{A1} is the partial pressure of A at the liquid surface, P_{A2} is the partial pressure of A in the gas B, P_A is the density, and M_A is the molecular weight of the liquid.

Diffusivity in Liquids

When measuring diffusivity in liquids, it is important to be aware that D_{AB} depends on the concentration of the diffusing substance in solution. Constant D_{AB} is only an approximation and may be used in cases of very dilute solutions where only small changes in the concentration occur. It should also be noted that D_{AB} indicates the diffusivity of A through B, and it is different from D_{BA} (diffusivity of B through A). For gases, $D_{AB} = D_{BA}$. A simple method to determine D_{AB} when A diffuses through B is shown in Figure 5. Two large containers, 1 and 2, are connected by a long capillary tube with diameter $2r$ and length Y. A dilute solution (of A in B) in container 2 has an initial concentration C_{A2_0} and is slightly more concentrated than C_{A1_0} in

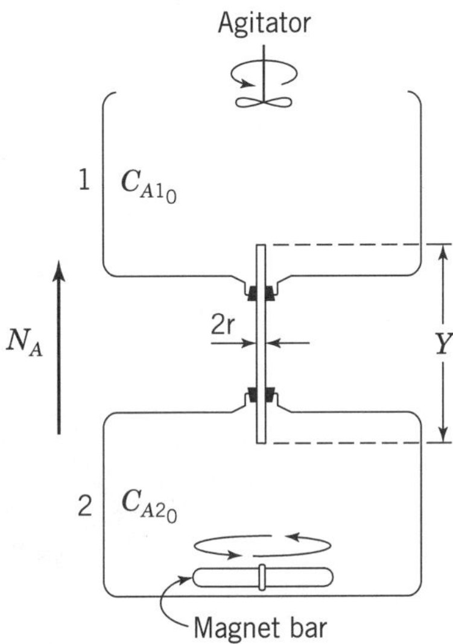

Figure 5. Arrangement for determination of diffusivity in liquids.

container 1. Both ends of the tube are gently stirred to gain uniform concentration. Diffusion takes place in the capillary tube and after a certain time, quasi-steady state is established. The total mass diffused can be calculated as

$$N_A = \frac{1}{\pi r^2}\frac{\Delta C_A}{\Delta t} = \frac{1}{\pi r^2}\frac{\bar{C}_A(X_{A2} - X_{A1})}{\Delta t} \tag{65}$$

where $\Delta C_A = \bar{C}_A(X_{A2} - X_{A1})$ is the concentration increase of the diffusing substance A in the upper container 1 during time Δt. Substituting equation 65 into equation 25 gives

$$\frac{1}{\pi r^2}\frac{(X_{A2} - X_{A1})}{\Delta t} = \frac{D_{AB}}{Y}\ln\left(\frac{1 - X_{A2}}{1 - X_{A1}}\right) \tag{66}$$

Solving for D_{AB} and accounting for the logarithmic mean of the molecular fractions gives

$$D_{AB} = \frac{Y(X_{A2} - X_{A1})}{\pi r^2 \Delta t}\ln\left(\frac{1 - X_{A1}}{1 - X_{A2}}\right) = -\frac{YX_{m\ln}}{\pi r^2 \Delta t} \tag{67}$$

Thus the diffusivity D_{AB} depends on the diffusion tube constant $Y/\pi r^2$ and the logarithmic mean $X_{m\ln}$ of the molecular fraction in solution.

Diffusivity in Solids

When a solid slab with some initial moisture content is brought into an airstream, the moisture inside the solid slab will diffuse through the bulk of the solid and evaporate from its surfaces. Assuming that the slab has a thickness much smaller than its length and width, the one-dimensional Fick's second law (equation 46) can be applied. If the system is maintained at equilibrium and both surface moisture contents are equal, the initial and boundary conditions for the system can be written as

$$\begin{aligned} &\text{at } t = 0,\ C = C_0, \text{ for } 0 < y < Y \\ &\text{at } y = 0,\ C = C_e, \text{ for } t > 0 \\ &\text{at } y = Y,\ C = C_e, \text{ for } t > 0 \end{aligned} \tag{68}$$

Equation 46 can be solved for the conditions of equation 68 (4) to give

$$\frac{C - C_e}{C_0 - C_e} = \frac{8}{\pi^2} \sum_{n=0}^{\infty} \frac{1}{(2n + 1)^2} \exp\left[-(2n + 1)^2 \pi^2 \frac{D_{\text{eff}} t}{Y^2} \right] \tag{69}$$

For approximation purposes, equation 69 can be simplified by keeping only the first term ($n = 0$) on the right-hand side. Taking the logarithm of both sides will give

$$\ln\left(\frac{C - C_e}{C_0 - C_e}\right) = \ln\left(\frac{8}{\pi^2}\right) - \frac{\pi^2}{Y^2} D_{\text{eff}} t \tag{70}$$

where C_0 is the initial concentration, C_e is the surface concentration at equilibrium, C is the average concentration in the solid at time t, Y is the thickness of the slab, and D_{eff} is the effective diffusivity of A in the solid. A plot of $\ln(C - C_e/C_0 - C_e)$ versus t will produce a line with $\lambda = -(\pi^2/Y^2)D_{\text{eff}}$. Thus

$$D_{\text{eff}} = -\frac{\lambda Y^2}{\pi^2} \tag{71}$$

For structure-dependent diffusion, it should also be recalled that $D_{\text{eff}} = f(D_{AB})$. Therefore, the diffusivity of a substance in solids is a function of concentration, structural parameters, and other system characteristics.

Diffusivity in Nonideal Systems

An experimental apparatus for measuring the diffusivity in nonideal systems is shown in Figure 6. Two chambers of equal volume, 1 and 2, are separated by a porous membrane and contain a true solution of a food system, whose concentration is slightly higher in 1. Both bulk concentrations are maintained uniform by agitation (rotating magnet), and diffusion takes place only through the membrane. Under quasi-steady-state conditions, the amount of A diffusing through the membrane in a small time interval can be found by taking the mass balance in chambers 1 and 2 separately. For chamber 1:

$$V dC_{A1} + \frac{\epsilon D_A}{Y}(C_{A1} - C_{A2})dt = 0 \tag{72}$$

For chamber 2:

$$V dC_{A2} + \frac{\epsilon D_A}{Y}(C_{A2} - C_{A1})dt = 0 \tag{73}$$

Combining equations 72 and 73, rearranging, and integrating gives

$$\ln\left(\frac{(C_{A1} - C_{A2})_t}{(C_{A1} - C_{A2})_0}\right) = -\frac{2\epsilon D_A}{YV} t = -\beta D_A t \tag{74}$$

where V is the volume of each chamber; $(C_{A1} - C_{A2})_0$ and $(C_{A1} - C_{A2})_t$ are the concentration differences between the two chambers at the initial and subsequent times, respectively; t is time; Y is the thickness of the membrane; ϵ is the effective porous fraction of the membrane; and β is a cell constant ($\beta = 2\epsilon/YV$) that can be determined by experimenting with a substance of known diffusivity.

The diffusivity of a nonideal system D_A is different from that of an ideal system D_{AB}. The relationship between D_A and D_{AB} is unpredictable and can be determined only by experiment. Furthermore, when binding reactions are present, the diffusivity (denoted by D_b) is different from D_A and D_{AB}. A semitheoretical equation has been presented (8) to estimate the diffusivity D_A of a globular-type protein in solution without binding action between the protein and the solute A as

$$D_A = D_{AB}(1 - 1.81 \times 10^{-3} C_p) \tag{75}$$

where C_p is the protein concentration in kg of protein/m^3. An equation that relates D_A, D_{AB}, and D_b during diffusion of sodium caprylate in bovine serum albumin solution was reported (9) as

$$D_A = D_{AB}(1 - \alpha\varphi_p)\left(1 - \frac{C_b}{C}\right) + D_b\left(\frac{C_b}{C}\right) \tag{76}$$

where α is the so-called diffusivity reduction shape factor for the protein, φ_p is the volume fraction of protein in solution, C_b is the concentration of protein-bound solute A in g/m^3, and C is the total concentration of solute A in g/m^3.

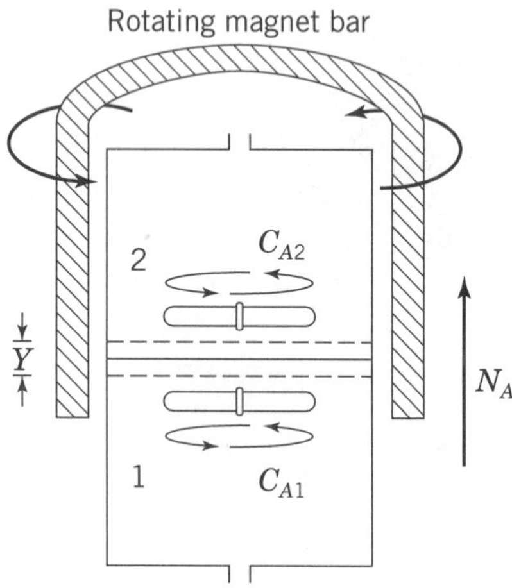

Figure 6. Diagram of a diffusion cell for determination of diffusivity in nonideal systems.

Overall Mass Transfer Coefficient

In a two-phase system, the mass transfer rate can be written in terms of the overall mass transfer coefficient and the bulk concentrations or partial pressures of both phases

$$N_A = K_g(P_A - P_A^*) \tag{77}$$

and

$$N_A = K_l(C_A^* - C_A) \tag{78}$$

where K_g and K_l are the overall mass transfer coefficients for the gas and liquid phases, respectively; P_A^* is the partial pressure of A at equilibrium with the bulk concentration of the liquid; and C_A^* is the concentration of A at equilibrium with the bulk partial pressure of the gas phase, as shown in Figure 3. For dilute solutions, Henry's law

$$P = HC \tag{79}$$

can be employed, where H is Henry's constant. It is also possible to write

$$(P_A - P_A^*) = P_A - P_{Ai} + (P_{Ai} - P_A^*) \tag{80}$$

Using equations 79 and 80, equation 77 becomes

$$N_A = K_g(P_A - P_{Ai}) + K_gH(C_{Ai} - C_A) \tag{81}$$

If these terms are rearranged, the following is obtained

$$\frac{1}{K_g} = \frac{P_A - P_{Ai}}{N_A} + \frac{H(C_{Ai} - C_A)}{N_A} \tag{82}$$

By combining equations 52, 53, and 82, the final expression is obtained

$$\frac{1}{K_g} = \frac{1}{k_g} + \frac{H}{k_l} \tag{83}$$

Hence, the overall mass transfer coefficient can be evaluated from the film coefficients. In a similar manner, an expression for K_l can be derived

$$\frac{1}{K_l} = \frac{1}{k_l} + \frac{1}{Hk_g} \tag{84}$$

Using equation 78 and noting that $P_A = H_{CA}^*$ and $P_A^* = H_{CA}$ gives

$$N_A = K_gH(C_A^* - C_A) \tag{85}$$

Comparison of equation 78 with equation 85 gives

$$K_l = K_gH \tag{86}$$

so that the overall mass transfer coefficient of the liquid phase can be related to that of the gas phase by Henry's constant.

APPLICATIONS OF DIFFUSION AND MASS TRANSFER IN FOOD PROCESSING

Absorption, distillation, evaporation, concentration, extraction, leaching, dehydration, and food packaging are some food-processing operations that involve diffusion or mass transfer. Among these, concentration deals with the removal of water from liquid foods and is normally dominated by heat transfer. Food packaging involves the diffusion of gases (O_2, CO_2, C_2H_4, etc.), water vapor, and volatile aroma components through one or more layers of polymeric films. The complex physical structure and chemical composition of foods, which may vary even within the same food and may change during processing or storage (10), makes the direct application of mass transfer and diffusion principles to food processing difficult. In this section some applications of mass transfer and diffusion principles are presented.

Diffusion in Gases

There are few cases of direct application of this principle to food processing, because convection effects are usually dominant. Corrections to the diffusivity are often necessary to account for such effects. The bulk flow of gases (convection) is caused by differences in total pressure and temperature created during processing. In many cases, mechanical equipment such as agitators, fans, or blowers are used to increase convection and, therefore, the transfer rate. Typical examples of diffusion in gases are (*1*) the evaporation of liquids from containers (e.g., benzene during cleaning of glass containers or hydrogen peroxide during sterilization of plastic containers for use in aseptic processing and packaging); (*2*) the escape of volatile-aroma food components from open storage tanks, mixing equipment, or concentration vessels; and (*3*) the loss of carbon dioxide from carbonated beverages.

Diffusion in Liquids

The phenomenon of diffusion in liquids is common in food processing, and its applications are extensive. A typical example is the extraction of sucrose from sugarcane or sugar beets (11), whereas other examples include the extraction of caffeine from coffee beans and edible oil from soybeans. Distillation, another application of diffusion in liquids, is used to remove or recover volatile components from foods (12), to produce ethanol and alcoholic beverages from fermentation liquids, and to purify solvents (10).

Diffusion in Solids and Nonideal Systems

The principle of diffusion in solids is used widely in food processing. Moisture diffusing through the porous skeleton of low-moisture foods during dehydration, gases or vapors diffusing through polymeric packaging films, small solutes diffusing through biomembranes during chemical peeling of fruits and vegetables (13), and oil leaching through the capillary pores of soybeans are some typical examples. The mechanism of diffusion in solids is more complicated than that in gases and liquids because solids usually have a heterogeneous structure. The concept of effective diffusivity is often employed to simplify the calculations. In some cases, interactions and binding actions of macromolecules affect the diffusion and movement of small molecules through food and other biological systems. Some examples include the diffusion of glucose in an agar gel and sucrose in gelatin (1), sorbic acid in agar (14), sodium chloride in meat muscle (15), and nitrite in beef (16).

NOMENCLATURE

A	Area
C	Concentration
C^*	Concentration at equilibrium with the bulk partial pressure
$\bar{C}$	Average concentration
D	Diffusivity
H	Henry's constant
j	Diffusion rate
k	Film coefficient
K	Overall mass transfer coefficient
l	Length of a solid cylinder
L	Time lag for unsteady-state mass transfer
M	Molecular weight
n	Mole number
N	Mass flux per unit area
$\dot{N}$	Total mass flux
P	Pressure
P^*	Pressure at equilibrium with the bulk concentration
Q	Total amount of substance being diffused in time t
r	Radius
$\bar{r}$	Mean capillary radius
R	Gas constant
S	Solubility
t	Time
T	Temperature
v	Velocity
V	Volume
X	Molecular fraction
y	Small distance of transfer
Y	Total distance of transfer

Subscripts

0	Properties at initial time
1,2	Properties at positions 1 and 2, respectively
A	Properties of matter A
AB	Properties of A when transferred through B
Ai	Properties of A at the boundary layer
b	Properties of bulk, or when binding effect is present
B	Properties of matter B
BA	Properties of B when transferred through A
d	Properties due to diffusion
e	Properties at equilibrium
g	Properties in the gas phase
i	Properties at the interface
l	Properties in the liquid phase
m	Arithmetic mean value
$m\ln$	Logarithmic mean value
p	Properties of protein
t	Properties at time t

Greek

α	Diffusivity reduction factor
β	Diffusion cell constant
δ	Thickness
ϵ	Porous fraction of solid or membrane
κ	Knudsen's constant
Λ	Molecular mean free path
λ	Slope of equation (70)
μ	First-order chemical reaction constant
ρ	Density
τ	Tortuosity of solid
φ	Volume fraction

BIBLIOGRAPHY

1. C. J. Geankoplis, *Transport Processes and Unit Operations*, 2nd ed., Allyn & Bacon, Boston, Mass., 1983.
2. R. M. Barrer, *Diffusion in and Through Solids*, Cambridge University Press, London, United Kingdom, 1951.
3. W. Just, *Diffusion in Solids, Liquids, Gases*, Academic Press, Orlando, Fla., 1960.
4. J. Crank, *The Mathematics of Diffusion*, 2nd ed., Clarendon Press, Oxford, United Kingdom, 1975.
5. J. D. Floros and M. S. Chinnan, "Determining the Diffusivity of Sodium Hydroxide through Tomato and Capsicum Skins," *J. Food Eng.* **9**, 129–141 (1989).
6. H. S. Mickley, T. K. Sherwood, and C. E. Reed, *Applied Mathematics in Chemical Engineering*, 2nd ed., McGraw-Hill, New York, 1957.
7. J. Crank and G. S. Park, *Diffusion in Polymers*, Academic Press, Orlando, Fla., 1968.
8. C. K. Colton et al., "Diffusion of Organic Solutes in Stagnant Plasma and Red Cell Suspensions," *AIChE Symposium Series* **66**, 85–100 (1970).
9. C. J. Geankoplis, E. A. Grulke, and M. R. Okos, "Diffusion and Interaction of Sodium Caprylate in Bovine Serum Albumin Solutions," *Industrial and Engineering Chemistry, Fundamentals* **18**, 233–237 (1979).
10. G. D. Saravacos, "Mass Transfer Properties of Foods," in M. A. Rao and S. S. H. Rizvi, eds., *Engineering Properties of Foods*, Marcel Dekker, New York, 1986, pp. 89–132.
11. D. R. Heldman and R. P. Singh, *Food Process Engineering*, 2nd ed., AVI, Westport, Conn., 1981.
12. J. L. Bomben et al., "Aroma Recovery and Retention in Concentration and Drying of Foods," in Chichester et al., eds., *Advances in Food Research*, Vol. 20, Academic Press, Orlando, Fla., 1973, pp. 2–111.
13. H. G. Schwartzberg and R. Y. Chao, "Solute Diffusivities in the Leaching Processes," *Food Technol.* **36**, 73–86 (1982).
14. I. J. Pfiug et al., "Diffusion of Salt in the Desalting of Pickles," *Food Technol.* **21**, 1634–1638 (1975).

15. G. Dussap and J. B. Gros, "Diffusion-Sorption Model for the Penetration of Salt in Pork and Beef Muscle," in P. Linko et al., eds., *Food Process Engineering*, Elsevier Applied Science, Barking, United Kingdom, 1980.
16. J. B. Fox, "Diffusion of Chloride, Nitrite and Nitrate in Beef and Pork," *J. Food Science* **45**, 1740–1744 (1980).

GENERAL REFERENCES

J. D. Babbit, "On the Differential Equations of Diffusion," *Can. J. Res.* **28**, 449–474 (1950).

A. Leniger and W. A. Beverloo, *Food Process Engineering*, D. Reidel, The Netherlands, 1975.

J. Lewis, *Physical Properties of Foods and Food Processing Systems*, Ellis Horwood, Chichester, United Kingdom, 1987.

Loncin and R. L. Merson, *Food Engineering: Principles and Selected Applications*, Academic Press, Orlando, Fla., 1979.

A. Rao and S. S. H. Rizvi, *Engineering Properties of Foods*, Marcel Dekker, New York, 1986.

T. Watson and J. C. Harper, *Elements of Food Engineering*, Van Nostrand Reinhold, New York, 1988.

John D. Floros
Jeff Rattray
Hanhua Liang
Purdue University
West Lafayette, Indiana

MEAT AND ELECTRICAL STIMULATION

Electrical stimulation of muscle from slaughtered animals hastens the process of rigor mortis. It does this through an initial pH fall that is then followed by a change in the rate of pH fall. The combined effect is that the muscles enter rigor mortis before the temperature falls sufficiently to result in *cold shortening* and toughening as a result of rapid chilling that may occur in present-day commercial situations. Electrical stimulation can be applied early postslaughter, in which cases relatively low voltages are effective as they operate via the nervous system. However, as the delay increases before stimulation, high voltages are then required. The applied electrical parameters generally used must consider the appropriate waveform and pulse frequency, duration, the prestimulation delay, the chilling rate, and the type of species involved. While electrical stimulation ensures that cold shortening is avoided, aging also starts at a high temperatures and is consequently more rapid. However, evidence suggests that other mechanism for tenderization are also taking place. Electrical stimulation must be considered as part of a total process from slaughter through chilling to final sale, considering electrical inputs and rates of pH fall and temperature fall to optimize tenderness and juiciness. Electrical stimulation has particular advantages for hot boning where the shortening and hence toughening conditions for unstimulated muscles are normally exacerbated by the improved chilling. When the stimulation conditions and chilling rates are tuned, hot boned meat is as tender as normal cold boned meat, and there can be significant improvements in other meat quality attributes such as drip and color.

HISTORY

The association between meat and electricity dates back to some of the earliest muscle physiology experiments (1). From Galvani's time, the use of electricity to study muscle function increased. The earliest reported use of electricity for meat improvement is its purported use by Benjamin Franklin in 1749 to electrocute turkeys with the result that they were "uncommonly tender" (2). In the 1950s the use of electrical stimulation as a means of improving tenderness of meat was investigated, but no commercial application of the process occurred then (3,4).

Electrical stimulation has been extensively used since the 1950s to hasten rigor mortis and to modify steps of the glycolytic pathway. Stimulation of horse muscle was used to facilitate a trial of microbial growth on prerigor and rigor muscles from the same animal (5). Others since have used it to modify and study steps of the glycolytic pathway (6). Landrace pigs were used to demonstrate that electrical stimulation accelerated the postmortem pH fall in normally slow glycolyzing muscles but produced no change in faster glycolyzing muscles (7). The principle was also used to predict the time course of rigor mortis (8).

The revived interest in electrical stimulation as an element of meat processing stems from the facts that it hastens development of rigor mortis (9–11), increases the rate of tenderization by postmortem aging (12,13), improves meat color (14), and provides an increased stiffness to muscles that can be of value for early boning.

DESCRIPTION

Electrical stimulation is a process that involves passing an electric current through the body or carcass of freshly slaughtered animals. This current causes the muscles to contract, increasing the utilization of muscle energy reserves, and leads to acceleration of glycolysis and subsequent rigor development. Electrical stimulation as a means of attempting to increase tenderness has been used experimentally on pigs, deer, goats, sheep, cattle, and various poultry species and, until recently, used commercially on all of these except pigs. This has changed as it has been shown that if stimulation is applied at 20 min postslaughter and the chilling rate is sufficiently fast (eg, deep leg to 10°C in 5–10 h), then the meat in those animals susceptible to pale, soft, and exudative (PSE) does not show it, and the remaining meat avoids cold shortening (15).

There are numerous physical methods by which electrical stimulation could be applied many different possible electrical specifications, and in reality many different perceptions of the response. Regardless of species, stimulation can be applied immediately after slaughter or at any point in time thereafter until the muscles become unresponsive. The time until muscles fail to respond is related to the natural rate of glycolysis and the voltage being applied, the duration of stimulation, and the type of response expected.

Most commercially used electrical stimulation systems employ the conveying rail as ground, and a live electrode contacts some other point of the body, carcass, or side. In the most basic systems, the live electrode contact is a clip manually applied to the head end of the animal body that is suspended by one or both hind legs, resulting in a current flow to the grounded rail support. More sophistication and protection is required as voltages increase and as application of the electrode becomes automated.

Procedures range from stimulating stunned but not bled animals, whole bodies, skinned bodies, carcasses, or sides. The electrical characteristics of the waveforms used are often poorly described, defying anyone to reproduce the results. Voltages used vary from 32 to 3,600 V and may be described in various ways. The value specified might be that of the peak or the rms (root mean square) voltage or in some cases the average over the total time. The rms voltage is the effective value or heating capacity of a waveform. For a sine wave, the rms value is the peak voltage divided by √2 (ie, 1130 V peak 50 Hz is 800 V rms), but for derived waveforms it may be quite different. For the MIRINZ waveform, which uses every seventh half-sine wave of a 50 Hz sine wave, the rms voltage is the peak voltage divided by √14.

Figure 1 sets out to illustrate the meaning of the different terms used to describe voltages and waveforms. Defining a waveform with a frequency (Hz) is likely to lead to confusion unless the waveform is defined in terms of shape, duration, and pulse spacing. Figure 2 shows some different waveforms. The applied waveforms may be unipolar or bipolar and applied as discrete pulses or as pulse trains. Again, their description has often defied interpretation. Some of the waveforms and pulse shapes used have been well researched and described, whereas others seem to have been determined merely by extending the use of available equipment, for example, electric stunning tongs.

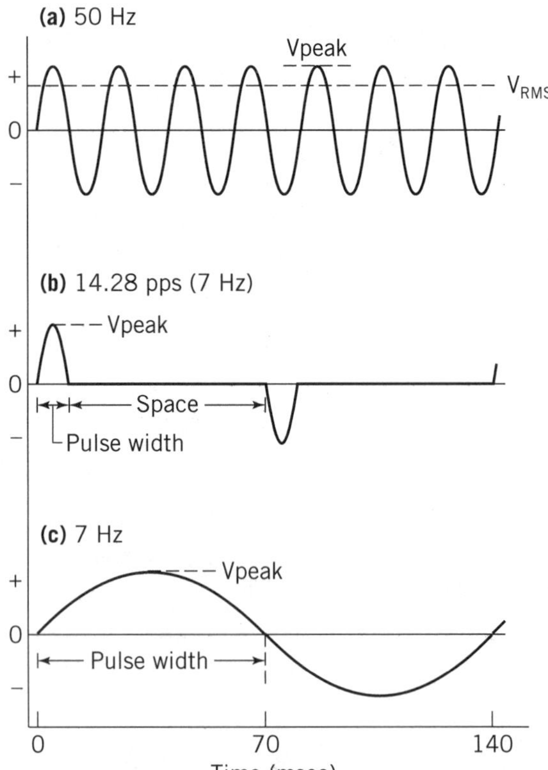

Figure 1. Terminology used to describe pulses and waveforms illustrated by sinusoidal (**a**) and (**c**) and derived waveforms (**b**). It is important to give the peak height in terms of either current or voltage while a carcass is being stimulated. The pulse shape must be specified either by description of the shape (eg, sinusoidal or square) or by specification of rise and fall times. The pulse width (mark) and space between pulses are also needed. The mark-to-space ratio helps specify a single cycle, which is the period from start to finish of the repetitive unit. The polarity of pulses is also necessary. The number of cycles per second then completes the description. The waveforms (**b**) and (**c**) both have the same period (inverse of wavelength) and peak but have different shape characteristics. The mark-to-space ratio for (**b**) is 1:7, whereas for (**c**) is 1:0. The time scale is given as an example only. The same-shaped waveforms could be given with different time scales.

Safety has been of utmost importance during experimentation and implementation of electrical stimulation in New Zealand, Australia, the UK, and France, to the point that in some instances safety concerns have effectively prevented commercial adoption of the process. Although not considered of overriding importance in some countries, safety of personnel can always be assured if the normal electrical safeguards are applied.

In New Zealand, practical on-line sheep operations stimulating up to 56 carcasses/min are a common operation. Less than 30 min after slaughter, dressed carcasses are suspended by metal skids and gambrels from a grounded rail and moved through a stimulation tunnel to make contact, at shoulder level, with an electrode supplied with high voltage pulses (16,17).

Similar systems can be applied for stimulation of beef bodies and sides of carcasses, and a novel system has been developed whereby sides contact a moving chain electrode curtain that maintains contact as the side curls (18) (Fig. 3). A range of other systems has also been developed for beef stimulation (19) covering both batch and continuous operations. The batch systems may involve manually inserted electrodes or electrode bars that move out to make contact with the body or carcass. In these systems the carcass or carcasses are enclosed within a shielded cabinet during stimulation. Continuous systems consist of stationary rubbing electrodes, or, where the stimulation is applied to carcasses prior to inspection, the electrode system consists of a moving series of electrodes that are sterilized between carcasses. In most instances electrical stimulation is applied with the aim of ultimately improving tenderness, but in the United States considerable attention has been devoted to color changes that improve grading of carcasses (20).

High or low voltages can be used for stimulation. Low voltages (less than 80 V peak) are commonly used within 5 min of slaughter in small operations where electrodes are manually applied and removed. There is often an overlap in use of electrical currents to immobilize carcasses and the use of electrical currents to accelerate glycolysis—the name is different but the process is the same. Although, in general, plants stimulate with an electrode on the nose or stick wound and ground via the rail from which the carcass is suspended, the resistance of the hind leg can be very high, due to the high bone and tendon but low muscle content of the lower narrow portion of the leg. This high

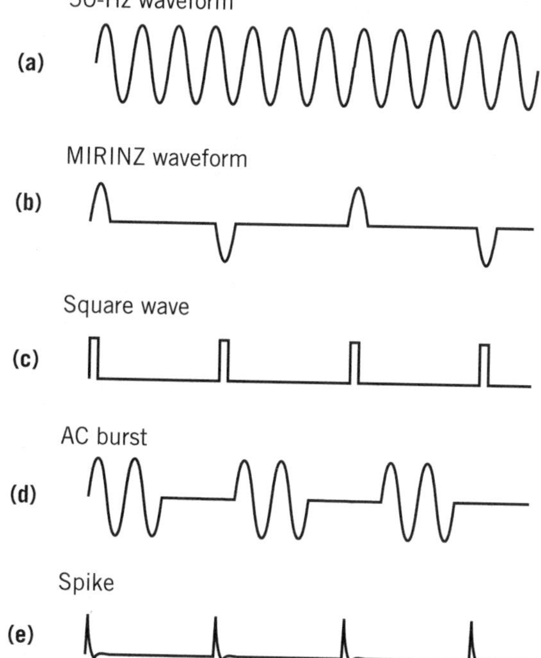

Figure 2. Common electrical waveforms that have been used for carcass electrical stimulation. All are shown at the same peak amplitude. (In the United States, a 60-Hz waveform is the basic supply.) (**a**) 50-Hz alternating current waveform (100 pulses/s). (**b**) MIRINZ waveform, derived by selecting every seventh pulse to give 14.28 pulses/s. (**c**) Square wave pulse train (14.28 pulses of 5-ms duration per second). (**d**) Bursts of the 50-Hz waveform. Common on-off times are 2 s on; 2 s off. (**e**). Spike pulses (commonly with base widths of 0.2–0.3 ms).

resistance can severely limit current flow. To solve this problem, there has been considerable effort to develop alternative procedures. The Australians (21) have used an anal electrode to provide good contact near the hind quarter and using this system have achieved effective low-voltage stimulation.

High voltages (greater than 200 V peak) are usually used for automated and high throughput systems to obtain maximum effect in the shortest period of time and to overcome variability of stimulation response that exists with low-voltage systems (22). Because high voltages are potentially lethal, considerable attention is paid to safety when they are being used.

MEAT TENDERNESS-TOUGHNESS

Meat tenderness has been regarded by many (23) as the prime determinant of consumer satisfaction with meat purchases. Although the ultimate measure of tenderness (or its converse, toughness) lies with the consumer, objective assessments can be made with a wide variety of mechanical devices, because these devices can reliably indicate differences attributable to animal and processing factors. Tenderness is the term used to describe the assessment by consumers (high scores being desirable), and the force to shear the meat, that is, shear force, gives an indication of the toughness (low values being desirable). The factors influencing meat quality, in particular tenderness, are covered the article MEAT SCIENCE.

Meat tenderness-toughness depends on both the myofibrillar strength and the connective tissue strength. Electrical stimulation seems mainly to modify the myofibrillar strength, although some work (24) suggests that electrical stimulation could also have an impact on the connective tissue component of meat tenderness. Most studies of electrical stimulation have been concerned with myofibrillar toughness as affected by cold shortening as muscle goes into rigor (termed conditioning). The basic tenets of much of meat science come from work that relates the degree of muscle shortening during the early postmortem period to the tenderness of cooked meat (25). The force required to shear a sample of cooked muscle across the grain rises steeply as the degree of muscle shortening exceeds 20%, reaching a peak at 40% shortening (see MEAT SCIENCE). The shear force of a sample shortened by 40% may be up to four times that of the nonshortened muscles.

If rigor (ie, conditioned) muscle is subsequently aged before cooking, it usually becomes appreciably more tender; however, the degree of tenderization is drastically reduced if the muscle has shortened during the conditioning (prerigor) phase (see MEAT SCIENCE). Even with extensive aging, there seems to be a limit beyond which any given muscle cannot be tenderized without loss of all structure. This background toughness is largely a result of the structure and composition of the connective tissues within the muscle. Aging reduces the myofibrillar strength (26,27) but is generally considered to have little effect on connective tissue, although some work questions this view (24).

Meat aging refers primarily to the tenderness improvement that takes place while postmortem muscle is held above freezing temperatures. Although there is no clear evidence as to when aging commences, much past evidence suggests that most of the aging commences with completion of rigor mortis (28). Other work (29) suggests that aging changes may commence prior to rigor completion, but they are relatively small. The changes in intramuscular connective tissue that have been noted (24) can be considered as aging changes.

The ultrastructural changes that occur during aging consist of a disappearance of Z disks and a disruption of the myofibrils at the junction of the I bands and Z disks. The ultrastructural changes are accompanied by parallel biochemical changes revealed by changes in the extractability of myofibrillar protein (30), changes to myofibrillar ATPase activity, and breakdown of myofibrillar proteins, for example, loss of troponin T (31,32). Proteinases termed calpains, with the inhibitor calpastatin, are considered to have a role (33). The ability of electrical stimulation to influence aging is disputed. However, it has been recently shown that rigor temperature affects tenderization with the optimum being 15°C (34), and this needs to be considered for optimum tenderization (Fig. 4) (36). Low level exercise affects nitric oxide (NO) in muscle (37), and recent work has shown that NO affects meat tenderizing, suggesting that electrical stimulation can have a similar role (38). Further work is still required to establish the effect of electrical stimulation on aging. The accelerated devel-

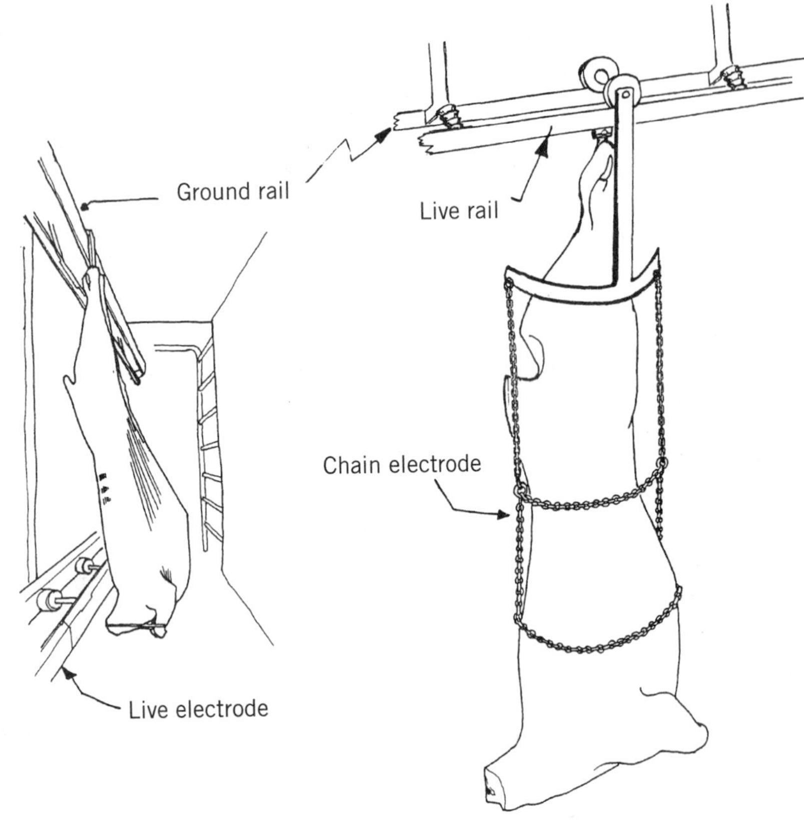

Figure 3. Diagrams showing electrode configuration used for high-voltage stimulation of sheep carcasses (only one is shown) and a novel chain electrode arrangement for beef sides. Stimulation area is enclosed in both situations and personnel entry controlled by floor switches and light beams. *Source:* Drawings by P. Hanara.

opment of rigor mortis means that even if aging does not start until after rigor is achieved, aging will commence at higher temperatures and, therefore, be more rapid (39).

SCIENTIFIC BASIS

When muscles of freshly slaughtered animals are electrically stimulated, they contract. There is a concomitant increase in biochemical reactions in the muscle cells leading to an accumulation of lactate and resulting in an immediate drop in the muscle pH (ΔpH). After stimulation, the rate of pH fall (dpH/dt) as the muscles go into rigor is increased (Fig. 5) (40). If a muscle is free, it will contract during stimulation. When stimulation ceases, the muscle will partially relax but unless the muscle is subjected to some force, either restrained or tensioned, it will remain in a shortened state. Stimulation of suspended bodies or sides with a balanced arrangement of opposing muscle systems ensures that there is sufficient physical force to pull most muscles back to rest length. If muscles have been freed from their natural attachments soon after stimulation, however, as occurs with hot boning, there may be no restraining force to return stimulated muscles to their natural length. The same situation applies to stimulation of isolated muscles. Electrical stimulation of muscle influences the binding of many enzymes associated with the glycolytic pathway, and this is exhibited in the accelerated rate of pH fall and of ATP disappearance and the extremely rapid loss of creatine phosphate (41) after stimulation.

Although there are various theories as to the mechanism of how electrical stimulation promotes meat tenderness, most meat scientists tend to favor the view that stimulation accelerates glycolysis so that the time when muscles will contract on exposure to cold is reduced, thereby avoiding cold shortening.

There is also strong support for stimulation increasing meat tenderness by accelerating postmortem aging, because rigor is being achieved while temperatures are still high. It has been claimed (42) that much of the effect of electrical stimulation, under conditions where cold shortening is unlikely, results from mechanical damage and is not due to rapid aging at high temperatures. Interestingly, it has been found (43) that supercontracture bands occur in both stimulated and nonstimulated muscles.

Cold shortening occurs when muscles of freshly slaughtered beef and sheep are subjected to temperatures below 8°C while there is sufficient energy supply for contraction to occur. Cold shortening will occur in pigs and poultry, but the critical temperatures are lower. In normal beef and sheep muscles, this equates to a tissue pH of about 6.0. A rule of thumb in prevention of cold shortening is to maintain temperature above 10°C until muscle pH falls below 6.0 (44). For a discussion on cold shortening, see the article MEAT SCIENCE. Electrical stimulation, therefore, can have an immediate benefit under commercial conditions by causing the pH to fall and thereby shortening the time before carcasses can be chilled without causing toughness. Although many meat processors use electrical stimulation to reduce the time when muscles are susceptible to cold

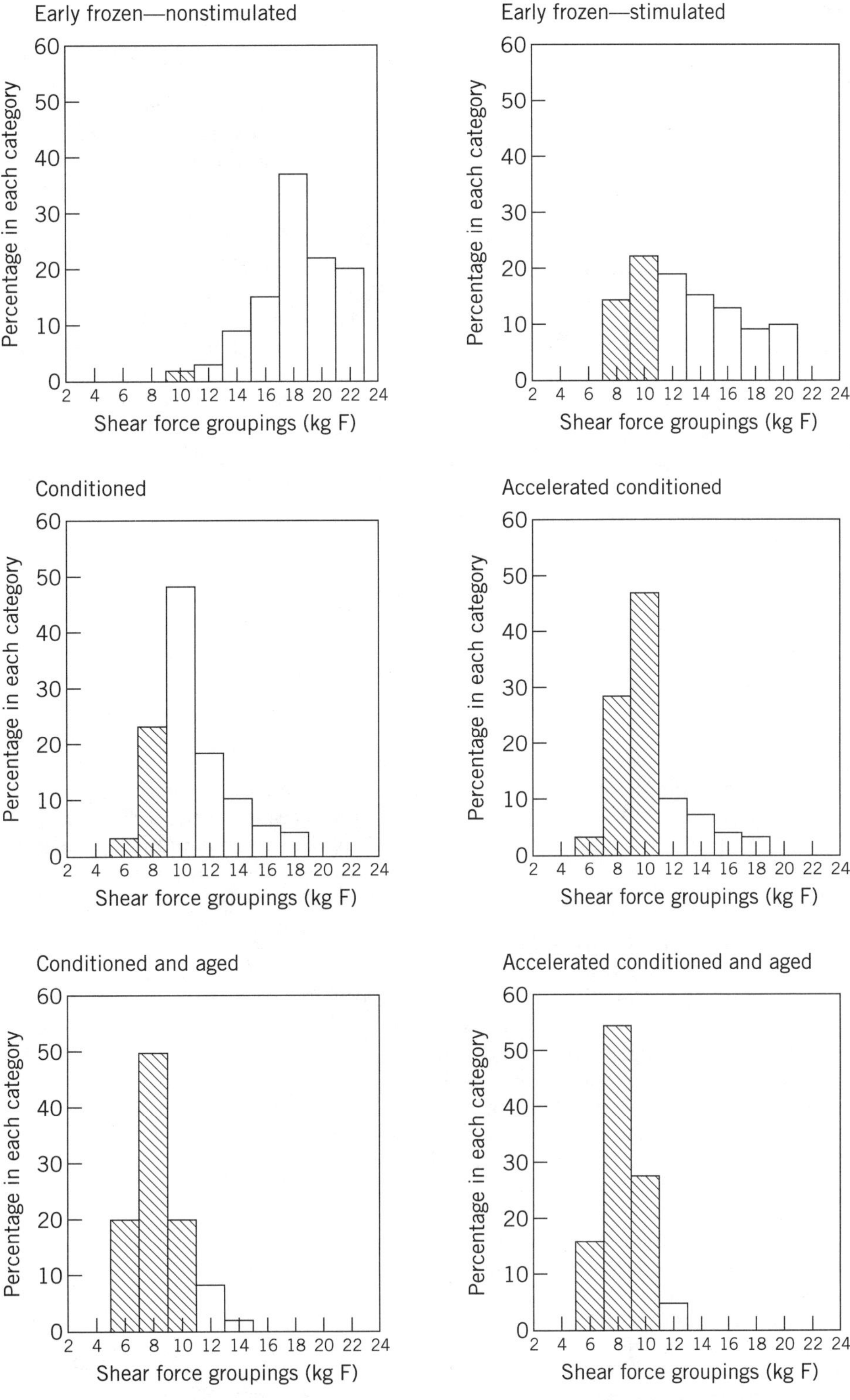

Figure 4. The advantage of stimulation can be seen in the following series of commercial situations. Distribution of force score values (kg F) from a MIRINZ pneumatic tenderometer with a wedge-shaped tooth, for loins from nonstimulated and electrically stimulated lamb carcasses. The early frozen product entered the freezer 40 min after slaughter, the conditioned carcasses were held at 10°C for 24 h before freezing, and the conditioned and aged product was held for 48 h at 10°C before freezing. The accelerated conditioned product was electrically stimulated at 30 min postmortem, then frozen 90 min later; the accelerated conditioned and aged product was electrically stimulated and held for 8 h at 7°C before freezing. All carcasses were frozen to a deep-leg temperature of −4°C over a 12-h period, then reduced to −12°C (66). *Source:* Ref. 35, used with permission.

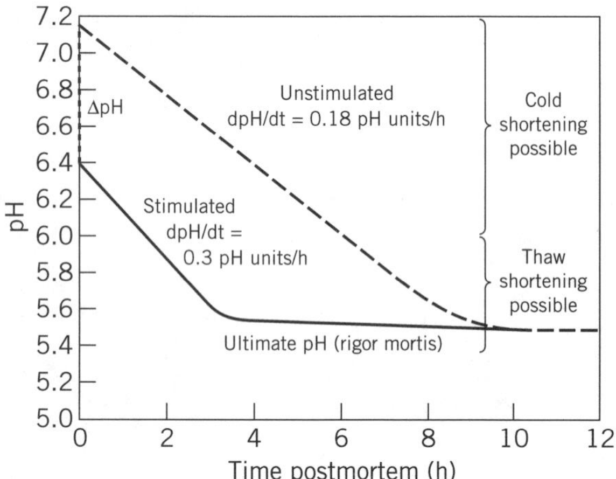

Figure 5. Postmortem pH fall in muscles after electrical stimulation. During stimulation the muscle pH falls (ΔpH). After stimulation the rate of pH fall (dpH/dt) is increased. At 35°C, unstimulated beef muscle dpH/dt = 0.18 pH units/h. With 120-s stimulation duration, the dpH/dt = 0.3 pH units/h.

shortening and get muscles into rigor so that aging commences while muscles are still warm, others have used electrical stimulation to produce color changes that can influence grading and even have used it for carcass immobilization. Electrical stimulation tends not to be used for pigs because early work (45) suggested that acceleration of rigor mortis through electrical stimulation increased the incidence of pale, soft, and exudative pork. However, no effect was found in the shoulders, but there was an increased paleness in the legs, unless rapid chilling was used (46), it is possible to use electrical stimulation if chilling is rapid enough (15).

Although remote from the actual changes produced by electrical stimulation, muscle pH has been used extensively as an indicator of progress into rigor. Muscle pH falls after stimulation provides a reasonable guide in normal muscles but is misleading in abnormal situations such as in the muscles from stressed animals. This is one case where muscle pH does not give a true indication of the stage of rigor development, because although the pH remains high, the ATP level may be zero and the muscle in rigor. The time to reach pH 6.0, or time to full rigor mortis, will be influenced by the magnitude of the ΔpH and the dpH/dt (Fig. 5) and provides a useful measure with which to evaluate the effectiveness of electrical stimulation systems in inducing early rigor. It does not provide an exact relationship with tenderness changes. Temperature will exert a substantial influence on both ΔpH and the dpH/dt.

FACTORS AFFECTING ΔPH

The magnitude of ΔpH is governed by the muscle fiber type, initial glycogen stores within the muscle, the electrical characteristics (current, frequency, pulse shape, and stimulation duration), the temperature of muscle, and the time after death that stimulation was applied.

Delay Postmortem

Postmortem delay before stimulation can have a threefold effect: muscle temperature can fall, greatly reducing the magnitude of ΔpH; glycolysis can have progressed so that muscle pH has fallen, reducing the ΔpH that can be achieved; and the nervous system can decay and become unresponsive so that its stimulation cannot elicit any muscle response. It has been shown that the ΔpH produced by any given stimulation decreases as prestimulation pH decreases (40,47). The ΔpH is 0 with a prestimulation pH of about 6.3.

The influence of temperature on the rate of pH fall has been considered in many reports (48–50) (Fig. 6). Since the advent of stimulation as a tool for meat science, the temperature dependence of the pH drop during stimulation and of the subsequent pH fall has been determined. When the muscle temperature falls, the magnitude of ΔpH is reduced. For example, in beef *m. sternomandibularis*, ΔpH ranges from 0.6 pH units at 35° to 0.018 units at 15°C. The energy of activation of ΔpH in stimulated beef *m. sternomandibularis* is calculated to be 97 kJ/mol (51), or very similar to that for calcium-activated actomyosin ATPase (44,50,52).

Muscle Type

Muscle type can influence ΔpH, for example, the fast-twitch beef *m. cutaneous trunci*, largely composed of white muscle fibers; and gives high values for ΔpH (and dpH/dt), whereas in the slow-twitch *m. masseter*, composed of red fibers, there is neither a distinct ΔpH nor an acceleration of dpH/dt, which is naturally rapid (0.4 pH units/h) (53). It has also been shown that ΔpH is minimal in predominantly slow-twitch muscles of beef (54). The greatest ΔpH

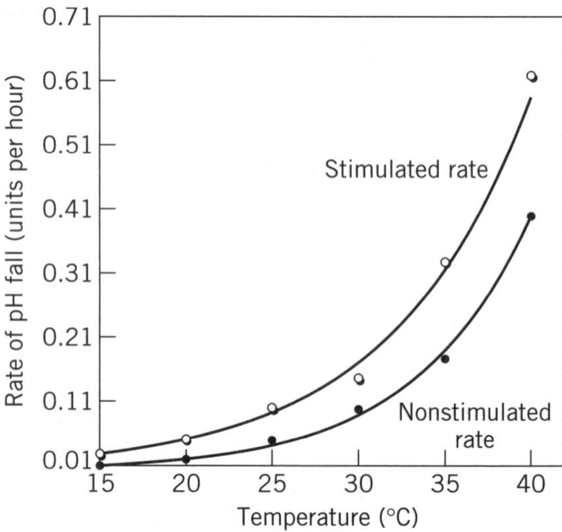

Figure 6. The various postmortem dpH/dt values for stimulated and nonstimulated muscle are plotted to show that the differences exist at all temperatures and illustrates how important it is to stimulate at high temperatures for full effectiveness. These results were obtained from beef *m. sternomandibularis* stimulated and held at the temperatures indicated.

values have been shown in beef *triceps brachii* and the least in the *m. semimembranosus* (55). It was reported (56) that within the *m. sternomandibularis*, fast fibers with strong ATPase and weak succinic dehydrogenase (SDH) reactions showed most glycogenolysis in response to electrical stimulation, while slow fibers with weak ATPase and strong SDH reactions showed the least.

Muscles also differ in their optimum response frequency. Beef *m. longissimus* is more responsive to 14.28 pulses/s than to 40 pulses/s (40). The converse has been shown for the beef *m. semimembranous* (57). Rat muscles have a much shorter twitch time than beef muscles, with optimum responses to stimulation with a waveform of 33 pulses/s (58).

Electrical Characteristics

For any given muscle, the pH response will be governed by the electrical characteristics of voltage (current), and pulse frequency, shape, and polarity. In general, the higher the current (at a constant resistance, current increases with increased voltage), the greater will be the effect. This response will be asymptotic to some maximal value, however, so continually increasing the current will not lead to a continuing increase in effect (40,55). If the current level becomes too high, the muscle can cook due to resistive heating, this is especially true for high-frequency input. A practical consequence of this was the melting, within 45 s of the tendons of the *m. gastrocnemius* when used to suspend lamb carcasses during stimulation with 800 V rms at 50 Hz in some early studies.

For all of the species tested, the pulse shape does not seem to be a critical factor, with equivalent responses being achieved with 10 ms duration half-sine wave pulses and 5 ms duration square-wave pulses at the same frequency (40). The pulse frequency, however, can exert a profound effect, because of the interaction with muscle type and response frequency. If a single muscle is considered, for example, beef *m. sternomandibularis*, the magnitude of ΔpH is maximal with a pulse rate of between 9 and 16/s. At frequencies between 10 and 20 pulses/s, ΔpH values are, respectively, 40 and 75% greater than at 50 and 100 pulses/s (40). Pulse polarity has not received much attention, although it was indicated (35) that for low-voltage stimulation the polarity of pulses influenced the magnitude of ΔpH and the movement of the carcass during stimulation. The greatest responses were obtained when the cranial end of the animal was positive relative to the rest of the body. The physical response to low-voltage alternating polarity waveforms used for stimulation of beef cattle can be movement at half the waveform frequency, which disappears as the voltage is raised.

Many different voltages, frequencies, and pulse shapes have been used, and these are sometimes inaccurately described in scientific publications. In the United States, many groups have pulsed the 60 Hz AC waveform with 1- or 2-s periods of current flow, then a similar period of rest (19,46,59), whereas groups in Australia, the UK, Sweden, and New Zealand have used discrete pulses. An unusual, but apparently effective, combination of high- and low-voltage pulses has been used by the Dutch (60). In many instances the reasons for the chosen frequency do not appear to have a scientific basis.

FACTORS AFFECTING DPH/DT

Almost any stimulation can affect the dpH/dt. Rates of pH fall in curarized muscles are much slower than those in muscles of animals slaughtered without curarization (61). Vigorous muscle movement during slaughter of normal animals can induce the apparent maximal twofold increase in dpH/dt (61). Sometimes there can be unexpected electrical stimulation arising from the stunning (62) and immobilization of carcasses and even effects from current application from downward hide pulling (63).

Attempts have been made to relate increases in dpH/dt to phosphorylase a and myofibrillar ATPase stimulation (47). The muscle temperature has a major effect on dpH/dt (Fig. 6). In nonstimulated beef *m. sternomandibularis*, the energy of activation is 40–45 kJ/mol, whereas that of stimulated muscle approaches 70 kJ/mol (51). This high energy of activation means that any cooling of the muscles will have a marked effect on the time for rigor to be achieved. Because the achievement of ultimate pH is asymptotic, any reduction in maximum rate can greatly prolong the period when muscle is above the ultimate pH and still has a measurable level of ATP. The rate of pH fall and ATP loss is affected not only by the temperature during stimulation, but also by the temperature history of the muscle after stimulation during the period of study (64).

PRACTICAL RESULTS OF STIMULATION

The most dramatic practical result of stimulation is the tenderness difference between nonstimulated and stimulated early frozen lamb carcasses (10). Loins and legs from nonstimulated lamb carcasses put into a blast freezer (−25°C) within 2 h of slaughter and later cooked from the frozen state are exceedingly tough, with only about 1% of shear force determinations on loins being below a level considered to be the maximum for acceptably tender meat. In contrast, loins from similarly processed but high-voltage stimulated lamb carcasses give about 75% of shear force values in the acceptable range (65) (Fig. 4). Lamb carcasses, because of their small size, can be chilled or frozen rapidly and, therefore, the muscles are more liable to cold shortening than are those of larger nonstimulated animals. However, there is a considerable body of evidence that electrical stimulation of beef under conditions where the carcasses are rapidly chilled also results in protection of tenderness (11,66).

Under slower cooling conditions, the differences are much less pronounced, and, in fact, if the evaluation is made after three weeks of aging, nonstimulated product not subjected to rapid chilling may not differ in tenderness from rapidly chilled stimulated product. Under these conditions, stimulation merely facilitates rapid processing. Rapidly chilled, nonstimulated product will, however, be significantly tougher than that from stimulated bodies (Table 1), as cold-shortened muscles have a decreased aging potential. Clearly it is more costly to hold nonstimulated

Table 1. Mean Shear Force Results (kg F) from Rapidly and Slowly Chilled Beef after Three Weeks of Holding Under Vacuum at 1°C

Stimulation	7°C chiller	2°C chiller
High-voltage stimulated	4.58	4.66
Low-voltage stimulated	5.99	8.2
Nonstimulated	8.1	12.2

Note: One side of each animal was chilled in 2°C air at 0.3 m/s, the other in 7°C air at 0.3 m/s. The lower the shear force valuee, the more tender the meat. Values above 11 kg F indicate unacceptable toughness. These results are for a MIRINZ tenderometer and correspond to a Warner Bratzler tenderometer if multiplied by 0.7 (67).

carcasses in a chiller for long durations to age to achieve the same desired tenderness of stimulated carcasses that occurs much earlier; thus, stimulation can be a cost-effective process.

The degree of tenderness improvement that researchers attribute to stimulation varies because of the widely different methods of assessing tenderness and the innumerable processing variations; comparison of results is, therefore, difficult.

Stimulation does not always result in tenderization, but neither does it in itself toughen meat. Studies showed that electrical stimulation of stressed animals caused a slight toughening of the *m. longissimus* in beef and sheep (68,69), although it was hypothesized (68,69) that the toughening was a consequence of muscles going into rigor during stimulation and remaining contracted. It has been suggested (70) that under the conditions present, the rapid reduction of the time that muscle pH is high reduces the aging that occurs, but additionally the lower tenderizing rate of intermediate pH meat may be important (71).

Electrical stimulation is also claimed to improve the color of lean and influence beef grading under the U.S. system (20,67,72). One study found no effect of low-voltage electrical stimulation on meat color as measured by reflectance but did find that postmortem aging significantly reduced reflectance values (73).

Comparisons of low- and high-voltage stimulation under comparable conditions have shown that if environmental conditions are such that cold shortening is unlikely, low- and high-voltage stimulation both result in an increase in tenderness (74), and both processes improve the youthful appearance and marbling scores of beef (75). Under conditions where chilling rates are more rapid, low-voltage stimulation does not give the same degree of protection against cold shortening as high-voltage stimulation (67). The interaction between three moderate chilling conditions and two electrical stimulation treatments were examined (76); it was shown that the temperature regime to which the muscles were exposed became more critical when low-voltage stimulation was used. When all things are considered, the use of electrical stimulation must be seen as part of a total process. The tuning of the amount of stimulation with chilling rate to reach rigor at 15°C results in optimum tenderization (Fig. 7) (36).

Not all muscles of a carcass are equally affected by stimulation. One study (11) showed that although the beef striploin and rump improved in tenderness, there was no change in the topside. Similarly a dramatic improvement in the tenderness of beef *m. longissimus* and *m. semimembranosus* was shown but not of the *m. triceps brachii* and *serrates ventralis* after high- or low-voltage stimulation. Some of the difference has been attributed to uneven distribution of the current during stimulation (59), but obviously some of the difference must be biochemically and physiologically based.

Because of the rapid fall in pH of pork and the tendency for PSE conditions of high temperatures and low pH values to be present, it was considered that there would be no advantage in stimulation. However, it has been shown that if electrical stimulation is applied at 20 min postslaughter and the chilling rate is sufficiently fast under ideal commercial conditions (eg, deep leg to 10°C in 5–10 h), then the meat in those animals susceptible to PSE does not show it and the remaining meat avoids cold shortening (15). There is an improvement in tenderness and drip loss with pelvic suspension (77). Further work has shown that there appears to be no advantages with electrical stimulation and slow chilling (77).

Cold shortening in poultry does not seem to cause the same toughness problems as in other species, as the maximum shortening occurs around 2°C (79), which is generally lower than the temperatures the birds reach following chilling by ice water and air chilling. However, rapid processing would be advantageous because the toughness that arises from portioning the birds early and removing the breast muscles in particular could be avoided. A stimulation system, part of the minimum time process system, was developed and patented by Campbell Institute for Research and Technology (80). The process proved ideal for further process applications with the meat at 3 h being the same as that of normal processing at 24 h, although meat was slightly tougher mainly because longer aging still would be beneficial. There are problems if the birds are either not chilled rapidly enough (81,82), or chilled rapidly and held at low temperatures (83), when it appears that there is no advantage in stimulation. If early boning is desirable, then stimulation is necessary. In a system developed and now in use in New Zealand with spin-chilling, when rigor is achieved at approximately 6°C, the breast muscles of such birds were able to be boned out in just over an hour poststun without any loss of tenderness (84). (D. J. C. Wild, C. E. Devine, and H. Reed, unpublished observations, 1988).

It is important to consider all aspects of stimulation and processing comparing results. It is almost impossible to compare results from one laboratory with those from another, because of the many variables introduced during processing. Not only are stimulation parameters different but also the processing conditions and the methods of evaluation can sometimes introduce artifacts; for example, freeing a normally restrained muscle can allow shortening, which can mask any other changes. Various stimulation treatments cause different degrees of internal heating and will produce different degrees of mechanical damage; for example, 2 Hz is claimed to cause no internal damage (42), whereas 60 Hz causes much more damage (12). Muscle

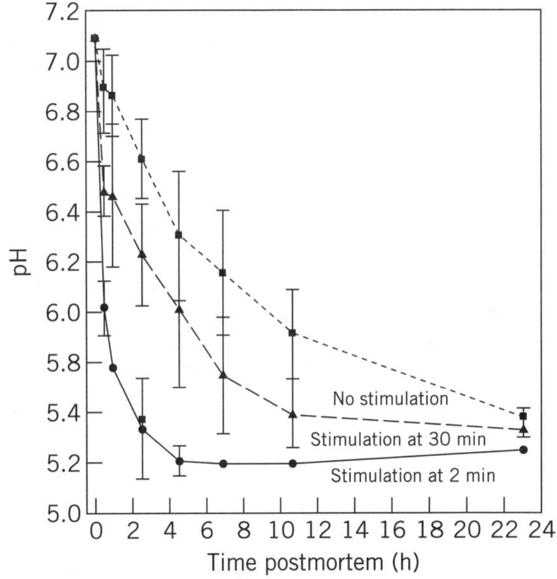

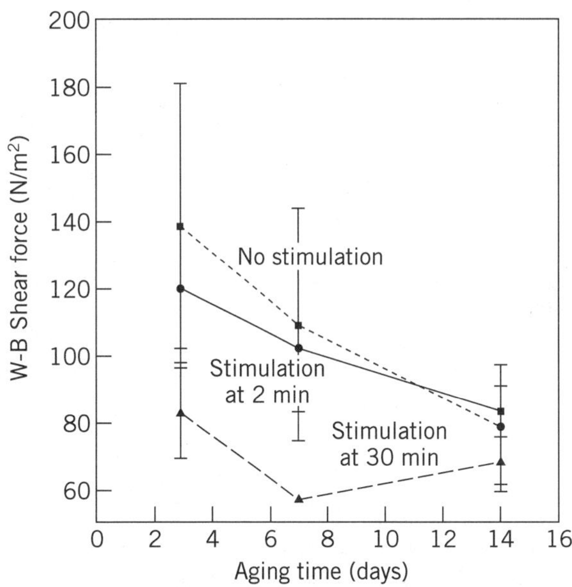

Figure 7. These two graphs show how the tenderness is affected by rigor temperature in a practical situation achieved through two different stimulation applications (37). The graph on the left-hand side shows the falls in pH. The fastest fall in pH arises from stimulation within 2 min of slaughter, the next fastest fall in pH arises from stimulation at 30 min postslaughter, and the slowest rate of pH fall arises from no stimulation at all. When the tenderness is measured at three days, the meat with the medium rate of pH fall is the most tender, the meat stimulated within 2 min is next, and the unstimulated meat is the toughest. In time, after aging, the meat all becomes tender, but the meat stimulated at 30 min is still the most tender. Temperature records showed that the meat stimulated at 30 min attained rigor close to 15°C. This study suggests that meat attaining rigor close to 15°C will be more tender than rigor at other temperatures, and future studies will determine the precise processing situations to achieve this.

damage seems unlikely to be the explanation for many of the effects of electrical stimulation.

ROLE OF ELECTRICAL STIMULATION IN HOT BONING

Hot boning, widely practiced in the past and common in many underdeveloped countries, is now experiencing revival in industrialized countries because of its economic advantages (savings in energy, space, labor, and time). The major constraints to the use of hot boning have been the contraction of prerigor excised muscles, the slippage of one muscle relative to another, and the extra shortening of excised muscles subjected to rapid chilling and in the interest of microbiological control. It has been suggested (84) that hot-boned beef should be chilled below 8°C within 4 h if boned at 37°C. However, it has been noted (85) that a reduction to below 21°C within 9 h would be acceptable. When muscle shortens by approximately 20%, it can still age and be tender (86), so hot boning is feasible if temperatures are controlled (see the article MEAT SCIENCE).

The contraction and consequent toughening of most prerigor excised muscles can be minimized by adequate holding at above 10°C (conditioning). In an early study (87), conditioned muscles were boned at 2 h postmortem, and then the vacuum-packaged primals were aged at 15°C. This gave a product that was as tender as, or more tender, than control samples from normally chilled and boned sides. However, boning at 2 h postmortem does not readily fit with a continuous operation; therefore, commercial processors bone earlier. These processors must use controlled chilling and tight packaging to minimize contraction and toughening; thus, there is interest in being able to reduce processing time and restraints without influencing product quality.

One study considered the physics of rapid chilling and cold shortening and concluded that cold shortening in rapidly chilled beef cannot be prevented without electrical stimulation (88). The situation was even more critical with hot boning because of the more rapid chilling of isolated product.

Electrical stimulation has been used to minimize or eliminate potential tenderness problems in hot-boned beef. The tenderness of product boned from stimulated carcasses at 1 h postmortem has been compared with product boned after 24 h of chilling (89). It was found that the stimulated hot-boned meat, even though chilled rapidly to 7°C, was as tender as that from sides conventionally chilled and boned at 24 h. If the electrically stimulated sides were held for 5 h so that they had entered rigor before boning, the product was also tender (90). Vacuum-packaged primals were prepared for studies of hot boning with and without electrical stimulation (91). It was found that cold-shortening toughness was avoided by either a delayed and slow chilling (21 h to 7°C) or by electrical stimulation of the carcass before boning followed by rapid chilling (17 h to 7°C).

Some attempts have been made to use high-temperature conditioning (eg, holding for 3 h at 37°C) as a means of ensuring that hot-boned beef does not toughen. It was shown that this high-temperature conditioning of hot-boned cuts gave the same results as did prior electrical stimulation of the carcass (92). The effects differed between the three muscles tested, with improvement noted in the *m. longissimus*, no change in the *m. semimembranosus*, and a decrease in tenderness of the *m. semitendinosus* compared with nonstimulated controls. It was theorized that the high-temperature conditioning caused increased proteolysis in the *m. longissimus* but connective tissue shrinkage in the *m. semitendinosus*. This theory does not seem to be completely substantiated by the results.

In general, hot boning has resulted in an increase in the toughness of beef *m. psoas major* muscles, which are normally stretched when carcasses are supported by the Achilles tendon; but if the product was aged for a day at 20°C, the differences between hot and cold boning were removed (93). The sarcomere length of cold-boned product was 3.3 μm (stretched), whereas that of the hot-boned product was 1.95 μm, close to that of rest length of most muscles in the body. Other cuts showed a lesser difference. High-temperature conditioning is not necessary to ensure tenderness if an effective electrical stimulation system was employed (92,93). It should be noted that the effect of a small degree of passive shortening can be overcome by additional aging. The observation that rigor at 15°C produces the most tender meat whether hot boned or restrained (34), suggests that further temperature controls will result in significant tenderness improvements.

Hot boning of pork for whole-hog sausage gives lower cooking losses and increased juiciness scores than does postrigor processed pork (94). For primal cuts, hot boning is reputed to give poorer initial color and appearance, but the differences from cold-boned product disappear over the following 120 h (95). Some studies have considered the use of electrical stimulation to aid hot processing of pork, but the results are conflicting due to the PSE condition, which is increased in susceptible pigs (46,96). Electrically stimulated pork hot boned at 1 h postmortem and then rapidly chilled produces cuts that give less purge and are more juicy than slowly chilled product (15,97). If the chilling is slower in the case of hot-boned pork, stimulation resulted in no tenderness advantage and increased drip and was not be desirable (98).

The role of electrical stimulation in hot-boning applications is clearly to hasten the onset of rigor mortis, so that cold shortening is minimized in cuts that will cool more rapidly than the corresponding cuts on the carcass. Rapid cooling, with its greater control of microbial proliferation can then be used without irrevocably toughening the product. The sooner and more rapid the chilling, the more efficient the electrical stimulation must be to be effective. Very early boning and rapid chilling, even after electrical stimulation, will result in cold shortening and toughness. If a little stimulation is good, a lot is not necessarily better. Future work will refine the temperature conditions to produce the most tender meat enabled by precisely controlled stimulation for each carcass. Future work should be aimed at refining the whole process to increase the uniformity of tenderness across muscles and animals.

BIBLIOGRAPHY

1. D. M. Needham, *Machina Carnis*, Cambridge Univ. Press, Cambridge, UK, 1971.
2. C. A. Lopez and E. W. Herbert, *The Private Franklin: The Man and His Family*, Norton, New York, 1975.
3. U.S. Pat. 2,544,681 (March 13, 1951), A. Harsham and F. E. Deatherage (to Kroger Co.).
4. U.S. Pat. 2,544,724 (March 13, 1951), H. C. Rentschler (to Westinghouse Electric Corp.).
5. M. Ingram and G. C. Ingram, "The Growth of Bacteria on Horse Muscle in Relation to the Changes after Death Leading to Rigor Mortis," *J. Sci. Food Agric.* **6**, 602–611 (1955).
6. G. T. Cori, "The Effect of Electrical Stimulation and Recovery on the Phosphorylase a Content of Muscle," *J. Biol. Chem.* **158**, 333–339 (1945).
7. O. Hallund and J. R. Bendall, "The Long Term Effect of Electrical Stimulation on the Postmortem Fall of pH in Muscles of Landrace Pigs," *J. Food Sci.* **30**, 296–299 (1965).
8. J. C. Forrest et al., "Prediction of the Time Course of Rigor Mortis through Response of Muscle Tissue to Electrical Stimulation," *J. Food Sci.* **31**, 13–21 (1966).
9. W. A. Carse, "Meat Quality and the Acceleration of Post-Mortem Glycolysis by Electrical Stimulation," *J. Food Technol.* **8**, 163–166 (1973).
10. B. B. Chrystall and C. J. Hagyard, "Electrical Stimulation and Lamb Tenderness," *New Zealand J. Agricultural Res.* **19**, 7–11 (1976).
11. C. L. Davey, K. V. Gilbert, and W. A. Carse, "Carcass Electrical Stimulation to Prevent Cold Shortening Toughness in Beef," *New Zealand Journal of Agricultural Research* **19**, 13–18 (1976).
12. J. W. Savell et al., "A Research Note: Structural Changes in Electrically Stimulated Beef Muscle," *J. Food Sci.* **43**, 1606–1607 (1978).
13. A. R. George, J. R. Bendall, and R. C. D. Jones, "The Tenderizing Effect of Electrical Stimulation of Beef Carcasses," *Meat Sci.* **4**, 51–68 (1980).
14. R. R. Riley et al., "Improving Appearance and Palatability of Meat from Ram Lambs by Electrical Stimulation," *J. Anim. Science* **52**, 522–529 (1981).
15. A. A. Taylor and M. Z. Tantikov, "Effect of Different Electrical Stimulation and Chilling Treatments on Pork Quality," *Meat Sci.* **31**, 381–395 (1992).
16. C. J. Hagyard, R. J. Hand, and K. V. Gilbert "Lamb Tenderness and Electrical Stimulation of Dressed Carcasses," *New Zealand J. Agricultural Res.* **23**, 27–33 (1980).
17. B. B. Chrystall and C. E. Devine, "Electrical Stimulation in New Zealand," in K. R. Franklin and H. R. Cross, eds., *Proc. Int. Symp. Meat Sci. Technol.*, National Livestock and Meat Board, Chicago, Ill., 1982, pp. 115–136.
18. Meat Industry Research Institute of New Zealand, *Annual Report*, MIRINZ, Hamilton, New Zealand, 1981.
19. J. W. Savell, "Industrial Applications of Electrical Simulation," in A. M. Pearson and T. R. Dutson, eds., *Electrical Stimulation*, Advances in Meat Research, Vol. 1, AVI, Westport, Conn., 1985.

20. G. C. Smith, "Effects of Electrical Stimulation on Meat Quality, Color, Grade, Heat Ring, and Palatability," in A. M. Pearson and T. R. Dutson, eds., *Electrical Stimulation*, Advances in Meat Research Vol. 1, AVI, Westport, Conn., 1985.

21. D. J. Walker, "Electrical Stimulation in Australia," in *Proc. Int. Symp. Meat Sci. Technol.*, National Livestock and Meat Board, Chicago, Ill., 1982, pp. 137–146.

22. V. Powell, "Quality of Beef Loin Steaks as Influenced by Animal Age, Electrical Stimulation, and Ageing," *Meat Sci.* **30**, 195–205 (1991).

23. P. V. Harris and W. R. Shorthose, "Meat Texture," in R. Lawrie, ed., *Developments in Meat Science—4*, Elsevier Applied Science, Barking, U.K., 1988, pp. 245–296.

24. E. W. Mills, S. H. Smith, and M. D. Judge, "Early Post-Mortem Degradation of Intramuscular Collagen," *Meat Sci.* **26**, 115–120 (1989).

25. R. H. Locker, "Degree of Muscular Contraction as a Factor in Tenderness of Beef," *Food Res.* **25**, 304–307 (1960).

26. R. H. Locker and D. J. C. Wild, "Yield Point in Raw Beef Muscle: The Effects of Ageing, Rigor Temperature, and Stretch," *Meat Sci.* **7**, 93–107 (1982).

27. R. H. Locker, D. J. C. Wild, and G. J. Daines, "Tensile Properties of Cooked Beef in Relation to Rigor Temperature and Tenderness," *Meat Sci.* **8**, 283–299 (1983).

28. C. E. Devine and A. E. Graafhuis, "The Basal Tenderness of Unaged Lamb" *Meat Sci.* **39**, 285–291 (1994).

29. D. J. Troy and P. V. Tarrant, "Changes in Myofibrillar Proteins from Electrically Stimulated Beef," *Biochem. Soc. Trans.* **15**, 297–298 (1987).

30. D. E. Goll and R. M. Robson, "Molecular Properties of Post-Mortem Muscle, 1: Myofibrillar Nucleosidetriphosphatase Activity in Bovine Muscle," *J. Food Sci.* **32**, 323–329 (1968).

31. I. F. Penney and E. Dransfield, "Relationship between Toughness and Troponin T in Conditioned Muscle," *Meat Sci.* **3**, 135–141 (1979).

32. R. H. Locker and D. J. C. Wild, "The Fate of the Large Proteins of the Myofibril during Tenderising Treatments," *Meat Sci.* **11**, 89–108 (1984).

33. D. E. Goll et al., "Role of Muscle Proteinases in Maintenance of Muscle Integrity and Mass," *J. Food Biochem.* **7**, 137–177 (1983).

34. C. E. Devine, N. M. Wahlgren, and E. Tornberg, "The Effects of Rigor Temperatures on Shortening and Meat Tenderness," *Proc. 42nd Int. Congr. Meat Sci. and Technol.*, Lillehammer, Norway, September 1–6, 1996.

35. B. B. Chrystall and C. E. Devine, "Electrical Stimulation: Its Early Development in New Zealand," in A. M. Pearson and T. R. Dutson, eds., *Electrical Stimulation* Advances in Meat Research, Vol. 1, AVI Publishing, Westport, Conn., 1985, pp. 73–119.

36. N. M. Wahlgren, C. E. Devine, and E. Tornberg, "The Influence of Different pH-Courses during Rigor Development on Beef Tenderness," *Proc. 43rd Int. Congr. Meat Sci. and Technol.*, Auckland, New Zealand, July 27–August 1, 1997.

37. J. Jungersten et al., "Both Physical Fitness and Acute Exercise Regulate Nitric Oxide Formation in Healthy Humans," *J. Appl. Physiol.* **82**, 760–764 (1997).

38. C. J. Cook, S. M. Scott, and C. E. Devine, "Measurements of Nitric Oxide and the Effect of Enhancing and Inhibiting It on the Tenderness of Meat," *Meat Sci.* **48**, 85–89 (1998).

39. C. L. Davey and K. V. Gilbert, "The Temperature Coefficient of Beef Aging," *J. Sci. Food Agric.* **27**, 244–250 (1976).

40. B. B. Chrystall and C. E. Devine, "Electrical Stimulation, Muscle Tension, and Glycolysis in Bovine Sternomandibularis," *Meat Sci.* **2**, 49–58 (1978).

41. D. J. Morton, J. F. Weidemann, and F. M. Clarke, "Enzyme Binding in Muscle," in R. Lawrie, ed., *Developments in Meat Science—4*, Elsevier Applied Science, Barking, UK, 1988, pp. 37–61.

42. B. B. Marsh et al., "Effects of Early-Postmortem pH and Temperature on Beef Tenderness," *Meat Sci.* **5**, 479–483 (1980–1981).

43. S. Fabiansson and R. Libelius, "Structural Changes in Beef Longissimus Dorsi Induced by Postmortem Low Voltage Electrical Stimulation," *J. Food Sci.* **50**, 39–44 (1985).

44. J. R. Bendall, "Cold Contracture and ATP-Turnover in the Red and White Musculature of the Pig," *J. Sci. Food Agric.* **26**, 55–71 (1975).

45. J. C. Forrest and E. J. Briskey, "Response of Striated Muscle to Electrical Stimulation," *J. Food Sci.* **32**, 482–488 (1967).

46. D. D. Crenwelge et al., "Effect of Chilling Method and Electrical Stimulation on Pork Quality," *J. Anim. Sci.* **59**, 697–705 (1984).

47. R. P. Newbold and L. M. Small, "Electrical Stimulation of Post-Mortem Glycolysis in Semitendinosus Muscle of Sheep," *Meat Sci.* **12**, 1–16 (1985).

48. E. C. Bate-Smith and J. R. Bendall, "Factors Determining the Time Course of Rigor Mortis," *J. Physiol.* **110**, 47–65 (1949).

49. B. B. Marsh, "Rigor Mortis in Beef," *J. Sci. Food Agric.* **5**, 70–75 (1954).

50. R. E. Jeacocke, "The Temperature Dependence of Anaerobic Glycolysis in Beef Muscle Held in a Linear Temperature Gradient," *J. Sci. Food Agric.* **28**, 551–556 (1977).

51. B. B. Chrystall and C. E. Devine, "Electrical Stimulation Developments in New Zealand," *Proc. 26th Eur. Meat Res. Workers Conf.*, Colorado Springs, Colo., August 31–September 5, 1980.

52. J. R. Bendall, *Muscles, Molecules, and Movement*, Heinemann, London, 1969.

53. C. E. Devine, S. Ellery, and S. Averill, "Responses of Different Types of Ox Muscle to Electrical Stimulation," *Meat Sci.* **10**, 35–51 (1984).

54. A. Houlier, C. Valin, and P. Sale, "Is Electrical Stimulation Muscle Dependent?" *Proc. 26th Eur. Meat Res. Workers Conf.*, Colorado Springs, Colo., 1980.

55. H. Specht and J. Kunis, "Kalteverkurzung und elektrostimulation: Auswirkungen auf bei beschaffenheit von Schaf- und Rindfleisch," *Fleischwirkehaft* **69**, 1275–1280 (1989).

56. H. J. Swatland, "Cellular Heterogeneity in the Response of Beef to Electrical Stimulation," *Meat Sci.* **5**, 451–455 (1980–1981).

57. P. E. Bouton, R. R. Wesle, and F. D. Shaw, "A Research Note: Electrical Stimulation of Calf Carcasses: Response of Various Muscles to Different Waveforms," *J. Food Sci.* **45**, 148–149 (1980).

58. B. B. Chrystall and C. E. Devine, "Electrical Stimulation of Rats: A Model for Evaluating Low Voltage Stimulation Parameters," *Meat Sci.* **8**, 33–41 (1983).

59. M. B. Solomon, "Response of Bovine Muscles to Direct High Voltage Electrical Stimuli: A Research Note," *Meat Sci.* **22**, 229–235 (1988).

60. F. G. M. Smulders and G. Eikelenboom, "Electrical Stimulation of Carcasses: Development of Fully Automated Equipment," *Fleishwirtschaft* **65**, 1356–1358 (1985).

61. C. E. Devine, B. B. Chrystall, and C. L. Davey, "Studies in Electrical Stimulation: Effect of Neuromuscular Blocking Agents in Lamb," *J. Sci. Food Agric.* **30**, 1007–1011 (1979).
62. C. E. Devine et al., "Differential Effects of Electrical Stunning on the Early Post Mortem Glycolysis in Sheep," *Meat Sci.* **11**, 301–309 (1984).
63. P. E. Petch and K. V. Gilbert, "Interaction of Electrical Processes Applied during Slaughter and Dressing with Stimulation Requirements," *Proc. 43rd Int. Congr. Meat Sci. Technol.*, Auckland, New Zealand, July 27–August 1, 1997.
64. A. C. Kondos and D. G. Taylor, "Electrical Stimulation and Temperature on Biochemical Changes in Beef Muscle," *Meat Sci.* **19**, 207–216 (1985).
65. B. B. Chrystall, "Trends and Developments in Meat Processing," in R. Purchas and B. Hogg, eds., *The Production and Processing of Meat*, Animal Production Society, Hamilton, New Zealand, 1989.
66. J. W. Savell, T. R. Dutson, and Z. L. Carpenter, "Beef Quality and Palatability as Affected by Electrical Stimulation and Cooler Aging," *J. Food Sci.* **43**, 1666–1669 (1978).
67. A. E. Graafhuis et al., "Tenderness of Different Muscles Cooked to Different Temperatures and Assessed by Different Methods," *Proc. 37th Int. Congr. Meat Sci. Technol.*, Kulmbach, Germany, September 1–6, 1991.
68. T. R. Dutson, J. W. Savell, and G. C. Smith, "Electrical Stimulation of Antemortem Stressed Beef," *Meat Sci.* **6**, 159–161 (1982).
69. B. B. Chrystall et al., "Tenderness of Exercise-Stressed Lambs," *New Zealand J. Agricultural Res.* **25**, 331–336 (1982).
70. B. B. Marsh, "Effects of Early-Postmortem Muscle pH and Temperature on Tenderness," *Proceedings of the Reciprocal Meat Conference* **36**, 131–135 (1983).
71. A. Watanabe, C. C. Daly, and C. E. Devine, "The Effects of Ultimate pH of Meat on the Tenderness Changes during Ageing," *Meat Sci.* **42**, 67–78 (1995).
72. C. Calkins et al., "Quality-Indicating Characteristics of Beef as Affected by Electrical Stimulation and Postmortem Chilling Time," *J. Food Sci.* **45**, 1330–1340 (1980).
73. S. A. Pommier, L. M. Poste, and G. Butler, "Effect of Low Voltage Electrical Stimulation on the Distribution of Cathepsin D and the Palatability of Longissmus Dorsi from Holstein Veal Calves Fed a Corn or Barley Diet," *Meat Sci.* **21**, 203–218 (1987).
74. A. A. Taylor, N. F. Down, and E. Dransfield, "Effect of High and Low Voltage Stimulation on Tenderness of Muscles from Slowly Cooled Beef Sides," *Proc. 30th Eur. Meat Res. Workers Conf.*, Bristol, UK, September 9–14, 1984.
75. D. M. Stiffler et al., "Comparison of the Effects of High and Low Voltage Electrical Stimulation on Quality-Indicating Characteristics of Beef Carcasses," *J. Food Sci.* **49**, 863–866 (1984).
76. K. C. Koh et al., "Effects of Electrical Stimulation and Temperature on Beef Quality and Tenderness," *Meat Sci.* **21**, 189–201 (1987).
77. A. A. Taylor and A. M. Perry, "Improving Pork Quality by Electrical Stimulation or Pelvic Suspension of Carcasses," *Meat Sci.* **39**, 327–337 (1995).
78. P. D. Warriss et al., "Potential Interactions between the Effects of Preslaughter Stresses and Post-Mortem Electrical Stimulation of the Carcasses on Meat Quality in Pigs," *Meat Sci.* **41**, 55–68 (1995).
79. Yu B. Lee and D. A. Rickansrud, "Effect of Temperature on Shortening in Chicken Muscle," *J. Food Sci.* **43**, 1613–1615 (1978).
80. U.S. Patent 4,675,947 (June 30, 1987), K. A. Clatfelter and J. E. Webb (to Campbell Soup Company).
81. G. W. Froning and T. G. Uijttenboogaart, "Effect of Post-Mortem Electrical Stimulation on Colour, Texture, pH, and Cooking Losses of Hot and Cold Deboned Chicken Broiler Breast Meat," *Poult. Sci.* **67**, 1536–1544 (1988).
82. C. E. Lyon et al., "Effects of Electrical Stimulation on the Post-Mortem Biochemical Changes and Texture of Boiler Pectoralis Muscle," *Poult. Sci.* **68**, 249–257 (1989).
83. D. K. Wakefield et al., "Influence of Postmortem Treatments on Turkey and Chicken Meat Texture," *Int. J. Food Sci. Technol.* **24**, 81–92 (1989).
84. L. S. Herbert and M. G. Smith, "Hot Boning of Meat: Refrigeration Requirements to Meet Microbiological Demands," *C.S.I.R.O. Food Research Quarterly* **40**, 65–70 (1980).
85. D. Y. C. Fung et al., "Initial Chilling Rate Effects on Bacterial Growth on Hot-Boned Beef," *J. Food Prot.* **44**, 539–544 (1981).
86. C. L. Davey, H. Kuttel, and K. V. Gilbert, "Shortening as a Factor in Meat Aging," *J. Food Technol.* **2**, 53–56 (1967).
87. G. R. Schmidt and K. V. Gilbert, "The Effect of Muscle Excision before the Onset of Rigor Mortis on the Palatability of Beef," *J. Food Technol.* **5**, 331–338 (1970).
88. J. Kunis and H. Specht, "Technologische kriterien fur da auftreten und die verhinderung von cold shortening in Rindfleisch," *Nahrung* **32**, 433–438 (1988).
89. K. V. Gilbert, C. L. Davey, and K. O. Newton, "Electrical Stimulation and Hot Boning of Beef," *New Zealand J. Agricultural Res.* **20**, 139–143 (1976).
90. K. V. Gilbert and C. L. Davey, "Carcass Electrical Stimulation and Hot Boning of Beef," *New Zealand J. Agricultural Res.* **19**, 429–434 (1976).
91. A. A. Taylor, B. G. Shaw, and D. B. MacDougall, "Hot Deboning Beef with and without Electrical Stimulation," *Meat Sci.* **5**, 109–123 (1980–1981).
92. S. C. Seideman, J. D. Crouse, and H. R. Cross, "High-Temperature Conditioning of Hot-Boned Subprimals," *J. Food Quality* **12**, 145–153 (1989).
93. R. L. J. M. van Laack, *The Quality of Accelerated Processed Meats: An Integrated Approach*, Rijksuniversitat, Utrecht, The Netherlands, 1989.
94. H.-S. Lin, D. G. Topel, and H. W. Walker, "Influence of Prerigor and Postrigor Muscle on the Bacteriological and Quality Characteristics of Pork Sausage," *J. Food Sci.* **44**, 1055–1057 (1979).
95. N. G. Marriott et al., "Acceptability of Accelerated-Processed Pork," *J. Food Prot.* **43**, 756–759 (1980).
96. R. L. Swasdee et al., "Processing Properties of Pork as Affected by Electrical Stimulation, Post-Slaughter Chilling, and Muscle Group," *J. Food Sci.* **48**, 150–151, 162 (1983).
97. J. O. Reagan and K. O. Honikel, "Weight Loss and Sensory Attributes of Temperature Conditioned and Electrically Stimulated Hot Processed Pork," *J. Food Sci.* **50**, 1568–1570 (1985).
98. R. L. J. M. van Laack and F. J. M. Smulders, "The Effect of Electrical Stimulation, Time of Boning, and High Temperature Conditioning on Sensory Quality Traits of Porcine *Longissimus dorsi* muscle," *Meat Sci.* **25**, 113–121 (1991).

B. B. CHRYSTALL
Massey University, Albany Campus
Auckland, New Zealand

C. E. DEVINE
Technology Development Group, HortResearch
Hamilton, New Zealand

MEAT, MODIFIED ATMOSPHERE PACKAGING

Chilled meat is packaged under modified atmospheres to preserve it against both microbial spoilage and nonmicrobial deterioration. Fresh meat is always contaminated by a variety of bacteria, which may include both spoilage and pathogenic strains. Meat is an ideal growth medium for many microorganisms, which have potential for growth to numbers that result in spoilage. Thus meat is a highly perishable food. However, altering the atmosphere can prevent the growth of some organisms and slow the growth of others, thus extending the time before microbial spoilage becomes evident or other microbial problems occur.

An important consideration for packaging and marketing is the retention of an attractive, fresh, salable appearance of meat in display. This includes most importantly muscle color, but also color of fat, bones, and purge in the package and includes prevention of excessive purge. During storage or display of meat, undesirable appearance gradually to rapidly occurs because of dulling, darkening, browning (even to tan or green) (1), and this color deterioration usually becomes a problem more rapidly than microbial spoilage. Dehydration of meat surfaces can also add to the appearance problem. Modified atmosphere packaging (MAP) packages are sealed and composed of films that are excellent barriers to water vapor transmission. MAP packages for red meats must maintain an atmosphere that prevents early deterioration of meat color. Nonmicrobial deterioration of meat flavor or texture will generally be significant only when microbial spoilage can be delayed for very long periods. These could include off-odors, flavors, or both caused by oxidative rancidity.

TYPES OF MAP PACKAGES

Three types of MAP packages include high oxygen, low oxygen, and ultra-low oxygen. Ultra-low oxygen is a more appropriate name for what some identify as controlled-atmosphere packaging (CAP). CAP implies control of in-package gas composition, which may not happen, but some systems definitely limit oxygen within the package and could fit into the ultra-low-oxygen category. Another relevant term is dynamic modified atmosphere packaging (2), sometimes called active packaging, and these systems may include a system to influence gas composition other than the influence of muscle metabolism.

Producing the initial gas atmosphere may involve an evacuation of air from the package followed by fill with the desired gas mix or, less likely, a displacement of air in the package with the gas mix. Gas mixes may be preformulated by the supplier or proper proportions from supply tanks of each desired gas mixed and controlled by a gas controller (mixer). Gas-mixture composition may be affected by the relative flow rates of the component gases during package filling. Given some variations in gas flows and variation in the amount of air removed by evacuation or displaced, the concentrations of the component gases of the modified atmosphere can be expected to vary by as much as 5% at the time packages are sealed.

High-oxygen MAP uses up to 80% O_2, 20 to 30% CO_2, and 0 to 20% N_2. Complete air evacuation is not as critical here as for low and ultra-low MAP. High-oxygen packaging machines may be dangerous, and a spark can result in an explosion. Absolutely complete insulation of electrical wires and a choice of safe lubricants (not petroleum-based) is essential.

Low and ultra-low-oxygen MAP packages are gassed with CO_2 or a CO_2-N_2 mixture (3,4). The composition of the package atmosphere at the time of closure will depend on the extent to which air was removed from the package. With simple gas-displacement systems, the level of residual oxygen may be up to 10%. With chamber evacuation systems, oxygen levels of 0.1 to 1.0% could be achieved, depending on time of each cycle. The push in industry toward high production rate pushes this figure toward the higher percentage. If pouches are evacuated through snorkels while subject to atmospheric pressure, or to a system to put outside pressure on packages, this may result in a within-package oxygen level as low as 0.1% (1,000 ppm). However, package collapse closer to the seal end before complete package evacuation can trap pockets of air in the package. These distribute themselves into the in-package environment and result in higher-than-desired levels of oxygen.

CAP implies that an atmosphere of a known, desired composition is established within a pack and remains unchanged during the life of the package. The only type of packaging in current use that comes close to this description is a gas-impermeable pouch filled with an atmosphere of unmixed CO_2 (5). Attainment of reasonably stable gas composition depends on rapid placement of cuts in the package after cutting and low holding temperatures to minimize muscle and microbial metabolism. Packages of that type can be prepared using snorkel systems but without high assurance that residual oxygen is always reduced to very low concentrations at the time of package closure. CAP packages are preferably prepared using a system in which pouches are evacuated and gassed by snorkel within a chamber that is simultaneously evacuated, to prevent locking off of air-filled regions, then partially pressurized before the pouch is gassed to gently collapse the pouch and thus expel much of the remaining atmosphere. When such a system incorporates a high-volume pump to produce a high vacuum ($\leq$2 torr, 29.90 in. of vacuum, a 99.7% vacuum), residual oxygen at package closure of less than 0.05% (500 ppm) can be attained providing the CO_2 gas contains 0.03% or less oxygen.

Packaging Materials

To adequately maintain modified atmospheres during storage, MAP packages must be constructed of materials that offer a high barrier to transmission of gases and water vapor. However, packages are constructed largely of materials that inevitably permit some degree of gas permeation. The amount of any gas actually transmitted by a film will depend on the film material, thickness of the materials, temperature of the film, and the difference between the partial pressure of the gas within and outside the pack. Most barrier packages are made of three layers of film. The outside layer must be scuff- and abrasive-resistant, the middle layer provides the gas barrier, and the inner layer is a sealant layer (6). Some barrier layers lose the barrier properties when wet and therefore must be protected from moisture on both sides (7,8). The rates at which different gases are transmitted through different films also vary widely, partially dependent on the chemical nature of the film polymer.

Gases that contribute to product life in MAP are CO_2 and O_2, and their relative rates of transmission depend on the film. As well as appropriately low gas transmission characteristics, films used for MAP packaging should have good sealing qualities. Although clipping or crimping is used in some systems for closing MAP packs, such closures are unreliable in that channels allowing excessive gas exchange may traverse the seals. To maintain a modified atmosphere, packaging materials that can be reliably sealed by heating are required. Tray and package design must consider film thickness at the thinnest points. An accordion-fold design meant to improve physical strength of a tray could result in thinning of gas barrier walls if the quantity of film material is limited.

Important physical characteristics needed include resistance to abrasion, as handling and transport are likely to impose abrasive stress; puncture resistance, particularly for bone-in products; appropriate thermoforming characteristics when tray forming is part of a packaging system; and, for display top-film layers particularly, good clarity and antifogging characteristics.

The critical gas-transmission characteristics are not well defined for many MAP purposes. In general, films with oxygen permeabilities of <100 $cm^3/m^2/24$ h at standard atmosphere, measured for the film at a specified temperature and exposed to a specified humidity (commonly, 25°C and 75% RH), are considered possible materials for constructing MAP packages for low-oxygen MAP (9). However, such data give only an approximate indication of the oxygen transmission in actual use conditions and not a real indication of the CO_2 transmission that could occur. Consequently, trials are necessary to establish the utility of films for particular MAP purposes.

For high-oxygen MAP display, appearance of the package is a major commercial concern. That aspect of the package may therefore be optimized in practice, even at the cost of some reduction in the preservative capabilities of the packaging.

The atmosphere of a high-oxygen MAP display pack is dynamic. The contained meat respires, converting O_2 to CO_2; O_2 and CO_2 are transmitted at different rates through the film, at least through the lid, which must have good optical qualities; and the highly soluble CO_2 will dissolve in the meat to an amount determined by the meat mass, area of exposed meat surface, the muscle tissue pH, the ratio of fat to muscle tissue, the temperature, and the partial pressure of the gas in the within-package atmosphere (10). To maintain an atmosphere of adequate composition against the meat-related factors tending to change it requires a gas-volume-to-meat weight ratio of about 3. When free space within the package is inadequate, an excessive change in the O_2 concentration may occur during storage. Package collapse can also occur because a rising CO_2 concentration leads to a reduction in the atmosphere volume as a result of CO_2 dissolving in the meat. To guard against collapse upon initial or progressing dissolution of CO_2, it is common practice to include some nitrogen, which is physiologically inert, as ballast in the package atmosphere. Even when the package is properly constructed and filled to an appropriate gas-volume-to-meat weight ratio, changes in the atmosphere composition occur during storage.

The inevitable economic (cost) pressures lead to reducing the in-package headspace-to-meat ratio, especially when transportation for a long distance is involved. In the high-oxygen package, the concentrations of both O_2 and CO_2 will decline and that of N_2 will increase proportionally with decreasing concentrations of the other gases (11).

Pouches for low-oxygen MAP packages are usually composed of either a high or very high gas-barrier laminate (<100 or <2 cm^3 $O_2/m^2/24$ h/atm). The pouch is evacuated and refilled with either CO_2 or a CO_2-N_2 mixture. Depending on the evacuation system used, residual oxygen may range from 10 to 0.1% (1,000 ppm) and sometimes as low as 300 ppm, although the latter, low level is not attained consistently.

After package closure, CO_2 will dissolve in the meat. When only CO_2 is added, if the quantity of CO_2 is insufficient to both saturate the meat at atmospheric pressure and fill the inevitable voids between meat pieces, the pack will collapse tightly on the product, compacting and deforming it. The partial pressure of CO_2 within the package will then be less than atmospheric, and the film may be prone to puncture by the contents, if bone-in cuts are used. Levels of CO_2 may also fall below the most effective antimicrobial levels. About 1.5 L of CO_2/kg of meat is generally required to avoid pack collapse (10).

Residual oxygen levels in low and ultra-low MAP packages may increase above initial levels for approximately 3 days, from an initial 2% to about 7% for a five–pork loin gas pack, then will decrease (12). The increase is partly due to oxygen that has diffused into the meat during exposure to air after cutting and before packaging. Some of this oxygen apparently diffuses out, increasing oxygen concentration of the in-package gas. Oxygen concentrations also rise because CO_2 dissolves into the meat. A shorter interval between cutting and packaging should decrease this short-term oxygen increase. After this initial increase, the concentration of residual oxygen within the MAP packages will tend to fall, because of reaction of the gas with the meat. However, the rate of such reactions will decline with the oxygen concentration and an equilibrium will ultimately be established between oxygen ingress through the film and scavenging of the gas from the package atmosphere. The equilibrium O_2 concentrations may reach about 1% in packs composed of high gas-barrier film but will be lower in packs composed of very high gas-barrier film (13). The subsequent lowering of oxygen concentration is affected by muscle metabolism (mitochondrial consumption and reaction with heme pigment), absorption by tissue fluids, and microbial metabolism.

MUSCLE SCAVENGING OF OXYGEN

Muscle metabolism can serve to scavenge oxygen within a MAP package, but such scavenging is slow at chilled meat temperatures and also slowed by lower concentrations of oxygen in the MAP headspace. Muscles differ in their ability to scavenge oxygen, with redder (more red fibers) muscle being more active. Scavenging rate may be influenced by pH.

Continued scavenging by muscle reduces its later ability to bloom (oxygenate) when exposed to air. Rapid reduction of in-package oxygen of ultra-low MAP packages seems important to retain substantial ability to bloom.

DYNAMIC MAP SYSTEMS

These systems (2) include active packaging, VSP (vacuum skin packaging) & MAP, CAPTECH, 2-phase MAP, static gas exchange, and dynamic gas exchange. VSP & MAP includes an inside vacuum skin of an oxygen-permeable film enclosed in a barrier bag into which a high-oxygen atmosphere (80% O_2, 20% CO_2) has been injected. The inner oxygen-permeable film allows oxygen at the meat surface while carbon dioxide controls microbial growth. The CAPTECH system has an inner oxygen-permeable film and the outer foil plus biaxially oriented nylon film that encloses a metered amount of carbon dioxide. The 2-phase MAP combines use of solid and gaseous carbon dioxide to avoid the package collapse that can result from CO_2 absorption by the meat.

Static gas exchange implies an inner package of oxygen-permeable film with an outer dome or master pack filled with a CO_2/N_2 mix. The outer dome or film is gas impermeable, and this system is dependent on a very good oxygen evacuation. This system may be vulnerable to oxygen that has diffused into meat after cutting and before packaging and later may partially diffuse out of the meat into the in-package atmosphere. Therefore the time interval between cutting and packaging should be very short.

Dynamic gas exchange implies mechanically replacing the original CO_2/N_2 gas mix with an 80% O_2/20% CO_2 or similar mix at the retail establishment. The original gas mix enables a longer storage and distribution product life of up to 3 weeks, and the gas exchange enables rapid bloom (oxygenation) of the myoglobin.

ACTIVE PACKAGING

Active packaging can use oxygen scavengers, carbon dioxide scavengers or emitters, chemically treated films, temperature-controlled packages, or time-and-temperature controlled packages (14). To insure early attainment and maintenance of continuous zero-oxygen concentrations in MAP, oxygen absorbers or scavengers can be added to the package to actively remove residual oxygen and oxygen that permeates the packaging film. The most common oxygen scavengers are reduced-iron powders mixed with acids, salts, or both to oxidize in the presence of oxygen. Most scavengers need to be activated by some wetted humectant (13). When a wetted humectant was added to the scavenger and placed in an aerobic atmosphere, oxygen scavenging occurred at near maximum rate (15). A scavenger used in an ultra-low-oxygen system can quickly lower oxygen levels below limits of detection.

Ageless® (Mitsubishi, Japan), a scavenger widely used by the food industry, has iron powder in a small packet that is highly permeable to oxygen. According to Nakamura and Hoshino (16), 1 g of iron can react with 0.0136 mol of oxygen, producing ferrous oxide. One gram or about 0.7 L of oxygen reacts with about 300 cm^3 of oxygen scavenger at standard pressure and temperature. Packets are available in sizes ranging from 20 to 2,000 cm^3 in oxygen-scavenging capacity. To remain within the scavenging ability of an oxygen scavenger, the barrier film permeability must be <20 $cm^3/cm^2/24$ hours, and the advantage of minimal oxygen to be scavenged is important.

An active packaging system (Cryovac OS 1000) released in 1998 (17) uses a polymeric oxygen-scavenging system that absorbs oxygen within the MAP package, and the polymer also serves as an oxygen barrier. The proprietary oxygen-scavenging layer consists of three primary components: an oxidizable polymer, a photoinitiator (PI), and a catalyst. The oxidizable polymer is the component responsible for binding the oxygen molecules. The PI absorbs the UV light and provides the energy to start the reaction. The catalyst helps to increase the rate of the scavenging reaction. The advantage of this system is that it is activated by UV light quickly after the package is sealed and does not require injection of an activating fluid.

An earlier stumbling block of sachet-type scavengers was a desire, especially by marketing people, not to have this "foreign material" in the final retail package. Some newer applications avoid this problem by attaching the iron-filled sachet to the outer barrier dome or bag, so it is disposed of and never appears with the final retail package. Most scavengers work more slowly at the low temperature required for chilled meat. This presents a problem as rapid scavenging of oxygen is a must for ultra-low-oxygen applications if rapid blooming is anticipated upon later exposure to air.

THE REACTIONS OF SPOILAGE FLORAS AND PATHOGENIC BACTERIA TO MODIFIED ATMOSPHERES

When lack of oxygen is not a factor, meat spoilage floras are inevitably dominated by strictly aerobic pseudomonads, at least at temperatures of 20°C or below. Those organisms will ultimately cause putrid spoilage of the meat. The presence of CO_2 in the atmosphere slows the rates of growth of both pseudomonads and competing species so that the pseudomonads retain the growth rate advantage that insures their predominance in the spoilage flora. Consequently, CO_2 in an aerobic atmosphere extends the time required for the spoilage flora to reach spoilage numbers but does not significantly alter the flora composition or the type of spoilage that finally occurs.

The inhibition of pseudomonads by CO_2 increases with decreasing temperature and with increasing CO_2 concentration. However, the increased inhibition is small for CO_2 levels >20%. Thus, provided the CO_2 concentration does not fall below that level, changes in CO_2 concentration will have an insignificant effect on the storage life of meat packaged in a high-oxygen MAP pack (18). At chiller temperatures, the growth rate of pseudomonads is approximately halved by 20% CO_2 in the atmosphere. Therefore, high-oxygen MAP packaging will at best double the time before microbial spoilage of meat develops. Microaerobic or anaerobic conditions severely inhibit or suppress the growth of pseudomonads. Under essentially anaerobic conditions, low-pH (<5.8) meat will develop a flora composed solely of anaerobic but aerotolerant lactobacilli. Lactobacilli produce acid or dairy spoilage flavors in meat, but only substantially after the bacteria have achieved their maximum numbers (19).

Under the same conditions, high-pH muscle tissue and meat with extensive fat cover, which is of neutral pH, will develop a flora containing facultatively anaerobic organisms in addition to the lactobacilli. Species of *Enterobacteriaceae* will inevitably be present and cause putrid spoilage as they approach their maximum numbers. Two other species, *Brochothrix thermosphacta* and *Alteromonas putrefaciens*, may also be present when processing hygiene is poor. When those organisms are present in a spoilage flora, they can cause early spoilage when their numbers are relatively low; *B. thermosphacta* produces sour-aromatic odors and flavors, and *A. putrefaciens* causes putrid spoilage with much H_2S.

High concentrations of CO_2 impose significant lags on the lactobacilli and *B. thermosphacta* before they can commence growth and thus halve the growth rates of these organisms. In addition, the minimum temperature for growth of *B. thermosphacta* is increased from about −3°C to above 0°C. The effect of high CO_2 concentrations on the enterobacteria is even more pronounced, because their lag phase is extended to very long times, and they finally commence only very slow growth. The effect of high CO_2 concentrations of *A. putrefaciens* is less certain, but available evidence suggests that its growth is effectively suppressed (20).

Increasing concentrations of O_2 in a nominally anaerobic CO_2 atmosphere will tend to accelerate the weak growth of the facultative anaerobes. When the O_2 concentration is sufficiently high, slow growth of pseudomonads will occur. Thus low-oxygen MAP atmospheres, with a range of O_2 concentrations that usually increase during storage, tend to allow growth of mixed spoilage flora in which facultative anaerobes are a substantial fraction and in which even pseudomonads may occur. In contrast, ultra-low-oxygen-packaged meat will develop a flora of lactobacilli, irrespective of the meat pH. Poor microbiological condition before packaging has minimal effect on the type of flora that develops when the product is stored at 0°C or below. However, at higher storage temperatures, *B. thermosphacta* will grow on ultra-low-oxygen-packaged meat that was heavily contaminated with that organism during processing, at a rate similar to that of the lactobacilli. The presence of *B. thermosphacta* in the flora will cause early spoilage of the product. On ultra-low-oxygen-packaged meat, the ubiquitous enterobacteria will finally initiate growth to cause putrid spoilage when they approach the relatively low numbers of $10^6/cm^2$. Such putrid spoilage may, however, be preempted by acid-dairy spoilage due to the lactobacilli that attain maximum numbers of about $10^8/cm^2$ well before the enterobacteria start to grow.

In addition to spoilage organisms, cold-tolerant pathogenic species may grow on chilled meat, and mesophilic pathogens may grow when the chilled meat experiences temperature abuse. As with the spoilage organism, pathogen growth is generally slowed by substantial concentrations of CO_2. However, the response of bacteria to CO_2 is highly varied, and under some circumstances, a CO_2-containing atmosphere may confer a growth-rate advantage on an individual pathogen over competing spoilage types. In that situation, the pathogen might reach hazardous numbers before the meat is rejected because of spoilage. Such a condition can develop with vacuum-packaged meat and might be expected in high- and low-oxygen MAP packs. However, in ultra-low-oxygen-packs, the combination of anoxia and high CO_2 concentration tends to raise the minimum growth temperature and extend the lag phase of pathogens so that their growth is disadvantaged relative to that of spoilage organisms (21).

MUSCLE PIGMENT REACTIONS IN MODIFIED ATMOSPHERE

When purchasing fresh meat, consumers judge the acceptability of the product largely on the appearance, mostly color, of the exposed muscle tissue. Dull and/or discolored fat and darkened cut bone surfaces can detract from meat cut appearance when the muscle tissue appears attractive, but good appearances of fat and bone cannot compensate for a degraded appearance of the muscle tissue (22).

Muscle tissue appearance is determined by the chemical state of the muscle pigment, myoglobin, as shown in Figure 1. In the absence of oxygen, the pigment is in the deoxymyoglobin state, which has a dark, purplish-red color (23). On exposure to air, the pigment is rapidly oxygenated to form oxymyoglobin, the bright red that consumers have been taught to expect and find attractive. Deoxymyoglobin and oxymyoglobin, which are both in the reduced state, can oxidize to metmyoglobin, which has a dull brown color associated with deterioration of quality. Oxidation of deoxymyoglobin can be more rapid than oxidation of oxymyoglobin. That results in the faster oxidation of myoglobin at low (highest at 4 mm partial oxygen pressure) than at higher concentrations of oxygen (24).

Myoglobin oxygenation is rapid, and the fraction of the pigment in the oxygenated form increases with increasing oxygen concentration. Metmyoglobin is more stable and is

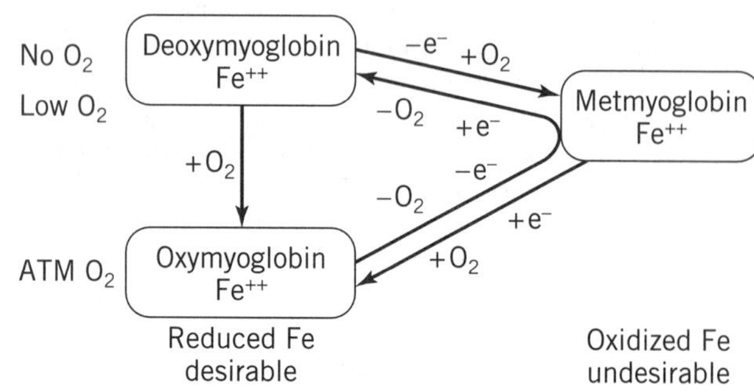

Figure 1. Fresh meat color triangle.

slowly converted to deoxymyoglobin by enzyme-mediated reactions termed metmyoglobin-reduction activity (25). Muscle tissue that is deficient in the enzymes that mediate metmyoglobin reduction or in the reduced cofactors necessary for the reduction reaction will be unable to reconvert metmyoglobin, which will persist once it is formed. Muscles vary widely in metmyoglobin-reduction activity. Those that tend to have a high activity, such as the longissimus dorsi, are more color-stable in air, their red color persisting for 3 or 4 times as long as color of unstable muscles of low metmyoglobin-reduction activity, such as the psoas. Metmyoglobin-reduction activity dissipates during storage of muscle. After lengthy storage, the color stability of initially color-stable muscles is similar to that of those muscles that were initially of relatively poor color stability (26).

Maintaining meat at as low a temperature as possible without freezing will slow pigment oxidation and increase the depth of the oxygenated surface layer. Two obvious means of preserving muscle color are to increase the proportion of oxidation-resistant oxymyoglobin by high-oxygen MAP or to largely exclude oxygen using ultra-low-oxygen MAP.

High-oxygen MAP is used mainly for retail product. If used for poultry, high-oxygen atmosphere may be inappropriate because of the meats' limited myoglobin level. The product may be packaged in gas barrier trays that contain the preservative atmosphere and are sealed with a lidding film of low gas permeability. Alternatively, several trays of product overwrapped with a film of high gas permeability may be master packaged in a bag of low gas permeability. The gas with which packs are filled may have a composition of about 65% O_2, 25% CO_2, and 10% N_2. The CO_2 is required to retard the growth of aerobic spoilage bacteria. Nitrogen has no preservative function for the meat and may be omitted, but it is often included to guard against the collapse of sealed trays that may result from the dissolution of CO_2 into the muscle.

The deoxymyoglobin chemical state predominates in normal muscle before cutting. Upon exposure to oxygen, usually in air, the oxygen diffuses further into the muscle with time, creating an oxymyoglobin layer on the meat surface. Diffusion into the muscle is more rapid in lower-pH muscle due to a more open muscle structure (27) and less scavenging of oxygen by muscle enzyme systems. Higher-oxygen tension at the meat surface, such as in high-oxygen MAP or by hyperbaric oxygen pressures, will drive the oxymyoglobin layer more deeply into the muscle, with the remaining muscle in the deoxymyoglobin state because of very low partial oxygen pressures in the meat-cut interior. At some variable time in display, metmyoglobin-reduction activity at the interface of the oxydeoxymyoglobin interface will be exhausted, so a layer of brown metmyoglobin begins to form, then becomes broader and begins to move toward the meat surface, diminishing the depth of the oxymyoglobin layer. At some point the brown metmyoglobin is close enough to the meat surface at some locations to contribute brown to the visual color as viewed on the meat surface, or as detected by a reflectance spectrophotometer. Because both the oxidative-reductive state and partial oxygen pressure may vary at different meat locations, the brownness is frequently spotty and not uniform, especially in ground meat where variable amounts of oxygen have been incorporated into the meat.

High-oxygen MAP systems often have atmospheres of 20% carbon dioxide and up to 80% oxygen and can produce and maintain a desirable red color in beef for up to 9 days and depress the formation of metmyoglobin by driving oxygen deep within the surface of the meat. Beef loin eye steaks in high-oxygen (70%) packages with 30% carbon dioxide had brighter, more desirable visual color scores and higher CIE a* (redness) values than vacuum controls (28) but with undesirable color beginning at 10 days of storage. Another study found that beef loin eye steaks were stable in high oxygen (70%) up to 14 days (29). Ground beef in a similar treatment was stable for only 6 days (30). Color life varies between species as well as muscle type. Pork loin chops in high oxygen had acceptable saturation indices and display color for 8 to 12 days with off-odor, not color, in beef steaks and pork the limiting factor to shelf life (31). High-oxygen-packaged beef loin eye steaks from loins in vacuum less than 7 days postmortem had much better display color stability than steaks packaged from loins vacuum aged for 14 or 28 days.

Carbon monoxide has been used in beef MAP systems to maintain a bright cherry-red color. Carbon monoxide binds strongly to myoglobin and hemoglobin, forming carboxymyoglobin and carboxyhemoglobin, respectively, pigments that are stable bright-red compounds (1). Round steak (pieces) stored at 0, 5, and 10°C in carbon monoxide from 0.5 to 1.0%, with the balance as nitrogen, had increased color and odor shelf life up to 20 days over that stored in air. Beef loin eye steaks at 4°C in a 0.4% carbon monoxide, 60% carbon dioxide, and 40% nitrogen system, compared with a high-oxygen (80%) system or vacuum packaging, were brighter red than the other two treatments and had better display color stability (28). Carbon monoxide at 0.4 to 0.5% is currently utilized safely in some domestic European fresh-meat markets. Low-oxygen packaging systems lack the benefit of high oxygen to cause oxymyoglobin to be formed more deeply into the meat. Low-oxygen systems often have too much oxygen for the meat to naturally scavenge and therefore may not discourage the formation of metmyoglobin.

In MAP systems with no oxygen in the gas mixture (ultra-low oxygen), a slight amount of residual oxygen frequently occurs due to small pockets of air not removed initially by the evacuation or flushing process. These small concentrations of oxygen frequently cause major discoloration, reduced color stability, or blooming ability problems. Initial oxygen concentrations $>0.15\%$ (1,500 ppm) seriously compromised the color stability of beef and lamb. Pork was affected by residual oxygen concentrations $>1.0\%$. At low oxygen partial pressure, oxidation of oxymyoglobin to metmyoglobin is highly favored. Oxygen concentrations between 0.5% and 2% most rapidly revert myoglobin to metmyoglobin (24). Beef or lamb stored in such atmospheres may be initially discolored by the residual oxygen, but fresh red meat retains a limited reducing capacity by which it can recover from this initial discoloration. Compared to pork, beef and lamb were more sensitive to residual oxygen. Provided the amount of oxygen did not exceed 0.5%, fresh meat will naturally scavenge most residual oxygen, but more muscle scavenging will reduce later ability to bloom (oxygenate). Constant exposure to low concentrations of oxygen will have detrimental effects on reblooming ability and color stability. Therefore, if the residual oxygen could be removed earlier, the meat would have less discoloration, better rebloom, and improved color stability.

Rapid scavenging of oxygen in an ultra-low-oxygen package is important if the meat is expected to bloom with acceptable speed and completeness upon later exposure to oxygen. Muscle oxygen scavenging may not be fast enough, and added chemical scavengers may be necessary.

SUMMARY

Storage life of MAP-packaged meat may be limited by microbial deterioration; loss of salable appearance; or, in limited circumstances, by development of oxidized odors, tastes, or both.

The meat industry is rapidly moving toward preparation of case-ready retail cuts at the slaughter-fabrication plants, and MAP is essential to make such a system viable.

The essential keys to success are:

1. Clean, high-quality product into the package
2. Rapid completion of packaging after fabrication
3. Cold temperature and good temperature control
4. Complete evacuation and fill with MAP gases
5. Sound package, good seal, and minimum of leakers
6. Final retail package not very different from what customers are used to
7. Package remains sealed until consumer opens it

BIBLIOGRAPHY

1. D. H. Kropf, "Effects of Display Conditions on Meat Color," *Proc. 33rd Annu. Reciprocal Meat Conf.*, West Lafayette, Ind., June 22–25, 1980.
2. Y. Zhao, J. H. Wells, and K. W. McMillin, "Application of Dynamic Modified Atmosphere Packaging Systems for Fresh Red Meats: Review," *J. Muscle Foods* **5**, 299–328 (1994).
3. E. H. Sanders and H. M. Soo, "Increasing Shelf Life by Carbon Dioxide Treatment and Low Temperature Storage of Bulk Pack Fresh Chickens Packaged in Nylon/Surlyn Film," *J. Food Sci.* **43**, 1519–1527 (1978).
4. F. M. Christopher et al., "Effect of CO_2-N_2 Atmospheres on the microbial Flora of Pork," *J. Food Prot.* **43**, 268–271 (1980).
5. C. O. Gill, "Packaging Meat for Prolonged Chill Storage: The CAPTECH Process," *Br. Food J.* **91**, 11–15 (1989).
6. W. H. Gehrke, "Film Properties Required for Thermo Formed and Thermal Processed Meat Packages," *Proc. 36th Annu. Reciprocal Meat Conf.*, Fargo, N.D., June 12–15, 1983.
7. M. Salame, "Barrier Polymers," in M. Bakker, ed., *The Wiley Encyclopedia of Packaging Technology*, Wiley, New York, 1986, pp. 48–54.
8. A. E. Lambden, D. Chadwick, and C. O. Gill, "The Oxygen Permeability at Sub-Zero Temperatures of Plastic Films Used for Vacuum Packaging of Meat," *J. Food Technol.* **20**, 781–783 (1985).
9. I. J. Eustace, "Some Factors Affecting Oxygen Transmission Rates of Plastic Films for Vacuum Packaging of Meat," *J. Food Technol.* **16**, 73–80 (1981).
10. C. O. Gill, "The Solubility of Carbon Dioxide in Meat," *Meat Sci.* **22**, 65–71 (1988).
11. G. C. Holland, "Modified Atmospheres for Fresh Meat Distribution," *Proc. Meat Industry Res. Conf.*, Chicago, Ill., March 27–28, 1980.
12. K. E. Warren et al., "Modified-Atmosphere Packaging of Bone-In Pork Loins," *J. Muscle Foods* **3**, 283–300 (1992).
13. D. J. Warburton, "The CAPTECH Process," *Proc. 25th Meat Industry Res. Conf.*, Hamilton, New Zealand, July 12–14, 1988.
14. T. P. Labuza and W. M. Breene, "Applications of 'Active Packaging' for Improvement of Shelf-Life and Nutritional Quality of Fresh and Extended Shelf-Life Foods," *J. Food Processing and Preservation* **13**, 1–69 (1989).
15. C. O. Gill and J. C. McGinnis, "The Use of Oxygen Scavengers to Prevent the Transient Discoloration of Ground Beef Packaged under Controlled, Oxygen-Depleted Atmospheres," *Meat Sci.* **41**, 19–27 (1995).
16. H. Nakamura and J. Hoshino, "Techniques for the Preservation of Food by Employment of an Oxygen Absorber," in *Sanitation Control for Food Sterilizing Techniques*, Sanyo, Tokyo, Japan, 1983.
17. P. Cook, "Stealth Scavenging Systems: The Scavenger Is the Polymer," *Inst. Food Technologists Annu. Mtg.*, Atlanta, Ga., June 20–24, 1998.
18. C. O. Gill and K. H. Tan, "Effect of Carbon Dioxide on Growth of *Pseudomonas fluorescens*," *Appl. Environ. Microbiol.* **38**, 237–240 (1979).
19. K. G. Newton and C. O. Gill, "The Development of the Anaerobic Spoilage Flora of Meat Stored at Chill Temperatures," *J. Appl. Bacteriol.* **44**, 91–95 (1978).
20. C. O. Gill and J. C. L. Harrison, "The Storage Life of Chilled Pork Packaged under Carbon Dioxide," *Meat Sci.* **26**, 313–324 (1989).
21. C. O. Gill and M. P. Reichel, "The Growth of the Cold-Tolerant Pathogens *Yersinia enterocolitica, Aeromonas hydrophila*, and *Listeria monocytogenes* on High-pH Beef Packaged under Vacuum or Carbon Dioxide," *Food Microbiol.* **6**, 223–230 (1989).
22. C. O. Gill, "Extending the Storage Life of Raw Chilled Meats," *Meat Sci.* **43**, S99–S109, (1996).
23. D. B. MacDougall, in G. G. Birch, J. G. Brennan, and K. J. Parker, eds., *Sensory Properties of Foods*, Applied Science, London, 1977.
24. J. Brooks, "The Oxidation of Hemoglobin to Metmyoglobin by Oxygen, II: The Relationship between the Rate of Oxidation and the Partial Pressure of Oxygen," *Proc. Rl. Soc. London Ser. B* **118**, 560–577 (1935).
25. D. A. Ledward, "Post-Slaughter Influences on the Formation of Metmyoglobin in Beef Muscles," *Meat Sci.* **15**, 149–171 (1985).
26. V. J. Moore and C. O. Gill, "The pH and Display Life of Chilled Lamb after Prolonged Storage under Vacuum or under CO_2," *New Zealand J. Agricultural Res.* **30**, 449–452 (1987).
27. M. C. Hunt and D. H. Kropf, "Color and Appearance" in A. M. Pearson and T. R. Dutson, eds., *Restructured Meat and Poultry Products*, Advances in Meat Research, Vol. 3, 1987, pp 125–159.

28. O. Sörheim, H. Nissen, and T. Nesbakken, "Shelf Life and Color Stability of Beef Loin Steak Packaged in a Modified Atmosphere with Carbon Monoxide," *Proc. 43rd Int. Congr. Meat Sci. Technol.*, 1997.
29. G. L. Nortjé and B. G. Shaw, "The Effect of Aging Treatment on the Microbiology and Storage Characteristics of Beef in Modified Atmosphere Packs Containing 25% CO_2 plus 75% O_2," *Meat Sci.* **25**, 43–58 (1989).
30. C. O. Gill and T. Jones, "The Display of Retail Packs of Ground Beef after Their Storage in Master Packages under Various Atmospheres," *Meat Sci.* **37**, 281–295 (1994).
31. T. J. Clark, "Rebloom and Display Color Stability of Beef and Pork Packaged in an Ultra-Low Oxygen Modified Atmosphere Packaging System," M.S. Thesis, Kansas State University, Manhattan, Kans., 1998.

Donald H. Kropf
Kansas State University
Manhattan, Kansas

MEAT PROCESSING: TECHNOLOGY AND ENGINEERING

The major processed meats consumed in the United States are made from beef, pork, turkey, or a combination of these. The term all-meat product usually refers to a mixture of beef and pork or beef, pork, and turkey. According to the type of process equipment used, the processed meats can be grouped into one of three categories: (1) muscle products, (2) coarse-ground products, and (3) emulsified products. Products in each group often share common process equipment. The process flowchart is illustrated in Figure 1. The goal of muscle product processing is to keep an original intact tissue appearance in the finished product. The products, such as ham, roast beef, and turkey breast, are produced from whole muscles or sectioned muscles. For coarse-ground products, the meat particles are substantially reduced and are then restructured back to a different physical form. The finished products, such as dry sausage and smoked sausage, still present the recognizable meat particles. Emulsified products are made from meats that are completely chopped and further reduced to a pastelike batter. The finished products are a smooth, homogeneous mass. The meat fibers are beyond recognition. Bologna and hot dogs belong to this category.

In the processed-meat industry, raw meats, either whole muscles or altered meat particles, are restructured to the predetermined product characteristics. The major processing equipment includes injectors and massagers for muscle products, grinders and mixers for coarse-ground products, and choppers and emulsifiers for emulsified products. Injecting and massaging processes facilitate brine distribution and protein extraction. Grinding reduces meat particle size, and mixing assures uniformity of chemical composition as well as protein extraction. The chopping process also reduces particle size and obtains salt-soluble protein. The emulsifying process forms a matrix in which the fat particle is encapsulated with the protein membrane (1).

Raw meats, after these preparation processes, are stuffed into casings or molds to form a defined geometric shape and size. During these processes, rheological properties are critical to equipment performance and final product quality (2–4). Regardless of the product category, the most important missions of process equipment are to extract salt-soluble protein for binding meats together and to hold moisture and fat inside the finished product.

INJECTING

Curing with a high concentration of salt was the sole method used for meat preservation in ancient times. With the availability of refrigeration today, mild curing is employed to improve eating quality more than for preservation. The common ingredients used in a pickle solution, or brine, are salt, nitrite, phosphates, sodium erythorbate, sugar, and starch. Because of the different solubility of each ingredient, the hard-to-dissolve ingredients should be mixed and dissolved first, and then the other ingredients should be added. The following preparation procedure is recommended (5).

1. Dissolve the sodium phosphates in cold water (40°F).
2. Add the sodium erythorbate and dissolve.
3. Add the remaining ingredients (salt, nitrite, sugar, etc).

Agitation in the brine tank is usually needed to prevent precipitation.

Accurate pumping and uniform brine distribution in meat muscle are two major process control points. The amount of pumping determines the profitability and legality of products. Uneven distribution of pickle can cause color, flavor, and texture problems. Small-diameter needles, closely spaced, provide more uniform injecting and less muscle damage. Because small-diameter needles are used for pickle injection, a filter system to prevent particles from plugging up the hole of the needles is vital for accuracy and uniformity. The pickle solution must be kept free of meat particles or other nondissolved ingredients before coming into the needle manifold. The injection needle can be single-hole or multihole. The former can create a large brine pouch inside the meat muscle and cause a leakage problem when the needle pulls out. The multihole needle is preferred for better distribution and less leakage. The amount of pumping is usually controlled by pressure, belt speed, and stitch sequence.

MASSAGING AND TUMBLING

After pickle injection, muscle products require a massaging or tumbling process to facilitate the salt-soluble protein extraction and pickle diffusion (6–8). The desired result is to produce a product with an original intact tissue appearance after heat processing. Before the massaging process, mechanical tenderization or maceration is often used to increase the surface available for protein extraction and to improve binding during cooking and pliability during stuffing. However, this step is not absolutely necessary. Tenderization or maceration has a tendency to disturb the

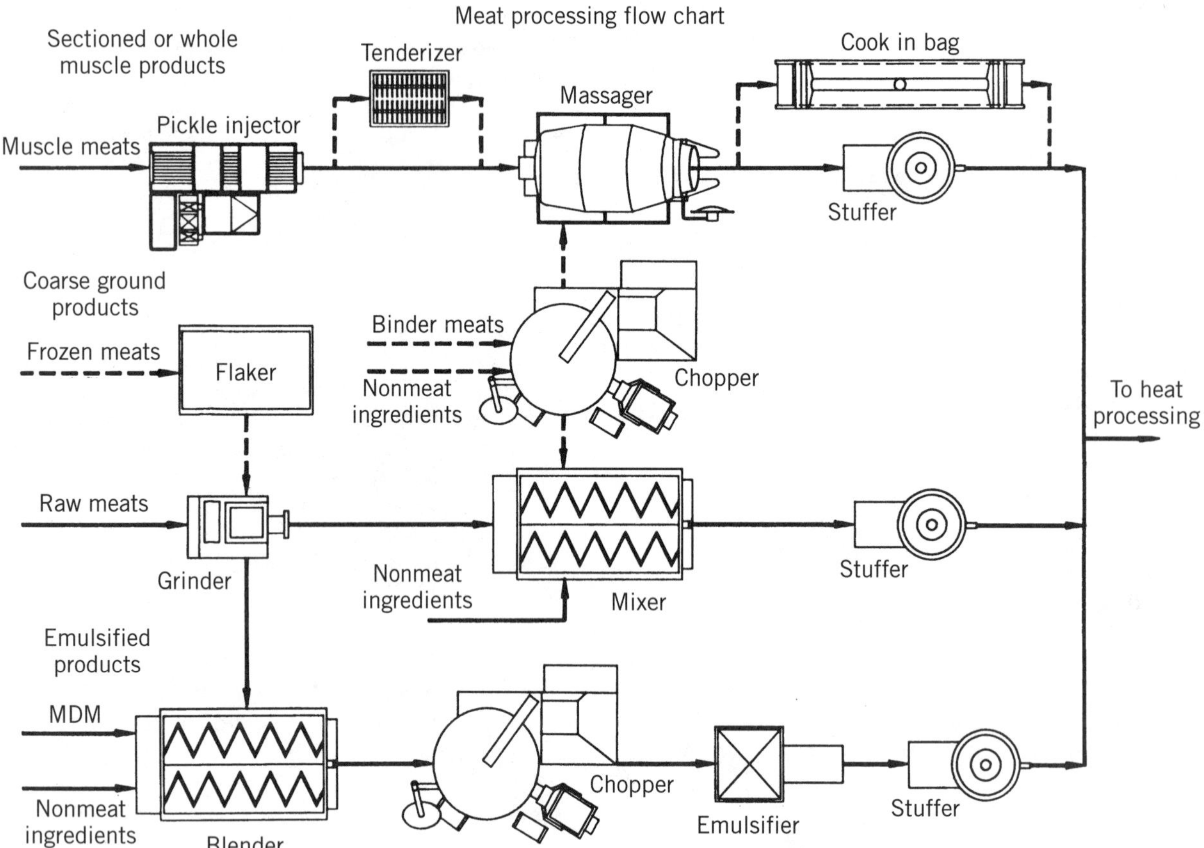

Figure 1. Meat processing flow chart.

original tissue appearance. The excessive pickle carried over from the injecting process is usually massaged into the product in order to attain a desirable brine level.

A massager is comprised of a stationary vat having one or two shafts extending therein with a plurality of paddles. The meats in the vat are kneaded by a slow-rotation mixing of paddles, which forces the meats to rub against one another. A tumbler is a rotating drum with blades built longitudinally on the inner surface. As the drum rotates, the meats inside the drum are lifted by the blades and fall on top of one another. If the blades have an angle with rotation direction, they force the meats to move against themselves.

A tumbler provides not only meat-on-meat friction but also falling-impact action. The most advanced tumbler has been improved to produce a gentle effect, similar to a massage machine. It consists of a rotating drum with two integral helical flights. The flights do not rotate independently from the drum and hence do not cause tearing and shearing effects as does a mixer. In addition, the flights are in a helical arrangement; they do not lift and drop meat like straight blades. There is no falling impact. As the drum rotates, the helical flights force the product into an elliptic motion and produce meat-on-meat massaging action. In industry, the massager and the tumbler are used equally because they provide the same function.

From a scanning electron micrograph study, fiber destruction was evident after a few hours of massaging (8). Further massaging resulted in longitudinal disruption. Intermittent tumbling resulted in more alteration in the cell structure than did continuous tumbling based on the constant tumbling time (9). Because the destruction of the cells on the surface and also deep inside the muscle provides more available salt-soluble protein, the massaging or tumbling process produces a better colored, more tender, and higher yield product. The application of a vacuum during massaging can further improve the product color, cohesion, and sliceability. Applying a vacuum during the massaging also gives a 2–3% higher yield. The optimum massaging temperature is suggested to be 43–46°F (10).

GRINDING

The grinder is the most commonly used piece of equipment in the meat industry for reducing particle size. The size of the meat particle is determined by the diameter of the hole in the plate and the revolutions per minute of the cutting knife. The grinding temperature is critical for obtaining clear particle definition and is found to be at or near the freezing point of meat, 28°F. The sharpness of cutting knives and the grinding plate is also vital to produce a product with a less smeared appearance. Products with high fat content can usually be ground at a colder temperature than can lean products. A double-plate grinding system can often be adapted to the regular grinder with adequate horsepower to reduce the capital and operation costs as well as to maximize use of space.

One quality concern of coarse-ground products is bone chips and gristle in the finished products. An automatic bone elimination system is often built into a grinder to remove bone, gristle, and any other hard-to-grind materials. There are two general types of removal systems: central removal and peripheral removal. The latter system passes the least amount of bone to the finished product and also has less of a plug-up problem for a prolonged period of use (11).

In poultry plants, a low-pressure grinding system has been used increasingly to separate meat from bone and sinew. Raw meat is fed into a precrusher, which consists of dual crushing rollers. The crushed meat is then carried and forced against a perforated drum by a flexible conveyor. Meat is pressed through the perforations into the inside of the drum. The deboned meat inside the drum is then discharged at the drum opening by a static product scraper. The bones, sinew, etc are carried around the outside surface of the drum and are then removed by a scraper knife. The low-pressure grinder functions very effectively for poultry products, but ineffectively for red meats, owing to the toughness of the muscle.

CHOPPING AND EMULSIFYING

A chopper is the other piece of equipment commonly used to reduce meat particle size. The major function of a grinder is solely to reduce particle size with clear particle definition. However, a chopper is employed to reduce particle size, to extract salt-soluble protein, to disperse fat, and to mix meats and spices uniformly. Products that go through a chopping process do not possess a particle definition as clear or as uniform as that for ground products. The chopper is usually used as the first phase of size reduction before the emulsifying process. Chopping is one of the most important processes to control emulsified product quality (12,13). The chopping process determines the amount of available protein and the size of fat particles. If the meats are not adequately chopped, there is a lack of soluble protein to coat the fat particles. If the meats are overchopped, the fat particle size is reduced and the surface area increases enormously. Eventually, the fat surface becomes so large that the protein solution cannot coat all the fat particles. In both cases, fat will render from emulsion during heat processing. To determine the optimum chopping condition, emulsion temperature is generally used as the process control criterion. An emulsion temperature of 55°F after chopping was found to be most stable (14). The variables that can effect emulsion stability in a chopper are chopping time, bowl speed, knife speed, bowl size, knife size, number of knives, and the mass of emulsion inside a chopper (2).

The emulsifier used in the meat industry is similar to a grinder. Because products that require an emulsifying process are usually prechopped or preground to a very fluid mass, an emulsifier normally does not have a feed screw to deliver product to the emulsifying plate. The cutting knives of the emulsifier function as impellers to suck meats from the hopper and to force them through the emulsifying plate simultaneously. The hole in the emulsifying plate ranges from 1 to 3 mm. The meat temperature out of the emulsifier should be kept below 70°F to assure emulsion stability.

In recent developments, one single piece of equipment combining chopping and emulsifying has been developed. The ground product is loaded in a bowl where a vacuum can be created and is chopped by two counter-rotating knives. Cutting time, temperature, and vacuum can be controlled by a microprocessor. A vertical augur turns clockwise gently while an agitator stirring arm lifts the product off the wall of the bowl. When the chopping cycle ends, the augur reverses and begins feeding the chopped meat into the emulsifier. The emulsifier consists of double plates. The product is forced through the plates by the impeller. The other new development is to eliminate chopping by combining chopping and emulsifying operations in one emulsion mill. The finely ground meat in a storage hopper is transferred to a deaerator. The deaerator plays dual functions; it removes air and accumulates meat temporarily. At the bottom of the deaerator, a positive pump forces meat through cutting knives and an emulsifying plate. The particle size and emulsion temperature are controlled by flow rate. Both systems can keep the whole process under vacuum conditions and also can save on unit operation cost.

Some physical and chemical properties of comminuted meats have been studied. The research on protein–protein interaction shows that the reaction among proteins results in molecular aggregation that is reversible between 39 and 86°F (15). A study on structural change during chopping and cooking has indicated that muscle fibers were reduced in size but some remain intact; fat particles were reduced in size with only a portion of them being surrounded by a protein interface, and collagen fibers were somewhat dispersed but otherwise unaltered (16). The application of a vacuum during the chopping process does promote cured color development (17).

MIXING

The mixing process is a necessary procedure for coarse-ground products. The only purpose of grinding is to reduce particle size. Lean and fat meat extruded through the grinder must be thoroughly mixed in order to assure the uniformity of chemical composition and to extract salt-soluble protein for fat-holding and water binding (18). A mixer consists of a hopper and a pair of agitators. Ribbon, paddle, or solid flight agitators are commonly used for mixing meat products. Agitators are designed to tumble and to circulate meat slowly inside the mixer. The agitators usually rotate in opposite directions to perform overlapping action. The optimum mixing temperature is suggested to be below 40°F to prevent smear problems.

Oxygen, which exists in meat products, accelerates the spoilage of finished product. A vacuum is often applied during the mixing process to evacuate air. To keep product temperature cold during the mixing operation, ice, chilled water, or liquid gas are added into the mixer. Meat is an excellent carbon dioxide (CO_2) absorbant. If dry ice is used for chilling, a vacuum should be applied to remove dissolved CO_2 at the end of the mixing cycle. Otherwise, the dissolved CO_2 will be released in heat processing or vacuum packaging and cause quality and shelf life problems.

STUFFING

Meat product is usually stuffed into natural casings, cellulose casings, collagen casings, cook-in-bags, or molds to form the final geometric shape. Several types of pumps are used in stuffers, such as piston, gear, vane, or worm gear types. For small link products, a continuous stuffing, linking, and looping machine is most commonly employed. The machine can produce 3,000–4,000 lb/h. For a large-diameter product, such as bologna, the meat dough is stuffed into the casing with a casing sizer to ensure a uniform diameter from one end to the other. A metal mold is often used in producing large-diameter products to replace expensive casing.

Sausage manufacturers spend about 5 percent of the cost on casing to form product and then peel it off and throw it away. To save the cost, a highly automatic coextrusion system has been developed. The system consists of two meat ingredients; one is sausage meat and the other a collagen paste that is derived from hide. Two meat ingredients are fed through a co-extrusion nozzle simultaneously. The nozzle consists of a central tube surrounded by two cone-shaped elements between which there is a narrow gap ending in an orifice circling the tube. The cones spin in opposite directions. The collagen is forced through the gap and spun onto an endless rope of sausage meat. This rope is then passed through a brine bath to coagulate the collagen and then through a crimper that separates and forms the individual links. The diameter of sausage is determined by the size of the nozzle. The co-extruded products are then transferred onto a conveyor to the drying tower. The co-extruded product has bite characteristics similar to those of the natural-casing product.

HEAT PROCESSING

Meat products, after being restructured from raw materials, are stuffed into casings, bags, or molds and are subjected to heat processing—except for the fresh-product category. The basic functions of smoking and heating processes are to inhibit bacterial growth, to provide desirable color, and to impart specific flavor and texture. Heat processing in the meat industry covers smoking, drying, cooking, cooling, and freezing (19). The smokehouse is the major heat-processing equipment. When meat products are subjected to heat processing in a smokehouse, not only does simultaneous heat and mass transfer occur, but also microbial and biochemical reactions are induced. The theoretical approach to process and engineering design is limited. According to material handling criteria, there are two types of smokehouses available. One is a batch-type smokehouse and the other is a continuous smokehouse. Batch-type smokehouses provide a cold-water shower for precooling. The final chilling is conducted at the separate blast-air chilling or brine chilling room. For a continuous smokehouse, the product is loaded on a conveyor and automatically carried through smoking, drying, cooking, and chilling processes. Products, after leaving the continuous oven, are ready for peeling, slicing, and packaging. Detailed heat-processing flowcharts for both batch and continuous smokehouses are shown in Figure 2.

Smoking

The smoking of meat is an ancient practice to preserve meat products in addition to developing color and flavor. The role of smoke as a meat preservative has declined owing to the availability of refrigeration. Today, the smoking process is used to enhance flavor and surface color. Several hundred chemical compound have been isolated from wood smoke. The most important compounds are acids, phenols, and carbonyls. Phenols and carbonyls are major contributors of smoke flavor and color, respectively. The acidic substances accelerate the curing reaction and contribute to the pink color of cured meats. For processing convenience, natural smoke is, for some uses, condensed to liquid form. It can be added directly to a product or applied on a product surface by dipping, spraying, atomizing, or regenerating.

Natural Smoke Generation

Natural smoke is produced from thermal decomposition of wood and is known as pyrolysis. The major constituents of wood are cellulose, hemicellulose, and liqnin. During thermal decomposition, a temperature gradient exists between the outer and inner core of wood chip or sawdust. The outer surface is being oxidized, and the inner surface is being dehydrated before it can be oxidized. During the dehydration process, the outer surface is at about water boiling temperature. Carbon monoxide, carbon dioxide, and some volatile short-change organic acids are released during the dehydration process. When the internal moisture is dried out, the temperature rapidly rises to 300–400°C (570–750°F). As long as the internal temperature reaches this range, the decomposition occurs and smoke is given off. The hemicellulose fraction is the first to undergo degradation. Cellulose and liqnin degradation follow accordingly.

Under normal smoking conditions, the smoking temperature ranges from 100 to 400°C (212–750°F) or higher. This results in the generation of more than 400 smoke compounds. Smoke composition can vary substantially with smoke generation temperature as well as with different varieties of wood. The effective smoke compounds—phenols, acids, and carbonyls—change with generation temperature, as shown in Table 1 (20); wood used here is maple sawdust.

Thermal decomposition of wood is induced by high temperature. Heat sources can be hot plate, self-burning, or friction. Sawdust is normally used in a smoke generator with hot plates. Sawdust is loaded in a hopper, and the feeding rate is controlled by the rotation speed of the feed plate. Under the feed plate there are three hot plates with electric heaters under each plate. On top of each hot plate there is a wiper to sweep sawdust across the hot-plate sur-

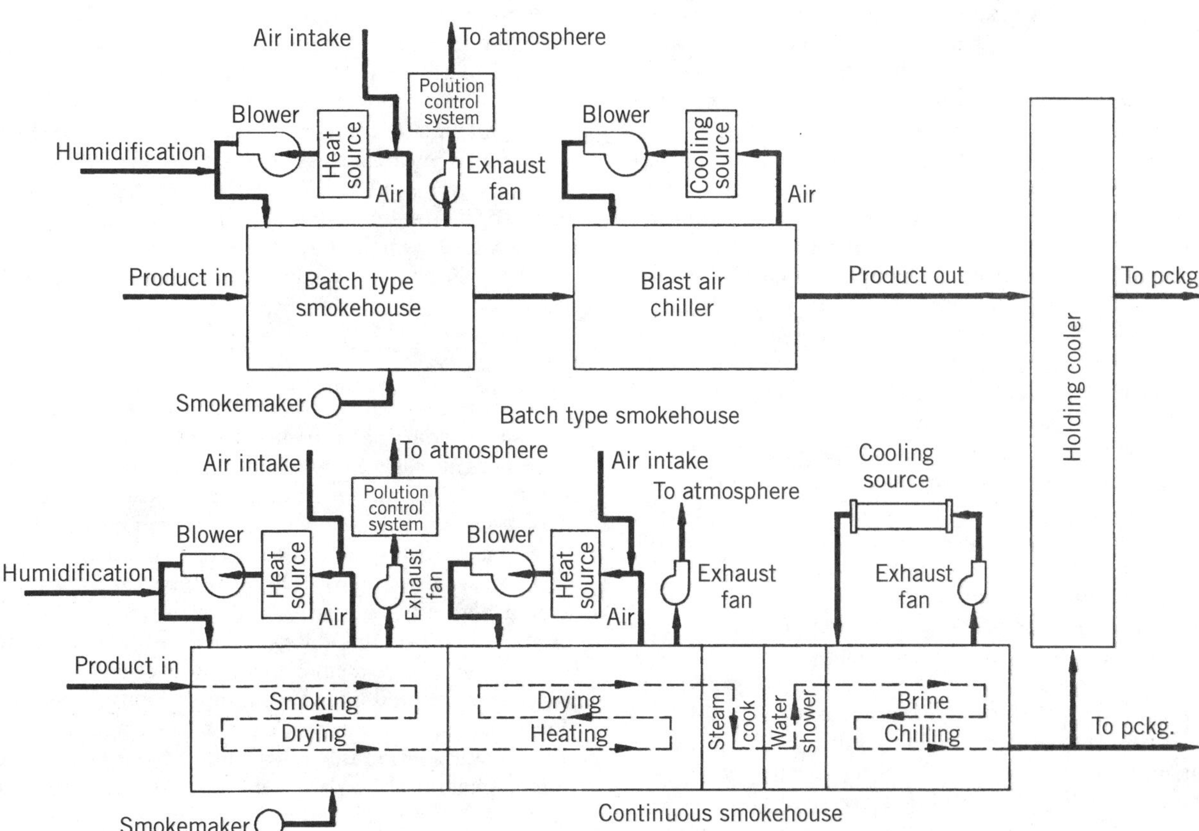

Figure 2. Heat processing flow chart.

Table 1. Smoke Composition Change Subjected to Smoke Generator Temperature

Generator Temperature (°F)	Acids (wt%)	Phenols (wt%)	Carbonyls (wt%)
575	41	2.3	57
600	40	1.7	59
650	32	1.7	66
725	27	1.6	72
780	24	1.5	84

face. The final ash is discharged into a cold-water stream and carried to an ash collector.

In the smoldering smoke generator, smoking materials—sawdust or wood chips—are ignited by an electric heating element. The ignition time can be preset. When the set time is expired, the ignition automatically switches off. Air is injected at the bottom by a blower and is heated by the burning sawdusts or wood chips. The heated air is moving upward, and sawdust is supplied from the top, functioning as a countercurrent flow. Because of the heat of the burning sawdust beneath, smoke is generated as fresh sawdust travels from top to bottom. Smoke produced with the smoldering principle often has high air velocity, leading to excessively high smoldering temperature. The smoke-generation temperature can be lowered by reducing the air supply and slightly moistening the sawdust or wood chips to a moisture content of 30%.

The friction smoke generator produces smoke by pressing a piece of wood log tightly against a spinning disc. Heat is produced by fierce friction. This results in pyrolysis of the wood. Friction smoke generators produce less amounts of smoke than does the sawdust burning type of smoke generator. Another method used for smoke generation is the superheated steam method. The low-pressure steam is mixed with air and conducted to the superheater. The temperature of the steam/air mixture can be adjusted within the prescribed range of 243–400°C (470–750°F). The amount of air can be varied as required and can thus have a varying effect on smoke production. A screw conveyor brings sawdust and superheated steam/air mixture into a smoke-generating chamber to induce pyrolysis. Among these smoke-generation methods, the hot-plate smoke generator and the smoldering smoke generators are most popular in the United States.

Smoke Deposition

Smoke deposition on a product involves several physical and chemical processes linked closely with one another. An important chemical process is the carbonyl-amino reaction, which produces a desired golden brown color. As for the physical process, the dominant mechanism of the smoking process has been identified as vapor absorption. During the normal smoking process, vapor smoke and the liquified smoke particulates coexist in an equilibrium condition. The smoke particulates function as a reservoir of smoke constituents. When the equilibrium between smoke vapor and particulates is changed by dilution of air because of absorption from the product or as a result of increased smoke temperature, the reservoir releases a part of its contents to maintain the new equilibrium condition. Because the smoking process is an absorption, the major physical parameters effecting the rate of smoking deposition are smoke density, air velocity, relative humidity, and the surface condition of products. It is clear that a wet surface absorbs smoke faster than a dry surface and that smoke absorption should follow the first order of kinetics. The higher the concentration gradient between the ambient and surface, the faster the absorption rate. The rate of smoke deposition is also faster at high air velocities, although it is difficult to maintain high smoke density and high air velocity simultaneously. Smoke density and air velocity should be optimized according to smoke-house pressure as well as the exhaust system.

The smoking process based on air temperature can be classified as cold smoking, warm smoking, and hot smoking. The temperature ranges are below 27°C (80°F), between 22 and 54°C (80–130°F), and above 54°C (130°F), respectively. The relative humidity during cold smoking is not as critical as during hot smoking because of the low drying rate. In the hot-smoking process, the relative humidity should be precisely controlled to keep the product surface damp. The optimum relative humidity (RH) in a smokehouse for hot smoking is 60% RH and 71.1°C (160°F) (21).

Smoking of meat products results in altering the protein composition, such as lowering myofibrillar and sarcoplasmic protein nitrogens while increasing stromal protein nitrogen. These protein changes result in a cross-linking of surface protein or tough skin (22).

Liquid Smoke

To find a way of better controlling smoke reaction, decrease the chance of carcinogenesis, reduce smoke effluent, and make the application easier, liquid smoke has emerged. Liquid smoke is generally produced from natural smoke by condensation or water scrubbing. The most prevalent method used in the United States is the water-scrubbing method. This method includes the smoldering of sawdust, controlling oxidation condition, and absorption of smoke in water. In this process, the smoke solution is recycled until a given concentration is developed. The other method is the pyrolysis of hardwood sawdust, where the generated smoke is condensed to a liquid-form solution through a series of condensers. Liquid smoke can be applied directly into raw products, thereby producing smoke flavor and color simultaneously. Liquid smoke is often applied to a product surface. The application technique can be classified as dipping, spraying, atomization, and regeneration. Dipping or spraying is implemented immediately after stuffing or before thermal processing. Atomization is using high liquid pressure to form a cloud of liquid smoke to which products are then exposed. Atomization can be applied either before or during heat processing. Regeneration of liquid smoke is converting the liquid smoke from a liquid phase to a gas phase by heat. The regenerated smoke can have the same application as for natural smoke.

Liquid smoke has been available for a while. However, it cannot replace natural smoking because of an unsatisfactory flavor profile. In order to impact sufficient smoke color on the product surface, liquid-smoked products often develop bitter smoke flavor. Recently, liquid smoke flavor has been improved by adjusting the phenol and carbonyl ratio.

Pollution Control

The natural wood smoke that can cause irritation and odors is discharged to the atmosphere with the exhausted air that is necessary for humidity control inside the smokehouse. To eliminate or reduce air pollution, afterburners, closed-system smokehouses, water-scrubbing systems, or electrostatic precipitators have been employed. The afterburner uses heat, above 1,000°F, to burn the contaminants. It is a high-efficiency air-pollution control unit. High energy cost has prohibited its application. The basic principle of the closed-system smokehouse is to circulate humid air through a dehumidifier for humidity control without discharging the contaminated air into the atmosphere. However, a substantial amount of smoke must be carried into the smokehouse. Some air must be released in order to attain the material and pressure balance. Therefore, a complete closed-system house cannot be achieved for a natural smoke system (23). For a water-scrubbing system, air with smoke contaminants passes through a venturi. In the venturi, air with contaminants come in close contact with the scrubbing water. The collection of particulates on scrubbing is accomplished by inertia impaction, interception, and diffusion. The water droplets are collected from the stream by a baffle and a mist eliminator before the scrubbed air is discharged to the atmosphere. The water scrubber can achieve up to 80% efficiency. The electrostatic precipitator draws the contaminated air through prefilters that collect the larger particles and then the air passes through ionizers that electrically charge all the particles in the airstream. Ionized particles enter the collecting cells, where ground plates remove the particles. The electrostatic precipitator can effectively remove the particulate material in the air-stream but cannot eliminate smoke odors in the gas phase. To improve efficiency, a chemical absorption tower is installed after the electrostatic filter. The alkaline solution is circulated inside the tower to dissolve smoke odors as well as to keep the tower clean. The new system can provide up to 97% efficiency at a flow rate of 1,500 ft^3/min.

Drying and Heating

The smokehouse is the major heat-processing equipment for producing processed meats. The modern smokehouse is equipped with instruments for programming air temperature and humidity during the process cycle. The air can be heated with direct gas-firing, steam coil, or electric heating elements. The humidity of air in a smokehouse is controlled by the amount of air intake and exhaust and the injection of steam when necessary. The conditions of returned air (dry-bulb and wet-bulb temperatures) are usually used as the process control references. The conditioned air is then introduced into the smokehouse by a blower. The air circulation rate is normally greater than 10 air changes a minute. One air change is equivalent to one smokehouse volume in cubic feet. The airflow coming into the smokehouse is alternated from one side to the other to provide more uniform heat processing. Heat-processing schedules are specifically designed according to product quality requirement as well as microbial kill. Because there are many different products and quality specifications, heat-processing schedules also differ substantially. In designing a smokehouse, the production rate, product drying rate, and product heating rate with a specific set of heat-processing schedules should be predetermined. By setting up material and energy balance, the equipment capacity for air-handling, heating, and cooling can be determined.

The quality of processed meats is significantly affected by the heat-processing schedule (24–26). High humidity in a smokehouse can cause surface grease and poor color problems. These problems are common for batch-type smokehouses. The batch-type smokehouse usually starts at low temperatures because the cold-water shower of the previous lot. A wet-bulb temperature setting higher than the house temperature at the beginning of heat processing often calls an excessive amount of steam into the smokehouse. It is common practice to set the wet bulb at zero degrees in order to prevent the excessive humidity. Because the wet-bulb temperature is set at zero position, which is way below smokehouse temperature, the intake and exhaust dampers are controlled at a wide-open position, disregarding the humidity condition in the smokehouse. There is substantial variation in year-round weather conditions. If the exhaust capacity is determined under hot, humid summer conditions and a good quality product is produced, it is not necessary to run the exhaust at maximum capacity in cold, dry winter weather. Therefore, the exhaust dampers should be controlled according to the dryness of intake air (23). Microprocessors have increasingly been used to control the smokehouse system. The exhaust and intake air, steam injection, and exhaust fan can be controlled and operated independently without interlocking problem as are seen in current control systems. Microprocessor control provides much more flexibility for heat-processing optimization and energy conservation.

Cooling

From microbiological research, it has been illustrated that product shelf life strongly depends on product temperature. By lowering product temperature, the onset of microbial growth can be substantially delayed. To reduce product temperature after smoking and cooking, air cooling and liquid chilling are two of the more commonly used methods in the meat industry. Salt brine and propylene glycol are two common cooling mediums. However, glycol solution can only be used for products with impermeable casing. There are several disadvantages with air cooling such as slow cooling rate, nonuniform cooling, and moisture loss. Air cooling is normally used to chill products in impermeable casing or metal molds. Because a continuous-processing system requires more rapid cooling, liquid cooling is used in almost every system. In addition, to compensate for the disadvantages of air cooling, liquid cooling with high chemical concentration can also retard the microbial growth. In general practice, the economical way to apply liquid cooling can be achieved by two steps:

1. Precool products with regular water.
2. Further cool the product with a refrigerated cooling medium such as brine.

Qualities that can be affected by brine chilling are shrink, flavor, and color. When moisture is extracted by brine, product weight loss and darker color occur. As long as brine concentration is higher than the salt concentrations in the product, salt migrates into the product and develops a salty flavor. Therefore, salt concentrations is the most critical condition in the brine-chilling process. The optimum salt concentration should be determined for each product. The best operation conditions should be at the equilibrium conditions of salt migration.

For the last decade, poultry products have taken a significant market share of processed meats. According to USDA regulation, frozen whole turkey must be chilled down to 0°F core temperature before it can be shipped. Freezing whole-bird turkey is a time-consuming process. The poultry industry is looking for a rapid-chilling method to increase productivity. Cryogenic-freezing can provide a fast freezing rate. However, it is a high-cost operation. Blast-air chilling is an economical process but takes a long time to achieve 0°F core temperature. Liquid freezing is an effective and also economical method. The common chemicals used to depress the freezing point below 0°F are calcium chloride, ethanol, and propylene glycol. However, calcium chloride contributes to bitterness at as low as 1% of concentration and ethanol causes medicinal aftertaste (27). Propylene glycol usually does not develop objectionable flavor. The high concentration of glycol at low temperature causes excessive viscosity of fluid and develops a pumping problem. Hence, the glycol solution is limited in the immersion system. Because of flavor and viscosity problems, liquid freezing has not been widely accepted. The most common practice in the poultry industry is blast-air freezing or salt brine prechilling followed by blast-air freezing. It usually takes 24–48 hours to reach 0°F core temperature.

PACKAGING

Products after cooking and cooling are usually stored in a holding cooler before packaging. Sometimes holding or tempering is a necessary process for sliced or cellulose-casing products. Rapid cooling can cause slicing problems due to fat crystallization. Blast-air chilling enhances the cohesion of meat and casing. It is difficult to separate casing right after blast-air chilling. The tempering process can improve product sliceability as well as peelability.

The packaging area is normally under refrigeration and kept around 40°F. General packaging procedures are illustrated in Fig. 3.

Peeling and Slicing

Products must have the casing peeled off or be ejected from a mold; except for natural casing, edible collagen casing, or cook-in-bag products. The mold products are ejected with compressed air or mechanical plungers. For casing products, there are various types of machinery to remove cellulose casing. The most commonly used peeling machine for link products is that the casing is first moistened with steam and is then slit longitudinally by a sharp razor blade. When the slit casing is moved forward, an air jet aims at the knife cut location to blow casing off the product. Afterward, the casing is stripped away with a vacuum suction drum. The peeled products are randomly collected in a hopper. A collating machine is needed to rearrange the link products in an orderly form for automatically loading into the packaging packet. The sliced products can be arranged in shingled or stacked configuration. A slicer with a rotating blade is the most commonly used slicing machine. Emulsion products require slower rotation speeds for clean and smooth appearance. Muscle products need more spinning action to protect product integrity and produce uniform slices. The sliced products are often manually loaded into a packaging cavity because of stacking or shingling characteristics. The slicing temperature should be kept below 32°F.

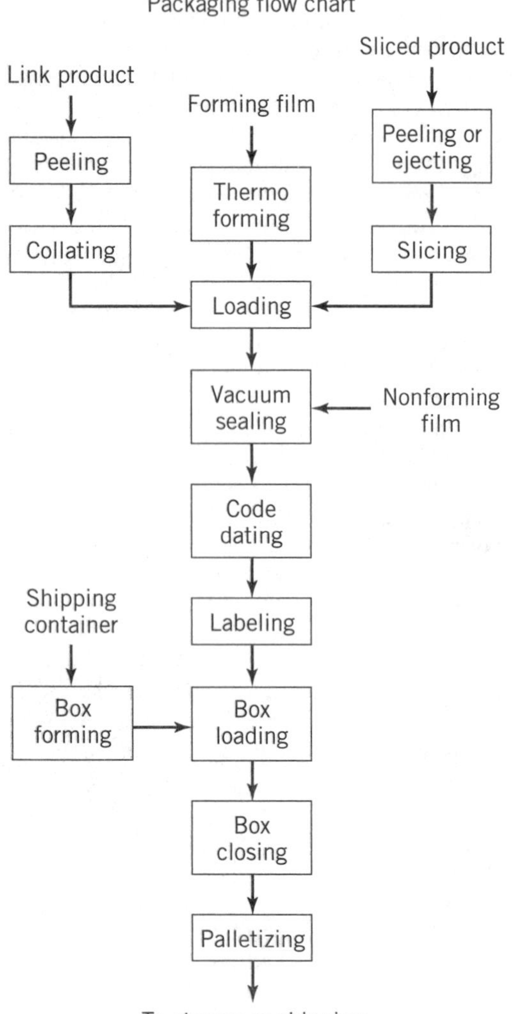

Figure 3. Packaging flow chart.

Packaging Machine

The most widely used packaging machine is a form-fill-seal machine. This machine requires two different types of film; a forming film and a nonforming film. The forming film is preheated and moved forward to the forming die where the desired shape and size can be formed by vacuum suction, forming plug, or both simultaneously. The formed cavity is indexed to the loading station. Products that come from a collator or slicer are automatically or manually placed into the formed cavity. After product loading, the nonforming film is brought in place over the formed film containing the product as the package moves to the final seal station. At the sealing station, air is evacuated from the cavity at a vacuum level higher than 27 in. and then both forming film and nonforming film are hermetically sealed together by heat-seal bars that melt the two sealant layers and weld them together under pressure. For some applications, such as a control atmosphere package, an inert gas or gas mixture is backflushed after the vacuumization before final sealing. The sealed packages are then trimmed and boxed. Oxygen causes processed meat to spoil and must be evacuated during mixing, chopping, emulsifying, and stuffing processes. Packaging film and a packaging machine alone cannot eliminate the residual oxygen inside the package. The performance of the packaging machine is extremely important. It can become out of adjustment easily, owing to its extreme high operating speed. A strict maintenance program is needed.

BIBLIOGRAPHY

1. D. M. Theno and G. R. Schmidt, "Microstructural Comparisons of Three Commercial Frankfurters," *Journal of Food Science* **43**, 845–848 (1978).
2. A. V. Gorbatov and V. M. Gorbatov, "Advances in Sausage Meat Rheology," *Journal of Texture Studies* **4**, 406–437 (1974).
3. P. Y. Wang, "Rheology of Sausage Dough," *American Society of Agricultural Engineering Paper* 79-6540 (1979).
4. R. Toledo, J. Cabot, and D. Brown, "Relationship Between Composition, Stability and Rheological Properties of Raw Comminuted Meat Batters," *Journal of Food Science* **42**, 725–727 (1977).
5. J. Cordray and D. Huffman, "The Art of Turkey Luncheon Meat," *Meat & Poultry* December, 16 (1987).
6. R. J. Krause et al., "Influence of Tumbling, Tumbling Time, Trim and Sodium Tripolyphosphate on Quality and Yield of Cured Hams," *Journal of Food Science* **43**, 853–855 (1978).

7. R. D. Cassidy et al., "Effect of Tumbling Method, Phosphate Level and Final Cook Temperature on Histological Characteristics of Tumbled Porcine Muscle Tissue," *Journal of Food Science* **43**, 1514–1518 (1978).
8. D. M. Theno, D. G. Siegel, and G. R. Schmidt, "Meat Massaging: Effect of Salt and Phosphate on Ultrastructure of Cured Porcine Muscle," *Journal of Food Science* **43**, 488–492 (1978).
9. R. R. Motycka and P. J. Bechtel, "Influence of Pre-Rigor Processing, Mechanical Tenderization, Tumbling Method and Processing Time on the Quality and Yield of Ham," *Journal of Food Science* **48**, 1532–1536 (1983).
10. D. Schied, "Cooked Ham Manufacturing," *Fleischwirtsch* **66**, 1022–1026 (1986).
11. P. H. Huebner, J. G. Sebranek, and D. G. Olson, "Effects of Meat Temperature, Particle Size, and Grinding Systems on Removal of Bone Chips from Ground Meat," *Journal of Food Science* **54**, 527–531 (1989).
12. K. W. Jones and R. W. Mandigo, "Effects of Chopping Temperature on the Microstructure of Meat Emulsions," *Journal of Food Science* **47**, 1930–1935 (1982).
13. D. Brown and R. Toledo, "Relationship Between Chopping Temperatures and Fat and Water Binding in Comminuted Meat Batters," *Journal of Food Science* **40**, 1061–1064 (1975).
14. K. M. Ladwig, C. L. Knipe, and J. G. Sebranek, "Effects of Collagen and Alkaline Phosphate on Time of Chopping, Emulsion Stability and Protein Solubility of Fine-Cut Meat Systems," *Journal of Food Science* **54**, 541–544 (1989).
15. J. C. Deng, R. T. Toledo, and D. A. Lillard, "Protein-Protein Interaction and Fat and Water Binding in Comminuted Flesh Products," *Journal of Food Science* **46**, 1117–1121 (1981).
16. R. L. Swasdee, "Ultrastructural Changes During Chopping and Cooking of a Frankfurter Batter," *Journal of Food Science* **48**, 1039–1041 (1982).
17. K. Tantikarnjathep et al., "Use of Vacuum During Formulation of Meat Emulsion," *Journal of Food Science* **48**, 1039–1041 (1983).
18. L. W. Solomon and G. R. Schmidt, "Effect of Vacuum and Mixing Time on the Extractability and Functionality of Pre- and Post-Rigor Beef," *Journal of Food Science* **45**, 283–287 (1980).
19. P. Y. Wang, "Smoking and Heating Meat," in C. W. Hall, A. W. Farrall, and A. L. Rippen, eds., *Encyclopedia of Food Engineering*, 2nd ed., AVI Publishing Co., Westport, Conn, 1987.
20. S. Simon, A. A. Rydinski, and F. W. Tauber, "Water-Filled Cellulose Casing as Model Absorbents for Wood Smoke," *Food Technology* **20**, 1494–1498 (1966).
21. W. S. Chan, R. T. Toledo, and J. Deng, "Effect of Smokehouse Temperature, Humidity and Air Flow on Smoke Penetration into Fish Muscle," *Journal of Food Science* **40**, 204–243 (1975).
22. J. A. Maga, "Smoke in Food Processing," CRC Press, Boca Raton, Fla., 1988.
23. P. Y. Wang, "Reduction of Energy Consumption and Air Pollution from Smokehouse," *American Society of Agricultural Engineering Paper* 76-6518, 1976.
24. G. S. Mittal, C. Y. Wang, and W. R. Usborne, "Smokehouse Process Conditions for Meat Emulsion Cooking," *Journal of Food Science* **52**, 1140–1144 (1987).
25. C. W. Monagle, R. T. Toledo, and R. L. Saffle, "Effect of Smokehouse Temperature, Humidity and Air Velocity on Rate of Heating and Quality of Frankfurters," *Journal of Food Science* **39**, 602–604 (1974).
26. I. Stech, W. R. Osborne, and G. S. Mittal, "Influence of Smokehouse Air Flow Patterns, Air Changes and Cook Cycle on Texture and Shrinkage of Wieners," *Journal of Food Science* **53**, 421–424 (1988).
27. G. H. Cipolletti, Robertson, and D. F. Farkas, "Freezing of Vegetables by Direct Contact with Aqueous Solution of Ethanol and Sodium Chloride," *Journal of Food Science* **42**, 911–916 (1977).

GENERAL REFERENCES

S. W. Ahmed, K. Sushil, and P. K. Mandal, "Tumbling and Massaging—An Emerging Technology for the Meat Industry," *Indian Food Industry* **14**, 37–41 (1995).

F. M. Aramouni, E. A. E. Boyle, and L. R. Vogt, "Introduction to the Hazard Analysis Critical Control Point (HACCP) Concept in a Small Meat-processing Plant," *Dairy Food and Environmental Sanitation* **16**, 431–439 (1996).

T. Backers and B. Noll, "Dietary Fiber for Meat Processing," *Food Marketing & Technology* **11**, 4–6, 8 (1997).

S. Barbut, "Effects of Three Chopping Methods on Bologna Characteristics," *Canadian Institute of Food Science and Technology Journal* **23**, 149–153 (1990).

Z. Bem, B. Zlender, and H. Hechelman, "Desirable and Undesirable Effects of Smoking Meat Products," *Fleischerei* **46**, III–VII (1995).

H. J. Buckenhueskes, "Functions of Sugars in Meat Processing," *Fleischwirtschaft* **78**, 1271–1275 (1998).

B. Carpentier, "Sanitary Quality of Meat Chopping Board Surfaces: A Bibliographical Study," *Food Microbiology* **14**, 31–37 (1997).

S. F. Chang, T. C. Huang, and A. M. Peason, "Control of the Dehydration Process in Production of Intermediate-Moisture Meat Products: A Review," *Advances in Food and Nutrition Research* **39**, 71–161 (1996).

K. J. Debrowski, S. Gwiazda, and A. Rutkowski, "Non-meat Proteins as Functional Ingredients in Meat Processing," *International Food Ingredients* No. 6, 26–30 (1991).

R. M. Estrada, E. A. E. Boyle, and J. L. Marsden, "Liquid Smoke Effects on *Escherichia coli* O157:H7 and Its Antioxidant Properties in Beef Products," *J. Food Sci.* **63**, 150–153 (1998).

M. Fode et al., "First Results on Experimentation of Solar Dryers in Tropical Ambient," *Rivista di Ingegneria Agaria* **28**, 65–72 (1997).

U. Gerhardt, "Starch and Plant Proteins in Meat Processing," *Fleischerei* **48**, III–VIII (1997).

H. G. Glenk, "Technical Gases in Meat Processing," *Fleischerei* **45**, 20, 22, 25–26 (1994).

M. D. Guillen and M. L. Ibargoitia, "New Components With Potential Antioxidant and Organoleptic Properties, Detected for the First Time in Liquid Smoke Flavoring Preparations," *J. Agric. Food Chem.* **46**, 1276–1284 (1998).

M. D. Guillen et al., "Effects of Smoking on Meat Products," *Alimentacion Equipos y Tecnologia* **17**, 104–114 (1998).

B. Hermey and H. Patzelt, "Using Liquid Smoke," *Fleischwirtschaft* **75**, 445–447 (1995).

P. Iascone, "Packaging of Meat Products," *Italia Imballaggio* **2**, 18–25 (1995).

A. T. Kamdem Kenmegne and J. Hardy, "Influence of Various Conditions on Meat Grinding Characteristics," *J. Food Eng.* **25**, 179–196 (1995).

E. Karmas, *Sausage Products Technology*, Noyes Data Group, Park Ridge, N.J., 1977.

C. Kuraishi, J. Sakamoto, and T. Soeda, "Application of Transglutaminase for Meat Processing," *Fleischwirtschaft* **78**, 657–658, 660, 702 (1998).

I. Lisse and A. L. Wack Raoult, "Drying of Meat Materials (Lean and Fat) by Deep-fat Frying in Animal Fat," *Sciences des Aliments* **18**, 423–435 (1998).

J. A. Maga, *Smoke in Food Processing*, CRC Press, Boca Raton, Fla., 1988.

D. Marque et al., "Safety Evaluation of an Ionized Multilayer Plastic Film Used for Vacuum Cooking and Meat Preservation," *Food Additives and Contaminants* **15**, 831–841 (1998).

J. D. Meyer, J. B. Luchansky, and J. G. Cerveny, "Inhibition of a Psychotrophic *Clostridium* Species by Sodium Diacetate and Sodium Lactate in a Cook-in-the-Bag, Refrigerated Turkey Breast Product," *J. Food Prot.* **58** (Suppl.), 9 (1995).

"Mincing, Chopping and Cutting: The Alpha and Omega of Meat Processing," *Fleischerei* **43**, 630–632 (1992).

S. A. Muller, "Packaging and Meat Quality," *Canadian Institute of Food Science and Technology Journal* **23**, 22–25 (1990).

A. M. Pearson and F. W. Tauber, *Processed Meats*, 2nd ed., AVI, Westport, Conn., 1987.

J. P. G. Piette et al., "Influence of Processing on Adherence of a Highly Extended Ham to its Cooking Bag," *Meat Science* **48**, 101–113 (1998).

"Processing and Packaging," *Food Review* **23**, 25, 27, 29, 31 (1996).

A. Rutkowski, *Advances in Smoking of Foods*, Pergamon Press, New York, 1976.

F. J. M. Smulders and R. L. J. M. van Laack, "Meat Quality. Ante Mortem Handling of Animals and Fresh Meat Processing. An Update," *Fleischwirtschaft* **72**, 1385–1389 (1992).

A. van der Leest, "Trends in the Modern Meat Industry," *Food Marketing & Technology* **8**, 64–66 (1994).

P. Y. Wang, "Meat Processing," in C. W. Hall, A. W. Farrall, and A. L. Rippen, eds., *Encyclopedia of Food Processing*, 2nd ed., AVI Publishing Co., Westport, Conn., 1987.

H. Weinberg, "Computer Controlled Manufacture of Meat and Sausage Products. Weight as a Factor in Manufacture of High-Quality Sausages," *Fleischerei* **44**, 200–203 (1993).

M. Weisenfels, "Low-Pollution Smoking in Batch Processes," *Fleischwirtschaft* **69**, 1111–1113, 1131 (1989).

M. Wilms, "Smoking of Meat Products. Development of Modern Smoking Processes From Economic and Ecological Viewpoints," *Fleischwirtschaft* **77**, 704–708 (1997).

R. Wittkowski, W. Baltes, and W. G. Jenning, "Analyses of Liquid Smoke and Smoked Meat Volatiles by Headspace Gas Chromatography," *Food Chem.* **37**, 135–144 (1990).

Y. Zhao and J. G. Sebraneck, "Technology for Meat Grinding Systems to Improve Removal of Hard Particles From Ground Meat," *Meat Science* **45**, 389–403 (1997).

PIE YI WANG
Armour Swift-Eckrich
Downers Grove, Illinois

See also MEAT AND ELECTRICAL STIMULATION; MEAT SLAUGHTERING AND PROCESSING EQUIPMENT.

MEAT PRODUCTS

Meat has been processed into various products for thousands of years. The purposes of processing have included preservation for future use, increased palatability and variety, enhanced convenience and complete utilization of the carcass and organs. Generally, people in warmer parts of the world processed meat to preserve it by salting and drying, and populations in the colder parts of the world processed meat to increase the quantity and variety to eat by adding such nonmeat ingredients as oatmeal or potatoes. Obviously, all the historic reasons for processing meat still remain today, but flavor, convenience, safety, and variety are key to the continued growth of processed meat sales in modern societies.

In 1997, *Meat Marketing & Technology* trade magazine published a report of a study sponsored by the National Pork Producers Council and the National Cattlemens Beef Association entitled *Consumer Purchase Behavior of Processed Meats at Retail* (1). The material collected in 1995 reported that 4.25 billion pounds of processed meats were purchased for $9.83 billion at retail stores. The major types of processed meats purchased are shown in Table 1. Lunch meat refers to a wide variety of products that are usually sliced to be utilized in making sandwiches. The penetration of lunch-meat-type products is shown in Table 2. A growing category of processed meats is dinner sausages, which provide a variety of flavors and convenience of preparation. The leading dinner sausage types are shown in Table 3.

Table 1. Processed Meats Purchase at Retail

Product	$ Billion	Billion pounds
Lunch meat	5.32	1.73
Hot dogs	1.14	0.75
Bacon	1.23	0.73
Dinner sausage	1.09	0.56
Breakfast sausage	0.55	0.29

Table 2. Penetration of Lunch Meat Types

Product	Percent of households purchasing per month
Ham	30
Turkey	21
Bologna	18
Sliced beef	11
Dry salami	5
Loaves	5
Chicken breast	3
Pepperoni	3
All other salamis	3
Liver sausage	3
Variety packs	3
Cotto/cooked salami	2

Table 3. Top Dinner Sausage Types

Product	Percent of total dinner sausage pounds purchased
Smoked	28
Polish	23
Italian	19
Bratwurst	10
Hot	9
Mild	5

FORMULATIONS

The quality of meat products is a reflection of the materials that have gone into making those products. Meat, water, salt, spices, casings, packaging materials and other components all have an impact on the quality of the final product. Since meat is a biological material that varies naturally in composition, it is difficult to ensure that the composition of the meat source used for further processing is similar from batch to batch. Establishing a program to monitor the selection and quality of raw materials is important to obtain a consistently made, quality product. Jones (2) and Rust and Olson (3) described the significance of raw material quality for processed meat products.

It is important to define purchase specifications for raw materials. Specific criteria representing the type or kind of product needed from a supplier should be established. It is up to the processor to determine how detailed the purchase description should be. The amount of detail specified and the level of monitoring that is applied contribute to the quality of the finished product. For example, specifications could be as rigid as stipulating the acceptable range of volatile oil content of garlic in a spice blend or merely stating the amount of garlic a spice blend must contain.

It is generally not recommended to base raw material purchases solely on price. A working relationship with suppliers should be developed to become familiar with their capabilities and reliability. In addition, suppliers need to be able to provide materials that meet purchase specifications at an agreed-upon price. To control raw material quality, the following steps are recommended: (*1*) establish well-defined purchase specifications, (*2*) set up testing procedures to verify that purchase specifications are met, (*3*) determine sampling frequency for testing procedures, and (*4*) accept or reject raw materials based on meeting purchase specifications.

The characteristics of processed meat products depend not only on the quality of the raw material used to formulate the product but also on the type of meat used. Before purchase specifications are developed, the type of meat used in processed meat products should be selected. Meat cuts vary in moisture, fat, and protein content; in the amount of pigmentation (redness); and in the ability to bind fat and water. These values can be obtained from tables; however, to more accurately determine the composition of a meat source, an analytical method should be used.

Knowing the moisture/protein ratio of a meat source can provide a guide to predict a meat product's final composition. In general, meat with lower moisture protein ratios perform better in sausage formulations. Fat contributes to palatability, tenderness, and juiciness of a processed meat product. Knowledge of the fat and moisture content is important in order to comply with meat inspection regulations that place limitations on their use.

Meat color is related to the concentration of myoglobin, a pigment found in muscle. Meat becomes redder as myoglobin concentration increases. The concentration of myoglobin in meat is dependent on several factors. An older animal has more myoglobin than a younger animal. Beef contains more myoglobin than lamb, and lamb contains more myoglobin than pork. Muscles used for locomotion have more myoglobin than support muscles such as the ribeye (*longissimus dorsi*). Genetics, nutrition, and environment also play a role in influencing myoglobin concentration in meat. To help maintain a consistence appearance from batch to batch, a color value ranking should be used in the formulation of meat products. This is particularly useful if a least-cost formulation program is used.

In the meat processing industry, the term *bind* has several meanings. It generally refers to (*1*) the ability of contractile protein to entrap water and fat such as in a sausage batter or (*2*) the surface cohesion of meat chunks to each other. Raw meat materials vary in their binding ability. Meats that have a high contractile protein content and low collagen content have high binding ability and are used to form stable batters. Beef skeletal muscle is an example of meat with high binding ability. Meats that have a high collagen and fat content and low contractile protein content have low binding ability. Heart and pork jowls are examples of meats with low binding ability. Bind value tables are available for a variety of meat sources. Tables providing proximate composition, color, and bind values for many raw meat sources, as well as an excellent overview of least-cost formulations, can be found in Pearson and Gillett (4).

Given the natural variation of raw materials, producing a quality product is both an art and a science. To succeed, it is extremely important to use meat that has been handled properly. Meat provides an excellent environment for bacterial growth. Since the meat used for processed meat products undergoes considerable handling during processing procedures, there are many opportunities for meat to become contaminated with bacteria. It is essential that all equipment be clean and sanitized prior to processing, and the meat must be kept cold at all times.

The development of off-odors and off-flavors in fat is known as rancidity. Over time, meat becomes rancid even if stored in a freezer. To produce quality products, meat that has become rancid should not be used. The rancid flavor and odor cannot be diluted out by combining rancid meat with fresh meat. If raw materials have become rancid, they should not be used as a base for processed meat products.

Purchase specifications are written descriptions describing raw materials. Each raw material should have a purchase description. When purchasing meat, in addition to species and portion cut, the description may specify grade (quality and/or yield), state of refrigeration for delivery of product (chilled or frozen), fat thickness (maximum average thickness and maximum at any one point), weight and thickness tolerances, muscling, trimming, netting or tying, cutting, or other material requirements. *The Meat Buyers Guide* (5) is a good resource for developing purchase specifications for meat cuts.

RESTRUCTURED MEAT PRODUCTS

The term *restructured* has the general meaning of binding smaller pieces of meat together to generally give the impression that a larger meat cut has been created that can produce slices of desired size for sandwiches or steak for plating at dinner. Processors have used various technologies to restructure ham since the 1950s. Mixing pieces of meat with ingredients, but especially appropriate levels of salt and alkaline phosphates, results in the osmotic movement of the ingredients throughout the meat particles and the extraction of myofibrillar proteins. The extracted myofibrillar proteins within the particles and on the surfaces of the particles form a heat set gel when the products are cooked. This gel is a very effective binder of water and fat and results in the binding of the meat pieces together and the retention of water and fat during cooking. This basic reaction is the primary system for the production of the very popular lunch meat types such as ham, turkey, and sliced beef. These products may be prepared from whole muscles or pieces of muscles or combinations of the two. The essential ingredients of salt and alkaline phosphates are added in a pickle that is usually injected into the meat pieces with large multineedle injectors. Sometimes the particles are slashed at frequent intervals to increase the surface area for protein extraction. This is called maceration or tenderizing. To further enhance protein extraction, the injected meat pieces are subjected to mechanical action in mixers, blenders, massagers, or tumblers. Portions of the mixture are then placed in appropriate casings or forms to create the final product shape and size. Often the shape of the final product is devised to mimic the whole muscle counterpart such as a ham, a beef roast, or a turkey breast. Cooking sets the protein gel, and the product can be uniformly sliced to create slices free of visible fat and connective tissue. Key references in this area are Pearson and Dutson (6) and Franklin and Cross (7).

Although the salt/phosphate-based binding system has obtained very widespread application, it has several limitations. Products must be marketed either precooked or frozen as the material will not cohere in the uncooked refrigerated state. Salt discolors the raw meat and accelerates oxidative rancidity. In recent years, several systems have been developed to bind meat without cooking to make products that look like whole tissue meats. The products can be marketed raw and have enhanced flavor and color stability. The five cold binding systems to be discussed are based on alginate/calcium, fibrinogen/thrombin, surimi, transglutaminase, or egg/milk/soy proteins. Although restructuring systems make comminuted meat appear to be an intact cut of meat, the restructured steak or roast has bacteria distributed throughout the interior of the product that is different from an unprocessed intact steak or roast. Therefore, it is imperative the restructured product be stored and, especially, cooked properly. Otherwise, food safety concerns for restructured products are the same as for any other comminuted meat.

Alginate forms a chemical bond with calcium ions to form a cold-set but heat-stable gel. High guluronic acid alginate is mixed with meat, and then a moderately slow release calcium source such as calcium carbonate or encapsulated calcium lactate is added along with an acidulant such as glucono-δ-lactone. After the ingredients are mixed with the meat pieces, the product is formed and chilled for a day to permit the chemical binding to continue to form (8).

Fibrinogen/thrombin binding system is based on the mechanism used to clot blood (9). The materials are usually prepared from beef blood collected in a sanitary manner. Approximately 5% of a fibrinogen/thrombin mixture is added to meat pieces to be bound together, mixed, and formed into the final desired shape. Once all the materials are mixed, it is important to quickly form the product and let it set for about 12 h. The time to gel is fairly rapid, and the process must be done in relatively small batches to avoid forming delays. In addition, the materials are natural, biological materials obtained from beef cattle during the slaughtering process. Special care must be taken to minimize microbial contamination and growth during binder preparation, storage, and utilization.

Surimi is a wet concentrate of the proteins of fish muscle produced by water washing of minced fish. Cryoprotectants such as 4% sucrose and 4% sorbitol permit surimi to be stored frozen for more than a year with little loss of gel forming ability (10). Research has shown surimi can be used in red meats as a cold binding system. There has been no commercial application of this process, probably due to flavor effects.

Transglutaminase is an enzyme that catalyzes the formation of a chemical cross-link between peptides containing glutamine and lysine. Transglutaminase can be used to increase the gelation capacity, physical strength, viscosity, thermal stability, and binding capacity of protein foods. A commercial binder based on this enzyme is used at the 1% level to bind meat, poultry, and fish pieces. This is a fast gelling binder and must be used in a batch system to mix and form in 30 min or less and sit for at least 2 h to gel. It is used in conjunction with sodium caseinate, sugar fatty acid ester, and dextrin or with sodium polyphosphate, sodium pyrophosphate, sodium ascorbate, and lactose (Ajinomoto U.S.A., Inc. Teaneck, NJ).

A commercial blend of proteins is sold as a cold-setting meat binder. The blend called PEARL MEAT contains egg white, casein, lactalbumin, gelatin, hydrolytic egg white extractive, hydrolytic casein extractive, soybean protein, and hydrolytic flour protein extractive. The meat is coated with the powder and quickly pressed and formed as the binding will be completed in 30 min (Chiba Flour Milling Company, LTD. Japan).

All of the binding systems require that the particles of meat being bound together be naturally tender, lean, have good color and flavor, and react well during subsequent cooking. Also, as mentioned earlier, the surfaces of the meat pieces that are subsequently bound on the inside of a restructured roast or steak must be assumed to contain some pathogenic bacteria such as *Salmonella*. Thus, restructured meats must be cooked sufficiently to kill these bacteria present in the center of the restructured cut. Thus, many of the cold binding systems are limited by the quality and cost of the raw materials and the need for more extensive cooking than intact cuts.

CURING

Historically, meat was preserved using salt. Early sausage makers recognized that using certain salts produced a distinct color and flavor in meat products. It is believed that these salts contained an impurity called potassium nitrate, better known as saltpeter. It wasn't until the late 1800s that scientists began to understand the role that saltpeter played in meat curing. In 1891, it was identified that the nitrate in saltpeter was chemically changed to nitrite by the action of bacteria found normally on meat. At the turn of the century, it was established that the reddish-pink color of cured meat was due to nitrite and not nitrate. These and other discoveries led to the use of nitrite as a meat curing ingredient.

Typical meat curing ingredients include salt, sodium nitrite, ascorbates, phosphates, and seasonings. Salt is used as a seasoning and a preservative, and it functions by extracting myofibrillar proteins to bind a product together. Nitrite imparts several important qualities to cured meat. The most visible characteristic is the pink color of cured meat. This color results from nitrite reacting with myoglobin, which, when denatured by heat, forms a stable pink color. Nitrite also provides a characteristic flavor and acts as a potent antioxidant to prevent the development of off-flavors in cured meat during storage. One of the most important reasons nitrite is used in cured meats is to inhibit the growth of food poisoning and spoilage microorganisms, especially *Clostridium botulinum*, the bacteria that causes botulism. Ascorbates, including sodium ascorbate and erythorbic acid, are used to improve and maintain the color of processed meats. Phosphates are used to enhance juiciness and texture and to help prevent fat from becoming rancid in products such as ham, bacon, and cooked sausages. The amount of phosphate that can be used in meat products is limited to a maximum of 0.5%.

There has been concern over potential health risks from nitrosamines in cured meat products. Nitrosamines are compounds that can form when nitrites combine with amines, a natural component resulting from the breakdown of proteins. Most cured meat products contain approximately 10 ppm residual nitrite (11). This is an 80% reduction in residual nitrite content in cured meats compared to mid-1970s levels. As a result, a very limited amount of nitrite is available in cured meats to react with amines to form nitrosamines. Previously, traces of nitrosamines had been detected in bacon fried at high temperatures until it was crisp and very well done. To minimize the risk of nitrosamine formation in fried bacon, the amount of nitrite that may be used to cure bacon has since been reduced. Cassens (11) concluded that health risks from nitrosamine formation in cured meat should be reevaluated.

Meats can be cured by directly mixing cure ingredients into comminuted products, or by using a pickle or applying a dry rub to primal or subprimal cuts of meat. To make a pickle, cure ingredients are dissolved into water. The resulting pickle can be incorporated into meat by injection, tumbling or immersion, or a combination of these techniques. The most widely used method by the industry is injection curing (12). Multineedle injection machines, handheld stitch devices, and artery pumping are mechanisms for injection curing. Alternatively, meat may be immersed in a container or mechanically manipulated in a tumbler containing pickle, or meat may be injected, then immersed, tumbled, or massaged to enhance pickle pickup and promote myofibrillar extraction. Application of salt, sodium nitrate, and possibly sodium nitrite directly onto the surface of raw meat, then holding under controlled conditions for defined time periods is used for dry curing meats. This technique is most often used to produce regional specialty products.

ACIDIFICATION

Fermented sausages are acidified by the deliberate or natural inoculation with specific microorganisms, incubated to ferment a sugar source to lactic acid, and subjected to heating and/or drying to alter pH and the moisture protein ratio, and to prolong shelf life. An update of the technology of meat fermentation with a list of key references on the topic is provided by Prochaska, Ricke, and Keeton (13). Fermented sausages have been produced for centuries. Salamis, mettwurst, thuringer, cervelat, teewurst, Lebanon bologna, and some summer sausages are produced by fermentation and drying without any cooking step. Others such as cooked salami, cooked summer sausage, and mortadella are cooked. Traditionally, fermentation, drying, and smoking of these products are done at 12 to 25°C (54–129°F) during times ranging from weeks to several months. Fermentation is done at 20 to 25°C (68–129°F) with a relative humidity of 92 to 95%. The subsequent smoking and drying are done at less than 20°C (68°F) and gradually decreasing relative humidity to about 78% (14). In the United States, fermented sausages are often produced by the addition of *pediococcal* starter cultures, fermented at 24 to 43°C (75–108°F) and greater than 85% relative humidity to a pH of approximately 4.8 and then dried at about 13°C (55°F) and 65% relative humidity to a moisture protein ratio of 1.6:1. However, emerging pathogenic bacteria such as *Escherichia coli* O157:H7 are not destroyed by this treatment (15). Acidification can be accomplished in meat via direct addition of acids. Citric acid is used at 0.001 to 0.01% together with antioxidants to prevent oxidative rancidity in dry sausage, fresh pork sausage, and dried meats. Citric and lactic acids are also encapsulated in coatings of lipids for direct addition to sausage products to lower the pH and mimic the flavor and preservation effects of natural fermentation. Glucono-δ-lactone, a neutral ester of gluconic acid crystallized through dehydration, is a mild acidulant and is used for the same effect. Direct acidification is most widely used on the less expensive snack sticks, which are also cooked and are shelf stable.

SMOKING

Meat is smoked to produce a desirable color, flavor, and aroma. Of the more than 390 individual chemical compounds that have been detected in wood smoke, more than 70 of these compounds have been found in smoked foods. The type of wood used to generate smoke, the moisture content of the wood, and the temperature and method of producing smoke all influence what types of compounds will be generated from smoke.

Phenols, carbonyls, and acids form the three basic categories of compounds important to smoked meat. Smoked meat color is primarily formed by carbonyls in wood smoke reacting with components of meat proteins to produce the characteristic reddish-brown appearance. Phenols contribute to smoked color development and are largely responsible for the smoky flavor imparted to meat. On the surface of smoked meats, phenols have a preservative effect by retarding rancidity and inhibiting bacterial growth. Acids play an important role in coagulating proteins on the surface of smoked meats, which is essential to achieving good skin formation.

Smoke may be generated using a variety of fuels including cord wood, corn cobs, wood shavings, or sawdust or may be applied in the form of liquid smoke. Hardwoods such as oak and hickory are commonly used to generate smoke because of their low resin content. Although softwoods can be used, they can produce a strong, bitter flavor.

Commonly, smoke is applied during the initial phases of cooking or even before a product is cooked. The deposition of smoke on a meat product is affected by the smoke density, the humidity and air movement in a smoker, and the surface conditions of the meat to be smoked. Dampened fuel is heated to generate smoke. High humidities during smoking will increase smoke deposition but may cause a muddy brown appearance on the meat surface. If the meat surface is wet when smoke is applied, streaking can occur. Too little humidity during smoking can cause a meat surface to become very dry, leading to poor smoke adherence and a pale color. It is recommended to apply smoke while the meat surface is tacky to the touch.

Liquid smoke, derived from wood using a series of extraction and refinement procedures, contains the same functional compounds that are found in vaporous smoke. Meat products can be dipped or sprayed in liquid smoke, the smoke can be added directly into pickles or sausage batters, or liquid smoke can be atomized onto meat surfaces during thermal processing.

The theory of smoking and its application to meat products was described by Hanson (16). Recommendations for optimizing processing variables and reducing process variation in the cooking and smoking process also are discussed by Hanson (17).

PRESERVATION METHODS

Meat is preserved to extend product shelf life. This is accomplished by applying techniques to delay or prevent microbial, enzymatic, chemical, and/or physical changes from occurring that would render a product unacceptable. Historically, meat was preserved for future consumption by curing, smoking, and drying. Chilling, freezing, thermal processing, dehydration, and irradiation are preservation methods currently employed by the meat industry. van Laack (18) presented an overview of spoilage mechanisms and techniques for preservation of muscle foods.

Refrigeration and freezing are the most common methods used to prolong the shelf life of meat. Refrigerated meats are stored between −2 and 5°C (28–41°F). Chilling meat typically begins immediately after slaughter. Carcasses are placed into coolers with forced-air movement to facilitate carcass cooling, while poultry and fish are chilled by immersion in ice water. Following fabrication, refrigerated shelf life of meats is dependent on the initial microbial load, temperature, and humidity conditions during storage, packaging, species, and type of product stored. Although the freezing point of muscle is approximately −2°C (28°F), it is not until −30°C (−22°F) that nearly 100% of the water in meat is frozen. The quality of frozen meat depends on freezing rate conditions and subsequent frozen storage, length of frozen storage, packaging, lighting, and type of product frozen. The rate of freezing impacts the size of ice crystals formed within a product. If meat is rapidly frozen, many small ice crystals form resulting in minimal muscle fiber shrinkage and distortion, and less drip loss upon thawing. Improperly packaged meat is susceptible to freezer burn, which is caused by a loss of moisture on the surface of meat. Freezer-burned meat will have a dry, discolored surface that, when cooked, will be tough and taste bland or rancid.

Canned meat products may be pasteurized or commercially sterile. Pasteurized products are heated to 58 to 75°C (136–167°F), which results in the destruction of all pathogenic bacteria; however, spores and some thermoresistant spoilage organisms may survive. Canned, pasteurized products must be stored under refrigeration. Pasteurized canned hams and picnics are examples of this type of product. Commercially sterile products generally are heated to an internal temperature of at least 107°C (225°F) to render the product free of microorganisms capable of growing at nonrefrigerated conditions (above 10°C/50°F). Canned, commercially sterile products may contain spores of thermophilic bacteria that do not germinate below 43°C (109°F). Meats are canned using a steel tank, called a retort, in which metal crates or baskets containing filled, sealed cans are placed for cooking and subsequent cooling. The amount of heat, time, and temperature required for a given degree of sterility depends on the nature of the product, pH, ingredients such as salt and nitrite, shape and size of the can, and the type of retort used. Footitt and Lewis (19) and Pearson and Gillett (6) provide considerable information on canning meat products.

The preservative effects of dehydration are due to reduction of water activity (a_w). The lean portion of freshly cut meat has an a_w of 0.99. When meat is dehydrated, the a_w is reduced to a level that inhibits microbial growth, allowing products to be shelf stable without refrigeration. According to Leistner and Rodel (20), factors most important in influencing a_w in processed meat products are the addition or removal of water, the addition of salts, and the amount of fat. Nieto and Toledo (21) demonstrated that both soluble and insoluble components affected a_w in processed fish products. The fat content of meat products has an indirect effect on a_w due to its low water-binding properties. In general, a_w increases with increasing fat content in a meat product because fat has little effect in depressing a_w (22).

Meat is dehydrated by using hot air drying or freeze drying. Jerky is commonly dried using hot air drying. Freeze drying involves the removal of water from meat by sublimation where water, in the form of ice, is directly transformed into water vapor without going through a liquid phase. In a conventional freeze drying process, meat is frozen, then dried under vacuum with pressures of 1.0 to 1.5 mm of mercury while it is in a frozen state to a residual moisture content of less than 2%. Because freeze-dried meat products are susceptible to enzymatic changes, rancidity development, nonenzymatic browning, and protein denaturation, oxygen and moisture impermeable packaging is necessary.

Irradiation involves treating a food item with energy from electrons or γ-rays to prevent foodborne illnesses, spoilage, and insect infestations. Food does not become radioactive when treated with irradiation. This process is not a substitute for good sanitary practices, and it will not make a "dirty" food clean. Food irradiation makes meat and food of good quality safer, plus increases the length of time food can be stored before it becomes spoiled. Organizations such as the American Medical Association and World Health Organization have endorsed the safety of irradiation for food. The Food and Drug Administration (FDA) has approved irradiation of wheat and wheat flour to control insects; of white potatoes to control sprouting; of spices to kill insects and control bacteria; of pork to control trichinosis; of fruits, vegetables, and grains to control insects and growth and ripening; and of uncooked poultry to control bacteria, particularly *Salmonella*. In 1997, the FDA approved irradiation for fresh or frozen red meats. A list of materials approved by the FDA that can be used to package foods before irradiation is found in 21 *CFR* 179.45. A scientific status summary on irradiation, particularly muscle foods, was prepared by Olson (23).

HOME MEAL REPLACEMENT

Ready-to-cook, partially cooked, or ready-to-eat food products prepared and sold at a wide range of establishments, then consumed at home, are commonly referred to as home meal replacements (HMR), or meal solutions. Sociocultural and socioeconomic changes have driven the demand for convenient, fresh (preferably not frozen), appealing, high-quality food products that require minimal preparation by the consumer. This trend was described by Hoogenkamp (24) and Farquhar (25).

HMRs commonly include chilled or hot meals marketed through supermarkets, restaurants, and nontraditional retail outlets. These products are prepared and packaged at U.S. Department of Agriculture Food Safety and Inspection Service (USDA FSIS) or FDA inspected plants, by in-house supermarket or restaurant chefs, or prepared as made-to-order meals on-site at retail establishments. Johnson (26) summarized food safety hazards and risks associated with production and handling of HMRs.

BY-PRODUCTS

Two excellent references on the topics of edible and inedible meat by-products are Pearson and Dutson (27,28). Although values may vary considerably due to such factors as international trade and changing consumer patterns of preferences, by-products account for about 10% of the value of products from the slaughter of domestic animals (29).

Edible meat by-products that are consumed directly or used in processed products are parts such as tongues, livers, hearts, tails, and brains. Other materials that are classified as by-products are meat trimmings from the carcass, head, neck, and viscera. These materials include cheek meat, weasand meat, giblet meat, head meat, salivary glands, and diaphragm meat. A list of these by-products, their potential uses and world production figures are given by Goldstrand (29). More recent export values of meat and meat by-products are shown in Table 4.

Mechanically separated meat is a continuously changing edible meat by-product. Engineering developments and regulatory requirements greatly affect the production methods, quality, and utilization of these products. There are no up-to-date comprehensive reviews of these products, because much of the methodology of processing and engineering is proprietary. The USDA FSIS regulates the composition and labeling of meat produced by advanced meat/bone separation machinery and recovery systems (30). The development of meat/bone separators have advanced such that bones no longer need to be ground or crushed by the separating machine. Rather, the bones emerge from the process in a manner consistent with hand-deboning operations that use knives. The modern systems produce distinct whole pieces of skeletal muscle tissue with a well-defined particulate size consistent with ground meat. There is no powdered bone or bone marrow in the product. Systems for producing these products use compression, heavy-duty sieve cylinders, warming, and centrifugation to separate meat from bone and fat. These materials are widely used in comminuted meat products, with separated beef being used in ground beef patties and mechanically separated poultry being widely used in sausage products. The mechanically separated tissue is usually sold at a lower price than hand-deboned tissue. As it is often finely comminuted, it is especially prone to oxidation and must be kept frozen or refrigerated for a minimum time before being incorporated into processed products.

Edible tallow and lard are still produced in abundance as by-products of the slaughter, fabrication, and retail cutting and processing of cattle and hogs. The 1986 production of these two materials was 2.4 billion pounds in the United States alone (31).

Inedible by-products are produced in abundance by the meat industry. The market forces that cause changes in the demand for inedible by-products are the supply of live-

Table 4. U.S. Export Values for Selected Animal Products for Calendar Year 1997

Product	Export values ($ million)
Animal fats	531
Hides and skins	1618
Red meats, fresh, chilled, frozen	4090
Red meats, prepared or preserved	401
Pet foods, dog and cat food	735
Poultry meat	2423

Source: USDA Foreign Agricultural Service Statistics.

stock, feed supplies, substitute products, product safety and health concerns, trade barriers, and politics (32). The raw materials processed into inedible by-products are hides, skins, pelts, hair, feathers, hooves, horns, feet, heads, bones, toenails, blood organs, glands, intestines, and fatty tissues. By recycling these inedible products, not only are very useful materials created for human use and animal feeds, but the material from a meatpacking plant that is classified as waste is greatly reduced. Two major recent developments that affect by-product values were the outbreak of bovine spongiform encephalopathy (BSE) and the growth of the biotechnology industry for synthesizing substitutes for natural gland extracts. The BSE outbreak in the United Kingdom resulted in the prohibition of feeding ruminant meat and bone meal to other ruminants and may eliminate the feeding of much meat and bone meal to meat animals (33).

Pet food literature is limited, and much of the work is proprietary. As shown in Table 4, pet food is a large volume and value export from the United States. According to a 1998 survey by the American Pet Products Association, 59% of U.S. households own a pet and Americans spent more than $21 billion on their pets in 1997 (34). A complete discussion of the use of meat products in pet foods is presented by Corbin (35).

REGULATIONS AND STANDARDS

Meat inspection is regulated by the government to ensure that meat is wholesome and safe to eat. Hale (36) reviewed major meat inspection laws and events that have impacted meat and poultry inspection. All meat that is sold must, by law, be inspected. Regulations for the meat industry are administered by the USDA FSIS. Currently, it is estimated that meat inspection costs each consumer 13 cents per month.

In 1906, the Federal Meat Inspection Act mandated inspection of all meat sold for interstate and foreign commerce. Prior to 1967, inspection of meat produced for intrastate commerce varied depending on the city and state where the plant was located. To standardize inspection programs, the Wholesome Meat Act of 1967 was passed. Individual states were given the opportunity to develop inspection programs that met or exceeded regulations established by the federal government. If a state chose not to set up an inspection program, the USDA FSIS assumed all meat inspection responsibilities within the state.

More recently, USDA FSIS (37) mandated implementation of Hazard Analysis and Critical Control Point (HACCP) programs for all inspected meat and poultry establishments. HACCP is a systematic, science-based process control system for food safety. This concept forms the basic structure for a preventative system for the safe production of meat products. Note that the key to this system is that it is a preventative approach to producing the safest possible meat products for human consumption. This means that potential biological, physical, or chemical food safety hazards—whether they naturally occur in food, are contributed by the environment, or are generated by a deviation in the production process—are prevented, eliminated, or reduced to produce safe meat products.

HACCP began in 1959 when the Pillsbury Company cooperated with the U.S. Army Natick Laboratories, the National Aeronautics and Space Administration (NASA), and the U.S. Air Force Space Laboratory Project Group to ensure the safety of food to be used for the space program. They used a system of analysis that had been developed by Natick called the "Modes of Failure," which was adapted and has since evolved into the concept now understood as HACCP. In 1971, HACCP was first presented to the public at the National Conference of Food Protection. Following the publication of a report in 1985 by the National Academy of Sciences, *An Evaluation of the Role of Microbiological Criteria for Foods and Food Ingredients*, HACCP received more recognition by the industry as a food safety concept. The seven principles of HACCP, as defined by the National Advisory Committee on Microbiological Criteria for Foods (38) are as follows:

1. Conduct a hazard analysis.
2. Determine the critical control points (CCPs).
3. Establish critical limits.
4. Establish monitoring procedures.
5. Establish corrective actions
6. Establish verification procedures.
7. Establish record-keeping and documentation procedures.

The USDA FSIS Final Rule on pathogen reduction mandated requirements in efforts to reduce the occurrence and numbers of pathogens on meat and poultry products, reduce the incidence of foodborne illness associated with consuming these products, and provide a framework for modernization of the meat and poultry inspection system. The new regulations mandated establishment of four new programs. The first program required that each establishment develop and implement written sanitation standard operating procedures (Sanitation SOPs). Second, regular microbial testing was required for slaughter establishments to verify the adequacy of a plant's process controls for the prevention and removal of fecal contamination and associated bacteria. All slaughter plants and plants producing raw ground products were required to meet pathogen reduction performance standards for *Salmonella* for the third program. Last, all meat and poultry plants were required to develop and implement HACCP programs. Pearson and Dutson (39) is a resource describing the application of HACCP principles to meat, poultry, and fish processing.

Animals are inspected for signs of disease before and after slaughter. The inspector relies on a veterinarian's judgment in those instances where there is a health or wholesomeness question. Extreme attention is paid to produce a carcass free of contamination. To ensure the sanitary handling of meat and meat products, the sanitation of equipment, buildings, and grounds are also inspected. An inspector has the authority to have condemned products destroyed to prevent their use for human food.

An inspection stamp is put on large cuts of meat or meat products to assure consumers that the product was wholesome when it was shipped from the plant where the meat was inspected. Every plant that is inspected has a unique, individual identification number. This number can be found as part of the round inspection stamp on meat labels. An edible ink is used to apply inspection stamps on fresh meat. USDA also regulates labeling of meat and poultry products.

Standards of identity have been established for some meat and poultry products that describe the use of certain meat and nonmeat ingredients and/or the amount of fat, moisture, and protein a product must contain. Standards of preparation have also been defined for some products. These regulations can be found in the *Code of Federal Regulations*. Additional regulatory information can be found in USDA FSIS policy memos and directives, and the *Standards and Labeling Policy Book*.

BIBLIOGRAPHY

1. B. Salvage, *Meat Marketing and Technology* **5**, special supplement 1–28 (1997).
2. K. W. Jones, "Quality Control of Ingredients," *1989 AMSA-AMI Intercollegiate Processed Meat Clinic*, Chicago, Ill., September 21–23, 1989.
3. R. Rust and D. Olson, "Processing Workshop, 1981–1986," *Meat and Poultry Magazine*, Oman Publ., Mill Valley, Calif., 1987.
4. A. M. Pearson and T. A. Gillett, *Processed Meats*, 3rd ed., Chapman & Hall, New York, 1996.
5. *The Meat Buyers Guide*, North American Meat Processors Association, Reston, Va., 1997.
6. A. M. Pearson and T. R. Dutson, *Restructured Meat and Poultry Products*, Advances in Meat Research, Vol. 3, Van Nostrand–Reinhold, New York, 1987.
7. K. R. Franklin and H. R. Cross, *Meat Sci. and Technol. Int. Symp. Proc.*, National Live Stock and Meat Board, Chicago, Ill., November 1–4, 1982.
8. U.S. Pat. 4,603,054 (July 29, 1986), G. R. Schmidt and W. J. Means (to Colorado State University Research Foundation).
9. U.S. Pat. 4,741,906 (May 3, 1988), E. J. C. Paardekooper, S. Michielsgestel, and D. R. Gerrit Wijngaards (to Nederlandse Centrale Organisatie Voor Toegepast-Natuurwetenschappelijk Onderzoek).
10. D. D. Hamann, "Rheological Studies of Fish Proteins," in K. Nishinari and E. Doi, eds., *Food Hydrocolloids: Structures Properties and Functions*, Plenum, New York, 1993.
11. R. G. Cassens, "Residual Nitrite in Cured Meat," *Food Technol.* **51**, 53–55 (1997).
12. J. R. Claus, J.-W. Colby, and G. J. Flick, "Processed Meats/Poultry/Seafood," in D. M. Kinsman, A. W. Kotula, and B. C. Breidenstein, eds., *Muscle Foods*, Chapman & Hall, New York, 1994.
13. J. F. Prochaska, S. C. Ricke, and J. T. Keeton, *Food Technol.* **52**, 52–57 (1998).
14. G. R. Schmidt, "Processing," in H. R. Cross and A. J. Overby, eds., *Meat Science, Milk Science, and Technology*, World Animal Science B3, Elsevier Science, New York, 1988.
15. N. G. Faith et al., "Viability of *Escherichia coli* O157: H7 in Pepperoni during the Manufacture of Sticks and the Subsequent Storage of Slices at 21, 4, and −20°C under Air, Vacuum, and CO_2," *International Journal of Food Microbiology* **37**, 47–54 (1997).
16. R. E. Hanson, *Meat Business* **54**, 9, 20–21 (1993).
17. R. E. Hanson, *Reciprocal Meat Conference Proceedings* **50**, 33–42 (1997).
18. R. L. J. M. van Laack, "Spoilage and Preservation of Muscle Foods," in D. M. Kinsman, A. W. Kotula, and B. C. Breidenstein, eds., *Muscle Foods*, Chapman & Hall, New York, 1994.
19. R. J. Footitt and A. S. Lewis, eds., *The Canning of Fish and Meat*, Chapman & Hall, New York, 1995.
20. L. Leistner and W. Rodel, "The Significance of Water Activity for Micro-Organisms in Meats," in R. B. Duckworth, ed., *Water Relations of Foods*, Academic Press, New York, 1975.
21. M. B. Nieto and R. T. Toledo, *J. Food Sci.* **54**, 925–930 (1989).
22. H. K. Leung, "Significance of Water Activity in Meat Products," *Proc. Meat Industry Res. Conf.*, American Meat Institute, Washington, D.C., 1984, pp. 142–147.
23. D. G. Olson, *Food Technol.* **52**, 56–62, 1998.
24. H. Hoogenkamp, *National Provisioner* **212**, S40, S42 (1998).
25. J. W. Farquhar, *Reciprocal Meat Conference Proceedings* **50**, 19–20 (1997).
26. J. L. Johnson, *Reciprocal Meat Conference Proceedings* **50**, 20–24 (1997).
27. A. M. Pearson and T. R. Dutson, *Edible Meat By-Products*, Advances in Meat Research, Vol. 5, Elsevier Applied Science, New York, 1988.
28. A. M. Pearson and T. R. Dutson, *Inedible Meat By-Products*, Advances in Meat Research, Vol. 8, Elsevier Applied Science, New York, 1992.
29. R. E. Goldstrand, "Edible Meat Products: Their Production and Importance to the Meat Industry," in A. M. Pearson and T. R. Dutson, eds., *Edible Meat By-Products*, Advances in Meat Research, Vol. 5, Elsevier Applied Science, New York, 1988.
30. U.S. Department of Agriculture, Food Safety and Inspection Service (USDA FSIS), *Fed. Regist.* **63**, 17959–17966 (1998).
31. I. D. Morton, J. I. Gray, and P. T. Tybor, "Edible Tallow, Lard, and Partially Defatted Tissues," in A. M. Pearson and T. R. Dutson, eds., *Edible Meat By-Products*, Advances in Meat Research, Vol. 5, Elsevier Applied Science, New York, 1988.
32. R. E. Goldstrand, "An Overview of Inedible Meat, Poultry, and Fish By-Products," in A. M. Pearson and T. R. Dutson, eds., *Inedible Meat By-Products*, Advances in Meat Research, Vol. 8, Elsevier Applied Science, New York, 1992.
33. R. M. Anderson et al., "Transmission Dynamics and Epidemiology of BSE in British Cattle," *Nature* **382**, 779–788 (1996).
34. D. E. Pszczola, "Poly Wants a Neutraceutical," *Food Technol.* **52**, 66–87 (1998).
35. J. E. Corbin, "Inedible Meat, Poultry, and Fish By-Products in Pet Foods," in A. M. Pearson and T. R. Dutson, eds., *Inedible Meat By-Products*, Advances in Meat Research, Vol. 8, Elsevier Applied Science, New York, 1992.
36. D. S. Hale, "Inspection," in D. M. Kinsman, A. W. Kotula, and B. C. Breidenstein, eds., *Muscle Foods*, Chapman & Hall, New York, 1994.
37. U.S. Department of Agriculture, Food Safety and Inspection Service (USDA FSIS), *Fed. Regist.* **61**, 38806–38989 (1996).
38. National Advisory Committee on Microbiological Criteria for Food, *J. Food Prot.* **61**, 762–775 (1998).
39. A. M. Pearson and T. R. Dutson, eds., *HACCP in Meat, Poultry, and Fish Processing*, Advances in Meat Research, Vol. 10, Blackie Academic and Professional, New York, 1995.

GENERAL REFERENCES

American Association of Meat Processors, URL: *http://www.aamp.com/*.

American Meat Institute, URL: *http://www.meatami.org/index.htm.*
American Meat Science Association, URL: *http://www.meatscience.org/.*
Code of Federal Regulations, URL: *http://www.access.gpo.gov/nara/cfr/cfr-table-search.html.*
Institute of Food Technologists, URL: *http://www.ift.org.*
International HACCP Alliance, URL: *http://ifse.tamu.edu/haccpall.html.*
Meat and Poultry Association Links, URL: http://www.meatpoultry.com/assn.asp.
Meat and Poultry Magazine, URL: *http://www.meatpoultry.com/.*
The Meating Place, URL: *http://www.mtgplace.com/.*
National Archives and Records Administration; Federal Register Online via GPO Access, URL: http://www.access.gpo.gov/su_docs/aces/aces140.html.
National Meat Association, URL: *http://www.nmaonline.org/.*
National Pork Producers Council, URL: *http://www.nppc.org/.*
North American Meat Processors Association, URL: *http://www.namp.com/.*
U.S. Department of Agriculture, URL: *http://www.usda.gov/.*
U.S. Department of Agriculture, Agricultural Research Services Nutrient Data Laboratory, URL: *http://www.nal.usda.gov/fnic/foodcomp/.*
U.S. Department of Agriculture/Food and Drug Administration Foodborne Illness Education Information Center, URL: *http://www.nal.usda.gov/fnic/foodborne/foodborn.htm.*
U.S. Department of Agriculture/Food and Drug Administration HACCP Training Programs and Resources Database, URL: *http://www.nalusda.gov/fnic/foodborne/haccp/index.shtml.*
U.S. Department of Agriculture Food Safety and Inspection Service, URL: *http://www.fsis.usda.gov/.*
United States Meat Animal Research Center, URL: *http://www.marc.usda.gov/.*
U.S. Meat Export Federation, URL: *http://usmef.org/.*
Worldmeat, URL: *http://www.worldmeat.com.au/.*

Elizabeth A. E. Boyle
Kansas State University
Manhattan, Kansas
Glenn R. Schmidt
Colorado State University
Fort Collins, Colorado

MEAT SCIENCE

WHAT IS MEAT SCIENCE?

Meat is a food prepared from skeletal muscle. The quality of the meat is affected by various live animal factors as well as the processing conditions. Meat science is a discipline that has evolved to understand the changes that occur from the live animal until the meat is ready for consumption. It should be distinguished from animal science, where meat yield and growth characteristics are considered to be the main aim rather than meat quality itself. This chapter considers all those factors that affect the main meat quality attributes—its tenderness, juiciness, flavor, color, and storage life—considering mainly meat from cattle, sheep, and pigs, with minor references to deer and poultry. Neither processed meats, in which meat and fat tissue are used as a raw material for derived foods, nor meat hygiene or meat packaging are considered. The freezing of meat will be considered, however, as this process effectively locks up the characteristics of the tissue that exist immediately before freezing. Therefore, the resultant product depends on the time postmortem when the meat was frozen, the rate of freezing, and frozen storage conditions.

PRINCIPLES OF MEAT SCIENCE

The major principle of meat science discussed in this chapter is that shortened muscle will be tough when cooked. Meat can age and become more tender, with the aging factors being affected by the amount of shortening and conditions at rigor. This gives rise to a second principle that, since shortening and degree of aging cannot be determined visually, the appearance of a piece of fresh meat cannot predict how tender or tasty it will be. The consequence of these two principles is that the consumer has to have faith that the meat has been processed properly. Such faith is often misplaced. The ultimate aim of meat science is to develop and encourage procedures in the production and processing of animals so that the full potential of tenderness, juiciness, and flavor, as well as good hygiene, can be achieved.

MUSCLE STRUCTURE AND FUNCTION

Definition of Meat and Muscle

Meat comes from the striated muscle of vertebrates. Some of the characteristics of the live muscle underlying meat quality, such as energy stores, connective tissue, and fat, are present at the time of death (see "Animal Factors Affecting Meat Quality," later in this article) and will affect the final meat quality. Other aspects of meat quality, particularly toughness, are affected by processing conditions that interact with some preslaughter attributes.

The muscles of a carcass, whatever the species, perform different functions while the animal is alive and are reflected in differences in physiology, structure, and biochemistry. The variation in these components underlies the differences between the various meat cuts, the use to which they are put, and the way they are cooked. The two major contributors of the muscle that have a bearing on meat quality and underpin meat science are the myofibrillar and the connective tissue proteins. To understand what happens when muscle changes into meat, including differences between the various meat cuts, it is necessary to consider the structure, biochemistry, and physiology of these proteins in detail.

Muscle tissue contracts in the living animal and can also contract after slaughter. The amount of shortening either before or after death is governed by skeletal restraint and the effect of antagonistic muscles. Figures 1 and 2 (1) show a muscle with its attachment to bones and a sequence of pictures of increasing magnification to show the structure of a single muscle fibril, 50 to 100 μm in diameter, at the electron microscope level.

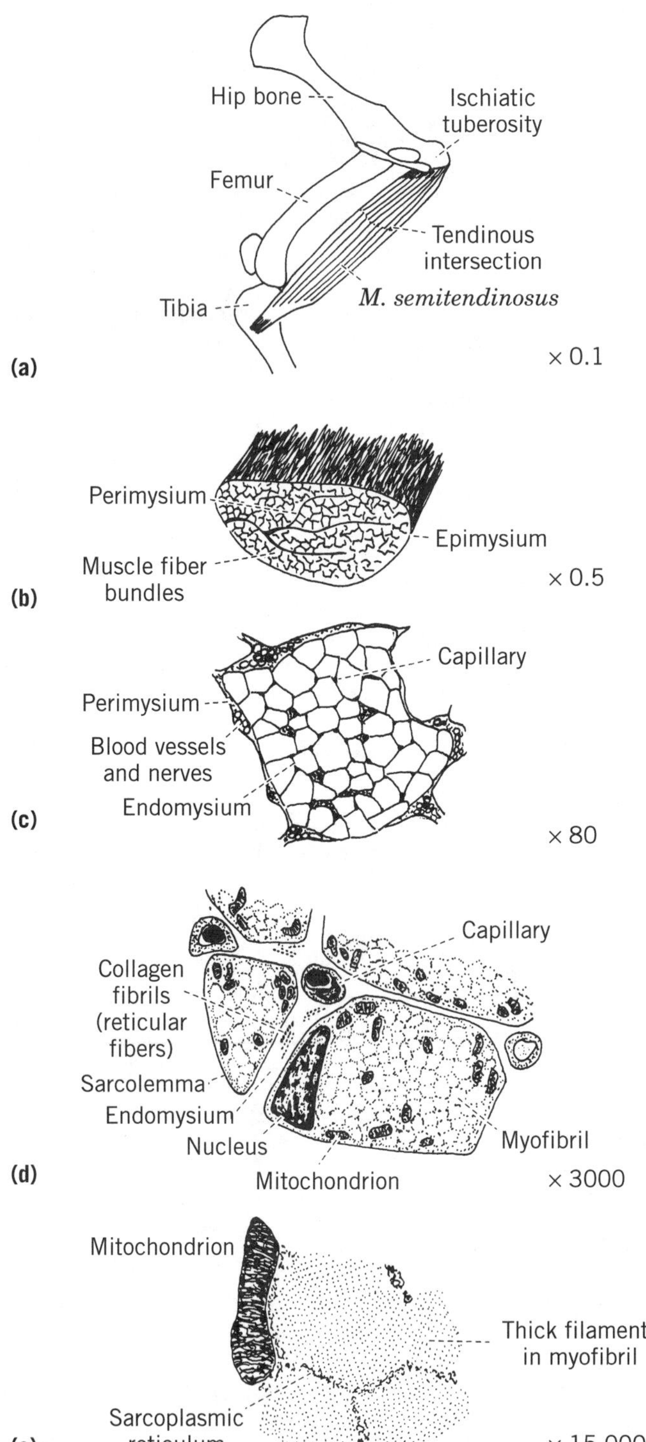

Figure 1. Gross and ultrastructural anatomy of the sheep semitendinosus muscle. (**a**) The semitendinosus muscle lies caudally in the thigh region. (**b**) Seen grossly in transverse section, perimysial strands course through the muscle and enclose muscle fiber bundles. (**c**) Within a bundle, an endomysial connective tissue network is revealed by light microscopy to surround individual muscle fibers. (**d**) Electronmicroscopic examination shows a network of sarcoplasmic reticulum enclosing myofibrils within the fibers. The two complete fibers drawn sectioned here are smaller in scale than actual fibers in the sheep semitendinosus muscle. (**e**) At a still higher magnification, a pattern of thick filaments is visible within each myofibril. *Source:* Ref. 1, used with permission.

Muscle Connective Tissue and Collagen Types

Each bundle of muscle fibers is surrounded by a connective tissue sheath, the epimysium, which makes the body of the muscle look like it is made up of strands. Smaller units of connective tissue associated with blood vessels and nerves within the epimysium are termed perimysium, and the feltlike connective tissue sheath around a single fiber (the muscle cell) is termed the endomysium. The types of collagen and their arrangement around muscle fibers have been described (2), and the different arrangement underlies the differences in the various meat cuts. Collagen is discussed later in this article.

Muscle Cell Structure

The single muscle fiber consists of a sarcolemma (termed cell membrane or plasmalemma in nonmuscle cells), mi-

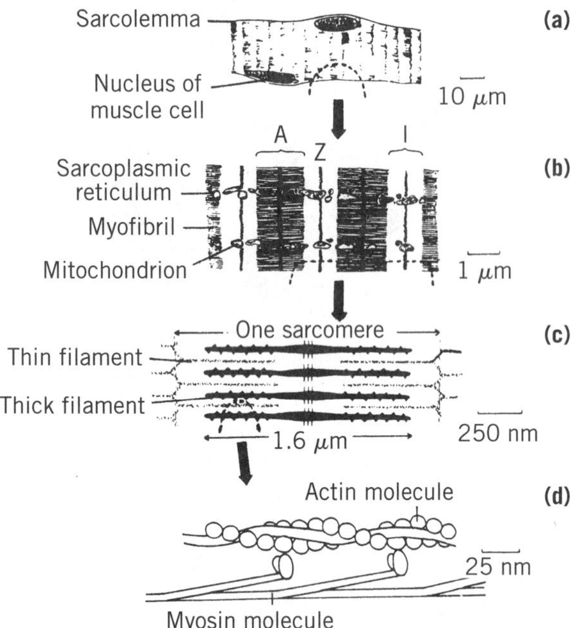

Figure 2. Fibrillar and molecular structure of the contractile elements of muscle. The skeletal muscle fiber in longitudinal section (**a**), as visualized by the light microscope; (**b**) and (**c**), by the electron microscope; and (**d**), as reconstructed from X-ray crystallographic evidence. *Source:* Ref. 1, used with permission.

tochondria, and the normal complement of cell organelles (Fig. 2). In particular, the cell is dominated by a unique arrangement of filamentous proteins, sarcoplasmic reticulum (termed endoplasmic reticulum in nonmuscle cells), and glycogen particles. In an electron micrograph, longitudinal sections show that the filaments consist of both thick (17 nm wide) and thin (8 nm wide) filaments corresponding mainly to the proteins myosin and actin, respectively. The thin filaments are attached to densely staining structures termed Z-lines, whereas the myosin filaments lie between the actin filaments and are joined to each other at the middle portion by an M-line. These structures, which together form a repeating unit from Z-line to Z-line termed a sarcomere, can be seen together with a further explanation in Figure 2. The part of the sarcomere containing myosin filaments is termed the A-band, and the part containing actin filaments is termed the I-band. Other filament structures, including filaments found in the gap of overstretched muscle (3,4), which have been now termed titin filaments, and proteins such as desmin and nebulin, are found with certain isolation, extraction, and staining procedures (see "Muscle to Meat" as follows). In transverse sections, the filaments appear as various size dots, and the Z-lines appear as smudges. When tissue is prepared from muscle at various stages of contraction, longitudinal sections show various degrees of interdigitation and overlap of thick and thin filaments. Transverse sections made at various levels of the sarcomere show corresponding patterns of dots associated with the overlap of the thick and thin filaments. Muscle and meat biochemistry have been reviewed (5).

Lying at the edge of the actin and myosin filament overlap, there is a system of tubules, termed T-tubules, derived from invaginations of the sarcolemma. The other tubular system, that is, the sarcoplasmic reticulum, lies within the cytoplasm of the muscle cell. The sarcoplasmic reticulum stores calcium ions. It is the controlled release of the calcium from the sarcoplasmic reticulum, raising the cytoplasmic calcium levels from 10^{-8} to 10^{-4} *M*, that causes a contraction (a shortening) in living muscle. The calcium ions are then retaken up into the sarcoplasmic reticulum and the muscle cell relaxes. The release of calcium from the sarcoplasmic reticulum is triggered by a wave of depolarisation via the T-tubules. Contraction occurs by the actin and myosin filaments sliding with respect to each other by attachment, release, and reattachment of portions of the myosin molecules (namely, myosin heads, also termed crossbridges), with specific regions of the actin filaments. This contractile activity, initiated via release of calcium from the sarcoplasmic reticulum, is activated and sustained by adenosine triphosphate (ATP). In the presence of calcium and absence of ATP (<1 μmol/g [6]), the muscles are in rigor, usually as a consequence of death. Eventually, ATP is completely transformed to inosine monophosphate, one of the flavor compounds of meat.

Muscle Fiber Types and Their Physiological Role

Although all striated muscles contract, the time course of contraction in the living animal varies with the muscle fiber type (fast twitch, slow twitch), substrate (ATP, creatine phosphate, protein, fatty acid, or glycogen), and the amount of oxygen available. Slow-twitch fibers (Type I fibers or red fibers) are mainly postural and are dominant in aerobic or endurance activity, whereas fast-twitch fibers (Type IIA and IIB fibers) are mainly fast acting and are involved in anaerobic activity. Type IIB muscles can take on a greater aerobic role than Type IIA muscles. Most body muscles consist of different proportions of all types, although instances of pure Type IIA (*m. cutaneous trunci*) or Type I (*m. masseter*) do exist in various parts of the body (7).

MUSCLE TO MEAT

The changes that occur when muscle in a living animal becomes meat are initiated from the moment the circulation stops, that is, slaughter. This stoppage of the circulation generally is preceded by preslaughter stunning and sticking. These events can affect meat quality in many different ways because they are extraordinarily variable. Slaughter by throat cutting can occur without a prior stun, can take place after a shooting with a penetrative captive bolt or nonpenetrative percussion head, or can take place after electrical stunning and be followed by electrical immobilization with important implications that are covered later. From the moment the blood supply stops, nutrients and oxygen are no longer available to the muscle from outside sources, so the energy stores present within the muscle start to be used up. Glycogen particles (see following) lying between the myofilaments and at various locations beneath the cell membrane are slowly depleted as the muscle's energy requirements are maintained, which may continue for a considerable period after slaughter. The depletion of muscle energy stores leads eventually to rigor mortis and a change in status; that is, the muscle is now meat. The pattern and extent of these changes are not the same for every muscle or species of animal and are influenced by a variety of physiological and physical interventions that have a major bearing on the ultimate quality of the meat.

The sliding of the myofilaments relative to each other can occur in muscle from living animals with the contraction initiated by the nervous system, as well as in muscle from animals that have been slaughtered, with the contraction initiated by physical and chemical changes. For example, if the carcasses of freshly slaughtered animals, particularly cattle and sheep, are exposed to cold, and as the normal metabolic arrangements fall down, the leakage of calcium into the cytoplasm (from mitochondria and sarcoplasmic reticulum) can cause irreversible cold contracture (a shortening with no relaxation). This ability to contract can last for many hours after death, depending on postslaughter conditions such as temperature and whether processes such as postmortem electrical stimulation were used.

The theory of muscle contraction based on a sliding of filaments with respect to each other is pivotal to the understanding of changes in muscle and the resulting effects on meat quality. Not only are the contractile filaments important, so too are a recently discovered set of filaments, originally termed gap filaments. These filaments, which contain a protein initially named connectin but now called titin, appear to contribute to the integrity of the myofibril (4,5,8–10) as well as play an important part in the postmortem changes that increase meat tenderness. Other minor proteins such as nebulin have an as-yet-unknown effect on meat tenderness, although desmin is susceptible to proteolysis and is likely to be involved in meat tenderness.

Muscle Energy Supply and Changes During Rigor

In a living muscle, or a muscle just at animal death, the immediate source of energy for contraction comes from ATP. Muscle ATP is quickly used up in two to five contractions, whereupon creatine phosphate immediately becomes dominant. Upon depletion of this source, glycogen takes over. Depending on the intensity and duration of muscle activity, and on the presence of blood circulation, either anaerobic or aerobic glycolysis occurs to sustain activity. Although protein or fat can be useful energy sources in sustained aerobic activity, glycogen is the only fuel that can be used for anaerobic activity. Therefore, glycogen is the only available energy source in postmortem muscle. It is metabolized when the muscle is triggered to contract via physical effects such as cold, and it is also metabolized slowly in noncontracting muscle. Without an intact circulation, lactic acid, a by-product of anaerobic glycolysis, accumulates in the tissue, causing the pH to fall (for reviews, see Ref. 6). At death, muscle tissue from rested, unstressed well-fed animals tends to contain 80 to 90 μmol of glycogen per gram of muscle tissue, but this is highly variable depending on muscle type and animal species. This glycogen allows the muscle to "survive" and contract until the pH falls from around 7.1 to below 6.0 to 6.2 (depending on animal species) and will still provide energy until the muscle goes into rigor at a pH of about 5.5 (Fig. 3) (11).

Immediately after slaughter, the muscle has an initial pH of 7.1 and is floppy and extensible. Upon attainment of rigor, the muscle is inextensible, and the pH has fallen to approximately 5.5. These changes have been explored in detail elsewhere (6,12). If the glycogen stores are less than optimum, the muscle pH cannot fall to the same degree. The consequences of this on meat quality are discussed in the section "Animal Factors Affecting Meat Quality."

Effect of pH on Myofibrillar Proteins and Water-Holding Capacity

Myosin, a highly charged molecule, undergoes some changes related to the pH of the muscle. At the pH of living tissue (approximately pH 7.1), the negative charges dominate, but at a pH of 5.5, which is closer to the muscle protein's isoelectric point, there is a similar number of negative and positive charges, and the capacity to bind water is least. When the muscle is in rigor, the bond between actin and myosin also causes the myofilament lattice to shrink, expelling water (13–15). Thus, for muscle in rigor that has a higher than normal ultimate pH, the lattice will not shrink so much, and the meat will bind water better than at a lower pH. Such meat with a pH greater than 6.0 is termed DFD (dark, firm, and dry). It has been suggested that changes involving the myosin heads by the temperature and pH conditions that exist (namely, rapid fall to pH 6.0 and below, when muscle temperature is still about 35°C) triggered by animal factors, result in pale, soft, exudative (PSE) meat that is caused by an even greater lattice contraction and greater water exudation than normal

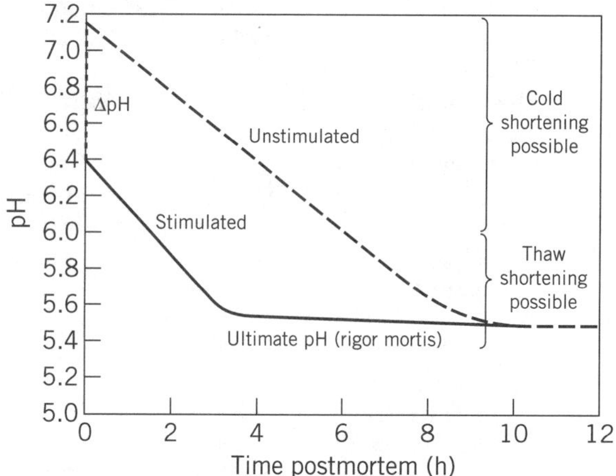

Figure 3. Fall in pH of unstimulated and stimulated beef sternomandibularis (neck muscle) held at 35°C. Unstimulated muscle reaches rigor and ultimate pH (approximately 5.5–5.6) in eight hours. Electrical stimulation for 120 s results in a concurrent fall in pH (ΔpH) to about 6.5. After stimulation, muscle pH falls at an increased rate. The combined effect of these pH falls is a decreased time for the muscle to enter rigor (approximately three to four hours). *Source:* Ref. 11, used with permission.

(15). Both DFD and PSE meat are discussed in more detail in the section "Animal Factors Affecting Meat Quality."

The water content of meat (approximately 70%) is important because it affects consumer acceptability as well as affecting the functional properties of meat and subsequent processed meat products (13). When meat pieces are treated with salt solutions, sometimes with added phosphate, weight gains of up to 40% may occur. Salting muscle before rigor can increase the water-holding capacity over that of meat salted postrigor (13). In salt solutions the fibers swell transversely, and with the addition of pyrophosphate, swelling occurs at lower salt concentrations. Recent work (16) has shown that the presence of the sarcolemma inhibits swelling, and this may cause considerable variation in the ability of different muscles to swell.

Muscle Glycolysis, pH Fall, and Temperature

The rate of production of lactic acid from glycogen in anaerobic muscle, which causes the postmortem fall in pH depends on muscle temperature (17). The rate in beef sternomandibularis muscle is 0.6 pH units per hour at 40°C and 0.15 pH units per hour at 15°C. As the temperature falls below 15°C, the rate increases; this can be attributed to the increased rate of glycogen used during cold shortening. The rate of muscle glycogen depletion can also be accelerated by preslaughter stunning (18), struggle (19), electrical stimulation by a factor of approximately two (20) (Fig. 3) (these effects are important and are covered by the article MEAT AND ELECTRICAL STIMULATION), and mincing (21). Under practical conditions, the rate of glycolysis slowly decreases due to the fall in muscle temperature in the cooling environments usually employed in chilling of carcasses. The rate of glycolysis also varies with muscle fiber type, being faster in beef *m. masseter* (0.4 pH units per hour at 35°C), which contains slow-twitch fibers, compared with the fast-twitch *m. cutaneous trunci* (0.2 pH units per hour at 35°C) (22). After electrical stimulation, the rates do not increase proportionately for particular muscles; the *m. cutaneous trunci* increases by a factor of two, but there is little change in the *m. masseter* (22). Differences in rates of glycolysis exist between species, with pigs having considerably faster rates than cattle. The rates of glycolysis are related to the basic metabolic rate of the animal, being very fast in small animals such as rats (23) and poultry. Chicken carcasses go into rigor in approximately four hours even when a fast chilling regime is used during processing (24). Cold shortening in poultry does not seem to cause the same toughness problems as in other species, as the maximum shortening occurs around 2°C (25), which is generally lower than the temperatures the birds reach following chilling by ice water and air chilling. However, rapid processing would be advantageous, because the toughness that arises from portioning the birds early and removing the breast muscles in particular could be eliminated. Electrical stimulation followed by rapid chilling of the birds to 2 to 6°C has been used to achieve rapid processing (see the article MEAT AND ELECTRICAL STIMULATION.)

The process whereby muscle goes into rigor is termed conditioning by New Zealand workers, and subsequent holding periods are termed aging. These terms are used throughout this article. Other countries use the term conditioning, or alternatively aging, for the whole process of going into rigor together with further postmortem holding. In general, the difference in terminology is unimportant, except when processes, such as electrical stimulation, are used that profoundly affect the conditioning stage. In this article, electrical stimulation is considered to produce its effects merely through acceleration of glycolysis to ensure that cold shortening is avoided and aging will start at high temperatures. This is not necessarily true in every instance, see the article MEAT AND ELECTRICAL STIMULATION.

MYOFIBRILLAR ASPECTS OF MEAT QUALITY

Muscle Shortening and Meat Quality

During normal conditioning procedures, as the muscle goes into rigor there is shortening, which depends on the temperature of the muscle, being greatest at high and low temperatures in unrestrained muscle, and differs with species. Early classical studies (26) showed that the minimum shortening for beef occurred at about 15°C (Fig. 4), and other studies showed that the greater the amount of shortening, the greater the toughening of the cooked meat (27) (Fig. 5). Increases in shortening at rigor temperatures above 15°C are as expected, as such contractures would increase with temperature in a normal way. Increases in shortening with temperature falls below 15° to 0°C are unexpected and are likely to arise from a combination of reduced calcium binding by cell organelles with the fall in temperature, causing an increase in intracellular calcium. The increased contracture by the muscle through increased calcium levels is more than offset by the concomitant falloff in muscle responses at the lowered temperature. A more quantitative analysis of the effect of

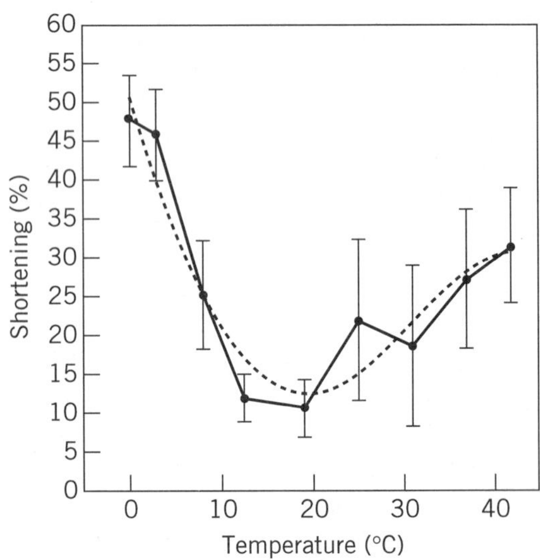

Figure 4. Mean ultimate shortenings of excised beef *m. sternomandibularis* stored at various temperatures during rigor development. A curve of best fit has been included. *Source:* Ref. 26, used with permission.

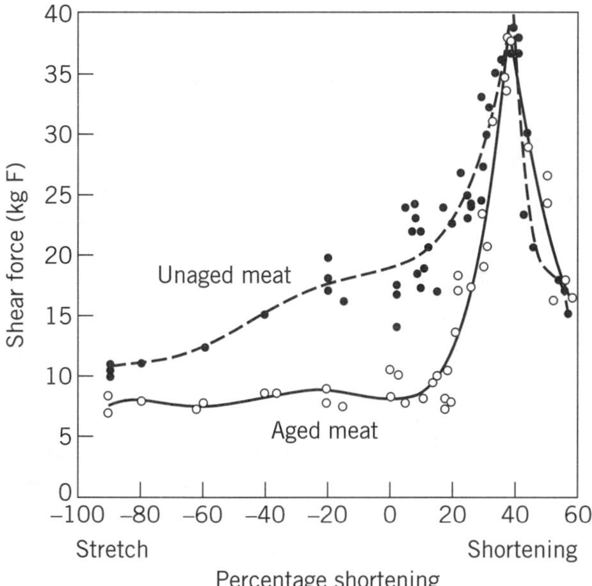

Figure 5. Influence of prerigor muscle shortening on the force score values of unaged and aged beef *m. sternomandibularis*. ●, rigor muscle; ○, muscle aged for three days at 15°C after setting in rigor at different degrees of shortening.[7] At shortenings close to rest, lengths the meat can tenderize, but as the shortenings increase, tenderization becomes less. The tenderization that occurs after the peak is as a result of shortenings being so large, the muscle tears itself apart. *Source:* Ref. 27, used with permission.

temperature on shortening and ATP utilization has been made (28).

Although shortening at temperatures higher than 15°C occurs, it is not strong and can be counteracted at these temperatures by postural alteration (29). With processing temperatures normally experienced, extreme toughness of the cooked muscle does not seem to be a problem except in poultry. Therefore, for most meat species, it seems likely that enzymes involved in postrigor proteolytic breakdown are more effective at higher temperatures and counteract any toughness due to shortening, although there are some problems with rigor at high temperatures where tenderizing enzymes are reduced (see "Variability in Tenderness"). The apparent anomalous shortening at temperatures below 15°C, which increases with colder temperatures, applies to most muscles and is termed cold shortening. Cold shortening is not a theoretical issue; it is a practical problem that results in tough meat, especially when carcasses are subjected to the rapid rates of chilling or freezing that currently take place in meat processing operations. When carcasses are treated in this way, and muscle temperatures fall below 10°C before the pH falls below pH 6.0, the muscles can shorten markedly. With extremes of shortening (arising from cold shortening, thaw shortening, or heat from prerigor cooking) the muscle fibers will supercontract in some regions with a consequent disruption in adjacent regions, leading to a tenderization (30) that corresponds to the tenderness beyond the peak of toughness in Figure 5.

Shortening can be partially prevented by hanging carcasses by the aitch bone (31), termed tenderstretch, or by other maneuvers of the carcass (29), or by electrical stimulation to accelerate rigor and to shorten possible exposure times to low temperatures while shortening can occur. If meat is frozen before rigor mortis is complete, the energy stores ensure that contraction will take place as the meat thaws, that is, thaw shortening (6,32). There are likely to be other as-yet-unknown effects that take place as muscles go into rigor. For example, even cold-shortened muscle becomes somewhat tender after being held at 37°C (33), and meat subjected to extremes of pressure will become tender (34). Even explosives will cause this (35).

The effects of cold shortening are greater when the low temperatures are attained soon after slaughter and seem to progressively disappear at a muscle pH between 6.2 and 6.0. Animals with a high ultimate pH generally reach rigor before shortening is manifest.

Cold shortening does not occur in all muscles of all species under the same conditions of temperature, and perhaps this is the underlying reason for its relatively late discovery. With the masseter, which consists of red (slow-twitch Type I) fibers, cold shortening readily occurs, yet in the cutaneous trunci (fast-twitch Type II A) cold shortening does not occur until muscle temperature reaches 0°C (22). Cold shortening is a problem with sheep and cattle carcasses, but not so much with pigs and poultry. With pig muscle, the rate of rigor is sufficiently fast and the tendency to shortening appears to be less in any event, so cold shortening does not appear to be a problem unless extremely high rates of cooling are involved. With poultry, cold shortening is also not a problem, as maximum shortening occurs at around 0°C and rigor occurs in approximately five hours. However, heat shortening at temperatures above 20°C can occur (25), and meat toughening occurs with stressed birds going into rigor early at high temperatures.

Prerigor muscle excised from the bone is free to shorten large amounts, the amount depending on the restraint of adjoining muscle and connective tissue and its temperature, and electrical stimulation is beneficial. If there is enough shortening, the meat will become tough. The practice of hot boning, therefore, has to be examined with respect to the end use of the meat. If the end use is for manufacturing processed meats, muscle shortening may not be a problem. Cold shortening does not cause changes in water-holding capacity (13), but thaw shortening changes the water-holding capacity, with some disadvantages in further processing. Muscle shortening, and hence toughening of the meat, cannot be predicted by meat appearance. Electrical stimulation can reduce the toughening of meat associated with hot boning for certain cuts (36).

Aging of Meat

Unshortened muscle at rigor has a basal level of tenderness that is enhanced by subsequent aging of the muscle. Severely cold-shortened meat does not age to any degree (Fig. 5) (27). Aging involves the breakdown of the muscle proteins (eg, troponin T, troponin I, tropomyosin, C-protein, and M-protein and cytoskeletal proteins, titin, nebulin, and desmin), by endogenous enzymes, termed calpains (37–39), although minimally involving the major proteins actin and myosin. Calpains are members of the cysteine class of proteases and are inhibited by calpastatins. They exert their effect by altering structural integrity at the junction of the A-I band, as well as fragmentation of the Z-lines. Calpains exist in two forms, μ- and m-calpains, that differ in their calcium requirements. μ-Calpain requires 1–30 μmol of calcium, and m-calpain requires 270 to 750 μmol of calcium for half-maximal activity. Binding of calcium also induces the autolysis of the calpains that results in loss of activity. In the presence of calcium sufficient to activate calpains, calpastatin complexes with calpains to inhibit activity; thus, the way calpain exerts its effects is still unclear. Even so, among the enzymes involved, the calpains seem to be the best candidates to tenderize meat, and the inhibitor calpastatin prevents their activity in normal tissues (40–43). Indeed there is more than enough calpastatin in any muscle to inactivate the calpains. A synergistic contribution by lysosomal (cathepsins) and calcium-dependent proteinases (calpains) has been proposed (44), with no direct relationship between aging rate and protease content of muscles found. The aging rate of meat increases dramatically as the temperature rises and has a temperature coefficient of 2.4, and aging even takes place during cooking but ceases dramatically at a temperature of 66°C (Fig. 6) (45).

As aging is temperature dependent and carcasses are in a cooling environment, processes such as electrical stimulation that produce an early rigor while the carcasses are still relatively warm ensure maximum aging in the shortest possible time (apart from also avoiding the effects

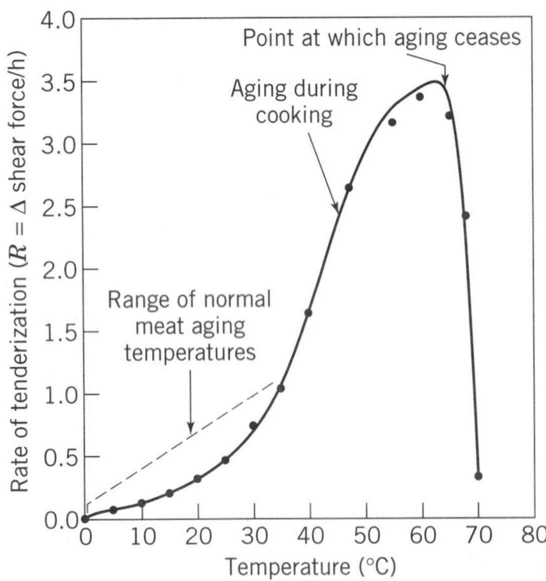

Figure 6. This diagram illustrates how the rate of tenderization of meat, through aging, increases with holding temperature. The rate of tenderization at temperatures of 0°C is relatively slow, and long holding periods are required at this temperature. As the temperature is raised during cooking, the rate of aging has an important part to play in the final tenderness. However, at temperatures above approximately 60°C, aging ceases. *Source:* Ref. 45, used with permission.

of cold shortening). High temperature tends to suggest that conditions are present to allow increased bacterial growth as well as meat aging, but in practical terms, surface drying prevents microbial growth on the meat surfaces. Rapid aging is not required when there is sufficient time available between slaughter and retail for aging to occur. Vacuum packaging or modified atmosphere packaging can ensure that bacterial growth is low under these circumstances. Aging also occurs when frozen meat is thawed and continues during subsequent holding. In certain species such as venison and other game animals, extreme aging is often regarded as desirable.

Variability in Tenderness

The initial tenderness of meat is not a constant and the tenderness of unaged meat and the extent of aging depends on muscle shortening (27). When one considers the variations of early tenderness of unaged meat, it is clear that there are large shortening-related changes in tenderness in the range close to rest length, and these changes will be influenced mainly by processing conditions, that effectively vary across a carcass. The sarcomere length is significantly changed when the carcass is hung from the pelvis rather than the Achilles tendon, and there are dramatic changes in tenderness with the increased sarcomere lengths (46–49).

Injections of various salts such as calcium (50) and phosphate (51) will increase tenderness and can overcome some effects of shortening. Marination of meat with acids does not dramatically improve tenderness (52), although the flavors added can enhance consumer appreciation. Making the meat alkaline (eg, sprinkling with bicarbonate) does increase tenderness, but with concomitant texture and small flavor changes that make it more useful in various ethnic dishes.

The endogenous enzymes responsible for tenderization are active throughout life in other roles, and by implication their action continues throughout the rigor process. With all inhibiting mechanisms in place effectively until rigor changes in tenderness are small; thus, aging effectively starts at rigor mortis (53).

Tenderization of meat is influenced by shortening, and in particular Locker and Wild (54) showed that proteolysis occurs in both cold-shortened and unshortened meat, but the cold-shortened meat was persistently tough. This suggests that myofibrillar fragmentation length, widely regarded as a measure of proteolysis, also must be considered with regard to processing conditions (53).

Dransfield et al. (56) suggested that there was some inhibition of aging at high rigor temperatures, which was shown by Devine et al. (53) to be greatest at 35°C and least at 15°C, and the mechanism was shown to involve an inactivation of calpains (56). Furthermore, it was found that by tuning electrical stimulation and obtaining the appropriate chilling rate, the greatest tenderness of meat could be achieved rapidly, when rigor occurs at at 15°C (57). Such mechanisms may also explain some aspects of tenderization of slow glycolyzing muscles.

Rapid chilling during hot-boning processing, when meat is removed prerigor, would exacerbate shortening; cold shortening in particular needs to be avoided. However, at 15°C, shortening is minimized and aging is maximized (58), thus making hot boning an effective option under such conditions (see the article MEAT AND ELECTRICAL STIMULATION). Because compounds such as endogenous nitric oxide affect the rate of tenderization (59), tenderness may be influenced by factors such as exercise that affect nitric oxide levels (60).

With so many factors affecting both the initial and ultimate tenderness of meat—animal factors such as breed, genetics, and sex—finishing regimes on meat tenderness must be considered only when rigorously controlling processing conditions (ie, rigor temperature and shortening).

ROLE OF CONNECTIVE TISSUE IN MEAT QUALITY

Closely associated with the contractile myofibrils in muscle tissue are connective tissue components, which form a three-dimensional network to support the muscle cells. The net of connective tissue encompasses complete muscle fiber bundles and even single muscle cells. Embedded in the connective tissue are also intramuscular fat, blood vessels, and nerves. The connective tissue elements contribute in a large way to meat texture as well as modifying meat tenderness.

Composition of Collagen

The major component of connective tissue is the protein collagen, with minor components elastin, glycoproteins, and proteoglycans (61). Connective tissue comprises less than 2% of most skeletal muscles (2). Such a small amount of protein, however, has a major impact on the tenderness of the meat (when other factors such as cold shortening are avoided).

Extensive studies have shown that it is not only the amount of collagen but also the degree to which cross-links in collagen are solubilized during cooking that determine tenderness. Collagen has a high mechanical strength, brought about through the formation within the collagen fibers of intramolecular cross-links (2,61), which are covalent in nature and result from a precise organization in the fiber. Two particular types of cross-links—an aldimine-type bond and a ketoimine-type bond—are replaced by more stable, nonreducible structures with increasing age. As the collagen molecules possess an extremely long biological half-life compared with most other proteins, changes in the numbers and type of cross-links tend to increase, causing the properties of the collagen to change with animal age. Thus, the fibers become progressively stronger and more rigid and less susceptible to enzymatic degradation and swelling by acids as the animals grow older. This maturation is achieved partly by further reactions to form multivalent cross-links that link several collagen molecules laterally, thus dramatically increasing the fiber stability.

Collagen Turnover

Collagen turnover varies considerably between tissues, and the relative growth rates of different muscles result in different levels of maturity of the collagen in a given animal (2). Similarly, differences in growth rates of bulls, steers, and double-muscled animals result in differences in both the amount of collagen and its maturity in terms of stable cross-links (2). Although the faster growth rate of bulls compared with steers would suggest that bull meat should be more tender, the rate of collagen turnover may well be different in bulls and allow earlier maturation of the bull collagen. On the other hand, the reported toughness of bull meat may arise from other factors, such as absence of intramuscular fat, which changes perceptions of tenderness during actual chewing, and sex effects such as an increased stress susceptibility and consequent elevation of muscle pH (see the section "Animal Factors Affecting Meat Quality"). Rapid finishing of animals appears to increase tenderness, which could be accounted for by a higher proportion of newly synthesized and consequently immature collagen being laid down. Possibly, larger amounts of intramuscular fat may also modify taste panel perceptions, by causing a smaller cross-sectional area of myofibrillar and collagen protein in any given bite.

Collagen Solubility

Animal age, due to a progressive increase in interfiber cross-links, has the most significant effect on meat texture, even with decreasing proportions of collagen, as the myofibrillar proteins increase in quantity during growth. The increase in toughness in cattle due to animal age was most clearly demonstrated (62) where some cuts had a fivefold increase in toughness in 5-year-old animals. The study also showed that the tenderness of grilling cuts was least influenced by animal age. Presumably this is because the amount of collagen is constant and low and any adverse effects of cross-linking cannot be expressed, as solubilization does not have time to occur during the brief cooking. This result has been confirmed in a study, using a sensory panel and instrumentally, which showed that toughness in *m. longissimus* did not increase and in *m. semitendinosus* increased in cows aged from 2 to 9 years (63).

For sheep *m. semimembranosus*, collagen concentration is highest in newborn animals, but decreases and becomes constant from 40 to 365 days. Collagen solubility was shown to be greatest in newborns, and as a consequence the meat is more tender (64). While the collagen solubility decreases with age, from approximately 50% in newborns to 31% at 42 days, down to 12% at one year (64), the decreases from then on clearly are not important in terms of tenderness. So although it is true that very young animals are tender from this component, it is not relevant in terms of the age at which animals are normally slaughtered. The studies also showed that high or low values for collagen levels and solubility in the *m. semimembranosus* were also true for the *m. biceps femoris* (64).

Studies on groups of steers and bulls slaughtered at various ages through to maturity also failed to show any major differences in terms of collagen solubility (65) and masculinity traits of bulls in meat quality (66).

Postmortem effects on collagen solubility are not fully explored. Researchers (67) showed that some proteolytic cleavage does occur but with little detectable change in solubility. More recently, it was shown that significant changes occur in the immediate postslaughter period (68), and it may be these early changes that are most important in any investigation of collagen solubility and its effect on tenderness. The effect of electrical stimulation on collagen solubility is unexplored, and the high rates of pH fall and consequent activation of various proteases that occur with electrical stimulation could profoundly affect the collagen molecules. Other connective tissue contributors to toughness may arise from changes in the collagen lattice, which alters significantly with muscle shortening and thus could affect texture (69).

COOKING AND MEAT TENDERNESS

Meat Tenderness/Texture

For consumers, the most important eating quality for meat is tenderness, which is generally rated on a consumer scale where high values are good. If the meat is evaluated by an instrument, the concept of tenderness is reflected in shear force values, where the lower it is, the better the meat. The word *tenderness*, however, does not completely describe the quality attribute, because various processed meats such as frankfurters are tender, yet do not have the texture associated with whole tissue meats. The tenderness/texture quality, a rather individual characteristic, is governed by a complex relationship between the myofibrillar structure and connective tissue structure, the response to which in turn is probably modified by associated juiciness parameters arising from fat and retained water. As a broad generalization, tenderness is more closely allied to myofibrillar factors, and texture is related to the connective tissue components. Texture is often noticed as the wad of tissue that remains after chewing. Ideally, to make a proper appraisal of meat, trained taste panels should always be used, but this is generally not feasible. An objective measuring device such as a tenderometer to measure the shear force is therefore desirable. It should be remembered, however, that many tenderometers measure only myofibrillar tenderness components to any degree of accuracy, although the measurements are influenced by extreme connective tissue variations. To some extent, texture issues are of lesser importance in cuts used for grillings, as only low connective tissue cuts are used.

Tenderometers

When meat is sheared in a tenderometer, the toothlike device or blade deforms the meat, essentially producing yield data for the cooked myofibrillar proteins with some contribution from the collagen. For all practical purposes, changes in the tenderness of meat from the same muscle correspond to changes in myofibrillar tenderness. A wide range of tenderometer devices have been developed and all have faults, but they yield useful information even though not accounting for all possible contributions to tenderness. The devices have been reviewed (70,71).

Only a few major basic designs of tenderometers are in current use (72). In the Warner Bratzler device, cylindrical or rectangular samples are placed in a triangular hole in a shear blade (some modifications use square-sectioned material in a square hole). There are a variety of tooth-shaped biting devices [Volodkevitch type toothed jaws, and a pneumatic tenderometer with toothlike jaws (MIRINZ tenderometer)] that are very similar in concept. In an Allo-Kramer shear press, there is a compression cell with an array of blades to penetrate the meat. Other measurements can also be made that indicate the contribution of connective tissue to toughness, such as the measurement of adhesion between fibers (72).

One difficulty of comparing tenderometer data is that no one has standardized the readings from the various types of tenderometers; thus, at present, each research group is unique in its presentation of data, although some attempt has been made to standardize them (73). All tenderometers work on cooked meat, and no group has yet devised a completely successful instrument for measuring raw meat characteristics that can be extrapolated to predict cooked meat values. There is potential in the use of near infrared spectroscopy where there is a correlation between measured tenderness values (74) and connective tissue solubility and concentration (75), from raw meat and predicted instrumental values, but such measurements only will estimate the tenderness at the time of measurement. From the preceding comments on meat conditioning, aging, and connective tissue changes during cooking, such correlations related to tenderness would need to consider time delays before consumption, as meat tenderness depends on various unpredictable aging treatments once product has left the meat plant. Any control of product is ultimately lost due to unpredictable handling by the consumer, but it should be acceptable at purchase. A noninvasive device applied to raw meat would be extremely valuable at this point.

Cooking of Meat

Tenderness/shear force and cooking have a complex interrelationship and are best discussed together. Cooking is the preferred method of preparing meat for serving. Cooking not only safeguards health by destroying bacteria and parasites, but the heat also coagulates the proteins, transforming a bland, chewy product to a juicy, flavorsome, and beautifully textured food.

The final outcome of cooking depends on factors such as the rate of heat transfer. Grilled products have a flavor development on the surface and a variable texture from the seared surface to the softer juicy, bland interior, whereas casserolling gives a complex effect arising from browned surfaces, which cause even deepening of the flavor throughout the whole meat. Roasting allows a slow heat transfer and aging of the meat during cooking.

The juiciness of the product and the contribution of the connective tissue influence the final outcome. Thus, products to be grilled must have low amounts of connective tissue (however, readily hydrolyzable cross-links will not be affected rapidly enough), otherwise they will be tough. The long, slow cooking of a casserole (or a roast) allows aging to take place and also breaks the collagen cross-links. This allows consumption of certain cuts that would be relatively unsatisfactory for grilling. In essence, the tenderness of a piece of meat is a function of the postslaughter handling, including conditioning and aging procedures, the rate of heat transfer with the various denatured muscle cooking methods, and the amount and type of connective tissue present.

The choice of cooking procedure depends first on the amount of connective tissue present and the age of the animal and, second, on whether the meat will benefit from aging by prolonged cooking procedures. Clearly, microwave cooking of meat has few advantages, other than convenience, as the temperature rises too fast in an uncontrollable manner. There is not enough time to break down connective tissue, and the myofibrillar proteins also will not age significantly above 66°C (45) (Fig. 6). If meat has severely cold shortened prior to any cooking procedure, it will never become tender, even with slow cooking. Such meat often is fed to pets, after the consumer finds it almost inedibly tough.

Cooking and Tenderometer Measurements

For tenderness or shear force measurements, and many meat quality assessments, the meat must be cooked under a specified set of conditions, so that either only meat processing or animal-related factors are compared. Meat is cooked by subjecting it to heat before it is either eaten or assessed in other ways. Cooking denatures the myofibrillar proteins as well as the connective tissue proteins. There is a complex interplay between the myofibrillar and connective tissue proteins in the meat as the temperature is progressively raised, including aging of the meat, which is inhibited at temperatures above 66°C (45). Only meat that has not been cold shortened will be discussed.

Changes of Myofibrillar and Connective Tissue Proteins Due to Temperature

The following description (2) of the processes occurring during cooking emphasizes the relationship between the myofibrillar and connective tissue proteins (neglecting aging effects during cooking). As the temperature of a piece of meat is raised, the myofibrillar and connective tissue proteins denature at different rates, and it is the properties of the denatured proteins that determine the texture of the meat. At temperatures between 40 and 65°C there is an increase in toughness (Fig. 7) (76) as determined by shear value, which is caused by aggregation of the denatured myofibrillar proteins. At this stage, there is a loss of fluid and shrinkage of the muscle fibers within the endomysial sheath. The collagen in this sheath is not affected at this temperature. However, because the endomysial sheath was under tension in the raw meat, the shrinking of the myofibrillar proteins releases the tension, which forces fluid out of the meat. As the temperature continues to rise from 63 to 80°C, the shear values increase further due to additional shrinkage as the collagen in the endomysium and perimysium denatures and the water is squeezed out. Complete shrinkage of the sheath is prevented by the presence of the myofibrillar proteins. With further increases in temperature above 80°C, there is a reduction in shear values (increase in tenderness), possibly due to peptide bond cleavage and/or cross-link rupture of denatured collagen and breakdown of myofibrillar proteins. The increased tenderness on prolonged heating is almost certainly primarily due to degradation of the denatured collagen.

During cooking, meat collagen is denatured and, because of its partially crystalline nature, it shrinks at about 65°C (depending on animal age and species) to form insoluble gelatin. The thermally stable cross-links modify the shrinkage characteristics so that when some residual tension is maintained after denaturation, there is considerable remaining strength. The tension generated during cooking varies considerably depending on the heat stability of the perimysium, which in turn is determined by the nature and extent of the cross-linking. Thus, the older the animal, the higher the proportion of heat-stable cross-links and the greater the tension generated on shrinkage. The total amount of collagen in itself is no indicator of strength or texture in the cooked meat. Instead, the solubility of collagen is important, as that changes with animal age and thus affects the consequent cross-linking in a complex way. It is the heat stability of the cross-links that affects the strength/texture components of tenderness. The changes in collagen can cause noticeable effects during normal cooking, as they are responsible for the extensive curling of chops or steaks that occurs during grilling and the amount of cook loss.

The end point of cooking is of interest to the consumer. A rare steak is very pink and has an internal temperature less than 60°C, and a well-done steak is grey to brown at internal temperatures greater than 75°C. Dissatisfaction often arises if the color and "degree of doneness" do not match (77). Meat proteins, especially myoglobin, are more stable to temperature as the ultimate pH rises, and the degree of doneness at the same internal temperature often appears to be insufficient (ie, redder). For institutional cooking, variable ultimate pH values give a variable appearance to the steaks.

Moisture Loss in Cooking

The moisture loss upon cooking is predictable. Whereas frozen meat initially loses moisture on thawing, and drip exudes in vacuum-packaged meat, the total water loss is an inevitable consequence of the cooking process and depends on the degree of cooking (and amounts of connective tissue) as previously described, being least in a very rare steak.

Consumer Aspects of Meat Texture and Tenderness

The overall tenderness of meat to the palate involves three aspects: (*1*) the initial ease of penetration of meat by the teeth (tenderness), (*2*) the ease with which the meat breaks into fragments (tenderness and texture), and (*3*) the amount of residue remaining after chewing (texture). Clearly, some of these attributes are modified by the amount of fat and the temperature at which the meat is served as well as the cooking conditions used. Because consumers clearly reserve tougher cuts of meat for casseroles, there is a perception that certain levels of tenderness are desirable. The less desirable cuts increase dramatically in toughness with animal age due to connective tissue changes (62) leaving more mouth residue, but frying cuts

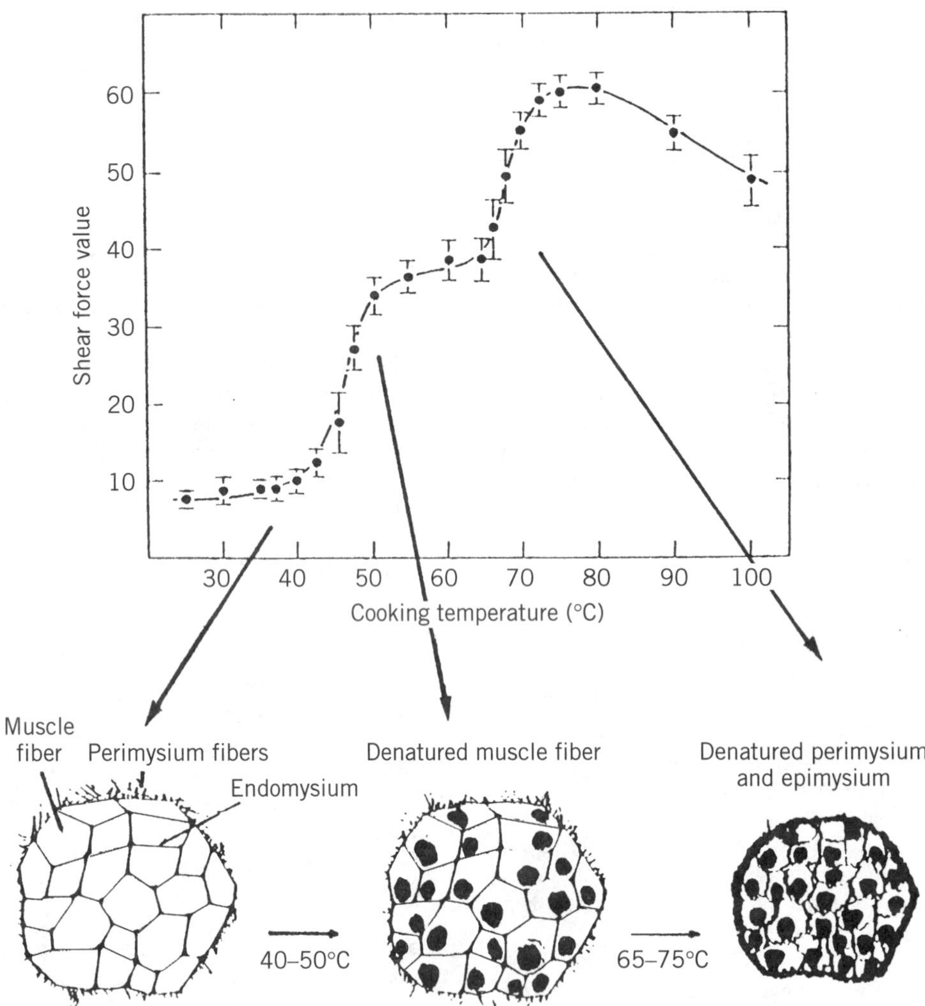

Figure 7. Diagrammatic representation of the changes in the muscle fibers and connective tissue during cooking in a cross section of muscle related to the change in shear value. The lower the shear values, the more tender the meat. The shear force units in this figure are on a different scale and are approximately four times those in other figures. The cooking temperatures for meat are very rare (55°C), rare (60°C), medium rare (65°C), medium (70°C), medium well done (75°C), well done (80°C) and very well done (85°C). *Source:* Ref. 2, used with permission.

are virtually unaffected by animal age if cooked to a comparatively rare state. The amount of subcutaneous fat appears to affect tenderness mainly by reducing the chilling rate, and fat within the muscle (ie, marbling) is a poor predictor of tenderness and palatability once there is a minimum amount. It requires large increases in marbling to produce detectable changes in palatability (78,79). Tenderness increases (ie, shear force reduction) and palatability improvement of the cooked meat with excessive marbling are likely because there is less meat, but there is often consumer resistance if they see the meat raw first!

ANIMAL FACTORS AFFECTING MEAT QUALITY

Preslaughter and Slaughter Components

This section in general considers the meat quality changes from the live animal through slaughter, but only considers briefly factors such as breed (ie, genetics), growth, and sex. Apart from stress factors that may be breed or sex related (such as in the case of bulls where pH effects occur—see later) and animal age (which has an obvious influence only at extremes—discussed earlier) and fat cover, which affects chilling rates, the influence of these is minor. This is not to suggest that the preceding factors do not affect meat quality, but rather the effects are minor in comparison with processing effects and are perhaps overemphasized. Cold-shortening capacity appears to be greater in older animals, where the tension generated and shortenings are greater by a factor of three (80). There was shown to be a reduced tenderness in old bulls compared with young ox (81). The muscle used was beef *m. sternomandibularis*, which has a higher connective tissue content than other muscles. We may merely be considering the expected age-related changes in the connective tissue net in a relatively atypical muscle (see age-related toughness changes covered earlier

[62]). Similar studies have not been made on other muscles. Much work in other studies has merely revealed either no statistical difference or a minor difference between breeds and tends to focus on differences between *Bos taurus* and *B. indicus* (82). For pigs, genetic effects on meat quality exemplified by different breeds seem to be more important than for ruminants (83). In general, breed effects would pale into insignificance compared with the processing-related effects on meat quality.

Animal welfare issues for all meat species are not only consumer issues but also need to be addressed for optimum product quality and production efficiency. The effect on meat quality is subtle and is of particular important for pigs, fish, and poultry (84). Preslaughter effects commence on the farm and may arise from simple factors such as insufficient feed. From this point, mustering or rounding up, transport, and moving through races to the slaughter area contribute to bruising and stress, much of which is unnecessary. Different problems arise for pig raising facilities or feedlots for cattle than for range-fed animals. Careful design of pens, races, trucks, ramps, and other animal handling facilities can reduce many of the problems (85), but further research is needed. Washing may reduce stress in the case of pigs or can be stressful in the case of swim washing for sheep.

Slaughter procedures depend on the type of animal involved (84,86). For sheep and cattle, two types of electrical stunning can be used: head-only stunning, which allows the heart to continue beating; and head-back stunning, which causes cardiac arrest. Bleeding out and meat quality are the same with either stunning technique for lambs (87) or with delayed bleeding for beef (88). As head-only electrically stunned animals will recover within a few minutes if they are not slaughtered, the animals actually die through loss of blood initiated by the slaughterman. This type of preslaughter stunning is therefore acceptable for halal slaughter of sheep or cattle. Cattle are generally captive-bolt stunned, although percussion stunning and either type of electrical stunning can be used with no deleterious effects on bleeding out. Carbon dioxide stunning and electrical stunning are used for pigs. Carbon dioxide stunning appears to be associated with a reduced incidence of both blood splash (hemorrhages in muscle) and pale, soft, exudative meat (89), but it also seems likely that this relationship is partially a result of animal selection. Recent studies indicate that there are no welfare or issues of humaneness with current slaughter if properly applied (84,90).

High Ultimate pH (DFD Condition)

The process of anaerobic conversion of muscle glycogen to lactic acid after slaughter reaches an end point when the muscle pH falls to about pH 5.5, at which point complete inhibition of the enzymes responsible occurs even if glycogen is still present. The pH reached when glycolysis ceases is termed the ultimate pH. The ultimate pH will be higher if the initial (preslaughter) stores of muscle glycogen are low. In the worst instances of extreme stress, the muscle ultimate pH does not change from immediate postslaughter values. Such high pH values have important consequences for meat quality. The meat is darker in color, firmer in texture, and dry (DFD) and does not keep as well as normal pH meat in vacuum packaging (91).

Depletion of Muscle Glycogen

The levels of muscle glycogen at slaughter are a result of the cumulative stressors that occur from the farm or feedlot, through transport and lairage to the point of slaughter. Stressors can be classified as psychological if the levels of glycogen are affected by increased levels of adrenaline, producing a concomitant increase of corticosteroids, or physical if they involve physical exercise, sickness, or inadequate nutrition. In the case of physical exercise such as rounding up sheep and cattle, mounting behavior in cattle, or swim washing for sheep, the muscle glycogen is quickly converted to lactic acid, which is transported away from the muscles by the circulation. This lactic acid ultimately can be resynthesized to muscle glycogen, given sufficient time. Feeding immediately preslaughter helps some species, such as pigs, to replenish their muscle glycogen, as the ingested high energy carbohydrate can rapidly be reincorporated into muscle glycogen. In ruminants, however, the process is much slower, because the rumen organisms convert carbohydrates into short-chain acids such as propionic, butyric, and acetic acids, which in turn have to be synthesized into glycogen. Even in conditions of complete rest, for sheep it takes 17 h for muscle glycogen to be resynthesized (92), and in stressful slaughter conditions this is slowed (93). Long-term malnutrition also has the effect of reducing initial muscle glycogen, but this in itself is usually not enough to cause meat with an excessively high ultimate pH, unless it is combined with other factors (94). It is the cumulative effect of many stressors that is important, and malnutrition may merely be the first in the series (95). Psychological stressors, mediated through adrenaline and cortisol release, are related to factors such as transport, contact with dogs in the case of sheep, and interaction with unfamiliar peers in a strange environment for cattle and pigs.

There is probably never a single cause of high-pH meat. It is this multiplicity of causes—many of which are not recognized—that makes identification of causes of high-pH meat in most circumstances difficult. In addition, the response to individual stressors is also different for each species. For example, dogs are identified as a predator by sheep, but they are a mere annoyance to cattle; swim washing for sheep is inordinately stressful because it involves severe physical exercise (96), but washing of pigs and cattle before slaughter can relieve stress. Mixing of cattle (97,98) and mixing of pigs is stressful but does not appear to affect sheep. Mounting behavior in cattle is also stressful (but can be reduced by an overhead electrified grid). Thus, elimination of high-pH animals requires a coordinated effort from the producer right through to the processor.

PSE Condition

Pale, soft, and exudative (PSE) meat appears to be restricted to pigs for all practical purposes and is a consequence of a particular muscle response to animal stress.

The outcome is a very rapid fall in postmortem pH while the meat is hot, which denatures muscle proteins in such a way that water binding is reduced, and the muscle drips and is soft to the touch (14). The denatured proteins scatter light, giving a pale appearance. Rapid chilling postslaughter reduces the incidence of PSE; although the meat is initially tough, it tenderizes with aging (99). Animal agitation and electrical stimulation increase the occurrence of PSE in certain circumstances (100). However, if electrical stimulation is applied at 20 min postslaughter and the chilling rate is sufficiently fast (eg, deep leg to 10°C in 5–10 h), then the meat in those animals susceptible to PSE does not show it and the remaining meat avoids cold shortening (101). The situation is further improved by pelvic suspension (102). Hot boning of pork can allow sufficient cooling to prevent PSE (103). Pale, wet muscle can occur in muscles from very large cattle when chilling is suboptimal and is not a result of stress susceptibility as in the case of pork, but rather of high temperatures and low pH. It is never as severe as it is in pigs.

Interrelationship of DFD and PSE in Pigs

Not only do certain handling situations tend to trigger PSE, but the genetic makeup of a pig may predispose the animal to PSE, and some European breeds are especially susceptible (83,84). In some particularly stress-susceptible animals, there is sudden death, especially during mating. Animals may be tested using exposure to the gaseous anesthetic halothane. Susceptible animals respond by a rise in temperature, which, if not checked, can lead to death (malignant hyperthermia) (see Ref. 104).

In Europe, studies have shown that for pigs there is a complex relationship between PSE and DFD (89). These studies indicated that feeding prior to short transport gives high PSE and low DFD. Fighting lowers the PSE incidence but raises DFD. Short transport times gives high PSE and reduced DFD, but as the transport distance lengthens, the PSE becomes less and the DFD becomes greater. The contribution of the nervous system to PSE is very significant, and even though the effects can be controlled by drugs (105), such treatment is not accepted for food animals. Under ideal conditions, the two extremes in meat appearance, PSE and DFD, can be measured by various electronic grading probes.

Effect of Muscle Ultimate pH on Meat Tenderness

It was shown (106) that mutton with a very high ultimate pH—greater than 6.5—was more tender than mutton of a normal low ultimate pH, and that meat with a moderately elevated ultimate pH was tougher than normal. The latter observation has also been consistently found with bulls and steers (107). Studies using lamb that had been electrically stimulated showed a similar pattern of toughness in moderate-pH animals (Fig. 8) (108). In the range pH 5.5 to pH 5.9, toughness increases sharply. Therefore, small differences in ultimate pH can easily result in "rogue" tenderness results in any experimental situation.

As ultimate pH values are rarely measured routinely during meat processing, and then only with the aim of excluding those animals with pH values over 6.0, the significance of factors that increase the ultimate pH even slightly has not been generally considered. It was suggested (107) that short sarcomere lengths as a consequence of moderate-pH effects were related to increasing toughness. When aging of meat was prevented by injections of zinc salts, the initial toughness of high medium and low pH meat was the same (109). The rate of tenderization of (non-zinc-injected) intermediate pH meat was slowest, suggesting that aging rate rather than toughness is an intrinsic property of high-pH meat (108). This observation is supported by the reduced disappearance of nebulin and titin at these intermediate-pH values (110). This suggests that high-pH meat can tenderize if aging is long enough, but there were still some outliers. It appears that other factors still influence tenderness, and the enzymes responsible for tenderization may be affected by *rigor* temperature and pH in some way (53).

MEAT COLOR

The appearance of raw meat gives no indication of its tenderness and little indication of its flavor. Nevertheless, consumers often use meat color at point of purchase to assure themselves that the meat is not spoiled and will be of good eating quality. This approach raises color to a position far beyond its real value. It is true that color does indicate something about the meat, but only in terms of prior exposure to oxygen or drying conditions; and in extreme instances, it may be a guide to flavor changes. Consumers expect poultry to be white, veal and pork to be a pale pink, lamb and beef to be cherry red. The darkest meat comes from bulls, deer, and horses, even under normal ultimate pH conditions, since these meats have high concentrations of the pigment myoglobin, which gives rise to their dark color.

Color Changes During Processing

In cattle and sheep, the normal color of living muscle tissue is purplish. This is a consequence of both the pigment myoglobin being at a high pH and any oxygen present carried by the blood being rapidly used up by the living tissues, effectively ensuring an anoxic state. Immediately after slaughter, the situation continues, so that for prerigor meat in an unoxidized anoxic state, the myoglobin is dark purple. It changes to a paler, bright red when at rigor with oxygen present. From this point, the color is determined by the proteins myoglobin and hemoglobin, which are affected by muscle temperature and pH postslaughter (111–113). Changes in muscle proteins during rigor development change the light scattering properties of the muscle tissue, so that it becomes paler as the muscle pH falls to pH 5.5, that is, as the pH approaches the isoelectric point of the proteins (114). Translucent meat, which appears dark, occurs when the pH is greater than pH 6.0 and light scattering diminishes (111,114). Thus, antemortem depletion of muscle glycogen results in meat of high ultimate pH and hence darkening in color—so-called dark, firm, and dry (DFD) meat. With high-pH meat as well, there is a closer packing of the myofibrils, with a resultant greater translucency, so the meat is darker in color (14). The as-

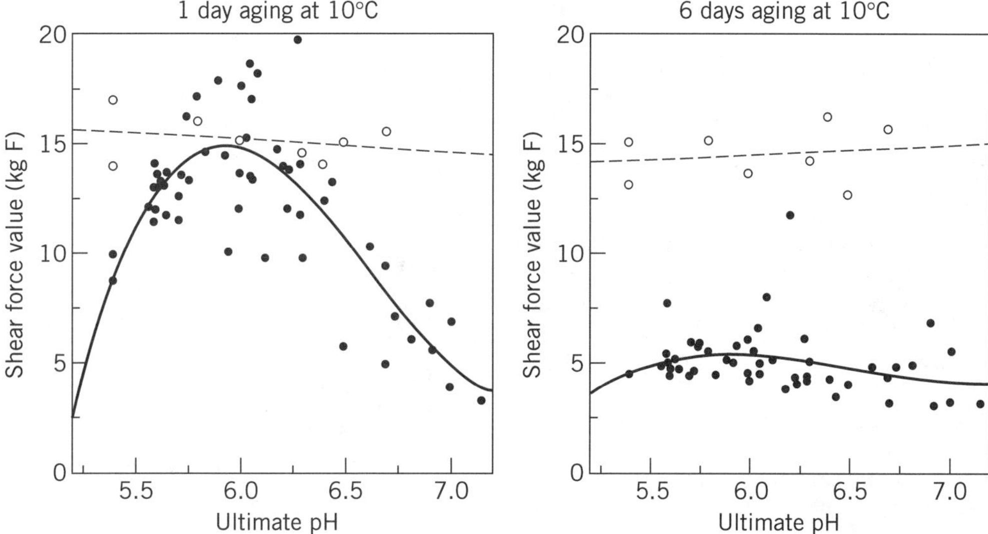

Figure 8. These results, at two aging times, were obtained from an experiment where the aging of some samples was inhibited by injection of zinc before rigor mortis. The shear forces of these muscles were high, were of similar values, and did not change with ultimate pH (○). There is a clear influence of ultimate pH on meat tenderness, where both high and low ultimate pH values result in tender meat, but meat with a moderate ultimate pH is tougher (●). After six days aging at 10°C, the meat with aging inhibited did not change, and all meat reached approximately the same levels of tenderness. The tenderness changes resulting from ultimate pH appear to relate to differences in aging rate, although there are still outliers arising from other factors. In the range of ultimate pH from 5.5 to 5.9, a small increase in pH results in large increases in toughness. *Source:* Ref. 108, used with permission.

sociated physical conditions of firmness and dryness are also associated with closer spacing of the myofibrillar lattice.

With normal ultimate pH values of around pH 5.5, the rate of cooling will also affect the muscle proteins. Portions cooled quickly scatter less light than those cooled slowly, such as regions in the muscle center (eg, beef hind quarter). In addition, aging of the meat also lightens meat color (111). The appearance of the meat is modified by electrical stimulation, as the rapid pH fall results in a more youthful appearance of the muscle, which improves the quality grade in the United States (115).

The most dramatic effect on meat color occurs in pork from stress-susceptible pigs (116). When these animals are stressed before slaughter, there is an extremely rapid fall in pH to pH 5.4 while the muscle is hot. The end result is extreme light scattering via denaturation of myofibrillar and sarcoplasmic proteins, to produce PSE meat.

Color Changes During Storage and Packaging

Red meats vacuum-wrapped in oxygen-impermeable films maintain a dark purple color. Consumer education has not reached the stage whereby this supposedly unattractive color is associated with the inherent advantages of either vacuum or anaerobic packaging for chilled meat, and gas-impermeable packaging for frozen meat. When vacuum or anaerobically packaged meats are removed from their package and exposed to air, the myoglobin oxygenates and the meat color changes from purple to bright red. When this color change occurs, the meat is said to *bloom*.

When fresh meat is exposed to high concentrations of oxygen—for example, as occurs in air, in a high oxygen modified atmosphere pack, or in a pack with an oxygen permeable wrapping—the myoglobin is converted to the bright red oxymyoglobin, which is resistant to further oxidation. However, with packaging in materials of low, but measurable, oxygen permeability, the low partial pressure of oxygen facilitates the formation of another more stable oxidized compound, metmyoglobin. Metmyoglobin is brown, giving the meat an unattractive color. In effect, the myoglobin of meat exposed to either high oxygen or no oxygen remains in a satisfactory state.

Even with exposure to high oxygen levels, the meat will finally turn brown. Metmyoglobin is slowly formed first in the deeper tissues where the critical low levels of oxygen occur. Eventually enough browning is produced that it can no longer be masked by the bright red oxymyoglobin lying above. Presumably, consumer experience with long storage of chilled meats, with eventual browning, has led to the association of dark meat color with "old meat" and spoilage.

Color Stability During Display

The important factors affecting the color stability of fresh meat are temperature, the packaging environment, and ultimate pH, the oxygen consumption and the reducing capacity of the meat. Thus, different muscles exhibit different rates of metmyoglobin formation, a change of temperature from 3 to 5°C in display will double the discoloration rate, and exposure to light also results in a decrease in display life (111–113). Exposure to light also accelerates rancidity development.

Aged meat can have a brighter color after blooming than unaged meat, not only because of increased light scatter and deeper penetration of the light, but also because of decreased catalytic oxidative capacity. However, aged meat has a poorer color stability (111–113).

Under certain conditions, microbial spoilage can affect meat color. Meat can turn green when sulfmyoglobin is formed. This occurs when certain spoilage organisms cause sulfur components to be released. Sulfmyoglobin formation is generally restricted to high-pH meat in vacuum packaging, where the environment is more conducive to sulfur-reducing bacteria (91). The addition of vitamin E to the diet improves the color stability at retail by reducing the metmyoglobin formation (117). This is more important in animals fed in a feedlot than for pasture-fed animals, which, in general, obtain sufficient from their daily food intake.

Frozen Meat Color

The freshly cut surfaces of frozen red meats are a dark purple but are lighter than the surfaces of chilled anaerobic meat because light transmission is inhibited and reflected light is scattered. In addition, only surface pigments can reoxygenate and affect frozen meat appearance, providing a redness overlying greyer myoglobin. Meat that has been frozen slowly is darker than meat that was rapidly frozen, for example, under liquid nitrogen; the difference being due to the smaller size of the ice crystals with rapid freezing.

Because of the low temperature of frozen meat, the deleterious effects of oxidation at the surface occur more slowly. However, various physical effects such as recrystallization occur at the meat surface. These reduce display life by affecting consumer acceptance. For this reason, it is much better to cut frozen meat into cuts immediately prior to display rather than to cut, package, and store the frozen cuts ready for retail display (118). As spoilage or flavor changes are minimal during frozen storage, cutting immediately prior to display is merely for consumer appeal and bears no relationship to actual product quality.

MEAT STORAGE AND FLAVOR CHANGES

Meat Odor and Flavor

Depending on species, meat has a characteristic odor/flavor, but this can be modified in various ways and significantly influences palatability (119). For nonruminants such as pigs and poultry, the type of food eaten (eg, fish meal) can influence the flavor of the meat more than in ruminants, where many flavor compounds are modified by the ruminant organisms. Even so, there are still some quality attributes that are different in pasture-finished and grain-finished animals that depend on the type of food eaten. For example, there are more desirable antioxidants and carotene in pasture-finished ruminant animals, but the flavor profile is different from the grain-finished animal, which has a higher glycogen content that could affect ultimate pH, a different fatty acid profile, and whiter fat. Preference of one over the other is mainly cultural. Animals with a high ultimate pH, via differences in the chemical reactions involving proteins and carbohydrates that take place during the cooking, may have a slightly different flavor than normal-pH animals.

Lipid Oxidation

When fresh meat is held in conditions free from bacterial spoilage, the meat will age and become more tender. In the presence of even small amounts of oxygen, oxidation will also take place at the double bond of the polyunsaturated fatty acids, leading to formation of peroxides that react chemically to produce secondary reaction products (120). These at first have a bland flavor but later result in a typical rancid flavor reminiscent of "old socks." The fats concerned are mainly the phospholipids associated with the cell membrane rather than the depot fatty tissues.

The lipids of raw meat are initially protected from oxidation by endogenous antioxidants, but eventually these antioxidants are inactivated and the rancidity process starts. Exposure to light and heavy metals promotes the onset of rancidity. Rancidity, being initiated by a free-radical reaction, continues even in the absence of additional oxygen, so conditions to minimize its initiation are most important. Vacuum packaging is one option; freezing is another. As all chemical reactions proceed more slowly at lower temperatures, meat should be stored at as low a temperature as practical.

One study on lamb (121) showed that at $-5°C$, rancidity was initiated, and it continued to progress even when the meat temperature was reduced to $-35°C$; rancidity did not appear to be initiated at the $-35°C$ temperature, however. This study illustrates that, when considering rancidity development, the entire history of the meat should be considered, rather than mere storage duration at various temperatures.

Although oxidative rancidity changes occur slowly in frozen meat, they occur extremely rapidly in meat after cooking and produce a characteristic "warmed-over flavor." Although the development of warmed-over flavor is reduced if cooked meat is chilled rapidly, this flavor development makes commercial storage of cooked meat for more than a few days a practical impossibility, unless oxygen is prevented from reaching the meat by storing and cooking meat in its own oxygen-impermeable package. The stability of meat lipids to oxidation is lowest in pork and poultry, intermediate in beef, and highest in lamb.

Frozen Storage

Freezing slows down any deterioration of the meat, but changes still occur. Color change (see earlier) and the onset of rancidity are slowed by low temperatures, although rancidity will continue once initiated. When meat freezes, it does so in such a way that extracellular ice crystals first form so that water moves out of the cells. Over time in frozen storage, recrystallization of ice occurs, and the ice so formed can cause structural damage. Fast freezing results in smaller ice crystals and less damage (122). In frozen storage, meat surfaces will sublime water unless pro-

Table 1. Composition of Lean Cooked Meat

	Beef			Lamb		Mutton	Pork		Chicken		
Cut	Rump steak	Fillet steak	Topside roast	Loin chops	Leg	Loin	Loin chops	Leg shank	Whole bird	Legs	Breast
					Cooking method						
	grill	grill	roast	grill	roast	roast	grill	roast	roast	broil	broil
					Cooked lean (% raw cut)						
	60.3	64.0	61.7	36.0	45.4	25.2	28.9	42.6	37.9	31.0	39.2
					Composition of cooked lean tissue						
Energy (kcal)	165.0	210.0	169.0	216.0	206.0	211.0	236.0	174.0	179.0	197.0	166.0
					Proximates						
Fat (%)	5.9	10.7	4.9	12.7	11.1	10.8	11.7	7.1	9.4	11.1	4.6
Moisture (%)	64.0	59.0	61.9	60.7	60.7	59.8	52.5	63.6	65.9	63.1	61.8
Protein (%)	27.8	28.3	31.2	25.5	26.5	28.4	32.7	27.5	23.7	24.2	31.2
					Minerals						
Calcium	6.0	5.6	4.2	9.8	5.4	10.4	22.6	8.5	6.9	25.0	13.6
Copper	0.12	0.10	0.06	0.04	0.10	0.12	0.06	0.08	0.02	0.04	0.02
Iron	3.6	4.4	3.6	3.6	4.6	4.0	2.7	2.2	3.0	2.6	1.9
Magnesium	20.5	23.0	23.5	20.5	20.5	21.8	22.5	19.5	18.0	15.0	22.5
Phospherous	230.0	260.0	260.0	220.0	240.0	190.0	270.0	240.0	210.0	200.0	260.0
Potassium	400.0	440.0	450.0	380.0	350.0	247.0	440.0	400.0	310.0	270.0	390.0
Sodium	60.0	66.0	59.0	109.0	78.0	50.0	89.0	82.0	71.0	93.0	65.0
Zinc	5.7	4.5	5.9	4.1	5.1	3.4	3.5	3.6	1.4	2.5	1.0
					Vitamins						
B1	0.14	0.15	0.09	0.20	0.11	0.14	1.01	0.69	0.09	0.11	0.10
B2	0.14	0.14	0.06	0.08	0.13	0.07	0.06	0.10	0.09	0.17	0.04
Niacin	4.9	5.1	6.0	6.0	5.8	5.5	4.8	4.1	3.8	3.5	8.1
B6	1.04	0.73	0.93	0.52	1.63	0.9	0.41	0.45	0.61	0.70	0.59
B12	3.4	3.9	2.3	1.3	2.1	1.4	0.6	0.5	0.5	1.3	0.6
					Cholesterol						
	84.0	106.0	81.0	115.0	109.0	101.0	118.0	100.0	109.0	148.0	112.0

Source: These results were obtained from unpublished studies by MIRINZ for the New Zealand Beef and Lamb Marketing Bureau, Inc., and from Ref. 125.
Note: Units are mg/100 g lean tissue unless otherwise indicated.

tected, and the effect on such surfaces is termed *freezer burn*. Sites of freezer burn allow rancidity to penetrate deeper into the meat. When frozen meat is thawed, the reverse of freezing takes place so that extracellular water is produced, but this can be resorbed during postthawing storage if this is possible by the nature of its packaging. The complex aspects of freezing and thawing meat and its effect on quality have been reviewed (122,123).

NUTRITIONAL ASPECTS OF MEAT

The major nutritional components of the meat of all species from fish through poultry to pork and so-called red meats are protein and iron. Although the protein contribution is obvious, the iron contribution is often underrated. Iron deficiency is the most common nutrition disorder, and nonanemic iron deficiency is far more insidious in its effects than normally considered. Even contributes to an impairment of cognitive function (124). Meat is one of the main natural sources of iron (Table 1). The fat component of meat provides desirable energy in certain malnourished communities, but it is possibly erroneously becoming increasingly regarded as a useless by-product in our overfed Western societies. Meat should be regarded as a valuable component of a balanced diet, rather than as a food item providing unwanted calories. The fat component may be either nonseparable fat, as is usually found in the lean of meat, or separable fat that is visible and can be trimmed off. As separable fat can be removed, it comes into the same food category as the skin of vegetables like potatoes or carrots, and therefore the consumer has a choice as to whether to consume it. Diets of meat animals with a high energy content result in the animals having extensive fat deposits; this even extends to the lean in substantial amounts, to give marbling. The contribution of meat to health has been reviewed (126,127).

Nutrient Content of Typical Servings

The nutrient content of some typical cooked 100-g servings of lean from several meat species is shown in Table 1 (125) and would provide approximately 50% of the recommended adult daily requirement for protein, zinc, niacin, and vitamin B_{12} and 15% or more of the requirement for iron, riboflavin, and vitamin B_6. In particular, the iron in meat is important because it is heme iron, which is more bioavailable than inorganic iron, and meat itself enhances the absorption of iron from other sources. Some meats, such as fish and poultry, have less iron than others. Cooked meats lose water, resulting in a proportional increase in protein, fat, and other nutrients per unit weight. The presentation of cooked meat data, therefore, gives a much more realistic indication of the actual food value of meat.

Cholesterol and Fats

Of particular topical interest is the amount of cholesterol in meat, ranging from 80 to 150 mg/100 g of cooked lean meat (Table 1). Phospholipids containing cholesterol are a necessary and expected component of animal cell membranes, with the greatest amounts being found in the brain. In fatty tissues from ruminants, the amount of cholesterol increases above that of the lean by approximately 50%, but even so, values are less than that present in the lean of some fish and poultry. Even considerable detailed research does not show conclusively that cholesterol in meat has adverse effects on health, and the studies giving a bad image to cholesterol have been under attack from many quarters with considerable justification (128). The dietary intakes of cholesterol only supplement cholesterol produced naturally by the body, and extremes of reduction of dietary cholesterol are required to lower serum cholesterol. Hypercholestemic people are at risk, however, but it is unclear whether the high cholesterol levels arise only from diet or from endogenous cholesterol production. Low blood cholesterol levels are implicated in another set of disorders, including cancer. There is increasing evidence that lipid oxidation products rather than cholesterol per se is one of the initiators of arterial injury and the train of events leading to coronary heart disease (129). This would suggest that absolute levels of cholesterol are not as important as other factors.

The saturated fats in meat from ruminants are mainly straight-chain C16:0 and C18:0 compounds, but in poultry and pork, saturated fats contain an increased proportion of branched chains, depending on the diet. Very lean meats such as venison are very low in both depot fats and separable fat. In effect, the fats in venison are primarily those concerned with cell structure and integrity. The lean of most species contains substantial amounts of monounsaturated or monoenoic (C18:1) fatty acids, which are implicated in reductions of low-density lipoprotein cholesterol in the blood, thus having potentially favorable effects on controlling heart disease (130). Sheep and cattle lean also contain over 3% of polyunsaturated or polyenoic C18 and C24 fatty acids, which are predominantly $\omega 3$ and $\omega 6$ fatty acids. Fish are often promoted as beneficial due to their being a rich source of the $\omega 3$ fatty acids (131), whereas meat is erroneously regarded as a poor source of these compounds, despite the fact that meat has been shown to have considerable proportions of both $\omega 6$ and $\omega 3$ fatty acids (132,133). Because of the relatively greater proportions of meat eaten than fish in many societies, the daily requirements of $\omega 3$ fatty acids can easily be met, and, in addition, the nutritionally necessary $\omega 6$ fatty acids, virtually absent in fish, can be supplied.

These comments on nutritional aspects are not exhaustive and clearly show a field in flux, as new information is sought to establish further the place of meat in the diet.

BIBLIOGRAPHY

1. A. S. Davies, "The Structure and Function of Meat Tissue in Relation to Meat Production" in R. W. Purchas, B. Butler-Hogg, and A. S. Davies, eds., *Meat Production and Processing*, Occasional Publication No. 11, New Zealand Society of Animal Production Hamilton, New Zealand, 1989, pp. 43–59.
2. A. Bailey, "Connective Tissue and Meat Quality," *Proc. Int. Congr. Meat Sci. and Technol.* **34**, 152–160 (1988).

3. R. H. Locker, "A New Theory of Tenderness in Meat, Based on Gap Filaments," in *Proceedings of the Reciprocal Meat Conference* **35**, 92–100 (1983).
4. R. H. Locker, "The Role of Gap Filaments in Muscle and in Meat," *Food Microstructure* **3**, 17–32 (1984).
5. A. M. Pearson and R. B. Young, *Muscle and Meat Biochemistry*, Academic Press, San Diego, Calif., 1989.
6. J. R. Bendall, "Post Mortem Changes in Muscle," in G. H. Bourne, ed., *The Structure and Function of Muscle*, Vol. 2, Academic Press, New York, 1973, pp. 243–309.
7. O. A. Young and C. L. Davey, "Electrophoretic Analysis of Proteins from Single Bovine Muscle Fibers," *Biochem. J.* **195**, 317–327 (1982).
8. K. Wang, R. Ramierez Mitchell, and D. Palter, "Titin Is an Extraordinary Long Flexible and Slender Myofibrillar Protein," *Proc. Natl. Acad. Sci. U.S.A.* **81**, 3685–3689 (1984).
9. K. Maruyama, "Myofibrillar and Cytoskeletal Proteins of Vertebrate Striated Muscle," in R. A. Lawrie, ed., *Developments in Meat Science—3*, Applied Science, London, 1985, pp. 25–50.
10. F. Parrish, B. C. Paterson, and J. M. Paxhia, "Titin: A Myofibrillar/Cytoskeletal Protein of Gigantic Proportions," *Proc. Reciprocal Meat Conf.* **41**, 1520–1546 (1988).
11. C. E. Devine, "Electrical Stimulation of Meat: Long-Term Research Vindicated," *Chemistry in New Zealand* **44**, 89–93 (1980).
12. E. C. Bate-Smith and J. Bendall, "Factors Determining the Time Course of Rigor Mortis," *J. Physiol.* **110**, 47–65 (1949).
13. R. Hamm, "Post Mortem Changes in Muscle Affecting the Quality of Comminuted Meat Products," in R. A. Lawrie, ed., *Developments in Meat Science—2*, Applied Science, London, 1981, pp. 93–124.
14. G. Offer et al., "The Structural Basis of the Water Holding, Appearance, and Toughness of Meat and Meat Products," *Food Microstructure* **8**, 151–170 (1989).
15. G. Offer et al., "Myofibrils and Meat Quality," *Proc. Int. Congr. Meat Sci. and Technol.* **34**, 161–168 (1988).
16. P. Knight, J. Elsey, and N. Hedges, "The Role of the Endomysium in the Salt-Induced Swelling of Muscle Fibres," *Meat Sci.* **26**, 209–232 (1989).
17. R. E. Jeacocke, "The Temperature Dependence of Anaerobic Glycolysis in Beef Muscle Held in a Linear Temperature Gradient," *J. Sci. Food Agric.* **28**, 551–556 (1977).
18. C. E. Devine et al., "Differential Aspects of Electrical Stunning on the Early Post Mortem Glycolysis in Sheep," *Meat Sci.* **11**, 301–309 (1984).
19. C. E. Devine, B. B. Chrystall, and C. L. Davey, "Studies in Electrical Stimulation: Effect of Neuromuscular Blocking Agents in Lamb," *J. Sci. Food Agric.* **30**, 1007–1010 (1979).
20. B. B. Chrystall and C. E. Devine, "Electrical Stimulation, Muscle Tension, and Glycolysis in Bovine Sternomandibularis," *Meat Sci.* **2**, 49–58 (1978).
21. R. Hamm, "Post Mortem Breakdown of ATP and Glycogen in Ground Muscle: A Review," *Meat Sci.* **1**, 15–39 (1977).
22. C. E. Devine, S. Ellery, and S. Averill, "Responses of Different Types of Muscle to Electrical Stimulation," *Meat Sci.* **10**, 293–305 (1984).
23. B. B. Chrystall and C. E. Devine, "Electrical Stimulation of Rats: A Model for Evaluating Low Voltage Stimulation Parameters," *Meat Sci.* **9**, 33–41 (1983).
24. J. M. Jones and T. C. Grey, "Influence of Processing on Product Quality and Yield," in G. C. Mead, ed., *Processing of Poultry*, Elsevier Science, London, 1989, pp. 127–181.
25. Y. B. Lee and D. A. Rickansrud, "Effect of Temperature on Shortening in Chicken Muscle," *J. Food Sci.* **43**, 1613–1615 (1978).
26. R. H. Locker and C. J. Hagyard, "A Cold Shortening Effect in Beef Muscles," *J. Sci. Food Agric.* **14**, 787–793 (1963).
27. C. L. Davey, H. Kuttel, and K. V. Gilbert, "Shortening as a Factor in Meat Aging," *J. Food Technol.* **2**, 53–56 (1967).
28. K. O. Honikel, R. Roncales, and R. Hamm, "The Influence of Temperature on Shortening and Rigor Onset in Beef Muscle," *Meat Sci.* **8**, 221–241 (1983).
29. C. L. Davey and K. V. Gilbert, "Carcass Posture and Tenderness in Frozen Lamb," *J. Sci. Food Agric.* **25**, 923–930 (1974).
30. B. B. Marsh, N. G. Leet, and M. R. Dickson, "The Ultrastructure and Tenderness of Highly Cold Shortened Muscle," *J. Food Technol.* **9**, 141–147 (1974).
31. R. L. Hostetler and S. Cover, "Relationship of Extensibility of Muscle Fibres to Tenderness of Beef," *J. Food Sci.* **26**, 535–540 (1961).
32. B. B. Marsh, P. R. Woodhams, and N. G. Leet, "Studies in Meat Tenderness, 5: The Effects on Tenderness of Carcass Cooling and Freezing before Completion of Rigor Mortis," *J. Food Sci.* **33**, 12–18 (1968).
33. R. H. Locker and D. J. C. Wild, "Tensile Properties of Cooked Beef in Relation to Rigor Temperature and Tenderness," *Meat Sci.* **8**, 283–299 (1983).
34. P. E. Bouton et al., "Pressure Heat Treatment of Postrigor Muscle: Effects on Tenderness," *J. Food Sci.* **42**, 132–135 (1977).
35. M. B. Solomon et al., "Hydrodyne-Treated Beef: Tenderness and Muscle Ultrastructure," *Proc. 43rd Int. Congr. Meat Sci. Technol.*, Auckland, New Zealand, July 27–August 1, 1997.
36. B. B. Chrystall, "Hot Processing in New Zealand," in K. R. Franklin and H. R. Cross, eds., *Proc. Int. Symp. Meat Sci. and Technol.*, National Livestock and Meat Board, Chicago, Ill., 1983, pp. 211–221.
37. I. F. Penney, "The Enzymology of Conditioning," in R. A. Lawrie, ed., *Developments in Meat Science—1*, Applied Science, London, 1980, pp. 115–153.
38. I. F. Penney and E. Dransfield, "Relationship between Toughness and Troponin T in Conditioned Beef," *Meat Sci.* **3**, 135–141 (1979).
39. D. E. Croall and G. N. DeMartino, "Calcium Activated Neutral Protease (Calpain) System: Structure, Function, and Regulation," *Physiol. Rev.* **71**, 813–847 (1991).
40. D. E. Goll et al., "Role of Muscle Proteinases in Maintenance of Muscle Integrity and Mass," *J. Food Biochem.* **7**, 137–177 (1983).
41. M. Koohmarie, "The Role of Endogenous Proteases in Meat Tenderness," *Proc. Reciprocal Meat Conf.* **41**, 89–100 (1988).
42. D. E. Goll et al., "Role of Proteinases and Protein Turnover in Muscle Growth and Meat Quality," *Proc. Reciprocal Meat Conf.* **44**, 25–36 (1992).
43. D. E. Goll et al., "Properties and Biological Regulation of the Calpain System," in A. Ouali, D. I. Demeyer, and F. J. M. Smulders, eds., *Expression of Tissue Proteinases and Regulation of Protein Degradation as Related to Meat Quality*, ECCEAMST, Utrecht, The Netherlands, 1995, pp. 47–65.
44. A. Ouali, "Meat Tenderization: Possible Causes and Mechanisms, a Review," *J. Muscle Foods* **1**, 129–165 (1990).
45. C. L. Davey and K. V. Gilbert, "The Temperature Coefficient of Beef Aging," *J. Sci. Food Agric.* **27**, 244–250 (1976).

46. P. E. Bouton et al., "A Comparison of the Effects of Aging, Conditioning, and Skeletal Restraint on the Tenderness of Mutton," *J. Food Sci.* **38**, 932–937 (1973).
47. P. E. Bouton et al., "Changes in the Mechanical Properties of Veal Muscles Produced by Myofibrillar Contraction State, Cooking Temperature, and Cooking Time," *J. Food Sci.* **39**, 869–875 (1974).
48. H. K. Herring, R. G. Cassens, and E. J. Briskey, "Sarcomere Length of Free and Restrained Bovine Muscles at Low Temperatures," *J. Sci. Food Agric.* **16**, 379–384 (1965).
49. F. L. Hostetler, T. R. Dutson, and Z. L. Carpenter, "Effect of Varying Final Internal Temperature on Shear Values and Sensory Scores of Muscles from Carcasses Suspended by Two Methods," *J. Food Sci.* **41**, 421–423 (1976).
50. M. Koohmaraie, J. D. Crouse, and H. J. Mersmann, "Acceleration of Postmortem Tenderisation in Ovine Carcasses through Infusion of Calcium Chloride: Effect of Concentration and Ionic Strength," *J. Anim. Sci.* **67**, 934–942 (1989).
51. J. M. Stevenson-Barry and R. G. Kauffman, "Tenderisation of Beef Muscles by Injection of Salts," *Proc. 43rd Int. Congr. Meat Sci. and Technol.*, Auckland, New Zealand, July 27–August 1, 1997.
52. N. Gault, "Marinaded Meat," in R. A. Lawrie, ed., *Developments in Meat Science—5*, Elsevier Applied Science, London, 1991, pp. 191–246.
53. C. E. Devine, N. M. Wahlgren, and E. Tornberg, "Effect of Rigor Temperature on Muscle Shortening and Tenderisation of Restrained and Unrestrained Beef *m. longissimus thoracicus et lumborum*," *Meat Sci.* **51**, 61–72 (1998).
54. R. H. Locker and D. J. C. Wild, "Aging of Cold Shortened Meat Depends on the Criterion," *Meat Sci.* **10**, 235–238 (1984).
55. E. Dransfield, D. K. Wakefield, and I. D. Parkman, "Modelling Post Mortem: Texture of Electrically Stimulated and Non Stimulated Beef," *Meat Sci.* **31**, 57–74 (1992).
56. N. J. Simmons et al., "The Effect of Prerigor Holding Temperature on Calpain and Calpastatin Activity and Meat Tenderness," *Proc. 42nd Int. Congr. Meat Sci. Technol.*, Lillehammer, Norway, September 1–6, 1996.
57. N. M., Wahlgren, C. E. Devine, and E. Tornberg, "The Influence of Different pH-Courses During Rigor Development on Beef Tenderness," *Proc. 43rd Int. Congr. Meat Sci. Technol.*, Auckland, New Zealand, July 27–August 1, 1997.
58. C. E. Devine, N. M. Wahlgren, and E. Tornberg, "The Effects of Rigor Temperatures on Shortening and Meat Tenderness," *Proc. 42nd Int. Congr. Meat Sci. Technol.*, Lillehammer, Norway, 1996.
59. C. J. Cook, S. M. Scott, and C. E. Devine, "Measurements of Nitric Oxide and the Effect of Enhancing and Inhibiting It on the Tenderness of Meat," *Meat Sci.* **48**, 85–89 (1998).
60. J. Jungersten et al., "Both Physical Fitness and Acute Exercise Regulate Nitric Oxide Formation in Healthy Humans," *J. Appl. Physiol.* **82**, 760–764 (1997).
61. T. J. Simms and A. Bailey, "Connective Tissue," in R. A. Lawrie, ed., *Developments in Meat Science—2*, Elsevier Applied Science, London, 1981, pp. 29–59.
62. W. R. Shorthose and P. V. Harris, "Effect of Animal Age on the Tenderness of Selected Beef Muscles," *J. Food Sci.* **55**, 1–8 (1990).
63. N. Madsen, "Effect of Animal Age on Cull Cow Beef Tenderness," *Proc. 43rd Int. Congr. Meat Sci. Technol.*, Auckland, New Zealand, July 27–August 1, 1997.
64. O. A. Young et al., "Collagen in Two Muscles of Sheep Selected for Weight as Yearlings," *New Zealand J. Agricultural Res.* **36**, 143–150 (1993).
65. H. R. Cross, B. D. Schanbacher, and J. D. Grouse, "Sex, Age, and Breed Related Changes in Bovine Testosterone and Intramuscular Collagen," *Meat Sci.* **10**, 187–195 (1984).
66. T. Pietersen et al., "Secondary Sexual Development (Masculinity) of Bovine Males, 2: Influence on Certain Meat Quality Characteristics," *Meat Sci.* **31**, 451–462 (1992).
67. C. Stanton and N. Light, "Effects of Conditioning on Meat Collagen: Part 2—Direct Biochemical Evidence for Proteolytic Damage in Insoluble Perimysial Collagen after Conditioning," *Meat Sci.* **23**, 179–199 (1988).
68. E. W. Mills et al., "Early Post Mortem Degradation of Intramuscular Collagen," *Meat Sci.* **26**, 115–120 (1989).
69. R. Rowe, "Collagen Fibre Arrangement in Intramuscular Connective Tissue: Changes Associated with Muscle Shortening and Their Possible Relevance to Raw Meat Toughness Measurements," *J. Food Technol.* **9**, 501–508 (1974).
70. A. M. Pearson, "Objective and Subjective Measurements for Meat Tenderness" in *Proceedings of the Meat Tenderness Symposium*, Campbell Soup Company, Camden, N.J., 1963, pp. 135–160.
71. B. B. Chrystall, "Meat Texture Measurement," in A. M. Pearson and T. R. Dutson, eds., *Quality Attributes and Their Measurement in Meat, Poultry, and Fish Products*, Advances in Meat Research, Vol. 9, Blackie Academic and Professional, London, 1994, pp. 316–336.
72. P. V. Harris and W. R. Shorthose, "Meat Texture," in R. Lawrie, ed., *Developments in Meat Science—4*, Applied Science, London, 1989, pp. 245–296.
73. B. B. Chrystall et al., "Recommendation of Reference Methods for Assessment of Meat Tenderness," *Proc. 40th Int. Congr. Meat Sci. Technol.*, The Hague, The Netherlands, August 28–September 2, 1994.
74. K. I. Hildrum et al., "Prediction of Sensory Characteristics of Beef by Near-Infrared Spectroscopy," *Meat Sci.* **38**, 67–80 (1994).
75. O. A. Young, G. J. Barker, and D. A. Frost, "Determination of Collagen Solubility and Concentration in Meat by Near Infrared Spectroscopy," *J. Muscle Foods* **7**, 377–387 (1996).
76. C. L. Davey and K. V. Gilbert, "Temperature-Dependent Cooking Toughness in Beef," *J. Sci. Food Agric.* **25**, 931–938 (1974).
77. R. J. Cox et al., "The Effect of Degree of Doneness of Beef Sirloin Steaks on Consumer Acceptability of Meals in Restaurants," *Meat Sci.* **45**, 75–85 (1997).
78. M. E. Dikeman, "The Relationship of Animal Leanness to Meat Tenderness," *Proc. Reciprocal Meat Conf.* **49**, 87–101 (1996).
79. L. E. Jeremiah, "A Canadian/North American Overview of the Contribution of Fat to Beef Palatability and Consumer Acceptance," *Meat Focus Int.* **5**, 175–178 (1996).
80. C. L. Davey and K. V. Gilbert, "Cold Shortening Capacity and Beef Muscle Growth," *J. Sci. Food Agric.* **26**, 755–760 (1975).
81. C. L. Davey and K. V. Gilbert, "The Tenderness of Cooked and Raw Meat from Young and Old Beef Animals," *J. Sci. Food Agric.* **26**, 953–960 (1975).
82. G. Whipple et al., "Evaluation of Attributes That Affect Longissimus Muscle Tenderness in *Bos taurus* and *Bos indicus* Cattle," *J. Anim. Sci.* **68**, 2716–2728 (1990).
83. P. D. Warriss et al., "The Quality of Pork from Traditional Pig Breeds," *Meat Focus Int.* **5**, 491–494 (1996).
84. N. G. Gregory, *Animal Welfare and Meat Science*, CABI Publishing, Wallingford, Oxon, UK, 1998.

85. T. Grandin, "Designs and Specifications for Livestock Handling Equipment in Slaughter Plants," *International Journal Studies on Animal Problems* **1**, 178–200 (1980).

86. T. M. Leach, "Preslaughter Stunning," in R. Lawrie, ed., *Developments in Meat Science—3*, Applied Science, London, 1985, pp. 51–87.

87. B. B. Chrystall, C. E. Devine, and K. G. Newton, "Residual Blood in Lamb Muscles," *Meat Sci.* **5**, 339–345 (1981).

88. J. C. Williams et al., "Influence of Delayed Bleeding on Sensory Characteristics of Beef," *Meat Sci.* **9**, 181–190 (1983).

89. P. A. Barton-Gade et al., "Slaughter Procedures for Pigs, Sheep, Cattle, and Poultry," in H. R. Cross and A. J. Overby, eds., *Meat Science, Milk Science, and Technology*, World Animal Science B3, Elsevier Science, London, 1988, pp. 33–111.

90. C. E. Devine et al., "The Humane Slaughter of Animals: A Realistic Goal," *Proc. 39th Int. Congr. Meat Sci. Technol.*, Calgary, Alberta, Canada, August 1–6, 1993.

91. K. G. Newton and C. O. Gill, "The Microbiology of DFD Fresh Meats: A Review," *Meat Sci.* **5**, 223–232 (1981).

92. B. B. Chrystall et al., "Animal Stress and Its Effect on Rigor Mortis Development in Lambs," in D. E. Hood and P. V. Tarrant, eds., *The Problem of Dark Cutting in Beef*, Vol. 10, Martinus Nijhoff, The Hague, The Netherlands, 1981, pp. 269–282.

93. G. Petersen, "The Effect of Swimming Lambs and Subsequent Resting Periods on the Ultimate pH of Meat," *Meat Sci.* **9**, 237–246 (1983).

94. A. Howard and R. A. Lawrie, "Studies on Beef Quality: Part II, Physiological and Biochemical Effects of Various Preslaughter Treatments," Department of Scientific and Industrial Research (DSIR) Special Report Food Investigation Board, No. 63, London, 1956.

95. A. Bray, A. E. Graafhuis, and B. B. Chrystall, "The Cumulative Effect of Nutritional, Shearing, and Preslaughter Washing Stresses on the Quality of Lamb Meat," *Meat Sci.* **25**, 59–67 (1988).

96. C. E. Devine and B. B. Chrystall, "High Ultimate pH in Sheep," in S. W. Fabiansson, W. R. Shorthose, and R. D. Warner, eds., *Dark Cutting in Cattle and Sheep, Proceedings of an Australian Workshop*, Australian Meat and Livestock Research and Development Corporation, Sydney, Australia, 1989, pp. 55–65.

97. A. Lacourt and P. V. Tarrant, "Glycogen Depletion Patterns in Myofibres of Cattle during Stress," *Meat Sci.* **15**, 85–100 (1985).

98. P. V. Tarrant, "Animal Behaviour and Environment in the Dark Cutting Condition," in S. W. Fabiansson, W. R. Shorthose, and R. D. Warner, eds., *Dark Cutting in Cattle and Sheep, Proceedings of an Australian Workshop*, Australian Meat and Livestock Research and Development Corporation, Sydney, Australia, 1989, pp. 8–18.

99. F. Feldhusen and M. Kühne, "Effects of Ultrarapid Chilling and Ageing on Length of Sarcomeres and Tenderness of Pork," *Meat Sci.* **32**, 161–171 (1992).

100. O. Hallund and J. R. Bendall, "The Long Term Effect of Electrical Stimulation on the Post Mortem Fall of pH in the Muscles of Landrace Pigs," *J. Food Sci.* **30**, 296–299 (1965).

101. A. A. Taylor and M. Z. Tantikov, "Effect of Different Electrical Stimulation and Chilling Treatments on Pork Quality," *Meat Sci.* **31**, 381–395 (1992).

102. A. A. Taylor and A. M. Perry, "Improving Pork Quality by Electrical Stimulation or Pelvic Suspension of Carcasses," *Meat Sci.* **39**, 327–337 (1995).

103. R. L. J. M. van Laack and F. J. M. Smulders, "The Effect of Electrical Stimulation, Time of Boning, and High Temperature Conditioning on Sensory Quality Traits of Porcine *Longissimus dorsi* Muscle," *Meat Sci.* **25**, 113–121 (1991).

104. D. Lister, N. G. Gregory, and P. D. Warriss, "Stress in Meat Animals," in R. A. Lawrie, ed., *Developments in Meat Science—2*, Applied Science, London, 1981, pp. 61–92.

105. P. D. Warriss and D. Lister, "Improvement of Meat Quality in Pigs by Beta Adrenergic Blockage," *Meat Sci.* **7**, 183–187 (1982).

106. P. E. Bouton, P. V. Harris, and W. R. Shorthose, "Effect of Ultimate pH upon the Water Holding Capacity and Tenderness of Mutton," *J. Food Sci.* **36**, 435–439 (1971).

107. R. W. Purchas, "An Assessment of the Role of pH Differences in Determining the Relative Tenderness of Meat from Bulls and Steers," *Meat Sci.* **27**, 129–140 (1990).

108. C. E. Devine et al., "The Effect of Growth Rate and Ultimate pH on Meat Quality of Lambs," *Meat Sci.* **36**, 143–150 (1994).

109. A. Watanabe, C. C. Daly, and C. E. Devine, "The Effects of Ultimate pH of Meat on the Tenderness Changes during Ageing," *Meat Sci.* **42**, 67–78 (1995).

110. A. Watanabe and C. E. Devine, "The Effect of Meat Ultimate pH on Rate of Titin and Nebulin Degradation," *Meat Sci.* **42**, 407–413 (1996).

111. D. B. MacDougall, "Changes in the Colour and Opacity of Meat," *Food Chem.* **9**, 75–88 (1982).

112. D. A. Ledward, "Colour of Raw and Cooked Meat," in D. E. Johnston, M. K. Knight, and D. A. Ledward, eds., *The Chemistry of Muscle-Based Foods*, Royal Society of Chemistry, London, 1992, pp. 128–144.

113. C. Faustman and R. G. Cassens, "The Biochemical Basis for Discoloration in Fresh Meat: A Review," *J. Muscle Foods* **1**, 217–243 (1990).

114. B. B. Chrystall, "Detection of High pH Meat with an Optoelectronic Probe," *New Zealand J. Agriculture Res.* **30**, 443–448 (1987).

115. J. W. Savell, G. C. Smith, and Z. L. Carpenter, "Beef Quality and Palatability as Affected by Electrical Stimulation and Cooler Aging," *J. Food Sci.* **43**, 1666–1668 (1978).

116. J. R. Bendall and J. Wismer Pedersen, "Some Properties of the Myofibrillar Proteins of Normal and Watery Pork Muscle," *J. Food Sci.* **27**, 144–157 (1962).

117. R. N. Arnold et al., "Tissue Equilibrium and Subcellular Distribution of Vitamin E Relative to Myoglobin and Lipid Oxidation in Displayed Beef," *J. Anim. Sci.* **71**, 105–118 (1993).

118. V. J. Moore, "Factors Influencing Frozen Display Life of Lamb Chops and Steaks: Effect of Packaging and Temperature," *Meat Sci.* **27**, 91–98 (1990).

119. O. A. Young, T. J. Braggins, and G. A. Lane, "Animal Production Origins of Some Meat Colour and Flavour Attributes," in Y. L. Xiong, C. Ho, and F. Shahidi, eds, *Quality Attributes of Muscle Foods*, Plenum, New York (1999).

120. D. Mottram, "Lipid Oxidation and Flavor in Meat and Meat Byproducts," *Food Sci. and Technol. Today* **1**, 159–162 (1987).

121. C. J. Hagyard et al., "Frozen Storage Conditions and Rancid Flavour Development in Lamb," *Meat Sci.* **35**, 305–312 (1993).

122. A. Calvelo, "Recent Studies on Meat Freezing," in R. A. Lawrie, ed., *Developments in Meat Science—2*, Applied Science, London, 1981, pp. 125–158.

123. C. E. Devine et al., "Red Meats," in L. E. Jeremiah, ed., *Freezing Effects on Food Quality*, Marcel Dekker, New York, 1996, pp. 51–84.

124. A. B. Brunner et al., "Randomized Study of Cognitive Effects of Iron Supplementation in Non-Anaemic Iron Deficient Adolescent Girls," *Lancet* **348**, 992–996 (1996).
125. J. West and B. B. Chrystall, "Composition of Cooked Tissues from Selected Lamb, Beef, Pork, and Chicken Cuts," *Food Technology in New Zealand* **24**, 23–39 (1989).
126. A. M. Pearson, "Meat and Health," in R. A. Lawrie, ed., *Developments in Meat Science—2*, Applied Science, London, 1981, pp. 241–292.
127. K. R. Franklin and P. N. Davis, eds., "Meat in Nutrition and Health," *Proc. Int. Symp. National Livestock and Meat Board*, Colorado Spring, Colo., September 2, 1981.
128. T. J. Moore, *Heart Failure*, Random House, New York, 1989.
129. P. B. Addis, "Coronary Heart Disease: An Update with Emphasis on Dietary Lipid Oxidation Products," *Food Nutrition News* **62**, 7–10 (1990).
130. S. M. Grundy, "Monounsaturated Fatty Acids and Cholesterol Metabolism: Implication for Dietary Recommendations," *J. Nutr.* **119**, 529–533 (1988).
131. M. S. Feder, "Fatty Acids of Current Interest, Omega 3," *Food Eng.* **59**, 64–66 (1987).
132. A. J. Sinclair and K. O'Dea, "The Lipid Levels and Fatty Acid Composition of the Lean Portions of Australian Beef and Lamb," *Food Technol. Australia* **39**, 228–231 (1987).
133. J. West and B. B. Chrystall, "Intramuscular Polyunsaturated Fatty Acids in New Zealand Meats," Technical Report No. 861, Meat Industry Research Institute of New Zealand, Hamilton, New Zealand, 1989.

C. E. Devine
Technology Development Group, HortResearch
Hamilton, New Zealand
B. B. Chrystall
Massey University, Albany Campus
Auckland, New Zealand

MEAT SLAUGHTERING AND PROCESSING EQUIPMENT

Mechanization and automation of animal stunning, slaughter, dressing, and boning operations can offer many benefits, including improved carcass hygiene through reduced hand-hide cross-contamination, better hide and pelt quality, better quality of the deboned product, improved worker safety, and reduced labor costs. The degree to which these operations have become mechanized or automated varies, depending on the species of animal.

Before animals can be appropriately dressed, they have to be presented to the processing slaughter equipment and personnel in a reproducible way. Thus the most effective method of stunning will be different for each species. Table 1 gives an overview of the stunning techniques currently being used.

OVINES

The Need for Mechanization

As a major producer of sheep meat for export, the New Zealand meat industry has over the last decade invested heavily in the development of slaughter, dressing, and boning equipment for sheep and lambs. This equipment was developed with inputs from meat processing companies, farmer producer boards, government research agencies, commercial engineering companies, and the Meat Industry Research Institute of New Zealand (MIRINZ). The program was initiated in response to increased labor costs and the need to meet more stringent hygiene regulations that had decreased slaughter and dressing productivity.

Table 2 illustrates the effect of carcass size and of mechanization on the labor requirements for processing various meat species from stunning to evisceration. The first column gives the number of worker-hours required to dress 10,000 kg of carcass weight. In the second column these data are adjusted to take into account the fact that mutton, lamb, and beef processing produces two salable products (the carcass and the skin or pelt), whereas for chicken and pork, the feathers and hair are of little economic value. With traditional manual dressing systems, even when the worker-hours are adjusted to take pelt value into account, the labor input for sheep and lambs is two to four times that for the other species. Therefore, for sheep meats to be competitive with other meats, the labor requirement of traditional sheep and lamb slaughter and dressing had to be reduced through mechanization.

Beef and pork have a high average carcass weight, which reduces the labor requirement to produce 10,000 kg of carcasses, because the manning for many operations (ie, hock removal, evisceration) is not affected by carcass size. On the other hand, although chickens have a very small carcass size, the labor requirement is relatively low primarily because chicken processing is heavily mechanized.

Stunning

Two types of electrical stunning, the preferred method for sheep, have been developed: head only and head to body. Head-only stunning, which meets halal slaughter requirements, results in an initially still animal that starts to produce a paddling or running movement even after the throat has been cut. Such movement can be reduced by passing an electric current through the carcass, preferably by using rubbing electrodes after shackling. A head-to-body stun results in cardiac arrest, and the current through the body reduces subsequent movement. Both types of stunning have been easily adapted for automation.

All automatic sheep stunners developed to date have used a "V" restrainer system for controlling the location of the animal throughout the operation. The first automatic stunner for sheep and lambs evolved from a unit developed in Europe for pigs. This unit was then modified for sheep by New Zealand researchers. MIRINZ developed an automatic stunner that used a single "V" restrainer. With this system, each sheep is brought to a position where its head is adjacent to two grids of nozzle electrodes. Once the animal is in place, the two grids move inward until they contact each side of the animal's head. The nozzle electrodes then simultaneously administer electrical current to the head and emit water, which assists passage of current (Fig. 1). The design of this stunning system was improved by New Zealand's Alliance Freezing Company to cope with

Table 1. Stunning Methods

Method	Uses and limitations
Captive Bolt	
Involves the controlled penetration of a steel bolt (8-mm dia.) into the brain. Energy is transmitted by a shockwave. Projection of the bolt is powered by blank cartridges or air.	Used for calves, cattle, deer, sheep; could be used for most other species; not reliable for pigs. There is a limited effect with animals having large bony skulls. The method is considered reliable and humane, but the bolt must be accurately shot into the head. Brains are not edible.
Percussion	
Percussion stunning is similar to captive bolt in principle but the projectile ends in a large flat or mushroom-shaped head. Energy is transmitted by a shockwave, but because the skull is not penetrated, there is variable dissipation of the energy, and stunning is therefore not controlled.	Used for calves, cattle, deer, sheep; could be used for most other species; not reliable for pigs. With percussion stunning there is a fine line between a reversible stun and irreversible brain damage, leading to doubts about humaneness in every situation. Brains can be recovered. Fulfills halal requirements.
Head-Only Electrical Stunning	
Electric current (50–60 Hz) is passed only through the brain, producing immediate unconsciousness. The animal will recover from the stun if left unslaughtered. Slaughter is by throat cutting or sticking. Electrodes can be in the form of two steel pins spread 5 cm apart with a pistol grip, or in the form of scissorlike tongs for small animals. Restraint and semiautomatic electrode placement is required for large animals such as adult cattle.	Used for sheep, calves, cattle, pigs; could be used for deer. Head-only stunning is humane if sticking quickly follows the stun. Poststun animal movement can be overcome by passing a current through the carcass shortly after slaughter (electrical immobilization). Automatic versions are available. Blood splash and petechial hemorrhages may occur. Brains can be recovered. Fulfills halal requirements.
Head-to-Body Electrical Stunning	
Electric current (50–60 Hz) is passed through the brain with electrodes as above, but current is also passed through the body to stop the heart (causing death) and to produce carcass stillness. Electrodes for the body can be sited on the back, forelegs, or brisket.	Used for sheep, calves, cattle, pigs, and poultry; could be used for deer, rabbits. The method gives animal stillness, and humaneness is guaranteed by the heart being stopped. Sticking need not quickly follow the stun. Automatic versions are available. Petechial hemorrhages may occur. Brains can be recovered. Nonhalal.
Carbon Dioxide Stunning	
CO_2 stunning involves anesthetizing the animal in an atmosphere of 60–70% CO_2 in air or oxygen. Animals are lowered into a chamber. Apparatuses for high or low throughputs are available.	Used only for pigs at present. CO_2 stunning reduces petechial hemorrhages and is claimed to reduce pale, soft, exudative meat.

horned stock. This improved stunning system uses a dual V restrainer system, in which one conveyor feeds the other. The use of two conveyors allows controlled spacing of the animals. The success rate of the machine was also improved by minor changes to the way the grids of electrodes operated. The machine has also been adapted for head-to-body stunning.

Pelting

Most of the effort devoted to mechanized sheep slaughter and dressing has been in the area of pelt removal. Many research organizations, engineering companies, and other groups have developed and refined various concepts designed to reduce the manpower needed for pelting.

In the late 1970s, the New Zealand meat industry initiated and funded a project to mechanize the pelt removal operation. People began to recognize the hygiene and labor benefits of depelting sheep from the shoulders to the hind legs, a process more easily done by suspending the animal from the front legs; traditionally sheep were depelted while hanging from their hind legs. Systems using this new carcass orientation became known as inverted dressing systems. A six-head rotary pelt removal machine, developed for the inverted orientation, removed the pelt from the belly, lower back, and hind legs by driving a ring between the pelt and the carcass, removing the pelt as a "sock" (Fig. 2). The machine was completed and released in 1982.

MIRINZ took over the project in 1983 and developed a simpler two-stage pelting system as an alternative. This system consists of a shoulder puller (Fig. 3), which removes the pelt from the shoulder and back regions, and a final puller (Fig. 4), which pulls the pelt off the rear legs. MIRINZ has refined the manual input of the inverted dressing system by introducing new butchering techniques to enhance the advantages that the inverted dressing system has over the traditional system, that is, less labor required, lower levels of microbial contamination particularly in the most valuable hind quarter area, reduced skills level required, and lower worker injury risk.

Table 2. Worker-Hours Needed for Producing 10,000 kg Dressed Carcass Weight

	Hours per 10,000 kg carcass weight	Data adjusted for pelt and hide value: Traditional systems	Data adjusted for pelt and hide value: Mechanized sheep chain[a]	Average carcass weight, kg
Chicken	25	25		1.5
Lamb	83	66	32	14
Mutton	60	48	23	20
Pork	15	15		65
Beef	22	18		250

[a]Incorporating all of the automated operations listed in Table 3.

Figure 1. Automatic stunner for sheep.

The MIRINZ inverted dressing system, combined with the pelt removal machines developed by MIRINZ, yields pelts of a quality at least as good as those from traditional manual systems. Over the years, various types of hand tools have been developed to aid the removal of the pelt from the sternum or brisket. Recently, MIRINZ has developed a brisket clearing machine (Fig. 5) that simplifies brisket processing, particularly with inverted dressing systems.

MIRINZ has also developed machines to automatically remove the front and rear hocks from carcasses (Figs. 6 and 7). For transferring the forelegs from the wide spreader to the narrow hock holder, a special wide-to-narrow transfer machine was also produced (Fig. 8). The narrow hock holder grips both the radius and ulna bones of the forelegs. Therefore, the front hocks (or trotters) can be removed without losing carcass support. If the front trotters are removed before removing the pelt, the potential for the front trotters to contaminate the carcass is eliminated.

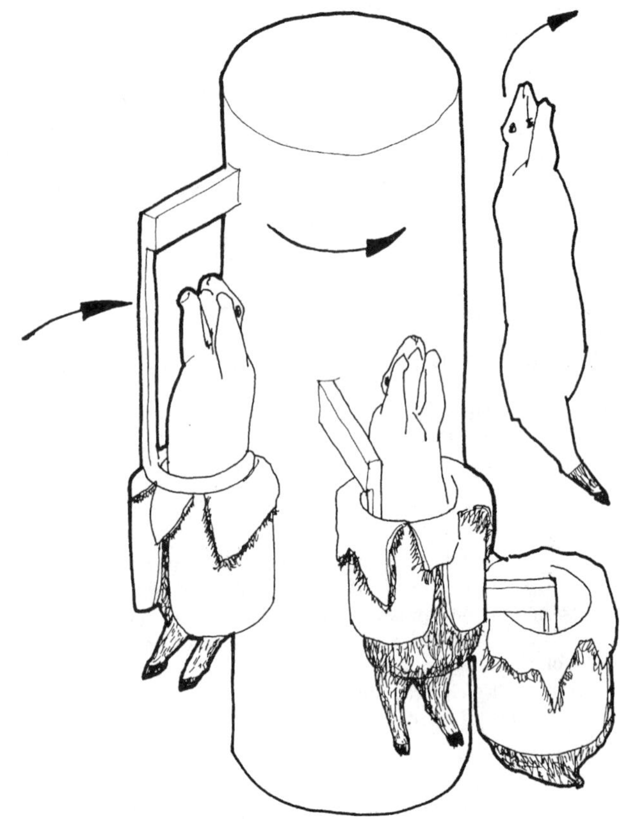

Figure 2. Rotary pelter.

Evisceration

Evisceration and offal (viscera) handling has the next greatest manpower requirement, after pelt removal. The Mechanical Dressing Project has undertaken several developments to partially mechanize evisceration and offal handling. Brisket cutting and belly opening was the first area studied. An early machine mechanically cut the brisket of inverted carcasses. A new machine has since been developed that not only cuts the brisket but also opens up the belly area. This machine is now awaiting production trials in a meat plant. Work is continuing on improved and mechanized methods of ovine viscera removal and handling.

Head Processing

In 1975 head skinning became necessary for sheep meats destined for export to the EEC. As part of carcass inspec-

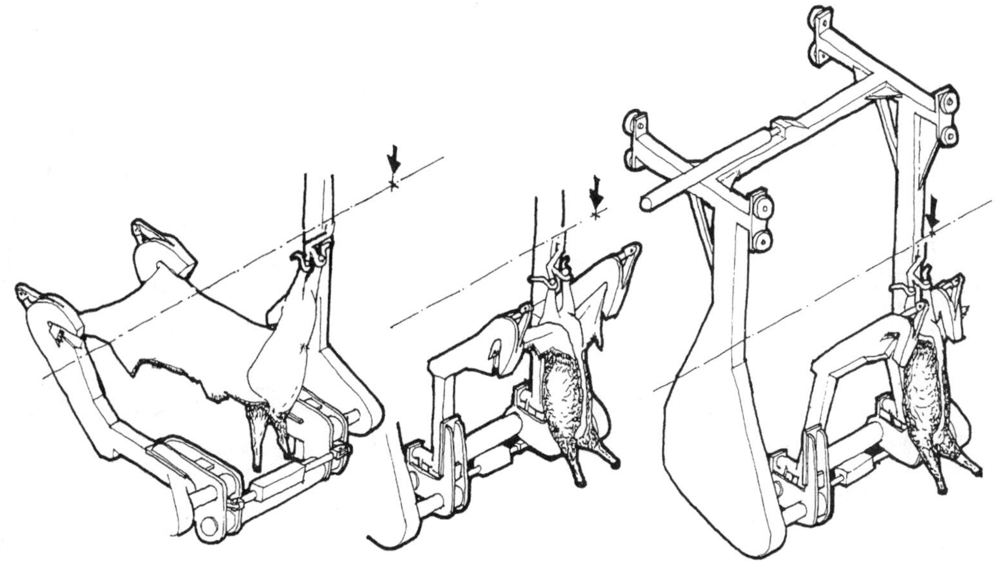

Figure 3. Shoulder puller.

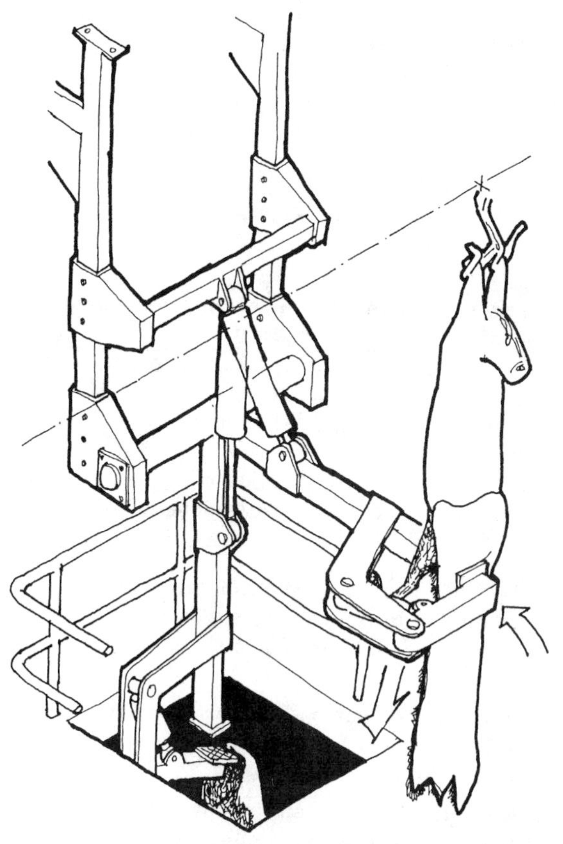

Figure 4. Final puller.

tion, the head, in a totally skinned state, had to be presented with the carcass for examination. In response to this requirement, several head-skinning machines were developed. The most successful was a machine that incorporated a small shaft that gripped a flap of skin near the nose and removed the skin by a rolling action. This machine was used in most meat plants throughout New Zealand and Australia.

The EEC regulation requiring heads to be presented with the carcass was partially relaxed in 1987, so that inspection was required only for those heads from which edible brains and tongues were to be saved. Therefore, the focus of developments in head processing (skinning and brain and tongue removal) has changed.

The Mechanical Dressing Project has recently developed several machines, including one that automatically severs the atlas joint at the base of the skull (Fig. 9), and one that automatically splits the head, followed by automatic extraction of the brain (Fig. 10).

Performance of a Mechanized Sheep-Dressing System

The potential manning levels for a mechanized sheep slaughter and dressing system in New Zealand, based on commercial developments so far and processing eight lambs per minute, are given in Table 3. This manning contrasts with that of the traditional manual system of 45 plus 15 assistants for the same throughput.

Carcass Hygiene

In 1988 a New Zealand meat plant replaced its traditional dressing system with an inverted system, along with four machines: wide-to-narrow transfer, front hock remover, shoulder puller, and final puller. This system processed 3500 lambs per day (one dressing chain). The plant, as part of its quality control program, regularly swabbed set places (flap, leg, brisket) on the surface of randomly selected carcasses, to give aerobic plate counts. These data reflect the hygienic condition of the carcasses processed in the old and new dressing systems.

With the new system, only 1% of the counts were above $10^4/\text{cm}^2$, whereas with the old system 11% were above that

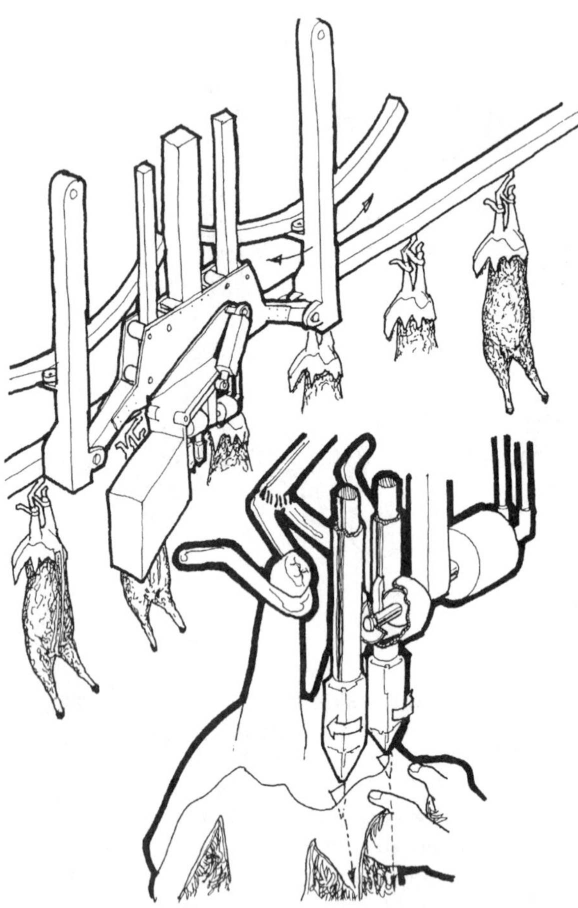

Figure 5. Brisket clearing machine.

Figure 6. Front hock remover.

level. The new system also gave a lower mean count than did the old manual system (Fig. 11).

Sheep and Lamb Boning and Cutting

Recent developments at MIRINZ in the mechanization and automation of sheep and lamb boning and cutting are showing a number of benefits, including the production of structurally intact whole-tissue meat, consistent and improved meat quality, greater meat recovery (higher yield), greater operator productivity, and reduced labor costs. Two developments are described here; one for whole mutton carcass boning, the other for boning short loins on both sheep (mutton) and lamb.

Whole Carcass Boning—The Frame Boner. The frame boner removes the two soft sides from mutton carcasses. Each soft side includes an intact form: the shoulder and foreleg bones (scapula, humerus, ulna, and radius), the leg bones (femur, tibia, patella, and fibula), and the muscle, connective tissue, and fat on the shoulders and legs, on the top and both sides of the neck bones (cervical vertebrae), and external to the backbone (thoracic and lumbar vertebrae), rib bones and aitch bones. A fully automated machine configuration is used, to minimize labor input, maximize yields, maximize processing rate, produce a product of consistent high quality, and accommodate and adjust to a range of carcass sizes.

The frame boner consists of four main components: the load station, the pedestal and carcass support, the linear drive and boning head, and the control cabinet. The physical relationship of each component to the machine as a whole is shown in Figure 12. The boning process performed by the frame boner consists of a sequence of five operations:

1. Lifting a carcass off the rail, removing the gambrel, loading onto the carcass support, and making carcass length measurements; performed at the load station.
2. Rotating the carcass support about the horizontal axis to present the carcass to the boning head; performed on the pedestal.
3. Clearing the pelvis on the upward travel of the boning head by grasping and pulling the rear legs; performed at the linear drive.
4. Boning off the soft sides on the downward travel of the boning head using front leg tensioners together with a combination of rotating knives, flexible discs, ploughs, and moving wires to do the cutting; performed at the linear drive.
5. Ejecting the skeletal frame during carcass support rotation; performed on the pedestal.

Figure 7. Automatic rear hock remover.

Figure 12 shows the machine with a single carcass support. A production machine would have two carcass supports fitted. With twin carcass supports, operation 1 would take place simultaneously with operations 3 and 4.

The boning process is controlled by programmable logic controllers (PLCs) interfaced to the human operator through a control panel. In this PLC-based system, individual components can be either cycled independently for checking and adjustment purposes or cycled together to perform the automated boning process. The machine can process over 200 thawed or chilled carcasses per hour (with twin carcass supports), and extra-heavy carcasses do not slow down the boning process. The yield is between 71 and 75% of the carcass weight for bone-in, untrimmed soft sides and is greater than 63% of the carcass weight for boneless, untrimmed soft sides. (These yields are based on average carcass weights of between 19 and 23 kg and assume that the shoulder and leg bones are 8% of carcass weight.)

Product quality is consistent within carcass grades, and the surface of backstraps (longissimus dorsi muscle) is smooth and even, with a complete absence of knife cuts. The machine gives significant labor savings. A production rate of 200 carcasses per hour is approximately equivalent to the work of five boners performing manually the equivalent breaking-down operation. Some workup on the carcasses prior to boning on the machine is necessary; for example, fillet removal, a V cut over the tail, and neck and pelvis underside clearing. This, together with machine supervision duties, is expected to require two people; thus giving a worker saving of three people.

Part Carcass Boning—The Loin Boner. The loin boner is designed to remove the two loin eye muscles (longissimus dorsi) from lamb and mutton short loins (short saddles) that contain the lumbar vertebrae. This machine, which is designed to fit existing boning rooms, is compact and robust and allows either one or two person operation. The loin boner consists of a horizontally moving carriage on which the loin support and clamp are mounted, a vertically moving assembly on which the knives and ploughs are located, a control system consisting of a PLC, pneumatic valves, motor speed controller and operator's console, a frame, and covers to provide hygienic machine protection.

In the boning process, a short loin is placed on the support and clamped manually. When the boning cycle is initiated by the operator, the support with the short loin saddle proceeds along the machine. Simultaneously, the knives and ploughs move downward to land on the support just prior to the loin passing under them. Boning is achieved during the passing of the loin under the knives and ploughs, the knives cutting either side of the vertical bones closely followed by the ploughs, which clear from the chine and horizontal bones. Thus, the muscles are separated from the bones. The knives and ploughs are then raised to allow the support to return to the operator for muscle and bone removal.

The machine can process up to six loins per minute with one operator and eight loins per minute with two operators. Compared with existing manual boning methods, there is a greater than 10% increase in muscle removed. At a production rate of three loins per minute, two fewer people are required than for the equivalent manual boning process. Product quality is excellent, as the boned loin has a consistent size and shape, and a smooth surface finish with a complete absence of knife cuts. This boning process is being extended to long saddles (6–8 thoracic vertebrae plus the lumbar vertebrae). Figure 13 shows the loin boner machine layout with a dual-purpose loin and long saddle support.

PIGS

Although the basic steps of the pig slaughter process have not changed in the last 50 years, considerable development has taken place to improve and optimize each of these steps. These steps include lairage, stunning, scalding, dehairing, singeing, and evisceration. To maintain the best possible hygiene level within the slaughter process, much attention has also been given to product flow, to ensure that material for human consumption, material for pet food, and waste material are kept appropriately separated.

Lairage and Handling

The conditions existing for the animal prior to stunning have been a subject of concern for the last 10 to 15 years. Preslaughter holding conditions and animal handling have been recognized as areas that can greatly contribute to the

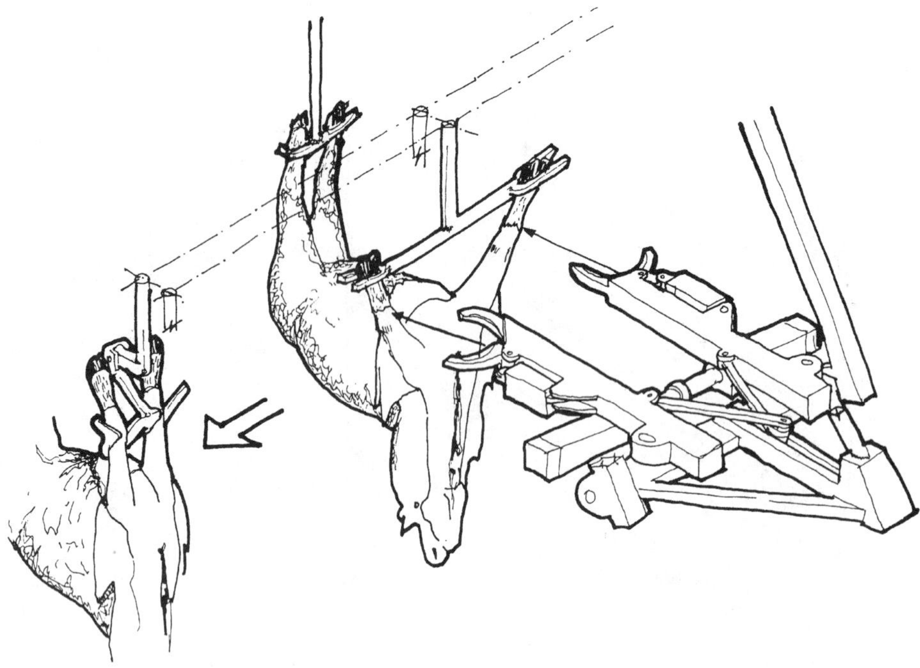

Figure 8. Wide-to-narrow transfer machine.

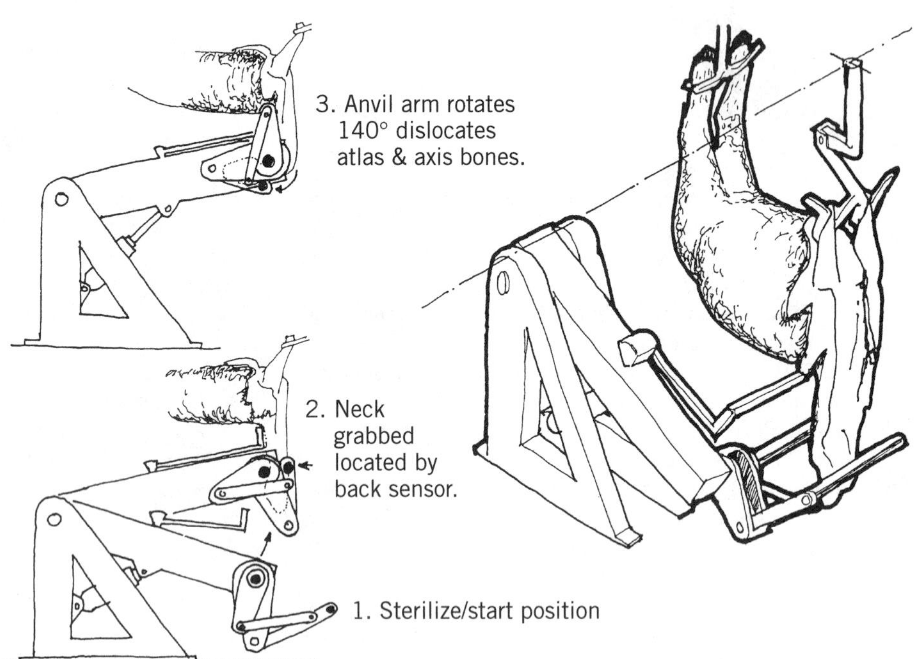
3. Anvil arm rotates 140° dislocates atlas & axis bones.
2. Neck grabbed located by back sensor.
1. Sterilize/start position

Figure 9. Automatic atlas joint severing machine.

stress level of pigs, which can affect meat quality. Here, traditionally, the animals have been forced into single file, a positioning that behavioral studies have shown is unnatural for these animals. Modern systems have been developed that use parallel single-file transport arrangements, which have greatly reduced the stress levels of the pig.

Stunning

The most common stunning method for pigs in the United States and Europe, particularly in larger plants, is electrical high-voltage stunning. The head-to-body version, causing cardiac arrest and reduction in poststunning movement, is preferred. More recently, studies in meat quality have shown that the occurrence of blood splash (small, visible hemorrhages within the muscle tissue) and breakage of the back bone and blade bone are attributable to electrical stunning methods. The incidence of these problems has been reduced by careful control of stun current and electrode position and by keeping the stun-to-stick time of the animal to a minimum.

CO_2 stunning is also very popular. Here the pig is lowered by a cradle into a pit containing a high concentration of CO_2 gas. The pigs are then raised out of the gas and

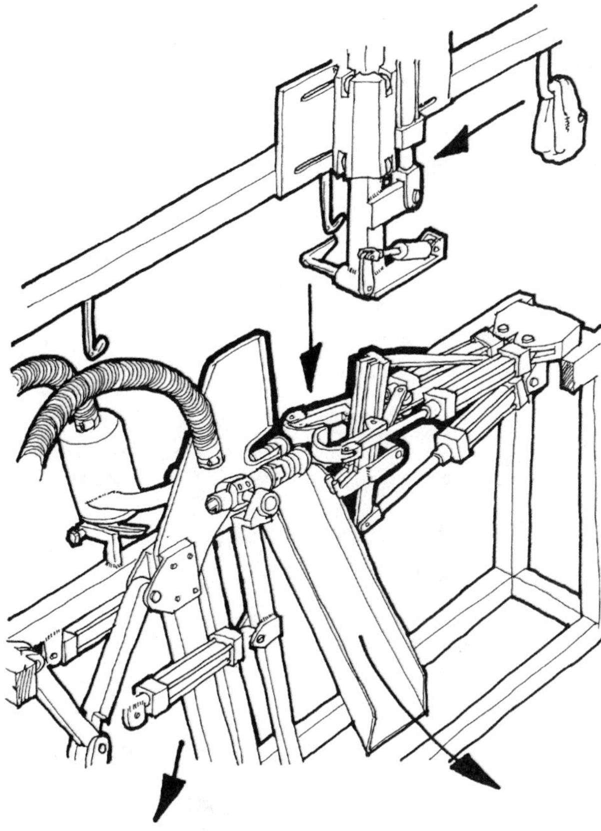

Figure 10. Automatic head splitting and brain extraction machine.

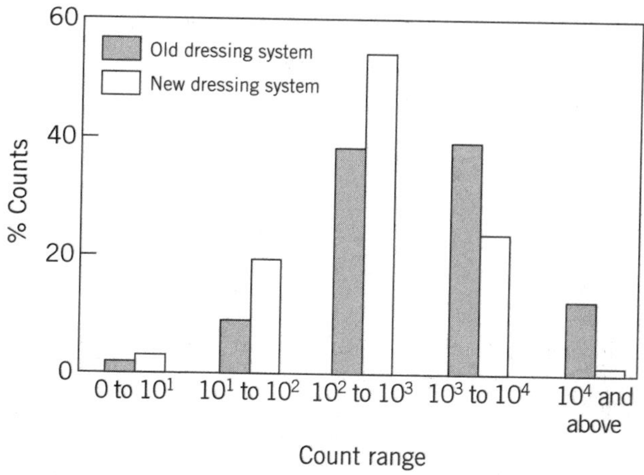

Figure 11. Aerobic plate counts (colony forming units/cm^2) for carcasses processed by a traditional dressing system and by an inverted dressing system in the same meat plant (n = 105 for each system).

Table 3. Potential Chain Manning for Mechanized Lamb Slaughter and Dressing, Processing Eight Animals per Minute

Task	Workers
Auto stun	—
Shackle, open, bleed	4
Auto neck break	—
Head cheek, remove	3
Y cut, push flap	3
Clear neck, lift shanks	2
Auto wide to narrow transfer	—
Auto front hock remover	—
Load brisket clearer	1
Pull brisket piece, Y-cut rear legs	3
Load shoulder puller	1
Clear breaks, punch tunnels	1
Auto final puller	—
Trim anus	1
Auto rear hock remover	—
Auto wash	—
Auto brisket cut	—
Gambrel	1
Evisceration	5
Total	*25 + 11 assistants*

discharged in a stunned state ready for sticking and further processing. There has been some criticism of CO_2 stunning of pigs, and this has been directed mainly toward the excitation phase that occurs after approximately 10 s exposure to the gas. In this phase, pigs tend to throw their heads and legs about with violent movements. It seems that no stunning method meets all the demands regarding meat quality, animal welfare, and worker safety. CO_2 stunning seems, however, to be one method that has potential for further development. Other stunning methods, such as the use of microwave energy, are also being assessed.

Scalding

Scalding is the most important part of the dehairing process. If the temperature of the scald is too high, there can be mechanical damage from the dehairing machinery due to skin softening. Also, too hot a scalding temperature and/or too long a time in the scalding tank has been associated with a higher level of PSE (pale, soft, exudative) meat.

Three type of scalding methods are used:

1. Scalding in a tank of water
2. Scalding by recirculating water sprinkled over the pigs
3. Individual scalding with temperate steam

There has been some debate as to whether water scalding is a clean operation or a contaminating one. That is, can water scalding cause contamination of the meat, or does meat contamination arise primarily by the sticking of the animal? Investigations have proved that neither shelf life nor flavor is affected negatively by the fact that the carcass has been scalded by either water or steam. The debate continues, however, and seems to suggest that scalding with steam will be the preferred option of the future. The problem with steam scalding is that heat transfer can be nonuniform. After scalding, the carcass is dehaired by means of a large number of rotating brushes.

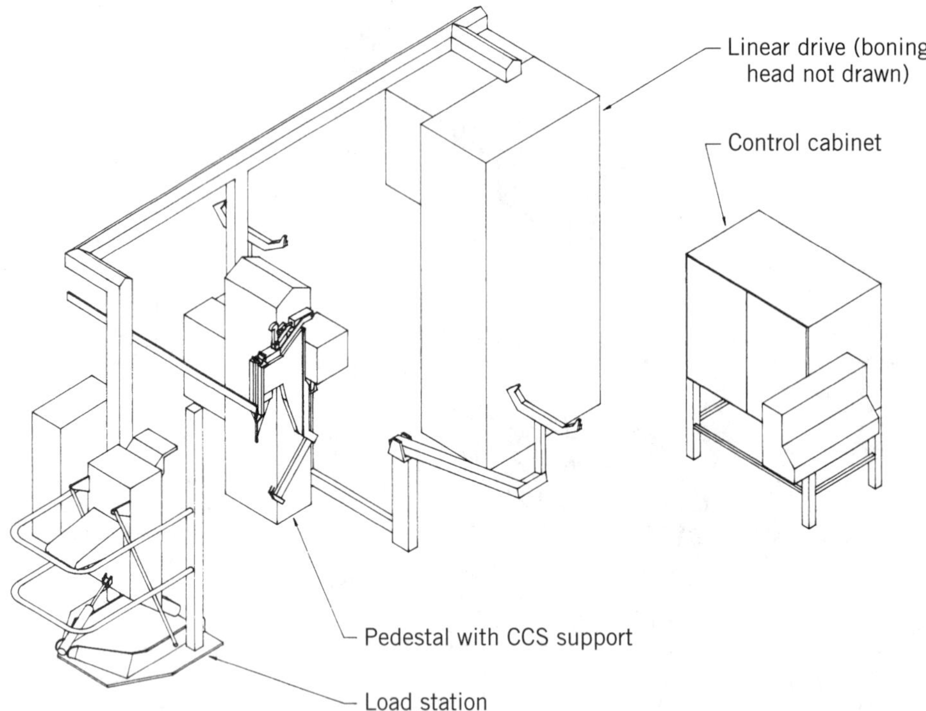

Figure 12. Frame boner layout.

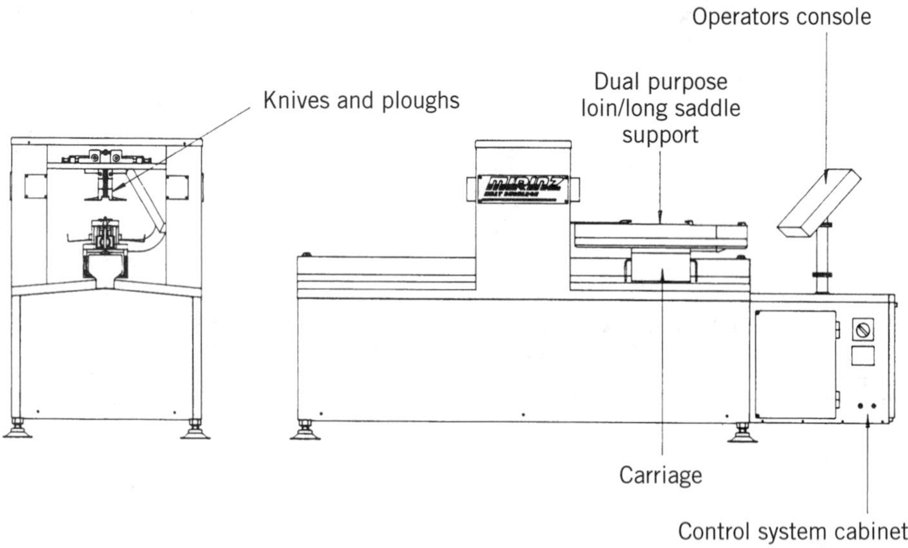

Figure 13. Loin boner machine.

Singeing

Carcasses are singed in cabinets, which are fired by either oil or gas, depending on the production setup. A newly developed ceramic material, coated on the inside of the furnace, reflects the energy of the flame and reduces the energy consumption. An exhaust gas heat exchanger retains half the energy that passes through the system. After singeing, the scraping treatment follows. Modern scraping machines use PLC technology. Scraper positions are hydraulically adjusted to an optimum position by sensing the height of the front legs. Hence, each pig receives optimal treatment from the scraper blades.

Dehiding

The method of dehairing just discussed consumes large quantities of energy and water. One alternative to dehairing is to dehide the pig. Dehiding pigs has the following advantages:

1. Longer shelf life for the meat (up to three days longer because of more hygienic slaughtering)
2. Less PSE because the hog carcass is not heated up by scalding and singeing
3. Lower energy and water costs

4. Reduced production area
5. An economic return from the hides

Disadvantages include:

1. Dehiding involves more work.
2. Grading circumstances are different.
3. Carcass stamping is more difficult.

Machinery Developments

A number of machines have been developed to assist the processing and slaughtering of pigs. One example is the fat end loosener, which drills out the fat end with the help of a vacuum. This machine was developed in Denmark and is now used in a large number of plants in that country. Prime quality fat ends have been increased from 75 to 95%. Another machine used in a number of abattoirs throughout Europe is an automatic pig carcass splitting machine. This machine replaces one operator and automatically splits the pig carcass into two halves. For hygienic operation, the cleaver mechanism that is used to split the carcass is sterilized between each pig. Considerable developmental resources are now being used in both Denmark and Holland to reduce the manpower involved in pig slaughtering. Some difficulties are being experienced with some of these machines when pigs are of an odd shape. Clearly, as robots become less expensive, there is a greater chance that these machines will succeed.

Automatic Grading

One of the great technical achievements in pig processing is the automatic carcass classification center, developed in Denmark. The center grades and stamps pigs automatically. These centers are now installed in a number of plants in Denmark. The system can cope with at least 400 pigs per hour. The main element of the center consists of an automatic measurement system. This system correctly positions 17 optical probes that measure the meat and fat thickness. The optical probes continuously measure the light reflectance value as they pierce the meat and fat.

Information from the probes is fed into a computer, which determines the fat and meat thickness. Subsequently, the carcasses are stamped automatically, with grading figures applied to each individual primal cut. The computer also transfers this information to a data sheet from which the farmer or supplier is paid.

POULTRY

The success of the poultry industry worldwide has been largely due to the automation of the slaughtering and dressing process. The development of this automation has taken place over the last 10 to 15 years.

Stunning, Slaughter, Dressing, Evisceration

The incorporation of mechanization into poultry processing lines has allowed the poultry meat industry to be cost competitive compared with other species. This mechanization started with the introduction of an automatic eviscerating machine. This machine started a revolution in methods of dressing chickens.

Today we have systems that involve overhead conveyors and a large number of machines—each capable of a line speed of around 4000 birds per hour. At present, the major manual effort into poultry processing lines occurs only where the birds are loaded into the processing conveyors. Once on these conveyors, the birds are transported through an electrical stunning area, a killing machine, a scalder, and a defeathering machine. Stunning generally occurs when the heads are immersed in a saline water bath, which causes cardiac arrest and also reduces post-stun movement. Problems reported to be associated with stunning can be eliminated. All machines on the killing line can be adjusted to suit the various sized birds being processed.

As with pig processing, the scalding process largely determines the defeathering result. Much work has been done to improve the temperature regulation of the water in the scalding tanks and to develop new systems for improved heat transfer from the water to the feathers.

The defeathering machine uses large rubber fingers which, when rotated in close proximity to the birds, remove the feathers. A number of other jobs are also carried out mechanically following this defeathering operation. The gullet is removed by machine, then a rotating knife removes the feet. In both cases, the design of the machine can cope satisfactorily with birds of various sizes.

After 10 years of effort, automation experts have succeeded in producing a fully automatic evisceration line for chicken. As with the killing lines, the principle of the evisceration line is to transport the birds on overhead conveyors, through a number of processing stations. These processing stations or machines include a vent cutter, an opener, an evisceration cut opener, and viscera removal. Machines are also being developed and manufactured for the automatic evisceration of other species, such as turkeys and ducks. Following evisceration, the birds are chilled, weighed, and packed. Much of this work has now been automated. For example, the birds are weighed on the conveyor and a trip system is used to grade the animals into certain precise weight ranges.

Cutting

During the last five years, the automated cutting of poultry into portions has attracted enormous interest, particularly in Western Europe. A number of manufacturers now produce automatic chicken portioning machines. Here again, these systems consist of an overhead conveyor, which transports the carcass through a number of cutting stations. Chickens can be cut into virtually any number of portions as desired, by switching cutting units off or on. Depending on the precision of the cutting, a throughput of around 2500 to a maximum of 3000 birds per hour is now achievable. The boning of chicken breasts is also now very popular. A number of machines are available to automatically fillet the breast. These machines operate at a rate of about 1500 birds per hour.

BEEF

Stunning and Slaughter

Cattle are normally stunned using a captive bolt or a percussion-type stunner and bled by a thoracic stick, but electrical stunning has been developed and used successfully. A head restraint system, which involves head capture and a chin lifter, is essential to allow reliable electrode placement. There are also advantages in using the same system for captive bolt or percussion stunning. If the current passes only through the head, the system is halal, although it is desirable to suppress poststun movement by electrical immobilization. If an additional set of electrodes makes contact with the brisket or other parts of the body of the animal, cardiac arrest occurs and poststun movement is reduced as well. A typical electrical stunning system is shown in Figure 14.

With electrical stunning systems, the stillness of the animal allows convenient typing off of the esophagus, preventing release of ingesta.

Dressing

After stunning, the carcasses are shackled and hoisted. The carcasses, suspended by their hind legs from overhead conveyors, pass along a number of platforms and units where the carcasses are dehided by means of hand tools and machines. On the first two platforms, both rear legs are dehided using knives and handheld dehiding tools. In this legging process, the hocks are removed by manually operated hock removers.

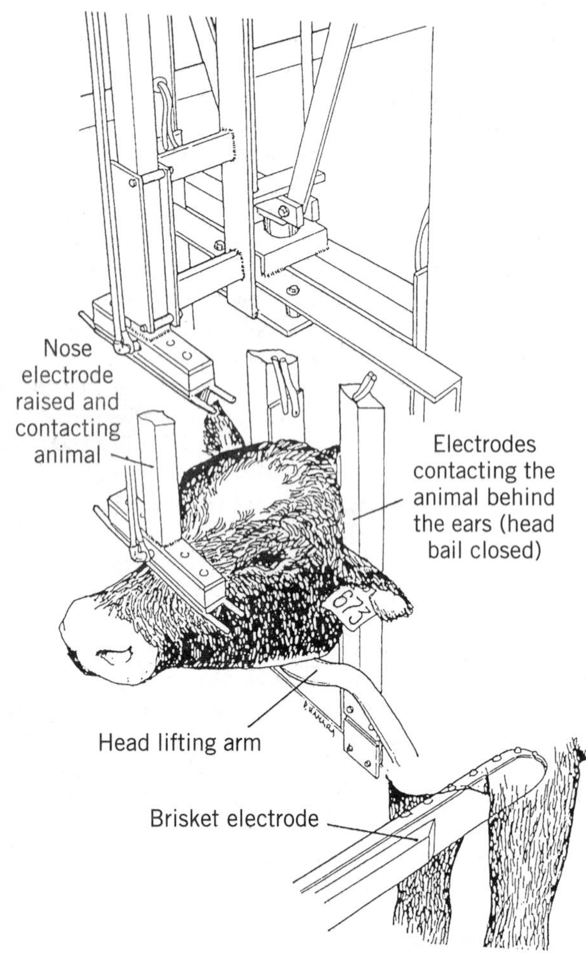

Figure 14. Electrode and head restraint for electrical stunning of cattle.

This preliminary dehiding is carried out in preparation for entry into the downward hide pulling machine. The dehiding machine consists of two operator platforms on either side of a powered rotating drum. The platforms and drum form an integrated assembly that moves up and down. Part of the hide that was removed during preliminary dehiding is fixed to the drum. As the platform-drum assembly is lowered, the hide is pulled off and rolled onto the drum. The two operators standing on each side of the carcass can assist the hide removal with handheld dehiding tools.

After complete removal of the hide, the drum is reversed and the pelt drops into a chute behind the machine. About 80 cattle per hour can be dehided with this system. After dehiding, the animal is eviscerated and all of the components of the viscera plus the carcass are inspected.

One machine development that has taken place in beef processing over the last 10 years has been automatic carcass splitting. This machine was developed in Europe and machines have been installed in the United States and Australia. The machine consists of a large circular saw that is controlled by a guiding system. The saw plus the guiding system is sterilized with hot water between each carcass. A large development program for automated beef dressing is nearing completion at the Commonwealth Scientific and Industrial Research Organization (CSIRO) Meat Research Laboratory in Queensland, Australia. This technology could have a significant impact on beef dressing in the future.

G. R. Longdill
Meat Industry Research Institute of New Zealand
Hamilton, New Zealand

MEAT STARTER CULTURES AND MEAT PRODUCT MANUFACTURING

The origin of fermented meat products remains unknown, although the history of fermented sausages probably can be dated back more than 2000 years to the Mediterranean (1). Traditionally, the chopped meat was either presalted to promote development of lactic acid bacteria, or the "back-slopping" method was used. Today, large-scale industrial productions require consistent and accelerated processing, which has led to the present general use of meat starter cultures. Around 1920, pioneering work was done in this field, but it was not until the 1950s that starter cultures for sausage fermentation were introduced on a commercial scale. In the United States, pure cultures of pediococci were introduced in 1957 (2), while pure cultures of micrococci were introduced in 1961 in Europe (3). The usage of starter cultures in the manufacture of dry-cured sausages became common in the 1980s. Apart from the United States, where pure pediococci are still preferred due to manufacturing conditions, the trend is now increasing toward the use of mixed cultures composed of staphylococci strains and strains of lactic acid bacteria. Starter cultures contribute to appearance, texture, flavor, and aroma. The usage of starter cultures for fermented meat products is largely confined to dry-cured sausages. Only a minor portion is employed for dry-cured ham production.

FERMENTED SAUSAGE MANUFACTURE

Fermented sausages are manufactured by chopping together a mixture of raw meat (usually a beef-pork combination), pork back fat, salt, fermentable carbohydrate, fillers (eg, caseinate, starch, or other carbohydrates), spices, nitrate or nitrite, and an acidifying agent such as starter culture, glucono-δ-lactone (GdL), or fermented meat. Briefly, the procedure is as follows: The meat, including the pork back fat, is usually chopped frozen to avoid smearing during the cutting process. Pork back fat is used due to its high melting point and low content of unsaturated fatty acids as opposed to beef fat, which tends to impart a characteristic tallowlike taste. In some cases, sausage manufacturers use chilled meat, which has been pretreated by salting it to the required level prior to cutting the meat in the bowl chopper. This addition of salt brings about an accelerated acidification process as the resulting reduction in water activity promotes the growth of the indigenous lactic acid bacteria. Prior to cutting, the presalted meat is usually held for two to three days at a temperature not to exceed 5°C. To ensure proper mixing, all nonmeat ingredients, including the starter culture, are added during chopping of the lean meat. To achieve as low a water-binding effect as possible, salt and pork back fat are the last ingredients to be added. The mixing stops when the particle size of the mixture is reduced to the required size. Following these operations, the mixture is stuffed into casings, which can be either natural or of synthetic material. At the point of stuffing, the temperature of the mixture should not exceed -2 to 0°C so as to avoid smearing of the fat during the stuffing process. Appropriate steps must be taken to ensure that all air is removed from the mixture. This is most often achieved by using a vacuum stuffer. Once stuffed, the sausages undergo a conditioning period. During this period, the temperature of the sausage equilibrates to ambient and the sausage surface is allowed to dry in order to remove condensate before the product is placed in the climate chamber. If the condensate is not entirely removed prior to this step, the sausages may appear streaked as a result of uneven coloration arising from smoke components that have deposited on the surface.

In Europe, initial fermentation conditions are temperatures between 20 and 24°C and a relative humidity of 95%. The humidity is lowered along with the temperature, and this parameter should always be adjusted to a value 2 to 4% lower than in the sausages. This ensures a proper drying out and minimizes the risk of dry rim, which may block water diffusion from the sausage core. The total drying time will usually be determined by the time a required water loss is reached. Sausages manufactured in Northern and Central Europe are often smoked. As well as imparting smoke color and flavor, compounds in the smoke have inhibitory properties that retard mold growth on the casing surface during fermentation and drying periods. If considered undesirable, mold growth on the surface of salami casings can be inhibited by resmoking during the drying period. With some sausages, such as traditional Hungarian types and southern European types, it is considered desirable, and in some markets essential, that the product is covered with white mold, and in these cases the product is often inoculated with mold cultures. North American type of sausages are fermented at temperatures between 40 and 43°C and a relative humidity of 95%. The fermentation time is short, usually between 12 and 20 hours, and is followed by a heating process and/or a drying period. The heating process is dictated for unfrozen pork meat not guaranteed free of trichiniae. Final products can be grouped in dry sausages (eg, pepperoni) with a moisture to protein ratio in the range of 0.75 to 2.3:1.0 and semidry sausages (eg, summer sausage) with a moisture to protein ratio of 2.5 to 3.7:1.0 (4).

THE INFLUENCE OF STARTER CULTURES IN FERMENTED SAUSAGES

The purpose of using meat starter cultures for dry-cured sausages can be summarized by the terms *product quality product safety*, and *technological advantages* (Table 1).

Acidification

A sufficient reduction in pH can only be obtained if carbohydrate is included in the sausage recipe, as the concentration of glucose and phosphorylated glycolytic intermediates in raw meat is low, usually <0.1% (5). Fermentable carbohydrate, most often glucose, is added at a level of approximately 0.5% or higher. Sucrose might substitute glucose, and more slowly metabolized carbohydrates such as lactose or dextrin may be added as an extra carbohydrate source or as fillers. Efficient fermentation is achieved by homofermentation, where the carbohydrate added is metabolized via glycolysis, with lactic acid virtually being the only end product. Lactic acid bacteria for sausage fermentation are facultatively heterofermentative (6). This means that under certain conditions (eg, carbohydrate depletion), the lactic acid bacteria may switch to heterofermentation, resulting in formation of acetic acid and carbon dioxide in addition to lactic acid. Such compounds lead to defects in dry-cured sausages because acetic acid imparts a bitter, vinegarlike taste and excessive carbon dioxide may result in gas pockets. The speed of lactate formation determines the course of the fall in pH, which again controls gel formation and texture. Initial pH of the sausage mixture is typically within the range from 5.8 to 6.0, and drying will, in practice, be initiated around pH 5.3. The degree of drying out has a significant influence on the final texture, whereas the type and the degree of acid formation determines whether the taste of the sausage is mild, acidic, or tangy. A rapid acid formation will, in conjunction with the salt concentration, ensure that pathogenic as well as spoilage bacteria are significantly inhibited. Acidification can, as mentioned, also be obtained chemically by usage by

Table 1. The Influence of Meat Starter Cultures on Dry-cured Sausage Manufacturing

Mode of action	*Lactobacillus / Pediococcus*	*Staphylococcus / Kocuria*
Product quality		
Reduction in pH	x	
Texture	x	
Color formation	x	x
Color stabilization		x
Flavor formation	x	x
Product safety		
Inhibition of pathogenic bacteria	x	
Inhibition of spoilage bacteria	x	
Technological advantages		
Shortened processing time	x	
Consistent production	x	x

GdL, added at a concentration around 0.5%. In the presence of water, GdL is instantly hydrolyzed to gluconic acid, thereby lowering the pH. Once hydrolyzed to gluconic acid, GdL can be further metabolized by lactic acid bacteria to form acetic acid and carbon dioxide (7), which may have a negative influence on the sensory properties.

Color Formation

The color of meat is directly related to the content of myoglobin, and the meat will thus appear more red at high myoglobin levels. Myoglobin is unstable and reacts easily with oxygen either under formation of the cherry red, and also unstable, oxymyoglobin or under formation of the stable, but brown, metmyoglobin. The red color in dry-cured meat products comes from the pigment nitrosylmyoglobin, which is formed in reactions between nitric oxide and myoglobin or metmyoglobin. The main nitric oxide source is the very reactive nitrite. Quantitatively, the reaction of nitrite with ascorbate is the most important with respect to formation of nitric oxide (8). Ascorbate or erythorbate (isoascorbate) is for this reason a typical ingredient in dry-cured sausage recipes. Nitric oxide is also formed via nitrous acid, the conjugated acid of nitrite, with a pK_a of 3.22. Nitrous acid is unstable and dissociates into nitric oxide and nitrate. The more acidic the environment becomes, a condition promoted by the activity of lactic acid bacteria, the more the equilibrium is shifted in direction of nitric oxide and nitrate.

Nitrite is the most common curing agent and is added as nitrite salt, in the form of sodium chloride containing 0.4 to 0.5% sodium nitrite (9). Although 30 to 50 ppm nitrite is enough for proper color formation (9), EU regulation instructs addition of 150 ppm sodium nitrite and permits 50 ppm sodium nitrite in ready-to-sale dry-cured fermented meat products (10). However, nitrate can be added instead of nitrite. The corresponding values for nitrate are 300 ppm and 250 ppm sodium nitrate, respectively. U.S. regulation instructs 150 ppm sodium nitrite and 1700 ppm sodium nitrate and permits a maximum level of 200 ppm sodium nitrite in the finished product (11).

Nitrate has no curing capability (9), and microbial reduction to nitrite is necessary to obtain curing effect. The reaction requires the nonendogenous enzyme nitrate reductase. The enzyme is produced by *Staphylococcus / Kocuria*, but even when nitrite is the only curing agent, *Staphylococcus / Kocuria* is reported to accelerate color formation (12). The reactions behind this formation are not known. It is hypothesized that because *Staphylococcus / Kocuria* utilizes nitrate as terminal electron acceptor for growth under anaerobic conditions, the reaction equilibria are shifted toward right, whereby the necessary nitric oxide for color formation is formed more rapidly.

Color Stabilization

Hydrogen peroxide is produced by the indigenous bacterial flora and can also be produced by lactic acid bacteria when oxygen is available. Even the more stable meat pigments are subject to oxidation and therefore also to reactions with hydrogen peroxide, which results in undesirable color pigments such as the green verdoheme and the yellow biliverdin (13). In combination with the red and brown meat pigments, the human eye will perceive this conglomerate of pigments as grey spots or even grey cores may be perceived to appear in the sausages (14). Both are typical defects associated with dry-cured sausages. Hydrogen peroxide is decomposed in presence of catalase, an enzyme produced by *Staphylococcus / Kocuria* (15).

Flavor Formation

The flavor and taste of cured meat products are to some extent defined by ingredients such as salt, nitrite/nitrate, and spices. The overall sensory quality is influenced by the choice of process (eg, smoking), as well as by microbial activity (16,17). The aroma of fermented meat products can be divided into the volatile compounds influencing flavor and the nonvolatile compounds determining the taste of the product. In both groups can be found a number of chemical compounds such as alkanes, alkenes, aldehydes, ketones, esters, organic acids, alcohols, phenols, furanes, terpenes, aromatic compounds, and compounds containing nitrogen, sulfur, or chloride (17,18). The aroma compounds derived from decomposition of carbohydrates, lipids, and proteins can be either of endogenous or microbial origin. Microbial activity transforms carbohydrate into lactic acid via glycolysis and provides dry-cured sausages with the characteristic tangy taste. If a part of the carbohydrate is metabolized via heterofermentation, the taste and the flavor is affected by the formation of compounds such as acetic acid, formic acid, and ethanol. The initial steps in protein breakdown are primarily facilitated by the endogenous enzymes (17). The degree of proteolysis is pH-dependent. Cathepsins as well as endopeptidases, active in protein degradation, have highest activity at low pH. As regional differences in processing technologies result in sausage types varying in final pH, the aroma will differ accordingly. Bacteria are involved in the later stages of protein degradation. Breakdown of peptides is partly caused by bacteria and they are also involved in amino acid catabolism (17,19). Aminopeptidase activity has been demonstrated in staphylococci and lactic acid bacteria. It is, however, remarkable that lactic acid bacteria apparently have not only a higher but also a more diverse aminopeptidase activity than staphylococci (20). The main bacterial impact on aroma lies probably in amino acid catabolism. The aldehydes 2-methylbutanal, 3-methylbutanal and 2-methylpropanal derived from branched-chained amino acids are important flavor components in dry-cured sausages (21). They can be of bacterial origin, and among starter culture strains, staphylococci seem to play a more important role than lactic acid bacteria (22). The lipids are decomposed via lipolysis and lipidoxidation. Lipolysis is caused by lipases and esterases, primarily of endogenous origin, and liberates free fatty acids of which particularly the short-chain acids affect taste. Staphylococcal lipases are reported but the activity may be limited at the conditions prevailing during fermentation and drying (23). Oxidation of fatty acids happens by chemical autoxidation or by microbial β-oxidation. There is evidence that the latter, resulting in methylketones and methylaldehydes as well as secondary alcohols, has a significant impact on the flavor generation in dry-cured sausages (19,24). Vice versa, synthesis catalyzed by lipases should also be considered. Fragrant esters, for example, are important flavor compounds in dry-cured sausages (25). The mechanism of their formation is not known; however, their presence and significance have been connected to sausages fermented with *Staphylococcus xylosus* (21).

Starter Cultures for Dry-Cured Sausages

Commercial starter cultures either consist of single-strain cultures or a mix of up to four or five strains. Single-strain cultures are most common in the United States with *Pediococcus acidilactici* being the typical species employed for high temperature fermentations (40–43°C). Under these conditions the decrease in pH will be so rapid that *Staphylococcus / Kocuria* are significantly inhibited. Mixed cultures consisting of lactic acid bacteria and *Staphylococcus / Kocuria* are normally chosen for fermentation at temperatures ranging from 18 to 24°C. To enhance aroma development, the cultures may also contain flavor-generating organisms such as yeast and *Streptomyces*. For surface treatment, single strain mold cultures or cultures composed of mold and yeast are available.

Lactic Acid Bacteria

The lactic acid bacteria, traditionally used for acidification of dry-cured sausages, are mesophilic species such as *Lactobacillus plantarum* and *Pediococcus pentosaceus*, and for American-type sausages, *P. acidilactici*. In recent years, psychrotrophic species such as *L. sakei* and *L. curvatus* have become more and more common in countries outside the United States. The lactic acid bacteria are added at a level of 5 to 10 $\times$ 10^6 cells/g meat mixture and reach approximately 10^8 cells/g in the final product. As described, lactic acid bacteria are responsible for acidification, for development of the acidic taste, and for development of texture. It has, however, also been demonstrated that some lactic acid bacteria possess enzymes normally attributed to *Staphylococcus / Kocuria*. It is reported that nitrate and nitrite reductases can be found in *L. plantarum* and *L. pentosus*, and a heme-dependent nitrite reductase has been demonstrated in *P. pentosaceus* (26). The heme-dependent nitrite reductase transforms nitrite into ammonia, whereas the end product from the heme-independent nitrite reductase is nitrous oxide, a precursor for the formation of nitric oxide. Results from practical experiments have, however, revealed that color formation, solely based on *Lactobacillus* strains possessing heme-independent nitrite reductase, was insufficient and caused color defects (26). Some lactic acid bacteria may have a color-stabilizing effect as they produce pseudocatalase, a peroxidase, which by nature is a flavoprotein, active under aerobic conditions (26).

Staphyloccous/Kocuria

A great part of meat starter culture preparations contain strains from the genera *Staphylococcus / Kocuria*. The genus *Staphylococcus* is more widely used than *Kocuria* due to their difference in oxygen requirement. The predominant species are *S. carnosus* and *S. xylosus*, whereas *Kocuria varians* is included only in a few preparations. *K. varians*, previously known as Micrococcus varians, was recently replaced with respect to genus (27). *Staphylococcus* is best suited for the sausage environment as they are facultatively anaerobic, whereas *Kocuria* by nature are aerobic. *Staphylococcus / Kocuria* are usually added at a level of approximately 5 $\times$ 10^6 cells/g meat mixture. They seldom increase in number during fermentation; however, they maintain a metabolic activity sufficient to ensure that the desired effects are achieved over a prolonged ripening period. *Staphylococcus / Kocuria* are, due to their nitrate and peroxidase reducing activity, involved in color development and color stabilization. Under anaerobic conditions, *Staphylococcus / Kocuria* utilize nitrate as an alternative electron acceptor in the respiration chain; that is, they gain metabolic energy via anaerobic respiration (28). *K. varians* does not reduce nitrite, but isolates among *S. carnosus* possess a dissimilatory nitrite reductase that results in increased growth under anaerobic conditions

(29,30). Acetate is the end product of anaerobic respiration, whereas lactate is the end product of fermentation. As described, *Staphylococcus* / *Kocuria* are also involved in flavor generation.

Molds

Raw ham and some fermented sausages are still being produced according to traditional methods, that is, inoculation with the *in-house mold flora*. However, by this inoculation method, toxinogenic strains may be among those colonizing the surface. Many of the mycotoxins that can be detected in culture media can also be detected when the same mold species grows on sausages or raw ham (31). Molded ham is generally regarded as more hazardous than molded sausages because the casing functions as a protecting barrier for the sausages. It is, however, not yet common to employ mold starter cultures for control of surface flora on dry-cured ham. In spite of this, there are multiple reasons to employ mold starter cultures. First, the starter culture provides the sausage surface with a white covering of a desirable shade, and, second, the starter culture dominates the flora and minimizes the risk for growth of mycotoxin-producing strains. In addition, the mold culture supplies the sausages with a characteristic flavor. Positive side effects such as reduced moisture loss due to mycelial covering, antioxidative effects because of catalase activity, oxygen consumption, and reduced oxygen penetration through the mycelium are factors mentioned among other advantages of a mold culture (32). Today, all mold strains used for surface treatment of dried sausages belong to the genus *Penicillium*. The most frequently used strain is *P. nalgiovense*, but also *P. chrysogenum* and *P. camembertii* are used (16,33,34). The preference of *P. nalgiovense* is linked to its growth ability on dry-cured meats as well as its toxicological safety. It should nonetheless be stressed that not all *P. nalgiovense* strains are nontoxic (35).

Yeasts

Commercial starter culture preparations contain only one type of yeast species, *Debaryomyces hansenii* or as the imperfect form *Candida famata*. Strains within this species are characterized by a high salt tolerance, no nitrate consumption, and high oxygen demands. The yeast is added directly to the meat at an inoculation level of 10^6 to 10^7 cells/g meat mixture. The activity is primarily in the periphery where the oxygen consumption accelerates the exterior color formation. The addition results in a characteristic flavor particularly desirable for Italian types of sausages (36).

PUBLIC HEALTH ASPECTS OF DRY-CURED SAUSAGES

Dry-cured sausages receive no heat-treatment prior to digestion, but they have a number of built-in microbiological hurdles such as low pH, low water activity, low redox potential, and nitrite. Despite these hurdles, there has been cases where severe illness could be associated to dry-cured sausages and outbreaks caused by bacteria have been reported. Mold-fermented sausages may encounter the risk of mycotoxin development, and chemical hazards comprise unacceptable levels of biogenic amines and nitrosamines.

Escherichia coli O157:H7

E. coli O157:H7 emerged as a food pathogen in 1982, and an outbreak was for the first time linked to dry-cured sausages in 1994. *E. coli* O157:H7 is more acid tolerant than other *E. coli*. It survives better at refrigeration temperature than at abuse temperature, and it has a low infectious dose, that is, <50 organisms (37). As a result of the outbreak, the U.S. Food Safety and Inspection Service issued guidelines for sausage manufacturers to validate processes to ensure a 5D-log unit reduction in counts of *E. coli* O157:H7 (38). In traditional processes for dry and semidry sausages, *E. coli* O157:H7 is reduced by 1 to 2 log units, and investigations have shown that the recommended 5D-log unit reduction can only be obtained by introducing a heating step following the fermentation (39,40). In European-type sausage manufacturing, similar results have been obtained. Thus, depending on inoculum, *E. coli* O157:H7 is reduced by 1 to 2 log units during normal processing (41). Subsequent storage results in further reduction, the reduction being more pronounced at ambient than at chill storage temperatures.

Salmonella

Nontyphoidal salmonellosis is described as a leading cause of foodborne illness in the United States (42), and in the mid-1990s, dry-cured sausages also caused outbreaks. In 1995, illness could be associated to consumption of Lebanon bologna, a semidry sausage, contaminated with *S. typhimurium*. In recent years, focus is more and more on phage type 104 strains of *Salmonella*, as they display a multiple resistance to antibiotics leading to increased hospitalization and increased mortality rate (43). To minimize the health hazard of *S. typhimurium* DT104 in dry-cured sausages in the United States, a more frequent sampling plan has been implemented for manufacturing plants not complying to the 5D-log reduction processes, described for *E. coli* O157:H7. *S. typhimurium* DT104 is more sensitive to manufacturing and storage conditions than *E. coli* O157:H7, but it is noteworthy that manufacturing of pepperoni is reported to result only in a 3D-log reduction (43). The survival is better in chill rooms than at ambient temperature.

Staphylococcus

Apathogenic strains of *Staphylococcus* are used as starter cultures, and, thus, the environmental conditions in dry and semidry sausages also allow *S. aureus* to compete successfully and grow. This organism is resistant to nitrite and salt and is capable of growing under anaerobic conditions. It has a minimum water activity for growth of 0.86, and the high fermentation temperatures used in the U.S. manufacturing process favor its rapid multiplication. At fermentation temperatures of 24 to 26°C, it has been demonstrated that the risk of food poisoning, resulting from the ingestion of enterotoxin produced by *S. aureus*, appears to be minor, provided the initial staphylococci count is low and provided an acidifying agent is used to ensure proper fermentation (44). Recent reports discuss pathogenicity of coagulase negative staphylococci, to which group the meat starter cultures belong (45). Hemolysins are demonstrated in strains of *S. carnosus* and *S. xylosus*, for example. In addition, enterotoxin production has been demonstrated in *S. xylosus*. Currently, there is no information available on the biological effect of hemolysins and enterotoxins. Neither is information available on parameters and factors affecting or controlling enterotoxin production (45).

Listeria monocytogenes

Studies have shown that *L. monocytogenes* may survive the combination of low pH, low water activity, and nitrite, which are the built-in hurdles in dry-cured sausages (46). Consequently, listeriosis should be regarded a potential health hazard in fermented meats, although no outbreaks so far have been associated with meat (47). *L. monocytogenes* has been shown to survive well in European as well as American-type sausages (41,48), the decrease in initial numbers being only one log unit. The survival of *L. monocytogenes* in stored, dry-cured sausages is better at chill than at ambient temperature. *In situ* produced pediocin, a bacteriocin effective against *L. monocytogenes* is, however, able to control the development of this pathogen in fermented sausages (48). Recently, a bacteriocin-producing starter culture, capable of controlling *L. monocytogenes*, has been launched for fermented sausages (49).

Yersinia enterocolitica

Y. enterocolitica is associated with pork and, as such should be regarded a potential health hazard in dry-cured meat products. No outbreaks or recalls have been associated with the organism. A reason for this could be that the nitrite level typically employed inhibits *Y. enterocolitica*, particularly in conjunction with a rapid decrease in pH. For example, *Y. enterocolitica* was eliminated in dry-cured sausages with added 80 ppm nitrite and fermented with a fast-acidifying starter culture (50).

Biogenic Amines

Biogenic amines in dry-cured sausages are of microbial origin. They are formed via decarboxylation of amines, by amination or transamination of aldehydes and ketones, or by hydrolysis of N-containing compounds. Due to their vasoactive properties, biogenic amines represent a health hazard. In retail dry-cured sausages, they are detected in varying levels and some of the biogenic amines are present in concentrations representing a risk for sensitive consumers (51). Many of the traditional starter cultures, such as *L. pentosus* and *L. plantarum*, are assumed not to form biogenic amines (52). Among the more recent meat starter cultures, *L. curvatus* has been shown to form biogenic amines, whereas the necessary decarboxylytic activity has yet to be demonstrated in *L. sakei* (53). High levels of biogenic amines in dry-cured sausages probably reflect usage of inferior raw materials, problems in recipe formulation or processing program, or choice of wrong starter culture (51).

N-Nitrosamines

The reaction of secondary or tertiary amines with nitric trioxide results in carcinogenic *N*-nitrosamines. The amount found in cured meat products is generally low (54), even though it has been demonstrated that GdL and starter cultures support the formation of *N*-nitrosamines, a phenomenon related to the fall in pH, which accelerates the reduction of nitrite to nitrous acid, the precursor for nitric trioxide. Ascorbate counteracts the reaction; however, it is claimed that more than 1000 ppm is needed to substantially limit the *N*-nitrosamine formation (55).

RAW HAM MANUFACTURE

Traditional dry-cured ham may be produced bone-in or boneless, the latter meaning that the aitchbone and shank is removed prior to salting. The next step is to rub a mixture of curing ingredients (sodium chloride and maybe nitrite and/or nitrate) onto the lean muscle surface of the ham. The ham is stored at temperatures below 5°C until the curing agents are evenly distributed throughout the ham. This period lasts from a few weeks to a few months. In between, the ham may be washed and resalted if necessary. The ham is then moved to the drying rooms and ripened at temperatures between 15 and 20°C and a relative humidity decreasing from 90 to 70%. In the last part of the ripening period, the ham may be covered with lard or pork fat to prevent excessive drying out. An in-house mold flora will usually grow on the ham surface. Some ham types, for example, country-style ham and German Westphalia ham, are smoked prior to the drying step. The raw material and the length of the ripening period characterize the type of ham. Thus the processing period of Serrano ham is 9 to 12 months, whereas it can be as long as 18 to 24 months for Iberian ham (56). Other important types of dry-cured ham are the Italian Parma ham and the French Bayonne ham.

Microbiology of Dry-Cured Ham

It can be debated whether dry-cured ham is a fermented product. The reason is that the microbial flora is seldom higher than approximately 10^4 cells/g meat, although the concentration can be as high as 10^8 cells/g on the surface (57,58). Growth is limited due to the initial low temperature and due to the low water activity throughout processing. The natural microflora, in the core as well as on the surface, consists of *Staphylococcus* / *Kocuria*. Initially, lactic acid bacteria, yeast, and the Gram-negative *Vibrio* may also be demonstrated, but the predominant organisms throughout processing are *Staphylococcus*, although their importance in the biochemistry of ham products is unknown (59). The characteristics of dry-cured ham are determined by type of raw material, processing conditions, and the biochemical changes taking place during manufacture. These are largely the result of muscle proteases and lipases (17,56). A recent study, however, suggests that microbial amino acid catabolism resulting in, among others, methyl-branched aldehydes, methylketones, and ethyl esters influences the flavor of dry-cured ham (58).

Particularly, staphylococci are capable of such reactions (22). Accelerated production processes have been developed, in which the dry-salting step is replaced by brine salting or brine injection pumping. The total processing time of such ham products is approximately one month as the drying and ripening period is significantly reduced, resulting in the final ham to deviate from the traditional type in texture and flavor. The application of starter cultures is an advantage for this production method. The starter cultures, most often pure cultures of staphylococci, but also occasionally mixed cultures of staphylococci (*S. carnosus, S. xylosus*) and lactic acid bacteria (*L. plantarum*), are added to the brine and support flavor development as well as color development and color stabilization.

MANUFACTURE OF MEAT STARTER CULTURES

Meat starter cultures are marketed as frozen or freeze-dried products, the latter being standardized with a carrier that is most often a carbohydrate. Meat starter cultures are produced as single-strain cultures that are mixed after fermentation and concentration (33). Starter cultures should be looked on as any other food additive; that is, they should be GRAS (Generally Recognized As Safe) approved. The legislation is not standardized and no official standard has been issued. However, requirements for meat starter cultures have been discussed in publications (33,60), and the following should be regarded as minimum requirements for a standard:

1. From a health perspective, only nontoxic and nonpathogenic strains should be used and the starter culture should be free of any impurity that could cause health risks.
2. Use of starter cultures should result in fermentations controlled with respect to acidification, texture, color, and aroma as well as microbiological stability of the meat product. Starter cultures should consist of phenotypically and genotypically stable as well as phage-resistant strains.

FUTURE PROSPECTS

In the future, lactic acid bacteria may play a more significant role as protective organisms in meat products, including those currently not treated with starter cultures. The mode of action comprises several activities, among which the production of bacteriocins is already in use for inhibition of *L. monocytogenes* in the manufacture of dry-cured sausages (49). The effect of a weak acid production combined with nutrient competition is utilized for control of the microflora of meat products packaged in vacuum or in modified atmosphere (61). The starter culture strain for this purpose is *L. alimentarius*. Probiotic bacteria such as *L. acidophilus, L. casei*, and *Bifidobacterium bifidum* are marketed for meat products to improve health (62). The idea is obviously adopted from dairy products but it is questionable whether the consumers will eat dry-cured sausages in such amounts that the probiotic effect is achieved. Finally, the use of genetically modified organisms (GMO) constructed by genetic engineering is another subject for the future. Legal approval may be difficult in certain regions, particularly in Europe, and consumer acceptance also remains to be obtained. It is, nonetheless, the most reliable way to develop strains contributing specific properties to the final products.

BIBLIOGRAPHY

1. C. S. Pederson, "Fermented Sausage," in *Microbiology of Food Fermentations*, 2nd ed., AVI, Westport, Conn. 1979, pp. 210–234.
2. U.S. Pat. 2,907, 661 (June 6, 1959), C. F. Niven, R. H. Deibel, and G. D. Wilson.
3. F. P. Niinivaara, M. S. Pohja, and S. E. Komulainen, "Some Aspects about Using Bacterial Pure Cultures in the Manufacture of Fermented Sausages," *Food Technol.* **18**, 147–153 (1964).
4. J. Bacus, *Utilization of Microorganisms in Meat Processing: A Handbook for Meat Plant Operators*, Wiley, New York, 1984.
5. R. A. Lawrie, *Meat Science*, 5th ed., Pergamon, New York, 1991.
6. O. Kandler and N. Weiss, "*Lactobacillus*," in P. H. A. Sneath et al., eds., *Bergey's Manual of Systematic Bacteriology*, Vol. 2, Williams & Wilkins, Baltimore, Md., 1986, pp. 1209–1234.
7. A. Kneissler et al., "Die wechselweise Beeinflussung von Glucono-delta-Lacton (GdL) und Starterkulturen bei der Rohwurstreifung," *Chem. Mikrobiol. Technol. Lebensm.* **10**, 82–85 (1986).
8. L. H. Skibsted, "Cured Meat Products and Their Oxidative Stability," in D. E. Johnston, M. K. Knight, and D. A. Ledward, eds., *The Chemistry of Muscle-Based Foods*, Royal Society of Chemistry, Cambridge, UK, 1992, pp. 266–286.
9. F. Wirth, "Restricting and Dispensing with Curing Agents in Meat Products," *Fleischwirtsch.* **71**, 1051–1054 (1991).
10. European Parliament and Counsils, Directive 95/2/EC, (on food additives other than colours and sweeteners), Luxembourg, 1995.
11. U.S. Department of Agriculture, Food Safety and Inspection Service, *Meat and Poultry Inspection Regulations*, Washington D.C., 1986.
12. G. Mogensen, "Starter Cultures," in J. Smith, ed., *Technology of Reduced-Additive Foods*, Blackie Academic and Professional, London, 1995, pp. 1–25.
13. E. Slinde, "The Color of Meat" (in Norwegian), in B. Underdal, ed., *Kjøttteknologi*, Landbruksforlaget, Oslo, Norway, 1984, 71–78.
14. K. Coretti, *Rohwurstreifung und Fehlerzeugnisse bei der Rohwurstherstellung*, Rheinhessischen Druckwerkstätte, Alzey, Germany, 1971.
15. W. E. Kloos and K. H. Schleifer, "*Staphylococcus*," in P. H. A. Sneath et al., eds., *Bergey's Manual of Systematic Bacteriology*, Vol. 2, Williams & Wilkins, Baltimore, Md., 1986, pp. 1013–1035.
16. F.-K. Lücke, "Fermented Sausages," in B. J. B. Wood, ed., *Microbiology of Fermented Foods*, Vol. 2, 2nd ed., Elsevier Applied Science, London, 1997, pp. 441–483.
17. A. Verplaetse, "Influence of Raw Meat Properties and Processing Technology on Aroma Quality of Raw Fermented Meat Products," *Proc. 40th Int. Congr. Meat Sci. Technol.*, The Hague, The Netherlands, August 28–September 2, 1994.
18. R. Dainty and H. Blom, "Flavour Chemistry of Fermented Sausages," in G. Campbell-Platt and P. E. Cook, eds., *Fermented Meats*, Chapman & Hall, London, 1995, pp. 176–193.

19. M. C. Montel, F. Masson, and R. Talon, "Bacterial Role in Flavour Development," *Proc. 44th Int. Congr. Meat Sci. Technol.*, Barcelona, Spain, August 30–September 4, 1998.
20. M. C. Montel et al., "Peptidasic Activities of Starter Cultures," *Proc. 38th Int. Congr. Meat Sci. Technol.*, Clermont-Ferrand, France, August 23–28, 1992.
21. L. Stahnke, "Dried Sausages Fermented with *Staphylococcus xylosus* at Different Temperatures and with Different Ingredient Levels: Part III, Sensory Evaluation," *Meat Science* **41**, 211–223 (1995).
22. J. L. Berdaguè et al., "Effects of Starter Cultures on the Formation of Flavour Compounds in Dry Sausage," *Meat Science* **35**, 275–287 (1993).
23. B. B. Sørensen and H. Samuelsen, "The Combined Effects of Environmental Conditions on Lipolysis of Pork Fat by Lipases of the Meat Starter Culture Organisms *Staphylococcus xylosus* and *Debaryomyces hansenii*," *International Journal of Food Microbiology* **32**, 59–71 (1996).
24. R. G. Berger et al., "Isolation and Identification of Dry Salami Volatiles," *J. Food Sci.* **55**, 1239–1242 (1990).
25. L. Stahnke, "Aroma Components from Dried Sausages Fermented with *Staphylococcus xylosus*," *Meat Science* **38**, 39–53 (1994).
26. W. P. Hammes, A. Bantleon, and S. Min, "Lactic Acid Bacteria in Meat Fermentation," *FEMS Microbiology Rev.* **87**, 165–174 (1990).
27. E. Stackebrandt et al., "Taxonomic Dissection of the Genus *Micrococcus: Kocuria* gen. nov., *Nesterenkonia* gen. nov., *Kytococcus* gen. nov., *Dermacoccus* gen. nov., and *Micrococcus* Cohn 1872 gen. emend.," *Int. J. Systematic Bacteriol.* **45**, 682–692 (1995).
28. C. Meisel, "Mikrobiologische Aspekte der Entwicklung von nitratreduzierende Starter-Kulturen für die Herstellung von Rohwurst und Rohschinken," Ph.D. Dissertation, Hohenheim University, Stuttgart, Germany, 1988.
29. S. Hartmann, G. Wolf, and W. P. Hammes, "Reduction of Nitrite by *Staphylococcus carnosus* and *Staphylococcus piscifermentans*," *System. Appl. Microbiol.* **18**, 323–328 (1995).
30. H. Neubauer and F. Götz, "Physiology and Interaction of Nitrate and Nitrite Reduction in *Staphylococcus carnosus*," *J. Bacteriol.* **178**, 2005–2009 (1996).
31. L. Leistner, "Toxinogenic Penicillia Occurring in Feeds and Foods: A Review," *Food Technol. Aust.* **36**, 404–406, 413 (1984).
32. F.-K. Lücke, "Microbiological Processes in the Manufacture of Dry Sausages and Raw Ham," *Fleischwirtsch.* **66**, 1505–1509 (1986).
33. B. Jessen, "Starter Cultures for Meat Fermentations," in G. Campbell-Platt and P. E. Cook, eds., *Fermented Meats*, Chapman & Hall, London, 1995, pp. 130–159.
34. F.-K. Lücke and H. Hechelmann, "Starter Cultures for Dry Sausages and Raw Ham: Composition and Effect," *Fleischwirtsch.* **67**, 307–314 (1987).
35. J. Fink-Gremmels, A. A. El-Banna, and L. Leistner, "Developing Mould Starter Cultures for Meat Products," *Fleischwirtsch.* **68**, 1292–1294 (1988).
36. L. Leistner and Z. Bem, "Vorkommen und Bedeutung von Hefen bei Pökelfleischwaren," *Fleischwirtsch.* **50**, 350–351 (1970).
37. D. C. R. Riordan et al., "Survival of *Escherichia coli* O157:H7 during the Manufacture of Pepperoni," *J. Food Prot.* **61**, 146–151 (1998).
38. C. A. Reed, "Approaches for Ensuring Safety of Dry and Semi-Dry Fermented Sausage Products," August 21, 1995, letter to plant managers, U.S. Department of Agriculture, Food Safety and Inspection Service, Washington, D.C., 1995.
39. J. C. Hinkens et al., "Validation of Pepperoni Processes for Control of *Escherichia coli* O157:H7," *J. Food Prot.* **59**, 1260–1266 (1996).
40. M. Calicioglu et al., "Viability of *Escherichia coli* O157:H7 in Fermented Semidry Low-Temperature-Cooked Beef Summer Sausage," *J. Food Prot.* **60**, 1158–1162 (1997).
41. H. Nissen and A. Holck, "Survival of *Escherichia coli* O157:H7, *Listeria monocytogenes* and *Salmonella kentucky* in Norwegian Fermented, Dry Sausage," *Food Microbiology* **15**, 273–279 (1998).
42. C. J. Sauer et al., "Foodborne Illness Outbreak Associated with a Semi-Dry Fermented Sausage Product," *J. Food Prot.* **60**, 1612–1617 (1997).
43. A. M. Ihnot et al., "Behavior of *Salmonella typhimurium* DT104 during the Manufacture and Storage of Pepperoni," *International Journal of Food Microbiology* **40**, 117–121 (1998).
44. J. Metaxopoulos et al., "Production of Italian Dry Salami: Effect of Starter Culture and Chemical Acidulation on Staphylococcal Growth in Salami under Commercial Manufacturing Conditions," *Appl. Environ. Microbiol.* **42**, 863–871 (1981).
45. S. A. A. Jassim and S. P. Denyer, "Coagulase-Negative Staphylococci: Useful Organism or Potential Problem for Food Processing?" *Food Quality* **3**, 31–35 (1997).
46. J. Junttila et al., "Effect of Different Levels of Nitrite and Nitrate on the Survival of *Listeria monocytogenes* during the Manufacture of Fermented Sausage," *J. Food Prot.* **52**, 158–161 (1989).
47. P. M. Foegeding et al., "Enhanced Control of *Listeria monocytogenes* by In Situ–Produced Pediocin during Dry Fermented Sausage Production," *Appl. Environ. Microbiol.* **58**, 884–890 (1992).
48. J. L. Johnson et al., "Fate of *Listeria monocytogenes* in Tissues of Experimentally Infected Cattle and in Hard Salami," *Appl. Environ. Microbiol.* **54**, 497–501 (1988).
49. B. Jessen, L. Andersen, and B. Jelle, "Bioprotection of Meat and Dairy Products" (in Danish), *Alimenta* **19**, 9–11 (1996).
50. K. Asplund et al., "Survival of *Yersinia enterocolitica* in Fermented Sausages Manufactured with Different Levels of Nitrite and Different Starter Cultures," *J. Food Prot.* **56**, 710–712 (1993).
51. R. Maijala, "Formation of Biogenic Amines in Dry Sausages with Special Reference to Raw Materials, Lactic Acid Bacteria, pH Decrease, Temperature, and Time," Ph.D. dissertation, National Veterinary and Food Research Institute, Helsinki, Finland 1994.
52. R. Maijala and S. Eerola, "Contaminant Lactic Acid Bacteria of Dry Sausages Produce Histamine and Tyramine," *Meat Science* **35**, 387–395 (1993).
53. W. P. Hammes and H. Knauf, "Starters in the Processing of Meat Products," *Meat Science* **36**, 155–168 (1994).
54. H. Schmidt et al., "Einfluss von Bakterien sowie Natriumascorbat und Glucono-delta-Lacton auf den Ab-, Um-, und Aufbau von Nitrosaminen," *Fleischwirtsch.* **65**, 1487–1489 (1985).
55. M.-L. Liao and P. A. Seib, "Selected Reactions of L-Ascorbic Acid Related to Foods," *Food Technol.* **41**, 104–107, 111 (1987).
56. F. Toldra, "The Enzymology of Dry-Curing of Meat Products," in F. J. M. Smulders et al., eds., *New Technologies for Meat and Meat Products*, ECCEAMST, Audet Tijdschriften bv, Nijmegen, The Netherlands 1992, pp. 209–231.
57. H. Silla et al., "A Study of the Microbial Flora of Dry-Cured Ham: 1, Isolation and Growth," *Fleischwirtsch.* **69**, 1128–1131 (1989).

58. L. Hinrichsen and S. B. Pedersen, "Relationship among Flavour, Volatile Compounds, Chemical Changes, and Microflora in Italian-Type Dry-Cured Ham during Processing," *J. Agric. Food Chem.* **43**, 2932–2940 (1995).

59. I. Molina et al., "Study of the Microbial Flora in Dry-Cured Ham: 2, *Micrococcaceae*," *Fleischwirtsch.* **69**, 1433–1434 (1989).

60. L. Leistner et al., *Anforderungen an Starterkulturen: Abschlussbericht zu einem Forschungsvorhaben*, Bundesanstalt Fleischforschung, Kulmbach, Germany, 1979.

61. L. Andersen, "Biopreservation with FloraCarn L-2," *Fleischwirtsch.* **75**, 1327–1329 (1995).

62. W. P. Hammes and C. Hertel, "New Developments in Meat Starter Cultures," *Proc. 44th Int. Congr. Meat Sci. Technol.*, Barcelona, Spain, August 30–September 4, 1998.

B. JESSEN
Danish Meat Research Institute
Roskilde, Denmark

MEMBRANE FILTRATION SYSTEMS

Membrane filtration has long been the method of choice for quantitative tests in water microbiology (1,2). Principally, this has been due to the ability to concentrate and detect low levels of indicator microorganisms by filtering large volumes of water. For this same reason, membrane filtration has been used to enumerate indicator organisms in beverages, including beer and wine. However, application of this tool in the beverage industry has been limited to those beverages that are low in particulate matter and that can easily pass through a membrane filter.

These same restrictions deterred food microbiologists from using membrane filtration as an analytical tool for many years. Early attempts to render food samples filterable met with limited success (3,4). In addition, the relatively narrow counting range of the standard 47-mm-diameter membrane filter (5) required a large number of sample dilutions to be filtered to ensure at least one countable filter. Despite these limitations, three very different approaches—membrane spread, direct epifluorescence, and hydrophobic grid membrane filter—have been used to adapt membrane filter techniques to food microbiology.

MEMBRANE SPREAD TECHNIQUE

A "membrane-spread" technique for enumerating *Escherichia coli*, first described in 1975, used a 90-mm-diameter membrane filter, laid on the surface of an agar plate (6). The sample aliquot was deposited on the membrane filter and spread over the entire surface with a bent glass rod. This 24-hour method, which used bile salts and a 44.5°C incubation temperature for selectivity and an indole reaction as the differential test, was highly specific for *E. coli* biotype 1. This method, later modified to include a nonselective repair step (7), is recognized as a valid method by the International Commission on Microbiological Specifications for Foods (8).

DIRECT EPIFLUORESCENT TECHNIQUE

A second membrane filter method, known as direct epifluorescent microscopy technique (DEFT), made use of fluorescent staining of cells trapped on the surface of a membrane filter followed by examination of the filter using a fluorescent microscope (9). This procedure overcame the limited counting range of standard membrane filtration by eliminating colony development. However, differentiation of viable and dead cells was not fully reliable, and reading the filters was tedious. Introduction of automated DEFT instruments addressed the latter problem (10,11). However, according to the developer of the technique, DEFT could not be applied to highly particulate food suspensions, nor could it be adapted to differential or selective enumerations (12).

Several researchers have improved on the original DEFT procedures by using prefiltration and enzyme digestion steps to reduce particle interference (13,14); adding short enrichment steps to enhance sensitivity (15); incubating the filters for a few hours after filtration to allow development of microcolonies, thus ensuring a true viable cell count and allowing for differential enumeration based on the culture media used (16,17); and using fluorescent antibody labeling in conjunction with DEFT to detect or enumerate specific target populations (18,19).

DEFT methods that do not incorporate enrichment or microcolony development steps offer the advantage of very fast results, with as little as 20 min required between analysis initiation and completion (14). Reliability of DEFT procedures has been evaluated for a variety of foods including raw minced meat (20), raw ham, ground beef and raw fish (21), raw lamb (22), and raw milk (14).

HYDROPHOBIC GRID MEMBRANE FILTRATION

The most widely adapted approach to membrane filtration of foods is the hydrophobic grid membrane filter (HGMF). The HGMF is a membrane filter on the surface of which a hydrophobic material is applied in a grid pattern. The hydrophobic material used to form the grid is a physical barrier to colony spread, largely restricting bacterial colonies to the confines of the individual grid squares in which they were first formed (see Fig. 1). Using this technique, the counting range of a single filter was extended to as high as 10^4 colony-forming units (23,24).

Initially, the HGMF was seen as a water microbiology and research tool. Beginning in 1978, its potential applicability to food microbiology was established when a series of studies on the filtering characteristics of a wide variety of food homogenates was published. It was reported that most food homogenates, if blended using a "Stomacher" and, where appropriate, if treated with a surfactant and/or an enzyme digestion, could be filtered (25). In a later study (26), a finely woven stainless steel cloth prefilter was used to remove particulate material from food homogenates immediately prior to filtration. The apparatus used to carry out this two-stage filtration is illustrated in Figure 2. This study established that neither prefiltration, nor surfactant treatment, nor enzyme digestions altered the bacterial counts obtained from food samples.

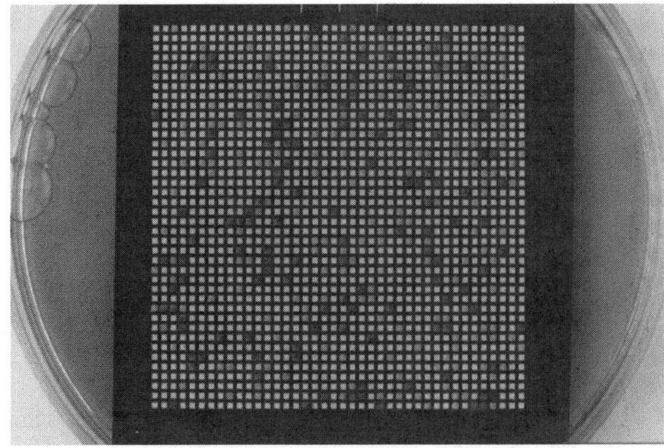

Figure 1. Coliform colonies growing on a hydrophobic grid membrane filter. The hydrophobic grid lines serve to confine colonies within the boundaries of the individual squares.

Figure 2. Filtration apparatus incorporating 5μ stainless steel prefilter used to inoculate hydrophobic grid membrane filters with food sample homogenates.

Since then numerous specific applications of the hydrophobic grid membrane filter to food microbiology have been introduced. These applications are listed in Table 1, and several of them are discussed in the following sections.

Table 1. Applications of Hydrophobic Grid Membrane Filter to Food Microbiology

Analysis	References
Aerobic plate count	27,41–43
Coliforms/*Escherichia coli*	28–31,54
E. coli O157:H7	32–36
Yeast and mold count	44–46,54
Staphylococcus aureus	54
Lactic acid bacteria	47,48,56
Fluorescent pseudomonads	57
Salmonella	37–40
Vibrio parahaemolyticus	49,50
Listeria monocytogenes	53

Coliforms, *Escherichia coli*, and *E. coli* O157:H7

The first HGMF method for total and fecal coliforms was an adaptation of the water microbiology fecal coliform test (27). Both counts were carried out using mFC Agar without rosolic acid. The total coliform test was incubated at 35°C; the fecal coliform test consisted of a preliminary resuscitation step followed by incubation of mFC Agar at 44.5°C. The companion test for *E. coli* was a direct adaptation of the Anderson and Baird-Parker membrane-spread method previously mentioned. These methods were subjected to validation studies in 1983 and were recognized as Official Methods by the Association of Official Analytical Chemists (now AOAC International) (28).

Developments in *E. coli* differential test procedures, notably the introduction of 4-methylumbelliferyl-β-D-glucuronide (MUG) as a differential reagent, prompted a redesign of the coliform and *E. coli* methods. In the newer method, a single HGMF was used for both the coliform and *E. coli* enumerations, eliminating the need for a 4-h resuscitation step. The entire analysis is complete in 24 h. This improved coliform/*E. coli* procedure was subjected to extensive in-house validation, followed by an AOAC-sponsored collaborative study (29,30). These studies demonstrated that the redesigned HGMF method produced quantitative total coliform and *E. coli* results that were not significantly different from the three-tube MPN method results over a wide range of food products. Furthermore, the confirmation rate of *E. coli* colonies on the HGMF method was in excess of 98% in the precollaborative study and more than 99% in the collaborative study. Based on combined results of both studies, the AOAC accorded "Official Action" status to the new procedure in 1990 (31).

The presence of *E. coli* O157:H7 in the food supply has become increasingly of concern. This *E. coli* serotype produces a negative MUG reaction and, therefore, is not detected by any *E. coli* enumeration procedures based on β-glucuronidase activity. It will also not grow in some culture media at elevated temperatures, making it more difficult to detect using the conventional three-tube MPN *E. coli* method. Several researchers have developed HGMF methods for *E. coli* O157:H7. The earliest of these described the development of HC Agar, a selective and differential culture medium for presumptive enumeration of *E. coli* O157:H7 (32). HC Agar relied on three differential reactions—MUG, indole, and sorbitol—to differentiate *E. coli* O157:H7 from other *Enterobacteriaceae*, including other *E. coli*. Two other methods employed enzyme-labeled antibody procedures to detect or enumerate *E. coli* O157:H7 (33,34).

Recently, SD-39 Agar was developed for use in conjunction with the HGMF to detect and enumerate presumptive *E. coli* O157:H7. The medium is based on three biochemical reactions that are read simultaneously: lysine decarboxylase, sorbitol fermentation, and β-glucuronidase production. Selectivity is provided by the presence of monensin and novobiocin and by incubation at 44 to 44.5°C. *E. coli* O157:H7 tolerates the elevated temperature because of the presence of NaCl in the culture medium. Confirmation is accomplished by subculturing presumptive positive colonies and carrying out O157 and H7 serological tests. This method was validated in a comprehensive precollaborative study and then in an AOAC-sponsored collaborative study (35,36). Results obtained by the HGMF method were either significantly higher than the reference method or not significantly different from the reference method, depending on the food product. The serological confirmation rates of presumptive positive *E. coli* O157:H7 were 99% in the precollaborative study and 94% in the collaborative study. Based on the combined results of both studies, the method was accorded First Action in 1997 (36).

Salmonella

In 1982 a rapid *Salmonella* detection procedure based on the HGMF was described (37). The method was subjected to collaborative study in 1984 and was granted official recognition by AOAC in that same year (38). Over time, it became evident that the plating media used in the original HGMF method—namely, Hektoen Enteric Agar and Selective Lysine Agar (SLA)—were not adequately specific to *Salmonella*. As a result, SLA was redesigned to increase its selective and differential properties. The improved medium, EF-18 Agar, relied on a dual biochemical differential reaction: lysine decarboxylation and sucrose fermentation. Alkaline colonies (ie, lysine positive and sucrose negative) were considered to be presumptive positive *Salmonella* and were subcultured for confirmation. Acid colonies (ie, lysine negative or sucrose positive organisms) were discarded.

EF-18 Agar was incorporated into the HGMF *Salmonella* method together with some relatively minor procedural changes. The improved method was subjected to extensive in-house validation against the conventional AOAC *Salmonella* method in a study comprising nearly 1000 naturally contaminated and inoculated food and feed samples (39). Overall, the HGMF method produced a false negative rate of 2.0% versus 1.9% for the reference method. The presumptive false-positive rate for the HGMF method was 0.3%, as compared with 8% for the reference method.

Following this successful validation, an AOAC-sponsored collaborative study of the method was organized. Thirty laboratories based in the United States and Canada took part, including both government and industry labs. The AOAC determined that the HGMF *Salmonella* method performance was statistically equivalent to that of the conventional AOAC *Salmonella* method, and awarded approval to the HGMF procedure (40).

Aerobic Plate Count

The HGMF aerobic plate count procedure has been an official method of the AOAC since 1985 (41) and is also a standard method of the American Public Health Association (42). The method consists of filtering a portion of sample homogenate through the HGMF and incubating on Tryptone Soy Agar supplemented with fast green dye at 35 to 37°C for 48 h. The method was subjected to extensive in-house validation (43), followed by an interlaboratory collaborative study (41). Results obtained by the HGMF method were at least statistically equivalent to the conventional pour plate method counts in both the in-house and collaborative validation studies while the colony-retaining properties of the hydrophobic grid allow the analyst to forgo duplicate plating and testing several serial dilutions without sacrificing accuracy or reproducibility.

Yeast and Mold Count

Historically, yeast and mold enumeration, with its requirement for a five-day incubation period, has been one of the slowest of the routine microbiology analyses to perform. Several attempts have been made over the years to shorten the incubation period by taking advantage of the ability of the HGMF to separate microorganisms from the food matrix and to confine the surface spread of colonies.

The first effort, reported in 1982, consisted of incubating the HGMF on Potato Dextrose Agar and staining the filter with safranin solution at the end of the incubation period to provide contrast between the colonies and the surface of the membrane filter (27). The method was refined by Lin et al. (44), who incorporated trypan blue dye into the Potato Dextrose Agar. The trypan blue was taken up by both yeasts and molds, producing blue colonies against the white background of the membrane filter. After determining that some molds would not develop visible colonies reliably within 48 h on Potato Dextrose Trypan Blue Agar, Entis and Lerner developed a new culture medium, YM-11 Agar, to optimize recovery of both yeasts and molds on the HGMF (45). YM-11 Agar enabled the enumeration of even slower growing yeasts and molds within a 50 ± 2 h incubation period at 25 ± 1°C. The performance of YM-11 Agar was evaluated in a precollaborative study and then further validated in an AOAC-sponsored collaborative study. The two-day HGMF method using YM-11 Agar performed equivalently to the five-day Potato Dextrose Agar pour plate method in both studies. As a result, AOAC recognized this HGMF method in 1995 (46).

Lactic Acid Bacteria

In 1988 a rapid method for differential enumeration of *Lactobacillus* and *Pediococcus* from meat starter cultures was described (47). An improved selective-differential culture medium was used together with the HGMF to achieve a differential count even when one genus was present in fairly large excess relative to the other. This method enabled confirmed enumeration of both genera in 48 h, as compared with a minimum of 72 h with the previous best method.

More recently, the development of an analogous procedure for enumerating *Lactobacillus bulgaricus* and *Streptococcus thermophilus* in yogurt was reported (48). Once again, the colony-separating properties of the grid, together with another newly designed culture medium, enabled direct differential enumeration of the two different bacterial species on a single filter even at cocci-to-rod ratios from 20:1 to 1:5.

Vibrio parahaemolyticus

An overnight HGMF enumeration method for *Vibrio parahaemolyticus* was first reported in 1983 (49) and later refined (50). The improved method consisted of first incubating the HGMF on a sucrose-containing selective differential agar for 16 to 18 h at 42°C. Sucrose-negative colonies (presumptive *V. parahaemolyticus*) were confirmed by removing the HGMF from the presumptive medium and placing it onto a secondary differential medium. Those squares producing typical reactions on both the primary and secondary media were counted as *V. parahaemolyticus*.

Enzyme Labeled Antibody and Colony Hybridization Methods

A number of researchers have taken advantage of the orderly positioning of colonies on the HGMF to develop either research or analytical tools based on either enzyme labeled antibody (ELA) or colony hybridization. ELA-based methods developed by Todd et al. (34) and by Doyle and Schoeni (33) for detecting or enumerating *E. coli* O157:H7 have already been mentioned. Peterkin et al. (51) developed a process for screening large numbers of DNA probes using HGMF. In addition, Kaysner et al. (52) reported on a DNA-DNA colony hybridization method to enumerate and differentiate *V. parahaemolyticus* and *V. vulnificus*, and Peterkin et al. (53) used a chromogen-labeled DNA probe to detect and confirm *Listeria monocytogenes* colonies on HGMF using a colony hybridization process.

Several characteristics of the HGMF contribute to its excellent performance. Principal among these are the colony-retaining and separating properties of the hydrophobic grid (23,27). The filter also serves as a passive support, enabling bacterial colonies to be transported undisturbed from one culture medium to another (32,54). Filtration often improves sensitivity of detection both by concentrating low levels of target organisms and by separating the organisms from any inhibitory materials (eg, preservatives, spice components) that might be present in the sample homogenate (27,54). The orderly grid array facilitates colony counting, thus reducing a significant source of between-analyst variation (23,27,55).

Numerous researchers have taken advantage of one or more of these HGMF characteristics to develop completely new procedures or to improve existing ones. It is inevitable that more uses will continue to be found for this highly versatile microbiological tool.

BIBLIOGRAPHY

1. A. Goetz and N. Tsuneishi, "Application of Molecular Filter Membranes to the Bacteriological Analysis of Water," *Journal of the American Water Works Association* **43**, 943–969 (1951).
2. H. F. Clark et al., "The Membrane Filter in Sanitary Bacteriology," *Public Health Reports* **66**, 951–977 (1951).
3. L. A. Nutting, P. C. Lomot, and F. W. Barber, "Estimation of Coliform Bacteria in Ice Cream by Use of the Membrane Filter," *Appl. Microbiol.* **7**, 196–199 (1959).
4. W. K. Kirkham and P. A. Hartman, "Membrane Filter Methods for the Detection and Enumeration of *Salmonella* in Egg Albumen," *Poultry Science* **41**, 1082–1088 (1962).
5. American Public Health Association, *Standard Methods for the Examination of Water and Wastewater*, 19th ed., American Public Health Association, Washington, D.C., 1995.
6. J. M. Anderson and A. C. Baird-Parker, "A Rapid and Direct Plate Method for Enumerating *Escherichia coli* Biotype I in Food," *J. Appl. Bacteriol.* **39**, 111–117 (1975).
7. R. Holbrook, J. M. Anderson, and A. C. Baird-Parker, "Modified Direct Plate Method for Counting *Escherichia coli* in Foods," *Food Technology in Australia* **32**, 78–83 (1980).
8. M. K. Rayman et al., "ICMSF Methods Studies. XIII. An International Comparative Study of the MPN procedure and the Anderson-Baird-Parker Direct Plating Method for the Enumeration of *Escherichia coli* Biotype I in Raw Meats," *Can. J. Microbiol.* **25**, 1321–1327 (1979).
9. G. L. Pettipher et al., "Rapid Membrane Filtration-Epifluorescent Microscopy Technique for Direct Enumeration of Bacteria in Raw Milk," *Appl. Environ. Microbiol.* **39**, 423–429 (1980).
10. G. L. Pettipher and U. M. Rodriques, "Semi-Automated Counting of Bacteria and Somatic Cells in Milk Using Epifluorescent Microscopy and Television Image Analysis," *J. Appl. Bacteriol.* **53**, 323–329 (1980).
11. G. L. Pettipher et al., "Preliminary Evaluation of COBRA, an Automated DEFT Instrument, for the Rapid Enumeration of Micro-organisms in Cultures, Raw Milk, Meat and Fish," *Lett. Appl. Microbiol.* **14**, 206–209 (1992).
12. G. L. Pettipher, R. J. Fulford, and L. A. Mabbitt, "Collaborative Trial of the Direct Epifluorescent Filter Technique (DEFT), a Rapid Method for Counting Bacteria in Milk," *J. Appl. Bacteriol.* **54**, 177–182 (1983).
13. I. Walls et al., "Separation of Micro-organisms From Meat and Their Rapid Enumeration Using a Membrane Filtration-Epifluorescent Microscopy Technique," *Lett. Appl. Microbiol.* **10**, 23–26 (1990).
14. C. P. Champagne et al., "Determination of Viable Bacterial Populations in Raw Milk Within 20 Minutes by Using a Direct Epifluorescent Filter Technique," *J. Food Prot.* **60**, 874–876 (1997).
15. M. T. Rowe, and G. J. McCann, "A Modified Direct Epifluorescent Filter Technique for the Detection and Enumeration of Yeast in Yoghurt," *Lett. Appl. Microbiol.* **11**, 282–285 (1990).
16. U. M. Rodrigues, and R. G. Kroll, "Rapid Selective Enumeration of Bacteria In Foods Using a Microcolony Epifluorescence Microscopy Technique," *J. Appl. Bacteriol.* **64**, 65–78 (1988).
17. U. M. Rodrigues and R. G. Kroll, "Microcolony Epifluorescence Microscopy for Selective Enumeration of Injured Bacteria in Frozen and Heat-Treated Foods," *Appl. Environ. Microbiol.* **55**, 778–787 (1989).
18. M. L. Tortorello and S. M. Gendel, "Fluorescent Antibodies Applied to Direct Epifluorescent Filter Technique for Microscopic Enumeration of *Escherichia coli* O157:H7 in Milk and Juice," *J. Food Prot.* **56**, 672–677 (1993).
19. L. Restaino et al., "Antibody-Direct Epifluorescent Filter Technique and Immunomagnetic Separation for 10-h Screening and 24-h Confirmation of *Escherichia coli* O157:H7 in Beef," *J. Food Prot.* **59**, 1072–1075 (1996).
20. F. Boisen et al., "Quantitation of Microorganisms in Raw Minced Meat Using the Direct Epifluorescent Filter Technique: NMKL Collaborative Study," *J. AOAC Int.* **75**, 465–473 (1992).
21. B. Abgrall and C. M. Bourgeois, "Dénombrement de la Flore Totale de Produits Alimentaires par la Technique DEFT [Counting of Total Microflora of Food Products by Direct Epifluorescent Filter Technique (DEFT)]," *Sciences des Aliments* **9**, 713–724 (1989).

22. M.-L. Sierra, J. J. Sheridan, and L. McGuire, "Microbial Quality of Lamb Carcasses During Processing and the Acridine Orange Direct Count Technique (a Modified DEFT) for Rapid Enumeration of Total Viable Counts," *International Journal of Food Microbiology* **36**, 61–67 (1997).
23. A. N. Sharpe and G. L. Michaud, "Hydrophobic Grid-Membrane Filters: New Approach to Microbiological Enumeration," *Appl. Microbiol.* **28**, 223–225 (1974).
24. A. N. Sharpe and G. L. Michaud, "Enumeration of High Numbers of Bacteria Using Hydrophobic Grid-Membrane Filters," *Appl. Microbiol.* **30**, 519–524 (1975).
25. A. N. Sharpe, P. I. Peterkin, and I. Dudas, "Membrane Filtration of Food Suspensions," *Appl. Environ. Microbiol.* **37**, 21–35 (1978).
26. P. Entis, M. H. Brodsky, and A. N. Sharpe, "Effect of Prefiltration and Enzyme Treatment on Membrane Filtration of Foods," *J. Food Prot.* **45**, 8–11 (1982).
27. M. H. Brodsky et al., "Enumeration of Indicator Organisms in Foods Using the Automated Hydrophobic Grid Membrane Filter Technique," *J. Food Prot.* **45**, 292–296 (1982).
28. P. Entis, "Enumeration of Total Coliforms, Fecal Coliforms and *Escherichia coli* in Foods by Hydrophobic Grid Membrane Filter: Collaborative Study," *J.—Assoc. Off. Anal. Chem.* **67**, 812–823 (1984).
29. P. Entis and P. Boleszczuk, "Direct Enumeration of Coliforms and *Escherichia coli* by Hydrophobic Grid Membrane Filter in 24 Hours Using MUG," *J. Food Protect.* **53**, 948–952 (1990).
30. P. Entis, "Hydrophobic Grid Membrane Filter/MUG Method for Total Coliform and *Escherichia coli* Enumeration in Foods: Collaborative Study," *J.—Assoc. Off. Anal. Chem.* **72**, 936–950 (1989).
31. "AOAC Official Method 990.11. Total Coliform and *Escherichia coli* Counts in Foods. Hydrophobic Grid Membrane Filter/MUG (ISO-GRID®) Method. First Action 1990. Final Action 1993," in *Official Methods of Analysis*, 16th ed., Vol. 1, AOAC International, Gaithersburg, Md., 1997.
32. R. A. Szabo, E. C. D. Todd, and A. Jean, "Method to Isolate *Escherichia coli* O157:H7 From Food," *J. Food Protect.* **49**, 768–772 (1986).
33. M. P. Doyle and J. L. Schoeni, "Isolation of *Escherichia coli* O157:H7 From Retail Fresh Meats and Poultry," *Appl. Environ. Microbiol.* **53**, 2394–2396 (1987).
34. E. C. D. Todd et al., "Rapid Hydrophobic Grid Membrane Filter-Enzyme-Labelled Antibody Procedure for Identification and Enumeration of *Escherichia coli* O157 in Foods," *Appl. Environ. Microbiol.* **54**, 2536–2540 (1988).
35. P. Entis and I. Lerner, "24-Hour Presumptive Enumeration of *Escherichia coli* O157:H7 in Foods Using the ISO-GRID® Method with SD-39 Agar," *J. Food Protect.* **60**, 883–890 (1997).
36. P. Entis, "Direct 24-Hour Presumptive Enumeration of *Escherichia coli* O157:H7 in Foods Using Hydrophobic Grid Membrane Filter, Followed by Serological Confirmation: Collaborative Study," *J. AOAC Int.* **81**, 403–418 (1998).
37. P. Entis et al., "Rapid Detection of *Salmonella* spp. in Food by Use of the ISO-GRID® Hydrophobic Grid Membrane Filter," *Appl. Environ. Microbiol.* **43**, 261–268 (1982).
38. P. Entis, "Rapid Hydrophobic Grid Membrane Filter Method for *Salmonella* Detection in Selected Foods: Collaborative Study," *J.—Assoc. Off. Anal. Chem.* **68**, 555–564 (1985).
39. P. Entis and P. Boleszczuk, "Rapid Detection of *Salmonella* in Food Using EF-18 Agar in Conjunction With the Hydrophobic Grid Membrane Filter," *J. Food Protect.* **54**, 930–934 (1991).
40. P. Entis, "Improved Hydrophobic Grid Membrane Filter Method, Using EF-18 Agar, for Detection of *Salmonella* in Foods: Collaborative Study," *J.—Assoc. Off. Anal. Chem.* **73**, 734–742 (1990).
41. P. Entis, "Hydrophobic Grid Membrane Filter Method for Aerobic Plate Count in Foods: Collaborative Study," *J.—Assoc. Off. Anal. Chem.* **69**, 671–676 (1986).
42. American Public Health Association, *Standard Methods for the Examination of Dairy Products*, 16th ed., American Public Health Association, Washington, D.C., 1992, pp. 237–239.
43. P. Entis and P. Boleszczuk, "Use of Fast Green FCF With Tryptic Soy Agar for Aerobic Plate Count by the Hydrophobic Grid Membrane Filter," *J. Food Protect.* **49**, 278–279 (1986).
44. C. C. S. Lin, D. Y. C. Fung, and P. Entis, "Growth of Yeast and Mold on Trypan Blue Agar in Conjunction With the ISO-GRID® System," *Can. J. Microbiol.* **30**, 1405–1407 (1984).
45. P. Entis and I. Lerner, "Two-Day Yeast and Mold Enumeration Using the ISO-GRID® Membrane Filtration System in Conjunction With YM-11 Agar," *J. Food Protect.* **59**, 416–419 (1996).
46. P. Entis, "Two-Day Hydrophobic Grid Membrane Filter Method for Yeast and Mold Enumeration in Foods Using YM-11 Agar: Collaborative Study," *J. AOAC Int.* **79**, 1069–1082 (1996).
47. R. A. Holley and G. E. Millard, "Use of MRSD Medium and the Hydrophobic Grid Membrane Filter Technique to Differentiate Between Pediococci and Lactobacilli in Fermented Meat and Starter Cultures," *International Journal of Food Microbiology* **7**, 87–102 (1988).
48. G. E. Millard, R. C. McKellar, and R. A. Holley, "Simultaneous Enumeration of the Characteristic Microorganisms in Yogurt Using the Hydrophobic Grid Membrane Filter System," *J. Food Protect.* **53**, 64–66 (1990).
49. P. Entis and P. Boleszczuk, "Overnight Enumeration of *Vibrio parahaemolyticus* in Seafood by Hydrophobic Grid Membrane Filtration," *J. Food Protect.* **46**, 783–786 (1983).
50. A. DePaola, L. H. Hopkins, and R. M. McPhearson, "Evaluation of Four Methods for Enumeration of *Vibrio parahaemolyticus*," *Appl. Environ. Microbiol.* **54**, 617–618 (1988).
51. P. I. Peterkin, E. S. Idziak, and A. N. Sharpe, "Screening DNA Probes Using the Hydrophobic Grid-Membrane Filter," *Food Microbiology* **6**, 281–284 (1989).
52. C. A. Kaysner et al., "A Research Note: Enumeration and Differentiation of *Vibrio parahaemolyticus* and *Vibrio vulnificus* by DNA-DNA Colony Hybridization Using the Hydrophobic Grid Membrane Filtration Technique for Isolation," *J. Food Protect.* **57**, 163–165 (1994).
53. P. I. Peterkin, E. S. Idziak, and A. N. Sharpe, "Detection of *Listeria monocytogenes* by Direct Colony Hybridization on Hydrophobic Grid-Membrane Filters by Using a Chromogen-Labeled DNA Probe," *Appl. Environ. Microbiol.* **57**, 586–591 (1991).
54. M. H. Brodsky et al., "Enumeration of Indicator Organisms in Foods Using the Automated Hydrophobic Grid-Membrane Filter Technique," *J. Food Protect.* **45**, 292–296 (1982).
55. A. N. Sharpe et al., "Colony Counting on Hydrophobic Grid-Membrane Filters," *Can. J. Microbiol.* **29**, 797–802 (1983).
56. J. L. Chu et al., "La Quantification des Populations de Bactéries Lactiques par la Methode HGMF," *Belgian Journal of Food Chemistry and Biotechnology* **42**, 65–74 (1987).
57. S. J. Knabel, H. W. Walker, and A. A. Kraft, "Enumeration of Fluorescent Pseudomonads on Poultry by Using the Hydrophobic Grid Membrane Filter," *J. Food Sci.* **52**, 837–841, 845 (1987).

GENERAL REFERENCES

B. J. Dutka, ed., *Membrane Filtration—Applications, Techniques, and Problems*, Marcel Dekker, New York, 1981.

P. Entis, "Membrane Filtration Systems" in M. D. Pierson and N. J. Stern, eds., *Foodborne Microorganisms and Their Toxins. Developing Methodology*, Marcel Dekker, New York, 1986.

A. N. Sharpe and P. I. Peterkin, *Membrane Filter Food Microbiology*, Research Studies Press Ltd., Taunton, Somerset, United Kingdom, 1988.

PHYLLIS ENTIS
QA Life Sciences
San Diego, California

MEMBRANE PROCESSING

Membrane technology was originally developed in 1960 for production of potable water from seawater and brackish water. The food industry has especially benefited from this technology because it is a gentle and efficient way of fractionating, concentrating, and clarifying components in liquid and gaseous streams. It is based on the use of semipermeable membranes (membranes that are permeable to some components but not to others) to separate molecules primarily on the basis of size and, to a certain extent, on shape and chemical composition. For example, as shown in Figure 1, reverse osmosis (RO) can be used to concentrate the solids in a liquid food, whereas nanofiltration (NF) membranes are designed to separate salts (primarily monovalent ions) from multivalent salts, sugars, and larger compounds. NF can also separate organic compounds by its degree of dissociation. Ultrafiltration (UF) can be used to fractionate components in the feed solution, whereas microfiltration (MF) is used to clarify slurries or remove suspended matter.

In addition, pervaporation (PV) separates liquid mixtures by partial vaporization through a membrane (Fig. 2). It is an enrichment technique similar to distillation, but it uses a membrane that is permeable either to water or selected organic compounds in vapor form. The driving force is maximized by application of low pressure to the permeate side of the membrane (eg, a vacuum), combined with immediate condensation of permeated vapors.

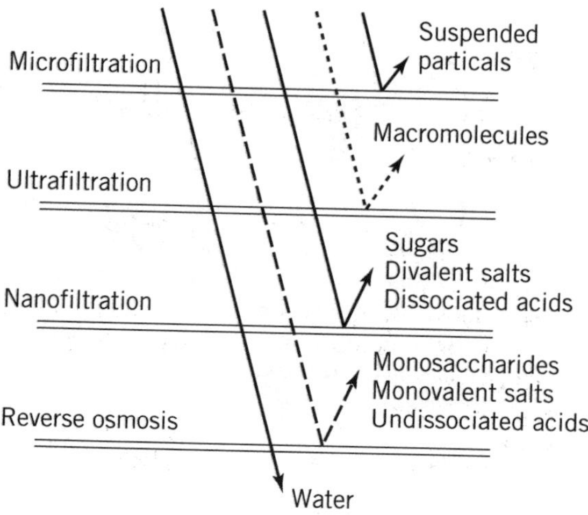

Figure 1. Definition of pressure-driven membrane processes.

Electrodialysis (ED) is the transport of ions through the membranes as a result of the application of direct electric current (Fig. 2). Only ionic species are transferred directly; thus ED can separate ionic species from nonionic components, so that both concentration and purification are possible. It is used mainly as a desalting technique.

Membrane technology requires less energy than many other dewatering techniques. Whereas open-pan evaporation may need over 600 kW h/ 1000 kg water removed and a five-effect evaporator requires 37–53 kW h/1000 kg, reverse osmosis for desalination requires 5–20 kW h/1000 kg water removed (1,2). In addition, no extremes of temperature are required as with evaporation and freeze concentration, thus preventing damage to heat-sensitive food components.

The technology is very simple. The membrane is assembled in a module, and the feed stream is pumped through the module over the membrane surface in a cross-flow mode (Fig. 3). The pressure gradient across the membrane forces solvent and solute molecules smaller than the pores on the membrane surface through the membrane into the permeate stream, while larger solutes are retained in the retentate stream.

IMPORTANT FACTORS

Several factors must be taken into account when considering a membrane process for a particular application: membrane material, membrane properties, module design, engineering factors, fouling and cleaning, and process design (1). Membranes have been made from more than 150 different polymers or inorganic materials, although fewer than 30 have achieved widespread commercial use. The majority of materials have been used for MF membranes and much fewer have worked successfully for RO, PV, and ED. Cellulose acetate and derivatives are still widely used for pressure-driven processes, despite their limitations of pH (generally usable only at pH 2–8) and temperature (not higher than 30°C). Thin-film composite membranes, containing a polyamide separating barrier on a polysulfone or polyethylene supporting layer, generally give better performance for RO and NF applications. Polysulfone, polyethersulfone, polyvinylidine fluoride, and polyacrylonitrile membranes are most common in UF and MF. MF membranes are also available in inorganic materials (alumina, zirconia/carbon composites, carbon/carbon composites, and stainless steel). Inorganic membranes have considerably widened the range of membrane applications, particularly in waste treatment, recovery and reuse of chemicals, and biotechnology applications where high temperature, acid and alkali stability, steam sterilizability, and cleanability are important.

Pore size is the most important property of a membrane. In reality, there is a distribution of pore sizes, which makes it difficult to get sharp or clean separations of molecules with membrane technology. Other factors affecting

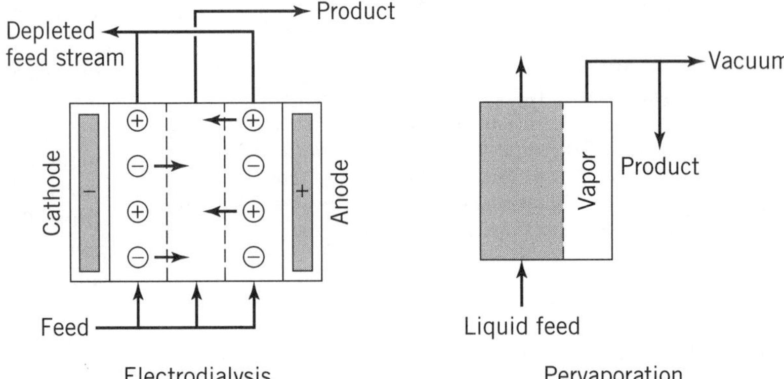

Figure 2. Schematic of electrodialysis and pervaporation.

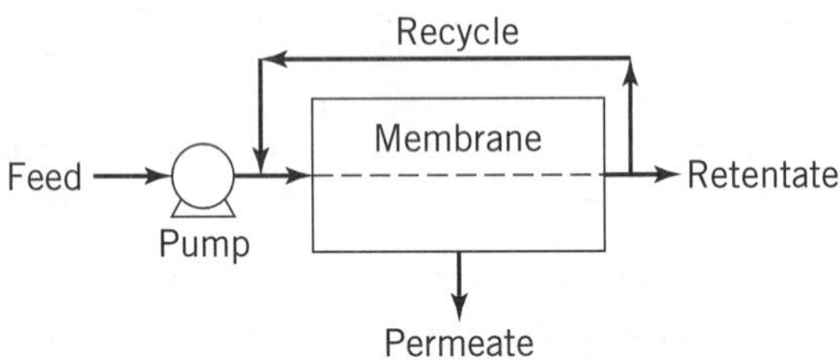

Figure 3. Typical flow diagram of a membrane process.

separation are the shape of the molecule; the presence of other solutes of similar size; operating parameters such as pressure, cross-flow rate, and temperature, and the microenvironment (1).

There are six different basic designs of membrane modules: tubular (with channel diameters >3 mm), hollow fiber or capillaries (self-supporting tubes, usually ≤2 mm i.d.), plates, spiral-wound, pleated sheets, and moving modules (1). The selection of a particular design depends on (1) the physical properties of the feed stream and retentate, especially viscosity and osmotic pressure; (2) particle size of suspended matter in the feed; (3) fouling potential of the feed stream, and (4) sanitary design, cleanability, and sterilizability.

FOOD APPLICATIONS

On a worldwide basis, the dairy industry is probably the largest user of membrane technology in food processing. RO is used as a preconcentration step prior to evaporation, NF is used for desalting cheese whey, ED is used for desalting milk and whey, and UF is used for purifying whey proteins from cheese whey to produce whey protein concentrates. UF is also used in the manufacture of cheese by preconcentrating the protein and fat to a level close to that normally found in the finished cheese. This precheese is then converted to cheese by conventional cheesemaking methods with appropriate modifications. The advantages of the UF method are that higher yields are obtained (10–30% in some cases) due to retention of whey proteins in the final cheese; the amount of enzyme required may be lower; and volume of milk to handle in the cheese vats is reduced and less whey is produced (although the milk permeate still needs to be disposed of). The UF method has been especially useful with soft cheeses.

MF is used for defatting and clarifying whey prior to ultrafiltration. MF is also used to reduce the microbial load of milk to extend the shelf life of pasteurized milk and the quality of cheeses.

The second-biggest use of membrane technology is for fruit juices, where UF and MF have been used in the manufacture of clear juices such as apple, pear, pineapple, kiwi, grape, lemon, lime, and cranberry. The goal is to replace the conventional holding, filtration, and decanting steps and perhaps the final pasteurization step. This has resulted in higher yields; improved juice quality due to removal of pectins, polyphenol oxidase, and tannin-protein complexes; elimination of filter-aid and fining agents; and much lower processing costs. Concentration of fruit juices by reverse osmosis to high levels (>42 Brix) using a combination of high-rejection and low-rejection RO membranes has been commercialized but is not widespread.

Other food applications include production of vegetable protein concentrates and isolates (eg, from soybeans), concentration of eggs prior to dehydration, processing of animal products such as blood and gelatin, refining of sugars and syrups, removal of yeast from beer and wine, and degumming vegetable oils. It is increasingly being used in waste treatment and recovery and reuse of chemicals such as acid and alkaline cleaners that are used for cleaning food equipment.

BIBLIOGRAPHY

1. M. Cheryan, *Ultrafiltration and Microfiltration Handbook*, Technomic, Lancaster, Pa., 1998.
2. M. Cheryan, "Concentration of Liquid Foods by Reverse Osmosis," in D. B. Lund and D. R. Heldman, eds. *Handbook of Food Engineering*, Marcel-Dekker, New York, 1992, pp. 393–436.

MUNIR CHERYAN
University of Illinois
Urbana, Illinois

MICROBIOLOGY OF FOODS

Food microbiology is the study of all aspects of microbial actions on food and food products, both directly and indirectly related to the welfare of mankind. Topics included in food microbiology are history of food microbiology, number and kinds of microbes in foods, intrinsic and extrinsic parameters of foods, methodologies, food spoilage, food preservation, and food-borne pathogens.

All living things smaller than 0.1 mm in diameter are classified as microorganisms because they are too small to be seen by the naked eye. Although human beings suspected the existence of these microbes for a long time, the real beginning of the field of study of microorganisms started when Antonie van Leeuwenhoek (1632–1723), a Dutch amateur lens grinder, reported the observation using his lens of many animalcules. He reported his observations to the Royal Society of London and described the microbes accurately, including bacteria (rod, sphere, and spiral), protozoa, and yeasts. Because he did not train any students to follow his work, no one really mastered his technique of lens grinding for the observation of microbes. His work was confirmed by Robert Hooke in 1678 using a compound microscope. No noticeable advancement in microbiology occurred until the middle of the 1800s.

The first major development in microbiology centered around the controversy over the theory of spontaneous generation. In Greek mythology humans were able to create life. People believed that animals could originate from the soil, maggots could be produced by exposing meat to warm air, and so on. Francesco Redi (1626–1697) showed that when gauze was placed over a jar containing meat, no maggots formed because flies were not allowed to contact the meat. Gradually the idea of life being generated from nonliving things was discarded by those performing experiments on the subject. However, this topic still lingered in the minds of many.

With the discovery of microbes the controversy rekindled. People questioned where these microbes came from. One camp believed that microbes occurred spontaneously, and the other camp believed that microbes were developed from other microbes and that life cannot come from nonliving matter. In the process of proving or disproving the theory of spontaneous generation, several important issues concerning microorganisms were raised and some experimental procedures were developed. John Needham in 1749 observed microorganisms in putrefying meat and explained this as evidence of spontaneous generation. Lazzaro Spallanzani boiled beef broth, sealed the flasks, and observed that no microorganisms developed in the broth. The pro-spontaneous generation camp proposed that exclusion of air prevented spontaneous generation. The con-spontaneous generation camp indicated that microorganisms in the air were the source of contamination in faulty experiments. By heating the air or trapping the microorganisms with heat-treated cotton, no microorganism would develop in heated broth even if the air was allowed to come in contact with the broth. The pro–spontaneous generation camp insisted that heat destroyed some vital force that prevented the development of life forms. Finally Louis Pasteur, using his famous goosenecked flask, showed that even unheated air, when allowed to contact heated broth, will not result in the development of microorganisms as long as the particles in air were allowed to settle at the bottom of the U-shaped neck of the flask. Disproving the theory of spontaneous generation was essential for the development of microbiology because this allowed scientists to observe microbiological events in sterile media with known inoculated cultures free from the uncertainties of some unknown living agent that might spontaneously develop in the media.

With his disproving of the theory of spontaneous generation, Pasteur introduced the field of fermentations and food microbiology. Since ancient times people have produced wine, beer, bread, and other fermented products without knowing the exact reason for the development of such foods. Pasteur was involved with fermentation because of the occurrence of spoilage in wine at that time in France. He started the project by first proving that alcoholic fermentation of grapes, fruits, and grains was the result of organisms he called ferments. He showed that good wine batches had certain types of ferment and bad batches had other types of ferment. By heating juice at 63°C for 30 min he could kill the bad ferments, and after cooling the juice he could consistently produce satisfactory wine by inoculating ferments from good wine batches into the juice. Not only did he solve the problem of wine disease, but he also developed the process of pasteurization of food and drink. In the process of these studies he discovered that some organisms can grow in the absence of air. His work resulted in the scientific understanding of food fermentation and thus greatly improved the quality of fermented food products. The value of his study was that he showed the direct relationship between activities of a specific microorganism in the development of a specific product (i.e., good yeasts acting on pasteurized grape juice under proper conditions will result in wine).

The precise time at which humans started to realize the role of microbes in food products cannot be determined. However, it is safe to assert that humans noticed the results of microbial action such as spoilage of food and food poisoning early in the history of food-gathering and food-producing periods. Humans experienced and noticed the changes occurring in foods without knowing the reasons behind such activities until the development of the science of microbiology. The real beginning of food microbiology coincided with the development of microbiology, especially Pasteur's work on food fermentation processes. Early studies in food microbiology centered on dairy bacteriology. Only about 20 years ago did the field of food microbiology become a recognized field. Today it has a great impact on issues such as food safety, genetic engineering of new food products, methodologies in applied microbiology, food technology, and preservation technologies.

NUMBERS AND KINDS OF MICROBES IN FOODS

The number and kinds of microbes in foods depend greatly on the food products and the conditions under which they are stored or processed. The microbial profiles of raw food products and processed products are completely different and pose different sets of problems in terms of food-spoilage potential and food-safety issues. In general, raw food products tend to have a heterogeneous population, whereas processed food usually contains those organisms that can survive the processes and subsequent storage conditions. Microorganisms are very small, and when they occur in foods they may be in the millions per gram, milliliter, or square centimeter. A convenient guide for categorizing microbial loads on meat surfaces is as follows: microbial load on meat surface is considered low when the count is 0–2 log colony-forming units (cfu)/cm^2; intermediate, 3–4 log cfu/cm^2; high, 5–6 log cfu/cm^2; and very high, 7 log cfu/cm^2 (1). Using this guide, samples with 0–4 log cfu/cm^2 would be considered as acceptable; samples with 5–6 log cfu/cm^2 would be considered questionable; and samples with 7 log cfu/cm^2 and above would be considered unacceptable from a food-spoilage standpoint. Generally, 7 log cfu/g, mL, or cm^2 is considered the index of spoilage, because food will exhibit odor, slime, or both at counts above 7.5 log cfu/g, mL, or cm^2. This guide does not include identification of potential pathogens in foods such as *Salmonella*, *Staphylococcus aureus*, or *Clostridium perfringens*.

The number of microorganisms in food is usually monitored by putting a known volume of liquid, weight of solid, or surface area sample into a sterile liquid diluent (usually a buffered liquid) and then appropriately diluting the sample such that when the sample is placed in a petri dish the number of organisms in the petri dish is in the range between 25 and 250. The agar medium used to grow commonly occurring microorganisms is called the standard plate count agar. A count is always converted to per milliliter, gram, or cubic centimeter for ease of comparison of data from laboratory to laboratory. The result provides only a count of all organisms that can grow in this agar under the time and incubation temperature (usually 48 h at 32°C). A differential count can also be made. By using different agars specifically designed for specific organisms, one can make a specific count, such as a coliform count by using a violet-red bile agar. This agar contains inhibitory compounds such that other organisms cannot grow, and when coliform organisms grow they will exhibit a characteristic color, shape, and size in the agar. By developing a variety of differential agars, scientists can monitor the occurrence of different types of organisms in food products and can make a *Staphylococcus aureus* count, a lactobacillus count, a *Clostridium perfringens* count, and so on. These types of differential counts give the food microbiologist a clearer picture of the specific population of microbes in certain food products. One of the concerns of food microbiologists is the recovery of injured cells by heat, cold, acid, base, radiation, and the like. On a nonselective agar the injured cells can grow due to rich nutrients. On a selective agar, however, many injured cells cannot grow due to added stress of the selective agents. Recently in the author's laboratory D. H. Kang and D. Y. C. Fung developed a procedure called thin agar layer method where a layer of nonselective agar was placed on top of a solidified selective agar. When injured cells are plated on top of the nonselective agar, they can repair and later will not be affected by the selective agents of the selective agar. With this method injured *Salmonella*, *Escherichia coli* O157:H7, and *Listeria monocytogenes* were recovered efficiently using special selective agar, while other nontarget organisms, such as *Pseudomonas*, could be differentiated on the same plate.

To pinpoint the exact organism in a food product, a food microbiologist must take a colony and purify it, then perform a variety of morphologic, biochemical, and physiologic tests to ascertain the genus and species of the microorganism in question. The identification scheme would include morphology (rod, sphere, spiral, fruiting bodies), gram reaction (gram-positive or gram-negative), biochemical tests (carbohydrate fermentation, enzyme production), pigment production (yellow, red, blue, black, gray, green, etc.), nutritional requirements (organic, inorganic, complex, simple), temperature requirements (psychrophiles, 0–10°C; mesophiles, 10–45°C; thermophiles 45–75°C; and psychrotrophs, 0–30°C), pH requirements, fermentation products (acid, alcohols, etc.); antibiotic sensitivity, gas requirement (aerobic, anaerobic, facultative); pathogenicity, and serology (serotyping with specific antibodies).

After these tests are made, the food microbiologist can determine the exact nature of the organism in terms of genus and species. In many instances the number of specific pathogens in food is quite small. For example, there may be only 10 *Salmonella* among hundreds and thousands of other organisms in a food product, yet the food microbiologist must be able to detect them. To achieve this, food microbiologists have developed many elaborate preenrichment, enrichment, and selective-enrichment procedures so that target organisms (eg, *Salmonella*) will grow while other organisms are suppressed. For each target organism a separate scheme is developed. Consult the "General References" section at the end of this article for detailed information on methodologies in food microbiology.

The kind of microorganisms in food depend on the food and the conditions under which the food was made, processed, or stored. Microorganisms in food can be classified as:

Bacteria. Bacteria are unicellular organisms ranging in size from 0.1 to 2.0 μm and occurring in rod, spherical, or curved shapes. They divide asexually by binary fission. Although no sexual stage occurs, bacteria can exchange genetic information from one to another.

Yeasts. Yeasts are unicellular fungi occurring singularly with round or oval shape. Asexual reproduction is by budding, and sexual reproduction is by sexual spores.

Molds. Molds are multicellular filamentous fungi, highly structured and organized in morphology. Reproduction is by sexual and asexual stages. They usually grow so profusely that humans can see them on foods (fuzzy masses) without magnification.

Viruses. Viruses are submicroscopic entities that cannot reproduce without a living host. They occur with a protein coat enclosing a coil of DNA or RNA material. They can infect bacteria, plants, and animals. The shapes and sizes vary greatly among different groups of viruses.

BACTERIA

Bacteria of interest in food microbiology can be divided into the groupings discussed next.

Lactic Acid Bacteria

These are gram-positive, non-spore-forming bacteria producing lactic acid as the major or sole product of fermentation. As a group they are important in food spoilage because they cause souring and discoloration. However, they are also very important in pickling, cheese making, fermented dairy products, and silage technologies. The major genera of lactic acid bacteria include *Streptococcus*, *Lactococcus*, *Enterococcus*, *Pediococcus*, *Leuconostoc*, and *Lactobacillus*.

Streptococcus species such as *S. lactis*, *S. cremoris*, *S. thermophilus*, and *S. diacetilactis* are important starter cultures in the dairy industry. Some *Streptococcus* are pathogenic, such as *S. pyogenes* and *S. faecalis*. The dairy *Streptococcus* cultures are named *Lactococcus* and fecal *Streptococcus* cultures are named *Enterococcus*.

Pediococcus produces large quantities of lactic acid and is very important as a starter culture in the curing of meat.

Leuconostoc produces gas as well as slime in the presence of sugar. Although they are important spoilage organisms, they are also important producers of flavor compounds in dairy products.

Lactobacillus is a heterogeneous group of organisms consisting of slender, gram-positive rods. One group is homofermentative (producing large quantities of lactic acid) and another group is heterofermentative (producing acid and gas). *Lactobacillus* species are important in dairy, meat, and silage fermentation but are undesirable as spoilage organisms because of the production of large quantities of lactic acid in a variety of food products during storage.

Aerobic, Gram-Positive Catalase-Positive Cocci

Aerobic, gram-positive catalase-positive cocci occur in pairs, short chains, or clusters. They produce catalase (a very active enzyme) and form acid from carbohydrate. Some of them are quite heat resistant and salt tolerant and produce a variety of colors in food and culture media. Among them are important food pathogens and spoilage organisms.

Micrococcus species are aerobic, gram-positive cocci occurring widely in nature. Many species can grow under refrigeration and on inadequately sanitized equipment.

Staphylococcus is an important genus of gram-positive cocci. This is a facultative anaerobic organism that occurs widely in nature and on human skin. These bacteria produce a variety of extracellular enzymes and metabolites. The most important metabolite they produce is a group of heat-stable toxins called enterotoxins, which are the agents of staphylococcal intoxications. Once formed in food, these toxins are very difficult to destroy. Boiling for 1 h will not destroy the toxins but will kill the pathogens easily. The organism is salt tolerant and can spoil a variety of foods besides being an important food pathogen.

Spore-Forming Bacteria

The two important spore-forming bacteria in food microbiology are *Bacillus* and *Clostridium*.

Bacillus is a gram-positive, aerobic spore-forming bacteria. It occurs widely in nature and in soil. It forms large spores. Bacterial spores are highly resistant to all forms of food-processing techniques that usually kill vegetative cells. *Bacillus* possesses a wide range of physiologic activities, including fermentation of sugar, peptonization of protein, hydrolysis of starch, and rennin coagulation of milk. *Bacillus anthracis* is an important animal pathogen. Many *Bacillus* species are environmental contaminants and enter the food chain through air, water, and surface contacts. Some important *Bacillus* species are *B. subtilis*, *B. cereus*, and *B. stearothermophilus*. *B. cereus* has been implicated as a foodborne pathogen that produces an emetic toxin causing vomiting and a diarrheal toxin causing diarrhea in susceptible persons who consume contaminated food.

Clostridium species are gram-positive, anaerobic spore formers. Some are highly anaerobic and can be killed in the presence of molecular oxygen, whereas others are aerotolerant. Important species in food spoilage include *C. butyricum*, *C. putrefaciens*, and *C. sporogenes*. From the standpoint of food safety, *C. botulinum* is the most important because it produces a group of highly toxic protein toxins called botulin. These toxins are responsible for the often fatal disease called botulism. The canning industry has spent millions of dollars designing time and temperature treatments aimed specifically at killing the spores of this organism in canned goods. Fortunately, the toxin is heat sensitive. Boiling the toxin for 10 min will render it inactive.

Another species of *Clostridium* important in food microbiology is *C. perfringens*. Although this organism is less toxic than *C. botulinum*, *C. perfringens* accounts for about one-fifth of all food-poisoning cases in the United States annually. It produces a toxin that causes a mild diarrhea in human intestines after consumption of a large number (in the millions) of *C. perfringens* in foods such as meat and gravy.

Gram-Positive Irregularly Shaped Bacteria

Among the group of bacteria that have irregular shapes and are gram-positive is *Propionibacterium*, which is a small, anaerobic, pleomorphic rod (cell with irregular shapes). *P. shermanii* is the organism responsible in forming the eyes in Swiss-type cheeses. It also imparts desirable flavor in cheese fermentation. *Corynebacterium* is a pleomorphic rod arranged in Chinese letter morphology when observed under the microscope. Most of the species are environmental contaminants. *C. diphtheriae* is the agent responsible for diphtheria in humans. *Microbacterium* is a small, aerobic, heat-resistant rod. It can withstand 80°C for 10 min and is important in the spoilage of vacuum-packaged meat products.

Gram-Negative Polarly Flagellated Bacteria

From the standpoint of food spoilage, *Pseudomonas* is considered one of the most important organisms. This gram-negative organism is a prolific metabolizer of organic compounds. The organism can grow in refrigerated temperatures (psychrotrophic) and form slime, fluorescent compounds, and pigments in cold-stored foods. It is responsible for the spoilage of chicken, meat, fish, vegetables, and all kinds of foods kept in cold storage. Because it is aerobic it usually is not responsible for the spoilage of canned or vacuum-packaged foods.

Acetobacter is also a member of this group of gram-negative bacteria. This organism oxidizes alcohol to acetic acid. In making vinegar it is desirable; however, in making wine it is the most important organism causing souring. *Photobacterium* can cause phosphorescence of meat and fish when incubated in suitable conditions. *Halobacterium* can grow in salt concentrations as high as 30%. It can produce pigments and spoil salty fish.

Gram-Negative Short Rods

The heterogeneous group of organisms consisting of gram-negative small rods, when motile, possess peritrichous (all-over-the-cell) flagella. They are facultative anaerobes that are found in water, soil, human and animal environments, and the food chain. Some of the organisms are exceedingly important in food microbiology. Many of them are food pathogens and food-spoilage organisms. The most important family in this group is the Enterobacteriaceae. According to the newest classification there are 14 genera in this family: *Escherichia*, *Shigella*, *Salmonella*, *Citrobacter*, *Klebsiella*, *Enterobacter*, *Erwinia*, *Serratia*, *Hafnia*, *Edwardsiella*, *Proteus*, *Providencia*, *Morganella*, and *Yersinia*.

Escherichia is a true fecal coliform; the type species is the well-known *E. coli*. This organism has been used as an indicator of fecal contamination. Its presence indicates the potential presence of other, more pathogenic enteric organisms and is highly undesirable. Recently the serotype, *E. coli* O157:H7, has been the source of many major outbreaks involving hundreds of people and resulting in deaths. Major recalls of foods (up to 25 million lb of ground beef) were also made because of contamination by this organism. *E. coli* O157:H7 is now considered a food adulterant and is the cause of many food-safety regulations. *E. coli* is the most monitored organism in food and water microbiology.

Salmonella is an organism of great concern to the food industry. It is ubiquitous in the animal population and especially in poultry flocks. Some raw poultry products harbor *Salmonella*; thus all poultry products should be well cooked before consumption. The organism, when consumed in large number (1 million), can cause a disease called salmonellosis, which includes vomiting, nausea, diarrhea, chills, and fever. Mortality can occur in the very old, the very young, or the immunocompromised. More than 2,000 serotypes of *Salmonella* are reported. Many new rapid methods are now being developed to detect this organism in about 1 day. The conventional method takes about 5 to 7 days.

All members of the Enterobacteriaceae are potentially pathogenic. A more thorough discussion of bacterial pathogens is recorded in a different section of this volume (see the articles FOODBORNE DISEASES and RAPID METHODS OF MICROBIOLOGICAL ANALYSIS). It is important to note that occurrence of members of the Enterobacteriaceae in food is generally undesirable. Proper food handling will prevent or retard their growth and keep food safer for the consumers.

It should be noted that the naming of bacteria is a dynamic process in a constant state of revision and updating as new research demands. Consult *Bergey's Manual of Determinative Bacteriology*, 9th edition (2) for the current genus and species names of bacteria of interest.

YEASTS AND MOLDS

In general, yeasts and molds are considered spoilage organisms in food microbiology. When conditions are not favorable for the growth of bacteria, yeasts and molds will take over and spoil the food items. For example, in citrus foods the pH is too low for bacterial growth, so yeasts and molds are more active in spoiling those foods. Another example is dry goods. Bread is spoiled by mold more easily than by bacteria because molds can grow with much less water than can bacteria.

The only real food-poisoning concern of molds is the possible production of mycotoxins and aflatoxins by some molds such as *Aspergillus flavus* and *A. paraciticus*. Some of these compounds are carcinogenic and thus are of concern to regulatory agencies. In the area of food fermentation, yeasts and molds are of great importance because a large variety of fermented foods are produced by direct or indirect activities of yeasts and molds (see the article FOOD FERMENTATION).

INTRINSIC AND EXTRINSIC PARAMETERS OF FOODS

All foods possess a set of conditions called intrinsic parameters. These parameters can be influenced by another set of conditions called extrinsic parameters. Together, these two groups of parameters have great influence on the number and kinds of microorganisms occurring in and on a food and their physiologic activities. Intrinsic parameters of food include pH, moisture, oxidation-reduction potential (presence or absence of oxygen), nutrient content, occurrence of antimicrobial constituents, and biologic structures.

All microorganisms have a minimum, maximum, and optimal pH tolerance; a moisture requirement; an oxygen-tension requirement; and a nutrient requirement. By knowing these parameters, one can predict the presence and growth potential of specific microorganisms in certain types of foods. A pH of 4.5 is considered the demarcation line between acidic foods (<pH 4.5) and basic foods (>pH 4.5). Yeast and molds can grow down to pH 1 whereas bacteria cannot grow below pH 3. Thus acidic foods such as citrus fruits and carbonated soft drinks will be spoiled more by yeasts and molds than by bacteria. On the other hand, in a more basic food (>pH 4.5) bacteria will outgrow yeasts and molds owing to their higher metabolic rates in a favorable growth environment. Moisture content is another important parameter. This is usually expressed as water activity (A_w). Most moist foods are in the range of 0.95 to 1.00 A_w). When the A_w drops to 0.9, most spoilage bacteria reach their minimum level. Most spoilage yeasts have their minimum at 0.88, and molds have theirs at 0.80.

Thus in dry food products yeasts and molds grow much better than bacteria, and in moist food bacteria will outgrow yeasts and molds. The role of oxygen tension in and around food also has a great impact on the type of organisms growing there. Bacteria can be aerobic, anaerobic, or facultative anaerobic, so they can grow in a variety of oxygen levels (although different types will grow in different oxygen-tension environments). Yeast can grow both aerobically and anaerobically. Most molds, however, cannot grow anaerobically. In a properly vacuum-packaged food, for example, one should not find mold growing. The amount of oxygen measured in terms of oxidation-reduction potential also dictates the types of bacteria that can grow in the food. Disrupting the oxygen tension of a food (e.g., grinding a piece of meat to make ground beef from a steak) makes it easier for aerobic organisms to spoil the food.

Nutrient content (water; source of energy for metabolism; source of nitrogen, vitamins, and growth factors; and minerals) of different foods will support different types of microbes. In general, a food nutritious for human consumption is also a good source of nutrients for microbes. Some foods have natural antimicrobial compounds, such as eugenol in cloves, allicin in garlic, and lysozyme in egg, that can suppress the growth of some microbes. Biologic structures of some foods are also important for the prevention of microbial invasion. An example is the skin of an apple. When the apple is bruised, microbes can easily enter the fruit and spoil it.

Extrinsic parameters of food also play an important role in the activities of microbes. Temperature of storage greatly influences the growth of different classes of microbes. The amount of moisture in the environment (relative humidity) also influences the absorption of moisture or the dehydration of the food during storage and thus also influences the growth of different organisms. Varying the gaseous environment in storage will also change the types and growth rates of different organisms during storage of the food items. And last, the length of time of food storage also influences the spoilage potential by microbes in the food.

Thus intrinsic and extrinsic parameters of food are of great concern to food microbiologists. Skillful manipulation of these parameters by food microbiologists will result in more stable, more nutritious, fresher, and safer foods for the consumer.

FOOD-PRESERVATION TECHNIQUES

Food-preservation techniques can be grouped under drying, low-temperature freeze-drying, high-temperature, radiation, and chemical treatments. The technologies of these processes are recorded in other articles in this encyclopedia. In this article only their effects on microorganisms are discussed.

Drying is the most widely used method of food preservation in the world. Meat, fish, cereals, fruits, and vegetables are dried and preserved for a long time. Controlled dehydration and sun drying of foods removes water from foods so that microorganisms cannot grow. This process does not sterilize the food, so when water is reintroduced (rehydration), microbial growth may resume and may result in spoilage of food. Spores of bacteria and mold are known to survive for long periods (months and years) in dried foods.

Low-temperature preservation of food is based on retardation of enzymatic activity of microbes. As temperature decreases, the enzymatic activity of microbes also decreases and eventually stops at around freezing temperature and below. Freezing will kill approximately 10% of the initial microbial population, but the remaining population can survive for a long time (years). Refrigeration temperature (0–10°C) will retard growth of most organisms; however, a group of organisms called psychrotrophs will grow slowly under refrigeration temperature and may eventually spoil the cold-stored food. Psychrotrophic bacteria (*Pseudomonas*, *Micrococcus*, etc.), yeasts (*Candida*, *Debaryomyces*, etc.), and molds (*Penicillium*, *Mucor*, etc.) can spoil foods under prolonged storage. Although most psychrotrophs are nonpathogenic, some pathogens such as *C. botulinum* and *Listeria monocytogenes* have been found to grow at or around 4 to 6°C. Freeze drying or lyophilization of foods depends on the unique property of water—the triple point of water. Food is first frozen and then a vacuum is applied to sublime the water out of the food mass. Freeze-dried foods need not be refrigerated because microorganisms cannot grow without water. However, rehydrated food will have the same spoilage potential as the original food. Freeze drying does not kill microorganisms effectively. In fact, freeze drying is the best method to preserve bacterial cultures.

High-temperature preservation depends on heat coagulation of proteins and enzymes, thus killing microorganisms. Pasteurization, a form of high-temperature preservation, usually refers to treatment of food at 63°C for 30 min or 72°C for 15 s. These time and temperature combinations are designed to kill most vegetative cells in milk, especially *Mycobacterium tuberculosis* and *Coxiella burnetti*. Thermoduric organisms are those that can survive pasteurization and later grow and spoil the pasteurized food. To reach sterilization temperature, food must be cooked under pressure in sealed containers using 121°C for 1 min as 1 sterilization unit. This is the practice used in commercial canning. The purpose is to achieve time and temperature combinations such that the most heat-resistant spores of *C. botulinum* are destroyed in the specific food item being canned. Commercial canning has an excellent record of safety. Improper home canning is the major source of botulism due to toxins produced by *C. botulinum* that survive the improper heating and sealing procedures in canning.

Radiation treatment of food is of two types: ionizing radiation and nonionizing radiation. Ionizing radiation such as α, β, γ, and X ray kill microorganisms by breaking chemical bonds of essential macromolecules such as DNA (target theory) or by the ionization of water, which results in the formation of highly reactive free radicals such as HO^- or H_2O^- capable of splitting chemical bonds in the microorganisms. Ionizing radiation destroys microorganisms without generation of heat, and thus it is called cold sterilization. Interest in radiation preservation of food has been fluctuating in the past 30 years. Recently interest has resurged on an international scale in the use of radiation for food preservation. In 1999, U.S. approved levels of radiation in foods are 1.5 to 3.0 kGy for poultry (fresh or frozen) and 4.5 kGy max for fresh red meat and 7.0 kGy max for frozen red meat (1 Gy = 100 rad; 1 kGy = 1000 Gy).

Nonionizing radiation includes UV treatment and microwave treatment. UV has poor penetration and thus is used only for surface decontamination. Microwave treatment of food has gained great popularity. It is estimated that 75% of the homes in the United States have a microwave oven. The waves at 2540 MHz when applied to foods bound back and forth and create vibration of asymmetric, dielectric molecules (such as water), which generates heat. This heat cooks food as well as kills microorganisms. There may exist some as yet unexplained mechanism of microwave destruction of microorganisms besides purely the effect of heat (3). Microwave cooking will continue to be an important method for homes and institutions, and it is an effective means of destroying microorganisms in foods as long as the food is given enough microwave exposure time.

Chemicals are used to kill organisms (bactericidal) or to prevent them from growing (bacteriostatic). Many chemicals are used to treat equipment and the environment such as sanitizers, disinfectants, strong acids and bases, and halogens. In terms of food science, the subject of food additives is of great interest. Food additives are compounds added to foods for purposes such as improvement or modification of flavor, texture, rheology, color, pH change, water-holding capacity, and emulsification. Some of these compounds (lactic acid, acetic acid, propionic acid, sorbate, etc.) are used as preservatives because they can kill microorganisms in foods. Compounds approved for use in foods are controlled by the Food and Drug Administration and are listed on the generally recognized as safe (GRAS) list and periodically updated (4).

There are many recent developments in methodologies pertaining to food microbiology. Rapid methods and automation in microbiology have been the subject of many national and international symposia. The four major developments are miniaturization of conventional procedures and development of diagnostic kits, such as Fung's mini systems, API, Enterotube, Spectrum 10, IDS, Minitek, MicroID, and Biolog; modification of viable cell count procedures, such as 3M Petrifilm, Redigel, Isogrid, Spiral System, and DEFT test; development of alternative approaches for the estimation of microbial populations, such as the use of adenosine triphosphate, electrical impedance or conductance, microcalorimetry, or radiometry to indirectly measure biomass (bacteria, yeasts, and molds) in food products; and identification of microbes by novel and sophisticated instruments and procedures, such as Vitek system, DNA Probe (Genetrak), polymerase chain reaction (PCR), ribotyping, ELISA (Organon Teknika, Tecra), motility enrichment (BioControl), protein profiles, and fatty acid analysis.

The field of food microbiology is very important for food science and technology. It is one of the central disciplines in food science. Food microbiologists are called on to solve microbiologic problems related to other branches of food science and technology. The future of food microbiology is very bright indeed. See the article RAPID METHODS OF MICROBIOLOGICAL ANALYSIS for more details.

BIBLIOGRAPHY

1. D. Y. C. Fung et al., "Mesophilic and Psychrotrophic Bacterial Populations on Hot-Boned and Conventionally Processed Beef," *J. Food Prot.* **43**, 547–550 (1980).
2. J. Holt et al., *Bergey's Manual of Determinative Bacteriology*, 9th ed., Williams & Wilkins, Baltimore, Md., 1994.
3. D. Y. C. Fung and F. E. Cunningham, "Effects of Microwave Cooking on Microorganisms in Foods," *J. Food Prot.* **43**, 641–650 (1980).
4. B. L. Oser and B. K. Bernard, "Thirteen GRAS Substances," *Food Technol.* **38**(10), 66–84 (1984).

GENERAL REFERENCES

Associations of Official Analytical Chemists, *Official Methods of Analysis*, AOAC Arlington, Va., 1988.

J. C. Ayres, O. Mundt, and W. E. Sandine, *Microbiology of Foods*, W. H. Freeman, San Francisco, Calif., 1980.

G. J. Banwart, *Basic Food Microbiology*, AVI, Westport, Conn., 1979; 2nd ed., Chapman & Hall, New York, 1989.

C. M. Bourgeois, J. Y. Leveau, and D. Y. C. Fung, Microbiological Control of Food and Agricultural Products (English edition), VCH, New York, 1995.

J. E. L. Corry, D. Roberts, and F. A. Skinner, *Isolation and Identification Methods for Food Poisoning Organisms*, Academic Press, New York, 1982.

M. P. Doyle, L. R. Beuchat, and T. J. Montville, Food Microbiology: Fundamentals and Frontiers. ASM, Washington D.C., 1997.

Food and Drug Administration, *Bacteriology Analytical Manual*, 6th ed., Association of Official Analytical Chemists, Arlington, Va., 1984; 18th ed., Association of Official Analytical Chemists International, Arlington, Va., 1995.

D. Y. C. Fung, "Microbiology of Batter and Breading," in D. R. Suderman and F. E. Cunningham, eds., *Batter and Breading Technology*, AVI, Westport, Conn., 1983, pp. 106–119.

D. Y. C. Fung, "Rapid Methods for Determining the Bacterial Quality of Red Meats," *J. Environ. Health* **46**, 226–228 (1984).

D. Y. C. Fung, "Types of Microorganisms," in F. E. Cunningham and N. A. Cox, eds., *Microbiology of Poultry Meat Products*, Academic Press, New York, 1987.

D. Y. C. Fung, M. A. Buono, and L. E. Erickson, "Mixed Culture Interactions in Anaerobic Fermentations," in L. E. Erickson and D. Y. C. Fung, eds., *Handbook on Anaerobic Fermentations*, Marcel Dekker, New York, 1988.

D. Y. C. Fung and C. L. Kastner, "Microwave Cooking and Meat Microbiology," *Proceedings Reciprocal Meat Conference* **35**, 81–85 (1983).

D. Y. C. Fung and K. Vicheinroj, *Introduction to Food Fermentation*, Kansas State Univ. Press, Manhattan, Ks., 1998.

W. F. Harrigan and M. E. McCance, *Laboratory Methods in Food and Dairy Microbiology*, Academic Press, New York, 1976.

P. A. Hartman, *Miniaturized Microbiological Methods*, Academic Press, New York, 1968.

C. G. Heden and T. Illeni, eds., *Automation in Microbiology and Immunology*, Wiley, New York, 1975.

C. G. Heden and T. Illeni, eds., *New Approaches to the Identification of Microorganisms*, Wiley New York, 1975.

B. C. Hobbs and J. H. B. Christian, *The Microbiological Safety of Food*, Academic Press, New York, 1973.

J. G. Holt et al., *Bergey's Manual of Determinative Bacteriology*, 9th ed., Williams & Wilkins, Baltimore, Md., 1994.

International Commission on Microbiological Specifications for Foods, *Microorganisms in Foods*, Vol. 2, University of Toronto Press, Toronto, Canada, 1974.

International Commission on Microbiological Specification for Foods, *Microorganisms in Foods, vol. 5: Microbiological Specifications of Food Pathogens*, Blackie Academics and Professionals, New York, 1996.

J. M. Jay, *Modern Food Microbiology*, 3rd ed., Van Nostrand–Reinhold, New York, 1986; 5th ed., 1998.

F. V. Kosikowski and V. V. Mistry, *Cheese and Fermented Milk Foods*, 3rd ed. Vol. 1: *Origins and Principles*, Vol 2: *Procedures and Analysis*, Edwards Brothers, Ann Arbor, Mich., 1997.

R. T. Marshall, *Standard Methods for the Examination of Dairy Products*, 16th ed., American Public Health Association, Washington, D.C., 1992.

E. H. Marth, ed., *Standard Methods for the Examination of Dairy Products*, American Public Health Association, Washington, D.C., 1978.

B. M. Mitruka, *Methods of Detection and Identification of Bacteria*, CRC Press, Cleveland, Ohio, 1976.

D. A. A. Mossel et al., *Essentials of the Microbiology of Foods*, Wiley, New York, 1995.

J. T. Nickerson and A. J. Sinskey, *Microbiology of Foods and Food Processing*, American Elsevier, New York, 1972.

H. Rieman and F. L. Bryan, eds., *Food-Borne Infections and Intoxications*, Academic Press, New York, 1979.

T. A. Roberts and F. A. Skinner, *Food Microbiology Advances and Prospects*, Academic Press, New York, 1983.

A. N. Sharpe, *Food Microbiology: A Framework for the Future*, Charles C. Thomas, Springfield, Ill., 1980.

J. H. Silliker et al., *Microbial Ecology of Foods*, 2 Vols., Academic Press, New York, 1980.

R. C. Tilton, *Rapid Methods and Automation in Microbiology*, American Society of Microbiology, Washington, D.C., 1982.

C. Vanderzant and D. Splittstoesser, *Compendium of Methods for the Microbiological Examination of Foods*, American Public Health Association, Washington, D.C., 1990.

H. H. Weiser, G. J. Mountney, and W. A. Gould, *Practical Food Microbiology and Technology*, AVI, Westport, Conn., 1971.

DANIEL Y. C. FUNG
Kansas State University
Manhattan, Kansas

MICROWAVE SCIENCE AND TECHNOLOGY

Interest in microwave processing began shortly after World War II with the introduction of the first microwave oven by the Raytheon Co. (Waltham, Mass.) a manufacturer of magnetrons for radar. The first U.S. patent granted to Raytheon Co., which claimed the novelty of microwave cooking, was illustrated with a food product moving on a conveyor belt past a microwave source (1). Many other patents followed, research was carried out on the effects of microwave energy on food materials, microwave cooking technology was developed, and eventually a consumer microwave oven sales boom occurred so that today this appliance is commonplace in most homes in the United States and a similar trend is taking place in most of the developed countries of the world. Microwave energy use in food processing is showing a similar growth trend, although at a much more modest rate.

FUNDAMENTALS OF MICROWAVE HEATING

Microwaves are electromagnetic (EM) waves of very short wavelength. In the EM spectrum, microwaves lie between the television frequencies and infrared. In terms of wavelength, radio waves are measured in kilometers, television frequencies in meters, microwaves in centimeters, and infrared in microns. The wavelength and frequency are related by the expression:

$$\text{wavelength } (\lambda_0) = \frac{\text{speed of light } (c)}{\text{frequency } (f)}$$

Thus for a typical microwave oven frequency of 2,450 MHz, the wavelength (λ_0) is

$$\frac{3 \times 10^{10} \text{ cm/s}}{2.45 \times 10^9 \text{ cycles/s}} = 12.25 \text{ cm/cycle}$$

Microwaves, as with other wave energy, radiate from a source in all directions. These waves carry energy and the amount of energy they carry depends on the amount of energy imparted to them. The energy of microwaves comes from electrical energy that is converted by a power supply to high voltages that in turn are applied to the microwave power tube or generator. The most common power tube used in microwave ovens is the magnetron and it broadcasts its energy into the applicator, which may be an oven, a waveguide, or some other device containing the material to be heated.

Microwaves, like infrared and visible light, are reflected, transmitted, and absorbed. They are reflected from metal surfaces: the microwave oven is basically a metal box in which the waves reflect from the walls and create a resonant system. Microwaves are transmitted, that is, they pass through many materials including glass, ceramics, plastics and paper. Some materials are only partially transparent to microwaves; that is, they absorb some energy. When microwaves are absorbed their energy is converted to heat.

How Microwaves Produce Heat

Microwaves in themselves are not heat. The materials that absorb microwaves convert the energy to heat. In foods, it is the polar molecules that for the most part interact with microwaves to produce heat. Water is the most common polar molecule and is a component of most foods. In the presence of a microwave electric field, water molecules attempt to line up with the field in much the same manner iron filings line up with the field of a magnet. Because the microwave field is reversing its polarity billions of times each second, the water molecule, because it is constrained by the nature of the food of which it is a part only begins to move in one direction when it must reverse itself and move in the opposite direction. In doing so, considerable kinetic energy is extracted from the microwave field and heating occurs. The phenomenon is similar to the heating of the human body when exposed to the sun or any other heat source. Energy in the form of infrared rays from the sun, is not heat until it is absorbed by the body and the polar molecules in the surface layers of the body convert it into heat.

Space charge polarization is an equally important microwave heating mechanism. Ions, are caused by the microwave field to flow first in one direction then in the opposite direction as the field is reversed. The effect of ionic conduction can be observed in the microwave heating of salted water in that higher temperatures are found at the surface. Ionic conduction has a negative effect on microwave energy penetration, thus foods with a high salt content show greater surface heating.

Microwave Properties of Foods

The amount of energy that can be absorbed by a substance is expressed by the relationship:

$$P = \sigma' E^2 \text{ (W/cm}^3\text{)} \tag{1}$$

where P is the the power absorbed in watts/cm^3, σ' is the equivalent dielectric conductivity, and E is the voltage gradient in V/cm. The dielectric conductivity,

$$\sigma' = 2\pi\varepsilon_0\varepsilon'' f \tag{2}$$

where f is the frequency of the energy source, ε_0 is the dielectric constant of vacuum (8.85×10^{-14} F/cm), and ε'' is the dielectric loss factor of the substance. Substituting equation 2 into equation 1 gives

$$P = 55.61 \times 10^{-14} f\varepsilon'' E^2 \tag{3}$$

Dielectric Loss Factor

Because the field strength (E) and the frequency are essentially constant for the microwave oven being used, the loss factor (ε'') is the only variable. The term loss originated from the discovery that energy loss occurred when the electrical energy through a capacitor was cycled on and off. The term today, in microwave heating technology, is representative of a desirable condition, whereas it once was considered undesirable. In the context being used, the loss factor represents a property of the material being processed, and a lossy material is one that heats well, while a low loss material is one that heats poorly and is, therefore, more transparent to microwave energy. The loss factor is a measurable quantity and a considerable volume of dielectric loss data have been accumulated (2–6).

The loss factor is the product of two measurable properties: the dielectric constant (ε') and the loss tangent (tan δ). The loss factor varies with temperature and frequency as shown in Table 1 and indicates that the penetration of microwave energy decreases with increasing dielectric loss.

Table 1. Dielectric Loss of Raw Turkey Roll as a Function of Temperature and Microwave Frequency

	Frequency, MHz		
Temperature, °C	2,450	915	300
−40	0.13	0.17	0.29
−20	0.61	0.73	1.21
5	21.6	26.4	58.1
25	21.5	33.2	86.2
45	23.8	42.8	118
65	26.8	53.1	159
80	25.0	69.0	179
100	27.1	78.7	205
110	29.2	83.4	218
115	29.9	85.8	224
120	30.3	88.2	231

Source: Ref. 7.

Penetration

The penetration depth is that depth in a material at which the energy level is 37% (or $1/e$) of the surface value. The term half-power depth is also used and is that depth in a material at which the power level is one-half that at the surface. The equation for converting dielectric property data into penetration depth (d) is

$$d = 1/2\alpha$$

for $1/e$, and for half-power depth it is

$$d = 1/2.886\alpha$$

where α is the attenuation constant, which is calculated as:

$$\alpha = \frac{2\pi}{\lambda_0}\left[\frac{\varepsilon'}{2}\left(\sqrt{1 + \tan^2\delta} - 1\right)\right]^{1/2}$$

Thus the main difference between microwave heating and other heating methods is that microwave energy penetrates deeply into food materials and is converted to heat as it penetrates. The temperature profile shown in Figure 1 illustrates the effect of microwave heating on a large food mass and the effect of time after the energy is no longer being applied. It is clear in this case that conduction heating will always play a role in any microwave heating process where a temperature gradient exists.

EFFECT OF VARIOUS PHYSICAL FACTORS

In addition to the dielectric properties of foods there are a number of other factors that affect microwave heating performance. Among these factors are geometry (shape), surface-to-volume ratio, specific heat, density, thermal conductivity, and evaporative cooling. These are important factors in food processing as well as cooking and heating in consumer and institutional microwave ovens.

Geometry

The shape of food items is critical to good microwave heating results. The sphere is the ideal shape as energy tends to be focused to give heating at the center of the sphere. Obviously as the diameter is increased it may be impossible for center heating to occur except by conduction.

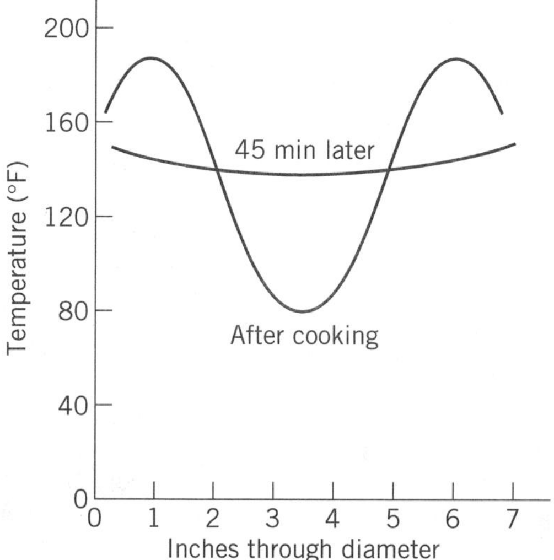

Figure 1. The effect of thermal conduction on the temperature profile of a large roast beef after 45 min standing out of a microwave oven.

Heating can be concentrated in the center of spheres with diameters (8) that measure between 20 and 60 mm. Such concentrated heating can be a disadvantage in that conduction heating may not be able to dissipate the temperature gradient and the mass would erupt, as in the heating of an egg in its shell. At lower microwave heating rates, more time is provided for conduction heating to take place. In the example of beef roasting the postmicrowave heating time could have been reduced if a lower heating rate were used, as the temperature gradients would not have been as great.

Computer simulation studies as well as actual heating of spheres and cylinders of various diameters have been carried out (8). A phantom food mixture with dielectric properties similar to food materials was used in these studies and temperature patterns were obtained by color thermography. For cylinders, thermography indicated maximum core heating occurred at diameters of 35 mm. At 50 mm, the core and surface heating occurred, while at 75 mm only surface heating was observed. Calculations support maximum center heating for diameters of 20 to 35 mm, while surface heating was more prominent at 40 to 50 mm and was dominant for larger cylinders.

While spherically shaped foods are commonplace among many natural foods eg, beets, onions, potatoes and some prepared foods (meatballs), the majority of foods are not spherical in shape. The cylinder is the next best shape in terms of heating performance.

For many foods it is the food container or dish that determines the shape and, therefore, affects the heating performance. A rectangular shape is common, but it is clear that food in the corners of a rectangular container will be overcooked before the remainder of the food. Overcooking in the corners also occurs when using a conventional oven, but because conduction cooking is much slower the temperature differences are not as pronounced. Where this shape is unavoidable, the corners can be shielded with aluminum foil to reduce the heating rate in these areas. Even when corners are unavoidable they should have generous radii to minimize the overheating effect. Also vertical or near vertical sides are preferable to sloping sides. Excessive sloping gives shallow areas where dehydration can occur.

Shielding and Shadowing

Shielding is the use of metal to reflect microwave energy and thereby reduce the heating rate in selected areas. Aluminum foil is usually used because of its availability, but also because it can be wrapped around the specific parts to be shielded. Shadowing can be described as a shielding effect by a food product that results in a reduced heating rate where the products are in contact. Thus it is a mutual effect with products shadowing each other. Food masses should always be spaced apart for best cooking or heating results.

Surface-to-Volume Ratio

As with conventional heating, the greater the surface area, the faster that cooking occurs. Thus potatoes and carrots can be cooked faster if they are diced. Vegetables such as corn and peas have a high surface-to-volume ratio and cook rapidly. This same condition also means that foods will cool more rapidly. This is an important consideration when developing new products, and means to reduce the cooling rate should be provided where possible, for example, by using insulated containers or preheated serving dishes.

Specific Heat

The specific heat of a material is the ratio of its thermal capacity to that of water. It is a measure of the energy required to raise product temperature by a specific amount. The specific heat of a food is closely related to its moisture content.

Density

Usually there is a clear relationship between density and moisture content. Thus bread, which has a low moisture content (about 35%) is not as dense as beef (65%); for example, 1 lb of bread requires about two-thirds as much energy as 1 lb of beef to heat to the same temperature. Thermal conductivity also comes into play in that the dimensions involved could mean substantial overheating of the surface of beef because of its density, while the more open structure of bread would not rely on thermal conductivity to the same extent.

Thermal Conductivity

Thermal conductivity is a measure of a material's ability to transfer heat in response to a temperature difference. Even in microwave cooking it plays an important role in spite of the penetrating nature of microwave energy. Without heat conduction unacceptable temperature differences would occur in most microwave heated or cooked foods. Such differences can be minimized by reducing the rate of microwave heating.

Starting and Final Temperature

Heating time depends not only on the power level but also on the temperature range that must be bridged to accomplish the desired result. Thus frozen foods take longer than refrigerated foods, which in turn take longer than shelf-stable food heated from room temperature assuming all were heated to the same final temperature. An additional quantity of heat is required to heat from the frozen state, ie, the heat of fusion.

Evaporative Cooling

This is a phenomenon that is much more evident when cooking with microwave energy. This effect is responsible for the lower surface temperature of some foods cooked in microwave ovens and, therefore, is responsible for the myth that microwaves cook from the inside out. It is not noticeable in a conventional oven because of heat radiation to the surface of foods.

FOOD-PROCESSING APPLICATIONS

A number of food-processing applications representative of current activity are listed in Table 2. Some are well-established commercial operations. Others have been demonstrated on a pilot scale. Still other have been discontinued for one reason or another.

Table 2. Microwave Food-Processing Applications

Process	Food products
Tempering	Frozen foods
Drying	Pasta, snack foods, onions, rice cakes, seaweed, fruit juices
Cooking	Bacon, meat patties, potatoes, poultry
Pasteurization	Fresh pasta, bread, meals
Sterilization	Meals
Rendering	Tallow
Roasting	Coffee, cocoa beans
Denaturation	Sausage emulsion
Baking	Bread

ADVANTAGES OF MICROWAVE PROCESSING

Speed

Because microwave heating occurs within the material being processed and not just at the surface the process can be greatly accelerated. This is particularly true for materials that are poor thermal conductors and for drying processes where the high dielectric loss properties of water result in selective heating of moist areas and rapid diffusion of water vapor to the surface.

Energy Savings

Microwave energy is not expended in heating the walls of the equipment or the environment. Heat is developed where it is needed, ie, in the material being processed. In drying processes, microwave energy is applied after conventional drying has reached the critical point where the drying rate falls off, thus the bulk of the heat of vaporization can be supplied by cheaper forms of energy.

Space Savings

The increase in processing rate makes it possible to design more compact equipment. In many cases plant capacity can be increased severalfold without the need for additional building space.

Fast Start-Up and Shutdown

There is practically no thermal lag with microwave processing so that the system can be operational almost immediately, adding in some cases extra hours of productive operation. Cleanup after shutdown can begin sooner in most cases because the equipment is cooler and can be worked on sooner. In a hybrid system, where conventional heat is used processing can be carried out at lower temperatures.

Precise Process Control

Heating adjustments are immediately effective and the operation can be fine-tuned easily. Systems respond instantaneously to feedback from sensors.

Improved Working Conditions

Because only the material being processed is heated there are negligible heat losses to the working environment. Worker productivity is enhanced and the requirement for ventilation and air-conditioning is reduced.

EXAMPLES OF MICROWAVE PROCESSING APPLICATIONS

Tempering

Conventional frozen food thawing methods are extremely inefficient taking many hours to several days depending on the product. The reason for this is that the thermal conductivity of the frozen product is about three times that of the thawed product. Thus as the surface thaws, heat is more slowly conducted toward the center.

Complete thawing of frozen foods by means of microwave energy also has its problems. It has been pointed out that as the frozen products approached 0°C, microwave penetration decreases rapidly and care has to be taken to prevent runaway heating (9). Furthermore, if frozen materials are held at ambient temperature too long before being exposed to microwave energy most of the energy will be absorbed at the surface and thawing to the core will not be possible.

Fortunately it is not necessary in most cases to completely thaw frozen materials. The tempering process stops short of thawing by several degrees; thus tempering has come to mean "raising the temperature of frozen food to a higher temperature that is still below the freezing point, but at which temperature the product is firm but no longer hard." Properly tempered frozen food is easily sliced, diced, separated, or otherwise handled for further processing.

Some data on the energy required to freeze foods are also pertinent to thawing foods and clearly illustrate the benefits of stopping the process prior to complete thawing (10). From Table 3 it can be seen that the energy required to raise the temperature of frozen beef from −17.7 to −4.4°C is about one-half that required to raise the temperature to −2.2°C. Thus, where possible, tempering should be terminated at the lowest acceptable temperature. It should also be noted that microwave tempering is uniform throughout the frozen mass being tempered.

Microwave tempering is one of the more economical processes in terms of energy consumption. The cost of tempering in terms of energy and magnetron replacement costs is less than $0.01/lb. Other factors that contribute to the return on investment include improved yield and quality, space and labor savings, simplified production scheduling, and increased plant productivity.

Most microwave tempering equipment today is operated at 915 MHz at power levels varying from 25 to at least 150 kW and are continuous. Batch units are available for smaller food-processing operations and usually operate with 40 kW of power at 915 MHz. Continuous equipment also is available at 2,450 MHz and varies in power from 33 to 132 kW. The microwave penetration at this frequency is less than at 915 MHz and requires the use of a refrigerated air circulation system to prevent surface thawing so that internal tempering can be accomplished. There are approximately three hundred microwave tempering systems in operation worldwide, with about one hundred eighty systems in the United States. These systems average 100 kW of microwave power. Annual tonnage of products tempered in such equipment is on the order of 5 billion lb.

Drying

Pasta drying is an excellent example of a process in which a relatively small amount of microwave energy is needed to generate maximum benefits. This process is accomplished in three stages: conventional hot air drying of the freshly extruded pasta followed by a microwave-hot air stage and a temperature equalization stage. The fresh pasta enters the first stage at about 30% moisture where it is exposed to hot air at 71–82°C. Rapid moisture loss to about 23% occurs in minutes and drying is continued for a total of about 35 min to reduce the moisture to about 18%. The product then falls by gravity into one end of the microwave thermal dryer where the combination of microwave energy at 915 MHz and hot air (82–93°C, 15–20% RH, 30.5 mpm) lowers the moisture level to 13–13.5% in about 12 min. The product finally falls once again by gravity to the final stage where it is conveyed slowly through a 70–80% RH zone to cool to temperatures low enough to prevent thermal stress cracking. About 1 h is required. An additional 0.5–1.0% moisture loss occurs in this stage. The major advantages of this process are time, space, and labor savings.

Table 3. Energy Requirement to Thaw Frozen Beef

J/g	°C
0	−40
50	−17.7
112.8	−4.4
173.3	−2.2

Source: Ref. 10.

Another drying process of interest is a system for the production of bread crumbs from bread dough. The fresh dough is extruded into one end of an 80 kW, 2450 MHz microwave-hot air tunnel where it is baked and dried so that the finished product can be cooled and ground into bread crumbs.

Some snack food items are being produced in a 240 kW, 915 MHz baking-drying unit. The dough is extruded as parallel strands onto a 5-ft-wide belt that passes through 12 independent sections, each with 25 kW of microwave power and an independent air system, where it is puffed, baked, and partially dried.

Microwave vacuum drying of fruit juice concentrates has been successfully demonstrated. In this process the concentrate is pumped into the vacuum chamber onto a conveyor belt where it foams under the effect of microwave energy and reduced pressure to a thickness of 80–100 mm. The product dries to a meringuelike condition, is crushed, scraped, and brushed from the belt into a receiver that is isolated from the vacuum chamber. A 48 kW, 2,450 MHz system will dry concentrate to 2% moisture in about 40 min at a rate of 50 kg/h.

Cooking Processes

A pair of microwave conveyor units has been used in the cooking of poultry parts. They operated at a frequency of 2,450 MHz: one with 80 kW of microwave power, the other with 50 kW of power. Cut-up breasts and thighs were processed in the 80 kW unit, while wings and legs were processed in the 50 kW unit. Saturated steam was introduced from a manifold of nozzles along the back walls of the microwave tunnels. As the cooked parts passed out of the ovens they were conveyed through several water chillers to lower their temperature. After the batter and breading operations the parts were blast frozen and packaged for the commercial food service trade. The processing rate was 2,500 lb/h.

Although in use for more than a decade in supplying precooked bacon to the institutional food service trade, there was a major escalation a few years ago in the requirement for precooked bacon by the fast-food business that now absorbs almost the entire production of this product. The major systems in use today are microwave-only systems, which operate at power levels up to 400 kW, and hybrid systems, which combine microwave energy (150 kW) and hot air.

An advantage of the microwave process is in the evenness of cooking such that the difference between microwave-cooked bacon 22–24 slices/lb and 16–18 slices/lb cannot be discerned. The microwave product is more tender and shrinks much less than the conventionally cooked product. To one fast-food firm, the greater number of slices per pound translates into substantial savings in shipping costs because 38% less raw bacon is required to meet their needs. Based on a monthly consumption of 600,000 lb of raw bacon, a 38% reduction amounts to a savings of almost $164,000 in raw bacon costs. In addition, the high quality of the rendered fat by-product commands a good price.

There are approximately fifty systems installed worldwide with an annual production of 8 billion slices of precooked bacon. Systems average 500 kW; however, some systems are 600 kW and at least one is 1,000 kW. Microwave cooking has the advantage over other methods in that nitrosamines are essentially absent.

A process was developed in Sweden for the cooking of meat patties; 30–45 kW, 2,450 MHz systems are typical. In practice, meat patties are formed by automated patty-making equipment at a rate of 1,000 kg/h (16,000 patties/h) and deposited on a conveyor belt. A few grams of margarine are deposited on each patty to improve heat transfer and surface browning as the patties are carried between Teflon belts that move between heated aluminum platens. The browned patties then pass through the microwave unit where in little more than a minute they are cooked to the proper degree. Typically, the temperature is raised from 40 to 70°C.

Pasteurization

Fresh pasta is a product that by its very nature is not easily pasteurized by conventional thermal methods. The loose condition of the product in the package makes conduction heat transfer impractical. Until recently the product was being marketed in modified atmosphere packaging that provides about 30 days of refrigerated shelf life. The combination of modified atmosphere packaging and microwave pasteurization extends the shelf life to 90–120 days.

The process is carried out by filling the fresh product into plastic trays, gas flushing, lidding, and passing the trays through a microwave tunnel oven to raise its temperature to 80°C, then cooling. Processing time is 3–5 min followed by a holding time at 80°C that varies with the product but is typically about 7 min. Cooling may take an additional 45 min. Thus far, this process has been carried out only at 2,450 MHz. A typical plant producing fresh gnocchi processes 3,000 lb/h. The microwave process is effective against molds, yeasts, and thermolabile microorganisms.

There are a number of full-scale systems in operation today for the microwave pasteurization of sliced packaged bread. The purpose of this process is twofold: (1) to increase the shelf life so that specialty breads, in particular, do not have to be baked as frequently and (2) to eliminate the need for chemical preservatives.

The most recently installed systems carry out pasteurization by conveying baskets of prepackaged bread through a microwave tunnel. The density of the product is sufficiently low that heating occurs deep within the product. As the baskets exit from the tunnel they are stacked to allow the temperatures to equilibrate. This may take 1–2 h. This practice appears to have no harmful effect on product quality and shelf life is substantially increased.

A wide variety of prepared foods are being pasteurized in package by two different microwave processes and sold at retail from refrigerated cabinets. One process involves hot filling the product in microwavable plastic dishes, lidding them, followed by a series of passes through a controlled-temperature tunnel and a microwave-hot air tunnel before being cooled. A refrigerated shelf life of six weeks at 4°C is claimed. The microwave system is operated at 2,450 MHz using 48–1.2 kW magnetrons. The other process involves microwave heating the products while they are immersed in a shallow, controlled-temperature water bath. This technique claims to provide more uniform control of product temperature.

FUTURE DEVELOPMENT

Food manufacturers are well aware of the acceptance of microwave cooking and heating by consumers because of the microwave oven sales boom over the last 20 years in the United States and in recent years in other nations throughout the world. There is already a large market for foods designed and packaged for microwave oven use. A substantial increase in this market can be expected if microwave pasteurization and sterilization of portion packed foods is successful. These two processes could account for a dominant share of the microwave equipment business and in the case of sterilization possibly cut deeply into the conventional equipment market for these processes.

Although practically all of the microwave sterilization of food has been carried out at 2,450 MHz and the first production installations have also been at this frequency, eventually major installations will shift to 915 MHz simply because of the power demand (i.e., hundreds of kilowatts of power). Experiments at this lower frequency indicate more uniform temperature distribution. All of the results are not yet in, but this has been the case with other major processes, including tempering and cooking bacon.

The first installations of microwave sterilization of ready-meals went on stream a few years ago. Total installed power is only a few hundred kilowatts, at 240 MHz. Process time is about 30 min, including cooling time. Interesting progress at the U.S. Army Natick Research and Development Laboratories (Natick, Mass.) has been made in development of a chemical method to assess process adequacy. Several chemical markers have been identified that are produced during normal processing of foods and correlate well with microbiological lethality. Thus, it is possible with a rapid analytical technique to determine the effectiveness of a process in minutes instead of the several days required for microbiological methods.

Microwaves may also find use in aseptic processing of flowable materials containing particulates, a process that has proven elusive because of the difficulty in ensuring that the fastest-moving particulates have received an adequate sterilization.

The adoption of microwave-based processes could influence the design of food plants of the future. Not only is space savings made possible by a reduced process time, but the reduced space requirements could have a powerful effect on process economics when one considers the cost of real estate. The labor requirement may also be reduced. A microwave tempering operation could be automated from freezer storage at least through the tempering stage. Pasta dryers already have reduced the space requirement three to four times with no reduction in production. Microwave bread baking could mean smaller bakeries located closer to their markets and have a significant impact on transportation costs. Microwave-assisted baking is another process that could find a niche in the supermarket bakeries to provide just-in-time baked products and thereby eliminate the daily loss of excess product to ensure latecoming shoppers of the same choice available to early shoppers.

Existing processes may be improved by including a microwave capability. Depending on the process, microwave heating could be applied as a preheating step, be built into existing equipment for simultaneous application of two or more forms of heating, or added on at the end of conventional equipment to increase the production rate.

Tools are available to evaluate new microwave processes without resorting to empirical methods alone. Microwave-compatible temperature-measuring systems are being used today. Infrared detectors are being used to follow surface temperature changes that can be translated into meaningful feedback data to control subsequent microwave input. Microwave radiometry, which has been demonstrated in measurement of temperature in human muscle tissue undergoing diathermy, could conceivably be used to noninvasively monitor product temperature during microwave pasteurization and sterilization to provide data for FDA approval. Continuous weighing systems have been used to monitor weight changes during microwave processing. There also is a large body of dielectric property data in the literature that can be useful in mathematical modeling and computer simulation in process development. All in all microwave processing has a promising future.

BIBLIOGRAPHY

1. U.S. Pat. 2,495,429 (Jan. 24, 1950), P. L. Spencer (to Raytheon Co.).
2. N. E. Bengtsson and P. O. Risman, "Dielectric Properties of Foods at 3 GHz as Determined by a Cavity Perturbation Technique. I. Measurements on Food Materials," *Journal of Microwave Power* **6**, 107–123 (1971).
3. R. E. Mudgett et al., "Prediction of Dielectric Properties in Solid Food of High Moisture Content at Ultrahigh and Microwave Frequencies," *J. Food Proc. & Pres.* **1**, 119–151 (1977).
4. T. Ohlsson et al., "The Frequency and Temperature Dependence of Dielectric Food Data as Determined by a Cavity Perturbation Technique," *Journal of Microwave Power* **9**, 129–145 (1974).
5. T. Ohlsson and N. E. Bengtsson, "Dielectric Food Data for Microwave Sterilization," *Journal of Microwave Power* **10**, 93–108 (1975).
6. E. C. H. To et al., "Dielectric Properties of Food Materials," *Journal of Microwave Power* **9**, 303–316 (1974).
7. D. I. C. Wang and S. A. Goldblith, "Dielectric Properties of Foods," Final Report, Contract DAAG17-72-C-0142, U.S. Army Natick R & D Command, Natick, Mass.
8. T. Ohlsson and P. O. Risman, "Temperature Distribution of Microwave Heating—Spheres and Cylinders," *Journal of Microwave Power* **13**, 303–310 (1978).
9. N. Meisel, "Tempering of Meat by Microwaves," *Microwave Energy Appl. Newsl.* **5**, 3–7 (1972).
10. L. Riedel, "Calorimetric Investigations of the Meat Freezing Process," *Kältetechnik* **9**, 38–40 (1957).

GENERAL REFERENCES

D. A. Copson, *Microwave Heating*, 2nd ed., AVI Publishing Co., Inc., Westport, Conn., 1975.

R. V. Decareau, *Microwaves in the Food Processing Industry*, Academic Press, Inc., Orlando, Fla., 1985.

A. C. Metaxas, *Foundations of Electroheat*, John Wiley & Sons, Chichester, United Kingdom, 1996.

A. C. Metaxas and R. J. Meredith, *Industrial Microwave Heating*, Peter Peregrinus Ltd., London, 1983.

R. V. Decareau and R. A. Peterson, *Microwave Processing and Engineering*, Ellis Horwood Ltd., Chichester, UK, 1986.

E. C. Okress, ed., *Microwave Power Engineering, Vol. 2, Applications*, Academic Press, Inc., Orlando, Fla., 1968.

ROBERT V. DECAREAU
Microwave Consulting Services
Amherst, New Hampshire

MILITARY FOOD

MILITARY SUBSISTENCE PROGRAM

Historical Background

To sustain troops under the operational demands and logistical constraints of modern warfare, military subsistence must possess characteristics markedly distinct from those of food products successful in the commercial marketplace. No program of research and development to diagnose and meet peculiarly military subsistence problems existed, or was even envisioned, until the then Office of the Quartermaster General (OQMG) established the Subsistence Research Laboratory at the Chicago Quartermaster Depot on July 24, 1936—less than six years before Pearl Harbor.

The World War I experience, without people knowledgeable in subsistence and unaided by systematic scientific investigation of military requirements, left procurement officials often dependent on almost miraculous improvisations by individuals within the food industry and the army. The troops were fed and the war won but the food margin was frequently so narrow that OQMG feared it could easily go the other way in the future if not addressed in time. The OQMG established the QM Subsistence School in 1919 toward generating a cadre of military experts in subsistence procurement, inspection, processing, transportation, and storage. Partly because of budgetary problems, the school was deactivated on June 30, 1936.

The new laboratory became heir to the school's research equipment, primarily a reasonably adequate kitchen, and was authorized $300 for supplies and equipment. It began operations with a staff of three: an officer-in-charge and two civilians who had worked at the school. Its official objectives were "to test foods and new style packaging of foods; prepare drafts of proposed specifications or reedit those which have become obsolete; to conduct studies of various reserve and emergency rations or components thereof; to prepare information bulletins and maintain liaison with other Government agencies." Its resources during the prewar period increased less rapidly than did its activities. By late 1941 its staff numbered 13. It had developed, field tested, and standardized the original C Ration, completed development of the D Ration, started work on the K Ration, essentially completed development of three-way frozen boneless beef, resumed the school's instructional functions, revised existing and prepared new specifications, and examined a melange of commercial products for potential military usefulness. This was done with no real advance knowledge, except possibly intuition, of the characteristics required for military suitability in the kind of conflict World War II proved to be. The type of food analyses the laboratory could then perform—mainly cooking tests, taste tests, and visual examination—was at best rough. It made a beginning in storage testing; laboratory members believed rations should be sufficiently palatable, as well as nutritional, to be eaten after long and adverse storage, but the implications of that belief were yet to be realized.

When World War II began there was virtually no experience in designing food or food packages for the conditions soon to be encountered overseas. A few separate agencies and individuals had done some research but without any attempt to translate results into food products meeting still-unknown military requirements. The food industry was concentrating on inexpensive and palatable foods that were stable enough to meet domestic distribution conditions. The small, poorly equipped, understaffed Subsistence Research Laboratory began its task of converting the immense industry to producing foods of a completely different type, despite labor and material shortages, as quickly as possible. Having no means of knowing precisely what was needed overseas, the laboratory began to formulate functional requirements as specifically as possible in the light of the best information it could obtain and to develop standard tests. Improvisation and trial and error were frequent in the evolution of products and rations. In general, the laboratory began to act as the hitherto missing military link between research groups and production groups—with a high degree of cooperation from both. That its work became effective was due largely to the notable cooperation it received from the industry.

For nearly three and a half years of the war, the laboratory concentrated on incorporating the most modern production procedures into specifications, enabling them to be put into operation in plants throughout the country. In 1944 it began to emphasize and accomplish development of new and better products and more acceptable rations. With the experience it had already gained, some overseas reports, increased facilities and additional technical personnel, it had a far clearer idea of what was needed and the means of obtaining it. Industry had also acquired experience and cooperation became more effective. With ability to define the broad types of military operation in terms of the subsistence problems distinguishing them, the laboratory could give clearer direction to its work by recognizing its goal as the development of food products and rations possessing the nutrition, stability, acceptability and utility properties required under the different operational conditions. By the end of the war it was recognized that science and technology would have to be expanded to meet known requirements more efficiently and to effectively respond to new demands.

Besides dealing directly with hundreds of food and food-packaging concerns and academic research groups, the laboratory had both formal and informal relationships with many government agencies, among them the U.S. Department of Agriculture, the Federal Food and Drug Administration, the Fish and Wildlife Service of the Department of the Interior, the other military services, and the Office of the Army Surgeon General. In recognition of its actual accomplishments and operations, it was redesignated the Quartermaster Food and Container Institute for the Armed Forces in 1946, and its mission, previously conducted without a specific charter, formalized. The new mission statement expanded relationships with other government agencies and added other military, federal, national, and international technical organizations. In 1950, as a result of Department of Defense (DOD) assignment to the army of primary responsibility for research and development of food, the institute was designated as the implementing agency for all DOD components, Following relocation of the institute to Natick, 1962–1963, DOD updated and expanded this responsibility in 1968 and added food service equipment.

LOGISTICS

Our national posture is defensive, requiring readiness to counter aggression wherever and whenever it occurs. The logistical demands inherent in this policy include need to stockpile enough food to ensure initial readiness for contingencies on a worldwide basis, transportability by any available means to and within operational areas, even when supply lines are threatened or cut by hostile action, and an economical production base to achieve mobilization objectives and to resupply troops once initial reserves are exhausted.

These demands present major and continuing challenges to our national food and packaging technical capability. They also impact the principles of nutritional adequacy, acceptability, stability, and military utility that the laboratory had formulated for operational subsistence during World War II.

Nutritional Adequacy

The primary purpose of military subsistence is to maintain the health and effectiveness of the persons subsisted. Military nutritional standards are based on the National Academy of Sciences Recommended Dietary Allowances (RDAs) as modified by military experience. As some of the known essential nutrients suffer loss in processing and preparation, raw, perishable foods would appear to be the best nutritional option, but the need for stockpiling, conservation of space and weight in storage and transport, and potential for worldwide distribution of the stored items obviously precludes reliance on the freezer facilities this would entail. In addition, availability of freezer and food service facilities and personnel on a worldwide basis during early days of emergencies is unlikely. To compensate for nutrients lost during processing and storage, it is necessary to fortify selected ration components that are stable carriers for the respective nutrients.

Acceptability

Nutrients are of no value unless consumed, and military experience has shown that troops will not necessarily eat "enough of anything" if they are hungry enough. At the least, unpalatable food is not a morale booster and can add to combat stress. The properties that make food desirable to eat—flavor, texture, odor, color and appearance—are sensory. Preservation to retain food wholesomeness over extended periods cannot be expected to improve these characteristics. A goal of food processing and preservation technology is to provide efficient, practicable preservation techniques that do minimal damage to initial food properties.

Consumer attitudes complicate acceptability problems. The people in the military services have generally formed their food habits, including likes and dislikes, before induction. Individual differences, changing food preferences, need to appeal to persons under extreme physical and emotional stress, for whom food may be the only break from the unpleasantness, discomfort, and monotony of fighting, are also part of the problem. So is the inherent American habit of griping about institutional foods.

Acceptability presents a major challenge to operational ration design. Capability of achieving it continues to challenge food science and technology.

Military Utility

A ration that cannot be used for the purpose intended will at best hamper the operation and the persons trying to conduct it. The criterion of utility affects the other attributes and ultimately dictates the form in which they appear. In general, all rations should be economical of weight and space in transport and storage, of facilities and labor in unloading, handling, and preparation, easy for the consumer to carry and eat, and should retain their configurations despite rough handling throughout the supply and distribution system.

Stability

Past experience has shown that operational rations must travel long distances, be exposed to wide temperature and humidity ranges, be handled roughly, and be stored where facilities are inadequate or none exist. Unless they retain the properties making them suitable and desirable for consumption, they are useless. Past emergencies have demonstrated both the economic waste and threat to preparedness of deteriorated, unserviceable supplies. Fielding of a ration entails capability of predicting its keeping quality under given time and temperature conditions, including prediction of the first date after manufacture at which stored rations must be opened to determine whether they should be issued, destroyed, or extended to another inspection period, along with the criteria for making these serviceability judgments.

Thus operational subsistence presents a continuing and open-ended demand on the nation's technical resources, but its demand on economic resources clearly cannot be open ended. Scale-up to volume production and continuing procurement on a competitive basis, within economic resources and without sacrificing a ration's essential characteristics, are major recurring hurdles. Unless they are overcome, the ration will be at best an academic curiosity with, perhaps, some aspects of knowledge or art it reflects applied to another, more workable item. There is irony in the recognition that products developed initially for specific military exigencies generally cannot be obtained for these unless they also prove to have some application in the civilian marketplace. Operational rations thus reflect compromises throughout their life cycle, but their essential characteristics cannot be compromised if the nation is to remain prepared.

CURRENT RATIONS

Military personnel located in fixed facilities are fed in dining halls not unlike those used for institutional feeding in the civilian sector. Because these situations are not unique, our focus in this article will be on foods used in field feeding situations that have no direct civilian counterpart.

Group feeding, where semifixed food preparation and serving facilities are possible, rations may be A Rations, B Rations, or T Rations, depending on the degree of sophistication of facilities, combat situation, availability of refrigeration, etc. A Rations are meals prepared from scratch, using primarily fresh or frozen ingredients, as well as many canned components. Although generally regarded as the most desirable, A Ration preparation requires facilities frequently unavailable in the field. In addition, trained cooks are needed to prepare and serve the food.

B Rations, prepared from shelf-stable ingredients such as canned foods, dehydrated foods, and packaged mixes also require trained cooks and equipment for frying, baking, broiling, etc, which may not be available, particularly in early stages of a conflict. T Rations consist of prepared foods thermally processed in tray packs, rectangular metal cans approximately the size of half-size steam table trays (10-1/16 × 12-3/8 × 2 in.). Sufficient tray pack entrées, starches, vegetables, and desserts have been developed to provide good menu variety for lunch/dinner meals, and improvements in breakfast item variety is currently a priority area of interest at Natick RD&E Center. T Rations offer the major advantage of requiring only heating, minimizing the need for equipment and trained food service personnel. Additionally, the tray can configuration allows processing of many popular foods such as lasagna, stuffed peppers and chicken breasts, which could not be thermally processed in conventional cylindrical cans.

For field-feeding situations where no preparation facilities are available and individuals must be self-sufficient, the basic ration is the Meal, Ready-to-Eat, Individual

(MRE). In addition, a number of special-purpose rations are stocked to meet specific, unique military needs, some of which will be described in this section.

The Meal, Ready-to-Eat, Individual has been in use since the early 1980s and has undergone major revision since first introduced. Based on retort pouch technology described in detail elsewhere, the MRE is a flexibly packaged, shelf-stable ration consisting generally of a casserole-type entrée or a meat/starch combination; fruit; crackers and spread; a cookie, cake, or brownie dessert item; as well as accessory items, condiments, candy, and beverage mixes. All MRE menus meet or exceed the recommended daily levels of vitamins, minerals, and calories prescribed for operational rations by the Office of the Surgeon General. Recent development of a highly acceptable packaged bread having a projected shelf life in excess of three years has enhanced the MRE acceptability and the technology developed for the bread is currently being applied to numerous other highly popular items such as shelf-stable packaged pizza, burritos, and a variety of new cake and snack items. Addition of these items as MRE components will add variety to the menu and further improve acceptability as a result of incorporation of more contemporary food items desired by today's service personnel.

To meet the unique needs of military personnel operating in the field during extremely cold weather where conventional hot meals cannot be provided, the Ration, Cold Weather (RCW) has been developed and is currently being procured. This ration, the first ever designed specifically for arctic/extreme cold conditions, is composed of dehydrated items that will not freeze. Salt and protein levels are adjusted to minimize water requirements, and several soup and hot drink mixes are included to help encourage adequate water intake because dehydration is a major problem in cold weather. A 4,500-calorie RCW weighs 2.7 lb. and has a volume of 225 $in.^3$ for a man-day supply of food. For comparison purposes, it would be necessary to carry four MREs to provide the 4,500 calories required for cold weather feeding. One RCW furnishes the total calorie requirement and saves 55% of the weight and approximately 40% of the cube when compared to the MRE.

The military inventory also contains survival subsistence items, generally in the form of small food packets. These are intended for use only for short periods of time in emergency situations. Since the space available for their storage aboard lifeboats and aircraft or in other survival gear is extremely limited, high-caloric-density foods are required. Also, because of the likelihood of little or no potable water available to personnel in such emergencies, foods low in protein and high in carbohydrate are required to provide the necessary calories while minimizing water requirements. Through the application of new technologies, a new survival ration is under development. This ration consists primarily of an assortment of high carbohydrate food bars including cornflake, shortbread, glucose, granola, and chocolate chip. The familiar flavor and texture of the new food bars will greatly increase acceptability, in comparison to current survival packet components.

TECHNOLOGY AND SCIENCE

Preservation Processes

The objective in processing or handling foods for military distribution is to retard or eliminate deteriorative processes that undermine food quality. Various biological and physicochemical processes, if uncontrolled, lead to spoiled and unsafe foods. Because microbiological contamination is the main concern, preservation technologies are used that retard microbial metabolism or destroy specific disease-causing or spoilage-inducing microbes.

One common strategy in retarding microbial growth is to limit the availability of water for utilization by the microorganism. It is particularly effective when combined with temperature lowering. Water can be made unavailable by freezing the food, by adding ingredients that tightly bind it, or by dehydrating the food either partially or completely.

The basic strategy in eliminating microbial growth is to destroy the microbes outright or to render them incapable of replication. When thermal energy or ionizing radiation energy is absorbed, cell rupture occurs, cellular processes are disrupted, or the DNA is irreversibly damaged. Thermoprocessing is very generally applied in stabilizing moist products for storage at ambient temperatures.

Dehydration

Freeze Drying. The primary dehydration technology used for certain ration components is freeze drying (1,2). By maintaining the product frozen well below 32°F and by removing the water through sublimation, highly desirable features are retained in the dried product. Freeze-dried components include fruits, vegetables, meat patties, and entrée items. All are blanched or precooked prior to processing.

Large-scale freeze drying by industry is done on a batch process basis following procedures that optimize the drying rate without compromising product quality. Pre-frozen products spread out on trays placed on radiant heating platens within a vacuum chamber are slowly processed to sublime away the moisture. Typically, the product temperature is maintained around 0– – 40°F and the chamber pressure is controlled to about 0.1–1.0 torr. As the process proceeds, the outer surface layer dries and the frozen ice core recedes. The energy for sublimation must transfer through the dried layer to the ice and the water vapor must migrate through the dry, porous layer to the surface. The transfer of this heat energy is driven by the difference between surface temperature and ice temperature and is affected by thermal conductivity of the porous layer. The migration of the moisture is driven by the difference between vapor pressure of the ice and the chamber pressure and is affected by the permeability of the porous layer. Typically, processors limit the slab thickness, adjust the platen temperature to maintain product surface temperature at or below 150°F, and try to achieve a suitable porosity in the product. Drying times are in the range of 10–15 h. The final product is packaged under vacuum or nitrogen in a water and oxygen impermeable material.

Reversible Compression. Because certain military rations must be compact, many freeze-dried components are reversibly compressed (3). A product that is "plastic" before compression recovers its original volume and textural properties upon rehydration. Such plasticity is achieved by spraying the freeze-dried material with enough moisture so that it has a uniform moisture content of about 12%. After compression, the product is redried and packaged in a protective container. Compression reduces the volume effectively, a ratio of 16:1 being achievable for green beans.

Microwave Freeze Drying. Microwave-assisted, freeze drying (MWFD) has been explored for producing products equivalent in quality to conventionally freeze dried products, but at lower cost. It has the advantage that the energy for sublimation is directly absorbed by the ice and that the remaining moisture can be continually redistributed throughout the sample. Control of microwave power and sample temperature is crucial, because the ice must not melt and the ice temperature must not be so low as to decrease the efficiency with which frozen water absorbs microwave radiation at 2,450 MHz. This advantage is evident in the decrease in drying time down to 5.5 h for green peas. Another advantage is the ability to dry the sample to a relatively uniform moisture level of 12%, which makes it directly compressible.

Centrifugal Fluidized-Bed Drying. Centrifugal fluidized-bed drying (CFBD) is another rapid drying technology. The fluidization is achieved by counterbalancing the centrifugal force exerted on diced or spherical samples contained in a rapidly rotating drum against the force of hot air streaming through small perforations in the drum wall. Air temperatures between 175–212°F are used. In 15 min, 50% of the moisture in green peas can be removed. Further drying would be inefficient and detrimental to quality. A combination of CFBD with MWFD could be used to optimize the dehydration process.

Thermoprocessing

Conventional Retorting. The primary thermoprocessing technology used in producing shelf stable tray-packed and pouch-packed ration components is retorting (4). By subjecting foods in these containers to high temperatures for specified times in steam-charged retort vessels, any contaminating disease-causing or spoilage-inducing microorganisms that might otherwise multiply during storage are destroyed. The minimum temperature–time requirement for the slowest heating point in a geometrically distinct container depends on the acidity of the product and on the heat resistance of the target microorganism. For most of the tray-packed and flexible pouch-packed components, the pH exceeds 4.5 and the targets are mesophilic spore-forming bacteria, which could grow between 50 and 105°F. It suffices then to ensure that the integrated temperature–time exposure is equivalent to 6 min at 250°F, which would reduce the population of the reference putrefying anaerobe, PA 3679, by 6 D (ie, six orders of magnitude) and which provides a large margin of safety for reducing *Clostridium botulinum* by about 24 D. For these products, the retorting temperature is set between 230 and 250°F, and the integrated lethality (ie, F_0) of the process takes into account the change in thermal resistance of PA 3679 as the product goes through the heat-up, cook, and cool-down cycles.

Industrial thermoprocessing of such different items as potatoes in butter sauce, frankfurters in brine, beef slices in barbecue sauce, lasagna with meat sauce, and spice cake is currently done in batch mode using either a still or rotating retort. The heating medium could be pressurized steam or water heated with steam. Tray packs or flexible pouches, after being either vacuum sealed or hot filled and purged with steam and then sealed, are loaded into the retort vessel. Once the process is begun, the heat is transferred from the surface of the container into the food. The flat shape of the 2-in.-deep tray-pack container and the 0.5-in.-thick profile of the pouch, in principle, allow the heat to penetrate quickly and uniformly, which should produce a high-quality product. If the food has a significant fluidity, the heat penetration is aided by natural convection; if it is solid, the penetration is exclusively conductive. In a rotating retort, the agitation introduced further facilitates convective heat transfer and would lead to a faster, more uniform heat treatment.

Before the process for a particular product in a specific retort is filed with the USDA or FDA, verification must be obtained that the cycle to be used ensures attainment of commercial sterility at the slowest heating point in the slowest heating container. For this purpose, thermocouples are used to obtain heat penetration data and calculations are made to specify the cycle time and temperatures needed to achieve the minimum F_0. Processors typically will add a safety factor to ensure that the processing is in excess of the requirement for microbial lethality. If an entirely new product or container has been developed, inoculation studies may be required.

Ration components so processed are not only safe, but show high retention of nutrients and score high in ratings of color, flavor, and textural attributes. However, some products might be overprocessed due to wide variation in treatment from container to container. Techniques are now being investigated to validate, noninvasively, that each container received a treatment that is greater than the minimum requirement and less than the maximum set as a high-quality limit. One promising technique uses a optically read, bar-coded label with a patch that becomes darker in color as the integrated temperature–time exposure increases.

Aseptic Processing. Another thermoprocessing technology that is receiving attention for possible future use is aseptic processing. It is a high-temperature, short-time exposure process that sterilizes the food product before it goes into a container. In this process, a fluid product with relatively large particulates, such as a stew with chicken cubes, potato dices, and carrot dices, is pumped through a scraped surface heat exchanger, then through a holding tube maintained at the specified high temperature, then through a rapid cool down system, and finally into a sterile filler unit utilizing surface-sterilized packaging material. Temperatures of 260–280°F can be used, and residence times of about 3 min in the holding tube are all that is needed to ensure that the center of the particulate has received an adequate lethal treatment. This technology has several inherent advantages including high quality and nutrient retention, lower overall energy input, continuous processing, and the option to package the product in any size container without the need for changing and verifying process parameters.

SENSORY ANALYSIS OF RATIONS

As described in preceeding sections, the development of today's rations is driven by the logistical requirements of light weight, low volume, minimal preparation, and long-term stability under climatic extremes. These requirements have forced dramatic changes in the ingredients and processing technologies for rations and have resulted in significant changes to the familiar appearance, flavor, and texture of the foods and beverages served to the soldier. These changes have the potential to dramatically affect the acceptability of the rations and, in turn, the soldier's nutritional status, because rations that are not acceptable will not be consumed.

To ensure that military rations retain high standards for quality and consumer acceptability, all rations are engineered to specific sensory criteria and levels of consumer acceptability. The process of evaluation used to insure these criteria includes a wide variety of sensory and psychophysical techniques. Many of these methods and their applications in military ration development have been published over the years (4–10).

Primary among the methods are those known as affective, ie, those that address the criteria of consumer liking for the product. In one of the largest in-house laboratory consumer-testing programs in the world, all military rations undergo extensive testing in the computerized sensory analytic laboratories at Natick, where they are evaluated by a volunteer panel of over 500 civilian and military personnel. The most commonly used rating scale in these tests is the nine-point hedonic scale developed in the 1950s at the Quartermaster Food and Container Institute in Chicago, Natick's predecessor laboratory. This simple scale with points labeled from "dislike extremely" to "like extremely," has been shown to be both a valid and reliable measure of consumer acceptability and is now employed throughout the food industry to assess likes and dislikes for foods and beverages.

In addition to hedonic testing, which can be conducted either in the laboratory or in the field, extensive attitudinal testing is conducted on rations and ration concepts. These tests frequently take the form of written surveys administered to subjects to assess their beliefs, feelings and attitudes toward foods, beverages, meal items/concepts, methods of packaging and preparation, food names, labels, condiment usage, etc. These data provide useful quantitative information for product and package developers to best meet the needs of the soldier in the field.

During the past few years greater emphasis has been placed on qualitative methods of consumer research, eg, focus group testing, for the assessment of consumer perceptions of ration and ration concepts. These qualitative procedures now supplement the quantitative survey and interview approaches. In addition, the concept of marketing military rations has been emphasized through closer examination of the importance of food names, packaging designs, and the role of consumer expectations in military feeding.

In contrast to the affective methods of ration assessment are the analytic methods, or those that use specialists with training in the lexicon and evaluation of the sensory characteristics of food. These methods include such approaches as flavor and/or texture profiling and food quality scoring. They are frequently used when (1) designing a ration item to meet an ideal or "target" sensory profile, (2) developing instrumental tests to correlate with specific sensory attributes, or (3) assessing the degradation of flavor or texture with time in storage.

Last, but no less important, are the wide variety of univariate and multivariate psychophysical methods that are applied to rations. These methods are most often used to analyze the combined effects on the soldier–consumer of the multiple sensory attributes of each ration item, and/or to determine the levels of each attribute required to achieve maximal acceptability of the item. These methods of analysis, coupled with the traditional methods described previously, have enabled military ration developers to more quickly achieve a desired target product than has previously been possible. Moreover, this sensory technology has been used to assist numerous other government agencies, such as the U.S. Departments of Commerce and Agriculture, who frequently engage Natick to help solve problems of national concern and attention. Such technology transfer from the military feeding program to the public sector is a spin-off of a military program designed to solve problems on the cutting edge of food technology.

In summary, the engineering of today's military rations is keyed on the criteria of sensory quality and soldier acceptability. Through detailed consumer and sensory analytic testing in the laboratory and subsequent field testing, it can be insured that fielded rations will be well liked and consumed with minimum waste, so that all of the available nutrient content will be used by the soldier to optimize performance and maintain readiness for combat.

RATION TESTING AND EVALUATION

The evaluation of military rations has traditionally focused on the sensory and hedonic responses to the food items that make up the ration (see above). In a typical sequence the food item is first tested under controlled conditions in a laboratory setting with civilian panelists. If the product is satisfactory, it is incorporated into the ration and subsequently tested for its acceptability to military consumers in a field-feeding context. If the ration is acceptable under field conditions, it is purchased and incorporated into the military feeding system.

Several converging lines of evidence have led to a reappraisal of this methodology. First a series of recent ration field tests that employed measures such as body weight, food consumption, water consumption, hydration status and consumer perceptions of the ration as well as the more traditional acceptance measures revealed that this more comprehensive profile is needed to adequately assess the effectiveness of operational rations for long-term use (11,12). These studies also revealed that, although the rations received high acceptance scores in the field, they were not consumed in sufficient quantity to meet the surgeon general's recommended daily allowance for calories. A decision to feed these rations over an extended time period based solely on acceptance measures would have been a mistake. A second shortcoming of relying solely on acceptance measures was revealed in a series of laboratory and field studies (13,14). This research showed that social and situational variables, can influence food acceptance and consumption as much as the intrinsic properties of the food. The third consideration that prompted a review of the way military rations are tested is that rations of the future are likely to emerge as unfamiliar and unconventional products in order to meet ever more stringent weight and volume requirements. Based on these factors, the U.S. Army Natick Research, Development and Engineering Center has developed a comprehensive seven-phase ration testing program.

Phase I—Consumer Marketing

The purpose of this initial step in the ration-development process is to identify potential items or ration concepts that match customer interests and requirements. It uses focus group testing, product profiling, marketing analyses, and food-preference surveys to define for the product developer what the military user would like in the ration. This phase of the testing program emphasizes that the military consumer is a customer who cannot be treated as a passive recipient if rations are to be successful products.

Phase II—Individual Item Sensory Testing

This phase is initiated when the product developer has fabricated prototype items and serves to insure that people find the food item acceptable.

Phase III—Consumer Meal Testing, Laboratory

Food items that passed through Phase II would be tested to determine their suitability as components of meals where item compatibility was evaluated. This work would be conducted in the laboratory in the context of a meal using civilian and military consumers where selection rate, item and meal acceptance, and consumption would be measured.

Phase IV—Meal Testing, Field

Phase IV would be identical to phase III except testing would occur with military consumers in a field environment.

Phase V—Prototype Ration Testing

When a sufficient number of food items have successfully passed through the preceding phases of testing, a new prototype ration would be assembled and compared to the ration it was meant to replace in a field test. The measures taken during this field test would include selection rate and consumption of all items, daily nutrient and fluid intake, body weight, hydration status, food acceptance, and consumer satisfaction with the utility of the ration. The testing would occur in all appropriate environments (hot, cold, temperate, and altitude) during field-training exercises. The data from these tests would serve as the basis for a precurement decision for rations that did not pose unusual nutritional problems or were designed for extended periods of use. Rations that fell into these latter categories would be more completely tested in Phase VI.

Phase VI—Extended Ration Use Validation

Some rations are designed to be used for extended time periods or are for special missions and do not provide adequate nutrition. Rations of this nature would be tested by the Military Nutrition Division of the U.S. Army Institute of Environmental Medicine to fully assess the ration's impact on health and performance. In addition to the measures employed in Phase V they would also monitor nutritional status, physical symptoms, as well as physical and mental performance.

Phase VII—Quality-Control Testing

This phase of the ration testing program is included to insure that the quality of rations that are part of the military feeding system are being maintained at their original, high level by the manufacturers. The sensory testing methods employed in Phase II would be used to answer this question.

At the present time this ration testing program is being introduced. Phases I, II and V and VI have been in operation for several years and have led to improvements in the operational rations (12). The program will be evaluated and modifications made as experience is obtained.

NUTRIENT STABILITY

As indicated above, nutritional adequacy of foods provided to military personnel is paramount. Because of the necessity to use processed foods with long storage lives, nutrient and microbiological stability are essential to insure that nutritional content is adequate and that foods are wholesome at time of consumption. An area of major emphasis, therefore, is focused on nutrient losses during processing and subsequent storage. Space constraints preclude a detailed discussion of all work ongoing in these areas. Following is a summary of several recent or current studies at Natick RD&E Center in the areas of nutrient and microbiological stability.

Proteins

Among the various degradative reactions that take place in foods, the most difficult to arrest are those between proteins and reducing carbohydrates. This situation is even more aggravated when foods, such as combat rations are stored for extended periods without refrigeration. Even when reducing carbohydrates are not part of product formulations, hydrolysis of the polysaccharides and oligosaccharides may occur during processing or during prolonged storage, and then protein quality loss will follow.

Maillard Reaction. Although it is generally recognized that protein quality losses occur during processing and storage through the Maillard reaction, the quantitative aspects of the protein-reducing sugar reaction, which leads to a decline in lysine availability, are subject to interpretation depending on the methodology employed (15). In the initial phase of the reaction, the free epsilon amino groups of lysine and *N*-terminal amino groups in the protein react with the reducing sugar to form a Schiff base, which cyclizes then rearranges to form a ketose sugar derivative, commonly referred to as the Amadori compound. This compound is not digested by mammalian proteolytic enzymes and, therefore, becomes biologically unavailable. Furthermore, the amino acids adjacent to the blocked lysine in food proteins may also be unavailable due to steric hindrance to enzymatic action. Acid hydrolysis of foods that have suffered quality loss due to Maillard reaction leads to the formation of a cyclic compound, furosine, in 20% yield, which may serve as an indicator of unavailable lysine (16). It has been used in Natick as a marker to determine protein quality losses in stored dairy bars.

Aminocarbonyl Interactions in Model Systems. Investigations focused on interactions in model systems between lysine, acetyllysine or albumin, and glucose at 40, 50, and 60°C and a low water activity, $a_w = 0.20$. The reaction was monitored by measurement of color, fluorescence, reducing capacity, furosine, glucose, and lysine. There was poor correlation of fluorescence and color increase with lysine loss. Both furosine and reducing capacity correlated highly with lysine loss (17).

Bioavailability of Essential Amino Acids in Food Products. The enzymatic availability of essential amino acids in beef stew processed using three technologies (retort pouch, retort can, and freeze-dehydration) were determined using three proteolytic enzymes (pronase, carboxypeptidase, and amino peptidase), followed by derivatization of amino acids with phenyl isothiocyanate and analysis by HPLC. As expected, because of less total heat input required to attain a comparable sterilization value in the retort pouch, lysine and methionine losses were slightly less in pouches than in # 21/2 cans. Accelerated storage at 125°F for 28 days did not further decrease available lysine.

Lipids

Autoxidation and Antioxidants. The central problem has been the radically different course of lipid autoxidation and its prevention by antioxidants in dispersed versus bulk systems. This was addressed by two systems: lipids coated as a monolayer on silica powder and lipids freeze dried onto carboxymethylcellulose. Oxygen uptake measurement was by gas partition chromatography. Among the findings from the monolayer work was the ordering of the relative effectiveness of antioxidants for lipids coated as monolayers on surfaces (18). The carboxymethylcellulose work revealed the relative effectiveness of many known and some previously unknown natural antioxidants. Their synergism with food-grade synthetic antioxidants was also proved (19).

Fluorescent Detection of Lipid Oxidation Products by Polyamide Surfaces. It was discovered that the volatiles from oxidizing lipids produce surface fluorescence on polyamide powder coated on glass or plastic (20). This enabled rapid vapor phase monitoring of oxidation either in tests of antioxidant effectiveness or in in-package quality monitoring.

Relative Effectiveness in Bulk Lipids vs Dispersed Lipids. Using cobalt-accelerated soy lecithin sonicated emulsions (liposomes) and dispersed and freeze-dried whole milk, a rationale to predict the effectiveness of polar versus nonpolar antioxidants in bulk versus dispersed systems has been formulated (21). In general, this rationale predicts that nonpolar antioxidants, eg, butylated hydroxytoluene (BHT), function best in polar lipid emulsions and membranes, while polar antioxidants, eg, gallic acid, are relatively more effective in nonpolar lipids and in bulk systems.

Rapid, Front-Face Fluorometric Methods of Quality Loss Assay on Comminuted Whole Tissue Foods. Experience with front-face, solid-surface fluorescence measurement extended the method to comminuted slurries of whole tissue foods. Rapid assays of the state of lipid oxidation, Maillard sugar-amine, and ascorbic acid-amine browning are possible with good discrimination between the types of quality loss. Dried whole milk and retorted wet-pack fruit and freeze-dried fruit have been studied (22).

Vitamins

Vitamins can be lost from foods during processing, storage, and preparation for consumption. Certain vitamins such as thiamin and ascorbic acid are more sensitive to processing conditions than are others such as riboflavin. However, riboflavin can be lost through exposure to light, so packaging becomes important in determining storage stability.

Ration Processing Effects. Studies on tray packs have concentrated on the effects of processing and preparation for serving on the vitamins thiamin, riboflavin, pyridoxine, and ascorbic acid (23–25). The products studied were beef stew (fortified at a level of two times the RDA per meal), sliced pork, and peas and carrots. It was found that riboflavin was stable to processing while thiamin content decreased by 53%, 68%, and 57%, respectively. Vitamin B_6 also showed losses due to processing. Ascorbic acid loss from peas and carrots due to processing was 72%, whereas the fortified beef stew showed only a 20% loss of this vitamin.

Ration Storage Studies. Both the Meal, Ready-to-Eat (MRE) (based on retort pouch technology) and the Long Range Patrol (LRP) (based on freeze drying technology) have been subjected to short- and long-term storage at 4°C, 21°C, and 38°C (26). The 1973 prototype MRE was nutritionally adequate after storage at 38°C for six months and at 4 and 21°C for four years. Losses of thiamin, riboflavin, pyridoxine, vitamin A, and ascorbic acid occurred at 21 and

38°C. Niacin and carotene were stable. In the 1980 procurement MRE storage study, the indicator vitamins thiamin and ascorbic acid were followed in individual meals and/or in composites of cases containing 12 meals. Case composites showed a loss of thiamin, but after 24 mo at 38°C this loss was only 12%. Ascorbic acid showed a 50% loss from cases after 24 mo at 38°C, but no loss after 60 mo at 4°C. As with the MRE, the LRP showed good stability of thiamin at 4°C. At 21°C losses did occur, being greatest for chicken stew (39%) and beef stew (46%) over 13 years of the study.

MICROBIOLOGICAL STABILITY

The role of microbiology in military subsistence systems is to ensure the safety and microbiological integrity of food-delivery systems and rations. This responsibility must respond to conventional feeding systems as well as those, present and future, that are considerably less conventional. These rations are prepared and delivered to their customers in a wide variety of facilities (dining halls, hospitals, submarines, ships, airplanes, in the field, etc) at widely scattered geographical locations and are either perishable or microbiologically stabilized rations produced commercially or under contract.

The methods currently used to stabilize military rations from microbiological deterioration and to render them safe are thermal processing, water activity control (dehydration, lowering osmotic pressure with humectant, salts, freezing, etc), low pH, microbial growth inhibitors (nitrites, benzoate, acids, etc), and refrigeration.

Inoculated pack studies on complex military rations that have been developed in-house are conducted to determine their thermal processing requirements. For these studies a simple, biphasic technique, which yields extremely heat resistant and clean spores of *Clostridium sporogenes* strain PA 3679, has been developed.

The problem of thermophiles, which are highly heat resistant organisms that grow above 45°C, is significant in military rations due to the increased chance of temperature abuse of rations during distribution and storage. Therefore, commercial sterility, based on minimum heat input necessary to provide stability and an acceptable safety margin for foods handled in a carefully controlled distribution environment, may be inadequate for military rations.

Experience has shown that to control or destroy microbial populations in a ration system of diverse foods, it is necessary to study each food group. For example, beef snacks, which are low-moisture foods designed for use in the field, must be processed so as to attain a water activity that will ensure that the organism of main concern for safety, in this case *Staphylococcus aureus*, will not grow. Water activity is the ratio of the water vapor pressure of the product to the vapor pressure of pure water equilibrated at a given temperature and is a measure of biological moisture available for microbial growth. Similarly, the minimum water activities for cakes, formulated for the tray pack container configuration, were also determined. The organism of concern in this case being *Clostridium botulinum*.

It is often necessary to use combinations of microbial inhibitory agents and environmental conditions to achieve microbial stability. The three-year shelf life required for canned bread is attainable by the combined use of a sufficiently low water activity to prevent the growth of *Clostridium botulinum* and the inclusion in the package of an oxygen absorber to prevent spoilage by fungal growth.

The military is also concerned with being able to predict and monitor the shelf life of perishable rations. This effort has resulted in the issuance of a guide designed to prevent the unnecessary discarding of temperature abused, although still safe and edible, perishable rations in commissaries.

To further ensure the safety of perishable foods a medium for quantitating *Listeria* organisms was developed which is particularly efficient in recovering heat-injured cells. This was accompanied by the development of a technique that can identify pathogenic *Listeria* in less than eight hours.

A program to evaluate microbial inactivation kinetics for the aseptic processing of particulates for application for future ration design is being actively pursued. For this purpose extremely thermoresistant *Bacillus stearothermophilus* spores have been produced. The application of modified atmospheres (carbon dioxide, nitrogen, oxygen removal, etc) to inhibit microbial growth and extend the shelf life of perishable foods is also being studied.

QUALITY MONITORING OF STORED FOODS

Rations formulated and processed to retain their nutritive value, to prevent microbial growth, and to appeal to the military man or women must be able to maintain such quality attributes throughout storage and distribution. Unfortunately, food cannot be preserved indefinitely and will slowly show some degradation. Several degradative processes take place, even in canned products, that are due to chemical, biochemical, or physical reactions. Browning in cream sauces and in fruits reflects the chemical reaction between sugars and proteins; decomposition of proteins or fats, which could affect texture or flavor, reflects an enzyme-catalyzed biochemical reaction; and the crystallization of starch in breads, which contributes to the perception of staling, reflects a physical process. All of these processes are influenced by the temperatures encountered in transportation to and from storage facilities and by the fluctuation in temperature during storage. Rapid, simple ways of accounting for or monitoring such degradation are needed to simplify the logistical burden of rotating products out of storage and into the hands of users before the quality of these products becomes significantly compromised.

Shelf Life Dating

Based on years of experience and extensive research by academia, industry, and government, the basic shelf lives of fresh or processed products stored at specified temperatures are known (27) and are used in establishing Pull Date, Sell by Date, or Best if Used by Date. These designations represent the dates at which a time that roughly corresponds to the product's shelf life has elapsed since the Date of Pack. One or another of these dates printed either explicitly or in coded form can be seen on many products intended for refrigerated, frozen, or room temperature storage. Although normally adequate for situations where temperature is relatively well controlled, this approach is inadequate to ensure quality if the storage equipment fails or if high average temperature is unavoidable.

Temperature Effects

The influence of the combined temperature–time effect can be illustrated with results on the acceptance of Meal, Ready-to-Eat components stored at different temperatures (28). Based on shelf life defined as the time it takes for an acceptance rating to decrease to 5 on a 1 to 9 hedonic scale, researchers found a wide variation in shelf lives among different components at any single temperature and in the sensitivity of shelf lives to temperature changes. At 80°F, the average shelf life for all entree items was found to be 140 mo, whereas pastry items last for only 40 mo. Most important, the pastry items are insensitive to changes in temperature, the shelf life dropping slightly to 32 mo as the temperature increases to 110°F, while in contrast, the fruits are extremely sensitive, their average shelf life dropping from 180 mo at 90°F to 20 mo at 100°F. The combined effect of temperature and time is such that, in terms of quality degradation, exposure for short times at high temperature is equivalent to long times at lower temperatures.

Time–Temperature Indicators (TTIs)

Research into several different time-temperature indicators that integrate this exposure and reflect the cumulative history of such stress has been undertaken. One promising indicator label made by I-Point Technologies is based on the enzymatic decomposition of fats, which produces an acid and causes a color change in the label from green to yellow to red. Another promising indicator made by LifeLines Technologies contains, in addition to the bar code, a patch with a substituted diacetylene monomer that is initially white and eventually becomes bluish and purplish as it polymerizes. Both can be made with different formulations to alter the sensitivity to temperature. The LifeLines TTI can be scanned with an optical reader to obtain product information and the percent reflectance of light from the patch. In studies of products with such labels and exposed purposely to different temperatures, data have been obtained indicating that the change in color reflectance correlates with changes in product quality. From such correlations, a scale can be developed from which an individual responsible for managing stock rotation could determine, from the percent reflectance of the patch, the time remaining at a given temperature before the shelf life limit is reached.

The suitability of such labels for stock management ultimately depends on how well the label temperature sensitivity matches the product's temperature sensitivity. If the label is more sensitive, it will overstate the loss in quality; conversely, if it is less sensitive, it will understate the loss. Moreover, the particular quality attribute in a product that most affects acceptance must be identified and the temperature sensitivity of this attribute should serve as the basis for selecting a TTI label. Current research indicates that sufficient flexibility exists in adjusting label temperature sensitivity to achieve the desired match.

Automated Stock Management

Other investigations are in process to establish an automated system for monitoring the integrated temperature stress on stored rations and for deciding on when to rotate the stock. In principle, it is possible to read such label with a handheld optical scanning wand, to store the data in a handheld computer system, and then to transmit the data over the phone lines to a central data collection and decision-making facility. Using the concept of Least Fresh—First Out, any product that experiences a high temperature stress would be stored for a shorter period than originally planned to ensure good acceptance when it is eventually delivered for use. The actual quality of the product becomes the determining factor instead of an assumed quality based on shelf life and date of pack.

FUTURE CONCEPTS

Calorie-Dense Rations

The need to reduce the size and weight of supplies soldiers must carry and to provide them with an eat-on-the-move capability has motivated research on calorie- and nutrient-dense rations. Provided that these rations are well accepted, their compactness and convenient shape offer significant benefits with respect to soldier performance and to ration storage and distribution. Consequently, some new concepts for ration design and tailoring are being explored.

Dehydration, Infusion, Compression, and Extrusion. Certain conventional and novel technologies, alone or in combination, can be considered for making calorie-dense ration components (29). Caloric density is defined as kilocalories per cubic centimeter (kcal/cc). One conventional approach involves dehydrating the product, which leads to a porous food, and then compressing it, which removes most of the air space. Caloric densities of about 1.1 kcal/cc can be obtained this way. One novel approach now being explored, even for civilian products, involves infusing a porous food with energy-rich ingredients, filling all of the air spaces with stable and nutritious constituents. Caloric densities between 5 and 6 kcal/cc are achievable. The technologies for producing the porous food could be either conventional dehydration or the more recently exploited cooking-extrusion process. Extrusion involves processing a flour and water mixture under high shear and high temperature within a confining extruder barrel so that, when the doughy mass is forced through a die opening, the superheated moisture rapidly vaporizes, blowing holes in the extrudate and forming a honey-combed structure. Commercial corn curls illustrate this type of product. Another way to achieve high density that does not require a structured, porous material is to directly compress dried powders.

Caloric Densities and Compactness. The volume reduction achievable through use of these technologies can be put into perspective by comparing some of the past and current rations. The Meal, Combat, Individual (MCI), with its moisture-laden food packed in metal cans, had a caloric density of 0.8 kcal/cc; the Meal, Ready-to-Eat (MRE), with similar foods packed in flexible pouches, weighs less but has a similar caloric density. The Long Range Patrol (LRP), with its dehydrated but noncompressed components, weighed less but had only a slightly higher density of 1.1 kcal/cc. The Food Packet Assault (FPA), with dry and compressed components, reached 1.5 kcal/cc, and the new Ration, Lightweight 30 Day (RLW30), with carefully selected and optimally processed dry and compressed components, reaches a high of 3.7 kcal/cc. Several experimental rations have reached as high as 7.1 kcal/cc. Assuming that the goals for acceptance, compactness, and cost are optimized, it should be possible to produce a ration with a caloric density of 4.8 kcal/cc. This sixfold greater density compared to the MRE means that, for an equivalent amount of calories, the proposed new ration would take up only one-sixth the space.

Current Thrusts. Research in this area is concentrating on refining the technologies so a ration can be produced with any prescribed nutritional or physical characteristics. The texture and infusibility of an extruded matrix are determined by the size and number of the air spaces as well as by the thickness and fragility of the walls defining those spaces. The nutritive value, the physical stability, and the textural attributes of the liquid with suspended particles that are infused into the extrudates are very much influenced by the size and shape of these particles and by their ability to penetrate the extrudate (30). Industry is currently infusing certain dried fruits with sugar solutions to increase their stability and appeal. Because one of the ultimate goals is to be able to tailor the ration composition to satisfy particular nutritional requirements for different climatic or tactical situations, the research should lead to an ability to produce the prescribed components by specifying the formulation and by adjusting the process parameters.

Self-Heating Rations

There has always been a need to heat certain foods to palate temperature to improve the taste, which in turn enhances the morale of the soldier in the field. Currently, the heating is done by burning a fuel tab under a canteen cup filled with water into which the soldier places a pouch-packed entrée item, such as beef in barbecue sauce. Heating food this way is relatively inefficient, limits the soldier to a fixed location, and, because of the open flame, sends out a visible signal. Newer developments aim for chemical methods of flamelessly heating the food within its packaging and while the soldier is on-the-move.

Chemical Heating Pad. A new product has been developed, through joint military—civilian cooperation, that uses a water-activated chemical system for efficiently generating and transferring heat (31). It contains a supercorrosive magnesium–iron alloy powder blended with sodium chloride and imbedded in a porous polymeric matrix. It is produced as a thin, rectangularly shaped pad enclosed in a perforated paperboard package and weighs 20 g. The exothermic reactions that take place with 2 oz of added water include the oxidation of magnesium, reduction of oxygen, and formation of gaseous hydrogen. Approximately 33 kcal of energy are released. The high ratio of energy released to the weight of the pad and the action of the streaming gas and steam in heating the packaged ration component make this product attractive.

Flameless Ration Heater (FRH). The initial application for this chemical heating pad will be as a flameless heater for MRE entrées. Whether packed with the MRE or issued separately, the FRH includes a heater pad in a polyethylene bag that is sized to accept the MRE pouch and is marked with a line to indicate a level of added water equivalent to 2 oz. It is capable of raising the temperature of an 8 oz entrée from 40 to 140°F. In practice, the entree is placed in the bag, the water added, the top of the bag folded over to avoid spillage, and the whole system placed back in the small carton originally used to protect the pouch. The carton is then set aside or carried (perhaps in the pocket of the soldier's uniform) for about 10 min while the heating takes place. If left in the carton, the entrée will stay warm for an hour.

Self-Heating Individual Meal Module. Another application of this chemical heating pad will be for use in an individual meal module designed as a laptop serving tray package containing all ration components, including utensils. The heater pad would be located in recessed sections of the tray just under the components to be heated. Water could be included in a puncturable reservoir or added when needed. In either case, channels would be included in the package to direct the water to the pads and to focus the hot gases around the compartments with the heatable components. Entrées, starchy items such as noodles, and vegetables such as carrots would be heated to the desired temperatures by judicious choice of the pad size and weight. This same general concept could be used to cool dessert items by choosing reagents whose reactions with water are endothermic.

Future Heating Concepts. Because this chemical heater concept is not practical for boiling water, other flameless heating concepts are being considered for supplementary use. A small device in which organic fuel could be catalytically combined with oxygen, without the propogation of a flame, would be especially attractive. Simple ways to initiate the catalytic combustion and to transfer the heat to a canteen would have to be included. Prospects for meeting these requirements are good.

BIBLIOGRAPHY

1. R. P. Singh and D. R. Heldman, *Introduction to Food Engineering*, Academic Press, Inc., Orlando, Fla., 1984.
2. M. Karel, O. R. Fennema, and D. B. Lund, *Physical Principles of Food Preservation. Part II*. Marcel Dekker, Inc., New York, 1975.
3. C. Andres, "New Class of Foods—Compressed Foods," *Food Processing* (Winter 1973).
4. A. V. Cardello and O. Maller, "Relationships Between Food Preferences and Food Acceptance Ratings," *Journal of Food Science* **47**, 1553–1561 (1982).
5. O. Maller and A. V. Cardello, "Ration Acceptance Methods: Measuring Likes and Their Consequences," *Nederlands Military Geneeskundig Tijshrift*, 91–96 (1984).
6. A. V. Cardello and O. Maller, "Psychophysical Bases for the Assessment of Food Quality," in J. Kapsalis, ed., *Objective Methods in Food Quality Assessment*, CRC Press, Inc., Boca Raton, Fla., 1987, pp. 61–125.
7. A. V. Cardello and O. Maller, "Sensory Texture Analysis: An Integrated Approach to Food Engineering," in H. R. Maskowitz, ed., *Food Texture*, Marcel-Dekker, Inc., New York, 1987, pp. 177–215.
8. A. V. Cardello, O. Maller, and M. V. Klicka, "Computer-aided Sensory Analysis and Evaluation: A Comprehensive System for the Assessment of Foods and Beverages," in H. R. Maskowitz, ed., *Applied Sensory Analysis of Foods*, CRC Press, Inc., Boca Raton, Fla., 1988, pp. 35–81.
9. H. L. Meiselman, R. Popper, and E. Hirsch, "Sensory, Hedonic and Situational Factors in Food Acceptance and Consumption," *Proceedings of the International Symposium on Food Acceptability*.
10. R. Popper, E. Risvik, H. Martens, and M. Martens, "A Comparison of Multivariate Approaches to Sensory Analysis and the Prediction of Acceptability," *Proceedings of the International Symposium on Food Acceptability*.
11. E. S. Hirsch et al., "The Effects of Prolonged Feeding of Meal-Ready-to-Eat (MRE) Operational Rations," *Technical Report NATICK / TR-85 / 035*, 1985.
12. R. D. Popper et al., "Field Evaluation of Improvement MRE, MRE VII, and MRE IV," *Technical Report NATICK / TR-87 / 027*, 1987.
13. H. L. Meiselman and E. S. Hirsch, "Situational and Sensory Factors Controlling Acceptance of Military Rations," *Proceedings of the 29th Conference of the Military Testing Association*, 1987, pp. 303–308.
14. H. L. Meiselman, E. S. Hirsch, and R. D. Popper, "Sensory, Hedonic and Situational Factors in Food Acceptance and Consumption," in D. M. H. Thompson, ed., *Food Acceptance*, Elsevier, 1988, pp. 77–87.
15. J. Mauron, "The Maillard Reaction in Food. A Critical Review from the Nutritional Standpoint," in C. Eriksson, ed., *Progress in Food and Nutrition Science. Maillard Reactions in Food*, Pergammon Press, Ltd., Oxford, UK, 1981.
16. R. F. Hurrel and K. J. Carpenter, "The Estimation of Available Lysine in Food Stuffs after Maillard Reaction," in Ref. 15.
17. K. A. Narayan and R. E. Andreotti, "Kinetics of Lysine loss in Compressed Model Systems Due to Maillard Reaction," *Journal of Food Biochemistry* **13**, 105–125 (1989).
18. W. L. Porter, L. A. Levasseur and A. S. Henick, *Journal of Food Science* **42**, 1533 (1977).
19. S. J. Bishov and A. S. Henick, *Journal of Food Science* **40**, 345 (1975).
20. W. L. Porter et al., "Analytical Use of Fluorescence-Producing Reactions of Lipid and Carbohydrate-Derived Carbonyl Groups with Amine End Groups of Polyamide Powder," in G. R. Waller and M. S. Feather, eds., *The Maillard Reaction in Foods and Nutrition*, ACS Symposium Series 215, American Chemical Society, Washington, D.C., 1983, p. 47.
21. W. L. Porter, E. D. Black and A. M. Drolet, *Journal of Agricultural and Food Chemistry* **37**, 615 (1989).
22. W. L. Porter et al., "Proceedings of the Second Natick Science Symposium," *Natick / TR-88 / 052*, June 1988, p. 175.
23. M. H. Thomas, B. M. Atwood, and K. A. Narayan, "Stability of Vitamins C, B_1, and B_2, and B_6 in Fortified Beef Stew," *Natick / TR-86 / 061*, Sept. 1986.
24. M. H. Thomas, B. M. Atwood, and K. A. Narayan, "Effect of Processing and Preparation for Serving on Vitamin Content in T, B, and A. Ration Pork," *Natick / TR-86 / 022*, Mar. 1986.
25. Y. Kim, "Effect of Processing and Preparation for Serving on Vitamin Content in T, B, and A Ration Peas and Carrots," *Natick / TR-86 / 047*, May 1986.
26. Natick RD&E Center, unpublished data.
27. T. P. Labuza, *Shelf-Life Dating of Foods*, Food and Nutrition Press, Inc., Westport, Conn., 1982.
28. E. W. Ross et al., "A Time-Temperature Model for Sensory Acceptance of a Military Ration," *Journal of Food Science* **52**, 1712 (1987).
29. J. L. Briggs et al., "A Calorically Dense Ration for the 21st Century," *Army Science Conference Proceedings*, Vol. 1, West Point, N.Y., 1986, p. 81.
30. A. H. Barrett et al., "Simulation of the Vacuum Infusion Process Using Idealized Components: Effects of Pore Size and Suspension Concentration," *Journal of Food Science* (in press).
31. U.S. Pat. 4,522,190 (June 11, 1985), W. E. Kuhn, K. H. Hu, and S. A. Black (to University of Cincinnati).

GERALD L. SCHULZ
ALICE MEYER
IRWIN TAUB
DANIEL BERKOWITZ
ARMAND CARDELLO
EDWARD HIRSCH
C. PATRICK DUNNE
ANANTH NARAYAN
BONNIE ATWOOD
WILLIAM PORTER
GERALD SILVERMAN
U.S. Army Natick
Natick, Massachusetts

MILK AND MILK PRODUCTS

By definition as mammals, humans are milk-drinking animals. Since at least from the beginning of recorded history, humans have used the milk of other mammals as a food source. Before urbanization, each family depended on its own animals for milk. Later, dairy farms were developed close to cities. The milk industry became a commercial enterprise when methods for the preservation of fluid milk were introduced. Like many other early industries—made up of many small producers and associated with a local production area—the dairy industry developed into large production units and large processing plants often located far from the production areas. The successful evolution of the dairy industry from small to large units of production, that is, from farm to dairy plant, depended on sanitation of product and equipment, cooking facilities, health standards for animals and workers, transportation systems, construction materials for process machinery and product containers, pasteurization methods, containers for distribution, and refrigeration for products in stores and in homes.

COMPOSITION AND PROPERTIES

Cow's milk consists of about 87% water and 13% total solids (Table 1). The latter are composed of solids-not-fat (SNF) and fat. Milk with higher fat content also has higher solids-not-fat content, with an increase of SNF of 0.4% for each 1% fat increase. The principal components of solids-not-fat are protein, lactose, and minerals (ash). Fat content and other constituents of milk vary with species, and, for the milk cow, vary with breed. Likewise, composition of milk may be influenced by many factors such as feed, stage of lactation, health and age of animal, and seasonal and environmental conditions. The SNF and moisture relationships are well established and can be used as a basis for detecting adulteration with water. The physical properties of milk are given in Table 2.

Fat

Milkfat is considered the most complex of all common fats. It is composed of triglycerides (96%) and contains more types of fatty acids than vegetable oils. Milkfat is mainly saturated (65%) but also contains monounsaturated (32%) and polyunsaturated (3%) fatty acids. Milkfat typically contains 7% short-chain fatty acids (C_4–C_8), 15–20% medium-chain fatty acids (C_{10}–C_{14}), and 73–78% long-chain fatty acids ($\geq C_{16}$). The fat is a carrier of numerous lipids (cerebrosides, sterols, and phospholipids) and the fat-soluble vitamins A, D, E, and K. Whole milk (3.3% fat) contains 14 mg cholesterol per 100 g milk, whereas milk of 0.18% fat contains 2 mg cholesterol per 100 g milk. Milk is an emulsion of fat in water (serum). The emulsion is stabilized by phospholipids that are absorbed on the fat globules. The emulsion is broken during treatments such as homogenization and churning. Current U.S. legal standards now allow skim milk (<0.5% fat) to use the terms fat free or nonfat, 1% milk can use light or lowfat, 2% milk can use reduced fat, and whole milk (3.25% minimum fat) remains unchanged.

Protein

Milk protein is composed of two major fractions, caseins and whey proteins. Milk proteins are important in human nutrition. They are also of more value today to dairy producers because they may be included in milk pricing structures, thus adding extra value to milk.

Nutritional Content

Milk is an excellent source of calcium, phosphorous, and riboflavin. Because raw milk is a poor source of vitamin D, vitamin D–fortified milk has been sold since the 1920s to prevent rickets in children. Originally milk was fortified with vitamin D by irradiation or by feeding irradiated yeast to the cows. Ergosterol is converted to vitamin D by ultraviolet irradiation. Currently, vitamin D is added directly to whole milk to provide 400 IU or 10 μg per quart.

Table 1. Gross Composition of Milks from Various Mammals

	Composition (g/100 g)						
Species	Water	Fat	Casein	Whey protein	Lactose	Ash	Energy (kcal/100 g)
Human	87.1	4.5	0.4	0.5	7.1	0.2	72
Rabbit	67.2	15.3	9.3	4.6	2.1	1.8	202
Rat	79.0	10.3	6.4	2.0	2.6	1.3	137
Guinea pig	83.6	3.9	6.6	1.5	3.0	0.8	80
Horse	88.8	1.9	1.3	1.2	6.2	0.5	52
Donkey	88.3	1.4	1.0	1.0	7.4	0.5	44
Pig	81.2	6.8	2.8	2.0	5.5	1.0	102
Camel	86.5	4.0	2.7	0.9	5.0	0.8	70
Reindeer	66.7	18.0	8.6	1.5	2.8	1.5	214
Cow	87.3	3.9	2.6	0.6	4.6	0.7	66
Zebu	86.5	4.7	2.6	0.6	4.7	0.7	74
Yak	82.7	6.5	5.8		4.6	0.9	100
Water buffalo	82.8	7.4	3.2	0.6	4.8	0.8	101
Goat	86.7	4.5	2.6	0.6	4.3	0.8	70
Sheep	82.0	7.2	3.9	0.7	4.8	0.9	102

Source: Ref. 1.

Table 2. Physical Properties of Milk

Property	Value
Density at 20°C of milk with 3–5% fat, average, g/cm^3	1.032
Weight at 20°C, kg/L[a]	1.03
Density at 20°C of milk serum, 0.025% fat, g/cm^3	1.035
Weight at 20°C of milk serum, 0.025% fat, kg/L[a]	1.03
Freezing point, °H[b]	−0.540
Boiling point, °C	100.17
Maximum density at °C	−5.2
Electrical conductivity, S $(=\Omega^{-1})$	$(45–48) \times 10^{-8}$
Specific heat at 15°C, kJ/(kg · K)[c]	
Skim milk	3.94
Whole milk	3.92
40% cream	3.22
Fat	1.95
Relative volume 4% milk at 20°C = 1, volume at 25°C	1.002
40% cream 20°C = 1.0010 at 25°C	1.0065
Viscosity at 20°C, mPa · s (=cP)	
Skim milk	1.5
Whole milk	2.0
Whey	1.2
Surface tension of whole milk at 20°C, mN/m (= dyn/cm)	50
Acidity, pH	6.3–6.9
Titratable acid, %	0.12–0.15
Refractive index at 20°C	1.3440–1.3485

[a]To convert kg/L to lb/gal, multiply by 8.34.
[b]Degrees Hortvet.
[c]To convert kJ/(kg · K) to Btu/(lb · °F), divided by 4.183.

Vitamin A must be added to lower-fat milk to provide 2,000 IU per quart. Multivitamin, mineral-fortified milk is available to meet recommended dietary requirements. The daily nutritional needs for an adult and the constituents of milk are given in Table 3. Average composition of selected dairy products is given in Table 4.

PROCESSING

Commercial dairy farm operations usually consist of a milking machine, a pipeline to convey the milk directly to the tank, and a refrigerated bulk milk tank in which the milk is cooled and stored. Milk at approximately 34°C from the cow is cooled rapidly to 4.4°C or below to maintain quality.

Processing operations for fluid milk or manufactured milk products include centrifugal sediment removal; cream separation; pasteurization; sterilization; homogenization; membrane separation; and packaging, handling, and storing.

Separation

Continuous-flow centrifugal cream separators using cone disks in a bowl were introduced in 1890. Originally, the cream separators were the basic plant equipment, and dairy plants were then known as creameries. Today's separators are pressure- or forced-fed, sealed airtight units. Separators develop 5,000 to 10,000 times the force of gravity to separate the fat (cream) from the milk. Cold incoming raw milk is generally filtered, clarified, or both in the plant to remove sediment. A clarifier is a special type of separator in which sediment is continuously removed from milk before further processing. Bactofugation is a specialized process of clarification in which high-speed centrifugal devices are used to remove most of the bacteria from milk. This process is used for sterile milk or cheese.

In today's large, automated fluid milk plants, milk is standardized automatically. A continuous separator splits the milk stream into fat-free milk and cream, the latter of which is added back to the milk stream to yield the desired fat content. Automatic (on-line) sampling and testing equipment along with air valves control the final tests of the desired product.

Homogenization

Homogenization is the process by which a mixture of components is treated mechanically to give a uniform product

Table 3. Reference Daily Intake[a] and Nutritional Content of Whole Milk

Nutrient	Reference daily intake	Supplied by 100 g	Supplied by 1 quart
Energy, kJ (kcal)[b]		257 (61)	2,510 (600)
Protein, g (daily reference value)	50	3.29	32.11
Calcium, g	1.0	0.119	1.165
Phosphorus, g	1.0	0.93	0.912
Magnesium, mg	400	13	131
Zinc, mg	15	0.38	3.71
Vitamin A, IU	5,000	31	303
Thiamine, mg	1.5	0.038	0.371
Riboflavin, mg	1.7	0.162	1.581
Niacin, mg	20	0.084	0.82
Ascorbic acid, mg	60	0.94	9.17
Vitamin D, IU	400	41[c]	400[c]

Source: Ref. 2.
[a]U.S. Food and Drug Regulations.
[b]To convert kJ to kcal, divide by 4.184; food calorie = 1 kcal.
[c]Fortified milk.

Table 4. Average Composition of Selected Dairy Products, %

Product	Moisture	Protein	Total fat	Total carbohydrate	Ash	Calcium	Phosphorous	Sodium
Fat-free milk	90.8	3.4	0.2	4.9	0.8	0.12	0.10	0.05
Whole milk	88.0	3.3	3.3	4.7	0.7	0.12	0.09	0.05
Half-and-half	80.6	3.0	11.5	4.3	0.7	0.10	0.09	0.04
Heavy whipping cream	57.7	2.0	37.0	2.8	0.4	0.06	0.06	0.04
Evaporated whole milk	74.0	6.8	7.6	10.0	1.5	0.26	0.20	0.10
Plain yogurt	87.9	3.5	3.3	4.7	0.7	0.12	0.10	0.05
Cultured low-fat buttermilk	90.1	3.3	0.9	4.8	0.9	0.12	0.09	0.10
Ice cream (economy)	60.8	3.6	10.8	23.8	1.0	0.13	0.10	0.09
Ice cream (premium)	58.9	2.8	16.0	21.6	0.7	0.10	0.08	0.07
Sherbet (orange)	66.1	1.1	2.0	30.4	0.4	0.05	0.04	0.05
Dried whole milk	2.5	26.3	26.7	38.4	6.1	0.91	0.80	0.40
Nonfat dry milk	3.2	36.2	0.8	52.0	7.9	1.26	0.97	0.53
Dried sweet whey	3.2	12.9	1.1	74.5	8.3	0.80	0.93	1.08
Dried acid whey	3.5	11.7	0.5	73.4	10.8	2.05	1.35	0.97
Creamed cottage cheese	79.0	12.5	4.5	2.7	1.4	0.06	0.13	0.40
Cheddar cheese	36.7	24.9	33.1	1.3	3.9	0.72	0.51	0.62
Swiss cheese	37.2	28.4	27.4	3.4	3.5	0.96	0.60	0.26
Mozzarella	54.1	19.4	21.6	2.2	2.6	0.52	0.37	0.37
Butter	15.9	0.9	81.1	0.1	2.1			
Butter oil	0.2	0.3	99.5	0.0	0.0			

Source: Ref. 1.

that does not separate. In milk, the fat globules are broken up into small particles that form a stable emulsion in the milk, that is, the fat globules do not rise by gravity to form a cream line. Today, most fluid milk products are homogenized.

A homogenizer is a high-pressure positive pump in which milk is forced through small passages under pressure of 14–17 MPa (2000–2500 psi) at velocities of approximately 183–244 m/s. The fat globules are broken up as a result of a combination of factors, namely shearing, impingement, distention, and cavitation. Fat globules in raw milk (1–15 μm in diameter) are reduced to 1 to 2 μm by homogenization.

Pasteurization

Pasteurization is the process of heating milk to kill yeasts, molds, and pathogenic and most other bacteria and to inactivate certain enzymes without greatly altering the flavor. The principles were developed by and named after Louis Pasteur and his work on wine in 1860 to 1864 in France. The basic regulations are included in the Pasteurized Milk Ordinance (3), which has been adapted by most local and state jurisdictions in the United States.

Pasteurization may be carried out by batch or continuous-flow processes. In the batch process, each particle of milk must be heated to at least 62.8°C and held continuously at or above this temperature for at least 30 min. In the continuous process, the milk is heated to at least 71.7°C for at least 15 s. This is known as high-temperature short-time (HTST) pasteurization.

Other continuous pasteurization processes using higher temperatures and shorter times called ultra-high-temperature (UHT) are commercially employed (Table 5). For milk products with a fat content above that of milk or with added sweeteners, additional time and temperature are required (see Table 5). Following pasteurization, the product should be cooled quickly to 7.2°C or less.

Membrane Separation

The separation of components of liquid milk products can be accomplished using semipermeable membranes by either ultrafiltration (UF), reverse osmosis or hyperfiltration (RO), microfiltration (MF), or nanofiltration (NF). The materials from which the membranes are made is similar for all processes and include cellulose acetate, polyvinyl chloride, nylon, polysulfone, and polyamide. Ceramic and mineral membranes are also widely used.

Ultrafiltration. Membranes capable of selectively passing molecules (molecular weights <10,000) are used. Pressures of 0.1 to 1.4 MPa (up to 200 psi) are exerted over the solution to overcome the osmotic pressure while still providing an adequate flow through the membrane for use. UF separates protein from lactose and mineral salts, the protein being the concentrate. Also, UF is used with skim milk to obtain a protein-rich concentrate from which cheese is made. The whey protein obtained by UF can be spray dried. UF is also used for removing minerals from whey and buttermilk.

Reverse Osmosis (Hyperfiltration). Membranes are used for separation of smaller components (molecular weights <500). They have smaller pore size than those used for UF. High-pressure pumps, usually of the positive-piston type or multistage centrifugal type, provide pressures up to 4.14 MPa (600 psi).

RO is used to remove only water from milk or whey; hence a concentrate similar to condensed milk is produced. RO has numerous applications in the dairy and food in-

Table 5. Pasteurization Times and Temperatures for Dairy Products

	Vat		HTST		UHT	
Product/Process	Time	Temp.[a]	Time	Temp.[a]	Time	Temp.[a]
Fluid milk	30 min	145°F	15 s	161°F	2.0 s	280°F
Milk products	Increase viscosity and added sweetener, stabilizers, etc.		15 s	166°F	2.0 s	280°F
	30 min	150°F				
Eggnog	30 min	155°F	25 s	175°F	2.0 s	280°F
Ice cream mix	Strictly on viscosity		15 s	180°F	2.0 s	280°F

Source: Ref. 3.
[a]To convert °C to °F, °C = (°F − 32) × 5/9.

dustries, among which are concentration of milk before cheese making and concentration of whey before drying.

Microfiltration. Membranes are used to selectively separate particles with molecular weights >200,000. This allows for the separation of proteins and bacteria. This technology has been used for the cold pasteurization of milk.

Nanofiltration. Nanofiltration falls between ultrafiltration and reverse osmosis. It removes particles with molecular weights of 300 to 1,000. It can therefore be used for demineralization as well as concentration of whey. It can also be used for the removal of salt from salty whey.

CLEANING SYSTEMS

In most plants today, the equipment surfaces are cleaned in place (CIP) by mechanical-chemical action. The results of CIP are influenced by equipment surfaces and time of exposure, mechanical action, and the temperature and concentration of the solution being circulated.

In the CIP procedure, a cold or tempered aqueous prerinse is followed by circulation of a cleaning solution for 10 min to 1 h at 54 to 82°C. A wide variety of cleaning solutions may be used, depending on the food product, the hardness of the water, and the equipment. Alkali or chlorinated alkaline cleaners are preferred. A chlorinated alkaline cleaner may be used separately or in combination with an acid detergent. The best combination of chemical, timing, and temperature is determined experimentally.

The CIP system is highly automated with valves, controls, and timers. The programmer controls the timing and airflow to the valves so that cleaning solutions are directed to the proper locations for a specific time, followed by a rinse and draining or air blow to clear the lines.

PACKAGING

Following pasteurization, milk and milk products are packaged for consumer use in a variety of package sizes and materials. The filling operation must be carried out without contamination of the product and as soon after pasteurization as possible. Following filling and closing, the packages are placed in a case. The containers are stacked on pallets and moved to a refrigerated area (2–4°C). These operations are mechanized, with continuous operator supervision.

Aseptic packaging has developed in conjunction with high-temperature processing and has continued to make sterile milk and milk products a commercial reality worldwide. In the United States, aseptic or UHT milk has been much slower to emerge, owing to well-developed refrigeration systems to handle dairy and other food products. As a result, most UHT systems are currently processing fruit juices and some cream and ice cream mixes.

ANALYSIS AND TESTING

Milk and milk products are subjected to a variety of tests to assure public safety and to meet composition standards (4).

Microbial Quality

The microbial quality of dairy products is related to the number of viable organisms present. A high number of microorganisms in raw milk suggests it was produced under unsanitary conditions or that it was not adequately cooled after removal from the cow. If noncultured dairy products contain excessive numbers of bacteria, in all likelihood postpasteurization contamination occurred, or the product was held at a temperature permitting substantial growth. Raw-milk and pasteurized-milk products are examined for the concentration of microorganisms by the agar plate method or the direct microscopic method.

The agar plate method consists of adding a known quantity of sample (usually ≤1.0 mL, depending on the concentration of bacteria) to a sterile petri plate and then mixing the sample with a sterile nutrient medium. After the agar medium solidifies, the petri plate is incubated at 32°C for 48 h, after which the bacterial colonies are counted and the number expressed in terms of standard plate count or colony-forming units per 1 mL or g of sample. This procedure measures the number of viable organisms present and able to grow under test conditions. An alternative method to the agar plate method is the Petrifilm™ aerobic count (PAC) method. This method uses a cold-water soluble gelling agent instead of agar in the nutrient medium. The methodology is similar to the agar plate count, but the Petrifilm™ plates require less labor in preparation and less incubator space.

The direct microscopic count determines the number of viable and dead microorganisms in a milk sample. The small amount (0.01 mL) of milk is spread over a 1.0 cm^2 area on a microscope slide and allowed to dry. After staining with an appropriate dye (usually methylene blue), the slide is examined with the aid of a microscope (oil-immersion lens). The number of bacteria cells and clumps of cells per microscopic field is determined and by proper calculations is expressed as the number of organisms per 1 mL of sample.

The keeping quality of milk and milk products is determined by several variations of the agar plate method. Coliform bacteria are detected by using the agar plate method and a selective culture media (violet-red bile agar). Because coliform bacteria do not survive pasteurization, they are an indication of recontamination or improper processing. Psychrotrophic bacteria are capable of growing at refrigerator temperatures, 2 to 10°C. They are responsible for fruity, putrid off-flavors in milk. These bacteria are measured by the agar plate method and incubation at 7°C for 10 days. Additional tests for milk quality can be found in Ref. 4.

Somatic Cells

Mastitis is an inflammation of the cow's udder that results in high levels of somatic cells in the milk. A level of somatic cells >750,000/mL is considered abnormal, and such milk must not be offered for human consumption. Treatment for mastitis with antibiotics may result in such compounds getting into the milk supply. When this occurs, starter culture growth and hence cheese making may be inhibited, or persons consuming such milk may have allergic reactions to these compounds. The presence of such chemicals is determined by a variety of very sensitive and rapid test procedures (4).

Other Tests

The phosphatase test (4) is a chemical method for measuring the efficiency of pasteurization. All raw milk contains phosphatase, and the thermal resistance of this enzyme is greater than that of pathogens over the range of time and temperature of heat treatments recognized for proper pasteurization. Underpasteurization as well as contamination of a properly pasteurized product with raw product can be detected by this test. Pesticides are often found in feed or water consumed by cows; subsequently, they may appear in the milk. At present, low-level residues of some of these chemicals are permitted in milk, and hence milk and its products must sometimes be tested for chlorinated hydrocarbon pesticides (4).

Milk Composition: Fat and Protein

Milk and milk products are purchased from the farmer and sold in stores based on content of fat, protein, or both for raw milk and according to federal and state standards. Milk was first analyzed for fat content by the Babcock test developed in 1890. Today, the Babcock test is used primarily to standardize newer, indirect fat-testing equipment such as the turbidimetric method, mid-infrared spectroscopic methods, and near infrared reflectance methods. Mid-infrared methods measure fat at specific wavelengths (3.48 and 5.723 μm) and are automated, capable of testing hundreds of samples per hour.

The protein content of milk can be determined by a variety of methods, including dye-binding, Kjeldahl, and colorimetric. Recently, procedures using infrared analyses have been adapted to the same equipment that measures fat, so that fat and protein can be measured at the same time using different wavelengths of light. Currently, much of the raw milk is being purchased based on fat and protein or fat and SNF contents. These fat and protein tests are described in detail in Ref. 5.

MANUFACTURED PRODUCTS

Cheese

Cheese making is based on the coagulation of casein from milk or, to a minor extent, of the proteins of whey. Casein is precipitated by acidification, which can be accomplished by adding bacteria that will produce lactic acid from lactose in milk. The procedures for making cheese vary greatly, and cheese products are countless.

Considerable art is involved in making the 400-plus existing cheese varieties. The composition and handling of the original milk, bacterial flora, and starter culture are the basic variables that along with heat treatments, flavoring, salting, and forming of the final product affect the final product. The composition of various cheeses is given in Table 4. Low-fat varieties can have higher moisture levels.

Cheddar Cheese. Pasteurized milk is inoculated with a lactic acid culture and rennet to coagulate casein. The coagulated milk is cut into cubes and cooked to remove whey. Whey is drained out, and the curd cubes are allowed to knit closely together by the cheddaring process. At the end of the cheddaring process, the curd is milled into smaller cubes and salted. The salted cheese is pressed overnight for further whey removal and aged up to a year for flavor development.

Cottage Cheese. Cottage cheese is made from skim milk. Compared to most other cheeses, cottage cheese has a short shelf life and must be refrigerated usually at 4.4°C or below to maintain quality and to provide a shelf life of 3 to 4 weeks. Cottage cheese is a soft, uncured cheese that contains not more than 80% moisture.

Cottage cheese is made by several procedures. In general, pasteurized skim milk is inoculated with lactic acid culture to coagulate the protein. A small amount of rennet may be added to firm the curd. The coagulated material is divided or cut, and the resulting curd is cooked to remove the whey. The whey is drained, and the curd is washed with water. Washed curd is mixed with cream. Mechanized operations are used for large-scale production.

Frozen Desserts

Ice cream is the principal frozen dessert produced in the United States. Known as the American dessert, it was first sold in New York City in 1777. The composition of various frozen desserts is given in Table 4.

Ice Cream. Ice cream is a frozen dessert prepared from a mixture of dairy ingredients (16–35%), sweeteners (13–20%), stabilizers, emulsifiers, flavoring, and fruits and nuts. Ice cream has 8 to 20% milkfat and 8 to 15% SNF with a total of 38.3% (36–43%) total solids. These ingredients can be varied. U.S. regulations also allow for reduced-fat ice cream (25% less fat than the reference ice cream), light ice cream (50% less fat or 1/3 fewer calories than the reference ice cream, as long as in the case of calorie reduction less than 50% of the calories are derived from fat), low-fat ice cream (not more than 3 g of fat per 1/2 cup serving), and nonfat ice cream (less than 0.5 g of fat per 1/2 cup serving). The dairy ingredients are milk or cream; milkfat, which is supplied by milk, cream, butter, or butter oil; as well as SNF, which is supplied by condensed whole, nonfat, or dry milk. The quantities of these products are specified by standards. Milkfat provides the characteristic texture and body in ice cream. Sweeteners are a blend of cane or beet sugar and corn syrup solids. Quantities of these vary, depending on the sweetness desired, cost, and effect on body and texture.

Stabilizers that improve the body of ice cream include gelatin, sodium alginate, sodium carboxymethyl cellulose, pectin, and guar gum. Emulsifiers such as lecithin, monoglycerides, diglycerides, and polysorbates assist in the incorporation of air and improve the whipping properties. Commercial ice cream stabilizers and emulsifiers are usually available as a blend. Use of stabilizers and emulsifiers is optional, and it is possible to produce good-quality ice cream without them. The liquid mixture of components for making ice cream is called ice cream mix.

Preceded by a blending operation, the ingredients are mixed in a freezer that whips the mix to incorporate air and freezes a portion of the water. Freezers may be of a batch type for a small operation or continuous type, in which most commercial ice cream is produced.

Incorporation of air decreases the density and improves the consistency of ice cream. If one-half of the final volume is occupied by air, the ice cream has 100% overrun and weighs 0.53 kg/L. The ice cream from the freezer is at about −5.5°C, with one-half of the water frozen, preferably in small crystals.

Packaged ice cream is hardened on plate-contact hardeners or by convection air blast as the product is carried on a conveyor or through a tunnel. Air temperatures for hardening are −40 to −50°C. The temperature at the center of the container as well as the storage temperature should be −23°C or below. Approximately one-half of the heat is removed in the freezer and the remainder in the hardening process. Although ice cream is by far the most important frozen dessert, other frozen desserts such as ice milk, frozen yogurt, sherbet, and mellorine-type products are also popular.

Frozen Yogurts. Frozen yogurts have become popular with health-conscious consumers. Product formulations vary widely. Frozen yogurt usually consists of 80% ice milk mix and 20% yogurt.

Sherbets. Sherbets have a low fat content (1–2%), low milk solids (2–5%), and a sweet (25–35% sugar) but tart flavor. Ice cream mix and water ice can be mixed to obtain a sherbet. The overrun in making sherbets is about 25 to 40%.

Mellorine. Mellorine is similar to ice cream except that the milkfat is replaced with a vegetable fat (6% minimum). It contains 10 to 12% milk solids.

Nondairy Frozen Dessert. Nondairy frozen dessert is a frozen product containing no dairy ingredients. It may contain vegetable fat, vegetable protein, sugar, eggs, and stabilizers.

Other Frozen Desserts. Other frozen desserts include parfaits, ice cream puddings, novelties, and water ice products. New low-fat and nonfat products and products containing high-intensity sweeteners and fat replacers are being introduced to the market.

Yogurt

Yogurt is a fermented milk product that is rapidly increasing in consumption in the United States. Milk is fermented with *Lactobacillus delbrueckii* spp. *bulgaricus* and *Streptococcus thermophilus* organisms that produce lactic acid and the characteristic yogurt flavor. Usually some cream or nonfat dried milk is added to the milk to obtain a heavy-bodied yogurt.

Yogurt is manufactured similarly to cultured buttermilk. Milk with a fat content of 1 to 5% and SNF content of 11 to 14% is heated to about 82°C and held for 30 min. After homogenization, the milk is cooled to 43 to 46°C and inoculated with 2% culture. The product is incubated at 43°C for 3 h in a vat or in the final container. The yogurt is cooled and held at ≤4.4°C. The cooled product should have a titratable acidity of 0.9 to 1.2% and a pH of 4.3 to 4.4. The titratable acidity is expressed in terms of percentage of lactic acid, which is determined by the amount of 0.1 *N* NaOH/9 grams required to neutralize the substance. Thus, 10 mL of 0.1 *N* NaOH represent 1% acidity. Yogurts with ≤2% fat are popular. Fruit-flavored yogurts are also common; 30 to 50 g of fruit is placed in the bottom of the carton (sundae style) or mixed with the yogurt (swiss style).

Cream

Cream is a high-fat product that is obtained by mechanical separation through differential density of the fat and the skim. Fat content may range from 10 to 40%, depending on use and federal and state laws. The U.S. *Code of Federal Regulations* (6) defines creams as products that contain not less than 18% milkfat. Whipping cream has a fat content of 36 to 40%; table, coffee, or light cream has a fat content of 18 to 30%. Half-and-half, a mixture of cream and milk, has not less than 10.5% milkfat and in some states up to 12%.

The sale of fresh cream for home use has decreased greatly over the past 20 years, primarily as a result of changing customer demand based on dietary concerns. A variety of cream and fat substitutes are available for spreads, toppings, whiteners, and cooking.

Dry Milk

Drying generally follows concentration in an evaporator. Clarification and homogenization precede evaporating and drying. Homogenization of whole milk at 63 to 74°C with pressures of 17 to 24 MPa (2500–3500 psi) is particularly desirable for reconstitution and the preservation of quality.

Dry milk is generally made by the drum-drying or spray-drying method. The spray process uses less heat than the drum process; hence the powder produced by the spray process is more soluble in water. Solubility of the powder can be further improved by coupling the agglomeration (or instantization) process with the spray process. In the agglomeration process, powder particles are agglomerated into larger particles that dissolve more easily than smaller particles.

Dry whole milk should be vacuum- or gas-packed to maintain quality under storage. Products containing milkfats deteriorate in the presence of oxygen, giving oxidation off-flavor. Antioxidants of many kinds have been used with various degrees of success, but a universally acceptable antioxidant that meets the requirements for food additives has not yet been found.

Dry milk reduces transportation costs, provides long-term storage, and supplies a product that can be used for food-manufacturing operations. It has been used primarily for manufactured products but is now used to a much greater extent for beverage products. The moisture content for nonfat dry milk, the principal dry milk product, is $\leq$5.0% for standard grade and $\leq$4.0% for extra grade. Dry whole milk contains $\leq$3.0% moisture.

Butter

Butter is defined as a product that contains 80% milkfat with not more than 16% moisture. It is made from cream containing 25 to 40% milkfat. The process of making butter is primarily a mechanical one in which the cream, an emulsion of fat in serum, is changed to butter, an emulsion of serum in fat. The process is accomplished by churning or, in recent years, by a continuous operation with automatic controls. Per capita annual sales of butter have remained steady at approximately 4.5 to 5.5 lb for the past 20 years.

Anhydrous Milkfat; Butter Oil

Butter oil is a high-milkfat material that is 99.7% milkfat. It is called anhydrous milkfat or anhydrous butter oil if $<$0.2% moisture is present. Although the terms are used interchangeably, anhydrous butter oil is made from butter and anhydrous milkfat is made from whole milk. For milk and cream there is an emulsion of fat in serum. For butter oil and anhydrous milkfat there is an emulsion of serum in fat, such as with butter. It is easier to remove moisture in the final stages to make anhydrous milkfat with the serum-in-fat emulsion.

Condensed Milk Products

Evaporated milk is produced by condensing milk to half its volume in a vacuum evaporator. The final product must have at least 6.5% milkfat and 23% total milk solids including fat. The process for evaporated fat-free milk is similar to that for whole milk except that the final product contains much less fat and total solids. The final product is packaged in cans and sterilized at 116 to 118°C for 15 to 20 min. It is subsequently cooled to room temperature within 15 min. Vitamins A and D and stabilizing salts such as sodium citrate and disodium phosphate may be added.

Condensed product made without sterilization is called condensed milk. Additionally, sugar may be added to the final product so that the final product contains 43 to 45% sugar, at least 8.5% milkfat, and 28% total milk solids. This product is called sweetened condensed milk. Owing to the high sucrose content, defects such as age thinning and age thickening may occur. These defects can be prevented by careful control of the manufacturing procedure. Condensed milk products are widely used in the manufacture of ice cream as well as bakery, confectionery, and other food products.

By-Products from Milk

Milk is a source for numerous by-products resulting from separation or alteration of the components. These components may be used in other nondairy manufactured foods, dietary foods, pharmaceuticals, animal feeds, and industrial products.

Lactose. Lactose or milk sugar makes up about 5% of cow's milk. Compared to sucrose, lactose has about one-sixth the sweetening strength. Because of its low solubility, lactose is limited in its application. However, it is soluble in milk serums and can be removed from the whey. Lactose is used in processed foods and pharmaceutical products.

Casein. Casein is the principal protein of milk, accounting for 2.5 to 4% of the weight. It possesses unique functional properties that enable its use as an ingredient in many food products. It can be separated from milk by acidification and removing the whey that is formed. The precipitate left behind, casein, is dried into a powder.

Casein is used to fortify flour, bread, and cereals. Casein also is used for glues and microbiological media. Calcium caseinate is made from pressed casein by rinsing, treating with calcium hydroxide, heating, and mixing followed by spray drying. A product of 2.5 to 3% moisture is obtained. Casein hydrolyzates are produced from dried casein. With appropriate heat treatment and addition of alkalies and enzymes, digestion proceeds. Following pasteurization, evaporation, and spray drying a dried product of 2 to 4% moisture is obtained.

Many nondairy products such as coffee creamers, toppings, and icings utilize caseinates. In addition to fulfilling a nutritional role, the caseinates impart creaminess, firmness, smoothness, and consistency to products. Imitation meats, soups, and the like use caseinates as an extender and to improve moistness and smoothness.

BIBLIOGRAPHY

1. N. P. Wong et al., *Fundamentals of Dairy Chemistry*, 3rd ed., Van Nostrand Reinhold, New York, 1988, p. 21.

2. U.S. Department of Agriculture, *Composition of Foods—Dairy and Egg Products*, USDA Agricultural Handbook No. 8-1, U.S. Government Printing Office, Washington, D.C., 1976.
3. U.S. Food and Drug Administration, *Pasteurized Milk Ordinance*, Publication No. 229, U.S. Department of Health and Human Sciences, Washington, D.C., 1995 edition, 1997 Revision.
4. R. T. Marshall, ed., *Standard Methods for the Examination of Dairy Products*, 16th ed., American Public Health Association, Washington, D.C., 1992.
5. P. Cunniff, ed., *Official Methods of Analysis*, 16th ed., Association of Official Analytical Chemists, Arlington, Va., 1995.
6. *Code of Federal Regulations*, Parts 100–169, Office of the Federal Register, National Archives and Records Administration, Washington, D.C., 1998.

GENERAL REFERENCES

References 1, 4, and 5 are good general references. Also refer to issues of *Dairy Science Abstracts* and *Food Science and Technology Abstracts* (England).

F. W. Bodyfelt, J. Tobias, and G. M. Trout, *The Sensory Evaluation of Dairy Products*, 3rd ed., Van Nostrand Reinhold, New York, 1988.

J. G. Brennan et al., *Food Engineering Operations*, 2nd ed., Applied Science Publishers, London, United Kingdom, 1976.

G. Bylund, *Dairy Processing Handbook*, Lund, Sweden, 1995.

J. R. Campbell and R. T. Marshall, *The Science of Providing Milk for Man*, McGraw-Hill, New York, 1975.

A. W. Farrall, *Engineering for Dairy and Food Products*, John Wiley & Sons, New York, 1963.

P. F. Fox, ed., *Advanced Dairy Chemistry, 1. Proteins, 2. Lipids, 3. Lactose, Water, Salts and Vitamins*, Elsevier Applied Science, New York, 1992–1996.

C. W. Hall, A. W. Farrall, and A. L. Rippen, *Encyclopedia of Food Engineering*, AVI, Westport, Conn., 1971.

C. W. Hall and T. I. Hedrick, *Drying of Milk and Milk Products*, 2nd ed., AVI, Westport, Conn., 1971.

W. J. Harper and C. W. Hall, eds., *Dairy Technology and Engineering*. AVI, Westport, Conn., 1976.

J. L. Henderson, *The Fluid Milk Industry*, 3rd ed., AVI, Westport, Conn., 1971.

H. G. Kessler, *Food Engineering and Dairy Technology*, Verlag A. Kessler, Freising, West Germany, 1981.

F. V. Kosikowski and V. V. Mistry, *Cheese and Fermented Milk Foods*, 3rd ed., Vol. I and II, F. V. Kosikowski, L.L.C., Westport, Conn., 1997.

M. Robinson, *Concentrated and Dried Milk Products (Evaporation) (Drying)*, Victorian College of Agriculture and Horticulture, Victoria, Australia, 1993.

P. Walstra and R. Jenness, *Dairy Chemistry and Physics*, John Wiley & Sons, New York, 1984.

JOHN G. PARSONS
ROBERT J. BAER
VIKRAM V. MISTRY
South Dakota State University
Brookings, South Dakota

See also CULTURED MILK PRODUCTS.

MINERALS: MICRONUTRIENTS

The minerals as nutrients are divided into two groups. Calcium, phosphorus, sodium, chloride, and magnesium comprise the macromineral group. The micromineral group is subdivided into a trace mineral group and an ultra trace mineral group. Iron, copper, and zinc comprise the former group and chromium, manganese, fluoride, iodide, cobalt, selenium, silicon, arsenic, boron, vanadium, nickel, cadmium, lithium, lead, and molybdenum comprise the latter. The basis for this subdivision lies with the magnitude of need. Whereas the human body needs the trace minerals in milligram amounts, the ultra trace minerals are needed in microgram (or less) amounts. For many of the microminerals the evidence of their essentiality is meager for humans, and yet reports exist on their essentiality to other species. Of the 18 microminerals thought to be essential, only four have been studied sufficiently such that a database exists to support a recommendation for a daily recommended dietary allowance (RDA). Table 1 gives the RDAs for these minerals (1). An additional five minerals have been studied sufficiently such that an intake recommendation that is generally recognized as safe and ade-

Table 1. Recommended Dietary Allowances (RDAs) for Iron, Zinc, Iodine, and Selenium

	RDA			
Age	Iron (mg)	Zinc (mg)	Iodine (mg)	Selenium (μg)
		Infants		
0–6 months	6	5	40	10
7–12 months	10	5	50	15
		Children		
1–3 years	10	10	70	20
4–7 years	10	10	90	20
8–11 years	10	10	120	30
		Males		
12–14 years	12	15	150	40
15–18 years	12	15	150	50
19–24 years	10	15	150	70
25–50 years	10	15	150	70
51+ years	10	15	150	70
		Females		
12–14 years	15	12	150	45
15–18 years	15	12	150	50
19–24 years	15	12	150	55
25–50 years	15	12	150	55
51+ years	10	12	150	55
		Pregnancy		
	30	15	175	65
		Lactation		
0–6 months	15	19	200	75
7–12 months	15	16	200	75

quate has been made. These are listed in Table 2. This figure is based on data extrapolated from animal studies and on data collected from human population studies. These studies of healthy humans may have included analysis of the food consumed as part of a regular diet and, where possible, tissue analysis. Blood values, especially those indicators of anemia; hair values; urine analysis; and fecal analysis may have been included as items in an epidemiological study. Although hair and blood analysis can reveal mineral nutrition status, these analyses are not without problems. The difficulty in using these analyses is that the mineral content can be transient—that is, blood levels of trace minerals represent the immediate mineral intake rather than long-term exposure, and hair mineral content can represent not only food or drink minerals but also airborne minerals. Hair minerals can be contaminated by shampoos and hair treatments such as coloring, bleaching, curling agents, and conditioning agents.

MICROMINERAL TOXICITY

Many of the microminerals are toxic when consumed in large amounts (2). Inadvertent exposure to a variety of minerals, whether via inhalation, absorption through the skin, or ingestion through food or drink, can elicit a toxic response. The gastrointestinal system offers the first line of defense against ingested excess mineral: vomiting and diarrhea. Through vomiting, contaminated food is expelled. Through diarrhea, excretion of excess consumed mineral is facilitated. This reduces the time that the intestinal tract is exposed to the mineral and thus reduces its subsequent uptake. The defense against excess exposure also involves the kidneys and the bone as well as the bilary tract. The kidney tubules will attempt to reduce the body load by increasing urinary excretion. Some minerals, however, namely, copper, iron, zinc, and lead, are not as subject to renal filtration as are other minerals, namely, magnesium, selenium, sodium, potassium, iodine, calcium, and molybdenum. Those minerals subject to renal filtration will appear in the urine; those that are not filtered out are excreted via the bile into the intestinal tract. When coupled with diarrhea, this provides a stimulated loss pathway. Last, reduction of the circulating toxic load of a micromineral is accomplished via deposit of the excess mineral in the bones. Bone mineral content has been used to document cases of suspected toxicity. Accidental or intentional poisoning can sometimes be masked by other nonspecific symptoms, but bone analysis can provide the documentation needed to support or deny a supposition of toxicity.

Table 2. Generally Recognized Safe and Adequate Intakes for Selected Minerals

Mineral	Intake
Copper	1.5–3.0 mg/day
Fluoride	1.4–4.0 mg/day
Manganese	2.0–5.0 mg/day
Chromium	50–200 μg/day
Molybdenum	75–250 μg/day

Note: These ranges are not age or gender specific.

The adverse effects of excess micromineral exposure are as diverse as the minerals themselves. Table 3 lists the characteristic signs and symptoms of micromineral toxicity as well as deficiency. Each mineral has its preferred target in the body. For some, the target is DNA (3). Certain minerals (copper, arsenic, nickel, chromium) bind to DNA in a cross-link fashion. The binding is covalent and is independent of a DNA binding protein. In this setting the excess mineral produces either a nonfunctional DNA or a DNA that cannot repair itself. Evidence of this cross-linking has been demonstrated *in vitro* using a variety of cell types. Chinese hamster cells have been used to show copper-induced, chromium-induced, and nickel-induced DNA cross-linking, and human fibroblasts and epithelial cells have been used to show arsenic-induced cross-linking.

The concept of chemically induced cross-linkage of DNA as a factor in carcinogenesis has been proposed to explain the role of asbestos and the development of mesothelioma (4). Mesothelioma is a malignant growth of the pleural and peritoneal cavities and develops as a response to the inhalation of asbestos fibers (pleural growth) or inadvertent ingestion (peritoneal growth) of the fibers. Asbestosis, another symptom of asbestos inhalation, results in impaired pulmonary function. This is due in part to iron-catalyzed free-radical damage to the lung tissue because asbestos contains iron as a contaminant. However, cytokines, growth factors, and proteases as well as effects of asbestos on gene expression are also involved. Malignant tumors are stimulated to grow by the presence of asbestos fibers that act as artificial linkers of DNA, resulting in mutations within the pleural and peritoneal cells (4). Changes in the CDKN2 gene that encodes protein 16 seem to be involved. This gene is either lost or mutated. Its gene product is a regulator of the phosphorylation of protein 105, a tumor suppressor. Unphosphorylated protein 105 can inhibit passage from the G1 to the S phase of the cell cycle, whereas phosphorylated protein 105 permits this passage. Passage inhibition is a common feature of cancer cell initiation. Any substance that interferes with this passage could be regarded as a carcinogen. Minerals in excess quantities can have this effect, and excess intakes of some have been linked with certain forms of cancer.

Excess exposure to certain minerals can invoke the stimulation of free-radical generation. Free radicals can damage membranes, altering their function, and can target numerous intracellular constituents (DNA, enzymes, transporters, signaling systems). Free-radical generation may be one of the responses to excess iron exposure and this in turn may link this mineral to cancer development (5). The mechanism whereby excess iron has its effect is far from clear. It may induce DNA cross-linking as described previously, but it may also act to stimulate free-radical formation. Free radicals can damage cell membranes and cell constituents as well as DNA. Free-radical attack of DNA could cause mutations and could explain an association between excess iron intake and cancer (6). At this time, more data are needed to support or deny such associations. Excess mineral intake either as a single mineral or as a mixture can have effects on metabolism as well

Table 3. Characteristic Signs of Micromineral Deficiency and Toxicity

Mineral	Deficiency	Toxicity
Iron	Anemia, ↓ amounts of hemoglobin, ferritin	Bloody diarrhea, vomiting occasionally, liver failure, hemorrhage, metabolic acidosis, shock
Zinc	Poor growth and sexual maturation, anemia, enlarged liver and spleen, rough skin, lethargy	Nausea, vomiting, epigastric pain, abdominal cramps, diarrhea; central nervous system deficits, copper deficiency
Fluoride	Excess tooth decay	Fluorosis; mottling of teeth
Copper	Rare; anemia, poor wound healing, muscle weakness; lethargy, depressed collagen synthesis	Rare
Iodine	Goiter (underproduction of thyroxine)	None
Manganese	Abnormal bone and connective tissue growth, decreased Mn superoxide dismutase	Rare
Molybdenum	No clear-cut symptoms	Rare; interferes with copper use
Selenium	Decreased glutathione peroxidase activity, fragile red cells; enlarged heart, skeletal muscle degeneration	Neuromuscular defects, trace mineral imbalance, liver and muscle damage

as on cellular respiration. Where the mineral is divalent (2^+), it can interfere with the actions of other similarly charged ions. Within the normal range of intakes, mineral–mineral interactions are common. However, when these ranges are exceeded, specific metabolic processes will be inhibited.

MICROMINERAL DEFICIENCY

Rarely are humans deficient in only one mineral. Usually the deficiency involves more than one, and these combined deficiencies have cumulative effects on health. As noted in Tables 3 and 4, micromineral interactions are common. Anemia usually due to iron deficiency can also be due to intake deficiency of zinc, copper, or both. Poor growth is typical of micromineral deficiency, either singly or collectively. Humans consuming a wide variety of foods therefore run scant risk of micronutrient deficiency. The inclusion of red meat, dairy foods, seafoods, and fresh fruits and vegetables should provide sufficient supplies of the microminerals to the average healthy adult. Of course, special health considerations, such as growth, pregnancy, excess blood loss, and the like, may call for adjustments in micromineral intake, and these adjustments might call for an oral supplement.

ANTAGONISMS AND INTERACTIONS AMONG TRACE MINERALS

No general discussion of trace minerals would be complete without mention of mineral interactions. Numerous antagonisms and synergisms have been reported. This should be expected, because many of the trace minerals have more than one charged state and living cells have preferences with respect to these states. For example, the uptake of iron is much greater when the iron is in the ferrous (2^+) state than when in the ferric (3^+) state. Minerals that keep iron in the ferric state will interfere with its absorption and use. Minerals that do the reverse will enhance iron uptake. Such is the beneficial action of copper on iron. The copper ion (Cu^{2+}) keeps the ferrous ion from losing electrons. Thus, copper prevents the conversion of the ferrous to ferric ion. The interactions of essential minerals are best illustrated in Figure 1 and listed in Table 4. Some of these interactions have important therapeutic value in

Table 4. Micronutrient Interactions

	Calcium	Phosphorus	Potassium	Sodium	Magnesium	Zinc	Iron	Copper	Iodine	Fluorine
Calcium	X	↑a	↓a	↓a↑m	↑m↓a					
Phosphorus	↑a	X	↑m	↑m	↓a					
Potassium	↑↓m	↑a	X	↑a	↓a, ↑m					
Sodium	↑↓m	↑a		X	↓a,↑m					
Magnesium	↓a	↑m		↓a,↑m	X				↑m	
Zinc						X		↓a, ↑m		
Iron	↑m	↓a				↓a	X	↓a, ↑m		
Copper						↓a	↓a,↑m	X		
Iodine									X	
Fluorine	↓a									X
Cobalt							↓a			
Chromium						↓a				
Manganese					↓a		↓a			
Molybdenum							↑m	↓a		
Selenium								↑m		

Note: ↑ = increase; ↓ = decrease; a = absroption; m = metabolism.

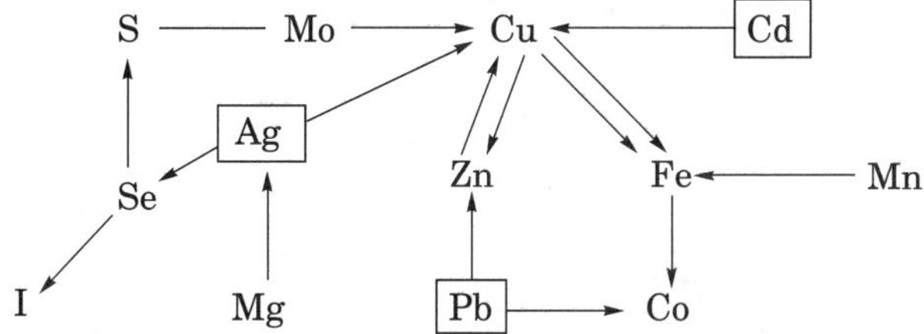

Figure 1. Toxicity and trace mineral interactions. Minerals that are important environmental toxins are in boxes. Abbreviations used: S, sulfur; Se, selenium; I, iodide; Ag, mercury; Mg, magnesium; Zn, zinc; Cu, copper; Pb, lead; Co, cobalt; Fe, iron; Mn, manganese; Cd, cadmium; Mo, molybdenum.

the treatment of inadvertent and toxic exposure to a single mineral. For example, lead toxicity can be minimized by consumption of a diet rich in calcium and phosphorus. The transport proteins of the enterocytes prefer calcium to lead. If the enterocytes are saturated with calcium, the lead will not be absorbed and will pass out of the body in the feces. A similar situation exists with iron and lead. Mineral interactions need to be recognized in the manufacture of special diets. Sometimes, so-called purified ingredients contain mineral contaminants. This is especially a problem in the protein portion of the diet. Proteins can bind minerals, and these minerals can be found in the food proteins. Soybean protein can contain phytate-bound phosphorus and magnesium; casein and lactalbumin, depending on origin, contain variable amounts of calcium, magnesium, and selenium. Unless the investigator determines the mineral content of the diet ingredients, an unexpected mineral imbalance could potentially occur, and this imbalance could affect the nutritional value of the product.

MINERAL ABSORPTION

Of all the elements mammals need, few enter the absorptive cell by passive diffusion. Several are carried into the body by proteins to which the minerals are loosely attached. Iron, zinc, and copper are imported via these transporters. Iodide and fluoride enter by way of an anion–cation exchange mechanism. Few of the minerals are 100% absorbable. For the others, the absorption process is dependent on a number of factors: the binding capacity of the transport protein (if needed); solubility; the composition of the diet; the mixture of elements present in the gut contents; and the presence of materials such as phytate and ethylenediaminetetraacetic acid (EDTA), which bind specific elements, thus changing the mineral mixture presented to the enterocyte. All of these factors contribute to the bioavailability of the essential micronutrients. In addition, physiological factors (eg, age, hormonal status, and health status) also influence absorption and subsequent use. Those minerals that are variable in terms of their absorption are those that are either divalent or multivalent ions (ie, ions with more than one charged state). Iron and chromium are multivalent ions whereas selenium, manganese, zinc, and molybdenum are divalent.

BIOAVAILABILITY

One concept that nutritionists have developed relates not only to absorption efficiency but also to mineral interactions at the site for absorption and the site of use. This concept is that of bioavailability (5). Bioavailability is defined as the percentage of the consumed mineral that enters via the intestinal absorptive cell, the enterocyte, and is used for its intended purpose. Thus, bioavailability includes not only how much of a consumed mineral enters the body but also how much is retained and available for use. An example might be the comparison of iron from red meat to the same amount of iron from spinach. Iron from red meat has a greater bioavailability than iron from spinach because it is an integral component of the protein heme. It is this form (heme iron) that is efficiently absorbed and used. The iron in the spinach is bound to an oxalate, and even though some of this iron can be released from the oxalate, it is in the ferric state and poorly absorbed.

APPARENT ABSORPTION

Another term referring to absorption is frequently used: apparent absorption. This term refers to the difference between the amount of mineral consumed and the amount that appears in the feces. Some minerals are recirculated via the bile whereas others are not. This recirculation, especially in a poorly absorbed mineral, can contribute to the mineral content of the feces, but there is no correction for the bilary contribution to the fecal mineral content. The term apparent absorption refers *only* to the difference between intake and fecal excretion.

MINERAL UPTAKE AND TRANSPORT

Because of their ionic nature, minerals can form electrovalent bonds to a variety of substances. Although ingested as salts, minerals ionize to their component parts, and it is the resultant ions that are absorbed, used, stored, or excreted. If absorbed, many of the microminerals are retained because there is little excretion. As ions, minerals react with charged amino acid residues of intact proteins and peptides. Table 5 provides a list of minerals and the amino acids with which they react. Depending on their valence state, these mineral–amino acid bonds can be very strong, moderate, or very weak associations.

Many minerals are carried in the blood by specific transport proteins. The reasons for this are not fully known. However, we do know that there are preferred ligand bonding groups as shown in Table 6. Other features of these

Table 5. Mineral–Amino Acid Interactions

Minerals	Amino acid
Copper	Histidine
Selenium	Methionine, cysteine
Zinc	Cysteine, histidine
Iron	Histidine

Table 6. Preferred Ligand Binding Groups for the Microminerals

Metal	Ligand groups
Mn^{2+}	Carboxylase; phosphate and nitrogen donors
Fe^{2+}	-SH, NH^2 > carboxylates
Fe^{3+}	Carboxylate, tyrosine, $-NH_2$, porphyrin (a nitrogen donor)
Co^{3+}	Similar to Fe^{3+} but is usually chelated by the NH groups as in vitamin B_{12}
Cu^+	-SH (cysteine)
Cu^{2+}	Amines ≫ carboxylates
Zn^{2+}	Imidazole, cysteine
Mo^{2+}	-SH
Cd^{2+}	-SH

transport proteins (eg, size, shape, amino acid sequence, tertiary and quatenary structures) may also determine specificity. As yet we lack this information for each of the specific proteins that bind the microminerals. In some instances a transport protein will carry more than one ion. An example is metallothionein, which will carry both zinc and copper. It will also carry some of the heavy metals, but its affinity for zinc and copper is greater than its affinity for heavy metals such as lead.

Many of the microminerals can be chelated by organic materials. Chelation is a common phenomenon in biological systems, and this process has many clinical applications. For example, EDTA is a potent chelator and is used to remove lead or other heavy metals from the body. It will also chelate calcium and magnesium, so the clinician using EDTA to treat lead overload will have to be aware of this feature as well. Penicillamine is another chelator of importance. It is used to remove excess copper, for example, in patients with genetically inherited Wilson's disease. 2,3-D-Dimercaptopropanol-1 is a chelator of lead, mercury, arsenic, copper, cadmium, tin, and other toxic metals. It solubilizes these metals and chelates them, allowing for their excretion in the urine. Chelation is important to the biological function of several minerals. Examples are the chelation of cobalt within the structure of vitamin B_{12} and the chelation of iron within the structure of hemoglobin, as shown in Figure 2. In the B_{12} structure, chelation is a function of the many nitrogen molecules within this very large molecule.

FUNCTIONS OF MICROMINERALS

The microminerals are widely dispersed in the living body. They serve a variety of functions, as summarized in Table 7, and most of these minerals can be found in trace amounts in the skeletal system. The bones and teeth are primarily calcium phosphate salts, yet within the bone matrix other minerals can be found as well. Some of the microminerals have only one function, for example, cobalt in vitamin B_{12} and iodide in thyroxine, but others serve at multiple levels, from influencing gene expression to serving as cofactors in enzymatic reactions to contributing to the stability and hardness of bone.

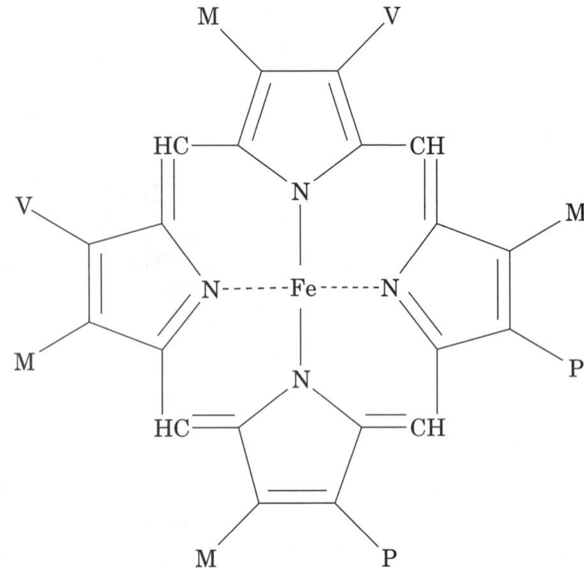

Figure 2. Heme as a chelator of iron. Abbreviations used: A, acetate (CH_2-COO_3H); M, methyl ($-CH_3$); P, proprionate ($-CH_2$-CH_2-COOH); V, vinyl ($-R$-$CH{=}CH_2$). These are the residues of the amino acids that project out from the heme protein structure.

Table 7. Microminerals and Their Functions

Mineral[a]	Function
Iron	Essential component of iron–sufur centers that function in redox reactions as well as in the cytochromes and hemoglobin; serves in the expression of the genes for the metallothioneins, ferritin and the transferrin receptor (7–9).
Zinc	Cofactor in more than 100 enzymatic reactions. Essential component of the zinc fingers that characterize many of the nuclear DNA-binding proteins. Serves in the expression of genes for metallothionein; has a vital role in apoptosis (7, 10–13).
Copper	Essential cofactor in several reactions concerning iron use, collagen synthesis, and free-radical suppression. Serves in the expression of the genes for metallothionein superoxide dismutase, and several mitochondrial enzymes (14–17).
Cobalt	Essential to the structure of vitamin B_{12}.
Iodine	Essential for thyroid hormone synthesis (18).
Manganese	Essential cofactor in many enzymatic reactions, especially those using ATP or UTP.
Molybdenum	Activates adenylate cyclase; cofactor in sulfite oxidase and xanthine oxidase (19).
Selenium	Essential to glutathione peroxidase and thyroxine deiodinase. Serves in the expression of the gene for glutathione peroxidase (20,21)
Fluorine	Increases hardness of teeth

[a]Arsenic, chromium, nickel, silicon, and vanadium appear to be essential for optimal growth and metabolism of several animal species, but their essentiality and function in humans has yet to be documented.

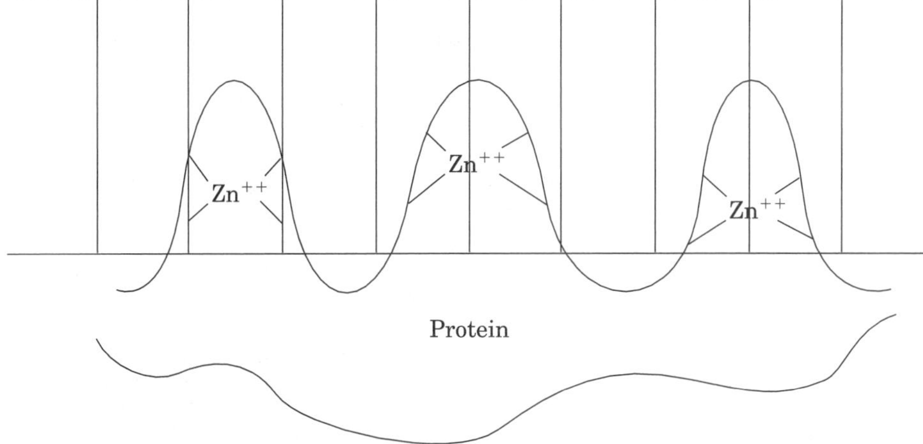

Figure 3. Zinc bound to cysteine and histidine residues from fingerlike projections that bind to DNA at specific regions.

MINERALS AND GENE EXPRESSION

As noted in Table 7, several of the microminerals influence the expression of genes encoding one or more important proteins (7–21). Doubtless we will discover many more of these gene products that require one or more of the microminerals as part of the regulation of the transcription and translation of their cognate genes. Perhaps the mineral that has been studied the most is zinc because it is part of a group of DNA-binding proteins called transcription factors. In this role, zinc is bound by the histidine and cysteine residues of these binding proteins, giving these regions fingerlike projections that in turn bind to DNA at specific regions. These are illustrated in Figure 3. These proteins are called zinc finger proteins and the zinc-bound regions, zinc fingers. A number of nutrients, namely, vitamin A and vitamin D, and hormones, namely, steroids, insulinlike growth factor I, growth hormone, and others, have their effects on the expression of specific genes because they can bind to very specific proteins that contain zinc fingers that in turn bind to very specific DNA sequences.

If there is a mutation in the genes that encode these DNA-binding proteins such that they lack the requisite two residues each of histidine and cysteine in the linear part of their structure, then the functional attributes of vitamins and hormones requiring these proteins at the genetic level will be ablated. Instances of such mutations have been published, as have instances where these zinc fingers have been purposefully modified as a therapeutic approach to disease control. As an example, a zinc-containing transcription factor, Zif268, has been modified such that its sequence-specific recognition by viral DNA results in an ablation of the viral invasion and takeover of specific target cells (22). Although this modification was done *in vitro* and not tested in whole animals, this approach has a potential use in battling viral infections.

Copper, zinc, and iron are all involved in the expression of genes for the metal-binding proteins, the metallothionines. These cysteine-rich proteins transport these divalent ions through the intestinal absorptive cells. Several genes are involved. Each has a metal response element that is specific for each of the minerals. In the case of copper, a copper-containing DNA-binding protein attaches to a specific DNA sequence called the CUP1 element. A zinc-containing DNA-binding protein is also required, and together these proteins stimulate the transcription of the metallothionine genes. The binding sites for these mineral-containing proteins are upstream from the promoter region of the genes, and the minerals by themselves (not incorporated into the DNA-binding proteins) do not directly stimulate transcription. In addition to the metallothionine genes, copper has been found to influence the expression of at least 10 other genes, all of which have a copper response element in their DNA sequence (17). Five of these are mitochondrial sequences, suggesting that copper has a function in mitochondrial gene expression as well as in nuclear gene expression.

Just as zinc, copper, and iron play multiple roles in gene expression, so too does selenium (20,21). Selenium is an essential cofactor to the activity of glutathione peroxidase in multiple forms. Selenium is also essential to the expression of the genes that encode these enzymes. In selenium-deficient animals, glutathione peroxidase messenger RNA falls to undetectable levels, indicating that this mineral functions in the transcription of this messenger RNA. How this occurs has not yet been elucidated, but likely the mechanism of action will be similar to that described earlier for copper.

Table 8. Genetic Diseases in Micromineral Metabolism

Disorder	Characteristics
Hemochromatosis	Excess iron absorption, resulting in damage to pancreas, liver, heart
Menkes disease	Defective intestinal copper absorption
Wilson's disease	Excessive copper accumulation due to impaired incorporation of copper into ceruloplasmin and decreased bilary excretion
Graves' disease	Autoimmune disease characterized by impaired thyroid gland function. Goiter (underproduction of thyroxine) or hyperthyroidism (overproduction of thyroxine) due to thyroid stimulating hormone receptor autoantibodies

Last, genetic errors or mutations in genes that encode specific proteins involved in micromineral metabolism have been described. These are listed in Table 8. These diseases are uncommon, yet their presence has allowed scientists to understand aspects of mineral metabolism previously unapproachable. Through the study of humans with Wilson's disease and Menkes disease, for example, an understanding of copper absorption and turnover was gained. Similar insights into the role of iron in the heme structure as well as iron–sulfur centers were gained through the study of hemochromatosis.

BIBLIOGRAPHY

1. Food and Nutrition Board, National Academy Press, Washington, DC, 1989.
2. D. Medeiros, R. Wildman, and R. Liebes, "Metal Metabolism and Toxicities," in *Handbook of Human Toxicology*, CRC Press, Boca Raton, Fla. 1997.
3. D. J. Paustenbach, B. L. Finley, and S. Kacew, "Biological Relevance and Consequences of Chemical or Metal Induced Cross Linking," *Proc. Soc. Exp. Biol. Med.* **211**, 211–217 (1996).
4. D. W. Kamp and S. A. Weitzman, "Asbestosis: Clinical Spectrum and Pathogenic Mechanisms," *Proc. Soc. Exp. Biol. Med.* **214**, 12–26 (1997).
5. B. Fairweather-Tait, "Bioavailability of Trace Elements," *Food Chem.* **43**, 213–217 (1992).
6. R. Yip, D. F. Williamson, and T. Byer, "Is There an Association between Iron Nutrition Status and the Risk of Cancer?" *Am. J. Clin. Nutr.* **53**, 30–34 (1991).
7. C. D. Berdanier and J. L. Hargrove, *Nutrition and Gene Expression*, CRC Press, Boca Raton, Fla., 1993.
8. O. S. Chen, L. Schalinske, and R. S. Eisenstein, "Dietary Iron Intake Modulates the Activity of Iron Regulatory Proteins and the Abundance of Ferritin and Mitochondrial Aconitase in Rat Liver," *J. Nutr.* **127**, 238–248 (1997).
9. G. T. Gassner et al., "Structure and Mechanism of the Iron-Sulfur Flavoprotein Phthalate Dioxygenase Reductase," *FASEB J.* **9**, 1411–1418 (1995).
10. S. R. Davis, R. J. McMahon, and R. J. Cousins, "Metallothionein Knockout and Transgenic Mice Exhibit Altered Intestinal Processing of Zinc with Uniform Zinc Dependent Zinc Transporter-1 Expression," *J. Nutr.* **128**, 825–831 (1998).
11. P. J. Fraker and W. G. Telford, "A Reappraisal of the Role of Zinc in Life and Death Decisions of Cells," *Proc. Soc. Exp. Biol. Med.* **215**, 229–236 (1997).
12. N. P. Koritschoner et al., "A Novel Human Zinc Finger Protein That Interacts with the Core Promoter Element of a TATA Box-less Gene," *J. Biol. Chem.* **272**, 9573–9580 (1997).
13. R. T. Nolte et al., "Differing Roles for Zinc Fingers in DNA Recognition: Structure of a Six-Finger Transcription Factor IIIA Complex," *Proc. Natl. Acad. Sci. U.S.A.* **95**, 2938–2943 (1998).
14. V. C. Culotta et al., "The Copper Chaperone for Superoxide Dismutase," *J. Biol. Chem.* **272**, 23469–23472 (1997).
15. P. Wang et al., "Overexpression of Human Copper, Zinc-Superoxide Dismutase (SOD1) Prevents Post Isechemic Injury," *Proc. Natl. Acad. Sci. U.S.A.* **95**, 4556–4560 (1998).
16. Y. Chen, J. T. Saari, and Y. J. Kang, "Copper Deficiency Increases Metallothionein mRNA Content Selectively in Rat Liver," *J. Nutritional Biochem.* **6**, 572–576 (1995).
17. Y. R. Wang et al., "Enhanced Expression of Hepatic Genes in Copper Deficient Rats Detected by the Messenger RNA Differential Display Method," *J. Nutr.* **126**, 1772–1781 (1996).
18. S. Boyages, "Iodine Deficiency Disorders," *J. Clin. Endocrinol. Metab.* **77**, 587–591 (1993).
19. K. V. Rajagopalan, J. L. Johnson, and B. E. Hainline, "The Pterin of the Molybdenum Co-Factor," *Fed. Proc.* **41**, 2608–2612 (1982).
20. S. Weiss and R. A. Sunde, "Selenium Regulation of Classical Glutathione Peroxidase Expression Requires the 3^1 Untranslated Region in Chinese Hampster Ovary Cells," *J. Nutr.* **127**, 1304–1310 (1997).
21. W-H. Cheng et al., "Cellular Glutathione Peroxidase Knockout Mice Express Normal Levels of Selenium-Dependent Plasma and Phospholipid Hydroperoxide Glutathione Peroxidases in Various Tissues," *J. Nutr.* **127**, 1445–1450 (1997).
22. H. Wu, W-P. Yang, and C. F. Barbas, "Building Zinc Fingers by Selection: Toward a Therapeutic Application," *Proc. Natl. Acad. Sci. U.S.A.* **92**, 344–348 (1995).

CAROLYN BERDANIER
University of Georgia
Athens, Georgia

MOLECULAR MODELING: COMPUTER MODELING OF FUNCTIONAL PROPERTIES

One of the principal goals of food science and technology is to be able to modify the measurable functional properties of a food by specific changes to the structures of the component molecules. Examples of this type of manipulation of molecular properties include direct chemical modification of polymers, as in modified starches, and the preparation of mutant proteins through site-directed mutagenesis (1). Although this goal of control of functional properties can be occasionally achieved as the serendipitous result of random experimentation, in principle, successful design of novel functional properties requires an understanding of the way in which these functional properties depend on molecular structure.

In general, such an understanding involves a detailed knowledge of the atomic and molecular-level interactions responsible for macroscopic bulk properties. Unfortunately, this knowledge is often unavailable, because few experimental techniques can directly probe chemical processes in the necessary detail. Even in those cases where experiment can directly examine molecular structure, such as X-ray diffraction experiments, it generally remains unclear how the molecular architecture functions or how it might be modified to successfully alter its workings. Fortunately, recent developments in theoretical chemistry now allow molecular events to be directly simulated on fast computers. These developments not only permit us to learn how biopolymer systems function but also present the possibility of modeling numerically the effects of changes in a polymer structure, such as a series of active site mutations in a protein, without the time and expense of actually producing each of these mutants in the laboratory. Experimental studies can thus be made more efficient by only making those mutants that the simulations suggest might actually possess the desired functional properties. This article provides an introduction to these theoretical simulation methods and their application to biopolymers of the types important in the physical properties of foods.

COMPUTER MODELING OF BIOPOLYMERS

The observable physical properties of any molecule are ultimately the result of the physical laws that determine the behavior of individual molecules and atoms. At the most fundamental level, of course, the properties of molecules are governed by quantum mechanics, and modeling the behavior of molecules numerically requires the direct solution of Schrödinger's equation for all of the nuclear and electronic degrees of freedom for all molecules in the system. Unfortunately, it is not practical to solve this time-dependent equation at the necessary level of accuracy for molecules containing more than a few heavy (nonhydrogen) atoms. It is thus not possible to make a direct quantum mechanical simulation of large and complex systems such as the mixtures of proteins, fats, water, salts, and other components that make up a typical food. However, in many cases, part of the solution of the quantum mechanical problem is known by other means. Typically, important features of the molecular structure, such as equilibrium bond lengths and angles, are known approximately from studies of less complicated molecules, and the primary challenge is to understand the conformational behavior of polymers, how different molecules interact, and the connection between basic structure and macroscopic properties. The link between basic structure and bulk properties is statistical mechanics, and the most common theoretical studies of biopolymer physical properties are various attempts to model this statistical mechanical connection.

Theoretical calculations that exploit a knowledge of molecular energies as a function of nuclear coordinates are called molecular mechanics calculations (2–4). Because it is impossible to accurately solve Schrödinger's equation for large polymer systems, it is common in theoretical simulations to approximate the quantum mechanical energy with continuous energy functions that have approximately the same behavior. By invoking the Born-Oppenheimer approximation (3), which says that the electronic motions of a molecule are far faster than the nuclear motions, it is possible to separate the inherently quantal problem of the electronic motions from the much more classical problem of the motions of the nuclei, by treating the electronic energy for each value of the nuclear coordinates as a potential energy function that governs the motion of the nuclei. It is then possible to use continuous, analytic functions to approximate this nuclear potential energy function. These semiempirical energy equations have theoretically reasonable functional forms that are parameterized to the results of various experiments and theoretical calculations such that properties calculated using these functions are those found in the experiments. Energy functions of this type usually contain terms to represent bond stretching, bond angle bending, hindered torsional rotations about chemical bonds, and nonbonded and electrostatic interactions. A typical example of such a function of the internal coordinates $\mathbf{q}$ might be

$$V(\mathbf{q}) = \sum k_b(b - b_0)^2 + \sum k_\theta(\theta - \theta_0)^2 + \sum k_\varphi[1 + \cos(n\varphi - \delta)] + \sum_{i>j}\left(\frac{A_{ij}}{r_{ij}^{12}} - \frac{B_{ij}}{r_{ij}^{6}} + \frac{q_i q_j}{r_{ij}}\right)$$

where the b's are the various bond lengths, which have equilibrium values b_0, the θ's are the bond angles with equilibrium values θ_0, the φ's are the torsional angles about each bond, and q_i and q_j are the partial atomic charges of each atom, separated by a distance r_{ij}. This equation treats bond stretching and bending as harmonic oscillator terms with force constants k_b and k_θ, and also includes a periodic hindered torsional energy term in the angle φ with the force constant k_φ, periodicity n, and phase δ. van der Waals interactions are treated as a typical 10–12 Lennard-Jones-type interaction, and the electrostatic interactions are represented by Coulomb's law. The various adjustable constants that appear in such equations (k_b, k_θ, k_φ, A_{ij}, B_{ij}, q_i) must be selected with great care in order for the results to be physically reasonable. This parameterization is accomplished by matching experimental properties of small molecules (equilibrium bond lengths and angles, IR and Raman spectral frequencies, etc) to the values calculated using the energy equation as a function of the adjustable parameters (2–4).

Energy Minimization

The most common types of molecular mechanics calculation are conformational energy studies (2–4). Conformational energy calculations involve the search for the molecular geometry with the lowest potential energy, on the assumption that the structure with the lowest mechanical energy (as calculated from the semiempirical energy functions) will be the form primarily found in nature. This search is accomplished using automatic energy minimization algorithms, which minimize the gradient (derivative) of the potential energy. Predictions of the structure-dependent physical properties of the biopolymer system are then based on those calculated from the lowest energy conformation. Unfortunately, the major problem with this approach is that due to the enormous complexity of large systems, it is difficult to identify the lowest energy structure. The problem arises because for large systems, there will, in general, be many local minimum energy geometries that are locally stable, with a zero gradient of the potential energy, but that are higher in energy than the structure with the lowest energy, which is called the global energy minimum. There is no general mathematical procedure for finding this global energy minimum. Most minimization techniques will simply locate the nearest local stable geometry for which the gradient is zero and then be unable to proceed over any energy barriers that separate this local minimum from any lower energy structures. This multiple-minimum problem (5) is the reason that it is not yet possible to predict the tertiary conformation of a protein given its primary sequence, even though in principle the sequence for most proteins contains all of the information necessary for correct folding (see later). Even if the global minimum-energy structure is identified, there are other possible problems with this type of approach. Polymers such as proteins are not static, but actually fluctuate considerably, sampling a number of nearby thermally accessible conformations (6) (which are usually quite similar to the lowest energy structure), and the experimental, or physical, structure is actually an appropriately weighted average of all these structures. Energy minimization studies, however, generally cannot include such conformational averaging, or its entropic effects (7). Perhaps most important, energy minimization calculations cannot directly include the effects of solvation, which can be quite crucial in many biological systems.

In spite of their limitations, energy minimization calculations can be of great utility in studying certain limited questions, such as predicting the preferred binding geometries of substrates in enzyme active sites or the local effects of a single point mutation in a protein. This type of calculation can be especially useful when combined with computer graphics representations of molecular structures, as is now frequently done in the pharmaceutical (8) and food industries in binding studies. Visual inspection of computer-generated three-dimensional color images of interacting molecules such as a target protein and a drug molecule allow researchers to use their chemical intuition and experience to identify promising approximate binding configurations, which can subsequently be refined by the application of energy minimization calculations. In this way, the human-guided input can avoid calculations for many of the unpromising minima that blind energy minimizations would have wasted much computer time exploring needlessly.

Molecular Dynamics Simulations

More theoretically realistic models of the behavior of molecular systems can be generated by recognizing that macroscopic properties are averages over large numbers of fluctuations of the system. Statistical mechanical calculations that explicitly average over these different states will thus give a more physical picture of the behavior of such systems. Among these types of calculations are Monte Carlo calculations (9) and molecular dynamics simulations (3,4,10). Molecular dynamics studies are a type of molecular mechanics calculation in which atomic motions are simulated numerically using a large digital computer. In this type of calculation, the classical Newton's equations of motion are solved numerically for every atom in the system as they respond to the forces arising from the empirical force field. Because the negative gradient, or derivative, of the potential energy function V with respect to the atomic coordinates is equal to the force $\mathbf{F}_i$ along that coordinate acting on the particle,

$$\mathbf{F}_i = -\nabla_i V$$

the derivatives of this analytic function can be used to numerically integrate the coupled Newton's equations of motion (11)

$$\mathbf{F}_i = m_i \mathbf{a}_i$$

to give a complete description of the motion of every atom in the molecule as a function of time, given the initial conditions (initial positions and velocities for each atom). For those molecules for which some plausible structure is known, either from reasonable chemical conjecture or from experiments such as X-ray diffraction from crystals, a random selection of initial atomic velocities from a Boltzmann distribution allows the researcher to produce a theoretical trajectory describing the way in which this molecule would move if it were possible to view it at the atomic level in such detail. Within the limitations imposed by the approximations (such as the use of empirical energy functions), this trajectory contains everything that can be known about the system, and all of its properties can be calculated as the appropriate time average over the trajectory, if it is of sufficient length. This absolute knowledge is the source of the great power of molecular dynamics simulations.

Unfortunately, the main limitation of this method is contained in the caveat that the trajectory must be of sufficient length. The statistical nature of the macroscopic properties requires averaging over either a very large number of systems identical in all respects except starting conditions or averaging over a single trajectory of very great duration. The numerical integration techniques used to solve the classical equations of motion require very small time steps, on the order of a femtosecond, to maintain an acceptable level of accuracy. Each time step in the integration is costly, requiring the evaluation of the forces arising from the extremely large number of nonbonded interactions, which varies roughly as N^2, where N is the number of atoms in the system. As a result, practical considerations concerning the amount of computer time available limit the trajectory time that can actually be simulated. The calculation of thermodynamically converged properties can thus be frustrated by large statistical inaccuracies. Nonetheless, such simulations have become routine in recent years and have been of great value in understanding a variety of physical systems. For example, much knowledge of the details of aqueous solutions has been based on such simulations (12). Calculations have now been reported for a variety of small molecules; for water and aqueous solutions (12,13), including those containing ions (14); and for proteins (15–18), carbohydrates (19–23), lipids, and nucleic acids (24,25).

A major advantage of this type of calculation over simple energy minimization calculations is that the trajectories contain entropic effects automatically "built in" to their behavior. Furthermore, systems do not necessarily become trapped in higher-energy local minimum geometries, because the momentum of the moving atoms can carry them over energy barriers and down the opposite side into other local minima. Thus, in principle, a trajectory of sufficient length would not suffer from the multiple minimum problem. Another important advantage of molecular dynamics simulations is that this type of calculation easily allows the modeling of solutions or crystal systems through the explicit inclusion of solvent water molecules or crystal neighbors in the calculations. Because solvation can play a critical role in determining the conformational behavior of polymers, and particularly proteins, the inclusion of water molecules directly in theoretical simulations can significantly increase their ability to model nature. Quite recently, a successful simulation of the complete folding of a very small globular protein in aqueous solution was reported (26). This goal, the prediction of a protein tertiary conformation from sequence alone, is the "holy grail" of protein modeling, and although the present simulation, which required more than two CPU months on one of the world's largest computers, is not necessarily the full realization of that goal, it offers renewed hope that such predictions might actually be possible in the future.

APPLICATION OF MOLECULAR MECHANICS CALCULATIONS TO FOOD SYSTEMS

Although molecular mechanics calculations have been widely used to study all major types of biological polymers, such calculations have only recently been attempted for typical food molecules. However, the potential for future use is great. This section simply indicates a few of the types of problems where such simulations have been profitably employed. Many other types of problems exist that could be usefully studied with theoretical calculations, however, and calculations of this type on food systems will probably become commonplace in the near future.

Because techniques already exist to modify the sequence of a protein almost at will by site-directed mutagenesis, perhaps the area with the greatest potential for practical applications for molecular mechanics is in studying possible protein mutants with the objective of modifying or improving their functionality. Such studies are already common in the biotechnology industry (8). However, because of the extreme difficulty or practical impossibility of making a prediction of protein conformation based solely on sequence information (see earlier), it is currently necessary to study those proteins whose tertiary structures have been previously determined by X-ray diffraction experiments. Fortunately, such diffraction studies are becoming more common for important food proteins.

One food protein whose tertiary conformation is known is the whey protein β-lactoglobulin (27). The structure of this 162-residue protein has been reported at 2.8-Å resolution. From this work it was determined that β-lactoglobulin has approximately the same tertiary structure as its sequential homologue, retinol-binding protein (28). Both proteins exist as a β-barrel in a calyx-shaped structure that contains an inner binding pocket that strongly binds retinol (β-lactoglobulin in fact binds retinol even more tightly than does retinol-binding protein). β-Lactoglobulin has already been used in food formulations as a carrier for additives immiscible in water. Among the modifications to the β-lactoglobulin molecule that could potentially be useful would be to increase its thermal stability, because this molecule denatures at a relatively low temperature, exposing a buried free cysteine to cross-linking with other proteins, including the free cysteine of κ-casein. Conformational energy calculations are currently being used to identify which mutants are possible that would allow the formation of an additional, third disulfide bond by including the free Cys-121 residue and another residue mutated into cysteine. Several such mutants identified by this procedure have been modeled by molecular dynamics simulations to determine whether the third disulfide bond does in fact introduce greater stability (29). Conformational energy calculations and docking studies are being used to model alternate substrates into the binding pocket of various additional mutants with substitutions on the inner surface of this pocket in an attempt to identify mutants that would have a high binding constant for these alternate substrates and that would be specific for these molecules. In this way, novel commercial molecular carriers for various substances might be developed that could function in somewhat the same way as the cyclodextrins, but with greater specificity. Such proteins could also potentially be designed to function as molecular scavengers for undesirable substances such as off-flavor compounds.

Other food-related proteins that have been modeled include the collagen of gelatin (30–32); various caseins (33); the enzyme xylose (glucose) isomerase (34), which is potentially of great commercial value in the corn syrup industry; and various cellulases (35). Because these enzymes are slow acting, and their mechanisms are still a matter of debate (36,37), simulations are being used not only to investigate the actual enzymatic mechanisms but also to suggest mutants that might have increased activity or specificity for other substrates. Similar studies could in principle be performed for any protein of interest whose tertiary conformation is known. The rate at which protein conformational determinations are reported continues to increase, and the list of known food proteins is growing. Notable recent examples include turkey troponin and the intensely sweet proteins monellin and thaumatin (38,39).

Proteins, of course, are not the only molecules to which molecular modeling calculations can be applied. Simulations of food lipids have in the past been rare, but there have been a number of molecular dynamics studies of membranelike phospholipid bilayers (40), and of a few milkfat triglycerides (41,42). The example of the sweet proteins mentioned in the previous paragraph serves to introduce another broad area of potential application of theoretical techniques to food molecules, which is the development of sweeteners. Theoretical studies might help not only to elucidate what makes these proteins sweet but also to identify the important functional characteristics of other sweeteners (or any other flavorant) as well. In general, the interaction of a flavorant molecule with its receptor is the same problem as in drug binding. Currently such studies are hampered by a nearly complete lack of information about the receptor proteins. This situation also exists for many medical problems, and theoretical studies under such circumstances employ techniques called quantitative structure-activity relationship (QSAR) studies (43) to identify structural features in the drug or flavorant that might give rise to the desired functional property.

Perhaps the most extensive use has been made of various types of modeling in the study of carbohydrates. Molecular mechanics conformational energy studies have long been used to assist in the interpretation of X-ray fiber diffraction patterns of polysaccharides (44,45). These polymers do not form a single crystal as do proteins, and the limited information provided by fiber diffraction experiments must be augmented by energy calculations to identify possible structures. Recently molecular dynamics studies of carbohydrates have begun to appear, including studies of carbohydrate solvation (20–23,46). Simulations have been used to understand the solution conformation of table sugar (47), and whether or not it is different from the known crystal structure, and to simulate the solution properties of the components of carrageenans (48–50). A recent simulation of the putative double-helical structure of carrageenan in water found the individual strands remained in the helical conformation proposed from fiber diffraction, but the double helix unraveled, indicating that the juncture zone in carrageenan gels may not be a double helix as previously thought (50). Simulations of water structuring around sugars (51) have not only explained the anomeric preference in sugars such as glucose and xylose (23,52), but have more generally revealed how these solutes affect their solvent water by imposing new structural arrangements on it and how this structuring depends on the stereochemistry of the sugars. Such information could be most useful in guiding the design of polymers with specific viscosities, gelation capacities, freezing points, and so on.

CONCLUSIONS

The growth of genetic engineering in the biotechnology industry offers the promise of the next industrial revolution in the coming decades as artificial proteins are designed for a host of applications and as the properties of wild-type proteins are altered through recombinant experiments. However, the ability to create improved protein products will require considerable guidance from chemical theory to select those mutants most likely to possess the desired functional properties. Molecular mechanics studies of possible mutant proteins will thus become increasingly common in biotechnology. Theoretical molecular design studies, particularly of molecular binding, have already become the principal methods of drug design in the pharmaceutical industry, and they will undoubtedly be widely used in the design of flavorants in the food industry. Other more complicated systems that might be susceptible to molecular design calculations as computers become larger and faster are gelation and textural properties. Computer simulations cannot currently be applied to every practical problem of interest, so care must be exercised in selecting for study those problems for which a well-defined objective might reasonably be accomplished with the resources available. Nevertheless, for those systems for which they are appropriate, theoretical calculations can provide information that is not available by any other technique and can be quite valuable in making the production of functional mutants more efficient.

BIBLIOGRAPHY

1. L. Perry and R. Wetzel, *Science* **226**, 555–557 (1984).
2. U. Burkert and N. L. Allinger, *Molecular Mechanics*, American Chemical Society, Washington, D.C., 1982, Vol. 177.
3. C. L. Brooks, M. Karplus and B. M. Pettitt, "Proteins: A Theoretical Perspective of Dynamics, Structure, and Thermodynamics," in I. Prigogine and S. A. Rice, eds., *Advances in Chemical Physics*, Wiley-Interscience, New York, 1988, Vol. 71.
4. J. A. McCammon and S. C. Harvey, *Dynamics of Proteins and Nucleic Acids*, Cambridge University Press, Cambridge, U.K., 1987.
5. H. A. Scheraga, *Biopolymers* **22**, 1–14 (1983).
6. R. Elber and M. Karplus, *Science* **235**, 318–321 (1987).
7. M. Karplus and J. N. Kushick, *Macromolecules* **14**, 325–332 (1981).
8. L. Regan and W. F. DeGrado, *Science* **241**, 976–978 (1988).
9. N. Metropolis et al., *J. Chem. Phys.* **21**, 1087–1092 (1953).
10. A. Rahman, *Phys. Rev.* **136**, A405–A411 (1964).
11. L. Verlet, *Phys. Rev.* **159**, 98–103 (1967).
12. F. H. Stillinger, A. Rahman, *J. Chem. Phys.* **60**, 1545–1557 (1974).
13. P. J. Rossky and M. Karplus, *J. Am. Chem. Soc.* **101**, 1913–1937 (1979).
14. G. Pálinkás, W. O. Riede, and K. Heinzinger, *Z. Naturforsch.* **32A**, 1137–1145 (1977).
15. J. A. McCammon, B. R. Gelin, and M. Karplus, *Nature* **267**, 585–590 (1977).
16. W. F. van Gunsteren et al., *Proc. Natl. Acad. Sci. USA* **80**, 4315–4319 (1983).
17. M. Levitt, *J. Mol. Biol.* **168**, 595–620 (1983).
18. M. Levitt, *J. Mol. Biol.* **168**, 621–657 (1983).
19. C. B. Post et al., *J. Mol. Biol.* **190**, 455–479 (1986).
20. J. W. Brady, *J. Am. Chem. Soc.* **108**, 8153–8160 (1986).
21. J. W. Brady, *J. Am. Chem. Soc.* **111**, 5155–5165 (1989).
22. J. W. Brady and R. K. Schmidt, *J. Phys. Chem.* **97**, 958–966 (1993).
23. R. K. Schmidt, M. Karplus, and J. W. Brady, *J. Am. Chem. Soc.* **118**, 541–546 (1996).
24. S. C. Harvey, M. Prabhakaran, and J. A. McCammon, *Biopolymers* **24**, 1169–1188 (1985).
25. U. C. Singh, S. J. Weiner, and P. Kollman, *Proc. Natl. Acad. Sci. USA* **82**, 755–759 (1985).
26. Y. Duan and P. A. Kollman, *Science* **282**, 740–744 (1998).
27. M. Z. Papiz et al., *Nature* **324**, 383–385 (1986).
28. M. E. Newcomer et al., *The EMBO Journal* **3**, 1451–1454 (1984).
29. Y. Cho et al., *Protein Eng.* **7**, 263–270 (1994).
30. H. A. Scheraga, *Biopolymers* **20**, 1877–1899 (1981).
31. G. Némethy and H. A. Scheraga, *Biochemistry* **25**, 3184–3188 (1986).
32. J. R. Grigera and H. J. C. Berendsen, *Biopolymers* **18**, 47–57 (1979).
33. C. Holt and L. Sawyer, *Protein Eng.* **2**, 251–259 (1988).
34. H. L. Carrell et al., *Proc. Natl. Acad. Sci. USA* **80**, 4440–4444 (1989).
35. J. S. Taylor et al., *Protein Eng.* **8**, 1145–1152 (1995).
36. C. A. Collyer, K. Henrick, and D. M. Blow, *J. Mol. Biol.* **211**, 211–235 (1990).
37. G. K. Farber et al., *Biochemistry* **28**, 7289–7297 (1989).
38. A. M. de Vos et al., *Proc. Natl. Acad. Sci. USA* **82**, 1406–1409 (1985).
39. S.-H. Kim et al., *Protein Eng.* **2**, 571–575 (1989).

40. K. Tu, M. L. Klein, and D. J. Tobias, *Biophys. J.* **75**, 2147–2156 (1998).
41. S. B. Engelsen, J. W. Brady, and J. W. Sherbon, *J. Agric. Food Chem.* **42**, 2099–2107 (1994).
42. Z.-Y. Yan et al., *J. Agric. Food Chem.* **42**, 447–452 (1994).
43. C. Hansch et al., *J. Med. Chem.* **25**, 777–784 (1982).
44. V. S. R. Rao et al., "Conformational Studies of Amylose," in G. N. Ramachandran, ed., *Conformation in Biopolymers*, Vol. 2, Academic Press, London, 1967, pp. 721–737.
45. D. A. Brant, *Ann. Rev. Biophys. Bioeng.* **1**, 369–408 (1972).
46. L. M. J. Kroon-Batenburg and J. Kroon, *Biopolymers* **29**, 1243–1248 (1990).
47. S. B. Engelsen and C. Herve du Penhoat, S. Peréz, *J. Phys. Chem.* **99**, 13334–13351 (1995).
48. K. Ueda and J. W. Brady, *Biopolymers* **38**, 461–469 (1996).
49. K. Ueda and J. W. Brady, *Biopolymers* **41**, 323–330 (1997).
50. K. Ueda, A. Imamura, and J. W. Brady, *J. Phys. Chem. A* **102**, 2749–2758 (1998).
51. Q. Liu and J. W. Brady, *J. Am. Chem. Soc.* **118**, 12276–12286 (1996).
52. S. Ha et al., *J. Am. Chem. Soc.* **113**, 1553–1557 (1991).

J. W. Brady
Cornell University
Ithaca, New York

MONOSODIUM GLUTAMATE (MSG). See Flavor chemistry.

MUSCLE AS FOOD

There are four types of biological tissue—epithelial, connective, muscular, and nervous. Muscle tissue is a major component of the animal and is valued as a food because of the nutritional value afforded as well as satiety. Skeletal muscle is voluntarily controlled and identified histologically because it is striated and multinucleated. In mature muscle the nuclei are located peripherally in the cell, and they do not divide. Cardiac muscle cells are also striated, but they are mononucleated and have, in addition, a distinguishing feature called the intercalated disc. The intercalated discs are found at opposing ends of the cardiac muscle cells and help to maintain cohesion of one cell with another. Smooth muscle cells are spindle shaped, involuntary, nonstriated, and mononucleated. Smooth muscle is located in such places as blood vessels, the digestive tract, and ducts of glands. Smooth muscle controls the internal environment of an organism by moving or controlling movement of substances through various routes by peristaltic movement. Cardiac muscle is responsible for pumping blood around, and striated muscle moves the organism or parts of it.

COMPOSITION OF MUSCLE TISSUE

Proteins are the important constituents that allow for the physiological functioning of meat animals. Meat provides a major source of high-quality proteins for the human diet. The proteins in muscle are categorized as sarcoplasmic, myofibrillar, and connective tissue proteins. Numerous other proteins are found in other parts of the animal, such as in the organs, bones, and skin.

Sarcoplasmic proteins, at a concentration of about 55 mg/mL, are found in the sarcoplasm or fluidlike substance that surrounds and bathes the contractile apparatus and other organelles of the muscle cell. The sarcoplasmic proteins are primarily metabolic enzymes, and the protein responsible for the color of meat is known as myoglobin. The function of myoglobin is to bind oxygen; it has a molecular weight of 17,500 and occurs in muscle at a concentration of 0.1 to 20 mg/g muscle, depending on function of the muscle and specie and age of the animal. The heme portion of myoglobin binds the oxygen, which is transported to the muscle cell via the circulatory system and bound in blood to a similar protein, hemoglobin. Other sarcoplasmic proteins include nucleoproteins, which are a complex of proteins and nucleic acids and function in protein synthesis, and lyosomal proteins, which are hydrolytic enzymes.

The contractile apparatus of muscle consists of thick and thin filaments that interdigitate and slide past each other as muscle shortens; these filaments are composed of myofibrillar proteins. The major protein of the thick filaments is myosin, and for thin filaments it is actin. Myosin has a molecular weight of 480,000 and consists of two heavy subunits and four light subunits. Two fragments are formed when myosin is subjected to digestion with trypsin. The larger fragment has a molecular weight of 350,000 and is called heavy meromyosin. It is soluble at low ionic strength and contains both actin binding and ATPase activity. The smaller fragment has a molecular weight of 150,000 and is called light meromyosin. Heavy meromyosin can be further digested into S-1 and S-2 fragments. The monomeric form of actin is called G actin and has a molecular weight of 42,000. G actin polymerizes to form the filament form of F actin. Tropomyosin and troponin are other muscle proteins that bind to actin and are responsible for regulation of contraction.

Connective tissue consists of a variety of cell types and proteins filaments that are secreted by them. It is the latter that are of interest in meat, since they not only hold the cells together and in place but are responsible for toughness. Muscle has three layers of connective tissue—one surrounding the entire muscle, known as the epimysium; another layer that segments the muscle into bundles of fibers, known as the perimysium; and a layer surrounding each individual cell or myofiber, known as the endomysium. Collagen and elastin are the two connective tissue proteins. Collagen is an abundant protein in muscle and is unique because it has a high complement of the amino acids hydroxyproline and hydroxylysine. It is converted to soluble gelatin at the usual cooking temperatures of meat. Elastin is, in fact, a minor component of connective tissue, contains the unique amino acids desmosine and isodesmosne, and does not solubilize upon heating, nor does it swell by acid or alkaline treatment.

Fats comprise 10 to 30% of the carcass weight of meat animals and are primarily triglyceride. The majority of fat is located in fat cells, but lipid also occurs in cellular membranes. The fat cells within muscle are located in the perimysial connective tissue, and their accumulation depends on genetics and nutrition. Also, species is important, as, for example, when beef or pork is compared with chicken or fish, with the latter having a much lower fat content. The composition of fat varies likewise. Softer or more unsaturated fat is more prone to autooxidation and the associated "off" or rancid odors or flavors. Other lipid components, although occurring in small amounts, are phospholipids, sterols, and free fatty acids.

Glycogen is the major carbohydrate in muscle and is normally present at a concentration of 0.5 to 1.0%. Ash is normally 1% or less and results from minerals such as iron, zinc, and calcium.

CONVERSION OF MUSCLE TO MEAT

When an animal is stunned and exsanguinated, oxygen supply to the muscle ceases and a series of biochemical changes commences. The end result depends on the condition of the animal and preslaughter handling. The main metabolic events are a depletion of the ATP energy store and a degradation of the glycogen, which is converted anaerobically to lactic acid. Accumulation of lactic acid results in a decline of pH from near 7.0 to the range of 5.5 to 5.9; this is a postmortem condition easily monitored to establish rate and status of changes in the muscle as it is converted to meat. Rigor mortis also occurs as the muscle is converted from a soft, pliable, and extensible condition to one of rigidity. This is explained by a locking together of the myosin to actin filaments as ATP depletes. The time for rigor mortis differs with species, with beef and lamb being longer than pork and poultry. Some shortening of muscle occurs during the onset of rigor mortis. There are also two other shortening conditions of interest to meat science. Cold shortening results when muscle is exposed to a low temperature prior to ordinary development of rigor mortis, and it may result in considerable toughening. If meat is frozen prior to the onset of rigor mortis, it may then shorten greatly when thawed. This is known as thaw rigor and results in severe toughening. The postmortem changes also influence color and water-holding capacity of the resulting meat. This is quite important in pork, where a rapid pH decline results in so-called pale, soft, and exudative (PSE) meat. In beef, a dark, firm, and dry (DFD) condition may result if the pH remains high. These undesirable conditions are due largely to the genetic predisposition of the animal. However, proper handling both pre- and postmortem is important. Preslaughter stress to the animal can greatly influence postmortem events. Likewise, postmortem handling, such as proper temperature control during the chilling process, influences the final outcome. Other procedures, such as electrical stimulation, employed during postmortem time can be used to control tenderness and quality of meat.

PROCESSING AND FABRICATION

Two types of meat are made available to consumers. These are fresh and processed, which are also known as not-ready-to-eat and ready-to-eat, respectively. With fresh meat, the carcass is normally broken down to smaller portions and then shipped to retailers for final cutting and packaging. Interest is strong in being able to send retail ready packages directly from the processing plants. Only a small amount of meat is sold in frozen form, although freezing is an excellent means of preservation. Vacuum packing of fresh meat during transfer to the retailer is routinely practiced, and more interest is being shown in other controlled atmosphere packaging. Numerous forms of processed meats are made, including canned products. The majority, however, are cured, which means they have been treated with nitrite, heated, and then vacuum packaged. The shelf life of these products is normally in weeks as compared with fresh meats, which last only a few days.

Nitrite and heating require special mention. Nitrite curing results in typical cured color and also imparts specific preservation qualities to the meat. The heating associated with curing not only gives desirable palatability changes but also essentially pasteurizes the product so that it is ready to eat and, if packaged properly, will have a very long shelf life.

Processed meat is often equated with sausage. Sausages have had a very long history and are often named after origins, as, for example, frankfurters from Frankfort. They are usually ground or chopped and made into specific shapes with characteristic flavor. They may be fresh or cured, and there are unique variations such as fermented sausages. Another variation of processed meats is restructured meat, in which smaller pieces of meat are formed into larger portions by bonding them together with extracted meat proteins. During the past 10 years, a new generation of processed meats has appeared that are low or no fat. In general, these are made by lowering fat content and adding water, which is held within the product by gums, carbohydrates, or other substances.

GENERAL REFERENCES

R. G. Cassens, *Meat Preservation, Preventing Losses and Assuring Safety*, Food & Nutrition Press, Inc., Trumbull, Conn., 1994.

P. J. Bechtel, ed., *Muscle as Food*, Academic Press, New York, 1986.

J. F. Price and B. S. Schweigert, eds., *The Science of Meat and Meat Products*, W. H. Freeman and Company, San Francisco, Ca., 1987.

Robert G. Cassens
University of Wisconsin
Madison, Wisconsin

MUSHROOMS: CULTIVATION

DESCRIPTION

The golden rule for testing edible mushrooms is never to experiment with fungi that you have not identified with certainty as harmless. It is only fair to point out that there are many fungi that are both edible and excellent, and in many countries they have formed part of the human diet for centuries (1,2). They are highly valued as a food source worldwide, and more than 30 species are sold commercially. Edible mushrooms contain more protein than any vegetable and are rich in vitamin B (3). The most delicious mushrooms include *Boletus edulis, Clitopilus prunulus, Macrolepiota procera, Cantharellus cibarius*, and *Tricholoma matsutake*. All edible fungi should be picked when young and fresh and cooked as soon as possible. The commercially grown products are much safer because growers have been aware over the years of the problems that exist. Also, the strains marketed today are those with a long history of reliability as edible species (3,4).

The observation that certain edible mushrooms grow naturally on certain decomposing organic matter and the desire to produce a tasty source of food have led to the development of various techniques for the cultivation of mushrooms, and in some cases they have been fairly profitable. The cultivated species, however, have never been the most prized by gourmets. The favorite types are still the boletes, specific species of *Amanita*, and truffles; all of them are mycorrhizal fungi, and their carpophores cannot be developed without the right symbiotic plant (1). To date, the only successes have been with the saprophytic species, that is, those that feed on dead organic matter. Among these the best known is undoubtedly the champion, *Agaricus bisporus*, which has given rise to a thriving agricultural industry. Each year millions of tons of these mushrooms are produced for direct consumption, for the canning industry, and for the preparation of soups and sauces.

In recent decades the microbiological industry has been enormously improved, and the production of mushrooms in a controlled environment has become an example of highly advanced technology. Today horse dung still constitutes the basis of the culture soil (5). With a given quantity of straw and droppings, the dung initially undergoes a process of natural though controlled fermentation, first out of doors and then in an enclosed area where it is enriched with nitrogenous sugar substances and vitamins required for the development of the field mushroom. After a variable period of time, depending on the composition of the substrate, now called compost, sowing takes place, using cereal seeds covered with *Agaricus bisporus* mold. Two main varieties of field mushrooms are used: white and brown. The white are better suited to canning, whereas the brown, with their hazel-colored cap and small brown scales, are more widely used in Europe for direct consumption because the flesh is firmer and tastier. The American preference for the white variety is an expression of the same prejudice reflected in the choice of white eggs, bread, and sugar (4).

The cultivation of the oyster mushroom, *Pleurotus ostreatus*, and closely related species is also highly developed, especially in the Orient. They are also found in western markets in various varieties. *Volvariella volvacea* (paddy straw mushroom) (6) and *Lentinula edodes* (shiitake mushroom) (5), common in Oriental cuisine, can now be obtained in speciality shops in North America. For other species, such as the morels, production techniques had been developed (5,7). Production lines are being launched for *Stropharia rugosoannulata* and *Agrocybe aegerita*.

All the saprophytic species, especially those growing on wood, can easily be cultivated, even though in many instances they do not command a large market because of their meager amount of flesh. However, they can be cultivated on a small scale or at home by infecting unhealthy plants or dead stumps with fragments of the cap. The species that lend themselves best to this practice are *Flammulina velutipes* and *Pleurotus ostreatus*.

The cultivation of mycorrhizal fungi, which need to live with a host plant in order to produce their fruiting bodies, is a demanding undertaking. This is known as indirect cultivation. The mycorrhizal symbiotic type that is cultivated quite extensively in France is the legendary French or Perigord truffle, the high quality and price of which amply justify the cost of installing artificial truffle beds and the long years of waiting before the carpophores can be picked. In centers of modern truffle cultivation, production is followed extremely closely and nothing is left to chance. The acorns are sterilized on the outside in a chemical bath or dip and are then put in direct contact exclusively with truffle spores. The seedings, which develop in small bags filled with sterilized earth, will have the truffle as the only symbiotic organism with which to associate. Once planted out in soil with the right characteristics and due care, they will start to produce their first truffles after about 10 years.

CONSERVATION OF CULTURE

Most mushroom growers have to rely entirely on commercial spawn producers and the spawn plants of larger mushroom farms to provide reliable products. Successful mushroom production depends on the proper maintenance of pure culture and spawn that is capable of producing fruiting bodies of high productivity, excellent flavor, texture, color, and resistance to pests, disease, or both (5). There is no satisfactory way to check and evaluate the qualities of spawn by any rapid on-the-spot examination. A method of preserving selected strains that have been thoroughly tested and proved desirable is of primary importance.

Stock cultures are often maintained in an actively growing state under optimum laboratory conditions on a suitable solid substrate, using a system of periodic transfers at reasonable intervals of time. Alternatively, spawn cultures are reestablished under field cultivation by using single or mass basidiospore isolation or tissue culture techniques from freshly harvested fruiting bodies. At present, the most effective methods for preservation of edible mushrooms are freezing and storing in liquid nitrogen (8–10). Both sporulating and nonsporulating fungi can be preserved by freezing with liquid nitrogen. If sporulation or production of aerial mycelia is abundant, fungi are cultured on agar slants or millet that mycelia had grown on.

If sporulation is sparse, or if the culture is nonsporulating and the production of aerial mycelium is sparse, cultures are grown in broth and fragmented in a sterile Waring blender (5). Some strains must be concentrated by centrifugation to obtain sufficient material for freezing. Strains such as the yeastlike *Tremella fuciformis* yield colonies in liquid culture small enough to be pipetted and are grown in flasks containing a suitable broth, either stationary or on a shaker. Survival of nonsporulating cultures or fragments of mycelium by freezing at subzero temperatures has been reported at −196°C (11). The cryogenic storage temperatures now commonly used are those of liquid nitrogen (−196°C) and liquid nitrogen vapor (−150°C and below). With the commercial availability of cryogenic materials and equipment, this process has been extensively used throughout the world on a routine basis by various laboratories with widely differing objectives. The advantages of liquid nitrogen for conservation of microorganisms have been emphasized.

The experimental results and experience with edible mushrooms of the Culture Collection and Research Center (CCRC) at the Food Industry Research Development Center in Taiwan have proved that storage of mushroom spawn cultures in the frozen state at liquid nitrogen temperatures has great value for the spawn-making and mushroom industries. In addition, the production performance of CCRC cultures of cultivated mushrooms of *Agaricus bitorquis* (Quélet) Saccardo and *Pleurotus cystidiosus* O.K. Miller, which have been frozen in liquid nitrogen at CCRC, has been tested in the Mushroom Research Laboratory at the Taiwan Agricultural Research Institute (12). *Pleurotus cystidiosus* has recently been brought into cultivation on a commercial scale in Taiwan for more than 20 years.

Three agar discs containing the hyphal tips or six to eight spawn grains are placed in polyvinyl vials of 1.5-mL capacity, and 0.8 mL of 10% (vol/vol) glycerol or 5% (vol/vol) dimethyl sulfoxide (DMSO)–water menstruum as a cryoprotective agent is added. The vials are precooled to 5°C to prevent overheating during the operation and attached to prelabeled canes and placed in the freezing chamber of a controlled-rate freezer. The initial cooling is carried out at a rate of 1°C/min from room temperature to −70°C; subsequent cooling to −120°C is rapid and uncontrolled. The vials are then immediately transferred to a liquid nitrogen refrigerator and stored in the liquid phase at −150°C and below.

Freezing injury to biological systems is very complicated and may occur physically, chemically, or both, depending on the cryoprotective agents and the cooling rate (13). Changes include ice crystal formation, pressure, solute concentrations, availability of water temperature, pH, and other factors. Best results for the survival and recovery of fungi from the frozen state generally have been achieved by slow cooling at approximately 1°C/min. Good recoveries can be expected with most fungi following freezing by either rapid or slow cooling if the frozen material is thawed rapidly in a 37°C water bath until the last trace of ice is dissipated (14,15).

Spawn maintenance by cryogenic freezing rather than periodic transfers of cultures eliminates degeneration and mutation, prevents losses due to contamination, and reduces the necessity for repeated production tests. Mushroom growers should now be assured of the ready availability of reliable, high-quality spawn.

CULTIVATION

Cultivation of the Common White Mushroom

The methods used for the cultivation of *Agaricus bisporus* are essentially similar to those pioneered by French horticulturists (9), but the old practices have been gradually refined and modified to the point where mushroom production is a closely controlled and scientific process. Mushroom spawn for each cultivation was simply transferred to an existing mushroom bed. A problem with this straightforward approach was the tendency for the spawn to gradually lose its vigor with repeated transfers from one bed to the next. Other improvements in the process have mainly concerned the development of specialized mushroom houses in which ventilation, temperature, and humidity can be carefully controlled to provide the best conditions for fruiting. In fairly deep caves the temperature remains remarkably stable, and for certain stages of mushroom development this advantage is obvious. Specially built mushroom houses are, however, more efficient, especially when it comes to preparing the mushroom beds or harvesting the crop.

The first stage in cultivating *Agaricus bisporus* is the germination of spores (5). This is carried out by taking spores or fragments from a mature fruiting body selected for desirable qualities such as color, taste, and vigor. The spores or fragments are then germinated on sterile nutrient medium to give a master culture. The master culture is used to prepare the spawn, which is then distributed to growers. This spawn was produced by inoculating a suitable sterile medium, such as horse-manure medium, but today a variety of spawn types are available, including grain spawn. Grain spawn is now almost universally used in houses; the grain may be wheat, rye, sorghum, or millet, in decreasing order of size. Whatever the medium, spawn production is merely a process of bulking up the initial mycelium to the point where it can be conveniently introduced into the culture beds themselves (Fig. 1, Fig. 2).

The fermentation process removes most of the simple sugars from the medium and leaves insoluble substances such as cellulose and lignin, which are readily utilized by *Agaricus bisporus*. The point of the composting step is to create a medium that both favors the mushroom mycelium and discourages its natural rivals. The composting process is usually carried out in the open, with the piles being regularly turned to ensure an even fermentation that results in the whole pile becoming hot. The length of time required to get the compost into the right condition varies, depending on the age of the manure, but normally it is at least 2 weeks. Composts for growing in mushroom houses vary according to the nature of the locally available materials and the type of pasteurization employed. Most trays now in use have short legs, which require the addition of spacing blocks for cropping. However, some farms use longer legs, so that the trays are always stacked at the full dis-

Figure 1. Gram spawn of *Agaricus bisporus*, with the fungus growing on sterilized wheat grains.

Figure 2. *Agaricus bisporus* growing in a commercial bed.

tance apart, as for cropping. Although the principle involved is the same in all cases, different growers often have their own minor modifications designed to improve the end product because their different ingredients and preparations require adjustments to the control of temperature and aeration. Correct timing is all-important. Undercomposted manure tends to overheat in the beds and leads to poor mycelial growth. On the other hand, overcomposting will remove many of the nutrients essential for production of a good crop. Once the compost has reached the correct stage of fermentation, the beds themselves can be prepared.

It has been known since mushrooms were first cultivated that the beds must be covered with a casting of soil or other suitable material to induce the change from vegetative growth of mycelium to fruiting. The virtues of different types of soil as casing materials are frequent subjects of argument among growers. Clay soils are reputedly better than sandy ones, but interest has also been extended to artificial substitutes, such as mixtures of peat and vermiculite (an artificial potting medium).

For many years the reason for casing was completely unknown. Recently research has revealed some of the effects produced by the soil. The change in growth conditions accompanying the addition of soil induces the formation of tiny primordia, which subsequently mature into mushrooms, but the exact action by which casing stimulates fruiting is still unknown.

The final stages in mushroom cultivation are cropping and packing. Following the first crop of mushrooms, several additional periods of fruiting usually occur at intervals of 2 or 3 weeks. The time at which the mushrooms are picked is important. It is agreed that commercial growers usually harvest their mushrooms before the veil has broken. The major consideration is the longer life of slightly immature mushrooms once they have reached the shop. In common with many other food items, the prevalent belief that a clean, uniform product is somehow superior to a more varied and possibly tattered product has no doubt contributed to the monopoly of the button mushroom. Once the beds are finished, the composting is removed and sold as manure for various horticultural activities.

Mushrooms are picked by gripping the cap and bending or twisting to break off the stem at the base. This is easily done with closed mushrooms; open ones are fragile, and more care is needed to ensure that the stem breaks at the bottom, not at the top.

The stage of growth at which mushrooms are picked is important. All open mushrooms, whatever their size, are picked. In many countries open mushrooms are considered extremely low grade, and in most they are cheaper in the market than closed cups or buttons. After opening, the stem grows but the weight of the cap does not increase appreciably, so that even disregarding quality and price, there is no further increase in yield. Up to the point of opening, the weight of the mushroom increases very rapidly, doubling in 24 h at its peak rate, so leaving closed mushrooms to grow as long as possible raises their picked weight. With tight picking, the more frequent flushes compensate for the loss of weight on the individual mushrooms. Closed mushrooms also have a better shelf life in the shop, particularly in warm weather.

Picking mushrooms has always been a major factor in wage costs. With closely packed trays or shelves, picking is carried out under cramped conditions, usually with poor lighting, so that selection and grading of mushrooms are difficult (5).

Two recent developments have promised to improve and accelerate the picking operation. One is development picking lines, in which the trays are transported to the pickers, where mushrooms can be picked under ideal conditions. The other is carrying of trays by forklift truck from the cropping room to a destacker at the input end of the line. They are automatically destacked and passed along by a conveyor system at a controlled speed that can be adjusted to run fast if there are not many mushrooms and slower for a heavy flush. Pickers stand on both sides of the line with baskets handy and pick as the trays pass in front of them. The speed of the conveyor is adjusted so that the trays are cleared just before they reach the end of the line. There may be conveyor belts for stumps and the like and another conveyor for filled baskets of mushrooms. There is also an overhead spray line at the end of the conveyor so that trays can be automatically watered as they pass along.

Figure 3. Cultivated *Pleurotus eryngii* growing in reused plastic bottles.

The other development is mechanical harvesting. With pickers' wages becoming an increasingly important factor in production costs, the possibility of harvesting mushrooms by machine has interested many growers. A machine that travels over the bed, cuts mushrooms with a reciprocating blade just over the casing, draws the cut mushrooms back over the blade, and delivers them to one side behind the machine has been described. Patents have also been filed for a machine that locates mushrooms by a number of sensor units, grips them on the top of the cap by suction, and plucks them off the bed (5). Mushrooms on a bed must all, or nearly all, be at the same stage of development, because they must all be cut at once. Any mushrooms that are higher or lower in the bed will be cut with too much stalk or through the cap,and any that are crooked will also suffer. Harvesting machines already in existence can pick from shelves or could be mounted over picking lines. It is necessary to provide for the machine to run on the beds or side boards or to be fitted with sensors that control its height above the casing. This may require the provision of bare strips of bed at each side of the strip to be harvested. A harvesting machine produces ungraded mushrooms. Postharvest grading for size is most easily done in the water flumes used in canneries so that the first application of these machines may be postharvest.

Cultivation of the Shiitake Mushroom

The shiitake grows naturally on logs of deciduous trees, in particular, beech, chestnut, hornbeam, and oak. Cultivation is therefore carried out on specially cut large branches from suitable trees, more substantial logs being in demand for construction purposes. Today a more scientific system is used (16). The logs are soaked in water and pounded to break the bark, or holes are made with a broad drill or a specially designed hammer. The logs are then inoculated with a spore emulsion prepared from mature fruiting bodies. Alternatively, a spawn consisting of mycelium grown on wood chips or sawdust is introduced into the holes. Then the infected logs are placed in a carefully selected site in

Figure 4. New developed and cultivated mushroom: *Tricholoma cystidiosus*.

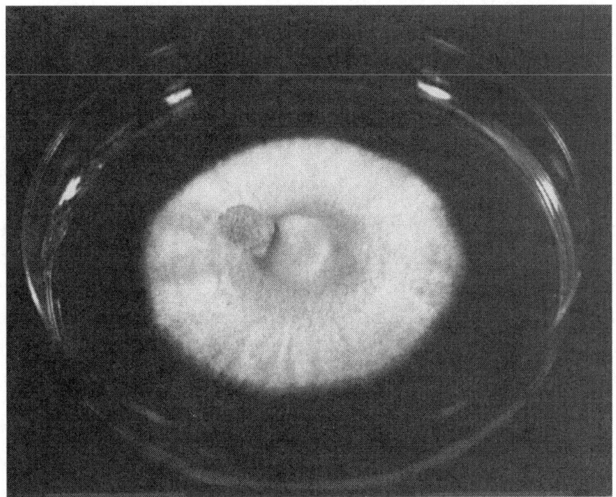

Figure 5. Mushroom primordium of fruiting body, *Fistulina hepatica*, of mycelia growing on potato dextrose agar (PDA) medium.

Figure 6. A fruiting body of cultivated *Fistulina hepatica*.

the forest, known as the laying yard. Choice of site is very important; if it is too moist, the natural wood-decaying competitors are favored and the crop is reduced or even lost altogether. The best laying yards are usually in well-ventilated clearings or positions at the edge of the forest. The logs themselves are placed crosswise at a slight angle to the ground.

Another method is cultivation in a polypropylene bag or space bag containing sterilized sawdust and rice bran (9:1 vol/vol) with a moisture of ca. 65%. The primordia are produced after one week when the plastic bags are opened and the humidity is kept at ca. 80 to 90%. This method is efficient for commercial cultivation and is now the major method in Taiwan and Japan (5).

Cultivation of Other Mushrooms

In addition to the aforementioned species, certain other fungi have been or still are subject to some form of primitive cultivation. The method used to grow cultivated paddy straw mushrooms (*Volvariella volvacea*) is similar to that used for the common white mushroom (17–21). Both methods are based on the preparation of special beds and the introduction of mushroom spawn, but the paddy straw mushroom is mainly cultivated on a small scale using traditional, much less scientifically refined methods.

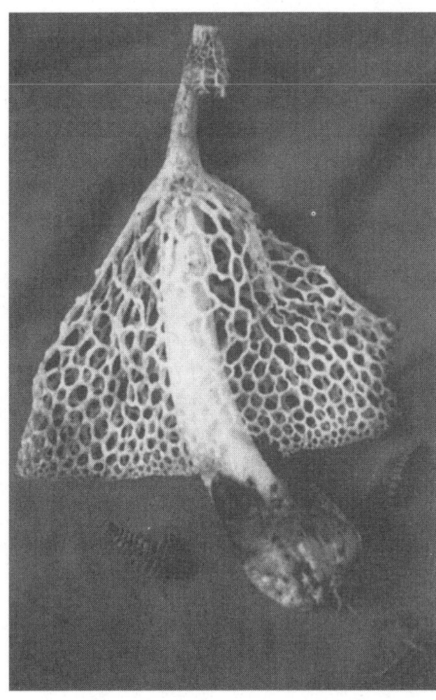

Figure 7. A longitudinal section of *Fistulina hepatica*. Its common names are beefsteak fungus, ox-tongue, and poor man's beefsteak.

The jelly fungus (*Auricularia polytricha*) is quite popular in the Far East and is produced in large quantities in China. A proportion of this crop is grown on logs or polypropylene bags set up with the specific intention of encouraging the fungus. A similarly haphazard process has been used to cultivate *Pleurotus ostreatus*, the oyster mushroom. Several other fungi have been grown deliberately on a small or experimental scale. These include *Pholiota nameko, Pleurotus eryngii* (Fig. 3), *Stropharia rugosoannulata, Auricularia* spp., *Tremella fuciformis, Tuber* spp., *Tricholoma matsutake, Tricholoma titans* (Fig. 4), *Flammulina velutipes, Morchella esculenta* (7), *Fistulina hepatica* (Fig. 5–7), *Agrocybe aegerita*, *Armillaria mellea*, and *Dictyophora indusiata*. Though some sort of systematic cultivation is probably feasible for many fungi, the major obstacle to large-scale production is the difficulty of obtaining economically worthwhile yields.

BIBLIOGRAPHY

1. G. H. Lincoff, *Simon & Schuster's Guide to Mushrooms*, Simon & Schuster, New York, 1981.
2. R. Philips, *Mushrooms and Other Fungi of Great Britain and Europe*, Pan Books, London, 1981.
3. G. H. Lincoff, *The Audubon Society Field Guide to North American Mushrooms*, Alfred A. Knopf, New York, 1984.

4. C. Dickinson and J. Lucas, *The Encyclopedia of Mushrooms*, Orbis Publishing Limited, London, 1979.
5. S. T. Chang and W. A. Hayes, *The Biology and Cultivation of Edible Mushrooms*, Academic Press, New York, 1978.
6. B. A. Oso, "Mushrooms and the Yorba People of Nigeria," *Mycologia* **67**, 311–319 (1975).
7. U.S. Pat. 4,594,809 (June 17, 1986), R. D. Ower et al. (to Neogen Corporation).
8. A. Dietz, "Nitrogen Presservation of Stock Cultures of Unicellular and Filamentous Microorganisms," in A. P. Rinfred and B. LaSalle, eds., *Round Table Conference on the Cryogenic Preservation of Cell Cultures*, National Academy of Sciences, Washington, D.C., 1975, pp. 22–36.
9. D. J. N. Hossack, "Liquid Nitrogen Frozen Inocula—A Year's Experience," *Proc. Soc. for Analytical Chem.* **9**, 36–38 (1972).
10. H. T. Meryman, "Review of Biological Freeying," in H. T. Merymann, ed., *Cryobiology*, Academic Press, New York, 1966, pp. 1–114.
11. S. W. Hwang, "Effects of Ultra-Low Temperature on the Viability of Selected Fungus Strains," *Mycologia* **52**, 527–529 (1960).
12. S. C. Jong and J. T. Peng, "Identity and Cultivation of a New Commercial Mushroom in Taiwan," *Mycologia* **67**, 1235–1238 (1975).
13. A. U. Smith, *Biological Effects of Freezing and Super-cooling*, Edward Arnold, Publishers Ltd., London, 1961.
14. R. D. Goos, E. E. Davis, and W. Butterfield, "Effect of Warming Rates on the Viability of Frozen Fungus Spores," *Mycologia* **59**, 58–66 (1967).
15. P. Mazur, "Survival of Fungi after Freezing and Desiccation," in G. C. Ainsworth and A. S. Sussman, eds., *The Fungi*, Vol. 3, Academic Press, New York, 1968, pp. 325–395.
16. P. J. Wuest, D. J. Royse, and R. B. Beelman, *Cultivating Edible Fungi*, Elsevier, Amsterdam, The Netherlands, 1987.
17. S. T. Chang, *The Chinese Mushroom*, The Chinese University of Hong Kong, Hong Kong, 1972.
18. S. T. Chang, "Production of Straw Mushroom (*Volvariella volvacea*) from Cotton Wastes," *Mushroom J.* **21**, 248–354 (1974).
19. S. E. Chua and S. Y. Ho, "Fruiting on Sterile Agar and Cultivation of Straw Mushrooms (*Volvariella* species) on Padi Straw, Banana Leaves and Sawdust," *World Crops* **25**, 90–91 (1973).
20. K. Ramakrishan et al., "A Simple Technique for Increasing Yield of Straw Mushroom, *Volvariella diplasta* (Berk. and Br.) Sacc.," *Madras Agric. J.* **55**, 194–195 (1968).
21. D. J. Pegler, *Mushrooms and Toadstools*, Mitchell Beazley, London, 1983.

CHEE-JEN CHEN
Food Industry Research and Development Institute
Hsinchu, Taiwan

MUSHROOMS: PROCESSING

PROCESSING

Canning

The mushroom processed in large quantities by canning, either in cans or in glass jars, is *Agaricus* spp. It is the common mushroom and called champignons in French. This is the most common edible mushroom cultivated and processed in the United States and in France. It is the only mushroom used for processing and more than two-thirds of the crop goes to processors in the United States (1).

Raw Material. Mushrooms which are destined to be sold canned in brine must be picked at the tight-cap stage of maturity, ie, the veil of the mushroom is closed and the gills are not visible, while more mature specimens may be utilized in other types of pack, such as chopped or sliced mushrooms, in soup, or in butter sauce. Mushrooms of large strain are cultivated and flat mushrooms with the cap fully opened are used for sliced mushrooms in butter sauce in Australia.

Mushrooms are sent to the cannery with the soil or bottom end of the stem either cut or uncut in the United States. It is believed that a mushroom with the soil end of its stem uncut is protected from dehydrating or losing its moisture. To obtain maximum quality, freshly harvested mushrooms to be packed in brine must be processed as quickly as possible. Otherwise, they must be cold stored until they are ready to be processed.

Washing and Sorting. Mushrooms with the soil end attached must be trimmed mechanically at the cannery. Workers place the mushrooms in the V-shaped grooves of the machine, which revolves and cuts the soil end from the stem and then cuts the stem from the cap if the mushrooms are to be packed as buttons or sliced buttons. For button mushrooms not more than 1/8 in. of the stem stub should be attached to the cap.

Generally, the trimmed mushrooms are dumped into a large vat filled with water and allowed to soak in water for a while to losen the adhering dirt. They are then moved through a spray washer. Cull mushrooms are removed either before or after spray washing. To permit more uniform blanching, size grading of the washed mushrooms before passing them to the blancher is recommended.

Post-Harvest Treatments. Mushrooms lose weight and shrink during blanching and thermal processing, mostly during blanching. This loss of canned product weight ranges between 30 and 40% (2). Since mushrooms are an expensive commodity and the canned product is sold on the drained weight basis, excessive shrinkage will result in considerable economic loss to the processor. A 5% reduction in the shrinkage could increase the revenue of mushroom processors by as much as 20% (3).

Postharvest storage and soaking of mushrooms in water prior to blanching could increase the yield of canned mushrooms. The PSU-3S process has been recommended (4). The process is a combined soaking and storage process, ie, fresh mushrooms are soaked in water [50–60°F (10–16°C)] for 20 min, followed by 18 h of cold storage at 35°F (2°C) and a 2-h second soaking prior to blanching. Soaking in water for 2 h prior to blanching could increase the yield of canned product by about 2%, and the increase could reach 2.9–4.4% when mushrooms are held at 35°F (2°C) for 24–48 h prior to processing (2). However, the PSU-3S process demonstrated yield increases as high as 9.4%. This process was made more feasible for the canning industry by replacement of the second soaking by vacuum soaking, in which mushrooms were held at 2-mmHg pressure for 5 min and kept in water for another 10 min after vacuum release. Vacuum soaking alone would result in a yield increase of 5%, but when it is combined with storage treatment the yields would be greater than those obtained with the PSU-3S process. The yield increases were the same (12%) regardless of the order in which the vacuum soaking and storage were carried out. The vacuum soaking also offers a significant advantage to commercial processors by reducing the time required for optimum water absorption by the product from 2 h to 10 min (5,6).

Mushrooms stored following harvest for 18 h at 35°F (2°C), 53°F (12°C), and 72°F (22°C) increased in yield, compared to the control, by as much as 3.6%, 7.5%, and 9.5%, respectively (5). Although the higher temperature and longer storage time resulted in a higher canned product yield, it also was accompanied by quality deterioration. The optimum process involved soaking mushrooms in water for 20 min, storing at 53°F (12°C) for 18 h, and applying vacuum soaking prior to blanching. The yield increase, compared to the control, was 15.3% and the quality was comparable to those processed after 18-h storage at 35°F (2°C) (6). When mushrooms were stored at 53°F (12°C) with 95% RH for 72 h, canned product yield increased up to 19% and the free amino acid content of the tissues increased proportionately. It has been confirmed that the increase of canned product yield caused by postharvest storage was due to qualitative protein changes that alter the water binding capacity and water holding capacity of the proteins (7).

The Food Industry Research and Development Institute of Hsinchu recommended an enzyme vacuum soaking process to the industry in Taiwan. In this process fresh mushrooms are soaked in water containing papain (0.02%) for 3 min at 50 mmHg pressure, followed by storage at room temperature for 2 h prior to blanching. The enzyme vacuum soaking process alone could result in an increase of the canned product yield by 5.2%, but when the process and 2-h storage at room temperature were combined, the yield increase was 10.1%. When an enzyme activator (such as L-cysteine) was added to the vacuum soaking water containing papain, the yield increase was even more significant (8).

Blanching. The precooking of fresh material is referred to as blanching in canning, freezing, or dehydration processes. Boiling water or steam may be used in the blanching of mushrooms, and the blanching times are adjusted accordingly to bring the center temperature of mushrooms to about 180°F (82°C). The blanching process removes the gases in the mushroom tissue, inactivates the polyphenoloxidase enzyme, improves the texture for slicing, and reduces the bacterial loads. Blanched mushrooms should be immediately cooled by potable water spray and soaking in running water unless the containers can be filled and thermal processed in a short time after blanching, as in the cases of chopped or sliced mushrooms in butter sauce or soup packs.

Size Grading and Filling. The blanched and cooled mushrooms next go through a size-grading machine,which is designed so that certain sizes of mushrooms will drop through holes of a particular size as the machine revolves. The mushrooms which do not drop through the holes will then be passed on to the next larger holes in succession. The machine sorts mushrooms so that uniformly sized mushrooms can be put into cans. Mushrooms may be placed in cans by hand or by machine. Before being put into cans, mushrooms are reinspected and defective ones removed. For sauce-type pack, mushrooms are chopped or sliced immediately after blanching and then cooked with other ingredients before being packed into cans. For mushrooms packed in brine, the cans filled with mushrooms are filled up with hot brine or hot water and a salt tablet. The salt concentration in the finished product is preferably about 1%. Ascorbic acid may be used up to 37.5 mg for each ounce of drained weight of mushrooms, and must be labeled. The filled cans, if not sufficiently hot, must be exhausted before being closed.

Thermal Processing. Since mushrooms are a low-acid food, they must be thermal-processed under pressure, either in a continuous processing machine or in a batch-type retort. In 1973 the Food and Drug Administration (FDA) published its findings on the presence of botulinum toxin in some cans of United States and foreign origin (but not from Taiwan). Many mushroom canneries in the United States were investigated and intensive studies on heat penetration rate were conducted. From the studies it is evident that a relatively wide variety of process times and temperatures were in use for canned mushroom products. This is a result of new styles of pack and new filling and processing methods used in the industry. The heat resistance of the spores from *Clostridium botulinum* isolates obtained from commercially canned mushrooms was determined and the D values for all of the *C. botulinum* spore crops overall were reported to be slightly higher in the buffer than in mushroom puree. The mean D-value (240°F) in buffer for the 10 spore crops was 0.24 min compared to a value of 0.19 min for spores in mushroom puree (9).

Good manufacturing practices for thermally processed low-acid foods packaged in hermetically sealed containers and an emergency permit control were promulgated in 1973 and in 1974. The processors must list their establishments and the scheduled processes required and used in their plants for their particular styles of pack with the FDA. Critical factors associated with each process must be specified in the scheduled process. Initial temperature is one of the critical factors for most of the processes. Other critical factors for still retort processing include slice thickness and the size of chopped or diced pieces, the proportion of small chips or fine pieces, the fill weight, the method of filling and brining, and the can position during the process (eg, horizontal or vertical). Other control factors determined for individual packing operations include the mushroom size and maturity as indicated by veil condition, the duration and temperature of storage before canning, the uniformity and degree of shrinkage during the blanching, and any condition or operation affecting the degree of compactness of the product within the can. Production and quality control records must also be kept on file to provide confirmation that the critical factors are under satisfactory control.

Sulfide spoilage of canned mushrooms was once a big problem encountered in the canneries located in the southern part of Taiwan. This has resulted in more severe thermal processing practices for the canned product since 1965. Later, however, the processors began to apply a higher temperature, shorter time, process to compensate for the deterioration in the quality of the product caused by the prolonged thermal processing (10,11). The yield, color, and texture decrease with increasing process time at all temperatures. Recently, activation energies and Z values during thermal processing were reported (12). The processed products must be cooled rapidly and not be allowed to stand at the higher temperatures.

Canning of Other Mushrooms. To a much lesser extent other mushrooms, such as *Volvaricella volvacea*, straw mushrooms; *Pleurotus ostreatus*, abalone mushrooms and *Flammulina velutipes*, called Enokitake in Japan are canned and consumed in the Orient. The canning processes for these are similar to those applied to Agaricus, described above. *Tremella fuciformis*, a white jelly fungus, is mainly utilized for cooking and for medical purposes by the Chinese. It is usually canned with sugar syrup and the net drained weight is always very light compared to the net weight. The consumption of *Tremella fuciformis* seems to be limited to the Chinese. The canned form is much as popular than the dried form.

Freezing and Freeze-Drying

Only a small portion of common (Agaricus) mushrooms produced are frozen or freeze-dried, and a large part of the frozen mushrooms are sold to reprocessors, such as freeze-dryers, or used in specialized frozen food products.

Washing and Blanching. Lower microbial loads are the basic requirements for these products, and vigorous washing and utilization of some disinfectants have been found effective for this purpose. High levels of chlorine, up to 50 ppm in the wash water, have been found beneficial for reducing the microbial load.

Mushrooms contain a highly potent polyphenoloxidase. It is this enzyme which causes the undesirable discoloration that appears in bruised or cut mushrooms. During the washing operation, soluble oxidized phenolic compounds resulting from the enzymatic action are washed off, and in some cases the washing water may turn a distinct red. Blanching adequate to destroy the enzyme also produces an undesirable gray color. Therefore, mushrooms for freezing are usually given a short blanching in hot water, primarily to help reduce microbial contamination rather than for enzyme inactivation. It is necessary to follow such a blanching immediately with a cold-water quench. A short-time dip in sodium bisulfite solution containing 200 ppm SO_2 or in 1% salt and 0.09% sodium bisulfite solution is an effective alternative for prevention of this enzyme discoloration.

Freezing and Freeze-Drying. A rapid freezing rate is essential for mushroom freezing. It is desirable to freeze mushrooms to a center temperature of -10°F (-23°C) in 3 to 5 min for diced or sliced mushrooms and in about 20 min for button or whole mushrooms. I.Q.F. (individually quick frozen) freezing and refrigerant dip freezing are recommended.

Freeze-drying is carried out in a high-vacuum chamber in which the ice on or in the mushroom tissue disappears by sublimation. Uniform loading is essential, otherwise burning or incomplete drying of a part of the product may occur. When the ice on the surface of the mushrooms disappears, application of heat can be started. The end point of the freeze-drying is the point at which the moisture content of the product reaches about 3%. From 12 kg of raw mushrooms, 1 kg of freeze-dried product is obtained. After reconstitution, 1 kg of the freeze-dried product will yield 8 kg of reconstituted product.

Packaging. Vacuum packaging inhibits oxidative discoloration (13) and a sauce coating may also serve the same purpose. The freeze-dried product is usually packed in airtight and moistureproof containers under nitrogen gas.

Dehydration

Among the cultivated mushrooms, the common mushroom (Agaricus) and Shiitake mushroom (*Lentinus edodes*) have enjoyed a good reputation internationally. While Agaricus is mainly canned and consumed by Western people, *Lentinus edodes*, of which the common name is Shiitake in Japanese, is mainly dried and consumed by Orientals. It is interesting to note that Japanese consumption of Shiitake per capita is about the same as that of Agaricus in European or North American countries. Japan is by far the world's largest producer of this mushroom. In Japan consumption of dried Shiitake is always higher than that of fresh Shiitake by about 20%.

Flavor and Aroma of Shiitake Mushrooms. The 5′-ribonucleotides have been shown to enhance the flavor of foods. As with L-glutamate (MSG), 5′-guanylate and 5′-inosinate are flavor potentiators, which are the compounds that amplify the effects of other flavor agents. While vegetables are generally low in 5′-nucleotides, ranging 1 to 10 μmol per 100 g fresh weight, mushrooms contain large amounts of these nucleotides. Common mushroom (Agaricus) contains about 50 μmol and fresh Shiitake mushroom from 182 to 235 μmol of 5′-nucleotides per 100 g fresh weight. Its content in the extract of dried Shiitake mushrooms is twice that in the extract from fresh ones. The process of drying not only makes it possible to preserve Shiitake for a long time but also enhances the flavor with a unique taste. Guanylic acid has been identified as the main constituent of the particular good taste of Shiitake extract. An extract containing 1–2 mg of guanylic acid can be obtained from 1 g of dried Shiitake, but a little free guanylic acid can be detected in fresh or dried Shiitake. The increase of guanylic acid content is due to the decomposition of ribonucleic acids by ribonuclease during cooking at 140°F (60°C)–158°F (70°C). Lenthionine ($C_2H_4S_5$) has been identified in the aroma from Shiitake mushrooms (14).

The Chinese believe that Shiitake is effective in the prevention of cerebral hemorrhage. Recently the medical effects of the Shiitake mushroom have been studied by many Japanese researchers, and its remarkable ability to remove serum cholesterol and its antiviral or antitumor activities have been shown through biological tests (15,16).

Drying of Shiitake. Nowadays harvested Shiitake are dried by artificial heating to maintain good flavor and the luster of the cap. Cabinet dryers are generally used for artificial heating. Revolving dryers are suitable for mass production. The mushrooms are dehydrated starting at 86°F (30°C) and the temperature is then increased 2°F (1°C)–4°F (2°C) per hour until a temperature of about 122°F (50°C) is reached (usually in 12–13 h). Finally they are heated to 140°F (60°C) and held there for 1 h to enhance the flavor and bring out the luster of the cap.

Drying of Other Mushrooms. In Europe and South America the most important dried mushroom is *Boletus edulis* (or *B. luteus*), a wild mushroom. In Chile, *B. luteus* is predried in the sun to about 15–20% moisture, held for a short time in temporary warehouses, and then taken to a central drying plant. Here the mushrooms are fumigated, cleaned, and sorted to remove dirt and foreign matter. The mushrooms are then sliced, spread on trays, and tunnel-dried. The drying cycle in Chile is reported to begin with a relatively low air temperature of about 120°F (49°C) which is gradually increased to a final drying temperature of 160°F (71°C). The final moisture content is approximately 10%.

Auricularia auricula-judae, which is referred to as Mu-Erh in Chinese and as Jew's-ear in English, is easy to grow and is preserved mostly in the dried form. The fruiting body can be dried under sunlight or with artificial heating systems. Since the water content is about 90% of a fresh fruiting body, but only 9–15% in the dried state, the weight of the dried Auricularia is only 10–12% of the fresh one. The duration of sun drying depends on the sunlight intensity and usually takes one to several days. If the fruiting bodies are dried in drying sheds with an artificial heating system, such as an electric heater or stove, the duration can be shortened to less than 24 h.

Tremella fusiformis, the white jelly fungus, is also called silver ear by the Chinese. They not only use this mushroom as a drug but consume it as a precious food as well. Because of its high cost it was considered a luxury only for the tables of the rich. The canned form is much less popular than the dried form. It is not necessary to avoid premature opening or discoloration of the fruiting bodies since they do not turn dark. The drying is usually done only by the sun or by an artificial heat source. Drying under the sun alone would not give good results in tropical and subtropical climates. Modern management calls for specially constructed drying sheds with a variety of arrangements, like those used for drying Shiitake. The drying process takes up to 8 h and the temperature is first held at 122°F (50°C), then gradually decreased until it reaches 104°F (40°C). During drying, a strong air flow should be maintained to supply hot air and to maintain an optimal temperature. Weak air circulation and high temperatures will affect product quality. Generally the dried white jelly mushrooms weigh only about 6–8% as much as fresh ones. Some instant-dried silver ear products have been produced recently. The fungus is cooked with sugar syrup prior to drying. This product can be rehydrated with hot water. The best of the dried fruiting bodies are packed in plastic bags and kept in airtight containers for storage. Otherwise there is always a danger of deterioration caused by insects or molds.

BIBLIOGRAPHY

1. W. B. Van Arsdel and M. J. Copley, *Food Dehydration, Vol. II, Products and Technology*, AVI Publishing Co., Inc., Westport, Conn., 1964.
2. F. J. McArdle and D. Curwen, "Some Factors Influencing Shrinkage of Canned Mushrooms," *Mushroom Science* **5**, 547 (1962).
3. C. W. Coale and W. T. Butz, "Impact of Selected Economic Variables on the Profitability of Commercial Mushroom Processing Operations," *Mushroom Science* **8**, 231 (1972).
4. R. B. Beelman, G. D. Kuhn, and F. J. McArdle, "Influence of Post-Harvest Storage and Soaking Treatments on the Yield and Quality of Canned Mushrooms," *J. Food Sci.* **38**, 951–953 (1973).
5. F. J. McArdle, G. D. Kuhn, and R. B. Beelman, "Influence of Vacuum Soaking on Yield and Quality of Canned Mushrooms," *J. Food Sci.* **39**, 1026–1028 (1974).
6. R. B. Beelman and F. J. McArdle, "Influence of Post-Harvest Storage Temperatures and Soaking on Yield and Quality of Canned Mushrooms," *J. Food Sci.* **40**, 669–671 (1975).
7. D. L. Eby, F. J. McArdle, and R. B. Beelman, "Post-Harvest Storage of the Cultivated Mushrooms (*Agaricus bisporus*) and Its Influence on Quality Changes Related to Canned Product Yield," *J. Food Sci.* **42**, 22–24 (1977).
8. "Feasibility Study on Mushroom Pretreatment with Vacuumed Solution (VS) in Canning Industry," *1988 Annual Report*, Food Industry Research and Development Institute, Hsinchu, Taiwan, 1989, pp. 36–40.
9. T. E. Odlaug, I. J. Pflug, and D. A. Kautter, "Heat Resistance of *Clostridium botulinum* Type B Spores from Isolates from Commercial Canned Mushrooms," *J. Food Protection* **41**, 351–353 (1978).
10. B. K. Wu, K. C. Lin, and K. M. Chiou, "Studies on Color Improvement and Spoilage Control of Canned Mushrooms. I. Control of Sulfide Spoilage," *Food Industry Research and Development Institute Research Report* **69**, 1–20 (1974).
11. B. K. Wu, Y. H. Lee, K. M. Chiou, K. L. Chang, and T. Y. Liu, "Studies on Color Improvement and Spoilage Control of Canned Mushrooms, II. Improvement in Color of Canned Mushrooms," *Food Industry Research and Development Institute Research Report* **69**, 21–42 (1974).
12. R. C. Anantheswaran, S. K. Sastry, R. B. Beelman, A. Okereke, and M. Konanayakam, "Effect of Processing on Yield, Color and Texture of Canned Mushrooms," *J. Food Sci.* **51**, 1197–1200 (1986).
13. E. Steinbunch, "Quality Retention of Unblanched Frozen Vegetables by Vacuum Packing," *J. Food Technol.* **14**, 321–323 (1979).
14. S. Wada, H. Nakatani, and K. Morita, "A New Aroma-Bearing Substance from Shiitake, an Edible Mushroom," *J. Food Sci.* **32**, 559–561 (1967).

15. T. Mizuno, "Shiitake, *Lentinus edodes*: functional properties for medicinal and food purposes," *Food Reviews International* **11**, 111–128 (1995).
16. T. Kawazoe and K. Yuasa, "Producing vitamin D_2-fortified shiitake mushroom powder in ultraviolet irradiation device," *Journal of Japanese Society of Food Science and Technology* **44**, 442–446 (1997).

BIH KENG WU
Food Industry Research and Development Institute
Hsinchu, Taiwan

MUSLIM DIETARY LAWS: FOOD PROCESSING AND MARKETING

Muslim dietary laws are a set of rules and requirements governing the lives of Muslims, or followers of the Islam religion. The food industry, like any industry, responds to the needs and requirements of consumers. Today's consumers are not only concerned with health and nutrition, they are also conscious of what goes into their bodies as food. Some of these restrictions by consumers are self-imposed philosophical or nutritional choices, whereas others are mandated by their religious beliefs. Muslim dietary laws are based on the Muslim scriptures, the Quran and the Hadith.

Muslims constitute almost 20% of the world population (1). With 1.2 billion adherents, Islam is the world's second-largest religion, after Christianity (2). In a area extending from the Atlantic Coast of Africa to Pakistan and from Central Asia to the Sahara, Islam is the religion of 90% of the population (2), as it is also in Bangladesh and Indonesia. Although the Muslims are concentrated in Asia and North Africa, they are also present as minorities throughout Europe (32 million), North America (5.5 million), Latin America (1.4 million), and even Oceania (0.4 million) (1). The basic tenets of Islam or the primary duties of Muslims are the belief in the oneness of God (Allah) and His messenger (Muhammad); performing the prayers, five times a day; almsgiving, or contributing to the welfare of the poor; fasting (abstaining from eating, drinking, having sex) from dawn to dusk during the 30 days of the month of Ramadan; and a pilgrimage to Makka in Saudi Arabia once in a lifetime, if physically and financially capable (3).

MUSLIM DIETARY LAWS

In addition to the basic duties, a set of guidelines direct the daily life of a Muslim, and these include dietary laws. The dietary laws are composed of the concepts of Halal, Haram, Mashbooh, and Makrooh (4).

What Is Halal?

Halal means lawful or permitted for the Muslims. According to the Quran, all good and clean foods are Halal. Consequently almost all foods from the sea, plants, and animals are considered Halal except those that have been specifically prohibited.

What Is Haram?

Haram means unlawful or prohibited for consumption. The prohibited categories mentioned in the Quran include the following:

- Carrion, or the meat of dead (unslaughtered) animals
- Blood
- Swine
- Animals blessed to others than God (Allah) (ie, to idols)
- Intoxicants, including alcoholic drinks

Additionally, carnivorous animals, birds of prey, and land animals without external ears are prohibited to the Muslims (5). Any products contaminated with Haram items also become Haram.

What Is Mashbooh?

Mashbooh means suspect, in doubt, or questionable. If the origin of a certain food item is in doubt, or there is uncertainty about its permission or prohibition under Islamic laws, then the product is considered Mashbooh. A wide range of products in today's marketplace fall under this category, which forms the gray area between what is permitted and what is prohibited (5).

What Is Makrooh?

Makrooh means detested or discouraged. Products that are discouraged by God or His messenger, Muhammad; are offensive to one's psyche; or may be otherwise harmful fall under this category, including stimulants and smoking (4).

Although the range of Haram and Makrooh foods in Islam is quite narrow, emphasis on observing the prohibitions is very strong. Islam is not oblivious to the exigencies of life, however. Islam permits a Muslim, under the compulsion of necessity, to eat a prohibited food in sufficient quantities to survive (4).

HALAL FOOD PROCESSING

There are in excess of 8,500 grocery items on the shelves of even a small supermarket in North America and Europe, and many new products are being added daily (6). Due to the complexity of food-product development and food processing, an uncertainty exists about the nature and kinds of ingredients present in a particular food. Muslims may either abstain from purchasing such foods or may end up eating what is prohibited. Food processors would need to use the following key points in the production of Halal products (7):

- Products must be from naturally Halal animals, such as cattle, goats, chicken, and the like.
- The animals must have been slaughtered properly, as described later.
- Products must be processed and packed using utensils, equipment, and machinery that has been properly cleaned.

- Products must be free from ingredients that are Haram, Makrooh, or Mashbooh.
- All raw materials used must be Halal.
- Products must be free from cross-contamination and must not come in contact with Haram substances.

Proper Method of Slaughter

There are strict requirements for slaughtering animals: the animal must be a Halal species; the animal must be slaughtered by a Muslim of proper age; the name of Allah must be pronounced at the time of slaughter; and the slaughter must be done by cutting the throat of the animal in a manner that induces rapid, complete bleeding and results in the quickest death (5). Slaughtering should be done with a sharp knife in one swift cut. Certain other conditions should also be observed to ensure humane treatment of animals before and during slaughter. It is required that land animals be slaughtered according to the preceding method. There is no such requirement to slaughter fish or other sea creatures. The meat of animals thus slaughtered is called Zabiha. The term Zabiha, or Dhabiha, implies that the animal has been slaughtered by a Muslim.

Ingredients

For the formulation and manufacture of Halal products, special consideration should be given to the following ingredients (6):

Alcohol: Ethyl alcohol is an intoxicant and is prohibited as such. Many liquid flavors, such as natural vanilla, contain alcohol to satisfy U.S. legal requirements. Use of alcohol in the food and flavor industries is quite common for a number of technical reasons, such as flavor extraction and standardization, ingredient precipitation, or use as a solvent. Although Halal production permits the use of synthetic or grain alcohol in their manufacture, it would be prudent for manufacturers to minimize the levels of alcohol present in the final products through evaporation or other methods. Though there is a small allowance for industrial alcohol, the tolerance is zero for grape wine and other liquors, both in manufacturing and the food-service industries. Many Muslim countries have set up quite elaborate analytical laboratories to test for alcohol in products at their points of entry.

Gelatin: Gelatin is generally made from pork skins, cattle bones, calf or cattle skins, or fish skins. Generally product labels make no mention of the source of gelatin. For Halal products, gelatin must not be porcine. Fish gelatin is Halal, and gelatin from cattle and calves is considered Halal when the animals are slaughtered by Muslims.

Lard: Lard is pork fat and is prohibited to Muslims. Products for Muslim markets must not be formulated with lard.

Enzymes: Enzymes can be from animals, plants, or microbes or produced through biotechnology. The use of enzymes is common in the manufacture of cheese and other foods. Use of porcine enzymes is not permitted in the manufacture of Halal foods. The use of enzymes extracted from animals that are not slaughtered by Muslims is also discouraged by most Muslim countries. Enzymes from plants, microbial sources, and biotech sources are generally considered Halal (see "Ingredients from Biotechnology").

Emulsifiers: Emulsifiers such as mono- and di-glycerides are made from vegetable oils, beef fat, or lard. Since vegetable mono- and di-glycerides are readily available, emulsifiers from beef fat which are Mashbooh and from lard which are Haram can easily be avoided by the food manufacturers.

Ingredients from Biotechnology

Biotechnology covers a wide range of activities and therefore leads to a large number of different applications in the food industry. For Muslims these new technologies open up opportunities for expanding their food supply. At the same time, these new technologies may create some difficulties in making Halal determinations for food ingredients, food products, food materials, and even modified species of animal and plants. Information should be made available to Muslim leaders on concepts and practices in food biotechnology and genetic engineering so they can properly evaluate their Halal implication for foods. In general, biotechnology ingredients and enzyme cultures are accepted by Muslims, with some reservations about certain products (eg, a porcine-derived biotech-produced enzyme) that need to be reviewed on a case-by-case basis (8).

Packaging

The Halal status of packaging materials is also questionable. Though a plastic container may appear acceptable, the source of some of the ingredients used to create the plastic may be of concern. In many cases, stearates from animal sources are used in the production of plastic containers. The formation and cutting of metal cans may require the use of oils, which may be derived from animals (9). Steel drums could have been used to carry foods containing pork or pork fat, and despite rigorous cleaning, these could retain small amounts and contaminate otherwise Halal products.

Marketing

With the global emphasis on food marketing, both food service and retail, people are becoming more accustomed to buying prepared or partially prepared meal items. Presently, two of the strongest import markets for Halal food products are Southeast Asia and the Middle East, accounting for more than 400 million Muslim consumers. A strong emphasis on Halal certification of products in Malaysia, Singapore, Indonesia, Thailand, and the Middle East is shaping the ways in which the exporting countries of Europe, North America, Australia, and New Zealand are modifying their production and marketing practices.

BIBLIOGRAPHY

1. B. Brunner, ed., "Religious Population of the World, 1996," *The TIME Almanac*, Information Please LLC, Boston, Mass., 1998.

2. C. I. Waslien, "Muslim Dietary Laws, Nutrition, and Food Processing," in Y. H. Hui, ed., *Encyclopedia of Food Science and Technology*, Vol. 2, John Wiley and Sons, Inc., New York, 1992, pp. 1848–1850.
3. S. Twaigery and D. Spillman, "An Introduction to Moslem Dietary Law," *Food Technol.* **7**, 88–90 (1987).
4. M. M. Hussaini, *The Islamic Dietary Concepts and Practices*, Islamic Food and Nutrition Council of America, Chicago, Ill., 1993.
5. M. M. Chaudry, "Islamic Food Laws: Philosophical Bases and Practical Implications," *Food Technol.* **12**, 92–104 (1992).
6. M. N. Riaz, "Halal Food—An Insight into a Growing Food Industry Segment," *Int. Food Marketing and Technol.* **12**, 6–9 (1998).
7. M. M. Chaudry et al., *Halal Industrial Production Standards*, My Own Meals, Inc., Deerfield, Ill., 1996.
8. M. M. Chaudry and J. M. Regenstein, "Implications of Biotechnology and Genetic Engineering for Kosher and Halal Foods," *Trends Food Sci. Technol.* **5**, 165–168 (1994).
9. M. M. Chaudry, "Islamic Foods Move Slowly Into Marketplace," *Meat Processing* **36**, 34–38 (1997).

MUHAMMAD MUNIR CHAUDRY
Islamic Food and Nutrition Council of America
Chicago, Illinois

JOE M. REGENSTEIN
Cornell University
Ithaca, New York

MYCOTOXIN ANALYSIS

Mycotoxins are a heterogeneous class of fungal toxins that contaminate virtually all agricultural commodities. They play an important role in foodborne disease in humans and animals, with toxic effects as variable as their composition. Although it is difficult to prevent mycotoxin contamination in food and feed, it may be possible to reduce exposure through rigorous monitoring programs that require accurate and precise analytical methods. In addition to monitoring and surveying, analytical methods are needed to determine compliance with tolerances and guidelines and for research purposes. A great deal of research has been devoted to developing sensitive, specific, and accurate methods for quantifying mycotoxins in food and feed. The progress of such efforts can be seen in the numerous recent reviews on mycotoxin analysis (1–9) and in the General Referee Report on Mycotoxins that appeared in the January/February issues of the *Journal of the Association of Official Analytical Chemists* (AOAC) *International* (10–14). Several excellent reference books have been written on mycotoxin analysis (15,16).

GENERAL ANALYTICAL PROCEDURES

Analysis of mycotoxins in foods and feeds requires trace analytical techniques because the mycotoxins are typically present in agricultural commodities at levels ranging from ng/g to μg/g. Mycotoxins vary greatly in their structural properties, and thus also their physical properties. Consequently, it is impossible to develop methods that are applicable to all mycotoxins. In addition, analytical difficulties exist since mycotoxins are not evenly distributed in food, and food matrices contribute interferences to sample extracts.

The procedure used to estimate the concentration of mycotoxins in food and feed consists of three steps. First, a random sample is taken from the lot. Second, the entire test sample is ground in a mill or grinder, and a subsample is removed from the ground sample. Finally, mycotoxins in the subsample are extracted with solvent and quantified. An accurate and precise estimate of the true concentration of a mycotoxin in a food depends on all three parts of the analysis (17). Of the three steps, initial sampling is the most critical because it contributes the most to variability in the analytical result. The analytical aspects will usually contribute the least error while sample preparation will usually have an error in between the two (18).

SAMPLING AND SUBSAMPLING

Much literature has been published on procedures for sampling agricultural commodities for contaminants in general (19) and for mycotoxins specifically (20–27), since sampling is one of the most critical aspects of contaminant analysis. Van Egmond (28) tabulated official sampling plans that are used by regulatory agencies throughout the world for monitoring mycotoxin levels in food and feed. Special sampling procedures for aflatoxins have been worked out by the Food and Agriculture Organization (29) and are under preparation by the Codex Alimentarius Commission (30) for control purposes (7).

The major goal of sampling is to obtain a portion that accurately represents the concentrations of individual mycotoxins in a given lot of food or feed. Traditional sampling methods for agricultural crops are usually not adequate for mycotoxin analyses because mycotoxins are rarely evenly distributed within lots of food and feed. Pockets of contamination may be found in bulk storage of grains, oilseeds, oilseed cakes, flours, or ground mixed feeds. These "hot spots" may be due to mold contamination and proliferation in a small portion of the field where the commodity was grown. It also may occur when small containers of contaminated material are added to loads of uncontaminated food. Pockets of contamination can form in stored commodities, especially in areas of high moisture (18).

It is important to design a sampling procedure that will allow the lowest variability that resources will allow. In general, increasing the sample size will result in more reliable analytical results. However, the amount of material collected should be cost effective and manageable for sample analysis. Success at obtaining representative samples of food or feed for analysis depends on several factors: (*1*) the physical characteristics of the product sampled, (*2*) the distribution of the contaminant, and (*3*) the sampling procedure (17). In general, it is easier to obtain representative samples of commodities that are liquids (ie, milk) or can be made into pastes or powders by grinding (ie, peanut butter, flour) than particulate foods such as grains, whole nuts, and mixed feeds (17). Liquids and samples with small particle sizes require little sample preparation because they can be readily made homogeneous by stirring before the sample is removed for analysis. In general, more heterogeneous samples with large particle sizes require a larger sample size (eg, peanuts) than samples with small particle size (grains). The nature and distribution of the mycotoxin is another important factor in obtaining representative samples (17). Several mycotoxins are very heterogeneously distributed in foods, making sampling difficult. For example, aflatoxin has been shown to contaminate only several kernels in an ear of corn or in a batch of peanuts (21).

Samples may be taken from crops in the field, during handling, during storage, and at other points in the production and processing system (31). The most effective sampling plans involve gathering equal portions at random points throughout a lot of material (17). Of the two methods that exist for collecting samples of contaminated food, the best sampling method is continuous stream sampling. Portions are collected at specific time intervals from a continuous stream of material and combined into a single sample. Often it is necessary to sample commodities held in bins, sacks, rail cars, barges, trucks, or other large containers. Portions are taken from these static populations using hand-operated or mechanical probes (18), then combined to obtain a representative sample of the entire bin of material.

Once a sample has been drawn from a lot, it is often too large to be analyzed. A smaller subsample must be taken from which the mycotoxin is extracted. The same heterogeneity that existed in the lot also occurs in the sample. Therefore, the samples should be ground to proper size and blended before mycotoxins are extracted (22). Ideally, analyses should be completed immediately after sampling. If this is not possible, samples must be stored properly to prevent further fungal growth and formation of mycotoxin. Warm, moist conditions should be avoided. Samples should be frozen or refrigerated in a dry state.

EXTRACTION

After sampling, the next phase of mycotoxin detection involves extraction of the toxin from the food matrix. Officially recognized extraction methods for foods are prescribed by the AOAC INTERNATIONAL Official Methods of Analysis (32) and for feeds by the International Standards of Organization (ISO) publication or the European Economic Community (EEC) directives (5,33–35). The main purposes of extraction are to transfer toxin from the sample to a solvent and to remove unwanted contaminants and interferences (2). The efficiency of extraction of mycotoxins depends on the physiochemical properties of the food matrix as well as those of the toxin.

Mycotoxins are often extracted from food or agricultural products by blending or shaking the sample with appropriate amounts of solvent (6). High-speed blending has the advantage of reducing the sample particle size, which may lead to better extraction (5). The selection of solvent type and ratio of sample to solvent volume depends on the mycotoxin of interest and the source material. The solvent-to-sample ratio is usually 2–5 mL:1 g sample. Most mycotoxins are readily soluble in organic solvents but sparingly soluble in water (5). Typical extraction solvents include chloroform, ethyl acetate, methanol, acetone, acetonitrile, and mixtures of these organic solvents and water. The water wets the substrate and increases penetration of the solvent mixture into the hydrophilic matrix (16). The aqueous phase can be acidified to release the toxin from interactions with proteins in the food. Sodium chloride or other inorganic salts are often added to the aqueous phase to reduce emulsion formation. In addition to liquid solvents, supercritical fluids (eg, CO_2 with methanol or acetonitrile as modifiers) have been successfully used to extract aflatoxin (36) and trichothecenes (37,38) from grain and feed. Nonpolar solvents such as hexane or 2,2,4-trimethylpentane can be used to defat samples of high-lipid content before extraction.

After extraction, filtration is often necessary to remove unwanted solids. When using highly nonpolar solvents, a common practice is to add a diatomaceous earth filter aid to assist in filtration. Centrifuging extracts prior to cleanup is a common practice when filtration difficulties arise (5).

CLEANUP METHODS

Some analytical methodologies, that is, immunoassays, require minimum sample cleanup for quantitation of mycotoxins. However, in most cases, extensive cleanup of agricultural commodities is needed before the final analysis is feasible. Removal of interfering substances is a key element in obtaining accurate quantitative results and is normally the most time-consuming stage in the analysis of mycotoxins. The choice of sample cleanup method depends on the type of mycotoxin and matrix, the expected concentration of the mycotoxin, and the available final analytical method used for detection and quantitation (5,6). Four varieties of cleanup methods have been used: dialysis, anionic precipitation, liquid–liquid partitioning, and column chromatography. For some samples one cleanup step is enough, but others may require a combination of two or more cleanup procedures before quantitation.

Dialysis cleanup procedures in combination with liquid–liquid partitioning have been developed for a variety of mycotoxins (39,40). The aqueous anionic precipitation method of cleanup is used mainly for extracts in polar organic solvents containing more than 55% water (5). Anionic precipitation involves coprecipitating plant pigments and proteinaceous substances with various reagents; lead, zinc and ammonium salts, phosphotungstic acid, iron, and copper compounds. Liquid–liquid partitioning in conjunction with column cleanup has been used to separate complex mixtures of mycotoxins having different physiochemical properties. Acids or bases can be added to the aqueous phase to facilitate separation of mycotoxins possessing strong functional groups such as the phenolic zearalenone and zearalenol (41), or acidic mycotoxins penicillic acid and ochratoxin (5). Use of liquid–liquid partitioning is declining because it requires large sample sizes and large volumes of potentially hazardous solvents.

Column chromatographic methods are the most widely used cleanup procedure. These methods include classical open column chromatography and chromatography using small disposable prepacked columns, also known as solid-phase extraction columns. In both types of column chromatography, packing materials can be normal or reverse-phase silica, alumina, polyamide, XAD-2, ion exchange, Sephadex®, charcoal, Florisil® and immunoaffinity. Available bonded phases for reverse-phase columns include ethyl, octyl, octadecyl, cyclohexyl, and phenyl.

Classical column chromatography has been used for purifying mycotoxins present in food and feed. Different modes of chromatography such as adsorption, partition, ion exchange, size exclusion, and affinity chromatography have been used (42). Due to the large amounts of solvent needed and the cost of waste solvent disposal, classical column chromatography is not commonly used today. One of the most widely used screening methods for mycotoxins (aflatoxins, ochratoxin A, zearalenone) uses prepacked cartridges containing one or more layers of different adsorbents to purify toxins before analysis (32,43–46). Prepacked, disposable solid-phase extraction (SPE) columns (cartridges) are increasingly being used for sample cleanup. The use of these cartridges saves time, uses less solvent than traditional liquid–liquid extraction or classical column chromatography, and is ideally suited to the analysis of large numbers of samples. The prepacked cartridges are reported to be more consistent than laboratory-packed columns, although batch-to-batch variations can occur. Consequently, recovery trials should be carried out on all batches of columns (5). SPE cleanup procedures have been developed for ochratoxin (47), trichothecenes (48,49), aflatoxins (50,51), fumonisins (52,53), and other mycotoxins. Recently, SPE disks requiring even less solvent than SPE cartridges have been developed (54) and were used for isolating zearalenol and zearalenone from corn extracts (55).

An important new cleanup technique is the use of antibody-based immunoaffinity columns (IAC). Advantages of IAC over other cleanup methods include specificity, speed, low solvent use, possibility of automation, and possibility of column reuse (56). Specific antibodies to a number of mycotoxins (aflatoxins B_1, B_2, G_1, B_2, M_1; fumonisins B_1 and B_2; ochratoxin A; deoxynivalenol; and zearalenone) have been developed (56) and are the basis of commercially available IAC. Among the commercial IAC, the Aflatest P is finding widespread use in the analysis of corn, peanuts, and complex foods and feeds (56). The Aflatest P test kit, which includes a fluorometer for quantitation, has been certified by the AOAC as an official method (32,57). Mycotoxins are extracted from a sample with aqueous methanol or acetonitrile. The extract is diluted further with water or buffer, and a portion of the diluted sample is passed through the column. The column is then washed with water to remove interferences. The purified mycotoxin is then eluted using methanol or acetonitrile. Quantification of the toxin can be accomplished using fluorometry or a variety of chromatographic methods.

Several studies have shown that IAC could be reused without loss of efficiency (58–60). Currently, multiple antibody IAC are under development to allow simultaneous separation of chemically diverse toxins from the same extract (56). Scudamore et al. (61) linked IAC in series to determine aflatoxin and ochratoxin levels in a variety of dry cereal-based pet foods. Maragos et al. (62) developed an IAC that isolated fumonisins B_1, B_2 and hydrolyzed forms of these toxins.

After cleanup procedures are complete, mycotoxins are present at low concentrations in large volumes of solvent. Therefore, concentration is required before analysis. This is usually done by heating at low temperatures in a water bath, temperature-controlled heating block, or a rotary evaporator (5). Evaporation under nitrogen is recommended for six mycotoxins listed in the *AOAC International Official Methods of Analysis* book (32), and exposure to light should be minimized.

ANALYSIS: SEPARATION, DETECTION, QUANTITATION

General Considerations

Choice of Analytical Method. The choice of analytical method used to separate, detect, and quantify mycotoxins depends on the physiochemical properties of the mycotoxin, concentration of the mycotoxin in the food or feed, cost and time constraints, and availability of analytical methodology and instrumentation. Early studies used biological methods, for example, animal toxicity assays, to detect the presence of mycotoxins in food and feed. With the development of sophisticated instrumentation, chemical methods became the method of choice for mycotoxin analysis. For mycotoxins that contain a chromophore having high molar absorptivity and/or high fluorescent properties, such as aflatoxin, zearalenone, patulin and ochratoxin, sample extracts are usually subjected to either thin-layer chromatography (TLC) or high-performance liquid chromatography (HPLC). For both methods, the compound of interest is identified by comparing the R_f (a measure of migration distance) or retention time to that of an analytically pure standard of the compound. Quantitation is accomplished by determining the fluorescence intensity or UV absorbance of an elution spot or an HPLC peak. Mycotoxins that do not contain a distinctive chromophore, for example, fumonisin and some of the trichothecenes, can be derivatized and analyzed by HPLC or gas chromatography (GC). Alternatively, these compounds can be analyzed directly by gas chromatography–mass spectrometry (GC/MS) or liquid chromatography–mass spectrometry (LC/MS).

Analytical Standards. Essential to mycotoxin analysis is obtaining pure analytical standards for the compound of interest. While standards for the most common mycotoxins are commercially available, newly discovered or uncommon toxins are often available only from the researchers who originally purified and identified the compounds. Analytical criteria of important mycotoxins were published by the Food Chemistry Commission of the International Union of Pure and Applied Chemistry (63). In addition, several extensive tabulations of chemical, physical and spectra characteristics of mycotoxins have been published by Cole and Cox (64) and Savard and Blackwell (65). Purity of primary standards should be determined by at least two analytical methods. If possible, concentration of standards in solution should be determined spectrophotometrically rather than gravimetrically. Some mycotoxins can be chemically labile, consequently, stability of the pure standard in solvent should be determined. Proper storage of the standard is essential, and periodic check of stability during storage should be determined. Since all mycotoxins are potentially hazardous, care should be taken while working with the pure compounds, and all manipulations should be performed in a hood or glove box (5,32).

Confirmation Methods. Interfering compounds can be present in sample extract even after exhaustive cleanup procedures are employed. Consequently, confirmation methods are needed to establish the true identity of the analyte. There are two main techniques for confirmation of identity: chemical derivitization and mass spectrometric analysis. Pre- and postcolumn derivatization of mycotoxins can be used to confirm the identity of mycotoxins analyzed by HPLC. For example, fumonisin previously analyzed by derivitizing with σ-phthaldialdehyde (OPA) can be analyzed using another derivitizing agent, for example, naphthalene dicarboxaldehyde (NDA). Diode array detectors are commonly available HPLC detectors that can be used to confirm the identity of mycotoxins having characteristic UV spectra. Using several detectors in tandem is another excellent confirmatory method (5). Numerous chemical methods for confirming the identity of aflatoxins and other mycotoxins exist with TLC. Various reagents can be sprayed onto developed chromatograms, and the reactions can be compared to those with pure standards (66,67).

The most powerful method for confirming the identity of mycotoxins is by mass spectrometry (MS), and numerous methods exist for the identification of mycotoxins by GC/MS (68–71) or LC/MS (72–77). The mass spectra of compounds eluting from a GC or HPLC column can be used to aid in the structural determination of unknowns or provide unequivocal identification of peaks (6). Selectivity can be increased by using select ion monitoring (SIM) that is unique to the analyte of interest. The affordability of GC/MS and LC/MS systems allows most modern laboratories to be equipped with this powerful analytical tool.

CHEMICAL METHODS

Chromatography

Introduction. In the past ten years, chromatography has been the most studied technique for detection, analysis, and characterization of mycotoxins. Chromatographic methods employed for these purposes include TLC, GC, HPLC, and supercritical fluid chromatography (SFC). A book by Betina (16) and reviews by Shephard (78) and Dorner (6) are excellent references on the use of chromatography for mycotoxin analysis.

Thin-Layer Chromatography. Despite the fact that many other techniques have been increasingly applied in the determination of mycotoxins in food and feed, TLC is still one of the most popular because of its relative simplicity and low operating cost. Other advantages are that it allows simultaneous analysis of many extracts, offers great selectivity, and makes multitoxin analysis possible (6). As with other chromatographic techniques, TLC requires extensive sample cleanup to remove interferences. Several excellent reviews have been published on TLC methods used for mycotoxin screening and quantitation (67,79–83).

The choice of type of TLC plate adsorbent and mobile phase depends on the chemical characteristics of the analyte as well the matrix from which it was extracted. Adsorbents used for normal-phase TLC include silica gel, aluminum oxide, and cellulose, whereas reverse-phase TLC plates have a layer of silica gel bonded to various nonpolar functional groups including ethyl, octyl, octadecyl, and phenyl. The majority of TLC plates used for mycotoxin analyses are glass or plastic supports coated with silica gel. Other substances, such as oxalic acid, sulfuric acid, or glycolic acid, can be incorporated into the silica gel layer to aid in the separation of acidic mycotoxins (citrinin, cyclopiazonic acid) (8). TLC plates can be prepared in the laboratory, but most frequently, they are purchased as precoated plates since they are more uniform and reproducible.

Sample extracts are applied to the TLC plate with a micropipet or syringe, and the plates are then placed in a mobile phase for chromatographic development. During development, components of the extract partition between the stationary and mobile phases. Mobile phases depend on the mycotoxins to be separated, the type of adsorbent, and the presence of compounds in the extracts that interfere with the analyte (6). In normal-phase TLC, the major component of the mobile phase is usually a relatively nonpolar solvent (chloroform, ethyl ether, or ethyl acetate). Polar solvents (acetone, alcohols, water, or organic acids) are usually added to achieve the desired separation of the analytes. With reverse-phase TLC, the mobile phase is usually composed of water in combination with methanol or acetonitrile. Development of plates is usually carried out with one mobile phase. However, use of more than one solvent can be used to improve resolving power (6).

When sample extracts contain few interfering substances, one-dimensional TLC is sufficient for separation and quantitation of analytes. Using this method, many sample extracts can be simultaneously analyzed (81). When cleanup is insufficient to remove interferences, the use of two-dimensional TLC (2D-TLC) can improve resolving power. In 2D-TLC, the plate is dried after developing in the first mobile phase, then it is rotated 90° and developed in a different solvent. 2D-TLC has been applied to the analysis of aflatoxins, ochratoxin A, and citrinin (81).

To visualize the mycotoxin spots on TLC plates, the plates are either examined under UV light for fluorescent mycotoxins such as aflatoxins, zearalenone, sterigmatocystin, penicillic acid, citrinin, and ochratoxin A, or they are sprayed with a chemical reagent that reacts with the mycotoxins to produce a colored or fluorescent derivative (83). Under long-wave UV light (365 nm), aflatoxins appear as blue, while zearalenone, citrinin, sterigmatocystin, and penicillic acid appear as blue-green, yellow, red, and purple spots, respectively (81). Aluminum chloride has been used to make fluorescent derivatives of deoxynivalenol (84) and sterigmatocystin (85), while ammonia was used to increase the fluorescence of ochratoxins (86) and patulin (87). Nonfluorescent mycotoxins are detected by chemical derivatization. For example, T-2 toxin can be detected by spraying plates with a mixture of sulfuric acid and methanol, then heating the plate for several minutes at 110°C (81). Mycotoxins are quantified by comparing the intensity of the fluorescence of the sample spot with that of a series of standards. Visual comparison has been widely employed for mycotoxin screening. More accurate quantitation is achieved instrumentally with a densitometer. Detection limits for TLC are in the range of pg/g to ng/g depending on the analyte, source of contamination, and cleanup method (83).

The development of high-performance TLC (HPTLC) in recent years has kept TLC competitive with other more sophisticated techniques. Plates are smaller than conventional TLC plates, and they are uniformly coated with a 0.1- to 0.3-mm layer of small particle size (2–10 μm) adsorbents (82). The small particle size results in rapid separation of analytes. HPTLC uses automatic sample application equipment for spotting small volumes of test sample on plates and a densitometer to quantify analytes. Detection limits for aflatoxins using HPTLC have been reported in the low picogram range (82).

High-Performance Liquid Chromatography. One of the first publications describing the use of HPLC to analyze mycotoxins in food was written by Seiber and Hseih (88). Since then HPLC has become the method of choice for quantifying mycotoxins in food and feed. HPLC is similar to TLC in principle, but it offers the analyst greater resolution, sensitivity, accuracy, and precision (25). HPLC allows ultratrace analysis of mycotoxins; some analytes can be determined at the nanogram or even picogram level. Other advantages of HPLC are its ability to analyze thermally labile, poorly volatile, polar, and ionic compounds. The introduction of autosamplers, computerized data retrieval systems and a variety of sensitive detectors make HPLC very useful for large-scale analyses. Important limitations of HPLC include the high cost of instrumentation and the lack of a sensitive universal detector. Like most of the other chromatographic methods, rigorous cleanup is usually required to achieve accurate, reproducible results. Reviews on the use of HPLC to analyze mycotoxins have been written by Shephard (78), Coker and Jones (89), Frisvad and Thrane (90), Shepherd et al. (91) and Kuronen (92).

The versatility of HPLC makes it ideal to analyze chemically diverse compounds such as the mycotoxins. This is mainly due to the advances made in the past decade in the chemistry of adsorption materials for column packings. Although use of normal-phase silica columns was initially favored for analysis of some of the mycotoxins, reverse-phase chromatography is currently being used most often for aflatoxins (93–97), zearalenone (41,98,99), nivalenol and deoxynivalenol (100), citrinin (101), ochratoxin (102), fumonisins (52,53), *Alternaria* toxins (103–105), patulin (106,107), and sterigmatocystins (108). There have been some reported use of ion-exchange chromatography (IEC) in the determination of moniliformin (109).

Several detectors are available for use with HPLC, which provides for great selectivity and sensitivity. Unlike GC, there is no truly effective universal detector for HPLC. Refractive index (RI) detectors lack sensitivity, which limits their use in mycotoxin analysis. An evaporative light-scattering detector (ELSD) is another universal detector that can be used for the detection of low-volatility components in purified extracts. As with RI detectors, ELSDs lack the sensitivity needed for detection of most mycotoxins. A literature search found only one published method for fumonisins (110).

As most mycotoxins absorb UV light, UV detectors are commonly used detectors for mycotoxin analysis. The photodiode array (PDA) detector has proven to be a powerful HPLC detector for detecting and quantifying mycotoxins. The advantage of PDA detectors is that they provide both multiwavelength and spectra information in a single chromatographic run (42). In combination with computer searching capabilities, it is possible to monitor the spectra of eluting components, allowing further identification of mycotoxins.

The fluorescence detector is commonly used to detect mycotoxins that fluoresce or fluoresce when derivitized. Fumonisins, a recently discovered group of mycotoxins, contain an amino group and several carboxylic acid groups that can be derivitized with fluorogenic reagents such as σ-phthaldialdehyde (OPA) (52), naphthalene dicarboxaldehyde (111), or *N*-hydroxysuccinimidylcarbamate (112). The trichothecenes and sterigmatocystin are other examples of mycotoxins that are derivatized with fluorescent compounds (113,114). Fluorescence detectors, which are highly selective and more sensitive than UV detectors, are the detectors of choice for aflatoxin and fumonisin analyses. In addition to UV and fluorometry, electrochemical detectors have been used for the detection of several mycotoxins (115–117).

Development of LC/MS techniques has been the subject of considerable study since its development in the late 1960s (118). Since that initial work LC/MS interfaces have undergone considerable development and improvement. First (direct liquid introduction, moving belt, particle beam) and second generation (thermospray, electrospray) interfaces have led the way to the generation of so-called atmospheric-pressure ionization (API) interfaces. The API techniques that comprise chemical ionization and other ionization techniques (electrospray and ionspray), have opened a new window in LC/MS, allowing analysis of either slightly polar analytes of low to medium molecular mass or more polar analytes and ions in solutions of medium to high molecular mass (75). Instruments have become smaller (benchtop size), less expensive, more rugged, and compatible with commonly used HPLC mobile phases and flow rates (75). Reports exist on the use of LC/MS for determination of such mycotoxins as aflatoxins (72), trichothecenes (73,74), zearalenone (75), ochratoxin (76), and fumonisins (77,119).

Gas Chromatography. GC is the method of choice for some mycotoxins that exhibit little or no UV absorption or fluorescence. GC methods have been published for virtually all mycotoxins, although their greatest utility lies in the analysis of the trichothecenes (6,119). With GC, extracts containing the analytes are first subjected to vigorous cleanup procedures. The next phase involves volatilizing the analytes. Most mycotoxins are not volatile at GC temperatures (30–350°C), and must be derivatized to a volatile form. The functional group in mycotoxins allowing derivatization is in most cases a hydroxyl group, and the derivatives formed are usually trimethylsilyl (TMS) ethers or heptafluorobutyryl (HFB), pentafluoropropionyl (PFP), or trifluoroacetyl (TFA) esters (120). Onji et al. (121) have described a method for analyzing several *Fusarium* mycotoxins including deoxynivalenol, 3-acetyldeoxynivalenol, fusarenon-X, diacetoxyscirpenol, 15-monoacetylscirpenol, T-2 toxin, scirpentriol and zearalenone without derivatization.

Packed and capillary GC columns are used for the separation of analytes. Packed columns are usually borosilicate glass tubes that are packed with an inert support that is coated with the polymer to be used as the stationary phase. With capillary columns, fused silica tubing is coated with thin films of stationary phase. Capillary columns offer great efficiency, which explains why the majority of methods published for the determination of trichothecenes used capillary columns (6).

Several types of detectors are available for GC, but for mycotoxin analysis, the flame ionization detector (FID), the electron capture detector (ECD), and the mass spectrometer (MS) are the most common. The FID detector is considered the universal GC detector because it detects all organic compounds. Although rugged and sensitive, FID detectors are not at all selective and require rigorous cleanup of extracts to obtain reliable results (6). The ECD is a more selective detector than the FID, but it is not necessarily more sensitive. ECD gives strong signals for halogenated compounds but not for hydrocarbons (6). MS is the ultimate detector for use with GC; it is sensitive, selective, and enables confirmation of the identity of eluting compounds. Selectivity can be enhanced by selecting SIM mode to monitor a single ion mass. GC/MS is a powerful tool especially for the analysis of trichothecene mycotoxins (69–71). Several recent publications (70,122) describe the use of GC/fourier transform infrared (FTIR) spectroscopy to aid in the characterization and identification of *Fusarium* mycotoxins. The use of GC/FTIR together with GC/MS provides a very powerful combination for the identification of a variety of mycotoxins that represent diverse structure types in food or feed extracts.

Supercritical Fluid Chromatography. Supercritical fluid chromatography (SFC) uses a mobile phase with the solvating power of a liquid and the diffusivity and viscosity of a gas to separate nonvolatile or thermally labile compounds (123). The combination of SFC with MS allows the analysis of compounds which require derivatization prior to GC/MS. Young and Games (124) used SFC on packed HPLC columns combined with UV and a moving belt MS to separate and identify some *Fusarium* mycotoxins (deoxynivalenol, isodeoxynivalenol, 3-acetyldeoxynivalenol, 3,15-diacetyldeoxynivalenol, calonectrin) in liquid culture extracts. The SFC separations were rapid, and detection limits for a variety of mycotoxin standards analyzed by MS were from 10 to 250 mg. Roach et al. (123) described a method for analyzing trichothecenes by SFC-negative ion chemical ionization MS. A disadvantage of SFC/MS is that the sensitivity is lower than GC/MS or LC/MS. In all, SFC is in its infancy due to high instrument costs compared with other analytical equipment.

CAPILLARY ELECTROPHORESIS

Capillary electrophoresis (CE) is capable of separating several charged and water-soluble molecules in a single analysis. The method often has similar cleanup requirements as available HPLC methods. An advantage of CE over HPLC is the elimination of organic solvents for the determinative step (125). Using a wide range of methods, CE has been used to separate and quantify aflatoxin, fumonisins, ochratoxin A and B, zearalenone, moniliformin, and α-cyclopiazonic acid with varying success (125–128). Maragos (127) and Maragos and Greer (125) described procedures for analyzing aflatoxin B_1 and fluorescein isothiocyanate-labeled fumonisin in corn using CE with laser-induced fluorescence detection. They reported that the limits of detection for these two toxins obtained by CE were as good as if not better than, the detection limits obtained by HPLC.

IMMUNOCHEMICAL METHODS

General Aspects

Within the past decade numerous reports have been published concerning the use of immunoassays (IAs) for mycotoxin analysis. IAs are rapid, simple, specific, sensitive, and portable and require little analyst training or skill (129). Because IAs are extremely sensitive, often in the picogram/milliliter range, simple dilution of the sample can often remove the need for extensive sample cleanup. However, with most food matrices, some form of extraction of the sample is needed. Three IAs that have been developed for mycotoxin analysis include the radioimmunoassay (RIA), the enzyme-linked immunosorbent assay (ELISA), and the IAC assay. The RIA and ELISA analyses are based on the competition of binding between unlabeled mycotoxin in the sample and labeled mycotoxin in the assay for the specific binding sites of antibody molecules (130). The IAC assay involves the use of antibody columns that specifically trap the mycotoxins in a sample. Trapped mycotoxin can then be eluted and quantified. Some IAs require specialized equipment, such as ELISA readers, tube readers, scanners, or fluorometers for quantitation, while others rely on visual quantitation (129).

Chu (131) listed four main criteria that need to be considered when choosing an IA for mycotoxin analysis. First, a well-labeled mycotoxin (as a marker) is needed in the assay system in addition to a specific antibody. Second, a good method for the separation of free and bound forms of mycotoxin is needed. Third, the degree of specificity of the antibody in the assay should be known. Finally, the sample matrix should be tested before the assay to determine if structurally related compounds in the sample react with the antibody.

Various commercial kits using immunochemical techniques such as ELISA and IAC assays are now available for several mycotoxins including aflatoxins, deoxynivalenol, fumonisins, ochratoxin A, and zearalenone (129). Such commercial tests will find universal acceptance if officially listed to be at least comparable with present analytical methods. Such assessment is currently under way with the AOAC International in the United States and the Ministry of Agricultural, Food and Fisheries (MAFF) in the United Kingdom (132). Several quantitative IA methods have been adopted as AOAC International Official methods for mycotoxins (10,129,130,133).

The development of recombinant antibodies and the making of reusable IAC should reduce the cost of IA and make it one of the more important techniques in the analysis of mycotoxins. In addition, recently published methods (134,135) have indicated that the possibility exists for quantifying several mycotoxins in the same IA. Several excellent reviews have been written on IA in general (132,136) and on the use of IA in the analysis of mycotoxins (56,58,129,137–140).

Radioimmunoassay (RIA)

Although the RIA is not used routinely for mycotoxin analysis, it is currently used by some laboratories for research purposes. RIA has been used to detect aflatoxins (140), nivalenol (141) deoxynivalenol (142), ochratoxin A (143), and T-2 toxin (144,145) in food.

In RIA, sufficient radiolabeled analyte (^{125}I, ^{14}C or ^{3}H) is added to an antibody such that 50% of the label becomes antibody bound. A proportion of this label is then displaced when a known concentration of standard analyte or an unknown amount of sample analyte is added. Activated charcoal is then used to remove the antibody-free fraction. The amount of analyte can be calculated by comparing the concentration of isotope in either the supernant fluid or the residue to a standard curve (146). Detection limits for purified mycotoxin are 0.25 to 0.5 ng when tritiated mycotoxins are used as markers (131). The sensitivity of the method can be improved by using radioactive markers of high specific activity or using a simple cleanup step after extraction. RIA has several drawbacks, including the instability and expense of some radioisotopes as well as the hazards in handling isotopes (146).

Competitive Enzyme Linked Immunosorbent Assay (ELISA)

Of the several available IA techniques, the most popular is the ELISA. Two different ELISA formats have been used for the analysis of mycotoxins, both of which involve the separation of free toxin in one phase from the bound toxin in another phase (130). In direct competitive (DC) ELISA, the mycotoxin-specific antibody is immobilized in the wells of a microassay plate or on the wall of an assay tube. Sample extracts and mycotoxin standard solutions are pipetted into the antibody-coated wells or tubes with a fixed amount of enzyme-labeled mycotoxin. The plates or tubes are incubated between five minutes and more than an hour, which allows the labeled and unlabeled mycotoxin to compete for binding sites on the antibodies. The unreacted material is then washed away and the amount of labeled analyte bound by the immobilized antibody is quantified by addition of an enzyme substrate that forms a colored product. The amount of color formed is inversely proportional to the amount of unlabeled mycotoxin in the sample (136). To avoid sample matrix interferences that can occur with DC ELISA, the sample can either be diluted or subjected to a cleanup step.

In indirect competitive (IDC) ELISA the mycotoxin is coupled to a carrier protein such as bovine serum albumin and bound to the wells of a microtiter plate or assay tube (136). The sample extract is added to the microassay well followed by addition of a fixed amount of mycotoxin specific antibody. The antibody in solution partitions between the free mycotoxin in the sample and the toxin bound by the microassay plate. The antibody not bound to the microassay plate is washed away and quantified by addition of a second antibody that specifically binds the first antibody. An enzyme substrate is added, and the amount of color formed is proportional to the amount of analyte in the sample extract. A major advantage of IDC ELISA is that it requires much less antibody than DC ELISA (130). One drawback of IDC ELISA is that it requires a longer analysis time than DC ELISA (131).

Several studies have investigated the accuracy and precision of ELISA by comparing the technique with other analytical methods. In general, excellent correlations have been found between results obtained by ELISA and TLC, HPLC, or GC for aflatoxins (147,148), deoxynivalenol (149), and fumonisins (150,151). However, fumonisin results obtained by ELISA can be higher than those determined by chromatographic methods (151–153). This phenomenon is due to the presence of structurally related compounds that cross-react with the antibodies. Use of antibodies of higher affinity to the analyte may prevent this problem (131).

In recent years, ELISA kits have been made available from commercial sources for quantitative and semiquantitative analysis of aflatoxins, deoxynivalenol, fumonisins, ochratoxin, and zearalenone in food and feed (129). The kits usually are self-contained and include all the reagents: standards, controls, buffers, and substrates. Several commercial ELISA kits have been approved as AOAC official methods for aflatoxin and zearalenone (32,129). A membrane-based visual dipstick enzyme immunoassay was developed for analyzing up to five mycotoxins simultaneously (134).

Immunoaffinity Columns

IACs are prepared by adsorbing immunospecific antibodies onto a gel support contained in a plastic column or cartridge (131). The sample extract is loaded onto the column, and as the extract is forced through, the analyte is captured by the antibodies. Impurities not bound to the antibodies are washed from the cartridge. Mycotoxins are then desorbed with methanol and quantitated by HPLC or by fluorometry with derivitization, if needed. Advantages of IAC over ELISA methods include selectivity, the ability to trap toxins from large volumes of sample extract, and the ability to combine with different analytical techniques. The cost of columns can be high (>$10.00 each), which can make analysis of many samples expensive (129). However, several reports have indicated that IAC could be reused, in some cases more than 100 times (58–60,129). Several investigators used IAC for analyzing several toxins simultaneously. Maragos et al. (62) developed a single IAC that simultaneously isolated fumonisin B_1 and B_2 and hydrolyzed forms of these toxins. In contrast, Scudamore et al. (61) linked two IACs in series to purify aflatoxins and ochratoxin A.

IACs are available from commercial sources for cleanup of aflatoxins B_1, B_2, G_1, G_2 and M_1, deoxynivalenol, fumonisins B_1 and B_2, ochratoxin A, and zearalenone. Antibodies against other mycotoxins (citrinin, patulin, cyclopiazonic acid, kojic acid, sterigmatocystin, and several others) have been produced. However, no reports have indicated the use of IAC for purification of these mycotoxins (129). IAC purification and SPE gave generally similar results for aflatoxins (154) in milk, corn, and sorghum, but higher recoveries were obtained when IACs were used to isolate fumonisin B_1 from canned and frozen corn (155).

BIOLOGICAL METHODS

General Concepts

Historically, biological assays have played a vital role in the detection and isolation of unidentified fungal metabolites in food and feed. In the classic studies of the early 1960s (156), a duckling bioassay was an instrumental tool in the isolation and purification of aflatoxin from groundnut meal. Since then biological assays have aided in the characterization of most mycotoxins. In general, chemical assays are preferred over biological assays for mycotoxin analysis because they are more rapid, reproducible, and specific. However, chemical analysis cannot be used until the toxin has been chemically characterized, standards purified, and appropriate methodology developed (157).

Before using a bioassay as a routine procedure for monitoring mycotoxins, several factors should be considered (157). First, the bioassay should be able to respond to a diverse range of mycotoxins. Second, the sensitivity and repeatability of the assay must be acceptable. Third, the cost of the assay should be relatively low. Fourth, the results should be obtained in a short time (preferably less than 24 h). Fifth, blanks should be included to determine if toxic effects occur with solvents or carrier reagents. Finally, the occurrence and frequency of false positives should be minimal to none (157). Bioassays can be classified by the organism used—microorganisms, animals, and plants. Vertebrate animals are the most dependable for detection of vertebrate toxins and they include intact animals, skin tests, and chick embryo assays. Bioassays can be highly specific or general, and the choice of type depends on the application. Excellent reviews on the use of biological assays for screening mycotoxins were written by Yates (158) and Cole et al. (157).

Microbiological Assays

The discovery that mycotoxins have antimicrobial properties has led to the development of mycotoxin screen assays using microorganisms (bacteria, fungi, and yeast). Surveys were carried out on numerous microorganisms to demonstrate which were sensitive to mycotoxins. Of those microorganisms tested, *Bacillus* spp. have been shown to be well suited for the detection of aflatoxin B_1 (159,160), patulin (161), penicillic acid (162), ochratoxin (163), and roquefortin (164). One of the most extensive testings of mycotoxins on a bacterial species was carried out by Boutibonnes et al. (165) on *B. thuringiensis*. Virtually all of the mycotoxins tested induced some toxic effect, indicating that the organism may be a good candidate for mycotoxin screening. Mycotoxin screening methods for mycotoxins were developed using *B. megaterium* (166), *B. cereus* (163), *B. stearothermophilus* (164,167), and *B. subtilis* (161). Yeasts such as *Kluyveromyces fragilis* and *Rhodotorula rubra* have been used to assay trichothecene mycotoxins (168,169).

Many of the mycotoxins have been subjected to the Ames mutagenicity assay (170) of which the aflatoxins, fusarin C, and some *Alternaria* toxins are reported to be mutagenic to at least one strain of *Salmonella typhimurium* (171). Other organisms that have been used to screen mycotoxins for mutagenicity include *B. subtilis* mutants, *Saccharomyces cerevisiae*, DNA polymerase deficient *Escherichia coli* and *Neospora crassa* (158). Advantages of mutagenicity assays include their ability to yield semiquantitative data and their sensitivity and simplicity.

Animal Assays

Vertebrates. Laboratory animals, especially mice and rats, have been successfully used for the detection and isolation of mycotoxins including aflatoxins, zearalenone, rubratoxins, and aflatrem (157). Advantages of using mice are that they are small, commercially available, and relatively inexpensive. Ducklings and chickens have been used successfully to detect ochratoxin A (172), α-cyclopiazonic acid (173), tremorgenic mycotoxins (174–176), moniliformin (177), and some trichothecenes (178). One of the most sensitive bioassays for aflatoxin is the production of liver carcinoma in rainbow trout (179). However, the test is difficult to run and is impractical as a screening method because it takes more than four months to complete (157).

Several specific animal tests exist for the trichothecene mycotoxins. In the skin assay, a test solution is applied to the back skin of an experimental animal such as a rabbit, rat, or guinea pig, and the area is monitored for skin lesions. Sensitivities for this bioassay have been reported to be 0.1 μg using T-2 toxin (180). Another trichothecene bioassay involves the rejection or acceptance of drinking water by mice (181). The test sample is dissolved in distilled water that is presented to mice. The volume of the contaminated water consumed by the animals is compared with the volume of control water consumed. A positive test occurs when the mice refuse to drink, indicating the presence of trichothecene.

A specific bioassay for zearalenone is based on the hyperestrogenic effects of the mycotoxin (157). Swine are the most sensitive animals, although dairy cattle, lambs, chickens, turkeys, and laboratory animals (rats, mice, guinea pigs, and monkeys) are also affected. The rat uterotopic bioassay has been used as an assay for zearalenone if other analytical methods are unavailable (182).

The only vertebrate bioassay adopted by the AOAC is the chick embryo assay also known as the CHEST (chick embryo toxicity screening test) assay (157). The assay involves injecting test solutions into the air cell of fertile chicken eggs. Toxicity of the test solution is determined by comparing the mortality of the embryos injected with the test substance with that of the undosed controls. The CHEST assay has been used to determine toxic effects of individual mycotoxins (183,184) as well as combinations of several toxins (185). Overall, the method is simple, inexpensive, and sensitive, but it is nonspecific and prone to false positives (157). The reproducibility of the CHEST assay depends on the type and volume of carrier solvent, the site of injection, and the observation period (186). The sensitivity of the CHEST assay is in the 0.1- to 100-μg range (158).

Factors that must be considered when using whole animal bioassays include the method of administration and the vehicle for administration (157). Oral routes of administration are the most valid since they simulate the processes that occur during ingestion, digestion, and absorption of the mycotoxins. Ideally, the test material is administered by feeding ad libitum in an inert carrier. Because of solubility problems, some toxins cannot be readily formulated in inert carriers. In addition, carriers can be toxic themselves or affect the absorption process (157).

Invertebrates. Several invertebrates (brine shrimp, protozoa, planaria, mollusks, and insects) have been screened for their ability to detect mycotoxins. The brine shrimp (*Artemia salina*) bioassay is one of the first and most widely used biological screening methods for mycotoxins and has been used to screen aflatoxin (187), trichothecenes (188), ochratoxin A (189), and fumonsin (184). The sensitivity range of the assay for 10 trichothecenes ranged from 0.04 to 0.4 μg/mL, with T-2 toxin being the most toxic (188). Attractive features of the assay include its sensitivity, simplicity, rapidity, and low cost. A disadvantage of the brine shrimp assay is that this organism is sensitive to many compounds present in normal foods and feeds, resulting in a high percentage of false-positive bioassay results (157).

Tissue Cultures and Cytotoxicity. Numerous vertebrate tissue and organ cultures have been used for bioassays for mycotoxins, including calf kidney cells; embryonic lung cells; rabbit reticulocytes; HeLa cells; human fibroblasts; and rat kidney, liver, and muscle cells (157). Cytotoxicity studies involve incubating the test substance with cell cultures for various periods of time. Cytological, morphological, and biochemical methods are used to evaluate toxicity (157). A mutagenicity/carcinogenicity assay using primary rat hepatocytes has been useful for determining if mycotoxins and other compounds induce chromosomal aberrations and micronuclei (190). Advantages of tissue culture and cytotoxicity tests are that they are sensitive, relatively inexpensive, and readily available. Several drawbacks include a lack of specificity and the fact that they do not take into account ingestion, digestion, and absorption of the toxin (157).

Plant Assays

Use of plants (whole, tissues) to detect mycotoxins has had limited application. However, there are several cases where plants have been excellent sensors for mycotoxins. Plant bioassays have an advantage over whole animal studies in that they are easier and less expensive to conduct. Plant bioassays have been used in the past to detect and isolate plant growth regulators. Since some mycotoxins also affect plant growth, whole plants and plant tissues are also used for mycotoxin screening. Malformin A, moniliformin, diacetoxyscirpenol, fusaric acid, and *Alternaria alternata* toxins are examples of mycotoxins that have been screened using bean, corn, rice, wheat, jimsonweed, and tobacco plants (191–193). The wheat coleoptile bioassay has been a useful tool for detecting chaetoglobosin K (194) and is especially sensitive to the 12,13-epoxytrichothecenes (157).

RAPID SCREENING METHODS

One of the simplest and most rapid mycotoxin screening methods uses black or long-wave UV light (365 nm) to examine cracked or coarsely ground corn. The characteristic bright greenish yellow fluorescence (BGYF) observed against a dark background is used to identify specific lots of corn contaminated with aflatoxin (133). When one BGYF glower is observed in 1 kg of ground corn, the corn must be analyzed by more quantitative methods (ie, TLC or HPLC). The main disadvantage of this test is that it gives a high rate of both false-positive and -negative results (133). Visual tests are also used in the United States to inspect peanuts for *Aspergillus flavus* conidial heads, and if present, suspected lots are not allowed into commerce for human consumption (95).

TLC has been an effective and fairly rapid method for qualitative/quantitative analysis of agricultural commodities for aflatoxin. The only supplies that are needed include aflatoxin standards, solvent, silica-coated TLC plates, a developing tank, and UV light. Samples are ground in a mill, extracted with chloroform:water, then spotted on a silica gel–coated plate. Plates are developed in tanks containing anhydrous diethyl ether and then examined under long-wave UV light. Identity and quantitation are accomplished by comparing unknown spots to those made by pure standards. This method can detect as low as 20-ng aflatoxin/g (133).

Minicolumn chromatography is a rapid screening method for aflatoxin, ochratoxin A, and zearalenone in foods before examination by other analytical techniques. Minicolumn screening methods include the following steps: extraction, purification, concentration, and development on a minicolumn for detection under UV light. The methods are simple, require little experience and no sophisticated equipment, and are useful for field analyses. The minicolumn method of Romer (44) has been approved by the AOAC as an official method for the detection of aflatoxin in mixed feeds, corn, peanuts, and cottonseed. Similar methods have been developed for ochratoxin A (45) and zearalenone (46). Detection limits for these methods range from 5 to 20 ng/g.

Widely used rapid screening methods use commercially available IA kits. By shortening the incubation time and adjusting antibody and enzyme concentration, microtiter ELISA screening kits can rapidly detect aflatoxin, deoxynivalenol, and zearalenone levels as low as 20 ng/g, 1 μg/g and 100 ng/g, respectively (32). Rather than use microtiter plates, some commercial kits have antibody mounted to a paper disk or membrane, a plastic cup, or polystyrene beads (130). Generally, these tests indicate mycotoxin levels above or below certain target levels. When no color develops, the test sample is above the target level. These tests, which are well suited for field analysis, can provide results in less than 10 min. All samples giving a positive response by the ELISA screening test must be analyzed by another procedure to verify the ELISA results. The ELISA procedure can reduce the laborious chemical testing of the majority of samples, which are negative for mycotoxin, thus saving time and expense (3). Rapid screening methods have been developed using IAC to clean up sample extracts containing fluorescent mycotoxins (aflatoxin, ochratoxin) followed by detection and quantitation using a fluorometer (2). Thompson and Maragos (195) designed a fiber-optic immunosensor that has the potential for screening corn for fumonisins in 6 min.

FUTURE DEVELOPMENT

Much research has been devoted in recent years to developing sensitive, accurate, and precise methods for measuring mycotoxin levels in food and feed. The efforts are continuing and keeping in pace with progresses in analytical chemistry. Considerable work needs to be done in reducing the time and expense required for the analysis of mycotoxins. In addition, research is needed to develop methods for analyzing several mycotoxins in the same assay. With the rapid development of new immunochemical methodologies, these needs may be fulfilled in the near future.

BIBLIOGRAPHY

1. D. W. Wilson et al., "Mycotoxin Analytical Techniques," in K. K. Sinha and D. Bhatnagar, eds., *Mycotoxins in Agriculture and Food Safety*, Marcel Dekker, New York, 1995, pp. 135–182.
2. F. S. Chu, "Recent Progress on Analytical Techniques for Mycotoxins in Feedstuffs," *J. Anim. Sci.* **70**, 3950–3963 (1992).
3. J. L. Richard et al., "Analysis of Naturally Occurring Mycotoxins in Feedstuffs and Food," *J. Anim. Sci.* **71**, 2563–2574 (1993).
4. J. Gilbert, "Recent Advances in Analytical Methods for Mycotoxins," *Food Additives and Contaminants* **10**, 37–48 (1993).
5. P. S. Steyn, P. G. Thiel, and D. W. Trinder, "Detection and Quantitation of Mycotoxins by Chemical Analysis," in J. E. Smith and R. S. Henderson, eds., *Mycotoxins and Animal Foods*, CRC Press, Boca Raton, Fla., 1991, pp. 165–221.
6. J. W. Dorner, "Chromatographic Analysis of Mycotoxins," in T. Shibamoto, ed., *"Chromatographic Analysis of Environmental and Food Toxicants,"* Marcel Dekker, New York, 1998, pp. 113–168.
7. W. Langseth and T. Rundberget, "Instrumental Methods for Determination of Nonmacrocyclic Trichothecenes in Cereals, Foodstuffs and Cultures," *J. Chromatogr. A* **815**, 103–121 (1998).
8. D. J. Webley, K. L. Jackson, and J. D. Mullins, "Mycotoxins in Food: A Review of Recent Analyses," *Food Australia* **49**, 375–379 (1997).
9. P. M. Scott, "Mycotoxin Methodology," *Food Additives and Contaminants* **12**, 395–403 (1995).
10. M. W. Trucksess, "Mycotoxins, General Referee Reports, Committee on Natural Toxins," *J. AOAC INT.* **78**, 135–141 (1995).
11. M. W. Trucksess, "Mycotoxins, General Referee Reports, Committee on Natural Toxins," *J. AOAC INT.* **79**, 200–205 (1996).
12. M. W. Trucksess, "Mycotoxins, General Referee Reports, Committee on Natural Toxins," *J. AOAC INT.* **70**, 119–125 (1997).
13. M. W. Trucksess, "Mycotoxins, General Referee Reports, Committee on Natural Toxins," *J. AOAC INT.* **81**, 128–137 (1998).
14. M. W. Trucksess, "Mycotoxins, General Referee Reports, Committee on Natural Toxins," *J. AOAC INT.* **82**, 488–495 (1999).
15. R. J. Cole, ed., *Modern Methods in the Analysis and Structure Elucidation of Mycotoxins*, Academic Press, Orlando, Fla., 1986.
16. V. Betina, ed., *Chromatography of Mycotoxins*, Elsevier, Amsterdam, The Netherlands, 1993.
17. D. L. Park and A. E. Pohland, "Sampling and Sample Preparation for Detection and Quantitation of Natural Toxicants in Food and Feed," *J. AOAC* **72**, 399–404 (1989).
18. A. D. Campbell et al., "Sampling, Sample Preparation, and Sampling Plans for Food Stuffs and Mycotoxin Analysis," *Pure and Appl. Chem.* **58**, 305–314 (1986).
19. F. M. Garfield, "Sampling in the Analytical Scheme," *J. AOAC* **72**, 405–411 (1989).
20. T. B. Whitaker and J. W. Dickens, "Variability Associated with Testing Corn for Aflatoxins," *J. AOAC* **56**, 789–794 (1979).
21. J. W. Dickens and T. B. Whitaker, "Sampling and Sample Preparation Methods for Mycotoxin Analysis," in R. J. Cole, ed., *Modern Methods in the Analysis and Structural Elucidation of Mycotoxins*, Academic Press, New York 1986, pp 29–49.
22. T. B. Whitaker, J. W. Dickens, and F. G. Giesbrecht, "Testing Animal Feedstuffs for Mycotoxins: Sampling, Subsampling, and Analysis," in J. E. Smith and R. S. Henderson, eds., *"Mycotoxins and Animal Foods,"* CRC Press, Baca Raton, Fla., 1991, pp. 153–164.
23. T. B. Whitaker et al., "Effects of Sample Size and Sample Acceptance Level on the Number of Aflatoxin-Contaminated Farmers' Stock Lots Accepted and Rejected at the Buying Point," *J. AOAC INT.* **77**, 1672–1680 (1994).
24. T. B. Whitaker et al., "Variability Associated with Sampling, Sample Preparation, and Chemical Testing for Aflatoxin in Farmers' Stock Peanuts," *J. AOAC INT.* **77**, 107–116 (1994).
25. T. B. Whitaker et al., "Variability Associated with Analytical Methods Used to Measure Aflatoxin in Agricultural Commodities," *J. AOAC INT.* **79**, 476–485 (1996).
26. T. B. Whitaker et al., "Variability Associated with Testing Shelled Corn for Fumonisin," *J. AOAC INT.* **81**, 1162–1168 (1998).

27. T. B. Whitaker, W. M. Hagler, Jr., F. G. Giesbrecht, "Performance of Sampling Plans to Determine Aflatoxin in Farmer's Stock Peanut Lots by Measuring Aflatoxin in High-risk Grade Components," *J. AOAC Int.* **82**, 264–270 (1999).
28. H. P. Van Egmond, "Current Status on Regulations for Mycotoxins, Overview of Tolerances and Status of Standard Methods of Sampling and Analysis," *Food Additives and Contaminants* **6**, 139–188 (1989).
29. "Sampling Plans for Aflatoxin Analysis in Peanuts and Corn," Technical Report 55, Food and Agriculture Organization, Rome, 1993.
30. Codex Alimentarius Commission, "General Guidelines in Sampling," World Health Organization, Geneva, Switzerland, 38-MAS, 1996.
31. N. D. Davis et al., "Protocols for Surveys, Sampling, Post-Collection Handling, and Analysis of Grain Samples Involved in Mycotoxin Problems," *J. AOAC* **63**, 95–102 (1980).
32. P. M. Scott, "Natural Toxins," in P. Cunniff, ed., *AOAC Official Methods of Analysis*, 16th ed., Vol. 2, AOAC INTERNATIONAL, Gaithersburg, Md., 1997, pp. 49-1 to 49-51.
33. Anonymous, "Determination of Aflatoxin B_1," *Official Journal of the European Communities* **L102/9** (1976).
34. Anonymous, "Animal Feedstuffs—Determination of Aflatoxin B_1 Content," *International Standard 6651*, 1st ed., International Organization for Standardization, Geneva, Switzerland, 1983.
35. Anonymous, "Animal Feedstuffs—Determination of Zearalenone Content," *International Standard 6870*, 1st ed., International Organization for Standardization, Geneva, Switzerland, 1985.
36. M. Holcomb et al., "Determination of Aflatoxins in Food Products by Chromatography," *J. Chromatogr.* **624**, 341–352 (1992).
37. R. P. Huopalahti, J. Ebel, and J. D. Henion, "Supercritical Fluid Extraction of Mycotoxins from Feeds with Analysis by LC/UV and LC/MS," *J. Liq. Chromatogr. Rel. Technol.* **20**, 537–540 (1997).
38. E. P. Jarvenpaa et al., "The Use of Supercritical Fluid Extraction for the Determination of 4-Deoxynivalenol in Grains: The Effect of the Sample Clean-up and Analytical Methods on Quantitative Results," *Chromatographia* **46**, 33–38 (1997).
39. B. A. Roberts and D. S. P. Patterson, "Detection of Twelve Mycotoxins in Mixed Animal Feedstuffs Using a Novel Membrane Clean-up Procedure," *J. AOAC* **58**, 1178–1181 (1975).
40. D. S. P. Patterson and B. A. Roberts, "Mycotoxins in Animal Feedstuffs; Sensitive Thin Layer Chromatographic Detection of Aflatoxin, Ochratoxin A, Sterigmatocystin, Zearalenone and T-2 Toxin," *J. AOAC* **62**, 579–585 (1979).
41. G. A. Bennett, O. L. Shotwell, and W. F. Kwolek, "Liquid Chromatographic Determination of α-Zearalenol and Zearalenone in Corn: Collaborative Study," *J. AOAC* **68**, 958–961 (1985).
42. P. Kuronen, "High-Performance Liquid Chromatographic Screening Method for Mycotoxins Using New Retention Indexes and Diode Array Detection," *Arch. Environ. Contam. Toxicol.* **18**, 336–348 (1989).
43. C. E. Holaday, "Minicolumn Chromatography: State of the Art," *J. AOAC* **58**, 931A–934A (1981).
44. T. R. Romer, "Screening Method for the Detection of Aflatoxins in Mixed Feeds and Other Agricultural Commodities with Subsequent Confirmation and Quantitative Measurement of Aflatoxins in Positive Samples," *J. AOAC* **58**, 500–506 (1975).
45. T. R. Romer and A. D. Campbell, "Collaborative Study of a Screening Method for the Detection of Aflatoxin in Mixed Feeds, Other Agricultural Products and Foods," *J. AOAC* **59**, 110–117 (1976).
46. T. R. Romer, N. Ghouri, and T. M. Boling. "Minicolumn Screening Methods for Determining Aflatoxins: State of the Art," *J. Am. Oil Chem. Soc.* **56**, 795–797 (1979).
47. A. Biancardi and A. Riberzni, "Determination of Ochratoxin A in Cereals and Feed by SAX-SPE Clean Up and LC Fluorimetric Detection," *J. Liq. Chromatogr. and Rel. Technol.* **19**, 2395–2407 (1996).
48. B. R. Malone et al., "One-Step Solid-Phase Extraction Clean-up and Fluorometric Analysis of Deoxynivalenol in Grains," *J. AOAC Int.* **81**, 448–452 (1998).
49. R. Krska, "Performance of Modern Sample Preparation Techniques in the Analysis of *Fusarium* Mycotoxins in Cereals," *J. Chromatog. A* **815**, 49–57 (1998).
50. H. Cohen and M. R. Lapointe, "High Pressure Liquid Chromatographic Determination and Fluorescence Detection of Aflatoxins in Corn and Dairy Feeds," *J. AOAC* **64**, 1372–1376 (1981).
51. J. E. Thean et al., "Extraction, Cleanup and Quantitative Determination of Aflatoxins in Corn," *J. AOAC* **63**, 631–633 (1980).
52. G. S. Shephard et al., "Quantitative Determination of Fumonisins B_1 and B_2 by High-Performance Liquid Chromatography with Fluorescence Detection," *J. Liq. Chromatogr.* **13**, 2077–2087 (1990).
53. L. G. Rice et al., "Evaluation of a Liquid Chromatographic Method for the Determination of Fumonisins in Corn, Poultry Feed, and *Fusarium* Culture Material," *J. AOAC Int.* **78**, 1002–1009 (1995).
54. S. Chiron and D. Barcel, "Identification of Trichothecenes by Thermospray, Plasmaspray and Dynamic Fast-Atom Bombardment Liquid Chromatography-Mass Spectrometry," *J. Chromatogr.* **645**, 125–134 (1993).
55. G. M. Ware et al., "Preparative Method for Isolating α-Zearalenol and Zearalenone Using Extracting Disk," *J. AOAC Int.* **82**, 90–94 (1999).
56. P. M. Scott and M. W. Trucksess, "Application of Immunoaffinity Columns to Mycotoxin Analysis," *J. AOAC Int.* **80**, 941–949 (1997).
57. Anonymous, "Vicam's Aflatest®," *J. AOAC Int.* **78**, 17A–19A (1995).
58. J. D. Groopman and K. F. Donahue, "Aflatoxin, A Human Carcinogen: Determination in Foods and Biological Samples by Monoclonal Antibody Affinity Chromatography," *J. AOAC* **72**, 861–867 (1988).
59. J. I. Azcona, M. M. Abouzied, and J. J. Pestka, "Detection of Zearalenone by Tandem Immunoaffinity-Enzyme-Linked Immunosorbent Assay and Its Application to Milk," *J. Food Prot.* **53**, 577–580 (1990).
60. A. Farjam et al., "The Determination of Aflatoxin M_1 Using a Dialysis-Based Immunoaffinity Sample Pretreatment System Coupled On-Line to Liquid Chromatography Reusable Immunoaffinity Columns," *J. Chromatogr.* **589**, 141–149 (1992).
61. K. A. Scudamore et al., "Determination of Mycotoxins in Pet Foods Sold for Domestic Pets and Wild Birds Using Linked-Column Immunoassay Clean-up and HPLC," *Food Additives and Contaminants* **14**, 175–186 (1997).
62. C. M. Maragos, G. A. Bennett, and J. L. Richard, "Affinity Column Clean-up for the Analysis of Fumonisins and Their

Hydrolysis Products in Corn," *Food and Agric. Immunol.* **9**, 3–12 (1997).

63. A. E. Pohland et al., "Physiochemical Data for Some Selected Mycotoxins," *Pure Appl. Chem.* **54**, 2219–2284 (1982).
64. R. J. Cole and R. H. Cox, *Handbook of Toxic Fungal Metabolites*, Academic Press, New York, 1981.
65. M. E. Savard and B. A. Blackwell, "Spectral Characteristics of Secondary Metabolites from *Fusarium* Fungi," in J. D. Miller and H. L. Trenholm, eds., *Mycotoxins in Grain, Compounds Other Than Aflatoxin*, Eagan Press, St. Paul, Minn., 1994, pp. 59–257.
66. C. P. Gorst-Allman and P. S. Steyn, "Screening Methods for the Detection of Thirteen Common Mycotoxins," *J. Chromatogr.* **175**, 325–331 (1979).
67. L. S. Lee and D. B. Skau, "Thin Layer Chromatographic Analysis of Mycotoxins: A Review of Recent Literature," *J. Liq. Chromatogr.* **4**, 43–62 (1981).
68. R. F. Vesconder and W. K. Rohwedder, "Gas Chromatographic-Mass Spectrometric Analysis of Mycotoxins," in R. J. Cole, ed., *Modern Methods in the Analysis and Structural Elucidation of Mycotoxins*, Academic Press, Orlando, Fla., 1986, pp. 335–357.
69. C. J. Mirocha et al., "Analysis of Deoxynivalenol and Its Derivatives (Batch and Single Kernel) Using Gas Chromatography/Mass Spectrometry," *J. Agric. Food Chem.* **46**, 1414–1418 (1998).
70. M. M. Mossoba, S. Adams, and J. A. G. Roach, "Analysis of Trichothecene Mycotoxins in Contaminated Grains by Gas Chromatography/Matrix Isolation/Fourier Transform Infrared Spectroscopy and Gas Chromatography/Mass Spectrometry," *J. AOAC Int.* **79**, 1116–1123 (1996).
71. M. Schollenberger et al., "Determination of Eight Trichothecenes by Gas Chromatography-Mass Spectrometry after Sample Clean-up by a Two-Stage Solid-Phase Extraction," *J. Chromatogr.* **815**, 123–132 (1998).
72. W. J. Hurst, R. A. Martin, Jr., and C. H. Vestal, "The Use of HPLC/Thermospray MS for the Confirmation of Aflatoxins in Peanuts," *J. Liq. Chromatogr.* **14**, 2541–2545 (1991).
73. T. Krishnamurthy et al., "Mass Spectral Investigations on Trichothecene Mycotoxins, VII: Liquid Chromatographic-Thermospray Mass Spectrometric Analysis of Macrocyclic Trichothecenes," *J. Chromatogr.* **469**, 209–222 (1989).
74. R. Kostiainen, N. Matsuura, and K. Njima, "Identification of Trichothecenes by Thermospray, Plasmaspray and Dynamic Fast-Atom Bombardment Liquid Chromatography-Mass Spectrometry," *J. Chromatogr.* **562**, 555–562 (1991).
75. E. Rosenberg et al., "High-Performance Liquid Chromatography-Atmospheric-Pressure Chemical Ionization Mass Spectrometry as a New Tool for the Determination of the Mycotoxin Zearalenone in Food and Feed," *J. Chromatogr. A* **819**, 277–288 (1998).
76. M. Becker "Column Liquid Chromatography-Electrospray Ionization-Tandem Mass Spectrometry for the Analysis of Ochratoxin," *J. Chromatogr.* **818**, 260–264 (1998).
77. E. D. Caldas et al., "Electrospray Ionization Mass Spectrometry of Sphinganine Analog Mycotoxins," *Anal. Chem.* **67**, 196–207 (1995).
78. G. S. Shephard, "Chromatographic Determination of the Fumonisin Mycotoxins," *J. Chromatogr. A* **815**, 31–39 (1998).
79. V. Betina, "Thin-Layer Chromatography of Mycotoxins," *J. Chromatogr.* **334**, 211–276 (1985).
80. D. Abramson, T. Thorsteinson, and D. Forest, "Chromatography of Mycotoxins on Precoated Reverse-Phase Thin-Layer Plates," *Arch. Environ. Contam. Toxicol.* **18**, 327–330 (1989).
81. R. D. Coker, A. E. John, and J. A. Gibbs, "Techniques of Thin Layer Chromatography," in V. Betina, ed., *Chromatography of Mycotoxins*, Elsevier Press, Elsevier, Amsterdam, The Netherlands, 1993, pp. 12–35.
82. S. Nawaz, R. D. Coker, and S. J. Haswell, "HPTLC—A Valuable Chromatographic Tool for the Analysis of Aflatoxins," *J. Planar Chromatogr.* **8**, 4–9 (1995).
83. L. Lin et al., "Thin-Layer Chromatography of Mycotoxins and Comparison with Other Chromatographic Methods," *J. Chromatogr. A* **815**, 3–20 (1998).
84. R. M. Eppley et al., "Thin Layer Chromatographic Method for Determination of Deoxynivalenol in Wheat: Collaborative Study," *J. AOAC* **69**, 37–40 (1986).
85. O. J. Francis, Jr., et al., "Thin Layer Chromatographic Determination of Sterigmatocystin in Cheese," *J. AOAC* **68**, 643–645 (1985).
86. F. S. Chu, "Studies on Ochratoxins," *CRC Crit. Rev. Toxicol.* **2**, 499–524 (1977).
87. T. F. Salem and B. G. Swanson, "Fluorodensitometric Assay of Patulin in Apple Products," *J. Food Sci.* **41**, 1237–1238 (1976).
88. J. W. Seiber and D. P. H. Hseih, "Application of High-Speed Liquid Chromatography to the Analysis of Aflatoxins," *J. AOAC* **56**, 827–830 (1973).
89. R. D. Coker and B. D. Jones, "Determination of Mycotoxins," in R. Macrae, ed., *HPLC in Food Analysis*, Academic Press, London, 1988, pp. 335–375.
90. J. C. Frisvad, and U. Thrane, "Liquid Column Chromatography of Mycotoxins," in V. Betina, ed., *Chromatography of Mycotoxins*, Elsevier, Amsterdam, The Netherlands, 1993, pp. 253–372.
91. M. J. Shepherd. "High-Performance Liquid Chromatography and its Application to the Analysis of Mycotoxins," in R. J. Cole, ed., *Modern Methods in the Analysis and Structural Elucidation of Mycotoxins*, Academic Press, Orlando, Fla., 1986, pp. 293–315.
92. P. Kuronen, "Techniques of Liquid Chromatography," in V. Betina, ed., *Chromatography of Mycotoxins*, Elsevier, Amsterdam, The Netherlands, 1993, pp. 36–77.
93. P. G. Thiel, S. Stockenstrom, and S. Gathercole. "Aflatoxin Analysis by Reverse Phase HPLC Using Post-Column Derivatization for Enhancement of Fluorescence," *J. Liq. Chromatogr.* **9**, 103–108 (1986).
94. D. M. Wilson, "Analytical Methods for Aflatoxins in Corn and Peanuts," *Arch. Environ. Contam. Toxicol.* **18**, 308–314 (1989).
95. R. W. Beaver, "Determination of Aflatoxins in Corn and Peanuts Using High Performance Liquid Chromatography," *Arch. Environ. Contam. Toxicol.* **18**, 315–320 (1989).
96. M. W. Trucksess et al., "Immunoaffinity Column Coupled with Solution Fluorometry of Liquid Chromatography Post-column Derivatization for Determination of Aflatoxins in Corn, Peanuts and Peanut Butter: Collaborative Study," *J. AOAC* **74**, 81–88 (1991).
97. A. L. Patey, M. Sharman, and J. Gilbert, "Liquid Chromatographic Determination of Aflatoxin Levels in Peanut Butters Using an Immunoaffinity Column Clean-up Method: Internal Collaborative Trial," *J. AOAC* **74**, 76–81 (1991).
98. H. L. Chang and J. W. DeVries, "Short Liquid Chromatographic Method for Determination of Zearalenone and α-Zearalenol," *J. AOAC* **67**, 741–744 (1984).

99. T. Tanaka et al., "Sensitive Determination of Zearalenone and α-Zearalenol in Barley and Job's-Tears by Liquid Chromatography with Fluorescence Detection," *J. AOAC* **76**, 1006–1009 (1993).

100. P. J. Martin et al., "Chromatography of Trichothecene Mycotoxins," *J. Liq. Chrom.* **9**, 1591–1602 (1986).

101. C. M. Franco et al., "Simple and Sensitive High-Performance Liquid Chromatography-Fluorescence Method for the Determination of Citrinin: Application to the Analysis of Fungal Cultures and Cheese Extracts," *J. Chromatogr. A* **723**, 69–72 (1996).

102. B. I. Vazquez et al., "Simultaneous High-Performance Liquid Chromatographic Determination of Ochratoxin A and Citrinin in Cheese by Time-Resolved Luminescence Using Terbium," *J. Chromatogr. A* **727**, 185 (1996).

103. E. G. Hiesler et al., "High-Performance Liquid Chromatographic Determination of Major Mycotoxins Produced by *Alternaria* Molds," *J. Chromatograph.* **194**, 89–92 (1980).

104. M. E. Stack et al., "Liquid Chromatographic Determination of Tenuazonic Acid and Alternariol Methyl Ether in Tomatoes and Tomato Products," *J. AOAC* **68**, 640–642 (1985).

105. T. Delgado, C. Gomez-Cordoves, and P. M. Scott, "Determination of Alternariol and Alternariol Methyl Ether in Apple Juice Using Solid-Phase Extraction and High-Performance Liquid Chromatography," *J. Chromatogr. A* **73**, 109–114 (1996).

106. J. Prieta et al., "Determination of Patulin by Reversed-Phase High-Performance Liquid Chromatography with Extraction by Diphasic Dialysis," *Analyst* **118**, 171–174 (1993).

107. A. R. Brause et al., "Determination of Patulin in Apple Juice by Liquid Chromatography: Collaborative Study," *J. AOAC INT.* **79**, 451–455 (1996).

108. D. Abramson and T. Thorsteinson, "Determination of Sterigmatocystin in Barley by Acetylation and Liquid Chromatography," *J. AOAC* **72**, 342–344 (1989).

109. M. J. Shepherd and J. Gilbert, "Method for the Analysis in Maize of the *Fusarium* Mycotoxin Moniliformin Employing Ion-Pairing Extraction and High-Performance Liquid Chromatography," *J. Chromatogr.* **358**, 415–422 (1986).

110. J. G. Wilkes et al., "Determination of Fumonisins B_1, B_2, B_3 and B_4 by High Performance Liquid Chromatography with Evaporative Light Scattering Detection," *J. Chromatogr.* **695**, 319–323 (1995).

111. G. A. Bennett and J. L. Richard, "Liquid Chromatographic Method for Analysis of the Naphthalene Dicarboxaldehyde Derivative of Fumonisins," *J. AOAC INT.* **77**, 501–506 (1994).

112. C. Velazquez et al., "Derivation of Fumonisins B_1 and B_2 with 6-Aminoquinolyl N-Hydroxylsuccinimidylcarbamate," *J. Agric. Food Chem.* **43**, 1535–1537 (1995).

113. D. R. Lauren and M. P. Agnew, "Multitoxin Screening for *Fusarium* Mycotoxins in Grain," *J. Agric. Food Chem.* **39**, 502–507, 1991.

114. W. L. Childress, I. S. Krull, and C. M. Selavka, "Determination of Deoxynivalenol (DON, Vomitoxin) in Wheat by High-Performance Liquid Chromatography with Photolysis and Electrochemical Detection (HPLC-hv-EC)," *J. Chromatogr. Sci.* **28**, 76–82 (1990).

115. G. M. Ware et al., "Determination of Zearalenol and Zearalenone Using Electrochemical Detection," *Anal. Lett.* **22**, 2335–2339 (1989).

116. B. T. Duhart et al., "Determination of Aflatoxins B_1, B_2, G_1 and G_2 by High-Performance Liquid Chromatography with Electrochemical Detection," *Anal. Chimica Acta* **208**, 343–346 (1988).

117. F. Palmisano "Determination of *Alternaria* Mycotoxins in Foodstuffs by Gradient Elution Liquid with Electrochemical Detection," *Chromatographia* **27**, 425–429 (1989).

118. M. Careri, A. Mangia, and M. Musci, "Applications of Liquid Chromatography-Mass Spectrometry Interfacing Systems in Food Analysis: Pesticide, Drug and Toxic Substance Residues," *J. Chromatogr.* **727**, 153–184 (1996).

119. R. D. Plattner, "Detection of Fumonisins Produced in *Fusarium moniliforme* Culture by HPLC with Electrospray MS and Evaporative Light Scattering Detector," *Natural Toxins* **3**, 294–300 (1995).

120. P. M. Scott, "Gas Chromatography of Mycotoxins," in V. Betina, ed., *Chromatography of Mycotoxins*, Elsevier, Amsterdam, The Netherlands, 1993, pp. 373–425.

121. Y. Onji et al., "Direct Analysis of Several *Fusarium* Mycotoxins in Cereals by Capillary Gas Chromatography-Mass Spectrometry," *J. Chromatogr. A* **815**, 59–65 (1998).

122. J. C. Young, and D. E. Games, "Analysis of *Fusarium* Mycotoxins by Gas Chromatography-Fourier Transform Infrared Spectroscopy," *J. Chromatogr. A* **663**, 211–218 (1994).

123. J. A. G. Roach et al., "Capillary Supercritical Fluid Chromatography/Negative Ion Chemical Ionization Mass Spectrometry of Trichothecenes," *Biomed. Environ. Mass Spec.* **18**, 64–70 (1989).

124. J. C. Young and D. E. Games, "Analysis of *Fusarium* Mycotoxins by Supercritical Fluid Chromatography with Ultraviolet or Mass Spectometric Detection," *J. Chromatogr. A* **653**, 374–379 (1993).

125. C. M. Maragos and J. I. Greer, "Analysis of Aflatoxin B_1 in Corn Using Capillary Electrophoresis with Laser-Induced Fluorescence Detection," *J. Agric. Food Chem.* **45**, 4337–4341 (1997).

126. B. C. Prasongsidh et al., "Analysis of Cyclopiazonic Acid in Milk by Capillary Electrophoresis," *Food Chem.* **61**, 515–519 (1998).

127. C. M. Maragos, "Capillary Zone Electrophoresis and HPLC for the Analysis of Fluorescein Isothiocyanate-Labeled Fumonisin B_1," *J. Agric. Food Chem.* **43**, 390–394 (1995).

128. B. Bohs, V. Seidel, and W. Lindner, "Analysis of Selected Mycotoxins by Capillary Electrophoresis," *Chromatographia* **41**, 631–637 (1995).

129. M. W. Trucksess and G. E. Wood, "Immunochemical Methods for Mycotoxins in Foods," *Food Testing and Analysis* **3**, 24–27 (1997).

130. F. S. Chu, "Current Immunochemical Methods for Mycotoxin Analysis," in M. Vanderlaan, ed., *Immunoassays for Monitoring Human Exposure to Toxic Chemicals in Food and Environment*, ACS Symposium Series, American Chemical Society, Washington, D.C., 1990, pp. 140–157.

131. F. S. Chu, "Recent Studies on Immunoassays For Mycotoxins," in R. C. Beier and L. H. Stanker, eds., *Immunoassays for Residue Analysis, Food Safety*, American Chemical Society, Washington, D.C. 1996, pp. 294–313.

132. A. A. G. Candlish, "Immunological Methods in Food Microbiology," *Food Microbiol.* **8**, 1–14 (1991).

133. M. W. Trucksess, "Methods Used in Testing of Aflatoxins, Deoxynivalenol, Ochratoxin A, and Zearalenone in Grains and Grain Products," in J. W. DeVries et al., eds., *Food Safety from a Chemistry Perspective, Is There a Role for HACCP?*, Analytical Progress Press, Minneapolis, Minn., 1996, pp. 101–112.

134. E. Schneider et al., "Multimycotoxin Dipstick Enzyme Immunoassay Applied to Wheat," *Food Additives and Contaminants* **12**, 387–393 (1995).

135. M. M. Abouzied and J. J. Pestka, "Simultaneous Screening of Fumonisin B_1, Aflatoxin B_1, and Zearalenone by Line Immunoblot: A Computer-Assisted Multianalyte Assay System," *J. AOAC Int.* **77**, 495–501 (1994).

136. L. H. Stanker and R. C. Beier, "Introduction to Immunoassays for Residue Analysis," in R. C. Beier and L. H. Stanker, eds., *Immunoassays for Residue Analysis, Food Safety*, American Chemical Society, Washington, D.C., 1996, pp. 2–15.

137. F. S. Chu, "Immunoassay for Analysis of Mycotoxins," *J. Food Protect.* **47**, 562–569 (1984).

138. J. J. Pestka, "Enhanced Surveillance of Foodborne Mycotoxins by Immunochemical Assay," *J. AOAC* **71**, 1075–1081 (1988).

139. J. J. Pestka, "Application of Immunology to the Analysis and Toxicity Assessment of Mycotoxins," *Food and Agricultural Immunol.* **6**, 219–234 (1994).

140. F. S. Chu and I. Ueno, "Production of Antibody Against Aflatoxin B_1," *Appl. Environ. Microbiol.* **33**, 1125–1128 (1977).

141. R. Teshima "Radioimmunoassay of Nivalenol in Barley," *Appl. Environ. Microbiol.* **56**, 764–768 (1990).

142. Y.-C. Xu, G.-S. Zhang, and F. S. Chu, "Radioimmunoassay of Deoxynivalenol in Wheat and Corn," *J. AOAC* **69**, 967–969 (1986).

143. D. M. Rousseau et al., "Detection of Ochratoxin A in Porcine Kidneys by a Monoclonal Antibody-Based Radioimmunoassay," *Appl. Environ. Microbiol.* **53**, 514–518 (1987).

144. F. S. Chu et al., "Production and Characterization of Antibody Against Diacetoxyscirpenol," *Appl. Environ. Microbiol.* **48**, 777–780 (1979).

145. J. J. Park, and F. S. Chu, "Assessment of Immunochemical Methods for the Analysis of Trichothecene Mycotoxins in Naturally Occurring Moldy Corn," *J. AOAC Int.* **79**, 465–471 (1996).

146. A. P. Wilkinson, C. M. Ward, and M. R. A. Morgan, "Immunological Analysis of Mycotoxins," in H. F. Linskens and J. F. Jackson, eds., *Modern Methods of Plant Analysis*, Springer-Verlag, Berlin, 1992, pp. 185–225.

147. F. S. Chu et al., "Improved Enzyme-Linked Immunosorbent Assay for Aflatoxin B_1 in Agricultural Commodities," *J. AOAC* **70**, 854–857 (1987).

148. J. W. Dorner and R. J. Cole, "Comparison of Two ELISA Screening Tests with Liquid Chromatography for Determination of Aflatoxins in Raw Peanuts," *J. AOAC* **72**, 962–964 (1989).

149. R. C. Sinha and M. E. Savard, "Comparison of Immunoassay with Gas Chromatographic Methods for the Determination of the Mycotoxin Deoxynivalenol in Grain Samples," *Canad. J. Plant Path.* **18**, 233–236 (1996).

150. E. W. Sydenham et al., "Polyclonal Antibody-Based ELISA and HPLC Methods for the Determination of Fumonisins in Corn: A Comparative Study," *J. Food Prot.* **59**, 893–897 (1996).

151. M. M. Sutikno et al., "Detection of Fumonisins in *Fusarium* Cultures, Corn, and Corn Products by Polyclonal Antibody-Based ELISA: Relation to Fumonisin B_1 Detection by Liquid Chromatography," *J. Food Protect.* **59**, 645–651 (1996).

152. F. S. Chu, "Immunochemical Methods for Fumonisins," in L. S. Jackson, J. W. DeVries and L. B. Bullerman, eds., *Fumonisins in Food*, Plenum Publishing, New York, 1996, pp. 123–133.

153. M. W. Trucksess and M. M. Abouzied, "Evaluation and Application of Immunochemical Methods for Fumonisin B_1 in Corn," in R. C. Beier and L. H. Stanker, eds., *Immunoassays for Residue Analysis*, American Chemical Society, Washington, D.C., 1996, pp. 358–367.

154. S. Dragacci et al., "Use of Immunoaffinity Chromatography as a Purification Step for the Determination of Aflatoxin M_1 in Cheeses," *Food Additives and Contaminants* **12**, 59–65 (1995).

155. M. W. Trucksess et al., "Immunoaffinity Column Coupled with Liquid Chromatography for Determination of Fumonisin B_1 in Canned and Frozen Sweet Corn," *J. AOAC Int.* **78**, 705–710 (1995).

156. K. Sargeant et al., "The Assay of Toxic Principle in Certain Groundnut Meals," *Vet. Res.* **73**, 1219–1223 (1961).

157. R. J. Cole, H. G. Cutler, and J. W. Dorner, "Biological Screening Methods for Mycotoxins and Toxigenic Fungi," in R. J. Cole, ed., *Modern Methods in the Analysis and Structural Elucidation of Mycotoxins*, Academic Press, New York, 1986, pp. 1–28.

158. I. E. Yates, "Bioassay Systems and Their Use in the Diagnosis of Mycotoxicoses," in C. Richard and T. Thurston, eds., *Diagnosis of Mycotoxicoses*, Martinus Nijhoff Publishers, Boston, 1986, pp. 331–378.

159. N. L. Clements, "Note on a Microbiological Assay for Aflatoxin B_1: A Rapid Confirmatory Test by Effects on Growth of *Bacillus megaterium*," *J. AOAC* **51**, 611–612 (1968).

160. N. L. Clements, "Rapid Confirmatory Test for Aflatoxin B_1 Using *Bacillus megaterium*," *J. AOAC* **51**, 1192–1194 (1968).

161. J. Reiss, "*Bacillus subtilis*: A Sensitive Bioassay for Patulin," *Bull. Environ. Contam. Toxicol.* **13**, 689–691 (1975).

162. F. J. Olivigni and L. B. Bullerman, "A Microbiological Assay for Penicillic Acid," *J. Food Prot.* **41**, 432–434 (1978).

163. D. Broce et al., "Ochratoxins A and B Confirmation by Microbiological Assay with *Bacillus cereus* Mycoides," *J. AOAC* **53**, 616–619 (1970).

164. G. Koppe, and H.-J. Rehm, "A Biological Assay for Quantitative Determination of Roquefortine," *Z. Lebesm.-Unters. Forsch.* **169**, 90–91 (1979).

165. P. Boutibonnes et al., "Mycotoxin Sensitivity of *Bacillus thuringiensis*," *IRCS Med. Sci. Biochem. Pharmacol.* **11**, 430–431 (1983).

166. L. Viitasalo and H. G. Gyllenberg, "Toxicity of Aflatoxins to *Bacillus megaterium*," *Lebesm.-Wiss. Technol.* **12**, 113–114 (1968).

167. J. Reiss, "Mycotoxin Bioassays Using *Bacillus stearothermophilus*," *J. AOAC* **58**, 624–625 (1975).

168. K. T. Schappert and G. G. Khachatourians, "A Yeast bioassay for T-2 Toxin," *J. Microbiol. Meth.* **3**, 43–46 (1984).

169. P. E. Stone, T. Rubidge, and K. D. MacDonald, "*Rhodotorula rubra* Bioassay for T-2 Toxin: Increased Sensitivity and Wider Application," *J. Microbiol. Meth.* **5**, 59–64 (1986).

170. D. M. Maron and B. N. Ames, "Revised Methods for the Salmonella Mutagenicity Test," *Mutation Res.* **113**, 173–215 (1983).

171. J. W. ApSimon, "The Biosynthetic Diversity of Secondary Metabolites," in J. D. Miller and H. L. Trenholm, eds., *Mycotoxins in Grain, Compounds Other Than Aflatoxin*, Eagan Press, St. Paul, Minn., 1994, pp. 3–18.

172. K. J. Van der Merwe, P. S. Steyn, and L. Fourie, "Mycotoxins, Part II: The Constitution of Ochratoxins A, B, and C, Metabolites of *Aspergillus ochraceus* Wlh," *J. Chem. Soc.* **74**, 7083–7088 (1965).

173. C. W. Holzapfel, "The Isolation and Structure of Cyclopiazonic Acid, a Toxic Metabolite of *Penicillium cyclopium* Westling," *Tetrahedron* **24**, 2101–2119 (1968).
174. R. J. Cole, J. W. Kirksey, and J. M. Wells, "A New Tremorgenic Metabolite from *Penicillium paxilli*," *Can. J. Microbiol.* **20**, 1159–1162 (1974).
175. R. J. Cole et al., "Tremorgenic Toxin from *Penicillium verruculosum*," *Appl. Microbiol.* **24**, 248–257 (1972).
176. R. J. Cole et al., "Papalum Staggers: Isolation and Identification of Tremorgenic Metabolites from Sclerotia of *Claviceps paspali*," *J. Agric. Food Chem.* **25**, 1197–2011 (1977).
177. R. J. Cole et al., "Toxin from *Fusarium moniliforme*: Effects on Plants and Animals," *Science* **179**, 1324–1326 (1973).
178. J. A. Lansden et al., "A New Trichothecene Mycotoxin Isolated from *Fusarium tricinctum*," *J. Agric. Food Chem.* **26**, 246–249 (1978).
179. R. O. Sinnhuber "Trout Bioassay of Mycotoxins," in J. V. Rodricks, C. W. Hesseltine, and M. A. Mehlman, eds., *Mycotoxins in Human and Animal Health*, Pathotox, Park Forest South, Ill., 1977, pp. 731–744.
180. R.-D. Wei, E. B. Smalley, and F. M. Strong, "Improved Skin Test for Detection of T-2 Toxin," *Appl. Microbiol.* **23**, 1029–1030 (1972).
181. H. R. Burgmeister, R. F. Vesconder, and W. F. Dwolek, "Mouse Bioassay for *Fusarium* Metabolites: Rejection or Acceptance when Dissolved in Drinking Water," *Appl. Environ. Microbiol.* **39**, 957–961 (1980).
182. C. J. Mirocha, S. V. Ptre, and C. M. Christensen, "Zearalenone," in J. V. Rodricks, C. W. Hesseltine, and M. A. Mehlman, eds., *Mycotoxins in Human and Animal Health*, Pathotox, Park Forest South, Ill., 1977, pp. 345–364.
183. M. J. Verrett, J.-P. Marliac, and J. McLaughlin, Jr., "Use of the Chicken Embryo in the Assay of Aflatoxin Toxicity," *J. AOAC* **47**, 1003–1006 (1964).
184. J. J. Hlywka, M. M. Beck, and L. B. Bullerman, "The Use of the Chicken Embryo Screen Test and Brine Shrimp (*Artemia salina*) Bioassays to Assess the Toxicity of Fumonisin B_1 Mycotoxin," *Food Chem. Toxicol.* **35**, 991–999 (1997).
185. C. W. Bacon, J. K. Porter, and W. P. Norred, "Toxic Interaction of Fumonisin B_1 and Fusaric Acid Measured by Injection into Fertile Chicken Egg," *Mycopathologia* **129**, 29–35 (1995).
186. D. B. Prelusky et al., "Optimization of Chick Embryotoxicity Bioassay for Testing Toxicity Potential of Fungal Metabolites," *J. AOAC* **70**, 1049–1055 (1987).
187. R. F. Brown, J. D. Wildman, and R. M. Eppley, "Temperature-Dose Relationships with Aflatoxin on the Brine Shrimp, *Artemia salina*," *J. AOAC* **51**, 905–906 (1968).
188. R. M. Eppley, "Sensitivity of Brine Shrimp (*Artemia salina*) to Trichothecenes," *J. AOAC* **57**, 618–620 (1974).
189. R. F. Brown, "The Effect of Some Mycotoxins on the Brine Shrimp *Artemia salina*," *J. Am. Oil Chem. Soc.* **46**, 119 (1969).
190. S. Knasmuller et al., "Genotoxic Effects of Three *Fusarium* Mycotoxins, Fumonisin B_1, Moniliformin and Vomitoxin in Bacteria and in Primary Cultures of Rat Hepatocytes," *Mutation Res.* **391**, 39–48 (1997).
191. H. K. Abbas, T. Tanaka, and W. T. Shier, "Biological Activities of Synthetic Analogues of *Alternaria alternata* Toxin (AAL-Toxin) and Fumonisin in Plant and Mammalian Cell Cultures," *Phytochem.* **40**, 681–689 (1995).
192. H. K. Abbas, C. D. Boyette, and R. E. Hoagland, "Phytotoxicity of *Fusarium*, Other Fungal Isolates, and of the Phytotoxins Fumonisin, Fusaric Acid, and Moniliformin to Jimsonweed," *Phytoprotection* **76**, 17–25 (1995).
193. C. W. Bacon and D. M. Hinton, "Fusaric Acid and Pathogenic Interactions of Corn and Non-Corn Isolates," in L. S. Jackson, J. W. DeVries, and L. B. Bullerman, eds., *Fumonisins in Food*, Plenum Publishing, New York, 1996, pp. 175–191.
194. H. G. Cutler et al., "Chaetoglobosin K: A New Plant Inhibitor and Toxin from *Diplodia macrospora*," *J. Agric. Food Chem.* **28**, 139–142 (1980).
195. V. S. Thompson and C. M. Maragos, "Fiber-Optic Immunosensor for the Detection of Fumonisin B_1," *J. Agric. Food Chem.* **44**, 1041–1046 (1996).

LAUREN JACKSON
FDA-NCFST
Argo, Illinois

MYCOTOXINS

Mycotoxins are chemicals that are produced by filamentous fungi that affect human or animal health. By convention, this excludes mushroom poisons. These fungi are called *toxigenic* fungi. All of these species are deuteromycetes (asexual) some of which have a known ascomycetous (sexual) stage. All of the mycotoxins discussed here are secondary metabolites of the fungi concerned, that is, compounds that are produced after one or more nutrients become limiting (1–3). The occurrence of mycotoxins is entirely governed by the existence of conditions that favor the growth of the fungi concerned. Under environmental conditions, different fungal species are favored as diseases of crop plants or as saprophytes on stored crops and sometimes other materials. When the conditions favor the growth of toxigenic species, it is an invariable and unfortunate rule that one or more of the compounds for which the fungus has the genetic potential are produced. Modern methods of agriculture appear to be selecting for increased prevalence of toxigenic strains (4).

With modern studies of toxigenic fungi, it has been appreciated that they can produce many compounds, often from different biosynthetic families. The ecological significance of the occurrence of these mixtures has been a fertile area of study in recent years. As it relates to human and animal toxicology, the potency of the contaminated material is due to the mixtures present (5,6). At the time of the discovery of penicillin and the first wave of fungal-derived antibiotics, the researchers believed that these compounds were active in nature. In the postwar period, secondary metabolites were characterized as everything from waste products to the consequences of "displacement activities" of the fungi. The current view is that these compounds are important as virulence factors and as mediators of interference competition; that is, they exclude competing microbes and animals from the food source (4,7,8).

Mycotoxins have affected human populations since the beginning of organized crop production. Ergotism is discussed in the Old Testament of the Bible. Some claim that the ancient Chinese used ergot for obstetrical purposes 5000 years ago. Many epidemics of ergotism were reported in western Europe from about A.D. 800. The screams of the victims, the stench of rotting flesh, extremities falling off, and death all feature in the descriptions of the disease. Large-scale mortalities persisted into the eighteenth century, when governments and the church promoted methods for the removal of sclerotia (9).

Outbreaks of ergotism have been reported in developing countries in modern times. However, because the cause of the problem—the consumption of ergot-infested grain—is something easily seen by the naked eye, ergotism is now uncommon. During the thirteenth century, rye was replaced by wheat in western Europe. The former species is resistant to *Fusarium* diseases and, as noted, the latter typically susceptible, leading to the accumulation of trichothecene mycotoxins. These compounds are prevalent in small grains in western Europe (10). In her book *Poisons of the Past* (11), the American historian Mary Matossian has analyzed the effect of weather and food consumption patterns during the plague epidemics in Europe. She found that plague epidemics took place when there were surpluses of small grains, the favourite food of rats. The occurrence of plague in the Middle Ages is very strongly associated with rainy and humid crop years. She found that the two years prior to the pandemic in Europe in 1378, the weather was extraordinarily rainy.

Although there are hundreds of fungal metabolites that are toxic in experimental systems, only five are of major agricultural importance: deoxynivalenol, aflatoxin, fumonisin, zearalenone, and ochratoxin (5). In grains, the toxins are concentrated into bran fractions during milling (12). All these toxins are stable in the processes typical of food and feed processing (12–15). Animal products can be a minor dietary source of ochratoxin and fumonisin; for the remaining three toxins, animal sources are not important under normal circumstances (16). Milk can contain aflatoxin M1 a derivative of aflatoxin B1 with much lower toxicity (17).

A recent report by the U.S. National Academy of Sciences notes that even with the high-quality food system in the United States, the carcinogenic mycotoxins in American diets may increase cancer rates. This is absolutely the case in many developing countries, where mycotoxins are a major population health problem (18,19). Information on regulations and guidelines for mycotoxins can be found in Van Egmond (20) and Kuiper-Goodman (21), and was discussed at the 1999 FAO Conference on Mycotoxins, in Tunis (*http: / /www.fao.org*).

A number of mycotoxins that occur from time to time in food in certain parts of the world will be considered. Human diseases associated with uncommon mycotoxins on rice and other crops are described in Pitt (22) and Beardall and Miller (23). There are also fungal toxins that occur in pastures that are not considered in this treatment. Inhalation exposure to mycotoxins is also not covered in this treatment but is a problem in farming and grain handling and in buildings with appreciable mold growth (24–27).

DEOXYNIVALENOL AND ZEARALENONE

These toxins occur when wheat, barley, corn, and sometimes oats and rye are infected by *Fusarium graminearum* and *F. culmorum*. These species cause *Fusarium* head blight in small grains, a major agricultural problem worldwide. These species cause a similar disease in corn called *Gibberella* ear rot. Disease incidence is most affected by moisture at flowering, and most cultivars and hybrids used today lack genetic resistance to the disease. *F. graminearum* is common in wheat from North America and China. *F. culmorum* is the dominant species in cooler wheat growing areas such as Finland, France, Poland, and The Netherlands (28). Oats, rye, and triticale have also been reported to contain deoxynivalenol (10,17,29–31). Wheat, corn, and barley comprise two-thirds of the world production of cereals; hence, deoxynivalenol is the mycotoxin to which the greatest number of humans are exposed.

A third species, *F. crookwellense* can also cause head blight or corn ear rot and produces nivalenol and zearalenone and many of the same families of minor metabolites as the related species (32). Nivalenol was first isolated from grains contaminated by *F. sporotrichioides* and misidentified as "*Fusarium nivale* Fn2B" (33,34). The toxic profile of nivalenol is thought to be similar to that of deoxynivalenol, although nivalenol is more potent in assays for some acute effects.

Different populations of *F. graminearum* and *F. culmorum* produce different toxins. In 1983, it was recognized that Japanese strains produce deoxynivalenol via 3-acetyldeoxynivalenol, but Canadian strains produced the 15-acetate. This has implications for consumers of grain contaminated with one or other of the populations because the toxicities of the two acetates are different (24), and there is always some of the acetate present. North and South American strains of these two species produce the 15-acetate, and most strains from Asia and Europe produce the 3-acetate (32,35,36). Authentic strains of *F. graminearum* have been shown to produce nivalenol instead of deoxynivalenol in Japan, New Zealand, Australia, and Italy (37–39). This has not been shown in North American isolates to date (the nivalenol in North American grains is presumed to come from *F. crookwellense*). *F. graminearum* and *F. culmorin* produce a wide variety of metabolites from several biosynthetic families (31). This includes trichothecenes, apotrichothecenes (40), zearalenone, calonectrins (41,42), sambucinol and sambucoin (42), culmorins (43), and butenolide.

Red mold poisoning was reported in rural Japan throughout the 1950s (44). Eventually, Japanese researchers discovered deoxynivalenol in grain that had made humans ill (45). The same chemical was subsequently re-reported as *vomitoxin* from *F. graminearum*–contaminated corn in 1973 (46). Deoxynivalenol was a widespread contaminant of wheat in the northeast of the United States and in eastern Canada in 1979 to 1981 and then again in 1993 to 1996 in the Great Lakes area and Red River Valley. Much of wheat crop of Ontario and several states could not be used in 1996 because of deoxynivalenol contamination. Large-scale acute human toxicoses from deoxynivalenol have occurred in modern times in India (47), China, and Korea, among other countries (23,48).

Trichothecenes are potent low-molecular-weight inhibitors of protein synthesis (49). In addition, they cause physical damage to membranes resulting in cell lysis. Red blood cells are a compartment for trichothecene metabolism and these cells will lyse in the presence of excess circulating toxin. The amount of toxin required to lyse red blood cells varies according to animal species (50).

Swine are the domestic animal species most sensitive to the effects of trichothecenes (6). Experiments feeding pure deoxynivalenol to swine suggested that the diets containing <2 mg/kg deoxynivalenol would have little impact on growth. However, many experiments using naturally contaminated grains often demonstrated that such grain was more toxic than indicated from the deoxynivalenol content (51,52). Several of the minor toxins from *F. graminearum* were present in such grain, sometimes at concentrations similar to deoxynivalenol (53). These co-occurring toxins were shown to increase the toxicity of deoxynivalenol when fed in combination to insects (54). This might be also true in swine (55). Toxicological interactions between trichothecenes were also discovered in yeasts (56) and in animals (57,58). The metabolism of deoxynivalenol and other trichothecenes at subacute doses results in rapid elimination in swine, other domestic animals, and rodents (16,17,25).

The basis for the feed refusal is the impressive neurotoxicity of dexoynivalenol. Experiments involving the dosing of the toxin by a continuous exposure osmotic pump implanted intraperitoneally resolved that the effects could not be due to taste or learned responses (52). A single dose of 0.25 mg/kg (IV) changed neurotransmitter concentrations in the hypothalamus, frontal cortex, and cerebellum up to 8 days postdosing (59). A very low dose (10 μg/kg) IV resulted in changes in cerebral spinal fluid neurotransmitters (60). Based on acute human exposure-emesis data, humans are not less, and are probably more, sensitive to deoxynivalenol than swine (21). Feed refusal also occurs in mice and in lifetime studies; this reduced the incidence of spontaneous liver tumors due to calorie restriction (61).

Because deoxynivalenol is less acutely toxic than the potent trichothecenes, the immunotoxic and neurotoxic properties of trichothecenes were recognized. Changes in immune system function in male mice occur at dietary concentrations often encountered by humans (62). As with other trichothecenes, high exposures increase susceptibility to facultative pathogens such as *Listeria*. Deoxynivalenol exposure produces prolonged elevations in serum IgA and mesangial IgA leading to hematuria (63). The calonectrins produced by *F. graminearum* also affect immune functions *in vitro* (64). Human IgA disregulation (Berger's disease) is common, and the only agents so far demonstrated to reproduce this condition in experimental animals are trichothecenes (63).

Cattle and cows are tolerant to deoxynivalenol, and milk production is not affected at typical field concentrations of deoxynivalenol (ie, dietary concentrations less than 5 mg/kg). Trichothecenes are detoxified by rumen bacteria. Poultry species are also tolerant to typical field concentrations of deoxynivalenol (6). Deoxynivalenol is not a carcinogen (17). The occurrence of deoxynivalenol in diets affects uptake of sugars and minerals (65). A review of the toxicology of deoxynivalenol is found in Rotter et al. (66).

As noted, crops that are contaminated by deoxynivalenol can often contain zearalenone, albeit at a lower frequency. Zearalenone is more common in maize than small grains (17). Zearalenone is an estrogen analogue and causes hyperestrocism in female pigs at low levels; the dietary no-effect level is less than 1 mg/kg. Cows and sheep are also sensitive to the estrogenic effects of this toxin with depressed ovulation and lower lambing percentages. The no-effect dietary levels are not clearly known (6). Nonhuman primates are also very sensitive to the estrogenic effects of zearalenone (20). Zearalenone has been implicated in several incidents of precocious pubertal changes in girls in Europe and South America (67,68); thus, this is a true environmental estrogen (69). There is limited evidence for its rodent carcinogenicity (17). A detailed description of the toxicology of zearalenone is found in Kuiper-Goodman et al. (70).

AFLATOXIN

Aspergillus flavus and *A. parasiticus* (Aflatoxins)

Aflatoxin is a problem in many commodities; however, as far as grains are concerned, it is primarily a problem in maize. This is because maize is colonized in the field depending on environmental conditions, whereas other grains are not. Of the other grains, rice is an important dietary source of aflatoxin in circumstances of poor storage in tropical and subtropical areas. The character of the problem varies by region. In the United States, storage systems are very good (71) and the problem is preharvest contamination of maize and peanuts (4,72). In tropical countries, such as Thailand and the Philippines, crop storage is a substantial problem (73). Aflatoxin contamination is managed by the development of systems to detect and segregate contaminated kernels and better storage systems. For corn and peanuts, all manner of efforts have been made to prevent aflatoxin contamination, including plant breeding and biological control to little effect (74).

In the United States, Mexico, and South America, *A. flavus* and *A. parasiticus* infect corn, although *A. parasiticus* is relatively uncommon (72). In the environmental circumstances prevalent in the corn belt of the United States, *A. flavus* contamination of corn occurs in two basic ways: (*1*) airborne or insect-transmitted conidia contaminate the silks and grow into the ear when the maize is under high temperature stress, or (more commonly) (*2*) insect- or bird-damaged kernels become colonized with the fungus and accumulate aflatoxin. In either case, drought-, nutrient-, or temperature-stressed plants are more susceptible to colonization by *A. flavus* (75). This is also the case for peanuts, which are mainly colonized by *A. parasiticus*. Aflatoxin contamination can be limited by preventing late season drought by irrigation where this is possible (74).

Insects and arthropods readily become contaminated with *A. flavus*. Soil-inhabiting mites feed on the germinated sclerotia and hence acquire conidia. Nitidulid beetles feed on moldy ears of maize, perhaps preferentially. Nitidulids are attracted to damaged ears, including those caused by corn ear worms and the European corn borer spreading the fungus into damaged kernels (4,76). In drought years, insects are attracted to peanuts and both wound the plant and bring the fungus.

The ecology of *A. flavus* in corn in subtropical regions of Asia appears to be different from that already described. In subtropical Asia there is a rapid rise in aflatoxin concentrations immediately postharvest. American studies have shown that, typically, *A. flavus*–infested kernels are randomly distributed in ears after wound inoculation (77). The kernels that were the sites of initial infection have very high aflatoxin contents. In studies done in Thailand, approximately 19% of kernels from 130 samples of maize collected from farmers fields throughout Thailand contained *A. flavus* (73).

There are many detailed reviews of the toxicology of aflatoxin, including that of the International Agency for Research on Cancer (IARC) (17). Aflatoxin B1, the most toxic of the aflatoxins, causes a variety of adverse effects in different animal species, especially chickens. In poultry, these include liver damage, impaired productivity and reproductive efficiency, decreased egg production in hens, inferior eggshell quality, inferior carcass quality, and increased susceptibility to disease (78). Swine are somewhat less sensitive than poultry species, with the LD_{50} being perhaps half of that of chickens. Aflatoxin is hepatotoxic, and its acute and chronic effects in swine are largely attributable to liver damage (79). In cattle, the primary symptom is reduced weight gain as well as liver and kidney damage. Milk production is reduced (80). Aflatoxin is also immunotoxic in domestic and laboratory animals with oral exposures in the ppm range. Cell-mediated immunity (lymphocytes, phagocytes, mast cells, and basophils) is more affected than humeral immunity (antibodies and complement) (63). The effects of aflatoxin on laboratory animals has been exhaustively reviewed by IARC (17).

Naturally occurring mixtures of aflatoxins were classified as class 1 human carcinogens, and aflatoxin B_1 is also a class 1 human carcinogen. There was inadequate evidence of the human carcinogenicity of aflatoxin M_1, the metabolite of aflatoxin B_1 found in human and animal milk (17). Many people in developing countries are seropositive for hepatitis B and C, which are also liver carcinogens. Although aflatoxin is a potent chemical carcinogen, its ability to alter response to the hepatocarcinogenic viruses is perhaps of greater importance. The relative rates of liver cancer in hepatitis B–positive populations are an order of magnitude greater (60X) when exposed to aflatoxin. This is because the toxin interferes with the processing of the virus (81,82).

The immunotoxicity of aflatoxin is also being increasingly studied; some think that it would have to be regulated for this toxicity regardless of its carcinogenicity. In one study, serum aflatoxin-lysine adducts were higher in protein energy malnourished (PEM) children compared with control children. Aflatoxin metabolism was affected, with relatively higher serum concentrations in PEM children. A second study compared PEM children with high and low serum aflatoxin concentrations. The serum aflatoxin positive group of PEM children showed a significantly lower hemoglobin level ($p = 0.02$), longer duration of edema ($p = 0.05$), an increased number of infections ($p = 0.03$), and a longer duration of hospital stay ($p = 0.008$). This finding was echoed in another study, which suggested that malaria infections in children were increased in children exposed to aflatoxin (18).

FUMONISINS

Fumonsisins are produced by *F. verticilloides* (formerly *moniliforme*), *F. proliferatum*, and several uncommon fusaria (83,84). Fumonisins were discovered in 1988 by two groups working independently. One was investigating the cause of human esophageal cancer in parts of southern Africa (85). The other was attempting to find the cause of a disease of horses known since 1850, equine leucoencephalomalacia (ELEM) (86). There are at least three naturally occurring fumonisins—B_1, B_2, and B_3; FB_1 occurs at highest concentration followed by B_2 and B_3. Fumonisins have been found as a very common contaminant of corn-based food and feed in the United States, China, Europe, southern Africa, South America, and Southeast Asia (5,17,87). In addition, there are a number of minor fumononisins. *F. verticilloides* also produces fusarin C, heat- and light-unstable compounds, and fusaric acid. *F. proliferatum* produces fumonisins, fusarins, fusaric acid, and moniliformin (88,89). Moniliformin is toxic to poultry species (6) and can occur in food (90).

F. verticilloides and *F. proliferatum* are the most common fungi associated with corn. For many years, *F. moniliforme* has been known to occur systematically in leaves, stems, roots, and kernels (91). These fungi can be recovered from virtually all corn kernels including those that are healthy, which suggests that it may be an endopyte that is, a mutualistic relationship (28). *F. verticillioides* and *F. proliferatum* cause a "disease" called fusarium kernel rot. In parts of the United States and lowland tropics, this is one of the most important ear diseases and is associated with warm, dry years and insect damage and fumonisin (92,93). Corn plant disease-stress also promotes the growth of *F. verticillioides* and fumonisin formation (94).

Fumonisins are toxic in all types of cells (yeast, plant, animal, human) due to their effects on sphinglolipid synthesis (95). Alteration in sphingolipid base ratios occurs almost immediately after exposure because fumonisin inhibits ceraminde synthetase. There are also many changes in the amounts and ratios of complex ceramides. In addition, fumonisins induce apoptosis leading to cell proliferation, which may explain their carcinogenic properties (96,97). Fumonisins B_2 and B_3 have similar toxic properties to B_1 but are less potent (98).

Pure fumonisin was demonstrated to cause equine leucocencephalomalia (ELEM) in 1988 (99). ELEM involves a massive liquefactive necrosis of the cerebral hemispheres; hence, the disease involves neurological manifestations including abnormal movements, aimless circling, lameness, and so on. At high exposures, death can occur within hours after the onset of visible symptoms. Damage to liver and kidneys in horses, features of fumonisins exposure in other animals, is poorly characterized in horses (6).

As in equine species, alterations of sphingolipid base ratios are indicative of fumononisin exposure in swine

(95). At high exposures, porcine pulmonary edema (PPE) has been shown to be caused by both pure fumonosin and *F. moniliforme* culture material and maize containing fumonisin (6). This is thought to be caused by fumonisin-induced heart failure (100). At lower exposures, both liver and kidney damage has been reported in swine (101). Fumonisin causes feed refusal and changes in carcass quality at dietary concentrations in the low mg/kg range (102,103). Feeder calves were reported to be unaffected by fumonisin (6).

Exposure to *F. verticilloides*–contaminated maize has been linked to the elevated rates of esophageal cancer in the Transkei for 25 years, and this has since been directly linked to fumonisin exposure (17,87,104). Fumonisin B_1 has been demonstrated to exhibit cancer-promoting activity in diethylnitrosamine-initiated rats (105). Fumonisin B_1 has been also shown to be hepatotoxic and hepatocarcinogenic in rats fed 50 mg/kg (90% purity) (106). Very pure fumonisin produced tumors in male and female mice and rats in the U.S. National Toxicology two-year bioassay (NTP TR 496; *http://www.ntp.gov*). In Fisher 344 rats, fumonisin B_1 exposure resulted in tumors at doses about 10 times less than aflatoxin. IARC (17) examined the human carcinogenicity of grain contaminated with *F. verticillioides* containing fumonisins and fusarin C and found them to be possible human carcinogens. There is an enormous amount known about the rodent toxicities of fumonisins from the just-noted NTP assay. An exhaustive treatment of the toxicology of fumonisins can be found in WHO (87).

OCHRATOXIN

Despite many contrary reports, ochratoxin is known to be produced by only one species of *Penicillium, verrucosum. Aspergillus ochraceous* and several related species also produce ochratoxin on grapes and coffee (17,107,108; M. Frank, unpublished data, 1999). A small percentage of surface-disinfected wheat and barley kernels collected at harvest in the UK and Denmark were contaminated by *P. aurantiogriseum* and *P. verrucosum*; this was similar in studies done in western Canada over many years. Infestation of some kernels by the ochratoxin-producing fungus *P. verrucosum* occurs from anthesis, and surface contamination is common at harvest. The absolute level of preharvest infestation varies according to site and season (5).

Ochratoxin is a potent nephratoxin in swine and causes kidney cancer in male Fisher 344 rats. Pigs are affected at low exposures in terms of kidney damage, but typically there are no overt signs of biochemical/hematological changes. At higher concentrations (>2 μg/g), decreased weight grains occur (6). Poultry are similarly affected, with reduced growth rate and egg production at low ochratoxin concentrations > 2 μg/g. Higher dietary ochratoxin concentrations are often fatal. Cattle are resistant to ochratoxin concentrations found in naturally contaminated grain (6). Ochratoxin is often found with other toxins such as citrinin and the naphthaquinone mycotoxins from *P. aurantiogriseum* (17,109). Citrinin mimics the effects of ochratoxin, although it is less potent (109). The naphthaquinones xanthomegnin and viomellein from *P. aurantiogriseum* are nephrotoxic (110). Interactions between ochratoxin and citrinin have been demonstrated in some experiments (6).

Ochratoxin is suspected as the cause of urinary tract cancers and kidney damage in areas of chronic exposure in parts of eastern Europe (17,111; M. Frank unpublished data, 1999). Human exposure to ochratoxin primarily occurs from whole-grain breads. Some exposure comes from the consumption of animal products, especially pork and pig-blood-based products (14). There is a great deal known about human serum ochratoxin concentrations in Europe (111,112). Despite considerable effort, no satisfactory conclusion has been reached regarding the linkage of ochratoxin with urinary tract cancers in humans (19,82,113). The last IARC evaluation of ochratoxin determined it to be a possible human carcinogen (17). As of 1999, The Joint Expert Committee for Food Additives & Contaminants of the WHO/FAO has a tolerable human daily intake for this toxin based on its renal toxicity. Ochratoxin exposure in the United States and Canada are considered to be low based on testing by Health Canada and the Food & Drug Administration.

OTHER MYCOTOXINS THAT CAN OCCUR IN FOOD

T-2 toxin was isolated from "strain T-2" of *F. sporotrichioides* misidentified as *F. tricinctum* isolated from corn associated with cow mortalities (33,114). This compound has been the subject of considerable toxicological study (17) because it is easy to isolate and purify. During World War II, there were large-scale poisoning of the rural population in the former Soviet Union caused by the consumption of grains left in the field over winter (estimates range to 1,000,000 victims). The disease was called alimentary toxic aleukia. Samples of extracts made at the time have been shown to contain the trichothecenes T-2 and HT-2 toxin. Shortly after consuming food prepared with contaminated grain, people reported a burning sensation in their mouths, vomiting, weakness, fatigue, and tachycardia. After a period of time, affected individuals felt better, but there was a progressive leucopenia, anemia, and decreased platelet count, lowering "the resistance of the body to bacterial infection." As consumption of toxic grains continued, petechial hemorrhages on the upper part of the body appeared together with necrotic lesions in the mouth and face. Bacterial infections were common. Patients who reached this stage almost always died (115). The toxicology of T-2 toxin is similar in character to that previously described for deoxynivalenol and is reviewed extensively in Ref. 17.

Patulin (116) is primarily found in apple and grape juices where is occurs from the growth of *Penicillium expansum*. Many other species of molds produce patulin. The presence of this toxin in juice is a sign that damaged fruit was present. The sparse toxicological data on this compound have been reviewed by the California Environmental Protection Agency in 1998 (*http://www.oehha.org*). This compound has low acute toxicity, and there is insufficient evidence to conclude that it is carcinogenic in rodents (IARC category 3). Human exposure in some countries has been high, and industry-government efforts have been made in North America and Europe to reduce the occurrence of patulin in juice.

Ergot alkaloids seldom appear in meaningful concentrations in food samples in the North America or Europe because the presence of *Claviceps* sclerotia in grains is a grading factor. However, infections by ergot alkaloid-producing fungi remain common in the sense that plants on the edges of rye, barley, and wheat fields are often infected. Thus the majority of products made from wheat or rye contain traces of ergot alkaloids (117). Grains downgraded to animal feed can contain ergot. Cattle are more sensitive than mice, sheep, or swine and sclerotial contents of about 0.1% appear to be tolerable. In cattle, the symptoms include lameness, and in other domestic animals reduced weight gain can be expected (6). Ergotism in humans has been reported in India and parts of Africa in modern times (29).

BIBLIOGRAPHY

1. J. D. Bu'Lock, "Secondary Metabolism," in J. E. Smith and D. R. Berry, eds., *The Filamentous Fungi*, Academic Press, New York, 1975, pp. 33–58.
2. J. D. Bu'Lock, "Mycotoxins as Secondary Metabolites," in P. S. Steyn, ed., *The Biosynthesis of Mycotoxins. A Study of Secondary Metabolism*, Academic Press, New York, 1980, pp. 1–16.
3. J. D. Miller, M. E. Savard, and S. Rapior, "Production and Purification of Fumonisins From a Stirred Jar Fermentor," *Natural Toxins* **2**, 354–359 (1994).
4. D. R. Wicklow, "Maize Cultivation Selects for Mycotoxin-Producing Ability Among Kernel Rotting Fungi," in C. H. Chou and G. R. Waller, eds., *Phytochemical Ecology: Allelochemicals, Mycotoxins and Insect Pheromones*, Academia Sinica Monograph Series 9, Taipei, China, 1989, pp. 263–274.
5. J. D. Miller, "Fungi and Mycotoxins in Grain: Implications for Stored Product Research," *J. Stored Product Res.* **31**, 1–6 (1995).
6. D. B. Prelusky, B. A. Rotter, and R. G. Rotter, "Toxicology of Mycotoxins," in J. D. Miller and H. L. Trenholm, eds., *Mycotoxins in Grain: Compounds Other than Aflatoxin*, Eagan Press, St. Paul, Minn., 1994, pp. 359–404.
7. B. B. Jarvis and J. D. Miller, "Natural Products, Complexity and Evolution," in J. T. Romeo, J. A. Saunders, and P. Barbossa, eds. *Phytochemical Diversity and Redundancy in Ecological Interactions*, Plenum Press, New York, 1996, pp. 265–294.
8. D. R. Wicklow, "Interference Competition and the Organization of Fungal Communities," in D. T. Wicklow and G. C. Carroll, eds., *The Fungal Community*, Marcel Dekker, New York, 1981, pp. 351–375.
9. S. J. Van Rensburg and B. Altenkirk, "*Claviceps purpurea*-Ergotism", in I. F. H. Purchase, ed., *Mycotoxins*, Elsevier, New York, 1974, pp. 69–96.
10. H.-M. Muller et al., "*Fusarium* Toxins in Wheat Harvested During Six Years in an Area of Southwestern Germany," *Natural Toxins* **5**, 24–30 (1997).
11. M. K. Matossian, *Poisons of the Past*, Yale University Press, New Haven, Conn., 1989.
12. J. C. Young et al., "Effect of Milling and Baking on Deoxynivalenol (Vomitoxin) Content of Eastern Canadian Wheats," *J. Agric. Food. Chem.* **32**, 659–664 (1984).
13. L. A. Goldblatt, ed., *Aflatoxin, Scientific Background, Control and Implications*, Academic Press, New York, 1969.
14. T. Kuiper-Goodman and P. M. Scott, "Risk Assessment of the Mycotoxin Ochratoxin A," *Biomedical and Environmental Sciences* **2**, 179–248 (1989).
15. M. M. Castelo, S. S. Sumner, and S. S. Bullerman, "Stability of Fumonisins in Thermally Processed Corn Products," *J. Food Protection* **61**, 1030–1033 (1998).
16. D. B. Prelusky, "Residues in Food Products of Animal Origin," in J. D. Miller and H. L. Trenholm, eds., *Mycotoxins in Grain: Compounds Other than Aflatoxin*, Eagan Press, St. Paul, Minn., 1994, pp. 405–420.
17. *Some Naturally Occurring Substances: Food Items and Constituents, Heterocyclic Aromatic Amines and Mycotoxins*, International Agency for Research on Cancer, Monograph 56, Lyon, France, 1993.
18. National Academy of Sciences, *Carcinogens and Anticarcinogens in the Human Diet*, National Academy Press, Washington, D.C., 1996.
19. J. D. Miller, "Global Significance of Mycotoxins," in M. Miraglia et al., eds., *Mycotoxins and Phycotoxins—Developments in Chemistry, Toxicology and Food Safety*, Alaken, Fort Collins, Colo., 1998, pp. 3–16.
20. H. P. Van Egmond, "Current Situation on Regulations for Mycotoxins, Overview of Tolerances and Status of Sampling Methods of Sampling and Analysis," *Food Additives and Contaminants* **6**, 139–188 (1989).
21. T. Kuiper-Goodman, "Prevention of Human Mycotoxicosis Through Risk Assessment and Risk Management," in J. D. Miller and H. L. Trenholm, eds., *Mycotoxins in Grain: Compounds Other than Aflatoxin*, Eagan Press, St. Paul, Minn., 1994, pp. 439–470.
22. J. I. Pitt, "*Penicillium* Toxins," *Australian Centre for International Agricultural Research Proc.* **36**, 99–103 (1991).
23. J. M. Beardall and J. D. Miller, "Diseases in Humans with Mycotoxins as Possible Causes," in J. D. Miller and H. L. Trenholm, eds., *Mycotoxins in Grain: Compounds Other than Aflatoxin*, Eagan Press, St. Paul, Minn., 1994, pp. 487–540.
24. J. L. Autrup et al., "Exposure to Aflatoxin B in Animal Feed Production Plant Workers," *Scand. Environ. Health. Perspect.* **99**, 195–197 (1993).
25. V. R. Beasley, ed., *Trichothecene Mycotoxins: Pathophysiologic Effects*, Vols. 1 and 2, CRC Press, Boca Raton, Fla., 1989.
26. B. Flannigan and J. D. Miller, "Health Implications of Fungi in Indoor Environments—An Overview," in R. Samson et al., eds., *Health Implication of Fungi in Indoor Environments*, Elsevier, Amsterdam, 1994, pp. 3–28.
27. E. L. Hintikka, "Human Stachybotrytoxicosis," in T. D. Wyllie and L. G. Morehouse, eds., *Mycotoxigenic Fungi, Mycotoxins, Mycotoxicoses: An Encyclopedic Handbook*, Vol. 3, Marcel Dekker, New York, 1978, pp. 87–89.
28. J. D. Miller, "Epidemiology of *Fusarium* Ear Diseases," in J. D. Miller and H. L. Trenholm, eds., *Mycotoxins in Grain: Compounds Other than Aflatoxin*, Eagan Press, St. Paul, Minn., 1994, pp. 19–36.
29. J. M. Beardall and J. D. Miller, "Natural Occurrence of Mycotoxins Other than Aflatoxin in Africa, Asia and South America," *Mycotoxin Res.* **10**, 21–40 (1994).

30. P. M. Scott, "The Natural Occurrence of Trichothecenes, in V. R. Beasley, ed., "*Trichothecene Toxicosis: Pathophysiological Effects*," Vol. 1, CRC Press, Boca Raton, Fla., 1989, pp. 2–26.

31. M. W. Trucksess et al., "Determination and Survey of Deoxynivalenol in White Flour, Whole Wheat Flour and Bran," *J. AOAC Int.* **79**, 883–887 (1996).

32. J. D. Miller et al., "Trichothecene Chemotypes of Three *Fusarium* Species," *Mycologia* **83**, 121–130 (1991).

33. W. F. O. Marasas, P. E. Nelson, and T. A. Toussoun, *Toxigenic Fusarium Species*, Pennsylvania State University Press, University Park, Pa., 1984.

34. T. Tatsuno et al., "Nivalenol, A Toxic Principle of Fusarium nivale", *Chem. Pharm. Bull.* **16**, 2519–2520 (1968).

35. C. J. Mirocha, "Variation in Deoxynivalenol, 15-Acetyl Deoxynivalenol, 3-Acetyl Deoxynivalenol and Zearlalenone Production by *Fusarium graminearum* Isolates," *Appl. Environ. Microbiol.* **55**, 1315–1316 (1989).

36. M. S. Piñeiro and G. E. Silva, "*Fusarium* Toxins in Uruguay: Head Blight, Toxin Levels and Grain Quality," *Cereal Res. Comm.* **25**, 805–806 (1997).

37. M. Ichinoe et al., "Chemotaxonomy of *Gibberella zea* with Special Reference to the Production of Trichothecenes and Zearalenone," *Appl. Environ. Microbiol.* **46**, 1364–1369 (1983).

38. D. R. Lauren, S. T. Sayer, and M. E. di Menna, "Trichothecene Production by *Fusarium* Species Isolated from Grain and Pasture Throughout New Zealand," *Mycopathologia* **120**, 167–176 (1992).

39. A. Logrieco, A. Bottalico, and C. Altomare, "Chemotaxomonic Observations on Zearalenone Production by *Gibberella zea* from Cereals in Southern Italy," *Mycologia* **80**, 892–895 (1988).

40. R. Greenhalgh et al., "Apotrichothecenes, Minor Metabolites of the *Fusarium* Species," in S. Natori, T. Hashimoto, and Y. Ueno, eds., *Proc. 7th Int. Symp. on Mycotoxins and phycotoxins*, Tokyo, 1989, pp. 223–232.

41. R. Greenhalgh et al., "Production and Characterization of Deoxynivalenol and Other Secondary Metabolites of *Fusarium culmorum* (CMI 14764, HLX 1503)," *J. Agric. Food Chem.* **34**, 98–102 (1985).

42. R. Greenhalgh et al., "Isolation and Characterization by Mass Spectrometry and NMR Spectroscopy of Secondary Metabolites of Some *Fusarium* species," in P. S. Steyn, ed., *Proc. 6th IUPAC Int. Symp. on Mycotoxins and Phycotoxins*, August 16–19, 1988, pp. 137–152.

43. G. C. Kasitu et al., "Isolation and Characterization of Culmorin Derivatives Produced by *Fusarium culmorum* CMI 14764," *Can. J. Chem.* **70**, 1308–1316 (1992).

44. S. Udegawa, "Mycotoxicoses—The Present Problem and Prevention of Mycotoxins," *Asian Med. J.* **31**, 599–604 (1988).

45. N. Morooka et al., "Studies on the Toxic Substances in Barley Infected with *Fusarium* Species," *J. Food Hygiene Soc. Jap.* **13**, 368–375 (1972).

46. R. Vesonder, A. Ceigler, and A. H. Jensen, "Isolation of the Emetic Principle from *Fusarium graminearum*-infected Corn," *Appl. Microbiol.* **26**, 1008–1010 (1973).

47. R. V. Bhat et al., "Outbreak of Trichothecene Mycotoxicosis Associated with Consumption of Mould-Damaged Wheat in Kashmir Valley, India," *Lancet* **7**, 35–37 (1989).

48. J. D. Miller, "Contamination of Food by *Fusarium* Toxins: Studies from Austria–Asia," *Proc. Jap. Assoc. Mycotoxicol.* **32**, 17–24 (1990).

49. B. Feinberg and C. S. MacLaughlin, "Biochemical Mechanism of Action of Trichothecene Mycotoxins," in V. R. Beasley, ed. *Trichothecene Mycotoxins: Pathophysiologic Effects*, Vol. 1, CRC Press, Boca Raton, Fla., 1989, pp. 27–36.

50. G. C. Khachatourians, "Metabolic Effects of the Trichothecene T-2 Toxin," *Can. J. Physiol. Pharmacol.* **68**, 1004–1008 (1990).

51. B. C. Foster et al., "Evaluation of Different Sources of Deoxynivalenol (Vomitoxin) Fed to Swine," *Can. J. Animal. Sci.* **66**, 1149–1154 (1986).

52. D. B. Prelusky, "Effect of Intraperitoneal Infusium of Deoxynivalenol on Feed Consumption and Weight Gain in the Pig," *Natural Toxins* **5**, 121–125 (1997).

53. B. C. Foster et al., "Fungal and Mycotoxin Content of Slashed Corn," *Microbiol. Aliment. Nutr.* **4**, 199–203 (1986).

54. P. F. Dowd, J. D. Miller, and R. Greenhalgh, "Toxicity and Interactions of Some *Fusarium graminearum* Metabolites to Caterpillars," *Mycologia* **81**, 646–650 (1989).

55. R. G. Rotter et al., "A Preliminary Examination of Potential Interactions Between Deoxynivalenol and Other Selected *Fusarium* Metabolites in Growing Pigs," *Can. J. Animal Sci.* **72**, 107–116 (1992).

56. H. A. Koshinski and G. C. Khachatourians, "Trichothecene Synergism, Additivity and Antagonism: The Significance of the Maximally Quiescent Ratio," *Natural Toxins* **1**, 38–47 (1992).

57. H. B. Schiefer, D. S. Hancock, and A. R. Bhatti, "Systemic Effects of Topically Applied Trichothecenes. I. Comparative Study of Various Trichothecenes in Mice", *J. Vet. Med.* **33A**, 373–383 (1986).

58. T. N. Bhavanishankar, H. P. Ramesh, and T. Shantha, "Dermal Toxicity of *Fusarium* Toxins in Combinations," *Arch. Toxico.* **61**, 241–244 (1988).

59. D. B. Prelusky et al., "Effect of Deoxynivalenol on Neurotransmitters in Discrete Regions of Swine Brain," *Arch. Environ. Contam. Toxicol.* **22**, 36–40 (1992).

60. D. B. Prelusky, "The Effect of Low-Level Deoyxivalenol on Neurotransmitter Levels Measured in Pig Cerebral Spinal Fluid," *J. Environ. Sci. Health.* **B26**, 731–761 (1993).

61. F. C. Iverson et al., "Chronic Feeding Study of Deoxynivalenol in B6C3F1 Male and Female Mice," *Teratogenesis, Carcinogenesis and Mutagenesis* **15**, 283–306 (1995).

62. D. M. Greene et al., "Role of Gender and Strain in Vomitoxin-Induced Dysregulation of IgA Production and IgA Nephropathy in the Mouse," *J. Toxicol. Environmental Health* **43**, 37–50 (1994).

63. J. J. Pestka and G. S. Bondy, "Mycotoxin-Induced Immune Modulation," in J. H. Dean et al., eds., *Immunotoxicology and Immunopharmacology*, Raven Press, New York, 1994, pp. 163–182.

64. G. S. Bondy et al., "Murine Lymphocyte Proliferation Impaired by Substituted Neosolaniols and Calonectrins, *Fusarium* Metabolites Associated with Trichothecene Biosynthesis," *Toxicon.* **29**, 1107–1113 (1991).

65. G. K. Hunder et al., "Influence of Subchronic Exposure to Low Dietary Deoxynivalenol, A Trichothecene Mycotoxin, On Intestinal Absorption of Nutrients in Mice," *Food Chem. Toxicol.* **29**, 809–814 (1991).

66. B. A. Rotter, D. B. Prelusky, and J. J. Petska, "Toxicology of Deoxynivalenol (Vomitoxin)," *J. Toxicol. Environ. Health* **48**, 101–134 (1996).

67. G. Falkay et al., "Affinity of *Fusarium* Toxins on Human Myometrial Estradiol Receptors," in S. Gorog, ed., *Proc. 5th*

Symp. on the Analysis of Steroids, Szombathely, Hungary, 1993, pp. 25–32.

68. H. B. Schiefer, "Mycotoxins," in Y. H. Hui, ed., *Encyclopedia of Food Science & Technology*, John Wiley and Sons, New York, 1990, pp. 1862–1869.
69. K. R. Price and G. R. Fenwick, "Naturally Occurring Oestrogens in Foods, A Review," *Food Additives and Contaminants* **2**, 73–106 (1985).
70. T. Kuiper-Goodman, P. M. Scott, and H. Watanabe, "Risk Assessment of the Mycotoxin Zearalenone," *Regulatory Toxicol. and Pharmacol.* **7**, 253–306 (1987).
71. D. B. Sauer and J. F. Tuite, "Conditions That Affect Growth of *Aspergillus flavus* and Production of Aflatoxin in Stored Maize," in M. S. Zuber, E. B. Lillehoj, and B. L. Renfro, eds., *Aflatoxin in Maize*, CIMMYT, Mexico, 1987, pp. 41–50.
72. G. A. Payne, "Aflatoxin in Maize," *Curr. Rev. Plant Sci.* **10**, 423–440 (1992).
73. P. Siriacha, P. Tanboon-Ek, and D. Buangsuwon, "Aflatoxin in Maize in Thailand," *Australian Centre for International Agricultural Research Proc.* **36**, 187–193 (1991).
74. R. J. Cole, J. W. Dorner, and P. D. Blankenship, "Management Strategies for Prevention and Control of Mycotoxins," in M. Miraglia et al., eds., *Mycotoxins and Phycotoxins—Developments in Chemistry, Toxicology and Food Safety*, Alaken, Fort Collins, Colo., 1998, pp. 189–202.
75. U. L. Diener, R. J. Cole, and R. A. Hill, "Epidemiology of Aflatoxin Formation by *Aspergillus flavus*," *Ann. Rev. Phytopathol.* **25**, 249–270 (1987).
76. W. W. McMillian, "Maize Plant Resistance to Insect Damage and Associated Aflatoxin Development," in M. S. Zuber, E. B. Lillehoj, and B. L. Renfro, eds., *Aflatoxin in Maize*, CIMMYT, Mexico, 1987, pp. 250–253.
77. M. G. Smart, D. T. Wicklow, and R. W. Caldwell, "Pathogenesis in *Aspergillus* Ear Rot of Maize: Light Microscopy of Fungal Spread from Wounds," *Phytopathol.* **80**, 1287–1294 (1990).
78. R. D. Wyatt, "Poultry," in J. E. Smith and R. S. Henderson, eds., *Mycotoxins and Animal Foods*, CRC Press, Boca Raton, Fla., 1991, pp. 553–606.
79. B. H. Armbrecht, "Aflatoxicosis in Swine," in T. D. Wyllie and L. G. Morehouse, eds., *Mycotoxic Fungi, Mycotoxins, Mycotoxicoses*, Vol. 2, Marcel Dekker, New York, 1978, pp. 227–235.
80. A. C. Keyl, "Aflatoxicosis in Cattle," in T. D. Wyllie and L. G. Morehouse eds., *Mycotoxic Fungi, Mycotoxins, Mycotoxicoses*, Vol. 2. Marcel Dekker, New York, 1978, pp. 9–27.
81. G.-S. Quian et al., "A Follow-up Study of Urinary Biomarkers of Aflatoxin Exposure and Liver Cancer Risk in Shanghai, Peoples' Republic of China," *Cancer Epidemiol. Biomarkers and Prevention* **3**, 3–10 (1994).
82. C. P. Wild, A. W. Daudt, and M. Castegnaro, "The Molecular Epidemiology of Mycotoxin Related Disease," in M. Miraglia et al., eds., *Mycotoxins and Phycotoxins—Developments in Chemistry, Toxicology and Food Safety*, Alaken, Fort Collins, Colo., 1998, pp. 213–232.
83. K. O'Donnell, E. Cigelnik, and H. I. Nirenberg, "Molecular Systematics and Phylogeography of the *Gibberella fujikuori* Species Complex," *Mycologia* **90**, 465–493 (1998).
84. P. E. Nelson, A. E. Desjardins, and R. D. Plattner, "Fumonisins, Mycotoxins Produced by *Fusarium* Species: Biology, Chemistry and Significance," *Ann. Rev. Phytopathol.* **31**, 233–252 (1993).
85. S. C. Bezuidenhout et al., "Structure Elucidation of the Fumonisin Mycotoxins from *Fusarium moniliforme*," *J. Chem. Soc. Chem. Commun.*
86. D. Laurent et al., "Etude en RMN ^{1}H et ^{13}C de la Macrofusin, Toxine Isolee de Mais Infeste par *Fusarium moniliforme*," *Sheld. Analusis* **18**, 172–179 (1990).
87. *International Program for Chemical Safety, Fumonisins*, World Health Organization, Geneva, 1999.
88. J. D. Miller et al., "Production of Fumonisins and Fusarins by *Fusarium moniliforme* from Southeast Asia," *Mycologia* **85**, 385–391 (1993).
89. J. D. Miller et al., "Mycotoxin Production by *Fusarium moniliforme* and *Fusarium proliferatum* from Ontario and Occurrence of Fumonisin in the 1993 Corn Crop," *Can. J. Plant. Pathol.* **17**, 233–239 (1995).
90. M. Sharman, J. Gilbert, and J. Chelkowski, "A Survey of the Occurrence of the Mycotoxin Moniliformin in Cereal Samples from Sources Worldwide," *Food Additives and Contaminants* **8**, 459–466 (1991).
91. D. C. Foley, "Systemic Infection of Corn by *Fusarium moniliforme*," *Phytopathol.* **68**, 1331–1335 (1962).
92. C. De Leon and S. Pandey, "Improvement of Resistance to Ear and Stalk Rots and Agronomic Traits in Tropical Maize Gene Pools," *Crop Sci.* **29**, 12–17 (1989).
93. H. Lew, A. Adler, and W. Edinger, "Moniliformin and the European Corn Borer (*Ostrinia nubilalis*)," *Mycotoxin Res.* **7**, 71–76 (1991).
94. A. W. Schaafsma et al., "Ear Rot Development and Mycotoxin Production in Corn in Relation to Inoculation Method and Corn Hybrid for Three Species of *Fusarium*," *Can. J. Plant Pathol.* **15**, 185–192 (1993).
95. R. T. Riley et al., "Evidence for Disruption of Sphingolipid Metabolism as a Contributing Factor in the Toxicity and Carcinogenicity of Fumonisins," *Natural Toxins* **4**, 3–15 (1996).
96. W. H. Tolleson et al., "The Mycotoxin Fumonisin Induces Apoptosis in Cultured Human Cells and in Livers and Kidneys of Rats," *Adv. Exp. Med. Biol.* **392**, 237–250 (1996).
97. W. H. Tolleson et al., "Apoptotic and Antiproliferative Effects of Fumonisin B_1 in Human Keratinocytes, Fibroblasts, Esophageal Epithelial Cells, and Hepatoma Cells," *Carcinogenesis* **17**, 239–249 (1996).
98. K. A. Voss et al., "*In vivo* Effects of Fumonisin B_1-producing and Fumonisin B_1-nonproducing *Fusarium moniliforme* Isolates are Similar: Fumonisins B_2 and B_3 Cause Hepato- and Nephrotoxicity in Rats," *Mycopathologica* **141**, 45–58 (1998).
99. W. F. O. Marasas et al., "Leucoencephalomalacia in a Horse Induced by Fumonisin B_1 Isolated from *Fusarium moniliforme*," *Onderspoort J. Vet. Res.* **55**, 197–203 (1988).
100. G. V. Smith et al., "Effects of Fumonisins on Cardiovascular Function in Swine," *Fund. Appl. Toxicol.* **31**, 169–172 (1996).
101. Riley et al., "Alteration of Tissue and Serum Sphinganine to Sphinganine Ratio: An Early Biomarker of Exposure to Fumonisin-containing Feeds in Pigs," *Toxicol. Appl. Pharmacol.* **118**, 105–112 (1993).
102. B. A. Rotter et al., "Response of Growing Swine to Pure Fumonisin B_1 During an 8 Week Period: Growth and Clinical Parameters," *Natural Toxins* **4**, 42–50 (1996).
103. B. A. Rotter et al., "Impact of Pure Fumonisin B_1 on Various Metabolic Parameters and Carcass Quality of Growing–Finishing Swine, Preliminary Findings," *Can. J. Animal Sci.* **77**, 465–470 (1997).

104. P. G. Thiel et al., "The Implications of Naturally Occurring Levels of Fumonisins in Corn for Human and Animal Health," *Mycopathologia* **117**, 3–9 (1992).

105. W. C. A. Gelderblom et al., "Fumonisins—Novel Mycotoxins with Cancer-Promoting Activity Produced by *Fusarium moniliforme*," *App. Environ. Microbiol.* **54**, 1806–1811 (1988).

106. W. C. A. Gelderblom et al., "Toxicity and Carcinogenicity of the *Fusarium moniliforme* Metabloite, Fumonisin B_1 in Rats," *Carcinogenesis* **12**, 1247–1251 (1991).

107. T. Battaglia, T. Hotzold, and R. Kroes, "Occurrence and Significance of Ochratoxin A in Food," *Food Additives and Contaminants* **13**, 1–57 (1996).

108. K. Jorgenson, "Survey of Pork, Poultry, Coffee, Beer and Pulses for Ochratoxin A," *Food Additives and Contaminants* **5**, 550–554 (1998).

109. P. Krogh, "Porcine Nephropathy Associated with Ochratoxin A," in J. E. Smith and R. S. Henderson, eds., *Mycotoxins and Animal Foods*, CRC Press, Boca Raton, Fla., 1991, pp. 627–646.

110. W. W. Carlton, M. E. Stack, and R. M. Eppley, "Hepatic Alterations Produced in Mice by Xanthomegnin and Viomellein, Metabolites of *Penicillium viridicatum*," *Toxicol. Appl. Pharm.* **38**, 455–459 (1976).

111. Scientific Committee for Food Reports, URL: *http://www.europa.com.*

112. M. Castegnaro et al., eds., *Mycotoxins, Endemic Nephropathy and Urinary Tract Tumours*, IARC Scientific Publication 115, Lyon, France, 1991.

113. D. Cvoriscec et al., "Endemic Nephropathy in Croatia," *Clin. Chem. Lab. Medicine* **36**, 271–277 (1998).

114. Y. Ueno, "Trichothecenes—Chemical, Biological and Toxicological Aspects," in Y. Ueno, ed., *Developments in Food Science*, Vol. 4, Elsevier, Amsterdam, The Netherlands, 1983.

115. A. Z. Joffe, "Toxicity of *Fusarium poae* and *F. sporotrichioides* and Its Relation to Alimentary Toxic Aleukica," in I. F. H. Purchase, ed., *Mycotoxins*, Elsevier, New York, 1974, pp. 229–262.

116. P. M. Scott, "*Penicillium* and *Aspergillus* Toxins," in J. D. Miller and H. L. Trenholm, eds., *Mycotoxins in Grain: Compounds Other than Aflatoxin*, Eagan Press, St. Paul, Minn., 1994, pp. 261–286.

117. P. M. Scott et al., "Ergot Alkaloids in Grain Foods Sold in Canada," *J. AOAC Int.* **75**, 773–779 (1992).

J. DAVID MILLER
Carleton University
Ottawa, Ontario
Canada

See also TOXICANTS, NATURAL.

N

NITROSAMINES

The chemistry of nitrosation of amines in foods is a complex process in which oxides of nitrogen (in oxidation states +3 and +4) formed during food processing, preservation, and preparation can react with amino compounds and other nucleophiles to produce N-, C-, O-, and S-nitroso compounds. The two major sources of oxides of nitrogen (nitrosating agents) result from (*1*) the addition of nitrate and/or nitrite to foods and (*2*) the heating and/or drying of foods in combustion gases in which molecular nitrogen can be oxidized to oxides of nitrogen. The only common feature of all *N*-nitroso compounds is the presence of the N-N=O functional group. Consequently, a wide range of chemical and physical properties exist for *N*-nitroso compounds, depending on the substitution on the amine/amide nitrogen.

The occurrence of *N*-nitroso compounds in foodstuffs probably represents the most comprehensively researched exposure situation for any class of genotoxic carcinogenic compounds in the human diet due to the possible link between various human cancers and exposure to *N*-nitroso compounds (1). Almost 80% of all *N*-nitroso compounds tested induce cancer in experimental animals, and some representative compounds of this class induce cancer in at least 40 different animal species including higher primates (2). Tumors induced in experimental animals resemble their human counterparts with respect to both morphological and biochemical properties. Extensive experimental and some epidemiological data suggest that humans are susceptible to carcinogenesis by *N*-nitroso compounds (1,3).

CHEMISTRY OF NITROSATION

The classic method for producing a nitrosating agent is the reaction between nitrite ions (NO_2^-) with protons (H^+ or H_3O^+) to give nitrous acid (HONO). Neither nitrite or nitrous acid per se are nitrosating agents, but are intermediates in the formation of the nitrosating agents dinitrogen trioxide (N_2O_3), dinitrogen tetraoxide (N_2O_4) and the nitrous acidium ion (H_2O^+NO) as shown in Figure 1. The acidity of the aqueous medium determines the relative proportions of the nitrosating species. Under moderately acidic conditions (pH 2–5) all three nitrosating species are present. At moderate pH (ca 3.0), N_2O_3 is the predominant nitrosating agent active in the nitrosation of secondary amines. At low pH (<2.0), H_2O^+NO is the predominant agent for nitrosation reactions involving weakly basic amines and amides.

$$NO_2^- \underset{+H^+}{\overset{-H^+}{\rightleftharpoons}} HONO \underset{+H^+}{\overset{-H^+}{\rightleftharpoons}} H_2O^+NO$$

$$HONO \rightleftharpoons N_2O_3 + H_2O$$

$$NO^\bullet + {}^\bullet NO_2 \rightleftharpoons N_2O_3 + H_2O$$

$$NO^\bullet \rightarrow NO_2 \qquad {}^\bullet NO_2 \rightleftharpoons N_2O_4$$

Figure 1. Equilibrium reactions of nitrite in aqueous media.

Primary amines are generally not considered to be precursors of *N*-nitrosamines, as nitrosation reactions generally proceed via diazotization and nucleophilic replacement of the amino group. However, nitrosation of the simplest aliphatic amine, methylamine, results in a complex mixture of products that include *N*-nitrosodimethylamine (NDMA). Nitrosation of higher primary aliphatic amines does not result in the formation of *N*-nitroso compounds.

Nitrosation of secondary amines is second order in terms of the concentration of nitrous acid and first order in terms of the amine concentration in which only the unprotonated amine reacts. Therefore, the reaction is strongly pH dependent. Strongly basic amines (eg, morpholine) are nitrosated to the full extent at ca pH 3.4. Introduction of ionizable groups near to the nitrosatable amino moiety reduces the basicity of the amine and the pH maxima for nitrosation is reduced to ca pH 2.5, as in the case of amino acids such as proline, sarcosine and hydroxyproline. For weakly basic amines (eg, *N*-alkylaromatic amines), nitrosation by N_2O_3 is very slow but proceeds more rapidly with the more reactive nitrosating species H_2O^+NO and NO^+ at low pH.

Tertiary amines present a rather complex situation in which simple trialkylamines react slowly with nitrous acid; more complex tertiary amines with substituents other than simple alkyl groups are readily nitrosated (eg, gramine and hordenine in malt). In most cases there appears to be no clear systematic approach in predicting which alkyl substituent will be replaced by the nitroso group.

In amides, the presence of the carbonyl group next to the nitrogen considerably reduces the basicity of the nitrogen. Nitrosation appears to involve an electrostatic interaction between the positively charged nitrosating species (eg, H_2O^+NO and NO^+) and the carbonyl oxygen, followed by rearrangement of the nitroso group onto nitrogen. The initial step (*O*-nitrosation) is readily reversible, but for rearrangement to the *N*-nitroso product, relatively strong acidic conditions are required, making this process unfavorable in food matrices.

Nitrosation reactions can be either catalyzed or inhibited by many factors. Nucleophilic anions such as I^-, Br^-, Cl^-, CNS^-, acetate, phthalate, weak acids, and some carbonyl compounds can exert a catalytic effect. The catalytic role of carbonyl compounds is related to their structure; acetone is inactive while formaldehyde and chloral are effective catalysts for the nitrosation of dialkylamines even under alkaline conditions. Other carbonyl catalysts include benzaldehyde and pyridoxal. Essentially any chemical that reacts with nitrite can inhibit the formation of N-nitroso compounds. Examples of such compounds include primary amines, sulfhydryl compounds, and certain aromatic compounds (eg, phenols). Several naturally occurring compounds in foods (eg, tannins in milk) can inhibit nitrosation via the formation of *C*-nitroso and *S*-nitroso compounds. In plant-based foods, phenolic compounds can catalyze and inhibit nitrosation, depending on their structure, the relative concentrations of the phenolic compound and the nitrosating agent, and the pH. High nitrite-to-phenol ratios favor catalysis while low ratios favor inhibition; 1,3-dihydroxyphenol (and related natural derivatives) are catalytic while 1,2- and 1,4-dihydroxyphenols inhibit nitrosation. Nitrosation can also result from the direct or indirect transfer of a nitric oxide radical (NO·) from N-, S-, O-, and C-nitroso compounds to a second nitrosatable precursor. The most important inhibitors of nitrosation reactions in food systems are redox compounds such as ascorbate and vitamin E. Ascorbate inhibits the formation of nitrosamines by competing with nitrosatable precursor compounds for the available nitrosating agent. On reaction with dinitrogen, tri- and tetraoxides nitric oxide is formed, which is not a nitrosating agent. However, this inhibition is not complete as oxidation of nitric oxide by molecular oxygen regenerates the nitrosating species. Complete inhibition is only effective in the presence of a large excess of ascorbate and/or under anaerobic conditions.

FORMATION OF NITROSAMINES IN FOODS

N-Nitroso compounds can be divided into either *N*-nitrosamines or *N*-nitrosamides. *N*-Nitrosamines can be subdivided into two groups depending on their physical properties: (*1*) Volatile *N*-nitrosamines (VNA), which are *N*-nitrosated analogues of simple low molecular weight dialkylamines and cyclic compounds such as NDMA, *N*-nitrosodibutylamine (NDBA), *N*-nitrosopiperidine (NPIP), *N*-nitrosopyrrolidine (NPYR), *N*-nitrosothiazolidine (NTHZ), and *N*-nitrosomorpholine (NMOR), as shown in Figure 2; and (*2*) nonvolatile *N*-nitrosamines (NVNA), which include a wide range of *N*-nitrosated hydroxylated or polyfunctional compounds, as shown in Figure 3. The most commonly encountered NVNA are the *N*-nitrosated amino acids *N*-nitrosoproline (NPRO), *N*-nitrososarcosine (NSAR), *N*-nitrosohydroxyproline (NHPRO), and a wide range of *N*-nitrosated heterocyclic carboxylic acids formed by nitrosation of the condensation products of the amino acids cysteine, serine, threonine, and tryptophan with simple aldehydes such as formaldehyde and acetaldehyde. The condensation of cysteine and formaldehyde to yield thiazolidine-4-carboxylic acid produces the amine precursor to *N*-nitrosothiazolidine-4-carboxylic acid (NTCA), which is the most frequently encountered *N*-nitrosated heterocyclic carboxylic acid formed by this type of reaction. Other examples of *N*-nitrosated heterocyclic carboxylic acids are *N*-nitroso-2-(hydroxymethyl)thiazolidine-4-carboxylic acid (NHMTCA) formed by condensation of cysteine with glycolaldehyde and the oxazolidines *N*-nitrosooxazolidine-4-carboxylic acid (NOCA) and *N*-nitroso-5-methyloxazolidine-4-carboxylic acid (NMOCA) formed by the condensation of formaldehyde with serine and threonine, respectively.

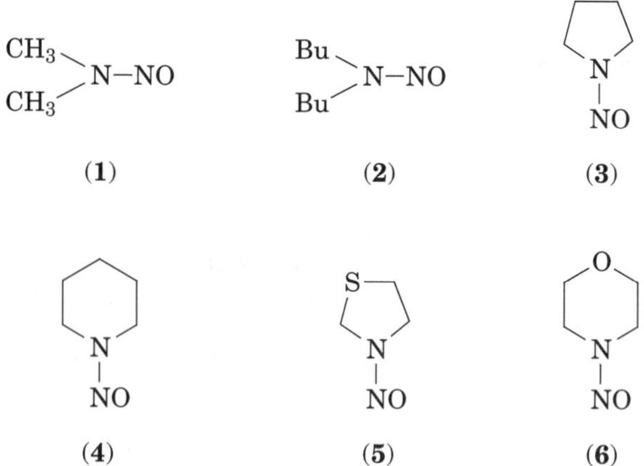

Figure 2. Chemical structures of volatile nitrosamines (VNA) occurring in foods: (**1**) NDMA, *N*-nitrosodimethylamine; (**2**) NDBA, *N*-nitrosodibutylamine; (**3**) NPYR, *N*-nitrosopyrrolidine; (**4**) NPIP, *N*-nitrosopiperidine; (**5**) NTHZ, *N*-nitrosothiazolidine; and (**6**) NMOR, *N*-nitrosomorpholine.

Several other nonvolatile *N*-nitroso compounds are probably present in certain foods and beverages; however, their presence remains uncertain (4). Examples of such compounds include *N*-nitrosated dipeptides or polypeptides with *N*-terminal proline or hydroxyproline residues, *N*-nitrosated 3-substituted indoles, *N*-nitrosated derivatives of certain pesticides (eg, *N*-nitrosocarbaryl, *N*-nitrosoatrazine, and *N*-nitrosoglyphosate), nonvolatile hydroxylated nitrosamines formed by the nitrosation of spermine and similar compounds, *N*-nitrosated glycosylamines and Amadori compounds.

Nitrosation of peptides and polypeptides to form stable *N*-nitroso compounds has been considered as a major potential source of NVNA in cured meats. The amide linkage is relatively unreactive towards nitrosation by nitrite in aqueous solution. However, nitrosation of dipeptides *N*-terminal in proline produces stable *N*-nitrosoprolyl dipeptides while higher peptides with *N*-terminal proline or hydroxyproline form stable derivatives under identical conditions.

Indoles occur in several plant species, particularly in the *Brassica* family, and are biosynthesized during the germination of grains. Naturally occurring indoles can be readily nitrosated under mild conditions; when carbon-3 is unsubstituted, C-3 nitrosation occurs in the first instance, where C-3 is already substituted, then nitrosation occurs preferentially at nitrogen unless the C-3 substituent is a powerful electron donor, in which case C-2 nitrosation results. Gramine, an indole alkaloid in barley malt acts as a precursor to NDMA found in beer during direct-drying of malt with hot flue gases containing nitrogen oxides. Nitrosation of gramine under aqueous acidic conditions yields NDMA and several nonvolatile reaction products, the major one of which is *N*-nitroso-3-nitromethylindole (5). Other indoles found in foods that

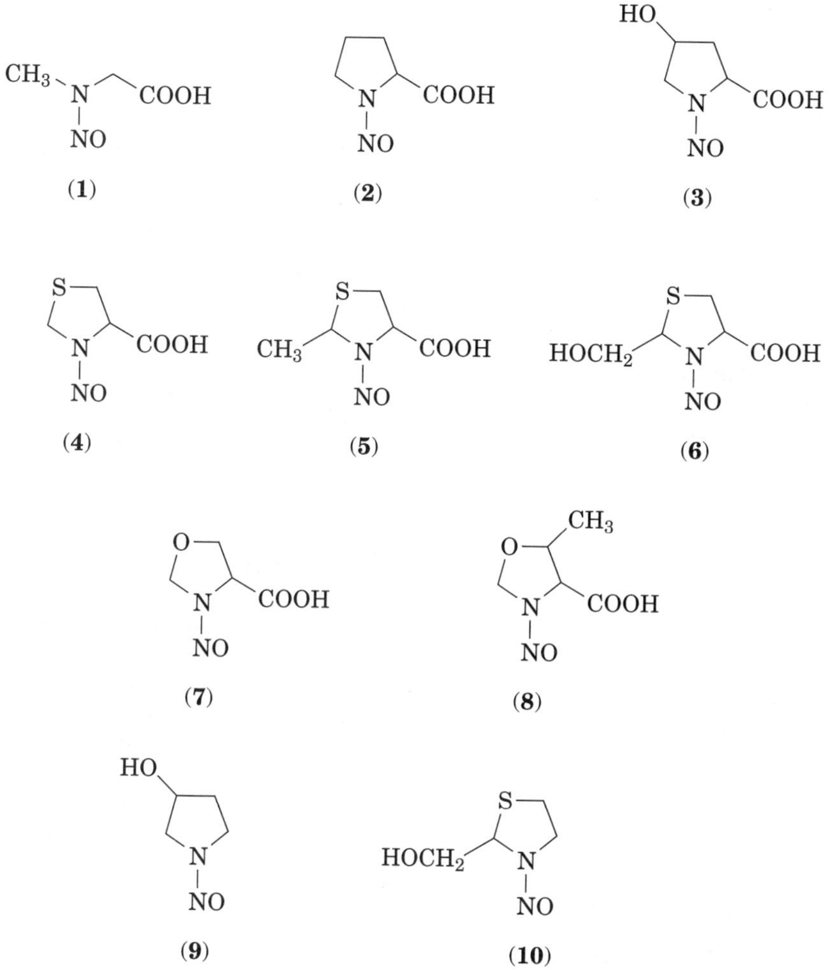

Figure 3. Chemical structures of nonvolatile nitrosamines (NVNA) occurring in foods: (**1**) NSAR, *N*-nitrososarcosine; (**2**) NPRO, *N*-nitrosopoline; (**3**) NHPRO, N-nitrosohydroxyproline; (**4**) NTCA. *N*-nitrosothiazolidine-4-carboxylic acid: (**5**) NMTCA, *N*-nitroso-2-methylthiazolidine-4-carboxylic acid: (**6**) NHMTCA, *N*-nitroso-2-hydroxymethylthiazolidine-4-carboxylic acid; (**7**) NOCA, *N*-nitrosooxazolidine-4-carboxylic acid; (**8**) NMOCA, *N*-nitroso-5-methyloxazolidine-4-carboxylic acid; (**9**) NHPYR, *N*-nitrosohydroxypyrrolidine; and (**10**) NHMTHZ. *N*-nitroso-2-hydroxymethylthiazolidine.

react with nitrite under acidic aqueous conditions include indole-3-acetonitrile, 4-methoxyindole-3-acetonitrile and 4-methoxyindole-3-acetaldehyde found in Chinese cabbage and 4-chloro-6-methoxyindole in fava beans (*Vicia faba*) All of which produce powerful direct-acting bacterial mutagens after nitrosation. The presence of nitrosated indoles have not yet been reported in foods or beverages.

During the cooking and baking of foods, nonenzymic browning (Maillard) reactions between aldoses or ketoses with amine compounds can result in low molecular weight volatile compounds (pyrazines, furans, etc) and high molecular weight compounds (melanoidines) such as glycosylamines and Amadori compounds. D-Fructose-L-amino acids formed by the Maillard reaction are readily nitrosatable under mild conditions, and a range of *N*-nitroso glycosylamines and Amadori compounds have been synthesized and shown to be direct-acting bacterial mutagens. Suitable methods for their analysis are still not available.

N-Nitrosamide precursors in foods and beverages include *N*-alkylureas, *N*-alkyl, carbamates, simple *N*-alkylamides, cyanamides, guanidines, amidines, hydroxylamines, hydrazones and hydrazides. At present, their is no direct evidence available to support the presence of *N*-nitrosamides in foodstuffs.

N-NITROSAMINES IN FOOD

Over the last decade, more than 500 publications on the occurrence of *N*-nitroso compounds in foodstuffs have been published, and several reviews have recently been published (4,6,7). The chemical stability of volatile nitrosamines, and the simplicity of their determination by distillation techniques followed by gas chromatography has resulted in the analysis of this group of *N*-nitroso compounds in almost all Western foods. In the first comprehensive dietary survey (8), 2,826 food samples on the German market were analyzed, revealing that only three volatile nitrosamines, namely NDMA, NPYR, and NPIP, were regularly present in foods; all other volatile nitrosamines were present only in rare cases. NDMA was detected in 30% of all samples, 6% of which contained concentrations above 5 μg/kg. NPYR and NPIP at concentrations above 0.5 μg/kg (ppb) were found in only 3 and 2% of all foods, respectively. Over the last decade, several major food surveys have been made in more than 10 different countries from which it is possible to evaluate basic trends in VNA contamination. The results from these surveys show that foodstuffs most commonly contaminated with *N*-nitroso compounds can be classified into several broad groups:

Foodstuffs preserved by either addition of nitrate and/or nitrite.

Foodstuffs preserved by smoking.

Foodstuffs subject to drying by combustion gases.

Pickled and salt-preserved vegetables.

Foodstuffs stored under humid conditions favoring fungal contamination, particularly the growth of *Fusarium moniliforme*.

Foodstuffs contaminated by food contact materials.

Foodstuffs with Added Nitrate and Nitrite

The addition of nitrite and/or nitrate salts to meat, poultry, fish, and to a lesser extent cheese at low concentrations (up to a few 100 ppm) has been a common method of preservation for centuries. Nitrite addition to meat products results in the formation of characteristic pink-red colorations in cured meats, prevents rancidity occurring under normal storage conditions, and inhibits *Clostridium botulinum* growth. As an alternative to direct nitrite addition, nitrate addition is used in several products as a continuous direct source of nitrite by reduction involving microbial enzymes.

The chemistry of interactions involving nitrite in food products is complex, owing to the highly reactive nature of nitrite. In cured meat products, a rapid decrease to about 50% of the initially added nitrite concentration occurs during processing, and a steady further decline occurs during storage. The decline in nitrite may result through several different mechanisms, of which reduction by ascorbic acid to nitric oxide and the formation of an equilibrium state between nitrite, nitrous acid, and oxides of nitrogen such as dinitrogen trioxide are the most important (refer to Fig. 1). Dinitrogen trioxide, which is a chemically active nitrosating species, reacts with several constituents present in meat. The fate of nitrite can be roughly summarized as follows: 1–5% is lost as nitric oxide gas or becomes bound to lipids, 1–10% is oxidized to nitrate, 5–10% remains as free nitrite, 5–15% reacts with sulfhydryl compounds or myoglobin, and 20–30% becomes bound to proteins. In addition, nitrosation of amino compounds (nitrosamine formation) may occur as well as the formation of organic nitrites and nitrates. To prevent the formation of *N*-nitroso compounds, nitrite cured meats are required to be formulated with 500 mg/kg sodium ascorbate in the United States and in most other countries.

Foodstuffs Preserved by Smoking

Smoking under traditional smokehouse conditions results in prolonged direct contact of foodstuffs with elevated concentrations of oxides of nitrogen present in wood smoke. Consequently, smoked meat products, fish, and cheese normally contain *N*-nitroso compounds. In cured meat products, several nonvolatile heterocyclic *N*-nitroso compounds have been identified that are not usually present in unsmoked products (4,9). Wood smoke contains several carbonyl compounds including formaldehyde, acetaldehyde, propionaldehyde, glyoxal, methylglyoxal, and glycolaldehyde. Formaldehyde and acetaldehyde readily condense with bifunctional food components containing thiol, hydroxyl, and free amino groups to form heterocyclic compounds that are subject to rapid nitrosation by oxides of nitrogen present in wood smoke. In smoked meats, poultry, and fish it is generally accepted that the condensation of formaldehyde with cysteamine and cysteine to form thiazolidine and thiazolidine-4-carboxylic acid, followed by nitrosation, results in the formation of NTHZ and NTCA (10). The potential routes of NTHZ and NTCA formation are summarized in Figure 4. Other simple carbonyl compounds in wood smoke may also undergo condensation reactions with amino acids such as cysteine, lysine, serine, threonine, and tryptophan to yield a wide range of nitrosatable heterocyclic amino acid derivatives. The corresponding *N*-nitroso derivatives, of which several examples are known, are commonly found in smoked meats and to a lesser extent in poultry and fish (reviewed in Ref. 4).

Foodstuffs Dried by Combustion Gases

By an analogus process, food-processing operations envolving the drying of foodstuffs in combustion gases (hot flue gases) containing oxides of nitrogen (NO_X) can result in the formation of N-nitroso compounds. The kilning (drying process) used in preparing malt for the brewing industry is directly responsible for the levels of NDMA in beer (11). Furthermore, specific types of malt produced by characteristic drying methods contain varying concentrations of NDMA as well as NPRO and NSAR. Amine precursors of NDMA in malt include dimethylamine, trimethylamine, and the alkaloids gramine and hordenine, which are biosynthesized in green malt during germination. In model experiments, both gramine and hordenine undergo nitrosation to yield 76 and 11% NDMA, respectively, at pH 4.4. In addition, gramine also yields the nonvolatile *N*-nitroso compound *N*-nitroso-3-nitromethylindole (5); the latter has not yet been detected in beer or malts. The identified nitrosation products from the alkaloid gramine are shown in Figure 5. Modification of kilning methods, in particular the introduction of indirect heating or low-temperature gas burners, produces lower NO_X concentrations and hence results in reduced levels of nitrosamines in malts. Malts treated with sulfur dioxide during the kilning process, either by burning sulfur or sulfur-containing compounds in the combustion flame, or by direct addition of sulfur dioxide gas into the combustion gas flow of direct drying kilns, generally contain lower VNA concentrations but have little or no effect on the reduction of NPRO in malt (12).

Foodstuffs produced by drying processes include dried soup mixes, tea, spices, coffee, soy protein isolates, cereal products, infant formulations, and dairy products such as cheese and nonfat dried milk powder. All these products have at one time been shown to be regularly contaminated with traces of VNA (mainly NDMA) whose formation was attributed to the drying process. However, with the exception of brewing malt, the occurrence of specific NVNA has not been demonstrated to occur in products dried with combustion gases.

Pickled and Salt-Preserved Vegetables

Nitrate, and to a lesser extent nitrite, are natural components of several plant species and vegetables. Fermen-

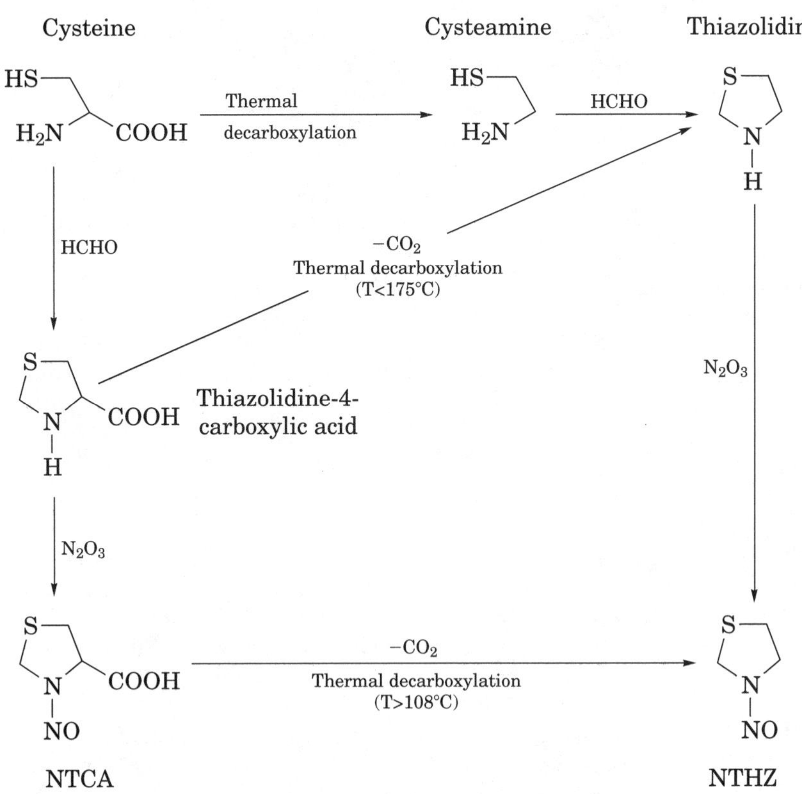

Figure 4. Proposed mechanisms for the formation of *N*-nitrosothiazolidine-4-carboxylic (NTCA) acid and *N*-nitrosothiazolidine (NTHZ).

N_2O_3 → NDMA + (indole-3-carboxylic acid)

N-Nitroso-3-nitromethylindole + Unidentified products

Figure 5. Nitrosation products of the alkaloid gramine.

tation by microorganisms (and surface bacteria) with nitrate-reductase activity can result in a large increase in nitrite content. On prolonged storage, nitrosamine formation may occur.

Fungal Contamination of Foods

The catalysis of nitrosamine synthesis from secondary amines and nitrate by several microorganisms and fungal biosynthesis of nitrosamines from primary amines and nitrite occurs under certain conditions in a limited range of foods. In certain areas in China where widespread contamination of foods by *Fusaria* occurs, low concentrations (<2 μg/kg) of *N*-nitroso-*N*-methylacetonyl-*N*-2-methylpropylamine (NMAMPA) and *N*-nitroso-*N*-methylacetonyl-*N*-3-methylbutylamine (NMAMBA) have been identified in moldy millet and wheat flour contaminated by *Fusaria* (13). A nonenzymic reductive condensation of isoamylamine, a decarboxylation product of leucine, and 3-hydroxy-2-butanone (acetoin) preset in cereal grains have been suggested as a possible route to the formation of the amine precursor to NMAMBA. The structures of NMAMBA and NMAMPA are shown in Figure 6.

Contact Materials

The contamination of amines by their corresponding *N*-nitroso analogues and the use of dialkylamine-derived accelerators and stabilizers in the production of rubber results in inadvertant *N*-nitrosamine contamination of rubber products. As a result, direct contact of foodstuffs with nitrosamine-contaminated food-packaging materials, in particular rubber, can result in nitrosamine contamination. Rubber nettings for the packaging of cured meats results in contamination of meat products with NDEA, *N*-nitrosodibutylamine (NDBA), and *N*-nitrosodibenzylamine (NDBzA), three *N*-nitroso compounds not generally encountered in foods (14), due to both active migration of the preformed nitrosamine into the food matrix and the interaction of nitrite in cured meats with amine additives (eg, zinc dialkyl or diaryldithiocarbamates) used as accelerators during rubber vulcanization. The use of rubber baby bottle nipples and pacifiers presents a more serious problem in which both nitrosatable amines and nitrosamines migrate from the rubber matrix into baby formulations. Government regulations in most countries have limited the permissible levels of VNA to 10 μg/kg and nitrosatable amines following extraction to 200 μg/kg (expressed in terms of nipple weight). Monitoring programs in most countries currently show negligible nitrosamine concentrations in rubber nipples and pacifiers produced after 1984. In most cases, nitrosamine free products are available.

Commercial wax preparations used for coating food-packaging materials, and liquid waxes used for coating apples and citrus fruit present a potential source of *N*-nitrosomorpholine (NMOR) contamination (15). Morpholine (which normally contains NMOR as an impurity) is used as a boiler additive for steam used in the manufacture of paper. Both morpholine and morpholineoleate are used as solvents for waxes used to treat paper for water resistance and to coat fresh fruits. NMOR contamination of milk packed in waxed cartons and dairy products in waxed paper wrappers was reported in several countries about a decade ago. However, this is not a current problem because

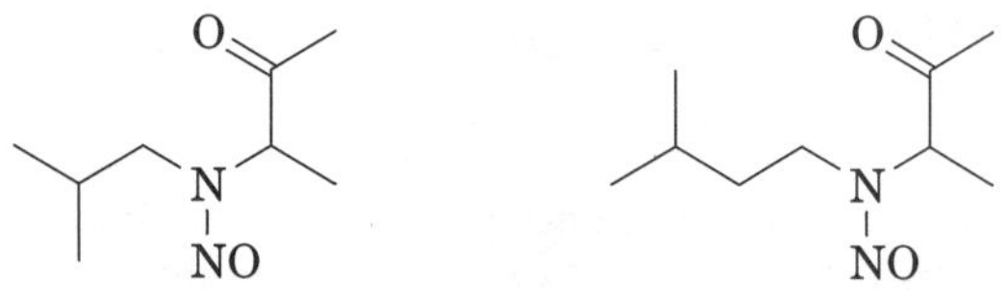

NMAMPA NMAMBA

Figure 6. Nitrosamines produced from fungal metabolites in *Fusarium*-contaminated foods. Nonvolatile nitrosamines abbreviated as NMAMPA, *N*-nitroso-*N*-methylacetonyl-*N*-2-methylpropylamine, and NMAMBA, *N*-nitroso-*N*-methylacetonyl-*N*-3-methylbutylamine.

waxed packaging has been replaced with plastic-coated surfaces.

TRENDS

Over the last decade, nitrosamine contamination trends in most Western food products have drastically decreased as a result of improved production technology, reduction in the use of nitrates and/or nitrites for curing, and changing from direct to indirect drying methods. Theses changes have resulted in a mean ten-fold reduction in the levels of VNA in cured meats and almost a 100-fold reduction in the level of NDMA in beer. The following data are representative of the current nitrosamine situation in food commodities in which a nitrosamine problem still exists.

Meat Products

Only nitrite- and nitrate-preserved meat products are contaminated with *N*-nitrosamines, the concentration and range of different *N*-nitroso compounds is increased following smoking. Early studies in the late 1970s in several countries showed that almost a third of all cured meat products contain VNA. In Germany (8), 32% of all cured meat samples analyzed contained VNA: <5 μg/kg NDMA was detected in 30% of all samples, with concentrations above 5 μg/kg in 2.1% of the samples analyzed; 6.8% contained <5 μ/kg NPYR (6.8% containing above 5 μg/kg NPYR); and 4% contained up to 6.4 μg/kg NPIP. *N*-Nitroso-3-hydroxypyrrolidine (NHPYR), a carcinogenic NVNA, was reported at about the same concentration levels. Similar trends were reported in the United States, Canada, and the United Kingdom. Fried bacon was identified as consistently having the highest concentrations of nitrosamines. NTHZ, a new nitrosamine, was identified in 1983 at concentrations under 5 μg/kg in fried bacon and at concentrations between 1.6–31.9 μg/kg in other cured meat products (16). Since these earlier reports, changes in production technologies, reduction in the levels of nitrate and nitrite added during curing, and the inclusion of nitrosation inhibitors such as ascorbic acid and α-tocopherol to the curing mixture have resulted in significant reductions in the levels of volatile nitrosamines, particularly for NPYR in bacon. During cooking and frying of cured meat products, the concentration of NDMA usually remains constant while the concentration of NPYR increases considerably. The volatility of NDMA causes losses during heating that are partially balanced by thermally induced synthesis. The lower volatility of NPYR decreases its losses during heating while at the same time synthesis of NPYR occurs from both reactions of pyrrolidine with residual nitrite and the thermal decarboxylation of NPRO and *N*-terminal *N*-nitrosoproline containing peptides. NPYR formation is increased during high-temperature frying for short periods as compared to low-temperature frying for prolonged periods. VNA exposure from consumer bought and cooked at home fried bacon has been calculated (ng/person/d) to be 18 ng NDMA, 6.6 ng NPIP, 76 ng NPYR, and 40 ng/person/d NTHZ. Microwave cooking produces almost undetectable (<0.1 μg/kg) levels of VNA. The thermal decarboxylation of NTCA during cooking appears to be the most important factor involved in the formation of NTHZ in fried bacon. Thiazolidine, the amine precursor to NTHZ, has not been reported in foods but may be formed by the condensation of cysteamine and formaldehyde or in cysteamine/glucose browning systems. The use of elastic rubber nettings for cured meat packaging provides an additional source of mainly NDEA, NDBA, NDBzA, and, to a lesser extent, NMOR in cured meat products (14).

Considerably higher concentrations of NVNA occur in nitrite cured and/or smoked meat products: Concentration ranges of $<$ 410 μg/kg NSAR, 20–360 μg/kg NPRO, and 10–560 μg/kg NHPRO occur in unsmoked meat products while condensation reactions between carbonyl compounds (aldehydes) and amino acids during smoking produces concentration ranges of up to 1620 μg/kg NTCA and 2,100 μg/kg NHMTCA (4,9). Some traditional smoked foods, eg, Icelandic smoked mutton (a Christmas delicacy that is first cured and then smoked over sheep dung) contain 40–70 μg/kg NOCA and 30–120 μg/kg of its 5-methyl derivative (NMOCA). Thermal decarboxylation of *N*-nitrosated amino acids during cooking yields the corresponding volatile *N*-nitrosamines; NPRO, NHPRO, NSAR and NTCA decarboxylate to NPYR, NHPYR, NDMA, and NTHZ, respectively. No evidence has been found for the presence of the decarboxylated analogues of NOCA and NMOCA in foods.

Fish and Sea Foods

Fresh fish and seafoods that are naturally rich in amines do not contain *N*-nitroso compounds; however, nitrosamines are occasionally found after processing. Early reports from England, the United States, and Canada showed 1–10 μg/kg NDMA concentrations in almost 80% of processed fish products. Today, NDMA is still present in very rare cases, but at levels seldom above 2 μg/kg. Processed products based on shrimp are nearly always contaminated with 5–30 μg/kg NDMA and in exceptional cases with several hundred μg/kg NDMA in shrimp sauces. The use of sea salt for fish drying can lead to nitrate contamination of fish and provides a potential source of nitrite for nitrosamine formation. Salt-dried fish in Japan contains about 1.0–34 μg/kg NDMA. Coalgas broiling, the traditional way of cooking fish in Japan, can increase the concentrations of NDMA in salted fish to 300 μg/kg and the concentration of NDMA in dried squid from 15–84 to 24–310 μg/kg. The formation of 2.4–13 μg/kg NPYR and up to 94 μg/kg NPRO on broiling squid has also been reported. Chinese and Cantonese salt-preserved fish often contains concentrations up to 130 μg/kg NDMA while dried fish from Greenland also contains up to 38 μg/kg NDMA.

Smoked fish and seafood contain high concentrations of NTHZ and NTCA: Smoked oyster has been reported to contain 109 μg/kg NTHZ and 167 μg/kg NTCA, while concentrations of 5–10 μg/kg NTHZ and 10–350 μg/kg NTCA in commercially smoked fish are not uncommon. Other *N*-nitrosamines including NPYR ($<$2 μg/kg) and NPRO ($<$10 μg/kg) are occasionally found in smoked fish.

Limited data show that most sun-dried fish, which forms a large part of the staple diet in several developing countries, is also subject to widespread NDMA contamination. The effect of traditional cooking methods such as smoking fish wrapped in grass or leaves on the concentration of *N*-nitroso compounds has not been investigated.

Cheese and Dairy Products

Current evidence shows that cheese is only a very minor dietary source of NDMA. No correlation exists between nitrate addition (used to prevent microbiological defects) and nitrosamine contamination. Concentrations of 5 μg/kg NDMA in cheese reported a decade ago are now reduced to $<$1 μg/kg in the limited number of samples still found to contain NDMA. A nationwide American survey (carried out in 1981) of non-fat dried milk powders produced by directly heated combustion gas driers showed 0.1–3.7 μg/kg (average 0.6 μg/kg) NDMA in 48 out of 57 dried-milk samples, with occasional traces of NPYR (0–0.8 μg/kg, average 0.1 μg/kg) and NPIP (0–0.5 μg/kg, average 0.1 μg/kg). Similar studies on dried-milk products produced in Europe and New Zealand using indirectly heated driers gave negative results, emphasizing the role of production technology on the formation of *N*-nitroso compounds. Changes in drying methods have since been introduced in the United States and contamination of dried-milk products is not a current problem.

Vegetables and Cereal Products

Volatile *N*-nitroso compounds have been detected rarely at trace levels in grain crops stored under humid conditions. The reported contamination of cereal products in New Zealand with levels of up to 1.2 μg/kg NDMA in grain, 2.8 μg/kg NDMA in maize meal, and 4.2 μg/kg NDMA in wheaten cornstarch is not representative of the general situation in cereal products where good agricultural practice (avoidance of wet harvesting and storage under humid conditions) does not favor mold growth and the subsequent possibility of nitrosamine contamination. Japanese fermented vegetables have been reported to contain $<$5 μg/kg NDMA and NPYR. Widespread contamination of homepickled vegetables with 1–15 μg/kg NDMA, 1–3 μg/kg *N*-nitrosoethylmethylamine (NEMA), 1–5 μg/kg NDEA, and $<$100 μg/kg NPYR have often been reported from China, Japan, and several Asian countries. Commercial products contain far lower concentrations. In some areas of China, contamination at levels of less than 2 μg/kg NMAMBA and NMAMPA in millet and wheat flour containing molds, particularly *Fusarium moniliforme*, is not uncommon (13). Indian and Tunisian foods containing red and black peppers, paprika, and spices contain various levels of VNA. Touklia, a stewing base that is the main ingredient of Tunisian food, contains relatively high concentrations of 12 μg/kg NDMA, 43 μg/kg NPIP, and 5.8 μg/kg NPYR. Contamination of most dried spices with NDMA and dried chillies with NPYR is a fairly widespread problem. However, the amounts of spices used in most Western cooking practices are not likely to lead to a significant exposure to the compounds.

Beer and Alcoholic Beverages

In the late 1970s, NDMA contamination of beer was reported in the former West Germany showing that almost

60% of bottled, canned, and draft beers contained a mean concentration of 2 μg/L NDMA with a maximum value of 68 μg/L NDMA. Similar concentrations were also reported in the United States, Canada, and most other European countries. Modifications in malt drying and kilning techniques, shown to be the stages involving the formation of NDMA, have resulted in considerable reductions in the levels of NDMA in beer. The current mean levels of NDMA reported in beer are 0.07 μg/L in the United States, 0.10 μg/L in Canada and West Germany, with most other European beers containing less than 0.3 μg/L. The mean concentration of NPRO, the only NVNA identified in beer, is less than 2 μg/L in most European, Australian, and American beers produced from indirectly kilned malt. Although NSAR and NPYR have been reported in malts, no evidence is available to show their presence in beer.

The only other commercially produced alcoholic beverage that contains nitrosamines is whiskey, in which current levels are generally below 0.3 μg/L. NDMA formation occurs during malt production, and no other *N*-nitroso compounds have been reported in whiskey. Contrary to earlier reports, *N*-nitrosamines are not present in wine, sherry, port, apple brandy, liqueurs, or fortified and distilled spirits such as rum, gin, and vodka. There is some evidence to support the presence of VNA, mainly NDMA and NDEA, in some African fermented beverages and both African and Indian illicitly distilled spirits.

Drinking Water

Early reports on the possible presence of nitrosamines in drinking water drawn from wells containing high nitrate concentrations have never been substantiated. Although industrial waste water and sewage sludge occasionally contain traces of NDMA, NDEA, and possibly other VNA at total concentrations of less than 5 μg/L, their incorporation into domestic water supplies has never been confirmed.

Other Food Products

In systematic studies carried out in several countries, only sporadic contamination of trace quantities of individual VNA have been found in other foods and beverages. In most cases, contamination occurred in single samples and could not be verified in other countries, or at a later date. Food contamination by direct contact materials has nearly always been assumed in unexplained nitrosamine contamination cases such as NMOR contamination of margarine packed in waxed paper wrappers.

DIETARY EXPOSURE TO *N*-NITROSO COMPOUNDS

The dietary exposure to NDMA (the most commonly occurring VNA in the diet) has been calculated in a number of food surveys and is summarized in Table 1. It should be taken into consideration that exposure estimates of this kind suffer from uncertainties in food consumption trends averaged over a population. Over the last decade, reductions in the use of nitrates and nitrites used for curing meats to the minimum amount required to inhibit bacterial growth, and modification of malting techniques in the brewing industry have resulted in significant reductions in the levels of NDMA. In most dietary surveys, cured meats and beer have been implicated as the major dietary sources of NDMA. As a direct consequence, NDMA exposure over the last decade has probably decreased from about 1 μg/d to ca 0.3 μg/d NDMA in most Western countries. An exposure estimate of between 10–100 μg/d for currently identified NVNA would not seem unreasonable. In developing countries, particularly China and other Asia countries, the occurrence and concentrations of VNA in common dietary items are considerably higher than in Western foods and a far higher dietary exposure to nitrosamines would be expected.

Table 1. Food Surveys and Daily NDMA Intake

Country	Year	NDMA intake (μg/d)	Major NDMA source
UK	1978	0.53	Cured meats (81%)
UK	1987	0.60	Beer, cured meats
Holland	1980	0.38	Beer (71%)
Holland[a]	1990	0.10	Not evaluated
The former FRG	1980	1.02 (men)	Beer (65%), cured meats (10%)
		0.57 (women)	
The former FRG	1983	0.53 (men)	Beer (40%), cured meats (18%)
		0.35 (women)	
Japan	1980	1.80	Dried fish (91%)
Japan	1984	0.50	Fish products (88%)
Sweden	1988	0.12	Cured meats (61%), beer (32%)
Finland[b]	1990	0.08 (adult)	Beer (75%), smoked fish (25%) for adults
		0.02 (child)	Smoked fish
China	1988	No data	Marine foods
Italy	1988	No data	Cured meats
The former USSR	1990	No data	Meat and fish

Source: Ref. 3.
Note: Beer not included in the survey.
[a]Determined by 24-h duplicate diet analysis.
[b]Based on limited data.

NITROSAMINE EXPOSURE AND HUMAN CANCER

Experimental studies provide evidence that the biological activity of *N*-nitroso compounds in humans is not substantially different from that in experimental animals (1). In contrast to animal experiments, in which exposure (normally at high concentrations) to single *N*-nitroso compounds may induce cancer (2), human exposure (3,6,7,17) results via several different sources (eg, diet, occupational exposure, and tobacco consumption) at a wide range of different concentrations. Dose-response studies using experimental animals show that NDEA, NMOR, and NPYR continuously administered in drinking water at exposure levels of 0.075 mg/L (0.075 ppm), 0.07 mg/L, and 0.01 mg/kg body weight/d, respectively, are sufficient to induce a significant incidence of tumors. In animal carcinogenicity experiments, the absence of a lower no-effect threshold and the syncarcinogenic activity of low concentration combinations of *N*-nitroso compounds at concentrations at which individual *N*-nitrosamine concentrations alone would not be expected to induce carcinogenesis suggest that multiexposure to several different *N*-nitroso compounds in the diet may present a potential carcinogenic risk to humans (1,3,18).

Furthermore, micronutrient deficiencies may modify nitrosamine-induced carcinogenesis with respect to both organotropism and cancer incidence (2). These confounding factors are partially responsible for the failure of epidemiologic studies in identifying exposures to individual compounds and foodstuffs as potential risk factors for human carcinogenesis despite the fact that there is sufficient evidence to support the hypothesis that humans are susceptible to *N*-nitroso compound induced carcinogenesis (1,3,18). Thus, continuous exposure to low concentrations of several different *N*-nitroso compounds in the diet would be expected to be an etiologic risk factor for certain human cancers. Epidemiologic case-control studies indicate that dietary nitrosamine exposure is an important risk factor for cancers of the esophagus, stomach, and nasopharynx (18).

BIBLIOGRAPHY

1. H. Bartsch and R. Montesano, "Relevance of Nitrosamines to Human Cancer," *Carcinogenesis* **5**, 1381–1393 (1984).
2. R. Preussmann and B. W. Stewart, "*N*-Nitroso Carcinogens," in E. Searle, ed., *Chemical Carcinogens. ACS Monograph 182*, American Chemical Society, Washington, D.C., 1984, pp. 643–828.
3. A. R. Tricker and R. Preussmann, "Carcinogenic *N*-Nitrosamines in the Diet: Occurrence, Formation, Mechanisms, and Carcinogenic Potential," *Mutations Research* **259**, 277–289 (1990).
4. A. R. Tricker and S. J. Kubacki, "Review of the Occurrence and Formation of Nonvolatile *N*-Nitroso Compounds in Foods," *Food Additives and Contaminants* (1991).
5. M. U. Ahmad, L. M. Libbey, J. F. Barbour, and R. A. Scanlan, "Isolation and Characterization of Products from the Nitrosation of the Alkaloid Gramine," *Food Chemistry and Toxicology* **23**, 841–847 (1985).
6. A. R. Tricker and R. Preussmann, "*N*-Nitroso Compounds and Their Precursors in the Human Environment," in M. J. Hill, ed, *Nitrosamines—Toxicology and Microbiology*, Ellis Harwood, Chichester, UK, 1988, pp. 88–116.
7. J. A. Hotchkiss, "A Review of Current Literature on *N*-Nitroso Compounds in Foods," *Advances in Food Research* **31**, 53–115 (1987).
8. B. Spiegelhalder, G. Eisenbrand, and R. Preussmann, "Volatile Nitrosamines in Food," *Oncology* **37**, 211–216 (1980).
9. A. R. Tricker et al., "Incidence of Some Nonvolatile *N*-Nitroso Compounds in Cured Meats," *Food Additives and Contaminants* **1**, 245–252 (1984).
10. J. W. Pensabene and W. Fiddler, "Effect of *N*-Nitrosothiazolidine-4-Carboxylic Acid on Formation of *N*-Nitrosothiazolidine in Uncooked Bacon," *Journal of the Association of Official Analytical Chemists* **68**, 1077–1080 (1985).
11. B. Spiegelhalder, G. Eisenbrand, and R. Preussmann, "Contamination of Beer with Trace Quantities of *N*-Nitrosodimethylamine," *Food and Cosmetics Toxicology* **17**, 29–31 (1979).
12. P. Johnson et al., "An Investigation into the Total Apparent *N*-Nitroso Compounds in Malt," *Journal of the Institute of Brewing*, **93**, 319–321 (1987).
13. C. Ji et al., "A new *N*-Nitroso Compound, *N*-(2-Methylpropyl)-*N*-(1-Methylacetonyl)Nitrosamine, in Moldy Millet and Wheat Flour," *Journal of the Agricultural and Food Chemistry* **34**, 628–632 (1986).
14. N. P. Sen, S. W. Seaman, P. A. Baddoo, and D. Weber, "Further Studies on the Formation of Nitrosamines in Cured Pork Products Packaged in Elastic Rubber Nettings," *Journal of Food Science* **53**, 731–734, 738 (1988).
15. N. P. Sen, "Migration and Formation of *N*-Nitrosamines from Food Contact Materials," in J. H. Hotchkiss, ed., *Food and Packaging Interactions. ACS Symposium Series No. 365*, American Chemical Society, Washington, D.C., 1988, pp. 146–158.
16. J. W. Pensabene and W. Fiddler, "*N*-Nitrosothiazolidine in Cured Meat Products," *Journal of Food Science* **48**, 1870–1874 (1983).
17. A. R. Trickler, B. Spiegelhalder, and R. Preussmann, "Environmental Exposure to Preformed Nitroso Compounds," *Cancer Surveys* **8**, 251–272 (1989).
18. P. N. Magee, "The Experimental Basis for the Role of Nitroso Compounds in Human Cancer," *Cancer Surveys* **8**, 207–239 (1989).

GENERAL REFERENCES

H. Biaudet, T. Malvelle, and G. Debry, "Mean Daily Intake of *N*-Nitrosodimethylamine from Foods and Beverages in France in 1987–1992," *Food Chemical Toxicol.* **32**, 417–421 (1994).

A. I. Burykin et al., "Effects of Drying Method on Carcinogen Content of Dairy Products," *Molochnaya Promyshlennost* **4**, 14–18 (1992).

R. G. Cassens, "Use of Sodium Nitrite in Cured Meats Today," *Food Technol.* **49**, 72–79, 115–116 (1995).

Lim Yung Choo et al., "The Formation of *N*-Nitrosamines During Storage of Salted Mackerel, *Scomber japonicus*," *J. Korean Soc. of Food Sci. and Nutrition* **26**, 45–53 (1997).

J. Cornee et al., "An Estimate of Nitrate, Nitrite and *N*-Nitrosodimethylamine Concentration in French Food Products of Food Groups," *Sciences des Aliments* **12**, 155–197 (1992).

T. P. Coulate, "Contaminants from Food Processing," *Food Sci. and Technol. Today* **9**, 44–49 (1995).

A. Dellisanti, G. Cerutti, and L. Airoldi, "Volatile *N*-Nitrosoamines in Selected Italian Cheeses," *Bull. Environ. Contamination and Toxicol.* **57**, 16–21 (1996).

W. Fiddler et al., "Nitrosamine Formation in Processed Hams as Related to Reformulated Elastic Rubber Netting," *J. Food Sci.* **63**, 276–278 (1998).

W. Fiddler et al., "*N*-Nitrosodibenzylamine in Boneless Hams Processed in Elastic Rubber Netting," *J. AOAC INTERNATIONAL* **80**, 353–358 (1997).

W. Fiddler et al., "Alaska Pollock (*Theragra chalcogramma*) Mince and Surimi as Partial Meat Substitutes in Frankfurters: *N*-Nitrosodimethylamine Formation," *J. Food Sci.* **58**, 62–65 (1993).

M. B. A. Gloria et al., "Influence of Nitrate Levels Added to Cheesemilk on Nitrate, Nitrite, and Volatile Nitrosamine Contents in Gruyere Cheese," *J. Agric. Food Chem.* **45**, 3577–3579 (1997).

M. B. A. Gloria, J. F. Barbour, and R. A. Scanlan, "*N*-Nitrosodimethylamine in Brazilian, U.S. Domestic and U.S. Imported Beers," *J. Agric. Food Chem.* **45**, 814–816 (1997).

M. B. A. Gloria, J. F. Barbour, and R. A. Scanlan, "Volatile Nitrosamines in Fried Bacon," *J. Agric. Food Chem.* **45**, 1816–1818 (1997).

R. Goutefongea, "Nitrosamines: Formation and Occurrence in Food Products," *Comptes Rendus de l'Academie d'Agriculture de France* **80**, 53–62 (1994).

J. S. Helmick and W. Fiddler, "Thermal Decomposition of the Rubber Vulcanization Agent, Zinc Dibenzyldithiocarbamate, and Its Potential Role in Nitrosamine Formation in Hams Processed in Elastic Nettings," *J. Agric. Food Chem.* **42**, 2541–2544 (1994).

M. Izquierdo-Pulido, J. E. Barbour, and R. A. Scanlan, "*N*-Nitrosodimethylamine in Spanish Beers," *Food Chem. and Toxicol.* **34**, 297–299 (1996).

H. Katoaka, M. Kurisu, and S. Shindoh, "Determination of Volatile *N*-Nitrosamines in Combustion Smoke Samples," *Bull. Environ. Contamination and Toxicol.* **59**, 570–576 (1997).

J. G. Kim, S. J. Lee, and N. J. Sung, "Influence of Nitrite and Ascorbic Acid on *N*-Nitrosamine Formation during Fermentation of Salted Anchovy," *J. Korean Soc. Food Sci. and Nutrition* **26**, 606–613 (1997).

D. Kuchne, "Nitrosamines in Meat Products—The Present Situation," *Mitteilungsblatt der Bundesanstalt fuer Fleischforschung Kulmbach* **34**, 220–225.

C. Lintas, G. Boccia Lombardi, and S. Nicoli, "Effect of Cooking on Availability and *In vitro* Nitrosation Precursors of Volatile *N*-Nitroso Compounds in Seafood," *Food Additives and Contaminants* **7**, 37–42 (1990).

R. N. Loeppky and C. J. Michejda, *Nitrosamines and Related N-Nitroso Compounds: Chemistry and Biochemistry*, American Chemical Society, Washington, D.C., 1994.

J. Marsden and R. Pesselman, "Nitrosamines in Food-Contact Netting: Regulatory and Analytical Challenges," *Food Technol.* **47**, 131–134 (1993).

R. C. Massey "Volatile, Nonvolatile and Total *N*-Nitroso Compounds in Bacon," *Food Additive Contamination* **8**, 585–598 (1991).

T. Mavelle, B. Bouchikhi, and G. Debry, "The Occurrence of Volatile *N*-Nitrosamines in French Foodstuffs," *Food Chem.* **42**, 321–338 (1991).

C. N. Mendoza et al., "Level and Occurrence of *N*-Nitrosodimethylamine, *N*-Nitrosodiethylamine and *N*-Nitrosopyrrolidine in Cured Meat Products," *Alimentos* **18**, 15–19 (1993).

J. W. Pensabene and W. Fiddler, "Effect of Carbohydrate Cryoprotecting Agents on the Formation of *N*-Nitrosodimethylamine in Surimi-Meat Frankfurters," *J. Food Safety* **13**, 125–131 (1993).

J. W. Pensabene, W. Fiddler, and R. A. Gates, "Nitrosamine Formation and Penetration in Hams Processed in Elastic Rubber Nettings," *J. Agric. Food Chem.* **43**, 1919–1922 (1995).

S. Raoul et al., "Rapid Solid-Phase Extraction Method for the Detection of Volatile Nitrosamines in Food," *J. Agric. Food Chem.* **45**, 4706–4713 (1997).

R. Rywotycki, "The Concentration of Nitrosamines in Pasteurized Beef Ham as Influenced by Heat Treatment and Functional Additives," *Medycyna Westerynaryina* **54**, 554–558 (1998).

R. A. Scanlan, "Volatile Nitrosamines in Foods—An Update, Food Flavors, Generation, Analysis and Process Influence," *Proc. 8th Int. Flavor Conf.: Developments in Food Science* **37**, 685–704 (1995).

R. A. Scanlan et al., "*N*-Nitrosodimethylamine in Nonfat Dry Milk," in R. N. Loeppky and C. J. Michejda, eds., *Nitrosamines and Related N-Nitroso Compounds, Chemistry and Biochemistry*, ACS Symposium Series, Washington, D.C., 1994, pp. 34–41.

R. A. Scanlan, J. E. Barbour, and C. I. Chappel, "A Survey of *N*-Nitrosodimethylamine in U.S. and Canadian Beers," *J. Agric. Food Chem.* **38**, 442–443 (1990).

N. P. Sen, P. A. Baddoo, and S. W. Seaman, "Nitrosamines in Cured Pork Products in Elastic Rubber Nettings: An Update," *Food Chem.* **47**, 387–390 (1993).

N. P. Sen et al., "Trends in the Levels of *N*-Nitrosodimethylamine in Canadian and Imported Beers," *J. Agric. Food Chem.* **44**, 1498–1501 (1996).

F. Shahidi, J. Synowiecki, and N. P. Sen, "*N*-Nitrosamines in Nitrite-Cured Chicken-Seal Salami," *J. Food Prot.* **58**, 446–448 (1995).

Nak Ju Sung, Jung Hye Shin, and Soo Jung Lee, "*N*-Nitrosamine in Korean Beer," *J. Food Sci. Nutrition* **1**, 6–9 (1996).

B. A. Tomkins and W. H. Griest, "Determinations of *N*-Nitrosodimethylamine at Part-Per-Trillion Concentrations in Contaminated Groundwaters and Drinking Waters Featuring Carbon-Based Extraction Disks," *Analytical Chem.* **68**, 2533–2540 (1996).

A. R. Tricker and S. J. Kubacki, "Review of the Occurrence and Formation on Nonvolatile *N*-Nitroso Compounds in Foods," *Food Additives and Contaminants* **9**, 39–69 (1992).

A. B. Tricker et al., "Mean Daily Intake of *N*-Nitrosoamines from Foods and Beverages in West Germany in 1989–1990," *Food and Chemical Toxicol.* **29**, 729–732 (1991).

C. Trova, G. Gandolfo, and G. Cossa, "*N*-Nitrosodimethylamine Levels in Beer," *Industrie delle Bevande* **22**, 528–530 (1993).

I. Vittozi, "Toxicology of Nitrates and Nitrites," *Food Additives and Contaminants* **9**, 579–585 (1992).

E. Wiegler, H. Kolb, and C. Ruehl, "Studies on the Nitrosamine Content of Pizza and Toast Products," *Fleischwirtschaft* **74**, 1296–1298 (1994).

C. P. Wild and R. Montesano, "Immunological Quantitation of Human Exposure to Aflatoxins and *N*-Nitrosamines," American Chemical Society Series 451, Washington, D.C., 1991, pp. 215–228.

A. R. THICKER
Institute for Toxicology and Chemotherapy
Heidelberg, Germany

See also TOXICANTS, NATURAL.

NUCLEOTIDES

Proteins and nucleic acids function as the key components to maintain life for plants and animals in terms of genetic heritage, self-duplication of genetic information, and regulation of metabolites for growth. It is, moreover, interesting that the components of these proteins and nucleic acids—such as L-glutamate in amino acids and 5′-ribonucleotides in nucleic acids—play a key role in making meals, which are indispensable for human life, very tasteful and joyful.

In the long history of humankind, umami was only first added to the restricted idea of four "basic" or primary tastes by Ikeda (1). He extracted monosodium L-glutamic acid from seaweed and identified the taste-active ingredient obtained as umami.

Not only must we consider the unique taste of L-glutamate, but also the remarkable effect of certain 5′-ribonucleotides in causing a synergistic taste effects in combination with L-glutamate. The culinary practices of Japanese cuisine have long taken advantage of the taste effects of certain combinations of natural products used as condiments, especially the seaweed sea tangle (*Laminaria*), in combination with either black mushroom (*shiitake*) or dried bonito (*katsuo-bushi*). The taste-active ingredients of black mushroom and bonito were identified as inosine 5′-monophosphate by Kodama (2) and guanosine 5′-monophosphate by Kuninaka (3), respectively. Kuninaka (4) discovered that monosodium glutamate (MSG) and certain 5′-ribonucleotides added together to foods have a synergistic effect in enhancing food flavor.

Nucleotides are now used not only as umami flavor enhances but also as the raw material of pharmaceuticals. Industrial production of nucleotides started in Japan in 1961, using the RNA degradation method. Nowadays, nucleotides are mostly produced by a fermentation method using microorganisms, which was developed mostly in Japan. The main purposes are to produce two umami flavor enhancers inosine 5′-monophosphate (5′-IMP) and guanosine 5′-monophosphate (5′-GMP).

SYNERGISM OF 5′-NUCLEOTIDES FOR UMAMI TASTE

When glutamic acid and 5′-nucleotides (free forms or salts) coexist, umami is dramatically enhanced. This phenomenon, well known as the synergistic effect of umami substances, has already been investigated and reported (5) mainly with regard to human taste sensations.

Ikeda (6) confirmed the synergistic effect in actual food. He conducted a discrimination test of MSG and IMP in a large variety of stocks or broth and compared the results with those of MSG and IMP in simple aqueous solutions. The thresholds of MSG and IMP were reported to be 0.03 and 0.025%, respectively. The nucleotide and glutamic acid contents in some of the stocks used are popular and commonly used in cooking to enhance the palatability of foods. Although the contents of umami substances are surprisingly low, these stocks are effective in foods apparently because of the synergism among the umami substances they contain.

The detection thresholds for five taste substances in single aqueous solutions and in 5 mM MSG (0.094%) or IMP (0.26%) solution are shown in Table 1. The presence of MSG or IMP did not lower the thresholds of the four basic tastes, although IMP raised that of quinine sulfate, and MSG and IMP raised that of tartaric acid. This means that umami does not improve sensitivity toward the four basic tastes. Only IMP remarkably lowered the threshold of MSG by more than 50-fold. This result suggests that umami neither enhances the four tastes nor is itself enhanced by the basic tastes. Only the synergistic effect between the two kinds of umami substances enhances the umami taste. Electrophysiological studies on the synergism have been carried out using the cat, rat, and hamster (7,8). In these studies, the synergism was examined under conditions in which both L-glutamate and 5′-ribonucleotides induced the taste responses. Hence, quantitative analysis of the data obtained was rather difficult, and it was not known by what mechanism the synergism occurs.

The recording in Figure 1 shows the responses of the rat chorda tympani nerve to 20 mM MSG in the absence and the presence of 0.1 mM 5′-GMP. As seen from the baseline of the recording, the response to 0.1 mM GMP itself is very small. The response to MSG is so greatly enhanced by the presence of 5′-nucleotides that it may derive mostly from the synergism between two substances.

The effect of various species of nucleotides on the response to L-glutamate is shown in Table 2. 5′-Nucleotides with a purine base, such as GMP, deoxy-GMP, guanosine diphosphate (GDP), guanosine triphosphate (GTP), adenosine 5′-monophosphate (AMP), and IMP, have the ability to enhance the response to L-glutamate, whereas prymidine nucleotides carrying the pyrimidine base, such as uridine 5′-monophosphate (UMP) and cytidine 5′-monophosphate (CMP), have only a small or no enhancing effect. Cyclic GMP, guanosine, and sodium phosphate also have only a small or no enhancing effect.

The preceding results demonstrate that only the nucleotides having a specific structure exhibit a synergistic effect on the responses to L-glutamate. Whether the monophosphate, diphosphate, or triphosphate is used does not affect the synergistic effect. Moreover, the nucleotides with a deoxy sugar have a small amount of synergistic effect.

Behavioral and electrophysiological studies on several mammalian species indicated that L-glutamate is a taste stimulus and that the synergism with 5′-ribonucleotides occurs in several species of mammals (9). Feeding studies have shown that weanling calves respond to L-glutamate positively, as indicated by their higher consumption of the diet with 0.2% MSG. Torii and Cagan (10) demonstrated that in the presence of low levels of certain 5′-ribonucleotides, the binding activity of the taste receptor prepared from bovine circumvallea papillae (containing approximately 90% of the tastebuds) for L-glutamate increased by severalfold, as shown in Figure 2. There was no effect on the low level of binding to the control epithelium. This was the first demonstration that the taste synergism of glutamate and 5′-ribonucleotide is a peripheral biochemical event, occurring at the level of the taste receptor membrane, and does not require neural integration. The specificity of the effect with respect to the nucleotide was similar, though not identical, with that for humans. In humans, the order of effectiveness is GMP > IMP > XMP, and AMP

Table 1. Detection Thresholds of Five Taste Substances

	Sucrose	NaCl	Tartaric acid	Quinine	MSG
Water	0.086	0.0037	0.00094	0.000049	0.012
5 mM MSG solution	0.086	0.0037	0.0019	0.000049	—
5 mM IMP solution	0.086	0.0037	0.03	0.0002	0.00019

Note: Expressed in percent (w/v).

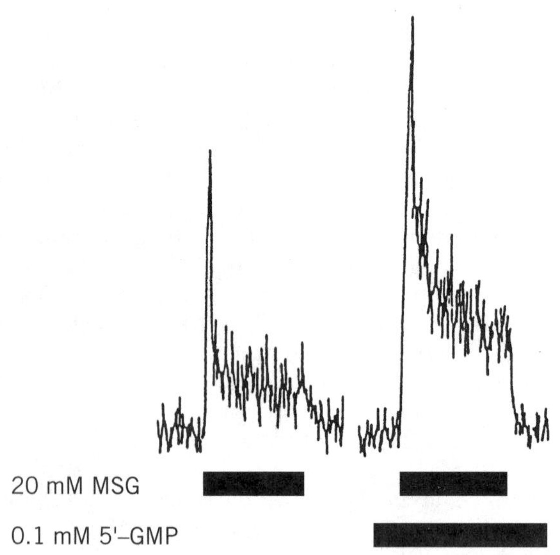

Figure 1. The summated responses of the rat chorda tympani nerve to 20 mM MSG in the absence and the presence of 0.1 mM 5′-GMP. The bars at the base of the recordings represent the duration of the application of the chemical stimuli.

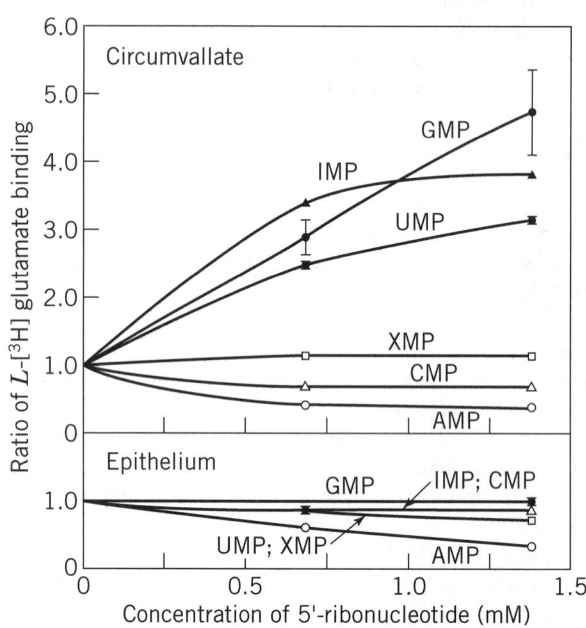

Figure 2. Effects of 5′-ribonucleotides on binding of L-glutamate to bovine taste receptors.

Table 2. Relative Effect of Various Nucleotides, Guanosine, and Sodium Phosphate on Rat Taste Nerve Response to 10 mM of Glutamate (taken as 100)

Nucleotide	Magnitude of responses
GMP	232 ± 13
dGMP	260 ± 24
GDP	192 ± 6
GTP	188 ± 22
cGMP	126 ± 31
Guanosine	109 ± 9
AMP	255 ± 27
IMP	171 ± 26
UMP	125 ± 25
CMP	124 ± 16
Sodium phosphate	108 ± 9

Note: Concentration, 10 mM; and pH 6.0.

is ineffective (11). The pyrimidine nucleotides are reportedly not effective (3). With the bovine system, the order of effectiveness was GMP ≥ IMP > UMP, and XMP, CMP, and AMP were effective (Fig. 2). Additional structure–activity considerations were studied by examining the effects of the free bases, the nucleoside, and the di- and triphosphates of GMP and AMP. The data in Figure 3 show that only 5′-GMP was effective in increasing binding of L-glutamate. The regions of the molecule that appear to be critical in the interaction with the regulatory site are shown in Figure 4. The keto function on the aromatic ring appears critical. With an amino group at this position, there is no synergistic activity. The data from human studies show that 5′-GMP and 5′-IMP are potent synergistically, 5′-XMP is less effective, but 5′-UMP is not effective. The binding studies agree with the earlier human psychophysical results showing the importance of the 4- or 6-position on the ring, the 2-position on the ring, and the 5′-monophosphate group on the ribose. There are two points of difference in the bovine and human data: (*1*) XMP did not enhance binding but is synergistic in humans, and (*2*) UMP increased binding but is not effective in humans.

The characteristics of the chemical structure of nucleotides attributed to the taste of umami for humans are summarized as follows: (*1*) a purine base on which the OH-bond is located at the 6-position carbon, and (*2*) a pentose on which the phosphate bond locates at 5′-position of the OH-bond.

When the total concentration of L-glutamate and IMP is kept constant (0.015 g/dL), the intensity of umami increases with the increasing ratio of IMP and goes to maximum when the ratio of IMP is 50%; it then decreases with the increasing ratio of IMP, which is shown in Figure 5.

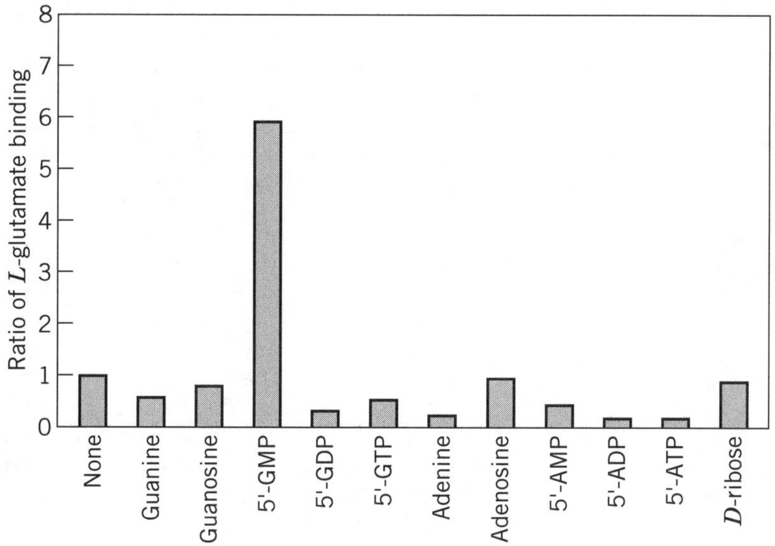

Figure 3. Effect of 5′-ribonucleotides, nucleosides, and free bases on binding of L-glutamate to bovine taste receptors.

IMP GMP UMP

XMP AMP CMP

Figure 4. Structure of 5′-ribonucleotides investigated for synergistic biochemical effect with L-glutamate.

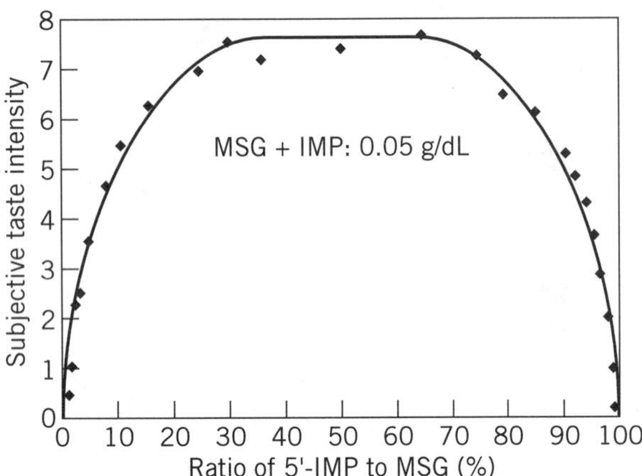

Figure 5. The synergistic effect shown as the relative intensity of umami taste to MSG, changing the ratio of 5′-IMP to MSG while the total concentration of MSG + 5′-IMP remained constant as 0.05 g/dL.

This intensity of umami of L-glutamate when IMP coexists is expressed by the following relation (5):

$$y = \mu + 1{,}200\mu\ v$$

where y is the concentration of L-glutamate (%) that shows the same intensity of umami as that shown by the mixture of L-glutamate and IMP; μ is the concentration of L-glutamate in mixture (%); and v is the concentration of IMP in mixture (%), respectively.

When using 5′-guanylic acid (GMP) in place of IMP in the mixture, the relation of intensity of umami was summarized as follows:

$$y = \mu + 2{,}800\mu\ v'$$

where v' is the concentration of GMP in mixture (%).

The concentration–response relationship for L-glutamate does not follow the simple taste equation obtained under the assumption that there exists only one type of receptor site for L-glutamate. Assuming that two types of receptor sites exist—with different dissociation constants for L-glutamate—in rat taste cell membrane we used the following equation (12):

$$R = R_H/K_H/S + 1 + R_L/K_L/S + 1$$

Here, R is the magnitude of the response; R_H and R_L are the magnitudes of the maximal responses brought about by occupation of high- and low-affinity sites, respectively; and S is the concentration. The best-fitted curves for the data were obtained with a computer utilizing a program for the linear square iterative fitting of the preceding equation. The parameters for the curves were determined as follows:

	L-glutamate alone	L-glutamate + 0.1 mM GMP
K_H (mole)	0.01	0.0015
K_L (mole)	0.32	0.17

The values of R_H = 1.18, and R_L = 15.59 were unchanged by the presence of 0.1 mM GMP.

There seem to be specific binding sites for purine-based 5′-nucleotides in the taste receptor membranes, because only the specific nucleotides have the ability to show the synergism. The binding sites for the nucleotides seem to be part of or closely associated with L-glutamate receptors.

STRUCTURE OF NUCLEIC ACIDS

The intrinsic parts of nucleic acids are more complex, being threefold aggregates of purine and pyrimidine bases, sugar, and phosphoric acid. The number of "building blocks," however, is much smaller, being six altogether, and in the great majority of cases, only four. Depending on the characteristic of the sugar and on the presence of uracil in place of thymine, two kinds of nucleic acid—ribonucleic acid (RNA) and deoxynucleic acid (DNA)—are defined.

The component unit, which can be thought of as analogous to an amino acid in a peptide chain, is

base—sugar—phosphate

and the linkage is by the phosphate to two sugars in a stepwise pattern, as follows:

```
—sugar—phosphate—sugar—phosphate—
   |               |
  base            base
```

A combination between a base and a sugar (ribose or deoxyribose) is called a nucleoside. When this is combined with a phosphate, it becomes a nucleotide. The important separate bases, six in number, are listed in Table 3, with their structural formulas. These, together with their sugar and phosphate complements, are linked into enormous molecules having molecular weights in the millions. The structure of nucleic acids is becoming understood and bringing with it an advance in molecular biology. The actual coupling among base sugar, and phosphate is shown in terms of the structural formula in Figure 6. This coupling permits the formation of a simplified double helix.

THE STRUCTURE OF NUCLEIC ACID POLYMERS

DNA

A firm knowledge of the structure of DNA is of crucial importance. The gross dimensions of DNA were found by light-scattering or viscosity measurements, which indicate that DNA is a long, thin, relatively rigid structure whose molecular weight is ~6×10^6 and density is 1.6 to 1.75 g/cm^3. Chemical evidence indicates that the phosphate (PO_4^-) groups are very reactive, but the amino ($-NH_2$) and CO groups are not readily titratable. The analysis of Wyatt and Chargaff (13) indicated that the ratio of thymine to adenine is unity, and the ratio of cytosine or its derivatives to guanine is also about unity. The absorption of polarized infrared light indicates that the -NH_2 and -C=O groups of the bases are oriented perpendicular to the molecular axis.

Table 3. Bases Involved in Nucleic Acids

Pyrimidines	*Purines*
Cytosine	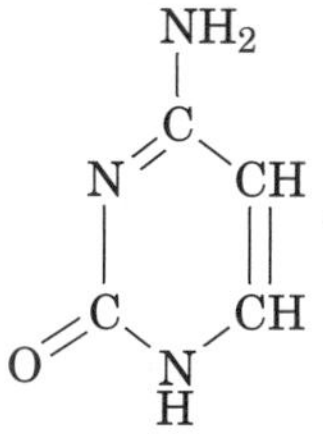 Adenine 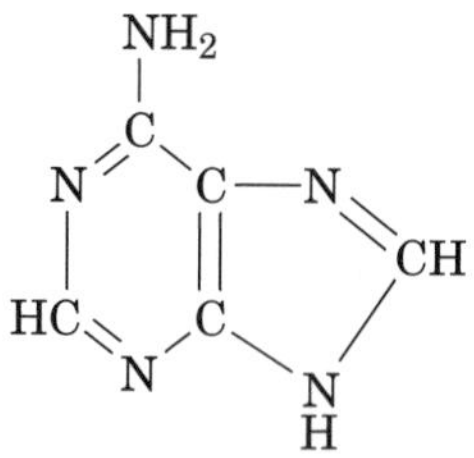
Alpha hydroxymethyl cytosine (found in virus DNA in some cases)	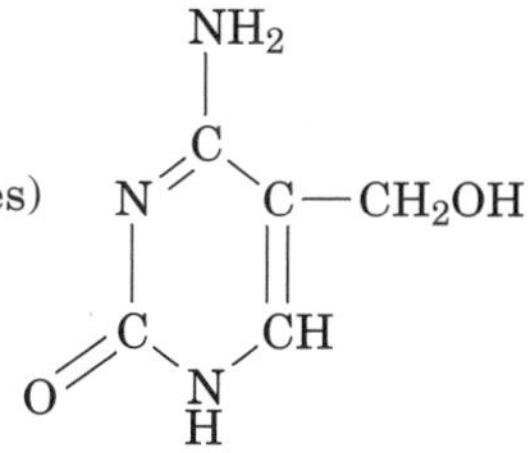 Guanine 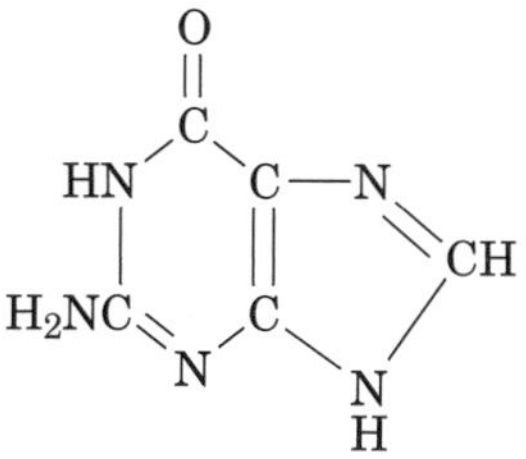
Uracil (in RNA only)	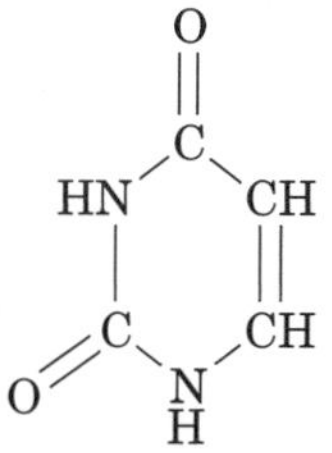 *Sugars* Ribose 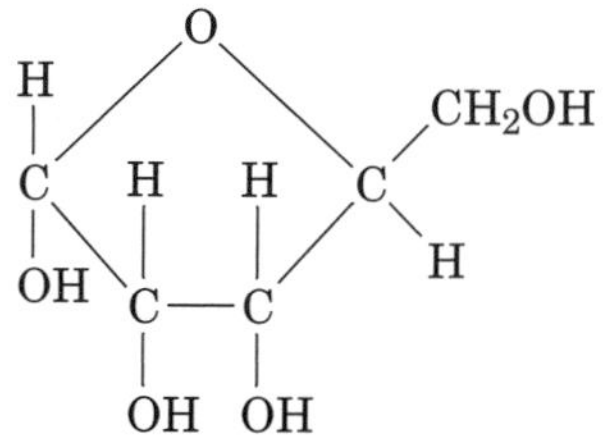
Thymine (in DNA)	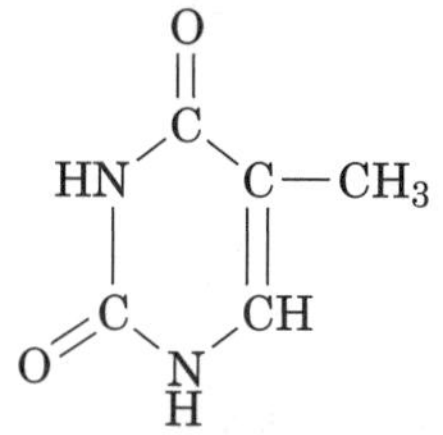 Deoxyribose 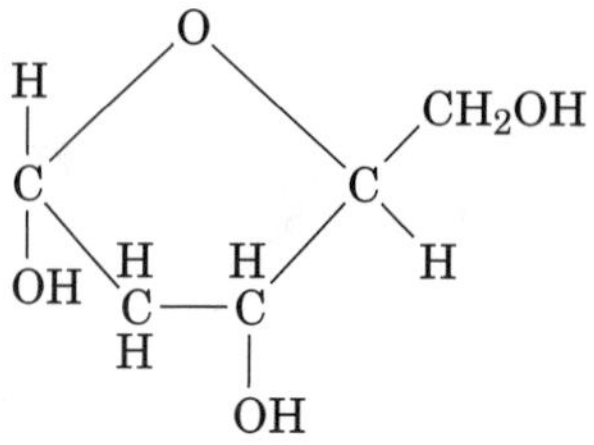

The P=O groups, on the other hand, are inclined an angle of ~55° to the axis. X-ray diagrams indicated that the planes of the bases were separated by a distance of 3.4 Å. Impressive advances in knowledge have come from the recognition by Watson and Crick (14,15) of the specific base pairing in DNA. Helical DNA has a pitch of 34 Å with 10 repeat units per turn, and its fundamental structure is not one helix but two right-handed helices of radius about 8.5 Å, spaced about one-half the pitch apart. The pairs of purines and pyrimidines that occur in equal amounts in DNA, namely, thymine–adenine and cytosine–guanine, have complementary structures that easily form hydrogen bonds. Watson and Crick found that the bases would be in a proper position for strong hydrogen bonds if the two helical chains had sequences of atoms in opposite directions. The helical chains are the phosphate–sugar backbones. The bases extend toward the center of the helix and hydrogen bond across the axis of the double helix. Thus the DNA spirals are held together by interchain bonding, whereas protein helices keep their configuration by means of intrachain hydrogen bonds.

The double spiral for the structure of DNA has very practical properties concerned with information storage and the duplication of genetic material.

RNA

The primary structure of RNA is similar to that of DNA, but with a few notable exceptions. First, in RNA, instead of thymine, the pyrimidine base uracil occurs, forming a complementary base pair with adenine in regions of double-stranded RNA. Also, a wide variety of ribonucleotides having modified or minor bases are found in naturally occurring RNA, one of the commonest of which is pseudouridine. In human tRNAs, as many as 25% of the bases are nonstandard. Although the role of base modification is not clear, it may be important for biological recognition. The other important feature of the primary structure of RNA is the presence of the 2′-hydroxyl group in ribose. Although this hydroxyl group is never involved in phosphodiester linkages, it does impose restriction on the helical conformations accessible to double-stranded RNA.

RNAs are usually single-stranded molecules that are stable allowing different regions of the ribonucleotide to form distinct secondary structural elements. When self-complementary regions of the RNA strand are aligned, duplex regions, which may have Watson–Crick base pairs, are formed.

RNA has a variety of functions within a cell; for each function, a specific type of RNA is required. Messenger

Figure 6. The detail of coupling between base, sugar, and phosphate in deoxyribonucleic acid.

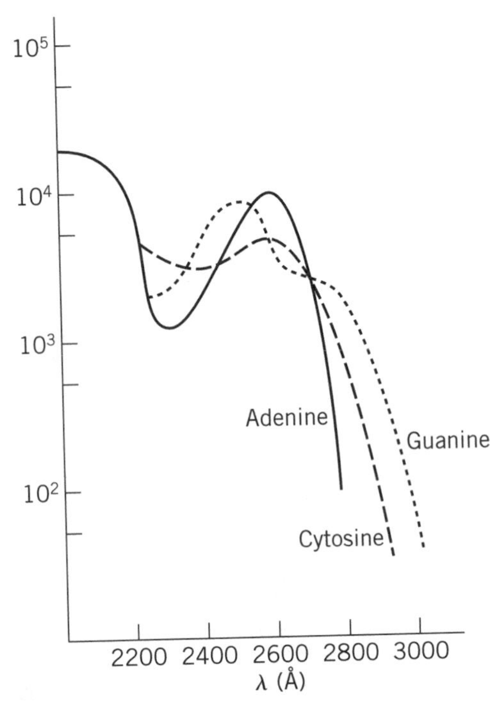

Figure 7. Absorption UV spectra of nucleoside that constitutes DNA at neutral pH in the dry state.

RNA (mRNA) serves as an intermediary for carrying genetic messengers from the DNA to the ribosome where protein synthesis takes place. Ribosomal RNA (rRNA) serves both structural and functional roles in the ribosome; it is diverse, both in terms of its size and structure. Transfer RNA (tRNA) is a group of small molecules that have a central role in protein synthesis. The multifunctional character of RNA, particularly the involvement of RNA in enzymatic processes, had led to the hypothesis that life on earth evolved from RNA, and that RNA had both the genetic and catalytic functions commonly associated with DNA and protein, respectively.

The functional diversity of RNA is directly related to its structural diversity. In contrast to DNA, RNA molecules are synthesized as single-stranded polynucleotides that fold to give complex tertiary structures. These structures, which incorporate hairpins, loops, bulges, and junctions between single-stranded and double-stranded regions, exhibit long-range interactions within the folded tertiary structure.

OPTICAL PROPERTIES OF NUCLEIC ACID

Ultraviolet Absorption

Figure 7 shows the absorption spectra of some bases of nucleic acids. From this figure one can see that the absorption spectra for nucleic acids would have a maximum at 260 μm. The optical densities shown in the figure were calculated from data obtained from thinner and more dilute solutions on the assumption that the Beer–Lambert law holds. At the longer ultraviolet wavelengths, nucleic acids will have about 20 times the optical density of an equal mass concentration of protein. Most of the amino acids are nonaromatic and contribute practically nothing to the protein absorption spectrum at wavelengths greater than 210 μm. The high and characteristic absorption by nucleic acids makes it relatively simple to detect them by ultraviolet spectroscopy and to localize them in cells by ultraviolet microscopes.

In Table 4, the wavelengths of maximum absorbency and the intensity (ϵ) for various bases, nucleosides, and nucleotides are summarized, which are tentative values compiled by the National Academy of Science.

Infrared Absorption Spectra

Infrared (IR) absorption methods are used mainly for two different purposes: one is as a “fingerprint” to identify a particular compound and to detect it in the mixture, the other is for structural studies. The logical approach to a spectroscopic study of the nucleic acids is to examine separately the base, sugar, and phosphate moieties; their combination in nucleosides and nucleotides; and their comparison with nucleic acids. The approximate vibrational frequency ranges for the components of functional groups related to nucleic acids are shown in Table 5.

Bases and Nucleosides

The fundamental modes associated with base residues occur throughout the IR spectrum and thus overlap bands

Table 4. The Maximum Absorbance Intensity (ϵ) for Various Nucleic Acids

Nucleic acid	The maximum absorbance			Ratio of observance between two wavelengths	
	Intensity ($\epsilon \times 10^3$)	Wavelength (m μ)		250/260	280/260
Adenine	13.1	262	0.1 N HCl	0.76	0.38
Adenosine	15.4	259	pH 7	0.98	0.15
5′-AMP	15.4	259	pH7	0.80	0.15
Cytosine	10.4	275	0.1 N HCl	0.48	1.50
Cytidine	13.0	280	0.1 N HCl	0.45	2.10
5′-CMP	13.0	280	pH2	0.45	2.10
Guanine	14.4	248	0.1 N HCl	1.37	0.84
Guanosine	13.6	252.5	pH 7	1.15	0.67
5′-GMP	13.7	252	pH 7	1.16	0.66
Hypoxanthine	10.6	250	pH 7	1.32	0.09
Inosine	12.2	248.5	pH 6	1.73	0.24
Uracil	8.2	260	pH 7	0.83	0.20
Uridine	10.1	262	pH 7	0.74	0.36
5′-UMP	10.0	262	pH 7	0.74	0.38
Thymine	7.89	264.5	pH 7	0.67	0.53
Xanthosine	11.4	248.5	pH 8	1.31	1.13

Table 5. Approximate Frequency Ranges for Characteristic Vibrations of Nucleic Acids and Related Compounds (infrared)

Frequency (cm^{-1})	Mode	Origin of vibration
2800–3500	ν OH	Water, sugar
	ν NH	Base residue
	ν CH	Sugar and base residue
1800–1500	ν C=O	Base residues, mixed bodies
	ν C=N	
	ν C=C	
	δ NH	
	δ HOH	Water
1250–950	νPO_2^-	Antisymmetric stretch
	νPO_2^-	Symmetric stretch
	ν CO	Sugar
1000–700	ν PO	Phosphate
	ν CO	Sugar
	τ NH	Base residue

Note: Modes: ν, stretching; δ, in-plane bending; τ, out-of-plane bending.

from the other entities in nucleic acids. Between 1500 and 1750 cm^{-1} several strong bands occur that are free from interference by phosphate and sugar groups and are related to the structure of the purine and pyrimidine rings in nucleosides and nucleotides, and nucleic acids. These bands also arise from strongly coupled vibration of the C=O (1725 cm^{-1} stretch), C=N (1600 cm^{-1} stretch), and C=C types with in-plane NH (1680 cm^{-1} bend) deformation.

On the basis of protonation studies of the base–sugar, the following are concluded (16):

1. The acidic form of adenosine has a proton attached to one of the ring nitrogens, not to $-NH_2$.
2. Guanosine exists in acidic, neutral, and alkali forms, with two possible neutral forms. Each form has a different IR spectrum. The alkali form is deprotonated in position 1, and the acidic form in position 7.
3. Deprotonation of inosine also causes π-electron delocalization with concomitant decrease in C=O stretching frequency from 1675 cm^{-1} to 1595 cm^{-1}.

Phosphate Group

Linking the sugars in the nucleotide backbone, the phosphate diester linkage can exist in two different states, protonated and charged:

$$\mathrm{RO{-}\overset{\overset{OR'}{|}}{\underset{\underset{O}{\|}}{P}}{-}OH} \leftrightarrow \left[\mathrm{RO{-}\overset{\overset{OR'}{|}}{\underset{\underset{O}{|}}{P}}{-}O}\right] + H+$$

In the IR spectrum the P-O modes are considerably stronger than those for the P-O-R linkage: 1230 cm^{-1} for PO_2^- asymmetric stretch, 1080 cm^{-1} for PO_2-symmetric stretch, and 980 cm^{-1} for PO_3^{2-}.

Sugar Moiety

The most characteristic group of bands occurs in a broad distribution in the range of 1000 to 1100 cm^{-1} and arises from C-O-C and C-C(OH)-C modes. The deoxy derivatives exhibit a sharp band at 975 cm^{-1}, which is absent in the ridonucleosides. There is some evidence that DNAs and RNAs may be distinguished in this manner.

DNA and RNA

As expected, the spectra of DNA and RNA show features in common with their simpler analogs. In the 1500 to 1750 cm^{-1} region there are broad, poorly resolved bands corresponding to the complex ring modes. The strong phosphate (asymmetric) mode is in the 1240 cm^{-1} region, and the broad and complex band is in the 1000 to 1100 cm^{-1} region for the sugar vibrations. The strong band in the 3400 cm^{-1} region, which contains the NH stretching mode, is influenced by the presence of water. One of the major difficulties of studying even solid-state nucleotides is that water almost inevitably crystallizes with the nucleotides.

Dichroic Ratios

The measurement of dichroic ratios of biological systems to determine their structure requires polarized light and oriented structures. Polarized IR light may be produced by reflection from or transmission through films of selenium 4 μm thick, set at the polarizing angle to the incident light. This technique has been applied to the analysis of nucleic acid structure (DNA). The position of the N-H and O-H frequencies indicates that these groups participate in hydrogen bonds. The dichroism of the N-H, O-H, and C=O groups indicates that the hydrogen bonds are preferentially oriented perpendicular to the molecular axis, and so they probably represent intermolecular regions of attraction. The P-O-C groups also have a component perpendicular to the molecular axis, but the absence of any dichroism in the P=O absorption band at 1250 cm^{-1} implies that they are oriented at angles of 55° to the molecular axis, which is in good agreement with the structural model of DNA discussed previously.

CONTENTS OF NUCLEOTIDES IN THE INGREDIENTS OF FOODS

Biochemical Character of Muscles

The greater part of muscle fiber is made up of a set of specialized protein molecules that are in some way concerned with the process of muscular action. There is far less nucleic acid and very little ribonucleic acid in muscle tissue. The specialized molecules constitute about two-thirds of the protein, while the remaining one-third is concerned with the enzymatic use of the substances brought in by the blood capillaries that surround the fiber in the whole muscle.

In 1939 Engelhaldt and Ljubimowa discovered that myosin is an enzyme that catalyzes the conversion of ATP to the diphosphate ADP, with consequent release, or transfer, of configuration energy. More thorough studies by Banga and Szent-Gyorgyi (17) showed that two proteins were involved in myosin, namely, actin and myosin, which form a combination known as actomyosin, a combination capable of changing shape in a manner very reminiscent of the behavior of muscle itself. Additional studies have brought out that its molecular weight is 450,000, with an axial ratio of about 50 and a total length of 1600 Å. That of actin is 60,000, with an axial ratio of 12 and a length of 290 Å.

In the biochemistry of muscle, ATP, whose general structure is a triphosphate of adenine and ribose, is characterized by a phosphate grouping that, on hydrolysis, can yield a diphosphate and orthophosphoric acid, resulting in a more stable total configuration of the same atoms. This increase in stability means that energy is released when the process takes place.

Fish Muscle

In the muscle of fish, nucleotides such as adenosine ATP, ADP, AMP, and IMP have been detected. The changes in these nucleotides in muscle between the conditions of *at rest* and *in fatigue* are shown in Table 6, which summarizes the analysis of the codfish (18). In the condition of *at rest*, a large amount of ATP and quite a bit of adenosine and inosinic acid were found. When *in fatigue*, conversely, contents of inosinic acid increased. As already mentioned, the muscle released energy in the active state by the release of phosphate bonds from ATP, converting it into ADP, then AMP. This AMP is then changed to IMP by the aid of the enzyme deaminase. The procedure is as follows:

$$\text{ATP} \rightarrow \text{ADP} + \text{Pi by ATPase}$$

$$\text{ADP} \rightarrow \text{AMP} + \text{Pi by myokinase}$$

$$\text{AMP} \rightarrow \text{IMP} + \text{NH3 by deaminase}$$

The autodigestion process occurs even after the death of animals; the change of ATP to AMP proceeds, and finally IMP is accumulated in the muscles. Therefore, meats in fatigue contain a high level of ATP in muscles so that, even after death, no conversion to IMP, yet occurred the taste of this supposed to be short of umami.

Fish meat after death contains mainly IMP, followed by inosine, which is decomposed to inosinic acid by the release of phosphoric acid; quite a bit of ADP and AMP remain. But *Crustacea* and *Mollusca* such as robster, crab, octopus, oysters, and cuttlefish kept in good condition for use as fresh meat have, on the contrary, a large amount of ADP or AMP and a trace of IMP because of a lack of the enzyme adenylic acid deaminase. Hence, there are two pathways of decomposition of adenylic acid in fish, shown as follows:

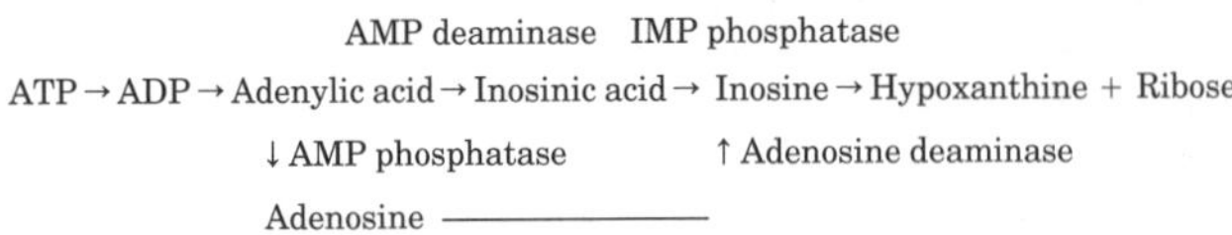

For *Crustacea* and *Mollusca*, adenylic acid is converted by AMP phosphatase in place of AMP daminase into adeno-

Table 6. Nucleotide Contents of Rested and Exhausted Codfish Muscle

	Content in muscle (μmoles/g)	
Nucleotide	Rested	Exhausted
DPN	0.107	0.338
TPN	0.0008	0.0005
AMP	0.69	0.57
ADP	0.576	0.426
ATP	5.32	0.26
IMP	1.26	5.86
GTP	0.0025	0
UTP	0.0036	0

Note: Values are mean determinations on six fish. DNP, diphosphopyridine nucleotoide; TPN, triphosphopyridine nucleotide; UTP; uridine 5′-triphosphate.

sine and then converted by adenosine deaminase into inosine. So far, in the *Mollusca*, AMP deaminase is not found; therefore, no inosinic acid is produced. Some kinds of *Crustacea* retain AMP deaminase in muscles; after killing, in a small amount of inosinic acid is detected in *Crustacea*.

As shown in Table 7, the IMP in fish meat is in the range of 100 to 200 mg%, and is regarded as the prime ingredient for taste of umami with L-glutamate. Base components constituted of nucleotides such as hypoxanthine, adenine, and guanine are found in the muscles of aquatic animals; in particular, a large amount of hypoxanthine is sometimes detected in many fish meals.

Meat

The contents of 5′-nucleotides in the meat of animals are summarized in Table 8. 5′-Inosinic acid is a prime ingredient in 5′-nucleotide for every animal and is found at a level of 100 mg% or so. The changes in the 5′-nucleotides during the maturing process have been investigated. Chicken (Fig. 8) has an active adenylic acid deaminase; in the case of storage in a refrigerator at 4°C, the concentration of inosinic acid in meat reached a maximum of 200 to 280 mg% in 4 to 10 h after killing, through the process of ATP → ADP → adenylic acid → inosinic acid conversion. Then after further storage in a refrigerator, inosinic acid began to decompose gradually to hypoxanthine, through inosine, and decreased to a half the original amount after 5 to 6 days. For pork, 3 days after slaughter the contents of inosinic acid reached a maximum, 100 to 170 mg%, and then gradually decomposed to hypoxanthine; after 10 days kept in a refrigerator, the contents of inosinic acid remained at 60 mg%, which proceeded through the conversion process inosinic acid → inosine → hypoxanthine, as well. In storage, no other 5′-nucleotides such as GMP and CMP were detected.

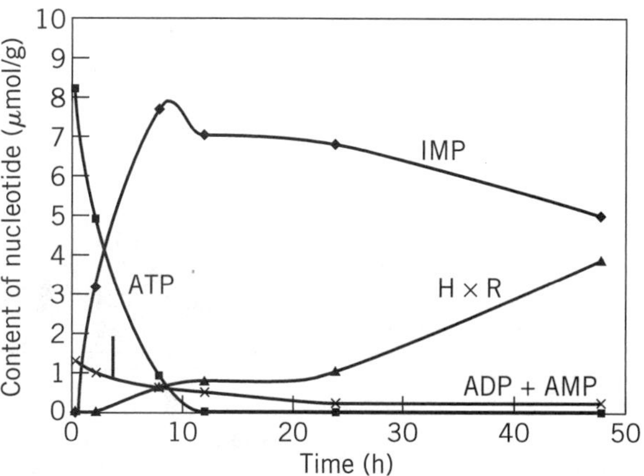

Figure 8. Change in the concentration of nucleic acid–related substances in chicken muscle after slaughter when kept at 4°C.

Table 7. Nucleotide Contents of Fish Meat

Fish	5′-AMP	5′-IMP
Saurel	6.4	212.6
Ayufish	7.2	189.0
Sea bass	8.4	124.9
Sardine	0.7	188.7
Sea bream	11.0	277.1
Saury	6.7	149.5
Mackerel	5.7	188.1
Salmon	6.9	154.5
Tuna	5.2	188.0
Swellfish	5.6	188.7
Eel	17.6	108.6
Sagittated calamari	163.2	0
Octopus	23.3	0
Lobster	72.5	0
Crab	10.1	0
Squilla	32.6	17.4
Ormer	71.7	0
Traugh shell	86.5	0
Scallop	102.3	0
Short-necked clam	10.8	0

Note: Content, mg %.

Table 8. Nucleotide Contents of Meat

Meat	5′-AMP	5′-IMP	5′-GMP	5′-UMP	5′-CMP
Beef	6.6	106.9	2.2	1.6	1.0
Pork	7.6	122.2	2.5	1.6	1.9
Chicken	11.5	75.6	1.5	1.3	2.6
Whale	2.1	214.5	3.6	1.9	—

Note: Content, mg %.

Vegetables and Mushrooms

In vegetables, a small amount of 5′-adenylic acid (10.4 mg% for tomato; 8.4 mg% for green soybeans; 6.5 mg% for corn), 5′-uridylic acid, and 5′-cytidylic acid were detected, while 5′-inosinic acid and 5′-guanylic acid were below the limit of measurement, as shown in Table 9. The taste of umami of vegetables mainly derives from the combination of L-glutamate and 5′-adenylic acid, which shows a weak synergistic effect with L-glutamate.

In mushrooms, as shown in Table 10, the order of qualitative amounts of nucleotides are as follows: 5′-adenylic acid > 5′-guanylic acid > 5′-uridylic acid, and 5′-inosinic acid was not detected. The 5′-guanylic acid with L-glutamate is regarded as the main umami component in mushrooms.

Table 9. Nucleotide Contents of Vegetables

Vegetable	5′-AMP	5′-IMP	5′-GMP	5′-UMP	5′-CMP
Asparagus	3.8	—	±	1.9	1.9
Welsh onion	0.9	—	—	0.4	—
Lettuce	0.9	±	±	0.5	—
Tomato	10.4	—	—	2.2	0.5
Field pea	1.9	—	—	1.3	±
Cucumber	0.5	—	—	0.6	±
Japanese radish	1.3	±	—	1.4	±
Onion	0.8	±	—	0.5	±
Bamboo shoot	1.1	—	—	1.3	0.5

Note: Content, mg %.

Table 10. Nucleotide Contents of Mushrooms

Mushroom	5′-AMP	5′-IMP	5′-GMP	5′-UMP	5′-CMP
Shiitake, *Cortinellus shiitake*	154.9	0	70.1	37.6	29.4
Dried shiitake	106.9	0	146.7	111.2	55.6
Mushuroom	11.3	0	—	6.4	—
Dried mushroom	167.7	0	—	59.7	—
Matsutake, *Armillarid matsutake*	99.3	0	64.6	65.2	35.6
Truffle	13.9	0	5.8	6.8	4.2
Hatsutake, *Latarius hatsutake*	51.4	0	58.1	41.5	7.4
Fly agaric, *Amanitamuscavia*	0.2	0	0	1.6	0.3

Note: Content: mg %.

PROPERTIES OF NUCLEIC ACIDS

Chemical Properties

Deamination Reaction. This reaction is the most important one with regard to production of the seasoning; it converts the amino bond to a hydroxyl bond on the purine base via nitrous acid, which may occur at any stage in nucleotides, nucleosides, and bases. Inosinic acid is produced by this process from adenylic acid (19).

Acylation Reaction. The hydroxyl bond on the sugar site of nucleosides has the ability of acylation. The order of the reaction ability is as follows: 5′-site > 3′-site > 2′-site.

When reacted with $POCl_3$ a product containing a mixture of three kinds of monophosphate esters was obtained; although the ratio of these products changes with the mole-ratio of water, no selective condition for 5′-phosphate was found.

Hydrolysis by Acid or Alkali. Both pyrimidine bases and purine bases are relatively resistant to the attack of acid or alkali under ordinary conditions, except that those with an amino group have a tendency to convert into a hydroxyl group when reacted with acid. Any *N*-glycoside bond of a nucleoside is relatively stable under alkali conditions, but not under acid conditions. The *N*-glycoside of pyrimidine-riboside is, contrarily, stable under acid conditions, which is related to the existence of the double bond between the 4- and 5-positions of the pyrimidine, for when this double bond is cleaved by an addition of hydrogen or bromide, *N*-glycoside is easily hydrolyzed with acid. The *N*-glycoside of a purine-riboside is easily hydrolyzed by acid as in the case of ordinary *N*-glycoside, but in the case of reaction of nitrous acid to adenosine or guanosine, no change in the site of the *N*-glycoside bond occurs, these two are converted to inosine or xanthosine, respectively. On the other hand, the *N*-glycoside of purine-deoxyriboside is extremely unstable under the reaction of nitrous acid.

In acid or alkali, the phosphate bond, conversely to the *N*-glycoside bond, is easily hydrolyzed by alkali and comparatively resistant to acid. Hence, when nucleotide was treated with alkali, both nucleoside and phosphoric acid were produced; with acid, on the other hand, base and pentose phosphoric acid ester were produced. But under conditions in the vicinity of pH 4, it is reported that the cleavage of a P=O bond occurred by a quite different mechanism.

DNA is stable with alkali but is unstable with acid, as mentioned earlier; the purine base is released by a cleavage of the *N*-glycoside bond of purine-nucleotide sites. RNA is easily hydrolyzed both with alkali and with acid; when in 1 N alkali solution for one day at room temperature, RNA decomposed to the mixture of mononucleotide and 2′- and 3′-nucleotide. When in 1 N HCl for 1 h at 100°C, RNA decomposed qualitatively into base, phosphoric acid, and ribose. Nevertheless, it is impossible to obtain 5′-nucleotides by using a reaction of the acid or alkali. For this purpose, an enzymatic hydrolysis method is utilized to obtain 5′-nucleotide from RNA.

Physical Properties

Tautomerism of the Bases. The pyrimidine base has two OH bonds that exist tautomerically (20), in the enol and lactam forms, respectively. Both in the solid state and in solution, pyrimidine is found to exist in the lactam form, as found from the IR spectrum (solid state) and the UV absorption spectrum (in solution) compared with relative compounds. Hydrogen in the NH= bond, which is located at the 3-position of pyrimidine, reacted with pentose, which constitutes a nucleoside. The ionic form of these pyrimidine bases changes with changing pH or solvent as cationic form (NH^+ at the 3-position of the ring), anionic form (=O at the 4-position, changeable to an $-OH^-$ bond), which shows the same amphoteric substance as amino acids. A purine base has both OH^-= and NH_2^-= bonds in the ring and shows the same both tautomerism and amphoteric character that a pyrimidine base does—lactam form for oxypurine (inosine, guanine) and amino form for aminopurine (adenine).

Nucleoside is also an amphoteric substance. No substantial change of physical properties is observed, except for the optical rotation caused by the induction of pentose, which changes with changes in pH in solution.

Localization of the Attachment and Detachment of Protons in Nucleosides and Nucleotides. It is natural to expect that a proton would be attached to those atoms in the molecule on which the highest net negative charge due to π electrons is concentrated, to the nitrogen atoms forming the two σ bonds. The possible sites of protonation (shown inside the circle in Fig. 9) in the predominant tautomeric forms of the bases are N3 in cytosine; N1, N3, and N7 in adenine; N3 and N7 in guanine and hypoxanthine; and N7 in xanthine. From the theoretical calculation, which takes into account

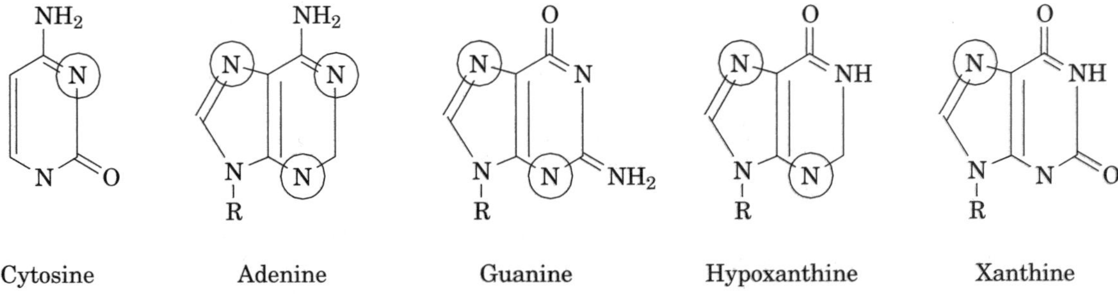

Figure 9. Chemical structure of various nucleosides; R denotes a hydrogen atom or various radicals.

the lone pair electrons, the position of the attachment of the proton is predicted to be the N1 position of adenine and the N7 of guanine and hypoxanthine. The results of X-ray structural analysis confirm the correctness of these conclusions regarding the site of proton attachment to the basis moieties of crystalline protonated bases, nucleosides, and nucleotides. On the basis of the UV spectra of bases, the location of protons in the ionized compounds can be established precisely.

Arguments in favor of protonation of adenosine at N1 and of guanosine at N7, as well as evidence of the protonation of cytidine at N3, at least in nonaqueous solvents, are given by the results of NMR spectroscopy (21).

Ionization Constants of the Bases of Nucleic Acids. Ionization constants are expressed in the form $pK_a = -\log K_a$.

Some of the bases of nucleic acids possess strong basic properties and are protonated in a weakly acidic medium but are deprotonated only in a strongly alkaline medium; other bases are weak acids and, although they form anions in a weakly alkaline medium, they are protonated only in a strongly acid medium. The first group includes cytosine and adenine; the values of pK_a are associated with

$$BH^+ \leftrightarrow B + H^+$$
$$K = (B)(H^+)/(BH^+) \quad (1)$$

The second group includes thymine and uracil; the values of pK_a are

$$BH \leftrightarrow B^- + H^+$$
$$K = (B^-)(H^+)/(BH) \quad (2)$$

Hypoxanthine, xanthine, and guanine show an intermediate position; they are protonated at comparatively high pH values for these compounds in accordance with equation 1 and are deprotonated in accordance with equation 2 at fairly low pH values in the alkaline region. Accordingly, their acid–base properties are described by two values of pK_a. Phosphate groups in nucleotides exert an appreciable influence on the pK_a value of the base. Values of the corresponding ionization constants pK_a for base, nucleosides, and nucleotides are given in Table 11.

The reason for the increase in pK'_a in series observed—nucleoside < nucleoside-2′, and 3′-phosphates < nucleoside-5′-phosphate—is evidently interaction of the ionized base with the phosphate group of the nucleotide.

It can be concluded from the conformation of ribose and deoxyribose that the possibility of such interaction is particularly great in the case of nucleoside-5′-phosphates, the phosphates groups of which may be in spatial proximity to the ring of the base. On dissociation of uracil derivatives, interaction between the negative charge on the ionized base and the negative charges of the phosphate group must lead to a decrease in the stability of the ionized form of the base and the shift of equilibrium toward the neutral form; pK'_a must increase in this case (22). These pK'_a values are utilized for the separation of these bases by ion exchange chromatography or electrophoresis.

Characteristics of Inosinic Acid and Guanylic Acid

Regarding 5′-nucleotides, both inosinic acid and guanylic acid are used as umami seasoning. The characteristics of these substances are summarized here (23).

Inosinic Acid. The common forms are 5′-inosinic acid, inosine 5′-phosphate, and inosine 5′-monophosphoric acid (5′-IMP). Molecular formula: $C_{10}H_{13}O_8N_4P$; molecular weight 348.2; C: 34.9 (%), H: 3.76 (%), O: 36.76 (%), N: 16.09 (%), P: 8.90 (%).

5′-IMP, which was the first nucleotide discovered by von Liebig in 1847, was isolated from the broth extracted from muscle as a salt of barium. It is said that von Liebig recognized that this inosinic acid related to the taste of the meat extract broth. In 1913, 5′-inosinic acid histidine salt was isolated from the extract of dried bonito as a umami component by Kodama. Later, the contribution of histidine to the umami taste was denied.

Inosinic acid, which is produced by a conversion of 5′-AMP through enzymatic deamination reaction, is found in the muscle of any kind of animals. In the metabolic pathway of nucleotides of microorganisms, 5′-IMP is produced first and then is converted to 5′-AMP by an amination reaction, and then proceed to 5′-XMP by oxidation with the aid of NAD; it then goes to 5′-GMP by a second amination reaction. Hence, 5′-IMP in microorganisms was observed when any enzymatic reactions in the pathway were blocked.

Table 11. pK'_a Values for Bases, Deoxyribonucleosides, and Ribonucleosides

				Nucleotides				
Derivatives	Base	Deoxyribonucleosides	Ribonucleosides	2′-Phosphate	2′(3′)-Phosphate	3′-Phosphate	5′-Phosphate	2′,3′-Cyclic phosphate
Uracil	9.5	9.3	9.25		9.96		10.06	9.47
Cytosine	4.46	4.3	4.1		4.43		4.54	4.12
Thymime	9.9	9.8	9.68	—		—	10.47	—
Guanine	—	9.33	9.22	9.87		9.84	10.00	—
Hypoxanthine	8.94	—	8.75	—		—	9.62	—
Xanthine	7.53	—	5.50					
Adenine	4.1	3.8	3.6	3.81		3.70	3.88	—

Note: Values determined spectrotometrically at 20°C.
For base and deoxyribonucleosides and ribonucleoside of guanine and adnine in solution of ionic strength = 1; in other cases in 0.1 M glysine or 0.025 M acetate buffers. For nucleotides: at zero ionic strength when pK'_a for deoxyguanosine was 9.50, pK'_a for deoxyguanosine 5′-phosphate was 10.0.

5′-IMP Characteristics

When condensed to the 5′-IMP solution, transparent syrup was obtained; this syrup, dried with 98% H_2SO_4, changed to a glassy solid.

Easily soluble in water; amorphous sediment was obtained with alcohol; sparingly soluble in organic solvent.

pK_1 = 1.54, pK_2 = 6.04, pK_3 = 8.88, acidic in solution; several kinds of metallic salts are formed; metallic salts with NH_4, Na, and K are transparent crystals.

Specific rotation in neutral or acidic solution; $[\alpha] = 18.5$ (0.1 N HCl solution). UV absorption spectrum changes with pH in solution; maximum wavelength is 248.5 nm at pH 6.0; molar absorbency coefficient: $\epsilon = 12.2 \times 10^3$.

In acidic solution, when heated it is hydrolyzed to ribose-5′-phosphoric acid and hypoxanthine; in alkali solution, when heated, phosphoric acid is released and inosine is produced.

Sodium Salt of Inosinic Acid

5′-IMP, when neutralized with 2 moles of NaOH and then crystallized, produces Na_2-5′-IMP-7.5H_2O.

These crystals are usually used as umami seasoning with L-glutamate.

Crystal group: orthorhombic.

Chemical formula: $C_{10}H_{11}N_4O_8PNa_2 \cdot 7.5H_2O$; molecular weight 527.32; C = 22.4 (%), H = 2.07 (%), N = 10.45 (%), O = 23.87 (%), P = 5.79 (%), Na = 8.58 (%).

From X-ray diffraction analysis: orthorhombic; space group $P2_1\ 2_1\ 2_1$, a = 11.63 Å, b = 21.84 Å, and c = 8.704 Å; Crystal density: 1.6063 g/cm^3; Z = 4.

Limited humidity at 20°C is more than 90% (less hygroscopy).

Easily soluble in water; about 24 g/p 100 mL in water; in acid (such as vinegar) about 30 g/100 mL.

Threshold value of umami is 0.025% in solution; thermal stability with a change of pH is summarized as follows and shows that Na_2 5′-IMP salts are stable for thermal and pH conditions in the range of ordinary cuisine conditions.

	Thermal stability		Residual ratio (%)	
pH	Temp.	Initial con.(g/L)	After 1 h	2 h
2.0	100	2.5	55.5	43.2
5.6	100	2.5	94.9	92.8
7.0	110	2.1	97.0	92.8
9.3	100	2.5	96.3	94.3

K and NH_4 salts of Inosinic Acid

Two moles ion are combined with 5′-IMP, similar to Na salt.

5′-Guanylic Acid. Guanosine 5′-monophosphate (GMP); chemical formula $C_{10}H_{14}O_8N_5P$; molecular weight 362.5; C: 33.06 (%), H: 3.88 (%), O: 35.24 (%), N: 19.28 (%), P: 8.53 (%).

Contrary to 5′-IMP, almost no distribution of 5′-GMP is reported in meats. Shiitake mushrooms contain 0.1 to 0.2% of 5′-GMP (as in dry base). In 1961 Kuninaka found that 5′-GMP was the third umami ingredient extracted from shiitake mushrooms.

Characteristics

Fine transparent needle crystals and no hydrate attached, decomposed at 190 to 200°C.

Sparingly soluble in cold water; comparatively strong acidic compound. Dissociation constants as follows: pK_1 = 0.8 (phosphoric acid), pK_2 = 2.4 (amino bond), pK_3 = 6.1 (phosphoric acid), pK_4 = 9.4 (enol).

Specific rotation is levo in solution.

UV spectrum changes with pH; at pH 2, maximum absorbency wavelength is 256 nm (ϵ = 12.3 × 103), and minimum absorbency wavelength is 228 nm (ϵ = 2.5 × 103), 261 nm is an isobestic point (ϵ = 11.6 × 103).

Sodium Salt of Guanylic Acid

With 2 moles of Na, 5′-GMP results in Na_2-5′-GMP · 8H_2O. This salt is difficult to obtain as a crystal form from aqueous solution and tends to white powder.

From the X-ray analysis: orthorhombic; space group $P2_12_12_1$; a = 21.33 Å, b = 21.18 Å, and c = 9.05 Å; Crystal density: 1.643 g/cm^3; Z = 8.

A little bit of hygroscopy (to the extent of 20 ~ 30% of water adsorbed), but no deliquescence observed.

Easily soluble in water (about 25 g/100 mL in water at 20°C); sparingly soluble in alcohol and acetone.

Threshold value for umami: 0.0125%, two- to threefold the synergistic effect of Na_2IMP with L-glutamate reported.

Industrial Production

The industrial production of IMP and GMP was started in 1961 by the enzymatic method; the cleavage of RNA by 5′-nuclease was produced by *Penicillium crysogenum* (24) and *Streptomyces aureus* (25).

Enzymatic Degradation of RNA. RNA distributed mainly in the cytoplasm of cells accounted for about 5 to 10% of the total in plants and animals, with only a limited amount in the nucleus. Yeast, especially, contains abundant, RNA—more than 10% on a dry basis—and a limited amount of DNA.

The separation of yeast is performed by centrifugation after the fermentation using cane molasses or pulp waste liquor as a carbon source. RNA can be extracted from yeast with hot sodium chloride solution and precipitated by acidifying the extract. The precipitate is neutralized and used for preparing 5′-nucleotides. An aqueous solution of yeast RNA is incubated with the heat-treated crude enzyme 5′-nuclease P1 (26). After incubation, the four 5′-nucleotides formed are separated by means of anion exchange resin column chromatography and purified. AMP is converted to IMP with *Aspergillus* adenyl deaminase.

Fermentation Method. Following the successful industrial production of amino acids by fermentation, extensive research was performed to establish the fermentation method for nucleosides and nucleotides. However, several different characteristics of amino acid fermentation are as follows: (*1*) in the metabolic pathway of nucleotides, there are both de novo synthesis and salvage synthesis, and moreover, mutual transformations are involved in purine nucleotide biosynthesis; (*2*) enzymes capable of degrading nucleotides to nucleosides, and to base and phosphoric acid, are widely distributed in microorganisms; and (*3*) nucleotides produced in a cell have less permeability through the cell membrane and do not easily secrete across the permeability barrier into broth.

For IMP, several methods are reported to overcome the preceding problems: (*1*) a direct fermentation method using a mutant; (*2*) a semifermentation method in which inosine, which accumulates more easily in broth, than IMP, was fermented by the fermentation method, to which phosphoric acid was added with a chemical or enzymatic method; (*3*) hypoxanthine added as a precursor was transformed into IMP by the ability of salvage synthesis of the microorganism; and (*4*) adenine (adenosine obtained by a chemical or fermentation method) was transformed by a chemical or enzymatic method into IMP. Methods (*1*) and (*2*) are regarded as more economical and were adopted in industrial production in line with enzymatic degradation of RNA.

Metabolic Pathway. The metabolic pathway of the nucleotide is shown in Figure 10 (27).

Regulation of Enzymatic Activity. PRPP-amidotransferase is an important initial enzyme for AMP, and GMP synthesis is specifically completely inhibited by AMP and ADP, but not by ATP, IMP, and GDP. SAMP (succino AMP)-lyase, which also takes part in the synthesis of AMP, and GMP are strongly inhibited by AMP and antagonistically inhibited by the substrate SAMP. IMP-dehydrogenase, the enzyme related to the IMP-to-GMP pathway, is strongly inhibited by the final product, GMP, and especially strongly by intermediate XMP as well. The enzyme SAMP-synthetase, related to the pathway from IMP to AMP, is strongly inhibited by AMP, followed by ADP and then GDP, but not by ATP and GMP (28,29).

Regulation of Enzyme Production. PRP-amidotransferase, IMP-trasformylase, and SAMP-lyase are released from repression under the conditions of deficiency of purinenucleotides and show high levels of enzyme activity but are repressed completely by addition of either adenine derivatives or guanine derivatives (see Fig. 11). IMP-dehydrogenase, which is important for the synthesis of GMP, is repressed completely by guanine derivatives, but not by adenine derivatives (30).

Bacillus subtilis (31) which has a characteristic high activity both of phosphatase and of 5′-nucleotidase, seems to be appropriate for the production of nucleoside; on the other hand, *Brevibacterium ammoniagenes* (32), having a phosphatase with low enzymatic activity is appropriate for the production of nucleotide.

Production of Nucleoside. For effective secretion of nucleoside through the cell membrane, the following properties seem to be necessary (33): (*1*) the activity of 5′-nucleotidase or phosphatase that catalyzes the cleavage of nucleotide should be absent or be at a very low level; (*2*) the regulation of enzyme involved in nucleotide formation, such as PRPP amidotransferase, by AMP and related nucleotides should be released, and (*3*) both SAMP and GMP reductase of IMP dehydrogenase should be lacking or present at a very low level.

Inosine. The appropriate strains for the accumulation of inosine are *B. subtilis* and *B. ammoniagenes*.

The adenine-requiring strain of *B. subtilis* was used as a parent strain and mutated by UV treatment to be deficient in the activity of both AMP deaminase and GMP reductase; moreover, from this strain an 8-azaguanine-resistant mutant was derived, one of the xanthine auxotrophic mutants that strengthens the ability of PRPP amidotransferase, and shows the result of accumulation of inosine at 16 to 18 g/L in broth (34).

Furuya et al. (35) derived 6-mercaptoguanine-resistant mutants from an adenine-leaky, IMP-producing strain of *B. ammoniagenes*. Furthermore, they induced the mutant that is devoid of IMP dehydrogenase and requires guanine for growth. Addition of the guanine-requiring character was most effective for accumulation of inosine. Under optimal conditions, this strain shows high accumulation of inosine: about 30 g/L from 17% of inverted cane molasses, which is equivalent to more than a 17% yield of inosine based on the consumed sugar.

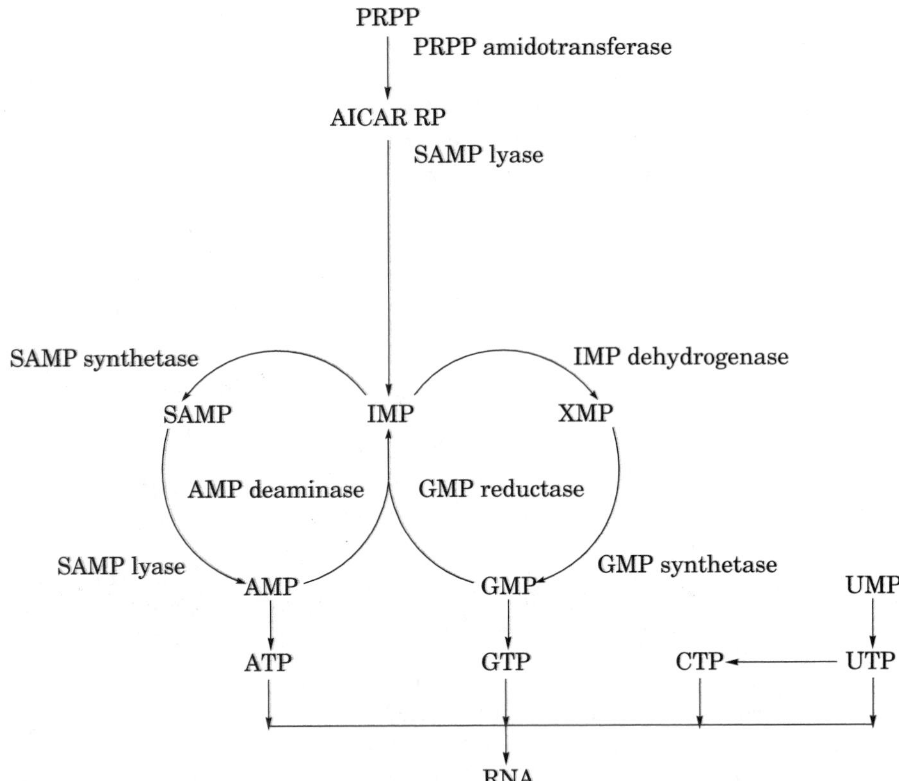

Figure 10. Metabolic pathway of nucleotide by de novo synthesis.

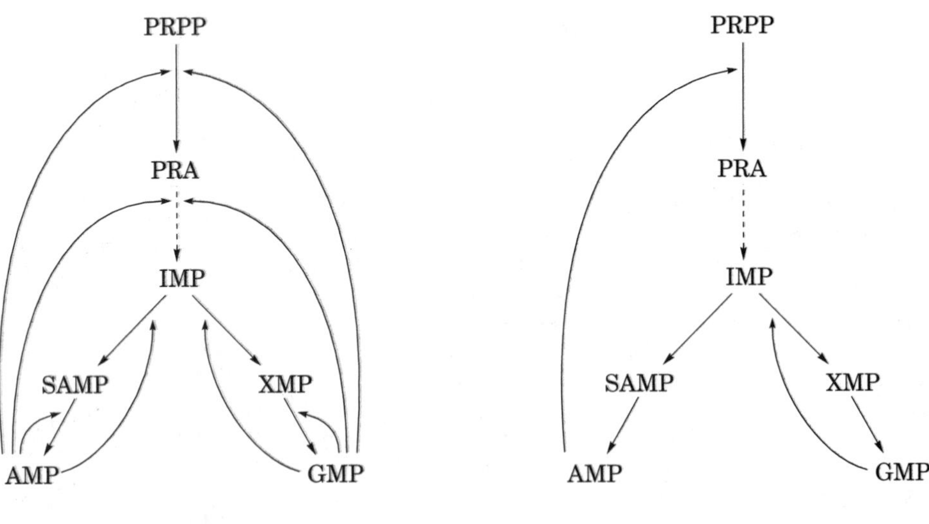

Figure 11. Regulation of metabolic pathway.

Guanosine. From *B. subtilis*, which is capable of producing inosine, in which PRPP amidotransferase IMP dehydrogenase (IMP → XMP), and GMP synthetase (XMP → GMP) are inhibited or repressed by GMP and guanine, methionine sulfoxide–, psicofuranine-, and decoyinine-resistant mutants were derived (36). The key to increasing guanosine is the loss of repression and feedback inhibition of IMP dehydrogenase, GMP synthetase, SAMP lyase, and PRPP amidotransferase. This strain showed the accumulation of 16 g/L of guanine and both hypoxanthine and inosine in the culture, glucose 80 g/L and L-histidine 0.3 g/L (72 h at 34°C).

Phosphorylation by the Chemical Method. For nucleosides, direct reaction with $POCl_3$ resulted in phosphorylation of not only a 5′-OH bond but also 2′- and 3′ OH bonds, simultaneously. So it is difficult to selectively obtain 5′-nucleotide. Obtaining 5′-nucleotide, 2′- and 3′-OH in the ribose of nucleosides was protected previously by acetone or benzaldehyde to form an isopropiridene bond or benziridene bond; the product was then reacted with $POCl_3$ to substitute the phosphate bond for 5′-OH on the riboside, at lower temperature (0 to 5 °C), and then a removal reaction of protected groups for the 2′- and 3′-bonds on the riboside was performed. The improved process is reported as follows: in the appropriate solvent with a small amount of water, $POCl_3$ was added directly to nucleoside; the protection of 2′- and 3′-OH bonds and 5′-phosphorylation occurred, simultaneously, and a highly pure 5′-nucleotide solution was obtained. To this solution, a large amount of water was added to stop the excess phosphorylation reaction, and then sodium hydroxide was added to crystallized the disodium 5′-nucleotide, producing, for example, Na_2GMP and Na_2IMP crystals.

Direct Fermentation

5′-Inosinic Acid. The adenine-leaky auxotrophic strain of *B. ammoniagenes* showed the ability to accumulate IMP; 12.8 g/L in broth was obtained (37), which is the first report exhibiting the possibility of direct fermentation of IMP. This strain showed a high sensitivity for Mn^{2+} ions and optimum concentration; an extremely low and narrow range of value (10 to 20 μg/L; and a change in cell shape and a decrease in vital cell number when it contained Mn^{2+} at a concentration suitable for IMP accumulation during cultivation. By adding the nitrosoguanizine-resistance ability, this strain mutated to Mn^{2+}-nonsensitive, and no change in cell shape was observed during cultivation, which accompanied the change in the mechanism of leakage of IMP, due to the improvement of the permeability of IMP through cell membrane associated with inhibition of the conjugated decomposition system (IMP → Hx) by glucose in the cell membrane.

These results lead to the conclusion that the leakage of IMP through the cell membrane attributes to the energy-dependent reaction.

5′-Guanylic Acid. So far GMP is not successfully produced by the direct fermentation method, mainly because of the intrinsic problem concerning the metabolic pathway; that is, no base corresponding to GMP for salvage synthesis exists, such as hypoxanthine, xanthine for IMP, or guanine for XMP (xanthosine monophosphate). Therefore, it is impossible to obtain the mutant-blocked pathway to GMP, and GMP itself is a cell-constituted component, different from IMP and XMP; then there exists a strict barrier in the cell membrane that inhibits GMP from passing through (38).

From the industrial viewpoint, the following three-step fermentation method is employed. First, XMP was produced by using the mutant of *B. ammoniagenes*, and then it was transformed into GMP by the enzymatic method using the other mutant of *B. ammoniagenes* (39).

1. *XMP fermentation*. In the case of XMP, *B. ammoniagenes* mutated to have both an adenine requirement (leakage type) and a guanine requirement. Nucleotide activity weakened, showing high productivity of XMP; also, no cell membrane barrier existed for XMP, which is different from IMP.
2. *Enzyme for conversion of XMP to GMP*. The nucleotide weakened *B. ammoniagenes* with docoynine resistance, which strengthens the activity of the GMP synthetase that was obtained. This strain showed an increased ability to convert XMP to GMP, with no cleavage of GMP and XMP. The cell membrane permeability of XMP and GMP was improved by adding a cationic surface-active agent, polyethylene stearylamine (40).
3. *Conjugated reaction with ATP energy system*. The XMP aminase reaction is a one-step reaction for conversion of XMP to GMP, conjugated with energy regenerated by the conversion of ATP to AMP and pyrophosphoric acid (PPi), which required an indispensable cofactor, the Mg^{2+} ion:

$$\text{XMP} + \text{NH}_3 + \text{ATP} \rightarrow \text{GMP} + \text{AMP} + \text{Ppi XMP aminase } (\text{Mg}^{2+})$$

Brevibacterium sp. treated with an active surfactant have an increased ability to synthesize ATP through the Krebs cycle from glucose. By using this process, in place of ATP, glucose was used as an energy supply source; conjugated with this process, the conversion of XMP to GMP was achieved economically. The enzymatic process was as follows:

$$\text{PPi} \rightarrow 2\text{Pi (pyrophosphatase)}$$

$$\text{AMP} + \text{ATP} \rightarrow 2\text{ADP (AMP kinase)}$$

$$\text{ADP} + \text{Pi} + \text{Glucose} \rightarrow \text{ATP (ATP-generating enzyme system)}$$

These enzymes, contained in a cell, are must be regulated appropriately to prevent the subreaction from occurring. Controlling the temperature to repress the activity of nucleotide monophosphate kinase and the concentration of phosphate ion ($PO^{3-}{}_4$) increases the ability of the ATP-generating system. If Mg^{2+} and PPi form insoluble magnesium pyrophosphate (Mg_2PPi), the preceding reactions cannot proceed smoothly. Phytic acid, a chelating agent, was able to prevent the formation of this Mg_2PPi.

When using purified XMP as raw material, overall conversion to GMP was more than 80% after 22 h (41).

Increasing the activity of XMP aminase is expected to reduce the number of cells required for conversion when whole cells are used as the enzyme sources, while reducing to a minimum the degradation of products and by-products GDP and GTP. The proliferation of the XMP aminase strain of *Escherichia coli* using genetic engineering techniques was attempted, which involved the subcloning of the *guaA* gene for XMP aminase into pBR322, followed by the construction of a plasmid utilizing the trp promoter. The amount of XMP aminase produced by the strain carrying the *gauA* gene accounted for 10% of total cellular protein, and the *E. coli* K294/XAR33 strain showed an increase in activity of about 80 times compared with *E. coli* K294. The conversion ratio from XMP to GMP was more than 90% after 7 h (42).

BIBLIOGRAPHY

1. K. Ikeda, *J. Tokyo Chem. Soc. Japan* **30**, 820 (1909).
2. S. Kodama, *J. Tokyo Chem. Soc. Japan* **34**, 751 (1913).

3. A. Kuninaka, M. Kibi, and K. Sakaguchi, *Food Technol.* **18**, 287 (1964).
4. A. Kuninaka, *J. Agric. Chem. Soc. Japan* **34**, 487 (1960).
5. S. Yamaguchi et al., *J. Food Sci.* **36**, 846 (1971).
6. S. Ikeda, *New Food Ind.* **7**, 41 (1965).
7. A. Adachi, M. Funakoshi, and Y. Kawamura, *Olfaction Taste* **2**, 411 (1967).
8. M. Sato, S. Yamashita, and H. Ogawa, *Olfaction Taste* **2**, 399 (1967).
9. R. H. Cagan, in M. R. Kare and O. Maller, eds., *The Chemical Sense and Nutrition*, Academic Press, New York, 1977, pp. 175–203.
10. K. Torii and R. H. Cagan, *Biochim. Biophys. Acta* **627**, 313 (1980).
11. A. Kuninaka, in W. H. Schluz, E. A. Day, and L. M. Libbey, eds., *The Chemistry and Physiology of Flavor*, AVI, Westport, Conn., 1967, pp. 515–535.
12. J. H. Hara, in J. H. Hara, ed., *Chemoreception in Fishes*, Elsevier, Amsterdam, The Netherlands, 1982, pp. 135–180.
13. E. Chargaff, *Essay on Nucleic Acids*, Elsevier, Amsterdam, The Netherlands, 1963.
14. J. D. Watson and F. H. C. Crick, *Proc. R. Soc. London, Ser. A* **223**, 80 (1954).
15. J. D. Watson and F. H. C. Crick, *Nature* **171**, 737 (1953).
16. T. Shimanouchi, M. Tsuboi, and T. Kyogoku, *Biochim. Biophys. Acta* **15**, 1 (1962).
17. A. Szent-Gyorgy, *Chemistry of Muscular and Contraction*, Academic Press, New York, 1951.
18. N. R. Jones and J. Murray, *Biochem. J.* **77**, 567 (1960).
19. Seasoning: Its Science and Production, in R. Takada, ed., Koseido, Tokyo, Japan, 1966, pp. 73–86.
20. N. K. Kochetkov and E. I. Budovskii, eds., *Organic Chemistry of Nucleic Acids*, Part A, L. Todd and D. M. Brown, trans., eds., Plenum, London and New York, 1971, pp. 134–000.
21. B. Pullman and A. Pullman, *Quantum Biochemistry*, Wiley, New York, 1963.
22. P. O. P. Ts'o, S. A. Rappaport, and F. J. Bollum, *Biochemistry* **5**, 4153 (1966).
23. A. Sumida, *New Food Ind.* **6**, 78–81 (1964).
24. A. Kuninaka et al., *Bull. Agrc. Chem.* **23**, 239 (1959).
25. K. Ogata et al., *Agree. Biol. Chem.* **27**, 110 (1963).
26. A. Kuninaka, in K. Ogata et al., eds., *Microbial Production of Nucleic Acid Substances*, Kodansha, Tokyo, Japan, 1976, pp. 75–86.
27. J. M. Buchanan and S. C. Hartman, in F. F. Nord, ed., *Advances in Enzymology*, Vol. 21, Interscience, New York, 1959.
28. K. Ishii and I. Shiio, *J. Biochem.* (Tokyo) **63**, 661 (1968).
29. K. Ishii and I. Shiio, *J. Biochem.* (Tokyo) **68**, 171 (1970).
30. H. Nishikawa, H. Momose, and I. Shiio, *J. Biochem.* (Tokyo) **62**, 92 (1967).
31. I. Shiio and H. Ozaki, *J. Biochem.* (Tokyo) **83**, 409 (1978).
32. M. Misawa, T. Nara, and K. Nakayama, *J. Agric. Biol. Chem. Japan* **38**, 167 (1964).
33. H. Enei, in T. Tochikura, ed., *New Biotechnology*, Japan Bioindustry Association, Tokyo, Japan, 1988, pp. 135–147.
34. H. Matsui et al., *Agric. Biol. Chem.* **46**, 2347 (1982).
35. A. Furuya, S. Abe, and S. Kinoshita, *Appl. Microbiol.* **20**, 263 (1970).
36. H. Matsui, *J. Agric. Biol. Chem. Japan* **58**, 175 (1984).
37. A. Furuya, S. Abe, and S. Kinoshita, *Appl. Microbiol.* **16**, 981 (1968).
38. T. Nara et al., *Agric. Biol. Chem.* **33**, 739 (1969).
39. M. Misawa et al., *Agric. Biol. Chem.* **33**, 370 (1969).
40. A. Furuya, S. Abe, and S. Kinoshita, *Biotechnol. Bioeng.* **13**, 229 (1971).
41. T. Fujio, K. Kotani, and A. Furuya, *J. Ferment. Technol.* **62**, 131 (1984).
42. A. Maruyama, T. Fujio, and S. Teshiba, *Agric. Biol. Chem.* **50**, 1879 (1986).

GENERAL REFERENCES

E. Chargaff and J. N. Davidson, eds., *The Nucleic Acids, Chemistry and Biology*, Vol. 1, Academic Press, New York, 1955.

D. O. Jordan, ed., *The Chemistry of Nucleic Acids*, Butterworth, London, 1960.

A. Kuninaka, in H. Pape and H.-J. Rehm, eds., *Nucleic Acids, Nucleotides, and Related Compounds*, Biotechnology, Vol. 4, VCH Verlargsgesellschaft mbH, Weinheim, Germany, 1986, pp. 71–114.

J. Rehmann, in *Encyclopedia of Chemical Engineering*, Vol. 17, *Nucleic Acids*, Wiley, 1982, pp. 507–543.

R. B. Setlow and E. C. Pollard, eds., *Molecular Biophysics*, Addison-Wesley, London, 1962.

Tetsuya Kawakita
Ajinomoto Company, Inc.
Kawasaki, Japan

NUTRITIONAL LABELING

Americans love their food. Their interest in all aspects of food has been steadily growing since the 1950s. Public demand for more and more information on food-product labels has been growing right along with this interest. Today, labels on food packages provide a wealth of facts about the products inside. It wasn't always this way, however. This chapter summarizes the development of food labeling in the United States and covers key elements of current labeling requirements. Emphasis is on the requirements of the U.S. Food and Drug Administration (FDA), which regulates the labeling of most packaged foods, except for meat and poultry products. These are the responsibility of the U.S. Department of Agriculture (USDA). Differences between what FDA and USDA require on product labels are minor.

SOME KEY MILESTONES IN FOOD LABELING

Before 1906, there were no requirements to include any information at all on food labels. In that year, the Federal Food and Drugs Act and the Federal Meat Inspection Act were first enacted. These groundbreaking laws formed the base for regular additions and refinements, right up to today's detailed labeling requirements.

The two 1906 acts authorized the federal government to regulate the safety and quality of food and assigned the responsibility for doing so to the U.S. Department of Agriculture and its Bureau of Chemistry. They became separate agencies when the Bureau of Chemistry was spun off in 1927, eventually to become the Food and Drug Administration. The USDA retained the responsibility for regulating meat and poultry products.

The 1906 acts set no detailed labeling requirements. At the time, it was expected these would be worked out based on enforcement actions taken against goods considered mislabeled. In 1913, an amendment to the 1906 acts established the first specifics by requiring labels to state the net quantity of contents in each food package. In 1938, the Federal Food, Drug and Cosmetic Act replaced the 1906 Food and Drugs Act and added the requirement that each package label state the name and address of the manufacturer or distributor. This act also required that labels list ingredients, including artificial colors, flavors, and preservatives, for most products and forbade false or misleading statements on the label. The amounts of vitamins and minerals were required only for products intended for special dietary uses.

The Fair Packaging and Labeling Act, passed in 1966, authorized and directed FDA to require more complete information on food product labels. Under this Act, FDA regulations specifically defined several key food label elements, for example, "label," "package," and "principal display panel." The regulations also established specific details as to the way the statement of identity of the food; the name and address of the manufacturer, packer, or distributor; and the net quantity of contents should be shown, such as location on the label and type style and size.

Another major milestone came in 1973 when FDA issued the first nutrition information labeling regulations, which became effective in 1975. These rules required nutrition information on the labels of only those products for which nutrition claims were made or to which nutrients were added. Many food companies took the opportunity to add nutrition information voluntarily to the labels of many products for which no claims were made, and as a result nutrition information became relatively common.

As interest in nutrition grew, especially in reducing calorie intake, FDA proposed regulations in 1978 that defined claims such as "low calorie" and "reduced calorie." They became final in 1980 after some debate. Low-calorie foods could contain only 40 cal per gram of food, whereas the calories in a serving of a reduced-calorie food had to be at least one-third lower than in a serving of the regular food. Products that met the qualifications for reduced-calorie were sold without being labeled as "imitation." Unfortunately, a loophole soon became obvious. Because the regulations did not define other terms that implied calorie reduction, a number of products that could not meet the requirements for reduced-calorie without a perceived sacrifice in taste were marketed as "light" or "lite," with relatively smaller reductions in calories. In 1984, a regulation was proposed to add sodium to the list of nutrients required to be included in the nutrition information. The rule was made final in 1985 and also defined claims such as "low sodium" (<140 mg per serving) and "sodium free" (<5 mg per serving).

A comparable regulation that defined claims about cholesterol in food products was proposed in November 1986. "Low cholesterol" was defined as <20 mg per serving, and "cholesterol free" was set at <2 mg per serving. Although this proposal was never made final, a number of food-product labels made claims based on the requirements, with FDA's official approval. Closely related claims for fat and fatty acid content were also discussed during this period but never formally proposed. Food-product labeling was about to undergo its next major revision.

THE NUTRITION LABELING AND EDUCATION ACT OF 1990

Several forces led Congress to pass the Nutrition Labeling and Education Act of 1990 (NLEA). As public interest in foods and nutrition, especially the relationship between diet and health, continued to grow, a number of food manufacturers had begun in the mid 1980s to make health and nutrition claims on certain product labels. Probably the best known of these was the claim that fiber in the diet was believed to reduce the risk of cancer. Because government regulations did not provide for this and other claims, questions and controversies arose that led to pressure on both Congress and FDA for labeling reform. In 1990, in what amounted to a race with Congress, FDA was the first to propose extensive labeling revisions. These were all superseded when, on its own, Congress passed the NLEA in October. The president signed NLEA in November 1990.

The key provisions of the NLEA, prescribing exactly what Congress wanted on food labels, are as follows:

Nutrition Labeling Required

All food products had to provide nutritional labeling, with very few exceptions. Exceptions included foods served in restaurants and food-service operations, foods containing insignificant levels of nutrients, and products from small manufacturers. Produce sold at retail, including fresh fruits, vegetables, and seafood, was expected to provide nutrition information voluntarily. Requirements would become mandatory if there was not "substantial compliance," as defined by FDA, by 30 months after the signing of the NLEA.

Government to Establish Serving Sizes

Serving sizes were to be established by government regulation, derived from amounts that were "commonly consumed" by the public. This provision was based on a perception that manufacturers often set unrealistically small serving sizes for high-fat/high-calorie products, or, conversely, unrealistically large serving sizes for foods with low levels of nutrients.

Specific Nutrient Requirements

The nutrition information had to include specific nutrients that were considered to represent major public concerns and interest. These were calories, calories from fat, total fat, saturated fat, cholesterol, sodium, total carbohydrates, sugar, dietary fiber, and protein. The door was left open for FDA to require additional nutrients on labels.

Prescribed Label Format

The information was to be presented in a form the public could easily understand. In effect, this provision required the government to establish a clear and consistent format for the nutrition information.

Government to Define Which Health and Nutrition Claims Would Be Permitted

The government was required to set detailed and specific requirements for any health or nutrition claims on labels. No claim could be made that was not government-approved. NLEA further required FDA to define terms such as "free," "low," "light" or "lite," "reduced," "less," and "high."

Preemption of State Labeling Regulations

Federal labeling regulations preempted any state or local regulations with respect to nutritional labeling but not with respect to safety. This added to clarify and ease of understanding label information across the country. It also prevented individual states from setting different labeling requirements, which might result in manufacturers having a different label for each state. This provision did not apply to specific warning statements, which states could still require individually.

Expanded Ingredient Labeling

Labels had to provide more details on certain ingredients, such as listing all certified Food, Drug and Cosmetic (FD&C) artificial food colors by name. NLEA also required declaration of the amount of juice in products that contain fruit juice.

"Hammer" Deadline

NLEA set a deadline of November 8, 1992, for all regulations based on the requirements of the act to become final. Any proposals not formally finalized would automatically become final on that date. This hammer provision was designed to make sure that the responsible government agencies published the detailed rules that would be required without undue delay.

Action by FDA and USDA

NLEA officially amended the federal Food, Drug and Cosmetic Act. Technically this required only FDA to develop new regulations for packaged foods under its authority. USDA, however, also developed NLEA-based regulations for meat and poultry products that are essentially the same as those of FDA.

FDA and USDA began to harmonize the earlier proposals with NLEA requirements. Both agencies issued the first new proposals in late November 1991, a little over a year after the signing. In July 1992, FDA followed up by publishing several different format designs. USDA proposed its own slightly different formal in August. Allowing time for review of public comments, final rules were scheduled for publication early in November to meet the hammer deadline. This was delayed at the last minute due to differences of opinion between the two agencies over several issues. Agreement was reached in December, and the final rules were formally published in the *Federal Register* of January 6, 1993. But this was not quite the end. On April 2, 1993, FDA published corrections to the final rules. These were followed on August 18 by additional technical revisions based on comments from the public. At this point, the major labeling provisions based on NLEA were finally complete.

The original target date for compliance was May 1993. Because of the delays, however, FDA extended the deadline to apply to all foods labeled on or after May 8, 1994. Exceptions were meat and poultry products, for which USDA extended the compliance date to July 6, 1994.

CURRENT U.S. FOOD-PRODUCT LABELING REQUIREMENTS

Today's food-product labels in the United States incorporate the requirements of the NLEA along with those established by the earlier laws and regulations. The provisions are many and very detailed, so only key elements are covered here. The discussion focuses mainly on FDA requirements, but those of the USDA are essentially similar. There is one major difference between the two agencies. USDA must review and approve labels before they go on the package. FDA does not have the authority for prior approval.

Labeling Panels on Food Packages

Labeling regulations define two key panels that carry required information:

1. The front label panel, known as the principal display panel (PDP)
2. The information panel, usually the panel just to the right of the PDP

The location of the information panel may vary depending on the size and shape of the package, but the PDP must always be the panel that is seen by the consumer at the point of purchase, almost always the front of the package.

Required Label Information

All food-product labels are required to carry the following information:

Statement of Identity (Product Name) This is a "common or usual" name that accurately describes the product. If the product is one that is regulated by a standard of identity, such as ice cream, the name established by the standard is the common or usual name. A brand name by itself is usually not enough. Although percentage labeling is not a general requirement, certain foods that contain characterizing ingredients, for example, apples in applesauce, may be required to include the percentage of that ingredient as part of the name.

Net Quantity Statement (Net Contents) Usually, solid foods are declared by weight, whereas liquid foods are declared by volume, but there are many exceptions. The United States still uses the avoirdupois system—pounds, ounces, pints, quarts, and so on. Use of metric units is optional for the time being; however, most package labels do carry metric equivalents.

Both the statement of identity and the net quantity statement *must* appear on the PDP.

Name and Address of the Manufacturer, Packer, or Distributor If the product is not actually made by the company whose name is on the label, the company name must be qualified by terms such as "manufactured for" or "sold by."

List of Ingredients All ingredients in a food product must be listed by name in descending order of predominance by weight:

- Type of oil, such as corn, soy, or canola, must be stated.
- Flavor enhancers such as monosodium glutamate and disodium inosinate, yeast extract, and certain vegetables that are commonly used as seasonings—garlic and onion, for example—must be listed by name. They may not legally be included in a general term such as "natural flavors" or "flavorings."
- Sulfiting agents such as sulfur dioxide and potassium metabisulfite that may be used as preservatives must be listed.
- Ingredients that are contained in other ingredients must be listed individually. For example, if cheese is an ingredient in a food, it is not enough to simply declare "cheese." All the ingredients in the cheese must be listed, for example, "cheese (milk, cheese cultures, enzymes, salt)."
- It is especially important that the most common allergens be listed in the ingredient statement, whether they are added separately or as a component of another ingredient. These are peanuts, tree nuts, soy, milk and dairy products, eggs, wheat, fish, and shellfish (crustacea and mollusks).

The NLEA and associated new FDA regulations added several more specific ingredient-labeling requirements:

- All packaged food products must carry a complete list of ingredients. Exemptions from full ingredient labeling for a few products covered by standards of identity, such as macaroni products, were removed.
- All artificial colors subject to certification, FD&C colors must be listed by name, for example, Blue 2 or Red 3.
- Hydrolyzed proteins, often used to enhance savory flavors, must now be listed by source, for example, "hydrolyzed soy protein" rather than "hydrolyzed vegetable protein." This is required to inform those with allergies to specific foods. Like monosodium glutamate, hydrolyzed proteins may not be listed simply as "natural flavors."
- Labels of drinks that contain fruit juice must now state the percentage of juice.

A few types of ingredients may still be declared by general rather than specific terms:

- Spices, except those that add color such as paprika, turmeric, and saffron. They may be declared either by name or as "spice and coloring."
- Flavors may be declared simply as "natural flavor(s)" or "artificial flavor(s)" without naming each individual flavoring material. Some countries recognize a "nature-identical" category of flavorings, which are made synthetically but are chemically identical to the natural flavoring, but FDA considers these artificial.
- Colorings that do not require certification may be declared simply as "artificial color" or "color added." Many companies voluntarily declare these by name and function, for example, "beta-carotene for color," "caramel color," and "colored with beet juice."

Warning Statements Warnings are required in only a very limited number of instances, such as cautions against overheating aerosol containers, saccharine and cancer, and so on.

Nutrition Facts Nutrition labeling is required by the NLEA for nearly all packaged foods sold at retail in the United States. The Nutrition Facts box is the most noticeable addition to food labels. The format and contents of the Nutrition Facts box and related nutritional labeling requirements are discussed in detail in a separate section.

The NLEA regulations did provide for a few exemptions from the requirements for nutritional labeling:

- Foods sold by businesses with less than $500,000 in gross annual sales
- Foods served in restaurants and other establishments for immediate consumption; ready-to-eat foods
- Products sold exclusively for food service or other institutional use rather than directly to consumers. Any of these products that may end up in warehouse or club stores must carry nutrition labeling.
- Foods with insignificant amounts of nutrients, such as spices, coffee, and tea
- Fresh produce and fish, meat, and poultry
- Products in very small packages with <12 in.^{2} of total surface area available to bear labeling. These products must list an address or phone number where consumers can obtain nutrition information.
- It is very important to note the requirement that any products that make any nutrition or health claims lose their exemption and must carry nutritional labeling.
- Medical foods, infant formulas, and dietary supplements have specific labeling requirements that are different from those for foods and are not discussed here.

The company name and address, ingredient list, nutritional labeling, and any warning statements may appear on either the PDP or the information panel, but the information panel is the most common location. There must be no intervening material, such as directions, recipes, logos, or the like, between these required items.

Figure 1, adapted from FDA's *A Food Labeling Guide*, shows a typical food label for a box of cereal, with the location of the required information on the PDP and the information panel.

Optional Label Information

Food-product labels carry much information in addition to that required by FDA or USDA. This can include pictures of the product; preparation directions; serving suggestions; recipes; sell-before, buy-before, or use-before dates; and claims for the product. Some of this optional information—health or nutrition claims, for example—may in fact trigger additional requirements.

The Universal Product Code (UPC) is an optional element that has become essential. The UPC is the familiar series of bars that stores scan at the checkout register. It is used to keep track of information such as prices and inventory. Although it is not officially required by any government agency, it is required for practical purposes on any product that is sold in today's market.

Food labels must contain all the required information in the correct places as well as all the optional information that makes a label attractive and helpful to consumers. Developing a new food label can thus present a real challenge.

NUTRITION INFORMATION BASED ON NLEA

NLEA focused on two key areas of food-product labels in the area of nutrition: how the nutrition information should be presented, and how claims should be regulated. Although nutrition information is required for most food labels, claims are optional, but making them may trigger additional requirements. Claims must be only those defined by NLEA regulations.

Nutrition Facts

According to the regulations, the nutritional values in the Nutrition Facts box are to be for the food "as packaged," that is, the product that is in the package. The regulations allow for providing nutritional values "as prepared," but this is optional.

FDA considered several approaches to how food-product labels should provide nutrition information. The final regulations followed NLEA requirements to establish what information must appear and prescribed the Nutrition Facts box as the exact format.

Figure 2 shows the Nutrition Facts box for the food label as it appears in FDA's regulations, with all the required details. This is known as the standard format.

Figure 3 is from a presentation by FDA and explains some of the considerations used to develop the Nutrition Facts box. The features were designed to help consumers apply the label information to their own diets.

Figures 2 and 3 show the conversion guide for calories per gram of fat, protein, and carbohydrate at the bottom of the Nutrition Facts box. The August 1993 amendments to the original rules made this optional, so few labels in the market actually carry it.

The serving size for each food product must be based on the established reference amounts using methods specified in the regulations. The purpose is to have consistent serving sizes for similar products, making it easier for consumers to make comparisons. Serving sizes must be expressed both in common household measures, such as teaspoons and cups, and in metric amounts, such as grams or milliliters. The exact amounts of each household measure are specified, for example, 1 teaspoon = 5 mL, 1 tablespoon = 15 mL, and 1 cup = 240 mL. The regulations list reference amounts for more than 120 foods and food categories, typically expressed both as common household measures and as metric amounts that are considered appropriate. It was hoped that providing the information in both forms would be another way to help make serving sizes easier for consumers to understand and use.

Servings per container seems straightforward at first, but it can get complicated in some situations. For example, the reference amount for carbonated beverages is 8 fluid ounces, but 12-oz cans are common. The rules require a container that contains up to twice the reference amount to be considered 1 serving. It can get even more complicated for products sold in large discrete units, such as pies or cakes of different sizes, in which the size of a common fraction of the unit—a slice, for example—may not match the official reference amount exactly. In such cases, serv-

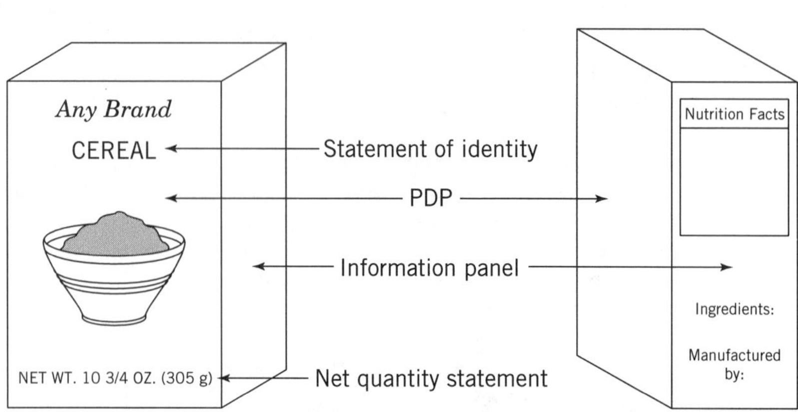

Figure 1. Typical placement of required label information on a package. *Source:* Ref. 1.

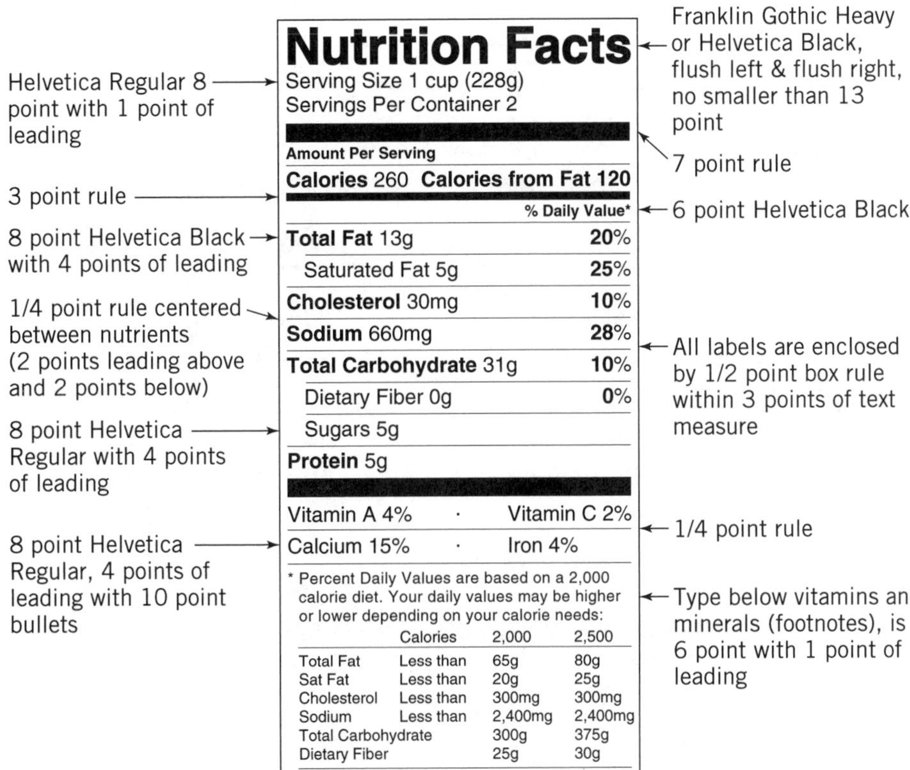

Figure 2. The new food label (graphic enhancements by FDA). *Source:* Ref. 1.

ings per container must be expressed in approximate terms—"about 6," for instance.

The nutrients shown in Figures 2 and 3 are those required by the regulations. For labeling purposes, they are divided into two groups: macronutrients, calories through protein (those between the two heaviest lines in the box), and micronutrients, the vitamins and minerals below the lower heavy line.

The regulations define each component and prescribe exactly how the numerical values shown in the nutrition facts box are to be expressed and rounded:

Serving size (common household measure and metric units): Metric units (grams or milliliters) must be rounded to the nearest 0.2 up to 2.0, the nearest 0.5 up to 5.0, and to the nearest whole unit thereafter

Calories and calories from fat: 5-cal increments up to 50 cal; 5-cal increments thereafter

Total fat and saturated fat: 0.5-g increments up to 5 g; 1-g increments thereafter

Cholesterol: 5-mg increments

Sodium: 5-mg increments up to 140 mg; 10-mg increments thereafter

Total carbohydrate, dietary fiber, sugars, and protein: nearest whole gram

Sugars: For labeling purposes, sugars are defined as the sum of monosaccharides (e.g., dextrose and fructose) and disaccharides (e.g., sucrose, or table sugar).

Percent daily value (DV) is rounded to the nearest 1% DV for total fat, saturated fat, cholesterol, sodium, total carbohydrate, and dietary fiber. There is no DV for sugars. The % DV for protein is optional and is usually not shown.

Vitamins and minerals: The % DV are rounded to the nearest 2% DV up to 10% DV, the nearest 5% DV from 10 to 50% DV, and the nearest 10% DV thereafter.

FDA expects the label values to be correct within the limits specified in the regulations. Nutrients that occur naturally in foods may vary by as much as 20% from the label value and still be within compliance limits. On the other hand, the amounts of any nutrients that are added must be at least 100% of the value stated on the label.

For example, suppose the label for a flavored milk drink declares "Protein 7 g" for a serving. If the protein is derived only from the milk ingredient, it would be "naturally occurring." The actual amount could be slightly less than 6 g in some individual products, and they would still be in compliance. However, if the product contains any added protein, such as that from soy or sodium caseinate and the label declares "Protein 9 g," the product would be required to contain no less than 9 g per serving.

A food product typically contains significant amounts of a few or several nutrients and smaller or even insignificant amounts of others. The regulations established levels that are considered "nutritionally insignificant" for each nutrient. If a serving of a food product contains less than this nutritionally insignificant amount, it may be declared as 0.

The specific nutritionally insignificant amounts are:

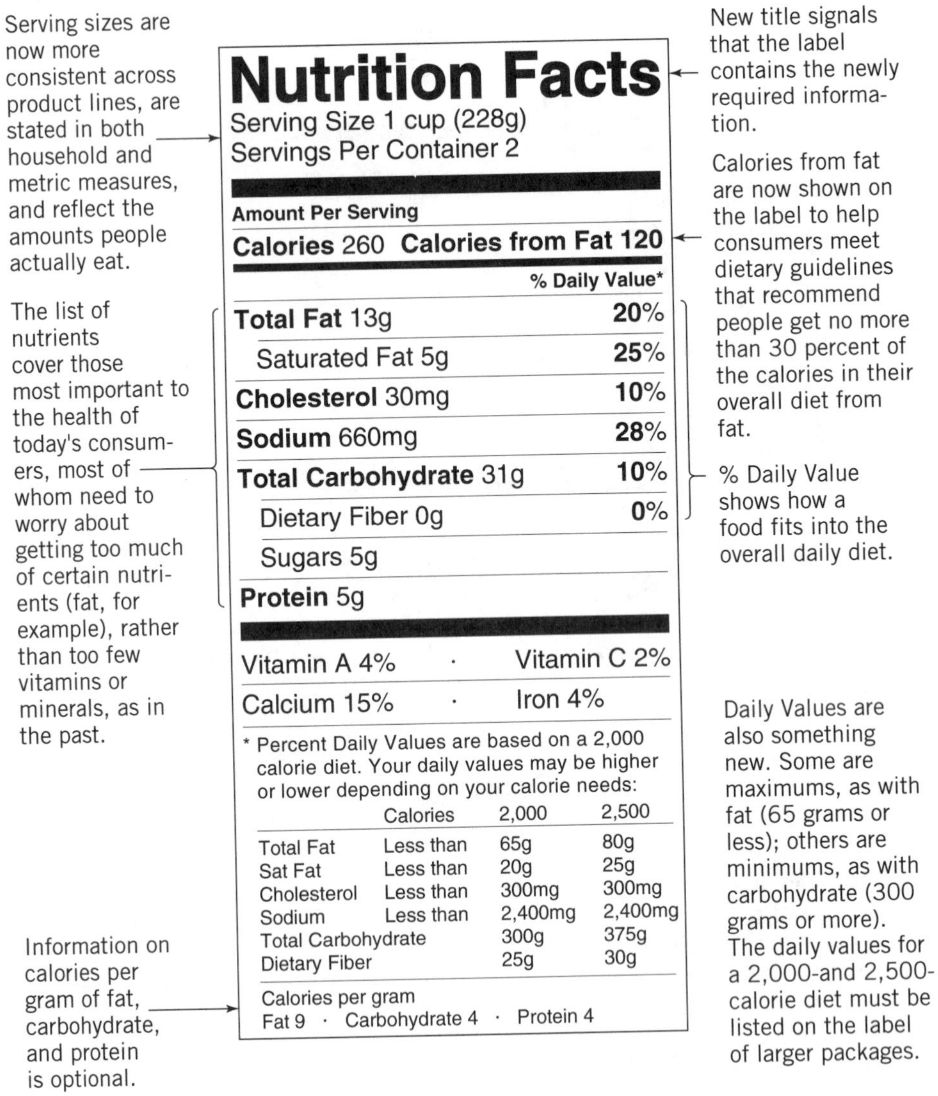

Serving sizes are now more consistent across product lines, are stated in both household and metric measures, and reflect the amounts people actually eat.

The list of nutrients cover those most important to the health of today's consumers, most of whom need to worry about getting too much of certain nutrients (fat, for example), rather than too few vitamins or minerals, as in the past.

Information on calories per gram of fat, carbohydrate, and protein is optional.

Nutrition Facts
Serving Size 1 cup (228g)
Servings Per Container 2

Amount Per Serving
Calories 260 **Calories from Fat 120**

	% Daily Value*
Total Fat 13g	**20%**
Saturated Fat 5g	**25%**
Cholesterol 30mg	**10%**
Sodium 660mg	**28%**
Total Carbohydrate 31g	**10%**
Dietary Fiber 0g	**0%**
Sugars 5g	
Protein 5g	

Vitamin A 4% · Vitamin C 2%
Calcium 15% · Iron 4%

* Percent Daily Values are based on a 2,000 calorie diet. Your daily values may be higher or lower depending on your calorie needs:

	Calories	2,000	2,500
Total Fat	Less than	65g	80g
Sat Fat	Less than	20g	25g
Cholesterol	Less than	300mg	300mg
Sodium	Less than	2,400mg	2,400mg
Total Carbohydrate		300g	375g
Dietary Fiber		25g	30g

Calories per gram
Fat 9 · Carbohydrate 4 · Protein 4

New title signals that the label contains the newly required information.

Calories from fat are now shown on the label to help consumers meet dietary guidelines that recommend people get no more than 30 percent of the calories in their overall diet from fat.

% Daily Value shows how a food fits into the overall daily diet.

Daily Values are also something new. Some are maximums, as with fat (65 grams or less); others are minimums, as with carbohydrate (300 grams or more). The daily values for a 2,000-and 2,500-calorie diet must be listed on the label of larger packages.

Figure 3. The new food label at a glance. *Source:* Ref. 2.

Calories and calories from fat: 5

Total fat, saturated fat: 0.5 g

Cholesterol: 2 mg (between 2 and 5 mg may be expressed as less than 5 mg)

Sodium: 5 mg

Total carbohydrate, dietary fiber, sugars, and protein: 0.5 g (values between 0.5 and 1 g may be expressed as "less than 1 g")

Vitamins and minerals: less than 2% DV

If a serving of a food contains insignificant amounts of 7 or more of the required 13 (total fat and calories from fat count as 1), several of the nutrients may be left off the label, and the Nutrition Facts may be in a simplified format (Fig. 4). The regulations set five "core nutrients"—total calories, total fat, sodium, total carbohydrate, and protein—that must be shown even if the amount present is 0.

The regulations list a number of nutrients that are optional. These include polyunsaturated fat, monounsaturated fat, potassium, soluble fiber, insoluble fiber, sugar

Nutrition Facts
Serving Size 1 can

Amount Per Serving
Calories 140

	% Daily Value*
Total Fat 0g	**0%**
Sodium 20mg	**1%**
Total Carbohydrate 36g	**12%**
Sugars 36g	
Protein 0g	

* Percent Daily Values are based on a 2,000 calorie diet.

Figure 4. Simplified format (example: soft drink). *Source:* Ref. 1.

alcohols (such as xylitol or sorbitol), and other carbohydrates (including starches).

Claims may result in some of the optional nutrients being required. For example, a "cholesterol free" claim requires monounsaturates and polyunsaturates to be shown in the nutrition facts box in addition to saturates.

In addition to the four required micronutrients, the regulations list 21 more vitamins and minerals, for a total of 25. Any of those listed that are present at levels of 2% DV or more may also be included in the Nutrition Facts. Any that are added by the manufacturer to foods such as breakfast cereals or fortified drinks must be included. Also, claims related to a vitamin or mineral, such as "high in niacin," requires the particular nutrient to be included in the nutrition facts. No vitamin or mineral that is not on the list in the regulations may be shown.

The information below the vitamins and minerals that shows the DV is called the footnote. The DV are based on recommendations agreed to by several well-known health groups, such as the American Medical Association and the American Heart Association. The amounts for total fat, saturated fat, and total carbohydrate were set to approximate the amounts of these nutrients in an ideal diet with about 30% of calories from total fat, 10% of calories from saturated fat, and 50 to 60% of calories from total carbohydrate. The amounts for the 2,000-calorie diet are used to calculate the "% Daily Values" in the macronutrient portion of the Nutrition Facts box.

The % DV for the vitamins and minerals are based on the Recommended Daily Intakes (RDIs) set by FDA regulations. These in turn are derived from the Recommended Dietary Allowances (RDAs) established by the National Academy of Sciences and the National Research Council (NAS/NRC).

The agency clearly wants the Nutrition Facts box to have a consistent look on all products. A few variations in the format of the Nutrition Facts box may be permitted under certain conditions, especially for packages that are small or have a shape that might restrict the area for labeling. (The simplified format mentioned earlier is based on nutrient content, not labeling area.) For example, some packages that are horizontal rather than upright may display the footnote to the side of the nutritional values instead of directly under them. Packages that have an area available for labeling of less than 40 in.^{2} may leave off the footnote or use a tabular display. Very small packages with labeling areas of less than 12 in.^{2} may use a linear display. Figure 5 shows examples of the tabular and linear displays.

Other format variations that are allowed include one for bilingual labeling, large Nutrition Facts boxes that show nutritional values for each different food in a multiproduct pack, and a format for showing products "as prepared" as well as "as packaged." Details of all formats must be exactly as prescribed in the regulations.

Claims

There are two categories of nutrition-related claims: nutrient content claims and health claims. The regulations define the claims that are permitted and the required qualifications quite strictly; they also provide for the review and possible approval of new claims.

Nutrient Content Claims/Descriptors Nutrient content claims and descriptors refer to elements of the composition of a food. "Reduced in calories" is one example. The new regulations officially expanded the earlier rules for reduced-calorie foods. Nutrient content claims are now permitted in the areas of sodium, fat and cholesterol, and light/lite.

The regulations set very detailed and specific requirements for each claim. For example, a "fat free" food must contain <0.5 g of fat per serving. "Cholesterol-free" means <5 mg of cholesterol per serving. A "low fat" food may contain no more than 3 g of fat in a serving, or, if the serving <2 tablespoons, no more than 3 g of fat in 50 g. This, together with established serving sizes, is supposed to help prevent manufacturers from reducing serving sizes to unrealistically small amounts just to make a claim.

The requirement for a "reduced calorie" food was changed from the earlier 33% (1/3) minimum reduction in calories to a 25% (1/4) reduction compared to the regular food. This was to make it easier for manufacturers to provide foods with reductions in calorie content. Higher reductions were applied to "light" or "lite" products, which must be reduced in calories by at least 33%, or reduced in fat by at least 50% compared to the regular food.

The new rules further extended the previous regulations for reduced-calorie foods to provide important incentives for manufacturers to develop foods with improved nutrition. Under old regulations, a food that was formulated to contain less fat or fewer calories might need to be labeled as "imitation." The specific provisions of the new rules make it clear that most foods that qualify for a "reduced calorie" claim, for example, but are nutritionally equivalent to the regular food in all other aspects, need not be called "imitation." Thus, we now see "reduced calorie" instead of "imitation" on products with at least 25% fewer calories.

The regulations also define the requirements for claims about vitamins and minerals. A food that is a "good source" of vitamin C, for example, must have at least 10% DV in a serving. To be "high" or "rich in" a nutrient, a serving of a food must have at least 20% DV. "More" means 10% or more of the DV, so that a drink that has "more calcium than milk" must contain at least 40% of the daily value of calcium compared to the 30% in a glass of whole milk.

Nutrient content claims are summarized in Table 1. The NLEA regulations list several synonyms for each claim. Examples are "zero," "no," and "without" for "free"; "little" or "contains a small amount of" for "low;" and "lower" or "fewer (calories)" for "reduced." No nutrient content claims or synonyms are allowed other than those that are specifically permitted, defined, and listed.

The regulations also define "healthy." A "healthy" food must at least meet the requirements for a food that is low in fat and saturated fat and must contain at least 10% DV, without fortification, of one or more of vitamin A, vitamin D, calcium, iron, dietary fiber, or protein. The 10% DV requirement in the original proposal might have prevented calling some fresh fruits and vegetables "healthy" because of their high moisture content. The final rules, however, specifically exempted fresh fruits and vegetables from this requirement, permitting them to be described as "healthy."

Tabular Display

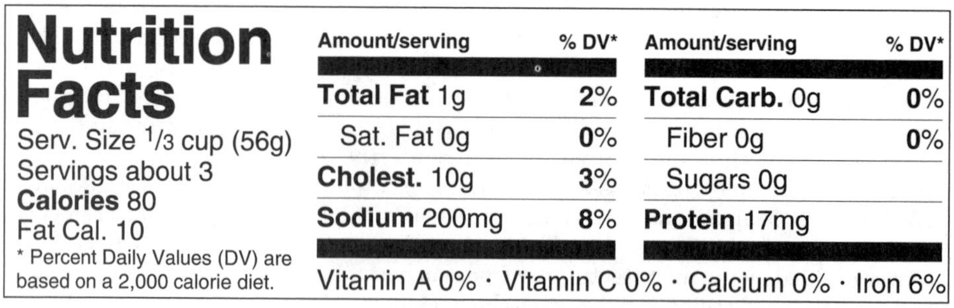

Nutrition Facts	Amount/serving	% DV*	Amount/serving	% DV*
Serv. Size 1/3 cup (56g)	**Total Fat** 1g	**2%**	**Total Carb.** 0g	**0%**
Servings about 3	Sat. Fat 0g	**0%**	Fiber 0g	**0%**
Calories 80	**Cholest.** 10g	**3%**	Sugars 0g	
Fat Cal. 10	**Sodium** 200mg	**8%**	**Protein** 17mg	
* Percent Daily Values (DV) are based on a 2,000 calorie diet.	Vitamin A 0% · Vitamin C 0% · Calcium 0% · Iron 6%			

May be used when the area available for labeling is less than 40 square inches.

Linear Display

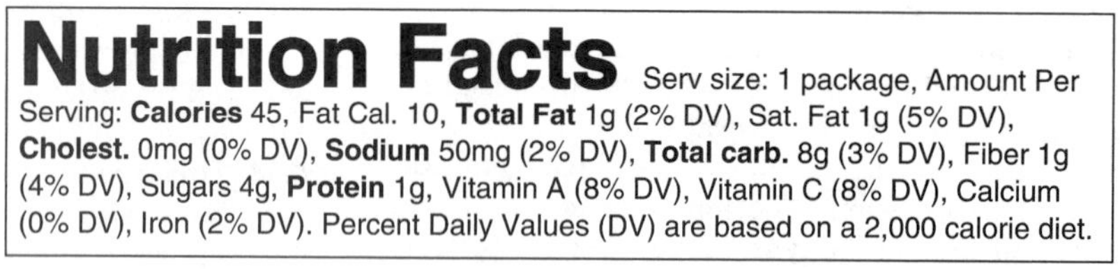

Nutrition Facts Serv size: 1 package, Amount Per Serving: **Calories** 45, Fat Cal. 10, **Total Fat** 1g (2% DV), Sat. Fat 1g (5% DV), **Cholest.** 0mg (0% DV), **Sodium** 50mg (2% DV), **Total carb.** 8g (3% DV), Fiber 1g (4% DV), Sugars 4g, **Protein** 1g, Vitamin A (8% DV), Vitamin C (8% DV), Calcium (0% DV), Iron (2% DV). Percent Daily Values (DV) are based on a 2,000 calorie diet.

May be used when the area available for labeling is less than 12 square inches.

Figure 5. Examples of permitted format variations. *Source:* Ref. 1.

Table 1. Summary of Nutrient Content Claims and Descriptors that May Be Used on Food Labels

Descriptor[a]	Definition[b]
Free	A serving contains no or a physiologically inconsequential amount: <5 cal, <5 mg of sodium, <0.5 g of fat, <0.5 g of saturated fat, <2 mg of cholesterol, or <0.5 g of sugar
Low	A serving (and 50 g of food if the serving size is small) contains no more than 40 calories, 140 mg of sodium, 3 g of fat, 1 g of saturated fat and 15% of calories from saturated fat, or 20 mg of cholesterol; not defined for sugar; for "very low sodium," no more than 35 mg of sodium
Lean	A serving (and 100 g) of meat, poultry, seafood, and game meats contains <10 g of fat, <4 g of saturated fat, and <95 mg of cholesterol
Extra lean	A serving (and 100 g) of meat, poultry, seafood, and game meats contains <5 g of fat, <2 g of saturated fat, and <95 mg of cholesterol
High	A serving contains 20% of more of the daily value (DV) for a particular nutrient
Good source	A serving contains 10–19% of the DV for the nutrient
Reduced	A nutritionally altered product contains 25% less of a nutrient or 25% fewer calories than a reference food; cannot be used if the reference food already meets the requirement for a "low" claim
Less	A food contains 25% less of a nutrient or 25% fewer calories than a reference food
Light	1. An altered product contains one-third fewer calories or 50% of the fat in a reference food; if 50% or more of the calories come from fat, the reduction must be 50% of the fat); or 2. The sodium content of a low-calorie, low-fat food has been reduced by 50% (the claim "light in sodium" may be used); or 3. The term describes such properties as texture and color, as long as the label explains the intent (e.g., "light brown sugar," "light and fluffy")
More	A serving contains at least 10% of the DV of a nutrient more than a reference food. Also applies to fortified, enriched, and added claims for altered foods
% Fat Free	A product must be low-fat or fat-free, and the percentage must accurately reflect the amount of fat in 100 g of food. Thus, 2.5 g of fat in 50 g of food results in a "95% fat-free" claim
Healthy	A food is low in fat and saturated fat, and a serving contains no more than 480 mg of sodium and no more than 60 mg of cholesterol
Fresh	1. A food is raw, has never been frozen or heated, and contains no preservatives (irradiation at low levels is allowed); or 2. The term accurately describes the product (e.g., "fresh milk" or "freshly baked bread")
Fresh frozen	The food has been quickly frozen while still fresh; blanching is allowed before freezing to prevent nutrient breakdown

[a]See the regulations for acceptable synonyms.
[b]These definitions have been simplified for this table; see the regulations for specific restrictions and additional requirements.
Source: Ref. 3.

Similar exemptions are expected for canned and frozen fruits and vegetables, as well as for bread and other grain products that have been fortified to meet the requirements of a standard of identity.

Health Claims Health claims refer to the possible effects of a food or the diet on health. The regulations as published in 1993 permitted health claims in only seven specific areas:

- Calcium and osteoporosis
- Dietary fats and cancer
- Sodium and high blood pressure
- Saturated fat, cholesterol, and heart disease
- Fruits, vegetables, and grain products that contain dietary fiber and cancer
- Fruits, vegetables, and grain products that contain dietary fiber and heart disease
- Fruits and vegetables (components other than fiber) and cancer)

Other permitted areas have since been added:

- Folic acid and neural tube defects in infants
- Sugar alcohols (such as xylitol) and dental caries (cavities)
- Soluble oat fiber in oats (oat bran) and heart disease

Direct claims for dietary fiber and cancer are not allowed at this time. This is ironic because the fiber versus cancer claims were among the first to appear on foods in the mid 1980s and helped reinforce the current interest in nutrition. Claims must be carefully worded to indicate that disease is the result of many factors and to position the food and its beneficial component within a healthy diet and lifestyle.

Several products on the market have taken advantage of the ability to make health claims. For example, consumers are familiar with product claims that link the fiber in oats with lessening the risk of heart disease, the calcium in dairy products with preventing osteoporosis, and the xylitol in chewing gum with cavity prevention.

Other Label Claims Labels may make other claims that are not considered related to nutrient content or health. "Fresh" and "fresh frozen" (see Table 1) are not nutrient content claims but were defined because of ongoing, existing prior differences in opinion as to their exact meaning as applied to specific foods.

FDA also considers statements such as "milk free" (related to allergy or religious concerns), "contains no preservatives" (no nutritive function), and "oat bran muffins" (statement of identity) not to be nutrient content claims. On the other hand, the statement "high in oat bran" is considered an implied nutrient content claim such that the product would have to meet all the necessary requirements, such as a minimum soluble fiber content of 0.75g/serving.

The Influence of Claims on Other Labeling Requirements Nutrient content or health claims typically lead to additional labeling requirements. For example, originally the label of any food that carried a nutrient content claim such as "low-fat" also had to carry a referral statement that directed consumers to the nutrition facts box. The line "See back panel for nutrition information" was typical. This requirement was made optional in 1998 in the belief that consumers are now familiar with the presence and location of the Nutrition Facts box.

The regulations also set disclosure levels, or levels that were considered high, for fat (13 g), saturated fat (4 g), cholesterol (60 mg), and sodium (480 mg). If a serving of a "reduced calorie" food, for example, contains more than the disclosure level of 480 mg of sodium, the label must carry a disclosure statement. The typical line "See nutrition information for sodium content" meets that requirement.

For some foods, the level of one nutrient may exclude a claim based on another. For example, peanut butter is naturally cholesterol free, but a serving also contains about 3 g of saturated fat. The regulations prohibit cholesterol-free claims for foods that contain more than 2 g of saturated fat in a serving, so the peanut butter label may not claim that the product is cholesterol free, although it may list "Cholesterol 0 mg" in the Nutrition Facts box.

Foods that exceed the same disclosure levels that trigger the disclosure statement for nutrient content claims may not make health claims at all.

FUTURE ISSUES AND CONCERNS

The public's interest in what is in their foods continues to be high, and the pressure to provide more information will surely continue. A few matters are left over from NLEA issues.

For the present, FDA considers the voluntary nutrition-labeling programs for fresh produce and fish to be adequate and has not proposed rules making nutrition labeling mandatory for these products.

Pressure continues from some quarters for the agency to require higher-profile nutrition labeling for all foods served in restaurants, not just those that make claims. This question is of particular concern to the operators of individual restaurants because of the difficulty involved in trying to make sure menu items that are prepared and served individually are consistent enough in composition for nutritional values to be meaningful.

Trans fats are unsaturated fats that form during partial hydrogenation of oils. They are different in structure from the *cis* forms that occur naturally. Products that contain significant amounts of partially hydrogenated fats or oils, such as shortenings and stick margarines, may be high in *trans* fats. Some recent research reports have suggested that *trans* fats may raise cholesterol levels in the body; however, whether the research findings have any significance in actual diets is not yet fully understood. The questions of whether and how to declare *trans* fats must eventually be decided.

The number of nutrient content claims and health claims that will be allowed will increase as more knowledge becomes available. The current interest in dietary supplements and functional foods will likely lead to the development of new products. Labeling will be complicated by questions of which products are foods and which are supplements.

In other areas, labels of foods that have been irradiated must disclose that fact. FDA has published guidelines for label statements on foods that require refrigeration and may eventually determine that detailed instructions for storage, preparation, and handling of certain foods should be required rather than optional. Labeling that calls specific attention to the presence of common allergens is beginning to appear. Standards for organic foods are being developed, and appropriate labeling will undoubtedly be required.

Within the public's interest in foods, a strong right-to-know element has been growing. Several consumer interest groups have asked the FDA to require percentage declaration of the major ingredients in foods. Specific groups have petitioned the agency to require the quantitative declaration of food components such as monosodium glutamate and caffeine. Groups have demanded that foods produced using biotechnology be labeled accordingly.

All this leads to the basic question of what kind of information should go on the label. Because label space is finite, a prevailing position has been that food-product labels should be required to carry only significant information about the composition and safety of the product in the package. It may be time for FDA and USDA to take positive steps to help avoid too much label clutter. It is worth remembering and realizing that there are many different ways to provide information to satisfy the public's right to know that do not necessarily involve the label itself.

ADDITIONAL INFORMATION

The FDA's *A Food Labeling Guide* provides detailed, practical information on food labeling in a question-and-answer format. It is available from the following:

Center for Safety and Applied Nutrition
Food and Drug Administration
5600 Fisher's Lane
Rockville, Maryland 20847

Several publications that describe the new food labels and are designed to help consumers use the information in planning their diets are available at this address as well.

The *Food Labeling Guide* and the other publications are also available on-line at *http://www.fda.gov/*. A wealth of additional information on labeling, nutrition, food safety, and related regulatory issues, including many of the materials used as references for this chapter, can be found at this site.

Nutrition and labeling information from the U.S. Department of Agriculture is available on-line at *http://www.usda.gov/*.

The magazine *FDA Consumer* is an excellent source of accurate and readable information on not only foods and labeling, but also other regulatory, health, and safety issues. FDA publishes it six times a year. Subscriptions are available from the following:

Superintendent of Documents
P.O. Box 371954
Pittsburgh, PA 15250-7954

BIBLIOGRAPHY

1. Food and Drug Administration, *A Food Labeling Guide*, Washington, D.C., 1994.
2. Food and Drug Administration, "21 CFR Part 1, et al.: Food Labeling; General Provisions; Nutrition Labeling; Label Format; Nutrient Content Claims; Health Claims; Ingredient Labeling; State and Local Requirements; and Exemptions; Final Rules," *Fed. Regist.* **58**, 2066–2941 (1993).
3. N. H. Marmelstein, "A New Era in Food Labeling," *Food Technol.* **47**, 81–96 (1993).

GENERAL REFERENCES

M. B. Browne *Label Facts for Healthful Eating: Educator's Resource Guide*, National Food Processors Association in cooperation with the Food and Drug Administration, U.S. Department of Health and Human Services, and the Food Safety and Inspection Service, U.S. Department of Agriculture, Washington, D.C., 1993.

Food and Drug Administration, "Focus on Food Labeling: Read the Label, Set a Healthy Table," *FDA Consumer* (Special Report) (1993).

Food and Drug Administration, "The New Food Label," *FDA Backgrounder, BG* 92–94 (1992).

Food and Drug Administration, *Notice to Manufacturers: Label Declaration of Allergenic Substances in Foods*, Washington, D.C., June 10, 1996.

Food and Drug Administration, "21 CFR Parts 1 and 101: Food Labeling; Mandatory Status of Nutrition Labeling and Nutrient Content Revision, Format for Nutrition Label; Correction," *Fed. Regist.* **58**, 17328–17340 (1993).

Food and Drug Administration, "21 CFR Parts 5 and 101: Food Labeling; Nutrient Content Claims; General Principles, Petitions, Definition of Terms; Definitions of Nutrient Content Claims for the Fat, Fatty Acid, and Cholesterol Content of Foods; Correction," *Fed. Regist.* **58**, 17341–17346 (1993).

Food and Drug Administration, "21 CFR Part 5, et al.: Food Labeling for Human Consumption; Rules, Proposed Rule, and Notice," *Fed. Regist.* **58**, 44019–44096 (1993).

Food and Drug Administration, "21 CFR Part 101: Food Labeling; Nutrient Content Claims, General Provisions," *Fed. Regist.* **63**, 26978–26980 (1998).

Food and Drug Administration and U.S. Department of Agriculture, Food Safety and Inspection Service, *The New Food Label: Check It Out*, Washington, D.C., 1994.

M. S. Meskin, "Regulating Organic Foods," *Food Technol.* **52**(7), 144 (1998).

Nutrition Labeling and Education Act, 101st Congress, November 8, 1990.

U. S. Department of Agriculture, Food Safety and Inspection Service, "9 CFR Parts 317, 320, and 381: Labeling of Meat and Poultry Products; Final Rule," *Fed. Regist.* **58**, 631–691 (1993).

B. A. Watkins, "*Trans* Fatty Acids: A Health Paradox?" *Food Technol.* **52**(3), 120 (1998).

PHILLIP WELLS
Bestfoods North America
Somerset, New Jersey

NUTS. See NUTS in the Supplement section.

NUTRITIONAL QUALITY AND FOOD PROCESSING. See FOOD PROCESSING: EFFECT ON NUTRITIONAL QUALITY.

OILSEEDS AND VEGETABLE OILS

Most oils of vegetable origin are derived from plant seeds, hence the term oilseed. However, some vegetable oils, such as olive and palm oil, come from the plant's mesocarp, the fleshy pulp covering the seed. This article will discuss only the more economically significant vegetable oils. The terminology is reviewed and then the processing of oilseeds to make various products is discussed. Information about specific oils is also presented. Tables 1–3 show the makeup of the oilseeds, the fatty acid composition of the oils, and the amino acid composition of the proteins. Figure 1 shows the worldwide production for the last 10 yr of the most economically important vegetable oils. Although the relationship between various human diseases and fats and oils in the human diet is of great concern, space limitations prevent a discussion of this topic and a variety of sources are available (14).

A vegetable oil is made up of monoglycerides, diglycerides and triglycerides with the latter predominating. A triglyceride is composed of a glycerol molecule backbone with three fatty acid groups branched off it. Technically the term fatty acid applies to the series of compounds referred to as $Cn{:}n'$, where n is the number of carbon atoms and n' is the number of double bonds. The simplest fatty acid is formic acid (C1:0). However, fatty acids that occur in vegetable oils start with caproic (C6:0) and go up to behenic (C22:0) and erucic (C22:1). A saturated fatty acid has no double bonds between any of the carbon atoms. An unsaturated fatty acid has one (monounsaturated) or more (polyunsaturated) double bonds between the carbons. Oils having a higher percentage of saturated fatty acids have higher melting points. For example, shortening and margarine have a higher proportion of saturated fatty acids than salad or cooking oils. Occasionally the terms fat and oil are used interchangeably. However, in common usage, a fat refers to a triglyceride mixture that is solid at room temperature, whereas an oil is liquid at room temperature. Different degrees of solidification or plasticity can be achieved by mixing a saturated fat, such as palm or coconut, with an unsaturated oil such as soybean. An unsaturated oil can also be made solid through a reaction called hydrogenation. In this process, hydrogen gas, in the presence of a catalyst, attaches to the double bonds of the unsaturated fatty acids and transforms them into saturated fatty acids. This large-scale industrial process is used extensively to manufacture margarine and shortening.

When iodine monobromide or monochloride reacts with an unsaturated fatty acid, the bromine or chlorine add to the double bonds. This reaction forms the basis of an oil characteristic known as iodine number. This quantity, defined as the number of grams of iodine absorbed under standard conditions by 100 g of oil, is a measure of the unsaturation of the oil, ie, higher numbers denote greater unsaturation. Saponification is the reaction of a fat or oil and an alkali to yield glycerol and a salt of the alkali metal. This process is used in the manufacture of soap. The saponification number is defined as the number of milligrams of potassium hydroxide required to saponify 1 g of oil. Once the composition of a pure oil is known its saponification number can be calculated. This number can be compared with the experimentally determined saponification number of an oil of unknown purity to determine the amount of unsaponifiable, ie, nonoil, material in the latter oil. The refractive index is a measure of the degree that incidental light is bent in the oil. Because this index is a function of the oil's molecular structure and impurities, it provides an easy and quick method to identify an oil and determine its purity.

PROCESSING OILSEEDS

Once harvested, oilseeds must be transported to processing plants and may be stored before being processed. Preventing a reduction in oilseed quality during transportation and storage is a major problem. Although weather conditions throughout the growing season affect the quality of oilseeds, rain at harvesting causes particular problems. Wet seed must be dried and cooled before storage to prevent mold growth and free fatty acid formation; the latter is caused by enzymatic hydrolysis of the triglycerides. Wet seed is typically dried and cooled by passing air through the seed piles in the storage houses. Oil is removed in two ways: mechanically pressing it from the seeds and soaking the seeds in a solvent, usually hexane, that dissolves the oil. Because of its efficiency, this latter process, called solvent extraction, is used in almost all commercial operations (15). A combination of these two processes, prepress solvent extraction, is sometimes used for high oil content oilseeds.

Figure 2 is a simplified schematic of the oilseed solvent extraction operation. First the dried seeds are cleaned, to remove stones, metals, and other objects that would damage the processing equipment. They are then dehulled and flaked. The flakes, typically about 0.025 cm thick, provide a more efficient extraction medium. The flakes are either prepressed or sent directly to the extractor. Extracted flakes are heated in a desolventizer to vaporize the solvent, which is recycled to the extractor. The meal from the desolventizer can be used as is or processed further. The oil-solvent mixture coming from the extractor, called miscella, is sent to an evaporator, where the solvent is driven off and recycled. The crude oil is sent to refining to be treated with caustic to remove most of the nontriglyceride components. The oil is then bleached and deodorized to give it a light color and bland odor. If it is to be used as a salad oil, it is also winterized. In the winterization process the oil is cooled to a low temperature and any crystallized material is removed from the liquid oil, which is now a suitable salad oil.

Table 1. Composition of Various Oilseeds and Characteristics of Their Oils

	Coconut[a]	Corn[b]	Cottonseed[c]	Olive	Palm[d]	Palm Kernel	Peanut	Rape	Soybean	Sunflower
Seed composition										
Oil, %	34	3.6	36	17	70	40	49	40	20	47
Protein, %	3.3	8.0	33	0.9		8.4	26	23	36	23
Crude fiber, %	4.3	2.5	2.0			5.8	4.9	6.7	5.0	4.2
Moisture, %	47	16	4.7	44		8.4	6.5	7.5	8.5	5.4
Ash, %	1.0	1.2	4.6	2.2		1.8	2.3	4.5	4.9	
Number of seeds g	0.0005	3	9	0.5	0.03	0.3	0.4	180	6	11
Oil										
Saponification number	250–284	187–193	189–198	185–200	195–206	242–255	188–195	170–180 190[e]	189–195	188–194
Iodine number	7.5–10.5	103–128	99–115	77–94	51–58	10–23	82–106	97–108 112–131[e]	120–141	125–136
Refractive index	1.448–1.450	1.470–1.474	1.468–1.472	1.469–1.470	1.453–1.456	1.449–1.452	1.470–1.472	1.470–1.474	1.470–1.476	1.466–1.684
Specific gravity	0.91–0.919	0.915–0.920	0.916–0.918	0.912–0.913	0.857–0.860	0.856–0.874	0.910–0.915	0.906–0.914 0.916–0.917[e]	0.917–0.921	0.894–0.899
References	1,3	2,3	1,3	3,4	5	5	6,7	8	3,6	1,9

[a]Coprs.
[b]Corn contains more than 60% starch.
[c]Delinted seed.
[d]Does not include the kernel.
[e]The second value is for canola (low–erucic-acid rapeseed).

Table 2. Fatty Acid Content of Various Oilseed Oils in Weight Percent

	Canola	Coconut	Corn	Cottonseed	Olive	Palm	Palm kernel	Peanut	Rape	Soybean	Sunflower
					Saturates						
Caproic C6:0[a]							0.2				
Caprylic C8:0		7.1					4.0	0.1			
Capric C10:0		7.3		0.5			3.9	0.1			
Lauric C12:0		54.0		0.4			50.4	0.6		0.1	
Myristic C14:0	0.1	17.4		0.8		2.5	17.3	0.3	0.1	0.3	
Palmitic C16:0	5.7	6.1	11.0	19.9	11.0	40.8	7.9	13.3	2.9	10.8	6.0
Stearic C18:0	2.1	1.6	1.8	3.1	2.2	3.6	2.3	2.1	1.4	3.2	4.0
Arachidic C20:0	0.2		0.2					1.2		0.1	
Behenic C22:0	0.2							2.9	0.5	0.1	
Total saturates	*8.3*	*93.5*	*13.0*	*24.7*	*13.2*	*46.9*	*86.0*	*20.6*	*4.9*	*14.6*	*10.0*
					Unsaturates						
Oleic C18:1	57.7	5.0	25.3	25.7	72.5	45.2	11.8	47.8	33.0	24.0	18.0
Linoleic C18:2	24.6	1.3	60.1	48.5	7.9	7.9	2.1	29.2	15.4	54.4	70.0
Linolenic C18:3	7.9		1.1	0.1	0.6				6.2	6.8	
Gadoleic C20:1	1.0				0.3			1.2	12.2		
Erucic C22:1	0.2							0.1	25.5		
Total unsaturates	*91.4*	*6.3*	*86.5*	*74.3*	*81.3*	*53.1*	*13.9*	*78.3*	*92.3*	*85.2*	*88.0*
References	10	11	12	11	13	3	3	11	10	11	9

[a]C6:0 means fatty acid has 6 carbons and 0 double bonds, C18:3 means fatty acid has 18 carbons and 3 double bonds, etc.

SPECIFIC OILS

Coconut

Description. There is one species of coconut palm, *Cocos nucifera* L., but there are several geographical varieties. These are divided into two groups: the tall typica and the dwarf nana: The tall coconut palms can live longer than 80 yr. After an initial flowering period of 5–8 yr, they continuously produce from 80 to 150 nuts per year. The coconuts from this variety have high oil yields and good-quality fiber and copra (dried coconut meat). The dwarf varieties live for about 40 yr, producing fruit after their third year. These colorful palms produce coconuts with a tough copra unsuitable for commercial purposes and so are planted mainly as an ornamental. Coconuts consist of an outer husk covering a thin, hard shell that in turn covers the copra. The inner cavity of the copra contains a watery substance called milk. A ripe nut weighs 2–3 kg and is about 25 cm long.

Origin and Cultivation. Records indicate that the coconut palm has existed for more than 3,000 yr. Although some theories place its origin in Central or South America, it most likely originated in Southeast Asia (16). The coconut palm grows in tropical areas throughout the world; in the western hemisphere, it is found as far north as Florida and as far south as Brazil. India, Indonesia, and the Philippines are the main coconut-producing countries. Africa, Latin America, and Oceania also are significant producers.

Composition and Uses. The husk consists of fibers that have a variety of uses and comprise more than 50% of the nut's mass. The shell and the milk each make up about 12% of the mass and the copra accounts for about 18%. More than 60% of the copra is a highly saturated oil. Because a large percentage of the oil's fatty acids is lauric (C12:0), coconut oil is known as a lauric acid oil. Although the oil has the greatest economic value, all parts of the coconut are used (16). The fibers from the husk, called coir, are used to make mats, nets, bags, ropes, and similar items. The shells are used to make activated carbon. The oil, extracted from the copra by crushing, is used in baking and in a variety of prepared foods. Of all the edible oils, coconut oil has the most nonedible uses. It is used in cosmetics, toiletries, and in the manufacture of soap by saponification. The glycerol by-product from soap manufacture is used in pharmaceuticals and in the manufacture of explosives. Coconut oil is used in the production of plasticizers, resins, detergents, and as a lubricant and fuel.

Corn

Description. *Zea mays* L. is a tall annual plant belonging to the grass family Graminese. Although it is commonly called maize throughout most of the world, it is referred to as corn in the United States and Canada. The female inflorescence, the ear, is where corn kernels are produced. Typically an ear of corn has about 800 kernels attached to its inner cylinder, called the cob (17). Kernels at either end of the ear are rounded whereas those at the center are flattened by pressure from adjacent kernels. An average kernel from the center of the ear is about 0.4 × 0.8 cm thick and about 1.2 cm long. Dent and sweet are two types of corn grown extensively in the United States. When drying, the center part of the dent corn kernel collapses making a distinct indentation. The sweet variety has mutant genes that retard the conversion of sugar to starch in the kernel.

Table 3. Amino Acid Composition of the Protein in the Various Oilseed Meals in Weight Percent

	Coconut	Corn	Cottonseed	Peanut	Rape	Soybean	Sunflower
Ala	0.2	0.8	1.5	1.0	1.7	1.7	1.1
Arg	0.6	0.4	4.4	3.1	2.3	2.8	2.4
Asp	0.3	0.7	3.5	3.2	3.1	4.6	2.5
Cys	0.1	0.2	0.9	0.3	0.4	0.6	0.5
Glu	0.8	1.8	8.2	5.4	6.3	7.1	5.6
Gly	0.2	0.4	1.6	1.6	1.9	1.7	1.5
His[a]	0.1	0.3	1.0	0.7	1.1	1.0	0.6
Ile[a]	0.1		1.2	0.9	1.5	1.8	1.1
Leu[a]	0.3	1.1	2.2	1.7	2.7	3.0	1.7
Lys[a]	0.2	0.2	1.7	0.9	2.3	2.4	0.9
Met[a]	0.1	0.2	0.5	0.3	0.7	0.5	0.5
Phe	0.2	0.4	2.0	1.3	1.5	1.9	1.2
Pro	0.1	0.8	1.4	1.1	2.7	2.1	1.2
Ser	0.2	0.4	1.6	1.3	1.7	2.1	1.1
Thr[a]	0.1	0.3	1.2	0.9	1.7	1.6	0.9
Trp[a]	0.04	0.1	0.5	0.3	0.4	0.5	0.4
Tyr	0.1	0.3	1.2	1.1	0.9	1.4	0.7
Val[a]	0.2	0.4	1.7	1.1	1.9	1.8	1.3
References	1	2	1	6	8	6	1

[a]Essential amino acid.

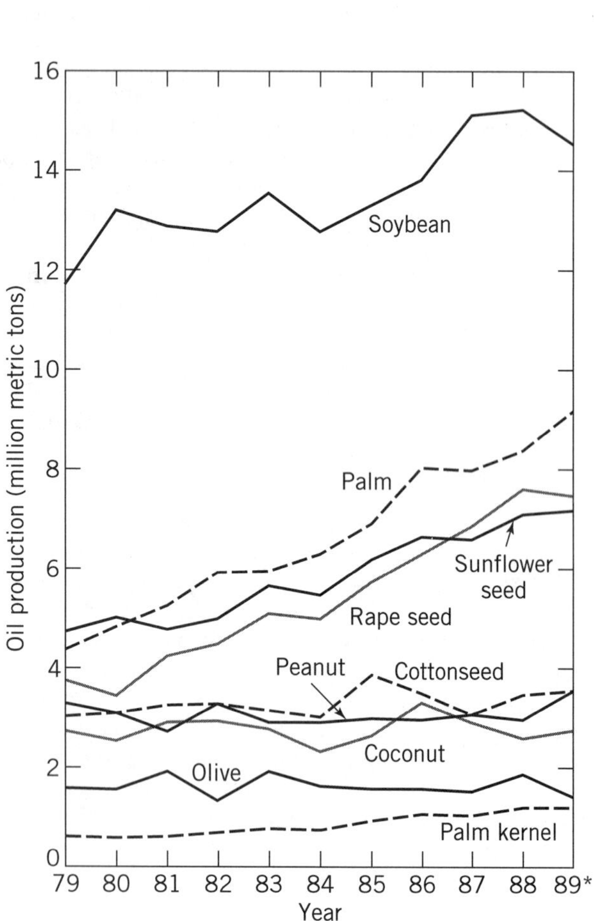

Figure 1. Graph of oil production for the last 10 yr for individual oils. Asterisk denotes projection.

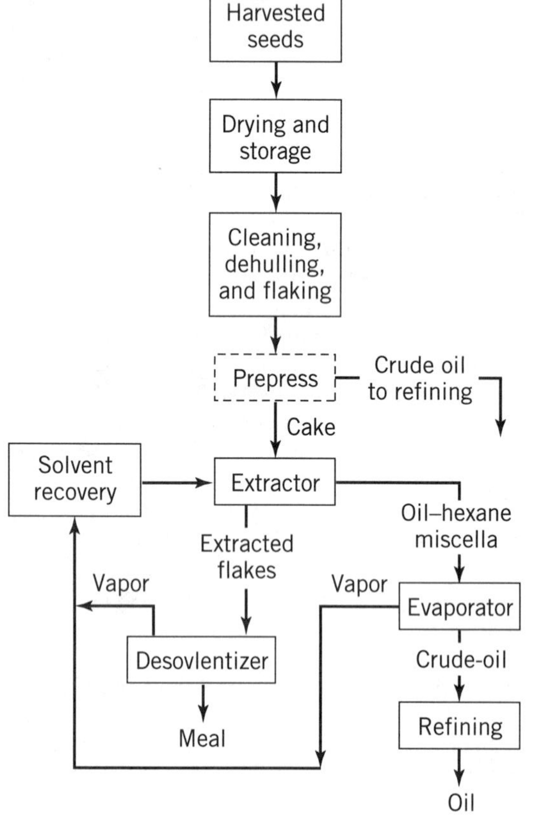

Figure 2. Oilseed solvent extraction flow sheet.

Origin and Cultivation. Archaeological evidence of earliest corn places its age at greater than 7,000 yr. The evidence further shows that this corn was domesticated, ie, natural seed dissemination was impossible. Corn probably originated in Mexico and spread north and south. Early

European explorers brought it to Europe, from where it spread to Africa and Asia (18). Every continent except Antarctica produces corn.

Today, the U.S. Corn Belt is the largest area of corn cultivation in the world. Through years of managed hybridization, Corn Belt corn is also the world's most productive variety. In fact, depending on climatic conditions, the per acre yield of corn is greater than any other crop. Over the last 60 yr, U.S. corn-growing area has decreased by half while production has doubled because of increased yields (18). Corn grows in the middle latitudes of both hemispheres. The growing season lasts through the freeze-free period. Corn grows from seed to a mature plant, 2–4 m tall, in about four months. Weather changes cause corn yield to vary across the Corn Belt by as much as 20% in a given year.

Composition and Uses. Grain makes up more than 40% of the dry matter of the corn plant. Stover, the nongrain parts of the plant, make up the rest of the dry matter. Because of its low oil content, corn is not generally considered an oilseed. More than 60% of the kernel is starch; oil comprises less than 4% (Table 1). In the United States, corn is primarily an animal feed crop, with less than 15% used for human consumption or various industrial purposes. For nonfeed purposes, corn is processed in one of two ways. Dry milling involves separating the germ from the whole kernel with an abrading action. The germ is then pressed or solvent extracted to remove the oil, which is used primarily as a salad and cooking oil. The rest of the kernel is then milled to produce grits, cornmeal, and flour. These products are used in a variety of applications, including snack products, breakfast cereals, brewing, pharmaceuticals, and building products. In wet milling, the corn is first softened by soaking it in a steeping liquor. The softened corn is then drained, coarsely ground, and combined with water to make a slurry. The slurry is then separated, in either hydroclones or floatation tanks, into an oil-bearing germ overflow and a starch underflow. The germ is further processed to extract oil. Most of the starch is processed into sweeteners and ethyl alcohol.

Cottonseed

Description. Four species of cotton are cultivated: *Gossypium arboreum*, L. *G. herbaceum* L., *G. hirsutum* L., and *G. barbadense* L. These are divided into two groups: the Old World (or Asiatic) and the New World. The Asiatic group, consisting of *G. arboreum* L. and *G. herbaceum* L., has short, harsh fibers that limit their use in fabrics. Their yield is comparatively low. The New World group consists of *G. hirsutum* L. and *G. barbadense* L. Their fibers are long, from 2 to 4.5 cm, and fine, making them well suited for a variety of uses. The seeds of the former are egg shaped and about 0.8–1.2 cm in length. They are densely covered with short cotton fibers, called linters, which remain after the fiber has been removed by ginning. The seed coat, or hull, is quite strong. The seeds of *G. barbadense* are similar in shape and size but have no adhering linters after ginning.

Origin and Cultivation. Archaeological evidence suggests that cotton was used for string and fabric in India around 3000 B.C. Although there have been claims that cotton fabric was used earlier than this in Egypt and Peru, the most reliable information suggests that this earlier material was actually linen (flax). The earliest reference to cotton in recorded literature appears in a Hindu hymn around 800 B.C. Evidence exists that the oil from cottonseed was also used by early Hindus and Chinese (19,20). However, throughout much of history, cultivation of the cotton plant was for fiber; the seeds were generally left to decay and yield fertilizer. In modern history, one of the earliest uses of cottonseed as a by-product of the cotton plant was in 1665, when inhabitants of the British West Indies made oil from the cotton flower to use for medicinal purposes. In 1769, a group of Pennsylvanians exhibited oil expressed from cottonseed and presented a sample of it to the American Philosophical Society to promote its use (21). The value of the residue of the seed after removal of the, oil, ie, cottonseed cake, soon began to generate interest. The Royal Society of Arts of London discussed the value of cottonseed cake as cattle feed and the potential of using sugar mill equipment for expression of the oil. In 1783, to spur the development of this technology, the Society offered a gold medal to any planter in the British West Indies who could produce 1 t of oil and 500 lb of cake within one year. Although this offer was renewed annually for six years no one ever succeeded; the large quantity required was, apparently, too great an obstacle. The difficulty of manually separating seed and fiber caused cotton, and consequently cottonseed, production to be relatively small. However, with the invention of the cotton gin in 1793, this manual operation became mechanized and production of cotton rapidly increased. The problem of disposing the large amount of seed being produced soon became apparent. Although methods did exist for crushing seeds to extract oil, cottonseed presented a particular problem. The kernel, which contains most of the oil, is surrounded by a tough hull making, it difficult to grind. Furthermore, this hull is covered with short fibers that absorb the oil being expressed, thus reducing yields. This changed in 1829 when the first cottonseed hulling machine was patented. That same year, the first cottonseed oil mill was put into production in Petersburg, Va. Because these early hullers applied a grinding action to the seed, screening did not completely separate the hulls, fibers, and meats. In 1857 an improved huller was patented that cut the seed open so that the kernels could fall cleanly out of the hulls. This huller, which uses the same principles as modern hullers, was instrumental in the rapid expansion of the cottonseed crushing industry following the Civil War.

During the crushing industry's early years, oil was not much in demand. It was not well suited for illumination, lubrication, or drying purposes and was not particularly desired as an edible oil in the United States. However, a market for edible cottonseed oil did develop in Europe that led to a significant export trade with the United States. In the early 1870s, 9 million gal were being produced in the United States, mostly for export. Later, the discovery that an acceptable substitute for lard could be made by mixing cottonseed oil with certain animal fats greatly increased the domestic market for the oil. By 1883 yearly production of cottonseed oil had reached 15 million gal, most of which was consumed domestically. By the turn of the century, annual production was 115 million gal.

The Asiatic group is grown mainly in China, India, and Pakistan, although cultivars from the New World group are replacing them. *G. hirsutum*, commonly referred to as American Upland Cotton, probably originated in southern Mexico and Guatemala and eventually spread to the southeastern United States. American Upland accounts for most of the U.S. production and 75% of the cotton produced worldwide. Sea Island cotton, the common name for *G. barbadense*, originated in Peru and spread to the Caribbean Islands. It adapted to islands off the coast of Georgia and the Carolinas and to nearby coastal areas of the mainland. Because of its limited growing regions and eventual devastation from the boll weevil, its production in the United States has been abandoned. A variety of *G. barbadense*, called Pima or Egyptian, is still produced in the United States although it accounts for only a small percentage of the U.S. production, its demand has increased in recent years due to its long, silky fibers. This variety, probably developed by crossing *G. barbadense* with *G. hirsutum*, was transferred from the Americas to Egypt and is the only variety of cotton grown in that country. The Egyptian government is very protective of the seed's germplasm, prohibiting export of the seed or import of other varieties.

Composition and Use. The cottonseed kernel makes up about 55% of the seed's weight. The surrounding hull and linters comprise about 32% and 13%, respectively. After processing, the seed yields about 16% crude oil, 45% meal, 9% linters, and 26% hull by weight (22). Most of the oil is in the kernel, which contains 35% oil and 39% protein. The hull contains less than 1% oil and 4% protein. Figure 3 shows a cross section of seeds from two varieties of *G. hirsutum*. The dark specks distributed over the glanded cross section (Fig. 3a) are pigment glands, which contain various material that impart the characteristic yellow-red color to the oil and extracted meal. The major constituent of these glands is gossypol, a yellow polyphenolic pigment found in all parts of the cotton plant, although it is concentrated mostly in the seed. Gossypol is insoluble in water but is soluble in oil, which accounts for the dark color of extracted crude oil. The amount of gossypol in a moisture-free kernel varies for different varieties and growing conditions, generally ranging between 0.39 and 1.7% (23). The presence of gossypol is a major problem in the use of cottonseed as an animal feed. Although cattle, sheep, and other ruminating animals can consume large quantities of raw cottonseed with no ill effects, rabbits and swine are very sensitive and can die from the gossypol's interference with the oxygen-carrying capacity of the blood. Chickens are moderately sensitive; green egg yolks result when chickens are fed too much cottonseed.

The gossypol in ginned seed before any processing treatment is said to be free. Once moist heat is applied to the seed, as is done in expeller processing or in the cooking step before solvent extraction, the gossypol reacts with the protein and is largely inactivated. This form of gossypol is said to be bound. Once bound, the gossypol's toxicity to nonruminating animals and its ability to discolor hen eggs is, for the most part, negated. However, because gossypol binds to the lysine, the quality of the protein is reduced.

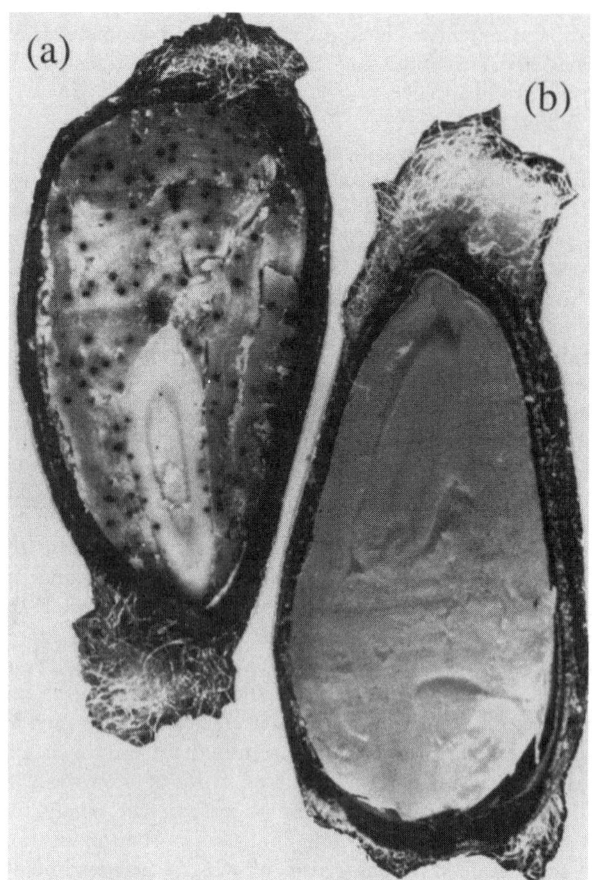

Figure 3. Photograph of a cross section of (**a**) glanded and (**b**) glandless cottonseed.

In 1953 a variety of *G. hirsutum* was discovered being grown by the Hopi Indians in Arizona that was free of gossypol glands on the leaves and bolls. The seeds, however, were still glanded. When these Hopi strains were crossed with cultivated Upland cotton, selections with glandless seed were found. After several generations of selecting for a reduced number of glands a true breeding glandless type was produced (24,25) (Fig. 3b). The glandless variety has a number of advantages. The crude oil is much lighter in color and consequently easier and less costly to refine. The meal is completely edible to both ruminants and monogastric animals. A variety of human food products, such as breads and snack items, can be produced from glandless cottonseed flour (26). However, despite the advantages, glandless varieties have not been a commercial success because the glandless varieties increase the potential value of only the seed, not the fiber. Because cottonseed represents less than 10% of the value of the cotton crop, farmers see no reason to shift to new cultivars (with all of the attendant uncertainties) for only a small potential increase in profits.

Of all the oilseeds, cottonseed probably has the most uses. The three major constituents of the seed, ie, kernel, hulls, and linters, all have economic value (27). The crude oil extracted from the kernel is refined and used in an assortment of food products. Snack and fast-food frying oil, salad and cooking oil, salad dressing, shortening, and margarine are major uses. One nonfood use for the refined oil has been as a carrier for agricultural sprays. By-products from the refining process are used in soap manufacture and as a source of fatty acids, which in turn have innumerable industrial uses. The meal is used as feed for beef and dairy cattle, swine, poultry, and fish and as fertilizer. The hulls are used in animal feed, as poultry litter, and as a mulch for soil conditioning. The linters are used in the preparation of high-quality bond paper and as felts for pads, cushions, comforters, and mattresses. They can be made into yarns or used as absorbent cotton for medical purposes. They can be processed so that the cellulose can be used in such diverse products as plastics, food casings, x-ray film, and many other industrial uses.

In recent years, the feeding of whole cottonseed to dairy cows has become increasing popular among dairy farmers (28). In 1981, 72% of the cottonseed production was crushed and 22% was fed as whole seed. In 1988, 64% was crushed and 35% was fed as whole seed (29). The whole seed is high in fat that, after digestion, increases the fat content, and hence the value, of the cow's milk.

Although gossypol has not yet been used commercially, much work has been done to find economically viable uses for it (30). Its antioxidant properties are well known. It also has significant antimicrobial properties. Possibly the area receiving the most attention is gossypol's potential use as a male contraceptive. In 1957 a Chinese researcher reported that a particular village had not had a single childbirth for a 10-year period between 1930s and 1940s. Before and after this time there did not seem to be a birthing problem. The researcher found that because of poor economic times, the villagers had switched to crude cottonseed oil for cooking. He postulated that its use might have caused female infertility. In the early 1970s researchers found it is the male who is actually affected. The gossypol initially reduces sperm motility and subsequently blocks sperm production (31). Much current research is being conducted on gossypol's potential contraceptive uses.

Olive

Description. The olive tree, *Olea europaea* L., is the only member of the Oleaceae family (which contains trees such as ash and shrubs such as lilacs) that is an important food source. The tree is a perennial evergreen, pyramidal in shape, that can grow to 20 m, but under cultivation is pruned to under 5 m. The fruit is fleshy, oblong or crescent in shape, and has a center stone. Olives change in color from green to red to black as they ripen.

Origin and Cultivation. The age of the olive tree is uncertain, but olives have been crushed for their oil for more than 6,000 years. The tree probably originated in Lebanon and Syria. It is believed to have been brought to Italy, through Greece, by Phoenician traders before the end of the fifth century B.C. (4). Olive trees grow in subtropical areas having dry summers and mild winters. For centuries, olives and olive oil have been among the most important agricultural products of Italy, Spain, Greece, and northern Africa.

Composition and Uses. The components of the olive are the edible pulp, which makes up 65–85% of the ripe fruit; the pit which consists of 13–23%; and the seed inside the pit, which is about 2–3% of the fruit. Olives have a high oil content but little protein. Green table olives are unripe and must be soaked in lye to remove oleuropein, a bitter principle, before pickling in brine. Ripe black olives do not need caustic treatment.

Oil is removed from olives by pressing. The oil obtained from the first pressing is called virgin. Virgin olive oil, which is used with no refining, is considered to be the highest quality salad and cooling oil. A second pressing of the olive yields an oil of lesser quality that must be refined, bleached, and deodorized. The residue from pressing can be solvent extracted to remove the remaining oil. The remaining olive cake has no feed use, because of its low protein content, and consequently is used as a soil conditioner.

Palm and Palm Kernel

Description. The oil palm, *Elaeis guineensis* Jacq., has a single stem that grows to 35 m in height and is topped by leaves typically 7 m in length. About six months after pollination, the fruit matures. It consists of a reddish orange oily pulp surrounding a kernel. The fruit, which is oval or pear shaped, about 3 × 5 cm, and weighs up to 30 g, grows in bunches in the axil of the leaves. These bunches can contain 1,500 fruits and weigh 20 kg. An adult palm tree is capable of producing 12 bunches per year. The trees are commercially useful for about 30 years.

Origin and Cultivation. *E. guineensis* Jacq. originated in west Africa along the Guinea coast. It was spread to other tropical regions by 15th-century Portuguese explorers. Palm oil, however, did not enter world trade until the end of the 18th century, making it one of the youngest of the major vegetable oils. Malaysia and Indonesia are the major producers of palm oil.

Composition and Uses. Both the pulp and kernel yield oil; each has a different fatty acid composition (Table 2). The pulp makes up 60–90% of the fruit's weight. On a dry weight basis, more than 70% of the pulp and 40% of the kernel consists of oil. A fruit bunch will yield about 20% palm oil and 2% palm kernel oil. Because palm oil is a relatively saturated oil, its food uses are mostly as shortening and frying oil. It is also used in the manufacture of soaps and fatty acids. Palm kernel oil is similar in composition and use to coconut oil.

Peanut

Description. *Arachis hypogaea* L., called peanut or groundnut, is a member of the family Leguminosae. Many varieties exist, but two general types are grown commercially. One is an upright plant with a single central stem reaching 30 cm in height and numerous upright branches. The second is a recombinant type, reaching 20 cm in height, with numerous creeping branches. The upright plant is better suited to mechanical harvesting. In all varieties of the peanut plant, small flowers form at the end of auxiliary branches called pegs. A few days after fertilization, the pegs push the flowers into the ground, where the peanut pod develops. At maturity, the pod contains one to three nuts or kernels.

Origin and Cultivation. The peanut originated in eastern South America. Evidence exists of its cultivation in Peru around 2000–3000 B.C.; it probably has a much longer history of domestication (9). By the time of the Columbus expedition in 1492 the peanut plant had spread to Central America and the Caribbean. Early European explorers probably brought the plant from Brazil to West Africa, from where it spread to India. Many varieties of peanuts now grown in the United States come from stocks developed in Africa. The plant requires sunshine and high temperatures. It can grow as far north as Canada and as far south as southern Argentina.

Composition and Uses. Most of the world production of peanuts is processed for recovery of the oil for edible use. Peanut oil, typically produced by solvent extraction, has a high smoke point, about 227°C, making it a good frying oil. However, peanuts are one of the few oil crops that can be used directly as food. They are among the highest protein content oilseeds (Table 1) and thus are nutritionally desirable. Eaten whole after shelling and oil- or dry-roasting (the terms used for deep-fat frying or hot-air cooking, respectively), they are one of the most popular snack foods in the United States. Peanuts are also used in a variety of confections, candy bars, and baked goods. About 25% of the U.S. supply of peanuts is processed into peanut butter, which is recognized as a nutritious, satisfying, and stable food and snack item for people in all age groups.

Rapeseed–Canola

Description. Rape refers to the oilseed forms of *Brasica napus* L. and *Brassica campestris* (32), which are in the same family as mustard seeds and are closely related to cabbage, cauliflower, broccoli, and other cole vegetables. The round seeds, about 0.1 cm in diameter, contain approximately 40% oil and yield a meal with more than 40% protein. The meal is used primarily as a feed supplement for livestock and poultry.

Origin and Cultivation. This family of oilseeds appears to have evolved separately in the Himalayan region of Asia and in the Mediterranean area of Europe. Sanskrit writings of 2000–1500 B.C. specifically mention the use of rapeseed oil for cooking and illumination; Greek, Roman, and Chinese writings between 500 and 200 B.C. ascribe medicinal value to them. It has been cultivated in Europe since about A.D. 1200. The Canadian rapeseed industry, the world's largest, started after World War II. By 1955 rapeseed oil was being used in Canada by major processors for salad oils (33).

Because rapeseed can survive at relatively low temperatures, they are one of the few vegetable oil sources that can be successfully cultivated in the colder temperate regions. For this reason they have become a major crop in Canada and throughout Europe. This characteristic also makes it possible for them to be cultivated as a winter crop in the subtropics. The winter form of rapeseed, typically found in Europe and Asia, is normally sown in August to September and is harvested in July. The summer form, found in Canada and Europe, is sown in April to May and harvested in September.

Composition and Use. Rapeseed oil differs from other vegetable oils because it contains significant quantities of eicosenoic and erucic fatty acids (Table 2). Studies since 1949 showed that erucic acid was poorly metabolized by rats and caused much physiological damage. In July 1956 the Canadian Department of National Health and Welfare ruled that rapeseed oil was not approved for edible purposes in Canada and all sales of oil for edible purposes were to cease immediately. In October of that same year the ban was rescinded, pending a thorough review, because of lack of evidence of harm to humans. Even though the ban was never implemented, the perception remained that health problems could arise from rapeseed oil consumption. Because of this, breeders began work to produce new varieties with reduced erucic acid content. In 1968 Oro, a variety of *B. napus*, was the first of several low-erucic-acid rapeseeds to be released. Industry began changing to these new rapeseed cultivars in the early 1970s. To differentiate the new oil composition, the terms low-erucic acid rapeseed (LEAR) and Canbra have been used. Sinola is the term used in the FGR for rapeseed containing less than 2% erucic acid. Most countries currently use the name canola for low-erucic acid rapeseed (Table 2).

Because rapeseed oil is increasing in importance for both edible and industrial uses in the United States, the U.S. Department of Agriculture is currently formulating standards for rapeseed. The proposed standards require that rapeseed oil with an erucic acid content of greater than 40% be used for industrial purposes only. Rapeseed oil with no greater than 2% erucic acid is edible and is called canola oil. Rapeseed oil containing between 2 and 40% erucic acid has little known commercial value. Although canola is used exclusively as an edible oil, rapeseed oil's high viscosity has traditionally made it a favorable lubricant for metal surfaces (34). However, where it was once used extensively for lubrication of steam locomotives and marine engines, synthetic derivatives are now used. Long-chain fatty amides of erucic acid are good plasticizers for vinyl chloride resins.

Soybean

Description. The cultivated soybean, *Glycine max* (L.) Merr., is an annual legume. It is usually erect, about 75 cm in height, and produces pods throughout its many branches. These pods, typically 2–10 cm in length, contain three yellow round seeds each about 0.5–1 cm in diameter.

Origin and Cultivation. *G. max* probably evolved from the wild species *Glycine soja* Sieb. and Zucc. Soy is sometimes referred to as one of the oldest cultivated crops. However, the earliest written record of it was 11th century B.C., long after the domestication of sesame in the Middle East. Although soybeans were first domesticated in northeastern China, the most recent evidence suggests that the plant originated in Australia (9). Seeds obtained from China were planted in Europe in the 1700s. In the early 1800s soybeans were introduced to the United States and later that century to Brazil. Not until the early 20th century was oil expressed from domestic soybeans in the United States, now the world's largest producer. Soybeans have only recently become important in Brazil; that country, the United States, and China are the world's major producers. Although soybean is a warm-temperature plant, it adapts well to temperate climates. Canada, the former USSR, South Africa, and Australia are some of the 35 countries that commercially produce soybeans.

Composition and Uses. Soybeans have the highest protein content of all the oilseeds. Processing the whole bean for food use is particularly popular in Asia. Soymilk, made by grinding beans that have been soaked in water, is consumed directly or used as a base for making tofu. Soy sauce is prepared by fermenting soybeans. The oil, obtained by solvent extraction, has many uses. Most of the production is consumed as salad oil, cooking oil, and margarine. It is also used in a variety of prepared foods such as frozen desserts, confections, and coffee whiteners (creamers). Because of its high linolenic acid content, soybean oil is considered a semidrying oil and has a variety of industrial uses. The major nonfood markets for the oil are in paints, resins, and plastics. Even though soybean oil is the dominant vegetable oil worldwide, the major economic value of soybean is in the protein. Defatted soybean meal is used mostly as animal feed; some, however, is also processed into flour, protein concentrate, and protein isolate, used in the manufacture of prepared foods.

Sunflower

Description. *Helianthus annuus* L. is the most common of the 67 species of sunflower and the only one grown commercially for oilseed production. A member of the family Compositae, the sunflower, is related to daisies, asters, marigolds, and dandelions. Although sunflower stems may reach heights of 5 m and have heads as large as 0.75 m in diameter (34), they typically are 1–3 m high with heads 0.3 m in diameter. The diameter of the stem is 3–6 cm. The head is actually a composite of 1,000–4,000 little flowers (florets) and will produce several hundred seeds. The head of the sunflower is heliotropic, ie, it turns to face the sun, until the majority of the flowers are fertilized, then it remains facing east. The easterly exposure minimizes the plant's temperature, which helps the seeds to set (9). The seed, more properly called an achene, is oblong and flattish, about 1.0–2.5 cm long, 0.75–1.5 cm wide, and 0.3–0.75 cm thick.

Origin and Cultivation. Sunflower seeds have been found at several archaeological sites in Mexico and southern and western United States. Although evidence exists of early precontact native Americans gathering wild sunflowers and making meal from the seeds, there is no evidence that they cultivated the plant. Even though corn, squash, and beans were cultivated, the prolific nature of wild sunflowers may have obviated the need to cultivate them. Around the 1500s, several native American nations were cultivating sunflower, although not nearly as much as they were cultivating corn. This is reasonable because corn is one of the best food plants known to humans.

The sunflower, like many other indigenous American plants, was taken back to Europe by early explorers. The first study of the sunflower, written by a Belgian herbalist (35), was published in 1568, and by 1616 sunflowers were common in England. Europeans did not at first know the uses for sunflower. In 1783 it was reported that not only are the seeds an excellent poultry feed but are also easily expressed to yield a good-quality oil. When sunflowers spread to Russia in the 1800s, the seeds became immensely popular there as a food source. This was because, during the forty days before Easter and the forty days before Christmas, the Russian Orthodox Church observed strict dietary guidelines. Almost all foods rich in oil were not to be eaten during these periods. Because the sunflower had only recently entered the country, it was not one of the prohibited foods. The people took advantage of this loophole and Russia became the foremost producer of sunflowers.

Although important in Europe, sunflowers were little cultivated by early American settlers. American seed companies did not offer sunflower seeds until Russian varieties were introduced in the 1880s. Even then, they never achieved great popularity. Only recently has the sunflower become an important oil crop in the United States. Ironically, even though the sunflower originated in the western hemisphere, USSR-cultivated varieties are the ones used almost exclusively throughout the world. Sunflower is commercially produced in warm to temperate regions. It grows well in the 20–28°C range. The former USSR is the major producer of sunflowers, although in recent years it has gained popularity in many countries. In the United States it is grown mainly in Minnesota and North Dakota.

Composition and Use. The oil content of the seed can vary from a low of 24%, caused by adverse growing conditions, to a high of 65% for experimental strains. The commercial seed typically contains about 40% oil. Its protein content is usually between 15 and 20% (Table 1). Probably the oldest use of sunflower seeds is in whole form as human food. In the former USSR they are eaten as commonly as peanuts are in the United States. In recent years, whole sunflower seeds have become increasingly popular in the United States because of the general increased interest of health food conscious consumers in whole grains and dietary fiber. Most of the commercial production of sunflower seeds goes into the manufacture of oil and animal feed. Because of its relatively high iodine number (ca 130) sunflower is considered a semidrying oil. As such, it can be used in the formulation of paints and for other industrial uses. It is, however, much more popular as a food and is considered by some as desirable a salad oil as olive. It is also used in cooking, frying, and in the manufacture of margarine and shortening. The meal left after oil removal is usually used as animal feed. In Canada and the former USSR the hulls are pressed into logs and used as fuel. The hulls are used in the manufacture of ethanol and furfural, as plywood filler, and in the production of yeast.

BIBLIOGRAPHY

1. U.S. Department of Agriculture, *Composition of Foods: Nut and Seed Products*, Agricultural Handbook No. 8–12, U.S. Government Printing Office, Washington, D.C., 1984.
2. K. N. Wright, "Nutritional Properties and Feeding Value of Corn and Its By-products," in S. A. Watson and P. E. Ramstad, eds., *Corn: Chemistry and Technology*, American Association of Cereal Chemists, St. Paul, Minn., 1987.
3. N. O. V. Sonntag, "Composition and Characteristics of Individual Fats and Oils," in D. Swern, ed., *Bailey's Industrial Oil and Fat Products*, Vol. 1, 4th ed., Wiley-Interscience, New York, 1979.
4. J. M. Martinez Suarez and J. A. Mendoza, "Olive Oil Processing and Related Aspects. Extraction and Refining: General Aspects," in A. R. Baldwin, ed., *Proceedings World Conference on Emerging Technologies in the Fats and Oils Industry, Cannes, France, November 3–8, 1985*, American Oil Chemists' Society, Champaign, Ill., 1986.
5. A. L. Winton and K. B. Winton, *Structure and Composition of Foods*, John Wiley & Sons, Inc., New York, 1932.
6. U.S. Department of Agriculture, *Composition of Foods: Legumes and Legume Products*, Agricultural Handbook No. 8–16, U.S. Government Printing Office, Washington, D.C., 1986.
7. E. M. Ahmed and C. T. Young, "Composition, Quality, and Flavor of Peanuts," in H. E. Pattee and C. T. Young, eds., *Peanut Science and Technology*, American Peanut Research and Education Society, Inc., Yoakum, Tex., 1982.
8. W. J. Pigden, "World Production and Trade of Rapeseed and Rapeseed Products," in J. K. G. Kramer, F. D. Sauer, and W. J. Pigden, eds. *High and Low Erucic Acid Rapeseed Oils*, Academic Press, Inc., Orlando, Fla., 1983.
9. E. A. Weiss, *Oilseed Crops*, Longman Inc., New York, 1983.
10. J. K. G. Kramer and F. D. Sauer, "Cardiac Lipid Changes," in Ref. 8.
11. R. Sreenivasan, "Component Fatty Acids and Composition," *Journal of the American Oil Chemists' Society* **45**, 259–265 (Apr. 1968).
12. F. T. Orthoefer and R. D. Sinram, "Corn Oil: Composition, Processing, and Utilization," in Ref. 2.
13. U.S. Department of Agriculture, *Composition of Foods: Fats and Oils*, Agricultural Handbook No. 8-4, U.S. Government Printing Office, Washington, D.C., 1978.
14. C. Galli and E. Fedeli, eds., *Fat Production and Consumption*, NATO ASI Series, Plenum Press, New York, 1987.
15. F. A. Norris, "Extraction of Fats and Oils," in D. Swern, ed., *Bailey's Oil and Fat Products*, Vol. 2, 4th ed., Wiley-Interscience, New York, 1982.
16. K. Satyabalan, "Coconut," in G. Röbbelen, R. K. Downey, and A. Ashri eds., *Oil Crops of the World*, McGraw-Hill, Inc., New York, 1989.
17. S. A. Watson, "Structure and Composition," in Ref. 2.
18. G. O. Benson and R. B. Pearce, "Corn Perspective and Culture," in Ref. 2.
19. A. N. Gulati and A. J. Turner, *A Note on the Early History of Cotton*, Indian Central Cotton Committee, Bulletin No. 17, Bombay, India, 1928.
20. M. R. Cooper, "History of Cotton and U.S. Cottonseed Industry," in A. E. Bailey, ed., *Cottonseed and Cottonseed Products*, Interscience Publishers, Inc., New York, 1948.
21. R. C. Curtis, *Cottonseed Meal: Origin, History, Research*, Robert S. Curtis Co., Raleigh, N.C., 1938.
22. J. P. Cherry, "Cottonseed Oil," *Journal of the American Oil Chemist's Society* 60, 360–367 (Feb. 1983).
23. L. C. Berardi and L. A. Goldblatt, "Gossypol," in I. E. Liener, ed., *Toxic Constituents of Plant Foodstuffs*, 2nd ed., Academic Press, Inc., Orlando, Fla., 1980.
24. S. C. McMichael, "Hopi Cotton, A Source of Cottonseed Free of Gossypol Pigments," *Agronomy Journal* **51**, 630 (1959).
25. P. A. Miller, "Genetic Basis and Breeding Procedures for Glandless Seeded Cotton," in *Glandless Cotton: Its Significance, Status, and Prospects: Proceedings of a Conference, ARS-USDA and National Cottonseed Products Association, Dallas, Tex., Dec. 13–14, 1977*. Agricultural Research Service-Southern Region, USDA, New Orleans, La., 1978.
26. D. C. Blankenship and B. B. Alford, *Cottonseed: The New Staff of Life*, TWU Press, Denton, Tex., 1983.
27. National Cottonseed Products Association, *Cottonseed and Its Products*, 9th ed., NCPA, Inc., Memphis, Tenn., 1989.
28. A. J. Kutches, W. Chalupa, and J. Trei, "Delinted Cottonseed Improves Lactational Response," *Feedstuffs* 16–17 (Aug. 17, 1987).
29. U.S. Department of Agriculture, *Oil Crops Situation and Outlook Report*, Economic Research Service, USDA, OCS-21, 37 (Apr. 1989).
30. R. J. Hron, Sr., S. P. Koltun, J. Pominski, and G. Abraham, "The Potential Commercial Aspects of Gossypol," *Journal of the American Oil Chemists' Society* **64**, 1315–1319 (Sept. 1987).
31. S. Z. Qian and Z. G. Wang, "Gossypol: A Potential Antifertility Agent for Males," *Annual Reviews of Pharmacology and Toxicology* **24**, 329–360 (1984).
32. R. K. Downey, "The Origin and Description of the *Brassica* Oilseed Crops" in Ref. 8.
33. J. M. Bell, "From Rapeseed to Canola: A Brief History of Research for Superior Meal and Edible Oil," *Poultry Science* **61**, 613–622 (1982).
34. J. S. R. Ohlson, "Rapeseed Oil," *Journal of the American Oil Chemists' Society* **60**, 337A–338A (Feb. 1983).
35. C. B. Heiser, *The Sunflower*, University of Oklahoma Press, Norman, Okla., 1976.

GENERAL REFERENCES

B. Alford, G. U. Liepa, and A. D. VanBeber, "Cottonseed Protein: What Does the Future Hold?," *INFORM* **5**, 1151–1155 (1994).

Anonymous, "Modern Practices in Cottonseed Oil Extraction," *Oils and Fats Int.* **5**, 32–33 (1989).

Anonymous, "Soybean-Based Tofu Production Develops in U.S.," *Oil Mill Gazetter* **103**, 20 (1997).

S. Berot et al., "Cottonseed Protein-Rich Products from Glandless African Varieties, 1. Pilot-Plant Scale Production of Protein Concentrates," *Sciences des Aliments* **15**, 203–215 (1995).

J. L. Bourely, "Development of Glandless Cottonseed as a Food Plant, Technological and Nutritional Aspects," *Sciences des Aliments* **10**, 485–514 (1990).

E. C. Canapi et al., "Coconut Oil," in Y. H. Hui, ed., *Bailey's Industrial Oil and Fat Products*, Vol. 2, 5th ed., John Wiley & Sons, New York, 1996, pp. 125–158.

Y. B. M. Che et al., "Aqueous Enzymatic Extraction of Coconut Oil," *J. American Oil Chemists' Society* **73**, 683–686 (1996).
G. Cole et al., "New Sunflower and Soybean Cultivars for Novel Vegetable Oil Types," *Fett / Lipid* **100**, 177–181 (1998).
M. G. Conception et al., "Chemical Characterization of Commercial Soybean Products," *Food Chem.* **62**, 325–331 (1998).
H. E. Davidson et al., "Sunflower Oil," in Y. H. Hui, ed., *Bailey's Industrial Oil and Fat Products*, Vol. 2, 5th ed., John Wiley & Sons, New York, 1996, pp. 603–689.
D. D. Duxbury, "Genetic Engineering Alters Fat Composition of Canola Oil," *Food Processing* **53**, 45–46 (1992).
N. A. M. Eskin and B. E. McDonald, "Canola Oil," *Br. Nutrition Foundation Nutrition Bull.* **16**, 138–146 (1991).
N. A. M. Eskin et al., "Canola Oil," in Y. H. Hui, ed., *Bailey's Industrial Oil and Fat Products*, Vol. 2, 5th ed., John Wiley & Sons, New York, 1996, pp. 1–95.
R. A. Ferrari, W. Esteves, and G. E. Plonis, "Triglyceride Composition of Corn Oil," *Ciencia e Tecnologia de Alimentos* **13**, 184–193 (1993).
L. Forman, S. Matthews, and C. C. King, "New Food, Feed Uses for Glandless Cottonseed," *INFORM* **2**, 737–739 (1991).
A. S. Guemueskesen and T. Cataloz, "Chemical and Physical Changes in Cottonseed Oil During Deodorization," *J. American Oil Chemists' Society* **69**, 392–393 (1992).
E. Hernandez and E. W. Lusas, "Trends in Transesterification of Cottonseed Oil," *Food Technol.* **51**, 72, 74–76 (1997).
J. Hounhouigan et al., "Revival of Coconut Oil Production Through the Hot Oil Immersion Drying Technique," *Plantations Recherche Development* **5**, 111–118 (1998).
L. A. Jones and C. C. King, "Cottonseed Oil," in Y. H. Hui, ed., *Bailey's Industrial Oil and Fat Products*, Vol. 2, 5th ed., John Wiley & Sons, New York, 1996, pp. 159–240.
A. P. Kiritsakis, *Olive Oil: From the Tree to the Table*, Food and Nutrition Press, Trumbull, Conn., 1998.
D. M. Klockman, R. Toledo, and K. A. Sims, "Isolation and Characterization of Defatted Canola Meal Protein," *J. Ag. and Food Chem.* **45**, 3867–3870 (1997).
D. Kritchevsky, S. A. Tepper, and D. M. Klurfield, "Lectin May Contribute to the Athergenicity of Peanut Oil," *Lipids* **33**, 821–823 (1998).
M. Naczk et al., "Current Developments of Polyphenolics of Rapeseed/Canola: A Review," *Food Chem.* **52**, 489–502 (1998).
J. R. K. Niemela, I. Wester, and R. M. Lahtinen, "Industrial Frying Trials with High Oleic Sunflower Oil," *Grasas y Aceites* **47**, 1–4 (1996).
J. M. Rouanet et al., "Comparative Study of Nutritional Qualities of Defatted Cottonseed and Soybean Meals," *Food Chem.* **34**, 203–213 (1989).
E. E. Sipos and B. E. Szuhaj, "Soybean Oil," in Y. H. Hui, ed., *Bailey's Industrial Oil and Fat Products*, Vol. 2, 5th ed., John Wiley & Sons, New York, 1996, pp. 497–601.
K. D. Tano and Y. Ohta, "Aqueous Extraction of Coconut Oil by an Enzyme-Assisted Process," *J. Sci. of Food and Ag.* **74**, 497–502 (1997).
M. Tsimidou, "Polyphenols and Quality of Virgin Olive Oil in Retrospect," *Ital. J. Food Sci.* **10**, 99–116 (1998).
R. E. Wilson, "New Commodity Products from Soybean Through Biotechnology," *Oil Mill Gazetter* **104**, 27–33 (1998).
C. T. Young, "Peanut Oil," in Y. H. Hui, ed., *Bailey's Industrial Oil and Fat Products*, Vol. 2, 5th ed., John Wiley & Sons, New York, 1996, pp. 337–392.

G. Abraham
R. J. Hron
U.S. Department of Agriculture
Southern Regional Research Center

OLESTRA. See Fats and oils: substitutes.

OLIVES AND OLIVE OIL

The cultivated olive (*Olea europaea*, Oleaceae) is a long-lived evergreen tree native to the Mediterranean basin. It is valued for its fruit and oil. The world production of olives in 1998 was 12,780,055 t (see Table 1). The Mediterranean countries, with ~9,000,000 ha dedicated to this cultivar, accounted for 95% of the production. In the Mediterranean, 90% of the olive trees are grown for oil. World olive oil production is currently increasing; the commercial crop during 1998 was 2,527,929 t (see Table 2), with the European Union (EU) producing around 1,876,000 t (Spain 51%, Italy 24%, and Greece 23%), Tunisia 170,000 t, and Turkey 190,000 t of oil. World table-olive production in

Table 1. World Production of Olives by Country for the Year 1998

Country	Production (t)	% World production
Afghanistan	900	0.01
Albania	40,000	0.31
Algeria	124,060	0.97
Argentina	91,940	0.72
Australia	1,000	0.01
Azerbaijan	300	<0.01
Brazil	4	<0.01
Chile	12,500	0.10
China	2,890	0.02
Croatia	10,405	0.08
Cyprus	9,000	0.07
Egypt	210,000	1.64
El Salvador	3,500	0.03
France	13,500	0.11
Gaza Strip	3,000	0.02
Greece	1,879,430	14.71
Iran	24,782	0.19
Iraq	13,000	0.10
Israel	18,700	0.15
Italy	2,232,000	17.46
Jordan	75,000	0.59
Kuwait	18	<0.01
Lebanon	98,000	0.77
Libya	58,000	0.45
Malta	10	<0.01
Mexico	7,205	0.06
Morocco	450,000	3.52
Peru	31,138	0.24
Portugal	287,000	2.25
Slovenia	235	<0.01
Spain	3,800,000	29.73
Syria	763,186	5.97
Tunisia	1,000,000	7.82
Turkey	1,300,000	10.17
United States	86,180	0.67
Uruguay	3,000	0.02
West Bank	130,000	1.02
Yugoslavia	172	<0.01
World	12,780,055	100.00

Source: FAOSTATS, Statistical database, Food and Agriculture Organization of the United Nations, 1998, *http: / / apps.fao.org*.

Table 2. World Production of Olive Oil by Country for the Year 1998

Country	Production (t)	% World production
Afghanistan	73	<0.01
Albania	2,500	0.10
Algeria	46,000	1.82
Argentina	11,500	0.45
Australia	90	<0.01
Chile	1,500	0.06
Croatia	1,572	0.06
Cyprus	1,200	0.05
El Salvador	525	0.02
France	2,000	0.08
Greece	430,000	17.01
Iran	1,650	0.07
Israel	23	<0.01
Italy	450,000	17.80
Jordan	14,100	0.56
Lebanon	6,500	0.26
Libya	10,400	0.41
Malta	2	<0.01
Mexico	400	0.02
Morocco	70,000	2.77
Portugal	44,000	1.74
Slovenia	470	0.02
Spain	950,000	37.58
Syria	110,000	4.35
Tunisia	170,000	6.72
Turkey	190,000	7.52
United States	490	0.02
West Bank	12,904	0.51
Yugoslavia	30	<0.01
World	2,527,929.24	100.00

Source: FAOSTATS, Statistical database, Food and Agriculture Organization of the United Nations, 1998, *http://apps.fao.org*.

Table 3. World Production of Table Olives for 1991–1992

Country	Production (thousand t)	% World production
Spain	230.0	25.4
Italy	130.0	14.3
Turkey	110.0	12.1
Morocco	90.0	9.9
Greece	85.0	9.4
Syria	50.0	5.5
United States	50.0	5.5
Argentina	30.0	3.3
Portugal	16.0	1.8
Israel	15.0	1.7
Tunisia	14.0	1.5
Jordan	13.0	1.4
Algeria	11.0	1.2
Peru	10.0	1.1
Egypt	9.0	1.0
Chile	9.0	1.0
Mexico	8.0	0.9
Cyprus	7.0	0.8
Other	7.0	0.8
Lebanon	4.0	0.4
Libya	3.5	0.4
Australia	2.0	0.2
France	2.0	0.2
Brazil	1.0	0.1
Yugoslavia	0.5	0.1
Total	907.0	100.0

Source: FAOSTATS, Statistical database, Food and Agriculture Organization of the United Nations, 1998, *http://apps.fao.org*.

1992 was 907,000 t. According to 1992 statistics, Spain is the world's largest producer (25.4%) and exporter of table olives, followed by Italy (14.3%), Turkey (12.1%), Morocco (9.9%), Greece (9.4%), Syria (5.5%), and the United States (5.5%). In Australia, Chile, China, Mexico, New Zealand, and South Africa, olives are considered a new crop (see Table 3).

The olive oil and table-olive industries play an important role in the agricultural and processing sectors of the major olive-producing countries. Most olive oil is consumed within the Mediterranean countries; only 18% of production enters world trade. On average, the world olive oil market represents ~6% of the quantity and ~23% of the value of the world trade in fluid edible oils. To a large extent olive oil does not compete with other vegetable oils but occupies a specialty niche market.

THE OLIVE TREE

Origin and Historical Evolution

The olive tree has a wide range of adaptability. It requires a mild climate with warm summers and cold winters. The tree needs substantial chilling for good fruiting (1) but is injured when temperatures fall below −10°C. Olive is considered a drought-resistant species because it thrives in areas where water stress is frequent, such as Mediterranean climates. It has been postulated that the minimum water requirement for olive is 2,000 m^3/ha per year, mainly during flowering and fruit setting in late spring and again in the summer as the fruit increases in size (2). Olive trees will grow on poor soils and rocky hillsides, but deep soils produce the best-quality fruit. They tolerate saline or alkaline soils and those with a high lime content. Their root system is relatively shallow and will not tolerate waterlogged soils.

The first reference to the existence of the olive tree is found in the Old Testament Book of Genesis, where the flight of the dove with an olive branch announces the end of the Flood. The origin of the cultivation of the olive tree is known through legends and tradition. It can be situated within a wide strip of land in the Mediterranean area (3) and adjacent zones comprising Asia Minor and parts of India, Africa, and Europe. However, the botanical ancestor of the olive tree is not precisely known (4). The two possible candidates are the oleaster *Olea sylvestris* and *O. chrisolphylla*, both of which might have had a common ancestor that covered most of the Sahara desert before the last glaciation. There also different hypotheses regarding the spread of its cultivation in the Mediterranean basin. This might have involved originally the Phoenicians and later the Greeks and the Romans. At the end of the Roman Empire, the olive trees were cultivated throughout the Arab and Roman worlds (5).

During the fifteenth and sixteenth centuries, colonizers and Spanish Franciscan monks extended the planting of the olive tree to various parts of the New World, finally reaching California during the eighteenth century. In the same manner, Italian and Spanish emigrants and missionaries spread the olive tree to Australia, South Africa, and Japan, completing its extension in both hemispheres. It is now distributed approximately between lat. 25 and 45° N and 15 and 35° S.

The Fruit

Composition and Changes during Growth and Maturation. The fruit of the olive tree is an edible, fleshy drupe, more or less oblong according to the variety, of a green color that changes to purple or black when mature and reaches a weight range of 1.5 to 12 g. The length of the fruit is generally between 2 and 3 cm and its transverse diameter between 1 and 2 cm. The specific weight is close to unity. The percentage of flesh, intensely bitter mainly when still green, varies between 70 and 90% of the fruit. The pit or stone represents 10 to 30% of the fruit, according to the variety, extent of growth, and maturity. The seed it contains accounts for less than 10% of the weight of the stone.

All the physical characteristics mentioned herein as well as the chemical composition of the edible flesh depend on several factors, among which the predominant are variety and degree of growth and ripeness when the fruits are harvested (6,7). Table 4 gives an overview of the composition of the fruit.

The main quantitative constituents of the flesh are water and oil, which show an inverse relationship for the same degree of maturity. The moderate degree of saturation of their fatty acids, and specifically, the high content of oleic acid, contributes to the fact that olive oil and table olives are considered high in biological and nutritive value. Water represents the main constituent of the olive fruit, accounting for up to 70% of its weight. Water serves as a solvent for water-soluble substances, including organic acids, tannins, and oleuropein.

Table 4. Composition of the Fruit of the Olive Tree and of the Fresh Pulp

			Weight %
	Fruit		
Pericarp	Epicarp Mesocarp	Pulp	70–90
Endocarp		Stone	9–27
		Seed	1–3
	Pulp		
Moisture			50–75
Lipids (oil)			6–30
Reducing sugars, soluble			2–6
Nonreducing sugars, soluble			0.1–0.3
Crude protein (N × 6.25)			1–3
Fiber			1–4
Ash			0.6–1
Organic acids and their salts			0.5–1
Phenolic compounds			1–3
Pectic substances			0.3–0.6
Other components			3–7

Oil is dispersed within the fruit cells in droplets that vary from 40 to 60 μm in diameter. The amount of oil increases through autumn and winter, reaching its maximum between late November and January when the fruit displays a reddish-bluish color, an indicator that it has reached the optimum-maturity stage. The oil content is lower in colder climates.

Soluble reducing and nonreducing sugars are the most important compounds for the fermentation and preservation stages of the process involved in the preparation of table olives. Glucose as a major component, followed by fructose and to a minor extent by sucrose, have been quantified; small amounts of xylose and rhamnose are also present, as well as mannitol (in the range of 0.5–1%), which is a poliol also important as fermentative matter. The evolution of all these components during the growth and maturation of the fruits has been studied for different varieties by gas-liquid chromatography (8).

Fiber is the fundamental support for the structure of the fruits. Its major components for fresh green fruits, in order of decreasing percentage, are cellulose, lignine, and hemicelluloses, with cellulose accounting for more than 50%, depending on the variety. However, when the degree of maturity advances, enzymatic degradation may produce changes in the relationship of the components. Thus, for Hojiblanca variety at different stages of ripeness, cellulose percentage goes from 40 to 45% down to 23%; lignine remains almost steady, between 33 and 38%; and hemicelluloses goes from 22 to 23% up to 41% when fruits are completely ripe (9).

Hemicellulose isolated from fruits of the Gordal and Manzanilla varieties have been studied (10,11). Cellulolytic enzymes were detected in the flesh of the Hojiblanca variety (12). These enzymes partially hydrolyze the cellulose fraction of the fiber to glucose and contribute to the softening of the fruits. Different factors influencing their action, such as incubation time, presence of sodium chloride, temperature, storage time, and degree of maturity when fruits are harvested, have been studied (13,14). Characterization and partial purification of the enzymatic complex cellulases and their inhibitors have been carried out (15,16).

Protein content is relatively low, between 1 and 3%, and remains almost constant during growth and ripening of the fruits. Hydrolysis shows that all essential amino acids are present (17). Ash percentage varies from 0.6 to 1, and the ash includes, in order of decreasing importance, K, Ca, P, Na, Mg, S, and, to a lesser extent, Fe, Zn, Cu, and Mn. The importance of the presence of organic acids and their salts in the juice of the fruits must be emphasized because of their buffering action during the fermentation stage. They range between 0.5 and 1%, based on the weight of the flesh.

Phenolic compounds, ranging from 1 to 3%, are responsible for color changes; pectic substances (0.3–0.6%) and pectic enzymes are related to texture (18); oleuropein, the most abundant phenolic glycoside found in olives (up to 2% in immature olives), is responsible for the bitterness of the fruit. It diffuses into the aqueous phase during the processing of the olive fruit and is hydrolyzed by alkaline solutions. Finally, certain vitamins, such as carotene, thiamine, and riboflavin, complete the known picture of the composition of the fresh fruit (19,20).

The chlorophyll and carotenoid presence in fruits of Hojiblanca and Manzanilla olive varieties have been studied (20,21). The qualitative composition is the same for both and does not change with maturation time. However, during the growth and development of the fruit, a gradual, homogeneous decrease is observed in the individual concentration of both chlorophylls and carotenoids (22). Hojiblanca variety always shows a greater amount of pigments than does Manzanilla.

As a general rule, water content decreases. Conversely, oil percentage, weight, and volume of the fruits, and flesh-to-pit ratio increase during growth and maturation. Soluble reducing and nonreducing sugars also decrease in a continuous manner. On the contrary, protein, ash, and total fiber remain almost stable, although qualitative changes may be produced in the total fiber.

Main Varieties The chemical composition and physical properties of the fruit that are closely related to variety and harvesting time are decisive factors in the quality of the final product. Varieties of olive trees in different countries may be either autochthonous or imported. Autochthonous varieties, because of their nature, and imported varieties, because of changes in climate, soil, and methods of cultivation, may yields products with diverse characteristics (23–27). Olive tree varieties are not well known in many cases, so national or local names rather than botanical classification are used to identify them.

Among available varieties, it is necessary to select the most suitable for the specific use (olive oil vs. table olives), the type of processing, and a definite style. For that selection, different factors must be carefully analyzed: (*1*) geographical situation of planting age and distribution on the soil; (*2*) agricultural characteristics, such as productivity of the trees, ripening cycle, and pest resistance; (*3*) type of culture (irrigated or nonirrigated system); (*4*) pruning methods; and (*5*) harvesting procedure (manual or mechanical). All these variables have an important influence on composition and hence on parameters to be considered in further processing.

Physical properties of fruits also have a relevant importance at the time a certain variety is selected for a definite type of processing. Among them, the most important are size, shape, flesh-to-pit ratio, ease of pitting, color, and texture.

Varieties used for oil production have a ratio of flesh to pit ranging from 4:1 to 8:1 and a ripe fruit that contains 15 to 40% of oil, whereas table oil varieties have a ratio of 7:1 to 10:1 and a lower upper range of oil content.

Spanish Varieties Among the 22 dominant Spanish varieties (Seed and Plant Genetic Resources Service B AGPS. Olive Germplasm: Cultivars and World-Wide Collections. Food and Agriculture Organization of the United Nations), the major cultivars used for oil production include Picual, Hojiblanca, and Lechin de Sevilla, whereas those for processing table olives are the Gordal (or Sevillana), Manzanilla, Morona, and Cacereña.

Picual. Picual has the widest spread of any single cultivar in Spain and is well suited to oil production. It is the main tree for the Jaen area (Andalucia), the largest olive-growing region in the world. It grows well in hilly to mountainous areas and is able to withstand cold winters. The fruit is elongated and nearly symmetrical, with a slight bend at the apex. At green maturation it has a somewhat dark shade. Blackening starts from the apex, and the ripe fruit is uniformly dark black. Fruit size varies from 3 to 5g according to yield and growing conditions. The mesocarp is light colored, smooth, and relatively firm. The oil content is medium-high, 21 to 25%, on a fresh-weight basis. The flesh-to-pit ratio oscillates between 3.8 and 5.9:1.

Hojiblanca (O. europaea arolensis). Hojiblanca ripens later than Gordal and Manzanilla and has a higher oil content, between 23 and 29%. It is the second most common Spanish variety with regard to market volume, being the most appreciated for processing either as natural black olives in brine or as pickled black olives in brine, although it is also used as pickled green olives in brine. The shape of the fruits is regular, the pit is straight and the olives vary greatly in size, from 230 to 700 fruits/kg. Smaller-size grades are used for oil extraction. The flesh-to-pit ratio oscillates between 4.9 and 6.6:1.

Lechin. Lechin has its origin between Cordoba and Seville and is cultivated in the area of Seville and Granada. The fruit is ellipsoidal and slightly bulged on the back. The size ranges from 3.58 to 3.80 g, with an oil content between 23.5 and 26.8%. The flesh-to-pit ratio ranges between 4.0 and 6.1:1.

Gordal or Sevillana (O. europaea Regalis Clemente). Gordal or Sevillana is a big, ellipsoid hearth-shaped fruit, with an average size of 100 to 120 fruits/kg and a relationship of flesh to pit of 7.5:1. The form is ellipsoidal with an incision in the area of the peduncle that gives it a shape resembling a heart. It has a fine epicarp, a mesocarp with good texture, and a right and regular endocarp. The color is deep green that changes to purple-black when the fruit is fully mature. Oil content is low, generally below 10%, and the sugar content is around 4 to 6%. Early ripeness is characteristic.

Manzanilla (O. europaea pomiformis). Different subvarieties, depending on the region, are known by different local names, including Fina, Serrana, and Carrasqueña. The fruits are the most appreciated variety in the international market, mainly because of their prominent organoleptic properties. It is the most important variety in Spain and together with Gordal and Morona is almost exclusively destined for the preparation of Spanish-style pickled green olives in brine. Average size is 200 to 280 fruits/kg; the flesh-to-pit ratio is 6:1. Manzanillas are apple-shaped, with fine skin and flesh with an excellent texture. The color is light green with mottled white spots, the fruit turns black-violet when fully mature. The pit is right and presents a very smooth surface. This variety reaches ripeness later than Gordal and shows a higher oil content, sometimes up to 15% by weight of the fruit, and a lower sugar content.

Morona. The Morona variety is very close to Manzanilla, it is perhaps the same variety. Its production is relativey small and is located around Moron de la Frontera (Sevilla province).

Cacereña. Although quite close to Manzanilla, Cacereña generally presents a lower average size and rougher texture. It is most suitable for processing pickled black olives in brine.

Greek Varieties. The major cultivars used for oil production include Koroneiki, Megaritiki, and Tsounati. Those for processing table olives include Conservolea, Kalamata, and Halkidiki.

Megaritiki (O. europaea argentata). Variety with fruits of small size (2–5 g) with a curved, cylinder-conic shape. About half of the production is dedicated to oil and half to dry salt black olives.

Conservolea (O. europaea media rotunda). Conservolea is the most important Greek variety, representing 80 to 85% of olive production in the country. Fruits, varying in shape between round and oval, show certain characteristics similar to those of Manzanilla and Hojiblanca, although closer to the latter. Average size is 180 to 200 fruits/kg; flesh-to-pit ratio is around 8:1. The flesh is fine and the texture consistent. Oil content varies from 22 to 25%.

Halkidiki. Some authors consider the not-well-defined Halkidiki variety a subvariety of Conservolea. However, its shape is more elongated, its pit is slightly curved, and the flesh does not present as good a texture as does Conservolea. Average size is 120 to 140 fruits/kg, and flesh-to-pit ratio is 10:1. Used mainly for pickled green olives in brine, it is the second-most-produced Greek variety on the market. The oil content ranges from 19 to 20%.

Kalamata (O. europaea ceraticarpa). This excellent variety, of limited culture, is third in production in Greece. Fruits are cylindric-conic shaped and curved, showing a prominent tip at the end. Average size is 180 to 360 fruits/kg; the flesh-to-pit ratio is analogous to that of Conservolea (8:1). It reaches a nice natural black color on late ripeness, and the oil content is high (25.5%). It is used mainly for special styles of natural black olives in brine.

Italian Varieties. Italy is the second-largest producer of olive oil, with cultivars such as Leccino and Frantoio, but lags behind in table olive varieties.

Nocellara di Belice. Considered the prime Italian table olive. The fruit is medium size, round or oval, and similar to Manzanilla or Conservolea. Flesh-to-pit ratio is 6.5:1 to 8:1.

Ascolana Tenera. This is the most common variety in Italy and is also cultivated in Argentina, California, Israel, and Mexico. The average size is 115 fruits/kg, with an ellipsoidal, slightly asymmetric shape. For the production of Spanish-style olives, the fruit is harvested when the skin color is yellowish green. The flesh is soft and the skin has little resistance to alkaline treatment (soaking the olives in a strong alkaline solution for several days). Oil content varies from 17 to 18%.

Cucco. A resistant variety exclusive to Italy. The fruit maintains the green color for a longer time than other varieties and changes to a violet black when it reaches full maturation. Oil content is about 17%.

Sant' Agostino. The fruits are clustered in groups of two or three with an oblong ellipsoidal form. The average size is 135 fruits/kg with an oil content of 14 to 15%.

Santa Caterina. Variety from central Italy with an average size of 120 fruits/kg with an asymmetric ellipsoidal shape. The oil content is about 17%.

Bella di Spagna or Cerignola. Variety from central Italy with fruit of large size (110 fruits/kg). The shape is oblong ellipsoidal and the mature color is black with white spots.

Moroccan Varieties. The main cultivated variety for pickling in Morocco is denominated Picholine. It comes from Argelia, although its origin is probably French. Average size is around 280 fruits/kg. It is oval shaped, with an elongated, slightly curved pit. The flesh-to-pit ratio is around 5.1:1. It is a resistant variety, well adapted to different kinds of soil. The flesh is flavorful.

Argentinian Varieties. Argentina prepares and exports mainly the Arauco variety, also called Criolla, of Spanish origin. This olive has flesh-to-pit ratio of 7:1 and high oil content (22–24%); the fruits are elongated with a tip at the end. Pits are slightly curved. It is used mainly as pickled green olives in brine.

Turkish Varieties. The most appreciated Turkish varieties are Domat, for pickled green olives in brine, and Gemlik, for natural black olives in brine or in dry salt. Domat averages 180 to 190 fruits/kg; Gemlik 270 to 280 fruits/kg. Oil content is high for both, oscillating between 22 and 24%.

Californian Varieties. The United States is the main importer of olives, but it also produces a significant amount of table olives (see Table 3): 82,000 tons in 1988–1989 and 86,500 tons in 1991–1992. Five varieties are cultivated in California: Manzanilla, Mission, and Gordal or Sevillana, of Spanish origin; Ascolana from Italy; and Barouni, from Tunisia, the last being the least important in production. The Mission variety is the oldest in California. The oil content is about 22%; average size, 240 to 260 units/kg; and the ratio of flesh to pit is 6.5:1 for processed fruits. This is a very tasty variety that shows a nice black color when ripes.

Ascolana variety averages 110 to 120 fruits/kg, with a flesh-to-pit ratio of 8.2:1. Oil yield is not very high (19%), and the flesh is very delicate. It may be used as pickled green or black olives in brine.

French Varieties. France is the smallest European producer, with nearly 40,000 ha planted with olive trees. French production during 1993–1994 was 2 million t of olive oil.

Table olives and olive oil from France are products of high quality. Olive growing represents an important economic and social factor in the Mediterranean region. The major varieties used in France for olive oil are Aglandau, Bouteillan, Dcayet roux, Cayon, Salonenque, Brun, and Ribier. For table olives the varieties most commonly used are Belgentieroise, Picholine, and Lucque.

OLIVE OIL

Atherosclerosis and coronary heart disease (CHD) as its main clinical manifestation have a multifactorial origin.

An individual's susceptibility to CHD is determined by both genetic and environmental influences. Among the latter, diet plays a central role. Dietary factors exert their influence largely through their effects on blood lipids and lipoproteins, but also through their great influence on other established, modifiable risk factors.

The dietary factors most directly implicated in CHD are dietary fats. Numerous comparisons between populations have shown a strong correlation between the intake of saturated fatty acids (SFA) and CHD morbidity and mortality. A diet high in SFA, such as is customary in the U.S. and Northern European countries, is associated with high levels of CHD. On the other hand, the Mediterranean countries, where the traditional diet derives the majority of its fat calories from olive oil, display a low incidence of CHD.

The Seven Countries study (28) gave scientific proof of the association between a diet low in animal products and saturated fat and low mean population levels of serum cholesterol with low incidence and mortality from CHD. It also documented a negative correlation between the intake of monounsaturated fatty acids (MUFA) and the MUFA-to-SFA ratio as well as CHD. All-cause and CHD death rates were low in cohorts with MUFA-rich olive oil as the main fat, underscoring the favorable role of olive oil in the diet. A body of indirect evidence from interventional studies shows that the traditional Mediterranean diet with its abundance of plant foods, preferential and regular intake of olive oil, and low to moderate consumption of animal foods efficiently protects against CHD.

Recent findings indicate that olive oil and the Mediterranean diet yield their benefits not only through their effects on established CHD risk factors such as hyperlipemia, hypertension, diabetes, and obesity, but also through directly protective effects, particularly their antioxidative properties. In addition, it has been documented that a Mediterranean diet adapted from the traditional Cretan diet rich in monounsaturated fat, is efficient in the secondary prevention of coronary events and death (29).

Constituents of Olive Oil

Olive oil is a complex compound made of fatty acids, vitamins, volatile components, water-soluble components, and microscopic bits of olive (26). Primary fatty acids are oleic and linoleic acids. Oleic acid is monounsaturated and constitutes 55 to 85% of the fatty acids in olive oil. Linoleic acids is polyunsaturated and represents about 9%. The polyunsaturated linolenic acid is present in small amounts ranging from 0 to 1.5%. The major vitamins are vitamin E and carotene. The specific fatty acid composition, the vitamin content, and the presence of several other minor components such as phenolic compounds may all contribute to the multiple health benefits associated with the consumption of olive oil (30–32).

The unique organoleptic characteristics of olive oil are attributed to a number of volatile components (26,32–35). Aldehydes, alcohols, esters, hydrocarbons, ketones, furans, and other compounds have been quantitated and identified by gas chromatography-mass spectrometry, nuclear magnetic resonance (NMR), and other techniques (6,36). The presence of flavor compounds in olive oil is closely related to its sensory quality. Hexanal, trans-2-hexenal, 1-hexanol, and 3-methylbutan-1-ol are the major volatile compounds of olive oil. Volatile flavor compounds are formed in the olive fruit through an enzymatic process. Olive cultivar, origin, maturity stage, storage conditions, and processing (6,26,27,34,37,38) influence the flavor components of olive oil and therefore its taste and aroma. The components octanal, nonanal, and 2-hexenal, as well as the volatile alcohols propanol, amyl alcohols, 2-hexenol, 2-hexanol, and heptanol, characterize the olive cultivar. Some slight changes in the flavor components in olive oil are obtained from the same cultivar grown in different areas. The highest concentration of volatile components appears at the optimal maturity stage of the fruit. During storage of olive fruit, volatile flavor components, such as aldehydes and esters, decrease. Phenolic compounds also have a significant effect on olive oil flavor. There is a good correlation between aroma and flavor of olive oil and its polyphenol content (39). Hydroxytyrosol, tyrosol, caffeic acid, coumaric acid, and p-hydroxybenzoic acid influence mostly the sensory characteristics of olive oil. Hydroxytyrosol is present in good-quality olive oil, whereas tyrosol and some phenolic acids are found in olive oil of poor quality. Various off-flavor compounds are formed by oxidation, which may be initiated in the olive fruit. Pentanal, hexanal, octanal, and nonanal are the major compounds formed in oxidized olive oil, but 2-pentenal and 2-heptenal are mainly responsible for the off-flavor (26,40–42).

Extraction of Olive Oil

A good olive oil is obtained from healthy, ripe olives that are processed without delay because the fermentation processes begin right after harvest and deteriorate the quality of the product, elevating the acidity and affecting the aroma and flavor of the oil (7).

The steps involved in the extraction of olive oil involve cleaning, milling, mixing, and the actual extraction, and these are described in detail. An scheme showing a traditional pressing process is shown in Figure 1.

Cleaning. The first operations to which the olives are subjected before oil extraction are the removal of leaves and washing. Leaves are removed using an automated aspiration system, and washing takes places using forced water that removes leftover leaves, branches, dust, soil, pebbles, and pesticides. The removal of the leaves is important for the final organoleptic characteristic as leaves can give the oil a bitter taste.

Milling. This operation breaks the flesh cells to induce the release of the oil from the vacuoles and their coalescence into larger oil droplets that can be separated from the other components.

A traditional milling machine consists of a bowl in which several heavy wheels turn, crushing the olives. Traditionally, the bowl and the wheels were made of stone and the oil stones, between 2 and 6 per machine, were either conical or cylindrical. The average machine could handle up to 500 kg of olives, turning at speeds of 15 rpm and taking between 15 and 30 min. This procedure has been used for several centuries and has many advantages:

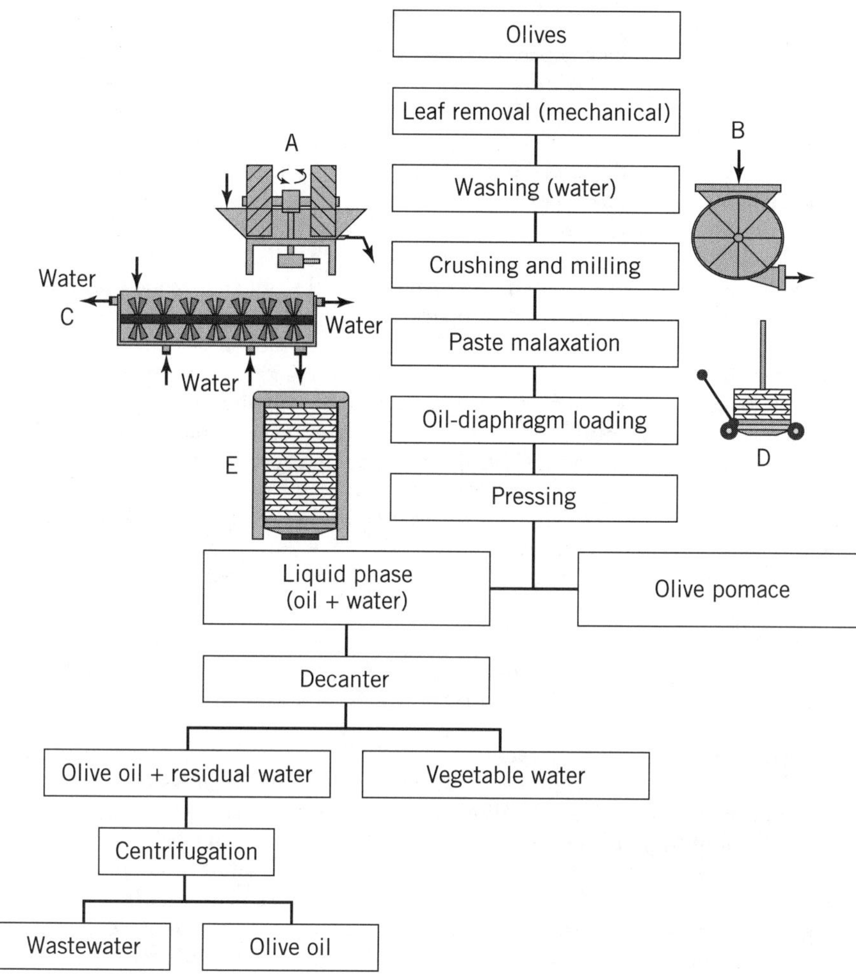

Figure 1. Scheme showing a typical pressing process. The crushing and milling can be done with (**a**) a traditional mill with cylindrical stones, or with the most modern apparatus, (**b**) steel hammer, (**c**) mixer, (**d**) oil diaphragms on trolley, or (**e**) hydraulic press.

- It does not produce emulsions.
- It does not raise the temperature.
- There is no contamination from metals.
- It crushes the pits to the proper size and achieves malaxation.

However, it also has certain disadvantages:

- High cost
- Low efficiency
- Discontinuous processing
- Need for highly skilled workers
- Increased oxidation due to lengthy air exposure of the paste

More recently, metal olive-crushing equipment, either roller, cylindrical, disc, toothed, or hammer, has been used. These work at much higher speeds (3,000 rpm) within a chamber that moves in the opposite direction at about 80 rpm. The advantages of these mechanical systems are:

- Continuous operation
- Lower cost
- Higher capacity
- Smaller size

Some drawbacks include:

- The high operating speed may leave a paste that is not properly prepared because the crushing is carried out quickly and may not be complete.
- Emulsions are created that are difficult to break.
- Alteration of the organoleptic characteristics of the oil induces bitterness.
- Wear and tear of the metallic parts results in contamination of the product.

Mixing (Malaxation). The mixing or malaxation of the paste resulting from the crushing involves stirring the mash slowly (~20 rpm) and continuously for 15 to 60 min. This process takes place in hemicylindric or semispheric water-heated (<30°C) stainless-steel vats with double walls and rotating blades. The mixing time depends on the crushing technique. When the traditional stone cones are used, 15 min at room temperature are enough, but when hammers are used, the mixing needs to be extended up to 60 min and the temperature increased to ~30°C. This temperature acts as an antiemulsionant and facilitates the activity of enzymes in the olives, which disrupt the membranes that surround small oil droplets and prevent self-aggregation. Modern techniques include the addition of inert coadjuvants (talc powder) during the mixing step to facilitate the emulsion separation and of enzymes to disrupt the membranes.

The purpose of this operation is to break the oil–water emulsion and promote the fusion of the small oil droplets into droplets of a diameter greater than 30 μm, the minimal dimension required for oil separation in continuous phase. Even after optimal crushing, only 40 to 45% of oil droplets have a diameter >30 μm. This percentage increases to 80 to 85% following good malaxation. Therefore, this operation is essential to increase the yield of the extraction regardless of the techniques utilized. Droplets of smaller size (<30 μm) remain as emulsion, and they stay in the subproducts.

Temperatures >30°C, lengthy mixing times, or both will have a negative effect on the aroma of the oil (43), as well as its antioxidant and vitamin content.

Extraction. The olive oil is separated from the olive paste using selective filtration (partial extraction), pressure (traditional system), or centrifugation (continuous system).

Selective Filtration. Selective filtration is based on the lower superficial tension of the oil compared to that of the water. Therefore, when both liquids are put in contact with the pores of the filtering surface, oil will pass through whereas water will be retained. The objective of selective filtration is not the total extraction of the oil in the paste, but the extraction of the oil naturally separated during the mixing step. Therefore, the factors that influence the amount of oil extracted by this system depend on:

- The amount of "free" oil in the paste
- Duration of the process, with an optimal time of about 30 min
- The characteristics of the equipment (filtering surface, rpm of the extractors, oil/water and oil/solids ratios)

The oil obtained by this procedure has a humidity content of ~1% and has to be centrifuged immediately.

Up to 60% of the oil in the paste can be extracted by this method. The product obtained from selective filtration maintains excellent organoleptic characteristics, low levels of acidity, and greater resistance to rancidity. Moreover, it has some other operational advantages, including:

- Low installation cost and maintenance
- Possibility of inclusion in previously established production lines (traditional or continuous)
- Low requirements in terms of labor and energy
- Facilitation of subsequent processing of the paste, thus improving the final recoveries

To understand the high cost of this product, is important to consider that the production of 1 L of olive oil by this process requires 11 to 12 kg of highly select olives, whereas only 5 kg are required for the extraction of 1 L of olive oil by the pressure processes described next.

Pressure. Pressure extraction the oldest procedure used to obtain olive oil. Originally, oil separation was achieved with pressure applied by humans or animals. Today, traditional olive oil mills use hydraulic presses. The olive paste is placed in thin layers (2–3 cm) over disks of filtration material (oil diaphragms). These diaphragm–paste layers are piled on top of each other and placed over a cart or trolley fitted with a central shaft to provide even distribution and support when these turrets are subjected to hydraulic pressure. This combination of trolley, shaft, oil paste, and diaphragms is subjected to pressure (300–400 kg/cm^2), the oil is extracted, and the apparatus is dissembled. Therefore this is a discontinuous process. The factors affecting the pressure process include:

- The characteristics of the oil diaphragms, which depend in part on the olive kernels in the paste and the degree of dispersion and concentration of the colloid constituents of the olive paste
- The humidity of the paste, the size and shape of the particles, and the physical characteristics of the olive (type, variety, ripeness, and temperature)

It is common to include a second step. Once the maximum pressure has been achieved, the pressure is partially released, allowing sponging of the paste and the diaphragms and reopening of channels that after pressure is reapplied allows the extraction of additional oil.

This system produces oils of excellent quality due to the low temperatures used during the process; however, it has elevated costs due to labor, the discontinuity of the process, and the use of optimal filtering materials.

Centrifugation. The use of centrifugation for oil extraction is relatively recent and is based on the different densities of the oil, the water, and the solids (pomace). Some of the first practical experiences with this technique were carried out with the Corteggiani system, a centrifuge with the capacity to process about 100 kg of paste spinning at 900 rpm. Water was added to facilitate the oil separation, and the process had to be stopped for removal of the solids.

Currently the process is carried out with a horizontal centrifuge (decanter). This system allows continuous solid–liquid separation and consists of a cylindrical–conical bowl that spins at speeds of 3,000 to 4,000 rpm. The interior consists of a hollow component of the same shape containing helical blades. There is a small difference between the speeds at which the bowl spins and the inner screw gyrates. The faster speed of the latter results in the movement of the pomace to one end of the decanter and the olive and water to the other end. The oily must is then fed to a vertical centrifuge revolving at 6,000 to 7,000 rpm for the final separation of the oil.

The separation of the solid and liquid phases by centrifugation requires the addition of water to the olive paste. The amount of water and the temperature influence the oil yield, and it is necessary to adjust both factors for each type of equipment.

Some of the advantages of the centrifugation process are:

- Small size of the equipment
- Automated process with semicontinuous cycles
- Reduced need for highly skilled labor
- Lower acidity of the oil
- Similar yields to those obtained using more-traditional systems
- Stainless-steel materials decrease the risk of contamination by other metals
- No diaphragms are used, thus improving the hygiene of the process and lowering the risk of contamination

Some of the disadvantages are:

- The initial capital investment is high.
- The process requires large amounts of water, which pollutes the environment and removes a significant amount of natural antioxidants.
- The energy consumption is high.
- The organoleptic characteristics of the oil may be adversely affected.

A variant consisting of a two-phase centrifugal continuous system reduces the use of water and eliminates the production of waste water (alpechin) (see Fig. 2).

Classification of Olive Oil Grades

Virgin Olive Oil. Oil extracted from olives by mechanical or other methods that do not modify its basic properties is called virgin olive oil. It is a completely natural product that maintains the taste as well as the chemical and biological characteristics of the olive. Within the virgin grade are three recognized quality levels:

Extra. Oil with the best organoleptic characteristics and with an acidity level not exceeding 1% is classified as extra virgin. The highest-quality extra virgin olive oils have an acidity level of at least 0.4 to 0.5% to maintain the organoleptic characteristics and lower than 0.7 to 0.8% so as not to exceed the maximum legal level. Its production requires the greatest care and attention from the cultivation stage all the way through to final processing.

The olives, unaffected by parasites, need to be harvested at the ideal point of maturation. If the fruit is still unripe, the extracted oil will be too sour. If the fruit is too ripe, the oil will be too sweet. The olives need to be harvested by hand to avoid any damage from mechanical processes. They should not be piled but should be arranged in layers of 15 to 30 cm in rigid containers. They should not be stored; pressing should take place immediately after removal of leaves and other detritus. If storage is required, it should be done only for the briefest time at low relative humidity (50%) and temperature (10–15°C). The use of water needs to be limited, and any rise in temperature should be avoided during the extraction process to ensure the highest-quality oil. Extra virgin oil is extracted during the first phase of pressing and therefore comes directly from the pulp of the fruit, whereas any oil that comes out later will also contain oil from the pit of the fruit. The oil extracted from the press needs to be separated immediately from the sludgy dregs and should be stored in a cool place away from direct light. Laboratory analysis is insufficient to detect the organoleptic qualities of extra virgin olive oil; taste tests are necessary. Olive oil tasters are highly qualified experts who use strict and sophisticated methods (26,44–55).

The cost of production is higher than that of other vegetable oils. This is due to the high amount of manual labor required for the cultivation and harvest of the olives. A person can pick from 60 to 100 kg of olives in a day, and from these about 13 to 20 L of oil can be extracted. The high cost is more than compensated, however, by the high yield and the oil's distinctly superior properties.

Average. Average oil has a good taste and acidity levels not exceeding 3.3%. This may be classified as:

- *Superfine virgin olive oil*: Obtained by mechanical extraction from olives and having undergone washing, sediment removal, and filtering with no chemical manipulations. It should not contain more than 1.5% acidity.
- Fine virgin olive oil: Obtained by mechanical extraction from olives and having undergone no chemical manipulations but only washing, sediment removal, and filtering. It should not contain more than 3.3% acidity.

Strong (Lampante). Strong virgin olive oil has inadequate taste, or acidity levels above 3.3%.

Refined Olive Oil. Refined olive oil is obtained by refining virgin oil whose taste, acidity levels, or both make it unsatisfactory for direct consumption. This is a healthy and perfectly acceptable food product, but it does not have the full taste of virgin olive oil.

Olive Oil. Olive oil is made by blending both refined and virgin olive oil. This type of oil is very much a standard in the marketplace; its properties fall somewhere between those of its two components.

Adulteration of Olive Oil

The high market price and the well-known health benefits attributed to olive oil make this product a target for different types of fraud, such as adulteration and mislabeling. This is especially true for extra virgin olive oil, which in the international trade market reaches a much higher price than any other vegetable oil (see Table 5).

Several physical, chemical, chromatographic, and spectroscopic tests are in use to detect adulteration of high-quality olive oil with low-grade olive oil (56,57), hazelnut oil (57), and seed oils (58,59). These have classically included determination of iodine value, saponification value, viscosity, density, refractive index, ultraviolet absorbance, fluorescence, and colorimetric reactions. More recently, FT-Raman spectra (60,61), mid-infrared spectroscopy (62), CO_2 laser infrared optothermal spectroscopy (58), reversed-phase liquid chromatography coupled to gas chromatography (63), and NMR (36) have been reported to detect the presence of $<1\%$ of other oils in extra virgin olive oil. These tests rely on the detection in the olive oil of unusual amounts of certain fatty acids or other compounds

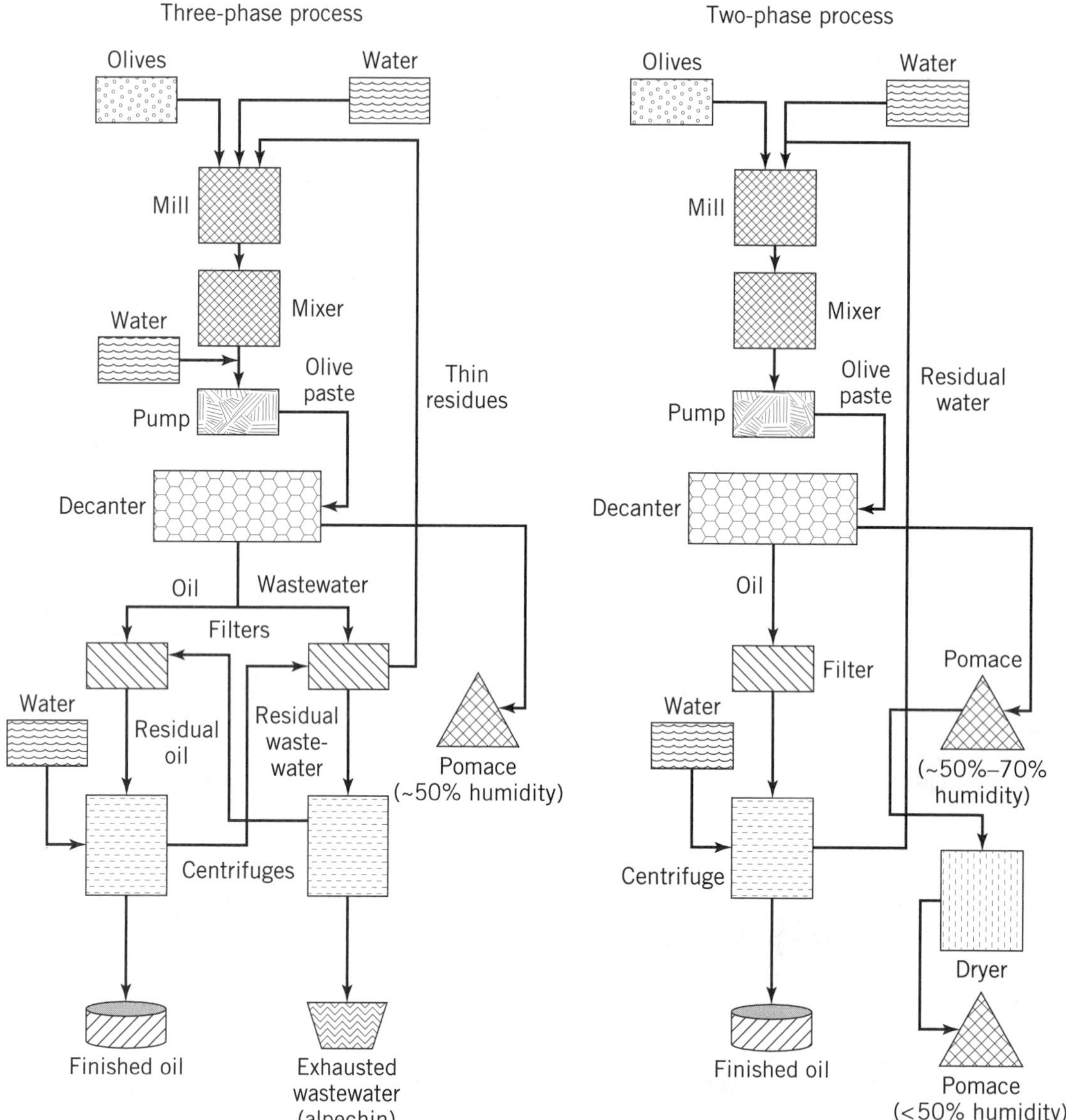

Figure 2. Scheme of three- and two-phase centrifugal processes for olive oil extraction.

such as fatty acids, triacylglycerols, squalene, sterols, tocopherols, alkanes, aliphatic alcohols, waxes, erythrodiol and uvaol, and stigmastadiene.

Fatty Acids. The International Olive Oil Council (IOOC) has established the normal ranges for the different fatty acids in olive oil. Gas chromatography has been used to measure these fatty acids, and this is the most common analytical technique for the assessment of the quality of olive oil (64). Among the different types of fatty acids, special attention should be paid to *trans* fatty acids. *Trans* fatty acids are not present in crude olive oils; however, they are formed during bleaching and deodorization. In addition to gas liquid chromatography, other techniques, such as infrared and nuclear magnetic resonance spectroscopy, high-performance liquid chromatography (HPLC), and supercritical fluid chromatography, have been applied to the detection of *trans* fatty acids (64).

Triacylglycerols. The triacylglycerol profile can provide information about the origin of the olive oil and its adulteration with other oils. Several methods are available for the determination of triacylglycerols, including high-temperature capillary gas chromatography and reversed-phase HPLC (65,66).

Squalene. Squalene content is very high in olive oil. Adulteration of olive oil decreases its content. Moreover, the refinement of olive oil induces squalene isomerization. Therefore, adulteration of olive oil with refined oil can be detected by the presence of these isomers using gas liquid chromatography (67) or HPLC (68).

Sterols. Large quantities of stigmasterol suggest adulteration with soybean oil. The ratio of beta-sitosterol (high in virgin olive oil) to campesterol + stigmasterol has been used for the detection of adulteration with seed oils.

Table 5. Price Ranges for Different Edible Vegetable Oils

Oil	Price ($/kg)
Extra virgin olive oil	2.98–3.25
Refined olive oil	2.10–2.17
Refined pomace olive oil	1.08–1.11
Peanut oil	1.08–1.09
Sunflower oil	0.79–0.80
Corn oil	1.01–1.02
Soybean oil	0.75–0.76
Canola oil	0.75–0.76

Note: Wholesale prices for edible oils quoted by the Roman Chamber of Commerce on September 17, 1998.

These compounds are analyzed using gas chromatography (69–71).

Tocopherols. Seed oils tend to contain higher quantities of beta, gamma, and delta tocopherol (soybean oil) and gamma tocopherol (cottonseed oil) than olive oil. These compounds can be detected using HPLC (72).

Alkanes. Olive oil contains primarily alkanes with chain lengths of C_{23}, C_{25}, and C_{27}, whereas other edible oils contain predominantly C_{27}, C_{29}, and C_{31} (73).

Aliphatic Alcohols. Olive oil contains lower levels of long chain alcohols and mono- and diunsaturated derivatives than does olive-pomace oil. This measurement could be used to assess contamination of virgin olive oil with low-grade olive oils (63,70).

Waxes. Extra virgin olive oil does not contain waxes; however, these are found in refined and olive-pomace oils. Therefore this determination could be used for detecting adulteration of extra virgin olive with olive-pomace oil (68,74,75).

Erythrodiol and Uvaol. Both erythrodiol and uvaol are increased in solvent-extracted olive oils (54); however, due to the variability of these compounds in olive oils obtained from different cultivars, these determinations are not commonly accepted as a measure to distinguish between pressed and solvent-extracted olive oils.

Stigmastadiene. Stigmasta-3,5-diene is formed by the removal of H_2O from beta-sitosterol during the refinement of olive oil. The determination of stigmastadiene has been proposed to detect refined oils in crude olive oil (68).

TABLE OLIVES

Pickled foods, which include table olives, are defined as those vegetables in which the preparation and preservation is carried out by fermentation, acidification, or both. This system of processing leads to end products with very special and well-defined sensorial characteristics, largely accepted by a wide consumer market; permits the preservation of perishable raw materials over a long time until they can be definitively processed; constitutes a manufacturing method that consumes little energy because the product is generally preserved under the final conditions of low pH value and relatively high acidity, thus not requiring any thermal treatment or requiring at most a simple pasteurization; and, finally, as a consequence of the previous considerations, retains more easily the natural nutrients, some of which are heat labile, and maintains important physical properties, such as color and texture.

In general, fermentation is carried out by homo- and heterofermentative lactic bacteria yeasts, or both. Both germ groups, in the natural sequence of events, are preceded by gram-negative aerobic bacteria and, if the ambient conditions are not adequate in certain moments of the sequence, can be accompanied by other genera, such as *Clostridium, Propionibacterium, Bacillus*, oxidative yeasts and molds, and other producers of diverse alterations.

The fundamental role that each microorganism is assigned in nature can be modified or substituted, partially, by technologically varying the related ambient conditions. These depend on a limited number of factors, including pH value, acidity, sodium chloride content, buffering capacity, temperature, degree of aerobicity or anaerobicity, and use of pure cultures.

To apply these modified technologies correctly, all investigators require the following: (*1*) knowledge, as wide-ranging as possible, of the chemical composition of the fruit or vegetable under study as well as the natural sequence, specific for each product, of the developing microorganisms; (*2*) selection of appropriate varieties as raw materials; and (*3*) determination of the optimum harvesting time.

These three points are evidently interdependent, but the natural germ sequence, whether maintained or modified, is perhaps what best determines the importance of the distinct components of the raw material and the physical and chemical modifications that are produced and must be produced in the composition. In other words, the aforementioned sequence frequently will mark the steps that must be the focus of basic research oriented toward technological advances. It is in this form that research on table olives has developed, initiated in the 1930s by Cruess (76) and continued mainly by the research of Vaughn et al. (77) de la Borbolla et al. (78), and Balatsouras (79).

Definition

According to the Unified Qualitative Standard Applying to Table Olives in International Trade, "Table olives are the sound fruit of specific varieties of the cultivated olive tree (*Olea europaea* sativa) harvested at the proper stage of ripeness and whose quality is such that, when they are suitably processed as specified in this standard, produce an edible product and ensure its good preservation as marketable goods. Such processing may include the addition of various products or spices of good table quality." The main purpose of processing is the removal, at least partially, of the natural bitterness of the fruit, which makes it suitable as either a food or an appetizer.

Harvesting, Transport, and Quality Evaluation

Harvest for table olives is fundamentally dependent on the type for which the fruits are destined and is influenced by distinct factors such as variety, climate, and cultivation systems.

Olives for green type are harvested when fruits have reached a green–straw yellow color and when the flesh is easily freed from the pit by finger torsion after the fruit is cut around the transverse diameter.

For turning-color type, fruits are harvested when the epicarp presents a purple color and the mesocarp is still white. For black-olive type, harvesting is at almost the same time; however, because fruits are further oxidized after lye treatment, the permissible range of color at harvest is wider than for the turning-color type.

Olives for natural black type must be harvested when ripeness is completed. Superficial color is not a good criterion in this case. It is necessary to cut along and close to the pit to obtain a representative sample of fruit and observe the flesh color. It is considered to be at a good stage of maturity if the purple color reaches at least 2 mm from the pit.

Generally, fruits are transported to the factory in holed plastic boxes of about 20- to 22-kg capacity. Parallelepipedic metallic bins with lateral plastic network walls of approximately 500-kg capacity are also used for this purpose. Bulk transportation in trucks is not used so as to avoid damaging the olives.

When the fruits reach the factory, different factors must be controlled, at arrival and before processing, to identify lots and sublots of the raw material. Among these the most important are variety, supplier, origin, cultivation system, date of harvesting, data of arrival at the factory, date of processing, stage of maturity, moisture content, and net weight.

To fix price and quality, it is necessary to determine, through a suitable sampling, the average weight and size distribution as well as the percentage of defects.

World Production and Distribution

According to recent data from the IOOC, the current world production of table olives is about 980,000 t. This figure, when compared to the 789,000-t figure for the 1987–1988 crop and the 450,000 t produced in 1967–1968, represents an increase of approximately 120% during the last 30 years. Spain is the major producer at 320,000 t, with 50% of this production consumed internally. Overall the European Economic Community (EEC) produces and takes to the market about 50% of world production.

About 277,000 t, 28% of world production, are exported yearly, with Spain the major exporter (125,000 t). The major importers are the United States and Canada (43.3%), the European Union (36.1%), and the Arab countries (7.2%). According to market trends during the last three decades, an average annual increment of at least 5% in production and consumption may be foreseen for the future; the industry could easily increase this figure by upgrading quality, preparing new derived items, and developing new presentation forms. This represents a challenge for food technologists.

Types of Table Olives

Differences among types of table olives are based on the degree of ripeness when they are harvested and the color of the fruit. According to Spanish standards, there are four established types.

Green Olives. Green olives are obtained from fruits harvested during the ripening period, before coloring and when they have reached normal size. Such olives must be firm, sound, resistant to a slight pressure between the fingers, and without marks other than the natural pigmentation, subjected to the tolerances set out in the standards. The color of the fruit may vary from green to straw yellow.

Turning-Color Olives. Turning-color olives are obtained from rose, wine-rose, or brown fruits harvested before the stage of complete ripeness is attained. They may or may not have been subjected to alkaline treatment and are ready for consumption. Californian green ripe olives are included in this type.

Black Olives. Black olives are obtained from fruits harvested when not fully ripe that have been blackened by oxidation and have lost their bitterness through alkaline treatment. These olives must be canned in brine and preserved by heat sterilization. Californian ripe olives are included in this type.

Natural Black Olives. Natural black olives are obtained from fruits harvested when fully ripe or slightly before full ripeness is reached. According to production region and time of harvesting, they may be reddish black, violet black, deep violet, greenish black, or a deep chestnut color.

IOOC standards consider only three types: green, turning-color, and black olives. However, under the title of commercial preparation, both the Spanish and IOOC standards agree and differentiate clearly between alkali-treated (black) and untreated (natural black) olives.

Commercial Preparation

The denomination of commercial preparation, or its equivalent, trade preparation, is used for all products destined for consumers. These preparations are diverse and include not only traditional methods of processing but also those derived from them and improved by new technologies. To facilitate better understanding, it is useful to first explain some other concepts.

Removal of the fruit's natural bitterness due to the presence of oleuropein glucoside may be carried out completely by alkaline hydrolysis, treating olives with an aqueous solution of sodium hydroxide before fermentation (80). In those cases, denomination in IOOC standards includes the word treated. The removal of oleuropein may also be achieved, slowly and partially, during the acid fermentation of fruits put directly into the brine or kept in dry salt. In those cases, the word untreated appears in the name.

Generally, the complete name for a specific commercial preparation includes information on (*1*) the type of raw material (green, turning-color, or black olives); (*2*) the procedure used to eliminate bitterness (treated or untreated);

and (3) the method of preservation (brine or dry salt). Different preparations described in the standards mentioned previously are as follows.

Treated Green Olives in Brine. The olives are treated in an alkaline lye and then packed in brine in which they undergo natural lactic fermentation, that is, either complete fermentation (Spanish style) or partial fermentation. Should the olives not undergo complete natural fermentation, their subsequent preservation at a pH value within the limits specified in the standards may be ensured by sterilization or pasteurization, the addition of preserving agents, refrigeration, or inert gas without brine.

Untreated Green Olives in Brine. Untreated green olives are placed directly in brine and preserved by natural fermentation.

Turning-Color Olives Treated in Brine. Turning-color olives are obtained after alkaline treatment and preserved by natural fermentation in brine or by heat treatment.

Untreated Turning-Color Olives in Brine. Directly treated in brine, turning-color olives are preserved by natural fermentation and are ready for consumption.

Olives Darkened by Oxidation. Olives darkened by oxidation are obtained from fruits that, when not fully ripe, have been darkened by oxidation and whose bitterness has been removed by treatment in alkaline lye. They have to be packed in brine and preserved by heat sterilization. Californian-style ripe olives correspond to this group.

Treated Black Olives. Treated black olives are obtained after alkaline treatment and are preserved by natural fermentation through one or a combination of the following procedures: brine, sterilization or pasteurization, or addition of a preserving agent.

Untreated Black Olives. Untreated black olives are placed directly in brine. They have a fruity flavor that is more marked than that of the treated black olives, and they usually retain a slightly bitter taste. They are preserved by natural fermentation through one or a combination of the following methods: brine, sterilization or pasteurization, or addition of a preserving agent. Greek-style natural black olives in brine correspond to this group.

Shriveled Black Olives. Shriveled black olives are harvested just before they are fully ripe and, after a brief immersion in a weak alkaline solution, are preserved by sprinkling salt in wooden casks, which are rotated daily until the olives are prepared for sale.

Untreated Naturally Shriveled Black Olives. Untreated naturally shriveled black olives are harvested when fully ripe, after they have become shriveled on the tree, and are subsequently treated directly in brine.

Treated Black Olives in Dry Salt. Treated black olives in dry salt are obtained from firm, practically ripe fruits that, following slight alkaline treatment, are preserved in alternating layers of olives and dry salt or by sprinkling dry salt over the olives.

Untreated Black Olives in Dry Salt. Untreated black olives in dry salt are obtained from fruits harvested when fully ripe and treated immediately or after partial drying in alternating layers of olives and dry salt or by sprinkling dry salt over the olives. They retain a degree of bitterness and a fruity flavor more marked than those of treated olives in dry salt.

Untreated Naturally Shriveled Black Olives in Dry Salt. Untreated naturally shriveled black olives in dry salt are obtained from fruits harvested when fully ripe that, after they have become shriveled on the tree, are preserved in alternating layers of olives and dry salt or by sprinkling dry salt over the olives.

Untreated Pierced Black Olives in Dry Salt. Untreated pierced black olives in dry salt are obtained from fruits harvested when fully ripe that, after the skin has been pierced, are preserved in alternating layers of olives and dry salt or by sprinkling dry salt over the olives.

Dehydrated Black Olives. Dehydrated black olives are obtained from ripe fruit that, after having been blanched and partially dehydrated in salt, undergo very gentle heating.

Bruised Olives. Bruised olives are obtained from whole fruit, fresh or previously treated in brine, subjected to a process whereby the flesh is opened without breaking the stone, which remains whole and intact within the fruit. They may be treated in weak lye and are preserved in brine, possibly flavored, with or without the addition of vinegar. They are prepared from either green or turning-color type.

Split Olives. Split olives are green olives, turning-color olives, or black olives split lengthwise by cutting into the skin and part of the flesh and then placed in brine, with or without vinegar; olive oil and aromatic substances may be added. They may be untreated or treated previously with alkali.

Specialties. Olives may be prepared by means distinct from those specified in the standards. Such specialties retain the name "olive" as long as the fruit used complies with the general definitions of those standards. The names used for these specialties should be sufficiently explicit to prevent any confusion as to the origin and nature of the products and, in particular, with respect to the designations in the standards.

Presentation Forms

According to their type and commercial preparation, olives may be offered to the consumer in one of the following styles, also denominated presentation forms.

- Whole olives: Olives that have their natural shape and from which the stone or pit has not been removed
- Without stem: Whole olives from which the stem has been removed
- With stem: Whole olives retaining the stem
- Pitted olives: Olives from which the pit has been removed and that basically retain their natural shape
- Stuffed olives: Pitted olives stuffed either with one or more suitable products (pimiento, onion, almond, celery, anchovy, olive, orange, or lemon peel, caper, etc.) or with their prepared natural pastes
- Halved olives: Pitted or stuffed olives sliced into two approximately equal parts, perpendicular to the major axis of the fruit
- Quartered olives: Pitted olives split into four approximately equal parts along the major axis of the fruit and perpendicular to the same
- Divided olives: Pitted or stuffed olives cut lengthwise into more than four approximately equal parts
- Sliced olives: Pitted or stuffed olives sliced into segments of fairly uniform thickness
- Chopped or minced olives: Small pieces of pitted olives of no definite shape and practically devoid of identifiable stem-insertion units as well as of slice fragments
- Olive paste: Exclusively olive flesh, finely crushed. Ingredients or additives may be incorporated for its preservation.
- Broken olives: Olives accidentally broken while being pitted or stuffed. They normally contain pieces of the stuffing material.
- Salad olives: Broken or broken-and-pitted olives with or without capers, plus stuffing material. The olives predominate in the total product marketed in this style
- Olives with capers: Whole or pitted olives, usually small in size, with capers and with or without pimiento. The olives predominate in the total product marketed in this style

According to the manner in which they are placed in the container, the olives may be presented as follows:

- Place-packed olives: Fruits are placed in transparent, rigid packs either in orderly, symmetrical fashion or forming geometrical shapes. Whole, pitted, stuffed, and halved olives may be exported in this manner.
- Random (thrown) packed olives: Fruits are not placed in an orderly fashion in the packs. All the olive styles may be exported in this manner.

Main Processes

The three preparations of table olives that represent high economic and commercial importance in international trade are:

1. Spanish- (or Sevillan-) style pickled green olives in brine
2. Californian-style pickled black olives in brine
3. Greek-style natural black olives in brine

The three preparations are similar in that the final products are prepared and preserved in diluted water solutions of sodium chloride, at different concentrations. For each style, fruits are harvested at a different stage of ripeness; green, black, and natural black, respectively. In each case, the adjective in the style name points out the country or state in which the process was first developed and used. In the first two styles, bitterness is removed by an alkaline treatment; for that reason, the word pickled appears in the name. In the case of the Greek-style process, fruits are placed directly into the brine, and removal of oleuropein is slow and only partial.

Sensory properties for each preparation are very different, although this variation may also occur even within the same preparation when parameters of the final product are adapted by technology to consumer preferences, which depend on the region or country where the product is going to be marketed.

For Spanish-style green olives in brine, a lactic fermentation constitutes the fundamental process (81). For Greek-style natural black olives, the main process responsible for the spontaneous fermentation are yeasts; lactic cocci and bacteria coexist in a smaller proportion.

Californian-style pickled black olives in brine do not necessarily require a fermentation process; they may be directly treated with lye and oxidized, then washed, placed in brine, canned, and heat sterilized. However, a fraction of the crop may be placed directly into brine for a variable preservation period and further processed. That process produces a kind of fermentation quite similar to that of natural black olives in brine.

Processed Olives

Regulations. Because of its organoleptic properties and its composition, the table olive is identified in many countries with a high standard of living. It is widely used as an appetizer to accompany all kinds of beverages, and as a decorative or nutritional element of various dishes. It also constitutes a basic food for the majority of people in some countries, mainly those situated in the Mediterranean area.

Table olives, prepared by different processes, are traded internationally by both developed and less-developed countries. For those transactions the IOOC through a committee of experts has evolved the Unified Qualitative Standard Applying to Table Olives in International Trade, mentioned previously, which is revised periodically. In the same manner, the Joint Program Food and Agriculture Organization/World Health Organization (FAO/WHO), with the cooperation of the IOOC has developed nutritional standards for table olives. The two sets of standards agree quite well in all their content and constitute an important aid in commerce and in the protection of consumers.

Spain, the main producer of table olives, has developed its own standards, which are even more detailed and more restrictive in certain aspects than those mentioned previously, although they are in agreement with them. In the same way, other exporter countries have developed their own standards, too.

Although the IOOC recommends that international standards apply to all transactions, it is also necessary to consider special regulations of importer countries in each particular case. Therefore Spain, the main supplier for the United States, regulates its exports of table olives according to U.S. regulations, which are gradually revised and published in the *Code of Federal Regulations* and *Federal Register*, Title 7, Chapters IX (Agricultural Marketing Service) and XXVIII (Agricultural Food Safety and Quality Service); and Title 21, Chapter I (Food and Drug). The EEC is taking actions to adopt the IOOC standards as official for its members.

Composition and Nutritive Value. Among the different factors that may affect the composition of the final product, the most important are the treatment of the fruit before fermentation or conditioning; the fermentation or conservation in brine; the storage after processing, bottling, or canning; and final conservation until consumption (shelf life) (82).

Table 6 shows the main components for some Spanish varieties that were processed by the three methods studied in this article. For Spanish-style pickled green olives in brine, ranges indicated in the table include four varieties: Gordal, Manzanilla, Hojiblanca, and Verdial. For Greek-style natural black olives in brine and Californian-style pickled black olives in brine, typical values for Hojiblanca variety are presented.

The concentration of moisture and oil in the flesh shows an inverse relationship to the concentration in fresh fruit. The wide range in the values is influenced by variety, maturity of the raw material, and processing conditions. The concentration of proteins as well as of fiber is slightly reduced, probably due to solubilization during the lye treatment and fermentation. Fiber is well balanced, showing a good digestibility because the ratio of lignin to cellulose is generally less than 0.5.

The percentage of ash increases as a consequence of the alkaline treatment, fermentation, and storage in brine. The majority of essential amino acids as well as a noticeable amount of vitamins and mineral elements are also present.

The caloric value is quite variable, depending mostly on the lipid content of different varieties. The quality of the oil fraction is good because of its low content of saturated fatty acids (12–19%) and the presence of an important fraction of linoleic acid (5–8%). For all these reasons, it can be said that table olives provide good food value (83,84). For Spanish-style pickled green olives in brine, soluble sugars disappear during fermentation and are absent from the final product. The same happens to the primary source of bitterness, oleuropein, which is mostly hydrolyzed during the alkaline treatment (80). Organic acids and their salts represent 1.5% of the weight of the pulp; they consist partly of the organic acids present in the fresh fruit and partly of those formed during the lye treatment and further fermentation.

Chlorophyll degradation during the processing of Spanish-style pickled green olives in brine can take place by two different mechanisms, resulting in a mixture of pheophytins and pheophorbides in the final product (85). In relation to the carotenoid fraction, only β-carotene and lutein are resistant. Phytofluene and ζ-carotene disappear, and violaxanthin, luetoxanthin, and neoxanthin give rise to their isomers auroxanthin and neochrome. The degradation taking place, however, does not give rise to uncolored final products with pigmentation loss, and the total balance of pigments remains constant with time.

In the case of Greek-style natural black olives in brine, prepared without a previous lye treatment, the concentration of soluble sugars remains relatively high (≧0.3%) even after long storage in brine. The same happens to oleuropein and tannins, as a consequence of a very slow fermentation process. For that reason, sensory properties of this

Table 6. Composition of Processed Table Olives: Concentration Ranges and Typical Values in the Pulp

Component	Spanish style	Californian style	Greek style
	Major component (weight %)		
Moisture	61.00–80.56	68.98	60.34
Lipids	9.05–28.19	21.24	23.43
Protein	1.00–1.45	1.12	1.15
Fiber	1.40–2.06	2.35	2.18
Ash	4.19–5.46	2.04	6.88
	Vitamins		
Carotene (mg/100 g)	0.02–0.23	0.09	
Vitamin C (mg/100 g)	1.44–2.87		
Thiamin (μg/100 g)	0.40–3.37	0.21	14.50
Riboflavin (μg/100 g)		96.30	91.00
	Essential Amino Acids (mg / 100 g)		
Valine	55–157	119	88
Isoleucine	43–121	118	37
Leucine	82–227	147	44
Threonine	5–64		
Methionine	13–79	32	10
Phenylalanine	39–111	89	47
Lysine	5–31		5
Tryptophan	13–18	15	11
	Minerals (mg / 100 g)		
Phosphorus	7–21	15	3
Potassium	34–109	12	29
Calcium	35–86	68	28
Magnesium	6–40	18	8
Sodium	1,313–1,753	634	3740
Sulfur	14–38	8	6
Iron	0.58–1.16	7	0.30
Manganese	0.06–0.12	0.14	0.02
Zinc	0.25–0.41	0.49	0.25
Copper	0.42–0.82	0.29	0.06
	Caloric value (cal / 100 g)		
	102–280	222	249

final product are quite different from those of the other two types and are also very appreciated by the consumer.

The main difference between Californian-style pickled black olives in brine and those previously described is due to the oxidation in an alkaline medium. If storage in brine before processing is used, the characteristics of the fruit at that stage are similar to those processed as natural black olives in brine, with only the difference in color due to the stage of maturity at which they are harvested. As a result of the processing conditions, tannin and oleuropein content, as well as total acidity of the final product, are very low, and it must be sterilized by heat as a consequence of the relatively high pH value.

In summary, the changes that occur during processing may be divided into three groups: (*1*) compounds that are simply lost during the different stages of the process; (*2*) compounds that are transformed by simple chemical reaction or by the action of enzymes originally present in the fruit or produced by microorganisms; and (*3*) compounds produced during fermentation by different microorganisms. Table 7, Table 8 and Table 9 show the main changes produced in the three types of processing mentioned in this article.

Main Industrial Processes

Spanish-Style Pickled Green Olives in Brine. Within the green-olive type, which accounts for ca. 42% of the world production of table olives, the most important commercial preparation is that of Spanish- or Sevillan-style, which in turn accounts for the highest percentage of the international trade. They generally may be preserved as is; however, if the fermentation is only partial, they keep the denomination pickled green olives in brine but do not enter within the Spanish style. In that case, their further preservation, at a pH value within the limits specified in that standard, may be ensured by (*1*) sterilization or pasteurization, (*2*) the addition of preserving agents, (*3*) refrigeration, or (*4*) inert gas, without brine.

Other green olives are prepared without a prior alkaline treatment and are placed directly into the brine, either to undergo a natural fermentation or to be preserved by any of the procedures mentioned earlier. In those cases, they keep the denomination green olives and the word pickled disappears. Their commercial importance is minor, although final products of excellent quality may be obtained by those procedures. Table 7 shows the different phases for the processing of Spanish-style pickled green olives in brine.

The fruits are picked when their color is still green or yellowish green and are harvested by hand, although numerous attempts have been made to harvest them mechanically. Tree shakers have not been well accepted yet because of the damage they cause to the trees and fruits.

Fresh fruits are then transported to the factory either in crates of perforated plastic material that allow access of air, with a 20 to 25-kg capacity, or in bins with perforated mesh walls of higher capacity. The fruits remain in their containers for a period varying from several hours to 3 or 4 days, depending on the variety. After sorting and an optional size grading, the fruits proceed to the lye treatment.

Fruits are treated with a diluted solution of sodium hydroxide to eliminate the greater part of the bitter glucoside oleuropein. In this treatment, the lye penetrates the pulp to a depth of two-thirds to three-fourths the distance between the skin and the stone. The lye concentration used varies generally between 1.3 and 2.6% (wt/vol), depending on temperature, variety, and stage of maturity of the fruits,

Table 7. Spanish-Style Pickled Green Olives in Brine: Scheme of the Process and Compositional Changes

Phases	Operations	Flow chart	Changes in composition
Prior phase	None	Fresh olives ⇓ Harvesting ⇓ Transport to the factory ⇓ Sorting (optional size grading) ⇓	None, under normal conditions
Fundamental phase	Alkaline treatment and washing with water	Lye treatment ⇓ Washing ⇓	Hydrolysis of oleuropein Loss of sugars and organic acids Formation of organic acids from sugars
	Fermentation in brine, mainly lactic acid bacteria Secondary action of other microorganisms	Brining ⇓ Lactic fermentation ⇓	Lactic acid formation from sugars and other fermentable compounds Formation of other organic acids Degradation of pigments
Final phase		Sorting and size grading ⇓ (optional pitting and stuffing) ⇓	
	Conservation in brine Bottling	Packing	None, under normal conditions

Table 8. Californian-Style Pickled Black Olives in Brine: Scheme of the Process and Compositional Changes

Phases	Operations	Flow chart	Changes in composition
Prior phase	None (optional fermentation in brine, by lactic acid bacteria and yeasts)	Fresh olives ⇓ Harvesting ⇓ Transport to the factory ⇓ Sorting (optional size grading) ⇓ (optional storage in brine) ⇓	None, under normal conditions (if optional fermentation is used: slow loss of sugars, tannins, and oleuropein. Formation of organic acids and possibly ethanol and aromatic compounds)
Fundamental phase	Alkaline treatment Washing with water Oxidation by air Fermentation in brine, mainly lactic acid bacteria Secondary action of other microorganisms Canning Heat sterilization	Lye treatment and air oxidation ⇓ Washing ⇓ Color fixing and brining ⇓ (Optional pasteurization) ⇓ Sorting and size grading ⇓ (Optional pitting, slicing, splitting, etc.) ⇓ Canning and heat sterilization ⇓	Hydrolysis of oleuropein Loss of sugars and organic acids Formation of organic acids from sugars Degradation of pigments
Final phase	Storage of the sealed and sterilized product	Storage	None, under normal conditions

Table 9. Greek-Style Pickled Black Olives in Brine: Scheme of the Process and Compositional Changes

Phases	Operations	Flow chart	Changes in composition
Prior phase	None	Fresh olives ⇓ Harvesting ⇓ Transport to the factory ⇓ Sorting (rough size grading) ⇓	None, under normal conditions
Fundamental phase	Washing with water Spontaneous fermentation in brine Yeast predominate; lactic acid bacteria sometimes present	Washing ⇓ Brining ⇓ Fermentation ⇓ Sorting and size grading (air exposure) ⇓	Slow loss of sugars, tannins, and oleuropein. Formation of organic acids, ethanol, acetaldehyde, and ethyl acetate.
Final phase	Storage in brine Canning	Packing and storage	None, under normal conditions

and it is regulated in such a way that the treatment takes a determined number of hours to reach the suitable penetration for each variety. As a general rule, time for the majority of the varieties is 5 to 7 h, except for Gordal and Ascolano, which require a slower treatment, around 9 to 10 h, with more diluted lyes because of the texture and composition of the skin and flesh. A good control of the factors mentioned previously (lye concentration, lye penetration, and time of the treatment) is essential for the quality of the final product. When the fruits are treated with too much of a low-concentration lye and hence over a longer time, the further developed color may be only fairly acceptable and the fermentation will be poor. Conversely, a high-lye concentration may produce texture deficiencies and a high loss of fermentable matter, which is important to further fermentation.

After the alkaline treatment, the fruits are washed with water to eliminate the major portion of the lye that remains in the flesh. The duration and number of necessary washings are also important factors. An excessive number of washings can deplete the fermentable matter and nutrients in such a way that it will be necessary to add more of those compounds to complete the fermentation process. An excessive amount of organic salts may also be lost, producing as a consequence a lack of buffer capacity of the medium. Finally, a long period of washing may lead to serious bacterial contamination before the fruits are placed into the brine. On the contrary, excessively short washings produce a high concentration of organic salts (frequently called residual lye) in the fruits, which further prevents the attainment of suitable pH values during fermentation (81).

In summary, for each variety, stage of maturity, and temperature, a compromise must be found between lye treatment and the washing system that permits the fruits, when placed in brine, to retain enough fermentable compounds to reach a suitable degree of free acidity during fermentation and a concentration of organic salts (normally around 0.1 N) that will maintain the right pH value for safe storage of the fruits in brine. If the lye treatment is correct, a quick rinse after the alkaline treatment, followed by a first washing of 2 to 3 h and a second washing of 10 to 12 h, is the schedule normally used in the factories.

After washing, fruits are placed into a brine of sodium chloride in which lactic acid fermentation takes place. The brine, by osmosis of the components of the fruit, is transformed in a rich culture medium for microorganisms that are responsible for fermentation. The speed of that transformation depends on the variety, lye treatment and washings, ratio of fruit to brine, salt concentration, temperature, and so on. The order of appearance of growth of different microorganisms is dependent on their nutritive requirements.

The concentration of initial brine is quite important. If too low, the also low osmotic pressure can lead to spoilage by sporulating microorganisms of the *Clostridium* type during the first stage of fermentation, if the pH value remains too high. On the contrary, if the concentration of salt is too high, fruits may become irreversibly wrinkled (86). Depending on the variety and the stage of maturity of the fruit, initial concentration must be regulated between 9 and 11 Beaume degree units.

After the initial decrease in the salt concentration, it must be gradually increased to maintain a good texture in the fruit and allow good fermentation and storage (86). However, the increments of salt must be slow enough to permit the right growth of the lactic acid bacteria.

A good approach is to maintain the salt concentration between 5 and 6% during the greater part of the fermentation phase, rising to 7% at the end of that part of the process. Furthermore, it may be increased up to 8% or even more during the storage phase to avoid the growth of *Propionibacteria* during the last stage of fermentation, which may produce a depletion of lactic acid.

At present, the containers used for all the fundamental phases of the process (treatment with lye, washing, fermentation, and further storage of the fermented product) in the majority of the producer countries are fermentors made of polyester and fiberglass. They can be fully closed to create anaerobic conditions and exclude the growth of yeast films on the surface of the brine. The containers have a large opening on the top and valves at the top and the bottom that facilitate the unloading of the fruits and the circulation of the brine. The most common capacity for these is about 10 t of fruit, some even 15 t. The same material is used for underground fermentors. In those cases, the opening at the top is used for all operations—loading, unloading, and recirculation of the brine—by use of suitable pumps.

The four phases of spontaneous fermentation of pickled green olives in brine have been described in detail. Each of them is characterized by a different development of the microbial population as well as by changes in the physical chemical characteristics of the culture medium, which is constituted by the initial brine and the soluble compounds of the fruit that pass to it by osmosis.

The duration of each phase as well as the relative importance of the growth of certain microorganisms depend on a series of factors, among which the following must be emphasized.

- Prior treatments of the olives (lye treatment, washing operations, and initial concentration of the brine)
- Capacity of the fermentor and, as a consequence, the ratio of fruit weight to brine volume
- Climatic and environmental conditions in the factory
- Variety of the olives
- Sanitary conditions of the plant and equipment

In all cases, the natural sequence of appearance, growth, and disappearance of microorganisms may be established as follows:

- Gram-negative bacteria, the majority of which belong to the Enterobacteriaceae family. They are the first to appear and grow, at the relatively high pH values of the initial brine.
- Lactic-acid-producing cocci, mainly from the genus *Pediococcus* (homofermentative) and *Leuconostoe* (heterofermentative). Their major or minor growth may have, together with other factors and the action of gram-negative bacteria, an influence on *Lactobacilli*, mainly from the species *L. plantarunz*, which, under normal conditions, become predominant in the third phase and are responsible for the typical fermentation of Spanish-style pickled green olives in brine.
- *Propionibacterium* genus, which, as indicated earlier, may grow during the fourth phase with depletion of lactic acid if salt concentration in the brine is not suit-

ably controlled during the storage of the fermented fruits.

A variable and nonabundant fermentative yeast population coexists with *Lactobacilli* throughout the whole fermentation period and probably partially contributes to the sensory properties of the final product. However, film-forming oxidative yeasts must necessarily be avoided with a suitable anaerobic closing of the fermentors. Each genus and species gradually adapts the characteristics of the medium to the requirements of the microorganism that succeeds it within the natural sequence, until the last one becomes dominant.

The correct control of a reduced number of factors, such as pH value, free and combined acidity, salt concentration, temperature, degree of aerobiosis or anaerobiosis, and use of a pure culture as starter, permits food technologists to vary the duration and relative importance of each of the phases and avoid the microorganisms that can have negative effects on the quality of the final product.

Acidification or injection of CO_2 into the brine during the first stage to prevent the development of gram-negative bacteria, and a further inoculation with a pure culture of *L. plantarum* or a well-fermented brine with a good population of active *Lactobacilli*, is a very efficient method to reduce to a minimum the effect of the first and second phases and accelerate the start of the fundamental third phase. A well-controlled fermentation process produces a final product that, in general, shows pH values of 3.8 to 4.2 units and a free acidity between 0.8 and 1.2%, expressed as lactic acid and mainly composed of that acid. The figures for combined acidity, also improperly called residual lye, oscillate between 0.09 N and 0.11 N, approximately. The final salt concentration should be about 7%, if the storage period is expected to be relatively short and room temperature is not too high. However, for longer periods of preservation or high summer temperatures, it is better to raise the sodium chloride concentration to at least 8%, to avoid the growth of *Propionibacterium* species. Only *Lactobacilli* in their declining phase and some fermentative yeasts constitute the microbial population of the fermented product.

Green olives to be sold are bottled or canned in small, hermetically sealed containers, which generally present the characteristics summarized in Table 10. The right selection of those parameters, the most important of which are fixed by the international standard of the IOOC, permits the bottled or canned product to be kept safely without pasteurization, and no sediment is formed in the container (87). According to those standards, salt concentration may be reduced up to 2% and pH value raised up to 4.3 if the containers are pasteurized.

Table 10. Typical Ranges for the Main Physical Chemical Parameters of Pickled Olives in Brine

Parameter	Spanish-style green olives	Californian-style black olives	Greek-style black olives
pH value	3.6–4.2	5.8–8.0	4.3–4.5
Free acidity (as lactic acid)	1%		0.3–0.5%
Sodium chloride	2–8%	1–5%	6–10%

The conditions for bulk pasteurization of green olives and pasteurization of different sizes and shapes of cans and bottles filled with different sizes of pickled green olives have been studied for Gordal and Manzanilla varieties (88). An objective spectrophotometric method by measurements of reflectance has been developed to follow the color evolution of the fruits during fermentation and further treatments (89).

Californian-Style Pickled Black Olives in Brine. Within the turning-color type, which represents about 20 to 23% of the world production of table olives, the most important commercial preparation is that of Californian-style pickled black olives in brine (called ripe olives in the United States). The main producer countries are the United States (46–48% of the world production of these olives) and Spain (25–32%). Table 8 shows the different phases for the process of this style. The process may be carried out either directly with the fresh fruits or after a holding period of the olives in brine, which varies from 2 to 6 months. The reason for the optional storage in brine is that not all the fruit available at the time of the harvest can be directly processed, because of the capacity of the processing plants and because it is not practical to keep large stocks of canned product for a long time. In the first procedure, sorted fresh olives are successively treated with sodium hydroxide aqueous solutions for varying periods to get a gradual penetration of the lye into the flesh. After each lye treatment, the fruits are placed in plain water and oxidized by air injection under pressure to transform the polyphenolic compounds and so permit a complete blackening of the skin and a uniform color in the flesh.

The color of the skin of the fruit at the time of harvest varies greatly from yellowish green to purple, and the taste, color, and texture of the final product will be highly dependent on the maturity stage and the variety. Mission and Manzanilla in the United States and Hojiblanca and Cacerena in Spain are the best varieties for this type of processing.

The number of lye treatments has traditionally been between 3 and 5, although at the present time industry tends to reduce this number to a minimum. In general, a first treatment just to penetrate the skin, a second one to penetrate 1 mm of the flesh, and a third one that permits the lye to reach the pit constitute a typical industrial schedule. The concentration of the lye depends on the maturity stage of the fruit, the variety, the temperature, and the desired penetration. It generally varies between 1 and 2%, the highest being used to penetrate the skin. Air oxidation periods, between lyes, are variable, frequently around 12 h.

Tanks for lye treatment, air oxidation, and washings have different sizes and shapes and are made of concrete, stainless steel, or polyester and fiberglass and must be suitably arranged for uniform distribution of pressurized air throughout the whole mass of fruits and liquid.

The number and duration of washings must be regulated to get the washed olives to reach a final pH value around 7 units. Washing water must be changed frequently to avoid the growth of spoiling aerobic microorganisms.

Generally, 0.1% of ferrous gluconate is added to the last washing water to fix the color of the fruits, obtained by oxidation. The effects of ferrous gluconate and ferrous lactate, at the same level of iron content, have been compared, and similar results with the two compounds have been reported (90). In addition, olives are placed and equilibrated into a brine containing 3% sodium chloride, canned, and heat sterilized. Varnished tinplate is generally used for the inside of the containers. The use of a needle board to puncture the skins of certain varieties of olives before the brining to avoid shriveling and a pasteurization before the canning operation to minimize the action of some aerobic bacteria during brining are optional parts of the process.

The fruits that are not directly treated after harvesting and that remain for a variable period in holding brines may undergo a spontaneous fermentation process that is similar in almost all aspects to the one followed by natural black olives in brine (Greek style), which will be described next. At the end of this holding period, fruits are processed in the same way as indicated previously for fresh fruits. However, lye concentration and time for each treatment will be influenced by the conditions used during the previous storage and must be carefully controlled.

Although traditionally this previous storage in brine has been undertaken under relatively close to anaerobic conditions to avoid films of yeasts and molds, which may spoil olives, an alternative aerobic procedure has been developed, with excellent results in Manzanilla and Hojiblanca varieties (91,92).

The final product for Californian-style pickled black olives in brine shows typical pH values between 5.8 and 7.9 units and salt concentrations between 1 and 3% sodium chloride (wt/vol). However, for certain consumer preferences, salt concentrations close to 5% are also used.

Greek-Style Natural Black Olives in Brine. Within the black-olive type, which represents about 33 to 40% of the world production of table olives, the most significant commercial preparation is that of Greek style. The main producing countries are Turkey (24–27% of the world production of this type) and Greece (18–21%). Table 9 shows the different phases for the processing of this style.

Fruits are harvested when fully ripe or slightly before full ripeness is reached, and they may, according to production region, variety, and time of harvesting, be reddish black, violet black, deep violet, greenish black, or a deep chestnut color. The olives are transported to the factory as described previously for the Spanish-style pickled green olives in brine. After a sorting to separate damaged fruits, and occasionally after a rough grading to eliminate small sizes, the fruits are quickly washed with water to remove superficial dirt and then placed into a brine with a salt concentration between 8 and 10 Beaume degree units, or slightly lower.

At present, the polyester and fiberglass fermentors described for green olives are also used for this type of fruit. They are generally buried underground to avoid high temperatures, because of the high degree of ripeness of the olives. Traditionally, the processors have tried to maintain strict anaerobiosis in the tanks to avoid the growth of film-forming yeasts and molds, which affect the texture and flavor of the olives. The fermentation process is very slow, because diffusion of soluble components of the fruit to the brine is slow due to the fact that it has not been treated with lye. A complex microflora develops during fermentation: gram-negative bacteria, gram-positive cocci of lactic acid, yeasts, and occasionally *Lactobacilli*—the last in the case of relatively low concentrations of sodium chloride, below 7 to 8%. Yeasts constitute the dominant population, and the most representative species are *Saccharomyces oleaginosus* and *Hansenula anomala*.

Under those conditions, the final product has pH values between 4.5 and 4.8 units, and free acidity ranges from 0.1 to 0.6%, expressed as lactic acid. For these reasons, there is a tendency to gradually raise the salt concentration, which frequently is about 10% or even higher at the end of the fermentation process.

Afterward, fermented olives are exposed to air to improve the skin color, classified by size; and packed with a new brine. Containers are either wooden or plastic barrels of about 130 to 150 kg of fruit, cans of 10 to 15 kg, or smaller plastic bags. During this traditional process, the spoilage, called *alambrado* or fish eyes, frequently produces big losses in certain varieties of olives. This effect is attributed to different causes, such as the fruit's metabolism, influenced by the anaerobic conditions of the medium and the maturity stage of the fruit itself, and the action of the most representative yeasts previously mentioned.

A new controlled aerobic fermentative process, which avoids spoilage, has been developed (93,94). It injects air through the mass of fruits and brine during fermentation. The system may be applied in either a continuous or discontinuous manner. The best conditions are volume of air, 0.1 to 0.5 L/h/L capacity; sodium chloride concentration, 10%; and pH value of 4.5 units, controlled by the addition of food-grade acetic acid. Under these circumstances, a population of yeasts of oxidative and facultative metabolism is responsible for fermentation, with *Debaryomyces hansenii* and *Toritlopsis candida* the most representative species. The system for polyester and fiberglass tanks, which are most frequently used in the industry, have been further optimized (94). After finishing the phase of active fermentation and interrupting the airflow, the investigators indicate that the safest method to avoid further spoilage is the addition of the necessary amount of potassium sorbate to reach 0.05%, expressed as sorbic acid, in the equilibrium (95).

Further studies indicate that the use of a low concentration of sodium chloride, around 2% at the beginning of the aerobic fermentative process, and a gradual increase of salt up to the final value of 9% is satisfactory (96). Some other species of representative yeasts have been isolated, such as *Pichia membranaefaciens*, *Hansenula mrakii*, and *Candida boldinii*. The effect of different levels of aeration and its mode, initial concentration of salt, maturity stage of the olives, and control of the pH value have been studied using a factorial design experiment (97).

The aerobic procedure is being gradually and successfully adopted by the processors of this style of black olive. The best conditions for packing the olives in small containers destined for the consumer have been studied (98). Two methods are equally effective: a pasteurization at 80°C for 4 min, and the addition of potassium sorbate 0.05%, expressed as sorbic acid, at the equilibrium. Typical values for these products, after 11 months of storage, are pH values of 4.40 to 4.45 units and sodium chloride between 7.2 and 7.4%.

REDUCTION AND REUSE OF WASTES

Most food industries that transform natural products, of either animal or vegetable origin, contribute to ambient pollution. So is the case with olive oil mills and olive-pickling plants. This situation has constituted an important concern for scientists and technologists in the olive oil and table olives industries for the last 35 years (99).

By-products of Olive Oil Production

Olive pomace and wastewaters are the major by-products resulting from the elaboration of olive oil (100–102). Olive pomace is the material remaining after most of the oil is removed from the olive paste. It contains fragments of skin, pulp, fragments of kernels, and some oil. In chemical terms, the major constituents are cellulose, protein, and water. The minor components are represented by a complex mixture of polyphenols (gallic acid, protocatechuic acid, p-hydroxybenzoic acid, vanillic acid, and caffeic acid). Olive pomace has some commercial value depending upon its oil and water content. The olive pomace can be further processed to extract the remaining oil. The exhausted pomace (kernel wood) can be used for several applications: (*1*) Its combustion can generate heating water in the olive oil mills; (*2*) kernel wood, after the removal of the stones, can be used as livestock feed; and (*3*) its ashes can be used as fertilizers.

The wastewater represents the water content of the olive fruit plus the water added during the processing of the olives in the olive mill. Carbohydrates are the major organic substances in wastewater; these are followed by proteins, organic acids, phenols and polyphenols, and some emulsified oil. Among the inorganic components, the most abundant are phosphates, sulphates, and chlorides, followed by carbonates and silicates. Sodium, potassium, and calcium are also found in large quantities. Wastewater has been used to grow yeast, to produce butanol using microorganisms, to isolate anthocyanin compounds for use in the food industry, and to produce steam. Efforts are being made to reduce wastewater by recycling in the milling process and to decrease its environmental pollution by treatment with biological or physical processes prior to its discharge (103–105).

By-products of Olive Pickling

Some important progress in solving the problem of olive-pickling by-products has been made in two complementary ways (102,106):

1. Reducing the volume of industrial wastes and reusing them in processing operations
2. Regenerating the fermentation brines to be further used in packaging the final product

For the first point, a series of internal-control regulations have been adopted, such as reusing lyes for the treatment of olives; reducing the lye concentration up to a minimum effective value; and reducing, and even completely eliminating the washing operations by neutralizing the residual free lye after the alkaline treatment (99). With regard to the regeneration of the fermentation brines, to be used in bottling and canning operations, the most successful procedures include ultrafiltration and a previous treatment with active coal followed by a filtration (107,108).

At present, it can be said that this is a problem that is technically solved, although it requires an economic adjustment to be developed industrially and generally applied.

THE INTERNATIONAL OLIVE OIL COUNCIL

Establishment and Functions

The IOOC was established in 1959 to administer the International Agreement on Olive Oil of 1956. This agreement has evolved over the last 40 years through a series of amendments approved at several United Nations conferences on olive oil. The most comprehensive International Agreement on Olive Oil and Table Olives was negotiated at the U.N. Conference on Olive Oil held in Geneva from June 18 to July 1, 1986, and extended with amendments until December 31, 1998. The chief objectives of the 1986 agreement are to facilitate international cooperation on world problems affecting olive products; to promote research and development for the modernization of olive cultivation and the industries processing its products by setting up technical and scientific programs to improve the quality and reduce the prices of the products obtained; to facilitate the study and application of measures devised to expand international trade in olive products; to seek to balance production and consumption, especially by increasing the latter; to reduce handicaps brought about by fluctuating market supplies; to forestall and oppose unfair competition practices and ensure delivery of goods that comply with the contracts signed; to foster the coordination of policies covering olive products; to enlarge market accesses; to ensure reliability of supplies and devise trade structures, especially by improving consultations and the supply of information; and to continue to develop the tasks undertaken within the context of the previous olive oil agreements.

The council undertakes, in conjunction with United Nations Development Programme (UNDP), FAO, and other international organizations, to collect technical information and circulate it to all members; to promote action to coordinate technical improvement activities among members; to assist national planning relating to technical improvements, in particular in the developing olive-growing countries; and to encourage the transfer of technology to developing olive-growing countries; and to encourage the transfer of technology to developing olive-growing countries from countries highly advanced in olive cultivation and oil-extraction techniques. The council provides a special fund for this purpose and, as part of the development of international cooperation, it procures financial and technical assistance from competent international, regional, and national organizations.

The agreement provides for a joint promotion fund set up every year with compulsory contributions from the mainly producing members. Mainly importing members may contribute to the fund under special agreement. The council is entitled to receive, for promotional purposes, voluntary contributions from a governments or other sources. The council also gives special attention to any scientific research on the biological value of the products obtained from the olive that might prove to be of use to consumption throughout the world.

Composition

The contracting parties to the International Agreement on Olive Oil and Table Olives, 1986, as amended and extended, are as follows:

- Mainly producing and mainly importing members: European Union (Austria, Belgium, Denmark, Finland, France, Germany, Greece, Ireland, Italy, Luxembourg, the Netherlands, Portugal, Spain, Sweden, United Kingdom)
- Mainly producing members: Algeria, Cyprus, Israel, Lebanon, Morocco, Syrian Arab Republic, Tunisia, Turkey, Yugoslavia
- Mainly importing member: Egypt
- Observers: Argentina, Australia, Brazil, Bulgaria, Canada, Chile, China, Colombia, Costa Rica, Croatia, Cuba, Dominican Republic, Ecuador, India, Islamic Republic of Iran, Iraq, Japan, Jordan, Libya, Mexico, New Zealand, Norway, Pakistan, Panama, Peru, Poland, Romania, Russian Federation, Saudi Arabia, Slovak Republic, Slovenia, Thailand, United States, Uruguay, Venezuela, the Palestinian National Authority.

General Publications

International Olive Oil Council Information Sheet (twice monthly); *OLIVAE* (5 times per year); *National Policies for Olive Products* (annual); various leaflets giving information on the International Agreement on Olive Oil and Table Olives, the IOOC, and IOOC activities; market research, technical handbooks, recipe books

SUMMARY

According to Greek mythology, thousands of years ago, Poseidon, god of the sea, and Athena, goddess of wisdom, were arguing over who was to going to control the destiny of a newly built city. They decided that the one who would present the city with the most valuable gift would be the one honored. Poseidon struck a rock with his trident, and a spring appeared. As the water started to flow, a horse emerged, symbol of strength and power. Athena stuck her spear in the ground, and it turned into an olive tree, symbol of peace, wisdom, and prosperity. Athena's gift was deemed the most valuable, and the city was named Athens in her honor. Today, an olive tree stands where the legend says Athena placed her spear. It is said that all the olive trees in Athens were descended from that first olive tree offered by Athena. The reality is that the olive tree and its fruits have played an integral part in the history of the people of the Mediterranean. Olive oil has been considered the cornerstone of the so-called Mediterranean diet. This diet has been associated with multiple health benefits, and current researchers are investigating the molecular basis for these effects. This chapter describes briefly some of the more technical aspects of the production of olive oil and table olives. Several monographs go into more detail about these aspects of olive oil; among the most recent and comprehensive is one published by A. Kiritsakis (109).

BIBLIOGRAPHY

1. G. C. Martin, L. Ferguson, and V. S. Polito, "Flowering, Pollination, Fruiting, Alternate Bearing, and Abscission," in L. Ferguson, G. Steven Sibbett, and G. C. Martin, eds., *Olive Production Manual*, Univ. of California, Div. Agr. Natural Resources, Oakland, Calif., 1994, pp. 51–56.
2. G. Bongi and A. Palliotti, "Olive," in B. Schaeffer and P. C. Andersen, eds., *Handbook of Environmental Physiology and Fruit Crops*, CRC Press, Boca Raton, Fl. 1994, pp. 165–188.
3. M. J. Fernandez-Diez, "Olives," in H. J. Rehm and G. Reed, eds., *Biotechnology*, Verlag Chemie, Weinheim, Germany, 1983, pp. 379–397.
4. N. Liphschitz et al., "The Beginning of Olive *Olea Europaea* Cultivation in the Old World: A Reassessment," *Journal of Archaeological Science* **18**, 441–454 (1991).
5. R. B. Hitchner and D. J. Mattingly, "Fruits of Empire: The Production of Olive Oil in Roman Africa," *Research and Exploration* **7**, 36, 38–48, 50–55 (1991).
6. R. Aparicio and M. T. Morales, "Characterization of Olive Ripeness by Green Aroma Compounds of Virgin Olive Oil," *J. Agric. Food Chem.* **46**, 1116–1122 (1998).
7. J. M. Garcia, S. Seller, and M. C. Perez Camino, "Influence of Fruit Ripening on Olive Oil Quality," *J. Agric. Food Chem.* **44**, 3516–3520 (1996).
8. J. Fernandez-Bolanos et al., "Sugars and Polyols in Green Olives: III, Quantitative Determination by Gas-Liquid Chromatography," *Grasas y Aceites* **34**, 168–171 (1983).
9. R. Guillen et al., "Dietary Fiber in Olives: Characterization of fractions," *European Journal of Clinical Nutrition* **49**, S224–S225 (1995).
10. A. Gil-Serrano, M. P. Tejero-Mateo, and J. Fernandez-Bolanos, "Polysaccharides in Olives: III, Partial Hydrolysis and Acetolysis of a Hemicellulose B Isolated in Olives of the Manzanilla Variety," *Grasas y Aceites* **35**, 224–227 (1984).
11. A. Jimenez et al., "Cell Wall Composition of Olives," *J. Food Sci.* **59**, 1192–1196 (1994).
12. A. Heredia-Moreno and J. Fernandez-Bolanos, "Cellulases in Olives and Its Possible Influence in the Changes of Texture: II, Activated Cellulolytic Activity in the Hojiblanca Variety," *Grasas y Aceites* **36**, 130–133 (1985).
13. A Heredia-Moreno and J. Fernandez-Bolanos, "Cellulases in Olives and Its Possible Influence in Texture Changes: III, Study of Some Factors That Could Modify Its Activity," *Grasas y Aceites* **36**, 171–176 (1985).

14. A. Heredia-Moreno, J. Fernandez-Bolanos, and R. Guillen-Bejarano, "Cellulolytic Activity in Olives during Maduration and Softening," *Alimentaria* **198**, 49–52 (1988).
15. A. Heredia-Moreno, J. Fernandez-Bolanos, and R. Guillen-Bejarano, "Characterization and Partial Purification of Cellulolytic Enzymes in Olives," *Grasas y Aceites* **40**, 190–193 (1989).
16. A. Heredia-Moreno, J. Fernandez-Bolanos, and R. Guillen-Bejarano, "Inhibitors of Cellulolytic Activity in Olive Fruits (*Olea europaea*, Hojiblance var.)," *Z. Lebensm. Unters.-Forsch.* **189**, 216–218 (1989).
17. M. Nosti-Vega, R. Castro-Ramos, and R. Vazquez-Ladron, "Composition and Nutritive Value of Spanish table olives: VI, Changes Due to Elaboration Processes," *Grasas y Aceites* **35**, 11–14 (1984).
18. M. I. Minguez-Mosquera, "Evolution of Pectic Constituents and Pectinolytic Enzymes During the Maturation and Storage of the Hojiblanca Olive Variety," *Grasas y Aceites* **33**, 327–333 (1982).
19. P. Manzi et al., "Natural Antioxidants in the Unsaponifiable Fraction of Virgin Olive Oils from Different Cultivars," *J. Sci. Food Agric.* **77**, 115–120 (1998).
20. B. Gandul Rojas and M. I. Minguez-Mosquera, "Chlorophyll and Carotenoid Composition in Virgin Olive Oils from Various Spanish Olive Varieties," *J. Sci. Food Agric.* **72**, 31–39 (1996).
21. M. I. Minguez-Mosquera and J. Garrido-Fernandez, "Chlorophyll and Carotenoid Presence in Olive Fruit (*Olea europaea*)," *J. Agric. Food Chem.* **37**, 1–7 (1989).
22. M. I. Minguez-Mosquera and L. Gallardo Guerrero, "Disappearance of Chlorophylis and Carotenoids during the Ripening of the Olive," *J. Sci. Food Agric.* **69**, 1–6 (1995).
23. L. Ferreiro and R. Aparicio, "Influence of Altitude on Chemical Composition of Andalusian Virgin Olive Oil Mathematical Classification Equations," *Grasas y Aceites* **43**, 149–156 (1992).
24. L. Cinquanta, M. Esti, and E. La Notte, "Evolution of Phenolic Compounds in Virgin Olive Oil during Storage," *J. Am. Oil Chem. Soc.* **74**, 1259–1264 (1997).
25. I. T. Agar et al., "Quality of Fruit and Oil of Black-Ripe Olives Is Influenced by Cultivar and Storage Period," *J. Agric. Food Chem.* **46**, 3415–3421 (1998).
26. A. K. Kiritsakis, "Flavor Components of Olive Oil: A Review," *J. Am. Oil Chem. Soc.* **75**, 673–681 (1998).
27. J. Tous et al., "Chemical and Sensory Characteristics of 'Arbequina' Olive Oil Obtained in Different Growing Areas of Spain" (in Spanish), *Grasas y Aceites* **48**, 415–424 (1997).
28. A. Keys et al., "The Diet and 15-Year Death Rate in the Seven Countries Study," *Am. J. Epidemiol.* **124**, 903–915 (1986).
29. M. de Lorgeril et al., "Mediterranean Diet, Traditional Risk Factors, and the Rate of Cardiovascular Complications After Myocardial Infarction: Final Report of the Lyon Diet Heart Study," *Circulation* **99**, 779–785 (1999).
30. F. Visioli and C. Galli, "The Effect of Minor Constituents of Olive Oil on Cardiovascular Disease: New Findings," *Nutrition Reviews* **56**, 142–147 (1998).
31. F. Visioli and C. Galli, "Oleuropein Protects Low Density Lipoprotein from Oxidation," *Life Sci.* **55**, 1965–1971 (1994).
32. F. Visioli, S. Bellosta, and C. Galli, "Oleuropein, the Bitter Principle of Olives, Enhances Nitric Oxide Production by Mouse Macrophages," *Life Sci.* **62**, 541–546 (1998).
33. F. Angerosa et al., "Characterization of Seven New Hydrocarbon Compounds Present in the Aroma of Virgin Olive Oils," *J. Agric. Food Chem.* **46**, 648–653 (1998).
34. F. Caponio, V. Alloggio, and T. Gomes, "Phenolic Compounds of Virgin Olive Oil: Influence of Paste Preparation Techniques," *Food Chem.* **64**, 203–209 (1999).
35. A. Bianco et al., "Microcomponents of Olive Oil: III, Glucosides of 2(3,4-dihydroxy-phenyl)ethanol," *Food Chem.* **63**, 461–464 (1998).
36. R. Sacchi et al., "Characterization of Italian Extra Virgin Olive Oils Using H-NMR Spectroscopy," *J. Agric. Food Chem.* **46**, 3947–3951 (1998).
37. A. Kiritsakis et al., "Effect of Fruit Storage Conditions on Olive Oil Quality," *J. Am. Oil Chem. Soc.* **75**, 721–724 (1998).
38. A. Ranalli, G. De Mattia, and M. L. Ferrante, "Comparative Evaluation of the Olive Oil Given by a New Processing System," *International Journal of Food Science and Technology* **32**, 289–297 (1997).
39. M. Esti, L. Cinquanta, and E. La Notte, "Phenolic Compounds in Different Olive Varieties," *J. Agric. Food Chem.* **46**, 32–35 (1998).
40. M. Rahmani and A. S. Csallany, "Role of Minor Constituents in the Photooxidation of Virgin Olive Oil," *J. Am. Oil Chem. Soc.* **75**, 837–843 (1998).
41. T. Gomes and F. Caponio, "Investigation on the Degree of Oxidation and Hydrolysis of Refined Olive Oils: An Approach for Better Product Characterisation," *Italian Journal of Food Science* **9**, 277–285 (1997).
42. M. T. Morales, J. J. Rios, and R. Aparicio, "Changes in the Volatile Composition of Virgin Olive Oil during Oxidation: Flavors and Off-Flavors," *J. Agric. Food Chem.* **45**, 2666–2673 (1997).
43. F. Angerosa et al., "Biogeneration of Volatile Compounds in Virgin Olive Oil: Their Evolution in Relation to Malaxation Time," *J. Agric. Food Chem.* **46**, 2940–2944 (1998).
44. R. Aparicio, J. J. Calvente, and M. T. Morales, "Sensory Authentication of European Extra-Virgin Olive Oil Varieties by Mathematical Procedures," *J. Sci. Food Agric.* **72**, 435–447 (1996).
45. N. Antoun and M. Tsimidou, "Gourmet Olive Oils: Stability and Consumer Acceptability Studies," *Food Research International* **30**, 131–136 (1997).
46. F. Angerosa et al., "Sensory Evaluation of Virgin Olive Oils by Artificial Neural Network Processing of Dynamic Head-Space Gas Chromatographic Data," *J. Sci. Food Agric.* **72**, 323–328 (1996).
47. R. Raoux, "Sensory Analysis," *Ocl-Oleagineux Corps Gras Lipides* **4**, 369–372 (1997).
48. R. Aparicio, M. T. Morales, and V. Alonso, "Authentication of European Virgin Olive Oils by Their Chemical Compounds, Sensory Attributes, and Consumers' Attitudes," *J. Agric. Food Chem.* **45**, 1076–1083 (1997).
49. P. C. Van Bruggen et al., "Robust Sensory Evaluation of Olive Oil by a Non-Parametric Scoring System," *J. Sci. Food Agric.* **67**, 53–59 (1995).
50. E. Pagliarini and C. Rastelli, "Sensory and Instrumental Assessment of Olive Oil Appearance," *Grasas y Aceites* **45**, 62–65 (1994).
51. M. Bertuccioli, "A Study of the Sensory and Nutritional Quality of Virgin Olive Oil in Relation to Variety, Ripeness, and Extraction Technology: Overview of the Three Year Study and Conclusion," *Grasas y Aceites* **45**, 55–59 (1994).
52. J. Mojet and S. De Jong, "The Sensory Wheel of Virgin Olive Oil," *Grasas y Aceites* **45**, 42–47 (1994).
53. R. Aparicio et al., "Relationship between the COI Test and Other Sensory Profiles by Statistical Procedures," *Grasas y Aceites* **45**, 26–41 (1994).

54. D. H. Lyon and M. P. Watson, "Sensory Profiling: A Method for Describing the Sensory Characteristics of Virgin Olive Oil," *Grasas y Aceites* **45**, 20–25 (1994).
55. M. A. Albi and F. Gutierrez, "Study of the Precision of an Analytical Taste Panel for Sensory Evaluation of Virgin Olive Oil Establishment of Criteria for the Elimination of Abnormal Results," *J. Sci. Food Agric.* **54**, 255–268 (1991).
56. G. P. Blanch, J. Villen, and M. Herraiz, "Rapid Analysis of Free Erythrodiol and Uvaol in Olive Oils by Coupled Reserved Phase Liquid Chromatography–Gas Chromatography," *J. Agric. Food Chem.* **46**, 1027–1030 (1998).
57. M. L. R. del Castillo et al., "Rapid Recognition of Olive Oil Adulterated with Hazelnut Oil by Direct Analysis of the Enantiomeric Composition of Filbertone," *J. Agric. Food Chem.* **46**, 5128–5131 (1998).
58. J. P. Favier et al., "CO2, Laser Infrared Optothermal Spectroscopy for Quantitative Adulteration Studies in Binary Mixtures of Extra-Virgin Olive Oil," *J. Am. Oil Chem. Soc.* **75**, 359–362 (1998).
59. A. S. Amr and A. I. Abu Al Rub, "Evaluation of the Bellier Test in the Detection of Olive Oil Adulteration with Vegetable Oils," *J. Sci. Food Agric.* **61**, 435–437 (1993).
60. V. Baeten et al., "Detection of Virgin Olive Oil Adulteration by Fourier Transform Raman Spectroscopy," *J. Agric. Food Chem.* **44**, 2225–2230 (1996).
61. R. Aparicio and V. Baeten, "Fats and Oils Authentication by FT-Raman," *Ocl-Oleagineux Corps Gras Lipides* **5**, 293–295 (1998).
62. N. A. Marigheto et al., "A Comparison of Mid-Infrared and Raman Spectroscopies for the Authentication of Edible Oils," *J. Am. Oil Chem. Soc.* **75**, 987–992 (1998).
63. F. Angerosa et al., "Carbon Stable Isotopes and Olive Oil Adulteration with Pomace Oil," *J. Agric. Food Chem.* **45**, 3044–3048 (1997).
64. O. Jimenez De Blas and A. Del Valle Gonzalez, "Gas Chromatographic Differentiation of Virgin, Refined, and Solvent-Extracted Olive Oils," *J. AOAC Int.* **79**, 707–710 (1996).
65. F. Santinelli, P. Damiani, and W. W. Christie, "The Triacylglycerol Structure of Olive Oil Determined by Silver Ion High-Performance Liquid Chromatography in Combination with Stereospecific Analysis," *J. Am. Oil Chem. Soc.* **69**, 552–556 (1992).
66. E. Salivaras and A. R. Mccurdy, "Detection of Olive Oil Adulteration with Canola Oil from Triacylglycerol Analysis by Reversed-Phase High-Performance Liquid Chromatography," *J. Am. Oil Chem. Soc.* **69**, 935–938 (1992).
67. A. De Leonardis, V. Macciola, and M. De Felice, "Rapid Determination of Squalene in Virgin Olive Oils Using Gas-Liquid Chromatography," *Italian Journal of Food Science* **10**, 75–80 (1998).
68. M. Amelio, R. Rizzo, and F. Varazini, "Separation of Stigmasta-Diene, Squalene Isomers, and Wax Esters from Olive Oils by Single High-Performance Liquid Chromatography Run," *J. Am. Oil Chem. Soc.* **75**, 527–530 (1998).
69. O. Jimenez De Blas and A. Del Valle Gonzalez, "Determination of Sterols by Capillary Column Gas Chromatography: Differentiation among Different Types of Olive Oil: Virgin, Refined, and Solvent-Extracted," *J. Am. Oil Chem. Soc.* **73**, 1685–1689 (1996).
70. A. Cert, W. Moreda, and J. Garcia-Moreno, "Determination of Sterols and Triterpenic Dialcohols in Olive Oils Using HPLC Separation and GC Analysis: Standardization of the Analytical Method," *Grasas y Aceites* **48**, 207–218 (1997).
71. F. Lanuzza, G. Micali, and G. Calabro, "On-Line HPLC-HRGC Coupling and Simultaneous Transfer of Two Different LC Fractions: Determination of Aliphatic Alcohols and Sterols in Olive Oil," *HRC-Journal of High Resolution Chromatography* **19**, 444–448 (1996).
72. E. Psomiadou and M. Tsimidou, "Simultaneous HPLC Determination of Tocopherols, Carotenoids, and Chlorophylls for Monitoring Their Effect on Virgin Olive Oil Oxidation," *J. Agric. Food Chem.* **46**, 5132–5138 (1998).
73. G. Bianchi and G. Vlahov, "Composition of Lipid Classes in the Morphologically Different Parts of the Olive Fruit, *cv. Coratina (Olea europaea Linn.)*," *Fett Wissenschaft Technologie* **96**, 72–77 (1994).
74. E. J. Birch et al., "Methods for Analysis of Esterified and Free Long-Chain Fatty Acids in High-Lipid Food Processing Wastes," *J. Agric. Food Chem.* **46**, 5332–5337 (1998).
75. W. Moreda, M. C. Perez-Camino, and A. Cert, "Determination of Some Purity Parameters in Olive Oils: Results from a Collaborative Study," *Grasas y Aceites* **46**, 279–284 (1995).
76. W. V. Cruess, "Pickling Green Olives," *California Agricultural Experimental Station Bulletin* **498**, 1–42 (1930).
77. R. H. Vaughn, H. C. Douglas, and J. R. Gililland, "Production of Spanish Type Green Olives," *California Agricultural Experimental Station Bulletin* **678**, 1–82 (1943).
78. J. M. R. de la Borbolla y Alcala and C. Gomez-Herrera, "La industria de aderezo de aceitunas verdes," *Revista de Ciencia Aplicada* **7**, 120–132 (1949).
79. G. D. Balatsouras, "The Chemical Composition of the Brine of Stored Greek Black Olives," *Grasas y Aceites* **17**, 83–88 (1966).
80. M. Brenes and A. de Castro, "Transformation of Oleuropein and Its Hydrolysis Products during Spanish-Style Green Olive Processing," *J. Sci. Food Agric.* **77**, 353–358 (1998).
81. J. E. De La Torre et al., "Physical, Chemical, and Microbiological Studies of the Fermentation of Arbequina Green Olives," *Grasas y Aceites* **44**, 274–278 (1993).
82. A. H. Sanchez, A. Montano, and L. Rejano, "Effect of Preservation Treatment, Light, and Storage Time on Quality Parameters of Spanish-Style Green Olives," *J. Agric. Food Chem.* **45**, 3881–3886 (1997).
83. V. Marsilio and B. Lanza, "Effects of Lye-Treatment on the Nutritional and Microstructural Characteristics of Table Olives (*Olea europaea L.*)," *Revista Espanola de Ciencia y Tecnologia de Alimentos* **35**, 178–190 (1995).
84. A. I. Yassa, A. B. S. S. Emmam, and R. T. Ahmed, "Effect of Different Methods of Processing on the Nutritional Value of Azizi and Picual Table Olives," *Annals of Agricultural Science* **35**, 775–786 (1990).
85. M. I. Minguez-Mosquera, J. Garrido-Fernandez, and B. Gandul Rojas, "Pigment Changes in Olives during Fermentation and Brine Storage," *J. Agric. Food Chem.* **37**, 8–11 (1989).
86. A. Jimenez et al., "Changes in Texture and Cell Wall Polysaccharides of Olive Fruit during 'Spanish Green Olive' Processing," *J. Agric. Food Chem.* **43**, 2240–2246 (1995).
87. J. M. R. de la Borbolla y Alcala and F. Gonzalez-Pellisso, "Studies on the Spanish-Style Packed Olives: XI, Sediment Inhibition," *Grasas y Aceites* **23**, 107–117 (1972).
88. F. Gonzalez-Pellisso, L. Rejano Navarro, and F. Gonzalez-Cancho, "The Pasteurization of Olives Sevillan Style," *Grasas y Aceites* **33**, 201–207 (1982).
89. A. Montano Asquerino, A. H. Sanchez Gomez, and L. Rejano Navarro, "Method for the Color Determination of Green Table Olives in Brine," *Alimentaria* **193**, 79–83 (1988).
90. P. Garcia-Garcia, M. Brenes Balbuena, and A. Garrido Fernandez, "The Use of Ferrous Lactate in the Elaboration of Black Table Olives," *Grasas y Aceites* **37**, 33–38 (1986).

91. A. Garrido Fernandez, M. C. Duran Quintana, and P. Garcia Garcia, "Study of Different Conservation Forms of Turning Color Olives of the Variety Manzanilla," *Grasas y Aceites* **37**, 1–7 (1986).

92. M. Brenes Balbuena et al., "Comparative Studies of the Conservation Systems of Black Table Olives," *Grasas y Aceites* **37**, 123–128 (1986).

93. P. Garcia Garcia, M. C. Duran Quintana, and A. Garrido Fernandez, "Aerobic Fermentation of Ripe Olives in Brine," *Grasas y Aceites* **36**, 14–20 (1985).

94. A. Garrido Fernandez, P. Garcia Garcia, and F. Sanchez-Roldan, "New Fermentation Aerobic Process for Natural Black Olives," *Alimentacion* **4**, 73–81 (1985).

95. A. Garrido Fernandez, P. Garcia Garcia, and M. C. Duran Quintana, "Conservation of Natural Black Olives Proceeding from Aerobic Fermentation," *Grasas y Aceites* **36**, 313–316 (1985).

96. M. C. Duran Quintana, P. Garcia Garcia, and A. Garrido Fernandez, "Aerobic Fermentation of Ripe Olives in Brine with Alternating Air Injection," *Grasas y Aceites* **37**, 242–249 (1986).

97. A. Garrido Fernandez, M. C. Duran Quintana, and P. Garcia Garcia, "Aerobic Fermentation of Natural Black Olives in Brine," *Grasas y Aceites* **38**, 27–32 (1987).

98. P. Garcia Garcia, M. C. Duran Quintana, and A. Garrido Fernandez, "Packing of Aerobically Fermented Natural Black Olives," *Grasas y Aceites* **37**, 92–96 (1986).

99. P. Garcia Garcia et al., "Physicochemical Depuration of the Waste Waters from the Green Table Olive Packing Industries," *Grasas y Aceites* **41**, 263–269 (1990).

100. S. Vitolo, L. Petarca, and B. Bresci, "Treatment of Olive Oil Industry Wastes," *Bioresour. Technol.* **67**, 129–137 (1999).

101. A. G. Vlyssides et al., "Olive Oil Processing Wastes Production and Their Characteristics in Relation to Olive Oil Extraction Methods," *Fresenius Environ. Bull.* **7**, 308–313 (1998).

102. G. C. Kopsidas, "Wastewater from the Table Olive Industry," *Water Res.* **28**, 201–205 (1994).

103. F. Cabrera et al., "Land Treatment of Olive Oil Mill Wastewater," *International Biodeterioration and Biodegradation* **38**, 215–225 (1996).

104. F. Flouri et al., "Decolorization of Olive Oil Mill Liquid Wastes by Chemical and Biological Means," *International Biodeterioration and Biodegradation* **38**, 189–192 (1996).

105. I. Chatjipavlidis et al., "Bio Fertilization of Olive Oil Mills Liquid Wastes: The Pilot Plant in Messinia, Greece," *International Biodeterioration and Biodegradation* **38**, 183–187 (1996).

106. B. Inigo Leal, "New Technological Trends in the Production and Preservation of Table Olives," *Alimentaria* **28**, 61–65 (1991).

107. A. Garrido Fernandez, M. Brenes Balbuena, and P. R. Garcia Garcia, "Treatment for Green Table Olive Fermentation Brines," *Grasas y Aceites* **43**, 291–298 (1992).

108. A. Garrido, P. Garcia, and M. R. Brenes, "The Recycling of Table Olive Brine Using Ultrafiltration and Activated Carbon Adsorption," *Journal of Food Engineering* **17**, 291–305 (1992).

109. A. K. Kiritsakis, *Olive Oil: From the Tree to the Table*, 2nd ed., Food and Nutrition Press, Trumbull, Conn., 1998.

JOSE M. ORDOVAS
Tufts University
Boston, Massachusetts

OMEGA-3 FATTY ACIDS.

See FATS AND OILS: CHEMISTRY, PHYSICS AND APPLICATIONS.

OPTIMIZATION METHODS IN FOOD PROCESSING AND ENGINEERING

THE BEST

One primary goal of food process engineering is to find and understand the mechanisms and phenomena that underscore a given food manufacturing process. Such basic knowledge allows the food engineer to grasp how a process works and thus operate an existing or design a new processing system. In today's rapidly changing, highly sophisticated, and extremely competitive world, however, such basic knowledge is not always enough. Often, it is important not only to know how a process works but also how it works "best." The meaning of best changes from system to system and depends on the criteria used. For example, if the chosen criterion is process yield, then the highest value (maximum) is best. On the other hand, if cost is the criterion, then the lowest value (minimum) is best. The desired extreme value (maximum or minimum) is optimum, and the values of the adjustable processing variables that produce this optimum are called the optimum conditions. The theory, methods, and techniques used to attain an optimum and to locate the optimum operating conditions are the subject of optimization. *Optimization* has been defined simply as "getting the best you can out of a given situation" (1). In addition, "it [optimization] opens up the possibility of achieving the very best, of reaching perfection, of choosing the superior course of action among all possible alternatives" (2).

The scope of this article is to (*1*) introduce the subject of optimization in simple terms; (*2*) clarify some of the terminology, theory, and methods used; and (*3*) cover some of the areas in food science and engineering where optimization is being applied or potential applications are likely to occur. The references provide more explicit and in-depth coverage of the subject.

THE ELEMENTS OF OPTIMIZATION PROBLEMS

In every case of optimization, the problem must first be formulated and then the system to be optimized must be defined. This requires specifying the variables that affect the system's behavior. For example, in making a hot cup of tea, an obvious variable that affects the quality of the tea is the length of time the tea bag stays in the hot water. Performing experiments to measure the quality of the tea as a function of time generates a curve like the one shown in Figure 1. Choosing the maximum quality value and tracing it back to the corresponding optimum time optimizes this oversimplified and trivial system. Accounting for the effect of the water temperature on the tea quality complicates the system. In such a case a series of constant quality curves might be obtained (Fig. 2), which produce a set of optimum conditions for temperature and time.

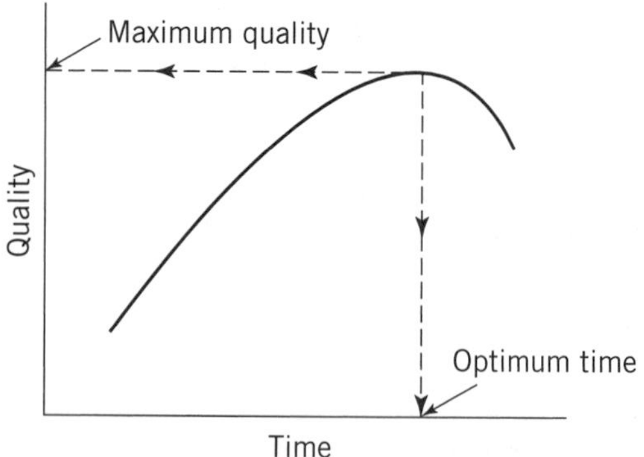

Figure 1. Quality of tea as a function of time the tea bag stays in hot water.

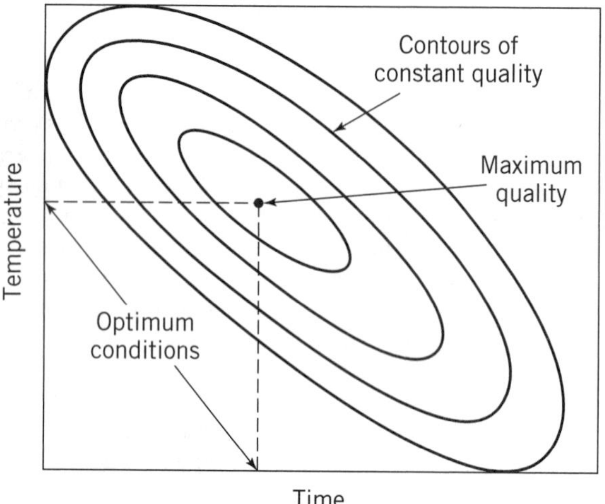

Figure 2. Quality of tea as a function of water temperature and processing time.

To make the preceding system more realistic, other variables could be included in the optimization problem: the amount of stirring, the amount of tea in the tea bag, the storage history of the tea leaves (temperature and humidity), and so on. All these factors will have some effect on the quality. Furthermore, it is difficult to measure and define quality, because it depends on the taste and preference of individuals, among other things. Despite the simplicity of the preceding example, it illustrates the basic problem of optimization and the need to formulate the problem, define the system, find a function (or mathematical model) that describes the system, and find the system's optimum.

More explicitly, all optimization problems must consider several important elements (2,3):

Selection of Decision Variables. The most critical element in optimization problems is to select all the appropriate variables that contribute to the decision-making process of attaining the optimum. These input, or independent variables affect the performance of the process or system and can be adjusted, changed, and controlled.

Definition of the Performance Function. Another important element of optimization problems is the process of defining which dependent variables require optimization (maximization or minimization). These measured variables, called performance or objective functions, determine and reflect the performance of the system or process.

Identification of Constraints. In most optimization problems, certain constraints may limit the values that dependent or independent variables can take, restricting, therefore, the region of allowable solutions. Such constraints may be imposed by physicochemical or other impossibilities (in the tea example, the temperature of the water cannot be more than 100°C). They may also represent restrictions endogenous to the system (eg, federal or state requirements on the composition of a food product, grading guidelines, and equipment restrictions). Whatever the case might be, such constraints must be taken into account, because they influence the solution of a given problem.

Development of a Mathematical Model to Describe the System. Attempts to solve a problem by trial and error require painstaking physical experimentation and application of statistical theory. Instead of this, a mathematical model is used for most cases of optimization. In some cases, mathematical models may already exist (eg, thermal processing and drying), but in other situations models must be developed based on available scientific and engineering knowledge. It is necessary to be aware of how to develop a model (4,5). Sometimes the complexity of the system might be such that a model cannot be developed or, if developed, cannot be solved. This is common in biochemical and biological systems such as foods, and usually an approximating (or graduating) function is developed, which represents the true nature of the system as closely as possible (6). Such functions are usually polynomials or transformations thereof, and a limited number of experiments are required to find their constant coefficients (7,8). Data collection methods (experimental design) and model fitting (linear, nonlinear, and other regression) techniques are important elements of finding the appropriate approximating function (3,7–11).

Attaining the Optimum. The final element of all optimization problems is to choose an appropriate method to optimize the system. It is imperative to realize that the scope of every optimization method is to adjust and readjust the values of the decision variables to locate the settings that maximize or minimize (optimize) the performance function(s), while ensuring that all variables (dependent or independent) satisfy the constraints of the system.

Figure 3 gives typical relationships between the important elements of an optimization problem and their role in its solution. The following section discusses the available optimization techniques and methods. This information

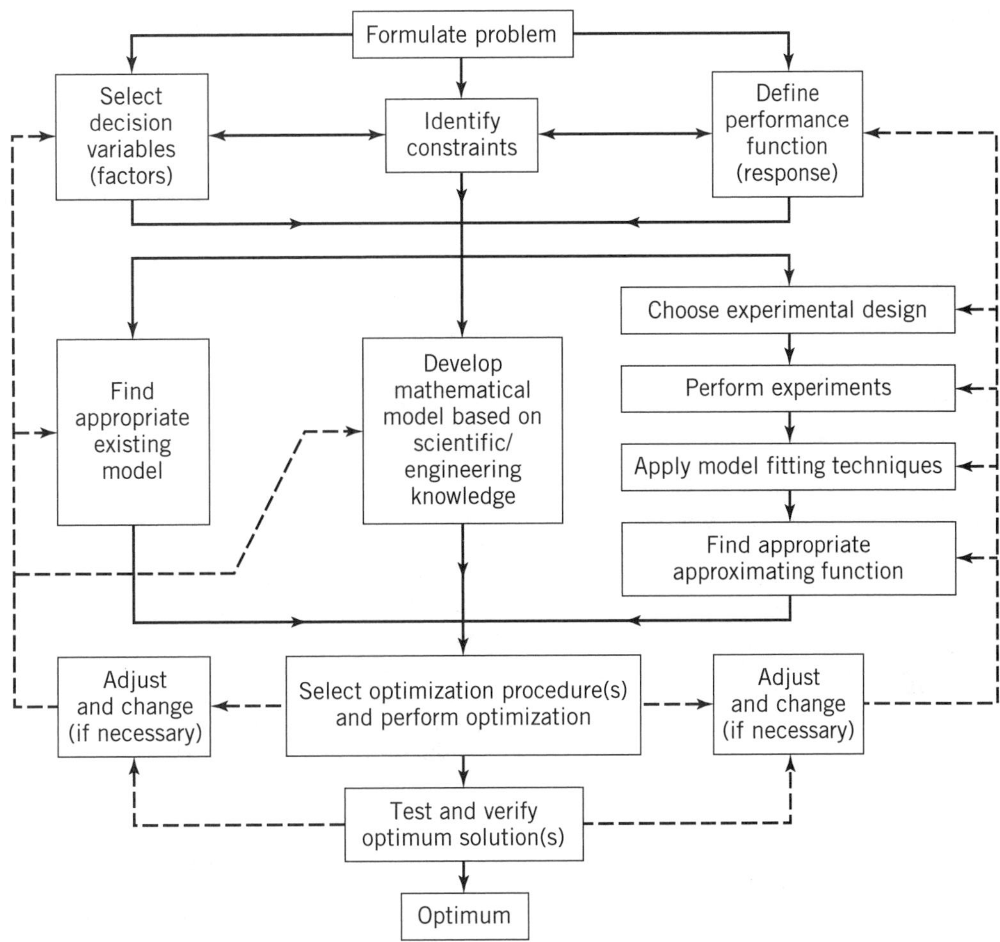

Figure 3. The important elements of an optimization problem and their interactions.

should be complemented with the more in-depth discussion given in the references cited therein.

OPTIMIZATION METHODS

Basic Theoretical Background

Among the various optimization techniques, each is best suited to operate on a particular type of mathematical model. Consequently, when formulating a problem and developing a specific model, the strong and weak points of the various optimization techniques must be kept in mind.

The basis for all optimization methods is the classic theory of maxima and minima. Mathematically, the theory is concerned with finding the minimum or maximum (extreme points) of a function of n variables, $f(x_1, x_2, \ldots, x_n)$, where n is any integer greater than zero. To simplify the concept, assume that f is a function of only one variable x (Fig. 4). In this example, point A is located at the boundary and it is a local minimum. Point B, where the first derivative f' is discontinuous, is a global maximum. Point C, where $f' =$ zero, is a local minimum, D ($f' = 0$) and G are local maxima, and E is the global minimum. Points C, D, E, and F are stationary points, because their first derivative f' is zero. A stationary point is a maximum (local or

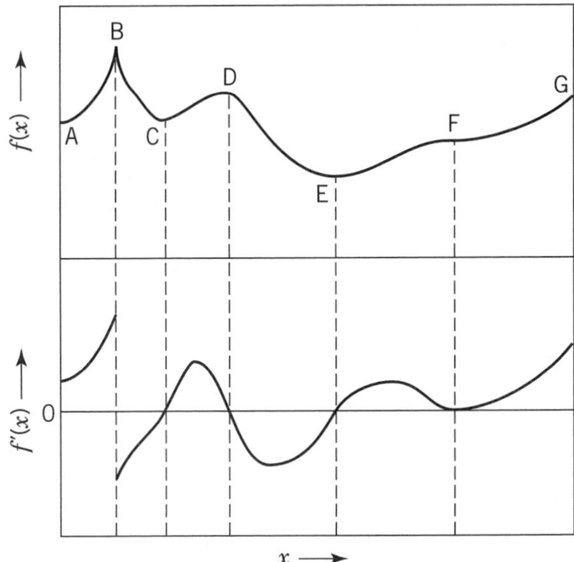

Figure 4. A schematic representation of a function $f(x)$ and its first derivative $f'(x)$.

global) if movement away from it causes the function $f(x)$ to decrease. A point is a minimum if movement away from it causes an increase. Mathematically, a point x^* is a local maximum of the function $f(x)$ if

$$f(x^*) \geq f(x) \tag{1}$$

and

$$|x^* - x| < \delta \tag{2}$$

where δ is a small positive number. The point x^* is termed a global maximum when the inequality (Equation 1) holds true for all values of x. Definitions of local and global minima are similar. Point F, however, whose first derivative is zero (a stationary point), is neither a maximum nor a minimum, but a mini-max, better known as a saddle point. In this case, movement away from point F will result in an increase or decrease in $f(x)$ depending on the direction of the movement.

The classic theory of maxima and minima is simply a search to find all local maxima or minima and then a comparison of the individual values to determine the global (absolute) maximum or minimum. The critical places to look for the extreme values are

1. at the stationary points (where $f'(x) = 0$),
2. on the boundaries of the domain of the decision variables, and
3. at points where the first derivatives are discontinuous.

When $f(x)$ and $f'(x)$ are continuous functions, the extreme points will most likely be at the stationary points (except for cases involving saddle points). Thus the problem becomes one of locating the points where the partial derivatives are zero. To accomplish this goal, the following n algebraic equations

$$\frac{\partial f}{\partial x_j}(x_1, x_2, \ldots, x_n) = 0 \quad (j = 1, \ldots, n) \tag{3}$$

must be simultaneously solved. Unfortunately, differential calculus does not always provide a method for solving such equations, particularly when real solutions do not exist. If some continuous and real functions $z_j(x)$ exist, an approximate solution to the equation

$$z_j(x) = 0 \quad (j = 1, \ldots, n) \tag{4}$$

can be obtained by minimizing the sum of squares of the residuals defined by

$$\sum_{j=1}^{n} (z_j(x))^2 \tag{5}$$

The calculus approach is useful when the equation 3 can be solved directly (when they are all linear, eg). It can also reduce the dimensionality, the number of variables required to solve the problem, in some cases.

These analytical methods may not always work. Optimization can still be performed, however, by iterative methods using a computer. Iterative methods are useful for solving simultaneous equations such as equation 4, but cannot guarantee optimization when used to solve equation 3, primarily because stationary points are not always optima (they may be saddle points). For that, it is usually better to apply the iterations directly to the function $f(x)$ and try to find the minimum (or maximum) by the following strategy (12):

1. Take a trial solution, x_k.
2. Find a direction from this trial solution in which $f(x)$ decreases (or increases).
3. Find a point x_{k+1} in this direction so that $f(x_{k+1}) < f(x_k)$ (or $f(x_{k+1}) > f(x_k)$).
4. Repeat the process from this new trial solution.

Even iterative methods might fail to produce an optimum, however. This might be due to a number of reasons, including the nature or structure of the system, slow convergence, rounding-off errors, and finding a local instead of a global extreme. With the preceding background, restrictions, and limitations in mind some basic optimization methods are presented next.

Unconstrained Optimization

Unconstrained optimization techniques are applicable when searching for a minimum (or maximum) of a function $f(x)$ that is not subject to any constraints. Sufficient conditions must be developed to allow evaluation of the nature of the stationary points and determine if they are minima, maxima, or other. Once all of the local maxima and minima of the function $f(x)$ are located, a comparison among them produces the global optimum (maximum or minimum).

If the function $f(x)$ has only one independent variable, a Taylor series expansion about the stationary point x^* can be performed.

$$f(x) = f(x^*) + f'(x^*)(x - x^*) + \frac{1}{2} f''(x^*)(x - x^*)^2 + \ldots + \text{higher order terms} \tag{6}$$

The first derivative vanishes at the stationary point ($f(x^*) = 0$). Also, if x is sufficiently close to x^*, the higher-order terms are negligible compared with the second-order terms, and equation 6 becomes

$$f(x) \approx f(x^*) + \frac{1}{2} f''(x^*)(x - x^*)^2 \approx \frac{1}{2} f''(x^*)(x - x^*)^2 \tag{7}$$

Because $(x - x^*)^2$ is always positive, the nature of $f(x^*)$ depends on the value of $f''(x^*)$, the second derivative of the function at the stationary point:

If $f''(x^*) > 0$, then $f(x^*)$ is a minimum.
If $f''(x^*) < 0$, then $f(x^*)$ is a maximum.

In the case of $f''(x^*)$ = zero, higher order derivatives must be examined. This method can be extended to systems with two or more variables by using multidimensional Taylor series expansions and Hessian matrices (5,13–15).

Another popular approach for finding local maxima (or minima) is an optimization algorithm known as Newton's method. It computes $f(x)$ and $f'(x)$ at an initial point x_i and then finds an improved estimate x_{i+1} by linear extrapolation (12). Algebraically, this is

$$x_{i+1} = x_i - \frac{f(x_i)}{f'(x_i)} \tag{8}$$

If the function $f(x)$ is either convex or concave, Newton's method should converge toward a local maximum (or minimum) very rapidly. If $f(x)$ is an S-shaped function, and the initial point x_i is poorly chosen, the method may oscillate with a continuously increasing amplitude. Newton's method can be extended and applied to multidimensional optimization (12). Other methods such as conjugate gradients, quasi-Newton methods, and various Newtonlike methods have also been successfully applied (12,16).

Finally, several other optimization methods work well with one-dimensional unconstrained systems. The bisection method and its variations, the method of false position and its modifications, the golden section search, the Fibonacci search, and so on (12,13) all have certain advantages and disadvantages.

When searching for a method to optimize a multidimensional system, the following questions deserve consideration (12):

1. Is it easy to calculate the values of the function?
2. Can the first derivatives be found easily?
3. Can the second derivatives be found easily?
4. Can the $(n \times n)$ Hessian matrix and its inverse be stored affordably?
5. Are many of the Hessian's elements zero? Can that be capitalized on and the calculation time reduced?
6. Is there anything special about the particular problem that might make the solution easier?

Because there are no easy answers to most of the preceding questions, it is advisable to further study and understand the strengths and weaknesses of each optimization method before applying it to solve specific problems.

Constrained Optimization

In a number of practical situations, optimization of a function occurs over a restricted domain of the independent variables. For example, flow rates, concentration, shelf life, and so on may never take negative values. Optimization techniques still locate the stationary points of the function, but this time the solutions for optima must be subject to equality or inequality constraints.

Optimization of Systems with Equality Constraints. For a function $f(x_1, x_2, \ldots x_n)$ of n independent variables and subject to $m < n$ constraint equations

$$g_i(x_1, x_2, \ldots, x_n) = 0 \quad (i = 1, \ldots, m) \tag{9}$$

three basic analytical methods can be used to locate the extreme points: direct substitution, solution by constrained variation, and Lagrange multipliers. The first method involves a simple substitution of the m constraint equation 9 directly into the function to be optimized. This results in an equation with $(n - m)$ unknowns subject to no constraints. The methods of unconstrained optimization may then be applied. Equation 9 is usually complicated, and it is not always possible to perform the substitutions.

Figure 5 illustrates the second method, solution by constrained variation. In this case, there are two independent variables x_1 and x_2, and the function $f(x_1, x_2)$ is subject to the constraint $g(x_1, x_2) = 0$. It is obvious that the maximum of the constrained system is point B and the maximum of the unconstrained system is point A. At point B the curve $f(x_1, x_2)$ = constant and the curve $g(x_1, x_2) = 0$ have the same slope and are tangent. Therefore, infinitesimal changes dx_1 and dx_2 affect the dependent variables $f(x_1, x_2)$ and $g(x_1, x_2)$ in a similar way. In other words

$$\left(\frac{dx_2}{dx_1}\right)_f = \left(\frac{dx_2}{dx_1}\right)_g \tag{10}$$

Because f = constant and $g = 0$, the total derivatives of both functions at point B are

$$df = \frac{\partial f}{\partial x_1} dx_1 + \frac{\partial f}{\partial x_2} dx_2 = 0 \tag{11}$$

$$dg = \frac{\partial g}{\partial x_1} dx_1 + \frac{\partial g}{\partial x_2} dx_2 = 0 \tag{12}$$

It should be recalled that in this case (constraint optimization) $\partial f/\partial x_1$ and $\partial f/\partial x_2$ are not zero (as they would have been in an unconstraint case). Equations 10 to 12 can be

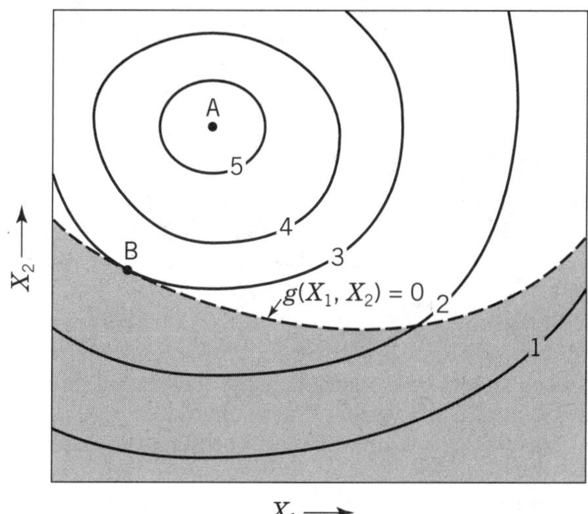

Figure 5. A schematic representation of a function $f(x_1, x_2)$ subject to constraint $g(x_1, x_2) = 0$, which restricts solutions to the shaded area only.

solved in combination with $g(x_1, x_2) = 0$ to locate the stationary point (5).

The preceding simple two-dimensional case generalizes to the case of n independent variables subject to $m < n$ constraints. Applying the same reasoning as before produces the so-called Jacobian determinants, which in turn can be used with the constraint equation 9 to obtain the stationary points (5).

The most commonly used of the three analytical methods mentioned is Lagrange multipliers. Its simplest form is the two-dimensional case. Multiplying equation 12 by a constant λ and adding the result to equation 11 yields

$$\left(\frac{\partial f}{\partial x_1} + \lambda \frac{\partial g}{\partial x_1}\right)dx_1 + \left(\frac{\partial f}{\partial x_2} + \lambda \frac{\partial g}{\partial x_2}\right)dx_2 = 0 \qquad (13)$$

Equation 13 transforms to

$$\frac{\partial}{\partial x_1}(f + \lambda g)dx_1 + \frac{\partial}{\partial x_2}(f + \lambda g)dx_2 = 0 \qquad (14)$$

The constant λ is the Lagrange multiplier of the system and is determined by equating the bracketed terms of equation 13 to zero. This produces (see equation 14) the augmented function Λ

$$\Lambda = f + \lambda g \qquad (15)$$

which guarantees that the stationary point(s) of the system will be located if λ is considered to be an independent variable and the following system of equations is solved.

$$\frac{\partial f}{\partial x_1} + \lambda \frac{\partial g}{\partial x_1} = 0 \qquad (16)$$

$$\frac{\partial f}{\partial x_2} + \lambda \frac{\partial g}{\partial x_2} = 0 \qquad (17)$$

$$g(x_1, x_2) = 0 \qquad (18)$$

In the general case of n independent variables with m constraint equations, the augmented function takes the form

$$\Lambda = f + \lambda_1 g_1 + \ldots + \lambda_m g_m \qquad (19)$$

and a system of $(n + m)$ equations must be solved to locate the stationary points. This is usually easier than a solution by direct substitution or constraint variation, unless the system is very simple.

Optimization of Systems with Inequality Constraints. Scientists and engineers understand that a profound difference exists between mathematical abstraction and physical reality. In the world of mathematics, an equality sign represents perfect equality, but in the real world equality requirements are physically impossible and unrealistic. For example, constraints such as net weight = 425 g or room temperature = 20°C are difficult to realize. However, targets such as contain no less than 425 g, or room temperature less than 20°C, are easier to achieve. Thus inequalities are an inherent part of the real world and require attention. With respect to optimization, several techniques exist to solve problems with inequality constraints.

If the system under consideration is linear, linear programming can be used to optimize it. The term *programming* here does not refer to computer programming, but rather to mathematical programming and scheduling. Several explanations of the mathematical relations involved in linear programming are available (12,16–21).

In some cases, it might be interesting to find out how a solution to a linear programming problem changes as the problem's data change in systematic and predetermined ways. Parametric programming can solve this problem. The algorithm for obtaining solutions of parametric programming problems has been published (12). In other cases, the objective function might not be linear, but quadratic instead. In such cases, quadratic programming can determine the optima (16). In more general terms, either the objective function or some (or all) of the constraints or both might be nonlinear in nature. Methods of nonlinear programming must then be applied (12,18). Other types of mathematical programming include (12,16) integer programming, which deals with optimization problems whose variables (some or all) are required to take integer values (ie, the number of machines or components required for some processing system). There is also geometric programming, which solves nonlinear programming problems containing special functions constructed from terms of the form

$$t_i = c_i \prod_{j=1}^{n} x_j^{a_{ij}} \qquad (20)$$

where t_i and c_i are positive coefficients and a_{ij} are constants (16).

Optimization of Dynamic Systems

A system is dynamic if it varies with time or distance. Optimization must account for such variation. An example of a dynamic system is the time-dependent behavior of batch and continuous fermentation vessels. The problem of optimizing dynamic systems can be stated as follows (19): given all the performance equations of a system (possibly a set of differential equations) and the initial-final values of some state variables, find the piecewise continuous decision control variable(s) that maximizes or minimizes the objective function.

Large dynamic systems with many decision variables can be broken down into stages and reduced to a series of interrelated systems, each containing only a few variables. The stages may be process components or equipment, units of time, or any other suitable entity. One of the methods to optimize such multistage dynamic systems is the so-called dynamic programming. Bellman (22) presented the basic theory and developed the principle of optimality. It states that an optimal policy has the property that whatever the initial state and initial decisions, the remaining decisions must constitute an optimal policy with regard to the state resulting from the first decision. The principle of optimality can be stated mathematically (5). In simple terms, the principle of optimality states that in a multistage serial system every component affects every downstream component. Figure 6 shows an example of a multistage process. Each stage has an input (I_n), an output (I_{n-1}), a decision (d_n), and a return (R_n). The output and return are dependent on the input and decisions. A dynamic programming analysis usually begins with the last stage, which affects no other stage in the system, and ends with the first stage, yielding an optimum for every input. For a multistage process, it is necessary first to find $f_1(I_1)$. With $f_1(I_1)$ known, then it is possible to find $f_2(I_2)$, then $f_3(I_3), \ldots, f_n(I_n)$.

Several other approaches can also solve dynamic optimization problems. Among the ones frequently applied are Pontyagrin's maximum principle (19,22–24), and the calculus of variations (25).

Experimental Optimization

Experimental optimization methods incorporate elements of statistical thinking into traditional scientific, engineering, and mathematical modeling and optimization. When the behavior of a system or process is unknown and a sufficiently simple deterministic (mathematical) model cannot be developed for further analysis, experimental optimization may be used to analyze the system and search for optimum conditions. The term *experimental* indicates that physical experimentation is involved. Well-designed experiments may produce statistically sound data, which in turn may result in reliable empirical models. Such models can predict or optimize, and often they provide the needed basis (information) for developing either more rugged or deterministic models. There are three well-known and widely used experimental optimization methods.

Response Surface Optimization. The term *response surface methodology* (RSM) refers to a group of mathematical and statistical techniques that, through limited physical experimentation, provide the means for attaining optimum operating conditions of complex systems. The theoretical basis of RSM (6) is a powerful tool for experimental optimization.

The basic idea of RSM is that for any given system, there must be a functional relationship φ that correlates the factors x_i (decision variables) to the response y (performance function)

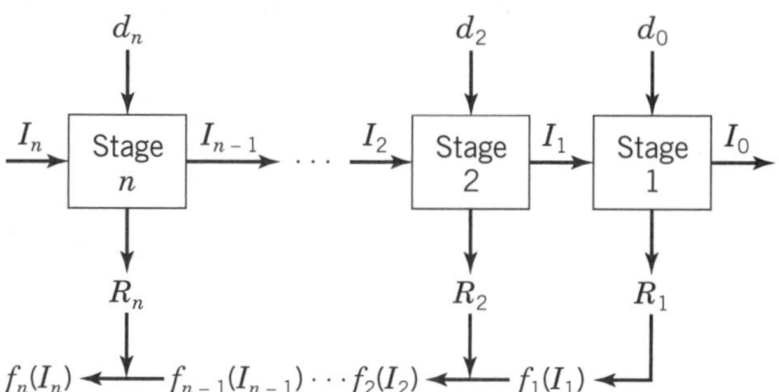

Figure 6. A multistage dynamic process.

$$y = \varphi(x_1, x_2, \ldots, x_n) \tag{21}$$

If the form of φ is explicitly known, then any of the methods discussed earlier can be used to optimize the system. If the function φ is unknown or very complex, another mathematically simpler function f must be found to approximate φ and describe the system. This new function

$$\hat{y} = f(x_1, x_2, \ldots, x_n) \tag{22}$$

estimates $\hat{y}$ rather than the true value y; it is called an approximating or graduating function and may take the form of practically any mathematical expression. The most commonly used expressions are polynomials of first or second order given by equations 23 and 24, respectively

$$\hat{y} = \beta_0 + \sum_{i=1}^{n} \beta_i x_i + \epsilon \tag{23}$$

$$\hat{y} = \beta_0 + \sum_{i=1}^{n} \beta_i x_i + \sum_{i=1}^{n} \beta_{ii} x_i^2 + \sum_{i=1}^{n-1} \sum_{j=i+1}^{n} \beta_{ij} x_i x_j + \epsilon \tag{24}$$

where β_0, β_i, β_{ij} are constant coefficients usually determined by least-squares methods and ϵ is the error involved in estimating the coefficients β from experimental data. At times, polynomials of higher order are used.

Polynomials are popular approximating functions for several reasons. They provide simple curvilinear relationships that can approximate practically any true continuous function within a specified range, and they usually possess a clearly defined optimum. They can be expanded to include any number of decision variables (x_i). Finally, a number of transformations such as logarithmic, exponential, inverse, power, and trigonometric, may be applied to the independent (x_i) or dependent (y) variables, which add some asymmetry and flexibility desirable in many cases of biological and biochemical systems. The weakness associated with the use of polynomials is that they are "smooth" functions without any biological or biochemical justification. Therefore, extrapolation beyond the experimental space (the region where data were collected to estimate β's) is usually not allowed.

The first step in a RSM study is to select an appropriate experimental design with a limited number of experimental runs k (where $k > n$), which will allow estimation of the coefficients β by minimizing the sum of squares of errors E

$$E = \sum_{i=1}^{k} \epsilon_i^2 \tag{25}$$

Ready-made designs already exist in terms of coded variables, which are linear functions of the actual decision variables, and minimize the overall error ϵ

$$\epsilon = \epsilon_v + \epsilon_b \tag{26}$$

where ϵ_v is the variance error, sometimes referred to as the experimental error, and is a result of the sampling variation, and ϵ_b is the so-called bias error (26), which refers to the failure of the approximating function (ie, polynomials) to exactly represent the true function $\varphi(x)$. Among the most commonly used designs for RSM studies are central composite designs (6,7), three-level designs (27), mixture designs (28–30), and several other fractional factorials (9,31,32).

After the coefficients β's have been determined (to satisfy equation 25), the stationary point(s) are determined. Taking the first partial derivatives of equation 23 or 24, equating them to zero, and solving the system of n equations (see equation 3 and related discussion) solves this issue. Matrix algebra can accomplish the same objective (7,8).

A canonical transformation of equation 23 or 24 will indicate the nature of the stationary point (8). If complicated ridge systems exist (when the stationary point is faraway from the origin, outside the experimental region), they may require further ridge analysis (33).

If the system has only one response (one performance function), then following the methods and analyses discussed achieves optimization easily. Often more than one response is involved, and sometimes the objectives are competing with each other. Under these circumstances, methods of multiresponse optimization (MRO) will locate the optimum operating conditions. The simplest form of MRO is the graphical approach. Equation 23 or 24, depending on the system, creates predictive models and contour plots. Superimposing contour plots of several responses identifies critical (optimum) regions. Examples of such graphical optimization approach exist in the literature (34–36).

An improved graphical method has been presented (3) that allows three (instead of two) independent variables to be represented continuously and simultaneously in a contour plot for one or more discrete values of the response y. It facilitates the process of locating optimum regions in multivariable systems, and requires some computer programming. An extended discussion and an application example of the improved graphical method are available (3,36).

Several other MRO techniques exist, such as the generalized distance approach (36–39), the extended response surface procedure (39), the so-called overall desirability function (7,40), and the normalized function approach (41). The application of these techniques requires caution, because they may lead to unanticipated and practically undesirable conditions (7).

Evolutionary Operation (EVOP). This is an experimental strategy for improving industrial processes on the plant floor during operation. The theoretical basis, methodology, and some application examples of this method have been presented (42,43). The basic idea of EVOP is to change a current manufacturing process slowly and methodically during production until no further improvement is possible. For example, if a process is presently producing an output y_p at factor settings x_{1p} and x_{2p} (Fig. 7), then according to EVOP, the factor setting should be changed to measure y_a, y_b, y_c, and y_d (in that order). These values are statistically compared to y_p. If significant differences exist, the best value becomes the center (the new y_p), and the

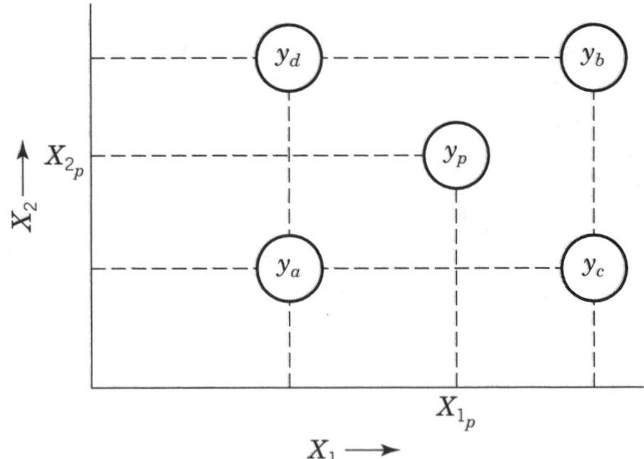

Figure 7. An evolutionary operation (EVOP) strategy.

process repeats until no further improvement occurs. Because experimentation takes place during production, it is important to avoid large changes in the factor settings, which may result in significant product variation or monetary loss. If applied wisely, EVOP provides a simple and straightforward technique for continuous process improvement. It works best when applied to simple systems with up to three independent variables (factors); therefore, its use should be limited to such systems.

Taguchi Methods. A broad spectrum of new ideas on applied statistics and engineering characterize this approach. Followers of these methods advocate a new philosophy on fundamental statistical theory, probability, experimental design, regression theory for modeling, quality and reliability, basic engineering design, and optimization. These ideas and philosophy are generally referred to as Taguchi methods (44,45).

To begin with, every quality characteristic (performance function or response) y relates to some loss L if the characteristic is not exactly on target (if it is not an optimum). Traditionally, the loss function $L(y)$ was thought of as being constant when y lay outside the specified limits (specifications, guidelines, or other quality-control requirements) and zero when y lay within those limits (Fig. 8). Expressed mathematically, this is

$$L(y) = \begin{cases} 0 & \text{for LL} \leq y \leq \text{UL} \\ c & \text{for } y < \text{LL or UL} < y \end{cases} \tag{27}$$

where c is a constant and LL and UL are the lower and upper specification limits, respectively. The Taguchi loss function is a quadratic expression (Fig. 8)

$$L(y) = k(y - a)^2 \tag{28}$$

where a is the target (optimum) value and k is a characteristic constant of the system. For a known loss value L corresponding to a given value of y, k is

$$k = L/(y - a)^2 \tag{29}$$

This loss function may be asymmetric, but it should approximate the true loss. By definition, this is the total loss

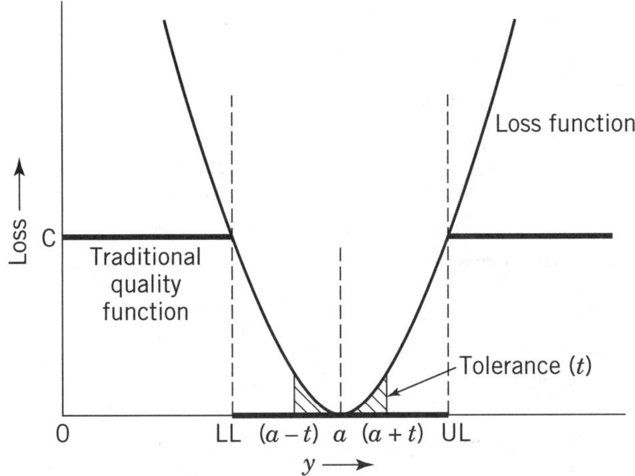

Figure 8. A comparison of Taguchi's loss function with the traditional quality function.

to society, not the loss to a company or an individual, when a product or process deviates from its target (optimum) value a. It has been further suggested that effort should be made to keep the process-product within a tolerance t from the target ($a - t \leqq y \leqq a + t$). For values of y outside the tolerance region (Fig. 8), the loss to society is greater than the amount of effort required to correct the process.

Another aspect of the Taguchi methods deals with the so-called factor space and states that all systems are affected (subject to changes) by some controllable and some uncontrollable factors. The controllable factors are the decision variables, which can be measured, controlled, and adjusted. The uncontrollable factors are either unknown variables or variables over which there is no real control. In either case the effect of uncontrollable factors is termed *noise* and should be minimized during optimization. Mathematically, this is

$$y = f(x_1, x_2, \ldots, x_n, z_1, z_2, \ldots, z_m) \quad (30)$$

where y is the response (performance function), $x_1, x_2, \ldots, x_n$ are the controllable factors (the normal decision variables), and $z_1, z_2, \ldots, z_m$ are the uncontrollable factors (noise). To optimize this system, the optimum values x_1^*, $x_2^*, \ldots, x_n^*$ must be found such that

$$y^* = f(x_1^*, x_2^*, \ldots, x_n^*, z_1, z_2, \ldots, z_m) \quad (31)$$

In this case, y^* is the value of y close to the ideal optimum, where the variability due to the effect of $z_1, z_2, \ldots, z_m$ is minimum. An example application in this area would be the formulation of a cake mix that produces consistently good cakes under uncontrollable variations in baking temperature and time.

The Taguchi methodology goes further and divides the uncontrollable factors into outer noise (ambient temperature, relative humidity, etc), inner noise (machine or equipment deterioration, component wear and tear, etc), and between product noise (product-to-product variation). Fractional factorial designs and orthogonal arrays are used to perform experiments and part of the analysis is the controversial signal to noise ratio.

Overall, the Taguchi methodology follows three basic design phases: (*1*) system design, the basic engineering design for plants, processes, and equipment; (*2*) parameter design, optimizing overall processing conditions; and (*3*) tolerance design, refining the system. This minimizes noise effects, reduces variation, and produces rugged and robust products, processes, and systems. Taguchi methods can also characterize and optimize multiresponse processes (46). Further information on this subject is available (43–46).

Artificial Intelligence. The increasing availability of low-cost computing power has made computationally intensive modeling methods more practical. These include neural network modeling (47), genetic algorithms, and fuzzy logic (48). These methods can provide models that conform more closely to the true response surface than traditional polynomials, but they require significantly more data to create such models.

APPLICATION AND IMPLEMENTATION

Several areas in food science and engineering research can benefit from the application of optimization methods. This is even more so in the broad area of food processing and food manufacturing. A general categorization of the types of problems that optimization methodology may assist solving has been attempted (2). Equipment design, which involves the design of individual components of a machine or a total system (eg, multiple effect evaporators), is a common area for application of optimization methods. Most equipment and food manufacturing companies use optimization techniques to design the best equipment possible.

Another area of application is plant design and layout, where the whole food manufacturing plant is considered as one system and may be optimized for overall efficiency, total operating cost, yield, total return, quality of output, and throughput. This is a far more difficult task than designing individual equipment, and thus its implementation has lagged behind. The optimization of food processing plants is more complex than other industrial plants because, except for the regular process optimization, changes in raw materials (input) and final product properties (output) must be accounted for (49). Thus, optimization of food processing plants is a subset of a much broader system called food chain optimization (49). This includes optimization of raw materials properties, process optimization, and product optimization. It is imperative to stress the importance of applying optimization methods in all of the previously mentioned areas.

Optimization also improves the management and control of processing plants (2). With the increased use of computers and computer control in food manufacturing, optimization should be an integral part of every sophisticated processing plant. However, routine application of optimization methods during operation (on- or off-line) exists in only a small fraction of industrial plants (2). This must change to increase efficiency, productivity, and quality and reduce energy use, product loss, and environmental pollution. The stiff competition in the world's markets will force an increased use of optimization on food manufacturers. Furthermore, population dynamics and changing consumer attitudes toward food and the environment will necessitate the adoption of optimization philosophy, ideas, and techniques throughout the food handling system. This will extend from the harvesting of raw materials to the transportation, processing, packaging, distribution, and consumption of the final products.

BIBLIOGRAPHY

1. R. Aris, *Discrete Dynamic Programming*, Blaisdell, New York, 1964.
2. L. B. Evans, "Optimization Theory and Its Application in Food Processing," *Food Technol.* **36**, 88–93 (1982).
3. J. D. Floros and M. S. Chinnan, "Computer Graphics-Assisted Optimization for Product and Process Development," *Food Technol.* **42**, 72–78, 84 (1988).
4. E. A. Bender, *An Introduction to Mathematical Modeling*, Wiley-Interscience, New York, 1978.
5. C. L. Smith, R. W. Pike, and P. W. Murrill, *Formulation and Optimization of Mathematical Models*, International Textbook, Scranton, Pa., 1970.
6. G. E. P. Box and K. B. Wilson, "On the Experimental Attainment of Optimum Conditions," *Journal of the Royal Statistical Society Ser. B* **13**, 1–45 (1951).
7. G. E. P. Box and N. R. Draper, *Empirical Model Building and Response Surface*, Wiley, New York, 1987.
8. R. H. Myers, *Response Surface Methodology*, Allyn and Bacon, Boston, Mass., 1971.
9. W. G. Cochran and G. M. Cox, *Experimental Designs*, Wiley, New York, 1957.
10. O. L. Davies, *Design and Analysis of Industrial Experiments*, Oliver & Boyd, Edinburgh, Scotland, 1954.
11. N. R. Draper and H. Smith, *Applied Regression Analysis*, Wiley, New York, 1981.
12. E. M. L. Beale, *Introduction to Optimization*, Wiley, New York, 1988.
13. I. Saguy, "Optimization Methods and Applications," in I. Saguy, ed., *Computer-Aided Techniques in Food Technology*, Marcel Dekker, New York, 1983.
14. T. N. Edelbaum, "Theory of Maxima and Minima," in G. Leitman, ed., *Optimization Techniques*, Academic Press, Orlando, Fla., 1962.
15. H. Hancock, *Theory of Maxima and Minima*, Dover, New York, 1960.
16. R. Fletcher, *Practical Methods of Optimization*, Wiley, New York, 1987.
17. V. Vemuri, *Modeling of Complex Systems*, Academic Press, Orlando, Fla., 1978.
18. G. P. McCormick, *Nonlinear Programming: Theory, Algorithms, and Applications*, Wiley, New York, 1983.
19. F. E. Bender, A. Kramer, and G. Kahan, "Linear Programming and Its Applications in the Food Industry," *Food Technol.* **35**, 94–96 (1982).
20. F. E. Bender and A. Kramer, "Linear Programming and Its Implementation," in I. Saguy, ed., *Computer-Aided Techniques in Food Technology*, Marcel Dekker, New York, 1983.
21. S. Vajda, *Problems in Linear and Nonlinear Programming*, Oxford Univ. Press, New York, 1975.
22. R. Bellman, *Dynamic Programming*, Princeton Univ. Press, Princeton, N.J., 1957.
23. I. Saguy, "Optimization of Dynamic Systems Utilizing the Maximum Principle," in I. Saguy, ed., *Computer-Aided Techniques in Food Technology*, Marcel Dekker, New York, 1983.
24. R. E. Kopp, "Pontyagrin Maximum Principle," in G. Leitman, ed., *Optimization Techniques*, Academic Press, Orlando, Fla., 1962.
25. G. Leitman, ed., *Optimization Techniques: With Applications to Aerospace Systems*, Academic Press, Orlando, Fla., 1962.
26. G. E. P. Box and N. R. Draper, "A Basis for Selection of a Response Surface Design," *J. Am. Stat. Assoc.* **54**, 622–654 (1959).
27. G. E. P. Box and D. W. Behnken, "Some New Three Level Designs for the Study of Quantitative Variables," *Technometrics* **2**, 455–475 (1960).
28. J. A. Cornell, "Experiments with Mixtures: An Update and Bibliography," *Technometrics* **21**, 95–105 (1979).
29. D. R. Thompson, "Designing Mixture Experiments: A Review," *Trans. ASAE* **24**, 1077–1086 (1981).
30. L. B. Hare, "Mixture Designs Applied to Food Formulations," *Food Technol.* **28**, 50–56, 62 (1974).
31. K. Mullen and D. M. Ennis, "Rotatable Designs in Product Development," *Food Technol.* **33**, 74–80 (1979).
32. K. Mullen and D. Ennis, "Fractional Factorials in Product Development," *Food Technol.* **39**, 90–103 (1985).
33. N. R. Draper, "Ridge Analysis' of Response Surfaces," *Technometrics* **5**, 469–479 (1963).
34. J. D. Floros and M. S. Chinnan, "Optimization of Pimiento Pepper Lye-Peeling Process Using Response Surface Methodology," *Trans ASAE* **30**, 560–565 (1987).
35. J. D. Floros and M. S. Chinnan, "Seven Factor Response Surface Optimization of a Double-Stage Lye (NaOH) Peeling Process for Pimiento Peppers," *J. Food Sci.* **53**, 631–638 (1988).
36. A. A. Guillou and J. D. Floros, "Multiresponse Optimization Minimizes Salt in Natural Cucumber Fermentation and Storage," *J. Food Sci.* **58**, 1381–1389 (1993).
37. A. I. Khuri and M. Conlon, "Simultaneous Optimization of Multiple Responses Represented by Polynomial Regression Functions," *Technometrics* **23**, 365–375 (1981).
38. A. I. Khuri and J. A. Cornell, *Response Surfaces, Designs, and Analysis*, Marcel Dekker, New York, 1987.
39. J. Fichtali, F. R. Van De Voort, and A. I. Khuri, "Multiresponse Optimization of Acid Casein Production," *J. Food Process Eng.* **12**, 247–258 (1990).
40. G. C. Derringer and R. Suich, "Simultaneous Optimization of Several Response Variables," *J. Qual. Technol.* **12**, 214–219 (1980).
41. J. D. Floros and H. Liang, "Multiresponse Optimization by a Normalized Function Approach," *Dev. Food Sci.* **37**, 2139–2150 (1995).
42. G. E. P. Box and N. R. Draper, *Evolutionary Operation*, Wiley, New York, 1969.
43. T. P. Ryan, *Statistical Methods for Quality Improvement*, Wiley, New York, 1989.
44. P. J. Ross, *Taguchi Techniques for Quality Engineering*, McGraw-Hill, New York, 1988.
45. G. Taguchi, *Systems of Experimental Design*, Vols. 1 and 2, American Supplier Institute, Dearborn, Mich., 1987.

46. N. Logothetis and A. Haigh, "Characterizing and Optimizing Multi-Response Processes by the Taguchi Method," *Qual. Reliab. Eng. Int.* **4**, 159–169 (1988).
47. V. Gnanasekharan and J. D. Floros, "Back Propagation Neural Networks: Theory and Applications for Food Science and Technology," *Dev. Food Sci.* **37**, 2151–2168 (1995).
48. I. G. Vradis and J. D. Floros, "Genetic Algorithms and Fuzzy Theory for Optimization and Control of Food Processes," *Dev. Food Sci.* **37**, 2169–2182 (1995).
49. P. Filka, "Optimization of Food Processing Plants," in M. Renard and J. J. Bimbenet, eds., *Automatic Control and Optimization of Food Processes*, Elsevier Applied Science, Barking, UK, 1986, pp. 405–415.

JOHN D. FLOROS
JEFF RATTRAY
Purdue University
West Lafayette, Indiana

ORANGES. See FRUITS, SEMI-TROPICAL.

OSTEOPOROSIS

Osteoporosis is a disease characterized by low bone mass and increased bone fragility and susceptibility to fracture. Adequate calcium, vitamin D, and other nutrients are critical to achieving optimal peak bone mass and modify the rate of bone loss with aging. After menopause, women experience rapid bone loss due to declining estrogen levels. Hormone replacement therapy, in conjunction with calcium, has been shown to be effective in reducing bone loss. Although less common in men, prolonged use of medications such as glucocorticoids as well as other factors may lead to osteoporosis in men.

DEFINITION

The National Institutes of Health (NIH) Consensus Development Conference defined osteoporosis as "A systemic skeletal disease characterized by low bone mass and microarchitectural deterioration of bone tissue, with a consequent increase in bone fragility and susceptibility to fracture" (1).

Osteoporosis affects more than 25 million people in the United States and is the major underlying cause of bone fractures in postmenopausal women and the elderly (2).

SYMPTOMS

Osteoporosis is often referred to as the silent disease because it can progress painlessly until a fracture occurs, typically in the hip, spine, or wrist. As the disease progresses, the vertebrae of the spine can become compressed leading to stooped posture and a loss of height. The abnormal curvature of the spine that results is referred to as kyphosis or a Dowager's hump.

The World Health Organization categorizes bone mass according to how it compares with a young adult reference range (see Table 1). The categories refer to the number of standard deviations that the patient's bone density differs from young, healthy women.

Table 1. World Health Organization Criteria for Osteoporosis in Women

Normal	BMD or BMC $\leq$ 1 SD below the young adult reference range.
Low bone mass	BMD or BMC 1–2.5 SD below the mean of young, healthy women.
Osteoporosis	BMD or BMC >2.5 SD below the mean of young, healthy women.
Severe osteoporosis	BMD or BMC >2.5 SD below the mean of young, healthy women and the presence of one or more fragility fractures.

Note: BMD = bone mineral density; BMC = bone mineral content.
Source: Ref. 3.

RISK FACTORS

The typical osteoporosis victim is a white or Asian postmenopausal woman with a thin body frame. Early menopause, either natural or surgical, combined with a diet low in calcium, family history of osteoporosis, sedentary lifestyle, smoking, and excessive alcohol intake further increase the risk. Numerous other factors increase the risk for osteoporosis. Table 2 provides a list of risk factors. Table 3 lists the types of osteoporosis.

BONE DEVELOPMENT OVER A LIFETIME

Bone is constantly in a state of flux. This state of flux is the result of two opposing forces: bone formation and resorption. Bone formation begins in utero, and the acquisition and development of bone mass continues throughout childhood. During growth and development of the skeleton, the cells lying on the surface of the developing bones, called *osteoblasts*, stimulate bone formation (Fig. 1). At the same time, a second group of surface cells, called *osteoclasts*, are involved in the process of resorption (breakdown) of bone. Both cell types are affected by physical and environmental factors (4,5).

At puberty, acquisition of bone mass is accelerated. This accelerated process is believed to be approximately 95% complete by age 18. Beyond this age, the process slows with the remaining 5% deposited in the bone matrix into the third decade of life (4,6,7). It is during these phases of growth and development that adequate intake of bone-related nutrients (calcium, vitamin D, magnesium, phosphorus, protein, vitamin K, boron, manganese, copper and zinc) is vital to a lifetime of strong and healthy bones (8–17).

During adulthood, from approximately the third decade of life to the onset of menopause in women at approximately 45 to 55 years of age, bone mass is considered stable and at its peak (18). With the onset of menopause, there occurs a rapid reversal of bone metabolism from acquisition to bone mass reduction. This is in response to declining estrogen levels, which initiate the menopausal process (4). This reduction process occurs in two phases. The first

Table 2. Osteoporosis Risk Factors

Genetic
Females
Asian or Caucasian individuals
Family history of osteoporosis
Small bone structure
Lifestyle
Inactive lifestyle
Prolonged bed rest
Never having been pregnant
Smoking
Nutritional
Excessive use of alcohol
Chronic low calcium intake
Vitamin D deficiency
High protein/phosphate ratio
High caffeine intake
Medical
Premenopausal oophorectomy
Premenopausal amenorrhea
Gastrointestinal dysfunction
Osteogenesis imperfecta
Parathyroid overactivity
Low testosterone levels in men
Medications
Prolonged glucocorticoid use
Prolonged use of some anticonvulsants
Chronic phosphate-binding antacid use
Diuretics producing calciuria
Miscellaneous
Late onset of menstruation
Early menopause, whether natural or surgical
Advanced age

stage lasts approximately 5 to 10 years and results in rapid reduction of bone mass, approximately 2 to 5% per year. In the second stage, bone loss is more gradual, approximately 0.2 to 0.5% per year. This loss continues unabated, in the absence of hormone replacement therapy, for the rest of the woman's life span, thereby increasing the risk of osteoporosis (4,19).

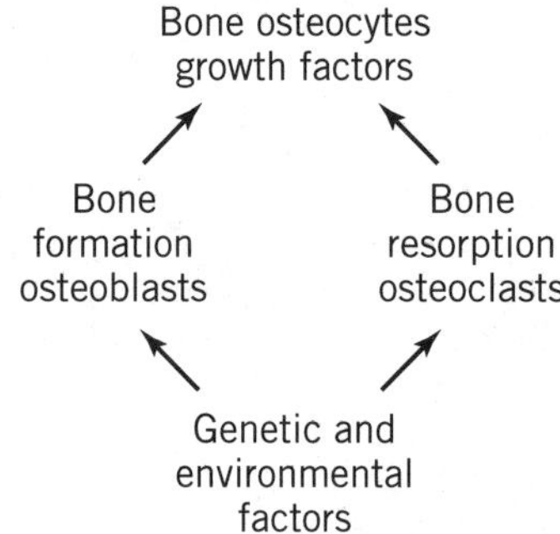

Figure 1. Bone turnover and remodeling.

As men age, the process of bone acquisition also converts to bone loss, but the process is much more gradual without the initial rapid losses created by the estrogen decline of menopause. The loss is consistent, however, and bone is continually lost as the male ages. Although this loss can lead to osteoporosis with age, over 75% of male osteoporosis is secondary to underlying chronic diseases, which accelerate bone loss similar to the effect of menopause in women (20). Other factors may also include age-related decline in testosterone, adrenal androgens, as well as other cellular factors (20–22).

As should be evident from the preceding, two critical factors determine the risk for osteoporosis. These are the peak bone mass attained and the level of bone loss after peak bone is attained (6). Peak bone mass is the total bone present in the final skeleton at maturity. Peak bone mass is affected by genetic factors, nutrition, hormonal factors, and exercise, among others (6,7).

The reduction in peak bone mass with aging is due to a number of interrelated factors. In women, estrogen decline after menopause is a major reason. In elderly men and women, other factors are important in age-related bone loss. With aging, there appears to be a general decline in osteoblast activity, as well as a reduction in calcium absorption from the GI tract due to a decrease in vitamin D intake and conversion to 1,25 $(OH)_2D$, the active form of vitamin D in the skin, kidney, and liver. In response to, and driven by this declining calcium uptake is hyperparathyroidism, which further drives the balance to resorption and bone loss.

Table 3. Types of Osteoporosis

	Type I, Postmenopausal	Type II, Senile	Type III, Secondary
Typical age	51–75	75–90	Any age
Gender	Female	Female: male ratio 2:1	Male or female
Risk factors	Estrogen deficiency	Advanced age	Certain medications, cancer, endocrine disorders, malnutrition or other conditions.
Bone involvement	Increased bone resorption without change in bone formation, particularly in trabecular bone	Bone formation and resorption are decreased, both trabecular and cortical bone are affected	Trabecular and cortical bone are affected equally
Fracture site	Vertebral, hip, wrist	Hip, spine, pelvis, and humerus	All types of fractures can occur

BONE COMPOSITION

Bone is composed primarily of calcium and phosphorus, in the form of hydroxyapatite crystals deposited in a collagen matrix (23). Adults have two types of bone: cortical and trabecular. Cortical bone provides rigidity and is the major component of the long bones of the skeleton. Cortical bone accounts for approximately 80 to 85% of the skeletal mass. Examples of cortical bone are the dense long bones of the hands, arms, and legs and bones of the feet. Trabecular bone is spongy in appearance and provides strength and elasticity. Trabecular bone makes up only about 15 to 20% of the skeletal mass. Examples of trabecular bone are the skull, the sternum, ribs, hip and shoulder girdle, and spinal column.

It would appear from the distribution of cortical and trabecular bone, that the cortical skeleton would be the most important area of bone turnover. However, bone turnover is a surface phenomenon, and trabecular bone has the greatest surface area. Therefore, 80% of bone turnover occurs in trabecular bone with only 20% occurring in cortical bone (4,5,19). This accounts for the greater fracture rate of trabecular bone versus cortical bone. A postmenopausal woman may lose 35% of her cortical bone and approximately 50% of her trabecular bone. A man of the same age, conversely, will lose about 12% of his cortical bone and approximately 17% of his trabecular bone (19).

GENETIC AND ENVIRONMENTAL FACTORS

Genetic and environmental factors are partners in the development of bone density and bone health over a lifetime (6). At present, the knowledge about genetic factors that are important in this process, and thus about the risk of osteoporosis, is meager at best. Genes, gene products, and nutrient–gene interactions are currently an evolving science (24,25). Ongoing research to understand the human genome, as well as many studies targeted at the environmental aspects of osteoporosis, will in time improve our knowledge. It is known now that hundreds of genes are involved in the growth, development, and maintenance of the skeletal system (6,26). A complete understanding of these genetic factors will help to clarify the underlying mechanisms of osteoporosis.

DIAGNOSIS

A number of techniques are available for measuring bone mass, each with its own advantages and disadvantages (see Table 4). Ideally, the technique should be rapid, reliable, accurate, inexpensive, and provide minimal radiation exposure. In addition, it should provide a prediction of the risk of subsequent fracture and have the ability to measure multiple sites (27).

Recommended Indications for Bone Density Studies

- Patients being treated for osteoporosis, to monitor bone mass
- Estrogen deficient women at risk for osteoporosis
- Patients with vertebral abnormalities, hip fractures, wrist fractures, or osteopenia

Table 4. Comparison of Bone Mass Measurement Techniques

Technique	Advantages	Disadvantages
Dual energy absorptiometry (DPA, DXA)	Best precision for measurement of spine or proximal femur Distinguishes between bone and soft tissue Multiple sites of measurement Low radiation exposure Moderate to low cost per test	Cannot distinguish between cortical and trabecular bone Limited mobility of equipment Moderate equipment cost
Peripheral DXA	Low equipment cost Low radiation exposure Equipment portable	Limited to study of wrist or heel Limited correlation to spine or hip
Quantitative Computed Tomography (QCT)	Distinguishes trabecular and cortical bone density Provides measure of volumetric BMD	More difficult to perform than DXA or DPA Higher radiation exposure than DXA Meticulous performance and calibration necessary Most costly of densitometric techniques Limited mobility of equipment Small body regions evaluated Accuracy can be severely affected by variations in fat content of marrow within the spine or femur
Radioabsorptiometry	Low equipment cost Equipment portable	Limited to study of phalanges Limited as a screening test Limited correlation to spine or hip
Single energy absorptiometry (SPA, SXA)	Useful for regions with low soft tissue (ie, forearm) Relatively inexpensive Equipment portable Low radiation exposure	Cannot measure bone mass near soft tissue (ie, vertebrae or femur) Limited correlation to spine or hip Limited as a screening test
Ultrasound	Low equipment cost Equipment portable No radiation exposure	Limited mesurement sites Limited as a screening test

- Patients receiving long-term glucocorticoid therapy
- Patients with primary hyperparathyroidism

PREVENTION OF OSTEOPOROSIS

Role of Calcium

Adequate calcium is critical to achieving optimal peak bone mass and modifies the rate of bone loss associated with aging (2). Numerous studies confirm the bone sparing effect of calcium. Reid et al. found that calcium supplementation significantly slowed axial and appendicular bone loss in normal postmenopausal women who received 1000 mg of calcium supplementation daily in addition to diet. The mean rate of loss of total body bone mineral density was reduced 43% in the calcium group compared with placebo (28). Recker et al. concluded that 1200 mg calcium per day reduced the incidence of spine fractures and halted measurable bone loss in elderly women with a history of spine fractures and self-selected calcium intakes of less than 1000 mg per day (29).

Recommended Calcium Intake

The NIH Consensus Development Panel on Optimal Calcium Intake defined optimal daily calcium intake as (2):

Infants birth to 6 months	400 mg
Infants 6–12 months	600 mg
Children 1–5 years	800 mg
Children 6–10 years	800–1200 mg
Adolescents and young adults 11–24 years	1200–1500 mg
Women 25–50 years	1000 mg
Pregnant or lactating women	1200 mg
Postmenopausal women on estrogen replacement therapy	1000 mg
Postmenopausal women not on estrogen replacement therapy	1500 mg
Men 25–65	1000 mg
Men and women over 65 years	1500 mg

The 1997 Dietary Reference Intakes (DRIs) for calcium recommend the following (30):

Males and females	9–13 year	1300 mg
	14–18 years	1300 mg
	19–50 years	1000 mg
	51–70+	1200 mg
Pregnancy and lactation	<18 years	1300 mg
	19–30 years	1000 mg
	31–50 years	1000 mg

National Osteoporosis Foundation Recommendations

The National Osteoporosis Foundation (NOF) advises all patients to obtain an adequate intake of dietary calcium (at least 1200 mg per day, including supplements if necessary) and vitamin D (400 to 800 IU per day for individuals at risk of deficiency).

Average Calcium Intake by Individuals

A large percentage of Americans fail to meet currently recommended guidelines for optimal calcium intake (2). According to the Third National Health and Nutrition Examination Survey (NHANES III), the average intake of calcium by females is only 744 mg per day. For women 50 years and older, calcium intake is even lower (626–711 mg per day). The average intake for males is considerably higher at 976 mg per day, although it is lower for men 70 and older (721–808 mg) (31). Since adequate calcium intake is critical to achieving optimal peak bone mass and modifies the rate of bone loss associated with aging, calcium supplementation may be indicated in individuals who cannot meet their need by ingesting conventional foods (2).

Sources of Calcium

Dairy products are the most concentrated food sources of calcium. While green, leafy vegetables can help contribute to calcium intake, these vegetables also contain oxalic acid that can bind with calcium and interfere with absorption. The following chart lists the calcium content of some common foods. Values listed are approximate.

Food	Calcium content
Plain lowfat yogurt, 1 cup	400 mg
Canned sardines (with bones), 3 oz.	400 mg
Skim, lowfat and buttermilk, 8 oz.	300 mg
Cheese (most hard cheeses), 1 oz.	200 mg
Tofu (processed with calcium), 4 oz.	150 mg
Broccoli (cooked), 1/2 cup	75 mg
Almonds, 1 oz.	75 mg
Cottage cheese, 1/2 cup	50 mg
Whole wheat bread, 1 slice	20 mg

Pharmacologic Interventions

A number of pharmaceutical approaches are currently available for the management of osteoporosis. These interventions are sometimes classified as either influencing bone formation or bone resorption. Major interventions are described briefly.

Hormone Replacement Therapy (ERT and HRT). Of the therapies currently available, estrogen replacement therapy (ERT) is the only one approved for both prevention and treatment of osteoporosis. Estrogen acts to reduce bone resorption and increase bone density. Increases in bone density vary from site to site, that is, cortical (1–3%) and trabecular (2–5%). The greatest benefits of ERT and hormone replacement therapy (HRT) are observed in the years immediately following menopause, and rapid bone loss is observed immediately following discontinuation of estrogen replacement therapy. A daily dose of 0.3 to 1.25 mg unopposed estrogen is effective in protecting against bone loss (32).

Epidemiologic data suggest a significant reduction in the incidence of all types of osteoporotic fractures in ERT users. Retrospective and prospective studies report a 70 to 80% reduction in vertebral fractures (33,34).

Combination hormone therapy, estrogen and progestin, are commonly prescribed for women with a uterus. A three-year, multicenter trial examining the impact of combination therapy on bone mineral density showed significant increases in the spine and hip while the placebo group exhibited a reduction in BMD (35). The optimal duration of HRT therapy for bone maintenance is still undefined. The ongoing Women's Health Initiative is expected to clarify this and other open questions about use of HRT.

A meta-analysis of the effect of calcium supplementation on the efficacy of estrogen and calcitonin on bone mass change suggests that a high calcium intake potentiates the positive effect of estrogen on bone mass at all skeletal sites and perhaps that of calcitonin on bone mass of the spine. An open study included in the analysis found that 0.3 mg estrogen was equally as effective as the higher standard estrogen dose of 0.625 mg when given in conjunction with a high calcium intake (1700 mg per day) (36).

In an accompanying editorial to the meta-analysis, Bess Dawson-Hughes indicates the meta-analysis supports the notion that the calcium intake requirement of women treated with estrogen may not be lower than that of average postmenopausal women. In fact, the studies do not exclude the possibility that women receiving antiresorptive treatments may benefit from even higher calcium intake than average women.

Raloxifene. Raloxifene is the most recent drug therapy to gain approval in the United States for prevention of osteoporosis in postmenopausal women and is considered a first choice when ERT or HRT is not indicated. Raloxifene is classified as a selective estrogen receptor modulator (SERM), meaning that it has both estrogen agonistic and antagonistic activities that vary by tissue type. A number of recent reviews detail the properties of Raloxifene and its safety and efficacy. Three large-scale clinical trials were conducted to look at Raloxifene's effect on osteoporosis. Delmas et al. (1997) reported that bone mineral density increased significantly over baseline in lumbar spine, total hip, femoral neck, and total body following 24-month treatment with 30, 60, or 150 mg Raloxifene per day (37). The other trials showed similar results (38–40). A number of shorter-term clinical studies, which compared Raloxifene to ERT in postmenopausal women, found comparable decreased bone turnover and resorption. To date, no data have been gathered on Raloxifene's effect on fracture rate. Other related compounds, that is, Droloxifene, Idoxifene, and Levormeloxifene, are currently under development.

Bisphosphonates. Bisphosphonates are synthetic analogs of naturally occurring pyrophosphates that are known to improve bone mineralization. The mode of action is not fully defined. Rogers et al. (1997) have reviewed the recent evidence (41). These compounds are viewed as effective treatments for osteoporosis. Long-term usage has been shown to increase bone density. Although bisphosphonates are effective, poor absorption and esophageal irritation limit compliance and effectiveness. Improved forms of bisphosphonates are currently being investigated. Two forms, etidronate and alendronate, have been shown to reduce fracture rate (42–44).

Calcitonin. Calcitonin is described as an antiresorptive peptide that acts by inhibiting osteoclast activity. It must be administered parenterally or nasally to preserve its activity and may be particularly suitable for women who cannot take HRT. A number of reviews have been published recently (45–47) that detail the pharmacology and clinical effectiveness of calcitonin. In a two-year placebo controlled trial conducted in early postmenopausal women, spinal bone loss was prevented in the group receiving 100 IU calcitonin (48). Calcitonin appears to be effective in increasing bone mass for approximately one to two years, and then effectiveness plateaus. Intermittent administration may enhance effectiveness of the therapy. A maximum of about a 10% increase in bone mass can be achieved under conditions of optimal use. This agent has been shown to have a good safety profile. At present, however, it is approved for use in a relatively small number of countries due to the limited portfolio of clinical data available. Data are not yet available to establish the effect of calcitonin on fracture rate.

Fluoride. Fluoride, an essential element in bone formation, has been used for decades in Europe in the management of osteoporosis. Supplemental fluoride, given in milligram quantities (50–75 mg/day), significantly enhances bone mineral density, by stimulating osteoblast activity. Conflicting data are currently available on fluoride's impact on fracture rate. A number of side effects have been reported with flouride usage; these include pain in the lower extremities and gastric irritation. Administration without adequate calcium intake impaired mineralization. A number of investigations are under way to evaluate the efficacy of other forms of fluoride as well as preferred dosage regimens (49).

Other Pharmacologic Therapies

A number of hormones such as PTH, growth hormone, and anabolic steroids have been studied in the management of osteoporosis. These agents and their utility have been effectively reviewed in recent publications by O'Connell. Although bone formation may be enhanced by a number of these agents, various side effects, for example, glucose intolerance, hypertension, and virulization, diminish their potential (49).

Role of Exercise in Prevention and Treatment

Exercise plays a pivotal role in building and maintaining bone mass throughout life. Bone formation is enhanced by mechanical loading and results in modest increases in bone mass and improvements in bone structure. The type and frequency of exercise appear to significantly influence bone health. Strength training and weight-bearing exercise appear to offer the greatest benefit. The American Academy of Sports Medicine recommends 50 to 60 min of weight-bearing exercise three times per week (50).

In the prepubescent and pubescent years, the years of greatest bone building, physical activity enhances bone mineralization (51). Physical activity at a young age is a known determinant of bone mass (52). However, excessive exercise leading to amenorrhea is detrimental to building bone mass (53). Studies conducted in young and middle-aged women indicate that bone mass may be enhanced in those who have undertaken training programs that include weight-bearing exercise (54–56).

A small number of intervention trials have examined the benefits of exercise on osteoporosis (57). Collectively, these trials indicate that regular exercise is beneficial in preventing and treating osteoporosis. The benefits of weight-bearing exercise have been reviewed by Reid (1996) (58). Exercise conferred benefits such as improved balance, strength, coordination, and flexibility, which appear to diminish the risk of falls and fall-related injury (59). Low muscle strength and mobility are recognized independent risk factors for hip fracture (60). In a meta-analysis of seven exercise intervention trials in postmenopausal women, a 10% reduction in fall frequency was found among exercising subjects (59).

Several studies have also examined the value of exercise when combined with other treatments. Prince et al. (1995) investigated the combined effects of calcium supplementation and HRT with exercise and found that consumption of approximately 1.8 g calcium/day coupled with a mild increase in exercise significantly reduced bone loss in the hip (61). An earlier study by Prince et al. (1991) demonstrated that both exercise plus HRT and exercise plus calcium supplementation were effective in slowing or stopping bone loss (62).

RESOURCES FOR MORE INFORMATION

National Osteoporosis Foundation
1150 17th Street, NW, Suite 500
Washington, DC 20036-4603
Phone: 202-223-2226
Web site URL: www.nof.org.

Osteoporosis and Related Bone Diseases–National Resource Center
1150 17th Street, NW, Suite 500
Washington, DC 20036
Phone: 202-223-0344 or 800-624-BONE
TTY: (202)466-4315
E-mail: orbdnrc@nof.org.

National Institutes of Health
9000 Rockville Pike
Bethesda, MD 20892

U.S. National Library of Medicine
8600 Rockville Pike
Bethesda, MD 20894
Phone: 800-272-4787 or 301-496-6308.

BIBLIOGRAPHY

1. National Institutes of Health Consensus Development Conference, "Osteoporosis," *American Journal of Medicine* **94**, 646–650 (1993).
2. NIH Consensus Development Panel on Optimal Calcium Intake, "Optimal Calcium Intake," *JAMA* **272**, 1942–1948 (1994).
3. "Assessment of Fracture Risk and Its Application to Screening for Postmenopausal Osteoporosis," WHO Technical Series 843, 1–129 (1994).
4. G. Birdwood, *Understanding Osteoporosis and Its Treatment*, Parthenon, New York, 1996.
5. J. Glowacki, "The Cellular and Biochemical Aspects of Bone Remodeling," in C. J. Rosen, ed., *Osteoporosis Diagnostic and Therapeutic Principles*, Humana Press, Totowa, N.J., 1996, pp. 3–14.
6. V. Matkovic et al., "Skeletal Development in Young Females: Endogenous versus Exogenous Factors," in P. Burckhardt, B. Dawson-Hughes, and R. Heaney, eds., *Nutritional Aspects of Osteoporosis*, Springer-Verlag, New York, 1998, pp. 26–41.
7. V. Matkovic, "Skeletal Development and Bone Turnover Revisited," *J. Clin. Endocrinol. Metab.* **81**, 2013–2016 (1996).
8. J. P. Bonjour et al., "Calcium Enriched Foods and Bone Mass Growth in Prepubertal Girls: A Randomized, Double Blind, Placebo-Controlled Trial," *J. Clin. Invest.* **99**, 1287–1294 (1997).
9. C. M. Weaver, G. P. McCabe, and M. Peacock, "Calcium Intake and Age Influence Calcium Retention," in P. Burckhardt, B. Dawson-Hughes, and R. P. Heaney, eds., *Nutritional Aspects of Osteoporosis*, Springer-Verlag, New York, 1998, pp. 3–10.
10. H. K. Kinyama et al., "Dietary Calcium and Vitamin D Intake in Elderly Women: Effect on Serum Parathyroid Hormone and Vitamin D Metabolites," *Am. J. Clin. Nutr.* **67**, 342–348 (1998).
11. B. Dawson-Hughes et al., "Rates of Bone Loss in Post Menopausal Women Randomly Assigned to One of Two Dosages of Vitamin D," *Am. J. Clin. Nutr.* **61**, 1140–1145 (1995).
12. R. P. Heaney, "Nutrition and Catch-Up Bone Augmentation in Young Women," *Am. J. Clin. Nutr.* **68**, 523–524 (1998).
13. D. Teegarden et al., "Dietary Calcium, Protein, and Phosphorus Are Related to Bone Mineral Density and Content in Young Women," *Am. J. Clin. Nutr.* **68**, 749–754 (1998).
14. D. Feskanich et al., "Vitamin K Intake and Hip Fractures in Women: A Prospective Study," *Am. J. Clin. Nutr.* **69**, 74–79 (1999).
15. H. P. Dimai et al., "Daily Oral Magnesium Supplementation Suppresses Bone Turnover in Young Adult Males," *J. Clin. Endocrinol. Metab.* **83**, 2742–2748 (1998).
16. C. D. Hunt, "Copper and Boron as Examples of Dietary Trace Elements Important in Bone Development and Disease," *Curr. Opin. in Orthopaedics* **9**, 28–36 (1998).
17. P. P. Saltman and L. G. Strause, "The Role of Trace Elements in Osteoporosis," *J. Am. College of Nutrition* **12**, 384–389 (1993).
18. V. Matkovic et al., "Timing of Peak Bone Mass in Caucasian Females and Its Implications for the Prevention of Osteoporosis," *J. Clin. Investigation*, 799–808 (1994).
19. G. R. Mundy, "Bone Remodeling and Mechanism of Bone Loss in Osteoporosis," in P. J. Meunier, ed., *Osteoporosis: Diagnosis and Management*, Mosby, New York, 1998, pp. 517–535.
20. P. Pilar and N. Guanabens, "Male Osteoporosis," *Curr. Opin. in Rheumatology* **8**, 357–364 (1996).
21. E. Seeman, "Osteoporosis in Men," *Bailliere's Clin. Rheumatology* **11**, 613–629 (1997).
22. M. E. Mussolino et al., "Risk Factors for Hip Fracture in White Men: The NHANES I Epidemiologic Follow-Up Study," *J. Bone Miner. Res.* **13**, 918–924 (1998).
23. A. Vies and B. Sabsay, "The Collagen of Mineralized Matrices," in W. A. Peck, ed., *Bone and Mineral Research*, Vol. 5, Elsevier, Amsterdam, The Netherlands, 1984.
24. E. Seeman et al., "Interaction between Genetic and Nutritional Factors," in P. Burckhardt, B. Dawson-Hughes, and R.

Heaney, eds., *Nutritional Aspects of Osteoporosis*, Springer-Verlag, New York, 1998, pp. 85–97.

25. S. Ferrari, R. Rizzoli, and J. P. Bonjour, "Genetics Dietary Calcium Interactions and Bone Mass," in P. Burckhardt, B. Dawson-Hughes, and R. Heaney, eds., *Nutritional Aspects of Osteoporosis*, Springer-Verlag, New York, 1998, pp. 99–106.
26. V. Matkovic et al., "Factors Which Influence Peak Bone Mass Formation: A Study of Calcium Balance and the Inheritance of Bone Mass in Adolescent Females," *Am. J. Clin. Nutr.* **52**, 878–888 (1990).
27. H. K. Genant et al., "Measurement of Bone Mineral Density: Current Status," *American Journal of Medicine* **91**, 49S–53S (1991).
28. I. R. Reid et al., "Effect of Calcium Supplementation on Bone Loss in Postmenopausal Women," *New Engl. J. Med.* **328**, 460–464 (1993).
29. R. Recker et al., "Correcting Calcium Nutritional Deficiency Prevents Spine Fractures in Elderly Women," *J. Bone and Mineral Res.* **11**, 1961–1966 (1996).
30. Institute of Medicine, *Dietary Reference Intakes: Calcium, Phosphorus, Magnesium Vitamin D, and Fluoride*, National Academy Press, Washington, D.C., 1997.
31. U.S. Department of Health and Human Services, *Advance Data: Dietary Intake of Vitamins, Minerals, and Fiber of Persons Ages Two Months and Over in the United States*, Third National Health and Nutrition Examination Survey, Phase 1, 1988–91, No. 258, 1994.
32. H. K. Genant, et al., "Low-Dose Esterified Estrogen Therapy," *Archives of Internal Medicine* **157**, 2609–2615 (1997).
33. B. Ettinger, H. K. Genant, and C. E. Cann, "Long-Term Estrogen Replacement Therapy Prevents Bone Loss and Fractures," *Ann. Intern. Med.* **102**, 319–324 (1985).
34. R. Lindsay et al., "Prevention of Spinal Osteoporosis in Oophorectomized Women," *Lancet* **2**, 1151–1154 (1980).
35. The PEPI Writing Group, "Effects of Hormone Therapy on Bone Mineral Density," *JAMA* **276**, 1389–1396 (1985).
36. J. W. Nieves et al., "Calcium Potentiates the Effect of Estrogen and Calcitonin on Bone Mass: Review and Analysis," *Am. J. Clin. Nutr.* **67**, 18–24 (1998).
37. P. D. Delmas et al., "Effects of Raloxifene on Bone Mineral Density, Serum Cholesterol Concentrations, and Uterine Endometrium in Postmenopausal Women," *New Engl. J. Med.* **337**, 1641–1647 (1997).
38. Eli Lily, Evista (raloxifene hydrochloride) package insert, Indianapolis, Ind., December 1997.
39. W. Khovidkhunkit and D. M. Shoback, "Clinical Effects of Raloxifene Hydrochloride in Women," *Ann. Intern. Med.* **130**, 431–439 (1999).
40. J. A. Balfour and K. I. Goa, "Raloxifene," *Drugs Aging* **12**, 335–341 (1998).
41. M. J. Rogers, D. J. Watts and R. G. Russell, "Overview of Bisphosphonates," *Cancer* **80**, (Supplement), 1652–1660 (1997).
42. T. Storm et al., "Effect of Intermittent Cyclical Etidronate Therapy on Bone Mass and Fracture Rate in Women with Postmenopausal Osteoporosis," *New Engl. J. Med.* **322**, 1265–1271 (1990).
43. N. B. Watts et al., "Intermittent Cyclical Etidronate Treatment of Postmenopausal Osteoprosis," *New Engl. J. Med.* **323**, 73–79 (1990).
44. D. M. Black et al., "Randomized Trial of Effect of Alendronate on Risk of Fracture in Women with Existing Vertebral Fractures: Fracture Intervention Trial Research Group," *Lancet* **348**, 1535–1541 (1996).
45. R. Eastell, "Treatment of Postmenopausal Osteoporosis," *New Engl. J. Med.* **338**, 736–746 (1998).
46. S. Silverman, "Calcitonin," *Am. J. Medical Sci.* **313**, 13–16 (1997).
47. C. Gennari and D. Agnusdei, "Calcitonins and Osteoporosis," *Br. J. Clin. Practice* **48**, 196–200 (1994).
48. K. Overgaard et al., "Effect of Calcitonin Given Intranasally on Early Postmenopausal Bone Loss," *Br. Med. J.* **299**, 477–479 (1989).
49. M. O'Connell, "Prevention and Treatment of Osteoporosis in the Elderly," *Pharmacotherapy* **19**, 7S–20S (1999).
50. American College of Sports Medicine, "ACSM Position Stand on Osteoporosis and Exercise," *Med. Sci. Sports Exer.* **27**, i–vii (1995).
51. C. W. Slemenda, et al., "Influences on Skeletal Mineralization in Children and Adolescents: Evidence for Varying Effects of Sexual Maturation and Physical Activity," *J. Pediatrics* **125**, 201–207 (1994).
52. P. Kanmus, et al., "Effect of Starting Age of Physical Activity on Bone Mass in the Dominant Arm of Tennis and Squash Players," *Ann. Intern. Med.* **123**, 27–31 (1995).
53. D. C. Cumming, "Exercise Induced Amenorrhea, Low Bone Density, and Extrogen Replacement Therapy," *Arch. Int. Med.* **156**, 2193–2195 (1996).
54. A. L. Friedlander et al., "A Two Year Program of Aerobic and Weight Training Enhances Bone Mineral Density of Young Women," *J. Bone Miner. Res.* **10**, 574–585 (1995).
55. T. Lohman et al., "Effects of Resistance Training on Regional and Total Bone Mineral Density in Premenopausal Women: A Randomized Study," *J. Bone Miner. Res.* **10**, 1015–1024 (1995).
56. B. P. Conroy et al., "Bone Mineral Density in Elite Junior Olympic Weight Lifters," *Med. Sci. Sports Exer.* **25**, 1103–1109 (1993).
57. E. Ernst, "Exercise for Female Osteoporosis: A Systematic Review of Randomized Clinical Trials," *Sports Medicine* **25**, 359–368 (1998).
58. I. R. Reid, "Therapy of Osteoporosis: Calcium, Vitamin D, and Exercise," *Am. J. Med. Sci.* **312**, 278–286 (1996).
59. M. A. Province, "The Effects of Exercise on Falls in Elderly Patients: A Preplanned Metaanalysis of FICsit Trials," *JAMA* **273**, 1341–1347 (1995).
60. S. R. Cummings, et al., "Risk Factors for Hip Fracture in White Women," *New Engl. J. Med.* **332**, 767–773 (1995).
61. R. Prince et al., "The Effects of Calcium Supplementation (Milk Powder or Tablets) and Exercise on Bone Density in Postmenopausal Women," *J. Bone Miner. Res.* **10**, 1068–1075 (1995).
62. R. Prince et al., "Prevention of Postmenopausal Osteoporosis: A Comparative Study of Exercise, Calcium Supplementation, and Hormone Replacement Therapy," *New Engl. J. Med.* **325**, 1189–1195 (1991).

RICHARD COTTER
CINDY DOMINGUEZ
SUSAN TRIMBO
Whitehall-Robins
Madison, New Jersey

See also MINERALS: MICRONUTRIENTS.

OXIDATION

Oxidation is one of the most important reactions occurring in food and food-related systems. The reaction affects not only the chemical nature but many interactions among food constituents, leading to both desirable and undesirable products.

Oxidation reactions can be categorized into chemical and enzymatic. Both processes can be naturally occurring or initiated and enhanced by food handling or processing. The former includes, for example, lipid oxidation and thermal oxidation, which are the frequent cause affecting the acceptability and qualities of many processed food products. Oxidation of proteins by radicals and disulfide exchange often results in cross-linking and subsequent changes in the physical and functional properties. Autoxidation of the heme iron in myoglobin is known to be the major cause of color change in meat products. Enzyme-catalyzed oxidations such as oxygenation of unsaturated fatty acids by lipoxygenase, the browning reaction catalyzed by polyphenol oxidase, and the conversion of glucose to lactone by glucose oxidase have been linked to various changes and interactions in food systems. The oxidative degradation of vitamins can lead to loss in nutrition, and similar types of reaction mechanisms are involved in the antioxidation effect of certain vitamins.

Lipid Oxidation

The reaction of oxygen with unsaturated lipids is one of the most extensively studied areas in food chemistry. Two different pathways have been identified- autoxidation and photosensitized oxidation (1).

Autoxidation

Autoxidation is a free-radical chain reaction involving initiation, propagation and termination steps (Scheme 1). Abstraction of the hydrogen from an unsaturated lipid is favored by resonance stabilization of the conjugated system. The initiation step is catalyzed by metal ion chelates or reducing agents present in biological systems, such as thiol groups in proteins, ascorbate, NADH, and $FADH_2$ (2,3). Both heme and nonheme irons present in muscle tissues are also implicated as initiator (4,5).

The peroxy radical formed by the reaction between the lipid alkyl radical and oxygen can react with another lipid molecule to form hydroperoxide and free radical. It is now well known that the breakdown of hydroperoxides constitutes the major source of rancid flavors in foods. The first step involves decomposition of the hydroperoxide to the alkoxy and hydroxy free radicals, followed by a carbon-carbon cleavage on either side of the alkoxy radical (Scheme 2). The alkoxy radical then reacts with other radicals. These reactions account for the formation of carbonyls, alcohols, esters, and hydrocarbons in peroxidizing lipid systems.

Photosensitized Oxidation

In systems containing a light sensitizer, such as chlorophyll in green vegetables, lipid oxidation proceeds via an "ene" reaction, in which the singlet oxygen molecule is added onto the double-bond (Scheme 3) (6). As free radicals are not involved in this reaction mechanism, the distribution of hydroperoxide products differs from that of autoxidation (1,7).

In peroxidized linolenic acid, cyclization of the hydroperoxide can lead to the formation of malonaldehyde, a dialdehyde that is known to cross-link proteins, enzymes, and nucleic acids. Malonaldehyde can form a Schiff base intermediate with the lysyl-ϵ-amino groups of proteins causing intermolecular cross-linking (7–9).

The radicals generated from oxidation of lipids can interact with some amino acid residues side chains in proteins to form (*1*) lipid-protein polymers, or (*2*) protein free radicals (P·), which may react further with other protein molecules to form cross-linked products (Scheme 4) (10,11). These subsequent polymerization reactions change the chemical structure and functionality of food proteins present in oxidized lipid systems. Cross-linked proteins are expected to possess very different nutritional properties, including reduced digestibility and loss of some of the most sensitive amino acids, such as histidine, methionine, lysine, and cysteine/cystine.

Oxidative Thermal Reactions

Thermal oxidation of saturated fatty acids generally occurs at the α, β, or γ position to form alkoxy radicals, followed

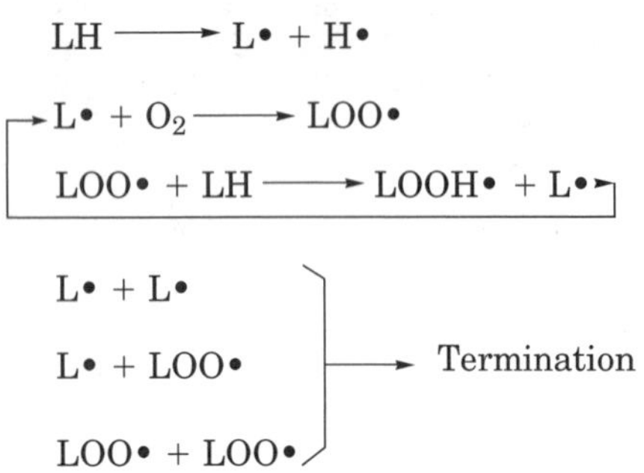

$$LH \longrightarrow L\bullet + H\bullet$$

$$L\bullet + O_2 \longrightarrow LOO\bullet$$

$$LOO\bullet + LH \longrightarrow LOOH\bullet + L\bullet$$

$$\left.\begin{array}{l} L\bullet + L\bullet \\ L\bullet + LOO\bullet \\ LOO\bullet + LOO\bullet \end{array}\right\} \longrightarrow \text{Termination}$$

Scheme 1.

$$\text{RCH(OOH)R'} \xrightarrow{\bullet OH} \text{RCH(O}\bullet\text{)R'} \xrightarrow{R''H} \text{RCH(OH)R'} + R\bullet''$$

$$\text{RCH(O}\bullet\text{)R'} \longrightarrow R\bullet + R'CHO$$

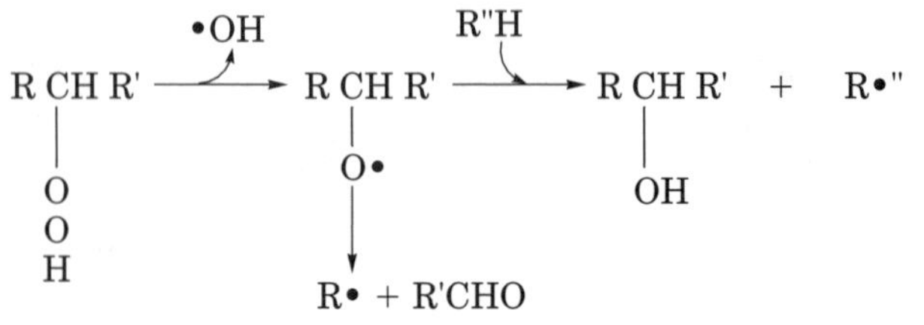

Scheme 2.

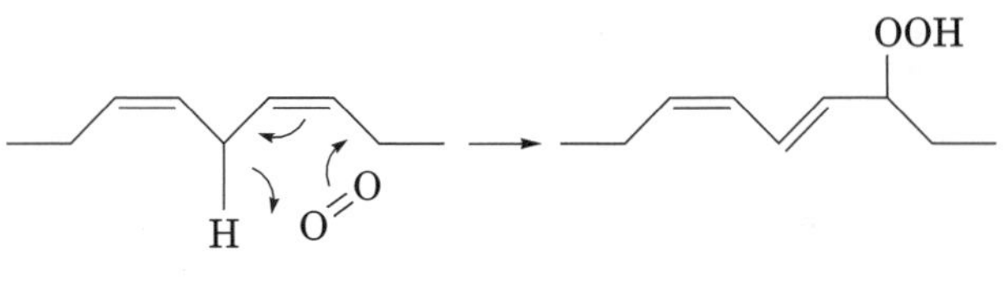

Scheme 3.

Formation of protein radical

$$L\bullet + P \longrightarrow LH + P\bullet$$

$$LO\bullet + P \longrightarrow LOH + P\bullet$$

$$LOO\bullet + P \longrightarrow LOOH + P\bullet$$

Liquid-protein cross-link

Termination

$$L\bullet + P\bullet \longrightarrow LP$$

$$LO\bullet + P\bullet \longrightarrow LOP$$

$$LOO\bullet + P\bullet \longrightarrow LOOP$$

Displacement

$$L\bullet + P \longrightarrow LP\bullet \xrightarrow{P} LPP\bullet \text{ etc}$$

$$LO\bullet + P \longrightarrow LOP\bullet \xrightarrow{P} LOPP\bullet \text{ etc}$$

$$LOO\bullet + P \longrightarrow LOOP\bullet \xrightarrow{P} LOOPP\bullet \text{ etc}$$

Protein-protein cross-link

$$P\bullet + P\bullet \longrightarrow PP$$

$$P\bullet + P \longrightarrow PP\bullet \xrightarrow{P} PPP\bullet \text{ etc}$$

Scheme 4.

by carbon-carbon cleavage producing various carbonyl compounds (12). Oxidative decomposition of unsaturated fatty acids yields dimers, trimers, and tetramers with polar groups. Radicals generated in these processes can enter into very complex combination reactions. Polymer formation causes increasing viscosity of the frying oil.

OXIDATION IN PROTEINS

Proteins are subjected to chemical reactions due to many reactive side-chain groups of the amino acids. The reactivity of these amino acids depends on their chemical nature, and very much on the environment in which they are placed. The pH of the system has a strong influence on the reaction. Depending on the particular pK_a, the side-chain group may be protonated or nonprotonated, and in general, it is the least protonated form that is most nucleophilic and reactive. The accessibility of the various groups also determines their differences in reactivity under a particular set of conditions.

The Hydroxy Radical

Proteins are very reactive toward reactive oxygen species generated by radiolysis of water in biological and food systems (11) (Scheme 5). Aromatic amino acids react with hydroxy radical (OH·), forming substituted cyclohexadienyl radicals (13). All sulfur-containing amino acids react at diffusion-controlled rates. The radical intermediates have been demonstrated to form cross-linked products between proteins and DNA (9,14).

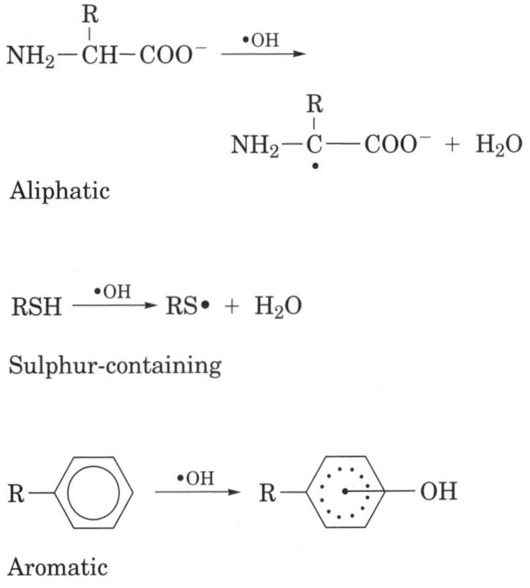

$$NH_2-CH(R)-COO^- \xrightarrow{\bullet OH} NH_2-\dot{C}(R)-COO^- + H_2O$$

Aliphatic

$$RSH \xrightarrow{\bullet OH} RS\bullet + H_2O$$

Sulphur-containing

$$R-C_6H_5 \xrightarrow{\bullet OH} R-C_6H_5(\bullet)-OH$$

Aromatic

Scheme 5.

Photosensitized Oxidation

In the presence of a suitable sensitizer such as flavin, cysteine is oxidized to cysteic acid and methionine to methionine sulfoxide (Scheme 6) (11,15). Interestingly, the same oxidation products are obtained using chemical agents such as hypochlorite or hydrogen peroxide. The treatment of soy proteins with 0.5% hydrogen peroxide oxidized half of the methionine to sulfoxide (16).

Oxidation of Cysteine

The oxidation reaction especially important to the functions of organized protein systems involves the reduction and reoxidation of disulfides. This type of sulfhydryl-disulfide reaction plays a major role in dough mixing (17–19). There is a direct correlation between the sulfhydryl-disulfide content and the rheological quality of doughs made from wheat flours of several varieties (20,21). However, the exact interpretation of disulfide exchange in the molecular model of dough forming remains unclear (22).

Another example of chemical and physical changes caused by reduction and reoxidation is exhibited by the whey protein, β-lactoglobulin. The protein is heat sensitive and starts to polymerize and aggregate when heated above 65°C. The primary reaction is believed to be the reduction of the two disulfides and reoxidation of the thiols forming intermolecular or intramolecular interchange with the free cysteine-121 (23). Sulfhydryl-disulfide interchange also occurs between β-lactoglobulin and κ-casein, producing a complex that is susceptible to heat-induced calcium phosphate precipitation and resistant to rennin action (24,25).

$$2RSH + H_2O \xrightarrow[2O_2,\ h\nu]{\text{Sensitizer}} RSO_3H + RS\bullet + H_2O_2$$

Scheme 6.

$$Fe^{2+} + O_2 \longrightarrow FeO_2^{2+}$$

$$FeO_2^{2+} + Fe^{++} \longrightarrow 2\,Fe^{3+} + O_2^{2-}$$

Scheme 7.

The loss in functionality limits the usefulness of this abundant whey protein as ingredients in food processing. The development of recombinant technology provides a novel procedure to clone the β-lactoglobulin gene, and specifically modify the free cysteine, with the ultimate aim of improving its thermal stability and functional properties (26–28).

Autoxidation of Heme Iron

Myoglobin, the protein that determines the color of meat products, is known to undergo autoxidation. The bright-red exterior color of fresh meat is derived from oxymyoglobin, which contains a heme iron in the +2 oxidation state with an oxygen molecule reversibly bound to the sixth coordination position. The Fe(II) is readily oxidized to Fe(III) to give ferrimyoglobin characteristic of the brown surface color of aged meat. The autoxidation of the heme iron has been shown not to occur by a simple one-electron transfer in which the iron is oxidized from +2 to +3 state and O_2 is reduced to $O^-{}_2$ (29,30). Instead, autoxidation in this case involves a two-electron reduction of the oxygen to peroxide via an iron-dioxygen complex (Scheme 7).

ENZYME-CATALYZED OXIDATION

Lipoxygenase-Catalyzed Peroxidation

The action of the enzyme lipoxygenase (EC1.13.11.12) in food and food-related systems leads to the formation of hydroperoxides in the aerobic pathway, and dimers and oxodienoic acids in the anaerobic reaction (31,32). The enzyme substrate must be unsaturated fatty acids containing *cis,cis*-1,4-pentadiene system, and the reaction is regiospecific and stereospecific in contrast to autoxidation. As the result of free-radical formation in the intermediate process, proteins present in the same system may be oxidized, and subsequent protein-protein polymerization yields many compounds similar to those outlined in autoxidation (33). Oxidized lipids also cause chemical changes in some sensitive amino acids in proteins, and cross-linking reaction between proteins and aldehydes generated from the breakdown of hydroperoxide.

Soybean lipoxygenase has been utilized to bleach flour by incorporating soy flour in wheat flour. The oxidative destruction of pigments in flour by the addition of lipoxygenase is believed to be a free-radical mediated reaction in which the alkyl or peroxy radicals are involved. Besides the bleaching effect, using soy flour in bread making often results in an improvement of the dough-forming properties (34–36). It has been proposed that in a peroxidizing system, the radicals enter into coupled oxidation of cysteine residues in the flour proteins and cause structural changes in the dough (Scheme 8).

The hydroperoxides formed from lipoxygenase-catalyzed oxidation of unsaturated lipids are converted to flavor compounds by the enzyme hydroperoxide lyase, which has been identified and purified from cucumber and tomato fruits, watermelon seedlings, pear, soybean seedlings and leaf tissues. The major flavor compound in cucumber, (*E,Z*)-2,6-nonadienal, is one of the products of this enzymatic rearrangement. Other aldehydes also originate from fatty acid through similar pathways.

Enzymatic Browning Reactions

Polyphenol oxidase (EC1.10.3.1) is an enzyme of great importance in that it not only causes the browning of cut or bruised fruits and vegetables but also plays a major role in the curing of tea, coffee, and tobacco. The enzyme functions as a monooxygenase in the ortho hydroxylation of monophenols to dihydroxyphenols, and a two-electron oxidase in the oxidation of the *o*-diphenols to *o*-quinone (Scheme 9) (37–39). The enzyme has broad substrate specificity. Some of the common substrates include catechol, 4-methylcatechol, dopamine, pyrogallol, catechin, caffeic acid, chlorogenic acid, *p*-cresol, tyrosine, and *p*-hydroxycinnamic acid. The product of the oxidation of diphenols is *o*-quinone, which polymerizes to form colored pigments (40–42).

In the oxidation of dopamine, the primary substrate found in banana, the dopamine quinone undergoes nonenzymatic rearrangement to indole-5,6-quinone that in turn is polymerized to form melanin (43,44).

The development of flavor (and color as well) in the fermentation of tea is related to the oxidation of phenolic compounds. The most important reaction is the oxidation of flavanols by polyphenol oxidase. The oxidized flavanols condense to form theaflavin, which is one of the major constituents of tea flavor (45,46). Similar type of desired activity is also evidenced in coffee cocoa, prunes, dates, and dark raisin production.

The oxidized flavanols can also undergo Strecker degradation with amino acids to form various aldehydes. Flavor compounds, such as isobutanal, 2-methylbutanal, isovaleraldehyde, and phenyacetaldehyde present in fermented tea aroma have been suggested to be the products of degradation of valine, isoleucine, leucine, and phenylalanine. The oxidation of flavanols to the *o*-quinone form

$$LH + O_2 \xrightarrow{\text{Lipoxygenase}} LOO\bullet$$

$$P{-}SH \longrightarrow PS\bullet + LOOH$$

$$P{-}S\bullet + P'{-}S\bullet \longrightarrow P{-}S{-}S{-}P$$

Scheme 8.

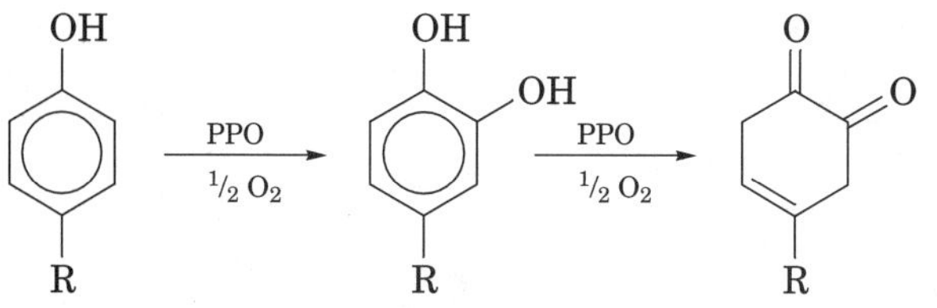

Scheme 9.

also causes the oxidative degradation of carotene to volatile products, such as ionone and 5,6-epoxy-β-ionone.

The polymerization of *o*-quinone to polyaromatic pigments requires water in the reaction components. The effort to stabilize the quinone for the purpose of regiospecific oxidation of the aromatic compounds has been attempted. Polyphenol oxidase has been found to function in chloroform whereby phenol substrates can be converted to stable quinone, which can be quantitatively reduced to catechols (47–49).

Glucose oxidase (EC1.1.3.4) is a food enzyme used for the removal of glucose to prevent the Maillard reaction in egg solids, dried meats, potatoes, and so on. When coupled with catalase, it is used for the removal of oxygen from the head space of packaged foods. The enzyme is also used in the production of gluconic acid, and for the quantitative determination of D-glucose in food, agricultural, and pharmaceutical products (50–52). The enzyme catalyzes the irreversible oxidation of a number of sugars to the corresponding lactones. Common substrates include glucose, deoxyglucose, mannose and galactose.

The enzyme is a dimer containing flavin adenine dinucleotide (FAD) as cofactor. In the reaction, the substrate is oxidized to the lactone while the FAD is reduced to $FADH_2$ (Scheme 10). In a subsequent step, the lactone is nonenzymatically hydrolyzed to gluconic acid, and the reduced enzyme is reoxidized (53).

OXIDATION OF VITAMINS

Loss of vitamins during food handling can be due to many types of degradations and one of these important reactions is oxidation.

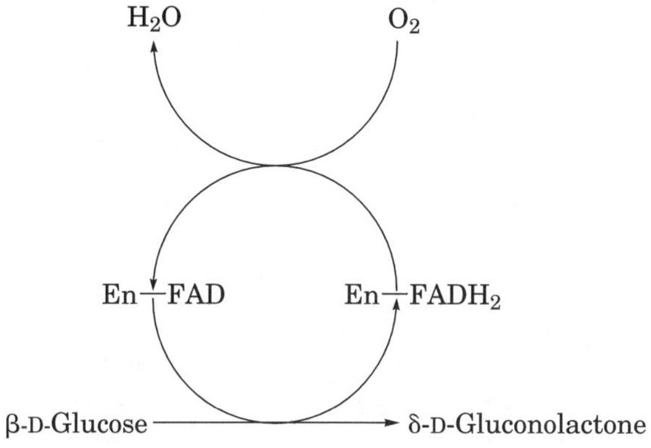

Scheme 10.

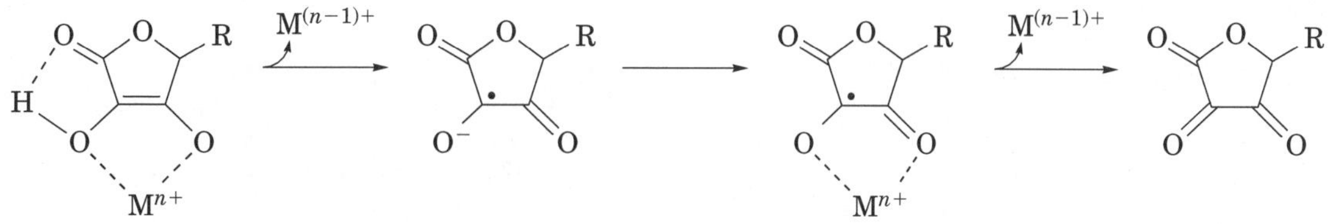

Scheme 11.

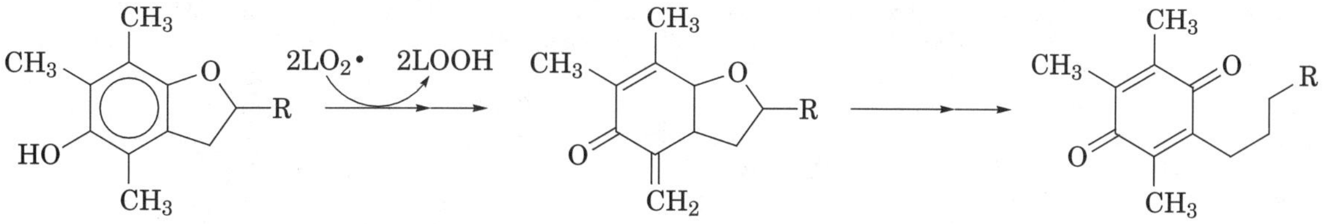

Quinone methine

Scheme 12.

Ascorbic Acid

Vitamin C is one of the least stable vitamins mainly due to oxidative degradations (54). Ascorbic acid can be oxidized by metal ions such as Fe(III) and Cu(II) in a two-sequential one-electron transfer to yield dehydroascorbic acid. Alternatively, metal ions can also catalyze the oxidation of ascorbic acid via the formation of an ascorbate-metal-dioxygen complex immediate (Scheme 11). In this reaction, the two-electron transfer occurs between the ascorbic and the dioxygen leading to the formation of dehydroascorbic acid and hydrogen peroxide (55).

Dehydroascorbic acid undergoes further degradation in acid medium (56), or reacts with amino acids via Strecker degradation. The latter reaction leads to the formation of scorbamic acid and, subsequently, polymeric compounds (57,58).

Tocopherol

Vitamin E also undergoes oxidative degradation via a free-radical pathway that accounts for its high efficiency as a chain-breaking antioxidant. The first step of the free-radical reaction involves the abstraction of two hydrogen atoms, forming the quinone methine that undergoes successive rearrangements to the stable quinone (Scheme 12). In the reaction, the tocopherol molecule consumes two peroxy radicals (LOO·) to form a nonradical product (59,60).

From the subject matter covered in the discussion, it is obvious that oxidation is a widespread reaction affecting every constituent in food and food-related systems. The chemistry involves free radical, singlet oxygen, coupled reduction-oxidation, oxidizing chemical, metal ion, enzyme catalysis, cofactor, and so on. Product interaction often leads to new degradation reactions, cross-linking, polymerization, and subsequently alteration in physical and chemical properties as well as functionality of the constituents. These reactions and interactions are complex and certainly should deserve more attention from food scientists and researchers.

BIBLIOGRAPHY

1. E. N. Frankel, "Lipid Oxidation," *Prog. Lipid Res.* **19**, 1–22 (1980).
2. L. Ernster and K. Nordenbrand, "Microsomal Lipid Peroxidation: Mechanism and Some Biomedical Implications," in K. Yagi, ed., *Lipid Peroxides in Biology and Medicine*, Academic Press, New York, 1982, pp. 55–79.
3. J. M. C. Gutteridge and B. Halliwell, "The Measurement and Mechanism of Lipid Peroxidation in Biological Systems," *TIBS* **15**, 129–135 (1990).
4. A. L. Tappel, "Hematin Compounds and Lipoxidase as Biocatalysts," in H. W. Schultz, E. A. Day, and R. O. Sinnhuber, eds., *Lipids and Their Oxidation*, AVI, Westport, Conn., 1962, pp. 122–138.
5. J. Kanner, J. B. German, and J. E. Kinsella, "Initiation of Lipid Peroxidation in Biological Systems," *CRC Crit. Rev. Food Sci. & Nutr.* **25**, 317–364 (1987).
6. F. D. Gunstone, "Reaction of Oxygen and Unsaturated Fatty Acids," *J. AOCS* **61**, 441–444 (1984).
7. E. N. Frankel, "Lipid Oxidation: Mechanisms, Products and Biological Significance," *J. AOCS* **61**, 1908–1917 (1984).
8. K. S. Chio and A. L. Tappel, "Inactivation of Ribonuclease and Other Enzymes by Peroxidizing Lipids and by Malonaldehyde," *Biochem.* **8**, 2827–2832 (1969).
9. M. Dizdaroglu and M. G. Simic, "Radiation-Induced Crosslinks Between Thymine and Phenylalanine," *Int. J. Radiat. Biol.* **47**, 63–69 (1985).
10. K. M. Schaich, "Free Radical Initiation in Proteins and Amino Acids by Ionizing and Ultraviolet Radiation and Lipid Oxidation, Part III: Free Radical Transfer from Oxidizing Lipids," *CRC Crit. Rev. Food Sci. & Nutr.* **13**, 189–244 (1980).
11. B. S. Berlett and E. R. Stadtman, Protein Oxidation in Aging, Disease, and Oxidative Stress," *J. Biol. Chem.* **272**, 20313–20316 (1997).
12. E. D. Cmjar, A. Witchwoot, and W. W. Nawar, "Thermal Oxidation of a Series of Saturated Triacylglycerols," *J. Agric. Food Chem.* **29**, 39–42 (1981).
13. P. S. Rao and E. Hayon, "Reaction of Hydroxyl Radicals with Oligopeptides in Aqueous Solutions, A Pulse Radiolysis Study," *J. Phys. Chem.* **79**, 109–115 (1975).
14. O. Yamamoto, "Radiation-Induced Binding of Methionine with Serum Albumin, Tryptophan or Phenylalanine in Aqueous Solution," *Int. J. Radiat. Phys. Chem.* **4**, 335–345 (1972).
15. S. F. Yang, H. S. Ku, and H. K. Pratt, "Photochemical Production of Ethylene from Methionine and its Analogues in the Presence of Flavin Mononucleotide," *J. Biol. Chem.* **242**, 5274–5280 (1967).
16. K. C. Chang, H. F. Marshall, and L. D. Satterlee, "Sulfur Amino Acid Stability, Hydrogen Peroxide Treatment of Casein, Egg White, and Soy Flour," *J. Food Sci.* **47**, 1181–1183 (1982).
17. A. Graveland et al., "Superoxide Involvement in the Reduction of Disulfide Bonds of Wheat Gel Proteins," *Biochem. Biophys. Res. Comm.* **93**, 1189–1195 (1980).
18. A. Graveland et al., "A Model for the Molecular Structure of the Glutenins from Wheat Flour," *J. Cereal Sci.* **3**, 1–16 (1985).
19. P. R. Shewry and A. S. Tatham, "Recent Advances in Our Understanding of Cereal Seed Protein Structure and Functionality," *Comments Agric. and Food Chem.* **1**, 71–94 (1987).
20. C. C. Tsen and W. Bushuk, "Reactive and Total Sulfhydryl and Disulfide Contents of Flours of Different Mixing Properties," *Cereal Chem.* **45**, 58–62 (1968).
21. R. Lasztity, *The Chemistry of Cereal Proteins*, CRC Press, Boca Raton, Fla., 1984.
22. D. K. Mecham, "Wheat Proteins—Observations on Research Problems and Progress, Part 1," *Food Technol. in Australia* **32**, 540–587 (1980).
23. M. Z. Papiz et al., "The Structure of β-Lactoglobulin and its Similarity to Plasma Retinol-Binding Protein," *Nature* **324**, 383–385 (1986).
24. P. F. Fox and M. C. T. Hoynes, "Heat Stability of Milk: Influence of Colloid Calcium Phosphate and β-Lactoglobulin," *J. Dairy Res.* **42**, 427–435 (1975).
25. D. M. Mulvihill and M. Donovan, "Whey Proteins and Their Thermal Denaturation—A Review," *Irish J. Food Sci. Technol.* **11**, 43–75 (1987).
26. L. K. Creamer, D. A. D. Parry, and G. N. Malcolm, "Secondary Structure of Bovine β-Lactoglobulin B," *Arch. Biochem. Biophys.* **227**, 98–105 (1983).
27. A. C. Jamieson et al., "Cloning and Nucleotide Sequence of The Bovine β-Lactoglobulin Gene," *Gene* **61**, 85–88 (1987).
28. C. A. Batt et al., "Expression of Recombinant Bovine β-Lactoglobulin in *Escherichia coli*," *Agric. Biol. Chem.* **54**, 949–955 (1990).

29. C. E. Castro, R. S. Wade, and N. O. Belser, "Conversion of Oxyhemoglobulin to Methemoglobin by Organic and Inorganic Reductants," *Biochem.* **17**, 225–231 (1978).
30. D. J. Livingston and W. D. Brown, "The Chemistry of Myoglobin and Its Reactions," *Food Technol.* **35**, 244–252 (1981).
31. K. J. Nelsen and S. P. Seltz, "The Structure and Function of Lipoxygenase," *Curr. Op. in Structural Biol.* **4**, 878–884 (1994).
32. D. W. S. Wong, *Food Enzymes: Structure and Mechanism*, Chapman & Hall, New York, 1995, pp. 237–270.
33. G. Matheis and J. R. Whitaker, "A Review: Enzymatic Cross-Linking of Proteins Applicable to Foods," *J. Food Biochem.* **11**, 309–327 (1987).
34. P. J. Frazier et al., "The Effect of Lipoxygenase Action on The Mechanical Development of Doughs From Fat-Extracted and Reconstituted Wheat Flour," *J. Sci. Food Agric.* **28**, 247–254 (1977).
35. P. J. Frazier, "Lipoxygenase Action and Lipid Binding in Breadmaking," *Baker's Digest* **53**, 8–20, 29 (1979).
36. R. Kieffer and W. Grosch, "Verbesserung der backeigenschaften von weizenmehlen durch die typ ii—lipoxygenase aus sojabohnen," *Z. Lebensm. Unters. Forseh* **170**, 258–261 (1980).
37. R. S. Himmelwright et al., "Chemical and Spectroscopic Studies of The Binuclear Copper Active Site of *Neurospora* Tyrosinase: Comparison to Hemocyanins," *J. Am. Chem. Soc.* **102**, 7339–7344 (1980).
38. D. E. Wilcox et al., "Substrate Analogue Binding to The Coupled Binuclear Copper Active Site in Tyrosinase," *J. Am. Chem. Soc.* **107**, 4015–4027 (1985).
39. M. A. Pavlosky and E. I. Solomon, "Near-IR CD/MCD Spectral Elucidation of Two Forms of the Non-Heme Active Site in Native Ferrous Soybean Lipoxygenase-1: Correlation to Crystal Structures and Reactivity," *J. Am. Chem. Soc.* **116**, 11610–11611 (1994).
40. C. R. Dawson and W. B. Tarpley, "On the Pathway of the Catechol-Tyrosinase Reaction," *Ann. New York Acad. Sci.* **100**, 937–948 (1963).
41. R. S. Phillips et al., "Oxygenation of Fluorinated Tyrosines by Mushroom Tyrosinase Releases Fluoride Iron," *Arch. Biochem. Biophys.* **276**, 65–69 (1990).
42. A. Nagai and H. Yamamoto, "Insolubilizing Studies by Water-Soluble Poly(LysTyr) by Tyrosinase," *Bull. Chem. Soc. Jpn.* **62**, 2410–2412 (1989).
43. J. K. Palmer, "Banana Polyphenoloxidase. Preparation and Properties," *Plant Physiol.* **38**, 508–513 (1963).
44. W. N. Vannesta and A. Zuberbuhler, "Copper-containing Oxygenases," in O. Hayaishi, ed., *Molecular Mechanisms of Oxygen Activation*, Academic Press, New York, 1974, pp. 371–404.
45. G. W. Sanderson, H. Co, and J. G. Gonzalez, "Biochemistry of Tea Fermentation: The Role of Carotenes in Black Tea Aroma Formation," *J. Food Sci.* **36**, 231–236 (1971).
46. G. W. Sanderson and H. N. Graham, "On the Formation of Black Tea Aroma," *J. Agric. Food Chem.* **21**, 576–585 (1973).
47. R. Z. Kazandjian and A. M. Klibanov, "Regioselective Oxidation of Phenols Catalyzed by Polyphenol Oxidase in Chloroform," *J. Am. Chem. Soc.* **107**, 5448–5450 (1985).
48. A. Zaks and A. M. Klibanov, "Enzymatic Catalysis in Nonaqueous Solvents," *J. Biol. Chem.* **263**, 3194–3201 (1988).
49. J. S. Deetz and J. D. Rozzell, "Enzyme-catalyzed Reaction in Non-Aqueous Media," *TIBTECH* **6**, 15–19 (1988).
50. M. Rohr, C. P. Kubicek, and J. Kominek, "Gluconic Acid," in H.-J. Rehm and G. Reed, eds., *Biotechnology*, Vol. 3, Verlag Chemie, Weinheim, Germany, 1983, pp. 455–465.
51. A. P. F. Turner, I. Karube, and G. S. Wilson, *Biosensors, Fundamentals and Applications*, Oxford University Press, Oxford, U.K. 1987.
52. F. Cioci and R. Lavecchia, "Effect of Polyols and Sugars on Heat-Induced Flavin Dissociation in Glucose Oxidase," *Biochem. Mol. Biol. Int.* **34**, 705–712 (1994).
53. D. W. S. Wong, *Food Enzymes: Structure and Mechanism*, Chapman & Hall, New York, 1995, pp. 308–320.
54. M.-L. Liao and P. A. Seib, "Chemistry of L-Ascorbic Acid Related to Foods," *Food Chem.* **30**, 289–312 (1988).
55. A. E. Martell, "Chelates of Ascorbic Acids, Formation and Catalytic Properties," in P. A. Seib and B. M. Tolbert, eds., *Ascorbic Acid: Chemistry, Metabolism, and Uses*, American Chemical Society, Washington, D.C., 1982, pp. 153–178.
56. T. Kurata and M. Fujimaki, "Formation of 3-Keto-4-Deoxypentosone and 3-Hydroxy-2-Pyrone by the Degradation of Dehydro-L-Ascorbic Acid," *Agric. Biol. Chem.* **40**, 1287–1291 (1976).
57. T. Hayashi, M. Namiki, and K. Tsuji, "Formation Mechanism of the Free Radical Products and Its Precursor by the Reaction of Dehydro-L-Ascorbic Acid with Amino Acid," *Agric. Biol. Chem.* **47**, 1955–1960 (1983).
58. T. Hayashi et al., "Red Pigment Formation by the Reaction of Oxidized Ascorbic Acid and Protein in a Food Model System of Low Moisture Content," *Agric. Food Chem.* **49**, 3139–3144 (1985).
59. E. H. Gruger, Jr., and A. L. Tappel, "Reactions of Biological Antioxidants: 1. Fe(III)—Catalyzed Reactions of Lipid Hydroperoxides with α-Tocopherol," *Lipids* **5**, 326–331 (1970).
60. G. W. Burton and K. U. Ingold, "Autoxidation of Biological Molecules. 1. The Antioxidant Activity of Vitamin E and Related Chain-Breaking Phenolic Antioxidants In Vitro," *J. Am. Chem. Soc.* **103**, 6472–6477 (1981).

DOMINIC W. S. WONG
USDA
Albany, California

OZONE AND FOOD PROCESSING

OZONE BACKGROUND

Mentioning ozone to most people in the 1990s evokes the thought of the ozone hole discovered by NASA's Landsat spacecraft over the South Pole or unhealthy air over some of our nation's cities. In both instances, the thoughts accompany a negative connotation of pollution by man-made chlorinated fluorocarbons destroying the ozone layer or a by-product of the photolysis of automobile exhaust. History has been much more favorable recognizing the positive attributes of ozone, and other countries have embraced the application of ozone for its environmentally safe biocidal and deodorizing properties for more than a century (1).

Ozone has been a natural part of our environment well before recorded history and indeed has enabled man to live today protected from dangerous ultraviolet rays from the sun. Ozone is created naturally by lightning and photochemically by ultraviolet (UV) light. Electrical discharge and UV light breaks the naturally occurring oxygen molecule, O_2, into two unstable atomic atoms, O, that can combine with other oxygen molecules and form ozone, O_3. Ozone produced by lightning is responsible for the fresh smell following a storm. If ozone were not formed in the upper atmosphere by O_2 absorption of the sun's UV rays, the radiation would be able to reach the surface of the earth where it can cause skin cancer in humans and other negative effects on plants and animals.

From the early writings of Homer, man has observed the smell that accompanied lightning as "full of sulphur" odor (2). In 1785 a Dutch physicist, van Marum (3), observed ozone's characteristic odor when electrical sparks were passed through air. From the electrolysis of water, another process to produce ozone, Cruickshank (3) observed in 1801 the same odor near the anode. However, it was Schöenbein who is credited with the discovery of ozone (4). In a memoir presented to the Academy of Munich in 1840, Schöenbein concluded that the odor during electrical discharge and electrolysis and after a flash of lightning were the same substance. He named the substance *ozone* after the Greek word *ozein* meaning "to smell."

Ozone exists naturally at ground level in low concentrations. It dissipates rapidly by releasing one of the three oxygen atoms and reverting to common molecular oxygen. Thus, ozone's unstable property necessitates that it be manufactured on-site where it will be used. Today, for most practical applications, electrical discharges, UV radiation, and electrolysis are used to produce ozone. Electrical discharges produce the highest ozone gas concentrations efficiently, up to 80 g O_3/kWh and up to 20% by weight in oxygen. UV radiation produces the lowest efficiencies, about 8 g O_3/kWh (5). Electrolysis can produce ozone directly in water, eliminating the need to dissolve the ozone gas water.

OZONE DISINFECTION PROPERTIES

Ozone is one of the strongest oxidizing reagents known. Its oxidation potential makes ozone effective as a biocide. Theories suggest that ozone attacks the lipid double bonds in the cell membrane and results in a change in cell permeability leading to lysis (6,7).

A demonstration of ozone's disinfection capability is its sterilization of medical instruments infected with the germicidal resistant microorganisms (8). Table 1 shows the spectrum of microorganisms encountered in the medical environment from the most resistant bacterial spores, *B. subtilis*, to the relatively sensitive lipid containing viruses, Herpes simplex 1. Ozone exposure conditions in the medical sterilizer were 12 to 14% ozone concentrations in oxygen, ambient temperature, and 85 to 95% relative humidity. Foodborne human pathogenic bacteria, molds, and viruses are represented by the spectrum of microorganisms listed in the table. In addition, ozone is effective in killing foodborne and waterborne parasites, such as *Giardia lamblia* and *Cyrptosporidum parvum*.

Table 1. Ozone is Effective in Killing Human Pathogens

Microorganism	20-min exposure	30-min exposure	90-min exposure
Bacillus subtilis	9/20[a]	16/20	0/20
Bacillus stearothermophilus	4/20	4/20	0/20
Clostridium sporogenes	3/20	1/20	0/20
Staphylococcus aureus	0/20	0/20	0/20
Salmonella choleraesuis	0/20	0/20	0/20
Pseudomonas aeruginosa	0/20	0/20	0/20
Mycobacterium phiel	0/20	0/20	0/20
Aspergillus niger	0/20	0/20	0/20
Polio 1	0/10	0/10	0/10
Herpes simplex 1	0/10	0/10	0/10

[a]The number of growers out of the number possible.

Reduction of the population of a specific microorganism by a given fraction depends uniquely on the product of the ozone concentration level, C, and exposure time, t. Different microorganisms' mortalities are easily compared by means the "C · t" product for the same reductions (9). For example, the C · t for *B. subtilis* is much greater than for Herpes simplex 1 for the same population reduction. Furthermore, comparison of disinfection in air and water can be made on a media mass fraction basis, that is, normalization of C · t to the mass of the fluid (10). Using the ratio of the mass of water to air, for example, indicates that a C · t would have to be 500 times greater in air for the same disinfection capability in water. For disinfection applications to food compared to medical sterilization, not only are the specific microorganisms anticipated quite different, the reduction objectives are much lower. In food disinfection, the objective is to reduce the foodborne microorganism populations to level preventing sickness and not to achieve clinical sterilization or twice the C · t required to achieve a zero population.

Besides destroying microorganisms, like mycotoxin-producing molds, ozone has the potential to destroy the mycotoxin itself. For example, patulin found at high levels in some natural apple juice and produced by several *Penicillium* species can be oxidized and decomposed by ozone rendering it harmless. Research in the oxidation of herbicide and pesticide residue by ozone has also demonstrated promise (11).

GRAS APPROVAL OF OZONE

In July 1997, a panel of experts concluded that there is sufficient scientific evidence to certify that ozone and its use is Generally Recognized as Safe (GRAS) when used at levels and by methods consistent with good manufacturing practices (12). This announcement fulfilled the requirements of a U.S. Food and Drug Administration (FDA) process that permits the use of ozone in those products for which it has jurisdiction; namely, fruits, vegetables, and seafood. Since then the U.S. Department of Agriculture (USDA) has approved the use of ozone on a case-by-case basis with submittal of a description of the process using ozone for meat and poultry. While the number of ozone applications is increasing in the United States, ozone seems to be a solution looking for a problem. The status quo is the use of chlorine and chlorine dioxide in various forms and commercial sanitizers, such as quaternary ammonium. The main problem ozone has in gaining acceptance is a lack of a track record and the high up-front cost associated with the generating equipment. Fortunately, there is continuing improvement on ozone's track record and equipment cost.

FOOD PROCESSING

Ozone is an effective biocide in air and oxygen gas and in water for applications from food storage to food processing and from fruits and vegetables to meat, poultry and seafood. Table 2 summarizes effective points where ozone has been introduced in the food supply chain. Like chlorine, ozone in water is more effective in disinfecting the water and preventing contamination of uninfected product than disinfecting the products directly (13). Unlike chlorine, ozone decomposes to oxygen and simple by-products, leaving no residue or buildup of compounds that requires the water to be discharged. Furthermore, ozonized water will have lower biological and chemical oxygen demand and lower fats, oil, and grease characteristics, reducing the environmental burden.

From the field to the grocery store, ozone is playing a more significant role in providing safe and environmentally sound food supply system as part of producers' Hazard Analysis and Critical Control Point program (14). Ozone plays a role in disinfecting wash water for fruits and vegetables before packing and storage. Potable water is generally not available for field processing in some areas and countries. Ozone is capable of disinfecting the water without leaving a chemical residue on the product or contaminating water that will find its way into the environment.

Ozone concentrations of 0.5 to 15 ppm in air applied to food storage have the capability to extend shelf life by inhibiting growth to destroying molds, viruses, and bacteria on food surfaces. In food processing plants, ozone treatment of the product transport flumes reduces the bacterial content of the final product by controlling the bacteria levels and cross-contamination. Ozonized water in place of chemical sanitizers is used in clean-in-place systems to disinfect process lines in food, dairy, and other beverage processing lines.

Animal slaughter plants can use ozone to wash down carcasses on the kill floor prior to cooldown and storage (15). Ozonized water spray and ozone gas applied in beef storage facilities controls mold growth. In poultry processing plants, filtration and ozonation of poultry chill water returns the water quality to a state that it can replace potable processing water (16). Not only is the water saved, but the energy cost of lowering the temperature to near 32°F is saved as well. The fishing industry is using ozone to wash freshly caught fish and storing them on ozonized ice on board ship to extend the shelf life.

Table 2. Successful Points for Ozone Intervention in the Food Supply Chain

Food	Successful ozone applications
Fruits and vegetables	Storage, field processing, plant processing
Grains	Storage
Milk and dairy products	Clean-in place, storage
Meat, poultry, and seafood	Postslaughter, poultry chiller, storage
Beverages	Clean-in place

OZONE APPLICATIONS FOR FOOD STORAGE

Ozone has found extensive use in extending the storage times for fruits, vegetables, eggs, cheese, meats, poultry, and fish (17). Many European countries have developed the technology and applications. Research began in this application in the early part of this century, but today's technology is based on research since World War II. Ozone's application continues to grow as the costs of generators, sensors, and control systems have come down. Today in the United States, fruit and vegetable storage and ripening rooms and farm storage facilities are using ozone gas to control molds and bacteria as well as ripening.

Fruits

Ripening rooms first gained popularity in the storage and forced ripening of bananas. Today, ripening rooms maintain temperature within 1.5°C and relative humidity to 85 to 95% with computer-controlled systems (18). Ozone offers the potential to control fungal spoilage on fruits, control or destroy pathogens, destroy ethylene gas produced as the fruits ripen, and improve the overall hygiene of the ripening room and air-handling equipment. By reducing ethylene levels, ripening of the fruit can be delayed until the fruit is ready to be shipped to market; then, introducing ethylene gas in the room accelerates the ripening.

Vegetables

From the relatively small volumes of ripening rooms (180 m^3) to vegetable storage facilities (25,000 m^3), ozone controls fungal spoilage on vegetables, reduces product shrinkage, and improves product margins. Most of the beneficial evidence of ozone in vegetable storage is anecdotal and difficult to compare because of the differences in storage facilities and differences from season to season. The owner of an onion storage facility with a Cyclopss ozone system reports that while neighboring storage facilities have black mold, his facility is mold free.

OZONE APPLICATIONS FOR FOOD PROCESSING

There are several points in a food processing plant where ozone can be used effectively. Many processing plants use water as part of the product washing or transporting between process steps. Ozone can effectively replace chlorine disinfectants by dissolving ozone gas in the water and controlling levels by dissolved ozone sensors or standard oxidation-reduction potential (ORP) sensors used to monitor chlorine levels. Ozone can be used to disinfect the water subsequently used in the finished product, as in the beverage industry. The FDA gave GRAS approval for bottled water in 1991 and today is used in over 95% of bottle water. Products that are processed dry can be treated with ozone

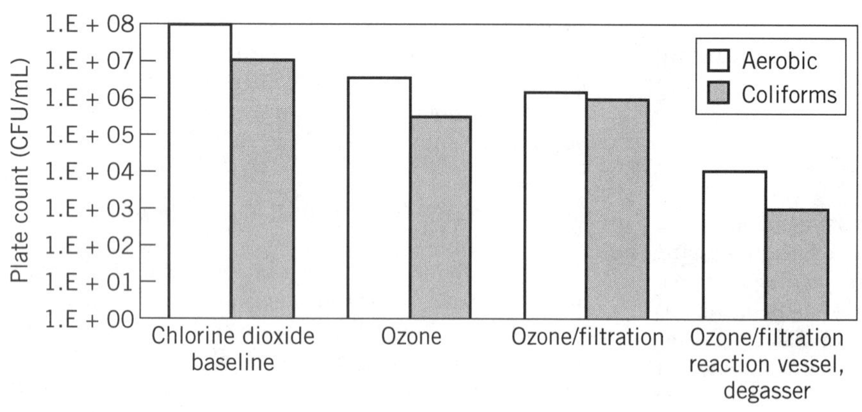

Figure 1. A comparison of ozone with chlorine dioxide shows ozone's disinfection capabilities with various application methods.

gas. For example, the dietary supplement ginger that has been partially dehydrated can be treated in ozone at 8% for 30 min to realize a 4-log reduction in total aerobic plate counts.

Vegetables

Ozone used in flume water for an onion processing line produced a 3.8-log reduction of total aerobic and coliform plate counts over chlorine dioxide. Ozone was used in the flume water between a peeler and a slicer on an onion line that processed an average of 6,000 kg/h. Figure 1 shows the comparison of chlorine dioxide and ozone with various water treatment steps. The first pair of columns represents the total aerobic and coliform plate counts with chlorine dioxide. The next three sets of columns shows ozone alone, ozone with a 100-μm filter prior to ozone injection, and ozone with the 100-μm filter, a 450-L reaction vessel, and a centrifugal gas separator. The quantity of ozone was adjusted to maintain ozone off-gassing below 0.05 ppm at the open water tanks for personnel safety. The amount of ozone used for each of the three conditions was 550, 330, and 500 g/h.

Figure 1 illustrates the effect of the organic load on the consumption of ozone. Comparison of ozone alone with ozone and filtration reduces the ozone required by over 20% for the same disinfection. To achieve the most efficient use of ozone for disinfection, filtration is required to reduce the organic load. Finally, one can also observed the benefit of increasing the contact time of the ozone with the flume water by the last pair of columns in the figure, where the plate counts were at their lowest levels. Similar results were obtained for potato processing lines where, again, ozone was used to replace chlorine dioxide. In the latter case, the organic load was from potato pulp, gratings, proteins, and starch content (as high as 15% by volume) in the process water. In both examples, ozone and filtration treatments lead to the extended reuse of the process water.

RECOMMENDATIONS

There are many successful applications of ozone to food processing as well as many disappointments. A user selecting knowledgeable ozone engineering companies is the main characteristic of the successes. Ozone engineering includes demonstrating the understanding of the $C \cdot t$ product to achieve a given objective, ozone safety, equipment specification, ozone compatible materials selection, installation, test, and support service. Ozone has great potential to address many of today's pressing food safety issues. Applications require significant investment in capital equipment up-front as opposed to pay as you go with chemicals. Be sure the application is done right by selecting a competent ozone engineering company so you will reap the rewards of a successful application.

BIBLIOGRAPHY

1. R. G. Rice and Aharon Netzer, eds., *Handbook of Ozone Technology and Applications*, Vol. 1, Ann Arbor Science Publishers, Ann Arbor, Mich., 1982.
2. Homer, *The Odyssey*, translated by W. H. D. Rouse, The New American Library, New York, 1937, p. 146.
3. E. K. Rideal, *Ozone*, Constable & Co., Ltd., London, 1920.
4. C. F. Schöenbein, "Recherches sur la Nature de l'Odeur Qui se Manifests dans Certaines Actions Chimiques," *Compte Rendus Hebd. Seances Acad. Sci.* **10**, 706–710 (1840).
5. R. Barker, *Improving the Efficiency of Ozone in Odor Treatment by Activation with UV Light*, Electricity Council Research Centre, Capenhurst, Chester, UK, 1979.
6. S. Farooq, E. S. K. Chian, and R. S. Engelbrecht, "Basic Concepts in Disinfection with Ozone," *J. Water Pollution Control Fed.* **49**, 1818–1831 (1977).
7. W. Riesser et al., "Possible Mechanisms of Poliovirus Inactivation by Ozone," in R. G. Rice and M. E. Browning, eds., *Forum on Ozone Disinfection*, International Ozone Association, Norwalk, Conn., 1977, pp. 186–192.
8. 510(k) Filing for Ster-O-Zone™ Medical Sterilizer to the U.S. FDA, Cyclopps Corporation, Salt Lake City, Utah, January 1995.
9. G. B. Wickramanayake and Otis J. Sproul, "Kinetics of the Inactivation of Microorganisms," in S. S. Block, ed., *Disinfection, Sterilization, and Preservation*, 4th ed., Lea & Febiger, Malvern, Pa., 1991, pp. 72–84.
10. W. J. Kowalski, W. P. Bahnfleth, and T. S. Whittam, "Bactericidal Effects of High Airborne Ozone Concentrations on

Escherichia coli and *Staphylococcus aureus*," *Ozone Sci. and Eng.* **20**, 205–221 (1998).

11. G. Crozes et al., "Evaluation of Ozone for Cryptosporidium Inactivation and Atrazine Oxidation in a Lime Softening Plant," *Ozone Sci. and Eng.* **20**, 177–190 (1998).
12. Electric Power Research Institute, *Expert Panel Report: Evaluation of the History and Safety of Ozone in Processing Food for Human Consumption*, Electrical Power Research Institute, Pleasant Hill, Calif., 1997.
13. R. E. Brackett, "Fruits Vegetables and Grains," in M. P. Doyle, L. R. Beuchat, and T. J. Montville, eds., *Food Microbiology*, ASM Press, Washington, D.C., 1997.
14. U.S. Department of Health and Human Services, *1997 Food Code, Annex 5, HACCP Guidelines*, Government Printing Office, Washington, D.C., 1997.
15. B. M. Gorman et al., "Evaluation of Hand-Trimming, Various Sanitizing Agents, and Hot Water Spray-Washing as Decontamination Interventions for Beef Brisket Adipose Tissue," *J. Food Protection* **58**, 899–907 (1995).
16. A. L. Waldroup, R. E. Hierholzer, and R. H. Forsythe, "Recycling of Poultry Chill Water Using Ozone," *J. Appl. Poultry Sci.* **2**, 330–336 (1993).
17. R. G. Rice, J. W. Farquhar, and L. J. Bollyky, "Review of the Applications of Ozone for Increasing Storage Times of Perishable Foods," *Ozone Sci. and Eng.* **4**, 147–163 (1982).
18. *Banana Handling Manual*, Dole Food Company, Inc., 1996.

DURAND SMITH
WILLIAM STODDARD
Cyclopss Corporation
Salt Lake City, Utah

P

PACKAGING: PART I—GENERAL CONSIDERATIONS

MARKETPLACE DEMANDS AND DRIVING FORCES

Packaging is critical in providing products that meet the consumer needs in a society. Packaging is even more critical in providing food products because they are more perishable and fundamental to the health and progress of the society.

Many driving forces result in increased marketplace demands for foods that offer quality, convenience, safety, and low cost. Some of these driving forces are lifestyle changes, demographic changes, and market globalization. The most noticeable lifestyle changes are influenced by smaller families, the increasing number of single-person households, and dual-income families. The major demographic change was the population spike that occurred after World War II ("Baby Boomers"), which resulted in an increased number of older consumers who demand more nutritious and healthier foods that are easy to prepare. Market globalization has resulted in increased importing and exporting of foods to and from many different and distant regions, creating strong consumer demands for regulations that ensure food safety and better product labeling. Some other driving forces are the macroeconomies in Europe and North America, the continued consolidation in industry, and the public concern over packaging disposal problems.

To meet the challenges presented by these driving forces, the packaging industry has made numerous technical innovations. Many innovations have come from the area of plastics packaging, because plastics are cost effective, lightweight, nonbreakable, heat sealable, microwavable, easily fabricated, and corrosion resistant. For example, significant advancements have been made in coextrusion, lamination, and coating technologies, which enable the design of plastic packaging materials with a wide range of properties (1). Another innovation is active packaging, including the technologies of modified atmosphere packaging for fresh produce, oxygen scavengers, time–temperature integrators, and antimicrobial films (2).

New concepts in food packaging technology are needed to meet the challenges of the Information Age. One of the concepts emerging is a greater emphasis on a systems approach in the delivery of better product value; one that addresses the integration of the product, the package, and the environment as a complete resource cycle, from raw material through use and recovery. Another concept is the use of information technology to enhance the communication role of packaging, allowing packaging to serve as a more intelligent messenger for information sharing in this resource cycle.

FUNCTIONS OF A FOOD PACKAGING SYSTEM

A major goal of food packaging is in the efficient delivery of products to the consumer. To accomplish this goal, the package must serve the important functions of containing the food, protecting the food, providing convenience, and conveying product information. The package protects the food against physical, chemical, and biological damages. It also acts as a physical barrier to oxygen, moisture, volatile chemical compounds, and microorganisms, which are detrimental to the food. The package provides the consumer with convenient features such as microwavability, resealability, and ease of use. The package conveys useful information such as product contents, nutritional values, and preparation instructions. These functions make food packaging an essential technology for maintaining the original food quality, with fewer chemical additives and less food waste.

The food package can function best when integrated into a food packaging system, which involves certain physical components and operations. The major physical components are the food, the package, and the environment (Fig. 1). It is useful to divide the environment into internal and external—the internal environment (referred to hereafter as the headspace) is inside the package and is in direct contact with the food, and the external environment is outside the package and depends on the storage and distribution conditions. The operations are the manufacturing, distribution, and disposal of the food package. In designing the food packaging system, these physical components and operations must be considered to prevent overpackaging and underpackaging, which result in higher costs, lower quality, and, in some cases, health risks.

GAS BARRIER PROTECTION AND SHELF LIFE

For foods that are sensitive to oxygen or moisture, gas barrier protection is the major function of the package in providing adequate shelf life, the time period during which the food maintains acceptable quality. For example, moisture can move from the external environment through the package into the headspace. This increase in relative humidity in the headspace can cause a moisture-sensitive food, such as potato chips, to have less crispness and a shorter shelf life. To ensure adequate shelf life, the package must help to reduce the moisture movement from the external environment to the headspace.

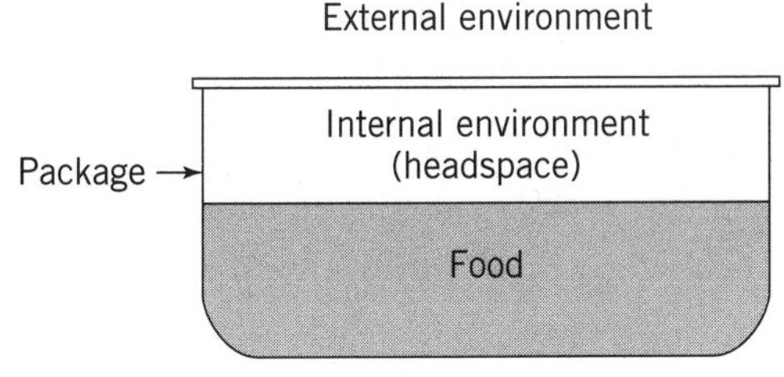

Figure 1. Physical components of packaging system.

Movement of gases between the external environment and the headspace through the package can occur by means of permeation and leakage. Gas permeation is an important consideration in packaging foods with plastics, because food packaging plastics are permeable to moisture, oxygen, carbon dioxide, nitrogen, and other gases. The gas-permeation rates of most interest are the oxygen transmission rate (OTR) and water vapor transmission (WVTR). Gas permeation is not an important consideration in packaging foods with metal or glass because these materials are not permeable.

Leakage can occur in packages made of plastics, metal, glass, or paper. Pinholes are sometimes found in thin metal foils and metallized films. Channel leaks are most commonly found in weak seals or contaminated seals. Pinholes and channel leaks can shorten the shelf life of a packaged food and, in some cases, allow microbial penetration, leading to a potential health risk.

Shelf life is best determined by actual field studies, although accelerated laboratory studies and computer simulations can sometimes provide useful shelf life predictions (1,3). The variables that affect the shelf life of packaged food are discussed next.

Shelf Life of Oxygen-Sensitive Foods

Equation 3 predicts the shelf life of packaged food that is oxygen sensitive. The equation assumes the package is made of a plastic material with uniform thickness and has complete integrity (no pinholes and channel leaks); hence, permeation is the only means of oxygen transfer. The equation can assist in understanding the relationship between the shelf life and the physical components of the food packaging system—the food, package, headspace, and external environment. Two useful terms, permeability ($\bar{P}$) and maximum allowable oxygen (O_{2max}), are introduced.

The OTR of a permeable package (eg, a plastic pouch) from the external environment to the headspace can be calculated using

$$\mathrm{OTR} = \frac{\bar{P}A}{L}(P_e - P_i) \tag{1}$$

where OTR is the oxygen transmission rate, cc O_2/day; A is the surface area of package, in.^{2}; L is the thickness of package, mil; $\bar{P}$ is the oxygen permeability, (cc $O_2 \cdot$ mil)/(100 in.$^2 \cdot$ day $\cdot$ atm); P_e is the partial oxygen pressure in external environment, atm; and P_i is the partial oxygen pressure in headspace, atm.

The shelf life (t_s) can be estimated using

$$t_s = \frac{O_{2max}}{\mathrm{OTR}} \tag{2}$$

where t_s is the shelf life, day and O_{2max} is the maximum allowable oxygen, cc O_2. Equation 2 assumes that O_2 permeation through the package is the major factor in limiting shelf life. This is the case for food packages of good oxygen barrier, in which the oxidative reaction rate of the food is greater than the oxygen transmission rate of the package.

Substituting equation 1 into equation 2 yields

$$t_s = \frac{O_{2max}L}{\bar{P}A(P_e - P_i)} \tag{3}$$

This equation is useful for evaluating many what-if scenarios. For example, if the thickness is decreased by 25% and the surface area is increased by 20%, then the equation predicts that the shelf life will be decreased by 37.5%. Equation 3 also shows that shelf life depends on the variables determined by the food, the package, and the environment.

The variable determined by the food is O_{2max}. Oxidation (especially lipid oxidation) is an important mode of deterioration for many foods (4). The incentive for studying oxidation is that it provides an objective measure of the food stability in addition to the subjective measure of sensory evaluation. The sensory acceptability of an oxygen-sensitive food can then be correlated to the extent of oxidation reaction. However, measuring the extent of oxidation reaction requires a great effort because it depends on the concentration and diffusion of oxygen in the food.

O_{2max} is sometimes used as an alternative measure of oxygen stability by the food packaging industry. It is the maximum amount of additional oxygen the food can absorb before becoming unacceptable. It is a simpler but less accurate measure of oxygen stability than the extent of oxidation reaction because O_{2max} usually ignores the effects of concentration and diffusion of oxygen in the food. Reference values for O_{2max} are available in the literature for some foods (3); for example, the O_{2max} for instant coffee is reportedly between 1 and 5 ppm. If the upper limit of 5 ppm is assumed, O_{2max} for a package containing 454 g of instant coffee $= 454\ (5 \times 10^{-6}) = 2.37 \times 10^{-3}$ g. Nevertheless, it is often necessary to determine O_{2max} experimentally, which involves correlating the food quality with the amount of oxygen absorbed in the food. O_{2max} is the difference between the critical oxygen level (at which the food is no longer acceptable) and the initial oxygen level (which depends on the initial condition of the food).

The variables determined by the package are L, A, and $\bar{P}$. L and A are specified by the dimensions of the package. The permeability $\bar{P}$ is a measure of the permeation of the gas through the packaging material—the lower the $\bar{P}$, the better the gas barrier. The values of $\bar{P}$ depend on the packaging material, permeant gas, temperature, and sometimes relative humidity. For example, the O_2 permeability of polyethylene terephthalate (PET) at 23°C is between 3 and 6 (cc $\cdot$ mil)/(100 in.$^2 \cdot$ day $\cdot$ atm). The relative humidity is not specified because it has little effect on the permeability of PET. $\bar{P}$ and gas permeation are further discussed in PACKAGING: PART III—MATERIALS.

The variables determined by the environment are P_e and P_i. P_e is the partial O_2 pressure in the external environment, which is about 0.21 atm for a normal storage environment. P_i is the partial O_2 pressure in the headspace, which is sometimes deliberately reduced using the techniques of vacuum packaging or modified atmosphere packaging for shelf life extension. (In vacuum packaging, the air in the headspace is removed before sealing. In modified atmosphere packaging, the air in the headspace is replaced by an inert gas such as nitrogen before sealing.) An important environmental variable not explicitly stated in equation 3 is temperature. An increase in temperature causes increases in the oxidation rate of the food and the package permeability.

The term $P_e - P_i$ in equation 3 is often called the concentration driving force—the larger the driving force, the higher the permeation rate. If the food is packaged in air, the external environment and the headspace have the same oxygen concentration, the driving force is $P_e - P_i = 0$, and there is no net flow of oxygen. If the food is packaged in a reduced oxygen environment, the driving force is $P_e - P_i > 0$. This causes a net flow of oxygen into the package, but at a slower rate as time passes, because $P_e - P_i$ continues to decrease.

Shelf Life of Moisture-Sensitive Packaged Foods

A more elaborate approach to predict the shelf life of moisture-sensitive packaged foods is described elsewhere (5). The approach requires the definition of critical limits of a_w, above or below which (depending on whether the food gains or loses moisture) the food is no longer considered acceptable. The moisture gain or loss through the package depends on the H_2O permeability and the temperature as well as the relative humidities in the headspace and the external environment.

Some foods are both oxygen and moisture sensitive. For example, potato chips can lose crispness because of moisture gain and become rancid because of oxygen absorption. In packaging these foods, oxygen and moisture protection are required. In packaging foods that are less sensitive to oxygen or moisture, physical protection is more important than gas barrier protection.

PHYSICAL PROTECTION

During distribution, the packages are exposed to physical abuses caused by shock, vibration, compression, or handling. The food packaging system must provide physical protection for the food package as well as facilitate safe and cost-effective product distribution. The important considerations in designing for distribution protection are the distribution environment and product fragility.

The distribution environment consists of all the events (including handling, storage, and transportation) that the food package encounters before consumer usage. The most important considerations in defining the distribution environment are shock, vibration, and compression (6). Severe shocks occur most likely when the package is dropped. Shock protection is often defined in terms of the most severe drop height to protect against, and this drop height is often selected based on the size and weight of the package as well as the probability of being dropped. Vibration occurs most likely during transportation, and the transportation vibration environment is often complex and random in nature. To protect the package from vibrational damage, the resonant frequencies of the package must be identified and protected against. Compression occurs most likely during warehousing and shipping. Static compression is determined by applying a load very slowly, and dynamic compression is determined by applying the load rapidly.

Product fragility describes the susceptibility of the product (the food and the package) to physical abuses, and it can be evaluated by product performance tests and packaging material tests (7). Product performance tests (such as drop test, incline impact test, vibration test, compression test, and burst test) are performed with the final package to ensure it can survive during distribution. The final package may be a food tray or a corrugated box containing several food trays. Packaging material tests (such as tensile test, Izod impact test, and creep test) are performed with the packaging materials. The properties of the packaging materials (such as tensile strength and tensile modulus) obtained from these tests are used to extrapolate the performance of the final package.

In addition to distribution, the manufacturing operations (such as forming, heat sealing, and retorting) can also cause physical stresses to the food package. During forming, the packaging material may be stretched into another shape (such as from a plastic sheet into a multicompartment tray). During heat sealing, the package is exposed to the heat of the sealing bar. During retorting of the packaged food product, the package is exposed to high temperature and pressure. These physical stresses may cause the package to break, deform, or develop leaks.

FOOD–PACKAGE INTERACTIONS

Food–package interactions are chemical and physical interactions that occur between the food and the package. Corrosion of a metal food can is an example of a chemical interaction between the food and the inner surface of the can. Mass transport between the food and the package is an example of a physical interaction—volatile compounds can move from the package into the food, a phenomenon called migration; conversely, volatile compounds or flavor from the food can be absorbed by the package, a phenomenon called scalping.

The first phenomenon, migration of volatile compounds into the food, is a greater concern because of the potential health hazard resulting from exposure to toxic migrants. The volatile compounds may arise from many sources, such as residual monomers and oligomers in the plastic packaging materials that cannot be easily eliminated during the polymerization process; residual processing aids and additives (lubricants, plasticizers, slip agents, antioxidants, light stabilizers, antistatic agents, etc.); and residual printing ink solvents used on the package. These volatile compounds may cause the food to acquire undesirable flavors or, in some cases, toxic components.

The migration of volatile compounds from microwave susceptors has received considerable attention in the past decade (8). A microwave susceptor is a metallized plastic or paper laminate that can convert the microwave energy into heat to rapidly achieve a high surface temperature (>400°F). It allows the food in contact with the hot surface to brown and crisp. The high temperature can cause degradation of the plastic layer in the microwave susceptor and result in migration of undesirable volatile compounds into the food.

The second phenomenon, scalping, may lead to diminished quality of the food. Flavor scalping is a major concern for aseptic fruit juice packages (9). Research has shown that absorption of flavor compounds may also adversely affect the barrier characteristics of the packaging material.

PACKAGING AND WASTE DISPOSAL

The disposal of postconsumer packaging has been a significant issue in the past two decades. The public concern about the disposal of food packaging heightened with the introduction of PET plastic beverage bottles because of their high visibility. The concern broadened to include the nonplastics, postconsumer packaging materials, and other materials destined for disposal through the traditional channels of landfill or incineration. There were numerous efforts by public and private groups to establish regulations for the control of packaging by such means as deposit laws, mandatory recycling, bans, and taxation.

The federal government and various industrial groups began to develop a more rational strategy to address the packaging disposal concerns. The Environmental Protection Agency (EPA) recommended a plan of integrated solid waste management, which consists of a prioritized strategy: source reduction, recycling, incineration, and landfill. The strategy provides useful guidelines for designing the food package.

REGULATIONS

Nearly every package or label is subject to some type of legislation that affects packing, shipping, selling, advertising, grading, standardizing, or marking. The regulations and compliance problems have been growing in number and complexity. Therefore, one should seek guidance of a law counsel when information on specific legal aspects of packaging is needed.

In general, the packaging-related regulations fall into three broad categories: weight and measure, adulteration, and public safety. Regulations related to weight and measure are designed to ensure that the consumer is not misled or deceived by the printing or appearance of the package. Regulations related to adulteration deal primarily with the wholesomeness of the product. These regulations prevent the use of any packaging material that might seriously affect the product by direct or indirect addition of foreign components. Regulations related to public safety, such as tamper-resistant packaging, also exist.

OTHER CONSIDERATIONS

In developing a food package, it is necessary to select the packaging machinery that is most suitable for the food product. The selection depends on the nature of the food product to be handled (size, form, flow behavior, moisture content, or fragility). The selection also depends on the speed and versatility of the packaging machinery. Speed is essential for high-volume production, and versatility is essential for production operations that require frequent changes in the package size or product.

In most cases, economic considerations determine the success or failure of the food package. The total cost is derived from material, processing, distribution, waste disposal, promotion, and research and development. Marketing research is needed to identify product life cycle, establish pricing and distribution policy, and project the potential market volume of new products. Good packaging graphics are also needed to attract customers, communicative value, and create the desire to purchase.

INFORMATION TECHNOLOGY AND PACKAGING

In the future, the integration of information technology and packaging technology is expected to deliver greater product value by providing more efficient and reliable decision making (10). Microwave reheating or cooking of prepared food has certain technical limitations that can be overcome by enabling the food, the package, and the microwave oven to share vital information. In the application shown in Figure 2, the package carries the vital informa-

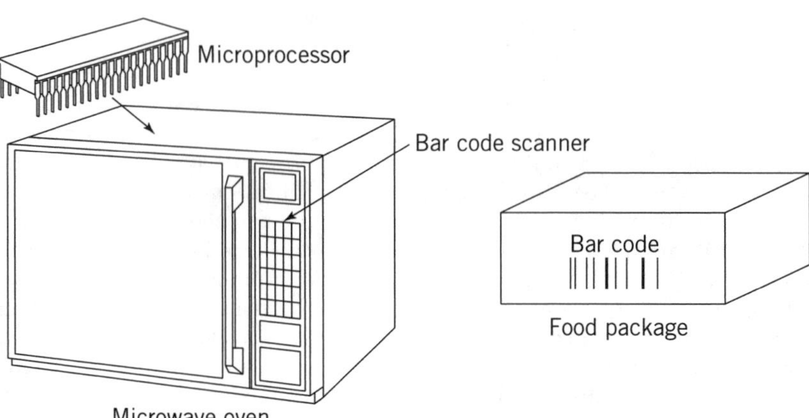

Figure 2. An application of information technology to packaging.

tion about the food and the package in a printed bar code, and the microwave oven is equipped with a bar code scanner and a microprocessor. The microprocessor, which is linked to the scanner, contains information about the microwave oven characteristics and also contains the logic to process the data from the scanner. Information sharing takes place when the bar code is scanned, and then the scanner transmits the information to the microprocessor, which in turn controls the magnetron and turntable (if applicable) in the microwave oven.

This application offers the advantages of convenience, accuracy, speed, and quality. The consumer has the convenience of automatic data entry that provides greater accuracy and speed, and it is particularly helpful with instructions involving multiple sequence cooking. The microprocessor allows the use of more sophisticated programming instructions to control the power and time of the microwave oven, which can improve the quality of the prepared food. The microprocessor can also be programmed to meet the food preferences of individual consumers.

The integrated technology can also help to facilitate handling, improve traceability, ensure safety, and reduce product loss. For example, other information devices can be incorporated into the package to enable the package to gather and share information on the time–temperature history, distribution pathway, and the quality or safety index of the product.

BIBLIOGRAPHY

1. W. E. Brown, *Plastics in Food Packaging*, Marcel Dekker, New York, 1992.
2. M. L. Rooney, *Active Food Packaging*, Blackie Academic & Professional, New York, 1995.
3. G. L. Robertson, *Food Packaging: Principles and Practice*, Marcel Dekker, New York, 1993.
4. O. R. Fennema, *Food Chemistry*, 3rd ed., Marcel Dekker, New York, 1996.
5. T. P. Labuza, "Moisture Gain and Loss in Packaged Foods," *Food Technol.* **36**, 92–97 (1982).
6. W. Soroka, *Fundamentals of Packaging Technology*, Institute of Packaging Professionals, Herndon, Va., 1995.
7. A. L. Brody and K. S. Marsh, eds., *Encyclopedia of Packaging Technology*, John Wiley & Sons, New York, 1997.
8. S. J. Risch and J. H. Hotchkiss, *Food and Packaging Interactions II*, American Chemical Society, Washington, D.C., 1991.
9. J. H. Hotchkiss, *Food and Packaging Interactions I*, American Chemical Society, Washington, DC, 1988.
10. K. L. Yam and R. G. Saba, "Intelligent Product Delivery: A Paradigm Shift in Packaging," *J. Technol. Eng.* **7**, 22–26 (1998).

Kit L. Yam
Raymond G. Saba
Y. C. Ho
Rutgers University
New Brunswick, New Jersey

PACKAGING: PART II—LABELING

Regulation of food labeling in the United States falls mainly under the jurisdiction of two federal agencies: the Food and Drug Administration (FDA) and the United States Department of Agriculture (USDA). USDA's Food Safety and Inspection Service (FSIS) oversees food labeling of products containing meat or poultry. All other food products, including certain products containing only small amounts of meat or poultry and certain products (eg, traditional sandwiches) not associated by consumers with the meat and poultry industry, fall under the jurisdiction of FDA's Center for Food Safety and Applied Nutrition (CFSAN). Although the two regulatory agencies approach many aspects of food labeling in a similar manner, occasionally they differ.

Other agencies may also have requirements that affect food labeling. For example, the U.S. Bureau of Alcohol, Tobacco and Firearms of the Treasury Department regulates labeling of alcoholic beverages. The Agricultural Marketing Service (AMS) within USDA administers grading of agricultural products.

KEY DIFFERENCES BETWEEN FDA AND USDA

FDA regulations are governed by the Federal Food, Drug, and Cosmetic Act and the Fair Packaging and Labeling Act (FPLA). (The Nutrition Labeling and Education Act of 1990 [NLEA] is an amendment to the Food, Drug and Cosmetic Act.) The Federal Meat Inspection and Poultry Products Inspection Acts define USDA's regulatory role and responsibilities. Because the two agencies are governed by separate laws, they have different missions, philosophies, and approaches to food labeling, resulting in subtle differences throughout the regulations. USDA plays a more "hands-on" role (ie, through plant inspections and label preapproval) in monitoring food production and packaging practices (Table 1).

Mission and Approach

FDA's regulatory history began with the passage in 1906 of the Pure Food and Drugs Acts, which instituted federal oversight of food labeling. The 1906 act was replaced in 1938 by the current law. In 1973, FDA initiated a voluntary nutrition-labeling program that represented a regulatory turning point—the inclusion of meaningful health information on food labels.

Under the Federal Meat Inspection and Poultry Products Inspection Acts, USDA's primary regulatory role has been to prevent public health hazards resulting from improper handling of meat and poultry during production and packaging. USDA has an extensive network of field offices responsible for conducting plant inspections, and the national office administers a label preapproval program.

Regulatory Documents and Guidance

Both FDA and USDA release regulatory changes in the *Federal Register (FR)*, published daily. Each year all fed-

Table 1. Key Differences Between FDA and USDA

FDA	USDA
Mission and approach	
Early emphasis on truth in labeling	Emphasis on prevention of public health hazards from improper handling
Broader span of regulatory oversight, limited resources	Extensive network of field offices available for consulting and monitoring
Regulatory documents and guidance	
Federal Register	*Federal Register*
Title 21 of the *Code of Federal Regulations*	Title 9 of the *Code of Federal Regulations*
Compliance policy guides	Policy manual and policy memos
Label preapproval	
No label preview or approval	Label designs reviewed and approved
Mistakes identified through market surveys, and warning letters issued	Plant inspections monitor application of approved labels
Nutrition labeling	
Mandated by NLEA	Exempted from NLEA
Spearheaded nutrition-labeling changes	Followed FDA regulations in general
Food categories regulated	
All packaged foods containing less than 2% cooked meat or poultry, or 3% or less raw meat or poultry	Food containing 2% or more cooked meat or poultry, or more than 3% raw meat or poultry
Traditional sandwiches	Fresh meats
Pizzas not containing meat or poultry	Fresh poultry
Dairy products	Processed meats and poultry
Eggs	Soups with meat or poultry
Fish and seafood	Salads with meat or poultry
Fruits	Pizzas with meat or poultry
Vegetables	Meat or poultry snacks
Soups not containing meat or poultry	Mixed dishes containing meat or poultry (eg, chili, lasagna, TV dinners, stews)
Salads not containing meat or poultry	
Nuts and legumes	

eral regulations are updated and compiled in the *Code of Federal Regulations* (*CFR*); FDA labeling regulations appear in Title 21 and USDA regulations in Title 9 of the *CFR*. Some USDA regulations cross-reference FDA regulations. FDA has issued Compliance Policy Guides, which reflect the agency's interpretations of the regulations and policies for enforcement. USDA publishes a standards and labeling policy manual and periodic labeling policy memos, which address similar issues. Both agencies provide consultation; however, manufacturers of USDA products obtain far more agency interaction and guidance through the label preapproval program and plant inspections.

Label Preapproval

USDA administers the label preapproval program in which food label designs traditionally have been reviewed and approved by the agency before they are printed. If manufacturers make mistakes, USDA tells them what is wrong so that manufacturers are able to learn the regulations and rely on the agency's guidance. Plant inspectors examine labels to make sure they correspond to the approved design.

FDA does not preview labels and offers no approval process. It places the responsibility on the manufacturer to understand the regulations and correctly apply the rules. The agency typically does not find mistakes until labels are printed and products are in the marketplace.

Nutrition Labeling

In 1990, Congress passed the NLEA, which made nutrition labeling mandatory on most packaged food products and mandated that FDA initiate extensive changes in the content and format of the nutrition label.

Nutrition labeling of USDA products was not included in the congressional mandate, but USDA decided to follow FDA to prevent consumer confusion arising from two different nutrition labels. In most significant aspects of the nutrition-labeling regulations, the two agencies have the same requirements; however, there are some differences.

USDA Special Requirements

USDA product labels are subject to additional requirements, some of which also pertain to select FDA products.

Handling Instructions. At times, FDA food products are required to carry warning statements or handling instructions. USDA product labels usually carry special instructions to ensure the safety and quality of the product (eg, keep refrigerated, keep frozen). In addition, raw meat and poultry products must carry safe-handling instructions with specified statements and graphics.

Inspection legends. All USDA products also must bear an official inspection legend in accordance with established inspection procedures.

Voluntary Grading Services

Voluntary grading services are available for certain products under both FDA and USDA jurisdiction. USDA's Agricultural Marketing Service (AMS) performs grading services for certain products (eg, dairy, egg, fruit, vegetable, meat, poultry) subject to FDA or USDA regulation. To use approved grade designations (eg, U.S. Grade A, U.S. No. 1, U.S.D.A. Prime, U.S. Fancy), a product must have been inspected and determined to comply with grade standards established by AMS.

Principal Display Panel

A principal display panel (PDP) must be included on all food sold in packaged form. The PDP is the part of the label that is most likely to be displayed for retail sale. The product identity statement and net quantity of contents declaration must be included on the PDP. At times, other required statements must also be displayed on the PDP.

Product Identity Statement. The product identity statement is intended to communicate important information to the consumer about the type and form of food contained in the product package. It should be incorporated into the PDP with sufficient prominence that the consumer can readily view it on a retail shelf. The terminology, as well as the placement and type size, of the identity statement are dictated by the regulations.

Net Quantity of Contents. The purpose of the net quantity of contents declaration is to tell the consumer the amount of product the package contains.

Information Panel

The information panel typically is contiguous to and immediately to the right of the PDP as viewed by the consumer. It includes detailed information about a food product that helps the consumer make knowledgeable purchasing decisions. The mandatory information required on the information panel consists of (*1*) the ingredient list, (*2*) the nutrition facts, and (*3*) the manufacturer identity statement. At times, other information may be required, such as warning statements, a percent-juice declaration, and claims-related statements. Mandatory information may be displayed on the PDP instead of on the information panel.

Ingredient List. A list of the ingredients of a multi-ingredient product must be included on the food label, typically on the information panel.

Nutrition Facts Information. With the introduction of the NLEA, the format of nutrition information has become much more strictly defined. A number of nutrition facts layouts are available, and specific criteria dictate their use.

Manufacturer Identity Statement. The manufacturer identity statement tells the consumer the name and place of business of the manufacturer, packer, or distributor. This information usually is placed on the information panel or on the PDP if there is no information panel.

Other Information. At times, additional statements may be required on the information panel, the PDP, or both. For example, if a PDP bears a comparative claim, such as "light" or "reduced," quantitative information comparing relevant nutrients in the product to those in a reference food must be included on the information panel or the PDP. Some products may require a warning statement, percent-juice declaration, or special handling instructions.

Exemptions

The regulations apply to retail primary packages, and so certain containers and wrappings are exempt from the labeling requirements. The regulations define the term "package" as any container or wrapping in which a food is enclosed for retail sale. Packages used for shipping, display, or other functions are exempt. In addition, some very small packages are exempt.

Inner Wrappings. When multiunit packages of food are not intended to be sold individually, their inner wrappings that bear no information are exempt from labeling requirements.

Shipping Cartons. Shipping cases for distribution of commodities in bulk are exempt from labeling requirements. If the shipping cartons will be sold as retail units (eg, in a warehouse club), however, all mandatory requirements apply.

Small Confections. Small confections (eg, "penny candy") weighing less than 1/2 oz (or 15 g) are exempt from all labeling requirements when the shipping container or retail package meets the labeling requirements.

PRINCIPAL DISPLAY PANEL

All retail food packages must bear a principal display panel (PDP) that includes the following mandatory label information: a product identity statement and a declaration of net quantity of contents. If there is no information panel, an ingredient list, a nutrition facts panel, and a manufacturer identity statement must also appear on the PDP. On some products, additional information may be required, such as claims-related information, a warning statement, or special handling instructions.

Location

The PDP is considered the area of the package that is most likely to be displayed, presented, or examined by the consumer at the point of sale. The PDP may be a spot label on a package surface, or it may be the whole surface. On jars and bottles, a spot label on the lid may be used for the PDP. A header strip attached across the top of a transparent or opaque pouch that contains no other printed or graphic material may serve as the PDP.

It is acceptable to include more than one PDP on a package design, but when this approach is used, all of the mandatory information must be duplicated in full.

Size

The PDP must be large enough to accommodate all mandatory label information with clarity and conspicuousness, and without obscuring or crowding by designs or vignettes. The size of the PDP determines the minimum letter height for some of the mandatory information contained in the

PDP. The area of the PDP is the same in similar-size packages, regardless of the size of the label. It is not the size of the label but the size of the container that determines the area of the PDP. For example, if a product carries a spot label, it is not the size of the label that determines the area of the PDP, but rather the area of the surface displayed to the consumer.

Rectangular Packages. For rectangular packages, one entire surface of the package is considered the PDP. Typically the PDP appears on the largest surface of the package. Multiply the height times the width of that surface.

$$\text{Area of PDP} = \text{height} \times \text{width}$$

Cylindrical Containers. For cylindrical (or nearly cylindrical) containers, such as jars, bottles, and cans, the PDP is considered 40% of the height multiplied by the circumference of the container. If a spot label is used, the area of the PDP is still determined by this formula.

$$\text{PDp area} = (\text{height} \times \text{circumference}) \times 0.40$$

Cone-Shaped Containers. For cone-shaped containers, such as jars, bottles, cans, and tubs, the PDP is considered 40% of the height multiplied by the circumference of the container. To determine the circumference, measure both the widest and the narrowest circumferences and average the two numbers.

$$\text{PDP area} = \text{height} \times \left(\frac{\text{widest circumference} + \text{narrowest circumference}}{2}\right) \times 0.40$$

Irregularly Shaped Packages. For containers of other shapes, 40% of the total surface of the container is considered the PDP. If there is an obvious surface to serve as the PDP (eg, top of a triangular or circular package of cheese), this entire surface should be measured. For extremely irregular containers, substitute an easily measured container of the same capacity.

$$\text{PDP area} = \text{total surface area} \times 0.40$$

PRODUCT IDENTITY STATEMENT

The product identity statement communicates what the product is. It includes the name of the food as well as any other defining characteristics, such as the form of the food, if it is an imitation food, or if it contains certain ingredients.

Terminology

The product name, typically a standard name or a common or usual name of the food, is used for the product identity statement (eg, orange juice). Sometimes an appropriately descriptive term (eg, taco seasoning) or a fanciful name (eg, Coca-Cola®) commonly used by the public for the food is allowed.

Form of Food. If a food is marketed in various forms (eg, whole, sliced, diced), the form of the food is a necessary part of the product identity statement. If the form of the food is visible through the container (eg, whole pickles in a clear jar) or it is depicted by an appropriate vignette, the particular form does not need to be stated.

Imitation Foods. If a food is an imitation food, the word "imitation" must be used as part of the product identity statement, immediately preceding the name of the food.

Juice-Containing Beverages. Special rules affect how beverages containing fruit or vegetable juices may be named. The following rules highlight some of the issues to consider:

- If a product contains less than 100% but more than 0% juice *and* the product name includes the word "juice," the name must also include a term such as "beverage," "cocktail," or "drink."
- If a product name specifically identifies a juice that has been reconstituted, the name must also include a qualifying term, such as "from concentrate" or "reconstituted."
- On a 100% juice *or* a diluted juice product, if specific juices are identified as part of the product name (or elsewhere outside of the ingredient list), either the juices must be listed in descending order of predominance by volume, or their relative predominance must be shown by other means. For example, a juice name could be combined with "flavored" to indicate a nonpredominant juice.
- When a label for a multiple-juice product represents that particular juices are present but does not identify all juices in the product, the label must reveal that other juices are present (eg, ". . . in a blend of two other juices").

In addition to the product name requirements, beverages that contain, or appear to contain, fruit or vegetable juice are subject to requirements for percent-juice labeling.

Characterizing Ingredients. Labels on some products may be required to include information about characterizing ingredients in the product identity statement. An ingredient is considered a characterizing ingredient when the proportion of an ingredient present in a food has an influence on the price or consumer acceptance of the product, or when labeling may create an erroneous perception (eg, pictorials depicting an ingredient).

The required information may be:

- The percentage of a characterizing ingredient
- The presence or absence of a characterizing ingredient
- The need for the consumer to add a characterizing ingredient

For example, foods packaged for use in preparing main dishes or dinners (eg, spaghetti dinner kit) require information about the need to add ingredients.

Characterizing Flavors. If the product labeling, advertising, or both make a direct or an indirect representation of any recognizable flavors through words or pictorial, the flavor is considered a characterizing flavor. A characterizing flavor must be incorporated into the product identity statement. For example, depiction of vanilla beans on an ice cream carton would trigger the inclusion of "vanilla" in the product name (eg, vanilla ice cream). The word "flavored" or "artificial" may have to accompany the name of the flavor in certain circumstances.

Design Elements

In addition to dictating the terminology, the regulations outline location, placement, and type-size requirements for the product identity statement. When other required information is incorporated into the product identity statement (eg, form of food, "imitation," "reconstituted," characterizing ingredients, characterizing flavors), additional design requirements apply.

Product Identity Statement. The product name must be in conspicuous type, located prominently on the PDP. The type size should be reasonably related to the largest printing on the PDP.

Form of Food. When language describing the form of the food is required, the letters should be in a type size bearing a reasonable relation to the type size of the other components of the product identity statement.

Imitation. If "imitation" is required, it must be in the same type size and prominence as the name of the food.

Reconstituted. When a term such as "reconstituted" or "from concentrate" is required as part of a juice beverage name, the letters may be no smaller than half the height of the letters in the name of the juice.

Characterizing Ingredients. Percent ingredient declarations must be in an easily legible, prominent type that distinctly contrasts with other printed or graphic material.

Characterizing Flavors. When a characterizing flavor must be included in the product name, the letters must be no smaller than half the height of the letters in the product name. If the word "natural," "artificial," or "flavored" is included, it must be no smaller than half the height of the letters in the characterizing flavor.

DECLARATION OF NET QUANTITY OF CONTENTS—FDA

A declaration of net quantity of contents, which generally must appear on all PDPs, provides information about the amount of product contained in the package. If there are alternate PDPs, this declaration must appear on each PDP.

Metric Labeling

The current regulations are based on the inch-pound system of measures (avoirdupois system) and also metric measures (International System of Units [SI]).

Selecting the Proper Unit. The net quantity of contents may be expressed as weight, a fluid or dry measure, or a numerical count. Units used in declaring the net quantity vary depending on the form of the food. If the food is sold in liquid form, a fluid measure must be used. If the food is solid, semisolid, viscous, or a mixture of solid and liquid, a weight measure must be used. Fresh fruits, vegetables, and other dry commodities may be labeled with a dry measure. If there is a firmly established consumer usage and trade practice for declaring the contents of a liquid product by weight—or solid, semisolid, or viscous product by fluid measure—it may be used. Similarly, if there is a firmly established consumer usage and trade practice of declaring net quantity by numerical count, linear measure, or area measure, it may be used and possibly augmented by a weight or fluid measure. It is *never* acceptable to use any adjective qualifying the unit (eg, jumbo quart, full gallon).

Using the Correct Term. When the net quantity is declared as a weight measure, the terms "net weight" and "net" are used. Net weight is used when the inch-pound declaration is first. If a fluid measure is used, then either net or net contents is used.

Converting to Metric. To assure uniformity and precision in net quantity declarations, the regulations provide conversion factors to use in declaring the inch-pound and metric amounts.

Largest Whole Unit. The net quantity of contents should be declared in the largest whole unit, with any remainder expressed as a decimal or a common fraction of the unit. Alternatively, the remainder may be expressed in terms of the next smaller whole unit and any decimal or common fraction of that unit. Common fractions are halves, quarters, eighths, sixteenths, or thirty-seconds. Common fractions must be reduced to the lowest (simplest) terms. Decimal fractions should be carried to no more than three places.

Dual Ounce-Pound Declaration. The current net quantity of contents regulations require a dual ounce-pound declaration on packages weighing between 1 and 4 lb or containing between 1 pt and 1 gal. This dual declaration requirement results in a statement of ounces followed by a parenthetical pound-ounce declaration (eg, 18 oz [1 lb 2 oz]) or a statement of fluid ounces followed by a parenthetical declaration in largest whole fluid unit with the remainder in fluid ounces (eg, 36 fl oz [1 qt 4 fl oz]). Under the metric labeling regulations, this dual ounce-pound declaration requirement is eliminated. Including a dual ounce-pound declaration would be optional; however, if used, it must appear on one line and may precede or follow the metric declaration.

Abbreviations. The regulations list the only abbreviations that may be used in the declaration of net quantity of contents (Table 2). Generally, periods and plural forms are optional.

Design Elements

The regulations specify a number of design requirements for the net quantity of contents declaration. It must appear as a distinct item in the lower portion of the PDP, be surrounded by white (ie, blank) space, and meet a minimum type-size requirement based on the area of the PDP.

Location. The net quantity of contents declaration must appear within the bottom 30% of the PDP. If the PDP is 5 square inches or less, any available space on the PDP may be used. The net quantity of Contents declaration must be placed in lines parallel to the base on which the package rests as it is displayed. It may appear on more than one line.

Copy-Free Area. The net quantity of contents declaration must be separated from other printed matter. Spacing requirements do not apply to pictorials or graphics, provided they do not render the declaration inconspicuous. The declaration may be placed closer to the extreme lower border than the space prescribed for below the statement.

Type Style. The net quantity of contents declaration must be in conspicuous type that is prominent and easily legible. The declaration must appear in distinct contrast (by typography, layout, color, embossing, or molding) to other matter on the package, *unless* it is blown, embossed, or molded on a glass or plastic surface, in which case the contrast is not mandatory.

Type Size. To ensure that all packages of the same size will have the same size net quantity of contents, the regulations define minimum type-size requirements based on the area of the PDP. To determine the minimum type size for the net quantity of contents, measure the area of the PDP (Table 3).

Table 2. Abbreviations for Net Quantity of Contents

Unit	Abbreviation
Weight	wt
Pint	pt
Ounce	oz
Quart	qt
Pound	lb
Fluid	fl
Gallon	gal
Kilogram	kg
Gram	g
Milligram	mg
Microgram	mg
Cubic Centimeter	cm^3
Liter	L or l
Milliliter	mL or ml

Special Labeling Provisions

The regulations allow for modification of the net quantity of contents declaration in certain circumstances. In some cases, a modified declaration is required (eg, multiunit packages). In many other cases, the requirements are simply relaxed to offer more flexibility in package design.

Multiunit Packages. Packages containing two or more individually packaged, identical units that may also be sold individually have special labeling requirements. The net quantity of contents declaration must be on the outside of the package if the individual labeling is obscured by the outer packaging. The net quantity declaration must include the number of individual units, the quantity of each individual unit, and, in parentheses, the total quantity of the multiunit package. The total quantity optionally may be preceded by "total" or "total contents."

Bulk Foods. Food sold in bulk containers at the retail level is exempt from declaring the net quantity of contents, provided it is accurately weighed, measured, or counted within the purchaser's view or in accordance with the purchaser's order.

Random-Weight Packages. Packages from one lot or shipment of the same commodity with no fixed weight pattern are random-weight packages (eg, cheese cut from a bulk block). If the product label contains the net weight, price per pound (or kilogram), and total price, the requirements for type size, metric declaration, location, and copy-free area are waived. In addition, the weight may be declared in decimal fractions of a pound even if the weight is less than 1 lb.

Individual Servings Not Intended for Retail Sale. Small, individual-serving containers for use in restaurants, institutions, and passenger carriers that are not intended for retail sale are exempt from declaring the net quantity. These packages must be less than 1/2 oz or 1/2 fl oz.

Shell Eggs. Cartons containing 12 shell eggs are exempt from placing the net quantity in the bottom 30% of the PDP.

Other Provisions. Special allowances are made for certain categories of food packaged in specific containers. These exceptions provide flexibility in package design and conformity to established trade practices.

Table 3. Type-Size Requirements for Net Quantity of Contents

Area of more than	The PDP less than or equal to	Minimum type size letter height not less than
	5 sq in	1/16 inch (1.6 mm)
5 sq in	25 sq in	1/8 inch (3.2 mm)
25 sq in	100 sq in	3/16 inch (4.8 mm)
100 sq in	400 sq in	1/4 inch (6.4 mm)
400 sq in		1/2 inch (12.7 mm)

Deviation from Declaration in Largest Whole Unit. Due to established trade practices that preceded the regulations, a variety of products may declare 8-fl-oz and 2-q volumes as 1/2 p and 1/2 gal, respectively, rather than following the rule for declaring the quantity in the largest whole unit. Products to which this deviation applies include bottled water, fruit juice beverages, ice cream and frozen desserts, and milk and fluid dairy products.

DECLARATION OF NET QUANTITY OF CONTENTS—USDA

Units and Terminology

As with FDA, units used in declaring the net quantity of USDA foods depend on the form of the food. Because most USDA foods are solid, semisolid, viscous, or a mixture of solid and liquid, a weight measure must be used. A food sold in liquid form requires a fluid measure (eg, chicken soup). However, firmly established consumer usage or trade practice for declaring the contents of a liquid product by weight—or solid, semisolid, or viscous product by fluid measure—may be used.

When the net quantity is declared in a weight measure, the term "net weight" or "net wt." is used. If a fluid unit is used, either "net contents" or "contents" should be used.

Dual Declaration

On certain-size packages, USDA requires that the net quantity be declared in ounces followed in parentheses by the quantity in pounds with any remainder expressed in ounces of a pound (similar requirements pertain to fluid measures). This is referred to as dual declaration. FDA eliminated its dual declaration requirement when it introduced metric labeling. USDA also requires metric declaration in addition to ounce-pound.

Abbreviations

USDA does not designate abbreviations that may be used in the net quantity declaration. Therefore, companies may use the FDA-defined abbreviations or other abbreviations.

Special Labeling Provisions

USDA allows for modification of the net quantity declaration in specific cases. USDA requires a modified declaration on multiunit packages. Some USDA requirements are relaxed to accommodate certain packaging situations and established trade practices.

Multiunit Packaging. USDA packages that contain two or more individually packaged, identical units that may also be sold individually have special labeling requirements. The net quantity declaration must be on the outside of the package if the individual labeling is obscured by the outer packaging. The declaration must include the number of individual units, the quantity of each individual unit, and, in parentheses, the total quantity of the multiunit package in ounces or fluid ounces. The dual declaration requirement is waived.

Random-Weight Packages. USDA packages from one lot or shipment of the same commodity with no fixed-weight pattern are considered random-weight packages (eg, ground beef). The net quantity declaration must be applied to random-weight consumer packages before retail sale; however, the declaration is exempt from type-size, dual declaration, and placement requirements, provided it appears conspicuously on the PDP.

Small Packages. If the shipping carton of individually wrapped, small packages (less than 1/2 oz) carries an accurate net quantity declaration, the USDA net quantity requirements are waived for the individual packages. When these small packages bear the net weight, price per pound, and total price, they are exempt from type-size, dual declaration, and placement requirements, provided the net quantity appears conspicuously on the PDP.

Margarine. Some margarine products contain animal fat and fall under USDA jurisdiction. Margarine in 1-lb rectangular packages (with the exception of whipped and soft margarine, or packages containing more than 4 sticks) are exempt from the requirements of placement in the bottom 30% of the PDP and the dual declaration. The net quantity declaration must appear as "1 pound" or "one pound" in a conspicuous manner on the PDP.

Sliced, Shingle-Packed Bacon. USDA exempts certain-size packages (ie, 8-oz, 1-lb, and 2-lb rectangular packages) of sliced, shingle-packed bacon from placement of the net quantity in the bottom 30% of the PDP and from dual declaration. These products are placed on boards that wrap around the top and serve as the PDP. Due to the space limitations of the PDP, these products have been granted more flexibility in incorporating the net quantity declaration. The declaration must appear conspicuously on the PDP.

In addition, sliced, shingle-packed bacon in any other size container (ie, other than 8-oz, 1-lb, and 2-lb) must show the net quantity declaration with the same prominence as the most conspicuous feature on the package.

CLAIMS-RELATED STATEMENTS

When a product includes a claim, a number of elements must be incorporated into the package design. The regulations do not specify where a claim must be placed; however, most claims appear on the PDP because it is the most conspicuous panel.

Types of Claims

The regulations define two categories of claims: nutrient content claims (sometimes referred to as "descriptors") and health claims. Nutrient content claims are statements about the level of a nutrient in a food (Table 4). Health claims, on the other hand, link the nutrient profile of a food to a health or disease condition (Table 5). Nutrient content claims are used widely by the food industry. Because the regulations governing health claims are more complicated and restrictive, health claims are not as common.

Nutrient Content Claims. Nutrient content claims characterize the level of a nutrient in a food. Only defined terms may be used on the label to describe a food's nutrient content. When these terms are used, a product must meet specific criteria. Nutrient content claims may be expressed or implied, comparative or absolute. These distinctions are important in package design because the type of claim dictates the information required.

Expressed Claim. Any direct statement about the level or range of a nutrient in a food is considered an expressed nutrient content claim. Specific terms are defined by regulation.

Implied Claim. An implied claim is any statement that leads the consumer to assume that a nutrient is absent or present in a certain amount or that the food may be useful in achieving dietary recommendations. Implied claims must follow the same requirements defined for expressed claims.

Comparative (or Relative) Claim. A claim comparing the level of a nutrient in one product to the level of that nutrient in another product or class of foods is considered a comparative claim (also called a relative claim). For example, "light," "reduced," "less," and "more" are comparative claims.

Absolute Claims. In contrast to comparative claims, absolute claims make a statement about the nutrient level in a food without stating or implying any comparison to another product. "Free," "low," "very low," "high," and "source of" are examples of absolute claims.

Table 4. Nutrient Content Claims

Comparative	Absolute
Light/Lite	Free
Reduced	Low
Less	Very Low
More	High
	Source of
	Healthy
	Lean
	Extra Lean

Table 5. Health Claims

Calcium and osteoporosis
Sodium and hypertension
Dietary saturated fat and cholesterol and risk of coronary heart disease
Dietary fat and cancer
Fiber-containing grain products, fruits, and vegetables and cancer
Fruits, vegetables, and grain products that contain fiber, particularly soluble fiber, and risk of coronary heart disease
Fruits and vegetables and cancer
Folic acid and neural tube defects
Sugar alcohols and dental caries

Health Claims. FDA has defined 9 health claims which may be used on labels (Table 5), and has created very strict criteria for using these claims. No other nutrient/disease associations may be made in food labeling. Written statements, third-party references, use of certain terminology in a brand name, symbols, and pictorials may be considered a health claim if the context in which they are presented either suggests or states a relationship between a nutrient and a disease. When a statement, symbol, pictorial, or other form of communication suggests a link between a nutrient and a disease, it is considered an implied health claim, and it is subject to all the requirements for health claims.

Nutrient Content Claims

When a product bears a nutrient content claim (whether it is expressed or implied), certain information must be incorporated into the design of the package. Because many claims appear on the PDP, this panel usually is affected. When a claim appears elsewhere on a package, however, certain information must appear on the panel with the claim.

Required Statement. Depending on the type of claim and whether the product is governed by FDA or USDA, the requirements for claims-related information vary. FDA mandates more extensive information than USDA (ie, the inclusion of either a referral or a disclosure statement). Both agencies require additional information on products bearing comparative claims.

Claims on FDA Products. All FDA products bearing nutrient content claims require either a referral or a disclosure statement to be included on each panel where a claim appears, except the nutrition facts panel. A referral statement, appearing next to the largest claim on each panel, directs the consumer to the nutrition facts statement. When a product contains excessive levels of key nutrients that are associated with health risks, the referral statement is replaced with a disclosure statement that flags the nutrient(s) of concern and directs the consumer to the nutrition facts panel.

Comparative Claims. Both FDA and USDA product labels that bear a comparative claim must include a nutrient claim clarification statement and quantitative information. On FDA products, this information is required in addition to a referral or a disclosure statement.

The nutrient claim clarification statement identifies the comparison food and states the percentage (or fractional) difference in the subject nutrient(s) between the product and its comparison food (eg, 50% less fat than [comparison food], 1/3 fewer calories than [comparison food]).

The quantitative information provides the absolute amounts of the subject nutrient(s) in the product and in the comparison food.

Design Elements. USDA and FDA differ slightly in their design requirements for claims-related information. USDA does not require referral or disclosure statements; FDA applies the type-size standards defined for these statements to other claims-related statements.

Claims Statements. Claims may not have undue prominence because of type style in comparison to the product identity statement. A claim may be no longer than two times the size of the product identity.

Referral or Disclosure Statement. On FDA products, a referral or a disclosure statement must be immediately adjacent to the claim. No intervening material may be placed between it and the claim, except for other claims-related statements, or a standard name that is modified by the claim, or both. It must appear on each panel where the claim is located, except for the panel that contains the nutrition facts (eg, if the nutrition facts and the claim are both on the PDP, the statement is not required). If multiple claims appear on a panel, the referral or disclosure statement must be adjacent to the largest claim.

It must be in easily legible, boldface type, in distinct contrast to other printed or graphic matter. It may be no smaller than the net quantity declaration, unless the claim is less than two times the size of the net quantity declaration.

Nutrient Claim Clarification Statement. The nutrient claim clarification statement must be placed in immediate proximity to the most prominent comparative claim. It must follow the same type-size requirements as the referral statement. If different comparative claims appear on the same label, the nutrient claim clarification statement must be placed in immediate proximity to the most prominent presentation of each claim.

Quantitative Information. Clear and concise quantitative information must appear adjacent to the most prominent claim or on the Information Panel.

Health Claims

When a product bears a health claim, very specific language must be used on the label. The regulations governing health claims are very complex; therefore, health claim situations should be handled on a case-by-case basis. The language associated with a health claim must conform to regulatory guidelines, which provide model health claim statements. Products bearing health claims must undergo careful review by legal experts, regulatory experts, or both to be certain the language is accurate and any graphics are acceptable.

Model Statements. The regulations pertaining to each health claim outline the assertions that may be made and any additional required statements. When a claim is implied through graphic representations, a complete claim statement must be included on the label.

Design Elements. All information required in a health claim must appear in one place, in the same type size, and without other intervening material. Because the required statements can be quite lengthy, the complete claim may appear on a back or side panel. When this approach is taken, a reference statement may be placed on the PDP, flagging the claim and directing the consumer to the location of the claim (eg, "See ______ for information about the relationship between ______ and ________"). The first blank contains the location of the health claim (eg, back panel, attached pamphlet), the second blank states the nutrient, and the third blank names the disease or health-related condition.

When any graphic material implying or expressing a health claim is used on the label or in accompanying labeling materials (eg, pamphlet), the entire claim statement or a reference statement must appear in immediate proximity to the graphics.

ADDITIONAL USDA LABELING REQUIREMENTS

The USDA has additional labeling requirements: official inspection legend and keep frozen/refrigerated statements.

Official Inspection Legend

The PDP of all USDA products must include an official inspection legend in accordance with established inspection procedures. The number of the official establishment (ie, the identity of the plant at which the product was manufactured) must also be included in one of the following manners:

- Within the official inspection legend
- Outside the official inspection legend, but elsewhere on the label (eg, on the lid of a can). When it is placed outside the inspection legend, the prefix "EST." must precede the establishment number. It must be shown in a prominent and legible manner in a size that ensures easy visibility and recognition.
- Off the exterior of the container (eg, on a metal clip used to close casings or bags) or on other packaging material in the container (eg, on aluminum pans and trays within the container). When it is placed on the exterior of the container or on other packaging material, a statement of its location must be printed continuous to the official inspection legend.
- On an insert label placed under a transparent covering, if it is clearly visible and legible and accompanied by the prefix "EST.".

Keep Frozen/Refrigerated Statements

Many USDA products require special instructions to ensure the safety and quality of the product. These products may include either of the following statements: "Keep Frozen" or "Keep Refrigerated." The statement must appear on the PDP in a conspicuous manner. There are no minimum-size requirements.

Other Required Information

On some products, additional information may be required on the PDP (Table 6). Products bearing claims have additional requirements that usually affect the design of the PDP. Other products require warning statements, some of which must be placed on the PDP. Certain products require special handling instructions to ensure the safety and

Table 6. Additional Information That May Be Required on the PDP

Information required on some products	FDA location	USDA location
Claims statements		
Disclosure or referral statement	Adjacent to largest claim on each panel	Not required
Nutrient claim clarification	Adjacent to most prominent claim	Adjacent to most prominent claim
Quantitative information	Adjacent to most prominent claim or on information panel	Adjacent to most prominent claim or on information panel
Safe-handling instructions for raw meat and poultry	Not required	Information panel or other panel
Percent-juice declaration	Information panel and/or PDP	Not required
Warning statements	Requirements vary	Requirements vary

quality of the product. When these statements are required, the regulations dictate the size and placement of the information.

INFORMATION PANEL

A package may be designed to include an information panel where some of the mandatory information may be placed (Table 7). If the ingredient list and manufacturer identity statement do not appear on the PDP, the package must have an information panel that includes these statements. Whenever practical, the nutrition facts statement should appear on the information panel with these statements. Other label statements required on some products (eg, a percent-juice declaration, some warning statements, special handling instructions) also may appear on the information panel.

Statements intended for the information panel may alternatively appear on the PDP if space on the information panel is insufficient. Also, it is permissible to omit an information panel, incorporating all mandatory statements on the PDP. Exemptions from labeling requirements for nonretail packages apply to the components of the information panel as well as to the PDP.

Location

The information panel generally is considered the part of the package immediately contiguous to and to the right of the PDP as observed by an individual facing the PDP. For situations when this arrangement is not possible, alternative locations are specified.

NUTRITION FACTS

Only when the nutrition facts statement (Figure 1) cannot be accommodated on the information panel or the PDP along with other mandatory information may it be placed elsewhere on the package. Incorporating it into the information panel (or PDP) should always be explored first.

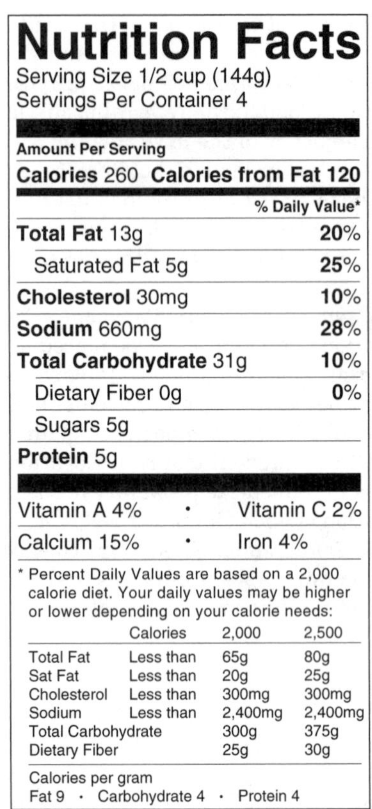

Figure 1. Typical nutrition facts label.

Table 7. Information Panel

Mandatory information	Additional required information
Nutrition facts	Percent-juice declaration
Ingredient list	USDA safe-handling instructions
Manufacturer identity	Special handling instructions
	Warning statements

INGREDIENT LIST

All products fabricated from two or more ingredients must include an ingredient list on either the PDP or the information panel.

Terminology

Ingredients are listed by their common or usual name in descending order of predominance by weight. Ingredients present in amounts of 2% or less may be listed at the end of the ingredient list following an appropriate qualifying statement (eg, "Contains ____ percent or less of ____, less than ____ percent of ____").

Design Elements

The ingredient list must be conspicuous and in a type size no smaller than 1/16 in.

Standardized Foods

With the inception of NLEA, standardized foods now typically must declare all ingredients.

Special Provisions for Bulk Foods

A food received in bulk containers at retail may provide the ingredient list by either displaying it on the bulk container or posting a counter card or sign with the required information.

MANUFACTURER IDENTITY

The manufacturer identity states the name and place of business of the manufacturer, packer, or distributor. The information panel typically includes this information.

Terminology

If the manufacturer is a corporation, the corporation name must be used and may be preceded or followed by the name of the particular division. If the manufacturer is an individual, a partnership, or an association, the name under which the business is conducted must be used. If the food is not manufactured by the person whose name is on the label, the name must be qualified with "Manufactured for ______," "Distributed by ______," or other wording that expresses the relationship.

The statement must include the street address, city, state, and postal code. If the name is in a current telephone directory, the street address may be omitted. The principal place of business may replace the actual place where the food was manufactured, packed, or distributed, assuming this would not be misleading.

PERCENT-JUICE DECLARATION

With certain exceptions, any beverage that purports or appears to contain a fruit or vegetable juice must declare the percentage of juice in the product. Any of the following situations could trigger this requirement: using the name (or variation) of a fruit or vegetable in advertising, labels, or labeling; depicting a fruit or vegetable in a vignette or other pictorial representation; or formulating a beverage to contain the color and flavor of a fruit or vegetable juice.

Percent-juice labeling (Table 8) is generally required if a beverage has the appearance and flavor of containing a fruit or vegetable juice, even if the product in fact contains no juice. Certain exceptions are made for products containing minor amounts of juice (typically less than 2%) that are labeled with a fruit or vegetable name and a term such as "flavored."

Terminology

The percentage of juice must be declared, and the type of juice may be declared. If a product does not contain juice but the labeling or color and flavoring suggest that it does, it must be declared as 0% (or a similar phrase).

Table 8. Juice Content Information

If beverage contains juice	If beverage does not contain juice
Contains __ percent (or %) juice	Contains 0% juice
Contains __ percent (or %) (*type*) juice	Contains 0% (*type*) juice
__ percent (or %) juice	Does not contain (*type*) juice
__ percent (or %) (*type*) juice	Contains no (*type*) juice
	Contains no fruit/ vegetable juice

USDA SAFE-HANDLING INSTRUCTIONS

USDA has regulations requiring safe-handling instructions (Table 9) on all raw meat and poultry products not intended for further processing at another USDA-inspected establishment. These safety instructions are intended to heighten consumer awareness of proper food-sanitation procedures and reduce the incidence and severity of foodborne illnesses.

Terminology

The regulations specify the exact wording that must be used.

Heading. The safe-handling information must be presented on the label under the heading "Safe Handling Instructions."

Rationale Statement. "This produce was prepared from inspected and passed meat and/or poultry. Some food products may contain bacteria that could cause illness if the product is mishandled or cooked improperly. For your protection, follow these safe handling instructions."

Safe Handling Statements. "Keep refrigerated or frozen. Thaw in refrigerator or microwave." Any portion of this statement that is in conflict with the product's specific handling instructions may be omitted (eg, instructions to cook without thawing).

"Keep raw meat and poultry separate from other foods. Wash working surfaces (including cutting boards), utensils, and hands after touching raw meat or poultry."

Table 9. Safe-Handling Instructions Model

Design elements
Set off by border
Prominent location, visible at point of purchase
Graphic illustrations next to safe-handling statements
Type no smaller than 1/16 in, except heading, which must be larger
One-color printing on single-color, contrasting background

"Cook thoroughly."

"Keep hot foods hot. Refrigerate leftovers immediately or discard.

OTHER LABELING REQUIREMENTS

In certain situations, other labeling requirements apply. The requirements may be triggered by voluntary declaration of related information (eg, stating number of servings, making a claim) or because a key substance is used in the product or packaging (eg, aspartame, saccharin, self-pressurized containers).

Labeling with Number of Servings

Although it is not required, some packages include a statement about the number of servings contained in the package. When this information is provided, the regulations require that the net quantity of each serving be stated adjacent to the number-of-servings declaration. The statement of net quantity must be in the same type size used in the declaration of number of servings. This net quantity information should be consistent with the serving size declaration in the nutrition facts statement, unless the product is exempt from nutrition labeling.

Quantitative Information for Comparative Claims

Products bearing a nutrient content claim that is considered a comparative claim must include quantitative information either adjacent to the most prominent claim or on the information panel.

Warning Statements

Certain products require warning statements that must be placed on the information panel (or the PDP). For example, products packaged in self-pressurized containers must warn consumers about health and environmental risks. Protein products in liquid, powder, tablet, capsule, or similar form must warn consumers about the risks of very-low-calorie diets. Products containing ingredients that have a health or safety concern (eg, aspartame, saccharin) are required to display warning statements or other notices.

Exemptions

Despite the goal of NLEA that virtually all packaged food products carry a nutrition label, some products are exempt (Table 10). Manufacturers of exempt products may voluntarily include nutrition labels on their products; however, all the requirements for nutrition labeling must be followed on such voluntary labels. Generally, if an otherwise exempt product carries a nutrition claim or other nutrition information on its label (or in labeling or advertising), nutrition labeling is required.

Foods Intended for Further Processing. Foods that are shipped to food manufacturers, distributors, or retailers for further processing, repacking, or labeling are generally exempt. The regulations detail the situations that qualify for exempt status. FDA and USDA take basically the same approach on this point; however, their criteria are outlined differently.

Foods Prepared in Restaurant, Food-Service, and Retail Settings. Restaurant and food-service foods are generally exempt from nutrition labeling requirements. Ready-to-eat foods from bakeries, delicatessens, and retail settings where the food is prepared on-site are also exempt.

Small Businesses. Both agencies exempt products of small businesses; however, each agency has its own criteria for defining a small business. Originally FDA used gross sales to define a small business, but due to a congressional mandate, FDA revised the small-business exemption, basing it on the number of full-time equivalent employees and units of product produced in a year. USDA's exemption criteria are based on the number of employees and pounds of product produced annually.

Other Exemptions. FDA exempts foods with no nutritional significance (eg, teas, coffee, spices); this provision is not applicable to USDA foods (ie, foods containing 2% or more cooked meat or poultry have significant nutrients). Foods intended for export and custom-slaughtered fish and game are exempt.

FDA oversees infant formula, medical foods, and dietary supplements. The labeling of these products is addressed under separate regulations; thus, they are exempt from the general nutrition-labeling regulations.

Special Labeling Provisions

Both FDA and USDA regulations have special labeling provisions for certain food categories. These products are not exempt from nutrition labeling, but some of the labeling requirements are more relaxed.

Small Packages. Both agencies allow food packed in small packages (less than 12 sq in of space available to bear labeling) that do not bear claims to omit the nutrition facts, but the label must include an address and/or a phone number where the consumer can obtain nutrition information. If nutrition labeling is required, there are special layout and design provisions. USDA exempts from nutrition labeling packages containing less than 0.5 oz, provided no nutrition-related claims are made or nutrition information is provided.

Gift Packs. For packages that contain a variety or assortment of foods intended to be used as a gift, FDA allows the nutrition labeling to be enclosed within the package. It may be provided on a composite basis for FDA-approved reasonable categories of foods that have similar dietary uses and nutritional characteristics. USDA also allows nutrition information to be provided by a label insert.

Raw Products. Raw fruits, vegetables, and fish are exempt from mandatory nutrition labeling, but FDA has outlined expectations for voluntary nutrition labeling by retailers at the point of sale. USDA has a similar program.

Foods for Infants and Children. Foods intended specifically for infants and children under 2 and under 4 years of

Table 10. Foods Exempt from Mandatory Nutrition Labeling

Exemption criteria[a]	FDA exempts	USDA exempts
Foods intended for further processing, labeling, or packaging, not for sale to consumers	Yes, even if bearing claims	Yes
Restaurant and food-service products	Yes	Yes
Food prepared at retail site	Yes	Yes
Food produced by small businesses[b]	Yes	Yes
Food with no nutritional significance (coffee, tea, spices)	Yes	N/A
Donated foods	Yes	No
Custom-slaughtered and processed products	Yes[c], even if bearing claims	Yes, even if bearing claims
Products intended for export	Yes, even if bearing claims	Yes, even if bearing claims
Infant formula, medical foods, and dietary supplements	Yes[d]	N/A

[a]Nutrition-related claims trigger mandatory nutrition labeling, except in a few situations in which the exemption is available regardless of the presence of claims.
[b]Both FDA and USDA provide for a small-business exemption; however, the agencies have different criteria for defining a small business.
[c]FDA oversees game meats, which have provisions for voluntary labeling.
[d]Addressed under separate regulations.

age have special format provisions, and certain nutrients are omitted.

Features of the Nutrition Label

Title. The label is entitled "Nutrition Facts" (Fig. 2).

Serving Size Information. Serving sizes are provided in common household measures (eg, cups), followed by the metric equivalent in parentheses. The regulations define standardized amounts based on the servings people typically consume (eg, 12 oz can of soda is 1 serving, not 2) and detail how to derive serving size declarations.

Nutrient List. Information on the following nutrients is mandatory: calories, total fat, saturated fat, cholesterol,

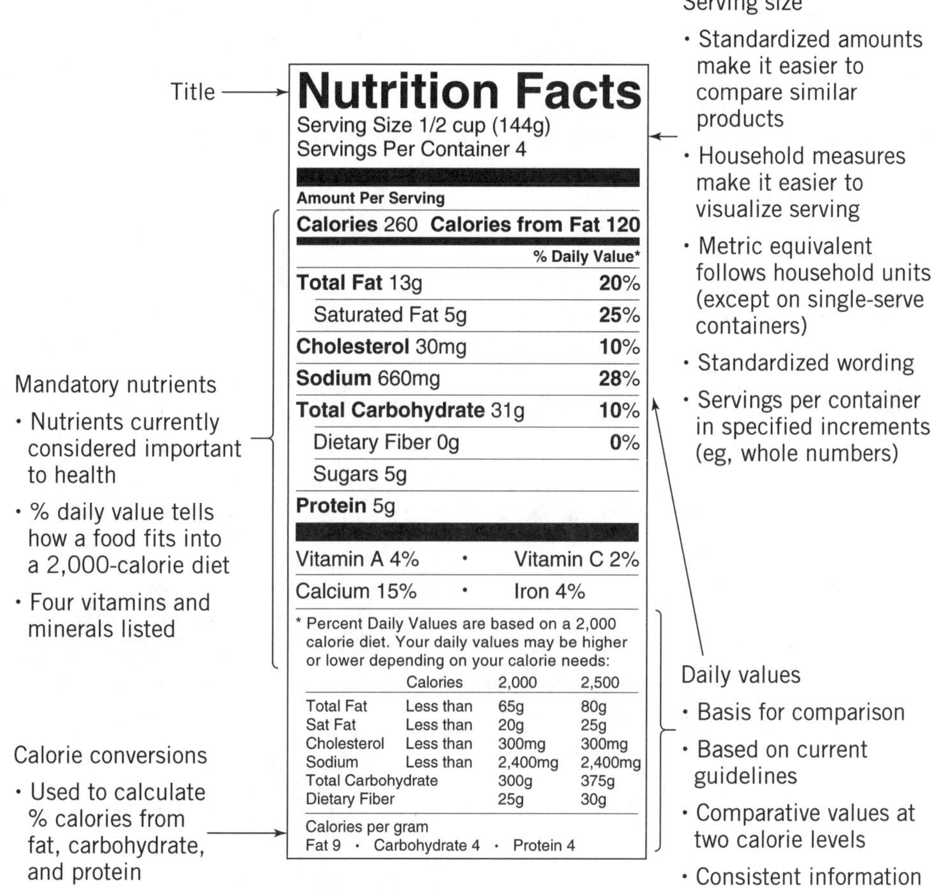

Figure 2. Nutritional label.

sodium, total carbohydrate, dietary fiber, sugars, protein, and four vitamins and minerals. In addition, certain nonmandatory nutrients may be included. The actual amount of each nutrient present in a serving is listed beside the nutrient. In the right-hand column, the amount of the nutrient is expressed as "% Daily Value," which shows how a food fits into a 2,000-calorie reference diet.

Daily Values Footnote. The daily values table at the bottom of the nutrition label presents the reference numbers used to calculate the % daily value. These numbers reflect current recommendations for health maintenance and disease prevention. They are provided at two calories levels and remain the same for all labels. On smaller packages and packages qualifying for the simplified format, this information may be omitted.

Calorie Conversion Footnote. Calorie conversions assist in calculating the percentage of calories from carbohydrate, protein, and fat. This feature is now optional.

Mandatory Design Elements

The regulations dictate a number of mandatory design elements that must be followed for all the nutrition facts layouts.

Hairline Box. The nutrition facts information must be set off from other printed material by a hairline box. The regulations are very precise on this point. No other information may be enclosed within the box. The box must be in the same color ink used for the type.

Typesetting. The type must be set in a single, easy-to-read type style. Helvetica is the recommended font; however, any sans serif type is permissible. Upper- and lowercase letters must be used. Minimum type sizes are specified for each component of the label.

The nutrients must be separated by 4 points of leading. All other text must be separated by at least 1 point of leading.

ACKNOWLEDGMENTS

This text is derived from *Food Label Design: A Regulatory Resource Kit*, published by the Institute of Packaging Professionals. Used with permission.

GENERAL REFERENCES

F. Olsson, and P. C. Weeda, *U.S. Food Labeling Guide*, The Food Institute, Fair Lawn, N.J., 1998.

R. Shapiro, *Nutrition Labeling Handbook*, Marcel Dekker, New York, 1996.

J. Storlie, *Food Label Design: A Regulatory Resource Kit*, Institute of Packaging Professionals, Herndon, Va., 1996.

AARON L. BRODY
Rubbright•Brody, Inc.
Duluth, Georgia

JEAN STORLIE
Nutrition Labeling Solutions
Maple Grove, Minnesota

PACKAGING: PART III—MATERIALS

The basic food-packaging materials can be divided into glass, metal, paper, and plastic. Each of these materials has both advantages and disadvantages (Table 1). The proper selection of these materials should be based on the functional requirements (see PACKAGING: PART I—GENERAL CONSIDERATIONS) and the economics of specific applications. To optimize the performance and cost, most food packages use more than one type of packaging material. For example, the packaging for a typical box of cereal consists of an inner plastic bag for moisture protection and an outer paperboard box for strength and rigidity.

GLASS

The glass container (bottle or jar) is one of the oldest food packages. It is an excellent barrier for protecting food from oxygen and moisture. It can provide food with excellent visibility and an image of cleanliness. Physically, glass is a very high viscosity supercooled liquid. The major constituents of container glass are silica (SiO_2), soda (Na_2O), and calcia (CaO), along with small amounts of other inorganic oxides. The properties of glass (such as strength, transparency, and moldability) can be modified by minor changes in the composition of these constituents.

Some important considerations in designing the glass container are mechanical, thermal, and optical properties. Mechanical properties include vertical load strength, internal pressure strength, impact strength, scratch resis-

Table 1 Advantages and Disadvantages of Food-Packaging Materials

Advantages	Disadvantages
Glass	
Excellent barrier against oxygen and moisture	Easily breakable
Chemically inert	Relatively heavier
Transparent	
Metal	
Excellent barrier against oxygen, moisture, and light	Susceptible to corrosion
Good mechanical strength and durability	Metal cans are generally more difficult to open and reseal
Good thermal stability	
Paper	
Relatively inexpensive	Poor gas and moisture barrier
Excellent printability	Greatly reduced mechanical strength when wet
Lightweight	
Plastic	
Most versatile packaging material	More susceptible to migration and flavor-scalping problems
Can be formed easily into many shapes	
Lightweight	

tance, and abrasion resistance (1). The strength of the glass container depends greatly on the condition of its surface, and the strength can be greatly reduced by a slight surface scratch. To reduce scratches and impact breakage, various lubricating coatings, such as special waxes and silicones, are often applied to the outer surfaces. Fairly thick glass is often needed to provide sufficient strength to prevent breakage due to internal pressure (especially for carbonated beverage bottles), impact, or thermal shock. This thicker glass, however, adds considerable weight to the package.

Thermal properties include thermal strength, which is a measure of the ability to withstand sudden temperature change. Stresses from sudden cooling are more damaging than those from sudden heating. A simple experiment to determine thermal shock resistance is to observe the number of breakages that occur when glass containers at a high temperature are suddenly transferred to a cold-water bath. The temperatures used in the experiment depend on the actual process temperatures that the container will experience, such as in pasteurization, hot pack, or retorting (1).

Optical properties include the degree of transmission to light and its effects on the quality of the product. Glass can be colored for decoration or light protection. Flint is basic clear glass used for most food-packaging applications. Amber is the familiar brown glass used for beer that can filter out ultraviolet light (300–400 nm). Emerald is bright-green glass used mostly in the beverage industry. Other blue, green, and opaque glasses are also available (2).

In recent years, thin layers of silica (SiO_x) are deposited onto films such as polyethylene terephthalate (PET), oriented polypropylene (OPP), and low-density polyethylene (LDPE). These silica-coated (or glass-coated) films have good gas barrier (3), and they are clear, recyclable, and retortable; however, the commercial applications of these films are rather limited because of the high cost and technical difficulties in ensuring good silica adhesion to certain base polymers (4).

METAL

Metal packages provide excellent protection against the intrusion and negative effects of oxygen, moisture, and light. Steel (with tin or chromium in trace amounts) and aluminum (with magnesium and manganese in trace amounts) are the most commonly used metals for food packaging. Cans and trays are the most common forms of metal packages.

Steel Can

The steel can (also referred to as tin can) is made of a low-carbon steel coated with a very thin layer of tin on both the inner and outer surfaces. Typically, the steel can is made of three pieces: the can body, the top lid, and the bottom lid. The can body may be mechanically seamed, bonded with adhesive, welded, or soldered (2).

The tin coating protects the steel from corrosion through reactions with water, oxygen, acids, and a host of other chemicals or foods. Most often, the inner tin surface is also coated with a nonmetallic lacquer to prevent undesired reactions of the tin with the foods.

Aluminum Can

Most beer cans and soft-drink cans are made of aluminum. Compared to the steel can, the aluminum can is lighter, more corrosion resistant, more formable, lower in structural strength, and more costly.

To a large degree, the lower structural strength has limited the use of the aluminum can to the bottling of carbonated beverages, where the can gains structural support from the internal gas pressure. However, the lower strength and greater formability also allow the two-piece aluminum can (can body plus the top lid) to be formed more readily. This is accomplished by forging an aluminum disk into a cup so that the bottom and the sides are all in one piece. The two-piece can eliminates the side and bottom joints, and thus the chance of leakers is also reduced.

The popularity of the aluminum can is partly attributed to its recyclability. The economics incentive for recycling of aluminum cans is high, and the aluminum industry has established an effective collection and reprocessing system.

Aluminum Foils and Metallized Films

Aluminum foil (between 0.0003 and 0.0015 in thick) is used in such food-packaging applications as pouches and liddings. The foil is usually laminated to one or more layers of plastic or paper. An example is the military meal-ready-to-eat pouch that uses a laminated structure of PET/adhesive/aluminum/adhesive/polypropylene, from outside to inside. The aluminum foil layer provides an excellent barrier to moisture and gases, unless pinholes exist.

The metallized film consists of a very thin layer of aluminum (compared to the foil) that is vapor deposited onto the surface of a plastic film, such as polyester or nylon. The aluminum improves the gas- and light-barrier properties of the film as well as its aesthetic appeal. However, the gas-barrier properties of the metallized film are easily reduced due to handling that results in the formation of pinholes. The metallized film is used in such applications as bags for potato chips or snacks. It can also be laminated to paperboard for such applications as liquid-box juice containers.

Susceptor is a special type of metallized film used in microwavable food packages (1). A typical susceptor is a laminate consisting of the structure PET/adhesive/aluminum/adhesive/paperboard. The aluminum in the laminate is a very thin, discontinuous layer of aluminum particles that can convert microwave energy into heat, causing the temperature of the susceptor to rise above 200°C rapidly. The high temperature is important for applications such as microwavable popcorn and pizza. However, the high temperature also raises safety concerns about thermal degradation of the PET and the migration of undesirable compounds from the susceptor to the food.

PAPER AND PAPERBOARD

Paper and paperboard are made from wood fibers composed of cellulose, hemicellulose, and polymeric residues.

Although there is no strict distinction between paper and paperboard, structures less than 0.012 in thick are generally considered to be paper. Because of their good mechanical strength, paper and paperboard are used mostly to protect the product from physical damage. However, paper and paperboard have poor gas-barrier properties, and their mechanical strength can decrease greatly in wet or humid conditions.

Many types of paper are used for food-packaging applications (1,2). Kraft paper is typically a coarse paper used in applications such as bags and wrapping where strength is required. Kraft paper is available either in unbleached brown form or in bleached form. Greaseproof paper consists of densely packed fine fibers that provide good oil and grease resistance. Greaseproof paper may be further treated with fluorochemicals or coated with polyvinylidene chloride (PVDC) to improve gas- and moisture-barrier properties. Glassine paper is produced by further processing greaseproof paper through a super-calendering operation, making the paper smoother, denser, and translucent. Waxed paper consists of a continuous layer of wax on a base paper such as greaseproof or glassine paper. The wax serves as both a moisture barrier and a heat-sealable layer.

There are also many types of paperboard used for food-packaging applications (5). Corrugated boxes are used mostly for shipping food packages. Folding cartons are used as outer boxes for packaging cereal products, cookies, snacks, frozen foods, and the like. The appearance and printing quality of paperboard can be enhanced by coating with clay and other minerals.

Paperboard is often coated or laminated with plastics or aluminum for better performance. For example, polyethylene (PE) is coated onto gable-top paperboard containers, which are commonly used to contain milk or juices. The coating protects the paperboard against moisture absorption, and it also provides the paperboard with heat sealability. Another example is the aseptic Tetra Pak juice carton, which is made from a paperboard laminate having the structure PE/paper/aluminum/PE. In this laminate structure, the inner PE layer protects the aluminum from the juice, the aluminum layer provides oxygen and a moisture barrier, the paperboard layer provides strength and rigidity, and the outer PE layer provides moisture protection for the paperboard.

PLASTIC

Plastics, relatively new in packaging materials, are used in many of the food-packaging applications that were once dominated by the more traditional packaging materials—glass, metal, and paper (6). Plastics are polymers or long-chain macromolecules that can be molded, extruded, and cast into various shapes such as films, sheets, and containers. Polymers are derived from petroleum, natural gas, or coal by polymerization processes. Most polymers used in food-packaging plastics have molecular weights between 50,000 and 150,000. In addition, most plastics contain small amounts of additives (such as plasticizers, lubricants, antioxidants, antistats, heat stabilizers, and ultraviolet stabilizers) that are used to facilitate processing or to impart desirable properties to the plastics. These additives can sometimes migrate into the food, causing the food to acquire undesirable flavors (see PACKAGING: PART I—GENERAL CONSIDERATIONS). Most packaging polymers are thermoplastics, which are characterized by the absence of cross-linking between polymer chains and can be softened repeatedly by heat and pressure to form different shapes (7–9).

Polymers can be formulated to provide a wide range of performance properties that are dependent on chemical structure, crystallinity, side-chain branching, molecular weight, molecular weight distribution, orientation, and so on (1). The chemical structures of some food-packaging polymers are shown in Figure 1. As a general rule, higher crystallinity results in greater strength, better gas-barrier properties, higher soften point, and lower clarity. The type of polymer used in a retail container can often be identified by the recycling code and lettering on the bottom.

The important considerations in selecting plastics for food-packaging applications are gas-barrier properties, strength, stiffness, chemical resistance, sealability, printability, formability, appearance, and cost. Plastics are unique in that they have a broad range of gas-barrier properties (Fig. 2), whereas glass and metal provide a total gas barrier, and paper provides virtually none. Thus plastics offer the versatility for packaging many different foods, including those that require a high gas barrier (eg, potato chips) and those that require a low gas barrier (eg, fresh produce).

The gas-barrier properties of polymers are often expressed in terms of permeability: the lower the permeability, the better the gas barrier. Permeability depends on the plastic material; permeant gas; temperature; and, sometimes, relative humidity. The permeabilities of some commonly used food-packaging polymers are listed in Table 2.

Following are general discussions of some commonly used food-packaging polymers. These polymers are used mostly as bottles, containers, bags, films, and so on. Because these polymers have some overlapping properties, more than one polymer can often meet the performance requirements of a food-packaging application. The specific properties of a polymer film, sheeting, or container are usually provided by the manufacturer.

Polyethylene

PE is the most frequently used polymer in food-packaging applications because of its low cost, easy processing, and good mechanical properties. PE also has the simplest chemical composition of all polymers, being essentially a straight-chain hydrocarbon. The major classifications of PE are high-density polyethylene (HDPE), low-density polyethylene (LDPE), and linear low density polyethylene (LLDPE). These classifications differ in density, chain branching, crystallinity, and so on. Within each classification are numerous grades that have different additives, molecular weight distributions, and other properties.

HDPE has a density typically between 0.94 and 0.97 g/cm^3. It is a linear polymer with relatively few side-chain branches (Fig. 3), and hence the macromolecules can fold and pack into an opaque, highly crystalline structure.

Polyethylene (PE)

Polypropylene (PP)

Polyvinyl chloride (PVC)

Polystyrene (PS)

Polyethylene terephthalate (PET)

Polyvinylidene chloride (PVDC) copolymer

Ethylene vinyl alcohol (EVOH) copolymer

An example showing the joining of repeating units to form polystrene polymer

Figure 1. Chemical structures of some common food-packaging polymers.

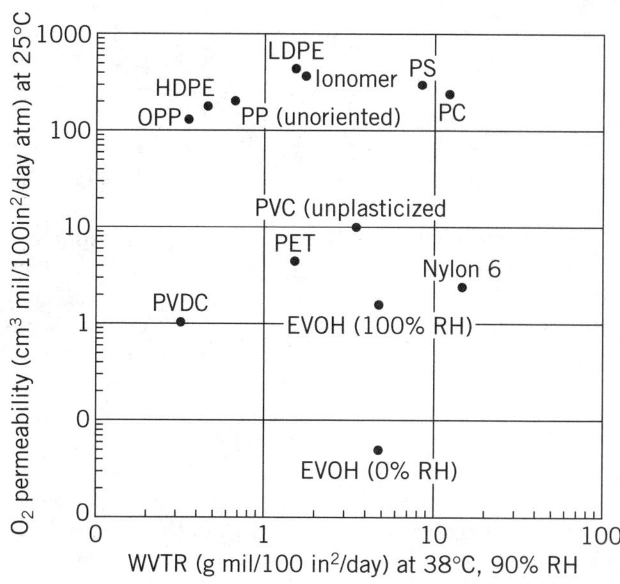

Figure 2. Gas barrier properties of some common food-packaging polymers.

Compared to LDPE, HPDE has a higher melting point (typically 135°C vs 110°C), greater tensile strength and hardness, and better chemical resistance. HDPE is used mostly to make blow-molded bottles for the packaging of products such as milk and water. It is also used for food containers, bags, films, extrusion coating, bottle closures, and the like.

LDPE has a density between 0.91 and 0.93 g/cm^3. It is a polymer with many long side-chain branches. LDPE is used mostly as film for packaging fresh produce and baked goods, as an adhesive in multilayer structures, and as waterproof and greaseproof coatings for paperboard packaging materials. The packaging film made from LDPE is soft, flexible, and stretchable. The film also has good clarity and heat sealability.

LLDPE has a density about the same as that of LDPE. LLDPE is a copolymer with many short side-chain branches. LLDPE has similar clarity and heat sealability to that of LDPE, as well as the strength and toughness of HDPE. With its superior properties, LLDPE has been replacing LDPE in many food-packaging applications. In recent years, many development efforts have focused on using single-site or metallocene catalyst to produce LLPDE films with superior heat seal and hot tack properties, clarity, abuse resistance, and strength (10,11).

Table 2 Barrier Properties of Some Common Packaging Plastics

Plastics	O_2 permeability[a]	CO_2 permeability[a]	WVTR[b]
PE			
Low density	300–600	1200–3000	1–2
High density	100–250	350–600	0.3–0.6
PP			
Unoriented	150–250	500–800	0.6–0.7
Oriented	100–160	300–540	0.2–0.5
PS	250–350	900–1050	7–10
PET	3–6	15–25	1–2
PVC			
Unplasticized	5–15	20–50	2–5
Plasticized[c]	50–1500	200–8000	15–40
PVDC	0.1–2	0.2–0.5	0.02–0.6
EVOH			
0% RH	0.007–0.1	0.01–0.5	—
100% RH	0.2–3	4–10	—
Ionomer	300–450	—	1.5–2
Nylon 6	2–3	10–12	10–20
PC	180–300	—	10–15

[a]Unit in (cm^3 mil)/(100 in^2/day atm) at 25°C.
[b]Unit in (g mil)/(100 in^2/day) at 38°C, 90% RH.
[c]Values depend greatly on plasticizer content.
Source: Data are taken from several sources.

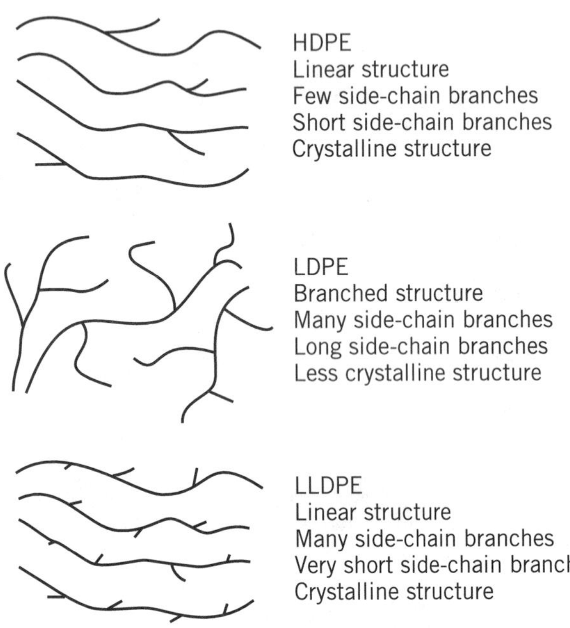

Figure 3. Molecular arrangements of HDPE, LDPE, and LLDPE.

Polypropylene

Polypropylene (PP) is a linear, crystalline polymer that has the lowest density (0.9) among all major plastics. Compared to PE, PP has higher tensile strength, stiffness, and hardness. It also has a higher melting temperature (165°C), and hence it is more suitable for hot filling and retorting applications.

A major application for PP is in packaging films. Commercial PP films are available in oriented and unoriented forms. Orientation can be achieved by stretching the film either uniaxially or biaxially during the film-forming process. Oriented PP film (OPP) has improved strength, stiffness, and gas-barrier properties; however, it is not heat sealable. Unoriented PP film has excellent clarity, good dimensional stability, and good heat-seal strength. Beside films, other major applications for PP are containers and closures.

Polystyrene

Polystyrene (PS) is an amorphous polymer that has excellent clarity. Solid PS is a clear, low-gas-barrier, hard, low-impact-strength (unless modified) material. It has a relatively low melting point (88°C) and can be readily thermoformed or injection molded into items such as food containers, cups, closures, and dishware. Because of its excellent clarity, PS film is often used as windows in paperboard boxes to display products (such as baked goods) that do not require a good gas barrier.

Expanded PS (EPS) of various bulk densities are manufactured by adding foaming agents in the extrusion process. EPS is used to make cups, bowls, plates, meat trays, clamshell containers, and egg cartons as well as protective packaging for shipping.

Polyethylene Terephthalate

PET (also abbreviated PETE) is the major polyester used for food packaging. The amorphous form of PET (APET) is used mostly as injection blow-molded bottles for carbonated soft drinks, water, edible oil, juices, and similar products. PET bottles are stronger and clearer and provide a better gas barrier than HDPE bottles, although they are more expensive.

APET is also used to produce films that have high strength, a high melting point (267°C), high scuff resistance, good clarity, good printing characteristics, and excellent dimensional stability. Because of their poor heat sealability, PET films are often laminated or extrusion coated with a heat-sealable layer such as PVDC and LDPE.

The crystallized form of PET (CPET) can withstand temperatures up to 220°C without deformation. Thus CPET food trays are used in microwave/conventional dual ovens. PETG is a copolymer of PET and 6% cyclohexane dimethanol that is used in thermoforming applications.

Polyvinyl Chloride

Polyvinyl chloride (PVC) is a clear, amorphous polymer used mostly for films and containers. Most often, plasticizers (organic liquids of low volatility) are added to the polymer to yield widely varying properties, depending on the type and amount of plasticizers used. Plasticized PVC films are limp, tacky, and stretchable, and the films are commonly used for packaging fresh meat and fresh produce. Unplasticized PVC sheets are rigid, and the sheets are often thermoformed to produce inserts for snacks such as chocolate and biscuits.

PVC bottles have better clarity, oil resistance, and barrier properties than those of HDPE. However, the use of PVC bottles in food packaging is relatively small due to poor thermal processing stability and environmental concerns with chlorine-containing plastics.

Polyvinylidene Chloride

PVDC commonly known under the trade name Saran, is a copolymer of vinylidene chlorine (85–90%) and vinyl chloride. The most notable advantages of PVDC are its excellent oxygen- and moisture-barrier properties. PVDC films have good clarity and grease and oil resistance.

PVDC is used in films, containers, and coatings. Monolayer PVDC films are used in household wraps. PVDC is often coextruded or laminated with other lower-cost polymers (such as OPP and PET) to form multilayer films or sheets. The multilayer films are used to package foods that require a good oxygen barrier, and the multilayer sheets are often thermoformed into semirigid containers. PVDC is also used in the form of latex for coating paper, film, and cellophane to achieve a better oxygen and moisture barrier, grease resistance, and heat sealability.

PVDC is more costly than most other commonly used food-packaging polymers. Similar to PVC, PVDC has poor thermal processing stability, and environmental concerns are associated with its use.

Ethylene Vinyl Alcohol Copolymer

Ethylene vinyl alcohol copolymer (EVOH) is a crystalline copolymer of ethylene (usually between 27 and 48%) and vinyl alcohol, and its properties are highly dependent on the relative concentrations of these comonomers. EVOH is best known for its exceptionally good oxygen-barrier properties, even compared to PVDC. The oxygen barrier increases with decreasing ethylene concentration and decreasing relative humidity. EVOH also has good resistance to hydrocarbons and organic solvents. EVOH is the most expensive of all commonly used food-packaging polymers (8).

EVOH is mostly used as an oxygen-barrier layer in multilayer structures, either laminated or coextruded. The multilayer structures containing EVOH are used as films or containers for packaging oxygen-sensitive foods. An example of a multilayer structure is PP/adhesive/EVOH/adhesive/PP, where the PP layers provide strength and a moisture barrier and the EVOH layer provides the oxygen barrier. This multilayer structure is suitable for hot filling and retorting applications.

Other Food-Packaging Polymers

Many other food-packaging polymers exist in addition to the major ones described previously. Ethylene vinyl acetate (EVA) is a low-cost copolymer used as an adhesive in coextrusion and as a heat-sealable layer. Compared to EVA, ionomers are more expensive adhesives that have better hot tack, adhesion to aluminum foil, and ability to heat seal through food contaminants (9). Nylon is a polyamide primarily used in applications that require good abrasion resistance and toughness; however, nylon has poor moisture-barrier properties and heat sealability. Polycarbonate (PC) is a relatively expensive polyester that has exceptional impact properties, and its food-packaging application is presently limited mostly to large returnable water bottles. Other packaging polymers include edible films (9), biodegradable films (9), oxygen-scavenging films (12), and temperature-compensating films (4).

BIBLIOGRAPHY

1. G. L. Robertson, *Food Packaging: Principles and Practice*, Marcel Dekker, New York, 1993.
2. W. Soroka, *Fundamentals of Packaging Technology*, Institute of Packaging Professionals, Herndon, Va., 1995.
3. E. Finson and R. J. Hill, "Glass-Coated Packaging Films Ready for Commercialization," *Packaging Technol. Eng.* 4, 36–43 (1995).
4. U. Stöllman, F. Johansson, and A. Leufvén, "Packaging and Food Quality," in C. M. D. Man and A. A. Jones, eds., *Shelf Life Evaluation of Foods*, Blackie Academic & Professional, New York, 1994, Chapter 4.
5. F. A. Paine, *The Packaging User's Handbook*, Van Nostrand Reinhold, New York, 1991.
6. W. A. Jenkins and J. P. Harrington, *Packaging Foods with Plastics*, Technomic, Lancaster, Pa., 1991.
7. W. E. Brown, *Plastics in Food Packaging: Properties, Design, and Fabrication*, Marcel Dekker, New York, 1992.
8. K. R. Osborn and W. A. Jenkins, *Plastic Films: Technology and Packaging Applications*, Technomic, Lancaster, Pa., 1992.
9. A. L. Brody and K. S. Marsh, eds., *The Wiley Encyclopedia of Packaging Technology*, 2nd ed., John Wiley and Sons, New York, 1997.
10. T. D. Stirling, "Solutions for Today's Food Packaging Challenges via Single-Site Catalyst Technology," in *Conf. Proc. of Future-Pak '97*, George O. Schroeder Associates, Appleton, Wisc., 1997.
11. J. deGroot et al., "ELITE Enhanced Polyethylene: A Performance and Cost Effective Alternative to EVA/LLDPE Blends," in *Conf. Proc. of Future-Pak '97*, George O. Schroeder Associates, Appleton, Wisc., 1997.
12. M. L. Rooney, *Active Food Packaging*, Blackie Academic & Professional, New York, 1995.

KIT L. YAM
RAYMOND G. SABA
Y. C. HO
Rutgers University
New Brunswick, New Jersey

PACKAGING: PART IV—CONTROLLED/ MODIFIED ATMOSPHERE/VACUUM FOOD PACKAGING

The shelf life of foods such as fresh and processed meat, eggs, fish, poultry, fresh fruits, fresh vegetables, and soft bakery goods is limited in the presence of atmospheric oxygen due to three important factors: the biochemical effect of atmospheric oxygen, the activity of oxidative enzymes, and the growth of aerobic spoilage microorganisms. A fourth factor of no small importance is attack by insects. Each of these factors, alone or in conjunction with one another, can result in changes in color, flavor, odor, and overall deterioration in food quality, and the hazard of microbiological safety. Technologies employed by food processors to retard these deteriorative changes include chilled storage, freezing, thermal processing, water removal, osmotic adjustment, pH change, and the use of chemical additives and preservatives. However, increasing energy costs associated with freezing and drying, quality changes imposed by the processes themselves, and growing consumer concerns about chemical additives has compelled the food industry to look for alternative methods of food preservation.

Controlled/modified atmosphere/vacuum packaging is a relatively new preservation technology used extensively for quality retention. Though these technologies have been in commercial use only since the 1960s, the volume of food preserved under them exceeds that of canned or frozen foods in North America—and is perhaps even proportionately higher in Europe.

The normal composition of air is 20.9% oxygen (O_2), 78% nitrogen (N_2), 0.9% argon, and 0.03% CO_2. A controlled-atmosphere process involves alteration of the gaseous environment in and around a food and its maintenance at a specified level throughout the preservation period. A modified atmosphere, as the name implies, is one in which the normal composition of air is changed or modified within a package, but the change is not constant due to continued product respiration and permeation of gases through the package. Any modification usually, but not always, results in a reduction of the O_2 content of the air in the package headspace while increasing the level of CO_2. This results in the potential for enhanced quality retention for food products without the use of covert chemical or physical treatments such as preservatives, freezing, and drying. Controlled- and modified-atmosphere preservation and packaging are almost always enhancements of refrigeration as a preservation technology.

Vacuum processing and packaging means the removal of oxygen from the foods' environment by mechanical methods. Oxygen removal may also be accomplished by displacement with an inert or nearly inert gas such as nitrogen.

The concept of controlled/modified atmospheres for shelf-life extension of food is not new in food preservation. During the nineteenth century scientists discovered that the elevation of CO_2 and reduction of O_2 retarded catabolic reactions in respiring foods and slowed the growth of aerobic spoilage microorganisms. Basic research on the use of modified atmospheres for shelf-life extension of fruit, vegetables, fish, and meat was performed during the 1920s and 1930s. By 1938, 26% of chilled-carcass beef shipped from Australia and 60% of that shipped from New Zealand was being shipped under a CO_2-enriched atmosphere.

In the United States and Europe, successful application of controlled-atmosphere storage is extensive for apples and pears, which can be stored for up to 9 months using the correct $CO_2/O_2/H_2O$ vapor mixture in conjunction with temperature, ethylene, and relative humidity control. This technology has been commercial for more than 35 years.

Distribution of food in retail and hotel/restaurant/institutional (HRI) units packaged under modified atmospheres, that is, modified-atmosphere packaging (MAP), is a major application of technology for a variety of fresh and minimally processed food products. MAP is defined as the packaging of food products in which the gaseous environment has been changed to slow respiration rates, reduce microbiological growth, retard enzymatic spoilage, and slow biochemical changes with the intent of prolonging quality retention. The objectives of this article are to review the reasons for the market growth of MAP technology, the roles of gases used in MAP, the methods of atmosphere packaging, and the applications of MAP for quality retention of specific food groups. The advantages and disadvantages of MAP technology and the public health concerns of this technology are also addressed.

GROWTH OF MAP TECHNOLOGY

Despite the paucity of visibility, MAP technology has emerged as the premier packaging technology of the last years of this century. Currently, the United States leads the way in MAP technology, followed by the United Kingdom, France, and Germany. Thousands of food processors around the world use MAP technology for shelf-life extension and food distribution. It is estimated that the production of CAP/MAP/vacuum-packaged foods in North America is well in excess of 30 billion lb annually. The growth of MAP technology, for both medium- and long-term preservation of food, is due to a number of interrelated factors: consumer desire for higher-quality, more nearly fresh, and higher nutritional quality foods; the development of better but still imperfect distribution systems; improved packaging technologies; and energy costs.

Developments in New Polymeric Barrier Packaging/ Materials

The success of any food-packaging technology as a means of extending the shelf life of food is dependent on the characteristics of the package materials surrounding a product. Developments in polymer chemistry have resulted in production of low-gas-permeability package films, such as polyvinylidene chloride (PVDC) and ethylene vinyl alcohol (EVOH), and high-gas-permeability materials such as styrene block copolymers. PVDC and EVOH films have excellent water vapor–oxygen-, and carbon dioxide–barrier characteristics and can be laminated to structural and/or water vapor–barrier polymers to impart the desired strength, heat sealability, and permeability characteristics for quality retention of packaged food products. In addition, developments in high-speed, continuous, vertical, horizontal, and thermoforming packaging equipment compatible with the machinability characteristics of these films have also promoted the growth of new packaging technologies.

Market Needs and Consumer Demands for Convenience

Over the past 30 years, many changes have occurred in consumers' lifestyles and food preferences. The fundamental change is in relation to the traditional roles of women, one of which in the past was meal preparation. With more than 60% of women in industrialized countries in the workplace, the time previously available for shopping and food preparation has decreased very substantially. The result is that today, with the need for convenience, many consumers are prepared to pay with their disposable incomes to have shelf-stable convenience foods that require a minimum of preparation. In fact, the consumer is indirectly asking the food industry to take over a part of, or in some cases all, of the more time-consuming steps associated with food preparation. The food industry has responded to these consumer demands by providing a variety of high-quality, more nearly fresh, easy-to-prepare, preservative-free foods packaged under a modified atmosphere. One surprising result has been that the proportion of family disposable income spent on food has been decreasing.

Increasing Energy Costs

Increasing energy costs associated with traditional methods of food preservation/storage, such as freezing, has resulted in the growth of less energy-intensive and more economical methods of short- and long-term preservation, such MAP. It has been estimated that MAP is 18 to 20% less energy intensive compared to freezing for shelf-life extension of bakery products. Thermal processing to achieve ambient temperature shelf stability is energy intensive, but not nearly as much as freezing, which requires removal of heat of fusion as well as temperature reduction and maintenance of low temperature. Further, it has been demonstrated repeatedly that the 0°F (−18°C) traditionally employed for frozen storage is well above the optimum temperature. Frozen foods are best stored at below their glass transition temperatures, a more costly and energy-intensive process being overtly resisted by commercial frozen food–distribution interests. These groups have failed to recognize the quality and hence economic benefits to be derived from optimum temperatures.

These same distribution channel members have also refused to reduce chill temperatures to their optima and have thus generated what might be described as the development of private chilled-food chains operated by food processors themselves to maintain safety and quality.

As a result of these interrelated factors, MAP is rapidly emerging as the packaging technology of choice for the preservation of minimally processed food products.

METHODS OF ATMOSPHERE MODIFICATION

Methods of atmosphere modification within a packaged food product may be subdivided into two main categories: passive modification and active modification. In commodity-generated or passive modification, the product is packaged in a film with the correct gas permeability characteristics, and the atmosphere within the packaged product is modified as a result of the consumption of O_2 and generation of CO_2 through respiration of the product, plus the permeation of gases through the package materials and structure. Passive modification is commonly used to modify the gas atmosphere of fresh respiring fruits and vegetables. However, to maintain the correct gas mixture within the packaged product, the gas permeabilities of the packaging films must be selected to allow O_2 to enter the package at a rate similar to its consumption by the product. Similarly, CO_2 must be vented from the package to offset the production of CO_2 by the product. Failure to achieve this gas balance will result in a depletion of O_2 and a buildup of CO_2, resulting in adverse changes in products. Depletion of oxygen to near 0 leads to anaerobic respiration or fermentation and the production of adversely flavored compounds. Most fresh vegetables and fruits are vulnerable to this type of injury under anoxic conditions.

Several methods can be used to actively modify the gas atmosphere within the packaged product. These include vacuum packaging, vacuum followed by gas, and injection of or sweep by gas mixtures.

Vacuum packaging is used extensively by the meat industry to extend the shelf life and the keeping quality of primal cut or wholesale cuts of fresh red meat. The product is placed in a package structure fabricated from film of low oxygen permeability, air is removed under vacuum, and the package is heat sealed. Under conditions of a good vacuum, headspace O_2 is reduced to $<1\%$. CO_2, produced from tissue plus microbiological respiration, may eventually increase to 10 to 20% within the package headspace. These conditions, that is, low O_2 and/or elevated CO_2 levels, extend the shelf life of meat by inhibiting the growth of aerobic meat-spoilage microorganisms, particularly *Pseudomonas* and *Alternaria* species.

A novel method of active modification is through the use of oxygen absorbents after mechanical oxygen removal. These consist of sachets that are placed inside the packaged product. Alternately, the oxygen scavenger may be incorporated into the film. Another method of active modification is the use of ethanol vapor generators that modify the gas atmosphere by producing ethanol vapor within the package headspace to suppress mold growth. Both oxygen absorbents and ethanol vapor generators have been used for shelf-life extension of food.

Active modification may also be achieved by gas packaging. Gas packaging is simply an extension of vacuum packaging technology and involves the evacuation of air followed by the injection of the appropriate gas mixture.

GAS-PACKAGING EQUIPMENT

The gas-packaging technique involves removing air from the pack and replacing it with a mixture of gases; the pressure of gas inside the package usually reaches about 1 atm, that is, equal to the external pressure. This is usually achieved by one of three types of packaging equipment: horizontal or vertical flexible packaging equipment, thermoform/fill/gas flush/seal or preformed tray fill/vacuum gas flush/seal equipment, and snorkel or related bulk packaging equipment.

In the continuous flow wrap/gas flushing technique, the machine creates a tube of flexible material that encloses the product either by itself or on a carrier tray. The appropriate gas mixture is introduced in a continuous countercurrent flow into the package to force the air out, the ends of the package are heat sealed, and the packages are cut from each other (Fig. 1).

In either the preformed tray or thermoform/fill/gas flush/seal technique, a compensated vacuum method is used to introduce the gas mixture. In this method, product is placed into a thermoformed tray and a vacuum is drawn to remove most of the air. The vacuum is broken by the appropriate gas mixture and the package heat sealed with a top web of film (Fig. 2). An example of a thermoformed gas-packaged product is shown in Figure 3. One advantage of the in-line thermoform method of gas packaging is the high efficiency of removing oxygen to residual levels of <1%. The in-line thermoforming type of equipment may also be used for vacuum packaging or vacuum skin packaging. The advantage of gas flushing versus vacuum plus back gas flush is its high production rate, with up to 120 packages/min being gas packaged.

The third type of equipment is the snorkel or bulk package equipment. The product, whether in the packaged form or without a package, is placed inside a large flexible pouch. The machine holds the pouch and inserts probes or snorkels that remove the air from inside the bag. The vacuum is broken by the addition of the appropriate gas mixture, the probes are removed, and the package is sealed. This type of equipment is used for bulk packaging and retains the product in the gas-packaged bag throughout storage and distribution of the product. Obviously, this type of equipment may also be employed for vacuum packaging, such as for primal cuts of fresh red meat or fresh-cut vegetables.

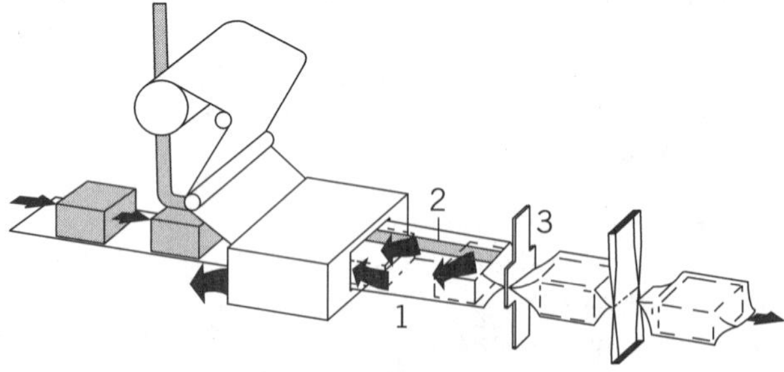

Figure 1. Horizontal flow wrap/fill/seal gas packaging equipment.

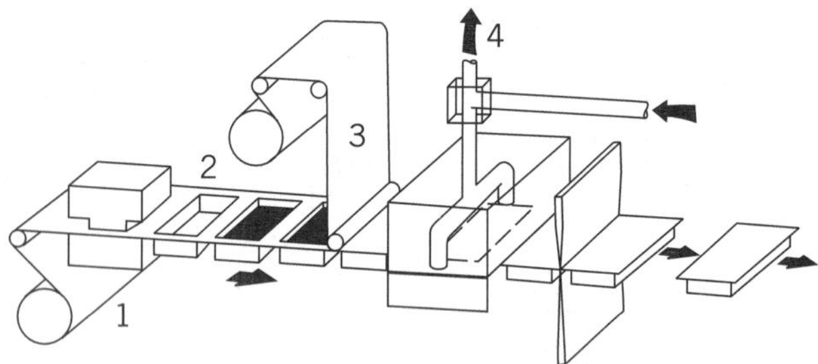

Figure 2. Thermoform/fill/vacuum/gas flush/seal packaging equipment. 1, Sheet; 2, formed cavity; 3, top closure web; 4, gas flush.

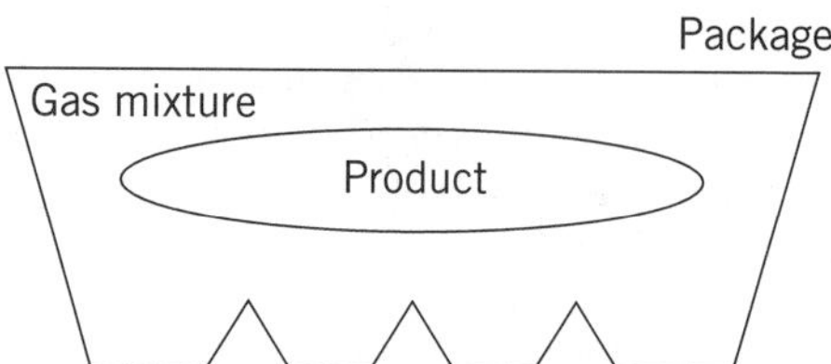

Figure 3. Example of a thermoformed gas-packaged meat product. The irregular base keeps the product above the package floor in order to have exposure to the modified atmosphere on all surfaces.

ROLES OF GASES USED IN MODIFIED ATMOSPHERE PACKAGING

Gases commonly used in gas packaging are nitrogen, oxygen, and carbon dioxide. Because the gases are those we breathe, the gases used in MAP are neither toxic nor dangerous, nor are they regarded as food additives.

Each gas plays a distinct and specific role in MAP foods. Nitrogen is an inert gas that has no effect on the food and has no antimicrobial properties. It is used mainly as a sweep gas to remove oxygen and as a filler gas to prevent package collapse in products that can absorb CO_2. It can also be used to replace oxygen in foods to retard biochemical spoilage of food, such as oxidations.

Oxygen is generally avoided in gas-packaging mixtures unless it is used to fulfill one of three functions: (*1*) it might be used with gas packaging of red meats to retain the desired oxymyoglobin red color or "bloom," apparently desired by many consumers; (*2*) it is used in low concentrations in packaging of products that respire, such as fresh-cut fruits and vegetables; (*3*) it may prevent anaerobic conditions and limit the growth of potentially harmful anaerobic microorganisms, specifically *Clostridium botulinum*, although this function has not been clearly demonstrated. Additionally, beneficial effects of high oxygen (above 40%) have been reported in retarding respiration rates of lettuce and on slowing microbiological growth on fresh nonfatty fish.

Carbon dioxide (CO_2) is both bacteriostatic and fungistatic, that is, it inhibits bacterial and mold growth. It can also be used to prevent insect growth in packaged and stored food products. Carbon dioxide is highly soluble in water and fats where it forms carbonic acid. Its high solubility may lower the pH, resulting in slight flavor changes in the food, and its absorption by the product may also cause package collapse. A summary of the salient properties of N_2, O_2, and CO_2 are shown in Table 1.

Table 1. Summary of Gas Properties Used in MAP

Oxygen	Maintains "bloom" (red color of fresh meat)
	Sustains basic metabolism of respiring foods
	Retards adverse effects of anaerobic respiration
	May prevent anaerobic microbiological growth
Nitrogen	Chemically inert
	Filler gas to occupy volume
	Retards oxidation by replacing oxygen
	Retards oxidative rancidity by replacing oxygen
Carbon dioxide	Inhibits bacterial and mold growth
	Fat and water soluble
	Prevents insect attack
	High concentrations can discolor products (meat) or injure produce (fresh fruits, vegetables)

Other gases that have antimicrobial or other functional properties include carbon monoxide, argon, nitrous oxide, ozone, sulfur dioxide, ethylene oxide, and ozone. Carbon monoxide, SO_2, ethylene oxide, and ozone generally are not included in MAP systems for a variety of reasons such as poor stability of the gas, limited approval for use in foods, or the formation of toxic or allergenic residues. Carbon monoxide (CO), however, is permitted in trace amounts (1–4%) in MAP of lettuce heads to prevent oxidative discoloration. Argon has been demonstrated to enhance MAP, perhaps by virtue of its molecular weight and size that may assist in removing oxygen more thoroughly than does nitrogen. Argon is being applied in fresh-cut vegetables and fish (and also in wine and salty snacks and chips that are not MAP applications).

ANTIMICROBIAL EFFECTS OF CO_2

Although the preservative action of CO_2 in foods has been known for many years, its mechanism of antimicrobial action has not been fully determined. Several theories have been postulated, however. One theory was that the displacement of O_2 was the main reason for the antimicrobial properties of CO_2. This theory was refuted by Coyne, who showed that aerobic spoilage organisms of fish grew well in 100% N_2 but not in 100% CO_2, indicating that displacement of O_2 was not the only reason for the antimicrobial effect of CO_2. Valley and Rettger (1) suggested that CO_2 acted by lowering extracellular pH as a result of the dissolution of CO_2 in the aqueous phase of the product. However, several studies have shown that when the pH is lowered by inorganic acid to values equivalent to those achieved under CO_2 atmospheres, bacterial and mold growth was less inhibited. Furthermore, CO_2 will inhibit microbial growth in buffered media and in naturally buffered foods such as meat. In response to these observations, Wolfe suggested that the inhibitory effect of CO_2 may be due to intracellular, rather than extracellular, pH changes that could interfere with enzymatic activities associated with cell metabolism. Several studies have shown that CO_2 inhibits oxaloacetate decarboxylase, succinate dehydrogenase, and cytochrome oxidase activity. However, King and Nagel (2) observed that CO_2 did not inhibit extracts of these enzymes. They observed that CO_2 specifically inhibited malic and isocitric dehydrogenase activity *in vitro* and concluded that the inhibitory effect of CO_2 may be due to its mass action effect on decarboxylases within the cell. Another theory suggests that CO_2 acts on the cell membrane, affecting the permeability characteristic of the membrane and its external environment by redistributing lipids at the surface. This has been demonstrated using a model system by Sears and Eisenberg (3) and has been proposed as the mechanism by which CO_2 may inhibit aerobic spore germination.

In conclusion, though many studies have been performed on the effect of CO_2 on microorganisms, there is little conclusive evidence of its mechanism of action. In a review on the effects of carbon dioxide on microbial growth and food quality by Daniels et al. (4) the following appear to be the salient points of investigations:

- The exclusion of oxygen by replacement with carbon dioxide may contribute to its overall antimicrobial effect by slowing the growth of aerobic spoilage microorganisms.
- The carbon dioxide/bicarbonate ion has an observed effect on the permeability of cell membranes.
- Carbon dioxide is able to produce a rapid acidification of the internal pH of the microbial cell with possible ramifications relating to metabolic activities.
- Carbon dioxide appears to exert an effect on certain enzyme systems.

Whatever the reason for its antimicrobial effect, CO_2 is effective in extending the shelf life of perishable foods by retarding microbiological growth. The overall effect of CO_2, in conjunction with refrigeration, is to increase both the lag phase and the generation time of spoilage microorganisms.

FACTORS INFLUENCING THE ANTIMICROBIAL EFFECT OF CO_2

Several factors influence the antimicrobial effect of CO_2, specifically, temperature, types and numbers of microorganisms, gas concentration, and package material properties. Each of these factors will be briefly reviewed.

Types of Microorganisms

The numbers and types of microorganisms present in a food product influence the antimicrobial effect of CO_2. Microorganisms differ considerably in their sensitivity to CO_2, and this sensitivity is related to the oxygen requirements of microorganisms (Table 2). It has been shown that CO_2 is most effective against aerobic spoilage microorganisms. Common aerobic spoilage organisms of meat, fish, and poultry (*Pseudomonads, Acinetobacter / Moraxella*) are inhibited by low concentrations of CO_2, a fact that is exploited in the gas packaging of muscle foods. Molds, most of which require oxygen for growth, are similarly inhibited by CO_2. Several studies have shown that low concentrations of CO_2 (10%) can be used to suppress mold growth. However, as with bacteria, mold species may vary in their sensitivity to the inhibitory effects of CO_2.

Table 2. Oxygen Requirements of Typical Food Spoilage Microorganisms

Aerobes	Require atmospheric oxygen for growth	Pseudomonads *Acinetobacter* / *Moraxella* *Micrococcus* Film yeasts Molds
Microaerophiles	Require low levels of oxygen	*Campylobacter* *Listeria* *Lactobacillus*
Facultative anaerobic microorganisms	Growing in the presence or absence of oxygen	*Staphylococcus* *Bacillus species* Enterobacteriaceae Vibrio Fermentative yeasts
Anaerobes	Inhibited (or killed) by oxygen	*Clostridium botulinum* *Clostridium perfringens*

Although CO_2 is effective against aerobic spoilage microorganisms, it has little or no antimicrobial effect against other spoilage microorganisms. Several studies have shown that CO_2 has little or no effect on the growth of facultative microorganisms in the Enterobacteriaceae or microaerophilic lactic acid bacteria. These organisms have been reported capable of growth in high concentrations of CO_2 (75–100%). Anaerobic bacteria, such as the food-poisoning organisms *Clostridium botulinum* and *C. perfringens*, are not affected by the presence of carbon dioxide, and the anaerobic conditions inside MAP foods may be conducive to their growth. There is concern that these organisms represent a potential public health hazard if present in gas-packaged foods, particularly if packaged under completely anaerobic conditions and stored under conditions of temperature abuse. Thus, although low concentrations of CO_2 (20–30%) may suppress the growth of aerobic microorganisms, they do little to inhibit the growth of facultative anaerobes, microaerophiles, and strict anaerobes. Under MAP conditions, the microbiological population shifts from a predominantly aerobic one to one comprising almost entirely CO_2-resistant anaerobic bacteria. In effect, the MAP suppresses competitive spoilage microorganisms and offers conditions conducive to pathogenic anaerobic microbiological growth. For this reason, prevention of contamination of MAP foods with pathogenic microorganisms is very important.

The age of the microbiological population also influences the inhibitory effect of CO_2. It has been shown that, as bacteria move from the lag phase to the log phase, the inhibitory effects of CO_2 are reduced. Thus, the earlier the product is gas packaged, the more effective CO_2 will be.

Concentration of CO_2

Early experiments clearly established that success in controlling aerobic spoilage deterioration of food was not simply dependent on the elimination of oxygen; rather, there was a definite requirement for CO_2 in the gas atmosphere. Coyne reported that the growth of *Achromobacter, Flavobacterium, Micrococcus, Bacillus*, and *Pseudomonas* was markedly inhibited by 25% CO_2 and completely inhibited by 50% CO_2 (5). In a later study, Coyne (6) reported that the optimal concentration for inhibition of aerobic spoilage microorganisms was 40 to 60% CO_2. No additional extension of shelf life was obtained by using higher concentrations of CO_2, and bacterial growth was less inhibited below these concentrations. Gill and Tan (7) examined the effect of various concentrations of CO_2, equivalent to pressures of 100 to 300 mm Hg, that is, 13 to 39% CO_2 in air, on the respiration rates of *Pseudomonas* species, *Acinetobacter, Alteromonas putrefaciens, Yersinia enterocolitica*, and *Enterobacter*. They reported that the respiration rates of most of the common spoilage organisms under investigation, with the exception of *Enterobacter* and *Brocothrix thermosphacta*, were affected by elevated levels of CO_2 in air. The level of CO_2 that resulted in maximum inhibition of the common spoilage organisms was approximately 200 mm Hg, or 26% CO_2 in air. However, the level of CO_2 investigated in this study had no antimicrobial effect on *B. thermosphacta*, which requires concentrations of 75% or more CO_2 for complete inhibition.

Several studies have shown that mold growth is also inhibited by low concentrations of CO_2. For example, many *Aspergillus, Rhizopus*, and *Cladosporium* species are completely inhibited by 5 to 10% CO_2 at 1°C. Other studies have shown that 20 to 30% CO_2 was sufficient to prevent the growth of meat-associated molds whereas 30 to 50% CO_2 was found to completely inhibit all mold species associated with the spoilage of bread and cakes. Again, this was not simply due to lower partial pressures of O_2 in the gas atmosphere, because it has been shown that many molds continue to grow normally when the O_2 concentration is maintained as low as 1%.

It is evident from these studies that the concentration of CO_2 in the gas mixture is very important in obtaining the desired extension of microbiological shelf life of the product. For most food products, with the exception of fruits and vegetables, a minimum of 20 to 30% CO_2 by volume is required to inhibit the growth of aerobic spoilage microorganisms and extend the shelf life of food, whereas for maximum shelf-life extension a concentration of 50 to 60% should be used. Though there is little or no increased antimicrobial effect or extension in shelf life at concentrations of CO_2 >50 to 60%, slightly higher concentrations are sometimes used to compensate for losses of headspace CO_2 through packaging films. However, too high a concentration of CO_2 may result in discoloration problems and drip loss in muscle foods. For fruits and vegetables, the maximum concentration of CO_2 is approximately 5 to 20% by volume because higher concentrations result in carbon dioxide damage.

Temperature

Temperature is a key variable in the antimicrobial activity of carbon dioxide. It has been shown that CO_2 is a very effective antimicrobial agent at low storage temperatures but less effective at higher temperatures. This increased inhibitory effect has been attributed to the greater dissolution of CO_2 in the aqueous phase of products at lower storage temperatures and resultant changes in intracellular pH and enzymatic activities of microorganisms. Therefore, any decrease in inhibition of spoilage and extension of shelf life at higher storage temperatures results from the lower solubility of CO_2 in the aqueous phase of the product. MAP should not be regarded as a substitute for proper storage temperature. Although MAP slows the deterioration of a food product, it never totally arrests deterioration. For respiring products, increasing the storage temperature also increases the rate of respiration, resulting in a decrease in shelf life.

The effects of temperature abuse are particularly important from the standpoint of safety. Temperature abuse of MAP muscle foods may result in the rapid growth of both spoilage and pathogenic bacteria. The minimum temperature for growth of *Salmonella* and *Escherichia coli* inoculated in ground meat and packaged in low- and high-permeability film is 0°C. Of major concern with respect to safety of MAP fish is the growth of and toxin production by *C. botulinum* type E, which has been demonstrated to be capable of growth at temperatures as low as 3.3°C. Proper refrigeration is therefore essential to assure the effectiveness of CO_2 as an antimicrobial agent and to prevent potential growth of pathogenic organisms that may not be suppressed by CO_2.

Package Material Permeability

The success or failure of MAP for respiring and nonrespiring foods depends on both the O_2 and CO_2 permeability of package materials to maintain the correct gas mixture in the package headspace. In addition, films used in gas packaging should also have low water vapor–transmission rates to prevent moisture loss or moisture gain, that is, to maintain the proper water vapor concentration because it enters into the MAP reaction. Polymers commonly used for MAP of food include polyester, polyamide (nylon), polypropylene (PP), polyvinylidene chloride (PVDC), ethylene vinyl alcohol (EVOH), ethylene vinyl acetate (EVA), and polyethylene (PE) (Table 3). Because all the desired characteristics of a package material, namely, structural strength, permeability, and heat sealability, are seldom found in one polymer, individual polymers may be coextruded or laminated to one another to produce films with the desired characteristics for MAP. Examples of laminated structures, some of which are shown in Table 4, for MAP of nonrespiring products include nylon/PE, nylon/PVDC/PE, EVA/EVOH/EVA, and nylon/EVOH/PE, all high-gas barrier structures. These composite structures have the desired characteristics for gas packaging of nonrespiring products, specifically strength, provided by the layer of gas and moisture-vapor impermeability (EVOH, PVDC, or nylon), and heat sealability, usually provided by PE.

For fruits and vegetables, the selection of the correct package material has been even more challenging for the packaging technologist because of the dynamic nature of the product. The ideal package material for gas packaging of fruits and vegetables must be able to keep a balance of low O_2 concentration (3–5%) within the package headspace and prevent buildup of high CO_2 concentrations (10–20%). Packaging films commonly used to achieve this balance include metallocene PE, EVA, and polypropylene.

Table 3. Examples of Food-Packaging Materials

Common name	Abbrev.	Subspecies
Polyethylene	PE	LDPE (low-density PE) LLDPE (linear LDPE) HDPE (high-density PE) Metallocene polyethylene
Polyvinyl chloride	PVC	UPVC (unplasticized PVC)
Polyvinylidene chloride	PVDC	
Polystyrene	PS	
Polyethylene terephthalate polyester	PET	
Ethyelene vinyl alcohol	EVOH	
Ethylene vinyl acetate	EVA	

Table 4. Examples of Laminated Films Used in MAP Foods

		Permeability ($cm^3/m^2/24h/1$ atm)		
Laminate	Gauge (μm)	O_2	CO_2	N_2
PET/PVDC/PE	12/3/50	8–10	30	8
UPVC/LDPE	400/75	15	30	4
Nylon/PVDC/PE	60/5/100	9	34	2.5
PVDC-coated PET/PE	15/60	2–4		
UPVC/PE	400/75	15		
Nylon/EVAL/Nylon/PE	25/10/25/100	5	20	1

APPLICATION OF GAS PACKAGING FOR SHELF-LIFE EXTENSION OF FOOD

The application of MAP using various gas mixtures has been used successfully by many food-processing and food-packaging companies around the world to extend the shelf life and retain quality of a variety of food products. Examples of food products currently gas packaged as well as the composition of gas mixtures used to extend the shelf life of each product are shown in Table 5. The optimum blend of gases for a specific product can be determined not simply by trial and error but only through a detailed, systematic study of the interdependent variables influencing product shelf life. These include the physical, chemical, and microbiological composition of the food product, the expected shelf life of the product under the normal storage conditions, and the choice of packaging film of correct gas- and moisture vapor–permeability characteristics. The application of MAP involving gas mixtures for shelf life extension of selected food groups is briefly reviewed.

Muscle Foods (Meat, Fish, and Poultry)

The shelf life of fresh beef, pork, fish, and poultry is limited in the presence of atmospheric oxygen by the growth and biochemical activities of gram-negative, psychrophilic (cold-loving) strains of microorganisms such as *Pseudomonas*, *Achromobacter*, *Flavobacterium*, and *Moraxella* species. Several studies have indicated that these spoilage

Table 5. Examples of Gas Mixtures for Selected Food Products

Product	Temp (°C)	O_2 (%)	CO_2 (%)	N_2 (%)
		Meat products		
Fresh red meat	0–2	40–80	20	Balance
Cured meat	1–3	0	30	70
Pork	0–2	40–80	20	Balance
Offal	0–1	40	50	10
Poultry	0–2	0	20–100	Balance
		Fish		
White fish	0–2	30	40	30
Oily fish	0–2	0	60	40
Salmon	0–2	20	60	20
Scampi	0–2	30	40	30
Shrimp	0–2	30	40	30
		Plant products		
Apples	0–4	1–3	0–3	Balance
Broccoli	0–1	3–5	10–15	Balance
Celery	2–5	4–6	3–5	Balance
Lettuce	<5	2–3	5–6	Balance
Tomatoes	7–12	4	4	Balance
		Baked products		
Bread	RT[a]		60	40
Cakes	RT		60	40
Crumpets	RT		60	40
Crepes	RT		60	40
Fruit pies	RT		60	40
Pita bread	RT		99	1
		Pasta and ready meals		
Pasta	4		80	20
Lasagna	2–4		70	30
Pizza	5		52	50
Quiche	5		50	50
Sausage rolls	4		80	20

[a]Room temperature; staling is accelerated at refrigerated temperatures.

organisms can be inhibited by packaging the product under a CO_2-enriched atmosphere. Under these packaging conditions, the growth of common spoilage microorganisms is inhibited, and microaerophilic strains of lactic acid bacteria, which are more tolerant of high CO_2 concentrations, become the dominant spoilage microorganisms. As a result of bacterial growth being suppressed, levels of chemical compounds, such as trimethylamine in fish, and total volatile nitrogen, which are chemical indicators of microbial spoilage of food, are also reduced.

It is evident from Table 5 that a variety of gas mixtures can be used to extend the shelf life of muscle foods. Although microbial quality is usually the primary concern of the processor, biochemical changes can take place over time within the food itself that adversely affect color, flavor, and texture. For example, with red meat, oxygen is necessary for the bright red color or "bloom" that many consumers associate with good-quality meat. However, oxygen also promotes microbial growth and oxidative rancidity. Carbon dioxide is a bacteriostatic agent, that is, it inhibits microbial growth, but in excess quantities it will discolor fresh meat. The problem of balancing these two separate effects can be overcome by use of a gas mixture incorporating CO_2, O_2, and N_2. The N_2 is needed to prevent the package structure from collapsing around the product as CO_2 dissolves in the flesh, the fat, or both. When intact fresh beef and pork are packed in an atmosphere of 40 to 80% O_2, 20% CO_2, and the balance N_2 and kept under chilled conditions, a shelf life of 10 to 12 days can be expected, providing the meat was of good microbiological condition at the time of packaging. Ground meat displays shorter microbiological shelf life due to its microbiological dispersion. High concentrations of O_2 may be favored because they have been demonstrated to exert an antimicrobial effect and also ensure that the desired bright red color will be maintained.

For cured meat products containing color-fixing agents, where O_2 is not necessary and is even detrimental to product color, it is necessary to package in an O_2-free vacuum—a mixture of CO_2:N_2, or 100% of either gas.

With fish, enzymatic and/or autooxidative changes lead to the formation of low molecular weight aldehydes, ketones, alcohols, and carboxylic acids. Here, the gas mix employed depends on the product's fat content, which varies from 1% to a maximum of 20% for mackerel. Low-fat fish can be packaged in 60% CO_2:40% O_2 whereas high-fat fish, such as mackerel and herring, should be packaged in an oxygen-free environment to prevent rancidity problems. In the United States, low-oxygen fish packaging is discouraged because of fear of growth of nonproteolytic anaerobic *C. botulinum*, capable of growth and toxin production at temperatures as low as 3.3°C.

For meat, fish, and poultry products, the main areas affected by packaging of product under a modified gas atmosphere are summarized in Table 6. It is evident that the major impact of gas packaging is shelf-life extension of product, which, for meat products, can be doubled or tripled compared to air-packaged products (Table 7). Similar extensions in shelf life have been attained for fish and poultry. However, these shelf-life extensions can be obtained only if MAP products are stored under refrigerated-temperature distribution conditions. The effect of distribution temperature is therefore critical to ensure

Table 6. Changes in Meat, Fish, and Poultry as Brought About by Modified Atmospheres

Enzymatic aging process	Unaffected
Microbial spoilage	Increased CO_2 reduces growth of aerobic spoilage pyschrotrophs and psychrophiles
Fat oxiation	Reduced O_2 reduces oxidation of fats, although some oxidation can still occur at low O_2 tensions
Oxidation of myoglobin	Increased CO_2 promotes metmyoglobin formation and color darkening
Enzymatic oxidation	Can be reduced
Color	Elevated O_2 can retain oxymyoglobin cherry red color

Table 7. Shelf-Life Extension of MAP Meat at Chilled Temperatures

		Shelf life (days)	
Product	Temperature (°C)	Air-packaged products	MAP products
Beef cuts	4	4	10–12
Pork buts	4	4	6–9
Ground beef	4	2	4
Offals (eg, liver)	4	1–2	6

maximum benefits of CO_2-enriched atmospheres for shelf-life extension of muscle foods.

Plant Products

MAP, in conjunction with proper temperature control, can be used to reduce spoilage and quality losses in fresh fruit and vegetables. The objectives of MAP of fresh produce are to:

- Inhibit microbiological spoilage, such as mold growth on product surface
- Reduce the respiratory activities of products, thereby delaying ripening and senescence
- Reduce enzymatic browning of plants

As a result of the biological nature of the stored products, the packaging of fruits and vegetables under MAP conditions is one of the most challenging problems facing the packaging industry. The major difference between plant products and other fresh foods is that plant products continue to respire after being harvested. This results in a depletion of O_2 and a buildup of CO_2. However, when O_2 supplies are too low or CO_2 levels too high, anaerobic respiration (fermentation) occurs, resulting in the production of alcohols, aldehydes, and ketones or other volatiles that impart undesirable off-odors and off-flavors to the product. To avoid oxygen starvation in MAP products, headspace oxygen generally should not fall below 2%. Some varieties of apples can withstand up to 10% CO_2, whereas strawberries can withstand 25% CO_2 for prolonged periods. Examples of gas mixtures for shelf-life extension of fruits and vegetables are shown in Table 5. Although carbon monoxide is not permitted in MAP of food, 4% by volume is permitted by the FDA in modified atmospheres to retard browning of lettuce cores. Currently, this is the only use of CO for food preservation permitted by regulatory authorities, although it is used rarely, if ever.

The key to successful MAP distribution of fresh and fresh-cut fruits and vegetables is to ensure the correct balance of gases within the package headspace. The challenge for the packaging industry in MAP of plant products is to develop package materials that allow the transfer of selected gases and moisture in a controlled fashion to reduce biochemical activity without a major reduction in headspace oxygen or a buildup of carbon dioxide. This is a very complex task, requiring the control of significantly more variables than with other products. Common films used are PE, PP, EVA, and styrene block copolymer. Microperforated films (eg, P-Plus from Sidlaw), films diluted with minerals (eg, FreshHold), and films containing special polymers responsive to temperatures (eg, Landec) are all commercially available to control oxygen concentrations within packages of respiring foods. The use of PE bags with silicone rubber windows for controlled permeation of O_2 and CO_2 have been used in France. These bags were capable of maintaining 3 to 5% O_2 and CO_2 at 3°C.

Using MAP in conjunction with temperature control, shelf-life extensions of 15 to 30 days are possible for fruits, vegetables, and prepared salads. Excellent reviews on MAP of fruits and vegetables can be found in the text by Brody and the paper by Zagory and Kader (8).

Bakery Products

The major problems limiting the shelf life of many bakery products are mold spoilage and staling. The bakery industry can use several methods to control mold spoilage, the most common being the use of preservatives, such as sorbates and benzoates. However, preservatives can lead to off-odors and off-flavors in certain products. Furthermore, with increasing consumer concerns about preservatives, the bakery industry has been actively searching for alternative methods to extend the mold-free shelf life of bakery products. One such method is MAP. Research on the use of CO_2-enriched atmospheres for shelf-life extension of bakery products has been performed by Seiler (9) at the Flour Milling and Bakery Research Association in England. He reported shelf-life extensions of 300 to 400% for bread and cakes stored in concentrations of 60% CO_2 or more. The shelf-life extension obtained was dependent on the water activity (a_w) of the product, with the most significant increases in shelf life occurring in products of lower water activities, that is, $a_w \leq 0.85$. This effect is due to the types of mold present. In low-moisture products, xerophilic molds (low moisture tolerant), such as *Aspergillus* species, were the predominant spoilage molds, while *Penicillium* species were more common in products with higher a_w values, such as crumpets and fruit pies. Ooraikul reported a mold-free shelf-life extension of 1 month or more for English-style crumpets packaged under 60% CO_2. Though slightly longer extensions were possible using higher CO_2 concentrations, the gas was absorbed by the product, creating a vacuum-packed effect. Bakery companies in Europe commonly use gas packaging for shelf-life extension of bread and cakes. In addition to extending the mold-free shelf life of products, CO_2-enriched atmospheres have also been reported to prevent staling in many bakery products. Examples of bakery products commonly gas packaged are shown in Table 5. Packaging films commonly used with these products are barrier films such as nylon/PE and PVDC-coated polypropylene laminated or extrusion coated with polyethylene or ionomer.

Fresh Pasta and Other Products

MAP is widely employed to extend the shelf life and keeping quality of fresh pasta and snack-food products. Major spoilage problems associated with moist (30% H_2O) pasta products are growth associated with bacteria and molds and rancidity problems from fat oxidation. These spoilage problems can be retarded by eliminating headspace oxygen in the packaged products by packaging under an appropriate CO_2:N_2 gas mixture or under zero oxygen using an oxygen scavenger (Table 5). For other products, such as snack foods, which generally have a lower water activity, specifically, a_w, <0.6, packaging in as close to 100% N_2 as possible is sufficient to minimize oxidative rancidity problems because these products are not susceptible to microbial spoilage due to their low a_w and therefore do not need CO_2 in the gas mixture.

ADVANTAGES AND DISADVANTAGES OF MAP

The main benefits associated with MAP of food products are better quality retention, extended product shelf life and associated increase in market area, improved product presentation and consumer appeal, and a reduction in energy costs associated with freezing and freezer storage costs (Table 8).

Some of the disadvantages of the technique (Table 9) include:

- The initial higher cost of packaging equipment
- Higher cost of package materials
- Secondary fermentation problems caused by CO_2-resistant microorganisms
- Production of acidic-type odors due to dissolution of CO_2, in certain products such as fish
- A reduction in water binding capacity and an increase in drip loss of muscle foods due to change in product pH
- Slight discoloration problems in muscle foods such as meat
- Package collapse in products using a high CO_2 concentration (100%)
- The potential for generating conditions favorable for anaerobic pathogenic microbiological growth

Table 8. Advantages of MAP of Food

Incrased shelf life
Superior quality retention
Incrased market area
Reduction in production and distribution costs
Improved presentation
Fresh appearance
Clear view of product when desired

Table 9. Disadvantages of Gas Packaging of Food

Initial high cost of packaging equipment, films, etc.
Discoloration of meat pigments
Leakage
Fermentation and swelling
Potential growth of anaerobic microorganisms of significance to public health

The high CO_2 problems can be inhibited by reducing the proportion of CO_2 in the gas mixture and replacing it with a filler gas (N_2).

MICROBIOLOGICAL SAFETY OF MAP FOODS

A major concern about MAP foods is that they may be a public health risk, particularly if subjected to temperature abuse during distribution and retail storage. Further, a higher risk exists for consumer temperature abuse, mishandling, and overextending the product's shelf life.

The major microbiological concern with MAP foods is the growth of and toxin production by *C. botulinum* types A, B, and E. These spore-forming microorganisms pose the greatest threat to consumer safety due to:

- The complete or partial reduction of aerobic microorganisms and potential indicators of incipient spoilage by CO_2-enriched atmospheres, and
- The presence of elevated levels of CO_2 and anaerobic packaging conditions conductive to the growth of, and toxin production by, *C. botulinum*.

Several studies have shown that spores of *C. botulinum* can outgrow and produce toxin in MAP fish and in other products such as nitrogen-packed sandwiches and vacuum-packaged potatoes, especially if stored under temperature-abuse conditions. Though several challenge studies have been done with *C. botulinum* in MAP food, there is little conclusive evidence that MAP represents a significantly greater hazard than packaging in air, particularly under conditions of temperature abuse where CO_2 is less effective. Some believe that the inclusion of oxygen in the package headspace may prevent the growth of *C. botulinum* in products that may be susceptible to contamination by this pathogen; however, recent studies have refuted this claim and have shown that the inclusion of O_2 in the package headspace may offer no additional protection against *C. botulinum*.

Non-spore-forming pathogens in MAP foods include *Salmonella, Staphylococcus, Listeria*, and *Yersinia* species. Several studies have concluded that MAP storage does not increase the microbiological hazards from *Salmonella* species and *Staphylococcus aureus*. Hintlian and Hotchkiss (10) reported that the growth of both *S. aureus* and *Salmonella* species is inhibited by high CO_2 concentrations and that the level of inhibition increased as storage temperature decreased. Other pathogens, such as *Y. enterocolitica* and *L. monocytogenes* can grow slowly at refrigerated storage temperatures and so should be controlled by other means. Little is known about the ability of these organisms to grow in MAP foods, however, and further research is needed in this area.

CONCLUSION

MAP involving gas mixtures is not a panacea and should not be regarded as a substitute for good manufacturing practices, HACCP programs, and proper distribution temperature conditions for shelf-life extension of food. When used in conjunction with these factors, however, substantial extensions in shelf life and economies in production and distribution are possible. Although the concept of MAP may appear to be simple, its implementation requires the careful evaluation of the chemistry, physiology, and microbiology of the food system in relation to the dynamics of the microenvironment and the packaging material.

BIBLIOGRAPHY

1. G. Valley and L. F. Rettger, "The Influence of Carbon Dioxide on Bacteria," *J. Bacteriol.* **14**, 101–137 (1927).
2. A. D. King and C. W. Nagel, "Influence of Carbon Dioxide Upon the Metabolism of *Pseudomonas aeruginosa*," *J. Food Sci.* **40**, 362–366 (1975).
3. D. F. Sears and D. F. Eisenberg, "A Model Representing a Physiological Role of Carbon Dioxide at the Cell Membrane," *J. Gen. Physiol.* **44**, 869–877 (1961).
4. J. A. Daniels, R. Krishnamurthi, and S. H. Rizvi, "A Review of the Effects of Carbon Dioxide on Microbial Growth and Food Quality," *J. Food Prot.* **48**, 532–537 (1985).
5. F. P. Coyne, "The Effect of Carbon Dioxide on Bacterial Growth with Special Reference to the Preservation of Fish, Part I," *J. Soc. Chem. Ind. (London)* **51**, 119T–121T (1933).
6. F. P. Coyne, "The Effect of Carbon Dioxide on Bacterial Growth with Special Reference to the Preservation of Fish, Part II," *J. Soc. Chem. Ind. (London)* **52**, 19–24 (1933).
7. C. O. Gill and K. H. Tan, "Effect of Carbon Dioxide on Meat Spoilage Bacteria," *Appl. Environ. Microbiol.* **39**, 317–324 (1980).
8. D. Zagory and A. A. Kader, "Modified Atmosphere Packaging of Fresh Produce," *Food Technol.* **42**, 70–77 (1988).
9. D. A. L. Seiler, "The Microbiology of Cake and Its Ingredients," *Food Trade Rev.* **48**, 339–344 (1978).
10. C. B. Hintlian and J. H. Hotchkiss, "The Safety of Modified Atmosphere Packaging: A Review," *Food Technol.* **40**, 70–76 (1986).

GENERAL REFERENCES

N. Y. Aboagye et al., "Energy Costs in Modified Atmosphere Packaging and Freezing Processes as Applied to a Baked Product," in M. LeMaguer and P. Jelen, eds., *Food Engineering and Process Applications*, Vol. 2, Elsevier Applied Science, Amsterdam, The Netherlands, 1986, pp. 417–427.

B. Blakistone, *Principles and Applications of Modified Atmosphere Packaging of Food*, 2nd ed., Blackie, Glasgow, Scotland, 1998.

A. L. Brody, "Integrating Aseptic and Modified Atmosphere Packaging to Fulfill a Vision of Tomorrow," *Food Technol.* **50**, 56–66 (1996).

A. L. Brody, *Controlled / Modified Atmosphere / Vacuum Packaging of Foods*, Food and Nutrition Press, Trumbull, Conn., 1990.

A. L. Brody, "Minimally Processed Foods Demand Maximum Research and Education," *Food Technol.* **52**, 62, 64, 66, 204, 206 (1998).

A. L. Brody, *Modified Atmosphere Food Packaging*, Institute of Packaging Professionals, Herndon, Va., 1994.

S.-O. Enfors and G. Molin, "The Influence of High Concentrations of Carbon Dioxide on the Germination of Bacterial Spores," *J. Appl. Bacteriol.* **4**, 279–285 (1978).

J. Farber and K. L. Dodds, *Principles of Modified Atmosphere and Sous Vide Product Packaging*, Technomic Lancaster, Pa., 1995.

G. Finne, "Modified- and Controlled-Atmosphere Storage of Muscle Food," *Food Technol.* **36**, 128–133 (1982).

C. A. Genigeorgis, "Microbial and Safety Implications of the Use of Modified Atmospheres to Extend the Storage Life of Fresh Meats and Fish," *Int. J. Food Microbiol.* **1**, 237–251 (1985).

K. E. Goodburn and A. C. Halligan, *Modified Atmosphere Packaging: A Technology Guide*, British Food Manufacturing Research Association, Leatherhead, England, 1988.

A. D. Lambert, J. P. Smith, and K. L. Dodds, "Combined Effects of Modified Atmosphere Packaging and Low-Dose Irradiation on Toxin Production by *Clostridium botulinum* in Fresh Pork," *J. Food Prot.* **54**, 94–101 (1991).

B. Ooraikul, "Gas Packaging for a Bakery Product," *Can. Inst. Food Sci. Technol. J.* **15**, 313–317 (1982).

S. A. Palumbo, "Is Refrigeration Enough to Restrain Food Borne Pathogens?," *J. Food Prot.* **49**, 1003–1009 (1986).

R. T. Parry, *Principles and Applications of Modified Atmosphere Packaging of Food*, Blackie, Glasgow, Scotland, 1993.

M. L. Rooney, *Active Food Packaging*, Blackie, Glasgow, Scotland, 1995.

J. P. Smith et al., Novel "Approach to Oxygen Control in Modified Atmosphere Packaging of Bakery Products," *Food Microbiol.* **3**, 315–320 (1986).

J. P. Smith et al., "Shelf Life Extension of a Bakery Product Using Ethanol Vapor," *Food Microbiol.* **4**, 329–337 (1987).

J. P. Smith, E. D. Jackson, and B. Ooraikul, "Microbiological Studies on Gas Packaged Crumpets," *J. Food Prot.* **46**, 279–283 (1983).

S. K. Wolfe, "Use of CO and CO_2 Enriched Atmospheres for Meats, Fish, and Produce," *Food Technol.* **34**, 55–58 (1980).

L. L. Young, R. D. Reviere, and A. B. Cole, "Fresh Red Meats: A Place to Apply Modified Atmospheres," *Food Technol.* **42**, 65–69 (1988).

AARON BRODY
Rubbright•Brody, Inc.
Duluth, Georgia

See also FRESH-CUT FRUITS AND VEGETABLES: MODIFIED ATMOSPHERE PACKAGING; CONTROLLED ATMOSPHERES FOR FRESH FRUITS AND VEGETABLES.

PALM OIL

Palm oil has now become the second most important vegetable oil after soybean oil in the world's oils and fats market. The dynamic growth of the palm-oil industry in Malaysia began around 1970. The pace of growth, in fact, has been so rapid that palm oil has now overtaken soybean oil and rapeseed oil as the growth leader. The oil palm is planted commercially in more than a dozen countries around the world, in West and Central Africa, Southeast Asia, Central and South America, China, and Papua New Guinea. However, the Southeast Asian nations of Malaysia and Indonesia are the major palm-oil-producing countries and between them account for about 80% of the world's palm-oil production. The overwhelming importance of Malaysia in world palm-oil exports can be seen from the fact that in 1989 it exported nearly 6 million tons and produced about 58% of the world palm-oil output (1,2).

THE OIL PALM AND ITS FRUITS

The oil palm *Elaeis guineensis* has its origin in West Africa. The world *Elaeis* is derived from the Greek word *elaion*, meaning oil, while *guineensis* refers to its origin in Guinea. *Elaeis guineensis* is the highest yielding oil-bearing plant and is now planted as a commercial crop throughout oil-palm growing countries of the world, thriving best within 10° of the equator (3,4).

The Central and South American oil palm, *Elaeis oleifera*, is another species in the genus, but it is of little economic importance owing to its poor oil yield. However, it is an important source of breeding material for oils of greater unsaturation.

Within the species *Elaeis guineensis*, several strains can be identified on the basis of variations of the endocarp of the fruits:

Dura: thick shell, comparatively less oil-bearing mesocarp.

Pisifera: very thin shell to shell-less, high proportion of oil-bearing mesocarp.

Tenera: hybrid of dura and pisifera, thin shell, more mesocarp, hence more oil.

Today, in Malaysia the commercially planted palms are mostly tenera.

The oil palm produces fruits by 2–3 years after planting. The mature palm produces 20–25 tons of fruit bunches per hectare annually, and the economic life span of the tree is about 25–30 years (see Fig. 1). The proportion of oil to bunch of tenera palms is around 25%, although the factory oil extraction ratio is about 20–22% owing to milling losses. The oil yield per hectare annually can average between 4–5 tons (3,4).

The palm fruits are borne on spikelets of the fruit bunch. The mature fruit bunch contains from a few hundred to a few thousand fruits and weighs as much as 40 kg. The loose or detached palm fruit is avoid in shape and varies from 2 to 5 cm in length, weighing 8–20 g. Unripe fruits of the tenera are mostly deep violet to black in color, ripening to a deep orange to bright red.

The palm fruit consists of the outer skin, the mesocarp (pulp), and the seed containing the endocarp (shell) and the kernel (see Fig. 2). Palm oil is extracted from the mesocarp and should be distinguished from palm kernel oil derived from the kernel.

Figure 1. The oil palm bearing fruit bunches.

Figure 2. Oil palm fruits and sections showing the oil-bearing mesocarp and the endocarp (kernel).

EXTRACTION OF CRUDE PALM OIL

Several stages of processing can be identified in the extraction of palm oil (1,5) from fresh fruit bunches:

Sterilization

Freshly harvested fruit bunches are brought to the mill and sterilized by high-pressure steam (120–140°C at 40 psi) with minimal delay so as to inactivate the lipolytic enzymes that cause oil hydrolysis and fruit deterioration. Sterilization also loosens the fruits from the bunch stalk and coagulates the protein and mucilaginous materials present in the palm fruit to prevent emulsion formation during oil recovery.

Bunch Stripping

After sterilization, the fruits are separated from the bunch stalks by mechanical stripping. The sterilized fruits go to the digester while the empty fruit bunches are incinerated, yielding an ash rich in potash that is used as a fertilizer, particularly in acid fields.

Digestion

The fruits are reheated (digested) by steam to a temperature of 80–90°C. The digestion process prepares the fruits for oil extraction by rupturing the oil-bearing cells in the mesocarp and loosening the mesocarp from the nuts.

Oil Extraction

Crude palm oil is extracted from the digested fruit mash by the use of the screw press without nut (kernel) breakage. Crude palm-oil liquor and fiber and nuts are discharged from the screw press.

Clarification and Purification

Crude palm oil extracted from the digested fruits by pressing contains varying amounts of water, solids, and dissolved impurities that must be removed. Fiber particles from the pressed crude oil are first removed by passing the oil over a vibrating screen, and sand and dirt are allowed to settle. Water is removed by settling or centrifuging and finally by vacuum drying. However the clarified crude oil still contains 0.1–0.25% of moisture, which helps to maintain oxidative stability and prevent the deposition of small amounts of soluble solids known as gums. The final oil is pumped to storage tanks pending despatch to the refinery.

The mill fibers (fibers from mesocarp after oil extraction) are partly burned to generate energy for mill-processing. The nuts are dried and cracked to obtain the kernels that can be further processed to yield palm kernel oil; the resultant cake left after oil extraction is used as animal feed.

REFINING AND FRACTIONATION

The minor components of crude palm oil are phospholipids, carbohydrates, trace metals, free fatty acids, mono- and diglycerides, sterols, carotenoid pigments, tocopherols, tocotrienols, and oxidized and odoriferous materials. Refining reduces those impurities of the crude oil that will adversely affect the quality of the end product, while retaining as much as possible of the tocopherols and tocotrienols because, as antioxidants, they confer oxidative stability to the end product. Crude palm oil is processed by two main methods, refining and fractionation (1,6).

Physical Refining

In Malaysia, physical refining, which subjects the oil to steam distillation under high temperature and vacuum, is the most widely adopted method for refining palm oil. The refining process consists of two major stages, pretreatment and distillation.

Pretreatment. Pretreatment involves conditioning with concentrated phosphoric acid, subsequent adsorption with activated bleaching earth under vacuum at a temperature of 90–130°C, followed by filtration. Pretreatment reduces the levels of phosphorus, trace metals, peroxides, carotenoid pigments, and their condensation products.

Deodorization and Distillation. The distillation step of physical refining involves steam distillation under vacuum at a temperature of 250–260°C to produce an oil of bland flavor and odor with oxidative stability. The process removes the volatile free fatty acids, aldehydes, and ketones (formed by oxidation of unsaturated fatty acids), degraded carotenoid pigments, hydrocarbons, sterols, tocopherols, and tocotrienols (vitamin E). However, not all the tocopherols and tocotrienols are lost in the distillate, as up to 62% of them is retained in the refined oil (7).

Fractionation

The triglyceride composition of palm oil is such that its major triglycerides include oleodipalmitin with a melting point of 37°C and palmitodiolein with a melting point of 19°C.

The aim of fractionation is to produce a liquid olein (slip melting point 21.6°C and cloud point 8.8°C) and a solid stearin. The olein is used mainly as a liquid vegetable oil for cooking and frying, while the stearin fraction is used in margarine, shortening, and frying fat.

In Malaysia, fractionation of palm oil into the low melting point olein and the high melting point stearin is done by two processes, referred to as dry fractionation and detergent fractionation.

During dry fractionation, the oil is first heated to 70–75°C to melt all crystal nuclei. The heated oil is then passed to a crystallizer equipped with a cooling jacket where it is cooled to 18–20°C for 4–8 h. The crystallization of stearin in liquid olein causes slurry formation. The liquid olein is separated from the solid stearin by membrane filtration of the slurry.

In detergent fractionation, the separation of olein and stearin is made possible by mixing the crystalline slurry with a detergent solution (sodium lauryl sulfate and magnesium sulfate). On centrifuging, the lighter olein is separated from the heavier aqueous phase containing the stearin. The liquid olein is washed, dried, and sent for storage while the stearin is recovered from the aqueous phase and the detergent is recycled. Special doubly fractionated olein (superolein) with a lower cloud point and higher iodine value and a softer stearin known as palm midfraction (PMF) can be obtained by a second fractionation. Crude palm oil can also undergo detergent fractionation, first to produce crude olein and crude stearin and thereafter refined to obtain the edible grades of palm olein and palm stearin.

THE CHEMICAL AND PHYSICAL CHARACTERISTICS OF REFINED PALM OIL AND ITS FRACTIONS

The chemical and physical properties of an oil are important determinants of its applications. Palm oil consists of more than 99% glycerides with less than 1% nonglyceridic minor components. Table 1 shows the fatty acid composition and other characteristics, including the solid fat con-

Table 1. Characteristics of Refined Palm Oil and its Fractionated Products

Characteristic	Refined palm oil	RBD palm olein	RBD palm olein (doubly fractionated)	RBD palm stearin	Palm midfraction
Fatty acid composition, %					
Lauric 12:0	0.2	0.2	0.1	0.3	0.1
Myristic 14:0	1.1	1.0	1.0	1.5	1.2
Palmitic 16:0	44.0	39.8	36.6	52.0	51.0
Palmitoleic 16:1	0.1	0.2	0		
Stearic 18:0	4.5	4.4	4.0	4.9	5.6
Oleic 18:1	39.2	42.5	44.8	33.3	34.0
Linoleic 18:2	10.1	11.2	12.6	7.6	7.4
Linolenic 18:3	0.4	0.4	0.2	0.1	0.1
Arachidic 20:0	0.4	0.4	0.7	0.3	0.6
Total saturated	50.2	45.8	42.4	59.0	58.5
Total monounsaturated	39.2	42.5	44.8	33.3	34.0
Total polyunsaturated	10.5	11.6	12.8	7.7	7.5
Iodine value	53.3	58.0	61.4	44	42.5
Slip melting point °C	36.0	21.6	19.0	44–56	
Solid fat content, %					
5°C	62.2	51.1		67.2	
10°C	50.3	37.0	16.7	61.2	76.1
15°C	35.2	19.2	5.0	53.1	66.2
20°C	23.2	5.9	3.1	43.4	52.1
25°C	13.7			33.9	21.9
30°C	8.5			27.0	18.3
35°C	5.8			22.2	13.4
40°C	3.5			17.3	7.9

tent at varying temperatures, of refined, bleached, and deodorized (RBD) palm oil and its fractions (8–10).

Table 2 shows the content of the principal minor components, namely, the vitamin E and sterols (11–13) in refined palm oil.

It is of interest to point out that palm oil is unusually rich in the vitamin E tocotrienols (unsaturated analogues of tocopherol), with the tocotrienols constituting between 70–80% of the total vitamin E content as opposed to other vegetable oils in which tocopherols are the predominant vitamin E species (12).

The sterols identified in palm oil consist of cholesterol, campesterol, stigmasterol, and sitosterol, but their levels are considerably reduced after refining. Palm oil, like other vegetable oils, is generally regarded as cholesterol-free, although a level of 2 ppm has been reported in the refined oil from the initial levels of 7–13 ppm in the crude oil (9).

Table 2. Vitamin E and Sterols in Refined Palm Oil

	Mean/range in ppm
Vitamin E	716 (559–902)
α-Tocopherol	158
α-Tocotrienol	143
γ-Tocotrienol	329
δ-Tocotrienol	86
Sterols	
Cholesterol	2
Campesterol	26–30
Stigmasterol	12–23
Sitosterol	68–114

Crude palm oil is one of the richest natural sources of β-carotene (12), containing an average of 700 ppm of carotenoids (giving it the red color) mainly in the forms of β-carotene (56%) and α-carotene (35%). However the carotenoids are absent in the refined oil as they are heat-destroyed and removed during the deodorization stage of oil refining.

Palm kernel oil (PKO), as distinct from palm oil, is extracted from the kernel of the oil-palm fruit. About 9 tons of kernel oil are produced for every 100 tons of palm oil. Although derived from the same palm fruit, the physical and chemical characteristics of the two oils are quite different. PKO is grouped in the same class as coconut oil. It is particularly rich in the short-chain and medium-chain fatty acids, viz, 12:0 lauric acid (48%) and 14:0 myristic acid (16%).

Like palm oil, PKO can also be fractionated into palm kernel olein and palm kernel stearin, which greatly extends its food applications, particularly in specialty fats that require a high solid fat content between 20 and 30°C.

USES OF PALM OIL AND ITS FRACTIONS

Refined palm oil, RBD palm olein, and RBD palm stearin are the principal forms of processed palm oil that are now traded in the world vegetable oils market. Palm oil, palm olein, and palm stearin are, respectively, semisolid, liquid, and solid at 25°C. Because of their wide plastic range, palm oil and its fractions lend themselves to a wide range of food uses from liquid oils to the semisolid fats such as shortening, margarine, and vanaspati (vegetable ghee). In temperate countries, palm olein can also be used as a liquid oil by blending with domestic sources of polyunsaturated oils up to a level of 30%. Such a blend greatly increases the oxidative stability of the polyunsaturates.

The extreme versatility of palm oil and its products for edible and nonedible uses (10,14) are shown as follows:

Food Uses

Cooking/frying oil
Shortening and cooking fats
Vanaspati
Margarine
Cocoa butter substitute
Ice cream
Coffee whitener
Bakery and biscuit fats
Instant noodles
Filled milk
Soup mixes

Nonfood Uses

Oleochemicals (fatty acids, fatty alcohols, fatty amines, glycerol and methyl esters)
Detergents and surfactants
Soap and metal soap
Candles
Lubricating grease
Cosmetics
Tin plating
Diesel substitute
Plasticizer for plastics

NUTRITIONAL AND HEALTH ASPECTS OF PALM OIL

Recent years have seen a considerable amount of anti-palm-oil publicity, claiming that palm oil is unhealthy because of its saturated fat content. The allegation, however, has been challenged by recent animal and human feeding experiments (15). New findings now indicate that not only has palm oil little effect on raising blood cholesterol levels, but, in many instances, its consumption has led to a reduction in blood cholesterol levels (Table 3). The rationale for palm oil's lack of a cholesterol-raising effect despite the fact that it is classified as a saturated fat may be attributed to the following:

- Its conspicuous lack of the cholesterol-raising 12:0 lauric acid and 14:0 myristic acid (the two acids accounting for less than 1.5% of total).
- Its principal saturated fatty acid, the 16:0 palmitic acid, is now regarded as neutral and certainly much less cholesterolemic than previously thought (19).
- Moderate abundance of the cholesterol-lowering monounsaturated 18:1 oleic acid plus adequate amounts of the essential 18:2 linoleic acid.

Table 3. Effects of Palm-Oil-Containing/Enriched Diets on Blood Cholesterol Levels

Effect	Reference
Chickens	
Palm oil lowered blood cholesterol compared to butter, lard, tallow, and coconut oil	16
Rats	
Palm oil lowered blood cholesterol compared to olive oil	17
Palm oil comparable to corn oil	18
Monkeys	
Palm oil lowered blood cholesterol compared to coconut oil and comparable to high-oleic safflower oil	19
Humans	
Palm oil lowered blood cholesterol compared to coconut oil	20
Palm oil lowered blood cholesterol compared to butterfat and vanaspati	21
Palm oil comparable to a Dutch Fat blend	22

- Presence of vitamin E tocotrienols that are known to suppress liver cholesterol synthesis, consequently reducing blood levels of cholesterol (23,24).

In addition, palm-oil derived tocotrienols have also been demonstrated to inhibit platelet aggregation and thromboxane production (24), which may well explain the observation of a favorable antithrombotic effect found in palm-oil-fed animals (25,26).

BIBLIOGRAPHY

1. *PORAM Technical Brochure*, 4th ed., The Palm Oil Refiners Association of Malaysia, 1989.
2. S. Mielke, "Past and Prospective World Production and Exports of Palm Oil," in F. D. Gunstone, ed., *Palm Oil*, John Wiley & Sons, New York, 1987.
3. B. J. Wood, "Growth and Production of Oil Palm Fruits" in Ref. 2.
4. C. W. S. Hartley, *The Oil Palm*, 3rd ed., Longman Scientific & Technical, UK, 1988.
5. J. H. Maycock, "Extraction of Crude Palm Oil" in Ref. 2.
6. F. V. K. Young, "Refining and Fractionation of Palm Oil," in Ref. 2.
7. M. L. Wong, R. E. Timms, and E. M. Goh, "Colorimetric Determination of Total Tocopherols in Palm Oil, Olein and Stearin," *Journal of the American Oil Chemists' Society* **65**, 258–261 (1988).
8. B. K. Tan and C. H. Flingoh, "Malaysian Palm Oil, Chemical and Physical Characteristics," *PORIM Technology* **3**, **4**, (May 1981).
9. C. L. Chong, "Chemical and Physical Characteristics of Palm Oil," in *Lecture Series, Seventh Palm Oil Familiarisation Programme*, September 3–11, 1987, Palm Oil Research Institute of Malaysia, 1987.

10. S. A. Kheiri, "End Uses of Palm Oil—Human Food," in Ref. 2.
11. A. Gapor, *Palm Oil Development*.
12. S. H. Goh, Y. M. Choo, and S. H. Ong, "Minor Constituents of Palm Oil," *Journal of the American Oil Chemists' Society* **62**, 237–240 (1985).
13. B. Tan, "Palm Carotenoids, Tocopherol and Tocotrienols," *Journal of the American Oil Chemists' Society* **66**, 770–776 (1989).
14. R. J. deVries, "End Uses of Palm Oil—Industrial Uses," in Ref. 2.
15. "New Findings of Palm Oil," *Nutrition Reviews* **45**, 205–207 (1987).
16. A. S. H. Ong, N. Qureshi, A. A. Qureshi, et al., "Effects of Palm Oil and Other Dietary Fats on Cholesterol Regulation in Chickens," *The Federation of American Societies for Experimental Biology Journal* **2**, A1541 (1988).
17. M. Sugano, "One Counter Argument to the Theory that Tropical Oils are Harmful," *Yukagaku* (Journal of the Japanese Oil Chemists' Society, in Japanese) **40**, 48–51 (1987).
18. K. Sundram, H. T. Khor, and A. S. Ong, "Effect of Dietary Palm Oil and its Fractions on Rat Plasma and High Density Lipoproteins," *Lipids*.
19. K. C. Hayes, A. Pronczuk, and S. Lindsey, "Dietary Palmitic Acid Lowers Cholesterol by Comparison to Lauric and Myristic Acids in Monkeys," *Circulation* **80** (Suppl. 2) (1989).
20. T. K. W. Ng, Khalid Hassan, J. B. Lim, "Non-Hypercholesterolemic Effects of a Palm Oil Diet in Malaysian Volunteers," *Abstract in Nutrition and Health Aspects of Palm Oil*, 1989 PORIM International Palm Oil Development Conference, September 5–9, 1989, Kuala Lumpur, Malaysia.
21. S. A. Khan, A. B. Chugtai, L. Khalid, and S. A. Jaffery, "Comparative Physiological Evaluation of Palm Oil and Hydrogenated Vegetable Oils in Pakistan," *Abstract in Nutrition and Health Aspects of Palm Oil*, 1989 PORIM International Palm Oil Development Conference, September 5–9, 1989, Kuala Lumpur, Malaysia.
22. G. Hornstra, K. Sundram, and A. Kester, "The Effect of Dietary Palm Oil on Cardiovascular Risk in Man," *Abstract in Nutrition and Health Aspects of Palm Oil*, 1989 PORIM International Palm Oil Development Conference, September 5–9, 1989, Kuala Lumpur, Malaysia.
23. A. A. Qureshi et al. "The Structure of an Inhibitor of Cholesterol Biosynthesis Isolated from Barley," *Journal of Biological Chemistry* **261**, 10544–10550 (1986).
24. A. A. Qureshi, N. Qureshi, Z. Shen, et al., "Lowering of Serum Cholesterol in Hypercholesterolemic Humans by Palm Vitee," *Abstract in Nutrition and Health Aspects of Palm Oil*, 1989 PORIM International Palm Oil Development Conference, September 5–9, 1989, Kuala Lumpur, Malaysia.
25. G. Hornstra, "Dietary Lipids and Cardiovascular Disease: Effects of Palm Oil," *Oleagineux* **43**, 75–87 (1988).
26. M. L. Rand, A. A. M. Hennissen, and G. Hornstra, "Effects of Dietary Palm Oil on Arterial Thrombosis, Platelet Responses and Platelet Fluidity in Rats," *Lipids* **23**, 1019–1023 (1988).

YOON HIN CHONG
Palm Oil Research Institute of Malaysia
Selangor, Malaysia

See also FATS AND OILS: CHEMISTRY, PHYSICS, AND APPLICATIONS.

PARASITIC ORGANISMS

Although long-recognized as an important public-health problem, foodborne diseases are receiving an increasingly greater share of the public's attention. The Centers for Disease Control estimates that 6.5–33 million Americans (3–14% of the population) contract a foodborne infection annually, with more than 9,000 associated fatalities. Foodborne pathogens also constitute a major cause of morbidity and mortality worldwide. The World Health Organization estimates that more than 1 billion cases of acute diarrhea occur annually in children 1–5 years of age in the developing countries. The most important microorganisms and parasites transmitted by food, particularly meat, are listed in Table 1. In the United States, concern over protozoan and helminth parasites lies mainly with *Toxoplasma gondii* (toxoplasmosis), *Sarcocystis* spp. (sarcocystosis), *Trichinella spiralis* (trichnellosis), larval and adult

Table 1. Foodborne Disease Agents

Viruses
Hepatitis A virus
Norwalk agent
Other Norwalklike viruses
Rotaviruses
Adenoviruses[a]
Astroviruses[a]
Echoviruses
Snow-mountain agent
Cockle agent
Coxsackie B viruses
Caliciviruses[a]
Bacteria
Salmonella spp.
Shigella spp.
Campylobacter spp.
Escherichia coli
Vibrio parahaemolyticus
Listeria monocytogenes
Yersinia spp.
Helminths
Anaskis spp.
Eustrongylides spp.
Pseudoterranova spp.
Trichinella spiralis
Angiostrongylus spp.
Paragonimus spp.
Diphyllobothrium latum
Taenia spp. (*cysticercus*)
Protozoa
Amoeba
Giardia
Isospora
Cryptosporidia
Sarcocystis
Toxoplasma

[a]Viruses that cause gastroenteritis and that may be foodborne.

infections of *Taenia saginata* and *Taenia solium* (taeniasis and cysticercosis), and the fishborne nematodes (eg, anasakiasis). The importance of these parasites relates not only to their direct clinical effect, but also to the high economic costs associated with prevention of their entrance into the domestic food chain and with medical care. Estimated annual costs for inspection and condemnation at animal slaughter are shown in Table 2. Note that losses for fascioliasis and ascariasis represent condemnations for reasons that are more esthetic than public health ones.

All these losses have an important indirect consequence, namely lowering consumer confidence in meat safety and quality. Although reliable figures are difficult to obtain, some economists and public-health workers believe that the cost of foodborne parasites, in terms of public-health expenditures and lost wages, is very high. Estimated annual costs for human taeniasis (from bovine cysticerocosis) is $100,000; for trichinosis, $1.5–2.2 million; and for congenital toxoplasmosis, up to $215–323 million in terms of medical care and work loss (1).

Consumer attitudes, even in the absence of firm facts, can also have a severe economic impact on an industry. A good example of this is trichinellosis. It is thought that public awareness that pork might be infected with *Trichinella spiralis* costs pork producers several hundred million dollars annually in suppressed consumer demand and loss in exports (2). This despite the fact that less than one hog per 1,000 might be infected and only about 50 to 100 human infections are diagnosed each year in the United States. Obviously, the potential consequences of increased public concern over pork-transmitted *Toxoplasma gondii* will also be economically important. It should be noted that firm facts on the role of meat in human toxoplasmosis are still lacking, and research is underway to clarify the question of the importance of pork vis à vis cat feces in transmission to humans. All this underscores the fact that producers have much at stake in controlling and, in some cases, eradicating foodborne parasites, regardless of their direct role in animal and human health.

The rising public concern over food safety presents new and difficult challenges for the research community. Many current procedures and strategies for ensuring a healthful and safe meat supply will require marked improvement in order to meet expected demands; this is especially true for foodborne parasites (3). For the three most serious meat-transmitted parasitic diseases, toxoplasmosis, cysticercosis, and trichinellosis, truly effective or practical control strategies or technologies are not yet available in or applicable for the United States.

Table 2. Estimated Losses Due to Slaughterhouse Condemnations

Commodity	Annual ($ millions)
Cattle	
Livers (fascioliasis)	7.1
Carcasses (cysticercosis)	?[a]
Carcasses (sarcocystosis)	2.0
Swine	
Livers (ascariasis)	50

[a]No economic estimates available; however, in 1988 approximately 11,000 carcasses required refrigeration or cooking before passing inspection.

In many countries, strategies for control emphasize detection of parasites at the abattoir. Of the 74 waterborne and foodborne parasites identified in a current review, virtually all must be diagnosed by tedious gross or microscopic observation for individual organisms (4). The biotechnological revolution, however, is having a significant impact on the efforts to develop rapid and more sensitive inspection techniques. Hybridoma technology has facilitated identification and isolation of antigens with superior immunodiagnostic value. In many instances, however, the availability of these antigens for application is limited. Advances in recombinant DNA technology, however, may solve this problem by permitting large scale *in vitro* production.

It is also becoming apparent that for many, if not all, foodborne pathogens, the ultimate solution is prevention of infection of food at the production or farm level. This strategy is dependent, however, on a thorough understanding of the epidemiology of the parasites. The improvement in diagnostic tools through application of hybridoma and recombinant DNA technologies will have a great impact here, also. Of particular value has been the development of molecular probes (DNA, monoclonal antibodies) to identify cryptic infections, to characterize parasite strains, and to determine transmission dynamics. Prevention of infection strategies may, in many instances, also require the use of vaccines. Again, recombinant DNA technology is playing a vital role in the development of candidate vaccines for parasites such as toxoplasmosis and cysticercosis.

This article identifies the themes common to the control strategies for these diseases and highlights the role of biotechnology in developing those controls.

TOXOPLASMOSIS

General Concepts and Unsolved Problems

Toxoplasmosis, caused by the protozoan parasite *Toxoplasma gondii*, is prevalent in humans and animals worldwide (5). *T. gondii* is transmitted by three routes: fecaloral, congenital, and carnivorism. Cats, the only definitive hosts, are the ultimate source of the infective oocyst stage, which is excreted in their feces. After a short period of exposure to air, oocysts sporulate (form sporozoites), becoming infectious for virtually all warm-blooded animals, including humans. Oocysts can survive in the environment for several months to a year. When oocysts from the environment are ingested the sporozoites, excyst from the oocyst in the intestine penetrate the wall then migrate to and grow inside st cells of the body. After several multiplication cycles, the parasite forms cysts in muscles, the liver, and the central nervous system (CNS). These tissue cysts can persist in humans and other animals virtually for life. The parasites within the tissue cysts are infectious to humans or animals if eaten with the meat. During pregnancy if a woman or animal acquires a primary infection with *Toxoplasma* there is a period when parasites circulate in the bloodstream, thereby rendering the fetus vulnerable to invasion. If cats ingest either infected meat or oocysts, the sexual stages develop in the small intestine. Oocysts eventually are produced and excreted.

Although 30–40% of adults in the United States have serum antibodies to *Toxoplasma*, most postnatally acquired *T. gondii* infections are asymptomatic or are manifested by mild flulike signs. However, toxoplasmosis can cause devastating illness in some adults and children. Loss of vision (chorioretinitis) and mental retardation are the two most important clinical symptoms in congenitally infected children. Although the child is infected before birth, such symptoms may not appear until adolescence. A fulminating, often fatal, illness may develop in patients with acquired immunodeficiency syndrome (AIDS) or in patients given immunosuppressive therapy while receiving organ transplants or treatment for malignancies. Toxoplasmosis is a widespread and important cause of abortion and neonatal death in sheep, goats, and pigs (5). Infection by live *T. gondii* parasites (live vaccine) induces a life-long immunity, thus there is optimism for developing a vaccine for animals, especially cats (6).

Control Strategies and Biotechnology

Strategies to control toxoplasmosis should include prevention of infection in livestock, identification of infected animals in the food chain and their removal or treatment to render the meat safe for human consumption, and prevention and treatment of toxoplasmosis in populations at risk (AIDS patients, transplant recipients, pregnant women and animals). Biotechnology can play an important role in all these strategies, as discussed below.

Identification and Characterization of Immunogens

Unlike the complex antigens of helminths (*Trichinella, Cysticercus*), *T. gondii* has only four major surface antigens (43, 30, 22, and 14 kDA); all strains of *T. gondii* so far investigated share these important antigens (6). Thus, there appears to be no potential strain-dependent vaccination or diagnostic problems. At least two proteins (P30 and P14) have been characterized. P30 is the major dominant antigen and constitutes 5% of the total surface protein (6,7). The gene encoding the P30 protein has been cloned, expressed in a vector, and the fusion protein is being tested for diagnosis. A preliminary study suggests that identification of high levels of specific IgA antibodies against P30 antigen is useful for diagnosing congenital toxoplasmosis in children (8).

DNA-Probes and Diagnosis

Cloned and amplified genes are proving to be a useful probe technology for diagnosis of infectious diseases. The B-1 gene probe of *T. gondii* has been cloned and amplified, and techniques have been developed to detect with it DNA from as little as one *Toxoplasma* organism. With refinements leading to mass production, the cost might be lowered so that this DNA probe can be used routinely for diagnosis (9).

Subunit Vaccines

Recent studies using mutant strains of *T. gondii* indicate that the persistence of live organisms in the host is not necessary for the maintenance of protective immunity (10). If the antigens that stimulate the protective response can be identified, cloned, and properly expressed in vectors, the expressed proteins can be tested as candidates for vaccination. The genes for two other major proteins, the P14 surface antigen and the P28 cytoplasmic antigen, have also been cloned and expressed in vectors. These fusion proteins are being tested as immunogens for possible inclusion in a *Toxoplasma* vaccine (10,11).

SARCOCYSTIS

Public-Health Significance

Two types of *Sarcocystis* infection (sarcocystis) are known in humans. One type is the intramuscular cyst stage probably acquired via feces of a carnivore, and the other type includes intestinal stages that develop after ingestion of raw infected meat. Because the actual sources of infection for the intramuscular cyst stage are unknown and because it is unlikely that human flesh is consumed often enough by carnivores to maintain a human–carnivore cycle in nature, it is hypothesized that such infections are accidental zoonoses, possibly existing in nature in a nonhuman primate–carnivore cycle (3). Because only about 40 such infections have been reported worldwide, most originating in tropical areas, this type of sarcocystosis does not represent a significant public-health problem. Two sources for the human intestinal infection are known: ingestion of raw beef and raw pork. However, little documentation exists regarding their public-health significance. There are several possible reasons for this lack of information. The life cycles and epidemiology of *Sarocystis* spp. have only recently been described. The sporocyst stage shed in the feces of infected humans is relatively small and often is overlooked in examination of feces. Finally, considering the excellent sanitation system used for disposal of human sewage in the United States, opportunities for infection are probably rare.

Biology and Life Cycle

Sarcocystis spp. are obligate two-host parasites, usually requiring a herbivorous intermediate host and a carnivorous final host. The specificity of these parasites is usually greater for the intermediate host than for the final host. Either host might be infected simultaneously with several species of *Sarcocystis*.

For *Sarcocystis hominis*, domestic cattle are the intermediate hosts; and for *Sarcocystis suihominis*, domestic swine are the intermediate hosts. Human and such nonhuman primates as cynomolgus and rhesus monkeys and chimpanzees serve as final hosts, which become infected by eating mature cysts containing zoites found in meat. The parasite undergoes development to the sexually reproducing stage in the lamnia propria of the small intestine. The offspring of the sexually distinct male and female gametes are called sporocysts (containing the infective sporozoite stage), which passes out in the feces. Cattle and pigs become infected when they ingest the sporocyst-containing sporozoites. After a complex migration and a sexual reproduction phase, a second-generation stage, termed the merozoite, invades the host's muscles where it develops to the cyst stage.

Diagnosis in Humans and Animals

In meat animals, the clinical signs of acute sarcocystosis provide a presumptive diagnosis. Animals may show excessive salivation or runny nose, loss of body hair, abortion, reduced milk production, or death. For a short time during peak acute illness the serum enzymes may be elevated. Serologic tests may be helpful in confirming the diagnosis; the indirect hemagglutination test and ELISA have been used in the laboratory but are not yet available commercially. Definitive diagnosis depends on finding parasites and lesions in sick animals. Hemorrhage and nonsuppurative inflammation are often found in tissues from accutely infected animals. The finding of large numbers of immature cysts in muscle biopsies or after death is indicative of a recent clinical infection.

Diagnosis of sarcocystosis in humans or other carnivores is based on finding sporocysts in the feces. This is best accomplished by standard laboratory fecal flotation procedures using saturated sugar solution or zinc sulfate followed by examination with bright-field microscopy.

Prevention and Control

The key to prevention and control of sarcocystosis in meat animals is the elimination from the environment of the sporocyst stage in carnivore feces. Any control measure designed to prevent carnivores from becoming infected and to prevent fecal contamination of feed and water will break the cycle. Such measures are also described for control of toxoplasmosis. In addition, every effort should be made to bury, incinerate, or otherwise remove dead livestock from farms, fields, or grazing areas to prevent domesticated and feral carnivores from acquiring infections.

To prevent human intestinal infection, meat should be cooked or frozen before it is eaten. In two studies of beef from retail food stores infectious *S. cruzi* (infectious for dogs but not humans) was present in various raw roasts, steaks, and hamburger but not in processed meats, such as frozen hamburger patties or minute steaks, beef bologna, frankfurters, or in hamburger cooked to 60°C (140°F) or higher.

TRICHINELLOSIS

Public Health Significance

Trichinellosis in the United States is not considered a major public-health problem in terms of number of human infections. But because the inspection of swine for *Trichinella spiralis* muscle larvae is not performed, as it is in Europe, the potential risk suppresses consumer demand for pork, resulting in considerable economic cost to the industry (2).

Life Cycle and Epidemiology

The most important features of this helminth's (nematode) life cycle is its obligatory transmission by ingestion of meat; there is no free-living stage, as exists in many other parasitic nematodes. *T. spiralis* is cosmopolitan in its distribution. Recent research has shown that the concept of only one species is wrong and that there are at least six to seven genetic types, some with distinct host specificities (12–14). The type that is normally associated with domestic swine is unique and appears to have evolved along with the domestication of *Sus scrofa* (pig) (12). Although there is some controversy as to the proper taxonomic status and host specificities of various isolates of *Trichinella* from domestic and wild animals, all appear to be infective for humans.

After ingestion, the larvae are digested out of the muscle capsule in gastric fluid and enter the small intestine, where within 4–6 days they develop into sexually mature males and females. Their offspring (newborn larvae) migrate via the circulatory system throughout the body, invade striated muscles, and eventually (17–21 days after initial infection) become infective, encapsulated larvae. Although larvae may invade smooth muscle and other tissue, they eventually die in these sites. The encapsulated larvae may persist for years in muscle, until they become calcified and die.

The degree of clinical disease in human trichinosis is somewhat dependent on the number of muscle larvae ingested. The ingestion of 500 or more larvae can produce moderate-to-severe and even life-threatening illness. During the first few weeks, if large numbers of worms are present, illness may be reflected by gastrointestinal signs such as nausea and abdominal pain. Subsequent to the production of the muscle-invading larval offspring, the acute muscle phase may be seen. This is usually characterized by muscular pain, facial edema, fever, and eosinophilia.

As pointed out above, the major source of *T. spiralis* for humans is the domestic pig, although game accounts for about a third of the annual cases. Periodic prevalence studies on swine have revealed a rather marked decline in infection rates among pigs in the United States during the past 85 years, although this varies considerably from one region to another. The prevalence in slaughter hogs has decreased from about 1.4% around the turn of the century to 0.63% in 1952 and 0.13% in 1970 (3). These rates are influenced by the preponderance, in the samples, of grain-fed Midwestern hogs, which account for the majority of production in the United States; prevalence surveys in the Midwest indicate only 0.001% of hogs are infected. Recent surveys (15) indicate that the rates in swine raised in the eastern United States, however, range from 0.5 to 1.0%; in this region swine husbandry includes more high-risk management practices, such a garbage feeding.

Control Strategies

There has been a resurgence of research on the epidemiology, biology, and control of *Trichinella spiralis* during the recent decade. A major ingredient in this renewed attention has been the biotechnological revolution and the new tools it has provided. Especially important has been the application of hybridoma technology to the development of better immunodiagnostic tests. The techniques of recombinant DNA have also been instrumental for providing a potential means to produce antigens and to develop probes for use in epidemiological research and systematics.

Because human infections derive mainly from the ingestion of infected pork, an immunodiagnostic test suitable for the abattoir has had high priority for both the Agricultural Research Service and the Food Safety and Inspection Service, USDA. The efforts of these agencies have been successful, and a commercial ELISA-type test is now available. The critical step in producing this test was the discovery of a simple parasite-cultivation procedure for obtaining the diagnostic antigen (16). The antigens have been purified using a monoclonal antibody in combination with immunoaffinity techniques. In field tests, the antigens proved to be highly specific and sensitive (17,18). Currently, research is underway to produce the antigens by recombinant methods, and preliminary results indicate these efforts will also be successful. This will enhance the commercial viability of the test.

This serological test has been valuable not only for national epidemiologic studies but also for the development of effective control strategies. An important aspect of the epidemiologic investigations has been the assessment of the role of the sylvatic trichinellosis in domestic pig infections. Considerable concern exists over the threat of wild-animal *Trichinella* infections as a source of infection for humans and domestic swine. Although many wild-animal isolates of *Trichinella* are poorly infective for swine, the highly infective pig-type strain has been isolated from wild animals (12). The high genetic variability among morphologically indistinguishable isolates of the *Trichinella* presents problems in interpreting epidemiologic data. However, these questions are being resolved through the application of such techniques as DNA restriction-enzyme fragment-length polymorphism analysis (RFLP) and the development of hybridization probes by cloning unique DNA sequences (12,19). It is clear that human encroachment on the habitat of wild animals has facilitated the introduction of the domestic pig type (*Trichinella spiralis sensu stricto*) into wild-animal populations and, hence, a sylvatic reservoir of parasites capable of infecting domestic swine has been established (20). Effective control strategies must take this potential threat into account.

Vaccines

A vaccine has been developed for swine trichinellosis (21). It will soon be subjected to field trials. However, the vaccine is based on whole, inactivated newborne larvae and can only be produced on a limited scale. If, however, the field trials are successful, efforts to produce the protective antigens by recombinant DNA methods can be anticipated.

CYSTICERCOSIS

Public-Health Significance

Two species of tapeworms may be transmitted to humans by ingestion of the larval or cysticercus stage encysted in the meat of livestock. These two species are *Taenia saginata*, the beef tapeworm, and *T. solium*, the pork tapeworm (3). The latter is of special significance because humans may also serve as an intermediate host for the cysticercus (larval) stage, where it often localizes in the brain, causing the disease termed neurocysticercosis. Neurocysticercosis is a major public-health problem in many areas of the world, especially Latin America (22). Neurocysticercosis, although an acute life-threatening disease in severe cases, is more often a long-lasting infection affecting the quality of the patient's life and social environment. The disease is of socioeconomic importance because 75% of patients with neurocysticercosis are at productive ages and are frequently unable to work soon after the onset of symptoms. Calculations of costs for medical care, such as hospitalization, chemotherapy, neurosurgery, and computed tomography, show that US$14.5 million were spent in Mexico during 1986 to treat only the 2,700 new hospitalized cases of neurocysticercosis. For these reasons, neurocysticercosis is recognized as being of major public-health significance. Furthermore, swine cysticerosis is considered a significant economic problem because of the condemnation of pig carcasses. In Mexico, more than US$43 million were lost in 1980, the equivalent of 68.5% of the total investment in pig stock production. In the southwest United States, human neurocysticerosis is increasingly encountered. The magnitude is suggested by the fact that in Los Angeles, fecal examinations revealed more than 60 positives out of 12,000 examined (23). In one neurologic service, 45 cases were reported over a 5-year period. High rates are also seen in other areas of the world. A 1984 survey in India revealed that 4% of humans were infected with adult *Taenia* and more than 17% of swine were infected with cysticerci of *T. solium* (24).

The bovine form, *T. saginata*, although less severe in humans because it is confined, as the adult worm stage, to the human intestine, does represent a considerable economic cost because inspection of beef carcasses for cysticerci is mandatory, although the incidence is very low. In the United States, the rate of bovine cysticercosis identified by carcass inspection has been less than 0.3/1,000 over the last 10 years. These rates frequently are higher in certain Western states because of sporadic outbreaks. For example, in 1984, the rate for Washington was 1.9/1000 (25).

Life Cycle and Epidemiology

The adult stage of *Taenia saginata* is a flat tapeworm that resides in the human small intestine (3). The tapeworm is composed of a chain (strobila) of proglottids or segments each of which contains male and female reproduction systems. As the proglottid matures, it produces a large number of eggs. Gavid proglottids detach from the strobila and pass out of the intestine through the host's anus. Generally, an infected person harbors one tapeworm, although in some regions of the world, such as, Asia Minor, multiple infections are frequent. The adult tapeworm normally localizes in the jejunum, about 50 cm below the duodenojejunal flexure, and rarely enters the gallbladder, appendix, or nasopharynx. The life span of *T. saginata* in humans may be as long as 30–40 years.

Taenia saginata is capable of producing a very large number of eggs, and an individual proglottid may contain about 80,000 eggs. Because each strobila can shed 6–9 gravid proglottids/day, a single infected individual may contaminate the pasture, lot, or barn with one-half million eggs or more each day. The mature egg contains the larval stage termed the onchosphere; immature eggs that are passed may complete their development outside the host. Ingestion of the egg by cattle is followed by disintegration of the outer embryophore and activation of oncosphere activity by gastric juice. Aided by its hooks and penetration glands, the onchosphere penetrates the bovine intestinal mucosa and within a couple of hours enters the blood or lymphatic system. The onchosphere utilizes the host's circulatory system to migrate throughout the host's body; however, successful development to the cyst or cysticercus stage usually occurs in skeletal muscle or the heart. The cysticercus, when fully developed, is composed of an invaginated scolex within a fluid-filled vesicle or bladder. When improperly cooked beef is eaten, the larvae are digested free in the gut and the scolex evaginates to attach by its suckers to the intestinal wall. The development of the stobila then commences, including proglottids with reproductive organs and eventually, eggs.

The life cycle of *T. solium* is similar but differs in several important respects (3). The gravid proglottids contain fewer eggs than do those of *T. saginata*. Another distinction is the lack of motility of *T. solium* proglottids, which results in their being shed in the feces as connected segments. The major difference from *T. saginata*, however, is the suitability of humans to serve as an intermediate host for *T. solium* cysticerci. As in pigs, the cysticerci are fairly evenly distributed throughout the liver, brain, central nervous system, skeletal muscle, and myocardium. The cysticercus of *T. solium* is bigger than that of *T. saginata* (5–20 mm in diameter). After 2–3 months of development, the cysticercus appears as a glistening pearly white cyst, with the scolex or head deeply invaginated into the fluid-filled bladder; the scolex bears four suckers and an apical crown of hooklets. When humans ingest infected pork, the larvae are digested out of the meat or tissue, the scolex evaginates from the bladder, attaches to the wall of the proximal portion of the ileum, and develops into a complete worm or strobila in about 3 months. The tapeworm begins to excrete proglottids after about 2 months and may live for many years.

Clinical Disease

Clinical disease associated with the intestinal adult form of *T. solium* is mild in humans and is similar to that of *T. saginata*. Far more serious may be infection with the larval cysticercus form, especially if the central nervous system is involved (3). Cerebral cysticercosis is usually classified according to its location: (1) ventricular-cisternal, (2) meningeal, (3) parenchymal, and (4) mixed. These localizations are often accompanied by a myriad of signs and symptoms, including seizures, hydrocephalus, headaches, dizziness, arterial thrombosis, loss of vision, and nausea. Cerebral cysticercosis is, on a worldwide basis, one of the most common parasitic diseases of the central nervous system (22).

Control Strategies

Current control strategies for these parasitic diseases are dependent on detecting infections in livestock in order to remove them from the food chain. However, procedures for this task are at present inadequate. For example, in the United States, the mandatory inspection of beef for cysticerci is known to be highly insensitive (26). Most current research on cysticercosis is aimed at developing rapid, more reliable immunodiagnostic tests for both livestock and humans.

Until recently, the prospects for an immunodiagnostic test for bovine cysticercosis were poor because of the low specificity of crude worm extracts as antigen reagents. Recently, biochemical fractionation approaches have greatly improved the specificity of these antigens. A genus-specific antigen from *T. hydatigena* (ThFAS) with a high degree of sensitivity and specificity for both bovine and porcine cysticercosis has been recently recovered (27,28). With this antigen, low infection levels in cattle could be detected as early as 3 weeks after exposure. Currently, efforts are underway to clone the gene for this antigen so that sufficient amounts of antigen for field tests can be made available.

The immunodiagnosis of human neurocysticercosis has also greatly improved. One of the most specific and sensitive test procedures is that described by Tsang and coworkers (29). This test is an enzyme-linked immunoelectrotransfer blot assay using glycoprotein antigens prepared by fractionation of *T. solium* cysticerci. In an evaluation using a large number of sera and cerebrospinal fluids, the test was 98% sensitive and 100% specific.

Recombinant DNA methods have had, to date, their greatest impact in human cysticercosis research in the area of taxonomy (30). As pointed out earlier, the systematics of the taeniid cestode group requires clarification because of its importance to understanding the epidemiology of these foodborne parasites. As various host and geographic isolates receive closer scrutiny, it is apparent that at least some of the species exist as complexes of intraspecific variants, frequently with variation in host specificity. Recently, a variety of geographic isolates of *T. solium*, have been analyzed with the aid of DNA probes (31). Considerable intraspecific genetic variation has been detected. Using a similar approach, distinct DNA differences between a *T. saginata*-like cestode from Taiwan and various other isolates of *T. saginata* have been observed (32). The human *Taenia* from Taiwan (and Korea) is unique in that it is more infective for swine than for cattle and has a predilection for the swine liver. The difference in DNA characteristics supports the interpretation that the Taiwan *Taenia* is genetically and biologically unique although closely related to *T. saginata*. It can be anticipated that these powerful methods will prove to be decisive in developing a greater understanding of the genetic variation and epidemiology of these important human parasites.

FISHBORNE PARASITES

The rising popularity of raw fish has increased the risk of a variety of new diseases, especially those caused by the larval stages of helminth parasites of fish. The majority of these seafood diseases occur in regions where seafood constitutes a major portion of the protein diet. The status of seafood zoonoses has recently been reviewed (33). The number of annual causes of anisakiasis (caused by *Anisakis* spp. and *Pseudoterranova* spp.) in Japan exceeded 3,100 in 1984. In the United States, although only about 50 cases have been reported since 1958, the annual number of new cases is increasing.

The most important freshwater fishborne parasites in the United States is the broad fish tapeworm *Diphyllobothrium latum*. Humans become infected by ingesting raw or inadequately cooked freshwater or andromous fish. Most commonly implicated are pike and walleys from the Great Lakes, where up to 50–70% may harbor the infective larval stage.

Currently, the control of these fishborne parasites relies on recommendations to the consumer for the preparation of safe fish meals and, in the case of marine fish, on some inspection by mechanical means. However, the pressure on government to develop mandatory, comprehensive inspection technologies for the inspection of marine and freshwater fish is increasing. Currently, accurate, rapid inspection methods do not exist. It can be expected, however, that if the need is sufficient, such inspection technology will be developed. Biotechnological tools may play an important role in this endeavor also.

BIBLIOGRAPHY

1. T. Roberts, "Microbial pathogens in raw pork, chicken and beef: benefit estimates for control using irradiation," *American Journal of Agric. Econ.* **67**, 957–965 (1985).
2. U.S. Department of Energy, "Trichina-safe pork by gamma irradiation processing: a feasibility study." Contract NO. DE-AC04-83AL-19411, Albuquerque, New Mexico, August 1983.
3. K. D. Murrell, R. Fayer, and J. P. Dubey, "Parasitic organisms," *Advances in Meat Research* **2**, 311–377 (1986).
4. R. Fayer, H. R. Gamble, J. R. Lichtenfels, and J. Bier, "Food-borne and waterborne parasites," in *APHA Compendium of Methods for the Microbiological Examination of Foods*, American Public Health Association. In press. 1990.
5. J. P. Dubey and C. P. Beattie, *Toxoplasmosis of Animals and Man*, CRC Press, Boca Raton, Fla., 1988.
6. A. M. Johnson, "Toxoplasma vaccines," in I. G. Wright, ed., *Veterinary Protozoan and Hemoparasite Vaccines*, CRC Press, Boca Raton, Fla., 1989, pp. 177–202.
7. L. H. Kasper, "Identification of stage-specific antigens of *Toxoplasma gondii*," *Infection and Immunity* **57**, 168 (1989).
8. A. DeCoster, F. Darcy, A. Capron, and A. Capron, "IgA antibodies against P30 as markers on congenital and acute toxoplasmosis," *Lancet* **2**, 1104 (1988).
9. J. L. Burg, C. M. Grover, P. Pouletty, and J. C. Boothroyd, "Direct and sensitive detection of a pathogenic protozoan, *Toxoplasma gondii*, by polymerase chain reaction," *Journal of Clinical Microbiology* **27**, 1787 (1989).
10. H. Waldland and J. K. Frenkel, "Live and killed vaccines against toxoplasmosis in mice," *Journal of Parasitology* **G9**, 60 (1988).
11. J. B. Prince, F. G. Araujo, J. S. Remington, J. L. Burg, J. C. Boothroyd, and S. D. Sharma, "Cloning of cDNAs encoding a 28 kilodalton antigen of *Toxoplasma gondii*," *Molecular and Biochemical Parasitology* **34**, 3 (1989).
12. J. B. Dame, K. D. Murrell, D. E. Worley, and G. A. Schad, "*Trichinella spiralis*: genetic evidence for synanthropic subspecies in sylvatic hosts," *Experimental Parasitology* **64**, 195 (1987).
13. E. Pozio, P. LaRosa, P. Rossi, and K. D. Murrell, "New taxonomic contributions to the genus *Trichinella* (Owen, 1835). I. Biochemical identification of seven clusters by gene-enzyme systems," in C. Tanner, ed., *Proceedings of the 7th International Conference on Trichinellosis*, CSIC Press, Madrid, Spain, 1989, pp. 76–82.
14. D. S. Zarlenga and K. D. Murrell, "Molecular cloning of *Trichinella spiralis* ribosomal RNA genes: application as genetic markers for isolate classification," in C. Tanner, ed., *Proceedings of the 7th International Conference on Trichinellosis*, CSIC Press, Madrid, Spain, 1989, pp. 35–40.
15. G. A. Schad, M. Kelly, D. A. Leiby, K. Blumrick, C. Duffy, and K. D. Murrell, "Swine trichinellosis in mid-Atlantic slaughterhouses: possible relationships to hog marketing systems," *Preventive Veterinary Medicine* **3**, 391–399 (1985).
16. H. R. Gamble, W. R. Anderson, C. E. Graham, and K. D. Murrell, "Diagnosis of swine trichinosis by enzyme-linked immunosorbent assay (ELISA) using an excretory-secretory antigen," *Veterinary Parasitology* **13**, 349 (1983).
17. H. R. Gamble and C. E. Graham, "Monoclonal antibody purified antigen for the immunodiagnosis of trichinosis." *American Journal of Veterinary Research* **45**, 67 (1984).
18. K. D. Murrell, W. R. Anderson, G. A. Schad, R. D. Hanbury, K. R. Kazacos, H. R. Gamble, and J. Brown, "Field evaluation of an ELISA test for swine trichinosis using an excretory-secretory antigen. *American Journal of Veterinary Research* **47**, 1046 (1986).
19. D. Boyd, T. deVos, G. Klassen, and T. Dick, "Characterization of the ribosomal DNA from *Trichinella spiralis*," *Molecular Biochemistry and Parasitology* **35**, 67 (1989).
20. K. D. Murrell, F. Stringfellow, J. B. Dame, D. Luiby, and G. A. Schad, "*Trichinella spiralis*: evidence for natural transmission of *Trichinella spiralis* from domestic swine to wildlife," *Journal of Parasitology* **73**, 103 (1987).
21. H. P. Marti, K. D. Murrell, and H. R. Gamble, "*Trichinella spiralis*: immunization of pigs with newborn larval antigens," *Experimental Parasitology* **63**, 68 (1987).
22. A. Flisser, "Cysticercosis: a major threat to human health and livestock production," *Food Technology* **37**, 61 (1985).
23. W. J. Brown and M. Voge, "Cysticercosis, a modern day plague," *Pediatric Clinica of North America* **32**, 953–968 (1985).
24. K. M. L. Pathak, S. N. S. Gaur, and D. Kumar, "The epidemiology of strobillar and cystic phases of *Taenia solium* in certain parts of Uttar Pradesh (India)," *Indian Journal of Veterinary Medicine* **4**, 1748 (1984).
25. D. D. Hancock, S. E. Wikse, A. B. Lichtenwalner, R. B. Wescott, and C. C. Gay, "Distribution of bovine cystercercosis in Washington," *American Journal of Veterinary Research* **50**, 564–570 (1989).
26. L. W. Dewhirst, "Antemortem diagnosis of bovine cysticercosis due to *Taenia saginata*," in *Proceedings of the 71st Annual Meeting, U.S. Livestock Sanitation Association*, 1968, pp. 540–545.
27. M. L. Rhoads, EIP Kamanga-Sollo, D. Rapic, K. D. Murrell, P. M. Schantz, and A. Beus, "Detection of antibody in humans and pigs to *T. solium* metacestodes using an antigenic fraction from *T. hydatigena* metacestodes," *Veterinary Archives* **57**, 143 (1987).

28. EIP Kamanga-Sollo, M. L. Rhoads, and K. D. Murrell, "Evaluation of an antigenic fraction of *Taenia hydatigena* metacestode cyst fluid for immunodiagnosis of bovine cysticercosis," *American Journal of Veterinary Research* **48**, 1206 (1987).
29. V. C. W. Tsang, J. A. Brand, and A. E. Boyer, "An enzyme-linked immunoelectrotransfer blot assay and glucoprotein antigens for diagnosing human cysticercosis (*Taenia solium*)," *Journal of Infectious Diseases* **154**, 50 (1989).
30. K. D. Murrell and M. L. Rhoads, "Application of biotechnology methods to the study of cestodes," *Southeast Asian Journal of Tropical Medicine and Health* **19**, 37 (1988).
31. A. K. Rishi and D. P. McManus, "Molecular cloning of *Taenia solium* DNA and characterization of taeniid cestodes by DNA analysis." *Parasitology* **97**, 161 (1988).
32. D. S. Zarlenga, D. P. McManus, P. C. Fan, and J. H. Cross, "DNA characterization of a prospective new species of *Taenia* from Taiwan using cloned ribosomal and repetitive DNA fragments," *Experimental Parasitology* **72**, 174–183 (1991).
33. T. L. Deardorff and R. M. Overstreet, "Sea-food transmitted zoonoses in the United States: the fishes, the dishes, and the worms," in C. R. Hackney and D. Ward, eds., *Microbiology of Marine Food Products*, Van Nostrand Reinhold, New York. In press.

GENERAL REFERENCES

Anonymous, "Issues in Pork Safety, Costs, Controls, and Incentives," *Agricultural Outlook* **201**, 28–32 (1993).

W. Buffolano et al., "Risk Factors for Recent Toxoplasma Infection in Pregnant Women in Naples," *Epidemiology and Infection* **116**, 347–351 (1996).

L. Bunetel et al., "Calibration of an in vitro Assay System Using a Nonadherent Cell Line to Evaluate the Effect of a Drug on *Toxoplasma gondii*," *Int. J. Parasitol.* **25**, 699–704 (1995).

D. Buxton and E. A. Innes, "A Commercial Vaccine for Ovine Toxoplasmosis," *Parasite Vaccines* S11–S16 (1995).

W. C. Campbell, "Meatborne Helminth Infections: Trichinellosis," in Y. H. Hui et al., eds., *Foodborne Disease Handbook, Diseases Caused by Viruses, Parasites and Fungi*, Vol. 2, Marcel Dekker Inc., New York, 1994, pp. 255–277.

A. Chapman et al., "Isolation and Characterization of Species-Specific DNA Probes from *Taenia solium* and *Taenia saginata* and Their Use in an Egg Detection Assay," *J. Clin. Microbiol.* **33**, 1283–1285 (1995).

J. L. De La Rosa et al., "Prevalence and Risk Factors Associated with Serum Antibodies Against *Trichinella spiralis*," *Int. J. Parasitol.* **28**, 317–327 (1998).

B. R. Dixon, "Prevalence and Control of Toxoplasmosis, A Canadian Perspective," *Food Control* **3**, 68–75 (1992).

J. P. Dubey, "Toxoplasmosis," *J. Amer. Veterinary Med. Assoc.* **205**, 1593–1598 (1994).

J. P. Dubey et al., "Oocyst-Induced Murine Toxoplasmosis: Life Cycle, Pathogenicity, and Stage Conversion in Mice Fed *Toxoplasma gondii* Oocysts," *J. Parasitol.* **83**, 870–882 (1997).

J. F. Edwards, R. B. Simpson, and W. C. Brown, "Bacteriologic Culture and Histologic Examination of Samples Collected from Recumbent Cattle at Slaughter," *J. Amer. Veterinary Med. Assoc.* **207**, 1174–1176 (1995).

K. L. Ehnert et al., "Cysticercosis, First 12 Months of Reporting in California," *Bull. of the Pan Am Health Organ.* **26**, 156–172 (1992).

Flair-Flow Europe, "Fish Parasites of Public Health Significance," *Flair Flow Rep.* **125**, 1 (1994).

A. A. Gajadhar, J. R. Bisaillon, and G. D. Appleyard, "Status of *Trichinella spiralis* in Domestic Swine and Wild Boar in Canada," *Can. J. Veterinary Res.* **61**, 256–259 (1997).

A. A. Gajadhar and W. C. Marquardt, "Ultrastructural and Transmission Evidence of *Sarcocystis cruzi* Associated with Eosinophilic Myositis in Cattle," *Can. J. Veterinary Res.* **56**, 41–46 (1992).

A. A. Gajadhar, W. C. Marquardt, and C. D. Blair, "Development of a Model Ribosomal RNA Hybridization Assay for the Detection of *Sarcocystis* and Other Coccidia," *Can. J. Veterinary Res.* **56**, 208–213 (1992).

H. R. Gamble, "Detection of Trichinellosis in Pigs by Artificial Digestion and Enzyme Immunoassay," *J. Food Protection* **59**, 295–298 (1996).

H. R. Gamble and I. V. Patrascu, "Whole Blood Serum and Tissue Fluids in an Enzyme Immunoassay for Swine Trichinellosis," *J. Food Protection* **59**, 1213–1217 (1996).

J. R. Isaac et al., "Detection of *Toxoplasma gondii* Oocysts in Drinking Water," *Appl. and Environmental Microbiol.* **64**, 2278–2280 (1998).

M. R. Lappin, "Diagnostic Testing for Toxoplasmosis," *Proc. North Amer. Veterinary Conf.* **9**, 221–223 (1995).

P. Mahannop et al., "Immunodiagnosis of Human Trichinellosis and Identification of Specific Antigen for *Trichinella spiralis*," *Int. J. Parasitol.* **25**, 87–94 (1995).

A. E. Marsh et al., "Sequence Analysis and Polymerase Chain Reaction Amplification of Small Subunit Ribosomal DNA from *Sarcocystis neurona*," *Amer. J. Veterinary Res.* **57**, 975–981 (1996).

A. Minniti et al., "Determination of Histamine in Common Sea Fish Parasites: *Anisakis simplex* and *Gymnorhynchus gigas*," *Ital. J. Food Sci.* **7**, 305–309 (1995).

K. D. Murrell, "Foodborne Parasites," *Int. J. Environmental Health Res.* **5**, 63–85 (1995).

Y. Ohnisni, T. Ono, and T. Shigehisa, "Histochemical and Morphological Studies on *Trichinella spiralis* Larvae Treated with High Hydrostatic Pressure," *Int. J. Parasitol.* **24**, 425–427 (1994).

K. S. Parmesh, D. W. Webert, and P. C. McCaskey, "Food Safety and Regulatory Aspects of Cattle and Swine Cysticercosis," *J. Food Protection* **60**, 447–453 (1997).

M. Pepin, P. Russo, and P. Pardon, "Public Health Hazards from Small Ruminant Meat Products in Europe," *Revue Scientifique et Technique Office International des Epizootes* **16**, 415–425 (1997).

J. H. Rim, "Field Investigation on Epidemiology and Control of Fish-Borne Parasites in Korea," *Int. J. Food Sci. and Technol.* **33**, 157–168 (1998).

F. Serrano et al., "Trichinella Strain, Pig Race and Other Parasitic Infections in the Reliability of ELISA for the Detection of Swine Trichinellosis," *Porositol.* **105**, 111–115 (1992).

L. Sloan, S. Schneider, and J. Rosenblatt, "Evaluation of Enzyme-Linked Immunoassay for Serological Diagnosis of Cysticercosis," *J. Clin. Microbiol.* **33**, 3124–3128 (1995).

H. J. Smith, K. E. Snowdon, and R. C. Finlay, "Serological Diagnosis of Cysticercosis by an Enzyme-Linked Immunosorbent Assay in Experimentally Infected Cattle," *Can. J. Veterinary Res.* **55**, 274–276 (1991).

J. L. Smith, "Documented Outbreaks of Toxoplasmosis Transmission of *Toxoplasma gondii* to Humans," *J. Food Protection* **56**, 630–639 (1993).

J. L. Smith, "*Taena solium* Neurocysticercosis," *J. Food Protection* **57**, 831–844 (1994).

J. L. Smith, "Long-Term Consequences of Foodborne Toxoplasmosis: Effects on the Unborn, the Immunocompromised, the Elderly, and the Immunocompetent," *J. Food Protection* **60**, 1595–1611 (1997).

A. M. Tenter, "Current Research on *Sarcocystis* Species of Domestic Animals," *Int. J. Parasitol.* **25**, 1311–1330 (1995).

F. van Knapen, "European Proposal for Alternative *Trichinella* Control in Domestic Pigs," *Fleischwirtschaft* **78**, 338–339 (1998).

K. DARWIN MURRELL
USDA/ARS/NRRC
Peoria, Illinois

PASTA. See CEREALS SCIENCE AND TECHNOLOGY; WHEAT SCIENCE AND TECHNOLOGY.

PATENTS

It is obviously not possible to discuss the complex subject of food patents in a few thousand words. It is hoped that this article will provide an introduction to the basic procedure of obtaining a patent for an invention in the United States. However, it must be emphasized that the service of a patent attorney is the best assurance for obtaining proper protection for an invention. This discussion will use the term patent instead of food patent throughout. There are obviously scientific differences between a patent for a motor and one for a food packaging material. However, the principles underlying the applications for both are the same.

THE LAW

There are four general areas of legal protection available for the exploitation of intellectual property. These are (1) patents, (2) copyrights, (3) trade secrets, and (4) trademarks. The area or areas available for protecting any particular type of intellectual property will depend on the specific type under consideration.

The Constitution of the United States (Article 1, Section 8) gave Congress the right or power to enact laws concerning patents and copyrights. In particular Article 1, Section 8 states the following:

> The Congress shall have the power . . . to promote the progress of science and useful arts, by securing for limited times to authors and inventors the exclusive right to their respective writings and discoveries.

The basic philosophy behind both the patent laws and copyright laws is to encourage authors and inventors to publicly disclose their respective writings and discoveries. This disclosure, in turn, is to add to the total sum of knowledge publicly available. To encourage such disclosures, patent and copyright laws were enacted to provide certain rights to authors and inventors. The patent laws have been codified as Title 35 of the United States Code (USC) and are referred to as 35 USC. The laws directed to copyrights are codified in Title 17 of the United States Code and are referred to as 17 USC.

Because the entire field of patent law has been preempted by Congress, individual states are excluded from any legal control over patent law. In addition, since January 1, 1978, Congress has preempted the copyright law and, therefore, all rights created from January 1, 1978, and thereafter are subject to federal law.

DEFINITION AND RIGHTS

A U.S. patent is actually a contract between the inventor and the people of the United States represented by the government and specifically by the U.S. Patent and Trademark Office. The right conferred on the inventor is the right to exclude others from making, using, selling, offering for sale, or importing that which is covered in the claims of the patent. Such rights are actually negative rights in the sense that the patentee is not given the right to do anything except exclude others from making, using, or selling the invention for a limited period of time. Typically, but with certain exceptions, U.S. patents expire 20 years from the filing date of the application for patent. One such exception relates to patents involving drugs or any medical device, food additive, or color additive subject to regulation under the federal Food, Drug and Cosmetic Act or the Virus-Serum-Toxic Act for veterinary biological products, where it is possible to recapture that portion of the patent term, up to a maximum of 5 years, that was lost because of delays resulting from the Food and Drug Administration's or the Animal and Plant Health Inspection Service's review prior to approving commercial marketing or use of the drug, medical device, food additive, color additive, or biological product.

As in any contract each party is to receive something of value and each party is to give up something of value in exchange. In exchange for the rights granted by the government, the inventor is required to provide a disclosure of the invention. This is the inventor's "consideration" or "value" given up as part of the contractual relationship. The disclosure of the invention is made available to the public when the patent issues (that is, is granted by the government). This disclosure of the invention by the inventor must be a full description of the invention.

The general subject matter to which patents are directed is considered to be the useful arts. In particular patents are granted for inventions directed to processes, machines, manufactures, compositions of matter, or any new and useful improvements thereof. Printed copies of granted patents can be purchased from the U.S. Patent and Trademark Office for $3.00 per copy.

JURISDICTION

The U.S. Patent and Trademark Office is a governmental agency within the U.S. Department of Commerce. For ease of reference, this office will be referred to as "the office" in this article. The head is referred to as the commissioner. The office has the responsibility for reviewing patent applications and determining whether or not to grant or issue a patent thereon. It does not have jurisdiction to determine whether a patent is infringed. Moreover, except to the limited extent authorized by the patent reexamination procedure, it has no jurisdiction over issued patents and cannot make determinations of patent validity or invalidity.

About 2,000 people with degrees in engineering or science are employed by the office to review and examine patent applications; they are referred to as patent examiners. Examiners conduct independent searches of the inventions described in the applications and then make decisions as to whether the invention described and claimed merits the granting of a patent.

The commissioner has the authority to promulgate rules and regulations for the purpose of administrating the patent and trademark laws so long as such rules are not inconsistent with these laws. In addition, the commissioner has the power to make rules that govern the recognition and conduct of those persons who represent applicants before the office. To be recognized as a patent agent or patent attorney and to advise and assist in the preparation or prosecution of applications and other business before the office, a person must comply with certain rules and regulations.

In the event an examiner refuses to allow or permit issuance of a patent for an application, the applicant has the opportunity at a particular stage of the proceedings to file an appeal from the examiner's decision to the U.S. Patent and Trademark Office Board of Patent Appeals and Interferences (hereafter referred to as "the board"). The board is composed of senior examiners, who are referred to as patent law administrative judges. Cases appealed to the board are considered by a three-member panel of the board.

A publication titled *Manual of Patent Examining Procedure (MPEP)* is available from the U.S. Government Printing Office. This publication is a guide to how the office interprets the laws, rules, and case law and how the patent examiners are to apply the same when examining and ruling on patent applications. Likewise, a similar publication available from the U.S. Government Printing Office titled *Trademark Manual of Examining Procedure (TMEP)* is a guide for trademark examiners in how to apply the laws, rules, and case law when ruling on trademark and service mark applications.

Another important publication available from the U.S. Government Printing Office is titled the *Official Gazette of the United States Patent and Trademark Office (OG)*. This is published each week in two separate sections. One is the patent section and the other is the trademark and service mark section. The patent section of the *OG* includes a short description of each patent granted during the week of publication, an index of the names of the patentees and the assignees for that week, notices of rule changes, and various other matters of interest with respect to patents.

Since 1929, published decisions in patent, trademark, and copyright cases have been reported in the *United States Patent Quarterly (USPQ and USPQ2d)*. This is published by the Bureau of National Affairs, Washington, D.C., which also publishes the weekly newsletter *Patent, Trademark, and Copyright Journal*. The latter contains news of matters of current development and interest in the patent, trademark, and copyright areas.

Decisions by the federal courts in trademark, copyright, and patent cases can also be found, depending on the particular court and the particular year in which the case was decided, in the *Federal Reporter* (F.), the *Federal Reporter* second series (F.2d), the *Federal Reporter third series* (F.3d), the *Federal Supplement* (F.Supp.), the *Supreme Court Reporter* (S.Ct.), and *The United States Reports* (U.S.).

PATENTABILITY

Specifications and Claims

To facilitate an understanding of the concept of patentability, some familiarity with the contents of patents and particularly with the specification and claims is desirable. The specification of a patent contains a detailed technical description of the invention involved. In addition, if an understanding of the invention would be facilitated by including drawings, the specification will do so and also provide an explanation of the drawings. The specification may also include a discussion of the background technology of the invention, the problems encountered in that technology, the state of the prior art, the objectives of the invention, and/or the advantages, if any, achieved by the invention.

The claim or claims of a patent appear at the end of the technical description of the patent. The function of the claim is particularly to point out and distinctly assert title to the subject matter that the applicant considers to be his or her contribution or invention for which he or she desires a patent grant. It is the claim or claims that are compared to the prior art to determine patentability. Likewise, it is the claim or claims that are compared to an accused activity to determine whether infringement actually exists.

What Can Be Patented

As required by 35 USC §101, an invention must be new and must be useful to be patentable. (Design patents need not be useful, however, to satisfy the requirements for patentability.) In addition, as required by 35 USC §101, the invention must be directed to a process, machine, manufacture, or composition of matter or to any new and useful improvement thereof. Patents are concerned with technology or the useful arts as contrasted to the liberal arts. A shorthand way to view the subject matter of a patent is that it represents the means by which a desired result is obtained.

The term process as used in the patent statute involves the treatment or manipulation of some material or materials or of information to cause some change to the material or information treated or manipulated by the process.

Novelty. Another requirement that an invention must satisfy to be patentable is that it must be new or novel (35 USC §102). Because patents are creatures of statutes, those conditions, referred to as prior art, that defeat the novelty of an invention are likewise defined in the patent laws. Only the more typically encountered ones are mentioned below.

The two most common categories of prior art are probably printed publications and patents. The publications and patents can be from any country and in any language. A patent or publication that describes an invention and is prior in time to it defeats the novelty of the invention. In the event a printed publication or patent describes an invention and is more than one year old when the patent application is filed in a patent office, such will render the invention of that patent application unpatentable no matter when the invention was actually made. In fact, the patent or publication could even be by the inventor and when such is more than one year before the actual filing of the patent application, granting of a patent is precluded. This situation is referred to as a statutory bar, because no matter whose patent or publication and no matter when the invention was made, a patent is barred. In essence the United States gives the inventor a one-year grace period after patenting or publication to file an application provided the invention was made prior to the patent or publication. However, most other countries do not give any grace period. Therefore, the patent application must already be on file before the invention is publically disclosed to be patentable in most other countries. Often an inventor desires to present or publish a paper describing some discovery. However, to preserve foreign rights, the paper should not be published until a patent application has been filed. After filing, it could be published without affecting rights.

Public use or public knowledge of an invention in the United States prior to the invention renders it unpatentable. Moreover a patent is precluded if the invention was in public use or on sale in the United States more than one year prior to filing the patent application. The public use or on sale activity could even be by the inventor and when it is more than one year before the filing of the patent application, such activity will prevent granting of a patent. This one-year period is an attempt to balance the desire for early disclosure of the invention to the public and the need to give the inventor some limited time to determine whether pursuing patent protection is justified.

Nonobviousness. The above-discussed novelty-defeating prior art items presuppose that the exact subject matter for which a patent is being sought is described by a single prior art item. However, even if the invention is not fully described by a single prior art item it still may not be patentable. In particular, the patent law also requires that the invention be nonobvious (35 USC §103). An invention is not patentable if the differences between the subject matter sought to be patented and the prior art are such that the subject matter as a whole would have been obvious at the time the invention was made to a person of ordinary skill in the technology of art that is the subject matter of the invention. Stated in other terms, the advance or difference between the subject matter sought to be patented and the prior art must not be so trivial to be readily apparent to persons working in that particular technology.

THE PATENT APPLICATION

The patent application, being a combination of a legal contract and a scientific paper, represents a unique written document. It is a scientific paper because it describes some technical achievement but differs from usual technical papers in that it must stand on its own to a greater extent than most technical papers. The patent represents a legal contract because it is actually a contract between the inventor and the people of the United States as represented by the government and, specifically, by the U.S. Patent and Trademark Office. In particular, the government gives the inventor the right to exclude others, for a limited period of time, from making, using, offering for sale, selling, or importing the invention as defined by the claims. The consideration given by the inventor in the contract is publicly disclosing information concerning the invention.

A patent application must include a specification, a drawing (in the event that the nature of the invention is such as to require a drawing for understanding), and a signed oath or declaration. The specification must contain both a written description of the invention and at least one claim. In addition, a government filing fee must be paid, which, of course, changes with time. The legal requirements for an application can be found in 35 USC §111. Moreover, by rule, an application must contain an abstract of the disclosure.

Abstract

The abstract is intended merely as a searching tool and should be a concise, informative statement that can give a quick and general idea of the nature and essence of an invention. The abstract has no legal effect on the interpretation of the claims of the patent or on the interpretation of the remainder of the disclosure of the patent.

Specification Description

The requirements for the specification of a patent are contained in 35 USC §112 and especially in ¶ 1 thereof. In particular, the requirements of the descriptive portion of the specification include the following:

1. A written description of the invention.
2. A written description of the manner and process of making and using the invention in such full, clear, concise, and exact terms as to enable any person skilled in the art to which it pertains or with which it is most nearly connected, to make and use the invention; a requirement usually referred to as the enablement requirement.
3. The best mode contemplated by the inventor of carrying out the invention; a requirement usually referred to as the best mode requirement.

This descriptive material is the consideration given by the inventor as the inventor's part of the bargain represented by the patent contract.

Written Description of the Invention

The requirements for a written description of the invention is that this specification must support the claims of the application. One way to insure this is to have the specification contain language that corresponds to the language employed in any of the claims or language that is its equivalent. Even if there is no explicit description of a generic invention in the specification, however, the mention of a representative component or a representative number of examples may provide an implicit description on which to base generic language. In particular, the specification, in addition to what it explicitly states, also implicitly discloses what would be apparent to those skilled in the art from a mere reading of it. It is not always safe to rely on what is implicit, however, because the outcome of having to convince another of the true import of an implicit disclosure is uncertain. In addition, it can be quite helpful, although not necessary, to include in the specification a discussion of the prior art or the efforts that have previously been made as well as a discussion of the advantages achieved by the invention or problems addressed by it. In fact, some courts have held that advantages that are not disclosed in the application to buttress the evidence of patentability cannot be relied on. In a sense the specification provides an opportunity to sell the patentability of the invention and, accordingly, can be used to that advantage.

Enablement

The specification must also include a written description that is sufficient to enable any person skilled in the art to make and use the invention. Although this enablement need only be addressed to those skilled in the art in explaining the technical aspects of the invention, it is desirable that it not be written in such a manner as to be understandable only by persons so skilled. This consideration may be ultimately important because the validity of a patent, if challenged, will initially be determined by a judge in a federal court who may have had little or no patent background and little or no technical background.

One way in the specification to teach persons skilled in the art how to make and use the invention is to include a number of representative examples. This represents only one way of teaching, however, and it is not the only means by which the requirements of the enablement portion of the statute can be satisfied. The specification need not convince those skilled in the art that its assertions are correct. There are instances, however, such as in certain chemical inventions where very broad claims are being advanced, that additional support may be requested to establish that the invention works as claimed.

A particular problem area involves inventions concerning biological materials such as microorganisms. Unless the biological material in question is known and available or can be readily produced by a known procedure without undue experimentation, it is usually not possible to explain how to obtain it merely by a written discussion. This problem of disclosure can be taken care of by depositing the material in an approved depository under conditions that will make it accessible once the patent issues. Making it accessible, however, does not give anyone the right to infringe the patent. In such a case, it is desirable for the biological material to be on deposit at the time the application is filed and for the application to refer to the access number given the microorganism by the approved depository. The description should also include any taxonomic information. Depositories in the United States include the following:

1. American Type Culture Collection, 12301 Parklawn Drive, Rockville, Md. 20852. Phone: 301-881-2600. Telex: 908-768.
2. National Research Culture Collection, U.S. Department of Agriculture, 1815 North University Street, Peoria, Ill. 61604. Phone: 309-685-4011. Fax: 309-671-7814.

A deposit in either of the foregoing depositories can also be used for satisfying deposit requirements for certain other countries, such as those observing the Budapest Treaty.

For applications involving a computer program, the disclosure should include the program itself or at least a flow diagram that sets forth the sequence of operations the program is to perform. Even with a flow chart, the disclosure might be questioned by the examiner, particularly if the operations stated are very general or very complex.

A specification must also teach how to use the invention it describes. In drug cases, the office refuses to accept general statements of utility as satisfying this teaching requirements. General statements that have been considered unacceptable include those purporting pharmacological or therapeutical purposes or biological activities. To satisfy the teaching requirements in such cases, it is necessary to present more specific uses, such as the treatment of a particular ailment or disease. It is also desirable to disclose precise dosages and treatment methods.

Besides requiring a statement of utility, the office may also require proof depending on the type of utility asserted. For instance, the highest degree of proof of utility is usually required for those inventions stated to cure or treat diseases, such as cancers, that are generally considered, at the present time, as being difficult to treat. On the other hand, compositions whose properties are usually predictable from a knowledge of their constituents, such as laxatives, antacids, and certain topical preparations, require little or no clinical proof.

Best Mode

The specification is required to include the best mode contemplated by the inventor for carrying out the invention. This does not mean that the best way of carrying out the invention in any absolute sense must necessarily be disclosed, but only the best way contemplated by the inventor. This requirement also pertains to the point in time when the application was filed and need not be updated if a new best mode is subsequently contemplated by the inventor.

This requirement for disclosure of the best mode is to prevent inventors from obtaining the benefits of patent protection while maintaining for themselves, by concealing from the public, the preferred ways in which to carry out the invention. The best mode disclosure, as well as the enablement requirements previously discussed, represents the consideration given by the inventor in exchange for the benefits of the rights granted by the patent (ie, the right to exclude others from making, using, selling, offering for sale, or importing the invention).

In the event a specification does not satisfy the requirements of 35 USC §112, a patent can be held invalid. In addition, if the best mode was deliberately concealed, the issue of inequitable conduct against the office may be involved.

Drawings

If the nature of an invention is such as to lend itself to illustration, drawings become a required part of the application. Inventions that require drawings are those directed to machines, articles of manufacture, and certain processes. Such drawings are not required to be drawn to scale. Therefore, if dimensions or spatial relationships are important, these should be explicitly stated in the specification or on the drawings themselves. The drawings must also be described in the specification. There are some very specific rules with respect to the size and type of paper, the margins, and the cross-hatching to be used in all such drawings.

Claims

The claim or claims of a patent represent the metes, or bounds, of the property to be protected. In other words, as in real property, a patent claim stakes out that territory that the patentee considers his or her own, and any encroachment on that particular territory, as in real property, constitutes an infringement. The claims of a patent can be viewed as the word fence surrounding the invention. It is important not to confuse such claims with technical claims that state benefits or advantages.

The second paragraph of 35 USC §112 states the purposes of claims. Their basic functions are to point out, define, and distinctly lay title to the subject matter that the applicant regards as his or her invention, but not necessarily to describe that invention in any great detail. It is the descriptive portion of the specification that performs this function. In fact, the descriptive portion of the specification and of the claims differ chiefly in the degree of detail presented by each. In particular, claims usually omit mentioning nonessential features and, whenever possible, attempt to use generic language to describe the particular elements of the claimed invention instead of the far more specific language found in specifications.

It is important to recognize, however, that despite the desirability of obtaining the most generic claim coverage possible, varying claims of lesser scope are also important if the invention is to be adequately protected. A generic claim is more difficult for potential infringers to design around than a very specific claim is, but the latter is more difficult to invalidate in terms of prior art. On the other hand, a specific claim is easier to design around, and thus avoid literal infringement, than a generic claim.

It is helpful to make claims of both specific and intermediate scope to hedge against the possible invalidation of the patent's generic claim or claims. In such cases, the ability to establish infringement of valid claims of lesser scope may be important because such claims are often the only ones of any commercial value. This is why it is recommended that specific claims recite the best or preferred embodiments of the invention that are more fully disclosed in the description portion of the specification.

A claim must be written as a single sentence. In actuality, the claim is really the predicate noun that completes the sentence, "What I(we) claim is . . . " or "What is claimed is . . . "

Generally speaking, a claim can be divided into three major components. The first may be referred to as the introductory phrase or preamble, followed by a transitional phrase, and then the body of the claim.

Preamble of the Claim. The introductory phrase, or preamble, of a claim sets the stage for the remainder of the claim. It may merely make a general statement concerning the invention, the title of the invention, or the general class into which the invention falls. The introductory phrase might also include a statement of the intended use, object, purpose, or advantages or the invention. Some examples of the range of introductory phrases are as follows:

A process.
A process for extracting oil from fish.
A process for extracting omega-3 fatty acids from fish.

A composition.
A broth composition.
A chicken broth composition.

An apparatus.
An apparatus for separating grains of different sizes.
An apparatus for separating corns of different sizes.

As can be appreciated from these samples, the preamble may be as general or as specific as desired, depending on the particular invention involved, the prior art, and the intent of the preamble.

Transitional Phrase of the Claim. The degree to which a claim is considered open (inclusive) or closed (exclusive) is determined by its transitional phrase. This is an important distinction because the addition of a constituent or a step not explicitly recited in a closed claim can avoid its literal infringement. In an open claim, on the other hand, the presence of an additional constituent or step not explicitly recited in the claim will not necessarily avoid its literal infringement.

Claim terms that are considered to be closed are "consisting of" and "composed of." These terms mean that the presence or addition of something other than that which is explicitly recited in the claim will avoid its literal infringement. Claim terms considered open are "comprising," "including," and "containing." When such language is used, the inclusion of steps or constituents not explicitly recited will not necessarily avoid a literal infringement of the claim.

It is not necessary for a claim to be entirely closed or entirely open. In particular, the presence of the phrase, "consisting essentially of," is usually interpreted to mean that the claim covers not only the recited constituents of the process, composition, article, or apparatus but also any additional ones so long as the latter do not significantly interfere with the primary function or interrelationship of those constituents explicitly recited. If the material or component or step that is added to those recited is of a type that is usually employed with the kind of subject matter in question, then infringement is usually easy to establish.

Body of the Claim. The structural elements of the claim are presented in what is known as its body. These elements include, for instance, the steps of a process, the components of a composition, or the parts of a machine or apparatus that constitute the prime subject matter of the invention. The body of the claim specifies the configuration, spatial relationship between elements, and relative amounts of components that may be important to point out the subject matter.

It is not necessary that the elements of the claim be recited in the form of the actual structure, material, or procedural acts involved. Alternatively, the words *means* or *step* may be used followed by a description of the function to be performed by the means or step.

For instance, in a claim directed to a bioreactor, reciting "means for supporting the bottom of a bioreactor at a suitable distance from the floor" could be substituted for a description of bioreactor legs.

It is suggested that terms such as "about" and "approximately" be employed whenever reciting amounts, distances, or spatial relationships, because such terms may provide a somewhat greater scope in the interpretation of the breadth of the claim.

Types of Claims. The subject matter of claims can take the form of a process, an apparatus, an article of manufacture, or a composition. Moreover, the subject matter of a claim often referred to as a hybrid claim can take the form of any combination of these. For instance, a process claim might include a recitation of a certain apparatus or an apparatus claim might include recitations concerning materials treated by the apparatus. Usually, the patentability of such hybrid claims will depend only on those differences from the prior art of the principal subject matter of the claim rather than on the differences of the secondary recitations. For instance, the patentability of a process claim with apparatus recitations will depend primarily on the process steps and not on the apparatus recitations. Likewise, the patentability of an apparatus claim containing recitations of the material being worked on will depend on the structure of the apparatus rather than the material in question.

A particular type of hybrid claim sometimes found in chemical cases is referred to as a product-by-process claim. It is used to define a composition of matter or a material by referring to the particular process employed to prepare the product in question. The patentability of such a claim depends on whether the product differs from prior-art products, not whether the process steps differ from prior-art processes. The product-by-process claim can be important when no other way seems convenient to define a product that differs from prior art products. Because of the nature of the product, its differences from the prior art may be difficult to define other than by reference to the process itself. Also, the product-by-process claim may be important when the possibility exists that the manner in which the product has been described is not entirely correct.

The inventions resulting from the development of a new technology need not be restricted to any one of the preceding categories of inventions; they may, in fact, be claimed in any number of different ways. For instance, the development of a new composition of matter may not only result in a patentable invention with respect to the composition itself, but also to the process for preparing it, the apparatus used to carry out the process of preparing it, the process or processes for using it, finished articles containing it, and so on. Accordingly, any number of patents may be obtainable for any particular development. In fact, it is desirable to attempt to claim a development by making as many different types of claims as possible. This is important because a variety of claims can encompass different classes of direct infringers, and it may be more desirable to sue one kind of infringer rather than another in view of the potential recovery or convenience involved. For instance, it would certainly be more desirable to be able to sue the manufacturer of a composition rather than the users, particularly, if the users are individual consumers or the inventor's own customers. Also, certain types of claims may be more difficult to invalidate in view of the prior art than others, thereby hedging against loss of all of the potential protection available.

Each claim of a patent or patent application is considered as a separate invention and is examined independently from the other claims with respect to its patentability and validity. During the examination of an application, it is not uncommon for an examiner to reject some claims in an application while allowing others. Likewise, a court can find some claims valid, while finding others invalid.

Oath or Declaration

To complete the requirements of the application, it is necessary to have a declaration or oath signed by the inventor(s). This declaration need not be filed along with the initial application papers but must be filed within one month after notice by the office, requiring such filing or within two months of the initial filing, whichever is later.

SUMMARY

The patent system serves to promote the progress of technology by encouraging innovation and by making available to the public the information concerning patented inventions. This article is intended to provide inventors and their managers with some basic familiarity and comfort level with the patent system. In view of the relative complexity of this subject matter, this article is not intended to teach patent law or how to best protect an individual invention. In view of the significant cost of present-day scientific research and development and the global nature of implementing new technology, it is suggested that a patent attorney or patent agent be consulted as early as practical in order that the fruit of the innovative labors be adequately protected.

GENERAL REFERENCES

This article is adapted from B. A. Amernick, *Patent Law for the Nonlawyer: A Guide for the Engineer, Technologist, and Manager*, Van Nostrand Reinhold, copyright, 1986. Used with permission.

BURTON A. AMERNICK
Pollock, Vande Sande, and Amernick, RLLP
Washington, D.C.

PEACHES. See FRUITS, TEMPERATE.

PECTIC SUBSTANCES

Pectic substances are a group of complex polysaccharides localized to the middle lamella, intercellular crevices, and primary cell walls of most if not all higher plants (1). They are significant contributors to the texture of fruits and vegetables and their processed products (2). Pectin and/or pectin fragments have been reported to possess pharmacological activities that include immunostimulation, antimetastasis, hypoglycemic, and cholesterol-lowering effects (3). In addition to the obvious structural role of pectic polysaccharides in plants, pectic fragments act as chemical messengers in the development, growth, senescence, and biochemical protection of plants (4).

Because the gel-forming properties of pectin substances (*pectin* is derived from a Greek word meaning to congeal or solidify) were discovered before the development of modern organic and macromolecular chemistry, common rather than structural names were applied to pectic polysaccharides, and these persist (5).

DISTRIBUTION IN NATURE

The percentage of pectic substances in plants varies with species, variety, anatomy, and maturity. The highest concentrations of pectic substances are found in young tissue and in fruit tissue. Typically, whole mature fruit contains 3 to 7% pectic substances on a dry-weight basis and 0.13 to 1.1% on a fresh-weight basis. The relative high pectic and low caloric content of citrus fruits make them a good source of soluble dietary fiber in that pectic substances are not digested until they reach the lower gastrointestinal tract. Apple pomace (the residue of pressed apples) may contain up to 20% pectic substances, and the albedo of oranges (the white, spongy material on the inside of the peel) may contain 40% or more pectic substances on a dry-weight basis.

ISOLATION AND PURIFICATION

The relative ease of the isolation and purification of large quantities of food-grade pectic substances account for their early discovery. Commercially, apple pomace or citrus peels (eg, orange, grapefruit, and primarily lemon or lime) are extracted with the goal of producing food-grade pectin for gels (6). Typically, pectin is acid extracted at low pH (1.5–3) to inhibit degradation by endogenous enzymes, ester saponification, and alkaline degradation by beta-elimination reactions. Extracts are filtered or centrifuged to remove insolubles and precipitated with alcohol or polyvalent salts such as those from aluminum or copper.

To obtain pectins with maximum retention of native structure and properties for research purposes, milder techniques have been adopted at the expense of product yield and with considerable lengthening of the isolation-purification process. In these procedures, relatively intact cell walls are isolated by macerating plant tissues in aqueous alcohol, washing to remove extraneous debris and deactivate endogenous enzymes, then sequentially extracting with chelating agent and mild alkali. Attempts to reduce size heterogeneity by gel filtration chromatography and chemical heterogeneity by ion exchange chromatography have been met with limited success.

CHEMICAL COMPOSITION AND STRUCTURE

Mildly extracted pectic substance is primarily a helical block copolymer of *D*-galacturonic acid and its methyl ester (7). These comonomers are (1→4) *O*-linked as poly(*α-D*-galactopyranosyluronic acid) and its methyl ester. Blocks are interrupted by (1→2) *O*-linked *α-L*-rhamnopyranosyl inserts. Some of these features are incorporated in Figure 1. Neutral sugars associated with pectin other than rhamnose include arabinose, galactose, glucose, xylose, and mannose. At least three of these neutral sugars, arabinose, galactose, and xylose, have been found in pectin as short or oligomeric side chains that themselves may be branched (8). Some portions of the pectin backbone have a high density of side chains and have been designated as hairy regions, whereas other portions of the pectin backbone are completely devoid of side chains and have been labeled as smooth regions. The hairy regions contain most of the neutral sugars found in pectin, namely rhamnose, arabinose, galactose, and xylose. A few sources of pectin (eg, sugar beets) contain acetyl and feruloyl esters. Acetyl esters are linked to ring hydroxyls in the glacturonate backbone, whereas feruloyl esters appear to be linked to neutral sugar side chains. Approximately 80 to 90% (by weight) of total sugars in commercial citrus pectins are residues of galacturonic acid and its methyl ester with the remaining sugars being neutral. Degree of methyl esterification (DM) in pectic substances will vary with plant source and method of extraction. Commercial citrus and apple pectins with DM ranging from 80 to 18% are available. X-ray diffraction patterns of sodium polygalacturonate indicate that water molecules are bound to each galacturonate residue in the backbone (9). Thus, bound water may account for about 8% of pectin samples by weight.

METHODS OF ANALYSIS

Because of its complex composition, several methods are necessary to analyze for pectin (10). A few of the more important methods will be discussed here. Of the several colorimetric methods developed to analyze for galacturonic

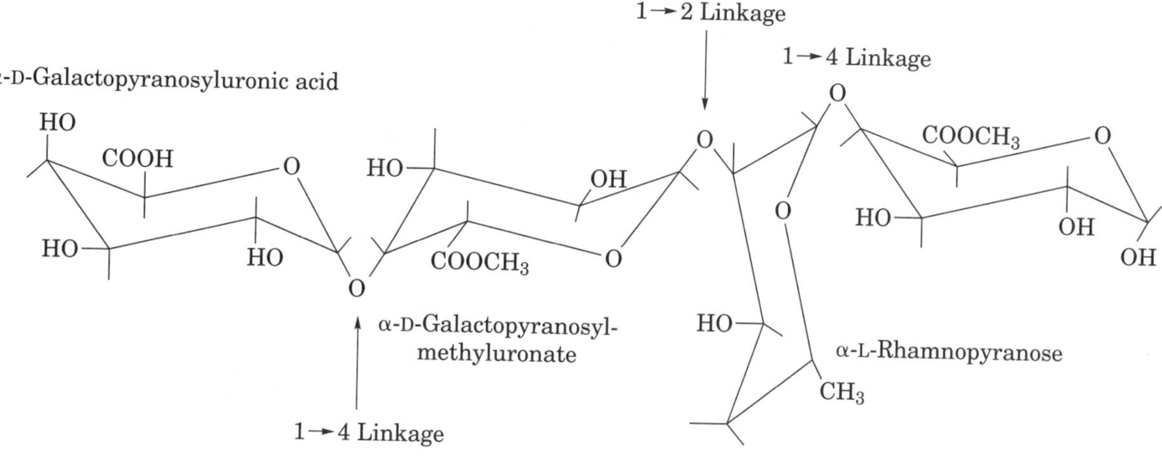

Figure 1. Portion of pectin molecule.

acid, the reaction of pectin with *m*-hydroxylbiphenyl in heated acids media to form a colored product is employed widely. Base titration before and after saponification is used for the simultaneous determination of DM and galacturonic content. Also, DM is determined by saponifying methyl ester groups and analyzing for free methanol by gas chromatography or by oxidizing the methanol to formaldehyde, which is determined colorimetrically. Carboxyl and methylester groups also may be determined by infrared spectroscopy in deuterated water provided there is a prior calibration against known standards. Often, neutral sugars are analyzed by gas chromatography after hydrolysis and conversion to volatile derivatives. Various derivatives and stationary phases have been employed. In one method, sugars are converted to their alditol acetates and chromatographed on a high-performance capillary column. Acetyl esters are analyzed by conversion to acetic acid, separation by steam distillation, and titration of the acid. Feruloyl esters are hydrolyzed to ferulic acid, which is analyzed by the Folin–Ciocalteu reagent.

CHEMICAL AND ENZYMATIC REACTIONS

The reactions of pectin are those characteristic of polysaccharides, esters, and organic acids (7). A select few of the more important reactions will be discussed briefly. Pectin undergoes acid-, base-, and enzyme-catalyzed depolymerizations. At acidic pH, glycosidic bonds other than those of (1→4) self-linked *α-D*-galacturonic acid are hydrolyzed preferentially. Thus commercial pectin is extracted from plant matrices by controlled acid hydrolysis. Prolonged acid hydrolysis can produce polygalacturonic acid with about 25 residues, whereas milder hydrolysis conditions have produced homogalacturonans with 70 to 100 galacturonan residues. Base-catalyzed depolymerizations occur at neutral and higher pH. These are beta-elimination reactions, which proceed with concurrent endodepolymerization, deesterification, and double bond formation (Fig. 2). The relatively high susceptibility to enzymatic and acid-catalyzed hydrolysis of the neutral sugar side chains and rhamnoglacturonan glycosidic bonds in the backbone of pectin may play an important role in plant metabolic processes. Evidence is accumulating that various pectic fragments act as biochemical messengers that initiate various biochemical reactions in plant development, senescence, and the defense of plants against pathogens (4). For example, evidence suggests that fragments from pectin-containing galactose may elicit ethylene production, which is important in the process of senescence. Endogenous endopolygalacturonases have been associated with fruits that ripen rapidly (eg, pears and freestone peaches), whereas those that contain only exopolygalacturonases (eg, apples and clingstone peaches) ripen more slowly. Polygalacturonases depolymerize deesterified pectins only. Most of the common pectin lyases depolymerize pectin in a fashion similar to beta-elimination reactions (Fig. 2). Commonly, pectin lyases are not found endogenously in plant cell walls but are produced by microorganisms. In addition to their biological importance, bacterial pectolytic enzymes have gained importance as probes to elucidate the neutral sugar side chain structure of pectin (11). For example, the action of *β*-(1,4)-galactanase has shown that apple pectin contains arabinogalactan side chains high in arabinose. More recently, fungal enzymes have been used to investigate the structure of the hairy regions in pectin (8). Pectin methyl esterase activity has been found in the plant cell wall, but the biological function of this enzyme is not clear. Reactions specific to the reducing sugar end group are important in that they permit determination of the number-average molecular weight and provide a method of following the course of depolymerization reactions. Chlorite oxidation of end groups has been used to determine pectin molecular weight, whereas several color reactions specific for reducing sugars have been used to assay for the galacturonic acid produced in depolymerization reactions (6). Ammonia partially amidates pectin through displacement of methyl ester groups. Amidated low methoxy pectic substances gel more readily than corresponding nonamidated low methoxy pectic substances (7). Pectin from sugar beet pulp has been gelled by intermolecular, oxidative coupling of feruloyl ester groups

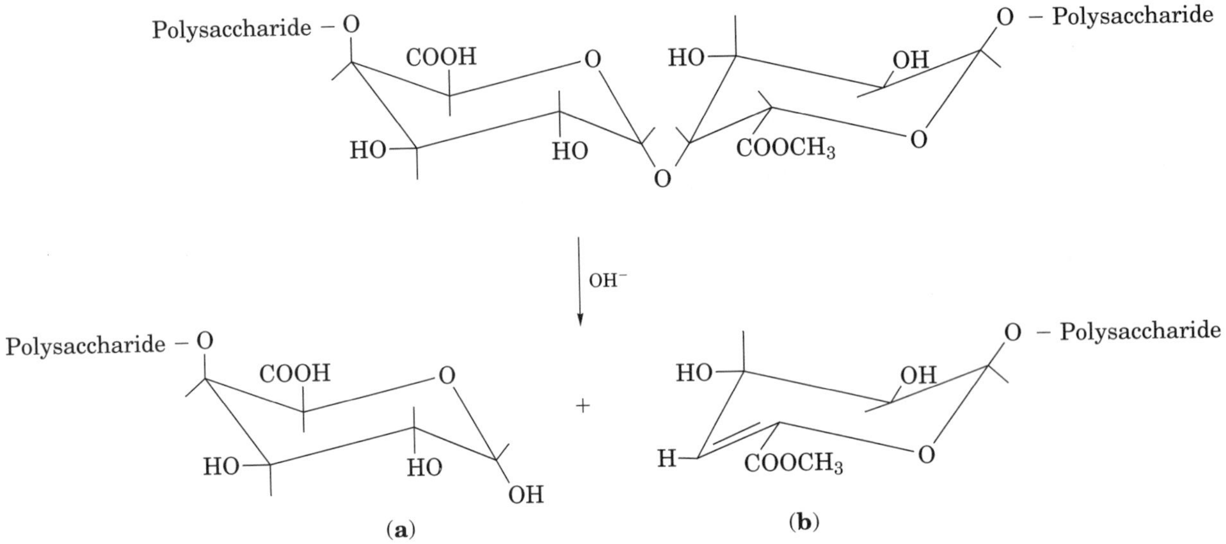

Figure 2. Schematic of β-elimination reaction: (**a**) is reducing sugar end group; (**b**) is unsaturated, nonreducing sugar end group.

through their aromatic rings. Coupling was achieved with a mixture of hydrogen peroxide and peroxidase (6). This reaction provides an opportunity for commercialization of beet pectin, because it is a poor gel former without oxidative coupling.

MOLECULAR WEIGHT, SIZE, AND SHAPE OF PECTINS

Over a period spanning more than 50 years of research, the molecular parameters of pectins have been measured extensively by numerous methods. Nevertheless, significant disagreements prevail concerning these parameters. The controversy may arise from the variability of pectin with plant source, whether dissolved pectin exists in solution as individual molecules or as aggregates, and the tendency of pectin to fragment. High methoxy (ca DM = 73) commercial citrus pectins give number-average molecular weights in the range 4–4.9 $\times$ 10^4 (12). A linear plot of reduced osmotic pressure against concentration appears to indicate a well-behaved unaggregated polymer. The second virial coefficient calculated from osmometry indicated a rodlike shape. Nevertheless, briefly heating pectin resulted in a slow concentration-dependent disaggregation that was also shown by osmometry. High-performance size-exclusion chromatography (HPSEC) with on-line viscometry detection on a series of commercial citrus pectins gave results consistent with the aggregated rod model (13). HPSEC studies of tomato cell wall pectins with computer-aided curve fitting of the chromatograms revealed that all pectins investigated were composed of a linear combination of five macromolecular subunits (14). The relative sizes of these subunits as obtained from their radii of gyration (R_g) were 1:2:4:8:16, with the smallest subunit having an R_g of about 25 Å. Dissociation of pectin into smaller-size fragments by dialysis against 0.05 M NaCl led to the conclusion that cell-wall pectin acted as if it were an aggregated mosaic, held together at least partially through noncovalent interactions. Superimposed and individual circular microgels in the micrometer size range were isolated from the cell walls of peaches and visualized by electron microscopy after rotary shadowing. Dilute NaCl and 50% aqueous glycerol disaggregated these microgels into rods, segmented rods, and kinked rods that collectively comprised the internal gel network of the microgels (15).

BINDING PROPERTIES

Many of pectin's unique properties result from its propensity to self-aggregate or to interact with cosolutes (6). A case in point is the pectin gels, which are three-dimensional networks capable of entrapping water. High-methoxyl pectins (HMP) with DM in the range 80 to 57% are induced to gel by adding sugars, which are thought to promote the self-aggregation of pectin by removing bound water, reducing ionic repulsions, and stiffening the polymer chain. These nonspecific interactions with sugars may be responsible for pectin's ability to reduce glucose intolerance in diabetics. For HMP at constant pH, gel strength and rate of gelling increase with DM, presumably through increased interchain hydrophobic interactions. HMPs of decreasing DM can be induced to gel by decreasing the pH, possibly through increased interchain hydrogen bonding. Low-methoxyl pectins (LMP), DM below 50%, gel with the addition of calcium or other divalent cations through interchain cross-links. Because decreasing DM increases the number of potential cross-links, LMP exhibits increased gel strength and higher gelling temperatures with decreasing DM. For HMP and LMP, gel strength increases with increased molecular size, possibly by cooperative self-interactions. It appears that all three major mechanisms exist for interchain interactions, namely divalent cation cross-links, hydrogen bonding, and hydrophobic interactions. Furthermore, for any mechanism to prevail, a critical number of functional groups must be present to promote cooperative interactions. Electron spin resonance studies support the concept that divalent cation binding to pectin in apple cell walls results in interchain bridges between adjacent smooth regions of pectin and that binding is a sequential cooperative process. Clinical studies suggest that pectin lowers blood cholesterol. This may be a case where the formation of divalent cation bridges is important in that calcium may form bridges between negatively charged pectin and bile acids. It has been suggested that binding and excretion of bile acids leads to a decrease of cholesterol in the bloodstream.

FOOD APPLICATIONS

A major application of pectin is in jams and jellies (16). A high-sugar jam contains 30 to 45% of fruit pulp and 0.20 to 0.4% pectin added as a gelling agent. Jams made with HMP must contain at least 60% soluble solids (sugars) to gel. Reduced-sugar or dietetic jams are manufactured with 55% or less soluble solids (even below 30%), by adding low methoxyl pectins (eg, in the range 0.75–1.0%). At very low soluble solids, a calcium salt often is added to aid gellation. Frequently, jellies are made from depectinized fruit concentrates with added pectin, water, and sugar. High-quality, tender confectionery jellies with excellent flavor-release characteristics contain pectin. Pectin is added to jams, fillings, and toppings as a gelling or thickening agent in the preparation of baked goods. HMP jams are useful in applications requiring resistance to the heat of baking such as occurs in producing tarts containing jam. Amidated low methoxyl pectins (ALMP) confer thermal reversibility to gels. These gels are useful as glazes for pastries or flans. Typically, these products are supplied as a paste containing ALMP, calcium diphosphate, and 65% sugar solids, which when diluted and melted can be reset to a clear, shiny glaze on cooling. In recent years, pectin has found increased application as an additive to dairy products. Yogurt containing fruit bases has been growing in popularity. Substituting pectin for modified starch as a thickening agent in yogurts will maintain a uniform distribution of fruit throughout the yogurt without masking delicate fruit flavors. Furthermore, unlike starches, pectin will not introduce a floury texture to yogurts. If the fruit bases contain 60% sugar, then HMP can be added. If the sugar content is lower than 60%, then ALMP is added. HMP stabilized casein against aggregation when heated at a pH less than 4.3. Thus it is added as a stabilizer in ultra-high-temperature-treated yogurt drinks and to milks blended with fruit juices. Pectin also stabilizes acidified soy milk drinks and whey products against protein precipitation. A low level of pectin is often added to low-calorie soft drinks to replace mouth-feel lost with the removal of sugar. Pectin is added to sorbet and ice pops to control ice crystal size; and to ice pops to prevent flavor and color from being sucked from the ice structure. Pectin is added to chutney and sauces to improve texture and batch-to-batch uniformity. LMP and ALMP gels can replace gelatin as a base in dessert jellies for the purpose of providing good flavor release. Pectin gels have higher melting points than gelatin gels and thus hold up better in warm weather.

Pectin has several applications in the pharmaceutical industry. It has been added to mixtures containing kaolin or bismuth compounds for the prevention of diarrhea. Pectin is added to maintain the viscosity of medicinal syrups. Recently, pectin has been employed as a filler in self-adhesive colostomy flanges and to promote healing by its addition to wound dusting powders and ulcer dressings.

GLOSSARY

Pectate. Pectate as in calcium or sodium pectate is pectic acid that is fully or partially neutralized with metal ions. Often referred to as polygalacturonate as in calcium or sodium polygalacturonate.

Pectic acid. Pectic acid is pure poly[(1-4)-*O*-linked (alpha-*D*-galactopyranosyl uronic acid)] and is often referred to as polygalacturonic acid.

Pectin. Pectin, as originally defined, is a water-soluble mixture of pectinic acids or partially neutralized pectinic acids capable of undergoing gel formation.

Pectinate. Pectinate as in calcium or sodium pectinate is pectinic acid that is partially or fully neutralized with metals ions.

Pectinic acid. Pectinic acid is a mixture of pectic substances, a significant portion of which is a copolymer of galacturonic acid and its methyl ester.

Protopectin. Protopectin refers to native, undissolved or insoluble plant tissue pectin.

DISCLOSURE STATEMENT

Reference to a brand or firm name does not constitute endorsement by the U.S. Department of Agriculture over others of a similar nature not mentioned.

BIBLIOGRAPHY

1. M. C. Jarvis, "Structure and Properties of Pectin Gels in Plant Cell Walls," *Plant Cell Env.* **7**, 153–164 (1984).
2. G. B. Fincher and B. A. Stone, "Metabolism of Noncellulosic Polysaccharides," in W. Tanner and F. A. Loewus, eds., *Plant Carbohydrates II*, Vol. 13, Springer-Verlag, New York, 1981, pp. 103–105.
3. H. Yamada, "Contributions of Pectins on Health Care," in J. Visser and A. G. J. Voragen, eds., *Pectins and Pectinases*, New York, 1996, pp. 173–180.
4. E. A. Nothnagel et al., "XXII A Galacturonic Acid Oligosaccharide from Plant Cell Walls Phytoalexins," *Plant Physiol.* **71**, 916–1026 (1983).
5. Z. I. Kertesz, *The Pectic Substances*, Wiley-Interscience, New York, 1951, pp. 3–9.
6. M. L. Fishman, "Chemical and Physical Properties of Pectin," *ISI Atlas of Science: Biochemistry* **1**, 215–219 (1988).
7. J. N. BeMiller, "An Introduction to Pectins: Structure and Properties," in M. L. Fishman and J. J. Jen, eds., *Chemistry and Function of Pectins*, ACS Symposium Series **310**, American Chemical Society, Washington, D.C., 1986, pp. 2–12.
8. H. A. Schols and A. G. J. Vorajen, "Complex Pectins: Structure Elucidation Using Enzymes," in J. Visser and A. G. J. Voragen, eds., *Pectins and Pectinases*, Elsevier, New York, 1996, pp. 3–19.

9. M. D. Walkinshaw and S. Arnott, "Conformations and Interactions of Pectins," *Molekulyarnaya Biologiya* **153**, 193–205 (1981).
10. L. W. Doner, "Analytical Methods for Determining Pectin Composition," in M. L. Fishman and J. J. Jen, eds., *Chemistry and Function of Pectins*, ACS Symposium, Series 310, American Chemical Society, Washington, D.C., 1986, pp. 13–21.
11. J. A. deVries et al., "Structural Features of the Neutral Sugar Side Chains of Apple Pectin Substances," *Carbohydr. Polym.* **3**, 193–205 (1983).
12. M. L. Fishman, L. Pepper, and P. E. Pfeffer, "Dilute Solution Properties of Pectin," in J. E. Glass, ed., *Water Soluble Polymers: Beauty with Performance*, Advances in Chemistry Series **213**, American Chemical Society, Washington, D.C., 1986, pp. 57–70.
13. M. L. Fishman et al., "Characterization of Pectins in Conjunction with Viscosity Detection," *J. Ag. Food Chem.* **37**, 584–591 (1989).
14. M. L. Fishman, D. T. Gillespie, and S. M. Sondey, "Macromolecular Components of Tomato Fruit Pectin," *Arch. Biochem. Biophys.* **274**, 179–191 (1989).
15. M. L. Fishman et al., "Progressive Dissociation of Pectin," *Carbohydrate Res.* **248**, 303–316 (1993).
16. C. D. May, "Industrial Pectin Sources, Production and Applications," *Carbohydr. Polym.* **12**, 79–99 (1990).

MARSHALL FISHMAN
USDA/ARS
Wyndmoor, Pennsylvania

PEPTIDES

Peptides have relevance to food science in various ways. Peptides contribute to the physical properties of food (physical aspect) and have taste (organoleptic aspect). Intestinal absorption of dipeptides or tripeptides is better than that of free amino acids (nutritional aspect). Biologically active peptides are found in food (physiological aspects). In this article the general aspects of peptides will be briefly summarized and various aspects of peptides as food constituents will be described.

GENERAL ASPECTS OF PEPTIDES

Definition

Peptide is a substance in which amino acids are bound together by peptide bonds, amide bonds between α-amino groups, and α-carboxyl groups of neighboring amino acids. Peptides containing fewer than 10 amino acids residues are called oligopeptides and those containing 10 amino acids or more are called polypeptides. Polypeptides containing more than 50 amino residues are called proteins.

Most peptides of natural origin are linear molecules containing L-amino acids. The structure of, for example, L-alanyl-L-threonyl-L-tyrosine is expressed as H-Ala-Thr-Tyr-OH or Ala-Thr-Tyr in three-letter abbreviations and as A-T-Y in simple one-letter abbreviations.

Usually, 20 different amino acids are found in proteins and peptides. Therefore, the number of possible peptides containing n residues of amino acid is 20^n.

Formation

Peptides are formed in four ways as described next.

1. Peptides are released by enzymatic digestion of proteins that have been synthesized on ribosomes according to the information coded in mRNA. This is the major route of peptide formation *in vivo*. Both the hydrolysis of food proteins in the digestive tract and also release of biologically active peptides in a variety of organs are classified in this category. The production of peptides by recombinant DNA technology essentially follows this route.
2. In living organisms there are peptide-synthesizing systems that are independent of mRNA and ribosomes. A few peptides (glutathione, kyotorphin, and cyclic peptide antibiotics) are synthesized in this way.
3. Under special *in vitro* conditions peptide bonds are also formed by protease-catalyzed reactions (1) (condensation and aminolyses of amides or esters).
4. Peptides are chemically synthesized by a condensation of amino acids in the presence of an activating reagent for carboxyl groups such as carbodiimides. If condensation between the carboxyl group of amino acid A and the amino group of amino acid B is to occur, the amino group of A and carboxyl group of B must be protected. Protection of other functional groups in the side chains of amino acids is also necessary. Stepwise syntheses of peptides are facilitated by a solid-phase synthesis (2). In the solid-phase synthesis developed by Merrifield, *t*-butyloxy-carbonyl (*t*-Boc) amino acid is coupled to chloromethyl polystyrene resin. After the acid-catalyzed removal of the *t*-Boc group from the resin, the next *t*-Boc amino acid is coupled to the resin in the presence of carbodiimide. Amino acids are thus coupled successively to the resin from the carboxyl terminus toward the amino terminus. Peptides are released from the resin in a strong acid such as anhydrous hydrogenfluoride. Protecting groups for amino acid side chains are also removed by this procedure. In a new procedure, the alkaline labile 9-fluorenyl-methoxycarbonyl (Fmoc) group is used as a blocking group for the α-amino group. In this case, recovery of the peptide from the resin is facilitated by trifluoroacetic acid.

CHEMISTRY AND ANALYSIS

The amino acid composition of a peptide is analyzed after acid- or base-catalyzed hydrolysis by an amino acid analyzer. The covalent structure of a linear peptide (primary structure) is analyzed by a stepwise chemical cleavage from the amino terminus of the peptide using phenyl isothiocyanate (Edman degradation) (3). This process is automatically done by a protein sequencer. The stepwise chemical cleavage method from the carboxyl terminus of peptide has also been reported (4). The primary structure of a peptide can be analyzed to some extent after the enzymatic cleavage by aminopeptidase or carboxypeptidase. The primary structure of a peptide can be also analyzed by X-ray crystallography. For the analysis of conformation of the peptides in solution, circular dichloism or nmr are available (5).

OCCURRENCE IN NATURE

A number of biologically active peptides occur in nature. Most of the biologically active peptides are synthesized as inactive precursor proteins and processed to an active form by a limited proteolysis. Acid- or base-catalyzed hydrolysis of proteins also produces peptides. Among heat-incoagulable peptides released by the digestion of proteins, those precipitable by the addition of ammonium sulfate are called proteoses and those soluble in saturated ammonium sulfate are called peptones.

Some peptides of microorganisms and plants have a cyclic structure and contain D-amino acid or amino alcohol or amino aldehyde. Those containing constituents other than amino acids are called depsipeptides; peptidic antibiotics belong to this category.

PEPTIDES IN PHARMACOLOGICAL USE

A number of biologically active peptides are used in pharmacology. These are classified into hormones, growth regulators, immunomodulators, neurotransmitters, substrates or inhibitors of enzymes, and so forth. Among them, some of the growth regulators and immunomodulators should be classified as proteins because of their molecular weights. The most widely used peptide drug is insulin. Not only agonists but also antagonists of these biologically active peptides show interesting pharmacological effects. Fragment peptides containing binding sites for receptors sometimes exhibit interesting biological effects as well. For example, Gly-Arg-Gly-Asp-Ser and Tyr-Ile-Gly-Ser-Arg are the fragment peptides of fibronectin and laminin, respectively. These peptides have been shown to prevent the metastasis of tumors. Asp-Ser-Asp-Pro-Arg (Hamburger peptide), a fragment peptide of human immunoglobulin E, shows an antiallergic effect. Peptide T (Ala-Ser-Thr-Thr-Asn-Tyr-Thr), a fragment peptide of the envelope glycoprotein of human immunodeficiency virus, inhibits the binding of the viral envelope to the CD-4 receptor of T-lymphocytes.

Inhibitor peptides for proteases show various pharmacological effects. Inhibitors for angiotensin I-converting enzyme or renin show antihypertensive effects, the aminopeptidase inhibitor bestatin shows an immunostimulating effect, and proryl endopeptidase inhibitor shows an antiamnesic affect.

Peptides are also promising as safe vaccines for the prevention of microbial or viral infections. Because of the difficulty of intestinal absorption and inactivation in the digestive tract, most peptide drugs are administered by injection. By suitable chemical modification, such as the incorporation of synthetic amino acids or acylation of the α-amino group or the amidation of the α-carboxyl group, some biologically active peptides are improved in specific activity, half-life, and proteolytic resistance. Some peptides become available for use through the oral or nasal route after such chemical modification or after physical modification such as emulsification and incorporation into liposomes. Another pharmacological use of peptide is as a solubilizer or emulsifier of insoluble drugs. Hydrolysates of gelation and casein have been shown to have such a potency (6).

PEPTIDES IN FOOD SCIENCE

Physical Aspect

Peptides contribute to the physical properties of food. For example, the smooth texture of cheese is dependent on the extent of proteolysis of casein during the maturation. Proteins and peptides have surface activities and show functional properties such as emulsifying and foaming. Amphiphatic properties, charge distribution, solubility, and the molecular weights of peptides are related to these functional properties. Usually the functional properties of food proteins are modified by proteolysis. In some proteins, like gluten, whose solubility is very low, the functional properties are improved by limited proteolysis. When a protein has good solubility, functional properties of the peptides derived from it tend to be lower than those of the parental proteins. Enzymatic digests of soybean proteins suppress thermal gelation and freezing-induced insolubilization of proteins. Soybean peptides protect starch from retrogradation.

Antioxidative Peptides

Amino acids such as tryptophan, tyrosine, histidine, methionine, proline, and branched-chain amino acids show antioxidative activity. Peptides containing these amino acids also show antioxidative activity (7,8). Antioxidative activities of peptides are sometimes higher than those of amino acids. This is partly because peptides have higher affinity for lipids than amino acids.

Taste of Peptides

Amino acids have taste. The taste of oligopeptides is sometimes more diverse and intense than that of amino acids. Oligopeptides elicit bitter, sweet, sour, umami, and salt (9).

Bitter Peptides. Hydrophobic acids (phenylalanine, tyrosine, tryptophan, leucine, valine, and isoleucine) of L-configulation taste bitter. Generally, oligopeptides consisting of these amino acids taste bitter. Bitter peptides are smaller than decapeptide in size and have two hydrophobic groups, or one hydrophobic group and one basic group, which are separated by 4.1 A (10). Hydrolysates of proteins, such as casein, soybean protein, and fish protein, sometimes taste bitter. Many bitter peptides have been isolated from hydrolysates of food proteins and their analogues have been synthesized. The bitterness of a protein hydrolysate is dependent on the types of protease used; formation of bitter peptide can be minimized by choosing protease (11). Carboxypeptidase W from wheat germ has been shown to be effective in debittering peptides. Oligopeptide rich in Glu is effective in masking a bitter taste.

Sweet Peptides. A typical sweet peptide (8) is aspartame (Asp-Phe-OMe), which is found during the synthesis of gastrin. Special arrangement of three functional groups (the basic α-amino group, the acidic β-carboxyl group of aspartyl residue, and the hydrophobic phenylalanyl residue) is required for sweetness. Phenyl-acetyl-glycyllysine and benzoyl-β-alanyllysine, which have a reverse arrangement of the three functional groups, have a sweetness that is one-fourth that of aspartame. Thaumatin, monellin, and curculin are sweet-tasting proteins isolated from tropical fruits. Curculin modifies sour taste into sweet taste. Miraculin, isolated from a plant native to western Africa, also modifies sour taste into sweet taste, although it is not sweet by itself. These proteins interact with the taste buds.

Sour Peptides. Glutamic acid and aspartic acid taste sour when their carboxyl groups are dissociated (8). Some oligopeptides containing Asp, Glu, or both also taste sour. Dipeptides formed by binding the γ-carboxyl group of Glu to the amino group of another amino acid have an astringent taste in addition to a sour taste.

Umami Peptides. Monosodium glutamate is a typical umami substance. Dipeptides containing Glu at the amino terminus elicit the umami taste in an aqueous solution at pH 6. The acidic oligopeptide fraction obtained from fish protein hydrolysates show an intense umami taste. An octapeptide, Lys-Gly-Asp-Glu-Glu-Ser-Leu-Ala, which was isolated from papin hydrolysate of beef, has an umami taste and has been named a delicious peptide (12).

Salty Peptides. Hydrochlorides of basic dipeptides, such as Orn-β-Ala and Orn-Tau, potentiate the salty taste of sodium chloride and therefore may be useful in decreasing the amount of salt in foods used for hypertensive patients (13).

Nutritional Aspects of Peptides

A fairly large part of digested protein is absorbed from the intestinal mucosa not as free amino acids but as dipeptides and tripeptides. A carrier-mediated transport system for dipeptides and tripeptides in the intestinal mucosa is operated by a proton gradient (14). For example, absorption rate of Gly-Gly is higher than that of Gly. Furthermore, Gly-Gly-Gly is absorbed more quickly than Gly-Gly (15). In a patient with Hartnup disease, whose ability to absorb free aromatic amino acid is defective, Phe-Phe could be absorbed well (16). Solubility of hardly soluble amino acids is improved by incorporation into peptides. The taste of oligopeptide mixtures is milder than that of free amino acids. The osmotic pressure of peptides in solution is lower than that of an equal amount of free amino acids. For these reasons an oligopeptide mixture is better than an amino acid mixture for an enteral diet. A low phenylalanine peptide (LPP) mixture for phenylketonuremic patients is prepared by the digestion of the protein by protease followed by the removal of phenylalanine by the absorption to Sephadex or charcoal (17).

Biologically Active Peptides

There are many biologically active peptides in foods. These peptides are classified into two groups: peptides found in food in active form from the beginning, and those that are released from inactive proteins by the limited proteolysis during food processing, storage, and digestion (see Table 1).

Biologically Active Peptides Found in Foods. Foods of animal origin, especially milk, contain large numbers of peptides belonging to the first group. Today, many biologically active peptides such as epidermal growth factor, nerve growth factor, insulin, prolactin, somatoatatin, thyroid-releasing hormone, thyroid-stimulating hormone, growth hormone-releasing factor, luteinizing hormone-releasing hormone, adrenocorticotropic hormone, erythropoietin, bombesinlike peptides, calcitonin, and delta-sleep-inducing peptide are detected in milk by sensitive radioimmunoassay. In neonates, the proteolytic activity in the gastrointestinal tract is usually low and the intestine shows considerable permeability to large peptides. Therefore, these biologically active peptides found in milk might have a physiological function in the development of the gastrointestinal tract and other organs. In fact, epidermal growth factor given orally to newborn rats has been shown to increase the weight of the intestine, liver, heart, and kidney (18).

Some peptides of plant origin also show biological activity for animal cells. Soybean trypsin inhibitor has growth-promoting activity for endothelial cells.

Biologically Active Peptides Released by Limited Proteolysis of Food Proteins. Many types of biologically active peptides are released from food proteins. These are classified as (*1*) peptides controlling intestinal absorption, (*2*) ligands for receptors, (*3*) enzyme inhibitors, (*4*) antimicrobial peptides, and (*5*) others including antioxidative peptides. Biologically active peptides are released not only from animal proteins but also from plant proteins.

Peptides Controlling Intestinal Absorption. **Casein phosphopeptide**. The bioavailability of calcium in milk is better than that in other foods. This is partly attributable to lactose. However, in cheese, which is low in lactose, the availability of calcium is also good. This is explained by the phosphopeptides that are released from α_{S1}- and β-casein released by trypsin (19). Similar peptides are found in the intestine. Phosphopeptides promote intestinal absorption of calcium by solubilizing calcium phosphate. This peptide also facilitates absorption of ferric ion by the same mechanism. The peptide itself is not absorbed from the intestine.

Cholesterol-lowering peptide. Serum cholesterol–lowering peptide was isolated from an enzymatic digest of soybean protein (20). Hydrophobic core peptides remaining after the digestion of the protein adsorb bile salt and destabilizes lipid emulsion (21). Consequently, intestinal reabsorption of bile salt and absorption of cholesterol are

Table 1. Biologically Active Peptides Derived from Food Proteins

Peptides	Structure
Phosphopeptides that promote calcium absorption	
α-CPP	α_{S1}-casein (43–79)
β-CPP	β-casein (1–25)
Peptide that lowers serum cholesterol	
Fragment peptide of soybean protein	Unknown
Opioid peptides	
β-casomorphin 7	YPFPGPI
α-casein exorphin	RYLGYLE
Gluten exorphin A	GYYPT
Gluten exorphin B	YGGWL
Gluten exorphin C	YPISL
Opioid antagonists	
Casoxin A	YPSYGLNY
Casoxin B	YPYY
Ileum-contracting peptides	
Oryzatensin	GYPMYPLPR
Casoxin C	YIPIQYVLSR
Casoxin D	YVPFPPF
Albutensin A	ALKAWAVAR
β-lactotensin	HIRL
Vasorelaxing peptide	
Ovokinin	FRADHPFL
Peptides promoting phagocytosis	
β-casein (63–68)	PGPIPN
β-casein (191–193)	LLY
Glycinin A_{1a} (279–282)	QRPR
Soymetide: β-conglycinin α'(322–334)	MITLAIPVNKPGR
Immunosuppressive peptides	
Proline-rich polypeptide fragment	YVPLFP
Human lactoferin (231–245)	CPDNTRKPVDKFKDC
Caseinomacropeptide	
Peptides that inhibit platelet aggregation	
κ-casein (100–105)	PHLSF
κ-casein (106–116)	MAIPPKKNQDK
Human lactoferrin (39–42)	KRDS
Inhibitor peptides for Angiotensin I-converting enzyme	
α_{S1}-Casein (23–34)	FFVAPFPEVFGK
α_{S1}-casein (194–199)	TTMPLW
Casein-derived peptide from fermented milk	IPP, VPP
Tuna peptide	PTHIKWGD
Bonito peptide	LKPNM
Antimicrobial peptides	
Lactoferricin B	FKCRRWQWRMKKLGAPSITCVRRAF

inhibited, and the serum cholesterol levels of animals fed with plant proteins are lower than those fed with animal proteins (22).

Ligands for Receptors. Biologically active peptides derived from food proteins bind to receptors for various types of endogenous biologically active peptides. Usually, their receptor affinities are weaker than those of endogenous peptides. However, some of them show physiological effects after oral administration because of their resistance to peptidase and smallness in molecular size.

Opioid peptides. Opioid peptides have an affinity for opiate receptors. There are more than 20 endogenous opioid peptides in the human body. Besides analgesic activity, these peptides have various physiological effects such as control of gastrointestinal functions and hormone secretion. Many opioid peptides are released from food proteins by limited proteolysis. β-casomorphin 7 is released from β-casein (23) and α-casein exorphin, from bovine α_{s1}-casein (24). Gluten exorphins A, B, and C are released from wheat gluten (25,26). β-casomorphin exhibits analgesic activity when administered intracerebroventricularly. β-casomorphin increases postprandial insulin and somatostatin levels. It also elevates the heart rate.

Opioid antagonist peptides are also released from milk proteins. Casoxin A and B are derived from κ-casein (27). Casoxin C is a functional antiopioid peptide of which activity is mediated by complement C3a receptor (27,28).

Ileum-contracting peptides. Many endogenous peptides such as angiotensin II, substance P, bradykinin, and neurotensin induce contraction of the ileum. Ileum-contracting peptides were isolated from digests of food proteins. β-lactotensin is released from β-lactoglobulin by chymotryptic digestion (29). Oryzatensin and casoxin C, which are released by trypsin from rice albumin and κ-casein, respectively, induce ileum contraction through receptor for complement C3a (28,30). Albutensin A, which is released from serum albumin by trypsin, also contracts ileum through receptors for both complements C3a and C5a (31). Casoxin D also induces contraction of isolated guinea pig ileum (32).

Vasorelaxing peptide. Ovokinin released from ovalbumin by the action of pepsin is a typical vasorelaxing peptide that has antihypotensive activity after oral administration (33). Some peptides isolated as ileum-contracting peptides also show vasorelaxing activity.

Immunostimulating peptides. Phagocytosis-stimulating peptides are released from various food proteins (34–36). Albutensin A stimulates phagocytosis as a complement C5a agonist. The complement C3a agonists, oryzatensin and casoxin C, also stimulate phagocytosis, but to a lesser extent than C5a agonist.

Immunosuppressive peptides. Proline-rich polypeptide from ovine colostrum and its fragment peptide YVPLFP show immunosuppressive activity (37). The human lactoferrin[231–425] has a homology to thymopentin and suppresses immune response (38). Caseinomacropeptide also has immunosuppressive activities (39).

Inhibitors for platelet aggregation. Peptides derived from food proteins inhibiting platelet aggregation have been reported (40). Some of them are homologous in structure to RGD, which is a recognition sequence for receptors of the integrin family.

Others. Peptides having affinity to calmodulin and inhibiting cyclic nucleotide phosphodiesterase are released from α_{s2}-casein (41). A peptide-stimulating proliferation of fibroblast and potentiating glucagon is released from β-casein (42). Caseinomacropeptide released from κ-casein by the action of chymosin inhibits gastric secretion (43).

Enzyme Inhibitors. The most typical enzyme inhibitor found in enzymatic digests of food protein is that for angiotensin I-converting enzyme; most of the digests show apparent inhibitory activity. Inhibitors for prolylendopeptidase derived from food proteins have been also reported.

Inhibitors for angiotensin I-converting enzyme. Angiotensin I-converting enzyme (ACE) is a dipeptidyl carboxypeptidase that catalyzes the conversion of angiotensin I to angiotensin II, a strong pressor. Inhibitors for this enzyme lower blood pressure in hypertensive animals. Peptidic ACE inhibitor was first isolated from snake venom. ACE inhibitor peptides were isolated from enzymatic digest of gelatin (44), casein (45,46), fish protein (47,48), and many other proteins. Some of these peptides have an antihypertensive effect after oral administration. Potentiation of hypotensive activity was observed when the ACE inhibitor peptide fraction from fish meal was orally administered as a emulsion in egg yolk (49).

Antimicrobial Peptides. Because of its iron-binding ability, lactoferrin shows bacteriostatic effect. Lactoferricin, released from the amino terminus region of lactoforrin by the action of pepsin, has bactericidal activity for various types of microorganisms (50). Immunostimulating activity of lactoferricin may also contribute to its antimicrobial effect *in vivo*. The carbohydrate chain attached to caseinomacropeptide is reported to be effective in preventing viral and bacterial infections (51).

Food Allergy

Many food materials, for example, eggs, milk, soybeans, and rice, are allergenic. Some food allergens are attributable to certain primary structures of the proteins. Allergens can be destroyed by limited proteolysis. Therefore, an oligopeptide mixture prepared by hydrolysis of proteins can be used as a low-allergenic food. In some other cases, allergenic epitopes are conformational. Allergenic structures can be destroyed by physical denaturation in this case.

CURRENT AND FUTURE STATUS

Bioreactor systems for the continuous production of peptides by a hydrolysis of proteins have also been developed. Release of peptides can be monitored by a peptide sensor, which is composed of aminopeptidase and amino acid oxidase (52). Peptides can be synthesized in a bioreactor system using synthetic reaction of proteases. Aspartame can be synthesized by this method.

Many biologically active peptides of food origin have been found by using sensitive *in vitro* assay systems. Although it is quite probable that milk-derived peptides may have teleological functions for newborns, only a few of these peptides have been demonstrated to be effective after oral ingestion. For those peptides having truly desirable functions for human health, methods to increase their availability, such as the enhancement of absorbability or half-life, should be explored. Although their specific activities are usually lower than those of endogenous biologically active peptides, those derived from food proteins sometimes show new and unexpected structure-activity relationships.

Recombinant DNA technology has been used to produce and improve pharmacologically useful peptides in microbial, plant, and animal cells. Production of heterogenous proteins in plants or milk by the transgenic technique is an especially promising method for the production of food protein. The amino acid composition and physical properties of food proteins can be improved by the site-directed mutagenesis technique. Furthermore, it will be possible to introduce biologically active peptide sequences into food proteins and to remove allergenic sequences from them.

BIBLIOGRAPHY

1. K. Nakanishi and R. Matsuno, *Synthetic Peptides in Biotechnology*, Alan R. Liss, Inc., New York, 1988, pp. 173–202.
2. J. M. Stewart and J. D. Young, *Solid Phase Peptide Synthesis*, Pierce Chemical Co., 1984.
3. P. Edman, *Acta Chemical Scandinavica* **4**, 277–282 (1950).
4. D. H. Hawke et al., *Analytical Biochem.* **166**, 298–307 (1987).
5. G. D. Rose et al., *Advances in Protein Chem.* **37**, 1–109 (1985).
6. T. Imai et al., *Chem. Pharm. Bull.* **37**, 2251–2252 (1989).
7. S. J. Bishov and A. S. Henick, *J. Food Sci.* **37**, 873–875 (1972).
8. H.-M. Chen et al., *J. Agric. Food Chem.* **44**, 2619–2623 (1996).
9. T. Nishimura and H. Kato, *Food Rev. Int.* **4**, 175–194 (1988).
10. N. Ishibashi et al., *Agric. Biol. Chem.* **52**, 819–827 (1988).
11. J. Adler-Nissen, *Protein Tailoring for Food and Medical Uses*, Marcel Dekker, Inc., New York, 1986, pp. 97–122.
12. Y. Yamasaki and K. Maekawa, *Agric. Biol. Chem.* **42**, 1761–1765 (1978).
13. M. Tamura et al., *Agric. Biol. Chem.* **53**, 1625–1633 (1989).
14. T. Hoshi, *Ion Gradient-Coupled Transport*, Elsevier, Amsterdam, The Netherlands, 1986, pp. 183–191.
15. D. M. Matthews and J. W. Payne, *Current Topics in Membrane and Transport*, Vol. 14, Academic Press, Inc., Orlando, Fla., 1980, pp. 331–425.
16. A. M. Asatoor et al., *Gut* **11**, 380–387 (1970).
17. S. Arai et al., *Agric. Biol. Chem.* **50**, 2929–2931 (1986).
18. O. Koldovsky, *J. Nutrition* **119**, 1543–1551 (1989).
19. R. Sato et al., "Casein Phosphopeptide (CPP) Enhances Calcium Absorption from the Ligated Segment of Rat Small Intestine," *J. Nutr. Sci. Vitaminol.* **32**, 67–76 (1986).
20. M. Sugano et al., *Atherosclerosis* **72**, 115–122 (1988).
21. K. Iwami et al., *Agric. Biol. Chem.* **50**, 1217–1222 (1986).
22. K. K. Carroll and M. W. Huff, *Nutrition and Food Science: Present Knowledge and Utilization*, Vol. 3, Plenum Publishing Corp., New York, 1980, pp. 379–385.
23. V. Brantl et al., *Hoppe-Seyler's Zeitschrift fur Physiologische Chemie* **360**, 1211–1216 (1979).
24. S. Loukas et al., *Biochem.* **22**, 4567–4573 (1983).
25. S. Fukudome and M. Yoshikawa, *FEBS Lett.* **296**, 107–111 (1992).
26. S. Fukudome and M. Yoshikawa, *FEBS Lett.* **316**, 17–19 (1993).
27. H. Chiba et al., *J. Dairy Res.* **56**, 363–366 (1989).
28. M. Takahashi et al., *Peptides* **18**, 329–336 (1997).
29. M. Yoshikawa et al., *Nippon Nogeikagaku Kaishi* **64**, 557 (1990).
30. M. Takahashi et al., *Peptides* **17**, 5–12 (1996).
31. M. Takahashi et al., *Lett. Peptide Sci.* **5**, 29–35 (1998).
32. M. Yoshikawa et al., *Peptide Chem. 1992*, ESCOM Science Publishers B. V., Leiden, The Netherlands, 1993, pp. 572–575.
33. H. Fujita et al., *Peptides* **16**, 785–790 (1995).
34. D. Migliore-Samour and P. Jolles, *Experientia* **44**, 188–193 (1988).
35. M. Yoshikawa et al., *Ann. New York Acad. Sci.* **685**, 375–376 (1993).
36. M. Tanaka et al., *Nippon Nogeikagaku Kaishi* **68**, 341 (1994).
37. M. Janusz et al., *Mol. Immunol.* **24**, 1029–1031 (1987).
38. I. Z. Siemion, *J. Peptide Sci.* **1**, 295–302 (1995).
39. H. Otani et al., *J. Dairy Res.* **62**, 349–357 (1995).
40. A.-M. Fiat et al., *J. Dairy Res.* **53**, 351–355 (1989).
41. K. Kizawa et al., *J. Dairy Res.* **62**, 587–592 (1995).
42. N. Azuma et al., *Agric. Biol. Chem.* **53**, 2631–2634 (1989).
43. G. K. Shlygin et al., *Byull. Eksptl. Biol. Med.* **72**, 9–13 (1971).
44. G. Oshima et al., *Biochimica et Biophysica Acta* **566**, 128–137 (1979).
45. S. Maruyama and Suzuki, *Agric. Biol. Chem.* **49**, 1405–1409 (1985).
46. Y. Nakamura et al., *J. Dairy Sci.* **78**, 777–783 (1995).
47. Y. Kohama et al., *Biochemical and Biophysical Res. Comm.* **155**, 332–337 (1988).
48. K. Yokoyama et al., *Biosci. Biotechnol. Biochem.* **56**, 1541–1545 (1992).
49. K. Suetsuna and K. Osajima, *Nippon Eiyou Shokuryou Gakkaishi (in Japanese)* **42**, 47–51 (1989).
50. W. Bellamy et al., *J. Appl. Microbiol.* **73**, 472–479 (1992).
51. Y. Kawasaki et al., *Biosci. Biotechnol. Biochem.* **57**, 1214–1215 (1993).
52. T. Tsuchida et al., *Hakkokogaku (in Japanese)* **67**, 499–507 (1989).

M. YOSHIKAWA
Kyoto University
Kyoto, Japan

H. CHIBA
Kobe Women's University
Kobe, Japan

See also PROTEINS: AMINO ACIDS; PROTEINS: STRUCTURE AND FUNCTIONALITY.

PESTICIDE RESIDUES IN FOOD

The term pesticide is used to describe any chemical agent that controls pests. In the United States, more than 20,000 pesticide products are registered (1). Traditionally, pesticides are used in agriculture, but nonagricultural uses are also common and include garden and household pest control, sanitation, wood preservation, and mosquito abatement. Pesticides contribute large economic and health benefits to society through minimizing crop losses, protecting the nutritional integrity of food, ensuring year-round storage, and providing appealing foods (2). Since as early as 1000 B.C. the Chinese used sulfur as a fungicide to control mildew on fruit; today sulfur remains an important fungicide. In the sixteenth century, arsenical compounds were popular insecticides; and in the seventeenth century, nicotine, rotenone, and *Chrysanthemum* extracts were introduced as insecticides and are still in use (3). In the United States, widespread utilization of pesticides in agriculture occurred after the 1920s with the introduction of mechanized power in farms. It has been estimated that 40% of the world's food supply would be at risk without pesticides (4).

Of the total amount of pesticides used in the United States during 1995 (4.5 billion pounds), less than 21% was used in the production of food and fiber products (5). The agricultural use of pesticides fluctuates yearly depending on climatic conditions, pest outbreaks, and planted acreage of pesticide intensive crops.

Food safety has been one of the driving forces behind pesticide legislation and regulation. Pesticide residues remain a major public concern today; consumer attitude surveys indicate that 72 to 82% of Americans consider pesticide residues to be a major concern (6). Nevertheless, in terms of overall food safety, the U.S. Food and Drug Administration (FDA) ranks pesticides as its fifth food safety priority and of less concern than (*1*) microbiological contamination of foods, (*2*) nutritional imbalance, (*3*) environmental contaminants, and (*4*) naturally occurring toxins (7).

Besides residues in foods, there are other pesticide concerns, such as potential acute and/or chronic toxicity to humans from occupational exposure. Other indirect risks include the destruction of susceptible crops and natural vegetation, reduction of natural pest enemies, effects on fish and wildlife populations, livestock losses, honeybee losses, evolved pesticide resistance, and creation of secondary pest problems.

PESTICIDE USE AND CLASSIFICATION

A wide variety of the types of pesticides is used in food production. Modern pesticides include synthetically produced organic chemicals, naturally occurring organic and inorganic chemicals, and microbial agents (natural or obtained through genetic manipulation). Table 1 lists many types of pesticides and their targets. Weeds, insects, and fungi are the major pests responsible for damage to agriculture. In terms of pounds applied, herbicides accounted for 55% of U.S. pesticide use in 1995, followed by insecticides (32%), fungicides (7%), and other categories (6%) (8).

Table 1. Pesticide Types and Targets

Pesticide type	Pest controlled
Insecticide	Insect
Herbicide	Weeds
Fungicide	Fungi
Nematicide	Nematodes
Acaricide	Mites
Defoliant	Leaves
Bacteriocide	Bacteria
Rodenticide	Rodents
Molluscicide	Snails
Algacide	Algae

Insecticides comprise compounds of various chemical classes; some of the most common are listed in Table 2. Insecticides have a variety of mechanisms of action in insects, including affecting the insects' metabolism (nerve poisons, muscle poisons, dessicants, and sterilants) or through a physical effect such as clogging air passages. The most common classes of insecticides are chlorinated hydrocarbons, organophosphates, and carbamates. The chlorinated hydrocarbons DDT and aldrin, among others (Table 2), were developed in the 1930s and 1940s. Such pesticides were very potent (high insect toxicity), but presented simultaneously chronic health and environmental effects because of their resistance to environmental and metabolic breakdown, which led to widespread environmental contamination and buildup of residues in a variety

Table 2. Some Pesticide Classes and Examples of Each Class

Pesticide	Examples
Insecticides	
Chlorinated hydrocarbons	Dicofol, methoxychlor, DDT,[a] aldrin,[a] dieldrin,[a] chlordane[a]
Organophosphates	Parathion, malathion, phosdrin, diazinon, chlorpyrifos, azinphos-methyl
Carbamates	Aldicarb, carbaryl, carbofuran
Pyrethroids	Permethrin, cypermethrin
Herbicides	
Triazine	Atrazine, cyanazine
Phenoxy	2,4 D
Quaternary ammonium	Paraquat
Benzoic acids	Dicamba
Acetanilides	Alachlor, metolachlor
Ureas	Linuron
Fungicides	
Inorganic	Sulfur
Ethylenebisdithiocarbamates	Maneb, mancozeb
Chlorinated phenols	Pentachlorophenol

[a]Banned in the United States.

of animals, including humans. Most chlorinated hydrocarbons are no longer permitted for use in the United States, although a few of the least persistent members of the family are still allowed. Historically, the carbamate and organophosphate insecticides served as replacements for many of the chlorinated hydrocarbons. These insecticides exert their toxicity on both insects and mammals through inhibition of cholinesterase enzymes that normally function to regulate nervous system activity. Although more acutely toxic to nontarget organisms, including mammals, the organophosphate and carbamate insecticides are less persistent in the environment. Another class of insecticides, the pyrethroids, were introduced in the 1970s. They are considered excellent broad-spectrum insecticides; they are effective at low doses, exhibit low toxicity to mammals, and break down quite rapidly in the environment. Pyrethroids continue to be used extensively in agriculture but also have the disadvantages of being relatively costly and environmentally labile, and they commonly lose their effectiveness due to the development of insect resistance.

Herbicides include a variety of chemical compounds that act upon weeds through different toxicological mechanisms. In some cases the toxicity results from direct plant contact by destroying leaf and stem tissues. Other herbicides inhibit seed germination or seedling growth; damage leaf cells, causing weeds to dry up; or affect the weed's ability to perform photosynthesis (2).

Fungi can cause damage or stress to food crops, and massive infestations can occur during storage of many foods, including grains, when storage occurs under conditions of high temperature and/or moisture. A serious consequence of fungal attack can be the production of mycotoxins, such as aflatoxins and fumonisins, by *Aspergillus flavus* and *Fusarium moniliforme*, respectively. Some fungicides can destroy fungi that have already invaded the plant while others may prevent fungal infestations.

PESTICIDE REGULATION

In the United States, the first legislation concerning pesticides was passed with the creation of the FDA in 1938, through the Federal Food, Drug and Cosmetic Act (FFDCA). The FFDCA established the requirement for pesticide tolerances when pesticide use could result in residues on food or feed crops.

In 1947, Congress passed the Federal Insecticide, Fungicide, and Rodenticide Act (FIFRA), which grouped all pesticide products under one law and mandated labeling and registration requirements. The U.S. Department of Agriculture (USDA) was given the initial responsibility to administer FIFRA. Several amendments to FIFRA were passed (1975, 1978, 1980, 1984, 1988, and 1996), which included, among others, provisions for use restriction, pesticide reevaluation and re-registration, and toxicological and environmental impact studies. In 1958, the Delaney Clause was approved as an amendment to the FFDCA. It stipulated that any food additive shown to cause cancer in humans or laboratory animals could not be used. In 1972, the responsibility for FIFRA administration was transferred to the newly created Environmental Protection Agency (EPA). EPA is authorized to grant pesticide registrations, to establish pesticide tolerances, and to regulate pesticide residues in food and feed under FIFRA.

In 1996, after legal disputes over the Delaney Clause, a new act, the Food Quality Protection Act (FQPA) was signed into law. The FQPA mandates a single, health-based standard for all pesticides, eliminating the Delaney paradox that arose due to inconsistencies resulting in different methods for regulating pesticides in raw commodities and those found in processed foods (9). The FQPA establishes standards that apply to all types of risks, including cancer risks and endocrine disruption, and guides the establishment for setting allowable levels (tolerances) for all pesticide residues on raw agricultural commodities and processed food. Important new provisions of the FQPA include additional protection for infants and children, consideration of aggregate risks from food, water, and domestic exposure, and consideration of cumulative risks from pesticides with common mechanisms of toxicity. Additionally, FQPA expedites approval of safer pesticides, requires periodic re-evaluation of pesticide registrations and tolerances, provides a consumer right-to-know provision and creates incentives for the development and maintenance of effective crop protection tools for farmers (1).

Three major U.S. regulatory agencies have the primary responsibility for regulating pesticides:

1. EPA, which has developed a series of guidelines for the toxicological testing of pesticides, registers pesticides for use, prescribes labeling, and establishes allowable levels (tolerances) of pesticides on food and feed crops;
2. FDA, which monitors domestic and imported foods for pesticide residues and enforces tolerances; and
3. USDA, which enforces tolerances on meat, poultry, and some egg products and also conducts the Pesticide Data Program (PDP) that provides important information on pesticide residue levels of fruits and vegetables in ready-to-eat form that may be directly used by EPA as a risk assessment tool.

At the international level, the Food and Agriculture Organization of the World Health Organization (FAO/WHO) is the regulatory authority on pesticides and develops, through the Joint FAO/WHO Meeting on Pesticide Residues, regulatory standards for maximum residue levels (MRLs). These standards are used widely throughout the world, but their use is not universal; several countries, including the United States, adopt their own standards for pesticide residue levels and enforce their sovereign standards on food entering their countries from foreign lands.

TOXICOLOGICAL EVALUATION OF PESTICIDES

Pesticides used in food production require detailed study in a series of toxicological tests. Laboratory animals (often rodents, dogs, and nonhuman primates) are used to study the acute, subchronic, and chronic toxicity of pesticides. In these studies, the metabolic fate, mutagenicity, carcinogenicity, and teratogenicity of pesticides are determined in addition to a large number of other toxicological effects. Additionally, EPA requires studies on the environmental fate of pesticides and their breakdown products, and the effects on nontarget organisms. All the collected data is reviewed by EPA for evaluation of the risks associated with the pesticide. This is a long and costly process, which frequently involves more than 10 years before a pesticide registration may be granted and at a cost in the tens of millions of dollars.

As FIFRA is primarily a risk-balancing statute, the EPA registers pesticides under a statutory standard that requires balancing the benefits of the use of the pesticide in question (such as increased crop yield, lower food cost, or public health protection) with potential risks such as health effects of consumers or agricultural workers and environmental damage. Thus the EPA grants a registration and specifies the commodity or commodities for which the pesticide may be used, as well as appropriate conditions for use and disposal. Failure to obey such legal requirements (printed on the pesticide labels) is a federal offense. Once federal registration is granted, the pesticide may still be subject to restrictions or denied use within individual states. More stringent use restrictions are frequently applied in some states, such as California.

Tolerance Setting

When the use of a pesticide may have the potential to leave a residue on a food in the United States, a tolerance, representing the maximum allowable residue permitted, is usually established. Tolerances represent the maximum expected residues of a pesticide on a specified commodity resulting from legal applications of the pesticide under established conditions for its use. When such conditions are followed, it is highly unlikely that residues in excess of tolerances would be detected. Tolerances are established for specific pesticide/commodity combinations; as such, the same pesticide may have different tolerance levels established for different commodities, and the tolerances of a variety of pesticides on a single commodity frequently vary considerably.

It is important to emphasize that the tolerance level is not established based on safety, but rather represents the maximum residue anticipated from the legal use of the pesticide (10). Nevertheless, before granting a tolerance, the EPA makes assessments of potential human exposure resulting from all registered (and proposed) uses of the pesticide to calculate the theoretical maximum residue contribution (TMRC). The TMRC assumes that the specified pesticide is always applied to all acreage planted, that it is used on all commodities for which it is registered, that residues are always present at the tolerance level, and that there is no reduction on pesticide levels from the plant to the postharvest stage, up to the table. It is a typical worst-case scenario using highly conservative assumptions that may overestimate exposures by factors of 100 to 100,000 times (2).

The TMRC value is compared with the reference dose (RfD), which is a daily exposure level not considered to represent any appreciable level of risk. A pesticide is considered to pose a negligible risk and tolerances are usually approved when the TMRC is below the RfD, provided that the carcinogenic risk at the TMRC is below the level of one excess cancer per million. When the TMRC is found to be higher than the RfD, or when exposure at the TMRC leads to a cancer risk greater than one excess cancer per million, the EPA may adopt a more refined risk assessment to more accurately calculate exposure estimates. Such refinements commonly represent the anticipated residue contribution (ARC) and may include adjustments of actual pesticide use, more realistic residue data, and consideration of potential pesticide levels reduction through washing, peeling, cooking, processing, and so on. The tolerance is established if the ARC is below the reference dose, and the carcinogenic risk at such an exposure is below the negligible risk of one excess cancer per million.

The passage of the FQPA in 1996 has served to make the process for establishing tolerances for pesticides more complicated. Prior to FQPA, tolerances were established on a chemical-by-chemical basis and considered only dietary exposure to the chemical. FQPA stipulates that the EPA may establish tolerances only when the EPA assesses that the risks posed by pesticides represent a "reasonable certainty of no harm" with respect to both carcinogenic and noncarcinogenic risks. In determining whether the pesticides satisfy the reasonable certainty of no harm criteria, the EPA considers the *aggregate* exposure to the chemicals from dietary, drinking water, and residential sources as well as *cumulative* exposure from pesticides possessing a common mechanism of toxic action (such as the organophosphates), meaning that determinations may be made on entire families of chemicals rather than on a chemical-by-chemical basis. In addition, the EPA is required to consider applying an additional 10-fold uncertainty factor in cases where infants and children may be more susceptible than adults to specific pesticides; this effectively reduces the RfD by a factor of 10. It is clear that the new FQPA requirements will require EPA scientists and others to develop significantly more robust models for assessing human pesticide risks and that the more stringent reregistration requirements posed by FQPA may significantly reduce the amounts and types of pesticides that may be used on food crops in the near future.

PESTICIDE RESIDUE MONITORING

It is the responsibility of the FDA to enforce tolerances in domestic and imported foods. Domestic samples are usually collected near the source of production or at the wholesale level, whereas imported foods are sampled at the entry point in the United States. The FDA has a regulatory monitoring program and a second program, the Total Diet Study, which estimates human dietary intakes of pesticides.

Regulatory commodity monitoring programs used by the FDA comprise both surveillance and compliance monitoring.

Surveillance Monitoring

The objective of surveillance monitoring is to identify violative residues in foods (11). Most regulatory samples analyzed by the FDA are taken in the surveillance monitoring program, and sampling is designed to maximize the chances of encountering illegal residues rather than providing a statistically representative look at residues in the U.S. food supply.

Compliance Monitoring

In compliance monitoring the FDA usually obtains follow-up samples after illegal residues have already been determined.

Although the FDA analyzes some processed foods, tests are primarily performed on raw commodities (prior to washing or peeling). The analytical methodologies adopted can detect more than 200 possible pesticides, and results are known shortly after samples are received by the laboratories. Clearly, results from the FDA's surveillance monitoring program indicate that residue levels rarely approach the tolerances. Although infrequently found, the vast majority of illegal residues represent cases where pesticides that are legally allowed to be used in the United States are detected on commodities for which a tolerance is not established (8). It is critical to realize that illegal residues should not be construed as "unsafe" residues and more properly serve as an indicator of erroneous application practices or the incidental contamination of commodities on which the pesticides were not directly applied but migrated to through the action of drift or uptake from contaminated soil (10).

Additionally, since 1961 the FDA has been carrying out its Total Diet Study annually. In this study, foods are collected in a "market basket" approach, using four geographical regions and three cities in each region each year. A total of 261 different food samples are collected to comprise each market basket. Each collection of foods is prepared for table-ready consumption and then analyzed for pesticide residues (8). Results from the FDA's Total Diet Study commonly indicate that the average human dietary exposure to pesticides is well below the established RfDs.

Since 1991 the USDA's Pesticide Data Program (PDP) has relied on cooperation with states in all regions of the United States to develop residue data that could be more useful for risk assessments than that collected by the FDA. In its initial years, PDP tested fresh fruits and vegetables, which were prepared for analysis simulating practices used by consumers such as washing and/or peeling. Since 1994 PDP monitoring has given more attention to foods frequently consumed by infants and children and has included samples of canned and frozen fruits and vegetables, wheat, soybeans, whole milk, fruit juices, and corn syrup. PDP's data on pesticides in selected commodities are used by the EPA to support its dietary risk assessment and pesticide registration programs, and by the FDA to refine sampling for tolerance enforcement (12).

At the state level, individual monitoring programs have been established by many states. Currently, the largest state program is conducted in California. California results from 1995 showed illegal residues in only 1.6% of the samples analyzed; from those only about one-quarter represented overtolerance violations, while three-quarters referred to pesticides detected on commodities for which they were not registered. The vast majority of samples analyzed revealed no detectable residue, and on samples containing detected residues, most were present below 10% of the tolerance level (13).

BIBLIOGRAPHY

1. Environmental Protection Agency, "The Food Quality Protection Act of 1996," EPA Office of Pesticide Programs, 1999, ⟨*http://www.epa.gov/oppfead1/fqpa*⟩ (September 17, 1999).
2. S. O. Archibald and C. K. Winter, "Pesticides in Our Food: Assessing the Risks," in C. K. Winter, J. N. Seiber, and C. F. Nuckton, eds., *Chemicals in the Human Food Chain*, Van Nostrand Reinhold, New York, 1990.
3. D. J. Ecobichon, "Toxic Effects of Pesticides," in C. D. Klaassen, ed., *Casarett and Doull's Toxicology: The Basic Science of Poisons*, 5th ed., McGraw-Hill, New York, 1996.
4. D. Pimentel et al., "Environmental and Economic Costs of Pesticide Use," *Bioscience* **42**, 750–760 (1992).
5. Environmental Protection Agency, "Pesticides and Toxic Substances," in *Pesticide Industry Sales and Usage, 1994 and 1995 Market Estimate*, EPA Office of Prevention, Washington, D.C., 1997.
6. C. M. Bruhn et al., "Consumer Response to Pesticide Food Safety Risk Statements: Implications for Consumer Education," *Dairy, Food, and Environmental Sanitation* **18**, 278–287 (1998).
7. C. K. Winter, "Lawmakers Should Recognize Uncertainties in Risk Assessment," *California Agriculture* **48**, 21–29 (1994).
8. Food and Drug Administration, "Pesticide Program Residue Monitoring 1997," 1998, ⟨*http//www.cfsan.fda.gov*⟩ (September 17, 1999).
9. National Research Council, *Regulating Pesticides in Food: The Delaney Paradox*, National Academy Press, Washington, D.C., 1987.
10. C. K. Winter, "Pesticide Tolerances and Their Relevance as Safety Standards," *Regul. Toxicol. Pharmacol.* **15**, 137–150 (1992).
11. D. V. Reed et al., "The FDA Pesticides Monitoring Program," *J. Assoc. Off. Anal. Chem.* **70**, 591–595 (1987).
12. U.S. Department of Agriculture, "USDA Pesticide Data Program: What Is PDP?" 1999, ⟨*http://www.ams.usda.gov/science/pdp/what.htm*⟩ (September 17, 1999).
13. California Department of Pesticide Regulation, "Residues in Fresh Produce," in *Executive Summary of 1995 Monitoring Program*, Sacramento, Calif., 1997.

ELISABETH L. GARCIA
CARL K. WINTER
University of California
Davis, California

See also FOOD SAFETY AND RISK COMMUNICATION.

PHENOLIC COMPOUNDS

The term "phenolic compounds" embraces a wide range of compounds that possess an aromatic ring bearing a hydroxyl substituent, including their functional derivatives. Phenolic compounds are present in many plants. They are directly related to food characteristics such as taste, palatability, nutritional value, pharmacological and toxic effects, and microbial decomposition. Among the natural phenolic compounds, of which approximately 8000 are known to occur in plants, the flavonoids and their relatives form the largest group with more than 5000 known structures (1). This is considered only a fraction of the total number that are likely to be present in nature, since only a small percentage of plant species has been properly examined for their phenolic compounds. Phenolic compounds range from structures that are very lipophilic (eg, tangeretin) to those that are very water soluble (eg, quercetin 3-sulfate). The size of molecule varies greatly, ranging from monomer, catechol with molecular weight of 110, to the complex heavenly blue anthocyanin pigment of *Ipomaea coerulea*, which has molecular weight of 1759 (2,3). Considerable numbers of simple monocyclic phenols, phenolic quinones, lignans, and xanthones, as well as polymeric materials such as lignins, melanins, and tannins are considered to be important phenolic compounds.

Only a relatively few polyphenols are considered to be important to foods and feeds. They are *p*-coumaric, caffeic, ferulic, sinapic, gallic acids and their derivatives, and the common flavonoids and their glycosides. Anthocyanins and flavonols are important pigments in a variety of fruits and vegetables. While many common polyphenols are not pigments that enhance quality, some are important food constituents because they are responsible for off-colors, usually browning, that develop during the storage and processing of fruits and vegetables. Many phenolic compounds participate in both enzymatic and nonenzymatic browning reactions. The retention of fruit and vegetable color has become a matter of great importance because color is one of the primary factors in consumer acceptance of plant foods. In this respect the need to prevent the formation of undesirable color during processing has always posed a challenge to food technologists.

In addition to color, polyphenols also contribute to food flavor and other qualities. For example, astringency of polyphenols and its ratio with sugar and acid are important and useful criteria for determining the overall quality of fresh fruits, fruit beverages, and wines. Some polyphenols, such as chalcones and related compounds found in citrus fruits, are exceedingly sweet or bitter. Both bitterness and astringency of wine are due to phenolic compounds present in grapes. There have been many claims for the adverse effect of polyphenol compounds on dietary proteins. Tannins have the potential to affect many aspects of digestion due to their affinity for proteins, and the tannin-protein complex decreases its digestibility (4). Another important function of polyphenol compounds in terms of the human health benefit is the growing evidence that suggests that polyphenol compounds in the diet have a long-term health benefit and may prevent or reduce the risk of some chronic diseases. Certain naturally occurring phenolic compounds, often referred to as flavonoids, seem to function via one or more biochemical mechanisms to interfere with, or prevent, carcinogenesis. Due to limited space, only those common phenolic compounds found in fruits and vegetables and other plant foods reported in recent years will be discussed here.

STRUCTURE

Hydroxycinnamic acids are widely distributed in relatively high concentrations in most fruits and vegetables. The important acids are *p*-coumaric, caffeic, ferulic, and sinapic acids (Table 1). They rarely occur in free state, but rather as simple esters. The most common is quinic acid, especially 5-caffeoylquinic or 3-caffeoylquinic acid, which are commonly known as chlorogenic acid and neochlorogenic acid. Hydroxycinnamic acids often occur in fruits and vegetables as esters of malic acid, tartaric acid, hydroxycitric acid, tartronic acid, shikimic acid, galactaric acid, gluconic acid, and methoxyaldaric acid (5).

Flavonoids constitute one of the most distinctive groups of higher plant secondary metabolites. The term flavonoid embraces all those compounds whose structure is based on that of flavone (2-phenylchromone) (Fig. 1).

It can be seen that flavone consists of two benzene rings (A and B) joined together by a three-carbon link that is formed into a γ-pyrone ring. The various classes of flavonoid compounds differ from each other only by the state of oxidation of this 3-C link. Various states of oxidation at the link produce a number of different compounds found in plants, such as flavan-3-ols (catechins), and 3-hydroxyflavones (flavonols, such as quercetin). Also included in the flavonoids are flavanones and anthocyanidins, chalcones, and isoflavones, as shown in Table 1. In the flavanones (naringenin), the double bond between C-2 and C-3 is reduced.

The individual compounds within each class are also distinguished by the number of hydroxyl, methoxyl, and other groups substituted in the two benzene rings (A and B in Fig. 1). Hydroxyl groups of the A ring are substituted at either both C-5 and C-7, or only at C-7, while those of the B ring are usually substituted by either one, two, or three hydroxyl or methoxyl groups. The hydroxylation pattern of the B ring thus resembles that found in the commonly occurring cinnamic acids and coumarins. Most flavonoid compounds except catechins occur in the plant as glycosides in which some of the phenolic hydroxyl groups (usually C-3) are combined with sugar residues such as galactose, arabinose, xylose, glucose, and rhamnose. More than 720 flavone and flavonol glycosides are known to occur in the plant kingdom (6).

Anthocyanins are the most important group of water-soluble plant flavonoid pigments visible to the human eye. With a few exceptions, they are universal plant colorants and largely responsible for the cyanic colors of flower petals and fruits. Most of the approximately 300 anthocyanins that have been characterized consist mainly of 17 known aglycones (7,8). Of these aglycones, six are important food colorants, due to their common occurrence (9). The sugars substituted on the aglycone are glucose, rhamnose, xylose,

Table 1. Common Phenolic Compounds Found in Fruits and Vegetables

Hydroxycinnamic acid derivatives

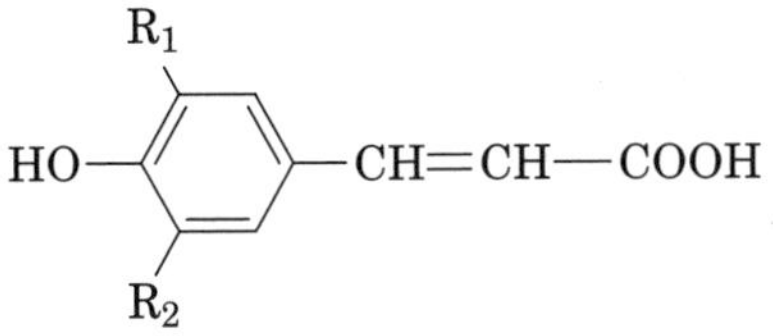

Cinnamic acids

R1 R2 = H, *p*-coumaric acid
R1 = OH, R2 = H, Caffeic acid
R1 = OMe, R2 = H, Ferulic acid
R1 R2 = OMe, Sinapic acid

Catechins

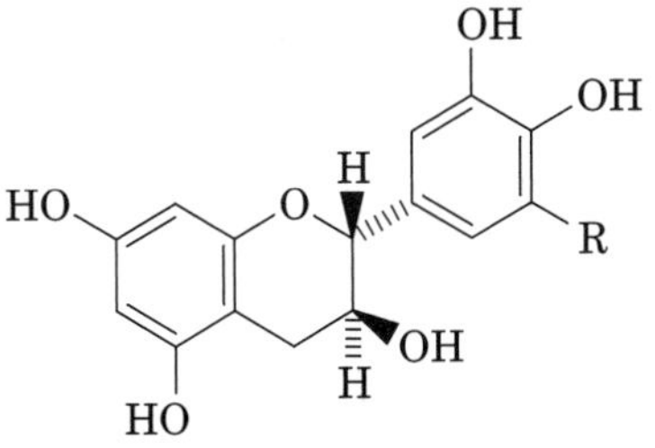

Catechins

R = H, (+) Catechin
R = OH (+) Gallocatechin
Opposite configuration at C-3
R = H (−) Epicatechin
R = OH (−) Epigallocatechin

Anthocyanidins

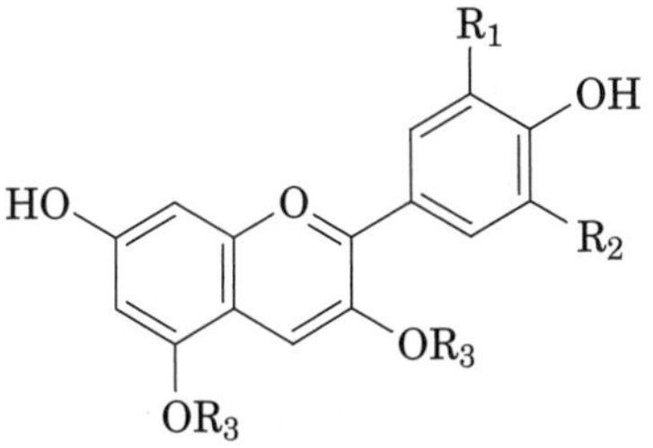

Anthocyanidins (R3 = H)

R1 R2 = H, Pelargonidin
R1 = OH, R2 = H, Cyanidin
R1 R2 = OH, Delphinidin
R3 = Sugar, Anthocyanins

Dihydrochalcones

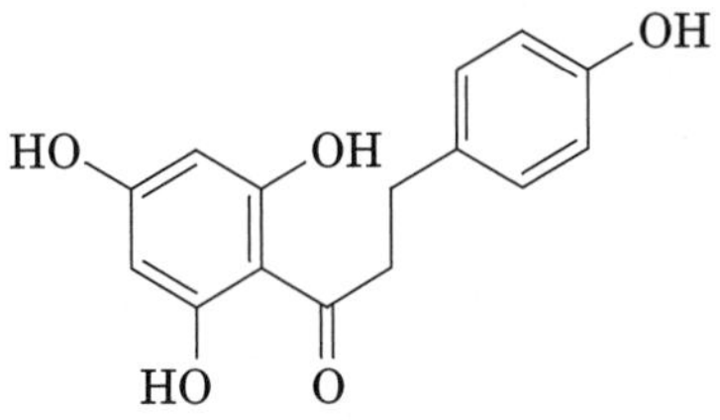

Phloretin

Flavanones

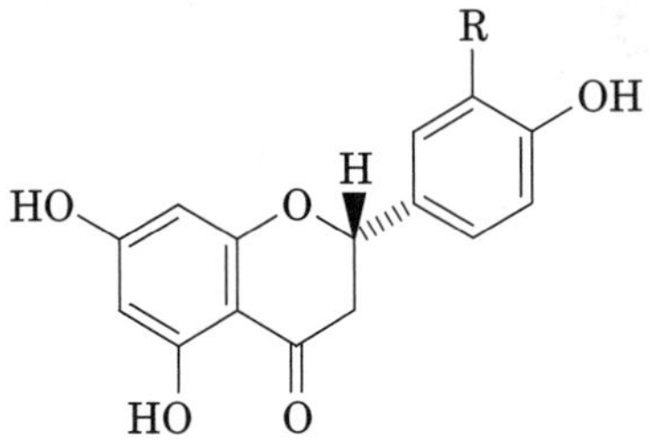

Flavanones

R = H, Naringenin
R = OH, Eriodictyol

Table 1. Common Phenolic Compounds Found in Fruits and Vegetables *(continued)*

Flavones

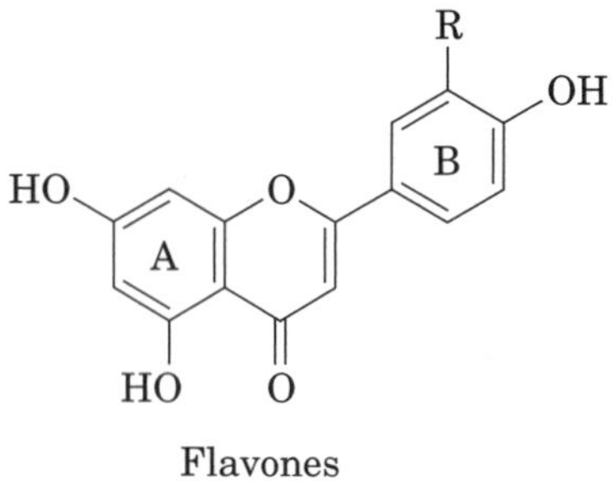

Flavones

R = H, Apigenin
R = OH, Luteolin
Common glycosides have the sugar at C-7 hydroxyl group

Flavonols

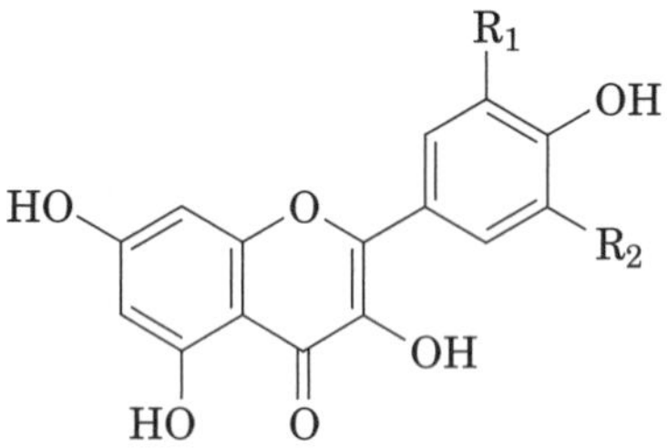

Flavonols (3-hydroxyflavones)

R1 R2 = H, Kaempferol
R1 = OH, R2 = H, Quercetin
R1 R2 = OH, Myricetin

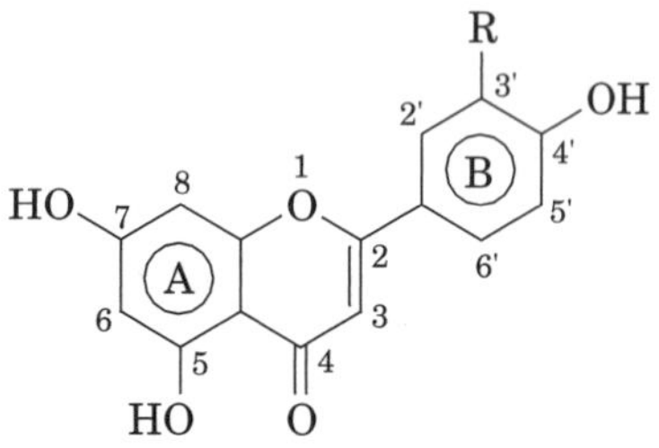

Figure 1. Flavones: the basic structure of flavonoids.

galactose, arabinose, and fructose. They occur as monoglycosides, diglycosides, and triglycosides. When the number of sugar residues is higher than three, they may attach to the basic molecule with alternating sugar and acyl acid linkages. The common acyl acids are coumaric, caffeic, ferulic, hydroxy benzoic, synapic, malonic, acetic, succinic, oxalic, and malic acids (9).

BIOSYNTHESIS

The biosynthesis of phenolic compounds occurs via various pathways. In recent years, considerable progress has been made in elucidating the biosynthesis of these compounds due to our rapidly developing knowledge of enzymology, and the improved quantification of substrates and products by high-performance liquid chromatography, mass spectrometry, and nuclear magnetic resonance spectroscopy. All flavonoids derive their carbon skeleton from two basic compounds, malonyl-CoA and the CoA ester of a hydroxycinnamic acid (10). The origins of the direct flavonoid precursors, 4-coumaroyl-CoA and malonyl-CoA are derived from carbohydrates (Fig. 2). Malonyl-CoA is synthesized from the glycolysis intermediate acetyl-CoA and carbon dioxide, the reaction being catalyzed by acetyl-CoA caboxylase. The synthesis of 4-coumaroyl-CoA involves the shikimate/arogenate pathway, which is the main route to the aromatic amino acids, phenylalanine, tryptophan, and tyrosine in higher plants (10,11). The condensation of three molecules of malonyl-CoA with 4-coumaroyl-CoA to the C_{15} chalcone is catalyzed by chalcone synthase. Transformation by chalcone isomerase produces naringenin (flavonone) and other flavonoids via various steps. Dihydroflavonols formed by direct hydroxylation of flavanones in the 3 position are biosynthetic intermediates in the formation of flavonols, catechins, proanthocyanidins, and anthocyanidins.

Modification by hydroxylation of the A- and B-ring, methylation of hydroxyl groups as well as glycosylation and acylation reactions result in the immense diversity of flavonoids found in nature (10). The detailed biosynthesis steps of various flavonoids have been extensively described (12). Recent advances on flavonoid biosynthesis were restricted to specific steps and filled some of the remaining gaps in the metabolic pathway. Yet the key reaction to the anthocyanins as well as the formation of epicatechin, the proanthocyanidins, and some minor flavonoid-related compounds, such as aurones and dihydrochalcones, still remains a matter of debate (12).

ANALYSIS

Quantitative analysis of phenolic compounds in biological extracts can be carried out in many different ways. The

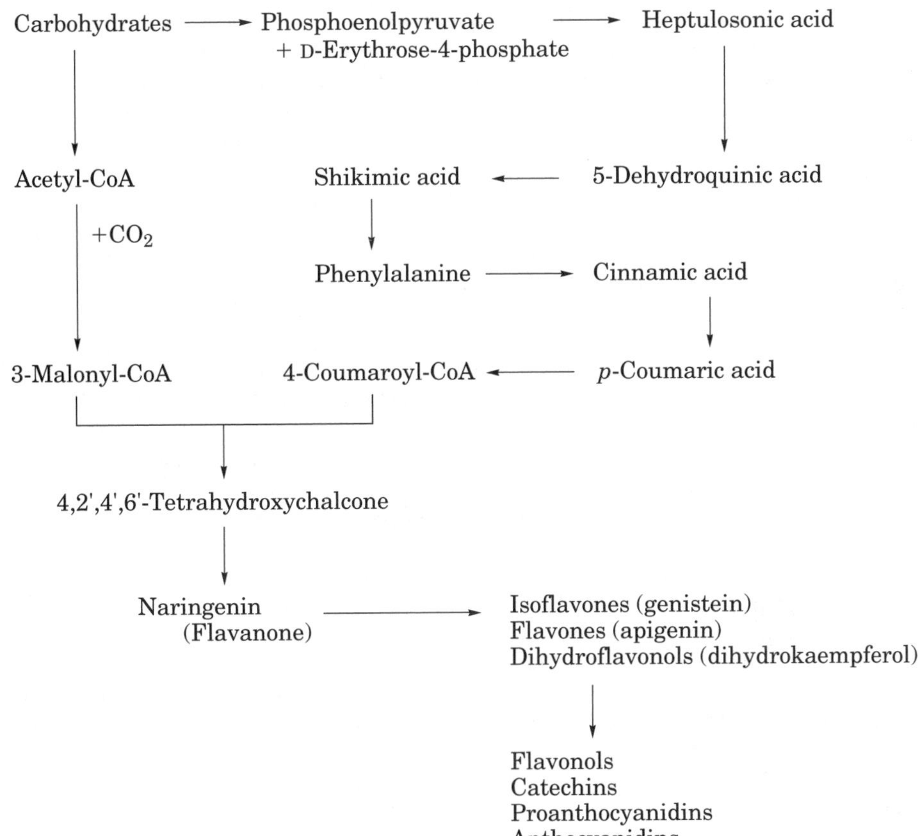

Figure 2. Biosynthetic pathways of phenolic compounds.

most reliable method for the determination of total phenols in food analysis laboratory is based on oxidation with the Folin-Ciocalteu reagent, which contains sodium phosphomolybdate and sodium tungstate. The intensity of the resulting blue complex can be estimated with a colorimeter or with a spectrophotometer (λmax at 725 nm). The colorless procoanthocyanidins can be estimated by heating them with acid and converting them to colored anthocyanidins. Individual phenolic compounds have been determined qualitatively by paper chromatography and semiquantitatively by densitometric analysis of the colored spots obtained by spraying two-dimensional chromatogram with a suitable reagent. Thin layer chromatography has been widely employed for polyphenols, because it is a highly effective, convenient, and inexpensive technique, especially for the separation of anthocyanin.

Today, it is possible to separate and quantify the individual polyphenols of fruits and vegetables by means of high-performance liquid chromatography (HPLC) with great success (13–15). HPLC has advantages of sensitivity, speed, and ease of use compared with other chromatographic procedures. When coupled with a diode array detector, HPLC provides an ideal procedure for accurately analyzing complex mixtures of polyphenols. Hydroxycinnamic acid esters in fruits and vegetables have been successfully separated by using a polyamide column (5). However, before using these procedures, the polyphenols must first be fractionated into several chemical groups to separate the individual polyphenols effectively (15,16). In addition, proton magnetic resonance, ^{13}C nuclear magnetic resonance (NMR), and mass spectrophotometry have been extensively used to determine the structures of polyphenols. Noteworthy achievements are the electron impact, chemical ionization, and fast atom bombardment (FAB) mass spectrometry that have been employed for the identification of various flavonoids, oligomeric hydrolyzable tannins, polyphenolic glycosides, and procyanidins (17). Previously, it had been difficult to obtain mass spectra of anthocyanins because they are not volatile and often present as salts. FAB mass spectrometry, however, gives molecular ions directly as $(M)^+$, and, therefore, its use has been extensive. The early use of CCl_4-soluble flavonoid trimethylsilyl ether derivatives for ^1H NMR spectroscopy and the advent of more sophisticated, higher-field spectrometers that have the capability to run ^{13}C as well as ^1H NMR spectra have been well reported (18,19). More recently, two-dimensional (2-D) homonuclear and heteronuclear spectra were produced for various flavonoids and its glycosides.

CONTENTS IN FRUITS AND VEGETABLES AND ITS PRODUCTS

The level of polyphenols in fruits and vegetables vary widely from species to species, cultivar to cultivar, season to season, and location to location. The accumulation of phenols in fruits may be higher or lower than in other parts of the plant, such as bark, leaves, or heartwood. The concentration of polyphenols often decreases as a fruit matures, but usually the amount per fruit increases. Fruits may contain considerable amounts of some types of polyphenols, such as anthocyanins, while other parts of the same plants, that is, leaves or bark, have very little or none. Concentrations of polyphenol compounds in various fruits and vegetables have been well documented (5,20) and some of them are presented in Table 2.

Among phenolic acids found in common fruits, chlorogenic acid is the major compound found in apples (62–385 ppm), pears (64–280 ppm), cherries (11–140 ppm), plums (15–142 ppm), peaches (43–282 ppm), and apricot (37–123 ppm). This is followed by neochlorogenic acid in cherries (73–628 ppm), plums (88–771 ppm), peaches (33–142 ppm), and apricots (26–132 ppm); and 3-*p*-coumaroylquinic acid in cherries (40–450 ppm), and plums (4–40 ppm). Among berries, blueberries are reported to contain the largest quantity of caffeoylquinic acids (1860–2080 ppm), followed by blackberries (45–53 ppm), and black currents (45–52 ppm) (5). The major phenolic acids in grapes are caffeoyltartaric acid, ranging from 46 to 397 ppm in white grapes (21) and from 50 to 435 ppm in the red cultivars (22).

Among vegetables, relatively large amounts of chlorogenic acid are found in eggplant (575–632 ppm), artichoke (433 ppm), endive (36–124 ppm), potatoes (22–71 ppm), carrots (23–121 ppm), lettuce (5–39 ppm), and tomatoes (12–71 ppm). The neochlorogenic acid content in Brussels sprouts is 70 to 120 ppm; red cabbage, 19 to 110 ppm; kale, 6 to 107 ppm; and broccoli, 58 ppm. Levels of sinapolyglucose have been reported in kale (16–273 ppm), red cabbage (28–84 ppm), Brussels sprouts (12–28 ppm), and broccoli (10 ppm). The levels of caffeoyltartaric acid in lettuce are 10 to 31 ppm while the dicaffeoyltartaric acid in endive is 38 to 334 ppm. Tomatoes contain a significant amount of caffeic acid-4-*o*-glucoside (15–48 ppm) and *p*-coumaric acid-*o*-glucoside (19–68 ppm) (5).

Among the various flavan-3-ols and their derivatives found in fruits and vegetables, apple skins mainly contain epicatechin (30–1010 ppm), procyanidins (30–910 ppm), phloretin glycosides (60–380 ppm), and quercetin glycosides (30–570 ppm) (23). A large quantity of epicatechin is also found in apricots (67–202 ppm), cherries (4–152 ppm), and pears (5–59 ppm) (5). Peaches are reported to contain mainly catechins (6–35 ppm) and its dimer, procyanidin B3 (4–85 ppm) (24). White grapes contain catechin gallates (3–105 ppm), epicatechin (7–63 ppm), and catechin (2–58 ppm) (21). As stated previously, the concentration of phenolic compounds in fruits and vegetables varies greatly depending on the tissue, cultivar, maturity, season, growing site, and various other factors such as cultivation practices.

Phenolic compounds constitute up to 35% of dry weight of tea, with major components such as epicatechin gallate, epigallocatechin, epigallocatechin gallate, catechin, quercetin, kaempferol, gallic acid, chlorogenic acid, and their derivatives. Green tea is known to contain more flavonols and their glycosides (25). Some fresh tea leaves (assamica variety) contain 9 to 13% epigallocatechin gallate, 3 to 6% epicatechin gallate, 3 to 6% epigallocatechin, and 1 to 2% epicatechin by dry weight as well as other flavonoids and their glycosides (26). Thearubigins, highly colored catechin oxidation products and their gallate are of major significance in determining the quality and flavor of tea. Black tea as consumed by humans contains about 36% thearubigins, 3% theaflavins, 5% epigallocatechin gallate, and 1% gallic acid by dry weight. Due to the large amounts of these phenolic compounds in tea, heavy drinkers of tea in Japan may consume 1 g of epigallocatechin gallate per day per person. Dry whole cocoa beans contain approximately 12 to 18% phenolic compounds and the major compound is epicatechin (27). Phenolic compounds in roasted coffee beans are produced during thermal processing from carbohydrates, chlorogenic acid, and lignins. Phenolic compounds in beer that contribute to bitterness, astringency, harshness, and the formation of haze are catechin and epicatechin (approximately 40 mg/L), gallocatechin (less than 15 mg/L), and hydroxycoumarins and anthocyanidins (less than 1 mg/L) (28). Phenolic compounds in wine derived from grape phenolics usually include derivatives of hydroxybenzoic and hydroxycinnamic acids, flavonoids such as flavan-3-ols, flavan-3, 4-diols, anthocyanins and anthocyanidins, flavonol, flavones, and condensed tannins (29). The composition of polyphenol compounds in wine depends on the type of grape used for vinification, extraction and wine making methods, and the chemical reactions that occur during the aging process. Individual phenolic compounds and their concentration in wine and grape are well documented (30).

EFFECTS OF FOOD PROCESSING

The phenolic compounds of fruit and vegetable products may be changed during storage and processing. From the view of the food processor, the tendency of polyphenols to undergo discoloration in fruits and vegetables is the main reason that they are significant constituents. They readily undergo color change because they serve as effective substrates for oxidation and react with other food components such as other polyphenols, thereby reducing sugars, metals and proteins. Phenolic compounds commonly found in plant foods are readily available substrates for polyphenol oxidase or peroxidase enzymes. Their reactivity lies not only in their molecular size and polyphenolic character, but also in their ability to complex strongly with protein, carbohydrates, nucleic acids, alkaloids, and minerals.

The oxidation of polyphenols during processing and storage of foods is accentuated by enzyme systems including catechol oxidase, laccase, peroxidases; also by alkali, especially in the presence of metal ions. The undesirable browning that occurs during processing of fruits and vegetables such as apples, apricots, banana, peaches, pears, and potatoes is due mainly to the oxidation of *o*-dihydroxyphenols by polyphenol oxidases and polymerization of the generated *o*-quinones (see the article BROWNING REACTION). The browning potential of many apple and peach cultivars correlates well with the quantity of certain phenolic compounds that is present. However, with certain cultivars, good correlation between browning potential and polyphenols is not always evident (31,32), presumably be-

Table 2. Concentration of Polyphenols in Fruits

Fruit	Hydroxybenzoic acid derivatives	Hydroxycinnamic acid derivatives	Anthocyanins	Flavonols	Falvon-3-ols	Tannins	Total phenolics
Apple		6.1–134	10–2160 (peel)	17.8–40 (peel)	0.2–16.3	40–350	50–1100
				154–285 (peel)	45.8 (peel)		78–120
							790–1160 (peel)
Avocado		2.8–46			3300 (peel)		2.8–28.7
				9 (pulp)		110–180	
Banana						0.6	150
Blackberry	0.9–3.0	6.2–8.8	82–180	0.85–2.15	7.3–17.7		
			234–326	20.7–31.4			
Blackcurrent	1.0–2.1	12.1–14.6	250	7.6–31.5	0.9–1.4	0.21–0.37%	
Cherry							
Sweet	0.05–4.1	28.5–140	350–450	1.1–3.1	2.0–7.0	70	360
Sour	0.1–0.8	27.0–67.1	28.8	4.2–6.0	15.2–20.2		200
Citrus							
Grapefruit	12.0–21.7						
Orange	13.6–16.3						
Gooseberry	1.1–2.1	2.5–14.8			1.7–3.3	0.06–0.1%	
Grape							
White	0.4 (skin, dw)	1.33–86.5 (skin)		0.81–8.19	1.4–52.7 (skin)	17.2 (skin)	350 (skin)
Red	24.7 (skin, dw)	10–109 (skin)	8–388	1.85–9.75	22–37 ppm	32.1–78.1 (skin)	900–950
Kiwi	60–100					35–60	
Peach		8.1–75		0.4–1.0	5.3–14.1		28–180
Pear		170	5–10 (peel)	4–160 (peel)	0.1–6.2	53	123–400
Persimmon					0.06% (dw)	0–2000	1–12% (dw)
Pineapple		5.3–14.9					
Plums	0.2–1.2	12.1–94	1.9–5.3	2–5.2	1.8–6.1	76–150	167–200
Raspberry, red	3.4–4.4	2–4	23–59	7.2–10.2	3.2–4.9	0.10–0.14	
Red currant	0.9–1.3	1–2.6	11.9–18.6	2–5.1	0.4–3.6		
Strawberry	0.7	1.4–3.1	28–70	2.7–17.4	2.4–9.6	110–150	
Tomato	0.15	7–23.2					45–57

Source: Ref. 20.
Note: All values are in mg/100 g fresh weight except where otherwise specified. dw:dry weight.

cause of the presence of other substances that can modify the reactions. The contributions of different polyphenols to the color of grape and apple juices depend on the amounts of certain polyphenols in the fruit.

In general, the significance of phenolic compounds in food can be related to ascorbic acid destruction, color formation with heavy metals, and antioxidant activity. When both active polyphenol oxidases and phenolic substrates are present in fruits and vegetables, tissue discoloration takes place. But this reaction does not begin until all of the ascorbic acid has been destroyed. The reason for this is that the oxidized polyphenols, in the course of oxidation, can reversibly transfer oxygen to ascorbic acid, being itself reduced to its original state.

Flavonoids affect anthocyanin color by copigmentation and by copolymerization. Polyphenols contribute most to browning when enzymatically oxidized to quinones, and the quinones polymerize to form relatively stable polymers. Phenolic compounds also give characteristic color reactions with a number of heavy metals, such as ferric salts and aluminum ions (33). For example, black spotting of cashew kernels is caused by iron interaction with catechin, and red-brown precipitates in wine are caused by the copper-procoanthocyanidin reaction (34). Since copper, iron, and other complexing ions catalyze oxidation of polyphenols, they can promote oxidative browning. Quercetin, chlorogenic acid, and gallic acid form highly colored complexes with cyanidin-3-glucoside in the presence of aluminum ions (35). The combination of anthocyanins with flavonoids of another types occurs in red wine during aging (36).

One of the discolorations caused by reactions of colorless plant polyphenols is the reddening or pinking that occurs when colorless anthocyanogens are converted to anthocyanidins. Acidic media and heat are important factors in the conversion of proanthocyanidins to anthocyandins; these conditions often occur in foods, particularly canned fruits. Pink color in canned pears is associated with high procoanthocyanidin content and failure to cool the cans quickly after thermal processing (37). Pink color development in banana, cabbage, and broad beans is also related to polyphenols (38–40).

Browning and blackening in fruits and vegetables during processing and storage involves the production of many overlapping chromophores that have absorbances in the visible region. Reactions of polyphenols to given colored products are not necessarily oxidative, particularly those that involve reaction with the phloroglucinol portion of flavonoids. Phenol-aldehyde condensations are an example of this. Carbonyl-amine reactions can result in browning if the carbonyl is in a quinone molecule that can be observed in the Maillard sugar-amino acid type reaction.

Certain phenolic compounds are antioxidants that suppress activity of the enzymes, lipoxygenase (41) and β-galactosidase (42). Lipoxygenase causes oxidation of carotenoids in certain vegetables and thus lowers the vitamin A value of these foods. Among phenolic compounds, flavans exhibit a higher inhibitory effect on lipoxygenase than flavonols and phenolic acids (43). Catechin, chlorogenic acid, and quercetin glycosides suppress β-galactosidase in apple and retard softening during cold storage (42).

Another important reaction of polyphenols in foods is the complexing with protein. It has long been known that the *o*-dihydroxy phenolic compounds and the carbonyl groups of protein interact by hydrogen bonding (44) and that quinone-protein reactions also participate in this interaction (45). This polyphenols-protein interaction affects the protein quality of certain foods. The presence of tannins has been shown to reduce the nutritive value of various grains such as sorghum and fava beans. Vegetable proteins that have been exposed to *o*-dihydroxyphenols in oxidizing condition would be expected to be less readily digested and to have less biologically available lysine and cystine (46). When polymerized tannins combine with the proteins of food, the complex is less likely to be absorbed. Condensed tannins also may pass unchanged through the digestive tract.

Phenolic compounds and proteins reduce apple juice quality by the formation of haze and sediment during storage. Phenolic compounds participate in formation of haze during processing of beer and wine. Gelatin treatments have been used to remove some of the tannins and thus minimize the problem (47). Oxidized polyphenols may also react with proteins; the mechanisms of these quinone-protein reactions are not completely understood.

Some beneficial effects of polyphenols in relation to color and flavor of food products are associated with tea and coffee. In the cultivation of the tea plant, conditions are designed to produce tea leaves that are rich in epicatechin, epigallocatechin, and their gallate esters (46). A large number of phenolic compounds are produced during the roasting process of coffee, while some phenolics such as chlorogenic acid (which is known to contain as much as 4% by weight of coffee beans) is destroyed significantly (48).

Since bound phenolic residues are components of unlignified cell walls of many plants, some polyphenols may contribute to the fiber content of the human diet. Both primary and secondary cell walls of wheat endosperm, spinach, potato, and other vegetables contain polyphenols such as ferulic and coumaric residues that are linked to hemicelluloses or peptides (49).

BIOLOGICAL AND CHEMICAL ACTIVITIES

Phenolic compounds are known to possess several biological and chemical properties, including antioxidant activity, the ability to scavenge active oxygen species, the ability to scavenge electrophiles, the ability to inhibit nitrosation, the ability to chelate metals, the potential for autoxidation, producing hydrogen peroxide in the presence of certain metals, and the capability to modulate certain cellular enzyme activities. Various phenolic acids and its esters are effective in the prevention of microbial growth. Phenolics such as gallic acid and *p*-hydroxybenzoic acid esters showed inhibitory effects against the growth and toxin production of *Clostridium botulinum* types A and B (50). *p*-Coumaric acid (>250 ppm) and ferulic acid (250 ppm) inhibit the growth of *Saccharomyces cerevisiae* (51). Proanthocyanidins, flavonols, and benzoic acid extracted from cranberry showed antimicrobial effects on *Saccharo-*

myces bayanus and Pseudomonas flurescens (52). Some phenolic compounds in red grapes, such as catechin gallate, are known to have antimicrobial activity on aciduric, aerobic, and spore-forming bacteria isolated from apple juice (53).

Natural antioxidants occur in all higher plants and in all parts of the plant—wood, bark, stems, pods, leaves, fruit, root, flowers, pollen, and seeds. These are usually polyphenol compounds. Typical compounds that possess antioxidant activity include tocopherols, flavonoids, cinnamic acid derivatives, and other compounds. The major function of flavonoids and cinnamic acids is in their primary antioxidant activity as free radical acceptors and as chain-breaker. This includes antioxidative activity against lipoxygenase catalyzed reactions. Spices and herbs also have antioxidant activity that originates mainly from their polyphenolic constituents. The antioxidant activity of rosemary depends primarily on the concentration of carnosic acid and rosmaric acid, a derivative of caffeic acid (54).

Several flavonoids have been found to be mutagens *in vitro* test, while other investigators have failed to observe any *in vivo* genetic toxicity of flavonoids. Quercetin and its glycosides found in many fruits and vegetables showed antiviral activities against various viruses (55). A variety of dietary flavonoids have been found to inhibit tumor development in experimental animal models. The hydroxylated flavonoids have been found to (*1*) inhibit the metabolic activation of carcinogenes by modulation of cytochrom enzymes; (*2*) inactivate ultimate carcinogens; (*3*) inhibit generation of active oxygen species and act as scavengers of active oxygen species; (*4*) inhibit arachidonic acid metabolism; (*5*) inhibit protein kinase C and other kinase activity; and (*6*) reduce the bioavailability of carcinogenes (56). Phenolic compounds such as caffeic acid and ferulic acid can block the nitrosation of amines by reducing nitrite to nitric oxide or by forming C-nitroso compounds. Epigallocatechin-3-gallate, isolated from green tea, was found to reduce the incidence of chemically induced tumors in experimental animals in the liver, stomach, skin, lungs, and esophagus (56). Ellagic and chlorogenic acid also showed chemopreventive activity against liver, colon, and tongue carcinogens (57). Epidemiological studies in Finland and Netherlands indicate an inverse relationship between flavonoid intake and coronary heart disease (58–60). Quercetin was reported to be the main flavonoid consumed in these studies. Since numerous reports related to the positive effects of polyphenol compounds in human health are being published, and the topic of polyphenol compounds in relation to functional foods has been the major subject among many scientific communities in recent years, we expect to know more about the bioactivity of polyphenol compounds soon.

COMMERCIAL APPLICATIONS

The most important group of polyphenols for commercial application has been anthocyanins. Commercial food colorants have been used for many years, especially in the wine trade. To enhance the color of red wine and also as a general food colorant, anthocyanin pigments from grape skins have been extensively utilized. Grapes are the largest single fruit crop grown in the world, and, therefore, the major source of phenolic compounds among the different fruits and vegetables. Grape pigments have significant commercial value; approximately 10,000 tons of grape anthocyanins are utilized annually. Since anthocyanins are natural colorants, many processors prefer anthocyanins to synthetic red dyes. However, anthocyanins are less stable in many foods because light and pH affect them. They are also readily bleached by sulfur dioxide, which is often used in beverage and food processing as a preservative. This instability can be counteracted by reacting anthocyanins with carbonyl compounds, such as acetaldehyde, to stabilize them (9). Also, it is possible to use acylated anthocyanins, which are more stable to light than the simple glycosides.

Some phenolic compounds are also valuable as antioxidants, especially the highly hydroxylated types. Butylated hydroxyanisole (BHA) and butylated hydroxytoluene (BHT) are synthetic phenolic compounds that are widely sued as antioxidant food additives, mainly in oils and food coating materials. Compounds such as alkyl gallates and nordihydroguaiaretic acid are well-known antioxidants. New phenolic compounds are currently being isolated and their chemical properties will be found. It is likely that some will have potent antioxidant properties that can be utilized in food processing.

Certain flavonoids have long been used as drugs in Eastern European and Southeastern Asia. The pharmacological activities of these polyphenols have been less specific and less pronounced than those of alkaloids or steroids. Therefore, doubt and controversy have always surrounded the therapeutic value of polyphenols. Many European countries and Japan have been leaders in the commercial production of herb phytomedicinals consisting mainly of polyphenol compounds. Germany accounts for the largest share of the European Union herbal drug market, with sales of $3 billion of an annual $6 billion total (61). Of the 10 best-selling herbs in the United States today, echinacea, garlic, ginseng, ginkgo, saw palmetto, and eleuthero have been popularized. This is primarily the result of European research, which takes place under a more favorable regulatory climate that permits reasonable evidence of efficacy. In the United States, demand for excess amounts of evidence regarding efficacy has discouraged research on those products and relegated them to the dietary supplement status (61). The Dietary Supplement Health and Education Act of 1994 permits manufacturers of supplements to make claims regarding the health benefits of the products, and numerous new products of dietary supplements that contain polyphenols are being produced. In the future, therefore, we expect to find more new health beneficial polyphenol compounds and will see the expanded use of natural phenolic compounds, both for the improvement of processed foods and also as new phytomedicines.

BIBLIOGRAPHY

1. J. B. Harborne, ed., *Flavonoids: Advances in Research since 1986*, Chapman & Hall, London, 1994.

2. J. B. Harborne, "Phenolics in the Environment: An Overview of Recent Progress," *Proc. 14th Int. Conf. Groupe Polyphenols*, Brock University, St. Catharines, Ontario, Canada, August 16–19, 1988.
3. J. B. Harborne, "Plant Polyphenols and Their Role in Plant Defense Mechanisms," in R. Brouillard, M. Jay, and A. Scalbert, eds., *Polyphenols 94*, INRA, Paris, 1995.
4. W. S. Pierpoint, "Reaction of Phenolic Compounds with Proteins and Their Relevance to the Production of Leaf Protein," in L. Telek and H. D. Graham, eds., *Leaf Protein Concentrates*, Avi, Westport, Conn., 1983.
5. K. Herrmann, "Occurrence and Content of Hydroxycinnamic and Hydroxybenzoic Acid Compounds in Foods," *Crit. Rev. Food Sci. Nutr.* **28**, 315–347 (1989).
6. J. B. Harborne and C. A. Williams, "Flavone and Flavonol Glycosides," in J. B. Harborne, ed., *The Flavonoids*, Chapman & Hall, London, 1988.
7. J. B. Harborne and R. J. Grayer, "The Anthocyanins," in J. B. Harborne, *The Flavonoids*, Chapman & Hall, London, 1988.
8. D. Strack and V. Wray, "The Anthocyanins," in J. B. Harborne, ed., *Flavonoids: Advances in Research since 1986*, Chapman & Hall, London, 1994.
9. F. J. Francis, "Food Colorants: Anthocyanins," *Crit. Rev. Food Sci. Nutr.* **28**, 273–314 (1989).
10. W. Heller and G. Forkmann, "Biosynthesis," in J. B. Harborne, ed., *The Flavonoids*, Chapman & Hall, London, 1988.
11. R. A. Jensen, "The Shikimate/Arogenate Pathway: Link between Carbohydrate Metabolism and Secondary Metabolism," *Physiol. Plant.* **66**, 164–186 (1985).
12. W. Heller and G. Forkmann, "Biosynthesis of Flavonoids," in J. B. Harborne, ed., *Flavonoids: Advances in Research since 1986*, Chapman & Hall, London, 1994.
13. B. Y. Ong and C. W. Nagel, "High-Pressure Liquid Chromatographic Analysis of Hydroxycinnamic Acid Tartaric Acid Esters and Their Glucose Esters in *Vitis vinifera*," *J. Chromatogr.* **157**, 345–341 (1978).
14. W. Brandl and K. Herrmann, "Analytical and Preparative High-Performance Liquid Chromatography of Hydroxycinnamic Acid Esters," *J. Chromatogr.* **260**, 447–451 (1983).
15. A. W. Jaworski and C. Y. Lee, "Fractionation and HPLC Determination of Grape Phenolics," *J. Agric. Food Chem.* **35**, 257–259 (1987).
16. J. Oszmianski and C. Y. Lee, "Isolation and HPLC Determination of Phenolic Compounds in Red Grapes," *Am. J. Enol. Vitic.* **41**, 204–206 (1990).
17. R. Self, "Fast Atom Bombardment Mass Spectrometry of Polyphenols," *Biomed. Environ. Mass Spectrom.* **13**, 449–469 (1986).
18. K. R. Markham, *Techniques of Flavonoid Identification*, Academic Press, London, 1982.
19. K. R. Markhan and H. Geiger, "^{1}H Nuclear Magnetic Resonance Spectroscopy of Flavonoids and Their Glycosides in Hexadeuterodimethylsulfoxide," in J. B. Harborne, ed., *Flavonoids: Advances in Research since 1986*, Chapman & Hall, London, 1994.
20. J. I. Macheix, A. Fleuriet, and J. Billot, *Fruit Phenolics*, CRC Press, Boca Raton, Fla., 1990.
21. C. Y. Lee and A. W. Jaworski, "Phenolic Compounds in White Grapes Grown in New York," *Am. J. Enol. Vitic.* **38**, 277–281 (1987).
22. V. L. Singleton, J. Zaya, and E. K. Trousdale, "Caftaric and Coutaric Acids in Fruit of Vitis," *Phytochemistry* **25**, 2127–2132 (1986).
23. S. Burda, W. Oleszek, and C. Y. Lee, "Phenolic Compounds and Their Changes in Apples during Maturation and Cold Storage," *J. Agri. Food Chem.* **38**, 945–948 (1990).
24. C. Y. Lee et al., "Enzymatic Browning in Relation to Phenolic Compounds and Polyphenoloxidase Activity among Various Peach Cultivars," *J. Agric. Food Chem.* **38**, 99–101 (1990).
25. D. A. Balentine, "Manufacturing and Chemistry of Tea," in C. T. Ho, C. Y. Lee, and M. T. Huang, eds., *Phenolic Compounds in Food and Their Effects on Health I*, American Chemical Society Symposium Series 506, ACS, Washington, D.C., 1992.
26. World Health Organization, International Agency for Research on Cancer, *Coffee, Tea, Mate, Methylxanthines and Methylglyoxal*, IARC Monographs on the Evaluation of Carcinogenic Risks to Humans, IARC Press, Lyon, France, Vol. 51, 1991, pp. 207–271.
27. H. Kim and P. G. Keeney, "Epicatechin Content in Fermented and Unfermented Cocoa Beans," *J. Food Sci.* **49**, 1090–1092 (1984).
28. I. McMurrough, G. P. Roche, and K. G. Cleary, "Phenolic Acids in Beers and Worts," *J. Institute of Brewing* **90**, 181–184 (1984).
29. V. L. Singleton and P. Essau, *Phenolic Substances in Grapes and Wine, and Their Significance*, Academic Press, New York, 1969.
30. J. J. Macheix, J. C. Sapis, and A. Fleuriet, "Phenolic Compounds and Polyphenoloxidase in Relation to Browning in Grapes and Wines," *Crit. Rev. Food Sci. Nutr.* **30**, 441–486 (1991).
31. M. Y. CoSeteng and C. Y. Lee, "Changes in Apple Polyphenoloxidase and Polyphenol Concentration in Relation to Degree of Browning," *J. Food Sci.* **52**, 985–989 (1987).
32. C. Y. Lee and A. W. Jaworski, "Phenolics and Browning Potential of White Grapes Grown in New York," *Am. J. Enol. Vitic.* **39**, 337–340 (1988).
33. E. C. Bate-Smith, "Flavonoid Compounds in Foods," in E. N. Mrak and G. F. Stewart, eds., *Advances in Food Research*, Vol. 5, Academic Press, New York, 1954, pp. 161–257.
34. V. L. Singleton, "Common Plant Phenols other Than Anthocyanins, Contributions to Coloration and Discoloration," in C. O. Chichester, ed., *The Chemistry of Plant Pigments*, Academic Press, New York, 1972.
35. J. Jurd and S. Asen, "The Formation of Metal and Copigment Complexes of Cyanidin-3-Glucoside," *Phytochemistry* **5**, 1263–1271 (1966).
36. T. C. Sommers, "Pigmentation Profiles of Grapes and of Wines," *Vitis* **7**, 303–320 (1968).
37. B. S. Luh, S. J. Leonard, and D. S. Patel, "Pink Discoloration in Canned Bartlett Pears," *Food Technol.* **14**, 53–56 (1960).
38. R. E. Guyer and F. B. Erickson, "Canning of Acidified Banana Puree," *Food Technol.* **8**, 165–167 (1954).
39. S. Ranganna and V. S. Govindarajan, "Leucoanthocyanins in Cabbage and Pink Discoloration," *J. Food Sci. Technol.* **3**, 155–158 (1966).
40. D. Dickson, M. Knight, and D. I. Rees, "Varieties of Broad Beans Suitable for Canning," *Chemistry and Industry*, 1503 (1957).
41. K. S. Rhee and B. M. Watts, "Effect of Antioxidants on Lipoxygenase Activity in Model Systems and Pea Slurries," *J. Food Sci.* **31**, 669–674 1966.
42. A. J. Dick et al., "Quercetin Glycosides and Chlorogenic Acid: Inhibitors of Apple b-Galactosidase and of Apple Softening," *J. Agri. Food Chem.* **33**, 798–800 1985.
43. J. Oszmianski and C. Y. Lee, "Inhibitory Effect of Phenolics on Carotene Bleaching in Vegetables," *J. Agri. Food Chem.* **38**, 688–690 (1990).
44. K. H. Gustavson, "Interaction of Vegetable Tannins with Polyamides as Proof of the Dominant Function of the Peptide Bond of Collagen for Its Binding of Tannins," *J. Polym. Sci.* **12**, 317–324 (1954).

45. W. D. Loomis and J. Battaile, "Plant Phenolic Compounds and Isolation of Plant Enzymes," *Phytochemistry* **5**, 423–438 (1966).
46. W. S. Pierpoint, "Phenolics in Food and Feedstuffs," in C. F. Van Sumere and P. J. Lea, eds., *Annual Proceedings of the Phytochemical Society of Europe*, Clarendon Press, Oxford, 1985, pp. 427–451.
47. A. Pollard and C. F. Timberlake, "Fruit Juice," in A. C. Hulme, ed., *The Biochemistry of Fruits and Their Products*, Vol. 2, Academic Press, New York, 1971.
48. I. Flament, "Coffee, Cocoa, and Tea," *Food Rev. Int.* **5**, 317–414 (1989).
49. S. C. Fry, "Incorporation of Cinnamate into Hydrolase-Resistant Components of the Primary Cell Wall of Spinach," *Phytochemistry* **23**, 59–64 (1984).
50. M. D. Pierson and N. R. Reddy, "Inhibition of *Clostridium botulinum* by Antioxidants and Related Phenolic Compounds in Comminuted Pork," *J. Food Sci.* **47**, 1926–1929 (1982).
51. J. D. Baranowski et al., "Inhibition of *Saccharomyces cerevisiae* by Naturally Occurring Hydroxycinnamate," *J. Food Sci.* **45**, 592–594 (1980).
52. A. G. Marvan and C. W. Nagel, "Microbial Inhibitors of Cranberries," *J. Food Sci.* **51**, 1009–1013 (1986).
53. D. F. Splittstoesser, J. J. Churey, and C. Y. Lee, "Growth Characteristics of Aciduric Sporeforming *Bacilli* Isolated from Fruit Juices," *J. Food Prot.* **57**, 1080–1083 (1994).
54. D. E. Pratt, "Natural Antioxidants from Plant Material," in M. T. Huang, C. T. Ho, and C. Y. Lee eds., *Phenolic Compounds in Food and Their Effects on Health II*, American Chemical Society Symposium Series 507, ACS, Washington, D.C., 1992.
55. I. Musci, "Combined Antiviral Effect of Quercetin and Interferon on the Multiplication of Herpes Simplex Virus in Cell Culture," in L. Farkas, M. Gabor and F. Kallay, eds., *Flavonoids and Bioflavonoids*, Elsevier, Amsterdam, The Netherlands, 1986.
56. M. T. Huang and T. Ferraro, "Phenolic Compounds in Food and Cancer Prevention," in M. T. Huang, C. T. Ho, and C. Y. Lee, eds., *Phenolic Compounds in Food and Their Effects on Health II*, American Chemical Society Symposium Series 507, ACS, Washington, D.C., 1992.
57. T. Tanaka et al., "Protective Effects against Liver, Colon, and Tongue Carcinogenesis by Plant Phenols," in M. T. Huang, C. T. Ho, and C. Y. Lee, eds., *Phenolic Compounds in Food and Their Effects on Health II*, American Chemical Society Symposium Series 507, ACS, Washington, D.C., 1992.
58. M. G. L. Hertog et al., "Dietary Antioxidant Flavonoids and Risk of Coronary Heart Disease: The Zutphen Elderly Study," *Lancet* **342**, 1007–1011 (1993).
59. P. Knekt et al., "Flavonoid Intake and Coronary Mortality in Finland: A Cohort Study," *Br. Med. J.* **312**, 478–481 (1996).
60. M. F. Muldoon and S. B. Kritchevsky, "Flavonoids and Heart Disease," *Br. Med. J.* **312**, 458–459 (1996).
61. V. E. Tyler, "Importance of European Phytomedicinals in the American Market: An Overview," in L. D. Lawson and R. Bauer, eds., *Phytomedicines of Europe*, American Chemical Society Symposium Series 691, ACS, Washington, D.C., 1998.

CHANG Y. LEE
Cornell University
Geneva, New York

See also COLORANTS: POLYPHENOLS; PHYTOCHEMICALS: BIOTECHNOLOGY OF PHENOLIC PHYTOCHEMICALS FOR FOOD PRESERVATIVES AND FUNCTIONAL FOOD APPLICATIONS.

PHOSPHATES AND FOOD PROCESSING

HISTORICAL DEVELOPMENT AND GENERAL USE

Phosphates exhibit functional properties in a wide variety of foods produced by all segments of the processed food industry. The availability of sodium, potassium, ammonium, calcium, and magnesium phosphates offers food technologists and food scientists formulation flexibility to control taste, nutritional, and other technical properties. The commercial use of food phosphates may be traced to 1864 when the first U.S. patent was granted for a phosphate-containing baking powder (1). The use of emulsifying salts in Europe for production of process cheese products began about 1895 and has been reviewed (2). Principal development in the United States began in 1916 (3). Since World War II new uses have developed in food products, including meats, poultry, seafood, beverages, dairy products, infant foods, cereals, desserts, produce, and nutritional supplements. Various detailed reviews have been published (4–6).

The largest market for food phosphates in the world is the United States where processing techniques and advanced food technology follow the consumer trends and demands for high-quality convenient foods. Phosphate use in food applications is an important segment of the overall industrial phosphate market. The U.S. consumption of food phosphates was estimated to be 87,000 t P_2O_5 in 1993, excluding phosphates sold into dentifrice applications and phosphoric acid sold directly into the food industry (7). The second largest market of food phosphates, Western Europe, is approximately 37,000 t P_2O_5 (1993), including the dentifrice market (7). The growth rate of food phosphates in the United States was predicted to be 1.5 to 2.3% annually (1993–1998, including phosphoric acid) while that predicted for Western Europe (1993–1997) was relatively flat (7). The bakery market segment represents approximately 40% of the food phosphates consumed in the United States while the meat, poultry, and seafood applications represent an estimated 30% of the phosphate consumed, followed by dairy (14%) and other (16%) (7).

NOMENCLATURE AND STRUCTURE

Of the phosphoric acids formed by the reaction between phosphorus pentoxide and water, orthophosphoric acid (H_3PO_4) is the simplest and most commonly encountered (8,9). One, two, or all three protons may be replaced by metal ions to form the orthophosphate salts. The acid orthophosphate salts can be dehydrated to form linear chains (polyphosphates) and rings (metaphosphates) wherein phosphate tetrahedrons share oxygen atoms. The polyphosphates have the general formula $M_{n+2}P_nO_{3n+1}$ (M equals one equivalent of hydrogen or metal ion), which approaches the formula of the cyclic metaphosphates as the chain length increases $(MPO_3)_n$.

Several systems of phosphate nomenclature exist, reflecting the historical changes in the understanding of the phosphate structures. Older names such as pyrophosphate and tripolyphosphate for the smallest of the polyphosphate anions are generally accepted in North American industry and commerce as opposed to the IUPAC names diphosphate and triphosphate, respectively, which are preferred in many other areas of the world. Some of the more common phosphates used in foods are listed in Table 1 along with their formulas.

MANUFACTURE

Phosphoric acid for food products can be produced either from elemental phosphorus or by the purification of wet-process phosphoric acid (9,11). Elemental phosphorus is produced by the reduction of phosphate rock with coke in the presence of silica in an electric furnace. Thermal (furnace) phosphoric acid is manufactured by the combustion and subsequent hydration of elemental phosphorus. For food use, thermal acid is treated with sulfide to remove traces of arsenic.

The North American production of technical- and food-grade phosphoric acid has shifted in the 1990s from almost exclusively thermal acid to a sizable proportion as purified wet-process acid. This shift was a continuation of a trend that had already taken place in many other parts of the world. In the wet process, phosphoric acid is made by reacting phosphate rock and sulfuric acid. The resultant acid contains percent levels of sulfate, fluoride, and metal impurities (Fe, Al, etc) and is therefore not suitable for food use. It is purified by a combination of solvent extraction and chemical treatment to yield a food-grade phosphoric acid. Phosphoric acid made by either the thermal or the wet acid purification route is typically available as 75 to 85% H_3PO_4.

The orthophosphate salts are made by neutralizing phosphoric acid with the appropriate base (Na_2CO_3, NaOH, KOH, NH_3, $CaCO_3$, $Ca(OH)_2$, $Mg(OH)_2$, etc) and separating the product by crystallization or drying. The condensed phosphate salts are made by thermal dehydration of the orthophosphate (9,11).

CHEMICAL PROPERTIES

Phosphates provide important functional properties for improving the processing and the final quality of foods. The use of a phosphate often improves more than a single property (4,8). The chemical effects of phosphates on foods are briefly outlined next. Physical properties of the phosphates (eg, particle size and rate of dissolution) are also important in many applications.

Acid-Base Properties

Various phosphates may provide acidity, alkalinity, or buffering, depending on the degree of neutralization and the structure of the phosphate. Orthophosphates and the short-chain polyphosphates have good buffering capacity, especially in the pH range of 6 to 8. Phosphoric acid is widely used as a beverage acidulant. Acidic phosphates are important leavening agents, and the basic phosphates such as sodium tripolyphosphate (STP) modify meat and dairy protein characteristics by altering their charge.

Table 1. Phosphates Commonly Used in Food Applications

Name	Formula
Orthophosphates	
Phosphoric acid	H_3PO_4
Monosodium phosphate (MSP)	NaH_2PO_4
Disodium phosphate (DSP)	Na_2HPO_4
Disodium phosphate dihydrate (DSPD)	$Na_2HPO_4 \cdot 2H_2O$
Trisodium phosphate (TSP)	Na_3PO_4
Trisodium phosphate dodecahydrate (TSPC)	$Na_3PO_4 \cdot 12H_2O \cdot 1/4NaOH$
Monopotassium phosphate (MKP)	KH_2PO_4
Dipotassium phosphate (DKP)	K_2HPO_4
Tripotassium phosphate (TKP)	K_3PO_4
Monoammonium phosphate (MAP)	$NH_4H_2PO_4$
Diammonium phosphate (DAP)	$(NH_4)_2HPO_4$
Monocalcium phosphate (MCPA)	$Ca(H_2PO_4)_2$
Monocalcium phosphate monohydrate (MCP)	$Ca(H_2PO_4)_2 \cdot H_2O$
Dicalcium phosphate (DCP)	$CaHPO_4$
Dicalcium phosphate dihydrate(DCPD)	$CaHPO_4 \cdot 2H_2O$
Tricalcium phosphate (TCP) (hydroxyapatite)	$Ca_{10}(PO_4)_6(OH)_2$
Dimagnesium phosphate (DMP)	$MgHPO_4 \cdot 3H_2O$
Sodium aluminum phosphates (SALP)	$NaAl_3H_{14}(PO_4)_8 \cdot 4H_2O$ $Na_3Al_2H_{15}(PO_4)_8$ $Na_{15}Al_3(PO_4)_8$
Polyphosphates	
Sodium acid pyrophosphate (SAPP)	$Na_2H_2P_2O_7$
Tetrasodium pyrophosphate (TSPP)	$Na_4P_2O_7$
Tetrapotassium pyrophosphate (TKPP)	$K_4P_2O_7$
Calcium acid pyrophosphate (CAPP)	$CaH_2P_2O_7$
Calcium pyrophosphate (TCPP)	$Ca_2P_2O_7$
Sodium tripolyphosphate (STP)	$Na_5P_3O_{10}$
Potassium tripolyphosphate (KTP)	$K_5P_3O_{10}$
Sodium potassium tripolyphosphate (SKTP)	$Na_3K_2P_3O_{10}$
Sodium hexametaphosphate (SHMP)[a] (Graham's salt)	$Na_{n+2}P_nO_{3n+1}$ ($n \sim 6$–20)
Insoluble sodium metaphosphate (IMP) (Maddrell's salt)	$(NaPO_3)_n$ ($n > 5{,}000$)
Potassium metaphosphate (KMP) (Kurrol's salt)	$(KPO_3)_n$ ($n \geq 1{,}000$)
Metaphosphate	
Sodium trimetaphosphate (STMP)	$(NaPO_3)_3$

Source: Refs. 9 and 10.
[a]Also known as sodium polyphosphates, glassy in *Food Chemicals Codex*.

Sequestration

Sequestration is the process of inactivating metal ions by forming soluble complexes. Polyphosphates chelate metal ions, thereby preventing the ions from entering into further (often undesirable) reactions. For example, polyphosphates inhibit the development of rancidity in oils and fats caused by iron (Fe^{3+}) catalyzed oxidation. The formation of precipitates and haze in beverages is eliminated by the addition of polyphosphates. Calcium ion sequestration aids in the removal of shells from shellfish and in the extraction of pectin from fruit.

Adsorption—Ionic Interaction

Phosphates (especially more basic condensed salts such as STP) may adsorb or complex with proteins and starches by virtue of their high charge. The charge and the extension of the protein is altered, thereby affecting its colloidal properties. Phosphates can, therefore, act to modify the processing characteristics or stability of proteins, for example, dispersion and peptization in dairy applications, improvement of water retention in meats, and emulsion stabilization in sausage and process cheese.

APPLICATIONS OF FOOD PHOSPHATES

Baking Industry Applications

The major function of phosphates in the baking industry is as leavening acids. Leavening is the process of making baked goods light by expanding or inflating the batter or dough during baking. Three general methods of leavening are employed: fermentation or yeast leavening, chemical decomposition, and chemical neutralization. In chemical neutralization, bicarbonate is neutralized by an acid salt such as monocalcium phosphate monohydrate (MCP). Although any acid component can be used to neutralize the bicarbonate, phosphate salts are preferred as they are manufactured to give controlled rates of reaction. The user can select the appropriate phosphate salt(s) so that carbon dioxide (CO_2) is released at the appropriate point in the baking process. Usually, CO_2 release will be desired during both mixing and while baking in the oven. Types of bakery product that are chemically leavened are listed in Table 2 along with typical use levels of sodium bicarbonate (10,12). The phosphate is used at a level to neutralize the bicarbonate. Other applications for phosphates in the baking industry include yeast food nutrients and dough conditioners.

Meat Industry Applications

The meat industry uses phosphate salts in processed meat, poultry, and seafood products to contribute a number of functionalities: pH change, buffering, sequestration of ions, and protein modification. The beneficial effects include an increase in water binding, which improves yield and reduces drip loss; emulsification of fats; solubilization of fibrous proteins, texture modification; flavor enhancement; color stabilization; and reduced rates of rancidity development. Increased pH and ionic strength modify protein properties to improve water-binding capacity. Applications in the meat, poultry, and seafood industries are summarized in Table 3.

Depending on the specific application, phosphates can be added to the muscle tissue in a number of ways. For most products, the phosphate is added in a pickling or curing solution by injection, vacuum tumbling, or through immersion of the tissue. It is important that the addition of the phosphate be followed by techniques to improve dispersion of the solution such as massaging or tumbling. This ensures uniformity and maximum yield and avoids localized pH effects, such as color change. U.S. Department of Agriculture (USDA) regulations do not allow phosphates to be added to fresh, uncooked meats or sausages. Phosphates are only permitted in frozen or processed meat products.

Dairy Industry Applications

In dairy processing, phosphates act as emulsifiers or melting salts accomplishing emulsification through protein modification. They help disperse ingredients in milk-based beverages and serve as acidifying and buffering agents in cheese production. Phosphates also prevent protein denaturation in dried and canned milks and coffee whiteners. In other applications, such as instant puddings or whey protein isolates, the ability of phosphates to promote the formation of protein complexes is used. By far the largest use of phosphates in the dairy industry is in the production of process cheese. In this application the phosphates modify proteins by sequestering calcium ions and adjusting pH. The modified proteins emulsify fats present in the cheese. As a result, the cheese is more uniform in flavor and exhibits controlled melt spread, and the fat does not separate when the cheese is melted. Some of the major applications for phosphates in the dairy industry are described in Table 4.

The Food and Drug Administration (FDA) standards of identity for process cheese, cheese food, and cheese spread allow phosphates to be added up to 3.0% by weight of the finished product. Phosphates permitted include disodium phosphate (anhydrous and dihydrate), trisodium phosphate (anhydrous and dodecahydrate), tetrasodium pyrophosphate, sodium hexametaphosphate, monosodium phosphate, and sodium acid pyrophosphate.

Miscellaneous Applications

Phosphoric acid and phosphates are useful as acidifying and buffering agents in beverages, preserves, and gelatin products. Powdered products may benefit from the flow-conditioning properties of tricalcium phosphate. Phosphates are used in a number of health-related applications where they contribute nutritional supplementation and fortification. Calcium and magnesium phosphates are used in infant foods for mineral supplementation. Calcium phosphates have also been used in several vitamin formulations. Both dicalcium phosphate (DCP) and tricalcium phosphate (TCP) have been used as excipients for tablets. In isotonic drinks, potassium orthophosphates are used to reestablish electrolyte balance. A summary of miscellaneous applications is listed in Table 5.

Table 2. Leavening Applications of Phosphates

Product	Typical $NaHCO_3$ level (%)	Leavening acid[a]
Baking powders	30–40	MCP, SAPP-28, SALP, DMP, MCPA
Biscuit doughs, frozen	1.5–2.0	SALP, SAPP-28, DMP
Biscuit and muffin mixes	1.5–2.0	SALP, DMP, MCPA, SAPP-28
Breading batter mixes	0.0–2.0	SAPP-40, SALP, MCP, DMP, SAPP-28, SAPP-RD
Cake mixes and bakery cakes	0.6–1.0	SALP, DMP, MCP, SAPP-28, DCPD, MCPA
Cookie mixes	0.0–2.0	SALP, MCPA, SAPP-28, SAPP RD
Doughnut mixes	0.5–1.0	SAPP-28, SAPP-40, SAPP-37, SALP, MCPA, SAPP-26, SAPP-43, MCP
Doughs, refrigerated	2.0–2.5[b]	SAPP RD, SAPP-26, SALP, DMP
Hush puppy mixes	1.5–2.0	SAPP-28, SAPP-40 MCPA, SALP
Pancake batters, frozen and refrigerated	1.7–2.2[b]	SALP, DMP, MCP, DCPD, SAPP-28
Pancake and waffle mix	1.5–2.0	SALP, DMP, MCP, MCPA, SAPP-28, SAPP-40
Phosphated flour		MCP, MCPA
Pizza mixes	0.3–1.2	SALP, DMP, MCP, SAPP-28, DCPD
Self-rising cornmeal	1.5–2.0	MCPA, SALP
Self-rising flour	1.2–1.5	SALP, MCPA, SAPP
Tortilla	0.2–1.0	SAPP, SALP, MCP

Source: Ref. 10.
[a]Refer to Table 1 for abbreviations; note that several grades of SAPP are produced as designated by the suffix -RD, -28, etc, which indicate different rates of CO_2 release.
[b]Percent of solids.

Table 3. Phosphates for Meat, Poultry, and Seafood

Food type	Phosphate[a]	Level (%)
Cooked sausages and other emulsion products	Sodium polyphosphate blends, TSPP	0.15–0.35
Ham	STP, sodium polyphosphate blends	0.4–0.5
Bacon	Sodium polyphosphate blends	0.4–0.5
Hamburger patties and ground beef patties (cooked or frozen)	Sodium polyphosphate blends, TSPP	0.25–0.4
Poultry, processed or frozen	STP, sodium polyphosphate blends	<0.5
Restructured poultry patties, rolls, nuggets	STP, sodium polyphosphate blends	0.15–0.35
Frozen shrimp	STP	<0.5
Frozen fish fillets	STP	<0.5
Surimi	STP, TSPP	0.1–0.5

Source: Ref. 10.
[a]refer to Table 1 for abbreviations

REGULATORY STATUS IN THE UNITED STATES AND OTHER COUNTRIES

The various food phosphates are classified as generally recognized as safe (GRAS) in the United States, with use limits set for foods not controlled by a Standard of Identity as that required for good manufacturing practice. Purity specifications have been defined (13). Regulations in all uses are defined by the FDA (14). Regulation of meats and poultry are administered by the USDA, which allows up to 0.5% of phosphate in all processed or frozen meats and poultry unless they are otherwise regulated (14). There are no regulated limits in seafood applications provided good manufacturing practices are followed. Process cheese products, phosphated flour, phosphated starch, and self-rising wheat and corn flours are foods with a Standard of Identity that defines and limits phosphate usage.

Many countries have regulations that are similar to those in the United States. There appears to be interest in standardized foods with regulatory limits for specific applications. Efforts are currently under way to harmonize the various national listings. This is being sponsored by the Food and Agricultural and World Health Organizations under UN sponsorship and is part of a broader program directed toward food phosphates in general. Similar efforts are under way in the European Economic Council.

TRENDS

The dynamics of the processed-food industry will continue to affect the need for food phosphates as consumers demand more convenient, nutritious, and healthy foods. Industry response to an aging population, more working parents, increased eating out, and other factors resulted in the introduction of 3400 new food products in the United States in 1988. Among these were many consumer products benefiting from the functionality of food phosphates, including frozen, microwaveable meat, poultry, and seafood entrees; dairy products and analogues; pet foods; instant desserts; and biscuits and other baked goods. Continued growth of fast foods and convenience foods should

Table 4. Phosphates for Dairy Applications

Application	Phosphate[a]	Level (%)
Evaporated or condensed milk or cream	DSP, DKP	0.02–0.10
Flavored milk powders	TSPP, TKPP	0.1–0.3
Nondairy coffee creamers	DKP, DSP	1.0–2.0 (dry) 0.1–1.0 (liquid)
Buttermilk	TSPP,TKPP, phosphoric acid	0.01–1.0
UHT concentrated milk	SHMP	0.1–1.0
Dried milk-drink products	DSP	2.0[b]
Sterile concentrated milk or cream	SHMP	0.1–1.0
Instant pudding or no-bake cheesecake	TSPP, DSP, DKP, MCP, SAPP, STP, TKPP	2–7
Milk foams	TSPP, TKPP, SHMP	1.5–2.0[b]
Imitation sour cream or chip dips	STP	0.05–0.2[c]
Spray-dried cheese	DSP, DKP, SHMP	1–3
Canned cream and cheese soups	DSP, DKP	0.2
Whipped toppings	TSPP, TKPP, DSP, DKP	0.025–1.0[d]
Process cheese	DSP or TSP blends with IMP, MSP, DSP, TSP, MKP, DKP, SAPP, TSPP, SHMP, TKPP	0.5–3.0
Cheese sauce	DSP, TSP, DKP, SHMP	0.5–3.0
Starter cultures	DSP, MAP, DAP, MKP, DKP	2–3
Frozen desserts and ice cream	DSP, TSPP, SHMP, DKP, TKPP	0.1–0.2
Direct-set cottage cheese	MCP, Phosphoric acid	0.03

Source: Ref. 10.
[a]Refer to Table 1 for abbreviations.
[b]Milk solids basis.
[c]Oil basis.
[d]Of liquid base.

Table 5. Miscellaneous Applications of Phosphates

Application	Phosphate[a]
Beverages	
Carbonated soft drinks	Phosphoric acid
Noncarbonated soft drinks	MCP
Fruit punch	TCP
Fruit juice	MKP
Produce	
Potatoes	SAPP, TSPP
Fruit	MCP, phosphoric acid, SHMP
Canned frozen vegetables	SHMP, TSPP, STP
Egg products (yolks, whole eggs, albumin, dried eggs)	STP, SHMP, MSP, MKP
Fats and oils	Phosphoric acid, TCP
Gelatin desserts	MSP, DSP
Salad dressings	Phosphoric acid
Jams and jellies	Phosphoric acid

[a]Refer to Table 1 for abbreviations.

result in growing demand for processed foods, many of which will benefit from the use of phosphates.

BIBLIOGRAPHY

1. U.S. Pat. 42,140 (March 29, 1864), E. N. Horford (to John H. Cheever).
2. F. Kosikowski, *Cheese and Fermented Foods*, 2nd ed., Edwards Bros., Ann Arbor, Mich., 1977.
3. U.S. Pat. 1,186,524 (June 6, 1916), J. L. Kraft.
4. R. H. Ellinger, *Phosphates as Food Ingredients*, CRC Press, Boca Raton, Fla., 1972.
5. R. H. Ellinger, "Phosphates in Food Processing," in T. E. Furia, ed., *Handbook of Food Additives*, 2nd ed., R. C. Press, Cleveland, Ohio, 1972, pp. 617–780.
6. J. M. Deman and P. Melnychyn, eds., *Phosphates in Food Processing*, AVI, Westport, Conn., 1971.
7. M. Smart, "Industrial Phosphates," CEH Marketing Research Report, February, 1995, Chemical Economics Handbook-SRI International, Menlo Park, Calif.
8. J. R. Van Wazer, *Phosphorus and Its Compounds*, Vol. 1, Interscience, New York, 1958.
9. D. R. Gard, "Phosphoric Acid and Phosphates," in J. I. Kroschwitz and M. Howe-Grant, eds., *Kirk-Othmer Encyclopedia of Chemical Technology*, Vol. 18, 4th. ed., John Wiley & Sons, New York, 1996, pp. 669–718.
10. "Food Phosphates by Monsanto," Solutia Inc. Publication No. 7779226, St. Louis, Mo., 1994.
11. D. R. Gard, "Phosphoric Acid and Phosphates," in J. J. McKetta, ed., *Encyclopedia of Chemical Processing and Design*, Vol. 35, Marcel Dekker, New York, 1990, pp. 429–495.
12. J. F. Conn, "Chemical Leavening Systems in Flour Products," *Cereal Foods World* **26**, 119–123 (1981).
13. Committee on Food Chemicals Codex, *Food Chemicals Codex*, 4th ed., National Academy Press, Washington, D.C., 1996.
14. *Code of Federal Regulations*, Title 21, Part 182 and Title 9, Parts 318 and 381, Office of the Federal Register, National Archives, Washington, D.C., 1997.

BARBARA B. HEIDOLPH
DAVID R. GARD
Solutia, Inc.
St. Louis, Missouri

See also MEAT PRODUCTS.

PHOTOSYNTHESIS

Photosynthesis is a process by which radiant energy in the form of visible light is captured by photosynthetic organisms (including plants, algae, cyanobacteria, and bacteria) and converted to stable chemical energy. Photosynthetic organisms harvest and use light energy to make carbohydrate (sugars) by fixing carbon dioxide. Animals must obtain photosynthetically generated chemical energy by ingesting plants or by consuming other animals that feed on plants. Photosynthetic efficiency places a ceiling on the amount of energy that can be stored in our food reserves, and thus, the properties of photosynthesis are important to the scientist working in food technology and production. This review deals mostly with the fundamentals of the process of photosynthesis as they occur in higher plants. Detailed reviews on other aspects of photosynthesis have been published elsewhere (1–5).

The complete process of photosynthesis can be summarized by the reaction:

$$CO_2 + H_2O \rightarrow C(H_2O) + O_2$$

That is, light drives the conversion of carbon dioxide and water to carbohydrate and oxygen. Oxygenic photosynthetic organisms use H_2O as an electron donor to convert light energy into chemical energy. However, anoxygenic photosynthetic bacteria use inorganic (eg, H_2S) or organic (eg, succinate) hydrogen donors instead of H_2O. Without the energetic input of light, the conversion of CO_2 to $C(H_2O)$ is thermodynamically uphill by 114 kcal/mol. In plants and algae, photosynthesis is confined to a special organelle called the chloroplast (1). Photosynthesis can be divided into two distinct sets of reactions: the light reactions (light harvesting and electron transport) and the dark reactions (carbon reduction and biosynthesis).

STRUCTURE AND ORGANIZATION OF CHLOROPLASTS

Chloroplasts are cytoplasmic organelles that are enclosed by a double membrane or envelope, which gives them an ellipsoidal or oval shape of about 1 by 5 (1,6–8). They are structurally sound entities that can be isolated in functional form from many plants (9). The interior of the organelle consists of an extensive membrane network (thylakoids), and an aqueous phase (stroma). Thylakoid membranes are interconnected series of disk-shaped, membrane-limited vesicles that encase a lumenal region. The pigments and proteins required for light harvesting and photosynthetic electron transport are associated with the thylakoid membrane (10).

The dark reactions of photosynthesis occur in the chloroplast stroma (1,11). This soluble phase contains the metabolic pathways that utilize the chemical energy (ATP) and reducing power (NADPH) generated by light reactions. The most conspicuous stromal process is the reductive pentose phosphate (RPP) cycle (Calvin-Benson cycle, photosynthetic carbon reduction cycle), which converts carbon dioxide to carbohydrate (11,12). Other important stromal biosynthetic pathways include amino acid synthesis (13), reduction and metabolism of nitrate and ammonia (14), and reduction of sulfate (13).

Photosynthetic Membranes and Photosystems

In oxygenic plants, algae and cyanobacteria, photosynthesis is carried out with the help of pigment–protein complexes associated with specialized structures called thylakoid membranes. Thylakoid membranes in plants and algae are located in the chloroplast, whereas in cyanobacteria these membranes arise from the inner cell membrane into the cytoplasm. Thylakoids from plants and algae, but not cyanobacteria, have two morphologies: stacked membranes (granal lamellae) and nonstacked membranes (stromal lamellae) (1,6–8). The thylakoid membrane contains four distinct complexes (Fig. 1)—photosystem II (PSII), photosystem I (PSI), cytochrome b6/f, and the ATP synthase—that are each composed of several protein subunits, as well as chromophores and cofactors (eg, the chlorophylls, carotenoids, and hemes) (10,15,16). The complexes are spatially organized in the thylakoid membrane such that PSII is in the granal lamellae, and PSI and the ATP synthase are located only in the stromal lamellae. The cytochrome b6/f complex is distributed into both morphologies of thylakoids. Contrary to oxygenic photosynthetic organisms, the anoxygenic photosynthetic bacterial pigment–protein complexes are located either directly in the cell membrane or in the invaginations that arise from the membrane.

All oxygenic photosynthetic organisms, including plants, algae, and cyanobacteria, contain two photosystems, PS I and PS II (Fig. 2a). These photosystems are the sites for the photosynthetic electron transport and are embedded in the thylakoid membrane. Each photosystem is composed of pigments, proteins, and inorganic and organic cofactors that participate in the photosynthetic electron transport. Nonoxygenic photosynthetic bacteria contain only one photosystem (Figs. 2b and 2c), also composed of pigments, proteins, and other cofactors that participate in the electron transport.

Photosynthetic Pigments and Pigment-Protein Complexes

In oxygenic photosynthetic organisms, photosynthesis is initiated by visible radiation (400–700 nm) that is absorbed by pigments. The most abundant photosynthetic pigment is chlorophyll (Chl), a cyclic tetrapyrrole with a bound Mg^{+2} (Fig. 3a). Chl is utilized for both light collection and initiation of electron transport. In higher plants and green algae there are two forms of Chl, Chla and Chlb (17,18). They absorb in the blue (400–480 nm) and red (620–700 nm) spectral regions (Fig. 3b). Chla is involved in light capture and electron transport, while Chlb only has a light-harvesting role (15,19). Carotenoids, which absorb in the blue and green spectral regions, also serve as photosynthetic pigments in higher plants (19). Carotenoids, however, only play a modest role in light capture, because they are in relatively low concentration. Carotenoids also play a key role in protecting the chloroplast from excess light via the xanthophyll cycle (for more details see Refs. 20 and 21). All Chls and carotenoids used as pigments are bound to integral membrane proteins. The in-

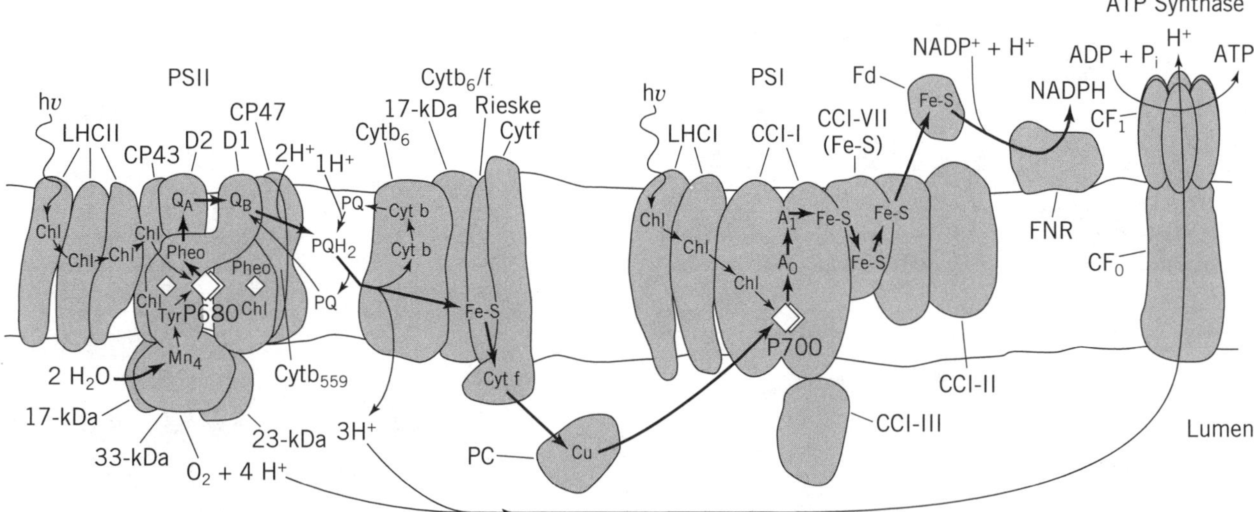

Figure 1. Organization of the four major thylakoid membrane complexes. PSII is constructed in two parts; the outer antenna (LHCII) and the core complex (the inner antenna [CP43 and CP47], cytochrome b559 and the D1/D2 reaction center). In PSII, electron flow is as indicated from water, through P680, to Q_B. The oxygen evolving Mn_4 cluster in the reaction center is stabilized by the 33-, 23- and 17-kDa polypeptides on the lumenal face of PSII. Although not depicted here, PSII is the stacked thylakoids. The cytochrome b_6/f complex is composed of four polypeptides. There are two electron pathways through this complex; one directly to PC and the other into the Q-cycle. PSI is also composed in two parts: the outer antenna (LHCI), and the core complex (CCI, which contains the inner antenna Chls, the reaction center Chls [P700], and electron acceptors). Electron flow is from PC to the Fd. CCI-III is a PC docking protein and CCI-II is an Fd docking protein. For both photosystems, light energy flows from the outer antenna, to the inner antenna, and finally to P680 or P700. The protons from water and PQH_2 oxidation drive ATP synthesis through the ATP synthase. Pigments and cofactors are as defined in Fig. 5 and the text.

teractions with protein widen absorbance bands of the pigments into the green spectral region (500–600 nm) (22).

Oxygenic photosynthetic cyanobacteria use Chla for light capture and initiation of electron transport. However, in place of Chlb-containing light-harvesting antennae, they possess special pigment–proteins, phycoerythrin, phycocyanin, and allophycocyanin, which serve a light-harvesting role. These pigment–proteins are organized into phycobilisomes and harvest red, blue, and green light (23). Contrary to the plant light-harvesting pigments that are embedded in the thylakoid membrane, the phycobilisomes are attached on the stromal surface of the thylakoid membrane.

All chlorophyll is bound exclusively to integral thylakoid membrane proteins (22,24). There are two classes of Chl-containing complexes: (*1*) antenna, which hold the vast majority of pigment (22,24,25) and (*2*) reaction centers, which have much less Chl but are the site of primary photochemistry initiated by special chlorophylls, P680 (PS II reaction center chlorophyll) and P700 (PS I reaction center chlorophyll) (16,26). Several antenna complexes combine with one reaction center to form a single photosystem.

The light-harvesting Chl-binding proteins of higher plants may be classified as either outer or inner antenna depending on their position relative to a reaction center (24–26). The outer antenna apoproteins, termed light-harvesting chlorophyll proteins (LHC) (Fig. 1), each bind a total of 7 to 10 Chla plus Chlb molecules. Those associated with PSII (LHCII) range in size from 24 to 29 kDa, while their PSI counterparts (LHCI) are generally smaller at 17 to 24 kDa (27,28). Each LHC polypeptide is proposed to contain three membrane spanning helices (29), and most LHCs share at least some primary amino acid sequence homology (30).

The inner antenna of PSII (part of the core complex [CC] of PSII [19,22,24,25]) is contained on two homologous polypeptides (CP43 and CP47, numbers refer to their molecular weight in kDa). These proteins are tightly bound to the PSII reaction center complex (Fig. 1). Each polypeptide binds approximately 25 Chla molecules (31). The inner antenna of PSI is composed of 40 to 50 Chla molecules, which are bound to the two reaction center apoproteins (CCI-subunit I) that contain P700 (25,32). The total antenna size (outer and inner antenna combined) of the respective photosystems is approximately 250 Chls per P680 and 150 Chls per P700 (15,25). The relative amount of excitation energy funneled to the two reaction centers from the antenna can be balanced by a process involving state I–state II transitions (1,33). Reversible binding of LHCII to PSII is associated with this phenomenon.

A photon of light is absorbed by an antenna Chl, and the energy is passed by resonance transfer from the outer antenna to the inner antenna and, finally, to P680 or P700 (Fig. 1). The movement of excitation energy is always

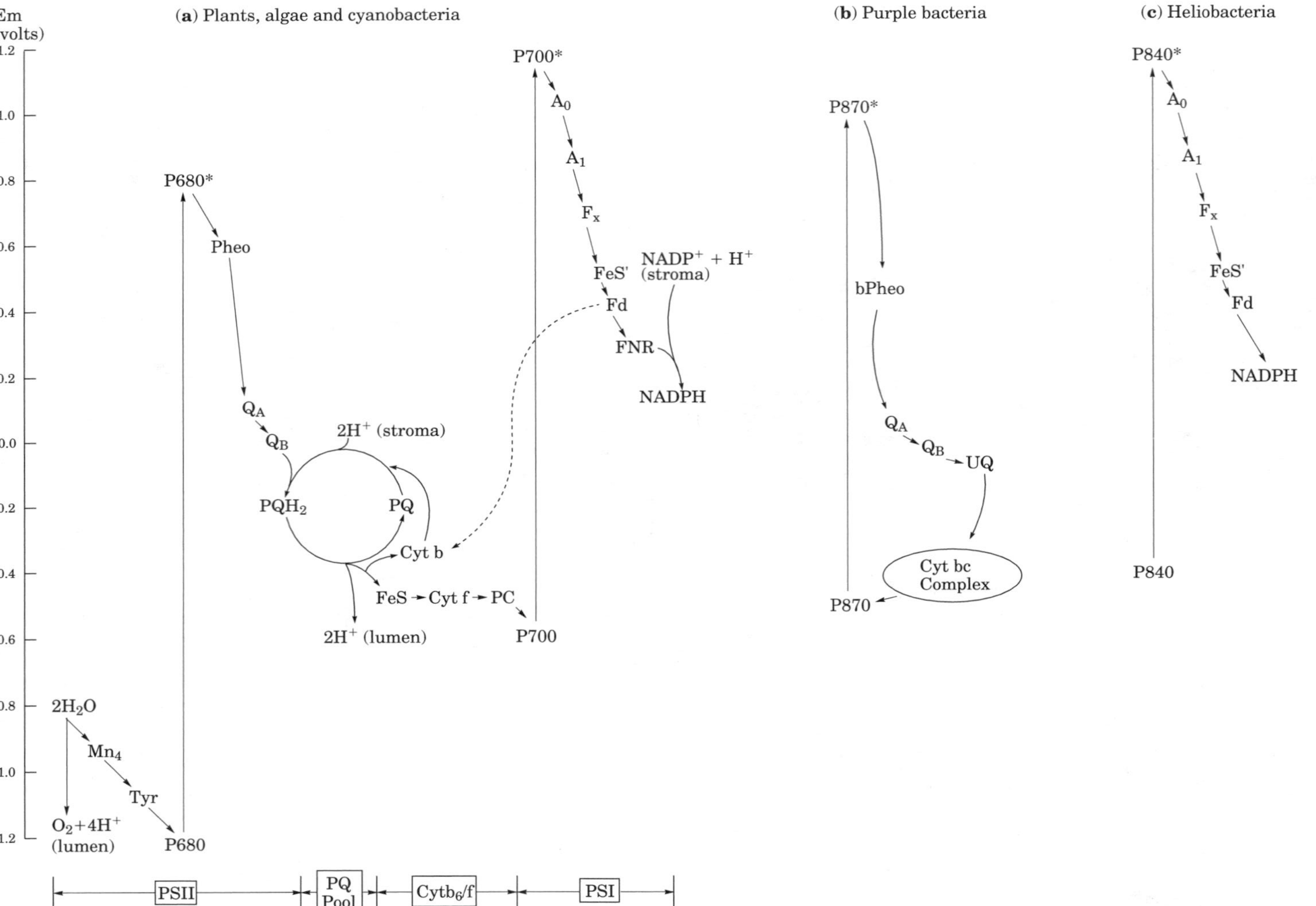

Figure 2. The schematic representation of photosynthetic electron transport. The ordinate represents the redox potential of the electron transport chain components. (**a**) Z-scheme of electron transport in plants, algae, and cyanobacteria. The electron flow from water to $NADP_+$ through two photosystems. The chlorophylls P680 and P700 are where light enters the system, raising the redox potential of the respective pigments. On the oxidizing side of PSII are water, the Mn_4 cluster and a tyr residue. On the reducing side of PSII are a pheophytin (Pheo) and two quinones (Q_A and Q_B). Mediating electron flow from PSII to PSI are the plastoquinone/plastoquinol (PQ/PQH_2) pool, the cytochrome b_6/f components and plastocyanin (PC). In the cytochrome complex, one electron goes directly to PC, while the other is cycled via cyt b back to the PQ pool (the Q-cycle). On the reducing side of PSI are a monomeric Chl (A_0), a phylloquinone (A_1), Fe-S centers and ferredoxin (Fd). Fd-$NADP^+$ reductase (FNR) catalyzes passage of electrons to $NADP^+$. Cyclic electron transport around PSI is connected by the dashed line. The protons consumed in the stroma and generated in the thylakoid lumen form the proton gradient used for ATP synthesis. (**b**) Photosynthetic electron transport in purple bacteria. There is only one photosystem involved in the electron transport which is analogous to PSII, because of the quinone type of electron carriers involved. (**c**) Photosynthetic electron transport in heliobacteria. Only one photosystem is involved in the electron transport that resembles the PSI, because of the iron-sulfur and ferrodoxins involved in the electron transport.

Chlorophyll a: R=CH_3
Chlorophyll b: R=CHO

CO_2-CH_2-CH=C(CH_3)−CH_2-(CH_2-CH_2-CH(CH_3)−CH_2)$_2$-CH_2-CH_2-CH(CH_3)$_2$

(a)

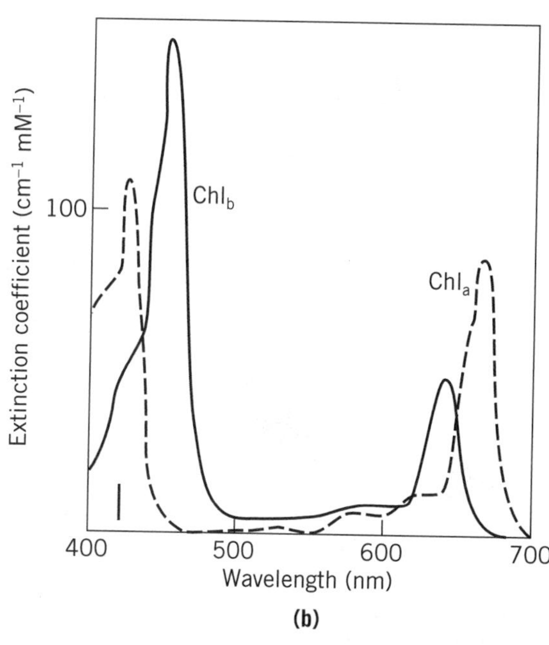

Figure 3. (**a**) Molecular model of chlorophyll a and b. (**b**) absorbance spectra of chlorophyll a and b.

downhill, progressing from pigments that absorb at shorter wavelengths (higher energy) to ones with longer wavelength absorbance bands (lower energy) (34). Photosynthetic electron transfer is initiated when the excitation energy reaches P680 or P700. The energy transfer from the light-harvesting complexes to the reaction centers is so efficient that more than 90% of the captured energy is passed on to the reaction center within a few hundred picoseconds.

The anoxygenic purple bacteria capture light with the help of bacteriochlorophyll (bChl) (22), which absorbs in the infrared (>700 nm), green (570 nm), and UVA (320–400 nm) spectral regions (17). These bacteria have different types of light-harvesting complexes and other mechanisms of regulating resonance transfer. However, their pigments still pass excitation energy down a gradient to the reaction center bChls. Like in plants, most anoxygenic bacteria contain two types of light-harvesting complexes: light-harvesting complex I (LH I) and light-harvesting complex II (LH II) (35).

Nonchlorophyll Thylakoid Complexes

The ATP synthase is the only major thylakoid membrane complex not directly involved in electron transport and does not contain chlorophyll (Fig. 1). It is a multisubunit protein complex consisting of two domains. CF0 is integral to the membrane and contains a proton channel. CF1 is extrinsic to the membrane and catalyzes ATP synthesis. ATP synthase utilizes the proton gradient across the thylakoid membrane during the light reaction to synthesize ATP (36). This chemiosmotic process of ATP synthesis in chloroplasts is very similar to the system mitochondria employ to generate ATP. This reaction in chloroplasts is called photophosphorylation and can be summarized as follows:

$$ADP + Pi \rightarrow ATP$$

THE LIGHT-DRIVEN PHOTOSYNTHETIC REACTIONS

The individual components of the light reactions, which include light-harvesting pigments, electron carriers, and integral thylakoid membrane proteins, work efficiently together to convert light energy into ATP and NADPH. This whole process is basically carried over the two morphologies in the thylakoid membrane using PS II and PS I to drive the process.

Reaction Centers

Photosynthetic reaction centers are Chl-protein complexes that convert the energy of a photon into photochemical work. The key property of a reaction center is its ability to stabilize the photoinduced charge separation between the primary donor (a special Chl) and an electron acceptor. Our knowledge of photosynthetic reaction centers was revolutionized when Michel and Deisenhofer (37) solved the crystal structure of an anaerobic purple bacterial reaction center (research for which they won the 1988 Nobel Prize in chemistry). In this reaction center, the special Chls (P870) are a pair of bChla molecules (Fig. 4). They are bound to a heterodimer of two homologous proteins (L and M, 32 and 34 kDa, respectively), which also hold two bacteriopheophytins (bPheo), two accessory bChls, two quinones, and a nonheme iron. P870 is an energy trap because it absorbs at a longer wavelength (870 nm) than the antenna (800–850 nm). Upon excitation, P870 passes an electron to bPheo (the primary electron acceptor), forming $bChl^+$ and $bPheo^-$. The electron is quickly advanced to a secondary electron acceptor, Q_A, and finally to a third species, Q_B (Fig. 4). Thus, the reaction center bChls transfer one electron to Q_B with the input of one photon. The charge separation between $bChl^+$ and Q_B^- is kinetically stabilized by the reaction center. This allows time for reduction of $P870^+$ by an electron donor (a cytochrome) back to P870. With the absorbance of a second photon, a second electron is transfered to Q_B^-, which then can diffuse out of the reaction center as a quinol (QH_2). Thus, Q_B is a constituent of the quinone pool that shuttles electrons from the reaction center to the cytochrome b-c complex (38).

The PSII of plants and purple bacterial reaction centers share several structural and functional properties (Fig. 4, see also Ref. 39). In PSII, the reaction center Chls (P680) are a special pair of Chla molecules (40,41), which absorb at a slightly longer wavelength than the antenna Chls. P680 is attached to a heterodimer of two homologous proteins (D1 and D2, 32 and 30 kDa, respectively), which also bind two pheophytins, two accessory Chls, Q_A, Q_B, and a nonheme iron (42). The D1 and D2 proteins and the L and M subunits are proposed to have similar structures in the photosynthetic membrane (Fig. 4, see also Refs. 39 and 43). D1 and D2 also share regions of important sequence homology with their bacterial counterparts (39,44). When P680 is excited, it reduces Pheo and the electron is rapidly transported across the membrane to Q_B via Q_A (Fig. 4). Concomitantly, $P680^+$ is reduced by a tyrosine residue that is part of the D1 protein (45). The electron hole in tyr^+ is

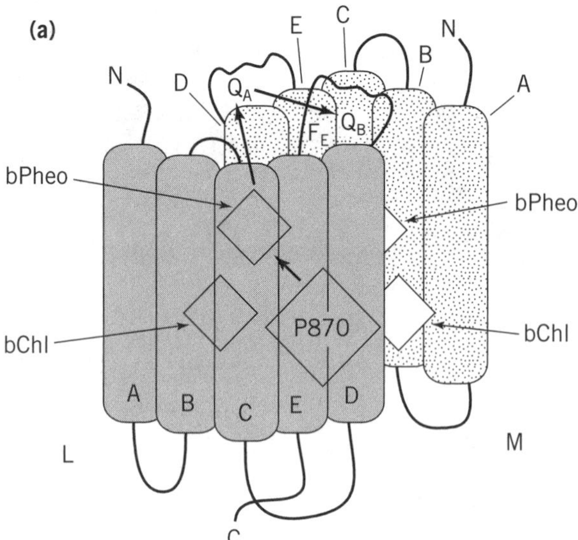

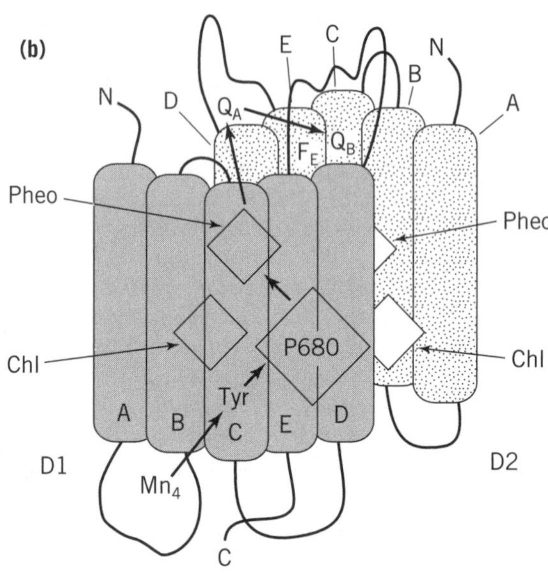

Figure 4. Structural comparison of the purple bacteria (**a**) and PSII (**b**) reaction centers. Each protein subunit (L, M, D1, and D2) has five membranes spanning helices labeled A through E. Both reaction centers contain the special pair Chls (P960 and P680), two Pheos, two accessory Chls, Q_A and Q_B, and a nonheme iron. The helices and cofactors are arranged with twofold rotational symmetry about an axis perpendicular to the membrane. The preferred pathway of electron transport through each complex is indicated.

filled by the oxygen evolving complex (a Mn_4 cluster). Similar to the bacterial reaction center, Q_B is a two electron gate (38). After this quinone is reduced to QH_2 by a second electron, it dissociates from the reaction center, blends in with the PQ pool, and migrates to the cytochrome b_6/f complex (Fig. 1).

A difference between the PSII and purple bacterial reaction centers is the rate of metabolism of the respective reaction center protein components (46). The D1 protein is subject to rapid, light dependent turnover, while all the proteins of the bacterial reaction center are kinetically stable. D1 degradation is one of the many effects associated with high light intensity damage to PSII (photoinhibition) (47). However, D1 turnover is also observed under low light intensities, including small amounts of UV radiation (48). Of environmental concern, PSII is a potential target of increased UV radiation from depletion of the ozone layer.

The PSI reaction center (Fig. 1) has a basic structure distinct from PSII (16,32,49). The reaction center Chls (P700) are a pair of Chla molecules that initiate primary photosynthesis (16,50). The primary electron acceptor (A_0) after P700 is a Chl (51), and the secondary acceptor (A_1) is a phylloquinone (52). These species and P700 are ligated to the heterodimer of 83 kDa apoproteins (CCI-subunit I) that also binds the inner antenna Chls (25,32). The electron passes from reduced phylloquinone through three Fe-S centers (one bound to CCI-subunit I and two bound to the 8 kDa CCI-subunit VII [53]) to ferredoxin (54). Ferredoxin in turn transfers the electron to $NADP^+$ (Figs. 1 and 2a).

Photosynthetic Electron Transport Chains

Electron transport begins in a photosystem. In oxygenic photosynthesis, two photosystems are used to oxidize water and reduce $NADP^+$ (26,55). Water oxidation is catalyzed by PSII, and reduction of $NADP^+$ is carried out by PSI (Fig. 2a). The formation of one oxygen molecule from two waters requires a four-photon input at PSII and pumps four electrons into the electron transport chain. Reduction of two $NADP^+$ molecules with these four electrons also depends on a four-photon input at PSI.

PSII, PSI, and the connecting electron carriers comprise the Z-scheme of photosynthesis (Fig. 2a). Electron transport is initiated when the energy of a photon excites a special Chl molecule, thereby increasing the reducing power of the pigment (34). In PSII, this species, P680, is probably a dimer of Chla molecules (40,41). P680 is excited to P680* by one photon. The latter transfers one electron to pheophytin (Pheo) which is a Chl without a bound Mg^{+2} (56). This results in $P680^+$ and $Pheo^-$. $P680^+$ is reduced by an electron from water. The electron on $Pheo^-$ is passed in series through two quinone (Q_A and Q_B) electron carriers (16,38). After Q_B is fully reduced, which requires two electrons (and thus two-photon events at P680), it leaves PSII as plastoquinol (PQH2). The electrons are transferred from PQH_2, in a semilinear fashion, through two cytochromes and a Fe-S center to plastocyanin (see the following and Fig. 2a). The two cytochrome proteins and the Fe-S protein collectively compose the cytochrome b_6/f complex (10). Plastocyanin in turn carries the electrons to PSI. PSI contains a special pair of Chla molecules (P700) (16,50) that harnesses the energy of a photon to elevate its redox potential. This drives an electron from the special Chls to A_0 (the primary acceptor) and subsequently to $NADP^+$.

Based on the Z-scheme, full-chain photosynthetic electron transport can be summarized by the reaction:

$$2H_2O + 2NADP^+ + 4ADP + 4Pi \xrightarrow{\text{8 photons}} O_2 + 2NADPH + 4ATP$$

As the final component of the Z-scheme, NADPH is produced directly by electron transport. ATP, however, is generated by an ATP synthase that utilizes a proton gradient established across the thylakoid membrane by photosynthetic electron transport (Figs. 1 and 2a). This pH gradient is generated by oxidation of water on the lumenal side of the thylakoid and reduction of $NADP^+$ on the stromal face of the membrane. Additionally, protons are carried across the membrane by PQH_2.

Electrons can also be driven around a contained loop (the Q-cycle) (15,26) involving PSI, ferredoxin, plastoquinone (PQ), cytochrome b_6/f, and plastocyanin (cyclic electron transport, Fig. 1a). For this reason, electron transport through the cytochrome b_6/f complex is said to be semilinear (one electron goes linearly to plastocyanin and the other to the Q-cycle). ATP is generated by this process, as protons are carried across the thylakoid membrane by PQ. However, $NADP^+$ is not reduced, since there is not a net input of electrons.

Nonoxygenic photosynthetic bacteria have a single photosystem (26). For example, in the purple nonsulfur bacteria, electrons are passed around a cyclic pathway from a special pair of bChls (P870) through bacteriopheophytin (BPheo), two quinones (Q_A and Q_B), a membrane-bound cytochrome b-c complex, and back to P870 (Fig. 2b). The quinone carries protons across the photosynthetic membrane generating a pH gradient used for ATP synthesis. Although this electron transport chain appears to resemble PSI cyclic electron transport, the purple nonsulfur bacterial photosystem is more analogous to PSII. However, there are other nonoxygenic photosynthetic bacteria (green sulfur bacteria and heliobacteria) that can use sulfur compounds and organic hydrogen as electron donors. Their electron transport chain does not use BPheo and quinones, instead iron–sulfur centers and ferredoxin are used to reduce NAD^+ or $NADP^+$ (Fig. 2c). Thus, this electron transport chain resembles that of PSI.

Oxygen Evolution

Water is oxidized by a Mn_4 cluster bound to the D1–D2 reaction center proteins on the lumenal face of the thylakoid membrane (Figs. 1 and 2a, and Ref. 40). This oxygen-evolving complex is stably maintained on the PSII reaction center by three extrinsic polypeptides that bind to the lumenal face of PSII. After the tyr residue on D1 is oxidized by $P680^+$ the Mn cluster gives up one electron to reduce tyr^+ back to tyr. The oxygen-evolving complex has a four-electron capacity, effectively allowing it to reside in five different redox states (S-states), ranging from S_0 (fully reduced) to S_4 (fully oxidized) (Fig. 5). A photon is required to extract an electron during each S-state transition from S_0 to S_4. The S_4 state is spontaneously returned to the S_0 state in a radiationless transition by releasing one O_2 molecule and binding two water molecules. The four protons from the two oxidized water molecules remain in the lumen, contributing to the pH gradient across the thylakoid membrane.

THE DARK REACTIONS

The RPP Cycle

Fixation of CO_2 takes place with the ATP and reduced NADP generated during the photosynthetic electron trans-

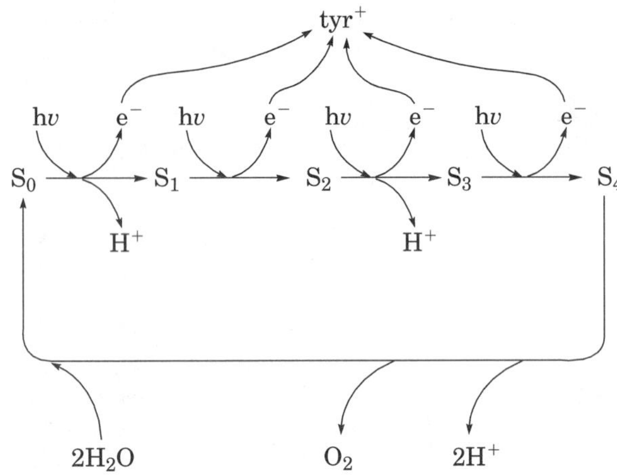

Figure 5. The S-state model for oxygen evolution in PSII. With the input of each photon, an electron is released and passed to tyr^+ (see text). Protons are placed in the lumen of the thylakoid during three of the five steps. Oxygen is released and water taken up during the S_4 to S_0 transition.

port. The reactions occur in the stroma of the chloroplast. The overall reaction of CO_2 fixation can be summarized as follows:

$$3CO_2 + 3RuBP + 9ATP + 6NADPH + 9H_2O + 6H^+ \rightarrow C_3(H_2O) + 3RuBP + 9ADP + 9Pi + 6NADP^+$$

The process is cyclic (the RPP cycle) (11,12), regenerating all the ribulose-bisphosphate (RuBP) consumed (Fig. 6). For each 3 CO_2s fixed, one extra C3 sugar is generated. ATP is utilized when 3-phosphoglycerate and ribulose-phosphate are phosphorylated. Diphosphoglycerate reduction to glyceraldehyde-3-phosphate is the only step in the RPP cycle involving NADPH.

Carboxylation of RuBP is carried out by RuBP carboxylase/oxygenase (Rubisco) (57), the most abundant enzyme in plants, if not the world. The enzyme is composed of eight large subunits (56 kDa) and eight small subunits (14 kDa). One catalytic domain is associated with each large subunit. Carboxylation of RuBP yields two C3 sugars (3-phosphoglycerate). Both 3-phosphoglycerates are reduced to glyceraldehyde-3-phosphate (GAP). GAP is necessary for the remaining metabolic reactions in the RPP cycle and those radiating from the cycle (11,12). Out of every six GAPs generated, five are converted back to RuBP (Fig. 6). The remaining GAP is either exported from the chloroplast, or it combines with another GAP and is stored as starch (58).

In addition to the carboxylation reaction, Rubisco can oxygenate RuBP (57). Although Rubisco has a lower affinity for O_2 than for CO_2, the O_2 concentration is high enough to make this oxygenation reaction significant. The products of RuBP oxygenation are 3-phosphoglycerate and phosphoglycolate. The process is called photorespiration because light drives the uptake of oxygen, and carbon dioxide is generated when phosphoglycolate is oxidized. A possible benefit of photorespiration is consumption of excess oxygen, which, if left unchecked, is potentially toxic (59). Also, it allows photosynthetic electron flow to continue at times when CO_2 is limiting.

Regulation of the RPP Cycle

Several key points in the RPP cycle are regulated in such a way that carbon fixation only proceeds in the light. For this reason, the "dark" reaction is a bit of a misnomer for the RPP cycle. Rubisco is activated in the light by Rubisco activase (60). For full activity, Rubisco must be stably carbamylated at a specific lys residue with a CO_2 that is distinct from the one used for carboxylation of RuBP. Rubisco activase probably plays a role in adding CO_2 to the lys. Rubisco is also inhibited by carboxyarabinitol 1-phosphate. This compound, which is produced in the dark, is a natural analogue of a carboxylation reaction intermediate. Activase probably catalyzes removal of carboxyarabinitol 1-phosphate from Rubisco in the light.

Four RPP cycle enzymes (glyceraldehyde-3-phosphate dehydrogenase, fructose-1,6-bisphosphatase, sedoheptulose-1,7-bisphosphatase, and phosphoribulokinase, reactions 3, 6, 9, and 13 in Fig. 6) are activated when key disulfide bridges in the proteins are reduced by thioredoxin (58). Thioredoxin is a 12-kDa peptide with two redox active cysteines. In the light, ferredoxin usually reduces $NADP^+$ (Fig. 1). Occasionally, however, it passes redox equivalents to thioredoxin instead. Thus, this system detects light dependant electron transport in the thylakoid membrane.

C-4 Pathway of CO_2 Fixation

The C-4 plants, many of which are common to hot tropical climates, utilize a distinctive method of concentrating CO_2 in the chloroplast (11,61). These plants contain two cell types: bundle sheath cells surrounded by mesophyll cells. Carbon dioxide is added to phosphoenolpyruvate by phosphoenolpyruvate carboxylase in the cytoplasm of mesophyll cells. The product, oxaloacetate, is reduced to malate and subsequently is transported to the chloroplasts of bundle sheath cells. There it is decarboxylated and the liberated CO_2 is utilized by the RPP cycle. Pyruvate, which results from malate decarboxylation, moves back to the mesophyll cell, where it is phosphorylated, regenerating phosphoenolpyruvate. The C4-pathway allows efficient delivery of CO_2 to the RPP cycle, and, by keeping oxygen concentration in bundle sheath cells to a minimum, dramatically diminishes photorespiration.

CONCLUDING REMARKS

Photosynthesis is the process driven by the absorbance of light by chlorophyll and ending with the conversion of CO_2 into carbohydrate. Although there is a diversity among the photosynthetic organisms, there appears to be a great similarity in the process of photosynthetic electron transport and CO_2 fixation. Many reactions are employed to orchestrate the transfer of radiant energy to reduced carbon (chemical energy). The light reactions, which occur on thylakoid membranes (plants, algae, and cyanobacteria) and cell membranes (anoxygenic bacteria), begin with charge

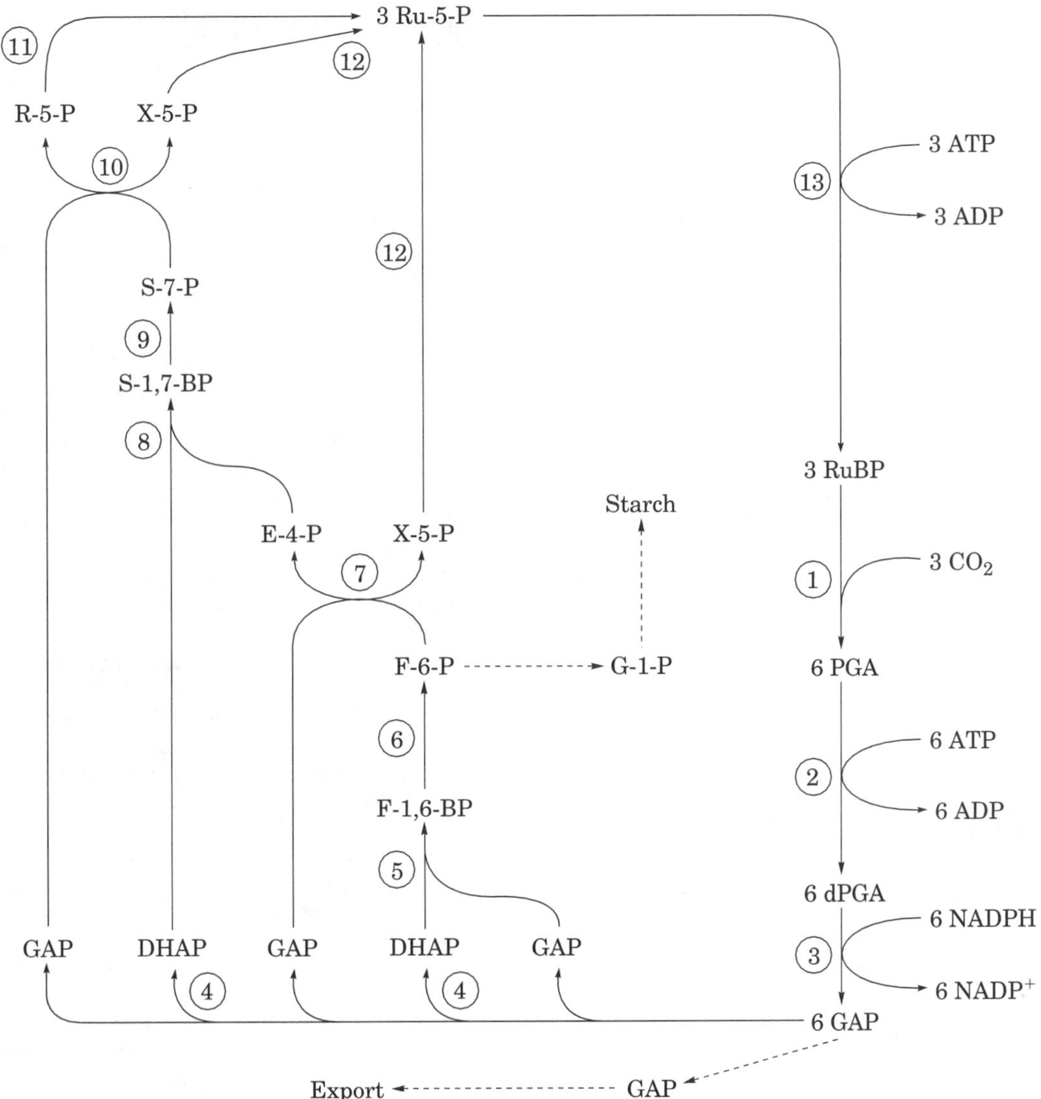

Figure 6. The reductive pentose phosphate cycle. The RPP cycle contains 13 reactions (see Robinson and Walker [Ref. 12] for a description of the enzymes). The carboxylation and regeneration of three ribulose-1,5-bisphosphate (RuBP) molecules are shown. One extra C3 sugar is generated, which is either exported from the chloroplast or incorporated into starch. PGA: 3-phosphoglycerate; DPGA: 1,3-diphosphoglycerate; GAP: Glyceraldehyde-3-phosphate; DHAP: dihydroxyacetone phosphate; F-1,6-BP: fructose-1,6-bisphosphate; F-6-P: fructose-6-phosphate; X-5-P: xyulose-5-phosphate; E-4-P: erythrose-4-phosphate; S-1,7-BP: sedoheptulose-1,7-bisphosphate; S-7-P: sedoheptulose-7-phosphate; R-5-P: ribose-5-phosphate; G-1-P: glucose-1-phosphate.

separation in a reaction center and result in temporary storage of chemical energy as ATP and NADPH. This process alone requires the coordinated operation of pigment–protein complexes, as well as numerous photochemical and redox reactions. The ATP and NADPH are then used in a complex, highly regulated metabolic cycle that fixes CO_2. Thus, photosynthesis entails a long chain of reactions, with each step playing an equally important role in sustaining life on our planet.

BIBLIOGRAPHY

1. L. A. Staehelin, "Chloroplast Structure and Supramolecular Organization of Photosynthetic Membranes," in L. A. Staehelin and C. J. Arntzen, eds., *Encyclopedia of Plant Physiology*, New Series, Vol. 19, Springer-Verlag, Berlin, 1986, pp. 1–84.
2. M. D. Hatch and N. K. Boardman, eds., *The Biochemistry of Plants*, Vol. 10, Academic Press, San Diego, Calif., 1987.
3. J. Amesz, ed., *Photosynthesis*, Elsevier Science, Amsterdam, The Netherlands, 1987.
4. R. P. F. Gregory, *Biochemistry of Photosynthesis*, Wiley, New York, 1989.
5. J. Amesz and A. J. Hoff, *Biophysical Techniques in Photosynthesis*, Kluwer Academic, Dordrecht, The Netherlands, 1995.
6. S. L. Wolfe, "Photosynthesis and the Chloroplast," in S. L. Wolfe, ed., *Introduction to Cell and Molecular Biology*, Brooks/Cole Wadsworth, Belmont, Calif., 1995, pp. 246–279.

7. A. Dyson, "Andy Dyson's Home Page, A Level Biology, Module 1: Ultrastructure of a Eukaryotic Cell," URL: *http://www.biology.demon.co.uk/Biology/mod1/organelles* (last accessed August 30, 1999).

8. P. Scott, "Biols Home Page," URL: *http://www.biols.susx.ac.uk/Home/Peter_Scott/Index.html* (last accessed August 30, 1999).

9. M. Edelman, R. B. Hallick, and N. H. Chua, eds., *Methods in Chloroplast Molecular Biology*, Elsevier Biomedical Press, Amsterdam, The Netherlands, 1982.

10. J. M. Anderson, "Molecular Organization of Thylakoid Membranes," in J. Amesz, ed., *Photosynthesis*, Elsevier Science, Amsterdam, The Netherlands 1987, pp. 273–297.

11. F. D. MacDonald and B. B. Buchanan, "Carbon Dioxide Assimilation," in J. Amesz, ed., *Photosynthesis*, Elsevier Science, Amsterdam, The Netherlands, 1987, pp. 175–197.

12. S. P. Robinson and D. A. Walker, "Photosynthetic Carbon Reduction Cycle," in M. D. Hatch and N. K. Boardman, eds., *The Biochemistry of Plants*, Vol. 8., Academic Press, San Diego, Calif., 1981, pp. 193–236.

13. B. Halliwell, "Nitrogen and Sulfer Metabolism in Photosynthetic Tissues," in *Chloroplast Metabolism*, Clarendon Press, Oxford, England, 1981, pp. 227–244.

14. D. H. Turpin et al., "Interactions between Photosynthesis, Respiration, and Nitrogen Assimilation in Microalgae," *Can. J. Botany* **66**, 2083–2097 (1988).

15. D. R. Ort, "Energy Transduction in Oxygenic Photosynthesis," in L. A. Staehelin and C. J. Arntzen, eds., *Encyclopedia of Plant Physiology*, New Series, Vol. 19, Springer-Verlag, Berlin, 1986, pp. 143–196.

16. P. Mathis and A. W. Rutherford, "The Primary Reactions of Photosystems I and II of Algae and Higher Plants," in J. Amesz, ed., *Photosynthesis*, Elsevier Science, Amsterdam, The Netherlands, 1987, pp. 63–96.

17. J. C. Goedheer, "Visible Absorption and Fluorescence of Chlorophyll and Its Aggregates in Solution," in L. P. Vernon and G. R. Seely, eds., *The Chlorophylls*, Academic Press, New York, 1966, pp. 147–184.

18. W. Rudiger and S. Schoch, "Chlorophylls," in T. W. Goodwin, ed., *Plant Pigments*, Academic Press, San Diego, Calif, 1988, pp. 1–59.

19. R. Cogdell, "The Function of Pigments in Chloroplasts," in T. W. Goodwin, ed., *Plant Pigments*, Academic Press, San Diego, Calif., 1988, pp. 182–230.

20. B. Demmig-Adams, A. M. Gilmore, and W. W. Adams, "Carotenoids 3: *In vivo* Function of Carotenoids in Higher Plants," *FASEB J.* **10**, 403–412 (1996).

21. A. M. Gilmore, "Mechanistic Aspects of Xanthophyll-Cycle Dependent Protection in Higher Plant Chloroplasts and Leaves," *Physiol. Plant.* **99**, 197–209 (1997).

22. H. Zuber, R. Brunisholz, and W. Sidler, "Structure and Function of Light-Harvesting Pigment-Protein Complexes," in J. Amesz, ed., *Photosynthesis*, Elsevier Science, Amsterdam The Netherlands 1987, pp. 233–271.

23. A. N. Glazer, "Phycobilisome: A Macromolecular Complex Optimized for Light Energy Transfer," *Biochim. Biophys. Acta* **768**, 29–51 (1984).

24. B. R. Green, "The Chlorophyll-Protein Complexes of Higher Plant Photosynthetic Membranes," *Photosynth. Res.* **15**, 3–32 (1988).

25. J. P. Thornber, "Biochemical Characterization and Structure of Pigment-Proteins of Photosynthetic Organism," in L. A. Staehelin and C. J. Arntzen, eds., *Encyclopedia of Plant Physiology*, New Series, Vol. 19, Springer-Verlag, Berlin, 1986, pp. 98–142.

26. N. K. Packham and J. Barber, "Structural and Functional Comparison of Anoxygenic and Oxygenic Organisms," in J. Barber, ed., *The Light Reactions*, Elsevier, Amsterdam, The Netherlands, 1987, pp. 1–30.

27. T. G. Dunahay and L. A. Staehelin, "Isolation and Characterization of a New Minor Chlorophyll a/b Protein Complex (CP24) from Spinach," *Plant Physiol.* **80**, 429–434 (1986).

28. A. Vainstein, C. C. Peterson, and J. P. Thornber, "Light-Harvesting Pigment-Proteins of Photosystem I in Maize," *J. Biol. Chem.* **264**, 4058–4063 (1989).

29. G. A. Karlin-Neumann et al., "A Chlorophyll a/b Protein Encoded by a Gene Containing an Intron with Characteristics of a Transposable Element," *J. Molecular and Applied Genetics* **3**, 45–61 (1985).

30. N. E. Hoffman et al., "A cDNA Clone Encoding a Photosystem I Protein with Homology to Photosystem II Chlorophyll a/b-Binding Polypeptides," *Proc. Natl. Acad. Sci. U.S.A.* **84**, 8844–8848 (1987).

31. B. A. Diner and F.-A. Wollman, "Isolation of Highly Active Photosystem II Particles from a Mutant of Chlamydomonas reinhardtii," *Eur. J. Biochem.* **110**, 521–526 (1980).

32. R. Nechushtai et al., "Photosystem I Reaction Center from the Thermophilic Cyanobacterium Mastigocladus laminosus," *Proc. Natl. Acad. Sci. U.S.A.* **80**, 1179–1183 (1983).

33. J. Bennett, K. E. Steinback, and C. J. Arntzen, "Chloroplast Phosphoproteins: Regulation of Excitation Energy Transfer by Phosphorylation of Thylakoid Membrane Polypeptides," *Proc. Natl. Acad. Sci. U.S.A.* **77**, 5253–5257 (1980).

34. K. Sauer, "Photosynthetic Light Reactions: Physical Aspects," in L. A. Staehelin and C. J. Arntzen, eds., *Encyclopedia of Plant Physiology*, New Series, Vol. 19, Springer-Verlag, Berlin, 1986, pp. 85–97.

35. X. Hu, A. Damajanovic, and K. Schulten, "Architecture and Function of the Light Harvesting Apparatus of Purple Bacteria," *Proc. Natl. Acad. Sci. U.S.A.* **95**, 5935–5941 (1998).

36. R. E. McCarty and C. M. Nalin, "Structure, Mechanism, and Regulation of the Chloroplast H+-ATPase (CF1-CF0)," in L. A. Staehelin and C. J. Arntzen, eds., *Encyclopedia of Plant Physiology*, New Series, Vol. 19, Springer-Verlag, Berlin, 1986, pp. 576–583.

37. J. Deisenhofer et al., "Structure of the Protein Subunits in the Photosynthetic Reaction Centre of *Rhodopseudomonas viridis* at 3 A Resolution" *Nature* **318**, 618–624 (1985).

38. P. R. Rich and D. A. Moss, "The Reactions of Quinones in Higher Plant Photosynthesis," in J. Barber, ed., *The Light Reactions*, Elsevier, Amsterdam, The Netherlands, 1987, pp. 421–445.

39. H. Michel and J. Deisenhofer, "Relevance of the Photosynthetic Reaction Center from Purple Bacteria to the Structure of Photosystem II," *Biochemistry* **27**, 1–7 (1988).

40. A. W. Rutherford, "Photosystem II, the Water-Splitting Enzyme," *Trends Biochem. Sci.* **44**, 227–232 (1989).

41. H. J. den Blanken et al., "High-Resolution Triplet-Minus-Singlet Absorbance Difference Spectrum of Photosystem II Particles," *FEBS Lett.* **157**, 21–27 (1983).

42. O. Nanba and K. Satoh, "Isolation of a Photosystem II Reaction Center Consisting of D-1 and D-2 Polypeptides and Cytochrome b-559," *Proc. Natl. Acad. Sci. U.S.A.* **84**, 109–112 (1987).

43. A. Trebst, "The Topology of the Plastoquinone and Herbicide Binding Peptides of Photosystem II in the Thylakoid Membrane," *Z. Naturforsch., C: Biosci.* **41**, 240–245 (1986).

44. D. C. Youvan et al., "Nucleotide and Deduced Polypeptide Sequences of the Photosynthetic Reaction-Center, B870 Antenna, and Flanking Polypeptides from *R. capsulata*," *Cell* **37**, 949–957 (1984).
45. R. J. Debus et al., "Site-Directed Mutagenesis Identifies a Tyrosine Radical Involved in the Photosynthetic Oxygen Evolving System," *Proc. Natl. Acad. Sci. U.S.A.* **85**, 427–430 (1987).
46. A. K. Mattoo, J. B. Marder, and M. Edelman, "Dynamics of the Photosystem II Reaction Center," *Cell* **56**, 241–246 (1989).
47. D. J. Kyle, I. Ohad, and C. J. Arntzen, "Membrane Protein Damage and Repair: Selective Loss of a Quinone-Protein Function in Chloroplast Membranes," *Proc. Natl. Acad. Sci. U.S.A.* **81**, 4070–4074 (1984).
48. B. M. Greenberg et al., "Separate Photosensitizers Mediate Degradation of the 32-kDa Photosystem II Reaction Center Protein in the Visible and UV Spectral Regions," *Proc. Natl. Acad. Sci. U.S.A.* **86**, 6617–6620 (1989).
49. B. D. Bruce and R. Malkin, "Subunit Stoichiometry of the Chloroplast Photosystem I Complex," *J. Biol. Chem.* **263**, 7302–7308 (1988).
50. K. D. Philipson, V. L. Sato, and K. Sauer, "Exciton Interaction in the Photosystem I Reaction Center from Spinach Chloroplasts: Absorption and Circular Dichroism Difference Spectra," *Biochemistry* **11**, 4591–4595 (1972).
51. M. R. Wasielewski, J. M. Fenton, and Govindjee, "The Rate of Formation of P700+-A0- in Photosystem I Particles from Spinach as Measured by Picosecond Transient Absorption Spectroscopy," *Photosynth. Res.* **12**, 181–190 (1987).
52. R. W. Mansfield, J. H. A. Nugent, and M. C. W. Evans, "ESR Characteristics of Photosystem I in Deuterium Oxide: Further Evidence That Electron Acceptor A1 Is a Quinone," *Biochim. Biophys. Acta* **894**, 515–523 (1987).
53. R. M. Wynn and R. Malkin, "Characterization of an Isolated Chloroplast Membrane Fe-S Protein and Its Identification as the Photosystem I Fe-SA/Fe-SB Binding Protein," *FEBS Lett.* **229**, 293–297 (1988).
54. W. Haehnel, "Plastocyanin," in L. A. Staehelin and C. J. Arntzen, eds., *Encyclopedia of Plant Physiology*, New Series, Vol. 19, Springer-Verlag, Berlin, 1986, pp. 547–559.
55. W. Haehnel, "Photosynthetic Electron Transport in Higher Plants," *Annual Rev. Plant Physiol.* **35**, 659–693 (1984).
56. V. V. Klimov et al., "Reduction of Pheophytin in the Primary Light Reaction of Photosystem II," *FEBS Lett.* **82**, 183–186 (1977).
57. T. J. Andrews and G. H. Lorimer, "Rubisco: Structure, Mechanisms, and Prospects for Improvement," in M. D. Hatch and N. K. Boardman, eds., *The Biochemistry of Plants*, Vol. 10, Academic Press, San Diego, Calif., 1987, pp. 131–218.
58. C. Cseke and B. B. Buchanan, "Regulation of the Formation and Utilization of Photosynthate in Leaves," *Biochim. Biophys. Acta* **853**, 43–63 (1986).
59. B. Halliwell, "Toxic Effects of Oxygen on Plant Tissues," in *Chloroplast Metabolism*, Clarendon Press, Oxford, England, 1981, pp. 179–205.
60. A. R. Portis, Jr., "Rubisco Activase," *Biochim. Biophys. Acta* **1015**, 15–28 (1990).
61. M. D. Hatch and C. R. Slack, "Photosynthetic CO2-Fixation Pathways," *Annual Rev. Plant Physiol.* **21**, 141–162 (1970).

T. Sudhakar Babu
Bruce Greenberg
University of Waterloo
Waterloo, Ontario
Canada

PHYTOCHEMICALS: ANTIOXIDANTS

The protective effects of fruit and vegetable consumption on diseases such as cancer and cardiovascular disease, among many other ailments, have been established for years. In recent history, the scientific community has been aggressively pursuing the basis of fruit and vegetable consumption and disease prevention to combat the prevalence of health disorders in today's society. Research studying the relationship between consumption of fruits and vegetables and cancer—and to a lesser extent cardiovascular disease—are the most abundant, because these conditions constitute the majority of health-related deaths in the industrialized world. Comprehensive reviews on the consumption of fruits and vegetables on the occurrence of cancer have resulted in 60 to 85% of the studies showing statistically significant associations in the decrease of cancer incidence (1). Individuals who consume the highest amount of fruits and vegetables have half the cancer rates as those who consume the least amount. A similar association has been seen with cardiovascular disease. Sixty percent of the studies reviewed by Ness and Powles (2) showed a statistically significant protective effect of fruit and vegetable consumption on coronary heart disease and stroke.

The consumption of an ample supply of fruits and vegetables (five to nine servings, as recommended by the U.S. Department of Agriculture Food Guide Pyramid) provides individuals with a wide variety of phytochemicals (both essential and nonessential nutrients from plants) that have been shown to have health benefits. A group of phytochemicals that have received significant attention for their health benefits are the antioxidants. These antioxidants include the nutritionally essential vitamins C (ascorbic acid), E (α-tocopherol), and β-carotene and also some nonessential compounds, such as the phenolics (most notably the flavonoids and isoflavones) and the carotenoids (Fig. 1). Because oxidative stress has been linked to many degenerative diseases, including cancer and cardiovascular disease, these antioxidants could potentially play a strong role in disease prevention. The fact that that only 9% of Americans consume at least five servings of fruits and vegetables per day is also substantial evidence that there is a large potential in preventing health problems through consumption of plant products (3).

Oxidative reactions that are harmful to health usually involve free radicals or reactive oxygen species that cause the oxidative modification of lipids, proteins, and nucleic acids (4). Free radicals can be produced as a by-product of normal metabolic reactions or from the decomposition of peroxides by transition metals or irradiation. In the case of lipid oxidation, the presence of free radicals that can oxidize unsaturated fatty acids will result in an autocatalytic reaction that involves the production of numerous free radical and lipid peroxide species. Reactive oxygen species are also produced as by-products of metabolic reactions and through the conversion of atmospheric oxygen to high-energy oxygen species by photosensitizers in the presence of light.

ANTIOXIDANT MECHANISMS OF PHYTOCHEMICALS

Phytochemical antioxidants can inhibit oxidative reactions by several different mechanisms. Both nutritionally non-

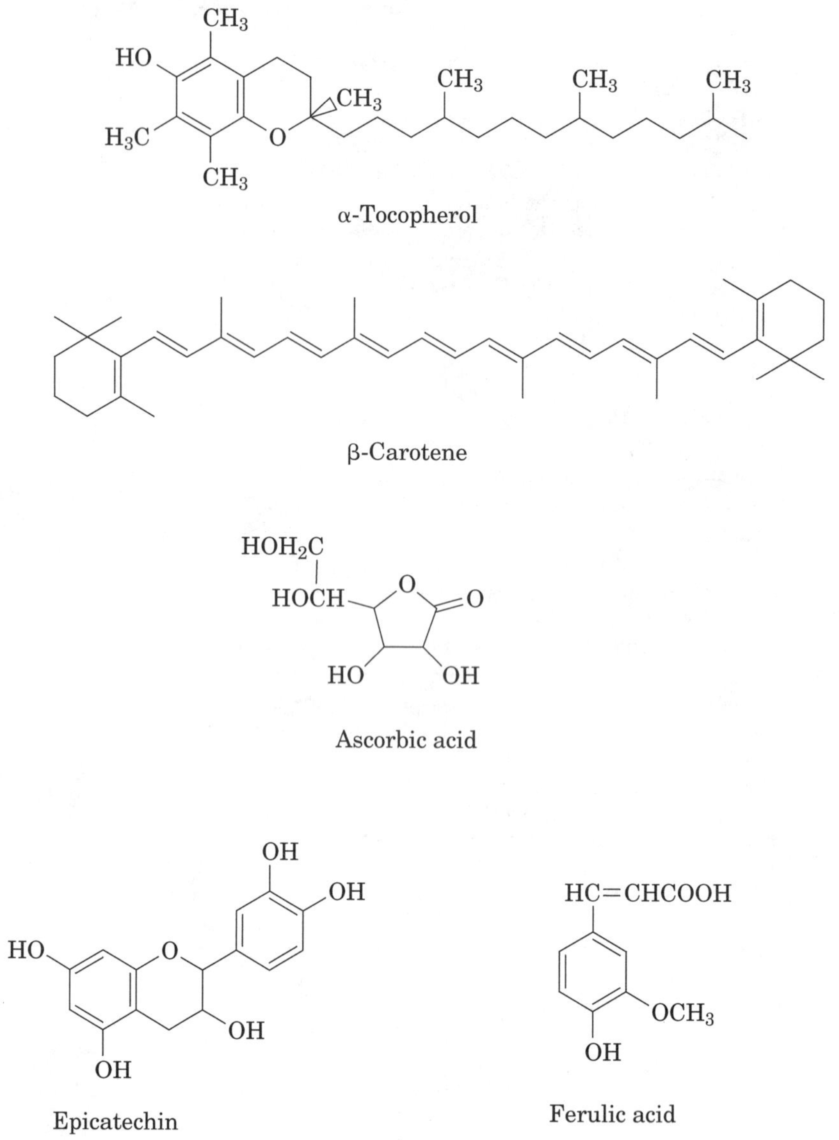

Figure 1. Structures of some common phytochemical antioxidants.

essential (eg, phenolic acids and isoflavones) and essential (α-tocopherol) phenolics inhibit oxidation by acting as chain-breaking antioxidants that are capable of inactivating free radicals and thus stopping the autocatalytic free-radical reaction (5). Phenolic compounds are efficient chain-breaking antioxidants because they are capable of scavenging free radicals, resulting in the formation of low-energy phenolic radicals that in turn do not have sufficient energy to further promote oxidative reactions. Ascorbic acid can also act as a chain-breaking antioxidant; however, it can also promote the reduction of transition metals (5). This metal-reducing activity will increase the pro-oxidant activity of metals because their reduced states rapidly promote the decomposition of peroxides into free-radical species.

Carotenoids can act as chain-breaking antioxidants when oxygen concentrations are low. However, in the presence of high oxygen concentrations, carotenoid radicals can actually increase oxidative reactions (6,7). Carotenoids are most effective at inhibiting oxidative reactions in the presence of systems that produce the reactive oxygen species, singlet oxygen. Singlet oxygen is produced from atmospheric oxygen when compounds known as photosensitizers (eg, chlorophyll, riboflavin, and myoglobin) are energized by light to an excited state. The excited state of the photosensitizer can then transfer its energy to oxygen to produce singlet oxygen that in turn oxidizes unsaturated fatty acids and amino acids (8). Carotenoids inhibit this reaction by physically absorbing energy from the excited photosensitizer or from singlet oxygen and then slowly returning the energy to the surrounding media through vibrational and rotational interactions. In order for a carotenoid to be able to inactivate singlet oxygen and activated photosensitizers through physical quenching, it must contain at least nine double bonds (9).

PHYTOCHEMICALS AND CANCER

Although there is abundant evidence implicating free radicals with cancer, the precise mechanism by which oxidative stress produces cancer is less clear. The free radicals (especially hydroxyl radicals), hydrogen peroxide, lipid peroxides, and singlet oxygen have been shown to cause DNA damage. The damage can occur through oxidative modification of the base and/or sugar moiety, single- and double-strand breaks, and DNA cross-linking (10). All this damage can be repaired through DNA repair enzymes, but problems may arise. Mutations following the incorrect repair or replication of DNA may lead to carcinogenesis. The body employs a series of enzymatic and structural antioxidant defenses along with exogenous dietary antioxidants to combat oxidative damage. However, these defenses are not perfect, and if the rate at which DNA becomes oxidized exceeds the rate at which it can be repaired, problems will arise. Over time, DNA damage becomes cumulative and is believed to account for the drastic increase in cancer rates seen with aging in humans and animals (3). Another possible mechanism may involve oxidative damage to critical proteins or enzymes associated with normal DNA repair. This damage may lead to erroneous repairs that may result in carcinogenic mutations (11). A third mechanism involves the enzymatic activation of compounds into carcinogens by free radicals, which is how many known procarcinogens are activated (10). These mechanisms are potential initiators of cancer and are believed to play a small role in the initiation stage of cancer. Free radicals are believed to play a larger role in the promotion of cancer through their ability to promote tumor development, stimulate oxygen radical production, and alter antioxidant defenses (10).

The ability of phytochemical antioxidants to act as free-radical scavengers *in vivo* demonstrates their potential as anticarcinogenic agents. By directly scavenging free radicals, oxidative stress in the body—and thus DNA damage—becomes diminished. Antioxidants have been shown to limit the chronic effects of smoking in animal models and reduce carcinogenic formation in chronic inflammation because smoking and inflammatory states both increase the exposure of tissues to free radicals (3). The protective effect of antioxidants is also related to their ability to decrease cellular division by limiting oxidants, which are known stimulants of cellular division (3). Rates of cellular division are strongly correlated to cancer rates because of increases in potential mutations with each replication of DNA. These mechanisms are the most important known effects of phytochemicals, but others that are either less understood or not as effective include bolstering enzymatic detoxification of carcinogens, directly hindering the formation of nitrosamines and other carcinogens, and maintaining normal DNA repair (1).

Despite the overwhelming protection seen in epidemiological studies relating fruit and vegetable consumption and cancer prevention, the results obtained by single or a combination of a few phytochemical antioxidants has been less encouraging. The next section presents an array of research, including epidemiological, clinical, animal, and *in vitro* studies in regard to phytochemical antioxidants and cancer. In-depth discussions of the research are impossible, and many limitations are inherent with each type of study. Animal studies often use extreme challenges on animals that do not relate to humans to obtain results. Epidemiological studies often ascertain nutrient values through questionnaires and dietary dairies that are difficult to interpret; therefore, it is often difficult to distinguish which phytochemicals are responsible for the observed effect. Intervention or clinical trials need to be very long and often run into problems of noncompliance, subjects are often at high risk (ie, smokers) for cancer development and may already be in the carcinogenic process before the study begins.

Carotenoids

Most of the epidemiological studies reviewed by Van Poppel and Goldbohm (12) that measure carotenoid intake from food and cancer risk have shown a strong protective effect, especially for lung, stomach, colorectal, and oral cancers (12). Intervention trials have not demonstrated such positive results. Recent studies observing lung and esophageal cancer have shown an increased risk or no effect with supplemental β-carotene. The α-Tocopherol, β-Carotene (ATBC) Cancer Prevention Study and the Carotene and Retinol Efficacy Trial (CARET) both demonstrated increases in lung cancer incidence and death in smokers among the β-carotene supplemented groups (13,14). A recent study that showed a protective effect was from Linxian, China, where esophageal and gastric cancers are prevalent (15). A combination of β-carotene, vitamin E, and selenium resulted in a 13% reduction in cancer deaths. A 6-year intervention trial in China, which was a follow up to the Linxian study, tested the ability of β-carotene to prevent esophageal and gastric cancers; it failed to show any reduction in cancer incidence (16). Studies on the recurrence of colorectal adenomas and skin cancer both failed to demonstrate any protection by β-carotene (17,18).

β-Carotene has been the predominant carotenoid studied, probably because of its availability in pill form, nontoxicity, and strong provitamin A activity, but other carotenoids have been shown to have stronger anticarcinogenic affects in animal models. In a study using chemical carcinogens as initiators and promoters of cancer, α-carotene was shown to be a more potent tumor suppressor in four locations (skin, lung, liver, and duodenum) compared to β-carotene (19). Palm carotene, which is a mixture of many carotenoids, was able to completely inhibit skin cancer development and was more effective in all sites, supporting evidence that a large mixture of phytochemicals are more potent in preventing cancer (19). Studies on high levels of carotenoids, such as lutein, lycopene, β-cryptoxanthin, and zeathin, in human blood have been linked to decreases in cancers of the prostate and breast (20,21), although other studies linking serum carotenoid levels with decreased risk of prostate and breast cancer have been less conclusive (22,23).

Ascorbic Acid

The ability of vitamin C to block carcinogenesis has been demonstrated in many animal models. Although its ability to promote or have no effect on cancer has also been demonstrated, overall vitamin C has shown anticarcinogenic activity. A study on female rats given chemical carcinogens resulted in the development of mammary tumors in the rats (24). When vitamin C only and vitamin C along with vitamin A, magnesium, and selenium were administered, tumorigenesis was reduced to 57.1% and 12%, respectively. A study by Pauling (25) that used hairless mutant mice irradiated with ultraviolet light resulted in cancer incidence being five times higher in rats not receiving vitamin C compared to rats receiving 10% vitamin C.

The epidemiological evidence supporting vitamin C has been promising for several types of cancer. From two comprehensive reviews of vitamin C and cancer incidence, the strongest association occurred with oral, esophageal, gastric, and pancreatic cancers (26,27). Of the 47 studies reviewed by Block (26) that reported vitamin C indexes or plasma ascorbate levels, 33 showed statistically significant protection, and none showed increased risk. Hormone-related cancers, such as breast, prostate, and ovarian cancers, have not been correlated to vitamin C intake. A study by Hunter et al. (28) of 89,494 woman concluded that large doses of vitamin C or E did not protect women from breast cancer. Intervention trials reviewed by Byers and Guerro (27) regarding supplemental vitamin C have been less encouraging, with most finding no or little protection. Clinical trials on the recurrence of cancer have been mixed. Survival times in cancer patients in Scotland receiving ascorbate supplements were twice as great as those of controls (29). In contrast, six intervention trials on the recurrence of colorectal polyps found no benefit from vitamin C supplementation (30).

Tocopherols

Animal studies evaluating the reduction of cancer with vitamin E (α-tocopherol) supplementation have been mildly supportive. A study using Wistar rats consuming diets with high vitamin E content found that rats developed fewer intestinal colorectal tumors induced by chemical carcinogens (31). Animal models testing skin and mammary glands have been inconclusive, but vitamin E has been shown to inhibit oral carcinomas (32). Observational epidemiological studies based on vitamin E consumed from the diet and intervention studies using supplemental vitamin E have been inconclusive (33). Colon cancer has received the most attention in regard to vitamin E and cancer prevention. There seems to be, at best, a weak beneficial effect with increased vitamin E intake. The Iowa Women's Health Study and the ATBC study, both large-scale trials with 35,000 and 29,000 participants, respectively, found vitamin E to be associated with a reduced risk of colon cancer (13,34). The Finnish Mobile Health Survey found a 1.5-fold increase in cancer incidence among individuals with the lowest serum α-tocopherol levels compared to the highest (33). Their findings supported a benefit only in the case of gastrointestinal cancers, with no effect seen with respiratory or hormone-related cancers. The Polyp Prevention Study Group, which assessed the benefit of vitamin E along with β-carotene in reducing colonic adenomas in 864 individuals, found no benefit from the supplementation (17).

Phenolics

Large-scale epidemiological or intervention studies regarding phenolics, as a group or individually, have yet to be performed. This may be due to the large diversity and abundance of phenolics found in our food supply or the limited availability of pharmaceutical sources of phenolics. Most of the research has been limited to *in vitro* and animal studies. The *in vitro* studies have proved the antioxidant effects of phenolics, and the animal studies have demonstrated their ability to block chemically induced carcinogenesis. Some examples of phenolics that have been shown to block carcinogenesis in animal models include quercetin, rutin, curcumin, caffeic acid, ellagic acid, genistein, and daidzein (35). Although phenolics have proved to be anticarcinogenic, studies have also demonstrated their ability to promote cancer. A study on rats that were fed high levels of caffeic acid, sesamol, and catechol resulted in the development of stomach cancer (36).

PHYTOCHEMICALS AND CARDIOVASCULAR DISEASE

Oxidative reactions have been linked to atherosclerosis through several different mechanisms. The most widely studied hypothesis of lipid oxidation and atherosclerosis involves the formation of oxidized, cytotoxic lipoproteins, particularly low-density lipoprotein (LDL) (37). During the oxidation of LDL, the lipoproteins become modified through either direct free-radical attack or formation of adducts between proteins and lipid oxidation products. The oxidized LDL can then be recognized and engulfed by macrophages, leading to the formation of foam cells that accumulate in arterial walls and form plaques (Fig. 2). Oxidized LDL has also been postulated to cause vascular inflammation and stimulate autoimmune reactions. Evidence supporting the relationship between LDL oxidation and cardiovascular disease is increasing; however, the importance of this mechanism to the development of cardiovascular disease has yet to be fully understood.

A second proposed mechanism for the link between oxidative reactions and atherosclerosis involves endothelium-derived relaxation factor (EDRF) (38). The endothelium, the cells that line the vascular lumenal surface, is a physiologically active organ that regulates vascular tone. The predominant regulator of vasodilation is EDRF, which

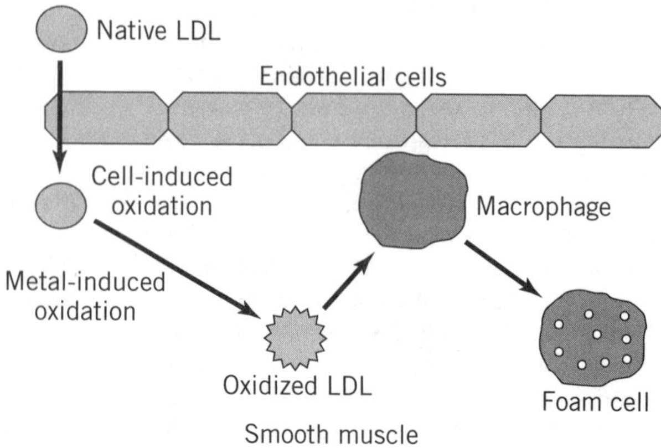

Figure 2. The proposed mechanism for promotion of coronary heart disease by oxidized LDL.

is thought to be identical to or similar to the free radical nitric oxide. Superoxide anion, another type of free radical, is believed to be vasoconstrictive owing to its ability to inactivate nitric oxide by converting it to peroxynitrite. Because peroxynitrite is a strong oxidizing agent, this can cause modification of proteins and DNA and thus damage endothelium function (Fig. 3). Therefore, the interaction of superoxide anion with nitric oxide leads to restriction in vascular dilation, which in turn can cause damage to the endothelium and lead to the development of atherosclerotic plaque formation. Theoretically, the introduction of antioxidants into the vasculature at suitable levels would negate the loss of dilation by inactivating superoxide anions and thus protect the vasodilatory activity of nitric oxide.

Carotenoids

Carotenoids, including β-carotene, are incorporated into LDL, but they only weakly inhibit LDL oxidation. Clinical and epidemiological studies have been largely inconclusive about the ability of carotenoids to inhibit atherosclerosis. These observations could be due to the weak antioxidant or even pro-oxidative activity of carotenoids (7). Recent research has suggested that foods high in carotenoids have been linked with decreased occurrence of atherosclerotic plaques, suggesting that carotenoids exert their protective effect later in the atherosclerotic process (39). However, this study was not able to specifically link dietary carotenoids with an antiatherogenic activity because the observed protective effects could be the result of other phytochemicals found in carotenoid-rich foods.

Ascorbic Acid

Vitamin C is an effective scavenger of superoxide anions and other free radicals, suggesting that it may protect endothelial function and inhibit the development of atherosclerosis. Ascorbate can inhibit the oxidation of LDL (40), and in studies on hypocholesterolemic individuals, ascorbic acid has been shown to improve vasodilation in subjects with chronic heart failure (41). However, epidemiological studies linking dietary vitamin C with prevention of atherosclerosis are inconclusive, and large-scale clinical trials with vitamin C alone have yet to be performed.

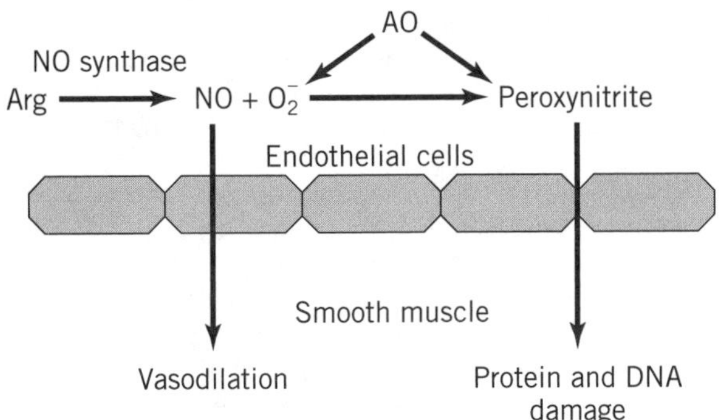

Figure 3. The proposed mechanism for the role of nitric oxide in the formation of coronary heart disease. Arg, arginine; AO, antioxidant; NO, nitric oxide; O_2^-, superoxide anion.

Tocopherols

Tocopherols, particularly vitamin E (α-tocopherol), have received much acclaim in the fight against atherosclerosis. Epidemiological evidence has shown a positive correlation between dietary α-tocopherol levels (>100 IU/day for 2 yr) and reduced incidence in cardiovascular disease (42,43). Clinical trials also support the idea that α-tocopherol is antiatherogenic, because 1000 IU of vitamin E for 2 yr decreased coronary lesions in men after bypass operations (44), and 400–800 IU vitamin E decreased heart attacks in 77% of men (45). The proposed mechanisms by which vitamin E protects against atherosclerosis includes its ability to inhibit LDL oxidation (46), maintain endothelial integrity (47), and inhibit monocyte adhesion to endothelial cells (48). In addition, vitamin E can protect the vasodilation activity of rat aortic strips (49).

Plant foods contain tocopherol isomers in addition to α-tocopherol, with γ-tocopherol being the most common. Tocopherol-binding protein is a liver protein responsible for tocopherol transport into very-low-density lipoprotein, which then carries tocopherol in the blood where it can be transported to peripheral tissues (50). The prevalence of α-tocopherol in most human tissues is due to the more than tenfold preference of tocopherol-binding protein for α-tocopherol compared to γ-tocopherol. Although γ-tocopherol inactivates most free radicals in a manner similar to α-tocopherol, it has been reported to have greater scavenging activity than α-tocopherol in the presence of peroxynitrite radicals and therefore has been suggested to be important in protecting endothelial function (51). This suggests that natural sources of tocopherols may be more beneficial to health because large doses of dietary α-tocopherol will result in very little γ-tocopherol being incorporated into tissue because of the large preference of tocopherol-binding protein for α-tocopherol.

Phenolics

Numerous plant phenolics, including those from tea, fruit, grains, and herbs, have been shown to inhibit the oxidation of LDL *in vitro* (52). The best studied of the plant phenolics are the catechins from tea, anthocyanins from grapes, and isoflavonoids from soybeans. Although many of these phenolics are absorbed from the diet into the blood, their retention time is normally very short (less than several hours). However, plasma metabolites of the plant phenolic quercetin have been shown to possess antioxidant activity, suggesting that the antioxidant effect of plant phenolics may be the result of a combination of the original compounds and their metabolites (53).

The ability of dietary catechins from tea to increase the antioxidant activity of blood plasma is unclear. Ingestion of green and black teas have been reported by Serafini et al. (54) and He and Kies (55) to increase the antioxidant activity of blood plasma, but McAnlis et al. (56) reported that black tea consumption did not affect plasma antioxidant capacity. Red grapes and red wine contain high amounts of anthocyanins that can inhibit both LDL oxidation (57) and platelet aggregation human subjects (58). The antiatherosclerotic benefits of grape anthocyanins have also been observed in clinical trials in which red wine was found to increase phenolics in blood and decrease lipid oxidation products in LDL (59) and in which grape juice consumption increased the antioxidant capacity of serum (60). Isoflavones can inhibit LDL oxidation and protect endothelial cells from the toxicity of oxidized LDL (61). Soy proteins containing isoflavones were found to decrease the occurrence of atherosclerotic lesions in monkeys (62).

CONCLUSION

Evidence continues to mount in support of the health benefits of phytochemicals. Unfortunately, the exact biochemical and physiological mechanisms for these health benefits are currently unclear. In fact, some of the phytochemicals, such as carotenoids and phenolics, have sometimes been shown to have pro-oxidative activity that could be harmful to health. In addition, very little is known about the absorption of the nonessential phytochemicals (eg, phenolics) through the gastrointestinal tract and their subsequent retention in tissues. Because of the lack of knowledge of the exact mechanisms and magnitude by which nutritionally nonessential antioxidants may affect health, consumption of large amounts of these compounds, in the form of either foods or supplements, may not be prudent until their bioactivity is better understood.

BIBLIOGRAPHY

1. K. A. Steinmetz and J. D. Potter, "Vegetables, fruit and cancer prevention: A Review," *J. Am. Dietetic Assoc.* **96**, 1027–1039 (1996).
2. A. R. Ness and J. W. Powles, "Fruit and Vegetables, and Cardiovascular Disease: A Review," *Int. J. Epidemiol.* **26**, 1–13 (1997).
3. B. N. Ames, M. K. Shigenaga, and T. M. Hagen, "Oxidants, antioxidants, and the degenerative diseases of aging," *Proc. Nat. Acad. Sci. U.S.A.* **90**, 7915–7922 (1993).
4. B. Halliwell and J. M. C. Gutteridge, "Role of Free Radicals and Catalytic Metal Ions in Human Disease: An Overview," *Methods Enzymol.* **186**, 1–88 (1990).
5. E. A. Decker, "Antioxidant Mechanisms," in C. C. Akoh and D. B. Min, eds., *Food Lipids, Chemistry, Nutrition, and Biotechnology*, Marcel Dekker, Inc., New York, 1998.
6. G. W. Burton and K. U. Ingold, "β-Carotene: An Unusual Type of Lipid Antioxidant," *Science* **235**, 1043–1046 (1987).
7. P. Palozza, "Prooxidant Actions of Carotenoids in Biologic Systems," *Nutr. Rev.* **56**, 257–265 (1998).
8. D. G. Bradley and D. B. Min "Singlet Oxygen Oxidation of Foods," *Crit. Rev. Food Sci. Nutr.* **31**, 211–236 (1992).
9. P. DiMascio, P. S. Kaiser, and H. Sies, "Lycopene as the Most Efficient Biological Carotenoid Singlet Oxygen Quencher," *Arch. Biochem. Biophys.* **274**, 532–538 (1989).
10. Y. Sun, "Free Radicals, Antioxidant Enzymes, and Carcinogenesis," *Free Radicals Biol. Med.* **8**, 583–599 (1990).
11. K. Frenler, "Oxidation of DNA Bases by Tumor Promoter Activated Processes," *Environ. Health Perspect.* **81**, 45–54 (1989).
12. G. Van Poppel and R. A. Goldbohm, "Epidemiological Evidence for Beta-Carotene and Cancer Prevention," *J. Am. Dietetic Assoc.* **62** (Suppl), 1393–1402 (1995).
13. The α-Tocopherol, β-Carotene Cancer Prevention Study Group, "The Effect of Vitamin E and Beta-Carotene on the Incidence of Lung Cancer and Other Cancers in Male Smokers," *N. Engl. J. Med.* **330**, 1029–1035 (1994).
14. G. S. Omenn et al., "The Beta-Carotene and Retinol Efficiency Trial (CARET) for Chemoprevention of Lung Cancer in High Risk Populations: Smokers and Asbestos Exposed Workers," *Cancer Res.* **54** (Suppl), 2038s–2043s (1994).
15. W. J. Blot et al., "Nutrition Intervention Trials in Linxian, China: Supplementation with Specific Vitamin/Mineral Combinations, Cancer Incidence, and Disease Specific Mortality in the General Population," *J. Nat. Cancer Inst.* **85**, 1483–1491 (1993).
16. J. Y. Li et al., "Nutrition Intervention Trials in Linxian, China: Multiple Vitamin/Mineral Supplementation, Cancer Incidence, and Disease Specific Mortality Among Adults with Esophageal Dysplasia," *J. Nat. Cancer Inst.* **85**, 1492–1498 (1993).
17. E. R. Greenberg et al. and the Polyp Prevention Study Group, "A Clinical Trial of Antioxidant Vitamins to Prevent Colorectal Adenoma," *N. Engl. J. Med.* **331**, 141–147 (1994).
18. E. R. Greenberg et al. and the Skin Cancer Prevention Group, "A Clinical Trial of β-Carotene to Prevent Basal Cell and Squamous Cell Cancer of the Skin," *N. Engl. J. Med.* **323**, 789–795 (1990).
19. H. Nishino, "Cancer Chemoprevention by Natural Carotenoids and Their Related Compounds," *J. Cell. Biochem.* **22** (Suppl), 231–235 (1995).
20. J. F. Dorgan et al., "Relationship of Serum Carotenoids, Retinol, α-Tocopherol, and Selenium with Breast Cancer Risk: Results from a Prospective Study in Columbia, Missouri (United States)," *Cancer Cases Control* **9**, 89–97 (1998).
21. E. Giovannucci et al., "Intake of Carotenoids and Retinol in Relation to Risk of Prostate Cancer," *J. Natl. Cancer Inst.* **87**, 1767–1776 (1995).
22. A. M. Nomura et al., "Serum Micronutrients and Prostate Cancer in Japanese Americans in Hawaii," *Cancer Epidemiol. Biomarkers Prev.* **6**, 487–491 (1997).
23. R. Jarvinen et al., "Diet and Breast Cancer Risk in a Cohort of Finnish Women," *Cancer Lett.* **114**, 251–253 (1997).
24. A. Ramesha et al., "Chemoprevention of 7,12-Dimethylbenz[a] Anthracene-Induced Mammary Carcinogenesis in Rat by the Combined Action of Selenium, Magnesium, Ascorbic Acid, and Retinyl Acetate," *Jpn. J. Cancer Res.* **81**, 1239–1246 (1990).
25. L. Pauling, "Effect of Ascorbic Acid on Incidence of Spontaneous Mammary Tumors and UV Light-Induced Skin Tumors in Mice," *Am. J. Clin. Nutr.* **54** (Suppl), 1252–1255 (1991).
26. G. Block, "Epidemiological Evidence Regarding Vitamin C and Cancer," *Am. J. Clin. Nutr.* **54** (Suppl), 1310s–1314s (1991).
27. T. Byers and N. Guerro "Epidemiological Evidence for Vitamin C and Vitamin E in Cancer Prevention," *Am. J. Clin. Nutr.* **62** (Suppl), 1385s–1392s (1995).
28. D. J. Hunter et al., "A Prospective Study of the Intake of Vitamins C, E, and A and the Risk of Breast Cancer," *N. Engl. J. Med.* **329**, 234–240 (1993).
29. E. Cameron and A. Campbell, "Innovation vs quality control: An 'unpublishable' clinical trial of supplemental ascorbate in incurable cancer," *Med. Hypotheses* **36**, 185–189 (1991).
30. S. N. Gershoft "Vitamin C (Ascorbic Acid): New Roles, New Requirements?" *Nutr. Rev.* **51**, 313–326 (1993).

31. M. G. Cook and P. McNamara "Effect of Dietary Vitamin E on Dimethylhydrazine-Induced Colonic Tumors in mice," *Cancer Res.* **40**, 1329–1331 (1980).
32. B. Toth and K. Patilk, "Enhancing Effect of Vitamin E on Murine Intestinal Tumorigenesis by 1,2-Dimethlylhydrazine Dihydrochloride," *J. Natl. Cancer Inst.* **70**, 531–536 (1983).
33. P. Knekt et al., "Vitamin E and Cancer Prevention," *Am. J. Clin. Nutr.* **53** (Suppl), 283s–286s (1991).
34. R. M. Bostizk et al., "Reduced Risk of Colon Cancer with High Intake of Vitamin E: The Iowa Women's Health Study," *Cancer Res.* **53**, 4230–4237 (1993).
35. H. L. Newmark "Plant Phenolics as Potential Cancer Prevention Agents," *Advances in Experimental Med. Biol.* **401**, 25–34 (1996).
36. M. Hirose et al., "Stomach Carcinogenicity of Caffeic Acid, Sesamol and Catechol in Rats and Mice," *Jpn. J. Cancer Res.* **81**, 207–212 (1990).
37. H. Esterbauer et al., "The Role of Lipid Peroxidation and Antioxidants in Oxidative Modification of LDL," *Free Radicals Biol. Med.* **13**, 341–390 (1992).
38. J. L. Mehta "Endothelium, Coronary Vasodilation, and Organic Nitrates, *Am. Heart J.* **129**, 382–391 (1995).
39. S. B. Kritchevsky et al., "Provitamin A, Carotenoid Intake and Carotid Artery Plaques: The Atherosclerosis Risk in Communities Study," *Am. J. Clin. Nutr.* **68**, 726–733 (1998).
40. A. Martin and B. Frei, "Both Intracellular and Extracellular Vitamin C Inhibits Atherogenic Modification of LDL by Human Vascular Endothelial Cells," *Arterioscler. Thromb. Vasc. Biol.* **17**, 1583–1590 (1997).
41. B. Hornig et al., "Vitamin C Improves Endothelial Function of Conduit Arteries in Patients with Chronic Heart Failure," *Circulation* **97**, 363–368 (1998).
42. M. J. Stampfer et al., "Vitamin E Consumption and the Risk to Coronary Disease in Women," *N. Engl. J. Med.* **328**, 1444–1449 (1993).
43. E. B. Rimm et al., "Vitamin E Consumption and the Risk of Coronary Heart Disease in Men," *N. Engl. J. Med.* **328**, 1450–1456 (1993).
44. H. N. Hodis et al., "Serial Coronary Angiographic Evidence That Antioxidant Vitamin Intake Reduces Progression of Coronary Artery Atherosclerosis," *JAMA* **273**, 1849–1854 (1995).
45. N. G. Stephens et al., "Randomized Controlled Trial of Vitamin E in Patients with Coronary Disease: Cambridge Heart Antioxidant Study," *Lancet* **23**, 781–786 (1996).
46. I. Jialal, "Evolving Lipoprotein Risk Factors: Lipoprotein (a) and Oxidized Low-Density Lipoprotein," *Clin. Chem.* **44**, 1827–1832 (1998).
47. B. Hennig, M. Toborek, and A. A. Cader, "Nutrition, Endothelial Cell Metabolism, and Atherosclerosis," *Crit. Rev. Food Sci. Nutr.* **34**, 253–282 (1994).
48. A. Martin et al., "Vitamin E Inhibits Low-Density Lipoprotein-Induced Adhesion of Monocytes to Human Aortic Endothelial Cells In Vitro," *Arterioscler. Thromb. Vasc. Biol.* **17**, 429–436 (1997).
49. C. Guarnieri et al., "α-Tocopherol Pretreatment Improves Endothelium-Dependent Vasodilation in Aortic Strips of Young and Aging Rats Exposed to Oxidative Stress," *Mol. Cell. Biochem.* **157**, 223–228 (1996).
50. M. G. Traber, "Determinants in Plasma Vitamin E Concentrations," *Free Radicals Biol. Med.* **16**, 229–239 (1994).
51. G. Wolf and D. Phil, "γ-Tocopherol: An Efficient Protector of Lipids Against Nitric Oxide-Initiated Peroxidative Damage," *Nutr. Rev.* **55**, 376–378 (1997).
52. J. A. Vinson et al., "Plant Flavonoids, Especially Tea Flavonals, Are Powerful Antioxidants Using an *In Vitro* Oxidation Model for Heart Disease," *J. Agric. Food Chem.* **43**, 2800–2802 (1995).
53. C. Morand et al., "Plasma Metabolites of Quercetin and Their Antioxidant Properties," *Am. J. Physiol.* **275**, R212–R219 (1998).
54. M. Serafini, A. Ghiselli, and A. Ferro-Luzzi, "Red Wine, Tea, and Antioxidants," *Eur. J. Clin. Nutr.* **50**, 28–32 (1996).
55. Y. H. He and C. Kies, "Green and Black Tea Consumption by Humans: Impact on Polyphenol Concentrations in Feces, Blood and Urine,"*Plant Foods Hum. Nutr.* **46**, 221–229 (1994).
56. G. T. McAnlis et al., "Black Tea Consumption Does Not Protect Low Density Lipoprotein from Oxidative Modification," *Eur. J. Clin. Nutr.* **52**, 202–206 (1998).
57. E. N. Frankel et al., "Inhibition of Oxidation of Human Low-Density Lipoprotein by Phenolic Substances in Red Wine," *Lancet* **341**, 454–457 (1993).
58. C. R. Pace-Asciak et al., "Wines and Grape Juices as Modulators of Platelet Aggregation in Healthy Human Subjects," *Clin. Chim. Acta* **246**, 163–182 (1996).
59. S. V. Nigdikar et al., "Consumption of Red Wine Polyphenols Reduces the Susceptibility of Low-Density Lipoproteins to Oxidation In Vivo," *Am. Soc. Clin. Nutr.* **68**, 258–265 (1998).
60. A. P. Day et al., "Effect of Concentrated Red Grape Juice Consumption on Serum Antioxidant Capacity and Low-Density Lipoprotein Oxidation,"*Ann. Nutr. Metab.* **41**, 353–357 (1997).
61. S. Kapiotis et al., "Genistein, the Dietary-Derived Angiogenesis Inhibitor, Prevents LDL Oxidation and Protects Endothelial Cells from Damage by Atherogenic LDL," **17**, 2868–2874 (1997).
62. M. S. Anthony et al., "Soybean Isoflavones Improve Cardiovascular Risk Factors Without Affecting the Reproductive System of Peripubertal Rhesus Monkeys," *J. Nutr.* **126**, 43–50 (1996).

ROSS TOMAINO
ERIC A. DECKER
University of Massachusetts
Amherst, Massachusetts

See also ANTIOXIDANTS.

PHYTOCHEMICALS: BIOTECHNOLOGY OF PHENOLIC PHYTOCHEMICALS FOR FOOD PRESERVATIVES AND FUNCTIONAL FOOD APPLICATIONS

Phytochemicals from herbs and fermented legumes are excellent dietary sources of phenolic metabolites. These phenolics are important for food preservation and increasingly for therapeutic and pharmaceutical applications. The functional food-related objectives of the food biotechnology program at the University of Massachusetts are to elucidate the molecular and physiological mechanisms associated with the synthesis of important health-related, therapeutic phenolic metabolites in food-related plants and fermented plant foods. Current efforts focus on the elucidation of the role of the proline-linked pentose phosphate pathway in regulating the synthesis of the anti-inflammatory compound rosmarinic acid (RA). Specific aims of current research efforts are (*1*) to develop novel tissue-based screening and selection techniques to isolate high RA-producing, shoot-based clonal lines from genetically heterogenous, cross-pollinating species in the family Lamiaceae and (*2*) to target genetically uniform, regenerated shoot-based clonal lines for characterization of key enzymes and genes associated with the pentose phosphate pathway and linked to phenolic synthesis. These research objectives have substantial implications for harnessing the genetic and biochemical potential of genetically heterogenous, food-related medicinal plant species. The success of this research also provides novel methods and strategies to gain access to metabolic pathways of pharmaceutically important metabolites from ginger, curcuma, chili peppers, melon, and other food-related species with novel phenolics.

PLANT PHENOLICS AS PHYTOPHARMACEUTICALS

Plant phenolics are an important group of secondary metabolites that have diverse medicinal applications. Examples of the use of specific phenolics as antioxidant and anti-inflammatory compounds are curcumin from *Curcuma longa* (1), *Curcuma mannga* (2), and *Zingiber cassumunar* (3), as well as rosmarinic acid from *Rosmarinus officinalis* (4). The use of phenolics as cancer chemopreventive metabolites has been also established with curcumin from *Curcuma longa* (1), isoflavonoids from *Glycine max* (5), and galanigin from *Origanum vulgare* (6). Other medicinal uses of plant phenolics include lithospermic acid from *Lithospermum* sp. as antigonodotropic (7), salvianolic acid from *Salvia miltiorrhiza* as antiulcer (8), thymol from *Thymus vulgaris* as anticaries (9), anethole from *Pimpinella anisum* as antifungal (10), and hellicoside from *Plantago asiatica* as antiasthmatic (8).

Species belonging to the family Lamiaceae are important sources of phenolic-type pharmaceuticals. Therapeutic use of these species can be observed in several countries. Examples are provided in Table 1.

GENETIC HETEROGENEITY IN FAMILY LAMIACEAE

The major problem in the use of phytopharmaceuticals from the family Lamiaceae is the plant-to-plant variability of specific metabolites due to the genetic heterogeneity common to all species in this family. Much of this genetic heterogeneity is the result of gynodioecy, resulting in breeding character being influenced by natural cross-pollination (15). Floral diversity and bee pollination also contribute to high cross-pollination. This gives rise to substantial variability in active ingredient levels and quality (16,17). Therefore, the biochemical characterization of pathways and genetic access to specific metabolites in all species in Lamiaceae is difficult. Each plant within a given sample extract originates from a different heterozygous seed. Also, this makes breeding of elite varieties targeting enhancement of specific metabolites very challenging. Current genetic improvements have been limited to random selection and in some cases vegetative propagation (18).

ROSMARINIC ACID AND MEDICINAL APPLICATIONS

RA (Fig. 1) is an important caffeoyl ester (phenolic depside) with proven medicinal properties and well-characterized physiological functions. RA is found in substantial quantities in several species in the family Lamiaceae with medicinal uses. *Salvia lavandulifolia* is used as choleretic, antiseptic, astringent, and hypoglycemic drug in southern Europe and contains high quantities of rosmarinic acid (19). RA-containing *Ocimum sanctum* (holy basil) is widely used to reduce fevers and against gastrointestinal disease in India. In Mexico, high RA-containing *Hyptis verticillata* is widely used by Mixtec Indians against gastrointestinal disorders and skin infections (11). In Indonesia and several other parts of the southeast Asia, RA-containing *Orthosiphon aristatus* is known for its diuretic properties and is also used against bacterial infections and inflammations of the urinary system (14). *Salvia cavaleriei*, a high RA-containing species, is used in China for treatment of dysentery, boils, and injuries (20).

PHARMACOLOGICAL FUNCTIONS OF ROSMARINIC ACID

The pharmacological effect of RA through inhibition of several complement-dependent inflammatory processes has been clearly proven (4). Therefore, it has tremendous potential as a therapeutic agent for control of complement-activation diseases (4,21). RA has been reported to have effects on both the classical C3-convertase and on the cobra-venom-factor-induced alternative convertase pathway (4). Other *in vivo* studies show that RA inhibits sev-

Table 1. Examples of Species of Lamiaceae Used as Medicine

Species	Key metabolite	Use	Reference
Hyptis verticillata	Rosmarinic acid	Gastrointestinal disorders	11
Lavandula spp.	Rosmarinic acid	Anti-inflammatory	12
Lithospermum erythrorhizon	Rosmarinic acid	Anti-inflammatory	13
	Lithospermic acid	Antigonodotropic	8
Origanum vulgare	Galangin	Antimutagen	6
Orthosiphon aristatus	Rosmarinic acid	Diuretic and anti-inflammatory	14
Rosmarinus officinalis	Rosmarinic acid	Anti-inflammatory	4
Salvia miltiorrhiza	salvianolic acid	Antiulcer	8
Thymus vulgaris	Thymol	Anticaries	9

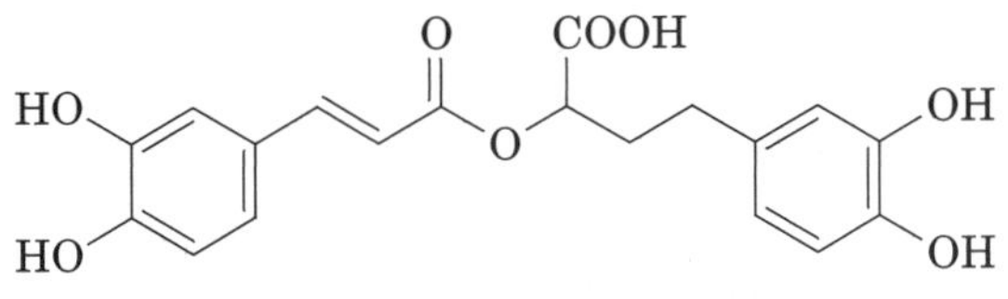

Figure 1. Structure of rosmarinic acid.

eral complement-dependent inflammatory processes, including paw edema induced by cobra venom factor and ovalbumin/antiovalbumin-mediated passive cutaneous anaphylaxis (21). It also inhibits prostacyclin synthesis induced by complement activation (22,23). It is also known to have complement-independent effects such as scavenging of oxygen free radicals (24) and inhibition of elastase. The relative safety of RA in relation to other methods of complement depletion is well documented (24). Among other actions of RA are antithyrotropic activity in tests with human thyroid membrane preparations, inhibition of complement-dependent components of endotoxin shock in rabbits, and the ability to react rapidly to viral coat proteins and so inactivate the virus (12,21). RA also inhibits Forskolin-induced activation of adenylate cyclase in cultured rat thyroid cells (25).

PRODUCTION OF ROSMARINIC ACID

RA has been targeted for production using undifferentiated cell suspension cultures of several species (12,14,26–30). The main purpose of cell suspension production of RA is the potential for large-scale production in bioreactors (31,32). Although large-scale production in bioreactors is feasible for RA, undifferentiated cell suspensions are not practical for metabolites produced in differentiated structures (eg, anethole in seeds, curcumin in rhizomes, and thymol in glandular cells of leaves). An additional disadvantage of undifferentiated callus-based suspension cultures is that the DNA is more error-prone; therefore, cell lines are genetically unstable (33). Even if problems of genetic stability and differentiation-linked metabolite production are solved, bioreactor-based production requires high initial operating costs and is not feasible in regions with poor industrial infrastructure. Gaining access to phenolic phytopharmaceuticals such as RA for all people at low cost can be best achieved by improving plant varieties for specific metabolites using modern biotechnology. Such improved, elite varieties could be incorporated into existing agricultural practices and systems. Keeping these perspectives in mind, the strategies outlined in this article are to isolate high RA-producing, genetically uniform, shoot-based clonal lines of several species belonging to the family Lamiaceae using tissue-culture techniques. Several species in this family are known to produce RA and are used in several parts of the world either as food additives, preservatives, or medicines. The high RA-producing, elite, shoot-based clonal lines will be screened based on resistance to a proline analogue and *Pseudomonas* spp. (34–36).

Selected elite, shoot-based clonal lines will be used to gain access to pathways important for biosynthesis of RA and for developing genetic transformation techniques. This will be used for subsequent engineering of RA biosynthesis using modern molecular biology techniques. Such a model for isolating genetically uniform, elite, shoot-based clonal lines from a heterogeneous and unknown genetic background can be extended for the metabolic engineering of other phytopharmaceuticals.

PATHWAYS ASSOCIATED WITH ROSMARINIC ACID BIOSYNTHESIS

It has been shown that two aromatic amino acids, phenylalanine and tyrosine, are precursors of RA biosynthesis (37). Using radioactive phenylalanine and tyrosine, it was established that they are incorporated into caffeic acid and 3,4-dihydroxyphenyllactic acid moieties, respectively (37). Steps in RA biosynthesis originating from phenylalanine and tyrosine have been characterized (Fig. 2) (13,14,30,38,39). In several cell cultures, the activity of phenylalanine-ammonia-lyase was correlated to RA (13,14). Using *Anchusa officinalis* cell suspension cultures, it was reported that tyrosine aminotransferase catalyzes the first step of the transformation of tyrosine to 3,4-dihydroxyphenyllactic acid. Several isoforms of tyrosine aminotransferase were found to be active in cell suspension cultures of *Anchusa officinalis* (13,39). Prephenate aminotransferase in *Anchusa* cell suspension cultures was found to be important; its activity was affected by 3,4-dihydroxyphenyllactic acid (40). Other enzymes of late steps in the RA biosynthesis pathway, such as hydroxyphenylpyruvate reductase and RA synthase, were isolated and characterized in cell cultures of *Coleus blumei* (41–43).

Recently, microsomal hydroxylase activities that introduce hydroxyl groups at positions 3 and 3′ of the aromatic rings of ester 4-coumaroyl-4′-hydroxyphenyllactate to give rise to RA were isolated (30). This led to the proposed complete biosynthetic pathway for RA biosynthesis originating from phenylalanine and tyrosine.

These reports point to good success in understanding RA biosynthesis from phenylalanine and tyrosine using various cell suspension cultures. However, several major issues have to be addressed to gain access and to control the interacting metabolic fluxes critical to RA biosynthesis for subsequent metabolic engineering. There are major gaps in understanding and significant questions to be answered: (*1*) What is the role of primary metabolism, particularly the pentose phosphate pathway? (*2*) How is the pentose phosphate pathway regulated during RA synthesis? (*3*) What is the role of light in regulating RA biosynthesis? (*4*) How can the problems associated with genetic instability of undifferentiated callus cultures be resolved? (*5*) How can the understanding of metabolic pathways and subsequent engineering of efficient RA biosynthesis be used to develop elite varieties for traditional and contemporary agricultural production systems?

MODEL AND RATIONALE OF CURRENT RESEARCH

Selection of High RA-Producing, Shoot-Based Clonal Lines

The focus of current research is to develop techniques to isolate genetically uniform, high RA-containing, shoot-

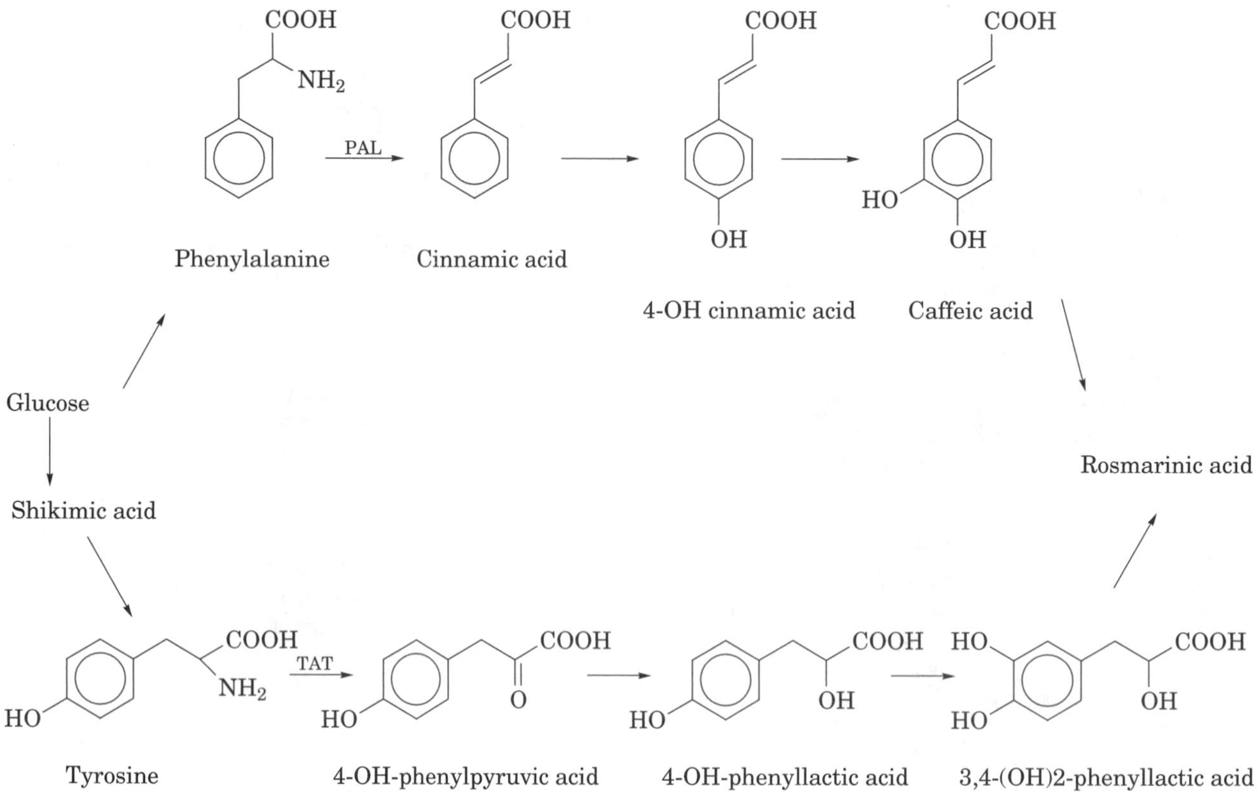

Figure 2. Rosmarinic acid biosynthesis from phenylalanine and tyrosine. PAL, phenylalanine ammonia lyase; TAT, tyrosine aminotransferase.

based clonal lines for metabolic pathway analysis and for developing gene transfer techniques for subsequent metabolic engineering. This will help provide excellent experimental systems and direction to fill gaps in knowledge of RA biosynthesis using several species used for food and medicinal applications. The basic strategy involves the isolation of genetically uniform, high RA-producing, shoot-based clonal lines from a heterogeneous genetic background. This heterogeneity is found in all species in the family Lamiaceae, and the breeding character is influenced by natural cross-pollination (15). The use of genetically uniform, shoot-based clonal lines envisioned in this article has the following advantages:

1. Shoot clones are genetically more stable than undifferentiated callus cultures.
2. Shoot clones allow the characterization of light-regulated pathways associated with RA synthesis.
3. Shoot clones can easily be targeted for large-scale greenhouse production of elite clonal lines or for incorporating into plant breeding programs to develop superior RA-producing seed varieties.
4. Shoot clones targeted for genetic engineering of metabolic pathways can be easily regenerated to whole plants for incorporation into plant-variety improvement programs.

Role of Proline-Linked Pentose Phosphate Pathway

High RA-producing, shoot-based clonal lines originating from a single heterozygous seed among a heterogeneous bulk-seed population are being selected based on tolerance to the proline analogue, azetidine-2-carboxylate (A-2-C) and a novel *Pseudomonas* sp. isolated from oregano. This strategy for selection of high RA clonal lines is based on the model that the proline-linked pentose phosphate pathway is critical for driving metabolic flux (erythrose-4-phosphate) toward shikimate and phenylpropanoid pathways (Fig. 3). Any clonal line with a deregulated proline synthesis pathway should have an overexpressed pentose phosphate pathway that allows excess metabolic flux to drive shikimate and phenylpropanoid pathways toward RA synthesis. Such proline-overexpressing clonal lines should be tolerant to A-2-C. If the metabolic flux to RA is overexpressed, it is likely to be stimulated in response to *Pseudomonas* sp. Therefore, such a clonal line is likely to be tolerant to *Pseudomonas* sp. Such a clonal line should also have high proline or proline oxidation (proline dehydrogenase) and RA content in response to A-2-C and *Pseudomonas* sp. In addition, in the presence of A-2-C or *Pseudomonas* sp., increased activity of key enzymes glucose-6-phosphate dehydrogenase (pentose phosphate pathway), pyrroline-5-carboxylate reductase (proline synthesis pathway), proline dehydrogenase (proline oxidation pathway), 3-deoxy-D-arabino heptulosonate-7-phosphate synthase (shikimate pathway), and phenylalanine ammonia-lyase (phenylpropanoid pathway) should be observed. The rationale for this model is based on the role of the pentose phosphate pathway in driving ribose-5-phosphate toward purine metabolism in cancer

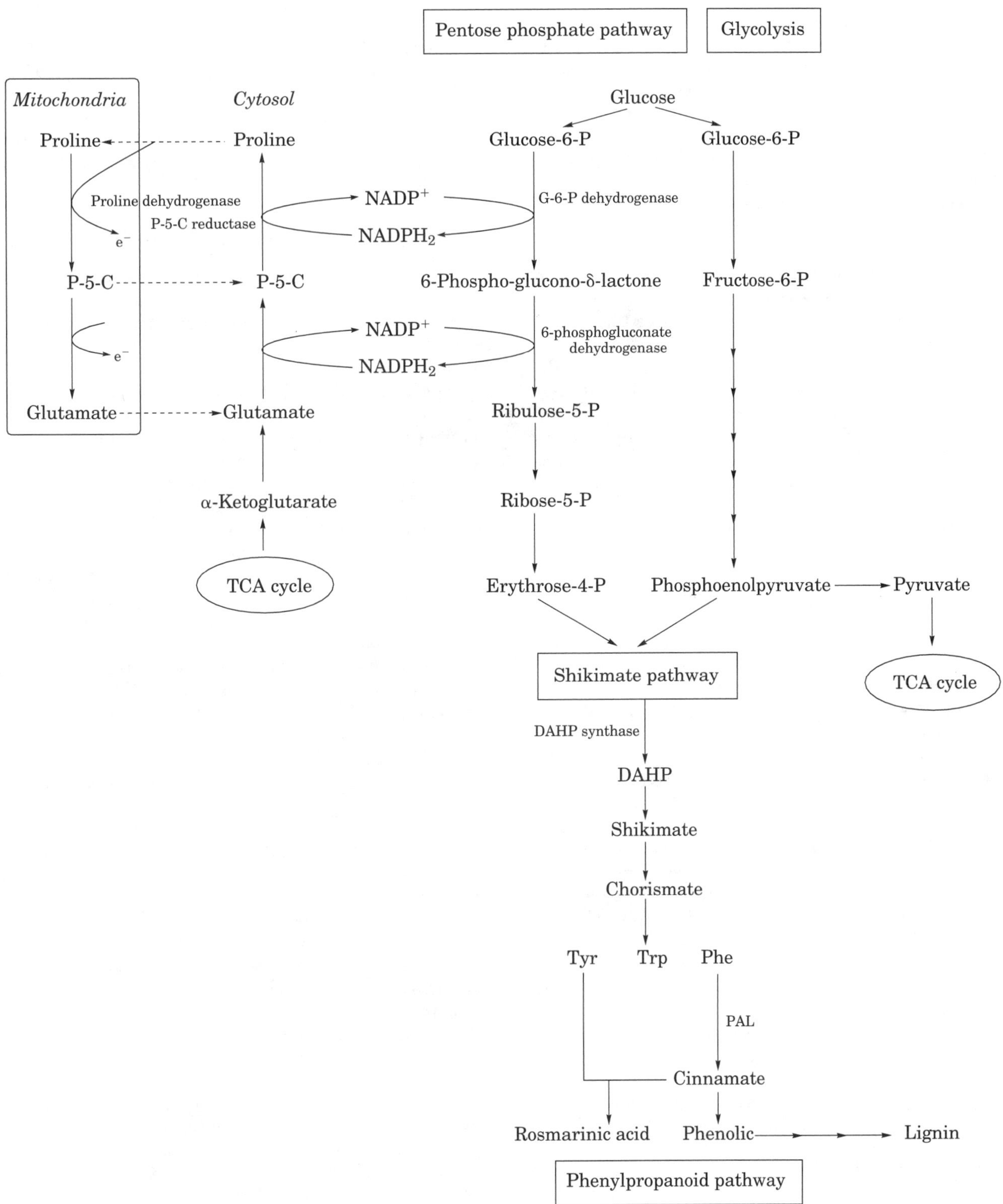

Figure 3. Proline-linked pentose phosphate pathway. TCA, tricarboxylic acid; PAL, phenylalanine ammonia lysase.

cells (44), differentiating animal tissues (45), and plant tissues (46). The hypothesis of this model is that the same metabolic flux from overexpression of the proline-linked pentose phosphate pathway regulates the interconversion of ribose-5-phosphate to the erythrose-4-phosphate-driving shikimate pathway. Shikimate pathway flux is critical for both auxin and phenylpropanoid biosynthesis, including RA. This hypothesis has been strengthened by preliminary results in which in several oregano clonal lines, RA biosynthesis was significantly stimulated by exogenous addition of A-2-C and ornithine (47). Such clonal lines are also tolerant to *Pseudomonas* spp. and respond to the bacterium by increasing RA and proline biosynthesis (34,48). High RA-producing clonal lines selected based on the model in this proposal will be targeted for preliminary characterization of key enzymes mentioned earlier. Such genetically uniform clonal lines will also be targeted for developing gene transformation techniques using *Agrobacterium* or particle gun bombardment. The success of this strategy will lead to access of critical interlinking metabolic pathways associated with RA biosynthesis. This will allow more detailed analysis that will lead to metabolic engineering for efficient RA biosynthesis. This strategy for RA biosynthesis can be the foundation for metabolic engineering of other phytopharmaceuticals from cross-pollinating, heterogeneous species.

BASIL AND FUNCTIONAL FOOD APPLICATIONS

The long-term goal of this project is to use the tools of biotechnology to develop improved clonal lines of dietary herbs and improved fermentation process for dietary legumes to generate consistent, nontoxic, and clinically relevant levels of phenolic metabolites for use as antimicrobials against chronic diseases caused by ulcer-associated *Helicobacter pylori* and urinary tract infection-associated *Escherichia coli*. Plant phenolic metabolites such as capsaicin from diet are known to be associated with low rate of ulcers through inhibition of *H. pylori* (49), and currently available synthetic drug treatments have significant side effects (50,51). Phenolics from cranberry have potential for use against urinary tract infections linked to *E. coli* (52). Use of dietary source of diverse antimicrobial-type plant phenolics could lead to reduced use of antibiotics and therefore reduce the potential increase in antibiotic-resistant, disease-causing bacteria. In addition, plant phenolic metabolites have antioxidant and anti-inflammatory properties (4) that contribute to cancer chemopreventive and immune-modulating potentials (eg, the French Paradox of reduced levels of high-fat diet-related diseases due to higher intake of phenolics from red wine and the reduced rate of breast cancer in Asian women due to high phenolics from a fermented soy-based diet). In order to overcome the problem of phytochemical inconsistency caused by genetic heterogeneity, plant-tissue-culture techniques have been developed to isolate a clonal pool of plants originating from a single seed (35). A single elite clonal line with superior phenolic profile (a combination of several bioactive phenolics) can then be selected based on tolerance to *Pseudomonas* spp. (34–36). In this project, this strategy will be used to select several elite clonal lines of basil (*Ocimum sanctum*), an herb species widely used in the diet and as medicine in several parts of the world. These elite lines (each clonal line originating from a different seed), following large-scale clonal propagation (micropropagation), will be targeted as dietary sources of phenolics (rosmarinic acid, eugenol, and methyl chavicol) to control chronic diseases associated with ulcer-linked *H. pylori* and urinary-tract-infection-linked *E. coli*. Additionally, the elite clonal lines will be screened for antioxidant activity.

THYME AND ANTIMICROBIAL APPLICATIONS IN FOOD

Therapeutic interventions for the treatment of *E. coli* 0157:H7 are limited, reinforcing the need to prevent its presence and reproduction in the food manufacturing process. The recent outbreaks of *E. coli* 0157:H7 through meat products has renewed interest in potential inhibitors of this organism (53,54). It is beyond the scope of this article to review all chemicals that are currently used for their antimicrobial activity. They include sorbates, benzoates, propionates, and sodium chloride. Others are added for color and flavor (nitrites) or for their antioxidant properties (BHA and BHT) but also possess antimicrobial activity. The food industry relies heavily on the use of synthetic antimicrobial agents to extend shelf life and preserve freshness. However, these synthetic preservatives in the food industry are coming under increasing scrutiny and reappraisal, resulting in the search for natural biochemicals from plants, including herbs and spices (55). Thus, there has been renewed interest in the antimicrobial properties of herbs and spices. In fact, about 20% of spices sold in the United States are used in the meat industry.

Most essential oils of spices and herbs are generally regarded as safe and are considered to contain the antimicrobial activity. The antimicrobial activity of plant extracts, including spices and essential oils, has been reviewed (56). Carvacrol and thymol are major volatile components of oregano, thyme, and savory. Generally, the thymol/carvacrol ratio in thyme is 10:1, whereas the carvacrol/thymol ratio in oregano is 20:1 (57). Carvacrol and thymol are known to have 1.5× and 20× the antimicrobial activity of phenol, respectively (8). In 1960, Katayama and Nagai (58) reported the antimicrobial activity of thymol and carvacrol against *Salmonella enteritidis, Staphylococcus aureus*, and *E. coli* (serotype not given). Similarly, Beuchat (59) in 1976 reported that the growth of *Vibrio parahaemolyticus* was delayed by the presence of 100 ppm of the essential oils of oregano and thyme. In 1989, Farag et al. (60) reported that of six essential oils examined (sage, rosemary, cumin, caraway, clove, and thyme), thyme oil was the most effective against three Gram-negative bacteria, with *E. coli* (serotype not given) being the most sensitive. The same workers (60) showed that thyme oil reduced the total bacterial count in butter stored at room temperature. This is the only report documenting the antibacterial effectiveness of essential oils in a food system.

The specific modes of action of plant extracts remain poorly defined, but there is some indication that they may cause a depletion of cellular energy (61). Because many of the components of essential oils, such as thymol, are similar in structure to phenolic antimicrobials, it seems reasonable that their modes of action would be similar.

Finally, environmental conditions may play a role in the effectiveness of natural antimicrobials. In particular, the antifungal properties of thymol and carvacrol were affected by pH (62). There is no similar information on its antibacterial properties, but its effectiveness in meat might be affected by such considerations. Our preliminary results demonstrating the effectiveness of thymol and carvacrol against *E. coli* in laboratory media were conducted at pH 6.5, near that of meat. This indicates that, at least with these two essential oils, pH alone would have a minimal adverse effect.

ELITE BASIL FOR POSTHARVEST PRESERVATION OF POTATOES

Preliminary studies conducted in several laboratories in the world indicate that naturally occurring plant extracts and natural compounds have demonstrated sprout inhibitory and antimicrobial properties (63–67). Several volatile monoterpenes that are primary components of essential oils inhibited sprouting when introduced as volatile into head space surrounding the potato tuber (66). Dimethylnapthalene (DMN) and diispropylnapthalene (DIPN) applied as thermal fog successfully suppressed sprouting under current storage conditions (68). DMN is reported to be a constituent of volatile released naturally by potatoes in storage (69,70).

Storage diseases of potatoes cause millions of dollars of loss each year. Fungicides are available to control some of the storage diseases, but none of the compounds currently registered for use function as sprout suppressants (63). *Helminthosporium solani*, the fungus that causes silver scurf on potatoes, caused a $8.6-million-dollar loss in 1992/1993 (71). *Fusarium* spp. (dry rot) and *Erwinia* spp. (soft rot) have become common ailments of the potato in storage, again causing loss of millions of dollars to the worldwide potato industry (72). Recent work from a few areas has demonstrated the dual advantage of natural products. The sprout suppressant properties and antimicrobial activity of ten essential oils were assessed on Spuntia (cv.) (73). Lavender (*Lavandula angustifolia*), sage (*Salvia fruticosa*), and rosemary (*Rosmarinus officinalis*) oils proved to be effective sprout inhibitors and also possesses antimicrobial activity (74). Observation during evaluation of naturally occurring aromatic compounds and monoterpenes for sprout suppression capabilities provided insight to this antimicrobial activity (66). Salicylaldehyde and 1,8-cineole-treated tubers were visibly free of storage diseases. Several *Fusarium* spp. strains tested in vivo were sensitive to DMN at ED_{50} values of 27 to 64 mL a.i L^{-1} and 78 to 124 mL a.i L^{-1} respectively (68). In the same study, salicylaldehyde suppressed growth and development of *H. solani*.

Natural products that potentially serve as sprout suppressants and disease-controlling agents will be an attractive proposition for postharvest use on potatoes. The potato industry would benefit from antimicrobial agent that could be fogged in or spray applied into a storage to disinfect exposed surfaces, including the surface layer of soil floor. Additionally, if seed potatoes could be sanitized of surface diseases such as silver scurf while in storage, the spread of inoculum in storage, to fields, and to subsequent crops could be decreased. Reduction or elimination of tuber surface defects caused by pathogens such as black scurf and silver scurf would be especially advantageous to the fresh market potato industry. The use of naturally occurring compounds for disease control would add to their acceptance by consumers. The priority in the industry today is to find a naturally occurring compound that (*1*) inhibits sprouting, (*2*) has no toxicity to humans at effective concentration, (*3*) has no adverse effect on potato quality, (*4*) requires little or no changes to existing postharvest methods, and (*5*) has additional benefits of disease suppression and additive health benefits that could be derived from the product itself.

Basil, which belongs to the family Lamiaceae, is an excellent culinary herb highly popular in gourmet food and in vegetable preparations such as pesto. The high phenolic content of basil, with chemicals such as eugenol methyl chavicol and rosmarinic acid, makes it an excellent natural preservative and a possible source of a natural disease-preventive phytochemical against chronic bacterial infections.

Development of new uses of crops such as basil through biotechnology for value-added preservation of potatoes will enhance the export potential of potatoes to the global markets. The current postharvest chemical used in potatoes, Chlorpropham™, is not permitted in Japan. Therefore, crops with preservative potentials, such as basil, are excellent targets for use of biotechnology to select elite lines with preservative qualities. The natural preservatives generated from basil will not only enhance the export potential of potatoes, but basil extracts can be potentially used for preservation of meats and seafood from oxidative deterioration (75). Basil-extract-treated food with consistent levels of specific phenolics like eugenol, methyl chavicol, and RA could also be a source of preventive medicinals against chronic infections such as ulcer-causing *H. pylori*.

BIBLIOGRAPHY

1. M. T. Huang, et al., "Inhibitory Effects of Curcumin on Tumor Promotion and Arachidonic Acid Metabolism in Mouse Epidermis," in L. Wattenberg et al., eds., *Cancer Chemoprevention*, CRC Press, Inc., Boca Raton, Fla., 1992, pp. 375–391.
2. A. Jitoe et al., "Antioxidant Activity of Tropical Ginger Extracts and Analysis of the Contained Curcuminoids," *J. Agric. Food Chem.* **40**, 1337–1340 (1992).
3. T. Masuda and A. Jitoe, "Antioxidative and Antiinflammatory Compounds from Tropical Gingers: Isolation, Structure Determination, and Activities of Cassumunins A, B and C, New Complex Curcuminoids from *Zingiber cassumunar*," *J. Agric. Food Chem.* **42**, 1850–1856 (1994).
4. P. W. Peake et al., "The Inhibitory Effect of Rosmarinic Acid on Complement Involves the C5 Convertase," *Int. J. Immunopharmacol.* **13**, 853–857 (1991).

5. A. M. Hutchins, J. L. Slavin, and J. W. Lampe, "Urinary Isoflavonoid Phytoestrogen and Lignan Excretion after Consumption of Fermented and Unfermented Soy Products," *J. Am. Dietetic Assoc.* **95**, 545–551 (1995).
6. K. Kanazawa et al., "Specific Desmutagens (Antimutagens) in Oregano against a Dietary Carcinogen, Trp-P-2, Are Galangin and quercetin," *J. Agric. Food Chem.* **43**, 404–409 (1995).
7. H. Winterhoff, H. G. Gumbinger, and H. Sourgens, "On the Antiogonadotropic Activity of *Lithospermum* and *Lycopus* Species and Some of Their Phenolic Constituents," *Planta Medica* **54**, 101–106 (1988).
8. J. B. Harbone and H. Baxter, "Phenylpropanoids," in J. B. Harbone and H. Baxter, eds., *Phytochemical Dictionary: A Handbook of Bioactive Compounds from Plants*, Taylor and Francis, London, 1993, pp. 472–488.
9. S. Guggenheim and S. Shapiro, "The Action of Thymol on Oral Bacteria," *Oral Microbiol. Immunol.* **10**, 241–246 (1995).
10. M. Himejima and I. Kubo, "Fungicidal Activity of Polygodial in Combination with Anethole and Indole against *Candida albicans*," *J. Agric. Food Chem.* **41**, 1776–1779 (1993).
11. M. Kuhnt et al., "Biological and Pharmacological Activities and Further Constituents of *Hyptis verticillata*," *Planta Medica* **61**, 227–232 (1995).
12. T. Lopez-Arnaldos et al., "Spectrophotometric Determination of Rosmarinic Acid in Plant Cell Cultures by Complexation with Fe^{2+} Ions," *Fresenius' J. Anal. Chem.* **351**, 311–314 (1995).
13. H. Mizukami and B. E. Ellis, "Rosmarinic Acid Formation and Differential Expression of Tyrosine Aminotransferase Isoforms in *Anchusa officinalis* Cell Suspension Cultures," *Plant Cell Rep.* **10**, 321–324 (1991).
14. W. Sumaryono "Induction of Rosmarinic Acid Accumulation in Cell Suspension Cultures of *Orthosiphon aristatus* after Treatment with Yeast Extract," *Phytochemistry* **30**, 3267–3271 (1991).
15. A. J. Richards, "Gynodioecy," in A. J. Richards, ed., *Plant Breeding Systems*, G. Allen and Unwin Publishers, London, 1986, pp. 89–331.
16. A. Fleisher and N. Sneer, "Oregano Species and *Origanum* Chemotypes," *J. Sci. Food Agric.* **33**, 441–446 (1982).
17. K. P. Svoboda and S. G. Deans, "A Study of the Variability of Rosemary and Sage and Their Volatile Oil on the British Market: Their Antioxidant Properties," *Flavour Fragrance J.* **7**, 81–87 (1992).
18. Y. P. S. Bajaj, M. Furmanowa, and O. Olszowska, "Biotechnology of Micropropagation of Medicinal and Aromatic Plants," in Y. P. S. Bajaj, ed., *Biotechnology in Agriculture and Forestry*, vol. 4, Springer-Verlag, Berlin, 1988, pp. 60–97.
19. S. Canigueral et al., "Phenolic Constituents of *Saliva lavandulifolia* ssp. *lavandulifolia*," *Planta Medica* **55**, 92 (1989).
20. H. J. Zhang and L. N. Li, "Salvianolic Acid I: A New Depside from *Salvia cavaleriei*," *Planta Medica* **60**, 70–72 (1994).
21. W. Engleberger et al., "Rosmarinic Acid. A New Inhibitor of Complement C3-Convertase with Anti-inflammatory Activity," *Int. J. Immunopharmacol.* **10**, 729–737 (1988).
22. H. Bult, A. G. Herrman, and M. Rampart, "Modification of Endotoxin-Induced Haemodynamic and Haemolytical Changes in Rabbit by Melthylprednisolone, $F(ab')_2$ Fragments and Rosmarinic acid," *Br. J. Pharmacol.* **84**, 317–327 (1985).
23. M. Rampart et al., "Complement-Dependent Stimulation of Prostacyclin Biosynthesis: Inhibition by Rosmarinic Acid," *Biochem. Pharmacol.* **35**, 1397–1400 (1986).
24. J. K. S. Nuytinck et al., "Inhibition of Experimentally Induced Microvascular Injury by Rosmarinic Acid," *Agents Actions* **17**, 373–374 (1985).
25. S. Kleemann and H. Winterhoff, "Rosmarinic Acid and Freeze-Dried Extract (FDE) of *Lycopus virginicus* Are Able to Inhibit Forskolin-Induced Activation of Adenylate Cyclase in Cultured Rat Thyroid Cells," *Planta Medica* **56**, 683 (1990).
26. M. H. Zenk, H. El-Shagi, and B. Ulbrich, "Production of Rosmarinic Acid by Cell Suspension Cultures of *Coleus blumei*," *Naturwissenschaften* **64**, 585–586 (1977).
27. W. De-Eknamkul and B. E. Ellis, "Rosmarinic Acid Production and Growth Characterization of *Anchusa officinalis* Cell Suspension Cultures," *Planta Medica* **50**, 346–350 (1984).
28. I. Hippolyte, et al., "Growth and Rosmarinic Acid Production in Cell Suspension Cultures of *Salvia officinalis*," *Plant Cell Rep.* **11**, 109–112 (1992).
29. H. Mizukami, Y. Tabira, and B. E. Ellis, "Methyl Jasmonate-Induced Rosmarinic Acid Biosynthesis in *Lithospermum erythrorhizon* Cell Suspension Cultures," *Plant Cell Rep.* **12**, 706–709 (1993).
30. M. Petersen "Proposed Biosynthetic Pathway for Rosmarinic Acid in Cell Cultures of *Coleus blumei* Benth," *Planta* **189**, 10–14 (1993).
31. B. Ulbrich, W. Wiesner, and H. Arens, "Large-Scale Production of Rosmarinic Acid from Plant Cell Cultures of *Coleus blumei* Benth," in K. Neumann, ed., *Primary and Secondary Metabolism of Plant Cell Cultures*, Springer-Verlag, Berlin, 1985, pp. 293–303.
32. W. Kreis and E. Reinhard, "The Production of Secondary Metabolites by Plant Cells Cultivated in Bioreactors," *Planta Medica* **55**, 409–416 (1989).
33. R. L. Phillips, S. M. Kaeppler, and J. Olhoft, "Genetic Instability of Plant Tissue Cultures: Breakdown of Normal Controls," *Proc. Natl. Acad. Sci. U.S.A.* **91**, 5222–5226 (1994).
34. K. Shetty et al., "Selection of High Phenolics-Containing Clones of Thyme (*Thymus vulgaris* L.) using *Pseudomonas* sp.," *J. Agric. Food Chem.* **40**, 3408–3411 (1996).
35. Y. Eguchi, O. F. Curtis, and K. Shetty, "Interaction of Hyperhydricity-Preventing *Pseudomonas* sp. with Oregano (*Origanum vulgare*) and Selection of High Phenolics and Rosmarinic Acid-Producing Clonal Lines," *Food Biotechnol.* **10**, 191–202 (1996).
36. R. Yang, O. F. Curtis, and K. Shetty, "Tissue-Culture-Based Selection of High Rosmarinic Acid-Producing Clonal Lines of Rosemary (*Rosmarinus officinalis*) using Hyperhydricity-Reducing *Pseudomonas*, " *Food Biotechnol.* **11**, 73–88 (1997).
37. B. E. Ellis and G. H. N. Towers, "Biogenesis of Rosmarinic Acid in *Mentha*, " *Biochem. J.* **118**, 287–291 (1970).
38. W. De-Eknamkul and B. E. Ellis, "Purification and Characterization of Tyrosine Aminotransferase Activities from *Anchusa officinalis* Cell Cultures," *Arch. Biochem. Biophys.* **257**, 430–438 (1987).
39. W. De-Eknamkul and B. E. Ellis, "Tyrosine Aminotransferase: The Entry Point Enzyme of the Tyrosine-Derived Pathway in Rosmarinic Acid Biosynthesis," *Phytochemistry* **26**, 1941–1946 (1987).
40. W. De-Eknamkul and B. E. Ellis, "Purification and Characterization of Prephenate Aminotransferase from *Anchusa officinalis* Cell Cultures," *Arch. Biochem. Biophys.* **267**, 87–94 (1988).
41. M. Petersen and A. W. Alfermann, "Two New Enzymes of Rosmarinic Acid Biosynthesis from Cell Cultures of *Coleus blumei*: Hydroxyphenylpyruvate Reductase and Rosmarinic Acid Synthase," *Z. Naturforsch.* **43c**, 501–504 (1988).

42. E. Hausler, M. Petersen, and A. W. Alfermann, "Hydroxyphenylpyruvate Reductase from Cell Suspension Cultures of *Coleus blumei Benth*," *Z. Naturforsch.* **46C**, 371–376 (1991).
43. M. S. Petersen, "Characterization of Rosmarinic Acid Synthase from Cell Cultures of *Coleus blumei*," *Phytochemistry* **30**, 2877–2881 (1991).
44. J. M. Phang, "The Regulatory Functions of Proline and Pyrroline-5-Carboxylic Acid," *Curr. Top. Cell Regul.* **25**, 91–132 (1985).
45. A. Jost et al., "Experimental Control of the Differentiation of Leydig Cells in the Rat Fetal Testis," *Proc. Natl. Acad. Sci. U.S.A.* **85**, 8094–8097 (1988).
46. D. H. Kohl et al., "Proline Metabolism in N_2-Fixing Nodules: Energy Transfer and Regulation of Purine Synthesis," *Proc. Natl. Acad. Sci. U.S.A.* **85**, 2036–2040 (1988).
47. R. Yang and K. Shetty, "Stimulation of Rosmarinic Acid in Shoot Cultures of Oregano (*Origanum vulgare*) Clonal Line in Response to Proline, Proline Analogue, and Proline Precursors," *J. Agric. Food Chem.* **46**, 2888–2893 (1998).
48. P. Perry and K. Shetty, "A Model for Involvement of Proline During *Pseudomonas*-Mediated Stimulation of Rosmarinic Acid Levels in Oregano Shoot Clones," *Food Biotechnol.* **13**, 137–154 (1999).
49. N. L. Jones, S. Shabib, and P. M. Sherman, "Capsaicin as an Inhibitor of the Growth of the Gastric Pathogen *Helicobacter pylori*," *FEMS Microbiol. Lett.* **146**, 223–227 (1997).
50. D. Y. Graham et al., "Factors Influencing the Eradication of *Helicobacter pylori* with Triple Therapy," *Gastroenterology* **102**, 493–496 (1992).
51. J. Labenz, F. Leverskus, and G. Borsch, "Omeprazole plus Amoxycillin for the Cure of *Helicobacter pylori* Infection: Factors Influencing the Treatment Success," *Scand. J. Gastroenterol.* **29**, 1070–1075 (1994).
52. A. E. Sobota, "Inhibition of Bacterial Adherence by Cranberry Juice: Potential Use for the Treatment of Urinary Tract Infections," *J. Urol.* **131**, 1013–1016 (1984).
53. M. Cohen and R. Giannella, "Haemorrhagic colitis associated with *E. coli* 0157: H7," *Adv. Intern. Med.* **37**, 173–189 (1992).
54. M. Neill, "*E. coli* 0157:H7—Current Concepts and Future Prospects," *J. Food Safety* **10**, 99–110 (1989).
55. W. H. Stroh, "New Biotechnologies Set to Impact Industrial Food Preservative Market," *Genet. Eng. News* **13**, 8 (1993).
56. D. Conner, "Naturally Occurring Compounds," in P. M. Davidson and A. Branen, eds., *Antimicrobials in Foods*, Marcel Dekker, Inc., New York, 1993, p. 441.
57. U. Salzer, "The Analysis of Essential Oils and Extracts (Oleoresins) from Seasonings—A Critical Review," *Crit. Rev. Food Sci. Technol.* **17**, 345–371 (1977).
58. T. Katayama and I. Nagai, "Chemical Significance of the Volatile Components of Spices from the Food Preservation Standpoint. IV. Structure and Antimicrobial Activity of Some Terpenes," *Nippon Suisan Gakkaishi* **26**, 29–34 (1960).
59. L. Beuchat, "Sensitivity of *Vibrio parahaemolyticus* to Spices and Organic Acids," *J. Food Sci.* **41**, 899–902 (1976).
60. R. Farag et al., "Antimicrobial Activity of Some Egyptian Spice Essential Oils," *J. Food Prot.* **52**, 665 (1989).
61. D. Conner et al., "Effects of Essential Oils and Oleoresins of Plants on Ethanol Production, Respiration and Sporulation in yeasts," *Int. J. Food Microbiol.* **1**, 63–67 (1984).
62. D. Thompson, "Influence of pH on the Fungitoxic Activity of Naturally Occurring Compounds," *J. Food Prot.* **53**, 428–431 (1990).
63. M. D. Lewis, Ph.D Thesis, University of Idaho, 1996.
64. D. F. Meigh, A. A. E. Filmer, and R. Self, "Growth-Inhibitory Volatile Aromatic Compounds Produced by *Solanum tuberosum* Tubers," Phytochemistry **12**, 987–993 (1973).
65. K. K. Shetty et al., "Studies on Alternative Methods of Sprout Suppression for Russet Burbank Potatoes" (abstract), *Am. Potato J.* **68**, 634 (1991).
66. S. F. Vaughn and G. F. Spencer, "Volatile Monoterpenes Inhibit Potato Tuber Sprouting," *Am. Potato J.* **68**, 821–831 (1991).
67. A. A. E. Filmer and M. J. C. Rhodes, "Investigation of Sprout Growth-Inhibitory Compounds in the Volatile Fraction of Potato Tubers," *Potato Res.* **28**, 361–377 (1985).
68. M. D. Lewis, M. K. Thornton, and G. E. Kleinkopf, "Commercial Application of CIPC to Storage Potatoes," CIS Newsletter, University of Idaho College of Agriculture, Moscow, Idaho.
69. J. L. Beveridge, J. Dalziel, and H. J. Duncan, "The Assessment of Some Volatile Organic Compounds as Sprout Suppressants for Ware and Seed Potatoes," *Potato Res.* **24**, 61–67 (1981).
70. J. L. Beveridge, J. Dalziel, and H. J. Duncan, "Dimethylnaphthalene as a Sprout Suppressant for Seed and Ware Potatoes," *Potato Res.* **24**, 77–88 (1981).
71. K. K. Shetty and P. Patterson, A. E. Research Series No. 93-9, University of Idaho, Moscow, Idaho, p. 11.
72. A. E. W. Boyd, "Potato Storage Diseases," *Rev. Plant Pathol.* **51**, 297–321 (1972).
73. D. Vokon, S. Vareltzidou, and P. Katinakis, "Effects of Aromatic Plants on Potato Storage: Sprout Suppression and Antimicrobial Activity," *Agric. Ecosys. Environ.* **47**, 223–235 (1993).
74. A. M. Jenssen, J. J. C. Scheffer, and A. Baerheim Svendsen, "Antimicrobial Activity of Essential Oils: A 1976–1986 Literature Review, Aspects of the Test Methods," *Planta Medica* **53**, 395–398 (1987).
75. H. L. Madsen and G. Bertelson, "Spices as Antioxidants," *Trends Food Sci. Technol.* **61**, 271–276 (1995).

KALIDAS SHETTY
University of Massachusetts
Amherst, Massachusetts

PHYTOCHEMICALS: CAROTENOIDS

Carotenoids are isoprenoids biosynthesized by plants, bacteria, molds, and yeasts. As integral components of photosynthetic reaction centers, carotenoids exist in all photosynthetic organisms and in many animals by virtue of their passage up the food chain. Of the roughly 600 structurally distinct carotenoids identified to date, only a few dozen are commonly found in human foodstuffs, and many of these circulate in plasma lipoproteins at concentrations that generally reflect dietary intake. This article discusses properties and features of carotenoids that relate to their occurrence and use in foods, including structure and reactivity, analytical methodology, nutritional and other functional properties, and the impact of food processing. Carotenoid biosynthesis, chemical synthesis, and nomenclature are adequately presented and still relevant in the 1992 edition of the *Encyclopedia* (1) and in a recent series detailing carotenoid chemistry and analysis (2–4).

STRUCTURE AND PHYSICOCHEMICAL PROPERTIES OF CAROTENOIDS

Carotenoids are oligoterpenic lipids most commonly possessing 40 carbons and an extensive conjugated double-bond system. They are classified into two main groups: the hydrocarbon carotenes and the xanthophylls, the latter of which possess one or more oxygen-containing polar functional groups. Carotenoids may be bicyclic, monocyclic, or acyclic. Several common carotenoids naturally occurring in foods or used as color additives are illustrated in Figure 1. The conjugated double-bond system is the most influential structural feature of the carotenoids, affecting both hue and chemical stability. Carotenoids with six or more conjugated double bonds absorb visible light, and absorption maxima generally increase with increasing number of double bonds. Selected physical characteristics of common carotenoids are listed in Table 1.

The two chemical reactivities of carotenoids most relevant to food systems are oxidation and isomerization. Reaction with oxygen, either autocatalytically or enzymatically, is the major cause of carotenoid degradation in foods and affects both color capacity and stability. Oxidation can disrupt conjugation and result in bleaching of endogenous or added carotenoid pigments and in formation of off-flavors (5). Oxidative degradation of carotenoids can be initiated by those factors that stimulate oxidation of unsaturated fatty acids, including light, transition metals, and lipoxygenases. Environmental factors that affect the rate of oxidation include oxygen tension, temperature, water activity, and pH (6–9). Linear (zero-order) kinetics of color loss is generally observed in dehydrated foods, whereas first-order kinetics commonly apply to aqueous systems (8,10,11), indicating that water plays a protective role in oxidative decolorization of carotenoids. Carotenoids in plant foods tend to be stabilized by complexation with proteins or polysaccharides, and processes that disrupt the microstructure of foods can result in decreased chemical stability, as discussed later.

Carotenoids are synthesized, either naturally or chemically, predominantly in the all-*trans* configuration. However, in solution, carotenoids undergo isomerization to yield a variety of *cis* isomers, and thermodynamic constants have in some cases been determined (12,13). The rate of isomerization increases with temperature and oxygen exposure. *Trans*-to-*cis* isomerization generally results in small decreases in absorption maxima, and consequently slight changes in hue.

ANALYSIS OF CAROTENOIDS

Methods of carotenoid analysis vary according the nature of the sample, but they usually involve an extraction step followed by chromatographic separation and quantification, most commonly by high-pressure liquid chromatography (HPLC). General precautions, analytical strategies, applications, and artifacts have been extensively reviewed (2). The choice of solvent(s) for extraction is determined in part by the sample matrix and may involve single solvents or solvent mixtures. Petroleum e+her is sufficient for the extraction of carotenes in oil-based products such as margarines (14). Complex matrixes such as plant and animal foods are often extracted with a mixture of solvents (14), although more recently, repeated tetrahydrofuran extraction has been successfully applied to many vegetables (15,16). In all cases, solvent purity, avoidance of oxygen and light exposure, and use of appropriate internal standards are essential. In many fruits, particularly citrus, carotenoids are present as fatty acid esters, and such samples must be saponified before chromatography since most HPLC methods assess only free carotenoids. For accurate determination of the provitamin A carotenoid content of foods, chromatographic systems that resolve provitamin A from nonprovitamin A carotenoids and the *cis* and *trans* isomers of β-carotene are required. Analytical approaches that fail to do so often overestimate the vitamin A potential of foods.

Given the large number of naturally occurring carotenoids and the lack of pure analytical standards for most, proper identification of carotenoids can be problematic. Photodiode array HPLC detectors are useful for routine structural confirmation of previously characterized carotenoids, but additional spectroscopic information is commonly required for structure elucidation of other carotenoids. Mass spectrometry, nuclear magnetic resonance, and circular dichroism are commonly employed for this purpose (3,17).

NUTRITIONAL PROPERTIES OF CAROTENOIDS

Carotenoids as Vitamin A Precursors

Although about 40 of the naturally occurring carotenoids can be metabolized to vitamin A, only 3 are quantitatively important with respect to total vitamin A intake in the United States. These are β-carotene, α-carotene, and cryptoxanthin, and of these, β-carotene is nutritionally most significant because of its prevalence in plant foods and subsequent higher frequency of consumption. Provitamin A carotenoids are estimated to account for 20–40% of total dietary vitamin A equivalents in the United States. Dietary carotenoids are freed from their food matrixes during digestion, diffuse into the intestinal mucosa with the aid of bile salt micelles, and undergo partial intracellular conversion to retinyl esters (18). Retinyl esters, along with nonmetabolized carotenoids and other lipids, are secreted in chylomicrons into the bloodstream, taken up by the liver, and either stored or complexed with retinol binding protein for transport to other tissues. The yield of vitamin A from any given food (actual vitamin A value) depends on the provitamin A carotenoid content of the food, the efficiency of digestion and uptake into the intestinal mucosa, and the extent of conversion to retinyl esters. At the present time, the true vitamin A value of specific plant foods is virtually unknown. Current values are based on the assumption that 6 mg of β-carotene consumed as food yields 1 mg retinol, or roughly 1 RDA. This equivalency, which assumes 33% absorption efficiency and 50% efficiency of conversion of β-carotene to vitamin A, is universally applied to all commodities, regardless of how they are prepared or consumed. It is generally held that oil solutions

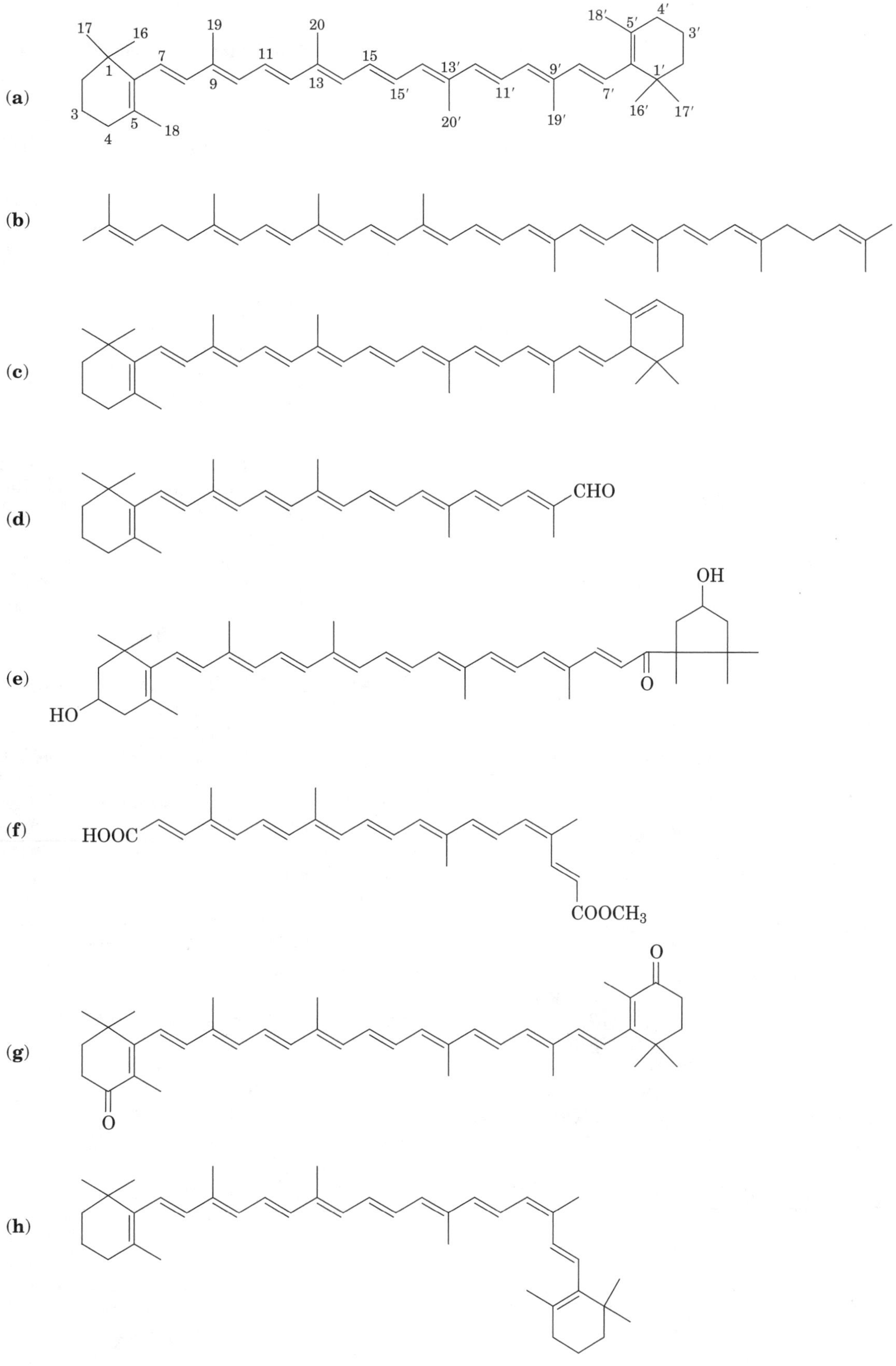

(a)

(b)

(c)

(d)

(e)

(f)

(g)

(h)

Figure 1. Chemical structure of common food carotenes and xanthophylls. (**a**) all-*trans*-β-Carotene; (**b**) lycopene; (**c**) α-carotene; (**d**) β-apo-8′-carotenal; (**e**) capsanthin; (**f**) bixin; (**g**) canthaxanthin; (**h**) 9-*cis*-β-carotene.

Table 1. Physical Characteristics of Several Common Food Carotenoids

Carotenoid	Color	λ_{max} (nm)[a]	$E^{1\%}\left(\frac{dL}{mg\ cm}\right)^a$
β-Carotene	Light yellow to violet-red	452	2560 (hexanes)
α-Carotene	Orange-red	444	2800 (pet ether)
		477	2180 (CS2)
		445	2710 (hexanes)
Lycopene	Dark-red	472	3450 (hexanes)
		470	3450 (pet ether)
		487	3370 (benzene)
Canthaxanthin	Red-orange to brown-violet	472	2110 (hexanes)
		466	2200 (pet ether)
		480	2092 (benzene)
Bixin (annatto)	Yellow-orange to dark red	456	4200 (pet ether)
Capsanthin (paprika)	Orange red to dark red	483	2072 (benzene)
β-apo-8′-carotenal	Light orange to purplish black	460	2640 (pet ether)

[a] λ_{max}, maximum wavelength; $E^{1\%}$, absorptivity.

of carotenoids are more efficiently absorbed than those of fruits or vegetables. Even in healthy, well-nourished individuals, a substantial proportion of absorbed β-carotene undergoes conversion to vitamin A, indicating that conversion is not inhibited by adequate vitamin A nutriture (18).

Physiological Effects of Carotenoids in Humans

A select few carotenoids have been investigated as potential preventative or ameliorative agents with respect to a wide variety of diseases, as recently reviewed (19). Numerous epidemiological studies have found inverse associations between dietary intake of carotenoids (or blood levels of carotenoids) and risk for various cancers, although a causal role for carotenoids cannot be ascertained from such data. Recently, lycopene intake has been linked with reduced risk of prostate cancer (20). Lycopene is a major dietary carotenoid in the United States (21), and its plasma concentration often exceeds that of β-carotene (22). Only β-carotene, and in some cases canthaxanthin, have been studied using pure supplements, owing to the lack of availability of sufficient quantities of other pure carotenoids. Clinical intervention trials with β-carotene in high-risk populations expected to develop cancer within a few years have yielded mixed results, with no evidence for a protective effect in cancers of the skin or lung. Protective effects of β-carotene have been reported with premalignancies of the oral cavity.

Carotenoid supplementation has been used successfully in the amelioration of certain inherited photosensitivity disorders, the best example of which is erythropoietic protoporphyria. Elevated skin levels of β-carotene or canthaxanthin have been shown to be effective in increasing light tolerance in such patients (19). The effect of β-carotene supplementation on various aspects of immune functions has been studied rather intensively, again with mixed results. In most cases it has not been possible to ascertain if observed effects were attributable to β-carotene per se or to its potential retinoid metabolites. Lutein and zeaxanthin are found in the macular pigment of the eye and are under investigation with respect to risk of macular degeneration.

Bioavailability of Carotenoids

Disparate estimates of carotene absorption efficiency, ranging from 1 to 99%, have been reported over the past three decades (23). Although it is generally recognized that the true vitamin A value of all plant foods is probably not the same, methods to more precisely determine the bioavailability of provitamin A carotenoids from foods in humans have been lacking. Plant foods contribute up to 90% of total dietary vitamin A equivalents in some developing nations where consumption of animal products is low, although the bioavailability of some plant foods has been questioned (24). Heat processing of plant food products may increase the bioavailability of at least some carotenoids (25), perhaps by weakening protein–carotenoid interactions, dissolution or dispersal of crystalline carotenoids, or improving digestibility of the matrix. The absorption efficiency of lycopene from heat-processed tomatoes (paste) is substantially greater than from an equivalent dose of fresh tomato (26). Attempts to compare the vitamin A yield from any plant food as a function of processing have not been reported.

OCCURRENCE AND FUNCTIONAL PROPERTIES OF CAROTENOIDS IN FOODS

More than 600 structurally distinct carotenoids have been identified in nature, of which only a few dozen commonly occur in human foodstuffs. Although the provitamin A and some nonprovitamin A carotenoids are important as nutrients, most carotenoids significantly influence food quality through the hues that they impart to fruits, vegetables, and animal foods. In most plant or animal tissues, only a small number of distinct carotenoids usually account for more than 80% of the total carotenoid content; examples of commodities with large numbers of carotenoids in significant quantities are relatively rare. Although some carotenoids, such as β-carotene, occur in many different fruits and vegetables, others, such as lycopene (tomato) and α-carotene (carrot), occur in appreciable quantities in only a few commodities.

NATURALLY OCCURRING CAROTENOIDS IN PLANT AND ANIMAL FOODS

In plant foods, carotenoids exist in various physical and morphological states depending on the carotenoid and the commodity. In leafy vegetables, carotenoids are associated primarily with chloroplasts as integral components of the photosynthetic apparatus. In fruits, roots, and tubers, carotenoids associate with chromoplasts or other intracellular structures, although the precise nature of their molecular orientation is largely uncharacterized (27). Most plant-associated carotenoids appear to exist in complexation with proteins or polysaccharides, rather than as true solutions in oil droplets. Although most common vegetable oils contain low concentrations of soluble carotenoids that impart yellow hues, a notable exception is the highly pigmented red palm oil, which is used sparingly in foods in some tropical countries.

Sources of Variation in Carotenoid Content of Plant Foods

Published data on the carotenoid content of fruits and vegetables reflect considerable variation for most species. For example, the β-carotene content (mg per 100 g fresh weight) has been reported to vary from 4 to 10 for carrot, 0.3 to 8.0 for tomato, and 0.1 to 5.2 for sweet potato (28). Variety, maturation, source, and analytical accuracy all contribute, in usually unknown proportions, to the reported variation. Recently, a carotenoid composition database was compiled by the USDA Food Composition Laboratory that contains updated analytical data for many fruits, vegetables, and processed foods available in the United States (29). The data were obtained predominantly by HPLC, in some cases with structural verification by mass spectrometry. A compilation of existing data on carotenoid content of foods available in other countries is also available (30), although much of these data were obtained by methods that do not clearly distinguish between different carotenoids or carotenoid isomers. Older analytical techniques often overestimate the provitamin A carotenoid content of foods, particularly fruits (31). Actual varietal differences in β-carotene content can be as high as 100-fold in carrots and tomato, and breeding techniques have increased the β-carotene content of sweet potatoes by threefold in the past 50 years (28). Several tomato varieties are now marketed with widely varying concentrations of lycopene and β-carotene.

Data concerning seasonal variation in carotenoid content have been reported for many plant foods, including carrot, kale, lettuce, snap beans, and turnip greens. In general, carrots harvested in summer contain higher levels of carotenoids than those harvested in winter months, whereas leafy and green vegetables exhibit the opposite effect of season (28). The effect of maturation on the carotenoid content of plant foods also varies by commodity. In general, carotenoid concentration of fruits and nonleafy vegetables increases with maturity, whereas that of leafy vegetables increases to a lesser extent or decreases (28). Postharvest losses of carotenoids in fruits and vegetables tends to be low. For example, the loss of β-carotene in fresh broccoli and green beans is very slight up to 9 days after harvest under typical refrigeration conditions (32). However, physical damage to fresh plant foods that disrupts normal compartmentalization can release lipoxygenases and increase exposure to oxygen or light, resulting in carotenoid oxidation and changes in product hue (5,33).

Carotenoids in Animal Foods

Dietary intake of carotenoids from animal foods is low compared to that from plant foods because of the relatively low rates of deposition of feed-associated carotenoids in most animal tissues. Tissue concentrations are entirely dependent on the carotenoid content of the feed. In animal foods, the carotenoids are localized predominantly in adipocytes, and thus generally accumulate in fatty tissues to a greater extent than in lean tissues. Animal food sources of carotenoids include egg yolks, chicken, certain fish such as salmon, and crustaceans. Xanthophylls tend to be the predominant carotenoids in animal foods, because their efficiency of tissue deposition is usually higher than that of the carotenes. Examples of the deliberate pigmentation of animal tissues using carotenoids include the use of yellow corn or marigold extracts in chicken feed and crab (or shrimp) shell waste or synthetic astaxanthin in shrimp and salmon feeds (34).

CAROTENOIDS AS COLORANT ADDITIVES IN FOODS

Carotenoids are useful to impart yellow, orange, and red hues to foods with no or minimal effects on flavor. Because of their high extinction coefficients, carotenoids used as colorants are generally added to foods in concentrations ranging from 1 to 25 ppm. Because carotenoids are generally stable across a range of unit processing conditions, they are often used as alternatives to the certified color additives (35). Carotenoid preparations used as colorants include natural isolates manufactured from plant products of high carotenoid content or synthetic carotenoids compounded into oil- or water-dispersible forms.

Natural Isolates

A variety of plant commodities rich in carotenoids have been used in the manufacture of powders or oil-based preparations suitable for use as colorants in foods. Intensely colored carotenoid isolates derived from paprika are available either as powders or oleoresins and impart orange or red hues to many products, including salad dressings, sauces, soups, beverages, snacks, and confections. A major carotenoid in these preparations is capsanthin (Fig. 1). Extracts of the annatto seed containing high concentrations of bixin are also commonly used in such products as cheeses, instant soups, dairy products, and desserts (36). Other natural isolates include products derived from carrot and tomato. Tomato oleoresins have recently become available and contain high concentrations of lycopene. These products are used in Europe but are not yet approved in the United States.

Synthetic Carotenoid Preparations

Three synthetic carotenoids, identical in chemical structure to their naturally occurring counterparts, are permanently listed by FDA as exempt from the certification requirements applied to the synthetic FD&C dyes. These carotenoids are β-carotene, β-apo-8′-carotenal, and canthaxanthin and are available in both oil-soluble and water-dispersible forms. The latter are prepared by emulsification into low-moisture colloidal beadlets. The carotenoids in these preparations are preserved by physical barriers to oxygen exposure and by the inclusion of hydrophilic and/or hydrophobic antioxidants such as ascorbic acid, ascorbyl palmitate, and α-tocopherol.

Physical Formulations of Synthetic Carotenoid Additives Used in Foods

β-Carotene is commercially available to food manufactures as vegetable oil dispersions of micronized crystalline β-carotene at concentrations ranging from roughly 10 to 30% by weight. Such concentrates, in which a portion of the β-carotene is in true solution, is diluted into warm oil to achieve a stock solution of desired concentration. These products are suitable for coloration of other oils, margarines, cheeses, ice cream, soups, or bakery products. Tocopherols naturally present or added to the vegetable oil carrier enhance stability. Several water-dispersible forms of β-carotene are commercially available to food manufacturers. Such products are produced as low-moisture beadlets 60 to 500 μ in diameter, using gelatin proteins, polysaccharides (eg, dextrin), or some combination thereof as the hydrophilic matrix material. Most are 1 to 10% β-carotene by weight and may contain antioxidants such as ascorbic acid, ascorbyl palmitate, or α-tocopherol. These products are used to color fruit juices, other beverages, pasta, cheeses, and baked goods. Usage levels are typically low, in the range of 2.5 to 3.5 mg per lb.

β-Apo-8′-carotenal is a synthetic monocyclic provitamin A carotenoid (Fig. 1) approved by the FDA for use in foods in 1963. This carotenoid is available as micronized dispersions or concentrated solutions in vegetable oil, stabilized with added α-tocopherol. It is commonly used in processed cheeses and colored salad dressings to impart orange hues. Canthaxanthin is a nonprovitamin A xanthophyll approved for food use in 1969, and it is used at concentrations ranging from 5 to 60 ppm, depending on the application and desired color intensity. Its primary advantage over the certified colors FD&C Red No. 2, FD&C Red No. 3, and paprika oleoresin is its ability to impart truer tomato-red hues in tomato-based products (35). Unlike FD&C Red No. 3, canthaxanthin will not precipitate at low pH. This carotenoid is available in water-dispersible beadlet forms and is used for coloring a variety of foods, including tomato products, simulated meat products, and surimi.

Regulatory Issues Related to Carotenoids as Food Colorants

β-Carotene, β-apo-8′-carotenal and canthaxanthin are pigments exempt from FDA certification. Carotenoid preparations may not be used to pigment foods for which a standard of identity exists unless the carotenoid has been authorized by the standard (37). Maximum legal use rates are 15 mg/lb for β-apo-8′-carotenal, and 30 mg/lb for canthaxanthin (21 CFR 101.22), although these levels are rarely approached owing to their high tinctorial properties. β-Carotene usage is governed by good manufacturing practices in which the use is restricted to that needed to accomplish the desired effect. However, since β-carotene is a vitamin A source, its use in some circumstances may be determined by a combination of nutritional and pigmentation effects. According to 21 CFR 101.22, carotenoid color additives may be labeled using an appropriate descriptor such as "artificial color or "color added" or may be identified specifically by name.

Carotenoids as Nutritional Additives

Fortification of foods with carotenoids for nutritional purposes is an uncommon but growing practice in the United States. Such products include fruit or vegetable-based drinks in which the added carotenoids augment that contributed by the ingredient commodities. Other foods, such as breakfast cereals or sports bars, are not usual sources of carotenoids. Because of the high tinctorial properties of carotenoids, the usage levels tend to be low in products that cannot tolerate intense yellow, orange, or red hues.

EFFECTS OF PROCESSING ON NATURALLY OCCURRING CAROTENOIDS AND CAROTENOID COLORANTS IN FOODS

As discussed earlier, carotenoids tend to be sensitive to oxygen, light, and heat. However, cooking-associated losses generally tend to be less than 30% (38), depending on circumstances and product. Heat-intensive processing of fruits and vegetables, such as commercial sterilization (canning), have in some cases been reported to result in significant carotenoid losses (38). However, such data have often not accounted for changes in moisture content, which generally increase with canning and other cooking procedures involving addition of water. Accounting for alteration in moisture content, carotenoid retention during canning is generally high. The major reactions which occur during canning are those involving isomerization rather than oxidation. Carotene losses during storage of canned fruits or vegetables at ambient temperatures are minimal, probably because of low oxygen tension and high water content. In most foods, an acid pH tends to stabilize carotenoid pigments by inhibiting oxidation and stereomutations.

Carotenoid retention during common moist cooking procedures is generally high. Because of their low water solubility, leaching losses are low, and moist heat is insufficient to induce significant isomerization. Thus, carotenoid retention in vegetables during boiling is usually 85 to 100% after taking into account changes in moisture content (38). There are relatively little data on the effect of home or commercial frying operations on carotenoid stability, but losses could be expected to be higher than during other unit thermal processes because of the combination of high temperature and oxygen exposure. Losses of carotenoids with freeze drying are generally low to moderate, ranging from 4 to 13% for freeze-dried orange juice and carrots, respectively, whereas tray-drying followed by explosion puffing can result in losses of up to 25 percent (39). Carotenoids commonly exhibit differential stabilities. During both processing and dark storage of paprika, β-carotene is lost more rapidly than capsanthin, with losses variable across variety of pepper (40).

Carotenoid Isomerization During Thermal Processing

Thermal processing conditions can result in isomerization of one or more *trans* double bonds to the *cis* configuration, and carotenes appear to undergo isomerization more readily than xanthophylls (5). Newer analytical methods now permit more precise quantification of *cis* carotenoid isomers in foods. Typical canning procedures can result in isomerization of 10 to 40% of carotenes in some products, including carrot, sweet potato, broccoli, and green leafy vegetables (41). Isomerization results in slight decreases in extinction coefficient and absorption maxima, but these generally have less impact on hue than reactions involving oxidation. Nutritionally, the vitamin A value of *cis* isomers of the provitamin A carotenoids is substantially less than that of their all-*trans* counterparts.

BIBLIOGRAPHY

1. K. L. Simpson, "Carotenoids," in *Encyclopedia of Food Science and Technology*, Y. H. Hui, ed., John Wiley and Sons, Inc., vol. 1A-D, pp. 293–310 (1992).
2. G. Britton, S. Liaaen-Jensen, and H. Pfander, *Carotenoids, Volume 1A: Isolation and Analysis*, Birkhauser Verlag, Basel, 1995.
3. G. Britton, S. Liaaen-Jensen, and H. Pfander, *Carotenoids, Volume 1B:Spectroscopy*, Birkhauser Verlag, Basel, 1995.
4. G. Britton, S. Liaaen-Jensen, and H. Pfander, *Carotenoids, Volume 2: Synthesis*, Birkhauser Verlag, Basel, 1995.
5. J. Davidek, J. Velisek, and J. Pokorny, *Chemical Changes During Food Processing*, Elsevier, Amsterdam, 1990.
6. H. T. Gordon and J. C. Bauernfeind, "Carotenoids as Food Colorants," *Crit. Rev. Food Sci. Nutr.* **18**, 59–96 (1983).
7. J. M. Dietz and W. A. Gould, "Effects of Process Stage and Storage on Retention of Beta-Carotene in Tomato Juice," *J. Food Sci.* **51**, 847–848 (1986).
8. M. I. Minguez-Mosquera and M. Jaren-Galan, "Kinetics of the Decolouring of Carotenoid Pigments," *J. Sci. Food Agric.* **67**, 153–161 (1995).
9. B. H. Chen, H. Y. Peng, and H. E. Chen, "Changes of Carotenoids, Color, and Vitamin A Contents during Processing Carrot Juice," *J. Agric. Food Chem.* **43**, 1912–1918 (1995).
10. L. A. Wagner and J. J. Warthesen, "Stability of Spray-Dried Encapsulated Carrot Carotenes," *J. Food Sci.* **60**, 1048–1053 (1995).
11. M. Goldman, B. Horev, and I. Saguy, "Decolorization of β-Carotene in Model Systems Simulating Dehydrated Foods. Mechanism and Kinetic Principles," *J. Food Sci.* **48**, 751–754 (1983).
12. W. E. Doering, D. Sotiriou-Leventis, and W. R. Roth, "Thermal Interconversions among 15-*cis*, 13-*cis* and all-*trans*-β-Carotene: Kinetics, Arrhenius Parameter, Thermochemistry, and Potential Relevance to Anticarcinogenicity of all-trans-β-Carotene," *J. Am. Chem. Soc.* **117**, 2747–2757 (1995).
13. M. Kuki, Y. Koyana, and H. Nagae, "Triplet-Sensitized and Thermal Isomerization of All-trans, 7-cis, 9-cis, 13-cis and 15-cis Isomers of β-Carotene: Configurational Dependence of the Quantum Yield of isomerization via the T1 State," *J. Phys. Chem.* **95**, 7171–7180 (1991).
14. E. De Ritter and A. E. Purcell, "Carotenoid Analytical Methods," in J. C. Bauernfeind, ed., *Carotenoids as Colorants and Vitamin A Precursers. Technological and Nutritional Applications*, Academic Press, Orlando, Fla., 1981, pp. 815–882.
15. F. Khackik, G. R. Beecher, N. F. Whittaker. "Separation, Identification, and Quantification of the Major Carotenoid and Chlorophyll Constituents in Extracts of Several Green Vegetables by Liquid Chromatography," *J. Agric. Food Chem.* **34**, 603–6 (1986).
16. F. Khachik and G. R. Beecher. "Application of a C-45-β-Carotene as an Internal Standard for the Quantification of Carotenoids in Yellow/Orange Vegetables by Liquid Chromatography," *J. Agric. Food Chem.* **35**, 732–738 (1987).
17. K. Schiedt and S. Liaaen-Jensen, "Isolation and Analysis," in G. Brotton, S. Liaaen-Jensen, and H. Pfander, eds., *Carotenoids, Volume 1A: Isolation and Analysis*, Birkhauser Verlag, Basel, 1995, pp. 81–83.
18. R. S. Parker, "Absorption, Metabolism and Transport of Carotenoids," *FASEB J.* **10**, 542–551 (1996).
19. J. E. Swanson and R. S. Parker, "Biological Effects of Carotenoids in Humans," in E. Cadenas and L. Packer, eds., *Handbook of Antioxidants*, Marcel Dekker, Inc., New York, NY, 1996.
20. E. Giovannucci, "Tomatoes, Tomato-Based Products, Lycopene, and Cancer: Review of the Epidemiologic Literature," *J. Natl. Cancer Inst.* **91**, 317–331 (1999).
21. L. C. Nebeling et al., "Changes in Carotenoid Intake in the United States: The 1987 and 1992 National Health Interview Surveys," *J. Am. Dietetic Assoc.* **9**, 991–996 (1997).
22. M. L. Nguyen and S. J. Schwartz, "Lycopene: Chemical and Biological Properties," *Food Technol.* **53**, 38–45 (1999).
23. S. de Pee and C. West, "Dietary Carotenoids and Their Role in Combating Vitamin A Deficiency: A Review of the Literature," *Eur. J. Clin. Nutr.* **50** (Suppl 3), S38–S53 (1996).
24. S. De Pee et al., "Lack of Improvement in Vitamin A Status with Increased Consumption of Dark-Green Leafy Vegetables," *Lancet* **346**, 75–81 (1995).
25. C. L. Rock et al., "Bioavailability of β-Carotene is Lower in Raw Than in Processed Carrots and Spinach in Women," *J. Nutr.* **128**, 913–916 (1998).
26. C. Gartner, W. Stahl, and H. Sies, "Lycopene is More Bioavailable from Tomato Paste than from Fresh Tomatoes," *Am. J. Clin. Nutr.* **66**, 116–122 (1997).
27. J. E. Bryant et al., "Isolation and Partial Characterization of α- and β-Carotene-Containing Carotenoprotein from Carrot (*Daucus carota* L.) Root Chromoplasts," *J. Agric. Food Chem.* **40**, 545–549 (1992).
28. A. Mozafar, *Plant Vitamins: Agronomic, Physiological and Nutritional Aspects*, CRC Press, Boca Raton, Fla., 1994.
29. A. R. Mangels et al., "Carotenoid Content of Fruits and Vegetables: An Evaluation of Analytic Data," *J. Am. Dietetic Assoc.* **93**, 284–296 (1993).
30. C. E. West and E. J. Poortvlilet, *The Carotenoid Content of Foods with Special Reference to Developing Countries*, Office of Nutrition, Bureau for Research and Development, U.S. Agency for International Development, Washington, D.C., 1993.
31. S. L. Booth, T. Johns, and H. V. Kuhnlein, "Natural Food Sources of Vitamin A and Provitamin A," *Food Nutr. Bull.* **14**, 6–19 (1992).
32. Y. Wu, A. K. Perry, and B. P. Klein, "Vitamin C and β-Carotene in Fresh and Frozen Green Beans and Broccoli in a Simulated System," *J. Food Quality* **15**, 87 (1992).
33. C. A. Pesek and J. J. Warthesen, "Photodegradation of Carotenoids in a Vegetable Juice System," *J. Food Sci.* **52**, 744–746 (1987).
34. J. C. Bauernfeind, *Carotenoids as Colorants and Vitamin A Precursors. Technological and Nutritional Applications*, Academic Press, Orlando, Fla., 1981.
35. J. D. Dziezak, "Applications of Food Colorants," *Food Technol.* **41**, 78–88 (1987).

36. D. E. Pszczola, "Natural Colors: Pigments of Imagination," *Food Technol.* **52**, 70–76 (1998).
37. FDA, Title 21.73.95, April, Food and Drug Administration, Washington, D.C., 1986.
38. E. Karmas and R. S. Harris, *Nutritional Evaluation of Food Processing*, 3rd ed., AVI Books, Van Nostrand Reihold Co., New York, 1988.
39. P. M. Bluestein and T. P. Labuza, "Effects of Moisture Removal on Nutrients," in E. Karmas and R. S. Harris, eds., *Nutritional Evaluation of Food Processing*, 3rd ed., AVI Books, Van Nostrand Reihold Co., New York, 1988, pp. 393–422.
40. M. J. Minguez-Mosquera, M. Jaren-Galan, and J. Garrido-Fernandez, "Effect of Processing of Paprika on the Main Carotenes and Esterified Xanthophylls Present in the Fresh Fruit," *J. Agric. Food Chem.* **41**, 2120–2124 (1993).
41. W. J. Lessin, G. L. Catigani, and S. J. Schwartz, "Quantification of *cis-trans* Isomers of Provitamin A Carotenoids in Fresh and Processed Fruits and Vegetables," *J. Agric. Food Chem.* **45**, 3728–3732 (1997).

ROBERT PARKER
Cornell University
Ithaca, New York

See also CAROTENOID PIGMENTS; COLORANTS: CAROTENOIDS.

PHYTOCHEMICALS: LIPOIC ACID

α-Lipoic acid is a naturally occurring compound present as a cofactor in a number of mitochondrial enzymes that are involved in metabolism and energy production. In its free form, lipoic acid is a powerful antioxidant, functioning as a free-radical scavenger and an antioxidant "protector." Recently, attention has focused on lipoic acid as a cellular redox regulator, having the capacity to interact at various stages in signal transduction pathways. Such properties make this compound of potential therapeutic importance in conditions in which oxidative stress is involved. These properties and their implications are discussed in the article.

NATURAL SOURCES OF LIPOIC ACID

Lipoic acid is an eight-carbon compound containing two sulfur atoms in a (dithiolane) ring structure. The naturally occurring form is present bound to a lysine residue in the four α-keto-acid dehydrogenase complexes of the mitochondria, which catalyze the oxidative decarboxylation of pyruvate, α-ketoglutarate, and branched-chain α-keto acids, respectively. These complexes all share similarities in both structure and function (1). The role of lipoic acid is to transfer an acyl group from the decarboxylated acid to coenzyme A. During this process, the lipoyl moiety is reduced and then reoxidized by the enzyme lipoamide dehydrogenase in order to keep the catalytic cycle in motion. The redox properties of lipoic acid make it ideal for such a function. As such, lipoic acid plays a crucial role in metabolism.

Lipoic acid was first isolated in 1951 by Reed and colleagues (2) and was tentatively described as a vitamin until it was later discovered to be synthesized in both plant and animal tissues. Recently, the levels of naturally occurring lipoic acid (lipoyllysine) in both plant and animal tissues have been determined (3). In animal tissues, the content was found to be tissue-specific, the highest levels being found in the kidney, heart, and muscular tissues (2.6, ~1, and ~1 μg/g dry weight, respectively), while lower amounts were found in the brain and lung (0.2 μg/g dry weight). Since natural source lipoic acid is present in mitochondrial enzymes, it is highly likely that lipoic acid content will correlate with metabolic activity of the tissue. In plant tissues, the highest concentrations were found in green tissues, especially spinach (3.6 μg/g dry weight). Plants are unique in that they possess a chloroplastic form of lipoic acid; therefore, lipoic acid content, dependent on the type of tissue, will be determined by the density of mitochondria and chloroplasts. Such low levels of lipoic acid (~0.1% total weight of a mitochondrion) express the need for supplementation if the antioxidant properties of lipoic acid are to be gained. The antioxidant properties of lipoic acid can rarely be achieved through a normal diet.

BIOAVAILABILITY

When taken orally as a supplement or administered *in vitro*, α-lipoic acid is rapidly absorbed and taken up into cells where it is reduced to the more active form, dihydrolipoic acid. The uptake of exogenously supplied α-lipoic acid has been studied in the perfused rat liver and isolated hepatocytes (4). Two different transport mechanisms were reported: carrier-mediated uptake, which is prominent below 75 μM, and passive diffusion, which is prominent at higher concentrations. This study also showed that the carrier-mediated uptake can be blocked by medium-chain fatty acids, suggesting that the same translocator was in use (4). In vitro studies have investigated the various cellular reduction pathways of lipoic acid (5). Two mechanisms exist that are both tissue- and stereo-specific. The natural R enantiomer is preferentially reduced in the mitochondria by lipoamide dehydrogenase, a nicotinamide adenine nucleotide (NADH) dependent enzyme. The S form is predominantly reduced in the cytosol by glutathione and/or thioredoxin reductase, which are NADPH dependent. The various contributions of each pathway are dependent upon the tissue: heart muscle almost entirely reduces the R form, while in the liver there is an equal distribution between the two pathways.

Reduction contributes one mechanism of lipoate metabolism. Other routes have been demonstrated using radiolabeled lipoic acid administered either orally (6) or via injection (7) into the rat. Several components were identified, including the short-chain homologues bisnorlipoic and tetranorlipoic acids (7). Interestingly, no evidence for the oxidation of the dithiolane ring was observed. Administration of [1,6-^{14}C] lipoic acid has shown that lipoic acid is rapidly absorbed in the gut and passed to various tissues for catabolism (6). The localization of administered lipoate was greatest in the liver, other intestinal organs, and skeletal muscle (6). The presence of both oxidized and reduced forms has also been demonstrated in various tissues of vitamin-E-deficient mice after lipoic acid supplementation (8).

ANTIOXIDANT PROPERTIES OF LIPOIC ACID

The strained-ring conformation of lipoic acid enables the molecule to readily undergo oxidation and reduction, and indeed this is reflected in the high negative redox potential of the oxidized and reduced form (-0.32 V), which makes the reduced form (dihydrolipoate) a very powerful reductant. Such properties enable the lipoate couple to interact with a number of reactive oxygen species.

Various cellular pathways like those involved in the electron transport pathways of cell respiration and drug detoxification or activation of neutrophils and macrophages result in the formation of reactive oxygen and nitrogen species. If overproduced and not scavenged or converted to a nonreactive form, they can be potentially damaging to the cell. These include superoxide ($O_2^{\cdot -}$), hydrogen peroxide (H_2O_2), hyperchlorous acid (HOCl), hydroxyl ($OH^{\cdot}$) and peroxyl ($ROO^{\cdot}$) radicals as well as nitric oxide ($NO^{\cdot}$) and peroxynitrite(ONOO). α-Lipoic acid and dihydrolipoic acid can scavenge both hydrogen peroxide and HOCl (9), while dihydrolipoic acid can also scavenge superoxide (9). Both α-lipoic and dihydrolipoic acid have been shown to scavenge hydroxyl radicals in a metal-catalysis system and in a metal-free reaction of ultraviolet irradiation induced by decomposition of the aromatic hydroperoxide model compound NP-III (9). Dihydrolipoic acid can also scavenge peroxyl radicals, formed from both lipophilic and hydrophilic peroxyl radical generators (9). Therefore, the lipoate couple represents a potent radical scavenging unit.

When antioxidants react with reactive oxygen species, the antioxidant is converted to a form that is no longer able to function and is said to be consumed. Therefore, this oxidized product needs to be recycled to this native form in order to function again. A number of antioxidants, including vitamin C, ubiquinols, and glutathione, can recycle vitamin E, the major chain-breaking antioxidant that protects biological membranes from lipid peroxidation. Dihydrolipoic acid has only a weak interaction with the tocopheroxyl radical, so the major recycling of vitamin E by dihydrolipoic acid occurs via the intermediary recycling of the ascorbyl radical by dihydrolipoic acid, which, in turn, recycles the vitamin E radical produced by oxidation. There is now evidence that lipoate supplementation increases tissue ubiquinol content, and ubiquinol can also recycle vitamin E. Therefore, there exists a network of antioxidants in which dihydrolipoic acid can interact and replenish in order to maintain both lipid and aqueous phase antioxidant status. This recycling by lipoate was shown as early as 1959 by Rosenberg and Culik (10), who demonstrated that lipoic acid supplementation could prevent the symptoms of both vitamin E and C deficiencies in guinea pigs. This work has since been repeated in vitamin-E-deficient mice (8).

Glutathione is a major intracellular antioxidant that acts as a sulfhydryl buffer, protecting cysteine residues in proteins from oxidation, the modulation of which has been discussed as a potential therapeutic strategy (11). α-Lipoate can interact with glutathione (GSH) both directly and indirectly. Dihydrolipoic acid (DHLA) can reduce oxidized glutathione (GSSG) to GSH, but GSH is incapable of reducing α-lipoate to DHLA (12). DHLA may also recycle vitamin E by reducing GSSG directly; the reduced GSH can then recycle vitamin E (13). Administration of α-lipoate to cells has been shown to cause an increase in intracellular GSH *in vitro* (14) and *in vivo* (15). This increase has been shown to be a consequence of increased GSH synthesis due to an improvement in cystine utilization (14).

SIGNAL TRANSDUCTION

Recently, attention has been focused on the role of oxidants and antioxidants in the regulation of intracellular signal transduction processes. Redox changes in cells trigger molecular responses, one such response being the activation of the transcription factor NF-κB. Most agents activating NF-κB tend to trigger the formation of reactive oxygen species or are oxidants themselves (eg, $O_2^{\cdot -}$, H_2O_2, or lipoxygenase products) (16). α-Lipoate can inhibit NF-κB activation induced by phorbol ester or tumor necrosis factor-α (TNF-α) in Jurkat T cells in a dose-dependent fashion (17), and both enantiomers of lipoate were found to be effective. It has recently been shown that the ability of α-lipoate to inhibit NF-κB was not dependent on increased intracellular glutathione levels (18). Following the activation of NF-κB, there is translocation into the nucleus and binding to specific areas of DNA, which elicits the response. α-Lipoate has been shown to inhibit the DNA binding of NF-κB (19); however, DHLA was found to enhance binding. Also, the inhibition of DNA binding by diamide can be overcome by the addition of DHLA. These observations suggest that two modes of redox regulation exist in cell signaling for NF-κB (20); the first, a requirement of oxidative processes in activating NF-κB; the second, a requirement for reductive processes in DNA binding.

THERAPEUTIC IMPLICATIONS

The chronic and degenerative diseases associated with aging that have been examined in detail have shown the involvement of free radicals or oxidants in their pathophysiology. Therefore, antioxidants may be important in slowing, and perhaps in the future will be of benefit in their treatment. All organ systems appear to be involved, from the eye, which is sensitive to photo-oxidative free-radical mediated reactions, to neurodegeneration, to initial stages of cancer, to cardiovascular diseases, and to viral activation, just to name a few. Redox and antioxidant regulation of physiological and pathophysiological processes are therefore assuming greater importance. These include redox-based antioxidants such as vitamin C and E; bioflavonoids and thiol antioxidants such as glutathione; and thioredoxin and α-lipoic acid, which are major substances that interact in the antioxidant network.

The properties of α-lipoate lead to the possibility that it may influence intracellular function via antioxidant actions and through affecting the redox status of thiol-containing proteins (9). Indeed, α-lipoic acid can be thought of as a metabolic antioxidant since it is a naturally occurring substance reduced by several cellular enzymatic systems. Beneficial effects of lipoic acid administration have been reported in diabetic complications (22,23), having been used in the treatment of diabetic neuropathy in Germany; ischemia-reperfusion injury (24); and liver disease (23). Furthermore, lipoic acid is be a good candidate for treatment in AIDS, neurodegenerative diseases (21), or heavy metal poisoning (9).

BIBLIOGRAPHY

1. M. S. Patel and T. E. Roche, *FASEB J.* **4**, 3224–3233 (1990).
2. L. J. Reed et al., *Science* **114**, 93–94 (1951).
3. J. K. Lodge et al., *J. Appl. Nutr.* **49**, 3–11 (1997).
4. J. Peinado, H. Sies, and T. P. M. Akerboom, *Arch. Biochem. Biophys.* **273**, 389–395 (1989).
5. N. Haramaki et al., *Free Radic. Biol. Med.* **22**, 535–542 (1997).
6. E. H. Haramaki and D. B. McCormick, *Arch. Biochem. Biophys.* **160**, 514–522 (1974).
7. J. T. Spence and D. B. McCormick, *Arch. Biochem. Biophys.* **174**, 13–19 (1976).
8. M. Podda et al., *Biochem. Biophys. Res. Commun.* **204**, 98–104 (1994).
9. L. Packer, E. H. Witt, and H. J. Tritschler, *Free Radic. Biol. Med.* **19**, 227–250 (1995).
10. H. R. Rosenberg and R. Culik, *Arch. Biochem. Biophys.* **80**, 86–93 (1959).
11. A. Meister, in D. Dolphin, R. Poulson, and O. Avramovic, eds., *Glutathione: Chemical, Biochemical, and Medical Aspects*, New York, John Wiley & Sons, 1989, pp. 367–373.
12. P. C. Jocelyn, *Eur. J. Biochem.* **2**, 327–331 (1967).
13. A. Bast and G. R. M. M. Haenen, *Biochim. Biophys. Acta* **963**, 558–561 (1988).
14. D. Han et al., *Biofactors* **9**, 1–18 (1997).
15. E. Busse et al., *Arzneim. Forsch.* **43**, 829–831 (1992).
16. P. A. Baeuerle and T. Henkel, *Annu. Rev. Immunol.* **12**, 141–179 (1994).
17. Y. J. Suzuki, B. B. Aggarwal, and L. Packer, *Biochem. Biophys. Res. Commun.* **189**, 1709–1715 (1992).
18. C. K. Sen and L. Packer, *FASEB J.* **10**, 709–720 (1996).
19. Y. J. Suzuki et al., *Biochem. Mol. Biol. Int.* **36**, 241–246 (1995).
20. L. Packer, S. Roy, and C. K. Sen, *Adv. Pharmacol.* **38**, 79–101 (1996).
21. L. Packer, H. J. Tritschler, and K. Wessel, *Free Radic. Biol. Med.* **22**, 359–378 (1997).
22. S. Jacob et al., *Exp. Clin. Endocrinol.* **104**, 284–288 (1996).
23. J. Bustamante et al., *Free Radic. Biol. Med.* **24**, 1023–1039 (1998).
24. M. Panigrahi *Brain Res.* **717**, 184–188 (1996).

JOHN K. LODGE
LESTER PACKER
University of California
Berkeley, California

PHYTOCHEMICALS: MELATONIN

N-acetyl-5-methoxytryptamine (melatonin) was first isolated from the pineal gland, a small organ in the brain of the mammal, 40 years ago. Since that time it has been characterized as the chief secretory product of the pineal gland of vertebrates (1) and linked to a variety of functions, including the regulation of seasonal reproduction, immune function stimulation, tumor inhibition, sleep induction, circadian rhythm regulation, and antioxidant function. In all vertebrates and most invertebrates where melatonin synthesis has been assessed, it is primarily produced in the pineal gland during the dark phase of the light/dark cycle. Because melatonin is normally released into the blood, circulating levels of melatonin are always higher at night than during the day.

Long before melatonin was found to exist in plants, the indoleamine was observed to have effects in plant cells similar to those in animals. Thus, in 1969 Jackson reported that melatonin increased the birefringence of the mitotic spindles in endosperm cells of *Haemanthus batherinae* just as it did in melanocytes obtained from a frog (2). Likewise, soon thereafter melatonin was reported to have similar cytoskeletal effects in onion root tips (3). Such cytoskeletal actions of melatonin may explain the effects of the indoleamine on the mitotic activity of cells from a variety of species.

IDENTIFICATION OF MELATONIN IN PLANTS

Melatonin was first detected and quantified in the dinoflagellate, *Gonyaulax polyedra* (4,5). In this species, melatonin was found to be at concentrations similar to those measured in the pineal gland of vertebrates. In addition to its obvious presence in this autotrophic organism, melatonin also exhibited a rhythm in these unicells like that observed in the pineal gland. Thus, melatonin levels measured either by high performance liquid chromatography or gas chromatography–mass spectrometry were higher at night than during the day. This finding stimulated research directed at identifying melatonin in higher plants, and in recent years, this indoleamine has been found in a remarkably large number of taxa (Table 1). Among angiosperms, melatonin has been identified in more than 30 species belonging to 19 different families, in both mono- and dicotyledons. Although melatonin has been reported in a rather small number of species, their diversity and the fact that almost all plants analyzed have been shown to contain melatonin suggests that this indoleamine may be ubiquitous in the plant kingdom. Likewise, within plant tissues, melatonin appears to be widespread, because it has been found in leaves, stems, roots, fruit, and seeds.

The concentration of melatonin in the plants studied varies widely, with the indoleamine being below detectable levels in potato tubers and very low in beet roots (2 pg/g weight) (12) to more than 5 ng/g weight in tall fescue (8). Even higher concentrations have been reported for some medicinal plants (Table 2). It is possible that some of the measurements actually underestimated the melatonin levels inasmuch as melatonin could easily have been de-

Table 1. Summary of the Species of Algae and Green Plants in Which Melatonin (*N*-acetyl-5-methoxytryptamine) Has Been Identified

Taxon	Species	Ref.
	Dinophyta	
	Gonyaulax polyedra	4,5
	Alexandruim lusitanicum	6
	Amphidinium carterae	6
	Phaeophyta	
	Pterygophora californica	7
	Chlorophyta angiospermae	
Rosaceae	*Malus domestica*	
	Fragaria magna	8
Brassicaceae	*Brassica oleracea*	
	Brassica compestris	
	Raphanus sativus	8
Apiaceae	*Angelica keiskej*	8
Actinidiaceae	*Actinidia chinensis*	8
Basellaceae	*Basella alba*	8
Chenopodiaceae	*Chenopoduim rubrum*	
	Beta vulgaris	9,10
Convolvulaceae	*Pharbitis nil*	11
Solanaceae	*Lycopersicon esculentum*	
	Lycospericon pimpinellifolium	
	Nicotiana tabacum	8,12
Cucurbitaceae	*Cucumis sativus*	8,12
Asteraceae	*Chrepanthemum coronarium*	
	Petasites japonicus	8
Liliaceae	*Allium cepa*	
	Allium fistulosum	
	Asparagus officinalis	8
Bromeliaceae	*Ananas comosus*	8
Poaceae	*Oryza sativa japonica*	8
	Hordeum vulgare	8
	Avena sativa	8
	Festuca arundinacea	8
	Zea mays	8
Zingiberaceae	*Zingiber officinale*	8
Musaceae	*Musa paradisiaca*	12
Araceae	*Colocasia esculenta*	8
Ateraceae	*Tanacetum parthenium*	13
Hypericaceae	*Hypericum perforatum*	13
Lamiaceae	*Scutellaria biacalensis*	13

Note: Melatonin was identified using a variety of techniques, including high-performance liquid chromatography, gas chromatography–mass spectroscopy, and radioimmunoassay.

graded during the extraction procedures used in these studies (14).

Among the normally consumed plant products that have been investigated to date, oats, rice, and barley products contain the largest amounts of measurable melatonin (8). When diets consisting of these plants were fed to chicks, levels of melatonin significantly increased in the circulation (Fig. 1) of the birds, and the authors of Reference 8 further showed that the ingested melatonin was bound to melatonin receptors in the animals. Clearly, the implication of these findings is that melatonin consumed in plant products is absorbed, enters the circulation, and could have physiological effects via specific receptors. Additionally, as will be discussed later, some of the melatonin's actions may be independent of receptors or binding sites so any ingested indoleamine may have a variety of actions, not only those that depend on its binding to receptors.

Table 2. Reported Melatonin Concentration in Commercially Obtained Medicinal Plants

Portion of plant	Melatonin (μg/g)
Tancetum parthenium	
Fresh green leaf	2.45
Fresh golden leaf	1.92
Freeze-dried green leaf	2.19
Freeze-dried golden leaf	1.61
Oven-dried green leaf	
Oven-dried golden leaf	1.69
Hypericum perforatum	
St. John's wort flower	4.39
St. John's wort leaf	1.75
Scutellaria biocalensis	
Huang-quin	7.11

Note: The three plants studied included feverfew (*Tanacetum parthenium*), St. John's wort (*Hypericum perforatum*), and Huang-qin (*Scutellaria biacalensis*).
Source: Reprinted with permission for Ref. 13.

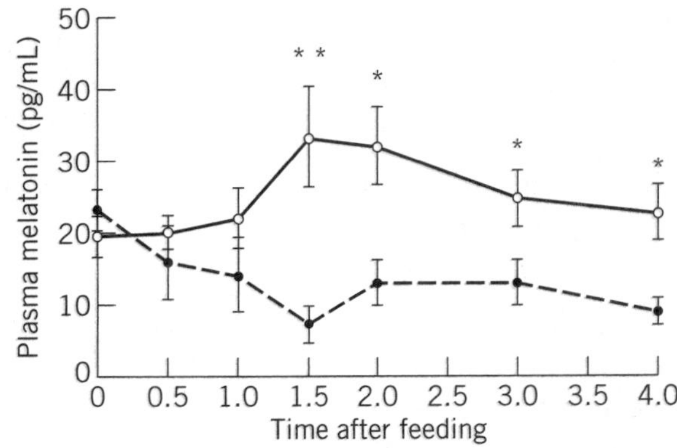

Figure 1. Blood levels of melatonin in chicks fed high-melatonin feed (3.5 mg/g) (open circle) compared to levels in chicks given low melatonin feed (<100 pg/g) (closed circle). This indicates that melatonin foods are absorbed and can increase circulating daytime levels of melatonin. The horizontal axis is in hours. $*p < 0.05$; $**p < 0.01$ is mean values in low-melatonin feed chicks. *Source:* Reprinted with permission from Ref. 8.

The lowest levels of melatonin are found in subterranean tubers, such as beet root and potato (12). According to R. Dubbels (personal communication), the concentrations of melatonin in tomatoes is inversely related to the degree of ripeness of the fruit. Thus, as fruit ripens, the concentration of melatonin seems to decrease accordingly.

TWENTY-FOUR-HOUR RHYTHMS IN PLANT MELATONIN

As already noted, a marked nocturnal rise in melatonin has been reported in the unicell, *Gonyaulax polyedra* (4).

The rise in melatonin in this organism begins with the onset of darkness, peaks near the middle of the dark phase of the light/dark cycle and then diminishes as morning approaches; this is highly reminiscent of the melatonin synthesis cycle in the pineal gland of mammals, including humans (15). A melatonin rhythm similar of that described in *G. polyedra* has also been measured in *Chenopoduim rubrum* (8). In this case, 15-day-old plants were maintained under a light/dark cycle of 12:12 (in hours). Melatonin concentrations were measured in the above-ground parts of the plant sampled at 2-hr intervals over a 24-hr period; during the dark period, the samples were collected with the aid of a dim green light. During the day, melatonin levels were uniformly low and in some cases undetectable. In contrast, during darkness, indoleamine concentrations began to increase to reach a peak at 4 to 6 hr after darkness onset; during the latter half of the dark period, melatonin levels diminished to day times values (Fig. 2). Whether the duration of elevated melatonin in either *G. polyedra* or *C. rubrum* would vary with the duration of darkness, as is the case in the mammalian pineal gland, has not been tested. Likewise, whether acute light exposure at night would alter the nocturnal increase in melatonin in plants is not known.

C. rubrum was specifically selected for the melatonin study because this obligate short-day plant exhibits pronounced and precise photoperiodic responses and rhythmic behavior; an endogenous rhythm in the flowering response, which is proportional to the duration of the dark period; and a sensitivity to night breaks with red light. Whether the melatonin rhythm relates to these cycles obviously remains unknown, as does whether other plants exhibit similar 24-hr rhythms in their melatonin content.

FUNCTION OF MELATONIN IN PLANTS

The functions of melatonin in animals are highly diverse and involve receptor and nonreceptor-related actions of

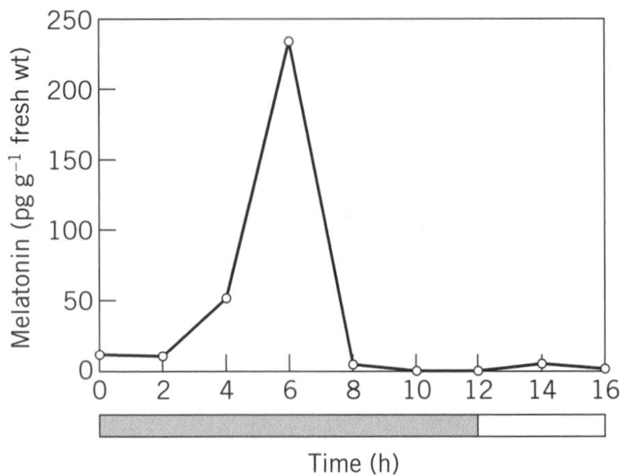

Figure 2. Nocturnal increase in the melatonin concentration in the above-ground parts of *Chenopodium rubrum* plants maintained under a light/dark cycle of 12:12. The dark bar at the bottom indicates the period of darkness. *Source:* Reprinted with permission from Ref. 10.

this chemical mediator (1,15,16). It seems possible that melatonin in plants has some of these same functions. Melatonin's action at the level of the biological clock of mammals is well documented (17). Based on the observations of the pronounced melatonin rhythm in *C. rubrum*, it was assumed that this cycle may relate to other 24-hr metabolic and morphogenetic rhythms (10). However, because so little is known of the circadian oscillator in plants, a mechanism for how the cycle of melatonin relates to the oscillator(s) remains uninvestigated.

A likely function of melatonin in plants relates to its ability to scavenge free radicals and function as an antioxidant (18–20). Melatonin is a direct scavenger of the hydroxyl radical (·OH) (21–23), the peroxynitrite anion ($ONOO^-$) (24,25), and possibly the peroxyl radical (LOO·) (26) and singlet oxygen ($^1O_2$) (27). Likewise, it either increases mRNA levels and/or the activity of several antioxidative enzymes in mammalian tissues (28,29). In view of these findings, it is not surprising that melatonin would also be tested for its ability to resist oxidative damage in plants as well.

The herbicide paraquat generates reactive oxygen species and free radicals in a variety of plants and animals, including *G. polyedra* (30). When this bioluminescent dinoflagellate is incubated with paraquat, the herbicide reduces, in a concentration-dependent manner, the nocturnal glow peak of the organism (Fig. 3) (31), indicating that a regulator molecule required for the expression of the bioluminescence, such as 5-methoxytryptamine and/or its precursor molecule melatonin, is oxidized by the free radicals generated by paraquat. The addition of supplemental melatonin to the incubation medium prevents the inhibitory action of paraquat on bioluminescence. These findings are consistent with the restoration of the glow peak being a consequence of melatonin's ability to scavenge free radicals generated by paraquat.

In another model of lethal oxidative stress in *G. polyedra*, melatonin was found to protect the unicell from death (32). While basal levels of melatonin in *G. polyedra* vary between roughly 0.1 and 1 μM, depending on the circadian phase (4), in response to low temperature melatonin levels can increase by orders of magnitude reaching the μM range (33). When *G. polyedra* were incubated with hydrogen peroxide (H_2O_2), which is readily converted to the ·OH, the cells exhibited a several-fold enhanced bioluminescence for several hours, indicating severe oxidative damage to the cells. Adding melatonin to the incubation medium 1 hr before the addition of H_2O_2 prevented the intense light emission and cell death. In this case, the melatonin concentrations that were added, although in the mM range, were physiological for this species. These findings, then, provide strong evidence that melatonin functions as a physiological antioxidant in the dinoflagellate.

It has also been speculated that melatonin functions as an antioxidant in higher plants as well (12). Ozone, which generates free radicals and reduces plant growth, causes visible foliar damage and reduces the fresh weight of fruits when applied at the time of flowering (34,35). *Lycopersicon pimpinelliform* is highly sensitive to ozone (36) and contains very low levels of melatonin (12). Similarly, leaves of different varieties of *Nicotiana tabacum* are differentially

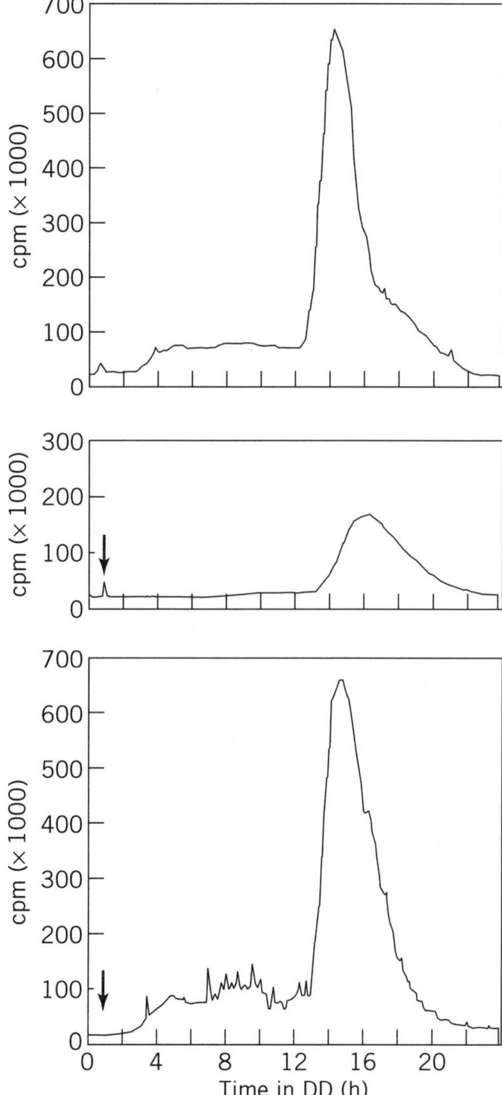

Figure 3. Spontaneous bioluminescence in *Gonyaulax polyedra* during darkness (DD). From top to bottom, control; middle, paraquat treatment that markedly suppresses to glow peak; bottom, combination of paraquat and melatonin showing that melatonin restored the glow peak. *Source:* Reprinted with permission from Ref. 31.

sensitive to ozone damage, and the sensitivity is lower in the leaves with the highest melatonin concentrations. This is consistent with the idea that melatonin in plants, as in animals, functions as an antioxidant.

Finally, considering melatonin's obvious function in animals as a modulator of circadian rhythms, the 24-hr fluctuations in this indoleamine in plants (9–11) and algae (4) indicates it may also be involved in the regulation of circadian rhythms and photoperiodicity in these species. Although there are no studies on these interactions in higher plants, in algae (18,37) the evidence is quite compelling for a role of melatonin in a variety of functions that vary over the light/dark cycle.

CONCLUSION

Relatively few plants have been examined as to their melatonin levels, but the preliminary data indicate the indoleamine is present and that its concentration may vary greatly over the light/dark cycle and between different species. Its presence in edible plants suggests that melatonin, like vitamin antioxidants, may be ingested in sufficient quantities to provide at least some protection against free radicals. In the plants themselves, melatonin is believed to function as a free-radical scavenger and in photoperiodism. This area of research is in its early stages of development, and the next several years will likely be associated with a substantial increase in the amount of information related to both the content and function of melatonin in plants.

BIBLIOGRAPHY

1. R. J. Reiter, "Pineal Melatonin: Cell Biology of Its Synthesis and of Its Molecular Interactions," *Endocrine Rev.* **12**, 151–180 (1991).
2. W. T. Jackson "Regulation of Mitosis. II. Interaction of Isopropyl *N*-Phenyl-Carbomate and Melatonin," *J. Cell Sci.* **5**, 745–755 (1969).
3. S. Banerjee and L. Margulis, "Mitotic Arrest by Melatonin," *Exp. Cell Res.* **78**, 314–318 (1973).
4. B. Poeggeler et al., "Pineal Hormone Melatonin Oscillates also in the Dinoflagellate *Gonyaulax polyedra*," *Naturwissenschaften* **78**, 268–269 (1991).
5. I. Balzer, B. Poeggeler, and R. Hardeland, "Circadian Rhythm of Indolemines in a Dinoflagellate, *Gonyaulax polyedra*: Persistence of Melatonin Rhythm in Constant Darkness and Relationships to 5-Methoxytryptamine," in Y. Touitou, J. Arendt, P. Pevet, eds., *Melatonin and the Pineal Gland*, Elsevier, Amsterdam, 1993.
6. I. Balzer and R. Hardeland "Melatonin in Algae and Higher Plants—Possible New Roles as a Phytohormone and Antioxidant," *Botanica Acta* **109**, 180–183 (1996).
7. B. Fuhrberg et al., "The Vertebrate Pineal Hormone Melatonin Is Produced by the Brown Alga *Pterygophora californica* and Mimics Dark Effects on Growth Rate in the Light," *Planta* **200**, 125–131 (1996).
8. A. Hattori et al., "Identification of Melatonin in Plants and Its Effects on Plasma Melatonin Levels and Binding to Melatonin Receptors in Vertebrates," *Biochem. Mol. Biol. Int.* **35**, 627–634 (1995).
9. J. Kolar et al., "Melatonin in Higher Plants Determined by Radioimmunoassay and Liquid Chromatography-Mass Spectrometry," *Biol. Rhythm Res.* **26**, 406 (1996).
10. J. Kolar et al., "Melatonin: Occurrence, and Daily Rhythm in *Chenopodium rubrum*," *Phytochemistry* **44**, 1407–1414 (1997).
11. D. L. Van Tassel, J. Li, S. D. O'Neil "Melatonin: Identification of a Potential Dark Signal in Plants," *Plant Physiol.* **102** (Suppl. 1), 659 (1993).
12. R. Dubbels et al., "Melatonin in Edible Plants Identified by Radioimmunoassay and by High Performance Liquid Chromatography-Mass Spectroscopy," *J. Pineal Res.* **18**, 28–31 (1995).
13. S. J. Murch, C. B. Simmons, and P. K. Saxena, "Melatonin in Feverfew and Other Medicinal Plants," *Lancet* **350**, 1598–1599 (1997).

14. B. Poeggeler and R. Hardeland, "Detection and Quantification of Melatonin in a Dinoflagellate, *Gonyaulax polyedra*, Solutions to the Problem of Methoxyindole Destruction in Nonvertebrate Material," *J. Pineal Res.* **17**, 1–10 (1994).
15. R. J. Reiter "The Melatonin Rhythm: Both a Clock and a Calendar," *Experientia* **49**, 654–664 (1993).
16. J. Arendt, *Melatonin and the Mammalian Pineal Gland*, Chapman and Hall, London, 1995.
17. A. J. Lewy, R. L. Sack, and C. M. Singer, "Bright Light, Melatonin, and Biological Rhythms in Humans," in J. Montplaisir and R. Godbout, eds., *Sleep and Biological Rhythms*, Oxford Press, New York, 1990.
18. R. Hardeland et al., "On the Primary Functions of Melatonin in Evolution, Mediation of Photoperiodic Signals in a Unicell, Photooxidation and Scavenging Free Radicals," *J. Pineal Res.* **18**, 104–111 (1995).
19. B. Poeggeler et al., "Melatonin and Structurally-Related, Endogenous Indoles Act as Potent Electron Donors and Radical Scavengers In Vitro," *Redox Rep.* **2**, 179–184 (1996).
20. R. J. Reiter et al., "Pharmacological Actions of Melatonin in Free Radical Pathophysiology," *Life Sci.* **60**, 2255–2271 (1997).
21. D. X. Tan et al., "Melatonin: A Potent, Endogenous Hydroxyl Radical Scavenger," *Endocrine J.* **1**, 57–60 (1993).
22. Z. Matuszak, K. J. Reszka, and C. F. Chignell, "Reaction of Melatonin and Related Indoles with Hydroxyl Radicals: EPR and Spin Trapping Investigations," *Free Radical Biol. Med.* **23**, 367–372 (1997).
23. P. Stasica, P. Ulanski, and J. M. Rosiak, "Melatonin as a Hydroxyl Radical Scavenger," *J. Pineal Res.* **25**, 65–66 (1998).
24. E. Gilad et al., "Melatonin Is a Scavenger of Peroxynitrite," *Life Sci.* **60**, PL169–174 (1997).
25. S. Cuzzocrea et al., "Protective Effect of Melatonin in Carrageenan-Induced Models of Local Inflammation," *J. Pineal Res.* **23**, 106–116 (1997).
26. C. Pieri et al., "Melatonin as an Efficient Antioxidant," *Arch. Gerontol. Geriatr.* **20**, 159–165 (1995).
27. C. M. Cagnoli et al., "Melatonin Protects Neurons from Singlet Oxygen-Induced Apoptosis," *J. Pineal Res.* **18**, 222–226 (1996).
28. M. O. Kotler et al., "Melatonin Increases Gene Expression for Antioxidant Enzymes in Rat Brain Cortex," *J. Pineal Res.* **24**, 83–89 (1998).
29. R. J. Reiter, "Aging and Oxygen Toxicity: Relation to Changes in Melatonin," *Age* **20**, 201–213 (1997).
30. C. B. Ramalko, J. W. Hastings, and P. Colepicolo "Circadian Oscillation of Nitrate Reductase Activity in *Gonyaulax polyedra* Is Due to Changes in Cellular Protein Levels," *Plant Physiol.* **107**, 225–231 (1995).
31. I. Antolin and R. Hardeland, "Suppression of the *Gonyaulax* Glow Peak by Paraquat and Its Restoration by Melatonin," in R. Hardeland, ed., *Biological Rhythms and Antioxidant Protection*, Cuvillier Verlag, Göttingen, 1997.
32. I. Antolin et al., "Antioxidative Protection in a High-Melatonin Organism: The Dinoflagellate *Gonyaulax polyedra* Is Rescued from Lethal Oxidative Stress by Strongly Elevated, but Physiologically Possible Concentrations of Melatonin," *J. Pineal Res.* **23**, 182–190 (1997).
33. B. Fuhrberg et al., "Dramatic Rises of Melatonin and 5-Methoxytryptamine in *Gonyaulax* Exposed to Decreased Temperature," *Biol. Rhythm Res.* **28**, 144–150 (1997).
34. A. Z. Tenga, B. A. Marie, and D. P. Ormrod, "Recovery of Tomato Plants from Ozone Injury," *Hortiscience* **25**, 1230–1232 (1990).
35. L. M. Mortensen, "Effects of Ozone Concentrations on Growth of Tomato at Various Light, Air Humidity and Carbon Dioxide Levels," *Sci. Hortic.* **49**, 17–24 (1992).
36. A. G. Gentile et al., "Susceptibility of *Lycopersion* ssp. to Ozone Injury," *J. Am. Soc. Hortic. Sci.* **96**, 94–96 1971.
37. R. Hardeland et al., "Chronobiology of Indoleamines in the Dinoflagellate *Gonyaulax polyedra*: Metabolism and Effects Related to Circadian Rhythmicity and Photoperiodism," *Braz. J. Med. Biol. Res.* **29**, 119–123 (1996).

Russel J. Reiter
Seok Joong Kim
The University of Texas Health Science Center
San Antonio, Texas

PHYTOCHEMICALS: *VACCINIUM*

The genus *Vaccinium* belongs to the health family, and members include the blueberries (bilberries), cranberries, and whortleberries. Total world production is estimated to about 1 million tons per year. These widespread, small fruits are found nearly everywhere in the Northern Hemisphere.

Knowledge of the chemical composition of the plant part of *Vaccinium* spp. is relevant because some phytochemicals (from the Greek *phyton*, plant) have biological activity, and berries are used as food and as medicine.

HISTORY

In the antiquity, Vergil and Ovid wrote verses on blueberries. Pliny the Elder and Vitruvius described the use of bilberries as a dye. The word *Vaccinium* derives from the Latin word *vacca*, ie, cow, or it is a misprint of *bacca*, berry.

Several plant parts (berries, leaves, flowers) of *Vaccinium* spp. have also been used in natural medicine dating back to ancient times. Dioscorides (first century A.D.) described the use of bilberries in the treatment of the diarrhea. This traditional folk remedy is still implemented today. St. Hildegarde of Bingen (twelfth century A.D.) recommended the use of these wild fruits to promote menstruation. Blueberry was entered in the pharmacopoeia in the sixteenth century.

Blueberries have been eaten in America since prehistoric times. Native American tribes not only ate bilberries, but also used the berries as a dye and as a medicine. Cranberry fruits, too acidic to be eaten raw, were generally mixed with honey or maple syrup or smoked. The main use for blueberries, however, was as a preservative. The high content of phenolic compounds and benzoic acid provides the bacteriostatic, antifermentative, and antioxidant properties. By adding blueberries to bison meat or deer fat, natives could preserve the food for the winter. This preparation was known as *pemmican*. Another was *venison*, meat filled with blueberries that was then smoked and dried. Several tribes used blueberries in medicine, using both berries and leaves.

Consumption of blueberries was introduced to the Pilgrims by the natives, but blueberry cultivation did not begin until the end of the nineteenth century. On the U.S. East Coast, July now is known as Blueberry Month, a time devoted to the appreciation of this most popular fruit.

Of curious interest is that the economic fortune of blueberries blossomed during wartime. During the American Civil War, blueberry juice and cranberries were consumed in high quantities by the troops. In the United Kingdom during World War I, blueberries were used as a substitute for synthetic aniline dyes. During World War II, pilots of the Royal Air Force, when flying in the night bombing raids over Nazi Germany, observed that when their diets included blueberry jam, they had great visual capacity and no night blindness.

PLANT COMPOSITION

Hundreds of compounds have been identified in *Vaccinium*. They are divided into several main classes, which are discussed in this section.

Sugars

Two main carbohydrates occur in *Vaccinium* berries, glucose (1–3%) and fructose (0.5–2%) (1,2). In cranberry, glucose occurs in a much higher amount than fructose; this is unusual because usually fructose and glucose occur in equal amounts.

Acids

Major organic acids in *Vaccinium* spp. are citric, malic, quinic, and benzoic acids (3,4). Quinic acid (1,3,4,5-tetrahydroxycyclohexanecarboxylic acid) and benzoic acid are the most important and characteristic acids of cranberry (*V. macrocarpom*). The mean benzoic acid content is 0.01%, occurring mainly as 6-benzoyl-D-glucose (vaccinin).

Phenolic Compounds

Phenolic compounds (4–6) are water-soluble compounds, sometimes present in glycosidic form, characterized by the occurrence of one or more hydroxyl residues in the molecule. They occur in monomeric or polymeric structures. They provide a high antioxidant potential in foods because they scavenge free radicals such as O_2^-, $HO^\cdot$, and $ROO^\cdot$. Phenolic compounds are also responsible for the color, taste, flavor, and astringency of foods.

Polymeric Phenols. The tannins comprise much of the phenolic content in *Vaccinium*. They have a molecular weight of up to 5000 and are characterized by their astringency, brown color, and ability to precipitate by bonding with proteins. Two types occur in vegetables, hydrolyzable and condensed. In *Vaccinium* they occur as condensed tannins.

Oligomeric Phenolic Compounds. Oligomeric (ie, of intermediate molecular weight that ranges from 500 to 1500) phenolic compounds (OPCs) (7,8) are dimeric, trimeric, or tetrameric forms of flavan-type phenols, such as catechin and epicatechin. They are called procyanidins or proanthocyanidins because they yield anthocyanidins after acid degradation. They occur as copolymer of all the possible combinations of the stereoisomer form of the monomeric flavans and are distinguished by an alphabetic letter (ie, B_1, B_2, B_3, and B_4 for the four-dimeric forms of [D,L]-catechin and [D,L]-epicatechin). With three monomers, the number of possible trimers would be 27, etc. A typical procyanidin, procyanidin B_1, that occurs in *Vaccinium* berries is represented in Figure 1.

Anthocyanins. The anthocyanins (5,6,9) are the pigments responsible for the red to dark blue color of the *Vaccinium* berries and their juice. Anthocyanins are flavonoids (a C_6-C_3-C_6 molecule) and occur in glycosidic form. The nature and the type of the attachment of the sugar may be different. The sugar can be also acylated, generally with an aromatic or an aliphatic acyl moiety. A typical anthocyanin is shown in Figure 2. The anthocyanin composition characterizes *Vaccinium* species with qualitative and quantitative differences. In highbush blueberry (*V. corymbosum* L.), the total anthocyanin content varies from 25 to 495 mg/100 g fresh weight (FW). In the lowbush blueberry (*V. angustifolium* L.), the anthocyanin content ranges from 120 mg/100 g in the variety Bloomingdon and 260 mg/100 g in the variety Chignecto. In *V. angustifolium*, it is important to observe the occurrence of up to 35% of anthocyanins acylated with acetic acid and the genetic relationship among varieties (10). In rabbiteye blueberry

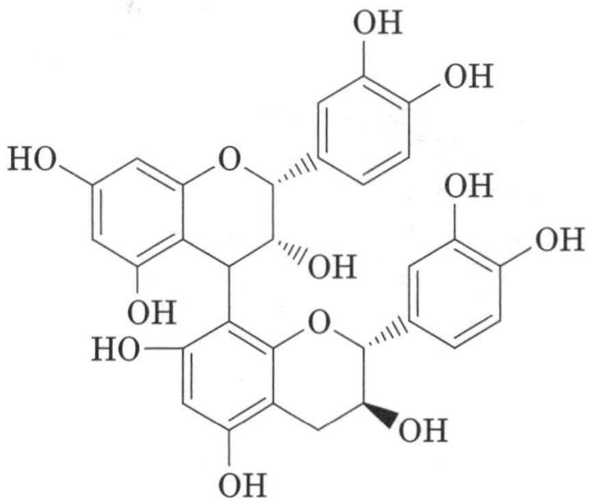

Figure 1. Procyanidin B_1, a catechin–epicatechin dimeric proanthocyanidin.

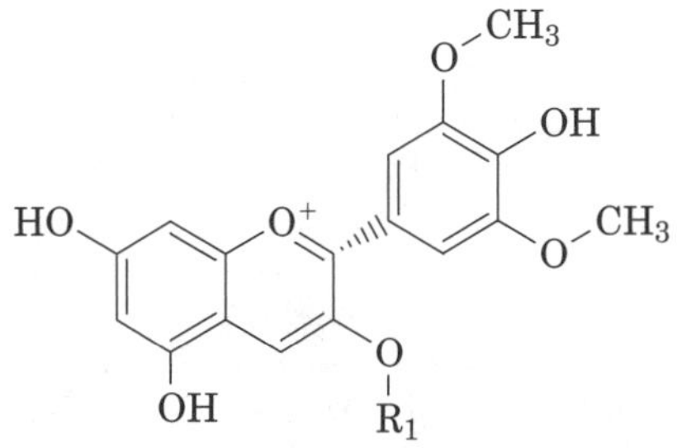

Figure 2. Malvidin, a typical *Vaccinium* anthocyanin. R_1, glucose, galactose, and arabinose.

(*V. ashei*), the total anthocyanin content is 210 mg/100 g in Tifblue berries and 272 mg/100 g in Bluegem berries.

Vaccinium elliotti, with an anthocyanin content of 760 mg/100 g, appears to be the species containing the highest amount of anthocyanins.

In *V. japonicum*, cyanidin 3-arabinoside (54%) and pelargonidin 3-arabinoside (39%) have been identified as two main pigments, they total 113 mg/100 g FW. This is the first report of pelargonidin glycosides in the genus *Vaccinium*.

In the fruit of *V. oxycoccus* L., or small cranberry, the main pigments present are peonidin 3-glucoside and cyanidin 3-glucoside, accounting for 41.9% and 38.3% of the total anthocyanin content, respectively. This anthocyanin pattern is rather different from that of the American cranberry (*V. macrocarpon*), which is rich in the 3-galactosides and the 3-arabinosides of peonidin and cyanidin; the total anthocyanin content of the fruit is about 78 mg/100 g FW. All possible combinations of cyanidin, delphinidin, petunidin, peonidin, and malvidin 3-galactosides, 3-glucosides, and 3-arabinosides have been found in the bilberry *V. myrtillus*. Quantitatively, the delphinidin glycosides are present in the largest quantities, and the peonidin glycosides are the least abundant. Fruit of bilberries from the Piedmont Alps in Italy contain 300–320 mg of anthocyanins/100 g FW. Bog whortleberry (*V. uliginosum*) contains 15 anthocyanins. The 3-glucoside of malvidin comprises the majority of these pigments (35.9%), and the total anthocyanin content is 256 mg/100 g FW. In Norwegian cowberries (*V. vitis-idaea* L.), the main anthocyanin is represented by cyanidin 3-galactoside (88.0%). The total anthocyanin content is 174 mg/100 g FW.

Phenolic Acids. *Vaccinium* berries contain a large number of phenolic acids as glycosides (6). The hydroxycinnamic acid derivatives (180–210 mg/100 g FW) are much higher in amount than the hydroxybenzoic derivatives (0.5–2.0 mg/100 g FW).

Simple Phenols. Arbutin (4-hydroxyphenyl-β-D-glucopyranoside) is a hydroquinone glycoside occurring in leaves and fruit (6). In the leaves of *V. vitis-idaea*, its content ranges from 40 to 90 mg g^{-1}. Its methyl ether, methylarbutin, occurs together with arbutin but in a lower amount.

Volatile Compounds

Several volatile compounds (11,12), including esters, hydrocarbons, aldehydes, ketones, cyclic ether, and sulfur-containing compounds, have been identified in different *Vaccinium* species, and there are qualitative and quantitative differences among them.

The impact compounds include *t*-2-hexenal, *t*-3-hexenol, and linalool in highbush blueberry. The impact character compounds in bilberry are *t*-2-hexenol, ethyl-3-methylbutanoate, and ethyl-2-methylbutanoate. Linalool contributes to the floral, rosy character unique to blueberry fruit, its concentration is less than 1 ppm. In cranberry, benzoic acid derivatives are responsible of the impact flavor character. As many as 50 volatile compounds have been identified in *Vaccinium* spp., but as many as 200 other compounds still remain unidentified.

Vitamins

Ascorbic acid is the vitamin occurring in highest amount in *Vaccinium*; its mean content in cranberry juice is 3 mg/100 mL (13,14). The folic acid content is 2.6 μg/100 g. The potential vitamin A content ranges from 50 to 100 IU/100 g. Trace amounts of thiamin, riboflavin, and niacin have also been detected.

Metals

The metal content includes potassium, sodium, calcium, rubidium, and magnesium, but *Vaccinium* berries are characterized by a high content of manganese, ranging from 28 to 250 ppm in *V. vitis-idaea* to 370 ppm in *V. myrtillus* (13,14).

NUTRITIONAL PROPERTIES

With their vivid color and their delicate flavor, *Vaccinium* berries stimulate the appetite. The energy supply of sugars, organic acids, minerals, and vitamins contained in them is important. However, an essential role is played by the phenolic compounds, particularly anthocyanins and OPCs. These phytochemicals are also responsible for the medicinal properties of bilberries (15–17).

An advantage of consuming large quantities of *Vaccinium* fruits may be an increase in antioxidative defenses in the body. The daily intake of phenolic compounds, including large quantities of anthocyanins, may prevent or lower the risk of a variety of ailments, particularly circulatory disease and stroke (17).

Flavonols and simple phenols retard the oxidation of ascorbic acid, tocopherols, and carotenoids, thus preserving their nutritional value and exerting a bacteriostatic action (18). Several scientific reports (19) have shown how phenolic substances, present in foods of plant origin, play an important role in nutrition. In *Vaccinium* berries, this fact is attributable to anthocyanins and procyanidins. They act as antioxidants in protecting the body from degenerative phenomena caused by the oxidation of lipids (cardiovascular diseases, atherosclerosis, etc.), and they have a chemopreventive action against the onset of cancer (20–31).

Manganese is an essential microelement for the synthesis of several enzymes, particularly superoxide dismutase (SOD), a very important enzyme that is the main front line defense against damaging free radicals (32).

PHARMACOLOGICAL PROPERTIES

The anthocyanins and OPCs are antioxidants; they are free radical scavengers and inhibitors of lipoxygenase, cyclooxygenases, hyaluronidases, and other oxidative enzymes, such as ascorbic oxidases. Anthocyanins also lower cholesterol levels and appear to have significant platelet antiaggregating activity as well as antithrombotic activity (33–39).

Ophthalmological studies done in the 1960s in France and Italy found that anthocyanins act directly on the speed of regeneration of the vision pigment. They have an indirect action by improving the blood flow to the strongly vascularized retina. Athletes and certain professionals (truck and bus drivers, video terminal operators), for whom good vision is critical, would benefit from high intake of *Vaccinium* anthocyanins (15).

A vast series of pharmacological and clinical studies conducted in Europe during the past 40 years has shown that the anthocyanin glucosides and OPCs have a marked action on the permeability and resistance of the capillary walls. In addition to the capillarotropic action, attributable to the so-called vitamin P factor, these compounds have a noticeable anti-inflammatory effect. They have been used successfully in capillary pathologies such as diabetes, atherosclerosis, arterial hypertension, and circulatory disturbances in superficial and deep veins and circulation problems during pregnancy (40–46).

Figure 3 shows some pharmaceutical formulations anthocyanin-containing from bilberries.

Other experimental and clinical studies have shown outstanding wound-healing, mucopoietic, capillaroprotective effects, not only on superficial wounds but also on the gastric mucosa and duodenum (40–46). Elderly persons, patients with diabetes, and persons with coronary disease, hypertension, or ulcers are among those who could greatly benefit from a diet rich in blueberries, cranberries, and whortleberries.

Recent clinical trials have also scientifically confirmed what has been known from popular experience. Cranberry juice and other *Vaccinium* berry juices prevent urinary tract infections in women and elderly individuals. The antiadhesion effect that impedes the proliferation of *Escherichia coli*, the bacteria responsible for the infection, has been proved to be due to the procyanidins contained in cranberry juice (47–50).

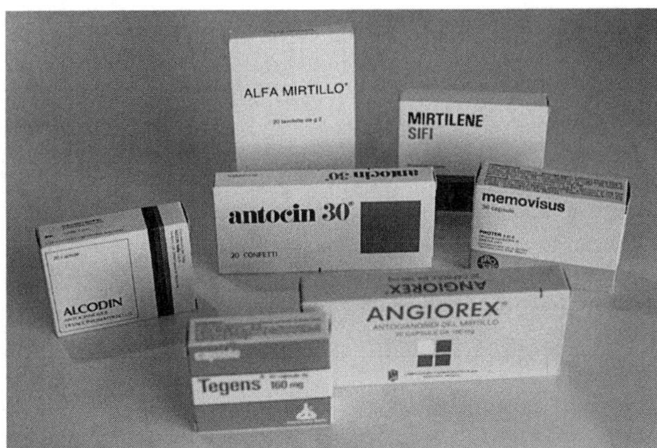

Figure 3. Pharmaceutical preparations based on bilberry anthocyanins.

CONCLUSION

Fresh and processed *Vaccinium* berries have been a part of our diet for a long time. In addition to their colorful appearance, they stimulate the appetite with flavorful and tasteful sensory properties and supply macronutrients and micronutrients to the body. The phytochemicals found in *Vaccinium* contribute significantly to cardiovascular disease prevention. They improve microcirculation, enhance eyesight, lower blood cholesterol levels, prevent lipid-oxidation-derived diseases (atherosclerosis, cancer, etc.), moderate the negative effects of aging, and have a synergistic action on all antioxidant vitamins (A, E, and C).

On the threshold of Third Millennium, Hippocrates' fifth-century B.C. statement is still topical:

> Let your food be your medicine, let your medicine be your food.

BIBLIOGRAPHY

1. W. Kalt and J. E. McDonald, "Chemical Composition of Lowbush Blueberry Cultivars," *J. Am. Soc. Hort. Sci.* **121**, 142–146 (1996).
2. A. B. Holmes and M. S. Starr, "Cranberry Juice," in S. Nagy, S. S. Chen, and P. E. Shaw, eds., *Fruit Juice Processing Technology*, Agriscience Inc., Auburndale, Fla., 1993, pp. 515–531.
3. E. D. Coppola, E. C. Conrad, and R. Cotter, "High Pressure Liquid Chromatographic Determination of Major Organic Acids in Cranberry Juice," *J. Assoc. Off. Anal. Chem.* **61**, 1490–1492 (1978).
4. V. Hong and R. E. Wrolstad, "Detection of Adulteration in Cranberry Juice Drinks and Concentrates," *J. Assoc. Off. Anal. Chem.* **69**, 208–213 (1986).
5. G. Mazza and E. Miniati, *Anthocyanins in Fruits, Vegetables and Grains*, CRC Press, Boca Raton, Fla., 1993.
6. J. J. Macheix, A. Fleuriet, and J. Billot, *Fruit Phenolics*, CRC Press, Boca Raton, Fla., 1990.
7. L. Y. Foo and L. J. Porter, "The Structure of Tannins of Some Edible Fruits," *J. Sci. Food Agric.* **32**, 711–716 (1981).
8. P. L. Wang, C. T. Du, and F. J. Francis, "Isolation and Characterization of Polyphenolic Compounds in Cranberries," *J. Food Sci.* **43**, 1402–1404 (1978).
9. J. Gross, *Pigments in Fruits*, Academic Press, London, 1987.
10. L. Gao and G. Mazza, "Quantitation and Distribution of Simple and Acylated Anthocyanins and Other Phenolics in Blueberries," *J. Food Sci.* **59**, 1057–1059 (1994).
11. W. D. Baloga, N. Vorsa, and L. Lawter, "Dynamic headspace gas chromatography-mass spectrometry analysis of volatile flavor compounds from wild diploid blueberry species," in R. L. Rousseff and M. M. Leahy, eds., *Fruit Flavors* American Chemical Society, Washington, D.C., 1994, pp. 235–247.
12. R. J. Horvath et al., "Comparison of Volatile Compounds from Rabbit Eye Blueberry (*Vaccinium ashei*) and Deerberry (*V. stamineum*) During Maturation," *J. Essent. Oil Res.* **8**, 645–648 (1996).
13. R. J. Bushway et al., "Mineral and Vitamin Content of Lowbush Blueberries (*Vaccinium angustifolium* Ait.)," *J. Food Sci.* **48**, 1878–1880 (1983).

14. R. R. Eitenmiller, R. F. Kuhl, and C. J. B. Smit, "Mineral and Water Soluble Vitamin Content of Rabbiteye Blueberries," *J. Food Sci.* **42**, 1311–1315 (1977).

15. E. Miniati and R. Coli, "Anthocyanins: Not Only Color for Foods," *Proc. 1st Int. Conf. on Natural Food Colorants*, University of Massachusetts, Amherst, Mass., 1994.

16. W. Kalt and D. Dufour, "Health Functionality of Blueberries," *Hort Technol.* **7**, 216–221 (1997).

17. P. Morazzoni and E. Bombardelli, "*Vaccinium myrtillus* L.," *Fitoterapia* **66**, 3–29 (1996).

18. C. A. Rice-Evans, N. J. Miller, and G. Paganga, "Structure–Antioxidant Activity Relationship of Flavonoids and Phenolic Acids," *Free Radiol Biol. Med.* **20**, 933–956 (1996).

19. M. T. Huang, C. T. Ho and C. Y. Lee, "Phenolic Compounds in Food and their Effects on Health. II. Antioxidants and Cancer Prevention," ACS Symposium Series 507, American Chemical Society, Washington, D.C., 1992.

20. J. A. Bomser et al., "Induction of NAD(P)H:Quinone Acceptor Oxidoreductase in Murine Cells (Hepa 1c1c7) by Fruit (*Vaccinium*) Extracts," *FASEB J.* **9**, A993 (1995).

21. R. L. Prior et al., "Antioxidant Capacity as Influenced by Total Phenolic and Anthocyanin Content, Maturity, and Variety of *Vaccinium* Species," *J. Agric. Food Chem.* **46**, 2686–2693 (1998).

22. H. Wang, G. Cao, and R. L. Prior, "Oxygen Radical Absorbing Capacity of Anthocyanins," *J. Agric. Food Chem.* **45**, 304–309 (1997).

23. H. Wang, G. Cao, and R. L. Prior, "Total Antioxidant Capacity of Fruits," *J. Agric. Food Chem.* **44**, 701–705 (1996).

24. Y. S. Velioglu et al., "Antioxidant Activity and Total Phenolics in Selected Fruits, Vegetables, and Grain Products," *J. Agric. Food Chem.* **46**, 4113–4117 (1998).

25. J. Bomser et al., "*In vitro* Anticancer Activity of Fruits Extracts from *Vaccinium* Species," *Planta Med.* **62**, 212–216 (1996).

26. M. T. Satue-Garcia, M. Heinonen, and E. N. Frankel, "Anthocyanins as Antioxidants on Human Low-Density Lipoprotein and Lecithin-liposome Systems," *J. Agric. Food Chem.* **45**, 3362–3367 (1997).

27. P. M. Laplaud, A. Leleubre, and M. J. Chapman, "Antioxidant Action of *Vaccinium myrtillus* Extract on Human Low Density Lipoproteins *In vitro*: Initial Observations," *Fundam. Clin. Pharmacol.* **11**, 35–40 (1997).

28. G. Cao et al., "Increases in Human Plasma Antioxidant Capacity After Consumption of Controlled Diets High in Fruit and Vegetables," *Am. J. Clin. Nutr.* **68**, 1081–1087 (1998).

29. M. T. Meunier, E. Duroux, and P. Bastide, "Activité Antiradicalaire D'Oligomères Procyanidoliques et D'Anthocyanosides Vis-à-vis de L'Anion Superoxyde et Vis-à-vis de La Lipoperoxydation," *Plant. Med. Phytothèr.* **23**, 267–274 (1989).

30. L. Costantino et al., "Activity of Polyphenolic Crude Extracts as Scavengers of Superoxide Radicals and Inhibitors of Xanthine Oxidase," *Planta Med.* **58**, 342–344 (1992).

31. T. Wilson, J. P. Porcari, and D. Harbin, "Cranberry Extract Inhibits Low Density Lipoprotein Oxidation," *Life Sci.* **62**, 381–386 (1998).

32. B. Frei, *Natural Antioxidants in Human Health and Diseases*, Academic Press, San Diego, Calif., 1994.

33. C. Ghiringhelli, F. Gregoratti, and F. Marastoni, "Capillarothropic Activity of Anthocyanosides in High Doses in Phebopathic Stasis," *Minerva Cardioangiologica* **26**, 255–276 (1978).

34. G. Spinella, "Natural Anthocyanosides in Treatment of Peripheral Venous Insufficiency," *Arch. Med. Int.* **37**, 21–29 (1985).

35. V. Bettini et al., "Effects of *Vaccinium myrtillus* Anthocyanosides on Vascular Smooth Muscle," *Fitoterapia* **55**, 265–272 (1984).

36. D. Colombo and R. Vescovini, "Controlled Clinical Trial of Anthocyanosides from *Vaccinium myrtillus* in Primary Dysmenorrhea," *G. Ital. Ostetr. Ginecol.* **7**, 1033–1038 (1985).

37. S. Bertuglia, S. Malandrino, and A. Colantuoni, "Effect of *Vaccinium myrtillus* Anthocyanosides on Ischaemia Reperfusion Injury in Hamster Cheek Pouch Microcirculation," *Pharmacol. Res.* **31**, 183–187 (1995).

38. A. Colantuoni et al., "Effects of *Vaccinium myrtillus* Anthocyanosides on Arterial Vasomotion," *Arzneim. Forsch.* **41**, 905–909 (1991).

39. G. Pulliero et al., "*Ex vivo* Study of the Inhibitory Effect of *Vaccinium myrtillus* Anthocyanosides on Human Platelet Aggregation," *Fitoterapia* **60**, 69–75 (1989).

40. J. C. Monboisse et al., "Non-enzymatic Degradation of Acid-soluble Calf Skin Collagen by Superoxide Ion: Protective Effect of Flavonoids," *Biochem. Pharmacol.* **32**, 53–58 (1983).

41. R. Boniface and A. M. Robert, "Effect of Anthocyanins on Human Connective Tissue Metabolism in the Human," *Klin. Monatsbl. Augenheilkd.* **209**, 368–372 (1996).

42. Z. Detre et al., "Studies on Vascular Permeability in Hypertension: Action of Anthocyanosides," *Clin. Physiol. Biochem.* **4**, 143–149 (1995).

43. M. J. Magistretti, M. Conti, and A. Cristoni, "Antiulcer Activity of an Anthocyanidin from *Vaccinium myrtillus*," *Arzneim. Forsch.* **38**, 686–690 (1990).

44. A. Cristoni and M. J. Magistretti, "Antiulcer and Healing Activities of *Vaccinium myrtillus* Anthocyanosides," *Farmaco. Ed. Prat.* **42**, 29–43 (1987).

45. A. Lietti, A. Cristoni, and M. Picci, "Studies on *Vaccinium myrtillus* Anthocyanosides. I. Vasoprotective and Antiinflammatory Activity," *Arzneim. Forsch.* **26**, 829–832 (1976).

46. A. M. Robert et al., "Action of Anthocyanosides of *Vaccinium myrtillus* on the Permeability of the Blood Brain Barrier," *J. Med.* **8**, 312–332 (1977).

47. I. Ofek et al., "Anti-*Escherichia coli* Adhesion Activity of Cranberry and Blueberry Juices," *New Engl. J. Med.* **324**, 1599 (1991).

48. J. M. T. Hamilton-Miller, "Reduction of Bacteriuria and Pyuria after Ingestion of Cranberry Juice," *J. Am. Med. Assoc.* **272**, 588–590 (1994).

49. J. Avorn et al., "Reduction of Bacteriuria and Pyuria Using Cranberry Juice," *J. Am. Med. Assoc.* **271**, 751–754 (1994).

50. A. B. Howell et al., "Inhibition of the Adherence of P-Fimbriated *Escherichia coli* to Uroepithelial-cell Surfaces by Proanthocyanidin Extracts from Cranberries," *New Engl. J. Med.* **339**, 1085–1086 (1998).

ENRICO MINIATI
Perugia University
Perugia, Italy

See also COLORANTS: ANTHOCYANINS.

PHYTOCHEMICALS: WINE

The understanding of the term *phytochemicals* (chemicals of plants) is likely to be related to the discipline in which a person has an interest and/or training (1). For example, plant biochemists are to be more apt to think of phytochemicals as secondary metabolites, that is, other natural chemicals than the primary plant components of carbohydrates, proteins, or lipids; toxicologists may define phytochemicals as natural compounds that cause or inhibit mutagenicity, carcinogenicity, or teratogenicity; food technologists mainly consider phytochemicals in relation to food colors and taste; nutritionists are now using the term *phytonutrients*—phytochemicals that enhance human health over and above vitamins and minerals. Whatever the definition, phytochemicals are natural compounds in plant foods and plant food products, such as wine. It is known that many of them contribute to the desirable qualities of the food and wine and, more recently, to optimal health. Many phytochemicals of wine may differ from the phytochemicals of the grapes used for vinification. The most-studied are red grape pigments that slowly turn into new, modified, more stable pigments during "wine aging" (2). These substances, among others, are known as polyphenols and can take months or years to evolve during the aging process. Various types of polyphenols are found in a variety of colored fruits and vegetables as well as in tea and coffee.

A phenolic index ($E_{280} - 4$) has been used as an arbitrary division between two divisions of red wines, that is, light red wines and robust red wines (3). A low phenolic index of the former is usually from young wines marketed after a few months of vintage, whereas the latter indicates prolonged maturation and further development during bottle aging.

PHENOLICS IN WINE

The quality and characteristics of wine depend on its phenol content. The natural phenols and their derivatives are crucial to wine's color, astringency, bitterness, resistance to oxidation, and other characteristics that make wine interesting and diverse (4). Although wines differ greatly in phenol content, a gross comparison between white and red young *Vitis vinifera* wines is presented in Table 1. Phenolic compounds are usually within several chemical classes but are usually separated into two general classes, the flavonoids and the nonflavonoids, as seen in Table 1. High-performance liquid chromatography (HPLC) is presently very useful in the analysis of wine phenols. However, wine chromatographs show very large numbers of components, many that are not identified and many that may not be phenols. For reliable quantitation, a known standard sample of each phenol to be determined must be chromatographed for comparison under exactly the same conditions. A direct HPLC method for separation of all classes of wine phenolic compounds, without prepurification steps, has been developed (5).

Table 1. Gross Estimated Phenol Content for Typical Young, Light Table Wines (mg L^{-1})

	White	Red
Nonflavodoids		
Volatile phenols	1	5
Tyrosol	14	15
Gallic and other C6C1 acids	10	40
Caffeic acid/related compounds	140	140
Total	*165*	*200*
Flavonoids		
Catechins (flavan-3-ols)	25	75
Anthocyanin and derivative	0	400
Other monomeric flavonoids	Trace	25
Oligomeric flavonoids	5	500
Total	*35*	*1000*
Total phenols	*200*	*1200*

Source: Ref. 4.

Nonflavonoid Phenols

The two most prevalent types of nonflavonoids, as observed in Table 1, are gallic acid and caffeic acid, as well as their related compounds (Fig. 1a,b). Gallic acid, the major wine hydroxybenzoate compound; caffeic acid, the primary hydroxycinnamate compound; and the stilbene derivative, *trans*-resveratrol, are the major, important nonflavonoids in wine. Other hydroxybenzoates and hydroxycinnamates appear in wine with substituted molecules for hydrogen in the final acid moiety of these compounds. All wines have quite similar amounts of nonflavonoids, and these compounds can contribute important sensory properties and color stability to wine. Other nonflavonoids in wines are short-chain aldehydes, caftaric acids, vanillin, syringaldehyde, tyrosol, 4-vinylguaiacol, acetovanillone, eugenol, and 4-ethylphenol. Some of these appear to be degradation products of other phenols and increases during red wine aging (4).

trans-Resveratrol (Fig. 1c) levels in wines, which are somewhat higher in red wines than whites, was first shown to be generally under 1 mg/L (6). This compound is considered to be a phytoalexin. Phytoalexins in grapes and other plants are formed in response to stress, especially microorganisms causing disease, and are part of the plant's response mechanism for disease resistance. The importance of *trans*-resveratrol in wine is its antioxidant effect on human low-density lipoprotein (LDL) as well as a possible cancer chemopreventive agent (7). Probably with the combination of other antioxidant flavonoid compounds in wine resveratrol may be effective in inhibiting the pathogenesis of atherosclerosis (see "The French Paradox").

Some red Spanish wines, such as pinot noir and merlot, have been reported to have average levels of *trans*-resveratrol of 5 and 4 mg/L, respectively (8). The *cis* isomer of resveratrol has not been reported as a natural product in wine, but it is present and probably occurs from light exposure during winemaking and/or during storage (8).

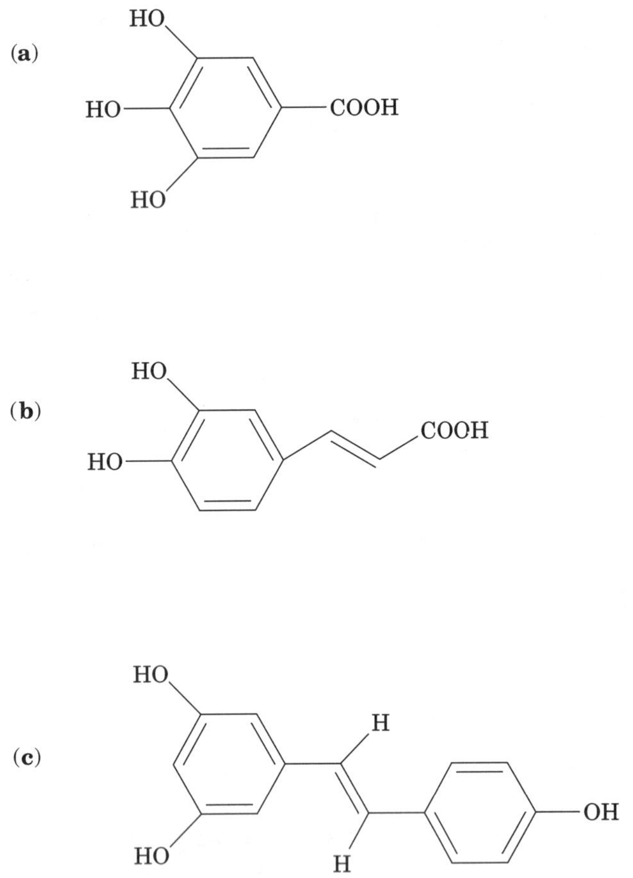

(a)

HO, HO—, —COOH, HO

(b)

HO, HO—, COOH

(c)

HO, H, OH, HO, H

Figure 1. Examples of wine nonflavonoids. (**a**) Gallic acid; (**b**) Caffeic acid; (**c**) *trans*-Resveratrol.

There is also a 3 β-glycoside of transveratrol noted in the skin of grapes called piceid (9). It is not known whether these resveratrol derivatives all have the attributed health effects reported for the *trans* isomer or how much is actually absorbed and metabolized.

Flavonoids

The highest level (Table 1) and most important phenols in wine are the flavonoids (Fig. 2). Nearly 5000 individual flavonoids have been characterized. In most cases, the various kinds of flavonoids arise by glycosylation or other substitution, and a very large number of different sugars have been found to substitute flavonoids, especially the flavonol class (10).

A significant flavonoid class are the flavan-3-ols. The monomer is catechin (Fig. 2a), and the flavan-3-ol oligomer is called procyanidin (Fig. 2b). An isomer of catechin, epicatechin, comprises one-half of the procyanidins and a good portion of the polymers of flavan-3-ols, also known as condensed tannins. These tannins, sometimes called flavolans (4), are a major component of grape seeds, skins, and stems. Tannins are astringent polyphenols, with molecular weights ranging from about 500 to 5000. These flavan-3-ol polymers are different from hydrolyzable tannins that appear in wine from wood contact, usually from barrel aging. Gallic esters of the catechins are also present in wine but at lesser amounts and are rather unstable to oxidation.

Levels of catechin, the most abundant monomeric phenolic in wine, vary from year to year and within wine varieties. As seen in Figure 3, the average catechin levels in selected California red wines by vintage show these large yearly variations. Wines made from pinot noir grapes appear to have quite low catechin levels in the 1986 vintage, yet possess the highest catechin levels, 300 mg/L, in 1989 and 1990 as compared to the cabernet sauvignon, zinfandel, and merlot wines made in these same years (10).

Quercetin (Fig. 2c) and other flavonols all contain a 4-keto group and occur in grapes as glycosides and in wines partly as aglycones. Quercetin is relatively high in grape leaves and may be controlled to a large extent by sun exposure. Flavanonols, such as dihydroquercetin and dihydrokaempferol in the form of their 3-rhamnosides, have been observed in grapes and wines at very low concentrations (4). Quercetin also appears to vary in wine by grape variety and vintage year, with cabernet sauvignon containing the relatively highest level compared to California pinot noir, zinfandel, and merlot, as seen in Figure 4. The largest variation of quercetin appears to be present in pinot noir wines. This flavonol was not detected in white wine varieties (10).

Anthocyanins are important phenols that give color to red wines primarily by extraction of these pigments from grape skins during vinification. These phenols become quite complex and seem to be of decreased importance to wine color during aging. Anthocyanins do not occur free but are glucosides in grapes and wines. Following the original work of P. Ribereau-Gayon (11) that showed the European grape *Vitis vinifera* contains only anthocyanins that are 3-monoglucosides, whereas other grape species are 3,5-diglucosides, there are more than six reviews considering analyses of anthocyanins in wine (4). Recent analytical methodologies include HPLC and polarography. However, analysis of anthocyanins in wines are quite difficult and complex. There are a number of nondefined polymeric forms and rapid and complex equilibria regarding the red forms of the anthocyanins (flavylium ion) with the purple anhydro base together with a colorless carbinol base. The acidity, temperature, interaction with other compounds (eg, catechins), concentration of the total amount of anthocyanins, and a very quick polymerization of the red pigment during winemaking and storage complicates anthocyanin analysis.

Three major anthocyanin structures are shown in Figure 5a–c. Malvidin 3-glucoside and its acylated derivatives account for more than 60% of total anthocyanins in *V. vinifera* wines. This 3,5-diglucoside, found in native American grapes such as Concord and Niagara, is termed *malvin*. Delphinidin (Fig. 5b) and peonidin (Fig. 5c) are usually the next two major anthocyanins in most wines. There are five major and about 15 minor anthocyanins believed to be present in all red wines.

Figure 6 shows the average levels of malvidin-3-glucoside in four varietal red wines. The older red wines from the 1989 vintage have only a portion of the malvidin levels than the newer 1990 and 1991 wines (10). This indicates,

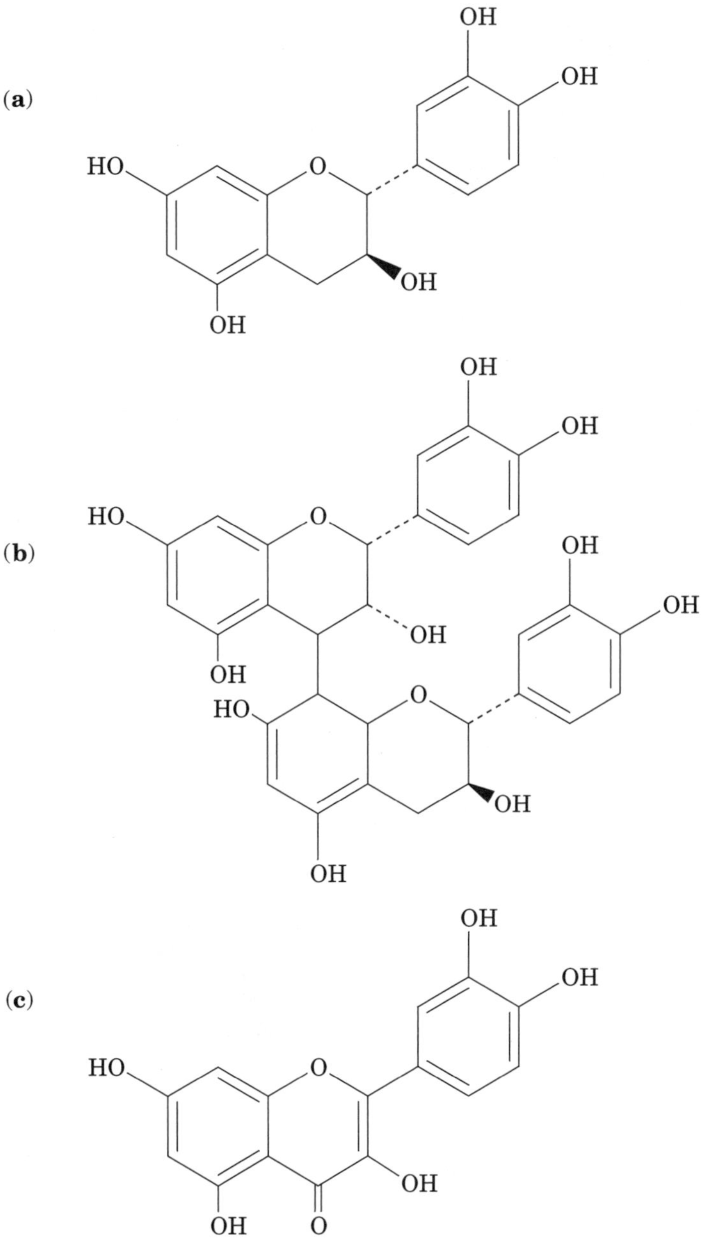

(a)

HO OH OH OH OH O

(b)

(c)

Figure 2. Examples of wine flavonoids. (**a**) Flavan-3-ol: catechin; (**b**) Flavan-3-ol oligomer: procyanldin; (**c**) Flavanol: quercetin.

as previously stated, that anthocyanins decrease with the age of wines.

VINIFICATION AND PHENOLICS

As seen in Figures 3, 4, and 6, the phenolic content of wines changes from vintage years and by grape varieties. Further phenolic changes in grapes have been noted from site of production, probably depending on a number of factors such as weather and soil conditions and grape degree of maturity. Catechins and procyanidins are observed at the highest level during the early stages of grape maturity and then decrease somewhat rapidly. Anthocyanins, developing in grape skins, steadily increase during maturity.

There are three main technological practices during vinification that change the flavonoid content of wines (12).

Destemming of Grape Clusters

After grape harvest, American winemakers usually first destem the grapes. Destemming in some European cellars is a current enological practice. Destemming grape clusters produces wines with less catechins and procyanidins. Some evidence indicates only a small reduction in epicatechins as compared to about a 25% reduction in catechin and procyanidins in wines from destemmed grapes.

Length of Maceration

Macerating red grapes with skins and seeds, and in some cases with stems, usually does not exceed 10 days because the maximum level of anthocyanin extraction then seems to occur. However, it has been noted that even with pro-

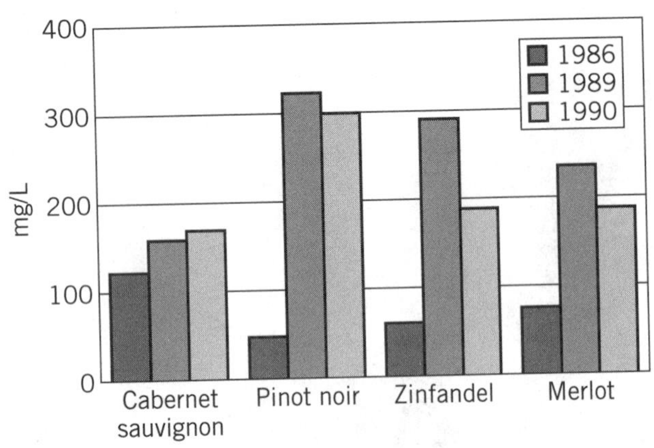

Figure 3. Average catechin levels in four red wines made in 1986, 1989, and 1990. *Source:* Reprinted with permission from Waterhouse and Teissedre, Ref. 10. Copyright 1997, American Chemical Society.

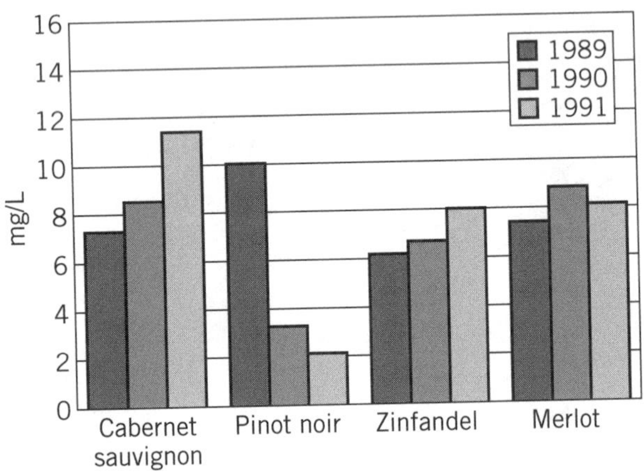

Figure 4. Average quercetin levels in four red wines made in 1989, 1990, and 1991. *Source:* Reprinted with permission from Waterhouse and Teissedre, Ref. 10. Copyright 1997, American Chemical Society.

Figure 5. Examples of three major wine anthocyanin flavonoids. (**a**) Malvidin; (**b**) Delphinidin; (**c**) Peonidin. R_1 and R_2 are glucose (3,5-diglucosides), except in *V. vinifera* grapes and wine, where R_1 is glucose and R_2 is H (3-monoglucosides).

longed skin contact and maceration, phenolic extraction rarely is more than 50% of the available grape amount. Maceration times, compared from 2 to 14 days, more than double the levels of catechin, epicatechin, and procyanidins in wine. These wines are more astringent because increased procyanidins and tannins causes high levels of astringency in young, red wines.

Fining of Wines

Fining or clarification of wines eliminates colloidal materials, usually involving phenolics and proteins, that cause turbidity and precipitates during aging. Clarification should be undertaken about 6 months after vintage, when both fermentation and malolactic fermentation have completed. The wine is then racked and filtered. Both red and white wines require fining, although some wines are now only filtered after centrifuging for faster marketing. Besides clarification of wine, certain fining agents stabilize wine against brown color development.

Bentonite, a common fining agent that is a clay consisting primarily of hydrated aluminum silicate, appears to be less effective in lowering catechins and procyanidins in wines than gelatin or poly(vinylpyrrolidone). However, the level of these and other phenols in wine as well as the amount of added fining agent used will affect the final amount of the finished wine's catechins and procyanidins.

WINE AND HEALTH

Wine is a rich source of polyphenolic antioxidants. Polyphenols are also widely present in fruits and vegetables.

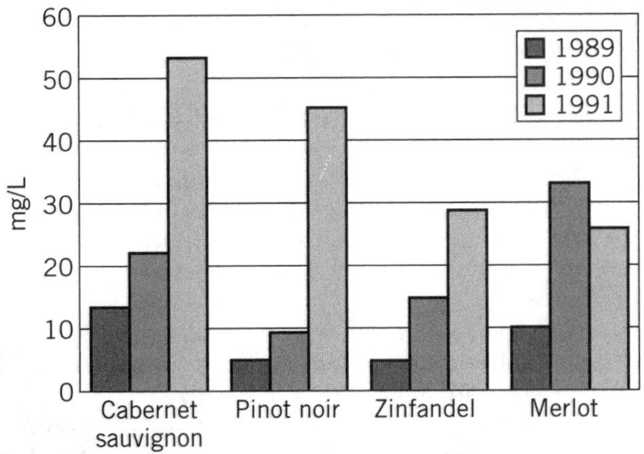

Figure 6. Average levels of malvidin-3-glucoside in four red wines made in 1989, 1990, and 1991. *Source:* Reprinted with permission from Waterhouse and Teissedre, Ref. 10. Copyright 1997, American Chemical Society.

Thus, healthy diets, as the Mediterranean diet, consists of large amounts of fruits and vegetables, with only limited levels of meat, and moderate amounts of red wine. Ethanol, the largest component of wine (<14%), and the polyphenols seem to be the healthful natural chemicals of wine as described in "The French Paradox."

Phytochemicals, such as the polyphenolics of food and wine that have disease-fighting and wellness-promoting qualities, have been recently termed *nutraceuticals*. The use of food as medicine was first presented in recorded history by the famous oath of Hippocrates, who stated he would use food first to treat disease and alleviate suffering in his patients. The use of wine in control of diseases of the heart and blood vessels has been advocated since the thirteenth century, but scientific studies published in 1904 first showed that consuming alcoholic beverages reduces the incidence of coronary artery diseases (13). Most physicians are hesitant to advise patients to drink moderate amounts of alcohol or wine, probably due to alcohol's potential abuse and known health problems in heavy drinkers. However, several population studies have shown that cardiovascular-related mortalities follow a U-shaped curve; light drinkers have lower cardiovascular-related mortalities than abstainers or heavy drinkers (14). Doll (15) has shown that drinking moderate amounts of alcohol daily, between one and four drinks for persons over the age of 45, reduces the risk of heart disease and premature death. There does not seem to be any positive effect from daily alcohol consumption for people under the age of 45.

The French Paradox

The French cardiologist S. Renaud first described a paradoxical situation in France, whereby there is a remarkably high intake of saturated fat but relatively low mortality from coronary artery disease. In most countries, a high intake of saturated fats is related to high mortality from heart disease. Renaud and de Lorgeril (16) indicated that the relatively high intake of wine in France explained this French paradox by the alcohol content of wine that inhibits blood platelet aggregation. Platelet aggregation is related to one of the conditions needed for the development of coronary disease. However, later studies by Frankel et al. (17) showed that the alcohol content of wine may not be the sole explanation for this protection. Red wine, containing antioxidant polyphenols, seemed to play an important role because *in vitro* studies showed that the red wine phenolics inhibited the oxidation of LDL. Continued studies by these investigators (18) showed that the two flavonoids, epicatechin and quercetin, had about twice the inhibiting potency as *trans*-resveratrol in LDL oxidation. The oxidation of LDL, commonly called the "bad cholesterol," is another condition usually required in the etiology of coronary disease.

Wine and Cancer

The study of food phytochemicals for cancer prevention and cure is a major research activity and will continue to expand as new knowledge develops (1). Most studies on chemical carcinogenesis in animals or with *in vitro* methodologies involve the pure phytochemical, including a number of polyphenols found in wine. Specifically, research on the flavonols, tannins (19), and resveratol (7) have shown anti-initiation, antipromotion, and/or antiprogression activities in chemical carcinogenesis studies with animal models.

The process of carcinogenesis is usually broadly divided into the three general stages of initiation, promotion, and progression. Initiation represents an alteration in the DNA of any cell by genetic mutations. Promotion is the process by which an initiated cell develops via changes of various enzyme systems and in intercellular communication patterns. Progression is the process by which a benign tumor progresses into a malignant one. The specific polyphenols, previously indicated, have shown antioxidant effects such as antimutagen mechanisms and the induction of phase II drug-metabolizing enzymes (anti-initiation effects), activation of membrane-stabilizing mechanisms (anti-inflammation responses), and inhibition of certain oxidase enzymes (antipromotion effects). Resveratrol inhibited tumorigenesis in a mouse skin cancer model and in carcinogen-treated mouse mammary glands in culture (antiprogression effects).

One recent study has shown a direct link to wine and cancer inhibition in a transgenic mouse model (20). These mice have a set of genetic disorders that cause tumors to grow along nerves similar to neurofibromatosis in humans. Dehydrated, dealcoholized red wine solids were fed as a supplement with a defined amino-acid-based diet to these mice successfully for three generations. Catechin was the major wine polyphenol, followed by gallic acid and epicatechin. These three phenols comprised about 93% of the total phenolics in the red wine used in this study. The mouse group fed the wine solids, compared to the unsupplemented controls, showed a significantly delayed tumor onset, and the major polyphenol, catechin, in the red wine was absorbed intact. This latter finding is important because if wine, or more specifically certain polyphenols in wine, has a role in the prevention of human cancer development, then future studies must show absorption, metab-

$$NH_2-\overset{\displaystyle O}{\overset{\|}{C}}-O-C_2H_5$$

Figure 7. Ethyl carbamate.

olism, organ distribution, and excretion patterns of these polyphenolics and their metabolites. In addition, ethanol could play an interactive role in the enhancement of these processes.

Ethyl Carbamate

Ethyl carbamate, also commonly known as urethane (Fig. 7), is produced at very low levels in many fermented foods and beverages, including bread, beer, soy sauce, yogurt, and red and white wine. It is developed in wine by the presence of urea with alcohol, by certain malolactic bacteria, or even by heating nonfermented grape juice with added ethanol (21).

Numerous wine surveys showed a range of ethyl carbamate levels from undetected to more than 100 μg/L. This compound is a water-soluble carcinogen as shown in numerous animal studies and is therefore considered a human dietary carcinogen. Thus, its content in foods and beverages should be as low as possible. In 1988, the Food and Drug Administration (FDA) established that American table wines should not have ethyl carbamate levels higher than 15 μg/L. The FDA also recommended that wine importers have some form of testing program so that foreign wines do not exceed this limit.

Carcinogenesis studies with mice have shown that the intake of ethanol in wine protects against the metabolite of ethyl carbamate, that is, the active or ultimate carcinogen, from forming through a competitive interaction mechanism. Also, lowered food and caloric intake with concomitant lowered body weights in mice seems to further reduce ethyl carbamate tumor development (22).

BIBLIOGRAPHY

1. M. Messina and V. Messina, "The Second Golden Age of Nutrition—Phytochemicals and Disease Prevention," in M.-T. Huang et al., eds., *Food Phytochemicals for Cancer Prevention I. Fruits and Vegetables*, ACS, Washington, D.C., 1994.
2. R. Brouillard, F. George, and A. Fougerousse, "Polyphenols Produced During Red Wine Ageing," *Biofactors* **6**, 403–410 (1997).
3. T. C. Somers and E. Verette, "Phenolic Composition of Natural Wine Types," in H. F. Linskens and J. F. Jackson, eds., *Wine Analysis*, Springer-Verlag, New York, 1988.
4. V. L. Singleton, "Wine Phenolics," in H. F. Linskens and J. F. Jackson, eds., *Wine Analysis*, Springer-Verlag, New York, 1988.
5. R. M. Lamuela-Raventos and A. L. Waterhouse, "A Direct HPLC Separation of Wine Phenolics," *Am. J. Enol. Vitic.* **45**, 1–5 (1994).
6. E. H. Slemann and L. L. Creasy, "Concentration of the Phytoalexin Resveratrol in Wine," *Am. J. Enol. Vitic.* **43**, 49–52 (1992).
7. M. Jang et al., "Cancer Chemopreventive Activity of Resveratrol, a Natural Product Derived from Grapes," *Science* **275**, 218–220 (1997).
8. R. M. Lamuela-Raventos et al., "Direct HPLC Analysis of *cis*- and *trans*-Resveratrol and Piceid Isomers in Spanish Red *Vitis vinifera* Wines," *J. Agric. Food Chem.* **43**, 281–283 (1995).
9. A. L. Waterhouse and R. M. Lamuela-Raventos, "The Occurrence of Piceid, a Stilbene Glucoside, in Grape Berries," *Phytochemistry* **37**, 571–573 (1994).
10. A. L. Waterhouse and P. -L. Teissedre, "Levels of Phenolics in California Varietal Wines," in T. R. Watkins, ed., *Wine Nutritional and Therapeutic Benefits*, ACS, Washington, D.C., 1997.
11. P. Ribereau-Gayon, "*Recherches sur les Anthocyanes des Vegetaux. Application au Genre* Vitis," Librairie Gen. de l'Enseignement, Paris, 1959.
12. E. Revilla, E. Alonso, and V. Kovac, "The Content of Catechins and Procyanidins in Grapes and Wines as Affected by Agroecological Factors and Technological Practices," in T. R. Watkins, ed., *Wine Nutritional and Therapeutic Benefits*, ACS, Washington, D.C., 1997.
13. R. T. Cabot, "The Relation of Alcohol to Arteriosclerosis," *JAMA* **43**, 774–775 (1904).
14. A. L. Klatsky, "The Epidemiology of Alcohol and Cardiovascular Diseases," in T. R. Watkins, ed., *Wine Nutritional and Therapeutic Benefits*, ACS, Washington, D.C., 1997.
15. R. Doll, "One for the Heart (Beneficial Effects of Alcohol Consumption on the Heart)," *Br. Med. J.* **315**, 1664–1668 (1997).
16. S. Renaud and M. de Lorgeril, "Wine, Alcohol, Platelets, and the French Paradox for Coronary Heart Disease," *Lancet* **339**, 1523–1526 (1992).
17. E. N. Frankel et al., "Inhibition of Oxidation of Human Low-Density Lipoprotein by Phenolic Substances in Red Wine," *Lancet* **341**, 454–457 (1993).
18. E. N. Frankel, A. L. Waterhouse, and J. E. Kinsella, "Inhibition of Human LDL Oxidation by Resveratrol," *Lancet* **341**, 1103–1104 (1993).
19. L. O. Dragsted, M. Strube, and J. C. Larsen. "Cancer-Protective Factors in Fruits and Vegetables: Biochemical and Biological Background," *Pharmacol. Toxicol.* **72** (Suppl. 1), 116–135 (1993).
20. A. J. Clifford et al., "Delayed Tumor Onset in Transgenic Mice Fed an Amino Acid-Based Diet Supplemented with Red Wine Solids," *Am. J. Clin. Nutr.* **64**, 748–756 (1996).
21. I.-M. Tegmo-Larsson and T. Henick-Kling, "Ethyl Carbamate Precursors in Grape Juice and the Efficiency of Acid Urease on Their Removal," *Am. J. Enol. Vitic.* **41**, 189–192 (1990).
22. G. S. Stoewsand, J. L. Anderson and L. Munson, "Wine and the Effect of Body Weight on Ethyl Carbamate-Induced Tumorigenesis in Mice," *J. Wine Res.* **7**, 207–211 (1996).

G. S. STOEWSAND
Cornell University
Geneva, New York

PINEAPPLES. See FRUITS, TROPICAL.

POTATOES AND POTATO PROCESSING

Potatoes are members of the Solanaceae family. Of the many tuber-forming *Solanum* species, the one that is most widely cultivated is *Solanum tuberosum*. The edible portion of potato is a tuber or an underground stem. While the potato is referred to as the "Irish potato," the center of origin of cultivated potatoes is in the Andes Mountains of South America (1,2). Long before the Spaniards arrived in South America, potatoes were a major portion of diet and a component of the culture of the ancient civilizations of that part of the world. Potatoes were introduced in Europe by the mid- to late-1500s by the Spanish. It took some time for them to be accepted as a food. However, potatoes became a vital part of the food supply of Ireland in the 1600s. Unfortunately, potatoes were the primary component of the Irish diet by the late 1840s when late blight (*Phytophthora infestans*) destroyed a major portion of the crop. The famine forced many Irish to immigrate to North America. The importance of potatoes in Ireland is responsible for the term Irish potato.

Potato tubers are composed of approximately 75 to 80% water, 20 to 23% carbohydrates, and about 2% protein. Approximately 80% by weight of potato carbohydrate is starch and is composed of amylopectin (75–79%) and amylose (21–25%) (3,4). The nutritional value of the crop was demonstrated by the Irish diet prior to the famine. Currently most individuals underestimate the nutritional value of potatoes. Potatoes are a good source of vitamin C, B_6, potassium, and dietary fiber (Fig. 1). Although the protein content of potatoes is relatively low, it is a high-quality protein, and as demonstrated by Irish history, if enough potatoes are consumed, the requirement for protein can be satisfied. Because of the potato's ability to produce high yields under a wide range of climatic conditions, it is an important crop worldwide (5). It ranks after wheat, rice, and maize as the fourth most important crop and is the most widely grown vegetable in the world. The largest potato-producing countries in 1997 were China (46 million t), Russia (40 million t), Poland (25 million t), and the United States (20 million t) (6).

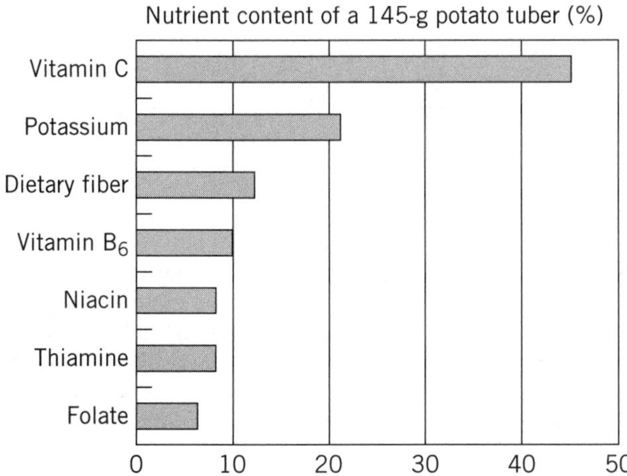

Figure 1. Nutritional contribution of a medium-sized potato to the daily requirements of 2000-calorie diet. *Source:* National Potato Board, Denver, Colo.

The per capita consumption of potatoes in the United States has increased from 46 kg in 1950 to 64 kg in 1997 (7). Approximately one-third of the United States crop is consumed in the fresh form while frozen products accounted for about 40% of the consumption (Fig. 2).

Nutritionally, potatoes are quite stable and are available year-round through modern storage and processing technology. Approximately 75% of the potatoes produced in the United States are stored for later consumption. Storage terms range from 2 to 11 months as required to match raw material supply with utilization. Because potatoes are highly perishable, potato storages must be designed and operated effectively and economically to limit losses in quality and quantity of the raw potatoes. Special storage environments are required to maintain potatoes for their intended end use (Table 1).

Potato varieties differ in their culinary characteristics. Potatoes with high dry matter (or specific gravity) are preferred for processing into frozen French fries, potato chips and dehydration, and as mealy, dry-textured baked potatoes. Potatoes used in potato salads or in canned products,

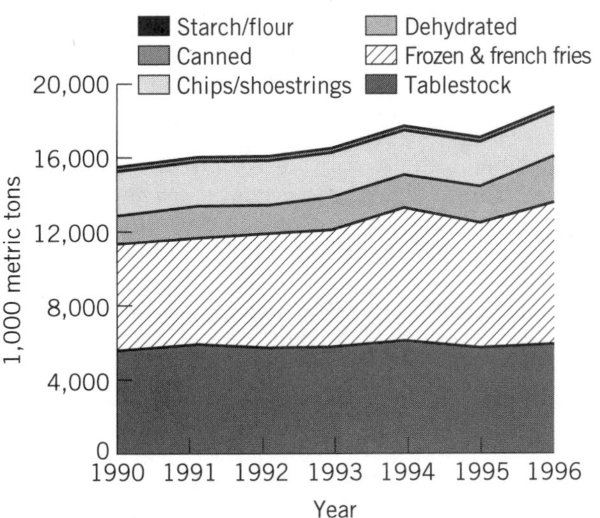

Figure 2. The fresh market equivalent of utilization of U.S. potatoes, 1990–1996.

Table 1. Environmental Requirements of Potatoes in Storage for Specific End Uses

End use	Temperatures, °C	Relative humidity, %	Airflow rate, m^3/kg-h
Seed	3–4	>90	0.02
Fresh	4	>90	0.02
French fries	7	>95	0.04
Chips	7–10	90–95	0.04

such as soup, should have low specific gravity. High-specific-gravity potatoes tend to fall apart when boiled.

The sugar content and composition of potatoes also determine how they are utilized in processing. Potatoes with a high reducing sugar level produce fried products that have a dark color.

PROCESSING OPERATIONS

Processing potatoes into finished food products involves a sequence of operations dependent on the product being prepared. Figure 3 illustrates major operations typically used for processing French fries, chips, and flakes. The initial stages of processing are common to the three major types of potato processing. Receiving, washing, peeling, inspection, and cutting/slicing potatoes are done in a similar manner, even though they may not be done in exactly the same. The steps outlined in Figure 3 are specific to the product being produced.

Receiving

Receiving operations are among the procedures that are common to all potato processing plants. Appropriate types and sizes of equipment and facilities are required for unloading (dump hoists, hoppers), handling (flumes, conveyors), and temporarily storing (bulk bins, pallet boxes) raw potatoes delivered to the plant by controlled-temperature highway trucks and railroad cars. Typical receiving operations consist of either wet fluming or dry conveying types of systems. Usually, some means of sampling the incoming potatoes is used to determine raw product quality.

Most potatoes are delivered to processing plants unwashed. Therefore, a system to remove soil and extraneous material is required at the plant. Forced water sprays combined with brushing are used to remove soil from the tubers. Destoners and trash elevators are normally incorporated into either the receiving operations or the initial washing operation at the plant.

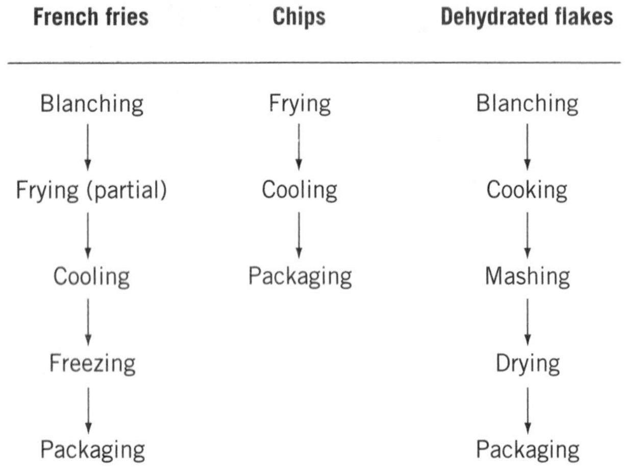

French fries	Chips	Dehydrated flakes
Blanching	Frying	Blanching
↓	↓	↓
Frying (partial)	Cooling	Cooking
↓	↓	↓
Cooling	Packaging	Mashing
↓		↓
Freezing		Drying
↓		↓
Packaging		Packaging

Figure 3. Flow diagram showing major operations in processing raw potatoes into French fries, chips, or dehydrated flakes.

Peeling

Peeling is a critical step in preparation for processing because the amount of material removed directly affects final product yield. Peel removal normally ranges from 1 to 10% by weight of raw product in chipping plants and from 10 to 25% in French fry plants. Steam peelers followed by abrasive peeling types are the most common. Abrasive peelers function by contact between tubers and an abrasive coating on moving rolls, drums, or disks to physically erode the peel from the tuber. The removed peel is then flushed away with flowing water. Steam peeling softens the peel at tuber surfaces via the action of heat and pressure in an enclosed vessel. Water sprays and brushes are used to remove the softened peel after the tubers exit the pressure vessel. Time and temperature settings are critical to controlling depth of peel softening and removal in the steam process.

Inspection

Removal of the peel often discloses blemishes within the potato tubers. Inspection for these defects is accomplished by visual means or electromechanical detection and removal. The number of inspectors or mechanical devices depends on product volume and/or defects. Defects such as bruised and discolored areas are commonly trimmed away or the entire tuber may be removed. Trimming of such defects from the tubers by hand is performed by the inspectors or by a photo/mechanical process. In some plants, a second inspection follows the cutting/slicing operation. Then optical detection and mechanical removal of only the defective portion of a French fry strip is performed.

Cutting/Slicing

Subsequent to the inspection operation, potatoes typically enter a cutting/slicing operation, which consists of equipment designed to produce cut potato pieces of appropriate size and shape from the raw tubers. A variety of sizes and shapes of slabs and strips may be obtained from a raw tubers by use of a basic industrial-type rotary vegetable slicer. In some plants, unique water knives, which cut peeled tubers into strips for French fries while being carried by flowing water in pipes, are used. Specially designed equipment provides for proper orientation of tubers relative to cutter blades and results in the longest possible cut strips. Multiple cutter units are used to meet plant capacity requirements.

Blanching/Cooking/Frying

Blanching is not used in chip processing but is a necessary step in French fry and flake production. Blanching (soaking in water at 70–80°C) accomplishes the gelatization of starch granules to improve product texture and limits subsequent cooking oil absorption. Additionally, blanching limits enzyme activity and leaches reducing sugars from the raw potato strips. The removal of reducing sugars is important in preventing dark-colored French fries. Cooling in water at approximately 10°C immediately following blanching is used in potato-flake processing to accomplish gelling of starch. This step is not needed in French fry production. To improve finished product quality (crispness and temperature retention), potato strips for French fries are dried after blanching and prior to frying. In some cases various coatings may be added prior to frying.

Although French fries and chips are both fried in cooking oil—times, temperatures, and equipment are different. Chips are fried as thin (0.125–0.175 cm) slices at times and temperatures that rapidly reduce their moisture content to less than 2%. Typical times and temperatures for chip frying might range from 1.5 to 3 min and from 155 to 190°C. French fries may be fried as strips ranging from 0.65 to 1.25 cm in thickness. Fry times are influenced by slice thickness and oil temperature, but range from 1 to 2.5 min in oil ranging from 160 to 190°C to accomplish an adequate par fry (partial frying), which removes moisture to less than 30 to 40%. Finish frying is done by the end-product user.

Cooling/Dehydrating

Products such as dehydrated potato flakes require cooking rather than frying. The cooking follows the blanching and cooling operations and use times and temperatures that best accomplish cell aggregation with minimal cell rupture. Typically, cooking involves the application of live steam (100°C) for times ranging from 20 to 30 min, dependent on slab thickness and potato solids content, to obtain a cooked product for mashing.

Potatoes are mashed immediately after cooking to minimize cell rupture. Additives such as emulsifiers that prevent gluiness/stickiness by complexing free starch (8), chelating agents that prevent after-cooking darkening by complexing metal ions (9), and antioxidants that prevent oxidation of potato components (10) are incorporated during the mashing process. Cooked potatoes are mashed in equipment that blends the potato tissue with additives until an appropriate consistency is obtained for application to the dehydrating drums for drying.

Dehydration to flakes consists of drying a thin layer of the mashed potato material to less than 6% moisture on a steam-heated rotating drying drum. This process results in a sheet of dried potato tissue 0.015 to 0.025 cm thick (11), which is then reduced to flakes via size-reduction/grinding equipment.

When processing potato chips or French fries, the frying operation is usually followed by a short cooling period before proceeding to the next operation. This cooling generally occurs on conveyor belts at ambient conditions in the plant (Fig. 4). French fries may undergo removal of excess surface oil in 70°C flowing air immediately following frying and before the cooling operation. Cooled French fries then enter a freezing operation. This operation rapidly reduces the product temperatures to −15 to −7°C with refrigerated air or by application of liquid nitrogen (−50°C) directly to the product surface.

Packaging

Packaging of all forms of processed potato products is accomplished by fully automated systems in most processing plants. Potato chips are packaged with form-fill equipment that currently uses special packing materials (foil laminates) that eliminate light and oxygen and provide for maintenance of injected nitrogen. This nitrogen serves both as an antioxidant for the contents and a cushion for the package. Frozen French fries are placed in retail or institutional packages via either form-fill or carton-fill equipment. Polyethylene materials and corrugated cardboard are commonly used for packaging frozen French fries. Potato flakes are filled into institutional or retail packages by automated fillers. Packages normally are made up of oxygen-barrier materials within corrugated boxes or polyethylene bags.

Figure 4. French fries cooling on conveyor belts. *Source:* Courtesy of J. R. Simplot Company.

Numerous types, sizes, and shapes of packages are used in the potato-processing industry to meet the myriad of marketing requirements. Individual packages are normally packed into cases that are further unitized into pallet loads for handling and transporting.

QUALITY

Recent additions of new processed products and heightened consumer awareness of quality aspects of existing processed products have drawn more attention to quality in all raw materials being processed. The primary raw material used in potato processing is, of course, the raw potato tuber. Finished product quality parameters such as color, flavor, texture, and even yield of finished product are directly related to specific quality characteristics of the raw tubers from which the products are processed (12).

Raw Product Quality

Physical characteristics of raw potatoes such as size, shape, smoothness, defects (greening, bruises), diseases (rots), and sprouts can influence the acceptance of potatoes for a particular processed potato product. For example, raw tuber length is extremely important for meeting the strip-length requirements of high-quality French fries. Conversely, smaller, round tubers are preferred for making potato chips, while neither tuber size nor shape is critical in flake processing.

Generally, chemical composition of the raw tuber is more critical across all forms of processed products. Dry-matter content (often measured as specific gravity) directly contributes to final yield of finished products in all processes and thus directly affects the economics of the processing plant (13). Typical ranges of acceptable specific gravity in raw potatoes for making various processed products are given in Table 2.

When the process involves frying of the potato tissue, a very high specific gravity is preferred. In addition to the concern for product yield, dry-matter content directly influences cooking oil absorption (15). In the production of French fries, however, specific gravities above 1.095 have been shown to contribute to poor texture and loss of desirable flavor in the final product.

The sugar content of raw potatoes for processing is another critically important chemical component, particularly when the process involves high-temperature frying. In such cases, the sugar content of potato tissue directly affects finished product color due to the Maillard reaction (16). In general, uniform light-colored finished products are desired (ie, chips, French fries); however, some products are required to exhibit some browning (hash browns, shoestrings). If the balance between sucrose and the reducing sugars, glucose and fructose, is unfavorable in the raw potatoes, the finished product will have poor color. Typically, a sucrose content less than 0.5% by weight and a glucose content less than 0.05% by weight are necessary for production of light-colored potato chips (17). Slightly higher concentrations of sugar are usable when producing French fries.

Finished Product Quality

Because raw-potato quality has a natural variability within any processing time interval, finished product quality will also exhibit variability. Therefore, monitoring of finished product quality versus some standard is practiced in most processing plants.

Potato-chip color can be either subjectively or objectively evaluated. Subjective methods generally use color photographs of chips to which sample chips can be visually compared. Objective methods, such as spectrophotometric instruments, use measurement of light reflected from the finished chip material. Visual comparison to photographs is common in French-fry processing. French fries are also checked against a cut-length standard by sorting the sample. Sensory analyses of finished product samples are conducted to monitor flavor, texture, and appearance.

Oil content of fried products is determined via rapid methods such as refractometry or slower methods such as Soxhlet extraction. The oil content of chips and French fries will vary according to raw product dry-matter content, cut sizes, fry times, and temperatures, but it will normally range from 30 to 45% by weight in chips and from 4 to 7% in partial fried French fries.

Table 2. Example Ranges of Specific Gravity for Selected Processed Products

Processed product	Specific gravity	Dry-matter content, %[a]
Chips	1.075–1.100	19.2–24.4
French fries	1.075–1.095	19.2–23.4
Flakes	1.075–1.100	19.2–24.4

Source: Ref. 14.
[a]% dry matter = 24.182 + 211.04 (sp. gr. − 1.0988)

BIBLIOGRAPHY

This article was initially written by Paul H. Orr, Red River Valley Potato Research Laboratory, East Grand Forks, Minnesota, and Jerry N. Cash, Michigan State University, East Lansing, Michigan.

1. R. N. Salaman, *The History and Social Influence of the Potato*, Cambridge University Press, 1987.
2. J. B. Sieczka and R. E. Thornton, "Commercial Potato Production in North America," in *Potato Association of America Handbook*, Potato Association of America, 1993.
3. B. B. Dean, *Managing the Potato Production System*, The Haworth Press, Inc., 1994.
4. A. van Es and K. J. Hartmaus, "Starch and Sugars During Tuberization, Storage and Sprouting," in A. Rostovski et al., eds., *Storage of Potatoes*, Pudoc Wageningen, 1987.
5. D. Horton, "Potato Nutrition and Consumption," in *Potatoes Production Marketing and Program for Developing Countries*, Westview Press, 1987, pp. 93–110.
6. National Potato Council, *Potato Statistical Yearbook*, Englewood, Co., 1997.
7. National Potato Council, *Potato Statistical Yearbook*, Englewood, Co., 1998.
8. E. C. Lulai, "Interrelationships of Quality Testing Methods and Relative Effectiveness of Surfactants on Potato Flake Quality," *Amer. Potato J.* **60**, 441–448 (1983).
9. O. Smith and C. O. Davis, "Potato Quality, XIV. Prevention of Graying in Dehydrated Potato Products," *Amer. Potato J.* **39**, 135–146 (1962).
10. G. M. Sapers et al., "Flavor Quality and Stability of Potato Flakes, Effects of Antioxidant Treatments," *J. Food Sci.* **40**, 797–799 (1975).
11. W. F. Talburt and O. Smith, *Potato Processing*, 4th ed., Van Nostrand Reinhold Co., New York, 1987.
12. W. A. Gould and S. Plimpton, "Quality Evaluation of Potato Cultivars for Processing," *Ohio Agricultural Experimental Station Bulletin* **1172**, 1985.

13. P. H. Orr and C. K. Graham, "A Generalized Model for Determining Least-Cost Sources of Processing Potatoes," *Trans. Amer. Soc. Agricultural Engineers* **26**, 1875–1878 (1983).

14. C. von Scheele, G. Svensen, and J. Rasmussen, "Du Bestamming au Patatiseus, Starelse och Torrsubstanshalt wed Tilhjap av Dess Specifka VKT," *Novd. Jordr. Forsk*, 12–37 (1935).

15. E. C. Lulai and P. H. Orr, "Influence of Potato Specific Gravity on Yield and Oil Content of Chips," *Amer. Potato J.* **56**, 379–390 (1979).

16. R. S. Shallenberger, O. Smith, and R. H. Treadway, "Role of the Sugars in the Browning Reaction in Potato Chips," *J. Ag. Food Chem.* **7**, 274–277 (1959).

17. J. R. Sowokinos and D. A. Preston, "Maintenance of Potato Processing Quality by Chemical Maturity Monitoring (CMM)," *Minnesota Agricultural Experimental Station Bulletin* AD-SB-3441, 1988.

Joseph B. Sieczka
Cornell University
Riverhead, New York

POULTRY FLAVORS

Poultry is among the most popular food products in the world. Its consumption has increased substantially in the last two decades. Consumer demand for poultry reflects the acceptability of the flavor of poultry products. Many factors influence the flavor quality of poultry foods, and numerous other factors affect the formation of flavors that evolve through complex chemical reactions during the heating process. This article summarizes pertinent production and processing factors associated with flavor variations and briefly reviews the chemical basis of poultry flavors.

For the purpose of this article, the flavor of poultry is defined as a blend of tastes and aromas of cooked poultry tissues, including lean dark and lean light meat, skin, and fat, primarily from chicken and turkey. In general, normal or conventionally produced and fresh-cooked poultry exhibit such flavor sensations as meaty, chickeny, chicken brothlike, and, to some extent, fatty-oily. On the other hand, any flavor notes that deviate from those characteristics normally expected from cooked poultry lead to the perception of off-flavors, such as fishy, stale, and rancid. However, there is no standard poultry flavor; sensory perception of a poultry food item with regular poultry flavor as judged by one panel in one region may be judged as weak or having no poultry flavor by another panel in a different region or as somewhat off in flavor by yet another panel. Hence, this article regarding flavor description and intensity is based mainly on individual reports in the literature.

EFFECT OF PRODUCTION AND PROCESSING METHODS

Production Factors

Production aspects that may influence the flavor of poultry include breed, sex, age, feed composition, and management. However, research findings regarding the effects of these factors are often inconsistent. An extensive investigation by the U.S. Department of Agriculture (USDA) in 1960 compared the flavor of 2,000 chickens grown under different production practices and concluded that modern birds (those grown in 1956) have as much flavor as old-style birds (1930s) regardless of the differences in breed and ration (1). On the contrary, some other reports indicated that broilers on rations containing dairy products produced meat with more flavor and that corn-fed poultry was more flavorful than poultry fed barley, oats, and wheat (2). Breast meat of older birds may have more flavor and odor intensity, whereas birds reared at low stocking density could produce thigh meat with stronger flavor (3).

Different species of poultry have different phospholipid profiles, and within one species the phospholipid profile is closely influenced by the fat composition of feed (4). One of the most significant findings concerning the effect of feed composition on poultry flavor was the determination of the responsible ingredients for the fishy off-flavor. It was found and confirmed that the fishy flavor was due to the presence of fish meal, fish oil, linseed oil, and other highly unsaturated fatty acids in the ration (2,5). The addition of alpha-tocopherol to feed reduces the fishy flavor to some extent. Dietary vitamin E also contributes to the oxidative stability of refrigerated and frozen turkey breast meat (6).

Processing, Preparation, and Cooking Effects

Many processing steps are required in the transformation of a live bird to a ready-to-cook, semiprepared (semicooked) or fully prepared (cooked) poultry product. Each processing step is also subjected to variations. Processing variables that have been investigated regarding the flavor of cooked poultry include chilling of eviscerated carcasses, application of phosphates, canning, refrigeration, and frozen storage of raw carcass or cooked meat.

The chilling method has little effect on cooked meat flavor provided that the meat is cooked within a short period after refrigeration or frozen storage. However, air-chilled chicken may not be as stable as immersion-chilled chicken over long storage periods (7). In general, frozen storage extends the storage time to 12 to 14 months without quality deterioration. Thawing and refreezing within a short period has no significant effect; the flavor of cooked meat was not affected when the raw carcass was thawed and frozen five times (8). Commercially processed fresh poultry items are usually distributed and marketed under refrigeration conditions. This short storage period before cooking facilitates the flavor development of cooked meat (9).

The flavor of cooked poultry also depends on the cooking and preparation methods. Flavor characteristics are different among boiled, roasted, and fried chicken, although all these items have similar chickeny flavor. Further-processed poultry products, such as refrigerated, precooked poultry meat, are susceptible to chemical deteriorations and the development of warmed-over flavor during storage. Frozen storage of cooked meats reduces the off-flavor problems to some extent. During chill storage, the flavor intensities of cardboard, warmed-over, and rancid/painty as well as overall off-flavor characteristics increased in broiler breast and thigh meat but not as much in skin (10). Mechanisms that have been suggested as being responsible for off-flavors include the breakdown of cell membranes during cooking and the autoxidation of polyunsaturated fatty acids of the membrane materials. Proper packaging and addition of antioxidants such as polyphosphate may retard off-flavor development (11). The stability of cooked, chill-stored poultry meat may also be affected by precooking temperature. Higher initial cooking endpoint temperatures accelerate the oxidative development during subsequent storage of cooked chicken (12,13).

CHEMICAL BASIS OF POULTRY FLAVORS

Development of Flavors

Raw meats, including poultry and red meats, generally have very little flavor except essentially a simple bloody aroma. During heating of the meat tissue, various chemical reactions take place, and the resulting numerous compounds are responsible for the cooked meat flavor. The precursors or the constituents in the raw tissue include proteins, peptides, free sugars, free amino acids, lipids, fatty acids, salts, minerals, nucleic acids, nucleotides, nucleosides, glycogen and amines, among others, with very few volatile compounds. The brothy taste of chicken is due to the presence of free amino acids and oligopeptides; these components are the flavor precursors (9). The removal of sulfur compounds from the volatile fraction of chicken meat and water slurries by passing the volatiles through mercuric chloride or mercuric cyanide solution resulted in an almost complete loss of "meaty odor." Removal of the carbonyl fraction via trapping in 2, 4-dinitrophenylhydrazine solution resulted in loss of "chickeny flavor," demonstrating a lipid role in species character (14).

The reactions and interactions of various constituents on heating, such as oxidation, decarboxylation, condensation, and cyclization, generate many volatiles. Major reactions involved in the development of cooked chicken flavor are the Maillard reaction and Strecker degradation, producing aldehydes, hydrogen sulfide, and ammonia. The Maillard reaction may result from the reaction of glycosidic hydroxyl groups of sugars and amino groups of amino acids, peptides, or proteins and can also occur between the hydroxyl groups of oxidized lipids and amino groups. Different tissues within the same species have different chemical compositions and undergo different chemical pathways (15). Hence, different volatiles and aqueous flavor compounds are generated on heating different raw tissues. Each compound may contribute a part of the blended flavor sensation. Numerous studies have attempted to qualitatively and quantitatively analyze all compounds related to poultry flavors and off-flavors.

Since the advent of analytical instrumentations, studies on the isolation and identifications of flavor compounds have increased rapidly. In 1972, a list of 178 compounds of raw and cooked poultry was compiled (16). Additional compounds were subsequently identified and 30 new compounds were found in chicken broth (17), including cyclic and acyclic hydrocarbons, alkylbenzenes, a terpene, and a nitrile. The number of total compounds of poultry flavors increased to more than 250 in 1982 (2). Later, 130 compounds, including many new ones, were isolated and identified in fried chicken that might originate from flour, oil, and/or the interaction products of chicken, flour, and oil (18). Eighty new compounds were reported in roasted chicken meat and skin (19) and 75 new compounds in fat drippings (20). Currently, more than 500 compounds, mostly volatiles, have been identified in heated poultry products. The major classifications and examples of compounds in each class are presented in Table 1.

Limited quantitative data have been reported. Although volatile components account for only a small fraction of chicken composition, these compounds are important to the flavor sensation. The aldehyde group is the predominant class of the total volatiles from roasted chicken meat and skin. Other important classes are the sulfur-containing compounds, alcohols, and ketones. The aldehydes are also abundant in the volatiles of fat drippings. The acid compounds are more important in fat drippings than in meat and skin. The identification of lactones in fat drippings deserve special attention in light of their absence in the meat and skin of roasted chicken. For a complete listing of all identified compounds, see Refs. 2 and 19 to 23. A brief discussion of major classes, their precursors, and the chemistry of flavor formation is presented in the following sections.

Hydrocarbons, Alcohols, and Carbonyl Compounds

Earlier studies of volatiles of cooked poultry identified a variety of hydrocarbons, alcohols, acids, aldehydes, and ketones. Of these classes, the carbonyl group (aldehydes and ketones) has often been suggested to be associated with meat flavor. Lipids are the principal precursors of these compounds, and many carbonyls probably originate from oleic, mevalonic, linoleic, and arachidonic acids (5,16). The flavor threshold values of some of the carbonyl compounds (up to C_9) are as low as a few parts per billion. The presence of highly unsaturated fatty acids, primarily in the phospholipids in chicken, is the basis for the development of the characteristic carbonyl profile that differentiates poultry flavor from the flavors of other species (14,24). At least nine compounds have been identified exclusively or predominantly in chicken, including 3-methyloctane, 2,2,6-trimethyloctane, 4-ethylbenzaldehyde, 1,2-dibutylcyclopentane, 2,6-bis(1,1-dimethylethyl)-4methylphenol, diethyl phthalate, 1,12-dodecanedial, 2(E), 4(E)-decadienal, and γ-dodecalactone (21,22). Aroma extract dilution analysis of volatiles obtained by the simultaneous distillation/extraction of chicken broth resulted in the identifications of nonanal, 2(E)-nonenal, 2(E), 4(E)-nonadienal, 2(E), 4(E)-decadienal, 2-undecenal, β-ionone, γ-decalactone, and γ-dodecalactone as primary chicken odorants. Alkanals, generated from turkey skin, contributed to meaty and turkeylike flavor, and the alk-2-enal group had a strong, oxidized, and brothlike flavor (25). Utilizing a supercritical carbon dioxide extraction technique followed by identification of compounds in isolated fractions, it was concluded that excess concentrations of enal aldehyde and ketones are related to the oxidized aroma and flavor of chicken broth (26).

In addition to the pathway of lipid oxidation, carbonyls can be derived from nonenzymatic browning (Maillard re-

Table 1. Representative Flavor Compounds in Poultry

Aliphatic Hydrocarbons (over 45 compounds)
n-decane
n-dodecane
n-heptane
1-octene
n-tridecane
Aromatic hydrocarbons (over 50 compounds)
benzene
n-propylbenzene
toluene
p-xylene
1,2,3-trimethylbenzene
Alcohols (over 45 compounds)
n-butanol
n-hexanol
n-heptanol
n-pentanol
2-methylpropanol-1-ol
Aldehydes (over 75 compounds)
acetaldehyde
n-butanal
2-methylbutanal
n-hexanal
trans-2-hexenal
Ketones (over 55 compounds)
acetone
2-butanone
2-heptanone
2,3-butanedione
3-octen-2-one
Amino acids, amines, and peptides (over 25 compounds)
cysteine
cystine
glutathione
methionine
dimethylamine
Fatty acids and esters (over 40 compounds)
pentanoic
ethyl acetate
hexanoic
2(3)-methylbutyric
ethyl lactate
Lactones (in fat drippings) (over 24 compounds)
4-hydroxybutanoic acid lactone
4-hydroxyhexanoic acid lactone
4-hydroxyoctanoic acid lactone
Pyrrols, pyridines, and pyrazines (over 50 compounds)
2-methylpyrazine
pyridine
pyrrole
3-ethylpyridine
2,6-dimethylpyrazine

Table 1. Representative Flavor Compounds in Poultry *(continued)*

Sulfur compounds (over 45 compounds)
methyl disulfide
thiophene
2-methyl thiophene
thiazole
ethane thiol
Furans (over 20 compounds)
2-butyl furan
2-hexyl furan
2-octyl furan
Miscellaneous (over 35 compounds)
hydrogen sulfide
ribose
inosine-monophosphate
chloroform
4,5-dimethyloxazole

action), Strecker degradation, and Amadori transformation reactions. The condensation of carbohydrates with amino acids during heating causes the formations of complex cyclic carbonyls. Carbonyl compounds could be important in secondary reactions; they resulted in furanones and mercapto compounds with roast meat aroma and meaty aroma (27).

Sulfur-Containing Compounds

Sulfur-containing compounds are also responsible for the species flavor (28), and they are important contributors for the meaty aroma in both dark and light chicken meat (14). The presence of hydrogen sulfide (H_2S) and its importance in chicken flavor have been confirmed in several studies (5). The low threshold level (10 ppb in water) makes H_2S an important factor in the cooked-meat aroma. H_2S may also contribute indirectly to the cooked poultry flavor by the formation of secondary products with other volatiles (29). During the heating process, sulfur compounds undergo decomposition and dissolve partially in the poultry lipid, giving the fat and the meat a characteristic sulfur flavor (30). It was found that 4,5-dimethylthiazole and 5-acetyl-2,4-dimethylthiazole have meaty, boiled poultry aroma (31). Also, using aroma extract dilution analysis, 2-methyl-3-furanthiol, 2-furfurythiol, methional, and 2,4,5-trimethylthiazole have been identified as the primary odorants of the chicken broth (22). The production of sulfur-containing volatiles and carbonyl compounds is also influenced by the cooking pH values; alkaline pH conditions increase the production of sulfur-containing volatile compounds but decrease the generation of carbonyls (32). The precursors of hydrogen sulfide in cooked poultry have been identified as mainly the cysteine and cystine of the proteins. Methional, dimethyl disulfide, and methanethiol are formed by Strecker degradation of methionine. But the identification of thiols and sulfides linked to alkyl groups indicate that sulfur-containing amino acids undergo breakdown by a more complex mechanism than the Strecker degradation (5).

Pyrazines and Furans

The pyrazine group has been regarded as one of the most important classes of flavor compounds (33). This group is usually produced by heating of appropriate precursors and exhibits a roasted or nutty aroma. The formation of pyrazines is probably due to the reaction of NH_3 or amino-containing compounds with sugars or other carbonyl compounds (25). It has also been shown that amino acid–N will react with sugar to produce different pyrazines and that degradation products of sugars can react with some amino acids to produce a number of pyrazines. At least one pyrazine, namely 5,7-dihydro-5,7-dimethylfuro *(3,4-b)*-pyrazine, has been incorporated in a chicken soup patent. Furans are derived in meat from the Amadori rearrangement, oxidation of unsaturated aldehydes, and the breakdown of 5′-ribonucleotides.

Other Compounds and Artificial Poultry Flavor

Many other classes of compounds, such as sugars, nucleotides, esters, fatty acids, and ammonia, also contribute to the total poultry flavor sensation. A number of them, for instance, inosinic acid, may contribute to mouth satisfaction and to the intensity of the flavor of other compounds. Other sulfur precursors that could possibly influence flavor include biotin, thiamine, and coenzyme A. A chicken flavor was developed by heating thiamine with S-containing polypeptides and alkanones or hydroxy-alkanones, then adding diacetyl and hexanal. Another formulation of artificial chicken flavor was generated from the combination of hexose, a bland fish or vegetable protein hydrolysate, cysteine, and arachidonic acid heated at 60°C for 10 min (2).

SUMMARY AND FUTURE DEVELOPMENT

Poultry flavors are composed of numerous chemical compounds. Some compounds, such as carbonyls and sulfur-containing compounds, have been shown to be indispensable in poultry flavors, but the literature has also suggested that other compounds play a role in the total flavor sensation of poultry. The flavors of various poultry products are affected by many factors: production, processing, and preparation conditions influence the tissue composition or the precursors of flavor compounds. The formation of flavor notes during heating of meat tissue involves complex chemical reactions; any variations in the precursors or heating conditions may affect the quality and quantity of the final reaction products—the flavor compounds. With the advances in analytical technologies and computer capabilities, it could be possible, in the near future, to identify those compounds critical to poultry flavor and to formulate poultry products with specific and desirable flavor characteristics.

BIBLIOGRAPHY

1. H. L. Hanson et al., "The Flavor of Modern and Old Type Chickens," *Poultry Sci.* **38**, 1071–1078 (1960).
2. H. S. Ramaswamy and J. F. Richards, "Flavor of Poultry Meat—A Review," *Can. Inst. Food Sci. and Technol. J.* **15**, 7–18 (1982).
3. L. J. Farmer et al., "Responses of Two Genotypes of Chicken to the Diets and Stocking Densities of Conventional U.K. and Label Rouge Production Systems. 2. Sensory Attributes," *Meat Sci.* **47**, 77–93 (1997).
4. G. Gandemer, "Muscle Lipids and Meal Quality—Phospholipids and Flavor," *Ocl-Oleagineux Corps Gras Lipides* **4**, 19–25 (1997).
5. E. L. Pippen, "Poultry Flavor," in H. W. Schultz, ed., *Symposium on Foods: The Chemistry and Physiology of Flavors*, AVI Publishing Co., Westport, Conn., 1967, pp. 251–266.
6. B. W. Sheldon et al., "Effect of Dietary Vitamin E on the Oxidative Stability, Flavor, Color, and Volatile Profiles of Refrigerated And Frozen Turkey Breast Meat," *Poultry Sci.* **76**, 634–641 (1997).
7. T. C. Grey et al., "The Effect of Chilling Procedure and Storage Temperature on the Quality of Chicken Carcasses," *Food Sci. Technol.* **15**, 362–365 (1982).
8. R. B. Baker et al., "Palatability and Other Characteristics of Repeatedly Refrozen Chicken Broilers," *J. Food Sci.* **41**, 443–445 (1976).
9. T. Nishimura et al., "Components Contributing to the Improvement of Meat Taste During Storage," *Agric. Biol. Chem.* **52**, 2323–2330 (1988).
10. C. Y. W. Ang and B. G. Lyon, "Evaluation of Warmed-Over Flavor During Chill Storage of Cooked Broiler Breast, Thigh and Skin by Chemical, Instrumental and Sensory Methods," *J. Food Sci.* **55**, 644–648, 673 (1990).
11. C. S. Rao et al., "Effects of Polyphosphates on the Flavor Volatiles of Poultry Meat," *J. Food Sci.* **40**, 847–849 (1975).
12. M. M. Mielche, "Development of Warmed-Over Flavour in Ground Turkey, Chicken and Pork Meat During Chill Storage, A Model of the Effects of Heating Temperature and Storage Time," *Zeitschrift fur Lebensmittel-Untersuchung und-Forschung* **200**, 186–189 (1995).
13. C. Y. W. Ang and Y. W. Huang, "Internal Temperature and Packaging System Affect Stability of Cooked Chicken Leg Patties during Refrigerated Storage," *J. Food Sci.* **58**, 265–269, 277 (1993).
14. L. J. Minor et al., "Chicken Flavor: The Identification of Some Chemical Components and the Importance of Sulfur Compounds in Cooked Volatile Fraction," *J. Food Sci.* **30**, 686–696 (1965).
15. C. Y. W. Ang, "Comparison of Broiler Tissues for Oxidative Changes After Cooking and Refrigerated Storage" *J. Food Sci.* **53**, 1072–1075 (1988).
16. R. A. Wilson and I. Katz, "Review of Literature on Chicken Flavor and Report of Isolation of Several New Chicken Flavor Components from Aqueous Cooked Chicken Broth," *J. Agric. Food Chem.* **20**, 742–747 (1972).
17. R. J. Horvat, "Identification of Some Volatile Compounds in Cooked Chicken," *J. Agric. Food Chem.* **24**, 953–958 (1976).
18. J. Tang et al., "Isolation and Identification of Volatile Compounds from Fried Chicken," *J. Agric. Food Chem.* **31**, 1287–1292 (1983).
19. I. Noleau and B. Toulemonde, "Quantitative Study of Roasted Chicken Flavour," *Lebensm.-Wiss. u.-Technol.* **19**, 122–125 (1986).
20. I. Noleau and B. Toulemonde, "Volatile Components of Roasted Chicken Fat," *Lebensm.-Wiss. u.-Technol.* **20**, 37–41 (1987).

21. N. Ramarathnam, L. J. Rubin, and L. L. Diosady, "Studies on Meat Flavor. 4. Fractionation, Characterization, and Quantitation of Volatiles from Uncured and Cured Beef and Chicken," *J. Agric. Food Chem.* **41**, 939–945 (1993).
22. U. Gasser and W. Grosch, "Primary Odorants of Chicken Broth," *Z. Lebensm Unters Forsch.* **190**, 3–8 (1990).
23. D. L. Taylor and D. K. Larick, "Investigation Into the Effect of Supercritical Carbon Dioxide Extraction on the Fatty Acid and Volatile Profiles of Cooked Chicken," *J. Agric. Food Chem.* **43**, 2369–2374 (1995).
24. M. Rothe, E. Kirova, and G. Schischkoff, "Sensory Profile Studies on Broth Quality Problems," *Die Nahrung* **25**, 543–552 (1981).
25. J. H. MacNeil and P. S. Dimick, "Poultry Product Quality. 3. Organoleptic Evaluation of Cooked Chicken and Turkey Skin Fractions as Affected by Storage Time and Temperature," *J. Food Sci.* **35**, 191–195 (1970).
26. D. L. Taylor and D. K. Larick, "Volatile Content and Sensory Attributes of Supercritical Carbon Dioxide Extracts of Cooked Chicken Fat," *J. Food Sci.* **60**, 1197–1200, 1204 (1995).
27. A. E. Wasserman, "Chemical Basis for Meat Flavor, A Review," *J. Food Sci.* **44**, 6–11 (1979).
28. J. Schliemann et al., "Chicken Flavour—Formation, Composition and Production. I. Flavour Precursors," *Die Nahrung* **31**, 47–56 (1987).
29. E. L. Pippen and E. P. Mecchi, "Hydrogen Sulfide, A Direct and Potentially Indirect Constituents to Cooked Chicken Aroma," *J. Food Sci.* **34**, 443–446 (1969).
30. E. L. Pippen, E. P. Mecchi, and M. Nonaka, "Origin and Nature of Aroma in Fat of Cooked Poultry," *J. Food Sci.* **34**, 436–440 (1969).
31. S. Fors, "Sensory Properties of Volatile Maillard Reaction Products and Related Compounds, A Literature Review," in G. R. Waller and M. S. Feather, eds., *The Maillard Reactions in Foods and Nutrition*, American Chemical Society Symposium Series 215, Washington, D.C., 1983, pp. 185–286.
32. C. S. Rao, E. J. Day, and T. C. Chen, "Effects of pH on the Flavor Volatiles of Poultry Meat During Cooking," *Poultry Sci.* **56**, 1034–1035 (1977).
33. S. S. Chang, and R. J. Paterson, "Recent Developments in the Flavor of Meat," *J. Food Sci.* **42**, 298–305 (1977).

Cathy Y. W. Ang
NCTR/FDA
Jefferson, Arkansas

Duane K. Larick
North Carolina State University
Raleigh, North Carolina

See also Poultry: meat from avian species.

POULTRY: MEAT FROM AVIAN SPECIES

The United States and China are the world leaders in meat animal production. The United States is the world's largest producer of beef, veal, broilers, and turkeys and is the second-largest producer of table eggs and pork. China is the world's largest producer of table eggs, swine, lamb and mutton, and horses, the second-largest in broilers, and third in beef and veal. None of that existed when Columbus first came to America (with the exception of possible domestication of turkeys in Mexico). Wild animals and birds were in abundance, and hunting easily supplied all the meat needs of the local inhabitants, and in future years a substantial part of the settlers' needs.

In his second voyage in 1493, Columbus brought to the West Indies livestock, which included chickens. In 1519 Cortez brought cattle and sheep to Mexico and brought turkeys back with him to Spain. De Soto brought horses and hogs to Florida in 1539, and later in the century missionaries brought these livestock to the Pacific Coast of North America. In the seventeenth century European settlers brought livestock and poultry to the United States. At that time pork was the main traditional meat source, partly because it could be well preserved without refrigeration. Chickens were used mostly for eggs and cockfights and less for meat. In 1641 the first meat-packing plant to produce salt pork was opened in Springfield, Massachusetts, by William Pynchon. Chickens were raised in the majority of households in small numbers. A hen laid about 60 eggs per year, mainly in the spring. Chickens provided the household with meat and eggs, and the surplus was bartered or sold in open markets.

In 1998 there were almost 9 billion meat-producing farm animals commercially raised on U.S. farms (Table 1). Many other animals were grown as specialty items for food, sport, and pleasure, mostly in low volume. Surprisingly, about 98% of all farm animals are birds. However, by amount of meat produced (in tons) and by revenue (in dollars), beef is still king, not only in the United States but also worldwide. Yet more people eat lamb, mutton, and sheep than any other animal flesh. The change of consumer preferences in meat consumption started slowly in the United States at the turn of the twentieth century and accelerated rapidly toward its end. The doubling of the American population since World War II from 132.1 million in 1940 to 269 million in 1998 strongly fueled the demand for poultry, resulting in the explosive growth in poultry production, meat consumption, and the emergence of a highly efficient vertically integrated poultry industry. Dramatic changes in lifestyle further increased the demand for poultry as a low-fat, convenient food. The majority of fat in poultry is deposited under the skin. Therefore substan-

Table 1. Major Meat-Producing Farm Animals On U.S. Farms (1998)

Broilers and roasters[a]	8,004,000,000
Layers	259,000,000
Turkeys	290,200,000
Ducks[b]	22,490,000
Beef and dairy cattle	89,485,000
Calves	10,016,000
Hogs	59,920,000
Sheep and lamb	7,616,000
Horses[c]	5,500,000
Total	*8,748,227,000*

Source: United States Department of Agriculture.
[a]Life span of a broiler is 6 weeks and of a roaster is 10 weeks.
[b]1996 USDA figures.
[c]Most horsemeat is exported.

tial fat reduction, sometimes to 2%, can easily be achieved through skin removal.

Chicken and turkey consumption more than tripled during the twentieth century, and today chicken is the most-consumed meat when calculated on retail weight (Table 2). Beef and pork subsequently declined, mainly during the 1980s and early 1990s. Yet total red meat (beef, pork, veal, lamb, and mutton) consumed by Americans in 1997 was 52.7% of all meats, compared to poultry (broilers, roasters, turkeys, ducks, and geese) at 40.6% and fish at 6.7%. Another method to calculate meat consumption proposed by the beef industry and now in use is on a boneless basis. According to that method, poultry lost out, as the majority of its retail parts contain significant amounts of bone (Table 2). However, the gap is closing again as 80 lb of broilers and 64 lb of beef per American were consumed in 1998.

The worldwide picture is similar to that in the United States. Since the early 1960s, the number of chickens slaughtered worldwide rose about sixfold (from 6.5 billion in 1961 to 39 billion in 1997). Increased production of turkeys, ducks, and geese was also phenomenal but confined to certain world regions such as China, the Pacific Rim, and Europe. Strong world population growth (4.8 billion in 1985, 5.7 billion in 1995, 6.0 billion in 1999) and the doubling of per capita annual income even in developing countries (to $600 in 1996) fueled this growth, as a large portion of this income in developing countries and a smaller proportion in developed countries was spent to buy more meat. In the United States total meat consumption has increased by 14 lbs per capita since 1970.

Poultry has numerous advantages over ruminants in intensive production:

1. A very low feed conversion range, between 1.75 and 3.0 lb of feed/1 lb of live weight
2. Short production period
3. Vertically integrated industry
4. Highly automated and fast processing lines
5. Small requirement for land
6. Low product price

As a result of these advantages, self-sufficiency in poultry production became a goal in many countries. However, intensive poultry production requires a large volume of grain and oil seeds that are not available in many countries or must be used for human food. Soybeans and corn are the major feed components used in poultry husbandry. The majority of soybeans are produced by five countries (Table 3). World corn production is less concentrated, and 81% is produced by 20 countries. Yet the United States, China, and Brazil are the largest corn producers, with 69% of world production (Table 4). Not surprisingly, the United States, China, and Brazil are also the largest poultry meat and egg producers (Table 5), and India, the fifth largest soybean producer, is the fastest-growing egg producer. Moreover, these countries have some of the world's lowest

Table 2. Per Capita Consumption of Meat in the United States (1996)

	Retail Weight (lb)	Boneless equivalent (lb)
Chicken	73.0	50.9
Beef	68.0	63.8
Pork	49.0	45.6
Turkey	18.5	13.9
Veal	1.0	0.9
Lamb and mutton	1.0	0.8
Total meat and poultry	*210.5*	*175.9*

Table 3. World Soybean Production (1994)

Country	Quantity (million metric tons)	%
World soybean production	113.1	100.00
United States	49.2	43.50
Brazil	23.8	21.04
China	13.0	11.49
Argentina	12.2	10.79
India	4.5	3.98
Rest of the world	10.4	9.20
Total of top 5	*102.7*	*90.80*

Table 4. World Corn Production and Largest Producers (1998)

	Quantity (metric tons)[a]	Percent of world production (%)
World	594,709,169	100
United States	247,932,436	42
China	123,995,428	21
Europe	34,099,568	6
Brazil	33,497,587	6
Mexico	17,998,476	3
Argentina[b]	13,500,127	2
India	9,499,619	1
Rest of the world	114,218,948	19

Source: United States Department of Agriculture.
[a]To convert to bushels, multiply by 39.37.
[b]In 1999 Argentina will dramatically increase its corn production.

Table 5. The Largest Chicken Meat Producers (1997)

Country	Quantity (metric tons)	Percent of world production (%)
United States	12,574,000	24.3
China	8,582,000	16.6
Brazil	4,340,000	8.4
Mexico	1,442,000	2.8
Japan	1,235,000	2.4
France	1,215,000	2.4
United Kingdom	1,160,000	2.2
Indonesia[a]	1,053,000	2.0
Thailand	955,000	1.8
Spain	870,000	1.7
Ten largest		*64.6*

[a]In 1998 poultry production in Indonesia was drastically reduced after civil unrest and the departure of many Chinese who raised poultry.

meat and egg prices because feed is the costliest item in poultry production. Worldwide grain trading, as well as poultry meat imports, has dramatically increased during recent times. Moreover, more countries are banning or reducing local grain production for animals in favor of crops for human consumption. This has made the poultry industry in many developing countries vulnerable to the instability of financial markets. A major financial crisis started in 1994 in Thailand and spread throughout the Pacific Rim. Later it reached Russia, some of its former states, and several South American countries. The crisis had a devastating effect on poultry production in these countries as well as grain producers and poultry meat exporters around the world. Strong and rapid currency devaluation prohibited or strongly reduced these countries' capability to buy feed or meats.

In contrast, most meat-producing ruminants can be grown on pastures that do not compete with human food. The majority of the feeding is done by grazing on land that is not suitable for extensive grain or other crop production. The amount of land in the world that is fit for extensive crop production (arable land) is relatively small and represents 11% of the world land area (Table 6). Furthermore, the arable land available for crop production is reduced by urban development. The arable land is not equally distributed. Although the United States enjoys 20% arable land, a densely populated country like China has only 8% and must depend on feed importation and innovative feeding systems. A vast continent such as Africa has only 7% arable land. Land topography such as high mountains and very large lakes affects rainfall; the angle of the land surface area to the sun, geological history, and land conservation practices are among the parameters that strongly affect the amount of arable land in each region and crop yield. In contrast, 26% of the world's land can be used as pasture, which sustains ruminants without competing with other human food production. Pasture area continues to grow as people clear forests, such as the rain forest in Brazil, or convert wild animal–grazing land into pasture for domestic animals at the cost of endangering numerous plant and animal species and negatively impacting the ecology.

There is still room for increases in all segments of meat production, but the future balance between poultry and other species will be determined by land use, environmental considerations, human population growth rate, available capital, and the stability of financial markets.

CHICKEN

Chickens, especially broilers, are the backbone of the poultry industry. In 1997 more than 39 billion broilers were slaughtered worldwide, and a continuous increase is predicted for the foreseeable future (Table 7). To process this huge number of birds, fast and highly automated processing lines have been developed. A modern plant that can process up to 3.5 million birds per week is the new standard for the industry. This plant can also further process the broilers by cutting up the whole bird into four or eight pieces and reassembling the pieces back into a cut-up whole chicken. Because many of today's consumers prefer specific parts, the modern plant provides a large selection of fresh or cooked products. This significantly increases its profit margins. The plant also performs a retail packaging service to individual supermarket chains that includes printing the chain's specific prices for that week on each package. This enables retail markets to reduce or even eliminate from their meat departments butchers who used to perform the same job manually and at higher cost. This has made retail meat departments more profitable.

In comparison, the standard size of a processing plant during the 1960s that produced mostly ice-packaged whole birds was 50,000 birds per day. This type of plant and smaller ones still operate around the world in areas with lesser chicken production. To supply a processing plant with capacity of 3.5 million birds per week without interruption requires a complicated infrastructure. This includes parent breeder farms, hatcheries, broiler production barns, feed storage and feed mills, railroad access, and a transportation fleet. Railroad access is needed as major feed ingredients are purchased by trainload. The transportation fleet moves fertile eggs, chicks, and market-ready broilers from one facility to another as well as distributes the finished products to retail, fast-food, and institutional distribution centers and warehouses, from which they are distributed to stores.

As environmental-control parameters tighten, waste management becomes a high-priority consideration. This includes manure handling, wastewater treatment of effluents from production and processing, dead birds, hatcheries refuse, feathers and other unwanted biological materials, and odor control. Food-safety programs to control the growth and spread of pathogens are also addressed. In recent years, the detection of human pathogens in finished products has resulted in painfully expensive major product recalls. When food poisoning outbreaks have occurred, recall costs, penalties, and lawsuit judgments run in some cases to hundreds of millions of dollars. In recent years, plant closures, bankruptcies, and the sale of an entire company have occurred in the United States.

Table 6. World Land Distribution and Usage

Land type	Size (in thousand hectares[a])	Percentage (%)
World land	13,041,713	100
Arable land (cropland)	1,441,573	11
Pasture	3,357,520	26
Forests and woodland	3,861,081	30
Other	4,381,539	33

[a]Hectare is equal to 2.47 acres or 0.00386 square miles.

Table 7. World and Continent Production of Chicken Meat (1997)

	Quantity (metric tons)
World	51,645,000
Asia	17,300,000
North and Central America	15,546,000
Europe	8,750,000
South America	7,206,000
Africa	2,240,000
Oceania	603,000

The coordination of all poultry production and processing activities is of highest priority in that plant downtime (being idle) or operation at less than full capacity is extremely expensive. The only way to strictly control and coordinate these complicated operations is by vertically integrating most aspects of the business. This requires a large capital investment. In 1900 5 million chicken farmers were reported and registered in the United States (minimum of 300 chickens). In 1998 only 341 egg producers (minimum of 70,000 hens) and several thousand broiler farmers remained. Clearly the poultry industry is becoming highly concentrated in the United States. The 50 largest companies produce 99% of all broilers, the 10 largest produce 66% of all broilers, and the largest company (Tyson) produces and processes 25% of all broilers. Most of the small broiler farmers became contract growers associated with integrated meat companies.

The disappearance of the majority of the small processing plants caused the broiler farmers that remained in the business to lose their independence. Today, many broiler-production facilities are under contract to the poultry meat companies. The company usually provides the farmers with chicks, feed, and veterinary and nutritional services. The farmer provides the labor, buildings, equipment, litter, light, heat, and water. The farmer is paid by the pound of live weight produced and gets a premium for better-performing flocks. There are few meat companies in each region, so the farmer has limited contract options.

As profit margins have narrowed during the years, larger production units have become vital. However, economical expansion or renovation is strongly related to interest rates. In many countries chronic high interest rates limit potential growth. Financing the expansion from profits and from depreciation has proved to be too slow a response in the poultry business. The low interest rates in effect since the late 1980s have fueled the dramatic expansion and the further concentration of the American broiler industry.

Producers of specialty birds such as Poussin (young spring chicken), petit poulet, the oriental black chicken, and others, or those providing live poultry for oriental markets are the last stronghold of the independent producers in that they do not depend on the large processing plants to take their birds. They produce low volume at higher prices in a family-style operation. In most parts of the world the poultry industry is less concentrated and the market is more regulated than in the United States, and producers there enjoy more freedom of choice and better returns. However, regulated markets generally need protection from lower-cost imports. According to the General Agreement on Tariffs and Trade (GATT), trade barriers for poultry meat and eggs should be removed by 2005. This is creating a lot of political and social concern in countries like France, Poland, Canada, and others where small farms are still common.

History and Breeds of the Chicken

Wild-chicken habitat was the jungles of Southeast Asia—Burma, Laos, Malaysia, Cambodia, Thailand, and Indonesia—as well as India, Pakistan, Sri Lanka, and China. The oldest chicken remains date from 31 million years ago. Four species of the wild jungle fowl are known: the red, gray, Java, and Ceylon jungle fowls. All of these species are good flyers, have beautiful feather colors and patterns, and can still be seen in the wild. The red jungle fowl (*Gallus gallus*) was the main source for domestication, which was done around 2000 B.C. Old Chinese records indicated that in China domestication took place around 1500 B.C. In Malaysia red jungle fowl were kept in captivity long before that. There, roosters were captured or hatched for timekeeping in the jungle and for cockfights. They also became part of religious practices as sacrifices to the gods and often replaced human sacrifices. The chicken was not domesticated for eggs or for meat because the red jungle fowl laid only eight small eggs in a clutch and their 500- to 1100-gram carcasses were stringy and inedible.

The domesticated chicken (*G. domesticus*) moved westward from India to Persia and South Russia, from there to central and western Europe, and then south to the Mediterranean and North Africa. The oldest Greek picture of a chicken on amphorae is from 500 B.C. The first Roman records of cockfights are from 200 B.C.; earlier Roman history makes no mention of the chicken. The New World (the Bahamas) got its first chickens from Columbus while the U.S. mainland got the chickens 150 years later from the settlers.

The domestic chicken went through intensive breeding, and as a result hundreds of breeds exist. Because the chicken is a beautiful and in many cases exotic bird, most breeds were bred for showmanship and sport. Cockfights were highly popular around the world, in particular in Rome and in Byzantium. In the United States cockfighting used to be very popular, but it is now prohibited in many states. Abraham Lincoln got his nickname “Honest Abe” because he was a reputable cockfight referee.

Egg consumption became more common after breeding increased egg size and the number of eggs laid. Yet egg laying remained seasonal (spring). Chicken meat started to be consumed after the Greeks on the island of Kos developed a feed formula that fattened chickens. The technology remained unknown for many years until the Romans spread it throughout the Old World after conquering Greece. Around 100 B.C. eggs and chicken meat were staple foods throughout the Mediterranean and Europe.

The first commercial breeds for eggs and meat were pure breeds. The most common breeds used are the White Leghorn, an Italian breed used for white shell eggs, and Rhode Island Red, used for production of brown eggs. New Hampshire Red, White Plymouth Rock, Cornish, and British Light Sussex are often used for meat production. At the beginning of the century the dual-purpose breeds were popular because they provided high-quality meat and eggs. As intensive production took place, however, specific breeds for meat and eggs were developed separately for greater efficiency. In broilers, fast growth rates, improved feed-to-meat conversion ratio, white feathers, white or yellow skin, and year-round production were the main breeding goals. In 1935, about 16 weeks were needed to raise a 2.8-lb broiler with a feed conversion ratio of 4.4 lb feed per 1 lb live weight. In 1994, 6.5 weeks were needed to produce a 4.65-lb broiler with 1.9 feed conversion ratio. Laying hens went from 131 eggs per year in 1925 to 260 to 300 eggs in 1998. These are small birds with substantially less meat than broilers. About two dozen commercial breeders now supply most of the commercial meat stocks.

Production

Production and processing of chicken for meat and eggs became an extremely intensive high-volume operation in the United States. Low profit margins and the low product prices do not leave much room for errors or inefficiencies. To operate under these constraints, companies must continually increase their scale and efficiency. This trend started in the 1920s and 1930s when successful breeding turned broilers and laying chickens from seasonal into year-round egg and meat producers. It exploded during the 1980s and 1990s, taking advantage of the marvelous developments in computers and automation and the dramatic increase in consumer demand. The twenty-first century is expected to bring even greater efficiencies.

To supply a modern processing plant with 3.5 million crossbred broilers per week, a huge production pyramid system is required. At the top of the pyramid is the breeding company, which develops the male and female lines of the grandparent breeder lines. The grandparent males and females produce fertile eggs or live chicks of the male and female lines of parent breeder lines. These are sold directly to parent breeder farms, which belong to very large broiler companies or to their contracted farmers. Fertile eggs are produced and shipped to hatcheries, which generally belong to the broiler companies.

The standard broiler is raised for about 6 to 7 weeks. Therefore 22–25 million broilers at different ages must be raised at the same time to provide a steady supply of 3.5 million birds per week to the processing plant (projected broiler mortality is included). About 275,000 grandparent chickens and 750,000 parent breeder chickens are needed to maintain the steady stream of broiler chicks needed. The grandparents and parent birds are replaced at around 15 months of age when egg production decreases. A standard production house (growing barn) maintains 20,000 to 25,000 broilers (Fig. 1). Fewer birds are housed in summer

Figure 1. Mother breeder chicken production barn (laying nests are on the right and left sides).

to reduce heat stress. Therefore, 840 to 1,050 production houses are needed to support one plant. Around 36 plants of this magnitude will be needed to process the 6.3 billion broilers slaughtered in the United States annually (1997), although today many processing plants are smaller. Countries with smaller production facilities incur inefficiencies in their production systems, which results in higher production costs and higher finished product prices. Low standard of living, low local currency exchange rate against U.S. money, and low feed cost can improve prices in these countries when converted to U.S. dollars for comparison. However, when calculated as percentage of income spent on food, these prices are enormously high. Only highly vertically integrated large companies can have megaplants because large capital investment is needed. Smaller companies that are partially integrated need to buy part of their services from independent hatcheries or feed mills, which makes their operation more expensive. As a result the larger companies keep buying smaller ones and integrating them into their megaoperations. They also buy smaller food companies that further expand their processing and marketing capabilities. In many countries the rate of annual disappearance of poultry companies is 5 to 10% of all existing companies.

Incubation is also a vital part of a modern megaproduction scale. The incubator was commercially developed around 500 B.C. by the Chinese, who improved the concept previously developed in what is now Malaysia. There, the jungle fowl eggs were gathered and incubated; because only cocks were wanted, all female pullets were sent free back to the jungle. The Chinese used a series of baskets that held about 1,200 eggs. Later, the Egyptians independently developed large hatcheries made from sun-dried mud bricks that held about 70,000 eggs at one time. The incubators were heated by burning coal or camel dung, and egg temperature was measured on the eyelid by an experienced operator. Controlling constant incubating egg temperature was an art passed down through families.

In the United States incubators started to be used in the late nineteenth century. During World War I a steep demand for eggs triggered the development of large incubators, as the hatching of eggs under sitting hens hit its limits. Among the most impressive developments was the million-egg room incubator developed in California. Incubation also substantially reduced the cost of chick production compared to mother hens. In 1928, 57% of all chicks were hatched in incubators; in 1959 96% of all eggs were incubated. Today incubators hold 80,000 to 90,000 eggs. They are fully automated and provide excellent control of temperature and other environmental parameters. Heating is done by electricity or hot water. Many modern incubators are of the tunnel type where the eggs are loaded into egg carts on one side and slowly move toward the hatcheries on the other. The chicks are hatched after 21 days. To supply the need of 7 million broilers per week, 10.5 million eggs must be incubated at any time. This requires 117 incubators. The efficiency of the hatchery is measured by hatchability percentage, which depends on parent breeding quality and also on the quality of the incubator operations. The average hatchability of broilers is 85 to 89%. The specific figures must be built into the production schedule of each operation by the addition of more incubators.

Harvesting

Harvesting is one of the most laborious jobs left in poultry production. Crews of bird catchers operate by night when the birds are immobilized due to their poor night vision. The catchers place the birds in shipping crates and load the crates on flatbed trucks. The accuracy and care of the catchers' operations strongly affect the processing plant yield in that bruises and broken limbs are trimmed off by the inspectors and used for pet food. A mechanical broiler loader was developed in England where a slow rotating drum with long rubber fingers carefully moves the birds into a tunnel equipped with conveyors that place the birds in the shipping crate. This equipment is operated by one person. Mechanical harvesters have not gained wide acceptance, but the diminishing labor force will require mechanical solutions in the future.

Other aspects of production are described in the turkey section that are similar in principle to those used for broilers. Processing is discussed in the article POULTRY MEAT PROCESSING AND PRODUCT TECHNOLOGY.

TURKEY

The wild turkey is native to North and Central America, and the discovery of America is associated with the discovery of the turkey. Turkeys roamed American forests long before humans crossed the Bering Strait. The oldest turkey fossils, which were found in California, are dated as over 10 million years old, late Pleistocene period (1). All turkeys belong to two species: the common turkey (*Meleagris gallopavo*), which is larger in size, and the ocellated turkey (*Agriocharis ocellata*), which is smaller and found mainly in Central America. All seven varieties still exist in the wild, however, the ocellated turkey is endangered.

The turkey used to be abundant in the United States during the Puritan era. In New England it was sometimes considered a pest to grain fields. At that time a whole wild turkey was sold for the very low price of 6 cents. Appreciation for the turkey drove Benjamin Franklin to propose it as the national bird, but it lost out to the eagle. Several factors could have led to the turkey's drastic population reduction: the disappearance of the chestnut, the turkey's main food source, due to the devastating chestnut blight; the excessive logging and drastic shrinking of the nation's forest areas, which provide the turkey's habitat; and excessive hunting. Fortunately, successful protection programs were implemented in time. The wild turkey was able to adjust to a different diet based on acorns, a variety of seeds, insects, and occasionally snakes. Today, population numbers are far below those at the time the Pilgrims landed in Plymouth, but there is no danger to the turkey's existence. About 84,000 wild turkeys are harvested annually in the United States.

The turkey can fly short distances up to a mile and uses this skill to reach higher tree branches for night resting after spending the day walking and running on the ground at speeds of up to 20 mph. The chicks, which are hatched on the ground, leave their nests after 1 day and are able to fly after 4 weeks. The mating season is in the spring, when a dominant male changes his external appearance by developing a thick red sac on his chest where fat is stored. He establishes a harem of several females and guards them and his territory from young challenging males. Fights for dominance may end with the death of the loser.

Domestication of the turkey was conducted mainly in Europe. The Native North American Indian did not domesticate the wild turkey, but it is documented that turkeys were kept in captivity in Mexico. The Aztec king Montezuma kept a large number of turkeys as well as other birds and animals for meat and for pleasure in zoological collections. However, the turkeys that were in abundance were also served as feed to other animals. Some believe that these turkeys were already domesticated. Montezuma's appreciation of the turkey was also expressed with turkey statues sculpted from pure gold. Columbus did not bring turkeys back with him to Spain, as turkeys did not exist in the West Indies. It is believed that Cortez brought them in the early 1500s from Mexico, followed by many other importations from all over the continent. Unlike tomatoes and potatoes, which weren't eaten until a century after their arrival due to suspicion that they were poisonous, turkeys were immediately well accepted. By 1570, domesticated turkeys were spread all over Europe and also returned to America with the settlers, where they were crossed with the eastern wild turkey (*Meleagris gallopavo silvestris*).

The meat of the early domesticated turkey was still tough, and most people did not know how to cook it to perfection. As a result, the literature of that time is split between negative opinions as well as praise-filled reports of turkey as a gourmet food. A major turning point was the wedding banquet of King Charles IX of France and Elizabeth of Austria in 1570. The cooked turkeys were so tasty that many of the noble guests started to raise turkeys on their estates. Even today, France remains the largest turkey producer in Europe. In the end, however, it was America that provided the turkey with eternal glory when the Pilgrims who landed in Plymouth, Massachusetts, celebrated Thanksgiving in November 1621 together with their Indian neighbors. In addition to giving thanks, the bountiful turkey was at the center of the celebration, which became an important holiday in the Puritans' life seeing they did not celebrate Christmas or other major holidays. Thanksgiving turned out to be a major civil holiday in the United States after proclamations by George Washington (1789), Abraham Lincoln (1863), and Franklin Roosevelt, who moved Thanksgiving to its present position on the calendar. Christmas, another holiday where turkey is a centerpiece food, had a tougher fight to be established as a holiday in the United States due to strong opposition. For 75 years, the U.S. Congress deliberately convened on Christmas Day. The glorification of the turkey as the holiday season bird placed extreme economic constraints on the turkey industry because most production was geared toward a short period of several meals per year. The marketing success of the chicken opened the door for the turkey. Many cut-up and value-added products with attractive nutritional profiles were developed, resulting in year-round consumption of 18 lb per capita annually in the U.S. (Table 8). In 1998 about 270 million turkeys were raised in the United States; 45 million were eaten at Thanksgiving, 22 million at Christmas, 19 million at Easter, and the remainder of 186 million were consumed year-round. However, 91% of the population ate turkey on Thanksgiving.

For a long time after the turkey was brought to Europe, it was thought that it was related to the guinea fowl, the West African bird that was brought to Europe by the Greeks from Egypt around 500 B.C. As a result both birds are called *meleagris* in Latin. The turkey was thought to be a cross between guinea fowl and male peacock, which is also reflected in its Latin name, *gallopavo* (Gallo = rooster, pavo = peacock). None other than Linnaeus, the father of scientific taxonomy, made this error. The English common name *turkey* also reflects the confusion between turkey and guinea fowl. Many exotic birds, including guinea fowls, as well as spices and other exotic goods came to Europe from India and frequently through Turkey. The turkey has many names in different languages. In Turkey the turkey is called "the American bird," whereas in Israel the name means "the rooster from India."

The 500 years of breeding has resulted in many varieties that exist all over the world. However, many of these varieties were developed for sport and showmanship and do not have much value as a commercial meat product. Commercial meat breeds were developed mainly in the twentieth century as the turkey industry was emerging from barnyard birds into vertically integrated, efficient commercial production and processing operations.

The following parameters are important to the turkey meat industry:

1. Rapid growth rate and efficient conversion ratio of feed into meat. Modern breeds can reach 40 lb in 20 weeks for the male (tom) and 20 lb for the female (hen). This is twice the weight in less time compared to wild or unimproved domesticated turkey growth (Table 9). The table demonstrates that hens eat less and gain less weight than toms but keep a similar feed conversion ratio. Hens are generally preferred when buying a whole bird. Europeans prefer smaller hens (12–15 lb) compared to Americans (15–18 lb).
2. Yield of breast meat. In the United States and in some European countries the white breast meat of turkey and chicken is overwhelmingly preferred over the dark thigh and leg meat and bears a much higher price. Therefore, effort was made to breed birds with a higher yield of white breast meat. The broad-breasted breed that was developed in England in the 1930s with no major impact on that market became a huge success in the United States and Canada. The wider skeletal structure allows larger breast muscles to be developed in a 40-lb tom. The huge breasts prevent the toms from mating, and artificial insemination is needed. Larger breast muscles can be further developed. However, to achieve this, selection for strains with good leg strength must occur simultaneously.
3. Uniformity. The need for uniform carcass size is most important for the meat processor, because the processing line and especially the evisceration machine cannot be adjusted to large differences in bird size. Breeding and management techniques are able to control size variation, although not to perfection.
4. Appearance. The external appearance of a processed whole bird is a key parameter for consumer acceptance and for marketing promotion. During processing, scalding and mechanical feather plucking can provide the desired appearance. However, small black feathers (pinfeathers) and black pigment spots on the follicles can strongly downgrade appearance due to their strong contrast with the yellowish white skin. Breeding of white-feathered turkeys solved this problem.

Table 8. Countries With Top Turkey Meat Consumption Per Capita (1998)

Country	Consumption (lb per capita)
Israel	27.8
United States	18.0
France	14.3
United Kingdom	11.5
Canada	9.7
Belgium-Luxemburg	7.5
Netherlands	3.9

Table 9. Weight Gain and Feed Conversion Standards for Tom and Hen Turkeys

Age	Tom average weight (lb)	Feed conversion[a]		Hen average weight (lb)
1	0.28	0.71	0.28	0.71
2	0.62	1.03	0.61	1.01
3	1.24	1.19	1.11	1.17
4	2.13	1.27	1.84	1.28
5	3.24	1.38	2.76	1.39
6	4.59	1.47	3.83	1.51
7	6.14	1.56	5.03	1.63
8	7.90	1.66	6.31	1.76
9	9.87	1.77	7.71	1.87
10	12.00	1.87	9.42	1.93
11	14.20	1.88	10.96	2.00
12	16.44	2.00	12.51	2.20
13	18.72	2.10	14.04	2.29
14	21.00	2.20	15.54	2.37
15	23.28	2.49	16.98	2.46
16	25.57	2.60	18.32	2.49
17	27.90	2.60	19.54	2.58
18	30.27	2.60	20.63	2.87
19	32.61	2.67	21.55	2.93
20	34.90	2.76	22.27	3.04
21	37.11	2.97	22.78	3.08
22	39.22	3.10	—	—

Source: Ref. 5.
[a]Feed conversion = pounds of feed/1 lb of live weight

Commercial Varieties

Commercial meat-producing turkey varieties are classified according to feather color, either bronze, black, or white, and by skeletal structure, the large broad breast and the smaller unimproved varieties.

The white varieties are now dominant in the United States and Canada. The White Holland and the Broad Breasted Bronze (BBB), were the most important source for breeding. Their descendant, the Broad Breast White, is today the main commercial breed in North America. The Beltsville Small White, with 15-lb toms and 10-lb hens, was popular in the 1950s, when many consumers preferred small birds. However, consumer preferences changed, and currently it holds a very small percentage.

The Bronze turkey is named after its unique color, a shimmering green-bronze that appears metallic in the sunlight. Two types exist, the Broad-breasted Bronze for commercial use and the unimproved variety, which is used in specialty small-scale production. The Broad-breasted Bronze was the world's main variety from the 1930s to 1950s but has been mostly replaced by the Broad-breasted White.

The Black turkey is a medium-size variety with a 27-lb tom and an 18-lb hen. In western Europe it is highly appreciated for its meat quality and therefore is intensely grown there, especially in France and England.

Figure 2. Broad-breast White turkey production barn.

Figure 3. Free-range chicken farm in France.

Production and Consumption

The turkey industry is composed of a small number of large vertically integrated companies. The entities own or contract out all major components of the businesses, such as hatcheries, production farms, feed mills, processing plants, and transportation fleets. In this way they can better control costs and quality and improve the efficiency of their operations by designing better production schedules.

In the commercial hatcheries eggs are placed in hatching trays, round edge up, and then secured in place. Environmental conditions and egg rotation are automatically controlled; chicks emerge after 28 days of incubation. The young poults go through sex determination, vaccinations against common diseases, and beak trimming to prevent cannibalism. Then they are placed in shipment packages, which are well ventilated and temperature controlled. Within 24 hours they are moved to brooder barns, where they live to about 6 weeks of age. During this time the poults need a special starter feed, supplemental heat from brooder stoves, adequate ventilation, and protection from exposure to diseases. Due to poult sensitivity to diseases, brooder barns are separated from hatcheries and growing barns. The growing barns are typically 500 feet long and 50 feet wide and hold about 10,000 hens or 7,000 toms (Fig. 2). Hens are grown for 14 to 16 weeks and toms 17 to 21 weeks before being marketed. In free-range operations, the birds can roam in a large fenced field adjacent to the barn, similar to free-range production of chickens (Fig. 3). In France, where the free-range system is well developed, other operational conditions besides outside space are required. These include no artificial lighting, large space allotment per bird inside and outside, no animal by-products in the feed, no antibiotics, no water chilling of carcasses during processing, and, most importantly, slow-growing breeds and a longer growth period for better flavor development. As a result, production costs are much higher, but devoted customers of free-range poultry are willing to pay this premium.

Feed Mill Operation

Turkey production is very extensive, and profit margins are low. Feed is the most expensive component in raising turkeys and has a major effect on the performance of the flock and finished product quality. Therefore, a strict economically and scientifically based diet is needed. The feed mill formulates various grains and other ingredients on a least-cost basis for sound diets that are the most economical for that day. Four to eight different formulas are used regularly during the bird's lifetime. However, the ingredient composition of each formula may be changed frequently.

When processing time arrives, the birds are placed in transport cages and transported by truck to the processing plant (see the article POULTRY MEAT PROCESSING AND PRODUCT TECHNOLOGY). The entire growing barn is evacuated at the same time and then is cleaned, sanitized, and prepared for the next flock. Turkeys are now produced worldwide. However, large-scale production is still conducted according to historical development lines in North America and Europe (Table 10). The United States,

Table 10. Worldwide Turkey Meat Production (1997)

	Metric tons
World production	4,740,050
North and central america	2,653,310
Europe	1,777,270
South America	144,594
Asia	111,984
Africa	34,856
Oceania	18,036

Source: World Health Organization.

France, the United Kingdom, Italy, and Germany are the top producers (Table 11). Consumption follows production patterns with a single surprise, as Israel tops world consumption with 27.8 lb per capita in 1998 (Table 8).

DUCKS

Origin of Species

Virtually every domesticated duck kept in captivity today can trace its origins back to the wild mallard duck (*Anas platyrhynchos*). The exception to this is the Muscovy duck, which originated in Brazil and can still be found wild in Central and South America.

Habitat. The natural habitat of the wild duck is in areas near rivers, lakes, and streams. Both tree-dwelling and ground-dwelling ducks exist. Over the years, domesticated ducks have become quite capable of living in a wide range of habitats and climates. Although ducks adapt quite readily to living in a variety of environments, unlike their chicken counterparts, they are seldom housed in cages.

Domestication. The duck was most likely first domesticated in India, China, or in both countries. Though records are inconclusive, it would seem that the duck was domesticated approximately 3,000 to 4,500 years ago.

History of Association with Humans. Since being domesticated, the duck has served a variety of purposes. For centuries, flocks of ducks have been kept to supply both meat and eggs. In addition to supplying food, large flocks of ducks were often employed to rid fields and rice paddies of unwanted insects and foliage. One of the most famous culinary dishes made from duck, Pekin duck (Bejing duck), made history in the United States. In preparing the dish, the skin of the fresh duck is pre-dried so it will become very crispy after roasting. The roasted duck is then hung for 6 to 8 hours at room temperature. In accordance with its regulations, the United States Department of Agriculture (USDA) called for refrigeration of the roasted duck. The proponents of the traditional Bejing duck fought back in court and won. Today Pekin duck is the only perishable cooked meat exempt from refrigeration.

Table 11. The 10 Largest Turkey Meat Producers (1997)

Country	Metric Tons
United States	2,499,000
France	745,000
United Kingdom	297,000
Italy	258,000
Germany	206,000
Canada	142,000
Brazil	110,000
Israel	76,000
Romania	44,000
Hungary	39,000

Production Statistics in the United States and Worldwide

World production of duck meat stood in 1997 was 2.73 million t, and 83% of it was produced in Asia (Table 12). China is the largest producer with 72% of world production, followed by France (6.7%), Thailand (4%), and the United States (1.7%) (Table 13). The United States produced 40,000 t duck meat in 1997. However, there has been a slow increase in the number of ducks marketed for meat in the United States over the past few years.

Major Breeds

Thousands of years of selective breeding have resulted in ducks that come in a wide variety of shapes and colors. The differences in these ducks have come to be recognized as distinct breeds as well as different varieties within a particular breed. Several breeds have been developed particularly for meat purposes. Pekin (Fig. 4), Rouen, Aylesbury,

Table 12. World and Continent Duck Meat Production (1997)

	Quantity (metric tons)	Percentage of world production (%)
World	2,734,417	100
Asia	2,274,740	83.3
Europe	296,148	10.8
North and Central America	74,180	2.7
Africa	52,189	1.9
South America	30,978	1.1
Oceania	6,182	0.2

Table 13. World's 10 Largest Duck Meat Producers (1997)

Country	Quantity (metric tons)
China[a]	1,973,000
France	184,000
Thailand	111,000
United States	46,000
United Kingdom	38,000
Eritrea	37,180
Malaysia	36,800
Germany	27,500
Brazil	22,100
Mexico	20,000

[a]Hong Kong included.

Figure 4. Heavy male Pekin duck (pure breed).

and Muscovy duck have all been selectively bred to provide a large carcass. Pekin duck is the most commonly marketed meat duck today. The Pekin is a large, pure white duck. It is typically marketed at 7 to 8 weeks of age. At this age, it generally weighs between 6.25 and 7.5 lb. The Muscovy duck has become more popular in recent years. Although it grows more slowly and therefore takes several weeks longer to reach optimum weight, it yields a much leaner carcass. The Muscovy also maintains a more efficient feed conversion ratio. The appearance of the Muscovy is quite unlike that of any other breed of duck. It is easily recognized by the large caruncles or fleshy red growths on its face. The Muscovy can be seen in a number of colors or varieties; however, the white Muscovy is most typically used for market purposes. Mule ducks, a sterile hybrid of the Muscovy and the Pekin (or occasionally another meat breed), are also becoming increasingly popular in the meat duck industry. The primary egg-producing breeds are Khaki Campbell and Indian Runner. The Campbell is primarily khaki colored, as the name implies. The Campbell has been bred for high production and exceeds even some of the best chicken breeds. It is not uncommon for some strains of Campbells to lay 300 or more eggs per year. The Indian Runner comes in a variety of colors, with white the most common. The average egg production of the Indian Runner is typically slightly less than that of the Khaki Campbell.

Production Practices

The production practices used in the duck meat industry follow the same general guidelines as those used in the chicken meat industry. The only significant difference is the picking process. The duck, with its thick feathers and down-covered body, is not as easily plucked as the chicken. To overcome this obstacle, many commercial processing plants have added a step to the processing line. After the birds are picked, they are run through hot wax. When the wax is cooled and hardened, it can be peeled away, thus removing any remaining feathers.

Products and By-products

In addition to supplying large quantities of meat and eggs, the duck industry provides numerous by-products. The feathers and down of ducks can be used to stuff clothing, pillows, sleeping bags, and comforters. It is quite common to use ducks, particularly the Muscovy, to produce liver pâté. In addition to this, a partially incubated duck egg is considered a delicacy in many countries.

Future Prospects

What does the future hold for the duck meat industry? Many would like to see an increase in the demand for duck meat and duck products. It is quite likely that artificial selection and well-managed breeding programs will provide us with a faster-growing duck that yields a larger carcass and requires less feed.

GOOSE

Origin of the Species

Habitat. The Romans referred to geese as *duplicis vitae*, or having two lives—being active both on land and in water. Indeed, geese are as comfortable on water as they are on land. Geese feed in both environments but are considered to be inefficient grazers because they cannot digest the cellulose of grasses.

Domestication. There is evidence that the Nile goose may have been domesticated by the end of the second millennium B.C. The Egyptians' utilization of the domesticated goose was short-lived; by the middle of the first millennium B.C. they had substituted domestic chicken meat for goose meat in their diet. The fact that wild waterfowl was so plentiful may have encouraged hunting of geese over farming of them. The domestic goose (*Anser domesticus*) is descended from the wild gray leg goose. This bird may have tended to linger, or lag, during its migrations, thus making it easier to catch and domesticate.

History of Association with Humans. As with other domesticated animals classified as household animals, the goose was probably domesticated by women and maintained in small numbers near the home. The ancient Greeks fattened geese, and goose meat was highly prized. Not only were geese kept by the Romans as a food source, they were also viewed as national heroes. The goose is said to have sounded the warning when the Gauls attacked Rome. The Romans consumed goose meat and eggs, learned how to produce fatty livers, used goose fat for medicinal purposes, and developed the goose quill pen. The Romans transported goose stock across Europe. Even after the fall of Rome, the goose was kept as a backyard/barnyard animal in western and central Europe. With the Europeans' colonization of North America came the goose as well. The domestic goose was kept in nearly every backyard through the early 1900s, for it offered more in return than any other animal or bird.

Production Statistics in the United States and Worldwide

A well-organized goose industry of any significant size does not currently exist in the United States. So relatively few geese are slaughtered under federal inspection that the USDA does not even have a separate listing for geese in its agricultural statistics report.

Europe and Asia continue to be the leaders in goose production. In Hungary, for example, geese represented 13.69% of all poultry slaughtered in 1991. The percentage for the rest of the world was 0.84%.

Major Breeds

Standardization of goose breeds has been ongoing since the late 1880s. The breed standards were set by poultry organizations whose members were concerned more with the "fancy" breeds (ie, breeding for show and competition) than with economic traits. Some standard breeds are Toulouse, the famous goose of France, and the blue-eyed German breed, Embden. Commercially raised meat geese in Europe are Rhenish and White Italians. The breeds used for the production of fatty livers, or foie gras, are Toulouse and Landaise. Both Rhenish and White Italians are white feathered, while Toulouse and Landaise are gray. A popular breed in China is Huoyan.

Production Practices

A variety of management systems are used for geese, depending on the country and the product of interest. It was once believed that the birds had to be at least partially maintained outdoors. A growing number of operations are raising their geese intensively, with no access to pasture. In Europe 10 to 20% of all the meat geese are being produced intensively. The birds may be kept on deep litter or slatted floors. A much smaller percentage of the geese kept for egg production are housed intensively with supplemental lighting. That figure is 5% or less in Europe (Metzer, personal communication, 1999). The birds kept for foie gras production are kept in a more restricted housing system during the period of intense feeding.

Products and By-products

Goose meat continues to have a place of prominence in many Asian and European menus. The ability of the goose, if properly managed, to produce an extremely large, fatty liver allows it to yield a product of extremely high value. The foie gras is sold whole or already cooked and made into pâté. Goose down is considered the premier down for the manufacture of clothing and bedding. Goose feathers are also utilized in similar products.

Future Prospects

Until recently, commercial goose production in the United States has been a relatively small-scale industry, with small flocks on many farms. The increasing diversity of the U.S. population, with many recent Asian and Eastern European immigrants, has already led to plans for and initiation of commercial goose operations in the United States. There appears to be room for growth and for opportunities to educate American palates about goose meat.

SQUAB (YOUNG PIGEON)

The squab is a young pigeon that is marketed at 26 to 30 days of age when it reaches 15 to 17 oz dressed weight. The meat has a distinctive delicate and unique flavor. Squab is preferred over adult pigeon because the meat is tender and has more flavorful fat. Squab meat is very light in color because the young bird has never flown. The squab is fed by both parents with pigeon milk, a thick, high-protein liquid produced in the parents' crop and fed directly to the squab. During the first several days of life, this is the only food the young bird receives. Later the pigeon milk is mixed in the crop with regular plant origin feed and, occasionally, insects. The feeding process continues until the squab leaves the nest in the wild or is sent to market.

Due to the unique feeding process, a pair of pigeons can have no more than two squabs at a time. In many cases one squab will become dominant, taking over the feeding or pushing its counterpart out of the nest, which could result in the death of the weaker squab. In captivity a pair of pigeons can produce up to 20 squabs per year, and modern breeding techniques have helped to increase the vitality of the weaker individuals and successfully bring most of them to market.

Pigeons and squabs have been eaten around the globe since ancient times. Wild pigeons could live almost everywhere and can be found from the Arctic Circle to the tip of South America, with the exception of Antarctica and the Hawaiian Islands. Pigeons and doves belong to the family Columbidae, which contains more than 300 species in the wild. The majority of the species nest in trees and bushes whereas other species reside in cavities in rocks or trees. Some of these species, such as the rock dove (*Columbia livia*), adapted to human surroundings and found shelter in barns, buildings, and the like. The rock dove, which originated in Europe, Africa, and Asia, is the bird most commonly seen in cities and towns around the world. This bird was domesticated and crossbred for many centuries, resulting in a variety of colors. The pure white dove became the symbol of purity and peace. Other breeds of this species were selected for racing, military and civil communication, and exhibitions. In searching for food, however, these birds became a pest in poultry farms and in food-processing plants. The birds can carry human pathogens such as *Salmonella* in their digestive systems and can recontaminate poultry feed and processed food with their feces. Another species, the mourning dove (*Zenidura macroura*), is found mainly in rural areas such as barns and is frequently seen sitting on telephone lines. Pigeons are good flyers, and some species migrate thousands of miles. The "homing" sense of certain species was used by humans for communication and racing. To support flying, the pigeon's breast muscles are large and dark and can reach up to one third of the bird's weight. A very large pigeon can weigh as much as 5 lb.

Pigeons are mentioned in early records of many nations from early biblical times (Noah and the ark), Greek mythology (the dove served ambrosia to Jupiter), and Roman records. Chinese records indicate that the Manchu emperors used to eat pigeons among five other game birds and their flesh had medicinal value. Until the twentieth century pigeons and squabs were hunted, not raised. As squabs became more popular, they were harvested from nests. As a result some species became extinct despite their initial huge numbers. One of the unlucky species was the passenger pigeon (*Ectopistes migratorius*), which at the beginning of the nineteenth century was the most abundant bird in North America. A report by the famous bird artist and naturalist John James Audubon describes a huge flock, estimated at 1.1 billion birds, in Kentucky (1813). When the flock became airborne, it eclipsed the sun. Flying at 60 mph the flock took 3 hours to pass him by, with a flock width estimated at 1 mile. When the birds nested in trees, more than 100 nests per tree were counted. The passenger pigeon was killed in huge numbers for food and hauled by trainloads to urban centers. The surplus was fed to hogs or rendered for fat to make soap. The expanding agriculture and the shrinking forests added to the disaster, turning the passenger pigeon into the pest of grain fields, and the last known free bird was killed in 1899. The bobwhite pigeon (*Colinus virginianus*) was also an important food source for settlers and Native Americans in the eastern United States and still exists in relatively small numbers.

Commercial breeding of pigeons was conducted first in France and Belgium, which developed the varieties of Bordeaux pigeon. In years to come these pioneer breeds were replaced by the French Mondain and the White Swiss Mondain. The United States followed in step by developing the White King, which produces a large full-breasted squab, and larger-scale production started in 1936. England used to be the traditional main market for squabs and pigeons and imported them from the United States and from the European continent. There the bird was served mainly to the well-to-do in restaurants and at parties and specialty events. In contrast, in China pigeons are the food of the common person.

Since the early 1980s California has become the largest center of squab production (Fig. 5). The climate in the central valley, a long, warm, and dry summer followed by a cool winter, is perfect for the locally grown King variants. The squab has a phenomenal growth rate and reaches almost 1 lb dressed weight in 1 month. Each parent pair produces up to 20 squabs per year.

Figure 5. Squab production pens. *Source:* Courtesy of the Squab Producers of California.

Squab production is a family operation conducted on a small to medium scale of 1,000 pairs on average. The pigeons are raised in a row of aviaries (pens), about 65 to 70 sq ft each, which hold 15 to 20 pairs in their individual nests. Management is relatively simple because the pigeons take care of each squab's needs. Feed and water are provided automatically; cleaning is conducted once anually. A pair's life span is about 5 years, and replacement of 20% of the flock annually is a management practice. In the United States about 1.6 to 2 million squabs are produced annually, mainly in California and South Carolina (about 1,000 short tons). Other major producers are France, Belgium, Switzerland, Canada, and China, but accurate production numbers are hard to come by. Most producers, with the exception of China, export a significant portion of their production.

In the United States squabs are sold mainly to white tablecloth restaurants and to oriental markets as well as exported frozen to many countries. The time that pigeons were a staple food in the United States are long gone, and the industry challenge is to introduce this meat again to a large customer base. Squabs bring the highest revenue and profit per bird than any other avian species, in part due to a year-round steady supply, control of surpluses, and devoted customers.

RATITES

Origin of the Species

The term *ratite* describes all the flightless birds and derives from the Latin word *ratis*, or raft. The birds in this grouping possess a sternum or breastbone that is flat, or shaped like a raft. This is in contrast to all other avian species, which are known as carinates. Carinates all have breastbones shaped like the keel (*carina* in Latin) of a boat. Though the ratite group is often described as large, flightless birds, this is partially incorrect. We commonly think of the ostrich (*Struthio camelus*) and emu (*Dromaius novaehollandiae*) as representative of the ratites, but the group also includes the much smaller rhea (*Rhea americana*, or greater rhea, and *Pterocnemi pennata*, or Darwin's rhea), cassowary (single wattled and double wattled), and the most diminutive, the kiwi (*Apteryx oweni, Apte haasti*, and *A. australis*). Only the ostrich, emu (Fig. 6), and rhea are raised commercially.

Habitat. Fossilized ostrich remains have been discovered in Europe, Asia, and Africa. All of the modern ostriches are native to sub-Saharan Africa. Emus are native to Australia, and both species of rhea are native to South America.

Domestication. Ostrich farming began in earnest ca 1863 in South Africa. Farmers in that area selected the wild-caught birds for certain traits associated with feather quality. After the turn of the century, these birds were crossed with imports from northern and eastern Africa, resulting in an ostrich with good feather quality and egg-laying capabilities. Although the emu had long been utilized by the aboriginal people of Australia and the rhea

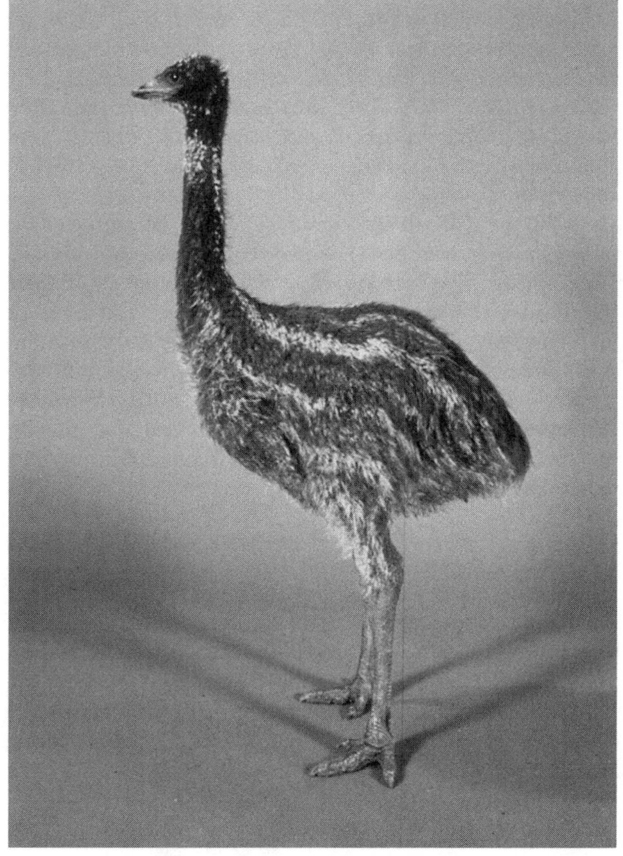

Figure 6. Emu, 2.5 months old.

had served as a meat source in South America, it is generally believed that neither the emu nor the rhea had been the subject of organized domestication efforts before 1988.

History of Association with Humans. The fact that humans have made use of ostrich feathers since ancient times is evidenced by biblical record, Egyptian hieroglyphs, and Greek writings. The efforts of South Africans as pioneers in ostrich farming are extensively documented. Though the South Africans were the first in the ostrich feather market, the Americans entered in the latter half of the nineteenth century. Ostrich farms in Pasadena and Los Angeles were southern California tourist attractions in the late 1800s and early 1900s.

With the collapse of the feather market in 1914, the South African industry experienced a great decline. No significant growth was observed until after 1960. The rise in environmental activism resulted in wildlife protection programs that prohibited the harvesting of rheas for meat in South America.

Production Statistics in the United States and Worldwide

Ratite production exploded onto the U.S. scene in the 1980s. Individuals with breeding stock touted what they considered the birds' strong selling points: valuable hides; a low-fat red meat; and by-products with unique characteristics, such as emu oil. Given the initially small number of ratite breeders but the extensive publicity, breeding stock commanded extremely high prices for several years. With the saturation of the breeder market, quick profits were no longer to be made by strictly selling birds. As the 1990s come to a close, ratite operators must make their money from selling product and not stock.

Major Breeds

Called the African Black by American producers, the farmed ostrich (*Struthio camelus* var. domedomesticus) are smaller than their wild counterparts. The farmed, blue-necked birds, however, have superior feather quality. The male and female have a height of 3.0 m and 2.5 m and a weight of 160 and 120 kg, respectively. The two-toed birds are sexually dimorphic. The male's coloration is black body, brown rump, and white wings and tail. The female is brownish gray. *Dromaius novaehollandiae* is the farmed emu. To date the commercial industry has not produced any specific emu strains or hybrids. The male and female have heights and weights of 1.5m, 36 to 38 kg and 1.8, 55 kg, respectively. The female is larger in these three-toed birds, but both sexes are grayish brown in color. Only the greater rhea is farmed in the United States and Europe. The male and female have heights of 1.7 m and 1.5 m, respectively. At 25 kg the male is slightly larger than the female. Both the male and female in these three-toed ratites are gray.

Production Practices

Traditional ostrich management in the Little Karoo in South Africa involved an extensive system, with the birds running on a natural field. The more intensive system in use today in South Africa involves taking the chicks from the incubator (42-day incubation period) and placing them in floor pens, either of concrete or on short lucerne pasture. From 6 weeks to 2 months the birds are placed in growing camps, where they do not have growing vegetation but rather are fed only a complete ration along with cut lucerne. From 6 months of age until slaughter, the birds are maintained in feedlots.

Management systems in the United States are much less uniform than in the well-established South African industry. As with South African operations, most producers in the United States use artificial incubation of the eggs and then place the newly hatched and dry chicks in chick pens. Concrete and sand are common substrates, and long narrow pens have been advocated to encourage exercise. Until the chicks can maintain their own body temperature, supplemental heat must be provided, and good ventilation is essential. In areas with a hospitable climate, the chicks can be turned out into enclosed pens during the warm part of the day. Chicks stay in the chick pens for 3 to 4 months. Juveniles are moved into larger pens that are well fenced. Breeding-stock birds are kept in pens in single pairs or in colonies of one male and several females. Breeder pens are 1/3 to 1/2 acre in size with a shelter. Again, good fencing is essential to prevent entry of predators, double fencing between adjacent pens can reduce male-to-male fighting. The feeding regimen involves diets containing 25.5% protein for chicks 0 to 2 months old. The protein content of the rations is decreased until the 10-month-old birds are eating a diet containing 8.596% protein. Ostriches are typically slaughtered at 10 to 14 months of age, at weights of 54 to 95 kg.

The majority of emu operators also use artificial incubation (50–52 day period) for their eggs. Housing systems for emus in the United States are similar to but smaller in scale than those for ostriches. A good percentage of the birds have been raised on small farms or large lots. The emu chicks are fed an 18 to 22% protein diet for the first 3 months. From that time on, they are placed on a 13% protein ration. Emus are typically slaughtered at 12 to 15 months of age, at a weight of 33.2 kg.

Rhea production is not an industry of any size or major significance. The incubation period for rhea eggs can vary from 20 to 43 days. Under natural conditions, the incubation and care of the young is performed by the male. Very little research has been conducted on the management of the rhea. For the most part, the birds have been raised more as a hobby than as a meat enterprise in the United States.

Products and By-products

Due to its reputed therapeutic value, emu oil is considered the emu's most valuable product. The oil must be rendered in a special manner before it can be used for medical and cosmetic purposes. Emu meat is another high-protein, low-fat red meat. However, the emu carcass does not yield large roasts and fillets such as can be cut from the ostrich carcass. Again, expert tanners are required to treat the emu hide. Bird size also imposes restrictions here. The hide is not large enough for the production of some leather goods but is well suited for the manufacture of small purses and accessories.

Those who have eaten rhea meat say it is a consistently flavorful product, and it has been used by South Americans for centuries. Some believe that rhea oil also has commercial possibilities.

All the ratites yield feathers at slaughter. Unless there is a rebirth of the millinery trade, the main outlet for the feathers will be costume and jewelry production. Producers often offer for sale blown-out eggs. The eggs are sought after by artists known as "eggers" who use eggs as their medium.

The growth of the ratite industries in the United States has been hampered by some of the same problems that face other young industries. The early years had too many unscrupulous promoters with no real dedication to seeing the industries survive as true agricultural enterprises. Those sincere producers who remain face developing markets, promotion, government-labeling restrictions, and an overall shortage of information. In addition to more publicity about the documented healthy nature of their meat products, the industries will also benefit from continuing research that seeks to verify the beneficial attributes of ratite oils.

GAME BIRDS

Game birds have long been consumed for food by hunters. In modern times certain species have become popular for domestic production (Fig. 7). The cost of raising these species in captivity varies with their reproductive biology, disease resistance, and adaptability to captive rearing. Artificial incubation and brooding (away from the mother) have dramatically reduced production costs. All of the popular species can be raised in confined houses, tiers of small wide floor brooders, or outdoor flight pens. Nutritional needs of the difference species vary, but game birds can generally be raised on feeds used for turkeys of a comparable age or breeding condition. Differences in behavior and disease susceptibility make it advisable for a novice to get professional advice before attempting to raise game birds commercially. Game birds, ratites, and squab have low fat levels, especially when they are skinless (Table 14). To overcome the effects of low fat content, they require special cooking techniques to produce a juicy and highly palatable entree.

Quail

Several species of quail are hunted for food throughout the world, but the Japanese quail (*Coturnix japonica*) and the bobwhite quail (*Colinus virginianus*) are two of the most commonly raised in North America for table use. The European quail (C. coturnix) has been widely consumed as food for centuries, but no accounts of its domestic rearing could be found.

Figure 7. Partridge, grouse, and other game birds in continental supermarket in Tours, France.

Table 14. Nutritional Analysis of Deboned Meat of Game Birds (per 100 g)

Species	Calories (kcal)	Fat (g)	Protein (g)	Iron (mg)	Vitamins (g)
Pheasant	181	9.3	22.7	1.15	177
Pheasant (skinless)	133	3.6	23.6	1.15	165
Squab	294	23.8	18.5	3.5	165
Squab (skinless)	142	7.5	17.5	4.5	94
Guinea fowl	158	6.5	23.4	0.8	92
Guinea fowl (skinless)	110	2.5	20.6	0.8	41
Quail	192	12.1	19.6	4.0	243

Japanese Quail. Japanese quail is used for the production of eggs and meat throughout the world. Imports into the United States were made in the 1950s, but these birds are thought to have been raised in captivity in eastern Asia for as long as domestic chickens. The stock imported into North America soon led to the development of several color mutations and strains selected for meat type or egg production. The body weight of imported Japanese quail was about 100 g, but the meat strains have been increased by genetic selection to a range of 200 to 400 g or more. The rapid reproduction of Japanese quail (fertile eggs are often obtained at 6 weeks of age; incubation requires 17 days) makes this species popular for egg and meat production and for use as an experimental bird by laboratories. Except for the small chick size, which requires special feeding and watering equipment, they can be raised much like chickens. These quail are often raised to maturity in chicken starting batteries, which have ample headroom for adults.

The meat of Japanese quail is prized for table use, and many are marketed to upscale restaurants in North America. The breast muscle of this quail is darker than that of a chicken or pheasant, but the breast meat is less likely to be dry in texture when cooked. The number of Japanese quail raised for food is unknown, but certainly several million are grown annually in Asia and North America. Processed birds are sold whole bodied or partially boneless for $1.50 to $2.50 per bird.

The eggs of Japanese quail are widely sold for food. This quail produces an egg that is 8 to 10% of its body weight in size. In contrast, a chicken egg is 3 to 4% of the hen's body weight. When the quail is raised on typical poultry diets, its eggs taste like chicken eggs. The shell membranes of the eggs are quite tough, making them difficult to open for fresh consumption. Eggs are often hard cooked and eaten fresh or pickled. The small size makes the eggs readily, absorb the flavor of pickle solutions and they make attractive hors d'oeuvres.

Bobwhite Quail. Bobwhite quail are native to North and Central America. The quail are 9.5 to 10.5 in long and weigh about 6 to 7 oz, although larger birds are probably used commercially in that quail respond rapidly to weight selection. Male and female quail differ in plumage. Males have a white stripe that extends from the beak past the eye to the neck. The females have buff-colored chins, upper throats and eye stripes. This species reaches sexual maturity at about 16 weeks. The eggs hatch after 23 to 24 days incubation. They are often grown in batteries or wire floor pens to prevent disease. Birds to be sold for hunting are grown in flight pens, but meat birds can be grown entirely in confinement. Like Japanese quail, bobwhite are sold as a specialty poultry item, but they are much less popular due to their higher cost (unprocessed young mature birds typically sell for $4 each). Eggs of this species are sometimes used for food, although the eggs are smaller than Japanese quail eggs and are more expensive to produce because of the bird's larger body size and later sexual maturity.

Pheasant

Most pheasants grown in North America for meat are either Chinese ring-neck (*Phasianus colchicus lorquatus*) or Mongolian (*P. colchicus mongolicus*) or commercial stocks derived from these. Pheasants are native to Asia and have been widely imported into North America as a game bird. The plumage of both species differs dramatically between males and females. Males have striking plumage color with a white ring around the neck. The Chinese male has a continuous white neck ring whereas the ring on the Mongolian male does not close under the beak. The Mongolian pheasant is about 10 to 20% larger in body weight than the Chinese. A white-feathered mutant pheasant is sometimes raised for meat in North America. This bird processes well, and no dark pinfeathers are visible. Under natural light pheasants produce about 50 eggs in the spring; reproduction then stops as the birds become refractory to the long days. The birds can be cycled to produce again in about 13 weeks by exposing them to a short day length for 9 to 10 weeks in a light-controlled facility. Pheasants are often sold in an eviscerated, whole-body package for $2.50 to $4 lb. They may be either young birds (called broilers) at 14 to 20 oz or adult birds at 2 to 4 lb. Some birds are sold smoked or fresh partially boned.

Partridge

Although several species of partridge can be reared in captivity, the chukar partridge (*Alectoris chukar*) is most commonly reared for hunting and table use. This bird, first imported into the United States in 1893, is native to Asia, where it inhabits mountains and valleys at elevations from 4,000 to 16,000 ft. Chukars have been established as a game bird in several western states. The plumage is predominately grey, with distinctive black-and-white markings on the head. Males and females are the same in appearance. Chukars are good fliers and are often released within private hunting clubs.

Chukars are widely raised throughout North America. Chukar eggs hatch in 23 days; hens reach sexual maturity at about 16 weeks and lay about 30 to 40 eggs in a season. Like the pheasant, the chukar hen becomes refractory to long days and can be cycled to lay again by exposure to a short day length for 8 to 10 weeks. Chicks can be brooded in wire floor or litter pens. Chukars are often raised on wire to prevent diseases. They are quite aggressive, and fighting and picking often injure pen mates. Beak trimming or use of antipicking devices may be necessary when the birds are raised in close confinement.

Chukars are excellent table birds and are often sold in an eviscerated, whole-body package for $3 to $4 per pound. They are often seen on the menus of upscale restaurants during the Christmas season.

Guinea Fowl

Guinea fowl (*Numida meleagris*) are native to Africa (Fig. 8). They are domestically raised and marketed as a specialty fowl in many parts of the world. Guinea females are reported to lay about 150 eggs in a 9-month season. Chicks are hatched artificially and raised much like meat chickens or turkeys. The meat of young guineas is reported to be excellent with a slightly gamy flavor. Eviscerated packaged birds typically weigh 3.5 to 4 lb and are often sold in an eviscerated, whole-body package for about $3 per pound.

Figure 8. Guinea fowl.

In France, guinea fowl are a staple food and are available in most supermarkets and many open markets around the country.

BIBLIOGRAPHY

1. G. Stocke, *Rancho La Brea: A Record of Pleistocene Life in California*, Los Angeles County Museum, Series 15, 5th ed., Los Angeles, Calif., 1953.
2. P. Ferket, "Turkey Growth Rate and Feed Consumption Standards: Turkey Performance Improves in 1998," *Turkey World*, 12–14 (January–February 1999).

GENERAL REFERENCES

U. Aldrovandi, *Aldrovandi on Chickens*, L. R. Lind, trans. and ed., University of Oklahoma Press, Norman, Okla., 1968.

A. W. Brant, "A Brief History of the Turkey World," *World's Poultry Sci. J.* **54**, 365–373 (1998).

The Broiler Industry, The Turkey Industry, and The Egg Industry: The Poultry Tribune Centennial Edition (1895–1995), Watt Mount Morris, Ill., 1995.

California Poultry Workgroup, *Turkey Care Practices*, Univ. of California, Cooperative Extension, Davis, Calif., 1998.

California Poultry Workgroup, *Broiler Care Practices*, 2nd ed., Univ. of California, Cooperative Extension, Davis, Calif., 1998.

Canadian Department of Agriculture, *Pictures Geese*, Publication No. 1345, Information Division of the Canadian Department of Agriculture, Ottawa, Canada, 1968.

E. Carbajo Garcia et al., "El sacrificio y la produccion de carne," E. Carbajo Garcia et al., eds., *Cria de Avestruces, Emues y Nandues*, Real Escuela de Avicultura, Barcelona, Spain, 1997, pp. 113–136.

E. Carbajo Garcia and F. Castello Fontova, "El Emu," E. Carbajo Garcia et al., eds., *Cria de Avestruces, Emues y Nandues*, Real Escuela de Avicultura, Barcelona, Spain, 1997, pp. 303–326.

E. Carbajo Garcia, E. F. Castello Fontova, and J. A. Castello Llobet, "Historia y Origen de la Produccion del Avestruz," in E. Carbajo Garcia et al., eds., *Cria de Avestruces, Emues y Nandues*, Real Escuela de Avicultura, Barcelona, Spain, 1987, pp. 19–40.

J. A. Crawford, "Foreword," in T. N. Tully, Jr. and S. M. Shane, eds., *Ratite Management, Medicine, and Surgery*, Krieger, Malabar, Fl., 1996.

R. D. Crawford, *Poultry Breeding and Genetics*, Elsevier, Amsterdam, The Netherlands, 1990.

A. Douglass, *Ostrich Farming in South Africa*, Cassell, Potter, Galpin, London, 1881.

Ducks and Geese, Reliable Poultry Journal, Quincy, Ill., 1910.

M. E. Ensminger, *Poultry Science*, Interstate, Danville, Ill., 1980.

I. Eric, *Geography of Domestication*, Prentice-Hall, Englewood Cliffs, N.J., 1970.

"Focus on the West," *Terra* **28**, 50–51 (1990).

F. D. Fowler, "Domesticated Geese," in *Ducks and Geese*, Reliable Poultry Journal, Quincy, Ill., 1910, p. 75.

M. E. Fowler, "Clinical Anatomy of Ratites," in T. N. Tully Jr. and S. M. Shane, eds., *Ratite Management, Medicine, and Surgery*, Krieger, Malabar, Fla., 1996, p. 1–10.

M. A. Hall, "Ratites," in M. A. Hall, ed., *National 4-H Avian Bowl Manual*, Clemson Univ., Clemson, S.C., 1998, pp. 171–173.

D. J. S. Hetzel, "Domestic Ducks: A Historical Perspective," in *Duck Production Science and World Practice*, University of New England Publishers, Armidale 1985.

F. W. Huchzermeyer, "Important Aspects of Ostrich Farming," in *Diseases of Ostriches and Other Ratites*, Agricultural Research Council, Ondersport, South Africa, 1998, pp. 77–119.

F. W. Huchzermeyer, "The Ostrich and Other Ratites," in *Diseases of Ostriches and Other Ratites*, Agricultural Research Council, Ondersport, South Africa, 1998, pp. 26–27.

L. M. Hurd, *Goose Raising*, Bulletin 823, State College of Agriculture, Ithaca, N.Y., 1951.

J. Kozak, "Changes of the Goose Stock and the Goose Production in Hungary," *Proceedings of the XX World's Poultry Congress* **3**, 764–767 (1996).

W. M. Levi, *The Pigeon*, Levi, Sumter, S.C., 1945, reprint 1992.

J. P. Mackenzie, *Game Birds*, North Word Press, Minocqua, Wisc. 1989.

J. Mesia Garcia, "La alimentacion del avestrucez," in E. Carbajo Garcia et al., eds., *Cria de Avestruces, Ernues y Nandues*, pp. 167–185.

Real Escuela de Avicultura, Barcelona, Spain, 1997, p. 167.

M. North and D. D. Bell, *Commercial Chicken Production Manual*, 4th ed., Chapman & Hall, New York, 1990.

C. G. Olentine, *Watt Poultry Statistics Yearbook*, Poultry International, Vol. 37, Watt Publishing Co., Mt. Morris, Ill., 1988.

M. E. Pennington, F. L. Platt, and C. G. Snyder, *Eggs*, Progress, Chicago, Ill., 1933.

W. Root, *Food: The Authoritative and Visual History and Dictionary of the Foods of the World*, Simon and Schuster, New York, 1980.

A. Rosinski et al., "Possibilities of Increasing Reproductive Performance and Meat Production in Geese," *Proceedings of the XX World's Poultry Congress* **111**, 724–735 (1996).

C. O. Sauer, *Seeds, Spades, Hearths, and Herds*, MIT Press, Cambridge, Mass., 1969.

A. W. Schorger, *The Wild Turkey: Its History and Domestication*, University of Oklahoma Press, Norman, Okla., 1966.

M. L. Scott, and W. F. Dean, *Nutrition and Management of Ducks*, M. L. Scott, Ithaca, N.Y., 1991.

D. J. v. Z. Smit, "Ostrich Farming Practice," in *Ostrich Farming in the Little Karoo*, Bulletin No. 358, Department of Agricultural Technical Services, Pretoria, South Africa, 1963, pp. 53–83.

A. Soyer, *The Pantropheon; or, A History of Food and Its Preparation in Ancient Times*, Paddington Press, London and New York, 1853; reprint 1977.

J. S. Stewart, "What Are the Kinds of Ostriches and Which One Is Best?" *Ostrich Report* **16**, 23 (May 1992).

L. Stromberg, *Poultry Oddities History and Folkore*, Stromberg, Pine River, Minn., 1982.

U.S. Department of Agriculture, *Agriculture Statistics*, Washington, D.C., 1998.

W. W. Van Wulffeten Palthe, *C. S. TH. Van Gink's Poultry Paintings, 1890–1968*, Dutch Branch of the World's Poultry Science Association, Beeklergen, The Netherlands, 1992.

D. W. Verwoerd, "Foreword," in F. W. Huchzermeyer, ed., *Diseases of Ostriches and Other Ratites*, Agricultural Research Council, Ondersport, South Africa, 1998.

J. R. Wade, "Restraint and Handling of the Ostrich," in T. N. Tully, Jr. and S. M. Shane, eds., *Ratite Management, Medicine, and Surgery*, Krieger, Malabar, Fla., 1996, pp. 37–45.

GIDEON ZEIDLER
University of California
Riverside, California

FRANCINE BRADLEY
RALPH ERNST
JENNIFER NEAR
University of California
Davis, California

POULTRY MEAT MICROBIOLOGY

CONTAMINATION OF POULTRY

The microflora of commercially processed raw poultry cannot be easily defined. Hundreds of different species of microorganisms have been isolated from live poultry and from fully processed, ready-to-eat carcasses and poultry parts. Nonpathogenic spoilage organisms and potential human pathogens are the two general categories of organisms that are of greatest concern to the poultry industry and the ultimate consumer. The following genera have been isolated from fully processed, raw poultry (1–4): *Pseudomonas, Achromobacter, Micrococcus, Flavobacterium, Alcaligenes, Proteus, Bacillus, Sarcina, Streptomyces, Penicillium, Oasora, Cryptococcus*, and *Rhodotorula*. These organisms originate from unlimited sources including the hatchery, the production environment, the feed or water, wild birds or animals, rodents, insects, domesticated farm animals, equipment, and even humans. In the United States and many other countries, almost all commercial broilers and turkeys are reared on the ground in large poultry houses. The outside surface of the live bird, including the feathers, feet, and skin, can support numerous different types of microorganisms. The contents of the intestinal tract of the bird is a constant source of contamination, which can easily be transferred to the live bird during production, transport, and processing. However, the skin of the carcass is more highly contaminated than the visceral cavity of freshly processed raw poultry (5,6). In the United States, the processing of almost all commercial poultry is extremely automated. This fact has caused considerable concern because birds move at considerable speed, up to 91 broilers/min. However, data suggest that commercially processed birds and birds processed by hand do not differ in the incidence or levels of microorganisms (7).

SHELF LIFE AND SPOILAGE

Freshly processed raw poultry carcasses or poultry parts typically harbor 1,000 to 10,000 microorganisms/cm^2 at the exit of the immersion chiller (5,8). At retail, total microbial counts can range from 10,000 to 10,000,000/cm^2 of skin surface area. Raw poultry is considered spoiled when total microbial counts exceed 10,000,000/cm^2 (9). At this level of bacterial growth, the product will be slimy and will have an uncharacteristic sweet smell (10). The two most important factors that influence shelf life are initial microbial levels and temperature of storage.

Pseudomonas, Moraxella, and *Alcaligenes* are the principal microorganisms on spoiled poultry carcasses and parts (1,2,11–13). At the time when spoilage becomes evident by slime formation and off-odor, *Pseudomonas* may also have produced a green fluorescent pigment that is visible under ultraviolet light. Any off-odors are the result of biochemical changes in the protein and fat. The time between day of processing and spoilage at retail for nonfrozen poultry carcasses or parts is typically 10 to 14 days at 4°C. The retail consumer should expect only 1 to 2 days of shelf life once the product is brought home and stored in the refrigerator. Thus, if fresh poultry is not to be used on the purchase day or the day after purchase, the product should be stored in the freezer. Frozen whole turkeys or chickens can be stored in the home freezer for 1 year; poultry parts can be frozen for 6 to 9 months (14). Frozen storage recommendations are based on oxidative stability, not microbial spoilage.

Bacterial levels increase when poultry carcasses are segmented, filleted, deboned, or comminuted. In the home, ground turkey or chicken will have a refrigerated shelf life of 1 to 2 days. Ground poultry products can be stored frozen for 3 to 4 months. Again, frozen shelf life is based on oxidative stability. Frozen chicken, fried chicken parts, patties, or nuggets are very susceptible to oxidative rancidity and should be stored in the home refrigerator for 3 to 4 days or up to 4 months in the home freezer.

Numerous methods have been used for evaluating the numbers of microorganisms on poultry carcasses. Among them are pressing metal or plastic dishes filled with solidified agar, using cut tissue sections sliced to known dimensions, using a blender or stomacher, swabbing the skin surface, and rinsing the product with known amounts of diluent. Although counts will not be identical, results from most of these methods are comparable (15).

Studies have been made to determine the effect of temperature on the growth of bacteria on poultry carcasses. A shelf life of 2 days at 15°C and 30 days at 1°C for uneviscerated chickens has been reported (16). It has been noted that broiler carcasses remained acceptable for 9 to 19 days at 1°C, depending on a number of factors, notably the mode of processing and the type of wrapping (17). One study showed that when carcasses were stored at 0°C, it took approximately 18 days for spoilage to occur, and at 20°C only 2 days were required (18). It has been reported that chicken meat with an initial bacterial count of approximately 6.0×10^4 cells/$in.^2$ spoiled after 16 days of storage at 0°C (9). Higher storage temperatures, such as 4.4 and 10°C, reduced the shelf life to 7 and 2 days, respectively.

A 10- to 12-day shelf life for immersion-chilled poultry meat and 18 days for hot-packaged poultry meat stored at 2 to 4°C been noted (3). Dry-chilling broiler carcasses by liquid nitrogen exposure resulted in a reduction of psychrophilic and memsophilic organisms initially and on storage at 2 to 4°C. An increase of approximately 2 days in shelf life of hot-packaged, liquid nitrogen—chilled cut-up parts was evident over the hot-packed parts (19). Lowering the storage temperature from 1.1 to 3.3°C to 2.2 to 1.1°C extended the shelf life for vacuum-CO_2—packed and the ice-packed broiler carcasses for approximately 5 and 4 days, respectively (4).

POTENTIAL HUMAN PATHOGENS

The microbiological population on poultry carcasses can, and often does, include potential human pathogens such as *Aeromonas* spp., *Salmonella* spp., *Campylobacter* spp., *Staphylococcus aureus, Clostridium perfringens, Yersinia enterocolitica*, and *Listeria* spp. Fortunately, most of these organisms are present only at very low levels and tend to remain at low levels due to the fact that they reproduce less vigorously than the many nonpathogens, which outnumber them from the beginning.

Aeromonas

To date, no fully confirmed outbreaks of foodborne illness have been attributed to *Aeromonas*. However, substantial data suggest that this organism is capable of causing foodborne illness in humans. Data from the United States and other developed countries report the incidence of this potential pathogen on raw poultry to be greater than 50%. Levels of this organism on raw poultry tend to exceed 100 per gram or per milliliter. Unfortunately, this organism can grow during normal refrigeration. However, this organism is very sensitive to heat and radiation and would primarily be a problem only if poultry was eaten raw or undercooked.

Salmonella spp.

In terms of the scientific literature and the popular press, the most common poultry pathogen is *Salmonella*. In the United States, incidence of this organism on raw poultry ranges from 8 to 50% (20). In other countries, incidence rates approaching 100% have been recorded (21). Regardless of country, levels of this organism on raw poultry are almost always extremely low, less than 30 organisms per carcass. Thus, in most cases of human salmonellosis due to consumption of poultry, contributing factors include cross-contamination, improper cooking, and temperature abuse. Currently, *Salmonella* is the only organism with a performance standard (20% positive) for raw poultry in the United States.

Campylobacter spp.

The incidence of *Campylobacter* on raw poultry is exceptionally high, greater than 80% (8,22,23). In the last 10 years the reported incidence of *Campylobacter* on poultry products has increased. In reality, the methodology for recovery of this organism has greatly improved, thus increasing the chance of recovery. In the few studies where this organism has been enumerated, data suggest that levels are very high in comparison to levels of other common poultry-associated pathogens (20). And even though the organism is unlikely to grow on raw poultry due to its fastidious nature, initial levels (>500/mL) are in the range to cause human illness. Most cases of campylobacteriosis are sporadic, and contributing factors typically include cross-contamination and lack of proper handwashing. There is particular concern with campylobacteriosis because of increasing antibiotic resistance and the organisms' connection with Guillain–Barré syndrome.

Staphylococcus aureus

Data indicate that most raw fresh and frozen poultry, both chicken and turkey, are contaminated with *S. aureus* (20). And in most studies, the numbers reported are rather high (>1000 organisms/g or cm^2). This organism must multiply to reach even higher levels in a food and produce toxin in order for a case of foodborne illness to occur. Staphylococci are not good competitors, so even if initial levels are high, as bacterial competition increases during refrigerated storage (even during temperature-abuse situations), the number of staphylococci on raw poultry tends to decrease considerably. This organism becomes a concern on raw poultry only when the natural bacterial population is significantly decreased by some type of bactericidal treatment, especially when followed by temperature-abuse conditions. Because normal heating or reheating will not destroy the toxin produced by this organism, the easiest precaution for preventing staphylococcal foodborne illness is to refrigerate cooked poultry quickly.

Clostridium perfringens

The incidence of *C. perfringens* on raw poultry ranges from 10% to 80% (20). In studies where counts have been reported, the numbers have been very low (<10 organism/g or cm^2). In virtually all cases of foodborne illness attributed to *C. perfringens*, the principal problem is trying to cool large batches of a particular food. The cooking process heat-activates spores, which can germinate during cooling. Thus, fast cooling of cooked poultry products is very important.

Yersinia enterocolitica

Depending on the reporting country, there is a wide range in the incidence of *Y. enterocolitica* organism on raw poultry. In the United States, incidence of this organism on raw broilers at retail is 27% (24). Unfortunately, there is very little information regarding the level of this organism on raw poultry. It is important to remember that not all serovars of this organism are pathogenic, but pathogenic serovars have occasionally been isolated from frozen chickens at the retail level. Unfortunately, *Yersinia* is a psychrotroph and can survive better than other gram-negative organisms under extreme alkaline conditions such as might be present on poultry treated with trisodium phosphate.

Listeria spp.

Few data are available concerning the incidence of this potential pathogen on raw poultry before the late 1980s. Two early European studies, conducted before 1980, suggested that the intestinal tract of broilers, freshly processed broilers, fresh broilers at the time of sale, and frozen broilers were contaminated with this organism. Since 1989 numerous studies have been designed to evaluate the incidence of this pathogen on raw poultry products. Depending on the country where such studies were conducted as well as the isolation techniques utilized, incidence rates range from 2 to 94%. Of course, not all isolates recovered have been *L. monocytogenes* (2–50%), but the presence of any *Listeria* spp. would lead to the conclusion that pathogenic species could be present. However, it is important to note that the serotype most often incriminated in cases of human listeriosis (serotype 4b) has not been a problem in raw poultry. Data regarding the numbers of *Listeria* on raw poultry are limited but suggest that levels are quite low.

FURTHER-PROCESSED POULTRY PRODUCTS

Further-processed poultry products or items of poultry meat processed beyond the eviscerated carcass stage are subjected to many subsequent handlings that are major sources of microbial contamination. However, information concerning the microbiological quality of precooked chicken products is limited. The destruction of foodborne microorganisms is an essential part of the commercial cooking operation.

The microbiological quality of further-processed turkey products has been reported. In the 38 cooked products examined, neither *Salmonella* spp. nor *Escherichia coli* were found, whereas 16 out of 38 contained *C. perfringens*, and only 1 out of 38 was positive for *S. aureus* (25). It was reported that bacterial counts of ready-to-eat turkey rolls increased following storage at 5°C (26). *Salmonella* and coagulase-positive *Staphylococcus* were isolated from some of these samples.

Frozen fried chicken is usually subjected to subsequent handling, which provides a potential opportunity for microbial contamination of the product. The microbiological quality of frozen fried chicken products was studied; it was reported that the log total microbial count ranged from 2.74 to 4.66/gm (27). Neither mold nor yeast was detected, and all samples were *Salmonella* negative. *Micrococci* and *Staphylococci* have been isolated from both hot water and microwaved precooked chicken parts and commercial ready-to-eat sliced chicken products (28).

The total aerobic and coliform counts in chicken patties during refrigerated storage were studied (29). *Escherichia* was the predominant genus at day 0, but *Enterobacter* was predominant after 10 days of storage. The quality of prefried chicken patties and nuggets was studied (30). It was indicated that the predominant microflora in raw and fried chicken patties were gram-positive, rod-type organisms. In addition, vacuum packaging did not inhibit the ultimate psychrotrophic microflora growth on patties and nuggets after refrigerated storage.

Chicken ham cooked to internal temperatures of 68.3 and 71.1°C had a longer lag phase microbial growth at 2 to 4°C than products cooked to an internal temperature of 65.0°C. The shelf life of chicken ham at 2 to 4°C with 68.3 and 71.1°C end point internal temperatures was 33 and 35 days, respectively. Chilling the products in ice slush immediately after cooking extended the shelf life approximately 7 days. Vacuum packaging was found to be an effective method for extending the shelf life of cooked ham (31).

An interaction between frying temperature and internal temperature was significant for microbial counts of deep-fat fried chicken patties. When patties were fried to an internal temperature of 48.9°C, there was a significant decrease in microbial counts as frying temperature increased. This effect was not significant when patties were fried to an internal temperature of 60.0 or 71.1°C. An internal temperature of 71.1°C is required to be labeled "fully cooked," "ready-to-eat," "baked," or "roasted" poultry according to the USDA (32). Raw chicken patties had a mean total psychrotrophic count of log 5.1 colony forming unit (CFU)/gm. Chicken patties fried to an internal temperature of 71.1°C had microbial counts ranging from 1.52 to 1.75 CFU/gm.

RETARDING THE GROWTH OF MICROORGANISMS ON POULTRY

Temperature Control

Of all the methods to delay microbial spoilage, temperature control remains the most dependable and widely used. According to the USDA, broiler carcasses must be chilled to at least 4.4°C after 4 hours; turkeys, after 8 hours (32). In most commercial facilities this is accomplished through immersion chilling. Freezing poultry destroys 96 to 99% of the total surface microflora (33). Numbers continue to decrease during prolonged frozen storage. Freezing does not completely free poultry from salmonellae or other potential pathogens.

The bacteria responsible for the spoilage of poultry are not notably xerotolerant. Freezing will subject these organisms to water stress at temperatures close to their minima for growth. The synergistic effects of these two factors, temperature and water activity, probably insure that spoilage bacteria do not usually grow on frozen poultry, although some strains could possibly grow at temperatures below 2°C. The preservation of food by freezing is based on the retardation of microbial growth to the point at which decomposition due to microbial action does not occur (34). Pathogenic microorganisms do not grow below 2°C.

Preservatives

Numerous efforts have been made to use additives to reduce the microbial load on fresh poultry. Some of the techniques used have been chilling and washing of carcasses with chlorinated water (35–45), 70% alcohol dipping (46), immersion of carcasses in solutions of poly(hexamethylenebiguanide hydrochloride) (47), dipping in various organic acids (19,48–51), and chilling in polyphosphate solutions (52,53).

It was reported that when *Lactobacillus* cells were added to chilled water at a concentration of 100 million/mL, shelf life was increased as much as 3 days. Carcasses immersed in 0.12% lactic acid had a total bacterial count of 400 compared with a control group that had 130,000 organisms/g (50). Sorbic acid was an effective preservative when a 7.5% solution was sprayed onto the chilled poultry parts at 140°F in a ratio of 70:20:10 propylene glycol to water to glycerin (54). Nine acids were studied, and it was concluded that adipic and succinic acid increased the shelf life of broilers by 6 days more than the water-chilled samples when the pH of the chilling solution was 2.5 (49). Another study revealed that dipping cup-up broiler parts in 1% ascorbic acid solution for 3 min retarded microbial growth and increased the refrigerated life for 6 to 7 days compared to that of the control (19). It has been suggested that the use of polyphosphate solutions was effective in controlling the growth of gram-positive micrococci and staphylococci (28).

Of the materials tested for use in preserving chilled poultry carcasses, antibiotics have received the most attention. The addition of antibiotics to the chilling water had been widely accepted in the 1950s and the early 1960s. Chlorotetracycline or oxytetracycline was added to the chilling bath through which the birds passed following evisceration; the results indicated that these antibiotics lengthened the keeping time by at least several days, but that bacteria resistant to the antibiotics (eg, pigmented pseudomonads and certain *Alcalilgenes* species) grew on the birds and built up in the plant (55). The antibiotics had no effect on yeasts and molds that were also a part of the microflora of poultry carcasses (56). Because of the fact that small residues, about 0.5 ppm, may be left after cooking and the effect that such residues might have on the intestinal flora of humans, permission to use such compounds in foods was withdrawn by the FDA (57).

Potassium sorbate has been extensively investigated as an antibacterial agent for extending shelf life and inhibiting pathogen growth. The effectiveness of a 10% potassium sorbate dip in inhibiting bacterial growth of fresh chicken breast meat inoculated with *Salmonella* was demonstrated (58). It has also been reported that dipping broiler carcasses in a 5% potassium sorbate dip for 1 min was effective in inhibiting *Salmonella* and *S. aureus* (59).

Preparation of germicidal ice incorporated with food additives such as calcium hypochlorite, chloramine benzoic acid, formaldehyde, hydrogen peroxide, sodium propionate, and sodium nitrite for ice-packing purposes have been studied (60). It was proven that ice-containing glycol diformate and sorbic acid were more effective against microbial growth than was common ice or the acronizing process. It has been reported that the average shelf life of broiler parts ice-packed with 0.075% sorbic acid ice and 0.1% potassium sorbate ice was 11.5 and 3.6 days longer than the ice-packed controls (61). Both sorbic acid ice and potassium sorbate ice decreased the incidence of gram-negative rod-type organisms and increased the incidence of gram-positive cocci on broiler parts after 4 days of storage at 2 to 4°C.

Packaging

A number of studies have been made to determine the effect of packaging on the shelf life of poultry carcasses. One study reported that packaging materials had a significant effect on the shelf life of fresh poultry (62). Storage time for poultry carcasses wrapped in cellophane and cellophane—vinylidene chloride copolymer sheets were the same under nonvacuum conditions, but when the air was evacuated, the storage time increased 4 days beyond that obtained without vacuum (63). Fewer pseudomonads were found, but *Microbacterium thermosphactum* and atypical lactobacilli were a significant part of the spoilage flora on turkeys wrapped in impermeable film at 1°C (64). It has been reported that while the pseudomonads were the main spoilage organisms on the carcasses wrapped in the permeable film, the *Alternomonas putrefacians* predominated on the carcasses wrapped in the impermeable film.

The influence of packaging materials seems to affect the quality of fresh poultry. Bacterial growth was inhibited by using a shrinkable poly(vinyl chloride) film with a permeability to oxygen of approximately 500 mL/cm^2, 24 h, 1 atm (65). It was also concluded that shrinkable poly(vinyl chloride) film was better for retaining the quality and shelf life of chicken than any other material that was not air evacuated.

Another effective way of packaging is by employing carbon dioxide treatment. Chickens stored in 10 and 20% carbon dioxide at 1°C effectively extended the shelf life (66). It has been suggested that packaging broiler carcasses in nylon—surlyn film with a carbon dioxide addition rate of 3.6×10^{-4} and 7.22×10^{-4} m^3/kg carcass extended shelf life at 1.1°C of storage to 22 and 27 days, respectively (67). Vacuum-CO_2 packaging extended the shelf life of broiler carcasses for approximately 7 days when compared to those of the ice-packed controls (4). It was also reported that the major microorganism groups isolated from the spoiled ice-packed samples were members of the *Pseudomonas* species (93.3%); however, *Lactobacillus* spp. (74.0%) were found to be the dominant microflora on the vacuum-CO_2 packaged broilers after 28 days of storage at 1.1 to 3.3°C (4).

Other Treatments

Ozone treatment also has been used for preservation of foods, especially of fruits and vegetables. Ozone-treated broiler parts have consistently lower microbial counts than air-treated control parts during the entire refrigerated observation period (68). Using log total microbial counts of 7.0/cm^2 as spoilage criterion, broiler parts treated with ozone had a shelf life that was extended for 2.4 days. It was also indicated that ozone-treated carcasses contained about 52.7% gram-positive cocci, while air-treated controls had 39.6% gram-positive cocci. Studies using microflora from spoiled poultry meat have also demonstrated that ozone treatment preferentially destroyed gram-negative, rod-type organisms.

Radurization processing of fresh eviscerated poultry with a dose of 5 kGy has been reported to extend the shelf life at 5°C by approximately 14 days (69). It has been reported that at 4°C the shelf life of chicken carcass was 3 days for nonirradiated samples, 13 days for those irradiated at 3 kGy, and greater than 30 days for those irradiated at 7 kGy (70). Chicken carcasses irradiated with 5 to 10 kGy at 3 to 4°C could extend the shelf life by 2 to 4 weeks (71). Chicken breast meats were irradiated with 3.7 kGy at 0°C; a satisfactory quality for about 3 weeks was reported (72). Recently, it was noted that irradiation increased the shelf life of chicken fillets (73). The time required to reach a log number of 6.5/g was 15 days for the controls and 35 days for the fillets treated with 2 kGy, all stored at 3°C. The combination of vacuum skin packaging and 4 kGy irradiation dose resulted in fillets with a shelf life of more than 45 days at 3°C. Irradiation at 3 kGy destroyed all inoculated *S. typhimurium* and *E. coli* in ground chicken meat.

Antagonisms of microorganisms on broiler carcasses have been observed. The growth-interfering effect of spoilage microorganisms by gram-positive cocci on broiler carcasses has been reported (74). This growth-interfering effect was greater at 2 to 4°C and 5 to 7°C than at 19 to 21°C. The higher the ratio of the gram-positive cocci to the rod-type spoilage microorganisms, the more the growth-interfering effect was observed. The cell-free filtrates from gram-positive broth cultures interfered with the growth of the spoilage microorganisms at 2 to 4°C, and the causative agent was heat-sensitive metabolites.

BIBLIOGRAPHY

1. J. C. Ayres, W. S. Ogilvy, and G. F. Stewart, "Post-Mortem Changes in Stored Meats. Microorganisms Associated with Development of Slime on Eviscerated Cut-Up Poultry," *Food Technology* **4**, 199–205 (1950).
2. A. S. Arafa and T. C. Chen, "Effect of Vacuum Packaging on Microorganisms on Cut-Up Chicken and in Chicken Products," *Journal of Food Science* **40**, 50–52 (1975).
3. A. S. Arafa and T. C. Chen, *Evaluation of Final Washing and Immersion Chilling on Carcass Microflora from Fourteen Mississippi Poultry Processing Plants*, MAFES Research Report **3**, 1977.
4. K. Viseshsiri and T. C. Chen, *Microbiological Characteristics of Vacuum-CO_2 Packaged Broiler Carcasses*, MAFES Bulletin **910**, Mississippi State, Miss., 1982.
5. H. W. Walker and J. C. Ayres, "Antibiotic Residuals and Microbial Resistance in Poultry Treated with Tetracyclines," *Food Research* **23**, 525–531 (1968).
6. M. Woodburn, "Incidence of Salmonellae in Dressed Broiler-Fryer Chickens," *Applied Microbiology* **12**, 492–495 (1964).
7. A. L. Izat, J. M. Kopeck, and J. D. McGinnis, "Research Note: Incidence, Number, and Serotypes of Salmonella on Frozen Broiler Chickens in Retail," *Poultry Science* **70**, 1438–1440 (1991).
8. A. L. Waldroup, B. M. Rathgeber, and R. E. Hierholzer, "Effects of Reprocessing on Microbiological Quality of Commercial Prechill Broiler Caracasses," *Journal of Applied Poultry Research* **2**, 111–116 (1993).
9. R. P. Elliott and H. D. Michener, "Microbiological Standards and Handling Codes for Chilled and Frozen Foods. A Review," *Applied Microbiology* **9**, 452–468 (1961).
10. A. A. Kraft, "Microbiology of Poultry Products," *Journal of Milk and Food Technology* **34**, 23–29 (1971).
11. J. C. Ayres, "The Relationship of Organisms of the Genus *Pseudomonas* as the Spoilage of Meat, Poultry and Eggs. *Journal of Applied Bacteriology* **23**, 471–486 (1960).
12. E. M. Barnes and C. S. Impey, "Psychrophilic Spoilage Bacteria of Poultry," *Journal of Applied Bacteriology* **31**, 97–107 (1968).
13. T. A. McMeekin, "Spoilage Association of Chicken Leg Muscle," *Applied and Environmental Microbiology* **33**, 1244–1246 (1977).
14. USDA Safety Inspection Service, *The Safe Food Book: Your Kitchen Guide*, Home and Garden Bulletin No. 241, Washington, D.C., 1985.
15. A. L. Izat et al., "Effects of Sampling Method and Feed Withdrawal Period on Recovery of Microorganisms from Poultry Carcasses," *Journal of Food Protection* **52**, 480–483 (1989).
16. E. M. Barnes and D. H. Shrimpton, "Causes of Greening of Uneviscerated Poultry Carcasses during Storage," *Journal of Applied Bacteriology* **20**, 273 (1957).
17. E. O. Essary, W. E. C. Moore, and C. Y. Kramer, "Influence of Scald Temperatures, Chill Times, and Holding Temperatures on the Bacterial Flora and Shell Life of Fresh Chilled, Tray-Packed Poultry, *Food Technology* **12**, 684–687 (1958).
18. L. E. Dawson and W. J. Stadelman, *Microorganisms and Their Control of Fresh Poultry Meat*, NCM-7 Technical Bulletin **278**, Michigan State University, East Lansing, Mich., 1960.
19. A. S. Arafa and T. C. Chen, "Ascorbic Acid Dipping as a Means of Extending Shelf-Life and Improving Microbial Quality of Cut-Up Broiler Parts," *Poultry Science* **57**, 99–103 (1978).
20. A. L. Waldroup, "Contamination of Raw Poultry with Pathogens," *World Poultry Science Journal* **52**, 7–25 (1996).
21. A. S. Kamat et al., "Hygienization of Indian Chicken Meat by Ionizing Radiation," *Journal of Food Safety* **12**, 59–71 (1991).
22. N. J. Stern and J. E. Line, "Comparison of Three Methods for Recovery of *Campylobacter* spp. from Broiler Carcasses," *Journal of Food Protection* **55**, 663–666 (1992).
23. A. L. Waldroup, B. M. Rathberger, and R. H. Forsythe, "Effects of Six Modifications on the Incidence and Levels of Spoilage and Pathogenic Organisms on Commercially Produced Postchill Broilers," *Journal of Applied Poultry Research* **1**, 226–234 (1992).
24. N. A. Cox et al., "The Presence of *Yersinia enterocolitica* and Other *Yersinia* Species on the Carcasses of Market Broilers," *Poultry Science* **69**, 482–485 (1990).
25. E. A. Zottola and F. F. Busta, "Microbiological quality of further-processed turkey products," *Journal of Food Science* **36**, 1001–1004 (1971).
26. A. J. Mercuri, G. J. Banwart, J. A. Kinner, and A. R. Sessums, "Bacteriological Examination of Commercial Precooked Eastern Type Turkey Rolls," *Applied Microbiology* **19**, 768–771 (1970).
27. P. L. Wang, E. J. Day, and T. C. Chen, "Microbiological Quality of Frozen Fried Chicken Products Obtained from a Retail Store," *Poultry Science* **55**, 1290 (1976).
28. T. C. Chen, J. T. Culotta, and W. S. Wang, "Effect of Water and Microwave Energy Precooking on Microbiology Quality of Chicken Parts," *Journal of Food Science* **38**, 155–157 (1973).
29. S. E. Craven and A. J. Mercuri, "Total Aerobic and Coliform Counts in Beef-Soy and Chicken-Soy Patties during Refrigerated Storage," *J. Food Prot.* **40**, 112–115 (1977).
30. Y. H. Yi, *Studies on the Quality of Prefried Chicken Patties During Refrigerated Storage*, Ph.D. dissertation, Mississippi State University, Mississippi State, Miss., 1986.
31. S. C. Yang and T. C. Chen, "Processing Yields and Shelf Life of Chicken Ham," *Taiwan Journal of Food Science* **15**, 133–141 (1988).

32. The U.S. Department of Agriculture, *Meat and Poultry Inspection Manual*, U.S. Department of Agriculture, Washington, D.C., 1973, Part 18, Subpart G, Article 37.

33. A. A. Kraft, J. C. Ayres, K. F. Weiss, W. W. Marion, S. L. Balloun, and R. H. Forsythe, "Effect of Method of Freezing on Survival of Microorganisms on Turkey," *Poultry Science* **48**, 128–136 (1963).

34. A. C. Peterson and R. E. Gunnerson, "Microbiological Critical Control Points in Frozen Foods," *Food Technology* **28**, 37–44 (1974).

35. E. E. Drewniak, M. A. Howe, Jr., H. E. Goresline, and E. R. Baucsh, *Studies on Sanitizing Methods for Use in Poultry Processing*, USDA Circular No. 930, 1954.

36. M. F. Gunderson, H. W. McFaddin, and T. S. Kyle, *The Bacteriology of Commercial Poultry Processing*. Burgess Publishing Co., Minneapolis, Minn., 1954.

37. L. E. Dawson, W. L. Mallmann, M. Frang, and S. Walters, "The Influence of Chlorine Treatments of Bacteria Population and Test Panel Evaluation of Chicken Fryers," *Poultry Science* **35**, 1140 (1956).

38. J. M. S. Dixon and F. E. Pooley, "The Effect of Chlorination on Chicken Carcass Infected with Salmonellae," *Journal of Hygiene* **59**, 343 (1961).

39. A. W. Kotula, J. E. Thompson, and J. A. Kinner, "Bacterial Counts Associated with the Chilling of Fryer Chickens," *Poultry Science* **41**, 818–821 (1962).

40. W. Mercer, "Product Protection by Chlorination and Other Safeguards," in *Proceedings of the Poultry and Egg Further-Processing Conference*, 1964, pp. 33–40.

41. M. D. Rankan, G. Clewlow, D. H. Shrimpton, and B. J. H. Stevens, "Chlorination in Poultry Processing," *British Poultry Science* **6**, 331–337 (1965).

42. J. E. Thomson, G. J. Banwart, D. H. Sanders, and A. J. Mercuri, "Effect of Chlorine, Antibiotics, Beta-Propolactone, Acids and Washing on *Salmonella typhimurium* on Eviscerated Fryer Chickens," *Poultry Science* **46**, 146–151 (1967).

43. C. J. Wabeck, D. V. Schwall, G. M. Evancho, J. C. Heck, and A. B. Rogers, "Salmonella Reduction, Chlorine Uptake, and Organoleptical Changes in Poultry Treated with Sodium Hypochlorite," *Poultry Science* **46**, 1333 (1967).

44. J. T. Patterson, "Bacterial Flora of Chicken Carcasses Treated with High Concentration of Chlorine," *Journal of Applied Bacteriology* **31**, 554–550 (1968).

45. R. M. Blood and B. Jarvis, "Chilling of Poultry: The Effects of Process Parameters on the Level of Bacteria in Spin-Chiller Waters," *Journal of Food Technology* **9**, 157–169 (1974).

46. K. N. Hall and J. V. Spencer, "The Effect of Ethanol on Shelf Life and Flavor of Chicken Meat," *Poultry Science* **43**, 573–576 (1964).

47. M. N. Islam and N. B. Islam, "Extension of Poultry Shelf Life by Poly(Hexamethylenebiguanide Hydrochloride), *J. Food Prot.* **42**, 416–419 (1979).

48. N. A. Cox, A. J. Mercuri, B. J. Juven, J. E. Thomson, and V. Chow, "Evaluation of Succinic Acid and Heat to Improve the Microbiological Quality of Poultry Meat," *Journal of Food Science* **39**, 985–987 (1974).

49. G. J. Mountney and J. E. O'Malley, "Acids as Poultry Meat Preservatives," *Poultry Science* **44**, 582–586 (1965).

50. U.S. Pat. (1962), J. F. Murphy and R. E. Murphy.

51. T. E. Minor and E. H. Marth, "*Staphylococcus aureus* and Staphylococcal Food Intoxications. A Review," *Journal of Milk and Food Technology* **35**, 228–241 (1972).

52. J. V. Spencer and L. E. Smith, "The Effect of Chilling Chicken Fryers in a Solution of Polyphosphates upon Moisture Uptake, Microbial Spoilage, Tenderness, Juiciness, and Flavor," *Poultry Science* **41**, 284–288 (1962).

53. R. P. Elliott, R. P. Straka, and J. A. Garibaldi, "Polyphosphate Inhibition of Growth of Pseudomonads from Poultry Meat," *Applied Microbiology* **12**, 517 (1964).

54. G. A. Perry, R. L. Lawrence, and D. Melnick, "Extension of Shelf Life of Poultry by Processing with Sorbic Acid," *Food Technology* **18**, 101–107 (1962).

55. H. W. Walker and J. C. Ayres, "Incidence and kids of microorganisms associated with commercially dressed poultry," *Applied Microbiology* **4**, 345–349 (1956).

56. R. H. Vaughn and G. F. Stewart, "Antibiotics as Food Preservatives," *Journal of the American Medical Association* **174**, 1308–1310 (1960).

57. J. T. Nickerson and A. J. Sinskey, *Microbiology of Foods and Food Processing*, Elsevier Publishing Co., Inc., New York, 1972.

58. M. C. Robach and F. J. Ivey, "Antimicrobial Efficacy of a Potassium Sorbate Dip on Freshly Processed Poultry," *J. Food Prot.* **41**, 284–288 (1978).

59. E. C. To and M. C. Robach, "Potassium Sorbate Dip as a Method of Extending Shelf Life and Inhibiting the Growth of *Salmonella* and *Staphylococcus aureus* on Fresh, Whole Broilers," *Food Technology* **37**, 107–110 (1980).

60. S. Kaloyereas, R. M. Crown, and C. S. McClesky, "Experiments on Preservation with a New Ice-Containing Glycol Diformate and Sorbic Acid," *Food Technology* **15**, 582–586.

61. Y. L. Lin and T. C. Chen, *Effect of Sorbic Acid Ice on Microorganisms of Broiler Meat*, MAFES Research Report **14**, Mississippi State, Miss., 1989.

62. J. V. Spencer, M. W. Eklund, E. H. Suter, and M. M. Hard, "The Effect of Different Packaging Materials on the Shelf Life of Antibiotic Treated Chicken Fryers," *Poultry Science* **35**, 1173 (1956).

63. F. E. Wells, J. V. Spencer, and W. J. Stadelman, "Effect of Packaging Materials and Techniques on Shelf Life of Fresh Poultry Meat," *Food Technology* **12**, 425 (1958).

64. E. M. Barnes and D. H. Shrimpton, "The Effect of Processing and Marketing Procedures on the Bacteriological Condition and Shelf Life of Eviscerated Turkeys," *British Poultry Science* **9**, 243–251 (1968).

65. J. M. Debevere and J. P. Voets, "Influence of Packaging Materials on Quality of Fresh Poultry," *British Poultry Science* **14**, 17–22 (1973).

66. C. J. Wabeck, C. E. Parmella, and W. J. Stadelman, "Carbon Dioxide Preservation of Fresh Poultry," *Poultry Science* **47**, 468–474 (1968).

67. E. H. Sander and H. M. Soo, "Increasing Shelf Life by Carbon Dioxide Treatment and Low Temperature Storage of Bulk Pack Fresh Chickens Packaged in Nylon/Surlyn Film," *Journal of Food Science* **43**, 1519–1523, 1527 (1978).

68. P. P. W. Yang and T. C. Chen, "Effect of Ozone Treatment on Microflora of Poultry Meat," *J. Food Proc. and Preserv.* **3**, 177–185 (1979).

69. E. S. Idziak and K. Incze, "Radiation Treatment of Foods, Radurization of Fresh Eviscerated Poultry," *Applied Microbiology* **16**, 1061 (1968).

70. H. E. Bok and W. H. Holzapfel, "Extension of Shelf Life of Refrigerated Chicken Carcasses by Radurization," *Food Review* **11**, 69–71 (1984).

71. H. O. Cho, M. K. Lee, M. W. Bynn, J. H. Kwon, and J. G. Kim, "Radurization of the Microorganisms Contaminating Chicken," *Korean Journal of Food Science Technology* **17**, 170–174 (1985).
72. R. C. Baker, D. Scott-Kline, J. Hutchison, A. Goodman, and J. Charvat, "A Pilot Plant Study of the Effect of Four Cooking Methods on Acceptability and Yields of Prebrowned Battered and Breaded Broiler Parts," *Food Technology* **65**, 1322–1332 (1986).
73. Y. Zhou and T. C. Chen, *Effect of Vacuum Skin Packaging and Irradiation on Quality of Marinated Chicken Fillets*, M.S. thesis, Mississippi State University, Mississippi State, Miss., 1989.
74. T. C. Chen and K. Tanteeratarm, "The Growth Interfering Effects of Spoilage Microorganisms by Gram-Positive Cocci on Broiler Carcasses," in *Proceedings of the Seventeenth World Poultry Congress*, 1984, pp. 681–683.

T. C. CHEN
Mississippi State University
Mississippi State, Mississippi

A. L. WALDROUP
University of Arkansas
Fayetteville, Arkansas

See also MICROBIOLOGY OF FOODS.

POULTRY MEAT PROCESSING AND PRODUCT TECHNOLOGY

Poultry meat is consumed all around the world. In the last few decades poultry meat has increased in popularity in many countries. Among the reasons for this increase are the relatively low growing costs, the rapid growth rate of poultry, the high nutritional value of the meat, and the introduction of many new further processed products. Overall, the poultry industry has drastically changed since the beginning of the century. In the early 1920s most poultry was produced in small flocks mainly to support small farm units, and the poultry was sold live in the local area markets. Most birds were used both as a source of meat and eggs. This is still the situation in some countries; however, in most countries the poultry industry has grown and specializes in meat production breeds and egg production breeds. Over the years the number of small flocks has decreased dramatically as large operations specializing in raising poultry have emerged. Today it is not uncommon to find farmers specializing in only one phase of the growing stage (eg, breeding, hatching, raising pullets, growing operations). The growing operations are usually fairly large (ie, a few hundred thousand to a few million birds) and must be efficient in order to compete. Vertical integration of poultry operations (from hatching to further processing) is common in quite a few countries and helps in making the production more cost effective and competitive. Genetic improvements and selection for higher feed efficiency have also contributed to the ability of the industry to improve its product and maintain a competitive price in comparison with other red meat sources. At the beginning of the century, poultry was mainly selected and bred for exhibitions. Later some breeders started selecting birds for economical characteristics. At that point the chickens selected were used for the general purpose of supplying both meat and eggs. Selection for meat-type and egg-type birds started 10 to 15 years later. Today we are at the point where in many areas cockerels from a meat-type breed are not raised because they are no longer a profitable source of chicken meat (1). Over the years, birds were selected for fast growing characteristics, and as a result the average time to reach marketing weight for broilers was significantly reduced (12 weeks 40 years ago compared with about 7 weeks today).

Changes in lifestyle and consumption patterns have also helped to make poultry a more popular source of meat. Today consumers in the Western world are much more concerned with the caloric and nutritional value of their food. In this context, poultry has gained popularity because it is viewed as a good source of lean meat and its fat as more unsaturated than red meat. The poultry industry has also taken the initiative to develop new further processed products and to respond to the increasing consumer demand for convenient food items (ie, semi/fully cooked). The changes in poultry meat consumption in the United States over the last 85 years are presented in Table 1. As can be seen, the total poultry meat consumption doubled for the first time between 1940 and 1960 and the second time between 1960 to 1980. This increase has been seen only in poultry meat, whereas red meat consumption has either stayed constant or slightly declined over the same period of time (3). This tremendous increase in poultry meat also represents a change in marketing strategies. The proportion of cut-up poultry increased for example, from 25% in 1970 to 60% in 1980 and the proportion of further processed products increased from 5 to 25%. Exporting of poultry meat has also become an important international trade. Some of the leading countries are the United States (2090 metric tons in 1996), Brazil (530 tons), France (381 tons), Netherlands (195 tons), and Thailand (165 tons; Ref. 3).

In the following section the different processes involved in converting live poultry into meat, the nutritional value of poultry meat, the equipment used for further processing, and the production of poultry products are discussed.

TYPES OF POULTRY

The poultry industry markets different types of birds (Table 2). On a worldwide basis the most popular species in-

Table 1. Civilian per Capita Consumption of Poultry in United States Between 1910 and 1996

Year	Broilers	Turkeys	Total poultry
1910	—[a]	—	15
1940	2	3	17
1960	23.4	6	34
1980	47	10	61
1990	61.5	17	81
1996	71	18	90

Note: Poultry meat consumption based on ready to cook weights in pounds. Only broiler and turkey meat are presented.
Source: Refs. 2 and 3.
[a]No data available.

Table 2. Types of Commonly Available Poultry and Their Average RTC Weight

Type	Age (weeks)	Weight RTC (kg)
	Chickens	
Broiler or fryer	6–8	1.2–1.7
Roaster	8–10	3.0
Rock cornish game	3–4	0.6
Hen/stewing fowl	>52	1.4
Cock or mature rooster	>30	3.0
	Turkeys	
Broiler hen	12	4.2
Young hen	16	7.0
Young tom	17–18	12.5
Spend breeder	>52	11.0
	Ducks	
Broiler or fryer	7	2.5–3.8
	Geese	
Mature	12–16	5.0
	Guineas	
Mature	12	1.5
	Pigeons	
	4–5	0.4

Note: RTC = Ready-to-cook weight (ie, body weight excluding feathers, blood, digestive tract, head, and feet).

clude chickens, turkeys, and ducks. However, other species such as geese and pigeons can be found marketed on a large scale in certain areas. The average marketing age of some of the most common species and their ready-to-cook weights (weight without blood, feathers, viscera, feet, and head) are provided in Table 2.

PROCESSING FRESH POULTRY

Special large-scale plants for processing poultry are common around the world. These plants are specifically designed to process poultry and usually include a slaughtering facility, defeathering, evisceration, inspection, chilling, and packaging. In some cases they are built adjacent to meat processing plants.

The steps involved in processing poultry are illustrated in Figure 1. Usually the process starts with a bulk weighing of the birds received on the truck. The live weight subtracted from the eviscerated weight (minus the weight of the condemned birds) provides the yield value and is often used, together with the grades assigned, to determine the payment to the farmer. Unloading the birds from the crates and placing them on the shackle line is often done manually. Special care should be exercised at this stage to minimize bruising of the excited birds. This is commonly followed by stunning the birds; however, stunning is not always used, especially when religious considerations are involved (eg, according to the Jewish laws known as kosher processing, stunning is prohibited). When stunning is used, it is done by electrical current passing through the bird; the amount of current and duration depends on the size of the birds. In the case of broilers, a current in the range of 60 V for 5 to 10 s is common. It is important not to use too much current, since it can completely stop the heart activity. Gas stunning (by CO_2, argon) is also possible (4) and has the advantage of reduced bruising of the birds.

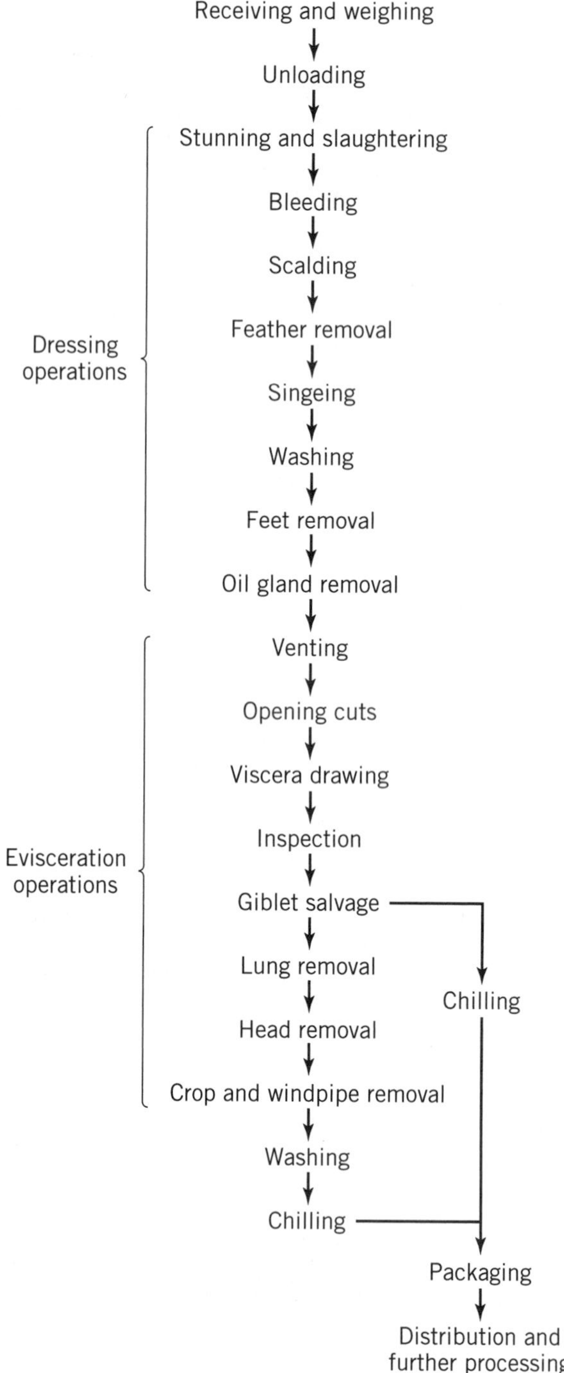

Figure 1. Unit operations in poultry processing.

Stunning, besides calming the birds, also helps in relaxing the muscles holding the feathers, which later facilitates feather removal.

The slaughtering itself can be done in different ways. The most commonly used method is the so-called modified kosher where the jugular vein is cut just below the jowls so that the esophagus and windpipe remain intact. According to the true kosher slaughtering procedure, the windpipe should pop out, and the whole procedure should be performed by a qualified person (*shochet*) and not by a machine. Decapitation and piercing through the roof of the mouth and brain are two other slaughtering methods that are seldomly used in Europe and North America today. After slaughtering, appropriate time should be allowed for bleeding since an excess amount of blood in the muscles will result in discoloration. About 35 to 50% of the total blood is removed during this stage, and it has been shown that the kosher or modified kosher slaughtering methods result in higher bleeding than decapitation.

Scalding is the process of immersing the birds in warm water to loosen the feathers. There are three commonly used temperatures for this purpose (Table 3), and they depend on the degree of difficulty in removing the feathers. The higher scalding temperature is the best for loosening feathers, but it is also the most harsh on the skin. In the case of the hard scalding, the outer layer of the skin (epidermis) is removed during plucking and the skin becomes discolored if dehydrated after processing. However, this method is the only satisfactory way to release the feathers of waterfowl. Relatively speaking, hard scalding does not cause as much discoloration in the thick skin waterfowl as it does in other poultry. Subscalding can remove part of the outer layer of the skin and leaves the skin sticky but will not result in excessive discoloration if the birds are kept in a moist environment. Semiscalding is used for young birds. It does not damage the outer layer of the skin while still allowing for relatively easy removal of the feathers in young birds. A true kosher processing, where scalding is prohibited, usually results in more skin tears because more force is required in the defeathering process. In large processing plants feather removal is done by large mechanical pickers equipped with rubber fingers that rub the feathers off the carcass. This is usually done while the carcass is hanging upside down and carried by a shackle line through the picker. The defeathering process can also be done in a batch-type operation by placing the carcasses in a drum equipped with rubber fingers. In small-scale operations, hand picking of the feathers is sometimes used. When pinfeathers are a problem (as with waterfowl), wax dipping after the picking process is common. Suspending the carcasses in hot wax, followed by cold water immersion, hardens the wax, which is later peeled in large pieces, pulling out the pinfeathers. The wax can be reused after reheating and filtering the feathers from the previous batch. When only minor problems of pinfeathers exist, singeing (the process of burning the small feathers) can be used. This is done by passing the carcass through a flame of a "clean" burning substance (eg, natural gas) that does not leave any off-odor or -flavor. The carcasses are then washed to remove all the soil left after the defeathering and singeing processes. Washing is usually followed by removing the oil gland. After this step, the carcasses are usually moved to another shackle line.

Table 3. Recommended Scalding Schedules for Defeathering

Scalding technique	Water temperature	Time (s)	Used for
Hard Scalding	>63°C	30–60	Waterfowl
Sub scalding	58–60°C	30–75	Mature birds
Semic scalding	52–54°C	30–75	Broilers, roasters, young turkeys

The evisceration process includes the opening of the body cavity and withdrawing the viscera. This process can be done manually by using a regular knife and a pair of scissors or automatically by using a circular saw and a scooplike arm to withdraw the viscera. In any case, special care should be taken not to pierce the viscera and contaminate the carcass. Such contamination can result, in some countries, in the condemnation of parts or the whole bird exposed to the spill. New automated equipment currently on the market allows viscera separation right after its withdrawal, which can improve the hygiene of the process. Once the viscera are exposed, the bird are inspected. The inspection is done at this point because certain diseases affect the intestine or liver. The inspection requirements differ among countries. In some countries it is required that each individual bird be inspected by a veterinarian; in other countries inspection is done on a whole flock basis and only a certain number of individual birds are inspected by a trained inspector. The viscera is removed after the inspection and giblets (liver, heart, and gizzard) are salvaged and washed in a separate line. This is followed by lung removal (manually or by a suction gun); head, crop, and windpipe removal; and a thorough washing of the carcass prior to chilling. The meat must then be chilled to minimize microbial growth. This step is mandatory in many countries, but where it is not, it is recommended that chilling take place as soon as possible. It is common to use long chillers with a counterflow of cold water or crushed ice to bring down the carcass temperature to about 4 to 5°C within 30 to 75 min. Air chillers are also used, but they might cause some surface dehydration if the relative humidity is not precisely controlled. In many countries, the amount of water pickup during the chilling process (when water is used) is regulated and is based on a certain percentage of the body weight. After chilling, the birds are usually packed or immediately deboned for further processing. Due to the relatively short time required for rigor mortis setup in poultry (as compared with red meat), no extra waiting time prior to chilling is required to prevent a phenomenon such as "cold shortening." Overall, the final meat quality is affected by different processing parameters (eg, stunning, chilling) as well as various growing and transporting parameters (1,5); all should be controlled for obtaining the best quality meat.

FRESH POULTRY MEAT

Composition and Nutritional Value

Poultry meat is widely accepted as a good source of high-quality protein, the B vitamins, and minerals such as iron.

The fact that poultry meat is considered to be a source of lean meat and has a higher level of unsaturated fat as compared with red meat has resulted in a significant increase in poultry meat consumption in North America (Table 1) and around the world. Overall, the composition of poultry meat is dependent on species, strain, sex, age, and diet. Table 4 illustrates that species vary in their fat content. Turkey meat is usually lower in fat than chicken, while goose and duck meat are higher in fat. In poultry fat is deposited either under the skin or in the abdominal cavity. Therefore it is easy to obtain very lean poultry meat by separating the skin from the flesh. This is different from red meat where marbling or intramuscular fat deposits are visible within a meat cut and are difficult to separate from the lean muscle. The diet of the monogastric birds can also significantly affect the composition of the meat. Carcass fat content and composition is particularly sensitive to the type of feed. In general, high-energy diets or low-protein diets have been shown to increase carcass fat. It is also possible to increase the proportion of unsaturated fat in poultry meat by manipulating the fat source in the diet.

The relatively high degree of fat unsaturation makes poultry meat susceptible to lipid oxidation during storage. Some processes, such as mechanical deboning, can further enhance the rate of lipid oxidation. In mechanically deboned poultry meat (MDPM), the release of heme from bone marrow, the aeration during the separation process, and extreme mechanical stress can further accelerate the oxidation process (7). The problem of oxidative rancidity can be partially controlled by the exclusion of oxygen by vacuum packaging, the addition of tocopherol to the diet of live birds (ie, vitamin E is deposited in the muscle tissue and later serves as an antioxidant), or the addition of synthetic or natural antioxidants to further processed products. It should be noted that synthetic antioxidants can be used only in countries where permitted.

The production of MDPM has increased substantially during the last three decades, mainly due to the development of equipment for harvesting the meat left on the skeletal frame after hand deboning and because of the increased demand for poultry meat. Today new aspects of mechanical deboning are investigated, such as the mechanical deboning of whole fowl and the subsequent washing of MDPM to obtain lighter meat. The latter is also less susceptible to lipid oxidation due to the removal of heme (eg, the process is similar to the production of fish-based surimi). MDPM is characterized by a pastelike texture and is readily available for the production of finely comminuted meat products (eg, frankfurters, bologna) where finely chopped muscle is traditionally used.

Cooking can also affect the final composition of the meat. This is basically due to the leaching of some of the meat components or the absorption of cooking media into the meat. Yields of cooked poultry average about 75%; the highest yields can be obtained by stewing (78%) followed by frying (77%) and roasting (69%) (6). However, since large variations exist in cooking methods, postslaughter practices (chilling and freezing), and preparation (with or without bone and skin), it is recommended that a reference to a specific cut and cooking procedure be sought.

Color of Fresh and Processed Poultry Meat

The basic color differences between muscles are a result of the relative amounts of white and red muscle fibers. These fibers have different characteristics, and the most noticeable difference to the consumer is their color. Meat color is largely dependent on the amount of meat pigment-myoglobin present in these fibers. Chicken breast muscle is predominantly composed of white fibers, which have a low level and therefore their color is light. On the other hand, thigh meat is mainly composed of red fibers and shows a darker color. Different poultry species also vary in the inherited amount of pigment in their muscles (chicken vs duck). The myoglobin is a complex molecule consisting of two major parts: the protein portion (called globin) and the nonprotein portion (called heme). The latter is capable of binding different compounds and, by that, changing the color of the meat. When the myoglobin is bound to oxygen, chicken leg meat will appear bright pink (Fig. 2). When the same fresh meat is packaged under vacuum it will develop a brownish color (the result of oxidizing the myoglobin into metmyoglobin). On cooking, the protein part of the meat pigment is denatured (similar to denaturation of egg proteins during cooking, which also results in a color and textural change) and gives the meat its typical greyish cooked color. When nitrite is added to the meat, prior to cooking, as is the case in many cured meat products (eg, turkey ham), a typical light pink color will develop. The difference between cured and noncured meat products can be clearly

Table 4. Composition and Nutritional Value of Four Raw Poultry Meats

Source of meat								
Species	Meat	Skin	Water (%)	Protein (%)	Fat (%)	Ash (%)	Iron (mg)	Calories (kcal)
Turkey	White	+	69.8	21.6	7.4	0.90	1.2	159
	White	–	73.8	23.5	1.6	1.00	1.2	115
	Dark	+	71.1	18.9	8.8	0.86	1.7	160
	Dark	–	74.5	20.1	4.4	0.93	1.7	125
Chicken	White	+	68.6	20.3	11.1	0.86	0.8	186
Broiler	Dark	+	65.4	16.7	18.3	0.76	0.98	237
Duck	All	+	48.5	11.5	39.3	0.68	2.4	400
Goose	All	+	50.0	15.9	33.5	0.87	2.5	371
	All	–	68.3	22.7	7.1	1.10	2.5	161

Source: Ref. 6. Expressed on a 100-g portion of meat and skin.

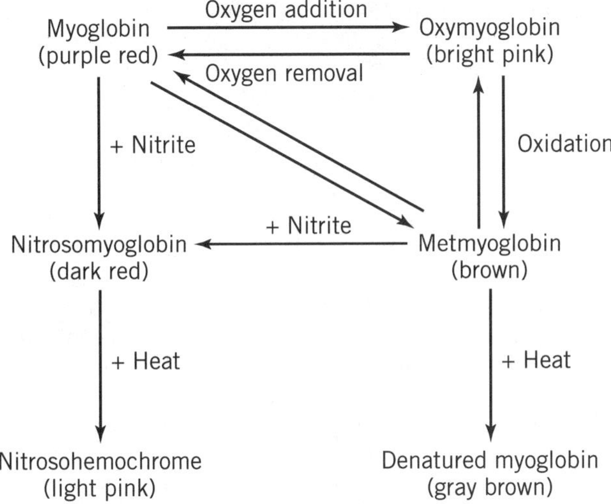

Figure 2. Colors of red poultry muscle (eg, chicken thigh meat).

illustrated when chicken leg meat is prepared at home (as a roast) resulting in a greyish color or when it is cured (as a turkey ham) in a processing plant resulting in the typical pink cured color.

Grading

Classifying poultry into ranges of quality is common in most countries around the world. The grades, based on certain standards, are usually developed by the government and can be voluntarily employed by the producers or imposed by a regulatory agency. Prior to the actual grading, poultry are divided into classes according to their species, sex, and age. This is done because each species has its own unique characteristics. The grading is usually done on the ready-to-cook bird; however, an inspection of the live birds can help in assessing the grade. Such an inspection usually includes: (*1*) general condition of the flock, (*2*) health of the flock, (*3*) feathering, (*4*) conformation, (*5*) flashing, and (*6*) lack of defects (1). The grading of the dressed ready-to-cook poultry is more accurate because the feathers have been removed. The five key areas considered in the grading process are (*1*) the conformation of the carcass, which is related to the presence of defects such as crooked bones, dented breast bones and swollen legs/wings; (*2*) fleshing, which refers to the amount of meat on the birds (well-fleshed birds are of the highest grade and the breast muscle is commonly used as the main indicator for the degree of fleshing); (*3*) fat covering, especially over the breast, back, and hips (although in young turkeys and chicken broilers only a moderate amount of fat is expected; (*4*) the presence of pinfeathers, which will lower the grade (ie, a grade A bird should be free of pinfeathers); and (*5*) the lack of defects such as bruises, broken bones, missing parts, tears, and discoloration of the skin. Additional information and precise details of the grading system used in the United States can be found in the U.S. Department of Agriculture (USDA) *Poultry Grading Manual* (8). In other countries consult the local inspection branch. All the grading systems include tolerances to compensate for interpretation variations and human error. It should be mentioned that in recent years a grading system based on meat quality attributes such as water-holding capacity and texture has been discussed. Such a system is of interest to further processors who are looking for meat that will hold added moisture and not fall apart during cooking regardless of skin tears or missing parts (5).

MEAT PROCESSING EQUIPMENT AND PROCEDURES

Different equipment is used in the fabrication of raw meat into further processed meat products. In the past most of the equipment was designed for a batch-type operation; however, today the trend is toward continuous operation. The latter speeds up the process, allows more volume to pass through a given plant, permits more automation, and, as a result, can save on operation costs. In this section the basic types of equipment used in meat processing plants and their operational principles will be reviewed.

Grinding or Size Reduction

Grinding or size reduction of the meat is the most common process in the manufacturing of meat products. Three major methods are used: grinding, flaking, and chopping.

Grinding. The meat is forced through a grinding plate that can have different opening sizes. An auger is used to convey the energy (manual or electric) to push the meat through the grinder plate (9).

Flaking. Partially frozen meat chunks or blocks are flaked by using a rotating cutting head. The machine has a circular cutting head and the meat is pushed close to the knives by centrifugal force. The size of the flakes is determined by the spacing in the cutting head. This eliminates the mechanical squeezing of the muscle fibers exerted in a conventional grinder, which results in higher moisture loss from the muscle (lean muscle contain about 70% moisture). The meat obtained from the flaker can be easily used in restructured meat products where a musclelike texture is reconstructed from small pieces of meat.

Chopping. The meat is passed through a set of cutting knives, and the degree of chopping is controlled by the length of cutting time. This process usually results in small particles mainly because of the small gap between the cutting knives and the bowl. Chopping is commonly used for fine comminution of meat products (frankfurters) and is not used to prepare products such as ground meat. The two popular designs include a bowl chopper and an emulsion mill. In a chopper, the meat is placed in a cutting bowl, which rotates at a relatively slow speed (15–30 rpm) while the meat is chopped by a set of semicircular knives (usually 3–15) at a speed of a few thousand revolutions per minute. In a mill, preground meat is passed through a fast rotating blade, positioned in a pretty similar way to a meat grinder, but operating at a much higher speed. Since emulsion mills operate at high speed and meat particles are subjected to considerable friction and relatively high temperature, special care should be given to the operating parameters.

Mixing

Mixing is another common step in further processing used to achieve four major goals: (*1*) provide uniformity when different meat sources are used, (*2*) mix nonmeat ingredients (eg, salt and spices) into the meat batter, (*3*) enhance absorption brine solution into the muscle structure, and (*4*) assist in extracting salt-soluble proteins from the muscles. Salt is usually added at the mixing stage because its uniform distribution is extremely important in manufacturing high-quality products (10,11). Briefly, the three functions of salt in meat processing are (*1*) extract salt-soluble proteins (mainly myosin and actin) required for binding and water holding, (*2*) provide flavor, and (*3*) serve as an antimicrobial agent.

Different mixer designs are available on the market. The paddle mixer uses a paddle to mix the meat in a stationary bowl, whereas the ribbon mixer has blades mounted on a horizontal shaft. With any mixer, blending should be precisely controlled to ensure uniform mixing and prevent overmixing, which can result in too much muscle fiber separation.

Tumbling and massaging are two newer procedures designed to achieve the same goals as mixing but in a gentler manner. In these processes the meat is subjected to a mild agitation in (*1*) a rotating drum (tumbling) and (*2*) a mixing vessel equipped with a slow moving paddle (massaging). This agitation, which is usually applied for a much longer period of time than in conventional mixing, helps disrupt some of the tissue structure, helps in distributing of the brine solution, and develops a protein exudate that binds meat chunks during cooking. These functions are further enhanced if the mechanical action is imposed in a vacuum (9).

Injection Machines

Injection machines are used to introduce the brine solution (water, salt, and flavorings) into large whole muscle chunks. In the past, the salt and spices were introduced by rubbing the dry ingredients onto the muscle surface or by immersing the muscle in a brine solution. These processes are time-consuming, especially when dealing with large muscles (eg, whole turkey breast). Today, mechanical injectors are commonly used in the industry. The injectors can consist of a single needle operated manually or have a few dozen needles automatically controlled to deliver a precise volume of brine. The needles must be narrow enough not to cause any damage to the appearance of the muscles. Special care should also be given to the uniformity of the injection process to prevent high salt and/or nitrite concentrations in localized areas. High concentrations of salt and nitrite can cause flavor and color defects (eg, nitrite burns) in the final product. It is common to tumble the meat after injecting the brine. This is done to ensure an even distribution of the brine, enhance brine absorption, and facilitate the extraction of salt soluble proteins (11).

Stuffing Machines

Machines are used to stuff the raw meat batter into the casings. Stuffers vary in size and degree of automation but generally can be divided into two basic types: piston and pump stuffers (12). In the piston stuffers, a piston driven by manual, hydraulic, or electrical energy forces the meat through a horn. The diameter of the horn, stuffing speed, and pressure are controlled by the operator and should match the size and type of the casings used. The piston-type stuffer is recommended for coarsely ground sausages and those having large particles such as fat and pickles, because they might be damaged by an impeller-type pump. In pump stuffers the meat is passed through an impeller-type pump that usually has feedback and pop-off connectors so a vacuum can be drawn to evacuate trapped air from the meat batter. Vacuum stuffers are more expensive but are advantageous in providing high-quality products since air pockets that might be filled with gelatin (melted connective tissue) or melted fat during the cooking process are eliminated. It is important to minimize these pockets because products with an excessive amount of pockets are rejected by the consumer.

Various types of casings are used for different types of meat products. They include:

1. Natural casings obtained from the gastrointestinal track of sheep, hogs, etc. They are digestible, permeable to moisture and smoke, and can shrink with the product when a drying step is employed. However, they are not uniform and are relatively difficult to handle especially when fast automated equipment is used.
2. Manufactured collagen casings that are made from natural collagen extracted from hides. They are uniform in size and usually stronger than natural casings.
3. Cellulose casings made from cotton fibers. They are also permeable to moisture and smoke, uniform, and very strong, but nondigestible. Overall, they are very popular where high-speed stuffing machines are used.
4. Plastic casings are moisture proof and used for either water or steam cooking. They are strong and can appear with printing on the outside surface; however, they are nondigestible and have to be removed prior to consumption.

Other casings are combinations of the preceding materials such as collagen reinforced with cotton, or cotton fibers coated with plastic. Rigid metal molds are used mainly for baking loaves.

Forming Machines

Forming machines are used to form products such as patties and nuggets. These products are made from a mixture of ground and emulsified meat and do not hold together very well in the raw state. The forming machine is basically a press with different templates that can be used to form any shape desired. To maintain the shape of the newly formed product it has to be either frozen immediately or cooked.

Battering and Breading Machines

Battering and breading machines are used to uniformly apply the batter and breading. These machines usually consist of a series of conveyor belts that carry the product (formed or whole cuts) through a series of ingredients used to coat the product. Some of the belts (usually a stainless steel mesh) can be shaken automatically to control the thickness of the breading layer.

Cooking

Most meat products are consumed after they have been cooked. This can be done in water, oil, by hot air, infrared, or microwave energy. Cooking results in distinct textural changes in the meat due to denaturation of different muscle proteins. Three major phases can be observed during cooking. The first increase in toughness (45–55°C) is due to myosin denaturation, the second (60–70°C) is due to sarcoplasmic protein and some connective tissue denaturation, while the third (above 75°C) actually results in reducing the toughness, which is the result of collagen transformation into gelatin (13). Differences exist in the denaturation profiles of different meats, and various muscles within the same animal. An example of this is the higher rigidity developed in white poultry meat (or products containing white meat) as compared with dark meat (14).

Hot air ovens, with/without a smokehouse attached, are very popular in the meat industry. The meat products are placed in a chamber and the heat generated from a gas burner or an electrical element is transferred into the products. The amount of moisture in the air, air flow, and temperature difference between the product and air will determine the rate of heating. Of particular importance is the relative humidity in the air (expressed as amount of moisture in the air at a given temperature). Water is a good conductor of heat and its presence helps in delivering the heat; however, a moist product surface results in a cooling effect due to evaporation. Therefore, a balance between the two factors should be always maintained.

Cooking in Water. Cooking in water is a faster way of transferring the heat into the product compared with hot air. Steam kettles and water baths are used for this purpose. Commercial meat products are usually stuffed into a moisture-proof casing to eliminate cooking losses (consisting of protein, fat and moisture) from the product. However, some meat cuts are not packaged and are immersed directly in water/soup. This is done where moist heat is required to tenderize the tough connective tissue (eg, in boiling mature hen meat).

Frying. Frying is a very efficient way of transferring heat into the meat since the temperature of the fat can be raised well above 100°C. Frying in oil also provides a crisp texture on the outside of the product, which is desired in products such as fried chicken and breading on chicken nuggets.

Infrared Heating. Infrared heating is achieved by the use of an infrared lamp that heats up the surface of the product. The heat is then slowly transferred by conduction into the center of the product. This type of heating is mainly used for warming up cooked products, keeping products hot on a display counter, and in combination with microwave heating when surface browning is desired.

Microwave Heating. Microwave heating is a fast way of cooking whereby heating results from converting microwave energy to heat by friction of water molecules rotating due to rapid fluctuation in the electromagnetic field (915 and 2450 MHz are used commercially). Since cooking is very fast, there is usually not enough time to develop the typical brown color on the surface of the product. Therefore, other cooking methods, such as infrared, should be employed if a typical brown surface is required. Low microwave energy is also used to defrost meat. However, special care should be taken since there is a big difference between the heating profile of water and ice.

Smoking. Smoking is the application of wood-burning smoke compounds onto the product. Overall, more than 200 individual compounds have been identified in wood smoke (15). In the past, the only way of smoking was to expose the product to smoke derived from burning wood. Today, the processor can choose different preparations of liquid smoke extracts that can be directly added to the product, used as a dip, or be sprayed onto the product prior to cooking. In the past, smoking was mainly used as a means of preserving the meat that was also dried at the same time over wood fire. Some of the smoke components exert strong antimicrobial activity (eg, phenols) and therefore help in preserving the product. However, since these compounds can only penetrate a few millimeters into the product, they are basically protecting the surface. Today, smoking is mainly used to provide typical flavors to the product (eg, hickory smoke). The compounds that contribute to flavor include carbonyls, organic acids, and some of the phenols.

Smoking and cooking are considered to be two separate processes; however, they are usually discussed together because they often occur in immediate succession or simultaneously. It is important to realize that to achieve the best smoke penetration, the product should not be cooked because the denatured protein film, formed during cooking, will prevent smoke migration into the product. Therefore, smoking is done at low temperatures even though some heat is often applied to partially dry the surface. The latter is done to ensure that the smoke will not be washed off the product. When liquid smoke (smoke extracts) is used, the product is dipped or sprayed prior to cooking. In all cases, casings permeable to smoke (eg, collagen, cellulose) should be used. If liquid smoke is to be added directly into the raw batter, a special preparation (eg, pH adjusted) and low concentration should be used; and in this case there is no need to use permeable casings (9).

SANITATION IN POULTRY PROCESSING PLANTS

Meat is a perishable food item (Fig. 3) because it contains all the nutrients required for microorganism to grow, and

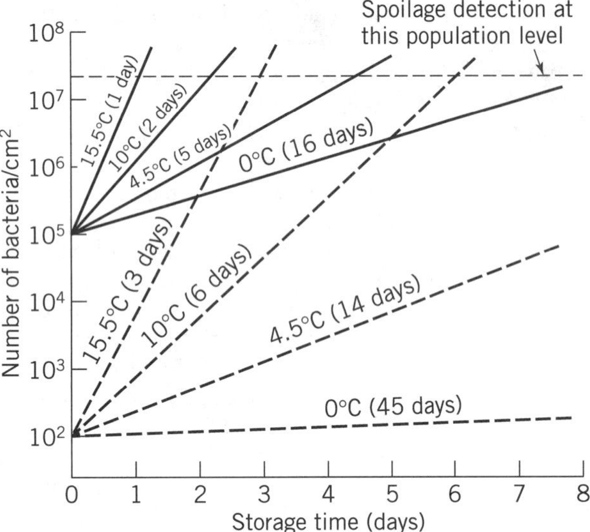

Figure 3. The time required for the spoilage of frankfurters that were contaminated with high and low levels of psychrophilic bacteria. Spoilage detection (by slime formation) was at a population level of 150 million bacteria/cm^2 of surface area. The high level of contamination was 1 million/cm^2, shown by the solid line; the low level was 100 bacteria/cm^2, shown by the broken line. *Source:* Ref. 16.

its pH (5.5–7.0) is not inhibitory to most microorganisms. The extensive fabrication and distribution of raw and processed meat further increases exposure to microbial contamination. It was reported that ground meat can be handled 10 to 15 times before it gets to the consumer.

One of the principal contamination sources during processing is slaughtering. The live healthy muscle is essentially free of microorganisms; however, after slaughtering the natural defense mechanisms no longer function. During slaughtering the sticking knife, which cuts through the skin, transfers microorganisms into the bloodstream. Because blood circulation is not immediately stopped, the microorganisms can still be distributed throughout the carcass. It is important to realize that 1 g of soil (dirt or manure) attached to the skin or feathers can contain 1 billion microorganisms. Another point of contamination is evisceration. The digestive tract harbors high numbers of microorganisms (about 200 million microorganisms per gram). If the gut contents are spilled on the carcass, high contamination levels can be expected. The process of defeathering can assist in the introduction of microorganisms into the skin. The scalding temperature used is insufficient to kill all the microorganisms, and the process of rubbing feathers off the skin actually helps to embed the microorganisms into the skin. Other potential contamination sources include people handling the meat, air, water used to rinse the carcasses, contact with equipment, and insects and rodents (17).

The method used for cleaning a food processing plant is based on the soil material present. Because meat contains mainly protein, fat, and moisture, alkaline solutions are the most common cleaning solutions used in the meat industry. An alkaline solution such as 1.5% sodium hydroxide can be used to saponify the fat and also to dipeptidize the protein deposits. Various alkaline phosphates and synthetic detergents are also used in meat processing plants. A new approach to cleaning involves the use of enzymes in a cleaning solution. A solution usually containing proteases (to break down protein deposits) is often used in a mild alkaline solution (to saponify the fat deposits). The main advantage of using enzymes is a significant reduction in corrosion in the plant. Because the alkalinity of the solution cannot be too high (otherwise will inactivate the enzymes), corrosion problems are minimized. However, it should be noted that enzyme solutions are more expensive to use, at least on a short-term basis.

The cleaning procedure in a meat plant includes:

1. *Removal of Heavy Soils from the Surface.* This step is usually done manually (scrapers) and can help in reducing waste loads and save on cleaning compounds.
2. *Rinse with Water.* High-pressure hoses can be used to facilitate this step. The water temperature should be below 55°C to prevent cooking the meat onto the surface.
3. *Wash with an Alkaline Solution or a Synthetic Detergent.* It is important to allow sufficient time for the chemical cleaning reaction(s) to take place. Usually a contact times of 5 to 10 min is recommended at a water temperature of 50 to 55°C. When vertical surfaces are cleaned, foaming agents are used to keep the cleaning solution in close contact with the surface. Enzyme solutions can also be applied at this stage; however, lower water temperature should be used to prevent enzyme inactivation.
4. *Rinse with Clean Water to Remove All the Alkaline and Detergent Solutions.*
5. *Acid Wash to Remove Scale Deposits.*
6. *Inspect to Ensure the Removal of All the Soil.*
7. *Sanitize the Plant.* It is crucial to start this step only after all the equipment is thoroughly cleaned, otherwise the sanitizer could not be in close contact with the surface and its activity is diminished. A chlorine solution (100–200 ppm), iodine (20–30 ppm), or quaternary ammonium solution (150–200 ppm) are commonly used.
8. *Rinsing and Drying.*
9. *Oiling, Only Areas Subject to Corrosion.*

Cleaning in place (CIP) is not very popular in meat plants because of plant outlay and design; however, where applicable (eg, a closed system such as a smokehouse) heavy duty detergents are used to effectively remove the soil deposits without exposing employees to harsh chemicals.

POULTRY PRODUCTS

Within the last 20 to 30 years many new processed poultry products have been introduced on the market. The poultry industry has taken the initiative to develop new products and also adopted some red meat recipes to increase consumption and to move away from seasonal demand. In the past whole turkeys were sold in the North American market mainly prior to Thanksgiving and Christmas. This kind of marketing significantly limited the increase in sales of poultry. Realizing these limitations (sale of whole birds and demand concentrated within 1–3 months) the poultry industry started to move methodically into further processing. At the beginning, red meat recipes were modified in order to manufacture poultry products (frankfurters), and later new technologies were developed exclusively for poultry (chicken nuggets). Chicken frankfurters were unheard of 25 years ago; however, after their introduction, they gained a significant market share, currently about 20% of the North American market. Overall, the poultry industry has moved into further processed poultry products, responding to consumer demand for more convenient food items such as semiprepared or fully prepared items. The increase in poultry meat consumption (Table 1) has been the result of aggressive marketing, the favorable nutrition profile of the meat, and the competitive price of the meat.

Today consumers can choose their meat products from a wide variety consisting of a few hundred different products. Becoming a knowledgable consumer can be a challenge. To assist the consumer, various systems have been suggested for classifying different meat products. One of the most common ways to classify the products is to group them into: (*1*) fresh: fresh poultry breakfast sausage, (*2*) uncooked and smoked: polish sausage, (*3*) cooked and smoked: frankfurter, bologna, (*4*) cooked: liver sausage, cooked salami, (*5*) dry/semidry or fermented: summer sausage, dry salami, and (*6*) cooked meat specialities: luncheon meats, jellied products, and loaves (9). However, a few recently developed technologies represent two additional categories: (*7*) restructured meat products: where small bits and pieces of meat are made into a steaklike product by using freezing and high pressure, and (*8*) surimi-like products: where the meat is minced, washed (to remove pigments and enzymes) and extruded to obtain a fiber muscle-like texture. In this section, examples of the major processed poultry products will be discussed (Table 5). Detailed information has been published elsewhere (1,2,18).

Table 5. Examples of Various Further Processed Poultry Products

Product	Comments
Smoked turkey breast	Whole muscle, without nitrite
Poultry roll	White and dark meat
Turkey ham	Cured dark meat
Turkey bacon	Layers of light and dark meat
Summer sausage	Fermented product
Poultry frankfurter	Fine emulsion, in small diameter casings
Poultry bologna	Fine emulsion, in large diameter casings
Poultry patties	Hamburger type, not cured
Breakfast sausage	Ground product, sold fresh or frozen
Chicken nuggets	Whole muscle or restructured
Fried chicken	Battered and breaded, sold un- or precooked
Roasted, barbecued	Prepared with dry heat, crisp skin
Jellied chicken loaf	Cooked meat held in a gelatin matrix

Smoked Turkey Breast

Smoked turkey breast is a premium product produced from the whole breast muscle. A brine solution (ie, water, salt, spices, and sometimes gums such as carrageenan) is added to the muscle, often by direct injection, prior to smoking and cooking. The meat is massaged or tumbled after brine injection to assist in absorbing the brine and achieving an even distribution within the muscle. The breast muscle is then stuffed into a netting (the meat can be covered with skin) and smoked and cooked in a smokehouse until an internal temperature of at least 70°C is reached. No nitrite is added to this product because it is desirable to maintain the white color of the cooked meats.

Poultry Rolls

Poultry rolls are made from dark meat, white meat, or their combination. The meat is obtained from whole chicken or turkey muscle, trimmings, and skin. In this product the pieces of meat are "glued" together to form a coherent product. Salt is added to extract the salt-soluble protein (mainly actin and myosin), which assists in binding the meat pieces and retaining the moisture and fat within the product. The skin and some of the trimmings are usually finely chopped (the term "emulsified" is used in the industry even though no true emulsion is formed) to facilitate the binding of the fat. Seasonings are added to provide flavor, and moisture is added to compensate for cooking losses and to improve the juiciness of the product. If added moisture exceeds a certain percentage (varies in different countries) of the raw meat moisture, the product should be labeled as a water-added product. The meat pieces are mixed together with the nonmeat ingredients until the meat becomes sticky (which is used as an indication of good protein extraction) and all the added moisture is absorbed. The mix is then stuffed into casings and the product is cooked either in a water bath or an oven.

Turkey Ham

Turkey ham is typically manufactured from turkey thigh meat. This product is much lower in fat content than the traditional pork ham. In the initial manufacturing step, a brine solution (ie, water, salt, phosphates, flavorings, and nitrite) is introduced either by injecting or marinating the meat. Tumbling is commonly used when a brine injection is employed and helps to get an even distribution of the curing ingredients and to extract the salt-soluble proteins to the surface. The meat is then stuffed into fibrous casings. The shape (round or oblong) and size are determined by the casings. The ham is smoked (smoke flavorings can also be added to the raw batter) and cooked to at least 68°C.

Summer Sausage and Salami

Summer sausage and salami are fermented poultry products made from dark poultry meat, skin, and fat. In this process microorganisms are used to reduce the pH of the product and add some typical flavor notes. In the past, microorganisms from a previous successful batch were introduced into a new batch and the product was allowed to ferment. Today the industry mainly uses starter cultures with a known composition of microorganisms (predominantly lactic acid bacteria). The producer can select from cultures that grow at different temperatures and produce specific flavors. Recently, a lot of progress has been made in the field of starter cultures, mainly through the use of genetic engineering where various characteristics can be included in one strain. The use of a starter culture is highly recommended because it ensures that lactic acid bacteria will dominate the fermentation, suppress pathogens, and produce the desired flavors. The degree of fermentation can be controlled by the quantity of the carbohydrate added (an energy source for the microorganisms) or by continuous pH monitoring. After the fermentation (1 to 2 days with a starter culture) the product is either smoked and cooked or only dried. The final product is usually shelf stable due to the low pH (4.6–5.1) and/or low water activity ($A_w < 0.90$).

Frankfurters and Bologna

Frankfurters and bologna are examples of finely comminuted meat products where the final product has a very homogeneous appearance. Dark muscle chunks and/or mechanically deboned meat are usually emulsified with the fat. A bowl chopper or an emulsion mill are used to achieve an efficient particle size reduction. Salt is added to extract the meat proteins, which is essential in binding the small meat particles and stabilizing the small fat globules within the protein matrix (10). Nitrite is added to inhibit *Clostridium botulinum* growth and to provide the typical cured meat color (Fig. 2). The meat batter is then stuffed into cellulose casings and smoked and cooked in a smokehouse. Since frankfurters are such a high volume item, some processors have dedicated an entire continuous line for this product. As with other meat products, low microbial contamination and refrigerated temperature can help prolong the shelf life (Fig. 3).

Chicken Nuggets

Chicken nuggets are one of the most successful poultry products introduced in the 1980s in North America. Originally, the product was prepared from a single piece of slightly marinated breast meat that was battered and breaded. Later, nuggets made from trimmings, dark meat, skin, mechanically deboned meat, and their combinations started to appear on the market. To prepare the products the meat pieces are marinated and mixed with a brine solution (containing salt and flavorings) until all the solution is absorbed by the meat. The meat is then put through a forming machine that creates the desired product shape, followed by battering and breading and deep fat frying. Frying is done after forming the product to preserve the product shape and to provide the typical crispy texture of the breading.

FUTURE DEVELOPMENT

In light of the material presented in the preceding sections and especially the continuous increase in poultry meat consumption (Table 1), it would seem that poultry meat is going to remain popular and might even gain a greater market share. Some predict that consumption level(s) will stabilize in this decade, and some forecast a further increase. One thing is certain, however; the poultry industry must be competitive to maintain its current market share. Poultry meat is also becoming more popular in many countries because of the ability of poultry to adapt to most areas of the world, the low economic value per unit and rapid growth rate (1,5). Therefore, it can be expected that on a worldwide basis, poultry consumption will increase.

Improvement in disease control will play an even more important role in the production of poultry as the trend toward larger growing units continues to increase. In addition, improvement in vaccination and feeding programs will further affect the competitiveness of the poultry industry. In the area of further processing, more automation and continuous operations will replace manual and batch operations. The meat industry is currently moving toward more automation and robotics to replace repetitive manual labor. Automated equipment for evisceration and cutting up poultry is already available on the market today. In the future, better cooperation between breeding and production personnel will help to produce a more uniform bird that will be easier to process in an automated plant.

Responding to emerging consumer demands, such as more convenient food items (eg, fully prepared in meat plants), microwave-ready food items and reduced-salt and -fat meat products will play a more important role in the future. Based on past responses of the poultry industry to consumer demands (eg, marketing fresh poultry, developing new products), it seems that the industry is well equipped to respond to future market trends.

BIBLIOGRAPHY

1. G. J. Mountney, *Poultry Products Technology*, Food Products Press, Binghampton, N.Y., 1989.
2. U.S. Department of Agriculture. *Labelling Policy Book*, USDA Food Safety Inspection Service, Meat and Poultry Inspection Technical Services, Washington, D.C., 1981.
3. American Meat Institute, *Meat and Poultry Facts*, Watt, Mt. Morris, Ill., 1997.
4. A. B. M. Raj, "Carcass and Meat Quality in Broilers either Killed with a Gas Mixture or Stunned with an Electric Current under Commercial Conditions," *British Poultry Sci.* **38**, 169–175 (1997).
5. S. Barbut, "Estimating the Magnitude of the Pale, Soft, and Exudative (PSE) Problem in Poultry: A Review," *J. Muscle Food* (1998).
6. U.S. Department of Agriculture. *Composition of Foods, Poultry Products: Raw, Processed, Prepared*, Agriculture Handbook 8-5, Washington, D.C., 1979.
7. G. W. Froning, "Mechanical Deboning of Poultry and Fish," *Advances Food Res.* **27**, 109–145 (1981).
8. U.S. Department of Agriculture. *Poultry Grading Manual*, Agricultural Marketing Service, Handbook No. 31, Washington, D.C., 1989.

9. H. B. Hedrick, et al., *Principles of Meat Science*, Kendall/Hunt, Dubuque, Iowa, 1994.
10. S. Barbut, and C. J. Findlay, "Sodium Reduction in Poultry Products: A Review," *CRC Crit. Rev. in Poultry Biology* **2**, 59–95 (1989).
11. S. Barbut, "Determining Water and Fat Holding," in G. M. Hall, ed., *Methods of Testing Protein Functionality*, Chapman & Hall, New York, 1996, pp. 186–225.
12. A. M. Pearson and F. W. Tauber, *Processed Meats*, AVI, Westport, Conn., 1984.
13. C. L. Davey and K. V. Gilbert. "Temperature Dependent Cooking Toughness in Beef," *J. Sci. Food Agric.* **25**, 931–937 (1974).
14. G. S. Mittal and S. Barbut, "Effects of Salt Reduction on the Rheological and Gelation Properties of White and Dark Poultry Meat," *J. Texture Stud.* **20**, 209–222 (1989).
15. J. A. Maga, *Smoke in Food Processing*, CRC Press, Boca Raton, Fla., 1989.
16. E. A. Zottola, *Introduction to Meat Microbiology*, American Meat Institute, Chicago, Ill., 1972.
17. N. G. Marriott, *Principles of Food Sanitation*, Chapman & Hall, New York, 1994.
18. W. J. Stadelman, *Egg and Poultry Meat Processing*, Ellis Horwood, Chichester, England, 1988.

SHAI BARBUT
University of Guelph
Guelph, Ontario
Canada

PROCESSED CHEESE

Processed cheese is manufactured by blending shredded natural cheeses of different types, sources, and degrees of maturity with emulsifying agents, adding water and heating the blend under partial vacuum with constant agitation until a homogeneous mixture is obtained. High shear mixing or homogenization could be used in addition. Besides natural cheeses, other dairy and nondairy ingredients may be included in the blend.

A number of natural cheese varieties were known and appreciated as a valuable food component in ancient Roman times, more than 2000 years ago. There are also quotations about cheeses in the Bible and in old Greek and Latin poetry. Compared with these data, processed cheese seems to be a brand new product, originating in this century, although based on natural cheeses.

Initially, processed cheese was manufactured without an emulsifying agent. The first attempt was as early as 1895, but only after the introduction of citrate (in Switzerland in 1912) as a calcium-sequestering agent did the industrial production of processed cheese become feasible. Actually, Swiss inventors first succeeded in changing the state of casein from coarsely dispersed calcium paracaseinate in the raw cheese, by using heat and sodium citrate as a peptizing agent, into a homogeneous, free-flowing condition: the sol state. Phosphates, which were introduced a few years later, enabled the real expansion of processed cheese manufacture. Quite independently, the production of processed Cheddar cheese started in the United States, developed by Kraft in Chicago in 1917, by using the mixture of phosphates and citrates as emulsifying agents.

The idea of processing was first meant to make use of natural cheeses that otherwise would be difficult or impossible to sell (cheeses with mechanical deformations and localized molds and trimmings produced during cheese formation, pressing, packaging, etc) and to obtain a product of better keeping quality. While soft cheese and semihard cheese could be canned, pasteurized, and consequently stored for a long time or exported overseas (even to the Tropics), such a procedure for hard cheeses has not been developed. All experiments with the aim to heat treat hard cheeses (eg, Emmentaler) failed, resulting in the breakdown of the cheese structure with exclusion of fat and water. Proteolysis and lipolysis continue beyond the stage where the optimum flavor has been developed in any given cheese. Advanced decomposition of proteins/lipids leads to a nutritionally inferior product, causing detrimental changes in sensory characteristics, ending up with an unsuitable cheese for consumption. The invention of processed cheese was the perfect solution for long storage of all native cheese varieties. Even longer storage could be achieved with processed cheese powder or sterilized processed cheese. The assortment of processed cheese was further expanded due to numerous possible combinations of various types of cheese and the inclusion of other dairy and nondairy components, which make it possible to produce processed cheese differing in consistency, flavour, size, and shape.

The main characteristics of processed cheese are composition, water contents, and consistency. According to these criteria, three main groups may be distinguished: processed cheese blocks, processed cheese foods, and processed cheese spreads (1–5) (Table 1). More recent subtypes are processed cheese slices and smoked processed cheese. Processed cheese analogues are imitation, simulated, or alternative groups of processed cheese that contain no cheese and are usually based on vegetable fat and casein blends.

Processed cheeses are advantageous compared with natural cheese primarily because of the following:

- Reduced refrigeration costs during storage and transport, especially important in hot climates
- Better keeping quality with less alterations during prolonged storage
- Great versatility of type and intensity of flavor, for example, from mild to sharp native cheese flavor or specific spices
- Milk replacement: excellent source of nutritively high valuable milk components for children or people who dislike milk
- Adjustable packaging for various usage, economical and imaginative.
- Suitability for home use as well as for fast-food restaurants, for example, in cheeseburgers, hot sandwiches, spreads, and dips

MANUFACTURING: PRINCIPLES AND TECHNIQUES

The manufacturing procedure for processed cheese consists of operations carried out in the following order: selec-

Table 1. Some Characteristics of Processed Cheese Types

Type of product	Ingredients	Cooking temperature, °C	Composition	pH	Ref.
Processed cheese block	Natural cheese, emulsifiers, NaCl, coloring	71–80	Moisture and fat contents correspond to the legal limits for natural cheese	5.6–5.8	1
		80–85		5.4–5.6	3
		74–85	≤45% moisture	5.4–5.7	2
Processed cheese food	Same as above plus optional ingredients such as milk, skim milk, whey, cream, albumin, skim milk cheese, organic acids	79–85	≤44% moisture, <23% fat	5.2–5.6	1
Processed cheese spread	Same as processed cheese food plus gums for water retention	88–91	≥44% and ≤60% moisture	5.2	1
		85–98		5.7–5.9	3
		90–95	≤55% moisture	5.8–6.0	2

Source: Ref. 4, courtesy of *Food Structure and Scanning Electron Microscopy.*

tion of natural cheese, computation of the ingredients, blending, shredding, addition of emulsifying agents, processing, homogenization (optional), packaging, cooling, and storage. A schematic representation of the processing procedure, including the possible facilities for particular operations, has been presented in Figure 1.

Selection of Natural Cheese

Proper selection of natural cheese is of special importance for the successful production of processed cheese. In some countries, processed cheeses manufactured from only one variety of cheese of different degrees of maturity are popular, for example, processed Cheddar cheese in the UK and Australia, Cheddar and Mozzarella in the United States and Canada, and Emmental in Western Europe. More frequently, processed cheeses are produced from a mix of various natural cheese types. Criteria for raw material selection are as follows (3):

- The type of flavour required
- The consistency and structure
- The fat content
- The variety and grade of fat in the raw material
- The maturity of the raw material
- The quantity of raw material in stock
- The nature and character of the additives
- The legal requirements
- The potential market
- Other, mostly economic, points

The combination of native cheeses, the processed cheese of proper maturity, is of special importance for the processed cheese quality, in terms of taste and flavor. Intact casein, present more in young cheese, has an emulsifying effect on milkfat and stabilizes the emulsion ("long structure"). This effect is not expressed by partially hydrolyzed protein, predominating in matured cheese ("short structure"). So, for example, high-fat spreadable processed cheese requires a larger proportion of young cheese in the blend with a correspondingly higher intact casein content.

Since it is possible to correct certain physical properties by skillful blending, some defective, but not harmful, cheeses can be used in processed cheese manufacture. Natural cheeses with microbial defects should not be selected for processing; spore-forming, gas-producing, and pathogenic bacteria are particularly hazardous. However, proper selection of good-quality natural cheeses is not, by itself, a guarantee that the processed cheese will have the desired high quality, if treated by improper process parameters.

Computation of Ingredients

The computation of the ingredients is based on fat and dry matter contents of natural cheese components as well as the composition of the final product. Formulation of the material balance of fat and dry matter, including all blend constituents, added water, and condensate from live steam used during processing, must be made in such a way as to yield a product with the desired composition. Additional adjustments of fat and dry matter during processing may be possible but this depends on the kind of equipment used.

Blending

Blending, which could be defined as designing the proper processed cheese blend composition and concerning, above all, selected cheese varieties, is highly influenced by the characteristics desired in the final product. However, there are no strict rules proscribed for designing the particular ratio among cheeses of different maturity in the blend. Only the general recommendations are suggested (1–3). For processed cheese blocks, mostly young to medium ripe cheese is used, whereas to produce processed cheese spread, a combination of young, semiripe and ripe cheese is preferred.

The main advantages of a high content of young cheese in the blend are as follows (2,5):

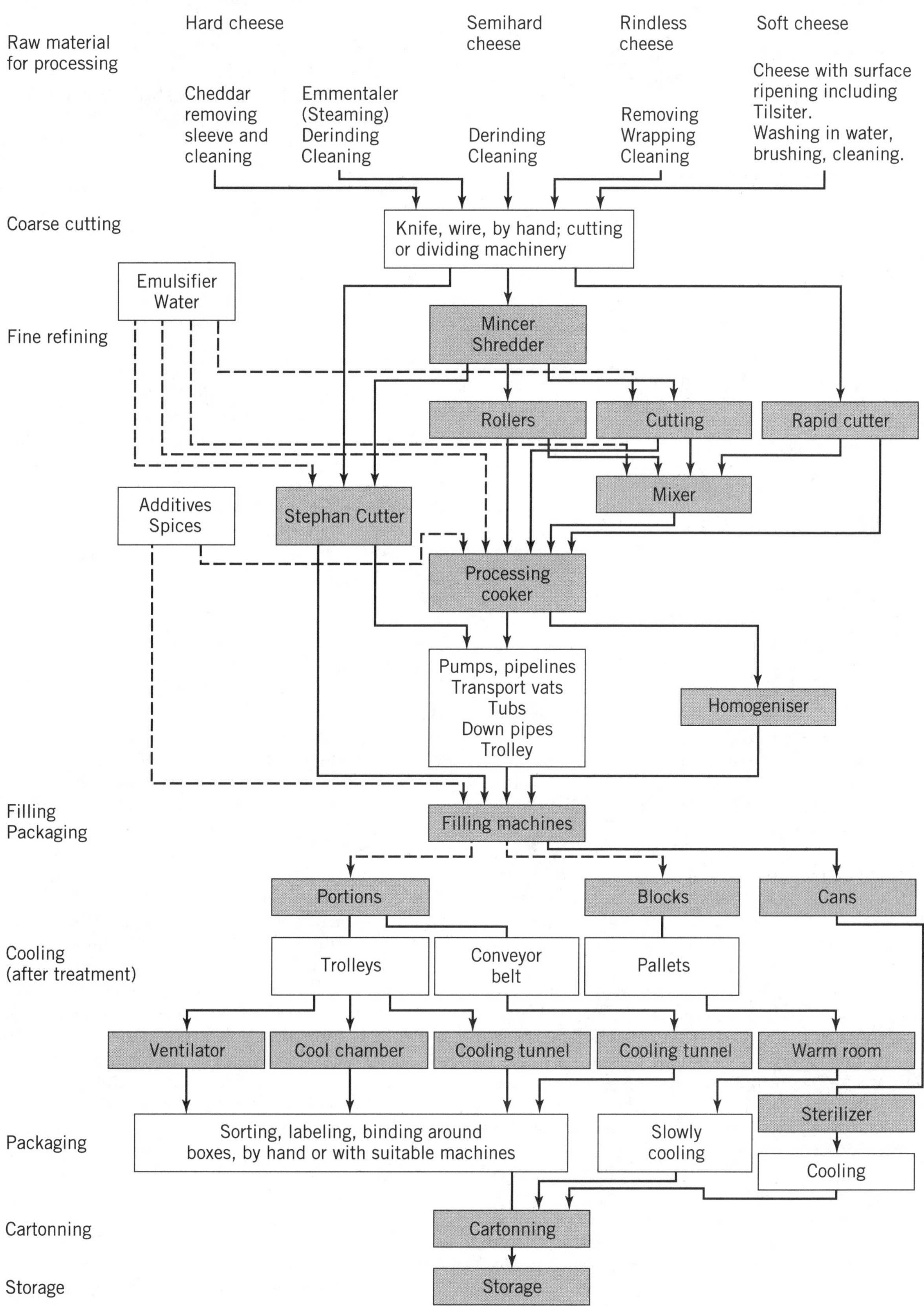

Figure 1. Diagrammatic representation of the processing procedure. *Source:* Ref. 3, courtesy of Food Trade Press.

- The reduction of raw material costs
- The possibility of using cheeses with poor curing properties immediately after manufacture
- The formation of a stable emulsion with high water-binding capacity
- The production of a firm body, with good slicing properties of the finished product

The main disadvantages are

- the production of a tasteless cheese,
- an emulsifier-like off-flavor,
- excessive swelling,
- the tendency to harden during storage, and
- the presence of small air bubbles (developed due to the high viscosity of the blend).

The advantages of a high contents of extramature cheese in the blend include (2,5)

- the development of a full flavor,
- good flowing properties, and
- a high melting index (processed cheese melts easily).

The disadvantages are

- the possibility of sharp flavor development,
- A low emulsion stability, and
- a soft consistency.

To ensure permanent production, the processed cheese factory needs a good stock of native cheese varieties of a constant quality (physicochemical and microbial). If the native cheese is not rindless (packed and ripened in plastic wrapping), the rind must be removed.

In addition to natural cheeses, various other dairy and nondairy ingredients are used in the production of processed cheese products, as shown in Table 2. The list of ingredients varies from country to country and some restrictions may apply. Because the quality of the final product is influenced considerably by all the components present in the blend, the noncheese components must also fulfill certain qualitative requirements. The most frequently used dairy, but noncheese ingredients are concentrated skim milk or skim milk powder, casein, whey protein concentrates, coprecipitates, various whey products, milkfat products, and so on.

Skim milk powder promotes the creaming properties and improves the spreadability and stability of processed cheese, but, if used in quantities exceeding 12% of the total mass, it may adversely affect the consistency or may remain undissolved. However, skim milk powder may be reconstituted first, its casein precipitated by citric acid or proteolytic enzymes and the resulting curd added to the blend (2,5). Discoloration of processed cheese due to the Maillard reactions is excluded if total lactose content is not over 6% in the final product. Skim milk powder could also be used in processed cheese manufacture by recombination and native cheese production (destined for processing). Such a procedure was developed and published in Iraq in 1987 (6). Fat contents of the cheeses were standardized by using corn oil. Processed cheese produced from reconstituted skim milk cheese had a crumbly structure, which was overcome by blending reconstituted skim milk cheese with corn oil prior to processing. In addition to being low in price, this product is low in cholesterol and is recommended to heart disease patients.

Milk protein coprecipitates, characterized with high emulsifying capacity, if added to the blend, increase the stability of the cheese emulsion and improve the physical characteristics of the finished product. Acting as an emulsifying agent, they even enable the reduction of the amount of emulsifying salt added. This is important particularly for dietary and special food products, where limitation of the sodium content may be desirable. Milk protein coprecipitates should not exceed 5% in processed cheeses (5).

Whey products incorporated in processed cheese blends favorably influence both nutritive and economic characteristics of the finished product. Although ordinary whey powder is the most common whey product used in processed cheeses, in concentrations ranging up to 7% in the blend (7), whey protein products with lower mineral and lactose contents are preferable because they yield processed cheeses with better flavor characteristics. However, some other whey products could be successfully used in the processed cheese blend as well, such as whey concentrate (2–4%), precipitated whey proteins (up to 25% with flavor correction), and native whey protein concentrates obtained by ultrafiltration (5–20%) (7).

All milkfat ingredients (Table 2) used to adjust the fat content of the processed cheese to the desired level must be of high quality and free from off-flavors.

Attempts have recently been made to develop processed cheese blends with improved characteristics that can be produced at a lower cost. Egyptian authors (8) have produced processed cheese spreads with good spreadability by partially substituting calcium caseinate for natural cheese in the blend. Although full replacement, with cheese flavor added, failed to yield a spread with good characteristics, partial replacement improved spreadability. The best results were obtained using a blend composed of 6 to 8% skim milk powder, 5 to 7% calcium caseinate, 15% mature Cheddar cheese, 14% butter oil, and 3% emulsifying agent.

Numerous procedures have been developed for producing economically favorable cheese or cheese bases intended exclusively for processing. One of these processes for production of a cheese base by ultrafiltration (UF) and diafiltration of whole milk was developed and patented in Germany (9). The retentate (40% dry matter, 4.17% lactose) obtained after UF was pasteurized (high temperature–short time [HTST]), cooled to 30°C, inoculated with lactic starter, and, after 2 h, evaporated at 42°C to 62% dry matter. It was further incubated at 25°C until the pH reached 5.2 and was then packaged in plastic bags under vacuum. The vacuum-packaged cheese base can be stored and later used in processed cheese production in combination with ripe cheese, in a 80:20 ratio.

A group of authors from Utah State University produced a cheese for processing from ultrafiltered whole milk adjusted to pH 5.2 to 6.6. The melting properties of the

Table 2. Ingredients Used in the Manufacture of Processed Cheese

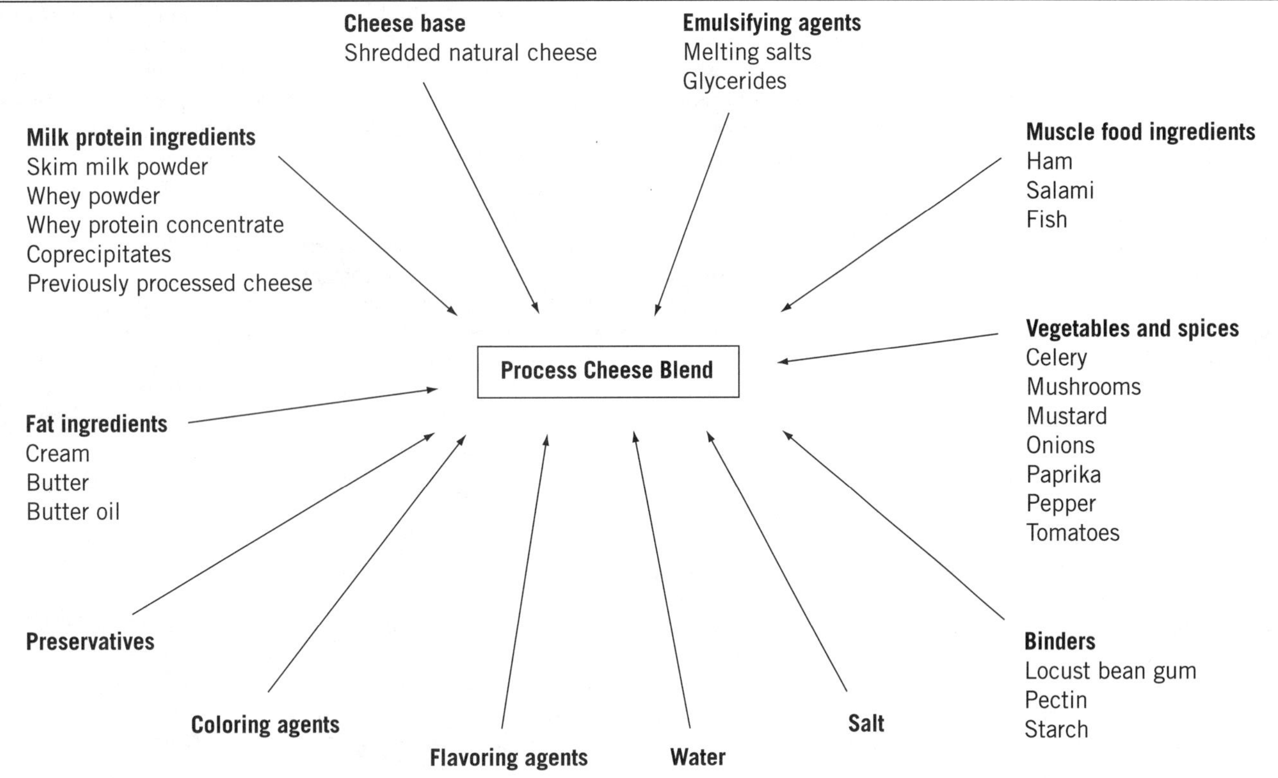

Source: Ref. 4, courtesy of *Food Structure and Scanning Electron Microscopy.*

product improved with decreasing pH of the milk before UF, caused by reduced calcium concentration in the retentate. Chymosin treatment of the UF retentate adversely affected the meltability of cheese produced (5).

Investigators at Cornell University (10) used ultrafiltered skim milk retentate to constitute up to 60% of the formulation base. The casein is mostly in the insoluble state; thus, the retentates cannot be used alone for processing. However, cheese containing up to 60% of retentate solids (treated with fungal protease and lipase preparations at 45°C, 24 h) had better sensory attributes than control cheese; a combination of sodium citrate (2.7%) and citric acid (0.3%) was the best emulsifier for retentate-containing cheese. It has been shown by calculation that the UF application in preparing raw material for a processed cheese blend pays off the investment costs after only two years of operation (11).

A kind of curd product, called Schmelzpack, is manufactured in Germany especially for processed cheese production. Containing 90 to 100% unhydrolyzed casein (original casein from milk curd), it can be blended with natural cheeses of diverse sources, types, and maturity (3,5). Good results on the use of chicken pepsin for the peptonization of curd for incorporation into processed cheese have also been reported (5).

Some more recent investigations (12) showed the possibility of successful processed cheese manufacture from Cheddar cheese and skim milk powder cheese base. Cheese base was produced from reconstituted skim milk powder after UF. In one experiment cheese base was subjected to accelerated ripening by adding commercial proteolytic enzyme. All cheeses were of a good quality, except cheeses containing a large proportion of cheese base (more than 40%). The microstructure of all finished processed cheeses stored at 10°C were similar to one another.

Enzymic methods for accelerating the ripening of natural cheeses are developed for commercial use and have been reviewed in detail (13). In general, the need to accelerate cheese ripening originates from the desire to decrease storage costs during ripening in cheese manufacture. Currently, there are three methods of proteinase addition in milk: direct blending with milk, encapsulated enzyme added to milk, and direct application to cheese curd. However, the investigations on accelerated cheese ripening are still being carried out and there is still the quality problem of maintaining the natural flavor balance and cheese texture. The importance of this problem diminishes if the cheese is manufactured exclusively for processing. For example, in Cheddar cheese manufacture, it was possible to reduce the duration of ripening from a few months to two to three weeks by proteinase addition (5). Pasteurized milk was coagulated at 36°C by chymosin and high quantities of starter culture (up to 4%). After cutting, cooking, and whey separation, the curd (at pH 5.1–5.3) was milled, mixed with an emulsifying agent to form a plastic mass, poured into containers, paraffined on top, and ripened for only a few weeks.

Authors from India (14) developed a processed cheese spread using accelerated-ripened Cheddar curd slurries.

The effect of pH; agitation; and the addition of reduced glutathione, cobalt, manganese, riboflavin, and diacetyl on flavor development in the slurries, as well as in processed cheese spread were investigated. Agitation and addition of glutathione had a positive effect on flavor development, whereas blending the slurries with 25% solids from fully ripened Cheddar cheese improved flavor, body, and texture of the finished product.

The relatively new possibility of achieving the wanted flavor in processed cheese is the addition of so-called modified cheese flavors (enzymatically modified cheese [EMC]). EMC are produced by inoculation and incubation of particular enzymes with specially designed media, followed by their heat inactivation. Flavor intensity of EMC is 10 to 30 times greater than the corresponding traditional cheese. EMC recently found application in various food products (confectionery, snack food, soups, sauces, etc). One study successfully substituted 15% ripe cheese in a processed cheese blend with a mixture of young cheese and whey powder by using Cheddar or Parmesan cheese flavor (7). Headspace analysis showed that experimental processed cheese produced with EMC, which had less ripened cheese in the blend, resulted in higher total aroma content than control samples. Thus, EMC application in processed cheese blends would be economically reasonable, resulting in a product of excellent quality. Egyptian authors (15) proved that the addition of modified Cheddar and Swiss cheese flavor affected the fatty acid composition of the produced processed cheeses. EMC caused the increase of short-chain fatty acid level, whereas Cheddar cheese flavor increased palmitic acid content as well.

A patent for chemical hydrolysis of cheese blends was granted in 1983 in the USSR. The trimmings were washed with hot water, centrifuged, comminuted, and hydrolyzed with 1.4 to 1.6 M HCl at 110 to 116°C for 3 to 5 h. A hydrolysate (about 20% TS with about 80% peptides in dry matter and 20% essential amino acids of total) was successfully incorporated in the blend, up to a maximum of 15% depending on the type and maturity of the natural cheese used (5).

As early as 1956, a rapid procedure for the production of raw material for processing based on curd acidification was patented in the United States (16). Milk was coagulated; the curd was mixed with lactic acid (0.2–1.5% acidity) and heated, with agitation, to 43°C for 6 to 40 min. The lactic acid solution was drained from the curd, which was washed with water, salted, and pressed. After two days' storage at 5°C, it was used in processed cheese manufacture. The same idea was explored in a recent French patent (17) in which the procedure for producing sliceable processed cheese of acidified milk or concentrated milk is described. The resulting curd is mixed with emulsifying agents, fat, and texture-modifying polysaccharides and further processed.

In addition to all the mentioned dairy-based products that could be included in the blend for processing, precooked cheese (also called *rework*) is an important blend component. Rework, or precooked, cheese is formerly processed cheese, or often it is the processed cheese from the previous charge. It is used to improve the texture and stability of the finished product, especially when very young or very mature raw cheese is used for processing. The percentage used in the blend varies between 1 and 30%, depending on the quality of the rework and the type of processed cheese wanted. Usually precooked cheese is used in the manufacture of processed cheese spreads to increase the creaming properties of the blend. Some dairies produce precooked cheese, which is stored in large containers. The presence of rework in processed cheese food has been detected by using electron microscopy (18).

The nondairy components of cheese blends can include spices, meat products, vegetables, and other ingredients. They all must be sterile and of the highest quality, with typical flavor. Their quantities must be properly prescribed for blending. The possible nondairy ingredients used in processed cheeses are shown in Table 2. Some attempts have been made to incorporate cottonseed flour (19) and dried vegetables as well (20).

Recent work from Alexandria University in Egypt (21) describes the texture and sensory characteristics of processed cheese made with palm oil (instead of butter) and Solva Complete B (instead of skim milk powder). The results were not encouraging.

Walnut is an attractive processed cheese additive; either shredded or cut in half, the quantity varies in the range of 8 to 10% (22,23). Additional sweet nondairy ingredients in processed cheese include fruit syrup, cocoa, honey, vanilla, hazelnut, and coffee extract (24).

Shredding (Grinding, Milling)

Shredding enables the emulsifying agent to come into close contact with the blend components during processing.

Addition of Emulsifying Agents

Addition of emulsifying agents is the last step in preparing the blend for processing. The quality of emulsifying agent added into the processed cheese blend depends on the type and age of cheese used in the blend (proportion of water and calcium) but is also determined by the final product group, which determines the sort of emulsifying agent as well. The amount of emulsifying salts that can be added to the cheese base are regulated by many countries and usually do not exceed 3 or 4%. Since the effects of emulsifying agents are responsible for the unique features of processed cheese production, their type and role will be treated in a separate section.

Processing

After all preparation treatments, the shredded, minced, and weighted raw material is transported to the cooker, where, by interaction with an emulsifying agent and water, processing is performed. Processing involves heat treatment of the blend with direct and/or indirect steam under partial vacuum. The product is constantly agitated through a continuous or batch method. If processing is carried out discontinuously, (ie, in a kettle), the temperature can reach 71 to 95°C for a period of 4 to 15 min (Table 1), depending on various parameters (3); this heating also provides pasteurization. In newly developed cookers it is also possible to reach the temperatures up to 140°C. A kettle, or cooker (Fig. 2), consists of two double-jacketed, round, stainless steel pans of various sizes (2 to 100 L), fitted with corresponding lids, three-stage switchable stirring equipment, and fittings for direct steam injection and vacuum draw. Double jackets enable indirect steam heating as well. There are specially designed units, similar to cutters used in meat processing, where cutting is completed prior to processing by the aid of rapidly rotating knives with simultaneous heating and homogenization of the product. In addition to this most common round design, the newer, horizontal, tube-shaped processing unit is also popular, particularly in the United States and Canada. This installation is fitted with one or two mixing arms (up to 4 m long); it is fed at one end and the final product is discharged at the opposite end, within 4 to 6 min. This type of unit is constructed for the batch operation, but it can be continuous and can process large capacities. A new programmed jacketed processor has been successfully developed (5), which is used to grind, mix, and process natural cheeses with other blend components, water, and emulsifiers using steam injection and vacuum, at 75°C for 5 min. The processed blend is discharged by either tilting the processor or by aseptic pumping to a packaging machine. This programmed batch processor acts via a punch card for blend formulation and cleaning in place.

By continuous processing, the blend is sterilized at temperatures of 130 to 145°C for 2 to 3 s in a battery of stainless steel tubes (1). Although continuous cheese cookers were developed as early as 1920, they are not commonly used in the processed cheese industry. The main reason is the great versatility in processed cheese products; therefore, changing over to a different product in continuous process, which requires intermediate cleaning, cannot be considered economical. A German patent describes a continuous process for simultaneous melting, homogenization, and sterilization in processed cheese production without application of pressure (25). A Japanese patent (26) describes a new method for the postprocessing heat treatment (to 100°C) of packed processed cheese, produced in the usual way.

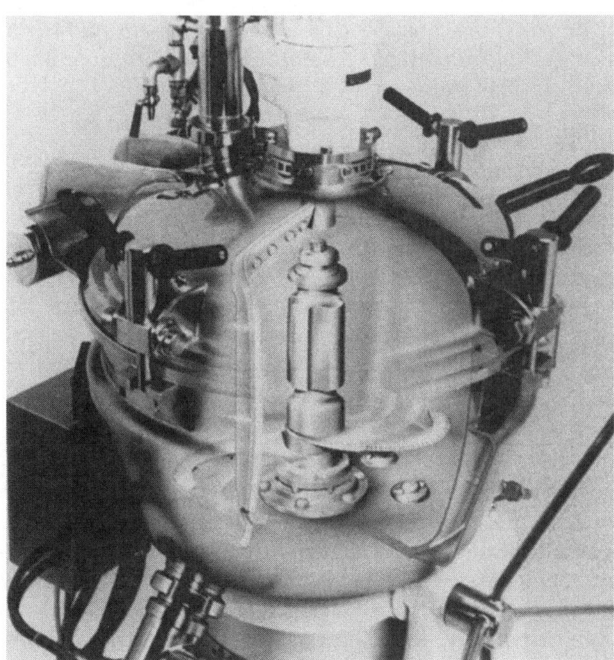

Figure 2. Universal machine for processed cheese production with cross section. *Source:* A. Stephan & Sons. Hameln, Germany.

A group of French authors (27) recently investigated the effects of blend variants and process conditions on protein–lipid interactions made by batch or extrusion cooker methods. Added emulsifying agents or premelted cheese mix increased lipid binding. Final cooling with slow mixing increased lipid binding in the extrusion cooker but not in the batch method. Proteolysis was greater in extruded than in the batch samples.

Regardless of the kind of facilities and technique of processing used, the emulsification of milkfat takes place. Transmission electron microscopy was used to observe the emulsification process (Fig. 3) (28).

The main chemical, mechanical, and thermal parameters in the cheese processing procedure are listed in Table 3. As evident from Table 3, the most important working conditions, which affect the processing and thus the quality of the final product, are as follows:

1. Temperature (heat induced by direct or indirect steam).
2. Duration of processing (depending on size and construction of the cooker, quality of raw material and blend composition, mechanical treatment, emulsifying agent used, desired keeping quality, etc).
3. Agitation (slow, at lowest speed of 60–90 rpm when producing processed cheese block, or fast, at 120–150 rpm for processed cheese spreads).
4. Acidity (pH) (a rather limited pH range; the increase of pH value, decrease of H+, causes better peptization of casein but can spoil keeping quality and flavor, whereas a decrease in pH value introduces

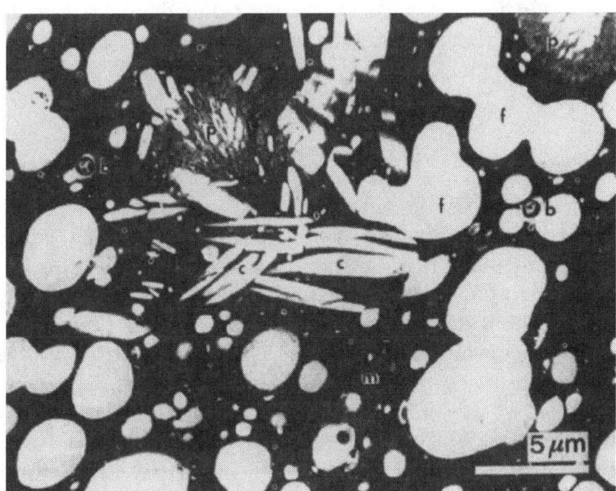

Figure 3. Transmission electron microscopy (TEM) of milkfat emulsification during cheese processing: f—fat; m—protein matrix; c—crystalline sodium citrate; p—calcium phosphate crystals; b—bacterium. *Source:* Ref. 28.

Table 3. Chemical, Mechanical, and Thermal Forces as Regulating Factors in the Cheese Processing Procedure

Process conditions	Firm slicing processed cheese, block cheese	Spreadable processed cheese, processed cheese spread
Raw material		
Average age	Young to medium ripe, predominantly young	Combination of young, medium ripe, overripe
Relative casein content	75–90%	60–75%
Structure	Predominantly long	Short to long
Emulsifying salt	Structure building not creaming, eg, high molecular polyphosphate C, SE, S7, PZ also citrate	Creaming, eg, lower and medium molecular polyphosphate, S9, S9 special, S10, S90
Water	10–25%	20–45%
Addition of water	All at once	In portions
Temperature	80–85°C	85–98°C (−150°C)
Duration of processing	4–8 min	8–15 min
pH	5.4–5.6	5.7–5.9
Agitation	Slow	Rapid
Precooked cheese	0–2.0%	5–20%
Milk powder or whey powder	0	5–10%
Homogenizing	None	Advantageous
Filling	5–15 min	10–30 min
Cooling	Slowly (10–20 h) at room temperature	Rapidly (15–30 min) in cool air
Treatment	Very light and carefully	Intensive treatment

Source: Ref. 3, courtesy of Food Trade Press.

thickening and solidifying of cheese structure). Higher pH values also favor more rapid product deterioration, in the event of postpasteurization contamination.

Homogenization (Optional)

Homogenization improves the stability of the fat emulsion by decreasing the average fat globule size. It also improves the consistency, structure, and appearance of the processed cheese. The flavor intensity is, however, reduced. If certain additives or spices are to retain their original form, they must be mixed into the product after homogenization. Because homogenization involves unnecessary additional capital, operational, and maintenance costs, and prolongs the production schedule, it is recommended only for blends with high fat contents. It is important to take care disinfecting the homogenizer, which could otherwise cause the reinfection of the final product.

Packaging

Processed cheese is usually packed and wrapped in laminated foil; in cardboard or plastic cartons; in tubes, cups, cans, and plastic containers; in sausage form; and occasionally in glass jars. Processed cheese, in sausage form, can be subjected to a smoking. A relatively new development is continuous slicing (Fig. 4) and packing of the cheese slices, suitable for sandwiches. Slices may also be obtained by mechanically slicing of rectangular processed cheese blocks. The common appearance of processed cheese in the European market is in a triangular portion (20–30 g), wrapped in heat-sealed laminated aluminum foil and assembled in round cardboard cartons. However, these packagings are quite rare in North America, where huge quantities of processed cheese block are packed in plastic foil, in the form of slices, for fast-food restaurants.

To avoid recontamination, finished processed cheese is transferred by means of stainless steel pipes to feed filling machines.

Cooling

The intensity and method of cooling is highly influenced by the type of processed cheese. Cooling of processed cheese spreads should be as fast as possible, while processed cheese blocks are cooled slowly, some time after production (rapid cooling softens the product). Cooling stops the creaming action by processed cheese spread, thus retaining creamy consistency and short structure. However, slow cooling can intensify Maillard reactions and promote the growth of spore-forming bacteria (29,30).

Storage

Processed cheese should be stored at temperatures in the range of 5 to 10°C, although such low temperatures may induce formation of calcium diphosphate–calcium pyrophosphate crystals. These crystals usually occur on the surface of product and can produce a gritty texture, but they are not harmful to the consumer.

EMULSIFYING AGENTS

Emulsifying agents (melting salts) are of major importance in processed cheese production. They provide a uniform structure during the melting process and affect the chemical, physical, and microbial quality of the product. Emulsifying agents are not emulsifiers in true chemical sense; that is, they are not surface-active compounds, although they help in emulsifying fat and stabilizing the emulsion. True emulsifiers like mono- and diglycerides may be included in commercially produced emulsifying agents, which are mixtures of selected compounds. The most com-

Figure 4. Continuous slicing machine, with two large cooling band units, to produce two different types of processed cheese. *Source:* Ref. 3, Sandvik Steel Inc, Fair Lawn, N.J. Courtesy of Food Trade Press.

mon components of the commercial salt mixtures are phosphates, polyphosphates, and citrates (Table 4), although compounds such as sodium potassium tartrate, complex sodium aluminum phosphate, sodium potassium tartrate, trihydroxyglutaric acid, and diglycolic acid could be used as well. The mixtures are of a constant and guaranteed quality; their composition is a secret and protected by the producer. The world's best-known commercial emulsifying agents producers both came from Germany: Benckiser-Knapsack, GmbH (Joha salts) and Giulini Chemie, and GmbH (Solva salts), which recently fused together in a single company entitled B. K. Giulini Chemie, in Ladenburg. Smaller producers come from a few other countries, such as Yugoslavia (KSS emulsifying salts, produced by Koteksprodukt) and the Czechoslovakia Republic (CI-FO emulsifying salts). However, some huge food companies and processed cheese plants, like Kraft in the United States and Canada Packers in Canada, design and use their own emulsifying agents.

The essential role of the emulsifying agents in the manufacture of processed cheese is to supplement the emulsifying capability of cheese proteins and can be summarized as follows (4,5):

1. Removing calcium from protein system by sequestering
2. Peptizing, solubilizing, and dispersing the proteins
3. Hydrating and swelling the proteins
4. Emulsifying the fat and stabilizing the emulsion
5. Controlling pH and stabilizing it
6. Forming an appropriate structure after cooling

The most important function of emulsifying agents is the ability to sequester calcium. Casein in cheese may be viewed as a molecule with a nonpolar, lipophilic end, whereas the other end, which contains calcium phosphate, is hydrophilic. Because of this structure, casein molecules function as emulsifiers (31). The solubility of casein in water, and hence its emulsifying capacity, are increased by reducing the calcium phosphate content. Calcium in the calcium paracaseinate complex of natural cheese is removed by the ion-exchange properties of melting salts, thus solubilizing the paracaseinate usually as sodium caseinate. Chemically, cheese processing could be observed as presented in Figure 5.

During processing, when higher temperatures are applied further polypeptide bonds are broken. Polyvalent anions of the emulsifying agents (eg, small ions of sodium diphosphate) attach themselves to the modified proteins, increasing their hydrophilic character. Protein molecules become larger, adsorb additional water, and thus increase the viscosity of the colloidal mass. This phenomenon is known as *creaming*. The affinity, that is, sequestering ability, of common emulsifiers for calcium increases in the following order: NaH_2PO_4, Na_2HPO_4, $Na_2H_2P_2O_7$, $Na_3HP_2O_7$, $Na_4P_2O_7$, $Na_5P_3O_{10}$ (5). The affinity of protein

Table 4. Emulsifying Salts Used in the Processing of Cheese

Group	Emulsifying salt	Formula	Molecular mass	P_2O_5 content, percent	Solubility at 20°C, percent	pH Value (1% solution)
Citrates	Trisodium citrate	$2Na_3C_5H_5O_7{\cdot}11H_2O$	714.31		High	6.23–6.26
Orthophosphates	Monosodium phosphate	$NaH_2PO_4{\cdot}2H_2O$	156.01	59.15	40	4.0–4.2
	Disodium phosphate	$Na_2HPO_4{\cdot}12H_2O$	358.14	19.80	18	8.9–9.1
Pyrophosphates	Disodium pyrophosphate	$Na_2H_2P_2O_7$	221.94	63.96	10.7	4.0–4.5
	Trisodium pyrophosphate	$Na_3HP_2O_7{\cdot}9H_2O$	406.06	34.95	32	6.7–7.5
	Tetrasodium pyrophosphate	$Na_4P_2O_7{\cdot}10H_2O$	446.05	31.82	10–12	10.2–10.4
Polyphosphates	Pentasodium tripolyphosphate	$Na_5P_3O_{10}$		57.88	14–15	9.3–9.3
	Sodium tetrapolyphosphate	$Na_6P_4O_{13}$		60.42	14–15	9.0–9.5
	Sodium hexametaphosphate (Graham's salt)	$(NaPO_3)_2$		69.60	Infinite	6.0–7.5

Source: Ref. 4, courtesy of Food Structure and Scanning Electron Microscopy.

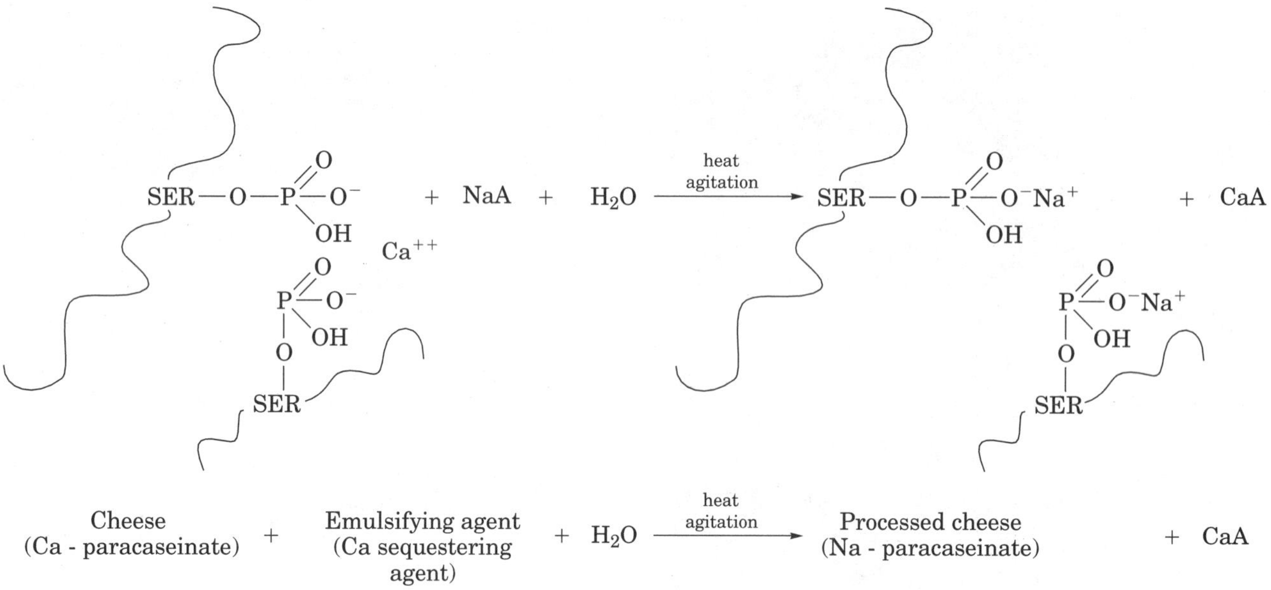

SER—O—P(=O)(OH)—O^- Ca^{++} ^-O—P(=O)(OH)—O—SER + NaA + H_2O $\xrightarrow{\text{heat agitation}}$ SER—O—P(=O)(OH)—O^-Na^+ + Na^+O^-—P(=O)(OH)—O—SER + CaA

Cheese (Ca - paracaseinate) + Emulsifying agent (Ca sequestering agent) + H_2O $\xrightarrow{\text{heat agitation}}$ Processed cheese (Na - paracaseinate) + CaA

Figure 5. Chemical reaction by cheese processing.

for the cations and anions of melting salts is determined by the valency of these ions (4,5).

The main types and roles of emulsifying agents and their characteristics and possible combinations are discussed in detail elsewhere (3,5,32,33).

Generally, the application of polyphosphates yields processed cheese with superior structure and better keeping quality than other emulsifying agents. This may be due to their ability to solubilize calcium paracaseinate and sequester calcium. Pyrophosphates and, in particular, orthophosphates contribute undesirable sensory attributes to the processed cheese. Trisodiumcitrate appears to be as efficient an emulsifier as polyphosphates but lacks the bacteriostatic effect possessed by the polyphosphates.

QUALITY DEFECTS OF PROCESSED CHEESE

A high-quality processed cheese should have a smooth, homogeneous structure; shiny surfaces; uniform color; and no gas holes due to fermentation. However, there are numerous possible defects, of physicochemical or microbial origin, caused either by (*1*) unsuitable blend, or (*2*) inadequate processing.

An unsuitable blend comprises poor-quality or contaminated natural cheese; a bad relationship of blend components; poor-quality proteins; an improper protein-to-fat ratio; irregular quality or quantity of emulsifying agent; and incorrect values for pH, moisture content, and quantity of reworked cheese.

Inadequate processing means unsuitable time-temperature regimes, inadequate agitation, improper cooling, or unsuitable storage. Fortunately, most of the problems in processed cheese technology, when once properly detected, can be corrected. So the first task is to identify the cause responsible for the defect. As soon as the cause is determined and eliminated, the defect showed disappears in further processing.

The most common quality defects of physicochemical and microbial origins in processed cheese, their causes, and suggestions for their correction are presented in Table 5. A number of the cited defects originate in the natural cheese used in the blend, some of which can be avoided by proper processing. Processed cheese with certain minor defects can be recovered by reworking small quantities of it into subsequent batches. More serious defects cannot be corrected and render the product unsuitable for human consumption (eg, microbial changes, *Clostridium botulinum* toxin, presence of metal ions, and excessive Maillard browning).

Crystal formation, sometimes visible, is a serious defect in processed cheese. Most often, the reason for crystal formation is a low solubility of the emulsifying agent used. The situation is accentuated with excessive amounts of emulsifying agent, high calcium content in the natural cheese, high pH of the finished product, and storage of the processed cheese at low temperatures. Some emulsifying agents react with calcium in the cheese, producing insoluble calcium salts. Using sophisticated instrumental methods, such as electron microscopy, energy dispersive spectrometric analysis, and Debye-Scherrer X-ray analysis, crystals in processed cheese have been characterized. They have been chemically identified as calcium phosphate, calcium citrate, sodium calcium citrate, and disodium phosphate (3–5,34–36). Apparently, the most frequently formed crystals are calcium phosphate. During processing, calcium from the natural cheese reacts with phosphates of the emulsifying agents, thus resulting in insoluble calcium phosphate crystals (1). The growth of calcium phosphate crystals in processed cheese, when sodium diphosphate is used as an emulsifying agent, has been shown (28). In Figure 6, calcium diphosphate crystals are visualized (a) in processed cheese, (b) as isolated from cheese, and (c) as a chemically pure form (spray dried). Sheetlike citrate

Table 5. Most Common Quality Defects in Processed Cheese

Defect	Cause	Correction
	Flavor	
Moldy	Air contamination, moldy raw cheese	Use hermetically sealed foils, eliminate all moldy cheeses from the blend
Acid	Excess phosphates	Reduce emulsifier
Salty	Salty raw cheese or other components, too much emulsifier	Add young, unsalted cheese or fresh curd to blend, decrease the quantity of emulsifier
Soapy	High pH value (>6.2)	Add younger cheese (with lower pH value), use emulsifying agent with lower pH value
Burned with browning	Maillard reaction (lactose and amino acids); usually when very young cheese or whey products are present	Use processing temperatures <90°C, cool processed cheese immediately after packaging, avoid large containers, store <30°C, avoid high pH values in final product
	Texture (body, consistency)	
Too soft	High moisture, improper emulsifier, insufficient emulsifier, high pH, fast cooling, excess ripe cheese in blend, prolonged processing, slow agitation	Reduce water content, use suitable emulsifier, increase emulsifier content, decrease pH, slow down cooling, increase proportion of young cheese in blend, reduce processing time, increase agitation speed
Too hard	Low moisture, improper or excess emulsifier, low pH, slow cooling, improper blend, excess creamed or overcreamed, reworked cheese	Increase water content, use proper emulsifier, decrease emulsifier content, increase pH, speed up cooling, change blend composition, avoid addition of creamed or overcreamed reworked cheese
Hard, with water separation	Colloidal change in cheese structure (overcreaming), bacteriological action leading to reduced pH	Remove all factors that affect excess creaming, choose blend components carefully, keep processing temperatures >85°C
Inhomogeneous (grainy)	Unsuitable blend, improper emulsifier, insufficient or excess emulsifier, low pH, short processing time, low processing temperature, improper amount of added water, inadequate agitation, colloidal or bacteriological changes caused by improper storage	Add younger cheese, use suitable emulsifier, correct emulsifier quantity, correct pH, prolong processing time, increase processing temperature >85°C, increase the ammount of added water, continue agitation during processing and filling; proper cold storage
Sticky (adhering to lid foil)	Sticky foil, insufficiently impregnated, excessively high pH, processed mass left hot too long without agitation	Change aluminum foil, decrease water addition and add in two portions, increase proportion of ripe cheese or cause better creaming, keep pH <6.0, continue agitation until packaging
	Appearance	
Holes (blown)	Bacteriological changes (growth of *Clostridia*, coliform or propionic bacteria); physical changes (occluded air, CO_2 from emulsifier mixture (citrates), holes filled up with fluid from emulsifying agent having low solubility); chemical changes (hydrogen from reaction between processed cheese and aluminum foil)	Select cheese blend components carefully, keep processing temperatures >95°C, use proper vacuum, preheat citrate emulsifier before processing, extend processing time, test porosity of aluminum foil and if necessary change it
Crystals	Calcium diphosphate and calcium monophosphate crystals (when phosphates are used in emulsifying agent), calcium crystals (when citrates are used in emulsifying agent), crystals due to undissolved emulsifying salt, large crystals due to excess emulsifier, lactose crystal formation, caused by excess whey concentrates or low water content, light coloured, grainy precipitate of tyrosine (very mature cheese in blend)	Avoid monophosphates and diphosphates as emulsifying agents, or combinations with higher phosphates and polyphosphates; exclude citrates from emulsifying agent; exclude sandy reworked cheese from blend; distribute emulsifying agent better; increase processing time; add emulsifying agent in solution; use prescribed quantity of emulsifier, reduce level of whey products, and increase water content; exclude raw cheese that contains tyrosine crystals

Source: Refs. 1–3, and 5.

crystals are shown in commercial processed Gruyère cheese in Fig. 7. Quite recently crystalline monoclinic calcium pyrophosphate dihydrate was also found in processed cheese when emulsifying agents containing pyro- and polyphosphates were used (37). Authors assume that this crystalline product either comes from migration in the protein matrix of calcium ions and pyrophosphate anions, which are either directly introduced, or results from the hydrolysis of the polyphosphates. In addition to the emulsifying agents, lactose and free tyrosine may de-

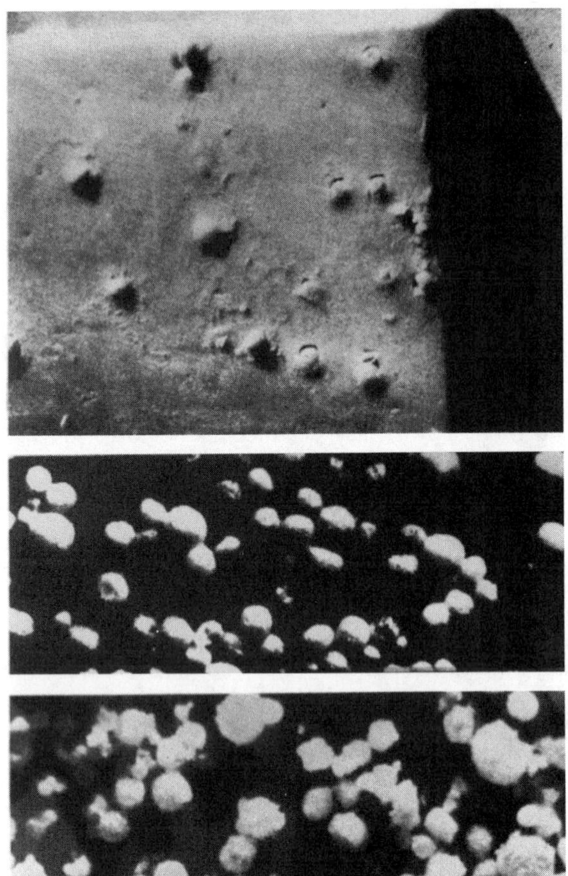

Figure 6. Calcium diphosphate crystals in (**a**) Gruyère processed cheese, (**b**) the isolated above diphosphate globular clumps, (**c**) chemically pure crystals (spray dried) (APV Anhydro A/S) *Source:* Ref. 3, courtesy of Food Trade Press.

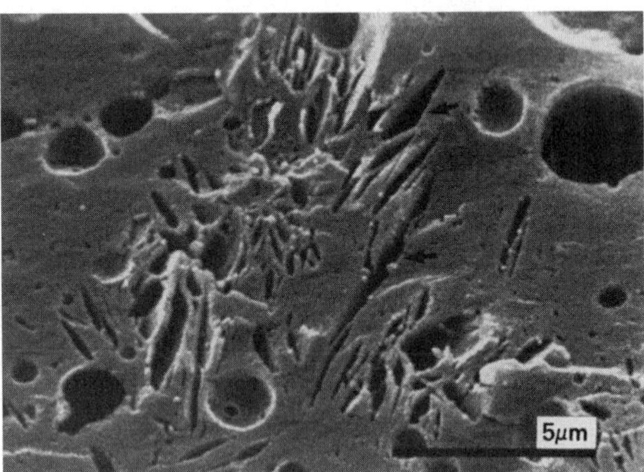

Figure 7. Imprints of sodium citrate crystals (arrows) in processed cheese. *Source:* Ref. 4, courtesy of Food Structure and Scanning Electron Microscopy.

velop crystals in processed cheese, if present at excessively high concentrations (1,2,5).

Discoloration or browning is a defect in processed cheese caused by the Maillard reactions (nonenzymic browning), when the product develops a dark brown or pink color. The exact mechanism and interrelations during all stages of Maillard reactions have been discussed in detail (30,38). Maillard reactions commence at elevated temperatures and continue autocatalytically. Because the main reactants in Maillard browning are amino acids and reducing sugars, the products most susceptible to these changes are blends containing high levels of young cheese; that is, high lactose concentration and other lactose-containing ingredients (particularly whey powder). Browning is more prevalent in processed cheese spreads because of higher processing temperatures, longer processing times, higher water and lactose contents, and higher pH. It has been shown that the intensity of the browning reaction can be reduced by using a galactose-fermenting strain of *Streptococcus salivarius* subsp. *thermophilus*, together with a mesophilic lactic starter culture, in curd production (39). A high NaCl content had the opposite effect, possibly by suppressing the activity of the lactic acid starter culture. Processed cheese that contains a high level of reworked cheese is exposed to more severe thermal treatment than usual and, consequently, the Maillard reactions are accentuated. Most noticeable are the levels of melanoidins, the main products of Maillard reactions, in sterilized processed cheese, even if cooled immediately after production, packed, and stored at low temperature (30).

Microbial defects in processed cheese are caused by spore-forming bacteria, which usually originate in the cheese milk and enter the process through the natural cheese used for blending. Other sources of microbial contamination include water supply, equipment, and additives (1–5). The normal thermal treatment during blend processing (Tables 1 and 3) does not eliminate viable spores from the product.

The spore-forming bacteria, which cause defects in processed cheese by producing gas, belong to the genera *Clostridium* (*C. butyricum*, *C. tyrobutyricum*, *C. histolyticum*, *C. sporogenes*, and *C. perfringens*) and *Bacillus* (*B. licheniformis* and *B. polymixa*). Especially hazardous is *C. botulinum*, producing a toxin that causes botulism. Whether produced continuously or in a batch cooker, processed cheese is not sterile. Germination of spores after processing is influenced by various factors; for example, blend composition, sodium chloride concentration, type and concentration of emulsifying agent, water level, pH, and the presence or absence of natural inhibitors. However, spore outgrowth can be prevented in a number of possible ways: preservatives in the cheese blend, sterilization of the processed cheese, or an increased redox potential of the blend. One of the most widely used methods of preventing spore outgrowth is the addition of preservative into the blend. There is, for example, a procedure patented in the United States where the outgrowth of *C. botulinum* spores and subsequent toxin formation is completely prevented in pasteurized processed cheese spread inoculated with 1000 spores per gram and incubated 48 weeks at 30°C, through the incorporation of nisin at the level of 250 ppm in the blend (40,41).

PROCESSED CHEESE ANALOGUES

There are two recognized groups of products not belonging to the classic range of dairy products: (1) modified dairy products, and (2) substitute (imitation or alternate) dairy products.

In modified dairy products, only one dairy component (eg, protein or fat) is substituted by a nondairy component for economic or nutritional reasons. Imitation products are based on novel components, frequently produced by newly developed technological procedures, using standard equipment but not containing the obligatory dairy component. In the United States, however, native cheeses are also labeled *imitation* if they do not meet the requirements proscribed by federal standards of identity for composition, which could result in confusion over terminology. Formulated or imitation products that closely resemble traditional dairy products are gaining in popularity. The main reason is that imitation products enable the utilization of milk components in combination with nonmilk ingredients, thus lowering production costs and resulting in the composition and nutritive characteristics as designed (42).

Substitution of one or more dairy macroconstituents in processed cheese manufacture usually does not cause any major technical problems. Processed cheese blends can easily be further enriched with desirable microcomponents (eg, vitamins and minerals). Blends can be tailored to yield less-expensive products. Essential components are casein and caseinates (Na, Ca, salts, etc), other proteins (soya, coconut, gluten, etc), suitable vegetable fats, flavorings, vitamins and minerals, food-grade acids (eg, lactic, citric) to correct the pH to 5.8 to 5.9, and emulsifying agents. A suitable blend, suggested for an imitation processed cheese product, is listed as follows (1):

Component	Quantity (%)
Water	100.00
Vegetable oil	3.50
Sodium caseinate	2.75
Dextrose or other sugar	3.80
Emulsifier	0.25
Stabilizer	0.40
Disodium phosphate	0.25
CaCl2	0.25
NaCl	0.15
Artificial color and flavor	Trace

Similar processing parameters and equipment are used as for conventional processed cheeses, processing temperature 79 to 85°C, giving a product of similar physicochemical and microbiological characteristics (1).

Numerous investigations have been carried out in the past decade to find new formulations (43–48). A method has been developed for a successful pilot plant production of a processed cheese based on dairy, noncheese components (43). Great similarity in components and in solubility, dispersibility, and colloidal stability of sodium and calcium caseinate dispersions and processed cheese have been found (44). Egyptian workers (45) included 20% Ras cheese, manufactured from recombined milk using a microbial enzyme, in a processed cheese blend containing butter oil as the lipid phase. Modification of processed cheese, by incorporating vegetable oil in the blend, improved cheese flavor and resulted in a 25 to 50% saving of butter (46). In a Russian patent, milkfat was substituted with a mixture of lard, beef fat, and sunflower oil up to 75% of the total fat content. The processed cheese obtained had clean flavor and mild aroma and was elastic with a uniform consistency (47). Various attempts have also been made to develop combined food products that contain processed cheese as a component. For example, a nutritious chocolate product, composed of 50% bitter chocolate and 50% processed cheese (emulsified, solidified, frozen, and coated with chocolate), was developed and patented in the UK (48). There are various processed cheese analogues in the U.S. market, even some with different colors and the look of a party cake.

NEW DEVELOPMENTS IN PROCESSED CHEESE PRODUCTS

In describing processed cheese analogues, constantly various new components and product modifications arise. However, new developments in classic processed cheese are targeted to obtain a product of better quality, to improve nutritive value of the product, or to widen the versatility of the production. The latest is achieved by introducing the nonconventional components in the blend or by developing a product of completely new characteristics by changing the technology. Some of these findings are still on a laboratory scale, while others are already patented and in industrial application.

In is well known that the addition of phosphates through emulsifying agents increases the phosphorus to calcium (P:Ca) ratio in the processed cheese, which is considered to be nutritionally detrimental. The P:Ca ratio has been reduced from 1.6 when sodium tripolyphosphate was used alone, to 1.1 when the same salt was used, but in combination with 1% surface active monoglyceride, which substituted half of the phosphate amount usually used (49). Another suggestion to correct the improper P:Ca ratio in processed cheese came from Japan (50). The idea of this patent is to fortify processed cheese with calcium by colloidal calcium carbonate addition.

On the other hand, there is a growing tendency to produce low-fat, low-sugar, and low-sodium food products. Consistent with this trend is the newly developed low-sodium processed cheese, manufactured with potassium-based emulsifying agent combined with δ-gluco-lactone (51) or low-fat sliced processed cheese with up to 50% less fat and no significant flavor loss, be it natural or herb-flavored processed cheese slices (41).

A recent review in processed cheese and related products (52) points to the development of a virtually fat-free processed cheese (53), virtually fat-free cheese analogues (54), as well as low-fat processed cheeses (55). The manufacture of virtually fat-free processed cheese (based on skim milk cheese, containing also skim milk powder, whey, and buttermilk powder) is covered by a U.S. patent (53).

The fat content of the finished cheese was approximately 1.67% (wt). Another U.S. patent (54) describes manufacture of a virtually fat-free cheese analogue that mimics the body, texture, and eating characteristics of ordinary processed cheese.

In the study from South Dakota State University (55), the composition and microstructure of some commercial full-fat and low-fat cheeses, including processed cheeses, have been investigated. The structure of low-fat cheeses was dominated by the protein matrix what explains the firmer texture of these cheeses. Commercial fat-reduced processed cheese spread in Germany contained 20% fat in total solids (TS), compared with classic processed cheese spread with 45% fat in TS (60% fat in TS in high-fat processed cheese) (56). The most recent U.S. patent (57) describes the processing of a fat-free high-moisture cheese sauce, based on a blend of skim milk cheese with milk protein and polyphosphate used as emulsifying agent.

However, some investigations at the University of Wisconsin have showed the advantages of particular milkfat components. It was reported that nine isomers of *cis*-9, *cis*-12-octadecadienoic acid (linoleic acid) were found in various processed cheese samples (58). The same synthetically prepared compound was effective in partial inhibition of mouse epidermal carcinogenesis.

There are numerous possibilities to enrich processed cheese with components of high nutritive value. One of these attempts conducted in Egypt (59) yielded a high-protein processed cheese in which 25, 20, 27.5, and 27.5% of nonfat solids were supplied from Cheddar cheese, whey protein concentrate, soy protein concentrate powder, and chickpea flour, respectively. It was reported that soy proteins, consumed at normal dietary levels, compare favorably with animal proteins for meeting human nutritional requirements. Chickpea seed lipids contain more polyunsaturated fatty acids, especially linoleic acid (62.44%). Blending of the selected vegetable proteins with whey proteins corrected their relative deficiency in sulfur-containing amino acids. The final product had fine consistency and high protein content and showed acceptable quality characteristics not only after production but also after two months of storage.

Excellent results were obtained in industrial-scale experiments, when isolated soy protein (Supro 710, Protein Technologies International) was included in the processed cheese blend (60). Compared with soy flour (about 56% proteins) or soy concentrate (about 72% proteins), soy protein isolate has remarkable nutritive, as well as functional, characteristics and does not adversely affect processed cheese products' flavor. The substitution rate of cheese dry matter was successful up to 15% in the blend, yielding products of high nutritive, sensory, and functional properties. Furthermore, if the flavoring agents are used in the processed cheese blend, the mentioned percentage could even be increased.

Additional new developments in processed cheese technology are predominantly targeted in the direction of obtaining processed cheese with natural cheese appearance. Some most recent findings of this type come from the well-known French cheese company Fromageries Bel and are covered by patents (61,62). Their process consists of melting a blend of cheeses, casein, skim milk, butter, and emulsifying agents; the mixture is agitated under an inert gas such as N_2, to give an expanded product, which is afterward partially defoamed, shaped, and rapidly cooled. The resulting processed cheese has an open texture similar to that of traditional cheeses with visible eyes approximately 0.5 mm in diameter.

Processed cheese, consisting mainly of a solution of dissolved proteins–peptides, has a high surface gloss, compared with natural cheese, consisting largely of undissolved suspended proteins–peptides with low surface gloss. By incorporation of milk concentrates obtained by membrane filtration, a new type of processed cheese was developed with the appearance (ie, low surface gloss) of natural cheese (63). The effect caused by undissolved suspended proteins–peptides was quantified by goniophotometric measurements. A German patent (64) prescribes the manufacture of foamed processed cheese, containing Cheddar, butter, yogurt, dried whey, acid casein, starch, emulsifying agents, and some optional flavorings, like fruits or meat products. The blend was processed in two stages, cooled, homogenized, and whipped, resulting in the completely new structure of the final product.

The U.S. processed cheese industry now produces colored, layered processed cheese tarts, nicely decorated and sweet in flavor. Several years ago the patent for continuous procedure for layered processed cheese tarts production was granted in Germany as well (65).

Processed cheese powder, convenient for application in the fast-food industry and snack foods such as popcorn, has been developed in the United States (66).

The demand for dried cheese product in the convenience foods industry is constantly increasing. For example, in Denmark the Cremo Cheese Company, MD FOODS, exports 97% of the 12,000 t annual production of cheese powder to be used as ingredients in biscuits and ready-made dishes and snacks (67).

The introduction of nonconventional components in the processed cheese blend, even the most recent attempts and findings such as modified cheese flavors, accelerated ripened cheese cheese produced of ultrafiltered or reconstituted milk, and various components other than cheese, are discussed in more detail in the section on manufacturing procedure ("Blending").

BIBLIOGRAPHY

Selected sections of this article have been adopted from Reference 5.

1. F. V. Kosikowski, *Cheese and Fermented Milk Foods*, Edward Brothers, Ann Arbor, Mich., 1982.
2. M. A. Tomas, *The Processed Cheese Industry*, Department of Agriculture Bulletin D44, Sydney, New South Wales, Australia, 1977.
3. A. Meyer, *Processed Cheese Manufacture*, Food Trade Press, London, 1973.
4. M. Caric, M. Gantar, and M. Kalab, *Food Microstructure* **4**, 297–312 (1985).

5. M. Caric and M. Kalab, "Processed Cheese Products" in *Cheese: Chemistry, Physics, and Microbiology*, Vol. 2, 2nd ed., Chapman & Hall, London, 1993, pp. 467–505.
6. R. M. Saleem and K. A. Al-Banna, *Iraqi Journal of Agricultural Sciences "Zanco"* **5**, 161–168 (1987).
7. Lj. Kulic, "Modification of Composition of Cheese Blend Intended for Processing by Using Natural Cheese Flavours," M.Sc. Thesis, University of Novi Sad, Cara Lazara 1, Yugoslavia, 1989.
8. A. Gouda et al., *Egyptian Journal of Dairy Science* **13**, 115 (1985).
9. Ger. Pat. Appl. DE 32 24 364 AL (1983), J. Rubin and P. Bjerre.
10. V. K. Sood and F. V. Kosikowski, *J. Dairy Sci.* **62**, 1713–1718 (1979).
11. R. F. Madsen and P. Bjerre, *Nordeuropais Mejeri Tidsskrift* **47**, 135–139 (1981).
12. A. Y. Tamime et al., *Food Structure* **9**, 23–37 (1990).
13. B. A. Law, *Dairy Industries International* **45**, 5 (1980).
14. J. M. Saluja and S. Singh, *Journal of Food Science and Technology (India)* **26**, 29–31 (1989).
15. E. H. El-Bagoury and F. R. Helal, *Egyptian Journal of Food Science* **15**, 219–223 (1987).
16. U.S. Pat. 2,743,186 (1956), N. Kraft and P. J. Ward.
17. Fr. Pat. Appl. FR 87-15458 (87 1106) (1989), L. R. Rizzotti and B. Villaudy.
18. M. Kalab, J. Yun, and S. H. Yin, *Food Microstructure* **6**, 181–192 (1987).
19. S. A. Abou-Donia, A. F. Salam, and K. M. El-Sayed, *Indian Journal of Dairy Science* **36**, 119 (1983).
20. P. Brezani and K. Herian, *Zbornik Prac Vyskumneho Ustavu Mliekarskeho v Ziline* **8**, 173 (1984).
21. A. F. Salam, *Alexsandria Science Exchange* **9**, 167–175 (1988).
22. Brit. Pat. 1,452,253 (1976), R. Invernizzi and G. Prella.
23. Swiss Pat. 584,515 (1977), Societe des Produits Nestle SA.
24. V. A. Samodurov, *Molochnaya Promyshlennost* **51**, 17–19 (1985).
25. Ger. Pat. Appl. DE 31 24 725 A1 (1983), F. Zimmermann.
26. Jpn. Exam. Pat. JP 57 380 B2 (1982), T. Hayashi, N. Shibukawa, Y. Yoneda, K. Musashi.
27. G. Blond, E. Haury, and D. Lorient, *Science des Aliments* **8**, 325–340 (1988).
28. A. A. Rayan, M. Kalab, and C. A. Ernstrom, *Scanning Electron Microscopy* **3**, 635 (1980).
29. R. K. Robinson, *Dairy Microbiology*, Vol. 2, *Microbiology of Dairy Products*, Elsevier Applied Science, Barking, UK, 1980.
30. B. Milic, M. Caric, and B. Vujicic, *Non-Enzimatic Browning Reactions in Food Products*, Naucna knjiga, Belgrade, Yugoslavia, 1988.
31. L. A. Shimp, *Food Technol.* **39**, 63 (1985).
32. M. Caric and S. Milanovic, *Processed Cheese*, Naucna knjiga, Begrade, Yugoslavia, 1997.
33. W. Berger et al., *Processed Cheese Manufacture*, BK Ladenburg, Ladenburg, Germany, 1993.
34. M. A. Thomas et al., *J. Food Sci.* **45**, 458 (1980).
35. H. A. Morris, P. B. Manning, and R. Jenness, *J. Dairy Sci.* **52**, 900 (1969).
36. H. Klostermeyer, G. Uhlmann, and K. Merkenish, *Milchwissenschaft* **38**, 582 (1983).
37. J. F. Pommert et al., *J. Food Sci.* **53**, 1367–1369, 1447 (1988).
38. H. D. Belitz and W. Grosch, *Food Chemistry*, Springer-Verlag, Berlin, New York, 1987.
39. M. E. Bley, M. E. Johnson, and N. F. Olson, *J. Dairy Sci.* **68**, 555 (1985).
40. U.S. Pat. 4,584,199 (1986), S. L. Taylor.
41. E. J. Mann, *Dairy Industries International* **52**, 11–12 (1987).
42. M. Caric, *Concentrated and Dried Dairy Products*, VCH, New York, 1994.
43. J. R. Rosenau, *Energy Management and Membrane Technology in Food and Dairy Processing*, American Society of Agricultural Engineers, Chicago, Ill., 1983.
44. O. Kirchmeier and F. X. Breit, *Milchwissenchaft* **38**, 80 (1982).
45. A. E. A. Hagrass et al., *Egyptian Journal of Food Science* **12**, 129 (1984).
46. USSR Pat. SU 971 216 A (1982), I. A. Snegireva et al.
47. USSR Pat. SU 1144 677 A (1985), V. A. Samodurov et al.
48. Brit. Pat. Appl. GR 2 113 969 A (1983), G. Vajda, L. Ravasz, B. Karacsonyi, and G. Tabajdi.
49. N. P. Zakharova, N. B. Gaurilova, and V. G. Dologoshchinova, *Trudy Uglich* **27**, 108–111, 121 (1979).
50. Jpn. Exam. Pat 5619967 (1981), Meiji Milk Products K.
51. R. C. Lindsay et al., *Proc. IDF Seminar, New Products via New Technology*, Atlanta, Ga., 1986.
52. E. J. Mann, *Dairy Industries International* **60**, 19–20 (1995).
53. U.S. Pat. 5,215,778 (1993), B. C. Davison.
54. U.S. Pat. 5,244,687 (1993), B. E. Rybinski.
55. V. V. Mistry et al., *Food Structure* **12**, 259 (1993).
56. B. K. Giulini Chemie Gmbh & Co, OHG, Ladenburg, *Technical Documentation* **6**, (1998).
57. U.S. Pat. 5,304,387 (1994), W. S. Hine.
58. Y. L. Ha, N. K. Grimm, and M. W. Pariza, *J. Agric. Food Chem.* **37**, 75–81 (1989).
59. A. A. El-Neshawy, S. M. Farahat, and H. A. Wahbah, *Food Chem.* **28**, 245–255 (1988).
60. M. Caric et al., *Mljekarstvo* **39**, 95–102 (1989).
61. Eur. Pat. Appl. EP 0 281 441 A1 (1988), J. Daurelles and J. Y. Bernard.
62. Fr. Pat. Appl. FR 2 610 794 A1 (1988), J. Daurelles and J. Y. Bernard.
63. K. H. Ney, *Alimenta* **26**, 123–124 (1987).
64. Ger. Pat. DE 33 14 551 C1 (1984), D. Bode.
65. Ger. Pat. Appl. DE 37 27 660 A1 (1989), F. Zimmermann.
66. U.S. Pat. 5,227,187 (1993), J. A. Wiser, S. J. Terhune, B. E. Jr. Gilmartin, and J. A. Kintner.
67. H. Mortensen, *Scandinavian Dairy Info* **8**, 24 (1994).

MARIJANA CARIC
Novi Sad University
Novi Sad, Yugoslavia

PROTEINS: AMINO ACIDS

Amino acids are fundamental to the pharmaceutical, food processing, and animal feed industries. Besides being integral to the health of humans, they are used in industry for their nutritive value, physiological metabolites, and taste.

In the food and animal feed industries, several essential amino acids, such as lysine, methionine, tryptophan, and threonine, are used in addition to L-glutamate, which accounts for more than 50% of total amino acid production. In the pharmaceutical industry, essential and nonessential amino acids are used, depending on the nutritional values desired. The demand for amino acids is expected to grow throughout the world as people seek improved lifestyles through diet.

L-Glutamate is used mainly as umami seasoning throughout the world. It was first produced in 1909 using an extraction method from the hydrolyzate of soybean with hydrochloric acid. A significant drawback to this method is that a large amount of by-products is produced, and it is necessary to deliver the solid matter as organic fertilizer and the amino acid moiety in solution as an alternative soybean source. In 1956, the fermentation method of L-glutamic acid from glucose using bacteria was reported, changing not only the production method but also the use of the amino acids themselves. The successful production of L-glutamic acid by fermentation was followed by production of other amino acids, using the same concept of regulation of the metabolic pathway of bacteria.

Methionine and glycine are produced by the chemical synthetic method. Methionine is mainly used as a feed additive in the DL-form because the D- and L-forms both have similar, nutritional values. Optical resolution is not necessary for glycine because it is achiral.

As the fermentation technology developed for L-glutamic acid production has been applied to other amino acids, the costs have become more reasonable. This has resulted in an increase in the distribution of L-glutamate, feed additives and L-lysine, and increased demand for amino acids by the pharmaceutical industry has been met.

USES OF AMINO ACIDS

Amino acids are used in animal feeds, food, parenteral and enteral nutrition, medicine, cosmetics, and raw materials for chemical industries.

Animal Feed

Although vegetable protein contains a high concentration of essential amino acids, it does not have a taste appreciated by many humans. When cooked with meat extract, the food's taste is improved, and its nutritive values for food animals are enhanced, thus providing an efficient concentration of essential amino acids that can be used to produce a tasty meat source throughout the world. The worldwide supply of free amino acids to increase the feed efficiency of feedstock has developed into a successful business in recent years.

Table 1 summarizes the limiting amino acids used as feed supplements (1). For grains such as maize, which is fed to both pigs and chickens, the first limiting amino acid is lysine, whereas in protein sources such as rapeseed meal, it is methionine, lysine, or tryptophan. There are three general methods of supplying lysine: adding soybean meal, which contains it in high quantity; adding a larger quantity of poor-quality protein such as corn gluten meal; or adding lysine directly. The first two approaches are more expensive and less economical in view of wasting nitrogen resources, so the direct addition of lysine has been adopted.

An optimal balance of amino acids is attained by adding deficient amino acids and reducing those in excess. When animals are fed corn–soybean meal (Fig. 1), about 2% of the feed protein contents can be saved without affecting the animal's growth response. One hundred kilograms of soybean meal contained in 2 tons of feed can be replaced by 97.5 kg of corn and 2.5 kg of lysine hydrochloride without lowering its nutritional value. When the price of the latter is lower than the former, the use of lysine increases.

Methionine. Fish meal and vegetable defatted meal are important protein sources of feed. The supply of fish meal has continuously decreased and cannot meet the demands for feed. Because of the decrease in fish meal for feed and the increased use of vegetable protein in vegetable oil meal, the demand for methionine has increased.

Lysine. As the demand for vegetable defatted meal has increased and the importance of amino acid balance in protein has been recognized, lysine has received increasing attention. Cereals are considered an energy source and comprise 60 to 80% of assorted feeds; they play an important role as protein sources but contain relatively small amounts of lysine. To compensate for this shortage, lysine has been used in place of fish meal.

When the ω-amino group of lysine reacts with a carbonyl group of a sugar derivative cmpound, which gives a positive Fehling reaction, a Schiff base is formed. These Schiff bases are not used as nutritional compounds. This reaction, which is accelerated by heat treatment, is usually performed to improve the protein efficiency.

Table 1. Limiting Amino Acids of Some Common Feeds for Pigs and Chickens

Ingredient	Crude protein (%)	Pig		Chicken	
		First	Second	First	Second
Maize	8.9	Lys	Trp	Lys	Trp
Sorghum	9.5	Lys	Thr	Lys	Arg
Barley	11.1	Lys		Lys	Met
Wheat	12.6	Lys	Thr	Lys	Thr
Soybean meal	46.2	Met	Thr	Met	Thr
Fish meal	64.3			Arg	
Rapeseed meal	35.3	Met		Lys	Arg
Peanut meal	47.4	Lys		Met	Lys
Sunflower seed meal	31.7	Lys		Met	Thr
Meat and bone meal	48.6	Lys		Trp	
Cottonseed meal	64.3	Lys	Thr	Lys	Met
Corn gluten meal	63.6	Lys	Trp	Lys	Trp

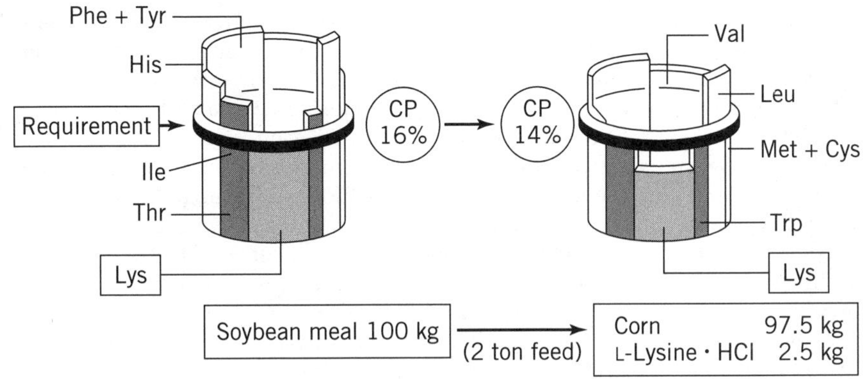

Figure 1. Balance of amino acids in feed for protein deficiency in a well-balanced diet.

When a group of animals received marginally deficient levels of lysine, the amount of weight gain was unchanged from the amount in the group receiving the required level of lysine. The lack of change in weight gain was due to the increase in fat deposition; nevertheless, protein retention decreased as the level of lysine decreased (2).

Fortification of Protein for Food

In southeast Asia, Africa, and South America, the population continues to increase, but no appropriate measures to improve the nutritional conditions in these societies have been implemented. The chronic shortage of protein, more than that of calorie intake, in these areas inhibits physical growth and mental development in children and deprives adults of the ability to perform physical labor. This makes impossible the development of a society with a sound economy. A 1957 report by the FAO to the United Nations stated that to make up the protein gap, it is necessary to provide a higher nutritional value of protein for the consumer (3). One concrete way to do this would be to fortify the amino acid content of cereal with seed protein, fish protein concentrate, and vegetable protein.

The principle of amino acid fortification is based on two factors. (*1*) Nutritionally efficient utilization of amino acids depends on the ratio of essential amino acids to all the amino acids contained in protein. (*2*) When the amino acid balance of a protein is corrected by the addition of certain amino acids, defined as limiting amino acids, significant improvements in the nutritional value of the protein are observed. The effects of fortification with limiting amino acids on the protein efficiency ratio (PER) for various cereals are summarized in Table 2 (4,5). The PER is defined as the ratio of the incremental weight gain in rats to the total amount of protein fed in certain intervals. The PER value for cereals such as rice, corn, barley, and wheat flour is approximately 1.0, which is lower than that of 2.5 for casein. The limiting amino acid for all cereals is lysine, in addition to threonine for rice, wheat, and barley flours and tryptophan for corn. When these limiting amino acids are added to cereals, the PERs for corresponding cereals are nearly the same as casein.

The limiting amino acid for soybean is methionine; for cotton seed, lysine and threonine; for sesame, lysine; and for peanuts, lysine, threonine, and methionine.

Table 2. The Effect of Adding Limiting Amino Acids on the Improvement of Protein Efficiency Ratio of Cereal

Protein	Limiting amino acid added (%)			Protein efficency ratio
	L-Lysine HCl	D,L-Threonine	D,L-Tryptophan	
Rice				1.50
	0.2	0.2	—	2.60
Wheat				0.65
	0.2	—	—	1.56
	0.4	0.3		2.67
Corn				0.85
	0.4	—	—	1.08
	0.4	—	0.07	2.55
Barley				1.66
	0.2	0.2	—	2.28
Kaoliang				0.69
	0.2	—	—	1.77

The idea of synthetic foods or chemical-defined foods composed of amino acids and other appropriate nutrients was investigated in the 1950s and applied to so-called space diets (6). The crystalline digestive diet, which contained crystalline amino acids as the nitrogen source as recommended by Greenstein, was the prototype of the space diet. It was investigated further in the U.S. space program, and from this the term *elemental diet* came into common use. When about 400 g of this synthetic food was administered to male adults for about half a year, no changes in pathophysiological or mental conditions were observed, except an extreme decrease in the amount of excretion, because no indigestible substances were contained in the synthetic food, and a decrease in several probiotic bacteria in the intestine (7,8).

Some infants exhibit lactose or cow's milk protein incompatibility. The formulas marketed for this condition often are based on isolated soybean protein and are supplemented with L-methionine to increase the nutritional value (9).

Umami Tastes

The specific tastes of foods are intimately related with their constituents, such as amino acids and peptides. The proteins contain about 20 kinds of amino acids, which are found in free form in dietary foods. As shown in Table 3, which summarizes the kinds of amino acids found in common foods, the differences in the amino acid patterns for each food are said to be key factors in their distinct tastes; the variety of tastes depends on the changes in the free amino acid components.

In urchin, amino acids such as glutamic acid, glycine, and alanine (and not so much methionine) contribute to specific sweet tastes. Green tea contains mainly glutamic acid and theanine; these two compounds are related to the umami tastes. These 20 or so amino acids found in food play an essential role in the harmonized mixed tastes of foods. The threshold value, which is defined as the limiting concentration distinguished from the intensity of specific taste, and the taste quality for various amino acids are summarized in Table 4 (10). L-Aspartic acid shows less intensity of umami taste than does L-glutamic acid; however, the sodium salt (monosodium aspartate), found in soybeans, enhances the umami tastes of soybeans usually associated with L-glutamate.

The specific taste of each amino acid generally remains unchanged when its concentration is changed. However, the specific tastes of L-alanine, L-arginine, L-glutamic acid, L-serine, and L-threonine, as shown in Table 5, change when their concentrations change (11,12).

Relationship between Umami Taste and Chemical Formula. The relationship between umami taste and the chemical structure of L-glutamate (Fig. 2) was investigated by Akabori and Kaneko (13–15). The threshold values for L-glutamate-related compounds for umami taste were as follows: 0.03% for L-glutamate, 0.16% for L-aspartate, 0.25% for sodium DL-α-amino adipic acid, 0.03% for sodium DL-threo-β-oxylglutamic acid, and 0.015% for sodium L-homocysteic acid.

Sodium L-aspartic acid has one less carbon than L-glutamate, and sodium L-amino pimeric acid has one more carbon than L-glutamate; these compounds as well as D-glutamate (Fig. 2b) show slight umami taste when compared with L-glutamate. When conversion of the α-amino bond occurs by acetylation of the carboxyl bond by esterification, or when the α-hydrogen bond is converted by methylation, these compounds lose the umami taste. When the hydrogen at the β-position is exchanged for an OH ion, the threo form keeps the umami taste, whereas the erythro form (Fig. 2f) loses the umami taste. When a γ-carboxyl bond was substituted for sulfonic acid (Fig. 2c), the umami taste was enhanced.

The intensity of umami taste depends on pH. When kept in the vicinity of pH 7, the intensity of umami taste is highest; when pH is acidic or basic, the intensity is lowered.

These variations are explained by Kaneko's five-membered-ring theory. For the umami taste to be present, the five-membered ring must have α-NH_3^+, and γ-carboxyl bonds that interact electrostatically. Moreover, it is important that three bonds α-NH_3^+, γ-COO^-, and α-H, are placed on the plane structure. D-Glutamate (Fig. 2b) and α-methyl L-glutamate (Fig. 2d) do not have this stereo structure and cannot interact with umami bud receptors because of this steric hindrance. For the stereo structure of β-hydroxyl L-glutamate, when the threo type (Fig. 2e) is formed, no hindrance of the OH bond for the formation of the five-membered-ring structure is observed, but in the erythro type (Fig. 2f), the umami taste is lost because the OH bond hinders the formation of the five-membered-ring structure.

In acidic regions, the γ-COO^- bond converts to COOH, and in basic regions, α-NH_3^+ bonds to NH_2, reducing the umami taste as a result of the reduction of the electrostatical interaction between COOH and NH_2 in both regions.

Flavor Enhancer

The free amino acids are used widely in food manufacturing for aromas and brown food colors (16). The flavors are

Table 3. Amino Acid Composition in Foods (mg%; dry base)

Amino acid	Pork	Beef	Urchin	Mackerel	Green tea	Seaweed
Alanine	4.19	11.28	261	37	25.2	1029
Arginine	—	—	316	6.1	142	90
Aspartic acid	1.37	0.28	4	9.8	136	230
Glutamic acid	1.95	4.63	103	20	668	640
Glycine	2.75	2.40	842	54	47	125
Histidine	2.55	4.10	54	563	6.7	0
Isoleucine	1.03	2.04	100	9.6	47	14
Leucine	2.68	3.81	176	14	34	20
Lysine	4.27	6.19	215	22	7.4	24
Methionine	0.69	2.01	47	7.3	0.6	0
Phenylalanine	0.51	1.36	79	9.2	10.1	15
Proline	0.64	—	26	5.4	18.3	62
Serine	2.95	7.53	130	6.9	81.1	53
Threonine	0.48	1.11	68	9.6	60.9	78
Tryptophan	—	—	39	2.2	11.6	0
Tyrosine	0.56	1.85	158	6.6	4.2	0
Valine	0.30	2.99	154	14	6.1	21
1-Methylhistidine	0.49	4.80	—	—	—	—
Theanine	—	—	—	—	1727	—

Table 4. Taste Profile of L-Amino Acids

L-Amino acid	Threshold value (mg/dL)	Taste quality				
		Sweet	Sour	Bitter	Salty	Umami
Gly	110	+ +				
Hyp	50	+ +		+ +		
β-Ala	60	+ +				+
Thr	260	+ +				
Pro	300	+ +		+ + +		
Ser	150	+ +				+
Lys HCl	50	+ +		+ +		+
Gln	250	+				+
Phe	150			+ + +		
Trp	90			+ + +		
Arg	10			+ + +		
Arg HCl	30	+		+ + +		
Ile	90			+ + +		
Val	150	+		+ + +		
Leu	380			+ + +		
Met	30			+ + +		+
His	20			+ +		
His HCl	5		+ + +	+	+	
Asp	3		+ + +			+
Glu	5		+ + +			+ +
Asn	100		+ +	+		
Monosodium glutamate	30	+			+	+ + +
Monosodium aspartate	100				+ +	+ +

Profiles of each basic taste intensity are expressed as follows: + + +, strongest; + +, stronger; and +, detectable. When a taste quality was not detectable, symbols and omitted.

Table 5. Change in Taste with Concentration of Amino Acids

Amino acid	Concentration		Concentration	
	Low (g/dL)	Taste	High (g/dL)	Taste
L-Alanine	0.5	Sweet	5.0	Sweet/umami
L-Arginine	0.2	Bitter/sweet	1.0	Bitter
L-Glutamic acid	0.025	Sour	0.2	Sour/umami
L-Serine	1.5	Sweet/sour	15.0	Sweet/umami/sweet
L-Threonine	2.0	Sweet/sour/bitter	7.0	Sweet/sour

formed mainly during food processing (eg, the fermentation of alcoholic beverages) or cooking (frying, roasting, boiling) by the Maillard reaction between amino acids and reducing sugars. The Strecker degradation of amino acids plays a central role in this process. It is often possible to assign certain aromas to specific amino acids. The sulfur-containing amino acid cysteine is primarily responsible for meaty flavors, proline produces the aroma of bread crust, and phenylalanine and branched amino acids provide the characteristic flavor of chocolate. Valine and leucine are involved in the aroma of roast nuts. Methionine plays a key role in the aroma of french fries. The flavors of such products as precooked foods, snacks, and spices may be improved by the addition of the proper Maillard aromas (17).

Other applications of amino acids in foods are listed in Table 6. Amino acids are used in the food manufacturing industry for purposes other than supplementation and flavoring. L-Cysteine is used in bread baking as a flour additive to relax wheat gluten proteins, improve the structure of the baked product, and allow shorter kneading times.

Melanoidines, which are formed through the Maillard reaction, are stronger antioxidants than amino acids themselves. Glycine apparently exhibits a special preservative effect.

Pharmaceuticals

Elman and Weiner (18) attempted the clinical application of intravenous casein hydrolyzates in humans in 1939, and the nutritional requirement of amino acids was reported by Rose (19) in 1949. Stimulated by these investigations, various types of protein hydrolyzates containing amino acids were applied in clinical practice throughout the world. But because of the allergic reaction and asymptomatic hyperammonemia that resulted when protein hydrolyzates in the form of administered, crystalline amino acid solutions that did not contain unnecessary components such as peptides were needed. However, it was not until

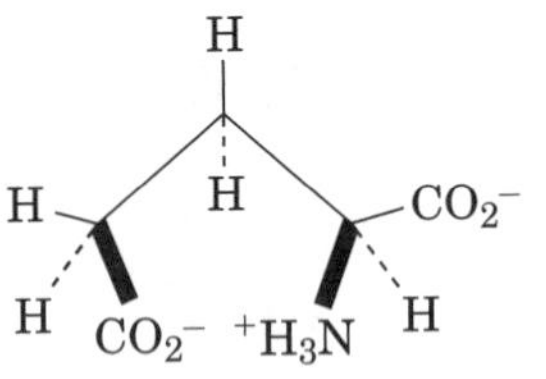

L-Glutamic acid (umami) (1)

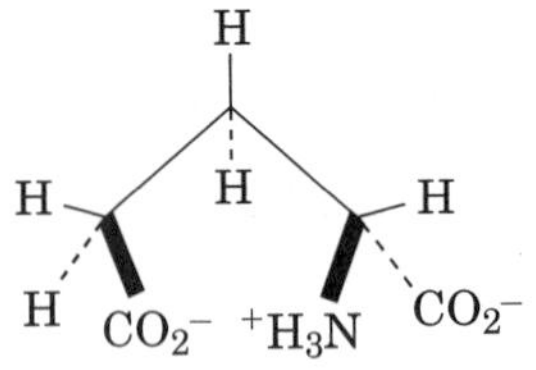

D-Glutamic acid (2)

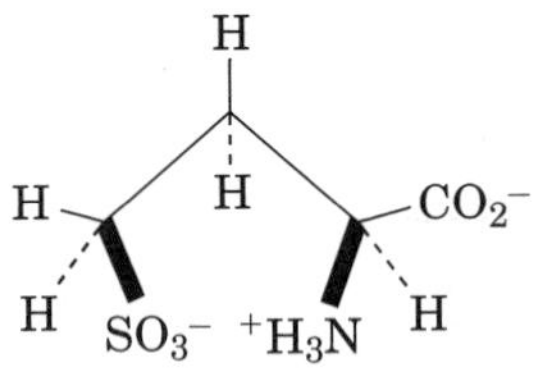

L-Homocysteinic acid (umami) (3)

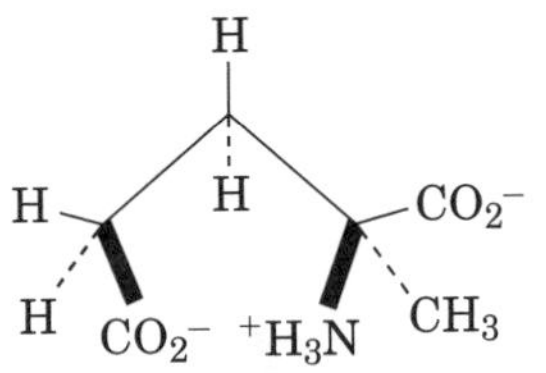

α-Methyl-glutamic acid (4)

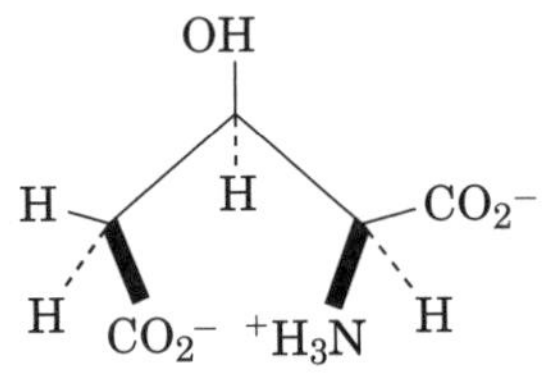

β-Hydroxyl L-Glutamic acid (threo type; umami) (5)

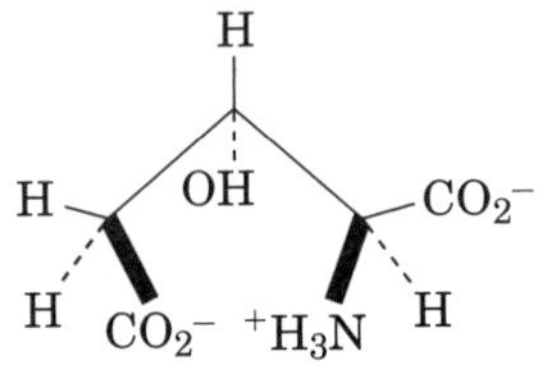

β-hydroxyl L-Glutamic acid (erythro type) (6)

Figure 2. Relationship between the chemical structure of glutamic-acid-related substances and the appearance of umami taste.

1959 that such solutions were commercially available in Japan and became broadly used in daily clinical practice. This type of solution contains amino acids but no peptides, making it possible to have the exact amount of each amino acid in the solution.

The composition of amino acids in each solution was essentially based on recommendations by the Joint FAO/WHO Expert Group (20), which recommended a ratio of 1:1 between essential and nonessential amino acids in mixtures administered intravenously and those administered orally or through a feeding tube.

Meeting the increased branched-chain amino acid requirement in patients with severe surgical injuries has been demonstrated as being a means of nutritional support and treatment of disease itself (21).

CHEMICAL PROPERTIES

Formation of Salt

Because amino acid molecules have carboxyl groups and amino groups, they are able to react with both acids and bases to form a corresponding salt. When the salts react with weak acid, such as acetic acid, and weak alkali, such as ammonia, they are generally unable to separate as crystals. When not only mineral acids but also benzoic sulfonic acid and picric acid reacts with amino acids, chemically stable and characteristic crystalline salts easily form, which are used as tools for separation and identification of various amino acids. With metallic ions, such as silver, lead, zinc, copper, and tungsten, amino acids form sparingly soluble complex salts, which can be used as tools for separation and colorimetric analysis to identify the respective amino acids.

In the copper complex, the α-carbon atom is activated to react with aldehyde, for example, the glycine–copper complex reacts with acetaldehyde to yield threonine.

Esterification

When heated with alcohol in the presence of anhydrous hydrochloric acid, COOH in the α-position on the amino acid converts to the hydrochloride of the ester salt, corresponding to the alcohol used. This ester bond is easily released by treatment with alkali solution. The free ester of an amino acid eliminates alcohol on standing or warming to form cyclic 2,5-diketopiprazine, which converts to the corresponding peptide by hydrolysis with weak acidic. Moreover, when this amino acid ester reacts with hydroxylamine, it changes to hydroxamic acid, which on colorimetric analysis exhibits a specific color with $FeCl_3$.

Acylation

With acid anhydride or acyl chloride compounds under alkaline solution conditions (Schotten-Baumann conditions), amino acids react to form *N*-acyl α-amino acids, which are used mainly for protection of the amino group for peptide synthesis. Amino acids acylated with naturally occurring fatty acid residues are used industrially as easily degradable surfactants.

Schiff Base Formation

An example of a typical amino acid reaction is that when the amino acid combines with an aldehyde to form a Schiff base, a ketoimine-type compound. This Schiff base is important in view of the physiology of amino acids. Vitamin B_{12} enzymes, which control the decarbonation of amino acid and the amino group transformation between amino acid and keto acid, react after the formation of the Schiff

Table 6. Use of Amino Acids in Foods

Amino acid	Food	Function
L-Alanine	Synthetic sake	Seasoning
Monosodium L-aspartate	Seasoning	Seasoning
L-Arginine, L-glutamate	Green tea	Seasoning
L-Cysteine HCl	Bread	Enrichment of nutrition
Glycine	Natural fruit juice	Seasoning
	Fish paste product	Microbiostasis
	Pickles	
	Synthetic sake	
	Synthetic vinegar	
L-Glutamic acid	Meat product	Seasoning
	Canned food	Prevention of struvite
L-Histidine	Patient diet	Enrichment of nutrition
	Baby food	
L-Isoleucine	Cereal grain	Enrichment of nutrition
L-Lysine	Cereal grain, bread	Enrichment of nutrition
	Noodles, rice	
L-Lysine, L-aspartate	Cereal grain	Enrichment of nutrition
		Seasoning
L-Lysine, L-glutamate	Cereal grain	Enrichment of nutrition
		Seasoning
L-Methionine	Bread	Promotion of fermentation
L-Phenylalanine	Bread	Promotion of fermentation
L-Threonine	Bread	Promotion of fermentation
L-Tryptophan	Bread	Promotion of fermentation
L-Theanine	Green tea	Seasoning
L-Valine	Cereal grain	Enrichment of nutrition

base with pyridoxal phosphate to form the coenzyme of vitamin B_{12}.

Racemization

Racemization is one of the most important reactions for the seasoning industry: optically active amino acids are transformed into racemized amino acids. In general, this reaction easily occurs in alkaline conditions, and proceeds by the following scheme: the formation of a carbanion releases the α-H from the α-C by a base catalytic reaction, and this then equilibrates with the α,β-unsaturated form (22).

Degradation of Amino Acids

Amino acids do not have a defined melting point. They decompose over the broad range of 200-to 300°C but are comparatively stable at ambient temperature and occur as transparent or white crystals.

In the solid state, decarbonation mainly occurs when the amino acid is heated. The amino acid converts to the amine that has the same carbon number as the amino acid.

When an α-amino acid reacts with nitrous acid, it yields α-hydroxycarboxylic acid, and nitrogen gas. The amount of gas that is measured, is equivalent to the amino-nitrogen. This is the basis for the van Slyke method, used for the quantitative determination of amino acids.

When ninhydrin is added to an amino acid solution, the CO_2 released causes the color of the solution to turn of blue-violet called Ruhemann's purple, which has a maximum absorbency at 570 nm. This reaction is often used as a quantitative analysis of amino acids.

PHYSICAL PROPERTIES

Amphoteric Compounds

A common characteristic of amino acids is to have a dipolar ion (Zwitterion) in molecular form—not an uncharged form of H_2N-CHR-COOH but a dipolar ion, H_3N^+-CHR-COO^-—in both neutral solution and in solid state.

Generally, this is recognized by the following physical properties. Amino acids have (*1*) high melting points; (*2*) high dipole moments; (*3*) extremely low volatile characters; and (*4*) in solution, changes in the value of the dielectric constant of water and electrical strain are determined directly by Raman spectra and infrared spectra.

Dielectric Constant

Because of the dipolar ion structure of the amino acids, they are able to considerably increase the dielectric constant of the solvent in which they are dissolved (23–25). This increase is on such a high an order of magnitude that it is explicable only on the basis of the possession by these compounds of a large permanent electric moment.

The electric moments of the amino acids themselves cannot be determined by Debye's procedure because of their nonpolar solvents. This fact suggests the highly polar character of the amino acids.

The moment can be calculated from the spatial considerations on the basis of a dipolar ion structure. If the center of the positive charge is located on the α-nitrogen atom, and the center of the negative charge is midway between the two oxygens of the carboxyl group, the dipole distance is 2.96 Å, and the electric moment is the product of the elementary charge on the electrons

$$4.8 \times 10^{-10} \text{ esu by } 2.96 \times 10^{-8} \text{ cm} = 14.2 \times 10^{-18} \text{ esu}$$

The Debye relation is $P = 4\pi N\mu^2/9$ kT for compounds with high permanent dipoles. P is the molar polarization and μ is the electric moment. The term P is a linear function of the dielectric constant of the medium, and the dielectric constant is related to the square of the electric moment per molecule of the compound.

The electric moment of a compound such as β-alanine would be expected from the Debye relation to be greater, and the effect on the dielectric constant of the medium would be expected to be greater than revealed by the isomeric compound α-alanine.

All α-amino acids produce an increase in the dielectric constant of water by about 23 per mol, which is independent of concentration. In the case of glycine, where the dielectric constant of the solution is 135, the increase is 2.5 mol/L. The dielectric increment per mole of solute, δ, is shown in the following equation:

$$\delta = D(\text{solution}) - D(\text{solvent})/\text{molarity}$$

For water at 25°C, the dielectric constant D_0 is 78.54; $\delta = 23$ for α-alanine and $\delta = 35$ for β-alanine.

The value of δ at a pH in the range 4.5 to 7.5 for glycine was measured. δ falls steeply on either side of this (isoelectric) pH range as the initial dipolar ions convert into either cations or anions with simultaneous loss of dipolar ion character.

The values of δ for all α-amino acids range from 22.6 to 23.2; the values for the β-amino acids are about 35.

Raman Spectra Studies

In Raman spectra, the peaks at 1730 cm^{-1} for COOH and at 3320 cm^{-1} for NH_2 are not found in amino acids, but a peak at1400 cm^{-1} for COO^- is detected. In the amino acid hydrochloride salt, a peak of 1730 cm^{-1} for COOH is detected, but the peak for NH_3^+ is not detected because of the overlap with the absorbency of water.

The unionized COOH group frequency near 1730 cm^{-1} is a particular feature of the spectra. This COOH group would be expected to be present in all relatively unionized carboxylic acids, and if the dipolar ion character of the amino acids is true, this group would also be expected to occur in the hydrochlorides of the amino acids. When the carboxyl group is ionized, the 1730-cm^{-1} line disappears, thus further suggesting the dipolar ion character of the amino acids under these conditions (26–31). RCOOH or $NH_3^+RCOOH + Cl^-$ shows a 1730-cm^{-1} line, but $RCOO^-$ or $NH_3^+COO^-$ does not show a 1730-cm^{-1} line.

Dissociation Constant

In aqueous solution, amino acids are in a state of dissociation equilibrium that is dependent on the pH. In acidic conditions, they exist as cationic forms (A^+); in basic conditions, as anionic forms (A^-); in between, as mainly zwitterions ($A^{\pm}$), where the cationic and anionic forms are equal.

When an amino acid solution is in a neutral state and acidic or alkaline agents are added slowly, an abrupt change in pH at the inflection is observed that plateaus at constant value. The more added, the move abrupt the change in pH at the infliction. Amino acids show a double buffer action in both acidic and basic regions, which are named pK_1 and pK_2, corresponding to the inflection point in the titration curve.

The pK_1, pK_2, and pI values are calculated from the titration curves with the Henderson–Hasselbalch equation.

The equilibrium constant K_1 in acidic solution is expressed as $K_1 = [A^{\pm}][H^+]/[A^+]$. Conversely, in basic solution, the equilibrium constant is $K_2 = [A^-][H^+]/[A^{\pm}]$. When 50% of amino acid is dissociated, $[A^{\pm}] = [A^+]$; $[A^-] = [A^{\pm}]$, ie, $K_1 = [H^+]$ and $K_2 = [H^+]$, then pK_1 and pK_2 each show the pH in solution.

When a direct current is applied to an amino acid solution and the pH in solution is lower than pK_1', the amino acids transfer toward the cathode; when pH is higher than pK_2', they transfer toward the anode. This phenomenon is called electrophoresis and is used as a effective tool for the separation of amino acids. In between pK_1 and pK_2, no migration of amino acids is observed because the apparent charge of amino acid in solution is zero, which assumes that the concentration of cations in solution is equal to the concentration of anions:

$$[A^+] = [A^-]$$

$$K_1K_2 = [H^+]^2[A^-]/[A^+]$$

It follows that $[H^+]$ at the isoelectric point $= \sqrt{K_1K_2}$, and in negative logarithmic terms, where pI is pH at the isoelectric point, $pI = pK_1 + pK_2/2$. For neutral amino acids, such as glycine, an amphoteric ion theoretically exists only at the point of pI, but in the broad range of pH between 4.3 and 7.7, no amino acid transfer in electrophoresis observed. This range is said to be the isoelectric region, in which more than 99% of amino acids are dissociated as amphoteric compounds.

For acidic amino acids such as L-aspartic acid, the amino group and carboxyl group on the α-position and an additional carboxylic group on the R-site are dissociated. This compound has three pK values: pK_1 (α-carboxylic group), pK_2 (β-carboxylic group), and pK_3 (α-amino group), named in the order from the acidic region. The isoelectric point is defined as $pI = pK_1 + pK_2/2$.

On the other hand, a basic amino acid such as lysine has an additional amino group on the R-site, corresponding to pK' values that exhibit pK_1 for the α-carboxyl group, pK_2 for the α-amino group, and pK_3 for the ω-amino group, respectively. At neutral pH 7, lysine exists in cation form, when moved toward an alkaline condition, the ratio of cation form gradually decreases and reaches an isoelectric point of $pI = pK_2 + pK3/2$.

The pK_1 values show that the amino acids are considerably stronger acids than acetic acid. Because of intramolecular protonation of the amino moiety by the carboxyl group, the amino acids solutions are weakly acidic.

The differences in acidity and basicits are used in the separation of amino acid mixtures by ion exchange chromatography and electrodialysis. The dissociation constant pK and the isoelectric point pI for various amino acids are summarized in Table 7 (32).

Optical Rotation

Optical rotation is an important characteristic for determining the purity of an amino acid. The value measures changes that occur with changes in temperature, pH, the kind of solvent, and other conditions that affect related optically active substances. For polarimetric measurement of the specific rotation, the sodium D line (wavelength, 589 nm) is used as a light source and expressed as $[\alpha]$D.

The optical rotation for various amino acids is minimum at neutral conditions. For acidic and alkaline conditions, the value of the optical rotation of L-amino acids rotates toward a positive direction. For D-amino acids, conversely, the value rotates toward a negative direction. This rule, called Lutz's law or the Clough–Lutz–Jirgenson law, is used to determine the stereo allocation of an amino acid molecule. The stereo allocation also is attributed to the difference in the ionic form of the amino acid. Therefore, when acid is added, a change in the optical rotation value is observed up to 2 Eq, but above 2 Eq, no change in the optical rotation is observed.

When the temperature is elevated, the dextrorotary value of the optical rotation of L-aspartic acid decreases. At 75°C, the value is 0; conversely, the levorotary value increases with increases in temperature.

Optical rotatory dispersion (ORD), is the phenomenon that changes in optical rotation are observed when the wavelength (λ) is changed. When $[\alpha]$ is plotted as a function of λ^2, it is the so-called rotatory dispersion curve, if it obeys the Drude monominal formula, it is called monorotation dispersion. $[\alpha] = A/\lambda^2 - \lambda_0^2$, where A is the rotation constant, λ is the wavelength of the incident light, and λ_0 is the dispersion constant that corresponds to the wavelength of the nearest optically active absorption band (33).

In general, amino acids show a normal dispersion, but tyrosine shows an anomalous dispersion due to two rotatory contributions of opposite sign, ie, with decreasing wavelength, the rotation is minimal, becomes zero, changes sign, and approaches $+\infty$ (an effect known as the Cotton effect). In the case of tyrosine, good agreement is obtained from three constant Drude equations.

Further methods for investigating the structure of amino acid enantiomers include circular dichroism, which shows differing absorption for left- and right-handed circularly polarized light.

Isoleucine, threonine, and hydroxyproline contain two chiral carbon atoms each; therefore, they appear in four stereoisomeric forms. Cystine, which also contains two chiral carbons, has only three stereoisomers: L-, D-, and *meso*-cystine. The *meso* form has a plane symmetry.

Absorbency Spectra

Aliphatic amino acids exhibit no absorption in the UV region above 220 nm, with the exception of cystine (240 nm). The aromatic amino acids, phenylalanine, tyrosine, and tryptophan, have a UV absorbency above 220 nm. The exact position of the maximum and the molar extinction coefficient ϵ are affected by the pH of the solution.

On the other hand, for infrared spectra, amino acids have common absorbency and specific absorbency. The common absorbencies of infrared spectra for amino acids are in the vicinity of 3070 cm^{-1} for NH_3^+, 1620 cm^{-1} for COOH, and 1560 to 1600 cm^{-1} for COO^-. The common absorbency nearly always shows a similar value irrespective of the state of the amino acid—solid or solution.

Table 7. Dissociation Constant and Isoelectric Point of Amino Acids

Amino acid	pK_1	pK_2	pK_3	pK_4	pI
Alanine	2.34	9.69			6.00
Arginine	1.82	8.99	12.48		10.76
Asparagine	2.02	8.80			5.41
Aspartic acid	1.88	3.65	9.60		2.77
Cysteine	1.92	8.35	10.46		5.07
Cysine	1.04	2.05 (-COOH)	8.20 ($-NH_2$)	10.25	5.02
Glutamic acid	2.19	4.25	9.67	3.22	
Glutamine	2.17	9.13			5.65
Glycine	2.35	9.78			5.97
Histidine	1.78	5.97	8.97		7.59
Isoleucine	2.36	9.68			6.02
Leucine	2.36	9.60			5.98
Lysine	2.20	8.90	10.28		9.78
Methionine	2.13	9.28			5.74
Phenylalanine	2.16	9.18			5.48
Proline	1.95	10.64			6.30
Serine	2.19	9.21			5.68
Threonine	2.15	9.12			6.16
Tryptophan	2.38	9.39			5.89
Tyrosine	2.20	9.11	10.07		5.66
Valine	2.32	9.62			5.96

The common absorbency in acidic solutions is in the vicinity of 3030 cm^{-1} (NH_3^+), 2130 cm^{-1} (NH_3^+), and 1730 cm^{-1} (COOH); in alkaline solutions, it is 3300 cm^{-1} (NH_2), 1610 cm^{-1} (COO^-), and 1400 cm^{-1} (COO^-). These absorbency bands show the ionic form of the amino acid in each solution, ie, in solid or in solution, of the isoelectric point, which is the evidence that the amino acid exists as a dipolar ion. Common absorbencies at 1450 cm^{-1}, 1370 cm^{-1}, and 1330 cm^{-1} are attributed to C-N and C-H and do not vary with change in pH.

Solubility

The solubility in water for different amino acids varies greatly. Cystine is the most sparingly soluble, followed by tyrosine. Aspartic acid, glutamic acid, and leucine belong to the category of comparatively insoluble amino acids. The easily soluble amino acids are proline and hydroxyproline, and other amino acids belong to the relatively soluble category. The optical isomers for each amino acid—L,D, and D,L—demonstrate slight difference in solubility.

In general, the solubility increases when the temperature increases. Acidic amino acids especially show a remarkable dependence on temperature, for example, aspartic acid: 2.09 g/L at 0°C and 68.87 g/L at 100°C.

The solubility is expressed using a function of temperature as follows:

$$\log S = a + bt + ct^2$$

The empirical coefficients a, b, and c for various amino acids are summarized in Table 8 (32,34,35).

The solubility of amino acids varies with the change in pH and is minimal at the isoelectric point. The solubility increases when the isoelectric point is deviated toward alkaline or acid in solution. The change in solubility is governed by the ratio of the ionic form corresponding to the pH in solution. Therefore, the monochloride salts of both neutral amino acids and basic amino acids are more soluble than the free forms of the corresponding amino acids. The sodium salts of amino acids are also easily soluble compared with the free forms of amino acids.

The solubility of an amino acid tends to increase when there are lower levels of inorganic substances present, and it gradually decreases when there are higher levels of inorganic substances present. This tendency is similar to the salt-out effect on the solubility of protein, which is the change in the molecular form caused by the formation of molecular compounds between amino acids and inorganic substances. By varying the concentration of ammonium sulfate, the solubility of cystine increases at low levels of ammonium sulfate and decreases when the ionic strength is increased. When amino acids exist together, the solubility of each amino acid increases.

In organic solvents, almost all amino acids, except amino acid, proline, or hydroxyproline, are sparingly soluble. Proline is soluble up to 1.6 g/dL in ethanol at 20°C.

Table 8. Parameter of Solubility for Various Amino Acids

Amino acid	a	$b \times 10^2$	$c \times 10^5$
Glycine	2.254	0.573	—
L-Alanine	2.1048	0.4669	—
D,L-Alanine	2.0830	0.5608	—
D,L-Valine	1.7749	0.2389	2.607
L-Leucine	1.3561	0.02233	3.727
D,L-Leucine	0.9013	0.2635	4.519
L-Isoleucine	1.5787	0.07682	2.594
D,L-Isoleucine	1.2616	0.2512	3.794
D,L-Serine	1.3432	1.520	−3.548
L-Aspartic acid	0.3194	1.519	—
D,L-Aspartic acid	0.4181	2.016	−5.000
L-Glutamic acid	0.5331	1.613	—
D,L-Glutamic acid	0.9317	1.523	—
L-Cystine	−1.299	1.357	—
L-Proline	3.1050	0.4206	—
L-Phenylalanine	1.2974	0.6982	—
L-Tryptophan	0.9456	0.4834	2.988

$\log S = a + bt + ct^2$ (S, g/L, t, °C).

INDUSTRIAL PRODUCTION

Amino acids have been produced by four methods: (*1*) extraction from protein hydrolysate, (*2*) chemical synthesis, (*3*) fermentation, and (*4*) enzymatic synthesis. The amount produced and the methods used for each amino acid in one year (1997) are summarized in Table 9 (36).

Extraction

For several amino acids (cystine, tyrosine, leucine), isolation from natural raw materials is still an economical pro-

Table 9. Production of Amino Acids (ton/year, 1997)

Amino acid	Method	World
D,L-Alanine	C	1,500
L-Alanine	Enz	500
L-Arginine	Enz, F	1,200
L-Aspartic acid	Enz	7,000
L-Aspargine	Enz	1,200
L-Citrulline	Enz	
L-Cysteine	Ext, Enz	1,500
L-Cystine	Ext, Enz	
Glycine	C	22,000
L-Glutamic acid	F	1,000,000
L-Glutamine	F	1,300
L-Histidine	F	400
L-Isoleucine	F	400
L-Leucine	Ext, F	500
L-Lysine hydrochloride	F	250,000
D,L-Methionine	C	350,000
L-Methionine	Enz	300
L-Ornithine	F	
L-Phenylalanine	C, F, Enz	8,000
L-Proline	F	350
L-Serine	F	200
L-Threonine	C, F	4,000
L-Tryptophan	C, Enz, F	500
L-Tyrosine	Ext, F	120
L-Valine	C, F	500

Note: F, fermentation method; C, chemical method, Enz, enzymatic method, Ext, extraction method.

duction method. The procedure for isolating an amino acid from a protein hydrolyzate consists of passing the hydrolyzate over an ion exchange resin. After elution with aqueous ammonia, aminoacids are collected in fractions.

Chemical Synthetic Production

The D,L forms of alanine, methionine, and tryptophan available for living organisms are chemically produced. Glycine is also produced chemically because there is no asymmetric carbon in its structure. Because of the increasing demand for D,L-cysteine in hairdressing products, the chemical process for production of this amino acid has been developed.

Fermentation

In fermentation, an amino acid–producing bacterium is cultured in a medium that contains sugars and ammonium compounds as carbon and nitrogen sources under aerobic conditions with aeration and agitation. Other growth factors are also added if required. Three groups of bacteria are used in the industrial production of amino acids: (*1*) *Brevibacterium* sp., the glutamic acid bacteria; (*2*) enteric bacteria, *Escherichia coli*; and (*3*) *Bacillus* spp. (37).

The optimal culture condition depends on the nature of the bacteria used and on culture conditions medium composition, temperature, pH, agitation, and aeration.

Mechanism of Production. The metabolic pathway used to produce amino acids is regulated by key enzyme regulatory mechanisms, feedback inhibition by the end product, and repression. In wild-type bacteria, these mechanisms prevent the overproduction of amino acids. A permeability barrier prevents the excretion of amino acids outside the cell wall.

The overproduction of amino acids is achieved by changing the bacterial nature through mutation to release these metabolic regulators and by selecting optimal culture conditions.

Induction by Controlling the Culture Conditions

Glutamic acid-producing bacteria such as *B. lactofermentum* require biotin for growth. When the concentration of biotin is controlled in a limited range of 3 to 5 μg/L, the amount of L-glutamic acid accumulated reaches 80 g/L or more if more than 50% sugar is added). If the culture contains more or less than this biotin level, almost no accumulation of L-glutamic acid is observed.

When a small amount of penicillin is added to a culture with an excess amount of biotin, L-glutamic acid production is induced as a result of the inhibition of peptide glycan synthesis by penicillin in cell walls. When cane molasses is used as a carbon source in cultures in which a sufficient amount of biotin exists, L-glutamic acid is produced by this penicillin method.

Auxotrophic Mutant Strains

Many kinds of amino acids are accumulated by auxotrophic strains that mutated to require a growth factor. L-Lysine is produced by a homoserine auxotrophic mutant of glutamic acid-producing bacteria; *B. lactofermentum* and *Corynebacterium glutamicum* and typical examples. In glutamic acid bacteria, L-lysine is formed by way of the branched pathway shown in Figure 3. The aspartate kinase, which is the first key enzyme of lysine biosynthesis, is inhibited by the combination of L-threonine and L-lysine, and L-lysine is among the amino acids not overproduced. In the homoserine auxotroph, if the supply of L-threonine is limited, L-lysine formed by *de novo* synthesis does not inhibit the first enzyme without the aid of L-threonine. In this mutant, the metabolic flux is genetically blocked to flow in the direction of threonine and methionine biosynthesis. This is one of the main reasons for the overproduction of L-lysine in mutant strains (38).

Regulatory Mutant Strains

When the feedback regulators in the biosynthetic pathway are genetically released, the overproduction of any amino acid is easily attained. These mutants are selected from among those resistant to the growth inhibition of amino acid analogues. *S*-(β-Aminoethyl)-cysteine (a lysine analogue)–resistant mutants of glutamic acid bacteria accumulate a large amount of L-lysine in the medium.

Cell Fusion

When protoplasts prepared from cells with two different genetic markers are mixed in the presence of poly(ethylene glycol) and $CaCl_2$, fused cells with different combinations of parental genetic markers are formed. By this method, a *B. lactofermentum* strain that rapidly produces L-lysine was obtained (39).

Gene Technology

The gene multiplication method with plasmid vectors has been applied to the breeding of amino acid producers. When a plasmid vector that can multiply within an amino acid producer is joined enzymatically to a DNA fragment carrying the genetic information from other strains of the same microorganism or other microorganisms, the resultant recombinant DNA is transferred to the amino acid producer to introduce new markers of the donor microorganisms. A plasmid such as pCG11 has sites sensitive to various restriction endonucleases and is easily joined to a DNA fragment prepared by digestion with corresponding restriction enzymes. Moreover, because it has the genes determining streptomycin resistance, it is easy to select strains with plasmids carrying these antibiotic resistance markers (40).

Enzymatic Process

Enzymatic catalysis is advantageous for cases where amino acids can be produced from relatively inexpensive chemicals with high specificity in a continuously operating process. The enzyme is recycled by binding it to a carrier and carrying out the reaction in a fixed-bed column.

Purification of Amino Acid from the Fermentation Broth

Ion exchange methods are usually used to recover amino acids from the fermentation broth. Because amino acids

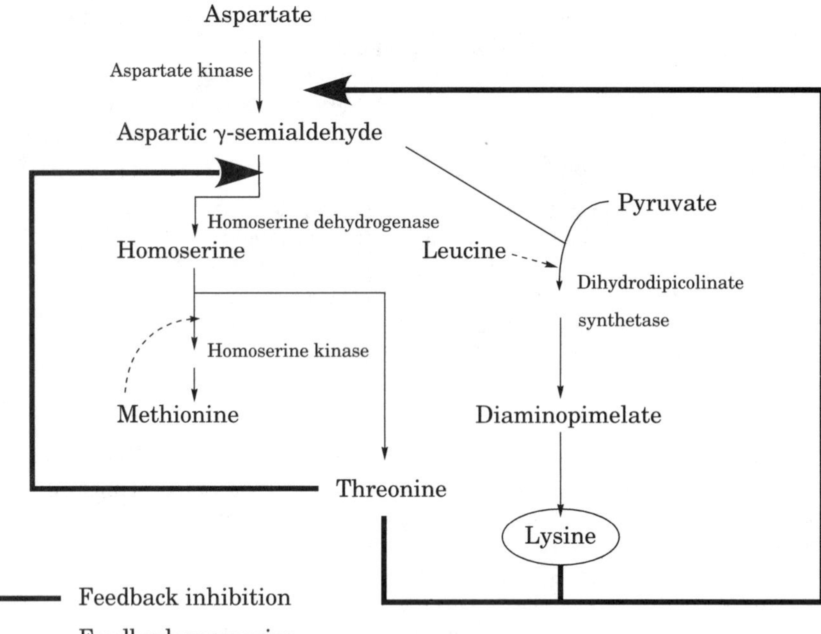

Figure 3. Regulation of lysine production in the metabolic pathway.

have an amphoteric character, when the pH is adjusted lower than pK_1, the amino acids change from zwitterion to cationic form. When a fermentation broth with a pH adjusted lower than pK_1 is passed through the strong cation exchange resin, the amino acid cation being exchanged with a counterion such as H^+ or ammonium ion is removed continuously until the concentration on the ion exchange resin reaches a value corresponding to equilibrium with that in the fermentation broth. The adsorbed amino acids are then eluted by ammonia water or other eluant. If ammonia water is used for the eluant, the counterion form of the ion exchange resin changes to the ammonium form simultaneously, and no generation process is necessary; moreover, almost no inorganic substances are contained in elution with the amino acid. The elution is condensed until the somewhat higher saturated concentration of amino acids is evaporated and crystallized to obtain amino acid crystals.

The crystallization method for an amino acid is selected on the basis of the characteristics of the amino acid and the degree of impurity contained in solution. The solubility minimum at the isoelectric point is useful for crystallizing amino acids. L-Aspartic acid and L-tyrosine, which show comparatively lower solubility, for example, are first dissolved in alkali or acid solution and then crystallized by neutralization to pI. On the other hand, L-lysine hydrochloride and L-glutamate, which show higher solubility, crystallize by the condensed crystallization method.

L-Leucine, L-isoleucine, and L-valine, branched-chain amino acids from which higher purity crystals are difficult to obtain, react with benzoic sulfonic acid derivatives to form corresponding complexes.

For pharmaceutical use, it is critical to prevent contamination by pyrogen substances and microorganisms. Pyrogen is a toxin consisting of polysaccharides and protein. It is excreted mainly by Gram-negative bacilli, and it is highly hydrophobic. The pyrogens are effectively adsorbed on activated carbon. Sterilization by heating and ultrafiltration membranes (molecular weight cutoff, 10,000 to 100,000) are used as well for rejection of pyrogens and microorganisms.

SAFETY

Physiological Effect of Imbalance

Excess amino acids are rapidly disposed of by increased metabolic degradation and are used to provide energy. The nitrogen is eliminated as urea.

When the component amino acids are imbalanced, alternation of the ribosome profile occurs in the liver, and ribonucleic acids are catabolized. Protein deficiency is usually coupled with calorie deficiency. The manifestation of chronic protein deficiency is known as marasmus and kwashiorkor. On the biochemical level, marasmus and kwaskiorkor result in a negative balance, indicating a reduction in protein inventory. The labile enzyme and plasma proteins are consumed, the greatest losses occurring first in the liver and then in the musculature. Brain, heart, and kidneys suffer minimal protein loss. The absence of a essential amino acid in the diet is more serious than protein deficiency because protein synthesis can then occur only by degradation of body protein.

Amino Acid Toxicity

When an excess amount of one or more amino acids is administered, some adaptive failure can lead to accumulation of the amino acids or certain metabolites in the organism, leading to anatomic or functional damage. The acute oral toxic levels have been studied by adding increasing quantities of individual amino acids to a protein diet (41)

Table 10. Acute Oral Toxicity of Amino Acids in Rats

Amino Acids	LD_{50} (g/kg weight of rat)
L-Alanine	>16
L-Arginine	~16
L-Asparagine	>16
L-Cysteine HCl	3.1 (2.7–3.6)
L-Cystine	11.2 (9.0–14.0)
L-glutamine	>16
Glycine	~16
L-Histidine	>16
L-Hydroxyproline	>16
L-Isoleucine	>16
L-Leucine	>16
L-Lysine HCl	10.6 (9.5–11.9)
L-Methionine	>16
L-Phenylalanine	~16
L-Proline	>16
L-Serine	14.0 (12.8–15.3)
L-Threonine	>16
L-Tryptophan	>16
L-Tyrosine	>16
L-Valine	>16
NaCl	3.75

(Table 10). The toxicity of individual amino acids depends on the total protein consumption.

The consumption of toxic amounts of amino acids increases their concentration in the plasma and brain. Because of the blood-brain barrier, however, the increase in the brain is not great (42).

BIBLIOGRAPHY

1. I. Chihata and K. Kawashima, in A. Yoshida et al., eds., *Nutrition: Proteins and Amino Acids*, Japan Scientific Societies Press, Springer-Verlag, New York, 1990, pp. 273–284.
2. H. W. Brown, B. G. Harmon, and A. H. Jansen, *J. Anim. Sci.* **37**, 708 (1973).
3. FAO Committee on Protein Requirements, *FAO Nutritional Studies* No. 16, FAO, Washington, D.C., 1957.
4. E. E. Howe, G. R. Jansen, and C. W. Gelfillan, *Am. J. Clin. Nutr.* **16**, 315 (1965).
5. E. E. Howe, G. R. Jansen, and M. L. Anson, *Am. J. Clin. Nutr.* **20**, 1134 (1967).
6. M. Winitz et al., *Nature* **205**, 741 (1965).
7. M. Winitz, D. A. Seedman, and J. Graff, *Am. J. Clin. Nutr.* **23**, 25 (1970).
8. M. Winitz et al., *Am. J. Clin. Nutr.* **23**, 546 (1970).
9. S. J. Fomon, *Am. J. Clin. Nutr.* **32**, 2460 (1979).
10. Y. Kobayashi, Y. Doi, and T. Takanami, in A. Yoshida et al., eds., *Nutrition: Proteins and Amino Acids*, Japan Scientific Societies Press, Springer-Verlag, New York, 1990, pp. 285–299.
11. Y. Komata, *Science of "Oishisa" and Taste* (in Japanese), Nippon Kougyou Shinbun Co., Ltd.
12. C. P. Berg, *Physiol. Rev.* **33**, 145 (1953).
13. S. Akabori, *Nippon Seikagaku Kaishi* **14**, 185 (1939).
14. T. Kanekò, R. Yoshida, and H. Katsura, *Nippon Kagaku Kaishi* **80**, 316 (1959).
15. U. K. Lee, T. Kaneko, *Bull. Chem. Soc. Jpn.* **46**, 3494 (1973).
16. W. J. Hertz and R. S. Shallenberger, *Food Res.* **25**, 491 (1960).
17. T. A. Rohan, *Food Technol.* **24**, 29 (1970).
18. R. Elman and D. O. Weiner, *J. Amer. Med. Assoc.* **112**, 795–802 (1939).
19. W. C. Rose, *Fed. Proc.* **8**, 546–552 (1949).
20. FAO/WHO Expert Group, *Protein Requirements*, WHO Technical Report Series No. 301, World Health Organization Geneva, Switzerland, 1965.
21. M. M. Meguid et al., in G. L. Blackburn, J. P. Grant, and V. R. Young, eds., *Amino acids, Metabolism and Medical Applications*, John Wright PSG Inc., 1983, pp. 147–154.
22. M. L. Anson, *Advances in Protein Chemistry* Vol. 4, 1948, p. 339.
23. J. Wyman, Jr., and T. L. McMeekin, *J. Am. Chem. Soc.* **55**, 908 (1933).
24. J. Wyman, Jr., *Chem. Revs.* **19**, 213 (1936).
25. K. J. Dunning and W. J. Shutt, *Trans. Faraday Soc.* **34**, 479 (1938).
26. T. Edsall, J. W. Otvos, and A. Rich, *J. Am. Chem. Soc.* **72**, 474 (1950).
27. F. Garfinkel and J. T. Edsall, *J. Am. Chem. Soc.* **80**, 3807 (1958).
28. E. Takeda et al., *J. Am. Chem. Soc.* **80**, 3813 (1958).
29. F. Garfinkel and J. T. Edsall, *J. Am. Chem. Soc.* **80**, 3818 (1958).
30. F. Garfinkel and J. T. Edsall, *J. Am. Chem. Soc.* **80**, 3823 (1958).
31. F. Garfinkel, *J. Am. Chem. Soc.* **80**, 3827 (1958).
32. J. P. Greenstein and M. Winitz, eds., *Chemistry of the Amino Acids*, Vol. 1, John Wiley & Sons, Inc., New York, 1966.
33. E. Brand, *J. Am. Chem. Soc.* **76**, 5037 (1954).
34. W. M. Hoskins, M. Randall, and C. L. A. Schmidt, *J. Biol. Chem.* **88**, 215 (1930).
35. M. S. Dunn, F. J. Ross, and L. S. Read, *J. Biol. Chem.* **103**, 579 (1933).
36. *Amino Acids* Japan Essential Amino Acids Association, Inc., 1997.
37. S. Kinoshita, in A. L. Demain and N. A. Solomon, eds., *Biology of Industrial Microorganisms*, Benjamin/Commings Publishing Co., Inc., London, 1985, pp. 115–142.
38. K. Nakayama, in G. Reed, ed., *Prescott & Dunn's Industrial Microbiology*, 4th ed., AVI Publishing Co., Inc., Westport, Conn., 1982, pp. 748–801.
39. O. Tosaka and K. Takinami, K. Aida et al., eds., *Biotechnology of Amino Acid Production*, Kodansha, Ltd., Tokyo, 1986, pp. 152–172.
40. R. Katsumata et al., in O. M. Neijsel, R. R. van der Meer, and K. Ch. A. M. Layben, eds., *Proc. 4th European Congr. Biotechnol.* Vol. 4, *Proceedings of the 4th European Congress on Biotechnology*, Elsevier, Amsterdam, 1987, pp. 767–776.
41. *Acute Oral Toxicity of Rat to Twenty Five Amino Acids*, Huntingdon Research Center, 1971.
42. Y. Peng, J. K. Tews, and A. E. Herper, *Am. J. Pysiol.* **222**, 314–321 (1972).

GENERAL REFERENCES

G. L. Blackburn, J. P. Grant, and V. R. Young, eds., *Amino Acids, Metabolism and Medical Applications*, John Wright PSG, Inc., 1983.

J. P. Greenstein and M. Winitz, eds., *Chemistry of the Amino Acids*, Vol. 1, John Wiley & Sons, Inc., New York, 1966.

T. Kaneko et al., eds., *Amino Acid Industry*, Kodansha, Japan, 1973.

J. I. Kroschwitz and M. H. Grant, eds., *Encyclopedia of Chemical Technology*, 4th ed., Vol. 2, John Wiley & Sons, Inc., New York, 1992.

A. Yoshida et al., eds., *Nutrition: Proteins and Amino Acids*, Springer-Verlag, New York, 1990.

TETSUYA KAWAKITA
Ajinomoto Company, Inc.
Kawasaki, Japan

PROTEINS: DENATURATION AND FOOD PROCESSING

Food proteins constitute a wide range of protein types (eg, structural, storage, catalytic) and essentially include all those that are palatable, digestible, nontoxic, and accessible to humans. Although contributing to the nutritional value of foods by supplying a dietary source of essential and nonessential amino acids, proteins also serve as integral food components or ingredients by virtue of their diverse technological or functional properties (Table 1). These properties affect the behavior of the protein in a given food system during preparation, processing, or storage as judged by the quality attributes of the final product (1–3). They reflect a variety of interactions between physicochemical properties of the proteins per se (eg, molecular weight, amphiphilicity, structural flexibility), other food components (eg, water, carbohydrates, lipids, vitamins, minerals), and the environment in which these are associated. A variety of commodities and by-products may serve as the source of food proteins, yet their inclusion in the human diet usually necessitates some form of processing and/or preparation because exploitation of technological attributes normally requires the destruction and/or modification of the native protein structure (ie, denaturation). Processing affects not only each individual food component, but also the nature and intensity of interactions between components. Because many processing treatments can result in undesirable changes in the functional properties of food proteins and, therefore, contribute to poor final quality of protein-containing food products, an understanding of the relationship between processing factors and the changes in proteins that they induce is imperative. This article discusses protein stability and folding, protein denaturation, factors that influence it, and specific examples of food proteins denaturated during processing. Thorough discussions of various aspects of protein denaturation can also be found in References 4 to 10.

Table 1. Functional Properties of Proteins in Foods

General classification	Functional property
Organoleptic	Color, odor, flavor, texture, turbidity
Hydration	Solubility, dispersibility, wettability, water-holding capacity, water absorption
Surface	Emulsification, emulsion stability, foamability, whippability, film formation, fat absorption, binding properties
Thermal	Thermocoagulability, heat stability
Rheological	Elasticity, viscosity, gelation, cohesion, extrudability, adhesion, coagulation, aggregation, hardness, chewiness
Other	Compatability with additives, enzyme activity, susceptibility to modification

NATIVE PROTEIN STRUCTURE AND STABILITY

The functional properties of food proteins ultimately arise from their unique molecular conformations, which derive from all four levels of structural hierarchy (ie, primary, secondary, tertiary, quaternary) (11). During synthesis on the ribosome, the polypeptide chain, pending subsequent post-translational processing (see section on protein folding, molten globules), adopts a unique molecular conformation traditionally referred to as the native conformation. This particular structure is dictated by the nature and sequence of amino acids in the polypeptide chain and is greatly influenced by solvent effects (12). The folding and resultant (native) conformation of a protein is largely governed by (equilibrium) thermodynamics. The third law of thermodynamics states that

$$\Delta G = \Delta H - T\Delta S \quad (1)$$

where ΔG is the change in Gibbs free energy, ΔH is the change in enthalpy, T is absolute temperature (Kelvin), and ΔS is the change in entropy. To obey this law, the polypeptide chain must fold into a conformation such that the least amount of free energy is expended in maintaining it. In terms of free energy, hydrophobic effects contribute most to protein folding and stabilization. Hydrophobic interactions are nonspecific and result from the strong hydrogen-bonding properties of water. Nonpolar amino acid residues generally cannot participate in hydrogen-bonding; thus, water molecules surrounding nonpolar residues hydrogen bond with each other to form a highly ordered icelike structure that is associated with unfavorable (ie, low) entropy. The native polypeptide chain, therefore, tends to bury its hydrophobic residues within the interior of the molecule with exclusion of water from this core and to orient its hydrophilic amino acids toward the protein exterior. Exclusion of water from the protein interior causes a large increase in the entropy of the previously structured surrounding water, while the close packing of residues in the protein interior generally leads to decreased enthalpy. The folding process may thus be regarded as an entropy-driven transition from a state of higher free energy (eg, random coil) to that of lower free energy (eg, native conformation). Although this transition is largely governed by hydrophobic effects, the particular folding patterns and final protein conformation(s) are governed by formation of (specific) hydrogen bonds, disulfide linkages, and electrostatic and van der Waals interactions within the protein. Maximization of these strongly interacting driving forces leads to the native conformation of proteins. The adopted conformation may not correspond to the global minimum of free energy because this structure may represent a metastable state (ie, local minimum). Therefore, in addition to thermodynamic constraints, protein folding and conformational stability are also governed by kinetic constraints.

The thermodynamic stability of the native protein is marginal, generally not exceeding 60 kJ/mol, or the equivalent strength of only 3 to 4 hydrogen bonds or a single electrostatic interaction (Table 2) (7,8). This lability suggests that such environmental and processing factors as pH, temperature, pressure, and solvent effects may readily alter protein conformation and, consequently, functional properties and product quality to varying degrees. The ability to predict and/or estimate protein stability and conformational potential ("the capacity of biopolymers to form intermolecular junction zones that generate the desired structural rheological and other physico-chemical properties of a given food system," [13]) is, therefore, of vital importance in protein isolation and in processing of protein-containing foodstuffs (9). For example, during isolation of food proteins (especially enzymes), it is generally desirable to avoid or to at least minimize disruption of native structure because native proteins possess higher conformational potential and, as a consequence, superior functional properties. During processing, knowledge of protein stability and of the extent of conformational (and functional) change resulting from manipulation of processing factors is advantageous and often necessary if high-quality protein-containing food products are to be produced and maintained throughout subsequent storage.

Protein Folding: An Overview

Prefolding State: The Molten Globule. In order for a protein to achieve a three-dimensional native state, it must undergo a large number of possible conformations rapidly, prior to achieving its final product. A problem for biochemists is understanding how a protein achieves its native state (N) or becomes folded. Anfinsen (14) discovered that a protein's amino acid sequence fully determines its three-dimensional structure and, thus, its biological function. Work from the late 1970s and early 1980s gave rise to the concept that the protein native (N) and unfolded states (U) were an all-or-none scenario ($U \leftrightarrow N$) (15,16). Yet, although this theory was generally accepted, strong experimental evidence of proteins treated with high guanidine hydrochloride (GdnCl) concentrations showed large differences in hydrodynamic and optical parameters when compared to little changes in the presence of high temperatures and/or low pH (plus addition of salts) (17). Evidence mounted for the existence of an intermediate equilibrium between the native and unfolded states. It was shown that temperature-denatured proteins are far from being completely unfolded and can undergo another cooperative transition when GdnCl or urea is added (18,19). Circular dichroism (CD) studies showed the existence of one or more equilibrium states that differed from the native state by the absence of the rigid aromatic side chains; at the same time, they differed from the unfolded states by the presence of secondary structure (18). The conclusion was made that these intermediates were unfolded, noncompact molecules with local secondary structure (20). These studies showed a physical state that existed between native and unfolded and was given the term *molten globule* (Fig. 1) (22).

Molten globule (MG), as a term, has been applied to all compact denatured states that have substantial secondary structure, but little or no tertiary structure. This can cover a wide range of ordered and disordered, partially folded proteins with and without disulfide bonds (23).

The molten globule conformation is frequently adopted rapidly where an unfolded protein is placed under refolding conditions before the appearance of the fully folded protein. Because of the prefolded state, occurrence of the molten globule (24) was thought to be an indicator of a kinetic intermediate and assigned the kinetic role of

$$U \leftrightarrow MG \leftrightarrow N$$

This equation would imply that the molten globule is required for the formation of the native state. From this equation, the initial rate of forming N should be zero upon

Table 2. Linkages and Interactions Involved in Stabilizing Protein Conformation

	Energy (kJ/mol)	Enthalpy contribution	Entropy contribution	Disrupting agents
Covalent bonding (–S–S–)	330–380	–	–	Reducing agents
Electrostatic interactions	42–84	+ or –	+	Salts, high or low pH
Hydrogen bonding	8–30	–	–	Urea, detergents, heat
Hydrophobic interactions	4–12	+	+	Detergents, organic solvents, urea, guanidine-HCl, cooling
van der Waals forces	1–9	–	–	

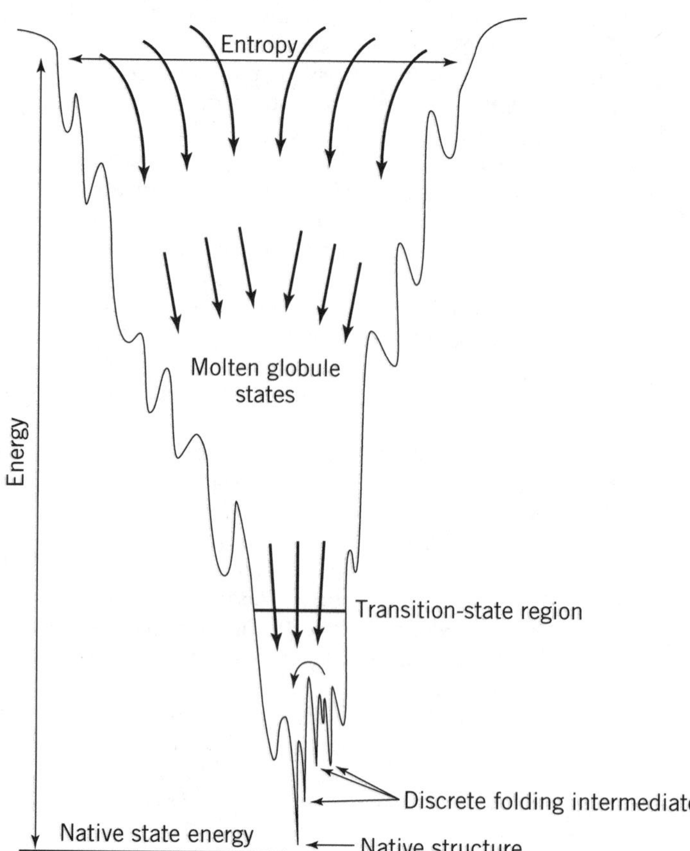

Figure 1. Molten globule state. *Source:* Ref. 21.

placement of the unfolded protein into refolding conditions. Only after the required molten globule intermediate has been generated should N appear. However, experimental data (25) have shown that under refolding conditions the equation should be rewritten to be

$$\begin{array}{rc} & \mathrm{U} \leftrightarrow \mathrm{N} \\ (\mathrm{Fast}) & \updownarrow \\ & \mathrm{MG} \end{array}$$

The molten globule state in itself does not produce rapid folding, but has been speculated to be an off-pathway, nonproductive species that is the energetically preferred form of the unfolded protein under refolding conditions (25). Research has shown that the primary role for the molten globule state is to only slightly increase the overall rate of disulfide formation and to favor those interactions between cysteine residues distant in the polypeptide chain (26).

Physiological Role of the Molten Globule: Molecular Chaperones and Foldases. The physiological role of the molten globule has been speculated greatly during the past few years. Because of the properties that define the molten globule, proteins *in vivo* must adjust themselves to a large set of different conditions, eg, in the cytoplasm and/or near membranes where a number of these conditions are denaturing ones, ie, low pH or differing salt concentrations. The molten globule state permits proteins to exist in this state of flux, which allows for more pronounced small-scale fluctuations than the native state, which protects it from occasional loss of folding pattern by large-scale thermal fluctuations (27). The molten globule state and the newer, more expanded molten globule, the premolten globule (28), may be important for a class of proteins, chaperones, to trap proteins after their biosynthesis for self-assembly, transmembrane transport, and other processes that use protein molecules in a semiflexible (rather than in a rigid) state (27).

Molecular chaperones have been identified as a family of proteins that bind to and assist in the folding of proteins into their functional states. They do not form part of the final protein structure, nor do they possess steric information specifying a particular folding or assembly pathway (29). By recognizing unfolded or partially denatured proteins, the predominant role of chaperones seems to consist in the prevention of incorrect intra-and intermolecular associations of polypeptide chains that would result in their aggregation (30).

Consistent with their role in the folding of newly synthesized translocated proteins, many molecular chaperones are constitutively expressed. Under conditions that compromise protein folding and cell physiology, eg, heat shock, the synthesis of most molecular chaperones is induced to higher levels. It is then not surprising that many of the molecular chaperones were first identified in one or more organisms as heat shock proteins (HSPs) (31). However, it is now recognized that the same closely related proteins are frequently essential components of normal cells (30).

Chaperones themselves have aids or accessory proteins called co-chaperones that are responsible for mediating the activity of specific chaperones (32). They were first identified in *Escherichia coli*, and some have been shown to stimulate the rate of ATP hydrolysis (DnaJ) or act as a nucleotide exchange factor (as in the case for DnaK) (33).

The final group of protein folding helpers includes the array of proteins that act as foldases. Foldases include enzymes such as protein disulfide isomerases (PDLs) (34) and the immunophiles or peptidylprolyl isomerases (PPIs or rotamase) (35). These proteins have demonstrated catalytic activities that increase the rate of protein folding (36,37).

To perform its biological function, a protein cannot remain as a linear string of amino acids that has just come off of the protein-making ribosome. It must fold into its three-dimensional native state. The molten globule state is widely considered to be an important intermediate in protein folding and to have a polypeptide backbone with a nativelike topology. The importance of the molten globule is not in rapid protein folding, but for aiding in the trapping of the protein by molecular chaperones that assist in protein folding and targeting. The notion that chaperones are needed to assist protein folding is an interesting concept that extends rather than negates Anfinsen's findings (14) that proteins fold spontaneously based on their sequence. Although Anfinsen's findings were based on *in vitro* experiments, the chaperone (GroEL, in this example) appears to form a large central cavity inside its structure that allows the molten globule state protein the ability to self-assemble within the confines of the chaperone environment (38). The role of the chaperone (GroEL) is believed to be through assisting in the repeated binding and releasing of unfolded or partially folded polypeptide. During each binding interval, the chaperone sequesters the polypeptide and prevents formation of nonnative conformations. Eventually, when the protein has achieved its native state, it dissociates from the chaperone to perform its function(s).

When proteins are not in their native conformation, they suffer a loss in functionality. This loss of functionality is an important aspect not only in food processes but in nature.

PROTEIN DENATURATION

Protein denaturation has traditionally been defined as any modification of conformation that is not accompanied by cleavage of peptide bonds involved in primary structure. This definition is interpreted differently by various researchers and disciplines owing to a general inability to recognize the phenomenon when it occurs (5,6,10). Many problems in recognizing denaturation of food proteins are due to the fact that they are rarely pure entities, often precluding direct measurement of changes in their conformation; rather, they are usually a heterogeneous mixture in isolates and in food. In addition, a conformational change may not be sufficient to effect a detectable or significant change in functional properties of a food protein. However, a significant change in functional properties is usually the result of structural alteration. Thus, from a food science perspective, protein denaturation may best be defined operationally as any modification of conformation not accompanied by alteration of primary structure that results in a change in one or more of the functional properties of the protein. Certainly, direct measurement of changes in protein conformation is preferable; however, measurement of functional properties (Table 1) as a function of denaturing conditions may provide a meaningful assessment of food protein denaturation. Denaturation may also be assessed by measuring other properties of proteins (Table 3). Because the transition from the native (N) to denatured (D) state of a protein is accompanied by a change in energy, manifested by absorption or liberation of heat (ie, enthalpy), one of the most useful and direct methods of measuring parameters associated with food protein denaturation is differential scanning calorimetry (DSC) (6). A comprehensive review of thermal analysis of proteins is given by Harwalker and Ma (39). The primary advantage of DSC is that from thermograms generated, the following parameters can be obtained directly: heat capacity (C_p) of N and D states, change in heat capacity (ΔC_p) associated with the N → D transition, enthalpy change (ΔH) associated with the transition, and the temperature of denaturation (T_d) (Fig. 2).

Denaturation may ultimately lead to a completely unfolded polypeptide structure (ie, random coil), to a more compact globular structure, or to any of a number of intermediary and often short-lived conformations (eg, molten globule). Despite the fact that the native conformation of various (purified) proteins have been characterized by X-

Table 3. Properties of Proteins to Assess Denaturation

Functional
Thermodynamic
Denaturation temperature
Enthalphy
Heat capacity
Hydrodynamic
Sedimentation behavior
Electrophoretic
Surface charge
Spectroscopic
Ultraviolet, visible absorption
Intrinsic fluorescence
Extrinsic fluorescence
Circular dichroism
Optical rotary dispersion
Light scattering
Infrared
Nuclear magnetic resonance
Electron microscopy
Biological
Catalytic/enzymatic
Immunological
Digestibility by proteases
Chemical reactivity of functional groups

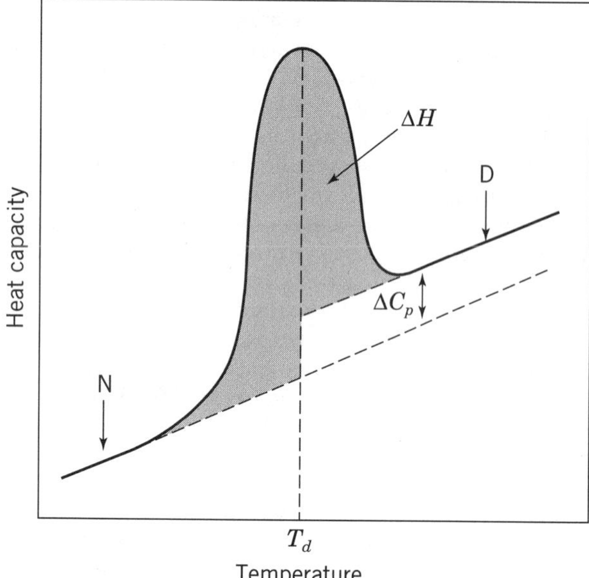

Figure 2. Parameters obtained from differential scanning calorimetry. N, Native state; ΔH, change in enthalpy; ΔC_p, change in specific heat capacity; T_d, denaturation temperature; D, denatured state.

ray crystallography and two-dimensional nuclear magnetic resonance, protein structures resulting from partial or complete denaturation are generally not well defined. The susceptibility of a protein to denaturation is dependent on the ability of the denaturant to break the bonds and/or interactions that stabilize the protein structure (Table 2). This requires energy input, ie, chemical or physical (Fig. 3). The relatively large values of denaturation constants in Table 4 (40) reflect the nature of denaturation (ie, N → D transition) in that many noncovalent interactions must be broken. Because every protein possesses a characteristic or unique structure, the effects of a denaturant are protein specific. For example, the denatured state of a protein may be transitional (eg, during thermal processing), after which it renatures with full recovery of its native (functional) properties (eg, many peroxidases). Conversely, the denatured state may be formed irreversibly and correspond to a new or technological state with predictable/

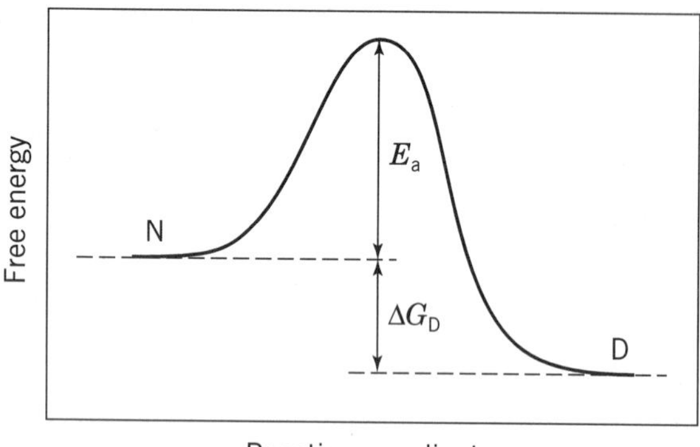

Figure 3. Energy output. N, Native state; ΔG_D, Gibbs free energy; E_a, activation energy; D, denatured state.

Table 4. Denaturation Constants for the Transition State of Various Food and Food-Related Proteins

Protein	°rH° (kJ/mol)	ΔS° (J/mol/°C)
Pepsin	232.8	474.4
Lipase (pancreatic)	190.1	285.5
Amylase (malt)	174.2	219.0
Trypsin	168.3	187.1
Chymosin	373.9	870.9
Hemoglobin	316.5	639.3
Egg albumin	552.7	1,321.8
Peroxidase (milk)	775.8	1,951.0
Invertase (yeast)		
pH 3.0	311.5	638.1
pH 4.0	462.2	1,099.0
pH 5.2	361.7	774.6
pH 5.7	219.4	354.6

Source: Ref. 40.

desirable functional properties that differ from those of the native protein (eg, texturized plant proteins). Denaturation is generally reversible if the denatured protein (ie, unfolded polypeptide chain) is thermodynamically stabilized by the denaturant; removal of the denaturant allows the native conformation of the protein to be reestablished. In addition, small molecular weight proteins are more likely to renature than large molecular weight proteins. Denaturation is irreversible if the unfolded polypeptide chain is stabilized by interactions with other denatured proteins, as in acid or heat-induced protein aggregation. If disulfide bonds contribute to protein conformation and these are broken, denaturation is often irreversible.

Protein denaturation is generally a cooperative multistate process (4) with many more-or-less unfolded intermediates (D_i) between the N and D states:

$$N \leftrightarrow D_1 \leftrightarrow D_2 \leftrightarrow D_3 \; . \; . \; . \; \leftrightarrow D_n$$

These intermediate structures correspond to progressive stages in the alteration of the protein conformation. A protein does not have a rigid or static conformation but, rather, a dynamic conformation, undergoing rapid thermodynamic fluctuations about a mean conformation. Thus, each of the states in the N → D_i transition may be regarded as a population of structures of similar energy. Their existence or identification as discrete states requires that their populations not overlap.

Thermodynamic analysis of protein denaturation and/or conformational stability according to the foregoing scheme is complicated. Although often an oversimplification, it is useful to consider the denaturation process in terms of a two-state equilibrium (Fig. 4) (4,6–8). Use of a two-state model is valid if the DSC enthalpy change (ΔH, Fig. 2) approximates the van't Hoff enthalpy change. The equilibrium or denaturation rate constant for a two-state equilibrium can be written

$$k_D = \frac{(1 - f_N)}{f_N} = \frac{[D]}{[N]} \qquad (2)$$

where k_D is the equilibrium constant and f_N is the fraction of protein molecules in the native state. Reversible and

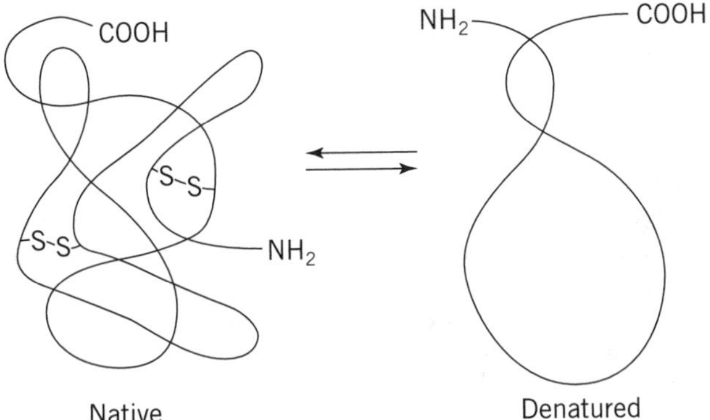

Figure 4. Two-state equilibrium model of the denaturation process.

irreversible denaturation are both initiated by reversible conformational change, eg, by unfolding. This initial conformational change is rate-limiting (41) and, in the case of irreversible denaturation, is followed by such secondary processes as aggregation or covalent modification (eg, proteolysis) resulting in a change in primary structure. If irreversible denaturation takes place, analysis of protein denaturation using the two-state model is precluded, because the two-state equilibrium is accompanied by an irreversible transition from the D state to an alternate structure(s) that cannot be readily characterized. Analysis of protein denaturation using the two-state model, therefore, requires that three criteria be met: (*1*) the N → D transition must be thermodynamically reversible; (*2*) an experimental technique must be used such that the transition be detected in the presence of the perturbing influence or denaturant; and (*3*) a reference conformational state, usually taken to be the N state, must be identified such that the stabilities of different proteins are compared. The limiting case for thermodynamic reversibility is an infinitely dilute solution. However, if thermodynamic reversibility can be assumed for the protein system in question, good approximation of the thermodynamic parameters ΔG°, ΔH°, ΔS°, and ΔC_p° can normally be obtained from

$$\Delta G^\circ_{app} = -RT \ln k_D \tag{3}$$

$$\Delta H^\circ_{app} = -R \frac{d(\ln k_D)}{d(1/T)} \tag{4}$$

$$\Delta C^\circ_{Papp} = \frac{d(\Delta H^\circ_{app})}{d(T)} = T \frac{d(\Delta S^\circ_{app})}{d(T)} \tag{5}$$

where ΔG°_{app} is the apparent change in standard free energy, ΔH°_{app} is the apparent change in enthalpy, ΔS°_{app} is the apparent change in entropy, ΔC°_{Papp} is the apparent change in heat capacity, k_D is the equilibrium or denaturation rate constant, R is the gas constant, and T is absolute temperature (Kelvin). Protein denaturation is normally accompanied by increases in (*1*) enthalpy, indicating that the N state possesses a lower free energy than the D state; (*2*) entropy, associated with the disorder that results from unfolding of the polypeptide chain; and (*3*) heat capacity, related to the transfer of apolar or hydrophobic groups to the aqueous environment during denaturation, with associated structuring of water. The stability of the N state relative to the D state, or the intrinsic conformational stability of the protein, is represented by ΔG_{app} (ΔG_D) (Fig. 3). A meaningful comparison of the stabilities of different proteins requires a knowledge of ΔG° and the derived thermodynamic quantities ΔH°, ΔS° and ΔC_p° (4). These quantities are normally obtained from temperature, pressure, or concentration dependencies of ΔG_{app} (or k_D).

FACTORS AFFECTING PROTEIN DENATURATION

As previously stated, many food processing operations to which protein-containing foods are subjected, can result in protein denaturation. Such factors as temperature, pH, pressure, shear, irradiation, and the presence of salts and oxidizing/reducing agents can be used by the processor to produce desirable and high-quality products. However, these factors may also serve as processing hazards that must be avoided in order to maintain native protein structure and associated functional properties. In the broadest sense, processes that result in protein denaturation can be divided into two major categories: physical and chemical.

Physical Denaturants

Heat. Application of thermal energy is the most common means by which food is processed. Among the major components in food, proteins may be the most sensitive to temperature extremes. Thermal denaturation of proteins may be detrimental to product quality, as reflected in their reduced functional properties (Table 1), or desirable, as in the heat processing of whey for use in confections. The denaturing effects of increased temperature are dependent on many factors, including protein type and concentration, water activity, pH, ionic strength, and the nature of ions present. For most chemical reactions, rate increases approximately twofold for each 10°C rise in temperature. However, the rate of protein denaturation may increase 600-fold. This large difference in reaction rate is attributable to the low-energy bonds/interactions stabilizing protein conformation (Table 2). The dependence of the rate of protein denaturation on temperature can be determined from (first-order) thermal denaturation curves derived using any of a number of different methods to monitor changes in protein conformation (Table 3). These curves provide information needed to calculate such kinetic/processing parameters as activation energy and z values. These values are related by

$$E_a = \frac{2.303RTT_1}{z} \tag{6}$$

where E_a is the activation energy (kJ/mol), R is the universal gas constant, T is temperature (Kelvin), T_1 is the temperature 10°C above T (Kelvin), $1/z$ is the slope of the denaturation curve, and z is the temperature change (°C) required to change the thermal denaturation rate by a factor of 10 (42,43). The energy required to denature the protein is represented by activation energy (Fig. 3, Table 4), which can be calculated from the Arrhenius law

$$E_a = RT^2 \frac{d(\ln k_D)}{d(T)} \quad (7)$$

where k_D is the equilibrium or denaturation rate constant and T is absolute temperature (Kelvin). The values for transition-state enthalpy (ΔH) of various proteins listed in Table 4 are related to E_a by equation 8:

$$E_a = \Delta H + RT \quad (8)$$

Values of E_a depend on the extent of denaturation, ie, the nature of the D state. For example, the E_a values associated with the thermal denaturation of proteins, from the N state to a completely unfolded state (eg, random coil), are large relative to other chemical reactions. Although covalent bonds other than disulfide cross-links are not broken during denaturation, a large number of low-energy noncovalent bonds and interactions are broken. The E_a values associated with thermal denaturation (ie, inactivation) of enzymes are relatively low, however, because the active sites of most enzymes are dependent on only a few low-energy bonds and/or interactions. Enzyme inactivation may be a direct consequence of thermally induced modification of protein conformation; however, the effects of temperature on substrate(s), activators, and inhibitors must also be considered.

Thermal denaturation of food and food-related proteins generally takes place between 55 and 80°C; enzymes tend to be more sensitive to the effects of heat and may begin to denature at a temperature as low as 45°C. Decreased solubility may accompany protein denaturation as heat-induced exposure of hydrophobic groups to solvent (ie, water) may result in the aggregation of unfolded protein molecules (7). A reduction or loss of biological activity and increased water absorption, susceptibility to protease digestion, and intrinsic viscosity may also accompany (thermal) denaturation. In addition, heat-mediated chemical alteration of proteins and their constituent amino acids can result (5,10). For example, dehydrogenation of serine, deamidation of glutamine and asparagine, and formation of new intra- and/or intermolecular covalent cross-links (eg, γ-glutamyl-ϵ-N-lysine) can significantly decrease the nutritional quality of proteins.

Cold. At some stage between postharvest/postslaughter and consumption, many foods are subjected to refrigeration and/or freezing so that product quality be maintained. However, just as the application of thermal energy can result in denaturation of proteins in foods, so too can the removal of thermal energy. Low-temperature denaturation of proteins is largely mediated by a reduction in hydrophobic interactions in conjunction with enhanced hydrogen bonding (Table 2). Thus, low temperatures can lead to aggregation and precipitation of proteins and/or to alteration of quaternary structure. For example, cold inactivation (ie, 10°C) of the glycolytic enzyme phosphofructokinase occurs by dissociation of the tetramer to two dimers as a result of weakened hydrophobic interactions between subunits.

The principal injurious effect of freezing is not low temperature per se, but the concomitant concentration of all soluble species as pure ice separates from the mixture. Concentration of acids and/or salts, resulting in large changes in pH and ionic strength, can have profound effects on proteins. At high subzero temperatures, the extent of protein denaturation is greatest, whereas at or below the eutectic temperature of the food system, minimal damage occurs (4). Fish proteins are particularly susceptible to destabilization by freezing temperatures. As a result, fish may become tough and exhibit excessive drip loss on thawing. Similarly, caseinate micelles of milk, which are relatively stable to heat, may be destabilized and coagulate during freezing. Not all proteins, however, are sensitive to freezing temperatures. In fact, several lipases and oxidases are resistant to freezing and remain active at subzero temperatures. In order to inactivate (ie, denature) these enzymes, foods are heat-treated (eg, blanching of vegetables) before frozen storage.

Pressure. Generally, proteins are not sensitive to pressure, and only when large pressures are applied do they exhibit changes. It appears that in most cases the denaturing effect is dependent on applied pressure, exposure time, pH, protein concentration, and temperature (44). Pressure-induced denaturation is thought to result from a decrease in protein volume accommodated by holes in the native protein structure and exposure of hydrophobic groups to solvent (45). The effects of shear on proteins are not unrelated to those for pressure. Shearing action during extrusion processing contributes to the texture of the final product; however, the denaturing effect of the shear plates is most likely secondary to the high temperatures and pressures used during texturization of the proteins. This subject has been reviewed (46). Shear denaturation can also occur during protein purification (eg, at pump heads and in chromatographic columns) and in immobilized enzyme reactors (4).

Interfaces. Gas–liquid, liquid–liquid, and liquid–solid interfaces commonly occur in food systems, eg, emulsions, foams, and aerosols. These interfaces are thermodynamically unstable; however, compounds may be present or added, with an affinity for each phase, that reduce energy at the interface and stabilize the system. Proteins, owing to their amphoteric nature and relatively high molecular weights, tend to migrate to interfaces and, in so doing, reduce interfacial tension between phases. The protein thereby adopts a higher energy state, ie, becomes denatured. This phenomenon is technologically exploited in the processing of such food products as milk, ice cream, butter, finely comminuted meats, cakes, and salad dressings. Yet denaturation of proteins at interfaces may also be detrimental, a phenomenon to be avoided or at least controlled during food processing.

The mechanism of denaturation at interfaces is a two-step process (Fig. 5), the first step involving rapid and diffusion-controlled sorption at the interface until a monolayer of concentration 2 to 3 mg/m^2 is attained. The pro-

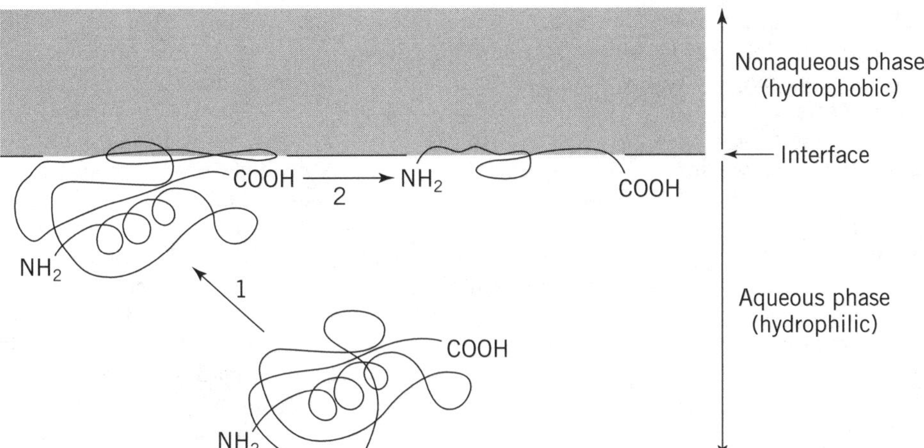

Figure 5. Two-step process of denaturation at interfaces.

pensity of different proteins to become sorbed at interfaces is dependent on their structures. Proteins that do not contain sizable hydrophobic or hydrophilic regions, or that possess disulfide cross-linked stabilized structures, tend not to adsorb at interfaces. When proteins are sorbed at interfaces, their sorbed structure depends on the stability of the native conformation. Once sorption has occurred, protein molecules reorganize (eg, unfold) and become denatured. This second step is rate-limiting, the rate being dependent on the conformational potential of the protein and its concentration. Detailed information on the denaturation behavior of β-casein, bovine serum albumin, and lysozyme at interfaces is reported in References 47 and 48.

Irradiation. Ionizing radiation (eg, γ, high-energy electrons) has been proposed for application in several areas of food processing, eg, inhibition of sprouting in onions, potatoes, and carrots; sterilization; and pasteurization. The effectiveness of irradiation as an antimicrobial agent has been well documented. However, at the levels of ionizing radiation required to destroy microorganisms (eg, 10 kGy), deterioration of sensory and nutritive quality may still result from autolysis (49). In general, the dose of ionizing radiation required for complete enzyme inactivation *in situ* is about 10-fold greater than that necessary to destroy microorganisms (50,51). Denaturation of proteins by irradiation is analogous to denaturation by other means; however, the specific effects are dependent on the wavelength and energy of the applied field. Other factors include the nature of the protein (enzyme), water activity, protein concentration and purity, oxygen tension, pH, and temperature (50). The structural alterations that irradiation can induce in proteins (eg, amino acid oxidation, ionization, free-radical formulation, polymerization) are often mediated by radiolysis of water (43). If the applied energy is sufficiently high, covalent bonds may be ruptured.

Chemical Denaturants

pH. Manipulation of pH is one of the oldest modes of food preservation, often used in combination with other preservation methods, eg, thermal processing, refrigeration, fermentation. pH is a critical determinant of protein conformation, stability, and function. Most proteins are stable over a characteristic pH range. Outside this range, denaturation can result from electrostatic repulsion of ionized groups within the protein molecule. The extent of denaturation is determined not only by the number of ionized groups but also by the location of these groups within the protein molecule (5,10). pH-induced denaturation of proteins is influenced by such factors as temperature, dielectric constant of the solvent, and ionic strength. In general, at high pH proteins remain soluble, largely because of the predominance of negative charges and their repulsion. In some instances, native conformation may be recovered when the pH is adjusted back to the characteristic range of the protein. Conversely, at low pH, proteins often aggregate, because positive charges rarely predominate within the pH ranges typically encountered during food processing.

Salts. Salts are used in many food processes and formulations for preservation and for development of flavor and texture. The denaturing effect of salts (ie, ions) on proteins in foods is difficult to assess: it is dependent on the nature of the protein, pH, and the size, charge, and concentration of the ion(s). Salts may react directly with proteins via electrostatic interaction, or they may alter the structure and orientation of water molecules around proteins. Ions may salt in (ie, solubilize) or salt out (ie, precipitate) proteins according to their location in the Hofmeister (lyotropic) series:

$$Ca^{2+} > Mg^{2+} > Li^{+} > Na^{+} > K^{+} > NH_4^{+}$$

$$PO_4^{3-} > SO_4^{2-} > \text{citrate}^{2-} > \text{tartrate}^{2-} > \text{acetate}^{-} > Cl^{-} > Br^{-} > NO^{3-} > I^{-} > ClO_4^{-} > SCN^{-}$$

Salting-in ions (eg, Cl^-, ClO_4^-) denature proteins, whereas salting-out ions (Ca^{2+} SO_4^{2-} PO_4^{3-}) stabilize native protein conformation. Salting out is believed to result from competition between proteins and ions for available water (4,5,10). Denaturation of proteins via salting-in arises from charge neutralization with concomitant reduction of hydrophobic effects.

Oxidation/Reduction. Various technological processes involve the use of oxidizing and reducing agents. Although not considered denaturants per se, these agents may lead to modification of amino acid residues (ie, cysteine, methionine, tryptophan, tyrosine, histidine), thereby resulting in protein denaturation. Food-grade oxidizing agents include peroxides (eg, H_2O_2, benzoyl peroxide), which are used for cold sterilization of milk and milk products and for bleaching of flour. Catalysts of lipid peroxidation (eg, light, heat, divalent metal ions, irradiation) also promote, directly and indirectly (ie, via reaction with lipid peroxides and their degradation products), oxidation of proteins. Bromates and oxidizing enzymes, which promote thiol-disulfide interchange through mild oxidation, are often used in the baking industry to improve the viscoelastic properties of gluten proteins. Reducing agents such as cysteine and ascorbic acid, which are commonly added to fruit juices, can similarly cause protein denaturation, but by disruption of disulfide cross-links. Alkaline conditions may also induce disulfide reduction in addition to lysinoalanine formation. Denaturation by reducing agents increases the susceptibility of the proteins to subsequent proteolysis and modification reactions.

DENATURATION OF FOOD PROTEINS

Controlled protein denaturation and associated reactions arising from processing of foods and isolated proteins are critical for some applications, eg, the formation of emulsions, foams, gels, and fibers. The following is a discussion of the effects of various environmental and processing factors on the major food protein classes (ie, meat and fish, milk, egg, plant).

Meat and Fish Proteins

The characteristics of proteins that affect their properties in comminuted meat products are not well defined (52). The ability of proteins to bind water and fat, as well as to retain these two components during heating and storage, is critical in the manufacture of processed meat products. Binding properties affect not only cook yield, but also final appearance and texture. Evaluation of individual muscle proteins suggests that the salt-soluble proteins (ie, myofibrillar proteins) are the major contributors in emulsification. Meat proteins generally display improved emulsification properties in the presence of increasing salt concentration, especially at pH values near or below their isoelectric points (pI). This apparent salt-induced shift in pI serves to increase or maintain protein solubility, an essential requirement for emulsification (53).

The emulsification properties of meat and fish protein are affected by changes in solubility, as induced by frozen storage, heating, and pH. However, solubility is not a good predictor of emulsification or other functional properties. One study reported that the amount of soluble protein from fresh meat sources was highly correlated with emulsification properties, irrespective of original meat source (54). For frozen or cooked meat, however, not only did the soluble protein content decrease, but the remaining noncoagulated soluble protein had lower emulsification properties compared to fresh meat sources. This observation was attributed to denaturation of soluble proteins that could have been caused by shearing during emulsification. Undenatured proteins are required for good emulsification properties since they possess greater conformational potential. Shear-induced protein denaturation may be followed by aggregation, this being intensified with increasing protein concentration. Dissociation of muscle protein complexes and subsequent protein denaturation without accompanying aggregation results in increased hydrophobicity of the salt-extractable proteins, with little change in their solubility or total sulfhydryl group content. Protein denaturation may be enhanced by hydrophobic association of the polypeptide chains at oil/water interfaces, resulting in a much larger available protein volume/surface area and increasing emulsifying capacity (1). Under moderate heating conditions (ie, $<50°C$), emulsification and functional properties of salt-extracted meat and fish proteins are often improved. At higher temperatures (ie, 50–70°C), functional properties are impaired, ie, proteins aggregate as reflected in decreased solubility and sulfhydryl group content. In addition, at temperatures above 70°C, proteins display decreased hydrophobicity (55).

The setting of a meat emulsion in comminuted products, in addition to binding of meat pieces in restructured or reformed products, is believed to be based on the establishment of a stable protein gel. Gelation arises from protein denaturation and subsequent association to form a three-dimensional matrix and is generally heat initiated, because raw meat pieces do not significantly bind to each other (56). An exception is the low-temperature (about 4°C) setting of sols from certain fish species that can occur on storage. During gel formation, exposure of sulfhydryl groups has been observed (55), in addition to shifts in pI caused by exposure of previously masked charged groups (57).

Perishability and compositional variations affect the use of fish muscle protein as a raw food material (eg, in the manufacture of kamoboko and surimi) (52). Aside from microorganisms and oxidative rancidity, the most conspicuous determinants of the quality of fish muscle proteins are alterations in functional properties, a direct reflection of protein denaturation (58). Formaldehyde is often produced during the storage of fish, especially gadoids, and has been implicated in the toughening of fish muscle during frozen storage. Through interaction with the side chain groups of fish muscle proteins, formaldehyde can increase the rate of protein denaturation, leading to aggregation of proteins and subsequent toughening (59). Denaturation of fish muscle proteins, especially on frozen storage, may also result from oxidation of free fatty acids and lipid peroxides or lipid–protein complex formation. When fish muscle proteins are thermally or surface denatured, the degree of denaturation is intensified in the presence of lipids. Yet, in other instances, intact lipids may function to stabilize and protect these proteins (53).

Milk Proteins

Milk proteins can be divided into two major groups: caseins and whey (or serum) proteins. Of these two, whey proteins, which are produced by heat precipitation and are used extensively as food ingredients, are extremely thermal labile (60). Thermal denaturation of whey proteins in fluid milk is manifested by the development of cooked flavor (61). Efficient use of whey as a functional food ingredient requires a knowledge of the denaturation behavior of individual whey proteins (62). Whey protein denaturation is considered a two-stage process: (*1*) disruption of tertiary and secondary structure, followed by (*2*) aggregation and coagulation, a phenomenon often associated with gel formation (5). Disulfide interchange reactions are largely responsible for aggregation and coagulation of β-lactoglobulin (a major component of whey) (63). The ability of whey proteins to form heat-induced gels is important in many food systems (eg, yogurt). Factors important in gelation include temperature, duration of thermal treatment, type and concentration of ions, state of the sulfur-containing amino acids, type of acidulant, and total solids concentration. In general, increases in total solids offer a protective effect by decreasing the rate and extent of whey protein denaturation (64). The presence of ions affects the thermal aggregation of β-lactoglobulin variably: phosphate and citrate inhibit, whereas calcium enhances, aggregation (65). Thermal denaturation of whey proteins modifies the course of milk coagulation and the rheological properties of the curd formed by acid or enzymes. In milk, thermally induced formation of intermolecular disulfide linkages occurs between unfolded β-lactoglobulin molecules and between unfolded β-lactoglobulin and casein micelles (ie, κ-casein and possibly α_{a1}-casein) (66). At temperatures in excess of 100°C, whey proteins may be extensively bound to casein micelles, thereby altering their surface properties.

Nonfat dry milk (NFDM) is used extensively in formulated foods and is produced from pasteurized skim milk, which is vacuum concentrated and spray-dried under conditions that result in either a low-heat or high-heat product. Low-heat NFDM is required for most applications that depend on a highly soluble protein (eg, emulsification) because it is processed under conditions that minimize whey protein denaturation and complexation with casein micelles (67). The casein proteins (ie, α-, β-, and κ-caseins) are generally not heat coagulable. In normal fluid milk, caseins resist coagulation for as long as 14 h at boiling temperatures or 1 h at 130°C. Thermocoagulation of casein in milk can occur as a result of compositional changes within the milk itself induced by sustained exposure to high temperatures (eg, increased acidity, a shift from soluble to colloidal forms of calcium and phosphates, denaturation, and hydrolysis of other milk proteins). Coagulation of milk is often attributed to the destabilization of casein micelles; however, this phenomenon may be the summation of many changes in the colloidal system (68).

Egg Albumin Proteins

The albumins (eg, ovalbumin, ovotransferrin, ovomucoid) are among the most widely used proteins in food formulation, owing to their exceptional gelling and whipping properties. Their functional diversity is directly attributable to their susceptibility to a variety of denaturants (69). Denaturation of albumins is a discrete phenomenon, the extent of which is dependent on temperature, pH, salt, and moisture content (69). When used in food formulations, egg protein gels that have been thermally set support and bind other ingredients within the matrix and contribute to the texture of the product (5,10). The extent of egg protein gelation (ie, denaturation) is influenced by the degree of oxidation of free sulfhydryls to form inter- or intramolecular disulfide linkages. For example, strong oxidizing agents, such as the metallic cations Fe^{3+} and Cu^{2+}, in addition to potassium iodate, enhance gel formation. However, hydrogen peroxide and potassium bromate, comparatively weak oxidizing agents, have little influence on gel strength. At alkaline pH values (eg, 9–10), reduction of disulfide linkages may occur that could also affect gel strength.

Egg albumin proteins are often pasteurized to facilitate their later use in formulated food products. Pasteurization generally does not affect their functional properties. However, ovotransferrin is more prone than ovalbumin to thermal denaturation. Ovalbumin is most stable at pH7.0, whereas ovotransferrin is least stable to thermal denaturation at this pH (5,10,69). At pH 6.8 and 50°C, up to 50% of the ovotransferrin is denatured in 4 to 5 min. In the presence of metals (eg, iron and aluminum), however, conformation of this protein is stabilized by a chelation-type mechanism. This protective effect is the impetus for the addition of aluminum and adjustment of pH to 7.0 with lactic acid before pasteurization of egg whites. Alternative pasteurization processes rely on lower temperatures and the addition of antibacterial agents (eg, hydrogen peroxide) (69). The native conformation of ovomucoid is resistant to extremes of pH and temperature. Prolonged exposure of ovomucoid to temperatures of 100°C does not alter its physicochemical properties. Its stability is attributed largely to intramolecular disulfide linkages (70). Denaturation of ovomucoid is suggested to be a three-stage process involving denaturation of each of three separate domains within the protein (71). If exposure to denaturing conditions is not prolonged, denaturation of ovomucoid is reversible (72).

Surface tension, important in such functional properties as foaming and emulsification, of egg and other proteins decreases markedly as denaturation proceeds. Thermal denaturation without coagulation improves surface properties (Table 1). This again implicates, as with meat proteins, the limited importance of solubility for functional (eg, surface) properties. Once sorbed at the interface, surface hydrophobicity plays a more dominant role in the foaming and emulsifying properties of egg proteins than does solubility.

Plant Proteins The impetus for utilization and development of (new) plant protein resources has been twofold: (*1*) to supplement, simulate, and/or replace muscle protein systems, and (*2*) to meet the increasing nutritional requirements of an increasing Third World population. Detailed reviews concerning denaturation of plant proteins in relation to their functional properties and food applications have been published (1,73).

Soybeans have been used in food systems, owing to their excellent functional and nutritional qualities. The major globulins of soy protein are conglycinin (7S) and glycinin

(11S). On heating, the subunits of both of these proteins can dissociate and reassociate in different ways. Formation of soluble aggregates of soy protein and the existence of sol, progel, gel, and metasol states and the roles of different forces in their formation have been discussed (74). The 11S globulin oligomer appears to undergo only minor changes on heating, retaining its quaternary structure and undergoing little obvious denaturation during formation of a gel when heated to 100°C (75). The importance of the overall hydrophobicity of unfolded proteins, rather than the surface hydrophobicity of undenatured proteins, for thermal functional properties has been noted for various proteins. High-heat treatment (121°C) of soy protein isolate at pH 5.5 led to a marked increase in overall hydrophobicity and a large reduction of solubility and emulsion stability (76). In contrast, the same heat treatment at pH 7.2 caused a moderate increase in overall hydrophobicity and a large increase in solubility and emulsifying properties. It is evident that environmental factors (eg, pH, heat) have a major impact on the functional properties of soy proteins.

Peanut and soy proteins that are heat denatured without any accompanying precipitation show enhanced foamability (77). Factors affecting foaming of these proteins include structural properties of the proteins per se (eg, flexibility, exposure of hydrophobic/hydrophilic groups, surface charge), ease of unfolding, and the Marangoni effect (ie, the ability to concentrate rapidly at a stress point). Also important are environmental factors such as temperature, pH, ionic strength, viscosity, and the presence of other components such as denaturants that could affect the intrinsic properties of the proteins (53).

The unique ability of wheat flour to form a cohesive and viscoelastic paste or dough when mixed and kneaded is due primarily to the properties of the two classes of principal storage proteins, gliadins and glutenins. These are collectively referred to as gluten proteins. Glutenins are responsible for the elasticity, cohesiveness, and mixing tolerance of the dough, whereas gliadins facilitate fluidity, extensibility, and expansion of the dough, thus contributing to loaf volume. A proper balance of both gluten proteins is essential for bread making. During the mixing and kneading of hydrated wheat, the gluten proteins orient, align, and partially unfold (ie, denature). Protein unfolding enhances both hydrophobic interactions and the formation of disulfide cross-links through disulfide interchange reactions, resulting in establishment of a three-dimensional matrix that serves to entrap starch granules and other dough components. The cleavage of disulfide cross-links by reducing agents such as cysteine destroys the cohesive structure of the hydrated dough. The addition of oxidizing agents such as bromate increases toughness and elasticity by promoting disulfide linkage formation. In addition to the gluten proteins, soluble proteins (ie, albumins and globulins) that are found in minor quantities denature and aggregate to aid in gel formation, thus contributing to setting of the bread crumb (78).

Drying, roasting, and cooking are thermal treatments routinely used to process plant proteins. These processes can cause substantial protein denaturation. Oat globulin, the major protein fraction in oats, has a quaternary structure similar to that of soy 11S globulin (glycinin), a heat-coagulable protein. Differential scanning calorimetry showed that oat globulin, heated under conditions inducing gelation, was not extensively denatured and exhibited highly cooperative transition characteristics (79). When 1% oat globulin was heated, aggregation and precipitation occurred. Ultraviolet and fluorescence spectra of soluble and insoluble fractions indicated no marked protein unfolding in the former fraction, but extensive denaturation in the insoluble aggregates. The insoluble fraction had significantly higher surface hydrophobicity than the soluble fraction and unheated protein (80).

Sonication has been used as a means to solubilize plant proteins and isolates/meals. Heat-treated, acid-precipitated soy proteins, intermediates in the commercial production of isolated proteins from defatted soybeans, have been dispersed by sonication (81). Changes in flow properties may have been derived from ultrasonic-induced dissociation of protein aggregates formed during thermal treatment. This was considered to be associated with partial cleavage of intermolecular hydrophobic interactions, but not cleavage of peptide or disulfide bonds. It was postulated that ultrasonically exposed hydrophobic regions are subsequently buried through rearrangement of the molecular structure, conferring greater hydrophilicity to the protein (81). However, it has been shown that sonication leads to agglomeration and aggregation, particularly of the 7S fraction (82). The ultrasonic action may have (*1*) promoted hydrophobic interactions between globular proteins; (*2*) induced formation of complex mixtures, as in the case of apolipoproteins; or (*3*) altered the equilibrium protein–protein and/or protein–lipid interactions, thereby favoring the formation of a cluster-type structure.

Thermoplastic extrusion technology has been used to texturize many plant proteins, especially soy, to produce fibrous structures that simulate meatlike products (83). The process begins with moistened, defatted soy flour, which is fed into an extruder where it is worked and heated, causing the protein molecules to denature and form new cross-linkages that result in a fibrous structure. The heated plasticized mass is forced through a die to form expanded texturized strands of vegetable proteins that have meatlike characteristics on rehydration (84). Extrusion of plant proteins has been reviewed (46).

STABILIZATION OF PROTEINS

The intrinsic instability of some proteins can be problematic for some food processing applications, eg, the use of a less-stable enzyme would require more time and larger amounts of enzyme to achieve the same processing goal as compared to a more stable enzyme. Following the introduction of site-directed mutagenesis in 1982 (85), research focused on the improvement of stability of proteins by site-specific mutations rather than random mutations. This approach was based on the hypothesis that the stability of proteins was governed by (*1*) bonds and interactions, (*2*) conformational factors, and (*3*) protection by modifications (86). To test this hypothesis, new disulfide bonds were introduced to stabilize various proteins, eg, T4 lysozyme

(87,88), subtilisin (89–91), dihydrofolate reductase (92), and λ repressor (93).

In a series of experiments, Matthews et al. stabilized T4 lysozyme using different strategies, eg, α-helix dipoles, entropy, conformation strain, and hydrophobic interactions (94,95). An 8°C increase in T_m (melting temperature) using a combination of strategic mutations was achieved (96–101). By changing the hydrophobic interaction between the subunits of L-lactate dehydrogenease, a 10°C increase in T_m was achieved (102). As an alternative type of hydrophobic interaction manipulation, site-specific glycosylation of hen egg white lysozyme resulted in a dramatic increase of thermostability (103). Suzuki et al. reported that the introduction of proline could stabilize protein by stabilizing the secondary structure (104). T4 lysozyme was stabilized by introducing a proline residue into an α-helix (105). Nakamura et al. also showed that *Bacillus* neutral protease was stabilized by introducting proline into a core α-helix (106).

Despite many successful reports on the stabilization of protein by site-directed mutagenesis, there is still no general theory to predict the site to be mutated. Gilis and Rooman used the solvent accessibility of the residues to determine the mutation site (107,108). The prediction of stabilization was based on a database containing the results of the mutation studies of various proteins.

Fersht's group, using barnase, conducted a series of mutation studies on their effects on stability and concluded the effects were varied even in a certain kind of the interactions, depending on the context (109,110). Furthermore, research based on the this hypothesis required that the three-dimensional structures be known in order to determine the mutation site.

Future research could provide a general theory on protein stabilization. Serrano et al. suggested one such possible theory. Systematic multiple mutation could stabilize proteins without the knowledge of the three-dimensional structures but would be dependent on two proteins with high sequence similarity (111). The reader is referred to the article by Jiminez-Flores and Bleck regarding recent advances in food protein biotechnology (112).

CONCLUSION

Various processing and environmental factors influence the functional properties of proteins; however, despite advances in molecular biology and analytical techniques the molecular basis for induced functional changes still remains ill defined. The reader is referred to articles by Jimenez-Flores and Bleck (112) and Yada et al. (113) for references regarding to recent advances in food biotechnology and analytical techniques, respectively. Although great gains have been made in the understanding of protein structure and its stabilization and, therefore, functionality, a need still exists to better define and characterize the denaturation of (specific) proteins if we are to control this phenomenon and take advantage of the technological attributes that proteins have to offer. Methodology by which to recognize denaturation when it occurs must be standardized. That food by its very nature is a multicomponent system in which a myriad of interactions can/have occurred makes this a foreboding task. Yet, in order to meet this mandate, the use of multicomponent or actual food systems is required. It must be stressed that although every protein has a unique structure and, therefore, unique conformational potential, the behavior of a protein in a given system is governed by thermodynamic and kinetic constraints.

ACKNOWLEDGMENTS

The financial support of the Natural Sciences and Engineering Research Council of Canada is gratefully acknowledged.

BIBLIOGRAPHY

1. J. E. Kinsella, "Functional Properties of Proteins in Foods," *CRC Crit. Rev. Food Sci. Nutr.* **7**, 219–280 (1976).
2. J. E. Kinsella, "Relationships between Structure and Functional Properties of Food Proteins," in P. F. Fox and J. J. Condon, eds., *Food Proteins*, Applied Science Publishers, New York, 1982, pp. 51–103.
3. S. Damodaran, "Functional Properties," in S. Nakai and H. W. Modler, eds., *Food Proteins: Properties and Characterization*, Wiley-VCH, New York, 1996, pp. 167–234.
4. F. Franks, "Conformational Stability: Denaturation and Renaturation," in F. Franks, ed., *Characterization of Proteins*, Humana Press, Clifton, N.J., 1988, pp. 95–126.
5. A. Kilara and T. Sharkasi, "Effects of Temperature on Food Proteins and Its Implications on Functional Properties," *CRC Crit. Rev. Food Sci. Nutr.* **23**, 323–395 (1986).
6. C. N. Pace, "The Stability of Globular Proteins," *CRC Crit. Rev. Biochem.* **3**, 1–43 (1975).
7. C. N. Pace, "Protein Conformations and Their Stability," *J. Am. Oil Chem. Soc.* **60**, 970–975 (1983).
8. R. H. Pain, "The Conformation and Stability of Folded Globular Proteins," in F. Franks, ed., *Characterization of Protein Conformation and Function*, Symposium Press, London, 1979, pp. 19–36.
9. V. B. Tolstoguzov, "Some Physico-Chemical Aspects of Protein Processing into Foodstuffs," *Food HydrocolLoids* **2**, 339–370 (1988).
10. A. Kilara and V. R. Harwalkar, "Denaturation," in S. Nakai and H. W. Modler, eds., *Food Proteins: Properties and Characterization*, Wiley-VCH, New York, 1996, pp. 71–165.
11. R. D. Ludescher, "Physical and Chemical Properties of Amino Acids and Proteins," in S. Nakai and H. W. Modler, eds., *Food Proteins: Properties and Characterization*, Wiley-VCH, New York, 1996, pp. 23–70.
12. G. E. Schulz and R. H. Schirmer, *Principles of Protein Structure*, Springer-Verlag, New York, 1979.
13. V. Tolstoguzov, "Structure–Property Relationships in Foods," in N. Parris et al., eds., *Macromolecular Interactions in Food Technology*, American Chemical Society, Washington, D.C., 1996, pp. 2–14.
14. C. B. Anfinsen, "Principles that Govern the Folding of Protein Chains," *Science* **181**, 223–230 (1973).
15. P. L. Privalov, "Stability of Proteins: Small Globular Proteins," *Adv. Protein Chem.* **33**, 167–241 (1979).
16. P. L. Privalov, "Stability of Proteins. Proteins Which Do Not Present a Single Cooperative System," *Adv. Protein Chem.* **35**, 1–104 (1982).

17. C. Tanford, "Protein Denaturation," *Adv. Protein Chem.* **23**, 121–282 (1968).
18. K. Kuwajima et al., "Three-State Denaturation of alpha-Lactalbumin by Guanidine Hydrochloride," *J. Mol. Biol.* **106**, 359–373 (1976).
19. M. Nozaka et al., "Detection and Characterization of the Intermediate on the Folding Pathway of Human alpha-Lactalbumin," *Biochemistry* **17**, 3753–3758 (1978).
20. K. Kuwajima, "A Folding Model of alpha-Lactalbumin Deduced from the Three State Denaturation Mechanism," *J. Mol. Biol.* **114**, 241–258 (1977).
21. J. N. Onuchic et al., "Towards an Outline of the Topology of a Realistic Protein-Folding Funnel," *Proc. Natl. Acad. Sci. USA* **92**, 3626–2630 (1995).
22. M. Ohgushi and A. Wada, "Molten Globule State: A Compact Form of Globular Proteins with Mobile Side-Chains," *FEBS Lett.* **164**, 21–24 (1983).
23. J. J. Ewbank et al., "What Is the Molten Globule," *Nature Struct. Biol.* **2**, 10–11 (1995).
24. L. C. Wu, Z. Peng, and P. S. Kim, "Bipartite Structure of the alpha-Lactalbumin Molten Globule," *Nature Struct. Biol.* **2**, 281–286 (1995).
25. T. E. Creighton, "How Important Is the Molten Globule for Correct Protein Folding?" *Trends Biochem. Sci.* **22**, 6–10 (1997).
26. Z. Peng, L. C. Wu, and P. S. Kim, "Local Structure Preferences in the alpha-Lactalbumin Molten Globule," *Biochemistry* **34**, 3248–3252 (1995).
27. V. E. Bychkova, R. H. Pain, and O. B. Ptitsyn, "The Molten Globule State Is Involved in the Translocation of Proteins across the Membranes?" *FEBS Lett.* **238**, 231–234 (1988).
28. O. B. Ptitsyn, V. E. Bychkova, and V. N. Uversky, "Kinetic and Equilibrium Folding Intermediates," *Philos Trans. R. Soc. London B, Biol. Sci.* **348**, 35–41 (1995).
29. J. Ellis, "Proteins as Molecular Chaperones," *Nature (London)* **328**, 378–379 (1987).
30. J. Martin, "Protein Folding Assisted by the GroEL/GroES Chaperonin System," *Biochemistry (Moscow)* **63**, 374–381 (1998).
31. E. A. Craig, B. D. Gambill, and R. J. Nelson, "Heat Shock Proteins: Molecular Chaperones of Protein Biogenesis," *Microbiol. Rev.* **57**, 402–414 (1993).
32. C. Georgopoulos and W. J. Welch, "Role of the Major Heat Shock Proteins as Molecular Chaperones," *Ann. Rev. Cell Biol.* **9**, 601–634 (1993).
33. K. Liberek et al., "*Escherichia coli* DnaJ and GrpE Heat Shock Proteins Jointly Stimulate ATPase Activity of DnaK," *Proc. Natl. Acad. Sci. USA* **88**, 2874–2878 (1991).
34. R. B. Freedman, "Protein Disulfide Isomerase: Multiple Roles in the Modification of Nascent Secretory Proteins," *Cell* **57**, 1069–1072 (1989).
35. F. X. Schmid et al., "Prolyl Isomerases: Role in Protein Folding," *Adv. Protein Chem.* **44**, 25–66 (1993).
36. C. T. Walsh, L. D. Zydowsky, and F. D. McKeon, "Cyclosporin A, the Cyclophilin Class of Peptidylprolyl Isomerases, and Blockade of T Cell Signal Transduction," *J. Biol. Chem.* **267**, 13115–13118 (1992).
37. H. F. Gilbert, "Protein Chaperones and Protein Folding," *Current Opinions Biotechnol.* **5**, 534–539 (1994).
38. K. Braig et al., "The Crystal Structure of the Bacterial Chaperonin GroEL at 2.8 Å," *Nature* **371**, 578–586 (1994).
39. V. R. Harwalkar and C.-Y. Ma, "Thermal Analysis: Principles and Applications," in S. Nakai and H. W. Modler, eds., *Food Proteins: Properties and Characterization*, Wiley-VCH, New York, 1996, pp. 405–427.
40. A. E. Stearn, "Kinetics of Biological Reactions with Special Reference to Enzymatic Processes," *Adv. Enzymol.* **9**, 25–74 (1949).
41. D. P. Goldenberg and T. E. Creighton, "Energetics of Protein Structure and Folding," *Biopolymers* **24**, 167–182 (1985).
42. D. B. Lund, "Heat Processing," in O. R. Fennema, ed., *Principles of Food Science Part II. Physical Principles of Food Preservation*, Marcel Dekker, New York, 1975, pp. 31–92.
43. S. Schwimner, *Source Book of Food Enzymology*, AVI, Westport, Conn., 1981.
44. M. Joly, *A Physico-Chemical Approach to the Denaturation of Proteins (Phenomenological Aspects of Denaturation)*, Academic, New York, 1965.
45. W. Kauzmann, "Some Factors in the Interpretation of Protein Denaturation," *Adv. Prot. Chem.* **14**, 1–63 (1959).
46. D. W. Stanley, "Protein Reactions during Extrusion Processing," in C. Mercier, P. Linko, and J. M. Harper, eds., *Extrusion Cooking*, American Cereal Chemists, St. Paul, Minn., 1989, pp. 321–341.
47. D. E. Graham and M. C. Phillips, "Proteins at Liquid Interfaces. I. Kinetics of Adsorption and Surface Denaturation," *J. Colloid Interface Sci.* **70**, 403–413 (1979).
48. D. E. Graham and M. C. Phillips, "Proteins at Liquid Interfaces. II. Adsorption Isotherms," *J. Colloid Interface Sci.* **70**, 415–439 (1979).
49. R. Zender et al., "Aseptic Autolysis of Muscle: Biochemical and Microscopic Modifications Occurring in Rabbit and Lamb Muscle during Aseptic and Anaerobic Storage," *Food Res.* **23**, 305–326 (1958).
50. T. Richardson and D. B. Hyslop, "Enzymes," in O. R. Fennema, ed., *Food Chemistry*, 2nd ed., Marcel Dekker, New York, 1985, pp. 371–476.
51. J. R. Whitaker, "Enzymes," in O. R. Fennema, ed., *Food Chemistry*, 3rd ed., Marcel Dekker, New York, 1996, pp. 431–530.
52. E. A. Foegeding, T. C. Lanier, and H. O. Hultin, "Characteristics of Edible Muscle Tissues," in O. R. Fennema, ed., *Food Chemistry*, 3rd ed., Marcel Dekker, New York, 1996, 879–942.
53. S. Nakai and E. Li-Chan, *Hydrophobic Interactions in Food Systems*, CRC Press, Boca Raton, Fla., 1988.
54. T. A. Gillett et al., "Parameters Affecting Meat Protein Extraction and Interpretation of Model System Data for Meat Emulsion Formulation," *J. Sci.* **42**, 1606–1610 (1977).
55. R. Hamm, "Changes of Muscle Proteins during the Heating of Meat," in T. Hoyem and O. Kvale, eds., *Physical, Chemical and Biological Changes in Food Caused by Thermal Processing*, Applied Science Publishers, London, 1977, pp. 101–134.
56. A. Ashgar, K. Samenjima, and T. Yasui, "Functionality of Muscle Protein in Gelation Mechanisms of Structured Meat Products," *CRC Crit. Rev. Food Sci. Nutr.* **22**, 27–106 (1986).
57. R. Hamm, "Heating of Muscle Systems," in E. J. Briskey, R. G. Cassens, and J. C. Trautman, eds., *The Physiology and Biochemistry of Muscle as a Food*, University of Wisconsin Press, Madison, 1966, pp. 363–385.
58. J. Spinelli and J. A. Dassow, "Fish Proteins: Their Modification and Potential Uses in the Food Industry," in R. E. Martin et al., eds., *Chemistry and Biochemistry of Marine Food Products*, AVI, Westport, Conn., 1982.
59. J. F. Ang and H. O. Hultin, "Denaturation of Cod Myosin during Freezing after Modification with Formaldehyde," *J. Food Sci.* **54**, 814–818 (1989).

60. H. E. Swaisgood, "Characteristics of Milk," in O. R. Fennema, ed., *Food Chemistry*, 3rd ed., Marcel Dekker, New York, 1996, pp. 841–878.

61. P. Walstra and R. Jennes, *Dairy Chemistry and Physics*, Wiley Interscience, New York, 1984.

62. F. Dannenberg and H.-G. Kessler, "Reaction Kinetics of the Denaturation of Whey Proteins in Milk," *J. Food Sci.* **53**, 258–263 (1988).

63. R. M. Hillier, R. L. J. Lyster, and C. C. Cheeseman, "Thermal Denaturation of α-Lactalbumin and β-Lactoglobulin in Cheese Whey: Effect of Total Solids Concentration and pH," *J. Dairy Res.* **46**, 103–111 (1979).

64. C. A. Zittle et al., "The Binding of Calcium Ions by β-Lactoglobulin Both before and after Aggregation by Heating in the Presence of Calcium Ions," *J. Am. Chem. Soc.* **79**, 4661–4666 (1957).

65. K. Watanabe and H. Klostermeyer, "Heat-Induced Changes in Sulfhydryl and Disulfide Levels of β-Lactoglobulin A and the Formation of Polymers," *J. Dairy Res.* **43**, 411–418 (1976).

66. S. I. Shalabi and J. V. Wheelock, "Effect of Sulfhydryl-Blocking Agents on the Primary Phase of Chymosin Action on Heated Casein Micelles and Heated Milk," *J. Dairy Res.* **44**, 351–355 (1977).

67. C. V. Morr, "Emulsifiers: Milk Proteins," in J. P. Cherry, ed., *Protein Functionality in Foods, ACS Symposium Series 147*, American Chemical Society, Washington, D.C., 1981, pp. 201–215.

68. R. M. Parry Jr., "Milk Coagulation and Protein Denaturation," in B. H. Webb, A. H. Johnson, and J. H. Alford, eds., *Fundamentals of Dairy Chemistry*, AVI, Westport, Conn., 1974, pp. 603–661.

69. D. V. Vadhera and K. R. Nath, "Eggs as a Source of Protein," *CRC Crit. Rev. Food Technol.* **4**, 193–309 (1973).

70. H. F. Déutsh and J. I. Moerton, "Physical-Chemical Studies of Soy Modified Ovomucoids," *Arch. Biochem. Biophys.* **931**, 654–660 (1961).

71. M. A. Baig and A. Salahuddin, "Occurrence and Characterization of Stable Intermediate State(s) in the Unfolding of Ovomucoid by Guanidine Hydrochloride," *Biochem. J.* **171**, 89–97 (1978).

72. T. Matsuda, K. Watanabe, and Y. Sate, "Temperature-Induced Structural Changes in Chicken Egg White Ovomucoid," *Agric. Biol. Chem.* **45**, 1609–1614 (1981).

73. Y. V. Wu and O. E. Inglett, "Denaturation of Plant Proteins Related to Functionality and Applications: A Review," *J. Food Sci.* **39**, 218–225 (1974).

74. N. Catsimpoolas and E. W. Meyer, "Gelation Phenomena of Soybean Globulins. I. Protein–Protein Interactions," *Cereal Chem.* **47**, 559–570 (1970).

75. T. Mori et al., "Differences in Subunit Composition of Glycinin among Soybean Cultivars," *J. Agric. Food Chem.* **29**, 20–23 (1981).

76. L. Voutsinas, S. Nakai, and V. P. Harwalker, "Relationships between Hydrophobicity and Thermal Functional Properties of Food Proteins," *Can. Inst. Food Sci. Technol. J.* **16**, 185–190 (1983).

77. J. P. Cherry and K. H. McWatters, "Whippability and Aeration," in J. P. Cherry, ed., *Protein Functionality in Foods, ACS Symposium Series 147*, American Chemical Society, Washington, D.C., 1981, pp. 149–176.

78. J. C. Cheftel, J.-L. Cuq, and D. Lorient, "Amino Acids, Peptides, and Proteins," in O. R. Fennema, ed., *Food Chemistry*, 2nd ed., Marcel Dekker, New York, 1985, pp. 245–369.

79. C.-Y. Ma and V. R. Harwalkar, "Chemical Characterization and Functionality Assessment of Oat Protein Fractions," *J. Agric. Food Chem.* **32**, 144–149 (1984).

80. C.-Y. Ma and V. R. Harwalkar, "Study of Thermal Denaturation of Oat Globulin by Ultraviolet and Fluorescence Spectroscopy," *J. Agric. Food Chem.* **36**, 155–160 (1988).

81. T. Furukawa and S. Ohta, "Ultrasonic-Induced Modification of Flow Properties of Soy Protein Dispersion," *Agric. Biol. Chem.* **47**, 745–750 (1983).

82. L. C. Wang, "Soybean Protein Agglomeration: Promotion by Ultrasonic Treatment," *J. Agric. Food Chem.* **29**, 177–180 (1981).

83. K. C. Rhee, C. K. Kuo, and E. W. Lusas, "Texturization," in J. P. Cherry, ed., *Protein Functionality in Foods, ACS Symposium Series 147*, American Chemical Society, Washington, D.C., 1981, pp. 51–88.

84. J. M. Harper, "Food Extrusion," *CRC Crit. Rev. Food Sci. Nutr.* **11**, 155–215 (1979).

85. M. J. Zoller and M. Smith, "Oligonucleotide-Directed Mutagenesis Using M13-Derived Vectors: An Efficient and General Procedure for the Production of Point Mutations in Any Fragment of DNA," *Nucl. Acids Res.* **10**, 6487–6500 (1982).

86. Y. Nosoh and T. Sekiguchi, "Protein Engineering for Thermostability," *Trends Biotechnol.* **8**, 16–20 (1990).

87. L. J. Perry and R. Wetzel, "Disulfide Bond Engineered into T4 Lysozyme: Stabilization of the Protein toward Thermal Inactivation," *Science* **226**, 555–557 (1984).

88. R. Wetzel et al., "Disulfide Bonds and Thermal Stability in T4 Lysozyme," *Proc. Natl. Acad. Sci. USA* **85**, 401–405 (1988).

89. J. A. Wells and D. B. Powers, "In vivo Formation and Stability of Engineered Disulfide Bonds in Subtilisin," *J. Biol. Chem.* **261**, 6564–6570 (1986).

90. M. W. Pantoliano et al., "Protein Engineering of Subtilisin BPN': Enhanced Stabilization through the Introduction of Two Cysteines to Form a Disulfide Bond," *Biochemistry* **26**, 2077–2082 (1987).

91. C. Mitchinson and J. A. Wells, "Protein Engineering of Disulfide Bonds in Subtilisin BPN'," *Biochemistry* **28**, 4807–4815 (1989).

92. J. E. Villafranca et al., "An Engineered Disulfide Bond in Dihydrofolate Reductase," *Biochemistry* **26**, 2182–2189 (1987).

93. R. T. Sauer et al., "An Engineered Intersubunit Disulfide Enhances the Stability and DNA Binding of the N-Terminal Domain of λ Repressor," *Biochemistry* **25**, 5992–5998 (1986).

94. B. W. Matthews, "Genetic and Structural Analysis of the Protein Stability Problem," *Biochemistry* **26**, 6885–6888 (1987).

95. B. W. Matthews, "Structural and Genetic Analysis of Protein Stability," *Ann. Rev. Biochem.* **62**, 139–160 (1993).

96. B. W. Matthews, H. Nicholson, and W. Becktel, "Enhanced Protein Thermostability from Site-Directed Mutations That Decrease the Entropy of Unfolding," *Proc. Natl. Acad. Sci. USA* **84**, 6663–6667 (1987).

97. X.-J. Zhang et al., "Enhancement of Protein Stability by the Combination of Point Mutations in T4 Lysozyme Is Additive," *Protein Eng.* **8**, 1017–1022 (1995).

98. S. Dao-Pin, W. A. Baase, and B. W. Matthews, "A Mutant T4 Lysozyme (Val 131 → Ala) Designed to Increase Thermostability by the Reduction of Strain within an α-Helix," *Proteins* **7**, 198–204 (1990).

99. H. Nicholson et al., "Analysis of the Interaction between Charged Side Chains and the α-Helix Dipole Using Designed Thermostable Mutants of Phage T4 Lysozyme," *Biochemistry* **30**, 9816–9828 (1991).
100. H. Nicholson, W. J. Becktel, and B. W. Matthews, "Enhanced Protein Thermostability from Designed Mutations That Interact with α-Helix Dipoles," *Nature* **336**, 651–656 (1988).
101. M. Matsumura, W. J. Becktel, and B. W. Matthews, "Hydrophobic Stabilization in T4 Lysozyme Determined Directly by Multiple Substitutions of Ile3," *Nature* **334**, 406–410 (1988).
102. H. K. W. Kallwass et al., "Single Amino Acid Substitutions Can Further Increase the Stability of a Thermophilic L-Lactate Dehydrogenase," *Protein Eng.* **5**, 769–774 (1992).
103. S. Nakamura et al., "Hyperglycosylation of Hen Egg White Lysozyme in Yeast," *J. Biol. Chem.* **268**, 12706–12712 (1993).
104. Y. Suzuki et al., "A Strong Correlation between the Increase in Number of Proline Residues and the Rise in Thermostability of Five *Bacillus* Oligo-1,6,-glucosidase," *Appl. Microbiol. Biotechnol.* **26**, 546–551 (1987).
105. T. Herning et al., "Role of Proline Residues in Human Lysozyme Stability: A Scanning Calorimetric Study Combined with X-Ray Structure Analysis of Proline Mutants," *Biochemistry* **31**, 7077–7085 (1992).
106. S. Nakamura et al., "Improving the Thermostability of *Bacillus stearothermophilus* Neutral Protease by Introducing Proline into the Active Site Helix," *Protein Eng.* **10**, 1263–1269 (1997).
107. D. Gilis and M. Rooman, "Stability Changes upon Mutation of Solvent-Accessible Residues in Proteins Evaluated by Database-Derived Potentials," *J. Mol. Biol.* **257**, 1112–1126 (1996).
108. D. Gilis and M. Rooman, "Predicting Protein Stability Changes upon Mutation Using Database-Derived Potentials: Solvent Accessibility Determines the Importance of Local versus Non-Local Interactions along the Sequence," *J. Mol. Biol.* **272**, 276–290 (1997).
109. A. R. Fersht, A. Matouschek, and L. Serrano, "The Folding of an Enzyme. I. Theory of Protein Engineering Analysis of Stability and Pathway of Protein Folding," *J. Mol. Biol.* **224**, 771–782 (1992).
110. L. Serrano et al., "The Folding of an Enzyme. II. Substructure of Barnase and the Contribution of Different Interactions of Protein Stability," *J. Mol. Biol.* **224**, 783–804 (1992).
111. L. Serrano, A. G. Day, and A. R. Fersht, "Step-wise Mutation of Barnase to Binase. A Procedure for Engineering Increased Stability of Proteins and an Experimental Analysis of the Evolution of Protein Stability," *J. Mol. Biol.* **233**, 305–312 (1993).
112. R. Jiminez-Flores and G. T. Bleck, "Biotechnology," in S. Nakai and H. W. Modler, eds., *Food Proteins: Properties and Characterization*, Wiley-VCH, New York, 1996, pp. 505–534.
113. R. Y. Yada et al., "Analysis: Quantitation and Physical Characterization," in S. Nakai and H. W. Modler, eds., *Food Proteins: Properties and Characterization*, Wiley-VCH, New York, 1996, pp. 333–403.

Rickey Y. Yada
Robert L. Jackman
Jeff L. Smith
Kenneth G. Payie
Takuji Tanaka
University of Guelph
Guelph, Ontario
Canada

PROTEINS: STRUCTURE AND FUNCTIONALITY

This discussion of protein structure begins with amino acids as building blocks of proteins, followed by different levels of protein structure. Chemical forces that maintain the protein structure or change the structure level are then explained. Interactions of proteins, especially protein–protein, protein–lipid and protein–water interactions, are briefly reviewed. Finally, structure–function relationships are discussed. For more details, readers are referred to the work by R. D. Ludescher (1).

Four levels of structural organization in proteins can be distinguished: primary, secondary, tertiary, and quaternary structures. These terms refer to the amino acid sequence, the regular arrangements of the polypeptide backbone, the three-dimensional structure of the globular protein, and the structures of aggregates of globular proteins, respectively. In addition to this classification, the supersecondary structure is frequently described, which refers to association products of secondary structure, eg, coiled-coil α-helix in which two α-helixes are wound around each other, α–β mixed structure, and βαβ-unit structure (2). Furthermore, domain refers to a structurally independent unit that has the characteristics of a small globular protein (3). A domain is composed of supersecondary structures. Domains often have a specific function. A functional motif (sequences shorter than 25 amino acids) or domain (larger than 50 amino acids) is part of a protein that serves a particular function (4).

AMINO ACIDS

Twenty standard amino acid residues occur in proteins. These are listed in Table 1 in the order of their general abundance in proteins in nature (2). The three-letter and one-letter abbreviations that are commonly used are included.

Amino Acid Analysis

The amino acid composition of proteins is analyzed after hydrolysis of the proteins. Acids or proteases are used for the hydrolysis. Because of the lability of some amino acids (eg, tryptophan, especially in the presence of carbohydrates, and sulfur-containing amino acids), enzymatic hydrolysis is preferable. However, because of the difficulty in achieving complete hydrolysis and high costs of the required enzyme mixtures, acid hydrolysis is more popular. To avoid the destruction of those amino acids during hydrolysis, methanesulfonic acid (7), mercaptoethanesulfonic acid (8), or thioglycolic acid (9) may be used for hydrolysis instead of the otherwise commonly used 6 N hydrochloric acid.

High-performance liquid chromatography (HPLC), especially using a cation exchange column, has been the most popular method for amino acid analysis of the resultant protein hydrolysate. However, reverse-phase (RP) chromatography with a C_{18} or C_8 column before or after derivatization of amino acids with *o*-phthaldialdehyde is convenient for detection using a fluorometric detector (10). This is because of simpler solvent systems required for

Table 1. Properties of the Amino Acid Residues of Proteins

Amino acid	Three-letter symbol	One-letter symbol		Molecular weight[a]	pK value of side chain	Hydrophobicity[b]	Bulkiness[c]
		Symbol	Mnemonics				
Alanine	Ala	A	Alanine	71		0.06	11.50
Glutamate	Glu	E	GluEtamate	128	4.3	−0.10	13.57
Glutamine	Gln	Q	Q-tamine	128		0.31	14.45
Either E or Q	Glx	Z					
Aspartate	Asp	D	AsparDate	114	3.9	−0.20	11.68
Asparagine	Asn	N	AsparagiNe	114		0.25	12.82
Either D or N	Asx	B					
Leucine	Leu	L	Leucine	113		3.50	21.40
Glycine	Gly	G	Glycine	57		0.21	3.40
Lysine	Lys	K	Before L	129	10.5	−1.62	15.71
Serine	Ser	S	Serine	87		−0.62	9.47
Valine	Val	V	Valine	99		1.59	21.57
Arginine	Arg	R	ARgining	157	12.5	−0.85	14.28
Threonine	Thr	T	Threonine	101		0.65	15.77
Proline	Pro	P	Proline	97		0.71	17.43
Isoleucine	Ile	I	Isoleucine	113		3.48	21.40
Methionine	Met	M	Methionine	131		0.21	16.25
Phenylalanine	Phe	F	Fenylalanine	147		4.80	19.80
Tyrosine	Tyr	Y	TYrosine	163	10.1	1.89	18.03
Cysteine	Cys	C	Cysteine	103		0.49	13.46
Tryptophan	Trp	W	tW0 rings	186		2.29	21.61
Histidine	His	H	Histidine	137	6.0	2.24	13.67
Undetermined or nonstandard		X					

[a]Weighted mean of relative molecular weight of all residues: 108.7.
[b]Hydrophobicity coefficients (**X**) derived from RP-HPLC of peptides, $\mathbf{X} = (\mathbf{A}^T\mathbf{b})(\mathbf{A}^T\mathbf{A})^{-1}$ where **A** is a sequence matrix and **b** is a mole fraction matrix (5).
[c]Bulkiness (Å) used for computation of protein compressibility (6).

eluting the amino acids from the column and quicker elution (less than 30 min) compared to ion exchange chromatography. The fluorometric RP chromatography can accurately quantify 5 pmol of 50-kDa protein, 2–3 times more sensitive than ion exchange/ninhydrin detector systems. Gas–liquid chromatography (GLC) is also being used for amino acid analysis after derivatization of amino acid. The purpose of this derivatization is to make the amino acids volatile by changing them to, for instance, *N(O,S)*-isobutoxycabonyl methyl esters among many other derivatization techniques. All serum amino acids were measured by gas chromatography (GC) with flame ionization detector using a DB017 capillary column in 9 min. Overall recoveries of amino acids added to serum samples were 88–108% (11). Recently, isotope dilution GC/electron capture ionization mass spectrometry was reported for less than 100 fmol amino acids in protein hydrolysates (12).

Progress has been made recently also in the RP/HPLC after precolumn derivatization. Rapid derivatization and direct injection without cleaning is now feasible by using 6-aminoquinolyl-*n*-hydroxysuccinimidyl carbamate for derivatization for fluorometric detectors (13). By using phenylisothiocyanate for derivatization, a UV detector that is more popular on HPLC instruments can be used with similar detectability as fluorometric detectors (14).

PRIMARY STRUCTURE

The backbone of polypeptide chains is described by two dihedral angles per residue: φ and ψ at the C^α atom. In a peptide linkage -NH-C^αH(R)-CO-NH-, the angles of -NH-C^α-, -C^α-CO-, and -CO-NH- are angles φ, ψ, and ω, respectively (Fig. 1). The two broken lines in Figure 1 show the location of peptide linkages with an amino acid residue -NH-C^αH(R)-CO- between these two linkages.

The genetic code involving the sequence of purine and pyrimidine bases in the DNA strand determines the se-

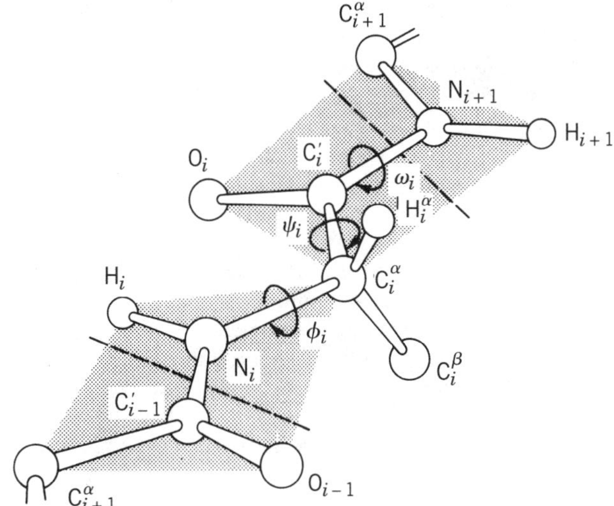

Figure 1. Polypeptide chain linking two peptide units. The chain is shown in a fully extended conformation with all φ_i, ψ_i, and ω_i angles greater than 180°. *Source:* Ref. 15.

quence of amino acids along the protein chain. Each group of three bases along the messenger RNA strand specifies a particular amino acid, and the sequence of these triplet groups dictates the sequence of the amino acids in the proteins. The genetic code is, however, degenerate; that is, most amino acids are coded by more than one codon. Therefore, a nucleotide sequence cannot be derived from the colinear amino acid sequence. But an unknown amino acid sequence can be deduced from the colinear nucleotide sequences. Thus, protein sequence can be determined by analyzing the underlying nucleotide sequence.

Sequence Analysis

For peptide sequence analysis, solid-phase Edman degradation has been used. Lately, gas-phase microsequencing has become popular. First, a phenylthiocyanate or other similar isocyanate solution is added, and then lower alkylamines, eg, triethylamine, in vapor are delivered by a stream of argon gas to convert the peptide sample on a porous support into a phenyl thiocarbamylated peptide. Then trifluoroacetic acid or another similar fluoride-containing organic acid in vapor form is delivered to liberate the phenylthiohydantion (pth) amino acid derivatives. The resultant pth-amino acids are extracted and analyzed by RP/HPLC. This cycle is repeated until it reaches the COOH terminus of the peptide chain (16). In the case of larger protein molecules, fragmentation to small peptides is needed since there are limits on the chain length that can be analyzed.

The advent of capillary electrophoresis in conjunction with nanoelectrospray mass spectrometry enabled the Edman sequencing of peptides and proteins using 5–10 picomoles of materials (17).

Recent developments in nucleic acid cloning and sequencing techniques have made the determination of the amino acid sequence of a protein straightforward, inexpensive, accurate, and rapid (18). For analysis of DNA sequences, two methods are widely used: the enzymatic dideoxy method and the chemical method. The difference is primarily in the technique used to generate the ladder of oligonucleotides. In the enzymatic dideoxy sequencing method, a DNA polymerase is used to synthesize a labeled, complementary copy of a DNA template. In the chemical sequencing method, a labeled DNA strand is subjected to a set of base-specific chemical reagents. A set of radiolabeled single-stranded oligonucleotides is generated in four separate reactions, either enzymatically or chemically. In each of the four reactions, the oligonucleotides have one fixed and one end that terminates sequentially at each A, T, G, or C, respectively. The products of each reaction are fractionated by electrophoresis on adjacent lanes of a high-resolution polyacrylamide gel. After autoradiography, the DNA sequence can be read directly from the gel.

Characteristic Properties of Amino Acids in Protein Structure

Each amino acid residue has its own characteristic property that cannot be replaced by another residue. All amino acid residues, except glycine, have characteristic side chains on the main chain linked through peptide linkage (2). Glycine increases flexibility to the main chain because it has no side chain but with only two hydrogens. A polypeptide chain at a glycine residue has considerably more conformational freedom than at any other residue. Alanine is the smallest nonpolar residue that does not have much preference on whether it is located inside or on the surface of a protein molecule. The nonpolar side chains of valine, isoleucine, and leucine are branched, thus restricting internal flexibility. Phenylalanine has the largest nonpolar side chain. Because of the single methylene group on C^{β} as with the other two aromatic side chains (tyrosine and tryptophan), the side chain flexibility is restricted. Tyrosine has by far the most reactive side chain of the three aromatic residues because of its hydroxyl radical.

The side chain of proline is characteristic because the last atom of the side chain is bonded to the main chain N atom, forming a ring structure. This prevents the N atom from participating in hydrogen bonding and also provides a steric hindrance to the α-helical conformation (19). Consequently, proline has the smallest degree of conformational freedom of all the amino acids, and a polypeptide chain at a proline residue has appreciably less conformational freedom. This characteristic property of proline can be used for stabilizing protein structure; however, its location should be carefully chosen so that the new residue should neither create volume interferences nor destroy stabilizing noncovalent interactions (20). Proline introduction to the N-terminal end of active-site helix of *Bacillus stearothermophilus* neutral protease improved thermostability. The glycine residues on the *N*-terminal side of proline residue relaxed the possible strain that resulted from proline introduction, thereby increasing molecular rigidity (21). Methionine has a rather flexible side chain with a sulfur in a thioether bond. This sulfur introduces an electrical dipole moment.

All the larger nonpolar residues, namely valine, isoleucine, leucine, phenylalanine, proline, tryptophan, and, to a lesser extent, methionine, are predominantly in the inside of protein molecules. Polar (uncharged) side chains form hydrogen bonds, for instance, serine and threonine have hydroxyl groups that form hydrogen bonds (1). Cysteine plays a special role by forming disulfide bridges between different parts of the main chain. A group can be activated in protein through specific hydrogen bonds, such as serine acting as a donor to an unprotonated imidazole group of a histidine. Such charge relay systems form an essential part of the active site of serine proteases. In both serine and threonine, the hydroxyl group can react with acids to form esters via enzyme-catalyzed reactions. Both are common sites for phosphorylation, fatty acid esterification, and glycosylation in proteins. The acid amides of asparagine and glutamine can also form hydrogen bonds. The amido groups function as hydrogen donors and the carbonyl groups as acceptors. The side chain of glutamine is more flexible than that of asparagine because of its extra methylene group. The polar hydroxyl group of tyrosine forms relatively strong hydrogen bonds.

Histidine has a heterocyclic aromatic side chain with a pK value of 6.0. In the physiological pH range, its imidazole ring can be either uncharged or charged. This chemical equilibrium is suitable for catalyzing reactions. This is one of the reasons why histidine is found in several of the active sites of enzymes. Aspartate and glutamate, usually located at the molecular surface, are negatively charged at physiological pH. Because of the short side chain, the carboxyl groups of aspartate are relatively rigid. This may be a reason why the carboxyl groups of active sites of enzymes, eg, aspartyl proteinases, are mainly provided by aspartates and not glutamates.

Most of the positively charged lysine and arginine residues are also at the molecular surface. They are long and flexible and do not usually adopt a defined conformation. The surface net charges of these residues in counterbalance with surface hydrophobicity increase the solubility of globular proteins (22). Sometimes they participate in internal salt bridges or in catalysis. The ϵ-amino group of lysyl residues, and to a lesser extent the guanidinium groups of arginyl residues, are the target of enzyme action, which either modifies the side chain or cleaves the peptide chain at the carboxyl end of lysyl and arginyl residues of substrates. The ϵ-amino group of lysine is considered to be the second-most reactive group in proteins, second to the cysteine sulfhydryl group. The most famous reaction of the ϵ-amino group of lysine is the formation of a Schiff base by the so-called Maillard reaction with aldehydes.

The folding process of a polypeptide chain depends on the hydrophobicity of the side chains, because the formation of a hydrophobic core in the globule seems to be one of the essential driving forces in folding. The hydrophobicity of amino acid residues is dependent on water-accessible surface area and dipole content.

SECONDARY STRUCTURE

Secondary structures are regular arrangements of the backbone of the polypeptide chain. They are stabilized by hydrogen bonds between peptide amide (>N-H) and carbonyl (>C=O) groups. α-Helices and β-sheets make up about half the secondary structures of globular proteins, with the remainder occurring as tight turns, small loops, and random coils.

Helices

The backbone of a polypeptide chain forms a linear group, if its dihedral angles ($\varphi \cong -60°$ and $\psi \cong -50°$) are repeated. An α-helix with 3.6 amino acid residues per turn has a rise/residue ratio of 0.15 nm. The C=O group of each residue at position i is hydrogen bonded to the NH group at position $i + 4$. The polypeptide backbone forms a cylinder of 0.23 nm radius in which the backbone atoms are close-packed in van der Waals contact with each other inside the helix. The hydrogen-bonding pattern generates a loop of connectivity through covalent and hydrogen bonds that includes 13 atoms in four turns of the helix (a 4_{13} helix).

In addition to the α-helix, a shorter helix, a 3_{10} helix, is sometimes observed in protein molecules. Its energetically disadvantageous geometry restricts frequent occurrence of the 3_{10} helix in proteins; usually pieces of about one turn are observed. These pieces tend to be at the N and C termini of α-helixes.

A third helical structure is the polyproline II helix found in collagen (gelatin). This is an extended, left-handed helical structure with 3.3 residues per turn generated by dihedral angles of $\varphi \cong -80°$ and $\psi \cong +150°$. Stability of this structure relies on the specific conformational constraints of the proline residues by locking into $\varphi \cong -80°$ and not by hydrogen bonds.

β-Pleated Sheet

Extended polypeptide chains can associate by hydrogen bonding to form sheetlike structures in parallel β-sheets, with all β-strands pointed in the same direction, and in antiparallel β-pleated sheets, with alternate strands pointed in opposite direction. Individual β-strands or β-pleated sheets have a structure generated by a regular sequence of dihedral angles ($\varphi \cong -120°$ and $\psi \cong +140°$) that orient the NH and C=O bonds nearly perpendicular to the long axis of the polypeptide chain. Also, the idealized β-structure based on 2.0 residues per turn originally proposed was a flat pleated sheet structure; actual β-pleated sheets found in proteins usually have a left-handed twist. Globular proteins contain about 15% sheet structures. The average length of β-structure is about 6 residues, which is equal to 20 Å in length, corresponding approximately to the diameters of domains.

Reverse Turns (β-Turns)

An energetically economical and space-saving way of changing direction in a polypeptide chain involves four amino acid residues, which are often joined to chains in the β-sheet conformation with a 180° chain reversal. Reverse turns (or loops) consist mostly of hydrophilic residues. Reverse turns are common structural elements; approximately one-third of the amino acid residues of proteins is found in this conformation. The best-known examples are β-hairpins that link two adjacent strands of an antiparallel β-pleated sheet. They are stabilized by hydrogen bonds connecting the C=O on residue i to the NH on residue $i + 2$ or $i + 3$. At least, 15 different variants of reverse turns have been described and categorized (23). Because of the geometric constraints required for the turn, certain classes of turns have preferences for specific amino acids (glycine or proline) at specific positions. β-Turns linking contiguous β-strands are of considerable interest to biochemists because they are usually at the protein surface and often are involved in molecular recognition and antigenicity.

Examples of secondary structures are shown in Figure 2; the ribbon-like coils at both right and left sides are α-helices; parallel β-sheets are at the center; a β-turn links the antiparallel β-sheet pair at the upper-right-hand corner (this is the reason for naming of β-turn).

DETERMINATION OF SECONDARY STRUCTURE

A variety of methods have been used to determine secondary structures of proteins. This section outlines those methods.

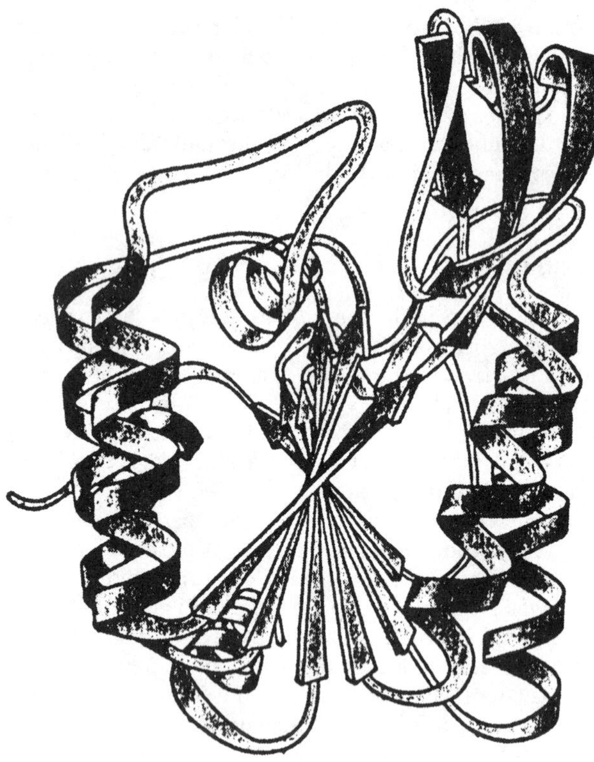

Figure 2. Secondary structure of phosphoglycerate kinase domain 2. *Source:* Ref. 24, with permission.

Infrared Spectroscopy

Assignments of the amide I band to various secondary structures in H_2O and D_2O are feasible by infrared spectroscopy. Amide I–III bands are caused by a vibration in the plane of the amide group of the polypeptide backbone; C=O stretch, N-H bend, and C-N stretch and N-H bend are the main vibrations for amide I, II, and III, respectively. The individual secondary structure fractions yield distinct maxima. However, the resolution of these structures in a globular protein that contains several of these structures is difficult because of the large half-widths of these structures, thereby interfering with their resolution. Least-squares optimization and Fourier deconvolution procedures have been used for the analysis of overlapping bands. Derivative spectroscopy is an alternative method for assessing the number and peak frequencies of the component peaks. The second derivative infrared spectra of various proteins have shown peaks associated with α-helical, β-sheet, and β-turn to be resolved and information pertaining to some amino acid side chains to be obtained.

Infrared spectra of 13 globular proteins in water were obtained in 1800–1480 cm^{-1}. High correlation coefficients of 0.80–0.99 were obtained between the infrared and X-ray estimates of ordered helix, disordered helix, ordered β-structure, disordered β-structure, turns, and remainder (25).

Laser Raman Spectroscopy

Raman and infrared techniques are complementary. Raman lines, derived-by a light-scattering effect measured by the emission of radiation from a vibrational exited state, are related to molecular polarizability originating from induced dipole moment, whereas infrared absorption is associated with the permanent dipole moment during molecular vibration. The amide I and II bands are strong in the infrared spectrum, whereas the amide I and III bands are prominent in the Raman spectrum. By analyzing the amide I Raman band directly as a linear combination of amide I bands of proteins whose secondary structures are known from X-ray data, excellent correlation between Raman and X-ray diffraction estimates was obtained for helix, β-strand, and β-turn composition (26).

Nuclear Magnetic Resonance Spectroscopy

The nuclear magnetic resonance (NMR) method for studying protein conformation is based on the fact that "nuclei of the same element in different chemical environments give rise to distinct chemically shifted spectral lines." The use of chemical shifts for the identification of specific secondary structures, however, has not been successful in determining secondary structures other than α-helix. The introduction of the nuclear Overhauser effect (NOE) into NMR has provided a network of short, intramolecular distance constraints between distinct locations along the polypeptide chain, which has allowed the characterization of other secondary structures. NOE is the enhancement in amplitude of the spectral line of nucleus observed during irradiation of one of a pair of coupled nuclear spins; irradiation of nuclei can cause changes in the population of the spin-state of the nuclei. Proton–proton NOE distance information was used for identification of polypeptide secondary structures in noncrystalline proteins. The combined information on all of the distances obtained from visual inspection of the two-dimensional NOE (NOESY) spectra was sufficient for determination of helical and β-sheet secondary structures in small globular proteins.

Although for larger proteins magnetic relaxation becomes a limiting factor, the benefits of using uniform high-level (>96%) deuteration was shown to inhibit relaxation processes (27). Complete deuteration provides significant signal-to-noise enhancement in heteronuclear NMR assignment and structure determination experiments, which uses the amide proton for detection. NMR pulse sequences lose sensitivity as the size of the protein under study increases above 25 kDa, mainly because of fast ^{13}C transverse relaxation via the strong dipolar coupling between a ^{13}C nucleus and its directly bonded protons. Since the gyromagnetic ratio of ^{2}H is 6.5 times smaller than that of ^{1}H, per deuteration dramatically reduces this relaxation.

A recent news release stated that "A technique called TROST (transverse relaxation–optimized spectroscopy) now greatly reduces NMR line broadening with increasing molecular size, permitting NMR analysis of molecules far beyond 100 kDa in size" (28).

Optical Rotatory Dispersion and Circular Dichroism

It is generally accepted that optical rotatory dispersion (ORD) and circular dichroism (CD) are a direct reflection of protein secondary structure. CD uses circularly polarized light (180–260 nm) to illuminate the specimen instead of plane polarized light in the case of ORD. The inherent problem of ORD curve analysis (ie, ORD absorption bands are infinitely broader compared to CD bands) has resulted in increasingly popularity of CD spectra analysis. Computer programs have been developed to compute the contents of α-helix, parallel and antiparallel β-sheets, β-turn, and random coil. Efforts have been made to expand the spectral range into the far ultraviolet, as low as 180 nm, for more detailed characterization of the CD spectra; thus, more accurate information can be obtained for the five forms of secondary structure. However, mechanical difficulty of maintaining high vacuum during measuring CD spectra restricts broad use of this approach. Instead, spectral analysis using advanced computer-aided classification (eg, artificial neural networks) is increasingly more popular (29).

PREDICTION OF SECONDARY STRUCTURE AND MOLECULAR MODELING

Many attempts have been made to predict protein structures from their primary sequences using a variety of parameters such as amino acid frequency, energy calculation, and φ and ψ values. As a result of these studies, it was possible to identify and quantify the secondary structure of proteins based on sequence data alone, with some degree of accuracy. In one of the most common of the secondary structure prediction methods, a statistical survey was conducted for 15 proteins in which the α-helix, β-sheet, and β-turn conformational potentials of all 20 amino acids were established (30). A set of empirical rules was then derived that allowed for the determination of the folding of the secondary structure regions in the proteins. Computer programs for this prediction, including programs for personal computers, have been published.

Although numerous prediction methods, in addition to the above Chou-Fasman approach (30), have been developed, the accuracy of these methods has not been very high, which is undoubtedly a result of the probabilistic nature of the methods (31). Moreover, protein structure hierarchy is not so rigorous as to assume that the secondary structure formation of a certain segment is entirely dependent on the sequence information alone; other segments at a distance along the chain also exert an influence.

To improve the correctness of prediction, attempts have been made by using sequence homology, multivariate analysis, and especially energy minimization. It is generally agreed that the maximum correctness is expected to be far less than 80%. In contrast, molecular modeling is based on known three-dimensional structures of homologous molecules with the expectation of maximum accuracy in prediction of about 80% (32).

TERTIARY AND QUATERNARY STRUCTURES

Tertiary structure refers to the three-dimensional structure of a polypeptide chain. The three-dimensional structure of multisubunit proteins is then described by the term quaternary structure, which refers to that structure resulting from the interactions between polypeptide chains, frequently through so-called self-association. However, in general, the three-dimensional structure includes all types of structure above the level of primary structure.

Three-dimensional structures of several hundred proteins have been determined by X-ray crystallography, yielding the location and bonding of all the atoms in the polypeptide chains. Detailed three-dimensional protein structure can also be determined by NMR methods (33), which are advantageous because the structure of proteins in solution can be determined in contrast to the requirement of protein crystals in X-ray diffraction measurement. It is now possible to determine the three-dimensional structure of proteins with molecular weights up to 20,000 routinely using NMR, provided that the protein can be obtained in large-enough quantities and is soluble and stable at room temperature over a period of days. This limit of molecular sizes is quickly increasing as described above (28).

Oligomeric proteins and multienzyme complexes such as pyruvate dehydrogenase are representative of the lowest level of macromolecular structural organization. "Supramolecular structures," such as ribosomes or the membranous components of the electron-transport chain, are examples of higher levels of macromolecular organization, which is the structural basis of life (3).

MOLECULAR FORCES

With the exception of disulfide bonds, molecular forces to maintain integrity of the structure are noncovalent, which are one to three orders of magnitude smaller than covalent energy.

Dispersion Forces

Dispersion forces occur between any pair of atoms. Each atom behaves like an oscillating dipole generated by electrons moving in relation to the nucleus of the atom. In a pair of atoms, the dipole in an atom polarizes the opposing atom. As a result, the oscillators are coupled, giving rise to an attractive force between the atoms. The theory of the attractive interaction between atoms is attributed to London (34), and such forces are often called London forces, or dispersion forces. They are long-range forces, arising from attractive forces between atoms over an extended distance.

It was shown by London, following the appearance of quantum theory, that this relationship is also applicable to the dispersion energy between two nonpolar molecules due to polarization of the electron density in one molecule by charge fluctuation in the other. Except in highly polar materials, it is the sum of London dispersion forces that contributes most to the total van der Waals attraction between macroscopic bodies.

Repulsive Forces

Short-range forces are repulsive, having their quantum mechanical origin in the overlap of electron density of adjacent molecules. Attractive and repulsive forces act simultaneously and cannot be isolated.

Electrostatic Interactions

Because covalent bonds between different types of atoms lead to an asymmetric bond electron distribution, most atoms of a molecule carry partial charges. Because a neutral molecule has no net charge, it contains only dipoles or higher multipoles. These multipoles interact with each other according to Coulomb's law.

$$U = kq_1q_2/Dr \qquad (1)$$

where U is the energy of association of two electric charges, q_1 and q_2, that are separated by the distance r. D is the dielectric constant of the medium.

Salt Bridges

The association of two ionic groups of opposite charge is known as an ion pair or salt bridge with a short r such as 4 Å in equation 1. There are only a few salt bridges in proteins, with the exception of phosphoproteins and some glycoproteins. These salt bridges amount to about 10–20 kcal/mol for adjacent carboxylate and ammonium groups. This value is higher than 3 kcal/mol of hydrogen bonds (35). Salt bridges are formed by about one-third of the charged residues; only 20% of these bridges are buried.

van der Waals Potentials

It is customary to combine all three noncovalent forces (ie, electron shell repulsion, dispersion forces, and electrostatic interactions) into a single simple potential function, or force field, that is called the van der Waals potential for historical reasons.

Hydrogen Bonds

The interatomic distance of monovalent atomic contacts are significantly shorter than the value calculated as the sum of the corresponding van der Waals radii when one of the atoms is hydrogen. Because the entire electron shell of hydrogen is appreciably shifted onto the atom to which hydrogen is covalently bound, the shell repulsion between contact partners is small, and the attracting charges can approach each other more closely. Such a short distance approach gives rise to a high attractive Coulomb energy and, therefore, a high dispersion energy.

Hydrophobic Interaction

The hydrophobic effect arises from the unfavorable interactions between water molecules and the nonpolar residues of a protein. When a hydrocarbon molecule is introduced into water, it induces changes in water structure, frequently decreasing its entropy. To minimize this unfavorable entropy change, the nonpolar molecules are forced to coalesce into dipoles or globules, reducing their surface of contact with water. As a result, the protein chain is forced to fold into a micellar structure with the hydrocarbon moiety on the inside of the globule and the polar groups on the outside. The removal of the nonpolar residues from contact with water makes a major contribution to the free energy of conformational stabilization. Thus, in this theory, the hydrophobic effect is purely the result of the phobia of water when contacting with hydrocarbons.

Disulfide Bonds

Disulfide bonds between pairs of cysteine residues can cross-link different chains of a protein, which gives rise to covalent chain assembles. Disulfide bonds occur between different segments within a single polypeptide chain as well. These bridges play an important role in rheological properties of food proteins. For instance, the cohesive elastic character of wheat flour doughs is based on the disulfide bonds of glutenin, and the three-dimensional network of disulfide bonds in glutenin is responsible for the difficulties encountered during wet milling of corn. Rearrangement of disulfide bonds in a protein by sulfhydryl–disulfide interchange reactions plays an important role in the regulation of activities of enzyme and immunoglobulins. By the catalytic action of sulfhydryl groups in low molecular weight compounds, eg, cysteine, glutathione and mercaptoethanol, proteins can be linked to each other to form large aggregates through a chain reaction. However, in the presence of higher concentrations of these reducing compounds, disulfide bridges in proteins are reduced to sulfhydryl groups, thereby decreasing the molecular weights as well as viscosity.

The measurement of sulfhydryl and disulfide contents of proteins is a very useful tool to assess protein structure. Raman spectroscopy is advantageous because sulfhydryl and disulfide groups appear as peaks in the spectrum, and even solid samples such as gels can be used for analysis with minimum pretreatment (36).

INTERACTIONS

Protein interactions are generally classified as the interactions of proteins with macromolecules and with small molecular weight compounds; a typical example of the former is a protein–protein association and that of the latter is a protein–lipid or protein–water interaction. They are an important property in elucidating the mechanisms of food protein functionality.

Protein–Protein Interaction

Gel filtration chromatography and gel electrophoresis are the most popular methods to determine changes in the molecular weight of proteins caused by interaction. However, to study chemical equilibrium, methods in which zonal separation occurs should be avoided. For this reason, frontal gel filtration (37), sedimentation equilibrium ultracentrifugation (38), light-scattering spectroscopy (39), and fluorescent polarization (40) are appropriate to use. Association constants of food proteins are in the order of $10^4\ M^{-1}$ as reported for κ–α_{s1}-casein and lysozyme–ovalbumin interactions (41). These values are lower than the order of 10^6–$10^7\ M^{-1}$ for inverse Michaelis-Menten constant ($1/K_m$) of enzymes vs substrates, 10^6–$10^9\ M^{-1}$ for antibodies vs antigens, and $10^{14}\ M^{-1}$ for avidin–biotin interaction, which is the strongest noncovalent interaction found in nature (38,42). The specific, strong avidin–biotin interaction has been used for chemical analysis, especially in immunoassay.

Protein–Lipid Interaction

Lipoproteins exist naturally in food systems such as egg yolk. However, lipids can also interact with proteins during food processing, thereby affecting the quality of foods such as bread and frozen fish. Protein–lipid interaction is the most important mechanism of emulsion formation. Analytical methods to measure this interaction include density gradient ultracentrifugation, fluorometry (43), gel filtration chromatography, and ultrafiltration. More sophisticated methods are electron spin-resonance spectroscopy using spin-labeling techniques with probes (44) and Raman spectroscopy (45).

Protein–Water Interaction

It has long been recognized that protein–water interactions play an important role in the determination and maintenance of the three-dimensional structure of proteins. Water in proteins involves stability, dynamics, and function of the proteins (46). Important states of water molecules surrounding protein molecules are structural water and monolayer water. The structural water refers to the water molecules that are part of the protein structure bound through hydrogen bonding. These 10–20 water molecules per protein molecule are unavailable for chemical reactions. The monolayer water refers to the water molecules tightly bound to the protein surface via dipole-induced dipole (hydrophobic hydration), ion–dipole (ionic hydration), and dipole–dipole (hydrogen bonding) interactions. At the saturated monolayer coverage, most proteins absorb about 0.3–0.5 g of water per gram of protein. This bound water is unfreezable and does not participate in any chemical reaction. Unlike the monolayer water, the multilayer water is available for chemical reactions, but some of these water molecules are unfreezable (47).

Most proteins exhibit the least water-binding capacity and solubility at isoelectric pH, presumably because of protonation of the carboxyl groups and enhanced hydrophobic interaction between protein molecules (47).

CHANGES IN PROTEIN STRUCTURE DURING FOOD PROCESSING

Physical treatments that modify protein structure include the use of thermal energy, mechanical energy, or pressure. Common examples of physical processes that alter food proteins are heating, freezing, radiating, extrusion, salting, fiber spinning, sonication, whipping, emulsification, and storage. These processes result in denaturation of the proteins. Denaturation is usually accompanied by unfolding of the protein molecules, without any apparent loss in solubility as long as they are monomeric. The unfolding step is frequently followed by aggregation, which may lead to loss of solubility. Thus, changes in solubility are frequently used as an indicator of protein denaturation. Determination of molecular weights by gel filtration chromatography and gel electrophoresis has been commonly used also for investigating protein denaturation (48).

The presence of a partially folded conformation called a molten globule state, that is, a kinetic intermediate of reversible denaturation equilibrium, was reported in many globular proteins (49). The molten globule has a native-like backbone secondary structure, and the side chain's environment undergoes a denaturation-like alteration. Hydrophobic clusters are exposed, as reflected in increasing binding of a hydrophobic fluorescent probe. It is postulated that molten globules are involved in the functional properties (eg, emulsification, foaming, and gelation) of food proteins (49).

In emulsions of oils in protein solutions, the proteins change from their native conformations to forms that depend largely on the hydrophobic interactions between the surface and the amino acid side chains in the protein. Ideally, hydrophobic side chains will be close to the oil surface and hydrophilic residues will favor the aqueous phase, but this will be constrained by the distribution of the amino acid residues in the protein. The conformation that is adopted also depends on the surface area that the protein is required to cover (47,50). Dickinson (51) classified milk proteins into two distinct classes: the disordered casein and the globular whey proteins. Substantial differences exist between these two classes in terms of adsorbed layer structure and surface rheological properties at the oil–water interface. Computer simulation showed promise for modeling the behavior of hypothetical proteins and peptides (51).

In the case of foam formation, the adsorption of proteins at the air–water interfaces has often been described in terms of three processes. They are transportation from bulk solution to the interface, penetration into the surface layer and reorganization of structure (surface denaturation) of the protein in the adsorbed layer (47,52). This process is similar to the case of emulsion formation, but the detail in structure changes at the air–water interface is unclear, mainly because of difficulty in analysis.

Conformational changes of proteins when the protein solutions are gelled upon heating are reviewed by Matsumura and Mori (53). It is interesting to note that β-sheet structure, which is mostly association units for gel formation, seems to be due to the formation of an intermolecular β-sheet structure induced by association between denatured molecules. That is, β-sheet formation itself probably has no direct relation to the denaturation step of globular proteins.

A similar phenomenon of aggregation of β-structure-forming (ie, cross β-fold) is observed in amyloid fibrinogenesis. Tissue deposition of soluble autologous proteins as insoluble amyloid fibrils is associated with serious diseases, including systemic amyloidosis, Alzheimer's diseases, and transmissible spongiform encephalopathy. Two naturally occurring human lysozyme variants were both amyloidogenic (54).

Modeling has been attempted to study the relationships between processing conditions and the resultant structural changes, using a variety of multivariate analysis techniques (55). The most advanced methods may be partial least squares regression (PLS), artificial neural networks, and genetic algorithm. Many commercial computer programs for these modeling methods are available.

For assessing the folding stability of protein molecules, in addition to the traditional spectroscopies (UV, fluorescence, and CD) calorimetry, solvent stabilization (56), a novel method of hydrogen exchange using NMR (57), and electrospray mass spectrometry (58) are being used. The exchange of the protons of buried backbone NH groups with those of the solvent is measured. Peptide NH groups that are exposed to solvent exchange rapidly, but those that form stable hydrogen bonds within the protein, or on occasion are simply buried, exchange slowly. The rate of $H \rightleftharpoons D$ exchange of individual groups can be measured conveniently using either normal protein dissolved in D_2O or D-labeled protein dissolved in H_2O.

STRUCTURE–FUNCTION RELATIONSHIPS

Structural changes of food proteins usually take place during processing. Because the structure closely relates to food quality, optimization of processing conditions to obtain the best quality is one of the most important objectives of food protein research. From the aspect of structure–function relationships of food proteins, there are three categories of structure changes. (*1*) The purpose of processing is not for changing structures, but the processing results in structure changes. An example is pasteurization or sterilization. (*2*) The structural changes are the main purpose of processing, such as in breadmaking. (*3*) A new type of structure change to be introduced, including chemical, enzymatic, and genetic changes of protein structure, optimizes these changes to achieve the best functions. The functions include all the chemical, physical, and biological properties of protein molecules. The use of bioactive peptides or proteins found in natural resources as ingredients in food is quickly becoming a new trend for contributing to human health. A new concept of chemopreventive agents is being introduced into diet to prevent diseases (59).

Meanwhile, a new paradigm is being discussed for protein research. A novel concept of protein substructure based on a knot-matrix construction principle revealed a much more powerful paradigm for protein research than that based on protein secondary structures (60). For macromolecular crystal structure analysis, the simple Debye-Waller model has been widely used. In this treatment of atomic motion, the probability of finding an atom in a given distance x from its equilibrium position x_0 is Gaussian. If it is assumed that the motion is isotropic, the model states that the motion in any direction can be characterized in terms of a mean-square vibration amplitude, $\langle x^2 \rangle$, also termed the mean-square displacement. The X-ray scattering from each atom is modified by a Gaussian function that is related to the mean-square displacement of that atom. The form of the Gaussian is

$$\exp(-B \sin^2\theta/\lambda^2)$$

where θ is the Bragg angle, λ is the wavelength of the incident radiation, and B is related to the mean-square displacement by

$$B = 8\pi^2\langle x^2 \rangle \quad (2)$$

B is called Debye-Waller factor (59). The B factor provides useful information about conformational dynamics, which is calculated for each nonhydrogen atom.

Three groups of B factors are distinguished: group I, from 2 to 8 Å^2; group II, 8–14 Å^2; and group III, all higher values. Functional domains consist of one group I substructure (knot) to which is tethered most of the group II atoms (matrices) of the domain. Group III atoms are restricted to surface. In the knot-matrix construction principle, which explains properties of protein molecules, especially functions, knots determine palindromic B factor patterns and matrices put them to work. Palindromic patterns can be found on B factor vs atom-number plots as two functional domains against the residue lying halfway between them. The palindromic patterns control specificity of proteins such as enzymes and antibodies (60).

Optimization of Structural Changes

When structural changes are directly related to functions, such as loaf volume of bread, the model yielded from the correlation computation can be used to predict the structure parameters. Based on the response surface graphically illustrated, the optimum may be able to be located.

However, when the processing data are inadequate to conduct the modeling computation, such as creating new processing procedures or new products, computer-aided optimization could be the best solution. Again, there are many commercial computer programs available for optimization. If it is difficult to select the best method for achieving the objectives, contacting Network Enabled Optimization Center (NEOS) is recommended. NEOS is the optimization technology center that is a joint enterprise of Argonne National Laboratory (Argonne, Ill.) and Northwestern University (Evanston, Ill.).

Recently, a unique experimental optimization technique was proposed (62). Random-centroid optimization (RCO) uses simple algorithms since there is no need for complicated advanced algorithms. The method has been successfully applied to chemical synthesis as well as food processing projects.

Quantitative Structure–Activity Relationship Computation

Since the introduction of computer-aided study of quantitative structure–activity relationships (QSAR) to explain the functions of chemical compounds (63), many computer programs have been used for modeling. Instead of using processing conditions as independent variables, hydrophobic, steric, and electronic parameters were used in *QSAR* at the beginning.

In the case of food proteins, hydrophobicity and solubility were first used for *QSAR* computation (55) because of unavailability of data on structure and electronic parameters. This was one step ahead of the solubility–function relationship in the 1970s. Although the electronic parameter can be replaced by charge density, the steric parameter is still difficult to define for food proteins.

Chemical and Enzymatic Modification for Improving Functionality

To improve functions of proteins, modification of charge density, hydrophobicity, and molecular size has been conducted (64). Cross-linking and hydrolysis are two major techniques for modifying the size of protein molecules.

Genetic Engineering to Improve Protein Functions

Genetic engineering is quickly becoming an important technique in food processing for using engineered products as bioactive ingredients. This is because of successful developments of high-level expression techniques, thereby yielding near 1 g of recombinant proteins per liter of culture media (65). To avoid possible allergenicity or side effects, human peptides or proteins can be produced by genetic engineering to use in diet at reasonable costs. Examples are human lysozyme in infant formula and human cystatins as chemopreventive ingredients. The random-centroid optimization program modified for genetic study (RCG) was intended for use in site-directed mutagenesis (66). The RCG can be used to improve biological functions or to eliminate adverse effects. Structural changes could be a result of these mutations. Study on changes in molecular structure should be conducted to elucidate the mechanism of function of protein molecules under investigation. Since the difficulty in defining structure parameters is involved, the strategy to change structure to optimize function is usually hard to establish, despite the effort being made as described above (60). An advantage of RCG is that there is no need of *a priori* information on the structure–function relationships. Therefore, experimental optimization such as RCG can be a useful approach at the present stage of progress in protein chemistry.

ACKNOWLEDGMENTS

The author is greatly indebted to Dr. John R. Whitaker, University of California at Davis, for his invaluable advice. Without Dr. Whitaker's profound knowledge in this area, there would not be the present form of this article.

BIBLIOGRAPHY

1. R. D. Ludescher, "2. Physical and Chemical Properties of Amino Acids and Proteins," in S. Nakai and H. W. Modler, eds., *Food Proteins: Properties and Characterization*, VCH Publishers, Weinheim, Germany, 1996.
2. G. E. Schulz and R. H. Schimer, *Principles of Protein Structure*, Springer-Verlag, New York, 1979.
3. D. Voet and J. G. Voet, *Biochemistry*, John Wiley & Sons, New York, 1990.
4. H. R. Matthews, R. Freedland, and R. L. Miesfeld, *Biochemistry, A Short Course*, Wiley-Liss, New York, 1997.
5. M. C. J. Wilce, M.-I. Aguilar, and M. T. Heam, "Physicochemical Basis of Amino Acid Hydrophobicity Scales: Evaluation of Four New Scales of Amino Acid Hydrophobicity Coefficients Derived from RP-HPLC of Peptides," *Analytical Chem.* **67**, 1210–1219 (1995).
6. M. M. Gromiha and P. K. Ponnuswamy, "Relationship between Amino Acid Properties and Protein Compressibility," *J. Theoretical Biol.* **165**, 87–100 (1993).
7. S. H. Chiou and K. T. Wang; "Simplified Protein Hydrolysis with Methanesulfonic Acid at Elevated Temperature for the Complete Amino Acid analysis of Proteins," *J. Chromatogr.* **448**, 404–410 (1988).
8. J. Csapo et al., "Determination of the Cysteine Content of Foods and Feeds by Mercaptoethanesulphonic Acid Hydrolysis," *Acta Alimentaria Academiae Scientiarum Hungaricae* **15**, 227–235 (1986).
9. Y. Yokote et al., "Recovery of Tryptophan from 25-Minute Acid Hydrolysates of Protein," *Analytical Biochem.* **152**, 245–249 (1986).
10. J. Haginaka and J. Wakai, "Fluorimetric Determination of Amino Acids by High Performance Liquid Chromatography Using a Hollow-Fiber Membrane Reactor," *J. Chromatogr.* **396**, 297–305 (1987).
11. S. Matsumura, H. Kataoka, and M. Makita, "Determination of Amino Acids in Human Serum by Capillary Gas Chromatography," *J. Chromatogr.* **681**, 375–380 (1996).
12. M. W. Duncan and A. Poljak, "Amino Acid Analysis of Peptides and Proteins on the Femtomole Scale by Gas Chromatography/Mass Spectrometry," *Analytical Chem.* **70**, 890–896 (1998).
13. M. Ward, "Amino Acid Analysis Using Precolumn Derivatization with 6-Aminoquinolyl-*N*-Hydroxysuccinimidyl Carbamate," in J. M. Walker, ed., *The Protein Protocols Handbook*, Humana Press, Totowa, N.J., 1996, pp. 461–465.
14. G. B. Irvine, "Amino Acid Analysis Using Precolumn Derivatization with Phenylisothiocyanate," in J. M. Walker, ed., *The Protein Protocols Handbook*, Humana Press, Totowa, N.J., 1996, pp. 467–472.
15. A. G. Walton, *Polypeptides and Protein Structure*, Elsevier Science Publishing, New York, 1981.
16. M. Haniu and J.E. Shively, "Microsequence Analysis of Peptides and Proteins. IX. Manual Gas-Phase Microsequencing of Multiple Samples," *Analytical Biochem.* **173**, 296–306 (1988).
17. M. D. Bauer, S. Yiping, and F. Wang, "Nano-electrospray Mass Spectrometry and Edman Sequencing of Peptides and Proteins Collected from Capillary Electrophoresis," in D. R. Marshak, ed., *Techniques in Protein Chemistry VIII*, Academic Press, New York, 1997, pp. 37–46.
18. F. M. Ausubel et al., eds., "DNA Sequencing," in *Current Protocols in Molecular Biology*, John Wiley and Sons, New York, 1998, pp. 7.01–7.7.31.
19. C. Branden and J. Tooze, *Introduction to Protein Structure*, Garland Publishing, New York, 1991.
20. C. Vieille and J. G. Zeikus, "Thermozymes: Identifying Molecular Determinants of Protein Structural and Functional Stability," *Trends in Biotechnol.* **14**, 183–190 (1996).
21. S. Nakamura et al., "Improving the Thermostability of *Bacillus stearothermophilus* Neutral Protease by Introducing Proline into the Active Site Helix," *Protein Eng.* **10**, 1263–1269 (1997).
22. S. Hayakawa and S. Nakai, "Relationship of Hydrophobicity and Net Charge to the Solubility of Milk and Soy Proteins," *J. Food Sci.* **50**, 486–491 (1985).
23. T. E. Creighton, *Protein: Structures and Molecular Properties*, 2nd ed., Freeman, New York. 1993.
24. J. King, "Deciphering the Rules of Protein Folding," *Chemical Eng. News* **67**(15), 32–54 (1989).
25. N. N. Kalnin, I. A. Baikalov, and S. Y. Venyaminov, "Quantitative IR Spectrophotometry of Peptide Compounds in Water Solutions. III. Estimation of the Protein Secondary Structure," *Biopolymers* **30**, 1273–1280 (1990).

26. R. W. Williams, "Protein Secondary Structure Analysis Using Raman Amide I and Amide III Spectra," *Methods in Enzymol.* **130**, 311–331 (1986).

27. R. A. Venters et al., "Strategies for NMR Assignment and Global Fold Determinations Using Perdeuterated Proteins," in D. R. Marshak, ed., *Techniques in Protein Chemistry VIII*, Academic Press, New York, 1997, pp. 605–615.

28. S. Borman, "Advance in NMR of Macromolecules, Technique Makes It Possible to Obtain Spectra of Much Larger Proteins," *Chemical and Eng. News* **76**, 55–56 (1998).

29. N. J. Greenfield, "Methods to Estimate the Conformation of Proteins and Polypeptides from Circular Dichroism Data," *Analytical Biochem.* **235**, 1–10 (1996).

30. P. Y. Chou and G. D. Fasman, "Empirical Prediction of Protein Conformation," *Annu. Rev. Biochem.* **47**, 251–276 (1978).

31. R. D. King, "Prediction of Secondary Structure," in M. J. E. Sternberg, ed., *Protein Structure Prediction*, Oxford University Press, Oxford, U.K., 1996, pp. 79–99.

32. N. Srinivasan, K. Gurprasad, and T. L. Blundell, "Comparative Modelling of Proteins," in M. J. E. Sterberg, ed., *Protein Structure Prediction*, Oxford University Press, Oxford, U.K., 1996, pp. 111–140.

33. D. G. Reid, ed., *NMR in Detail: Protein NMR Techniques*, Humana Press, Totowa, N.J., 1997.

34. F. London, " Über einige Eigenshaften und Anwendungen der Molekularkräfte," *Zeitschrift für Phisikalische Chemie, Abteilung B* **11**, 222–251 (1930).

35. K. Hamaguchi, *The Protein Molecule, Conformation, Stability and Folding*, Japan Scientific Societies Press, Springer-Verlag, Tokyo, 1992.

36. M. Nonaka, E. Li-Chan, and S. Nakai, "Raman Spectroscopic Study of Thermally Induced Gelation of Whey Proteins," *J. Agric. Food Chem.* **41**, 1176–1181 (1993).

37. L. W. Nichol et al., "Evaluation of Gel Filtration Data on Systems Interacting Chemically and Physically," *Arch. Biochem. Biophysics* **121**, 517–521 (1967).

38. S. Nakai and M. Nonaka, "Computation of Molecular Weight and Weight Fraction of Five and Six Components in Mixtures from Model Equilibrium Ultracentrifugation Data," *J. Agric. Food Chem.* **40**, 824–829 (1992).

39. G.-M. Wu, D. Hummel, and A. Herman, "Analysis of the Solution Behavior of Protein Pharmaceuticals by Laser Light Scattering Photometry," in ACS Symposium 675, *Therapeutic Protein and Peptide Formulation and Delivery*, American Chemical Society, Washington, D.C., 1997, pp. 168–185.

40. K. L. Bentley et al., "Fluorescence Polarization: A General Method for Measuring Ligand Binding and Membrane Microviscosity," *Biotechniques* **3**, 356–366 (1985).

41. S. Nakai and C. M. Kason, "A Fluorescence Study of the Interactions between κ- and α_{s1}-Casein and between Lysozyme and Ovalbumin," *Biochimia et Biophysica Acta* **351**, 21–27 (1974).

42. F. M. Ausubel et al., eds., "Analysis of Protein Interactions," in *Current Protocols in Molecular Biology*, John Wiley and Sons, New York, 1998, pp. 20.01–20.5.9.

43. L. A. Sklar et al., "Conjugated Polyene Fatty Acids as Fluorescent Probes: Binding to Bovine Serum Albumin," *Biochem.* **16**, 5100–5108 (1977).

44. B. H. Robinson et al., "Application of EPR and Advanced EPR Techniques to Study of Protein Structure and Interactions," *EPR and Advanced EPR Studies of Biological Systems* **1985**, 183–255 (1985).

45. S. Nakai et al., "Quantification of Hydrophobicity for Elucidation of the Structure–Activity Relationships of Food Proteins," in *Interactions of Food Proteins*, ACS Symposium 454, American Chemical Society, Washington, D.C., 1991, pp. 42–58.

46. R. B. Gregory, "Protein Hydration and Glass Transfer Behavior," in R. B. Gregory, ed., *Protein-Solvent Interactions*, Marcel Dekker, New York, 1995.

47. S. Damodaran, "Functional Properties," in S. Nakai and H. W. Modler, eds., *Food Proteins: Properties and Characterization*, VCH Publishers, Weinheim, Germany, 1996, pp. 167–234.

48. A. Kilara and V. R. Harwalkar, "Denaturation," in S. Nakai and H. W. Modler eds., *Food Proteins: Properties and Characterization*, VCH Publishers, Weinheim, Germany, 1996, pp. 71–165.

49. M. Hirose, "Molten Globule State of Food Proteins," *Trends in Food Sci. Technol.* **4**, 48–51 (1993).

50. D. D. Dalgleish, "Adsorption of Protein and the Stability of Emulsions," *Trends in Food Sci. Technol.* **8**, 1–6 (1997).

51. E. Dickinson, "Proteins at Interface and in Emulsions, Stability, Rheology and Interactions," *J. Chemical Soc. Faraday Trans.* **94**, 1657–1669 (1998).

52. P. J. Wilde and D. C. Clark, "Foam Formation and Stability," in G. M. Hall, ed., *Methods of Testing Protein Functionality*, Blackie Academic and Professional, London, 1996.

53. Y. Matsumura and T. Mori, "Gelation," in G. M. Hall, ed., *Methods of Testing Protein Functionality*, Blackie Academic and Professional, London, 1996.

54. D. R. Booth et al., "Instability, Unfolding and Aggregation of Human Lysozyme Variants Underlying Amyloid Fibrillogenesis," *Nature* **385**, 787–793 (1997).

55. S. Nakai, G. E. Arteaga, and E. C. Y. Li-Chan, "Computer-Aided Techniques for Quantitative Structure Activity Relationships Study of Food Proteins," in N. S. Hetiarachchy and G. R. Ziegler, eds., *Protein Functionality in Food Systems*, Marcel Dekker, New York, 1994, pp. 121–145.

56. B. A. Shirley, *Boiled Eggs and Molten Globules—Protein Stability and Folding: Theory and Practice*, Humana Press, Totowa, N.J., 1995.

57. J. Clarke, L. S. Itzhaki, and A. R. Fersht, "Hydrogen Exchange at Equilibrium: A Short Cut for Analysing Protein-Folding Pathways?," *Trends in Biological Sci.* **22**, 284–287 (1997).

58. Z. Zhang et al., "Higher-Order Structure and Dynamics of FK506-Binding Protein Probed by Backbone Amide-Hydrogen/Deuterium Exchange and Electrospray Ionization Fourier Transform Ion Cyclotron Resonance Mass Spectroscopy," in D. R. Marshak, ed., *Techniques in Protein Chemistry VIII*, Academic Press, New York, 1997.

59. P. Greenwald, "Chemoprevention of Cancer," *Scientific American* **275**, 96–99 (1996).

60. R. Lumry, "The New Paradigm for Protein Research," in R. B. Gregory, ed., *Protein–Solvent Interactions*, Marcel Dekker, New York, 1995, pp. 1–141.

61. G. A. Petsko and D. Ringe, "Fluctuations in Protein Structure from X-Ray Diffraction," *Annu. Rev. Biophysics and Bioeng.* **13**, 331–371 (1984).

62. S. Nakai et al., "Optimization of Site-Directed Mutagenesis. 1. A New Random-Centroid Optimization Program for Windows Useful in Research and Development," *Agric. Food Chem.* **46**, 1642–1654 (1998).

63. A. J. Stuper, W. E. Bugger, and P. G. Jurs, eds., *Computer Assisted Studies of Chemical Structure and Biological Functions*, John Wiley and Sons, New York, 1979.

64. S. Nakai, "An Overview," in S. Nakai and H. W. Modler, eds., *Food Proteins: Properties and Characterization*, VCH Publishers, Weinheim, Germany, 1996, pp. 1–21.
65. J. M. Cregg, T. S. Vedvick, and W. C. Raschke, "Recent Advances in Expression of Foreign Genes in *Pichia pastoris*," *Bio/Technol.* **11**, 905–910 (1993).
66. S. Nakai, S. Nakamura, C. H. Scaman, "Optimization of Site-Directed Mutagenesis. 2. Application of Random-Centroid Optimization to One-Site Mutation of *B. stearothermophilus* Neutral Protease to Improve Thermostability," *J. Agric. Food Chem.* **46**, 1655–1661 (1998).

S. Nakai
University of British Columbia
Vancouver, British Columbia
Canada

PULSED LIGHT PROCESSING

A new processing technology uses intense flashes of broad-spectrum white light to kill microorganisms on packaging and food surfaces and in air and water. The flashes have a duration of 100 to 300 millionths of a second and effectively kill all microorganisms (bacteria, fungi, spores, cysts, protozoa, and viruses) in less than a second. The process is effective for sterilization when the light can reach all surfaces (such as the surfaces of packaging materials) or the total volume of a product such as water and other transmissive fluids. Microbial levels on foods, including pathogens, are reduced to extend shelf life and reduce public health risk. The process is marketed under the trade name of PureBright® by PurePulse Technologies (a subsidiary of Maxwell Technologies Inc.), 4241 Ponderosa Avenue, San Diego, California 92123.

PULSED LIGHT

Electrical energy stored in a capacitor can be rapidly released to produce short, intense high-power pulses (Fig. 1). Such pulses are used to generate the intense flashes of PureBright light by electrically ionizing a xenon gas lamp. The broadband emitted light flash has wavelengths from far-UV (200–300 nm) through the near-UV (300–380 nm) and visible spectra (380–780 nm), to the infrared (780–1,200 nm). Approximately 25% of the light is in the UV wavelength range.

Flash duration is typically about 300 μs and the light intensity is greater than 20,000 times that of sunlight. A few flashes applied in a fraction of a second kills all exposed microorganisms, spores, and viruses.

Although the peak power of each flash is very high, the total energy consumed per flash is relatively low and the average operating power requirement modest.

All PureBright systems have a treatment-monitoring capability. Each pulse of light is monitored for total fluence and UV content. A feedback circuit is used to assure that proper treatment is provided and controlled. The process is essentially fail-safe.

MICROBICIDAL EFFECTS

PureBright antimicrobial mechanisms are based on the sensitivity of microorganisms to UV wavelengths that are readily absorbed by microbial nucleic acids and proteins, and the very high power and intensity of the flash.

The antimicrobial effects of one to several flashes are dramatic. PureBright treatment at 1.5 J/cm^2 kills 7 logs of *Staphylococcus aureus* ATCC 27661 suspended in 25-mL droplets dried on the surface of microbiological media.

Figure 2 shows the pulsed light kill kinetics obtained using the microdrop assay for a number of different bacteria and spores. The results show 7 to 9 logs/cm^2 reduction using a few flashes.

PureBright kill dynamics are superior to conventional ultraviolet light from a high-intensity mercury vapor lamp. Using relatively UV-resistant *Aspergillus niger* spores inoculated and dried on a packaging surface, conventional continuous wave UV kills 3.5 to 4.5 logs in 6 to 10 s with no appreciable increase in kill with additional treatment time. By comparison, PureBright kills more than 7 logs of *A. niger* spores with a few flashes applied in a fraction of

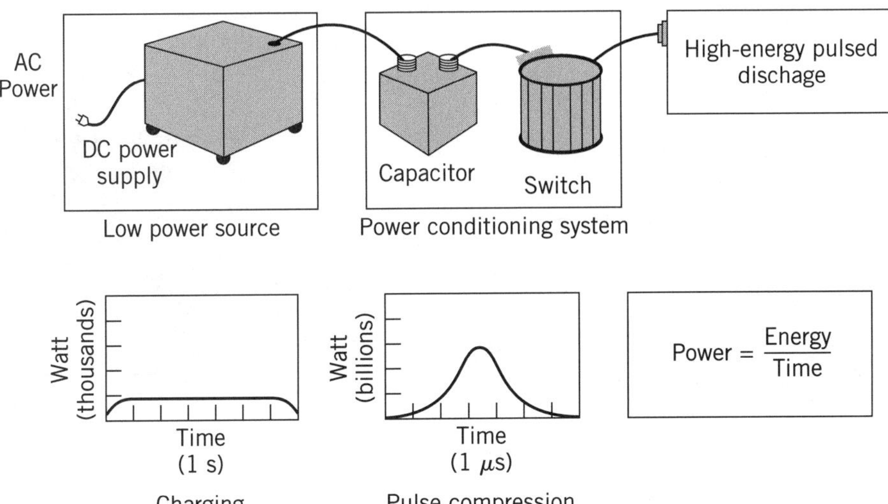

Figure 1. Pulsed energy processing.

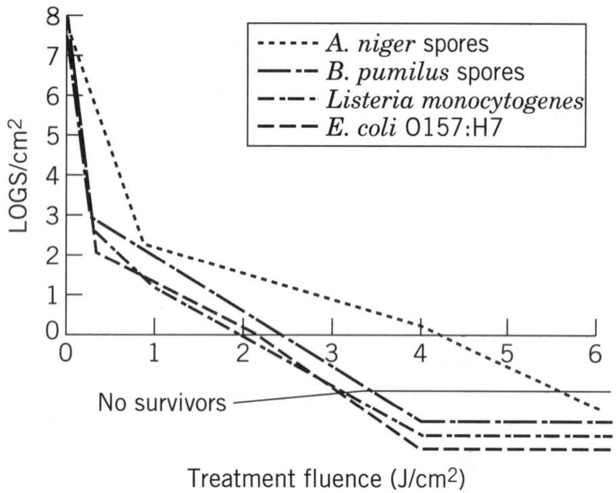

Figure 2. Pulsed light kill kinetics.

a second. This dynamic provides the effectiveness required for high sterility assurance (Fig. 3).

TREATMENT OF PACKAGING

PureBright can be used to sterilize packaging material surfaces, thus replacing hydrogen peroxide in many aseptic packaging applications. On most package surfaces, PureBright kills greater than 7 logs of *Bacillus pumilus* spores (a radiation-resistant bacterial spore) in challenge tests.

PureBright is also effective for package applications involving perishable, refrigerated, extended–shelf life food products, such as containers for yogurt, cottage cheese, or milk. Figure 4 shows a typical cup-and-lid-packaging line configuration that can be used for extending the shelf life of refrigerated desserts and other products.

Pulsed light is transmitted through many plastics, so certain packages can be sterilized on the inside by flashing the outside. Also, the light can be used to treat a filled and sealed product through the package if the product has good transmission properties such as water or other clear solutions. In general, colors will inhibit the UV transmission and microbial kill.

The olefins (polyethylene, polypropylene), acrylics, many polyamides and nylons, ethylene acrylic acid, EVA, EVOH, and similar plastics that transmit in the ultraviolet are good candidates for PureBright treatment through the package. Plastics with aromatic hydrocarbon backbones, side groups, or additives generally absorb UV too strongly to permit treatment through the package.

No adverse chemical or functionality effects on plastic packaging materials have been found using a variety of plastics treated with PureBright at levels in excess of those needed to provide appropriate microbial kill.

PureBright is also effective for treating metal or plastic bottle caps. Tests using four different types and configurations of caps have shown total kill of 4 to 5 logs of inoculated *A. niger* conidiospores using 2 to 6 flashes.

TREATMENT OF FOOD

PureBright reduces microbial loads on perishable foods such as baked goods, fresh and processed meats, poultry, fish, fresh fruit, and vegetables. Due to the generally opaque and irregular surfaces of foods, lower kill levels are realized than on packaging; however, significant shelf life extension is achieved and pathogenic microorganism levels reduced.

Microorganisms on fresh meat surfaces are reduced 1 to 3 logs by PureBright, resulting in extended refrigerator shelf life of 2 or more days. Similarly, the shelf life of baked goods is increased by the elimination of surface mold (Fig. 5).

The FDA has issued a food additive regulation that approves the use of PureBright on all foods for the control of bacteria, molds, yeast, and other microorganisms.

TREATMENT OF WATER

PureBright kills a wide range of microorganisms in clear water without the use of heat, chemicals, or fine filtration.

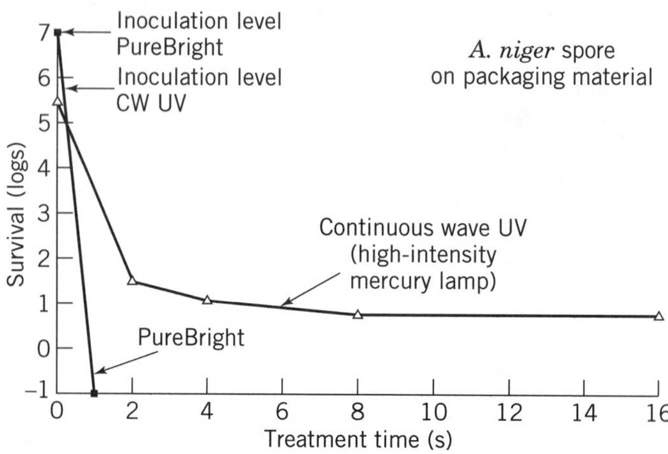

Figure 3. PureBright vs. high-power continuous-wave UV.

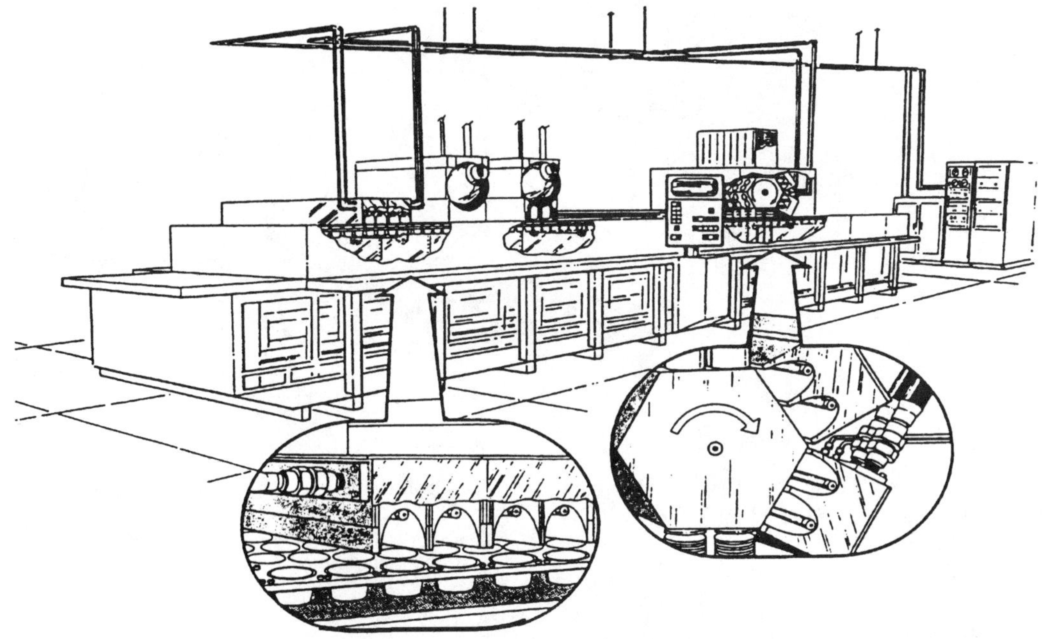

Figure 4. PureBright kills bacteria on packages in food-production line.

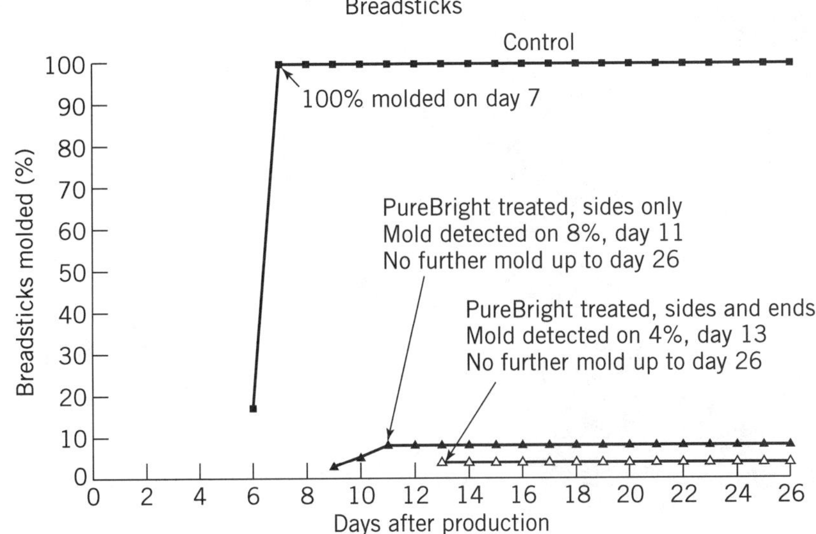

Figure 5. Extended shelf life of baked goods.

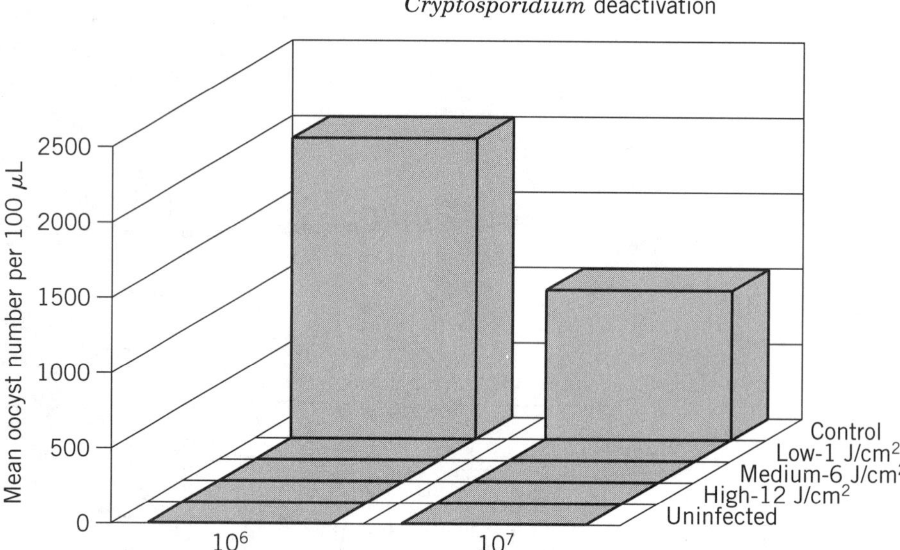

Figure 6. PureBright effect on *Cryptosporidium*.

Unlike other treatment methods, PureBright, at a relatively low level, has demonstrated efficient kill of *Cryptosporidium* oocyts and viruses (Fig. 6).

GENERAL REFERENCES

J. Dunn, T. Ott, and W. Clark, "Pulsed Light Treatment of Food and Packaging," *Food Technol.* **49**, 95–97 (1995).

J. Jagger, *Ultraviolet Photobiology*, Prentice Hall Inc., Englewood Cliffs, N.J., 1967.

L. Koller, *Ultraviolet Radiation*, John Wiley and Sons, Inc., New York, 1977.

K. Smith, *The Science of Photobiology*, Plenum Press, New York, 1977.

THOMAS M. OTT
Purepulse Technologies
Westwood, New Jersey

Q

QUALITY ASSURANCE

Americans spent about $337 billion on food purchased in retail food stores in 1994 and about $303 billion on food purchased in foodservice establishments (1). A variety of factors are responsible for the changes in U.S. consumption patterns, including changes in consumer preferences, relative prices, increases in real (adjusted for inflation) disposable income, and more food assistance to the poor. An expanded scientific base relating diet and health, new dietary guidelines for Americans designed to help people make food choices that promote health and prevent disease, and burgeoning interest in nutrition and food safety also influence marketing and consumption trends.

Consistent with dietary and health recommendations, Americans in 1996 consumed two-fifths more grain products and a fifth more fruits and vegetables per capita than they did in 1970, ate leaner meat, and drank lower-fat milk (2).

Though consumers do not want to spend a lot of time on food preparation, they also are not willing to scrimp on health and nutrition. The trend to lowering fat consumption has contributed to gains in poultry consumption and declines in red meat. Consumer concerns about food additives, chemicals, and preservatives have translated into a growing market for pesticide-free products, organic produce, and vegetarian choices. As consumer concerns about food safety have increased, so have the demands that government regulators increase efforts to assure food safety and guarantee consumers better and more accurate information about the food they eat. We have seen some major initiatives by the government such as the Food Code 1997, the Nutrition Labeling and Education Act (NLEA), United States Department of Agriculture (USDA) and Food and Drug Administration (FDA), Hogard Analysis Critical Control Point (HACCP) initiatives, and the President's National Food Safety Initiative. Environmental awareness has spawned a wave of recyclable, biodegradable, and nonpolluting products. Consumers are demanding "environmentally friendly" products. Manufacturers are using packaging and other environmental characteristics to appeal to consumers in advertising campaigns. From farm to table, technologies are being developed that contribute to sustainability of agriculture and environment (3).

Food service has shown tremendous growth. Industry growth is attributable in large part to the bundling of food products with convenience and quality.

A variety of factors are fueling the trends toward the use of more prepackaged and prepared items. In the commercial food service sector, the conflicting goals of expanding menu offerings and reducing operating costs are demanding that operators explore new types of products and packaging. It is expected that increased variety and greater speed will be the future direction of this industry. Increased variety will exist in many alternatives in the service and packaging methods. Speed will be reflected in the short time between when the food is bought and when it is consumed. Speed will also play a role in functions such as ordering and food preparation. Electronic systems will be increasingly critical to the operating process. Sophisticated heating and cooling mechanisms will be demanded. In the future, doing business will require science-based quality systems and the increased use of food technology. What will not change is the consumer's desire for good-tasting and healthy products that satisfy both nutritional and social needs. This is the service in food service that cannot be replaced by the increasingly sophisticated activities that are behind what is ultimately served.

THE FOOD QUALITY SYSTEM

The significance of any food quality system has to be discussed within the context of the food supply in its entirety. The quality of our food supply is of supreme importance to us all: consumers, farmers, processors, distributors, food service operators, and the regulatory agencies. Increasing affluence has given birth to a new breed of consumer, one accustomed to eating away from home frequently. This new consumer shows greater reliance on convenience and expects safe, high-quality foods for consumption (4).

The U.S. food supply is abundant. Our preservation processes have changed dramatically and include such developments as irradiation, genetic engineering, and controlled atmospheres. Distribution includes air transport, refrigerated railroad cars, trucks, freezers, and shipments by sea. We have a global system of distribution that makes, for example, a variety of out-of-season produce available year-round, with improved taste. This complexity has brought about concerns about food safety (5).

The consumer drives the food quality system. No quality program can be successful if it does not address consumer concerns, real or perceived. F. J. Francis discusses public perception of food safety in a detailed report and urges the scientific community to stand up and speak out when something is scientifically absurd. Epidemiologists have estimated the potential deaths due to pesticides as close as zero. The FDA and National Cancer Institute have repeatedly stated that risks from pesticides are insignificant. Yet USDA appropriated $40 million for pesticide monitoring, not to increase public safety but to reassure the public. It emphasizes the need for a science-based quality system that incorporates risk assessment and a mechanism to explain to the public the benefits of such a system through a core of knowledgeable food scientists (6).

A recent survey revealed that freshness is the most important quality consumers look at when purchasing refrigerated foods (7). In the writer's view, consumers associate freshness with quality, safety, and health. A USDA Economic Research Service (ERS) study (8) indicated that in 1995, 5.4 billion lb of food were lost at the retail level, and 91 billion lb were lost by consumers and food service. Fresh fruits and vegetables accounted for nearly 20% of consumer and food service losses. It was estimated that the total annual cost of unsalables to U.S. packaged goods manufactured in the 1996 supermarket channel was $1.96 billion (9). There is a need to improve the quality of our food system, maintain food safety, and return the benefits to the consumer. The various components of the food system are interdependent. The private sector, regulators, consumers, and universities need to work closely to insure that the goals of a science-based quality system are achieved.

SYSTEM DESIGN

Definitions

- Quality system: The organizational structure, responsibilities, procedures, and resources for implementing quality management
- Quality management: That aspect of overall management function that determines and implements the quality policy
- Quality policy: The overall quality intentions and direction of an organization, as regards quality, formally expressed by top management
- Quality control: The operational techniques and activities that are used to fulfill requirements for quality
- Quality assurance: All those planned and systematic actions necessary to provide adequate confidence that a product will satisfy given requirements

These terms are used to define quality systems and their implementation (10,11).

The elements of putting a system together are detailed in the publication *Food Processing Industry Quality System Guidelines* (12). These guidelines were written by a committee of experts and adapted from American National Standards Institute/American Society for Quality Control (ANSI/ASQC) Q9004-3-1993, *Quality Management and Quality System Elements—Guidelines for Processed Materials*. The monograph includes definitions of terms used in quality systems; management responsibility; quality systems principles; economics—quality-related cost considerations; quality in marketing, specification, design/development, procurement, and production; product verification; measuring and control of test equipment; nonconformity, corrective action, handling, and postproduction functions; quality documentation and records; personnel; product safety and liability; and use of quality methods.

The International Organization for Standardization (ISO) in Geneva, Switzerland, has developed a series that provides a framework in the development of quality plans. They are:

- ISO 9000 Quality Management and Quality Assurance Standards: Guidelines for Selection and Use
- ISO 9001 Quality Systems: Model for Quality Assurance in Design, Development, Production, Installation, and Servicing
- ISO 9002 Quality Systems: Model for Quality Assurance in Production, Installation, and Servicing
- ISO 9003 Quality Systems: Model for Quality Assurance in Final Inspection and Test
- ISO 9004 Quality Management and Quality System Element Guidelines
- ISO A 8402 Management and Quality Assurance Vocabulary

The Malcolm Baldridge National Quality Award is presented by the President of the United States to companies in the manufacturing and service industries. The criteria consist of seven categories: leadership, strategic planning, customer and market focus, information and analysis, human resources focus, process management, and business results. It is a good model and provides yardsticks by which companies can be recognized for both business excellence and quality achievements.

Although models provide a good framework in the development and implementation of quality plans, the models need to be adapted to individual business segments.

QUALITY COSTS

It is important to determine the impact of quality costs on company profitability. Cost assessment can be useful in evaluating not only the effectiveness of the quality system, but also criteria for internal improvement.

Costs can be divided into those for operating quality costs and those for external assurance. Operating quality costs are incurred by the company to attain and insure a specified quality level. They consist of prevention and appraisal costs (or investments) and failure costs (or losses). Prevention costs are incurred to prevent failures; some examples include ingredient sampling and analysis, equipment maintenance and design to maintain quality and food safety, and integrated pest management. Appraisal costs are incurred during testing and inspection to insure product quality; examples are laboratory testing and internal verification audits. External assurance costs are related to the demonstration and proof required as objective evidence by customers or regulatory agencies. Additional costs are opportunity costs, loss of business due to poor quality, and costs of exceeding requirements (11,12).

FOOD SAFETY

The Centers for Disease Control and Prevention (CDC) maintains statistics on foodborne outbreaks. From 1988 to 1992 (13), among outbreaks for which the etiology was determined, bacterial pathogens caused the largest percentage of outbreaks (79%) and the largest percentage of cases (90%); chemical agents caused 14% of outbreaks and 2% of cases; parasites, 2% of outbreaks and 1% of cases; and viruses, 4% of outbreaks and 6% of cases.

The Center for Science in the Public Interest compiled an inventory of 225 foodborne illness outbreaks (14) that occurred between 1990 and 1998. Ground beef caused 37 of 65 outbreaks; fruits and salads were the second most likely foods to be linked with an outbreak, with a total of 48 outbreaks. Seafood, both finfish and shellfish, was responsible for 32 outbreaks. Multi-ingredient foods, which included desserts, sauces, egg dishes, pasta dishes, and stuffing, were responsible for 63 outbreaks. In three-quarters of these outbreaks, the cause of illness was *Salmonella enteridis*. It is possible that emerging and reemerging pathogens contributed to the large number of outbreaks of unknown etiology (14).

The Council for Agriculture and Science Technology (CAST) concludes that microbial pathogens in foods cause 6.5 to 44 million cases of human illnesses and up to 9000 deaths each year (15). It has been estimated that the costs of human illness for six bacterial pathogens—*Salmonella*, *Campylobacter*, *E. coli* 0157:H7, *Listeria*, *Staphylococcus aureus*, and *Clostridium pefigus*—are estimated to be $9.3 to $12.9 billion. Of these costs, $2.9 to $6.7 billion are attributed to foodborne bacteria (16). The estimated annual cost of *Campylobacter*-associated Guillain-Barré syndrome was $0.2 to $8.0 billion, with an estimation of $0.8 to $5.6 billion from food sources (17). Estimates are subject to errors because most of the comprehensive population-based studies have not attempted to determine which proportion of the reported illnesses are from food consumption and which are from other sources; also, foodborne illnesses can cause clinical conditions, not characteristically gastrointestinal symptoms (5,18).

Foodborne Diseases Active Surveillance Network (FoodNet) is a collaborative project among CDC; the health departments of California, Connecticut, Georgia, New York, Maryland, Minnesota, and Oregon; FDA; and USDA. The following bacteria have been targeted: *Salmonella*, *Shigella*, *Campylobacter*, *E. coli* 0157:H7, *Listeria*, *Yersinia*, and *Vibrio* (19). FoodNet reports that *Campylobacter* was the most frequently isolated foodborne bacterium pathogen (49.4%), followed by *Salmonella* (27.4%), *Shigella* (15.7%), *E. coli* 0157:H7 (4.2%), *Yersinia* (1.7%), *Listeria* (1%), and *Vibrio* (0.6%) (20). FoodNet estimated that there were 8 million cases of these bacterial infections in 1997 in the United States (21).

EMERGING FOODBORNE DISEASES

Emerging and reemerging infections have been defined as new, recurring, or drug-resistant infections whose incidence in humans has increased in the past two decades, or whose incidence threatens to increase in the near future (22). This subject has been reviewed by Altekrusse (23). A committee of the Institute of Medicine recognized as early as 1992 the microbial threats to health in the United States of emerging infections and made recommendations (22). A number of factors contribute to the emergence of foodborne diseases (24).

A growing segment of the population is immunocompromised because of advancing age, chronic diseases, or infection with human immunodeficiency virus (HIV). Advances in medical sciences have extended life expectancies of persons with chronic diseases such as cancer, thus increasing their susceptibility to foodborne diseases.

Changes have occurred in industry, technology, agriculture, and lifestyles. There has been an increase in the consumption of fruits and vegetables. Produce is susceptible to microbial contamination during growth, harvest, and distribution. There has also been an increase in the demand for fresh-cut produce. The larger surface area makes this type of product more susceptible to contamination. Outbreaks have been associated with alfalfa sprouts, unpasteurized cider, fresh-squeezed orange juice, and frozen strawberries. Consumers have increased their spending on foods eaten away from home. Food service is a growing market. Improper cooking of items such as hamburgers, holding foods at improper temperatures, and cross-contamination are some of the causes of outbreaks. Extensive use of antibiotics has led to the emergence of multidrug-resistant *Salmonella typhomurium* (DT) 104. Bovine spongiform encephalopathy (BSE) has emerged as an infectious agent, mainly in the form of adult dairy cattle in the United Kingdom. There is evidence that this prion-derived disease was a result of feeding cattle meat and bonemeal derived from dead ruminants such as sheep and slaughterhouse products infected with scrapie as a protein supplement. The problem was compounded by reduction of temperature used in heat treatment, and elimination of solvent extraction of animal tissue during meat and bonemeal processing led to a failure to inactivate BSE prion and thus allowed its contraction into the cattle population (24).

Allergenic mites are emerging as a food safety issue. Four species of food-contaminating mites have been reported to have caused allergenic reactions, including anaphylaxis in persons who consumed mite-contaminated foods. The mite species are the American house dust mite *Dermatophagoides farinae* Hughes (Acarina: Pyroglyphidae), the scaly grain mite *Suidasia* sp prob. *pontifica* Oudemans (Acarina: Suidasiidae), an acarid mite *Thyreophagus entomophagus* Portus and Gomez (Ascarina: Acaridae), and the mold mite *Tyrophagus putrescentiae* (Schrank) (Acarina: Acaridae) (25).

FDA guidance is delineated in Compliance Policy Guide 585.500, *Mushrooms, Canned or Dried (Freeze Dried or Dehydrated)—Adulteration Involving Maggots, Mites, Decomposition*. These levels are based on the assumption that the contaminants are harmless and unavoidable. The true level of safety to allergenic mites is unknown (25).

Miller mentions a number of pathogens that might potentially emerge, including prior induced variant of Kreutzfeld-Jakob disease, viruses, *Citrobacter freundii*, the newer genes of *Salmonella* such as *S. enteriditus* PT4 and *S. typhomurium* DT104, and parasites such as *Cyclospora*. Miller recommends strategies to anticipate the emerging pathogen challenge. They include intelligence, contingency, and strategic planning. Intelligence consists of gathering information through a global surveillance program. Contingency is rapid resource mobilization and assures that resources are available to quickly characterize the emerging pathogen. Strategic planning requires futuristic thinking to anticipate what may happen, taking into consideration the various factors such as society, economics, and technology.

NUTRITIONAL QUALITY

With increasing consumer awareness of nutrition, it becomes important to insure the proper handling of food products to maintain the claimed nutrition content. The Nutritional Labeling and Education Act (NLEA) defines the labeling requirements for food products, and quality systems need to be in place to insure that these requirements are met. Advances in food technology will result in minimal nutrient losses during harvest, processing, storage, and distribution. The challenge remains in educating the consumer about the importance of proper food handling (26).

HAZARD ANALYSIS: CRITICAL CONTROL POINT SYSTEMS

The scientific community accepts that HACCP systems, when used properly, will substantially enhance food safety. HACCP is defined as a management system in which food safety is addressed through the analysis and control of physical, chemical, and microbiological hazards from raw material production, procurement, and handling to manufacturing and consumption of finished product. The successful implementation of HACCP requires a firm commitment from top management (27). Bryan (28) provides a historical perspective on HACCP. The principles of HACCP were endorsed in the 1980s by the National Research Council (NRC) (29). NRC has issued a number of reports on meat, poultry, and seafood very supportive of the HACCP concept (30–32). A number of reports have been issued by the National Advisory Committee on Microbiological Criteria for Foods (33–38). The HACCP principles have gained recognition by industry and regulatory agencies. USDA has mandated the use of HACCP in meat and poultry (39), and FDA in seafood (40). Guidelines for fresh fruits and vegetables have been issued by FDA (41).

The FDA Food Code (42) incorporates the principles of HACCP. There have been a number of guidelines, manuals, and generic plans in the literature. The National Advisory Committee on Microbiological Criteria for Foods standardized the hazard principles as follows:

1. Conduct a hazard analysis
2. Determine the critical control points
3. Establish critical limits
4. Establish monitoring procedures
5. Establish corrective actions
6. Establish verification procedures
7. Establish record-keeping and documentation procedures

To facilitate the development of a HACCP plan, certain prerequisite requirements are suggested. The importance of education and training of management and employees cannot be overemphasized. An in-depth discussion on the preparation and implementation of HACCP plans is beyond the scope of this article. The HACCP steps, however, are briefly discussed here:

Conduct a Hazard Analysis. A hazard analysis requires identification of the hazards that might reasonably be expected to occur. The hazards must be critical to assure food safety. Examples include physical hazards (metal, glass), chemical hazards (pesticide residues, undeclared allergenic ingredients), and microbiological hazards (pathogens, parasites). Regulatory agencies have defect action levels such as pits and stems. They do not present a health hazard. Physical hazards have been reviewed.

Determine Critical Control Points (CCPs). Critical control points are defined as the steps that are applied to prevent, eliminate, or reduce a hazard to an acceptable level. Examples would be cooking, metal, or allergen controls.

Establish Critical Limits. Establishing critical limits involves determining the maximum and/or minimum values to which a physical, chemical, or microbiological hazard must be controlled at a determined CCP to prevent, eliminate, or reduce to an acceptable level the occurrence of a food safety hazard. Examples include cooking and cooling time and temperature limits, and metal detection limits.

Establish Monitoring Procedures. Monitoring procedures assess whether CCP is under control. They provide a record for future use in verification.

Establish Corrective Actions. Should deviations occur from the critical limits, then corrective action is needed to bring the critical limits into compliance. The steps include determining the cause and, of course, correcting it; disposing of the noncompliant product; and recording the corrective actions that need to be taken to prevent reoccurrence. These actions should be developed and be a part of the HACCP plan.

Establish Verification Procedures. Verification procedures are developed to determine whether the HACCP plan is valid. A well-designed, functional HACCP plan has enough safeguards built in so that end-product testing should be minimal.

Establish Record-Keeping and Documentation Procedures. Records are important in that they provide a history of the HACCP system performance and should include, among other information, results of the monitoring procedures, corrective actions, and validation records.

RISK ANALYSIS

Risk analysis is an important tool. It yields information on estimating and analyzing costs and benefits of various alternatives, provides direction on what should be allocated for long- and short-term benefits, and serves as a source of resource identification. The components of risk analysis are risk assessment, risk management, and risk communication. It is important to understand that risk analysis will not yield a process that will lead to the development of zero risk to the end user. There is an inherent risk in all food products. HACCP systems do use risk analysis. HACCP is a risk-management system. Microbiological risk assessments are complex because of the many complexities of the food system from farm to table (15).

PHYSICAL HAZARDS

Foods, as grown, can become naturally contaminated with rocks, stones, glass, metal, and so on. Additional contamination can occur throughout the food chain. Physical hazards are controlled by processes that use exclusion such as magnets, metal detectors, and screens. Plant policies such as preventive maintenance, glass control, and good manufacturing practices assist the process. Federal agencies have delineated defect action levels that include stems, pits in vegetables, and insect fragments. These levels are generally considered a health hazard.

CHEMICAL HAZARDS

Chemical hazards can be introduced into foods either directly or indirectly. These are also naturally occurring toxicants. Chemical hazards include pesticides used to control pests in crops, cleaning compounds and sanitizers used on food equipment surfaces, and environmental contaminants such as mercury in fish and polybrominated biphenyls in feed. Certain foods, such as peanuts, eggs, milk, fish, and wheat, contain allergens that can be harmful to an allergen-sensitive individual. There are questions as to the adequacy of current pesticide safety assessments for children (43). Consumer interest in dietary supplements has raised concerns about adverse reactions (44). Chemical hazards present a continuing challenge to the scientific community.

BIOLOGICAL HAZARDS

Bean and Griffin (45) summarized CDC statistics and reported that 66% of the outbreaks and 87% of cases were bacterial in nature. Some of the more common biological agents are *Salmonella*, *Sa. Aureus*, *Campylobacter*, *E. coli* 0157:H7, and viruses such as hepatitis A and Norwalk. There are concerns about some of the emerging pathogens such as *Listeria*. Hazard assessment will require an assessment of why they occur and survive in a particular food, and the determination of CCPs that would prevent these organisms from becoming a potential health hazard.

QUALITY MANAGEMENT

In a recent survey, 7 out of 10 (69%) consumers cited spoilage-related concerns, including bacterial contamination, as the most significant threat to food safety. Consumers are also taking action; almost half of survey respondents (45%) said they are doing something different as a result of the safe-handling labels on meat products. More respondents are washing their hands, cooking properly, not leaving meat out to thaw, and washing meat. More than 90% of the consumers surveyed indicated that the nutritional content of the foods they eat was important. Fat, salt, and cholesterol remain the top nutritional concerns (46,47). The consumer expects his or her food to be of high quality, safe, and nutritious. The consumer drives the food system, which in turns drives the industry to meet consumer expectations; hence the need for quality management.

The literature is replete with books and papers featuring quality buzzwords such as CP (critical path), PC (process control), SPC (statistical process control), and TQM (total quality management). Colleges and junior colleges offer differing curricula, as do associations, trainers, and consultants who offer courses and seminars.

The systems used for the implementation of quality will vary depending on the nature of the product, the process, and the end user. Agricultural commodities require implementation of good agricultural practices along with proper controls during distribution, such as temperature and humidity to maintain freshness and nutritional content. Similar precautions need to be taken with seafood products. Programs cannot be successful without management commitment and the appropriate allocations of funds to successfully implement programs. Food service offers unique challenges because of its complexity. Challenges will vary depending on the type and size of the company.

It is becoming increasingly evident that quality needs to be controlled in all its aspects, which includes all parameters that will directly or indirectly affect quality, namely, the raw materials, agricultural practices, manufacturing process, environment, finished product, product handling, and consumer guidance. Monitoring of these parameters needs to be directed toward minimizing product failures, prevention of substandard product, and production of uniform quality product that will meet defined criteria—organoleptic, physical, chemical, and microbiological.

Ingredients and packaging should be purchased based on specifications under a continuing food guarantee. The specifications should take into consideration the capabilities of the supplier to meet the specifications and the ultimate use of these ingredients in the finished product. The specification should define the physical, chemical, and microbiological requirements. The capabilities of the supplier should be taken into consideration, including the adequacy of HACCP and quality programs. The raw materials should be statistically sampled (48).

The increased production needs have increased demands for automated processes—complex equipment that requires the utilization of sophisticated techniques to sense and correct malfunctions—hence the need for statistical process controls. Statistical process control will increase the effectiveness of HACCP systems and assist in product improvement.

Finished products need to be tested on a periodic basis. An adequate process and HACCP controls will greatly minimize the need to test finished products.

It is important that environmental sanitation in the plant be taken seriously. A clean plant instills pride, increases productivity, and is conducive to the production of consistent-quality products. Continuous follow-up and inspections by trained personnel will determine the program's success or failure. The program should include equipment design and maintenance, cleaning, pest control, and personnel practices.

Equipment should be designed so that it is easily cleaned. Hard-to-clean spots can be focal points for microbial growth. Improperly maintained equipment is conducive to accidents and foreign material contamination. Through integrated pest management control, pesticides must be used judiciously. The extent of the programs depends on the nature of the food and the location of the manufacturing facility. Equipment must be properly cleaned and sanitized. The adequacy of cleaning and sanitizing needs to be monitored using a microbiological control program. Plants must maintain the highest standards of hygiene and follow good manufacturing practices.

The quality functions need to be working in an interdisciplinary manner with company functions such as research and development, operations, sales, marketing, financial, and human resources. All of the quality considerations driving manufacturing, packaging, and distribution will be rendered worthless if the product is mishandled by the consumer or the food service operator. Handling instructions need to be adequate and explicit.

The final link in the food chain is the consumer. Assuming that each American eats four times a day, there could be one billion opportunities each day for someone to contract or transmit a foodborne illness (49) or consume a substandard product. Therefore, a concerted effort to educate the consumer is very important. The successful quality program will require the implementation of a system approach from farm to fork (4).

CONCLUSION

The food industry has a long history, beginning when humans first served a meal from home. The industry has seen that event refined to its current state of a multisegmented domestic and international business. In addition to having to meet customer expectations in quality and price, the food industry must satisfy a number of requirements such as labeling, standards of identity, food safety, and so on. We have seen spiraling ingredient, labor, and regulatory costs; such increased costs lead to the pursuit of more efficient methods of manufacture and optimum formulations (50).

Thomas Jefferson prescribed the need for a wise and frugal government. We have initiatives undertaken by a number of presidents—Kennedy, Nixon, and most recently Clinton, with the Food Safety Initiative. The Center for the Study of American Business at Washington University estimates that the private sector will pay $688 billion in "hidden compliance costs" related to federal regulations in 1997, roughly 9% of the gross domestic product (51). The consumer will ultimately pay the price; yet government regulations are necessary. It is important, therefore, that the desired food quality be achieved through engineering quality into the process at minimum regulatory costs through such programs as HACCP and quality systems, and communicating and educating the consumer on how to handle the food product for optimum taste and nutrition.

BIBLIOGRAPHY

1. U.S. Department of Agriculture, *Food Marketing Review 1994–95*, Agricultural Economic Report No. 743, Washington, D.C., 1996.
2. J. Putnam, *Food Consumption, Prices, and Expenditures, 1970–75*, USDA Economic Research Service Report No. SB-939, Washington, D.C., 1997.
3. K. L. Lipton, W. Edmondson, and A. Manchester, *The Food and Fiber System: Contributing to the US and World Economics*, USDA Economic Research Service Agriculture Information Bulletin No. 742, Washington, D.C., 1998.
4. M. P. de Figueiredo, "Quality Assurance of Food Safety," *Food Technol.* **44**, 58–59 (1981).
5. Institute of Medicine, *Ensuring Safe Food from Production to Consumption*, National Academy of Press, Washington, D.C., 1998.
6. F. J. Francis, "Public Perception of Food Safety Priorities," *American Council on Science and Health* **3**, 29–32 (1991).
7. S. Dowdell, "Looking for a Date: The Vital Statistics in a Recent Survey Reveal That Freshness Is the Most Important Quality Consumers Look For in a Perishables Department," *Supermarket News* **46**, 27–28 (1996).
8. L. S. Kantor et al., "Estimating and Addressing America's Food Losses," *Food Review* **20**, 2–12 (1997).
9. S. R. Matthews, "Is Damage Done," *Progressive Grocer* **76**, 63–68 (1997).
10. American Society for Quality, *ANSI / ISO / ASQC Q9000 Series Quality Standards*, ASQ Quality Press, Milwaukee, Wisc., 1994.
11. R. Early, *Guide to Quality Management Systems for the Food Industry*, Blackie Academic and Professional, Glasgow, Scotland, 1995.
12. American Society for Quality Food, Drug and Cosmetic Division, *Food Processing Industry Quality System Guidelines*, ASQ Quality Press, Milwaukee, Wisc., 1998.
13. N. H. Bean et al., "Surveillance for Foodborne Disease Outbreaks; United States 1988–1992," *Morbidity, Mortality Weekly Report* **45**, 16–20 (1996).
14. C. Smith de Waal, L. Aderton, and M. Jacobson, *Outbreak Alert: Closing the Gaps in Our Federal Food Safety Net*, Center for Science in the Public Interest, Washington, D.C., 1999.
15. Council for Agricultural Science and Technology, *Foodborne Pathogens*, CAST Task Force Report No. 122, Ames, Iowa, 1994.
16. J. Buzby et al., *Bacterial Foodborne Disease: Medical Costs and Productivity Losses*, USDA Agricultural Economic Report No. 741, Washington, D.C., 1996.
17. J. Buzby, T. Roberts, and B. M. Allos, *Estimated Annual Costs of Campylobacter-Associated Guillain-Barre Syndrome*, USDA Agricultural Economic Report No. 756, Washington, D.C., 1997.
18. J. G. Morris and M. Potter, "Emergence of New Pathogens as a Function of Changes in Host Susceptibility," *Emerging Infectious Diseases* **3**, 435–441 (1997).
19. U.S. Department of Health and Human Services, Centers for Disease Control and Prevention, *FoodNet: CDC's Emerging Infections Program*, Washington, D.C., 1998.
20. U.S. Department of Agriculture, Food Safety and Inspection Service, Report to Congress, *FoodNet: An Active Surveillance System for Bacterial Foodborne Diseases in the United States*, Washington, D.C., 1998.
21. U.S. Department of Health and Human Services, Centers for Disease Control and Prevention, Foodnet, CDC/USDA/FDA Foodborne Diseases Active Surveillance Network, *CDC's Emerging Infections Program: 1997 Surveillance Results*, Washington, D.C., 1997.

22. National Academy of Sciences, *Emerging Infections: Microbial Threats to Health in the US*, National Academy Press, Washington, D.C., 1993.

23. S. F. Altekrusse, M. L. Cohen, and D. L. Swerdlow, "Emerging Foodborne Diseases," *Emerging Infectious Diseases* **3**, 285–292 (1997).

24. A. J. Miller, J. L. Smith, and R. L. Buchanan, "Factors Affecting the Emergence of New Pathogens and Research Strategies Leading to Their Control," *Journal of Food Safety* **18**, 243–263 (1998).

25. A. R. Olsen, "Regulatory Action Criteria," *Regul. Toxicol. Pharmacol.* **28**, 190–198 (1998).

26. E. Karmas and R. S. Harris, *Nutritional Evaluation of Food Processing*, Van Nostrand–Reinhold, New York, 1988.

27. National Advisory Committee on Microbiological Criteria for Foods, "Hazard Analysis and Critical Control Point Principles and Application Guidelines," *J. Food Prot.* **61**, 1246–1259 (1998).

28. F. Bryan, "HACCP Approach to Food Safety: Past, Present, and Future," *Food Testing Analysis* **5**, 13–19 (1999).

29. Subcommittee on Microbiological Criteria, Committee on Food Protection, Food and Nutrition Board, National Research Council, *An Evaluation of the Role of Microbiological Criteria for Foods and Food Ingredients*, National Academy Press, Washington, D.C., 1985.

30. Committee on Scientific Basis of Meat and Poultry Inspection, Food and Nutrition Board, National Research Council, *Scientific Basis of Meat and Poultry Inspection*, National Academy Press, Washington, D.C., 1985.

31. Committee on Evaluation of USDA Streamland, Food and Nutrition Board, Institute of Medicine, National Academy of Sciences, *Inspection System Guide for Cattle (515-C)*, National Academy Press, Washington, D.C., 1990.

32. Committee on Evaluation of the Safety of Fishery Products, Institute of Medicine, National Academy of Sciences, *Seafood Safety*, National Academy Press, Washington, D.C., 1991.

33. National Advisory Committee on Microbiological Criteria for Foods, "Outstanding Symposia in Food Science and Technology," *Food Technol.* **54**, 142–162 (1991).

34. "Microbiological Criteria for Raw Molluscan Shellfish," *J. Food Prot.* **55**, 463–480 (1992).

35. "Listeria Monocytogenes," *International Journal of Food Microbiology* **14**, 185–246 (1991).

36. "The Role of Regulatory Agencies and Industry in HACCP," *International Journal of Food Microbiology* **21**, 187–195 (1994).

37. "Generic HACCP for Raw Beef," *Food Microbiology* **10**, 449–488 (1993).

38. "Hazard Analysis and Critical Control Point System," *International Journal of Food Microbiology* **16**, 1–23 (1992).

39. U.S. Department of Agriculture, Food Safety and Inspection Service, "Pathogen Reduction: Hazard Analysis Critical Control Point Systems, Final Rule," *Fed. Regist.* **6**, 38805–38855 (1996).

40. U.S. Department of Health and Human Services, Food and Drug Administration, "Procedures for the Safe and Sanitary Processing and Importing of Fish and Fishery Products, Final Rule," *Fed. Regist.* **60**, 65095–65202 (1995).

41. U.S. Department of Health and Human Services, Food and Drug Administration, "Guidance for Industry to Minimize Food Safety Hazards," *Fed. Regist.* **63**, 58055–58056 (1998).

42. U.S. Department of Health and Human Services, Food and Drug Administration, *FDA Food Code 1997*, National Technical Information Service, Springfield, Va., 1997.

43. National Research Council, *Pesticides in the Diets of Infants and Children*, National Academy Press, Washington, D.C., 1993.

44. Institute of Food Technologists, "IFT Says Flexibility Is Key to Food Safety," *IFT Voices* **1**, 1 (1999).

45. N. H. Bean and P. M. Griffin, "Foodborne Disease Outbreaks in the United States 1973–1987: Pathogens, Vehicles, and Trends," *J. Food Prot.* **53**, 804–817 (1990).

46. International Food Information Council, *Newsbites: 97 Consumer Trends*, Washington, D.C., 1998.

47. Food Marketing Institute, *Trends in the United States Consumer Attitudes and the Supermarket*, Washington, D.C., 1997.

48. C. Stannard, "Development and Use of Microbiological Criteria for Foods," *Food Science and Technology Today* **11**, 137–177 (1997).

49. A. M. Coulston, "President's Page: Personal Responsibility and Food Safety," *J. Am. Dietetic Assoc.* **99**, 236 (1999).

50. S. A. Goldblith, "Consumer, the Product the Premise," in S. R. Tannenbaum ed., *Nutritional Safety Aspects of Food Processing*, Marcel Dekker, New York, 1979, pp. 1–5.

51. Center for the Study of American Business, *1997 Annual Review*, Washington University, St. Louis, Mo., 1998.

MARIO P. DE FIGUEIREDO
Consultant
Chesterfield, Missouri

RAPID METHODS OF MICROBIOLOGICAL ANALYSIS

Rapid methods and automation in microbiology are dynamic fields of study that address the utilization of microbiological, chemical, biochemical, biophysical, immunological, and serological methods for the study of improving isolation, early detection, characterization, and enumeration of microorganisms and their products in clinical, food, industrial, and environmental samples. In the past 10 years, food microbiologists have started to adapt rapid and automated methods in their laboratories (Fig. 1). Conventional methods of detection, enumeration identification and characterization of microbes, are described in reference books such as *Compendium of Methods for the Microbiological Examination of Foods* (1), *Official Methods of Analysis of the AOAC* (2), *Bacteriological Analytical Manual* (3), *Standard Methods for the Examination of Dairy Products* (4), and *Modern Food Microbiology* (5). A comprehensive treatment of all areas of food microbiology was recently published by Doyle, et al. (6).

Important publications on the subject of rapid methods for medical specimens, water, food, industrial, and environmental samples are in a series of papers by Fung and colleagues (7–11) and books such as *Instrumental Methods for Quality Assurance in Foods* (12) and *Rapid Analysis Technique in Food Microbiology* (13). Hartman et al. (14) had an excellent chapter on rapid methods and automation in *Compendium* (1). Swaminathan and Feng (15) also provided updated materials in rapid methods.

The purpose of this article is to review the basic principles and practical applications of a variety of instruments and procedures directly and indirectly related to improved methods for microbiology in quality assurance and research in food microbiology.

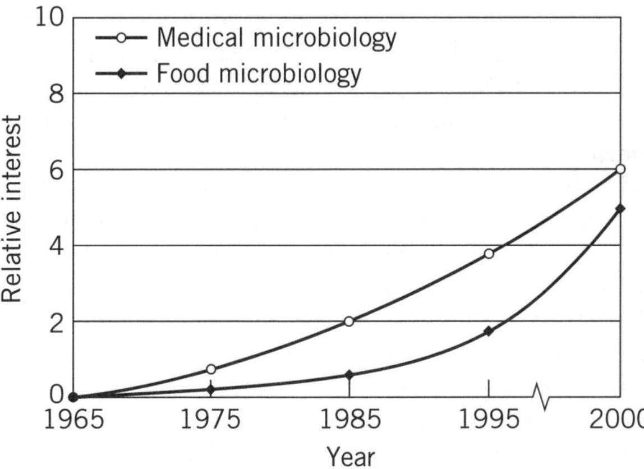

Figure 1. Relative interest in rapid methods among medical microbiologists and food microbiologists.

IMPROVEMENTS IN SAMPLING AND SAMPLE PREPARATION

One of the most useful instruments developed for sample preparation is the Stomacher (Tekmar, Cincinnati, Ohio). This instrument is designed to massage food samples in a sterile bag. The food sample is first placed in the sterile disposable plastic bag, and appropriate sterile diluents are added. The bag with the food is placed in the open chamber. After the chamber is closed, the bag is then massaged by two paddles for a suitable time, usually from 1 to 5 min. No contact occurs between the instrument and the sample. During massaging, microorganisms are dislodged into the diluent for further microbiological manipulation. Massaged slurries are then used for microbiological analysis. More than 30,000 units of the Stomacher have been sold worldwide since its introduction in 1975. A similar system named Masticator Homogenizer is marketed by IUL Instruments, Germany. Recently Sharpe invented a new instrument called Pulsifier that can dislodge bacteria from food by pulsification in a bag. Fung et al. (16) evaluated the Pulsifier versus the Stomacher and found that both instruments provided similar microbial counts from food samples. The Pulsifier resulted in less food debris in the sample, which is advantageous for pathogen detection using technology such as enzyme-linked immunosorbent assay (ELISA), Gene Probe, and polymerase chain reaction (PCR).

An instrument for sample preparation is the Gravimetric Diluter (Spiral Biotech, Bethesda, Maryland). One of the routine procedures in food microbiological work is to aseptically measure a sample of food (eg, 5 g of meat) and then aseptically add an exact amount of sterile diluent (eg, 45 mL) to make a desired dilution (1:10). With the Gravimetric Diluter, the analyst needs only to aseptically place an amount of food (eg, 5.3 g), into a Stomacher bag, set a desired dilution (1:10), and set the instrument to deliver the appropriate amount of sterile diluent (eg, 47.7 mL). Thus, the dilution operation can be automatically done. The dilution factor can be chosen by the analyst to satisfy the need (1:10, 1:50, 1:100, etc) simply by programming the instrument. Manninen and Fung (17) evaluated the Gravimetric Diluter and found that depending on the volume tested the accuracy of delivery for most samples was found to be in the range of 90 to 100%. A new version of this instrument is called Diluflo and has been in use satisfactorily in the author's laboratory since 1992. Recently Pbi of Italy marketed a similar instrument named Dilumacher for automatic microbiological dilution.

ALTERNATIVE METHODS FOR VIABLE CELL COUNT PROCEDURE

The conventional viable cell count or standard plate count method is time-consuming in terms of both operation and collection of data. Several methods have been explored to improve the efficiency of operation of the viable cell count procedure.

The spiral plating method is an automated system for obtaining viable cell counts (Spiral Biotech, Bethesda, Md.). The instrument can spread a liquid sample on the surface of agar contained in a petri dish in a spiral shape (the Archimedes spiral) with a concentration gradient starting from the center and decreasing as the spiral progresses outward on the rotating plate. The volume of liquid deposited at any segment of the agar plate is known. After the liquid containing microorganisms is spread, the agar plate is incubated overnight at an appropriate temperature for the colonies to develop. The colonies appearing along the spiral pathway can be counted either manually or electronically. Table 1 demonstrates that the spiral plating method provides the same counts as the conventional method. A skilled operator can perform 10 times more samples using the spiral system compared with conventional methods.

New versions of the spiral plater have recently been introduced as Autoplater (Spiral Biotech, Bethesda, Md.) and Whitley Automatic Spiral Plater (Bioscience International, Inc., Rockville, Md.). With these instruments an analyst need only present the liquid sample and the instrument completely and automatically processes the sample, including resterilizing the unit for the next sample.

Many instruments are now available for rapid enumeration of bacterial colonies in agar plate. Spiral Biotech (Bethesda, Md.) markets a laser system called CASBA II that has been in use in the author's laboratory for many years with satisfactory results. Recently the same company introduced CASBA III, which includes a high-resolution scanner with a high-performance personal computer and Windows-based colony image analysis software. Protos colony counter (Bioscience International, Inc., Rockville, Md.) is another automatic colony counter that can count 600 plates/hour at 0.5 s/plate. Many other companies also market automatic plate counters. Table 1 also shows that counting manually and counting by a laser counter provided the same number, but the laser counter can complete the task in a few seconds.

Another alternative method for viable cell count is the Isogrid system (QA Laboratories Ltd., San Diego, Calif.).

Table 1. Comparison of Pour Plate and Spiral Plate Counted Manually and by Laser for Bacterial Cultures

	Log_{10}cfu/ml			
	Pour plate		Spiral plate	
Test cultures	Manual	Laser	Manual	Laser
Escherichia coli	8.86	8.85	8.73	8.85
Salmonella enteritidis	8.76	8.66	8.78	8.92
Pseudomonas aeruginosa	8.00	8.00	8.00	8.00
Staphylococcus aureus	8.04	7.78	8.18	8.18
Lactobacillus plantarum	9.48	9.40	9.60	9.69
Streptococcus sp.	7.73	7.66	8.00	8.08
Bacillus cereus	7.26	7.15	7.15	7.26
Micrococcus luteus	7.40	7.32	7.51	7.57

This system consists of a square filter with hydrophobic grids printed on the filter to form 1,600 squares for each filter. Food samples are weighed, blended, and enzyme treated before passage through the membrane filter containing the hydrophobic grids by vacuum. The filter is then placed on agar containing a suitable nutrient for growth of the bacteria, yeasts, or molds. The hydrophobic grids prevent colonies from growing further than the square grids; thus all colonies have a square shape. This facilitates counting of the colonies both manually and electronically. This method has been successfully used to make viable cell counts for a variety of foods, including milk, meat, black pepper, flour, peanut butter, mushrooms, rice, fish, shrimp, and oyster (18–20). Other applications of the Isogrid system include determination of total coliforms, fecal coliforms, *Escherichia coli*, *Salmonella* spp., and other microorganisms (21).

Rehydratable nutrients are embedded into a series of films in the Petrifilm (3M Co., St. Paul, Minn.) system. The top layer of the protective cover is lifted and 1 mL of liquid sample introduced to the center of the unit, and the cover is then replaced. A plastic template is placed on the cover to make a round mold. The rehydrated medium with nutrient will support the growth of microorganisms after suitable incubation time and temperature. The colonies can be counted directly in the unit. The unit is about the size and thickness of a plastic credit card, thus providing great savings of space in storage and incubation. Petrifilm units have been developed for total bacterial count, coliform count, fecal coliform count, yeast and mold counts, and hemorrhagic *E. coli* 0157:H7. Table 2 shows excellent correlation between Petrifilm and standard plate count method for the food.

Redigel system (RCR Scientific, Inc., Goshen, Ind.) is another convenient viable cell count system. This system consists of sterile nutrients with a pectin gel in a tube. The tube is ready to be used any time and no heat is needed to "melt" agar. A 1-mL food sample is first pipetted into the tube. After mixing, the entire content is poured into a special Petri dish previously coated with calcium. When liquid comes in contact with the calcium, a Ca-pectate gel is formed, and the complex swells to resemble conventional agar. After an appropriate incubation time and temperature, the colonies can be counted exactly like conventional standard plate count method. Besides total count, Redigel also has systems for coliform, fecal coliform, yeast and molds, lactic and bacteria, *Staphylococcus aureus*, and *Salmonella*.

The four aforementioned methods have potential as an alternative to the conventional agar pour plate method. Chain and Fung (23) made a comprehensive analysis of all four methods against the conventional method on seven different foods (skinless chicken breast, fresh ground beef, fresh ground pork, packaged whole shelled pecans, raw milk, thyme, and whole wheat flour, 20 samples each) and showed that the new systems and the conventional method were highly comparable and exhibited a high degree of accuracy and agreement ($r = 0.95+$). It should be noted that those performing these newer methods need some training and experience before satisfactory results can be obtained consistently. These are good methods if performed under careful operation.

In the direct epifluorescent filter technique (DEFT) method, the liquified sample is first passed through a filter that retains the microorganisms. The filter is then stained with acridine orange, and the slide is observed under ultraviolet microscopy. "Live" cells usually stain orange-red, orange-yellow, or orange-brown, whereas "dead" cells fluoresce green. The slides can be read by the eye, or by a semiautomated counting system marketed by Bio-Foss. A viable cell count can be made in less than an hour. With the use of an image analyzer, an operator can count 50 DEFT slides per hour (24). This method has been used satisfactorily for counting viable cells in milk and other food samples such as fish (25). Recently the Nordic countries used this method for quality assurance of ground beef. Tortorello and Gendel (26) further developed this procedure by using fluorescent antibodies in conjunction with DEFT to enumerate *E. coli* 0157:H7 in milk and juice.

INSTRUMENTS FOR ESTIMATION OF MICROBIAL POPULATIONS AND BIOMASS

Many methods have been developed in recent years to estimate the total number of microorganisms by parameters other than the viable colony count as described in the previous section. For a new method to be acceptable, it should have some direct correlation with the total viable cell count. Thus, standard curves correlating parameters such as adenosine triphosphate (ATP) level, detection time of electrical impedance or conductance, generation of heat, radioactive CO_2, and so on, with viable cell counts of the same sample series must be made. In general, the larger the number of viable cells in the sample, the shorter the detection time of these systems. A scattergram is then plotted and used for further comparison of unknown samples. The assumption is that as the number of microorganisms increase in the sample, these physical, biophysical, and biochemical events will also increase accordingly. When a sample has 10^5 to 10^6 organisms/mL, detection can be achieved in about 4 to 6 h.

All living things utilize ATP. In the presence of a firefly enzyme system (luciferase and luciferin system), oxygen, and magnesium ions, ATP will facilitate the reaction to generate light. The amount of light generated by this reaction is proportional to the amount of ATP in the sample; thus, the light units can be used to estimate the biomass of cells in a sample. The light emitted by this process can be monitored by a variety of fluorimeters. These procedures can be automated for handling large numbers of samples. Some of the instruments can detect as little as 10^2 to 10^3 fg. The amount of ATP in one colony-forming unit has been reported as 0.47 fg with a range of 0.22 to 1.03 fg. Using this principle, many researchers have tested the efficacy of using ATP to estimate microbial cells in foods and beverages.

Lumac (Landgraaf, the Netherlands) markets several models of ATP instruments and provides customers with test kits with all necessary reagents, such as a fruit juice kit, hygiene monitoring kit, and so on. The reagents are injected into the instrument automatically, and readout is reported as relative light units (RLUs). By knowing the number of microorganisms responsible for generating known RLUs, one can estimate the number of microorganisms in the food sample. In some food systems, such as wine, the occurrence of any living matter is undesirable; thus, monitoring of ATP can be a useful tool for quality assurance in the winery. Recently, much interest has been expressed in using ATP estimation not for total viable numbers but as a sanitation check by companies such as Lumac, BioTrace (Plainsboro, N.J.), Lightning (IDEXX, Westbrook, Maine), Hy-Lite (Glengarry, Biotech, Cornwall, Canada), Charm 4000 Luminometer (Charm Sciences, Malden, Mass.), and others.

As microorganisms grow and metabolize nutrients, large molecules change to smaller molecules in a liquid system and cause a change in electrical conductivity and resistance in the liquid as well as at the interphase of electrodes. By measuring the changes in electrical impedance, capacitance, and conductance, scientists can estimate the number of microorganisms in the liquid because the larger

Table 2. Comparison of the Standard Plate Count Method and the Petrifilm Method for Viable Cell Counts of Shrimp, Perch, Cod, and Whiting

	Colony-forming units/gm	
Sample	Standard plate count	Petrifilm SM
	Shrimp	
1	2.1×10^4	0.9×10^4
2	9.3×10^3	6.8×10^3
3	1.5×10^4	1.2×10^4
4	4.4×10^4	2.0×10^4
5	2.0×10^4	0.9×10^4
	Perch	
1	1.4×10^3	1.2×10^3
2	5.4×10^2	3.7×10^2
3	8.0×10^2	4.8×10^2
4	5.0×10^2	3.2×10^2
5	1.0×10^3	1.2×10^3
	Cod	
1	1.2×10^2	1.1×10^2
2	2.2×10^2	2.4×10^2
3	1.0×10^4	1.0×10^4
4	2.1×10^4	2.3×10^4
5	1.2×10^4	1.2×10^4
	Whiting	
1	3.0×10^2	3.1×10^2
2	5.8×10^2	5.6×10^2
3	1.2×10^3	1.1×10^3
4	5.4×10^2	5.2×10^2
5	8.8×10^2	8.4×10^2

Note: Samples were massaged in a Stomacher for 1 min in sterile diluent. Viable cells counts were made according to standard method. Incubation time was 48 h at 32°C. All samples were done in duplicate. Correlation coefficient between the two methods is r = 0.99.
Source: Ref. 22.

the number of microorganisms in the fluid, the faster the change in these parameters that can be measured by sensitive instruments. A detailed analysis on the subject of impedance, capacitance, and conductance in relation to food microbiology has been made by Eden and Eden (27).

The Bactometer (bioMerieur Vitek, Inc., Hazelwood, Mo.) is an instrument designed to measure impedance changes in foods. Samples are placed in the wells of a 16-well module. After the module is completely or partially filled, it is plugged into the incubator unit to start the monitoring sequence. At first, there is a stabilization period for the instrument to adjust to the module, then a baseline is established. As the microorganisms metabolize the substrates and reach a critical number (10^5–10^6 cells/mL), change in impedance increases sharply, and the monitor screen shows a slope similar to the log phase of a growth curve. The point at which the change in impedance begins is the detection time, and this is measured in hours from the start of the experiment. The detection time is inversely proportional to the number of microorganisms in the sample. By knowing the number of microorganisms per milliliter in a series of liquid samples and the detection time of each sample, one can establish a standard curve. From the curve one can decide the cutoff points to monitor certain specifications of the food products.

Impedance methods have been used to estimate bacteria in milk, dairy products, meats, and other foods. This method has been used for determining the shelf life potential of pasteurized whole milk by Bishop et al. (28).

The Malthus system (Crawley, UK) works by measuring the conductance of the fluid as the organisms grow in the system. It also generates a conductance curve similar to the impedance curve of the Bactometer, and it also uses detection time in monitoring the density of the microorganisms in the food.

The Malthus system has been used for microbial monitoring of brewing liquids and hygiene monitoring. Gibson and colleagues (29–31) have done considerable work using the Malthus system to study seafood microbiology.

Besides estimating viable cells in foods, both the Bactometer and the Malthus systems can detect specific organisms by the use of selective and differential liquid media. An automatic instrument for measuring direct and indirect impedance has been developed in Europe and named RABIT (Rapid Automated Bacterial Impedance Technique).

An instrument called the Omnispec bioactivity monitor system (Wescor, Inc., Logan, Utah) is a tri-stimulus reflectance colorimeter that monitors dye pigmentation changes mediated by microbial activity. Dyes can be used that produce color changes as a result of pH changes, changes in the redox potential of the medium, or the presence of compounds with free amino groups. Samples are placed in microtiter wells or other types of containers and are scanned by an automated light source with computer interface during the growth stages (0–24 hours). The change of color or hue (a*, b*, L) can be monitored similar to impedance curve and conductance curve. Manninen and Fung (32) evaluated this system in a study of pure cultures of *Listeria monocytogenes* and food samples and found high correlation coefficients (r) of 0.90 to 0.99 for pure bacterial cultures and 0.82 for minced beef between the colony counts predicted by the colorimetric technique and the results of the traditional plate count method. They also showed that detection times for bacterial cultures such as *Enterobacter aerogenes, E. coli, Hafnia alvei*, and several strains of *L. monocytogenes* were substantially (2–24 h) shorter using the instrument than using the traditional method and concluded that the colorimetric detection technique employed by the Omnispec system simplifies the analyses, saves labor and materials, and provides a high sampling capacity. Tuitemwong (33) completed an extensive study using Omnispec 4000 to monitor growth responses of food pathogens in the presence or absence of membrane-bound enzymes. This instrument is highly efficient in large-scale studies of microbial interaction with different compounds in liquid and food.

The catalase test is another rapid method for estimation of microbial populations in certain foods. Catalase is a very reactive enzyme. Microorganisms can be divided into catalase positive and catalase negative. Both groups are important in food microbiology; however, under certain food-storage conditions, a certain group predominates. Most perishable foods (commercial as well as domestic) are cold stored under aerobic conditions. The organisms causing spoilage of these foods are psychrotrophs. The predominant psychrotrophic bacteria are *Pseudomonas* spp., which are strongly catalase positive. Other important psychrotrophs such as *Micrococcus, Staphylococcus*, and a variety of enterics are also catalase positive. Thus, one can make use of the presence of catalase to estimate the bacterial population. Catalase activity can be detected by a simple capillary tube method (34). Recently Binjasass and Fung (35) completed an extensive study using the capillary catalase tube method for monitoring microbial load and endpoint cooking temperature of fish and found the method to be reliable and simple to use.

Ang et al. (36) also showed that heating poultry meat to 71°C (a legal requirement for these products) will destroy both bacterial and animal catalase. The test is 99% accurate and is simple and inexpensive to perform.

MINIATURIZED MICROBIOLOGICAL TECHNIQUES

Identification of microorganisms is an important part of quality assurance and control programs in the food industry. The author has developed many miniaturized methods to reduce the volume of reagents and media (from 5 to 10 mL to about 0.2 mL) for microbiological testing in microtiter plates. The basic components of the miniaturized system are the microtiter plates for test cultures, a multiple inoculation device, and containers to house solid media (large petri dishes) and liquid media (another series of microtiter plates). The procedure involves placing liquid cultures (pure cultures) to be studied into sterile wells of a microtiter plate to form a master plate. Each microtiter plate can hold up to 96 different cultures, 48 duplicate cultures, or various combinations as desired. The cultures are then transferred by a sterile multipoint inoculator (96 needles protruding from a template) to solid or liquid media.

Sterilization of the inoculator is by alcohol flaming. Each transfer represents 96 separate inoculations in the conventional method. After incubation at an appropriate temperature, the growth of cultures on solid media or liquid media can be observed and recorded, and the data can be analyzed. These miniaturized procedures save a considerable amount of time in operation, effort in manipulation, materials, labor, and space. These methods are ideal for studying large numbers of isolates or for research involving challenging large numbers of microbes against a host of test compounds.

The miniaturized methods have been used to study large numbers of isolates from foods (37–39) and to develop bacteriological media and procedures (40–43). Many useful microbiological media were discovered through this line of research. For example, an aniline blue *Candida albicans* medium was developed and marketed by Difco under the name Candida Isolation Agar. The sensitivity and specificity were 98.0 and 99.5%, respectively, with a predictive value of 99.1% (44).

On the commercial side, many diagnostic kits to identify microorganisms have been developed and marketed since the 1970s. Currently, API, Enterotube, R/B, Minitek, MicroID, and IDS are available. Most of these systems were first developed for the identification of enterics (*E. coli, Salmonella, Shigella, Proteus, Enterobacter* spp., etc). Later many of the companies expanded the capacity to identify nonfermentor, anaerobes, gram-positive organisms, and even yeasts and molds. Most of the early comparative analyses centered around evaluation of these kits for clinical specimens. Comparative analysis of diagnostic kits and selection criteria for miniaturized systems were made by Fung et al. (45) and Cox et al. (46). They concluded that these miniaturized systems are accurate, efficient, labor saving, space saving, and cheaper than the conventional procedure. Their usefulness in clinical and food microbiological laboratories will continue to be important.

TECHNIQUES INVOLVING INSTRUMENTS, IMMUNOLOGY, AND GENETIC TESTS

Many sophisticated instruments have been developed to identify isolates from clinical specimens such as Sensititre (Radiometer Amer, Westlake, Ohio) and Biolog (Hayward, Calif.). One of the most automated systems for the identification of isolates (clinical and foods) is the Vitek system. The system depends on the growth of target organisms in specially designed media housed in tiny chambers in a plastic card. The card is then inserted into the incubation chamber. The instrument periodically scans the wells of the cards and sends information to the computer, which then matches the database and identifies the unknown cultures in the cards. The system is entirely automated and computerized and provides hard copies for record keeping. The system is capable of identifying enterics, yeast, *Bacillus*, selected gram-positive pathogens, and other organisms.

The DNA probe (Genetrak, Farmingham, Mass.) is a sensitive method to detect pathogens such as *Campylobacter, Salmonella, Listeria*, and *E. coli*. At first, the system utilized radioactive compounds for assay. The second generation of probes uses enzymatic reactions to detect the presence of pathogens. Another major change in this area is the development of probes to detect target RNA. In a cell there is only one copy of DNA; however, there may be 1,000 to 10,000 copies of ribosomal RNA. Thus, the new generation of probes are designed to probe target RNA. After enrichment of cells (either in liquid or colonies on solid agar), DNA or RNA of target cells can be extracted and released into the liquid and then detected by the appropriate DNA and/or RNA probes. These methods are currently being automated. Currently, kits are available for *Salmonella*, *Listeria*, *Campylobacter*, *Yersinia*, and others. As the need arises more organisms will be added to the list.

Polymerase chain reaction (PCR) systems are the latest development in DNA amplification technology and recently have gained much attention in food microbiology. Originally the procedures are highly complicated, and a very clean environment is needed to perform the test. Recently, much research has been directed to simplifying the procedure for laboratory analysts. DuPont recently commercialized a system called BAX for PCR.

The BAX screening system is designed to work from overnight enrichment broths. In the case of the BAX system for *Salmonella*, the food is enriched overnight in any standard nonselective broth (lactose broth, buffered peptone water, etc). A 1:10 dilution into BHI is performed followed by a 3-h grow-back incubation. An aliquot is taken into a tube and treated with lytic enzymes and heat. An aliquot of the cell lysate is used to rehydrate the PCR reagents housed in a tube, and a control tablet and the sample is placed into the thermal cycler along with the corresponding control tube. Thermal cycling, UV visualization of the gel, and photography to document the results are then performed to detect PCR products.

PCR technology is now steadily moving into the food laboratories. Some obstacles of PCR technology include the occurrence of inhibitors in the food samples and the fact that DNA from dead cells can also be amplified, thus giving a false positive result for a food that may be safe to eat. For pure cultures this technology is very useful. Dilution of samples (1:10) will reduce the inhibitors, and enrichment of target culture will ensure amplication of live cells even in the presence of other organisms.

Another important development, also by Qualicon, is the RiboPrinter characterization system. The RiboPrinter is designed to accept isolated colonies of bacteria as the sample. A sample is prepared as follows:

1. A colony of bacteria is picked from an agar plate using a sterile plastic stick (provided in the sample kit).
2. Cells from the stick are suspended in a buffer solution by mechanical agitation.
3. An aliquot of the cell suspension is loaded into the sample carrier to be placed into the instrument. Each sample carrier has space for eight individual colony picks.

The patterns of organisms would allow an analyst to pinpoint the source of the pathogen. For example, even the pathogen *L. monocytogenes* has many different patterns. By matching the pattern of an isolate from food and patterns from isolates from a slicer or drain, the analyst can determine exactly the source of the *L. monocytogenes* in the finished product. This system received the Industry Award given by the Institute of Food Technologists in 1997.

Modification of the basic PCR procedure includes reverse transcriptase PCR, nested/multiplex PCR, randomly amplified polymorphic DNA (RAPD), fluorescent probes in PCR, among others. These methods are slowly finding their way to the food microbiology laboratory.

The enzyme-linked immunosorbent assay (ELISA) systems method commercialized by Organon Teknika (Durham, N.C.) utilizes two monoclonal antibodies specific for *Salmonella* detection. In a comparative study involving 1,289 samples, Eckner et al. (47) found no significant difference between the conventional method and the ELISA method for food samples except cake mix and raw shrimp. Another ELISA system, the Tecra system (International BioProducts, Redmond, Wash), was developed in Australia and uses polyclonal antibodies to detect *Salmonella*. These methods have also been used to detect *Listeria* and *E. coli*. Many companies are providing a host of monoclonal and polyclonal antibodies for a variety of diagnostic tests, some including food pathogens (48).

One of the drawbacks of the ELISA test is the many steps necessary for adding reagents and washing test samples. In the Vitek ImmunoDiagnostic Assay System (VIDAS™) system, all intermediate steps are automated. VIDAS is a multiparametric immunoanalysis system that utilizes the enzyme-linked fluorescent immunoassay (ELFA) method. The end result of the test protocol is a fluorescent product, and the VIDAS reader utilizes a special optical scanner that measures the degree of fluorescence. From the moment the solid phase receptacles and the reagent strips are placed in the instrument, the VIDAS is fully automated. Many automated ELISA instruments are on the market now, including TECRA OPUS (International BioProducts, Redmond, Md.), Bio-tek Instruments (Highland Park, Vt.), Automated EIA Processor (BioControl, Bethell, Wash.), among others.

In a related development, several self-contained small units (REVEAL, VIP, etc) have been marketed recently. After enrichment (with or without an Oxyrase type of stimulation) an analyst need only apply a small aliquot (boiled or unboiled) to the kit. Reaction occurs in a few minutes. These kits have been used to rapidly screen *E. coli* 0157:H7 and *Salmonella* in ground beef.

UNIQUE system for *Salmonella* is another way of using immunocapture technology. A dipstick with antibody against *Salmonella* is applied to the pre-enriched liquid. The antibody captures the *Salmonella* if present. This charged dipstick is then placed in a fresh enrichment broth and the cells are allowed to multiply for a few hours. After the second enrichment step, the dipstick, with a much larger population of *Salmonella* attached to it, will be subject to further ELISA procedures. The entire test is housed in a convenient self-contained plastic unit. This type of method is very useful for a small laboratory with low-volume testing of pathogens.

Immunomagnetic capture methods have attracted much attention lately. VICAM and DYNAL developed magnetic beads coated with a variety of antibodies to capture target cells or cellular components in foods. After the magnetic beads have a chance to interact with potential target cells in a tumbling apparatus for an hour, the magnetic beads are physically separated from the food or liquid by a powerful magnet applied to the side of the test tube. Further microbiological procedures such as direct plating or ELISA tests can be made on these charged beads. Many test systems are now incorporating immunomagnetic capture technology to shorten the enrichment steps by at least 1 day. This is especially important for foods with very low number of target pathogen such as *Listeria* or *Salmonella*.

Motility enrichment is a very useful concept in rapid isolation and identification of food pathogens. Fung and Kraft (49) described a motility flask system for rapid detection and isolation of *Salmonella* spp. from mixed cultures and poultry products. The system involves a flask with a side arm that contains several agar layers. Lactose broth is placed in the flask, and then a sample (with or without salmonellae) is inoculated into the lactose broth. When salmonellae are present, they will swim through the first level of agar, which contains selenite cysteine and sodium lauryl sulfite that inhibits other organisms but allows salmonellae to pass through. Once salmonellae pass the first layer, they can grow and metabolize compounds in the second and third layers. By looking at the color changes in the second and third agar layers, one can make an assumption that salmonellae were in the sample that was put into the lactose broth. Further serological tests can then be performed to confirm the presence of *Salmonella*.

In more recent years, a commercial system called *Salmonella* 1-2 test (BioControl, Bothell, Wash.) was developed that utilizes motility as a form of selection. The food sample is first preenriched for 24 h in lactose broth, and then 0.1 mL is inoculated into one of the chambers in an L-shaped system. The chamber contains selective enrichment liquid medium. There is also a small hole connecting the liquid chamber with the rest of the system, which has a soft agar through which salmonellae can migrate. An opening on the top of the second chamber allows the analyst to deposit a drop of polyvalent anti-H antibody. If the sample contains salmonellae from the lower side of the L unit, salmonellae will migrate through the hole and up the agar column. Simultaneously, the antibody against flagella of salmonellae will move downward by gravity. When the antibody meets the salmonellae, they will form a visible immunoband. The presence of an immunoband in this system is a positive test for *Salmonella* spp. The system is easy to use and has gained popularity because of its simplicity.

Oxyrase (Mansfield, Ohio), a membrane fraction of *E. coli*, was found in the author's laboratory to stimulate the growth of a large number of important facultative anaerobic food pathogens. In the presence of a hydrogen donor such as lactate, Oxyrase can convert O_2 to H_2O, thereby reducing the oxygen tension of the medium and creating anaerobic conditions that favor the growth of facultative anaerobic organisms. In a medium containing 0.1 units/

mL of the enzyme, the growth of *L. monocytogenes*, *E. coli* 0157:H7, *S. typhimurium*, *S. faecalis*, and *Proteus vulgaris* were greatly enhanced; colony counts were greater by 1 to 2 log units, depending on the initial count and the strain studied, after incubation in the presence of the enzyme for 5 to 8 h at 35 to 42°C compared with control without Oxyrase.

By combining the oxyrase enzyme and a unique U-shaped tube, Yu and Fung (50–52) developed an effective method to detect *L. monocytogenes* and *Listeria* spp. from laboratory cultures and meat systems. Niroomand and Fung (53) studied the effects of oxyrase in stimulation of growth of *Campylobacter* from foods. Tuitemwong et al. (54) also found that these membrane fragments and those obtained from *Acetobacter* and *Gluconobacter* can stimulate growth of starter cultures in food fermentation.

There are many other systems that involve modern biochemistry, chemistry, and immunology. For example, one can use protein profiles for microbial fingerprinting (AMBIS system, San Diego, Calif.) or cell composition as a way to identify bacterial cultures (Hewlett-Packard, Palo Alto, Calif.).

CONCLUSIONS

This article describes a variety of methods that are designed to improve current methods, explore new ideas, and develop new concepts and technologies for the improvement of applied microbiology. Although many of these methods were first developed for clinical microbiology, they are being used for food microbiology. This field will certainly grow, and many food microbiologists will find these new methods very useful in their routine work in the immediate future. Many methods described here are already being used by applied microbiologists nationally and internationally.

ACKNOWLEDGMENTS

This material is based on work supported by the Cooperative State Research Service, U.S. Department of Agriculture, under Agreement No. 8890341874511, Agricultural Experiment Station, Kansas State University, Manhattan, Kansas 66506.

BIBLIOGRAPHY

1. C. Vanderzant and D. Splittstoesser, eds., *Compendium of Methods for the Examination of Foods*, American Public Health Association, Washington, D.C., 1992.
2. Association of Official Analytical Chemists, *Official Methods of Analysis of the AOAC (Association of Official Analytical Chemists)*, Vols. 1 and 2, Arlington, Va., 1990.
3. Food and Drug Administration, *Bacteriological Analytical Manual*, 6th ed., Association of Official Analytical Chemists, Food and Drug Administration, Arlington, Va., 1996.
4. American Public Health Association, *Standard Methods for the Examination of Dairy Products*, Washington, D.C., 1992.
5. J. M. Jay, *Modern Food Microbiology*, 4th ed. Van Nostrand–Reinhold, New York, 1992.
6. M. P. Doyle, L. R. Beuchat, and T. J. Montville, *Food Microbiology Fundamentals and Frontiers*, ASM Press, Washington, D.C., 1997.
7. D. Y. C. Fung, "Rapid Methods and Automation for Food Microbiology," in D. Y. C. Fung and R. F. Matthews, eds., *Instrumental Methods for Quality Assurance in Foods*, Marcel Dekker, New York, 1991.
8. D. Y. C. Fung, "Historical Development of Rapid Methods and Automation in Microbiology," *Journal of Rapid Methods and Automation in Microbiology* **1**, 1 (1992).
9. D. Y. C. Fung, "Rapid Methods and Automation in Food Microbiology: A Review," *Food Rev. Int.* **10**, 357–375 (1994).
10. D. Y. C. Fung, "What's Needed in Rapid Detection of Food Borne Pathogens," *Food Technol.* **44**, 64–67 (1995).
11. D. Y. C. Fung et al., "Rapid Methods and Automation: A Survey of Professional Microbiologists," *J. Food Prot.* **52**, 65 (1989).
12. D. Y. C. Fung and R. F. Matthews, eds. *Instrumental Methods for Quality Assurance in Foods*, Marcel Dekker, New York, 1991.
13. P. D. Patel, *Rapid Analysis Techniques in Food Microbiology*, Chapman & Hall, New York, 1994.
14. P. A. Hartman et al., "Rapid Methods and Automation," in C. Vanderzant and D. Splittstoesser, eds., *Compendium of Methods for the Examinations of Foods*, American Public Health Association, Washington, D.C., 1992.
15. B. Swaminathan and P. Feng, "Rapid Detection of Food-Borne Pathogenic Bacteria," *Rev. Microbiol.* **48**, 401 (1994).
16. D. Y. C. Fung et al., "The Pulsifier: A New Instrument for Preparing Food Suspensions for Microbiological Analyses," *Journal of Rapid Methods and Automation in Microbiology* **6**, 42–50 (1998).
17. M. T. Manninen and D. Y. C. Fung, "Use of the Gravimeter Diluter in Microbiological Work," *J. Food Prot.* **55**, 59 (1992).
18. P. Entis, "Enumeration of Coliforms in Nonfat Dry Milk and Canned Custard by Hydrophobic Grid Membrane Filter Method: Collaborative Study," *J. Assoc. Off. Anal. Chem.* **66**, 897 (1983).
19. P. Entis, "Enumeration of Total Coliforms and *Escherichia coli* in Foods by Hydrophobic Grid Membrane Filters Collaborative Study," *J. Assoc. Off. Anal. Chem.* **67**, 812 (1984).
20. A. W. Sharpe and P. I. Peterkin, *Membrane Filter Food Microbiology*, Research Studies Press, Letchworth, U.K., 1988.
21. A. W. Sharpe, "Rapid Methods: Consideration to Adoption by Regulatory Agencies," *Proc. 105 AOAC Annu. Int. Mtgs*, Phoenix, Ariz., August 12–15, 1991.
22. D. Y. C. Fung, R. A. Hart, and V. Chain, "Rapid Methods and Automated Procedures for Microbiological Evaluation of Seafood," in D. E. Kramer and J. Liston, eds., *Seafood Quality Determination*, Elsevier, Amsterdam, The Netherlands, 1987, pp. 247–253.

23. V. S. Chain and D. Y. C. Fung, "Comparison of Redigel, Petrifilm, Spiral Plate System, ISOGRID, and Standard Plate Count for the Aerobic Count on Selected Foods," *J. Food Prot.* **54**, 208 (1991).

24. G. L. Pettipher, "Review: The Direct Epifluorescent Filter Technique," *J. Food Technol.* **21**, 535 (1986).

25. G. L. Pettipher, "The Direct Epifluorescent Filter Technique," in M. R. Adams and C. F. A. Hope, eds., *Rapid Microbiological Methods*, Elsevier, New York, 1989.

26. M. Tortorello and S. M. Gendel, "Fluorescent Antibodies Applied to Direct Epifluorescent Filter Techniques for Microscopic Enumeration of *Escherichia coli* 0157:H7 in Milk and Juice," *J. Food Prot.* **56**, 672 (1993).

27. R. Eden and G. Eden, *Impedance Microbiology*, Research Studies Press, Letchworth, U.K., 1984.

28. J. R. Bishop, C. R. White, and R. Firstenberg-Eden, "A Rapid Impedimetric Method for Determining the Potential Shelf-Life of Pasteurized Whole Milk," *J. Food Prot.* **47**, 471 (1984).

29. D. M. Gibson and G. Hobbs, "Some Recent Developments in Microbiological Methods in Seafood Quality," in D. F. Kramer and J. Listen, eds., *Seafood Quality Determination*, Elsevier, Amsterdam, The Netherlands, 1987, pp. 283–298.

30. D. M. Gibson and I. D. Ogden, "Assessing Bacterial Quality of Fish by Conductance Measurement," *J. Appl. Bacteriol.* **49**, 12 (1980).

31. I. D. Ogden, "Use of Conductance Methods to Predict Bacterial Counts in Fish," *J. Appl. Bacteriol.* **61**, 36 (1986).

32. M. T. Manninen and D. Y. C. Fung, "Estimation of Microbial Numbers from Pure Bacterial Cultures and from Minced Beef Samples by Reflectance Colorimetry with Omnispec 4000," *Journal of Rapid Methods and Automation in Microbiology* **1**, 41 (1992).

33. K. Tuitemwong, "Characteristics of Food Grade Membrane Bound Enzymes and Applications in Food Microbiology and Food Safety," Ph.D. Dissertation, Kansas State University Library, Manhattan, Kans., 1993.

34. D. Y. C. Fung, "Procedures and Methods for One-Day Analysis of Microbial Loads in Foods," in K.-O. Habermehl, ed., *Rapid Methods and Automation in Microbiology and Immunology*, Springer-Verlag, Berlin, 1985, pp. 656–664.

35. F. M. R. Binjasass and D. Y. C. Fung, "Catalase Activity as an Index of Microbial Load and End-Point Cooking of Temperature of Fish," *Journal of Rapid Methods and Automation in Microbiology* **6**.

36. C. Y. W. Ang et al., "Sensitive Catalase Test for End-Point Temperature of Heated Chicken Meat," *J. Food Sci.* (1994).

37. D. Y. C. Fung and P. A. Hartman, "Miniaturized Microbiology Techniques for Rapid Characterization of Bacteria," in C. G. Heden and T. Illeni, eds., *New Approaches to the Identification of Microorganisms*, Wiley, New York, 1975, pp. 347–370.

38. C. Y. Lee, D. Y. C. Fung, and C. L. Kastner, "Computer-Assisted Identification of Bacteria on Hot-Boned and Conventionally Processed Beef," *J. Food Sci.* **47**, 363 (1982).

39. C. Y. Lee, D. Y. C. Fung, and C. L. Kastner, "Computer-Assisted Identification of Microflora on Hot-Boned and Conventionally Processed Beef: Effect of Moderate and Slow Initial Chilling Rate," *J. Food Sci.* **50**, 553 (1985).

40. S. P. Chein and D. Y. C. Fung, "Acriflavin Violet Red Bile Agar for the Isolation and Enumeration of *Klebsiella pneumoniae*," *Food Microbiology* **7**, 73 (1991).

41. D. Y. C. Fung and C. Liang, "A New Fluorescent Agar for the Isolation of *Candida albicans*," *Bulletin d'information des Laboraores des Service Veterinaries (France)* **29/30**, 1–2 (1989).

42. C. C. S. Lin and D. Y. C. Fung, "Effect of Dyes on Growth of Food Yeast," *J. Food Sci.* **47**, 770 (1985).

43. C. C. S. Lin and D. Y. C. Fung, "Critical Review of Conventional and Rapid Methods for Yeast Identification," *CRC Crit. Rev. in Microbiol.* **14**, 273 (1987).

44. M. C. Goldschmidt, et al., "New Aniline Blue Dye Medium for Rapid Identification and Isolation of *Candida albicans*," *J. Clin. Microbiol.* **29**, 1098–1099 (1991).

45. D. Y. C. Fung, M. C. Goldschmidt, and N. A. Cox, "Evaluation of Bacterial Diagnostic Kits and Systems at an Instructional Workshop," *J. Food Prot.* **47**, 68 (1984).

46. N. A. Cox, "Selecting a Miniaturized System for Identification of *Enterobacteriaceae*," *J. Food Prot.* **47**, 74 (1984).

47. K. F. Eckner et al., "Comparison of *Salmonella* Bio-EnzaBead Immunoassay Method and Conventional Culture Procedure for Detection of *Salmonella* in Foods," *J. Food Prot.* **50**, 379 (1987).

48. D. Y. C. Fung, R. Bennett, and G. C. Lehleitner, "Rapid Diagnosis in Bacteriology: Contribution of Polyclonal and Monoclonal Antibodies," in M. M. Gaeteau, J. Henry, and G. Siest eds., *Biologie Prospective, le Colloque de Pont-e-Mousson*, John Libbey, London, 1988, pp. 21–26.

49. D. Y. C. Fung and A. A. Kraft, "A Rapid and Simple Method for the Detection and Isolation of *Salmonella* from Mixed Cultures and Poultry Products," *Poult. Sci.* **49**, 46 (1970).

50. L. S. L. Yu and D. Y. C. Fung, "Oxyrase Enzyme and Motility Enrichment Fung-Yu Tube for Rapid Detection of *Listeria monocytogenes* and *Listeria* spp.," *Journal of Food Safety* **11**, 149 (1991).

51. L. S. L. Yu and D. Y. C. Fung, "Effect of Oxyrase Enzyme in *Listeria monocytogenes* and Other Facultative Anaerobes," *Journal of Food Safety* **11**, 163 (1991).

52. L. S. L. Yu and D. Y. C. Fung, "Growth Kinetics of *Listeria* in the Presence of Oxyrase Enzyme in a Broth Model System," *Journal of Rapid Methods and Automation in Microbiology* **1**, 15 (1992).

53. F. Niroomand and D. Y. C. Fung, "Effect of Oxygen Reducing Membrane Fragments on Growth of *Capmylobacter*," *Journal of Rapid Methods and Automation in Microbiology* **2**, 247 (1994).

54. K. Tuitemwong, D. Y. C. Fung, and P. Tuitemwong, "Acceleration of Yoghurt Fermentation by Bacterial Membrane Fraction Biocatalysis," *Journal of Rapid Methods and Automation in Microbiology* **3**, 127 (1994).

Daniel Y. C. Fung
Kansas State University
Manhattan, Kansas

REFRIGERATED FOODS: FOOD FREEZING AND PROCESSING

In high latitudes it is possible to freeze foods simply by leaving the product outside in winter. This familiarity with food freezing resulted in a general feeling that such a simple process needs little planning or care in product selection, processing, packaging, or subsequent storage and can be accomplished with the minimum of skill. The enormous amount of scientific research put into these three P's (intrinsic Product quality, Processing, and Packaging) by food scientists has resulted in frozen raw materials and products, now increasingly available across the world, meeting the needs of the most sophisticated consumers. Equally important are the three T's (Time, Temperature, and Tolerance), which govern the practical shelf life (PSL) of all frozen products.

The advent of mechanical refrigeration (1) led by the Linde air compressor in the 1850s (capable of maintaining temperatures of about −9°C—the temperature at which mold activity can be ignored over short periods) made it possible to emulate winter conditions in high latitudes across the temperate and tropical regions irrespective of season. A frozen food industry developed, with its capability of transporting products such as beef, lamb, fish, shellfish, and butter around the world and particularly across the Tropics.

The pioneer of the retail frozen food industry, an American named Clarence Birdseye, invented a multiplate froster that enabled commercial quantities of food to be frozen rapidly and brought down to −18°C, which he specified as the warmest temperature at which this new "frosted" food should be stored (2). Today, some 70 years after the launch of the first commercially frozen retail frozen products in Springfield, Massachusetts (3), there has been only little change in the basic precepts of this now large industry. The results of a massive research program, carried out at the U.S. Department of Agriculture (USDA) Western Research Laboratory in Albany, California, in the 1950s, known as the Time–Temperature Tolerance (T–TT) program, validated Birdseye's original prescription of −18°C as the accepted temperature for the industry and has been immensely influential in defining custom and practice throughout the industry (4). Later researchers (5) favored colder temperatures for long-term storage of fish, especially fatty fish such as salmon. Many cold stores have been built capable of holding temperatures of −30°C and, in the case of surimi (to preserve the white color) for the Japanese market, temperatures as cold as −60°C.

The first textbook on food freezing (2) omitted two products that, when developed later, now play a large role in the industry—French fries and frozen concentrated orange juice. Sources of information on food freezing technology include the International Institute of Refrigeration (IIR), the proceedings of which cover the whole gamut of refrigerants; refrigerating equipment; cold store construction; road, rail, and sea transport of frozen foods; and the characteristics of various products during freezing, storage, and thawing (6). In the United States the American Society of Heating, Refrigerating and Air-Conditioning Engineers, Inc. (ASHRAE) handbooks (7) summarize published research and provide essential background information on refrigerant characteristics. The IIR has also published a concise guide to the processing and handling of frozen foods (8).

Following the successful T–TT program in the 1950s, attention was drawn to the importance of other factors apart from time and temperature—the three P's (5). Subsequent publications (9,10) have elaborated on these developments and advanced our knowledge of the science and practice of food freezing, storage and transport.

T–TT

The concept that the rate of quality loss, as influenced by temperature, in most frozen foods during storage broadly follows the Arrhenius equation, as postulated in the T–TT studies (4), has stood the test of time. One refinement has been the investigation of the glass transition phenomenon in many frozen products (11). As temperature is reduced well below the freezing point, the ice crystals grow until the remaining unfrozen moisture, the freeze-concentrated unfrozen phase, becomes an increasingly highly concentrated and viscous solution. If temperature is lowered further, the point is reached (T_g) at which it becomes a "glass"—an amorphous solid, devoid of structure and totally unlike the highly ordered structure of ice crystals. Below the glass temperature (T_g) the rate of molecular mobility is markedly reduced and hence the rate of quality loss brought almost to a standstill. The kinetics of reactions associated with quality loss at temperatures close to the glass transition temperature (T_g) can be described by the Williams–Landel–Ferry (WLF) model, in which reaction rates (and hence quality loss) are a function of the temperature difference between the storage temperature and T_g (12). Shelf-life modeling of frozen foods is well described by Fu and Labuza (13). While the T_g of most frozen foods may be so cold as to render it too expensive (and unnecessary in view of the short storage periods—a few months—necessary to span the seasons or transport food across the world) to store at or near the T_g these findings open up new avenues of research to breed food plants and formulate prepared dishes with a "warm" T_g, thus increasing their PSL at conventional storage temperatures. Perhaps this research will even extend the range of products that can be successfully frozen.

PPP: PRODUCT QUALITY, PROCESSING, AND PACKAGING

Intrinsic Product Quality

Following Jul's work (5) that drew attention to the PPP factors, numerous researchers have confirmed and extended the understanding of the importance of initial product quality, principally in terms of freshness of the living tissue or formulated food. Perhaps the most dramatic claim showed that the PSL of squid held for only 24 h in ice between death and freezing was reduced by one-half compared with a similar squid frozen immediately on death (14). Cyroprotectants have an important part to play in modifying products to increase their PSL. Cryoprotectants commonly employed include sugars, amino acids, polyols, and so on. Indeed, the loss of PSL in the frozen state consequent upon loss of "freshness" in the raw material may be considered to be due to a loss of naturally occurring cryoprotectants. Antifreeze proteins have been found to depress the freezing point markedly (Antarctic fish can survive without freezing at −70°C), and the possibility that the use of antifreeze proteins may be able to improve the quality of frozen foods is being actively researched (15).

Processing

Any enzyme-mediated reactions will be greatly accelerated by freezing. The cryconcentration effect, caused by the removal of water in the form of ice from the medium, may be 5 or 10 times and is only feebly compensated for by the lowering of temperature. A cooked product, in which the enzymes have been destroyed, is more stable. This enhanced enzymic activity on freezing (which results in loss of color, flavor, and development of off-flavors in cold storage) calls for inactivating the enzymes in vegetables (such as sweet corn, peas, and green beans) by blanching; the choice of which indicator enzyme ensures adequate blanching is addressed by Whitaker (16).

Packaging

Oxygen is the great enemy of frozen food quality. Oxygen can be excluded by covering the product with sauce, by water glazing of fish or using packaging with a low oxygen permeability that can be reinforced by exhausting the package before sealing, by gas flushing, or by skin packaging (17), in which the preheated film is dropped onto the product, which is supported on a lower web of the same film. On withdrawing the air between the two films, a skin-tight package is formed, which is then heat sealed in a vacuum chamber. Figures 1 and 2 illustrate the effects of some common packaging materials on the stability of various products. Packaging may also serve as the cooking container, in a microwave oven or microwavable/dual ovenable packaging.

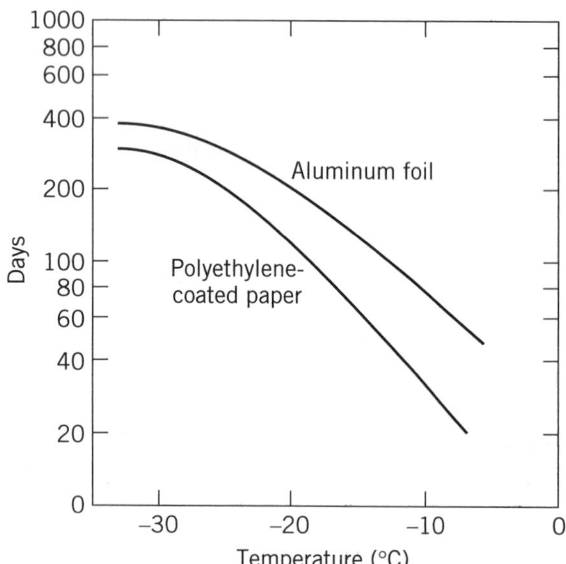

Figure 1. The effect of various packaging materials on shelf life of individually packaged hamburger meat.

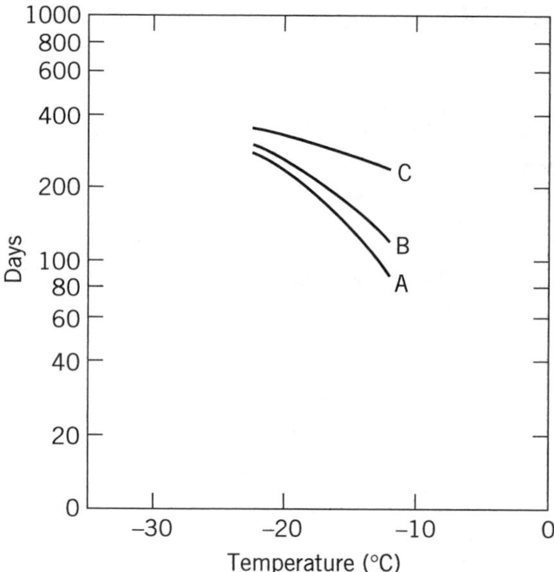

Figure 2. Acceptability time for pork chops. A wrapped in PE film and placed in master cartons; B same as A but a PA–PE laminate film; and C, vacuum packaged in PA–PE laminate (same material as in B).

BIBLIOGRAPHY

1. R. Thevenot, *A History of Refrigeration Throughout the World*, International Institute of Refrigeration, Paris, 1979.
2. D. K. Tressler and C. F. Evers, *The Freezing Preservation of Foods*, AVI Publishing, Westport, Conn., 1943.
3. E. W. Williams, *Frozen Foods: Biography of an Industry*, E. W. Williams Publications, New York, 1963.
4. W. B. Van Arsdel, M. J. Copley, and R. L. Olson, *Quality and Stability in Frozen Foods*, Wiley Interscience, New York, 1969.
5. M. Jul, *The Quality of Frozen Foods*, Academic Press, Orlando, Fla., 1984.
6. *Congress Proceedings of the International Institute of Refrigeration*, International Institute of Refrigeration, Paris, 1959–1995.
7. ASHRAE, *Refrigeration Systems and Applications*, Atlanta, Ga., 1986.
8. *Recommendations for the Processing and Handling of Frozen Foods*, 3rd. ed., International Institute of Refrigeration, Paris, 1986.
9. C. P. Mallet, ed., *Frozen Food Technology*, Blackie Academic and Professional, Glasgow, Scotland, 1993.
10. M. C. Erickson and Y.-C. Hung, eds., *Quality in Frozen Food*, Chapman and Hall, New York, 1997.
11. D. S. Reid, "Basic Physical Phenomena in the Freezing and Thawing of Plant and Animal Tissues," in C. P. Mallet, ed., *Frozen Food Technology*, Blackie Academic and Professional, Glasgow, Scotland, 1993.
12. H. D. Goff, "Measurement and Interpretation of the Glass Transition in Frozen Foods," in M. C. Erikson and Y.-C. Hung, eds., *Quality in Frozen Foods*, Chapman and Hall, New York, 1997.

13. B. Fu and T. P. Labuza, "Shelf-Life Testing: Procedures and Prediction Methods," in M. C. Erikson and Y.-C. Hung, eds., *Quality in Frozen Foods*, Chapman and Hall, New York, 1997.
14. J. Joseph, P. A. Perigreen, and M. R. Nair, "Effect of Raw Material Quality on the Shelf-Life of Frozen Squid (*Loligo duvaucelli*) Mantles," in *Storage Lives of Chilled and Frozen Fish and Fish Products*, International Institute of Refrigeration, Paris, 1984–1985.
15. G. A. MacDonald and T. C. Lanier, "Cryoprotectants for Improving Frozen Food Quality," in M. C. Erikson and Y.-C. Hung, eds., *Quality in Frozen Foods*, Chapman and Hall, New York, 1997.
16. D. C. Williams et al. "Blanching of Vegetables for Freezing—Which Indicator Enzyme to Choose?," *Food Technol.* **40**, 130–140 (1986).
17. V. M. Balasubramanian and M. S. Chinnan, "Role of Packaging in Quality Preservation of Frozen Foods," in M. C. Erikson and Y.-C. Hung, eds., *Quality in Frozen Foods*, Chapman and Hall, New York, 1997.

HUGH SYMONS
Kingston Upon Thames, England

REFRIGERATED FOODS: FOOD FREEZING AND WORLD FOOD SUPPLY

Freezing is widely regarded as the most elegant method of long-term preservation of a wide variety of highly perishable foods at economical cost. The list of products that cannot be frozen (salad vegetables, bananas, other whole fruits, etc) is much shorter than the list of products that are successfully frozen. Freezing enables perishable products (fish, shellfish, meats, vegetables, berries, and fruit juices) to be transported across the world and to span the seasons. It makes a multitude of high-quality perishable products available year-round anywhere in the world.

Freezing retains most nutrients, although a few thermolabile vitamins (Vitamin C, thiamine etc) are degraded by perhaps 10 to 30% over a year's storage at −18°C in most products (1).

MEAT

Most meats (beef, lamb, poultry, pork, and game) freeze excellently; that is, after freezing, frozen storage, and thawing, they are often indistinguishable from the original material. Quality retention during long-term storage is also very high, especially for meats containing highly saturated fat. Mammoth meat unearthed from the permafrost in Siberia after 20,000 years was found to be edible (2,3).

The frozen meat trade is vital to the major meat producing countries in the Southern Hemisphere (Argentina, Brazil, Australia, and New Zealand), as it enables them to export enormous tonnages of this vital export to the more densely populated countries in the Northern Hemisphere (North America and Europe). Before the advent of refrigerated transport, these countries could export only wool and tallow. Now butter, cheese, seafood, and vegetables have been added to the export of meat. At the dawn of the era of refrigerated transport the carrying temperature was −10°C, about the best that a Linde air compressor could achieve when relying on tropical seawater for condenser cooling. It is also the warmest temperature at which white and black molds fail to develop or only develop very slowly. This traditional "meat trade" carrying temperature has given way, largely due to containerization, to the customary frozen food temperature of −18°C. The impracticability of providing different temperatures in a container ship and the introduction of fish, frozen orange juice concentrate, vegetables, prepared meals, and so on, all of which demand at least −18°C, has led to standardization on this colder temperature.

SEAFOOD

Seafood remains the only substantial item of human diet a large part of which is still hunted, apart from small quantities of game. Aquaculture now provides substantial quantities of farmed salmon, tilapia, catfish, and trout to supplement the declining productivity of most fishing grounds. Wild seafood often occurs in greatest abundance at short seasons of the year and in areas remote from human populations desirous of consuming them. Several species (herring, cod, mackerel, and hake) are migratory, often present in concentrated shoals and in good eating condition for only a few months of the year. A preservation method is required to preserve them from the time of catching until ultimate consumption perhaps several thousands of miles away and several months later. Traditional methods such as salting, drying, and pickling (in either acid or alkali) have been largely supplanted by freezing.

The most heavily traded seafood commodity in world trade is the international fish block. Made from skinned and deboned fish, generally cod, pollack, or hake, which may be in fillets or sliced fish, often with tripolyphosphate added, this is the raw material from which portions, fish sticks, and fish fingers can be cut. These are generally battered and often breaded before sale to food service or retail outlets. Fish sticks, a relatively simple frozen fish product, developed in the United States in the early 1950s, found enthusiastic acceptance in many markets. Renamed fish fingers when launched in the rapidly growing frozen food market in the UK in 1955, this product became the second most popular item of diet in that country, second only to the traditional dish of fish and chips. Much whole, gutted fish is frozen at sea in upright plate freezers on board the trawler.

Aquaculture of salmon, catfish, trout, shrimp, mussels, and oysters relies almost exclusively on freezing to transport the catch after processing. Freezing has made seafood, often of hitherto exotic species, available to inland consumers to whom the product may have been unfamiliar. Seafood also makes a significant contribution to the nutritional status of the world. Species low in fat (cod, haddock, sole, pollack, hake, catfish, etc) constitute a rich source of animal protein coupled with a low caloric content. Species high in fat (salmon, herring, trout etc.) are very low in saturated fats, while the omega-3 fatty acids present are protective against cardiovascular disease. Surimi, a Japanese technology, depends on freezing for its production and distribution. This technology transforms otherwise disregarded species (eg, pollack in the Bering Sea) into washed, minced fish, which is then flavored and colored before binders are added; the material then simulates exotic species such as crab, lobster, King Crab, shrimp, and so on.

FRUIT AND FRUIT JUICES

Strawberries were the first popular frozen fruit in the United States. They were mixed with sugar in wooden barrels and placed in a cold store where the barrels were turned regularly to ensure proper mixing of the fruit/juice/sugar while being slowly frozen. Chilled whole fruit has eroded the market for this melange that is used in jams, jellies, ice cream, and so on. In the berry-producing countries of Eastern Europe the introduction of individually quick-frozen (IQF) berries, packed hygienically in tote bins each containing perhaps a thousand kilograms has superseded the sulfited pulps in wooden barrels used in the production of conserves and so on. Most whole fruit suffer a substantial loss of texture and mouth-feel during freezing and storage and are used in further manufacture: into pies, flans, jams, jellies, tortes, and so on.

The Florida citrus groves had to wait for the development of vacuum-concentrated juice technology before the enormously popular frozen citrus concentrates could be marketed. Frozen concentrated juice (FCOJ) now accounts for over 95% of the citrus crop in Florida, Israel, and Brazil. FCOJ is interesting from another point of view. In general the objective in freezing is to maintain the original quality of the raw material. Freezing does not seek to improve the raw material. In the case of FCOJ it is possible to blend, in the frozen state, the high acid–low sugar juice produced early in the season with low acid–high sugar juice produced later in the season to produce a consistent quality product. These juice concentrates are increasingly presented to the consumer as diluted to single strength, often as multiple blended chilled juices in single-serving containers.

FROZEN VEGETABLES

Development of a long-term preservation technique, such as freezing, coupled with economical temperature-controlled transport, has enabled the growing of each species of vegetable to be located in the area most conducive to obtaining high yields of high-quality crops often far removed from the urban communities that will consume the products. This is in contrast to the previous habit of clustering canneries around each conurbation.

The introduction of blast-freezing tunnels, and later the spiral tunnels (which occupy less factory floor space), enabled the products to be frozen in the IQF state. These bulk frozen vegetables can then be shipped across the world before being packed into consumer packages, blended, mixed with sauces, or forming one component of a multicomponent dish. In this way half a dozen different vegetables, grown in as many different continents, can be enjoyed year-round in any part of the world. Blanched and frozen within a few hours of harvesting, frozen vegetables often enjoy a superior nutritional profile—even after frozen storage for a year—than their fresh counterparts, which often spend several days before reaching the consumer and then may be stored in a refrigerator before being consumed. Not only is a year-round supply of nutritious vegetables made available but consumers can be introduced to exotic vegetables unfamiliar in their part of the world.

POTATOES

The pioneers of frozen foods in the 1930s and early 1940s, in the United States, knew that potatoes could not be successfully frozen. Maynard Joslyn is credited with developing the technology that enabled frozen French fries to be introduced into the United States market in 1947 (4). U.S. production alone now totals 8.4 billion pounds annually (Fig. 1). Much of this is consumed in fast-food outlets where four out of five customers order French fries, a greater proportion than those who order any other single product, even hamburgers! Instantly acceptable in every market to which they have been introduced, the quality standard for most of the world is the product processed from Russet Burbank potatoes in the northwestern states of Idaho, Washington, and Oregon in the United States.

BIBLIOGRAPHY

1. Harris and Karmes, *Nutritional Evaluation of Food Processing*, 2nd. ed., AVI Publishing Co. Inc., Westport, Conn., 1975.
2. B. Kurten, *How to Deep-Freeze a Mammoth*, Columbia University Press, New York, 1986.

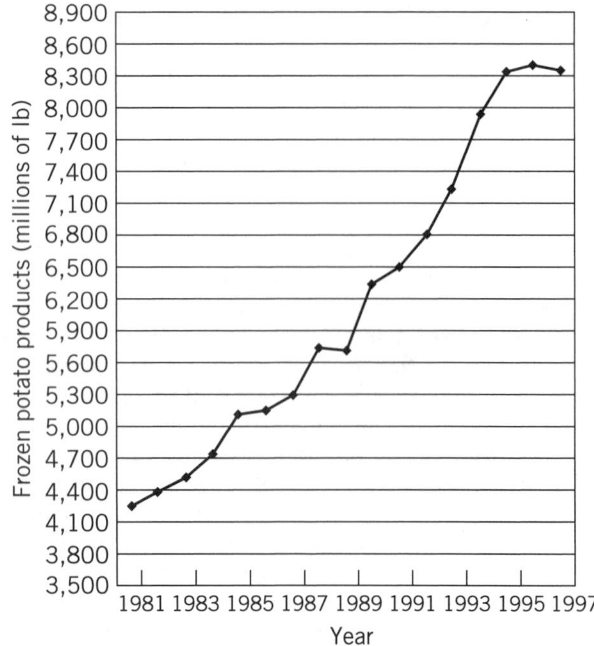

Figure 1. U.S. frozen potato production. *Source:* American Frozen Food Institute.

3. P. Mazur, "Frozen Neanderthals," *Nature* **342**, 23–52 (1989).
4. Talburt and Smith, *Potato Processing*, AVI Publishing Co., Inc. Westport, Conn., 1967.

HUGH SYMONS
Kingston Upon Thames, England

REFRIGERATED FOODS: HANDLING AND INVENTORY

POSTFREEZING OPERATIONS

The International Code of Practice for the Processing and Handling of Frozen Foods (1) elaborated by Codex Alimentarius forms the basis for any recommendations regarding these products. Product may be frozen before packaging, often in the individually quick-frozen (IQF) state, or go directly into the retail or food service package. IQF product may be stored in tote bins, containing around one ton of product, lined with high-density polyethylene liners of sufficient length to allow for sealing. As many postfreezing operations as possible should be carried out in cold store; any packaging operations carried out in chilled areas should be severely limited so that the frozen product spends as few seconds or minutes outside the cold store as possible. Two reports (2,3) produced under the EU FAIR program cover the cold chain and maximizing the quality of frozen foods.

CASING AND PALLETIZING

Casing is almost always carried out in a chilled area or ambient; mechanical palletizing is common, following which the pallet may be shrink-wrapped with polythene for stability during transport. Energy can be saved and product quality retained by introducing cases through a small aperture fitted with a guillotine door (which rises only high enough to admit the case) and palletizing in cold store. The use of forklift trucks, which necessitates two openings of a door sufficiently large to admit a truck for every pallet, incurs significant energy penalties. Vehicles are often sealed onto port doors opening into cold stores; making the interior of the vehicle contiguous with the cold store environment. Cold stores are often highly automated, using high-rise racking accessible with reach trucks fitted with closed cabins for driver comfort and radio communication informing the driver of his order-picking schedule (4).

INVENTORY CONTROL

Recent technological advances (bar coding, computerized control—Electronic Point of Sale, EPOS) have resulted in substantial changes in the management of the flow of product down the cold chain leading to Just-in-Time (JIT) inventory control. Instead of manufacturers producing stocks likely to be able to meet probable future sales to supermarkets or food service outlets and then awaiting sales orders, the sales outlets can now automatically call off the precise quantities needed, which in turn causes the manufacturers to schedule the required production. Reading the bar code of an individual package at a checkout automatically results in instructions passed back to the producing unit to replenish the stock.

OPEN SHELF-LIFE DATING

The EC Frozen Food Directive demands a minimum durability date on all frozen products. An evaluation of open-dating of foods (5) concluded:

> There is little evidence to support or to negate the contention that there is a direct relationship between open shelf-life dating and the actual freshness of food products when they are sold.

The International Institute of Refrigeration (IIR) (6) found a similar lack of evidence in support of open shelf-life dating when it stated:

> The antagonists of mandatory open shelf-life dating of frozen foods quote several disadvantages. The most important requirement for the retention of high quality in frozen foods is the maintenance of an adequately cold temperature. At sufficiently cold temperatures [at least −18°C (0°F) or colder] frozen foods possess a high quality life far in excess of the period such products spend in the cold chain between processing and consumption. Anything which detracts from good temperature management, such as unnecessary regard for product age, would be counterproductive to the objective of presenting the highest practicable quality to the ultimate consumer.

A food stability survey (7) gives a comprehensive account of the custom and practice in various countries and made a number of recommendations including adoption of the AFDOUS Code (8). Open shelf-life dating of frozen foods, which often have a long practical shelf life (PSL), leads to the erroneous concept that the quality is in some way guaranteed up to the date given irrespective of storage temperature. An intending purchaser when faced with open dating can react in one of three ways. The date can be ignored, in which case the costs associated with open dating will have been in vain. Alternatively the purchaser may rummage to find the "youngest" product, which turns First In First Out (FIFO) into Last In First Out, (LIFO) and puts pressure on the retailer to display product with only one date. This will lead to more out-of-life stock to be disposed of. Finally the purchaser can select the oldest product knowing that this will assist the retailer in managing his inventory on a FIFO basis. Regrettably this last purchaser does not exist!

TIME-TEMPERATURE INTEGRATORS (T-TI)

Since time elapsed since freezing is an imperfect indicator of product quality, a device that will indicate the integrated effects of time and temperature (a T-TI) is obviously to be preferred and forms part of intelligent or *active* packaging

(9). The deteriorative changes (quality loss) in frozen foods are complex, but the reaction kinetics can be simplified into either pseudo-zero-order or pseudo-first-order kinetics (10). A large number of T-TIs have been patented; these have been reviewed by Taoukis (11). Only three seem to be currently available; the 3M Monitor Mark (which relies on the travel of a colored ink along a wick), the Life Lines T-TI (which relies on a progressive color change in a polymer) and the VITSAB T-TI (which relies on a color change from green to yellow mediated by an enzyme). Life Lines is used on frozen turkey rolls in the United States (10) and also, since 1991, on some 100 chilled products by the supermarket chain Monoprix in France (12). Several studies have shown a close correlation between the quality loss in various frozen and chilled foods and the response of T-TIs (13,14), and an industry specification for evaluating T-TIs for use in food has been elaborated (15). A British standard (16) sets out performance specification and reference testing for T-TIs. The biggest obstacle to the wider use of these inexpensive devices, which present quality assurance to the consumer, is probably the mandatory requirement for open dating in most countries and the resulting confusion that would arise between this almost totally meaningless and potentially misleading date and the far more useful reading of the T-TI.

IN-HOME STORAGE

Most domestic refrigerators possess a frozen food storage compartment while many households also have a home freezer, capable not only of storing frozen foods but also of freezing food. In Europe, star-marked refrigerators are common, a classification system that indicates the temperature in the compartment. One star indicates −12°C; two stars, −15°C; and three stars (the most common), −18°C under specified test conditions. These stars are repeated on the packaging of frozen foods, coupled with an indication of the maximum time the product should be stored according to the number of stars on the storage compartment. For the vast majority of products, the storage time is one week in a one-star, one month in a two-star, and three months in a three-star storage compartment. Home freezers carry three stars while some carry an additional star to denote that they are capable of freezing a substantial quantity of food without unduly elevating the temperature of the already frozen food. The commonest refrigerator-freezer in the United States permits adjustment of the thermostat and louvers that apportion the air from the evaporator between the two compartments so that the body of the refrigerator is maintained at 5°C and the freezer compartment at −18°C under specified test conditions.

THAWING

Most frozen foods requiring cooking are best cooked from the frozen state. Bulky products, such as poultry, whole fish, and so on, that require thawing are best thawed in the body of the refrigerator or in a microwave oven. These ovens often have a defrost program that allows periods of low or no energy to allow thermal equilibration to occur since ice and water absorb microwave energy very differently. It is important to allow extra cooking time if end cooking from the frozen state.

BIBLIOGRAPHY

1. Codex Alimentarius, *International Code of Practice for the Processing and Handling of Quick-Frozen Foods*, U.N. Food and Agriculture Organization, Rome, 1964.
2. R. Fuller, ed., *A Practical Guide to the Cold Chain from Factory to Consumer*, European Union FAIR Programme Concerted Action (CT96-1180), Leeds, United Kingdom, 1997.
3. C. J. Kennedy and G. P. Archer, eds., *Maximising Quality and Stability of Frozen Foods*, European Union FAIR Programme Concerted Action (CT96-1180), Leeds, United Kingdom, 1997.
4. A. F. Harvey, "Mechanical Handling," in C. V. J. Dellino, ed., *Cold and Chilled Storage Technology*, 2nd ed., Van Nostrand Reinhold, New York, 1997, p. 249.
5. Office of Technology Assessment, *Open Shelf-Life Dating of Food*, U.S. Government Printing Office, Washington, D.C., 1989.
6. *Recommendations for the Processing and Handling of Frozen Foods*, 3rd. ed., International Institute of Refrigeration, Paris, 1985.
7. *Food Stability Survey*, Vol. 2, U.S. Government Printing Office, Washington, D.C.
8. *Frozen Food Code*, Association of Food and Drug Officials of the United States, 1961.
9. J. D. Selman, "Time–Temperature Indicators," in M. L. Rooney, ed., *Active Food Packaging*, Blackie Academic and Professional, London, 1995, p. 215.
10. B. Fu and T. P. Labuza, "Shelf-Life Testing: Procedures and Prediction Methods," in M. C. Erikson and Y.-C. Hung, eds., *Quality in Frozen Food*, Chapman and Hall, New York, 1997, p. 377.
11. P. S. Taoukis, B. Fu, and T. P. Labuza, "Time–Temperature Indicators," *Food Technol.* **45**, 70–72 (1991).
12. P. Toursel, "La Puce, un Saut dans la Chaine du Froid," Process No. 1130, Paris, 1997.
13. R. P. Singh and J. H. Wells, "Use of Time–Temperature Indicators to Monitor Quality of Frozen Hamburger," *Food Technol.* **39**, 42–50 (1985).
14. R. P. Singh and J. P. Wells, "Monitoring Quality Changes in Frozen Strawberries with Time–Temperature Indicators," *Int. J. Refrig.* **10**, 296–300 (1987).
15. R. M. George and R. Shaw, "A Food Industry Specification for Defining the Technical Standard and Procedures for the Evaluation of Temperature and Time–Temperature Indicators," *Technical Manual No. 35*, Campden Food and Drink Research Association, United Kingdom, 1992.
16. British Standard 7908, "Packaging—Temperature and Time–Temperature Indicators—Performance Specification and Reference Testing," British Standards Institute, London, 1999.

HUGH SYMONS
Kingston Upon Thames, England

REFRIGERATED FOODS: TRANSPORTATION

The transportation of refrigerated raw materials, ingredients, and food products is an essential link between the food industry and consumers. Each food product possesses characteristics that decide its practical storage life (PSL). Temperature control is a vital factor in determining the PSL of all refrigerated products. Good transport meets these requirements in the chilled or frozen food chains. Some shorter-life products spend more than half their PSL in transport. For others the transit time is a small interval from production to consumption. Apart from temperature, other important factors in the maintenance of quality during transport are packaging, humidity, and protection from damaging product integrity.

Commercial pressures and the advance of computer technology (especially Electronic Point of Sale [EPOS] and Just-in-Time [JIT] inventory control) are helping to reduce the time many items are held in storage or transport. Transportation by sea, land, and air is becoming intermodal, with users demanding the mix that provides them with the optimum quality and cost in an acceptable time. There is keen competition among transportation providers that is quickening the development of improved technology and customer service.

SPECIAL REQUIREMENTS BY TYPE OF FOODSTUFF

Sensitive Chilled Foodstuffs

This growing sector includes a variety of meat, poultry, dairy, fish, and prepared food products. Some are carried within about 1°C of their freezing points (many foods start to freeze at about −1.5°C). The longest sensitive chill chain carries meat from New Zealand to Europe by way of Cape Horn. This involves at least 35 days. These cargoes do not respire, so they can be in tightly packed loads. Precooling to the carrying temperature is essential. These products need to be held within a narrow temperature band to minimize the loss of PSL (1).

Fresh Fruits and Vegetables

Products of this type are living and need transporting in equipment that can remove their heat of respiration, water vapor plus gases such as carbon dioxide and ethylene. The aim is to slow the rate of ripening by moving them at the coolest workable temperature without causing chill injury. This varies between products but can occur at temperatures as warm as 11°C. Products may need protecting from cold ambient temperatures in northern latitudes during the winter.

Some products benefit from being carried under controlled or modified atmosphere when the percentage of oxygen in the vehicle, or container, is reduced (2). Carbon dioxide levels can be changed by supplying measured amounts of the gas from a cylinder. Humidity control will become more important in the future, reducing cargo weight loss and helping to maintain quality longer. An evolving technology is the use of modified atmosphere packaging around the product. This material will prevent oxygen from passing through it but will allow carbon dioxide to flow out. As the product respires, the level of oxygen falls and the process slows down.

Other Chilled and Cooled Products

Modern transport systems allow product carriage under controlled conditions as warm as +30°C. The range continues to expand with increasing volumes of sugar confectionery, chocolate, biscuits, cheese, and shelf-stable products that need protection from extremes of temperature and humidity. Dehumidifiers are being fitted to refrigeration machinery that can help to dry the air surrounding products or packaging. This is useful when a low relative humidity will improve a product's shelf life (when it is not in fully sealed packaging) (3).

Frozen Foods and Ice Cream

The colder the temperatures, the longer these foods will maintain their practical storage lives. The old rule of keeping them colder than −18°C is being replaced by colder temperature requirements. The set point for some modern integral containers can be as cold as −27°C. A few specialized containers can achieve −60°C and are mainly used for carrying tuna to Japan (M. S. Walker, unpublished data). The aim is to reduce temperature fluctuations that cause losses in flavor, texture, and moisture from the surfaces of the produce. Moisture is usually deposited as ice crystals on the packaging or the evaporator coils of the refrigeration machinery (4). Recent research into the "glassy state" of frozen foods may result in demands for colder temperatures for specific foods in transit (5).

Good-quality packaging is essential to protect products from dehydration when exposed to moving refrigerated air within transport. Packaging must also withstand the coldest temperatures, retaining the product's properties while subject to vibration and impact shocks (6).

Compatibility of Mixed Loads

Precautions are needed to ensure that mixed loads can be carried together. The following define compatibility: transit temperature, relative humidity, emission of active gases such as ethylene, odor-absorbing qualities, and modified atmosphere requirements (6).

MODES OF REFRIGERATED TRANSPORT

Sea

Temperature-controlled cargo carriage is by intermodal containers and conventionally refrigerated ships. Some conventional refrigerated ships also carry integral containers (see Fig. 1). The world's container capacity is about the same as conventional ships and continues to expand. Some trades also use porthole-insulated containers in ships with central refrigeration systems.

Containers manufactured to international standards (ISO) provide a secure door-to-door delivery system with landside movements by road, rail, or barges (7). This in-

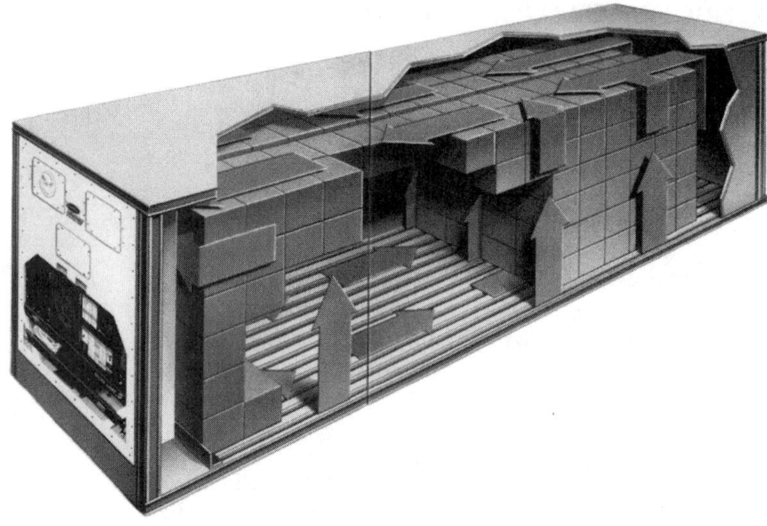

Figure 1. Cutaway view of a loaded integral container. *Source:* Carrier Transicold, reproduced with permission.

termodal chain depends on a supply of electricity from electrical mains or generators. Containers operate with air supplied from the refrigeration system along the T-section floor (see Fig. 2). It circulates around the cargo and in the space between the cargo and the doors. The air then returns along the space between the roof and the top of the cargo to the machinery section at the front (8).

In the chilled mode (warmer than −10°C), a sensing probe in the supply air controls the air temperature to the container. In the frozen mode (colder than −10°C), a sensing probe in the return air of the refrigeration unit controls the air temperature.

Porthole containers have air delivered through couplings to their bottom portholes. It passes around the container and returns through the top portholes. This older system continues to operate well, especially when high volumes of cargo are being carried. Clip-on units (see Fig. 3), towers, or total loss refrigerants such as liquid nitrogen or carbon dioxide provide refrigeration while ashore. This system is restricted to certain trade routes and ports.

Road Trailers

Most temperature-controlled trailers have mechanical refrigeration units powered by their own diesel engine. Major design advances have occurred in recent years due to noise restrictions, moves to more environmentally friendly refrigerants, microprocessor controls, and the need for smaller turning circles. They have resulted in better air temperature control, less defrosting, and lower power demands. Many countries have introduced less restricting maximum weights as road systems have improved, and vehicles have more axles.

Many trailers use refrigerated air blown over the top of the cargo (top-air delivery) and returned along the floor (see Fig. 4). In temperature climates, flat floors predominate, with return air passing through the pallet bases to the refrigeration unit (see Fig. 5). T-section floors are available for use in warmer climates. Air chutes, mounted under the roof, are sometimes used to provide better air distribution along the length of the unit (9). Some haulers use liquid nitrogen or carbon dioxide to provide refrigeration. These systems are almost noiseless in operation but need cargo close to the carriage temperature for efficient use.

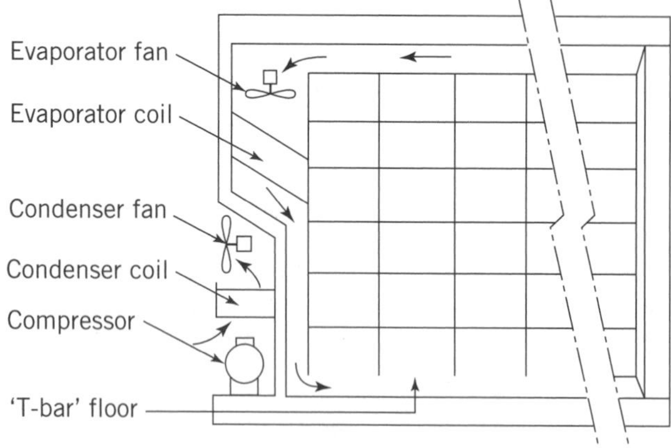

Figure 2. Schematic layout of a refrigerated container.

Figure 3. View of a porthole container with a clip-on unit.

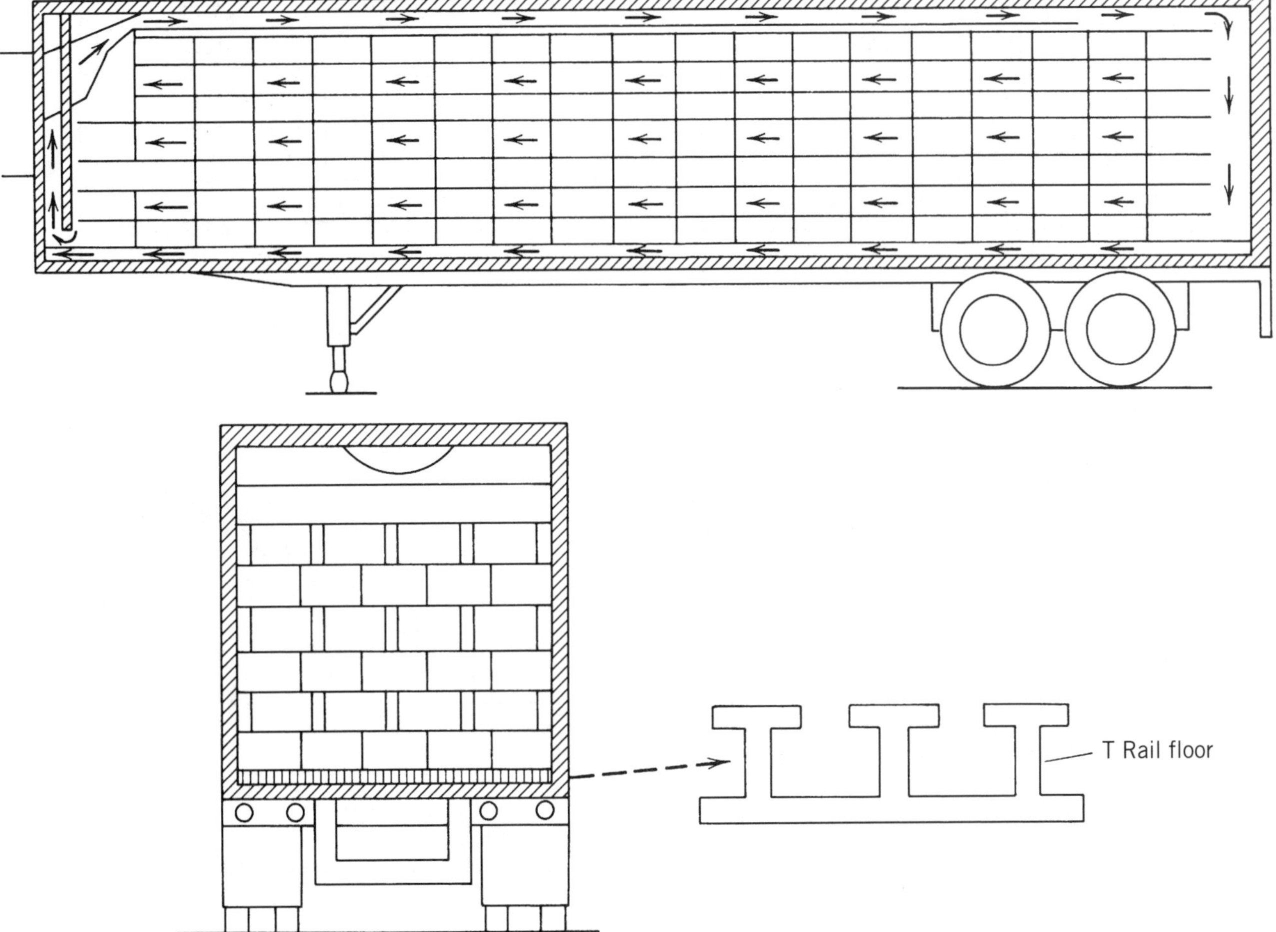

Figure 4. Pathways of air in a trailer equipped with a traditional top-air system and T-section floor.

Multicompartment vehicles are available for carrying products that need different temperatures. Movable bulkheads have also been designed to allow compartments to have variable volumes. If multiple deliveries occur, then it is important to ensure that the load on the tractor does not exceed legal limits. This can happen if the center of pressure of the trailer moves forward when products are taken away from the rear section.

Road Delivery Vehicles

There are many designs of rigid body insulated vehicles with refrigeration systems sized for the operational requirements. Multidelivery vehicles require adequate door protection such as plastic curtains to minimize heat entry when the driver is picking and removing products.

Some companies use a eutectic plate system built into the walls and ceiling. The eutectic is refrozen using an electrical main's electricity to power a small compressor when the vehicle returns to the depot. Ice can build up inside by freezing air that enters when the doors are open. It is removed when the vehicle body is allowed to warm up for routine cleaning.

Insulated pallet-sized containers with eutectic plates are used by some companies requiring small volumes of chilled, or frozen, products delivered to shops next to ambient commodities.

Air

Carriage by air provides a JIT service that can deliver perishable products to customers to meet market demands. It can also provide a quicker way of balancing stocks during times of peak demand. Air freight is very useful in supplying niche markets with products whose shelf life is too short for carriage by surface transport.

Cargoes need to be, tightly packed, and consignments insulated, if a suitable container is not available. Refrigerants such as dry ice may be added, providing they do not freeze the cargo and the gases emitted do not damage sensitive products or provide a health hazard. Ice is also used to keep some cargoes cool. Drip loss during melting needs to be contained to prevent damage to packaging and the aircraft. Airlines normally need to give permission for the use of these refrigerants.

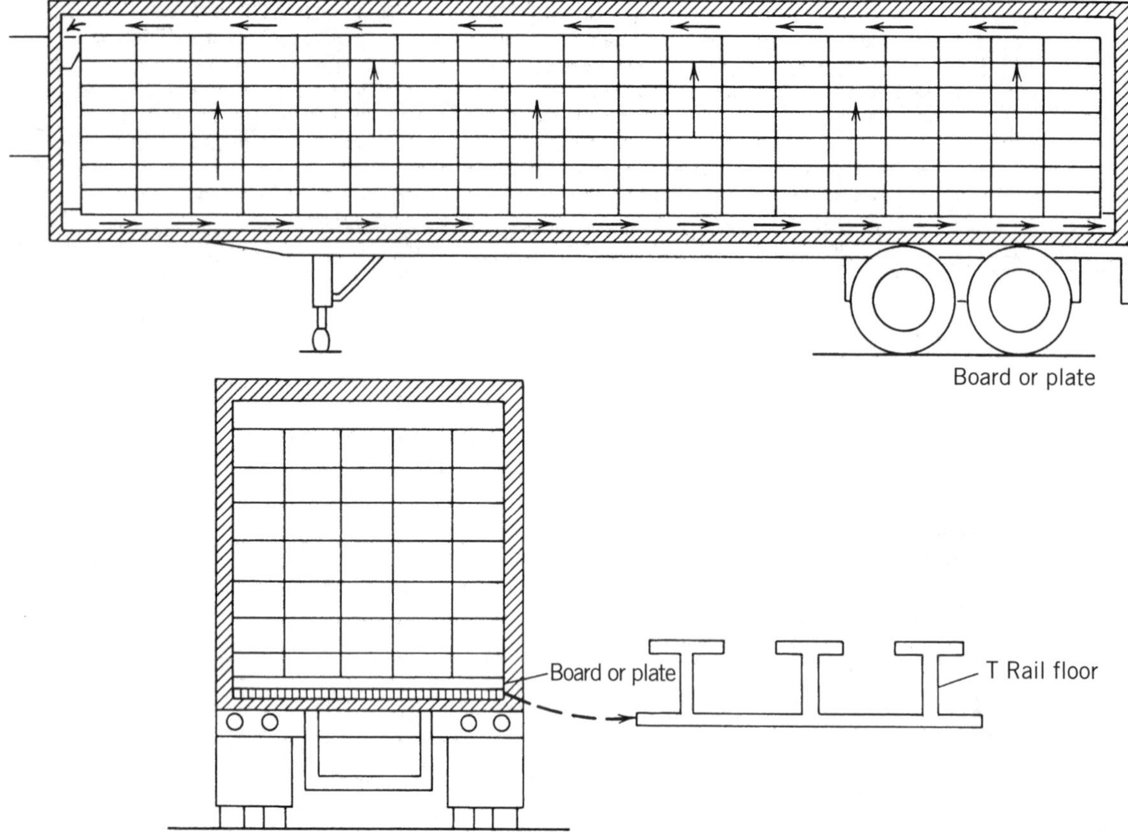

Figure 5. Pathways of air in a trailer equipped with a bottom-forced-air-supply system and a T-section floor.

Administration procedures must be efficient to avoid delays in cargo clearance. Completion of landside operations needs to be in time to prevent damage to the products.

Rail

The volumes of refrigerated cargo moved by rail varies greatly between various parts of the world. New Zealand and Australia have established routes for containers as part of an intermodal system. In the United States there are several routes where containers are stacked two high, giving improved utilization and cost economy.

Deregulation of rail freight services in many countries should lead in the future to more container movements by rail (10). The popularity of the hicube (9 ft 6 in) high container offers a challenge to the older rail systems with height limits. Special low loader flatbeds are being manufactured to contain them.

The former Soviet Union makes much use of refrigerated rail wagons. Trailers are carried on flat wagons in some areas. The Channel Tunnel between England and France requires fast trains that can carry complete road tractors and trailers. This quick drive on and off system has proved to be popular and efficient.

NEW TECHNOLOGY

Digital electronics is giving improved control of refrigeration systems and air temperatures. Data loggers are included in many systems, providing an accurate record of times on and off power plus supply and return air temperatures. Additional temperature-probe readings can also be supplied if required. Records of alarms caused by malfunctions in equipment can also be logged.

Integral containers fitted with modems can transmit data through the power cable to a host computer indirectly linked to the electrical supply (11). The technology exists for the transmission of data through a satellite to a central control room. Remote monitoring allows the checking of location and temperatures from many containers or vehicles continually. Alarms and events can also be transmitted. Routine adjustments of settings are also possible without reference to drivers or other staff. None of these systems obviate the need to check refrigeration systems periodically to ensure all mechanical parts are working.

ENVIRONMENT

Pressures to reduce environmental damage are leading to systems that consume less energy, have refrigerants and foam blowing agents that are more ozone friendly, and equipment with on/off options when reaching required carriage temperatures (9). These developments will continue.

QUALITY ASSURANCE

Demands for higher-quality foodstuffs, with a need for year-round supplies, are causing many retailers to source products worldwide. Food safety and quality requirements (12,13) plus a premium on "freshness" will continue to provide challenges to all sectors of the food transportation sector. Competition is leading to mergers and takeovers of companies with major transport companies operating globally.

BIBLIOGRAPHY

1. M. S. Walker, *Chilled Meat Technical Cold Chain Challenges*, Intermodal, Rotterdam, The Netherlands, 1998.
2. *Controlled Atmosphere in Marine Transport Refrigeration*, Cambridge Refrigeration Technology Seminar Cambridge, U.K., October 1, 1998.
3. Internal *Temperature Controlled Cargoes* P&O Nedlloyd Internal Procedures, London, 1997.
4. A. E. Hawkins, C. A. Pearson, and D. Raynor, "Advantages of Low Emissivity Materials to Products in Commercially Refrigerated Display Cabinets," *Proc. Inst. Refrig.* **69**, 54–67 (1973).
5. H. D. Goff, "Measurement and Interpretation of the Glass Transition in Frozen Foods," in M. C. Erickson and Y.-C. Hung, eds., *Quality in Frozen Foods*, Chapman and Hall, New York, 1997, pp. 29–46.
6. R. Heap, *Food Transportation*, Blackie Academic and Professional, London, 1998, pp. 75–96.
7. *International Standard for Thermal Containers*, ISO 1496 Part 2.2, October 1996.
8. J. P. Sheehan, "Meeting Diverse Reefer Cargo Demands with Marine Container Refrigeration Equipment," *Proc. Inst. Refrig.* **94**, 41–53 (1997–1998).
9. M. J. Heyward, *Food Transportation*, Blackie Academic and Professional, London, 1998, pp. 51–74.
10. T. C. Harris, *Inland Challenges Facing Deep Sea Carriers in Europe*, Intermodal, Rotterdam, The Netherlands, 1998.
11. *International Standard for Reefer Container Monitoring Using Power Cable Transmission*, ISO10368, 1992.
12. European Union General Food Hygiene Directive 94/43/EEC, *Off. J. European Community* **36**, 1–11 (1993).
13. Institute of Food Science and Technology, (U.K.) *Food and Drink Good Manufacturing Practice Guide*, London, 1998.

MICHAEL SANDERSON WALKER
P&O Nedlloyd
London, United Kingdom

RETORT POUCH

A retort pouch is a flat polymeric film laminate package designed to hermetically contain thermoprocessed food. Though some products are contained in pillow pack or bottom-gusseted designs, the predominant version has flat seals on all four sides. The laminate materials range from two-ply to four-ply depending on the desired shelf life, packaging and processing equipment, and product. The design as shown in Figure 1 resembles frozen boil-in-bag entrées but requires no refrigeration. In this context, the retort pouch is a shelf-stable item, the same as food products in hermetic metal cans or glass jars. Usually, a paperboard outer carton is used as an additional precautionary protective measure and to provide surfaces for graphics and print. Retort pouch is the predominant terminology for this concept, although synonyms are flex-pack, flex can, soft can, and pantry pack. Products packaged in the pouches are often referred to as ready meals; the term retort pouch usually encompasses products packaged in this manner.

The term retort correctly signifies that sterilization occurs after three traditional commercial canning steps: filling, headspace air removal, and sealing. This differentiates the retort pouch and its supporting technology from aseptic techniques and the term pouch from semirigid aluminum or thermoformable shallow polymeric tubs or trays for shelf-stable foods. The retort pouch postfilling thermoprocessing approach simplifies, to some extent, the sterilization operation but also imposes a heavier technological and performance burden on the basic package.

In the 1950s, the incentives for development were exploratory and the potential for a shelf-stable competitor for the then visible and strongly marketed frozen boil-in-bag concept. As feasibility approached reality, incentives beyond the convenience aspect of boil-in-bag included the potential for high-quality products, efficient retail shelf space utilization, and less energy usage for materials, processing, storage, and retail display.

The U.S. military, because of its unique combat ration requirements, saw potential advantages of the retort pouch (1): the flat shape fitted into field clothing and other gear conveniently; the pouch flexibility precluded injury if the soldier was falling, crawling, or crouching; opening the pouch required no separate tool or device; and cube and weight savings were possible. The thin cross-section, a nominal 0.75 in, would be suitable for high-quality products, especially those with conductive heating characteristics. Because the geometric center reached sterilization temperatures in a very short time, there would be less overprocessing in the pouch's peripheral areas. The ready-to-eat feature was appealing. The advent of the pouch

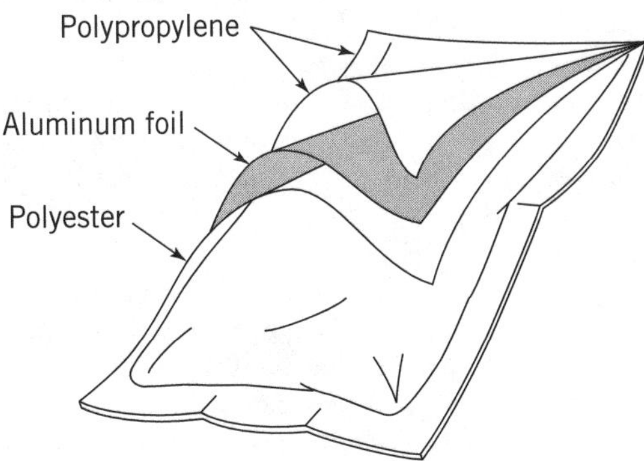

Figure 1. Retort pouch.

turned out to be fortuitous in that to meet the military's need for a long shelf life, procurement of suitable steel-plate cans was becoming problematical.

HISTORY

The history of the retort pouch covers initial curiosity-based exploratory experiments with polymeric films, recognition of the potential advantages as incentives to further delve into the concept, and performance of experiments to establish a database for the eventual design and production requirements of the pouch. The earliest formally recorded activity for applying flexible packaging to thermoprocessed foods is that of Hu and coworkers (2,3), which expanded on earlier graduate studies by Hu. Their results revealed that polymeric pouches of high- and low-acid foods could survive water processes using superimposed air to prevent bursting and also pointed out a need for stronger, heat-sealable materials with low oxygen- and water-permeability rates. Wallenberg and Jarnhall (4), showing European interest, confirmed the concept's feasibility. Limited production was reported in Italy as early as 1960 (5). In Japan, the growth of the retort pouch benefited from lack of competition from other preservation methods and from post–World War II development of pasteurized (up to 100°C) high-protein entrées (6).

During this period, up to approximately 1960, many fundamental questions were answered: materials were screened and their performance specifications established; durability of the basic package was confirmed, product formulations were optimized, processing techniques were further refined, and heat-sealing methods were studied and improved. Key contributions were made by film converters who developed and supplied the heat-resistant three-ply laminates. Equipment manufacturers proffered rotary and in-line equipment, ranging from use of preformed pouches to form, fill, and seal operations starting with roll stock film. Thermal processing centered on water cooks with superimposed air pressure.

Near the end of this early period of development, field tests performed by the military revealed that defects at the point of use occurred at a low frequency and were related to manufacturing deficiencies—mostly contaminated or leaking seals, but also body cuts from poor in-plant handling. In other words, if made and processed well, the pouch, as part of a ration meal, performed well. Similar manufacturing problems were reported in northern and southern Europe (7,8) and Japan (9).

The next sphere of activity, from 1961 through 1975, centered on transition of the retort pouch from a pilot plant operation and test production runs to commercial-level production systems. The approaches to success in this time frame universally reflected an acknowledgment that the production-oriented technology—as borrowed from the canning and frozen food industries—existed, and that standard equipment could be used, but that equipment and unit operation performance had to be at a high level. References 6 and 9–11 are reports on ensuing representative manufacturing system developments. If, for example, the pouch forming and filling operations started with roll stock and were performed on a standard intermittent motion machine, at least 14 additional functions need to be performed. A summary of production systems up to 1977 is presented in reference 12.

TECHNOLOGY BASE

Materials

During the earlier development phases, the film structure consisted of three plies, built to take advantage of and support the oxygen and water vapor barrier properties of aluminum foil. At that time—the mid-1950s to late 1980s—shelf life of over a year, and for the military, 3 to 5 years, was the given requirement. The three-ply materials consisted of:

1. The *outer ply*, offering some barrier and strength properties but mainly relied on for abrasion resistance, is usually polyethylene terephthalate (the polyester Mylar) in thicknesses of 9 to 12 μ. An adhesive system has been used to combine the outer ply with the aluminum foil.
2. The *center* material is dead soft aluminum foil in thicknesses of 9 to 12 μm. This ply is the barrier to oxygen and water vapor passage. Although some pinholes (foil breaks) are inevitable because of the foil's prelamination fragility, calculations of the percentage open area and storage tests both indicate that these breaks have no measurable adverse effect on performance. Neither do these breaks per se permit bacterial penetration; fractures must exist through all plies for bacteria to penetrate into the product.
3. The *inner ply* is the major performance contributor. It must be suitable for extrusion bonding to foil, where the bond must withstand thermoprocess temperatures as high as 270°F, transportation abuse, and exposure to a wide variety of troublesome food ingredients (lipids, salts, flavor oils) for long periods. It must be heat sealable on basically standard equipment; free of toxic or objectionable components, odor, or flavor; machinable; and capable of retaining its functional integrity at storage temperatures ranging from −25°F to over 110°F. The first inner food-contacting material used in any significant testing was polyvinyl chloride, adhered by an adhesive system to the foil layer.

Problems with delamination and off-odors with polyvinyl chloride led to two alternatives that prevailed until the late 1970s: a modified medium- to high-density polyolefin called C-79 by its producer (Continental Can Company) and a blend of ethylene–propylene copolymers. These two materials were also combined with the foil by an adhesive (a polyester–isocyanate system) (10). These materials, or minor modifications thereof, were used successfully worldwide.

In late 1974 and into 1975, there was a hiatus in retort pouch development while the U.S. Food and Drug Administration reassessed extractives attributable to the polyester–isocyanate adhesive system. Because of concern over the adhesive approach and the precautions necessary to produce an acceptable material, an extrusion lamination approach was perfected; the current universal inner film is a cast polypropylene that is then co-extruded with the foil.

The three-ply foil laminate was satisfactory where preformed pouches were used or where pouches were made vertically on rotary or in-line form, fill, and seal machinery. Because the successful initial performance characteristics of the pouch were based on products made via such production procedures (both sides of the pouch equally manipulated and/or stressed), the only other film structures used were those without foil and, occasionally, also with a polyamide in lieu of the outer polyester. These other films were for products where a shorter shelf life (8–18 months) was acceptable.

In the late 1980s and into the 1990s, as experience and confidence accumulated with production procedures, distribution, and user performance of the retort pouch, some materials changes were made to accommodate horizontal form, fill and seal equipment, and the greater range of materials requirements engendered by consumer needs. Relative to use of horizontally oriented equipment, the alternate use of drawn semirigid polymer trays and the desirability of microwavability were also pertinent factors. Currently, for shelf life of a minimum of 3 years, such as required for the U.S. military's meals ready to eat combat ration, the cast polypropylene/foil/polyester structure is used for any pouches still vertically formed and filled and as the top or cover layer when horizontal machinery is used. The bottom or drawn laminate in the horizontal mode consists of cast polypropylene as the sealant ply, then foil, and oriented polypropylene as the outer ply. In Japan and impending in the United States is a four-layer material—a polyamide ply will be inserted between the foil and the polyester for preformed pouches or between the cast polypropylene and the foil for horizontal form, fill, and seal operations.

Where a shorter shelf life (8–18 months) is adequate, that is, for practically all applications except the military, the central aluminum foil ply is replaced by a high-barrier polymer, ethylene vinyl alcohol (EVOH). The resulting polypropylene/EVOH/polyester film—on horizontal systems—presents an adequate 18-month shelf life and a microwave reheat feature. EVOH meets U.S. regulatory requirements for direct food contact (13).

PACKAGE DESIGNS

The flat four-seal design has dominated retort pouch usage from the beginning. In addition to favorable marketing characteristics, such as flat surfaces for graphics and efficient shelf space usage, this design, whether implemented on vertical or horizontal equipment, facilitates visual inspection of the seals, minimizes seal junctions that are potential leak sites, provides the thin cross section needed for shorter thermoprocesses, and results in fast heating in hot water for serving. Tear notches provide an easy opening feature.

Except for some institutional packs where a sturdy bulk shipping case is used, an outer carton is used to provide additional handling protection. Experience has shown that gluing the pouch to the carton is not necessary.

Sacharow (14), reviewing Japanese experience, reported that a stand-up pouch featuring a bottom gusset is used for stews, soups, boiled vegetables, and oriental-style foods. In the United States, a juice drink product uses this bottom gusset design. The paperboard carton is absent in both of these applications.

PRODUCTS

The number and variety of products successfully packaged in retort pouches is very large. Lampi (12) listed more than 80 brand-name products marketed worldwide up to the year 1976. Since then, variations and forms of entrees, but probably not total numbers, have increased. Items include ready meals (eg, stews, meatballs in tomato sauce), straight sauces, meats with minimal fluids, vegetable packs, fruits, soups, and bakery items. The U.S. military's meals ready to eat combat ration currently has 24 menus—24 retort-pouched entrées supplemented with retort-pouched fruits, rice, and noodles. Shelf-stable bread and meat-containing sandwiches, though not basically thermoprocessed, exist in pouches of the same three-ply material.

Retort pouches continue to hold a share of the processed food market in Japan and are used for selected items in Europe. In the United States, following several test-market ventures, the market has narrowed to the military; items in support of weight control and health programs; microwavable meals for children; campers and hikers' items; specialty, high-quality-image items; and entrées and adjuncts for direct-networking marketing systems.

One of the cited advantages of the retort pouch has been a quality improvement over cylindrical cans because of reduced heat exposure to achieve sterilization. In actuality, conflicting experiences make this postulation difficult to confirm. The U.S. military, before accepting the retort pouch as a replacement for cans in its operational rations, gathered taste test data that showed a higher acceptance for the retort pouch (15). In addition to formal, controlled acceptance assessments, a perhaps truer indication of military acceptance, though anecdotal, was the posttest, ad libitum selection of leftover rations by the test participants; 4 out of 5 selected the pouched over the canned items. Undoubtedly, commercial entities, before entering into expensive test-marketing programs, satisfied themselves that quality, in addition to convenience and other user attributes, was high. And where there has been minimal competition from frozen foods, such as in Japan, the pouch has done well (16).

Beverly (17), summarizing opinions on various aspects of retort pouch processing, reported a range of conclusions. One large U.S. food processor reported that consumers found no quality difference between pouches and retail cans for the same product. This report also cited the internal evaluations by a second large processor as revealing a pouched food quality halfway between frozen and canned and a third stating they can achieve frozen food quality. Commercial processes of 10 oz of pouched product receive 70 to 83% of the total heat exposure of same-weight cans, which raised the question as to whether this decrease was significant enough to result in detectable product quality improvement.

Retention of thiamine in thermoprocessed foods has been used as a quantitative measure of quality. With pouches of sweet potato puree processed at 250°F, 77% of the thiamine was retained as compared to 60.4% for equal-volume cans (18). A computer model calculation for a conductive heating product indicated 84.6% thiamine retention at 250°F for a 12-oz pouch and 64% for a 12-oz can (19). Data comparing thickness of the product and thiamine retention confirmed these findings (20).

In summary, product quality of retort pouches can be better than that of cans, possibly equal to that expected from some frozen products. Commercial realization has been hindered by production inabilities to get pouch-to-pouch uniformity and repeatability of product heat treatment and quality. Current indications are that this gap is being overcome.

PRODUCTION SYSTEMS

The design of the retort pouch that was initially subjected to acceptance, storage, handling, and distribution durability trials was a flat four-seal version where both sides of the pouch structure were subjected to equal stress during filling and processing. Accordingly, when manufacturing systems were investigated, these systems were geared to manipulating opposing pouch walls equally and filling through the fourth seal position that had been left open. Then, after air removal, a closure seal was made. This principle, tied to vertical pouch handling, was applied to all production systems, whether predominantly manual or automated, up to the late 1980s, and whether preformed pouches or roll stock was the starting point.

Early user tests had shown that retort pouch defects were related to manufacturing deficiencies, primarily the need to fill pouches without seal-area contamination, remove residual air, and form a closure seal of adequate strength and high integrity. A judgment was also made that standard equipment principles could be used. To prove this point, a comprehensive development program was completed by a consortium of companies headed by Swift and Company. Program results are summarized in reference 21. Reference 12 describes production systems that existed up to 1977.

Since then, horizontally oriented equipment as shown in Figure 2 (courtesy of Multivac, Inc.) has been perfected and has gained acceptance to the point where it now prevails. The intermittent motion horizontal form, fill, and seal machine starts with roll stock film, as defined previously in the section on materials, where a lower web is formed into drawn pockets in dies assisted by compressed air or vacuum (and heat and plugs, if necessary). Each index movement forwards a set of pouches, for example, six. Next the pouches are filled in a single step or in multiple steps, automatically or manually, and, along with the upper web, are passed through a chamber where after closing, a vacuum is drawn and the top lidding film is heat sealed to the bottom web, seals being located where the die outline shows solid support. Pouches then exit the chamber and are cut into separate units, inspected, and transferred to the thermoprocessing operation.

The horizontal configuration is currently preferred because it is faster—a bigger pouch pocket area allows faster filling of several pockets or pouches at one time, and because the same equipment can be converted from pouches to drawn, semirigid polymer trays.

Whether prepared on vertically or horizontally oriented equipment, the finished pouched item must meet rigid specifications. References 22–25 provide requirements and criteria.

Filling

The filling of retort pouches can be divided into three steps or functions (12): (*1*) the use of pumps, augers, conveyors, and the like to move the product to the pouch opening; (*2*) the positioning and feeding of the product into the pouch through a nozzle designed or fitted with a positive cutoff or suck-back antidrip feature; and (*3*) the simultaneous presentation of the opened pouch or drawn pouch cavity to the filling station. Filling for vertical systems is discussed in detail in reference 12. The same steps and precautions apply to horizontal drawn pouch procedures. Controlling fill weight is critical because of its effect on the thermoprocess (26).

Air Removal

Before closure sealing, as much of the noncondensible gases as possible should be removed from the pouch for product stability, avoidance of pouch bursting during retorting (without resorting to high air counterpressures), assurance of product sterilization, easier detection of spoilage (swelling), and easier cartoning and casing. The air-removal step is carried out on equipment that also performs other steps such as filling or, more frequently, the closure seal operation. Specific residual gas levels are a function of product, occluded air, air-removal technique, and production rates. Recently, to accommodate solid, placeable items, the permissible residual air in an 8-oz pouch has been increased to 20 cc.

Air-removal techniques for vertical packaging systems used vacuum chambers, snorkel suction tubes, steam-flush air displacement, and counterpressure via opposing metal plates or waterhead pressure. On a horizontal drawn pouch machine, air is removed in an on-line hoodlike chamber where the four-sided sealing operation also takes place.

Sealing

Next to the basic pouch material itself, retort pouch seals have received the most attention relative to their assured performance through the processing steps and over the postprocess distribution period. There are two performance aspects to retort pouch sealing. The first is the creation of an inherently sound seal when a pouch is being vertically formed or formed over a lower web of horizontal pockets; that is, when seal surfaces are clean; the second,

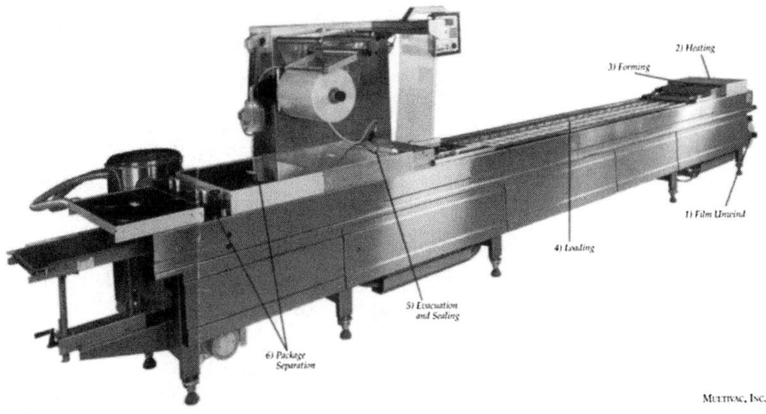

Figure 2. Horizontally oriented intermittent motion pouch form, fill, seal machine. *Source:* Courtesy of Multivac, Inc.

integral with the filling steps, is the creation of a sound seal free of contamination, severe distortion, or significant wrinkles, again true for vertical and horizontal approaches. These aspects of seals are covered by Lampi et al. (27). Young (28) presents a thorough discussion of the principles and options for sealing thermoplastics.

Both hot bar and impulse sealing techniques have been successfully used with retort pouches prepared on vertical systems. Improvements to bar design and control of sealing parameters have resulted in adequate pouch-forming seals, but most of the attention has been focused on the closure seal. This is where sealing occurs as an integral part of the air-removal operation and frequently has had to overcome minor product contamination or moisture (from steam flushing for air removal) in the seal area. Also, wrinkles have resulted from distortion during filling. Various tensioning grippers and forming bars have been used to maintain a taut, wrinkle-free seal area. Bars with a slightly transverse radiused cross-section used against a flat anvil bar can seal through water and grease contamination (29). In Japan, a superimposed triple hot bar technique where a second and a third bar flatten out blisters caused by vaporization of product from the initial hot bar sealing action has been used (9). With correct design and close surveillance of the filling, air-removal, and sealing operation, satisfactory closure seals can routinely be made.

On horizontal equipment, sealing dies are formed to match the outline and surface edges of the lower pockets for a set of pouches, for example, six. Once hot seals are made, there is a cooling pause or a movement to a second station where unheated bars provide cooling and stabilization of the seal area. Following that, the pouches are separated by a cutting die and moved on to the retorting operation.

RETORTING

Along with establishing consistently satisfactory materials and obtaining defect-free closure seals, the subject of retorting, the commercial sterilization step, has received considerable attention. During the early exploratory phases, enough studies, including inoculated packs, were performed so that apprehension over sterility was relieved and reliable cooking procedures were established. Then, as experience was accumulated and as productivity as well as quality became important for successful commercialization, retorting received more attention. For example, process times for retort pouches were established at 25 to 40% of those required for equivalent canned items; however, the total heat exposure in commercial processes was actually 70 to 83% of cans (26).

For determining thermoprocessing conditions (ie, cook time and temperature), the general and mathematical techniques for calculating cook times for traditional cylindrical cans based on heat-penetration data have been found to be generally suitable for retort pouches. The lethality levels, known as F_0, adequate for commercial cans were also found to be suitable for pouched products. Comparisons of the various process determination techniques and listings of precautions to note during their application to pouches (such as difficulty in locating the cold spot or occasional occurrence of nonlinear heating curves) have been published (12,19,30).

For successful retort pouch thermoprocesses, the following factors need to be closely controlled (19,26,31,32):

1. Pouch fill weight and residual gases
2. Pouch thickness during processing
3. Control and reproducibility of the entire cook cycle: come-up, process, and cool
4. The retort rack or other device that controls thickness during processing, and that must not adversely affect heat medium flow or temperature uniformity in the retort
5. Overriding or total pressure

A variety of processing techniques and equipment designs have been used with retort pouches, and each has its proponents. Still cooks under water with superimposed air pressure was the initial reliable procedure and is still used in some instances for some products. Pouches are positioned usually horizontally onto perforated pocketed trays and accumulated on carts that are wheeled into horizontal retort chambers. Most systems are designed so that rotation of the cart is possible. To conserve energy and reduce come-up time, the processing water is preheated to a high temperature in a separate tank and then fed into the retort chamber after loading. Processes are now computer controlled (33). Steam-air retorts are now proven, accepted, and more prevalent. They are basically the same design except that an air tank replaces the hot water preheating tank.

Other retort designs include a continuous cooker-cooler (trade-named Hydrolock) using steam-air as the heating medium, and a Universal Convenience Food Sterilizer (34), where pouches in retort carts are exposed to horizontally flowing hot water. Water is pumped from a hot water storage tank into each rectangular tanklike cart (as opposed to the entire chamber) fitted with a water flow distribution plate configured to assure uniform lateral water flow distribution.

RELIABILITY AND PERFORMANCE

Two initial concerns with the advent of interest in using flexible packaging for shelf-stable foods were:

1. Whether the pouch could be formed, filled, sealed, and processed at an acceptably low reject rate (i.e., be free of manufacturing defects), and
2. Whether the end product could withstand the rigors of postprocess handling and distribution (i.e., be free of leakers swells, and/or contamination). Although some technology and criteria were usable from metal and glass "canning" and from frozen and dry food flexible packaging, the retort pouch's characteristics were singular enough that new specifications and testing protocols were necessary.

The specific tests and the criteria required to assure that the material and the production line were in control and that items had been correctly packaged and to present a confidence that items would survive postprocess rigors changed over the years from the late 1950s to the present. These changes resulted from accumulated experience-based confidence with end-item performance; improvements in and institutionalization of manufacturing systems; improved materials converting; and the transition from vertical filling and sealing of pouches to horizontally oriented pouch drawing, filling, air removal, and sealing.

The initial performance standards from the mid-1960s to the late 1980s were built around a three-ply foil laminate, semiautomated packaging procedures, and the ability to resist military distribution abuse (considered the most rigorous). Lampi et al. (27) discuss these in detail. In summary, postretort pouch seals had to exhibit complete fusion, meet an internal pressure level of 20 psig for a 30-s hold time, meet a tensile strength of 12 lb/in for a 1/2-in sample width, and show no visual aberrations. Photographs of acceptable and nonacceptable seals were published as guides for visual inspection. Residual gases were initially kept at 5 cm^3 or less. An inspection plan suitable for monitoring production is shown and discussed by Lampi (12).

The key elements of the early test pack performance trials and supporting laboratory experiments and test have been incorporated into several documents:

1. National Food Processors Association Bulletin 41-L. *Flexible Package Integrity Bulletin.* (22).
2. *Classification of Visible Exterior Flexible Package Defects*. Poster. Association of Official Analytical Chemists (23). To be used in conjunction with NFPA Bulletin 41-L.
3. *Standard Guide for Use and Handling of Flexible Retort Pouches in the Manufacturing Environment.* ASTM F-1278-90 (24).
4. *Packaging and Thermoprocessing of Foods in Flexible Pouches.* Military Specification MIL-P-44073 (Latest version is E) (25).

The acceptance, and now prevalence, of horizontal form, fill, and seal machinery for pouches and the continuing compliance of films to specifications has resulted in emphasis on a burst test and two visual in-plant inspections as adequate in-plant control. The burst test is a simple puncture of the pouch with a needle, controlled pressure increase to a set level, and hold for 30 s. Visual inspections are pre- and postthermoprocessing.

The overall result of the maturation and stabilization of the processing technology, diligent use of key test(s), and trained visual inspection is that reports of retort pouch failures at the point of use are rare, so rare that the U.S. military reports "no problems" with their use of the pouch, and their usage rate is 24 million pouches per year. The absence of any failures is difficult to accept, but it most likely reflects that two visual examinations are more inclusive in picking out defects than anticipated and that field failures are so few that they are simply discarded without comment or complaint. Signs of postprocess failure parallel those of cans: obvious leakage, malodor, and swelling.

There has been interest in on-line defect-detection devices since the earliest establishment of a production procedure, primarily to replace visual examinations and their people-based frailties. Defects were defined as seal occlusions and leaks through seal areas or on the body of the pouch. Spencer and Bodman (35) surveyed leak detection methods, identified 23, and determined that helium gas as a tracer for "drier" foods and electrical conductivity changes in an external fluid for "wetter" foods were the best options to find a 10-μm hole on the body of the pouch. Lampi et al. (27) reported that the cost, size, and complexity of adaptation to a production scenario were prohibitive. Since then, interest has periodically resurfaced. The use of horizontal form, fill, and seal equipment provides the option, depending on the specific principle, of testing several pouches at once, favorably impacting on test equipment design. Floros and Gnanasekharan (36–38) discuss principles, technology, and applications to aseptic and flexible food packaging. In a symposium jointly sponsored by the Institute of Packaging Professionals and the Food Processors Institute (39), several aspects of package integrity and some candidate nondestructive defect-detection methods were discussed. To date (April 1998 at this writing) on-line defect detectors have not materialized, although some pressure differential profile procedures are promising, both technically and economically.

TRENDS

Retort-pouched foods, once they were proven viable, shelf-stable items, resulted in considerable commercial interest and several test-market trials by major food processors. Since the initial rush, commercial interest has subsided, and current applications are centered on specialty gourmet-image foods: liquid diets as used in hospitals; health and calorie-control entrées marketed in conjunction with health programs; and convenience entrées through direct-marketing network systems. Nearly all of these commercial pouches are foil free, with ethylene vinyl alcohol resins providing the oxygen and water vapor barrier. Shelf life is 18 months, and products are microwavable and hot-water heatable. The nature of these applications requires a high organoleptic quality, a major factor in their successes.

The U.S. military remains steadfast as one of the major users of retort-pouched foods; in the combat ration, the meals ready to eat, the retort pouch continues to meet the requirements of rigorous storage (including serving as war materiel reserves), distribution (which includes occasional free-fall air drops), climatic and geographical environmental extremes, and ultimate consumer preferences. Some variations among entrées have been introduced so that the ration can be used to alleviate indigenous food shortages in strife-torn areas. The meals ready to eat now feature 24 menus, and the ration includes a flameless ration heater (water-activated controlled electrochemical reaction). Through 1991, just under 400 million meals had been procured. Since then, consumption has hovered at 2 million cases (24 million meals) per year, and all indications are that these levels will remain constant.

Now that the retort pouch has been technically feasible for nearly 40 years and several attempts have been made at introducing it into major market segments, its future is, in all likelihood, to remain with the applications listed earlier. Some reasons for this are:

1. The frozen food distribution system in the United States from the manufacturer to the home freezer has been well established and has provided products that the retort pouch has had difficulty replacing. Conversely, the pouch has done well in Japan, where competitive processes do not exist to the same degree.
2. Consistency in quality (not quality per se) for some lines of flavor-sensitive products has not been satisfactory; that is, because of deficiencies (mostly related to inconsistent time variables with unit operations) in filling, sealing, and processing, some package-to-package product variations have been greater than the manufacturers felt was acceptable. In addition, equipment was not available for high-volume production.
3. The boil-in-bag approach, even with frozen foods, has not appealed to customers to the degree manufacturers expected.
4. The greatest deterrent has been an outgrowth of the pouch and its technology; namely, the shallow, semirigid, thermoprocessable polymer trays with barrier properties suitable for an acceptable 12- to 18-month shelf life. These packages have a thin cross section so that high-quality products are achieved with minimum thermoprocessing. They also provide structural protection for fragile, placeable items; easy, direct consumption from the trays (seals are peelable); and, most significantly, the capability to be heated in a microwave oven for serving.

The retort pouch has provided an essential package for the military and a viable one for selected commercial foods. Its applications have generally stabilized. Increased usage will depend on innovative products and marketing rather than on further technical improvements. Its development and technical acceptance has broadened shelf-stable food packaging by demonstrating that another materials class—polymers, in pouch, tray, or tub form—provides quality and performance at least equal to more traditional materials. The existence of these packaging options has resulted in the National Canners Association becoming the National Food Processors Association.

BIBLIOGRAPHY

1. F. J. Rubinate, "Flexible Containers for Heat Processed Foods," in *Proceedings of the Conference on Flexible Packages for Military Food Items*, National Academy of Science/Quartermaster Food and Container Institute for the Armed Forces, Washington, D.C., 1960, pp. 21–28.
2. K. H. Hu et al., "Feasibility of Using Plastic Film Packages for Heat-Processed Foods," *Food Technol.* **19**, 236–240 (1955).
3. A. I. Nelson, K. H. Hu, and M. P. Steinberg, "Heat Processable Food Films," *Modern Packaging* **20**, 173–179 (1956).
4. E. Wallenberg and B. Jarnhall, "Heat Sterilization in Plastics," *Modern Packaging* **31**, 165–167 (1957).
5. F. Nughes, E. Mantovani, and R. Merloni, "Nuova Serie di Controlli di Imballaggi per Cibi Pronti," *Imballaggio* **24**, 14–15 (1973).
6. Y. Tsutsumi, "Retort Pouch: Its Development and Application to Foods in Japan," *Journal of Plastics* **6**, 24–30 (1972).
7. S. F. T. Nieboer, "Flexible Vacuum Packs for Processed Vegetables," *Food Manufacturing* **45**, 60–64 (1970).
8. F. Nughes, "Perspective Report on Retortable Pouch Packaging in Europe," Report No. T-7313, Packaging Institute Seminar, St. Louis, Mo., 1973.
9. Y. Tsutsumi, "The Growth of Food Packed in Retortable Pouches in Japan," *Activities Report, Research and Development Associates for Military Food and Packaging Systems, Inc.* **27**, 149–153 (1975).
10. P. L. Goldfarb, "Pouch for Low Acid Foods," parts 1 and 2, *Modern Packaging* **43**, 70–76 (1970); **44**, 70–76 (1971).
11. D. D. Duxbury et al., "Reliability of Flexible Packaging for Thermoprocessed Foods under Production Conditions: Phase I, Feasibility," Technical Report No. 72-77-GP. U.S. Army Natick Laboratories, Natick, Mass., 1970.
12. R. A. Lampi, "Flexible Packaging for Thermoprocessed Foods," *Advances in Food Research* **23**, 305–428 (1977).
13. R. H. Foster, "Polymers for Packaging, EVOH," in R. B. Holmgren and J. R. Russo, eds., *Packaging Encyclopedia*, Cahners, Newton, Mass., 1987, pp. 70–72, 75.
14. S. Sacharow, "Retortable Food Packaging: The Japanese Experience," *Prepared Foods* **154**, 29–30 (1985).

15. R. A. Lampi, "The Current Prognosis for Flexible Packaging of Military Rations," *Annu. Packaging Inst. Forum*, Chicago, Ill., October 7, 1974.
16. Y. Saito, "The Retort Pouch Is a Way of Life in Japan," *Activities Report, Research and Development Associates for Military Food and Packaging Systems, Inc.* **35**, 8–12 (1983).
17. R. G. Beverly, "Retort Pouch in the 80's," *Food Engineering* **52**, 100–103 (1980).
18. S. H. Rizvi, "Nutritional Retention of Retort Pouched Foods," *Activities Report, Research and Development Associates for Military Food and Packaging Systems, Inc.* **33**, 112–115 (1981).
19. R. Greene, "Designing Thermal Processes for Optimizing Product Quality in Retort Pouches," *Activities Report, Research and Development Associates for Military Food and Packaging Systems, Inc.* **35**, 19–24 (1983).
20. T. Ohlsson, "Optimal Sterilization Temperatures for Flat Containers," *J. Food Sci.* **45**, 848–852 (1980).
21. R. A. Lampi and F. J. Rubinate, "Thermoprocessed Foods in Flexible Packages: Transition to Production," *Package Development* **3**, 12–18 (1973).
22. National Food Processors Association, *Flexible Package Integrity Bulletin*, NFPA Bulletin 41-L, Washington, D.C., 1989.
23. Association of Official Analytical Chemists, *Classification of Visible Exterior Flexible Package Defects* (poster), Arlington, Va. 1989.
24. American Society of Testing and Materials, *Standard Guide for Use and Handling of Flexible Retort Food Pouches in the Manufacturing Environment*, ASTM F-1278-90, Philadelphia, Pa. 1990.
25. *Packaging and Thermoprocessing of Foods in Flexible Packages*, Military Specification MIL-P-44073E, U.S. Army Natick Research, Development and Engineering Center, Natick, Mass., 1995.
26. R. G. Beverly, J. Strasser, and B. Wright, "Critical Factors in Filling and Sterilizing in Institutional Pouches," *Food Technol.* **34**, 44–48 (1980).
27. R. A. Lampi et al., "Performance and Integrity of Retort Pouch Seals," *Food Technol.* **30**, 38–48 (1976).
28. W. T. Young, "Sealing," *Packaging* **23**, 200, 203–207 (1987).
29. G. L. Schulz and R. T. Mansur, "Sealing through Contaminated Pouch Surfaces," Technical Report No. 69-76-GP, U.S. Army Natick Development Center, Natick, Mass., 1969.
30. S. H. Spinak and R. C. Wiley, "Comparisons of the General and Ball Formula Methods for Retort Pouch Process Calculations," *J. Food Sci.* **47**, 880–884 (1982).
31. A. F. Badenhop, ed., *Proc. Conf. Using Retort Pouch Worldwide, Food Sciences Inst.*, Purdue University, Indianapolis, Ind., March 14–15, 1979.
32. A. A. Kopetz, C. A. Prange, and R. J. Flessner, "The Future Is in Our Hands: Critical Factors in Retort Pouch Thermal Process Assurance," *Activities Report, Research and Development Associates for Military Food and Packaging Systems, Inc.* **31**, 49–53 (1979).
33. R. Lingle, "Ameri Qual Foods: At Ease with MREs," *Prepared Foods* **158**, 144–146 (1989).
34. T. L. Heyliger, "Temperature Distribution Universal Sterilizer (Preliminary Studies)," Technical Memorandum No. 333, Food Research Department, Food Processing Machinery Division, FMC Corporation, Santa Clara, Calif., 1985.
35. W. T. Spencer and H. A. Bodman, "Nondestructive Testing of Packages of Thermoprocessed Foods: Final Report, Phase I (Draft)," Contract DAAG17-69-C-0013, U.S. Army Natick Laboratories, Natick, Mass., 1970.
36. J. D. Floros and V. Gnanasekharan, "Principles, Technology, and Application of Destructive and Nondestructive Package Integrity Testing," in R. K. Singh and P. E. Nelson, eds., *Advances in Aseptic Processing Technology*, Elsevier Science, London, 1991, pp. 157–188.
37. V. Gnanasekharan and J. D. Floros, "Package Integrity Evaluation: Part I, Criteria for Selecting a Method," *Packaging Technology and Engineering* **3**, 44–48 (1994).
38. V. Gnanasekharan and J. D. Floros, "Package Integrity Evaluation: Part II, Criteria for Selecting a Method," *Packaging Technology and Engineering* **3**, 67–72 (1994).
39. B. A. Blackistone and C. L. Harper, eds., *Plastic Packaging Integrity Testing*, Institute of Packaging Professionals and the Food Processors Institute, Herndon, Va., 1995.

RAUNO LAMPI
Westboro, Massachusetts

RHEOLOGY

Rheology is the study of deformation and flow of matter (1). In food applications, deformation is usually involved with assessing texture, while flow is generally associated with the property of viscosity (2). Generally texture is more loosely defined as "the response of the tactile senses to physical stimuli that result from contact between some part of the body and the food . . . a group of physical properties that derive from the structure of the food . . . properties expressed in terms of mass, distance and time only." In cheese grading, texture may refer to appearance while body may be the term used to denote rheological properties. Here, texture is used as a rheological term. Viscosity can be more rigorously defined as "the internal friction of a fluid or its tendency to resist flow." Often texture is used when describing the rheological properties of solid or semisolid foods such as cheese, fruit, bread, or cold ice cream, while viscosity is generally applied to foods that flow, such as pourable salad dressings, vegetable oils, syrups, or melted ice cream. Texture can also be used with a broader meaning, synonymously with rheology, including viscous properties of the food; this meaning of texture is especially common in describing semisolid foods such as ketchup, mayonnaise, or purées. Of the many sensory food attributes of concern to a food scientist, texture is the least well described (3). It has been observed that the consumer may also have difficulty describing texture but this difficulty does not lessen the importance of this attribute:

> When first asked about food texture, the consumer appears to exhibit very little spontaneous awareness. Flavor overshadows texture at the conscious level. People simply take the texture of a food for granted. . . . An average consumer may have difficulty in visualizing the concept of texture per se. . . . If the texture of a food is the way people have learned to expect it to be, and if it is psychologically and physiologically acceptable, then it will scarcely be noticed. If, however, the texture is not as it is expected to be . . . it becomes a focal point for criticism and rejection of the food. Care must be taken not to underestimate the importance of texture just because it is taken for granted when all is as it should be (4).

The science of rheology made significant advances since 1940 in the determination and explanation of the viscous and viscoelastic properties of polymers; this part of the field had its origins in the time of Newton and now has a solid theoretical and experimental basis (2,5,6). Analysis of the texture of foods is a more recent application of rheology. Modern food texture analysis could be said to have begun with the development of texture profile analysis (TPA, a scheme for measuring and classifying textural properties) (3,4,7–12). A texturometer (9), which evolved from a denture tenderometer built at MIT (13), has largely been replaced by tensile–compression testing machines made by Instron Corp. (14) and other similar instruments.

FLUID FOODS

The force required to make fluids flow depends on how fast they are made to flow. Conversely, the speed at which fluids flow depends on the force that is causing them to flow. The relationship between force and flow rate for a spreadable food such as mayonnaise can often be expressed as

$$\sigma = \sigma_y + m\dot{\gamma}^n \qquad (1)$$

where σ is the shear stress (force per unit area) observed at a sheer rate $\dot{\gamma}$ (velocity gradient s^{-1}); σ_y (force per unit area) is a so-called yield stress, the minimum stress that must be applied to a material to initiate flow: and m and n are parameters characteristic of the material and are called the consistency coefficient and the flow behavior index, respectively. It has been argued that the concept of yield stress is not rigorously correct, that given enough time any material will flow under a small stress. In a real situation, as in consuming a food, time is not unlimited and many foods do exhibit a yield stress as will be seen in some of the following examples. The variables in equation 1 differ from simple force and flow rate in that the effects of geometry (of the instrument, pipeline, etc) have been taken into account.

Newtonian Flow

It is convenient to classify the rheological behavior of fluid foods according to their flow curves or rheograms, which are plots of shear stress against shear rate; that is, they are graphical portrayals of equation 1. Some liquid foods exhibit linear dependence of shear stress on shear rate, as shown in Figure 1 for 50 and 60% sucrose solutions. In the case of the sucrose solutions, there is no yield value and the flow curves are straight lines passing through the origin. Such fluids are called Newtonian after Newton who found that for such fluids equation 1 reduces to

$$\sigma = m\dot{\gamma}^1 \text{ or } \eta = \sigma/\dot{\gamma} \qquad (2)$$

where m has been replaced by η, the Newtonian viscosity. As also shown in Figure 1, the viscosity of a Newtonian fluid is a constant and does not depend on shear rate. For most fluid foods, the ratio of $\sigma/\dot{\gamma}$ is not a constant, but changes (usually decreases) with increasing shear rate.

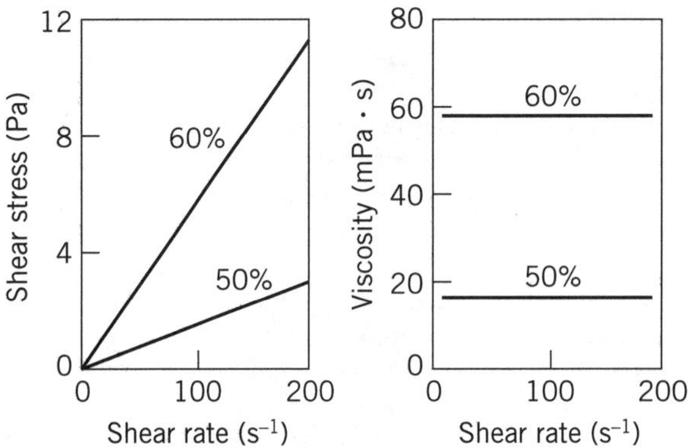

Figure 1. Rheograms for 50 and 60% sucrose solutions at 20°C showing ideal viscous flow.

Equation 2 may still be used to calculate an apparent viscosity of such materials at any shear rate.

Non-Newtonian Flow

Only a few fluid foods exhibit Newtonian flow behavior. Some syrups are approximately Newtonian over a limited shear rate range, but often their flow curves deviate from a straight line if the shear rate range is extended enough. Usually, syrups, oils, and most beverages do not have yield values even though they may not be Newtonian in their flow behavior. In the more general case, the flow behavior of a food is represented by equation 1 with a nonzero yield value. An example of such flow behavior is shown in Figure 2 for a tomato ketchup. This type of flow behavior is also shown by mayonnaise, viscous salad dressings, cheese sauces, and many other products. Products that exhibit decreasing apparent viscosity with increasing shear rate (ie, have a flow behavior index less than 1) are called pseudoplastic. If, as in Figure 2, the rheogram is curved and does not pass through the origin, the material is classified as pseudoplastic with a yield value. Non-Newtonian fluids that follow equation 1 (with or without yield values) are called power law fluids; however, many other equations have been developed to describe non-Newtonian flow. If the flow curve is a straight line intercepting the shear stress

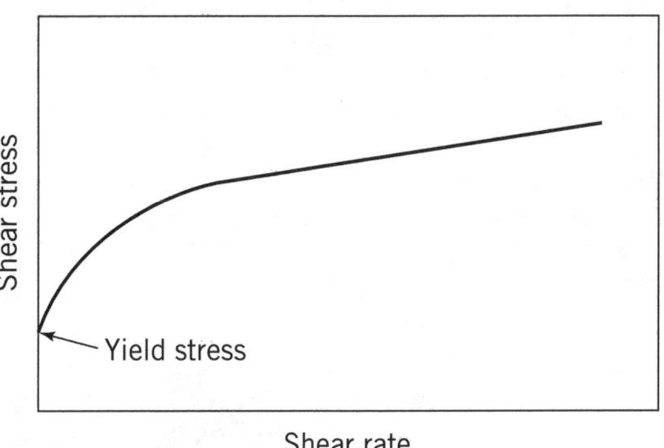

Figure 2. Rheogram for tomato ketchup plotted on arithmetic coordinates up to a shear rate of 500 s^{-1}.

axis at a nonzero value, then the material is said to be plastic, and if the rheogram is a linear function of shear rate, the material is said to show Bingham plastic behavior. Such behavior is unusual, although not unknown, in foods. Usually, the effect of shear is to break down the structure that may be present in a fluid at rest, so their apparent viscosity will decrease with increasing shear rate. This is especially evident in some gels that will flow like a normal fluid once the gel structure has been broken. Consumers commonly shake containers of products such as ketchup or Thousand Island salad dressing because the shearing makes such products more fluid.

A few products (some starch slurries, for example) show dilatant flow behavior, wherein the apparent viscosity increases with increasing shear rate. One explanation of non-Newtonian flow behavior is that long molecules or asymmetric particles may tend to align themselves so their long axes are parallel to the shear field and this orientation leads to a change in resistance to flow.

Some products are slow to recover from the effects of shearing, and under some circumstances some products may not recover at all. The time lag in recovering an initial structure after it has been broken by shear is the reason it is possible to pour ketchup following shaking the bottle. The shearing breaks the initial structure and it stays broken for a short time after the shearing has stopped. (If a bottle is completely filled, it may be difficult to apply much shear to the product and shaking may not initiate flow.) Such time dependence in recovering the initial structure is called thixotropy. If there is no recovery, the process is called rheodestruction. Dilatant fluids may also relax slowly to their unsheared (in this case, lower viscosity) initial state and that time dependence is called rheopexy. Shearing may be applied to a product at a fixed level for a variable amount of time, at levels that vary linearly as a function of time, or at any shear rate and time combination. While the process of applying shear to a product is somewhat arbitrary, for rigorous rheological analysis of the data, it is essential that the shearing history be exactly known; furthermore, certain controlled applications of shear to the product will give data that are easier to analyze, so in practice a specific, not arbitrary, shearing process is followed. Idealized rheograms for thixotropic and rheodestructive dispersions are shown in Figure 3 for processes in which a constant shear rate is applied for a fixed time and then stopped. Idealized rheograms for thixotropic and rheopectic fluids are shown in Figure 4 for shearing processes in which the shear rate is increased and then decreased linearly with time while recording the resultant shear stress. Instruments are commercially available for recording flow curves in this manner, using shear rate as the independent variable, or for recording shear rate as a function of shear stress. The choice of instrument type depends on the objective of the measurement.

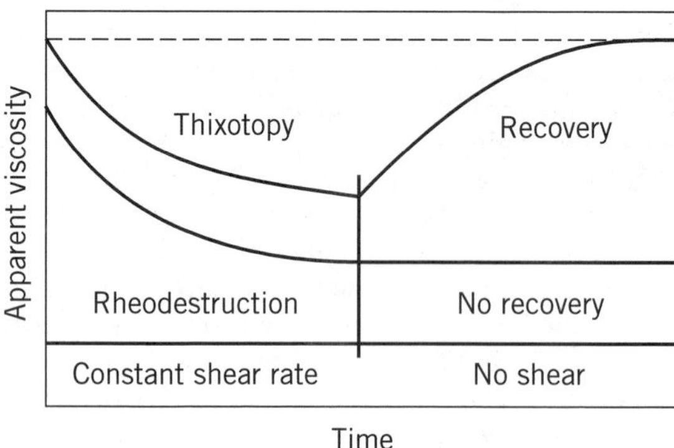

Figure 3. Idealized rheograms for thixotropic and completely rheodestructive dispersions.

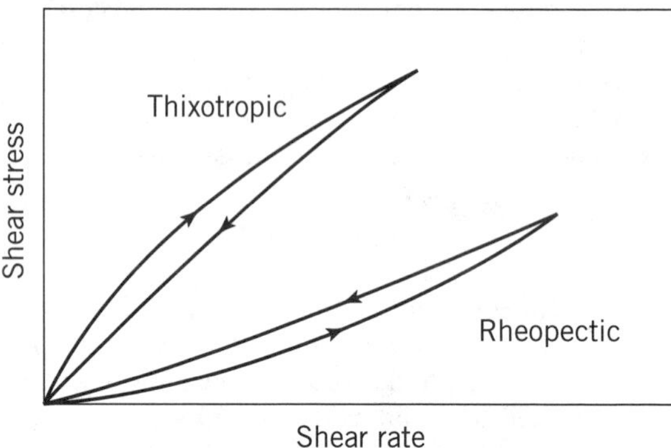

Figure 4. Idealized rheograms of time-dependent flow in continuous up curve and down curve experiments.

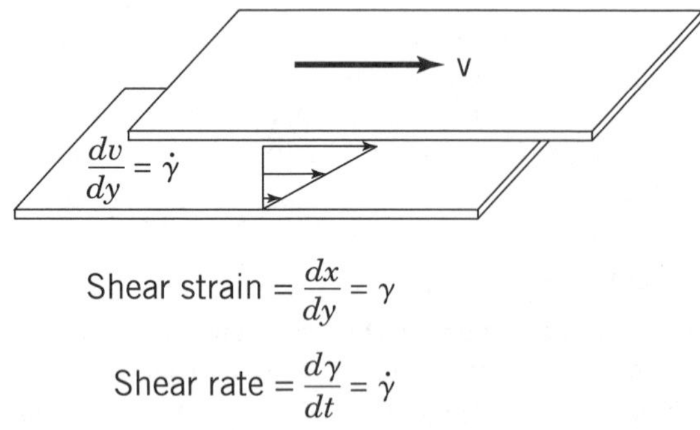

Figure 5. Velocity profile for an ideal viscous fluid in steady laminar shearing flow between two parallel plates.

Principles of Flow Measurement

The schematic diagram of a parallel plate viscometer shown in Figure 5 will be used to illustrate how rheological data are obtained. If a fluid is placed in the gap y between the parallel plates (each of area A) and a force F is applied to the (movable) upper plate, it will move with a velocity v relative to the (stationary) lower plate. The shear strain (γ) is a measure of the relative distortion of the sample (x/y, or in differential form, dx/dy); the shear rate, or more precisely the shear strain rate, is the change in strain as a function of time ($d\gamma/dt$ or dv/dy); and the shear stress is the force per unit area ($F/A = \sigma$); the apparent viscosity (defined earlier) is the ratio of shear stress to shear rate. The fluid next to the upper plate moves with it at a velocity v and the fluid next to the lower plate is stationary. Between the plates there exists a velocity gradient; for a Newtonian material the gradient is linear as shown in Figure 5 but the gradient will have a more complex shape for other types of flow behavior.

While parallel plate viscometers are available commercially, in practice this is sometimes not a convenient form of the instrument. Coaxial cylinder viscometers are more commonly used than are parallel plate viscometers. It is possible to imagine constructing a coaxial cylinder viscometer from a parallel plate viscometer by forming one of the plates into a cylinder and then shaping the second plate into a cylinder around the first one. Then one cylinder is rotated while the other is stationary and the fluid fills the gap between the cylinders. Parameters are obtained in basically the same way as discussed in the preceding paragraph for the parallel plate viscometer, except the factors that account for geometric effects are somewhat more complicated. A rotating bob immersion viscometer could be regarded as a simplified version of the coaxial cylinder viscometer in which the inner cylinder rotates and the outer cylinder has been replaced with the wall of the container. If the container is large enough, it can be assumed that the gap between the rotating bob and the container wall is infinitely large. In fact, with a 2-cm diameter bob, this assumption is fair even if the container is a medium size (say a 250-mL) beaker. If the assumption can be made, the data analysis is somewhat simplified. Such immersion viscometers are simple to use and very popular in industry. While it is more difficult to obtain rheologically rigorous parameters from immersion viscometers, their readings can be very useful in quality assurance and process control.

Temperature Effects on Flow

Temperature effects can be critical in rheological measurements. Some viscometers have a built-in means for controlling the sample temperature, such as a thermostatic jacket. Sometimes the measurements can be made in a room where the temperature is well controlled. At any rate, temperature control is necessary; this is illustrated in Figure 6, which shows the effects of temperature on the consistency coefficient of plain and salted liquid egg yolk. Note that the ordinate scale in Figure 6 is logarithmic, showing that the consistency coefficients for these samples decrease exponentially with increasing temperature, so a relatively small error in temperature control could lead to a much larger (proportionally) error in the rheological parameters. In general, the viscosities of Newtonian liquids decrease logarithmically as the temperature is raised, and this dependence can be expressed as an Arrhenius-type relationship:

$$\eta = Ae^{E/RT} \quad (3)$$

where E is the activation energy of flow (E is an energy barrier that must be overcome before the elementary flow processes can occur and is related to the coherence of the molecules in the liquid), A is a constant, R is the gas constant (8.314 J/K mole or 1.987 cal/K mole), and T is the absolute temperature (K) (6,15).

An Arrhenius-type plot is shown in Figure 7 for a 40% sucrose solution over a temperature range of 0–80°C. In this example, the activation energy of flow (E) is 23.8 kJ/mole. If data can be fitted in this manner, it is possible to use equation 3 to predict the viscosity of a Newtonian fluid at any temperature in the range. Moreover, because E represents molecular associations within a fluid, this may be a useful tool in the study of interactions in fluid systems, and their changes with the application of heat. (It should be remembered that equation 3 is an empirical equation and the activation energy is not derived from fundamental molecular theory, even though it is thought to represent molecular-level phenomena.)

For non-Newtonian fluids, temperature effects on flow behavior may be more complex. An approach to analyzing such effects was indicated by the egg yolk data (Fig. 6) where the temperature dependence of the consistency coefficient (m) was shown. Similarly, the temperature dependence of the flow index parameter (n) could be determined and, from equation 1, the apparent viscosity could be calculated at any temperature–shear rate combination; if the yield value were nonzero, it would also need to be determined as a function of temperature.

SOLID FOODS

Viscoelasticity

The rheology of solid foods is usually considered in the area of food texture (2,7,8,16,17). Most solid foods are viscoelas-

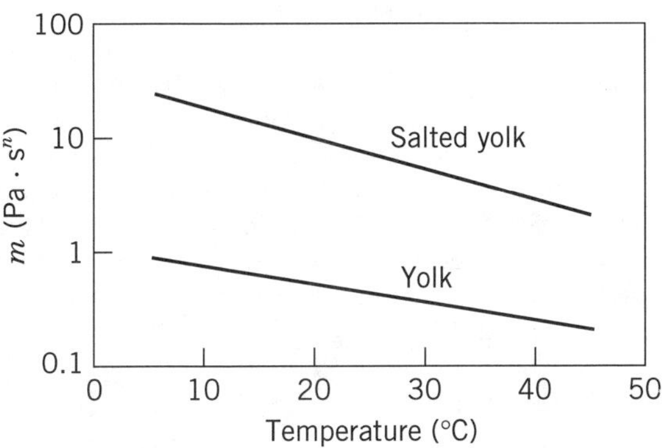

Figure 6. Effect of temperature on consistency coefficient for liquid egg yolk and salted yolk.

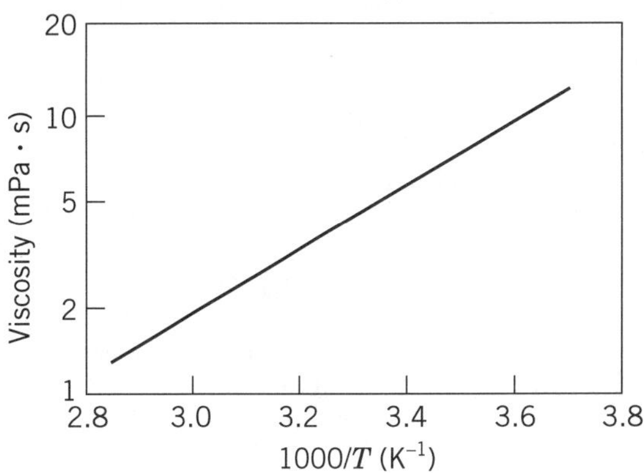

Figure 7. Arrhenius-type plot of viscosity for 40% w/w sucrose solutions between 0 and 80°C.

tic, that is, they exhibit viscous flow behavior (as discussed), but also behave somewhat as elastic solids. The viscoelastic aspect of rheological behavior is particularly relevant to material response in situations of unsteady motion, that is, when stresses or strains are changing with time. Examples of this are cutting, mashing, chewing, and swallowing of solid materials, and mixing, pouring, spreading, or pumping of fluid systems.

Elasticity can be defined as that property of a food by virtue of which, after deformation and on removal of stress, it tends to recover part or all of its original size, shape, or both (18). Clearly many products, such as cheese, are elastic if the stresses are small, but if the stress is high so that the product is crushed to only a fraction of its original height, for example, then there will be little recovery. The limit of stress, beyond which a food loses its elasticity, is called the elastic limit. On mastication the elastic limit of any food will be exceeded, but the elasticity of many foods (such as Mozzarella cheese, marshmallows, gelatin desserts) is an important sensory aspect of the food. Products, such as margarines, which tend to flow when the elastic limit is exceeded are viscoelastic in their rheological behavior. A food's viscoelasticity can be measured and expressed in terms of its storage and loss moduli. The storage modulus G' is a measure of elasticity, or energy stored, and the loss modulus G'' is a measure of viscosity, or energy lost, by a food when it is subjected to stress and strain (19). The ratio of the loss and storage moduli is called the loss tangent.

$$\tan\delta = G''/G' \tag{4}$$

Viscoelastic parameters depend on food composition and temperature, so, as with other rheological measurements, the sample and test conditions must be described in detail (8).

Principles of Measuring Viscoelastic Properties

There are different techniques for measuring viscoelastic properties of foods, but probably the most commonly used is dynamic testing wherein linear viscoelastic response to very small oscillatory shear is observed. Idealized dynamic responses of elastic, viscous, and viscoelastic systems to sinusoidal oscillatory shear are shown in Figure 8. In such tests, the strain γ is a sinusoidal function of time t:

$$\gamma = \gamma_0 \sin(\omega t) \tag{5}$$

where ω is the oscillatory frequency and γ_0 is the maximum strain amplitude. The strain rate, or shear rate, will be the first derivative of strain with respect to time:

$$d\gamma/dt = \dot{\gamma} = \omega\gamma_0 \cos(\omega t) \tag{6}$$

If the food behaves as an ideal elastic (Hookean) solid, stress is directly proportional to strain:

$$\sigma = k\gamma_0 \sin(\omega t) \tag{7}$$

but in an ideal viscous (Newtonian) fluid, the stress is directly proportional to shear rate:

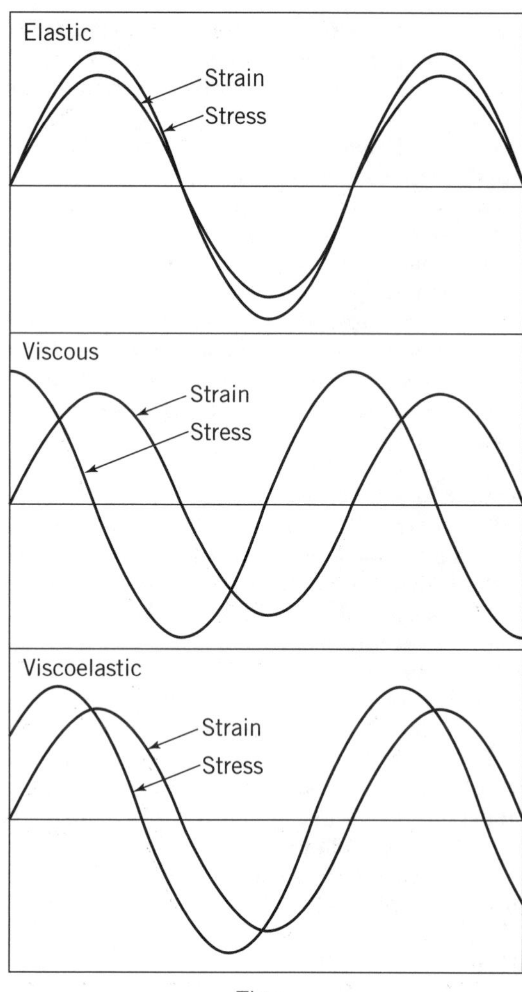

Figure 8. Idealized dynamic response of elastic, viscous, and viscoelastic systems to sinusoidal oscillatory shear.

$$\sigma = \eta\omega\gamma_0 \cos(\omega t) \tag{8}$$

Most foods exhibit both viscous and elastic properties at the same time and will have a viscoelastic stress response that combines the features of equations 7 and 8:

$$\sigma = \gamma_0[G' \sin(\omega t) + G'' \cos(\omega t)] \tag{9}$$

where the moduli G' and G'' are as defined previously. On softening a food (increasing the temperature of a cheese or increasing the water content of a dough, for example) the food will become more liquidlike and less solidlike and its loss tangent (G''/G', eq. 4) will increase. Such objective characterization, putting numbers on what is observed sensorily, can be very valuable in product development and in quality control.

Texture Profile Analysis

Dynamic testing, as described in the preceding paragraph, can give a rigorously correct rheological characterization of a food, and it may be possible to correlate viscoelastic parameters with fundamental, molecular-level properties. The analysis of dynamic testing data becomes highly complex if the elastic limit of the product is exceeded. As indicated earlier, mastication almost always exceeds the elastic limit of solid or semisolid foods, and so do many food processing operations. A rheological method was needed to imitate the mastication (chewing) process and still provide objective data; therefore, texture profile analysis (TPA) was developed in response to this need.

In a TPA measurement, a food sample of specific dimensions is compressed, the compressive force is removed and the sample is compressed again. This two-bite sequence is imitative of the chewing process. During the test, the amount of compression (distance) and compressive force are recorded. The resultant force vs distance plot (sometimes compressive stress, force per unit area, is used instead of force) is called a TPA curve. An Instron TPA curve for Cheddar cheese is shown in Figure 9. In that measurement, a cylinder of cheese 25 mm high and 21 mm in diameter was compressed 20 mm (80%) until it was only 5 mm high. Then the platen was lifted off the cheese and lowered again to the same point. Several TPA parameters may be derived from the TPA curve: the maximum force H, which occurs at the end of the first compression, is called the hardness; the force of the first maximum F is called the fracturability (not every food shows a fracturability peak); the work done to compress the sample on the "first bite" is given as the area A_1, and on the second bite, A_2, and the ratio A_2/A_1 is called the cohesiveness C; the distance S is called the springiness; the negative or tensile area A_D is the adhesion or stickiness; gumminess G is the product of hardness times cohesiveness; and chewiness is the product of hardness times cohesiveness times springiness. One advantage of the TPA method is that the parameters derived from it have been shown to correlate with sensory perception; it really is a way to replace or supplement subjective evaluation of foods with numbers and so has proven invaluable in product characterization and development.

FUTURE DEVELOPMENTS

TPA parameters depend on experimental conditions; compression rate, temperature, sample size, and nature of the platen surfaces. These conditions have not been standardized between laboratories and, until they are, it will remain difficult to compare TPA data obtained by different investigators. Such standardization is an urgent need.

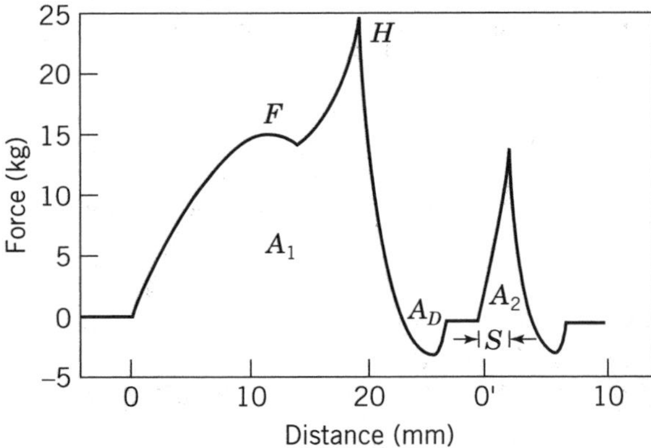

Figure 9. Instron TPA curve for cheddar cheese.

In recent years, significant progress has been made in correlating sensory and objective evaluation of food rheology; the texts cited in the bibliography all include information on sensory–instrument correlation. Much work is still needed in this area. The effects of temperature, composition, and other variables on food rheology can be extremely complicated, and consequently have not been extensively investigated. Work is needed in this area, especially with the objective of drawing general conclusions that can be applied to a variety of foods.

Rheological instruments for in-line measurement and control of food processes are generally not completely satisfactory. Instruments are needed in this area, particularly to handle non-Newtonian, viscoelastic systems, and the influence on results caused by temperature variations. Such instrumentation must be designed to permit measurement of complex rheological properties that will be consistent with the processing or sensory conditions of interest, and to comply with good manufacturing practice regulations.

Correlations between food rheology and microstructure and molecular interactions have rarely been entirely successful. Rheological behavior and microstructure are expressions of molecular and structural organization within a food that should be further explored by these methodologies to address both fundamental and applied challenges in food science. This is an area that should receive more attention.

BIBLIOGRAPHY

1. S. L. Passman, *J. Rheology* **28**, 663 (1984).
2. M. C. Bourne, *Food Texture and Viscosity: Concept and Measurement*, Academic Press, Inc., Orlando, Fla., 1982.
3. A. S. Szczesniak, *J. Food Sci.* **28**, 385 (1963).
4. A. S. Szczesniak and E. L. Kahn, *J. Texture Studies* **2**, 280 (1971).
5. J. D. Ferry, *Viscoelastic Properties of Polymers*, 3rd ed., John Wiley & Sons, Inc., New York, 1980.
6. J. R. Van Wazer, J. W. Lyons, K. Y. Kim, and R. E. Colwell, *Viscosity and Flow Measurement: A Laboratory Handbook of Rheology*, Wiley-Interscience, New York, 1963.
7. A. Kramer and A. S. Szczesniak, eds., *Texture Measurements of Foods*, Reidel Publishers, Dordrecht, The Netherlands, 1973.
8. P. Sherman, ed., *Food Texture and Rheology*, Academic Press, Orlando, Fla., 1979.
9. H. H. Friedman, J. E. Whitney, and A. S. Szczesniak, *J. Food Sci.* **28**, 390 (1963).
10. A. S. Szczesniak, M. A. Brandt, and H. H. Friedman, *J. Food Sci.* **28**, 397 (1963).
11. A. S. Szczesniak, *J. Food Sci.* **28**, 410 (1963).
12. A. S. Szczesniak, *J. Texture Studies* **6**, 5 (1975).
13. B. E. Proctor, S. Davison, G. J. Malecki, and M. Welch, *Food Technol.* **9**, 471 (1955).
14. M. C. Bourne, *J. Food Sci.* **32** (1967).

15. I. Tinoco, Jr., K. Sauer, and J. C. Wang, *Physical Chemistry: Principles and Applications in Biological Sciences*, Prentice Hall Press, Englewood Cliffs, N.J., 1978.

16. H. R. Moskowitz, ed., *Food Texture: Instrumental and Sensory Measurement*, Marcel Dekker, Inc., New York, 1987.

17. M. Peleg and E. B. Bagley, eds., *Physical Properties of Foods*, AVI Publishing Co., Inc., Westport, Conn., 1983.

18. J. M. deMan, P. W. Voisey, V. F. Rasper, and D. W. Stanley, eds., *Rheology and Texture in Food Quality*, AVI Publishing Co., Westport, Conn., 1976.

19. H. Faridi and J. M. Faubion, eds., *Fundamentals of Dough Rheology*, American Association of Cereal Chemists, St. Paul, Minn., 1986.

David N. Holcomb
Kraft General Foods Technology
Glenview, Illinois

Marvin A. Tung
Technical University of Nova Scotia
Halifax, Nova Scotia
Canada